SECOND EDITION

DEVELOPMENTAL and REPRODUCTIVE TOXICOLOGY

A Practical Approach

SECOND EDITION
DEVELOPMENTAL and REPRODUCTIVE TOXICOLOGY
A Practical Approach

Edited by
Ronald D. Hood

Taylor & Francis Group
Boca Raton   London   New York

A CRC title, part of the Taylor & Francis imprint, a member of the
Taylor & Francis Group, the academic division of T&F Informa plc.

Published in 2006 by
CRC Press
Taylor & Francis Group
6000 Broken Sound Parkway NW, Suite 300
Boca Raton, FL 33487-2742

© 2006 by Taylor & Francis Group, LLC
CRC Press is an imprint of Taylor & Francis Group

No claim to original U.S. Government works
Printed in the United States of America on acid-free paper
10 9 8 7 6 5 4 3 2 1

International Standard Book Number-10: 0-8493-1254-X (Hardcover)
International Standard Book Number-13: 978-0-8493-1254-0 (Hardcover)
Library of Congress Card Number 2005043988

This book contains information obtained from authentic and highly regarded sources. Reprinted material is quoted with permission, and sources are indicated. A wide variety of references are listed. Reasonable efforts have been made to publish reliable data and information, but the author and the publisher cannot assume responsibility for the validity of all materials or for the consequences of their use.

No part of this book may be reprinted, reproduced, transmitted, or utilized in any form by any electronic, mechanical, or other means, now known or hereafter invented, including photocopying, microfilming, and recording, or in any information storage or retrieval system, without written permission from the publishers.

For permission to photocopy or use material electronically from this work, please access www.copyright.com (http://www.copyright.com/) or contact the Copyright Clearance Center, Inc. (CCC) 222 Rosewood Drive, Danvers, MA 01923, 978-750-8400. CCC is a not-for-profit organization that provides licenses and registration for a variety of users. For organizations that have been granted a photocopy license by the CCC, a separate system of payment has been arranged.

Trademark Notice: Product or corporate names may be trademarks or registered trademarks, and are used only for identification and explanation without intent to infringe.

Library of Congress Cataloging-in-Publication Data

Developmental and reproductive toxicology : a practical approach / edited by Ronald D. Hood. -- 2nd ed.
   p. cm.
   First ed. published in 1997 under title: Handbook of developmental toxicology.
   ISBN 0-8493-1254-X
   1. Developmental toxicology--Handbooks, manuals, etc. 2. Reproductive toxicology--Handbooks, manuals, etc. I. Hood, Ronald D. II. Handbook of developmental toxicology.

RA1224.2.H36 2005
615.9--dc22
                                                  2005043988

Taylor & Francis Group
is the Academic Division of T&F Informa plc.

Visit the Taylor & Francis Web site at
http://www.taylorandfrancis.com

and the CRC Press Web site at
http://www.crcpress.com

CONTENTS

PART I
PRINCIPLES AND MECHANISMS 1

Chapter 1 Principles of Developmental Toxicology Revisited 3
Ronald D. Hood

Chapter 2 Experimental Approaches to Evaluate Mechanisms of Developmental Toxicity 15
Elaine M. Faustman, Julia M. Gohlke, Rafael A. Ponce, Thomas A. Lewandowski, Marguerite R. Seeley, Stephen G. Whittaker, and William C. Griffith

Chapter 3 Pathogenesis of Abnormal Development...61
Lynda B. Fawcett and Robert L. Brent

Chapter 4 Maternally Mediated Effects on Development..93
Ronald D. Hood and Diane B. Miller

Chapter 5 Paternally Mediated Effects on Development...125
Barbara F. Hales and Bernard Robaire

Chapter 6 Comparative Features of Vertebrate Embryology 147
John M. DeSesso

PART II
HAZARD AND RISK ASSESSMENT AND REGULATORY GUIDANCE.......................199

Chapter 7 Developmental Toxicity Testing — Methodology 201
Rochelle W. Tyl and Melissa C. Marr

Chapter 8 Nonclinical Juvenile Toxicity Testing 263
Melissa J. Beck, Eric L. Padgett, Christopher J. Bowman, Daniel T. Wilson, Lewis E. Kaufman, Bennett J. Varsho, Donald G. Stump, Mark D. Nemec, and Joseph F. Holson

Chapter 9 Significance, Reliability, and Interpretation of Developmental and Reproductive Toxicity Study Findings 329
Joseph F. Holson, Mark D. Nemec, Donald G. Stump, Lewis E. Kaufman, Pia Lindström, and Bennett J. Varsho

Chapter 10 Testing for Reproductive Toxicity 425
Robert M. Parker

Chapter 11 The U.S. EPA Endocrine Disruptor Screening Program: *In Vitro* and *In Vivo* Mammalian Tier I Screening Assays...489
Susan C. Laws, Tammy E. Stoker, Jerome M. Goldman, Vickie Wilson, L. Earl Gray, Jr., and Ralph L. Cooper

Chapter 12 Role of Xenobiotic Metabolism in Developmental and Reproductive Toxicity 525
Arun P. Kulkarni

Chapter 13 Use of Toxicokinetics in Developmental and Reproductive Toxicology 571
Patrick J. Wier

Chapter 14 Cellular, Biochemical, and Molecular Techniques in Developmental
Toxicology 599
Barbara D. Abbott, Mitchell B. Rosen, and Gary Held

Chapter 15 Functional Genomics and Proteomics in Developmental and
Reproductive Toxicology 621
Ofer Spiegelstein, Robert M. Cabrera, and Richard H. Finnell

Chapter 16 *In Vitro* Methods for the Study of Mechanisms of Developmental
Toxicology 647
Craig Harris and Jason M. Hansen

Chapter 17 Statistical Analysis for Developmental and Reproductive Toxicologists 697
James J. Chen

Chapter 18 Quality Concerns for Developmental and Reproductive Toxicologists 713
Kathleen D. Barrowclough and Kathleen L. Reed

Chapter 19 Perspectives on the Developmental and Reproductive Toxicity Guidelines 733
Mildred S. Christian, Alan M. Hoberman, and Elise M. Lewis

Chapter 20 Human Studies — Epidemiologic Techniques in Developmental and
Reproductive Toxicology 799
Bengt Källén

Chapter 21 Developmental and Reproductive Toxicity Risk Assessment
for Environmental Agents 841
Carole A. Kimmel, Gary L. Kimmel, and Susan Y. Euling

Chapter 22 Principles of Risk Assessment — FDA Perspective 877
**Thomas F. X. Collins, Robert L. Sprando, Mary E. Shackelford,
Marion F. Gruber, and David E. Morse**

Appendix A Terminology of Anatomical Defects 911

Appendix B Books Related to Developmental and Reproductive Toxicology 967

Appendix C Postnatal Developmental Milestones 969

Index 1131

Preface

The areas of developmental and reproductive toxicology are becoming increasingly important. *Developmental toxicology* encompasses the study of hazard and risk associated with exposure to toxicants during prenatal development and has been expanded in recent years to include effects on the developmental process until the time of puberty, i.e., until the completion of all developmental processes. It is well known that a considerable percentage of newborns have significant anatomical defects, that birth defects are a major cause of hospitalization of infants, that "spontaneous" abortion and perinatal death are common, and that numerous individuals suffer from congenital functional deficits such as mental retardation. Although considerable progress has been made in determining the cause of these defects, the etiology of the majority of birth defects is not yet known or only poorly established. We still must learn much about the mechanisms involved in eliciting congenital defects and the genetic and environmental factors and their interactions that trigger these mechanisms. *Reproductive toxicology* is the study of adverse effects of chemical or physical agents on the reproductive processes of both sexes and the causes of such effects. Concern has been voiced regarding reproductive issues, such as the possibility that human sperm counts have decreased in at least some geographic areas in recent times and that early menarche and precocious breast development may have environmental causes. Thus it is certain that both mechanistic studies of known developmental and reproductive toxicants and the toxicological assessment of pharmaceutical agents, food additives, pesticides, industrial chemicals, environmental pollutants, and the like to which humans may be exposed will be of importance for the foreseeable future. This situation points to the need for useful references in the field of developmental toxicology, teratology, and reproductive toxicology, and provides the impetus for the current volume.

The purpose of this book is to provide a practical guide to the practice of developmental and reproductive toxicology; inclusion in a single volume of material from these areas will be of value to the many individuals with professional responsibilities in both. Further, this book provides information in one source that is currently scattered through the literature or has not been readily available; it provides much of that information in considerable detail. Although the current work is primarily oriented toward research designed to establish the likelihood of harm to humans, it should also prove useful to those who are primarily interested in effects on other organisms. This book should be especially useful for individuals working in industry who are responsible for testing chemical agents for developmental or reproductive toxicity and for those who manage such endeavors. It should be helpful to regulatory scientists at all levels of government who must evaluate the adequacy of studies and who must interpret data on hazard and establish the potential risk from exposures. This volume will also be useful in training students and technicians, and for individuals active in other areas who find the need to become familiar with the principles and practice of developmental or reproductive toxicology.

Developmental and Reproductive Toxicology: A Practical Approach is intended to be a practical guide as well as informative, providing insights gained from hands-on experience along with a theoretical foundation. Although this book is intended to be a relatively comprehensive guide to the fields of developmental and reproductive toxicology, there were practical limits to the number and scope of areas that it could address. It should also be noted that mention of vendors, trade names, or commercial products does not constitute an endorsement or recommendation for use.

The editor wishes to especially thank the contributing authors, whose efforts and expertise made this project a success. Thanks also go to Taylor & Francis and the following individuals in its employ: Stephen Zollo, who proposed that the current work be published as a follow-up to the *Handbook of Developmental Toxicology*, which I had previously edited, and who has provided invaluable advice and encouragement on the project, and to Patricia Roberson, who has provided essential technical guidance during the process of bringing the book to publication.

The Editor

Ronald D. Hood, Ph.D., is Professor Emeritus of Biological Sciences, having retired from the faculty of the Cell, Molecular, and Developmental Biology Section of the Department of Biological Sciences of The University of Alabama, Tuscaloosa, Alabama, in June, 2000. Dr. Hood remains active in research and consulting, and has retained his office and laboratory at the university. He is also principal of Ronald D. Hood and Associates, Toxicology Consultants, and Adjunct Professor of Public Health in the School of Public Health of the University of Alabama at Birmingham. Dr. Hood received his B.S. and M.S. degrees from Texas Tech University and his Ph.D. in reproductive physiology from Purdue University (1969). He joined the faculty of The University of Alabama in 1968 as assistant professor, advanced to the rank of full professor in 1978, and served as interim department chair in 1996 to 1997. Dr. Hood was also Consultant in Environmental Medicine, U.S. Veterans Administration, Office of Medicine and Surgery, Agent Orange Special Projects Office (off site) during 1983 and Special Consultant, Science Advisory Board, U.S. Environmental Protection Agency, from 1983 through 1993. In addition, he has served as a consultant to industrial clients, trade associations, federal and state agencies, and law firms since 1978, and as a grant or document reviewer for the Environmental Protection Agency, the Agency for Toxic Substances and Disease Registry, the Congressional Office of Technology Assessment, ICCVAM, and the National Research Council.

Dr. Hood is a member of the Teratology Society, the Society of Toxicology, and the Reproductive and Developmental Toxicology Specialty Section of the Society of Toxicology (RDTSS), and he is the current editor of the *RDTSS Newsletter*. Dr. Hood was also a charter member of the Society for the Study of Reproduction and of the Neurobehavioral Teratology Society. He has been particularly active in the Teratology Society, where he has chaired the society's Membership, Education, and Constitution/Bylaws committees, was a member of the society's Ad Hoc Committees on Ethics, Warkany Lecturer Selection, Expert Testimony, and Web Site. Dr. Hood has served as a member of the editorial boards of *Fundamental and Applied Toxicology*, *Toxicological Sciences*, and *InSight*, and he is a current member of the EPA's Food Quality Protection Act Science Review Board (FQPA/SRB).

At the University of Alabama, Dr. Hood taught courses on teratology, developmental toxicology, both general and environmental toxicology, developmental biology, human embryology, reproductive physiology, endocrinology, and general physiology.

Dr. Hood's current research interests include investigation of mechanisms and development of assays for developmental toxicity. His recent research has involved assessment of the ability of indole-3-carbinol to protect against developmental toxicity, determination of whether the developmental toxicity of chromium picolinate is due to the picolinate alone, detection of toxicant-induced DNA strand breaks in organogenesis stage mammalian embryos, influence of maternal diet and biotransformation by methylation on arsenical-induced developmental toxicity, maternal stress and teratogen interactions, and mechanistic studies of mitochondrial poisons as teratogens. He has also investigated the teratogenic potential of environmental toxicants, such as mycotoxins and arsenicals; he has conducted research on teratogen-teratogen and gene-teratogen interactions, the use of developing *Drosophila melanogaster* as an "*in vitro*" screening system for developmental toxicants, and arsenic biotransformation *in vivo* and *in vitro*. Dr. Hood has participated in numerous workshops and expert review panels on developmental toxicity and risk assessment. He has also authored or edited some 92 publications in print or in press (research articles, reviews, books, and book chapters), including *Handbook of Developmental Toxicology,* the predecessor to the current volume, in addition to numerous unpublished reports.

Contributors List

Barbara D. Abbott
U.S. Environmental Protection Agency
Research Triangle Park, North Carolina

Kathleen D. Barrowclough
DuPont
Newark, Delaware

Melissa J. Beck
WIL Research Laboratories
Ashland, Ohio

Christopher J. Bowman
WIL Research Laboratories
Ashland, Ohio

Robert L. Brent
DuPont Hospital for Children
Wilmington, Delaware

Robert M. Cabrera
Texas A & M University System Health Science Center
Houston, Texas

James J. Chen
U.S. Food and Drug Administration
Jefferson, Arkansas

Mildred S. Christian
Argus Research
Horsham, Pennsylvania

Thomas F.X. Collins
U.S. Food and Drug Administration
Laurel, Maryland

Ralph L. Cooper
U.S. Environmental Protection Agency
Research Triangle Park, North Carolina

John M. DeSesso
Mitretek Systems
Falls Church, Virginia

Susan Y. Euling
U.S. Environmental Protection Agency
Washington, D.C.

Elaine M. Faustman
University of Washington
Seattle, Washington

Lynda B. Fawcett
Jefferson Medical College
Wilmington, Delaware

Richard H. Finnell
Texas A & M University System Health Science Center
Houston, Texas

Julie M. Gohlke
University of Washington
Seattle, Washington

Jerome M. Goldman
U.S. Environmental Protection Agency
Research Triangle Park, North Carolina

Earl Gray, Jr.
U.S. Environmental Protection Agency
Research Triangle Park, North Carolina

William C. Griffith
University of Washington
Seattle, Washington

Marion F. Gruber
U.S. Food and Drug Administration
Rockville, Maryland

Barbara F. Hales
McGill University
Montreal, Quebec, Canada

Jason H. Hansen
Emory University
Atlanta, Georgia

Craig Harris
The University of Michigan
Ann Arbor, Michigan

Gary Held
U.S. Environmental Protection Agency
Research Triangle Park, North Carolina

Alan M. Hoberman
Argus Research
Horsham, Pennsylvania

Joseph F. Holson
WIL Research Laboratories
Ashland, Ohio

Ronald D. Hood
The University of Alabama
Tuscaloosa, Alabama

Bengt Källén
Tornblad Institute
Lund, Sweden

Lewis E. Kaufman
WIL Research Laboratories
Ashland, Ohio

Carole A. Kimmel
U.S. Environmental Protection Agency
Washington, D.C.
Currently at
Kimmel and Associates
Silver Spring, Maryland

Gary L. Kimmel
U.S. Environmental Protection Agency
Washington, D.C.
Currently at
Kimmel and Associates
Silver Spring, Maryland

Arun P. Kulkarni
University of South Florida
Tampa, Florida

Susan C. Laws
U.S. Environmental Protection Agency
Research Triangle Park, North Carolina

Thomas A. Lewandowksi
University of Washington
Seattle, Washington

Elise M. Lewis
Argus Research
Horsham, Pennsylvania

Pia Lindström
Maxim Pharmaceuticals
San Diego, California

Melissa C. Marr
Center for Life Sciences and Toxicology
Research Triangle Park, North Carolina

Diane B. Miller
Centers for Disease Control and Prevention
Morgantown, West Virginia

David E. Morse
U.S. Food and Drug Administration
Rockville, Maryland

Mark D. Nemec
WIL Research Laboratories
Ashland, Ohio

Eric L. Padgett
WIL Research Laboratories
Ashland, Ohio

Robert M. Parker
Hoffmann-LaRoche Inc.
Nutley, New Jersey

Rafael A. Ponce
University of Washington
Seattle, Washington

Kathleen L. Reed
DuPont
Newark, Delaware

Bernard Robaire
McGill University
Montreal, Quebec, Canada

Mitchell B. Rosen
U.S. Environmental Protection Agency
Research Triangle Park, North Carolina

Marguerite R. Seeley
Gradient Corporation
Cambridge, Massachusetts

Mary E. Shackleford
U.S. Food and Drug Administration
College Park, Maryland

Ofer Spiegelstein
Texas A & M University System Health Science Center
Houston, Texas

Robert L. Sprando
U.S. Food and Drug Administration
Laurel, Maryland

Tammy E. Stoker
U.S. Environmental Protection Agency
Research Triangle Park, North Carolina

Donald G. Stump
WIL Research Laboratories
Ashland, Ohio

Rochelle W. Tyl
Center for Life Sciences and Toxicology
Research Triangle Park, North Carolina

Bennett J. Varsho
WIL Research Laboratories
Ashland, Ohio

Stephen G. Whittaker
University of Washington
Seattle, Washington

Patrick J. Wier
GlaxoSmithKline Pharmaceuticals
Upper Merion, Pennsylvania

Daniel T. Wilson
WIL Research Laboratories
Ashland, Ohio

Vickie Wilson
U.S. Environmental Protection Agency
Research Triangle Park, North Carolina

Part I
Principles and Mechanisms

CHAPTER 1

Principles of Developmental Toxicology Revisited

Ronald D. Hood

CONTENTS

I. Introduction 3
II. Basic Principles 5
    A. Some Basic Terminology 5
    B. Wilson's Principles 6
        1. Susceptibility to Teratogenesis Depends on the Genotype of the Conceptus
           and the Manner in Which This Interacts with Adverse Environmental Factors 6
        2. Susceptibility to Teratogenesis Varies with the Developmental Stage
           at the Time of Exposure to an Adverse Influence 7
        3. Teratogenic Agents Act in Specific Ways (Mechanisms) on Developing
           Cells and Tissues to Initiate Sequences of Abnormal Developmental
           Events (Pathogenesis) 8
        4. The Access of Adverse Influences to Developing Tissues Depends
           on the Nature of the Influence (Agent) 8
        5. The Four Manifestations of Deviant Development Are Death, Malformation,
           Growth Retardation, and Functional Deficit 9
        6. Manifestations of Deviant Development Increase in Frequency and Degree
           as Dosage Increases, from the No-Effect to the Totally Lethal Level 10
III. Who Will Conduct the Tests, and Who Will Interpret the Results? 11
IV. Where Do We Go from Here? 12
References 12

I. INTRODUCTION

Developmental toxicology has been evolving as a discipline for decades with only modest initial recognition despite the early knowledge that an excess of certain nutrients (e.g., vitamin A)[1] or administration of various chemicals could cause developmental defects in various animal species.[2–4] As has been stated many times, it took the revelation in the early sixties that thalidomide, a drug promoted as a relatively innocuous sedative and antiemetic, was a potent human teratogen[5] to arouse interest in testing for potential developmental toxicants[6] and to toughen the Food, Drug, and Cosmetic Act of 1938.[7] Since that time, testing protocols have slowly evolved, first under the guidance of regulatory agencies in individual countries and more recently as the result of joint efforts to harmonize test paradigms and reduce duplication of effort (cf. Chapter 19).

Relatively early in this process, a series of three protocols was designed to evaluate test agents for their effects on developing mammals, with the intent of protecting humans exposed to pharmaceuticals, food additives, pesticides, workplace chemicals, and environmental pollutants. These protocols were developed for the purpose of assessing (1) the outcomes on the conceptus of maternal exposures, beginning prior to mating and ending prior to implantation (Segment I: "Study of Fertility and General Reproductive Performance"), (2) exposures during major organogenesis (Segment II: "Prenatal Developmental Toxicity Study" or "Teratological Study"), and (3) exposures during late gestation, parturition, and lactation (Segment III: "Perinatal and Postnatal Study"). The current iterations of these protocols, descriptive terminology, and generated data are described and discussed elsewhere in this volume and particularly in Chapters 7, 9, and 19. It is of interest to note, however, that although the test procedures have evolved in specific aspects, they have not changed greatly since they were first recommended by the U.S. Food and Drug Administration (FDA).[8]

Changes in test protocols have typically been modest, such as increases in the numbers of test females required or in the duration of treatment. Another example is the requirement for neurobehavioral testing in certain cases.[9,10] Perhaps the greatest advances in developmental toxicity testing have come not from improvements in the standard protocols but from our increased knowledge of how to interpret test outcomes, and how and when to modify the protocols. And it must be kept in mind that the standard testing protocols are necessarily compromises between the demands of efficiency and cost effectiveness and the quality and completeness of the information the tests provide.[11] The need to keep costs at a bearable level is in conflict with the needs of regulatory agencies to obtain adequate data to serve as the basis for the required decisions. Thus, the Segment II test protocol compromises by calling for treatment throughout organogenesis (or even throughout gestation if the treatment is not expected to prevent implantation) until the day of palate closure or until the day prior to scheduled sacrifice at term, depending on the specific guideline. That is the case even though the use of several groups of mated females treated at each dosage level during brief, discrete periods of organogenesis would be more effective at revealing a compound's teratogenicity potential. Interestingly, the latter methodology had been proposed initially.[12] Conversely, although smaller test groups of rabbits than of rats were once allowed, primarily to contain costs, today the minimum number of rabbits has been increased to provide more meaningful data.

At some future date, cellular/molecular assays and quantitative structure-activity-relationships (QSAR) may become more routine to supplement (or even replace) the current whole-animal developmental toxicity tests. However, that eventuality is likely to require major increases in our understanding of both the mechanisms of developmental toxicity and the complex interplay of the molecular and physiological systems that govern and regulate both developmental processes and maternal physiology and homeostasis.

As is discussed in Chapter 2, developmental toxicants first act via specific mechanisms, i.e., the initial event(s) in the germ cells or in the cells of the embryo or fetus that begin the series of processes (i.e., pathogenesis, described in Chapter 3) leading to adverse outcomes. This is true of toxic insults to adults as well, but such occurrences in immature systems are made more complex by the constantly changing nature of the developing organism, especially during the period from conception through major organogenesis. Adding even further to the complexity in mammals is the interplay between the developing conceptus and the supporting maternal "environment," mediated during most of development by the extraembryonic membranes and with the eventual addition of the placenta.

Our understanding of specific incidences of events such as abnormal development, functional deficit, or prenatal demise is further confounded by the likelihood that such manifestations may, at times, merely be sporadic failures of complex systems. It is likely that the genetic "blueprint" for the development of complex organisms is not failure proof, and that even in the absence of deleterious mutations or chromosomal anomalies, development can fail. This might happen if certain critical gene alleles, which would ordinarily direct a robust developmental process, specified a more error prone process when present in a specific combination. Alternatively, some species or strains of animals seem to have "weak points" in their development, such that some percentage of offspring, even if they had identical genomes and similar environments, would manifest an anomaly. In other words, the genetically specified plan for development is seldom if ever perfect. The myriad events it specifies must occur at just the right time, in the right location, and in a reasonably correct manner, and in a small percentage of individuals one or more such processes may fail. This seems a likely mechanism in cases where an inbred strain of mice exhibits a high "spontaneous" incidence of some defect, such as cleft palate. The incidence of such anomalies can be further increased by exposure of the conceptus to a toxic agent or a compromised maternal environment, which presumably nudges borderline cases in the direction of abnormal development.

Although much remains to be learned about the causation of adverse effects on developing offspring, there are certain principles that should be considered by anyone seeking to plan, carry out, or interpret the results of tests for developmental toxicity, including epidemiologic studies. A number of these principles have been known for some time, and much of this chapter will be devoted to such basic principles.

II. BASIC PRINCIPLES

A. Some Basic Terminology

According to Wilson,[13] *teratology* is "the science dealing with the causes, mechanisms, and manifestations of developmental deviations of either structural or functional nature." He also defined teratology as "the study of adverse effects of environment on developing systems, that is, on germ cells, embryos, fetuses, and immature postnatal individuals." Although it is recognized that a portion of developmental defects have a genetic causation, Wilson reckoned that, "Even hereditary defects...were initiated as mutations at some time in the past," and thus, "It is probable that all abnormal development has its causation in some aspect of environment."

In descriptions of the harmful effects of chemical or physical agents on developing systems, terms such as *embryotoxicity* and *fetotoxicity* have often been used. These are legitimate terms, but it should be recognized that they are properly applied only to toxic insults occurring during the specified portion of the developmental process. They should not be used as all-encompassing terms to describe the effects of exposures throughout the entirety of development. *Developmental toxicity* can be defined as the ability of a chemical or physical agent to cause any of the manifestations of adverse developmental outcome (i.e., death, malformation, growth retardation, functional deficit), individually or in combination. *Teratogenicity* has been used to mean just the ability to produce "terata" or malformations. In the broader sense, it has been used in the same way as developmental toxicity, and the study of developmental toxicity is referred to as *developmental toxicology*. *Malformation* has been defined as "a permanent structural change that may adversely affect survival, development, or function," while a *variation* is "a divergence beyond the usual range of structural constitution that may not adversely affect survival or health."[14] Distinguishing between these two can be difficult, however, because they exist on a continuum between the normal and the abnormal.

Although basic principles pertaining to developmental toxicology and teratology are presumably known to practitioners in the field, it can be useful to review some of these principles and experimental support for them. Also, individuals new to the field can benefit from such accumulated wisdom as an aid in designing, carrying out, and interpreting the results of experimental and epidemiological studies.

B. Wilson's Principles[15]

1. Susceptibility to Teratogenesis Depends on the Genotype of the Conceptus and the Manner in Which This Interacts with Adverse Environmental Factors[15]

It must also be kept in mind that in some cases there is a purely genetic cause, e.g., Down syndrome in humans, and there can be purely environmental causes, such as x-rays at high doses. In still other cases, however, a combination of environmental insult(s) and a susceptible genome in the conceptus is required for adverse effects to appear, a scenario that Fraser termed *multifactorial causation*.[16]

It is now well established that the genetic makeup of the conceptus can significantly influence the outcome of exposures to developmental toxicants, especially if the level of exposure is near the threshold for causing a particular adverse effect. This has been readily observed in studies involving treatment of inbred mouse strains and crosses between them. However, one must be aware of the caveat that the developmental outcome may also be influenced by the genotype of the dam as a determinant of, for example, the rate or preferred pathway of biotransformation of the toxicant in question and its peak level in the maternal blood or its area under the curve (AUC, the area under the plasma [or serum or blood] concentration versus time curve). The outcome here can be influenced by whether it is the parent compound or a metabolite that is developmentally toxic, and of course in some cases *both* can be toxic. Fetal alcohol syndrome may be a case where the maternal and/or fetal genotype can interact with the toxic agent (and probably with other environmental influences, such as the maternal diet) to produce damage in some instances and few or no obvious effects in others.[17]

Recently, confirmation of the significant influence of specific genes has come from experiments in which knockout mice lacking a specific gene have been found to be either enhanced or diminished in susceptibility to exposure to a developmentally toxic agent.[18] Further, recent human studies have probed the potential influence of combinations of specific gene polymorphisms and dietary deficiencies, especially folate deficiency, on the incidence of neural tube defects.[19] And deficiency in the activity of a detoxifying enzyme, epoxide hydroxylase, may be an important determinant of the manifestation of fetal hydantoin effects.[20]

Species differences in response to developmental toxicants may be due to differences in inherent susceptibility of the conceptus to differences in maternal pharmacokinetics — including biotransformation — and maternal physiology, or to a combination of these. The same is true of strain and litter differences in response to toxic insult, and the basis for strain differences can be determined by use of reciprocal crosses and by embryo transfer. Typically, there are also individual differences within litters. This is probably most commonly due, at least in part, to genotypic differences among the individual fetuses, though this should be less often true when inbred strains are involved.

Differences in developmental stage at the time of exposure to toxic insults — those of short duration or those that begin during organogenesis — may explain some of the differences in individual response among fetuses within the same litter. In some cases, there are sex differences in the response, i.e., males and females may be affected differently. Differences in the intrauterine environment of each conceptus may also contribute to nonuniformity of response within litters. For example, placental blood supply varies somewhat according to location in the uterus[21,22] and may bring or remove greater or lesser amounts of a toxic substance or of nutrients, waste products, or respiratory gases. Female rodent fetuses that are found next to one male or (especially) between two male fetuses can be altered in certain attributes, such as sexual attractiveness and estrous cycle length, compared with those of females not found next to males.[23] Such effects are presumably caused by the transfer of androgen from the male to the adjacent female fetuses.

Even differences in fetal drug metabolizing capability may be influential, at least in fetuses in which enhanced levels of xenobiotic biotransforming enzymes had been induced. Nebert found this to be true for a strain of mice heterozygous at the *AHH* (aryl hydrocarbon hydroxylase) locus, where some fetuses lack the induction receptor.[24] This receptor responds to aryl hydrocarbons, such as TCDD (2,3,7,8-tetrachlorodibenzo-*p*-dioxin), forming a complex involved in activating the genes for cytochrome P-450 monooxygenases, a process thought to be involved in initiating the toxic effects of TCDD.

2. Susceptibility to Teratogenesis Varies with the Developmental Stage at the Time of Exposure to an Adverse Influence[15]

Development can be roughly divided into three periods with regard to susceptibility to toxic insult:

a. Predifferentiation

If damaged during this period, the embryo typically either dies or completely repairs the damage, an "all or none" effect. There are some exceptions, however, such as the effects of x-rays or certain highly potent genotoxic chemicals, and their effects may be genetically mediated.[25]

b. Early Differentiation or Early Organogenesis

This occurs roughly from days 9 through 15 in the rat and days 9 through 18 in the rabbit (day on which mating was confirmed is day 0), with the timing varying by species (see Chapter 6, Table 1 for others). Organ primordia, foundations for later development, are laid down at this time. The embryo is most susceptible to induction of malformations during this period of development because once the basic structures are formed, it becomes increasingly difficult to alter them structurally. Organs whose development is multiphasic, such as the eye and the brain, tend to have more than one susceptible period.

c. Advanced — or Late — Organogenesis

This period is largely occupied with histogenesis and functional maturation. Insults during this period mainly cause growth retardation, developmental delay, or functional disturbances, particularly neurobehavioral problems, as the brain matures relatively late. However, even at this time, interrupted blood supply to a localized area or structure (e.g., because of an amniotic band) can cause degeneration of that area or structure, resulting in a malformation.

A malformation may sometimes occur well after the initiating toxic insult, and this might be due to alteration of biochemical events, such as gene activation and mRNA synthesis, that may occur prior to organ differentiation. This might also be due to alteration of an earlier event in a sequence of events leading up to the formation of an organ — for example, interference with an induction early in a "chain of inductions."

Organisms tend to be significantly more sensitive to many adverse environmental influences during early developmental stages, although this differential may not be quite as universally applicable in mammals as was once thought. And even though the human gestation period is long, the portion of that time spent in early organogenesis is fortunately relatively short; thus, the human embryo is not at peak risk in comparison with its total gestation length. Increased sensitivity, especially during early organogenesis, apparently occurs because many complex and alterable events are taking place. Many tissues are undergoing rapid cell division, and the embryo, and to a considerable extent, the fetus, has much less capacity to metabolize xenobiotics than does the adult. Some embryos, such as those of rodents, however, can be induced to biotransform a significant amount of certain developmental toxicants. Thus, metabolism can, in some cases, actually *activate* a toxicant, increasing its effects on the conceptus.

3. Teratogenic Agents Act in Specific Ways (Mechanisms) on Developing Cells and Tissues to Initiate Sequences of Abnormal Developmental Events (Pathogenesis)[15]

One consequence of mechanistic specificity is the frequently observed "agent specificity" of malformations, malformation syndromes, and/or behavioral or other functional effects. Many teratogens produce characteristic patterns of malformations and other effects as a result of their particular mechanisms of action and/or their tissue distribution. Although this specificity is most often mentioned in the context of evaluating human data, it is also important to consider it when interpreting animal test outcomes that appear somewhat ambiguous.

It must also be appreciated that a given developmental toxicant may act by more than one mechanism at the same time in the same organism, although one or more such mechanisms may predominate. Unfortunately, there has often been the tendency to interpret the outcomes of toxicological testing of individual compounds in terms of a single mechanism proposed for the test agent. This has generally been the case even though it has long been known that many agents may have multiple actions in biological systems. In addition, most tests for developmental or reproductive toxicity of necessity employ only one test agent and carefully control for other variables that might influence test outcomes. This is true even though such conditions are never seen outside the laboratory, as humans and other organisms are continually subjected to a variety of biologically active agents and other environmental influences throughout the reproductive and developmental process.

With regard to pathogenesis, we must also consider that multiple defects are often seen in the same individual. Thus the same or different mechanisms may initiate abnormalities that are acting concurrently by pathogenetic pathways in different organs. Alternatively, as pointed out by Bixler et al.,[26] expression of a given pathway of dysmorphogenesis may secondarily result in a defect in some other structure of the embryo or fetus. Interpretation of experimental findings is made more challenging because, according to the concept of common pathogenetic pathways as proposed by Wilson,[27] even like defects do not necessarily share the same initial causal mechanism.

Mechanisms are discussed more fully in Chapter 2, and pathogenesis in Chapter 3, as these important topics require a more complete treatment than would be appropriate here.

4. The Access of Adverse Influences to Developing Tissues Depends on the Nature of the Influence (Agent)[15]

a. Placental Transfer

The layer of cellular and extracellular material between the maternal and fetal bloodstreams has sometimes been called the *placental barrier* because at one time it was believed to afford great protection to the embryo and fetus. We now know that the degree of protection is often modest at best, and that instead of being a barrier, the placental membrane acts more as an ultrafilter.[28] Also, a number of physiologically important substances, such as amino acids and glucose, are transferred across the placental membrane by specific mechanisms.[29] This can sometimes allow xenobiotics to be transferred as well, if they can take advantage of the physiologic systems.

A number of factors are known that can influence how readily a given substance crosses the placenta:

- Molecular size — smaller molecules cross more readily, especially those under about 1000 in molecular weight.
- Charge — uncharged molecules cross more readily than charged compounds of the same size; negatively charged molecules pass more readily than those that are positively charged.
- Lipid solubility — lipophilic compounds penetrate more quickly than more polar forms.

Degree of ionization — less ionized substances pass more readily.

Formation of complexes — complexed molecules are impeded in comparison with their "free" forms, especially if they are complexed with proteins.

Existence of concentration gradients — substances present in high concentrations in the maternal blood are likely to cross the placenta in larger amounts.

Placental metabolism — the placenta can metabolize certain substances, but the influence of placental metabolism is thought to be minor for most compounds.

From the foregoing, it can be seen that although the placenta only rarely acts as a complete barrier, it can greatly slow the passage of certain water soluble molecules, such as heparin and plasma proteins, that are large and/or highly charged. Also, certain physical agents, such as x-rays, gamma rays, ultrasound, and radiofrequency radiation, can readily reach the embryo or fetus, even from outside the mother. And maternal metabolism and excretion may eliminate a portion of the absorbed dose of a chemical agent or alter its nature prior to its reaching the conceptus.[29]

b. Relationship of the Placenta to Teratogenesis

It has been suggested that interference with the function of the yolk sac, particularly during development of rodents and lagomorphs, may result in developmental defects because these species depend for a considerable portion of gestation on an "inverted yolk sac placenta"[30] (see Chapter 6 for more discussion and for comparisons with other species).

c. Exposure of the Conceptus

If a potentially developmentally toxic agent reaches the embryo at a sufficiently high rate, it may reach a sufficiently high concentration to cause an adverse effect; on the other hand, if the rate is too low, it will not reach an effective concentration. This is because at very low exposure rates, cells may not be significantly affected, and at somewhat higher rates, damage that occurs may be repaired before irreparable harm is done. It should be noted that although mammalian embryos and fetuses can repair minor injuries, if they have developed past the blastocyst stage, they cannot regenerate lost parts. It is also of interest to note that in some cases a chemical may become more concentrated in the conceptus than in the mother, depending on the particular agent and exposure route.[31-33]

5. The Four Manifestations of Deviant Development Are Death, Malformation, Growth Retardation, and Functional Deficit[15]

In some cases, it is likely that malformation precedes and results in death. According to Kalter,[34] as the dose is increased in such a scenario, an inverse relationship between malformation and prenatal mortality would be seen. This would occur as the severity of the induced defects increased to the point of causing deaths among the offspring. A practical consequence is that an increased incidence of prenatal deaths may obscure our ability to determine if an agent has caused malformations because the most severely malformed offspring may not survive to be counted at cesarean section in a typical teratogenicity assay. However, malformation as the cause of prenatal demise can be revealed if examination of embryos or fetuses prior to birth reveals more malformations and fewer deaths than would be expected at term.[15] In other cases, death and malformation may be due, at least in part, to different causes, and thus it is sometimes possible to block or enhance one effect without blocking or enhancing the other. Independence of these two manifestations (malformation and death) would be suspected if their frequencies increased independently with increasing dose.[34] The two additional possible consequences of developmental toxicity, growth retardation and functional deficit, may also be caused by malformations, or they may be induced independently by either the same or different mechanisms from those resulting in malformation or death. Further, the consequences of exposure to a developmental toxicant are strongly influenced by the timing of exposure, as pointed out in Section II.B.2. and discussed elsewhere in this volume (e.g., Chapter 6).

6. Manifestations of Deviant Development Increase in Frequency and Degree as Dosage Increases, from the No-Effect to the Totally Lethal Level[15]

It is often assumed, although there are arguments on both sides of the issue,[35–37] that developmentally toxic agents typically have a "no-effect" threshold because of the repair and regulative abilities of the embryo. Existence of or lack of such a threshold is virtually impossible to prove experimentally, however. Even if large group sizes are employed, it is difficult to decide how to fit a dose-response curve to the data at the lowest doses, as there is always a background level of adverse outcomes, such as malformation and prenatal demise. An argument against the existence of developmental toxicity thresholds in some cases states that when a hormone or other endogenous chemical is responsible for adverse effects that make up a portion of the control (background) incidence, a threshold dose will not be observed for an exogenous chemical acting by the same mechanism. Even though a threshold may exist, it has already been exceeded by the endogenous chemical, and agents acting by the same mechanism merely add to the background incidence.[38]

The threshold concept has also been debated with regard to other forms of toxicity, such as carcinogenicity and mutagenicity, but the existence of a threshold seems more likely to hold generally true for teratogenicity, with a few possible specific exceptions as described above. And the assumption of thresholds, along with no-observed-adverse-effect levels (NOAELs), has proven useful for practical risk assessment, even in the absence of ability to confirm their existence absolutely.

Teratogenesis dose-response curves are often quite steep, so that merely doubling the dose (or less) may extend the range from minimal to maximal effects, but there are exceptions (e.g., thalidomide in humans). It is also possible to have a U-shaped (or J-shaped) dose response, where an *inadequate level* of a compound is harmful, a *higher level* is beneficial, and a *still higher level* is again harmful. Such responses are common in the case of vitamins and essential minerals. For example, vitamin A deficiency is teratogenic, appropriate levels are required for normal development, and excessive amounts are again teratogenic.[1] Another possibility is hormesis, which has been described as producing a U- or J-shaped dose-response curve. In this case, if a low dose results in a stimulatory effect on a normal function, e.g., growth, in comparison with no exposure, while higher doses are inhibitory, the response curve would appear as an "inverted U."[39,40] When the measured response shows some dysfunction (as would be expected with most toxicants), but low doses are beneficial relative to controls while higher doses result in increased levels of dysfunction, the dose-response is described as a "J" or a "noninverted U."[40] The occurrence of hormesis and mechanisms for such responses remain as areas of controversy, and whether hormesis occurs at low doses of developmental or reproductive toxicants in laboratory animals or in humans is as yet unclear.

It is not always appreciated that the current protocol for the developmental toxicity study is not terribly efficient at eliciting malformations,[41] and that, as stated by Palmer,[42] malformations can be a somewhat unreliable guide to developmental toxicity in Segment II tests. The practice of dosing throughout much or all of gestation does severely stress the maternal/embryonic/fetal system, however, and is likely to elicit other useful dose-related indicators of adverse effects on the conceptus, such as decreased survival, diminished fetal weight, or elevated dose-related incidences of developmental variations. But production of malformations often requires a relatively specific set of conditions, including the right species and appropriate treatment dose, mode, and timing. It can even be influenced by such factors as the vehicle for the test article. And it is well established that dosing a pregnant animal once or twice on just one or two gestation days is generally more effective at eliciting malformations (and other manifestations of developmental toxicity) than is dosing for much of gestation. That is largely because if the dam receives only one or a few doses, higher doses can generally be used without causing maternal demise. Further, such targeted treatments can be administered during peak periods of sensitivity of the conceptus.

The A/D ratio is a concept that was proposed by Marshall Johnson.[43,44] It is the ratio of the "adult toxic dose" to the "developmental toxic dose" (the dose harmful to the conceptus). If the conceptus is significantly more sensitive than the adult, the ratio is above 1, if the two are similarly sensitive, the ratio approximates 1, while if the mother is more sensitive, the ratio is less than 1. In practice, the teratogenic dose is often a level that is toxic or sometimes lethal to the mother, i.e., the A/D ratio is typically close to 1, but some teratogens can be effective at doses that are apparently harmless to the mother. These agents with high A/D ratios have often been considered to be especially significant in human risk assessment because exposures to toxicants at doses high enough to be obviously harmful to adults are commonly avoided. Agents not obviously harmful to adults, however, might fail to cause sufficient concern about potential exposures. Nevertheless, pregnant women have frequently been exposed to developmental toxicants at doses harmful to *both* mother and offspring. Obvious examples include alcohol and cigarette smoke. In such a scenario, the *likely level of maternal exposure* vs. the *margin of safety for the conceptus* can be of more importance than the relative sensitivities of adult and offspring.[45]

Other important considerations include the relative extent and duration of harm to the offspring vs. the mother. For example, a pregnant woman who is treated with Accutane during early pregnancy is not likely to experience a lasting harmful effect, while her baby may be irreparably harmed.[45] It is also of interest to note that recent tests comparing the A/D ratios for the same compounds in different species suggest that the ratio is not at all constant across species.[46,47] Such findings indicate that the A/D ratio is of relatively little use for risk extrapolation.

One last comment on the issue of dose vs. effect is that the span of time during development when a defect can be produced by a toxic insult tends to widen as the dose level increases.[15]

III. WHO WILL CONDUCT THE TESTS, AND WHO WILL INTERPRET THE RESULTS?

Graduate programs that are the source of the new generation of toxicologists have increasingly emphasized research and course work in cellular and molecular toxicology, areas that are of obvious importance (e.g., see Chapters 2, 11, 12, and 14 to 16). However, new personnel hired in other areas of our disciplines often start out with inadequate training for the positions they are expected to fill. For several years this state of affairs has become increasingly evident to many in the fields of developmental and reproductive toxicology, and it appears true of positions in both testing laboratories and regulatory agencies. Students are no longer being trained in significant numbers in what might be termed "classical" testing methodology (e.g., the types of tests outlined in Chapters 7 through 10). Lack of in-depth experience working with laboratory animals and insufficient knowledge of animal biology and principles of toxicology can hinder the ability of new professionals to design, carry out, and interpret results of safety evaluation studies unless they are given considerable "on the job" training.

The reasons for this apparent lack of classically trained graduates from institutions of higher learning, including graduate programs in "toxicology," seem clear. Research funding for academia rapidly shifted toward support of mechanistic studies conducted solely at the cellular, biochemical, and especially the molecular level, as investigators increasingly make use of the newly available tools of molecular biology. Not only has there been a major shift in research funding, there has also been a change in the way certain types of research are viewed by many. Molecular approaches are often seen as the only "real science," and work with whole animals may be viewed as outdated. New college students are told about the latest hot cellular or molecular research and often never hear about other possibilities throughout their undergraduate and graduate training. Thus, the typical student can complete a master's or doctoral degree without ever touching a live mammal or having to give much thought to the original source of the cells and molecules that he or she may encounter and manipulate in the laboratory. Nevertheless, despite advances in developmental and reproductive toxicology at the cellular and molecular levels, the need for animal testing remains and will likely persist for some time in the future. This situation poses a challenge to the developmental and reproductive toxicology profession. It puts increased pressure on the (hopefully) more knowledgeable supervisors of newly hired professionals to mentor and monitor the activities of the new hires until they can attain the appropriate levels of knowledge and experience. If this is not done, the result will inevitably be less well-designed studies, poorly presented and interpreted data, and ineffective or inappropriate regulatory decisions.

IV. WHERE DO WE GO FROM HERE?

Although, as stated above, developmental and reproductive toxicity testing in laboratory animals remains a major and essential enterprise, especially in industry and in regulatory agencies, scientific advances in related disciplines have opened doors to complementary areas of toxicological research.[18] Studies at the biochemical, cellular, and molecular levels (e.g., Adeeko et al.[48]) have brought the promise that we will increasingly understand the mechanistic bases of manifestations of toxicity. They also bring the hope that we will become better able to predict toxicity and to extrapolate findings in laboratory animals to likely outcomes of human exposure with increasingly greater accuracy. For example, as stated by MacGregor,[49] "It is clear that genomic technologies are already being used to develop new screening strategies and biomarkers of toxicity, to determine mechanisms of cellular and molecular perturbations, to identify genetic variations that determine responses to chemical exposure and sensitivity to toxic outcomes, and to monitor alterations in key biochemical pathways." Such advances in science bring on new regulatory challenges, however, in that considerable understanding and wisdom will be required to make the best use of the escalating wealth of new data. We must also take extreme care not to lose sight of our "roots." An understanding of whole animal biology and toxicology will remain essential regardless of advances in the study of mechanisms.

REFERENCES

1. Szabo, K.T., *Congenital Malformations in Laboratory and Farm Animals*, Academic Press, San Diego, 1989, chap. 2.
2. Gilman, J., Gilbert, C., and Gilman, G.C., Preliminary report on hydrocephalus, spina bifida, and other congenital anomalies in rats produced by trypan blue, *South Afr. J. Med. Sci.*, 13, 47, 1948.
3. Hoskins, D., Some effects of nitrogen mustard on the development of external body form in the rat, *Anat. Rec.*, 102, 493, 1948.
4. Ancel, P., *La Chimiotératogenèse. Réalisation des Monstruosités par des Substances Chimiques Chez les Vertébrés*, Doin, Paris, 1950.
5. Lenz, W., A short history of thalidomide embryopathy, *Teratology*, 38, 203, 1988.
6. Wilson, J.G., The evolution of teratological testing, *Teratology*, 20, 205, 1979.
7. Kelsey, F.O., Thalidomide update: Regulatory aspects, *Teratology*, 38, 221, 1988.
8. Goldenthal, E.I., *Guidelines for Reproduction Studies for Safety Evaluation of Drugs for Human Use*, Drug Review Branch, Division of Toxicological Evaluation, Bureau of Science, U. S. Food and Drug Administration, 1966.
9. Hass, U., Current status of developmental neurotoxicity: regulatory view, *Toxicol. Lett.*, 140–141, 155, 2003.
10. Kaufman, W., Current status of developmental neurotoxicity: an industry perspective, *Toxicol. Lett.*, 140–141, 161, 2003.

11. Hood, R.D., Tests for developmental toxicity, in *Developmental Toxicology: Risk Assessment and the Future*, Hood, R.D., Ed., Van Nostrand Reinhold, New York, 1990, chap. 15.
12. Wilson, James G., *Environment and Birth Defects*, Academic Press, New York, 1973, chap. 10.
13. Wilson, James G., *Environment and Birth Defects*, Academic Press, New York, 1973, chap. 1.
14. U.S. Environmental Protection Agency, Guidelines for health assessment of suspect developmental toxicants, *Federal Register*, 51, 34028, 1986.
15. Wilson, James G., *Environment and Birth Defects*, Academic Press, New York, 1973, chap. 2.
16. Fraser, F.C., Relation of animal studies to the problem in man, in Wilson, J.G. and Fraser, F.C., Eds., *Handbook of Teratology, Vol. 1., General Principles and Etiology*, Plenum, New York, 1977, chap. 3.
17. Davidson, R.G. and Zeesman, S., Genetic aspects, in *Maternal-Fetal Toxicology, A Clinician's Guide*, 2nd ed., Koren, G., Ed., Marcel Dekker, New York, chap. 23.
18. Committee on Developmental Toxicology, Board on Environmental Studies and Toxicology, Commission on Life Sciences, National Research Council, *Scientific Frontiers in Developmental Toxicology and Risk Assessment*, National Academy Press, Washington, DC, 2000.
19. Botto, L.D. and Yang, Q., 5,10-Methylenetetrahydrofolate reductase gene variants and congenital anomalies: A HuGE review, *Am. J. Epidemiol.*, 151, 862, 2000.
20. Finnell, R.H. Teratology: General considerations and principles, *J. Allergy Clin. Immunol.*, 103, S339, 1999.
21. McLaren, A., Genetic and environmental effects on foetal and placental growth in mice, *J. Reprod. Fertil.*, 9, 79, 1965.
22. Bruce, N.W. and Abdul-Karim, R.W., Relationships between fetal weight, placental weight and maternal circulation in the rabbit at different stages of gestation, *J. Reprod. Fertil.*, 32, 15, 1973.
23. Vom Saal, F.S. and Bronson, F.H., Variation in length of the estrous cycle in mice due to former intrauterine proximity to male fetuses, *Biol. Reprod.*, 22, 777, 1980.
24. Shum, S., Jensen, N.M., and Nebert, D.W., The murine Ah locus: In vitro toxicity and teratogenesis associated with genetic differences in benzo[a]pyrene metabolism, *Teratology*, 20, 365, 1979.
25. Hood, R.D., Preimplantation effects, in *Developmental Toxicology: Risk Assessment and the Future*, Hood, R.D., Ed., Van Nostrand Reinhold, New York, 1990, chap. 6.
26. Bixler, D., Daentl, D., and Pinsky, L., Panel discussion: Applied developmental biology, in Melnick, M. and Jorgenson, R., Eds., *Developmental Aspects of Craniofacial Dysmorphology*, Alan R. Liss, New York, 1979, p. 99.
27. Wilson, James G., *Environment and Birth Defects*, Academic Press, New York, 1973, chap. 5.
28. Faustman, E.M. and Ribeiro, P., Pharmacokinetic considerations in developmental toxicity, in *Developmental Toxicology: Risk Assessment and the Future*, Hood, R.D., Ed., Van Nostrand Reinhold, New York, 1990, chap. 13.
29. Glazier, J.D., Harrington, B., Sibley, C.P., and Turner, M., Placental function in maternofetal exchange, in *Fetal Medicine: Basic Science and Clinical Practice*, Rodeck, C.H. and Whittle, M.J., Eds., Churchill Livingstone, London, 1999, chap. 11.
30. Hood, R.D., Mechanisms of developmental toxicity, in *Developmental Toxicology: Risk Assessment and the Future*, Hood, R.D., Ed., Van Nostrand Reinhold, New York, 1990, chap. 4.
31. Ranganathan, S. and Hood, R.D., Effects of *in vivo* and *in vitro* exposure to rhodamine dyes on mitochondrial function of mouse embryos, *Teratogen. Carcinog. Mutagen.*, 9, 29, 1989.
32. Ranganathan, S., Churchill, P.F., and Hood, R.D., Inhibition of mitochondrial respiration by cationic rhodamines as a possible teratogenicity mechanism, *Toxicol. Appl. Pharmacol*, 99, 81, 1989.
33. Nau, H. and Scott, W.J., Drug accumulation and pH$_i$ in the embryo during organogenesis and structure-activity considerations. Mechanisms and models in toxicology, *Arch. Toxicol., Suppl.*, 11, 128, 1987.
34. Kalter, H., The relation between congenital malformations and prenatal mortality in experimental animals, in Porter, I.H. and Hook, E.B., Eds., *Human Embryonic and Fetal Death*, Academic Press, New York, 1980, p. 29.
35. Brent, R.L., Editorial comment. Definition of a teratogen and the relationship of teratogenicity to carcinogenicity, *Teratology*, 34, 359, 1986.
36. Gaylor, D.W., Sheehan, D.M., Young, J.F., and Mattison, D.R., The threshold question in teratogenesis, *Teratology*, 38, 389, 1988.
37. Giavini, E., Evaluation of the threshold concept in teratogenicity studies, *Teratology*, 38, 393, 1988.

38. Sheehan, D.M., Literature analysis of no-threshold dose-response curves for endocrine disruptors, *Teratology*, 57, 219, 1998.
39. Calabrese, E.J. and Baldwin, L.A., The frequency of U-shaped dose responses in the toxicological literature, *Toxicol. Sci.*, 62, 330, 2001.
40. Rodricks, J.V., Hormesis and toxicological risk assessment, *Toxicol Sci.*, 71, 134, 2003.
41. Johnson, E.M. and Christian, M.S., When is a teratology study not an evaluation of teratogenicity? *J. Am. Coll. Toxicol.*, 3, 431, 1984.
42. Palmer, A.K., The design of subprimate animal studies, in Wilson, J.G. and Fraser, F.C., Eds., *Handbook of Teratology, Vol. 4., Research Procedures and Data Analysis*, Plenum, New York, 1978, chap. 8.
43. Johnson, E.M., and Gabel, B.E.G., Application of the hydra assay for rapid detection of developmental hazards, *J. Am. Coll. Toxicol.*, 1, 57, 1982.
44. Johnson, E.M., and Newman, L.M., The definition, utility and limitations of the A/D ratio concept in considerations of developmental toxicity, *Teratology*, 39, 461, 1989.
45. Hood, R.D., A perspective on the significance of maternally-mediated developmental toxicity, *Regulat. Toxicol. Pharmacol.*, 10, 144, 1989.
46. Rogers, J.M., Barbee, B., Burkhead, L.M., Rushin, E.A., and Kavlock, R.J., The mouse teratogen dinocap has lower A/D ratios and is not teratogenic in the rat and hamster, *Teratology*, 37, 553, 1988.
47. Daston, G.P, Rogers, J.M., Versteeg, D.J., Sabourin, T.D., Baines, D., and Marsh, S.S., Interspecies comparison of A/D ratios: A/D ratios are not constant across species, *Fundam. Appl. Toxicol.*, 17, 696, 1991.
48. Adeeko, A., Li, D., Doucet, J., Cooke, G.M., Trasler, J.M., Robaire, B., and Hales, B.F., Gestational exposure to persistent organic pollutants: Maternal liver residues, pregnancy outcome, and effects on hepatic gene expression profiles in the dam and fetus, *Toxicol. Sci.*, 72, 242, 2003.
49. MacGregor, J.T., Editorial: SNPs and Chips: Genomic data in safety evaluation and risk assessment, *Toxicol. Sci.*, 73, 207, 2003.

CHAPTER 2

Experimental Approaches to Evaluate Mechanisms of Developmental Toxicity

Elaine M. Faustman, Julia M. Gohlke, Rafael A. Ponce, Thomas A. Lewandowski, Marguerite R. Seeley, Stephen G. Whittaker, and William C. Griffith

CONTENTS

I. Introduction 16
   A. Definitions 16
   B. General Mechanistic Considerations..16
   C. Guidelines for Evaluating Critical Events 18
   D. Levels of Mechanistic Inquiry..20
   E. Genomic Conservation 21
   F. Processes of Organogenesis and Implication of Three-Dimensional
      Context of Evaluation...23
      1. Cell-Signaling Pathways..23
      2. Cell-Cell Communication 27
II. Examples of Mechanistic Approaches..28
   A. Mitotic Interference 29
      1. 5-Fluorouracil 29
      2. Radiation 30
      3. Antitubulin Agents 30
      4. Methylmercury...30
   B. Altered Energy Sources..32
      1. Inhibitors of Mitochondrial Respiration..32
      2. Cocaine 33
      3. Rhodamine Dyes..33
   C. Enzyme Inhibition 33
      1. Methotrexate 33
      2. Chlorpyrifos 34
      3. 5-Fluorouracil 35
      4. Mevinolin 35
   D. Nucleic Acids..35
      1. Hydroxyurea 35
      2. Cytosine Arabinoside..36
   E. Mutations 37
      1. Alkylating Agents 38
      2. Aromatic Amines 39
      3. Radiation 39

        F. Alterations in Gene Expression 40
            1. Retinoic Acid 40
            2. Dioxin 41
            3. New Directions in "Omic" Research 42
        G. Programmed Cell Death 42
            1. Retinoic Acid 43
            2. Dioxin 43
            3. Ethanol 44
III. Conclusions 46
Acknowledgments 46
References 47

I. INTRODUCTION

Daedalus, an architect famous for his skill, constructed the maze, confusing the usual marks of direction, and leading the eye of the beholder astray by devious paths winding in different directions. Thanks to the help of the princess Ariadne, Theseus rewound the thread he had laid, retraced his steps, and found the elusive gateway…

Ovid

The purpose of this chapter is to review methodological approaches for elucidating the mechanisms by which chemical and physical agents cause or contribute to dysmorphogenesis and teratogenicity. Emphasis is given to the approaches rather than to agent-specific mechanisms, and the focus is on how molecular and cellular information is combined to evaluate mechanistic hypotheses.

A. Definitions

In order to develop this paper, a few common definitions must be discussed. *Mechanisms of toxic action* is used to refer to the detailed molecular understanding of how chemicals can impair normal physiological processes and hence, produce developmental toxicity. *Mechanistic information* can include biochemical, genetic, molecular, cellular, and/or organ systems information.[1] *Mode of action* for developmental toxicants is frequently used to refer to the identification of critical steps that can explain how an agent can produce developmental toxicity and usually refers to a less detailed but more comprehensive description of the overall process of developmental toxicity. This chapter will include a discussion of approaches used for understanding mechanisms for all four endpoints of developmental toxicity: lethality, growth retardation, morphological defects (teratogenicity), and functional impacts. Throughout this chapter, the general term *developmental dynamics* is used to describe the genetic, biochemical, molecular, cellular, organ, and organism level processes that change throughout development and that define and characterize the developing organism at each life stage.[2] The term *kinetics* is used to refer to the absorption, distribution, and metabolism of chemicals because many of our discussions of developmental dynamics directly relate to the amount and form of the environmental or pharmacological agent that reaches the developing organism.

B. General Mechanistic Considerations

Figure 2.1 shows an overall framework for assessing the effects of a toxicant on development.[1,3] This figure illustrates how both kinetic and dynamic considerations are needed to link exposure with developmental outcome. It provides a context for collecting mechanistic data and for ordering sequence of events data that structures a mode of action hypothesis. This framework has been

EXPERIMENTAL APPROACHES TO EVALUATE MECHANISMS OF DEVELOPMENTAL TOXICITY

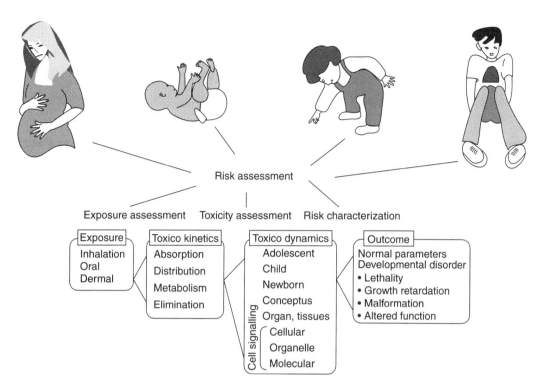

Figure 2.1 Overall framework for assessing the effects of a toxicant on development. (Adapted from National Research Council, *Scientific Frontiers in Developmental Toxicity Risk Assessment,* National Academy Press, Washington, 2000, p. 354 and from Faustman, E. et al., *IRARC Technical Report on Developmental Toxicity,* Institute for Risk Analysis and Risk Communication, University of Washington, Seattle, WA, 2003. With permission.)

Table 2.1 Example mechanisms for developmental toxicity

Mitotic interference	Altered nucleic acid synthesis
Altered membrane function or signal transduction	Mutations
Altered energy sources	Gene and protein expression changes
Enzyme inhibition	Alterations in programmed cell death

modified from the original National Academy of Sciences (NAS) framework to illustrate that the affects of exposure during development can occur *in utero*, in newborns, in childhood, and in early adulthood. Manifestations of early exposures sometimes cannot be observed until adulthood.[4]

The study of mechanisms of toxicity is of vital importance not only for the insights provided into the events underlying adverse developmental outcomes, but also for the information gained concerning the processes involved in normal development. Recently, there has been an increased interest in mechanistic information as a result of legislative actions. For example, in the Food Quality Protection Act, exposure to agents that have common mechanisms of action should be considered in a cumulative manner. This has led to the joint evaluation of organophosphates in regard to their developmental neurotoxicity.

Table 2.1 includes a list of potential mechanisms first proposed by Wilson[5] that included the following general categories: mitotic interference, altered membrane function or signal transduction, altered energy sources, enzyme inhibition, altered nucleic acid synthesis, and mutations. Because these processes play essential roles in embryogenesis and normal development, it is logical to expect that alterations may result in developmental toxicity, and the research literature is replete with proof of this assumption. With our increased understanding of the molecular mechanisms

Table 2.2 Guidelines for assessment of proposed mechanistic pathways in chemically induced developmental toxicity

Temporal association	Does developmental toxicity precede, occur simultaneously with, or follow the initial event?
Dosage relationship	Does the potential mechanistic event occur at or below those doses that result in the developmental toxicity?
	Is there a dose-response relationship between exposure and severity of the developmental outcome?
Structure-activity relationships	Do chemicals with similar structures cause similar developmental outcomes?
Strength of association	Is the proposed mechanistic process strongly or weakly linked to the appearance of the developmental outcome?
Consistency of association	Are the proposed mechanistic processes required for the appearance of the developmental outcome?
	Does modification of the mechanistic event, or of one step in the process, alter or eliminate the adverse developmental outcome?
Coherence	Is there a molecular basis for the proposed mechanism of action by the chemical or physical agent that elicits the initial event?

underlying normal development, we can now propose additional mechanisms, including perturbations in gene and protein expression and programmed cell death. However, even with inspection at this more basic level of action, only partial segments of the mechanistic path from initial insult to dysmorphogenesis are understood for even the most well-characterized developmental toxicants. In most circumstances, only phenomenological information is available.[1]

C. Guidelines for Evaluating Critical Events

This chapter emphasizes the need to consider the underlying causes of developmental toxicity rather than relying on phenomenology. To accomplish this, a series of guidelines has been developed to ascertain the significance of postulated critical events in the mechanistic pathway leading to an adverse developmental outcome. Such guidelines need to be considered when evaluating the validity of the relationship between an initial toxic insult and teratogenicity. These guidelines have been developed from basic pharmacological principles of drug action[6] and from epidemiological approaches, such as the Bradford Hill criteria of causality.[7] The general adaptability of these approaches is evidenced by their recent use in the International Programme on Chemical Safety (IPCS) Harmonization Approaches for Risk Assessment[8] and in EPA's Carcinogen Risk Assessment Guidelines.[9] Table 2.2 lists these assessment guidelines.

The first guideline is the issue of *temporal association*. In this assessment, the question is the temporal association between any altered development and the potential initial mechanistic event. Obviously, for this event to be a critical event in the process of developmental toxicity that event must precede or occur simultaneously with the pathology. Complex temporal relationships associated with development can often complicate analysis, and events that might be labeled as *nontemporally associated* are missed if the biology of developmental processes is not a prime factor in reviewing temporal associations.

Because of the hierarchical nature of tissue organization in the developing organism, patterns of affected tissue can provide important temporal mechanistic clues. For example, Figure 2.2 illustrates that if dysmorphogenic alterations are observed in cardiac cells and sensory and stomach epithelia (i.e., endoderm, mesoderm, and ectoderm), then events occurring during blastula and gastrula stages might be the first to be evaluated as potential critical events for the mechanism of dysmorphogenesis linking these three responses. Nevertheless, later events occurring separately in each of these tissue types (such as changes in cell proliferation or changes in receptor expression) could also explain the common responses in these three tissues. Thus, such temporal associations can be used as initial clues, but mechanistic investigation must always be open to multiple explanations for the same response.

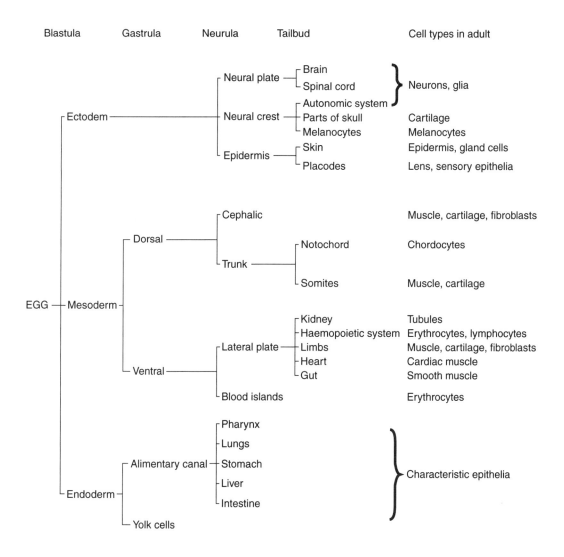

Figure 2.2 The hierarchical embryonic origins of tissues and cells within the vertebrate embryo. (From Slack, J., *From Egg to Embryo: Regional Specification in Early Development,* 2nd ed., Cambridge University Press, Cambridge, 1991, p. 348. With permission.)

The second guideline concerns the establishment of a *dose-response relationship* for the proposed critical mechanistic events. If exposure to a suspected developmental toxicant produces a dose-related increase in malfunctions, then the possibility that the chemical is a developmental toxicant, causing the adverse outcome, is strengthened. Lack of a dose-response relationship, however, does not rule out the possibility that the suspected toxicant is a developmental toxicant. For example, Selevan and Lemasters[10] have dramatically illustrated the concept of competing risk. Although there might not be an observable, dose-related increase in malformations, the effects of the suspected developmental toxicant may be to increase embryolethality. Thus, at higher doses, fewer embryos survive to manifest increased malformations.

The third guideline for assessing proposed mechanisms of action for developmental toxicants relates to *structure activity relationships* (SARs) and whether they exist for the compound under investigation. Good SAR examples for developmental toxicants are published in the literature for alkylating agents,[11–13] retinoic acid derivatives,[14,15] alkoxy acids,[16] short-chain carboxylic acids

Table 2.3 Levels of mechanistic inquiry for developmental toxicity

Intracellular events	Biochemical and molecular mechanisms of action define key intracellular events for both normal and abnormal developmental responses.
Intercellular events	Specific cell-cell interactions and activities define activities for specialized cell populations.
Organ level events	Specialized functions of organs define organ development and function.
Organism level events	Embryonic and fetal responses are defined by the collective responses of organ and intra- and intercellular events.
Litter responses	The combined embryo and fetal responses of a litter are defined within the single maternal unit.

(valproic acid derivatives),[17] and phenols.[18] Note that many of these SARs are determined using *in vitro* as well *as in vivo* investigations to establish relationships.

The fourth guideline concerns the *strength* and *consistency of occurrence* of the adverse outcome with the postulated critical mechanistic event. For example, if the proposed mechanisms of action for compound X is that it inhibits neuronal cell division by 50% in the brain during early brain formation, causing microcephaly, then to determine the strength of this mechanistic association, two types of model experiments could be planned. First, the investigator could determine if other agents that cause a comparable decrease in cell division at this time in development also cause microcephaly. Secondly, the investigator could see if blocking the effects of compound X on cell division would reverse the incidence of microcephaly. Cross-species extrapolation of results could also increase the consistency of these observations as key mechanistic processes. Although these example observations provide clues, failure to see these changes does not mean that the mechanistic hypothesis must be abandoned.

The last guideline relates to the importance of *coherence* in the overall mechanistic hypothesis. If a possible molecular or cellular basis can be described for the proposed mechanism of action, then this coherence provides a stronger degree of confidence in the postulated pathway. If no molecular basis is found, then the proposed mechanism may have a difficult battle for acceptance because it may be a mechanism whose conception may have outpaced related molecular experimentation.

D. Levels of Mechanistic Inquiry

Our current lack of understanding of the events underlying teratogenicity reflects the complexity of the developmental process. It may never be possible to describe every molecular or cellular event that ultimately leads to dysmorphogenesis. However, of the myriad of potential effects elicited by chemical and physical agents on embryonic development, it is probable that only a relative few represent critical events responsible for developmental toxicity. Therefore, it is essential to identify the key events, based on an understanding of the toxicological properties of the agent and the biological processes involved. To gain an understanding of developmental toxicity, investigations must focus on multiple biological levels. The initial molecular and subcellular events must be defined along with key processes occurring at the cellular, tissue, and organ level. Investigations at the organ system and organism level are then required. Table 2.3 lists these levels of mechanistic inquiry for developmental toxicity. Recognition of these levels is critical for several reasons. First, mechanisms are frequently defined at only a single level. Thus, a cellular mechanism of action will be defined in isolation from events occurring at higher levels of organization. Later investigators working only at the fetal level may dismiss these mechanistic observations because strict temporal or dose-response relationships may be unclear at the higher level of investigation. However, if both levels are examined, both observations can be confirmed and a broader appreciation for the mechanistic complexities involved can be realized. The possibility of multiple mechanisms should be recognized; an adverse developmental outcome will probably not be attributable to a single event but rather to a cascade of events.

EXPERIMENTAL APPROACHES TO EVALUATE MECHANISMS OF DEVELOPMENTAL TOXICITY

Table 2.4 Critical intercellular signaling pathways important for development[a]

Period during Development when Signaling Pathway Is Used	Pathway
Early (axis specification, germ layer specification, left-right asymmetry) and continued in all later stages.	1. Wnt pathway 2. Hedgehog pathway 3. TGF receptor (ser/thr kinase) pathway 4. Receptor tyrosine kinase (small G protein) pathway 5. Notch/Delta pathway 6. Cytokine receptor (cytoplasmic tyrosine kinases) JAK/STAT pathway
Middle (during organogenesis and cytodifferentiation) and continued in all later stages.	7. IL1/Toll NFkB pathway 8. Nuclear hormone receptor pathway 9. Apoptosis pathway 10. Integrin pathway 11. Receptor phosphotyrosine phosphatase pathway
Late (after cell types have differentiated). Used in fetal, larval, and adult physiology.	12. Receptor guanylate cyclase pathway 13. Nitric oxide receptor pathway 14. G-protein coupled receptor (large G protein) pathway 15. Cadherin pathway 16. Gap junction pathway 17. Ligand-gated cation channel pathway

[a] The mammalian fetus uses all 17 pathways.

Source: Adapted from National Research Council, *Scientific Frontiers in Developmental Toxicity Risk Assessment,* National Academy Press, Washington D.C., 2000, p. 354.

E. Genomic Conservation

Inherent in most toxicological studies is the premise that chemically induced effects in animals are predictive and instructive for understanding the potential for a chemical to alter development in humans. Recent advances in developmental and molecular biology make these assumptions even more important. One of the most exciting advances is the determination that most of development is controlled by approximately 17 cell-signaling pathways and that these signaling pathways are genomically conserved.[1] These pathways are listed in Table 2.4.

A generalized signal transduction pathway is the basic mechanism underlying each of these cell-signaling pathways. This multistep process is composed of a series of switchlike intermediates that are activated by a receptor-mediated signal, which ultimately activates a protein kinase. The target protein is hence phosphorylated and is either activated or inactivated. Target proteins in these signaling cascades include proteins that are integral to processes of transcription, translation, cell cycling, cell migration, differentiation, etc. Fourteen of the cell-signaling pathways shown in Table 2.4 involve transmembrane receptors and two involve intracellular receptors.[1]

These findings on genomically conserved pathways have had some important implications for improved approaches for mechanistic studies. First, the function of many of these pathways across model organisms has been determined by transgenic studies. Genes for the cell-signaling processes have been cloned, and a variety of transgenic technologies have been applied to evaluate their significance. Knockout and/or null mutations, overexpression, or miss-expression of genes can be studied and used to evaluate the role that these specific genes and signaling pathways may play in development. Table 2.5 shows examples of such transgenic studies, where the phenotypes of mouse mutants lacking components of specific cell-signaling pathways are evaluated. For example, if the *Wnt-1* pathway is knocked out, the offspring are viable to adulthood, but no midbrain, cerebellum, or rhombomere 1 is present. Mice obviously also display behavioral defects.[1,19] Likewise, if Sonic hedgehog (*Shh*) is knocked out, perinatal death is observed and embryos have evidence of cyclopia and spinal cord, axial skeleton, and limb defects.[1,20]

Table 2.5 shows results from transgenic animal studies for selected key cell-signaling pathways; however, there are several concepts that are not captured in this table. One issue is that

Table 2.5 Example phenotypes of mouse mutants lacking components of signaling pathways

Signaling Component	Variability of Null Mutant	Phenotype of Null Mutant	Ref.
Wnt Pathway			
Wnt-1	Adulthood	No midbrain, cerebellum and rhombomere 1; behavioral deficits	19, 320
Transforming Growth Factor β Pathway			
TGFβ 1	Adulthood	Immune defects, inflammation	297
TGFβ 2	Perinatal death	Defects of heart, lung, spine, limb and craniofacial and spinal regions	298,320, 321
BMP5	Adulthood	Thin axial bones; abnormal lung, liver, ureter, and bladder; like a short ear mutant	299
BMP7	Adulthood	Defects in eye and kidney; skeletal abnormalities; hindlimb polydactyly	300, 323
Hedgehog Pathway			
Sonic hedgehog	Perinatal death	Cyclopia, defects of spinal cord, axial skeleton, and limbs	20
Patched receptor	Homozygotes, early lethality	Open neural tube	301
	Heterozygotes, adulthood	Rhabdomyosarcomas, hindlimb defects, large size. Like Gorlin syndrome in humans	302, 324
Nuclear Receptor Pathway			
Progesterone receptor	Adulthood	Males normal; females: no ovulation, no mammary glands, uterine hyperplasia	297,303, 304
Retinoic acid receptor β	Adulthood	Fertile, small size, transformations of cervical vertebrae	305
RARα and RARβ	Perinatal death	Visceral abnormalities, reduced thymus and spleen	305

Source: Adapted from National Research Council, *Scientific Frontiers in Developmental Toxicity Risk Assessment,* National Academy Press, Washington D.C., 2000, 354. With permission.

as organisms become more complex, there is increasing redundancy in the downstream pathways of the 17 cell-signaling processes. If extensive redundancy exists, then interpretation of the significance of specific pathways in transgenic animal studies is complicated. For example, in many cases where high levels of redundancy exist in a process, knockout animals will have minimal phenotypic changes. These may result from the fact that in rodent models there may be multiple ligand genes (i.e., 24 TGFβ genes and 11 *Wnt* ligand genes in mice versus 3 to 5 TGFβ genes and 1 to 3 *Wnt* genes in *Drosophila*).[1] In addition, when a single gene is knocked out in mice, the developmental defect may be very subtle since that specific gene may only be expressed in a very narrow temporal and tissue-specific context where no related genes are expressed to provide redundant function.

These transgenic models are of methodological use for studying developmental toxicity in several ways. First, many researchers compare the phenotype of transgenic mouse models with the phenotype that can arise from treatment of animals with developmental toxicants. Such an approach was useful for studies of *Veratrum* alkaloids, where cyclopamine produced cyclopia in livestock that fed on plants containing *Veratrum* alkaloids. Molecular investigations of cyclopamine have revealed that it can interfere with *Shh* signaling.[21–23] Mouse knockout studies showed that genetic manipulation of the *Shh* pathway could result in the same types of defects.

Other ways that developmental toxicants have been found to act through conserved cell-signaling pathways are portrayed in Table 2.6, which shows a list of examples of receptor-mediated developmental toxicity. These include both environmentally relevant examples (i.e., TCDD and cyclopamine) and examples of interest from medicinal chemistry (retinoic acid and diethylstilbestrol [DES]). In all of these cases, knowledge about the normal role of the receptor aided the mechanistic studies of nonendogenous ligands.

In some cases, a developmental toxicant can interact directly or indirectly with a receptor to activate a receptor inappropriately (agonist) or inhibit signaling via the normal ligand (antagonist). If the response of a developmental toxicant or a receptor is to activate the receptor but to produce a less than maximal response, that agent is known as a partial agonist. If the developmental toxicant can cause a decrease in activity from baseline, then the agent can be considered a negative agonist. Hence, mechanistic approaches for evaluating receptor-ligand mediated developmental toxicity can include assessment of receptor binding, but more importantly, a quantitative characterization of the type of response when the developmental toxicant binds is usually more informative. Pioneering work by Nebert showed the importance of Ah receptors (AhR) in mediating polycyclic aromatic hydrocarbon-mediated developmental toxicity.[24,25] This was demonstrated with genetically different strains of Ah-responsive and nonresponsive mice.

Other examples included in this table are the myriad of studies on retinoic acid receptors where good structure-activity relationships are available for developmental toxicity because of the production of numerous candidate drugs. Depending upon receptor subtype, timing of exposure, and dose, almost all organ systems can be affected via this receptor interaction.[1]

F. Processes of Organogenesis and Implication of Three-Dimensional Context of Evaluation

1. Cell-Signaling Pathways

The importance of evaluating cell-signaling pathways within the overall context of organogenesis is elegantly illustrated in our knowledge of limb development from over 50 years of experimental embryology and recent intense molecular developmental biology studies.[1]

Figure 2.3 shows the development of limb buds in vertebrates (tetrapods) and shows the complex interactions of precise temporal and spatial signaling that is required for organ development. This figure also shows the multitude of signaling pathways controlling proliferation that are required to establish the three-dimensional morphology of this organ.[1,26,27] For example, to determine the possible mechanistic ramifications of chemically induced changes in *Shh* for limb bud development, one would need to understand the impacts this signal would have initially on *BMP2* signaling and cell proliferation activity in the bud mesoderm, and subsequently on the overall anterioposterior gradient that determines bud extension. The dorsal epidermis secretes Wnt-7a onto the mesenchyme, where it induces expression of the *LMX-1* gene and suppresses the *engrailed-1* gene. If Wnt-7a signaling is defective, double ventral limbs can form.[1]

These changes would then need to be evaluated in terms of possible changes in the dorsal ventral gradient and possible alterations in *Wnt-7a/engrailed* gene expression. Such three-dimensional evaluations are difficult to impossible to glean from simple cell experiments. They require model systems that have complex three-dimensional cell-signaling interactions. Mechanistic clues to examine *Shh* or *Wnt-7a* cell-signaling pathways can come from more simplified systems when relevant dose-response relationships are established. It has been postulated that thalidomide can act by reducing cell proliferation in the progress zone (PZ), and this can result in prolonged contact of cells with fibroblast growth factor (FGF) secreted by the apical ectodermal ridge (AER).[1,28] Other researchers have postulated that thalidomide can interfere with integrin gene expression, hence inhibiting angiogenesis and subsequently proliferation.[29] These are just two of the myriad of proposed hypotheses for the action of thalidomide; however, knowledge of normal developmental biology and of these complex signaling processes allows one to set up studies to methodologically and quantitatively determine the validity of such mechanisms. Using the guidelines presented in

Table 2.6 Examples of receptor-mediated developmental toxicity

Receptor (Official Name[a])	Endogenous Ligands	Developmentally Toxic Ligand and Modifier	Typical Effects	Recent References
Basic Helix-Loop-Helix Transcription Superfamily				
Aryl hydrocarbon AHR	Unknown	Agonists: TCDD and related polycyclics	Cleft palate, hydronephrosis	25, 31, 306, 325
Nuclear Hormone Receptors				
Androgen AR (NR3C4)	Testosterone Dihydro-testosterone	Agonists: 17 a–ethinyl-testosterone and related progesterones Antagonists:[b] Flutamide	Agonists: Masculinization of female external genitals Antagonists: Inhibition of Wolffian duct and prostate development and feminization of external genitals in males	307
Estrogen ERa, ERb (NR3A1 and 2)	Estradiol	Agonist: DES Antagonist: tamoxifen, clomiphene—weak	Agonist: various genital-tract defects in males and females	308, 326
Glucocorticoid GR (NR3C1)	Cortisol	Agonists: cortisone, dexamethasone, triamcinolone	Cleft palate	309, 327
Retinoic acid RARα, β, and γ (NR1B1, 2, and 3)	All trans and 9-cis retinoic acids	Agonists: numerous natural and synthetic retinoids	Almost all organ systems can be affected	226, 310–312, 328
RXRα, β, and γ (NR2B1, 2, and 3)	9-cis retinoic acid	Antagonists: BMS493, AGN 193109, and others		
Thyroid hormone TRα and β (NR1A1 and NR1A2)	Thyroxine (T4 and T3)	Antagonist: nitrophen	Lung, diaphragm, and harderian-gland defects	313

Hedgehog receptor

Patched	Sonic, Desert, and Indian Hedgehogs	Veratrum alkaloids: cyclopamine (mechanism unclear)	Cyclopia, holoprosencephaly	21, 22, 329

Membrane

Endothelin receptors A and B	Endothelins 1, 2, and 3	Antagonists: L-753, 037, SB-209670, SB-217242	Craniofacial, thyroid, and cardiovascular defects, intestinal aganglionosis (Hirschsprung's disease)	314–316, 330, 331

Cation Channels

Delayed-rectifying IKr	Potassium ion	Inhibitors: almokalant, dofetilide, d-sotalol	Digit, cardiovascular, orofacial clefts	317, 318

[a] Nuclear Receptors Committee 1999.
[b] Also, 5-alpha reductase inhibitors (e.g., finasteride) affect prostate and external genitals.[319]

Source: Adapted from National Research Council, *Scientific Frontiers in Developmental Toxicity Risk Assessment*, National Academy Press, Washington D.C., 2000, 354. With permission.

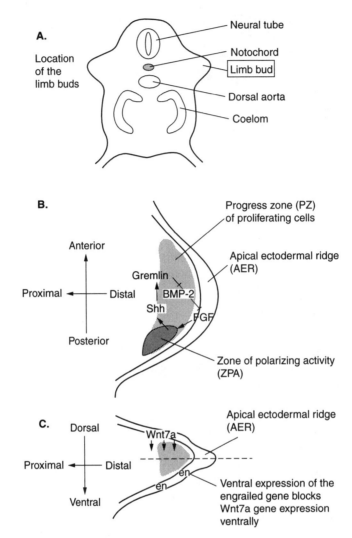

Figure 2.3 Three-dimensional development of vertebrate limbs (tetrapod). This figure shows the three axes of limb development for four legged vertebrates: anteroposterior, proximodistal and dorsoventral. (A) Location of limb buds in relationship to an overall cross-sectional view of vertebrate development. (B) Cross-section of a limb bud showing the anteroposterior and dorsoventral axis. The location of the zone of polarizing activity (ZPA) is shown, as is the apicalectodermal ridge (AER) and the progress zone (PZ) of proliferating cells. The signaling feedback between Sonic hedgehog (*Shh*), gremlin, BMP2, and FGF is illustrated. (C) Dorsoventral and proximodistal axis and the relationship of *engrailed* (en) gene and its inhibition of *Wnt7a* gene expression ventrally. (Adapted from National Research Council, *Scientific Frontiers in Developmental Toxicity Risk Assessment*, National Academy Press, Washington, D.C., 2000, p. 354. With permission.)

Table 2.2, researchers can determine, for example, if the temporal relationship, dose, and consistency in changes in cell proliferation of the PZ correlate with the incidence of phocomelia.

Likewise, researchers can determine the temporal relationship, dose, and consistency of specific limb abnormalities by integrating gene expression changes induced with thalidomide. Because of the cross-species similarity in appendage or limb patterning, other approaches for mechanistic studies could include looking at the consistency of these observations in *Drosophila* and chick embryos, as similar conserved gene domains exist. For example, similar domains exist for arthropods and chordates for *WG-HH-DPP* and *Wnt-Shh-BMP* as do similar expressions of *En*, *Ap*, and *LMX* selector genes.[1] One caution that must be kept in mind when performing cross-species experiments with thalidomide is that although the cell-signaling processes controlling the developmental dynamic processes may be conserved, dissimilar conservation of drug metabolism genes in these various species may result in toxicokinetic differences. Hence, the use of "active metabolites" and correction for dose-to-target concentrations of thalidomide and its metabolites may be very important for such mechanistic determinations. Such investigations would help to weave together significant mechanistic clues for thalidomide effects on limb development.

2. Cell-Cell Communication

The interaction between cells and their environment during fetal development plays a significant role in both cell differentiation and morphogenesis. This interaction leads to information transfer and can occur through direct cell-cell contact, through the activation of membrane-bound cellular receptors, or through the associations created between a cell and the surrounding extracellular matrix. Cell fate determination by specific cell interactions of the cell and its surrounding environment is conditional specification and is dependent upon extracellular factors.[30] If the extracellular signal causes a specific manifestation of one differentiation fate over others, the process is referred to as *instructive induction*. In contrast, the other form of conditional specification, *permissive induction*, results from a cell that already is committed to a specific differentiation path but only expresses this differentiation phenotype after exposure to a signal. A good example of this is seen in limb development (see Figure 2.3) where a complex three-dimensional morphogenic signaling gradient is established across the limb bud.[1,30]

Appositional induction results from tissue interactions where signaling and responding tissues come together and a common response is induced in the contact region. The fundamental importance to normal development of this system was recognized by Wilson[5] in a discussion on the role of altered cell membrane function as a contributor to teratogenesis. Under the most severe chemical exposure conditions, altered membrane integrity will likely result in cytolysis and cell death. This occurs primarily as a result of the inability of a cell to maintain a normal physiologic ionic balance and osmolarity due to an altered membrane permeability. Thus, at this extreme, monitoring cytotoxicity is an indirect and nonsubtle measure of altered membrane integrity. Less extreme examples can result in functional changes, such as the excess cell proliferation seen at the fusion points in TCDD-exposed palatal shelves.[31-33]

Although much mechanistic information is lacking, there is some understanding of the role of altered membrane function and the importance of intracellular communication, cell adhesion, cell migration, cell shape, and cellular receptors in developmental toxicology. Of particular interest are recent investigations into the role of adhesion molecules on normal membrane function in cell migration.

A key cellular process occurring during differentiation is the migration of cells to new locations. Examples of this include: neural crest cells, which develop into a variety of cell types; precardiac mesodermal cells, which form the heart; and neurons, which migrate to various regions of the cerebrum from the ventricular zones following cell division.[34] In general, cell migration relies heavily on cell-cell interaction, particularly cell adhesion. For example, the migrating neuron is believed to receive guidance cues from the extracellular environment and the extracellular matrix, as well as through contact between the neuron and supporting glial and neuronal cells.[35-46] When these cues are altered by changing the chemical composition of the extracellular matrix, dramatic changes in differentiation patterns occur; this is one technique that can be used to define the essential nature of this matrix for developmental processes.

Intracellular cytoskeletal components, such as microtubules and actin, provide a structural basis for migration,[47,48] while extracellular cell adhesion molecules provide support and guidance for the migrating cell by offering a preferred substrata.[35,49] Extracellular adhesion molecules are also likely to be involved in signal transduction, which results in directional guidance cues and cell motility.[48] The proper functioning of the intracellular cytoskeleton and extracellular adhesion molecules is crucial for normal nervous system development; alterations in either may result in a variety of cortical malformations.[43] For example, in cleft palate, faulty adhesion may underlie the failure of epithelial fusion even though the palatal shelves may be in close apposition.[35,50]

Adhesion molecules are involved in numerous neurodevelopmental processes, including neurite outgrowth (integrin, neural cell adhesion molecule [N-CAM], N-cadherin, L1), peripheral nerve regeneration (L1, N-CAM), nerve target adhesion (N-CAM), regulation of intracellular and extracellular ionic composition (adhesion molecule on glia, AMOG), cell-cell stabilization (N-CAM 180), and others.[48] Alterations to these proteins, therefore, may play a role in toxicant-induced cortical dysfunction. For example, N-CAM, an adhesion molecule involved in synaptogenesis and migration, undergoes maturation during development from a sialic-acid–poor form in early embryogenesis to a sialic-acid–rich form, and finally to a sialic-acid–poor form again in the mature adult. Alterations to the sialylation state of N-CAM occur during cell migration and synaptogenesis.[51] Low-level lead exposure has been associated with an inhibition of normal developmental N-CAM desialylation,[52] and it has been proposed based on dose and time exposure studies that this lead-induced alteration to the normal maturation of N-CAM results in the observed abnormal synaptogenesis in the developing central nervous system.[51] Similar alterations to N-CAM maturation have also been reported following exposure to other metals, such as methylmercury;[53] these findings support a role for faulty adhesion molecule function in methylmercury-induced neuronal ectopia.

Another developmental process that relies heavily on a normal membrane function is the extension of neurites by the young neuron during the formation of a synaptic network. Neuronal fasciculation is accomplished through the activity of a motile growth cone found on the extremity of the neurite. As the growth cone advances away from a relatively stationary cell body, the axon is developed. Termination of growth cone activity may occur through contact inhibition involving cell-to-cell communication, but this process is poorly understood.[54] Therefore, chemical agents that alter cell-cell communication may disturb synaptogenesis and perhaps other developmental processes, such as cell proliferation and differentiation. For example, Trosko and colleagues[55] have forwarded a series of arguments regarding the role of abnormal cell-cell communication in teratogenesis. In this work, they summarized evidence that gap junctional communication is involved in normal differentiation and development, and extend the hypothesis that a disruption in this communication during histo- or organogenesis will result in pathology. The identification of the integrin, cadherin, and gap junction pathways in the 17 key cell-signaling pathways that define development supports these concepts and highlights the importance of these processes in normal development.

In summary, the cell membrane is a key site where developmental toxicants may act. However, there is very little direct evidence implicating altered membrane function as the critical step in the etiology of developmental toxicity. Additional approaches to ascertain the role of cell communication are needed.

II. EXAMPLES OF MECHANISTIC APPROACHES

The following review is organized according to principal mechanisms by which chemical and physical agents are thought to elicit dysmorphogenesis and developmental toxicity. Examples are provided of chemical and physical agents that have been shown to be associated with the mechanism being discussed, and methodological approaches used to make those associations are highlighted. The purpose of this chapter is not to be all-inclusive in terms of the mechanistic knowledge for the developmental toxicants. Rather, the purpose is to highlight several of the proposed mechanistic hypotheses for developmental toxicity and to highlight examples of the types of experimental approaches employed and the results that were generated to test these hypotheses.

In this review, we will examine approaches used for evaluating mitotic interference, altered cell signaling, enzyme inhibition, mutation, alterations in gene expression, and programmed cell death. In a few cases where detailed kinetic and dynamic information is available, some explanation on how these considerations can add to our mechanistic evaluations has been included. Information is provided that illustrates the guidelines discussed in this introductory section. This approach will provide tools for the reader to critically evaluate mechanistic information.

A. Mitotic Interference

Normal fetal development is characterized by rapid and coordinated cell replication. Thus, it follows that mitotic interference, defined as a change in the rate of cell proliferation,[5] is a potential mechanism underlying chemically induced developmental defects. This rapid and specific cell proliferation confers a unique sensitivity of the fetus toward agents affecting cell division processes. Differential rates of cell division within developing tissues or organs may create specific subsets of cells that are especially sensitive to chemical exposure.[56] Mitotic interference is most commonly elicited by chemical or physical agents that delay or block cell cycling. However, an increase in the cell cycle rate due to compensatory repair following an exposure might also contribute to a teratogenic outcome.[57] The nature of the pathological outcome and the ability of the organism to compensate for the damage will depend upon, among other factors, the nature of the exposure, the developmental stage of exposure, and the affected tissue or cell types. Because normal morphogenesis depends upon a highly synchronized progression of events, a reduction in the total number of cells or a delay in the production of cells as a result of cell cycle arrest or inhibition can have long-term and irreversible consequences. For example, the rate of neuroblast proliferation may be an important determinant of cerebral organization and synaptic network formation,[58] and cell cycle delays may lead to altered synaptogenesis and an altered cortical cytoarchitecture. Additionally, the normal development of certain tissues may depend on attaining a critical cell mass for the progress of normal cell differentiation and organogenesis;[59] an inhibition or delay in the proliferation of these tissues may result in developmental toxicity. In both models, normal morphogenesis relies on the normal progression of cell division.

General mechanisms have been identified that result in an altered cell cycle rate. These include: (1) reduction of DNA synthesis; (2) interference in the formation or separation of chromatids; and (3) failure to form or maintain the mitotic spindle.[5] Critical cell-signaling pathways for controlling cell cycle pathways (i.e., p21 and p53) have also been identified. The mechanisms of mitotic inhibition for several example developmental toxicants have been described. Agents that affect cell cycling through these pathways are discussed.

1. 5-Fluorouracil

Developmental toxicants have been identified that may act through inhibition of DNA synthesis. These include hydroxyurea, cytosine arabinoside, and 5-fluorouracil. 5-Fluorouracil (5-FU) exposure is associated with morphologic abnormalities in several animal species.[60,61] The S-phase inhibition elicited by 5-FU exposure,[62] via inhibition of thymidylate synthetase, is well documented. However, because 5-FU is also incorporated into RNA, resulting in cell death, it is difficult to attribute 5-FU–induced teratogenicity solely to an inhibition of DNA synthesis. In fact, depletion of cytidine, rather than thymidine, may be the proximate cause of interrupted cell division.[63] It is unclear how the incorporation of 5-FU into DNA affects DNA fidelity or cell viability, and whether these effects are associated with teratogenicity (for a review of these mechanisms see Parker and Cheng[64]). The effects of 5-FU on cell cycling and viability are dependent on the cell type, cell cycle phase, and exposure conditions,[62] which implies a differential sensitivity among developing tissues. Coordinated biochemical, cellular, and morphological studies conducted by Shuey et al.[65] demonstrate successful mechanistic approaches to understanding 5-FU–induced developmental toxicity. These studies revealed that the levels of incorporation of 5-FU derivatives into nucleic acids and direct interactions with DNA were too low to explain the resultant developmental toxicity. Subsequent mechanistic studies of 5-FU have focused on inhibition of thymidylate synthetase and resultant cellular perturbations. (See discussion below on enzyme inhibition and how these mechanistic studies were incorporated into a biologically based dose-response model.) These mechanistic studies thus reveal the need for researchers to shift their investigations from one biological level — DNA and molecular level changes — to assessments at the cellular level. They might thus identify critical rate-limiting processes that could more effectively explain the overall complex processes of nucleotide pool alterations and subsequent DNA synthesis perturbations.

2. Radiation

The effects of radiation on fetal development, including the developmental effects of radiation-induced cell loss, have been well documented.[58,66–71] The effect of radiation on chromatid formation and cell cycling likely underlies radiation-induced cell cycle perturbations.[5] However, the varied effects of radiation on DNA fidelity make it difficult to attribute radiation-induced teratogenicity solely to effects on cycling cells. Among the DNA effects showing dose-response relationships following radiation exposure are chromosome instability, single strand breaks, double strand breaks, DNA-protein cross-links, apoptosis, p53 activity, and mitotic inhibition.[72–75] The most sensitive period for radiation-induced malformations is following day eight of organogenesis in the rat (corresponding to weeks 8 to 25 in the human fetus), during the period of maximal proliferation of neuronal precursors. Exposure (e.g., 100 Rd in the rat) prior to organogenesis can cause either fetal death or have no effect. Exposure during organogenesis results in decreased weight and thickness of cortical layers, formation of ectopic structures, and microcephaly. CNS effects are noted if exposure takes place late in gestation, reflecting the extended period of sensitivity of nervous system development (see reviews by Brent,[76] Beckman and Brent,[77] and Kimler[67]).

3. Antitubulin Agents

Perturbations in the mitotic spindle may result in cytoskeletal disruption, aneuploidy, micronuclei, alterations in cell division rate, cell cycle arrest, and/or cell death. The coincidence of these types of toxicological manifestations is indicative of antimitotic agents that affect tubulin.[78] Classic antitubulin agents, such as benzimidazoles, carbamates, and colchicine, have been demonstrated to elicit aneuploidy[79–86] and developmental toxicity both *in vivo* (reviewed in Delatour and Parish,[87] Ellis et al.,[88] and Van Dyke[89]) and *in vitro*.[78,90] These studies used dose-response relationships and structure-activity information to strengthen the support for mitotic perturbations as the mechanism of action by which these agents cause developmental toxicity. Other methodological approaches used in these studies were comparisons across species that revealed significant cross-species differences in tubulin binding affinities that were related to their differing potency as toxicants in different species.

4. Methylmercury

Numerous mechanisms have been proposed by which methylmercury (MeHg) may disrupt normal cell function, and these mechanisms are postulated to result in neurodevelopmental toxicity. Most, if not all, are associated with the exceptional affinity of MeHg for the thiol group, with the association constant of the Hg-SH pair being orders of magnitude greater than MeHg's interaction with any other ligand.[91] MeHg may interfere with the proper functioning of cells by disrupting the thiol bond–mediated structure and the function of key proteins or other molecules. While this idea is mechanistically a very simple concept, the numerous affected pathways may include (1) disruption of protein synthesis,[92–94] (2) disruption of cellular energy production,[95–97] (3) disruption of intracellular calcium levels,[98,99] particularly in the mitochondria, (4) disruption of microtubule assembly and cellular division and transport,[100–104] and (5) induction of oxidative stress, either

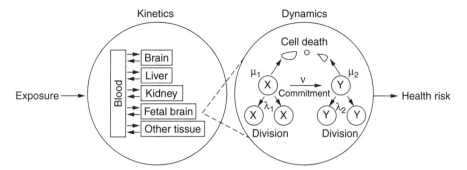

Figure 2.4 Example of a combined toxicokinetic and dynamic model for methyl mercury exposure during pregnancy. Exposure may occur through ingestion, inhalation, or dermal absorption, through which the toxicant rapidly enters the bloodstream. Distribution throughout the body (toxicokinetics) determines the dose to the conceptal brain, affecting cell proliferation, differentiation, and death (toxicodynamics) and the risk of developmental neurotoxicity. The structure of the kinetic model is shown in more detail in Faustman, E., et al. *Environ Toxicol and Pharmacol*: 2005. (Adapted from Faustman, E., et al. *Inhalation Tox,* 11, 101, 1999. With permission).

through the depletion of the intracellular redox agent, glutathione,[88,105–107] or through the generation of reactive species.[108–110] Each of these mechanisms has been studied in considerable detail, and it is unclear whether any one mechanism is predominant. Many of these mechanisms, however, will affect the cell cycle and appear as an antimitotic effect. To illustrate approaches for evaluating this type of common "synthesized" endpoint, we have chosen in this discussion to focus on an evaluation of these mechanisms at a higher level of biological organizations, namely, how does MeHg cause neuronal cell loss during brain organogenesis?

Studies in both humans and experimental animals have shown that MeHg causes developmental CNS abnormalities, notably decreases in brain cell number and improper neuronal alignment. As Burbacher et al.[111] noted, this phenomenon is observed consistently across a considerable dose range and across a variety of species, including humans,[112,113] rodents,[114–117] and monkeys.[111] Although MeHg is also known to produce necrosis and apoptosis in neuronal cells,[118–120] alterations in proliferative activity represent a more sensitive effect, and that is associated with low-dose human exposures. Both *in vivo* and *in vitro* studies reveal that MeHg exposure can affect the dynamics of cell cycling in the CNS.[101,114,121–123] In these assessments, DNA in actively proliferating cells is identified in two ways: (1) mitotic figures are determined morphometrically or (2) actively proliferating DNA is labeled with the thymidine analog, 5′ bromodeoxyuridine (BrdU), and BrdU incorporation and cell cycle progression is determined using bivariate flow cytometry. The *in vitro* studies revealed that a G2/M cell cycle arrest occurred in the absence of direct cytolethality, supporting the hypothesis that cell cycle effects may be a more sensitive endpoint than cell death.[122] *In vivo* studies have revealed that although both rats and mice are sensitive to cell cycle effects, mouse cells appear to be more sensitive.[123]

To put our *in vivo* and *in vitro* observations into context, we adapted our previous biologically based dose-response (BBDR) model to evaluate the effects of MeHg on CNS cell dynamics.[124,125] Figure 2.4 shows a toxicokinetic and dynamic model framework for MeHg developmental toxicity that builds from the general framework presented in Figure 2.1. Within the dynamic portion of this framework is a dynamic model that evaluates the impacts of MeHg on the normal developmental processes of proliferation, differentiation, and cell loss (apoptosis and necrosis). As shown in Figure 2.4, our dynamic model was linked with a toxicokinetic model for MeHg exposures during pregnancy to assess neuronal cell exposures at realistic environmental exposures.[126] To evaluate brain concentrations during development, we linked our kinetic model outputs in a stepwise manner to our dynamic model of midbrain development.[125,127]

The dynamic model framework was also used to evaluate the role that specific cell-signaling pathways, which control cell cycle checkpoints, might play in mediating MeHg cell cycle effects and hence developmental toxicity. In these studies, the response of wild-type and *p21* cell cycle gene knockout mouse cells to MeHg was evaluated. Whereas the G2/M accumulation induced by MeHg was independent of *p21* status (as was cytotoxicity), a greater proportion of *p21*(-/-) cells were able to complete one round of cell division in the presence of MeHg as compared with *p21*(+/-) or *p21*(+/+) cells. These data suggest an important role for *p21* cell checkpoint pathways in mediating MeHg's effects on the cell cycle. The importance of MeHg's effects on cell cycle in our analysis was significantly enriched by making comparisons across species and across both *in vivo* and *in vitro* assessments. The modeling framework was critical for placing these mechanistic clues into a larger, more environmentally relevant context.

In summary, agents that cause mitotic interference can be potent developmental toxicants. Dose-response and structure-activity relationship have strengthened our understanding of common mechanisms of these agents causing delays and complete blockage of rapidly proliferating cells within the conceptus. Transgenic models for key cell cycle checkpoint pathways are also proving useful in evaluating key signaling pathways.

B. Altered Energy Sources

The high replicative activity of cells during "biosynthesis and proliferation requires an uninterrupted source of intrinsic energy generated in the developing tissues."[5] Oxidative metabolism is essential for fetal development, and oxidative phosphorylation increases as gestation proceeds.[128,129] However, in only a few situations has altered mitochondrial function been associated with an adverse developmental outcome. Among these are achondroplasia and riboflavin deficiency.[130–134] In these studies, skeletal system malformations were associated with deficient mitochondrial activity. Chondrogenesis may be especially sensitive to agents that interfere with energy production because the growth plates of the long bones have "the lowest oxygen tension of any bodily organ undergoing active proliferation."[133] Although studies demonstrate that reduced mitochondrial function is associated with skeletal malformations, the relationship between reduced energy status and the appearance of teratogenesis remains unclear. For example, other studies carried out by Mackler and Shepard demonstrated that iron deficiency inhibited mitochondrial function (as measured by a 60% reduction in mitochondrial NADH oxidase activity) and produced a marked decreased fetal viability and size (but no congenital malformations).[135–137]

Few studies link chemically induced mitochondrial dysfunction with developmental toxicity. For example, classic inhibitors of mitochondrial respiration, such as rotenone or cyanide, have not been associated or are only weakly associated with teratogenic outcomes. While the high toxicity of these chemicals may preclude the observation of altered morphology because of fetal or embryonic death, there may be a narrow range of exposures where malformations are observed.[138] Investigations into the effects of chemically induced inhibition of mitochondrial respiration and the appearance of dysmorphogenesis have included studies of rhodamine dyes (rhodamine 6G and rhodamine 123),[139] diphenylhydantoin, chloramphenicol and sodium phenobarbital,[138] and cocaine.[140–142] More recent studies have investigated the role that mitochondria play in mediating apoptotic signals (see Section II.G for more details on this mechanism).

1. Inhibitors of Mitochondrial Respiration

Mackler et al.[138] investigated agents such chloramphenicol, phenobarbital, and malonate because of their known inhibitory effects on mitochondrial respiration. All of these agents were found to inhibit fetal growth, with the exception of malonate. Phenobarbital produced profound skeletal alterations, including cleft palate, edema, spinal retroflexion, and delayed ossification of the occiput and sternum. Diphenylhydantoin produced syndactyly and oligodactyly, and delayed ossification of the occiput and sternum. Chloramphenicol produced edema, wavy ribs, and fused ribs. The low percentage of effects following exposure to malonate and chloramphenicol may be attributable to the steep dose-response relationship between the production of dysmorphogenesis and lethality. While these chemicals were shown to decrease the specific activity of various enzymes involved in electron transport *in vitro*, the authors found little inhibition of these enzymes when studying homogenates prepared from the exposed fetuses, with the exception of those treated with phenobarbital. Therefore, the relationship between inhibition of mitochondrial function and the production of dysmorphogenesis was not determined.

2. Cocaine

The effects of cocaine on fetal dysmorphogenesis, while producing mitochondrial inhibition, have not been ascribed to mitochondrial dysfunction. Rather, cocaine-induced ischemia reperfusion has been hypothesized to result in the production of reactive oxygen species, leading to focal tissue damage.[143]

3. Rhodamine Dyes

The cationic rhodamine dyes, which include rhodamine 123 and rhodamine 6G, have been used as mitochondrial-specific markers. The strongly negative charge potential across the mitochondrial membrane causes accumulation of the positively charged dyes, while the neutral rhodamine dyes show no specific localization to mitochondria.[139] The cationic rhodamine dyes also interfere with mitochondrial respiration, which has been observed to result in low ATP production following either *in vivo* or *in vitro* exposure.[139] Hood et al.[144] investigated the teratogenic effects of rhodamine 123 in mice during gestation (7 to 10 days). Administration of rhodamine 123, in combination with 2-deoxyglucose, an inhibitor of glycolytic ATP generation, led to elevated levels of both gross and skeletal malformations, as well as an increased incidence of early fetal death.[144]

The few studies presented here demonstrate that a compromised energy production capacity has the potential to lead to adverse developmental outcomes that can range from relatively minor abnormalities to fetal death. However, at present, there is no clear understanding of the fundamental contribution of this pathway to teratogenesis. The investigations presented here support a model where skeletal development is at highest risk from exposures to chemical agents that inhibit oxidative respiration.

C. Enzyme Inhibition

The teratogenic effects of some compounds may be attributed to inhibition of specific enzymes. Enzymes critical for cell growth and proliferation, such as those involved in synthesis of DNA and RNA, are ones whose inhibition might have the greatest effect on developmental processes. Four model developmental toxicants, methotrexate, chlorpyrifos, 5-fluorouracil, and mevinolin, are highlighted as examples of agents that cause enzyme inhibition.

1. Methotrexate

Methotrexate (MTX), a cancer chemotherapeutic agent, is a competitive inhibitor of dihydrofolate reductase (DHFR), the enzyme that converts folate to tetrahydrofolate. Tetrahydrofolate is subsequently metabolized to various coenzymes that participate in one-carbon metabolism (OCM), which is critical for the synthesis of purines and amino acids and the conversion of deoxyuridylate to thymidylate. *In utero* exposure to MTX causes craniofacial defects, limb deformities involving reduction in size, and decreased fetal weights.[145]

To determine whether a chemical's teratogenicity reflects inhibition of a specific enzyme, the effects of administration of the inhibited enzyme's product may be determined. By administering leucovorin, a metabolic derivative of tetrahydrofolate, to animals treated with MTX DeSesso and Goeringer[146] demonstrated that MTX's teratogenic effects are specifically due to inhibition of DHFR: Treatment with leucovorin protected animals from the teratogenic effects of MTX.

While it seems plausible that the teratogenic effects of MTX are due to its interference with OCM, owing to the importance of this pathway, it is still conceivable that MTX's teratogenicity is a result of depletion of tetrahydrofolate, which may participate in an unidentified, yet crucial, metabolic pathway. This possibility was addressed by DeSesso and Goeringer[145] using 1-(p-tosyl)-3,4,4-trimethylimidazolidine (TTI), a functional analog of tetrahydrofolate, which participates in OCM. Since TTI is structurally dissimilar from tetrahydrofolate, yet still enables OCM to proceed as usual, the ability of TTI to prevent MTX-induced teratogenic effects can be attributed to its restoration of OCM. Results from this study provide strong evidence that the teratogenic effects of MTX are related to its interference with OCM, in that TTI dramatically reduced both the incidence and severity of MTX-induced malformations. Although TTI did not completely alleviate MTX-associated teratogenicity, this may be partly due to the dosing regimen used by DeSesso and Goeringer.

2. Chlorpyrifos

Chlorpyrifos (CP) and its active metabolite chlorpyrifos-oxon (CPO) act as nervous system toxicants through their ability to inhibit acetylcholinesterase (AchE) activity, which can explain pesticide action of CP and its ability to act as a poison at high doses in adults. Thus, a key question for developmental toxicologists is whether similar mechanisms underlie the neurodevelopmental toxicity of chlorpyrifos in children.

CP/CPO is capable of potent and irreversible inhibition of cholinesterase, and this inhibition results in subsequent accumulation of the neurotransmitter acetylcholine in the synaptic cleft.[147] This may lead to multiple toxic effects, including cholinergic crisis and disruption of neurodevelopment. Acetylcholine is responsible for cholinergic neurotransmission. If this ligand is not removed from the synaptic cleft, overstimulation may occur at high doses, leading to cholinergic crisis. Such poisoning effects have been well documented in agricultural workers,[148] and monitoring of cholinesterase levels is used as a method of exposure surveillance in California agricultural workers.

Researchers are particularly interested in fetal and neonatal exposure because recent studies have focused on adverse developmental effects associated with chlorpyrifos intake. Multiple mechanisms of chlorpyrifos toxicity have been proposed, and their relative importance appears to depend on developmental lifestage and dose. To identify critical modes of action that can explain specific adverse outcomes that can arise following CP exposures during development, consideration of dose, time of exposure, and target tissue is essential. Cholinesterases have been proposed to have distinct roles in multiple phases of neurogenesis, including neuronal differentiation, cell migration, neurite outgrowth, and synaptogenesis (reviewed by Small et al.[149]). *In vitro* midbrain micromass studies of differentiation suggest that many neuronal cells are cholinergic in nature.[150] There is also evidence that neurodevelopment is influenced by AchE expression. In neuroblastoma cells, the ability of neurites to extend and the appearance of other differentiation markers varied with AchE expression.[151] In rabbits, AchE transcripts have been identified on embryonic day 12,[152] indicating an early role for this enzyme. Thus, agents that disrupt these pathways would be hypothesized to have developmental impacts. *In vitro* studies with primary cultured chick neurons indicate that cholinesterase inhibitors induce growth cone collapse and inhibit neurite extension.[153] Thus, there is the potential for CP to affect a variety of neurodevelopmental processes via AchE inhibition and many *in vitro* and *in vivo* studies have been conducted that investigate these effects. However, it must also be noted that *in vivo* neurobehavioral effects by CP have been observed at levels below its effects on AchE activity[154] and at time points prior to AchE dependent neurodevelopmental processes. *In vitro* studies[155] suggest other mechanisms such as the production of reactive oxidative stress may be significant.[155–161] Thus, AchE inhibition is thought by many to be one of several modes of CP neurotoxicity.[162]

3. 5-Fluorouracil

Teratogenic effects of 5-FU include cleft palate and limb and tail defects.[164,165] As mentioned earlier, 5-FU can affect a variety of processes, including the direct incorporation of 5-FU into nuclear RNA, resulting in processing errors in forming mRNA and rRNA,[163] and incorporation of nucleotide base residues into DNA. However, the resulting mitotic inhibition is insufficient to explain 5-FU's developmental toxicity.[63] Hence, other effects of 5-FU have been examined.

Of particular interest for developmental effects is 5-FU's ability to inhibit thymidylate synthetase (TS), which methylates deoxyuridylic acid to form thymidylic acid. This inhibition can lead to an imbalance of nucleotide pools and alterations in cell proliferation and/or cell death.

Evidence that the toxicity of 5-FU may be due to inhibition of TS was provided in a study by Elstein et al.[62] in which 5-FU induced an accumulation of murine erythroleukemic cells (MELC) in early S-phase. Specifically, effects of 5-FU were greatest when cells are exposed during the S-phase of the cell cycle, when DNA synthesis occurs and when TS activity is highest. If 5-FU acted by misincorporation into RNA or DNA, effects would not necessarily be specific to the S-phase of the cell cycle. A study by Abbott et al.[60] suggests that the teratogenicity of 5-FU was specifically due to TS inhibition. In this study, inhibition of TS activity in palatal shelves of rat embryos was correlated with effects on growth and fusion of the palatal shelves.[60]

4. Mevinolin

Mevinolin is a competitive inhibitor of 3-hydroxy-3-methylglutaryl coenzyme A (HMG-CoA). HMG-CoA participates in the mevalonate pathway, a precursor for the synthesis of isoprenoids and cholesterol. Teratogenic effects of mevinolin include neural tube defects and rib and vertebral malformations.[170] These effects may be due to depletion of isoprenoids, which, among other things, are required for the posttranslational farnesylation of the p21^{ras} protein.

One indication that the teratogenicity of mevinolin results from inhibition of HMG-CoA reductase is that the teratogenic effects of mevinolin can be diminished by mevalonate, the product of HMG-CoA reductase. Additionally, the teratogenicity of mevinolin analogs correlates with their ability to inhibit HMG-CoA reductase.[170]

The approach used by Brewer et al.[171] to demonstrate that mevinolin's teratogenicity may be due to inhibition of HMG-CoA reductase is similar to that used for 5-FU by Abbott et al.[60] Using *in situ* hybridization with a cRNA probe, Brewer et al. demonstrated that expression was high in the neural tube, where mevinolin is known to produce developmental abnormalities. High expression levels of HMG-CoA reductase could indicate a requirement for high levels of mevalonate products and likewise a high degree of sensitivity to depletion of mevalonate by HMG-CoA inhibitors.

In summary, this section has highlighted three developmental toxicants that inhibit key enzymes critical to proper cellular function. As noted, inhibition of these enzymes critical to DNA and RNA synthesis has dramatic effect on normal development.

D. Nucleic Acids

Agents that interfere with the normal synthesis and functioning of DNA and RNA can be teratogenic because these processes are so vital to the rapidly proliferating cells of a developing embryo.

1. Hydroxyurea

Hydroxyurea (HU) blocks DNA synthesis by inhibiting ribonucleotide reductase, the enzyme responsible for reducing uridine, cytidine, adenosine, and guanosine diphosphate to their corresponding deoxyribonucleotides.[172] Teratogenic effects of HU in rats include malformations of limbs, palate, jaw, and tail, as well as mortality.[173] The morphological changes in embryonic murine cells transplacentally exposed to HU are similar to apoptotic cell death because the observed effects involve condensation of chromatin and shrinkage of cytoplasm.[174]

Evidence that HU depletes deoxyribonucleotides and interferes with DNA synthesis to elicit teratogenicity is provided by a study in which the teratogenic effects of HU in rats were eliminated by coadministration of deoxycytidine monophosphate (dCMP) and HU.[173] Similar results were obtained by Herken,[175] who found that coadministration of dCMP partially prevented cytotoxic effects of HU on murine neuroepithelial cells isolated from transplacentally exposed embryos. In addition to abrogating some of HU's cytotoxic effects, dCMP also reduced inhibition of DNA synthesis. Since dCMP can be converted to dTTP as well as dCTP, the ability of dCMP to provide protection from the effects of HU may indicate that the availability of pyrimidine deoxynucleotides is more of a limiting factor on DNA synthesis then the availability of the purine deoxynucleotides.

The inability of dCMP to provide complete protection from HU-induced cytotoxicity in the Herken[175] study may be due to a lack of the purine deoxynucleotides. However, simultaneous injection of all four deoxynucleotides into mouse fibroblasts failed to completely restore DNA synthesis.[176] This suggests the existence of an additional mechanism by which HU exerts effects on DNA synthesis and developmental toxicity.

The dual nature of HU's teratogenicity may be related to the presence of a hydroxylamine functional group on the molecule. This hydroxylamine group is capable of reacting with oxygen to form hydrogen peroxide, which can ultimately generate highly reactive hydroxyl free radicals (discussed in DeSesso and Goeringer[177]). To determine if any of the teratogenic effects of hydroxyurea can be attributed to generation of hydroxyl free radicals, DeSesso and Goeringer[177] pretreated rabbits with either ethoxyquin or nordihydroguaiaretic acid, both of which are antioxidants that can terminate free radical reactions. Since pretreatment with either of these compounds delayed the onset of embryonic cell death and lowered both the number of malformed fetuses and the incidence of specific malformations, while increasing body weight, DeSesso and Goeringer[177] suggested that the developmental toxicity of hydroxyurea can be at least partly attributed to the generation of reactive oxygen species. Similar results were obtained with propyl gallate, which delayed onset of embryonic cell death without disrupting hydroxyurea's inhibition of DNA synthesis.[177,178] Taken together, these data are significant in that all three of these antioxidants are structurally dissimilar. Thus, their ability to protect embryos from the teratogenic effects of hydroxyurea is probably due to their antioxidant properties, rather than an unknown mechanism related to their structure. Since these three antioxidants didn't completely prevent cell death completely or hydroxyurea-induced developmental toxicity, it is reasonable to postulate that other properties of hydroxyurea, such as inhibition of DNA synthesis, also contribute to its teratogenicity.

2. Cytosine Arabinoside

Cytosine arabinoside (Ara-C) inhibits DNA synthesis by functioning as a pyrimidine analog following its intracellular phosphorylation to Ara-CTP.[179] Teratogenic effects of Ara-C in mice include increases in resorptions, decreased fetal weight, cleft palate, defects of the long bones, and oligodactyly. Higher doses can also cause fusion of vertebrate bodies and ribs. These effects occur following treatment between gestation days (GD) 10.5 and 12.5. However, exposure to Ara-C after GD 13 produced no discernable malformations,[180] possibly because Ara-C is particularly toxic to rapidly proliferating cells in the S-phase of the cell cycle[181] while cells that have undergone some differentiation appear to be relatively insensitive to Ara-C.[182] One of the direct effects of Ara-C is its incorporation into DNA. DNA polymerase alpha can be moderately inhibited by Ara-C, but this mechanism probably does not account for all the observed inhibition of DNA synthesis. Ara-C also inhibits DNA ligase, the enzyme responsible for joining Okazaki fragments. Inhibition of DNA ligase could contribute to the inhibition of DNA synthesis and consequently the cytotoxicity of

Ara-C, and may account for the small size of DNA fragments seen with Ara-C treatment. However, cytotoxicity may be more related to formation of Ara-CTP.[183]

Ara-CTP may cause a deficiency in deoxycytidylic acid triphosphate (dCTP) by inhibiting the reduction of cytidylic acid diphosphate (CDP) to deoxycytidylic acid diphosphate (dCDP).[184] If this mechanism occurs, then a restoration of normal levels of dCTP should protect embryos from the effects of Ara-C. This protection was achieved in a study in which deoxycytidine was administered at a dose four times higher than that of Ara-C.[184] Deoxycytidine similarly protected mouse embryos from the cytotoxic and teratogenic effects of Ara-C when its dose was eight times greater than that of Ara-C.[180] However, these studies don't rule out the possibility that DNA synthesis is inhibited by incorporation of Ara-CTP into DNA, because the high doses of deoxycytidine used could outcompete Ara-C for incorporation.

The effect of HU and Ara-C on DNA synthesis have been highlighted in this section. Research on these two agents provides excellent examples of mechanistic studies using strength and consistency guidelines. (See also earlier comments on 5-FU for related discussions.)

E. Mutations

Mutations are alterations of DNA nucleotide sequence. Such changes in the DNA sequence can result from exchange of one base pair for another (transitions or transversions) or deletions or insertions of a few bases, as well as from inversions, deletions, and translocations involving changes in long segments of DNA following strand breaks and errors in repair. Mutations generally arise from agents that damage DNA, including ionizing radiation and highly electrophilic substances. Because DNA replication is not 100% accurate, a low rate of spontaneously occurring DNA damage also occurs. In addition, mutations can arise from inhibition or altered function of either DNA repair enzymes or the DNA polymerases involved in proofreading. Twenty percent of malformations in humans are attributable to known genetic transmissions, and up to five percent are due to chromosomal aberrations. This section will not focus on these known genetic birth defects but rather will focus on malformations that are chemically or physically induced.

There are a number of factors to consider when evaluating mutation as a mechanism of chemically induced teratogenesis. One factor is the location of the mutation within the genome. For example, a loss-of-function mutation in a housekeeping gene that is required for cell survival would probably be cytotoxic, and therefore would not be passed on to future generations of cells. For a mutation to persist in the developing embryo, it would have to occur in a gene that is not required for cell survival; otherwise, the mutation would result in cell death. Examples of genes that can be mutated and yet allow for cell survival are proto-oncogenes and tumor suppressor genes, many of which are involved with regulating cell growth and proliferation. Both proto-oncogenes and tumor suppressor genes are expressed at very specific times during the course of development. Additionally, mutated forms of these genes are found in a wide variety of tumors. Because proto-oncogenes and tumor suppressor genes have developmentally specific expression patterns, any mutation that alters either the timing or level of expression of one of these genes might be expected to alter normal developmental processes.

Another factor to consider is that the occurrence of a mutation in a gene that isn't required for cell survival is probably a very rare event. For such a rare event to have a significant effect on embryogenesis, it would probably have to occur early, for instance, at or before the 22 blastocyst stage, or would have to occur nonrandomly in the genome. Evidence of nonrandom distribution of chemically induced genetic alterations has been reported.[185,186]

In this section, alkylating agents, aromatic amines, and ionizing radiation will be discussed as examples of developmentally toxic agents that have both mutagenic and teratogenic properties. With all of these agents, it is not clear whether the teratogenic effects arise because of mutations or whether the cytotoxicity of the agents determines the teratogenic outcome.

1. Alkylating Agents

Monofunctional alkylating agents are highly electrophilic substances that form covalent alkyl bonds with nucleophilic sites on DNA and proteins. Commonly alkylated sites include the N-3 and N-7 positions on purines, the O-6 positions on guanine, and the O-2 and O-4 positions on thymine. The mutagenic potential of alkylating agents is primarily due to alkylation at the exocyclic oxygens of guanine and thymine.

Teratogenic effects in rodents following *in vivo* gestational exposure to alkylating agents include mortality, growth retardation, and cephalic, CNS, palatal, and limb malformations, as well as anophthalmia.[12] Effects following *in vitro* exposure are similar to those seen following *in vivo* exposure and include mortality, growth retardation, abnormal neurulation, abnormal flexure, and optic malformations.[11]

Although alkylating agents form DNA lesions that are both cytotoxic and mutagenic, studies by Bochert et al.[187] strongly suggest that the teratogenic potential of alkylating agents may be related to their mutagenic potential. In this study, the teratogenic potencies of three different alkylating agents (ethylmethane sulfonate, methylnitrosourea, and dimethylnitrosamine) in mouse embryos were related to adduct levels at the O-6 position of guanine, a promutagenic lesion. In contrast, there was no correlation between teratogenic potency and adduct levels at the N-7 position of guanine, which is considered to be primarily a cytotoxic lesion. The importance of promutagenic lesions in developmental toxicity is further supported by the observation that similar adduct levels at the O-6 position of guamine were observed at equally developmentally toxic levels of exposure of these three chemicals, despite significant physical and structural differences among these chemicals.

The relative importance of the O^6-alkylguanine adduct on cytotoxicity and inhibition of differentiation of primary embryonic rat midbrain (CNS) and limb bud (LB) cells was studied using O^6-benzylguanine (O^6-Bg), which is a potent and specific inhibitor of the protein that repairs O^6-alkylguanine DNA adducts.[188] In these modulation studies, O^6-Bg potentiated the effects of methylnitrosourea (MNU) to a greater extent than those of ethylnitrosourea (ENU), and for both compounds inhibition of differentiation was potentiated to a greater extent than cytotoxicity. These results provide further evidence that the promutagenic O^6-alkylguanine adduct may be of particular importance for developmental toxicity.

Bifunctional alkylating agents, such as the cancer chemotherapeutic agent cyclophosphamide (CP), can also have both teratogenic and mutagenic properties. CP is bioactivated to a teratogenic metabolite, 4-hydroperoxycyclophosphamide (4-OOH-CP). 4-OOH-CP spontaneously decomposes to phosphoramide mustard and acrolein, which is considered to be mutagenic. Exencephaly, cleft palate, abnormal prosencephalon, as well as limb malformations have been observed in embryos following *in utero* CP exposure.[189–192] CP has also been shown to inhibit DNA synthesis.[193]

Some investigators have hypothesized that CP-induced teratogenicity may be related to its mutagenic potential. For example, treatment of male rats with CP altered growth and development of both second and third generation offspring. Effects included increases in mortality and malformations and reductions in body weight, as well as learning disabilities.[194–196] CP-induced mutations were also detected in transgenic mice containing shuttle vectors for detection of mutagenicity.[197–199] However, the results of these studies do not exclude the possibility that CP could cause mutations indirectly through epigenetic mechanisms, such as alteration of cellular redox status.

To determine whether CP's teratogenicity reflects the mutagenicity of the metabolite acrolein, investigators exposed D10 rat embryos *in vitro* to an analog of CP that breaks down into acrolein and dechlorophosphoramide. Dechlorophosphoramide does not have the DNA alkylating properties of phosphoramide. Consequently, any DNA damage may be attributable to binding of acrolein to DNA. This study revealed that DNA damage occurred only if embryos were cultured in serum-free media with buthionine sulphoximine (BSO) depletion of glutathione, and only at concentrations lethal to the embryo. DNA damage was not detected at concentrations that produced malformations but minimal lethality.[200] Thus, mutagenicity did not appear to be the mechanism by which CP exerts developmental toxicity *in vitro*.

2. Aromatic Amines

Aromatic amines are a class of industrially important chemicals that include potent mutagenic and carcinogenic compounds. These agents have also been of interest as transplacental carcinogens and developmental toxicants.[201–205] In particular, this class of arylating agents was investigated to determine if mutagenic metabolites important in defining the carcinogenic potency of the arylating agents were also important in the etiology of their developmental toxicity. It was determined that metabolites of 2-acetylaminofluorene (AAF), such as 7-hydroxy-acetylaminofluorene, that were not important for carcinogenesis were important contributors to the developmental effects seen *in vitro*.[204] In addition, Faustman-Watts et al.[203] were able to separate the mutagenic, teratogenic, cytolethal, and embryolethal effects of AAF metabolites, thus minimizing the strength and consistency of any association between mechanisms of carcinogenesis and teratogenesis.

3. Radiation

Ionizing radiation is another teratogen that can cause mutations. Ionizing radiation's ability to cause mutagenicity has been studied in many mammalian cell types, including those employed in specific locus (such as the hypoxanthine guanine phosphoribosyltransferase locus) and shuttle vector systems.[206,207] These studies reveal that ionizing radiation causes both point mutations (base changes, frameshifts, and small deletions) and large deletions, with the relative proportion of each type dependent on the type of radiation (either high or low linear energy transfer focus) dose and both the cell type and the locus or vector used.[208–212] Detection of mutations in oncogenes and tumor suppressor genes in radiation-induced tumors suggests that ionizing radiation could cause mutations in specific genes.[213–215] However, ionizing radiation is also cytotoxic, and significant increases in mutations at specific genes are usually observed only at doses where there is also substantial cell killing.[215] Prenatal exposure to ionizing radiation can impair fetal growth and cause structural, physiological, and behavioral abnormalites.[213,214,216–219] However, it is not clear whether the effects seen following exposure to ionizing radiation are due to its cytotoxic potential, its mutagenic potential, or a combination of both.

Recent work on low dose effects of radiation has used microbeams, a unique type of tool that can irradiate either the nucleus or cytoplasm of selected cells in culture. These studies have shown that mutational effects can occur when only the cytoplasm is irradiated or in adjacent cells that were not irradiated, which has led to the term "bystander effects" to describe these phenomena.[206] Other types of special tools for radiation dose delivery have also demonstrated elevated mutation rates *in vivo* in nonirradiated tissues.[206] This suggests that such bystander effects may also provide an important mechanism by which development could be disrupted.

Techniques are now available to determine whether a mutation in a specific gene can alter normal developmental processes. One method that has been used to study effects of specific mutations on developmental processes involves transfection of cells with activated proto-oncogenes. Embryonal carcinoma cells, which are undifferentiated stem cells isolated from teratocarinomas and which can be induced to differentiate into a variety of cell types by manipulating culture conditions, can be used for this purpose.[220] Aberrant expression of proto-oncogenes alters normal differentiation of these cells. For example, transfection of v-*src* into P19 embryonal carcinoma cells alters morphology and causes loss of expression in stem cell markers, in addition to preventing normal induction of differentiation along neuronal and mesodermal pathways.[221] Ectopic expression of c-*jun* into P19 cells resulted in a cell population containing endodermal and mesodermal cells, whereas control cells remained as relatively undifferentiated, pluripotent cells.[222] The phenotype of differentiated cells can also be transformed by transfection with proto-oncogenes. For example, adult human pigment epithelial cells acquire characteristics of neuronal cells when transfected with the H-*ras* proto-oncogene.[223] Results from studies such as these suggest that mutations in developmentally important genes could result in developmental abnormalities.

Another useful method for studying effects of specific mutations is generation of transgenic mice carrying activated proto-oncogenes. Transgenic mice expressing high levels of the *mos* proto-oncogene transgene in brain tissue exhibit degeneration of neurons, axons, and spiral ganglia, as well as gliosis. These physiological abnormalities are accompanied by neurobehavioral anomalies, such as circling, head tilting, and head bobbing.[224] Expression of a mutated *WT-1* tumor suppressor gene in transgenic mice caused abnormal development of kidney, mesothelium, heart, and lungs.[225] Transgenic mice have been developed that lack functional HRas, KRas2, and N-Ras genes, resulting in tumors in all three mutants, with variable to midgestational deaths reported. In particular, KRas transgenic animals display CNS tissue effects.[1]

Although studies using transfected cells and transgenic mice are useful for demonstrating that mutations in specific genes can cause developmental abnormalities, it would also be interesting to determine if mutated proto-oncogenes are observed in animals or cells exposed to mutagens. In summary, the role of mutation in the mechanistic process of developmental toxicants still remains to be clarified. Questions regarding whether a significant level of mutation would occur in viable embryos and lead to teratogenic events are still at issue. Obviously, additional studies will be required to separate correlation and phenomenology versus true mechanistic paths.

F. Alterations in Gene Expression

1. Retinoic Acid

Development occurs according to very specific and well-orchestrated patterns of gene expression. Hence, alterations in these expression patterns can result in serious adverse developmental consequences. Retinoic acids (RAs), the biologically active metabolites of Vitamin A, play an important role in controlling these synchronized expression patterns. Exogenous retinoids can be equally effective in disrupting these processes.

Retinoids are well characterized developmental toxicants, with their potency being determined by both kinetic and dynamic factors, including the timing of exposure, dose, and structural form of the retinoid under evaluation.[1] Structure-activity relationships established for retinoids have demonstrated that an acidic polar terminus is essential for developmental toxicity[15,226] and that greater than a 1000-fold difference in potency between different retinoids can be observed. Many of these differences appear to be due to kinetic differences resulting from changed elimination rates and reduced affinity for cellular retinoid acid binding proteins.[227] Kinetic studies have shown that the area under the curve correlates better with RA teratogenicity than does the peak high dose.[228] RA can produce many dysmorphogenic effects, such as truncation of the forebrain and posteriorization of the hindbrain. Cross-species evaluations have revealed that these same effects are observed in mammals, birds, amphibia, and fishes.

There are minimal strain and species differences when specific metabolites of RA (e.g., all-*trans*-RA or 13-*cis*-RA) are evaluated. However, species differences in metabolism of RA can affect the form and type of retinoid at target sites and underscores the significance of kinetic studies in discussing mechanistic differences in cross-species RA responses.[229,230]

Our knowledge of retinoic acid receptors indicates that they belong to the nuclear hormone ligand–dependent transcription-factor superfamily. This information has greatly facilitated our mechanistic understanding of RA developmental toxicity[1]. Two classes of receptors have been identified: RARs and RXRs.[231] Each class of receptors has three receptor types: α, β, and γ; each is encoded by a separate gene, and multiple isomers can be formed by differential promoter usage and alternative splicing.[1] The isoforms, which differ in domains, are responsible for conferring cell-type and tissue specificity.

Much of our mechanistic knowledge about RA developmental toxicity has resulted from elegant transgenic animal studies. Knockout mice have been generated for each receptor (RAR and RXR) and for most of the isoforms. These studies have revealed that most of the developmental toxicity of RA is mediated by RAR-RXR heterodimers.[1] This was determined by evaluating similarities in phenotype and from treatment studies in these genetically engineered mice. Null mutant mice have been used to correlate specific receptors with specific RA-induced teratogenic effects. For example, RXR appears to be responsible for RA-induced limb defects,[232] whereas RAR appears to be responsible for RA-induced truncation of the posterior axial skeleton and partially responsible for cranial, facial, and neural tube defects.[1,233,234] Kochar and Kumawat[235] showed that RAR agonists were potent teratogens and that RXR agonists were relatively inactive. Mixed agonists revealed intermediate developmental effects. In total, these observations from receptor binding studies and from transgenic and knockout mouse studies reveal the power of these two approaches for providing mechanistic clues for developmental toxicants.

This research has been extended to molecular mechanisms of action by analyzing the DNA target sequences for RAR-RXR heterodimers. These RA response elements are found on *Hox* genes and are shown to be under transcriptional control by *ras*.[226,231] The *Hox* genes are downstream targets of RA in developmental toxicity. These transcription factors control developmental patterning of the CNS, limbs, skeleton, etc., and their expression encodes postural identity.[1] The *Hox* expression patterns following RA exposure can be expanded, reduced, or miss-expressed, resulting in abnormal cell fate and morphogenesis.[236] Treatment with RA can result in alternations in rhombomere expression in the brain and can result in altered patterning of expression domains.[236] Small changes in gene expression patterns in RA treated embryos can result in alterations in cell migration, differentiation, and proliferation.[226]

2. Dioxin

Dioxin (2,3,7,8-tetrachloro-dibenzo-*p*-dioxin, TCDD) also alters gene expression by binding with a nuclear receptor. Dioxin has been hypothesized to cause developmental toxicity by interacting with an endogenous cytoplasmic receptor, a basic helix-loop-helix DNA-binding receptor, and causing gene expression changes in a host of genes. These include genes that regulate proliferation, differentiation, and the stress response.[1,5] *In utero* exposure to TCDD causes a wide range of adverse developmental outcomes, including mortality, growth retardation, behavioral abnormalities, and structural defects such as cleft palate and hydronephrosis. Evidence from rodent studies using aryl hydrocarbon (Ah) responsive and nonresponsive strains supports the hypothesis that activation of the Ah receptor is a critical process mediating TCDD's developmental effects.[237–239] Both mRNA and protein expression levels for the Ah receptor correlate with sensitivity to TCDD-related developmental toxicity.[240] Abbott et al.[241,242] found that TCDD-induced gene expression changes are seen in levels of epidermal growth factor (EGF), transforming growth factor α (TGFα), EGF receptor, transforming growth factor β1 (TGFβ1), and TGFβ2 and also correlate with TCDD-induced cleft palate. These investigators have examined expression changes in *in vitro* models of palatogenesis (mouse and human palate organ cultures), and these studies have informed species differences in the TCDD response, as well as providing detailed mechanistic assessments. The induction of a TCDD responsive gene, CYP1A1, was used to compare mouse and human responsiveness to TCDD. Tissue-level dose and time-response assessments were made to quantitate species differences in Ah receptor and Ah receptor nuclear translocator. By comparing the differences in response patterns, these investigators were able to explain quantitative differences in species response to TCDD by identifying approximately 350 times fewer receptors in human tissues than in mouse tissues and have estimated that approximately 200 times higher levels of TCDD are required to produce equivalent responses in human tissues versus mouse tissues.

3. New Directions in "Omic" Research

Clearly, more research needs to be done in the area of teratogen-induced changes in gene expression. For instance, for genes whose expression is altered by exposure to teratogens, knowledge of their precise function during normal development would enable a prediction of whether alterations in their expression could result in the observed teratogenic outcome. This chapter has highlighted some of the recent advances in our understanding of 17 key cell-signaling pathways that can explain a significant portion of normal developmental processes (Table 2.4). A major focus of current molecular biology research is defining the downstream gene responses associated with these processes. Increases in our knowledge about these normal expression patterns and function will greatly aid our investigations of toxicant-altered gene expression patterns. Since this chapter was first written an "omic" revolution has taken place in which our ability to evaluate changes in gene expression is now at the level of simultaneously evaluating 10,000 genes or more. Such genomic analysis of gene expression patterns via microarray analysis is just part of these "omic" assessments. In conjunction with proteomics (study of protein expression patterns) and metabolomics (study of biochemical functional changes), analyses of the dynamics of expression data can be linked with functional assessments. Monitoring normal and toxicant-induced changes in these patterns requires an understanding of temporal, tissue, and even cell-specific expression changes. This need has driven the development of amplification techniques that allow for assessment of gene expression changes within a single cell,[243] and of laser capture *in situ* microdissection techniques that can allow for linkage of gene and protein expression patterns within an anatomical context. International and national efforts to develop developmentally relevant databases of expression data are underway,[244–247] as are efforts to develop databases for specific types of toxicant-induced expression changes (NIEHS Toxicogenomics Initiative). The generation of bioinformatic tools that will allow developmental toxicologist to assess toxicant-induced changes in genes and proteins with functional impacts within the context of developmental life stage, biological level of assessment, and functional consequence represents both the promise and challenge for interpretation of "omic" data of our mechanistic studies.[1] Additional material regarding such studies can be found in Chapter 15.

G. Programmed Cell Death

Programmed cell death (PCD), which is sometimes referred to as apoptosis, is a normal physiological process that occurs during development. PCD serves a number of very important functions, which include providing the embryo with the proper morphology and removing vestigial structures.[248,249] PCD is an integral component of the development of the central nervous system (CNS). It is estimated that as much as half of the original cell population in the CNS may be eliminated as a result of apoptosis.[250,251] Apoptosis is thought to optimize synaptic connections by removing unnecessary neurons. This is accomplished by the direct relationship between the extent of neural connections to a postsynaptic target and the survival of the trophic factor–dependent neuron.[252] PCD is morphologically and biochemically distinct from necrosis.

Although the exact sequence of events occurring during PCD depends on the cell type, a common feature of many cells undergoing PCD is activation of key intracellular cysteinyl-aspartate proteases know as caspases, which cleave specific target substrates, such as poly (ADP-ribose) polymerase (PARP). Hallmarks of PCD include condensation of chromatin at the periphery of the nucleus, resulting in a pyknotic nucleus and shrinkage of the cytoplasm. Throughout this process, the cell membrane normally remains intact, although blebbing may occur. Thus, the appearance of a cell undergoing PCD differs from that of a necrotic cell, where cell swelling and breakdown of cellular membranes is observed. PCD also differs from necrosis in that it is an active process, often requiring ongoing protein synthesis. Because PCD is critical for normal morphogenesis, alterations in normal patterns of PCD are therefore an important mechanism of teratogenicity. Observations that areas of the body with a high incidence of malformations coincide with areas where PCD[253] occurs support the idea that disruptions in PCD could be teratogenic.

1. Retinoic Acid

One example of a teratogen whose effects may be due to disruption of PCD is retinoic acid (RA). Under normal physiological conditions, RA appears to be involved in expression of homeobox-containing genes that regulate pattern formation and specify positional identity in developing embryos. Some of these genes, such as *GHox-8* (expressed in chick limb bud) may be involved in PCD, as this gene is expressed in zones of PCD.[254] A study by Coelho et al.[254] provides evidence that addition of exogenous RA may alter the normal pattern of PCD. In this study, bead implants coated with RA diminished cell death and inhibited expression of *GHox-7*, another chick limb bud homeobox gene that is normally expressed in necrotic zones of chick limb bud mesoderm but not in mutant limb buds, which lack such necrotic zones.

PCD was also inhibited in palatal shelves of mice exposed to RA on day 10 of gestation (GD 10). In exposed shelves, the medial epithelial cells continued to express EGF receptors and bind EGF at a time when EGFR expression and binding of EGF were decreasing in medial epithelial cells of control shelves. However, the effects of RA on palatal shelf cells may be secondary to other effects. The phenotype of medial epithelial cells depends on interactions with mesenchyme, basal lamina, extracellular matrix, and growth factors, all of which could be affected by RA. For example, RA decreases extracellular spacing between mesenchymal cells underlying the medial epithelium.[33]

Results from several additional studies suggest that under some conditions RA may increase the extent of PCD. For example, human neuroblastoma cells exposed to RA show growth inhibition, neurite outgrowth, and PCD.[255] Vitamin A, the naturally occurring form of RA, can increase the extent of PCD in the interdigital necrotic zones that appear during limb morphogenesis.[256]

Sulik et al.[257] suggest that RA may cause both craniofacial and limb malformations by increasing cell death in areas of PCD. These investigators used supravital staining with Nile Blue Sulfate (NBS) as well as scanning electron microscopy (SEM). According to Sulik, uptake of NBS is most intense in regions that have a high percentage of apoptotic, but not necrotic cells.[257,258] SEM plus NBS staining revealed that treatment of pregnant mice with a single oral dose of 13-*cis*-RA on GD 11 causes excessive cell death in the apical ectodermal ridge of fetal limbs 12 h after treatment. When observed on GD 14, limbs from treated mice exhibit malformations, including oligodactyly and polydactyly.[258] Treatment of pregnant mice with all-*trans*-RA on GD 11 caused mesomelic limb defects in fetuses observed on GD 18. Cell death patterns observed 12 h posttreatment suggested that cell death induced by RA was associated with zones of PCD as seen in control embryos.[259] Treatment of dams with 13-*cis*-RA at 8 d, 14 h or 9 d, 6 h caused malformations of the secondary palate in areas that coincide with the PCD that normally occurs in the first visceral arch.[260] Alles and Sulik[259] suggest that cells in the vicinity of those undergoing PCD may be induced to undergo PCD abnormally if perhaps an endogenous signal is stronger than usual or if an exogenous agent has a mechanism similar to that of the normal stimuli.

2. Dioxin

TCDD is another teratogen that may act by altering normal patterns of PCD. In at least one example of exposure during embryogenesis, TCDD prevented PCD.[241] Between GD 14 and 16, the medial peridermal cells of mouse embryonic palatal shelves normally stop incorporating ^{3}H-thymidine, and both EGF binding and EGF receptor expression decrease. In control shelves, medial epithelial peridermal cells detached from basal cells, and a high percentage of cells contained dense cytoplasm and pyknotic nuclei (two features that are characteristic of PCD). In contrast, shelves exposed to

TCDD *in vivo* or *in vitro* on GD 10 or *in vitro* on GD 12 continued expressing EGFR and binding EGF and showed no peridermal cell degeneration.[32,241]

3. Ethanol

The mechanism of action of ethanol-induced developmental toxicity has not been elucidated fully; however, oxidative stress, interference with growth factor regulation, and interference with retinoic acid synthesis are leading hypotheses.[261–263] The complex nature of ethanol-induced developmental neurotoxicity provides a unique opportunity to evaluate the potential integration of numerous proposed mechanisms into a common mode of action. Of particular relevance to this discussion are the relative effects of ethanol on proliferation versus those on apoptosis, as both have been proposed as key toxic impacts.[264–267]

Ethanol is a particularly prevalent and harmful developmental neurotoxicant. Numerous anomalies are characteristic of gestational ethanol exposure, including general growth retardation, abnormal brain development, microcephaly, mental retardation, and specific craniofacial malformations.[261,268] Data from human and rat embryological and morphological studies identifying targets of ethanol toxicity are consistent. Morphologically, the human and rat central nervous system are highly susceptible to ethanol-induced growth retardation as manifested in microcephalic children and microencephalic rodent models.[261,269,270] Also, dissections of human FAS brains show similar deformities to those seen in rats, including decreased cerebral cortex, hippocampus, and cerebellum size.

Many recent advances in design based on "unbiased" stereological methods in the field of neuroscience have lead to a more complete picture of cellular composition and its relationship to development and aging, and to a greater understanding of the effects of exogenous toxicants.[271] Before the development of three-dimensional, unbiased stereological techniques to determine particle number in a structure, the field relied on two-dimensional "assumption-based" counting methods to determine cellular density and on morphometrics for volume estimates. Earlier studies making relative comparisons based on only density or volume estimates may be misleading, because changes in just one of these parameters may not reflect changes in cell number. To account for variations in both of these parameters simultaneously, total cell number must be analyzed. Furthermore, earlier two-dimensional estimates of numerical density may be biased because of after-the-fact corrections for assumptions regarding particle size, shape, and orientation (e.g., Abercrombie, Weibul, and Gomez corrections). For example, larger nuclei will appear more often in sections than will smaller nuclei, and nuclei will appear more often in sections when they are cut across their long axis. Assumptions about the size, shape, and orientation accompany corrections for split nuclei, lost caps, and overprojection. Recent articles highlight this current controversy.[272–275]

The development of design-based stereological methods made possible statistically unbiased estimation of volume, area, surface boundary, length, population size, and density. This is accomplished by using an *a priori* combination of systematic random sampling and counting rules to eliminate the biases associated with size, shape, orientation, and spatial distribution of objects instead of applying corrections after the fact. This has provided developmental toxicologists with improved tools to investigate pathological differences.

However, it must be emphasized that although these new methods theoretically eliminate statistical biases, the methods have no way of eliminating biases associated with the methodologies (e.g., sectioning and embedding) or observations (e.g., correct identification of a particle or structure boundary by the investigator). For example, bias introduced by counting poorly stained, thick sections used in optical methods or by counting sections by trying to align two separate sections may well be much greater than the statistical bias introduced by two-dimensional methods. Furthermore, these new methods have not been rigorously compared with the two-dimensional estimates or validated by three-dimensional reconstruction studies. Therefore, no independent, quantified measures of the benefits, whether in accuracy or efficiency, of these new methods over the old methods has been done.[272–275]

Stereological methods provide invaluable tools for estimation of cell number in various structures. Effects caused by perturbations can be measured precisely and efficiently to determine if an exposure affects the volume of a structure or if it also influences particle number. Stereological methods provide quantitation of pathological states that can then be used in quantitative risk assessments.[271,276]

Investigations of ethanol-induced neuronal loss highlight the usefulness of stereology in toxicology. Investigators have documented that the most severe deficits in cell number may occur in the neocortex, the hippocampus, or the cerebellum, depending on the stage of CNS development in which exposure to ethanol occurs.[264–267,277] Not only are distinct regions of the brain affected differently by ethanol, but the timing and pattern of exposure plays a critical role in the final outcome of ethanol-induced cellular loss.[278,279] The neocortex is particularly sensitive to neuronal loss following a relatively low exposure (resulting in a peak blood ethanol concentration [BEC] of 150 mg/dl which is roughly equivalent to a BEC in humans after imbibing three to five standard drinks) during early developmental events, including neurogenesis and migration.

Decreases in cell numbers can be caused by a decrease in proliferation or an increase in cell death, and various studies have shown ethanol to be a potent inhibitor of cellular proliferation. A single dose of ethanol administered to female rats within 8 h after mating results in a dose-dependent retardation of cell division in the fertilized ova, which is sustained up to 42 h after the exposure.[280] Within the cerebral cortex, ethanol-exposed rat fetuses generate 30% fewer neocortical neurons between gestational day (GD) 12 and 19, the peak time of cortical neurogenesis in the rat.[281] When ^{3}H-thymidine incorporation into rat fetal brain and liver tissue after exposure to ethanol *in utero* on GD 16 and 20 was compared, the brain tissue showed decreased incorporation, suggesting increased susceptibility of proliferating neuronal and glial precursors as opposed to proliferating liver cells.[282]

In-depth *in vivo* research on neocortical neurogenesis in the mouse model has been performed, relating functional data (cell cycle rates and migration) within an anatomical context.[289,290] Ethanol-induced effects documented include a reduction in the proliferating cell population and an increase in the length of the cell cycle, both contributing to fewer numbers of neurons or glial cells generated.[283–286] The development of a cumulative BrdU incorporation technique allowed determination of the effect of moderate alcohol intake (peak BEC of 153 mg/dl) on the cell cycle length of the proliferating cells of the dorsal neocortices.[266,287] These studies have shown 30% increase in cell cycle length (18 h compared with 11 h) during early neocortical neurogenesis (GD 13 to 16); however, the increase was not constant throughout neurogenesis. As normal neurogenesis proceeds, the cell cycling rate naturally becomes longer, whereas the ethanol-exposed cells showed the same cell cycling rate throughout cortical neurogenesis. No increase in pyknotic cells was detected, suggesting again that the cycling cell is the target.[288]

The postnatal period of neocortical synaptogenesis for the rat, which includes the brain growth spurt, is a highly sensitive period. *In vitro* and *in vivo* studies suggest that ethanol neurotoxicity during this period may be orchestrated by mechanisms different from those governing the earlier period of neocortical neurogenesis. Recently, convincing evidence of increased cell death due to postnatal exposure to ethanol has been documented. The period of natural cell death for the cerebral cortex occurs between postnatal days (PD) 1 and 10 in the rat, with a peak on PD 7.[292] Ikonomidou[293] showed that by blocking N-methyl-D-aspartate (NMDA) glutamate receptors and activating γ-aminobutyric acid (GABA) receptors, ethanol triggers widespread apoptosis (as great as 30 times the baseline rate) in the synaptogenesis period of many brain regions, including the hippocampus, thalamus, and frontal, parietal, cingulate, and retrosplenial cortex. In this study, cell death was measured by DeOlmos silver staining and was confirmed with caspase 3 activation in a more recent study.[294] Furthermore, if blood concentrations exceeded 200 mg/dl for 4 h, apoptotic neurodegeneration was significantly increased compared with controls. If this threshold was exceeded for more than 4 h, the degenerative response became progressively more severe in proportion to the length of time the threshold had been exceeded. A discrete window of time, coinciding with the synaptogenesis period of each region tested and occurring anywhere from GD 19 to PD 14, depending on the region, was the most susceptible period.[293] However, in comparison with the peak BEC of 150 mg/dl shown to inhibit proliferation during the neurogenesis stage, the threshold of 200 mg/dl BEC for over 4 h is relatively high.

In light of the above evidence of ethanol-induced cell death, it is interesting that no stereological studies to date have shown reductions in neuronal cell number in the neocortex when exposure occurs during the synaptogenesis period. However, this exposure scenario has been shown to cause reduction in the volume and weight of cortical structures.[277,279] Conversely, Mooney et al.[277] showed that no decrease in adult cortical volume or in neuronal cell number occurred after rat pups were subjected to a bingelike exposure regimen from PD 4 to 9, acheiving a peak BEC of approximately 300 mg/dl. Furthermore, no differences from controls were detected in adult cerebral cortex DNA content after postnatal exposure to ethanol.[278] When comparing results from prenatal and postnatal exposure paradigms, one must keep in mind that the exposure scenario may result in different pharmacokinetics becoming a potential confounding factor.[295] The studies described in this section for ethanol-induced neuronal deficits show some of the approaches available as well as complexities in linking observation of these morphological impacts with developmental dynamics. The differential impacts of dynamic mechanisms, such as insults to proliferation compared with cell death, can be quantitatively evaluated using BBDR models.[296]

III. CONCLUSIONS

As our initial quote by Ovid suggests, the mazelike path to discernment of mechanisms of developmental toxicity is indeed challenging. The aim of this chapter has been to suggest guidelines for evaluating these paths and to provide approaches for new mechanistic investigations. We have emphasized the idea that mechanisms can be defined on many levels of organization, ranging from the molecular and cellular to the whole organism. We have also illustrated the importance of using an evaluation framework that allows for examination of both kinetic as well as dynamic responses in deciphering mechanistic clues. Although few complete mechanisms are known, recent advances in molecular and cellular biology have provided new mechanistic "threads" and like Theseus, the "elusive gateway" beckons.

A significant challenge for mechanism-based research is the need to identify critical rate-limiting changes that are associated with an adverse outcome. In many cases, our ability to measure changes (for example, microarray gene changes) outstrips our ability to interpret the significance of subtle changes. Thus, besides the criteria for causality that are discussed within this chapter; there is a need to link quantitatively dynamic changes at any and/or all the levels of biological outcome with the manifestations and the dose and temporal context of developmental toxicity. For example, in the section in this chapter that discusses gene expression changes as proposed mechanisms of developmental toxicity, there is a need to understand the significance of subtle, transient changes in overall developmental outcomes over time and in the context of rapidly changing morphology. Although faced with somewhat similar issues in the past, such as assessments of subtle changes in fetal body weight, the sensitivity and ready availability of genomic tools will press the need to frequently address and reevaluate this assessment issue. Presentation of a kinetic and dynamic framework for organizing our multilevel assessment is one step toward this evaluation.[1]

ACKNOWLEDGMENTS

This work was supported by the Institute for Risk Analysis and Risk Communication, the UW Center for Child Environmental Health Risks research (EPA R826886 and NIEHS 1PO1ES09601), the UW NIEHS Center for Ecogenetics and Environmental Health (5 P30 ES07033), the Institute for Evaluating Health Risks, DOE Low Dose Radiation Research Program Grant, and EPA Contract No. W-2296-NATA.

References

1. National Research Council, *Scientific Frontiers in Developmental Toxicity Risk Assessment,* National Academy Press (Committee on Developmental Toxicology, Board on Environmental Studies and Toxicology, National Research Council), Washington D.C., 2000, p. 354.
2. ILSI, I. L. S. I., Workshop to develop a framework for assessing risks to children from exposures to environmental agents, *Environ. Health Perspect.*, 2003 (in press).
3. Faustman, E.M., Griffith, W., Lewandowski, T., Gohlke, J., and Webb, T., *IRARC Technical Report on Developmental Toxicity,* Institute for Risk Analysis and Risk Communication, University of Washington, Seattle, WA, 2003a.
4. Barker, D., Fetal origins of coronary heart disease, *BMJ,* 311, 1995.
5. Wilson, J.G., Current status of teratology: general principles and mechanisms derived from animal studies, in *Handbook of Teratology: General Principles and Etiology,* Wilson, J.G. and Fraser, F.C., Eds. Plenum Press, New York, 1977.
6. Chabner, B., Allegra, C., Curt, G., and Calabresi, P., Antineoplastic agents, in *Goodman and Gilman's The Pharmacological Basis of Therapeutics,* Hardman, J., Limbird, L., Molinoff, P., et al., Eds., McGraw-Hill, New York, 1996, p. 1233.
7. Hill, A., The environment and disease: association or causation, *Proc. Roy. Soc. Med.,* 58, 295, 1965.
8. IPCS, I. P. o. C. S. IPCS workshop on developing a conceptual framework for cancer risk assessment, Lyon, France, 1999,
9. U.S. EPA, Draft final guidelines for carcinogen risk assessment, U.S. EPA, Washington, D.C., 2003.
10. Selevan, S.G. and Lemasters, G.K., The dose-response fallacy in human reproductive studies of toxic exposures, *J. Occup. Med.,* 29, 451, 1987.
11. Faustman, E.M., Kirby, Z., Gage, D., and Varnum, M., In vitro developmental toxicity of five direct acting alkylating agents in rodent embryo cultures: Structure — activity patterns, *Teratology,* 40, 199, 1989.
12. Platzek, T., Bochert, G., Schneider, W., and Neubert, D., Embryotoxicity induced by alkylating agents: 1. Ethylmethanesulfonate as a teratogen in mice — a model for dose response relationships of alkylating agents, *Arch. Toxicol.,* 51, 1, 1982.
13. Platzek, T., Bochert, G., Meister, R., and Neubert, D., Embryotoxicity induced by alkylating agents: 7. Low dose prenatal-toxic risk estimation based on NOAEL risk factor approach, dose-response relationships, and DNA adducts using methylnitrosourea as a model compound, *Teratog. Carcinog. Mutagen.,* 13, 101, 1993.
14. Kistler, A., Limb bud cell cultures for estimating the teratogenic potential of compounds. Validation of the test system with retinoids, *Arch. Toxicol.,* 60, 403, 1987.
15. Willhite, C.C., Wier, P.J., and Berry, D.L., Dose response and structure-activity considerations in retinoid-induced dysmorphogenesis, *Crit. Rev. Toxicol.,* 20, 113, 1989.
16. Rawlings, S.J., Shuker, D.E., Webb, M., and Brown, N.A., The teratogenic potential of alkoxy acids in post-implantation rat embryo culture: structure-activity relationships, *Toxicol. Lett.,* 28, 49, 1985.
17. Brown, N.A., Coakley, M.E., and Clarke, D.O., Structure-teratogenicity relationships of valproic acid congeners in whole-embryo culture, in *Approaches to Elucidate Mechanisms in Teratogenesis,* Welsch, F., Ed., Hemisphere Publishing Corp., Washington, D.C., 1987, p. 17.
18. Oglesby, L.A., Ebron-McCoy, M.T., Logsdon, T.R., Copeland, F., Beyer, P.E., and Kavlock, R.J., In vitro embryotoxicity of a series of para-substituted phenols: structure, activity, and correlation with in vivo data, *Teratology,* 45, 11, 1992.
19. McMahon, A.P., Joyner, A.L., Bradley, A., and McMahon, J.A., The midbrain-hindbrain phenotype of Wnt-1-/Wnt-1-mice results from stepwise deletion of engrailed-expressing cells by 9.5 days post-coitum, *Cell,* 69, 581, 1992.
20. Chiang, C., Litingtung, Y., Lee, E., Young, K.E., Corden, J.L., Westphal, H., et al., Cyclopia and defective axial patterning in mice lacking Sonic hedgehog gene function, *Nature,* 383, 407, 1996.

21. Cooper, M.K., Porter, J.A., Young, K.E., and Beachy, P.A., Teratogen-mediated inhibition of target tissue response to Shh signaling, *Science*, 280, 1603, 1998.
22. Incardona, J.P., Gaffield, W., Kapur, R.P., and Roelink, H., The teratogenic Veratrum alkaloid cyclopamine inhibits sonic hedgehog signal transduction, *Development*, 125, 3553, 1998.
23. Roelink, H., Porter, J.A., Chiang, C., Tanabe, Y., Chang, D.T., Beachy, P.A., et al., Floor plate and motor neuron induction by different concentrations of the amino-terminal cleavage product of sonic hedgehog autoproteolysis, *Cell*, 81, 445, 1995.
24. Nebert, D., Pharmacogenetics and pharmacogenomics: Why is this relevant to the clinical geneticist?, *Clin. Genet.*, 56, 247, 1999.
25. Couture, L.A., Abbott, B.D., and Birnbaum, L.S., A critical review of the developmental toxicity and teratogenicity of 2,3,7,8-tetrachlorodibenzo-p-dioxin: recent advances toward understanding the mechanism, *Teratology*, 42, 619, 1990.
26. Johnson, R. and Tabin, C., Molecular models for vertebrate limb development, *Cell*, 90, 979, 1997.
27. Ferretti, P. and Tickle, C., The limbs, in *Embryos, Genes and Birth Defects*, Thorogood, P., Ed., John Wiley & Sons, Chichester, 1997, p. 101.
28. Tabin, C., A developmental model for thalidomide defects, *Nature*, 396, 322, 1998.
29. Stephens, T.D. and Fillmore, B.J., Hypothesis: thalidomide embryopathy — proposed mechanism of action, *Teratology*, 61, 189, 2000.
30. Brown, N., *Normal Development: Mechanism of Early Embryogenesis*, 2nd Ed., Raven Press, New York, 1994, p. 15.
31. Abbott, B.D., Probst, M.R., Perdew, G.H., and Buckalew, A.R., AH receptor, ARNT, glucocorticoid receptor, EGF receptor, EGF, TGF alpha, TGF beta 1, TGF beta 2, and TGF beta 3 expression in human embryonic palate, and effects of 2,3,7,8-tetrachlorodibenzo-p-dioxin (TCDD), *Teratology*, 58, 30, 1998.
32. Abbott, B.D., Diliberto, J.J., and Birnbaum, L.S., 2,3,7,8-Tetrachlorodibenzo-p-dioxin alters embryonic palatal medial epithelial cell differentiation in vitro, *Toxicol Appl Pharmacol*, 100, 119, 1989.
33. Abbott, B.D., Harris, M.W., and Birnbaum, L.S., Etiology of retinoic acid-induced cleft palate varies with the embryonic stage, *Teratology*, 40, 533, 1989.
34. Gohlke, J., Griffith, W., Bartell, S., Lewandowski, T., and Faustman, E., A computational model for neocortical neuronogenesis predicts ethanol-induced neocortical neuron number deficits, *Develop. Neurosci.*, 24, 467, 2002.
35. Bernfield, M., Mechanisms of congenital malformations, in *The Biological Basis of Reproductive and Developmental Medicine*, Warshaw, J.B., Ed., Elsevier Biomedical, New York, 1983, p. 143.
36. Chow, I., Cell-cell interaction during synaptogenesis, *Physiol., Paris*, 84, 121, 1990.
37. Doherty, P. and Walsh, F.S., The contrasting roles of N-CAM and N-cadherin as neurite outgrowth-promoting molecules, *Cell Sci.*, 15, 13, 1991.
38. Grumet, M., Structure, expression and function of Ng-CAM, a member of the immunoglobulin superfamily involved in neuron-neuron and neuron-glia adhesion., *Neurosci. Res.*, 31, 1, 1992.
39. Hatten, M.E. and Mason, C.A., Mechanisms of glial-guided neuronal migration in vitro and in vivo, *Experientia*, 46, 907, 1990.
40. Lehmann, S., Kuchler, S., Theveniau, M., Vincendon, G., and Zanetta, J.P., An endogenous lectin and one of its neuronal glycoprotein ligands are involved in contact guidance of neuronal migration, *Nat.. Acad. Sci.*, 87, 6455, 1990.
41. Liesi, P., Extracellular matrix and neuronal movement, *Experientia*, 46, 900, 1990.
42. Pinto-Lord, M.C., Evrard, P., and Caviness, V.S., Jr., Obstructed neuronal migration along radial glial fibers in the neocortex of the reeler mouse: a golgi-EM analysis, *Brain Res.*, 256, 379, 1982.
43. Rakic, P., Defects of neuronal migration and the pathogenesis of cortical malformations, *Prog. Brain Res.*, 73, 15, 1988.
44. Tang, J., Landmesser, L., and Rutishauser, U., Polysialic acid influences specific pathfinding by avian motoneurons, *Neuron*, 8, 1031, 1992.
45. Walsh, F.S. and Doherty, P., Glycosylphosphatidylinositol anchored recognition molecules that function in axonal fasciculation, growth and guidance in the nervous system, *Cell Biol. Int. Rep.*, 15, 1151, 1991.
46. Wilkin, G.P. and Curtis, R., Cell adhesion molecules and ion pumps—Do ion fluxes regulate neuronal migration?, *Bioessays*, 12, 287, 1990.

47. Kirschner, M. and Weber, K., Cytoplasm and cell motility, *Current Sci.*, 1, 3, 1989.
48. LoPachin, R.M. and Aschner, M., Glial-neuronal interactions: Relevance to neurotoxic mechanisms, *Toxicol. Appl. Pharmacol.*, 118, 141, 1993.
49. Yamada, K., Cell Morphogenic Movements, in *Handbook of Teratology*, Wilson, J. and Fraser, F., Eds., Plenum Press, New York, 1977, p. 199.
50. Pratt, R.M., Goulding, E.H., and Abbott, B.D., Retinoic acid inhibits migration of cranial neural crest cells in the cultured mouse embryo, *J. Craniofac. Genet. Dev. Biol.*, 7, 205, 1987.
51. Regan, C. M., Neural cell adhesion molecules, neuronal development and lead toxicity, *Neurotoxicology*, 14, 69, 1993.
52. Cookman, G.R., King, W., and Regan, C.M., Chronic low-level lead exposure impairs embryonic to adult conversion of the neural cell adhesion molecule, *Neurochem.*, 49, 399, 1987.
53. Graff, R. and KR, R., Neural cell adhesion molecules and molecules and microtubles are distinct targets of methylmercury in vitro, *Toxicologist*, 15, 1386, 1995.
54. Bray, D. and Hollenbeck, P.J., Growth cone motility and guidance, *Ann. Rev. Cell Biol.*, 4, 43, 1988.
55. Trosko, J.E., Chang, C.C., and Netzloff, M., The role of inhibited cell-cell communication in teratogenesis, *Teratog. Carcinog. Mutagen.*, 2, 31, 1982.
56. Scott, W., Cell death and reduced proliferative rate, in *Handbook of Teratology: Mechanisms and Pathogenesis*, Wilson, J. and Fraser, F., Eds., Plenum Press, New York, 1977, p. 81.
57. Snow, M.H., Restorative growth in mammalian embryos, in *Issues and Reviews in Teratology*, Kalter, H., Ed., Plenum Press, New York, 1983, p. 251.
58. Bayer, S.A., Cellular aspects of brain development, *Neurotoxicology*, 10, 307, 1989.
59. Trasler, D.G. and Fraser, F.C., Time-position relationships, with particular reference to cleft lip and cleft palate, in *Handbook of Teratology*, Wilson, J.G. and Fraser, F. C., Eds., Plenum Press, New York, 1977, p. 271.
60. Abbott, B.D., Lau, C., Buckalew, A.R., Logsdon, T.R., Setzer, W., Zucker, R.M., et al., Effects of 5-fluorouracil on embryonic rat palate in vitro: fusion in the absence of proliferation, *Teratology*, 47, 541, 1993.
61. Chaube, S. and Murphy, M.L., The teratogenic effects of the recent drugs active in cancer chemotherapy, *Advances Teratol.*, 3, 181, 1968.
62. Elstein, K.H., Zucker, R.M., Shuey, D.L., Lau, C., Chernoff, N., and Rogers, J.M., Utility of the murine erythroleukemic cell (MELC) in assessing mechanisms of action of DNA-active developmental toxicants: application to 5-fluorouracil, *Teratology*, 48, 75, 1993.
63. Shuey, D.L., Setzer, R.W., Lau, C., Zucker, R.M., Elstein, K.H., Narotsky, M.G., et al., Biological modeling of 5-fluorouracil developmental toxicity, *Toxicology*, 102, 207, 1995.
64. Parker, W.B. and Cheng, Y.C., Metabolism and mechanism of action of 5-fluorouracil, *Pharmacol. Ther.*, 48, 381, 1990.
65. Shuey, D.L., Buckalew, A.R., Wilke, T.S., Rogers, J.M., and Abbott, B.D., Early events following maternal exposure to 5-fluorouracil lead to dysmorphology in cultured embryonic tissues, *Teratology*, 50, 379, 1994.
66. Kimler, B.F. and Henderson, S.D., Cyclic responses of cultured 9L cells to radiation, *Radiat. Res.*, 91, 155, 1982.
67. Kimler, B., Prenatal irradiation: a major concern for the developing brain, *Int. J. Radiat. Biol.*, 73, 423, 1998.
68. Brent, R.L. and Beckman, D.A., Environmental teratogens, *Bull N. Y. Acad. Med.*, 66, 123, 1990.
69. Jensh, R.P., Weinberg, I., and Brent, R.L., Teratologic studies of prenatal exposure of rats to 915 MHz microwave radiation, *Radiat. Res.*, 92, 160, 1982.
70. Rodier, P.M., Time of exposure and time of testing in developmental neurotoxicology, *Neurotoxicology*, 7, 69, 1986.
71. Schull, W., Norton, S., and Jensh, R., Ionizing radiation and the developing brain, *Neurotox. Teratol.*, 12, 249, 1990.
72. Bolaris, S., Bozas, E., Benekou, A., Philippidis, H., and Stylianopoulou, F., In utero radiation-induced apoptosis and p53 gene expression in the developing rat brain, *Int. J. Radiat. Biol.*, 77, 71, 2001.
73. Brooks, A.L., Khan, M.A., Jostes, R.F., and Cross, F.T., Metaphase chromosome aberrations as markers of radiation exposure and dose, *J. Toxicol. Environ. Health*, 40, 277, 1993.

74. Frankenberg-Schwager, M., Induction, repair and biological relevance of radiation-induced DNA lesions in eukaryotic cells, *Radiat. Environ. Biophys.*, 29, 273, 1990.
75. Iliakis, G., The role of DNA double strand breaks in ionizing radiation-induced killing of eukaryotic cells, *Bioessays*, 13, 641, 1991.
76. Brent, R.L., Radiations and other physical agents., in *Handbook of Teratology*, Wilson, J.G. and Fraser, F. C., Eds., Plenum Press, New York, 1977, p. 153.
77. Beckman, D.A. and Brent, R.L., Mechanism of known environmental teratogens: drugs and chemicals, *Clin. Perinatol.*, 13, 649, 1986.
78. Whittaker, S.G. and Faustman, E.M., Effects of albendazole and albendazole sulfoxide on cultures of differentiating rodent embryonic cells, *Toxicol. Appl. Pharmacol.*, 109, 73, 1991.
79. Albertini, S., Analysis of nine known or suspected spindle poisons for mitotic chromosome malsegregation using Saccharomyces cerevisiae D61.M., *Mutagenesis*, 5, 453, 1990.
80. Delatour, P. and Richard, Y., Proprietes embryotoxiques et antimitotiques en serie benzimidazole, *Therapie*, 31, 505, 1976.
81. Eichenlaub-Ritter, U. and Boll, I., Nocodazole-sensitivity, age-related aneuploidy, and alterations in the cell cycle during maturation of mouse oocytes, *Cytogenet. Cell Genet.*, 52, 170, 1989.
82. Generoso, W., Katoh, M., Cain, K., Hughes, L., Foxworth, L., Mitchell, T., et al., Chromosome malsegregation and embryonic lethality induced by treatment of normally ovulated mouse oocytes with nocodazole, *Mutat. Res.*, 210, 313, 1989.
83. Kappas, A., Georgopoulos, G., and Hastie, A.C., On the genetic activity of benzimidazole and thiophanate fungicides on diploid Aspergillus nidulans, *Mutat. Res.*, 26, 17, 1974.
84. Vanderkerken, K., Vanparys, P., Verschaeve, L., and Kirsch-Volders, M., The mouse bone marrow micronucleus assay can be used to distinguish aneugens from clastogens, *Mutagenesis*, 4, 6, 1989.
85. Whittaker, S.G., Moser, S.F., Maloney, D.H., Piegorsch, W.W., Resnick, M.A., and Fogel, S., The detection of mitotic and meiotic chromosome gain in the yeast Saccharomyces cerevisiae: effects of methyl benzimidazol-2-yl carbamate, methyl methanesulfonate, ethyl methanesulfonate, dimethyl sulfoxide, propionitrile and cyclophosphamide monohydrate, *Mutat. Res.*, 242, 231, 1990.
86. Whittaker, S.G., Zimmermann, F.K., Dicus, B., Piegorsch, W.W., Fogel, S., and Resnick, M.A., Detection of induced mitotic chromosome loss in Saccharomyces cerevisiae — an interlaboratory study, *Mutat. Res.*, 224, 31, 1989.
87. Delatour, P. and Parish, R., Benzimidazole anthelmintics and related compounds: Toxicity and evaluation of residues, in *Drug Residues in Animals*, Rico, A.G., Ed., Academic Press, Orlando, FL, 1986, p. 175.
88. Ellis, W.E., Roos, F.D., Kavlock, R.J., and Zeman, F.J., Relationship of periventricular overgrowth to hydrocephalus in brains of fetal rats exposed to benomyl, *Teratogen. Carcinog. Mutagen.*, 8, 377, 1988.
89. Van-Dyke, J. and Ritchey, M., Colchicine influence during embryonic development in rats, *Anat. Rec.*, 97, 375, 1947.
90. Whittaker, S.G. and Faustman, E.M., Effects of benzimidazole analogs on cultures of differentiating rodent embryonic cells, *Toxicol. Appl. Pharmacol.*, 113, 144, 1992.
91. Hughes, W., *Ann. N.Y. Acad. Sci.*, 65, 454, 1957.
92. Kajiwara, Y., Yasutake, A., Adachi, T., and Hirayama, K., Methylmercury transport across the placenta via neutral amino acid carrier, *Arch. Toxicol.*, 70, 310, 1996.
93. Brubaker, P., Kleing, R., Herman, S., Lucier, G., Alexander, L., and Long, M., DNA, RNA, and protein synthesis in brain, liver, and kidneys of asymptomatic methylmercury treated rats, *Exp. Mol. Pathol.*, 18, 263, 1973.
94. Syversen, T., Effects of methyl mercury on protein synthesis in vitro, *Acta. Pharm. Tox.*, 49, 422, 1981.
95. Salvaterra, P., Lown, B., Morganti, J., and Massaro, E., Alterations in neurochemical and behavioral parameters in the mouse induced by low doses of methyl mercury, *Acta. Pharmacol. Tox.*, 33, 177, 1973.
96. Massaro, E., The developmental cytotoxicity of mercurials, in *Toxicology of Metals*, Chang, L., Ed., CRC Press, Boca Raton, FL, 1996.
97. Lund, B.O., Miller, D.M., and Woods, J.S., Mercury-induced H_2O_2 production and lipid peroxidation in vitro in rat kidney mitochondria, *Biochem. Pharmacol.*, 42 Suppl, S181, 1991.
98. Denny, M.F. and Atchison, W.D., Mercurial-induced alterations in neuronal divalent cation homeostasis, *Neurotoxicology*, 17, 47, 1996.

99. Hare, M.F., McGinnis, K.M., and Atchison, W.D., Methylmercury increases intracellular concentrations of Ca++ and heavy metals in NG108-15 cells, *J. Pharmacol. Exp. Ther.,* 266, 1626, 1993.
100. Miura, K., Suzuki, K., and Imura, N., Effects of methylmercury on mitotic mouse glioma cells, *Environ. Res.,* 17, 453, 1978.
101. Sager, P., Doherty, R., and Rodier, P., Effects of methylmercury on developing mouse cerebellar cortex, *Exp. Neurol.,* 77, 179, 1982.
102. Vogel, D., Margolis, R., and Mottet, N., The effects of methyl mercury binding to microtubules, *Tox. Appl. Pharm.,* 80, 473, 1985.
103. Zucker, R., Elstein, K., Easterling, R., and Massaro, E., Flow cytometric analysis of the mechanism of methylmercury cytotoxicity, *Am. J. Pathol.,* 137, 1187, 1990.
104. Clarkson, T., Mercury — an element of mystery, *N. Engl. J. Med.,* 323, 1137, 1990.
105. Ali, S.F., LeBel, C.P., and Bondy, S.C., Reactive oxygen species formation as a biomarker of methylmercury and trimethyltin neurotoxicity, *Neurotoxicology,* 13, 637, 1992.
106. Woods, J. and Ellis, M., Upregulation of glutathione synthesis in rat kidney by methyl mercury: relationship to mercury-induced oxidative stress, *Biochem. Pharmacol.,* 50, 1719, 1995.
107. Ornaghi, F., Ferrini, S., Prati, M., and Giavini, E., The protective effects of N-acetyl-l-cysteine against methyl mercury embryotoxicity in mice, *Fund. Appl. Tox.,* 20, 437, 1993.
108. Yohana, M., Saito, M., and Sagai, M., Stimulation of lipid peroxidation by methyl mercury in rats, *Life Sci.,* 32, 1507, 1983.
109. Sarafian, T. and Verity, M., Oxidative mechanisms underlying methyl mercury neurotoxicity, *Int. J. Devl. Neurosci.,* 9, 147, 1991.
110. Verity, M.A., Torres, M., and Sarafian, T., Paradoxical potentiation by low extracellular Ca^{2+} of acute chemical anoxic neuronal injury in cerebellar granule cell culture, *Mol. Chem. Neuropathol.,* 15, 217, 1991.
111. Burbacher, T.M., Rodier, P.M., and Weiss, B., Methylmercury developmental neurotoxicity: a comparison of effects in humans and animals, *Neurotoxicol. Teratol.,* 12, 191, 1990.
112. Matsumoto, H., Koya, G., and Takeuchi, T., Fetal Minamata disease. A neuropathological study of two cases of intrauterine intoxication by a methyl mercury compound, *J. Neuropathol. Exp. Neurol.,* 24, 563, 1965.
113. Choi, B.H., Lapham, L.W., Amin-Zaki, L., and Saleem, T., Abnormal neuronal migration, deranged cerebral cortical organization, and diffuse white matter astrocytosis of human fetal brain: a major effect of methylmercury poisoning in utero, *J. Neuropathol. Exp. Neurol.,* 37, 719, 1978.
114. Chen, W.J., Body, R.L., and Mottet, N.K., Some effects of continuous low-dose congenital exposure to methylmercury on organ growth in the rat fetus, *Teratology,* 20, 31, 1979.
115. Geelen, J.A., Dormans, J.A., and Verhoef, A., The early effects of methylmercury on the developing rat brain, *Acta Neuropathol. (Berl.),* 80, 432, 1990.
116. Kakita, A., Wakabayashi, K., Su, M., Sakamoto, M., Ikuta, F., and Takahashi, H., Distinct pattern of neuronal degeneration in the fetal rat brain induced by consecutive transplacental administration of methylmercury, *Brain Res.,* 859, 233, 2000.
117. Sager, P., Aschner, M., and Rodier, P., Persistent, differential alterations in developing cerebellar cortex of male and female mice after methylmercury exposure, *Dev. Brain Res.,* 12, 1, 1984.
118. Castoldi, A.F., Barni, S., Turin, I., Gandini, C., and Manzo, L., Early acute necrosis, delayed apoptosis and cytoskeletal breakdown in cultured cerebellar granule neurons exposed to methylmercury, *J. Neurosci. Res.,* 59, 775, 2000.
119. Miura, K., Koide, N., Himeno, S., Nakagawa, I., and Imura, N., The involvement of microtubular disruption in methylmercury-induced apoptosis in neuronal and nonneuronal cell lines, *Toxicol. Appl. Pharmacol.,* 160, 279, 1999.
120. Nagashima, K., Fujii, Y., Tsukamoto, T., Nukuzuma, S., Satoh, M., Fujita, M., et al., Apoptotic process of cerebellar degeneration in experimental methylmercury intoxication of rats, *Acta Neuropathol. (Berl.),* 91, 72, 1996.
121. Rodier, P., Aschner, M., and Sager, P., Mitotic arrest in the developing CNS after prenatal exposure to methylmercury, *Neurobehav. Toxicol. Teratol.,* 6, 379, 1984.
122. Ponce, R., Kavanagh, T., Mottet, N., Whittaker, S., and Faustman, E., Effects of methyl mercury on the cell cycle of primary rat CNS cells in vitro, *Toxicol. Appl. Pharmacol.,* 127, 83, 1994.

123. Lewandowski, T., Ponce, R., Charleston, J., Hong, S., and Faustman, E., Effect of methylmercury on midbrain cell proliferation during organogenesis: Potential cross-species differences and implications for risk assessment, *Toxicol. Sci.,* 75(1), 124–133, 2003.
124. Leroux, B.G., Leisenring, W.M., Moolgavkar, S.H., and Faustman, E.M., A biologically-based dose-response model for developmental toxicology, *Risk Anal.,* 16, 449, 1996.
125. Faustman, E., Gohlke, J., Judd, N., Lewandowski, T., Bartell, S., and Griffith, W., Modeling developmental processes in animals: applications in neurodevelopmental toxicology, *Environ. Toxicol. Pharmacol.,* 19, 615, 2005.
126. Lewandowski, T., Pierce, C., Pingree, S., Hong, S., and Faustman, E., Methylmercury distribution in the pregnant rat and embryo during early midbrain organogenesis, *Teratology,* 66, 2002.
127. Faustman, E., Lewandowski, T., Ponce, R., and Bartell, S., Biologically based dose-response models for developmental toxicants: lessons from methylmercury, *Inhalat. Toxicol.,* 11, 101, 1999.
128. Fantel, A.G., Person, R.E., Burroughs, G.C., Shepard, T.H., Juchau, M.R., and Mackler, B., Asymmetric development of mitochondrial activity in rat embryos as a determinant of the defect patterns induced by exposure to hypoxia, hyperoxia, and redox cyclers in vitro, *Teratology,* 44, 355, 1991.
129. Mackler, B., Grace, R., Haynes, B., Bargman, G.J., and Shepard, T.H., Studies of mitochondrial energy systems during embryogenesis in the rat, *Arch. Biochem. Biophys.,* 158, 662, 1973.
130. Aksu, O., Mackler, B., Shepard, T.H., and Lemire, R.J., Studies of the development of congenital anomalies in the ribloflavin-deficient, galactoflavin fed rats. II. Role of the terminal electron transport system, *Teratology,* 1, 93, 1968.
131. Mackler, B., Davis, K.A., and Grace, R., Cytochrome a3 deficiency in human achondroplasia, *Biochim. Biophys. Acta,* 891, 145, 1987.
132. Mackler, B., Grace, R., Davis, K.A., Shepard, T.H., and Hall, J.G., Studies of human achondroplasia: oxidative metabolism in tissue culture cells, *Teratology,* 33, 9, 1986.
133. Mackler, B. and Shepard, T.H., Human achondroplasia: defective mitochondrial oxidative energy metabolism may produce the pathophysiology, *Teratology,* 40, 571, 1989.
134. Shepard, T., Lemire, R.J., Aksu, O., and Mackler, B., Studies of the development of congenital anomalies in the riboflavin-deficient, galactoflavin fed rats. I. Growth and embryologic pathology, *Teratology,* 1, 75, 1968.
135. Mackler, B., Grace, R., Person, R., Shepard, T.H., and Finch, C.A., Iron deficiency in the rat: biochemical studies of fetal metabolism, *Teratology,* 28, 103, 1983.
136. Sulik, K.K., Cook, C.S., and Webster, W.S., Teratogens and craniofacial malformations: relationships to cell death, *Development,* 103 (Suppl.) 213, 1988.
137. Shepard, T.H., Mackler, B., and Finch, C.A., Reproductive studies in the iron-deficient rat, *Teratology,* 22, 329, 1980.
138. Mackler, B., Grace, R., Tippit, D.F., Lemire, R.J., Shepard, T.H., and Kelley, V.C., Studies of the development of congenital anomalies in rats. III. Effects of inhibition of mitochondrial energy systems on embryonic development, *Teratology,* 12, 291, 1975.
139. Ranganathan, S., Churchill, P.F., and Hood, R.D., Inhibition of mitochondrial respiration by cationic rhodamines as a possible teratogenicity mechanism, *Toxicol. Appl. Pharmacol.,* 99, 81, 1989.
140. Fantel, A.G., Person, R.E., Burroughs, G.C.J., and Mackler, B., Direct embryotoxicity of cocaine in rats: effects on mitochondrial activity, cardiac function, and growth and development in vitro, *Teratology,* 42, 35, 1990.
141. Finnell, R.H., Toloyan, S., van-Waes, M., and Kalivas, P.W., Preliminary evidence for a cocaine-induced embryopathy in mice, *Toxicol. Appl. Pharmacol.,* 103, 228, 1990.
142. Plessinger, M.A. and Woods, J.R., Jr., Maternal, placental, and fetal pathophysiology of cocaine exposure during pregnancy, *Clin. Obstet. Gynecol.,* 36, 267, 1993.
143. Fantel, A.G., Barber, C.V., and Mackler, B., Ischemia/reperfusion: a new hypothesis for the developmental toxicity of cocaine, *Teratology,* 46, 285, 1992.
144. Hood, R.D., Ranganathan, S., Jones, C.L., and Ranganathan, P.N., Teratogenic effects of a lipophilic cationic dye rhodamine 123, alone and in combination with 2-deoxyglucose, *Drug Chem. Toxicol.,* 11, 261, 1988.
145. DeSesso, J.M. and Goeringer, G.C., Methotrexate-induced developmental toxicity in rabbits is ameliorated by 1-(p-tosyl)-3,4,4-trimethylimidazolidine, a functional analogue for tetrahydrofolate mediated one carbon transfer, *Teratology,* 45, 271, 1992.

146. DeSesso, J. and Goeringer, G., Amelioration by leucovorin of methotrexate developmental toxicityin rabbitss, *Teratology*, 43, 201, 1991.
147. Costa, L.G., Organophosphorus compounds, in *Recent Advances in Nervous System Toxicology*, Spencer, P.S., Ed., Plenum, New York, 1988, p. 203.
148. Levine, R., Recognized and possible effects of pesticides in humans, in *Handbook of Pesticides Toxicology*, Hayes, W. and Lawa, E., Eds., Academic Press, New York, 1991, p. 275.
149. Small, D.H., Michaelson, S., and Sberna, G., Non-classical actions of cholinesterases: role in cellular differentiation, tumorigenesis and Alzheimer's disease, *Neurochem. Int.*, 28, 453, 1996.
150. Ribeiro, P.L. and Faustman, E.M., Embryonic micromass limb bud and midbrain cultures: different cell cycle kinetics during differentiation, *Toxic. In Vitro*, 4/5, 602, 1990.
151. Koenigsberger, C., Chiappa, S., and Brimijoin, S., Neurite differentiation is modulated in neuroblastoma cells engineered for altered acetylcholinesterase expression, *J. Neurochem.*, 69, 1389, 1997.
152. Jbilo, O., L'Hermite, Y., Talesa, V., Toutant, J.P., and Chatonnet, A., Acetylcholinesterase and butyrylcholinesterase expression in adult rabbit tissues and during development, *Eur. J. Biochem.*, 225, 115, 1994.
153. Saito, S., Cholinesterase inhibitors induce growth cone collapse and inhibit neurite extension in primary cultured chick neurons, *Neurotoxicol. Teratol.*, 20, 411, 1998.
154. Jett, D.A., Navoa, R.V., Beckles, R.A., and McLemore, G.L., Cognitive function and cholinergic neurochemistry in weanling rats exposed to chlorpyrifos, *Toxicol. Appl. Pharmacol.*, 174, 89, 2001.
155. Crumpton, T., Seidler, F., and Slotkin, T., Is oxidative stress involved in the developmental neurotoxicity of chlorpyrifos?, *Dev. Brain. Res.*, 121, 189, 2000.
156. Levin, E.D., Addy, N., Baruah, A., Elias, A., Christopher, N.C., Seidler, F.J., et al., Prenatal chlorpyrifos exposure in rats causes persistent behavioral alterations, *Neurotoxicol. Teratol.*, 24, 733, 2002.
157. Song, X., Violin, J.D., Seidler, F.J., and Slotkin, T.A., Modeling the developmental neurotoxicity of chlorpyrifos in vitro: macromolecule synthesis in PC12 cells, *Toxicol. Appl. Pharmacol.*, 151, 182, 1998.
158. Qiao, D., Seidler, F.J., and Slotkin, T.A., Developmental neurotoxicity of chlorpyrifos modeled in vitro: comparative effects of metabolites and other cholinesterase inhibitors on DNA synthesis in PC12 and C6 cells, *Environ. Health Perspect.*, 109, 909, 2001.
159. Das, K. and Barone, S., Jr., Neuronal differentiation in PC12 cells is inhibited by chlorpyrifos and its metabolites: Is acetylcholinesterase inhibition the site of action?, *Toxicol. Appl. Pharmacol.*, 160, 217, 1999.
160. Chanda, S.M. and Pope, C.N., Neurochemical and neurobehavioral effects of repeated gestational exposure to chlorpyrifos in maternal and developing rats, *Pharmacol. Biochem. Behav.*, 53, 771, 1996.
161. Moser, V.C. and Padilla, S., Age- and gender-related differences in the time course of behavioral and biochemical effects produced by oral chlorpyrifos in rats, *Toxicol. Appl. Pharmacol.*, 149, 107, 1998.
162. Slotkin, T.A., Developmental cholinotoxicants: nicotine and chlorpyrifos, *Environ. Health Perspect.*, 107 Suppl, 1, 1999.
163. Pinedo, H.M. and Peters, G.F., Fluorouracil: biochemistry and pharmacology, *J. Clin. Oncol.*, 6, 1653, 1988.
164. O'Dwyer, P.J., King, S.A., Hoth, D.F., and Leyland-Jones, B., Role of thymidine in biochemical modulation: a review, *Cancer Res.*, 47, 3911, 1987.
165. Scheutz, J., Wallace, H., and Diasio, R., 5-Fluorouracil incorporation into DNA of CF-1 mouse bone marrow cells as a possible mechanism of toxicity, *Cancer Res.*, 44, 1358, 1984.
166. Shuey, D.L., Lau, C., Logsdon, T.R., Zucker, R.M., Elstein, K.H., Narotsky, M.G., et al., Biologically based dose-response modeling in developmental toxicology: biochemical and cellular sequelae of 5-fluorouracil exposure in the developing rat, *Toxicol. Appl. Pharmacol.*, 126, 129, 1994.
167. Shuey, D.L., Zucker, R.M., Elstein, K.H., and Rogers, J.M., Fetal anemia following maternal exposure to 5-fluorouracil in the rat, *Teratology*, 49, 311, 1994.
168. Lau, C., Mole, M.L., Copeland, M.F., Rogers, J.M., Kavlock, R.J., Shuey, D.L., et al., Toward a biologically based dose-response model for developmental toxicity of 5-fluorouracil in the rat: acquisition of experimental data, *Toxicol. Sci.*, 59, 37, 2001.
169. Setzer, R.W., Lau, C., Mole, M.L., Copeland, M.F., Rogers, J.M., and Kavlock, R.J., Toward a biologically based dose-response model for developmental toxicity of 5-fluorouracil in the rat: a mathematical construct, *Toxicol. Sci.*, 59, 49, 2001.

170. Minsker, D.H., MacDonald, J.S., Robertson, R.T., and Bokelman, D.L., Mevalonate supplementation in pregnant rats suppresses the teratogenicity of mevinolinic acid, an inhibitor of 3-hydroxy-3-methylglutaryl-coenzyme a reductase, *Teratology*, 28, 449, 1983.
171. Brewer, L.M., Sheardown, S.A., and Brown, N.A., HMG-CoA reductase mRNA in the post-implantation rat embryo studied by in situ hybridization, *Teratology*, 47, 137, 1993.
172. Krakoff, I., Brown, N., and Reichard, P., Inhibition of ribonucleoside diphosphate reductase by hydroxyurea, *Cancer Res.*, 28, 1559, 1968.
173. Chaube, S. and Murhph, M.L., Protective effect of deoxycytidylic acid (CdMP) on hydrosyurea-induced malformations in rats, *Teratoloy*, 7, 79, 1973.
174. Herken, R., Merker, H.J., and Krowke, R., Investigation of the effect of hydroxyurea on the cell cycle and the development of necrosis in the embryonic CNS of mice, *Teratology*, 18, 103, 1978.
175. Herken, R., The influence of deoxycytidine monophosphate (cDMP) on the cytotoxicity of hydroxyurea in the embryonic spinal cord of the mouse, *Teratology*, 30, 83, 1984.
176. Wawra, E., Microinjected deoxynucleotides for the study of chemical inhibition of DNA synthesis, *Nucleic Acids Res.*, 16, 5249, 1988.
177. DeSesso, J.M. and Goeringer, G.C., Ethoxyquin and nordihydroguaiaretic acid reduce hydroxyurea developmental toxicity, *Reprod. Toxicol.*, 4, 267, 1990.
178. DeSesso, J.M., Amelioration of teratogenesis. I. Modification of hydroxyurea-induced teratogenesis by the antioxidant propyl gallate, *Teratology*, 24, 19, 1981.
179. Kessel, D. and Shurin, S.B., Transport of two non-metabolized nucleosides, deoxycytidine and cytosine arabinoside, in a sub-line of the L1210 murine leukemia, *Biochim. Biophys. Acta*, 163, 179, 1968.
180. Kochhar, D.M., Penner, J.D., and McDay, J.A., Limb development in mouse embryos., *Teratology*, 18, 71, 1978.
181. Bhuyan, B.K., Scheidt, L.G., and Fraser, T.J., Cell cycle phase specificity of antitumor agents, *Cancer Res.*, 32, 398, 1972.
182. Kochhar, D.M. and Agnish, N.D., "Chemical surgery" as an approach to study morphogenetic events in embryonic mouse limb, *Develop. Biol.*, 61, 388, 1977.
183. Zittoun, J., Marquet, J., David, J.C., Maniey, D., and Zittoun, R., A study of the mechanisms of cytotoxicity of Ara-C on three human leukemic cell lines, *Cancer Chemother. Pharmacol.*, 24, 251, 1989.
184. Chaube, S., Kreis, W., Uchida, K., and Murphy, M.L., The teratogenic effect of 1--D- arabinofuranosylcytosine in the rat, *Biochem. Pharmacol.*, 17, 1213, 1968.
185. Faustman, E.M. and Goodman, J.I., Alkylation of DNA in specific hepatic chromatin fractions following exposure to methylnitrosourea or dimethylnitrosamine, *Toxicol. Appl. Pharmacol.*, 58, 379, 1981.
186. Ohgaki, H., Eibl, R.H., Wiestler, O.D., Yasargil, M.G., Newcomb, E.W., and Kleihues, P., p53 mutations in nonastrocytic human brain tumors, *Cancer Res.*, 51, 6202, 1991.
187. Bochert, G., Platzek, T., Rahm, U., and Neubert, D., Embryotoxicity induced by alkylating agents: 6. DNA adduct formation induced by methylnitrosourea in mouse embryos, *Arch. Toxicol.*, 65, 390, 1991.
188. Kidney, J. and Faustman, E., Modulation of nitrosourea toxicity in rodent embryonic cells by O^6-benzylguanine, a depletor of O^6-methylguanine-DNA methyltransferase, *Toxicol. Appl. Pharmacol.*, 133, 1, 1995.
189. Francis, B.M., Rogers, J.M., Sulik, K.K., Alles, A.J., Elstein, K.H., Zucker, R.M., et al., Cyclophosphamide teratogenesis: evidence for compensatory responses to induced cellular toxicity, *Teratology*, 42, 473, 1990.
190. Gebhardt, D.O., The embryolethal and teratogenic effects of cyclophosphamide on mouse embryos, *Teratology*, 3, 273, 1970.
191. Gibson, J.E. and Becker, B.A., The teratogenicity of cyclophosphamide in mice, *Cancer Res.*, 28, 475, 1968.
192. Mirkes, P.E., Fantel, A.G., Greenaway, J.C., and Shepard, T.H., Teratogenicity of cyclophosphamide metabolites: phosphoramide mustard, acrolein, and 4-ketocyclophosphamide in rat embryos cultured in vitro, *Toxicol. Appl. Pharmacol.*, 58, 322, 1981.

193. Chernoff, N., Rogers, J.M., Alles, A.J., Zucker, R.M., Elstein, K.H., Massaro, E.J., et al., Cell cycle alterations and cell death in cyclophosphamide teratogenesis. *Teratogen, Carinogen, Mutagen.*, 9, 199, 1989.
194. Auroux, M.R., Dulioust, E.J., Nawar, N.N., Yacoub, S.G., Mayaux, M.J., Schwartz, D., et al., Antimitotic drugs in the male rat. Behavioral abnormalities in the second generation, *J. Androl.*, 9, 153, 1988.
195. Hales, B.F., Crosman, K., and Robaire, B., Increased postimplantation loss and malformations among the F2 progeny of male rats chronically treated with cyclophosphamide, *Teratology*, 45, 671, 1992.
196. Dulioust, E.J., Nawar, N.Y., Yacoub, S.G., Ebel, A.B., Kempf, E.H., and Auroux, M.R., Cyclophosphamide in the male rat: new pattern of anomalies in the third generation, *J. Androl.*, 10, 296, 1989.
197. Hoorn, A.J., Custer, L.L., Myhr, B.C., Brusick, D., Gossen, J., and Vijg, J., Detection of chemical mutagens using Muta Mouse: A transgenic mouse model, *Mutagen.*, 8, 7, 1993.
198. Kohler, S.W., Provost, G.S., Fieck, A., Kretz, P.L., Bullock, W.O., Putman, D.L., et al., Analysis of spontaneous and induced mutations in transgenic mice using a lambda ZAP/lacI shuttle vector, *Environ. Mol. Mutagen.*, 18, 316, 1991.
199. Kohler, S.W., Provost, G.S., Fieck, A., Kretz, P.L., Bullock, W.O., Sorge, J.A., et al., Spectra of spontaneous and mutagen-induced mutations in the lacI gene in transgenic mice, *Proc. Natl. Acad. Sci. USA*, 88, 7958, 1991.
200. Little, J.B., Cellular, molecular, and carcinogenic effects of radiation, *Hematol. Oncol. Clin. North Am.*, 7, 337, 1993.
201. Faustman-Watts, E., Greenaway, J.C., Namkung, M.J., Fantel, A.G., and Juchau, M.R., Teratogenicity in vitro of 2-acetylaminofluorene: role of biotransformation in the rat, *Teratology*, 27, 19, 1983.
202. Faustman-Watts, E., Yang, H., Namkun, M., Greenaway, J., Fantel, A., and Juchau, M., Mutagenic, cytotoxic, and teratogenic effects of 2-acetylaminofluorene and reactive metabolites in vitro, *Teratogen. Carcinog. Mutagen.*, 4, 273, 1984.
203. Faustman-Watts, E.M., Greenaway, J.C., Namkung, M.J., Fantel, A.G., and Juchau, M.R., Teratogenicity in vitro of two deacetylated metabolites of N-hydroxy-2-acetylaminofluorene, *Toxicol. Appl. Pharmacol.*, 76, 161, 1984.
204. Faustman-Watts, E.M., Namkung, M.J., Greenaway, J.C., and Juchau, M.R., Analysis of metabolites of 2-acetylaminofluorene generated in an embryo culture system. Relationship of biotransformation to teratogenicity in vitro, *Biochem. Pharmacol.*, 34, 2953, 1985.
205. Juchau, M.R., Giachelli, C.M., Fantel, A.G., Greenaway, J.C., Shepard, T.H., and Faustman-Watts, E.M., Effects of 3-methylcholanthrene and phenobarbital on the capacity of embryos to bioactivate teratogens during organogenesis, *Toxicol. Appl. Pharmacol.*, 80, 137, 1985.
206. Morgan, W., Non-targeted and delayed effects of exposure to ionizing radiation: I. Radiation-induced genomic instability and bystander effects *in vitro*, *Rad. Res.*, 159, 567, 2003.
207. Morgan, W., Non-targeted and delayed effects of exposure to ionizing radiation: II. Radiation-induced genomic instability and bystander effects *in vivo*, *Rad. Res.*, 159, 2003.
208. Fuscoe, J.C., Zimmerman, L.J., Fekete, A., Setzer, R.W., and Rossiter, B.J., Analysis of X-ray-induced HPRT mutations in CHO cells: insertion and deletions, *Mutat. Res.*, 269, 171, 1992.
209. Grosovsky, A.J., de Boer, J.G., de Jong, P.J., Drobetsky, E.A., and Glickman, B.W., Base substitutions, frameshifts, and small deletions constitute ionizing radiation-induced point mutations in mammalian cells, *Proc. Natl. Acad. Sci. USA*, 85, 185, 1988.
210. Hutchinson, F., Molecular biology of mutagenesis of mammalian cells by ionizing radiation, *Semin. Cancer Biol.*, 4, 85, 1993.
211. Lutze, L.H. and Winegar, R.A., pHAZE: a shuttle vector system for the detection and analysis of ionizing radiation-induced mutations, *Mutat. Res.*, 245, 305, 1990.
212. Skandalis, A., Grosovsky, A.J., Drobetsky, E.A., and Glickman, B.W., Investigation of the mutagenic specificity of X-rays using a retroviral shuttle vector in CHO cells, *Environ. Mol. Mutagen.*, 20, 271, 1992.
213. Brachman, D.G., Hallahan, D.E., Beckett, M.A., Yandell, D.W., and Weichselbaum, R.R., p53 gene mutations and abnormal retinoblastoma protein in radiation-induced human sarcomas, *Cancer Res.*, 51, 6393, 1991.
214. Garte, S.J. and Burns, F.J., Oncogenes and radiation carcinogenesis, *Environ. Health Perspect.*, 93, 45, 1991.

215. Little, J.B., Cellular, molecular, and carcinogenic effects of radiation, *Hematol. Oncol. Clin. North Am.,* 7, 337, 1993.
216. Jensh, R.P. and Brent, R.L., Effects of prenatal X-irradiation on the 14th-18th days of gestation on postnatal growth and development in the rat, *Teratology,* 38, 431, 1988.
217. Kiyono, S., Seo, M., and Shibagaki, M., Effects of environmental enrichment upon maze performance in rats with microcephaly induced by prenatal X-irradiation, *Jpn. J. Physiol.,* 31, 769, 1981.
218. Norton, S., Behavioral changes in preweaning and adult rats exposed prenatally to low ionizing radiation, *Toxicol. Appl. Pharmacol.,* 83, 240, 1986.
219. Tamaki, Y. and Inouye, M., Avoidance of and anticipatory responses to shock in prenatally x-irradiated rats, *Physiol. Behav.,* 22, 701, 1979.
220. Rudnicki, M. and McBurney, M., Cell culture methods and induction of differentiation of embryonal carcinoma cell lines, in *Teratocarcinomas and Embryonic Stems Cells: A Practical Approach,* Robertson, E. J., Ed., IRL Press, Oxford, 1987, p. 19.
221. Boulter, C.A. and Wagner, E.F., The effects of v-src expression on the differentiation of embryonal carcinoma cells, *Oncogene,* 2, 207, 1988.
222. de Groot, R.P., Kruyt, F.A., van der Saag, P.T., and Kruijer, W., Ectopic expression of c-jun leads to differentiation of P19 embryonal carcinoma cells, *Embo. J.,* 9, 1831, 1990.
223. Dutt, K., Scott, M., Sternberg, P.P., Linser, P.J., and Srinivasan, A., Transdifferentiation of adult human pigment epithelium into retinal cells by transfection with an activated H-ras proto-oncogene, *DNA Cell Biol.,* 12, 667, 1993.
224. Propst, F., Rosenberg, M.P., Cork, L.C., Kovatch, R.M., Rauch, S., Westphal, H., et al., Neuropathological changes in transgenic mice carrying copies of a transcriptionally activated Mos protooncogene, *Proc. Natl. Acad. Sci. USA,* 87, 9703, 1990.
225. Kreidberg, J.A., Sariola, H., Loring, J.M., Maeda, M., Pelletier, J., Housman, D., et al., WT-1 is required for early kidney development, *Cell,* 74, 679, 1993.
226. Collins, M. and Mao, G., Teratology of retinoids, *Annu. Rev. Pharmacol. Toxicol.,* 39, 399, 1999.
227. Pignatello, M.A., Kauffman, F.C., and Levin, A.A., Multiple factors contribute to the toxicity of the aromatic retinoid TTNPB (Ro 13-7410): interactions with the retinoic acid receptors, *Toxicol. Appl. Pharmacol.,* 159, 109, 1999.
228. Tzimas, G., Thiel, R., Chahoud, I., and Nau, H., The area under the concentration-time curve of all-trans-retinoic acid is the most suitable pharmacokinetic correlate to the embryotoxicity of this retinoid in the rat, *Toxicol. Appl. Pharmacol.,* 143, 436, 1997.
229. Nau, H., Chaoud, I., Dencker, L., Lammer, E., and Scott, W., Teratogenicity of vitamin A and retinoids, in *Health and Diseases,* Blomhoff, R., Ed., Marcel Dekker, New York, 1994.
230. Kraft, J.C., Shepard, T., and Juchau, M.R., Tissue levels of retinoids in human embryos/fetuses, *Reprod. Toxicol.,* 7, 11, 1993.
231. Chambon, P., A decade of molecular biology of retinoic acid receptors, *FASEB J.,* 10, 940, 1996.
232. Sucov, H.M., Izpisua-Belmonte, J.C., Ganan, Y., and Evans, R.M., Mouse embryos lacking RXR alpha are resistant to retinoic-acid-induced limb defects, *Development,* 121, 3997, 1995.
233. Lohnes, D., Kastner, P., Dierich, A., Mark, M., LeMeur, M., and Chambon, P., Function of retinoic acid receptor gamma in the mouse, *Cell,* 73, 643, 1993.
234. Iulianella, A. and Lohnes, D., Contribution of retinoic acid receptor gamma to retinoid-induced craniofacial and axial defects, *Dev. Dyn.,* 209, 92, 1997.
235. Kochar, D.K. and Kumawat, B.L., Cerebral malaria or Plasmodium falciparum malaria with hypoglycaemia, *Lancet,* 347, 1549, 1996.
236. Marshall, H., Morrison, A., Studer, M., Popperl, H., and Krumlauf, R., Retinoids and Hox genes, *FASEB J.,* 10, 969, 1996.
237. Chang, C., Smith, D.R., Prasad, V.S., Sidman, C.L., Nebert, D.W., and Puga, A., Ten nucleotide differences, five of which cause amino acid changes, are associated with the Ah receptor locus polymorphism of C57BL/6 and DBA/2 mice, *Pharmacogenetics,* 3, 312, 1993.
238. Lambert, G.H. and Nebert, D.W., Genetically mediated induction of drug-metabolizing enzymes associated with congenital defects in the mouse, *Teratology,* 16, 147, 1977.
239. Poland, A. and Glover, E., 2,3,7,8,-Tetrachlorodibenzo-p-dioxin: segregation of toxocity with the Ah locus, *E. Mol. Pharmacol.,* 17, 86, 1980.

240. Abbott, B., Perdew, G., and Birnbaum, L., Ah receptor in embryonic mouse palate and effects of TCDD on receptor expression, *Toxicol. Appl. Pharmacol.,* 126, 16, 1994.
241. Abbott, B.D. and Birnbaum, L.S., TCDD alters medial epithelial cell differentiation during palatogenesis, *Toxicol. Appl. Pharmacol.,* 99, 276, 1989c.
242. Abbott, B.D. and Birnbaum, L.S., Retinoic acid-induced alterations in the expression of growth factors in embryonic mouse palatal shelves, *Teratology,* 42, 597, 1990.
243. Eberwine, J., Yeh, H., Miyashiro, K., Cao, Y., Nair, S., Finnell, R., et al., Analysis of gene expression in single live neurons, *Proc. Natl. Acad. Sci. USA,* 89, 3010, 1992.
244. Yu, Y., Zhang, C., Zhou, G., Wu, S., Qu, X., Wei, H., et al., Gene expression profiling in human fetal liver and identification of tissue- and developmental-stage-specific genes through compiled expression profiles and efficient cloning of full-length cDNAs, *Genome Res.,* 11, 1392, 2001.
245. Master, S.R., Hartman, J.L., D'Cruz, C.M., Moody, S.E., Keiper, E.A., Ha, S.I., et al., Functional microarray analysis of mammary organogenesis reveals a developmental role in adaptive thermogenesis, *Mol. Endocrinol.,* 16, 1185, 2002.
246. Stuart, R.O., Bush, K.T., and Nigam, S.K., Changes in global gene expression patterns during development and maturation of the rat kidney, *Proc. Natl. Acad. Sci. USA,* 98, 5649, 2001.
247. Ko, M.S., Kitchen, J.R., Wang, X., Threat, T.A., Hasegawa, A., Sun, T., et al., Large-scale cDNA analysis reveals phased gene expression patterns during preimplantation mouse development, *Development,* 127, 1737, 2000.
248. Shier, W.T., Why study mechanisms of cell death?, *Methods Achiev. Exp. Pathol.,* 13, 1, 1988.
249. Schwartz, L.M., The role of cell death genes during development, *Bioessays,* 13, 389, 1991.
250. Nijhawan, D., Honarpour, N., and Wang, X., Apoptosis in neuronal development and disease, *Annu. Rev. Neurosci.,* 23, 73, 2000.
251. Oppenheim, R., Cell death during development of the nervous system, *Annu. Rev. Neurosci.,* 14, 453, 1991.
252. Cowan, W.M., Fawcett, J.W., O'Leary, D.D., and Stanfield, B. B., Regressive events in neurogenesis, *Science,* 225, 1258, 1984.
253. Menkes, B., Sandor, S., and Ilies, A., Why study mechanisms of cell death?, *Methods Achiev. Exp. Pathol.,* 13, 1, 1970.
254. Coelho, C.N., Upholt, W.B., and Kosher, R.A., The expression pattern of the chicken homeobox-containing gene GHox-7 in developing polydactylous limb buds suggests its involvement in apical ectodermal ridge-directed outgrowth of limb mesoderm and in programmed cell death, *Differentiation,* 52, 129, 1993.
255. Melino, G., Stephanou, A., Annicchiarico, P.M., Knight, R.A., Finazzi, A.A., and Lightman, S.L., Modulation of IGF-2 expression during growth and differentiation of human neuroblastoma cells: Retinoic acid may induce IGF-2, *Neurosci. Lett.,* 151, 187, 1993.
256. Schweichel, J., The influence of oral vitamin A doses on interdigital necrosis in the limb bud of the rat, *Teratology,* 4, 502, 1971.
257. Sulik, K.K., Johnston, M.C., Smiley, S.J., Speight, H.S., and Jarvis, B.E., Mandibulofacial dysostosis (Treacher Collins syndrome): a new proposal for its pathogenesis, *Am. J. Med. Genet.,* 27, 359, 1987.
258. Sulik, K.K. and Dehart, D.B., Retinoic-acid-induced limb malformations resulting from apical ectodermal ridge cell death, *Teratology,* 37, 527, 1988.
259. Alles, A.J. and Sulik, K.K., Retinoic-acid-induced limb-reduction defects: perturbation of zones of programmed cell death as a pathogenetic mechanism, *Teratology,* 40, 163, 1989.
260. Sulik, K.K., Smiley, S.J., Turvey, T.A., Speight, H.S., and Johnston, M.C., Pathogenesis of cleft palate in Treacher Collins, Nager, and Miller syndromes, *Cleft Palate J.,* 26, 209, 1989.
261. Clarren, S.K. and Smith, D.W., The fetal alcohol syndrome, *New Eng. J. Med.,* 298, 1063, 1978.
262. Miller, M.W., *Development of the Central Nervous System: Effects of Alcohol and Opiates,* Wiley-Liss, New York, 1992.
263. Rodier, P., Developing brain as a target of toxicity, *Environ. Health Perspect.,* 103, 73, 1995.
264. Napper, R.M.A. and West, J.R., Permanent neuronal cell loss in the cerebellum of rats exposed to continuous low blood alcohol levels during the brain growth spurt: A stereological investigation, *J. Comp. Neurol.,* 362, 283, 1995.
265. Miller, M.W. and Potempa, G., Numbers of neurons and glia in mature rat somatosensory cortex: Effects of prenatal exposure to ethanol, *J. Comp. Neur.,* 293, 92, 1990.

266. Miller, M.W. and Kuhn, P.E., Cell cycle kinetics in fetal rat cerebral cortex: effects of prenatal treatment with ethanol assessed by a cumulative labeling technique with flow cytometry, *Alcohol Clin. Exp. Res.*, 19, 233, 1995.
267. West, J.R., Hamre, K.M., and Cassell, M.D., Effects of ethanol exposure during the third trimester equivalent on neuron number in rat hippocampus and dentate gyrus, *Alcohol Clin. Exp. Res.*, 10, 190, 1986.
268. Coles, C.D., Prenatal alcohol exposure and human development, in *Development of the Central Nervous System: Effects of Alcohol and Opiates*, Miller, M.W., Ed., Wiley-Liss, Inc., New York, 1992, p. 9.
269. Allebeck, P. and Olsen, J., Alcohol and fetal damage, *Alcohol Clin. Exp. Res.*, 22, 329S, 1998.
270. Samson, H.H., Microcephaly and fetal alcohol syndrome: human and animal studies, in *Alcohol and Brain Dev.*, West, J. R., Ed., 1986, p. 167.
271. Duffell, S.J., Soames, A.R., and Gunby, S., Morphometric analysis of the developing rat brain, *Toxicol. Pathol.*, 28, 157, 2000.
272. Benes, F.M., About assumptions in estimation of density of neurons and glial cells — Reply, *Biolo. Psych.*, 51, 842, 2002.
273. Benes, F.M. and Lange, N., Benes and Lange respond: reconciling theory and practice in cell counting, *Neurosci.*, 24, 378, 2001.
274. Benes, F.M. and Lange, N., Two-dimensional versus three-dimensional cell counting: a practical perspective, *Neurosci.*, 24, 11, 2001.
275. von Bartheld, C.S., Counting particles in tissue sections: Choices of methods and importance of calibration to minimize biases, *Histol. Histopath.*, 17, 639, 2002.
276. Skoglund, T., Pascher, R., and Berthold, C., Aspects of the quantitative analysis of neurons in the cerebral cortex, *J. Neurosci. Meth.*, 70, 201, 1996.
277. Mooney, S.M., Napper, R.M.A., and West, J.R., Long-term effect of postnatal alcohol exposure on the number of cells in the neocortex of the rat: A stereological study, *Alcohol Clin. Exp. Res.*, 20, 615, 1996.
278. Miller, M.W., Effect of early exposure to ethanol on the protein and DNA contents of specific brain regions in the rat, *Brain Res.*, 734, 286, 1996.
279. Maier, S.E., Chen, W.-J.A., Miller, J.A., and West, J.R., Fetal alcohol exposure and temporal vulnerability: Regional differences in alcohol-induced microcephaly as a function of the timing of binge-like alcohol exposure during rat brain development, *Alcohol Clin. Exp. Res.*, 21, 1418, 1997.
280. Pennington, S.N., Taylor, W.A., Cowan, D.H., and Kalmus, G.W., A single dose of ethanol suppresses rat embryo development in vivo, *Alcohol Clin. Exp. Res.*, 8, 326, 1984.
281. Miller, M.W., Effects of alcohol on the generation and migration of cerebral cortical neurons, *Science*, 233, 1308, 1986.
282. Dreosti, I.E., Ballard, J., Belling, G.B., Record, I.R., Manuel, S.J., and Hetzel, B.S., The effect of ethanol and actealdehyde on DNA synthesis in growing cells and on fetal development in the rat, *Alcohol Clin. Exp. Res.*, 5, 357, 1981.
283. Miller, M.W., Effects of prenatal exposure to ethanol on neocortical development: II. Cell proliferation in the ventricular and subventricular zones of the rat, *J. Comp. Neurol.*, 287, 326, 1989.
284. Guizzetti, M. and Costa, L.G., Inhibition of muscarinic receptor-stimulated glial cell proliferation by ethanol, *J. Neurochem.*, 67, 2236, 1996.
285. Miller, M.W., Effects of prenatal exposure to ethanol on cell proliferation and neuronal migration, in *Development of the Central Nervous System: Effects of Alcohol and Opiates*, Miller, M. W., Ed., Wiley-Liss, New York, 1992, p. 47.
286. Miller, M.W., Effect of pre- or postnatal exposure to ethanol on the total number of neurons in the principal sensory nucleus of the trigeminal nerve: Cell proliferation and neuronal death, *Alcohol Clin. Exp. Res.*, 19, 1359, 1995.
287. Nowakowski, R., Lewin, S., and Miller, M.W., Bromodeoxyuridine immunohistochemical determination of the lengths of the cell cycle and the DNA-synthetic phase for an anatomically defined population, *J. Neurocyt.*, 18, 311, 1989.
288. Miller, M.W. and Muller, S., Structure and histogenesis of the principle sensory nucleus of the trigeminal nerve: Effects of prenatal exposure to ethanol, *J. Comp. Neurol.*, 282, 570, 1989.

289. Caviness, V., Takahashi, T., and Nowakowski, R., Numbers, time and neocortical neuronogenesis: a general developmental and evolutionary model., *Trends Neurosci.,* 18, 379, 1995.
290. Takahashi, T., Nowakowski, R., and Caviness, V., The leaving or Q fraction of the murine cerebral proliferative epithelium: A general model of neocortical neuronogenesis., *J. Neurosci.,* 16, 6183, 1996.
291. Climent, E., Pascual, M., Renau-Piqueras, J., and Guerri, C., Ethanol exposure enhances cell death in the developing cerebral cortex: Role of brain-derived neurotrophic factor and its signaling pathways, *J. Neurosci. Res.,* 68, 213, 2002.
292. Ferrer, I., Bernet, E., Soriano, E., del Rio, T., and Fonseca, M., Naturally occurring cell death in the cerebral cortex of the rat and removal of dead cells by transitory phagocytes, *Neuroscience,* 39, 451, 1990.
293. Ikonomidou, C., Ethanol-induced apoptotic neurodegeneration and fetal alcohol syndrome, *Science,* 287, 1056, 2000.
294. Olney, J.W., Farber, N., Wozniak, D.F., Jevtovic-Todorovic, V., and Ikonomidou, C., Environmental agents that have the potential to trigger massive apoptotic neurodegeneration in the developing brain, *Environ. Health Perspect.,* 108, 383, 2000.
295. Light, K.E., Kane, C.J., Pierce, D.R., Jenkins, D., Ge, Y., Brown, G., et al., Intragastric intubation: Important aspects of the model for administration of ethanol to rat pups during the postnatal period, *Alcohol Clin. Exp. Res.,* 22, 1600, 1998.
296. Gohlke, J.M., Griffith, W.C., and Faustman, E.M., A mechanism based model for evaluating neocortical development: Applications for understanding ethanol neurodevelopmental toxicity, *Teratology,* 65, 335, 2002.
297. McCartney-Francis, N. L., Mizel, D. E., Frazier-Jessen, M., Kulkarni, A. B., McCarthy, J. B., and Wahl, S. M., Lacrimal gland inflammation is responsible for ocular pathology in TGF-beta 1 null mice, *Am. J. Pathol.,* 151, 1281, 1997.
298. Dunker, N. and Krieglstein, K., Targeted mutations of transfroming growth factor-beta genese reveal important roles in mouse development and adult homeostasis, *Eur. J. Biochem.,* 267(24), 6982, 2000.
299. Mikic, B., Van der Meulen, M. C., Kingsley, D. M., and Carter, D. R., Mechanical and geometric changes in the growing femora of BMP-5 deficient mice, *Bone,* 18, 601, 1996.
300. Dudley, A. T. and Robertson, E. J., Overlapping expression domains of bone morphogenetic protein family members potentially account for limited tissue defects in BMP7 deficient embryos, *Dev. Dyn.,* 208, 349, 1997.
301. Goodrich, L. V., Milenkovic, L., Higgins, K. M., and Scott, M. P., Altered neural cell fates and medulloblastoma in mouse patched mutants, *Science,* 277, 1109, 1997.
302. Hahn, H., Wojnowski, L., Zimmer, A. M., Hall, J., Miller, G., and Zimmer, A., Rhabdomyosarcomas and radiation hypersensitivity in a mouse model of Gorlin syndrome, *Nat. Med.,* 4, 619, 1998.
303. Lydon, J. P., DeMayo, F. J., Funk, C. R., Mani, S. K., Hughes, A. R., Montgomery, C. A., Jr., et al., Mice lacking progesterone receptor exhibit pleiotropic reproductive abnormalities, *Genes Dev.,* 9, 2266, 1995.
304. Lydon, J.P., DeMayo, F.J., Conneely, O.M., and O'Malley, B.W., Reproductive phenotypes of the progesterone receptor null mutant mouse, *J. Steroid Biochem. Mol. Biol.,* 56(1–6 Spec No), 67, 1996.
305. Wendling, O., Ghyselinck, N.B., Chambon, P., and Mark, M., Roles of retinoic acid receptors in early embryonic morphogenesis and hindbrain patterning, *Development,* 128(11), 2031–2038, 2001.
306. Moriguchi, T., Motohashi, H., Hosoya, et al., Distinct response to dioxin in an arylhydrocarbon receptor (AHR)-humanized mouse, *Proc. Natl. Acad. Sci. U.S.A.,* 100(10), 5652–5657, 2003.
307. Kassim, N. M., McDonald, S. W., Reid, O., Bennett, N. K., Gilmore, D. P., and Payne, A. P., The effects of pre- and postnatal exposure to the nonsteroidal antiandrogen flutamide on testis descent and morphology in the albino Swiss rat, *J. Anat.,* 190 (Pt 4): 577, 1997.
308. Cunha, G. R., Forsberg, J. G., Golden, R., Haney, A., Iguchi, T., Newbold, R., et al., New approaches for estimating risk from exposure to diethylstilbestrol, *Environ. Health Perspect.,* 107 Suppl 4, 625, 1999.
309. Abbott, B.D., Perdew, G.H., Buckalew, A.R., and Birnbaum, L.S., Interactive regulation of Ah and glucocorticoid receptors in the synergistic induction of cleft palate by 2,3,7,8-tetrachlorodibenzo-p-dioxin and hydrocortisone, *Toxicol. Appl. Pharmacol.,* 128(1), 138–150, 1994.
310. Chazaud, C., Chambon, P., and Dolle, P., Retinoic acid is required in the mouse embryo for left-right asymmetry determination and heart morphogenesis, *Development,* 126, 2589, 1999.

311. Kochhar, D. M., Jiang, H., Penner, J. D., Johnson, A. T., and Chandraratna, R. A., The use of a retinoid receptor antagonist in a new model to study vitamin A-dependent developmental events, *Int. J. Dev. Biol.*, 42, 601, 1998.
312. Elmazar, M., Ruhl, R., Reichert, U., Shroot, B., and Nau, H., RARalpha-mediated teratogenicity in mice is potentiated by an RXR agonist and reduced by an RAR antagonist: Dissection of retinoid receptor-induced pathways, *Toxicol. Appl. Pharmacol.*, 146, 21, 1997.
313. Brandsma, A. E., Tibboel, D., Vulto, I. M., de Vijlder, J. J., Ten Have-Opbroek, A. A., and Wiersinga, W. M., Inhibition of T3-receptor binding by Nitrofen, *Biochim. Biophys. Acta*, 1201, 266, 1994.
314. Spence, S., Anderson, C., Cukierski, M., and Patrick, D., Teratogenic effects of the endothelin receptor antagonist L-753,037 in the rat, *Reprod. Toxicol.*, 13, 15, 1999.
315. Treinen, K. A., Louden, C., Dennis, M. J., and Wier, P. J., Developmental toxicity and toxicokinetics of two endothelin receptor antagonists in rats and rabbits, *Teratology*, 59, 51, 1999.
316. Gershon, M. D., Lessons from genetically engineered animal models. II. Disorders of enteric neuronal development: insights from transgenic mice, *Am. J. Physiol.*, 277(2 Pt 1), G262, 1999.
317. Webster, W. S., Brown-Woodman, P. D., Snow, M. D., and Danielsson, B. R., Teratogenic potential of almokalant, dofetilide, and d-sotalol: drugs with potassium channel blocking activity, *Teratology*, 53, 168, 1996.
318. Wellfelt, K., Skold, A. C., Wallin, A., and Danielsson, B. R., Teratogenicity of the class III antiarrhythmic drug almokalant. Role of hypoxia and reactive oxygen species, *Reprod. Toxicol.*, 13, 93, 1999.
319. Clarke, C., Clarke, K., Muneyyirci, J., Azmitia, E., and Whitaker-Azmitia, P. M., Prenatal cocaine delays astroglial maturation: immunodensitometry shows increased markers of immaturity (vimentin and GAP-43) and decreased proliferation and production of the growth factor S-100, *Brain Res. Dev. Brain Res.*, 91, 268, 1996.
320. Brault, V., Moore, R., Kutsch, S., et al., Inactivation of the beta-catenin gene by Wnt1-Cre-mediated deletion results in dramatic brain malformation and failure of craniofacial development, *Development*, 128(8), 1253–1264, 2001.
321. Larsson, J., Goumans, M.J., Sjostrand, L.J., et. al., Abnormal angiogenesis but intact hematopoietic potential in TGF-beta type I receptor-deficient mice, *Embo. J.*, 20(7), 1663–1673, 2001.
322. Miettinen, P.J., Chin, J.R., Shum, L., et al., Epidermal growth factor receptor function is necessary for normal craniofacial development and palate closure, Nat. Genet., 22(1), 69–73, 1999.
323. Dudas, M., Sridurongrit, S., Nagy, A., Okazaki, K., and Kaartinen, V., Craniofacial defects in mice lacking BMP type I receptor Alk2 in neural crest cells, *Mech. Dev.*, 121(2), 173–182, 2004.
324. Hahn, H., Wojnowski, L., Specht, K., et al., Patched target Igf2 is indispensable for the formation of medulloblastoma and rhabdomyosarcoma, *J. Biol. Chem.*, 275(37), 28341–28344, 2000.
325. Peters, J.M., Narotsky, M.G., Elizondo, G., et al., Amelioration of TCDD-induced teratogenesis in aryl hydrocarbon receptor (AhR)-null mice, *Toxicol. Sci.*, 47(1), 86–92, 1999.
326. Tanaka, M., Ohtani-Kaneko, R., Yokosuka, M., Watanabe, C., Low-dose perinatal diethylstilbestrol exposure affected behaviors and hypothalamic estrogen receptor-alpha-positive cells in the mouse, *Neurotoxicol. Teratol.*, 26(2), 261–269, 2004.
327. Abbott, B.D., Schmid, J.E., Brown, J.G., et al., RT-PCR quantification of AHR, ARNT, GR, and CYP1A1 mRNA in craniofacial tissues of embryonic mice exposed to 2,3,7,8-tetrachlorodibenzo-p-dioxin and hydrocortisone, *Toxicol. Sci.*, 47(1), 76–85, 1999.
328. Elamazar, M.M. and Nau, H., Potentiation of the teratogenic effects induced by coadministration of retinoic acid or phytanic acid/phytol with synthetic retinoid receptor ligands, *Arch. Toxicol.*, 78(11), 660–668, 2004.
329. Nagase, T., Nagase, M., Osumi, N., et al., Craniofacial anomalies of the cultured mouse embryo induced by inhibition of sonic hedgehog signaling: an animal model of holoprosencephaly, *J. Craniofac. Surg.*, 16(1), 80–88, 2005.
330. Gershon, M.D., Endothelin and the development of the enteric nervous system, *Clin. Exp. Pharmacol. Physiol.*, 26(12), 985–988, 1999.
331. Brand, M., Kempf, H., Paul, M., et al., Expression of endothelins in human cardiogenesis, *J. Mol. Med.*, 80(11), 715–723, 2002.

CHAPTER 3

Pathogenesis of Abnormal Development

Lynda B. Fawcett and Robert L. Brent

CONTENTS

I. Introduction 61
II. Manifestations of Developmental Toxicity 62
   A. Structural Anomalies: Malformations, Deformations, and Disruptions 62
   B. Multiple Defects: Syndromes, Sequences, and Associations 63
III. Factors That Influence the Pathogenesis of Abnormal Development 63
   A. Stage of Development 63
   B. Tissue Specificity 64
   C. Influence of Dose 65
IV. Pathogenesis at the Cellular Level 67
   A. Cell Death 68
   B. Cell Proliferation 70
   C. Alterations in Cell-Signaling and Cell–Cell Interactions 71
   D. Cell Migration and Differentiation 74
V. Examples of Abnormal Pathogenesis 75
   A. Defects with Different Underlying Pathogenesis but Similar Morphology 76
      1. Cleft Palate 76
      2. Neural Tube Defects 77
   B. Abnormalities with Similar Pathogenesis but Different Morphology 78
VI. Overview and Perspective 80
References 81

I. INTRODUCTION

Vertebrate development proceeds in a sequence of carefully timed events that progress from the cellular level to the formation of tissues, organ systems, and morphologic structures. Alterations at any level may permanently alter later developmental processes. The initial adverse effects of an environmental agent on a developing organism most frequently occur at the molecular level. Such effects constitute the *mechanisms* by which the agent alters development. Subsequent events that include deviations at the cellular, tissue, and organism levels constitute the *pathogenesis* of abnormal development. It is this sequence of abnormal developmental processes, i.e., events downstream of the mechanism, that is the focus of this chapter. Unfortunately it is not possible within the scope of this chapter to include all that is known or suspected about the pathogenesis of the myriad agents that have been classified as developmental toxicants. Moreover, the pathogenesis involved in the genesis of many congenital anomalies has not yet been determined. Therefore, we instead cover the pathogenesis of selected toxicants and selected well-defined malformations as examples to illustrate the major considerations involved in the study of the pathogenesis of abnormal development.

II. MANIFESTATIONS OF DEVELOPMENTAL TOXICITY

Vertebrate development is hierarchical, composed of a series of events that are both temporally and spatially regulated. Perturbations in any of the normal developmental processes can potentially alter subsequent growth and morphogenesis and result in congenital malformations. Adverse fetal outcomes that result from exposure to developmental toxicants are not limited to congenital malformations but also manifest as functional deficits, growth retardation, and death. Functional deficits refer to physiological or system deficits (e.g., immunological and hormonal) and, more commonly, to neurological and behavioral deficits. The potential adverse fetal outcomes are not mutually exclusive. In fact, they are frequently associated. For instance, anatomical defects are often accompanied by growth retardation.[1]

The etiology and discrete pathogenesis responsible for the majority of congenital defects in humans is unknown.[2-4] For the most part, our understanding of the pathogenesis of congenital anomalies is based on studies that use as models animals exposed to developmentally toxic agents, and on studies of spontaneous mutants and transgenic mice. These studies have revealed that although there are some consistent features associated with the induction of a given anomaly, the underlying pathogenesis may be quite different. Thus, a particular type of anomaly, such as cleft palate, may be caused by diverse agents and mechanisms. While the early pathogenesis of the defect may differ, the pathogenic pathways ultimately converge. Likewise, a particular agent or resulting pathogenesis may result in very different outcomes depending on factors such as dose, embryonic age, and genetic susceptibility. Because of this complexity, it is extremely difficult to ascertain the underlying pathogenesis of a given developmental anomaly based on the final outcome, even to the extent of determining whether it is genetic or environmentally induced. Rather one must look earlier in the developmental process to determine which tissues were initially affected and what the effects or consequences on the target tissue or cell population were. This knowledge, coupled with an understanding of the normal embryological processes that would govern later recovery and development of the structure, will provide insight into the overall pathogenesis leading to the defect.

A. Structural Anomalies: Malformations, Deformations, and Disruptions

Dysmorphology resulting from developmental toxicants can be described as resulting from processes of malformation, deformation, or disruption.[5] Malformations occur when the normal developmental process is altered such that a given structure cannot form or forms improperly. In this case, the error is intrinsic to the morphogenetic process itself. Generally speaking, the embryo is most susceptible to the adverse effects of toxicants that produce malformations during organogenesis, when cells are rapidly proliferating and differentiating, and extensive patterning is taking place. Malformations such as limb defects, whether induced by chemical means during early organogenesis or due to genetic defect, are usually (but not always) bilateral.[6-8]

Unlike malformations, deformations and disruptions involve alterations to already existing structures. In these cases, the intrinsic developmental processes had proceeded normally, and extrinsic factors not related to the developmental processes lead to the dysmorphology observed at term. Deformations usually result from mechanical factors, such as uterine constraint or amniotic bands, and include alterations in body shape, form, or position. Well-known examples are congenital dislocation of the hip, plagiocephaly, clubfoot, and some facial anomalies. Disruptions differ from deformations in that they are usually more severe, often involving extensive destruction of tissue, disruption of tissue function, and/or prevention of subsequent tissue or organ formation. Like deformations, disruptions can arise from mechanical factors impinging on the fetal environment or fetal tissue directly, such as might occur in amniotic band syndrome, fetal vascular occlusion, or placental emboli. Additionally, disruption can result from alterations in fetal physiology, especially those resulting in localized or general hypoxia and vascular insults.[9] It is important to remember that these classifications are not mutually exclusive. This is especially true when multiple structural defects are present. Thus, disruptions to another tissue may be caused by mechanical factors related to a previous defect, such as occurs with hydronephrosis. Excessive buildup of urine due to ureter blockage leads to oligohydramnios, damage to the kidneys, and pulmonary hypoplasia (reviewed by Coplen[10]).

B. Multiple Defects: Syndromes, Sequences, and Associations

The majority of developmental toxicants do not produce only a single anomaly. The effects of a toxicant on development will typically manifest as a distinguishable pattern of multiple anomalies that vary with the timing of exposure and the differing tissue susceptibilities of the embryo. There are some exceptions to this general rule, such as glucocorticoid-induced cleft palate in mice.[11,12] Such outcomes can occur in the absence of other malformations and only the frequency of the defect, and not the defect itself, changes with timing of administration. However, growth retardation usually accompanies the experimentally induced malformation. When multiple anomalies are observed, it is often helpful to classify them according to whether the multiple defects arise in tissues independently as a result of a similar pathogenesis, are secondary to a single anomaly or pathogenic mechanism, or represent other associations. A recognizable pattern of multiple structural anomalies is generally referred to as a syndrome, a sequence, or an association.

Multiple anomalies are considered a *syndrome* when they result from the same or similar underlying mechanisms and thus are causally related.[5,13] Syndromes can result from genetic mutation or may be teratogen induced and thus usually have a known etiology. Examples of genetic syndromes in humans include Down's, Meckel, and Prader-Willi syndromes. Recognized teratogen-induced syndromes include thalidomide syndrome, fetal alcohol syndrome, retinoic acid embryopathy, and diphenylhydantoin syndrome. A *sequence* represents a pattern of defects that are secondary to a single defect, with a discrete known or suspected pathogenesis but often with an unknown etiology. The pathogenic events that result downstream from amniotic bands or from fetal hydronephrosis would represent a sequence.

A pattern of malformations that have not yet been classified as either a syndrome or a sequence, but that occur in a nonrandom fashion, is referred to as an *association*.[14] As the etiology and pathogenesis become more understood, associations are reclassified into one of the other two categories. Examples of associations in the human include the VATER association (*v*ertebral defects–*a*nal atresia–*t*racheoesophageal fistula–*e*sophageal atresia–*r*adial and renal dysplasia). When cardiac and limb defects are also present it is often termed the VACTERAL association.[14] Because these defects tend to occur as a pattern of anomalies, there is a high likelihood that there is a common, although as yet undiscovered, underlying mechanism and pathogenesis.

III. FACTORS THAT INFLUENCE THE PATHOGENESIS OF ABNORMAL DEVELOPMENT

A. Stage of Development

Developmental toxicity induced by environmental agents usually results in a spectrum of morphologic anomalies or intrauterine death that varies in incidence depending on stage of exposure and dose. The developmental period at which exposure occurs will determine which structures are most susceptible to the deleterious effects of the agent and to what extent the embryo can repair the damage. The period of sensitivity may be narrow or broad, depending on the environmental agent and the malformation in question. Limb defects produced by thalidomide have a very short period of susceptibility. Moreover, the type of limb defect that can be induced changes rapidly as embryonic age increases from 22 to 36 days postconception.[15,16] The mechanism of thalidomide's action on the cells, while still undetermined, very likely remains the same. Thus, the differing sensitivities presumably result from changes in the target cell population itself over time. Such changes could include alterations in proliferation rate, stage in cell cycle, commitment to differentiation, or expression of specific receptors. Other teratogens, and particularly those that cause a more nonspecific cytotoxicity, may have much longer periods of susceptibility. Radiation-induced microcephaly, which can be induced far into the fetal period, is an example.[17,18]

Numerous studies and observations have shown us that during the first period of embryonic development, from fertilization through the early postimplantation stage, the embryo is most sensitive to the embryolethal effects of drugs and chemicals. Surviving embryos have malformation rates similar to the controls, not because malformations cannot be produced at this stage, but because significant cell loss or chromosome abnormalities at these stages have a high likelihood of killing the embryo. Because of the omnipotentiality of early embryonic cells, surviving embryos have a much greater ability to have normal developmental potential. This trend of marked resistance to the malforming consequences of teratogens has been termed the "all-or-none phenomenon." Utilizing x-irradiation as the experimental teratogen, Wilson and Brent demonstrated that the all-or-none phenomenon disappears over a period of a few hours in the rat during early organogenesis.[19]

The term "all-or-none phenomenon" has been misinterpreted by some investigators to indicate that malformations cannot be produced at this stage. On the contrary, it is likely that certain drugs, chemicals, and other insults during this stage of development can result in malformed offspring.[20,21] But the nature of embryonic development at this stage will still reflect the basic characteristic of the all-or-none phenomenon, which is a propensity for embryolethality, rather than surviving malformed embryos.

The period of organogenesis encompasses rapid cell division, extensive patterning, and tissue differentiation. Thus, it is not surprising that this period represents the stage when the embryo is most susceptible to dysmorphogenesis from developmental toxicants and that the majority of congenital malformations can be produced by alterations in developmental processes during this period. Exceptions include malformations of the genitourinary system, palate, and brain, as well as abnormalities produced by deformations and disruptions.

During the fetal period, teratogenic agents may decrease the cell population by producing cell death, inhibiting cell division, or interfering with cell differentiation. The resulting effects, such as cell depletion or functional abnormalities, may not be readily apparent at birth. Major structural anomalies can be produced during the fetal period, and they usually result from disruptions or deformations due to factors such as uterine constraint, hypoxia, or the action of vasoactive substances such as cocaine. Disruptions induced at this stage may involve limbs, major organs, including the brain, or other systems. Severe growth retardation in the whole embryo or fetus may also result in permanent deleterious effects on many organs or tissues, especially the brain and reproductive organs.

B. Tissue Specificity

The pathogenesis associated with a developmental toxicant depends not only on the nature of the teratogen itself but also on the susceptibility of the embryonic tissues. In general, rapidly dividing cell populations will be the most sensitive to developmental toxicants that result in cell death or reduced cell proliferation. This is not because other cell populations receive no exposure, but because depletion of the cell population in areas of rapid cell division is more likely to interfere with the normal developmental process. Exceptions include agents that act specifically on dividing cells, such as ionizing radiation, 5-aza-2-deoxycytadine (5-AZA), and 5-fluorouracil. In this case only dividing cells will be affected because the agents affect only active DNA synthesis. Some tissues such as the heart appear to be quite resistant to cytotoxic agents such as 5-AZA and cyclophosphamide metabolites. Other tissues, such as those in the limb bud and neural tube, are quite sensitive to these toxicants. It is interesting that the chorioplacenta, which is a proliferative organ, is very resistant to cytotoxic agents.[22]

Some developmental toxicants act only on very specific subpopulations of cells that have unique, tissue specific receptors. Teratogens that have this characteristic will only affect the target organ and do not usually produce malformations at other sites. Examples of this class of toxicants include sex hormones, such as the progestins, estrogens, and androgens. In contrast to progesterone and 17α-hydroxyprogesterone caproate, high doses of some of the synthetic progestins have been reported to cause virilizing effects in humans. Exposure during the first trimester to large doses of androgen-related progestins, such as 17α-ethinyltestosterone, have been associated with masculinization of the external genitalia of female fetuses.[23,24] Similar associations result from exposure to large doses of 17α-ethinyl-nortestosterone (norethindrone) and 17α-ethinyl-17-OH-5(10)estren-3-one (Enovid-R).[24,25] The preparations with androgenic properties may cause abnormalities in the genital development of females only if present in sufficient amounts during the critical period of development.[23,24,26] Grumbach and colleagues[25] point out that labial fusion could be produced with large doses if the fetuses were exposed before the 13th week of pregnancy, whereas clitoromegaly could be produced after this period. Such findings demonstrate that a specific form of maldevelopment can be induced only when the embryonic tissues are in a susceptible stage of development. The synthetic progestins, like progesterone, can influence only those tissues with the appropriate steroid receptors. Because the steroid receptors that are necessary for naturally occurring and synthetic progestin action are absent from nonreproductive tissues early in development, the evidence is against the involvement of progesterone or its synthetic analogs in nongenital tissues.[6,27–31] Only cells that bear the appropriate sex hormone receptors can be affected by sex steroids. Therefore, it follows that only those tissues bearing cells with the appropriate receptors can be developmentally modified by the presence of these hormones.

C. Influence of Dose

The dose or magnitude of the exposure to a developmental toxicant may greatly affect the ensuing pathogenesis and final outcome. Agents that have more generalized mechanisms of action, such as antiproliferative or cytotoxic drugs, may result in intrauterine growth retardation at lower doses and structural anomalies or death at higher doses. On the other hand, agents that have receptor-mediated tissue specificity may not manifest fetotoxicity at higher doses because tissues not bearing appropriate receptors would not be affected at any dose. However, one might expect a dose responsive gradation in the severity of the insult to the specific tissues or organs affected.

An important consideration is not only the dose but also the overall magnitude of the exposure to the toxicant. Prolonged exposure to doses below cytotoxic levels or below levels that inhibit cellular function would not be expected to induce perceptible changes to developmental processes. When designing toxicity studies to assess human reproductive safety, investigators need to carefully consider the effects from chronic exposure because gestation in the human is much longer than for most animal species used for these studies. The pathogenesis of some agents, such as the angiotensin-converting enzyme (ACE) inhibitors, involves effects secondary to oligohydramnios in the fetal period, a consequence of fetal anuria.[32–39] Because of the much shorter gestation and fetal period of most experimental species, pathogenesis relating to these types of secondary effects may not be observed in animal toxicology studies. In some instances the ACE-inhibitor syndrome was not found because the experimental animals were only exposed during early organogenesis.

Table 3.1 Stochastic and threshold dose-response relationships of diseases produced by environmental agents

Phenomenon	Pathology	Site	Diseases	Risk	Definition
Stochastic	Damage to a single cell may result in disease	DNA	Cancer, germ cell mutation	Some risk exists at all dosages; at low doses, risk may be less than spontaneous risk	The incidence of the disease increases but the severity and nature of the disease remain the same
Threshold	Multicellular injury	Multiple; variable etiology affecting many cell and organ processes	Malformation, growth retardation, death, toxicity, etc.	No increased risk below the threshold dose	Both the severity and incidence of the disease increase with dose

Source: Modified from Brent, R.L., *Teratology*, 34, 359, 1986. With permission.

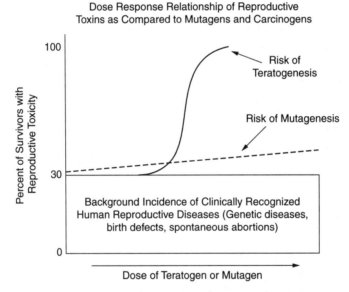

Figure 3.1 The dose response curve of environmental toxicants (drugs, chemicals, and physical agents) can reveal deterministic (threshold) and/or stochastic effects. Mutagenic and carcinogenic events are stochastic phenomena and theoretically do not have a threshold exposure below which no risk exists. At low exposures the risk still exists but is usually below the spontaneous risk of cancer and mutations. Whether the curve is linear or curvilinear for stochastic phenomena can be debated, but from a theoretical point of view it intersects zero. Toxicological phenomena, such as teratogenesis, that do not involve mutagenic and carcinogenic effects, usually follow an S-shaped curve with a threshold below which no effects are expected.

A final but extremely important consideration regarding pathogenesis and dose involves the concept of threshold. The threshold dose is the dosage below which the incidence of death, malformation, growth retardation, or functional deficit is not statistically greater than that of controls. For drugs and other chemicals, the threshold level of exposure is usually from less than one to three orders of magnitude below the teratogenic or embryopathic dose required to kill or malform half the embryos. A teratogenic agent therefore has a no-effect dose, as compared with mutagens or carcinogens, which have a stochastic dose-response curve (Table 3.1, Figure 3.1).[40] The severity and incidence of malformations produced by every exogenous teratogenic agent that has been appropriately tested have exhibited threshold phenomena during organogenesis.[2]

Table 3.2 Mechanisms of action of environmental teratogens

1. Cell death or mitotic delay beyond the recuperative capacity of the embryo or fetus (e.x., ionizing radiation, chemotherapeutic agents, alcohol)
2. Inhibition of cell migration, differentiation, and/or cell communication (e.x., retinoic acid, cyclopamine)
3. Alterations in programmed cell death
4. Interference with histogenesis by processes such as cell depletion, necrosis, calcification, or scarring
5. Biologic and pharmacological receptor-mediated developmental effects (e.x., etretinate, isotretinoin, retinol, sex steroids, streptomycin, thalidomide)
6. Metabolic inhibition (e.x., warfarin, anticonvulsants, nutritional deficiencies)
7. Deformations (physical constraint from: uterine myoma, multiple pregnancy, oligohydramnios, amniotic band syndrome, etc.)
8. Disruptions: vascular disruption, placental emboli, inflammatory lesions, amnionitis; (e.x., cocaine, chorionic villus sampling, misoprostol)

Source: Modified from Beckman, D.A. and Brent, R.L., Mechanism of known environmental teratogens: Drugs and chemicals, in *Clinics in Perinatology*, Brent, R.L. and Beckman, D.A., Eds., W.B. Saunders, Philadelphia, 1986, p. 649. With permission.

It is important to keep in mind that the threshold concept, as it is usually applied, concerns fetal outcome and malformation, not effects observed at the cellular level. Effects at the cellular level are probably still detectable below the observable threshold of an agent, but these effects may not be deleterious and may be completely reversible. Cellular effects most likely also manifest threshold phenomena because a certain number of receptors or ligands must be affected by a chemical or drug before cellular processes themselves are diverted via second messenger pathways or by other means. At the tissue level, a threshold number of cells within the tissue must be diverted from normal processes or killed to result in dysmorphology, dysfunction, or embryonic death. Thus, another way to view threshold is to say that below the threshold, a developmental toxicant has no discernable *final* pathogenesis. However, that should not imply that there are no discernable early cellular effects from the agent. Empirically, it means that the dose was low enough that the affected cells did not permanently alter the developmental process, and recent experimental evidence supports this concept. Francis, Rogers, and colleagues[41,42] observed that cyclophosphamide and 5-AZA, agents that result in significant cell death and resultant developmental toxicity and dysmorphology at higher doses, induced significant perturbations in the cell cycle at doses that did not produce overt teratogenesis. These studies illustrate the ability of the embryo to compensate or recover from damage of a certain level. It is expected that as techniques for studying early pathogenesis become more sensitive, we will have the increased ability to detect changes at the cellular level that do not translate into overt developmental toxicity or dysmorphogenesis.

IV. PATHOGENESIS AT THE CELLULAR LEVEL

There are myriad mechanisms by which developmental toxicants can affect cellular processes. Fortunately, from the standpoint of understanding pathogenesis, effects at the cellular level will usually manifest as one or more of the following: (a) cell death, (b) altered cell–cell interactions, (c) reduced biosynthesis, (d) impaired cell migration, or (e) reduced or impaired proliferation. The potential cellular outcomes from exposure to a developmental toxicant were first described by Wilson[2] and have been modified and expanded by other investigators (Table 3.2).[43] For instance, while it has long been accepted that excessive cell death may result in dysmorphology, it is now recognized that a failure to induce appropriate programmed cell death can also result in dysmorphogenesis (reviewed by Mirkes[44]). It is important to keep in mind that although alterations to cellular function can be described as a discrete outcome (e.g., cell death), the various cellular outcomes may often occur concurrently within the organism. Moreover, because of the hierarchical nature of pathogenesis, potentially all of these cellular events will eventually come into play as development proceeds.

A. Cell Death

A generalized mechanism of cell death is often suspected as an early event in the pathogenesis of abnormalities when growth retardation is also present. It should also be suspected when the agent causes manifestations of abnormal development in multiple tissues that, based on timing of exposure, correlate with the tissues' rapid growth or development. However, the only way to determine if cell death represents a major early pathogenic event is to monitor for cell death soon after exposure to the agent or event that results in the malformation. The toxicologist should then focus not only on those tissues known to be adversely affected by the agent but also on other tissues that are rapidly dividing and differentiating during that period of gestation.

Pathogenesis of malformations resulting from exposure to cytotoxic agents is highly dependent on the stage of development, the number of cells affected, and the inherent ability of the tissue to recover. The cells of very early embryos (preorganogenesis) retain a great deal of multipotentiality. Thus, division of the remaining cells can often overcome cell loss, with little or no consequences to later development. However, when the number of cells that die exceeds the capacity for the embryo to replace them, embryonic death or malformation may result. When organogenesis stage embryos are exposed to agents that cause cell death, there is a high likelihood of abnormal development. Pathogenesis will depend on both the nature and location of the cell type affected and the extent of the cell death in that tissue. Many agents that produce developmental abnormalities have been associated with an increase in cell death in the affected structure.[45]

It is generally accepted that cell death can take one of two forms, i.e., apoptosis, often referred to as programmed cell death, and necrosis. Necrosis usually results from a more rapid and severe insult to a cell, and has a greater potential for damage to surrounding cells and tissues. Cell death may take the form of necrosis when cells are irreversibly injured as a result of exposure to cytotoxic agents, extremes of pH or temperature, free radicals, or membrane disrupters. During the necrotic process, there is loss of cell membrane integrity, with concomitant leakage of cytoplasm and marked enlargement of mitochondria (reviewed by Zakeri and Ahuja[46]). Necrotic cell death culminates with cell lysis and release of the cellular contents into the surrounding environment, where nucleases and proteases then degrade the cellular constituents. This rapid disintegration of the cell has injurious effects on surrounding tissues because of the release of hydrolytic (lysosomal) enzymes and the activation of inflammatory mediators.

Unlike necrosis, apoptosis is a carefully regulated and tightly controlled process that has few or no direct adverse consequences on surrounding tissues. Apoptosis is a normal and essential component of development with well-known examples that include apoptosis in the vertebrate neural tube to facilitate neural tube fusion and removal of interdigital mesenchyme to form the digits of the limb.[44] Recent studies using *p53* knockout mice and 4-hydroperoxycyclophosphamide (4OOH-CPA) as the teratogen have provided evidence that whether the cellular response to a cytotoxic agent follows a necrotic or an apoptotic pathway may be influenced by levels of *p53*. Wild type mice responded to 4OOH-CPA with cell death characteristic of apoptosis, while cells in mice null for *p53* displayed characteristics of necrosis. Heterozygotes had cells that displayed characteristics of both apoptosis and necrosis.[47]

There are two major pathways for apoptosis: a receptor-mediated, or extrinsic, pathway, and a mitochondrial, or intrinsic pathway.[44] Extrinsic apoptosis occurs in response to the binding of ligands, such as TNFα, to cellular receptors, while intrinsic apoptosis is initiated by factors that ultimately cause the release of cytochrome c from the mitochondria. It is this latter type of apoptosis that is most frequently associated with exposure to developmentally toxic agents.[44] Both pathways are mediated by caspase cascades, especially the cleavage of procaspase 3 to caspase 3, and both result in similar dismantling of cellular constituents. However, in receptor-mediated apoptosis, caspases 8 and 10 are involved in the initiation of apoptosis, while in intrinsic apoptosis the cascade is initiated by caspase 9.[44]

Table 3.3 Methods to detect cell death

Technique	Methodology	Comments	Selected References
Acridine orange	Staining with vital dye	Stains phagolysosomes associated with cell death; visual inspection using fluorescence microscopy	49, 64, 226
Neutral Red	Staining with vital dye	Same as above	52
Nile Blue Sulfate	Staining with vital dye	Same as above	41, 42, 60, 61, 227
Apoptosis — DNA ladder	DNA electrophoresis	Care must be taken with DNA sample to prevent extraneous degradation	47, 49, 52
Apoptosis — TUNEL	Terminal transferase dUTP nick end labeling; labels 3′ ends of cleaved DNA	Detects DNA strand breaks associated with apoptosis	55, 65, 228, 229
Dual staining for apoptosis	Combines TUNEL with immunohistochemistry	Requires multiple wavelengths for fluorescence but can be modified for colorimetric substrates	48

Studies using a variety of agents and adverse stimuli have demonstrated reproducible differences in the sensitivity of embryonic tissues to cell death. For instance, regions including the prosencephalic neuroepithelium and surrounding mesenchyme are especially sensitive to cell death induced by agents such as hyperthermia, while the heart, mesencephalic mesenchyme, and yolk sac are resistant.[44,48] A similar pattern of tissue selectivity has been reported for cell death induced by ethanol, 4OOH-CPA, 2-deoxyadenosine, staurosporine, and other developmental toxicants.[44,49–53] This differential sensitivity appears to be related, at least in part, to the ability of cells in resistant tissues such as the heart to prevent the release of cytochrome c by the mitochondria and thus prevent the apoptotic cascade.[44,52–54] Recent evidence also indicates that the expression of p53 in response to cytotoxic agents may also differ among tissues and may contribute to differing tissue susceptibilities.[55]

Some agents, by their nature, target proliferating cells. This replication-associated cytotoxicity most frequently results from agents that affect DNA synthesis. For example, cyclophosphamide (or its activated analogs) causes alkylation of DNA during the S-phase of the cell-cycle.[56] This is initially detectable as an alteration of cell cycle, and eventually leads to cell death in the rapidly dividing cell populations of the developing embryo.[41,56] Similar findings have been reported with 5-AZA.[42,57]

Unlike the case of agents that target proliferating cells, the pathogenesis of some developmental toxicants involves quantitative increases in cell death in regions of tissues that are already undergoing programmed cell death. Retinoic acid (RA) is a good example of a teratogen that displays these properties. Retinoic acid exposure results in a variety of defects involving diverse tissues, depending on dose and stage of exposure. Studies have demonstrated that exposure to RA results in increases in cell death in regions of normally occurring cell death. Excessive cell death in these zones has been correlated to the formation of structural anomalies, including spina bifida, mesomelia, digit defects, and craniofacial malformations.[58–61]

It is important to differentiate between effects due to cell death and those caused by a failure of cellular proliferation or differentiation, as these have the potential to produce similar outcomes. In most instances, when excessive cell death is an initial component of pathogenesis, it occurs rapidly and is detectable within the first few hours following exposure. Numerous techniques have been developed to measure cell death and to differentiate between necrosis and apoptosis, and many of these have been applied to embryonic studies. These techniques are summarized in Table 3.3. One of the most frequently applied methods involves the use of vital dyes, such as Nile Blue Sulfate (NBS). NBS, like many vital dyes used to detect cell death, is accumulated in membrane bound acidic compartments (lysosomes) that occur when neighboring cells or macrophages engulf cellular debris from dead cells, and it also stains free apoptotic bodies.[62–64] More specific approaches used to detect apoptosis in embryonic studies include examination of DNA fragmentation and detection of the presence of activated factors in the apoptotic cascade. Because apoptosis results in the cleavage of DNA into discrete lengths, electrophoresis of DNA from tissues undergoing apoptosis will reveal a "DNA ladder" effect.[47,49,52] Terminal transferase d*U*TP *n*ick *e*nd *l*abeling (TUNEL) can also detect DNA fragmentation.[65] This technique is based on the use of terminal deoxynucleotidyl transferase (TdT) to label free 3-hydroxyl ends of DNA fragments with a labeled nucleotide. This technique is relatively flexible with regard to the label used on the dUTP and has included the use of peroxidase, alkaline phosphatase, biotin, and fluorescent markers. Cell death can also be examined by detecting specific molecules involved in apoptosis, such as cleaved caspase-3, by immunohistochemistry on histological sections or whole mounts. A recent and useful technique has been described that combines both immunostaining for cleaved caspase-3 and TUNEL,[48] and allows for a more complete assessment of the presence of apoptosis in embryonic tissues.

B. Cell Proliferation

Agents that produce alterations in cell proliferation can produce fetal outcomes that are similar to agents that cause cell death. Like cell death, reduced cellular proliferation reduces the number of cells available in a tissue for later growth and differentiation. Alterations in cellular proliferation may be suspected as an early pathogenetic event when excessive cell death is not readily apparent or is insufficient to account for ensuing dysmorphogenesis. Reduced cell proliferation can result from direct effects on progression through the cell cycle and mitosis, deficiencies in growth factors, deficiencies in signaling or induction factors, or impairment of the cellular response to these molecules.

Tissues most likely affected by agents that directly impair cell-cycle mechanisms are those that contain the most rapidly dividing cells at the time of exposure. Replication-associated cytotoxic agents that affect DNA replication or mitosis, such as cyclophosphamide and 5-AZA, have been shown to induce detectable but reversible alterations in the cell cycle at doses below those that result in cell death or dysmorphology.[41,42,56] Other teratogens, such as valproic acid (VPA), may alter the cell cycle without directly interfering with DNA replication. For instance, VPA treatment induced a cell-cycle arrest in the mid-G1 phase in the C6 glioma cell line *in vitro* at doses below those producing overt cytotoxicity.[66,67] *In vivo* studies in mice have reported evidence that the VPA-induced cell-cycle arrest may involve alterations to the enzyme ribonucleotide reductase in specific regions of the neuroepithelium associated with neural tube closure.[68] Alterations in C6 glioma cell proliferation *in vitro* have also been associated with the ability of anticonvulsant agents to induce neural tube defects (NTDs) *in vivo* and thus have the potential to be a predictive indicator for NTD pathogenesis.[69]

In addition to direct effects on cell replication, failure of cell proliferation can also occur as a result of a deficiency of the growth factors or signaling molecules that induce proliferation. Failure of cell proliferation by these means usually occurs downstream from other pathogenic events that affect the cell populations responsible for the synthesis of the growth factor or signaling molecule involved. In these cases, measurement of cell proliferation in affected regions may show decreases in the number of cycling cells but may not reveal discrete cell-cycle arrest. The effects of cell-signaling molecules on tissue growth and differentiation will be discussed in greater depth later in this chapter.

Numerous methods are employed to detect changes in cell proliferation and the cell cycle. Analysis of levels of cell proliferation can be performed on histological sections by analysis of the number of mitotic cells present or by assessment of the incorporation of labeled nucleotides. Some traditional methods, such as incorporation of radiolabeled nucleotides into cells undergoing DNA synthesis, have largely been replaced by similar techniques that utilize nonradioactive labels. For example, incorporation of bromodeoxyuridine (BrdU), a thymidine analog, can be detected by

Table 3.4 Methods to detect alterations in cell-cycle

Technique	Methodology	Comments	Selected References
Flow cytometry	Marker of cell cycle using propidium iodide (or other) fluorescent DNA stain	Requires use of expensive equipment	41, 42, 55, 56
Incorporation of DNA analogs	Incorporation of analogs into DNA of replicating cells	Most radioisotope techniques have been replaced by use of nonisotope labels, analogs such as BrdU	68, 70, 71, 228
Indicator cell lines; micromass culture	In vitro growth of cells or tissue	Follow by analyzing proliferation by use of flow cytometry, cell number, or DNA synthesis markers	66–69, 230–234

immunohistochemistry in tissue sections or by use of *in vitro* enzyme-linked immunosorbant assays (ELISAs) on cells following *in vitro* or *in vivo* exposure to potential toxicants.[68,70,71] A more quantitative measure for cell-cycle analysis utilizes flow cytometry on isolated nuclei stained with fluorescent labels, such as 4,6-diamidino-2-phenylindole (DAPI) or propidium iodide, that bind DNA. This type of analysis relies on distinguishing nuclei based on their DNA content and can thus differentiate cells in G0/G1, S, and G2/M phases.[41,42,56,72] It is important when using this technique to isolate the cell populations of interest, which may be a difficult task depending on the size or stage of the embryo employed. Flow cytometric analysis of the cell cycle is also useful when measuring the effects of a teratogen on indicator cell lines, such as C6 glioma cells, cells isolated from embryos and grown *in vitro* as monolayers, or cells from micromass or organ culture. Selected methods for assessing cell proliferation are given in Table 3.4.

C. Alterations in Cell-Signaling and Cell–Cell Interactions

Cell-signaling pathways form the basis for embryonic patterning and differentiation and thus are an important consideration in determining the pathogenesis of a developmental toxicant. Signaling pathways control morphogenesis by the secretion of cell products that are both temporally and spatially distributed. The downstream effect of these developmental signaling molecules on a tissue usually occurs at the level of transcription and results in alterations of gene expression in the target cells. The role of signaling molecules at the cellular and tissue level is discussed in the following text in the context of their involvement in both normal and abnormal development.

Numerous cell-signaling pathways are involved in embryogenesis. Perhaps the best understood involves the ligand Sonic Hedgehog (Shh), the receptors patched (Ptc) and smoothened (Smo), and the GLi family of transcription factors that are involved in the transduction of the Shh signal. This pathway has been highly conserved across organisms as diverse as drosophila, amphibians, and mammals (reviewed by Walterhouse et al.[73]). Genetic alterations in Shh signaling result in similar and reproducible phenotypes in humans and in mice. For example, altered expression of *Shh* in humans results in holoprosencephaly,[74–76] and in mutant mice lacking *Shh* (*shh -/-*) the resulting phenotype is also holoprosencephaly, with other axial patterning anomalies.[77] Human syndromes involving mutations in the *GLi3* gene include Greig cephalopolydactyly and Pallister–Hall syndrome, both of which include postaxial polydactyly and craniofacial malformations.[78,79] The phenotype of a mouse mutation of *Gli3*, extra-toes (*Xt*), also includes polydactyly and malformations similar to those observed in humans.[80] Such observations from genetic mutants and numerous embryological studies illustrate the importance of signaling molecules such as Shh for the development of numerous tissues and structures. For instance, in vertebrates the Shh pathway is crucial to dorsal-ventral patterning of the CNS, anterior-posterior specification in the limbs and somites, axial skeleton patterning, and gastrointestinal tract and lung development.[73,81–87]

The patterning or morphogenesis achieved by signaling pathways is based on the secretion of different signaling molecules from several signaling pathways in overlapping gradients and along multiple axis. Alterations in the strength of one signal, either by changes in the secretion or the transduction of the signal, will substantially alter morphogenesis. Changes in transduction may involve alterations in the expression of receptors, transcription factors, inhibitors, or downstream signaling factors, or in the distribution of gap junctions.[88–90] Cell signaling in the vertebrate limb has been extensively studied and provides a good example of the effects of signaling molecules on morphogenesis. In the vertebrate limb, as in other tissues, signaling molecules form overlapping gradients in three dimensions, such that cells within the limb bud attain a positional value.[86] Signaling molecules for each dimensional specification are produced by discrete tissues within the limb bud. Damage to these regions or alterations in the signaling pathways evoked from these regions results in characteristic phenotypes that give insight to the identity of the affected tissue or pathway responsible. For example, dorsal-ventral patterning of the limb involves signaling pathways that originate in the overlying ectoderm of the limb bud and involves the expression of *Wnt7a* for dorsal specification and *engrailed-1* for ventral patterning.[91] Mutations in *engrailed-1* result in both dorsal and ventral expression of *Wnt7a* and dorsal morphology on both dorsal and ventral sides, while mutations in *Wnt7a* give the reverse scenario.[92]

Anterior-posterior patterning in the limb largely involves Shh, produced by cells in the zone of polarizing activity (ZPA), and transduction of the Shh signal by other proteins, such as BMP-2.[93–95] Altered expression of *Shh* results in altered patterning of distal structures.[96,97] An increase in Shh signaling, produced by beads soaked with Shh and placed on the anterior margin of the developing limb bud, results in the development of additional digits.[95] In the mouse mutant extra-toes there is up-regulated signaling of Shh in the anterior regions of the limb bud and the addition of a supernumerary digit. Conversely, an absence of Shh results in the absence of distal structures (such as digits), but proximal structures are generally unaffected.[96,97]

Absence of distal structures can also occur because of altered signaling involved in apical-distal growth. Apical-distal outgrowth requires signals produced from the apical ectodermal ridge (AER) that overlies the rapidly dividing progress zone of the limb bud. The signals produced by the AER that appear responsible for apical-distal outgrowth are members of the FGF family of proteins.[86,98] Shh from the ZPA also forms a positive feedback loop with FGF-4 from the AER, and this feedback is also required for continued outgrowth of the limb bud and maintenance of the progress zone.[99–101] Damage to the AER results in truncation of the limb, with failure of distal structures to form,[86] a phenotype similar to that observed with lack of *Shh* expression. The effect is stage dependent. Earlier removal of the AER leads to loss of all structures and later removal allows formation of proximal features, such as long bones, but an absence of distal features, such as digits.[102,103]

In contrast to damage in the AER, depletion of cells in the progress zone by teratogens that kill or impede cell proliferation, such as x-irradiation and 5-AZA, results in formation of distal structures and failure to produce proximal structures,[71,104] a phenotype similar to that observed for thalidomide. Because the progress zone is a highly proliferative tissue, it is potentially susceptible to teratogens that target mitotic cells. Excessive cell death or reduced proliferation in the progress zone may reduce the number of cells below the threshold number required to produce apical-distal structures. Positional value is also influenced by the length of time cells spend in the progress zone, as well as by specification under the influence of *Hox* genes.[95,105,106]

Developmental toxicants can alter cell-signaling pathways involved in the morphogenesis of structures by causing a deficiency in either the appropriate inducer or target cells involved (due to failures of differentiation or proliferation, or to excessive cell death), by inappropriate biosynthesis of signaling molecules, or by interference with the transduction of the signal to the target cell population. It is probable that the pathogenesis of many defects induced by environmental agents includes altered cell-signaling patterns in the overall pathogenesis and that more than one signaling pathway will be altered, particularly by cytotoxic agents. However, some teratogens, and particularly those that are receptor mediated, may alter signaling pathways more specifically. This class of teratogens includes the Veratrum alkaloids, plant toxins long known to cause cyclopia in sheep, that inhibit tissue responsiveness to Shh.[107,108] In studies with chicks, jervine, an especially toxic Veratrum alkaloid, also caused defects that were characteristic of holoprosencephaly.[108]

Another developmental toxicant that causes changes in cell signaling pathways is retinoic acid (RA). RA is a natural morphogen that is essential to many normal developmental processes.[109] Investigators using animal models have shown that both a deficiency and an excess of RA can lead to malformations, including limb defects such as oligodactyly, polydactyly, and reduction defects, and in some cases, limb duplication. Differences in observed abnormalities depend on the timing of administration, dose, route, and species employed.[109–111] RA deficiency prevents normal AER–FGF-4–ZPA signaling in both chick and mouse embryos.[112,113] Studies using RA-deficient mice have demonstrated that RA is required for cells of the ZPA to acquire competence for Shh secretion, most likely via induction of *Hox* genes in the mesenchyme and by secretion of FGF-4 by the AER.[85,100] Mice deficient in RA lack normal ZPA patterning and exhibit oligodactyly.[109] In contrast to RA deficiency, direct application of RA to the anterior domain of the chick limb bud results in extra digits or in mirror image limb duplication via the formation of a second ZPA.[93,114–116] At higher doses of RA, skeletal elements are reduced or do not form.[110, 117]

Although RA is teratogenic in humans[118,119] and in animals,[120–123] there are notable species differences in the effects observed. While similar defects exist in humans and rodents for craniofacial, CNS, cardiac, and thymic development, limb defects are seldom reported in humans exposed to RA *in utero*.[124] Like other teratogens, the effects of exogenous RA on development are highly stage dependent. Late administration during limb development in the mouse causes skeletal defects,[125] whereas administration during preorganogenesis can induce the development of supernumerary hind limbs.[126,127] Differences also occur with different methods of administration. Direct application of RA leads to digit duplication, while systemic administration more often results in oligodactyly. Polydactyly can also occur, particularly at lower doses.[109,128,129] As is the case with several other teratogens, there appears to be a preferential loss of postaxial digits of the right forelimb,[116] a phenotype also prevalent in *Wnt7a* null mice that display reduced *Shh* expression.[91] Studies have also implicated alterations in the underlying mesenchyme and interference with ectodermal-mesenchymal interactions in loss of AER function and signaling in RA pathogenesis, and it is of no small interest that cells of the underlying mesenchyme express receptors for RA.[116,130–133]

It is important to keep in mind that the preceding overview of signaling pathways in the limb is necessarily greatly simplified for the present discussion. Although discrete morphogenetic functions can be assigned to regions such as the AER and ZPA, these regions are codependent for limb development, adding an additional layer of complexity to the system. For instance, maintenance of the AER requires a functional progress zone and ZPA, while signaling from the AER and the ZPA is required to maintain proliferation in the underlying mesenchyme and progress zone (reviewed by Johnson et al.[103]). The complexity of this system, as well as species differences and even differences between fore- and hind limb development within a species, will often make determining the initial pathological deviation from normal development quite difficult.

Perhaps the preferred technique for detecting alterations in cell-signaling patterns following toxicant exposure is *in situ* hybridization using labeled probes to the mRNAs of the molecules of interest. There are several good reviews covering this highly valuable technique.[134–137] Patterns of signaling molecules in tissues can also be detected by whole mount immunostaining or immunohistochemistry when appropriate antibodies are available. For instance, Shh is a protein that undergoes autocatalytic cleavage to yield a secreted amino (N) peptide and a carboxy (C) peptide that remains cell associated. Specific antibodies are available to both peptides for several species. Antibodies to cellular receptors, such as cellular retinoic acid binding proteins (CRABPs), are also useful in identifying changes to populations of cells responsive to signaling molecules.

D. Cell Migration and Differentiation

The dynamic process of morphogenesis involves not only cell proliferation and specification as already discussed but also cell migration and differentiation. Perhaps nowhere is the importance of cell migration and differentiation better illustrated than in the changes that occur to the group of cells collectively termed *the neural crest*. The neural crest initially develops from the neuroepithelium on the dorsal aspect of the neural folds early in embryogenesis. Neural crest cells detach from the neuroepithelium and migrate as single cells in a process referred to as *delamination and epithelial-mesenchymal transformation* (EMT). Delamination involves specific changes to extracellular matrix molecules that allow the neural crest cells to detach from the neuroepithelium. These changes include down regulation of neural cell adhesion molecule (N-CAM) and N-cadherin, and up regulation of cadherins 7 and 11 (reviewed by Nieto[138]). Neural crest cells then undergo extensive migration throughout the embryo, whereupon they eventually differentiate to form tissues and structures as diverse as the connective tissues of the face, elements of the peripheral nervous system, and melanocytes. Because of the rapid dynamics of this cell population and the involvement of neural crest cells in the development of so many organ systems and tissues, it is not surprising that alterations in neural crest cell migration and differentiation often play a major role in the pathogenesis of developmental toxicants.

It is generally accepted that early neural crest cells are multipotent and that specification of neural crest cell lineages is based on signals encountered during cell migration, interactions with the extracellular matrix, and tissue-specific gene expression of a variety of induction factors.[139] However, evidence also suggests that some neural crest cells, and particularly those originating in the hindbrain, may acquire some positional specification while the cells are still in the neural tube, possibly under the influence of *Hox* genes.[140,141] Neural crest cells that arise in the neuroepithelium of the hindbrain form three discrete streams of migrating cells that invade the branchial arches to form the cranial ganglia and components of the craniofacial skeleton. Specifically, neural crest cells adjacent to rhombomeres r2, r4, and r6 populate the first, second, and third branchial arches, respectively.[139] Misdirection of these discrete streams of cells in their allotted pathways, either by altered prespecification or altered signals from the tissues through which the cells migrate, will result in craniofacial anomalies.

Neural crest cells arising in the trunk follow two defined pathways of migration. The first migration pathway occurs ventrally and leads to formation of the neurons and glia of the dorsal root ganglia, the sympathetic ganglia, and the adrenal medulla. Later, trunk neural crest cells follow a dorsolateral pathway that gives rise to the melanocytes. The formation of these discrete pathways for neural crest cell migration is controlled to a large extent by interactions of the neural crest cells with the extracellular matrix. These allow or impede migration in the tissue, depending on the pattern of expression of cell surface receptors, such as the ephrin receptors, on the neural crest cells themselves.[142,143]

Because the migration and differentiation of the neural crest cell population is necessary for the development of so many tissues and organs, it is intuitive that drugs and chemicals that alter the mechanics of cell migration, for instance, those that cause alterations to the cytoskeleton, have the potential to profoundly affect embryonic development. Furthermore, altered cell proliferation or cell death, not only in the neural crest population but also in cells responsible for secretion of extracellular matrix components and signaling factors required for neural crest migration and specification, may alter the fate of the neural crest population and result in dysmorphology. For example, spinal nerve abnormalities in mouse embryos exposed to VPA coincided with alterations in somite morphogenesis, suggesting that the spinal nerve abnormalities resulted, at least in part, from anomalous patterns of migration and induction of the neural crest by the developing somites themselves.[144]

Altered patterns of neural crest cell migration may also arise from alterations in cell signaling that result in changes in neural crest fate specification. For example, RA is involved in the prepatterning and specification process of hindbrain neural crest cells,[145,146] presumably by altering

Table 3.5 Methods to detect neural crest cells and derivatives

Marker	Method	Comments	Examples of Use
Neurofilament protein	Immunostaining for protein found in neurons, craniofacial and dorsal root ganglia.	Can be used either as whole mount or on histological sections. A variety of labels is available for secondary antibody. Staining is time consuming for whole-mount technique.	144, 147, 148, 227, 235–237
HNK-1	Immunostaining for sialoprotein found on migrating NC and derivatives	Same as above	145, 238, 239
CRABP1	Immunostaining for RA binding protein found in certain neural crest cell populations and cells of the hindbrain	Same as above	133, 147, 148, 152, 153
DiI (1,1-dioctadecyl-3,3,3′,3′-tetramethyl indocarbocyanine)	Lipophilic fluorescent dye injected into cells of interest	Requires microinjection equipment and fluorescence microscopy	145, 147, 238, 240

Hox gene expression.[141] At high concentrations, RA appears to alter neural crest cell identity, resulting in altered migration patterns and subsequent dysmorphology in the target tissues.[145,147,148] Similar effects have been noted across several species; however, there appear to be notable differences with respect to the resultant structures affected.[145,147,148] Some of the differences observed may result from inherent differences between species in early neural crest cell development. In chick embryos, for example, closure of the neural tube is a requirement for neural crest induction, while in mice the neural crest is formed and can migrate from the cranial region in the absence of neural tube closure.[138,149,150] Differences in early neural crest development have also been noted in Xenopus and zebra fish.[138]

In some cases neural crest cell migration may proceed normally but later differentiation of the neural crest is perturbed, resulting in dysmorphology similar to that induced by lack of neural crest cell migration. For instance, migration of the neural crest proceeds normally in mouse embryos lacking the endothelin A receptor on those cells. However, later signaling from the postmigratory neural crest cells to surrounding ectomesenchymal cells in the arches for production of downstream inductive factors is impaired. This results in cardiac malformations and in defects in branchial arch derived craniofacial tissues.[151–153]

There are several markers that can be used to study the effects of a drug or chemical on the migration and fate of the neural crest cell population. Antibodies to the marker HNK-1 have proven exceptionally useful for detection of migrating neural crest cells in mammalian embryos. This cell surface antigen is present on greater than 90% of migrating crest cells but is not expressed by the mesectoderm.[154,155] Other markers include antibodies to neurofilament proteins that are expressed by neural crest derived neurons and ganglia, CD44 (a marker restricted to cranial neural crest in the chick),[156] and cellular retinoic acid binding protein (CRABP1), which is expressed in hindbrain regions and the neural crest from r4 to r6, but not r1 or r3.[147,157,158] Selected markers and their uses are presented in Table 3.5.

V. EXAMPLES OF ABNORMAL PATHOGENESIS

Deviations from normal developmental processes at the cellular level eventually result in alterations in tissue and organ development. Alterations at the tissue level can manifest as hypoplasia, retarded or arrested growth, altered patterns of cellular differentiation, altered patterns of growth, altered

Table 3.6 Morphological observations: methods for external, visceral, and skeletal examinations

Technique	Methodology	Comments	Ref.
Fresh visceral	Dissection at termination	Not appropriate to detect some malformations; cephalic examination requires fixation	241
Wilson's dissection	Serial sections through fixed tissue to visualize gross internal organ and tissue structures	Hazardous as Bouin's fixative contains picric acid	242, 243
Frozen cephalic sections	Frozen sectioning for cephalic examination combined with fresh visceral exam	Allows for faster analysis of cephalic morphology compared to fixation in Bouin's	244
Skeletal	Alizarin Red (bone) and Alcian Blue (cartilage) staining	Differential staining of cartilage and bone.	245–250
External	Staining of fixed specimen to reveal details of external morphology, especially somites	Can be done after whole mount immunostaining or other techniques. Requires fluorescence microscopy	144, 251

morphologic expression, tissue destruction, deformation, and/or altered morphogenetic movements (i.e., failure of elevation of palatal shelves or neural folds). Unlike alterations that occur on a cellular level, effects at the tissue level can be observed histologically or by gross observation. The remainder of this chapter focuses on features involved in the pathogenesis of several of the more common congenital defects to illustrate some of the ways in which alterations at the cellular level induced by a toxicant may translate into a congenital defect recognizable at birth. Selected methods for gross morphological observations of fetal anomalies are listed in Table 3.6.

A. Defects with Different Underlying Pathogenesis but Similar Morphology

1. Cleft Palate

One example of a discrete defect that can result from diverse agents and differing underlying pathogenesis is clefting of the secondary palate. Palatogenesis involves morphogenetic processes that include cell migration, proliferation, cell signaling, alterations in extracellular matrix synthesis, and cell specification and differentiation. Changes to the normal developmental sequence involving any of these processes can lead to a failure of palate fusion and a resulting cleft (reviewed by Young et al.[159]). Cleft palate can occur as an isolated morphological defect, as observed with corticosteroid administration in rodents,[11,12,160] as part of a multiple defect syndrome, as occurs with retinoic acid dysmorphogenesis,[111] or secondary to another anomaly, such as cleft lip.[161,162] Although the underlying pathogenesis of a cleft palate resulting from exposure to a toxicant or environmental insult can be quite different from that caused by another agent, the end result can be virtually indistinguishable in morphology.

As with other anomalies, a key to determining the underlying pathogenesis of cleft palate is to first determine at what stage the normal developmental processes were initially affected. For practical purposes, the development of the secondary palate can be divided into three stages.[163] The first of these is the formation and growth of the maxillary processes. These structures originate from neural crest–derived mesenchymal cells surrounded by an epithelium of ectodermal origin. The maxillary processes initially sit vertically in the oral cavity. Elevation of the shelves to a horizontal position is the second stage of palatogenesis. The third and final stage comprises the fusion of the two shelves, degeneration of the medial edge epithelium (MEE), and formation of a contiguous mesenchyme. Identifying a defect in one of these major developmental events of palate formation can provide important clues to the underlying pathogenesis of the anomaly.

Early changes to craniofacial morphogenesis that affect the formation of the maxillary processes tend to involve perturbations in cell signaling, migration, and subsequent patterning. The connective tissue of the various prominences that form the face and palate are derived from neural crest cells that originate from the neural tube in the midbrain and rostral hindbrain.[159] Failure of neural crest cell migration, differentiation, or proliferation, or extensive cell death in the developing primordium results in failure of the maxillary processes to grow to sufficient size for later palatal fusion. Patterning of the craniofacial structures may also be influenced by regional differences in the overlying ectoderm.[159] An interesting feature of the craniofacial prominences that has direct implications for ascertaining the pathogenesis of a teratogen is that growth and development of each of the facial primordia appear to be controlled by differing morphogenetic processes.[159] Thus, teratogens may affect a specific portion of craniofacial morphogenesis while having no effect on others. The differing susceptibility of the facial prominences to teratogens such as RA may result, at least in part, in differences in the expression of RA receptors and signaling pathways, such as the Shh pathway.[159]

Another frequent cause of cleft palate is a failure of palatal shelf elevation. Although the persistence of the palatal shelves in the vertical position would be quite evident to gross visual inspection, it is often a delay in shelf elevation, rather than a complete lack of shelf movement, that results in cleft palate. When the shelves elevate too late in the developmental sequence, the other structures of the head have grown and changed in size and shape such that the two shelves can no longer meet. Continued growth of the head further separates the shelves over time.[164] Although the processes that govern shelf elevation are not well understood, there is substantial evidence that alterations in the extracellular matrix are essential to the process, possibly through the interactions with matrix metalloproteinases (reviewed by Kerrigan et al.[163]). Studies also indicate that the composition of the extracellular matrix molecules synthesized in the shelves changes with respect to position along the shelf. This difference might impart a tension or force that is involved in the process of elevation.[163] Others have suggested that the shelves are initially held in the vertical position by the tongue. Movement of the tongue, which occurs developmentally at the same time as shelf elevation, releases the shelves to attain the horizontal position.[163]

The final series of events in palatogenesis that are subject to perturbations brought about by exposure to developmental toxicants involves the fusion of the palatal shelves, the breakdown of the MEE, and the consolidation and fusion of the underlying mesenchyme. The degeneration of the MEE and its associated basement membrane is essential for fusion of the palate. Failure of the MEE to degenerate provides a barrier to the mesenchymal cells within the shelves, and a contiguous mesenchyme cannot form. As the craniofacial structures grow, the seam formed at the MEE is not sufficient to hold the shelves together, and clefting occurs.[159,163] A complete failure of fusion of the mesenchyme of the two shelves following shelf elevation may leave a cleft that is virtually indistinguishable from that resulting from delayed shelf elevation.

Although it was initially thought that the MEE was removed via programmed cell death, current studies have failed to identify apoptotic cells in the MEE.[165,166] Other studies suggest two possibilities for the removal of the MEE. The first is that the MEE cells are transformed into mesenchymal cells that migrate into, and remain in, the fused shelves.[165–167] An alternative explanation for the fate of the MEE is that they migrate out of the seam and become incorporated into the oronasal epithelium.[168] Although little is known concerning the exact mechanisms involved in palate fusion, there is growing evidence for the involvement of growth factors, such as TGFβ3, in stimulating removal of the MEE and subsequent fusion. Mice that are null for TGFβ3 have incomplete palatal fusion, and antisense oligonucleotides or antibodies to TGFβ3 block palatal shelf fusion *in vitro*.[169–171]

2. Neural Tube Defects

The vast majority of neural tube defects (NTDs) arise from a failure of neural tube closure. Failed closure of the anterior neural tube results in anencephaly, while failed closure in the posterior neural tube region results in spina bifida. It should be noted, however, that central nervous system defects involving the neural tube may also result from improper closure, bifurcations in the neural tube or groove, changes involving the overlying ectoderm, and alterations in vesicle formation. Aspects of the pathogenesis of NTDs resulting from failed neural tube closure, the most frequent types of NTDs, will be considered below. The processes governing neural tube closure are complex and involve diverse cell types and morphogenetic mechanisms. Because of this complexity, the pathogenesis of NTDs can vary widely. Although the initial events that lead to failed closure may differ greatly, the morphologic defect that results, once fusion has failed to occur, will proceed through development along similar pathogenetic lines. The result is that NTDs resulting from differing underlying pathogenesis may ultimately appear morphologically indistinguishable from each other.

There are two major processes involved in formation of the neural tube. One is primary neurulation, in which the neural plate is formed ectopically and subsequently fuses to form a tube. Following primary neurulation, there is secondary neurulation, a process that proceeds caudally and forms the remainder of the tube by hollowing out the region below the ectoderm. NTDs such as anencephaly and most spina bifida arise from aberrations in primary neurulation. Studies on human embryos suggest that the pathogenesis of the majority of human NTDs is limited to those processes involved in neural tube closure itself.[150] Other elements of neural tube development, such as maturation of the neuroepithelium, appear to continue to occur despite lack of closure. Anencephaly and most myelomeningoceles result from failed neural tube closure. Other disorders such as encephaloceles and a small percentage of myelomeningoceles, appear to be related to subsequent development of the nervous system and its coverings, rather than being directly related to neural tube closure, and thus should be considered separately.[150]

Numerous *in vivo* and *in vitro* animal models have been used to study the process of neurulation, neural tube closure, and the pathogenesis of NTDs. Agents that are known or suspected to cause NTDs in humans, such as methotrexate, alcohol, hyperthermia, and others, also cause NTDs in animal models, regardless of the species employed. However, it should be noted that there are distinct differences between vertebrate species in the process of neural tube formation that may potentially alter the effects of a toxicant. For example, the zones of primary and secondary neurulation overlap in both humans and chicks, a feature not as evident in rodent embryos. This is of some significance, as observations of human infants with NTDs reveal that a large number of myelomeningoceles occur at the site of overlap.[150] Moreover, differences among species appear to exist in the number and position of closure points that may also influence the location and severity of NTDs. Despite these differences, studies using animal models, chemical agents, and genetic mutants have contributed a large body of information regarding the process of neural tube closure and the genesis of NTDs.[172] The majority of studies have demonstrated a strong link between inhibited cell proliferation and increased cell death and failed neural fold elevation and closure. Because cell proliferation is so crucial for neural tube closure, it is probable that almost any agent that induces cell death and is administered at the appropriate time and dose may result in open neural tube.

Studies have repeatedly demonstrated a strong link between nutritional factors and the occurrence of NTDs. In particular, it is now well established that supplemental folic acid lowers the risk of NTDs in the human population, a finding that may be true for other defects as well.[173-175] Folic acid is required for DNA synthesis, and thus cell replication, via its role in one-carbon metabolism. Methotrexate (MTX), a folic acid antagonist, causes NTDs in rodents, rabbits, and humans by inhibiting dihydrofolate reductase.[176] In rabbits, MTX-related defects were severely lessened when an alternate one-carbon donor was provided concurrently with MTX treatment.[177] Other nutrients that may alter the incidence of or susceptibility to NTDs in experimental animal models include zinc,[150,178-183] methionine,[184-191] and inositol,[192-194] but the role of these nutrients in the etiology and pathogenesis of NTDS in humans has not been clearly determined.[195-201]

B. Abnormalities with Similar Pathogenesis but Different Morphology

As discussed above, defects that are morphologically similar, such as closure defects, can be caused by different underlying pathologies. In the alternate scenario are those defects that differ morphologically at birth, and can involve dissimilar structures, but that have the same underlying pathogenesis. Many developmental toxicants induce different defects in an organism based on factors such as dose and timing of administration. In this case, the underlying pathogenesis may be similar but the tissue affected and the degree to which it is affected changes, resulting in different malformations at term. Disruption defects, and in particular vascular disruption defects, are another example of a group of defects that have a shared pathogenesis but have the potential to result in very different morphologic outcomes.

Unlike malformations, disruption defects result from the destruction of existing tissues or structures that may have previously developed normally. Vascular disruption defects refer to those involving the interruption or destruction of some part of the fetal vasculature.[9] The pathogenesis of vascular disruption is extremely important to developmental toxicology because it can cause dysmorphology at nearly any stage of gestation, affect almost any tissue or structure, and occur as the primary cause of dysmorphology or secondary to existing malformations. The latter situation can be extremely difficult to resolve because the secondary disruption can obliterate evidence for the primary malformation, thus hampering determination of the underlying cause of the congenital anomaly.

Vascular disruption defects usually occur as a consequence of localized or general fetal hypoxia resulting from direct effects on the fetal vasculature or secondary to changes to the maternal uterine or placental vasculature. Vasoconstrictive drugs, such as cocaine, may alter both maternal and fetal hemodynamics. This results in decreased uterine and placental blood flow resulting from constriction of the uterine arteries, and brings about increased fetal blood pressure and heart rate.[202] Vasodilating agents may decrease uterine blood flow secondary to diversion of blood to peripheral vasculature and yet have no direct effect on fetal hemodynamics.[203] In humans, defects consistent with a vascular disruption pathogenesis have been associated with exposure to cocaine,[204–206] antihypotensives,[203] and the abortifacient misoprostol.[207–209] Defects associated with such mechanical disturbances as chorionic villus sampling (CVS), uterine trauma, twin–twin transfusion syndrome, and amnion disruption sequence may also result from vascular disruption.[9,210–212] In animals, defects attributable to vascular disruption have been induced with a variety of treatments. These include physical methods, such as uterine vascular clamping in the rat,[213–215] and cocaine and other vasoactive substances.[202–204,216,217]

Studies have shown that the fetal response to hypoxia follows the same pattern of events regardless of the agent or mechanism responsible. Early histopathological changes in affected tissues reveal that fetal hypoxia leads to edema and an increase in vessel diameter, particularly in distal structures. Loss of fetal blood vessel wall integrity leads to vessel rupture, hemorrhage, and formation of subcutaneous blebs or blood-filled blisters.[204,214–217] Necrosis of affected tissues occurs within 24 to 48 h, and subsequent resorption results in distortion or loss of already formed structures. It has recently been shown that mechanisms involved in the rupture of fetal vessels may share similarities with mechanisms associated with ischemia-reperfusion injury in adult tissues and specifically may involve the detrimental action of oxygen free radicals.[218–220]

Like other teratogenic events, the types of anomalies associated with vascular disruption depend on both the stage of gestation and severity of the insult. However, vascular disruptive events, unlike most teratogens, have been shown to induce congenital defects to some degree during a broad range of developmental stages, including postorganogenesis.[9] Studies in humans who had been exposed to cocaine indicate that the fetus is sensitive to vascular disruptive defects throughout the second and third trimesters,[205,206,221] although genitourinary tract malformations and more severe malformations may be induced during the first trimester.[222]

Of the types of defects ascribed to vascular disruption pathogenesis, limb defects, including adactyly, transverse reduction defects, syndactyly, polydactyly, and club foot appear to be the most common.[9,205,206,213–217] The type of defect depends largely on the blood vessel affected and the severity of the ensuing hemorrhage. In animals and humans, the most distal parts of the skeleton are the most easily affected by fetal hypoxia; thus, abnormalities of the phalanges as well as facial abnormalities are the most frequent morphologic manifestations of fetal hypoxia.[9,216,223] Other defects often ascribed to vascular disruption include limb body wall complex, gastroschisis, micrognathia, anencephaly, limb reduction defects, syndactyly, and orofacial malformations, such as cleft palate.[9,212]

Discerning whether a given anomaly has an underlying pathogenesis of vascular disruption can be difficult. For instance, a hematoma occurring in the early progress zone of the limb bud can lead to a truncation defect similar to those observed with other pathogenic phenomena, such as x-irradiation-induced cell death. However, substances that induce overt vascular disruption, such as cocaine, unlike many toxicants that cause anomalous development, are less likely to result in a consistent pattern of malformations. With other developmental anomalies, such as RA-induced oligodactyly, one finds consistent, bilateral changes to the limb buds, with increasing severity that correlates to increasing dose. In contrast, agents that induce disruption defects produce defects that are often asymmetrical and sporadic. A hallmark of vascular disruption is the occurrence of subcutaneous blebs or hematomas on the fetus, and hemorrhage in the affected structure early in the pathogenetic sequence, features that are usually readily apparent if sought early following exposure.[203,213–215,217,224,225] Moreover, the ability to induce defects in structures at a period in gestation beyond the normal period of susceptible development of that structure is also a good indication that pathogenesis of the defects involves disruption.

VI. OVERVIEW AND PERSPECTIVE

Any discussion of the mechanisms that are operative and responsible for congenital malformations following exposure to toxic xenobiotics, including drugs, other chemicals, and physical agents, must include some aspects of the genetic control of normal development as well as the genetic changes that are responsible for abnormal development. In the past, we studied the genetics of abnormal development by studying genetic alterations that occurred spontaneously. But with the advances in molecular biology resulting in deciphering the human genome and many animal genomes, scientists have the ability to alter the genome of experimental animals to study the impact of individual genes or growth factors on development. It is not surprising that advances in genetics have assisted teratological investigations in many ways. We have known for a long time that environmental toxicants that are teratogenic can alter development and mimic some aspects of genetic malformations. In fact, clinicians have used the terms *phenocopy* and *phenotype* for almost a century to indicate that some environmentally produced malformations and genetic malformations cannot be distinguished from each other. Many genetic congenital malformations are due to an abnormality at one locus that results in a cascade of abnormal molecules or abnormal quantities of normal molecules that can affect many structures. Thus, genetic flaws can result in a single isolated anatomical defect or a recognizable syndrome of malformations. Environmental toxicants are less likely to produce a single malformation and more likely to result in a so-called teratogen syndrome. What is responsible for this difference between the effect of teratogens and gene mutations on development?

Joseph Warkany referred to teratogens as sledgehammer toxicants. In other words, their mechanism of action was due to their toxicity. Of course this label does not apply to the mechanisms of all teratogens. Warkany was probably referring to the many teratogens that are cytotoxic, e.g., ionizing radiation, many cancer chemotherapy drugs, and many mutagenic drugs and chemicals. Geneticists may look upon teratology as an unsophisticated methodology to produce congenital malformation compared to the preparation of knockout mice or site-directed mutagenesis. But unfortunately, human malformations can be produced by sledgehammer teratogens as well as by inherited genes, chromosome abnormalities, and new mutations that affect development. Only certain classes of teratogens can come close to the specificity of effect of an abnormal gene on development. The best examples are the receptor mediated teratogens, e.g., retinoic acid, sex steroids, and possibly thalidomide. If environmental toxicants produce malformations that affect the genome, in most cases it will be only indirectly. And some teratogens, such as deformations and vascular disruptive phenomena, are unrelated to any genetic alteration.

The methods of molecular biology are used to study the structure and nature of the genes, and their products that result in congenital malformations. After we achieve an understanding of the basic science of genes that produce congenital malformations, there is frequently a long and difficult road before this information can be utilized to treat or prevent birth defects.

Developmental toxicology is much different. If we can identify an environmental agent as a teratogen, our first task is to prevent exposure to the agent. In many instances that is a readily attainable goal, especially if the agent is a drug or chemical over which society has control. Yet knowing that alcohol is a teratogen does not and has not solved the problem. Similarly environmental contaminants, such as mercury, are not readily controlled. Maternal disease states that are teratogenic, such as diabetes or teratogenic infections, represent more difficult problems, but they are still solvable. The answer for the teratologist is epidemiological studies that identify developmental toxicants or animal studies that indicate or confirm the potential for harm. From the human standpoint, understanding the pathophysiology is irrelevant. Yet, as scholars it is important to study and understand the mechanisms of teratogenesis and the pathophysiology from a scientific and clinical perspective. Drugs and chemicals with similar pathophysiological effects may represent a risk. Greater knowledge might thus permit chemists to prepare drugs that have the therapeutic benefit sought, while eliminating the teratogenic risk, as has been attempted with valproic acid analogs.

The approach for solving the problem of genetically transmitted congenital malformations is much different. Basic science research into the nature of the normal and abnormal genes involved in the genetically transmitted malformations will be the key to preventing or treating the genetic defect. Since there is a cascade of gene products that can be unveiled once the gene is identified, understanding the totality of a particular gene's role in development may permit the replacement of the deficient products attributable to the abnormal gene. There is even the possibility of inserting the normal gene; although presently largely theoretical, it is a real possibility.

So developmental toxicologists have to examine their field and recognize that it consists of the following components:

Epidemiological studies to identify agents that are causally associated with the production of birth defects.
Basic science research dealing with the mechanisms of teratogenesis and the pathophysiology of abnormal development.
Ecological interests to identify potential new environmental toxicants.
Social and political actions to make certain that the information about environmental reproductive risks is acted upon promptly by governmental agencies.
Responsibility for educating the public about environmentally induced birth defects as well as to educate their scientific, clinical, and regulatory colleagues about these risks.

REFERENCES

1. Brent, R.L. and Jensh, R.P., Intrauterine growth retardation, *Adv. Teratol.*, 2, 139, 1967.
2. Wilson, J.G. *Environment and Birth Defects*, Academic Press, New York, 1973.
3. Brent, R.L., Environmental factors: miscellaneous, in *Prevention of Embryonic, Fetal and Perinatal Disease*, Brent, R.L. and Harris, M.J., Eds., DHEW Pub. No. (NIH) 76-853, Department of Health, Education, and Welfare, Bethesda, 1976, p. 211.
4. Brent, R.L., The magnitude of the problem of congenital malformations, in *Prevention of Physical and Mental Congenital Defects. Part A. Basic and Medical Science*, Marois, M., Ed., Alan R. Liss, New York, 1985, p. 55.

5. Spranger, J., Benirschke, K., Hall, J.G., Lenz, W., Lowry, R.B., Opitz, J.M., Pinsky, L., Schwarzacher, H.G., and Smith, D.W., Errors of morphogenesis: concepts and terms. Recommendations of an international working group, *J. Pediatr.*, 100, 160, 1982.
6. Wilson, J.G. and Brent, R.L., Are female sex steroids teratogenic?, *Am. J. Obstet. Gynecol.*, 114, 567, 1981.
7. Brent, R.L., Bendectin: Review of the medical literature of a comprehensively studied human nonteratogen and the most prevalent tortogen-litigen, *Reprod. Toxicol.*, 9, 337, 1995.
8. Brent, R.L., Review of the scientific literature pertaining to the reproductive toxicity of Bendectin, in *Modern Scientific Evidence: The Law and Science of Expert Testimony*, Faigman, D.L., Kaye, D.H., Saks, M.J., and Saunders, J.W., Eds., West Publishing Group, St. Paul, 1997, p. 373.
9. Van Allen, M.I., Structural anomalies resulting from vascular disruption, *Pediatr. Clin. N. Amer.*, 39, 160, 1992.
10. Coplen, D.E., Prenatal intervention for hydronephrosis, *J. Urol.*, 157, 2270, 1997.
11. Fraser, F.C. and Fainstat, T.D., Production of congenital defects in the offspring of pregnant mice treated with cortisone, *Pediatrics*, 8, 527, 1951.
12. Fraser, F.C., Kalter, H., Walker, B.E., and Fainstat, T.D., The experimental production of cleft palate with cortisone and other hormones, *J. Cell. Comp. Physiol.*, 43, 237, 1954.
13. Khoury, M.J., Moore, C.A., and Evans, J.A., On the use of the term "syndrome" in clinical genetics and birth defects epidemiology, *Am. J. Med. Genet.*, 49, 26, 1994.
14. Lubinsky, M., Vater and other associations: historical perspectives and modern interpretations, *Am. J. Med. Genet. Suppl.*, 2, 9, 1986.
15. Brent, R.L. and Holmes, L.B., Clinical and basic science lessons from the thalidomide tragedy: What have we learned about the causes of limb defects? *Teratology*, 38, 241, 1988.
16. Lenz, W., A short history of thalidomide embryopathy, *Teratology*, 38, 203, 1988.
17. Jensh, R.P. and Brent, R.L., The effect of low level prenatal x-irradiation on postnatal growth in the Wistar rat, *Growth Dev. Aging*, 52, 53, 1988.
18. Jensh, R.P. and Brent, R.L., The effects of prenatal x-irradiation in the 14th-18th days of gestation on postnatal growth and development in the rat, *Teratology*, 38, 431, 1988.
19. Wilson, J.G. and Brent, R.L., Differentiation as a determinant of the reaction of rat embryos to X-irradiation, *Proc. Soc. Exp Biol. Med.*, 82, 67, 1953.
20. Generoso, W.M., Rutledge, J.C., Cain, K.T., Hughes, L.A., and Downing, D.J., Mutagen-induced fetal anomalies and death following treatment of females within hours of mating, *Mutat. Res.*, 199, 175, 1988.
21. Pampfer, S. and Streffer, C., Prenatal death and malformations after irradiation of mouse zygotes with neutrons or X-rays, *Teratology*, 37, 599, 1988.
22. Brent, R.L., The indirect effect of irradiation on embryonic development. II. Irradiation of the placenta, *Am. J. Dis. Child.*, 100, 103, 1960.
23. Wilkins, L., Jones, H.W., Holman, G.H., and Stempfel, R.S., Masculinization of the female fetus associated with the administration of oral and intramuscular progestins during gestation: nonadrenal female psuedohermaphrodism, *J. Clin. Endocrinol. Metab.*, 18, 559, 1958.
24. Wilkins, L., Masculinization due to orally given progestins, *JAMA*, 172, 1028, 1960.
25. Grumbach, M.M., Ducharine, J.R., and Moloshok, R.E., On the fetal masculinizing action of certain oral progestins, *J. Clin. Endocrinol. Metab.*, 19, 1369, 1959.
26. Van Wyk, J. and Grumbach, M.M., Disorders of sex differentiation, in *Textbook of Endocrinology*, Williams, R.H., Ed., W.B. Saunders, Philadelphia, 1968, p. 537.
27. Briggs, M.H. and Briggs, M., Sex hormone exposure during pregnancy and malformations, in *Advances in Steroid Biochemistry and Pharmacology*, Briggs, M.H. and Corbin, A., Eds., Academic Press, London, 1979, p. 51.
28. Seegmiller, R.E., Nelson, B.W., and Johnson, C.K., Evaluation of the teratogenic potential of Delalutin (17a-hydroxyprogesterone caproate) in mice, *Teratology*, 28, 201, 1983.
29. Hochner-Celinkier, D., Marandici, A., Iohan, F. and Monder, C., Estrogen and progesterone receptors in the organs of prenatal Cynomolgus monkey and laboratory mouse, *Biol. Reprod.*, 35, 633, 1986.
30. Carbone, J.P., Figurska, K., Buck, S., and Brent, R.L., Effect of gestational sex steroid exposure on limb development and endochondral bone ossification in the pregnant C57B1/6J mouse. I. Medroxyprogesterone acetate, *Teratology*, 42, 121, 1990.

31. Carbone, J.P. and Brent, R.L., Genital and non-genital teratogenesis of prenatal progesterone therapy: the effects of 17 alpha-hydroxyprogesterone caproate on embryonic and fetal development and endochondral bone ossification in the C57B1/6J mouse, *Am. J. Obstet. Gynecol.*, 169, 1292, 1993.
32. Guignard, J.P., Burgener, F., and Calame, A., Persistent anuria in neonate: a side effect of captopril, *Int. J. Pediatr. Nephrol.*, 2, 1981.
33. Guignard, J.P., Effect of drugs on the immature kidney, *Adv. Nephrol. Neker Hosp.*, 22, 193, 1993.
34. Hanssens, M., Keirse, M.J.N.C., Vankelecom, F., and Van Assche, F.A., Fetal and Neonatal effects of treatment with angiotensin-converting enzyme inhibitors in pregnancy, *Obstet. Gynecol.*, 79, 128, 1991.
35. Piper, J.M., Ray, W.A., and Rosa, F.W., Pregnancy outcome following exposure to angiotensin converting enzyme inhibitors, *Obstet. Gynecol.*, 80, 429, 1992.
36. Martin, R.A., Jones, K.L., Mendoza, A., Barr, M., and Benirschke, K., Effect of ACE inhibition in the fetal kidney: decreased renal blood flow, *Teratology*, 46, 317, 1992.
37. Pryde, P.G., Sedman, A.B., Nugent, C.E., and Barr, M., Angiotensin-converting enzyme inhibitor fetopathy, *J. Am. Soc. Nephrol.*, 3, 1575, 1993.
38. Brent, R.L. and Beckman, D.A., Angiotensin-converting enzyme inhibitors, an embryopathic class of drugs with unique properties: information for clinical teratology counselors, *Teratology*, 43, 543, 1991.
39. Beckman, D.A. and Brent, R.L., Teratogenesis: alcohol, angiotensin converting enzyme inhibitors, and cocaine, *Curr. Opin. Obstet. Gynec.*, 2, 236, 1990.
40. Brent, R.L., Editorial comment: definition of a teratogen and the relationship of teratogenicity to carcinogenicity, *Teratology*, 34, 359, 1986.
41. Francis, B.M., Rogers, J.M., Sulik, K.K., Alles, A.J., Elstein, K.H., Zucker, R.M., Massaro, E.J., Rosen, M.B., and Chernoff, N., Cyclophosphamide teratogenesis: evidence for compensatory response to induced cellular toxicity, *Teratology*, 42, 473, 1990.
42. Rogers, J.M., Francis, B.M., Sulik, K.K., Alles, A.J., Massaro, E.J., Zucker, R.M., Elstein, K.H., Rosen, M.B., and Chernoff, N., Cell death and cell cycle perturbation in the developmental toxicity of the demethylating agent 5-aza-2′-deoxycytidine, *Teratology*, 50, 332, 1994.
43. Beckman, D.A. and Brent, R.L., Mechanism of known environmental teratogens: drugs and chemicals, in *Clinics in Perinatology*, Brent, R.L. and Beckman, D.A., Eds., W.B. Saunders, Philadelphia, 1986, p. 649.
44. Mirkes, P.E., 2001 Warkany Lecture: To die or not to die, the role of apoptosis in normal and abnormal development, *Teratology*, 65, 228, 2002.
45. Knudsen, T.B., Cell death, in *Drug Toxicity in Embryonic Development.*, Kavlock, R.J. and Daston, G.P., Eds., Springer-Verlag, New York, 1997, p. 211.
46. Zakeri, Z.F. and Ahuja, H.S., Cell death/apoptosis: normal, chemically induced, and teratogenic effect, *Mutation Res.*, 396, 149, 1997.
47. Moallem, S.A. and Hales, B.F., The role of p53 and cell death by apoptosis and necrosis in 4-hydroperoxycyclophosphamide-induced limb malformations, *Development*, 125, 3225, 1998.
48. Umpierre, C.C., Little, S.A., and Mirkes, P.E., Co-localization of active caspase-3 and DNA fragmentation (TUNEL) in normal and hyperthermia-induced abnormal mouse development, *Teratology*, 63, 134, 2001.
49. Gao, X., Blackburn, M.R., and Knudsen, T.B., Activation of apoptosis in early mouse embryos by 2′-deoxyadensosine exposure, *Teratology*, 49, 1, 1994.
50. Dunty, W.C., Chen, S.-Y., Zucker, R.M., Dehart, D.B., and Sulik, K.K., Selective vulnerability of embryonic cell populations to ethanol-induced apoptosis: implications for alcohol related birth defects and neurodevelopmental disorder, *Alcohol Clin. Exp. Res.*, 25, 1523, 2001.
51. Thayer, J.M. and Mirkes, P.E., PCD and N-acetoxy-2acetylamineofluorene-induced apoptosis in the rat embryo, *Teratology*, 51, 418, 1995.
52. Mirkes, P.E. and Little, S.A., Teratogen-induced cell death in postimplantation mouse embryos: differential tissue sensitivity and hallmarks of apoptosis, *Cell Death Differ.*, 5, 592, 1998.
53. Mirkes, P.E. and Little, S.A., Cytochrome c release from mitochondria of early postimplantation murine embryos exposed to 4-dydroperoxycyclophosphamide, heat shock and staurosporine, *Toxicol. Appl. Pharmacol.*, 162, 197, 2000.
54. Little, S.A. and Mirkes, P.E., Caspase-3-induced augmentation of the intrinsic, mitochondrial apoptotic pathway, *Teratology*, 63, 248, 2001.

55. Torchinsky, A., Ivinisky, I., Savion, S., Shepshelovich, J., Gorivodsky, M., Fein, A., Carp, H., Schwartz, D., Frankel, J., Rotter, V., and Toder, V., Cellular events and the patterns of p53 protein expression following cyclophosphamide-initiated cell death in various organs of the developing embryo, *Teratogen Carcinogen Mutagen.*, 19, 353, 1999.
56. Little, S.A. and Mirkes, P.E., Effects of 4-hydroperoxycyclophosphamide (4-OOH-CP) and r-hydroperoxydechlorocyclophosphamide (4-OOh-deCICP) on the cell cycle of post implantation rat embryos, *Teratology*, 45, 163, 1992.
57. Li, L.H., Olin, J., Fraser, T.J., and Bhuyan, B.K., Phase specificity of 5-azacytadine against mammalian cells in tissue culture, *Cancer Res.*, 30, 2770, 1970.
58. Sulik, K.K., Cook, C.S., and Webster, W.S., Teratogens and craniofacial malformations: Relationships to cell death, *Development*, 103, 213, 1988.
59. Sulik, K.K. and Dehart, D.B., Retinoic acid-induced limb malformations resulting from apical ectodermal ridge cell death, *Teratology*, 37, 527, 1988.
60. Alles, A.J. and Sulik, K.K., Retinoic acid-induced limb-reduction defects: perturbation of zones of programmed cell death as a pathogenic mechanism, *Teratology*, 40, 163, 1989.
61. Alles, A.J. and Sulik, K.K., Retinoic acid-induced spina bifida: evidence for a pathogenic mechanism, *Development*, 108, 73, 1990.
62. Saunders, J.W., Gasseling, M.T., and Saunders, L., Cellular death in morphogenesis of the avian wing, *Dev. Biol.*, 5, 147, 1961.
63. Pexieder, T., Cell death in the morphogenesis and teratogenesis of the heart, *Adv. Anat. Embryol. Cell Biol.*, 51, 1, 1975.
64. Abrams, J.M., White, K., Fessler, L.I., and Steller, H., Programmed cell death during drosophila embryogenesis, *Development*, 117, 29, 1993.
65. Ansari, B., Coates, P.J., Greenstein, B.D., and Hall, P.D., In situ end-labeling detects DNA strand breaks in apoptosis and other physical states, *J. Pathol.*, 170, 1, 1993.
66. Nau, H. and Hendrickx, A.G., Valproic acid teratogenesis: ISI atals of science, *Pharmacology*, 1, 52, 1987.
67. Martin, M.L. and Regan, C.M., The anticonvulsant valproate teratogen restricts the glial cell cycle at a defined point in the mid-G1 phase, *Brain Res.*, 554, 223, 1991.
68. Craig, J.C., Bennett, G.D., Miranda, R.C., Mackler, S.A., and Finnell, R.H., Ribonucleotide reductase submit R1: a gene conferring sensitivity to valproic acid-induced neural tube defects in mice, *Teratology*, 61, 305, 2000.
69. Regan, C.M., Gorman, A.M., Larsson, O.M., Maguire, C., Martin, M.L., Schousboe, A., and Williams, D.C., In vitro screening for anticonvulsant-induced teratogenesis in neural primary cultures and cell lines, *Int. J. Dev. Neurosci.*, 8, 103, 1990.
70. Miller, M.W. and Nowakowski, R.S., Use of bromodeoxyuridine-immunohistochemistry to examine the proliferation, migration and time of origin of cells of the central nervous system, *Brain Res.*, 457, 44, 1988.
71. Rosen, M.B. and Chernoff, N., 5-Aza-2'-deoxycytidine-induced cytotoxicity and limb reduction defects in the mouse, *Teratology*, 65, 180, 2002.
72. Chernoff, N., Rogers, J.M., Alles, A.J., Zucker, R.M., Elstein, K.H., Massaro, E.J., and Sulik, K.K., Cell cycle alterations and cell death in cyclophosphamide teratogenesis, *Teratogen Carcinogen Mutagen,.* 9, 199, 1989.
73. Walterhouse, D.O., Yoon, J., and Iannaccone, P.M., Developmental pathways: Sonic hedgehog-patched-GLl, *Environ. Hlth. Perspect.*, 107, 167, 1999.
74. Roessler, E., Belloni, E., Gaudenz, K., Jay, P., Berta, P., Scherer, S.W., Tsui, L.C., and Muenke, M., Mutations in the human Sonic hedgehog gene cause holoprosencephaly, *Nat. Genet.*, 14, 357, 1996.
75. Roessler, E., Belloni, E., Gaudenz, K., Vargas, F., Scherer, S.W., Tsui, L.C., and Muenke, M., Mutations in the C-terminal domain of Sonic Hedgehog cause holoprosencephaly, *Hum. Mol. Genet.*, 6, 1847, 1997.
76. Wallis, D.E. and Muenke, M., Molecular mechanisms of holoprosencephaly, *Mol. Genet. Metab.*, 68, 126, 1999.
77. Chiang, C., Litingtung, Y., Lee, E., Young, K.E., Corden, J.L., Westphal, H., and Beachy, P.A., Cyclopia and defective axial patterning in mice lacking Sonic hedgehog gene function, *Nature*, 383, 407, 1996.
78. Kang, S., Graham, J.M., Jr., Olney, A.H., and Biesecker, L.G., GLI3 frameshift mutations cause autosomal dominant Pallister-Hall syndrome, *Nat. Genet.*, 15, 266, 1997.

79. Wild, A., Kalff-Suske, M., Vortkamp, A., Bornholdt, D., Konig, R., and Grzeschik, K.H., Point mutations in human GL13 cause Greig syndrome, *Hum. Mol. Genet.,* 6, 1979, 1997.
80. Hui, C.C. and Joyner, A.L., A mouse model of Greig cephalopolysyndactyly syndrome: the extra-toes J mutation contains an intragenetic deletion of the Gli3 gene, *Nat. Genet.,* 3, 241, 1993.
81. Bumcrot, D.A. and McMahon, A.P., Sonic signals somites, *Current Biology,* 5, 612, 1995.
82. Roberts, D.J., Johnson, R.L., Burke, A.C., Nelson, C.E., Morgan, B.A., and Tabin, C.J., Sonic hedgehog is an endodermal signal inducing BMP-4 and Hox genes during induction and regionalization of the chick hindgut, *Development,* 121, 3163, 1995.
83. Motoyama, J., Liu, J., Mo, R., Ding, Q., Post, M., and Hui, C.C., Essential function of Gli2 and Gli3 in the formation of lung, trachea, and oesophagus, *Nat. Genet.,* 20, 54, 1998.
84. Litingtung, Y., Lei, L., Westphal, H., and Chiang, C., Sonic hedgehog is essential to foregut development, *Nat. Genet.,* 20, 58, 1998.
85. Pearse, R.V. and Tabin, C.J., The molecular ZPA, *J. Exp. Zool.,* 282, 677, 1998.
86. Wolpert, L., Vertebrate limb development and malformations, *Ped. Res.,* 46, 247, 1999.
87. Warburton, D., Schwarz, M., Tefft, D., Flores-Delgado, G., Anderson, K.D., and Cardoso, W.V., The molecular basis of lung morphogenesis, *Mechan. Dev.,* 92, 55, 2000.
88. Makarenkova, H. and Patel, K., Gap junction signaling mediated through connexin-43 is required for chick limb development, *Dev. Biol.,* 207, 380, 1999.
89. Coelho, C.N. and Kosher, R.A., A gradient of gap junctional communication along the anterior-posterior axis of the developing chick limb bud, *Dev. Biol.* 148, 529, 1991.
90. Coelho, C.N. and Kosher, R.A., Gap junctional communication during limb cartilage differentiation, *Dev. Biol.,* 144, 47, 1991.
91. Parr, B.A. and McMahon, A.P., Dorsalizing signal Wnt-7a is required for normal polarity or D-V and A-P axes of mouse limb, *Nature,* 374, 350, 1995.
92. Cygan, J.A., Johnson, R.L., and McMahon, A.P., Novel regulatory interactions revealed by studies of murine limb pattern in Wnt-7a and En-1 mutants, *Development Supplement* 124, 5021, 1997.
93. Riddle, R.D., Johnson, R.L., Laufer, E., and Tabin, C.J., Sonic hedgehog mediates the polarizing activity of the ZPA, *Cell,* 75, 1401, 1993.
94. Francis, P.H., Richardson, M.K., Brickell, P.M., and Tickle, C., Bone morphogenic proteins and a signaling pathway that controls patterning in the developing chick limb, *Development* 120, 209, 1994.
95. Wada, N., Kawakami, Y., and Nohno, T., Sonic hedgehog signaling during digit pattern duplication after application of recombinant protein and expressing cells, *Dev. Growth Differen.* 41, 567, 1999.
96. Chiang, C., Litingtung, Y., Harris, M.P., Simandl, B.K., Li, Y., Beachy, P.A., and Fallon, J.F., Manifestation of limb prepattern: limb development in the absence of Sonic Hedgehog function, *Dev. Biol.* 236, 421, 2001.
97. Litingtung, Y., Dahn, R.D., Li, Y., Fallon, J.F., and Chiang, C., Shh and Gli3 are dispensable for limb skeleton formation but regulate digit number and identity, *Nature,* 418, 979, 2002.
98. Hara, K., Kimura, J., and Ide, H., Effects of FGFs on the morphogenic potency of the AER-maintenance activity of cultured progress zone cells of the chick limb bud, *Int. J. Dev. Biol.* 42, 591, 1998.
99. Laufer, E., Nelson, C.E., Johnson, R.L., Morgan, B.A., and Tabin, C.J., Sonic hedgehog and FGF-4 act through a signaling cascade and feedback loop to integrate growth and patterning of the developing limb bud, *Cell,* 79, 993, 1994.
100. Niswander, L., Jeffrey, S., Martin, G., and Tickle, C., Positive feedback signaling in vertebrate limb development, *Nature,* 371, 609, 1994.
101. Tickle, C., Vertebrate limb development, *Curr. Opin. Gen. Dev.,* 5, 478, 1995.
102. Summerbell, D., A quantitative analysis of the effect of excision of the AER from the chick limb bud, *J. Embryol. Morphol.,* 32, 651, 1974.
103. Johnson, R.L., Riddle, R.D., and Tabin, C.J., Mechanisms of limb patterning, *Curr. Opin. Gen. Dev.,* 4, 535, 1994.
104. Wolpert, L., Tickle, C., and Sampford, M., The effect of cell killing by x-irradiation on pattern formation in the chick limb, *J. Embryol. Exp. Morphol.,* 50, 175, 1979.
105. Nelson, C.E., Morgan, B.A., Burke, A.C., Laufer, E., DiMambro, E., Murtaugh, L.C., Gonzales, E., L., T., Parada, L.F., and Tabin, C.J., Analysis of Hox gene expression in the chick limb bud, *Development Supplement,* 122, 1449, 1996.

106. Goff, D.J. and Tabin, C.J., Analysis of Hoxd-13 and Hoxd-11 misexpression in chick limb buds reveal that Hox genes affect both bone condensation and growth, *Development Supplement* 124, 627, 1997.
107. Bryden, M.M., Evans, H.E., and Keeler, R.F., Cyclopia in sheep caused by plant teratogens, *J. Anat.* 110, 507, 1971.
108. Cooper, M.K., Porter, J.A., Young, K.E., and Beachy, P.A., Teratogen-mediated inhibition of target tissue response to Shh signaling, *Science*, 280, 1603, 1998.
109. Niederreither, K., Vermot, J., Schuhbaur, B., Chambon, P., and Dolle, P., Embryonic retinoic acid synthesis is required for forelimb growth and anteroposterior patterning in the mouse, *Development*, 129, 3563, 2002.
110. Summerbell, D., The effect of local application of retinoic acid to the anterior margin of the developing chick limb, *J. Embryol. Exp. Morphol.*, 78, 269, 1983.
111. Hendrickx, A.G., Teratogenicity of retinoids in rodents, primates and humans, *J. Toxicol. Sci.*, 23, 272, 1988.
112. Power, S.C., Lancman, J., and Smith, S.M., Retinoic acid is essential for Shh/Hox signaling during rat limb outgrowth but not for limb initiation, *Dev. Dyn.*, 216, 469, 1999.
113. Stratford, Abnormal anteroposterior and dorsoventral patterning of the limb bud in the absence of retinoids, *Mech. Dev.*, 81, 115, 1999.
114. Tickle, C., Alberts, B., Wolpert, L., and Lee, J., Local application of retinoic acid to the limb bud mimics the action of the polarizing region, *Nature*, 296, 564, 1982.
115. Wanek, N., Gardiner, D.M., Muneoka, K., and Bryant, S.V., Conversion by retinoic acid of anterior cells into ZPA cells in the chick wing bud, *Nature*, 350, 81, 1991.
116. Bell, S.M., Schreiner, C.M., and Scott, W.J., Jr., Disrupting the establishment of polarizing activity by teratogen exposure, *Mech. Dev.*, 88, 147, 1999.
117. Kwasigroch, T.E. and Bullen, M., Effects of isotretinoin (13-cis-retinoic acid) on the development of mouse limbs *in vivo* and *in vitro*, *Teratology*, 44, 605, 1991.
118. Lammer, E.J., Chen, D.T., Hoar, R.M., Agnish, N.D., Benke, P.J., Braun, J.T., Curry, C.J., Fernhoff, P.M., Grix, A.W., Lott, I.T., Richard, J.M., and Sun, S.C., Retinoic acid embryopathy, *N. Engl. J. Med.*, 313, 837, 1985.
119. Rosa, F.W., Wilk, A.L., and Kelsey, F.O., Teratogen update: vitamin A cogeners, *Teratology*, 33, 355, 1986.
120. Kochhar, D.M., Limb development in mouse embryos. I. Analysis of teratogenic effects of retinoic acid, *Teratology*, 7, 289, 1973.
121. Fantel, A.G., Shepard, T.H., Newell-Morris, L.L., and Moffett, B.C., Teratogenic effects of retinoic acid in pigtail monkeys (*Macaca nemestrina*). 1. General features, *Teratology*, 15, 65, 1976.
122. Kochhar, D.M., Penner, J.D., and Tellone, C.J., Comparative teratogenic activities of two retinoids: effects of palate and limb development, *Teratogen Carcinogen Mutagen.*, 4, 377, 1984.
123. Webster, W.S., Johnston, M.C., Lammer, E.J., and Sulik, K.K., Isotretinoin embryopathy and the cranial neural crest: an *in vivo* study, *J. Craniofacial Genet. Dev. Biol.*, 6, 211, 1986.
124. Rizzo, R., Lammer, E.J., Parano, E., Pavone, L., and Argyle, J.C., Limb reduction defects in humans associated with prenatal isotretinoin exposure, *Teratology*, 44, 499, 1991.
125. Kwasigroch, T.E., and Kochhar, D.M., Production of congenital limb defects with retinoic acid: phenomenological evidence of progressive differentiation during limb morphogenesis, *Anat. Embryol.*, 161, 105, 1980.
126. Rutledge, J.C., Shourbaji, A.G., Hughes, L.A., Polifka, J.E., Cruz, Y.P., and Bishop, J.B., Limb and lower body duplications induced by retinoic acid in mice, *Proc. Natl. Acad. Sci., USA*, 91, 5436, 1994.
127. Niederreither, K., Ward, S.J., Dolle, P., and Chambon, P., Morphological and molecular characterization of retinoic acid-induced limb duplications in mice, *Dev. Biol.*, 176, 185, 1996.
128. Cusic, A.M. and Dagg, C.P., Spontaneous and retinoic acid-induced postaxial polydactyly in mice, *Teratology*, 31, 49, 1985.
129. Kwasigroch, T.E., Vannoy, J.F., Church, J.K., and Skalko, R.G., Retinoic acid enhances and depresses *in vitro* development of cartilaginous bone anlagen in embryonic mouse limbs, *Dev. Biol.*, 22, 150, 1986.
130. Maden, M., Ong, D.E., Summerbell, D., and Chytil, F., The role of retinoid-binding proteins in the generation of pattern in the developing limb, the regenerating limb and the nervous system, *Dev. Suppl.*, 107, 109, 1989.

131. Tickle, C., Crawley, A., and Farrar, J., Retinoic acid applications to chick wing buds leads to a dose-dependent reorganization of the apical ectodermal ridge that is mediated by the mesenchyme, *Development,* 106, 691, 1989.
132. Dolle, P., Ruberte, E., Leroy, P., Morris-Kay, G., and Chambon, P., Retinoic acid receptor and cellular retinoid binding proteins. I. A systematic study of their differential pattern of transcription during mouse organogenesis, *Dev. Suppl.,* 110, 1133, 1990.
133. Vaessen, M.J., Meijers, J.H., Bootsma, D., and Van Kessel, A.G., The cellular retinoic-acid-binding protein is expressed in tissues associated with retinoic-acid-induced malformations, *Dev. Suppl.,* 110, 371, 1990.
134. Wilkinson, D.G. *In Situ Hybridization: A Practical Approach*, IRL Press, Oxford, 1992.
135. Nieto, M.A., Patel, K., and Wilkinson, D.G., In situ hybridization analysis of chick embryos in whole mount and tissue sections, *Methods Cell Biol.,* 51, 219, 1996.
136. Lowe, L.A. and Kuehn, M.R., Whole-mount in situ hybridization to study gene expression during mouse development, in *Methods in Molecular Biology, Vol 137: Developmental Biology Protocols,* Tuan, R.S., and Lo, C.W., Eds., Humana Press, Totowa, NJ, 1997, p. 125.
137. Tuan, R.S., mRNA and protein co-localization on tissue sections by sequential, colorimetric *in situ* hybridization and immunohistochemistry, in *Methods in Molecular Biology, Vol 137: Developmental Biology Protocols*, Tuan, R.S. and Lo, C.W., Eds., Humana Press, Totowa, NJ, 1997, p. 117.
138. Nieto, M.A., The early steps of neural crest development, *Mech. Dev.,* 105, 27, 2001.
139. Trainor, P.A. and Krumlauf, R., Hox genes, neural crest and branchial arch patterning, *Curr. Opin. Cell. Biol.,* 13, 698, 2001.
140. Noden, D.M., Interactions and fates of avian craniofacial mesenchyme, *Development,* 108, 121, 1988.
141. Chisaka, O., Musci, T.S., and Capecchi, M.R., Developmental defects of the ear, cranial nerves and hindbrain resulting from targeted disruption of the mouse homeobox gene Hox 1.6, *Nature,* 355, 516, 1992.
142. Santiago, A. and Erickson, C.A., Ephrin-B ligands play a dual role in the control of neural crest cell migration, *Development,* 129, 3621, 2002.
143. Krull, C.E., Segmental organization of neural crest migration, *Mech. Dev.,* 105, 37, 2001.
144. Menengola, E., Broccia, M.L., Prati, M., and Giavini, E., Morphological alterations induced by sodium valproate on somites and spinal nerves in rat embryos, *Teratology,* 59, 110, 1999.
145. Gale, E., Prince, V., Lumsden, A., Clarke, J., Holder, N., and Maden, M., Late effects of retinoic acid on neural crest and aspects of rhombomere identity, *Development,* 122, 783, 1996.
146. Gale, E., Zile, M., and Maden, M., Hindbrain respecification in the retinoid deficient quail, *Mechan. Dev.,* 89, 43, 1999.
147. Lee, Y.M., Osumi-Yamashita, N., Ninomiya, Y., Moon, C.K., Eriksson, U., and Eto, K., Retinoic acid stage-dependently alters the migration pattern and identity of neural crest cells, *Development,* 121, 825, 1995.
148. Wei, X., Makori, N., Peterson, P.E., Hummler, H., and Hendrickx, A.G., Pathogenesis of retinoic acid-induced ear malformations in a primate model, *Teratology,* 60, 83, 1999.
149. Nichols, D.H., Neural crest formation in the head of the mouse embryo as observed using a new histological technique, *J. Embryol. Exp. Morphol.,* 64, 105, 1981.
150. Campbell, L.R., Dayton, D.H., and Sohal, G.S., Neural tube defects: A review of human and animal studies on the etiology of neural tube defects, *Teratology,* 34, 171, 1986.
151. Clouthier, D.E., Hosoda, K., Richardson, J.A., Williams, S.C., Yanagisawa, H., Kuwaki, T., Kumada, M., Hammer, R.E., and Yanagisawa, M., Cranial and neural crest defects in endothelin-A receptor deficient mice, *Development,* 125, 813, 1998.
152. Clouthier, D.E., Williams, S.C., Yanasigawa, H., Wieduwilt, M., Richardson, J.A., and Yanagisawa, M., Signaling pathways crucial for craniofacial development revealed by endothelin A receptor deficient mice, *Dev. Biol.,* 217, 10, 2000.
153. Yanagisawa, H., Hammer, R.E., Richardson, J.A., Williams, S.C., Clouthier, D.E., and Yanagisawa, M., Role of endothelin-1/endothelin-A receptor mediated signaling pathway in the aortic arch patterning in mice, *J. Clin. Invest.* 102, 22, 1998.
154. Abo, T. and Balch, C.M., A differentiation antigen of human NK and K cells identified by a monoclonal antibody (HNK-1), *Immunology,* 127, 1024, 1981.

155. Tucker, G.C., Aoyama, H., Lipinski, M., Turz, T., and Thiery, J.P., Identical reactivity of monoclonal antibodies HNK-1 and NC-1: Conservation in vertebrates on cells derived from neural primordium and on some leukocytes, *Cell Differ.,* 14, 223, 1984.
156. Corbel, C., Lehmann, A., and Davison, F., Expression of CD44 during early development of the chick embryo, *Mech. Dev.,* 96, 111, 2000.
157. Maden, M., Horton, C., Graham, A., Leonard, L., Pizzey, J., and Eriksson, U., Domains of cellular-retinoic-acid-binding protein 1 (CRABP 1) expression in the hindbrain and neural crest of the mouse embryo, *Mech. Dev.,* 37, 13, 1992.
158. Eriksson, U., Hansson, E., Nordlinder, H., Busch, C., Sundelin, J., and Peterson, P.A., Quantitation and tissue localization of the cellular retinoic acid binding protein, *J. Cell Physiol.,* 133, 482, 1987.
159. Young, D.L., Schneider, R.A., Hu, D., and Helms, J.A., Genetic and teratogenic approaches to craniofacial development, *Crit. Rev. Oral Biol. Med.,* 11, 304, 2000.
160. Fawcett, L.B., Buck, S.J., Beckman, D.A., and Brent, R.L., Is there a no-effect dose for corticosteroid-induced cleft palate? The contribution of endogenous corticosterone to the incidence of cleft palate in mice, *Ped. Res.,* 39, 856, 1996.
161. Hackman, R.M. and Brown, K.S., Corticosterone-induced isolated cleft palate in A/J mice, *Teratology,* 6, 313, 1972.
162. Kalter, H., The history of the A family of inbred mice and the biology of its congenital malformations, *Teratology,* 20, 213, 1979.
163. Kerrigan, J.J., Mansell, J.P., Sengupta, A., Brown, N., and Sandy, J.R., Palatogenesis and potential mechanisms for clefting, *J.R. Coll. Surg. Edinb.,* 45, 351, 2000.
164. Fraser, F.C., The multifactorial/threshold concept — uses and misuses, *Teratology,* 14, 267, 1976.
165. Griffith, C.M. and Hay, E.D., Epithelial-mesenchymal transformation during palatal fusion: carboxyfluoresence in traces cells at light and electron microscopic levels, *Development,* 116, 1087, 1992.
166. Shuler, C.F., Halpern, D.E., Guo, Y., and Sank, A.C., Medial edge epithelium fate traced by cell lineage analysis during epithelial-mesenchymal transformation *in vivo, Dev. Biol.,* 154, 318, 1992.
167. Fitchett, J.E. and Hay, E.D., Medial edge epithelium transforms to mesenchyme after embryonic palatal shelves fuse, *Dev. Biol.,* 131, 455, 1989.
168. Carette, M.J.M. and Ferguson, M.W.J., The fate of medial edge epithelial cells during palatal fusion *in vitro*: an analysis of Dil labeling and confocal microscopy, *Development,* 114, 379, 1992.
169. Brunet, C.L., Sharpe, P.T., and Ferguson, M.W.J., Inhibition of TGF-beta 3 (but not TGF1 beta 1 or TGF-beta 2) activity prevents normal mouse embryonic palate fusion, *Int. J. Dev. Biol.,* 39, 345, 1995.
170. Proetzel, G., Pawlowski, S.A., Wiles, M.V., Yin, M., Bovin, G.P., Howles, P.N., Ding, J., Ferguson, M.W.J., and Doetschman, T., Transforming growth factor beta 3 is required for secondary palate fusion, *Nat. Genet.,* 11, 409, 1995.
171. Kaartinen, V., Cui, X.M., Heisterkamp, N., Goffen, J., and Shuler, C.F., Transforming growth factor-beta 3 regulates transdifferentiation of medial edge epithelium during palatal fusion and associated degradation of the basement membrane, *Dev. Dyn.,* 209, 255, 1997.
172. Harris, M.J. and Juriloff, D.M., Toward understanding mechanisms of genetic neural tube defects in mice, *Teratology,* 60, 292, 1999.
173. Medical Research Council (MRC) Vitamin Study Research Group, Prevention of neural tube defects: Results of the Medical Research Council Vitamin Study, *Lancet,* 338, 131, 1991.
174. Czeizel, A.F. and Dudas, I., Prevention of the first occurrence of neural tube defects by periconceptional vitamin supplementation, *Arch. Dis. Child.,* 56, 911, 1992.
175. Smithells, R.W., Sheppard, S., and Schorah, C.J., Apparent prevention of neural tube defects by periconceptional vitamin supplementation, *Arch. Dis. Child.,* 56, 911, 1981.
176. Goldman, I.D., The mechanism of action of methotrexate: I. Interaction with a low-affinity intracellular site required for maximum inhibition of deoxyribonucleic acid synthesis in L-cell mouse fibroblasts, *Mol. Pharmacol.,* 10, 257, 1974.
177. DeSesso, J.M. and Goeringer, G.C., Methotrexate-induced developmental toxicity in rabbits is ameliorated by 1-(*p*-tosyl)-3,4,4-trimethylimidizoline, a functional analog for tetrahydrofolate-mediated one-carbon transfer, *Teratology,* 45, 271, 1992.
178. Warkeny, J. and Petering, H.G., Congenital malformations of the central nervous system in rats produced by maternal zinc deficiency, *Teratology,* 5, 319, 1972.

179. Keen, C.L., Taubeneck, M.W., Daston, G.P., Rogers, J.M., and Gershwin, M.E., Primary and secondary zinc deficiency as factors underlying abnormal CNS development, *Ann. New York Acad. Sci.,* 678, 37, 1993.
180. Meiden, G.D., Keen, C.L., Hurley, L.S., and Klein, N.W., Effects of whole rat embryos cultured on serum from zinc- and copper-deficient rats, *J. Nutrition,* 116, 2424, 1986.
181. Apgar, J., Effects of zinc deprivation from day 12, 15, or 18 of gestation on parturition in the rat, *J. Nutrition,* 102, 343, 1972.
182. Hurley, L.S., Gowen, J., and Swenerton, H., Teratogenic effects of short-term and transitory zinc deficiency in rats, *Teratology,* 4, 199, 1971.
183. Hurley, L.S., Teratogenic aspects of manganese, zinc, and copper nutrition, *Physiol. Rev.,* 61, 249, 1981.
184. Essein, F.B., Maternal methionine supplementation promotes the remediation of axial defects in Axd mouse neural tube mutants, *Teratology,* 45, 205, 1992.
185. Essein, F.B. and Wannberg, S.L., Methionine but not folinic acid or Vitamin B-12 alters the frequency of neural tube defects in Axd mutant mice, *J. Nutrition,* 123, 27, 1993.
186. Fawcett, L.B., Pugarelli, J.E., and Brent, R.L., Effects of supplemental methionine on antiserum-induced dysmorphology in rat embryos cultured *in vitro, Teratology,* 61, 332, 2000.
187. Flynn, T.J., Friedman, L., Black, T.N., and Klein, N.W., Methionine and iron as growth factors for embryos cultured on canine serum, *J. Exp. Zool.,* 244, 319, 1987.
188. Coelho, C.N.D., Weber, J.A., Klein, N.W., Daniels, W.G., and Hoagland, T.A., Whole rat embryos require methionine for neural tube closure when cultured on cow serum, *Teratology,* 42, 437, 1989.
189. Chambers, B.J., Klein, N.W., Nosel, P.G., Khairallah, L.H., and Romanow, J.S., Methionine overcomes neural tube defects in rat embryos cultured on sera from laminin-immunized monkeys, *J. Nutrition,* 125, 1587, 1995.
190. Coelho, C.N. and Klein, N.W., Methionine and neural tube closure in cultured rat embryos: morphological and biochemical analysis, *Teratology,* 42, 437, 1990.
191. Nosel, P.G. and Klein, N.W., Methionine decreases the embryotoxicity of sodium valproate in the rat: *in vivo* and *in vitro* observations, *Teratology,* 46, 499, 1992.
192. Cockcroft, D.L., Brook, F.A., and Copp, A.J., Inositol deficiency increases the susceptibility to neural tube defects of genetically predisposed (curly tail) mouse embryos *in vitro, Teratology,* 45, 233, 1992.
193. Greene, N.D. and Copp, A.J., Inositol prevents folate-resistant neural tube defects in the mouse, *Nat. Med,.* 3, 60, 1997.
194. Van Straaton, H.W. and Copp, A.J., Curly tail: a 50-year history of the mouse spina bifida model, *Anat. Embryol,.* 203, 225, 2001.
195. Mukherjee, M.D., Sandstead, H.H., Ratnaparkhi, M.V., Johnson, L.K., Milne, D.B., and Stelling, H.P., Maternal zinc, iron, folic acid and protein nutriture and outcome of human pregnancy, *Am. J. Clin. Nutr.,* 40, 496, 1984.
196. Velie, E.M., Block, G., Shaw, G.M., Samuels, S.J., Schaffer, D.M., and Kulldorff, M., Maternal supplemental and dietary zinc intake and the occurrence of neural tube defects in California, *Am. J. Epidemiol.,* 150, 605, 1999.
197. Shaw, G.M., Todoroff, K., Schaffer, D.M., and Selvin, S., Periconceptional nutrient intake and risk for neural tube defect-affected pregnancies, *Epidemiology,* 10, 711, 1999.
198. Bower, C., Stanley, F.J., and Spickett, J.T., Maternal hair zinc and neural tube defects: no evidence of an association from a case control study in western Australia, *Asia-Pacific J. Pub. Health,* 6, 156, 1992-19993.
199. Milunski, A., Morris, J.S., Jick, H., Rothman, K.J., Ulcickas, M., Jick, S.S., Shoukimas, P., and Willett, W., Maternal zinc and fetal neural tube defects, *Teratology,* 46, 341, 1992.
200. Shaw, G.M., Velie, E.M., and Schaffer, D.M., Is dietary intake of methionine associated with a reduction in risk for neural tube defect-affected pregnancies? *Teratology,* 56, 295, 1997.
201. Shoob, H.D., Sargent, R.G., Thompson, S.J., Best, R.G., Drane, J.W., and Tocharoen, A., Dietary methionine is involved in the etiology of neural tube defect-affected pregnancies in humans, *J. Nutr.,* 131, 2653, 2001.
202. Woods, J.R. and Plessinger, M.A., Maternal-fetal cardiovascular system: a target of cocaine, *NIDA Research Monograph* 108, 7, 1991.

203. Danielsson, B.R.G., Danielson, M., Reiland, S., Rundqvist, E., Denker, L., and Regard, C.G., Histological and *in vitro* studies supporting decreased uteroplacental blood flow as explanation for digital defects after administration of vasodilators, *Teratology*, 41, 185, 1990.
204. Webster, W.S. and Brown-Woodman, P.D.C., Cocaine as a cause of congenital malformations of vascular origin: experimental evidence in the rat, *Teratology*, 41, 689, 1990.
205. Jones, K.L., Developmental pathogenesis of defects associated with prenatal cocaine exposure: fetal vascular disruption, *Clin. Perinatol.*, 18, 139, 1991.
206. Gingras, J.L., Weese-Mayer, D.E., Hume, R.F., and O'Donnell, K.J., Cocaine and development: mechanisms of fetal toxicity and neonatal consequences of prenatal cocaine exposure, *Early Hum. Dev.*, 31, 1, 1992.
207. Gonzalez, C.H., Vargas, F.R., Perez, A.B.A., Kim, C.A., Marques-Dias, M.J., Leone, C.R., Neto, J.C., Llerena, J.C., and Cabral de Almeida, J.C., Limb reduction defects with or without mobius sequence in seven Brazilian children associated with misoprostol use in the first trimester of pregnancy, *Amer. J. Med. Genet.*, 47, 59, 1993.
208. Brent, R.L., Congenital malformation case reports: The editor's and reviewer's dilemma, *Am. J. Med. Genet.*, 47, 872, 1993.
209. Genest, D.R., Di Salvo, D., Rosenblatt, M.J., and Holmes, L.B., Terminal transverse limb defects with tethering and omphalocele in a 17 week fetus following first trimester misoprostol exposure, *Clin. Dysmorphol.*, 8, 53, 1999.
210. Burton, B.K., Schultz, C.J., and Burd, L.I., Limb anomalies associated with chorionic villus sampling, *Obstet. Gynecol.*, 79, 726, 1992.
211. Holmes, L.B., Chorionic villus sampling and limb defects, *Prog. Clin. Biol. Res.*, 383A, 409, 1993.
212. Stoler, J.M., McGuirk, C.K., Lieberman, E., Ryan, L., and Holmes, L.B., Malformations reported in chorionic villus sampling exposed children: a review and analytic synthesis of the literature, *Genet. Med.*, 1, 315, 1999.
213. Brent, R.L. and Franklin, J.B., Uterine vascular clamping: New procedure for the study of congenital malformations, *Science*, 132, 89, 1960.
214. Franklin, J.B. and Brent, R.L., The effect of uterine vascular clamping on the development of rat embryos three to fourteen days old, *J. Morphol.*, 115, 273, 1964.
215. Leist, K.H. and Grauwiler, J., Fetal pathology in rats following uterine-vessel clamping on day 14 of gestation, *Teratology*, 10, 55, 1974.
216. Danielson, M.K., Danielsson, B.R.G., Marchner, H., Lundin, M., Rundqvist, E., and Reiland, S., Histopathological and hemodynamic studies supporting hypoxia and vascular disruption as explanation to phenytoin teratogenicity, *Teratology*, 46, 485, 1992.
217. Danielsson, B.R.G., Danielson, M., Rundqvist, E., and Reiland, S., Identical phalangeal defects induced by phenytoin and nifedipine suggest fetal hypoxia and vascular disruption behind phenytoin teratogenicity, *Teratology*, 45, 247, 1992.
218. Fantel, A.G., Barber, C.V., Carda, M.B., Tumbic, R.W., and Mackler, B., Studies of the role of ischemia/reperfusion and superoxide anion radical production in the teratogenicity of cocaine, *Teratology*, 46, 293, 1992.
219. Zimmerman, E.F., Potturi, R.B., Resnick, E., and Fisher, E.J., Role of oxygen free radicals in cocaine-induced vascular disruption in mice, *Teratology*, 49, 192, 1994.
220. Fantel, A.G. and Person, R.E., Further evidence for the role of free radicals in the limb teratogenicity of L-NAME, *Teratology*, 66, 24, 2002.
221. Brent, R.L., Editorial Comment: Relationship between uterine vascular clamping, vascular disruption syndrome, and cocaine teratogenicity, *Teratology*, 41, 757, 1990.
222. Chavez, G.F., Mulinare, J., and Coredero, J., Maternal cocaine use during early pregnancy as a risk factor for congenital urogenital anomalies, *JAMA*, 262, 795, 1989.
223. Gaily, E., Distal phalangeal hypoplasia in children with prenatal phenytoin exposure, *Am. J. Med. Genet.*, 35, 574, 1990.
224. Webster, W.S., Lipson, A.H., and Brown-Woodman, P.D.C., Uterine vascular trauma and limb defects, *Teratology*, 35, 253, 1987.
225. Fawcett, L.B., Buck, S.J., and Brent, R.L., Limb reduction defects in the A/J mouse strain associated with maternal blood loss, *Teratology*, 58, 183, 1998.

226. Menkes, B., Prelipceanu, O., and Capalnasan, I., Vital fluorochroming as a tool for embryonic cell death research, in *Advances in the Study of Birth Defects*, Persaud, T.V.N., Ed., University Park Press, Baltimore, 1979, p. 219.
227. Kohlbecker, A., Lee, A.E., and Schorle, H., Exencephaly in a subset of animals heterozygous for AP-2a mutation, *Teratology*, 65, 213, 2002.
228. Ivnitski, I., Elmaoued, R., and Walker, M.K., 2,3,7,8-tetrachlorodibenzo-p-dioxin (TCCD) inhibition of coronary development is preceded by a decrease in myocyte proliferation and an increase in cardiac apoptosis, *Teratology*, 64, 201, 2001.
229. Takagi, T.N., Matsui, K.A., Yamashita, K., Ohmori, H., and Yasuda, M., Pathogenesis of cleft palate in mouse embryos exposed to 2,3,7,8-tetrachlorodibenzo-p-dioxin (TCCD), *Teratogen Carcinogen Mutagen.*, 20, 73, 2000.
230. Slavkin, H., Nuckolls, G., and Shum, L., Craniofacial development and patterning, in *Methods in Molecular Biology, Vol 137: Developmental Biology Protocols*, Tuan, R.S. and Lo, C.W., Eds., Humana Press, Totowa, NJ, 1997, p. 45.
231. Abbott, B., Palatal dysmorphogenesis, palate organ culture, in *Methods in Molecular Biology, Vol 137: Developmental Biology Protocols*, Tuan, R.S. and Lo, C.W., Eds., Humana Press, Totowa, NJ, 1997, p. 195.
232. DeLise, A.M., Stringa, E., Woodward, W.A., Mello, M.A., and Tuan, R.S., Embryonic limb mesenchyme micromass culture as an *in vitro* model for chondrogenesis and cartilage maturation, in *Methods in Molecular Biology, Vol 137: Developmental Biology Protocols*, Tuan, R.S. and Lo, C.W., Eds., Humana Press, Totowa, NJ, 1997, p. 359.
233. Renault, J.Y., Caillaud, J.M., and Chevalier, J., Ultrastructural characterization of normal and abnormal chodrogenesis in micromass rat embryo limb bud cell cultures, *Toxicol. Appl. Pharmacol.*, 130, 177, 1995.
234. Flint, O.P., *In vitro* tests for teratogens: desirable endpoints, test batteries and current status of the micromass teratogen test, *Reprod. Toxicol.*, 7 Suppl 1, 103, 1993.
235. Van Maele-Fabry, G., Gofflot, F., Clotman, F., and Picard, J.J., Alterations of mouse embryonic branchial nerves and ganglia induced by ethanol, *Neurotoxicol. Teratol.*, 17, 497, 1995.
236. Van Maele-Fabry, G., Clotman, F., Gofflot, F., Bosschaert, J., and Picard, J.J., Postimplantation mouse embryos cultured *in vitro*, assessment with whole-mount immunostaining and *in situ* hybridization, *Int. J. Dev. Biol.*, 41, 365, 1997.
237. Mark, M., Lufkin, T., Vonesch, V.L., Ruberte, E., Olivo, J.C., Dolle, P., Gorry, P., Lumsden, A., and Chambon, P., Two rhombomeres are altered in Hoxa-1 mutant mice, *Development*, 119, 319, 1993.
238. Krull, C.E., Collazo, A., Fraser, S.E., and Bronner-Fraser, M., Segmental migration of trunk neural crest: time-lapse analysis reveals a role for PNA-binding molecules, *Development*, 121, 3733, 1995.
239. Bhattacharyya, A., Brackenbury, R., and Ratner, N., Axons arrest the migration of schwann cell precursors, *Development*, 120, 1411, 1994.
240. Serbedzija, G.N., Bronner-Fraser, M., and Fraser, S.E., A vital dye analysis of the timing and pathways of neural crest migration, *Development*, 106, 809, 1989.
241. Stuckhardt, J.L. and Poppe, S.M., Fresh visceral examination of rat and rabbit fetuses used in teratogenicity testing, *Teratogen Carcinogen Mutagen.*, 4, 181, 1984.
242. Wilson, J.G., Methods for administering agents and detecting malformations in experimental animals, in *Tertology: Principles and Technique.*, Wilson, J.G. and Warkany, J., Eds., University of Chicago Press, Chicago, 1965, p. 262.
243. Barrow, M.V. and Taylor, W.J., A rapid method for detecting malformations in rat fetuses, *J. Morphol.*, 127, 291, 1969.
244. Astroff, A.B., Ray, S.E., Rowe, L.M., Hilbish, K.G., Linville, A.L., Stutz, J.P., and Breslin, W.J., Frozen-sectioning yields similar results as traditional methods for fetal cephalic examination in the rat, *Teratology*, 66, 77, 2002.
245. Inouye, M., Differential staining of cartilage and bone in fetal mouse skeleton by alcian blue and alizerin red, *S. Cong. Anom.*, 16, 171, 1976.
246. Webb, G.N. and Byrd, R.A., Simultaneous differential staining of cartilage and bone in rodent fetuses: an alcian blue and alizarin red S procedure without glacial acetic acid, *Biotechnic Histochem.*, 69, 181, 1994.
247. Kelly, W.L. and Bryden, M.M., A modified differential stain for cartilage and bone in whole mount preparations of mammalian fetuses and small vertebrates, *Stain Technol.*, 58, 131, 1983.

248. Kimmel, C.A. and Trammell, C., A rapid procedure for routine double staining of cartilage and bone in fetal and adult animals, *Stain Technol.,* 56, 271, 1981.
249. Whitaker, J. and Dix, K.M., Double staining technique for rat foetus skeletons in teratological studies, *Laboratory Animals,* 13, 309, 1979.
250. Young, D.L., Phipps, D.E., and Astroff, A.B., Large-scale double-staining of rat fetal skeletons using Alizarin Red S and Alcian Blue, *Teratology,* 61, 273, 2000.
251. Zucker, R.M., Elstein, K.H., Shey, D.L., Ebron-McCoy, M., and Rogers, J.M., Utility of fluorescence microscopy in embryonic/fetal topographical analysis, *Teratology,* 51, 430, 1995.

CHAPTER 4

Maternally Mediated Effects on Development

Ronald D. Hood and Diane B. Miller

CONTENTS

I. Introduction 94
   A. Maternally Mediated Effects Defined 94
   B. Historical Background 94
   C. Causes of Maternal Stress 95
      1. Neurogenic Stress 96
      2. Systemic Stress 97
      3. An Alternative Categorization of Stressors 97
   D. Developmental Hazard Associated with Maternal Stress or Toxicity 97
      1. Supernumerary Ribs 98
      2. Other Variations and Malformations 98
      3. Prenatal Mortality and Growth Inhibition 99
   E. Concern about Maternally Mediated Effects 99
      1. Distinguishing Maternally Mediated from Direct Effects 99
      2. Influence on Developmental Toxicity Test Outcomes 100
      3. Influence on Interpretation of Developmental Toxicity Test Results 101
      4. Potentiation of Chemical Teratogenesis by Maternal Stress 102
      5. Possible Influence on Human Pregnancy Outcomes 103
II. Possible Mechanisms for Maternally Mediated Effects 103
III. Assessment of Maternal Stress or Toxicity 104
   A. Biochemical Measures 104
      1. Measures Related to the Stress Cascade 104
      2. Catecholamines 106
      3. Stress Proteins 106
      4. Measures Related to Zinc Metabolism 107
   B. Organ Weights 108
   C. Behavioral Measures 108
   D. Measures of Toxicity 109
   E. Overview of Stress and Toxicity Assessment 110
IV. Experimental Assessment of Possible Maternally Mediated Effects 110
   A. Factors to Be Considered in Experimental Design and Conduct 110
      1. Controlling the Animal's Environment 110

	2. Measurement of Stress 111
	3. Appropriate Controls 111
	4. Developmental Timing 111
	5. Consistency of Methodology.. 112
	6. Outcome Assessment and Interpretation of Results 112
	B. Induction of Maternal Restraint Stress as a Model System................................. 113
V.	Overview 115
References... 115	

I. INTRODUCTION

A. Maternally Mediated Effects Defined

In developmental toxicology, maternally mediated effects on development are those adverse consequences that occur secondarily, as a result of some effect on the pregnant mother. They differ from direct effects on the conceptus primarily in their immediate source, rather than their end result. Since likely mechanisms for maternally mediated effects are more limited in number than are direct-acting mechanisms, the range of consequences to the offspring may also be more limited. Nevertheless, as pointed out by Daston,[1] since there are several mechanisms by which maternally mediated effects may occur, it is unlikely that they would all result in only a single limited spectrum of effects on the offspring.

The potential for maternally mediated effects also makes the task of extrapolating from animal data to potential human outcomes more problematic.[1,2] As stated by Kimmel and colleagues,[3] "The developing mammal and its maternal support system present a special situation in toxicology and risk assessment." They contend that because of the dependence of the conceptus on the nurturing maternal environment, factors disturbing that environment may adversely affect the offspring's development. They also recognize, however, that the maternal organism offers a degree of protection against at least some environmental perturbations.

It is well established that fetal disruptions can be maternally mediated, but questions remain regarding the kinds of effects produced, their prevalence, and their biological significance.[1,4] It is likely that fetal variations, malformations, functional alterations, and deaths can result from direct effects on the fetus, indirect (maternally mediated) effects, or a combination of the two.[5] Nevertheless, there has often been insufficient concern on the part of researchers for the potential of maternally mediated effects to affect the outcome of developmental toxicity studies.[3] The converse can also happen, however, in that some authors consider any effects on the conceptus that are seen only at maternally toxic doses to be secondary to the maternal toxicity although there may be little or no evidence to substantiate that supposition. A more reasonable approach is that of Chernoff and coworkers,[2] who advocate concern about manifestations of developmental toxicity, whether or not there is concurrent maternal toxicity, unless there is unequivocal evidence that humans would not be similarly affected.

B. Historical Background

The concept that stress and/or toxicity to the pregnant mother could indirectly result in adverse effects on the developing conceptus is not new. Early studies investigated the potential of such "maternally mediated effects" on developing offspring. These studies generally used such stressors as restraint, hypothermia, electric shock, noise, visual stimuli, shipping, or crowding on pregnant rodents.[6–17] Some investigators[14,18] used maternal food and water deprivation as a stressor for mice, but it is not entirely clear if alterations seen in the offspring were due to indirect effects or, at least in part, to malnutrition of the conceptus. Nevertheless, food and water deprivation was correlated

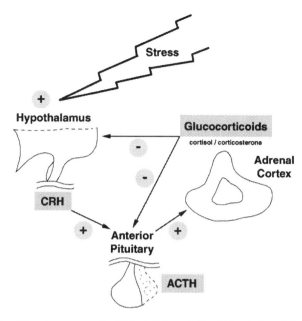

Figure 4.1 Schematic of the stress cascade. (Adapted from Colby, H.D., *J. Am. Coll. Toxicol.*, 7, 45, 1988.)

with increased maternal plasma corticosterone levels,[14,19] and resulted in an increased incidence of cleft palate. Additional early studies suggested that maternal restraint stress in the rat could potentiate the effects of chemical teratogens.[20,21]

Although investigations of the role of maternal effects in causing growth retardation, death, behavioral effects, biochemical alterations, or dysmorphogenesis in the embryo and fetus continued into the 1980s,[22–30] the major impetus for further research came as a consequence of publications by Khera.[31–34] Khera reviewed published studies and proposed that a number of effects on the offspring of mice, rats, and rabbits occurred merely as a consequence of maternal toxicity. Such putative maternally mediated fetal effects included decreased body weights, certain malformations and variations, and resorptions. Additional discussion of the significance of maternally mediated effects can be found in reviews by Hood,[4,35] Chernoff and colleagues,[2] and Daston.[1]

C. Causes of Maternal Stress

In the broadest sense "stress" is any change in the organism's environment that disturbs homeostasis. The resulting series of neural and endocrine adaptations is commonly referred to as the "stress response" or "stress-cascade" (Figure 4.1). Operationally defined, a "stressor" is any manipulation capable of disturbing homeostasis. Many of the procedures utilized in the collection of data in toxicology, and teratology is no exception, would qualify as stressors under this definition (Table 4.1). Thus, stress can be considered an adaptive mechanism that has evolved to protect the organism in times of crisis.[36]

The mammalian response to stress is a basic regulatory mechanism carried out in part by the limbic-hypothalamo-pituitary-adrenal (LHPA) axis.[36] Release of catecholamines from the adrenal medulla and the sympathetic nervous system; adrenocorticotropin (ACTH) from the anterior portion of the pituitary gland; corticotropin releasing factor (CRF) and arginine vasopressin (AVP) from the hypothalamus; and glucocorticoids from the adrenal cortex are all key components of this action. The stress cascade is considered an example of a classic negative feedback loop. When glucocorticoids, the terminal element in the cascade, reach a sufficiently high level in the circulation, they inhibit the release of the initiating components of the cascade, CRF and ACTH.

Table 4.1 Stress inducers in toxicology

Dosing

Route
    Inhalation (e.g., nose-only)
    Dermal (e.g., wearing jackets)
    Gavage (e.g., inexperienced technician)
Maximum tolerated dose (MTD)

Deprivation

Food or water
Maternal

Restraint

During dosing
During monitoring

Housing Conditions

Group
Single

The response of the body to stress can also result in alteration or release of other hormones and biochemical substances, including prolactin, growth factors, prostaglandins and other arachidonic acids, as well as proteinases, lymphokines, and peptides. Many body systems, including endocrine, neural, renal, and immune systems, can be activated.[36–40]

A wide variety of stressors have been employed in attempts to understand the effects of stress on adult animals. Many of these same stressors have also been used in developmental studies to elucidate both the prenatal and postnatal consequences of stress. In a practical sense the broad definition of a stressor as any disrupter of homeostasis may not be the most useful in delineating the role of stress in maternally mediated effects on development. It is obvious that events disruptive to homeostasis can result in a myriad of biochemical and physiological changes, and that events as diverse as moving an animal to a new cage or immobilizing it would qualify as stressors. It is, however, equally obvious that events categorized as stressors[41] do not all result in exactly the same spectrum of biochemical and physiological changes, nor do they result in the same spectrum of developmental changes when applied during the prenatal period.

The developmental alterations engendered by stress depend to a large extent on the species studied as well as the type and the duration of the stressor involved. For these reasons it may be more useful to further subcategorize events capable of disturbing homeostasis. The scheme developed by Allen and coworkers,[42] in which stressors are considered as either neurogenic or system stressors, may be useful in this regard. Stressors are subdivided into those directly affecting various components of the areas normally participating in the stress reaction (e.g., a test chemical directly activates the adrenal cortex)[43] and those where the action of the stressor is initiated at the neural level (e.g., an experimental procedure, such as injection, is interpreted as aversive by the organism, resulting in a release of CRF from the hypothalamus and initiation of the stress cascade). Of course, there may be instances in which a stressor could act as both a systemic and a neurogenic stressor.

1. Neurogenic Stress

"Neurogenic stress," according to Allen et al.,[42] is the type of stress that causes signals to be sent to the hypothalamus by neural pathways. For neurogenic stressors to be effective, the peripheral nervous system must interact with the central nervous system. Both aspects of the nervous system must thus be intact for a response, such as stimulation of ACTH production, to occur. The various inputs indicative of stress are presumed to converge via a final common path, the neurons of the brain's medial parvocellular division of the paraventricular nucleus of the hypothalamus.[36] Neurogenic stressors are apparently "psychological stressors" that presumably do not affect the pregnant mother directly through toxicity, physical effects, or direct physiological alterations. Instead, such stressors act by causing the animal to perceive its environment as potentially hazardous or annoying. These perceptions then result in changes in the pregnant animal's physiology and/or biochemistry that could potentially affect the conceptus. Examples of neurogenic stressors include noxious stimuli, such as trauma, electric shock, and vibration, as well as annoying stimuli, such as noise, light, and crowding. Allen and coworkers[42] also list "psychological" or "emotional" stress in the neurogenic category. When using such stressors, appropriate control groups must be included, and care must be taken to ensure that effects seen are not merely due to secondary consequences to the mother, such as decreased food or water intake.

2. Systemic Stress

The second major category of stress listed by Allen et al.[42] is "systemic stress." According to those authors, systemic stressors may be pharmacologic agents, such as ether or endotoxin, that appear to act following transport to the hypothalamus and pituitary by the circulation. Other chemical toxicants may act by similar means, or they may elicit at least part of their effects by causing pain and acting by neurogenic pathways. Allen et al.[42] listed other examples of systemic stressors, namely "hypotension, (forced) exercise, and starvation," which were said to act by "humoral and/or neural pathways," and environmental factors, such as heat or cold. They also listed "forced immobilization or restraint" as a systemic stressor, but it seems more likely that restraint stress is primarily psychologically mediated (i.e., a neurogenic stressor), and forced immobilization or restraint can be considered both a systemic and a neurogenic stressor. Other potential stressors, such as acidosis and alkalosis, would cause a physiologically based stress, but were not listed. It is not always clear *a priori* which stressors are primarily neurogenic, which are systemic, and which may act by both pathways to stimulate responses, such as stress hormone release.

3. An Alternative Categorization of Stressors

Pacák and Palkovits proposed a different classification of stressful stimuli.[41] They divided stressors into four categories: (1) physical stressors that have a negative (or, in some cases, a positive) psychological component and that include cold, heat, intense radiation, noise, vibration, etc., chemical toxicants, and pain induced by chemical or physical means, (2) psychological stressors that reflect a learned response to previously experienced adverse conditions, and may evoke responses that can be considered to reflect anxiety, fear, or frustration, (3) social stressors reflecting disturbed interactions among individuals, e.g., an animal being placed in the home cage of a dominant animal, or in the case of humans, unemployment or marital separation, and (4) stressors that challenge cardiovascular and metabolic homeostasis, for example, exercise, hypoglycemia, and hemorrhage. Some stressors would include components of more than one of the four categories of stressors. Examples include handling, restraint, and anticipation of a painful stimulus.

D. Developmental Hazard Associated with Maternal Stress or Toxicity

A number of developmental alterations have been suggested as possible maternally mediated effects. Only a few, however, have been experimentally shown to be commonly inducible by at least some types of maternal stress and/or toxicity, and these are discussed in later sections. They form a subset of the anomalies listed for rodents by Khera,[31,32] and even this subset is not invariably seen in offspring of stressed or intoxicated dams. Maternally mediated behavioral and physiological (especially endocrine) alterations in mammalian offspring, including those of humans, have also been noted by a number of investigators,[6,9,25,29,30,44–53] but these will not be specifically addressed in this chapter.

1. Supernumerary Ribs

According to Russell,[54] supernumerary (lumbar) ribs (SNR) result from "homeotic shifts" in the axial skeleton, i.e., they reflect "evolutionarily based capabilities for alternative development" that may involve serially segmented structures. Presumably SNR reflect altered expression of homeobox genes, although the details remain to be elucidated,[55,56] and they can occur in untreated animals.[57] In a study by Chernoff and collaborators,[58,59] fetuses from pregnant mice subjected to immobilization stress responded with an increased incidence of SNR, while the offspring of rats showed no such effect. These results indicated that extra rib production in mice may be a general response to maternal stress, and that stress alone (or at least stress in the presence of food and water deprivation) was adequate to induce such a response.

Lumbar ribs were the only anomalies consistently found by Kavlock and coworkers[60] when several test compounds were given to mice at maternally toxic doses; seven of the ten test agents increased the incidence of extra ribs. Kimmel and Wilson[61] differentiated between "extra" (supernumerary) ribs — at least half as long as the thirteenth rib — and "rudimentary" ribs (RR) — ossified structures shorter than extra ribs; they proposed that the former tended to be treatment related, while the latter were more variable in incidence and did not appear to be related to test agent dose. Kavlock et al.[60] did not make a distinction between the two "rib" types, as they found the incidence of both to be concurrently increased.

In a study[62] employing eight compounds at maternally toxic doses in the rat, only two test agents increased the incidences of SNR. Another study found that maternally toxic doses of the herbicide bromoxynil induced extra ribs in both mice and rats.[63] When Chernoff and coworkers treated both species, the incidence of SNR was elevated by treatment, but those seen in the rat were mainly rudimentary ribs, while mice exhibited both RR and SNR.[64] The persistence of the extra ossified structures through postnatal day 40 varied by rodent species and "rib" length.

The studies in which maternally toxic doses of chemicals were associated with increased incidences of SNR suggest that this endpoint may have occurred as a secondary effect of maternal toxicity-associated stress, at least in the mouse. However, the increase in SNR may represent a direct effect of the compounds on the fetus, an indirect effect due to disturbance of maternal homeostasis, or a combination of direct and indirect effects.

2. Other Variations and Malformations

Khera's literature surveys[31,32] identified a constellation of defects that often appeared in offspring of mice, rats, rabbits, and hamsters when their pregnant dams were given maternally toxic doses of chemicals. The studies[60,62] described above that were done to test Khera's hypothesis that maternal toxicity would commonly result in specific defects did not find a consistent relationship for defects other than SNR. Such data suggest that maternal toxicity, as usually defined, is not an effective or consistent inducer of most developmental defects.

Some data exist, however, that indicate a possible relationship between at least one type of maternal stress and exencephaly-encephalocele and fused ribs, in addition to the SNR discussed above. This was first suggested in 1986[58] and 1987[59] in related reports describing the developmental effects of maternal immobilization in mice. The particular method of restraint utilized, however, appears to be crucial in inducing the malformations, a finding supported in a mouse restraint study by Rasco and Hood.[65] The degree to which the movement of the dam is restricted may be the defining factor. When employed later in gestation, maternal restraint induced cleft palate in mice.[18]

Another maternal stressor, audiogenic stress (objectionable noise), has been said to cause adverse effects in mice. Defects such as exencephaly, encephalocele, and fused sternebrae have been seen in fetuses from dams subjected to such stress,[23,66] although Kimmel and coworkers[15] found no such effects in rats and only an increased incidence of resorptions in mice. More recently, gestational exposure to loud noise was reported to alter the development and responsiveness of immune system components in rats.[67]

3. Prenatal Mortality and Growth Inhibition

In the studies[58,59,65,68] done to date with mice, significant decreases in prenatal survival or prenatal weight gain have not generally been noted as a result of maternal restraint stress on only one day of gestation; the one exception was a statistically significant but numerically rather modest trend toward embryo or fetal mortality noted across treatment days by Chernoff and coworkers.[69] When pregnant rats were stressed on multiple gestation days, however, litter size[13] or birth weight[70] was decreased. In the latter study, the decreased birth weight may have merely been due to food and water deprivation as there was no mention of deprivation of the unstressed controls. In that study, restraint of young rats (70 to 120 days) both before and during gestation delayed parturition, but the same was not true for a group of older (11 to 13 months) rats.[70]

Other stressors have been said to have deleterious effects on survival to term or fetal growth. Audiogenic stress[8,15,17,22,23,66] and anticipation of electric shock[11] have been reported to cause embryo or fetal mortality in rodents and decreased fetal weight at term.[11,17,23,66,71] Stress-induced immunological imbalances have been proposed as causes of abortion in both mice[73] and women.[74]

E. Concern about Maternally Mediated Effects

1. Distinguishing Maternally Mediated from Direct Effects

Khera hypothesized that a number of common effects on the offspring of rodents and rabbits whose dams had received toxic doses of test agents during gestation were secondary effects of maternal toxicity.[31,32,34] Khera also stated that such effects often were not dose related, tended to be species specific, and were seldom seen at doses below the maternally toxic dose. Khera's proposal precipitated increased interest in the potential of maternally mediated effects to influence the outcomes of developmental toxicity studies and has been supported by Black and Marks.[72] Nevertheless, his proposal has also received criticism. His literature review approach indicates a possible association but does not establish a causal relationship.[59] Also, because of the retrospective nature of the studies analyzed by Khera and because negative data often remain unpublished, Khera's reviews could not avoid a degree of selection bias.[75] As pointed out by Schardein,[76] the endpoints used in literature reports to assess maternal toxicity are often ill defined and are sometimes ignored altogether.

There are plausible alternative explanations for Khera's findings. If both offspring and mother have similar inherent sensitivities to the same or different mechanisms of toxicity, the conceptus may be similar to the dam in susceptibility to many chemical agents.[3,59] Alternatively, lower absorbed dose or lack of activating enzymes in the embryo or fetus might increase its resistance to the toxicity of certain compounds to a level approximating or exceeding that of the mother. Many developmental toxicants appear to have cytotoxic effects, and some may produce maternal and developmental effects at similar doses because of a lack of selectivity. Also, a given species or strain is likely to have specific points in its program of development that are the most sensitive to toxicity. Thus, each species and strain is likely to exhibit a specific spectrum of common developmental defects in response to a variety of toxic insults,[75] whether a specific effect on the offspring is maternally mediated or direct.

A critical consideration in any attempt to understand the relative role of maternally mediated effects is the ability to determine which effects are direct effects on the conceptus and which, if any, are indirect (maternally mediated). As noted previously in this chapter, maternal stress alone can cause certain adverse fetal outcomes. Although such findings are of interest, the more critical question in interpreting results of developmental toxicity tests is how commonly is toxicity to the mother translated into adverse effects on the offspring and under what circumstances is such maternal mediation likely to occur?

In an attempt to test Khera's proposal,[31] Kavlock et al.[60] exposed groups of pregnant mice to maternally toxic doses of one of ten different chemicals. Although a number of effects on the offspring were seen, only supernumerary ribs (SNR) were observed in a majority of the test groups (7 of the 10), and no such association was found for any of Khera's other proposed maternally mediated effects. Expected effects, such as embryonic resorption, inhibited fetal growth, and gross malformations, were seen in only a minority of the test groups. The study by Kavlock and colleagues[60] does not rule out the possibility that tests with another set of compounds would yield results more compatible with Khera's predictions, but it does indicate that maternal toxicity is by no means invariably associated with Khera's predicted major effects. The authors concluded that "there clearly is no direct relationship between the induction of maternal toxicity and the production of major abnormalities."[60]

A more recent study with rats took another approach to examine maternal influence on the manifestation of developmental anomalies.[77] The timing of maternal toxicity, as indicated by clinical signs and effects on body weight, was correlated with the specific period of gestation during which observed fetal defects were most likely to have been elicited. In addition, maternal toxicity data from individual dams with and without affected litters were compared. Both approaches failed to suggest a role for maternal toxicity in causing the observed malformations (eye defects, such as anophthalmia and microphthalmia) associated with exposure to the herbicide cyanazine.

Examination of a representative sample of the large number of existing safety evaluation studies could provide valuable insight into the maternal toxicity issue and perhaps settle any remaining debate concerning its contribution to fetal defects. Another possible approach would be to conduct parallel experiments with chemicals known to produce teratogenic effects only in the presence of maternal toxicity. One set of pregnant dams, probably rats or mice, would be given the test agents by the usual routes, while another set would be treated via intraamniotic or intrauterine administration, with doses scaled to result in similar exposures to the conceptus. Presumably, the second method would still allow direct effects, but should allow use of doses low enough to avoid maternal toxicity. As stated by Schardein,[76] the literature has yet to show "an unequivocal relationship between specific maternal and developmental toxicities," and "developmental disruption appears not to result unconditionally from maternal toxicity."

2. Influence on Developmental Toxicity Test Outcomes

The possibility of maternally mediated adverse developmental effects in developmental toxicity testing is of concern because if such secondary effects occur, they might produce positive results in animal tests that would not be seen in man (or other species of interest) at expected exposure levels. According to Seidenberg and Becker,[78] testing for developmental toxicity at dose levels that are maternally toxic is a somewhat controversial issue. This is because at times, "Investigators may argue that the use of such high dose levels would lead to a large number of false positive results as a direct consequence of disturbance of the maternal-fetal homeostasis by the induced maternal toxicity." Although Seidenberg and Becker[78] were discussing interpretation of the results of a Chernoff-Kavlock screen, which differs from the traditional Phase II (developmental toxicity) tests, the same principles are involved. The authors further stated that their data from 55 chemical compounds failed to indicate that maternal toxicity alone causes effects on the offspring detectable in such a screen. A similar conclusion was drawn by Chahoud et al.,[79] who found that maternal body weight change, as an indicator of maternal toxicity, was not always associated with embryo or fetal toxicity. Critical analyses by others[80–82] support the concept that maternal toxicity is not invariably followed by adverse prenatal effects, such as decreased survival or fetal weight or an increased incidence of abnormal morphology. Black and Marks[72] note that in range-finding studies, where the range of doses typically includes some that are highly toxic to the dam, such doses do not often result in an increased incidence of malformation, "although dose-related increases in the incidence of [unspecified] fetal variations are almost always seen."

3. Influence on Interpretation of Developmental Toxicity Test Results

Whether animal test results accurately predict human hazard potential, and especially whether the predicted effects are biologically significant and irreversible, are important issues in developmental toxicology.[35] Thus, any potential confounders — such as maternal toxicity effects — are of importance, especially when they differ between the test species and humans. The argument that high-dose testing may produce misleading results has been forcefully expressed by Khera,[32] who stated that, "Teratology-testing studies usually include apparently maternotoxic dose levels, and the repetitious and quite often predictable fetal outcome have [sic] incriminated a large number of compounds as potential teratogens thus making the testing methods a meaningless exercise." It can also be argued, however, that it is their interpretation, rather than the results themselves, that is of concern. Since it has been shown experimentally that even severe maternal toxicity is not necessarily accompanied by developmental toxicity, it appears that many of the fetal outcomes seen concurrent with maternal toxicity are not necessarily secondary to maternal effects. A remaining and perhaps more important possibility is that in at least some cases the maternal stress induced by toxicity may *exacerbate the effects* of a chemical teratogen.

The consensus at a workshop convened by the U.S. EPA was that if developmental toxicity is observed it cannot routinely be ignored or discounted as secondary to maternal toxicity.[3] The workshop discussions also lead to the conclusion that hazard assessments should be conducted for all agents that elicit developmental effects, even if those effects "are seen only in the presence of maternal toxicity." The effective experimental doses should then be compared with likely human exposures to assess the risk to humans. The U.S. EPA *Guidelines for Developmental Toxicity Risk Assessment*[83] reflect this point of view. Those guidelines state that even if fetal effects are seen only when maternally toxic doses have been administered, "the developmental effects are still considered to represent developmental toxicity and should not be discounted as being secondary to maternal toxicity." They further state that, "Current information is inadequate to assume that developmental effects at maternally toxic doses result only from maternal toxicity." Similarly, the position of the U.S. FDA was reflected in comments by Collins et al.,[84] who stated, "Developmental effects that occur in the presence of minimal maternal toxicity are thus considered to be evidence of developmental toxicity, unless it can be established that the developmental effects are unquestionably secondary to the maternal effects. In situations where developmental effects are observed only at doses where there is a substantial amount of maternal toxicity, then the possible relationship between maternal toxicity and the developmental effects should be evaluated in order to make a proper assessment regarding the toxicity of the test substance."

The question of interpretation of fetal effects seen in the presence of maternal toxicity has also been addressed by Schardein.[76] He pointed out that statements have been made to the effect that a specific chemical was not actually "teratogenic," even though it induced malformations, because such effects occurred only in the presence of maternal toxicity. Schardein further mentioned comments such as, "the chemical was teratogenic, but not a teratogenic *hazard*." As he properly points out, an agent that causes malformations is teratogenic, regardless of whether the mechanism is direct or indirect; the important point is whether the teratogen is *selective* in its effects. A similar point of view was expressed by Hood,[35] who commented, "Once one ascertains whether there are effects on the offspring, then it is important to determine as much as possible about the mechanism(s) involved, but if the same mechanism(s) may occur in man, it does not matter whether the effect on the embryo/fetus is direct or indirect. All that truly matters in such a case is the final outcome," and "In reality, it is of little consequence to the embryo/fetus whether it was harmed by a chemical that acted directly or through maternal mediation." Thus, until we know much more about maternally mediated effects and their relation to developmental hazard, the critical factors should be the likelihood of toxicity to the pregnant woman and the relative margin of safety between likely maternal exposures and toxic doses.

4. Potentiation of Chemical Teratogenesis by Maternal Stress

Several studies have attempted to determine if maternal stress can interact with or potentiate the effects of teratogenic chemicals or radiation when given in combination with such agents. In an early study,[20] Hartel and Hartel subjected pregnant rats to both intermittent loud noises and bright lights or to immobilization during midgestation, in combination with vitamin A. Only immobilization increased the incidence of retinoid-induced malformations, primarily cleft palate, and prenatal mortality. In contrast, Ishii and Yokobori[66] found that loud noise increased the incidence of prenatal deaths and malformations in mice treated with trypan blue. Goldman and Yakovac[21] then coadministered restraint stress and salicylate to pregnant rats and observed an enhancement of the teratogenic effects of the chemical. The finding that two central nervous system (CNS) depressants, chlorpromazine and pentobarbital, attenuated the ability of maternal restraint to potentiate salicylate teratogenicity in rats suggested that the restraint effect is mediated via the CNS.[85] In a later study, pregnant mice of two strains were either briefly restrained, injected with lucanthone, or both.[86] Lucanthone treatment of NMRI mice caused a high incidence of malformations that was not increased by maternal restraint, but restraint appeared to have increased the incidence of resorptions. Conversely, in the F/A strain, the chemical alone did not significantly increase the malformation rate but its teratogenic effect was potentiated by restraining the pregnant dams, with no increase in resorptions.

More recently, combined exposure to cadmium and noise was assessed in mice.[87] The combination was said to have significantly increased the fetal malformation rate (gross plus skeletal), but the data appeared equivocal, and the specific defects seen were not listed. In another study, female Uje:WIST strain rats were dosed with lithium prior to and throughout pregnancy.[88] Their female offspring were then mated at maturity, with no further lithium exposure, and half of them were subjected to restraint on gestation days 6 through 20. The restrained females gained less weight during pregnancy, and their offspring weighed less at birth than was true of the (unrestrained) controls. The lack of an appropriate food- and water-deprived control group limits the interpretation of the data because the restrained group was food and water deprived as well as restrained.

Rasco and Hood exposed pregnant mice to restraint stress and low doses of teratogens to determine if the apparent enhancement of teratogenic effects seen by others when combining maternal immobilization stress and chemicals was a common response or an anomaly. They found that maternal restraint concurrent with either sodium arsenate or all-*trans* retinoic acid enhanced the teratogenicity of the chemical.[68,89] The timing of administration of the retinoid during the restraint period influenced the intensity of the potentiative effect.[90] Although Domingo's laboratory subsequently found relatively few apparent developmental effects of maternal restraint when combined with treatment with various metal salts and other chemicals, they most often used restraint periods of only 1 to 2 h/d, repeated for several days.[91,92] Brief restraint periods may not be effective, and there is also the possibility of habituation to the repeated stress.[93]

The mechanisms by which maternal stress potentiates the effects of chemical teratogens remain to be investigated. A number of possibilities come to mind, including altered biotransforming capability,[94] changes in gastrointestinal secretion and motility that may influence absorption of an orally administered compound, altered blood flow to the placenta and/or the maternal liver, altered levels of cell proliferation or thresholds for intracellular signaling due to effects of increased maternal serum hormone levels, altered target tissue receptor binding, altered gene or protein expression,[95] altered fetal programming,[96] altered lifestyle (in humans),[97,98] and altered body temperature. Note that some of these putative mechanisms are secondary effects of increases in maternal stress-related hormones and other endogenous compounds, while others may be caused, at least in part, by neurogenically induced physiological changes.

5. Possible Influence on Human Pregnancy Outcomes

Little is known of the potential of maternal stress or toxicity to cause adverse fetal outcomes or to potentiate chemical teratogenicity in humans, although several possibilities for such effects exist. According to early theories, exposure of a pregnant woman to a shocking, worrisome, or frightening situation could result in birth of a malformed child, but such theories have fallen into disfavor.[99] Also, no association was shown between maternal "emotional upsets" and giving birth to a baby with cleft lip or palate,[100] and similar results were reported for maternal exposure to "airport noise" and malformation.[101] However, three later studies have reported an apparent connection between maternal emotional stress and malformed offspring.[102–104] Although some studies have suggested a relationship between maternal stress and low birth weight, McAnarney and Stevens-Simon[105] found that the data were not definitive. More recently, Lou and coworkers[106] reported that maternal stress was associated with decreased offspring head circumference, although the number of individuals compared was small, and Chen et al.[107] described birth weight reduction following maternal occupational stress.

Workplace stress has not generally been found to increase the likelihood of adverse pregnancy outcomes,[108] but there are few studies of such possibilities, and epidemiologic studies typically can identify only relatively strong developmental hazards or those with unusual outcomes. In a few studies some apparent stressors, such as working long hours and "psychosocial stress in the workplace," have been linked to preterm birth[109] or reduced fetal growth,[110] and both workplace stress and "negative life events" have been associated with spontaneous abortion.[111–113] Some common sources of stress, such as marital conflict,[114] appear not to have been investigated for effects on fetal outcome since the work of Stott.[115] It should be kept in mind, however, that studies seeking effects of stressors in women have been hampered by the difficulty of measuring psychological parameters in ways that allow for comparisons among groups.[116]

As pointed out by Daston,[1] the increased likelihood of neural tube defects associated with maternal treatment with anticonvulsants might at least partly be a secondary effect of the concomitant drug-induced folate deficiency. He also comments that certain drug treatments or toxic exposures can adversely alter maternal zinc metabolism, which might affect offspring development, as may the vitamin A depletion seen in alcoholic women. Further, as stated by Brent and Holmes,[117] abnormal maternal metabolic states, such as diabetes mellitus and phenylketonuria, can contribute to abnormal development of the embryo or fetus. And a study of maternal stress response and fetal cardiac activity revealed that fetuses of high anxiety mothers had increases in heart rate during periods of maternal stress, suggesting that maternal stress can alter fetal physiology,[99a] providing another possible means for influencing birth outcome.

II. POSSIBLE MECHANISMS FOR MATERNALLY MEDIATED EFFECTS

Numerous possible mechanisms for secondary effects of maternal stress or toxicity have been proposed, and no attempt will be made to review them in general here, as this topic has been extensively reviewed by Daston[1] and by Khera.[33] Kalter and Warkany[119] and DeSesso,[5] also have reviewed the potential for maternal factors, including abnormal metabolic states, to influence developmental outcome, and Khera recently described damage to maternal placental circulation as a possible contributor to adverse developmental effects in the mouse.[120] An interesting recent study of rats found that late gestational blockade of opioid receptors by administration of naltrexone prevented reduction in male anogenital distance and altered certain behavioral outcomes in offspring of light and noise-stressed dams.[121] Further, the stress-induced reductions in pain induced by restraint and other stressors can be blocked by opioid antagonists.[122] These data suggest that opioid receptors may be involved in at least some maternally mediated effects of stress.

At least one potential maternally mediated factor has apparently not been discussed in the above-mentioned reviews. That is the possibility of inducing abnormal maternal biotransformation at high — often maternally toxic — doses of chemical agents that are typically metabolized to less toxic forms by the mother. Such high exposures may overwhelm the normal maternal biotransforming abilities and lead to metabolism by normally minor pathways, some of which may produce hazardous metabolites. If such metabolites are produced in significant quantities, some may have the potential to harm the conceptus. Further, maternal stress can inhibit normal xenobiotic biotransformation[123] and might result in exposure of the conceptus to higher levels of toxicants than would otherwise be the case.

III. ASSESSMENT OF MATERNAL STRESS OR TOXICITY

Multiple biochemical, physiological, and behavioral changes are engendered by stress and would appear to provide the toxicologist or teratologist with a variety of endpoints suitable for assessing maternal stress. In studies specifically designed to evaluate the impact of stress on development, there is no doubt that both systemic and neurogenic stressors,[42] when applied to pregnant animals in an experimental setting, will alter developmental outcome (see Chernoff et al.,[2,59,60,62,124] Joffe,[125] Hood,[4,35] and Weinstock et al.,[126] for studies relevant to this issue). These studies, as well as studies investigating the impact of stress in the adult, utilize a variety of methods to measure the many changes associated with exposure to a stressor. In the following section, we briefly catalog a variety of methods for assessing stress. However, the reader should be aware that their implementation in a standard teratology study may be quite difficult. It is one thing to determine the consequences of an operationally defined stressor; it is quite another to determine if stress plays a role in a particular developmental outcome. As Chernoff et al.[62] found in their investigation of overt maternal toxicity and adverse developmental effects, the relationship between maternal toxicity, maternal stress, and developmental outcomes is not easy to define. One should, however, be aware of how various manipulations can act as neurogenic stressors (e.g., nose-only inhalation exposure procedures) and the impact these manipulations may have on development.

A. Biochemical Measures

1. Measures Related to the Stress Cascade

The biochemical indicators that are nearly synonymous with stress are those associated with the stress cascade; namely glucocorticoids (corticosterone in rodents and cortisol in primates, guinea pigs, and hamsters). Selye[127] in his landmark studies first emphasized the use of this adrenal cortical product as a marker for stress. The levels of glucocorticoids in blood and their metabolites in urine are used much more often to gauge stress than are the levels of either ACTH or CRF. In most experimental studies that utilize circulating glucocorticoid levels as a stress marker, their elevation in blood is accepted as prima facie evidence that other components of the stress cascade have been activated. That is, CRF has been released from the hypothalamus into the hypophyseal portal vessels leading to the pituitary gland. Subsequent release of ACTH from the corticotropes in this structure results in the eventual stimulation of the adrenal cortex and release of glucocorticoids into the general circulation. Generally, ACTH and CRF are evaluated after elevated glucocorticoid levels are noted and usually serve to provide knowledge about the functioning of the axis as a whole. For example, the direct action of a chemical at the adrenal cortex leading to the sustained secretion of high levels of glucocorticoids over a period of weeks or months might be expected to result in a decreased secretion of ACTH (see Krieger and Liotta[38] for a discussion of ways to test the functioning of the HPA axis).

Studies too numerous to list have measured plasma or serum levels of total blood glucocorticoids as a primary indicator of stress. The total value includes free glucocorticoids as well as those bound to transcortin, the binding globulin found in blood.[128] Only unbound glucocorticoids are believed to have biological activity, but see Rosner and Hochberg[129] for a discussion of this issue. There has been a recent emphasis, especially in humans, on utilizing saliva as an easily accessed body fluid that can be repeatedly collected without trauma and contains only unbound glucocorticoids. Saliva can be used to evaluate event-related activation of the HPA axis, as well as its general function.[130] Saliva can be collected from small laboratory animals, but the limited volume may limit its use in gauging stress.

Glucocorticoids can cross the placenta. Thus, quite a few investigators have attempted to delineate the role of elevated glucocorticoids in various morphological developmental abnormalities, and they have found that the ability of the human fetus to respond to stressors with a release of cortisol matures around the 20th week of gestation.[14,124,131–137] The role of glucocorticoids in postnatal functional abnormalities, including altered sexual differentiation and sociosexual behavior[25,126] and the response to stress by the neonatal, juvenile, and adult organism,[29,138–141] has also been investigated. However, the relationships between stress, maternal elevations in glucocorticoids, and developmental outcomes have remained elusive (see Hansen and coworkers[133] and Miller and Chernoff[124] for a discussion of this issue). Glucocorticoid elevations usually accompany stress, but this does not necessarily mean they are directly responsible for producing developmental alterations, or that they play any role at all. Abnormalities may occur due to secondary alterations in maternal and fetal physiology affected by elevated glucocorticoids. These abnormalities could include altered uterine and placental blood flow, transient hypoxia due to limited oxygen or other substances in the blood, alterations in uterine contractility, and decreased production of essential hormones (e.g., estrogen, progesterone, or DHEA)(see Schneider et al.[142] for a discussion).

While elevated glucocorticoid levels may signal activation of other components of the axis, this is not always the case, as a chemical may act directly on the adrenals.[43] Developmental toxicology studies commonly involve the testing of chemicals, and a chemical may act as a systemic stressor, with direct activation of the HPA axis at various levels. Additional work would be needed to determine if the elevation was due to an activation of the stress cascade or to a specific and direct release of glucocorticoids by the compound.

Many diverse agents can affect the adrenal glands, either through direct damage or by sustained activation, with a resultant increase in the circulating levels of its products (see Colby[143] for a discussion). The susceptibility of the adrenal gland to toxic insult is documented by Ribelin[144] in a survey of the literature, where he noted that more than 90% of the citations describing toxic actions of chemicals on endocrine organs concern the adrenal and testes. While extensive documentation is not available to indicate direct chemical effects at other levels of the feedback loop, it is certainly possible (e.g., Bestervelt and coauthors[145,146]).

A primary difficulty in implementing measures of glucocorticoids, ACTH, CRF, or the catecholamines as markers for stress is that blood must be collected. The collection process itself can induce stress (e.g., see Reinhardt et al.[147]), and to avoid this animals are anesthetized prior to collection or receive indwelling catheters to facilitate blood sampling.[147,148] In some studies, groups separate from those used to assess toxic effects are included to determine the degree of stress induced by various experimental conditions (e.g., Miller and Chernoff[124]). In particular, care must be taken not to stress animals at the time of blood collection. Elevations in blood glucocorticoid levels due to the manipulation of interest (whether chemical or procedural in nature) in subjects could be masked by increased levels in inadvertently stressed control subjects.

The small size of some species (e.g., mice) can preclude the collection of large amounts of blood, limiting the number of end or time points that can be evaluated in a given sample. Investigators should also be aware that there is a marked circadian pattern in the levels of blood glucocorticoids.[149] In the rodent, the nadir (~1 g/dl) occurs during the lighted period, with a peak (~20 g/dl) during the dark period. Thus, measuring blood glucocorticoids at a single time point can lead to difficulties in interpretation (see Kopf-Maier[135] and Hansen et al.[133] for a discussion of these issues). There are also multiple reports of strain differences in the responsiveness of the HPA axis in many species.[150–154] Because of the many factors that affect glucocorticoid secretion, investigators should not begin using glucocorticoids as markers for stress until they determine the basal circadian pattern of secretion obtained in their animal facility with the strains they normally use to evaluate chemicals for developmentally toxic effects. Finally, the investigator should be aware that in the pregnant mouse, as well as other species (e.g., rabbit, human), the total glucocorticoid level found in blood increases dramatically at about gestational day 9, with the peak level observed between days 12 and 15.[133,155,156] The day and degree of peak elevation are strain dependent and may be related to pregnancy-induced increases in the glucocorticoid binding protein in blood that occur in a number of species.[157–159]

2. Catecholamines

Activation of the HPA axis can also result in the release of catecholamines from the adrenal medulla, as well as from sympathetic nerve endings. Plasma norepinephrine is released primarily from sympathetic nerve terminals, while epinephrine is secreted exclusively from the adrenal gland.[160] These compounds have hemodynamic properties and can alter blood flow to the uterus. Exogenous treatment of the dam with these substances, as well as other vasoactive agents, such as vasopressin, can cause developmental abnormalities (e.g., Chernoff and Grabowski[161]). Thus, there has been some interest in determining their role in maternal toxicity and stress in experimental animals as well as in humans.[23,137] Cook and colleagues did not observe consistent alterations in blood or uterine catecholamine levels in response to noise, and the levels induced were much lower than would be obtained by exogenous treatment with these agents.[23] However, Gitau and colleagues have observed impaired uterine flow in pregnant humans with documented high anxiety and attributed the impairment to possible elevations in catecholamines.[137]

There is some suggestion that plasma levels of catecholamines are more useful than glucocorticoid levels for assessing stress because the former are more correlated with the intensity of the stressor. Indeed, blood catecholamine levels have been proposed as the best visceral indicator of stress (e.g., see Natelson and coworkers[162]). That is, catecholamine (rather than glucocorticoid) levels more accurately reflect levels of arousal. Both measures are similar in that a repeated exposure to the same neurogenic stressor results in a diminution of the response.[162] Again, these measures require collection of blood, and care must be taken to avoid stressing animals during the collection. In many studies indwelling catheters are preferred for sampling, as release of catecholamines occurs quickly in response to a stressor (e.g., within 5 min of the onset of restraint). Blood catecholamine levels can be determined by high performance liquid chromatography. These methods have become quite sophisticated in recent years and are able to detect quite small amounts (see Konarska and coauthors[163] and Kopin[164]).

3. Stress Proteins

In recent years there has been great interest in using those proteins known as "stress proteins" as cellular level markers for areas activated by stress or for monitoring the influence of different environmental factors on animals.[165–168] The designation "stress proteins" usually includes the immediate-early gene products (IEGPs) and the heat-shock proteins (HSPs). The HSPs were first discovered in experiments involving hyperthermia but can be induced in response to other disruptive conditions.[169–171] Stress proteins include c-FOS, c-JUN, and the HSPs, which are named and classified into different families on the basis of molecular weight (e.g., HSP 70, 90, or 100). Other proteins, such as ornithine decarboxylase and metallothionein, as well as acute phase proteins, are also sometimes included in the category of stress proteins. Stress proteins are highly conserved, rapidly inducible, and are synthesized in a variety of tissues (including brain, lung, heart, and lymphocytes) in response to conditions that alter homeostasis.[172–179] The areas where these proteins are induced and their patterns of induction are dependent on the type of stressor (e.g., see Ceccatelli et al.[173]). What is interesting is that these proteins are synthesized at a time when conditions (e.g., stress) dictate a reduced synthesis of most other proteins.

As stress proteins can be induced in response to chemical exposure, their measurement in the embryo has been suggested as a possible biomarker for teratogenic effects induced by hyperthermia[180] or more generally as a means of screening for or detecting teratogens in general.[181] German[165] has also proposed the embryonic stress hypothesis of teratogenesis, but in general studies have not supported his contention that induction of the heat-shock response is a common pathway capable of mediating developmental defects induced by diverse agents (see Finnell and colleagues[181] or Mirkes and coworkers[182]). It is more likely the induction of these proteins under certain conditions associated with developmental defects indicates the activation of particular cellular signaling pathways that are parallel to those involved in development. The dam is not normally screened for the induction of these proteins. However, as stress proteins appear to be induced in multiple organs and the pattern of their induction is dependent on the disrupting condition, they may serve as a means for identifying different kinds of stress (e.g., neurogenic vs. systemic). Conversely, they may identify stressors operating through common pathways because they are a biomarker for activation in a particular organ. Screening for the induction of stress proteins may not be feasible in standard teratology studies because tissue from the dam is required. However, in follow-up studies they may provide a way to better delineate the biochemical status of the dam when an anomaly is suspected to be due to stress.

Use of stress proteins as a means of identifying target organs or transynaptic pathways activated by certain conditions requires some means of identifying them. Some techniques that have been used are two-dimensional gel electrophoresis, treatment of histological sections with antibodies, immunoblots of tissue homogenates, antibody-based detection, or general detection procedures with the new proteomic platforms linked to mass spectrometry.[167,177,183–185] Because these proteins are rapidly and transiently induced, time course determinations will provide the most information.

4. Measures Related to Zinc Metabolism

Adequate maternal zinc levels are needed for the appropriate development of the fetus. Altered maternal zinc metabolism may be a mediating factor in the abnormal development caused by diverse toxicants.[1,186–188] Many of the same manipulations that are considered stressors (e.g., hyperthermia, food and water deprivation) will cause a redistribution of maternal zinc, copper, and iron. This redistribution is suspected to occur as a consequence of an acute-phase response to certain toxicants or manipulations, with a concomitant increased synthesis of certain serum and liver proteins.

One of the proteins believed to figure prominently in the redistribution of zinc is the liver protein, metallothionein (MT). Zinc, as well as copper, are bound to the newly synthesized MT and are therefore unavailable for distribution to the developing fetus. Glucocorticoids and catecholamines, both released in response to stress, can induce MT, and MT is sometimes classified as a stress protein.[189] As with other potential markers of maternal stress, time course considerations are important. For example, an acute injection of α-hederin to the pregnant rat will induce MT; this induction reaches its peak in approximately 12 to 24 h. Serum from these dams was able to support embryo development in culture at 2 h but not 18 h after treatment.[186]

Whether repeated exposure to compounds capable of inducing MT and altering zinc distribution would effectively lower blood zinc levels for a sufficient period to be detected at the usual tissue collection period used in a standard developmental toxicology testing protocol is unknown. Blood should be collected in zinc-free collection tubes to ensure accurate assessment of plasma zinc levels. Zinc concentrations in serum and other tissues as well as the concentration of other minerals can be determined by flame atomic absorption spectrophotometry. The investigator should also be aware there is a midgestational increase in the constitutive expressions of MT in mice but not in rats.[190] This may serve to increase the vulnerability of mice to developmental insults mediated through alterations in zinc metabolism.[186]

B. Organ Weights

Developmental toxicity protocols routinely evaluate maternal toxicity by monitoring weight and body weight gain, food consumption, clinical signs of toxicity, and morbidity or mortality. While organs are usually inspected at necropsy for gross toxicity, some protocols (e.g., reproductive toxicity evaluation) require recording of organ weights.[191] Changes in certain organ systems should alert the investigator to possible maternal stress or activation of the stress cascade. For example, altered thymus or adrenal weight may suggest chronic adrenal axis activation or direct toxicity at some level of the HPA axis.[192,193] As Akana and colleagues[192] have noted, both body and thymus weight are tightly regulated by glucocorticoids levels. Liver-to-body-weight ratio changes may signal alterations in the liver, such as those associated with altered zinc metabolism.[186]

Chernoff et al.[62] investigated whether organ weight differences at necropsy define commonalities between diverse chemicals capable of inducing maternal toxicity. Organs (adrenals, thymus, spleen) expected to be affected by HPA axis activation were weighed at necropsies conducted at gestational days 8, 12, 16, or 20 from dams exposed to diverse chemicals agents on days 6 to 15. Maternal toxicity was defined as alterations in body or organ weight and/or lethality. Consistent patterns in organ or body weight changes were sought. All the evaluated chemicals induced maternal toxicity as indicated by lethality, reduced weight gain, and decreased relative thymus and spleen weight, as well as altered relative adrenal weight. Further, the changes were obvious throughout the treatment period and persisted until term. As would be expected from the work of Akana and colleagues,[192] the most sensitive organ was the thymus; the spleen and adrenal were less affected. Although all compounds induced maternal toxicity and affected organs associated with the stress cascade, there were no consistent developmental effects associated with maternal toxicity. A further investigation of the usefulness of organ weights for delineating the presence and consequences of maternal toxicity and stress would be helpful.

C. Behavioral Measures

Most of the measures outlined above necessitate extra groups of subjects because tissue must be collected close to the time of exposure rather than at the necropsy period utilized in the standard developmental toxicology protocol. Thus, there is a place for noninvasive measures that will indicate if maternal stress is induced by certain conditions or compounds but that do not require termination of the dam.[2] Along with glucocorticoids and catecholamines, stressors may also cause the release of endogenous opiates. As is the case with medicinal opiates (e.g., morphine) that are used to block pain, one method of evaluating levels of endogenous opiates is to measure pain perception. Many putative stressors, including restraint and electric shock, can rapidly diminish sensitivity to noxious or painful events (i.e., stress-induced analgesia).[122] Altered pain perception has been utilized as a noninvasive gauge of stress in several studies evaluating the relationship between maternal stress and developmental outcome.[69,124]

Recently, Chernoff and colleagues[2] suggested that tail-flick assays, a measure of analgesia, may be useful in determining if a compound or condition is stressful for species known to be responsive to stress. It should be noted, however, that a test agent itself may have analgesic properties and be acting directly, rather than through the release of endogenous substances, to induce analgesia. It is true that analgesia assessment is a noninvasive procedure, but it does require skill as well as considerable handling of the animal and, at the minimum, additional groups of nonhandled subjects. Again, this procedure, like others mentioned previously, may be the most useful in situations where the investigator suspects on the basis of preliminary information (e.g., 90-d toxicity studies) that maternal toxicity or stress may be a factor and where the study is designed specifically to address this issue.

Analgesia can be detected with a variety of methods, but the most commonly used (e.g., in pharmaceutical studies) is the tail-flick assay.[194] This may have some benefits in testing pregnant animals for stress-induced analgesia. For example, it has been suggested that endorphin systems are activated during pregnancy, resulting in a form of endogenous analgesia, but there is some controversy regarding possible confounding of the analgesia measures employed.[195,196] Analgesia increased over the duration of pregnancy, but so did the weight of the rat, and the particular analgesia method used (flinch/jump test) can be influenced by the weight of the subject. The tail-flick assay involves little motor movement and is considered to be a polysynaptic reflex mediated within the spinal cord.[197]

Many manipulations considered to be stressful (e.g., restraint, shock, type of housing) can alter general activity for a significant period following their removal (see Miller[193] for a discussion of this in the rodent). Recently, Barclay and colleagues[198] suggested the use of the "Disturbance Index" (basically, movement in an open field) as a way of gauging distress in laboratory animals. Noticeable alterations in normal behaviors observed during preliminary toxicity tests or during cage-side observations during teratology testing should serve to alert the investigator that the compound under study may cause maternal distress or stress.

D. Measures of Toxicity

End points of maternal toxicity were discussed at some length in a 1989 review by Chernoff and coworkers[2] and more recently by the U.S. EPA.[83] The typical endpoints assessed during the course of developmental toxicity assays are listed in Table 4.2. In general, these endpoints were chosen

Table 4.2 End points of maternal toxicity

Mortality
Mating index [(no. with seminal plugs or sperm/no. mated) X 100]
Fertility index [(no. with implants/no. of matings) X 100]
Gestation length (useful when animals are allowed to deliver pups)
Body weight:
  Day 0
  During gestation
  Day of necropsy
Body weight change:
  Throughout gestation
  During treatment (including increments of time within treatment period)
  Posttreatment to sacrifice
  Corrected maternal
    (body weight change throughout gestation minus gravid uterine weight or litter weight at sacrifice)
Organ weights (in cases of suspected target organ toxicity and especially when supported by adverse histopathology findings):
  Absolute
  Relative to body weight
  Relative to brain weight
Food and water consumption (where relevant)
Clinical evaluations:
  Type, incidence, degree, and duration of clinical signs
  Enzyme markers
  Clinical chemistries
Gross necropsy and histopathology

Source: From U.S. Environmental Protection Agency, 1991.[83]

because they were readily assessable without adding greatly to the cost or length of studies; however, they are relatively crude, and their abilities to detect subtle or unusual toxicities are limited. Further, there has been some controversy over whether certain endpoints, such as enzyme induction or certain physiological changes, should be considered to be manifestations of toxicity or merely useful adaptive responses. As stated by Chernoff and colleagues,[2] "maternal toxicity," as it is currently determined, is often imprecisely defined, providing little that is of value to attempts to understand the potential impact of maternal effects on development.

E. Overview of Stress and Toxicity Assessment

It is clear from the preceding sections that there are a number of measures that may prove useful in determining if certain conditions or agents induce "stress" in a pregnant dam. However, these measures would be much more useful if there was a clearer understanding of the relationship between different stressors and fetal outcome. As Schwetz and Harris[199] noted in their comments on the field of developmental toxicology, there is as yet no complete mechanistic understanding of the actions of various developmental toxicants. Maternal toxicity and stress is certainly no exception to this statement.

IV. EXPERIMENTAL ASSESSMENT OF POSSIBLE MATERNALLY MEDIATED EFFECTS

A. Factors to Be Considered in Experimental Design and Conduct

A number of factors must be considered in designing experiments to elucidate mechanisms of developmental toxicity. Some are typically needed in investigations of other types of toxicity, but others are unique to developmental toxicity studies. This is true because in such a study one is assessing interactions between the maternal and the developing organism and because the conceptus is continually changing in its attributes and potential responses to toxicity over time. Thus, apparent findings may have been confounded by maternal and/or developmental factors that must be taken into account. Such factors may be unknown, and they may be difficult or impossible to control for, even if known. The remainder of this section deals with factors that should be considered in designing experimental investigations of the potential influence of maternal factors or in interpreting results of such studies.

1. Controlling the Animal's Environment

It is not widely appreciated that the typical environment of laboratory animals is somewhat stressful, as measured by endogenous levels of stress-related hormones and other endpoints. According to reviews by Rowan[200] and by Barnard and Hou,[201] mere routine handling of animals that are unaccustomed to being handled is stressful, and acclimation to handling and to experimental housing or a test apparatus may decrease stress. Laboratory animals typically have an acute sense of hearing. Even routine animal room activities, such as changing cages, moving cage racks and feed containers, and running large cage or rack washers, can generate considerable levels of stressful noise, as can the activities of some laboratory animals themselves (e.g., rabbits).[202] Barking of dogs housed in a nearby animal room may be heard by animals such as rats and causes stress. Even shipping, especially by air, is stressful. According to Brown et al.,[12] a 48-h transport of pregnant A/J strain mice increased the incidence of cleft palate in their offspring.

Merely moving the cages housing rats has been reported to cause altered hematologic values and heart function, as well as significant increases in serum levels of several hormones, such as corticosterone, prolactin, and thyroxin.[203] Changing cages and introducing animals into a clean cage is also stressful, as it removes familiar odors. Riley designed a "low-stress" mouse facility with decreased noise levels and less cage changing.[204] He found that mice in such an environment had plasma corticosterone levels no higher than 35 ng/ml, while conventionally housed mice had levels ranging from 150 to 500 ng/ml.

Also typically ignored is the possible stress from isolating experimental animals that are normally somewhat social. For instance, although female rodents to be used in developmental toxicity research are commonly group housed initially, once they are mated they are often housed individually. Whether this significantly affects the outcome of developmental toxicity tests is not known. Excessive crowding can also be stressful, but this is unlikely to occur because it would be a violation of modern animal care standards.

Further, few investigators consider that the effects of a given stressor may be quite complex. For example, restraint, in addition to inducing analgesia and activation of the HPA axis, can greatly reduce body temperature in mice (e.g., 2 to 3 degrees in certain strains).[205] Such complex effects may complicate interpretation of experimental outcomes.[206] Conversely, restraint does not appear to cause hypothermia in rats.[207]

The researcher should also consider that samples of blood or solid tissue constituents can be significantly influenced by stressful conditions and by hypoxia prior to sampling. For example, levels of metabolites, such as adenosine monophosphate, in rat liver were affected by the time between sacrifice of the animal and freezing of the liver, as well as by even normal levels of daytime lighting and animal room background noises and by pentobarbital anesthesia.[208]

2. Measurement of Stress

Various measures of maternal stress, as described in Section III, can be employed and attempts made to correlate them with developmental outcome. An example is the use of the tail flick test by Chernoff et al.[69] to correlate stress and fetal outcome (i.e., incidence of supernumerary ribs) in mice. Although those authors achieved success in terms of measuring relative degrees of maternal stress, their measure of stress did not correlate significantly with fetal effects as manifested by increases in supernumerary ribs.

3. Appropriate Controls

Controlling for confounders in studies dealing with maternal effects can require a great deal of thought and consideration of more experimental factors than is the case with more straightforward experiments, e.g., standard safety evaluation tests. For example, maternal stress (such as restraint) can prevent the dam from consuming food or water, thus requiring that controls be deprived of food and water. In studies of interactions of stress and chemical toxicants, controls should also include dams that are subjected to stress alone, the toxicant alone, and the toxicant plus food and water deprivation. A clever method of determining the relative role of corticosterone in mediating maternal effects was employed by Barbazanges and coworkers who compared effects of stressors on offspring of rats that were either intact or were adrenalectomized and given substitutive corticosterone therapy.[209]

4. Developmental Timing

Experiments dealing with the mechanistic bases for adverse effects on development are greatly influenced by aspects of timing. As outlined in Chapter 6 and the related tables, the embryo (and to a lesser extent the fetus) undergoes profound changes in its anatomy and extraembryonic membranes as development progresses. These morphological alterations are accompanied by equally striking changes in biochemical, physiological, and genomic attributes. Thus, when subjecting the conceptus to a toxic insult, one is always dealing with a "moving target," and a difference of only a few hours can be critical in determining the outcome. The practical consequences of this are critically important. Obviously, treatments must be given to each test animal at the same time, or (if known) at as nearly the same developmental stage as possible. Since most experiments dealing with maternally mediated effects will involve pregnant animals, treatment timing is generally used as a surrogate for developmental stage. There are typically differences in developmental stage between individuals in the same litter in the common species of laboratory animals.[210] Thus, not all animals in the study are exposed at exactly the same stage in development, with a resultant increase in the variability of the results. Also, the timing of fertilization in relation to treatment timing tends to be somewhat variable as well and might result in between-litter variability. This may not be a major factor, however, as Ishikawa et al.[211] have found that timing of ovulation is a more important determinant of developmental stage than is time of fertilization. Some researchers have advocated methods for reducing variability in developmental timing by restricting the time span in which mating is allowed.[212] However, the most certain way to decrease intralitter stage variability is to use embryo culture and select embryos at as nearly the same developmental stage as possible, an approach often incompatible with maternal effects research.

A further consideration is the choice of controls for evaluating the effects of treatment timing on development. As the ability of the embryo or fetus to respond to a toxic insult changes over time, the inappropriate choice of controls may result in data that cannot be interpreted.

5. Consistency of Methodology

Experimental variability is inherent in all animal experimentation but is held to a minimum by requiring a high degree of consistency and control in experimental technique. It is not always appreciated, however, that even seemingly minor differences in methodology can result in significantly different experimental outcomes. This is especially the case with highly complex interacting systems such as the mammalian mother and conceptus as they respond to stress or toxicity.

Adult rodents respond to diverse stress paradigms to different degrees. The rat responds to restraint stress by producing gastric ulcers, and ulcer production is further facilitated by cold, water immersion, or starvation. When more intense stress or multiple stressors are employed, less time of exposure is needed for the same effect.[213] The intensity and nature of the stressor and the number of episodes of stress determine the degree of stress perceived by the animal (or human) and thus the type and degree of response. Such responses are mediated by a complex interplay of neural and humoral factors, and it is not surprising that what seem to be modest changes in the application of stress can have significant effects on biological outcome. An example is seen in the results of a restraint stress study of pregnant mice by Rasco and Hood,[65] in which seemingly minor alterations in the restraint method resulted in consistent differences in the incidence of restraint-induced rib fusion.

6. Outcome Assessment and Interpretation of Results

Traditional maternal and fetal endpoints — such as maternal clinical signs or body weight changes and fetal incidence of fused or supernumerary ribs — are in some cases adequate for establishing maternally mediated effects of a test compound. More frequently the cause of a defect is obscure and more novel approaches are needed for determining the mechanism(s) responsible for the observed defect. Often, in the case of chemical toxicity to the dam, it is difficult or impossible to determine if specific effects seen in the offspring were direct or indirect.

There are numerous ways in which the maternal physiology or biochemistry can be perturbed, and these can in turn adversely influence prenatal development. Thus it is essential to consider clues, such as knowledge of the target organs or possible mechanisms of maternal toxicity, that may suggest the proper endpoints to assess. Indeed, there have already been a number of cases where alert researchers were able to pinpoint such mechanisms and show that they were specific to the species involved or only occurred as a result of maternal toxicity.[24,26–28,132,214–217] Future mechanistic studies may be aided by the availability of transgenic animals and knockout mice with defined defects in various components of pathways or systems activated by stressors or toxic substances (e.g., the LHPA axis[218]). Use of such animal models could allow inferences to be made regarding the possible influences of alterations in maternal physiology on offspring development and survival.

B. Induction of Maternal Restraint Stress as a Model System

Physical restraint, the partial or complete limitation of movement, has been used extensively as an inducer of stress in laboratory animals, and Paré, Glavin, and their colleagues have reviewed its use.[213,219] Restraint has advantages as a stress inducer: It can be applied relatively consistently, it can be used in a manner that is not likely to cause pain to the experimental subjects, it does not require elaborate or expensive equipment, and it does not involve toxicity, so purely stress-related effects can be separated from maternal toxicity effects. Nevertheless, results seem to differ depending on the exact methodology employed for restraint (as this determines the degree to which movement is restricted) and the length of the restraint period.

Chernoff and coworkers have established maternal restraint as a relatively consistent inducer of supernumerary ribs (SNR) in the CD-1 strain mice.[58,59,69] Exact timing is critical, as fetuses from mice restrained on gestation day 8 (copulation plug day is day 0 throughout this chapter) from 9:00 a.m. to 9:00 p.m. had a significantly elevated incidence of SNR, while those from dams restrained during the following 12 h did not. When pregnant mice were restrained for 12 h on one of gestation days 6 through 14, increased incidences of SNR were found only in fetuses from dams restrained on days 7 or 8.[69] Rib fusion was seen in a few fetuses from mice immobilized (i.e., completely restricted in their movement) in a supine position by use of Johnson and Johnson Elastikon® surgical tape, but not in those from dams restrained in padded conical holders, which allow some movement. This suggestion that the exact method of restraint may be important was confirmed by Rasco and Hood,[65] who found that a slight modification of the taping procedure could consistently influence the incidence of fused ribs in CD-1 mice.

The CD-1 mouse is an appropriate animal model for the investigation of effects of maternal immobilization stress on offspring development. The restraint method must be applied consistently, and it should be employed on gestation day 7 or 8 if SNR are to be used as an endpoint. The initial studies generally employed 12-h restraint periods; shorter restraint periods of 8 or 9 h may not be consistently effective. The most acute stress effects on the conceptus appear to result from maternal restraint in the supine position by some means such as immobilization with surgical tape, but other methods may prove to be adequate.

Plastic conical tipped screw cap 50-ml centrifuge tubes, such as the Falcon brand produced by Becton Dickinson Labware, act as inexpensive, easily obtained restraint devices. They have frosted areas useful for marking numbers on the tubes for identification of the restrained animals and assignment to treatment protocols. Holes for breathing should be drilled in the closed end and elsewhere as desired. The holes can be made using an ordinary electric power drill, but the bit should be as sharp as possible to minimize melting the plastic from friction during the drilling, as this leaves rough edges. The location of the holes should be kept consistent among different tubes. This can be facilitated by designating the appropriate locations with a marking pen prior to drilling. However, investigators should take care to use tubes from the same manufacturer to maintain consistency within a given study or laboratory, as 50-ml centrifuge tubes from various suppliers can vary in their inner diameter.

Pregnant mice can be introduced to the open end of the tube, and with a little encouragement will generally enter it. They will quickly begin to attempt to back out, but they can be kept in by use of the screw cap. Other means of keeping mice in the tubes include drilling two holes on opposite sides near the open end, placing a bolt through both holes, and securing it in place with a nut. Similarly, one can place a long hypodermic needle through both holes and stick the sharp end into a cork to secure the needle in place. Use of the screw cap allows the mouse to move back and forth in the tube, while use of the bolt or needle can either allow such movement or, if placed nearer to the closed end, can inhibit such movement. The size and weight of the mouse are also determinants of how much movement is afforded in the restraint method chosen. The latter two methods utilizing the bolt or needle to prevent escape also allow access to the mouse for tail flick tests, obtaining blood samples, or measuring rectal temperatures with a suitable electronic thermometer and miniprobe. It should be kept in mind, however, that such additional manipulations may also stress the animals and may make experimental design and/or interpretation more difficult. If necessary, additional mated females, treated identically to the unsampled mice, can be used for collecting such additional data, but their use does not allow for as accurate a correlation between the data obtained and the fetal outcome.

An alternative tube closure is the Identi-Plug plastic foam stopper, in the 35 to 45 mm size, made for closing *Drosophila* culture vials and similar containers. These are made by Jaece Industries and marketed by major laboratory supply vendors. These resilient stoppers can be pushed into the tube behind the mouse and will typically stay in place, preventing the animal from moving back and forth. Care should be taken to avoid trapping the mouse's tail between the stopper and the side of the tube; twisting the stopper into place helps to avoid this.

The ability of restraint to induce developmental effects in mice can be validated under one's chosen experimental conditions by findings of an increased incidence of SNR. Since most restraint methods prevent access to food and water, use of a food- and water-deprived control group is of value, although employment of appropriate controls can result in a relatively high number of test groups. An example can be seen in a study by Rasco and Hood[68] that assessed the ability of maternal restraint to enhance sodium arsenate teratogenicity. In addition to the experimental group given arsenate alone, five control groups were used. Control mice were given one of the following: the vehicle, restraint, arsenate, food and water deprivation, or arsenate plus food and water deprivation. Similarly, in another study by the same authors,[90] timing of administration of all-*trans*-retinoic acid (tRA) during a period of maternal restraint was investigated by use of six control groups, along with five experimental groups. Controls included vehicle control, food and water deprived, restrained, tRA treated plus food and water deprived, and tRA given at one of two times concurrent with the timing of restraint for the experimental (restrained plus tRA-treated) mice. Several additional controls could have been used but were not considered to be critical.

A further consideration in tests using incidence of SNR as one of the endpoints is the process of obtaining an accurate count of the fetal ribs. Thus, adequate clearing of the soft tissues is important. This is especially true for visualizing cervical ribs and extra thoracic ribs, which are induced by some teratogens, but is somewhat less critical for seeing the lumbar ribs brought about by stress. The ribs may be counted beginning at the most caudal if one is concerned only with detecting lumbar ribs. In this case, anything over 13 ribs is assumed to be a supernumerary rib. If detection of cervical or thoracic ribs is desired, the cervical vertebrae (normally, seven) should be counted, beginning at the cranial end of the spinal column, to determine if there are any "ribs" associated with them. Then, the count can be continued through the thoracic vertebrae and their associated ribs, and finally, presence of lumbar ribs can be noted.

Certain teratogens, e.g., all-*trans*-retinoic acid, may induce anteriorization of lumbar vertebrae, transforming them into replicas of thoracic vertebrae, with their associated ribs. These transformations can be distinguished from typical lumbar ribs in that the transformed ribs are more similar in appearance to normal ribs. That is, tRA-induced ribs and normal ribs are similarly wide and uniform in width, with blunt ends, whereas lumbar ribs are generally more slender, have tapered ends, and tend to point more ventrally. It is also possible for a fetus to have both one or two teratogen-induced pairs of extra thoracic ribs and uni- or bilateral lumbar ribs (either stress induced or "spontaneous," which of course may possibly have been induced by the "normal" stress of life in a typical noisy, well-lit laboratory animal facility). Finally, so-called rudimentary ribs[64] or "ossification sites" associated with lumbar vertebrae may be seen. These are also inducible by stress, but they are generally small and delicately formed and are commonly lacking in associated cartilage. Thus rudimentary ribs are usually readily distinguishable from supernumerary ribs.

V. OVERVIEW

The potential influence of changes in the maternal compartment on developmental outcome is not well understood and is somewhat controversial, and continued interest in the potential for maternal toxicity to affect the outcome and interpretation of developmental toxicity studies is evidenced by the recent workshop dealing with the topic.[220] It is well established that maternal stress alone can result in morphological or behavioral developmental effects on the offspring of laboratory rodents. Whether the outcomes of many developmental toxicity studies are greatly influenced by maternal toxicity and/or stress is uncertain, although there are apparent examples of such effects in the literature. Nevertheless, we must beware of improperly interpreting animal test results seen at maternally toxic doses, and we should keep in mind that human exposures also sometimes occur at levels that are maternally toxic. It appears that maternal stress can influence the developmental toxicity of a variety of chemical agents, but it is again not known if this significantly influences the outcome of typical safety evaluation tests. It is also not understood whether the incidences of developmental defects, pre- or perinatal mortality, or growth retardation in human beings are significantly influenced by maternal stress. If such is the case, it would be useful to know if animal models can provide reliable information that can be extrapolated to the human situation.

In brief, we know relatively little about maternally mediated effects on developing offspring, but we know they can occur, at least in certain laboratory animals. We need much more information about how this relates to human development, and we must be cautious about either over- or underestimating the influence of maternal effects on the outcome of developmental toxicity assays.

References

1. Daston, G.P., Relationships between maternal and developmental toxicity, in *Developmental Toxicology*, 2nd ed., Kimmel, C.A. and Buelke-Sam, J., Eds., Raven, New York, 1994, p. 189.
2. Chernoff, N., Rogers, J.M., and Kavlock, R.J., An overview of maternal toxicity and prenatal development: considerations for developmental toxicity hazard assessments, *Toxicology*, 59, 111, 1989.
3. Kimmel, G.L., Kimmel, C.A., and Francis, E.Z., Implications of the Consensus Workshop on the Evaluation of Maternal and Developmental Toxicity, *Teratogen. Carcinog. Mutagen.*, 7, 329, 1987.
4. Hood, R.D., Maternal vs. developmental toxicity, in *Developmental Toxicology: Risk Assessment and the Future*, Hood, R.D., Ed., Van Nostrand Reinhold, New York, 1990, p. 67.
5. DeSesso, J.M., Maternal factors in developmental toxicity, *Teratogen. Carcinog. Mutagen.*, 7, 225, 1987.
6. Thompson, W.R., Influence of prenatal maternal anxiety on emotionality in young rats, *Science*, 125, 698, 1951.
7. Geber, W.F., Developmental effects of chronic maternal audiovisual stress on the rat fetus, *J. Embryol. Exp. Morph.*, 16, 1, 1966.
8. Geber, W.F. and Anderson, T.A., Abnormal fetal growth in the albino rat and rabbit induced by maternal stress, *Biol. Neonat.*, 11, 209, 1967.
9. Hutchings, D.E. and Gibbon, J., Preliminary study of behavioral and teratogenic effects of two "stress" procedures administered during different periods of gestation in the rat, *Psych. Reports*, 26, 239, 1970.
10. Ward, C.O., Barletta, M.A., and Kaye, T., Teratogenic effects of audiogenic stress in albino mice, *J. Pharmaceut. Sci.*, 59, 1661, 1970.
11. Smith, D.J., Heseltine, G.F.D., and Corson, J.A., Pre-pregnancy and prenatal stress in five consecutive pregnancies: Effects on female rats and their offspring, *Life Sci.*, 10, 1233, 1971.
12. Brown, K.S., Johnston, M.C., and Niswander, J.D., Isolated cleft palate in mice after transportation during gestation, *Teratology*, 5, 119, 1972.
13. Euker, J.S. and Riegle, G.D., Effects of stress on pregnancy in the rat, *J. Reprod. Fertil.*, 34, 343, 1973.
14. Barlow, S.M., McElhatton, P.R., and Sullivan, F.M., The relation between maternal restraint and food deprivation, plasma corticosterone, and induction of cleft palate in the offspring of mice, *Teratology*, 12, 97, 1975.
15. Kimmel, C.A., Cook, R.O., and Staples, R.E., Teratogenic potential of noise in rats and mice, *Toxicol. Appl. Pharmacol.*, 36, 239, 1976.

16. Barlow, S.M., Knight, A.F., and Sullivan, F.M., Prevention by diazepam of adverse effects of maternal restraint stress on postnatal development and learning in the rat, *Teratology*, 19, 105, 1979.
17. Fanghanel, J., and Schumacher, G.-H., Environmental effects on normogenesis and teratogenesis, with special regard to noise and vibration, in *Teratological Testing*, 2, Persaud, T.V.N., Ed., MTP Press, Lancaster, England, 1979, p. 325.
18. Rosenzweig, S. and Blaustein, F.M., Cleft palate in A/J mice resulting from restraint and deprivation of food and water, *Teratology*, 3, 47, 1970.
19. Hemm, R.D.L., Arslanoglau, L., and Pollock, J.J., Cleft palate following prenatal food restriction in mice: Association with elevated maternal corticosteroids, *Teratology*, 15, 243, 1977.
20. Hartel, A., and Hartel, G., Experimental study of teratogenic effect of emotional stress in rats, *Science*, 132, 1483, 1960.
21. Goldman, A.S. and Yakovac, W.C., The enhancement of salicylate teratogenicity by maternal immobilization in the rat, *J. Pharmacol. Exptl. Therapeut.*, 142, 351, 1963.
22. Nawrot, P.S., Cook, R.O., and Hamm, C.W., Embryotoxicity of broadband high-frequency noise in the CD-1 mouse, *J. Toxicol. Environ. Health*, 8, 151, 1981.
23. Cook, R.O., Nawrot, P.S., and Hamm, C.W., Effects of high-frequency noise on prenatal developmental and maternal plasma and uterine catecholamine concentrations in the CD-1 mouse, *Toxicol. Appl. Pharmacol.*, 66, 338, 1982.
24. Clark, R.L., Robertson, R.T., Minsker, D.H., Cohen, S.M., Tocco, D.J., Allen, H.L., James, M.L., and Bokelman, D.L., Diflunisal-induced maternal anemia as a cause of teratogenicity in rabbits, *Teratology*, 30, 319, 1984.
25. Ward, I.L. and Weisz, J., Differential effects of maternal stress on circulating levels of corticosterone, progesterone and testosterone in male and female fetuses and their mothers, *Endocrinology*, 114, 1635, 1984.
26. Clark, R.L., Robertson, R.T., Peter, C.P., Bland, J.A., Nolan, T.E., Oppenheimer, L., and Bokelman, D.L., Association between adverse maternal and embryo-fetal effects in norfloxacin-treated and food-deprived rabbits, *Fund. Appl. Toxicol.*, 7, 272, 1986.
27. Moriguchi, M. and Scott, W.J., Prevention of caffeine-induced limb malformations by maternal adrenalectomy, *Teratology*, 33, 319, 1986.
28. Ugen, K.E. and Scott, W.J., Acetazolamide teratogenesis in Wistar rats: potentiation and antagonism by adrenergic agents, *Teratology*, 34, 195, 1986.
29. Fride, E. and Weinstock, M., Prenatal stress increases anxiety related behavior and alters cerebral lateralization of dopamine activity, *Life Sci.*, 42, 1059, 1988.
30. Herrenkohl, L. R., Ribary, U., Schlumpf, M., and Lichtensteiger, W., Maternal stress alters monoamine metabolites in fetal and neonatal rat brain, *Experientia*, 44, 457, 1988.
31. Khera, K.S., Maternal toxicity — a possible factor in fetal malformations in mice, *Teratology*, 29, 411, 1984.
32. Khera, K.S., Maternal toxicity: a possible etiological factor in embryo-fetal death and fetal malformation of rodent-rabbit species, *Teratology*, 31, 129, 1985.
33. Khera, K.S., Maternal toxicity of drugs and metabolic disorders — a possible etiologic factor in the intrauterine death and congenital malformation: a critique of the human data, *CRC Critical Rev. Toxicol.*, 17, 345, 1987.
34. Khera, K.S., Maternal toxicity in humans and animals: effects on fetal development and criteria for detection, *Teratogen. Carcinog. Mutagen.*, 7, 287, 1987.
35. Hood, R.D., A perspective on the significance of maternally mediated developmental toxicity, *Regulat. Toxicol. Pharmacol.*, 10, 144, 1989.
36. Lopez, J.F., Akil, H., and Watson, S.J., Role of biological and psychological factors in early development and their impact on adult life, *Biol. Psychiatry*, 46, 1461, 1999.
37. Hanoune, J., The adrenal medulla, in *Hormones: From Molecules to Disease*, Baulieu, E.-E. and Kelly, P.A., Eds., Chapman & Hall, New York, 1990, p. 309.
38. Krieger, D. and Liotta, A., ACTH and related peptides, in *Hormones: From Molecules to Disease*, Baulieu, E.-E. and Kelly, P.A., Eds., Chapman & Hall, New York, 1990, p. 229.
39. McEwen, B.S., DeKloet, E.R., and Rostene, W., Adrenal steroid receptors and actions in the nervous system, *Physiol. Rev.*, 66, 1121, 1986.

40. Munck, A., Guyre, P.M., and Holbrook, N.J., Physiological functions of glucocorticoids in stress and their relation to pharmacological actions, *Endocrin. Rev.*, 5, 25, 1984.
41. Pacák, K. and Palkovits, M. Stressor specificity of central neuroendocrine responses: implications for stress-related disorders, *Endocrine Rev.*, 22, 502, 2001.
42. Allen, J.P., Allen, C.F., Greer, M.A., and Jacobs, J.J., Stress-induced secretion of ACTH, in *Brain-Pituitary-Adrenal Interrelationships*, Brodish, A. and Redgate, E.S., Eds., S. Karger, Basel, Switzerland, 1973, p. 99.
43. Brunetti, L., Preziosi, P., Ragazzoni, E., and Vacca, M., Effects of vindesine on hypothalamic-pituitary-adrenal axis, *Tox. Lett.*, 75, 69, 1995.
44. Herrenkohl, L.R., Prenatal stress may alter sexual differentiation in male and female offspring, *Monogr. Neural Sci.*, 9, 176, 1983.
45. Peters, D.A.V., Maternal stress increases fetal brain and neonatal cerebral cortex 5-hydroxytryptamine synthesis in rats: a possible mechanism by which stress influences brain development, *Pharmacol. Biochem. Behav.*, 35, 943, 1990.
46. Lopez-Calderon, A., Ariznavarreta, C., and Chen, C.C.-L., Influence of chronic restraint stress on pro-opiomelanocortin mRNA and β-endorphin in the rat hypothalamus, *J. Mol. Endocrinol.*, 7, 197, 1991.
47. Ohkawa, T., Takeshita, S., Murase, T., Kambegawa, A., Okinaga, S., and Arai, K., Ontogeny of the response of the hypothalamo-pituitary-adrenal axis to maternal immobilization stress in rats, *Endocrinol. Japon.*, 38, 187, 1991.
48. Armario, A., Marti, O., Gavalda, A., Giralt, M., and Jolin, T., Effects of chronic immobilization stress on GH and TSH secretion in the rat — response to hypothalamic regulatory factors, *Psychoneuroendocrinology*, 18, 405, 1993.
49. Schmitz, C., Rhodes, M.E., Bludau, M., Kaplan, S., Ong, P., Ueffing, I., Vehoff, J., Korr, H., and Frye, C.A., Depression: reduced number of granule cells in the hippocampus of female, but not male, rats due to prenatal restraint stress, *Mol. Psychiat.*, 7, 810, 2002.
50. Frye, C.A. and Orecki, Z.A., Prenatal stress produces deficits in socio-sexual behavior of cycling, but not hormone-primed, Long-Evans rats, *Pharmacol. Biochem. Behav.*, 73, 53, 2002.
51. Williams, M.T., Davis, H.N., McCrea, A.E., Long, S.J., and Hennessy, M.B. Changes in the hormonal concentrations of pregnant rats and their fetuses following multiple exposures to a stressor during the third trimester, *Neurotox. Teratol.*, 21, 403, 1999.
52. Koenig, J.I., Kirkpatrick, B., and Lee, P, Glucocorticoid hormones and early brain development in schizophrenia, *Neuropsychopharmacology*, 27, 309, 2002.
53. Avishai-Eliner, S., Brunson, K.L., Sandman, C.A., and Baram, T.Z., Stressed-out, or in (utero)? *Trends Neurosci.* 25, 518, 2002.
54. Russell, L.B., Sensitivity patterns for the induction of homeotic shifts in a favorable strain of mice, *Teratology*, 20, 115, 1979.
55. Gossler, A. and Balling, R., The molecular and genetic analysis of mouse development, *Eur. J. Biochem.*, 204, 5, 1992.
56. Kessel, M., Respecification of vertebral identities by retinoic acid, *Development*, 115, 487, 1992.
57. Lansdown, A.B.G., Supplementary rib formation: spontaneous anomaly or teratogenic effect? *Lab. Animals*, 10, 353, 1976.
58. Beyer, P.E. and Chernoff, N., The induction of supernumerary ribs in rodents: Role of maternal stress, *Teratogen. Carcinog. Mutagen.*, 6, 419, 1986.
59. Chernoff, N., Kavlock, R.J., Beyer, P.E., and Miller, D., The potential relationship of maternal toxicity, general stress, and fetal outcome, *Teratogen. Carcinog. Mutagen.*, 7, 241, 1987.
60. Kavlock, R.J., Chernoff, N., and Rogers, E.H., The effect of acute maternal toxicity on fetal development in the mouse, *Teratogen. Carcinog. Mutagen.*, 5, 3, 1985.
61. Kimmel, C.A. and Wilson, J.G., Skeletal deviation in rats: malformation or variation? *Teratology*, 8, 309, 1973.
62. Chernoff, N., Setzer, R.W., Miller, D.B., Rosen, M.B., and Rogers, J.M., Effects of chemically induced maternal toxicity on prenatal development in the rat, *Teratology*, 42, 651, 1990.
63. Rogers, J.M., Francis, B.M., Barbee, B.D., and Chernoff, N., Developmental toxicity of bromoxynil in mice and rats, *Fund. Appl. Toxicol.*, 17, 442, 1991.
64. Chernoff, N., Rogers, J.M., Turner, C.I., and Francis, B.M., Significance of supernumerary ribs in rodent developmental toxicity studies: postnatal persistence in rats and mice, *Fund. Appl. Toxicol.*, 17, 448, 1991.

65. Rasco, J.F. and Hood, R.D., Differential effect of restraint procedure on incidence of restraint-stress-induced rib fusion in CD-1 mice, *Tox. Lett.*, 71, 177, 1994.
66. Ishii, H. and Yokobori, K., Experimental studies on teratogenic activity of noise stimulation, *Gunma J. Med. Sci.*, 9, 153, 1960.
67. Sobrian, S.K., Vaughn, V.T., Ashe, W.K., Markovic, B., Djuric, V., and Jankovic, B.D., Gestational exposure to loud noise alters the development and postnatal responsiveness of humoral and cellular components of the immune system in offspring, *Environ. Res.*, 73, 227, 1997.
68. Rasco, J.F. and Hood, R.D., Effects of maternal restraint stress and sodium arsenate in mice, *Reprod. Toxicol.*, 8, 49, 1994.
69. Chernoff, N., Miller, D.B., Rosen, M.B., and Mattscheck, C.L., Developmental effects of maternal stress in the CD-1 mouse induced by restraint on single days during the period of major organogenesis, *Toxicology*, 51, 57, 1988.
70. Glöckner, R. and Karge, E., Influence of chronic stress before and/or during gestation on pregnancy outcome of young and old Uje:WIST rats, *J. Exptl. Animal Sci.*, 34, 93, 1991.
71. Smith, D.J., Joffe, J.M., and Heseltine, G.F.D., Modification of prenatal stress effects in rats by adrenalectomy, dexamethasone, and chlorpromazine, *Physiol. Behav.*, 15, 461, 1975.
72. Black, D.L. and Marks, T.A., Role of maternal toxicity in assessing developmental toxicity in animals: a discussion, *Regulat. Toxicol. Pharmacol.*, 16, 189, 1992.
73. Arck, P.C., Merali, F.S., Stanisz, A.M., Stead, R.H., Chaouat, G., Manuel, J., and Clark, D.A., Stress-induced murine abortion associated with substance P-dependant alteration in cytokines in maternal uterine decidua, *Biol. Reprod.*, 53, 814, 1995.
74. Arck, P.C., Rose, M., Hertwig, K., Hagen, E., Hildebrandt, M., and Klapp, B.F., Stress and immune mediators in miscarriage, *Human Reprod.*, 16, 1505, 2001.
75. Palmer, A.K., Kavlock, R.J., et al., Consensus workshop on the evaluation of maternal and developmental toxicity. Work Group II Report, *Teratogen. Carcinog. Mutagen.*, 7, 311, 1987.
76. Schardein, J.L., Approaches to defining the relationship of maternal and developmental toxicity, *Teratogen. Carcinog. Mutagen.*, 7, 255, 1987.
77. Iyer, P., Gammon, D., Gee, J., and Pfeifer, K., Characterization of maternal influence on teratogenicity: an assessment of developmental toxicity studies for the herbicide cyanazine, *Regulat. Tox. Pharm.*, 29, 88, 1999.
78. Seidenberg, J.M. and Becker, R.A., A summary of the results of 55 chemicals screened for developmental toxicity in mice, *Teratogen. Carcinog. Mutagen.*, 7, 17, 1987.
79. Chahoud, I., Ligensa, A., Dietzel, L., and Faqi, A.S. Correlation between maternal toxicity and embryo/fetal effects, *Reprod. Toxicol.*, 13, 375, 1999.
80. Francis, E.Z. and Farland, W.H., Application of the preliminary developmental toxicity screen for chemical hazard identification under the Toxic Substances Control Act, *Teratogen. Carcinog. Mutagen.*, 7, 107, 1987.
81. Hardin, B.D., A recommended protocol for the Chernoff-Kavlock preliminary developmental toxicity test and a proposed method for assigning priority scores based on results of that test, *Teratogen. Carcinog. Mutagen.*, 7, 85, 1987.
82. Palmer, A.K., An indirect assessment of the Chernoff/Kavlock assay, *Teratogen. Carcinog. Mutagen.*, 7, 95, 1987.
83. U.S. Environmental Protection Agency, Guidelines for developmental toxicity risk assessment, *Federal Register*, 56, 63798, 1991.
84. Collins, T.F.X., Sprando, R.L., Shackelford, M.E., Hansen, D.K., and Welsh, J.J., Food and Drug Administration proposed testing guidelines for developmental toxicity studies, *Regulat. Toxicol. Pharmacol.*, 30, 39, 1999.
85. Goldman, A.S. and Yakovac, W.C., Prevention of salicylate teratogenicity in immobilized rats by certain central nervous system depressants, *Proc. Soc. Exptl. Biol. Med.*, 115, 693, 1964.
86. Michel, C. and Fritz-Niggli, H., Radiation-induced developmental anomalies in mammalian embryos by low doses and interaction with drugs, stress and genetic factors, in *Late Biological Effects of Ionizing Radiation*, II, International Atomic Energy Agency, Vienna, 1978, p. 397.
87. Murata, M., Takigawa, H., and Sakamoto, H., Teratogenic effects of noise and cadmium in mice: Does noise have teratogenic potential? *J. Toxicol. Environ. Health*, 39, 237, 1993.

88. Glöckner, R., Schwarz, S., and Jähne, F., Enhanced effect of chronic stress on pregnancy outcome in Uje:WIST rats by prenatal treatment with lithium, *Exp. Toxicologic Pathol.*, 45, 35, 1993.
89. Rasco, J.F. and Hood, R.D., Maternal restraint stress-enhanced teratogenicity of all-trans-retinoic acid in CD-1 mice, *Teratology*, 51, 57, 1995.
90. Rasco, J.F. and Hood, R.D., Enhancement of the teratogenicity of all-trans-retinoic acid by maternal restraint stress in mice as a function of treatment timing, *Teratology*, 51, 63, 1995.
91. Colomina, M.E., Sanchez, D.J., Sanchez-Turet, M., and Domingo, J.L., Behavioral effects of aluminum in mice: influence of restraint stress, *Biol. Psychol./Pharmacopsych.*, 40, 142, 1999.
92. Torrente, M., Colomina, M.T., and Domingo, J.L., Effects of prenatal exposure to manganese on postnatal development and behavior in mice: influence of maternal restraint, *Neurotox. Teratol.*, 24, 219, 2002.
93. Viau, V. and Sawchenko, P.E., Hypophysiotropic neurons of the paraventricular nucleus respond in spatially, temporally, and phenotypically differentiated manners to acute vs. repeated restraint stress, *J. Comp. Neurol.*, 445, 293, 2002.
94. Konstandi, M., Marselos, M., Radon-Camus, A.M., Johnson, E., and Lang, M.A., The role of stress in the regulation of drug metabolizing enzymes in mice, *Eur. J. Drug Metab. Pharm.*, 23, 483, 1998.
95. Cianfarani, S., Geremia, C., Scott, C.D., and Germani, D., Growth, IGF system, and cortisol in children with intrauterine growth retardation: Is catch-up growth affected by reprogramming of the hypothalamic-pituitary-adrenal axis? *Pediat. Res.* 51, 94, 2002.
96. Edwards, C.R., Benediktsson, R., Lindsay, R.S., and Seckl, J.R., Dysfunction of placental glucocorticoid barrier: link between fetal environment and adult hypertension? *Lancet*, 341, 355, 1993.
97. Vogel, W.H., The effect of stress on toxicological investigations, *Human Exptl. Toxicol.*, 12, 265, 1993.
98. Slotkin, T.A., Lau, C., McCook, E.C., Lappi, S.E., and Seidler, F.J., Glucocorticoids enhance intracellular signaling via adenylate cyclase at three distinct loci in the fetus: a mechanism for heterologous teratogenic sensitization?, *Toxicol. Appl. Pharmacol.*, 127, 64, 1994.
99. Warkany, J. and Kalter, H., Maternal impressions and congenital malformations, *Plastic & Reconstr. Surg.*, 30, 628, 1962.
99a. Monk, C., Fifer, W.P., Myers, M.M., Sloan, R.P., Trien, L., and Hurtado, A., Maternal stress responses and anxiety during pregnancy: Effects on fetal heart rate, *Dev. Psychobiol.*, 36, 67, 2000.
100. Fraser, F.C. and Warburton, D., No association of emotional stress or vitamin supplement during pregnancy to cleft lip or palate in man, *Plastic & Reconstr. Surg.*, 33, 395, 1964.
101. Edmonds, L.D., Layde, P.M., and Erickson, J.D., Airport noise and teratogenesis, *Arch. Environ. Health*, 34, 243, 1979.
102. Nimby, G.T., Lundberg, L, Sveger, T., and McNeil, T.F., Maternal distress and congenital malformations: Do mothers of malformed fetuses have more problems?, *J. Psychiat. Res.*, 33, 291, 1999.
103. Hansen, D., Lou, H.C., and Olsen, J., Serious life events and congenital malformations: a national study with complete follow-up, *Lancet*, 356, 875, 2000.
104. Carmichael, S.L. and Shaw, G.M., Maternal life event stress and congenital anomalies, *Epidemiology*, 11, 30, 2000.
105. McAnarney, E.R. and Stevens-Simon, C., Maternal psychological stress/depression and low birth weight. Is there a relationship? *Am. J. Dis. Child.*, 144, 789, 1990.
106. Lou, H.C., Hansen, D., Nordentoft, M., Pryds, O., Jensen, F., Nim, J., and Hemmingsen, R., Prenatal stressors of human life affect fetal brain development, *Develop. Med. Child Neurol.*, 36, 826, 1994.
107. Chen, D., Cho, S.-I., Chen, C., Wang, X., Damokosh, A.I., Ryan, L., Smith, T.J., Christiani, D.C., and Xu X., Exposure to benzene, occupational stress, and reduced birth weight, *Occupat. Environ. Med.*, 57, 661, 2000.
108. U.S. Congress Office of Technology Assessment, *Reproductive Health Hazards in the Workplace*, U.S. Government Printing Office, Washington, DC, 1985.
109. Luke, B., Mamelle, N., Keith, L., Munoz, F., Minogue, J., Papiernik, E., and Johnson, T.R.B., The association between occupational factors and preterm birth: a United States nurses' study, *Am. J. Obstet. Gynecol.*, 173, 849, 1995.
110. Hatch, M., Ji, B.-T., Shu, X.O., and Susser, M., Do standing, lifting climbing, or long hours of work have an effect on fetal growth? *Epidemiology*, 8, 530, 1997.
111. Fenster, L., Schaefer, C., Mathur, A., Hiatt, R.A., Pieper, C., Hubbard, A.E., Von Behren, J., and Swan, S.H., Psychologic stress in the workplace and spontaneous abortion, *Am. J. Epidemiol.*, 142, 1176, 1995.

112. Neugebauer, R., Kline, J., Stein, Z. Shrout, P, Warburton, D., and Susser, M., Association of stressful life events with chromosomally normal spontaneous abortion, *Am. J. Epidemiol.*, 143, 588, 1996.
113. Luke, B., Avni, M., Min, L., and Misiunas, R., Work and pregnancy: the role of fatigue and the "second shift" on antenatal morbidity, *Am. J. Obstet. Gynecol.*, 181, 1172, 1999.
114. Malarkey, W.B., Kiecoltglaser, J.K., Pearl, D., and Glaser, R., Hostile behavior during marital conflict alters pituitary and adrenal hormones, *Psychosomatic Med.*, 56, 41, 1994.
115. Stott, D.H., Follow-up study from birth of the effects of prenatal stresses, *Develop. Med. Child. Neurol.*, 5, 770, 1973.
116. Scialli, A.R., Is stress a developmental toxin? *Reprod. Toxicol.*, 1, 163, 1988.
117. Brent, R.L. and Holmes, L.B., Clinical and basic science lessons from the thalidomide tragedy: What have we learned about the causes of limb defects? *Teratology*, 38, 241, 1988.
118. Monk, C., Fifer, W.P., Myers, M.M., Sloan, R.P., Trien, L, and Hurtado, A., Maternal stress responses and anxiety during pregnancy: effects on fetal heart rate, *Devel. Psychobiol.*, 36, 67, 2000.
119. Kalter, H. and Warkany, J., Congenital malformations: Etiologic factors and their role in prevention, *New Engl. J. Med.*, 308, 424, 1983.
120. Khera, K.S., Mouse placenta: hemodynamics in the main maternal vessel and histopathologic changes induced by 2-methoxyethanol and 2-methoxyacetic acid following maternal dosing, *Teratology*, 47, 299, 1993.
121. Keshet, G.I. and Weinstock, M., Maternal naltrexone prevents morphological and behavioral alterations induced in rats by prenatal stress, *Pharmacol. Biochem. Behav.*, 50, 413, 1995.
122 Miller, D.B., Restraint-induced analgesia in the CD-1 mouse: interactions with morphine and time of day, *Brain Res.*, 473, 327, 1988.
123. Matamoros, R.A. and Levine, B.S., Stress response and drug metabolism in mice, *Fundam. Appl. Toxicol.*, 30, 255, 1996.
124. Miller, D.B. and Chernoff, N., Restraint-induced stress in pregnant mice — Degree of immobilization affects maternal indices of stress and developmental outcome in offspring, *Toxicology*, 98, 177, 1995.
125. Joffe, J.M., Hormonal mediation of the effects of prenatal stress on offspring behavior, in *Studies on the Development of Behavior and the Nervous System: Early Influences*, Gottlieb, G., Ed., Academic Press, New York, 1978, p. 108.
126. Weinstock, M., Fride, E., and Hertzberg, R., Prenatal stress effects on functional development of the offspring, in *Progress in Brain Research,* Boer, G.J., Feenstra, M.G.P., Mirmiriam, M., Swaab, D.F., and van Haaren, F., Eds., Elsevier, Amsterdam, vol. 73, 319, 1988.
127. Selye, H., The general adaptation syndrome and disease of adaptation, *J. Clin. Endocrinol. Metab.*, 6, 117, 1946.
128. Hammond, G.L., Molecular properties of corticosteroid binding globulin and the sex-steroid binding proteins, *Endocrine Rev.*, 11, 65, 1990.
129. Rosner, W. and Hockberg, R., Corticosteroid-binding globulin in the rat: isolation and studies of its influence on cortisol action in vivo, *Endocrinology*, 91, 626, 1972.
130. Kirschbaum C. and Hellhammer, D.H., Salivary cortisol in psychoneuroendocrine research: recent developments and applications, *Psychoneuroendocrinology*, 19, 313, 1994.
131. Barlow, S.M., Knight, A.F., and Sullivan, F.M., Diazepam-induced cleft palate in the mouse: the role of endogenous maternal corticosterone, *Teratology*, 21, 149, 1980.
132. Eldeib, M.M.R. and Reddy, C.S., Role of maternal plasma corticosterone elevation in the teratogenicity of secalonic acid D in mice, *Teratology*, 41, 137, 1990.
133. Hansen, D.K., Holson, R.R., Sullivan, P.A., and Grafton, T.F., Alterations in maternal plasma corticosterone levels following treatment with phenytoin, *Toxicol. Appl. Pharmacol.*, 96, 24, 1988.
134. Iida, H., Kast, A., Tsunenari, Y., and Asakura, M., Corticosterone induction of cleft palate in mice dosed with orciprenaline sulfate, *Teratology*, 38, 15, 1988.
135. Kopf-Maier, P., Glucocorticoid induction of cleft palate after treatment with titanocene dichloride? *Toxicology*, 37, 111, 1985.
136. Uno, H., Lohmiller, L., Thieme, C., Kemnitz, J.W., Engle, M.J., Roecker, E.B., and Farrell, P.M., Brain damage induced by prenatal exposure to dexamethasone in fetal rhesus macaques. I. Hippocampus, *Devel. Brain Res.*, 53, 157, 1990.
137. Gitau, R., Fisk, N.M., and Glover, V., Maternal stress in pregnancy and its effect on the human foetus: An overview of research findings, *Stress*, 4, 195, 2001.

138. Ader, R. and Plaut, S.M., Effects of prenatal handling and differential housing on offspring emotionality, plasma corticosterone levels, and susceptibility to gastric erosion, *Psychosomatic Med.*, 30, 277, 1968.
139. Clarke, A.S., Wittner, D.J., Abbott, D.H., and Schneider, M.L., Long-term effects of prenatal stress on HPA axis activity in juvenile rhesus monkeys, *Devel. Psychobiol.*, 27, 257, 1994.
140. Fride, E., Dan, Y., Feldon, J., Halevy, G., and Weinstock, M., Effects of prenatal stress on vulnerability to stress in prepubertal and adult rats, *Physiol. Behav.*, 37, 681, 1986.
141. Peters, D.A., Prenatal stress: Effects on brain biogenic amine and plasma corticosterone levels, *Biochem. Behav.*, 17, 721, 1982.
142. Schneider, M.L., Coe, C.L., and Lubach, G.R., Endocrine activation mimics the adverse effects of prenatal stress on the neuromotor development of the infant primate, *Devel. Psychobiol.*, 25, 427, 1992.
143. Colby, H.D., Adrenal gland toxicity: chemically induced dysfunction, *J. Am. Coll. Toxicol.*, 7, 45, 1988.
144. Ribelin, W.E., Effects of drugs and chemicals upon the structure of the adrenal gland, *Fund. Appl. Toxicol.*, 4, 105, 1984.
145. Bestervelt, L.L., Yong, C., Piper, W.N., Nolan, C., and Pitt, J.A., TCDD alters pituitary-adrenal function. I. Adrenal responsiveness to exogenous ACTH, *Neurotoxicol. Teratol.*, 15, 365, 1993.
146. Bestervelt, L.L., Pitt, J.A., Nolan, C.J., and Piper, W.N., TCDD alters pituitary-adrenal function II. Evidence for decreased bioactivity of ACTH, *Neurotoxicol. Teratol.*, 15, 371, 1993.
147. Reinhardt, V., Cowley, D., Scheffler, J., Vertein, R., and Wegner, F., Cortisol response of female rhesus monkeys to venipuncture in home cage versus venipuncture in restraint apparatus, *J. Med. Primatol.*, 19, 601, 1990.
148. Wiersma, J. and Kastelijn, J., A chronic technique for high frequency blood sampling/transfusion in the freely behaving rat which does not affect prolactin and corticosterone secretion, *J. Endocrinol.*, 107, 285, 1985.
149. Dallman, M.F., England, W.C., Rose, J.C., Wilkinson, C.W., Shinsako, J., and Siedenburg, F., Nyctohemeral rhythmic adrenal responsiveness to ACTH, *Am. J. Physiol.*, 235 (Regulatory Integrative Comp. Physiol. 4), R210, 1978.
150. Clarke, A.S., Mason, W.A., and Moberg, G.P., Differential behavioral and adrenocortical responses to stress among three macaque species, *Am. J. Primatol.*, 14, 37, 1988.
151. Dhabhar, F.S., McEwen, B.S., and Spencer, R.L., Stress response, adrenal steroid receptor levels, and corticosteroid-binding globulin levels — a comparison between Sprague Dawley, Fischer 344, and Lewis rats, *Brain Res.*, 616, 89, 1993.
152. Dhabhar, F. S., Miller, A. H., McEwen, B. S., and Spencer, R. L., Differential activation of adrenal steroid receptors in neural and immune tissues of Sprague Dawley, Fischer 344, and Lewis rats, *J. Neuroimmunol.*, 56, 77, 1995.
153. Griffin, A.C. and Whitacre, C.C., Sex and strain differences in circadian rhythm fluctuation of endocrine and immune function in the rat: Implications for rodent models of autoimmune disease, *J. Neuroimmunol.*, 35, 53, 1991.
154. Sternberg, E.M., Glowa, J.R., Smith, M.A., Calogero, A.E., Litwak, S.J., Aksentijevich, S., Chrousos, G.P., Wilder, R.L., and Gold, P.W., Corticotropin releasing hormone related behavioral and neuroendocrine responses to stress in Lewis and Fisher rats, *Brain Res.*, 570, 54, 1992.
155. Barlow, S.M., Morrison, P.J., and Sullivan, F.M., Plasma corticosterone levels during pregnancy in the mouse: the relative contributions of the adrenal gland and foeto-placental units, *J. Endocrinol.*, 60, 473, 1974.
156. Salomon, D.S., Gift, V.D., and Pratt, R.M., Corticosterone levels during midgestation in the maternal plasma and fetus of cleft palate-sensitive and resistant mice, *Endocrinology*, 104, 154, 1979.
157. Ballard, P.L., Delivery and transport of glucocorticoids to target cells, in *Glucocorticoid Hormone Action, Monographs on Endocrinology, 12,* Baxter, J.D., and Rousseau, G G., Eds., Springer-Verlag, Berlin, 1979, p. 25.
158. Doe, R.P., Zinneman, H.H., Flink, E.B., and Ulstrom, R.A., Significance of the concentration of non-protein bound plasma cortisol in normal subjects, Cushing's syndrome, pregnancy and during estrogen therapy, *J. Clin. Endocrinol. Metab.*, 20, 1484, 1960.
159. Seralini, G.-E., Smith, C.L., and Hammond, G.L., Rabbit corticosteroid-binding globulin: primary structure and biosynthesis during pregnancy, *Mol. Endocrinol.*, 4, 1166, 1990.
160. McCarty, R. and Kopin, I.J., Stress-induced alterations in plasma catecholamines and behavior of rats: effects of chlorisodamine and bretylium, *Behav. Biol.*, 27, 249, 1979.

161. Chernoff, N. and Grabowski, C., Responses of the rat foetus to maternal injections of adrenaline and vasopressin, *Br. J. Pharmacol.*, 43, 270, 1971.
162. Natelson, B.H., Creighton, D., McCarty, R., Tapp, W.N., Pitman, D., and Ottenweller, J.E., Adrenal hormonal indices of stress in laboratory rats, *Physiol. Behav.*, 39, 117, 1987.
163. Konarska, M., Stewart, R.E., and McCarty, R., Predictability of chronic intermittent stress: effects on sympathetic-adrenal medullary responses of laboratory rats, *Behav. Neural Biol.*, 53, 231, 1990.
164. Kopin, I. J., Catecholamine metabolism: basic aspects and clinical significance, *Pharmacol. Rev.*, 37, 333, 1985.
165. German, J., Embryonic stress hypothesis of teratogenesis, *Am. J. Med.*, 76, 293, 1984.
166. Kochevar, D.T., Aucoin, M.M., and Cooper, J., Mammalian heat shock proteins: an overview with a systems perspective, *Tox. Lett.*, 56, 243, 1991.
167. Kohler, H.-R., Triebskorn, R., Stocker, W., Kloetz, P.-M., and Alberti, G., The 70 kD heat shock protein (hsp 70) in soil invertebrates: a possible tool for monitoring environmental toxicants, *Arch. Environ. Contam. Toxicol.*, 22, 334, 1992.
168. Sagar, S.M., Sharp, F.R., and Curran, T., Expression of c-fos protein in brain: metabolic mapping at the cellular level, *Science*, 240, 1328, 1988.
169. Dienel, G.A., Kiessling, M., Jacewicz, M., and Pulsinelli, W.A., Synthesis of heat shock proteins in rat brain cortex after transient ischemia, *J. Cerebral Blood Flow Metabol.*, 6, 505, 1986.
170. Munro, S. and Pelham, H., What turns on heat shock genes? *Nature*, 317, 477, 1985.
171. Schlesinger, M.J., Ashburner, M., and Tissieres, A., *Heat Shock: From Bacteria to Man*, Cold Spring Harbor Laboratory, Cold Spring Harbor, 1982.
172. Brown, I.R. and Rush, S.J., Induction of a "stress" protein in intact mammalian organs after the intravenous administration of sodium arsenite, *Biochem. Biophys. Res. Commun.*, 120, 150, 1984.
173. Ceccatelli, S., Villar, M., Goldstein, M., and Hokfelt, T., Expression of c-Fos immunoreactivity in transmitter-characterized neurons after stress, *Proc. Natl. Acad. Sci. USA*, 86, 9569, 1989.
174. Dragunow, M. and Robertson, H.A., Generalized seizures induce *c-fos* protein(s) in mammalian neurons, *Neurosci. Lett.*, 82, 157, 1987.
175. Hammond, G.L., Lai, Y.-K., and Markert, C.L., Diverse forms of stress lead to new patterns of gene expression through a common and essential metabolic pathway, *Proc. Natl. Acad. Sci. USA*, 79, 3485, 1982.
176. Ewing, J.F., Haber, S.N., and Maines, M.D., Normal and heat-induced patterns of expression of heme oxygenase-1 (HSP32) in rat brain: Hyperthermia causes rapid induction of mRNA and protein, *J. Neurochem.*, 58, 1140, 1992.
177. Kononen, J., Honkaniemi, J., Alho, H., Koistinaho, J., Iadarola, M., and Pelto-Huikko, M., Fos-like immunoreactivity in the rat hypothalamic-pituitary axis after immobilization stress, *Endocrinology*, 130, 3041, 1992.
178. White, F.P., The induction of "stress" proteins in organ slices from brain, heart, and lung as a function of postnatal development, *J. Neurosci.*, 1, 1312, 1981.
179. Yang, G., Koistinaho, J., Zhu, S., and Hervonen, A., Induction of c-fos-like protein in the rat adrenal cortex by acute stress — immunocytochemical evidence, *Mol. Cell. Endocrinol.*, 66, 163, 1989.
180. Kimmel, C.A., Kimmel, G.L., Lu, C., Heredia, D.J., Fisher, B.R., and Brown, N.T., Stress protein synthesis as a potential biomarker for heat-induced developmental toxicity, *Teratology*, 43, 465, 1991.
181. Finnell, R.H., Ager, P.L., Englen, M.D., and Bennett, G.D., The heat shock response: potential to screen teratogens, *Tox. Lett.*, 60, 39, 1992.
182. Mirkes, P.E., Doggett, G., and Cornel, L., Induction of a heat shock response (HSP 72) in rat embryos exposed to selected chemical teratogens, *Teratology*, 49, 135, 1994.
183. Nowak, T.S., Jr., Synthesis of a stress protein following transient ischemia in the gerbil, *J. Neurochem.*, 45, 1635, 1985.
184. Goering, P.L., Fisher, B.R., and Kish, C.L., Stress protein synthesis induced in rat liver by cadmium precedes hepatotoxicity, *Toxicol. Appl. Pharmacol.*, 122, 139, 1993.
185. Sawyer, T.K., Proteomics — structure and function. *BioTechniques,* 31, 156, 2001.
186. Daston, G.P., Overmann, G.J., Baines, D., Taubeneck, M.W., Lehman-MeKeeman, L., Rogers, J.M., and Keen, C.L., Altered Zn status by alpha-hederin in the pregnant rat and its relationship to adverse developmental outcome, *Reprod. Toxicol.*, 8, 15, 1994.

187. Keen, C.L. and Hurley, L.S., Zinc and reproduction: effects of deficiency on fetal and postnatal development, in *Zinc and Human Biology*, Mills, C.F., Ed., Springer-Verlag, New York, 1989, p. 183.
188. Taubeneck, M.W., Daston, G.P., Rogers, G.P., and Keen, C.L., Altered maternal zinc metabolism following exposure to diverse developmental toxicants, *Reprod. Toxicol.*, 8, 25, 1994.
189. Klaassen, C.D. and Lehman-McKeeman, L.D., Induction of metallothionein, *J. Am. Coll. Toxicol.*, 8, 1315, 1989.
190. Quaife, C., Hammer, R.E., Mottett, N.K., and Palmiter, R.D., Glucocorticoid regulation of metallothionein during murine development, *Devel. Biol.*, 118, 549, 1986.
191. Francis, E.Z., Testing of environmental agents for developmental and reproductive toxicity, in *Developmental Toxicology*, 2nd Ed., Kimmel, C.A. and Buelke-Sam, J., Eds., Raven Press, New York, 1994, p. 403.
192. Akana, S.F., Cascio, C.S., Shinsako, J., and Dallman, M.F., Cotricosterone: narrow range required for normal body and thymus weight and ACTH, *Am. J. Physiol.*, 249, R527, 1985.
193. Miller, D.B., Caveats in hazard assessment: stress and neurotoxicity, in *The Vulnerable Brain and Environmental Risks. Vol. 1. Malnutrition and Hazard Assessment*, Isaacson, R.L. and Jensen, K.F., Eds., Plenum Press, New York, 1992, p. 239.
194. D'Armour, F.E. and Smith, D.L., A method for determining loss of pain sensations, *J. Pharmacol. Exptl. Therapeut.*, 72, 74, 1941.
195. Gintzler, A.R., Endorphin-mediated increases in pain threshold during pregnancy, *Science*, 210, 193, 1980.
196. Dahl, J.L., Silva, B.W., Baker, T.B., and Tiffany, S.T., Endogenous analgesia in the pregnant rat: an artifact of weight-dependent measures?, *Brain Res.*, 373, 316, 1986.
197. Kirby, M.L., Alterations in fetal and adult responsiveness to opiates following various schedules of prenatal morphine exposure, in *Neurobehavioral Teratology*, Yanai, J., Ed., Elsevier, New York, 1984, p. 235.
198. Barclay, R.J., Herbert, W.J., and Poole, T.B., *The Disturbance Index: A Behavioural Method of Assessing the Severity of Common Laboratory Procedures on Rodents*, Universities Federation for Animal Welfare, Potters Bar, UK, 1988.
199. Schwetz, B.A. and Harris, M.W., Developmental toxicology — status of the field and contribution of the National Toxicology Program, *Environ. Health Perspect.*, 100, 269, 1993.
200. Rowan, A.N., Refinement of animal research technique and validity of research data, *Fund. Appl. Toxicol.*, 15, 25, 1990.
201. Barnard, N. and Hou, S., Inherent stress. The tough life in lab routine, *Lab. Animal*, 17, 21, 1988.
202. Milligan, S.R., Sales, G.D., and Khirnykh, K., Sound levels in rooms housing laboratory animals—An uncontrolled daily variable, *Physiol. Behav.*, 53, 1067, 1993.
203. Gärtner, K., Büttner, D., Döhler, K., Friedel, R., Lindena, J., and Trautschold, I., Stress response of rats to handling and experimental procedures, *Lab. Animal*, 14, 267, 1980.
204. Riley, V., Psychoneuroendocrine influences on immunocompetence and neoplasia, *Science*, 212, 1100, 1981.
205. Miller, D.B. and O'Callaghan, J.P., Neurotoxicity of d-amphetamine in the C57BL/6J and CD-1 mouse: interactions with stress and the adrenal system, *Ann. N.Y. Acad. Sci.*, 801, 148, 1996.
206. Johnson, E.A., Sharp, D.S., and Miller, D.B., Restraint as a stressor in mice: Against the dopaminergic neurotoxicity of d-MDMA, low body weight mitigates restraint-induced hypothermia and consequent neuroprotection, *Brain Res.*, 875, 107, 2000.
207. Wright, B.E. and Katovich, M.J., Effect of restraint on drug-induced changes in skin and core temperature in biotelemetered rats, *Pharmacol. Biochem. Behav.*, 55, 219, 1996.
208. Faupel, R.P., Seitz, H.J., Tarnowski, W., Thiemann, V., and Weiss, C., The problem of tissue sampling from experimental animals with respect to freezing technique, anoxia, stress and narcosis. A new method for sampling rat liver tissue and the physiological values of glycolytic intermediates and related compounds, *Arch. Biochem. Biophys.*, 148, 509, 1972.
209. Barbazanges, A., Piazza, P.V., Le Moal, M., and Maccari, S., Maternal glucocorticoid secretion mediates long-term effects of prenatal stress, *J. Neurosci.*, 16, 3943, 1996.
210. Thiel, R., Chahoud, I., Jurgens, M., and Neubert, D., Time-dependent differences in the development of somites of four different mouse strains, *Teratogen. Carcinog. Mutagen.*, 13, 247, 1993.
211. Ishikawa, H., Omoe, K., and Endo, A., Growth and differentiation schedule of mouse embryos obtained from delayed matings, *Teratology*, 45, 655, 1992.

212. Endo, A. and Watanabe, T., Interlitter variability in fetal body weight in mouse offspring from continuous, overnight, and short-period matings, *Teratology*, 37, 63, 1988.
213. Paré, W.P. and Glaven, G.B., Restraint stress in biomedical research: a review, *Neurosci. Biobehav. Rev.*, 10, 339, 1986.
214. Terry, R.D., Marks, T.A., Hamilton, R.D., Pitts, T.W., and Renis, H.E., Prevention of tilorone developmental toxicity with progesterone, *Teratology*, 46, 237, 1992.
215. Sadler, T.W., Denno, K.M., and Hunter, E.S., Effects of altered maternal metabolism during gastrulation and neurulation stages of embryogenesis, in *Maternal Nutrition and Pregnancy Outcome (Series: Ann. New York Acad. Sci., 678)*, Keen, C.L., Bendich, A., and Willhite, C.C., Eds., New York Academy of Sciences, New York, 1993, p. 48.
216. Tritt, S.H., Tio, D.L., Brammer, G.L., and Taylor, A.N., Adrenalectomy but not adrenal demedullation during pregnancy prevents the growth-retarding effects of fetal alcohol exposure, *Alcoholism: Clin. Exper. Res.*, 17, 1281, 1993.
217. Marks, T.A., Black, D.L., and Terry, R.D., Counteraction of the embryolethal effects, but not the maternal toxicity, of bropirimine and tilorone by coadministration of indomethacin, *J. Am. Coll. Toxicol.*, 13, 93, 1994.
218. Bornstein, S.R., Böttner, A., and Chrousos, G.P., Knocking out the stress response, *Mol. Psychiatry*, 4, 403, 1999.
219. Glavin, G.B., Paré, W.P., Sandbak, T., Bakke, H.K., and Murison, R., Restraint stress in biomedical research: an update, *Neurosci. Biobehav. Rev.*, 18, 223, 1994.
220. ECETOC, *Influence of Maternal Toxicity in Studies of Developmental Toxicity, Workshop Report No. 4*, European Centre for Ecotoxicology and Toxicology of Chemicals, Brussels, 2004.

CHAPTER 5

Paternally Mediated Effects on Development

Barbara F. Hales and Bernard Robaire

CONTENTS

I. Role of the Male in Mediating Developmental Toxicity 125
   A. Evidence from Epidemiological Studies 126
   B. Evidence from Animal Experimentation 127
II. Potential Mechanisms Involved in Male-Mediated Developmental Toxicity 128
   A. Drugs or Toxicants in Semen 128
   B. Drugs or Toxicants Affecting the Male Germ Cell 128
      1. Germ Cells in the Testis or the Posttesticular Excurrent Duct System 129
      2. Reversibility 131
      3. Heritability 132
III. Methodological Approaches in Male-Mediated Developmental Toxicity 132
   A. Effects of Toxicants in Seminal Fluid 132
   B. Effects on Sperm Quantity and Characteristics 132
   C. Effects on Sperm Quality 134
      1. Sperm Chromatin Packaging and Function 134
      2. Sperm Genetic Integrity 136
      3. Epigenetic Changes 138
IV. Summary and Conclusions 138
Acknowledgments 138
References 139

I. ROLE OF THE MALE IN MEDIATING DEVELOPMENTAL TOXICITY

It is well established that there are risks to the progeny if the mother is exposed to a variety of chemicals during pregnancy. The extent to which paternal exposures contribute to infertility and pregnancy loss is less evident. There is growing concern that paternal exposure to drugs, radiation, or environmental toxicants may result in a decrease in sperm count and an increase in male infertility, spontaneous abortions, birth defects, or childhood cancer, adversely affecting reproduction and progeny outcome. Clear evidence that the exposure of males to xenobiotics can result in adverse effects on progeny outcome has accumulated over the past two decades.[1] The recent publication of the proceedings of a multidisciplinary international conference on male-mediated developmental toxicity has highlighted the need for more research in this area. The goal of this research is to

Table 5.1 Paternal exposures associated with adverse progeny outcomes

Agent	Pregnancy Loss[a]	Birth Defects[a]	Childhood Cancer[a]
Radiation	0.9–1.5	1.4–5.6	1.8–6.7
Solvents	0.9–2.3	NA[b]	1.7–7
Anesthetic gases	1.5–1.8	NA	NA
Heavy metals	0.9–2.3	1.5–249	3.5–7
Cigarette smoke	0.6–1.4	1.3	1.2–3.9
Herbicides/pesticides	NA	5.7–405	2.4–7.1
Anticancer drugs	NA	4.1	NA

[a] Values represent the range of OR/RR (odds ratios/relative risk) found by different studies
[b] NA = not available

Source: Adapted from Aitken, R. J. and Sawyer, D., The human spermatozoon – not waving but drowning, in *Advances in Male-Mediated Developmental Toxicity, Advances in Experimental Medicine and Biology, Vol. 518,* Robaire, B. and Hales, B.F., Eds. Kluwer Academic/Plenum Press, New York, 2003. With permission.

elucidate the extent to which the father contributes to the approximately 60% of all birth defects that are of unknown cause.[1,2]

Studies of the mechanisms by which paternal exposures adversely affect progeny outcome are essential to provide the scientific basis for the development of biomarkers for risk assessment. The two major mechanisms by which exposure of a male to a drug or environmental toxicant may adversely affect his progeny are: (1) direct exposure of the conceptus during mating to a chemical present in the seminal fluid and (2) toxic effects on male germ cells.

Two major approaches have been taken to identify instances in which paternal exposure to a xenobiotic adversely affects progeny outcome, namely, epidemiological studies and animal experiments.

A. Evidence from Epidemiological Studies

Epidemiological studies have focused largely on determining the consequences of paternal occupational exposures on progeny. The effects of paternal occupational exposures on the offspring range from early spontaneous abortion, which may be perceived as infertility, to delayed spontaneous abortions and stillbirth, malformations, preterm delivery, delivery of a small-for-gestational age infant, childhood cancer, altered postnatal behavior, or changes in reproductive function.[1,3–9]

A variety of different paternal occupations have been associated with adverse progeny outcomes (Table 5.1).[3–9] Paternal exposures to radiation, solvents, heavy metals (e.g., mercury), pesticides, and hydrocarbons were associated with an increased incidence of spontaneous abortion or miscarriage, birth defects, or childhood cancer.[3–9] Paternal occupations that have been associated with an increased risk of having a liveborn child with a birth defect include janitor, woodworker (forestry or logging, sawmill, carpenter), firefighter, electrical worker, and printer.[6,7] Men employed in occupations associated with solvent exposures (with painters having the highest risk) were more likely to have offspring with anencephaly.[5] Further, an increased risk of stillbirth, preterm delivery, or delivery of a small-for-gestational age infant was associated with paternal employment in the art (painters) or textile industries.[8,9]

Concerns about the potential consequences to offspring of exposure to herbicides were brought to the forefront by male Vietnam veterans. The data from some of the studies conducted to address these concerns suggested that there was an increased incidence of birth defects among the children fathered by veterans, while other inquiries did not find a positive relationship.[10–13] Arbuckle and colleagues reported recently that adverse progeny outcomes, including an increased risk of early abortions, were associated with the exposure of male farmers to pesticides that included carbamates, organophosphates, and organochlorines.[14]

Quantitative exposure estimates are available in very few studies. In one European study of exposure levels and adverse effects, median sperm concentrations were reduced almost 50% in men with a blood lead concentration ≥50 µg/dl.[15] Furthermore, in most studies, the specific chemical exposures for each occupational group were not identified. Depending on the profession and the study, the likelihood that paternal occupational exposure is a factor in abnormal progeny outcome has ranged from 1.5-fold to as high as 4.5- to 5-fold. In addition, a wide range of paternal professions either were not associated or were associated with a negative likelihood of abnormal progeny outcome.[4,7]

The possibility that "life style" or "recreational" exposures of the father to cigarette smoke, alcohol, caffeine, methadone, cannabinoids, cocaine, and other illicit agents may affect progeny outcome also deserves consideration. Although some of these exposures have received at least limited attention, most studies have not shown any definitive evidence that paternal smoking or alcohol consumption causes birth defects in the offspring.[16] Several studies have established that paternal smoking is associated with increased miscarriages, while others have reported a decrease in birth weights.[17,18] Adverse effects of cigarette smoking on male reproductive functions include a decrease in sperm count, poor sperm motility, abnormal morphology, including retention of cytoplasm, and ultrastructural abnormalities, as well as an increase in DNA adducts and reduced fertility.[19-23] Increased chromosome aneuploidy has been reported in the sperm of men who smoke compared with the sperm of those who do not.[19] At least four constituents in cigarette smoke have been shown to cause aneuploidy in different test systems.[24]

Therapeutic drug exposures are more readily documented than are "lifestyle" exposures. Despite this, few studies have focused on the consequences to the progeny of paternal treatment with a variety of commonly used drugs, such as antihypertensives or antidepressants. One group of drugs that has received some attention is the anticancer drugs. Treatment of men with these drugs is often associated with transient or permanent infertility. In those instances when treated men have fathered children, the limited studies available to date have found that the proportion of malformed children in the treatment group was not different from that in the control group or the general population.[25-27] There was also no significant increase in cancer or genetic disease among the offspring of male cancer survivors.[28] At present, however, it is still difficult to assess the impact of the treatment of men with anticancer drugs because the number of patients in most studies is very low. Additional cohorts of thousands of patients would be necessary to rule out a relative risk on the order of 1.5 for a germ cell mutation. Thus, the true impact of the latest combination chemotherapeutic regimens on reproductive health, gamete genetic integrity, and more importantly, the genetic risks to future generations, remains to be evaluated.

To date, epidemiological studies have largely used reproductive outcome as the measure of a paternally mediated effect. In the future, as the knowledge gained from basic research is applied to humans, it may be possible to analyze changes in spermatozoal DNA, gene expression, protein products, or chromatin packaging as biomarkers of the impact of exposure to a male-mediated developmental toxicant.

B. Evidence from Animal Experimentation

Animal studies have demonstrated that paternal exposures to a wide range of environmental chemicals (e.g., carbon disulfide, lead, dibromochloropropane) and drugs (e.g., cyclophosphamide, chlorambucil, melphalan, ethanol) can result in abnormal progeny outcome.[1] Drugs or environmental chemicals to which the male is exposed may be present in his seminal fluid and thus may have direct effects on the sperm or ovulated egg, on the process of fertilization, or on the embryo itself. Alternatively, drugs or other chemicals may have adverse effects on the number of male germ cells by affecting the hypothalamo-pituitary-testicular axis. Thus, they may alter the hormonal milieu required to maintain spermatogenesis. The numbers of male germ cells may also be affected by inhibiting mitosis or meiosis, by triggering germ cell apoptosis, or by affecting the shedding of the germ cells from the seminiferous epithelium. Insufficient numbers of functional sperm will result in male infertility. Alternatively, the male germ cell may be functional but altered, with the consequence that genetic defects or mutations are transmitted to the progeny, resulting in early or late conceptal loss, malformations, or serious diseases after birth. Finally, epigenetic changes in noncomplementary or imprinted regions of the male genome may also adversely affect embryo development *in utero*. Anticancer drugs are among the best studied male-mediated developmental toxicants. These drugs are present in the seminal fluid and are transferred to the female during mating; they affect germ cell numbers and alter germ cell quality.

II. POTENTIAL MECHANISMS INVOLVED IN MALE-MEDIATED DEVELOPMENTAL TOXICITY

A. Drugs or Toxicants in Semen

A drug in the seminal fluid may be absorbed and distributed throughout the female, and may thus affect the conceptus. A wide range of compounds has been shown to enter semen.[29–32] From animal experiments, methadone, morphine, thalidomide, and cyclophosphamide are all examples of drugs reported to adversely affect progeny outcome by this mechanism.[33–36] Cyclophosphamide and/or its metabolites can enter all tissues of the male reproductive tract, including the seminal vesicle fluid; this drug is transmitted to the female partner, where it is absorbed through the vagina and distributed to a large array of tissues.[36] When male rats were given a single dose (10 to 100 mg/kg) of cyclophosphamide immediately prior to cohabitation with females in proestrus, there was a significant dose-dependent increase in preimplantation loss (the number of resorption sites in the uterus minus corpora lutea in both ovaries).[36] No significant increases in postimplantation loss (the total number of resorbed or dead fetuses per pregnant female) or abnormal fetuses were noted.

The increase in preimplantation loss after the acute treatment of males with cyclophosphamide may be due either to the presence of the drug in the seminal fluid or to an effect of the drug on spermatozoa in the testicular excurrent duct system. To distinguish between these possibilities, females in proestrus mated to control males were remated within 2 h to azoospermic, vasectomized males treated acutely with cyclophosphamide or saline. Once again, preimplantation loss was increased significantly in both drug-treated groups, suggesting that the increase in preimplantation loss was due to the presence of the drug in seminal fluid, rather than to an effect on spermatozoa stored in the testicular excurrent duct system.[33] Thus, the presence of drugs in semen can modify progeny outcome; these effects are likely to occur early during gestation.

B. Drugs or Toxicants Affecting the Male Germ Cell

The second major mechanism by which drugs given to the male may affect progeny outcome is that they alter male germ cell numbers or quality, either during spermatogenesis in the testis or during spermatozoal maturation in the epididymis. Examples of male-mediated developmental toxicants thought to act via an effect on the male germ cell include lead, dibromochloropropane, vinyl chloride, 1,3-butadiene, acrylamide, and anticancer drugs.[1,37,38]

Lead exposure impairs reproductive capacity by inducing testicular toxicity, effects on spermatogenesis, and altered androgen metabolism.[39,40] Chronic low level exposures may compromise fertility in the absence of demonstrable effects on endocrine function and semen quality.[41,42] Effects on reproductive and learning behavior in the F_1 generation of rodents whose fathers were exposed to lead have been reported after exposure to low levels of lead, even in the absence of infertility or overt testicular damage.[43,44]

Exposure to dibromochloropropane (DBCP), a nematocide, has been associated in men with azoospermia and oligospermia, as well as infertility, and in male rats with hepatic and renal effects, as well as testicular atrophy.[45,46] DBCP increased sister chromatid exchanges and chromosomal aberrations and tested positive in the dominant lethal test,[47] which involves mating treated and control males with untreated females. The females are examined to determine the numbers of corpora lutea in the ovaries (an indication of the number of potential embryos) and implantations in the uterus. Implants are classified as normal fetuses, dead fetuses, or early or late resorptions, and fetal weights may be monitored. Usually both pre- and postimplantation embryonic losses are considered to be indicative of dominant lethality. Vinyl chloride is a mutagen and carcinogen in many test systems, but most studies found that it did not induce dominant lethality.[48] Dominant lethal mutations, congenital malformations, and heritable translocations were induced by the exposure of mice to butadiene.[49,50] Carbazole and cypermethrin were also positive in the dominant lethal test when male germ cells were exposed to these insecticides.[51,52] The exposure of male rats to acrylamide in the drinking water for 10 weeks induced pre- and postimplantation loss in their progeny, as well as axonal fragmentation and/or swelling in the adult F_1 male progeny.[53]

Animal data reveal that anticancer drugs have a plethora of effects on male germ cells. That anticancer drugs affect germ cell numbers is not surprising, as they are almost without exception toxic to rapidly dividing cells. Exposure to anticancer drugs may affect germ cell numbers by inhibiting mitosis or meiosis or by triggering germ cell death by apoptosis.[54] Multiple measures of germ cell quality (morphology, motility, DNA damage and genetic integrity, chromatin packaging) are also affected by anticancer drugs.[1,37] Alkylating agents, such as mechlorethamine, dacarbazine, or cyclophosphamide, were among the most potent germ cell mutagens, inducing heritable translocations and dominant lethal and specific locus mutations; the predominant effects are usually in spermatids and early spermatozoa.[55,56] Cisplatin-induced dose-related increases in (1) DNA adducts and (2) chromatid and isochromatid breaks in leptotene and preleptotene spermatocytes and differentiating spermatogonia; these germ cell effects were expressed as an increase in pre- and postimplantation loss, as well as an increase in malformed and growth retarded fetuses.[57,58] Effects on stem cells represent the most serious threat to subsequent generations; these effects may differ from those seen in post-stem cells.[59] Importantly, several alkylating agents (cyclophosphamide, procarbazine, melphalan, mitomycin C) induce chromosomal translocations in stem cells.[60] There was a significant increase in abnormal fetuses among the progeny of male mice mated to control females 64 to 80 days after treatment with genotoxic chemicals, such as mitomycin C, ethylnitrosourea, or procarbazine.[60]

Doxorubicin and etoposide both inhibit topoisomerase II. Doxorubicin is cytotoxic at early meiotic stages and at all spermatogonial stages, and it has been demonstrated to affect sperm motility, sperm head structure, and sperm numbers, resulting in an increase in preimplantation loss.[61] Etoposide inhibits premeiotic DNA synthesis and induced micronuclei in spermatids after exposure of diplotene-diakinesis and preleptotene spermatocytes. Like doxorubicin, bleomycin is an intercalating agent; it is postulated to induce DNA damage by generating reactive oxygen species. Interestingly, bleomycin induced specific locus mutations in spermatogonia but not in postspermatogonial stages.[62] The *Vinca* alkaloids vinblastine and vincristine interfere with the spindle apparatus, resulting in an arrest of mitosis and meiosis followed by cell death, as well as an increase in aneuploidy in surviving germ cells. Large doses of both drugs also affect Sertoli cells, destroying microtubules and mitochondria.[63] Antimetabolites, such as 6-mercatopurine or 5-fluorouracil, induce dominant lethality and chromosomal aberrations in differentiating spermatogonia at stages during which rapid DNA synthesis is taking place. Thus, many drugs and other chemicals have been identified in animal experiments as male-mediated developmental toxicants.

1. Germ Cells in the Testis or the Posttesticular Excurrent Duct System

Spermatogenesis in the testis is a tightly regulated process. Spermatogonia undergo mitotic proliferation and then meiosis to form primary and secondary spermatocytes (spermacytogenesis). After completion of the second meiotic division, spermatocytes become spermatids and differentiate into spermatozoa (spermiogenesis), primarily by condensing nuclear elements, shedding most of the

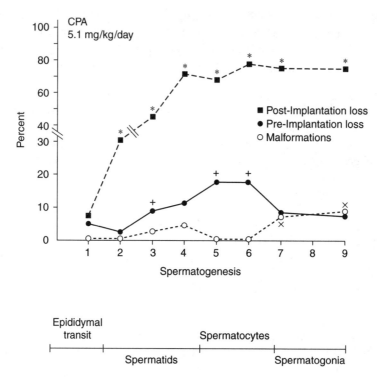

Figure 5.1 Adverse progeny outcomes after mating of control females to male rats treated by gavage with cyclophosphamide (5.1 mg/kg/d) for the indicated number of weeks. The observed effect depends on the phase of spermatogenesis in the germ cells when they are first exposed to cyclophosphamide. (Adapted from Trasler, J.M., Hales, B.F., and Robaire, B., *Biol. Reprod.*, 34, 275, 1986. With permission.)

cytoplasm, and forming an acrosome and a flagellum. From the timing of the effect of an exposure, one can assess the stage specificity of the susceptibility of the germ cells proceeding through spermatogenesis. For example, in mice, an effect on progeny outcome after exposure of males to a drug or x-rays for 1 to 5 days would represent an effect on spermatozoa, most probably those residing in the epididymis.[64,65] Exposure to a toxicant 10 to 18 days prior to conception would represent an effect on spermatids, while long exposures (35 days or more) prior to conception should represent an effect on spermatogonia. Germ cells at the spermatogonial stage in both mice and humans are very susceptible to x-ray exposure; sperm numbers are reduced, and the surviving sperm are morphologically abnormal.[64,65] Following exposure of mice to chlorambucil, a peak in mutation yield is observed when offspring are conceived from germ cells exposed as spermatids.[64,65] In contrast, spermatozoa are the germ cells that are maximally sensitive to the specific locus mutations induced by acrylamide monomer.[65] Ethylnitrosourea has been shown to result in a high frequency of embryo death (dominant lethality) in the offspring of mice after short treatment periods in which only spermatozoa were exposed.[64,65]

Cyclophosphamide, a commonly used anticancer drug, remains one of the best studied examples of a drug that affects male germ cells in a stage-specific manner (Figure 5.1).[66–70] Increased postimplantation loss was found after 2 weeks of chronic low dose cyclophosphamide treatment of male rats. This postimplantation loss rose dramatically to plateau at a level dependent on drug dose by 4 weeks of treatment and was reversed within 4 weeks of the termination of drug treatment.[67,71] Thus, cyclophosphamide-induced postimplantation loss was associated primarily with germ cell exposure during spermiogenesis. Postmeiotic germ cells were also most susceptible to effects leading to the induction of learning abnormalities in the progeny after paternal exposure to cyclophosphamide.[72–73] Heritable translocations were found in mice after exposure of spermatids and spermatozoa to cyclophosphamide.[74]

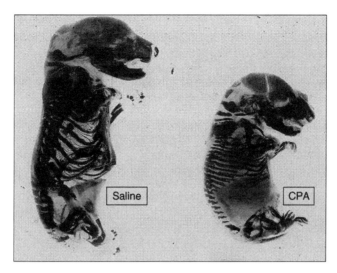

Figure 5.2 Day 20 rat fetuses stained to reveal skeletal malformations were sired by control (left) and cyclophosphamide-treated (right, 5.1 mg/kg/day for 9 weeks) males.

Interestingly, exposure of rat spermatocytes to cyclophosphamide resulted in increased preimplantation loss, as well as in synaptic failure, fragmentation of the synaptonemal complex, and altered centromeric DNA sequences.[75] An increase in malformed (hydrocephaly, edema, micrognathia) and growth retarded fetuses was observed after 7 to 9 weeks' treatment of male rats with a chronic low dose of cyclophosphamide (Figure 5.2), representing progeny sired by germ cells first exposed to cyclophosphamide as spermatogonia.[66,67] However, spermatogonia have been reported to be at low risk for the induction of heritable translocations by cyclophosphamide.[74,76]

It is noteworthy that the malformations produced by exposure of male mice to urethane or x-rays or of male rats to cyclophosphamide are very similar; these malformations include dwarfism, open eyelids, and tail anomalies.[64,67] More than one mechanism may be involved in the developmental toxicity of such drugs. This is suggested by the spectrum of effects that are produced, ranging from pre- and postimplantation effects to growth retardation, malformations, and behavioral abnormalities. It is also indicated by the specificity of the susceptibility of germ cells at different stages of development to insult with these agents.

Many laboratories have concentrated their efforts on the effects of drugs on the male germ cell during its differentiation in the testis. But drugs may also affect spermatozoa during their transit through the epididymis and vas deferens, or they may initiate drug-sperm interactions during ejaculation. As sperm transit through the epididymis, they become mature, i.e., able to fertilize an egg.[77] The administration of methyl chloride caused an increase in dominant lethal mutations as a consequence of the selective inflammatory action of this agent on the epididymis; this increase in embryo loss was reversed by administering an anti-inflammatory agent.[78,79] Treatment of male rats with cyclophosphamide resulted in an increase in postimplantation loss when the exposed spermatozoa originated in the caput or corpus, but not the cauda epididymidis.[80] The cysteine-rich protamines become progressively more tightly compacted by the formation of disulfide bonds during epididymal transit; the result of this may be that cauda epididymal spermatozoa are relatively inaccessible to drugs such as cyclophosphamide.

2. Reversibility

A major question that arises from studies on the consequences of exposure to male-mediated developmental toxicants is the degree to which these effects are reversible. To determine whether the effects of chronic exposure of male rats to cyclophosphamide were reversible, investigators treated adult male rats with saline or cyclophosphamide for 9 weeks and then mated them at various intervals posttreatment.[71] The high level of postimplantation loss observed after 9 weeks of exposure to cyclophosphamide was markedly decreased by 2 weeks post-treatment and had returned to the control range by 4 weeks posttreatment. Thus, in rats, the male-mediated developmental toxicity of cyclophosphamide was reversible. Moreover, the rate of onset for the actions of cyclophosphamide on progeny outcome was parallel to the rate of reversal.

3. Heritability

Are there adverse effect(s) of paternal drug exposure that persist to subsequent generations in surviving offspring? Significantly, after paternal cyclophosphamide treatment, increased postimplantation loss and malformations persisted to the F_2 generation; moreover, the malformations observed in the offspring of rats whose fathers were exposed to cyclophosphamide were similar to those found in the F_1.[66,67,81] Behavioral abnormalities caused by paternal cyclophosphamide treatment also persisted in subsequent generations.[72,73] Thus, this drug appears to affect spermatogonial stem cells, as well as postmeiotic germ cells, because effects are transmissible to the next generation. There is evidence that this is also true for certain other male-mediated developmental toxicants. A number of the viable congenital anomalies in the F_1 generation of urethane-treated males were expressed in the F_3 generation.[64] X-ray induced anomalies have also been shown to be heritable.[64] Thus, germ-line alterations causing malformations can be transmitted to the next generation as dominant mutations, often with reduced penetrance. No specific chromosomal aberrations were associated with these malformations.

While it is clear from animal experiments that exposure of the father to a variety of chemicals can result in a heritable alteration in his surviving "apparently normal" F_1 progeny, the limited studies done to date do not permit us to conclude whether it is valid to extrapolate between species, e.g., from rodents to humans.

III. METHODOLOGICAL APPROACHES IN MALE-MEDIATED DEVELOPMENTAL TOXICITY

A. Effects of Toxicants in Seminal Fluid

The measurement of drugs or chemicals in the seminal fluid can be accomplished fairly readily with the variety of chemical and physical techniques currently available. We know that drugs can be transmitted to the female through the semen. One of the tasks facing investigators is the assessment of the consequences to progeny of the presence of drugs or chemicals in seminal fluid. Unless they are bound to spermatozoa, drugs or chemicals in the seminal fluid that enter the female are likely to be extensively diluted in the female reproductive tract before they reach the oocyte. Drugs bound either reversibly or irreversibly to spermatozoa may have greater access to the conceptus. Vasectomy experiments are useful in determining whether any effect on progeny outcome is due to the presence of drug in the seminal fluid or the physical binding of the drug to the spermatozoon that fertilizes the egg. Females can be sequentially mated to a control (fertile) male, and then a drug-treated vasectomized male. Using this approach, it was shown that the effect of cyclophosphamide after acute administration to male rats was mediated by metabolites of the drug in the seminal fluid, rather than by drug bound to spermatozoa.[33]

B. Effects on Sperm Quantity and Characteristics

Ideally, we should like to use the male germ cell itself to evaluate the effects of toxicants with the potential to cause developmental toxicity to the progeny. Most commonly, toxicants affect male germ cell numbers, structure, motility, viability, or ability to fertilize the oocyte.[82]

The relationship between sperm numbers in the ejaculate and fertility and progeny outcome has not been studied extensively. The administration of sustained release testosterone capsules to rats reduced epididymal sperm reserves to less than 5% of control before fertility was affected, and no abnormal progeny outcome was noted with reduced sperm production.[83] While such direct studies have not been done in humans, contraceptive development studies indicate that sperm counts may need to be reduced to less than 3 million per milliliter to produce effective contraception.[84] Based on a large number of clinical studies of men seeking treatment for infertility, it would appear that in spite of highly variable counts within a given individual, there is no correlation between sperm count and the quality of progeny outcome.[85]

One of the striking characteristics of the spermatozoon is its potential for rapid progressive motion. Computer assisted semen analysis allows for the determination of various parameters of sperm movement, including forward progression rate and lateral head displacement.[86] Many drugs and environmental chemicals have been shown to modify sperm motility characteristics, while at least one male-mediated developmental toxicant, cyclophosphamide, has been reported not to affect rat sperm motility.[86–88] Sperm motility may be an early and sensitive endpoint for the assessment of the male reproductive toxicity of chemicals such as cadmium.[89] As this method becomes accepted as a standard for the quantification of the effects of drugs on sperm motility, the impact of male-mediated developmental toxicants on this parameter of sperm function will become clearer.

Sperm motility is needed not just for aiding the movement of sperm through the female reproductive tract to the egg but also to allow the spermatozoon to "drill" its way through the various layers surrounding the egg. The first event in this process is the recognition by the zona pellucida (ZP-3 receptor protein) of specific proteins on the capacitated sperm membrane.[90] The acquisition of this protein, as well as the ability of spermatozoa to undergo the acrosome reaction, can provide information about spermatozoal function. To date, this approach has not been used to assess the action of drugs on sperm function.

It is perceived of as normal in humans that as many as 50% of spermatozoa have an abnormal appearance, including such features as misshapen nuclei and abnormal arrangements of axonemal elements in the tail. One of the consequences of such structural abnormalities may be the inability to fertilize. Among those sperm that are capable of fertilization, the consequences to progeny outcome have not been well defined. However, with the introduction of intracytoplasmic sperm injection (ICSI), it has become possible to show that injection of abnormally shaped spermatozoa can result in progeny that produce spermatozoa with even greater defects.[91] The full range of potential effects of using ICSI remains to be determined, but this procedure does seem to be associated with a higher rate of abnormal progeny outcome. Additional studies to ascertain the impact of sperm with an abnormal appearance on pronuclear formation and early embryo development will help to shed light on this possibility.[92]

The profile of proteins found in the sperm membrane changes almost continually during the transit of spermatozoa through the epididymis.[93] Several of these proteins have been postulated to play key roles in the process of sperm maturation and in the development of the components that will recognize the sperm receptor on the zona pellucida.[94] Simple tools, such as immunocytochemistry using monoclonal antibodies or polyacrylamide gel electrophoresis, may be useful in identifying drug-dependent changes in sperm membranes and determining whether such changes have significant effects on sperm function and progeny outcome.

During spermiogenesis, histones are removed and replaced with protamines, which are small, highly basic, sulfhydryl-rich proteins. This transition allows for a remarkably tight packaging of chromatin in the sperm nucleus, the shape of which is species specific.[95] Interestingly, in humans the absence of protamine 2 in sperm was associated with infertility, abnormal sperm penetration rates, abnormal morphology, and decreased progressive motility, although fertilization after ICSI and early embryo development were unaffected.[96] Acrylamide is a male-mediated developmental toxicant that has been hypothesized to act, at least partially, by reacting directly with sperm nuclear protamines.[97,98] Interaction with protamine may lead to chromosome breakage, possibly during sperm chromatin remodeling after fertilization. Underprotaminated sperm bind to a fluorochrome, chromomycin A_3 (CMA_3).[99,100] Interestingly, the percentage of CMA_3 positive sperm was positively correlated with the presence of endogenous nicks in sperm DNA and with low sperm counts, high percentages of abnormal sperm, and low *in vitro* fertilization rates.[99,101] As spermatozoa progress from the testis to the cauda epididymidis, the extent to which chromatin is cross-linked by disulfide bonds increases remarkably.[102] The ability of drugs to interfere with the disulfide cross linkage of chromatin may correlate with the loss of the ability of spermatozoa to produce normal offspring.

C. Effects on Sperm Quality

To date, studies of male reproductive toxicity and male-mediated developmental toxicity in animal models have focused generally on "gross" abnormalities, such as pregnancy loss, growth retardation, or malformations, and effects on germ cell numbers, motility, morphology, or DNA. Some of the more subtle measures, such as changes in chromatin packaging, specific changes in the DNA structure, imprinting errors, altered methylation patterns, altered template function, or transcription in the early embryo may affect progeny outcome and may serve as biomarkers for the assessment of sperm quality or even toxicant exposure. The recent establishment of primordial germ cell–derived permanent female and male murine embryonic cell lines presents an interesting avenue to explore the establishment of an *in vitro* method to assess germ cell damage and quality.[103] Nevertheless, the use of an "isolated germ cell" model system to assess male-mediated developmental toxicity would presuppose that the male germ cell is the exclusive target. It assumes that the surrounding cellular organization and environment (germ cells at different stages of differentiation, and Sertoli, myoid, and Leydig cells) and the role of the oocyte are unimportant in translating any male germ cell "hit" into developmental toxicity.

1. Sperm Chromatin Packaging and Function

Exposure to a toxicant may affect sperm chromatin packaging and function; such an effect may be detected as a change in gene expression profile, in the accessibility of chromatin to specific dyes, in the ability to decondense, or in pronuclear formation or template function. Interestingly, cyclophosphamide exposure affected male germ cell gene expression in a cell-, stage-, and treatment-specific manner. Exposure to a single high dose most affected round spermatids, especially their expression of heat shock proteins, their cochaperones, and genes associated with DNA repair. Chronic low dose treatment resulted in an overall decrease in gene expression in both pachytene spermatocytes and round spermatids (Figure 5.3).[104,105]

The ability of the sperm nucleus to decondense and serve as a template is another parameter that could be very useful in determining the site of action of a toxicant on the germ cell. Decondensation of the chromatin of the mammalian spermatozoon takes place after fertilization, restoring the paternal genome to an active conformation.[106] *In vivo*, the complete decondensation process is characterized by reduction of the disulfide bonds of the protamines, followed by degradation of these nuclear proteins and their replacement with histones.[107,108] Spermatozoa can be decondensed *in vitro* by incubating them with a reducing agent, such as dithiothreitol, and a protease, such as proteinase K.[102]

Although there is little in the literature on the effects of most drugs or toxicants on the ability of spermatozoal nuclei to decondense, exposure to cyclophosphamide has been shown to alter *in vitro* decondensation.[102] Spermatozoa from rats treated with cyclophosphamide for 1 week showed the same decondensation pattern as did those from the control group. Conversely, while the decondensation pattern of spermatozoa from rats exposed to cyclophosphamide for 6 weeks was similar to that of control for the first 60 min, marked chromatin dispersion was noted in the next 30 min. The cell area, perimeter, curvature, length, and straight length were all significantly less than those of control spermatozoa. We speculate that other drugs may also alter the decondensation

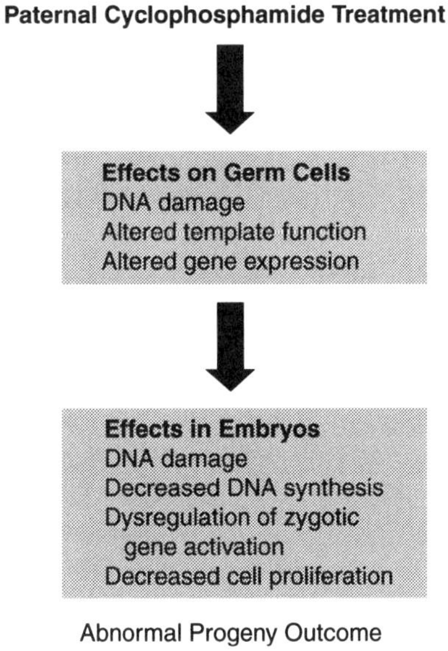

Paternal Cyclophosphamide Treatment

⬇

Effects on Germ Cells
DNA damage
Altered template function
Altered gene expression

⬇

Effects in Embryos
DNA damage
Decreased DNA synthesis
Dysregulation of zygotic
 gene activation
Decreased cell proliferation

Abnormal Progeny Outcome

Figure 5.3 Diagrammatic representation of the effects of paternal cyclophosphamide exposure on male germ cells and embryos.

pattern of spermatozoa, perhaps by affecting the cross-linking of protamines, and that this may have consequences on the ability of the male genome to be activated in the early embryo.

Spermatozoal DNA template function is a measure of sperm nuclear decondensation. By incubating rat sperm *in vitro* with cytoplasmic extracts of *Xenopus laevis* eggs and assessing chromatin decondensation and DNA synthesis, Sawyer and colleagues investigated how chemical damage affected nuclear activation.[109] Whereas exposure to a cross-linking agent blocked decondensation, treatment with a DNA base modifier, hydroxylamine, enhanced decondensation, induced gross chromatin abnormalities, and increased [^{3}H]TTP incorporation into activated sperm nuclei.[109] The availability of spermatozoal DNA for template function was not affected by 1 week of treatment with cyclophosphamide, but was markedly affected after 6 weeks of treatment with this drug.[110]

Pronucleus formation by human sperm has been studied extensively in denuded hamster eggs as an endpoint in fertility assessment. Using this approach, we demonstrated that the decondensation of spermatozoa from cyclophosphamide-treated males was more rapid than that of control spermatozoa.[111] In addition, male pronucleus formation was early in rat oocytes sired by drug-treated males. One possibility is that cyclophosphamide-induced chromatin damage prevents "normal" condensation during spermiogenesis.

A disturbance in male germ cell chromatin condensation, remodeling, and pronucleus formation may result in dysregulation of zygotic gene activation, leading to adverse effects on embryonic development (Figure 5.3). In fact, transcription is turned on earlier in the male pronucleus than in the female pronucleus.[112] The assessment of RNA synthesis in embryos sired by control and cyclophosphamide-treated males revealed that total RNA synthesis ([^{32}P]UTP incorporation) was constant in one to eight cell embryos sired by drug-treated fathers, while in control embryos RNA synthesis increased fourfold, to peak at the four-cell stage.[111] Moreover, BrUTP incorporation into RNA and Sp1 transcription factor immunostaining were increased and spread over both the cytoplasmic and nuclear compartments in two-cell embryos sired by cyclophosphamide-treated males.[111] In contrast, both BrUTP incorporation and Sp1 immunostaining were in the nuclear compartment only in embryos fertilized by control spermatozoa.

The profile of expression of specific genes was altered in embryos sired by drug-treated males, even as early as the one-cell stage.[111,113,114] In the one- and two-cell stage embryo, the relative abundance of transcripts for candidate DNA repair, imprinted, growth factor, and cell adhesion genes was elevated significantly above control in embryos sired by cyclophosphamide-treated males; a peak in the expression of many of these genes was not observed until the eight-cell stage in control embryos.[111,113,114] Thus, paternal drug exposure temporally and spatially dysregulated rat zygotic gene activation, altering the developmental clock.

2. Sperm Genetic Integrity

Genetic damage of the male germ cell is of major concern; such damage may be transmitted to the offspring and lead to abnormal progeny outcome, not only in the F_1 generation but also in subsequent generations. Even fairly severely DNA-damaged sperm may be capable of fertilization. Moreover, some germ line alterations appear to cause phenotypic malformations. Measurements of genetic damage to the male germ cell include dominant lethal and specific locus mutation tests as well as cytogenetic, fluorescent *in situ* hybridization (FISH), and polymerase chain reaction (PCR) based methods for the detection of DNA damage or mutations.

Nonmammalian test systems have been used extensively in screening banks of drugs and chemicals for mutagenicity. These tests have been invaluable in selecting chemicals for further *in vivo* genetic mutagenicity testing. One example of such a system involves the Japanese medaka. Use of arbitrarily primed PCR and fingerprinting in fish treated with γ-irradiation allows changes in the genomic DNA of individual progeny to be detected as bands lost or gained.[115] By taking advantage of suitable reporter genes, one can use male germ cells of *Drosophila* to detect a spectrum of genetic damage, ranging from recessive lethal (or visible) mutations, deletions, reciprocal translocations, chromosome loss, or dominant lethal mutations to aneuploidy.[116]

A number of *in vivo* animal tests rely heavily on dominant characteristics in inbred mice. Dominant lethal and specific locus mutation tests are examples of *in vivo* animal tests that have been used extensively to identify the chemicals that are capable of mutating germ cells.[65,117–119] In these tests, male rodents are treated either acutely or chronically with the chemical to be investigated and mated to control females. For the dominant lethal test, the outcome measured is embryolethality, usually postimplantation. Thus, the spermatozoa are capable of fertilizing an egg, but the conceptus fails to develop normally, either at the time of implantation or shortly thereafter. The specific locus mutation test uses mice with mutations in a number of loci coding for "visible" features, such as coat color, and evaluates the ability of the chemical in question to cause a mutation in the male germ cell at these loci. This approach has been very valuable in detecting a number of mammalian germ cell mutagens. Both of these approaches use progeny outcome to measure the effects of the drug on the male germ cell. More recently, transgenic mice have been used as a model with which to study gene mutations during different phases of spermatogenesis.[120,121] The results of specific-locus mutation studies have suggested that exposure of spermatogonia to chemicals or radiation yields few large lesions, while large lesions are common after exposure of postspermatogonial germ cells.

At the chromosome level, it may be possible to detect bulky deletions, aneuploidy, or chromosomal duplications by use of cytogenetic approaches.[122] The inherent difficulty in cytogenetic analysis of spermatozoa has been the lack of mitosis and hence of chromosomal structure. It was only by fertilizing denuded hamster eggs with the sperm in question that chromosomal structures in the male pronucleus could be analyzed. This approach was used to identify the effects of age, x-irradiation, and drugs on the chromosomal banding pattern of human sperm.[123] More recently, FISH was developed for the analysis of aneuploidy in the sperm genome.[124] This method involves the staining of specific chromosomal regions with fluorescent-labeled complementary DNA sequences, allowing the identification of sperm with chromosomal abnormalities, such as trisomies or aneuploidies. FISH has been used extensively to study human sperm and was adapted for the rat by Wyrobek's group to apply this powerful tool to toxicology.[125]

Other tests that have detected DNA damage in male germ cells include the unscheduled DNA synthesis assay, the DNA alkaline elution assay, the single cell gel electrophoresis (SCGE or comet) assay, and the sperm chromatin structure assay (SCSA™). In the unscheduled DNA synthesis assay, the repair of chemically induced DNA damage is assessed in postmitotic male germ cells by measuring the incorporation of radiolabeled thymidine into their DNA.[126] Because of their active DNA replication, unscheduled DNA synthesis cannot be determined in spermatogonia or preleptotene primary spermatocytes. Unscheduled DNA synthesis does not occur in later stage spermatids or in spermatozoa because of the loss of the enzymes involved in DNA repair; both of these phases of spermatogenesis are thus susceptible to damage leading to developmental toxicity. Therefore, this test is selectively useful for assessing DNA repair in late spermatocytes and early spermatids.

In the alkaline elution assay, germ cells that have been exposed to a test drug or chemical are lysed and decondensed on a filter prior to the addition of a highly alkaline buffer.[110,127,128] Under alkaline conditions, the DNA unwinds and is eluted through the filter at a rate reflecting the extent to which the test chemical has resulted in DNA single-stand breaks or cross-links. One week of treatment with cyclophosphamide caused DNA single-strand breaks that could be detected only in the presence of proteinase K in the lysis solution, but no DNA cross-links were observed.[110] In contrast, 6 weeks of treatment with cyclophosphamide induced a significant increase in both DNA single-strand breaks and cross-links in spermatozoal nuclei; the cross-links were due primarily to DNA–DNA linkages.[110] In general, there was a close correlation between the DNA damage responses of human and rat testicular cells as assessed with alkaline elution.[129] The detection and localization of drug adducts within the genome of the male germ cell would add additional information about the target of such DNA damaging agents.

Although alkaline elution provides a powerful test of the interaction of a drug with chromatin in a large population of cells, it cannot be used on an individual cell basis. In the comet assay, individual cells, lysed and decondensed, are electrophoresed in agar after treatment with alkali to assess double-strand breaks or under neutral conditions to identify single-strand breaks.[130] The electric field causes the migration of fragments of DNA. The smaller the fragment, the greater is the migration or "comet." The DNA can be visualized with a fluorescent dye, and the relative amount of DNA that migrated, as well as the distance it migrated, provides quantitative data on DNA integrity after drug exposure. There is a highly significant correlation between DNA fragmentation as detected by the comet assay in ejaculated spermatozoa and infertility.[130] When the alkaline comet assay was used to assess DNA damage in one-cell embryos sired by cyclophosphamide-treated males, a significantly higher percentage (68%) of the embryos fertilized by drug-exposed spermatozoa displayed a comet indicative of DNA damage compared with embryos sired by control males (18%).[114]

The sperm chromatin structure assay is an indirect indicator of DNA damage because it measures the amount of single-stranded DNA after treatments that normally do not denature sperm DNA (heat or acid pH).[131,132] The test employs the unique metachromatic and equilibrium staining properties of acridine orange, a dye that fluoresces green when intercalated into double-stranded "native" DNA and red when bound to single-stranded "denatured" DNA (or RNA). Alternate approaches to assess sperm genetic integrity include the terminal deoxynucleotidyl transferase (TDT)-mediated d UTP nick-end labeling (TUNEL) and *in situ* nick translation (NT) assays.[133] In these assays, sperm are labeled with fluorescently tagged DNA precursors at sites of single- and/or double-stranded breaks in DNA. Endogenous DNA breaks in human spermatozoa have been demonstrated with the NT assay.[99]

There is evidence in most genetic mutation test systems for "hot spots" or loci that are more susceptible to mutations. This specificity has also been observed for the specific locus mutation test.[117] The importance of "specific" genes or chromosomes as targets in mediating the adverse effects of chemicals on male germ cells is not known. Dubrova and colleagues have suggested the use of hypervariable tandem repeat loci to evaluate germ line mutation induction in mice and humans.[134,135] These loci are capable of detecting changes in mutation rates in samples from relatively small populations because they have a very high spontaneous mutation rate in both humans and mice.[136–139]

The position of specific genes on DNA loops attached to the nuclear matrix is constant.[95] Hence, we can speculate that specific gene loci may be targeted even by "nonspecific" exposures. Any selectivity of the effect of male-mediated developmental toxicants on sperm genetic integrity may be based on the DNA sequence and/or conformation in chromatin, as well as on epigenetic modifications.

3. Epigenetic Changes

There is evidence that the male and female genomes are not equivalent during development.[140,141] The presence of both the maternal and paternal genome is essential for normal development in the rodent. Surani and his co-workers have suggested that chromosomes are imprinted in a germ-line specific manner during gametogenesis.[142,143] Thus, the absence of the male genome in a zygote leads to embryos with poor development of the extraembryonic tissues, while the absence of the maternal genome results in embryos having markedly reduced embryonic tissues. The exposure of male germ cells to a developmental toxicant could be embryolethal if the toxicant targeted genes that are paternally imprinted. A number of studies have implicated the methylation status of various regions of the gene in imprinting.[144] Maternally imprinted genes, such as *Snrpn*, *Mest*, and *Peg3*, are unmethylated in sperm and 100% methylated in mature oocytes, whereas the 5′ region of *H19*, a paternally imprinted gene, is completely methylated in sperm and unmethylated in oocytes.[145] The treatment of male rats with 5-azacytidine, a drug that blocks DNA methylation, resulted in an increase in preimplantation loss when germ cells were first exposed as spermatogonia or spermatocytes.[146,147]

The male genome is required for normal development of the trophectoderm. In embryos sired by cyclophosphamide-treated male rats, cell death occurred selectively in those tissues derived from the inner cell mass, while the trophoblast-derived trophectoderm cells appeared morphologically normal.[148] Thus, exposure of the male rat to cyclophosphamide may affect paternal genes essential for the development of inner cell mass–derived tissues in the embryo, sparing those genes required for normal trophectoderm development. This is not what would have been predicted from the nuclear transplantation experiments cited above. However, this is consistent with the tissue specificity of the effects of radiation on early embryos.[149] Transgenic mouse experiments have shown that inner cell mass–derived tissues are eliminated in mice deficient in fibroblast growth factor-4.[150] Thus, inner cell mass cells have different growth factor requirements than the trophectoderm. Moreover, the male genome is essential for the development of the inner cell mass as well as trophectoderm-derived cells.

IV. SUMMARY AND CONCLUSIONS

Over the past few decades it has become clear that drugs given to the father may affect his progeny's outcome. The range of effects that can occur encompasses infertility and reduced fertility, as well as malformations, growth retardation, and behavioral alterations in the progeny. Furthermore, it is apparent that the germ cell line of the progeny may be affected. Technological advances in the methods available to study changes in DNA at the molecular level help to elucidate the molecular mechanisms mediating these consequences of paternal drug exposure for the progeny. Although few of these observations have been extended to the human, it seems essential that carefully designed clinical and epidemiologic studies be undertaken to establish the extent to which such effects may contribute to infertility and/or abnormal progeny outcome in humans.

ACKNOWLEDGMENTS

The studies from our laboratories were done with the support of grants from the Canadian Institutes of Health Research.

References

1. Robaire, B. and Hales, B.F., Eds., *Advances in Male-Mediated Developmental Toxicity. Advances in Experimental Medicine and Biology, Vol. 518*, Kluwer Academic/Plenum Press, New York, 2003.
2. Shepard, T.H., Ed., *Catalog of Teratogenic Agents*, 7th ed., Johns Hopkins University Press, Baltimore, MD, 1992.
3. Sawyer, D.E. and Aitken, R.J., Male mediated developmental defects and childhood disease, *Reprod. Med. Rev.*, 8, 107, 2001.
4. McDonald, A.D., McDonald, J.C., Armstrong, B., Cherry, N.M., Nolin, A.D., and Robert, D., Fathers' occupation and pregnancy outcome, *Br. J. Ind. Med.*, 46, 329, 1989.
5. Brender, J.D. and Suarez, L., Paternal occupation and anencephaly, *Am. J. Epidemiol.*, 131, 517, 1990.
6. Olshan, A.F., Teschke, K., and Baird, P.A., Birth defects among offspring of firemen, *Am. J. Epidemiol.*, 131, 312, 1990.
7. Olshan, A.F., Teschke, K., and Baird, P.A., Paternal occupation and congenital anomalies, *Am. J. Ind. Med.*, 20, 447, 1991.
8. Savitz, D.A., Whelan, E.A., and Kleckner, R.C., Effect of parents' occupational exposures on risk of stillbirth, preterm delivery, and small-for-gestational-age infants, *Am. J. Epidemiol.* 129, 1201, 1989.
9. Savitz, D.A., Sonnenfeld, N.L., and Olshan, A.F., Review of epidemiologic studies of paternal occupational exposure and spontaneous abortion, *Am. J. Ind. Med.*, 25, 361, 1994.
10. Friedman J.M., Does Agent Orange cause birth defects? *Teratology*, 29, 193, 1984.
11. Field, B. and Kerr, C., Reproductive behaviour and consistent patterns of abnormality in offspring of Vietnam veterans, *J. Med. Genet.*, 25, 819, 1988.
12. Kaye, C.I., Rao, S., Simpson, S.J., Rosenthal, F.S., and Cohen, M.M., Evaluation of the chromosome damage in males exposed to Agent Orange and their families, *J. Craniofac. Genet. Dev. Biol. Suppl.*, 1, 259, 1985.
13. Stellman, S.D., Stellman, J.M., and Sommer, J.F. Jr., Health and reproductive outcomes among American legionnaires in relation to combat and herbicide exposure in Vietnam, *Environ. Res.*, 47, 150, 1988.
14. Arbuckle, T.E., Lin, Z., and Merv, L.S., An exploratory analysis of the effect of pesticide exposure on the risk of spontaneous abortion in an Ontario farm population. *Environ. Health Perspect.*, 109, 851, 2001.
15. Bonde, J.P., Joffe, M., Apostoli, P., Dale, A., Kiss, P., Spano, M., Caruso, F., Giwercman, A., Bisanti, L., Porru, S., Vanhoorne, M., Comhaire, F., and Zschiesche, W., Sperm count and chromatin structure in men exposed to inorganic lead: lowest adverse effect levels, *Occup. Environ. Med.*, 59, 234, 2002.
16. Little, J. and Vainio, H., Mutagenic lifestyles? A review of evidence of associations between germ-cell mutations in humans and smoking, alcohol consumption and use of 'recreational' drugs, *Mutat. Res.*, 313, 131, 1994.
17. Savitz, D.A., Schwingl, P.J., and Keels, M.A., Influence of paternal age, smoking, and alcohol consumption on congenital anomalies, *Teratology*, 44, 429, 1991.
18. Davis, D.L., Paternal smoking and fetal health, *Lancet*, i, 123, 1991.
19. Robbins, W.A., FISH (fluorescence in situ hybridization) to detect effects of smoking, caffeine, and alcohol on human sperm chromosomes, in *Advances in Male-Mediated Developmental Toxicity. Advances in Experimental Medicine and Biology, Vol. 518*, Robaire, B. and Hales, B.F., Eds., Kluwer Academic/Plenum Press, New York, 2003, chap 6.
20. Wong, W.Y., Thomas, C.M.G., Merkus, H.M.W.M., Zielhuis, G.A, Doesburg, W.H., and Steegers-Theunissen, R.P.M., Cigarette smoking and the risk of male factor subfertility: minor association between cotinine in seminal plasma and semen morphology, *Fertil. Steril.*, 74, 930, 2000.
21. Vogt, H.J., Heller, W.D., and Borelli, S., Sperm quality of healthy smokers, ex-smokers, and never-smokers, *Fertil. Steril.*, 45, 106, 1986.
22. Zavos, P.M., Correa, J.R., Karagounis, C.S., Ahparaki, A., Phoroglou, C., Hicks, C.L., and Zarmakoupis-Zavos, P.N., An electron microscope study of the axonemal ultrastructure in human spermatozoa from male smokers and nonsmokers, *Fertil. Steril.*, 69, 430, 1998.
23. Zenzes, M.T., Bielecki, R., and Reed, T.E., Detection of benzo(*a*)pyrene diol epoxide-DNA adducts in sperm of men exposed to cigarette smoke, *Fertil. Steril.*, 72, 330, 1999.

24. Dellarco, V.L., Mavournin, K.H., and Waters, M.D., Aneuploidy data review committee: Summary compilation of chemical data base and evaluation of test methodology, *Mutat. Res.*, 167, 149, 1986.
25. Senturia, Y.D., Peckham C.S., and Peckham, M.J., Children fathered by men treated for testicular cancer, *Lancet*, ii, 766, 1985.
26. Fried, P., Steinfeld, R., Casileth, B., and Steinfeld, A., Incidence of developmental handicaps among the offspring of men treated for testicular seminoma, *Int. J. Androl.*, 10, 385, 1987.
27. Mulvihill, H.J., Reproductive outcomes among men treated for cancer, in *Male-Mediated Developmental Toxicity*, Olshan, A.F. and Mattison, D.R., Eds., Plenum Press, New York, 1994, p. 197.
28. Olshan, A.F., Breslow, N.E., Daling, J.R., Falleta, J.M., Grufferman, S., Robison, L. L., et al. Wilm's tumor and paternal occupation, *Cancer Res*, 50, 3212, 1990.
29. Eliasson, R. and Dornbusch, K., Secretion of metronidazole into the human semen, *Int. J. Androl.*, 3, 236, 1980.
30. Malmborg, A.S., Antimicrobial drugs in human seminal plasma, *J. Antimicrobe Chemother.*, 6, 483, 1978.
31. Mann, T. and Lutwak-Mann, C., Passage of chemicals into human and animal semen: mechanisms and significance, *CRC Crit. Rev. Toxicol.*, 11, 1, 1982.
32. Forrest, J.B., Turner, T.T., and Howards, S.S., Cyclophosphamide, vincristine and the blood testis barrier, *Invest. Urol.*, 18, 443, 1981.
33. Robaire, B. and Hales, B.F., Post-testicular mechanisms of male-mediated developmental toxicity, in *Male-Mediated Developmental Toxicity*, Mattison, D.R. and Olshan, A.F., Eds., Plenum Press, New York, 1994, p. 93.
34. Soyka, L.F., Peterson, J.M., and Joffe, J.M., Lethal and sublethal effects on the progeny of male rats treated with methadone, *Toxicol. Appl. Pharmacol.*, 45, 797, 1978.
35. Lutwak-Mann, C., Observations on the progeny of thalidomide-treated male rabbits, *Br. Med. J.*, 1, 1090, 1964.
36. Hales, B.F., Smith, S., and Robaire, B., Cyclophosphamide in the seminal fluid of treated males: Transmission to females by mating and effects on progeny outcome, *Toxicol. Appl. Pharmacol.*, 84, 423, 1986.
37. Hales, B.F. and Robaire, B., The male-mediated developmental toxicity of cyclophosphamide, in *Male-Mediated Developmental Toxicity*, Mattison, D.R. and Olshan, A.F., Eds., Plenum Press, New York, 1994, p. 105.
38. Ratcliffe, J.M., Paternal exposures and embryonic or fetal loss: the toxicologic and epidemiologic evidence, in *Male-Mediated Developmental Toxicity*, Mattison, D.R. and Olshan, A.F., Eds., Plenum Press, New York, 1994, p. 185.
39. Winder, C., Reproductive and chromosomal effects of occupational exposure to lead in the male, *Reprod. Toxicol.*, 3, 221, 1989.
40. Alexander, B.H., Checkoway, H., van Netten, C., Kaufman, J.D., Vaughan, T.L., Mueller, B.A., and Faustman, E.M., Paternal occupational lead exposure and pregnancy outcome. *Int. J. Occup. Environ. Health*, 2, 280, 1996.
41. Foster, W.G., McMahon, A., and Rice, D.C., Sperm chromatin structure is altered in cynomolgus monkeys with environmentally relevant blood lead levels, *Toxicol. Ind. Health*, 12, 723, 1996.
42. Johansson, L. and Wide, M., Long-term exposure of the male mouse to lead: effects on fertility, *Environ. Res.*, 41, 481, 1986.
43. Stowe, V.M. and Goyer, R., Reproductive ability and progeny of F1 lead-toxic rats. *Fertil. Steril.*, 22, 755, 1971.
44. Brady, K., Herrera, Y., and Zenick, H., Influence of paternal lead exposure on subsequent learning ability of offspring, *Pharmacol. Biochem. Behav.*, 3, 561, 1975.
45. Slutsky, M., Levin, J.L., and Levy, B.S., Azoospermia and oligospermia among a large cohort of DBCP applicators in 12 countries, *Int. J. Occup. Environ. Health*, 5, 116, 1999.
46. Leone, M., Costa, M., Capitanio, G.L., Palmero, S., Prati, M., and Leone, M.M., Dibromochloropropane (DBCP) effects on the reproductive function of the adult male rat, *Acta Eur. Fertil.*,19, 99,1988.
47. Whorton, M.D. and Foliart, D.E., Mutagenicity, carcinogenicity and reproductive effects of dibromochloropropane (DBCP), *Mutat. Res.*,123, 13, 1983.
48. Thornton, S.R., Schroeder, R.E., Robison, R.L., Rodwell, D.E., Penney, D.A., Nitschke, K.D., and Sherman, W.K., Embryo-fetal developmental and reproductive toxicology of vinyl chloride in rats, *Toxicol. Sci.*, 68, 207, 2002.

49. Anderson, D., Genetic and reproductive toxicity of butadiene and isoprene, *Chem. Biol. Interact.*, 135–136, 65, 2001.
50. Pacchierotti, F., Adler, I.D., Anderson, D., Brinkworth, M., Demopoulos, N.A., Lahdetie, J., Osterman-Golkar, S., Peltonen, K., Russo, A., Tates, A., and Waters, R., Genetic effects of 1,3-butadiene and associated risk for heritable damage, *Mutat. Res.*, 397, 93, 1998.
51. Shukla, Y. and Taneja, P., Mutagenic potential of cypermethrin in mouse dominant lethal assay, *Environ. Pathol. Toxicol. Oncol.*, 21, 259, 2002.
52. Jha, A.M. and Bharti, M.K., Mutagenic profiles of carbazole in the male germ cells of Swiss albino mice, *Mutat. Res.*, 500, 97, 2002.
53. Tyla, R.W., Friedman, M.A., Losco, P.E., Fisher, L.C., Johnson, K.A., Strother, D.E., and Wolf, C.H., Rat two-generation reproduction and dominant lethal study of acrylamide in drinking water, *Reprod. Toxicol.*, 14, 385, 2000.
54. Cai, L., Hales, B.F., and Robaire, B., Induction of apoptosis in the germ cells of adult male rats after exposure to cyclophosphamide, *Biol. Reprod.*, 56, 1490, 1997.
55. Adler, I.D., Kliesch, U., Jentsch, I., and Speicher, M.R., Induction of chromosomal aberrations by dacarbazine in somatic and germinal cells of mice, *Mutagenesis*, 17, 383, 2002.
56. Witt, K.L. and Bishop, J.B., Mutagenicity of anticancer drugs in mammalian germ cells, *Mutat. Res.*, 355, 209, 1996.
57. Hooser, S.B., van Dijk-Knijnenburg, W.C., Waalkens-Berendsen, I.D., Smits-van Prooije, A.E., Snoeij, N.J., Baan, R.A., and Fichtinger-Schepman, A.M., Cisplatin-DNA adduct formation in rat spermatozoa and its effect on fetal development, *Cancer Lett.*, 151, 71, 2000.
58. Seethalakshmi, L., Flores, C., Kinkead, T., Carboni, A.A., Malhotra, R.K., and Menon, M., Effects of subchronic treatment with cis-platinum on testicular function, fertility, pregnancy outcome, and progeny. *J. Androl.*, 13, 65, 1992.
59. Russell, L.B. and Rinchik, E.M., Structural differences between specific-locus mutations induced by different exposure regimes in mouse spermatogonial stem cells, *Mutat. Res.*, 288, 187, 1993.
60. Nagao, T. and Fukikawa, K., Genotoxic potency in mouse spermatogonial stem cells of triethylene-melamine, mitomycin C, ethylnitrosourea, procarbazine, and propyl methanesulfonate as measured by F_1 congenital defects, *Mutat. Res.*, 229, 123, 1990.
61. Kato, M., Makino, S., Kimura, H., Ota, T., Furuhashi, T., and Nagamura, Y., Sperm motion analysis in rats treated with adriamycin and its applicability to male reproductive toxicity studies, *J. Toxicol. Sci.*, 26, 51, 2001.
62. Russell, L.B., Hunsicker, P.R., Kerley, M.K., Johnson, D.K., and Shelby, M.D., Bleomycin, unlike other male-mouse mutagens, is most effective in spermatogonia, inducing primarily deletions, *Mutat. Res.*, 469, 95, 2000.
63. Parvinen, L.M., Soderstrom, K.O., and Parvinen, M., Early effects of vinblastine and vincristine on the rat spermatogenesis: analyses by a new transillumination-phase contrast microscopic method, *Exp. Pathol. (Jena)*, 15, 85, 1978.
64. Nomura, T., Male-mediated teratogenesis: ionizing radiation/ethylnitrosourea studies, in *Male-Mediated Developmental Toxicity*, Mattison, D.R. and Olshan, A.F., Eds., Plenum Press, New York, 1994, p. 117.
65. Russell, L.B., Effects of spermatogenic cell type on quantity and quality of mutations, in *Male-Mediated Developmental Toxicity*, Mattison, D.R. and Olshan, A.F., Eds., Plenum Press, New York, 1994, p. 37.
66. Trasler, J.M., Hales, B.F., and Robaire, B., Paternal cyclophosphamide treatment of rats causes fetal loss and malformations without affecting male fertility, *Nature*, 316, 144, 1985.
67. Trasler, J.M., Hales, B.F., and Robaire, B., Chronic low dose cyclophosphamide treatment of adult male rats: effect on fertility, pregnancy outcome and progeny, *Biol. Reprod.*, 34, 275, 1986.
68. Trasler, J.M., Hales, B.F., and Robaire, B., A time course study of chronic paternal cyclophosphamide treatment in rats: effects on pregnancy outcome and the male reproductive and hematologic systems, *Biol. Reprod.*, 37, 317, 1987.
69. Qiu, J., Hales, B.F., and Robaire, B., Adverse effects of cyclophosphamide on progeny outcome can be mediated through post-testicular mechanisms in the rat, *Biol. Reprod.*, 46, 926, 1992.
70. Anderson, D., Bishop, J.B., Garner, R.C., Ostrosky-Wegman, P., and Selby, P.B., Cyclophosphamide: review of its mutagenicity for an assessment of potential germ cell risks, *Mutat. Res.*, 330, 115, 1995.

71. Hales, B.F. and Robaire, B., Reversibility of the effects of chronic paternal exposure to cyclophosphamide on pregnancy outcome in rats, *Mutat. Res.*, 229, 129, 1990.
72. Fabricant, J.D., Legator, M.S., and Adams, P.M., Post-meiotic cell mediation of behavior in progeny of male rats treated with cyclophosphamide, *Mutat. Res.*, 119, 185, 1983.
73. Auroux, M., Dulioust, E., Selva, J., and Rince, P., Cyclophosphamide in the F_0 male rat: physical and behavioral changes in three successive adult generations, *Mutat. Res.*, 229, 189, 1990.
74. Generoso, W.M., Cattanach, B., and Malashenko, A.M., Mutagenicity of selected chemicals in mammals; the heritable translocation test, in *Comparative Chemical Mutagenesis*, de Serres, F.J. and Shelby, M.D., Eds., Plenum Press, New York, 1981, p. 681.
75. Backer, L.C., Gibson, M.J., Moses, M.J., and Allen, J.W., Synaptonemal complex damage in relation to meiotic chromosome aberration after exposure of male mice to cyclophosphamide, *Mutat. Res.*, 203, 317, 1988.
76. Sotomayor, R.E. and Cumming, R.B., Induction of translocations by cyclophosphamide in different germ cell stages of male mice: cytological characterization and transmission, *Mutat. Res.*, 27, 375, 1975.
77. Orgebin-Crist, M-C. and Olsen, G.E., Epididymal sperm maturation, in *The Male in Farm Animal Reproduction*, Courot, M., Ed., Martinus Nijhoff, Amsterdam, 1984, p. 80.
78. Chellman, G.J., Bus, J.S., and Working, P.K., Role of epididymal inflammation in the induction of dominant lethal mutations in Fischer 344 rat sperm by methyl chloride, *Proc. Natl. Acad. Sci. USA*, 83, 8087, 1986.
79. Chellman, G.J., Morgan, K.T., Bus, J.S., and Working, P.G., Inhibition of methyl chloride toxicity in male F-344 rats by the anti-inflammatory agent BW755C, *Toxicol. Appl. Pharmacol.*, 85, 367, 1986.
80. Qiu, J., Hales, B.F., and Robaire, B., Adverse effects of cyclophosphamide on progeny outcome can be mediated through post-testicular mechanisms in the rat, *Biol. Reprod.*, 46, 926, 1992.
81. Hales, B.F., Crosman, K., and Robaire, B., Increased postimplantation loss and malformations among the F_2 progeny of male rats chronically treated with cyclophosphamide, *Teratology*, 45, 671, 1992.
82. Wyrobek, A.J., Methods and concepts in detecting abnormal reproductive outcomes of paternal origin, *Reproductive Toxicol.*, 7, 3, 1993.
83. Robaire, B., Smith, S., and Hales, B., Suppression of spermatogenesis by testosterone in adult male rats: effect on fertility, progeny outcome and pregnancy, *Biol. Reprod.*, 31, 221, 1984.
84. Handelsman, D.J., Farley, T.M., Pregoudor, A., and Waites, G.M., Factors in non-uniform induction of azoospermia by testosterone enanthate in normal men. World Health Organization Task Force for Methods for the Regulation of Male Fertility. *Fertil. Steril.*, 63, 125, 1995.
85. Robaire, B., Pryor, J., and Trasler, J.M., Eds., *Handbook of Andrology*, American Society of Andrology, Allen Press, Inc., Lawrence, KS, 1995.
86. Slott, V.L., Suarez, J.D., Poss, P.M., Linder, R.E., Strader, L.F., and Perreault, S.D., Optimization of the Hamilton-Thorn computerized sperm motility analysis system for use with rat spermatozoa in toxicological studies, *Fund. Appl. Toxicol.*, 21, 298, 1993.
87. Linder, R.E., Strader, L.F., Slott, V.L., and Suarez, J.D., Endpoints of spermatotoxicity in the rat after short duration exposures to fourteen reproductive toxicants, *Reprod. Toxicol.*, 6, 491, 1992.
88. Higuchi, H., Nakaoka, M., Kawamura, S., Kamita, Y., Kohda, A., and Seki, T., Application of computer-assisted sperm analysis system to elucidate lack of effects of cyclophosphamide on rat epididymal sperm motion, *J. Toxicol. Sci.*, 26, 75, 2001.
89. Xu, L.C., Wang, S.Y., Yang, X.F., and Wand, X.R., Effects of cadmium on rat sperm motility evaluated with computer assisted sperm analysis, *Biomed. Environ. Sci.*, 14, 312, 2001.
90. Bleil, J.D. and Wasserman, P.W., Identification of a ZP-3 binding protein on acrosome-intact mouse sperm by photoaffinity crosslinking, *Proc. Natl. Acad. Sci. USA*, 87, 5563, 1990.
91. Akutsu, H., Tres, L.L., Tateno, H., Yanagimachi, R., and Kierszenbaum, A.L., Offspring from normal mouse oocytes injected with sperm heads from the *azh/azh* mouse display more severe sperm tail abnormalities than the original mutant, *Biol. Reprod.*, 64, 249, 2001.
92. Hansen, M., Kurinczuk, J.J., Bower, C., and Webb, S., The risk of major birth defects after intracytoplasmic sperm injection and in vitro fertilization, *N. Engl. J. Med.*, 346, 725, 2002.
93. Robaire, B. and Hermo, L., Efferent ducts, epididymis, and vas deferens: structure, functions and their regulation, in *The Physiology of Reproduction*, Vol. 1, Knobil, E. and Neill, J.D., Eds. Raven Press, New York, 1988, p. 999.

94. Robaire, B. and Hinton, B.T., Eds. *The Epididymis: From Molecules to Clinical Practice — A Comprehensive Survey of the Efferent Ducts, the Epididymis and the Vas Deferens.* Kluwer Academic/Plenum Publishers, New York, 2002.
95. Ward, W.S. and Coffey, D.S., Specific organization of genes in relation to the sperm nuclear matrix, *Biochem. Biophys. Res. Commun.*, 173, 20, 1990.
96. Carrell, D.T. and Liu, L., Altered protamine 2 expression is uncommon in donors of known fertility, but common among men with poor fertilizing capacity, and may reflect other abnormalities of spermiogenesis, *J. Androl.*, 22, 604, 2001.
97. Sega, G.A., Alcota, R.P.V., Tancongco, C.P., and Brimer, P.A., Acrylamide binding to the DNA and protamine of spermiogenic stages in the mouse and its relationship to genetic damage, *Mutat. Res.*, 216, 221, 1986.
98. Dearfield, K.L., Douglas, G.R., Ehling U.H., Moore, M.M., Sega, G.A., and Brusick, D.J., Acrylamide: a review of its genotoxicity and an assessment of heritable genetic risk, *Mutat. Res.*, 330, 71, 1995.
99. Manicardi, G.C., Bianchi, P.G., Pantano, S., Azzoni, P., Bizzaro, D., Bianchi, U., and Sakkas, D., Presence of endogenous nicks in DNA of ejaculated human spermatozoa and its relationship to chromomycin A3 accessibility, *Biol. Reprod.*, 52, 864, 1995.
100. Bizzaro, D., Manicardi, G.C., Bianchi, P.G., Bianchi, U., Mariethoz, E., and Sakkas, D., In-situ competition between protamine and fluorochromes for sperm DNA, *Mol. Hum. Reprod.*, 4, 127, 1998.
101. Lolis, D., Georgiou, I., Syrrou, M., Zikopoulos, K., Konstantelli, M., and Messinis, I., Chromomycin A3-staining as an indicator of protamine deficiency and fertilization, *Int. J. Androl.*, 19, 23, 1996
102. Qiu, J.P., Hales, B.F., and Robaire, B., Effects of chronic low dose cyclophosphamide exposure on the nuclei of rat spermatozoa, *Biol. Reprod.*, 52, 33, 1995.
103. Klemm, M., Genschow, E., Pohl, I., Barrabas, C., Liebsch, M., and Spielmann, H., Permanent embryonic germ cell lines of BALB/cJ mice — an *in vitro* alternative for *in vivo* germ cell mutagenicity tests, *Toxicol. In Vitro*, 15, 447, 2001.
104. Aguilar-Mahecha, A., Hales, B.F., and Robaire, B., Acute cyclophosphamide exposure has germ cell specific effects on the expression of stress response genes during rat spermatogenesis, *Mol. Repro. Dev.*, 60, 302, 2001.
105. Aguilar-Mahecha, A., Hales, B.F., and Robaire, B., Chronic cyclophosphamide treatment alters the expression of stress response genes in rat male germ cells, *Biol. Reprod.*, 66, 1024, 2002.
106. Naish, S.J., Perreault, S.D., and Zirkin, B.R., DNA synthesis in the fertilizing hamster sperm nucleus: sperm template availability and egg cytoplasmic control, *Biol. Reprod.*, 36, 245, 1987.
107. Perreault, S.D., Naish, S.J., and Zirkin, B.R., The timing of hamster sperm nuclear decondensation and male pronucleus formation is related to sperm nuclear disulfide bond content, *Biol. Reprod.*, 36, 239, 1987.
108. Perreault, S.D. and Zirkin, B.R., Sperm nuclear decondensation in mammals: role of sperm-associated proteinase in vivo, *J. Exp. Zool.*, 224, 253, 1982.
109. Sawyer, D.E., Hillman, G.R., Uchida, T., and Brown, D.B., Altered nuclear activation parameters of rat sperm treated in vitro with chromatin-damaging agents, *Toxicol. Sci.*, 44, 52, 1998.
110. Qiu, J., Hales, B.F., and Robaire, B., Damage to rat spermatozoal DNA after chronic cyclophosphamide exposure, *Biol. Reprod.*, 53, 1465, 1995.
111. Harrouk, W., Khatabaksh, S., Robaire, B., and Hales, B. F., Paternal exposure to cyclophosphamide dysregulates the gene activation program in rat pre-implantation embryos, *Mol. Reprod. Dev.*, 57, 214, 2000.
112. Worrad, D.M., Ram, P.T., and Schultz, R.M., Regulation of gene expression in the mouse oocyte and early preimplantation embryo: developmental changes in Sp1 and TATA box-binding protein, TBP, *Development*, 120, 2347, 1994.
113. Harrouk, W., Robaire, B. and Hales, B.F., Paternal exposure to cyclophosphamide alters cell-cell contacts and activation of embryonic transcription in the pre-implantation rat embryo, *Biol. Reprod.*, 63, 74, 2000.
114. Harrouk, W., Codrington, A., Vinson, R., Robaire, B., and Hales, B.F., Paternal exposure to cyclophosphamide induces DNA damage and alters the expression of DNA repair genes in the rat preimplantation embryo, *Mutat. Res. — DNA Repair*, 461, 229, 2000.
115. Kubota, Y., Shimada, A., and Shima, A., DNA alterations detected in the progeny of paternally irradiated Japanese medaka fish (*Oryzias latipes*), *Proc. Natl. Acad. Sci. USA*, 92, 330, 1995.

116. Vogel, E.W. and Nivard, M.J.M., Model systems for studying germ cell mutagens: from flies to mammals, in *Advances in Male-Mediated Developmental Toxicity, Advances in Experimental Medicine and Biology, Vol. 518*, Robaire, B. and Hales, B.F., Eds., Kluwer Academic/Plenum Press, New York, 2003, p. 99.
117. Favor, J., Specific-locus mutation tests in germ cells of the mouse: an assessment of the screening procedures and the mutational events detected, in *Male-Mediated Developmental Toxicity*, Mattison, D.R. and Olshan, A.F., Eds., Plenum Press, New York, 1994, p. 23.
118. Ehling, U.H., Dominant mutations in mice, in *Male-Mediated Developmental Toxicity*, Mattison, D.R. and Olshan, A.F., Eds., Plenum Press, New York, 1994, p. 49.
119. Hales, B.F. and Cyr D.G., Study designs for the assessment of male-mediated developmental toxicity, in *Advances in Male-Mediated Developmental Toxicity*, Robaire, B. and Hales, B.F. Eds., Kluwer Academic/Plenum Press, New York, 2003, p. 271.
120. van Delft, J.H., Bergmans, A., and Baan, R.A., Germ-cell mutagenesis in lambda lacZ transgenic mice treated with ethylating and methylating agents: Comparison with specific-locus test, *Mutat. Res.*, 388, 165, 1997.
121. Douglas, G.R., Gingerich, J.D., Soper, L.M., and Jiao, J., Toward an understanding of the use of transgenic mice for the detection of gene mutations in germ cells, *Mutat. Res.*, 388, 197, 1997.
122. Allen, J.W., Collins, B.W., Cannon, R.E., McGregor, P.W., Afshari, A., and Fuscoe, J.C., Aneuploidy tests: cytogenetic analysis of mammalian male germ cells, in *Male-Mediated Developmental Toxicity*, Mattison, D.R. and Olshan, A.F., Eds., Plenum Press, New York, 1994, p. 59.
123. Martin, R.H., Rademaker, A., Hildebrand, K., Barnes, M., Arthur, K., Ringrose, T., Brown, I.S., and Douglas, G., A comparison of chromosomal aberrations induced by in vivo radiotherapy in human sperm and lymphocytes, *Mutat. Res.*, 226, 21, 1989.
124. Wyrobeck, A.J., Weier, H.-U., Robbins, W., Mehraein, Y., and Pinkel, D., Detection of sex-chromosomal aneuploidies in human sperm using two-color fluorescence in situ hybridization, *Environ. Molec. Mutagen.*, 19, 72, 1992.
125. Lowe, X.R., de Stoppelaar, J.M., Bishop, J., Cassel, M., Hoebee, B., Moore, D. 2nd, and Wyrobek, A.J., Epididymal sperm aneuploidies in three strains of rats detected by multicolor fluorescence in situ hybridization, *Environ. Mol. Mutagen.*, 31, 125, 1998.
126. Bentley, K.S., Sarrif, A.M., Cimino, M.C., and Auletta, A.E., Assessing the risk of heritable gene mutation in mammals: Drosophila sex-linked recessive lethal test and tests measuring DNA damage and repair in mammalian germ cells, *Environ. Molec. Mutagen.*, 23, 3, 1994.
127. Kohn, K.W., Erickson, L.C., Ewig, R.A.G., and Friedman, C.A., Fraction of DNA from mammalian cells by alkaline elution, *Biochemistry*, 15, 4629, 1976.
128. Sega, G.A., Sluder, A.E., McCoy, L.S., Owens, J.G., and Generoso, E.E., The use of alkaline elution procedures to measure DNA damage in spermiogenesis stages of mice exposed to methyl methanesulfonate, *Mutat. Res.*, 159, 55, 1986.
129. Bjorge, C., Brunborg, G., Wiger, R., Holme, J.A., Scholz, T., Dybing, E., and Soderlund, E.J., A comparative study of chemically induced DNA damage in isolated human and rat testicular cells, *Reprod. Toxicol.*, 10, 509, 1996.
130. Irvine, D.S., Twigg, J.P., Gordon, E.L., Fulton, N., Milne, P.A., and Aitken, R.J., DNA integrity in human spermatozoa: relationships with semen quality, *J. Androl.*, 21, 33, 2000.
131. Evenson, D.P., Alterations and damage of sperm chromatin structure and early embryonic failure, in *Towards Reproductive Certainty: Fertility and Genetics Beyond 1999*, Janssen, R. and Mortimer, D., Eds., Parthenon Publishing Group Ltd, New York, 1999, p. 313.
132. Evenson, D.P., Larson, K., and Jost, L.K., The sperm chromatin structure assay (SCSA™): clinical use for detecting sperm DNA fragmentation related to male infertility and comparisons with other techniques. Andrology Lab Corner. *J. Androl.*, 23, 25, 2002.
133. Perreault, S.D., Aitken, R.J., Baker, H.W.G., Evenson, D.P., Huszar, D., Irvine, S., Morris, I.O., Morris, R.A., Robbins, W.A., Sakkas, D., Spano, M., and Wyrobek, A.J., Integrating new tests of sperm genetic integrity into semen analysis: breakout group discussion, in *Advances in Male-Mediated Developmental Toxicity, Advances in Experimental Medicine and Biology, Vol. 518*, Robaire, B. and Hales, B.F., Eds., Kluwer Academic/Plenum Press, New York, 2003, p. 253.
134. Dubrova, Y.E., Jeffreys, A.J., and Malashenko, A.M., Mouse minisatellite mutations induced by ionizing radiation, *Nature Genet.* 5, 92, 1993.

135. Dubrova, Y.E., Nesterov, V.N., Krouchinsky, N.G., Ostapenko, V.A., Neumann, R., Neil, D.L., and Jeffreys, A.J., Human minisatellite mutation rate after the Chernobyl accident, *Nature*, 380, 683, 1996.
136. Jeffreys, A.J., Royle, N.J., Wilson, V., and Wong, Z., Spontaneous mutation rate to new length alleles at tandem-repeat hypervariable loci in human DNA, *Nature*, 332, 278, 1988.
137. Vergnaud, G. and Denoeud, F., Minisatellites: mutability and genome architecture, *Genome Res.*, 10, 899, 2000.
138. Kelly, R., Bulfield, G., Collick, A., Gibbs, M., and Jeffreys, A.J., Characterization of a highly unstable mouse minisatellite locus: evidence for somatic mutation during early development, *Genomics*, 5, 844, 1989.
139. Gibbs, M., Collick, A., Kelly, R., and Jeffreys, A.J., A tetranucleotide repeat mouse minisatellite displaying substantial somatic instability during early preimplantation development, *Genomics*, 17,121, 1993.
140. Barton, S.C., Surani, M.A.H., and Norris, M.L., Role of paternal and maternal genomes in mouse development, *Nature*, 311, 374, 1984.
141. McGrath, J. and Solter, D., Completion of mouse embryogenesis requires both the maternal and paternal genomes, *Cell*, 37, 170, 1984.
142. Surani, M.A.H., Barton, S.C, and Norris, M.L., Development of reconstituted mouse eggs suggests imprinting of the genome during gametogenesis, *Nature*, 308, 548, 1984.
143. Surani, M.A.H., Evidences and consequences of differences between maternal and paternal genomes during embryogenesis in the mouse, in *Experimental Approaches to Mammalian Embryonic Development*, Rossant, J. and Pedersen, R. A., Eds., Cambridge University Press, New York, 1986, p. 401.
144. Bartolomei, M.S., Webber, A.L., Brunkow, M.E., and Tilghman, S.M., Epigenetic mechanisms underlying the imprinting of the mouse H19 gene, *Genes Dev.*, 7, 1663, 1993.
145. Lucifero, D.L., Mertineit, C., Clarke, H.J., Bestor, T.H., and Trasler, J.M., Methylation dynamics of imprinted genes in murine germ cells, *Genomics*, 79, 530, 2002.
146. Doerksen, T. and Trasler, J.M., Developmental exposure of male germ cells to 5-azacytidine results in abnormal preimplantation development in rats, *Biol. Reprod.*, 55,1155, 1996.
147. Doerksen, T., Benoit, G., and Trasler, J.M., DNA hypomethylation of male germ cells by mitotic and meiotic exposure to 5-azacytidine results in altered testicular histology, *Endocrinology*, 141, 3235, 2000.
148. Kelly, S.M., Robaire, B., and Hales, B.F., Paternal cyclophosphamide treatment causes postimplantation loss via inner cell mass-specific cell death, *Teratology*, 45, 313, 1992.
149. Goldstein, L.S., Spindle, A.L., and Pedersen, R.A., X-ray sensitivity of the preimplantation mouse embryo in vitro, *Radiat. Res.*, 62, 276, 1975.
150. Feldman, B., Poueymirou, W., Papaioannou, V.E., DeChiara, T.M., and Goldfarb, M., Requirement of FGF-4 for postimplantation mouse development, *Science*, 267, 246, 1995.

CHAPTER 6

Comparative Features of Vertebrate Embryology

John M. DeSesso

CONTENTS

I. Introduction 147
II. Comparative Placental Characteristics 148
III. Embryological Processes 154
IV. Comparative Embryological Milestones...159
Acknowledgments 162
References 189
References for Tables 3 to 12..191

I. INTRODUCTION

Multicellular animals have limited life spans. Consequently, for a species to survive, a mechanism must exist for the successive production of new generations. The solution to this problem lies in the process of reproduction. This process typically involves the presence of two sexes, the production by each sex of specialized cells called gametes, and a complicated series of events resulting in the joining of two gametes to form a new individual. Gametes are referred to as haploid cells because they contain one-half the number of chromosomes found in somatic cells of the particular species. Male gametes (spermatozoa) are generated in the testes and are small, motile cells, millions of which are produced daily. In contrast, female gametes (ova) are large, nonmotile cells that develop in the ovaries. Relatively few ova are produced, and only a few hundred mature during the reproductive lifetime of female mammals.

Fertilization is the union of a single spermatozoon and an ovum. It occurs in the female reproductive tracts of birds and mammals and produces a new single celled organism, the zygote. Fertilization restores the diploid number of chromosomes, so that the zygote has the same amount of genetic material as did the somatic cells of its parents. Fertilization in mammals and birds also determines the sex of the zygote and initiates the process of cleavage. Cleavage is a rapid series of mitotic divisions that allows the relatively large amount of cytoplasm contributed by the ovum to be divided into progressively smaller cells.

In mammals, fertilization occurs in the uterine tube (oviduct). Cleavage divisions occur as the zygote progresses to the uterus, where it will become attached to the maternal uterine wall. During this time, the zygote is surrounded by the zona pellucida, an acellular mucopolysaccharide layer that prevents the zygote from implanting prematurely. When the zygote reaches the uterus, it is a

Table 6.1 Gestational milestones for mammals

	Gestational Milestone[a]				
	A[b]	B	C	D	E
Species	Implantation	Primitive Streak	Early Differentiation	Partial Closure[c]	Usual Parturition
Rat	5–6	8.5	10 *organogenesis begins*	15	21–22
Mouse	5	6.5	9	15	19–20
Rabbit	7.5	7.25	9	18	30–32
Hamster	4.5–5	7	8	13	16
Guinea pig	6	12	14.5	~29	67–68
Monkey	9	17	21	~44–45	166
Human	6–7	13	21	~50–56	266

[a] In gestational days; day of confirmed mating is gestational day 0.
[b] Letters refer to positions on Figure 6.4 (conceptual roadmap of embryonic development).
[c] Marks the end of major organogenesis for most organ systems.

cluster of small cells surrounded by the zona pellucida; this cluster is called a *morula*. Subsequently, the zona pellucida thins, ruptures, and eventually disappears, while the morula cavitates to become a sphere of cells surrounding a fluid-filled cavity. At this stage, the zygote is termed a *blastocyst*.

In most mammals that are used in experimental studies, the blastocyst arises between days 5 and 8 of gestation and attaches to the uterine wall during this time (Table 6.1). Two populations of cells are recognized in a blastocyst. They include the outer sphere of cells, called the *trophoblast* that gives rise to the placenta and fetal membranes, and a small cluster of cells on the inside, the *inner cell mass* that gives rise to the embryo proper.

II. COMPARATIVE PLACENTAL CHARACTERISTICS

One of the earliest tasks of the blastocyst is the establishment of a mechanism for maintenance of nutrient supply and disposal of metabolic wastes. This is accomplished through the development of a placenta from the trophoblast. The placenta and fetal (extraembryonic) membranes are temporary organs that form early in development and exist for a brief period compared to the life span of the organism. Because of their importance to embryonic development, however, they will be described in some detail before we return to further discussion of the embryo proper.

The extraembryonic membranes provide nutrition, respiration, metabolic waste elimination, and protection to the embryo and fetus, in addition to assisting in the establishment of embryonic vascularity. The four fetal membranes of vertebrates are the amnion, chorion, allantois, and yolk sac. Not all vertebrate species exhibit all four membranes; for instance, animals that lay eggs in water (*anamniota*) do not possess an amnion.

A placenta is an organ composed of fetal and parental tissues that are intimately apposed for the purpose of physiological exchange.[1] The fetal tissues of the placenta include one or more of the extraembryonic membranes (e.g., yolk sac, allantois), whereas the parental tissue is usually part of the uterus. The types of placentas can be described by the fetal membranes that participate in the apposition of fetal to maternal tissues. In general, the definitive placentas of eutherian mammals are formed from the outermost membrane of the embryonic vesicle (the avascular chorion), which is augmented by, and receives vascularization from, the allantois. This type of placenta is a chorioallantoic placenta. Most marsupial species develop placentas from the chorion, which is vascularized by the yolk sac. This type of placenta is called a *choriovitelline* (or *yolk sac*) placenta. (See discussion of the rodent "inverted" yolk sac placenta below.)

The placenta and fetal membranes are tissues with diverse structures and functions. In addition, these tissues are dynamic and modify both their structures and functions during gestation. Consequently, when assessing the role of the placenta in developmental toxicity, one must be aware not

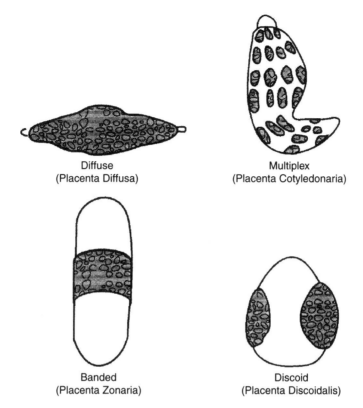

Figure 6.1   Types of placentas, classified by shape. The images depict the outer surface of the chorion, the fetal membrane that is apposed to the maternal reproductive tract. The white territories represent the smooth portions (chorion laeve); the gray regions represent the part of the chorion (chorion frondosum) that is modified to increase the surface area between embryo and mother. Note that banded placentas may exist as complete bands (illustrated) or as partial belts. Note also that discoid placentas may exist as single placentas, as in humans, or as paired structures, as in rhesus monkeys (illustrated).

only of interspecies differences in placental structure and function, but also of differences in the structure and function of the same placenta at different times in gestation.

Placentas are classified according to their gross appearance, their mode of implantation, their type of modification of the chorionic surface to increase surface area, and the intimacy of embryonic invasion into maternal tissues.[2,3] Because these differences can influence the efficiency or rate of transfer of materials between mother and embryo, they will be described briefly.

The outermost fetal membrane is the chorion. To the naked eye, it appears either as a smooth membrane (chorion laeve) or as a roughened or fuzzy membrane (chorion frondosum). The distribution of the villous areas of the chorion may take on one of four shapes (see Figure 6.1).

1. Diffuse (placenta diffusa): Villi are maintained over the entire chorion (e.g., pigs,* horses, humans [early in gestation], lemurs).
2. Multiplex (placenta cotyledonaria): Villi are grouped in discrete rosettes (cotyledons) that are separated by regions of smooth chorion (e.g., cattle, sheep, deer, ruminants).
3. Banded (placenta zonaria): Villi assume a girdle-like configuration around the middle of the chorionic sac (e.g., carnivores — dogs, cats [complete band], bears [less than half]).
4. Discoid (placenta discoidalis): Villi are grouped into one or two disk-shaped regions (e.g., insectivores, bats, rodents, nonhuman primates, definitive human placenta).

---

* Villi in the pig are actually plicate elevations.

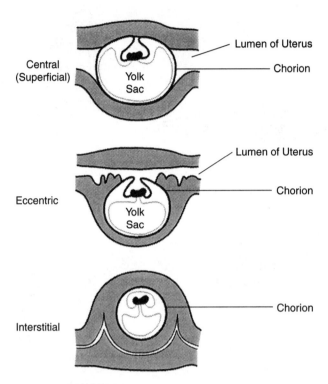

Figure 6.2 Types of placentas, classified by mode of implantation. The depth of implantation into the uterine wall increases from central, in which the conceptus essentially lies in the uterine lumen, to interstitial, in which the conceptus resides completely within the uterine wall and the uterine lumen is obliterated.

The relationship of the chorionic sac to the uterine wall and lumen can be described in terms of the extent or the depth of embryonic implantation into the uterine wall, and three general types of implantation can be distinguished[4] (see Figure 6.2).

1. Central (superficial): The chorionic sac remains in contact with the main uterine lumen (e.g., ungulates, carnivores, monkeys).
2. Eccentric: The chorionic sac lies in a pocket or fold that is partially separated from the uterine lumen (e.g., rodents — early in gestation).
3. Interstitial: The chorionic sac penetrates the uterine mucosa and loses contact with the uterine lumen (e.g., guinea pigs, human beings, rodents — late in gestation).

In rodents, the uterine lining of a pregnant female assumes a characteristic topography while awaiting the arrival of the blastocysts. The uterine mucosa appears scalloped, with evenly spaced indentations (or implantation chambers) along the long axis of each uterine horn. One blastocyst will come to occupy each implantation chamber in such a manner as to make the relationship between the chorion and the uterine lining eccentric. With further development, the rodent embryo will completely embed itself into the uterine wall, making the relationship interstitial.

The modifications of the chorionic surface to increase the area of contact between the chorion frondosum of the embryo and the maternal reproductive tract also demonstrate species differences.[3,5]

Plicate: The surface of the chorion exhibits elevated ridges or folds (e.g., pigs).
Villous: The chorionic surface exhibits fingerlike projections of embryonic tissue that project into maternal blood. The maternal circulatory pattern is described as entering lacunae, or pools, in which the villi are bathed (e.g., primates).

Labyrinthine: The chorionic surface exhibits anastomosing cords or trabeculae of embryonic tissue through which maternal blood flows. The maternal circulatory pattern is described as labyrinthine (e.g., insectivores, rodents, bats).

Great species differences also exist with respect to the layers of embryonic and maternal tissues that are interposed between their respective circulations. The invasiveness of the trophoblast can be gauged by the amount of maternal tissue that is eroded.[2,3,6,7] The three most common placental types, as classified by extent of invasiveness, are described below (see Figure 6.3).

1. Epitheliochorial: The least invasive type of placenta. No maternal tissue is destroyed. The six layers separating the maternal bloodstream from the embryonic bloodstream are maternal capillary endothelium, maternal uterine connective tissue, uterine epithelium, trophoblast, embryonic connective tissue, and embryonic capillary endothelium (e.g., pigs, horses).
2. Endotheliochorial: The trophoblast invades the endometrium and connective tissue, allowing the trophoblast to approach the maternal capillaries. The four layers interposed between maternal and embryonic circulations are maternal capillary endothelium, trophoblast, embryonic connective tissue, and embryonic capillary endothelium (e.g., dogs, cats).
3. Hemochorial: The trophoblast eliminates all maternal tissue, allowing the trophoblast to come into direct contact with the maternal blood. The three layers that separate the maternal from the embryonic circulation are trophoblast, embryonic connective tissue, and embryonic capillary endothelium (e.g., lagomorphs, bats, rodents, primates).

The gestational periods of rodents and lagomorphs (rabbits and guinea pigs) are brief (16 to 68 days). Consequently, development in these species occurs rapidly. Because the definitive chorioallantoic placenta is not established until a competent embryonic circulatory system is operative (at about the 20 somite stage), these species develop an early placenta that uses the membranes of their rather large yolk sacs. This early placenta is frequently termed the "inverted" yolk sac placenta because portions of the outer yolk sac membranes attenuate and become discontinuous at the plane of apposition to the uterine wall, leaving the epithelium of the inner yolk sac membrane in virtual contact with the uterine lumen and epithelium. The inverted yolk sac placenta ferries nutritive substances to the embryo by a *histiotrophic* process that entails the pinocytosis by yolk sac epithelial cells of maternally derived macromolecules found in uterine secretions and the subsequent breakdown of those macromolecules within lysosomal vacuoles, followed by diffusion into the embryo. In contrast to the inverted yolk sac placenta, the chorioallantoic placenta accomplishes the exchange of nutrients, gases, and metabolic wastes between mother and embryo by means of a *hemotrophic* interchange of solutes between the respective circulations. Such a direct exchange can move materials between mother and embryo more efficiently and rapidly in the chorioallantoic placenta than can the multistep, lysosome-dependent process of the inverted yolk sac placenta.

While the importance of the rodent inverted yolk sac placenta is greatly diminished after the establishment of the chorioallantoic placenta, it does remain functional throughout gestation. For purposes of temporal comparison, the rat inverted yolk sac placenta develops about gestational day 6.5 to 7, whereas the rat chorioallantoic placenta is established at approximately gestational day 11 to 11.5. Inverted yolk sac placentas do not develop in humans or other primates. Table 6.2 summarizes the placental, uterine, and other gestational characteristics of humans and six commonly used experimental mammals.

Since the placenta is the interface between the embryo and the maternal environment, it is the site of absorption, transfer, and metabolism of nutrients and foreign compounds. In the not so distant past, the placenta was believed to be a barrier that prevented the movement of all unwanted xenobiotic (foreign) compounds into the embryo. The thalidomide tragedy of the late 1950s and early 1960s dispelled that idea. The placenta was reconceptualized as a sieve that retarded or eliminated the transfer of molecules that weighed greater than around 1000 Da or were highly charged, highly polar, or strongly bound to (serum) proteins. Currently, however, it is recognized

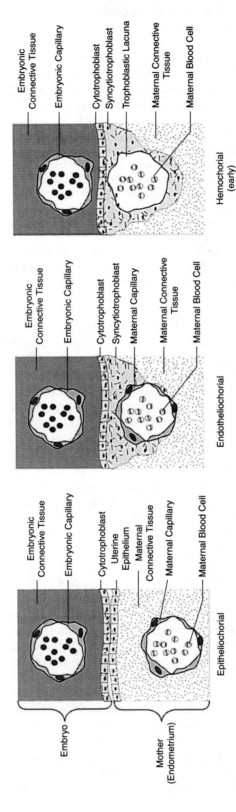

Figure 6.3 Types of placentas, classified by extent of invasiveness. The three most common placental types are depicted in a series that illustrates the progressive loss of tissue layers between the maternal and embryonic vascular systems. Six layers are interposed between the two circulations in the epitheliochorial placenta, four layers in the endotheliochorial placenta, and three layers in the hemochorial placenta.

Table 6.2 Comparative reproductive and placental features in selected experimental mammals and humans

Feature	Rat	Mouse	Rabbit	Hamster	Guinea Pig	Rhesus Monkey	Human
Estrus cycle (days)	4–6	3–9	None	4	16	28 (menstrual cycle)	28 (menstrual cycle)
Ovulation stimulus	Spontaneous	Spontaneous	Coitus	Spontaneous	Spontaneous	Spontaneous	Spontaneous
Uterus	Bicornuate	Bicornuate	Duplex	Bicornuate	Bicornuate	Simplex	Simplex
Usual no. of offspring	6–14	8–16	6–9	5–10	3–4	1	1
Gestation (days)	22	19	30–32	15–16	67–68	166	266
Implantation type	Eccentric — early Interstitial — late	Eccentric — early Interstitial — late	Superficial	Interstitial	Interstitial	Superficial	Interstitial
Classification by fetal membranes that contribute to placenta	*Early* Inverted yolk sac *Definitive* Chorioallantoic	*Early* Inverted yolk sac *Definitive* Chorioallantoic	*Early* Inverted yolk sac *Definitive* Chorioallantoic	*Early* Inverted yolk sac *Definitive* Chorioallantoic	*Early* Inverted yolk sac *Definitive* Chorioallantoic	Chorioallantoic	Chorioallantoic
Placental shape	Discoid	Discoid	Discoid	Discoid	Discoid	Bidiscoid	Discoid
Internal placental structure	Labyrinthine	Labyrinthine	Labyrinthine	Labyrinthine	Labyrinthine	Villous	Villous
Placental relation to maternal tissues	Hemotrichorial	Hemotrichorial	Hemodichorial	Hemotrichorial	Hemomonochorial	Hemomonochorial	Hemomonochorial

that there exists a broad diversity of mechanisms for transporting molecules through the placenta. The transport mechanisms[8-10] include both simple diffusion for most molecules (e.g., urea, oxygen, carbon dioxide) and carrier-mediated transport. The carrier-mediated mechanisms include active transport (e.g., for sodium/potassium, calcium, amino acids), facilitated diffusion (e.g., for D-glucose), and receptor-mediated endocytosis (e.g., for immunoglobulins, vitamin B-12). Thus, given the multiplicity of available transport mechanisms, when any substance is presented to the placenta, the question concerning entry into the embryo should not be whether placental transfer occurs, but rather by what mechanism and at what rate will transfer occur. The closest phenomenon to a barrier function is the expression in trophoblast cells of the multidrug resistance (*mdr*) gene family, which encodes a p-glycoprotein on the surface of the trophoblast membrane of conceptuses[11-13] exposed to certain xenobiotics. This phenomenon serves to limit the exposure of embryos to selected molecules.

In addition to transferring nutritive molecules to the embryo, the placenta may metabolize substances, whether they are nutrients or xenobiotic compounds.[9,10,14] For example, in cattle and sheep, the placental trophoblast converts maternally delivered glucose to fructose, which is in turn transferred to the embryo. In those species, an intravenous dose of glucose to the pregnant female causes a dramatic rise in fetal plasma fructose concentrations, rather than a rise in fetal plasma glucose. This illustrates the concept that placentas are not merely sieves but have the ability to alter some of the types of molecules that traverse them.

Placentas also contain various enzymes that are capable of metabolizing xenobiotics.[15-17] These enzymes include reductases, epoxide hydrases, cytochrome P-450 monooxgenases, glucuronidases, and others. These enzymes are not present at all times during gestation but make their appearances as the placenta (and embryo) mature. The presence (or absence) of these enzymes reflects the genotype of the embryo rather than that of the mother. Placental enzymes can be induced by inducers of monooxygenases, such as phenobarbital, benzo(*a*)pyrene, and 3-methylcholanthrene. In addition, the formation of reactive intermediates from xenobiotic compounds by placental enzyme preparations has been demonstrated *in vitro* (e.g., see Chapter 12 of this volume).

Placental toxicity, per se, is rarely cited as a primary mechanism for developmental toxicity. This does not mean that the importance of the placenta in development is not recognized, nor does it mean that placental dysfunction can be discounted as playing a critical role in development.[10,14,17-20] That the placenta plays a role in developmental toxicity is not in dispute; rather, it has proved difficult to determine whether developmental toxicity arises as a result of direct placental toxicity or from combined effects on the materno-feto-placental unit. Examples of developmental toxicity that have been ascribed to some combination of mother, fetus, and placenta include reductions in utero-placental blood flow subsequent to hydroxyurea,[21,22] altered transport of nutrients by azo dyes,[23,24] immunotoxicants,[25,26] lectins,[71] and hemoglobin-based oxygen carriers,[72] as well as pathological changes observed in the trophoblast after exposure to placental toxicants, such as cadmium.[27,28]

III. EMBRYOLOGICAL PROCESSES

Development from zygote to embryo to fetus to independent animal is a dynamic and carefully orchestrated phenomenon that involves numerous simultaneous processes that occurs in specific sequences and at particular times during both gestation and the postnatal period. This is especially true for rodents, wherein many of the organ systems of neonates have attained only the state of maturation found in late second or early third trimester human fetuses.[29] While it is imperative that developmental schedules be maintained, each embryo develops at its own rate, and there is some room for adjustment to the schedules. That is, some developmental events may be delayed to a certain extent without adverse consequences. Thus, the gestational ages given for developmental events are merely averages of the observed events. Embryos within the same litter of polytocous species are frequently at different developmental stages, especially during early embryogenesis. This may have resulted from different times of fertilization as well as from differences in the rate at which each embryo progresses through its own developmental schedule.

While many of the details concerning the development of embryos from various species (e.g., length of gestation, size of fetuses, time at which developmental landmarks appear) differ, the sequence of developmental events and many other features and processes are remarkably consistent across species. The following paragraphs will provide an overview of the consistencies and similarities of processes that take place in all embryos.

The blastocyst is a small organism, the cells of which have relatively few distinguishing morphological characteristics when observed with a microscope. There are two geographically distinct areas, the trophoblast and the inner cell mass. Not only does the blastocyst grow larger in size, but also the cells that comprise the blastocyst must become different in structure and in function as development of the individual progresses. As mentioned previously, the trophoblast forms the fetal membranes, whereas the inner cell mass gives rise to the embryo proper.

The cells of the inner cell mass quickly segregate into a two-layered disk that unequally transects the blastocyst cavity. One layer of cells (epiblast) is associated with the developing amniotic cavity; the other layer (hypoblast) is associated with the developing yolk sac cavity. The epiblast in turn rearranges by a process of cellular migration (variously called *invagination*, *ingression*, or *gastrulation*)[30–33] into the three germ layers (ectoderm, mesoderm, and endoderm), as well as the notochord. The hypoblast gives rise to the epithelial lining of the yolk sac, but does not contribute to the embryo proper.

Specific tissues of the body are derived from each germ layer. The ectoderm will give rise to the nervous system, skin, and adnexal dermal organs, including teeth, nails, hair, and both sweat and mammary glands. The derivatives of the mesoderm include cartilage, bone, muscle, tendons, connective tissue, kidneys, gonads, and blood. The endoderm gives rise to the linings of the alimentary, respiratory, and lower urinary tracts. The notochord serves as a primitive supporting tissue for the embryo and actively participates in the organization of the embryo. It degenerates, leaving no derivative except for the nucleus pulposus of each intervertebral disk.

The primordia of the organ systems are formed from combinations of tissues derived from the germ layers. To execute this process efficiently and accurately, many controls operate to maintain embryonic schedules and to control the fates of populations of cells, although there is some room for flexibility in these schedules and fates.

To help understand how the development of the embryo proper unfolds in an orderly fashion, two important concepts will be explained. These relate to the potential fate of a given cell (embryonic cellular potency) and its state of differentiation. Briefly, *embryonic cellular potency* is the total range of developmental possibilities (i.e., all possible adult tissues) that an embryonic cell is capable of manifesting under any conditions. In contrast, *differentiation* is the process whereby an embryonic cell attains the intrinsic properties and functions that characterize a particular tissue. Differentiation is a progressive, continuous phenomenon that involves at least three steps: (1) determination, during which stable biochemical changes occur within cells, but the changes are not apparent microscopically, (2) cytodifferentiation, during which those biochemical changes manifest themselves, resulting in the characteristic cytological and histological features that distinguish cell or tissue types from one another, and (3) functional differentiation, during which the cell or tissue begins to act in a physiologically mature role (e.g., insulin synthesis and release by pancreatic islet cells).

An embryonic cell's potency and its state of differentiation are reciprocal characteristics. Cells in the early stages of development, such as the blastomeres, the cells of the morula, or those of the inner cell mass of the blastocyst, are not differentiated; they are morphologically similar, and they have the potential to become nearly any type of embryonic cell. The pluripotent cells of the inner cell mass constitute the population of embryonic stem cells, which hold the promise of cures for many debilitating diseases.[73] As development proceeds, however, developmental decisions are made concerning the fate of each cell. Thus, at later periods of gestation the cells have become different from one another. One cell may have become an endoderm cell lining the liver parenchyma, while another may be a mesoderm cell that is providing smooth muscle in the walls of a blood vessel. The possible ultimate fates available to an endoderm cell are not the same as those of a mesoderm cell. Thus, the cells have restricted their potential. Also, they look different from one another because their state of differentiation has increased.

Cells become increasingly differentiated with increasing gestational age, and their embryonic potential decreases, as depicted in Figure 6.4. For both of these processes to occur in the proper sequence to result in a well-formed, normal individual, mechanisms must be available to keep populations of cells on schedule. A primary means for accomplishing this is the process of tissue interactions. As an example of tissue interactions, we discuss embryonic induction.[34]

Embryonic induction requires two populations of cells of developmentally dissimilar origin that attain proximity to one another. Developmental information of a directive nature is released or transferred from one population of cells (the inducer) for a finite period. The receiving population of cells must be competent (i.e., able to react to the directive message) for a limited period. The change that is evoked in the competent tissue must be progressive, stable, and maturational. One important thing to note about this process is that the ability of one population of cells to send a message and the second to receive and respond to the message is limited to a finite period (or window) that is intrinsic to each cell population. It is the transfer of developmental information through these open "windows" that maintains the embryo on its schedule. Genetic control over the timing and location of inducing and competent tissues is likely related to the sequence of expression and spatial delimitation within the embryo of genes controlling the synthesis of transcription factors and developmental control genes, such as the homeobox genes.[35–37]

The nature of the message substance or inducer has been investigated for many years. It is not known what the medium of the message is in all cases. In some cases, the message appears to require direct contact between the cells; in others it appears to be the release of a chemical substance into the extracellular space. In still other cases, a combination of the two appears to be required. For our purposes, however, the nature of the message is not as important as the fact that appropriate communication between the populations of cells has occurred in a timely fashion.

It is important to recognize that, for a given cell, the information required to direct its differentiation (e.g., manufacture of cellular structural proteins, receptor molecules, and extracellular matrix molecules) resides within the genetic material of its own nucleus, whereas the information required to maintain developmental schedules usually comes from environmental stimuli (e.g., inducer molecules as well as permissive and instructive signal molecules that are manufactured and released by other embryonic cells). Successful development of an organism requires timely interactions between (normal) environmental stimuli and embryonic genes as they are expressed or repressed throughout development.[38] It should not be surprising, then, that abnormalities in either an embryo's genetic material (i.e., mutations or chromosomal aberrations) or its environment can lead to developmental anomalies.

The subcellular and molecular interactions that direct or contribute to the execution of these developmental processes (differentiation, induction, pattern formation), as well as to their control by gene expression, are active areas of research that are beyond the scope of this brief overview. The reader is referred to more detailed texts and articles that capture this information.[32,33,39]

To respond to challenges external to the embryo or to untoward intrinsic influences on development, the embryo can possibly undergo internal rearrangements of its schedule or populations of cells, thus maintaining normal, orderly development. This process has been termed *embryonic regulation*. Regulation is an important concept because it demonstrates that the process of embryonic development is a dynamic progression that is able to adjust to changing conditions.

When an embryo is challenged by an environmental agent, many components contribute to the eventual outcome. Some of these are extraembryonic in nature, whereas others are embryological components. Although the extraembryonic components are not the main thrust of this discussion, they will be addressed briefly because they can affect the rate and quality of embryonic development. First, the nature of the environmental agent itself must be considered. For instance, is it a physical agent or a chemical agent? If it is a chemical agent, then its structure, polarity, and lipid solubility are all important properties to be considered as they affect the amount of uptake of the chemical

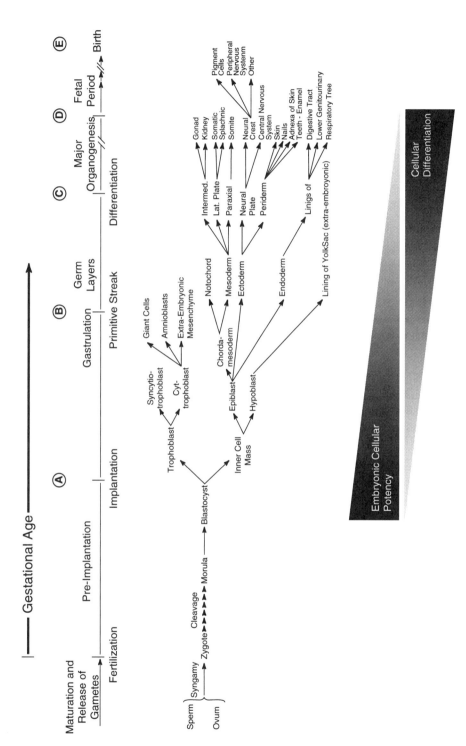

Figure 6.4 Diagrammatic representation of embryonic development. Fertilization is depicted on the left, and developmental maturation proceeds to the right. The dashed arrow represents possible differentiative pathways; the series of arrowheads denotes the rapidly occurring cell divisions during cleavage. Diverging arrows represent developmental decisions made by tissues as they differentiate. With each succeeding developmental decision, a cell's developmental potential decreases, while its state of differentiation increases. The circled letters denote the gestational milestones for the species listed in Table 6.1.

into the mother and the amount that will ultimately reach the embryo. In addition, each environmental agent may be thought to act in a specific way on some aspect of embryonic metabolism, and this specificity may help to determine how the agent interferes with embryonic development.

A second extraembryonic component is the dosage of the particular agent. The dosage is not as simple a component as one might assume; not all doses of proven teratogens cause birth defects. Typically, there is a lower dose range that allows most, or all, embryos to proceed through normal development, whereas higher doses may kill the embryo (and perhaps the mother as well). In between those two doses, there is usually a rather narrow teratogenic range in which sufficient damage is elicited in the embryo to disrupt developmental events without destroying it entirely. In addition, the dosage may be administered either acutely or chronically, and this also will affect the nature of any interference with embryonic development that may occur.

A third extraembryonic component is the physiological state of the mother because she provides the physical environment of the embryo. The state of the mother's nutrition and her general state of health are important, as is her ability to metabolize chemical agents and thereby change the nature of the compound to which the embryo may be exposed.[40] A fourth extraembryonic component is the previously discussed efficiency of the maternal-fetal exchange through the placenta. In summary then, the major nonembryonic considerations are the nature of the teratogenic agent, the dosage of the agent and timing of exposure, the maternal organism, and the effectiveness of maternal-embryonic exchange.

There are also important embryological components that affect embryonic outcomes.[41–43] The first of these components is the embryonic genotype and its expression, the theoretical basis of which has been discussed elsewhere[38] and which is the topic of ongoing, in-depth research.[31,33,39] In simplest terms, the embryonic genotype is an important embryonic consideration because it determines the inherent susceptibility of the embryo to exogenous agents at any given time during development. Alterations in a cell's DNA are the cause of mutations. Throughout most of the life span of mammals, DNA is replicated with great fidelity, and alterations to nonreplicating DNA, caused by environmental agents such as irradiation or chemicals, are rapidly repaired. There are two periods, however, when mammals are rather vulnerable to permanent changes in their DNA. One of these periods occurs during cleavage, when cell cycle times are shortest and extremely rapid synthesis of DNA is required. The fidelity of DNA replication diminishes with the continued rapidity of its synthesis. The other period is during the postmeiotic stage of gamete development. The greatest sensitivity occurs in males during spermiogenesis, when spermatozoa are maturing. The maturation of spermatozoa involves a process that drastically decreases the cytoplasm of the cells. In concert with the reduction of nonessential cytoplasm, the enzymes required for DNA repair are lost, leaving the maturing gametes unable to repair DNA damage. These topics have been discussed at length by others.[44,45]

A second important embryonic component is the stage of development of the embryo. In general, the time at which an agent acts on an embryo determines which tissues will be susceptible to the effects of the agent. This means that susceptibility to a particular agent will vary greatly during the course of gestation. Agents that are applied, even at high doses, during the predifferentiation period (from the time of fertilization through formation of the blastocyst) typically produce no teratogenic response, although exposure of females to mutagens within a few hours of mating has been reported to induce malformed offspring in some instances.[46] The reason why young embryos appear to be resistant to the effects of teratogens is not well understood; however, that resistance is probably a result of either the lack of specialization by the cells of the zygote to form specific parts of the organism, thereby retaining their embryonic potency, or a large number of stem cells. As long as all or many cells of the zygote retain a high degree of potency, the destruction or damage of some of those cells can be tolerated because the embryo can still undergo sufficient regulation to allow normal development to proceed. Although it appears that the destruction of a small number of undifferentiated cells in the embryo does not necessarily result in a structural malformation, there does appear to be a critical limit beyond which damaging even nonspecialized cells cannot be tolerated if the embryo is to live; if that critical limit is exceeded, the zygote will die. Further, nonlethal damage to the genome of pluripotent cells can result in a mosaic of tissues that exhibit increased likelihood of disease or other pathology.[44]

During the period of early organogenesis (when the embryo begins to undergo differentiation and the establishment of the germ layers), the onset of greatest susceptibility to teratogenesis occurs. This is coincident with the processes of gastrulation or invagination. For mammals, this occurs approximately five days postconception in small rodents (e.g., hamsters and mice), and up to 10 to 12 days postconception in primates. Not only is the onset of susceptibility to teratogenesis sudden, but also the majority of teratogenic agents produce their highest incidences of malformations at about this time.[41] Although there are no indications of the definitive organs in the embryo at that time, the cells of the germ layers have become determined (i.e., the morphologically undetectable aspect of differentiation has occurred) and have, therefore, lost some of their embryonic potency. Thus, cells that have become determined are susceptible to teratogenic agents even though their ultimate morphology is not yet evident. For example, rat embryos that have been exposed to x-rays on gestational day 10 exhibit malformations of the kidney at term.[47] This is of interest because the definitive kidney of the rat develops from the metanephros, which does not appear until day 12 of gestation.

This illustrates the concept that it is the stage of development at which an agent is effective, rather than the time at which it is administered, that determines the embryo's susceptibility.[48] This concept is important for those agents that might be stored in adipose tissues of the body. By way of example, this has been used as a basis to allege that the vitamin A derivative, etretinate, may have caused malformations in the offspring of a woman who had terminated its use several months prior to conception.[49]

Not only do embryos themselves have a sudden onset of susceptibility to teratogenesis, but also each organ of an embryo has a sensitive period for teratogenesis.[41,45] This sensitive (or critical) period is the time during which a small dose of a teratogen produces a great percentage of fetuses that will exhibit malformations of the organ in question. The critical period coincides with the early developmental events and tissue interactions that occur within the organ. In general, the susceptibility to teratogenesis decreases as differentiation and organogenesis proceed. This is because the proliferative and morphogenetic activities that characterize the early stages of the formation of tissues and organs become less prominent as the organ develops.

As an embryo progresses through the period of organogenesis, and as differentiation continues, production of a given teratogenic effect requires increasingly higher doses of the teratogen. This means that as organ systems and the embryo itself become progressively more differentiated, they become increasingly resistant to teratogenesis. Most of the organ systems have been laid down by the period of late organogenesis and the early fetal period, and the critical events involved in their formation have been completed. What remains to be accomplished during the remainder of prenatal and postnatal development is the progressive growth and functional maturation of each organ system. Strictly speaking, the majority of gross malformations become increasingly less problematic, although malformations of late-developing organs (e.g., kidneys, genitalia, and brain), altered histodifferentiation, growth retardation, and postnatal functional deficits (including neurobehavioral problems) may still be caused.

IV. COMPARATIVE EMBRYOLOGICAL MILESTONES

Because the primordia of the organ systems of an embryo are laid down in sequence, and not concomitantly at any given time in gestation, each organ system is likely to be at a unique stage of differentiation. For this reason, agents given acutely during a particular period of gestation may cause malformations of one organ system but not of another, or they may cause different malformations of the same organ system. Thus, the pattern of defects caused by any particular teratogen may change if the time at which the agent is applied or if the time at which the agent is effective occurs successively later in gestation. This has led to the construction of developmental schedules for embryos and, subsequently, both the use and misuse of those embryonic timetables.[50] It is important to realize that embryonic timetables can be used to determine at what developmental time a given organ is formed. From such a table, it is possible to ascertain the earliest and latest gestational times at which a particular organ system is likely to be grossly malformed by a noxious agent. Such embryonic schedules are useful for determining whether an embryo was exposed during or prior to the time of development for a given organ; they cannot identify the exact date at which an embryo was exposed to a particular agent because, as mentioned previously, some agents have delayed effects.

The differences among species, especially with respect to the timing of prenatal developmental events, are the subjects of Tables 6.3 to 6.12. The tables present the times of appearance for events in the embryology of various organ systems for selected laboratory animal species. The timing of such events is important if one wishes to investigate the normal development of a particular organ system. Timing is also crucial to studies of the genesis of malformations of an organ, using an animal model, or if one wishes to determine whether treatment with a specific agent is capable of eliciting the malformation of a given organ system. In cases such as the latter, the investigator must know at what gestational time the organ system in question is undergoing organogenesis in the appropriate animal model. Thus, the reader is referred to Tables 6.3 to 6.12 for interspecies comparisons of embryonic events related to development in general (Table 6.3); the circulatory system (Table 6.4); the digestive system (Table 6.5); selected endocrine glands (Table 6.6); the respiratory system (Table 6.7); the nervous system (Table 6.8); selected organs of special sense, i.e., eye, ear, and olfactory region (Table 6.9); the muscular, skeletal, and integumentary systems (Table 6.10); the excretory system (Table 6.11); and the reproductive system (Table 6.12). It is important to reemphasize that development does not end at birth and that it may be important to study events in postnatal animals. When the developmental phases of an organism's life are scaled according to the appearance of developmental landmarks (rather strict chronology) so that the developmental schedules of different species are congruent, they are being compared according to physiologic time. This concept is explained in greater detail elsewhere.[29]

Although avian models are not considered to be relevant for the assessment of human developmental toxicity, data for the chicken are included because of its long-standing use in embryological studies and because of its possible usefulness in assessment of the developmental toxicity of environmental pollutants toward wildlife. More complete texts and monographs should be consulted for the detailed embryology of particular species (e.g., human,[32,33,51,52] rat,[53,54] mouse,[55–57] hamster,[58] rabbit,[53,59] guinea pig,[55,60] rhesus monkey,[61,62] and chicken[63–65]).

Determination of the precise times in gestation for each species at which developmental events take place is difficult for a number of reasons. Even though the process of prenatal development proceeds sequentially, the rate at which it proceeds is neither standardized nor constant, even among offspring within the same litter. Development is based upon the expression of information contained within the genome of embryonic cells, and the timing of that expression is both triggered by and permitted by signals in the environment of those cells. Thus, there can be substantial variation in the time of appearance of rudimentary embryonic structures. This is especially true for those species with longer gestational periods.

The timing of embryonic events is further complicated by the fact that the starting point for timing (the instant of fertilization) is not known precisely. In most cases, the time of copulation is used as a surrogate for the time of fertilization. By convention, gestational age is measured from the time that mating is either observed (rabbits) or deduced from evidence of mating (such as observation of a copulatory plug in mice or rats or finding sperm in a vaginal smear in rats, mice, or hamsters). When mating is deduced, the time of fertilization is usually considered to have occurred at 9:00 a.m. of the day that the observations were made. Thus, by embryological convention, 9:00 a.m. of the day that the observations are made is set as day 0, hour 0 of gestation.

In humans and other primates, gestational age is estimated by ovulation age (or, at times, menstrual age — which is about 14 days longer than ovulation age). This means that the actual time of fertilization may be miscalculated by as much as 12 h in rodents and much longer in primates. For avian embryos, including chickens, development is initiated approximately 24 h prior to laying.

A final impediment to establishing the timing for embryonic events is caused by the sample size (or number of examined specimens) from which the times have been derived. In some species, particularly humans and other primates, the number of available specimens is quite small, leading to variations in the timing of events reported by the source documents. Thus, to group or classify embryos by their stage of development rather than by the time after fertilization, some investigators report other embryonic characteristics, such as crown-rump length, number of somites, or external features, as surrogates for gestational age.[52,53,63,66–70]

The aforementioned challenges to determining timing have led to the inclusion of several entries for many developmental events in Tables 6.3 to 6.12. These tables present the timing of developmental events of seven mammalian species and the chicken, organized by organ system. The entries include the estimated time during gestation and (where appropriate and available) the surrogate descriptors somite number or crown-rump length. Where source data have diverged, the entries are given as ranges. It should be emphasized, however, that even though the timing for the developmental events may be somewhat imprecise for certain events, the order of developmental events within a given organ system rarely changes.

ACKNOWLEDGMENTS

The author is grateful to Ms. Sue Walter for her diligence and patience in extracting data and preparing multiple revisions of the tables for this manuscript. The author also wishes to thank Ms. Judith Pals, Mrs. Pauline Kapoor, and Mr. Michael Yang for their assistance with the figures.

Table 6.3 Comparative early developmental milestones

Description	Rat Age[a] (d)	Rat Size (mm)	Rat Somites	Rat Ref.	Mouse Age (d)	Mouse Somites	Mouse Ref.	Rabbit Age (d)	Rabbit Somites	Rabbit Ref.	Hamster Age (d)	Hamster Somites	Hamster Ref.	Guinea Pig Age (d)	Guinea Pig Somites	Guinea Pig Ref.	Rhesus Monkey Age (d)	Rhesus Monkey Somites	Rhesus Monkey Ref.	Human Age (d)	Human Size (mm)[b]	Human Somites	Human Ref.	Chicken Age (d/h)	Chicken Somites	Chicken Ref.
One cell (in oviduct)	1	0.07		1–4,6,9,12	1		5,7–9				1		42				1		33	1	0.13		61,62			
Two cells (in oviduct)	2	0.08		1–4,6,9,12	1		5,7–9,12	0.33		12,28	1–1.5		42	0.96		49,56	1		12	2	0.12		62,63	3h[c]		12
Four cells (in oviduct)	3	0.08		1–4,6,9,12	2.25		12	0.46		12,28	1.67		12,42	1.25		50	1.5		12	1.5–2	0.12		62,63	3.25h[c]		12
Eight–twelve cells	3.25	0.08		1–4,6,9,12	2		5,7–9							3.5		56				3	0.1		63			
Morula (in uterus)	3.5	0.08		1–4,6,9,12	3		5,7–9				3		42				4		177	4	0.1		41			
Free blastocyst (in uterus)	5	0.12		1–4,6,9,12	4		5,7–9,12				3.5–4		42	4.75		56	5–8		33	5	0.1		63			
Implantation	5.5–6	0.28		10–12	4.5–5		10,34	7–7.5		10,12	4.5–6		10,12,42,43	6–6.5		10,51–53,56	9		10,12,46,60	6–7.5				NA		
Shell membrane formed in oviduct (chick)																								3.5–4.5h[c]		86
Shell of egg formed in uterine portion of oviduct (chick)																								4.5–24h[c]		86
Hypoblast formed	6			1–4,6,9	4.5		5,7–9,12				~5		42	6–6.5		51	9		46	7–8	0.5		65			
Primitive streak	8.5–9			1–4,6,9,10,12			5,7–9,10,35	7.25		10,57	6.5–7		10,42	12–13		10,15	15–17		10,33	13.5–17	0.3–1.2		9,10,35,66	7–19h		10,86
Neural folds	9	1		1–4,6,9,11	7.5		5,7–9	7.75–8.25	1–4	27	7.75	5	44	14–14.5		15,55	20–21	3	33	18–21	1.5–2	1–4	36,37,41,67,68,70	22–26h	1	86,87
First myocardial contractions	9.5	1.5	1–4	1,2,9,13,14,18,24,54	8		24	8.5	9	24,57	8–8.25	12–13	42,43,44	16–16.5	23	56				21–24	2–3.5	4–12	59,71	1.5d		30,86
Yolk sac; exocoelem	9.5	1.5	1–4	1–4,6,9,12							7		42				12		46	11–13	0.15		9,72	2d		87
Head process/notochord					7		36				7.5		44	11–18		55	16–18		33	18			36	19–22h		86

COMPARATIVE FEATURES OF VERTEBRATE EMBRYOLOGY

Event																					
Amniotic cavity	9.5	1.5	1–4	1–4,6, 9,12	7.75	5,7–9				6.5–7.5	12,42, 44	9	51,58	10	46,178	8			41	2d	87
Start of somite phase	9.5–10	1.5	1–4	11,12	7.75–8	5,7–9, 34,37	7.75–8.25	1–4	27	7.5	42	14.5	59	20–21	46,60	19–21	1.5–2	1–4	37,64, 67,70, 73	2d	1–4	87
Allantois arises	10	2	1–4	1–4,6, 9, 12	7.25–7.75	5,7–9, 35	8–10		99	7.75	42,44	11.75 –13	15,55, 56			16.5– 19			35,41	2–3d	30–36	86, 87
Oralmembrane perforates	10	2	5–12	1,2,9, 12,18, 22–26	9–9d 2h	5,7,26, 38	10		10	8.5	44			27–28	10,33	26–30	3.3–4	17	12,26, 39,74, 75,76, 77,78	2.2–3d	29– 32	26, 88
Ten somites	10.5			10	19±		19±	10	10	10	10,42	15		10	10	25		10	10	1.5d	10	10, 12
Fusion of neural folds (early)	10.5– 10.75			11,27			8.5–9	10– 14	27, 99	8.25	42,44	14– 15	15,55	21–23	33	22–24	2–3.5	4–12	66–68, 71,73, 74, 79–82	26–29h	3–4	28
Anterior neuropore closed	10.5– 10.75	2.4	17	1,2,4, 9,10, 11,18, 23,27	9d 1h	5,7,8, 39	4	9–9.5	10, 27	8.5	10,44	15.25		25	10	24–26	2.5– 4.9	13– 20	10,39, 71,73, 83	2.3d		10
Both neuropores closed	10.5– 11.5			10,11	9–9.5	10	9.5– 10.5		10, 99	8.5–9	10	15.25 –15.5	10	28–31	10,33	25–28			10			
Dorsal flexure disappears; embryo curves ventrally	10.5– 11.5	2.4	13– 20	1–4, 6,9, 12,27	8.5–9	5,7–9, 39				8.5– 8.75	42,44			25–27	33	26±			39			
Anterior limb bud appears	11	3.3	21– 25	1–4, 6,9,12	9.5– 9.75	5,7–9, 39	10.5– 11		27,57, 99	8.75	42,44, 46,47	16.5	56	25–28	33,46	26	3–5	21– 29	39,73	51–56h	26– 28	30, 86
Tail bud	11– 11.5	3.3	21– 25	1–4, 6,9,12	9.5	5,7–9	9.5		12	8–8.5	12,44		23	26	12	29	3.8		12	ca. 50– 52h	20– 21	86
Hind limb bud appears[d]	11.5– 12	3.8	26– 28	1–4,6, 9,10, 12	10– 10.3	5,7–10	11– 12	10, 14	10, 27, 57, 99	9	44	17.5– 18.5	29	28–30	10,33, 60	28–32	4–6	30– 32	9,10, 60,71, 73,84, 85	2.2–3d	29– 32	10, 86
Hand (forepaw) rays	13.5– 14			3,4, 10–12, 27, 30–32	12.3	7,10, 29,40	14.5		10, 27, 29	10.25– 11	10,42, 44,48	22– 23.75	10,56	34–35	10,46	35–37	8–11		10	4.75d		30, 86

[a] Age is measured in hours and days from the time of evidence of intromission. For rats, mice, hamsters, and guinea pigs age is counted from 9:00 am on the morning of discovery of either sperm in the vaginal lavage or a copulatory plug. In rabbits, it is measured from time of observation of mating. For primates, age is measured from the midpoint of the cycle (14 d after onset of last menses). In chickens age is generally given as "incubation age" or time after laying. The actual age of the chicken embryo is approximately 24 to 25 h older than the incubation age.
[b] Crown–rump length.
[c] Preincubation age.
[d] Hindlimb bud forms earlier in rodents than in primates.

Table 6.4 Comparative gestational milestones in circulatory system development

Description	Rat Age[a] (d)	Rat Size (mm)	Rat Somites	Rat Ref.	Mouse Age (d)	Mouse Somites	Mouse Ref.	Rabbit Age (d)	Rabbit Somites	Rabbit Ref.	Hamster Age (d)	Hamster Somites	Hamster Ref.	Guinea Pig Age (d)	Guinea Pig Somites	Guinea Pig Ref.	Rhesus Monkey Age (d)	Rhesus Monkey Somites	Rhesus Monkey Ref.	Human Age (d)	Human Size (mm)[a]	Human Somites	Human Ref.	Chicken Age (d/h)	Chicken Somites	Chicken Ref.
Vascularization of yolk sac	7.25–8.5	0.6		1,2,9, 13–22	ca 8d 8h	4	5,7,8, 90				7.5		44	13.5		56	16–18		33	19–20	1.5		74,84, 96–103	20–29h	4	12, 86
Bilateral heart primordia in ventral wall of coelom; two dorsal aortae								8.25		10	7.75	5	42,44	14.5		15,56						2–6	90			
Fusing heart tubes	9–9.5	1	6	1,2,7–10,12, 13–22, 24	7		5,7,8, 10	8.5–9	12	10, 24	8–8.25	12–13	10, 42–44	15		10,15	22		10,33	21	2		10,24	1.2d		10
First myocardial contractions	9.5	1.5	1–4	1,2,8, 9,13, 14,24, 54	8		5,7,8, 24	8.5	9	24	8–8.25	12–13	42–44	16.5	23	15,56				21–24	2–3.5	4–12	59,71	1.5d		30, 86
Dorsal mesocardium disappears					8d 21h	15	90				8.75–9.25		44									16–17	90			
Aortic arch arteries forming	10–12			10,24	8.5–11		10,24	9–11		10, 24	8–9.5		10	15.5–21.5		10	22–30		10	22–32	2–4		10,12, 24	1.5–4.5d		10, 24
Aortic arch I	10	2	5–12	2,8,9, 13,14, 18,24	8d 13h	9	7,81, 24	9.25		24	8		44			56	21–23		33	22	2	7	7,12, 81,24	33–38h	9–10	88
Sinus venosus; umbilical vessels; cardinal veins; endocardium	10.5	2.4	16–20	2,8,9, 13,14, 18							8.5		44	17.5	29	56	24–26	13–20	33	24–27	2.5–4.5	13–20	12,71			
S-shaped heart	10	2	10–12	2,8,9, 13,14, 18,24	8.5		10,24	9.5	21	10,24	8.5		10,44	16		10	25		10,33, 44	25–27	3.3		10,12, 24	48–54h	24–27	10, 24
Anterior cardinals	10.5	2.5	16–20	2,8,9, 13,14, 18	8d 14h	9–10	91				8		44	15.5	13	56						14	91	40h	13	12
Aortic arches I & II	10.5–11	2–2.4	11–20	2,8,9, 13,14, 18,89	9d 4h	20±	92,24	9.5		24	8.5		44	15.5	13	56	21–23		33	30	4	28	12,92, 24	50–55h	20–26	88, 24
Dorsal aortae fuse					9d 4h	20±	92,24				9		44	16.5	23	56				27	3.3	20	92,24	56h	24–27	24

COMPARATIVE FEATURES OF VERTEBRATE EMBRYOLOGY

Feature																							
Posterior cardinal veins established	10.5	2.5	16–20	2,9,13,14,18			20±	7,39									26±		7,10	7,39		10	10, 12
Ten-somite stage	10.5		10	10		7,10		10		10			10		23	10	23–25			10	1.5d	10	24
First circulation	10–10.5	2.4	16–20	2,9,13,14,18,176	8.5			5,7,24									26	3		71	2d	18	88
Duct of Cuvier (common cardinal vein)	12.5	6.2		12							8.5	44	16.5	23	56		24		14	83	ca. 45h	15	88
Aortic arch III	11	3.3	21–25	2,9,13,14,18	10	27±	7,93		9.75		9	44	16.5	23	56	27–28	28–31	4			50–55h	26	88, 24
Posterior cardinal; branches of anterior cardinal on mesencephalon	13.5	8.5		12							9	44	88,24 17.5	29	56		24		14	83	ca 45h	15	88
Septation occurring	11.5–12.5			10,12, 24	10.5–11.5		10,40, 24		13	34	9.25	10	19.5		10	28	28–37	3.5–6	16	10,24, 74,90, 99, 104–111	2.3d	26	10, 24
Four aortic arches; I regressing, IV still small	11.75–13	4.2	29–31	2,9,13,14,18,24	10		24		11.75		9.25	44	17.5	29	56	28–29	31–34	4–4.6		12,24, 83	3d	36	88
Atrioventricular canal	12	5.1	34–35	2,9,13,14,18													25			83	3–4d	ca. 30–44	88
Endocardial cushions appear	12–12.375	5.1–6	35–40	2,9,13,14,18,24	9d 18h	26±	39,24		11			10					26–28			39,71	3–4d	ca 30–34	88, 24
Aortic arches III, IV, & VI	12.125	5.2	36	2,9,13,14,18	10.5		7,24, 93		11		9.5	44	19	35	56	29–30	32–35	6		24,93	4d		88, 24
Beginning interventricular septum	12.125–13	5.2	36	2,9,13,14,18,24	9–10.5		76,24		12		8.5	44					29–35	6		24,76	4d		88, 24
Vitelline veins anastomose with liver plexus	12.5	6.2		12	10.5		39				9.25	44							26	39			

Table 6.4 Comparative gestational milestones in circulatory system development (continued)

Description	Rat Age (d)	Rat Size (mm)	Rat Somites	Rat Ref.	Mouse Age (d)	Mouse Somites	Mouse Ref.	Rabbit Age (d)	Rabbit Somites	Rabbit Ref.	Hamster Age (d)	Hamster Somites	Hamster Ref.	Guinea Pig Age (d)	Guinea Pig Somites	Guinea Pig Ref.	Rhesus Monkey Age (d)	Rhesus Somites	Rhesus Ref.	Human Age (d)	Human Size (mm)[a]	Human Somites	Human Ref.	Chicken Age (d/h)	Chicken Somites	Chicken Ref.
Intersegmental artery supplying anterior limb bud																								4d		88
Primordium of atrioventricular valve																				31–35	5		83			
Septum primum	12–12.5	5.1	34–35	2,9,13,14,18,24	11–11.5		7,24,90		34	24	9.25		44	23.75	35	56				28–37	4.3–5.4		74,84,95,104–111	50–55h	26	88,24
Pulmonary vein enters left atrium	12.375	6	39–40	2,9,13,14,18	11.5		7,90							18.5	31	56				28–37	3.5–6		7,24,71,74,84,90,95,104–111	ca 50h	20	88
Endocardial cushions fused	12.5	6.2	41–42	2,9,13,14,18	11		77,24	14		24				20.75	39	56				35–40	6–8		7,74,84,90,95,104–111	5.5–6d		24
Subcardinal veins formed					11.5		94				9.5		44							35–40	8–10		24,74,77,84,96–98,100–103			
Initiation of aortic–pulmonary septum					11.5		77,24				10		44							28	4		94	70h	30–36	88
Ostium secundum	13.25			24	11		24	12.75		24	13		44							32–35	7–8		24,71,77	4d		24
Septum secundum	14			24	11–12			13		24										40	8–10		24	5d		24
Foramen ovale present	13–14	8	46–48	2,9,13,14,18,24	11–12		10,24	14		10,24	10–13		10,44	19.25–21	38	10,56	34		10	40	8–10		24	5d		10,24
Subcardinal anastomosis					12.5		7,94				11.5		44							41–44	8–10		10,24			
Inferior vena cava enters heart					12.5		7,24,94				9.75		44							45–51	11		7,94	6–6.5d		24

COMPARATIVE FEATURES OF VERTEBRATE EMBRYOLOGY

Feature																			
Posterior cardinal degeneration	15			12.5	94											94			
Right dorsal aorta between arches III and IV disappears	15		24	13	7,24,93	14	24	11	44						41–42	12–14	7,24,93	7–7.5d	24
Interventricular septum complete	15[c]	12	2,9,13,14,18,24	13	7,24,95	16.5	24	11	44						43–46	13–17	7,24,71,95	8d	24
Bursa of Fabricius; posterodorsal wall of cloaca (chick only)																		5–6d	88
Truncal septation complete	15.5		10	13–14	10,77	16.5	10	15	10	22		10		36	35–46	11–14	10,71,77	5–7d	10, 88
Initiation of aortic and pulmonary semilunar values				13.5	77,24										35–38		77,24	5.5–6d	24
Fetal circulatory system is established[b]	15.5	14.2	64	2,9,13,14,18															

[a] Crown-rump length.
[b] Differs from that in humans mainly by persistence of a capacious vitelline circuit in addition to allantoic (umbilical) circuit.
[c] Completion of rat membranous interventricular septum may occur as late as postnatal day 7 (179, 180).

Table 6.5 Comparative gestational milestones in digestive system development

Description	Rat Age[a] (d)	Rat Size (mm)	Rat Somites	Rat Ref.	Mouse Age (d)	Mouse Somites	Mouse Ref.	Rabbit Age (d)	Rabbit Somites	Rabbit Ref.	Hamster Age (d)	Hamster Somites	Hamster Ref.	Guinea Pig Age (d)	Guinea Pig Somites	Guinea Pig Ref.	Rhesus Monkey Age (d)	Rhesus Monkey Somites	Rhesus Monkey Ref.	Human Age (d)	Human Size (mm)[a]	Human Somites	Human Ref.	Chicken Age (d/h)	Chicken Somites	Chicken Ref.
Foregut and oral plate	9.5	1.5	1–4	1,2,9,10,18,22–26	7.8	0	10,26	8.5		10	7.75–8	5	10,44	14–14.5		10,56	20.5		10	20.5–22	2.1	2	10,26,71,115	23h–1.1d		10,26,88
Pharyngeal pouches appear; stomodeum	10	2	5–12	2,9,18,25,26	8.3–8.8	4	26				8		44				24–26		33	24–27	3.3	7	26,39,76,77,81,96,116	36–39h	12	26,88
Oral membrane perforates	10	2	5–12	1,2,9,18,22–26	9–9d 2h	19±	26	10		10	8.5		44				27–28	21–29	10,33	26–30	3.3–4	17	12,26,39,76–78,96,116	2.2–3d	29–32	26,88
Liver primordium	10.5–11	2.4	13–20	1,2,9,10,18,22–26	8.8	14–15	10,26	9.5		10	8.5		10,44	16		10	24–26	13–20	10,33	21–27	2–3.3		10,26,39,78,116–118	50–56h	22	10,12,26,88
Hindgut	11	3.3	21–25	1,2,9,18,22–26	8–8.5	6–7	10,26	9		10	8		10,44	15.5		10,56	21		33	21.5		7	10,26,81	50–53h	21	10,12,26,88
Second pharyngeal pouch	11	3.3	21–25	2,9,18,25	8d 19h	14	26	9.5		57	8.5		44	15.5		56	25–26		33			14	91			
Gallbladder[b]	NA				9.625–9.7	25±	10,26	11.5		10	8.7		10	19		10	28–29		10,33	26–30	3.3–4		10,26,39,71,78	2.8–3.5d		10,26,88
Pancreas, ventral	11			26	9.7–11		7,26										29–30		33	31–35	4.3–7.5		26,39,71,96,117,119,120	3d		26
Pancreas, dorsal	11			26	9.7		26										28–29		33	26	3.5		26,71	4	35	26,88
Vitelline duct closes	11	3.3	21–25	1–4,6,9,18,22–25	9.5		5,7–9																			
Cloaca	11	3.3	21–25	1,2,9,18,22–25																27	3.3		39,76,77,96,116			
Liver epithelial cords	11.5			10,26	9.5–9.625	25±	10,26	10.5		10	8.75–9		10,44	16.5		10				26		25	10,26,39	3		10

COMPARATIVE FEATURES OF VERTEBRATE EMBRYOLOGY

Event																				
Stomach appears	11.5		10,26	11.5	10,26	10.5	10	8.5	10	16.5	10	28–29	10,33	28–32	3.5–4	10,26,39,76,77,84,96,116	3d		10,26
Third pharyngeal pouch; laryngotracheal groove	11.5	3.8	26–28	2,9,18,25				8.75	44	19	57	27–28	21–24	33	28	4.5	71	50–55h	23–24	12
First and third pharyngeal pouches touch ectoderm, second ruptured into visceral groove	12.125	5.2	36	1,2,9,18,22–25				8.75	44						25		83		14	
Umbilical hernia begins	12.25–14			9	11–12.3	12.5–14.5	10	9.3–11	10	22–23	10	33–35		10	36–45	8	10,26,84,121	4.75–6d		10
Urorectal septum appears	12.375–17	6	39–40	2,9,10,26,112,113		15	10	9.5	10,44						28–48	4.3–6	10,39,76,77,84,96,114,116			
Trachea separates from esophagus	12.5			1,2,9,18,22–26	11			9.75	44						29–31		26,71	4–4.5d		12
Primordium of bile duct	12.5			1,2,9,18,22–26				9.5	44		57				31–37	4.3–6	96,122	68h		88
Anal plate posterior to genital tubercle	12.5	6.2		12				10	44	20.75	40								35	
Tongue primordium; tuberculum impar	12.5	6.2	41–42	1,2,9,18,22–25				9.75	44	20.75	40				36–40	6–10	39,76,77,84,123	4		88
Fusion of dorsal and ventral pancreas	13	8	46–48	1,2,9,18,22–25	11.5		7,26	11.5	44			35–36		33	35–44	8–14				
Tip of tongue free	14.5	10.5	56–60	1,2,9,18,22–25				10.5	44	23.75	57									
Dental lamina; upper and lower incisor buds forming	14.5	10.5	26–60	1,2,9,18,22–25				11.5	44	20.75	40				40–48	8–15.6	125–128	6d		88

Table 6.5 Comparative gestational milestones in digestive system development (continued)

Description	Rat Age[a] (d)	Rat Size (mm)	Rat Som-ites	Rat Ref.	Mouse Age (d)	Mouse Som-ites	Mouse Ref.	Rabbit Age (d)	Rabbit Som-ites	Rabbit Ref.	Hamster Age (d)	Hamster Som-ites	Hamster Ref.	Guinea Pig Age (d)	Guinea Pig Som-ites	Guinea Pig Ref.	Rhesus Monkey Age (d)	Rhesus Monkey Som-ites	Rhesus Monkey Ref.	Human Age (d)	Human Size (mm)[a]	Human Som-ites	Human Ref.	Chicken Age (d/h)	Chicken Som-ites	Chicken Ref.
Fusion of mandibular and lower jaw elements completed	14.5–15	10.5–12	56–63	1,2, 9,18, 22–25										19.75	38	57	37–38		33	38	8		39,76, 77,84, 123			
Mandibular glands; mucosa near mandibular symphysis to base of tongue																				40–50	8–17		96, 116, 124	8		88
Anal membrane perforates	15	12		10,12, 26	14		10	10		10	13		10				39–40		33	45–50	16.5		10,26, 39,76, 77,84, 96,116	6		10,26
Maximal size of umbilical hernia	15.5	14.2	64	9	14.5		5,7–9				12.5		44							9wk		167				
Palatal folds uniting (not all at the same stage of union)[c]	17			10,11	15		10	19.5		10	12		10	26		10	45–46		10,33	56–63			10,12, 83, 116, 167	N/A		
Umbilical hernia reduced	17–18.5	16–20		1–4, 6,9, 10–12, 27	16–16.5		5,7–10	20		10	13		10				45–48		10,33	8.5–10wk	26–45		9,10, 39,66, 76,77, 84,96, 116, 121	18		10

[a] Crown-rump length
[b] Rats do not have gallbladders.
[c] In the chick palatal folds do not fuse.
[d] Transient structure; teeth do not form.

Table 6.6 Comparative gestational milestones in endocrine system development

Description	Rat Age (d)	Rat Size (mm)	Rat Somites	Rat Ref.	Mouse Age (d)	Mouse Somites	Mouse Ref.	Rabbit Age (d)	Rabbit Somites	Rabbit Ref.	Hamster Age (d)	Hamster Somites	Hamster Ref.	Guinea Pig Age (d)	Guinea Pig Somites	Guinea Pig Ref.	Rhesus Monkey Age (d)	Rhesus Monkey Somites	Rhesus Monkey Ref.	Human Age (d)	Human Size (mm)[a]	Human Somites	Human Ref.	Chicken Age (d/h)	Chicken Somites	Chicken Ref.
Pharyngeal Pouches																										
Two pharyngeal pouches	11	3.3	21–25	2,9,18,25	8.5–8.8	6–14	91,115										24–26		33	27	3.3	12–14	91,115			
Three pharyngeal pouches	11.5	3.8	26–28	2,9,18,25	9d 15h	25±	135										27–28	21–29	33	28	3.5	22	135			
Four to six pharyngeal pouches (ultimobranchial complex)	12.125	5.2	36	2,9,18,25							11.5		44				28–29		33		13.5		83	68h		12,88
Epiphysis — Pineal Gland																										
Pineal; epiphyseal evagination	14–14.5	9.5–10.5	52–60	1,2,4,9,18,23,25,129–134	11.5		77				11		44				32–38		60	33–48	13–17		60,77	52–64h	30–35	88
Adrenal Gland																										
Adrenal gland, cortical component; coelomic epithelium	12.5			10	11		10	18		10	10		10	23		10				34			10	3.25d	37–40	10,88
Adrenal gland, medullary component; migratory cells of neural crest and sympathetic ganglia	13.5	8.5	49–51	2,9,18,25,129–134							11–12.5		44							44	16		83	4–7d		88
Pancreas																										
Pancreas — dorsal	11–12.125	5.2	36	2,9,10,18,25,129–134	9.5		10	10		10	9.5		10	17.5		10	28–29		10,33	28			10	3d	35	10,88

Table 6.6 Comparative gestational milestones in endocrine system development (continued)

Description	Rat Age (d)	Rat Size (mm)	Rat Som-ites	Rat Ref.	Mouse Age (d)	Mouse Som-ites	Mouse Ref.	Rabbit Age (d)	Rabbit Som-ites	Rabbit Ref.	Hamster Age (d)	Hamster Som-ites	Hamster Ref.	Guinea Pig Age (d)	Guinea Pig Som-ites	Guinea Pig Ref.	Rhesus Monkey Age (d)	Rhesus Monkey Som-ites	Rhesus Monkey Ref.	Human Age (d)	Human Size (mm)[a]	Human Som-ites	Human Ref.	Chicken Age (d/h)	Chicken Som-ites	Chicken Ref.
Pancreas — ventral	11.5–12.125	5.2	36	2,9,10,18,25,129–134	9.7		10	11.5		10	9.5		10				29–30		10,33	31–32			10	4d		10,88
Islets of Langerhans within pancreatic diverticula	13	8	46–48	2,9,18,25,129–134																8–9 wk	40–50		96, 117, 137	8–9d		88
Pancreas fused	13	8	46–48	2,9,10,18,25,129–134	11.5		10	14		10	11.5		10				35–36		10	40–44				6d		10
Thymus																										
Thymus	12.375–12.5	6	39–40	2,9,10,18,25,129–134	12		10	12.5		10	8.75–9		10,44				*			30–40			10	6–8d		10,88
Thymus and parathyroid detaching from 3rd pouch	13	8	46–48	2,9,18,25,129–134							11.5		44								23		83			
Thyroid																										
Thyroid	10.5–12	2.4	13–20	2,9,10,18,25,129–134	8.5		10	9.5		10	8.5		10	16.5	23	10,57	28–29		10,33	24–27			10	40h		10,88
Thyroid shows open diverticulum from floor of mouth	11.75	4.2	29–31	2,9,18,25,129–134										17.5	29	57										
Vesicular ultimobranchial body (lateral thyroid) detaching	13	8	46–48	2,9,18,25,129–134	9.5	24±	136				12		44									20	136			
Ultimobranchial vesicles detached from pharynx	13.5	8.5	49–51	2,9,18,25,129–134																						

COMPARATIVE FEATURES OF VERTEBRATE EMBRYOLOGY

Parathyroid

Feature																						
Parathyroids	12.5	6.2	41–42	2,9,10,18,25,129	11	5,7,10		8.75–9	10,44								35–38			10	6–8		10,88
Parathyroids attached to left and right wings of thyroid	17–18	16–20		2,9,18,25,129		5,7,132																	

Hypophysis — Pituitary Gland

Feature																						
Rathke's pouch appears	10.5			10	8.5–9	23+	10,39	9.5	10	8.5	10	15.5	13	10,57	28–32	10,60	28–34		10,39,60	50–52h	20	10,88
Neural hypophyseal evagination	11.5–11.75	4.2		10,12,60	11.5	10,77	12	10	10	10	18.5		10	30–34	10,60	30–42	8–11	10,60,77	3		10	
Rathke's pouch closed off, connected to oral ectoderm	13.5–14	8.5	49–51	2,9,11,18,25,129	12	60			9	44	19.75	38	57				14	83				
Stalk of Rathke's pouch detached from stomodeal epithelium	15.5	14.3	64	2,9,11,18,25	12.5	138			12	44				36–42	60		19	138				
Pars intermedia thin-walled; pars distalis — trabecular and secondary vesicles					14	60								40–44	60	53–54	22–24	60				

[a] Crown-rump length.

Table 6.7 Comparative gestational milestones in respiratory system

Description	Rat Age (d)	Rat Size (mm)	Rat Somites	Rat Ref.	Mouse Age (d)	Mouse Somites	Mouse Ref.	Rabbit Age (d)	Rabbit Somites	Rabbit Ref.	Hamster Age (d)	Hamster Somites	Hamster Ref.	Guinea Pig Age (d)	Guinea Pig Somites	Guinea Pig Ref.	Rhesus Monkey Age (d)	Rhesus Monkey Somites	Rhesus Monkey Ref.	Human Age (d)	Human Size (mm)[a]	Human Somites	Human Ref.	Chicken Age (d/h)	Chicken Somites	Chicken Ref.
Laryngotracheal groove	11	3.3	21–25	1,2,9,18,22–25										18–18.5		56								ca. 50–54h	23	88
2nd pharyngeal pouch	11	3.3	21–25	2,9,18,25	8.75		7				8.5		44				24–26		33							
Primary lung diverticulum	11.5–12	3.8	26–28	2,9,10,18,25	9.5–9.75	25±	7,10,76	10.5		10	9		10	16		10	27		10	26–28	3.3		10,71,76,84,140–142	3d		10
Primary bronchi											9		44	18.5		56	29–30		33	29	4.5		71	96h		88
Trachea separated from esophagus	11.75–12.5	4.2	29–31	2,9,18,25	11		76				9		44	16.5		56				29–32	6		71,76	96h		88
Secondary bronchi	12.75	7	43–45	2,9,18,25	12		7,77							18.5		10				35–38	9		71,77			
Asymmetric lung buds; 3 bronchial areas in right lung bud	12.75–13	7	43–45	2,9,10,18,25	10.5–11.5		10,76	12		10	9.5–10		10,44	18.5–21.5		10,56	29		10	32			10	NA[b]		10
Bucconasal membrane ruptured					13		60				15–15.25		60				36–42		60	48–51	16–18		60			
Major bronchial divisions	15.5			10	13		10	15		10	10.5–11		10,44	21.5		10,56	36		10	46			10	NA[b]		10
Palatal shelves uniting (Not all at the same stage of union)	17			10	15		10	19.5		10	12–12.5		10,44	26		10	45–46		10,33	57			10	NA[c]		10
Developing alveoli					16.5		139														85		139			

[a] Crown-rump length
[b] Pattern of avian lung development is different from that of mammals.
[c] In the chick, palatal shelves do not fuse.

COMPARATIVE FEATURES OF VERTEBRATE EMBRYOLOGY

Table 6.8 Comparative gestational milestones in nervous system development

Description	Rat Age (d)	Rat Size (mm)	Rat Somites	Rat Ref.	Mouse Age (d)	Mouse Somites	Mouse Ref.	Rabbit Age (d)	Rabbit Somites	Rabbit Ref.	Hamster Age (d)	Hamster Somites	Hamster Ref.	Guinea Pig Age (d)	Guinea Pig Somites	Guinea Pig Ref.	Rhesus Monkey Age (d)	Rhesus Monkey Somites	Rhesus Monkey Ref.	Human Age (d)	Human Size (mm)[a]	Human Somites	Human Ref.	Chicken Age (d/h)	Chicken Somites	Chicken Ref.
Primitive streak	9			10	8		10	7.25		10	7		10	13		10	17		10	17			10	12h		10
Neural plate	9–9.5	1		1,2,4,9,10,18,23	7		5,7–10	8		10	7.5		10	13.5		10	20		10	18–19	1–1.5		10,39	1d		10
Elevated brain plate, neural folds	9.5	1.5	1–4	1,2,4,9,11,18,23				7.75–8.25	1–4	27				14.5	2	56				ca 19–20	1–2		115	23–26h	4	88
Neural crest for ganglia of IX and X; spinal flexure sometimes present	10	2	5–12	1,2,4,9,18,23																24±1	3–5		92,144	ca 45–49h	17	88
Trigeminal, neural crest component; neural crest at level of metencephalon	10	2	5–12	1,2,4,9,18,23							8.5	17–20	44											ca. 40–45h	13–14	88
Ganglia of VII and VIII; neural crest and posterodorsal epibranchial placode of first pharyngeal groove	10	2	5–12	1,2,4,9,18,23							8.5	17–20	44											ca 45h	15	88
Fusion of neural folds (early)	10.5–10.75			11,27				8.5–9	10–14	27	8.25	12–13	44	14–15		15,55				22–24	2–3.5	4–12	66–68, 71,73, 74, 79–82	26–28h	3–4	28
Anterior neuropore completely closed	10.5–10.75	2.4	17	1,2,4,10,11,18,23,27	9d 1h	18±	5,7–9, 39	9–9.5	10–14	10,27	8.5	17–20	10,44	15.25		10	25		10	24–26	2.5–4.9	13–20	10,39, 71,73, 83	2.3d		10
Otic pits; optic vesicles, and auditory pits appear	10.5–10.75			11,27	8d 22h	16	39	8.5–9	10–14	27	8.25	12–13	44									20	39			
Both neuropores closed — anterior first	10.5–11.5			10,11	9–9.5		10	9.5–10.5		10	8.5–9		10	15.25–15.5		10	25–31		10	25–28			10			

Table 6.8 Comparative gestational milestones in nervous system development (continued)

Description	Rat Age (d)	Rat Size (mm)	Rat Somites	Rat Ref.	Mouse Age (d)	Mouse Somites	Mouse Ref.	Rabbit Age (d)	Rabbit Somites	Rabbit Ref.	Hamster Age (d)	Hamster Somites	Hamster Ref.	Guinea Pig Age (d)	Guinea Pig Somites	Guinea Pig Ref.	Rhesus Monkey Age (d)	Rhesus Monkey Somites	Rhesus Monkey Ref.	Human Age (d)	Human Size (mm)[a]	Human Somites	Human Ref.	Chicken Age (d/h)	Chicken Somites	Ref.
Optic vesicle; optic pits	10.5–11	2	4–13	27,89				8.5–9	10–14	27,99	8.5	17–20	44	15–16		56	21–23		33	24±	2.5–5	13–20	71,73,92,144	29–33	7	86
Neural tube differentiates into the three primary brain vesicles; anterior neuropore closed	10.5–12	2.4	13–20	1,2,4,9,10,18,23	8		5,7–10	8.25–8.5		27	8.5	17–20	10,44	15–15.3		10,15	25–30		10,60	26±1	3.8–4.9		10,39	33–38	10	10,88
Five brain vesicles	11.5	3.8	26–28	1,2,4,9,18,23							8.75		44	17.5	29	56	30–33		60	33±	7–11		77,144,145			
Otic cyst and otic pit closed; endolymphatic appendage; deep cervical flexure	11.5–12			27	9.1–10	19±	39,60	10		99				16.75		56	28–30		60	28–32	4–6		39,60,146			
Posterior neuropore closing	11			11	9.5	24±	76	9+		99							27		60	26±			76			
Thickened lens disc; lens placode					9.625	25±	76	11		99							28–29		33	28–35	5.4	38	76,84,85			
Endolymphatic evagination					10		60				9		44	19		56	28–30		60							
Shallow olfactory pits					10		60	11–12		27,99	9		44				28–30		33,60							
Posterior neuropore closes[b]	11.5–12			11,27	10.5		60	10		99	9		44				28–30		60	26–27	3–6	21–29	71,73,76			
Cerebral hemispheres (early)	12–12.125	5.2	36	1,2,4,9,10,18,23	10	27±	5,7–9,77	11		10,99	9		10	17		10	29		10	29–33	5.5–9		10,73,76,77,144	3d	10	10,88
Pontine flexure	13.5	8.5	49–51	1,2,4,9,18,23	10.5		5,7–9,60	10		99							30–33		60	35–38	7–9		60			
Vomeronasal organ					11.5		77													37±			77			

COMPARATIVE FEATURES OF VERTEBRATE EMBRYOLOGY

	13.5	8.5	49–51	1,2,4,9,18,23	12.5–13	5,7–9,60,143			11	44			36–42	60	48–51	16–18	60,143				
Formation of choroid plexus of lateral and fourth ventricles																					
Cerebellum	14			10	12	10	15	10	11	10	19	10	30–36	10,60	37		10		4.5d		10
Lens cavity crescentic; primitive lens fibers present; primordial semicircular ducts; primordial cochlear duct; nasal fin					12	60							34–36	60	42–44	11–14	60				
Superior and inferior colliculi separate					12.5	77									37±		77				
Earliest reflex responses observed															41±1	18–23	144, 147, 148				
Neostriatum; telencephalic cortex																~17	83		8–12d		88

[a] Crown-rump length.
[b] Posterior neuropore closes earlier in primates than in rodents.

Table 6.9 Comparative gestational milestones in the development of sense organs

Description	Rat Age (d)	Rat Size (mm)	Rat Somites	Rat Ref.	Mouse Age (d)	Mouse Somites	Mouse Ref.	Rabbit Age (d)	Rabbit Somites	Rabbit Ref.	Hamster Age (d)	Hamster Somites	Hamster Ref.	Guinea Pig Age (d)	Guinea Pig Somites	Guinea Pig Ref.	Rhesus Monkey Age (d)	Rhesus Monkey Somites	Rhesus Monkey Ref.	Human Age (d)	Human Size (mm)[a]	Human Somites	Human Ref.	Chicken Age (d/h)	Chicken Somites	Chicken Ref.
Eye																										
Optic sulci	10	2		2	8.54	23±	39														14		39			
Optic vesicle forming	10.5		10	10	9.5	20±	10,39	9		10	8	10	10	15.5	13	10,56	23–25	10	10,33	21–24	2–2.8		10,74	26–33h		10,88
Optic bulbs	10.5	2.4		2		23±	39,60																		20–21	176
Lens placode	11.5	3.8		2	10	24±	39,60				9.5		44				28–30		33,60	27–35	3.3–5.4		84,149,151			
Invagination of lens placodes begins	11.75	4.2		2	10–10.5		60,76				9.25		44	17.5	29	56	28–32		60	27–32	4–7	30+	71,73,76	48h		88
Extrinsic premuscle masses appear																				31–44	4.3–14		152			
Lens vesicle forming; choroid fissure	12.125–12.375	5.2–6		2							10		44	16.5		56	35–38		60					48h		88
Lens separated	12.5	6.2		2,10	11.5		10	11.5		10	10		10,44	18		10	32		10	35–37	6		10,84,149,151,153,154	2.5d		10
Primordium of hyaloid artery	12.5	6.2		2							10		44							35–37	6		84,149,151,153,154			
Lens vesicle closed	13–14	6.2		2,11,27	11–11.5		60							19.75		56	32–34		60	34.5	7–9		71			
Definite retinal pigmentation					11–12	25±	39,60	14–15		27,57							30–36		60	33–44	7–14	25	39,60,73,77,84	3–3.5d	40–43	86,88
Differentiation of retina																				42–44	12–14		84,149,151	100h		88
Choroid fissure fusing; hyaloid canal	13.5	8.5		2	12		149				15–15.25		60				34–36		60	42–44	14.5		60,149			

COMPARATIVE FEATURES OF VERTEBRATE EMBRYOLOGY

Optic nerve fibers present	14	9.5		2,10	13		10	15	10	11	10	21.5	10	39	10	46–48	14.6–15.6		10,84,149,151	4d	10
Differentiation of cornea									60	15–15.25	60			36–42	60	48–51	16–18		60,84,149,151		
Anterior chamber differentiating	15	12		2	13			9						40–44	9	53–54	22–24		60	6d	88
Ciliary body differentiation; primordium of choroidea sclera																56–70	26–45		84,149,151	8d	88

Ear

Otic placodes	10	2		2	8.54	23±	39									27		14	39		
Otic cups	10.5	2.4		2,11		23±	39														
Otic vesicle forming	11.5			10	8.5–8.75	13	10,39	9		10	8.5	10	15.5	13	10,56		25	10	21–29	2–6	10,73,146	2.3d	10
Otocyst (closure)	11–11.5	3.3–3.8		2,11		24±	39				9	44				29		33	30	4	39,146	11d	88
Otic vesicles with short endolymphatic duct	11.75	4.2		2	10.5		76												29–32	5–7	73,76		12
Endolymphatic sac appears pinched off from otic vesicle	12–13			11,27	11		77										30		33–37	6.5	77,146,155		
Thickened, hollow epithelial primodia of semicircular canals					11.5		77												38–44	8–13	77,155,156	35±	
Cochlear and vestibular regions	12–13	6.2		2,27	11–11.5		60										30–34	60	35–42	7–9	60	6–7d	88
Separation of utricular and saccular regions	12.5	6.2		2	14.5		150													33	150		
Cochlea appearing	13.5			10	12		10	13		10	10	10	20.5		10		37	10	44		10	7d	10

180 DEVELOPMENTAL REPRODUCTIVE TOXICOLOGY: A PRACTICAL APPROACH, SECOND EDITION

Table 6.9 Comparative gestational milestones in the development of sense organs (continued)

Description	Rat Age (d)	Rat Size (mm)	Rat Som-ites	Rat Ref.	Mouse Age (d)	Mouse Som-ites	Mouse Ref.	Rabbit Age (d)	Rabbit Som-ites	Rabbit Ref.	Hamster Age (d)	Hamster Som-ites	Hamster Ref.	Guinea Pig Age (d)	Guinea Pig Som-ites	Guinea Pig Ref.	Rhesus Monkey Age (d)	Rhesus Monkey Som-ites	Rhesus Monkey Ref.	Human Age (d)	Human Size (mm)[a]	Human Som-ites	Human Ref.	Chicken Age (d/h)	Chicken Som-ites	Chicken Ref.
One or more semicircular canals formed	14	9.5	4–13	2,89	12		77										34–36		60	44–48	13.5–15	37±	77,84, 146, 149, 151, 155, 156	5–6d		88
Condensations of ocular muscles	14	9.5		2																						
Otic capsule cartilaginous	15			10	14.5		10	20		10	11		10	22		10	42		10	56			10	8d		10
Ocular muscles innervated	15	12		2																					6d		88
Ossification of anterior process of malleus																				56–70	28		160, 161			
Olfactory																										
Olfactory placodes	11–11.5	2–3.8		2,89	10	24–27	39,76				8.75		44				28–29		33	28–31	7–9	17–25	39,76, 77, 157, 158		28	88
Olfactory pit	12.125	5.2–5.6		2	10.5–11		60,76							18.5		56	30–33		60	33–38	7–8	29±	73,76, 84, 141, 159	52–64h	29–32	88
Olfactory nerve; olfactory epithelium	12.5	6.2		2	11	27±	39																	4–6d		88
Olfactory bulbs	13.5			10	11		10	14		10	11		10	23		10	38		10	37			10	7d		10
Partly cartilaginous nasal septum and capsule	15	12		2																				6d		88

[a] Crown-rump length.

COMPARATIVE FEATURES OF VERTEBRATE EMBRYOLOGY

Table 6.10 Comparative gestational milestones in muscular, skeletal, and integumentary system development

Description	Rat Age (d)	Rat Size (mm)	Rat Somites	Rat Ref.	Mouse Age (d)	Mouse Somites	Mouse Ref.	Rabbit Age (d)	Rabbit Somites	Rabbit Ref.	Hamster Age (d)	Hamster Somites	Hamster Ref.	Guinea Pig Age (d)	Guinea Pig Somites	Guinea Pig Ref.	Rhesus Monkey Age (d)	Rhesus Monkey Somites	Rhesus Monkey Ref.	Human Age (d)	Human Size (mm)[a]	Human Somites	Human Ref.	Chicken Age (d/h)	Chicken Somites	Chicken Ref.
Muscular System																										
Primitive streak	9			10	8		10	7.25		10	7		10	13		10	17		10	17			10	0.5 d		10
First pharyngeal arch	10	2	5–12	1–4,6,9	8–8.5		5,7–9	9.5		27										23	10		83			
Ten-somite stage	10.5		10	10	8.5		10	8.5	10	10	8	10	10	15	10	10	23	10	10	25		10	10	1.5 d	10	10
Pharyngeal arches I and II, clefts I and II								9.5–9.75		27	8.5	17–20	44							24–27	3.3	13–20	9,71, 83–85, 91	2–3 d	22	86,87
Anterior limb bud	11	3.3	21–25	1–4,6,9	9.5–9.75	23±	5,7–9,39	10.5–11		27,57	8.75	17–20	42,44,46,47	16.5	23	56	25–26		46	26	3–5	21–29	39,73	51–56 h	26–28	30,86
Three pharyngeal arches	11.5	3.8	26–28	1–4,6,9	10		5,7–9,60	9.75		27	8.75		44	16.5	23	56	28–30		60	28–32	4–6	21–24	60,84, 92,166	53–55 h	24–27	86,87
Hind limb bud[b]	11.5–12	3.8	26–28	1–4,6,9,10	10–10.3		5,7–9,10	11–12		10,27,57	9		44	17.5–18.5	29	10,56	28–30		10,60	28–32	4–6	30–32	9,10, 60,71, 73,84, 85	2.2–3 d	29–32	10,86
Appearance of 4th pharyngeal arches	11.75	4.2	29–31	1–4,6,9	10.25		5,7–9	9.75–10.5		27	9		44	16.5	23	56	28–29		33	32	4.6–5	33–34	83,84	3.5 d	43–44	86
Maxillary processes meet nasolateral and medial processes	12–13	7	43–45	1–4,6,9,27				9	12–13	27	9.25		44				37–38		33	42–45	12–13		83	23–26 h	4–5	88
Subdivision of forelimb bud	12.75–14	7	43–45	1–4,6,9,27	10.5		60	12–13		27	10.5		44	20.75	39	56	30–32		60	31–35	5–7		60			
Cervical sinus obliterated	13–13.5	8–8.5	46–51	1–4,6,9,18,25	12		5,7–9	13–14		27	10		44	23.75		56	33–34		33	40–44	8–15		83,84			
Digital rays (forepaw)	13.5–14			3,4,10,27,29–32	12.3		7,10,29,40	14.5		10,27,29	10.25–11		10,42,44,48	22–23.75		10,56	34–35		10,46	35–37	8–11		10	4.75 d		30,86
Pleuroperitoneal canal closed; complete diaphragm	15	12	61–63	1–4,6,9										26		56					20		83			

Table 6.10 Comparative gestational milestones in muscular, skeletal, and integumentary system development (continued)

Description	Rat Age (d)	Rat Size (mm)	Rat Somites	Rat Ref.	Mouse Age (d)	Mouse Somites	Mouse Ref.	Rabbit Age (d)	Rabbit Somites	Rabbit Ref.	Hamster Age (d)	Hamster Somites	Hamster Ref.	Guinea Pig Age (d)	Guinea Pig Somites	Guinea Pig Ref.	Rhesus Monkey Age (d)	Rhesus Monkey Somites	Rhesus Monkey Ref.	Human Age (d)	Human Size (mm)[a]	Human Somites	Human Ref.	Chicken Age (d/h)	Chicken Somites	Chicken Ref.
Skeletal System																										
Subdivision of forelimb bud	12.75–14	7	43–45	1–4,6, 9,27	10.5		60	12–13		27	10.5		44	20.75	39	56	30–32		60	31–35	5–7		60			
Subdivision of hindlimb bud					11		60										30–33		60	37	8–11		73			
Mesenchymal condensation for ribs					12		162				10.5		44							31–33			162	5d		88
Auditory ossicles; mesoderm above dorsal extremity of tubotympanic cavity											11		44											5–6d		88
Digital rays (forepaw)	13–14.5	8.5	19–51	1–4,6, 9,10, 11,27	12–12.5		5,7–9, 10,60	14.5		10	11–14		10,44, 60	22		10	34–38		10,60	37–48	11–17		10,60, 73,84, 121	4.75d		10
Anlagen of centra and neural arches	14	9.5	52–55	1,2,9, 13–15, 17–22							9.5		44	18.5	31	56				28			167			
First sign of elbow					13		60										36–42		60	48–51	16–18		60	4.5–5d		86
Distinct finger rays; rim of hand plate crenated; primitive palatine processes					13		60				14		60				35–38		60	44–48	13–17		60			
Chondrification centers in ribs					13		162				11		44	23.75		56					15		162			
Primordial Meckel's cartilages					13		60				14		60				35–38		60	44–48	13–17		60			
Interdigital notches in hand plate	15–15.5			27	13.5		77	16–17		27							36–42		60	37±			77			
First sign of wrists					14		60										40–44		60	53–54	22–24		60			

COMPARATIVE FEATURES OF VERTEBRATE EMBRYOLOGY

Distal phalanges of fingers separated	16.5–17					60	14						40–44	60	53–54	22–24	60
First seven ribs chondrified and in contact with sternum							14.5	162								30–35	162
Digital separation of hindpaws	15.5–16.5			27									40–44	60	45	16–18	60
Complete separation of digits of forelimb	16.5–17			11,27					17–17.5	27			46–50	60	56–60	27–31	60
Palatal folds uniting (not all at the same stage of union)	17–18			1–4,6, 9,10, 11	15–16.5	5,7–9, 10	19.5	10	12	10	26	10	44–48	10,60	54–57	23–28	10,60
Primary ossification center in humerus with trabeculae					15.5	123							43–44	33		50	123
Ossification centers present in all ribs					15.5	163										68	163	NA[c]

Integumentary System

Milk line appears	12.125–13	5.2–8	8–36	1–4,6, 9, 13–15, 17–22	12	5,7–9	14–15		27	10	44	21.75	56			7	83
Nasolacrimal groove	13–14			27	11.5	60								30–34	60	37–42	8–11	60
Periderm present					12	164											16	164
Skin differentiated into stratum germinativum and stratum intermedium					13.5	165											22	165

Table 6.10 Comparative gestational milestones in muscular, skeletal, and integumentary system development (continued)

Description	Rat Age (d)	Rat Size (mm)	Rat Som-ites	Rat Ref.	Mouse Age (d)	Mouse Size (mm)	Mouse Som-ites	Mouse Ref.	Rabbit Age (d)	Rabbit Som-ites	Rabbit Ref.	Hamster Age (d)	Hamster Som-ites	Hamster Ref.	Guinea Pig Age (d)	Guinea Pig Som-ites	Guinea Pig Ref.	Rhesus Monkey Age (d)	Rhesus Monkey Som-ites	Rhesus Monkey Ref.	Human Age (d)	Human Size (mm)[a]	Human Som-ites	Human Ref.	Chicken Age (d/h)	Chicken Som-ites	Chicken Ref.
Stratum granulosum					15.5			164														60		164	N/A		
Vibrissary papillae appear on maxillary process	14–14.5	9.5	52–55	1,2,9, 11, 13–15, 17–22, 27					14–15		5,7, 8,27	10.5		44	23.75		56										
Hair follicle primordia appearing												12		44											N/A		
Distinct auricular hillocks	14–14.5	9.5	52–55	1–4,6, 9,27	11.5			60	14–15		27	14		60				32–34		33,60	37–48	8–17		60,73			
Eyelids — small ectodermal folds	15	12	61–63	1,2,9, 13–15, 17–22	13			5,7–9, 60,154				15–15.25		60				36–42		60	48–51	16–18		60,154			
Feather germs (chick)																									6.5–7d		86
First trunk hair papillae appear	15.5–16.5			11,27					17.5–18.75		27	12.5		44											NA		

[a] Crown-rump length.
[b] Hindlimb bud forms earlier in the rodents than in primates.
[c] In the chick, palatal folds do not fuse.

COMPARATIVE FEATURES OF VERTEBRATE EMBRYOLOGY

Table 6.11 Comparative gestational milestones in excretory system development (continued)

Description	Rat Age[a] (d)	Rat Size (mm)	Rat Somites	Rat Ref.	Mouse Age (d)	Mouse Somites	Mouse Ref.	Rabbit Age (d)	Rabbit Somites	Rabbit Ref.	Hamster Age (d)	Hamster Somites	Hamster Ref.	Guinea Pig Age (d)	Guinea Pig Somites	Guinea Pig Ref.	Rhesus Monkey Age (d)	Rhesus Monkey Somites	Rhesus Monkey Ref.	Human Age (d)	Human Size (mm)[a]	Human Somites	Human Ref.	Chicken Age (d/h)	Chicken Somites	Chicken Ref.
Intermediate mesoderm thickens nephrogenic cord	10	2	5–12	2,9,112,113	8.75		7				7.75		44	15.5	13	56				22	2.1		71			
Pronephros appears	10			10,26																22			10,26	1.4–1.5d		10,26
Ten-somite stage	10.5	10		10	8.5	10	10	8.5	10	10	8	10	10	15	10	10	23	10	10	25		10	10	1.5d	10	10
Nephrogenic cord with mesonephric tubules and duct	11	3.3	21–25	2,9,112,113	8d 21h–9.5	15–24	5,7–9,38,79				8.75		44	16.5		56				25–28	3.5	10	71,79	1.75d		30,86
Mesonephros appears	11.5		10	10,26	9.5		10,26													24–25	3.5	10	26,38			
Kidney: mesonephric duct enters urogenital sinus or cloaca	11.75–12	4.2	29–31	2,9,10,26,112,113	11		5,7–9,10,26,76	11.5		10	9–9.5		10,44	17.5–19	29–35	10,56	28–29		10,33	28±	4.5		10,26,71,76	3d	10	10
Primitive mesonephric tubules, mostly solid; Wolffian duct discontinuous	11.75–12.125	4.2–5.2	29–36	2,9,112,113										17.5–21.75	29–35	56				28	3.5		71	1.75d	19–32	88
Germinal epithelium (testis) appearing	12			10,26	11.5–12.5		10,26	13		10	9		10				33–34		10	38–40			10,26	4d	10	10,26
Kidney:ureteric bud	12.3			26	11–11.5		76,114	11.5		10	9.25–9.3		44	19–20		56	29–30		33	28–29			26,76	4d		114
Urorectal septum dividing cloaca	12.375–17	6–6.2	39–42	1,9,10,18,22–25,112,113				15		10	9.5		10							28–48			10			
Ureteric bud with metanephric "cap"	12.5			2,9,10,26,112,113				13		10	9.5		44	21		10	31–32		10	32	6		10,26,71	5d		10

Table 6.11 Comparative gestational milestones in excretory system development (continued)

Description	Rat Age[a] (d)	Rat Size (mm)	Rat Somites	Rat Ref.	Mouse Age (d)	Mouse Somites	Mouse Ref.	Rabbit Age (d)	Rabbit Somites	Rabbit Ref.	Hamster Age (d)	Hamster Somites	Hamster Ref.	Guinea Pig Age (d)	Guinea Pig Somites	Guinea Pig Ref.	Rhesus Monkey Age (d)	Rhesus Monkey Somites	Rhesus Monkey Ref.	Human Age (d)	Human Size (mm)[a]	Human Somites	Human Ref.	Chicken Age (d/h)	Chicken Somites	Chicken Ref.
Kidney: metanephros	12.5–12.8			10,26	11		10	23		10	10		10	23		10	38–39		10	35–37			10,26	6d		10,26
Paramesonephric duct appears	13.5	8.5	49–51	2,9,10, 26, 112, 113	10		5,7–9, 10,26	15		10	11		10,44	23.75		56	35–36		10	42–44			10,26	4d		10,26
Testes histologically differentiated	13.5			10,26, 114	12		10,26	16.5		10, 26	12		10,26	26		10, 26	36–39		10,26	46–48			10,26	13d		10,26
Paramesonephric duct reaches cloaca	15.5			10,26	14		10	20		10	13		10	26		10	37–38		10	49–56			10,26	7d		10
Rectum and urogenital sinus completely separated	17	16		2,9, 112, 113							15		44							43			59	N/A		

[a] Crown-rump length.

COMPARATIVE FEATURES OF VERTEBRATE EMBRYOLOGY

Table 6.12 Comparative gestational milestones in reproductive system development

Description	Rat Age[a] (d)	Rat Size (mm)	Rat Somites	Rat Ref.	Mouse Age (d)	Mouse Somites	Mouse Ref.	Rabbit Age (d)	Rabbit Somites	Rabbit Ref.	Hamster Age (d)	Hamster Somites	Hamster Ref.	Guinea Pig Age (d)	Guinea Pig Somites	Guinea Pig Ref.	Rhesus Monkey Age (d)	Rhesus Monkey Somites	Rhesus Monkey Ref.	Human Age (d)	Human Size (mm)[a]	Human Somites	Human Ref.	Chicken Age (d/h)	Chicken Somites	Chicken Ref.
Primitive streak	9			10	8		10	7.25		10	7		10	13		10	17		10	17			10	0.5d		10
Germ cells in yolk-sac epithelium	9.5–10	1.5–2	1–12	1–4,6, 9,18, 112, 131, 134, 168–171	8–8.5		5,7–9													27	3.3	13–20	9,84, 85,91			
Pronephros appears	10			114																22			114	1.4–1.5d		10,26
Ten-somite stage	10.5			10	8.5	10	10	8.5	10	10	8	10	10	15	10	10	23	10	10	25			10	1.5d	10	10
Germ cells in mesentery	10.5–11.5	2.4–3.8	13–28	1–4,6, 9,18, 112, 131, 134, 168–171	9.5		5,7–9													29	3.8	25–27	84,85, 116			
Mesonephros appears	11.5			114	9.5		114													24			114	3d		114
Germ cell migration reaches borders of mesonephric ridges	11.75	4.2	29–31	2,9,18, 112, 131, 134, 168–171										20.75	39	56				31	4.3	30–32	9,84, 85			
Germinal epithelium	12			10, 114	11.5–12.5		10,114	13		10	9		10				33–34		10	38–40			8	3–4d	38	10, 88, 114
Mesonephric duct enters urogenital sinus	12			10, 114	11		76,114	11.5		10	9.25–9.3		10,44	19	35	10,56	28–29		10,33	28	4.5		10,26, 71,76, 114			
Indifferent gonadal folds; rapid increase in number of germ cells	12.125	5.2	36	2,9, 112, 113													33–34		33							
Germ cells in genital ridges, end of migration	12.75	7	43–45	2,9,18, 112, 131, 134, 168–171																35–37	6		9,84, 121, 174			

Table 6.12 Comparative gestational milestones in reproductive system development (continued)

Description	Rat Age (d)	Rat Size (mm)	Rat Som-ites	Rat Ref.	Mouse Age (d)	Mouse Som-ites	Mouse Ref.	Rabbit Age (d)	Rabbit Som-ites	Rabbit Ref.	Hamster Age (d)	Hamster Som-ites	Hamster Ref.	Guinea Pig Age (d)	Guinea Pig Som-ites	Guinea Pig Ref.	Rhesus Monkey Age (d)	Rhesus Monkey Som-ites	Rhesus Monkey Ref.	Human Age (d)	Human Size (mm)[a]	Human Som-ites	Human Ref.	Chicken Age (d/h)	Chicken Som-ites	Chicken Ref.
Paramesonephric duct appears	13.5	8.5	49–51	2,9–10,112–114	10		5,7–9,10,114	15		10	11		10,44	23.75		56	35–36		10,33	42–44			10,114	4d		10,88,114
Gonads begin sexual differentiation	13.5	8.5	49–51	2,9,18,25,129–134																	17		83			
Histologic differentiation of testes	13.5–14.5	10.5	56–60	2,9,18,112,114,131,134,168–171	12.5		5,7–9,172				12		11,173				38–39		33	46–48	14–16		85,114,16,175	5.5d		114
Gonad, rete cords; in stroma between genital primordium and mesonephros					12.5		172							21.75		56					17		83	5d		88
Oogonia; germ cells in secondary sexual cords of ovarian cortex	14.5	10.5	56–60	2,9,18,112,131,134,168–171																	17		83	7–8d		88
Paramesonephric ducts reach urogenital sinus	15.5	14.2	64	2,9,18,112,114,131,134,168–171																49–56			114	7d		88
Indifferent external genitalia	19			10																37			10			
Differentiation of male and female external genitalia																				56–70	26–45		9,66,84,121			

[a] Crown-rump length.

References

1. Mossman, H.W., Comparative morphogenesis of the foetal membranes and accessory uterine structures, *Contrib. Embryol. Carneg. Inst.*, 26, 129, 1937.
2. Beck, F., Comparative placental pathology and function, *Environ. Health Perspect.*, 18, 5, 1976.
3. Ramsey, E.M., *The Placenta: Human and Animal*, Praeger, New York, 1982.
4. Enders, A.C., Mechanisms of implantation of the blastocyst, in *Biology of Reproduction: Basic and Clinical Studies*, Velardo, J.T. and Kasprow, B A., Eds., III Pan American Congress of Anatomy, New Orleans, LA, 1972, p. 313.
5. Reynolds, S.R.M., On growth and form in the hemochorial placenta: an essay on the physical forces that shape the chorionic trophoblast, in *Biology of Reproduction: Basic and Clinical Studies*, Velardo, J.T. and Kasprow, B.A., Eds., III Pan American Congress of Anatomy, New Orleans, LA, 1972, p. 7.
6. Freese, U.E., The maternal-fetal vascular relationships in the human and rhesus monkey, in *Biology of Reproduction: Basic and Clinical Studies*, Velardo, J.T. and Kasprow, B.A., Eds., III Pan American Congress of Anatomy, New Orleans, LA, 1972, p. 335.
7. Harris, J.S.W. and Ramsey, E.M., The morphology of human uteroplacental vasculature, *Contrib. Embryol.*, 38, 43, 1966.
8. Wild, A.E., Endocytic mechanisms of protein transfer across the placenta, *Placenta*, 1, 165, 1981.
9. Miller, R.K., Koszalka, T.R., and Brent, R.L., Transport mechanisms for molecules across placental membranes, in *Cell Surface Reviews*, Poste, G. and Nicholson, G., Eds., Elsevier/North-Holland, Amsterdam, 1976, p. 145.
10. Miller, R.K., Placental transfer and function: The interface for drugs and chemicals in the conceptus, in *Drug and Chemical Action in Pregnancy: Pharmacologic and Toxicologic Principles*, Fabro, S. and Scialli, A.R., Eds., Marcel Dekker, New York, 1986, p. 123.
11. Arceci, R.J., Baas, F., Raponi, R., Horwitz, S.B., Housman, D., and Croop, J.M., Multidrug resistance gene expression is controlled by steroid hormones in the secretory epithelium of the uterus, *Mol. Reprod. Dev.*, 25, 101, 1990.
12. Nakamura, Y., Ikeda, S., Furukawa, T., Sumizawa, T., Tani, A., Akiyama, S., and Nagata, Y., Function of P-glycoprotein expressed in placenta and mole, *Biochem. Biophys. Res. Commun.*, 27, 235, 1997.
13. Smit, J.W., Huisman, M.T., van Tellingen, O., Wiltshire, H.R., and Schinkel, A.H., Absence or pharmacological blocking of placental P-glycoprotein profoundly increases fetal drug exposure, *J. Clin. Invest.*, 104, 1441, 1999.
14. Miller, R.K., Levin, A.A., and Ng, W.W., The placenta: relevance to toxicology, in *Reproductive and Developmental Toxicity of Metals*, Clarkson, T., Nordberg, G., and Sager, P., Eds., Plenum Press, New York, 1983, p. 569.
15. Juchau, M.R., Mechanisms of drug biotransformation reactions in the placenta, *Fed. Proc.*, 31, 48, 1972.
16. Juchau, M.R., Drug biotransformation in the placenta, *Pharmacolog. Therapeut.*, 8, 501, 1980.
17. Juchau, M.R. and Rettie, A.E., The metabolic role of the placenta, in *Drug and Chemical Action in Pregnancy: Pharmacologic and Toxicologic Principles,* Fabro, S. and Scialli, A. R., Eds., Marcel Dekker, New York, 1986, p. 153.
18. Kelman, B.J., Effects of toxic agents on movements of material across the placenta, *Fed. Proc.*, 38, 2246, 1979.
19. Goodman, D.R., James, R.C., and Harbison, R.D., Placental toxicology, *Food Chem. Toxicol.*, 20, 123, 1982.
20. Juchau, M.R., The role of the placenta in developmental toxicology, in *Developmental Toxicology*, Snell, K., Ed., Praeger, New York, 1982, p. 187.
21. Millicovsky, G. and DeSesso, J.M., Uterine versus umbilical vascular clamping: differential effects on the developing embryo, *Teratology*, 22, 335, 1980.
22. Millicovsky, G., DeSesso, J.M., Clark, K.E., and Kleinman, L. I., Effects of hydroxyurea on maternal hemodynamics during pregnancy: a maternally mediated mechanism of embryotoxicity, *Am. J. Obstet. Gynecol.*, 140, 747, 1981.
23. Beck, F., Lloyd, J.B., and Griffiths, A., Lysosomal enzyme inhibition by trypan blue: a theory of teratogenesis, *Science*, 157, 1180, 1967.
24. Williams, K.E., Roberts, G., Kidston, M.E., Beck, F., and Lloyd, J.B., Inhibition of pinocytosis in rat yolk-sac by trypan blue, *Teratology*, 14, 343, 1976.

25. New, D.A.T. and Brent, R.L., Effect of yolk-sac antibody on rat embryos grown in culture, *J. Embryol. Exp. Morphol.*, 27, 543, 1972.
26. Brent, R.L., Beckman, D.A., Jensen, M., Koszalka, T.R., and Damjanov, I., The embryopathologic effects of teratogenic yolk sac antiserum, *Trophoblast Res.*, 1, 335, 1983.
27. Parizek, J., Vascular changes at sites of oestrogen biosynthesis produced by parenteral injection of cadmium salts: the destruction of placenta by cadmium salts, *J. Reprod. Fertil.*, 7, 263, 1964.
28. Dencker, L., Possible mechanisms of cadmium fetotoxicity in golden hamsters and mice: uptake by the embryo, placenta and ovary, *J. Reprod. Fertil.*, 44, 461, 1975.
29. Morford, L.L., Henck, J.W., Breslin, W.J., and DeSesso, J.M., Hazard identification and predictability of children's health risks from animal data, *Environ. Health. Perspect.*, 112, 266, 2004.
30. Browder, L.W., Erickson, C.A., and Jeffrey, W. R., *Developmental Biology*, 3rd ed., W. B. Saunders, Philadelphia, 1991.
31. Gilbert, S.N., *Developmental Biology*, 5th ed., Sinauer Associates, Sunderland, MA, 1997.
32. Larsen, W.J., *Human Embryology*, 3rd ed, Churchill Livingstone, New York and London, 2001.
33. Carlson, B.M., *Human Embryology and Developmental Biology*, 3rd ed., C.V. Mosby, St. Louis, 2004.
34. Hay, E. D., Embryonic induction and tissue interaction during morphogenesis, in *Birth Defects*, Littlefield, J.W. and De Grouchy, J., Eds., Excerpta Medica, Amsterdam, 1978, p. 126.
35. Holland, P.W.H. and Hogan, B.L.M., Expression of homeobox genes during mouse development: a review, *Genes Dev.*, 2, 773, 1998.
36. Kessel, M. and Gruss, P., Murine developmental control genes, *Science*, 249, 374, 1990.
37. DeRoberts, E.M., Oliver, G., and Wright, C.V.E., Homeobox genes and the vertebrate body plan, *Sci. Am.*, 7, 46, 1990.
38. Edelman, G., *Topobiology: An Introduction to Molecular Embryology*, Basic Books, New York, 1988.
39. National Research Council, *Scientific Frontiers in Developmental Toxicology and Risk Assessment*, National Academy Press, Washington, DC, 2000.
40. DeSesso, J.M., Maternal factors in developmental toxicity, *Teratogenesis, Carcinog. Mutagen.*, 7, 225, 1987.
41. Wilson, J.G., *Environment and Birth Defects*, Academic Press, New York, 1973.
42. Beckman, D.A. and Brent, R.L., Basic principles of teratology, in *Medicine of the Fetus & Mother*, Reece, E.A., Hobbins, J.C., Mahoney, M.J., and Petrie, R.H., Eds., J. B. Lippincott, Philadelphia, 1992, p. 293.
43. DeSesso, J.M. and Harris, S.B., Principles underlying developmental toxicity, in *Toxicology and Risk Assessment*, Fan, A.M. and Chang, P., Eds., Marcel Dekker, New York, 1996, p. 37.
44. Mohrenweiser, H. and Zingg, B., Mosaicism: the embryo as a target for induction of mutations leading to cancer and genetic disease, *Environ. Mol. Mutagen.*, 25(Suppl. 26), 21, 1995.
45. Drost, J.B. and Lee, W.R., Biological basis of germline mutation: comparisons of spontaneous germline mutation rates among Drosophila, mouse, and human, *Environ. Mol. Mutagen.*, 25(Suppl. 26), 48, 1995.
46. Generoso, W.M., Rutledge, J.C., Cain, K.T., Hughes, L.A., and Downing, D.J., Mutagen-induced fetal anomalies and death following treatment of females within hours after mating, *Mutat. Res.*, 199, 175, 1988.
47. Wilson, J.G., Embryological considerations in teratology, in *Teratology: Principles and Techniques*, Wilson, J.G. and Warkany, J., Eds., University of Chicago Press, Chicago, 1965.
48. Ritter, E.J., Scott, W.J., and Wilson, J.G., Relationship of temporal patterns of cell death and development to malformations in the rat limb: possible mechanisms of teratogenesis with DNA synthesis inhibitors, *Teratology*, 7, 219, 1973.
49. Lammer, E.J., A phenocopy of the retinoic acid embryopathy following maternal use of etretinate that ended one year before conception, *Teratology*, 37, 472, 1988.
50. Warkany, J., Sensitive or critical periods in teratogenesis: uses and abuses of embryologic timetables, *Congenital Malformations*, Year Book Medical Publishers, Chicago, 1971, p. 49.
51. Hamilton, W.J. and Mossman, H.W., *Hamilton, Boyd and Mossman's Human Embryology*, 4th ed., Williams and Wilkins, Baltimore, 1972.
52. O'Rahilly, R. and Muller, F., *Developmental Stages in Human Embryos*, Publication No. 637, Carnegie Institute, Washington, DC, 1987.

53. Edwards, J.A., The external development of the rabbit and rat embryo, in *Advances in Teratology*, Woolam, D.H., Ed., Academic Press, New York, 1968, p. 239.
54. Hebel, R. and Stromberg, M.W., *Anatomy and Embryology of the Rat*, BioMed Verlag, Worthsee, Germany, 1986.
55. Nelsen, O.E., *Comparative Embryology of the Vertebrates*, McGraw-Hill, New York, 1953.
56. Snell, G.D. and Stevens, L.C., The early embryology of the mouse, in *Biology of the Laboratory Mouse*, Green, E.L., Ed., Blakiston, Philadelphia, 1966, p. 205.
57. Rugh, R., *The Mouse: Its Reproduction and Development,* Burgess, Minneapolis, 1968.
58. Boyer, C.C., Chronology of development of the golden hamster, *J. Morphol.*, 92, 1, 1953.
59. Waterman, A.J., Studies on the normal development of the New Zealand White strain of rabbit, *Am. J. Anat.*, 72, 473, 1943.
60. Scott, J.P., The embryology of the guinea pig. I. Table of normal development, *Am. J. Anat.*, 60, 397, 1937.
61. Heuser, C.H. and Streeter, G.L., Development of the macaque embryo, *Contrib. Embryol.*, 29, 15, 1941.
62. Hendrickx, A.G. and Sawyer, R.H., Embryology of the rhesus monkey, in *The Rhesus Monkey*, Vol. 2, Bourne, G.H., Ed., Academic Press, New York, 1975, p. 141.
63. Hamburger, V. and Hamilton, H.L., A series of normal stages in the development of the chick embryo, *J. Morphol.*, 88, 49, 1951.
64. Patten, B.M., *Early Embryology of the Chick*, 4th ed., McGraw-Hill, New York, 1951.
65. Arey, L.B., *Developmental Anatomy,* 5th ed., W.B. Saunders, Philadelphia, 1965.
66. Streeter, G.L., Developmental horizons in human embryos (XI-XII), *Contrib. Embryol.*, 30, 211, 1942.
67. Streeter, G.L., Developmental horizons in human embryos (XIII-XIV), *Contrib. Embryol.*, 31, 27, 1945.
68. Streeter, G.L., Developmental horizons in human embryos (XV-XVIII), *Contrib. Embryol.*, 32, 133, 1948.
69. Wanek, N., Muneoka, K., Holler, D.G., Burton, R., and Bryant, S.V., A staging system for mouse limb development, *J. Exp. Zool.*, 249, 41, 1989.
70. Witschi, E., *Development of Vertebrates*, W.B. Saunders Company, Philadelphia, 1956.
71. DeSesso, J.M., Niewenhuis, R.J., and Goeringer, G.C., Lectin teratogenesis II. Demonstration of increased binding of concanaualih A to limb buds of rabbit embryos during the sensitive period, *Teratology,* 39, 395, 1989.
72. Holson, J.F., Stump, D.G., Pearce, L.B., Watson, R.E., and DeSesso, J.M., Mode of action: yolk sac poisoning and impeded histiotrophic nutrition. HBOC-related congenital malformations, *Crit. Rev. Toxicol.,* in press, 2005.
73. DeSesso, J.M. and Lavin, A.L., Stem cell research. *Sigma,* Fall edition, 2002, p. 11. Available online at http://www.mitretek.org/pubs/sigma/fall2002/chap3.pdf.

References for Tables 3 to 12

1. Butcher, E.O., The development of the somites in the white rat (*Mus norvegicus albinus*) and the fate of the myotomes, neural tube, and gut in the tail, *Am. J. Anat.,* 44, 381, 1929.
2. Henneberg, B., *Normentafel zur Entwicklungsgeschichte der Wanderratte (*Rattus norvegicus *Erxleben),* G. Fischer, Jena, Germany, 1937.
3. Huber, G.C., The development of the albino rat, *Mus norvegicus albinus, J. Morphol.*, 26, 1, 1915.
4. Long, J.A. and Burlingame, M.L., The development of the external form of the rat with some observations on the origin of the extraembryonic coelom and fetal membranes, *Univ. Calif., (Berkeley) Mem. Zool.*, 43, 143, 1938.
5. MacDowell, E.C., Allen, E., and MacDowell, C.G., The prenatal growth of the mouse, *J. Gen. Physiol.,* 11, 57, 1927.
6. Nicholas, J.S. and Rudnick, D., Development of rat embryos of egg-cylinder to head-fold stages in plasma cultures, *J. Exptl. Zool.,* 78, 205, 1938.

7. Otis, E.M. and Brent, R., Equivalent ages in mouse and human embryos, *Anat. Record,* 120, 33, 1954.
8. Snell, G.D., *The Early Embryology of the Mouse,* Blakiston, Philadelphia, 1941.
9. Witschi, E., *The Development of Vertebrates,* W.B. Saunders, Philadelphia, 1956.
10. Hoar, R.M. and Monie, I.W., Comparative development of specific organ systems, in *Developmental Toxicology,* Kimmel, C.A. and Buelke-Sam, J., Eds., Raven Press, New York 1981, p. 13.
11. Christie, G., Developmental stages in somite and post-somite rat embryo, based on external appearance, and including some features of the macroscopic development of oral cavity, *J. Morphol.,* 114, 263, 1964.
12. Altman, P., and Dittmer, D.S., Eds., *Growth Including Reproduction and Morphological Development,* Federation of American Societies for Experimental Biology, Washington, D.C., 1962, 304.
13. Burlingame, P.L. and Long, J.A., The development of the external form of the rat, with some observations on the origin of the extraembryonic coelom and foetal membranes, *Univ. Calif. (Berkeley) Publs. Zool.,* 43, 249, 1939.
14. Goss, C.M., Development of the median coordinated ventricle from the lateral hearts in rat embryos with thirteen somites, *Anat. Record,* 112, 761, 1952.
15. Harman, M.T. and Prickett, M., The development of the external form of the guinea-pig (Cavia Cobaya) between the ages of eleven days and twenty days of gestation, *Am. J. Anat.,* 49, 351, 1932.
16. Krehbiel, R.H., Cytological studies of the decidual reaction in the rat during early pregnancy and the production of deciduomata, *Physiol. Zool.,* 10, 212, 1937.
17. Marit, C., Les premiers stades de la formation du myocarde chez le rat, *Bull. Acad. Roy. Med. Belg.,* 24, 451, 1959.
18. Witschi, E., unpublished, *Embryology of the Rat,* State Univ. Iowa, Iowa City, 1961.
19. Aoto, T., *J. Fac. Sci. Hokkaido Univ.,* 13, 364, 1957.
20. Herrmann, J., Nicholas, J.S., and Vosgian, M.E., *Proc. Soc. Exptl. Biol. Med.,* 72, 454, 1949.
21. Johnson, M.L., Time and order of appearance of ossification centers in albino mouse, *Am. J. Anat.,* 52, 241, 1933.
22. Schour, I. and Massler, M., The teeth, in *The Rat in Laboratory Investigation,* Griffith, J.Q. and Farris, E.J., Eds., J.B. Lippincott, Philadelphia, 1942.
23. Adelmann, H.B., The development of the neural folds and cranial ganglia of the rat, *J. Comp. Neurol.,* 39, 19, 1925.
24. Sissman, N.J., Developmental landmarks in cardiac morphogenesis: comparative chronology, *Am. J. Cardiol.,* 25, 141, 1970.
25. Rogers, W.M., Development of pharynx and pharyngeal derivatives in white rat (*Mus norvegicus albinus*), *Am. J. Anat.,* 44, 283, 1929.
26. Hoar, R.M., Comparative developmental aspects of selected organ systems. II. Gastrointestinal and urogenital systems, *Environ. Health Perspect.,* 1, 61, 1976.
27. Edwards, J.A., The external development of the rabbit and rat embryo, in *Advances in Teratology,* Vol. 3, Woollam, D.H.M., Ed., Academic Press, New York, 1968, p. 239.
28. Nelsen, O.E., *Comparative Embryology of the Vertebrates,* Blakiston, New York, 1953.
29. Davies, J. and Hesseldahyl, H., Comparative embryology of mammalian blastocysts, in *Biology of the Blastocyst,* Blandau, R.J., Ed., University of Chicago Press, Chicago, 1971, p. 27.
30. Monie, I.W., Comparative development of rat, chick and human embryos, in *Teratologic Workshop Manual* (supplement), Pharmaceutical Manufacturers Associations, Berkeley, 1965, p. 146.
31. Nicholas, J.S., Experimental methods and rat embryos, in *The Rat in Laboratory Investigation,* Farris, E.J. and Griffith, J.Q., Eds., Hafner Publishing, New York, 1962, p. 51.
32. Witschi, E, Development of the rat, in *Growth Including Reproduction and Morphological Development,* Altman, P. and Dittmer, D.S., Eds., Federation of American Societies for Experimental Biology, Washington, D.C., 1962, p. 304.
33. Hendrickx, A.G. and Sawyer, R.H., Embryology of the rhesus monkey, *The Rhesus Monkey, Volume II: Management, Reproduction and Pathology,* Bourne, G.H., Ed., Academic Press, New York, 1975, p. 141.
34. Theiler, K., *The House Mouse. Development and Normal Stages from Fertilization to 4 Weeks of Age,* Springer, Berlin, Heidelberg, New York, 1972.
35. Heuser, C.H., Rock, J., and Hertig, A.T., Two human embryos showing early stages of the definitive yolk sac, *Contribs. Embryol. Carnegie Inst. Wash.,* 31, 85, 1945.

36. Heuser, C.H., A presomite human embryo with a definite chorda canal, *Contribs. Embryol. Carnegie Inst. Wash.,* 23, 251, 1932.
37. Ludwig, E., Uber einen operativ gewonnenen menschlichen Embryo mit einem Ursegmente (Embryo Dal), *Morphol. Jahrb,* 59, 41, 1928.
38. West, C.M., A human embryo of twenty-five somites, *J. Anat.,* 71, 169, 1937.
39. Streeter, G.L., Developmental horizons in human embryos. Description of age group XI, 13 to 20 somites, and age group XII, 21 to 29 somites, *Contribs. Embryol. Carnegie Inst. Wash.,* 30, 211, 1942.
40. Gruneberg, H., The development of some external features in mouse embryos, *J. Hered.,* 34, 88, 1943.
41. Tuchmann-Duplessis, H., David, G., and Haegel, P., *Illustrated Human Embryology: Embryogenesis,* Vol. 1, Hurley, L.S. Translators, Masson & Co., Paris, 1972.
42. Graves, A.P., Development of the golden hamster *Cricetus auratus* waterhouse, during the first nine days, *Amer. J. Anat.,* 77, 219, 1945.
43. Boyer, C.C., Development of the Golden Hamster *Cricetus Auratus* with Special Reference to the Major Circulatory Channels, doctoral dissertation, Duke University, Durham, NC, 1948.
44. Boyer, C.C., Chronology of the development for the golden hamster, *J. Morph.,* 92, 1, 1953.
45. Tanimura, T., Nelson, T., Hollingsworth, R.R., and Shepard, T.H., Weight standards for organs from early human fetuses, *Anat. Rec.* 171, 227, 1971.
46. Heuser, C.H. and Streeter, G.L., Development of the macaque embryo, *Contrib. Embryol.,* 29, 15, 1941.
47. Iffy, L., Shepard, T.H., Jakobovits, A., Lemire, R.J., and Kerner, P., The rate of growth in young embryos of Streeter's horizons XIII to XXIII, *Acta Anat.* (Basel), 66, 178, 1967.
48. Shenefelt, R.E., Morphogenesis of malformations in hamsters caused by retinoic acid: relation to dose and stage of treatment, *Teratology,* 5, 103, 1972.
49. Lams, H., Etude de l'oeuf de cobaye aux premiers stades de l'embryogenese, *Arch. de Biol., T.* 28, 229, 1913.
50. Squire, R.R., The living egg and early stages of its development in the guinea pig, *Carnegie Institution Contributions to Embryology,* 137, 225, 1932.
51. Selenka, E., *Studien uber Entwicklungsgeschichte der Thiere,* Bd. 3, Wiesbaden, 1884.
52. Maclaren, N., Development of Cavia: implantation., *Trans. Roy. Soc.* Edinburgh, 55, 115, 1926.
53. Sansom, G.S. and Hill, J.P., Observations on the structure and mode of implantation of the blastocyst of Cavia, *Trans. Zool. Soc.* London, 21, 295, 1931.
54. Hall, E.K., Further experiments on the intrinsic contractility of the embryonic rat heart, *Anat. Record,* 118, 175, 1954.
55. Huber, G.C., On the anlage and morphogenesis of the chorda dorsalis in mammalia, in particular the guinea pig (Cavia cobaya), *Anat. Rec.,* 14, 217, 1918.
56. Scott, J.P., The embryology of the guinea pig. I. Table of normal development, *Am. J. Anat.,* 60, 397, 1937.
57. Waterman, A.J., Studies of normal development of the New Zealand White strain of rabbit, *Am. J. Anat.,* 72, 473, 1943.
58. Hensen, V., Beobachtungen uber die Befruchtung und Entwicklung des Kaninchens und Meerschweinchens, *Zeitsch. f. Anat. u. Entwick.,* Bd.1, S.213, 353, 1876.
59. Shepard, T.H., *Growth and Development of the Human Embryo and Fetus. Endocrine and Genetic Diseases of Childhood,* Gardner, L.I., Ed., W.B. Saunders, Philadelphia, 1969.
60. Gribnau, A.A.M. and Geijsberts, L.G.M., Developmental stages in the Rhesus monkey (*Macaca mulatto*), *Adv. Anat. Embryol. Cell Biol.,* 68, 1, 1981.
61. Menkin, M.F. and Rock, J., *In vitro* fertilization and cleavage of human ovarian eggs, *Am. J. Obstet. Gynecol,* 55, 440, 1948.
62. Shettles, L.B., Observations on human follicular and tubal ova, *Am. J. Obstet. Gynecol.,* 66, 235, 1953.
63. Hertig, A.T., Adams, E.C., and Mulligan, W.J., On the preimplantation stages of the human ovum: a description of four normal and four abnormal specimens ranging from the second to the fifth day of development, *Contribs. Embryol. Carnegie Inst. Wash.,* 35, 199, 1954.
64. O'Rahilly, R., Early human development and the chief sources of information on staged human embryos, *Eur. J. Obstet Gynecol. Reprod. Biol.,* 9/4, 273, 1979.
65. Hertig, A.T. and Rock, J., Two human ova of the pre-villous stage, having a developmental age of about seven and nine days respectively, *Contribs. Embryol. Carnegie Inst. Wash.,* 31, 65, 1945.

66. Keibel, F., in *Manual of Human Embryology*, Vol. 1, Keibel, F. and Mall, F.P., Eds., Lippincott, Philadelphia, 1910, p. 59.
67. Heuser, C.H. and Corner, G.W., Developmental horizons in human embryos. Description of age group X, 4 to 12 somites, *Contribs. Embryol. Carnegie Inst. Wash.*, 36, 29, 1957.
68. Politzer, G. and Hann, F., Z. Uber die Entwicklung der branchiogenen Organe beim Menschen, *Anat. Entwicklungsgeschichte,* 104, 670, 1935.
69. Sensenig, E.C., The development of the occipital and cervical segments and their associated structures in human embryos, *Contribs. Embryol. Carnegie Inst. Wash.*, 36, 141, 1957.
70. Sternberg, H., Z., Beschreibung eines menschlichen Embryos mit vier Ursegmentpaaren, nebst Bemerkungen uber die Anlage und fruheste Entwicklung einiger Organe beim Menschen, *Anat. Entwicklungsgeschichte,* 82, 747, 1927.
71. Lemire, R.J., Loeser, J.D., Leech, R.W., and Alvord, E.W., Eds., *Normal and Abnormal Development of the Human Nervous System*, Harper and Row, Hagerstown, MD, 1975.
72. Hertig, A.T. and Rock, J., Two human ova of the pre-villous stage, having an ovulation age of about eleven and twelve days respectively, *Contribs. Embryol. Carnegie Inst. Wash.*, 29, 127, 1941.
73. O'Rahilly, R. and Muller, F., *Human Embryology & Teratology*, Wiley-Liss, New York, 1992.
74. Bartelmez, G.W. and Evans, H.M., Development of the human embryo during the period of somite formation including embryos with 2 to 16 pairs of somites, *Contribs. Embryol. Carnegie Inst. Wash.*, 17, 1, 1926.
75. Hochstetter, F., *Beitrage zur Entwicklungsgeschichte des menschilichen Gehirns*, F. Deuticke, Leipzig, 1929.
76. Streeter, G.L., Developmental horizons in human embryos. Description of age group XIII, embryos about 4 or 5 millimeters long, and age group XIV, period of indentation of the lens vesicle, *Contribs. Embryol. Carnegie Inst. Wash.*, 31, 27, 1945.
77. Streeter, G.L., Developmental horizons in human embryos. Description of age groups XV, XVI, XVII, and XVIII, being the third issue of a survey of the Carnegie collection, *Contribs. Embryol. Carnegie Inst. Wash.*, 32, 133, 1948.
78. Atwell, W., A human embryo with seventeen pairs of somites, *Contribs. Embryol. Carnegie Inst. Wash.*, 21, 1, 1930.
79. Corner, G.W., A well-preserved human embryo of 10 somites, *Contribs. Embryol. Carnegie Inst. Wash.*, 20, 81, 1929.
80. Ludwig, E., Embryon humain avec dix paires de somites mesoblastiques, *C. R. Ass. Anat.*, 24, 580, 1929.
81. Payne, F., General description of a 7-somite human embryo, *Contribs. Embryol. Carnegie Inst. Wash.*, 16, 115, 1925.
82. Bartelmez, G.W., The subdivisions of the neural folds in man, *J. Comp. Neurol.*, 26, 79, 1923.
83. *Hamilton, Boyd, & Mossman's Human Embryology*, 4th ed., Hamilton, W.J. and Mossman, H.W., Eds., The Williams & Wilkins Co., Baltimore, 1972.
84. Streeter, G.L., Developmental horizons in human embryos. Age groups XI-XXIII, *Contribs. Embryol. Carnegie Inst. Wash.*, Embryol. reprint vol. 2, 1951.
85. Witschi, E., Migration of the germ cells of human embryos from the yolk sac to the primitive gonadal folds, *Contribs. Embryol. Carnegie Inst. Wash.*, 32, 67, 1948.
86. Hamburger, V. and Hamilton, H.L., A series of normal stages in the development of the chick embryo, *J. Morphol.*, 88, 49, 1951.
87. Patten, B.M., *Early Embryology of the Chick*, McGraw-Hill, New York, 1951.
88. Hamilton, H.L., *Lillie's Development of the Chick*. H. Holt, New York, 1952.
89. Hebel, R. and Stromberg, M.W., *Anatomy and Embryology of the Laboratory Rat*, BioMed Verlag, Fed. Rep. Germany, 1986.
90. Wilson, J., *Embryology of the Human Heart*, Ward's Natural Science Est., Inc., Rochester, New York, 1945.
91. Heuser, C.H., A human embryo with 14 pairs of somites, *Contribs. Embryol. Carnegie Inst. Wash.*, 22, 135, 1930.
92. Davis, C.L., Description of a human embryo having twenty paired somites, *Contribs. Embryol. Carnegie Inst. Wash.*, 15, 1, 1923.
93. Congdon, E.D., Transformation of the aortic-arch system during the development of the human embryo, *Contribs. Embryol. Carnegie Inst. Wash.*, 14, 47, 1922.

94. McClure, C.F.W. and Butler, E.G., The development of the vena cava inferior in man, *Am. J. Anat.*, 35, 331, 1925.
95. Kramer, T.C., The partitioning of the truncus and conus and the formation of the membranous portion of the interventricular septum in the human heart, *Am. J. Anat.*, 71, 342, 1942.
96. Arey, L.B., *Developmental Anatomy*, 5th ed, W.B. Saunders, Philadelphia, 1946.
97. Bloom, W. and Bartelmez, G.W., Hematopoiesis in young human embryos, *Am. J. Anat.*, 67, 21, 1940.
98. Gilmour, J.R., Normal haemopoiesis in intra-uterine and neonatal life, *J. Pathol. Bacteriol.*, 52, 25, 1941.
99. Pitt, J.A. and Carney, E.W., Development of a morphologically-based scoring system for postimplantation New Zealand White rabbit embryos, *Teratology*, 59, 88, 1999.
100. Jageroos, B.H., On early development of vascular system; development of blood and blood-vessels in chorion of man, *Acta Soc. Med. Duodecim*, B, 19, 1, 1934.
101. Kampmeier, O.F., Hemopoietic foci in wall of thoracic duct, and cellular constituents of its lymph stream in human fetus, *Am. J. Anat.*, 42, 181, 1928.
102. Kling, C.A., Studien uber die Entwicklung der Lymphdrusen beim Menschen, *Arch. Mikroskop. Anat. u. Entwicklungsmech.*, 63, 575, 1904.
103. Sabin, F.R., in *Manual of Human Embryology*, Vol. 2, Edibel, F. and Mall, F.P., Eds., J.B. Lippincott, Philadelphia, 1912, p. 709.
104. Davis, C.L., Development of human heart from its first appearance to stage found in embryos of 20 paired somites, *Contribs. Embryol. Carnegie Inst. Wash.*, 19, 245, 1927.
105. Ebert, J., An analysis of the synthesis and distribution of the contractile protein, myosin, in the development of the heart, *Proc. Natl. Acad. Sci. U. S.*, 39, 333, 1953.
106. Goss, C.M., The physiology of the mammalian heart before circulation, *Am. J. Physiol.*, 13, 146, 1942.
107. Marcell, M.P. and Exchaquet, J.P., L'electrocardiogramme du foetus humain avec un cas de double rhythme auriculaire verifie, *Arch. maladies coeur et vasseaux*, 31, 504, 1938.
108. Odgers, P.N.B., The development of the pars membranacea septi in the human heart, *J. Anat.*, 72, 247, 1938.
109. Odgers, P.N.B., The development of the atrio-ventricular valves in man, *J. Anat.*, 73, 643, 1939.
110. Robb, J.S., Kaylor, C.T., and Trumers, W.G., A study of specialized heart tissue at various stages of development of the human fetal heart, *Am. J. Med.*, 5, 324, 1948.
111. Walls, E.W., The development of the specialized conducting tissue of the human heart, *J. Anat.*, 81, 93, 1947.
112. Hall, K., The structure and development of the urethral sinus in the rat, *J. Anat.*, 70, 413, 1936.
113. Ludwig, E., Uber die Anlage der Nach-niere beider Maus, *Acta Anat.*, 4, 193, 1947.
114. Hoar, R.M., Comparative female reproductive tract development and morphology, *Environ. Health Perspect.*, 24, 1, 1978.
115. Ingalls, N.W., A human embryo at the beginning of segmentation, with special reference to the vascular system, *Contribs. Embryol. Carnegie Inst. Wash.*, 11, 61, 1920.
116. Patten, B.M., *Human Embryology*, 2nd Ed., Blakiston, New York, 1953.
117. Hamilton, W.J., Boyd, J.D., and Mossman, H.W., *Human Embryology*, W. Heffer, Cambridge, 1952.
118. Sulik, K.K. and Sadler, T.W., Postulated mechanisms underlying the development of neural tube defects. Insights from in vitro and in vivo studies, *Ann N Y Acad. Sci.*, 678, 8, 1993.
119. Bremer, J.L., Description of a 4 mm. human embryo, *Am. J. Anat.*, 5, 456, 1906.
120. Simkins, C.S., *Textbook of Human Embryology*, F. A. Davis, Philadelphia, 1931.
121. His, W., *Anatomie menschlicher Embryonen. Atlas*, Vogel, Leipzig, 1880-1885.
122. Nelson, W., et al., Congenital absence of gall bladder, *Surgery*, 25, 916, 1949.
123. Streeter, G.L., Developmental horizons in human embryos (fourth issue): a review of the histogenesis of cartilage and bone, *Contribs. Embryol. Carnegie Inst. Wash.*, 33, 149, 1949.
124. Odgers, P.N.B., Some observations on the development of the ventral pancreas in man, *J. Anat.*, 65, 1, 1930.
125. Ahrens, H., Die Entwicklung der menshlichen zähne, *Anat. Hefte*, 48, 167, 1913.
126. Chase, S.W., unpublished, Western Reserve Univ., Cleveland, OH, 1955.
127. Churchill, H.R., *Histology and Histogenesis of the Human Teeth*, J.B. Lippincott, Philadelphia, 1935.
128. Orban, B., *Oral Histology and Embryology*, C.V. Mosby, St. Louis, 1953.
129. Harland, M., Early histogenesis of thymus in white rat, *Anat. Record*, 77, 247, 1940.

130. Jost, A., The role of fetal hormones in prenatal development, *Harvey Lectures*, 55, 201, 1961.
131. Price, D., Normal development of prostate and seminal vesicles of rat with study of experimental postnasal modifications, *Am. J. Anat.*, 60, 79, 1936.
132. Rogers, W.M., Fate of ultimobranchial body in white rat (Mus norvegicus albinus), *Am. J. Anat.*, 38, 349, 1927.
133. Sehe, C.T., Steriod hormones at early developmental stages of vertebrates, Ph.D. Thesis, State Univ. Iowa, Iowa City, 1957, p. 1.
134. Witschi, E. and Dale, E., *Gen. Comp. Endocrinol*, Suppl. 1, 356, 1962.
135. Girgis, A., Description of a human embryo of twenty-two paired somites, *J. Anat.*, 60, 382, 1926.
136. Weller, G.L., Development of the thyroid, parathyroid and thymus glands in man, *Contribs. Embryol. Carnegie Inst. Wash.*, 24, 93, 1933.
137. Frazer, J.E., *A Manual of Embryology*, Bailliere, Tindall, and Cox, London, 1940.
138. Atwell, W., The development of the hypophysis cerebri in man, with special reference to the pars tuberalis, *Am. J. Anat.*, 37, 159, 1926.
139. Bender, K., Uber die entwicklung der lungen, *Z. f. Anat. Entwicklungsgeschichte*, 75, 639, 1925.
140. Heiss, R., Zur Entwicklung und Anatomie der menschlichen Lunge, *Arch. Anat. Physiol. Anat. Abt.*, 1919, p. 1.
141. Keibel, F. and Elze, C., *Normentafeln zur Entwicklungsgeschichte des Menschen*, G. Fischer, Jena, Germany, 1908.
142. Puiggros-Sala, J.Z., *Uber die Entwicklung der Lungenanlage des Menschen, Anat. Entwicklungsgeschichte*, 106, 209, 1937.
143. Weed, L.W., The development of the cerebro-spinal spaces in pig and in man, *Carnegie Contrib. Embryol.*, 5, 3, 1917.
144. Bartelmez, G.W. and Dekaban, A.S., The early development of the human brain, *Contribs. Embryol. Carnegie Inst. Wash.*, 35, 12, 1962.
145. Hines, M., Studies in the growth and differentiation of the telencephalon in man. The fissura hippocampi, *J. Comp. Neurol.*, 34, 73, 1922.
146. Anson, B.J., The early development of the membranous labyrinth in mammalian embryos, with special reference to the endolymphatic duct and the utriculo-endolymphatic duct, *Anat. Record*, 59, 15, 1934.
147. Fitzgerald, J.E. and Windle, W.F., Some observations on early human fetal movements, *J. Comp. Neurol.*, 76, 159, 1941.
148. Hooker, D., Fetal reflexes and instinctual processes, *Psychosomat. Med.*, 4, 199, 1942.
149. Bach, L. and Seefelder, R., *Atlas zur Entwicklungsgeschichte des menschlichen Auges*, W. Englemann, Leipzig, 1911–1912.
150. Streeter, G.L., On the development of the membranous labyrinth and the acoustic and facial nerves in the human embryo, *Am. J. Anat.*, 6, 139, 1906.
151. Scammon, R.E. and Armstrong, E.L., On the growth of the human eyeball and optic nerve, *J. Comp. Neurol.*, 38, 165, 1924 or 1925.
152. Gilbert, P., The origin and development of the head cavities in the human embryo, *J. Morphol.*, 90, 149, 1952.
153. Keibel, F., in *Manual of Human Embryology*, Vol. 2, Keibel, F. and Mall, F.P., Eds., J.B. Lippincott, Philadelphia, 1912, p. 218.
154. Mann, I., *The Development of the Human Eye*, Cambridge University Press, England, 1928.
155. Bast, T.H. and Anson, B.J., *The Temporal Bone and the Ear*, C.C. Thomas, Springfield, IL, 1949.
156. Bast, T.H., Anson, B.J., and Gardner, W.D., The developmental course of the human auditory vesicle, *Anat. Record*, 99, 60, 1947.
157. Gilbert, M.S., The early development of the human diencephalon, *J. Comp. Neurol.*, 62, 81, 1935.
158. His, W., *Die Entwicklung des menschlichen Gehirns wahrend der ersten Monate*, S. Hirzel, Leipzig, 1904.
159. Schaeffer, J.P., The lateral wall of the cavum nasi in man with special reference to the various developmental stages, *J. Morphol.*, 21, 613, 1910.
160. Richany, S.F., Anson, B.J., and Bast, T.H., The development and adult structure of the malleus, incus, and stapes, *Quart. Bull. Northwestern Univ. Med. School*, 28, 29, 1954.
161. Richany, S.F., Anson, B.J., and Bast, T.H., The development and adult structure of the malleus, incus, and stapes, *Ann. Otol. Rhinol. Laryngol.*, 63, 399, 1954.

162. Sensenig, E., The early development of the human vertebral column, *Carnegie Contrib. Embryol.*, 33, 21, 1949.
163. Noback, C. and Robertson, G., Sequences of appearance of ossification centers in the human skeleton during the first five prenatal months, *Am. J. Anat.*, 89, 1, 1951.
164. Patten, B., *Human Embryology*, Blakiston, Philadelphia, 1946.
165. Hogg, I., Sensory nerves and associated structures in the skin of human fetuses of 8 to 14 weeks of menstrual age correlated with functional capability, *J. Comp. Neurology.*, 75, 371, 1941.
166. Johnson, F.P., A human embryo of twenty-four pairs of somites, *Contribs. Embryol. Carnegie Inst. Wash.*, 6, 125, 1917.
167. Arey, L.B., *Developmental Anatomy*, 7th ed., W.B. Saunders Company, Philadelphia, 1965.
168. Mintz, B., Embryological development of primordial germ cells in the mouse: influence of new mutation WJ_1, *J. Embryol. Exptl. Morphol.*, 5, 396, 1957.
169. Mintz, B., Continuity of the female germ line from embryo to adult, *Arch. Anat. Microscop. (Paris)*, 48, 155, 1959.
170. Raynaud, A., Recherches embryologiques et histologiques sur la différenciation sexuelle normale de la souris, *Bull. Biol. Fr. Belg.*, 29, 1, 1942.
171. Thomson, J.D., *Proc. Iowa Acad. Sci.*, 49, 475, 1942.
172. Gillman, J., The development of the gonads in man, with a consideration of the role of fetal endocrines and the histogenesis of ovarian tumors, *Carnegie Contrib. Embryol.*, 32, 81, 1948.
173. Ortiz, E., The embryological development of the Wolffian and Mullerian ducts and the accessory reproductive organs of the golden hamster (*Cricetus auratus*), *Anat. Rec.*, 92, 371, 1945.
174. Hamilton, W.J., Boyd, J.D., and Mossman, H.W., *Human Embryology*, 2nd ed., Williams and Wilkins, Baltimore, 1952, p. 87.
175. Witschi, E., in *The Ovary*, Grady, H.G. and Smith, D.E., Eds., Williams and Wilkins, Baltimore, 1962.
176. Holson, J.F., Relative Transport Roles of Chorioallantoic and Yolk Sac Placentae on the 12th and 13th Days of Gestation, Doctoral dissertation, University of Cincinnati, 1973.
177. Lewis, W.H. and Hartman, C.G., Early cleavage stages of the egg of the monkey (macacus rhesus), *Carnegie Contrib. Embryol.*, 24,187, 1933.
178. Wislocki, G.B. and Streeter, G.L., On the placentation of the macaque (*Macaca mulatta*) from the time of implantation until the formation of the definitive placenta, *Carnegie Contrib. Embryol.*, 27, 1, 1938.
179. Solomon, H.W., Wier, P.J., Fish, C.J., Hart, T.K., Johnson, C.M., Prosobiec, L.M., Gowan, C.C., Maleef, B.E., and Kerns, W.D., Spontaneous and induced alterations in the cardiac membranous ventricular septum of fetal, weaning, and adult rats, *Teratology*, 55, 185–194, 1997.
180. Fleeman, T.L., Cappon, G.D., and Hurtt, M.E., Postnatal closure of membranous ventricular septal defects in Sprague-Dawley rat pups after maternal exposure with trimethadione, *Birth Defects Research: Part B: Developmental and Reproductive Toxicology*, 71, 185–190, 2004.

PART II

Hazard and Risk Assessment and Regulatory Guidance

CHAPTER 7

Developmental Toxicity Testing — Methodology

Rochelle W. Tyl and Melissa C. Marr

CONTENTS

I.	Introduction and Objective	207
II.	Materials and Methods	208
	A. Test Substance	208
	1. Chemical Safety and Handling	209
	2. Dose Formulation and Analyses	209
	B. Test Animals	210
	1. Species and Supplier	210
	2. Live Animals and Species Justification	211
	3. Total Number, Age, and Weight	211
	C. Animal Husbandry	213
	1. Housing, Food, and Water	213
	2. Environmental Conditions	214
	3. Identification	215
	4. Limitation of Discomfort	215
	5. Mating	216
III.	Experimental Design	219
	A. Study Design	219
	B. Dose Selection	220
	C. Allocation and Treatment of Mated Females	220
	1. Allocation	220
	2. Treatment of Mated Females	221
	D. Observation of Mated Females	228
	1. Clinical Observations	228
	2. Maternal Body Weights	229
	3. Maternal Feed Consumption	229
	E. Necropsy and Postmortem Examination	234
	1. Maternal	234
	2. Fetal	236
	F. Statistics	252
	G. Data Collection	252
IV.	Storage of Records	253

V. Compliance with Appropriate Governmental and AAALAC International Regulations 254
VI. Reports 254
   A. Status Reports 254
   B. Final Report 254
VII. Personnel 255
VIII. Study Records to Be Maintained 255
Acknowledgments 256
References 256

Soon after the horror of the thalidomide disaster in the late 1950s and early 1960s, resulting in over 8000 malformed babies in 28 countries, the U.S. Food and Drug Administration (FDA) assumed regulatory responsibilities for requiring specific testing paradigms "for the appraisal of safety of new drugs for use during pregnancy and in women of childbearing potential."[1] A letter was sent from the Chief of the Drug Review Branch, FDA, to all corporate medical directors,[1] establishing what became known as the Guidelines for Reproductive Studies for Safety Evaluation of Drugs for Human Use. These guidelines (Figure 7.1) encompassed three test intervals.

1. Phase I (Segment I): Prebreeding and mating exposures for both sexes and exposure to pregnant dams until implantation on gestational day (GD) 6, to provide information on possible effects on breeding, fertility, preimplantation, and embryonic development to midgestation (GD 13 to 15 in the absence of maternal exposure), and implantation (Figure 7.1A).
2. Phase II (Segment II): Exposures during major organogenesis, to provide information on possible effects on *in utero* survival and morphological growth and development, including teratogenesis (Figure 7.1B).
3. Phase III (Segment III): Exposures from the onset of the fetal period through weaning of the offspring, to provide information on maternal parturition and lactation, on F_1 offspring late intrauterine and postnatal growth and development to reproductive maturity, and production of F2 offspring (Figure 7.1C).

The procedures for Segment II studies have essentially been followed by the U.S. Environmental Protection Agency (EPA),[2-4] Japan,[5] Canada,[6] Great Britain,[7] and the Organization for Economic Cooperation and Development (OECD).[8]

Recently, attempts have been made to make testing guidelines for reproductive and developmental toxicity studies consistent among major nations. The International Conference on Harmonisation (ICH), representing the FDA, the European Community (EC), and Japan, has promulgated testing guidelines for registering pharmaceuticals within the three regions.[9] These guidelines are presented graphically in Figure 7.2. They are similar to the original FDA guidelines (Figure 7.1) and assess exposure during prebreeding, mating, and gestation, until implantation on GD 6 (Study 4.1.1), exposures from implantation to weaning (Study 4.1.2), exposures only during organogenesis (Study 4.1.3), and combined single- and two-study designs. ICH Study 4.1.3 is, in fact, identical to the original FDA Phase II study, and ICH Study 4.1.2 is similar to the FDA Phase III study, except that exposures start at the beginning of organogenesis rather then at the end. Final testing guidelines from the EPA Office of Prevention, Pesticides and Toxic Substances (OPPTS)[10] and from the EPA Toxic Substances Control Act (TSCA)[11] for developmental toxicity studies have also been promulgated recently. The FDA has also recently revised its developmental toxicity testing guideline[12] as has the OECD.[13]

All previous governmental developmental toxicity testing guidelines specified exposure as beginning after implantation is complete and continuing until the completion of major organogenesis (closure of the secondary palate). This corresponds to GD 6 through 15 for rodents and GD 6 through 18 (FDA and TSCA) or GD 7 through 19 (Federal Insecticide, Fungicide, and Rodenticide Act; FIFRA) for rabbits, if the day of impregnation is designated GD 0. The 1994 ICH guideline[9]

DEVELOPMENTAL TOXICITY TESTING — METHODOLOGY

A. **Phase I: Fertility and General Reproductive Performance Study**

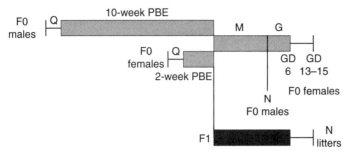

Information on: breeding, fertility, nidation, embryonic development

B. **Phase II: Developmental Toxicity Study**

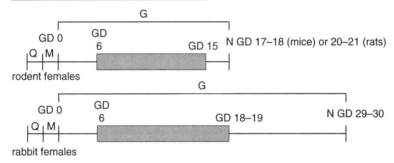

Information on: embryotoxicity, fetotoxicity, teratogenicity

C. **Phase III: Perinatal and Postnatal Study**

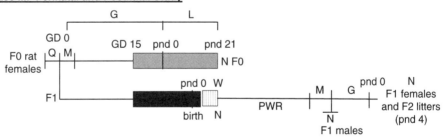

Information on: fetal development, parturition, lactation, peri-, neo-, and postnatal effects, adult offspring structures and functions

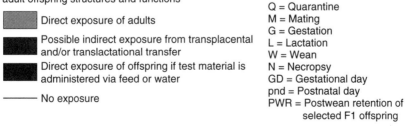

Q = Quarantine
M = Mating
G = Gestation
L = Lactation
W = Wean
N = Necropsy
GD = Gestational day
pnd = Postnatal day
PWR = Postwean retention of selected F1 offspring

Figure 7.1 FDA study designs.

A. Study of Fertility and Early Embryonic Development (4.1.1), Rodent (see Phase I)

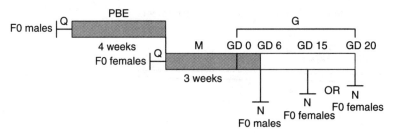

Assess: Maturation of gametes, mating behavior, fertility, preimplantation, implantation

B. Study for Effects on Prenatal and Postnatal Development, Including Maternal Function (4.1.2), Rodent (see Phase III)

Assess: Toxicity relative to nonpregnant females, prenatal and postnatal development of offspring, growth and development of offspring, functional deficits (behavior, maturation, reproduction)

C. Study for Effects on Embryo-Fetal Development (4.1.3), Rodent and Nonrodent (see Phase II)

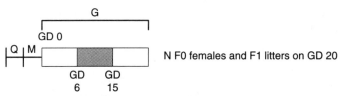

Assess: Toxicity relative to nonpregnant females, embryo/fetal death, altered growth of offspring, and structural changes of offspring *in utero*

Figure 7.2 International Conference on Harmonization (ICH) study designs.

has retained this duration of exposure only during major organogenesis. In a departure from the previous guidelines, the recently finalized developmental toxicity testing guidelines by U.S. EPA (OPPTS),[14] FDA,[9] and OECD[8] specify exposure during the entire gestational period (Figure 7.3), from GD 0 through scheduled sacrifice at term or from GD 6 to term (see Section C, Allocation and Treatment of Mated Females, 2b. Duration of Administration, for a discussion of the rationale for new start and end times of administration). There are other differences as well (see Table 7.1).

The Phase II study, the major topic for this chapter, had traditionally been termed a "teratology study," since the initial focus was on structural malformations (terata). It is currently more appropriately termed a "developmental toxicity study," as it evaluates (and the term "developmental toxicity" encompasses) a spectrum of possible *in utero* outcomes for the conceptus, including death, malformations, functional deficits, and developmental delays.[16–19] There has been discussion on whether this test, as structured, does assay developmental toxicity,[20] or whether it is even necessary

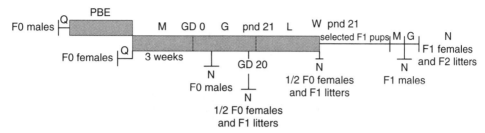

D. Single Study Design (4.2), Rodents (combine 4.1.1 and 4.1.2)

E. Two-Study Design (4.3), Rodents

4.1.1 with 4.1.2: 1/2 F0 females and F1 litters necropsied on GD 20
1/2 F0 females and F1 litters necropsied on pnd 21
(retained selected F1 pups followed through mating and gestation of F2 litters)

Q = Quarantine
PBE = Prebreed Exposure Period
M = Mating
G = Gestation
L = Lactation
W = Wean
N = Necropsy
GD = Gestational Day
pnd = Postnatal Day

▨ Direct exposure of adults

Figure 7.2 (continued)

for assessing developmental risk.[21] However, the consensus is that this study protocol, scientifically designed and performed, provides useful, critical information for human risk assessment of potential developmental toxicants.[14,15,22]

It is important to note that a Phase II Study evaluates only structural growth, development, and survival of offspring only during *in utero* development. The conceptuses are evaluated at term. The parameters evaluated are *in utero* demise (resorption, fetal death), fetal body weights, and the size and morphology of external, visceral, and skeletal structures. For example, if the organs are the right size, shape, and color and in the correct location, they are judged to be normal. There is no assessment for microscopic integrity or function and no way to assess structural and/or functional effects that might have occurred (or become evident) during postnatal life if the fetuses had been born.[23,24]

Because there was (and is growing) concern about postnatal sequelae to *in utero* structural and/or functional insult, as well as a recognition that exposure of a developing system may result in qualitatively or quantitatively different effects than exposure of an adult system, the Phase III study was designed to investigate postnatal consequences to late *in utero* exposures (Figure 7.1C).[1] In brief, the Phase III study consists of exposure of pregnant rats to the test agent from the end of organogenesis (GD 15), through histogenesis (during the fetal stage), through parturition (birth), and through lactation until the offspring are weaned (postnatal day [PND] 21). The offspring are "exposed" only via possible transplacental and/or lactational (via the milk) routes. There are usually three test material groups and a vehicle control group, with at least 20 litters per group; exposure of the dam is usually by gavage (to minimize disruption of the mother and her litter and to control the internal dose). During gestation, the dam is weighed periodically and feed consumption is measured. Dams and pups are weighed, sexed, and examined externally, with food consumption measured at birth (PND 0) and repeatedly during the lactation period (e.g., on PND 0, 4, 7, 14,

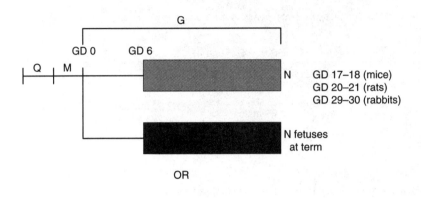

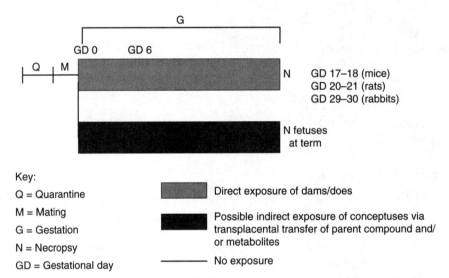

Figure 7.3 New EPA, OECD, and FDA prenatal developmental toxicology ("Phase II") study exposure durations.

and 21). Litters are culled to eight pups on PND 4. The time of acquisition of developmental landmarks, such as surface righting reflex, pinna (external ear) detachment, incisor eruption, eye opening (pups are born blind with eyes shut), auditory startle (pups are born deaf with the external auditory meatus [ear canal] closed), and midair righting reflex, is recorded. The age at testis descent may also be recorded, occurring in male rats late in lactation, typically on PND 16 to 20 in the CD® (SD) rat. If the pups are maintained after weaning, then vaginal patency (opening of the vaginal canal) and/or preputial separation are monitored, along with motor activity (initial exploratory behavior as well as habituated behavior); learning and memory may also be assessed. This test provides information on the last "trimester" of pregnancy, delivery, maternal-pup interactions and behaviors (such as pup retrieval, nursing, grooming, nest building, etc.), and pup postnatal growth and development.

At weaning, the dam is sacrificed and the number of uterine implantation scars counted to obtain information on prenatal (postimplantation) loss; pups can be necropsied at weaning or later, with target tissues examined histologically. The pups may also be raised to adulthood and mated to ascertain any effects of early indirect exposure on reproductive competence.

This chapter is designed to provide the methodology to perform a Phase II study according to current U.S. governmental testing guidelines and in compliance with Federal Good Laboratory

Table 7.1 Differences between old and new developmental toxicity test guidelines

Event/Parameter	Previous Requirements[a]	Current Requirements[b]
Maternal Evaluations		
Assignment to dose group	Not specified	Assignment by a body weight dependent random procedure
Definition of high dose level	Should induce some overt maternal toxicity, but not more than 10% maternal deaths	Should induce developmental and/or maternal toxicity, but not more than 10% maternal deaths
Test substance administration: period of dosing	During the major period of organogenesis: days 6–15 in rodents and 6–18 (or 7–19 FIFRA) in rabbits	Dose from implantation through termination (days 6–20 or 21 in rats, 6–17 or 18 in mice, and 6–29 or 30 in rabbits): option to begin on GD 0; ICH retains original dosing period
Test substance administration: dose adjustment	Dosage based upon the body weight at the start of test substance administration or adjusted periodically by body weight	Dosage adjusted periodically throughout the period of administration by body weight
Number of pregnant animals at termination (presumed pregnant animals assigned to study)	Rodents: 20 per group Rabbits: 12 per group	Rodents and rabbits: 20 litters per group (females with implantation sites at termination)
Maternal postmortem data: ovarian corpora lutea counts	Data required for all species except mice (TSCA only)	Data required for all species (including mice)
Fetal Evaluations		
Rodents: Assignment of fetuses for evaluation	One-third to one-half of each litter assigned for skeletal evaluation, the remainder for visceral evaluation	One-half of each litter assigned for skeletal evaluation, the remainder for visceral evaluation
Rabbits: Coronal sectioning	Not required	Required (50% serial sections, 50% coronal sections)
Ossified and cartilaginous skeletal evaluation	Only ossified specified (alizarin red S stain)	Both ossified and cartilaginous skeletal examination required (unspecified method of staining; usually alizarin red S for ossified bone and alcian blue for cartilaginous bone and other structures), all species

[a] Requirement under FIFRA, TSCA, and FDA.
[b] OPPTS (EPA TSCA and FIFRA) Draft Guidelines, Public Draft, February 1996,[10] and U.S. Environmental Protection Agency, OPPTS, Health Effects Test Guidelines, OPPTS 870.3700, Prenatal Developmental Toxicity Study, Public Draft, U.S. Government Printing Office, Washington, D.C., February, 1996 and U.S. Environmental Protection Agency, *Fed. Regist.* 62(158), 43832, August 15, 1997, and U.S. Environmental Protection Agency, *Fed. Regist.*, 56(234), 63798, 1991.

Practice (GLP) Regulations.[25-27] The rest of this chapter is therefore organized according to the headings (I through X) for a typical GLP-compliant study protocol, as currently followed in the authors' laboratory (Table 7.2). An additional useful reference for techniques and methods for and problems encountered in the performance of a Phase II study is a small book by Pamela Taylor.[28]

I. INTRODUCTION AND OBJECTIVE

The protocol introduction should indicate the reason for the study and the objective of assessing developmental toxicity (including teratogenicity) in the test animal species after *in utero* exposure from implantation to term.

Table 7.2 Typical protocol contents

Introduction and Objective
Materials and Methods
   Test substance:
      Characterization
         Identification (CAS No., lot/batch number, supplier)
         Chemical safety and handling
         Dosage formulation and analyses
   Animals:
      Species and supplier
         Justification for live animals and species
         Total number, age, and weight
   Quarantine
   Animal Husbandry:
         Housing, food, and water
         Environmental conditions
   Animal Identification
   Limitation of Discomfort
   Mating
Experimental Design
   Study design
   Dose selection
   Allocation and treatment of mated females
   Observation of mated females
   Maternal clinical observations
      Maternal body weights
      Maternal food consumption
      Necropsy and postmortem evaluation:
         Maternal
         Fetal
Statistics
Storage of Records
Compliance with Governmental and AAALAC Regulations
Reports
   Status reports
   Final report
Personnel
Study Records to be Maintained
References

ATTACHMENT I – Certificate of Analysis (of the specific wwwlot/batch number of test material
ATTACHMENT II — Material Safety Data Sheet
ATTACHMENT III — etc.

II. MATERIALS AND METHODS

A. Test Substance

The test material should be characterized as to sponsor designation, chemical name, CAS Registry number, molecular formula, molecular weight, supplier, lot (or batch) number, chemical purity, appearance, solubility, and storage conditions. All of these parameters should be supplied by the study sponsor or by the performing laboratory. Information on the vehicle selected and the amount of test article required should also be included.

1. Chemical Safety and Handling

If any relevant published toxicity information is available (e.g., from sponsor, RTECS, Toxline, Medline, etc.), it should be accessed and extracted. Any chemical-specific information pertaining to the toxic properties of the test material (i.e., eye irritation, skin irritation, sensitization, anticholinesterase activity, etc.) should be detailed in this section. The same is true of any chemical-specific handling information, such as "hygroscopic," "light sensitive," "temperature sensitive" (e.g., store frozen at 20 ± 5°C or –80 ± 5°C; refrigerated at 5 to 10°C; room temperature). A Certificate of Analysis and a Material Safety Data Sheet (MSDS), or Experimental Safety Data Sheet (ESDS), or other formal written safety information should be incorporated into the protocol (e.g., as attachments), and/or read and understood by all participating staff (with appropriate documentation).

2. Dose Formulation and Analyses

Prior to the start of the study, representative formulations of the test material, in vehicle at concentrations encompassing the range of dose levels to be employed in the study, must be assayed for homogeneity and stability. Samples for homogeneity testing should be obtained from representative locations (e.g., the top, middle, and bottom of a container of solution or suspension or from the left, right, and center of a V-shell diet blender). Stability of formulations should be ascertained under storage conditions (e.g., refrigerator or freezer) and at room temperature. The duration of storage stability assessments should allow for formulation and dose level verification analyses prior to use and time to reformulate and/or reanalyze, if necessary, before administration to the animals. The duration of room temperature stability assessments depends on the route of administration selected and should allow for bolus dose administration (gavage) in the animal room (usually 1 to 4 h/d maximum), for cutaneous application (usually 6 to 8 h/d), for dose administration in feed or water (usually 9 d, to allow for 7 d of presentation and a "safety net" if the next formulation is not appropriate for administration), etc. Dose level verification should be performed on all doses for each formulation if the formulation interval is reasonable (e.g., weekly or every 2 weeks) or on first, middle, and last formulation if formulations are frequent (based on stability data). For generation of exposure concentrations of materials such as gases, aerosols, or dusts, uniformity of concentration level within the chamber and actual (analytical) concentrations in the breathing zone of the test animals must also be established prior to the study's start.

To prepare oral or cutaneous doses, the following equation is useful:

$$\text{Concentration (mg/ml)} = \frac{\text{dose level (mg/kg)}}{\text{dose volume (ml/kg)}}$$

To prepare feed or water dosage formulations, the dose level may be expressed in ppm, percentage (weight/weight), or a constant level of intake (in milligrams per kilogram per day). For feed or water dosing, the actual intake in milligrams per kilogram per day can be calculated based on the amount of feed or water consumed per interval (converted to grams per day), the animal's average body weight over the feeding interval, and the percentage of test material in the diet or water. For example, for a 0.5% dietary dose to a rat weighing an average of 300 g (0.3 kg) over the interval, and eating 20 g feed/d, the actual intake is computed as follows:

$$\frac{20 \text{ g/d} \times 0.005}{0.3 \text{ kg}} = \frac{0.1}{0.3} = 0.333 \text{ g/kg/d} = 333 \text{ mg/kg/d}$$

When you want to provide a fixed intake in milligrams of the test agent per kilogram of the animal's body weight per day (mg/kg/d), the concentration in the diet or water must be adjusted, usually weekly, based on food or water intake and the projected test animal body weight for the next week (if possible by dose group, or based on historical control data). For example, if the targeted intake is 500 mg/kg/d, the projected daily feed consumption is 30 g/d, and the projected body weight at the midpoint of the next interval is 400 g, the calculation for the dietary concentration is:

$$\text{Conc (g/g)} = \frac{\text{intake (g/kg/d)} \times \text{body weight (kg)}}{\text{feed consumption (g/d)}}$$

$$= \frac{0.500 \times 0.4}{30} = 0.0067 \text{ g/g} \quad \text{equiv. to } 0.67\% \text{ in diet}$$

For oral doses (gavage), since the dosing volume is usually kept constant (in milliliters per kilogram), the concentration will vary by dose level; the dosing volume is usually adjusted based on each animal's most recent body weight. For example, if the animal weighed 350 g and the dosing volume was 5 ml/kg, then the dosing volume on that particular dosing day would be:

$$\frac{5 \text{ ml}}{1000 \text{ g}} = \frac{x}{350 \text{ g}} = 1.75 \text{ ml}$$

If the animal weighed 375 g at the next weigh day, the dosing volume would be 1.875 ml, etc. For cutaneous application, the volume may vary by dose if the chemical is administered "neat" (undiluted).

B. Test Animals

1. Species and Supplier

Mice, rats, and rabbits are the most commonly used species for developmental toxicity studies. Using mice or rats as well as rabbits satisfies the rodent/nonrodent FDA and EPA testing requirements. These three species also have the most extensive historical databases available. For rodents, both inbred and outbred strains are available; many strains are more or less sensitive to specific or general chemical insult during gestation, and this strain diversity in sensitivity is also seen with regard to the target organs that may be affected. Both types of strains are capable of changing over time as a result of genetic drift, founder effect, selection, and/or new mutations. Each performing laboratory must have a historical control database for each test species/strain used to submit guideline studies to governmental regulatory agencies. Recent control values from the authors' laboratory are listed in Table 7.3 of this chapter, and additional historical control databases are listed in the Appendix of this volume.

It is imperative that the animals come from a reputable supplier. Most reputable commercial breeders now have extensive health quality control programs as well as genetic monitoring (to ensure the genetic integrity of their animal strains). For a laboratory to maintain a reliable historical database, it is best if all animals come from the same supplier at the same location. In addition, as discussed later in this chapter, the laboratory should maintain a regular program of testing animals for common laboratory animal diseases, regardless of the supplier. There is no absolute certainty that the animals will be disease free as received or that they will remain so in one's own facility.

Table 7.3 Summary of historical control maternal endpoints[a]

Parameter	Gestational Days of Dosing	
	6–15	6–19
Number of dams	178	136
Maternal body weight (GD 0) (g)	247.56 ± 1.27	244.00 ± 1.29
Maternal body weight (GD 20 at sacrifice) (g)[c]	410.24 ± 2.54	377.71 ± 2.03
Maternal body weight change (gestation) (g)[c]	162.72 ± 1.87	133.71 ± 1.60
Maternal body weight change (corrected) (g)[c]	73.85 ± 1.27	52.95 ± 1.32
Gravid uterine weight (g)[b,c]	88.87 ± 1.22	80.76 ± 1.11
Maternal liver weight (g)[b,c]	17.97 ± 0.17	16.33 ± 0.13
Relative maternal liver weight (g)[a,d]	4.42 ± 0.03	4.32 ± 0.03

[a] Includes all pregnant control dams until terminal sacrifice on GD 20. Reported as the mean ± S.E.M.; GD = gestational day
[b] Weight change during gestation minus gravid uterine weight
[c] $p \leq 0.05$ significant difference between the two groups
[d] $p \leq 0.001$, significant difference between the two groups

Source: Marr, M.C., Myers, C.B., Price, C.J., Tyl, R.W., and Jahnke, G.D., *Teratology*, 59, 413, 1999 (tables provided by the authors).

2. Live Animals and Species Justification

The use of live vertebrate animals must be justified for federal GLPs, and the protocol must be submitted to each organization's Institutional Animal Care and Use Committee (IACUC). The justification usually presented is that the sponsor requested the animals, and that this test with live animals is required by the applicable governmental testing guidelines, e.g., preclinical testing for new drug development (FDA), for pesticide registration or Data Call-In (FIFRA), for a premanufacturing notice (PMN) of a commercial chemical under TSCA, for a mandated Test Rule or negotiated test agreement (TSCA), or for a significant new use registration (SNUR) for a commercial chemical (TSCA). It should be stated that alternative test systems are not available and/or currently accepted by the scientific community for the assessment of chemical effects on prenatal mammalian growth and development. This is usually the section where historical control data available in the performing laboratory are noted for the species/strain on test along with any published historical control databases.

3. Total Number, Age, and Weight

Rat and mouse females are sexually mature at approximately 50 days of age.[29] Male rodents do not have sperm in the cauda epididymis until 60 (mice) or 70 (rats) days of age. Most developmental toxicity studies use female rodents 8 to 10 weeks of age and males 10 to 12 weeks of age (if breeding is to be done in-house). An 8-week-old female CD® rat weighs approximately 200 g, an 8-week-old female CD-1® mouse weighs approximately 20 g (based on Charles River Laboratories' growth charts), and female rabbits should be between 4 and 6 months old (2.5 to 5 kg). Does mature earlier than bucks, which do not attain adult sperm levels until 6 to 7 months of age.[29] A prime consideration is to use the most reproductively sound age for any animal, but use animals as young as possible to avoid the costs of purchasing older animals.

Numbers are based on guideline requirements. The new guidelines require a minimum of 20 pregnant animals per dosage group for rodents and rabbits. The previous guidelines required 12 pregnant does per group for rabbits. Inseminated rats and mice typically have at least a 90% pregnancy rate. Therefore, putting 25 sperm- or copulation plug-positive females per dose group on study should be sufficient, taking into consideration the historical pregnancy rate in the laboratory for the species and strain to be used. Ordering 30 to 50% extra female rodents, if doing one-on-one in-house mating to obtain 100 sperm-/plug-positive rodents over a 3- to 4-d period, is a good guideline. A 10% increase over the number of pregnant rabbits desired (i.e., 22 mated/inseminated does per group), to obtain 20 litters at term per group, is also a reasonable guideline, as rabbits do occasionally spontaneously abort or deliver early without regard to treatment. In the authors' laboratory, pregnancy rates for naturally bred rabbits are typically over 95%.

a. Physical Examination

Upon arrival, while the animals are being uncrated, a well-trained laboratory animal technician should inspect each animal for external alterations of the head, trunk, appendages (limbs and tail), and orifices (mouth, anus, genitourinary tract) and for congenital defects, such as microphthalmia. The condition of the coat, eyes, ears, and teeth should also be evaluated. Any abnormal clinical observation should be brought to the attention of the investigator. Rabbits are commonly fed a rationed amount upon arrival and during quarantine to alleviate the onset of the mucoid enteritis (enteropathy) occasionally brought on by shipping stress. Covance Research Products has suggested no feed for the first 24 h (water *ad libitum)* and an increase of 25 g/d, up to a 125-g ration. For rabbits received timed-pregnant, a half ration is recommended for the first 24 h (65 to 70 g) and then full ration (120 to 150 g) or *ad libitum* feeding.

b. Pathogen Antibody Screen

Rodents and rabbits may be purchased certified pathogen antibody free (with documentation from the supplier), but additional quality control evaluations are commonly done by many laboratories. In fact, many testing protocols require in-house health quality control. Each shipment of animals should be quarantined on arrival, and quality control evaluation should be initiated within one day after receipt. On the day after receipt, five rats per sex should be randomly chosen from the shipment of animals. They should be sacrificed and their blood collected for assessment of viral antibody status. Commercial testing laboratories generally offer rat and mouse viral screening assays. A typical viral screen for rats, available from BioReliance Corporation (Rockville, MD), consists of evaluation for the presence of antibodies against the following: Toolan H-1 virus (H-1), Sendai virus, pneumonia virus of mice (PVM), rat coronavirus/sialoacryoadenitis (RCV/SDA), Kilham rat virus (KRV), CAR bacillus, *Mycoplasma pulmonis* (*M. Pul.*), and parvo. Thus, health status is assured from two independent sources other than the toxicology laboratory. Survival quality control (QC) is possible with rabbits because they may be bled for serum from the central ear artery.

c. Flotation and/or Tape Test for Intestinal Parasites

Most endoparasites can be detected by fecal examination. This may be accomplished by direct smear, in which a small amount of feces is mixed with a drop of saline on a slide and then cover slipped. The preparation is then examined for the presence of parasites or parasitic ova. Protozoan parasites will be motile. For fecal flotation, a larger specimen is mixed in a solution of zinc sulfate, sodium nitrate, or supersaturated sugar, with a specific gravity of 1.2 to 1.8. The mixture is centrifuged or simply allowed to settle. The tube is filled to the top with the flotation solution, and a clean coverslip is placed on top of the tube so that it touches the liquid. This should stand for 15 to 30 min. The cover slip is then removed and placed on a microscope slide. The slide should then be examined for the presence of parasite ova and coccidial oocysts. To simplify the process, a commercial kit may be used. The method of cellophane tape may be used by pressing a strip of tape to the animal's perianal area. The tape is then placed on a microscope slide and examined.[30]

d. Histopathology

Occasionally, extra animals are ordered and sacrificed upon arrival, and slides of likely target organs for disease are prepared for histological examination. These may include representative sections

Table 7.4 Recommended parameters for caging and environment for common laboratory animals

Animal	Body Weight	Cage Floor Area		Cage Height		Relative Humidity (%)	Dry-Bulb Temperature	
		in.²	cm²	in.	cm		°C	°F
Mouse	15–25 g	12.0	77.42	5	12.70	30–70	18–26	64–79
	>25 g	>15.0	96.78	5	12.70			
Rat	<100 g	17.0	109.68	7	17.78	30–70	18–26	64–79
	100–200 g	23.0	148.40	7	17.78			
	200–300 g	29.0	187.11	7	17.78			
	300–400 g	40.0	258.08	7	17.78			
	400–500 g	60.0	387.12	7	17.78			
	> 500 g	>70.0	451.64	7	17.78			
		ft²	m²					
Rabbit	< 2 kg	1.5	0.14	14	35.56	30–70	16–21	61–72
	2–4 kg	3.0	0.28	14	35.56			
	4–5.4 kg	4.0	0.37	14	35.56			
	>5.4 kg	>5.0	0.46	14	35.56			

Source: Data from National Research Council, *Guide for the Care and Use of Laboratory Animals,* Institute of Laboratory Animal Resources, Commission on Life Sciences, National Research Council. National Academy Press, revised 1996, Tables 2.1, 2.2, 2.3, and 2.4.

of the liver, spleen, kidneys, gastrointestinal tract (esophagus, stomach, duodenum, jejunum, ileum, cecum, colon, and rectum), lungs, lymph nodes (submaxillary and mesenteric), and reproductive organs (testes, epididymides, prostate, seminal vesicles, coagulating gland, vagina, corpus and cervix uteri, oviducts, and ovaries). This adds increased costs to studies and is generally unnecessary for the relatively short-term developmental toxicity studies if a reputable vendor is used. (For longer term studies, such as multigeneration, Phase I or III studies, this procedure is recommended.)

C. Animal Husbandry

1. Housing, Food, and Water

Animals should be singly housed (except during quarantine or mating for rodents) so that food consumption can be determined (as required by testing guidelines) and so that there will be no confounding factors from group dynamics. For example, a dominant female rat or mouse will eat more than others, a dominant female mouse may overgroom (to the point of "barbering") other cohabited mice, and stress levels of variously ranked females may vary, with consequences unrelated to chemical exposure.

Rodents can be housed in solid-bottom polycarbonate or polyethylene cages with stainless-steel wire lids (e.g., from Laboratory Products, Rochelle Park, NJ), using hardwood or other well-characterized bedding (see below). Alternatively, they can be maintained in stainless-steel, wire-mesh cages mounted in steel racks, with Deotized® paperboard (e.g., from Shepherd Specialty Papers, Inc., Kalamazoo, MI) placed under each row of cages to collect solid and liquid excreta (copulation plugs for rats will also be detectable on the paperboard if the animals are individually housed in hanging cages). Rabbits are housed in stainless-steel cages with mesh flooring (e.g., from Hoeltge, Inc., Cincinnati, OH), with a pan lined with paperboard beneath each cage to collect excreta. The dimensions of the cages as required by the NRC Guide[31] (update of the NIH Guide[32]) are presented in Table 7.4, by species and body weight range.

Feed must be certified and analyzed (usually by the supplier) and is usually available *ad libitum* throughout the study (after any initial adjustments in food presentation for rabbits; see previous section on quarantine). For rodents, feed can be pelleted and available on the cage lid or in feeders within the cage, or it can be ground and available in feeders within the cage. For rabbits, pelleted feed is presented in feeders attached to the cage, and it must be changed frequently. For rodents, a good feed source is Purina (PMI® Nutrition International) Certified Rat and Mouse Chow, No. 5002 (ground or pelleted); for rabbits, we have used Purina Certified Rabbit Chow (e.g., No. 5322 or high fiber diet No. 5325, for animals on restricted feed). A high-fiber diet that stimulates hindgut motility, reduces enteritis, and protects against fur chewing and formation of trichobezoars (hair balls) in the stomach is necessary for rabbits.[33] Additional analyses for possible contaminants and/or nutrient levels can be provided by commercial sources. Feeds should be stored at or below 15°C to 21°C (60°F to 70°F) and should not be used more than 6 months past the milling date.

Drinking water must meet EPA standards for potable water and must be analyzed for contaminants. If the water is taken directly from a municipal water supply, the supplier can provide analyses showing levels of contaminants to be within acceptable limits for human consumption. If the incoming water is further treated (e.g., acidified, deionized, filtered, deionized/filtered, distilled), then these procedures must be documented and posttreatment analyses recorded. The water, regardless of treatment, may also be analyzed by an independent testing laboratory (e.g., Balazs Laboratory, Inc., Sunnyvale, CA. The water can be presented in plastic (polypropylene, polycarbonate) or glass bottles with stainless-steel sipper tubes and rubber stoppers, or via an automatic watering system (e.g., from Edstrom Industries, Inc., Waterford, WI).

2. Environmental Conditions

a. Light Cycles

According to the NRC Guide (1996), light levels of 30 foot candles are adequate for most routine care procedures. Illumination of excessive intensity and duration may cause retinal lesions in albino rats and mice.[32,34] The color balance of light should approach that of sunlight to allow the most accurate observations of the conditions of the animals' eyes and other body parts, for which color is an important factor.[30] Light cycles of 12 h light/12 h dark seem to be adequate to promote breeding of rodents. Ovulation in mice generally occurs during the midpoint of the dark cycle. Continuous light will depress cycling; therefore, documented quality control of computer- or timer-controlled room light timing is essential. Light cycles for rabbits can vary from 8 to 10 h of light for males and 14 to 16 h of light for females. An intermediate compromise (i.e., 12:12) is adequate for breeding. Shortening of the lighted segment of the light cycle may bring on autumnal sexual depression.

b. Temperature and Relative Humidity

Temperature and relative humidity are two of the most important factors in an animal's environment because of their effects on metabolism and behavior. Consequently, they may have an effect on the animal's biological reactions to various test agents (Table 7.4). The range of temperatures suggested in the NRC Guide[31] is slightly lower than each species' thermoneutral zone, but allows optimal comfort, reactivity, and adaptability.[35] Hyperthermia is of concern in pregnant animals, but the core temperature of pregnant Sprague-Dawley (SD) rats is less affected by heat stress than that of nonpregnant rats.[36] Of more concern is the effect of elevated temperature on male fertility. Typically, sustained temperatures above 85°F (29.4°C) will result in temporary infertility. Therefore, careful monitoring of temperature deviations and immediate response to any equipment failure is essential in a facility engaged in reproductive and developmental toxicity studies if breeding is done in-house. Relative humidity levels, as suggested by the NRC Guide,[31] are 30% to 70% for rodents and rabbits (Table 7.4). Excessively low humidity in rodent rooms can cause "ring-tail" in neonates.

c. Ventilation

The suggested rate of ventilation in the NRC Guide[31] (10 to 15 air changes per hour) is designed to help create an odor-free environment. This, along with an adequate frequency of bedding changes, ensures an acceptably low level of ammonia (< 20 ppm) in the microenvironment of the cage.

d. Noise

The NRC Guide's[31] recommendations for limitation and reduction of noise inherent in the day-to-day operations of an animal facility include separating the animal rooms from the procedures that involve the most noise (such as cage washing and refuse disposal). Noisy animals, such as dogs and primates, should be separated from rodents and rabbits. Continuous exposure to acoustical levels above 85 dB can cause reduced fertility in rodents.[37]

e. Chemicals

Aromatic hydrocarbons from cedar shavings and pine bedding material can induce the biosynthesis of hepatic microsomal enzymes, so softwood shavings should be avoided for developmental toxicity studies. Excellent hardwood bedding materials include Ab-Sorb-Dri® hardwood chips (Laboratory Products, Inc., Garfield, NJ), Sani-Chip® cage litter (P.J. Murphy Forest Products Corporation, Montville, NJ), and Alpha-Dri® purified α-cellulose (Shepherd Specialty Papers, Inc., Kalamazoo, MI). Alpha-Dri may be more expensive but is more absorbent; its use may require fewer cage changes and therefore save time and money.

Bedding can be sterilized and certified, and should be changed as often as is necessary to keep animals dry and clean. Litter should be emptied from cages in areas other than the animal rooms to minimize exposure to aerosolized waste. Although rodents will successfully mate following recent cage changes, it is best to allow the male time to mark "his" cage with urine prior to introduction of the female for mating; male rabbits generally mate better if their bedding has not been changed immediately prior to mating. It is unwise to change a littering rodent's cage within 24 h of delivery; therefore, the last suggested cage change is usually on GD 20 (day of vaginal sperm or copulation plug is GD 0).

3. Identification

Rodents may be identified by ear tag, tail tattoo, ear notch, toe notch, or implant. An inexpensive and highly reliable method, if done properly, is the tail tattoo. This is accomplished with a tattoo gun. The tail is cleaned with soap and water or alcohol. The tattoo needle is dipped into the ink, and the individual ID is slowly "written" onto the tail. The animal is typically restrained in a rodent restrainer for this process. Ear tags are inexpensive, but if not applied properly, they may pose irritation problems. Multiply housed animals may also lose tags through grooming or playing.

Some commercial rabbit suppliers ear tag their kits or subcutaneously implant an identification chip (e.g., AVID microchips, AVID Microchip I.D. Systems, Folsom, LA) at weaning. This allows easy identification for particular clients, following genetic traits (such as excessive aggression, hair growth pattern, and litter size) and tracking individuals for mating (to preclude sibling matings). Rabbits' ears may also be tattooed, but this requires sedation. Toe clipping (once very common for mice) is not currently considered appropriate for humane reasons and for ease in "reading" numbers. Ear notching (usually in association with toe clips) is also not currently done; multiply-housed animals create new "notches" as they interact, and notches can be difficult to read.

4. Limitation of Discomfort

U.S. Department of Agriculture regulations require annual documentation of animals (currently large mammals, including dogs, cats, rabbits, nonhuman primates, etc., but it is anticipated that rats, mice, and birds may soon be included) that are subjected to more than momentary or slight pain or distress without the benefit of anesthetic, analgesic, or tranquilizing drugs. As part of the study protocol, it is necessary and appropriate to state that some adult toxicity may be caused by exposure to the high and/or mid doses. It should be anticipated that those oral doses employed (or the cutaneous doses applied) will not result in irritation or corrosion to the gastrointestinal tract (or the skin) of the test animals. Discomfort or injury to animals should be limited, so if any animal becomes severely debilitated or moribund, it should be immediately and humanely terminated after appropriate anesthesia. All necropsies should be performed after terminal anesthesia. If blood is to be collected, it should be collected after appropriate anesthesia. Animals should not be subjected to undue or more than slight or momentary pain or distress.

5. Mating

a. Rodents

Rats are continually polyestrous, with a 4- to 5-d estrous cycle (estrus lasts approximately 12 h). Breeding considerations for developmental toxicity studies include the number of pregnant females that can be processed in one day at sacrifice, the range of mating days, and the need to limit the number of females mated to any given male in a dosage group. Females are mated at between 8 and 10 weeks of age. Males may be mated at approximately 12 weeks of age and appear to breed successfully for 6 to 8 months after this. Monogamous (one female with one male) or polygamous (harem) mating may be employed. Disadvantages of the monogamous scheme include the need to house a larger population of males. Polygamous mating pairs two to three (rarely as high as six) females with one male. If the male is housed with multiple females and mates with more than one on the same night, the distinct possibility exists that females impregnated later will have reduced pregnancy rates and/or litter size (due to reductions in the number of ejaculated sperm in later matings). This can be disadvantageous since the number of females employed per male per dose group must be limited to reduce the likelihood that heritable malformations from a male will affect study outcome.

Females may be checked by means of vaginal lavage with 0.9% saline for determination of their estrous cycle stage (i.e., proestrus, estrus, metestrus, or diestrus) using fixed cells and Toluidine blue stain, if necessary. Only the females in estrus or late proestrus would be used for mating. Charles River Laboratories can now provide guaranteed rats with 4-d cycles from their Portage, MI, facility. One caveat is that the stress of shipment may shift the cycle (usually by one day). The females' estrous stage should be confirmed by the performing laboratory upon arrival. Conversely, females may be paired one-to-one with a male, regardless of estrous stage. Each morning following pairing, females are checked by vaginal lavage for the presence of sperm. If sperm negative, females remain with the male and are checked daily until found sperm positive. A medicine dropper containing approximately 0.1 ml of saline is inserted in the vaginal opening, and the fluid is expelled into the vaginal canal and then drawn back into the eye dropper. The contents are expelled onto a microscope slide and then examined for the presence of sperm using a light microscope (400× magnification). The presence of a copulation plug in the vagina (for mice) or a dropped copulation plug (for rats) is also a positive sign of mating. In solid-cage systems, vaginal lavage or manual checking for the presence of a copulatory plug is necessary. In a hanging cage system, the presence of a dropped plug on the shelf below the cage may be used as evidence of mating in monogamously paired rats (the plug dries, shrinks, and falls out within 6 to 8 h postcoitus). Once the female is found sperm- or plug-positive, she is typically weighed and removed to her own cage. This date is normally designated as GD 0 (rarely, GD 1).

One-to-one breeding of rats, without regard to the female's estrous cycle, will usually generate 25% to 30% sperm-positive females per day (i.e., in a 4- to 5-d cycle, theoretically 20% to 25% of the animals are in the appropriate stage for mating on any given day).

Mouse estrous cycles are also 4 to 5 d, with an evening estrous period of 12 h. Females may be primed prior to mating.[38] Group-caged female mice may enter a phase of anestrus, which is terminated by the odor or presence of a male (from a pheromone excreted in male urine). A male in a wire-mesh cage can be placed in a larger cage containing a group of females 3 d prior to the desired day of mating (or females can be housed in the presence of bedding taken from a male's cage). This synchronizes the estrous cycle to typically yield a 50% plug-positive rate when the females are mated.

Female mice are introduced into the male's cage (not vice versa and not into a clean cage; the male will more likely be successful if he has previously marked his territory) in the late afternoon, and the following morning the females are checked for the presence of copulation plugs in the vagina. Plugs are normally readily apparent, but if not, a pair of closed, rounded-point forceps are gently inserted into the vaginal opening and then opened to expose any plug present.

Timed-pregnant rodents are now available from most suppliers. This can be advantageous because of space limitations and the high cost of maintaining a male breeding colony, but the animals are more expensive, and there is no time for a quarantine period prior to GD 0 following receipt in the research laboratory. Also, shipping stress can result in implantation failure in mice and rats.

b. Rabbits

Rabbits are induced ovulators, releasing ova 9 to 13 h after copulation; ovulation may also be induced by an injection of luteinizing hormone (LH). The number of females to be mated each day is dependent, as with rodents, on the number that can be successfully processed at sacrifice.

1) Natural Mating — A receptive doe (characterized by a congested, moist, purple vulva) will raise her hindquarters (flagging) to allow copulation when placed in a buck's cage. Once copulation has occurred, the female may be introduced into a second male's cage to increase the fertility rate. A high success rate is obtained even with randomly mated females (i.e., a group of females chosen to breed on a given day regardless of receptivity). If a female shows no interest, another male may be tried. If the male shows no interest, the female's hindquarters may be placed over his nose to make sure he notices her. Some males may attempt to mate in inappropriate positions, but the technician can manually place the female in a better position.

Occasionally, a female will refuse to mate. This will have to be considered an unsuccessful mating, and an attempt can be made the next day. Only rarely will a doe fail to mate during the two or three days of mating for a study.

2) Artificial Insemination — Ovulation is induced by intravenous injection (marginal ear vein) of 20 to 25 IU of human chorionic gonadotropin (hCG). This may be done immediately prior to mating or up to 4 to 5 h prior to mating. Sperm is collected from bucks by use of artificial vaginas (AV), which have been previously described.[39,40] An AV can be assembled easily by use of 1-in. rubber latex tubing, two bored neoprene stoppers, K-Y Jelly™, and glass centrifuge tubes (see Figure 7.4A). The latex can be replaced as it becomes brittle with use and washing. Glass insemination tubes (see Figure 7.4B) have become fairly expensive to have custom made but will last for years with careful handling. The AVs are heated in an incubation oven to approximately 40°C to 45°C prior to use. Semen is collected by using a tubally ligated female rabbit (or one with a contraceptive implant). The technician holds the AV, the opening of which has been coated with K-Y Jelly™, between the teaser's hind legs. The male will normally mate and ejaculate within 1 min. (The teaser should be sterilized in case the male is faster than the technician and "misses" the AV.) Alternatively, a sleeve made from a female rabbit skin is placed on the arm of the technician. The buck mounts the sleeve and ejaculates in the AV held in the technician's hand. This eliminates the need for live, sterilized does.

A. Artificial Vagina and Collecting Tube

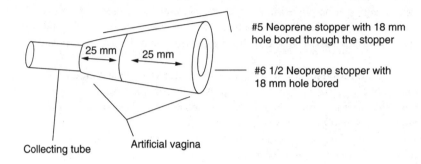

6" length of 1" diameter Penrose drain tubing is sufficient for 1 A.V.

The two stoppers are held together with an adhesive by 3M- Scotch Grip Rubber Adhesive #1300.

B. Artificial Insemination Tube

Figure 7.4 Artificial insemination equipment.

The advantage of artificial insemination is that the semen sample can be evaluated for sperm count, viability, and motility before insemination to assure that an optimal sample is introduced into the female. This can be accomplished manually by using a blood diluting pipette and a hemocytometer. The semen is drawn up to the 0.5 mark on the pipette, followed by 0.9% saline to the 1.01 mark. The sample is gently mixed, and a drop is released into the groove of the hemocytometer. The number of sperm in five 0.2×0.2 mm squares is counted at 400× magnification. If a count of greater than 10 per five squares is obtained (i.e., 10 million sperm), then the sample is acceptable for use. An additional advantage is that up to five females may be inseminated with one sample, thus reducing the number of bucks needed in the breeding colony. Precise records must be kept of the male's performance, including fetal outcomes; any males producing malformed control kits must be culled from the colony.

By use of a 1-cc syringe, a 0.25-ml sample of undiluted semen is drawn into the insemination tube. The female to be inseminated is placed upside down with her head between a seated technician's legs. The hind legs are grasped by the technician and spread. A second technician inserts the glass insemination tube until the pelvic brim is felt and then rotates the tube 180° and continues to insert it up to a depth of 7 to 10 cm. The semen is then injected and the tube withdrawn. If the doe urinates during insemination, another semen sample is injected 15 to 30 min later.

The day of successful natural mating or the day of artificial insemination is usually designated GD 0. One major consequence of hormonally priming the female (necessary for artificial insemination, less commonly used for natural breeding) is the possibility of "superovulation." A large number of eggs are ovulated, but many are not implanted, resulting in a large percentage of preimplantation loss.

Table 7.5 Summary of developmental endpoints in control litters

Parameter	Gestational Days of Dosing	
	6–15	6–19
All litters		
No. litters examined	178	136
No. corpora lutea per dam	16.93 ± 0.23	15.45 ± 0.18
No. implantation sites/litter[a,b]	15.79 ± 0.22	14.69 ± 0.20
Percent preimplantation loss/litter[a]	6.66 ± 1.04	5.47 ± 0.79
No. resorptions/litter (percent postimplantation loss/litter)[a]	2.99 ± 0.41	3.61 ± 0.67
No. (percent) litters with resorptions	56 (31.5)	52 (38.2)
No. late fetal deaths/litter[a]	0.0	0.0
No. adversely affected implants/litter[c,d,e]	1.10 ± 0.12	0.67 ± 0.08
No. (percent) litters with adversely affected implants[d]	94 (52.81)	63 (46.32)
Live litters[e]	4.42 ± 0.03	4.32 ± 0.03
No. live fetuses/litter[b,c]	15.30 ± 0.22	14.21 ± 0.21
Average fetal body weight (g)/litter (sexes combined)[a,e]	3.671 ± 0.026	3.566 ± 0.024

[a] Reported as the mean ± S.E.M.
[b] $p \leq 0.05$, significant difference between the two groups
[c] Includes all dams pregnant at terminal sacrifice on GD 20; litter size = No. implantation sites per dam
[d] Adversely affected = nonlive plus malformed
[e] $p \leq 0.01$; significant difference between the two groups
[f] Includes only dams with live fetuses; litter size = no. live fetuses per dam

Source: Marr, M.C., Myers, C.B., Price, C.J., Tyl, R.W., and Jahnke, G.D., *Teratology*, 59, 413, 1999 (tables provided by the authors).

$$\frac{\text{No. corpora lutea} - \text{No. implantation sites}}{\text{No. corpora lutea}} \times 100$$

For example, the value for percent preimplantation loss in the authors' laboratory for artificially inseminated does with hormonal priming is 30.74 ± 2.90% (88 does) versus the range of values for naturally bred does (not primed) of 5.82%–17.48% (224 does) (Table 7.6).

3) Vendor Supplied Timed-Pregnant Rabbits — Vendors (e.g., Covance Research Products, Denver, PA) are now offering timed-mated rabbits and have done extensive studies with numerous laboratories that show the fertility rate based on gestational day of shipment. Unless the toxicology facility doing the study is in close proximity to the vendor, it is not possible to measure food consumption from GD 0. The advantage of purchasing timed-mated rabbits is eliminating the upkeep of a large number of bucks and the technical time expended for mating procedures and recordkeeping, but there is no on-site quarantine period prior to GD 0 with these animals. In the authors' laboratory, we waive the prebreed quarantine. We monitor the purchased timed-mated does from GD 1, 2, or 3 (depending on when they arrive at our facility) until the start of dosing on GD 6 for body weight, weight changes, feed consumption, clinical signs, etc., so this time serves as a predosing "quarantine" period in lieu of a premating quarantine.

III. EXPERIMENTAL DESIGN

A. Study Design

This section should document the following:

- The number of mated females per group
- The number of days (and gestational days) of treatment
- The dose levels in mg/kg/day (as assigned), mg/ml (as formulated), ppm, dietary percentage, and in mg/m^3, etc.
- The dosing volume in ml/kg, if appropriate
- Projected in-life study performance dates should also be documented as follows:
  - Animals arrive at performing laboratory
  - Dates of GD 0 (for rodents, the end date is tentative since it will depend on performance of the breeding pairs)
  - Inclusive dates of treatment (the end date will depend on GD 0 dates)
  - Dates of scheduled sacrifice (the end date will depend on GD 0 dates)

The date of anticipated submission of the draft report may also be included in this section (or under the section on reports).

B. Dose Selection

All current governmental testing guidelines call for at least three dose or concentration levels plus a concurrent vehicle control (four groups total). The top dose level should be chosen to result in demonstrable maternal (and possibly developmental) toxicity. Maternal indicators can include reductions in body weight or weight gain, treatment-related clinical signs of toxicity, sustained reductions in food and/or water consumption, and maternal mortality (not to exceed 10%).[41,42] The middle dose should result in minimal maternal (and possible developmental) toxicity. The low dose should be a NOAEL (No Observable Adverse Effect Level) for both maternal and developmental toxicity. Optimally, the selection of doses should be based on results of range-finding studies in pregnant animals. A distant second choice on which to base doses would be a 14-day repeated dose study in the same species and strain by the same route in the same sex (although the animals were not pregnant, and pregnancy changes many physiologic and toxicologic parameters). Information on absorption, distribution, metabolism, and excretion in the test species (and sex) would be very useful, but it is rarely available for commercial chemicals and pesticides (most likely available for new drug preclinical work) and is almost never available for pregnant females. The spacing of the doses may be arithmetic or logarithmic; the low dose should be selected, if possible, to allow application of safety factors during risk assessment and still be above any expected human exposure levels. The doses selected and their justification should be documented in this section.

C. Allocation and Treatment of Mated Females

1. Allocation

The number of females assigned to dose groups is defined by the protocol. Animals can be assigned to dose groups totally randomly (by computer, by a table of random numbers, by numbered cards dealt randomly, etc.), or they can be assigned on the basis of body weight (i.e., stratified by body weight but randomly within body weight classes).

For mated females to be assigned to study by stratified randomization, GD 0 weights are arranged in ascending order, from the lightest to the heaviest on the first GD 0 date, and assigned in the same order or reverse order (i.e., descending order from heaviest to lightest) on subsequent days. Beginning with the lightest weight, animals are assigned to groups stratified by body weight (number of animals per group equals number of treatment groups). Within each stratified group, one animal will be randomly assigned to each treatment group. In the event that the total number of animals inseminated on a given day is not an even multiple of the number of treatment groups, the stratified group with the heaviest body weights (the last group filled by stratified randomization) will be assigned randomly (one animal per treatment group) until all animals have been assigned.

When consecutive breeding days are required, then female assignment to dose groups on each subsequent GD 0 will place females in sequence beginning with the dose group following the last dose group assigned on the previous day. In other words, the mated females will be assigned to complete the last incomplete stratification group (if necessary) and then assigned to the next stratification group until all GD 0 females for that day are assigned.

The objective at the end of the mating period is to have all groups with very similar mean body weights (not statistically significantly different) with very similar indicators of variance (e.g., standard deviation, standard error) and representatives from each mating date in each group. Obviously, if some of these females turn out not to be pregnant at scheduled sacrifice (and so are not included in summarized and analyzed study data, such as body weights, weight changes), the summarized GD 0 weights for reporting purposes (based on pregnant animals) will not be identical to the summarized weights initially run on GD 0 to guarantee homogeneity of study groups. This possibility should be explained in an Standard Operating Procedure (SOP) or memorandum to the study records. The values should still be very similar.

2. Treatment of Mated Females

a. Routes of Administration

Vehicles used for gavage dosing should be innocuous. Suspensions may be administered orally, but injections (i.v., i.p., or i.m.) require the compound be soluble in a physiologically compatible, innocuous vehicle such as water, dextrose solution, or physiological saline (0.9% NaCl). The vehicle may affect the outcome, e.g., corn oil may result in slower absorption than water, or in a different control embryofetal profile.[43]

1) Gavage — Commonly used vehicles include water, corn (or other) oil, and aqueous (up to 0.5% to 1.0%) methylcellulose, with or without polysorbate 800. The compound may be soluble in the vehicle, or a suspension may be created. The only limiting factor is the physical nature of the compound to be administered in the vehicle. The formulation should be easily drawn up into the dosing needle used (based on inner diameter of the needle) and be amenable to being expelled from the syringe without loss of homogeneity.

Rodents and rabbits may be gavaged with blunt-ended, stainless-steel feeding needles (e.g., from Popper and Sons, Inc., New Hyde Park, NY), 18 gauge and 1.5 in. long for mice and 16 or 18 gauge and 2 in. long for rats. A 13-gauge, 6-in. stainless-steel feeding needle is normally used for rabbits.[44] These needles are attached to appropriately sized syringes to deliver the entire dose to the animal while allowing accurate measurement of the dosing volume. Alternatively, a flexible, rubber catheter may be used for oral gavage dosing in rabbits. This involves using a Nelaton® French catheter from 8 to 12 in. long attached to a syringe adapter. The rabbit is put in a restrainer, and a plastic bit is inserted into the rabbit's mouth; the catheter is eased into the hole in the bit and down into the stomach. Rats may also be dosed by catheter. Dosing volumes are typically 5 to 10 ml/kg for mice; 2.5, 5, or 10 ml/kg for rats; and 1 to 5 ml/kg for rabbits (lower volumes should be employed for corn oil solutions or suspensions). These volumes allow administration of the compound into a full stomach. For rodents, the compound is drawn into an appropriately sized syringe with the stainless-steel feeding needle attached. Since the stainless-steel feeding tubes go only halfway down the esophagus to the stomach, care must be taken while inserting the needle to avoid inadvertently inserting the needle into the trachea. A well-trained technician who is alert to signs of a distressed animal can avoid such a misinsertion of the needle and a resultant "lung shot" (i.e., introduction of the dosing agent into the trachea and/or lungs). In rodents, a lung shot results in immediate demise; in rabbits, sequelae may not be detected until necropsy. A disadvantage of the steel feeding tube is efflux of the compound should it be injected too quickly, or as more often happens, the esophagus constricts, causing backwash. This will often result in aspiration pneumonia.

When catheters are used, the dosing compound is drawn directly into the syringe. Then the adapter with the catheter is attached to the syringe, and the catheter is fed through the bit completely into the stomach. If the tube is guided into the trachea, the rabbit will twitch its ears and cough, signaling the technician to withdraw the tube and try again. Steady movement of the nostrils should be apparent while the compound is being injected by use of a rubber catheter. With this method, it is necessary to "wash" the catheter with approximately 1 ml of vehicle after dosing, to administer the last milliliter of compound remaining in the catheter.

2) Dosing via the Diet — The test compound may be introduced directly into ground rodent or rabbit feed, depending on the characteristics of the compound, including its palatability and solubility. The test compound may be mixed with a solvent prior to mixing with the feed, or it may be microencapsulated to limit release to a particular part of the intestines or to ensure that unpalatable compounds will be eaten.

Ground feed for rodents is made available in the home cage in glass feed jars that are chemically resistant, easily sanitized, and transparent (allowing cage-side observation of the feeder). The jars are typically fitted with a stainless-steel lid. Mouse feed jars are also fitted with a wire cylinder that allows access to the feed but prevents nesting in the feed. Rats will generally use their paws to scoop out the feed and then lick their paws. The feed jar is weighed at intervals based on the size of the jar, the amount consumed, and the stability of the compound in the feed. This allows the calculation of grams of feed eaten per kilogram of animal body weight per time interval (usually daily), and therefore quantitates (in milligrams or gram per kilogram body weight per day) the actual amount of test article consumed. Most feed consumption occurs during the dark cycle, so the rodent is being "dosed" periodically at night. This differs from gavage, where the test material is generally given as a once-daily bolus dose in the morning.

Rabbits will not eat ground food;[33,45] therefore, any ground, feed-based dose formulation must be pelleted prior to administration to rabbits. This may be done in-house or by commercial laboratories. Because most performing laboratories do not have pelleting capabilities, many compounds that may be administered to rodents in the feed are given to rabbits by gavage.

Before glass feed jars are used for another study, they should be washed. Then, several jars that had been used for the high-dose feed should be rinsed, employing a solvent appropriate for the test compound they had contained. The jar rinse is then analyzed for the presence of the test compound. If the results of the analysis are negative, the jars are considered suitable for use in another study. If not, additional washing may be done, or the jars may be soaked with solvent and reanalyzed.

3) Dosing via Drinking Water — The major consideration for use of drinking water as a vehicle for compound administration is the palatability of the test compound in water. If the animals won't drink their water because of its smell or taste, you cannot successfully dose them by this route. Water bottles are used for the dosing solution. These may be polypropylene or polycarbonate if the test compound is nonreactive with plastic. Glass bottles may be used (either clear or amber), depending on the reactivity of the compound to light. In our experience, plastic bottles are unsuitable for rabbit studies because rabbits tend to play with the bottles to the extent that measurement of water consumption is almost impossible. Glass bottles in sturdy holders seem to be more reliable. Water consumption is determined by weighing the bottle at intervals determined by the stability of the compound and the volume of water consumed daily by the animal. Bottle washes should also be done prior to use for another study to verify absence of test compound.

4) Injection — **Intravenous** — The blood vessels used for intravenous (i.v.) injection in the rodent are normally the dorsal and ventral tail veins. Rodents are dosed i.v. at between 2.5 and 10 ml per kilogram of body weight. To dilate the vein, the rodent is placed in a warming box under a heat lamp for a few seconds, or the tail is placed in a warm water bath at 37°C. The rodent is placed in a Plexiglas box that has a narrow V opening in one side. The rodent's body remains in the box with the tail extending to the outside through the V opening. Alternatively, the rat can be suspended in a cone-shaped container (attached to a ring stand) with the tail hanging down in warm water.

To ensure that a length of vein will remain patent on subsequent days of dosing, the site of the first injection should be as near the tip of the tail as possible. Subsequent injection sites would be located more proximal to the body. Holding the tip of the tail, the technician inserts the needle (normally 26 gauge attached to a 1-cc syringe) into the vein at a slight angle and slowly injects the test article. A localized swelling at the injection site indicates the dose was not delivered i.v. Following withdrawal of the needle, gauze is placed over the injection site, and pressure is applied until bleeding stops.

Rabbits are dosed via the marginal ear vein, usually at 1 ml/kg of body weight. The rabbit is placed in a restrainer that allows full access to the ears. The hair from the ear to be dosed is plucked from the tip to the base of the ear at the ear margin to expose the vein. A 25- or 26-gauge needle is used with the appropriate size syringe. The vein is compressed proximal to the injection site. The needle is inserted into the vein (pointing toward the head), the pressure on the vein is released, and the dosing solution is injected slowly. The needle is then removed, and gauze is held firmly over the injection site until the bleeding stops. The ear used is alternated each day, and the injection sites should be selected beginning at the tip of the ear and subsequently moving toward the head. Syringes should be changed after each dosage group, and needles should be changed after each animal.

Subcutaneous — In rodents, subcutaneous (s.c.) injections are normally made in the interscapular area on the back of the animal. The volume used is generally up to 10 ml/kg of body weight, with needle size dependent on the size of the animal and the viscosity of the dosing formulation. The loose skin behind the neck is grasped between the thumb and forefinger. The needle is inserted through the fold of skin but not into the underlying muscle. Loose skin behind the neck or in the hip area of the rabbit is used for s.c. injections. Syringes should be changed between dosage groups, and needles should be changed after each animal is dosed.

Intraperitoneal — A rat is restrained by holding its head and thorax. A mouse is grasped by the scruff of the neck, and the tail is twined around a finger to control the hindquarters. The needle should be introduced rapidly into a point slightly left or right of the midline and halfway between the pubic symphysis and xiphisternum on the ventral abdominal surface. A 25- or 26-gauge needle is used, and the typical injection volume is 2.5 to 10 ml/kg of body weight.

Rabbits are held by the scruff of the neck by a seated technician, with the rabbit's hips resting in the technician's lap and the rabbit facing another technician who performs the dosing. The needle is inserted through the skin of the abdomen just lateral to the midline and just posterior to the area of the umbilicus, pointed toward the spine. The syringe is changed between dose groups, and the needle is replaced between injections. Intraperitoneal injections in late pregnancy run the risk of injection directly into the gravid uterus, since a great deal of the space in the peritoneal cavity is taken up by the uterus.

Intramuscular — The site of intramuscular dosing is usually the haunch or upper hind leg ("thigh" region), with the needle inserted into the muscle; repeated dosing should alternate dosing sites (right, left, and repeat).

5) Inhalation Exposures — Mated females can be exposed in their home cages (if they are wire-mesh hanging cages), with food and water sources removed (to prevent inadvertent exposure via ingestion of absorbed or dissolved test material), or the animals can be transferred to exposure cages. Automatic watering systems can be employed within the exposure chambers to provide drinking water *ad libitum* during the exposure periods. The amount of water exposed to the test atmosphere in the tip of the "nipple" is very small. Therefore, the amount of dissolved test material in the water available for drinking is also very small. Animals are usually exposed 6 to 8 h/d. This interval is determined from the time the desired concentration is reached at the start of exposure, using a calculated t_{99} (time required to attain 99% of the target concentration) until the exposure system is shut down, with subsequent exhaust until the calculated t_{01} (time to 1% of the target concentration) is achieved. Individual animal clinical observations should begin as soon as possible after the chamber is opened. General observations on animals viewed through the chamber during exposures should also be made (although not all animals will be visible, and the identity of a given animal may not be ascertainable) since the postexposure observations may not coincide with the time of maximal clinical response.

6) Nose- or Head-Only Exposures — Nose-only or head-only exposures are usually reserved for exposure to aerosols, dusts, or mists (which the animal can groom from the fur and ingest and/or absorb through the intact skin), or for radiolabeled, difficult to obtain, or expensive test materials (to reduce the amount of material needed for exposures). These exposures require restraint (and therefore cause stress); the confinement equipment must minimize compression of the rapidly expanding abdominal contents and prevent additional heat stress. Commercially available equipment for nose-only or head-only exposures can be purchased for rabbits, rats, and mice.[46] It may be appropriate to adapt the animal to the exposure apparatus for one to two days prior to the onset of exposures to minimize stress.

7) Subcutaneous Insertion — Subcutaneous insertion of test material, an implant, or an osmotic minipump (for continuous infusion) is done under anesthesia. An incision is made on the interscapular dorsum, and one or more subcutaneous pockets is created with blunt forceps in the connective tissue space above the muscular layer. The implant material is inserted, and the incision site is stitched or closed with wound clips. The implant material is usually inserted at the start of the dosing period, and pre- and postexposure weights of the implanted material can be used to ascertain actual "dose." Alternatively, the implanted material can be retrieved at the scheduled terminal gestational sacrifice. Concurrent control group animals undergo the same procedures (anesthesia, pocket formation, implantation of an implant, wound closure, etc.), with the implant empty or filled with the vehicle.

8) Cutaneous Application — The EPA has examined the acceptability and interpretation of cutaneous developmental toxicity studies.[47-49] The procedures for this type of administration are as follows. The dosing site on the interscapular dorsum and up to 10% of the total body surface are shaved or clipped prior to initial application, with repeated shaving or clipping as needed during the dosing period. For a mouse, the application site is 2.5 × 2.5 cm, and 0.1 ml of the test article is applied (i.e., 3 ml/kg for a 30-g animal); for a rat, the site is 5 × 5 cm or 7.5 × 7.5 cm, with 0.5 to 0.6 ml applied (1.5 to 1.8 ml/kg for a 300-g animal); for a rabbit, the site is 7.6 × 7.6 or 10 × 10 cm, with 1 to 2 ml applied (0.3 to 0.6 ml/kg for a 3-kg animal).

For a nonoccluded application, the animal is manually restrained, and the test material is usually applied by syringe over the prepared site. A rat or rabbit may be provided with a collar (see Figure 7.5A) to preclude access to the dosing site, but the collar should not prevent access to food or water. Collars sized for mice (approximately 2-in. outer diameter and ¾-in. inner diameter) are now available commercially (by special order from Lomir Biomedical, Inc., Quebec J7V7M4, Canada).

Occlusion (covering the dosing site) reduces any volatilization of test material, so more test material remains on the site, and the test animal does not inadvertently get exposed by inhalation of the volatilized material. Occlusion also prevents access to the dosing site during the animal's grooming, so more test material remains on the site and the test animal does not inadvertently get exposed by ingestion of test material contaminating the hair. It further maximizes and/or accelerates absorption through the skin by increasing both temperature and humidity at the dosing site.[49]

Methods of occlusion (Figure 7.5) include application of sterile gauze and Vetwrap™ (3M Animal Care Products, St. Paul, MN) over the dosing site, usually accompanied by use of an "Elizabethan collar" (Lomir Biomedical, Inc., Quebec J7V7M4, Canada). This technique is commonly used on rabbits and rats (Figure 7.5A).[49-51] A second method useful in rabbits and rats

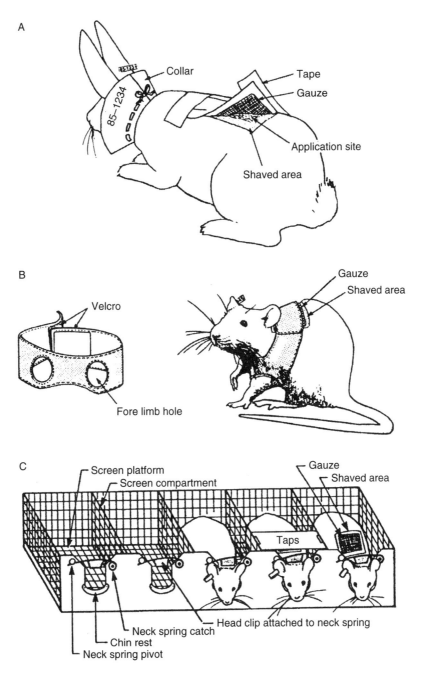

Figure 7.5 Methods of topical occlusion for pregnant animals. (From Tyl, R.W., York, R.G., and Schardein, J.L., Reproductive and developmental toxicity studies by cutaneous administration. In *Health Risk Assessment: Dermal and Inhalation Exposure and Absorption of Toxicants*, Wang, R.G.M., Knaak, J.B., and Maibach, H.I., Eds., CRC Press, Boca Raton, FL, 1992. With permission.)

involves use of a Spandex® elastic jacket, with forelimb holes, Velcro® closures on the back, and a plastic sheet on the dorsal underside to overlay the dosing site; these are available commercially (e.g., Lomir Biomedical, New York, NY). The jacket is placed on a flat surface, the test animal is placed with its forelegs in the holes, and the test article is applied. Sterile gauze is placed over the site (if appropriate), and the jacket flaps are brought up over the site and snugly attached by the Velcro strips. In most cases, a collar is not needed (Figure 7.5B).[49–52]

A method developed for occluded cutaneous application in mice (Figure 7.5C)[49,53] utilizes a stainless-steel mesh tray, compartmentalized by stainless-steel fences between each animal slot. Each compartment has an opening at one end for the animal's head, with a chin rest and a neck spring. Once the mouse is restrained, the dorsal dosing site is prepared, the test article is applied, and sterile gauze is placed over the site. Adhesive tape is secured across the dosing sites of all animals in the tray by clips on each fence.

In each case, the site is examined, and the test article and occlusion are applied at the start of each daily dosing period. At the end of each dosing period, the occlusion is removed, the gauze is discarded, and the site is gently washed and examined. A useful system of scoring the application site is provided by Draize et al.[54] for edema (swelling), erythema (redness), and any eschar (necrotic tissue) formation.

b. Duration of Administration

The recent governmental testing guidelines specify exposure beginning after implantation is complete (or on the day of insemination; see below) and continuing until the day before scheduled sacrifice at term. This corresponds to GD 6 (or 0) through 19 (or 20) for rats, GD 6 (or 0) through 17 (or 18) in mice, and GD 6 (or 0) through 28 (or 29) for rabbits, if the day sperm or a copulation plug is found (rodent) or mating is observed (rabbit) is designated GD 0. In the previous protocols, the guidelines specified exposure only from GD 6 to the end of major organogenesis (closure of the secondary palate), GD 6 through 15 for rodents, GD 6 through 18,[55] or GD 7 through 19[56] for rabbits.

The rationale for starting exposures after implantation is complete is based on two possible confounding scenarios:

- If the initial (parent) test material is teratogenic and the metabolite(s) is not, and if metabolism is induced by exposure to the parent compound, then exposure beginning earlier than implantation (with concomitant induction of enzymes and enhanced metabolism) will result in the conceptuses being exposed to less of the teratogenic moiety, and the study may be falsely negative.
- If the test chemical and/or metabolites interferes with implantation, then exposure prior to implantation will result in few or no conceptuses available for examination.

However, there are situations when initiation of exposure should begin prior to completion of implantation. These include:

- For exposure regimens that are anticipated to result in slow systemic absorption (e.g., cutaneous application, subcutaneous insertion, or dosing via feed or water, steady-state), maximal blood levels may not be attained until the very end of (or beyond) organogenesis if exposures begin on GD 6 or 7.
- For materials that are known to have cumulative toxicity (due to buildup of chemical and/or insult) after repeated exposures, exposure should begin on GD 0 (or earlier) so that the conceptuses are developing in a fully affected dam.
- For materials that are known to deplete essential components (such as vitamins, minerals, cofactors, etc.), exposures should begin early enough so that the dam is in a depleted state by the start of organogenesis (or by GD 0).
- For materials that are innocuous as the parent chemical but that are metabolized to teratogenic forms, exposures should begin early enough so that postimplantation conceptuses are exposed to maximal levels of the teratogenic metabolites.
- For test materials that are known not to interfere with implantation, exposure encompasses the entire gestational period and offspring are available for examination.

In the case of exposures that do not involve bolus dosing (i.e., gavage), such as dosing via the feed or water, and subcutaneous implants, the duration of each daily exposure is essentially continuous for the entire dosing period *ad libitum*. For cutaneous application and inhalation exposures, the daily duration is usually 6 to 8 h (corresponding to a human workday), although some studies have employed exposures of 22 h/d, with 2 h/d for housekeeping and allowing the animals to eat and drink unencumbered.

The termination of exposures was previously specified as the end of major organogenesis. It is signaled by the closure of the secondary palate and the change in designation of the conceptus from embryo to fetus. The cessation of exposure prior to term allowed for a postexposure recovery period for both the dam and the fetuses, and an assessment could be made regarding whether the observed maternal effects (body weights, clinical observations) are transient or permanent. However, fetal evaluations take place at term, and there is no commonly employed way to detect early adverse effects on the conceptus that resolve (are repaired, compensated for, or result in *in utero* demise) earlier in gestation. What was observed at term was the net result of the original insult and any repair or compensation that occurred subsequently in the conceptus, as well as any observable effects on the dam after a postexposure period. Thus, no detailed information on the dam was obtained during the exposure period, and there was no way to distinguish effects during exposure from those occurring afterward. The new developmental toxicity testing guidelines require exposure until term, which includes the fetogenesis period.

The consequences to exposures continuing until term are as follows:

- There is no postdosing recovery period so the outcome at term is not confounded by insult during dosing and compensation in the postdosing period. Example: There is no increase in maternal feed consumption or weight gain in postdosing period and no obscuring of effects on fetal body weight (since most fetal body weight gain is in the "last trimester"). There may be increased *in utero* fetal death due to continued direct or indirect toxic insult.
- The NOAEL may allow a lower dose (than if the dosing ended on GD 15), which may be a more appropriate value.
- The incidence of fetal malformation may be underestimated because more malformed fetuses may die with prolonged maternal dosing.
- Dams will require more test material as they gain weight in the "last trimester" (to maintain constant dose in milligrams per kilogram per day), so maternal and/or fetal toxicity may be greater at a given dose (i.e., the maternal liver will have to metabolize more test material).
- Fetal malformation rates will probably not increase because they are usually induced during the period of major organogenesis (GD 6 through 15), common to both dosing regimens.
- Maternal assessments at scheduled necropsy on GD 20 (e.g., toxicokinetic, biochemical, physiological endpoints) reflect effects of continued dosing (not after a postdosing recovery period). A satellite group of dams to be sacrificed at the end of dosing under the shorter dosing period for such assessments is, therefore, not necessary (this saves animals, time, and money).
- More test material is necessary (more dosing days and dams are heavier in the "last trimester," so more test material is administered to maintain a constant dose in milligrams per kilogram per day).
- The costs to perform a Phase II study with prolonged dosing are increased because more test chemical is required and there is increased labor to dose the dams on the additional days.

The rationale for continuing exposure until term includes:

- Maternal exposure until term is a better model for human exposure than exposure only during a portion of gestation, with abrupt cessation at the end of embryogenesis.
- Continued maternal responses at term (e.g., changes in organ weights, hematology, clinical chemistry, histopathology) can be better interpreted in terms of causality; there is no maternal postexposure period for compensatory changes to occur.
- Many systems continue to develop in the fetal period, both in terms of increases in cell size and the number and differentiation of specialized cells, tissues, and organs (e.g., central nervous, pulmonary, renal, gastrointestinal systems, etc.). The effects on these processes occurring in the presence of continual maternal exposure will be manifested at term (for most of the systems) and will not be confounded by compensatory processes that may occur in a postexposure period.

- The male reproductive system is established and differentiates internally and externally *in utero*, beginning on GD 13 to 14 in rodents, so effects from possible endocrine-active or reproductively toxic compounds with other mechanisms may be detected. However, at term, only the testes and epididymides can be reasonably assessed, and most effects are not discernible until weaning, puberty, or adulthood (such as morphological and/or functional effects on accessory sex organs, adult testicular and epididymal spermatogenesis, and sperm transit).

Recently in the authors' laboratory, a comparison was made of parameters of maternal and developmental effects in control CD® (SD) rats dosed from GD 6 through 15 versus from GD 6 through 19.[57] Although the GD 6 through 15 dosing was employed for earlier studies (1992 to 1997) and the GD 6 through 19 dosing was employed for more recent studies (1996 to 1998), there was an overlap in time between studies employing the different dosing durations.

As anticipated, control maternal body weights and weight gain, gravid uterine weight, absolute and relative liver weight, and maternal body weight change (GD 20 – GD 0, minus weight of the gravid uterus) were all significantly lower with the extended dosing period (the control dams received vehicle only, with no exposure to test chemical; Table 7.6). Litter size and fetal body weights were also reduced with the extended maternal dosing period (in the absence of any test chemical; Table 7.6). Interestingly and unexpectedly, uterine implantation sites per litter were also reduced as were the number of adversely affected fetuses per litter (Table 7.6). The same was true of percent fetuses per litter with visceral malformations, total (any) malformations, visceral variations, and total (any) variations (Table 7.7 and Table 7.8).

The authors concluded that the longer dosing regimen with no recovery period resulted in significant depression of maternal body weight and weight gain endpoints, as well as a reduction in fetal body weights. Presumably, the stress of handling and dosing is responsible for these differences. The three vehicles used (methylcellulose, corn oil, and water) were equally represented in both data sets. The concern is whether dosing with a potentially toxic test material will result in even further reductions due to a synergistic effect of the longer dosing period and the toxicity of the test material. The decrease in the number of implants and live fetuses may be due to the differences in times of performance of the two groups of studies. In the early 1990s, Charles River Laboratories selected all offspring from larger litters (i.e., rather than a set number per litter, regardless of litter size) as breeders, so that the average litter size rose. In the latter 1990s, a more balanced selection program was instituted (the CD®[SD] "international gold standard" IGS) to halt and reverse the increasing litter sizes. The decreased incidence of hydronephrosis (a common malformation), as well as enlarged lateral ventricles of the brain and rudimentary ribs on lumbar I (both variations), may represent genetic drift in this strain over the years evaluated. The relative developmental delay of the fetal skeleton, shown by the increased incidences of dumbbell cartilage and bipartite ossification centers in the thoracic centra, is likely due to decreased fetal body weights at term in the litters under the longer dosing regimen (i.e., the fetuses are delayed in late gestational development but are appropriate for their size).[58]

These considerations are the basis for the requirement in the new OPPTS,[10] FDA,[12] and OECD[13] final developmental toxicity testing guidelines that exposures continue until terminal sacrifice.[16]

D. Observation of Mated Females

The better and more complete the profile on maternal effects, the better the interpretation of embryofetal findings in the context of maternal toxicity, if any.[58–61]

1. Clinical Observations

Clinical observations of all animals should be made once daily during the pre- and posttreatment periods, at least once daily at dosing, and at least once after dosing for gavage studies (with timing of observation based on initial findings during the dosing period) throughout the dosing period. Observations should be made for (but not limited to) the potential findings listed in Table 7.9.

In the authors' laboratory, mated females are observed closely on the first day of dosing at a number of postdosing time points (e.g., 15 and 30 min and 1, 2, and 4 h). This is done to observe any possible treatment-related clinical signs, to identify the times at which such signs are maximally expressed (for determination of times of observation for subsequent dosing days), and to document whether the signs are transient (and the time frame if they are).

2. Maternal Body Weights

Mated females should be weighed on GD 0 and at least once more on GD 3 prior to the exposure period (if exposures are started on GD 6). Females should be weighed frequently during the exposure period, at least every three days (e.g., GD 6, 9, 12, 15, 17 [mice], 18, and 20 [rats]; GD 6, 9, 12, 15, 18, 21, 24, 27, and 29 or 30 [depending on the day of necropsy] for rabbits).[42] Assessment of maternal weight is very useful for detection of early consequences of dosing (e.g., taste aversion from gavage, which may become more profound over time, or palatability problems from treated food or water, which may resolve over time) and stress for other routes (e.g., inhalation, cutaneous application), compounded by any effects from the chemical. Maternal weights can also be used to ascertain whether early effects are transient and if there is complete litter loss prior to term. For example, CD® rats will usually gain 40 to 60 g from GD 0 to 6 if they are pregnant (F344 rats will gain at least 20 g, and CD-1® mice will gain 2 to 4 g). Subsequent weight loss to GD 0 levels may indicate total litter loss.

Maternal body weight changes should be calculated for the preexposure period if employed (GD 0 to 6 or 7), for time increments during exposure, and for the total exposure period (GD 6 to 9, 9 to 12, 12 to 15, 15 to 18, and 18 to 20, and 6 to 20 for rats; GD 6 to 9, 9 to 12, 12 to 15, 15 to 18, 18 to 21, 21 to 24, 24 to 27, 27 to 30, and 6 to 30 for rabbits). Weight changes (calculated per animal and summarized for each dosage group) are more sensitive indicators of effects than are body weights per se. Such measures can detect early, transient weight reductions during the dosing period and compensatory increases late in the dosing period.

Total gestational weight change (GD 0 to 17 or 18 for mice, GD 0 to 20 or 21 for rats, and GD 0 to 29 or 30 for rabbits) is also calculated. Gestational weight change corrected for the weight of the gravid uterus (weight change during gestation minus the weight of the gravid uterus) and corrected terminal body weight allow measurement of a weight change or terminal weight of the dam, with no contribution from effects on the conceptuses. This corrected parameter will provide information on maternal toxicity per se, unconfounded by reduced numbers and/or body weights of the litter. This correction can only be made using the weight of the term gravid uterus. The impact on maternal body weights of effects on the conceptuses (e.g., reduced number of live fetuses, reduced embryo/fetal body weights) during gestation cannot be ascertained unless extra females are included in the various treatment groups and are necropsied during the exposure period to provide input on body weight, gravid uterine weight, and status of the offspring.

3. Maternal Feed Consumption

Feed consumption of singly housed animals should be measured throughout gestation on the same gestational days on which the maternal animals are weighed. This is done by measuring the full feeder (or the feed alone) at the start of each interval and the "empty" feeder (or the remaining feed alone) at the end of each interval. Pretreatment, treatment, posttreatment, and gestational feed consumption should also be calculated.

Feed consumption should be presented as grams per animal per day and grams per kilogram per day, with the latter obviously factoring in the weight of the female during the consumption interval. For example, if a rat that weighed 300 g at the start and 350 g 2 d later ate 25 g/d during the interval, then the average food consumption is as follows:

$$\frac{25 \text{ g/d}}{\frac{(0.300 \text{ kg} + 0.350 \text{ kg})}{2}} = \frac{25 \text{ g/d}}{0.325 \text{ kg}} = 76.92 \text{ g/kg/d}$$

Table 7.6 Recent historical control data for developmental toxicity studies in common laboratory animals

Parameter	CD® (SD) Rats	CD-1® Mice	NZW Rabbits[a]	NZW Rabbits[b]
No. pregnant	672	71	227	88
No. live litters	672	70	224	83
No. fetuses	10,033	841	1800	558
No. corpora lutea/litter	14.91–18.74	12.43–13.86	8.31–10.38	10.48 ± 0.26
No. implants/litter	13.71–16.75	11.57–13.35	7.58–9.06	7.19 ± 0.31
Percent preimplantation loss/litter	0.55–14.76	5.12–9.25	5.82–17.48	30.74 ± 2.90
No. (%) litters with nonlive implants	5 (13.04)–17 (70.83)[d]	7 (30.43)–17 (68.00)	3 (14.29)–17 (42.11)	37 (42.0)
No. nonlive implants/litter	0.22–1.21	0.61–1.00	0.14–0.74	0.85 ± 0.15
Percent nonlive implants/litter	1.18–6.82	4.85–11.50	1.85–16.15	14.46 ± 2.80
No. live fetuses/litter	13.30–16.24	10.96–12.70	7.00–8.50	6.72 ± 0.31
Male	6.21–8.72	5.42–6.87	3.50–4.74	3.39 ± 0.23
Female	6.32–8.65	5.04–6.96	3.50–4.64	3.33 ± 0.21
Percent male fetuses/litter	41.83–57.06	44.19–54.87	43.04–51.51	49.68 ± 2.85
Fetal body weight/litter (g)[c]	3.430–3.866	1.007–1.042	48.09–53.30	50.23 ± 0.96
Males (g)	3.519–4.063	1.007–1.065	48.14–54.16	49.65 ± 1.00
Females (g)	3.424–3.791	0.999–1.021	46.96–52.79	48.86 ± 0.88
No. (%) fetuses with malformations	0 (0.00)–48 (10.84)	3 (1.01)–31 (12.30)	1 (0.64)–16 (10.53)	32 (5.7)
No. (%) litters with malformed fetuses	0 (0.00)–13 (56.52)	3 (12.50)–14 (60.87)	1 (5.26)–9 (42.86)	21 (25.3)
No. malformed fetuses/litter	0.00–1.66	0.13–1.35	0.05–0.80	0.39 ± 0.10
Male	0.00–0.97	0.08–1.00	0.00–0.40	0.25 ± 0.08

Female	0.00–0.69	0.004–0.39	0.00–0.40	0.16 ± 0.05
Percent malformed fetuses/litter	0.00–10.60	1.14–12.18	0.62–10.57	5.54 ± 1.48
Male	0.00–10.36	1.67–18.44	0.00–9.50	5.21 ± 1.61
Female	0.00–12.64	0.60–5.41	0.00–9.25	5.63 ± 2.00
No. (%) fetuses with variations	16 (4.95)–202 (45.60)	38 (13.01)–161 (54.21)	51 (42.02)–158 (65.60)	343 (61.5)
No. (%) litters with variant fetuses	10 (43.48)–30 (100.00)	17 (73.91)–23 (95.83)	11 (85.00)–45 (100.00)	77 (92.8)
No. variant fetuses/litter	0.70–6.96	1.65–6.71	3.22–5.96	4.13 ± 0.31
Male	0.39–3.72	1.04–2.88	1.30–3.22	2.24 ± 0.22
Female	0.30–3.24	0.61–3.83	1.83–2.89	2.19 ± 0.19
Percent variant fetuses/litter	5.22–46.55	12.87–54.70	42.53–62.70	60.74 ± 3.62
Male	6.27–42.60	16.11–55.52	38.40–59.45	57.66 ± 4.46
Female	3.82–49.57	8.86–54.63	47.74–82.36	63.24 ± 4.11

Note: For data involving ranges as No. (%)–No. (%), the specific percent is not necessarily based on the specific number. The range of numbers and the range of percentages are not necessarily the same since different studies had different numbers of litters/fetuses. The values are the lowest to the highest numbers and the lowest to the high percentages.

[a] Naturally bred, without hormonal priming.
[b] Artificially inseminated, with hormonal priming; values are presented as grand mean of study control group means ± SEM.
[c] Sexes combined.

Source: From the Reproductive and Developmental Toxicology Laboratory of the authors at RTI; data are presented as range of study control group means. For CD® rats, 28 studies; for CD-1® mice, 3 studies; and for naturally bred rabbits, 11 studies.

Table 7.7 Summary and statistical analysis of control rat fetal malformations and variations

Parameter	Gestational Days of Dosing	
	6–15	6–19
No. fetuses examined[a]	2723	1932
No. litters examined[b]	178	136
Malformations		
No. fetuses with external malformations/litter[c,d]	0.02 ± 0.01	0.02 ± 0.01
Percent fetuses with external malformations/litter[c,d]	0.16 ± 0.08	0.13 ± 0.08
No. (percent) fetuses with external malformations[d]	4 (0.1)	3 (0.2)
No. (percent) litters with external malformations[e]	4 (2.2)	3 (2.2)
No. fetuses with visceral malformations/litter[c,d]	0.55 ± 0.10	0.11 ± 0.04
Percent fetuses with visceral malformations/litter[c,d,f]	6.92 ± 1.36	1.41 ± 0.49
No. (percent) fetuses with visceral malformations[d]	98 (6.4)	15 (1.6)
No. (percent) litters with visceral malformations[e]	43 (24.2)	9 (6.6)
No. fetuses with skeletal malformations/litter[c,d]	0.04 ± 0.02	0.05 ± 0.02
Percent fetuses with skeletal malformations/litter[c,d]	0.89 ± 0.42	0.77 ± 0.03
No. (percent) fetuses with skeletal malformations[d]	8 (0.4)	7 (0.7)
No. (percent) litters with skeletal malformations[e]	7 (3.9)	7 (5.1)
No. fetuses with any malformations/litter[c,d]	0.61 ± 0.10	0.18 ± 0.05
Percent fetuses with malformations/litter[c,d,f]	4.73 ± 0.87	1.22 ± 0.30
No. (percent) fetuses with malformations[d]	109 (4.0)	25 (1.3)
No. (percent) litters with malformations[e]	53 (29.8)	17 (12.5)
Variations		
No. fetuses with external variations/litter[c,d]	0.02 ± 0.01	0.0 ± 0.0
Percent fetuses with external variations/litter[c,d]	0.10 ± 0.07	0.0 ± 0.0
No. (percent) fetuses with external variations[d]	3 (0.01)	0 (0.0)
No. (percent) litters with external variations[e]	2 (1.1)	0 (0.0)
No. fetuses with visceral variations/litter[c,d]	3.62 ± 0.23	1.15 ± 0.11
Percent fetuses with visceral variations/litter[c,d,f]	44.9 ± 2.89	16.04 ± 1.54
No. (percent) fetuses with visceral variations[d]	645 (41.9)	157 (16.3)
No. (percent) litters with visceral variations[e]	140 (78.7)	79 (58.1)
No. fetuses with skeletal variations/litter[c,d]	0.97 ± 0.10	1.00 ± 0.11
Percent fetuses with skeletal variations/litter[c,d]	10.87 ± 1.21	14.40 ± 1.58
No. (percent) fetuses with skeletal variations[d]	172 (9.4)	136 (14.1)
No. (percent) litters with skeletal variations[e]	94 (52.8)	74 (54.4)
No. fetuses with variations/litter[c,d]	4.46 ± 0.24	2.15 ± 0.015
Percent fetuses with variations/litter[c,d,f]	29.09 ± 1.50	15.26 ± 1.05
No. (percent) fetuses with variations[d]	793 (29.1)	293 (15.2)
No. (percent) litters with variations[e]	160 (89.9)	113 (83.1)

[a] Only live fetuses were examined for malformations and variations
[b] Includes only litters with live fetuses
[c] Reported as the mean ± S.E.M.
[d] Fetuses with one or more malformations or variations
[e] Litters with one or more fetuses with malformations or variations
[f] $p \leq 0.001$; significant difference between the two groups

Source: Marr, M.C., Myers, C.B., Price, C.J., Tyl, R.W., and Jahnke, G.D., *Teratology*, 59, 413, 1999 (tables provided by the authors).

Technicians should make notations regarding spilled feeders, feed in bottom of cage, etc., to account for "outlying" values not appropriate for inclusion in summarization and analysis (statistical tests for outliers can also be used to identify suspect values). Water consumption should be assessed in exactly the same way as feed consumption. If the animals are dosed via feed or water, the actual intake of test material in grams or milligrams per kilogram of body weight per day can be easily calculated. This is done by using the value for consumption in grams per kilogram per day multiplied by the percentage of test material in the diet or water.

Table 7.8 Summary of morphological abnormalities in control CD® rat fetuses: listing by defect type[a]

Parameter	Gestational Days of Dosing	
	6–15	6–19
Fetal Malformations and Variations		
Total no. fetuses examined externally[b]	2723	1932
Total no. fetuses examined viscerally[b]	1538[d]	966
Total no. fetuses examined skeletally[b]	1838[d]	965[e]
Total no. litters examined[c]	178	136
External Malformations		
Anasarca	1 (1)	
Cleft palate	1 (1)	
Anal atresia	1 (1)	2 (2)
Agenesis of the tail	1 (1)	
Short, thread-like tail		2 (2)
Short tail		1 (1)
Umbilical hernia	1 (1)	
Visceral Malformations		
Abnormal development of the cerebral hemispheres		1 (1)
Abnormally shaped heart	1 (1)	
Hydronephrosis:		
Bilateral	2 (2)	6 (4)
Left	3 (2)	3 (2)
Right	1 (1)	
Hydroureter:		
Bilateral	54 (28)	4 (3)
Left	36 (19)	2 (2)
Right	3 (3)	3 (3)
Skeletal Malformations		
Unossified frontals, parietals, and interparietals	1 (1)	
Discontinuous rib		1 (1)
Discontinuous rib cartilage	1 (1)	
Fused rib cartilage		1 (1)
Fused cartilage: thoracic centrum		1 (1)
Bipartite cartilage, bipartite ossification center: thoracic centrum	6 (5)	6 (6)
External Variations		
Hematoma:		
Neck	2 (1)	
Forelimb	1 (1)	
Visceral Variations		
Enlarged lateral ventricle of brain (full):		
Bilateral	100 (43)	18 (15)
Left	20 (13)	8 (7)
Right	19 (14)	4 (4)
Enlarged lateral ventricle of brain (half):		
Bilateral	129 (52)	48 (28)
Left	12 (9)	21 (15)
Right	11 (10)	7 (5)
Enlarged lateral ventricle of brain (partial):		
Bilateral	283 (63)	
Left	14 (11)	
Right	14 (13)	

Table 7.8 Summary of morphological abnormalities in control CD® rat fetuses: listing by defect type[a] (continued)

Parameter	Gestational Days of Dosing	
	6–15	6–19
Enlarged nasal sinus	1 (1)	2 (2)
Agenesis of the innominate artery	1 (1)	5 (5)
Pulmonary artery and aorta arise side by side from heart	1 (1)	
Extra piece of liver tissue on median liver lobe		1 (1)
Very soft tissue: liver	1 (1)	
Blood under kidney capsule: Right		1 (1)
Small papilla: Right	1 (1)	
Distended ureter:		
Bilateral	37 (21)	27 (18)
Left	20 (17)	15 (12)
Right	12 (9)	11 (10)
Skeletal Variations		
Misaligned sternebrae		2 (2)
Unossified sternebra (I, II, III, and/or IV only)		5 (4)
Rib on lumbar I:		
Bilateral full	3 (2)	
Left full	4 (4)	
Right full	1 (1)	1 (1)
Bilateral rudimentary	31 (20)	6 (4)
Left rudimentary	40 (29)	7 (6)
Right rudimentary	15 (12)	3 (3)
Short rib: XIII	5 (3)	2 (2)
Wavy rib	1 (1)	
Incomplete ossification, cartilage present: thoracic centrum	1 (1)	
Dumbbell cartilage, normal ossification center: thoracic centrum	1 (1)	
Dumbbell cartilage, dumbbell ossification center: thoracic centrum	33 (19)	3 (3)
Dumbbell cartilage, bipartite ossification center: thoracic centrum	10 (10)	40 (23)
Normal cartilage, bipartite ossification center: thoracic centrum	36 (22)	70 (32)
Normal cartilage, no ossification center: thoracic centrum	1 (1)	
Normal cartilage, unossified ossification center: thoracic centrum (VI-XIII only)		1 (1)

[a] A single fetus may be represented more than once in listing individual defects. Data are presented as number of fetuses (number of litters). See Table 7.7 for summary and statistical analysis of fetal malformation/variation incidence.
[b] Only live fetuses were examined.
[c] Includes only litters with live fetuses.
[d] For many studies, a single fetus was examined externally, viscerally, and skeletally.
[e] One fetus was lost during processing prior to skeletal examination.

Source: Marr, M.C., Myers, C.B., Price, C.J., Tyl, R.W., and Jahnke, G.D., *Teratology*, 59, 413, 1999 (tables provided by the authors).

E. Necropsy and Postmortem Examination

1. Maternal

Mated females that die during the course of the study should be necropsied in an attempt to determine the cause of death, with target tissues saved for possible histopathology. Females that appear moribund should be humanely euthanized (by CO_2 asphyxiation for rodents and i.v. euthanasia solution for rabbits) and necropsied to attempt to determine the cause of the morbidity, with target tissues saved for optional histopathology. Females showing signs of abortion or premature delivery should also be sacrificed, as described above, as soon as the event is detected. They should be subjected to a gross necropsy, and any products of conception should be saved in neutral buffered

Table 7.9 Some typical maternal clinical observations

Alteration of:
    Body position
    Activity
    Coordination
    Gait
Unusual behavior, such as:
    Head flicking
    Compulsive biting or licking
    Circling
    Rooting in bedding[a]
Presence of:
    Hemorrhages
    Petechiae
    Convulsions or tremors
    Increased salivation[a]
    Increased lacrimation or other ocular discharge
    Red-colored tears (chromodacryorrhea)
    Nasal discharge
    Increased or decreased urination or defecation (including diarrhea)
    Piloerection
    Mydriasis or miosis (enlarged or constricted pupils)
    Unusual respiration (fast, slow, gasping, or retching)
    Vocalization

[a] Salivation pre- or postdosing and/or rooting in bedding postdosing may indicate taste aversion (from gavage dosing). Palatability problems from dosed feed or water are most likely to be detectable from reduced feed or water intake initially, with "recovery" back to normal intake values over time.

10% formalin. Approximately 1.0 to 1.5 d before expected parturition (GD 17 or 18 for mice, 20 or 21 for rats, and 29 or 30 for rabbits), all surviving dams/does should be killed by CO_2 asphyxiation (rodents) or i.v. injection into the marginal ear vein (rabbits). They should be laparotomized, their thoracic and abdominal cavities and organs examined, and their pregnancy status confirmed by uterine examination. Uteruses that present no visible implantation sites or that contain one-horn pregnancies should be stained with ammonium sulfide (10%) to visualize any implantation sites that may have undergone very early resorption.[62] Immersion of the uterus in ammonium (or sodium) sulfide renders it useless for fixation and histopathologic examination. When it is to be subsequently fixed and examined, the authors stain the fixed uterus with potassium ferricyanide by a method they developed to detect resorption sites. The stain is then washed out of the uterus, and the uterus is replaced in fixative for subsequent histopathologic examination (the staining does not interfere with subsequent evaluations).

At sacrifice, the body, liver, any identified target organs, and the uterus of each mated female should be removed and weighed. Any maternal lesions should be retained in appropriate fixative for possible subsequent histologic examination. The ovarian corpora lutea should be counted to determine the number of eggs ovulated (to allow calculation of preimplantation loss and to put the number of implants in perspective). For rabbits, each corpus luteum is a round, convex body with a central nipplelike projection; for rats, each corpus luteum is a round, slightly pink, discrete swelling (counted with the naked eye or under a dissecting microscope). For mice, the ovarian capsule must be removed (peeled off) and the corpora lutea counted under a dissecting microscope; each appears as a rounded, slightly pink, discrete swelling. For partially resorbed litters, the number of corpora lutea may be less than the number of total implantation sites since corpora lutea involute (to become small corpora albicantia) when they no longer sustain live implants (note that the concordance between number of resorbed implants and the loss of corpora lutea is not exact). Corpora lutea can be discounted for summarization when the count is less than the number of implants, or the number of corpora lutea can be set to correspond to the number of implants to correct for involuting corpora lutea, typically in partially resorbed litters. The uterine contents (i.e., number of implantation sites, resorptions, dead fetuses, and live fetuses) should be recorded (see below). Dead fetuses should be weighed, examined externally, and saved in neutral buffered 10% formalin.

2. Fetal

a. Anesthesia and Euthanasia

Once fetuses are removed from the uterus, several options for anesthesia or euthanasia are acceptable, but one may be preferred over the other, depending on the constraints of each institution's Animal Care and Use Committee. Rodent and rabbit fetuses may be anesthetized or euthanized by i.p. (or sublingual) injection of a sodium pentobarbital solution. Alternatively, the USDA considers hypothermia an acceptable form of anesthesia for rodent fetuses. The fetus is placed on a wetted paper towel over ice, which induces anesthesia by lowering the core body temperature below 25°C.[63,64] This method does not interfere with the internal examination of organ systems, as sometimes occurs with injected anesthetics. Rabbit fetuses are routinely euthanized by intraperitoneal injection of 0.25 ml of sodium pentobarbital solution. Anesthesia and/or euthanasia must be achieved prior to any further examinations.

b. Implantation Detection, Designation, and Recording

Once the uterus is removed, weighed, and opened, the status of implantation sites can be evaluated. Beginning at a designated landmark specified by the individual laboratory's SOPs (e.g., starting at the left ovarian end and moving, in order, to the cervix and up to the right ovary, or alternatively beginning at the left ovary and moving to the cervix, then starting at the right ovary and moving to the cervix), the status of implantation sites is determined and the location of the cervix specified (since uterine position affects the reproductive outcome in unexposed and exposed conceptuses).[65] Implantation sites can be characterized as follows:

> Live: Fetus exhibits spontaneous or elicited movement.
> Dead: Nonlive fetus with discernible digits in any or all paws, body weight *at or greater than* 0.8 g for rats, 0.3 g for mice, or 10.0 g for rabbits.
> Full: Nonlive fetuses with discernible digits below the weights listed above for fetuses designated as "dead."
> Late: Some embryonic or fetal tissue from fetal remains to a full fetus with discernible limb buds but no discernible digits; fetuses may show signs of autolysis or maceration and may appear pale or white.
> Early: Evidence of implantation sites visualized only after staining fresh uterine preparation with ammonium sulfide; metrial glands and decidual reaction only; maternal placenta; maternal and fetal placenta.

Once implantation status is recorded, the umbilical cord of each fetus is gently pinched with forceps to occlude blood flow, and the cord is cut as close to the body as is practical. The placentas are examined and discarded (some laboratories weigh and/or retain placentas in fixative for possible subsequent histopathology). The ICH testing guideline[9] requires gross examination of the placentas. The OPPTS draft guidelines[10] called for closer examination of placentas (e.g., weight and retention in fixative), but this suggestion did not make it to the final guideline.[66]

Each live fetus is given a fetal number, while dead fetuses and early or late resorptions are designated by a letter (e.g., D for dead, L for late resorption, E for early resorption) since no further tracking will be done. To induce hypothermia, rodent fetuses may be placed in order (by uterine horn) on a wetted paper towel over ice, or they may be placed in a compartmentalized box over a bag of crushed ice. Alternatively, the fetuses may be euthanized with sodium pentobarbital and placed in a compartmentalized box. Each fetus may be individually identified by a plastic or paper tag (e.g., jewelry "price tag") on an alkali-resistant string (e.g., dental floss) secured around the fetus' middle, below the rib cage, and above the pelvis.

Rabbit fetuses may be numbered on the top of the head with a permanent marker once they are removed from the amniotic sac, or they may be individually identified by a tag as described above. This must be done quickly, as rabbit fetuses are active, and their position in the uterus can be uncertain if they are not marked immediately. The head is gently blotted dry, and a permanent marker is used to gently record the fetal number on the top of the head. Excessive pressure may cause intracranial swelling, thus creating artifacts observed during the craniofacial examination.

c. *External Examination*

All live fetuses are weighed, normally to a hundredth of a gram (thousandth of a gram for mice), in numerical order. Dead fetuses are also weighed. After weighing, each fetus is held under a magnifying lens and its sex (for rodents) is determined by gauging the anogenital distance (AGD; the distance between the anus and the genital tubercle [papilla]). In male CD® (SD) rat fetuses, the historical control range for AGD is approximately 1.8 mm; in females, the distance is about one-half that, approximately 0.85 mm). In CD-1® mouse term fetuses, the male AGD is approximately 1.2 mm, and the female AGD is approximately 0.6 mm. AGD can be accurately and precisely measured with an eyepiece diopter (accurate to 0.01 mm), attached to a stereo microscope, and a microscope stage grid. Alternatively, vernier calipers or a micrometer eyepiece on a dissecting microscope, calibrated with a micrometer stage, can be used. Rabbit fetuses are normally not sexed externally, as they have no distinct sex difference in AGD.

The fetuses are carefully examined externally.[67] The examination should proceed from head to tail in an orderly manner (with the aid of a magnifying lens for rats and/or a dissecting microscope for mice, if necessary). The contour of the cranium should be noted in profile and full-face view. The bilateral presence of the eye bulges should be noticed; they should be symmetrical and of normal size and position. The eyelids should be closed. The pinnae (external ear flaps) should not be detached from the head at this developmental stage (rodents only), and they should be checked for symmetry, size, shape, and location. The alignment of the upper and lower jaw is examined in profile. A clean angle should be formed with the upper jaw and nose protruding slightly further than the lower jaw. The upper jaw is checked face-on for notches, furrows, or distortions. The tongue and palate are checked by depressing the tongue while opening a pair of closed forceps that have been inserted between the upper and lower jaws. The status of the teeth is also checked in rabbit fetuses.

The skin covering the head and the rest of the body is checked carefully for continuity, and any abnormalities in color, texture, or tone are noted. Irregular swellings, depressions, bumps, or ecchymoses (subdermal hematomas) are recorded as well. The overall posture of the fetus should be observed at this time. The fore- and hindlimbs are checked carefully for normal size, proportions, and position. The number and disposition of the digits is noted (four digits plus a dewclaw on the forelimbs, five digits on the hind limbs), and the depth of the interdigital furrows is explored by gently pressing against the paws with the forceps to spread the digits.

Abnormalities of the general body form, failure of dorsal or ventral midline closure, and defects involving the umbilicus are typically quite obvious but must be examined carefully for accurate description. The anus, external genitalia, and tail are examined next. The anal opening must be present and in the proper location. The external genitalia are checked for general shape and size. The tail is also examined for normal length and diameter (to detect thread-like tail), as well as abnormal kinking or curling and localized enlargements or constrictions. Dead fetuses may be saved for visceral and skeletal examination or may be preserved in neutral buffered 10% formalin for possible subsequent examination.

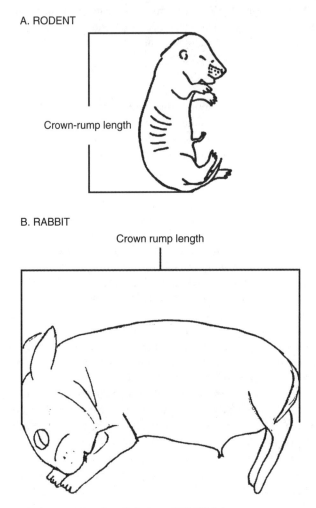

Figure 7.6 Measurement of crown-rump length in term (GD 20) fetuses.

Other evaluations, such as crown-rump length measurement, may be done prior to visceral examination. For measurement of crown-rump length, each fetus is placed on its side in a natural fetal position on a metric rule calibrated in millimeters (rodents) or centimeters (rabbits), or is measured with a vernier caliper or micrometer eyepiece. Length is measured from the crown tip (dorsal aspect of the head, corresponding to the hindbrain) to rump base (excluding the tail) and recorded (see Figure 7.6). For example, the crown-rump length for control CD® rat fetuses ranges from 30 to 40 mm (3.0 to 4.0 cm) and that for control New Zealand White (NZW) rabbit fetuses ranges from 8 to 12 cm.

d. *Visceral Examination—Staples's Technique*[68]

1) Selection — Current guidelines specify that 50% of the litter for rodents and 100% for rabbits must be evaluated for soft tissue alterations (malformations or variations). A random or arbitrary process of selection should be employed, such as evaluating every other fetus. In the case of 50% visceral evaluations, alternating even- or all odd-numbered fetuses from sequential litters could be selected (or one could select even-numbered fetuses from a dam with an even study number and odd-numbered fetuses from a dam with an odd study number, etc.). The designated rodent fetuses for soft tissue evaluation will also have their heads removed, with appropriate unique identification (e.g., individual, labeled scintillation vials for rodents; a large jar for each litter with individually numbered heads for rabbits) for subsequent fixation, decalcification, and soft tissue craniofacial examination. Since all rabbit fetuses will be examined viscerally, selection for which ones will have their heads removed and evaluated can be made, as above, for the selection of rodent fetuses for visceral examination.

The tools needed for rodents include microdissecting scissors, forceps, and ultra microdissecting or iridectomy scissors (for heart cuts in rodents). All visceral examinations should be performed under a dissecting microscope. Each fetus designated for visceral examination is placed in the supine position (on its back) on a wax block with a number corresponding to the fetal number, and the fetal limbs are secured by rubber bands or dissecting pins. A ventral midline incision is made from the umbilicus, cutting caudal to the genital tubercle and cutting cranially to the neck. The cut should first be made through the skin so that the sternebrae are visible, and then through the muscular abdominal wall and the ribcage, slightly lateral to and on the right of the sternebrae. Once the incision is completed, the ventral attachment of the diaphragm is located. Working to either side from this point, the technician checks the diaphragm for herniations or other abnormalities (i.e., thinness) and carefully clips it away from the rib cage. Once the diaphragm has been clipped away, the rib cage may be gently opened on either side and secured to the block with pins.

2) Thoracic Viscera — The fetal viscera are examined sequentially, generally beginning with the thoracic organs and moving caudally. The bilobed thymus is first examined for normal size, shape, and coloration, and removed. The lungs should be pink and frothy in appearance, with clearly visible alveoli. For rodents, there should be three lung lobes on the right side of the fetus, one on the left, and a small intermediate lobe that crosses the midline and lies slightly over to the left side. Rabbits have three lobes on the right side and two mediastinal lobes on the left side. The trachea and esophagus are checked for normal alignment and their relationship to the vessels of the heart, i.e., the aortic and pulmonary arches, should be ventral to (in front of) the trachea and esophagus.

The heart and great vessels should be examined carefully. Each vessel entering and leaving the heart should be the proper size and should course normally as far as it can be traced. The atria of the heart may be heavily engorged (this is not abnormal), but the heart should have a smooth, rounded appearance and a well-defined apex slightly left of center. A light indentation is usually visible over the area of the interventricular septum. The pericardium, a thin transparent membrane that surrounds and cushions the heart, should be carefully stripped away. The first heart cut (with ultra microdissecting scissors for rodents and sharp microdissecting scissors in rabbits) begins to the right of the ventral midline surface at the apex. It extends anteriorly and ventrally into the pulmonary artery. When opened, this incision permits examination of the tricuspid valve, papillary muscles, chordae tendineae, right side of the interventricular septum, atrioventricular septum, and semilunar valve of the pulmonary artery. The second heart cut originates slightly to the left of the ventral midline surface of the apex and extends anteriorly and ventrally into the aorta. Because the pulmonary artery lies ventral to the aorta, it is cut as this incision extends into the aorta. Following the second cut, these structures are visible: bicuspid valve, papillary muscles, chordae tendineae, left side of the interventricular septum, atrioventricular septum, and semilunar valve of the aorta.

3) Abdominal Viscera — The liver lobes (four including the median, right lateral, left lateral, and caudate) are counted and checked for fusion, texture, and normal coloration. Rats do not have a gallbladder. Rabbits' gallbladders vary more in shape and size than those of other species. The stomach plus the lower esophagus, spleen, pancreas, and small and large intestines are examined next. The stomach and intestines should be filled with a viscous green or yellowish fluid (bile plus swallowed amniotic fluid). The kidneys should then be examined; the left kidney should be located slightly more caudal than the right. The ureters are checked for normal size and location and for continuity from the renal hilus to the urinary bladder. Then, each kidney is cut transversely (or the left longitudinally and the right transversely) so that the renal papilla and renal pelvis may be examined. Any abnormal dilation of the ureters or the renal pelvis (more common in rats) is noted. The urinary bladder should be checked at this time.

The sex of each fetus is then determined by carefully inspecting the gonads. Rodent testes are defined by their round to bean shape and the tortuous spermatic artery that runs along the side of each testis, with associated epididymides (caput [head], corpus [body], and cauda [tail]) on the lateral side of the testes. The testes should be located low in the pelvis on either side of the urinary bladder. Additional reproductive structures can be seen in the term rat fetus with a magnifying lens or dissecting microscope. For males, these include the white, threadlike efferent ducts (vasa efferentia) from the testes to the caput epididymis, the single, thin ductus deferens (vas deferens, Wolffian duct) from the cauda epididymis to the urethra and the thick gubernaculums, which extends caudally from the cauda epididymis to the inguinal ring. In rabbits, the testes are in the lower quadrant; they are slightly oval, and their superficial blood vessels can be seen.

Ovaries are small and elongated. Their location in rodents is high in the pelvis, just inferior to the kidneys, and they are cupped by the funnel-shaped ends of the oviducts (uterine tubes, fallopian tubes). The bicornuate ("two horned") uterus is continuous with the proximal ends of the oviducts. For females, in addition to the ovaries, oviducts, and uterus, the cranial suspensory ligament can be seen (with magnification) extending from each ovary anteriorly to the diaphragm along the dorsal body wall. Rabbit ovaries are pinkish and elongated but, like the testes, are lower in the abdomen than is the case for rodents. Once the visceral examination is concluded, the viscera are removed *in toto*, and the eviscerated carcass is prepared for skeletal staining.

e. Visceral Examination—Wilson's Technique

An alternative method for visceral examination of rodent fetuses involves fixation and decalcification in Bouin's solution of fetuses selected for visceral examination.[17, 69–74] Each fetus is then serially cross-sectioned through the head (see head examination, below) and trunk regions by freehand cutting with a razor or scalpel blade. The advantages to this technique include examination of the fetuses at the convenience of the technical staff (since the fetuses are fixed) and retention of sections for documentation or subsequent histologic confirmation of a lesion. The disadvantages include: (1) inability to examine the same fetus for visceral and skeletal alterations (since Bouin's fixative decalcifies the skeleton and the serial sections preclude such examination),[75] (2) inability to use color changes in fresh specimens and/or blood flow through the heart and great vessels as aids in detection and diagnosis of circulatory alterations, (3) difficulty in sectioning each fetus in a litter and every litter in precisely the same way (e.g., through the heart) to assure comparability of sections, and (4) difficulty in visualizing morphological alteration from serial cross-sections. A comparison of Staples's versus Wilson's technique in rat fetuses indicated that Staples's technique identified more heart and great vessel malformations than Wilson's technique did.[76] An alternative microdissection technique after fixation has been presented by Barrow and Taylor.[77]

Surface staining of Wilson's sections can enhance detection of visceral alterations.[78] Since Bouin's fixative contains picric acid, which is reactive, unstable, explosive, and also a strong irritant and allergen (and therefore a safety and health hazard), and Bouin's fixative can result in fragile friable soft tissues, alternative fixatives that do not contain picric acid have been suggested. The use of a modified Davidson's fixative for fetuses has been reported[79] (and for fixation of testes and eyes[80]), with subsequent serial sectioning of fetuses by Wilson's technique. The authors[79] report that side-by-side comparisons of results with the two fixatives indicate Davidson's fixative produces superior contrast and definition of organs and vessels in the cranial, thoracic, and abdominal regions, with the tissues remaining moist for a longer time. The procedure reported is as follows. Fetuses are immersed in modified Davidson's fixative (14 ml ethanol, 6.25 ml glacial acetic acid, 37.5 ml saturated commercial grade 37% formaldehyde, and 42.25 ml distilled water for 100 ml of fixative) for 1 week, rinsed twice with tap water, and stored in 70% isopropyl alcohol. Another decalcifying fixative that does not contain picric acid is Harrison's, with similar benefits to Davidson's. Most laboratories utilize the Staples's technique[68] or modifications of the Staples's technique[77,81] for visceral examination of rodent fetuses. Wilson's technique cannot be used for rabbit fetuses since current guidelines require both visceral and skeletal examination of all fetuses of that species.

f. Craniofacial Examination

Prior to or after visceral examination (for each fetus scheduled for craniofacial examination), the head is removed and placed in Bouin's solution or another decalcifying fixative (see above) for fixation and decalcification. Rodent heads may be put individually in a scintillation vial approximately three-quarters full of fixative. Because rabbits have identification numbers on top of their heads, the heads from an entire litter may be put in a large container for fixation. (Some laboratories perform only a single midcoronal section on rabbit heads at sacrifice; this provides information on eye structure and on the status of the lateral ventricles of the cerebrum [e.g., is there hydrocephaly?] but does not provide information on other areas of the fore-, mid-, and hindbrain.) Rodent heads should remain in the fixative for approximately 72 h prior to examination, and rabbit heads should remain for about a week. The following recipe may be used to prepare 6 L of Bouin's fixative:

Saturated picric acid (57.13 g in 4200 ml distilled water)
37% Formaldehyde (1428 ml)
Glacial acetic acid (286 ml)

The equipment necessary for craniofacial evaluation includes a dissecting microscope, scalpels or razor blades, and forceps. The head is removed from the bottle and blotted dry. Up to seven cuts may be made using a sharp, clean blade. They should always be made in the sequence listed, but additional cuts may be added if a structure appears abnormal, is missed, or must be verified as missing. The cuts should be smooth and perpendicular to the cutting surface. The following descriptions and the head cuts are modified from Wilson[17,69] and van Julsingha and Bennett[82] (Figure 7.7).

Before the heads are cut, they are examined for any grossly apparent abnormalities that should be more carefully explored as the cutting proceeds. The first cut is made with the head turned nose upward (use a pair of blunt forceps to grip the head) and exposes the tongue, palate, upper lips, and lower jaw. It is a ventrodorsal section (i.e., horizontal section) beginning at the mouth and coursing immediately inferior to the ears. The tongue should be lifted from the palate after the cut is made, and the palate and upper lips examined for incomplete closure (clefting). The pattern of the rugae should be examined for correct closure of the palate (the rugae should not be misaligned where they meet in the midline). The nasopharyngeal opening, the cochlea of the ears, and the base of the brainstem (medulla oblongata) may also be visualized. The flat surface produced by this cut simplifies the remaining slices by stabilizing the head, which should now be turned over (anterodorsal side up).

The second cut is made about half way between the tip of the nose and the foremost corner of the eye slits. The nasal passages, nasal conchae, nasal septum, palate, and insertions for the vibrissae should be visible on either side of the cut.

The third cut is made through the eyes. Tooth primordia and the Harderian glands may be visible, in addition to the nasal septum, nasal passages, nasal conchae, and palate. Both eyes, including cornea, lens, and retina, are visible in cross-section. The cerebral hemispheres and the anterior-most portions of the two lateral ventricles are readily visualized, as are the mandibular rami on either side. Occasionally, the interventricular foramen may be bisected, and a small portion of the thalamus and third ventricle will be visible. Other structures that should be located in this section include the optic nerves and/or the optic chiasma, the nasopharynx, the soft palate, and cranial nerves V and VII.

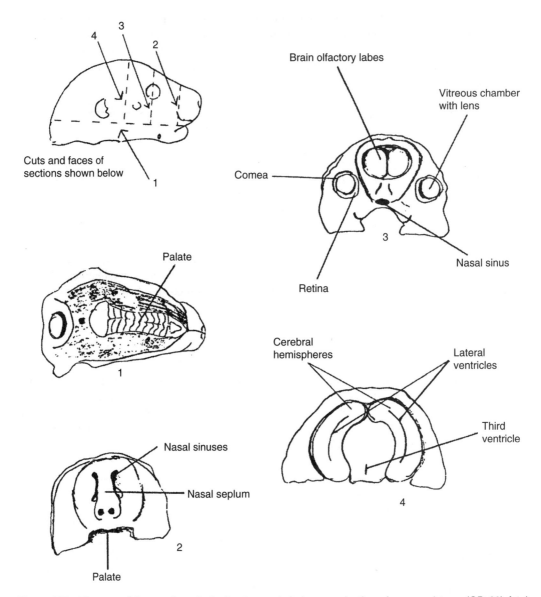

Figure 7.7 Diagram of five sections typically observed during examination of a normal term (GD 20) fetal rodent head.

The fourth cut is made through the largest vertical diameter of the cranium, well in front of the ear flaps. The thalamus, surrounding the slitlike third ventricle, will occupy a large central area, and the left and right cerebral hemispheres should surround it dorsolaterally. The lateral ventricles and cerebellum can be visualized, as can the pons and medulla oblongata (brainstem). After this cut, the final posterior section of the brain should be lifted from the cranial vault for a superficial observation of the cerebellum and surrounding structures.

After examination, the fetal head sections (rodents) may be stored in labeled cassettes (e.g., Tissue-Tek Uni-Cassette, Miles, Inc., Diagnostics Division, Elkhart, IN) and sealed in plastic bags (one per litter), with 70% ethanol as a preservative. Rabbit head sections from each fetus may be sealed in an individual, plastic, heat-sealable bag with 70% ethanol. All the bags from a litter can be stored in a single larger bag.

g. Skeletal Examination

Currently, the remaining 50% of the litter for rodents (intact, not decapitated) and 100% of the litter for rabbits are processed and examined for skeletal alterations. Fetal carcasses are eviscerated and prepared for staining with alizarin red S (for ossified bone)[83] or "double stained" with alizarin red S and alcian blue (for ossified and cartilaginous bone plus other permanent cartilaginous structures, such as the tip of the nose, external pinnae, etc.).[84,85] New guidelines require evaluation of both cartilaginous and ossified skeletal components but do not specify how (double staining is the preferred method and strongly recommended).

1) Single Staining for Ossified Bone — Two and one-half days is the minimum time required to hand stain rodent fetuses for skeletal evaluation. However, commercially available automatic stainers (e.g., the Sakura Teratology Processor from Sakura Finetek, USA, Inc., Torrance, CA) allow very rapid single or double staining of fetuses, with appropriate documentation for GLP compliance.

Once fetuses are eviscerated, they are immersed in 70% ethanol at least overnight in preparation for staining.[86,87] The ethanol is drained off the next morning and replaced with 1% potassium hydroxide (KOH) solution containing alizarin red S (25 mg/liter of solution) for 24 h. After 24 h, this is poured off and replaced with a fresh 1% KOH solution (without alizarin).[88] Six hours later, the 1% KOH is replaced with 2:2:1 solution (2 parts 70% ethanol to 2 parts glycerin to 1 part benzyl alcohol). The next morning the 2:2:1 solution is decanted and replaced with a 1:1 solution (1 part 70% ethanol to 1 part glycerin). The skeletons can remain in this solution indefinitely for evaluation and storage.

Seven to eight days is the minimum time required to process a rabbit fetus for skeletal examination. Once the fetus has been eviscerated, it is skinned, placed in a partitioned plastic box, and either air dried overnight or placed in 70% ethanol. The following morning, the ethanol is drained off and replaced with KOH solution containing alizarin red S as described above. Forty-eight hours later, this staining solution is decanted and replaced with a fresh 1% KOH solution. After up to 4 d, depending on the size of the fetus, the 1% KOH is replaced with the above-mentioned 2:2:1 solution. The next morning the 2:2:1 solution is decanted and replaced with 1:1 70% ethanol to glycerin for evaluation and storage.

2) Double Staining for Ossified Bone and Cartilaginous Tissue — **Rodents** — Fetuses should be skinned (or the skin should be separated from the underlying tissue at least) after evisceration to allow the alcian blue to penetrate. The fetus is immersed in a 70°C water bath for approximately 7 s (the fetus will appear to "unfold"). The epidermis is then peeled off the body, paws, and tail. This is best done by a gentle rubbing between the thumb and fingers, as the skeleton at this point is extremely fragile. The skeleton is placed in 95% ethanol overnight. The following morning, the ethanol is drained and replaced with alcian blue stain (15 mg alcian blue to 80 ml 95% ethanol to 20 ml glacial acetic acid). This is decanted after 24 h and replaced with 95% ethanol for 24 h. The following morning, the 95% ethanol is decanted, and the skeletons are placed in alizarin red S staining solution (25 mg/l of 1% KOH) for 24 to 48 h. The stain is drained and replaced with 0.5% KOH for 24 h; this step may not be necessary for small specimens. After the 0.5% KOH solution is decanted, it is replaced with 2:2:1 solution (70% ethanol to glycerin to benzyl alcohol). The following morning the fetal skeletons are placed in 1:1 70% ethanol to glycerin for evaluation and storage.[89,90]

Rabbits — The process is the same as for staining with alizarin red S, with the following alterations. After skinning, the fetal skeletons are placed in 95% ethanol. The following morning the ethanol is decanted and replaced with the alcian blue stain (see rodent staining for formulation). The specimens remain in the alcian blue for approximately 24 h and are then placed in 95% ethanol for 24 h. The following morning, the fetal skeletons are placed in the alizarin red S solution, and the remainder of the process is as outlined above for single staining of rabbit fetuses.

h. Skeletal Evaluation

The following description of a fetal skeletal exam is based on using a double-stained preparation. In a double-stained skeleton, the ossified bone will be stained red to purple and the cartilage blue. The staining of the cartilage allows the examiner to ascertain whether the underlying structure is present (i.e., is the bone merely not yet ossified, although the cartilage anlage is present, or is the underlying cartilage actually missing?). Similarly, a cleft in the cartilage plate of the sternum explains bipartite ossification sites of the sternebrae, and a misalignment of the sternebrae may be seen by the abnormal fusion of the cartilage plate. The normalcy of the cervical vertebrae can only be determined in a double-stained specimen, as these are not normally ossified in the term fetal skeletal preparation.

It is critical that the examiner be familiar with the normal appearance of bone structures in the fetus and with the degree of ossification that should be evident on the sacrifice date. Accurate skeletal descriptions enable the investigator to pinpoint accelerated or retarded ossification.[90-94]

The skeletal system consists of axial and appendicular sections. The axial skeleton includes the skull, vertebral column, sternum, and ribs. The pelvic and pectoral girdles and the appendages comprise the appendicular skeleton.

The paired bones of the skull (Figure 7.8), which must be identified during the skeletal examination, are described as follows. The exoccipitals and supraoccipital form the posterior wall of the cranial cavity. The supraoccipital is fused in rats on GD 19 and on GD 28 in rabbits. However, this bone has paired aspects in GD 18 mice. The interparietal forms the posterior portion of the cranial roof, and the more anterior parietals form part of the roof and sides of the cranial cavity. Frontals and nasals cover the anterior portion of the brain. The premaxillae, maxillae, zygomatics, and squamosals are the bones of the face and the upper jaw. The zygomatic bone, squamosal bone, and zygomatic process of the maxilla combine to form the zygomatic arch. The mandibles (unfused medially at this stage of development) form the lower jaw. An incisor originates from each mandible in the rabbit skeleton.

Usually, six ossified sternebrae are present (Figure 7.9). In rodents, sternebra 5 ossifies last, number 6 second to last, and number 2 (and/or 4) third to last. The sternebrae fuse to form the sternum or breastbone. The sternum consists of a cartilage plate, with the distal cartilage portion of the first seven ribs fused to it. The sternum is formed from two symmetrical halves that fuse during development. Incomplete fusion will result in a sternal cleft or perforation of the cartilage, as well as misalignment of the ossified sternebrae.

Vertebrae are the basic structural units of the vertebral column (Figure 7.10 and Figure 7.11). Each vertebra consists of a ventral centrum and paired lateral arches. The vertebral column articulates anteriorly with the exoccipital bones of the skull. The vertebrae are divided into the following groups: cervical (7 vertebrae), thoracic (typically 12 vertebrae in rabbits, 13 in rodents), lumbar (7 vertebrae in rabbits, 6 in rodents), sacral (4 vertebrae), and caudal (number of vertebrae varies with tail length).

The centrum consists of an ossified center surrounded by cartilage. The caudal centra are entirely cartilage. The examiner should note unossified centra (in a normal GD 20 rat, thoracic centra I to VII are normally unossified; all lumbar and sacral centra should be ossified) and misaligned, bipartite, dumbbell, or unilateral ossification centers. If the cartilage anlage is affected, it should be recorded along with any findings for the ossified portion. The ossified portion of the centra may or may not be bipartite or dumbbell when the cartilage portion is bipartite or dumbbell. The centra are also examined for fusion of the cartilage between centra. The vertebral arch is almost completely ossified except for a cartilage tip, which may or may not be present on the transverse process,

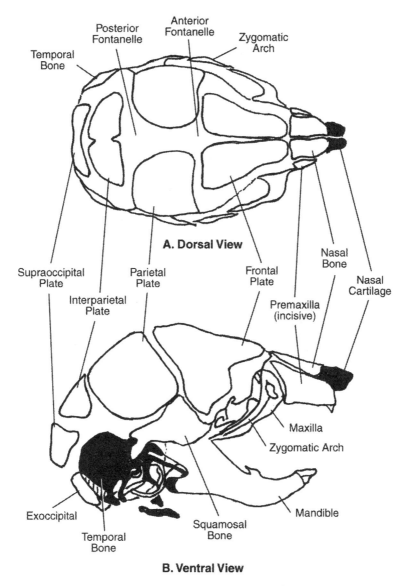

Figure 7.8 Dorsal and ventral views of the skull from a normal term (GD 20) rat fetus.

depending on the relative maturity of the fetus. The spinous process is entirely cartilaginous and can be examined only from the dorsal side. The cartilage of the transverse processes of sacral centra I, II, and III is normally fused, providing extra support for the pelvis. The two most anterior cervical vertebrae are the atlas and the axis. Each thoracic vertebra articulates dorsally with a pair of ribs.

Rabbits usually have 12 pairs of ribs, whereas rodents usually have 13 pairs (Figure 7.12). Each species often has a full rib, rudimentary rib, or an ossification site at lumbar I. In rodents, this may only be unilateral, but in rabbits, it is usually bilateral. Additional rib structures are typically classified as extra, full, or supernumerary ribs if they are at least one-half or greater than the length of rib I or XIII; as rudimentary ribs if they are less than one-half the length of rib I or XIII; or as ossification sites if they are small, round dots. Each rib consists of a cartilage tip proximal to the vertebral arch, an ossified middle portion, and a distal cartilage portion. The distal cartilages of

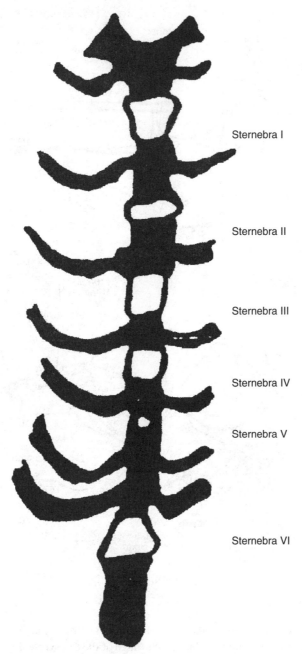

Ossified areas are open, cartilage is shaded.

Figure 7.9 Sternum of a normal term (GD 20) rat fetus.

ribs I to VII are attached to the sternum, and ribs VIII to XI curve upward, with their tips in close proximity. Ribs XII and XIII are free distally. Extra ribs may also be found on cervical arch VII and in rats are sometimes found only as cartilage; this extra cartilage is sometimes found to be fused to rib I of the same side.

Paired dorsal scapulae and ventral clavicles comprise the pectoral girdle (Figure 7.10). The scapulae are flat, trapezoidal bones with anterior-dorsal projections; the clavicles are elongated,

DEVELOPMENTAL TOXICITY TESTING — METHODOLOGY

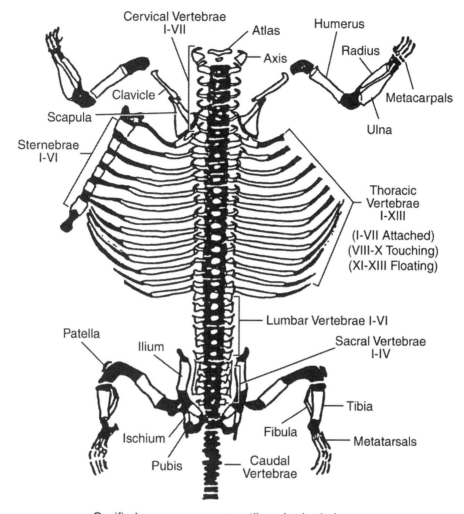

Ossified areas are open, cartilage is shaded.

Figure 7.10 View of axial and appendicular skeleton of a normal term (GD 20) rat fetus (decapitated).

curving, slender bones that articulate laterally with the scapulae and medially with the sternum (postnatally). The forelimb skeleton consists of the humerus in the upper foreleg, the radius and ulna in the lower foreleg, and carpals, metacarpals, and phalanges in the forefoot (Figure 7.13). In rodents, the only bones of the forefoot that have ossified by the time of sacrifice on GD 20 are the metacarpals, located between the carpals and phalanges (by GD 21, phalanges are visible). In mice, usually only metacarpals II to IV are ossified at term; in rats, metacarpals II to V are normally ossified at term.

Three pairs of bones are the basic units of the fetal pelvic girdle (Figure 7.14). Individually they are the ilium, ischium, and pubis. The femur, patella, tibia, and fibula are the bones of the thigh and hindleg, and the tarsals, metatarsals, and phalanges make up the skeletal support of the hindfoot. The five metatarsals ossify first. The hindfoot of the rabbit has only four digits.

If data are being collected by hand, rather than by use of computer software, there is the option of a checklist of bones that must be checked off as normal or any deviations recorded. The other option is recording any deviations from normal — any and all developmental delays that are usually

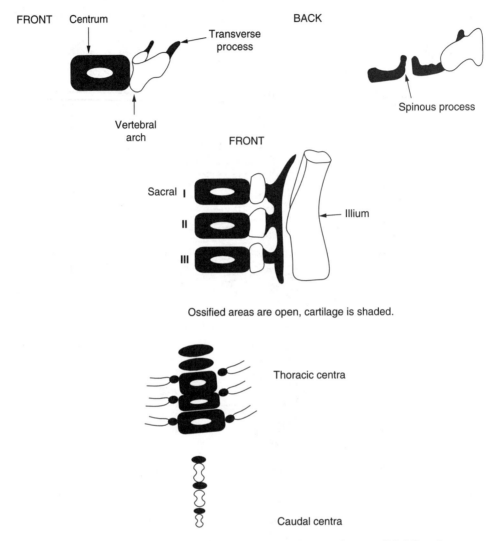

Figure 7.11 Enlarged views of selected elements of the axial skeleton of a term (GD 20) rat fetus.

manifested as lack of ossification of sternebrae (V, VI, and/or II in rats, VI in mice) and as bipartite or dumbbell-shaped cartilage or unossified centra (in rats). There is less variability in the development of the skeleton if the sacrifice day is GD 21 for rats and GD 18 for mice, but the risk of delivery prior to scheduled sacrifice is increased.

During the fetal examinations, all unusual (notable) observations are recorded or described by the examining technician. This is done for each individual fetus on the appropriate data collection sheet or in the appropriate locations on the computer terminal display. The observations will subsequently be coded as malformations (M) or variations (V) by a designated staff member (with input from the study director, laboratory supervisor, veterinarian, pathologist, etc., as appropriate), based on previously established criteria and designations (usually defined by the study director and laboratory supervisor). The determination of M or V status for each observation is made in the context of three salient factors:

- In the test animal and human literature on malformations, there is no case to date where a teratogenic agent causes a new, "never before seen" malformation. What is detected is an increase in the incidence of the malformation(s) above those seen in the general population. The current

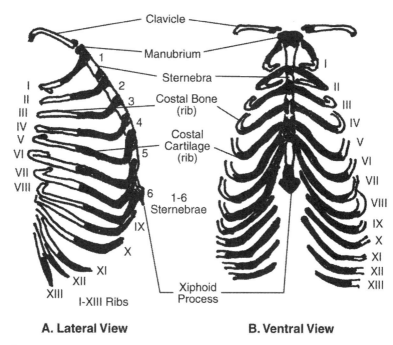

Figure 7.12 Detailed view of ribs from a term (GD 20) rat fetus.

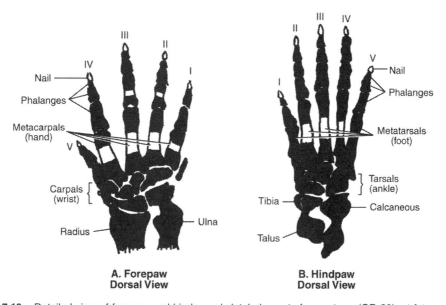

Figure 7.13 Detailed view of forepaw and hindpaw skeletal elements from a term (GD 20) rat fetus.

view is that teratogenic agents act on susceptible genetic loci or on susceptible developmental events. Therefore, the response seen is influenced by the genetic background and will vary by species, strain, race (in the case of humans), and individual (the last is more relevant to outbred strains and/or to genetically heterogeneous populations than to inbred strains).
- There is genetic predisposition to certain malformations that characterizes specific species, strains, races, and individuals. Historical control data are indispensable (along with concurrent controls) for determining the designation and occurrence of the present finding(s) in the context of the

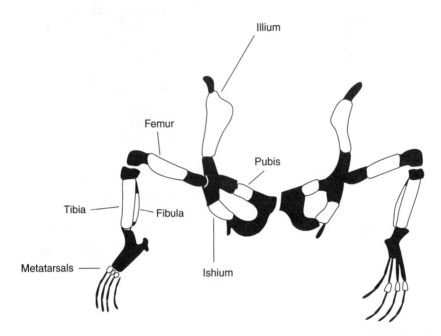

Figure 7.14 Detailed view of the pelvic and hind limb skeletal elements from a term (GD 20) rat fetus.

background "noise" of the population on test or at risk. For comparison purposes, the most recent MARTA/MTA compiled control data from several strains of rats, mice, and rabbits can be accessed at the following URL: http://hcd.org/search/abnormality.asp. Additional historical control databases or teratogenicity studies with control data exist for a number of commonly employed test animals, for example: CD® rat;[95–98] CD-1® mouse;[96,99,100] F-344 rat;[101–116] B6C3F1 mouse;[104,117] and New Zealand White rabbit.[118–124]

- The general considerations for designation of a finding as an M or a V are imprecise, may vary from study to study and teratologist to teratologist, may be relatively arbitrary, and are not necessarily generally accepted.[125]

The following sections present general classification criteria employed in the authors' laboratory to aid in the appropriate categorization of experimental findings.

i. Malformations

- Incompatible with or severely detrimental to postnatal survival, e.g., ventricular septal defect (VSD), diaphragmatic hernia, exencephaly, anencephaly, spina bifida, cleft palate
- Irreversible without intervention, e.g., hydroureter, hydronephrosis, VSD
- Involve replication, reduction (if extreme), or absence of essential structure(s) or organs, e.g., limbs, organs, major blood vessels, brain, heart chambers or valves, skeletal components unossified in an abnormal pattern
- Result from partial or complete failure to migrate, close, or fuse, e.g., cleft palate, cleft lip, ectopic organs, facial clefts, renal agenesis, lung agenesis, open neural tube
- May include syndromes of otherwise minor anomalies
- Exhibit a concentration- or dose-dependent increased incidence across dose groups, with a quantitative and/or qualitative change across dose groups, e.g., meningocoele → meningomyelocoele → meningoencephalocoele → exencephaly; foreshortened face → facial cleft → facial atresia; short tail → no tail → anal atresia; short rib → missing rib; brachydactyly (short digits) → oligodactyly (absence of some digits) → adactyly (absence of all digits); missing distal limb bones (hemimelia) → missing distal and some proximal limb bones → amelia (missing limb)

j. Transitional Findings

These may be upgraded to "malformation" or downgraded to "variation" status, depending on severity and/or frequency of occurrence.

- Nonlethal and generally not detrimental to postnatal survival
- Generally irreversible
- Frequently may involve reduction or absence of nonessential structures, e.g., innominate artery
- Frequently may involve reduction in number or size (if extreme) of nonessential structures or may involve their absence
- Exhibit a dose-dependent increased incidence

k. Variations

- Nonlethal and not detrimental to postnatal survival
- Generally reversible or transitory, such as wavy rib or the reduced ossification in a cephalocaudal sequence frequently seen associated with immaturity or delayed development as result of toxicity, e.g., reduced ossification in fore- and hindpaws, caudal vertebrae, pubis (but usually not ilium or ischium), skull plates, sternebrae (especially 5, 6, 2, or 4, in that order), cervical centra (especially 1)[126]
- May occur with a high frequency and/or not exhibit a dose-related increased incidence, e.g., reduced renal papilla and/or distended ureter in CD® rat fetuses at term,[127] dilated lateral cerebral ventricles (in rodents), extra ribs (on lumbar vertebra I) in rat and rabbit fetuses
- Detectable change (if not extreme) in size of specific structures (subjective), e.g., renal papilla one-half to one-quarter normal width; spleen greater than 1.5 times normal; kidney less than one-half normal size, kidney two times normal size; liver lobe less than one-half or up to 2 times normal size.

A profile of recent historical control data from the authors' laboratory of developmental toxicity studies on CD® rats, CD-1® mice, and New Zealand White rabbits is presented in Table 7.6. These test animal species exhibit characteristics that maximize their usefulness for such studies, such as:

- High pregnancy rate with very low levels of spontaneous full litter resorptions
- Large litters (total implants) with low preimplantation loss
- Large live litters with low postimplantation loss
- Approximately equal sex ratios in fetuses
- Consistent fetal body weights
- Low malformation rates
- Reasonable variation rates (high enough to provide sensitivity to test agents that increase the incidence of variations and/or exacerbate the variations to malformations; low enough to allow detection of treatment-related increases)

The interpretation of fetal findings requires a "weight of the evidence" approach (examples follow). If the incidence of fetal malformations is increased at the middle dose but not at the high dose, but the *in utero* mortality is highest at the high dose, then it is likely that the most affected conceptuses at the high dose died, especially with the prolonged dosing period specified in the new guidelines. In this case, the lack of a dose-response pattern is spurious, and an umbrella parameter like affected implants (nonlive plus malformed) will indicate the real dose-response pattern.

In the presence of reduced fetal body weights, it is typical to observe increased incidences of reduced ossification, especially in the skeletal areas that ossify last *in utero*, such as sternebrae V and VI of the prenatal sternum, bipartite or dumbbell ossification sites in the thoracic centra, and anterior and posterior phalangeal segments of the paws.[58] Reduced fetal body weights are commonly associated with dilated renal pelves (a delay in growth of the renal papilla;[98] and/or dilated (enlarged) lateral ventricles of the cerebrum without compression of the cerebral walls (again a delay in growth).[57]

Conversely, an increase in the incidence of an ossification site, partial or full rib on lumbar 1 (i.e., a 14th rib in rodents or a 13th rib in rabbits), considered a variation in the authors' laboratory, may indicate an alteration in developmental patterning (e.g., *Hox* genes), genetic drift over time, and/or may predict possible greater effects at higher doses.[125] In the authors' laboratory, cervical ribs (ossification site, partial, or full ribs) are considered malformations because of their rarity in historical controls and the possibility that they may have postnatal consequences, including alteration of blood flow to the head. Biological plausibility must also be considered. Kimmel and Wilson[125] conclude that "the best basis for rational interpretation of such anatomical variants is cumulative (historical) data on untreated controls, vehicle-treated controls, and various experimental groups in the strain or stock of animals used in a given teratogenicity study."

F. Statistics

As part of protocol development, the choice of statistical analyses should be made *a priori*, although specific additional analyses may be appropriate once the data are collected. The unit of comparison is the pregnant female or the litter and not the individual fetus, as only the dams are independently and randomly sorted into dose groups.[128] The fetus is not an independent unit and cannot be randomly distributed among groups. Intralitter interactions are common for a number of parameters (e.g., fetal weight or malformation incidence). Two types of data are collected: (1) ordinal/discrete data, which are essentially present or absent (yes or no), such as incidences of maternal deaths, abortions, early deliveries, or clinical signs, and incidences of fetal malformations or variations; and (2) continuous data, such as maternal body weights, weight changes, food and/or water consumption, organ weights (absolute or relative to body or brain weight), and fetal body weights per litter. For both kinds of data, three types of statistical analyses are performed. Tests for trends are available and appropriate to identify treatment-related changes in the direction of the data (increases or decreases), overall tests are performed for detecting significant differences among groups, and specific pairwise comparisons (when the overall test is significant) are made to the concurrent vehicle control group values. Pairwise comparisons are critical to identify statistically significant effects at a given dose relative to the concurrent vehicle control group. Continuous data are designated parametric (requiring "normally distributed" data) or nonparametric ("distribution free"), with different specific tests employed for the three types of statistical analyses, depending on whether the data are parametric or nonparametric. (See Chapter 17 for a more complete discussion of statistical approaches and appropriate tests.)

G. Data Collection

Historically, data have been collected by hand and subsequently entered into various software systems for statistical analyses and summarization (the latter was also typically done by hand in some laboratories). Summary tables were then constructed by hand or word processing. Currently, most laboratories are using (or acquiring) automated online, real-time data collection systems. These systems are either constructed in-house or purchased as a turnkey package available from a number of vendors (e.g., Provantis® (Instem), ISIS® BioComp, PathTox®, WIL®, TopCat, Teros®, etc.). In-house systems can be designed to meet the specific needs and procedures of the performing laboratory; purchased systems are standardized (with modifications available at additional cost). Regardless of the data collection method and analysis, the requirements under GLPs must be satisfied. Also, various regulatory agencies have provided guidance on how to purchase, validate, and acceptance test the automated systems (see Chapter 18 for details on GLPs). The advantages with automated systems include: more efficient use of technical staff, no need for back-entering data, less errors on data collection, elimination of paperwork, ease of QA auditing, and the possibility of electronically submitting the finished report to the appropriate cognizant federal or international agency (see Section V of this chapter for a discussion of electronic record keeping and reporting). One note: an audit trail of changes to a dataset is, if anything, even more important for electronic data capture than for hand-collected data. The audit trail (required, permanent, and accessible by appropriate staff) must contain the "who, what, when, and why" of each change.

IV. STORAGE OF RECORDS

All original study records, including all original data sheets, should be bound and stored in the secure archives of the performing laboratory, under the control of the quality assurance officer. Any biological samples collected during the course of the study (e.g., slides, blocks, wet tissues, fetal skeletons, sectioned fetal heads, dead fetuses, maternal gross lesions, etc.) should be placed in secure storage in the archives. Work sheets and computer printouts generated in the statistical analysis of data should also be stored in the archives. Copies of the final study report should be filed with the contract laboratory as well as with the sponsor. All study records, data, and reports should be maintained in storage in accordance with the appropriate federal guidelines, e.g., for the registration lifetime of the pesticide (FIFRA), storage is for 10 years after submission of the final report (FDA, EPA, TSCA), and in accordance with the appropriate governmental GLPs, e.g., EPA (FIFRA),[26] EPA (TSCA),[27] or FDA.[25] These records and samples may be released to the sponsor upon written request. See Table 7.10 for a list of typical study records to be maintained, and for further details regarding GLP compliance, refer to Chapter 18.

Table 7.10 Study records to be maintained

Protocol and any amendments
List of standard operating procedures
Animal requisition and receipt records
Quarantine records
Temperature and humidity records for the treatment room(s)
Animal research facility room log(s)
Deionized water analysis (if appropriate)
Feed type, source, lot number, dates used, certification, contaminants
Dose code records containing rx code (if used), color code (if used), and concentrations
Mating/insemination records
Randomization records
Assignment to study records
Bulk chemical receipt, storage, and use records
Dose formulation records
Analytical chemistry report(s)
Disposal of archival dose formulation samples
Poststudy shipment of bulk chemical to supplier
Dosing records
Clinical observations
Maternal body weights
Maternal food and/or water consumption
Necropsy sheets (in the event of maternal mortality)
Teratology sacrifice records: results of external, visceral, and skeletal examination sheets
Fetal carcasses and head sections
Wet tissues, blocks, and slides (if appropriate)
Computer printouts
Photographs (if taken)
Correspondence
Any deviations from the protocol
Draft report
Final report

V. COMPLIANCE WITH APPROPRIATE GOVERNMENTAL AND AAALAC INTERNATIONAL REGULATIONS

The performing laboratory must be operated in compliance with the appropriate governmental GLPs (see above). The animal research facility should be accredited by AAALAC International (Association for Assessment and Accreditation of Laboratory Animal Care International). Thus, the study should be conducted in compliance with appropriate GLPs and in compliance with the appropriate testing guidelines (e.g., FIFRA;[3,4,56] TSCA,[2,10] FDA,[55,129]) and AAALAC International accreditation standards. It may be useful if the performing laboratory is also approved by the Ministry of Agriculture, Forestry and Fisheries (MAFF), Japan, for toxicology studies on agricultural chemicals, so that work performed under EPA (FIFRA) GLPs and OPPTS or OECD testing guidelines will be acceptable to the Japanese regulatory agencies. The quality assurance unit of the performing laboratory must review the protocol and any amendments; inspect all in-life phases, necropsy, and fetal evaluations; audit the raw and summarized data and the final report; and provide a compliance statement to that effect for the final report. For further details regarding GLP compliance, refer to Chapter 18.

With the increasing use of automated data collection systems, the possibility and usefulness of submitting the final report *in toto* electronically to the appropriate federal or international agency has been recognized. To date, the FDA[130] has promulgated regulations for electronic records and electronic signatures (known affectionately as CFR, Part II), and the EPA has published[131] a proposed rule on establishment of electronic reporting and electronic records (known as CROMERRR, Cross Media Electronic Reporting and Recordkeeping Rule).

The objective of both the FDA and EPA initiatives is to "set forth the criteria under which the agency considers electronic records, electronic signatures, and handwritten signatures executed to electronic records to be trustworthy, reliable, and generally equivalent to paper records as handwritten signatures executed on paper."[130] The EPA proposal rule[131] would "allow electronic reporting to EPA by permitting the use of electronic document recovery systems to receive electronic documents in satisfaction of certain document submission requirements in EPA's regulations." The EPA acknowledges that "the electronic records criteria in [their] rule are not as detailed as that contained in FDA's 21 CFR Part II."[131] Laboratories providing studies for submission to the EPA or FDA should read the referenced documents in their entirety. One concern the authors of this chapter have is whether the electronic submissions would be limited to "read only" or would have to be sent in a format that allowed manipulation by regulators (as is apparently one objective of the FDA code). The integrity of the text and data submitted by the performing laboratory must be maintained. The study director will have no control over any alterations or manipulations to the data (e.g., deletions of an animal's data, a different interpretation of results, etc.) postsubmission.

VI. REPORTS

A. Status Reports

Status reports should be provided to the sponsor at a frequency specified by the sponsor. The content and format of these reports should remain flexible so as to accommodate unforeseen situations.

B. Final Report

The final report should include an abstract, objectives, materials and methods, results (narrative and summary tables), discussion, conclusions, references, any deviations from the protocol, a GLP compliance statement, a QA statement, and appendixes. The appendixes should include analytical chemistry reports (if appropriate), data from individual mated females, individual embryo and fetal data by litter and uterine location, histopathology and clinical chemistry reports (if appropriate), and the study protocol with any amendments. If the test material is a pesticide, the report must include the Statement of (No) Data Confidentiality (page 2 of report), the GLP compliance statement (page 3), and the FIFRA Flagging Statement (page 4), as well as the QA statement.

Data in the final report should be summarized in tabular form, showing for each dose group: numbers of animals at the start of test, pregnant animals, ovarian corpora lutea, and total uterine implantations; numbers and percentages of live fetuses; pre- and postimplantation loss; numbers of fetuses and litters with any external, visceral, or skeletal abnormalities; percentage of fetuses per litter with malformations and variations, by sex per litter (if possible). The findings should be evaluated in terms of the observed effects and the exposure levels (doses) producing effects. In addition, specific information on toxic responses should be reported by dose, species and strain employed, date of onset and duration of each abnormal sign and its subsequent course, food consumption, body weight and weight changes, uterine weight data, and fetal findings (number of live/dead, resorptions, fetal body weight, sex, and external, visceral, and skeletal alterations [malformations, variations]). Historical developmental toxicity control data, both published and from the performing laboratory, should also be considered, if appropriate. It is anticipated that a NOAEL would be identified for both maternal and developmental toxicities.

The final report for this study, in draft form, should be submitted to the sponsor within a certain interval (e.g., 3 months of the last sacrifice date). Within a specified interval (e.g., 30 days) of receipt of any comments from the sponsor's representative, at least one copy of the fully signed final report should be submitted to the sponsor. If the test material is evaluated under FIFRA testing guidelines, the final report must be in FIFRA format, according to EPA PR Notice 86-5.

VII. PERSONNEL

All senior staff involved in the study ("key personnel") should be identified by name and responsibility. Technical support staff are listed by name and responsibility or title in the authors' laboratory, but that is not required. The list should include:

- Study director (required)
- Sponsor's representative or study monitor (if appropriate)
- Animal research facility veterinarian and staff supervisor/manager
- Developmental toxicology laboratory staff supervisor/manager
- Data analyst
- Chemical handling supervisor (chemical receipt, disbursement, return to sponsor)
- Dosing formulation supervisor
- Supervisor for analytical chemistry (dosing formulations)
- Supervisor for clinical chemistry (if appropriate)
- Histology supervisor (if appropriate)
- Veterinary pathologist (if appropriate)
- Supervisor/manager of quality assurance unit

It is always useful to add a statement to the study protocol (prior to its initial submission to the sponsor when the study is initially commissioned) such as "Additional team members, if any, will be documented in the study records and listed in the final report." Otherwise, the protocol will have to be amended if new staff are added (it will also have to be amended if identified staff leave or do not work on the study).

VIII. STUDY RECORDS TO BE MAINTAINED

A typical list of study records to be maintained is presented in Table 7.10. Most of the entries are self-explanatory. The listing is there to provide a listing of all SOPs used by title and effective date (at least) or an actual copy of each SOP. Since SOPs evolve (new ones are written, old ones are retired, current ones are revised or amended), it is necessary to provide the specific SOPs that were in use at the time the study was performed.

ACKNOWLEDGMENTS

The authors wish to acknowledge with profound appreciation the dedication, expertise, and efforts of Dr. Tyl's staff at the University of Connecticut, Chemical Industry Institute of Toxicology, Bushy Run Research Center, and especially at RTI. Special thanks go to C.B. Myers, data specialist at RTI, for summarizing historical control data for Table 7.3 and to C.A. Winkie at RTI for her expert and patient typing (and retyping) of this chapter. The authors have learned a great deal and thoroughly enjoyed their long association with the science and art, as well as with the practitioners of developmental toxicity testing and research as the discipline has grown and evolved.[11,114] We cannot imagine a more exciting, rewarding, or relevant way to serve science and improve the human condition.

References

1. Goldenthal, E.I., *Guidelines for Reproduction Studies for Safety Evaluation of Drugs for Human Use*, U.S. Food and Drug Administration, Washington, D.C., 1966.
2. U.S. Environmental Protection Agency, Toxic Substances Control Act (TSCA) test guidelines: final rule, *Fed. Regist.*, 50, 39412, 1987.
3. U.S. Environmental Protection Agency, Pesticides Assessment Guidelines (FIFRA), Subdivision F. Hazard Evaluation: Human and Domestic Animals (Final Rule), NTIS (PB86-108958), Springfield, VA, 1984.
4. U.S. Environmental Protection Agency, Pesticide Assessment Guidelines, Subdivision F Hazard Evaluation: Humans and Domestic Animals, Series 83-3, Rat or Rabbit Developmental Toxicity Study, June, 1986 (NTIS PB86-248184), as amended in *Fed. Regist.*, 53(86), Sect. 158.340, May 4, 1988.
5. Ministry of Health and Welfare, *Japanese Guidelines of Toxicity Studies, Notification No. 118 of the Pharmaceutical Affairs Bureau, 2. Studies of the Effects of Drugs on Reproduction*, Yakagyo Jiho Co., Ltd., Tokyo, Japan, 1984.
6. Canada Ministry of Health and Welfare, Health Protection Branch, *The Testing of Chemicals for Carcinogenicity, Mutagenicity and Teratogenicity*, Ministry of Ottawa, Ottawa, 1973.
7. Department of Health and Social Security, *United Kingdom, Committee on Safety of Medicines: Notes for Guidance on Reproduction Studies*, Department of Health and Social Security, Great Britain, 1974.
8. Organization for Economic Cooperation and Development (OECD), *Guideline for Testing of Chemicals: Teratogenicity, Director of Information*, OECD, Paris, France, 1981.
9. U.S. Food and Drug Administration, International Conference on Harmonisation (ICH), Guideline for detection of toxicity to reproduction for medicinal products, Section 4.1.3, study for effects on embryo-fetal development, *Fed. Regist.*, 59(183), 48749, September 22, 1994.
10. U.S. Environmental Protection Agency, OPPTS, Health Effects Test Guidelines, OPPTS 870.3700, Prenatal Developmental Toxicity Study, Public Draft, U.S. Government Printing Office, Washington, D.C., February, 1996.
11. U.S. Environmental Protection Agency, TSCA, Toxic Substances Control Act Test Guidelines, Final Rule 40 CFR Part 799.9370, TSCA Prenatal Developmental Toxicity, *Fed. Regist.*, 62(158), 43832, August 15, 1997.
12. U.S. Food and Drug Administration, *Guidelines for Developmental Toxicity Studies*, Center for Food Safety and Applied Nutrition Redbook 2000, IV.C.9.b, July 20, 2000.
13. Organization for Economic Cooperation and Development, *OECD Guideline for the Testing of Chemicals; Proposal for Updating Guideline 414: Prenatal Developmental Toxicity Study*, pp. 1–11, adopted January 22, 2001.
14. U.S. Environmental Protection Agency, Guidelines for the health assessment of suspect developmental toxicants, *Fed. Regist.*, 51, 34028, 1986.

15. U.S. Environmental Protection Agency, Part V, Guidelines for developmental toxicity risk assessment; notice, *Fed. Regist.*, 56(234), 63798, 1991.
16. Wilson, J.G., The evolution of teratology testing, *Teratology*, 20, 205, 1979.
17. Wilson, J.G., Principles of teratology, in *Environment and Birth Defects*, Academic Press, New York, 1973, p. 11.
18. Johnson, E.M., Screening for teratogenic potential: Are we asking the proper questions? *Teratology*, 21, 259, 1980.
19. Johnson, E.M., Screening for teratogenic hazards: nature of the problems, *Ann. Rev. Pharmacol. Toxicology*, 21, 417, 1981.
20. Johnson, E.M. and Christian, M.S., When is a teratology study not an evaluation of teratogenicity? *J. Am. Coll. Toxicol.*, 3, 431, 1984.
21. Marks, T.A., A retrospective appraisal of the ability of animal tests to predict reproductive and developmental toxicity in humans, *J. Am. Coll. Toxicol.*, 10(5), 569, 1991.
22. U.S. Environmental Protection Agency, Proposed amendments to the guidelines for the health assessment of suspect developmental toxicants; request for comments; notice, *Fed. Regist.*, 54, 9386, 1989.
23. Daston, G.P., Rehnberg, B F., Carver, B., Rogers, E.H., and Kavlock, R.J., Functional teratogens of the rat kidney. I. Colchicine, dinoseb and methyl salicylate, *Fundam. Appl. Toxicology*, 11, 381, 1988.
24. Daston, G.P., Rehnberg, B.F., Carver, B., and Kavlock, R.J., Functional teratogens of the rat kidney. II. Nitrofen and ethylenethiourea, *Fundam. Appl. Toxicology*, 11, 401, 1988.
25. U.S. Food and Drug Administration, Good Laboratory Practice Regulations for Nonclinical Laboratory Studies, *Code of Federal Regulations* (CFR), 229, April 1, 1988.
26. U.S. Environmental Protection Agency, Federal Insecticide, Fungicide, and Rodenticide Act (FIFRA) good laboratory practice standards; final rule, *Fed. Regist.*, 54, 34051 (40-CFR-792) August 17, 1989.
27. U.S. Environmental Protection Agency, Toxic Substances Control Act (TSCA), good laboratory practice standards; final rule, *Fed. Regist.*, 54, 34033, August 17, 1989.
28. Taylor, P., *Practical Teratology*, Academic Press, New York, 1986.
29. Harkness, J.E. and Wagner, J.E., *The Biology and Medicine of Rabbits and Rodents*, 3rd ed., Lea and Febiger, Philadelphia, 1989.
30. American Association for Laboratory Animal Science, *Training Manual Series, Vol. 3, for Laboratory Animal Technologist*, Stark, D.M. and Ostrow, M.E., Eds., AALAS, Cordova, TN, 1991.
31. National Research Council, *Guide for the Care and Use of Laboratory Animals*, Institute of Laboratory Animal Resources, Commission on Life Sciences, National Research Council, National Academy Press, revised 1996.
32. National Institutes of Health, *Guide for the Care and Use of Laboratory Animals*, PHS NIH Pub. No. 86-23, revised, U.S. Department of Health and Human Services, Bethesda, MD, 1985
33. Cheeke, P.R., Nutrition and nutritional diseases, in *The Biology of the Laboratory Rabbit*, 2nd ed., Manning, P.J., Ringler, D.H., and Newcomer, C.E., Eds., Academic Press, New York, 1995, p. 321.
34. Bellhorn, R.W., Lighting in the animal environment, *Lab. Anim. Sci.*, 30, 440, 1980.
35. Besch, E.L., Environmental quality within animal facilities, *Lab. Anim. Sci.*, 30, 385, 1980.
36. Wilson, N.E. and Stricker, E.M., Thermal homeostasis in pregnant rats during heat stress, *J. Comp. Physiol. Psych.*, 93, 585, 1979.
37. Peterson, E.A., Noise and laboratory animals, *Lab. Anim. Sci.*, 30, 422, 1980.
38. Whitten, W.K., Modification of the oestrus cycle of the mouse by external stimuli associated with the male, *J. Endocrinol.*, 13, 399, 1956.
39. Hafez, E.S.E., Ed., *Reproduction and Breeding Techniques for Laboratory Animals*, Lea and Febiger, Philadelphia, 1970.
40. Bredderman, P.J., Foote, R.H., and Yassen, A. M., An improved artificial vagina for collecting rabbit semen, *J. Reprod. Fertil.*, 7, 401, 1974.
41. Chernoff, N., Rogers, J.M., and Kavlock, R.J., Review paper. An overview of maternal toxicity and prenatal development: considerations for developmental toxicity hazard assessments. *Toxicology*, 59, 111, 1989.
42. Schwetz, B.A. and Tyl, R.W., Consensus workshop on the evaluation of maternal and developmental toxicity workgroup iii report: low dose extrapolation and other considerations for risk assessment — models and applications. *Teratogen., Carcinogen., Mutagen.*, 7(3), 321, 1987.
43. Kimmel, C.A., Price, C.J., Sadler, B.M., Tyl, R.W., and Gerling, F.S., Comparison of distilled water (DW) and corn oil (CO) vehicle controls from historical teratology study data, *Toxicologist*, 5, 185, 1985.

44. Feldman, D.B., Simplified gastric intubation in the rabbit, *Lab. Anim. Sci.* 27, 1037, 1977.
45. American Association for Laboratory Animal Science, *Training Manual Series, Vol. 1, for Assistant Laboratory Animal Technician*, Stark, D.M. and Ostrow, M.E., Eds., AALAS, Cordova, TN, 1991.
46. Tyl, R.W., Ballantyne, B., Fisher, L.L., Fait, D.L., Dodd, D.E., Klonne, D.R., Pritts, I.M., and Losco, P.E., Evaluation of the developmental toxicity of ethylene glycol aerosol in CD-1® mice by nose-only exposure, *Fundam. Appl. Toxicol.*, 27(1), 49, 1995.
47. Francis, E.Z. and Kimmel, C.A., Proceedings of the workshop on the acceptability and interpretation of dermal developmental toxicity studies, *Teratology*, 39, 453, 1989.
48. Kimmel, C.A. and Francis, E.Z., Proceedings of the workshop on the acceptability and interpretation of dermal developmental toxicity studies, *Fundam. Appl. Toxicol.*, 14, 386, 1990.
49. Tyl, RW., York, R.G., and Schardein, J.L., Reproductive and developmental toxicity studies by cutaneous administration, in *Health Risk Assessment through Dermal and Inhalation Exposure and Absorption of Toxicants*, Wang, R.G., Knaak, J.B., and Maibach, H.I., Eds., CRC Press, Boca Raton, FL, 1993, p. 229.
50. Tyl, R.W., Fisher, L.C., France, K.A., Garman, R.H., and Ballantyne, B., Evaluation of the teratogenicity of bis(2-dimethylaminoethyl) ether after dermal application in New Zealand white rabbits, *J. Toxicol. Cutaneous Ocular Toxicol.*, 5, 263, 1986.
51. Tyl, R.W., Fisher, L.C., Kubena, M.F., Vrbanic, M.A., Gingell, R., Guest, D., Hodgson, J.A., Murphy, S.R., Tyler, T.R., and Astill, B.D., The developmental toxicity of 2-ethylhexanol (2-EH) applied dermally to pregnant Fischer 344 rats, *Fundam. Appl. Toxicol.*, 19, 176, 1992.
52. Fisher, L.C., Tyl, R.W., and Kubena, M.F., Cutaneous developmental toxicity study of 2-ethylhexanol (2-EH) in Fischer 344 rats, *Teratology*, 39(5), 452, 1989.
53. Tyl, R.W., Fisher, L.C., Kubena, M.F., Vrbanic, M.A., and Losco, P.E., Assessment of the developmental toxicity of ethylene glycol applied cutaneously to CD-1® mice, *Fundam. Appl. Toxicol.*, 27, 155, 1995.
54. Draize, J.H., Woodard, G., and Calvery, H.O., Methods for the study of irritation and toxicity of substances applied to the skin and mucous membranes, *J. Pharmacol. Exp. Therap.*, 82, 377, 1944.
55. U.S. Food and Drug Administration, *Toxicological Principles for the Safety Assessment of Direct Food Additives and Color Additives Used in Food* ("Redbook"), Appendix II, Guidelines for teratogenicity testing in rat, hamster, mouse, and rabbit, 108, 1982.
56. U.S. Environmental Protection Agency, OPP Guideline 83-3, *Teratogenicity Study, Pesticide Assessment Guidelines, Subdivision F, Hazard Evaluation, Human and Domestic Animals*, Office of Pesticides and Toxic Substances, Washington, D.C., 1982.
57. Marr, M.C., Myers, C.B., Price, C.J., Tyl, R.W., and Jahnke, G.D., Comparison of maternal and developmental endpoints for control CD® rats dosed from implantation through organogenesis or through the end of gestation, *Teratology*, 59, 413, 1999 (tables provided by the authors).
58. Kimmel, G.L., Kimmel, C.A., and Francis, E.Z., Evaluation of maternal and developmental toxicity, *Teratogen. Carcinogen. Mutagen.*, 7, 203, 1987.
59. Khera, K.S., Maternal toxicity — a possible factor in fetal malformations in mice, *Teratology*, 29, 411, 1984.
60. Khera, K.S., Maternal toxicity — a possible etiological factor in embryo-fetal deaths and fetal malformations of rodent-rabbit species, *Teratology*, 31, 129, 1985.
61. Clark, R.L., Robertson, R. T., Minsker, D. H., Cohen, S.M., Tocco, D.J., Allen, H.L., James, M.L., and Bokelman, D.L., Diflunisal-induced maternal anemia as a cause of teratogenicity in rabbits, *Teratology*, 30(3), 319, 1984.
62. Salewski, E., Färbemethode zum makroskopischen nachweis von implantationsstellen am uterus der ratte, Naunyn-Schmiedebergs *Arch. Exp. Pathol. Pharmakol.*, 247, 367, 1964.
63. Lumb, W.V. and Jones, E.W., *Veterinary Anesthesia*, Lea and Febiger, Philadelphia, 1973, p. 452.
64. Blair, E., Hypothermia, in *Textbook of Veterinary Anesthesia*, Soma, L., Ed., Williams and Wilkins, Baltimore, MD, 1979.
65. Woollam, D.H.M. and Millen, J.W., Influence of uterine position on the response of the mouse embryo to the teratogenic effects of hypervitaminosis A, *Nature*, 190, 184, 1961.
66. U.S. Environmental Protection Agency, Office of Prevention, Pesticides and Toxic Substances, Health Effects Test Guidelines OPPTS 870.3700, *Prenatal Developmental Toxicity Study*; final guideline, August 1998.

67. Edwards, J.A., The external development of the rabbit and rat embryo, in *Advances in Teratology*, Vol. 3, Woolam, D.H.M., Ed., Academic Press, New York, 1968.
68. Staples, R.E., Detection of visceral alterations in mammalian fetuses, *Teratology*, 9, 37, 1974.
69. Wilson, J.G., Embryological considerations in teratology, in *Teratology: Principles and Techniques*, Wilson, J.G. and Warkany, J., Eds., University of Chicago Press, Chicago, 1965, p. 251.
70. Wilson, J.G., Methods for administering agents and detecting malformations in experimental animals, in *Teratology: Principles and Techniques*, University of Chicago Press, Chicago, 1965, p. 262.
71. Wilson, J.G., *Environmental and Birth Defects*, Academic Press, New York, 1973.
72. Wilson, J.G., Reproduction and teratogenesis: Current methods and suggested improvements, *J. Assoc. Off. Anal. Chem.*, 58, 657, 1975.
73. Wilson, J.G., Critique of current methods for teratogenicity testing and suggestions for their improvement, in *Methods for Detection of Environmental Agents That Produce Congenital Defects*, Shepard, T., Miller, J.R., and Marois, M., Eds., American Elsevier, New York, 1975, p. 29.
74. Wilson, J.G. and Fraser, F.C., Eds., *Handbook of Teratology*, Plenum Press, New York, 1977.
75. Monie, I.W., *Dissection Procedures for Rat Fetuses Permitting Alizarin Red Staining of Skeleton and Histological Study of Viscera*, Supplement to Teratology Workshop Manual, Berkeley, CA, 1965, p. 163.
76. Fox, M.H. and Goss, C.M., Experimentally produced malformations of the heart and great vessels in rat fetuses, *Amer. J. Anat.*, 102, 65, 1958.
77. Barrow, M.V. and Taylor, W.J., A rapid method for detecting malformations in rat fetuses, *J. Morphol.*, 127, 291, 1969.
78. Faherty, J.F., Jackson, B.A., and Greene, M.F., Surface staining of 1 mm (Wilson) slices of fetuses for internal visceral examination, *Stain Tech.*, 47, 53, 1972.
79. Creasy, D.M., Jonassen, H., Lucacel, C., Agajanov, T., Gondek, H., and Schoenfeld, H., Modified Davidson's fixative as an alternative to Bouin's for evaluation of fetal soft tissue anomalies using the Wilson's serial sectioning technique, Society of Toxicologic Pathology, 21st Annual Symposium, meeting program abstract no. P35, 73, Denver, CO, 2002.
80. Latendress, J.R., Warbritton, A.R., Jonassen, H., and Creasy, D.M., Fixation of testes and eyes using a modified Davidson's fluid: Comparison with Bouin's fluid and conventional Davidson's fluid, *Toxicol. Pathol.* 30, 524, 2002.
81. Stuckhardt, J.L. and Poppe, S.M., Fresh visceral examination of rat and rabbit fetuses used in teratogenicity testing, *Teratogen. Carcinogen. Mutagen.*, 4, 181, 1984.
82. van Julsingha, E.B. and Bennett, C.G., A dissecting procedure for the detection of anomalies in the rabbit foetal head, in *Methods in Prenatal Toxicology*, Neubert, D., Merker, H J., and Kwasigrouch, T.E., Eds., George Thieme Publishers, Stuttgart, Germany, 1977, p. 126.
83. Dawson, A.B., A note on the staining of the skeleton of cleared specimens with alizarin red S, *Stain Tech.*, 1, 123, 1926.
84. Inouye, M., Differential staining of cartilage and bone in fetal mouse skeleton by alcian blue and alizarin red S, *Congenital Anomalies*, 16, 171, 1976.
85. Kimmel, C.A. and Trammel, C., A rapid procedure for routine double staining of cartilage and bone in fetal and adult animals, *Stain Tech.*, 56, 271, 1981.
86. Peltzer, M.A. and Schardein, J.L., A convenient method for processing fetuses for skeletal staining, *Stain Tech.*, 41, 300, 1966.
87. Staples, R.E. and Schnell, V.L., Refinements in rapid clearing technic in the KOH-alizarin red S method for fetal bones, *Stain Tech.*, 39, 62, 1964.
88. Crary, D.D., Modified benzyl alcohol clearing of Alizarin-stained specimens without loss of flexibility, *Stain Tech.*, 37, 124, 1962.
89. Marr, M.C., Myers, C.B., George, J.D., and Price, C.J., Comparison of single and double staining for evaluation of skeletal development: The effects of ethylene glycol (Eg) in CD® rats, *Teratology*, 37, 476, 1988.
90. Marr, M.C., Price, C.J., Myers, C.B., and Morrissey, R.E., Developmental stages of the CD® (Sprague-Dawley) rat skeleton after maternal exposure to ethylene glycol, *Teratology*, 46, 169, 1992.
91. Fritz, H., Prenatal ossification in rabbits as indicative of fetal maturity, *Teratology*, 11, 313, 1974.
92. Spark, C. and Dawson, A.B., The order and time of appearance of centers of ossification in the fore- and hindlimbs of the albino rat, with special reference to the possible influence of the sex factor, *Am. J. Anat.*, 41, 411, 1928.

93. Strong, R.M., The order, time and rate of ossification of the albino rat (*Mus norvegicus albinus*) skeleton, *Am. J. Anat.*, 36, 313, 1928.
94. Walker, D.G. and Wirtschafter, Z.T., *The Genesis of the Rat Skeleton*, Charles C. Thomas, Springfield, IL, 1957.
95. Banerjee, B.N. and Durloo, R.S., Incidence of teratological anomalies in control Charles River CD® strain rats, *Toxicology*, 1, 151, 1973.
96. Perraud, J., Levels of spontaneous malformations in the CD® rat and the CD-1® mouse, *Lab. Anim. Sci.*, 26, 293, 1976.
97. Charles River Laboratories, *Embryo and Fetal Developmental Toxicity (Teratology) Control Data in the Charles River Crl:CD® BR Rat*, Charles River Laboratories, Inc., Wilmington, MA, 1988.
98. Woo, D.C. and Hoar, R.M., Reproductive performance and spontaneous malformations in control Charles River CD® rats: a joint study by MARTA, *Teratology*, 19, 54A, 1979.
99. Fritz, H., Grauwiler, J., and Himmler, H., Collection of control data from teratological experiments in mice, rat and rabbits, *Arzneim-Forsch/Drug Res.*, 28, 1410, 1978.
100. Palmer, A.K., Sporadic malformations in laboratory animals and their influence in drug testing, *Drugs Fetal Dev. Adv. Biol. Med.*, 27, 45, 1972.
101. DePass, L.R. and Weaver, E. V., Comparison of teratogenic effects of aspirin and hydroxyurea in the Fischer 344 and Wistar strains, *J. Toxicol. Environ. Health*, 15, 297, 1982.
102. Wolkowski-Tyl, R., Jones-Price, C., Ledoux, T.A., Marks, T.A., and Langhoff-Paschke, L., Teratogenicity evaluation of technical grade dinitrotoluene in the Fischer 344 rat, *Teratology*, 23, 70A, 1981.
103. Wolkowski-Tyl, R., Jones-Price, C., Ledoux, T.A., Marks, T.A., and Langhoff-Paschke, L., Teratogenicity evaluation of aniline HCl in the Fischer 344 rat, *Teratology*, 23, 70A, 1981.
104. Wolkowski-Tyl, R., Phelps, M., and Davis, J.K., Structural teratogenicity evaluation of methyl chloride in rats and mice after inhalation exposure, *Teratology*, 27, 18, 1983.
105. Wolkowski-Tyl, R., Jones-Price, C., Marr, M.C., and Kimmel, C.A., Teratologic evaluation of diethylhexyl phthalate (DEHP) in Fischer 344 rats, *Teratology*, 27, 85A, 1983.
106. Jones-Price, C., Wolkowski-Tyl, R., Marks, A., Langhoff-Paschke, L., Ledoux, T. A., and Reel, J. R., Teratologic and postnatal evaluation of aniline hydrochloride in the Fischer 344 rat, *Toxicology Appl. Pharmacol.*, 77, 465, 1985.
107. Hayes, W.C., Cobel-Geard, S.R., Hanley, T.R., Murray, J.S., Freshour, N.L., Rao, K.S., and John, J.A., Teratogenic effects of vitamin A palmitate in Fischer 344 rats, *Drug Chem. Toxicology*, 4, 283, 1981.
108. Peters, M.A. and Hudson, P.M., Effects of totigestational exposure to ethchlorvynol on development and behavior, *Ecotoxicology Environ. Safety*, 5, 494, 1981.
109. Peters, M.A., Hudson, P.M., and Dixon, R.L., Effect of totigestational exposure to methyl n-butyl ketone on postnatal development and behavior, *Ecotoxicology Environ. Safety*, 5, 291, 1981.
110. John, J.A., Hayes, W.C., Hanley, T.R., Cobel-Geard, S.R., Quast, J.F., and Rao, K.S., Behavioral and teratogenic potential of vitamin A in the Fischer 344 rat, *Teratology*, 24, 55A, 1981.
111. Keller, W.C., Inman, R.C., and Back, K.C., Evaluation of the embryotoxicity of JP10 in the rat, *Toxicologist*, 1, 30, 1981.
112. Snellings, W.M., Pringle, J.L., Dorko, J.D., and Kintigh, W.J., Teratology and reproduction studies with rats exposed to 10, 33, or 100 ppm of ethylene oxide (EO), *Toxicol. Appl. Pharmacol.*, 48, A84, 1979.
113. Sunderman, F., Shen, S.K., Mitchell, J.M., Allpass, P.R., and Damjanov, I., Embryotoxicity and fetal toxicity of nickel in rats, *Toxicology Appl. Pharmacol.*, 43, 381, 1978.
114. Sunderman, F., Allpass, P.R., Mitchell, J.M., Baselt, R.C., and Albert D.M., Eye malformations in rats: induction by prenatal exposure to nickel carbonyl, *Science*, 203, 550, 1979.
115. Bus, J.S., White, E.L., Tyl, R.W., and Barrow, C.S., Perinatal toxicity and metabolism of n-hexane in Fischer 344 rats after inhalation exposure during gestation, *Toxicol. Appl. Pharmacol.*, 51, 295, 1979.
116. Olson, C.T. and Back, K.C., Methods for teratogenic screening of air force chemicals, US NTIS AD Report, AD-A052002, 1978.
117. Wolkowski-Tyl, R., Lawton, A.D., Phelps, M., and Hamm, Jr., T.E., Evaluation of heart malformations in B6C3F1 mouse fetuses induced by in utero exposure to methyl chloride, *Teratology*, 27, 197, 1983.
118. Cozens, D.D., Abnormalities of the external form and of the skeleton in the New Zealand White rabbit, *Food. Cosmet. Toxicology*, 3, 695, 1965.

119. DeSesso, J.M. and Jordan, R.L., Drug-induced limb dysplasias in fetal rabbits, *Teratology*, 15, 199, 1977.
120. Hartman, H.A., The fetus in experimental teratology, in *The Biology of the Laboratory Rabbit*, Weisbroth, S.H., Flatt, R.E., and Kraus, A.L., Eds., Academic Press, New York, 1974, p. 92.
121. McClain, R.M. and Langhoff, L., Teratogenicity of diphenylhydantoin in the New Zealand white rabbit, *Teratology*, 21, 371, 1980.
122. Palmer, A.K., The design of subprimate animal studies, in *Handbook of Teratology*, Vol. 4, Wilson, J.G. and Fraser, F.C., Eds., Plenum Press, New York, 1978, p. 215.
123. Stadler, J., Kessedjian, M.-J., and Perraud, J., Use of the New Zealand white rabbit in teratology: incidence of spontaneous and drug-induced malformations, *Food Chem. Toxicol.*, 21, 631, 1983.
124. Woo, D.C. and Hoar, R.M., Reproductive performance and spontaneous malformations in control New Zealand white rabbits: a joint study by MARTA, *Teratology*, 25, 82A, 1982.
125. Kimmel, C.A. and Wilson, J.G., Skeletal deviations in rats: malformations or variations? *Teratology*, 8, 309, 1973.
126. Aliverti, V., Bonanomi, L., Giavini, E., Leone, V.G., and Mariani, L., The extent of fetal ossification as an index of delayed development in teratogenic studies on the rat, *Teratology*, 20, 237, 1979.
127. Woo, D.C. and Hoar, R.M., Apparent hydronephrosis as a normal aspect of renal development in late gestation of rats: the effect of methyl salicylate, *Teratology*, 6, 191, 1972.
128. Weil, C.S., Selection of the valid number of sampling units and a consideration of their combination in toxicological studies involving reproduction, teratogenesis, or carcinogenesis, *Fd. Cosmet. Toxicol.*, 8, 177, 1970.
129. U.S. Food and Drug Administration, *Guidelines for Reproduction Studies for Safety Evaluation of Drugs for Human Use*, U.S. Food and Drug Administration, Washington, D.C., 1966.
130. U.S. Food and Drug Administration, Federal Food, Drug, and Cosmetic Act, Commissioner of Food and Drugs, Title 21, Chapter I of Code of Federal Regulations (FR), amended by addition of Part II: "Part II — Electronic Records, Electronic Signatures." *Fed. Regist.*, 62(54), 13464–13466 (March 20, 1997)
131. U.S. Environmental Protection Agency, Part V, 40 CFR parts 3, 51, et al., Establishment of Electronic Reporting: Electronic Records; Proposal Rule. *Fed. Regist.*, 66(170), 46162–46195, 2001.

CHAPTER 8

Nonclinical Juvenile Toxicity Testing

Melissa J. Beck, Eric L. Padgett, Christopher J. Bowman,
Daniel T. Wilson, Lewis E. Kaufman, Bennett J. Varsho,
Donald G. Stump, Mark D. Nemec, and Joseph F. Holson

CONTENTS

I. Introduction	264
A. Objectives	265
B. Background	265
1. Historical Perspective of Pediatric Therapeutics	266
2. Adverse Events	267
3. Agency Intervention/Regulatory Guidelines	267
II. Importance of Juvenile Animal Studies	272
A. Differences in Drug Toxicity Profiles between Developing and Developed Systems	272
B. Utility of Studies in Juvenile Animals	273
1. Clinical Considerations	273
2. Predictive Value	274
III. Design Considerations for Nonclinical Juvenile Toxicity Studies	274
A. Intended or Likely Use of Drug and Target Population	274
B. Timing of Exposure in Relation to Phases of Growth and Development of Target Population	275
C. Potential Differences in Pharmacological and Toxicological Profiles between Mature and Immature Systems	276
D. Use of Extant Data	277
1. Pharmacokinetic (PK) and Toxicokinetic (TK) Data in Adult Animals	277
2. Extant Adult Nonclinical Animal Toxicity Data	280
3. Stand-Alone Juvenile PK and TK Studies	280
IV. Juvenile Toxicity Study Design	280
A. Types of Studies	280
1. General Toxicity Screen	281
2. Adapted Adult Toxicity Study	281
3. Mode of Action Study	283

        B. Model Selection 284
            1. Species 284
            2. Sample Size 284
        C. Exposure 286
            1. Route of Administration 286
            2. Frequency and Duration of Exposure 290
            3. Dose Selection 290
        D. Organization of Test Groups 292
            1. Within-Litter Design 292
            2. Between-Litter Design 293
            3. Fostering Design 294
            4. One Pup per Sex per Litter Design 295
        E. Potential Parameters for Evaluation 295
            1. Growth and Development 295
            2. Food Consumption 299
            3. Serum Chemistry and Hematology 299
            4. Macroscopic and Microscopic Evaluations 301
            5. Physical and Sexual Developmental Landmarks and Behavioral Assessments 301
            6. Reproductive Assessment 318
        F. Statistical Design and Considerations 318
    V. Conclusions 319
Acknowledgments 319
References 319

I. INTRODUCTION

The use of animals to evaluate the potential adverse effects of xenobiotics (environmental chemicals or therapeutic agents) is required as an essential component of safety and risk assessment to ensure human health. However, many of the traditional study designs use direct dosing in adult animals to provide safety data to extrapolate for human risk assessment, a practice that provides less than adequate safety data for the pediatric population. The predictive value of the study designs for purposes of risk assessment following direct dosing of xenobiotics to juvenile animals has lagged far behind that of risk assessment based on direct dosing in adult animals or prenatal exposure. Although a considerable volume of material has been written on juvenile or pediatric pharmacokinetics,[1] less attention has been given to juvenile or pediatric toxicity studies that employ direct administration of xenobiotics during critical periods of development. Many nonclinical safety evaluations of therapeutic agents are conducted in adult animals, with the result that many drugs approved for use in the adult human population are not adequately labeled for use in the pediatric population. Physicians prescribe therapeutic agents off-label for children to treat ailments such as depression, epilepsy, severe pain, gastrointestinal problems, allergies, high blood pressure, and attention deficit hyperactivity disorder (ADHD). Even though many of the drugs prescribed for these indications have been extensively evaluated in adults, very little, if any, safety or efficacy data have been obtained from controlled studies in juvenile animals or clinical trials in children.

Historically, physicians have determined the dosage of therapeutics for use in children based upon professional experience and the child's body weight. The problems with this approach are that children are not just "little adults," and the differences in sensitivity between children and adults can be chemical specific. Developing organs or organ systems and metabolic pathways in children and fully developed adult systems can react very differently to drugs. Because of their small stature, infants and/or children are often assumed to be more susceptible to the toxic effects of drugs or chemicals in the environment. However, their susceptibility depends on the substance

Table 8.1 Proposed pediatric age classification

Preterm newborn infants	Born prior to 38 weeks of gestation
Term newborn infants	0 to 27 days of age
Infants and toddlers	28 days to 23 months of age
Children	2 to 11 years of age
Adolescents	12 to 16-18 years of age

Source: Modified from *Guidance for Industry: E11 Clinical Investigation of Medicinal Products in the Pediatric Population* (2000).[6]

and the exposure situation (e.g., timing of exposure during critical developmental periods). In some cases, there may be no difference in response between children and adults. In other cases, different physiological and metabolic factors, pharmacokinetics, and behavioral patterns can render children more or less sensitive than adults.

Results of studies in juvenile animal models indicate that exposure to certain environmental chemicals,[2] drugs,[3] and ionizing radiation[4] can lead to potential developmental and/or functional deficits. These deficits may range from immunomodulation (e.g., suppression of vital components of the immune system) to altered or poorly regulated processes that may be debilitating (e.g., behavioral impairment). Concern regarding potential toxic effects in children has increased in recent years with the recognition that children are a unique target population. However, the inherent risks, ethical issues, and costs associated with extensive human pediatric testing can be significant.

One way to address the concerns that children may be more sensitive than adults to drug or chemical exposures is to conduct safety evaluations designed to target critical periods of development following direct exposure in juvenile animals. Results obtained from testing at the appropriate developmental stage in animal models may be used to characterize and extrapolate both efficacy and safety in the pediatric population. More importantly, information from juvenile animal studies may support reducing the number of children required for pediatric clinical trials.

A. Objectives

The purpose of this chapter is to provide a brief introduction to the regulatory history of pediatric safety assessment and to present information for the design, conduct, and interpretation of nonclinical juvenile safety studies for extrapolation to the pediatric population. Examples of possible variations in study designs are presented. In addition, selection of the appropriate juvenile animal model, advantages and disadvantages of various animal species, and some mistakes to avoid in the practical aspects of conducting the studies are presented. Where appropriate, examples are given regarding the differences in organ system maturation between animal species and their relationship to human development. Other topics of importance for study design, including dose selection; organization of test groups (e.g., litter design); route, frequency, and timing of exposure; and endpoints to measure are discussed.

B. Background

Newborns (preterm or term), infants, toddlers, children, and adolescents are recognized as different from adults in many ways with respect to behavioral, developmental, and medical requirements. Equally important is the recognition that developmental differences exist within the pediatric population as well. Grouping the pediatric population into age categories is to some extent subjective, but classifications such as the one in Table 8.1 provide a basis for considering age groups within the pediatric population in which developmental differences may exist. Indeed, more than a century ago, Dr. Abraham Jacobi, father of American pediatrics, recognized the age differences between his pediatric patients, as well as the need for age-related drug therapy, when he wrote,

"Pediatrics does not deal with miniature men and women, with reduced doses and the same class of disease in smaller bodies, but...its own independent range and horizon."[5]

Traditionally, physicians have recognized that tissues and organs in children mature and function differently depending on the developmental period of life, and that anomalies in these early developmental processes could be associated with disease progression found specifically in the pediatric population.[6] Despite the known differences in the physiological ontogeny from early child development to adulthood, there was very little difference in the way in which drugs were prescribed between the pediatric and adult populations.[7]

Prior to the increased concern of off-label use of drugs in children, as well the consideration for using developmental pharmacology in the therapeutic evaluation process, there was a lack of appropriate drug-dosing guidelines to help in the determination of pediatric drug dose levels to be used in the clinic.[8] Typically, adjustments for drug dose in children were based upon rudimentary formulary calculations that used age or relative body size (e.g., Young's Rule, Cowling's Rule, or Clark's Rule), as well as labeling information derived from both nonclinical and clinical adult studies.[9,10] These practices were based upon default assumptions that there are no developmental differences for a drug's pharmacokinetic or pharmacodynamic characteristics between children and adults, and that children and adults have similar disease progression. However, pediatric growth and development are not one-dimensional processes, and age-associated changes in body composition and organ development and function are variables that can have an impact on drug toxicity and/or efficacy, as well as on disease progression.[9] Therefore, these prescribing practices were not considered sufficient for individualizing drug doses across the entire course of pediatric development.[8] As a result, the use of dose level equations is being replaced by normalization of the drug dose for either body weight or body-surface area, in combination with understanding the safety and pharmacokinetic data obtained from testing drugs in pediatric clinical trials and/or from juvenile animal studies.[8,11–13]

1. Historical Perspective of Pediatric Therapeutics

Many prescription drugs and biological therapeutics are marketed with little or no dosage information for administration to pediatric patients. This information, if available for use in the pediatric population, is provided to physicians in the product label (package insert). Therefore, the drugs without adequate labeling information which are given to children are unlicensed or prescribed off-label. The off-label prescription of a drug therapy approved by the U.S. Food and Drug Administration (FDA) to a patient outside the specification of the product license involves drugs being administered by an unapproved route, formulation, or dosage, or outside an indicated age range.[14] According to a report in the *FDA Consumer Magazine,* some classes of drugs and biologics, such as vaccines and antibiotics, generally have adequate labeling information for pediatric use.[15] However, pediatric labeling for other classes of drugs has been deficient. Examples include steroids to treat chronic lung disease in preterm neonates,[16] agents to treat gastrointestinal disorders,[17,18] prescription pain medications,[19] antihypertensives,[20] and antidepressants.[21] In addition, some age groups have less labeling information available to them than others. For example, children under 2 years of age have virtually no pediatric use information on drug products in several drug class categories.[22]

Despite regulatory efforts to increase the rate of labeling of pharmaceuticals for pediatric indications, little has changed in the way that these products are prescribed for off-label use in the past two decades. According to some literature, more than half of the drugs approved every year that are likely to be used in children are not adequately tested or labeled for treating patients in the pediatric population.[15] A survey of the 1973 *Physicians' Desk Reference* (PDR) showed that 78% of drugs listed contained either a disclaimer or lacked sufficient dose information for pediatric use.[23] A later survey of the 1991 PDR indicated that 81% of the listed medications contained information disclaiming use in children or limiting use to specific age groups in the pediatric population.[24] Another survey of new molecular entities (NMEs; an NME contains an active substance that has not previously been approved for marketing in any form in the United States) approved by the FDA from 1984 through 1989 showed that 80% of the listed drugs were approved without label information for the pediatric population.[25] In addition, the FDA identified the 10 drugs most commonly prescribed to children in 1994 that lacked sufficient pediatric labeling information.[15] Together, these drugs were prescribed more than five million times in the pediatric population. Between 1991 and 2001, the FDA approved 341 NMEs.[26] Of these NMEs, only 69 (20%) were labeled for use in children at the time they were initially approved. Although the percentages varied from year to year, there was no apparent trend toward an increase over time for NMEs with pediatric labeling. In addition, between 1991 and 1997, the FDA also gathered statistics on the number of NMEs that were potentially useful in the pediatric population. Of 140 such drugs, only 53 (38%) were labeled for use in children when they were initially approved. This relatively low percentage of NMEs developed and/or labeled for pediatric use is somewhat surprising, considering that some published data indicate that more children than adults in the United States are taking prescription medications.[27,28]

The lack of drug approval for pediatric use does not imply that a drug is contraindicated. It simply means that insufficient data are available to grant approval status and that the risks or benefits of using a drug for a particular indication have not been examined. This leads to the following basic issues concerning the off-label use of drugs in children: the lack of age-related dosing guidelines, the lack of age-related adverse-effect profiles, and the unique pediatric clinical problems that must be considered when prescribing off-label.[7] These issues create an ethical dilemma for the physician in that a decision must be made to either deprive a child of the potential therapeutic benefits of using a drug or to use it despite disclaimers or the lack of sufficient safety and efficacy data. The latter choice has a potential consequence of underdosing with a resulting lack of efficacy or overdosing with a resulting increase in toxicity.

2. Adverse Events

The history of pediatric medicine is well documented with progressive improvements in children's health care. However, interspersed throughout the medical successes have been many therapeutic tragedies of adverse drug reactions in the pediatric population (refer to Figure 8.1 for examples). Some of the tragedies include kernicterus in newborns from administration of sulfa drugs,[29] seizures and cardiac arrest from the local anesthetic bupivacaine,[30] neurological effects from the unexpected absorption through the skin of infants bathed with hexachlorophene,[31] fatal reactions in neonates exposed to benzyl alcohol used as a preservative in dosing formulations,[32] fatal hepatotoxicity to children 2 years or younger administered valproic acid as part of a multiple anticonvulsant therapy,[33] cardiotoxic effects of anthracyclines in the treatment of childhood cancer,[34] and cardiovascular collapse (i.e., gray-baby syndrome) in neonates administered chloramphenicol as a treatment for severe bacterial infections.[35]

All of these events have the common underlying theme of introducing treatments into the pediatric population without having adequately investigated their potential harm. It is these types of misfortunes that have been primarily responsible for the major changes in laws and regulations that guide the testing and marketing of new drugs.

3. Agency Intervention/Regulatory Guidelines

Many drugs marketed in the United States and used in the pediatric population lack safety and efficacy information derived from this population. In many cases, safety data from clinical trials in adults, supported by nonclinical studies in adult animals, have supported the use of a therapeutic agent in the pediatric population. In the case of a potential therapeutic agent or new environmental chemical, the question with respect to early safety studies is usually, "Has the compound been

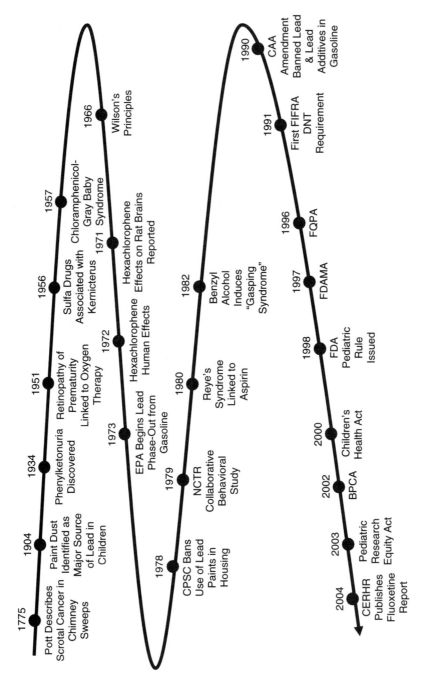

Figure 8.1 Influencing events and key times in guideline development and their changes for juvenile toxicology.

evaluated in adult animals?" Better questions would have been, "Has the substance been evaluated in developing animals?" or "Have the effects in adult animals been studied following direct exposure to developing offspring?" Despite the lack of pediatric and nonclinical juvenile toxicity studies, many of the therapeutic drugs on the market today have been used safely in children. For example, acetaminophen has been prescribed for many decades to children.[36] Although there has been some formal pharmacological testing of acetaminophen following direct administration to children, a great deal of information about its safety and efficacy has been acquired over time from anecdotal reports. An extrapolation from the clinical experience in adults was used to provide initial dosing information for children. In general, the risk of adverse events from acetaminophen administration appears to be lower in pediatric patients than adults.[37] However, regardless of the very low incidence of adverse effects, acetaminophen toxicity still remains a concern because it is widely used throughout the pediatric population.[38] In general, many drugs prescribed to the pediatric population have insufficient information, either anecdotally and/or from controlled studies, on which to base the appropriate dosage regimen.

The lack of appropriate safety or efficacy information for the use of therapeutic agents in children may have been due in part to the trepidation that physicians had about testing drugs in children. Many in the health community argued that it was unethical to put children at risk, especially since children could not give legal consent.[39] Despite these concerns, children were still prescribed drugs that the FDA had approved for use in adults only. Even though it was known that juveniles, especially neonates and infants, metabolized some drugs differently than adults, the therapeutic benefits were assumed to outweigh the unknown risks. In addition to the ethical concerns for testing drugs in children, pediatric trials and/or nonclinical juvenile studies would also add millions to the cost of drug development. Physicians can legally prescribe adult products to children off-label. Thus, the drug industry had little incentive to increase the number of drugs targeting the pediatric population or to evaluate existing drugs currently on the market that could be or were being administered to children. Instead, many companies opted to add a disclaimer on the label indicating that safety and effectiveness have not been established in the pediatric population. Even though off-label usage may have been supported by much professional experience, it was recognized by the lay, medical, scientific, and governmental communities that this could not continue to be the type of evidence on which to base safe and effective drug therapies for use in the pediatric population. Indeed, as illustrated in Figure 8.1, there already existed scientific evidence of pediatric toxicities that were not readily predicted from adult use of the same drugs.

With the increasing concern that pediatric patients should be recognized as a sensitive subpopulation in risk assessment, and with drug development targeting conditions specific to the pediatric population, new regulations were enacted to increase the extent to which drugs were tested prior to use in children (refer to Figure 8.1 for a brief history of pediatric drug labeling). Two of the most prominent U.S. regulations used to advance pediatric risk assessment for drug development include the Pediatric Rule[40] and the Best Pharmaceuticals for Children Act (BPCA).[41]

The Pediatric Rule established the presumption that all candidate drugs and biologics must be evaluated in the pediatric population. Under the Rule, pediatric safety and effectiveness data must be included for New Drug Applications (NDAs), Biologics License Applications (BLAs), supplemental applications for new active ingredients, and new indications, dosage forms, dosing regimens, or routes of administration. For currently marketed drugs and biologics, the Rule required pediatric studies for products that were used for a labeled indication in 50,000 or more pediatric patients, where the absence of adequate labeling could pose a significant risk to these patients. In addition, the Rule required pediatric studies for products that would provide a meaningful therapeutic benefit over existing treatments for pediatric patients for one or more of the claimed indications where the absence of adequate labeling information could pose significant risk in these patients. However, under the Rule, the FDA had no authority to require a manufacturer to conduct pediatric studies for products used for off-label indications.

In December of 2000, the Pediatric Rule was challenged in federal court.[42] The lawsuit stated that the Rule was not valid, arguing that the FDA had no statutory authority to promulgate the Rule. The court decided in favor of the lawsuit, struck down the Rule, and enjoined the FDA from its enforcement. However, despite the ruling, legislation was eventually introduced in Congress to write the Pediatric Rule into law. As a result, the Pediatric Research Equity Act of 2003 amended the Federal Food, Drug and Cosmetic Act to provide the FDA with the statutory authority to require the drug industry to submit assessments regarding the use of drugs and biologics in pediatric patients in certain defined circumstances.[43] In fact, the authority granted in the new legislation follows many of the same key elements of the former Pediatric Rule.[44]

The BPCA[41] reauthorized until 2007 the 6 months of additional drug patent protection originally enacted in the Food and Drug Administration Modernization Act (FDAMA) of 1997, which incorporated an incentive-based program for encouraging the drug industry to provide pediatric drug labeling information.[45] In addition, the BPCA made available alternative mechanisms, such as studies by the National Institutes of Health (NIH), to obtain information about generic or patented drugs that a manufacturer does not study in children. The BPCA authorized the FDA to request pediatric studies of marketed drugs upon determination that the information relating to their use in the pediatric population provides additional health benefits or because their use predisposes children to greater risk. The BPCA required the NIH (in consultation with the FDA) to develop, prioritize, and publish an annual list of approved drugs for which additional studies are needed to assess safety and effectiveness in the pediatric population. The FDA can issue written requests for the conduct of pediatric or nonclinical juvenile studies to manufacturers of approved NDAs (drugs with or without patent exclusivity). If there is no response from the manufacturers, the NIH must publish a request for contract proposals to conduct the studies. Following their completion, reports of the studies, including all data generated, must be submitted to the NIH and FDA. Following submission, the FDA has a 180-day period to negotiate label changes with the manufacturer and publish in the Federal Register a summary of the report and a copy of the requested label changes. If a manufacturer does not agree to the recommended labeling changes, then the FDA refers the request to the Pediatric Advisory Subcommittee for review. After the subcommittee makes its recommendations to the FDA, the manufacturer is requested to make the labeling changes. If the manufacturer fails to agree, the FDA may deem the drug misbranded.

The combination of mandatory (Pediatric Rule, now law under the Pediatric Research Equity Act of 2003) and incentive-based (FDAMA, now reauthorized under BPCA) legislative policies appears to have contributed to an increased number of drug studies involving children, which resulted in pediatric labeling changes[46] and an increase in the number of drugs being developed specifically for diseases in the pediatric population.[47]

Although these legislative policies advocate the collection of pediatric safety and effectiveness data for marketed drugs (Pediatric Rule and BPCA) and biological products (Pediatric Rule only), very little, if any, guidance was provided to indicate specifically what to do to obtain adequate safety information for pediatric labeling. There are, however, several guidance documents that provide outlines of critical issues in pediatric drug development and approaches to the safe, efficient, and ethical study of drug products in the pediatric population.[6,48] These guidance documents acknowledge that assessing the effects (safety and effectiveness) of drugs in pediatric patients is the primary goal of any pediatric clinical trial. However, pediatric clinical trials should be ethically conducted without compromising the well-being of pediatric patients or predisposing them to more than minimal risk in cases where there may be only slight, uncertain, or no beneficial returns.[49] Therefore, previously collected, relevant safety data from adult studies, as well as data from nonclinical repeated-dose toxicity, reproductive toxicity, and genotoxicity studies, should be considered in an effort to anticipate and reduce potential hazards in pediatric subjects before a clinical trial is initiated. Given that there are developmental differences between pediatric and adult patients that can affect risk assessment and that acceptable concordance has been shown for the predictability of drug toxicity between adult animals and humans,[50] nonclinical juvenile studies should be useful for predicting potentially adverse events prior to the initiation of pediatric clinical trials.

Recognizing the important benefits of juvenile toxicity studies, in February 2003 the FDA issued a draft guidance document on nonclinical safety evaluation of pediatric drug products.[51] This document provides general information on the nonclinical safety evaluation of drugs intended for the pediatric population. In addition, the guidance presents some conditions under which nonclinical juvenile studies are considered meaningful predictors of pediatric toxicity and recommends points to consider for the design of nonclinical juvenile studies. Because the document attempts to encompass most pediatric products (e.g., NMEs, drugs used off-label, or pediatric-only indications), there is no one conventionally accepted nonclinical juvenile study design that is recommended to address the need for, timing of, or the most appropriate parameters to include for assessment. This has led to the current practice for determining the need for nonclinical juvenile studies on a case-by-case basis. This approach should take into consideration the following issues for determining the need for conducting a nonclinical juvenile study: (1) the intended likely use of the drug in the pediatric population, (2) the timing and duration of dosing in relation to the growth and development in children, (3) the potential differences in pharmacological or toxicological profiles between children and adults, and (4) the evaluation of whether available adult data from nonclinical or clinical studies support reasonable safety for drug administration in pediatric patients. For specific study design considerations for a nonclinical juvenile study, the guidance document discusses appropriate selection of animal species, age at initiation of dosing, sex, sample size, route, frequency, and duration of drug administration, dose selection, toxicological end points, and timing of monitoring. The content of this chapter spans some of the same considerations recommended in the guidance document and attempts to further elaborate on specific examples of study design and implementation.

In regard to industrial and environmental chemicals, the U.S. Environmental Protection Agency (EPA), along with many in the medical community, recognize the necessity of hazard identification studies for pesticides, toxic substances, and environmental pollutants to address the potential for adverse effects to pediatric development.[52,53] Hazard and dose-response information are used in the risk assessment process and are typically derived from laboratory animal studies. These include the prenatal developmental toxicity study in two species,[54] the two-generation reproduction study in rodents,[55] and for some chemicals, the developmental neurotoxicity study in rats.[56] The reproduction and developmental neurotoxicity studies include juvenile animals that have been potentially exposed to a test chemical through placental transfer during *in utero* development and until weaning from lactational exposure. The juvenile animals may also be directly exposed through treated-diet consumption during the latter part of the preweaning period. The data used for risk assessment in the juvenile population from these types of animal studies commonly assume the following: (1) offspring exposure to the chemical has in fact occurred, (2) in the absence of kinetic data, levels of chemical exposure in the offspring are similar to those in adults, and (3) exposures of young rodents (prenatal and/or weanling animals) are similar to potential exposures in downstream developmental periods (e.g., neonates, infants, children, or adolescents). In recognition that these assumptions may not provide the most accurate data for risk assessment in infants, children, or adolescents (the presumed sensitive populations), the EPA recommends that consideration be given to direct dosing in immature animals to characterize potential adverse effects during critical developmental periods.[52] These studies can be conducted on a chemical-specific basis, in which modifications to existing guideline protocols may be used to specifically address the developing animal. This approach in study design has been pursued by the EPA Office of Pesticide Programs since 1999 to evaluate potential developmental neurotoxic effects and age-related sensitivity to cholinesterase inhibition by organophosphorous pesticides.[57] The EPA has begun advocating the expanded use of biomarkers and toxicokinetic data to help characterize direct exposure and/or aid in the design of targeted animal studies for risk assessment for the pediatric population.

II. IMPORTANCE OF JUVENILE ANIMAL STUDIES

A. Differences in Drug Toxicity Profiles between Developing and Developed Systems

Some drugs or environmental chemicals in the pediatric population show different toxicity profiles compared with those of adults. Intrinsic differences between developed and immature systems may result in an increase or decrease in drug toxicity in the pediatric population. From a physiological and anatomical perspective, many factors can contribute to these differences in toxicity profiles. Some of these factors include rapid cell proliferation during early pediatric development, age-related differences in body composition (e.g., fat, muscle, and bone composition, and water content and distribution), level of hepatic enzyme development, age-related differences in protein synthesis and function in the pancreas and gastrointestinal tract, composition and amount of circulating plasma proteins that bind to free drug, maturation of renal function, degree of immune system functional maturity, ontogeny of receptor expression, intestinal motility and gastric emptying, and maturation of the blood–brain barrier.

As stated previously, neonates (preterm and term), infants, children, and adolescents should not be considered simply as "little adults" when toxicological risk is evaluated. Instead, pediatric patients are a unique population for assessing risk associated with drug therapies or chemical exposures. From the viewpoint of pediatric medicine, there is a large potential for children to be more sensitive to drug effects during growth and development, which encompasses the period between birth and adulthood. The structural and functional characteristics of many cell types, tissues, and organ systems differ significantly between children and adults because of rapid growth and development. These developmental differences can lead to different outcomes for drug absorption, distribution, metabolism, and elimination, which can have an impact on toxicity and/or efficacy.

For example, as a result of reductions in both basal acid output and the total volume of gastric secretions during the neonatal period, the gastric pH is relatively higher (greater than 4) than the gastric pH in older children and in adults.[58,59] Subsequently, oral administration of acid-labile compounds to neonates can lead to greater drug bioavailability than in older children or adults.[60] In contrast, drugs that are weak acids may require larger oral doses during the early postnatal period to achieve the therapeutic levels observed in older children or adults.[61] In addition, age-dependent changes in body composition can alter the physiological compartments in which a drug may be distributed.[9] For example, compared to adults, neonates and infants generally have relatively larger extracellular and total-body water spaces associated with adipose stores, resulting in a higher ratio of water to lipid composition.[62] This could potentially lead to changes in the volume of drug distribution, affecting free plasma levels of the drug.

From a functional perspective, a fully mature immune system is lacking in the pediatric population, where adult response antibody levels for immunoglobulin G (IgG) and immunoglobulin A (IgA) are not achieved until about 5 and 12 years of age, respectively.[63] Experimental animal studies, and to a lesser extent human studies, describing adverse immunological effects in neonates presumed to be exposed to toxic agents during the prenatal or early postnatal period have been published.[64–66] Of particular concern is that aberrant effects on the immune system often appear more severe and/or persistent when the drug or chemical exposure occurred peri- or postnatally, compared with exposure in adults.

In regard to normal cognitive and motor function, several neurodevelopmental processes undergo relatively rapid postnatal change from birth to 3 years of age when compared with those of adults. These processes include migration of neurons in the cerebellum, synapse formation in the association cortex, closure of the blood brain barrier, and myelination in the somatosensory, visual, and auditory areas of the brain.[67] When administered to preterm infants for the treatment of bronchopulmonary dysplasia, inhaled steroids can affect postnatal development during the period in which these critical neurological processes occur.[68] The resulting toxic effects have been implicated as the cause of an increased association with the occurrence of cerebral palsy.[69]

Children metabolize some drugs in an age-related manner, and adverse events may not be predicted from adult experiences. For example, children may be more sensitive to adverse events despite exposures being equal to or less than those in adults because maturational differences in enzyme activities and/or concentrations may preclude a simple scale down from adult to pediatric doses. Administration of the antibiotic chloramphenicol in neonates (less than 1 month of age) leads to a toxic condition described as "gray baby syndrome."[35] This condition can be fatal via a form of circulatory collapse associated with excessive and sustained serum concentrations of the unconjugated drug. Neonates are susceptible because of the absence of Phase II enzymes responsible for the conjugation reaction converting the active drug to a biologically inactive, water-soluble monoglucuronide.[70] Because this is a dose-dependent toxicity, chloramphenicol must be administered to neonates at a lower dosage than is administered from infancy through adulthood.

In contrast, hepatotoxicity from acetaminophen-induced depletion of glutathione levels is less severe in neonates and young children than in adults.[71] The major pathway for the metabolism of therapeutic doses of acetaminophen is through conjugation reactions with sulfate and glucuronide. The minor metabolism pathway for the drug is oxidation by the mixed-function oxidase P450 system (mainly CYP2E1 and CYP3A4) to a toxic, electrophilic metabolite, *N*-acetyl-*p*-benzoquinoneimine (NAPQI).[72] Under conditions of excessive NAPQI formation or reduced glutathione stores (e.g., acetaminophen overdose), the toxic metabolite covalently binds to proteins and the lipid bilayer, resulting in hepatocellular death and subsequent liver necrosis.[36] Because young children have lower enzyme activity levels for the minor pathway and higher glutathione stores and rate of glutathione turnover, there is less toxic intermediate formed.

B. Utility of Studies in Juvenile Animals

1. Clinical Considerations

Traditionally, drugs have not been sufficiently evaluated in the pediatric population. The reason is multifactorial and relates to ethical issues, technical constraints, financial and practical concerns, and the potential for long-term adverse effects, as well as deficiencies in previous laws and regulations to require more labeling information.[73] It is considered unethical for a child to be prescribed medications that have not been adequately evaluated; however, studies involving procedures submitting children to more than minimal risk with only slight, uncertain, or no benefit are also regarded as unethical.[49] Under federal regulations, children are considered a vulnerable group and require additional protection as research subjects.[74] In addition, obtaining the assent of a child and the permission of a parent or guardian is not the same as obtaining informed consent from a competent adult.[39] Furthermore, the parameters utilized to monitor children's health and safety during clinical trials are of major technical and ethical concern.

The financial and practical challenges in drug development involving children include the small number of patients with certain age-related medical conditions, the expense of studying children in various stages of development, difficulties in developing formulations appropriate for children, difficulty in recruitment of pediatric patients, and the possibility that an approved drug will be used by only a small number of pediatric patients.[26] In addition, previous FDA regulations allowed the effectiveness of a drug treatment to be extrapolated from studies in adults to children if the progression of the medical condition being treated and the effects of the drug were sufficiently similar.[75] The adult data were usually augmented with pharmacokinetic studies obtained in children. However, the safety of a drug prescribed in children cannot always be extrapolated from data obtained in adults; therefore, the drug and course of treatment could potentially be more or less toxic in children. Because of these issues, governmental institutions and professional organizations have started to address the need to evaluate drugs for children in a similar manner as they are for adults.

2. Predictive Value

The traditional model for prediction of potential hazards in pediatric patients is to consider study data (safety, effectiveness, and exposure) from previous adult human studies, as well as data from nonclinical repeated-dose toxicity, reproductive toxicity, and genotoxicity studies.[48] However, this model inherently assumes (correctly or incorrectly) that a similar disease progression occurs between children and adults, that there will be a comparable response to the drug for a specific indication, and that there are no differences between adults and developing children that may affect the safety profile. Therefore, given that there are known differences between pediatric and adult patients that can affect risk-benefit analysis, and that acceptable concordance has been shown for the predictability of drug toxicity between adult animals and humans, nonclinical juvenile studies should be used to predict potentially adverse events prior to the initiation of pediatric clinical trials.[50,76]

In general, some juvenile animals (e.g., rodents, dogs, minipigs, nonhuman primates) exhibit age-related and developmental characteristics similar to those in the human pediatric population, thus making them suitable for toxicity testing.[77] Because of these similarities, nonclinical juvenile studies have been shown to be useful for identifying, evaluating, or predicting age-related toxicities in children. For example, the adverse effects of phenobarbital on cognitive performance in children are predicted by administration of the drug to the developing rodent during periods of critical neurodevelopment.[78,79] Increased sensitivity of human infants to hexachlorophene neurotoxicity was replicated and evaluated in juvenile rats of comparable developmental age.[80] Proconvulsant effects observed in developing rodents treated with theophylline could be predictive of risk to similar effects in children.[81] In addition, a juvenile rat model has been used to understand the toxic effects on craniofacial growth and development observed in young children administered prophylactic treatments (irradiation with or without chemotherapeutic drugs) to reduce the recurrence of childhood acute lymphoblastic leukemia.[4] Unfortunately, most of these examples of the predictive utility of nonclinical juvenile studies for identifying adverse events occurred after the adverse events were identified in children.

III. DESIGN CONSIDERATIONS FOR NONCLINICAL JUVENILE TOXICITY STUDIES

When designing nonclinical juvenile toxicity studies, various factors must be considered to obtain data that allow the assessment of the toxicological profile of a compound in young animals and provide information pertaining to specific endpoints deemed important based on existing clinical and/or nonclinical data. For example, the intended use of a drug and the intended target population must be addressed, along with the developmental period of exposure in the target population. The duration of clinical use of a drug must also be considered when determining the proper exposure period in animal studies. The test article exposure regimen in an animal model should correspond to the intended exposure period in the human target population. Finally, to the extent possible, inter- and intraspecies physiological, pharmacological, and toxicological profiles should be understood.

A. Intended or Likely Use of Drug and Target Population

Juvenile toxicity studies are often conducted in the rat, which is a species that reaches adult status much more rapidly than humans (as illustrated in Figure 8.2). This compression of physiological

NONCLINICAL JUVENILE TOXICITY TESTING

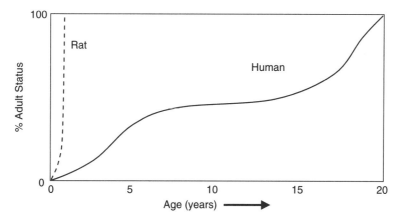

Figure 8.2 Time to develop adult characteristics. Rats, because of their shorter life span compared to humans, reach adult status much more rapidly. This compression of physiological time must be considered when determining the appropriate exposure periods in juvenile toxicity studies.

time in the rat and other nonhuman species relative to humans must be considered when determining the appropriate time of initiation and duration of exposure in the nonclinical juvenile study. For example, if a compound is administered for a number of years in children, the critical period of exposure in the rat may be merely weeks. Additional thought must be given to the ontogeny of specific organ systems in the animal versus the human. Maturation of several organ systems (e.g., the renal system) occurs *in utero* in humans but occurs in the early postnatal period in rodents. Therefore, the window of exposure in nonclinical juvenile studies must take into account the physiological age correlation between humans and the animal model, as well as differences in timing of development of specific organ systems of interest. As an example, Figure 8.3 illustrates the comparative age categories based on CNS and reproductive system development in several species.[82]

B. Timing of Exposure in Relation to Phases of Growth and Development of Target Population

There are multiple examples that illustrate the idea of physiological time and differences in development between humans and animals. Well-documented examples include the differences between humans and rats in the timing of formation of the functional units of the lung (alveoli) and those of the kidney (nephrons). Detailed reviews of the development of the kidneys, lungs, and other organ systems can be found in Appendix C of this text. In the human, alveolar proliferation is initiated during gestational week 36 (approximately). It is completed by 2 years of age, and the expansion stage is completed by 8 years of age. In the rat, however, alveolar development occurs strictly during the postnatal period, with proliferation occurring between postnatal days (PND) 4 and 14, and the expansion stage is completed by PND 28.[66] Based on this information, it would be appropriate to expose young rats during the first 4 weeks of postnatal life in order to study the effects of a compound on alveolar development. The rat, therefore, is a widely accepted model for studying the effects of compounds on lung development, while other species, such as the rabbit, sheep, pig, and monkey are not acceptable for a postnatal assessment of lung development because of the advanced stage of lung development seen in these species at birth.[83]

Nephrogenesis also occurs prenatally in the human but postnatally in the rodent.[84–86] Nephrogenesis is completed by gestational week 34 to 35 in humans,[85] while it has been reported that this process is completed by PND 11 in the rat, with further kidney maturation extending until postnatal week 4 to 6.[87] Along with considering the relative timing of morphological development of the kidney between humans and test species, functional development should also be considered. The

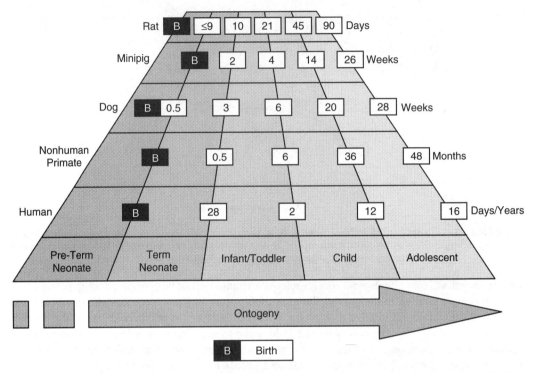

Figure 8.3 Comparative age categories based on overall CNS and reproductive development. (From Buelke-Sam, J., Comparative schedules of development in rats and humans: implications for developmental neurotoxicity testing, presented at the 2003 Annual Meeting of the Society of Toxicology, Salt Lake City, UT, 2003.) When designing juvenile toxicity studies, specific organ system development must be considered in conjunction with overall developmental age categories.

rat model is appropriate in this respect when investigating potential nephrotoxicants because several aspects of kidney function, including glomerular filtration rate, concentrating ability, and acid-base equilibrium, mature postnatally in both humans and rats.[87] Therefore, postnatal exposure in the rat would be appropriate when examining drug- or chemical-induced effects on nephrogenesis and functional kidney development.

Extensive investigations have also been performed to elucidate the ontogenies of immune system function, bone growth, and the reproductive and central nervous systems in various species, including humans (cf. Appendix C). As an example, the ontogeny of cholinesterase was examined in five compartments in the rat, including the plasma, red blood cells, brain, heart, and diaphragm.[88] On a whole-tissue basis, cholinesterase in the heart and diaphragm increased with age from gestation day 20 through PND 21 and then remained relatively constant through adulthood, while in the brain, cholinesterase increased until approximately 6 weeks of age. However, when normalized to the weight of the tissue, cholinesterase activity in the heart and diaphragm increased until PND 11 and decreased thereafter until adulthood, while it continued to increase with age and growth of the brain.

C. Potential Differences in Pharmacological and Toxicological Profiles between Mature and Immature Systems

The underlying physiology of juveniles and adults leading to the differences in metabolic clearance and/or activation of compounds should be identified, if possible, prior to initiating juvenile toxicity testing. Table 8.2 lists several factors responsible for the different responses to toxicant exposure in adults versus children. As mentioned previously in this chapter, the timing of exposure to a drug

Table 8.2 Factors responsible for differences in risk assessment between children and adults

Growth and development
Time between exposure and manifestation
Diet and physical environment
Parameters of toxicity assessment
Biochemical and physiological responses
Drug and chemical disposition
Exposure/behavior pattern

Source: Modified from Roberts, R.J., *Similarities & Differences Between Children & Adults: Implications for Risk Assessment,* Guzelian, P.S., Henry, C.J., and Olin, S.S., Eds., ILSI Press, Washington, 1992, p. 11.

or chemical is an important factor in the resulting effects seen in exposed individuals. During the neonatal and postnatal periods, a multitude of developmental processes occur, including CNS development, cardiovascular system development, and reproductive system maturation. Because young children are growing and maturing, they may express unique susceptibilities to toxicant exposure. For example, while a chemical or drug that delays sexual maturation may not have an effect on a sexually mature adult, profound adverse effects may occur in a child exposed prior to or during puberty. Similarly, if a drug or chemical affects long bone growth by disruption of the epiphyseal plates and subsequent endochondral bone formation, the resulting toxicity would be manifested to a greater degree in an exposed child compared with an adult. Since the growth plates have already fused in the adult, manifestations of toxicity due to growth plate disruption would be unlikely. This idea of critical exposure periods is one that resonates throughout this chapter. It is crucial to understand the developmental processes occurring in a test system and the temporal relationship to human development.

If a chemical causes alterations because of a long-term cascade of events, it is more likely that a child exposed to the material will develop symptoms while an adult who is exposed may not. Ultraviolet irradiation and the subsequent development of skin cancer is an example of a delayed response, which would be more likely to occur the earlier the exposure took place. Other examples of toxicants in which toxicity is manifested in increasing severity with age include methylmercury and triethyltin.[67,89–91] By disrupting neurodevelopmental processes in the neonate, these compounds cause life-long aberrations in motor and sensory function that worsen with age.

D. Use of Extant Data

1. Pharmacokinetic (PK) and Toxicokinetic (TK) Data in Adult Animals

Because of physiological differences between adult and juvenile systems, the bioavailability and biotransformation of xenobiotics may be difficult to predict accurately in juveniles when only adult data are available.[92] For example, young children and animals have the potential to be exposed to much higher levels of chemicals administered dermally (on a milligram per square meter) basis because of their large surface-area-to-body-weight ratio.[93]

Gastric pH in juvenile animals is generally higher than in adults, thereby increasing the absorption of basic molecules and decreasing absorption of acidic molecules. Differences in gastrointestinal (GI) absorption and motility in the juvenile model may also affect the toxicological profile of drugs and chemicals. Gastrointestinal motility in infants and neonatal children is low, whereas in older children GI motility is higher. For example, the average GI absorption rate of lead in infants is approximately fivefold higher than in adults.[94] The difference in absorption of lead between infants and adults, coupled with the rapid neurodevelopmental processes occurring during

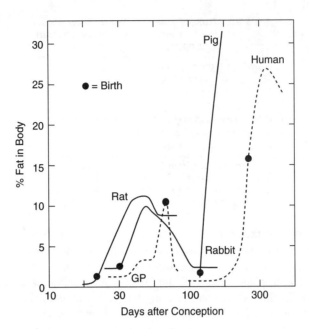

Figure 8.4 Comparative ontogeny of fat. The human and guinea pig deposit fat prior to birth. (After Adolph, E.F. and Heggeness, F.W., *Growth*, 35, 55, 1971.)

infancy (neural migration, brain topographical mapping), exacerbates the devastating results for those who are exposed during the early postnatal period.

Distribution of a drug throughout the body may also differ between adults and infants because of differences in plasma protein binding. Generally, newborns display less protein binding than adults because of a combination of factors. First, albumin from infants has less affinity for certain drugs, but adult binding capacity is developed during the first year of life. In addition, basic compounds may not be bound to a high degree because of low levels of alpha-1-acid glycoprotein in the neonate.[9] Lastly, neonatal serum can contain high levels of free fatty acids that may cause dissociation of drugs from albumin.[93]

Another factor that influences distribution of chemicals throughout the body is fat composition (see Figure 8.4 for comparative fat ontogeny).[95] Since many drugs are lipophilic, the body fat content (which changes during development) can influence the sequestration and availability of free drug. Total body water, extracellular fluid, and intracellular fluid levels also change during maturation and may lead to differences in drug distribution between developing and fully developed organisms.[94] See Figure 8.5 and Figure 8.6 for comparative perinatal water content. In general, species with longer gestational periods have lower water contents at birth.[95]

Drug receptor expression and binding may also vary between mature and immature organisms. It has been postulated that the differences in response to phenobarbital and methylphenidate in children compared with adults result from differences in drug-receptor interactions. Phenobarbital increases the effectiveness of GABA (an inhibitory neurotransmitter) and inhibits the release of glutamate (an excitatory neurotransmitter). Interestingly, phenobarbital is a sedative in adults but produces hyperactivity in children.[76] Conversely, methylphenidate, a dopamine reuptake inhibitor that also affects norepinephrine levels, is a sedative in children but a stimulant in adults.[76]

Metabolism of chemicals varies with age. Generally, metabolism in the human infant is somewhat slower than in the adult.[96] Although dependent on the specific enzyme, Phase I metabolism is fully developed in humans by 3 years of age, while the Phase II enzymes continue to develop later in life. The development of metabolic systems in laboratory animals follows patterns similar to those of humans, with lower enzyme activity at birth and a rapid increase in metabolic activity within the first few postnatal months.[94] Many enzymes have unique developmental expression

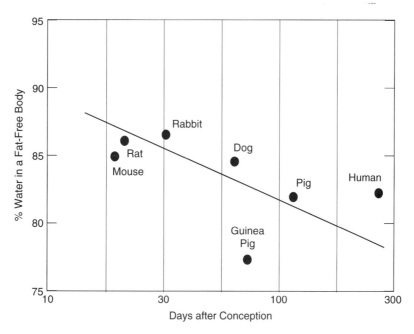

Figure 8.5 Comparative water content at birth. Longer gestations develop drier ("denser") animals. (After Adolph, E.F. and Heggeness, F.W., *Growth*, 35, 55, 1971.)

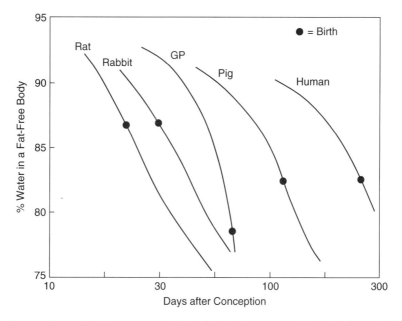

Figure 8.6 Comparative perinatal water content. Water fraction decreases with age in all species. (After Adolph, E.F. and Heggeness, F.W., *Growth*, 35, 55, 1971.)

patterns. Therefore, the toxicological or pharmacological profile of a xenobiotic often differs depending on age.

Lastly, differences in excretion rate exist between infants and adults.[97] Generally, the glomerular filtration rate is lower in young children,[87] which may increase the half-life of a compound that is readily excretable by the kidneys. Biliary excretion follows a similar time course of development as renal clearance, with biliary excretion of compounds being low in the neonate and increasing with age.[97] To illustrate this point, biliary excretion of methylmercury is approximately 10-fold lower in neonatal rats than in adults, which may contribute to the longer half-life and increased sensitivity observed in young rats.[98] This difference in excretion of methylmercury is a result of lower excretion of glutathione into the bile in neonatal animals. Since methylmercury is carried into the bile as a glutathione conjugate, the lack of glutathione excretion in turn causes lower excretion rates of methylmercury.

2. *Extant Adult Nonclinical Animal Toxicity Data*

Pediatric administration of pharmaceuticals has been primarily based on adult toxicity data, and doses have been adjusted for body weight or surface area without regard to differential sensitivity between children and adults. As previously discussed, one example of the differing responsiveness between adults and children to drugs is the lower toxicity of acetaminophen observed in children. Alternatively, children may be more sensitive than adults to certain compounds because of developmental processes occurring in the growing child. For example, corticosteroids cause a decrease in growth velocity in children but do not have a similar effect in adults, as their bone growth has already ceased.[99] Other examples of juvenile toxicants that are less toxic to adults include phenobarbital, hexachlorophene, and valproic acid. It is becoming increasingly obvious that there are many factors that must be considered when using adult data to evaluate the pharmacological and toxicological profiles of drugs administered to young children.

3. *Stand-Alone Juvenile PK and TK Studies*

As mentioned earlier in the chapter, the pharmacokinetic profile of a drug can differ dramatically between juveniles and adults and between animals and humans. Thus, the existing pharmacokinetic data of the drug from clinical and nonclinical adult studies may not be of value when predicting pediatric exposure to the drug, and the manufacturer may wish to conduct a stand-alone juvenile pharmacokinetic and toxicokinetic study prior to administration of the drug in the clinic. These stand-alone studies provide several advantages to the manufacturer, including the ability to compare pharmacokinetic data between the juvenile and adult animal to examine potential differences in exposure across ages that may require selection of different dosages for the different aged populations and for the pediatric population in the clinic. Additionally, by evaluating differences in exposure profiles across the species, such studies can aid the manufacturer in determining whether the animal model chosen is an appropriate species for use in the definitive nonclinical juvenile study. Finally, the studies can provide limited toxicity data and may guide the investigator to select additional endpoints for evaluation in the definitive study.

IV. JUVENILE TOXICITY STUDY DESIGN

A. Types of Studies

Juvenile toxicity studies are generally conducted using one of three designs: (1) adapted adult toxicity studies (usually reproductive study designs that are modified to address specific concerns in the juvenile), (2) general toxicity screens, and (3) focused or mode of action studies. Each of these study designs serves an important role in the nonclinical development of a drug that will be administered to pediatric patients and for those chemicals for which human adolescent exposure may be unavoidable. One or more of these study designs may be required to evaluate the juvenile safety profile of the test agent for risk assessment in the pediatric population. It cannot be stressed enough that researchers need to stay in contact with the agency to which the study will be submitted and receive their input prior to conducting the study to ensure the acceptability of the finished study.

1. General Toxicity Screen

The term "general toxicity screen" suggests that basic toxicity endpoints, such as clinical observations, body weight, food consumption, clinical pathology, and histopathology will be evaluated; however, these studies may encompass additional endpoints. While growth and pathology endpoints play a key role in the general toxicity screening study, they may be accompanied by more specialized evaluations, including assessments of reproductive performance, behavior, and immune competence. These study designs often initiate dose administration prior to or at the time of weaning, and measure growth and development into young adulthood. Typically, these studies are conducted in rodents and require many animals to address relevant concerns. However, it is not uncommon for these studies to be conducted in large animal models. A large animal model provides some advantages over the rodent, including the capability of longitudinal assessments of various physiological endpoints. The general toxicity screen also commonly includes a posttreatment (or recovery) period, in which persistent and/or latent effects of the test article can be examined in comparison to changes observed during treatment. An example of one such study design (the within-litter study design) is illustrated in Figure 8.7.

The general toxicity screen serves an important purpose in the nonclinical juvenile study battery because it may reveal target organs not identified in adult toxicity studies. In addition, the general toxicity screen may serve to highlight areas that require further research to define the mode of action of the test article.

2. Adapted Adult Toxicity Study

For the purposes of this chapter, the authors define an adapted adult toxicity study to be a reproductive toxicity study in which the design is modified to include a satellite juvenile phase (e.g., immunotoxicity, neurotoxicity, endocrine modulation screen, comet assay, or brain morphometric assessment). For some chemicals, the adapted adult design may be sufficient to address pediatric concerns in lieu of conducting a general toxicity screen. These modified studies have the advantage of a transgenerational exposure regimen, wherein the parental population is first exposed, followed by exposure of the juvenile population (either with or without concurrent parental exposure); this exposure model may more closely parallel the real-world situation for some chemicals. For example, the design for a two-generation reproductive toxicity study required by the EPA and the OECD encompasses a 10-week premating exposure period that begins, for the F_1 generation, immediately following weaning.[55,100] During this period, growth and sexual developmental landmarks are evaluated. If there is concern that the infant and/or toddler may be directly exposed to the test article, appropriate adjustments to the onset of dose administration for the F_1 generation may be considered. The adjustments may consist of oral intubation of the offspring (in a study where administration is by gavage), modified feeding canisters (see Figure 8.8 for a diagrammatic representation of a Lexan® rat creep feeder) that would allow measurement of whole-litter offspring food consumption (in a study where administration is via the diet), or specially designed inhalation chambers that allow litters to be exposed concurrently with dams (Figure 8.9). In the same way, for some pre- and postnatal development studies, concerns over the bioavailability of the test article in the neonate dictate that a modified approach for dose administration must be used. For these situations, pharmaceutical companies have developed a strategy wherein maternal animals are administered the test article during gestation, followed by a combination of maternal and offspring dose administration during the postnatal period.

In cases where inadequate offspring exposure has been demonstrated following delivered doses to the dam only, both the EPA and OECD have requested modifications to the developmental neurotoxicity study design to incorporate direct administration of the test article to the F_1 offspring prior to weaning.[52,167] These requests have been made to the registrants for certain classes of compounds and reflect heightened public concern for special populations at risk.

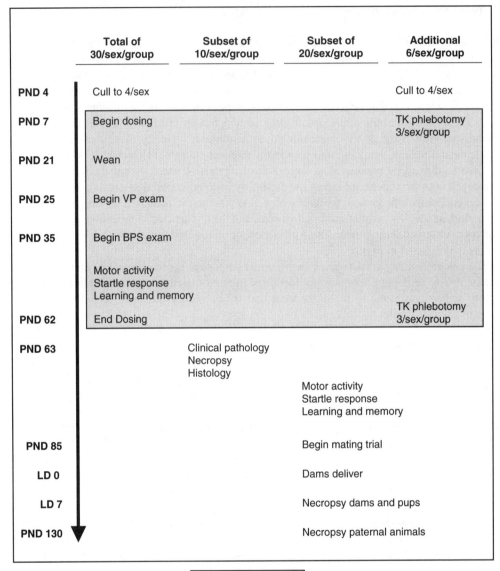

Figure 8.7 Within-litter study design. A robust within-litter design requires approximately 36 litters obtained via in-house mating or time-mated animals. VP, vaginal patency; BPS, balanopreputial separation.

A specific type of adapted adult toxicity study design that has been proposed as an option for evaluating postnatal toxicity is the transgenerational protocol. This design has been proposed to evaluate effects of chemicals on the development of the reproductive system that may not otherwise be detected by standard testing paradigms. This alternative design has been suggested because some endocrine-active chemicals were negative in standard multigeneration studies (likely because of postweaning selection of only one pup per sex per litter).[101] The dosing regimen for this transgenerational design is *in utero* and lactational exposure (with the option of direct dosing of the offspring), followed by necropsy of all offspring in adulthood.[102] The salient points to this design are that more offspring per litter are used (though fewer litters per group) and more endpoints related to endocrine-active compounds are evaluated. An important feature of this design is that multiple offspring per litter are examined following sexual maturity (since earlier examination may

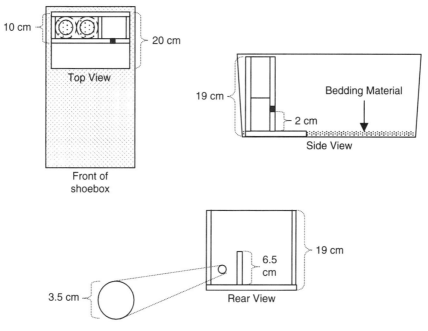

Figure 8.8 Construction of a Lexan® creep feeder shoebox caging insert for rats. This system allows rat pups access to treated feed but prevents access by the dam. Untreated feed for consumption by the dam must be offered in a separate cage to prevent preferential consumption of untreated diet by the pups. During the second half of the lactational period, dams can be separated from their pups for at least 8 h with minimal impact on either.

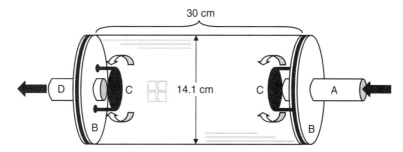

Figure 8.9 A 4.7-liter glass and anodized aluminum inhalation chamber. (A) Inlet pipe that carries air/test atmosphere at 2.5 to 3.0 l/min. (B) End cap with O-ring that seals chamber. (C) Dispersal plate that distributes atmosphere throughout chamber and prevents aerosols from affecting animals directly. (D) Outlet pipe that attaches to manifold.

not fully characterize effects on reproductive organs that are not yet fully developed), decreasing the potential for false negatives of low-incidence responses. An example of this type of study (although without direct dosing of pups) is the one-generation extension study sponsored by the EPA that evaluated vinclozolin and di(n-butyl) phthalate. That investigation demonstrated enhanced sensitivity in detecting low-incidence responses as statistically and biologically significant effects (thus reducing the chance of false negatives).[103]

3. Mode of Action Study

Mode of action studies are usually conducted when disruption of a specific organ system or endpoint has been demonstrated to be of concern in the juvenile and/or adult. Alternatively, mode of action studies may be undertaken when the registrant believes that the toxicity observed in a general toxicity screening study is species specific and not relevant to humans. These studies are designed to address specific issues and/or class effects, and do not necessarily involve large numbers of animals or long exposure periods. They do not generally include many endpoints; however, those parameters that are evaluated are typically much more closely examined than would be the case in a general toxicity screen. For example, in a study designed to examine the toxic potential of the angiotensin converting enzyme inhibitor, ramipril, 16 rat pups per sex per group were administered the test article on one of two discrete days during development (PND 14 or 21). Clinical chemistries were assessed together with macroscopic and microscopic examination of several organs 3 or 14 days later.[92] In addition, the pharmacokinetic profile of the parent and metabolite were evaluated on the days of dose administration. Although only limited numbers of animals were used (8 per sex per group per age at necropsy), statistically significant changes in renal structure and function were noted, and dramatic differences were observed in the exposure of the juvenile rats to both the parent molecule and metabolite when ramipril was administered on PND 14 or 21.

B. Model Selection

Selection of the animal model is based on the study objective, the available data for the test article in the model, the availability of historical control data, and the experience of the researcher and testing laboratory with the model chosen. Several key factors must be considered when selecting the animal model to be used in a juvenile toxicity study. These factors include the species, age, sex, sample size, and relevance to human development.[51]

1. Species

A majority of nonclinical juvenile studies utilize the rodent as the animal model of choice for a number of reasons. For adapted adult toxicity studies, the animal model is dictated by basic guideline requirements for the standard study, which in most cases tends to be the rodent. Moreover, rats and mice are well characterized with regard to growth and development, given the large number of reproductive, developmental, and general toxicity studies that have been conducted with these species. However, very little historical control data, such as the ontogenic profile of clinical chemistry parameters (refer to Table 8.3 for one example in rats) or the microscopic changes that occur in many organ systems during development, are available for more traditional toxicity endpoints in developing animals.

Despite the paucity of historical control data for traditional toxicity endpoints in the developing rodent, there is a greater breadth of literature regarding the developmental changes that occur in the rodent than in other species selected for juvenile toxicity testing (e.g., dog, swine, and nonhuman primate). Furthermore, if the intent is to administer the test article during the period of adolescent development, larger animal species will typically require much longer exposure periods than the rodent, given the longer developmental spans of larger species. Dog, swine, or nonhuman primate study designs also require more test article, housing space, technical involvement, and socialization than a comparable rodent study. However, large animal models afford some key advantages in a nonclinical juvenile study. For example, the larger circulating blood volume in these species provides the opportunity to evaluate changes in clinical chemistry parameters very early in development and repeatedly over time in the same animal; longitudinal assessments of standard clinical chemistry panels are extremely difficult, if not impossible, to conduct in rodents, because of the small circulating blood volume.

2. Sample Size

The number of animals required for a nonclinical juvenile toxicity study entails a balance between achieving the statistical power required to detect potentially adverse outcomes and the logistical

Table 8.3 Ontogeny of serum chemistry parameters in the Crl:CD®(SD)IGS BR rat

Parameter	PND 17	PND 24	PND 28	PND 35
Albumin (g/dl)				
M	3.2 ± 0.2	3.8 ± 0.1	3.9 ± 0.2	4.1 ± 0.2
F	3.4 ± 0.2	3.8 ± 0.2	3.9 ± 0.2	4.3 ± 0.2
Total protein (g/dl)				
M	4.7 ± 0.2	4.9 ± 0.1	5.1 ± 0.3	5.5 ± 0.3
F	4.9 ± 0.1	4.9 ± 0.2	5.1 ± 0.3	5.6 ± 0.4
Total bilirubin (mg/dl)				
M	0.5 ± 0.3	0.2 ± 0.1	0.2 ± 0.1	0.1 ± 0.1
F	0.4 ± 0.1	0.2 ± 0.1	0.1 ± 0.1	0.1 ± 0.1
Urea nitrogen (mg/dl)				
M	18.7 ± 3.9	11.7 ± 2.9	11.9 ± 2.6	12.5 ± 2.1
F	18.1 ± 3.4	12.8 ± 2.7	11.9 ± 2.4	12.8 ± 2.6
Alkaline phosphatase (U/l)				
M	304 ± 27	333 ± 74	332 ± 65	368 ± 74
F	302 ± 31	373 ± 92	342 ± 55	295 ± 89
Alanine aminotransferase (U/l)				
M	21 ± 9	59 ± 9	58 ± 11	73 ± 16
F	19 ± 6	56 ± 8	56 ± 13	62 ± 13
Glucose (mg/dl)				
M	277 ± 12	264 ± 26	215 ± 28	259 ± 51
F	284 ± 34	265 ± 42	235 ± 27	217 ± 30
Cholesterol (mg/dl)				
M	183 ± 13	74 ± 9	78 ± 10	71 ± 9
F	205 ± 22	76 ± 13	83 ± 14	76 ± 10
Phosphorus (mg/dl)				
M	13.2 ± 0.4	13.3 ± 1.2	13.9 ± 1.1	13.4 ± 0.5
F	13.6 ± 0.5	13.0 ± 0.7	13.1 ± 1.2	12.0 ± 1.1
Potassium (mEq/l)				
M	8.65 ± 1.03	7.77 ± 0.91	7.33 ± 0.85	7.73 ± 0.75
F	8.67 ± 1.17	7.57 ± 0.93	7.08 ± 0.55	6.9 ± 0.68
Sodium (mEq/l)				
M	136 ± 2	143 ± 2	144 ± 4	144 ± 1
F	137 ± 2	143 ± 2	145 ± 4	144 ± 2
Aspartate aminotransferase (U/l)				
M	104 ± 24	130 ± 24	111 ± 18	116 ± 18
F	109 ± 15	126 ± 15	119 ± 20	101 ± 20
Creatinine (mg/dl)				
M	0.1 ± 0.05	0.1 ± 0.05	0.1 ± 0.00	0.2 ± 0.08
F	0.1 ± 0.05	0.1 ± 0.05	0.1 ± 0.05	0.2 ± 0.05
Glutamyltransferase (U/l)				
M	A	0.9 ± 0.6	0.8 ± 0.5	0.8 ± 0.4
F	0.9 ± 0.6	0.5 ± 0.2	0.9 ± 0.3	0.7 ± 0.5
Globulin (g/dl)				
M	1.5 ± 0.1	1.1 ± 0.1	1.2 ± 0.2	1.4 ± 0.2
F	1.4 ± 0.1	1.1 ± 0.1	1.2 ± 0.1	1.3 ± 0.2

Data represent the means ± standard deviations of eight rats/sex/age, except for glutamyltransferase, which was often below instrument range for individual animals. (Collected at WIL Research Laboratories, LLC.)
A = Below instrument range; M = males; F = females.

difficulty involved in conducting the study. If the purpose of the study is to evaluate neurobehavioral deficits that may result from juvenile rodent exposure to the test article, it may be of benefit to use a sample size of 20 animals per sex per group to provide enough power to detect statistically significant changes in performance.[104] However, if the species of choice for the same study is the beagle, animal costs, test article requirements, and the feasibility of conducting quantitative behavioral assessments (e.g., locomotor activity) may limit the sample size, increasing the likelihood of Type II errors.

The availability of sufficient litters to produce the number of animals required for the study may be a limiting factor, and overselection from a small number of litters to yield the desired number of animals per group may compromise the study (a discussion of litter-based selection will follow). In more focused study designs, smaller numbers of animals may be adequate to accurately assess the toxic potential of the test article, as demonstrated in the ramipril study described previously.[92] In reproductive and developmental toxicity studies adapted to address specific concerns in the juvenile, the availability of an adequate sample size is of less concern because these studies are typically conducted in the rodent, with approximately 20 to 30 litters per group available for analysis. Selection of a subset of animals from these litters for juvenile testing is quite feasible in the adapted adult study designs, and the researcher is still capable of fulfilling the basic requirements of the standard study.

A critical issue that must be addressed when determining the appropriate number of animals for use in a study is the effect of the litter on the response to test article exposure. It was demonstrated in the Collaborative Behavioral Teratology Study (CBTS) that offspring from the same litter respond more similarly to a test article than do nonlittermates.[105] This response is known as the litter effect. Therefore, guidelines for developmental toxicity studies require that the litter be the experimental unit for statistical and biological analysis.[54,106–110] Unfortunately, many researchers fail to carry this requirement into the postweaning period. Instead, littermates are treated as though they are not siblings. Determination of the appropriate number of animals for use for a study should take into account the number of litters available for selection. Attempts must be made to select animals from as many litters as possible in order to create the required sample size for each group. This can be accomplished by conducting within-litter selections, fostering, or selecting only one animal per sex per litter (between-litter selection) for analysis. Between-litter designs should only be considered when culling is planned, when data from siblings are used to obtain a litter mean, or when siblings are selected for different endpoints of analysis. The issue of litter-based selection is addressed in more detail later in the chapter.

C. Exposure

When designing a juvenile toxicity study, the proper exposure paradigm must be determined. This involves selection of an acceptable route of administration, consideration of the frequency and duration of exposure and the timing of dose administration, and appropriate dose level selection. When feasible, the route of exposure should match the expected exposure scenario for the pediatric population.

1. Route of Administration

Choosing an acceptable route of administration may seem an easy task for most test articles, in that the most likely route of clinical administration or exposure to the human population would already have been determined. However, there are many more challenges when the route of administration is selected for a juvenile toxicity study. The first issue that must be considered is the feasibility by which the test article may be administered via a particular route. Although the common route of administration in preweaning animals is by gavage, oral intubation can be technically challenging at some ages (Figure 8.10 and Figure 8.11). For example, in a PND 4 mouse pup, the technical challenges involved in administering the test article can be extreme. When the test article must be administered as a viscous suspension through a small gauge cannula, these challenges may be impossible to overcome. While a rat pup is several times larger than a mouse pup of the same age (at PND 1, rats weigh approximately 6 g whereas mice weigh approximately 2 g), the size of the animal and the susceptibility of the esophagus to perforation early in postnatal life cannot be ignored. The staff involved in administering doses to these young animals must be thoroughly trained; never assume that an individual who is capable of dosing an adult animal will be able to dose a neonate of the same species. At the authors' laboratory, staff members are required

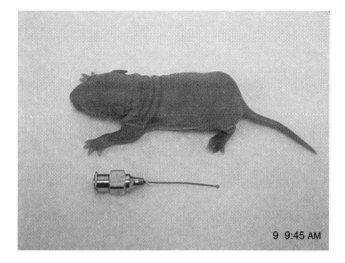

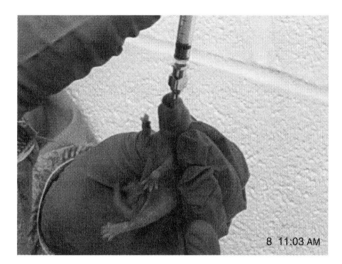

Figure 8.10 Example of gavage technique for PND 4 to 10 rat pups, using 24-gauge stainless-steel dosing cannula.

to intubate 30 pups once daily from PND 3 through PND 28, with no more than one death during the dosing period, to be qualified to gavage juvenile rodents on study. Based upon this training regimen, the authors have found that the success rate for oral intubation of juvenile rats is nearly 100% (Table 8.4).

In the authors' experience, a dosage volume of 5 to 10 ml/kg has been found to be ideal. Experimentation may be required to determine dosage volumes that are technically feasible for the vehicle.

Finally, the time required to administer a dose should be considered when choosing the route. This includes predosing activities, as well as the actual dosing of the animal. Since developing animals have difficulty maintaining core body temperature early in life and time away from the dam can have a negative impact on development, the time it takes to identify and segregate littermates prior to dosing should be minimized. If prolonged separation from the dam is unavoidable, the environment supporting the pups should be warmed appropriately.

Other routes of administration may also be considered, including dietary, dermal, inhalation, and intravenous. Each of these routes poses different technical challenges to the researcher, depending on the species and age of animal involved. For example, dietary administration may seem

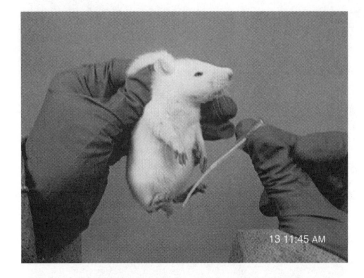

Figure 8.11 Example of gavage technique for PND 21 to 35 rat pups, using 21-gauge Teflon® dosing cannula.

appropriate if it is the likely route of human exposure, but the age at which pups start to consume solid food must be considered. If dietary exposure commences prior to weaning, specialized feeding apparatuses may be required to prevent maternal consumption of the test diet, while allowing offspring consumption (refer to Figure 8.8). In this scenario, untreated diet must be provided to the dam in a separate cage, and the dam must be removed from the litter each day for up to eight hours in order to prevent consumption of the untreated maternal diet by the offspring. Because rat pups do not consume appreciable quantities of diet prior to PND 14, the dietary route is not appropriate when exposure is required early in postnatal life. Furthermore, a within-litter testing scheme (with all groups represented in the litter) employing the dietary route of administration is not possible prior to weaning.

Inhalation exposure may be the most appropriate choice in some circumstances, but issues of animal welfare and inhalation capacity of the test species may make this choice difficult to implement. For example, the inhalation route has been used for modified developmental neurotoxicity studies, which typically use the rat model.[111] In this exposure paradigm, both the dam and

Table 8.4 Flexible cannula intubation error rates in juvenile Crl:CD®(SD)IGS BR rats

Age at Intubation	Number of Animals	Number of Doses Administered	Confirmed Intubation Errors[a]	Suspected Intubation Errors[b]	Intubation Error Rate, %
PND 4-31	256	7,168	4	0	0.056
PND 7-13	3681	25,767	0	5	0.019
PND 7-34	256	7,936	0	0	0.000
PND 14 or 21	376	376	0	0	0.000
PND 14-41	1800	50,400	0	0	0.000
PND 19-21	212	636	2	0	0.314
PND 21-34	520	14,560	0	0	0.000
PND 22-41	285	5,700	0	0	0.000
All ages combined	**7386**	**112,543**	**6**	**5**	**0.010**

[a] Intubation errors were confirmed at necropsy by observation of a perforated esophagus.
[b] Suspected intubation errors resulted in animal death, with reddened esophagus and/or foamy contents in trachea and/or lungs noted at necropsy.

Source: Data were collected at WIL Research Laboratories, LLC (9/22/00 – 6/10/04).

litter are exposed simultaneously to the test article during early postnatal life of the pups until weaning. Therefore, specialized whole-body inhalation chambers must be used for the animals during exposure (refer to Figure 8.9). If the pups are exposed without the dam, the temperature of the inhalation chamber must be maintained appropriately since neonatal rats are inefficient at regulating core body temperature, although this is not recommended early in lactation. The inhalation route of administration for juvenile rodents must be whole-body until the animal reaches sufficient size to make head- or nose-only inhalation possible. The whole-body route of exposure introduces the possibility that the juvenile and/or the maternal animal (if concurrently exposed) may also be exposed orally (during grooming) and dermally. Finally, neonatal animals in these modified inhalation chambers may be exposed to lower than desired concentrations of the test article as a result of filtration when nursing.[77] In contrast, exposure by the inhalation route presents fewer technical difficulties when using larger species because they may be exposed via head- or nose-only methodology, with some acclimatization.

Less common routes of administration, such as intravenous, present technical challenges that are related to animal size. It is technically more feasible to administer a test article intravenously to a puppy or a piglet than to a rodent pup. There are more sites into which the dose may be administered in a larger animal (e.g., jugular, saphenous, marginal ear, or femoral vein) than in a rodent. If intravenous administration is required and the animal of choice is the rodent, the investigator may have to compromise by using older animals at the onset of dose administration. Continuous infusion would not be possible in rodent juvenile studies that initiate prior to weaning. Moreover, the logistical challenges involved in continuous infusion in any species may be daunting given the potential need to provide new catheters at multiple intervals throughout the study to compensate for growth of the animal.

The dermal route is another option for administration of the test article to the juvenile, although International Life Sciences Institute (ILSI) does not recommend this route for juvenile rodents because of architectural differences between rodent and human skin.[77] The use of this route should be limited to the postweaning period to prevent unintended oral administration to the dam during grooming and/or maternal rejection of the offspring because of changes in olfactory cues caused by application of the material. Moreover, if siblings are gang-housed after weaning, a period of separation may be required for those studies requiring dermal applications. Finally, to prevent ingestion of the test article, Elizabethan collars (see Chapter 7, Figure 7.5A) may be required, and the size of these collars should be monitored closely and appropriately adjusted to accommodate the growth of the animal during the dose administration period.

Other routes of administration that are less commonly employed in nonclinical juvenile studies are subcutaneous, intramuscular, and intraperitoneal injection. These routes are adaptable to the juvenile animal, depending upon the age and species of the model.

2. Frequency and Duration of Exposure

The frequency of dose administration prior to weaning will most often be once daily, with dosages based on daily body weight determination. Previous studies in adult animals, which may include kinetic information, will aid in developing the most appropriate regimen for the test article. However, the researcher is cautioned not to overlook differences in absorption, distribution, metabolism, and elimination (ADME) profiles between juvenile and adult animals, such as differences in the concentration and/or activity of key metabolic pathways. Carefully designed pharmacokinetic studies in juvenile rodents have revealed marked differences in clearance rates, peak plasma concentrations, and area under the curve (AUC) values for test articles when administered at different ages in the same model.[92] These differences clearly demonstrate that developmental changes occurring in metabolic pathways may result in significant quantitative differences in systemic exposure and influence the selection of exposure frequency.

In addition, the treatment period must be considered when juvenile toxicity studies are designed. Consideration should be given to covering the entire period of growth, from the neonate to the sexually mature animal. In rodent models, test article administration from weaning until or beyond the age when mating would occur may be required. Alternatively, administration may begin at approximately PND 4 to 7, which is roughly equivalent to a preterm human infant, and continue until just after the males enter puberty, at approximately 6 to 7 weeks of age. At a minimum, this treatment period, coupled with the adult general and reproductive toxicity studies that are already required, would encompass all periods of the animal's life span. If there is concern that a test article may reveal oncogenic potential as a result of juvenile (or even prenatal) exposure, a subset of animals from a pre- and postnatal development study may be selected to become part of a carcinogenicity study, with dose administration beginning immediately following weaning. Dose administration throughout growth may be impractical in large animal species (longer duration and higher cost), for which the entire growth and development period may span several months to years. At a minimum, in situations where a large animal model is the species of choice, the period of dose administration for a general screening study should cover the period of target organ system development (if known).

A recovery (or posttreatment) period may be included in the design of the study. For some test articles, changes in functional parameters are expected because of the pharmacological action. Therefore, there may be less concern regarding acute effects of the test article (e.g., locomotor activity changes immediately following administration of a sedative) and more concern about withdrawal, persistent, and/or latent effects. In addition, when the posttreatment period contains the same assessments as those in the treatment period, comparisons can be made regarding the severity of the changes, persistence of these changes, and their relationship to the expected pharmacology of the test article. If functional assessments are not included in the treatment period, interpretation of findings during the posttreatment period becomes much more difficult. In this scenario, responses in the juvenile study must be compared with those that may have been observed in studies with adult animals.

3. Dose Selection

For the general toxicity screen, the draft FDA guidance document recommends that the high dose produce frank toxicity and that a no-observed-adverse-effect level be identified, if possible.[51] However, when enzymatic pathways involved in metabolism of the test article are immature, different sensitivities may be observed between the adult and the juvenile. Moreover, functional changes in developing organ systems can create periods of greater sensitivity in the juvenile, with resulting toxicities that would not be evident in the adult. Therefore, dose levels used for adult animals may not be applicable to juvenile animals. As an example, the benzodiazepines produce paradoxical responses in opposition to their anxiolytic properties, including convulsions.[112] Those convulsions were characterized in preweaning rats and were postulated to result from differences in the Type 1 and Type 2 benzodiazepine receptor sites expressed in immature and mature rats.

Another example of apparent differential sensitivity between adults and juveniles relating to enzyme system competence is that of the pyrethroids.[113] As an example, the LD_{50} for cypermethrin (a type II pyrethroid) is far lower in juvenile rats than in adults of the same strain. However, when tri-o-tolyl phosphate was used to inhibit drug metabolism prior to treatment in the adult to mimic the inherently lower esterase level in the juvenile rat, the observed LD_{50} at both developmental stages became similar.[114] An interesting age-related manifestation of toxicity of a type II pyrethroid evaluated in the authors' laboratory in the juvenile rat is a gait abnormality that was termed "carangiform ataxia." This abnormality is a fish-tailing gait wherein the forepaws pedal in a forward direction while the posterior third of the body (behind the rib cage) rapidly oscillates in a fishlike motion with maximal lateral abdominal flexion. Carangiform ataxia is temporally limited and generally corresponds to the time of locomotor function development in the rat during which pivoting is observed (approximately PND 6 to 11).[115] This age-specific manifestation of pyrethroid toxicity is speculated to result from the limited neural development in the hindquarters of the rat at this age; there is no known adult correlate. This gait abnormality was observed in juvenile rats at a dose level approximately 10-fold lower than the dose level that produced maternal toxicity, emphasizing both the apparent sensitivity of juveniles and the possibility of unique manifestations of toxicity based on physiological development. For current thinking and future directions in the study of developmental neurotoxicity of pyrethroid insecticides, the reader is referred to a recent review article by Shafer et al.[116]

Dose selection in a general sense requires careful consideration of adult nonclinical data, data from analogs, existing human exposure data, and knowledge of the organ systems that are developing during the proposed period of dose administration. In addition, well-designed range-finding studies are essential for the selection of acceptable dose levels. In these studies, gross changes in body weight, food consumption, and clinical chemistry, and macroscopic changes in organ structure and weight can be evaluated. Although less commonly included in dose range-finding studies in adult animals, pharmacokinetic profiling of the test article can be extremely valuable in the selection of dose levels for the definitive juvenile study. The pharmacokinetic profile (e.g., C_{max}, AUC) in the juvenile after direct administration can be compared with previously developed profiles in adult animals. Any potential differences that are identified can provide guidance as to the modifications in the dosage levels, dosage regimen, vehicle, etc., that may be required. Moreover, to date many nonclinical juvenile studies are being conducted after some initial data have been developed in the human. In these circumstances, inclusion of a pharmacokinetic phase in the dose range-finding study will guide the researcher to select dose levels that will more accurately reflect the human exposure scenario or that may even indicate the appropriateness of the species as a model for the human situation.

As stated previously, the FDA draft guidance document recommends that the high dose should produce frank toxicity.[51] One way to ensure that toxicity is observed is by reaching the maximum tolerated dose (MTD). Since clinical studies in the pediatric population will only characterize the efficacy of the drug, and perhaps some side effects, reaching an MTD in the nonclinical juvenile study will allow for an examination of the range of potential adverse outcomes following exposure. However, several challenges may be imposed by the use of the MTD. For example, reaching an MTD in an adult study will result in toxicity that may be limited to a single organ system. Conversely, administration of the test article to a juvenile animal population at a dose level at or above the MTD may not only result in an adverse outcome in a particular organ or organ system but may also result in confounding growth retardation. The authors have experience using dose levels that were far above the MTD for the juvenile. One example was a developmental neurotoxicity study in rats of methimazole, a drug used to treat hyperthyroidism.[117] Although not strictly a juvenile toxicity study, in this design the test article was administered in the drinking water from gestation day 6 through lactation day 21. Body weight gain was decreased early in the lactational period when pups did not consume water; this effect was exacerbated later in the lactational period when pups began to independently consume water containing the test article. Growth retardation can have a profound influence on the development of many organ systems and on the ability to interpret direct effects of the test article compared with secondary changes resulting from the growth retardation.

D. Organization of Test Groups

Several organizational models are available, each with advantages and disadvantages. These models are based on the principle of the litter as the experimental unit of testing and analysis and include the within-litter, between-litter, fostering, and one pup per sex per litter designs.

1. Within-Litter Design

The within-litter design is based on the idea that all dosage groups should be equally represented in each litter (refer to Figure 8.7 for a schematic representation of a study design using the within-litter approach). Depending on the number of dose levels in a particular study design or the size of the litter, a true within-litter design may be difficult to achieve. In these cases, a modified approach, the split-litter design, may be employed. In such designs, no two same-sex siblings are assigned to the same dosage group, and multiple litters are required to obtain a sample size of one per sex for each group. In either case, the litter effect is distributed across all dosage groups. Refer to Figure 8.12 for a depiction of the distribution of pups in within-litter and split-litter designs.

A	Each Dam			
	Group 1 male pup	Group 2 male pup	Group 3 male pup	Group 4 male pup
	Group 1 female pup	Group 2 female pup	Group 3 female pup	Group 4 female pup
B	Dam 1				Dam 2			
	Group 1 male pup	Group 2 male pup	Group 3 male pup	Group 4 male pup	Group 5 male pup	Group 6 male pup	Group 1 male pup	Group 2 male pup
	Group 1 female pup	Group 2 female pup	Group 3 female pup	Group 4 female pup	Group 5 female pup	Group 6 female pup	Group 1 female pup	Group 2 female pup
C	Each Dam			
	Group 1 male pup	Group 1 male pup	Group 1 male pup	Group 1 male pup
	Group 1 female pup	Group 1 female pup	Group 1 female pup	Group 1 female pup

Figure 8.12 Distribution of pups in various study designs. (a) For within-litter designs, each treatment group is represented in each litter by one male and one female pup. (b) For split-litter designs, each treatment group is represented by no more than one male and one female pup per litter, but all groups may not be represented in each litter. (c) For between-litter designs, all pups in the litter are assigned to the same treatment group.

Considering that the response to a test article is more alike in siblings than in offspring from different litters,[105] the within-litter approach provides the advantage of exposing genotypically similar offspring to different levels of the test article. Another advantage of the within-litter design is that litter-specific maternal and environmental factors are appropriately distributed across dosage groups. This design also provides ethical and practical advantages, in that it reduces the number of animals (dams with litters) required for the study and minimizes the time required for dose administration and data collection.

However, the within-litter design has some key disadvantages, including the possibility of cross-contamination because of exposure to the test article or metabolite when administration begins prior to weaning. For example, siblings assigned to the control or lower dosage groups may consume feces from siblings assigned to higher dose groups, or they may be dermally exposed to the test article through contact with contaminated bedding. Also, the dam may be inadvertently exposed to the test article while grooming the neonate (or if she has consumed one of her young), and the offspring may then be exposed to the test article while nursing.

2. Between-Litter Design

Rather than representing all dosage groups in the litter, the between-litter design assigns the entire litter to the same dosage group (refer to Figure 8.12). Compared with the within-litter design, this is a very straightforward design in principle. Between-litter designs have the advantage of decreased chances of cross-contamination of dosage levels and simplification of dose administration procedures. Figure 8.13 presents a schematic representation of a study design using the between-litter approach.

For the between-litter design, it is expected that the parental genetics and maternal care would be similar for all offspring in the litter. A disadvantage of this design is that it requires a large number of animals. In a rodent juvenile toxicity screen with a sample size of 25 per group, the number of progeny being dosed each day can exceed 1500, depending on whether standardization of litters is performed. This alone can create a logistically challenging study design. When the endpoints of analysis are also considered, these studies may become too demanding for many laboratories to conduct properly.

Between-litter designs are incorrectly implemented when the litter effect is ignored (i.e., the pup is considered the statistical unit instead of the litter). For a detailed exposition of the litter effect in statistical analysis, refer to Chapter 9 of this text. For example, when incorrectly applied, each sibling is assigned to the same dosage group as a separate unit for statistical analysis. Rather than decreasing the effect of the litter, incorrect use of the between-litter design can enhance the effect of the litter on the observed response. For this reason, between-litter designs should only be used when the researcher plans to cull the litter with increasing age, when data from littermates are used to obtain a litter mean at those times when the entire litter has been examined for a particular endpoint, or when littermates are selected for different endpoints.

Few studies have been designed to specifically evaluate the same endpoints by use of both the within-litter and between-litter designs. In one such study, pups treated with 9 mg triethyltin/kg from the within-litter design weighed less than those in the between-litter design from PND 14 to 20. In addition, the hyperactivity induced in young adult animals by triethyltin in the within-litter design was higher than that in the between-litter design.[118] Conversely, in another study, 6-hydroxydopamine induced hyperactivity was less in a modified within-litter design (half of pups treated) than it was in a between-litter design (all pups treated). In addition, 6-hydroxydopamine-treated pups in the within-litter design demonstrated better avoidance than those in the between-litter design.[119] Although these results do not conclusively demonstrate that the within-litter design is more scientifically relevant, they do illustrate that different results may be obtained depending on the organization of the test groups. No definitive regulatory guidance has been provided to indicate which approach should be used.

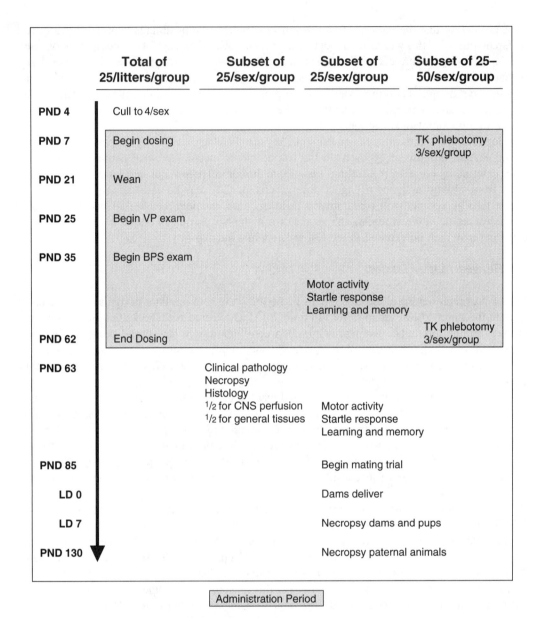

Figure 8.13 Between-litter study design. A robust between-litter design requires approximately 100 litters obtained via in-house mating or via time-mated animals.

3. Fostering Design

The fostering approach has been employed by some researchers in an effort to minimize the effect of the litter and genetic bias on the study. In this design, offspring are fostered to new mothers immediately after birth. Although on the surface this approach may seem relatively simple, it can become difficult to track and assess the effects of genetics or gestational influences on the observed responses to a test article. A true fostering study does not merely replace the litter of one female with that of another. Instead, it replaces the original litter of one female with one offspring from each of several litters of other females (Figure 8.14). Fostering also requires that a sufficient number of litters be born on the same day. If insufficient litters are available for fostering on a given day,

	A: Prior to Fostering			B: After Fostering	
Dam A	**Dam B**	**Dam C**	**Dam A**	**Dam B**	**Dam C**
A1 A2 A3 A4 A5 A6 A7 A8	B1 B2 B3 B4 B5 B6 B7 B8	C1 C2 C3 C4 C5 C6 C7 C8	B1 C2 D3 E4 F5 G6 H7 I8	C1 D2 E3 F4 G5 H6 I7 A8	D1 E2 F3 G4 H5 I6 A7 B8
Dam D	**Dam E**	**Dam F**	**Dam D**	**Dam E**	**Dam F**
D1 D2 D3 D4 D5 D6 D7 D8	E1 E2 E3 E4 E5 E6 E7 E8	F1 F2 F3 F4 F5 F6 F7 F8	E1 F2 G3 H4 I5 A6 B7 C8	F1 G2 H3 I4 A5 B6 C7 D8	G1 H2 I3 A4 B5 C6 D7 E8
Dam G	**Dam H**	**Dam I**	**Dam G**	**Dam H**	**Dam I**
G1 G2 G3 G4 G5 G6 G7 G8	H1 H2 H3 H4 H5 H6 H7 H8	I1 I2 I3 I4 I5 I6 I7 I8	H1 I2 A3 B4 C5 D6 E7 F8	I1 A2 B3 C4 D5 E6 F7 G8	A1 B2 C3 D4 E5 F6 G7 H8

Figure 8.14 Fostering procedure. (A) Prior to fostering; a total of at least nine litters of eight pups on one day are required for a fostering procedure that would create an entirely new litter. B. After fostering; no pups remain with original dams and no siblings are present in the fostered litter.

then all the litters born on that day cannot be fostered. Therefore, careful planning must be used to ensure efficient use of litters. Finally, fostering does not remove the effect of the litter entirely. Rather, the care provided by the foster dam is as important to the litter effect as that provided by the birth dam. Thus, using the fostering approach does not entirely alleviate concerns of using the between-litter design, although genetic factors should be evenly distributed. The authors have conducted several studies using the fostering design in rats and have observed that an environmental litter effect was imposed on the fostered litter such that pups in the fostered litter gained weight at a similar rate. If the within-litter design is to be used, the fostering approach is not recommended because it offers very little improvement in control of bias but increases the complexity of the study. Moreover, for logistical reasons, the fostering approach is not recommended for studies with large numbers of litters.

4. One Pup per Sex per Litter Design

The final organizational design that is used in juvenile toxicity studies is the selection of one pup per sex per litter. This design is the preferred approach for selecting a subset of offspring from a pre- and postnatal development study. This approach eliminates the chance for cross-contamination (if the animals are group-housed) and avoids the large number of animals that would be dosed in a between-litter design.

E. Potential Parameters for Evaluation

1. Growth and Development

The FDA's current draft guidance states that nonclinical juvenile studies should include appropriate growth measurements, such as body weight, growth velocity, crown-rump length, tibia length, and organ weights.[51] The utility of body weight and body weight change as sensitive measures of toxicity has been previously described in other texts and therefore is not discussed in detail here.

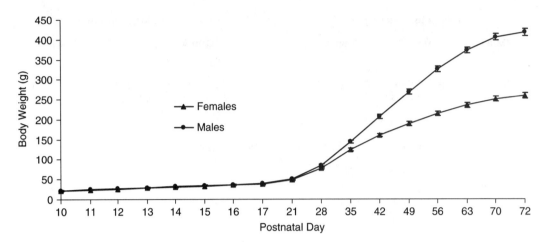

Figure 8.15 Body weight development of the Crl:CD®(SD)IGS BR rat. Each body weight interval represents the mean weight (± S.E.M.) of litters by sex (n = 20 to 25 litters per interval) collected at WIL Research Laboratories, LLC (2003–2004).

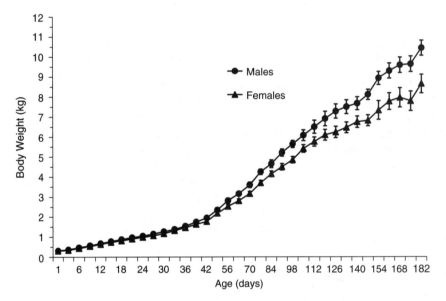

Figure 8.16 Body weight development of the beagle dog. Each body weight interval represents the mean weight (±S.E.M.) of litters by sex (n = 11 to 16 litters per interval) collected at WIL Research Laboratories, LLC (2001–2002).

An important factor regarding body weight change in juvenile study designs is the relatively rapid growth during the early postnatal period (Figure 8.15 and Figure 8.16 illustrate growth curves for rats and dogs, respectively). Body weight in the rat pup increases by approximately twofold between PND 1 and 7 and approximately threefold between PND 7 and 21. In the postweaning period, male rat body weight increases twofold between approximately PND 28 and 42 and again between PND 42 and 70, after which body weight gain is not so dramatic. Female rat postweaning body weight doubles between approximately PND 28 and 49, but increases only approximately 30 to 40% between PND 49 and 70. Consequently, body weight should be measured frequently prior to 10 weeks of age (at least twice weekly) to evaluate subtle changes that may not be detected by use of longer intervals. Furthermore, frequent body weight measurements will allow more accurate dosage calculations during rapid periods of growth.

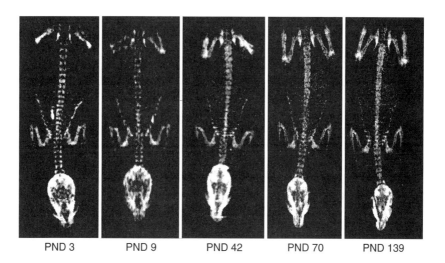

Figure 8.17 Examples of full-body DEXA scans of anesthetized beagle dogs from each postpartum developmental age category (WIL Research Laboratories, LLC, 2001–2002).

One example illustrating differential toxicity on growth and development with age and gender comes from a study of methylphenidate. When administered to female rats from PND 5 through 24 via subcutaneous injection (s.c.) once daily (35 mg/kg/day), the reductions in mean body weight, femur length, and pituitary weight were approximately 9%, 5%, and 16%, respectively, and were statistically significant compared with controls on PND 25.[120] After a 30-day recovery period, mean body weight, femur length, and pituitary weight in the methylphenidate-treated females were similar to those of control animals. These data in females were consistent with data from previous studies in male rats.[121,122] However, no such effects were observed in PND 55 female rats treated from PND 35 through 54.[120] When methylphenidate was administered similarly (35 mg/kg/day, s.c.) from PND 35 through 54 to male rats, a statistically significant decrease in mean body weight of approximately 10% by PND 55 was seen (no effect was observed on mean femur length or pituitary weight).

In addition, evaluation of the overall growth and/or function of organ systems that develop postnatally (e.g., skeletal, renal, pulmonary, neurological, immunological, and reproductive systems) is recommended by the FDA guidance document.[51] As suggested above, additional direct growth measurements, such as crown-rump length or long bone length (tibia or femur), can be assessed longitudinally. Bone growth can be evaluated by noninvasive techniques, such as dual energy x-ray absorptiometry (DEXA) or computerized tomography (refer to Figure 8.17 for an example of DEXA beagle full-body scans). A recent review comparing important developmental milestones demonstrated that patterns of postnatal skeletal growth and development between humans and some animal models are similar.[123] Additional measures of growth, including craniofacial dimensions, femur length, and body length following exposure to radiation and/or chemotherapeutic compounds have been described in an animal model.[4]

Changes in body composition may be considered in addition to any changes in body weight in juvenile animals (see Table 8.5 for beagle reference data). Total body composition (including bone mineral density and content, lean mass, and total body fat) can also be evaluated longitudinally through imaging techniques such as DEXA. Water content is very high during early prenatal development, decreasing perinatally and with advancing age (refer to Figure 8.5 and Figure 8.6).[95] Consequently, preweaning animals have higher water content and less fat than mature animals. This can result in a larger volume of distribution for water soluble compounds than in older animals.[1] With respect to fat content, only the guinea pig and humans begin depositing fat prenatally (Figure 8.4).[95]

Table 8.5 Total body composition by DEXA for beagle dogs

Parameter[b]		3	9	21	30	Postnatal Day[a] 42	70	115	139
BMD (g/cm^2)	M	0.23 ± 0.014 (9)	0.32 ± 0.016 (15)	0.38 ± 0.033 (14)	0.41 ± 0.036 (14)	0.47 ± 0.031 (14)	0.56 ± 0.033 (14)	0.67 ± 0.027 (14)	0.7 ± 0.028 (10)
	F	0.24 ± 0.011 (8)	0.33 ± 0.019 (16)	0.39 ± 0.025 (14)	0.42 ± 0.029 (14)	0.47 ± 0.023 (14)	0.55 ± 0.02 (14)	0.64 ± 0.018 (14)	0.66 ± 0.049 (10)
BMC (g)	M	3.2 ± 1.07 (9)	5.2 ± 2.37 (15)	15.6 ± 6.67 (14)	25.6 ± 9.13 (14)	44.9 ± 11.89 (14)	112 ± 25.96 (14)	197.2 ± 32.49 (14)	244.3 ± 35.17 (10)
	F	3.3 ± 0.87 (8)	5.2 ± 1.55 (16)	15.6 ± 5 (14)	24.6 ± 6.9 (14)	43.6 ± 8.56 (14)	104.1 ± 15.49 (14)	171.2 ± 24.76 (14)	213.9 ± 29.49 (10)
Soft tissue (g)	M	410 ± 65.7 (9)	658 ± 141 (15)	1066 ± 237.5 (14)	1374 ± 305.6 (14)	2123 ± 397.9 (14)	3881 ± 694.2 (14)	6227 ± 1059.6 (14)	6952 ± 991.8 (10)
	F	406 ± 49.7 (8)	654 ± 102.6 (16)	1077 ± 210.4 (14)	1319 ± 236.4 (14)	2027 ± 320.1 (14)	3599 ± 439.8 (14)	5512 ± 849.9 (14)	6227 ± 856.6 (10)
Lean (g)	M	376 ± 56.3 (9)	602 ± 120.8 (15)	965 ± 190.9 (14)	1258 ± 257.7 (14)	1881 ± 306.9 (14)	3215 ± 536.6 (14)	5607 ± 786.9 (14)	6263 ± 739.3 (10)
	F	374 ± 41.5 (8)	600 ± 94.7 (16)	966 ± 170.8 (14)	1172 ± 260.2 (14)	2511 ± 2831 (14)	3056 ± 348 (14)	5090 ± 685.1 (14)	5739 ± 698.6 (10)
Fat (g)	M	35 ± 17.4 (9)	57 ± 31.7 (15)	101 ± 65 (14)	117 ± 62.3 (14)	242 ± 128 (14)	686 ± 177.2 (14)	621 ± 323.2 (14)	689 ± 313.9 (10)
	F	32 ± 14 (8)	54 ± 22.5 (16)	111 ± 62.4 (14)	118 ± 68.2 (14)	231 ± 117.5 (14)	542 ± 128.8 (14)	422 ± 194.5 (14)	488 ± 210.7 (10)
Fat (%)	M	8.3 ± 3.3 (9)	8.2 ± 3.7 (15)	8.9 ± 4.06 (14)	8.1 ± 2.93 (14)	10.9 ± 4.37 (14)	17.4 ± 2.15 (14)	9.5 ± 3.38 (14)	9.5 ± 3.21 (10)
	F	7.9 ± 2.79 (8)	8.1 ± 3.29 (16)	9.9 ± 4.07 (14)	8.5 ± 3.52 (14)	11.1 ± 4.67 (14)	15 ± 2.27 (14)	7.3 ± 2.33 (14)	7.5 ± 2.54 (10)
DEXA BW (g)	M	414 ± 66.5 (9)	663 ± 142.9 (15)	1081 ± 243.5 (14)	1400 ± 313.9 (14)	2168 ± 408.5 (14)	3993 ± 718.8 (14)	6425 ± 1088.8 (14)	7196 ± 1024 (10)
	F	410 ± 50.4 (8)	659 ± 103.5 (16)	1093 ± 214.8 (14)	1343 ± 242.2 (14)	2071 ± 328.2 (14)	3703 ± 453.8 (14)	5683 ± 872.7 (14)	6441 ± 882.3 (10)
Scale BW (g)	M	351 ± 56.7 (9)	572 ± 118.2 (15)	990 ± 223.9 (14)	1232 ± 271.9 (14)	1951 ± 353.5 (14)	3595 ± 677.6 (14)	6492 ± 1190.8 (14)	7363 ± 1099.8 (10)
	F	347 ± 50.7 (8)	570 ± 87.2 (16)	993 ± 194.8 (14)	1215 ± 224.2 (14)	1817 ± 241 (14)	3265 ± 409.7 (14)	5782 ± 965 (14)	6641 ± 921.9 (10)
BW diff. (%)	M	15.2 ± 2.27 (9)	13.4 ± 5.68 (15)	8.4 ± 1.4 (14)	11.4 ± 4.78 (14)	9.2 ± 7.35 (14)	10.1 ± 3.93 (14)	−0.9 ± 2.79 (14)	−2.2 ± 1.75 (10)
	F	15.4 ± 5.3 (8)	13.3 ± 5.01 (16)	9.2 ± 1.38 (14)	9.6 ± 1.9 (14)	11.8 ± 4.7 (14)	11.8 ± 2.45 (14)	−1.6 ± 2.8 (14)	−3.2 ± 2.38 (10)

[a] Values are expressed as: mean ± SD (N = number of litters).
[b] BMD, bone mineral density; BMC, bone mineral content; Soft tissue, soft tissue [lean (g) + fat (g)]; DEXA BW, body weight calculated from the DEXA total body scan [soft tissue (g) + BMC (g)]; Scale BW, body weight recorded from a scale just prior to scanning; BW Diff., percent difference between the DEXA calculated body weight and the scale body weight [DEXA BW (g) − Scale BW (g)]×100, M, male; F, female.

Source: Data were collected at WIL Research Laboratories, LLC (2001–2002).

2. Food Consumption

Monitoring food consumption in juvenile studies is a general measure of animal health and a potential indicator of toxicity, just as it is in other bioassays. During the preweaning period, it is not typical to evaluate food consumption for individual offspring. If animals are single-housed after weaning, food consumption can be reliably evaluated. As shown in Figure 8.18, food consumption, starting on PND 28, increases throughout the growth phase in the rat, if calculated on a gram per animal per day basis (although females appear to plateau around PND 39). Interestingly, food consumption in the rat actually decreases over time if compared relative to body weight (refer to Figure 8.18), with remarkable similarity between males and females. This decrease in food consumption, when calculated on a gram per kilogram body weight basis, has also been observed in humans.[94]

Species-specific considerations should be made in the design and interpretation of food consumption data. One consideration is that feeding patterns in rodents are linked to circadian rhythms, thus most of the food intake occurs during the dark cycle. In contrast, humans, minipigs, nonhuman primates, and dogs are not influenced in this manner.

3. Serum Chemistry and Hematology

The current FDA draft guidance for juvenile studies states that clinical pathology can be useful,[51] but these evaluations may be limited by the technical feasibility of obtaining adequate sample volumes for analysis, particularly in the case of rodents. To evaluate a standard clinical chemistry panel prior to PND 28 in the rat, researchers must use terminal blood collections, and samples may have to be pooled to obtain adequate sample sizes in rats younger than PND 17. A review of rat and dog standard serum chemistry panels from juvenile animals reveals specific patterns of change throughout postnatal development (Table 8.3 and Table 8.6).[92,124] For example, in both the rat and the dog, cholesterol, total bilirubin, and serum urea nitrogen levels decreased markedly with age. However, alanine aminotransferase (ALT) increased by approximately 50% in beagles during the first 20 weeks of life, while in the rat, ALT increased by approximately 300% from 2 to 5 weeks of age.

Prior to investigating effects of a test article on clinical pathology parameters, as much information as possible about the ontogeny of these parameters should be gathered. For example, if a chemical inducing α-2 urinary globulin accumulation were under investigation in a juvenile male rat, the researcher should be cognizant that α-2 urinary globulin is not produced by the liver in substantial amounts until the time of puberty.[125–126] In addition to evaluation of standard enzyme panels, specific enzymes associated with a known mode of action may be evaluated. One example is cholinesterase activity, which increases dramatically from birth until PND 42 in the rat brain (Figure 8.19).[88] An understanding of the ontogeny of this type of enzyme system can be critical in the design and interpretation of nonclinical juvenile toxicity studies to evaluate compounds that target specific enzyme systems.

Reference hematology values at defined developmental intervals should also be considered when designing a juvenile study, especially if the class of compounds has known effects on hematopoeisis or the developing immune system. There is a lack of hematology reference values at different developing time points for small laboratory animals (e.g., mice and rats) because of the technical challenges associated with collecting enough terminal blood from individual animals to evaluate a full panel of hematology parameters for reasonable interpretation. For larger animals (e.g., dogs, nonhuman primates, and pigs) the technical challenges of obtaining the blood volumes required for a full hematology panel are not so great. A review of the standard hematology panel from juvenile dogs reveals some specific developmental differences from birth to adolescence (refer to Table 8.7).[124] At birth, reticulocytes are larger and more numerous than those of adults, as indicated by the higher mean reticulocyte counts and mean corpuscular volumes (MCV) during

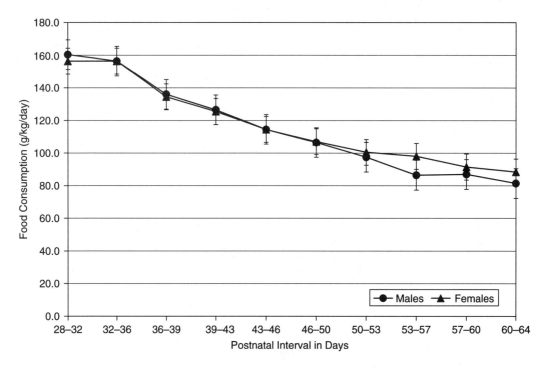

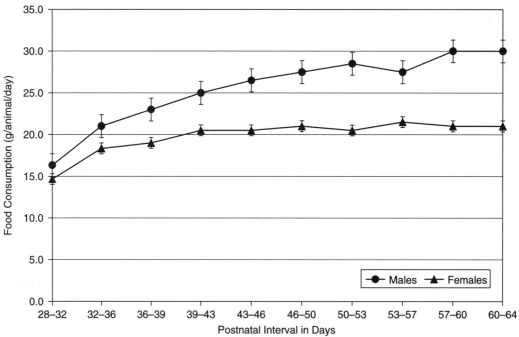

Figure 8.18 Food consumption in juvenile Crl:CD®(SD)IGS BR rats during development (PND 28 through PND 64).

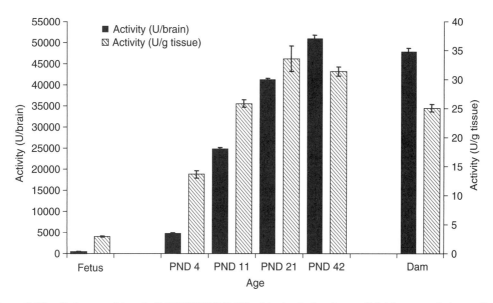

Figure 8.19 Ontogeny of female Crl:CD®(SD)IGS BR rat brain cholinesterase (fetal tissue pooled regardless of sex). Data were collected at WIL Research Laboratories, LLC.

the first month following birth. As neonatal dogs develop from birth to 3 months of age, these values decline to normal adult ranges. In contrast, many of the other hematology values (neutrophil, lymphocyte, and platelet counts) have similar values from birth through adolescent development.

4. Macroscopic and Microscopic Evaluations

Macroscopic and microscopic evaluations are also recommended by the FDA draft guidance document for nonclinical juvenile studies.[51] The importance of gross examinations is similar to that for standard toxicity studies, although with less emphasis on chronic lesions. Macroscopic changes consistent with hypoplasia may correlate with changes in growth in juvenile studies. One example is a change in the relative size of organs or structures such as the kidneys, liver, testes, or bones. If there are known target organs, based on data from adult animals or humans, these organs should be examined in the juvenile. However, histopathology varies with stage of organ maturity (e.g., reproductive organs, such as the testes), and these differences between developing and adult animals must be considered. As an example, basophilic tubules observed in regenerating tubular epithelium in an adult kidney are considered evidence of ongoing degeneration of the renal tubule and are considered an adverse finding.[127] However, in a juvenile study of the angiotensin converting enzyme (ACE) inhibitor, ramipril, the finding of basophilic tubules on PND 17 and 28 was noted for both well-developed tubules and for clusters of relatively undeveloped tubules in both control and treated animals.[92] In that study, the adverse change caused by ramipril was the increased incidence of basophilic tubules in the treated animals, potentially suggesting a delay in development of the kidneys of these animals. Another ACE inhibitor was also shown to cause arrested maturation of the developing kidney (immature glomeruli, distorted and dilated tubules, and relatively few and short, thick arterioles) when administered to newborn rats.[128]

5. Physical and Sexual Developmental Landmarks and Behavioral Assessments

The assessment of growth can involve evaluation of a variety of different parameters, including body weights and landmarks of physical and sexual development. Physical developmental landmarks are commonly included as endpoints in reproductive and developmental toxicity studies

Table 8.6 Serum chemistry values for beagle dogs

Parameter[b]	7	14	21	28	42	60	85	108	135
ALB (g/dl)									
M	2.3 ± 0.22 (8)	2.3 ± 0.13 (8)	2.5 ± 0.17 (9)	2.9 ± 0.17 (7)	3 ± 0.13 (7)	2.9 ± 0.13 (7)	3 ± 0.12 (7)	3 ± 0.12 (7)	3.1 ± 0.13 (7)
F	2.3 ± 0.1 (9)	2.3 ± 0.1 (9)	2.5 ± 0.16 (10)	2.9 ± 0.17 (8)	3 ± 0.15 (8)	3 ± 0.14 (8)	3.1 ± 0.09 (8)	3.1 ± 0.1 (8)	3.2 ± 0.13 (8)
ALP (U/l)									
M	223 ± 50.3 (8)	199 ± 38.2 (9)	168 ± 7.5 (7)	178 ± 12.3 (7)	232 ± 34.1 (7)	246 ± 55.3 (7)	253 ± 29 (7)	209 ± 27.4 (7)	167 ± 19.6 (7)
F	214 ± 45 (9)	183 ± 53.3 (10)	168 ± 16.7 (8)	167 ± 12 (8)	217 ± 41.8 (8)	247 ± 51.6 (8)	253 ± 22.5 (8)	213 ± 26.6 (8)	168 ± 25.9 (8)
ALAT (U/l)									
M	28 ± 17.2 (8)	15 ± 3.1 (9)	17 ± 2.6 (7)	24 ± 5.3 (7)	36 ± 11.9 (7)	33 ± 5.7 (7)	37 ± 6.7 (7)	39 ± 7.5 (7)	40 ± 3.2 (7)
F	24 ± 11.5 (9)	17 ± 6.4 (10)	17 ± 3.5 (8)	21 ± 7.5 (8)	32 ± 13.7 (8)	31 ± 6.3 (8)	38 ± 5.5 (8)	39 ± 7.4 (8)	41 ± 2.9 (8)
ASAT (U/l)									
M	53 ± 25 (8)	34 ± 11.7 (9)	29 ± 6.2 (7)	28 ± 6.4 (7)	30 ± 1.8 (7)	34 ± 8.3 (7)	35 ± 12.1 (7)	36 ± 7 (7)	30 ± 3.3 (7)
F	49 ± 26.6 (9)	31 ± 10.6 (10)	28 ± 8 (8)	23 ± 6.5 (8)	29 ± 4.6 (8)	32 ± 6 (8)	31 ± 7.9 (8)	34 ± 3 (8)	31 ± 4 (8)
TOT. BIL (mg/dl)									
M	0.7 ± 0.22 (8)	0.5 ± 0.28 (8)	0.4 ± 0.11 (9)	0.4 ± 0.13 (7)	0.2 ± 0.07 (7)	0.2 ± 0.06 (7)	0.2 ± 0.03 (7)	0.1 ± 0.05 (7)	0.1 ± 0.03 (7)
F	0.9 ± 0.49 (9)	0.5 ± 0.24 (9)	0.4 ± 0.06 (10)	0.3 ± 0.1 (8)	0.3 ± 0.11 (8)	0.2 ± 0.06 (8)	0.2 ± 0.08 (8)	0.2 ± 0.04 (8)	0.2 ± 0.03 (8)
UREA N (mg/dl)									
M	30 ± 17 (8)	21 ± 5.5 (9)	16 ± 1.6 (7)	15 ± 2.6 (7)	13 ± 3.3 (7)	11 ± 2.4 (7)	15 ± 4.7 (7)	15 ± 5.1 (7)	11 ± 2 (7)
F	26 ± 3.7 (9)	21 ± 4.6 (10)	16 ± 3.2 (8)	14 ± 1.4 (8)	13 ± 2.9 (8)	11 ± 3 (8)	13 ± 1.8 (8)	14 ± 3.5 (8)	12 ± 2.9 (8)
CHOL (mg/dl)									
M	285 ± 128.8 (8)	218 ± 46.1 (9)	233 ± 41.9 (7)	239 ± 46 (7)	156 ± 36.3 (7)	127 ± 14 (7)	143 ± 10.8 (7)	166 ± 10.5 (7)	183 ± 10.7 (7)
F	249 ± 39.4 (9)	241 ± 39.7 (10)	263 ± 49 (8)	259 ± 42.1 (8)	171 ± 37.5 (8)	126 ± 11.7 (8)	143 ± 19.4 (8)	166 ± 16.5 (8)	174 ± 14.7 (8)
CREAT (mg/dl)									
M	0.1 ± 0.061 (8)	0.06 ± 0.046 (9)	0.03 ± 0.039 (7)	0.04 ± 0.045 (7)	0.06 ± 0.059 (7)	0.14 ± 0.057 (7)	0.14 ± 0.073 (7)	0.19 ± 0.099 (7)	0.29 ± 0.078 (7)
F	0.08 ± 0.067 (9)	0.06 ± 0.066 (10)	0.05 ± 0.039 (8)	0.04 ± 0.048 (8)	0.07 ± 0.051 (8)	0.16 ± 0.044 (8)	0.13 ± 0.057 (8)	0.21 ± 0.065 (8)	0.32 ± 0.079 (8)

Parameter									
GLT (U/L)									
M	77.4 ± 44.5 (8)	16.7 ± 8 (8)	5.4 ± 0.91 (9)	3.3 ± 0.73 (7)	2.8 ± 0.73 (7)	3 ± 0.6 (7)	3.8 ± 0.35 (7)	3.2 ± 0.41 (7)	3.7 ± 0.58 (7)
F	58.8 ± 33.66 (9)	18.2 ± 7.68 (9)	6.1 ± 1.68 (10)	3.5 ± 1.05 (8)	2.6 ± 0.57 (8)	3.2 ± 0.76 (8)	7.4 ± 10.15 (8)	3.4 ± 0.5 (8)	3.5 ± 0.68 (8)
GLOB (g/dl)									
M	1.6 ± 0.08 (8)	1.4 ± 0.08 (8)	1.4 ± 0.09 (9)	1.3 ± 0.13 (7)	1.4 ± 0.1 (7)	1.9 ± 0.25 (7)	2.2 ± 0.13 (7)	2.1 ± 0.13 (7)	2 ± 0.16 (7)
F	1.5 ± 0.13 (9)	1.4 ± 0.14 (9)	1.4 ± 0.11 (10)	1.4 ± 0.13 (8)	1.4 ± 0.13 (8)	1.9 ± 0.26 (8)	2.1 ± 0.17 (8)	2.1 ± 0.16 (8)	2 ± 0.06 (8)
GLUC (mg/dl)									
M	120 ± 11.8 (8)	121 ± 7.5 (9)	128 ± 12 (7)	135 ± 13 (7)	131 ± 8.2 (7)	125 ± 10.1 (7)	113 ± 4.8 (7)	116 ± 7.1 (7)	109 ± 5.7 (7)
F	122 ± 13.9 (9)	123 ± 10.7 (10)	129 ± 11.9 (8)	136 ± 10.3 (8)	135 ± 10.5 (8)	126 ± 8.7 (8)	117 ± 5.5 (8)	109 ± 5.9 (8)	106 ± 6.3 (8)
K+ (mEq/l)									
M	5.7 ± 0.19 (8)	5.7 ± 0.44 (8)	5.8 ± 0.44 (9)	5.9 ± 0.49 (7)	6 ± 0.49 (7)	5.7 ± 0.52 (7)	5.4 ± 0.38 (7)	5.4 ± 0.3 (7)	5 ± 0.14 (7)
F	5.8 ± 0.48 (9)	5.7 ± 0.48 (9)	5.8 ± 0.42 (10)	5.9 ± 0.55 (8)	6 ± 0.53 (8)	5.5 ± 0.67 (8)	5.3 ± 0.34 (8)	5.2 ± 0.3 (8)	5.1 ± 0.3 (8)
Na+ (mEq/l)									
M	144 ± 5.7 (8)	144 ± 2.5 (9)	146 ± 1.8 (7)	146 ± 0.8 (7)	149 ± 2 (7)	149 ± 1 (7)	148 ± 1.4 (7)	149 ± 1.2 (7)	
F	142 ± 3.8 (9)	144 ± 3.1 (10)	145 ± 1.8 (8)	147 ± 1.3 (8)	149 ± 2.4 (8)	149 ± 0.8 (8)	149 ± 2 (8)	150 ± 1.8 (8)	
I. PHOS (mg/dl)									
M	9.8 ± 1.13 (8)	9.5 ± 0.5 (8)	9.8 ± 1.1 (9)	9.5 ± 0.72 (7)	9.6 ± 0.29 (7)	9.3 ± 0.37 (7)	9.2 ± 0.39 (7)	8.5 ± 0.34 (7)	7.9 ± 0.28 (7)
F	10.6 ± 1.23 (9)	10 ± 0.74 (9)	9.6 ± 0.57 (10)	9.5 ± 0.57 (8)	9.4 ± 0.32 (8)	9.1 ± 0.56 (8)	9.2 ± 0.36 (8)	8.6 ± 0.28 (8)	7.8 ± 0.41 (8)
PROT (g/dl)									
M	3.9 ± 0.26 (8)	3.6 ± 0.17 (8)	3.9 ± 0.2 (9)	4.2 ± 0.22 (7)	4.4 ± 0.11 (7)	4.8 ± 0.29 (7)	5.2 ± 0.19 (7)	5.1 ± 0.17 (7)	5.1 ± 0.21 (7)
F	3.7 ± 0.16 (9)	3.7 ± 0.2 (9)	3.9 ± 0.19 (10)	4.2 ± 0.26 (8)	4.4 ± 0.2 (8)	4.8 ± 0.23 (8)	5.2 ± 0.17 (8)	5.2 ± 0.14 (8)	5.2 ± 0.16 (8)

[a] Values are expressed as: Mean ± SD (N = number of litters).
[b] Serum Chemistry Key: P, parameter; ALB, albumin; ALP, alkaline phosphatase; ALAT, alanine aminotransferase; ASAT, aspartate aminotransferase; TOT. BIL, total bilirubin; UREA N, urea nitrogen; CHOL, cholesterol; CREAT, creatinine; GLT, glutamyltransferase; GLUC, glucose; I.PHOS; inorganic phosphorus; PROT, total protein; M, males; F, females.

Source: Data were collected at WIL Research Laboratories, LLC (2001–2002).

Table 8.7 Hematology values for beagle dogs

Parameter[b]	Postnatal Day[a]							
	7	14	21	28	42	60	85	108	135
WBC (thousands/µl)									
M	12.7 ± 2.12 (8)	10.4 ± 1.84 (8)	12.9 ± 3.43 (9)	12.6 ± 1.46 (7)	13.2 ± 2.56 (7)	13.3 ± 3.6 (7)	15.2 ± 3.33 (7)	12.2 ± 2.56 (7)	10.2 ± 1.88 (7)
F	13.7 ± 3.11 (8)	10.1 ± 2.95 (8)	10.5 ± 2.98 (10)	11.7 ± 1.13 (8)	12.5 ± 2.39 (7)	13 ± 3.09 (8)	13.8 ± 2.82 (8)	11.4 ± 3.22 (8)	10.1 ± 2.53 (8)
RBC (millions/µl)									
M	4 ± 0.41 (8)	3.6 ± 0.27 (8)	3.6 ± 0.22 (9)	3.6 ± 0.24 (7)	3.7 ± 0.36 (7)	4.5 ± 0.34 (7)	5.2 ± 0.39 (7)	5.7 ± 0.31 (7)	5.9 ± 0.49 (7)
F	4 ± 0.33 (8)	3.6 ± 0.45 (8)	3.6 ± 0.27 (10)	3.7 ± 0.23 (8)	3.9 ± 0.29 (7)	4.6 ± 0.47 (8)	5.2 ± 0.42 (8)	5.7 ± 0.34 (8)	6 ± 0.47 (8)
HB (g/dl)									
M	11.8 ± 1.18 (8)	9.8 ± 0.75 (8)	9.4 ± 0.42 (9)	8.8 ± 0.53 (7)	8.6 ± 0.86 (7)	10.2 ± 0.76 (7)	11.7 ± 0.65 (7)	12.9 ± 0.55 (7)	13.5 ± 0.98 (7)
F	11.5 ± 1.44 (8)	9.9 ± 1.23 (8)	9.5 ± 0.6 (10)	9.4 ± 0.59 (8)	9.1 ± 0.8 (7)	10.6 ± 0.92 (8)	11.9 ± 0.71 (8)	13.2 ± 0.68 (8)	13.9 ± 1.03 (8)
HCT, %									
M	32 ± 3.1 (8)	27 ± 1.7 (9)	26 ± 1.2 (7)	25 ± 1.6 (7)	24 ± 2.4 (7)	29 ± 1.8 (7)	33 ± 2 (7)	36 ± 1.7 (7)	37 ± 2.6 (7)
F	31 ± 2.8 (8)	27 ± 3.2 (10)	26 ± 1.6 (8)	26 ± 1.6 (8)	26 ± 2 (8)	30 ± 2.7 (8)	33 ± 2.1 (8)	37 ± 1.9 (8)	38 ± 3 (8)
MCV (fl)									
M	79 ± 2.5 (8)	75 ± 2.7 (9)	72 ± 1.6 (7)	69 ± 2.3 (7)	65 ± 1.6 (7)	64 ± 2.1 (7)	63 ± 2 (7)	63 ± 1.4 (7)	63 ± 1.4 (7)
F	76 ± 3.3 (8)	75 ± 2.5 (10)	73 ± 2.1 (8)	70 ± 2.1 (8)	66 ± 1.3 (8)	65 ± 1.7 (8)	64 ± 2 (8)	64 ± 1.7 (8)	63 ± 1.6 (8)
MCH (µµg)									
M	29 ± 1.1 (8)	27 ± 1.2 (9)	26 ± 0.5 (7)	25 ± 0.4 (7)	23 ± 0.9 (7)	23 ± 1.2 (7)	23 ± 0.7 (7)	23 ± 0.6 (7)	23 ± 0.6 (7)
F	29 ± 2.2 (8)	28 ± 1.2 (10)	26 ± 0.7 (8)	25 ± 0.5 (8)	23 ± 0.8 (8)	23 ± 0.6 (8)	23 ± 0.7 (8)	23 ± 0.6 (8)	23 ± 0.4 (8)
MCHC (g/dl)									
M	37 ± 1.1 (8)	37 ± 1.2 (9)	36 ± 0.4 (7)	36 ± 0.8 (7)	35 ± 0.7 (7)	36 ± 0.8 (7)	36 ± 0.9 (7)	36 ± 0.2 (7)	37 ± 0.4 (7)
F	38 ± 1.9 (8)	37 ± 1 (10)	36 ± 0.6 (8)	36 ± 0.6 (8)	35 ± 0.8 (8)	36 ± 0.6 (8)	36 ± 0.5 (8)	36 ± 0.3 (8)	37 ± 0.5 (8)
PLAT (thousands/µl)									
M	431 ± 102.2 (8)	361 ± 110.7 (9)	392 ± 128.8 (7)	394 ± 101.9 (7)	360 ± 86.3 (7)	331 ± 107.7 (7)	440 ± 77.3 (7)	442 ± 47.9 (7)	401 ± 90.6 (7)
F	491 ± 126.5 (8)	412 ± 72.7 (10)	442 ± 107.7 (8)	427 ± 91.4 (8)	413 ± 90.6 (8)	330 ± 103.9 (8)	409 ± 94.8 (8)	422 ± 42.1 (8)	386 ± 81.3 (8)
RETIC (%)									
M	6.7 ± 7.54 (7)	4.5 ± 5.77 (7)	1.5 ± 0.29 (7)	1.4 ± 0.31 (7)	2.1 ± 0.69 (7)	1.7 ± 0.66 (7)	1.7 ± 0.63 (7)	0.9 ± 0.18 (7)	0.7 ± 0.15 (7)
F	6.4 ± 7.92 (8)	4.1 ± 3.68 (8)	1.5 ± 0.24 (9)	1.4 ± 0.36 (8)	2 ± 0.52 (8)	1.4 ± 0.54 (8)	1.4 ± 0.28 (8)	0.9 ± 0.17 (8)	0.7 ± 0.21 (8)
RETIC AB (millions/µl)									
M	0.31 ± 0.363 (7)	0.16 ± 0.2 (7)	0.05 ± 0.011 (7)	0.05 ± 0.011 (7)	0.08 ± 0.027 (7)	0.08 ± 0.032 (7)	0.09 ± 0.035 (7)	0.05 ± 0.011 (7)	0.04 ± 0.011 (7)
F	0.25 ± 0.288 (8)	0.14 ± 0.126 (9)	0.05 ± 0.01 (8)	0.05 ± 0.014 (8)	0.07 ± 0.018 (8)	0.06 ± 0.027 (8)	0.07 ± 0.017 (8)	0.05 ± 0.012 (8)	0.04 ± 0.014 (8)

NONCLINICAL JUVENILE TOXICITY TESTING

P									
N (thousands/μl)									
M	7 ± 1.94 (8)	6.2 ± 1.27 (8)	7.7 ± 2.55 (9)	7.3 ± 0.66 (7)	7.6 ± 1.44 (7)	9.2 ± 3.47 (7)	10.3 ± 2.6 (7)	7.4 ± 1.53 (7)	5.8 ± 1.56 (7)
F	8.1 ± 2.34 (8)	5.6 ± 2.22 (8)	5.6 ± 2.29 (10)	6.5 ± 0.92 (8)	6.8 ± 1.42 (7)	8.7 ± 2.65 (8)	9.2 ± 1.88 (8)	6.4 ± 2.36 (8)	5.7 ± 1.29 (8)
L (thousands/μl)									
M	4.4 ± 0.66 (8)	3.2 ± 0.95 (8)	4 ± 1.06 (9)	3.9 ± 0.47 (7)	4.3 ± 1 (7)	3 ± 0.92 (7)	3.5 ± 0.79 (7)	4 ± 0.98 (7)	3.7 ± 1.04 (7)
F	4.2 ± 0.96 (8)	3.5 ± 0.97 (8)	4.1 ± 1.09 (10)	4 ± 0.51 (8)	4.4 ± 0.95 (7)	3.5 ± 1.31 (8)	3.3 ± 1.07 (8)	4.1 ± 1.42 (8)	3.8 ± 1.41 (8)
MO (thousands/μl)									
M	1.2 ± 0.41 (8)	1 ± 0.29 (8)	1 ± 0.7 (9)	1.3 ± 0.72 (7)	1.1 ± 0.77 (7)	0.9 ± 0.51 (7)	1.3 ± 0.85 (7)	0.8 ± 0.44 (7)	0.5 ± 0.24 (7)
F	1.3 ± 0.73 (8)	0.9 ± 0.6 (8)	0.8 ± 0.33 (10)	1 ± 0.54 (8)	1.2 ± 0.79 (7)	0.7 ± 0.5 (8)	1.2 ± 0.9 (8)	0.8 ± 0.47 (8)	0.6 ± 0.47 (8)
E (thousands/μl)									
M	0.15 ± 0.21 (8)	0.03 ± 0.1 (9)	0.06 ± 0.059 (7)	0.02 ± 0.06 (7)	0.09 ± 0.182 (7)	0.09 ± 0.114 (7)	0.09 ± 0.134 (7)	0.07 ± 0.09 (7)	0.16 ± 0.161 (7)
F	0.06 ± 0.177 (8)	0.07 ± 0.127 (10)	0.11 ± 0.159 (8)	0.04 ± 0.082 (8)	0.03 ± 0.041 (8)	0.13 ± 0.233 (8)	0.06 ± 0.119 (8)	0.05 ± 0.085 (8)	0.08 ± 0.113 (8)
B (thousands/μl)									
M	0.04 ± 0.052 (8)	0.03 ± 0.044 (9)	0.03 ± 0.055 (7)	0.12 ± 0.243 (7)	0.12 ± 0.104 (7)	0.04 ± 0.057 (7)	0.07 ± 0.093 (7)	0.03 ± 0.036 (7)	0.02 ± 0.026 (7)
F	0.04 ± 0.088 (8)	0.01 ± 0.016 (10)	0.01 ± 0.015 (8)	0.06 ± 0.051 (8)	0.11 ± 0.151 (8)	0.03 ± 0.069 (8)	0.04 ± 0.046 (8)	0.03 ± 0.054 (8)	0.05 ± 0.077 (8)
N (%)									
M	54 ± 8.3 (8)	59 ± 6.1 (9)	60 ± 6.3 (7)	58 ± 2.3 (7)	58 ± 5.9 (7)	68 ± 8.6 (7)	67 ± 7.6 (7)	60 ± 4.3 (7)	56 ± 7.4 (7)
F	59 ± 8.1 (8)	54 ± 10.5 (10)	51 ± 9.9 (8)	56 ± 4.6 (8)	54 ± 7.5 (8)	66 ± 8.1 (8)	67 ± 8.4 (8)	56 ± 9.3 (8)	56 ± 9.9 (8)
L (%)									
M	35 ± 6.2 (8)	31 ± 5.9 (9)	32 ± 6.1 (7)	31 ± 4.3 (7)	33 ± 4.7 (7)	23 ± 7.8 (7)	24 ± 5.7 (7)	33 ± 2.7 (7)	38 ± 5.2 (7)
F	31 ± 7.1 (8)	36 ± 8 (10)	39 ± 7.3 (8)	34 ± 5.3 (8)	35 ± 3.9 (8)	28 ± 9.1 (8)	25 ± 6.1 (8)	37 ± 10.4 (8)	37 ± 7.5 (8)
MO (%)									
M	9.4 ± 4.34 (8)	9.7 ± 3.44 (8)	7.7 ± 4.22 (9)	10.1 ± 4.98 (7)	7.9 ± 4.74 (7)	7.2 ± 4.28 (7)	8.1 ± 4.37 (7)	6.7 ± 3.48 (7)	5.1 ± 1.9 (7)
F	9.1 ± 4.13 (8)	9.1 ± 4.94 (8)	7.9 ± 3.62 (10)	8.6 ± 4.44 (8)	9.2 ± 5.09 (7)	5 ± 2.74 (8)	7.9 ± 4.77 (8)	6.6 ± 3.28 (8)	5.1 ± 4.38 (8)
E (%)									
M	1.2 ± 1.62 (8)	0.3 ± 1 (8)	0.6 ± 0.74 (9)	0.2 ± 0.53 (7)	0.6 ± 1.08 (7)	0.6 ± 0.63 (7)	0.7 ± 0.94 (7)	0.5 ± 0.62 (7)	1.4 ± 1.41 (7)
F	0.6 ± 1.77 (8)	0.7 ± 1.27 (8)	1.5 ± 2.41 (10)	0.4 ± 0.78 (8)	0.3 ± 0.31 (7)	1.1 ± 1.64 (8)	0.4 ± 0.79 (8)	0.6 ± 0.96 (8)	0.9 ± 1.38 (8)
B (%)									
M	0.27 ± 0.398 (8)	0.28 ± 0.441 (9)	0.23 ± 0.369 (7)	1.05 ± 1.932 (7)	0.82 ± 0.7 (7)	0.4 ± 0.572 (7)	0.39 ± 0.456 (7)	0.25 ± 0.348 (7)	0.18 ± 0.263 (7)
F	0.25 ± 0.535 (8)	0.05 ± 0.158 (10)	0.14 ± 0.147 (8)	0.55 ± 0.473 (8)	0.94 ± 1.054 (8)	0.22 ± 0.364 (8)	0.26 ± 0.355 (8)	0.28 ± 0.452 (8)	0.44 ± 0.729 (8)

[a] Values are expressed as: Mean ± SD (N = number of litters).
[b] Hematology Key: P, parameter; WBC, white blood cells; RBC, erythrocytes; HB, hemoglobin; HCT, hematocrit; MCV, mean cell volume; MCH, mean cell hemoglobin; MCHC, mean cell hemoglobin concentration; PLAT, platelets; RETIC, reticulocytes; AB, absolute; N, neutrophils; L, lymphocytes; MO, monocytes; E, eosinophils; B, basophils; M, males; F, females.

Source: Data were collected at WIL Research Laboratories, LLC (2001-2002).

Table 8.8 Developmental landmarks for beagle dogs

	Eyelid Separation		Incisor Eruption		Vaginal Patency	Testes Descent	Balanopreputial Separation
	M	F	M	F	F	M	M
Day of acquisition (PND)							
Mean	15.5	15.6	20.2	20.3	21.3	22.4	50.7
S.D.	1.62	1.81	2.09	3.30	4.36	4.72	7.34
N (litters)	16	17	16	17	16	16	14
Body weight (g)							
Mean	792	774	927	877	915	1057	2572
S.D.	123.7	169.5	157.2	162.1	247.9	338.6	607.6
N (litters)	16	17	16	17	16	14	14

M = male, F = female
Source: Data were collected at WIL Research Laboratories, LLC (2001–2002).

conducted for both environmental chemicals and drugs and are required endpoints in the OECD draft guideline for developmental neurotoxicity studies.[129] Recently, studies evaluating juvenile toxicity have included these parameters, as they provide key indicators of perturbations in growth and function, and because the dose administration period for these studies often encompasses ages at which many of these landmarks are acquired.

In addition to physical landmarks of development, the study of immediate onset and latent behavioral expression of CNS damage associated with exposures during prenatal and early postnatal development has been accorded a higher level of concern in recent years. Tests to evaluate these aspects of development are routinely performed during the International Conference on Harmonisation (ICH) pre- and postnatal development study,[106] as well as under the more robust and focused EPA and OECD developmental neurotoxicity protocols.[56,129] A plethora of standard testing paradigms have been proposed, validated, and/or used in the process of identifying and characterizing agents that may represent a threat to the developing nervous system. The authors have chosen representative examples of these tests, and the following discussion should not be construed to be a comprehensive treatment of the subject. The reader is referred to the indicated references for additional exposition.

a. Physical Landmarks of Development

There are several accepted measures of growth and functional development of the juvenile animal that have been employed in a variety of study designs, including the multigeneration studies submitted to the EPA and the pre- and postnatal development studies submitted to the FDA. Although the most robust historical control data exist for the rat, landmarks can and have been evaluated in a variety of species. An example data set for these parameters collected in the beagle dog is presented in Table 8.8. Inclusion of these landmarks of development may provide insight into potential developmental delays occurring as a result of direct administration of a test article. However, the use of these landmarks is always contingent on the age at onset of exposure and the duration of the study in general. It is also important to note that while these landmarks are still evaluated in tests for development, many are positively correlated with body weight, and developmental delays or accelerations are typically a function of neonatal weight.[130,131] Therefore, merely collecting body weights at key developmental stages may be as effective at elucidating changes in growth and development without requiring the addition of labor-intensive neonatal evaluations. Moreover, the OECD draft guideline for developmental neurotoxicity studies has suggested that these endpoints are "recommended only when there is prior evidence that these endpoints will provide additional information."[129] In addition, the ICH guideline for reproductive and developmental toxicity studies states that "the best indicator of physical development is body weight."[131]

Table 8.9 Example of developmental delay in surface righting reflex

Sex	Control	T1	T2	T3
A. Total Number of Pups Available for Assessment				
Males	5.1 ± 0.10	5.0 ± 0.08	5.1 ± 0.13	5.1 ± 0.19
Females	5.1 ± 0.15	5.1 ± 0.17	5.1 ± 0.19	5.2 ± 0.27
Combined	5.1 ± 0.10	5.1 ± 0.11	5.1 ± 0.14	5.2 ± 0.18
B. % Positively Responding (Number of Pups)				
Males				
Total pups	95	119	100	108
PND 5	95% (90)	98% (117)	94% (94)	85% (92)
PND 6	100% (95)	100% (119)	100% (100)	100% (108)
Females				
Total Pups	108	112	110	102
PND 5	86% (93)	89% (100)	88% (97)	78% (80)
PND 6	100% (108)	99% (111)	100% (110)	98% (100)
PND 7		100% (112)		100% (102)
Combined				
Total Pups	203	231	210	210
PND 5	90% (183)	94% (217)	91% (191)	82% (172)
PND 6	100% (203)	100% (230)	100% (210)	99% (208)
PND 7		100% (231)		100% (210)

Notes: Surface righting response evaluated in Crl:CD®(SD)IGS BR rat offspring, beginning on PND 5, with assessment continuing until positive evidence of righting was observed (WIL Research Laboratories, LLC). The same data are presented in A and B. Section A presents the mean ± S.D. day of acquisition of surface righting for males, females, and combined sexes. Section B presents the percentage (total number) of rat offspring acquiring surface righting on each day of assessment.

The ICH guideline specifically mentions several preweaning developmental landmarks, including pinna detachment, hair growth, and incisor eruption, and notes that these measures are highly correlated with body weight and postcoital age, especially "when significant differences in gestation length occur." Moreover, functional developmental landmarks, such as surface and air righting, were also noted in the ICH guideline to be dependent on physical development. While a complete discussion of the many developmental landmarks is beyond the scope of this chapter, the authors have chosen to address a few key evaluations, and we encourage the reader to review the extensive literature on this topic for more detail on specific evaluations.

One commonly evaluated landmark of functional development is surface righting. The surface-righting reflex evaluates the ability of the animal to regain a normal upright stance when placed in a supine position on a flat surface.[132] This reflex is present soon after birth in rats, although the time required to attain an upright stance decreases as the animal ages.[133] At the authors' facility, this test is performed beginning on PND 5 (one day after litter standardization) for the rat, with a 15-s window for test completion. Neonates with negative responses (no observed righting) are evaluated on sequential days until a positive response is noted. Under these testing conditions, the approximate average age at which righting occurs is PND 5. Because the age at onset of testing equals the age at which a positive response is observed, a delay in acquisition of this landmark is most easily noted as an increase in the number of animals (on a litter basis) with a negative response on the first day of testing. An example of such a delay is provided in Table 8.9. Although not evident in this example, it is important to note that a statistical difference from control in the mean age at acquisition for this test is a key indicator of a test article–related change and should not be ignored.

Hair growth, incisor eruption, and eyelid separation are well accepted landmarks of physical development, and both incisor eruption and eyelid separation were required endpoints of analysis in the CBTS.[105] For each of these tests, the neonate is evaluated from the first day of testing until a positive response is noted. Of course, evaluation of the acquisition of some landmarks is not possible in species with such long gestational lengths that they are typically born with hair (dog, pig) or with separated eyelids (pig).

Habituation is another functional reflex that is often assessed in a variety of study designs. Habituation is the progressive decrease in response to repeated stimulation that cannot be attributed to receptor fatigue or adaptation.[134] Glass and Singer characterized habituation as the most important mechanism by which humans adapt and survive in modern society.[135] It is ubiquitous in nature and most multicellular organisms have been shown to exhibit some form of habituation.[136] The regulatory agencies consider development of the habituation response to be a key endpoint for evaluation.[56,129] Habituation can also be used as a measure of basic learning, and in the CBTS, it was determined to be a consistent method for detecting toxicity caused by methyl mercuric chloride.[104] This reflex can be evaluated with a number of automated functional tests, such as the auditory startle habituation test and the locomotor activity test. More detailed information on this reflex is provided below as part of the discussion of each behavior.

As Lochry pointed out in her review, landmarks of physical and functional development are often highly correlated with body weight.[130] Therefore, when test article–related changes in landmarks of physical and functional development are observed, correlative changes in behavioral assessments may be noted as well. For example, delays in the acquisition of physical landmarks can indicate a general developmental delay that may influence the normal ontogeny of locomotor activity or the habituation to a startle stimulus. It is also critical to note that preweaning developmental landmarks may be affected by maternal behavior. If the mother becomes incapable of adequately rearing her young, growth of the offspring will be negatively impacted. This detrimental maternal influence can profoundly affect the offspring, should be carefully monitored during the course of the study, and must be considered during evaluation of the data.

b. Landmarks of Sexual Development

Two additional standard landmarks of sexual development that are included in both juvenile and adult reproduction studies are balanopreputial separation and vaginal patency (see Chapters 9 and 10 for descriptions). These landmarks are acquired by males or females, respectively, prior to the onset of puberty and are triggered by the presence of reproductive hormones.[137]

Evaluation of balanopreputial separation is typically conducted daily until there is evidence that the prepuce has separated from the glans penis.[138] There is some thought that preputial separation can be accelerated by repeated examinations; however, to ensure that the day of acquisition is not overlooked, many laboratories will begin the examinations between PND 35 to 40 in the rat.[137] In the authors' laboratory, the mean age at acquisition of balanopreputial separation in the Crl:CD®(SD)IGS BR rat is approximately PND 45. The authors have also developed limited data in the New Zealand White rabbit, which indicate that balanopreputial separation is achieved on approximately PND 72. In addition, data developed at the authors' laboratory in the beagle dog reveal that the mean age at acquisition of balanopreputial separation is approximately PND 51 (see Table 8.8). As with the other landmarks of development, delays and accelerations in the acquisition of balanopreputial separation can occur. Acquisition is correlated with body weight,[139,140] but other determining factors, such as the presence of reproductive hormone stimulation or endocrine-modulating chemicals, including 1,1-dichloro-2,2-bis(*p*-chlorophenyl)ethylene, can also disrupt normal development.[141] Testes descent is another landmark of sexual development in the male that can be assessed in a number of species and is preferred by some researchers when evaluating sexual maturation in mice. Recently published data from three investigations in Swiss albino mice suggest that testes descent is very consistent over time and across laboratories, and occurs at approximately day 23 or 24 of postnatal life.[142–144]

In females, the acquisition of vaginal patency is detected by examining the vagina to determine if the septum covering the vagina has ruptured.[145] Typically, the initial age at testing is between PND 25 to 30 in the rat. In the authors' laboratory, the mean age at acquisition of vaginal patency in the Crl:CD®(SD)IGS BR rat is approximately PND 33. In addition, in a limited data set developed in the authors' laboratory, the mean age at acquisition for vaginal patency in the New Zealand White rabbit was shown to be approximately PND 29. In comparison, data developed at the authors' laboratory for the beagle dog indicate that the mean age at acquisition of vaginal patency is approximately PND 21 (see Table 8.8). Swiss albino mice acquire vaginal patency at approximately day 23 or 24 of postnatal life.[142–144]

Vaginal patency is positively correlated with body weight, and substantial changes in growth can alter the day of acquisition.[139] Aside from effects on growth, estrogenic and antiestrogenic compounds can dramatically influence the day on which vaginal opening is first observed.[137] One potential confounder in the observation of vaginal patency is the somewhat rare occurrence of a thin strand of tissue remaining over the vaginal canal. The presence of the thin strand of tissue does not interfere with estrous cyclicity or mating, and it can be removed by the observer or during the mating process.[137] However, there seems to be some disagreement as to whether the rat should be considered as having acquired vaginal patency if this tissue strand is still present. Gray and Ostby noted that when animals with the thin strand of tissue were eliminated from the analysis, there was no effect of 2,3,7,8-tetrachlorodibenzo-p-dioxin (TCDD) on the mean age at acquisition.[146] As with many laboratories, the authors consider a rat to have acquired vaginal patency if only the thin strand of tissue spans the lumen.

Given the potential for acceleration in the acquisition of these sexual developmental landmarks by potent endocrine modulators, it is recommended that the age of initial evaluation should be well in advance of any normal age of acquisition. In the authors' laboratory, male and female rats are initially evaluated on PND 35 and 25, respectively. That timing is used because diethylstilbestrol, a potent estrogen, has been shown to accelerate vaginal opening in a female pubertal assay to 29.7 and 24.1 days of age at dose levels of 1 and 5 μg/kg, respectively.[147] The mean age at acquisition of vaginal patency in that study was prior to the initial evaluation age used by most laboratories, but PND 25 should be early enough for most cases, unless the test article is a suspected endocrine modulating compound.

Another key point to consider when evaluating vaginal patency is the variability of the test in a given test species. The mean age at acquisition of vaginal patency in mice has been shown to be affected by cohabitation with adult males and by the presence of bedding from male cages.[137] This potential confounding influence should be taken into consideration when designing a juvenile study in mice.

When evaluating effects on balanopreputial separation or vaginal patency, it is important to consider changes in other growth parameters to determine whether there is a pattern of developmental delay or acceleration. Changes in the mean age at acquisition of either of these endpoints in the absence of effects on either body weight or the mean age at acquisition of other physical landmarks can be a key indicator of endocrine modulation by a test article. However, when the mean age at acquisition of balanopreputial separation or vaginal patency is altered in the presence of clear changes in body weight, these effects are often considered secondary to overall changes in growth. In these cases, it is imperative to evaluate the entire data set for other similar patterns of effect.

Ashby and Lefevre reported a relationship between the age at acquisition of balanopreputial separation and group mean body weight.[140] They showed that in the absence of changes in body weight, changes in the age at acquisition of balanopreputial separation in Alpk:APfSD rats of more than two days would likely be test article–related. Similarly, Merck performed calculations using historical data to determine the power of a standard two-tailed test to detect changes in the age at acquisition of both balanopreputial separation and vaginal patency.[137] As an example, their calculations demonstrated that with a sample size of 20 Sprague-Dawley rats/group, a 2.5-day change in the mean age at acquisition of balanopreputial separation would have a probability of 0.87 of being a true change from control. Likewise, the data indicated that a difference in the mean age at acquisition of vaginal patency of 2.0 days would have a probability of 0.87 of being a true change from control.[137]

c. Behavioral Assessments

The decision to conduct behavioral testing as part of the juvenile study can be a difficult one because assessment of behavior may be required for other agents in addition to neuroactive compounds. Factors influencing the decision include the age at initiation of clinical dose administration, the duration of the dose administration period, whether the central or peripheral nervous systems are still developing in the species, and whether the test article has been shown to cause adverse changes in organ systems that are key to internal homeostasis in adult animals. As mentioned earlier, there are critical windows during which organ systems may be susceptible to test article administration. For example, the development of the human nervous system extends well beyond birth.[148] If the pediatric indication includes the period of nervous system development, regulatory agencies may recommend an evaluation of behavioral endpoints. In addition, functional deficits in key organ systems can produce detrimental effects on behavior that are separate from direct neurotoxicity. However, these potential secondary changes in behavior should not be ignored because they can provide important clues to the clinician regarding the types of side effects that pediatric patients may experience. The next several sections describe basic behavioral testing paradigms that may be employed to evaluate changes in motor function and coordination, sensory perception, and cognition. The list of endpoints discussed is not meant to cover every possible parameter, and the reader is invited to review the extensive literature on these topics for more detail.

1) Locomotor Activity — Locomotor activity can be generally quantified as total, fine, and ambulatory activities. There are several styles of test chambers for rats, and the authors' facility employs an open-field environment surrounded by four-sided black plastic enclosures to decrease the potential for distraction by extraneous environmental stimuli. Activity is measured electronically with a series of infrared photobeams surrounding the clear plastic rectangular open field environment. The animal's activity can be expressed as total and ambulatory activity (for rodents), although this is not a universally applied approach. Total activity is generally the sum of both gross and fine motor movements (any photobeam break during the testing paradigm). Ambulatory activity is a measure of only gross movements (two or more consecutive photobeam breaks, i.e., the animal is moving from one location to another). Testing is usually conducted over a 60- to 90-min session, although some laboratories conduct locomotor activity test sessions over a 23-h period. The length of the test session is determined by the desire to demonstrate habituation during a particular test. Many laboratories have based the session length on EPA guideline recommendations that "the test session should be long enough for motor activity to approach asymptotic levels by the last 20 percent of the session for nontreated control animals."[56] Appropriate validation and positive control studies must be conducted prior to the conduct of any animal study to determine the session length necessary to evaluate habituation, using the equipment available for testing.

For the rat, total activity will normally be relatively low at PND 13, will increase to a maximum preweaning level by PND 17, and will decrease by PND 21. This ontogenic profile of activity has been well characterized.[149] For rats and mice, locomotor activity is generally assessed longitudinally during early postnatal life (PND 13, 17, and 21 for rats) and again at young adulthood (PND 60 to 61) in standard developmental neurotoxicity studies.[56] More limited assessments are generally

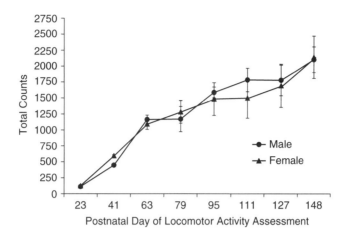

Figure 8.20 Locomotor activity development of the beagle dog. Mean number of total counts for litters by sex at each day of locomotor activity assessment (n = 7 to 8 litters per time point). Data were collected at WIL Research Laboratories, LLC.

conducted for the pre- and postnatal development studies,[131] although these assessments are generally conducted in a longitudinal fashion as well.

The first three time points listed above for rats examine the ontogeny of habituation, the development of motor activity coordination, and the presence or absence of the characteristic pattern of early rodent activity.[150–151] In more traditional reproduction and developmental neurotoxicity studies, the assessment at approximately PND 60 examines potential latent alterations in motor activity and the persistence of changes observed prior to weaning. Other ages may be substituted or added as appropriate for the age at which dose administration occurs and to assess potential recovery from or latency of motor activity decrements after juvenile exposure.

The use of different species, such as the dog, will require adjustments in the timing for evaluating changes in the ontogenic profile of activity. Figure 8.20 illustrates the development of locomotor activity in beagle puppies over a period of approximately 18 weeks.[124] If no data are available regarding the ontogeny of motor activity in the species selected for the juvenile study and if it is considered necessary to understand the potential effects of the test article on this profile, some species characterization will be required prior to study initiation or else selection of a different animal model should be considered. Many nonrodent species have substantially longer periods of motor development than those of rodents (refer to Figure 8.20 for ontogeny of motor function in the dog), requiring longer study durations to assess changes in the ontogeny of motor activity.[115]

When evaluating locomotor activity, the pattern of habituation should be examined within the control group at each age of assessment. In the rat on PND 13, there should be very little change in total activity across each test interval. A test interval is defined in this text as a period during which activity counts are collected and then presented. A test interval is typically between 10 and 15 min and is determined through validation of the equipment. Between PND 17 and 21, a graph of cross-interval activity should begin to show a near-hyperbolic shape. The first test interval should contain the most movement, with each subsequent interval showing an approximately 30 to 50% decrease in movement (when the total activity session is divided into four 15-min intervals) until the final test interval, when very little change should occur. In testing paradigms in which six or nine intervals (usually 10 min each) are used for data analysis, the magnitude of interinterval decreases will likely differ from that described above. Interval data from motor activity sessions at adulthood should demonstrate a well-defined habituation curve. There should be no consistent sex-related differences among young animals, although gender-specific differences may develop with age.

Potential test article–related effects observed in motor activity should be compared with those observed in other behavioral and/or physiological endpoints. For example, if a delay is suspected in the normal pattern of motor activity development, similar changes indicative of delay or decreased performance in age- or weight-sensitive parameters in a functional observational battery, such as grip strength, rotarod performance, or gait, may also be expected. In comparison with a change in multiple related behavioral endpoints, a change in only one motor activity parameter in the absence of any other change in behavior may carry less weight relative to test article exposure.

Some locomotor activity test systems evaluate activity in both a vertical plane and along a horizontal surface. These systems contain two levels of photobeams. The lower level detects interruptions in photobeams resulting from gross movements and grooming, while the upper level detects interruptions caused by rearing. Rearing is typically noted in aroused adult rats and provides an indication of exploratory behavior.[133] When examined in an open field environment, it has been demonstrated that rearing is not well developed in rats until PND 18 and continues to increase in frequency in response to a novel environment beyond weaning. Rearing decreases over time in a novel environment. An interpretation of this parameter in the locomotor activity assessment should first consider whether rearing should be present in the animal at the age tested. Also, if a reduction in rearing counts is not observed over the testing interval, this should be considered an important test article–related change.

2) *Auditory Startle Response* — The auditory startle test used most often at the authors' laboratory is based on the procedures used in the CBTS.[145] In this testing paradigm, the animal is placed in an enclosure atop a force transducer in a sound-attenuated chamber. Background noise is used to dampen external environmental stimuli and enhance the response to the startle stimulus.[152–153] The force applied to the transducer is measured for some period after a high-pitched noise of approximately 115 dB(A) is produced. Examples of the resultant data include the maximum response to the stimulus and the time to maximum response. The data are typically presented as blocks of trials for a given assessment. Other paradigms include prepulse inhibition, which provides a stimulus (auditory or tactile) prior to the startle burst to detect an attenuated response.[152] More complex startle testing paradigms, involving a combination of habituation and prepulse inhibition, have been successfully used at some testing facilities.[154]

As with the locomotor activity assessment, the auditory startle response paradigm is typically conducted longitudinally. The standard ages of assessment in most reproductive and developmental neurotoxicity studies are PND 20 and 60 for the rodent.[56] Within a particular assessment interval, the pattern of habituation should be examined. During preweaning assessments in the rodent at the authors' laboratory, there is typically an attenuated change in peak response across each trial block. At young adulthood, peak response to the startle stimulus should exhibit a characteristic habituation curve, with the greatest response observed during the first trial block. Each subsequent trial block should demonstrate a reduction in force, until the final intervals, when very little change is expected. Latency to peak response does not change remarkably over a given test session. Typically younger control animals will show a slight increase in latency to peak response over the five blocks of trials. No gender-specific differences in peak response should be observed for preweaning animals, although gender-related differences become apparent with increasing age and body weight (because males generally weigh more than females).

It is important to compare potential test article–related effects observed in the auditory startle response testing paradigm with those for other behavioral and/or physiological endpoints. For example, if deficits in peak response related to neuromotor fatigue or disruption are observed, the investigator might also expect to observe adverse effects on motor activity and grip strength. Severe deficits in body weight can confound interpretation of the effects of the test article on the auditory startle response and may mask subtle changes in neuromotor function in this testing paradigm.

For nonrodent species, alternative methods for evaluating auditory startle habituation may be required. The authors are aware of only two systems capable of evaluating auditory startle response

Table 8.10 Biel maze learning and memory paradigm

Day	Number of Trials	Path	Data Collected	Parameter Evaluated
1	4	Straight channel	Time to escape	Swimming
2	2 (Trials 1 and 2)	Path A (forward)	Time to escape and number of errors	Learning
3	2 (Trials 3 and 4)	Path A (forward)	Time to escape and number of errors	Learning
4	2 (Trials 5 and 6)	Path B (reverse)	Time to escape and number of errors	Learning
5	2 (Trials 7 and 8)	Path B (reverse)	Time to escape and number of errors	Learning
6	2 (Trials 9 and 10)	Path B (reverse)	Time to escape and number of errors	Learning
7	2 (Trials 11 and 12)	Path A (forward)	Time to escape and number of errors	Memory

in large animals, although basic startle habituation can be evaluated in some forms of cognitive testing, such as the conditioned eyeblink response in the rabbit. Instead, many testing facilities evaluate sensory perception in large animal models by use of standard assessments in a functional observational battery examination. These measures of sensory perception provide an indication of hearing and responsiveness but do not easily afford the opportunity for quantitative evaluation of the response magnitude or evaluation of habituation.

3) Learning and Memory — Although there are several forms of learning and memory testing paradigms in use, four accepted models will be discussed in this section, including the Biel water maze, the Morris water maze, active avoidance, and passive avoidance. Each model has its own advantages and limitations that have been presented in various forums and are not expounded upon in this chapter.

The model of learning and memory used at the authors' laboratory is the complex eight-unit T-maze (Biel water maze),[155] initially applied to neurotoxicity testing by Vorhees and colleagues.[156] Each animal is assessed for acquisition of learning and memory at a single stage of development; for standard reproduction and developmental neurotoxicity studies, the most common ages of onset of testing are approximately PND 22 or 62. According to the current EPA developmental neurotoxicity testing guideline,[56] it is not appropriate to test an animal at more than one stage by use of the same learning and memory paradigm because it may confound the results of both the learning and memory components of the test. However, if two different measures of learning are used in the same study, the same animals may be used at multiple ages because there would be no expected confounding influences of one testing period on the other. This would provide the additional benefit of a longitudinal assessment of learning and memory. Regardless, for the Biel maze procedure used at the authors' testing facility (refer to Table 8.10), the first day of testing is used to measure the animal's ability to navigate a straight channel, followed by 5 days of testing during which the animal is required to learn first the forward path and then the reverse path. On the final day of testing, animals are probed for the ability to remember the forward path. The first four trials (conducted on testing days 2 to 3) are designed to measure learning in the forward path and shorter-term memory of that path. The second six trials (conducted on testing days 4 to 6) are designed to measure learning and shorter-term memory in the reverse path. The last two trials (conducted on testing day 7) are designed to measure long-term memory of the forward path.

An alternative model of learning and memory used in many laboratories is a maze developed by Morris.[157] Like the Biel maze, the Morris maze involves immersion of the rodent in water, although the maze in this case is a circular tank divided into quadrants. The animal is placed in the tank and must learn the location of the submerged platform through a series of trials. After a sufficient number of trials, the platform is moved to a different quadrant, and the animal must learn the new location. For all trials, the time spent in each quadrant of the tank and the number of errors (entries into the wrong quadrant) are recorded.[158] Learning can be observed as an increase in the time spent in the correct quadrant, as well as by a decrease in the number of errors. Memory is assessed in one of two ways. If the animal is not tested for a period after learning the location of the platform, memory can be observed by relatively few errors and relatively long periods spent in the correct quadrant. Alternatively, if the platform is placed in a new location, memory is noted as a dramatic decrease in the time spent in the correct quadrant and a large increase in the number of errors, while the animal learns the new location.

One of the most popular tests of learning and memory is passive avoidance. This assessment involves the use of an aversive stimulus to shape the animal's behavior. One typical design involves placement of the rodent in a lighted chamber, with access to a dark chamber.[159] As rats prefer a darkened environment, the rat will cross into the darkened chamber. Upon entry, an electric shock is administered to the animal through the floor plate, ending the trial. The latency for the rat to cross into the darkened chamber is recorded. After one or more training trials, the rat is placed in the lighted chamber and the latency to cross into the darkened chamber is again recorded. An increase in the latency to cross is considered evidence of learning. To determine the retention of training (i.e., memory), the animal is returned to the testing apparatus after a period of days, and the latency to cross is recorded.[154]

In contrast to the passive avoidance paradigm, active avoidance requires that the animal perform a task to avoid the aversive stimulus.[159] In this evaluation, an animal is placed into a chamber and allowed access to another chamber. A conditional stimulus, such as a tone or light, is presented, whereupon the animal must cross into the opposite chamber within a given time frame. If the animal does not cross within the allotted time, an aversive stimulus (typically an electric foot shock) is applied until the animal moves to the other chamber, and an escape event is recorded. If the animal crosses within the allotted time, an avoidance is recorded. The typical measures of learning are the numbers of escapes and the numbers of avoidances over a series of trials.

A common assessment paradigm for active avoidance is the two-way shock avoidance, which alters the placement of the electric shock between the two chambers. In this paradigm, the rat must learn that the conditional stimulus is applied prior to the presentation of the shock and that to avoid or escape the aversive stimulus it must reenter a chamber in which it was previously exposed to the shock.[159] Another paradigm of active avoidance is the Y-maze.[160-164] In this assessment, three arms are used in place of the two chambers for the two-way shock avoidance. The rat is placed in an arm of the maze, the conditional stimulus is applied, and the rat must move to the appropriate arm to avoid the aversive stimulus. If the rat crosses to the correct arm within the allotted time, an avoidance is recorded. If the rat crosses to the incorrect arm within the allotted time, an error is recorded, and the aversive stimulus is applied. If the rat does not cross within the allotted time, an escape event is recorded when the rat crosses after being exposed to the aversive stimulus.

For water-based assessments of learning and memory, when comparing treated groups to the control group, one must first determine whether the test article–treated groups demonstrate difficulties in swimming ability. If there is a significant change in this parameter, all other endpoints in the assessment may be affected. Developmental delays can impair swimming ability. This usually lengthens the latency to escape within the first few days of testing during the early postweaning period, without concomitant effects on the number of errors committed. Increased time to escape throughout early postweaning testing and/or during later testing may indicate clear neuromotor impairment. Changes in learning in the Biel maze will be detected by alterations in the slopes of the curves (response versus time; Figure 8.21 and Figure 8.22) for the forward and reverse paths. In the authors' experience, the slope should appear somewhat similar to a habituation curve, although asymptotic levels are rarely achieved in either learning phase. Therefore, a potential change may be represented by a flattened curve (animal never learns or cannot remember) or a curve that reaches asymptotic levels (animal is a fast learner). Changes in memory may be detected in several ways. First, test article–related changes in shorter-term memory may be detected by alterations in the slope of the line between the two trials on the same day. Effects on shorter-term memory may also present as changes in the slope of the line between two sequential sessions on different days. Test article–related effects on long-term memory may be detected as significantly longer latencies to escape or as a greater mean number of errors in trial 11 than in trial 1, with relatively little change between trials 11 and 12.

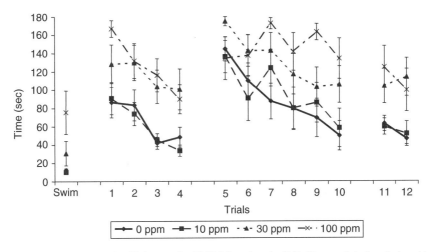

Figure 8.21 Times to escape the Biel maze for PND 22 male rats (WIL Research Laboratories, LLC).

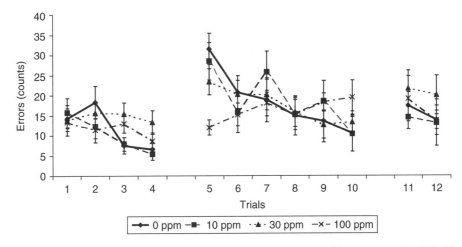

Figure 8.22 Number of errors committed by PND 22 male rats during Biel maze testing (WIL Research Laboratories, LLC).

When evaluating passive avoidance, the change in latency over time becomes the measure of both learning and memory. Since the aversive stimulus typically is only applied during the training session(s), learning can be observed as an increase in the latency to cross to the darkened chamber (when multiple training sessions are used). Upon completion of the training session(s), the aversive stimulus is no longer applied and memory can be examined by the latency to cross to the darkened chamber. An additional assessment of learning can be observed as a decrease in the latency to cross with repeated tests, as the animal learns that the aversive stimulus is no longer being applied to the floor plate in the darkened chamber.

Evaluation of the data for active avoidance depends on the paradigm being used. With two-way shock avoidance, learning can be detected as a decrease in the number of escapes and an increase in the number of avoidances for a series of trials. In the conduct of the Y-maze avoidance test, learning is observed as a decrease in the number of escapes and in the number of errors, while the number of avoidances increases with learning. Again, memory can be observed as a relatively small difference in the performance of the animal between the training and memory sessions.

In the overall assessment of the response, it is important to evaluate interrelated endpoints in other behavioral and/or physiological evaluations before drawing conclusions. For example, an increase in the mean latency to escape from a straight channel during Biel maze swimming evaluation might also be correlated with decreased motor activity, decreased peak response to an auditory stimulus, or reduced body weight. The Biel maze test evaluates a unique form of habituation. Immersion in water provides an aversive stimulus to rats. Therefore, rats must habituate to the test conditions while also learning and remembering the escape route from the maze. It is likely that a significant test article–related effect on habituation in locomotor activity or the auditory startle response may also manifest itself in the early phases of the learning component of the Biel maze test. Finally, changes in neuromotor function can negatively affect the ability to interpret all the described assessments of learning and memory. If an animal is not able to locomote or has impaired motor coordination, it may be unable to perform the required tasks adequately. Thus, these data should be examined in combination with data from other measures of neuromotor function, such as the locomotor activity assessment or grip strength, rotarod performance, and landing foot splay, in the functional observational battery assessment.

4) Functional Observational Battery Assessments — The current FDA draft guidance for juvenile toxicity studies[51] states that well-established methods should be used to monitor key functional domains of the central nervous system and these should include assessments of reflex ontogeny, sensorimotor function, locomotor activity, reactivity, and learning and memory. Many of these domains can be evaluated in the functional observational battery (FOB) assessment. The FOB, in combination with an automated measure of locomotor activity, was designed to quickly screen for neurobehavioral deficits in Tier 1 adult neurotoxicity studies.[165] It has been incorporated into a variety of study designs, including acute and subchronic neurotoxicity studies, developmental neurotoxicity studies, and more recently, nonclinical juvenile studies.[56,129,166,167] The FOB consists of a series of endpoints that assess six functional domains, as first defined by Moser.[165] At the authors' testing facility, the FOB includes the parameters listed in Table 8.11.

The FOB can be adapted to nonrodent species with some modifications. At the authors' laboratory, FOB testing paradigms have been developed for use in the dog and nonhuman primate, and the authors are aware of FOB tests designed for use in the rabbit.[168] Although evaluations cannot necessarily be conducted in an open-field arena for some large animal models (e.g., nonhuman primates and rabbits), deficits in neuromotor function that may affect gait can be characterized by modifying standard assessments used in rodents.

Regardless of the animal model chosen and the specific endpoints evaluated in the FOB, it is imperative that appropriate controls (e.g., training, interobserver reliability, validation) be implemented at the testing facility. First, the investigator must ensure that observers are appropriately trained in the assessments, and the observer should understand the types of abnormal neurobehavioral outcomes that may be encountered. Of equal importance, the observer must be able to recognize the normal behavior of the test species. In the authors' laboratory, new personnel are not permitted to be trained in the rat FOB until after a year of employment. It is our position that it takes at least a year to develop a good understanding of normal rat behaviors, and even then, FOB training requires that the individual evaluate 60 control animals in addition to approximately 20 animals that have been treated with various positive control agents.

Another important control in the conduct of the FOB is the use of blinded examinations. Since many of the endpoints included in the FOB are subjective in nature, it is possible for the observer to unintentionally bias the results by applying slightly harsher expectations to animals in higher dose groups than to control animals. Therefore, neurotoxicity testing guidelines require FOB assessments to be conducted without knowledge of the animal's dosage group assignment.[166,167]

The testing laboratory must also ensure that observer variability be minimized. This can be accomplished by using only one observer for the entire study, which may not always be possible, or by conducting specialized positive control studies designed to minimize interobserver variability.

Table 8.11 Functional domains as described by Moser and associated battery tests

Home Cage Observations	Sensory Observations	Handling Observations	Neuromuscular Observations	Physiological Observations	Open Field Observations
Biting - E	Air righting reflex - N	Ease of handling animal in hand - E	Grip strength - fore- and hindlimb - N	Body temperature - P	Arousal - E
Convulsions or tremors - E	Approach response - S	Ease of removal from cage - E	Hindlimb extensor strength - N	Body weight - P	Backing - C
Feces consistency - A	Eyeblink response - S	Eye prominence - A	Hindlimb foot splay - N	Catalepsy - P	Bizarre or stereotypic behavior - E
Palpebral closure - A,C	Forelimb response - S	Fur appearance - A	Rotarod performance - N		Convulsions or tremors - E
Posture - N	Hindlimb extension - S	Lacrimation or chromodacryorrhea - A			Gait - N
	Olfactory orientation - S	Mucous membranes, eye, or skin color - A			Gait score - N
	Pupil response - A	Muscle tone - N			Grooming - N
	Startle response - S	Palpebral closure - A,C			Mobility - N
	Tail pinch response - S	Piloerection - P			Rearing - C
	Touch response - S	Red or crusty deposits - A			Time to first step (seconds) - N
		Respiratory rate or character - P			Urination or defecation - A
		Salivation - A			

Note: Endpoints observed during the functional observational battery assessment conducted at the authors' testing facility. Functional domains, as defined by Moser (Moser, V.C., *J. Am. Coll. Toxicol.*, 10, 661, 1991) examined by each endpoint are as follows: A = autonomic, N = neuromuscular, S = sensorimotor, E = CNS excitability, C = CNS activity, P = physiological

These interobserver reliability studies should be conducted periodically to prevent drift in the observations by a single observer or across multiple observers.[166,169]

Finally, FOB assessments in large animal species should be conducted only after the observer has developed a rapport with the subject. For example, prior to conducting the FOB in nonhuman primates, the observer should enter the testing room and sit quietly for several minutes to allow the animals to become acclimated to the observer's presence.

In adult neurotoxicity studies conducted with the rat, pretest FOB assessments are performed to provide baseline data on the behavior and neuromotor function of the animal. Pretest evaluations are particularly important when conducting studies with dogs and nonhuman primates because these species have distinct personalities and behaviors that are intrinsic to the individual animal. It is, therefore, especially important to understand what the normal behavior for these animals is prior to administration of a drug that may influence that behavior. Unfortunately, pretest evaluations usually are not possible for nonclinical juvenile studies because animal behavior and neuromotor function change with age, and pretest evaluations of neonatal animals are unlikely to predict the pattern of responses that these animals will display as adults. For example, in a juvenile rat study in which dose administration is initiated on PND 7, gait cannot be evaluated during the predosing period because rats at this age are incapable of coordinated locomotion. For the same reason, posttreatment FOB assessments may not be comparable to assessments conducted during the treatment period because neurological development may have progressed to a point beyond which juvenile toxicities may be observed.

6. Reproductive Assessment

Although reproductive performance assessment is required in the standard battery of tests conducted for submission to the EPA and FDA,[55,100,131,170] it has also been included in many juvenile study designs for a number of reasons. As with other endpoints, differential sensitivity to the test article following juvenile or adult exposure may lead to different effects on reproduction. In addition, the period of exposure often encompasses sexual maturation of the test system. Therefore, reproductive performance may be affected after juvenile exposure, and such assessments should be considered as part of a juvenile toxicity program. Regardless of the reason for inclusion in the study, specific details of the design, conduct, and interpretation of these assessments are discussed in Chapters 9 and 10, and the reader is referred to those chapters for extensive overviews.

F. Statistical Design and Considerations

One of the most important elements in the design of any study is the statistical methodology. The statistical procedures employed in a study are critically dependent upon the other elements of the design, including the test species, sample size, animal assignment, experimental methodology, and data collected. Each of these elements is discussed in detail in earlier sections of the chapter. Therefore, this section addresses specific issues with regard to statistical approaches. It cannot be emphasized enough that the effect of the litter cannot be ignored in the statistical analysis of juvenile data. Regardless of the age of the animal, the litter has been shown to play a key role in the response of the animal to toxic insult, although it has been demonstrated that the effect of the litter decreases with increasing age.[105]

Many endpoints evaluated in a juvenile study are collected longitudinally. These endpoints, including body weights, food consumption, clinical pathology, and many of the behavioral tests, exhibit patterns of change over time. Often, laboratories statistically evaluate mean values at discrete intervals from test article–treated groups compared with the control group, while eyeballing alterations in the patterns over the assessment period. One method by which patterns are evaluated is through graphical presentation of the data. While it is clear that one can visually identify gross changes in body weight or food consumption by inspecting a graph, subtle changes in the pattern of growth may be overlooked. Moreover, altered patterns may not be identified by simple one-way analyses of variance (ANOVA), which are designed to evaluate snapshots in time. That is because the ANOVA proceeds by parsing the data into arbitrarily discrete blocks, which may have little to do with the actual pattern of change. However, one statistical approach that lends itself well to both a biological and statistical evaluation of the data is the repeated measures analysis of variance (RANOVA).

For a more complete discussion of the conduct and interpretation of the RANOVA, the reader is referred to the multitude of existing statistical texts. Briefly, however, the RANOVA is a means of detecting a main effect of treatment, a main effect of time, and an interaction effect of treatment by time.[171] Often, a main effect of treatment is a change in the cumulative mean of a particular data set in a treated group compared with the control group. Main effects of treatment, when presented graphically, would be evident by a shift in the overall data, without a change in the shape of the response curve. For many endpoints, main effects of time are not typically examined because the investigator expects to observe a change in the data with time (e.g., body weights change dramatically with age during the growth spurt of a juvenile animal). Interaction effects of treatment by time are observed when the difference from control changes over time, as evident by a difference in the slopes of the response curves for control and treated animals. More complex repeated measures analyses can also be conducted, using such factors in the analyses as sex and behavioral experience. An example of one such analysis was presented as part of the CBTS, and the reader is referred to these texts for excellent reviews of the analyses.[105,171]

V. CONCLUSIONS

Children are a unique population for health risk assessment; they require special considerations when exposed to environmental toxicants or when prescribed drugs with insufficient labeling information. Numerous tragedies affecting the pediatric population have shaped current scientific, medical, and public opinion, resulting in legislation that has encompassed both incentive-based voluntary participation and mandatory requirements for assessing the effects of xenobiotics in children. Although pediatric clinical trials for risk assessment are more common today, nonclinical juvenile animal studies are increasingly recommended or required for early hazard identification.

If nonclinical juvenile safety studies are undertaken, challenges inherent to their design include understanding the complexity of physiological and anatomic differences that exist between immature and mature organisms. The concept of physiological time is the basis for identifying the critical windows for exposure, and for selection of an appropriate animal model and the endpoints evaluated. Although guidance documents have been published, current nonclinical juvenile study designs are generally developed on a case-by-case basis. Factors influencing study design include the age of the target population, availability of safety and exposure data, relevancy of endpoints evaluated, and practicality of execution.

Regulatory guidelines and study designs will continue to evolve as more is learned about organ system maturation in various animal models, and that will lead to better predictive value of nonclinical juvenile studies to humans. Increased focus on safety assessment for pediatric populations should minimize the likelihood of future tragedies while improving the overall protection of children's health.

ACKNOWLEDGMENTS

The authors would like to acknowledge with deep appreciation the efforts of various individuals who contributed in tangible ways to the completion of this project. Special thanks go to Dr. Gerald Schaefer (WIL Research Laboratories, LLC) and Ms. Judy Buelke-Sam (Toxicology Services) for their thoughtful comments and suggestions regarding our discussion of behavioral assessments. Jon Hurley, Kim Rhodes, Robert Wally, and Matthew Coffee compiled and verified much of the WIL historical control data referenced in this chapter. The authors would also like to recognize the numerous staff members at WIL, too many to mention individually, who through their dedication, hard work, and meticulous data collection over the past 10 to 20 years have laid the foundation for this chapter.

References

1. Ginsberg, G., Hattis, D., Sonawane, B., Russ, A., Banati, P., Kozlak, M., Smolenski, S., and Goble, R., Evaluation of child/adult pharmacokinetic differences from a database derived from the therapeutic drug literature, *Toxicol. Sci.*, 66, 185, 2002.
2. Rice, D.C., Behavioral impairment produced by low-level postnatal PCB exposure in monkeys, *Environ. Res.*, 80, S113, 1999.
3. Rofael, H.Z., Turkall, R.M., and Abdel-Rahman, M.S., Immunomodulation by cocaine and ketamine in postnatal rats, *Toxicology*, 188, 101, 2003.
4. Schunior, A., Zengel, A.E., Mullenix, P.J., Tarbell, N.J., Howes, A., and Tassinari, M.S., An animal model to study toxicity of central nervous system therapy for childhood acute lymphoblastic leukemia: effects on growth and craniofacial proportion, *Cancer Res.*, 50, 6455, 1990.
5. Halpern, S.A., *American Pediatrics: The Social Dynamics of Professionalism, 1880-1980*, University of California Press, Berkeley, CA, 1988, p. 52.
6. U.S. Food and Drug Administration, *Guidance for Industry: E11 Clinical Investigation of Medicinal Products in the Pediatric Population*, U.S. Department of Health and Human Services, Food and Drug Administration, Rockville, Maryland, December 2000.
7. Cuzzolin, L., Zaccaron, A., and Fanos, V., Unlicensed and off-label uses of drugs in paediatrics: a review of the literature, *Fund. Clin. Pharmacol.*, 17, 125, 2003.
8. Kearns, G.L., Impact of developmental pharmacology on pediatric study design: overcoming the challenges, *J. Allergy Clin. Immunol.*, 196, S128, 2000.
9. Kearns, G.L., Abdel-Rahman, S.M., Alander, S.W., Blowey, D.L., Leeder, J.S., and Kauffman, R.E., Developmental pharmacology: drug disposition, action, and therapy in infants and children, *N. Engl. J. Med.*, 349, 1157, 2003.
10. Blumer, J.L., Off-label uses of drugs in children, *Pediatrics*, 104, 598, 1999.
11. Leeder, J.S. and Kearns, G.L., Pharmacogenetics in pediatrics: implications for practice, *Pediatr. Clin. North Am.*, 44, 55, 1997.
12. Brent, R.L., Utilization of animal studies to determine the effects and human risks of environmental toxicants (drugs, chemicals, and physical agents), *Pediatrics*, 113, 984, 2004.
13. Preclinical data may support single study for pediatric cancer drugs, *The Pink Sheet*, 66, 35, 2004.
14. Sutcliffe, A.G., Prescribing medicines for children: major problems exist, but there are some promising developments, *BMJ*, 319, 70, 1999.
15. Nordenberg, T., Pediatric drug studies: protecting pint-sized patients, *FDA Consumer Magazine*, 33(3), May–June 1999. Accessed April 21, 2004 at http://www.fda.gov/fdac/features/1999/399_kids.html.
16. Weiler, H.A., Paes, B., Shah, J.K., and Atkinson, S.A., Longitudinal assessment of growth and bone mineral accretion in prematurely born infants treated for chronic lung disease with dexamethasone, *Early Hum. Dev.*, 47, 271, 1997.
17. Evans, J.S. and Huffman, S., Update on medications used to treat gastrointestinal disease in children, *Curr. Opin. Pediatr.*, 11, 396, 1999.
18. Martin, R.M., Wilton, L.V., Mann, R.D., Steventon, P., and Hilton, S.R., Unlicensed and off label drug use for paediatric patients, *BMJ*, 317, 204, 1998.
19. Conroy, S. and Peden, V., Unlicensed and off label analgesic use in paediatric pain management, *Paediatr. Anaesth.*, 11, 431, 2001.
20. Shacter, E. and DeSantis, P.L., Labeling of drug and biologic products for pediatric use, *Drug Inf. J.*, 32, 299, 1998.

21. Antidepressant prescriptions by pediatricians have doubled: FDA analysis, *The Pink Sheet*, 66, 37, 2004.
22. Pina, L.M., Center ids top 10 drugs used off-label in out-patient setting, *News Along the Pike*, 3(1), 6–7, 17 January 1997. Accessed April 20, 2004 at http://www.fda.gov/cder/pike/jan97.pdf.
23. Wilson, J.T., Pragmatic assessment of medicines available for young children and pregnant or breast-feeding women, in *Basic and Therapeutic Aspects of Perinatal Pharmacology*, Morselli, P.L., Garattini, S., and Sereni, F., Eds., Raven Press, New York, 1975, p. 411.
24. Gilman, J.T. and Gal, P., Pharmacokinetic and pharmacodynamic data collection in children and neonates, *Clin. Pharmacokinet.*, 23, 1, 1992.
25. U.S. Food and Drug Administration, Center for Drug Evaluation and Research, *Office of Drug Evaluation Statistical Report*, U.S. Department of Health and Human Services publication 89–233530, Rockville, MD, 1989.
26. Steinbrook, R., Testing medications in children, *N. Engl. J. Med.*, 347, 1462, 2002.
27. Blassingame, K.M., Spending on childrens'[sic] Rx rises rapidly, *BenefitNews.com*, January 2003. Accessed May 11, 2004 at http://www.benefitnews.com/pfv.cfm?id=3882.
28. *Express Scripts Drug Trend Report*, Express Scripts®, 2001. Accessed May 10, 2004 http://www.express-scripts.com/other/news_views/outcomes2001/dt_report.pdf.
29. Andersen, D.H., Blanc, W.A., Crozier, D.N., and Silverman, W.A., A difference in mortality rate and incidence of kernicterus among premature infants allotted to two prophylactic antibacterial regimens, *Pediatrics*, 18, 614, 1956.
30. Gunter, J.B., Benefit and risks of local anesthetics in infants and children, *Paediatr. Drugs*, 4, 649, 2002.
31. Anderson, J.M., Cockburn, F., Forfar, J.O., Harkness, R.A., Kelly, R.W., and Kilshaw, B., Neonatal spongioform myelinopathy after restricted application of hexachlorophene skin disinfectant, *J. Clin. Pathol.*, 34, 25, 1981.
32. Lovejoy, F.H., Fatal benzyl alcohol poisoning in neonatal intensive care units, *Am. J. Dis. Child.*, 136, 974, 1982.
33. Dreifuss, F.E., Santilli, N., Langer, D.H., Sweeney, K.P., Moline, K.A., and Menander, K.B., Valproic acid hepatic fatalities: a retrospective review, *Neurology*, 37, 379, 1987.
34. Kremer, L.C.M. and Caron, H.N., Anthracycline cardiotoxicity in children, *N. Engl. J. Med.*, 351, 120, 2004.
35. Weiss, C.F., Glazko, A.J., and Weston, J.K., Chloramphenicol in the newborn infant: a physiological explanation of its toxicity when given in excessive doses, *N. Engl. J. Med.*, 262, 787, 1960.
36. American Academy of Pediatrics Committee on Drugs, Acetaminophen toxicity in children, *Pediatrics*, 108, 1020, 2001.
37. Lesko, S.M. and Mitchell, A.A., The safety of acetaminophen and ibuprofen among children younger than two years old, *Pediatrics*, 104, 1, 1999.
38. Kearns, G.L., Leeder, J.S., and Wasserman, G.S., Acetaminophen overdose with therapeutic intent, *J. Pediatr.*, 132, 5, 1998.
39. American Academy of Pediatrics Committee on Bioethics, Informed consent, parental permission, and assent in pediatric practice, *Pediatrics*, 95, 314, 1995.
40. U.S. Food and Drug Administration, Regulations requiring manufacturers to assess the safety and effectiveness of new drugs and biological products in pediatric patients: final rule, *Fed. Regist.*, 63, 66632, 1998 (codified at 21 CFR §231).
41. *Best Pharmaceuticals for Children Act*, Pub. L. 107–109, 107th U.S. Congress (January 4, 2002). Accessed March 17, 2004 http://www.fda.gov/opacom/laws/pharmkids/contents.html.
42. History of pediatric studies, rule, legislation and litigation, *Biotechnology Industry Organization*, Washington, DC, January 8, 2002. Accessed October 29, 2004 http://www.bio.org/reg/action/pedhist.asp.
43. *Pediatric Research Equity Act of 2003*, Pub. L. 108–155, 108th U.S. Congress (December 3, 2003). Accessed March 15, 2004 at http://www.fda.gov/opacom/laws/prea.html.
44. Congress gives FDA the power to require pediatric studies. *Arnold & Porter Update*, December 2003. Accessed April 4, 2004 at http://www.arnoldporter.com/pubs/files/Advisory_FDA_Pediatric_Studies.PDF.
45. *Food and Drug Administration Modernization Act of 1997*, Pub. L. 105–115, 105th U.S. Congress (November 21, 1997). Accessed April 10, 2004 at http://www.fda.gov/cder/guidance/105-115.htm.

46. U.S. Food and Drug Administration, Center for Drug Evaluation and Research, *Pediatric Exclusivity Labeling Changes as of May 7, 2004*. Accessed May 12, 2004 at http://www.fda.gov/cder/pediatric/labelchange.htm.
47. Drug companies are developing nearly 200 medicines for children, in *New Medicines in Development for Children,* The Pharmaceutical Research and Manufacturers of America (PhRMA), May 29, 2002. Accessed April 20, 2004 at http://www.phrma.org/newmedicines/resources/2002-05-29.57.pdf.
48. ICH Expert Working Group, ICH Harmonised Tripartite Guideline, *Maintenance of the ICH Guideline on Non-Clinical Safety Studies for the Conduct of Human Clinical Trials for Pharmaceuticals, M3(M)*, International Conference on Harmonisation of Technical Requirements for Registration of Pharmaceuticals for Human Use (ICH), Geneva, Switzerland, Finalized July, 1997.
49. American Academy of Pediatrics Committee on Drugs, Guidelines for the ethical conduct of studies to evaluate drugs in pediatric populations, *Pediatrics*, 95, 286, 1995.
50. Olson, H., Betton, G., Robinson, D., Thomas, K., Monro, A., Kolaja, G., Lilly, P., Sanders, J., Sipes, G., Bracken, W., Dorato, M., Van Deun, K., Smith, P., Berger, B., and Heller, A., Concordance of the toxicity of pharmaceuticals in humans and in animals, *Regul. Toxicol. Pharmacol.*, 32, 56, 2000.
51. U.S. Food and Drug Administration, Center for Drug Evaluation and Research, *Guidance for Industry: Nonclinical Safety Evaluation of Pediatric Drug Products*, U.S. Department of Health and Human Services, Rockville, MD, 2003 (Draft).
52. U.S. Environmental Protection Agency, *Toxicology data requirements for assessing risks of pesticide exposure to children's health: Report of the Toxicology Working Group of the 10X Task Force*, April 28, 1999 (Draft). Accessed October 13, 2004 http://www.epa.gov/scipoly/sap/1999/may/10xtx428.pdf.
53. Schettler, T., Stein, J., Reich, F., Valenti, M., and Wallinga, D., *In Harm's Way: Toxic Threats to Child Development*, Greater Boston Physicians for Social Responsibility (GBPSR), Cambridge, MA, 2000. Accessed October 29, 2004 at http://psr.igc.org/ihwrept/ihwcomplete.pdf.
54. U.S. Environmental Protection Agency, *Health Effects Test Guidelines: Prenatal Developmental Toxicity Study*, Office of Prevention, Pesticides and Toxic Substances (OPPTS), 870.3700, 1998.
55. U.S. Environmental Protection Agency, *Health Effects Test Guidelines: Reproduction and Fertility Effects*, Office of Prevention, Pesticides and Toxic Substances (OPPTS), 870.3800, 1998.
56. U.S. Environmental Protection Agency, *Health Effects Test Guidelines: Developmental Neurotoxicity Study*, Office of Prevention, Pesticides and Toxic Substances (OPPTS), 870.6300, August 1998.
57. Kimmel, C.A. and Makris, S.L., Recent developments in regulatory requirements for developmental toxicology, *Toxicol. Letters*, 120, 73, 2001.
58. Agunod, M., Yamaguchi, N., Lopez, R., Luhby, A.L., and Glass, G.B., Correlative study of hydrochloric acid, pepsin and intrinsic factor secretion in newborns and infants, *Am. J. Digest. Dis.*, 14, 400, 1969.
59. Rødbro, P., Krasilnikoff, P.A., and Christiansen, P.M., Parietal cell secretory function in early childhood, *Scand. J. Gastroenterol.*, 2, 209, 1967.
60. Huang, N.N. and High, R.H., Comparison of serum levels following the administration of oral and parenteral preparations of penicillin to infants and children of various age groups, *J. Pediatr.*, 42, 657, 1953.
61. Morselli, P.L., Antiepileptic drugs, in *Drug Disposition during Development*, Morselli, P.L., Ed., Spectrum, New York, 1977, p. 60.
62. Siber, G.R., Echeverria, P., Smith, A.L., Paisley, J.W., and Smith, D.H., Pharmacokinetics of gentamicin in children and adults, *J. Infect. Dis.*, 132, 637, 1975.
63. Miyawaki, T., Moriya, N., Nagaoki, T., and Taniguchi, N., Maturation of B-cell differentiation ability and T-cell regulatory function in infancy and childhood, *Immunol. Rev.*, 57, 61, 1981.
64. Barnett, J.B., Developmental immunotoxicology, in *Experimental Immunotoxicology*, Smialowicz, R.J. and Holsapple, M.P., Eds., CRC Press, Boca Raton, FL, 1996, p. 47.
65. Holliday, S.D. and Smialowicz, R.J., Development of the murine and human immune system: differential effects of immunotoxicants depend on time of exposure, *Environ. Health Perspect.*, 108, 463, 2000.
66. Dietert, R.R., Etzel, R.A., Chen, D., Halonen, M., Holladay, S.D., Jarabek, A.M., Landreth, K., Peden, D.B., Pinkerton, K., Smialowicz, R.J., and Zoetis, T., Workshop to identify critical windows of exposure for children's health: immune and respiratory systems work group summary, *Environ. Health Perspect.*, 108, 483, 2000.

67. Rice, D. and Barone, S., Jr., Critical periods of vulnerability for the developing nervous system: evidence from human and animal models, *Environ. Health Perspect.*, 108, 511, 2000.
68. Bos, A.F., Dibiasi, J., Tiessen, A.H., and Bergman, K.A., Treating preterm infants at risk for chronic lung disease with dexamethasone leads to an impaired quality of general movements, *Biol. Neonate*, 82, 155, 2002.
69. Barrington, K.J., The adverse neuro-developmental effects of postnatal steroids in the preterm infant: a systemic review of RCTs., *BMC Pediatr.*, 1, 1, 2001.
70. Gupta, A. and Waldhauser, L.K., Adverse drug reactions from birth to early childhood, *Pediatr. Clin. N. Am.*, 44, 79, 1997.
71. Miller, R.P., Roberts, R.J., and Fischer, L.J., Acetaminophen elimination kinetics in neonates, children, and adults, *Clin. Pharmacol. Thera.*, 19, 284, 1976.
72. Corcoran, G.B., Mitchell, J.R., Vaishnav, Y.N., and Horning, E.C., Evidence that acetaminophen and N-hydroxyacetaminophen form a common arylating intermediate, N-acetyl-p-benzoquinoneimine, *Mol. Pharmacol.*, 18, 536, 1980.
73. Coté, C.J., Kauffman, R.E., Troendle, G.J., and Lambert, G.H., Is the "therapeutic orphan" about to be adopted, *Pediatrics*, 98, 118, 1996.
74. U.S. Food and Drug Administration, Federal policy for the protection of human subjects: additional protections for children involved as subjects in research (technical amendment), *Fed. Regist.*, 56, 28032, 1991 (codified at 45 CFR §46.404-406).
75. U.S. Food and Drug Administration, Specific requirements on content and format labeling for human prescription drugs: revision of the "pediatric use" subsection in the labeling, *Fed. Regist.*, 59, 64240, 1994 (codified at 21 CFR §201.57).
76. Roberts, R.J., Overview of similarities and differences between children and adults: implications for risk assessment, in *Similarities & Differences Between Children & Adults: Implications for Risk Assessment*, Guzelian, P.S., Henry, C.J., and Olin, S.S., Eds., ILSI Press, Washington, D.C., 1992, p. 11.
77. ILSI Risk Science Institute Expert Working Group on Direct Dosing of Pre-weaning Mammals in Toxicity Testing, *Principles and Practices for Direct Dosing of Pre-weaning Mammals in Toxicity Testing and Research*, Zoetis, T. and Walls, I., Eds., ILSI Press, Washington, DC, 2003.
78. Diaz, J., Schain, R.J., and Bailey, B.G., Phenobarbital-induced growth retardation in artificially reared rat pups, *Biol. Neonate*, 32, 77, 1977.
79. Fonseca, N.M., Sell, A.B., and Carlini, E.A., Differential behavioral responses of male and female adult rats treated with five psychotropic drugs in the neonatal stage, *Psychopharmacologia*, 46, 253, 1976.
80. Steinschneider, M., Hexachlorophene, in *Experimental and Clinical Neurotoxicity*, Spencer, P.S. and Schaumburg, H.H., Eds., Oxford University Press, Oxford, 2000, p. 630.
81. Yokoyama, H., Onodera, K., Yagi, T. and Iinuma, K., Therapeutic doses of theophylline exert pro-convulsant effects in developing mice, *Brain Dev.*, 19, 403, 1997.
82. Buelke-Sam, J., Comparative schedules of development in rats and humans: implications for developmental neurotoxicity testing, presented at the 2003 Annual Meeting of the Society of Toxicology, Salt Lake City, UT, 2003.
83. Zoetis, T. and Hurtt, M.E., Species comparison of lung development, *Birth Defects Res. Part B Dev. Reprod. Toxicol.*, 68, 121, 2003.
84. Fouser, L. and Avner, E.D., Normal and abnormal nephrogenesis, *Am. J. Kidney Dis.*, 21, 64, 1993.
85. Gomez, R.A. and Norwood, V.F., Recent advances in renal development, *Curr. Opin. Pediatr.*, 11, 135, 1999.
86. Kleinman, L.I., Developmental renal physiology, *Physiologist*, 25, 104, 1982.
87. Zoetis, T. and Hurtt, M.E., Species comparison of anatomical and functional renal development, *Birth Defects Res. Part B Dev. Reprod. Toxicol.*, 68, 111, 2003.
88. Beck, M.J., Nemec, M.D., Schaefer, G.J., Stump, D.G., and Buelke-Sam, J., Validation of peripheral tissue cholinesterase activity assessment in rats administered chlorpyrifos by gavage [Abstract], *Teratology*, 61, P5, 2002.
89. Rice, D.C., Age-related increase in auditory impairment in monkeys exposed in utero plus postnatally to methylmercury, *Toxicol. Sci.*, 44, 191, 1998.
90. Rice, D.C., Evidence for delayed neurotoxicity produced by methylmercury, *Neurotoxicology*, 17, 583, 1996.

91. Barone, S., Jr., Stanton, M.E., and Mundy, W.R., Neurotoxic effects of neonatal triethyltin (TET) exposure are exacerbated with aging, *Neurobiol. Aging*, 16, 723, 1995.
92. Beck, M.J., Ching, S.V., Leung, E., Moorman, A.R., Lai, A.A., Radovsky, A., and Buelke-Sam, J., An acute toxicity and pharmacokinetic study of the ace inhibitor, ramipril (R), in juvenile rats [Abstract], *Birth Defects Res. Part A Clin. Molecular Teratol.*, 70, 294, 2004.
93. Snodgrass, W.R., Physiological and biochemical differences between children and adults as determinants of toxic response to environmental pollutants, in *Similarities and Differences Between Children and Adults: Implications for Risk Assessment*, Guzelian, P.S., Henry, C.J., and Olin, S.S., Eds., ILSI Press, Washington, DC, 1992, p. 35.
94. Plunkett, L.M., Turnbull, D., and Rodricks, J.V., Differences between adults and children affecting exposure assessment, in *Similarities and Differences Between Children and Adults: Implications for Risk Assessment*, Guzelian, P.S., Henry, C.J. and Olin, S.S, Eds., ILSI Press, Washington DC, 1992, p. 79.
95. Adolph, E.F. and Heggeness, F.W., Age changes in body water and fat in fetal and infant mammals, *Growth*, 35, 55, 1971.
96. Kacew, S., General principles in pharmacology and toxicology applicable to children, in *Similarities and Differences Between Children and Adults: Implications for Risk Assessment*, Guzelian, P.S., Henry, C.J. and Olin, S.S, Eds., ILSI Press, Washington DC, 1992, p. 24.
97. Radde, I.C., Mechanisms of drug absorption and their development, in *Textbook of Pediatric Clinical Pharmacology*, Macleod, S.M. and Radde, I.C., Eds., PSG Publishing Co., Littleton, MA, 1985, p. 17.
98. Ballatori N. and Clarkson T.W., Developmental changes in the biliary excretion of methylmercury and glutathione, *Science*, 216, 61, 1982.
99. Salvatoni, A., Piantanida, E., Nosetti, L., and Nespoli, L., Inhaled corticosteroids in childhood asthma: long-term effects on growth and adrenocortical function, *Paediatr. Drugs*, 5, 351, 2003.
100. Organization for Economic Cooperation and Development (OECD), *Guideline for Testing of Chemicals, Guideline 416, Two-Generation Reproduction Study*, Paris, France, 2001.
101. Gray, L.E., Jr. and Foster, P.M.D., Significance of experimental studies for assessing adverse effects of endocrine-disrupting chemicals, *Pure Appl. Chem.*, 75, 2125, 2003.
102. Gray, L.E. Jr., Wilson, V., Noriega, N., Lambright, C., Furr, J., Stoker, T.E., Laws, S.C., Goldman, J., Cooper, R.L., and Foster, P.M., Use of the laboratory rat as a model in endocrine disruptor screening and testing, *ILAR J.*, 45, 425, 2004.
103. George, J.D., Tyl, R.W., Hamby, B.T., Myers, C.B., and Marr, M.C., *One-Generation Extension Study of Vinclozolin and Di-n-butyl Phthalate Administered by Gavage on Gestational Day 6 to Postnatal Day 20 in Cd® (Sprague-Dawley) Rats*, presented at the Endocrine Disruptor Methods Validation Subcommittee (EDMVS) Seventh Plenary Meeting, June 5–6, 2003. Accessed October 29, 2004 at http://www.epa.gov/scipoly/oscpendo/docs/edmvs/onegenextensionfinalreport5803.pdf.
104. National Academy of Sciences, Sample size determination, in *Guidelines for the Care and Use of Mammals in Neuroscience and Behavioral Research*, National Academies Press, Washington, DC, 2003, p. 176, Appendix A.
105. Buelke-Sam, J., Kimmel, C.A., Adams, J., Nelson, C.J., Vorhees, C.V., Wright, D.C., Omer, V.St., Korol, B.A., Butcher, R.E., Geyer, M.A., Holson, J.F., Kutscher, C.L., and Wayner, M.J., Collaborative behavioral teratology study: results, *Neurotoxicol. Teratol.*, 7, 591, 1985.
106. U.S. Food and Drug Administration, ICH guideline on detection of toxicity to reproduction for medicinal products, *Fed. Regist.*, 59, 183, 1994.
107. Organization for Economic Cooperation and Development, *Guidelines for Testing of Chemicals. Section 4, No. 414, Teratogenicity*, adopted January 22, 2001.
108. Office of Food Additive Safety, IV.C.9.b., Guidelines for developmental toxicity studies, *Redbook 2000—Toxicological Principles for the Safety Assessment of Food Ingredients*, Office of Premarket Approval, U.S. FDA, CFSAN, College Park, MD, 2000.
109. U.S. Food and Drug Administration, Center for Biologics Evaluation and Research, *Guidance for Industry—Considerations for Reproductive Toxicity Studies for Preventive Vaccines for Infectious Disease Indications*, U.S. Department of Health and Human Services, Rockville, MD, 2000 (Draft).
110. Japanese Ministry of Agriculture, Forestry and Fisheries (MAFF), *Developmental Toxicity Testing Guidelines*, 1985.

111. Vitarella, D., James, R.A., Miller, K.L., Struve, M.F., Wong, B.A., and Dorman, D.C., Development of an inhalation system for the simultaneous exposure of rat dams and pups during developmental neurotoxicity studies, *Inhalation Toxicol.*, 10, 1095, 1998.
112. Barr, G.A. and Lithgow, T., Effect of age on benzodiazepine-induced behavioural convulsions in rats, *Nature*, 302, 431, 1983.
113. Sheets, L.P., A consideration of age-dependent differences in susceptibility to organophosphorous and pyrethroid insecticides, *Neurotoxicology*, 21, 57, 2000.
114. Cantalamessa, F., Acute toxicity of two pyrethroids, permethrin, and cypermethrin in neonatal and adult rats, *Arch. Toxicol.*, 67, 510, 1993.
115. Kallman, M.J., Assessment of motoric effects, in *Developmental Neurotoxicology*, Harry, G.J., Ed., CRC Press, Boca Raton, 1994, pp. 103–122.
116. Shafer, T.J., Meyer, D.A., and Crofton, K.M., Developmental neurotoxicity of pyrethroid insecticides: critical review and future research needs, *Environ. Health Perspect.* (in press, accessed October 29, 2004 at http://ehp.niehs.nih.gov/members/2004/7254/7254.pdf).
117. Schaefer, G.J., Nemec, M.D., Herberth, M.T., Pitt, J.A., Radovsky, A.E., and Knapp, J.F., Validation of developmental neurotoxicity endpoints in rats administered methimazole in drinking water [Abstract], *The Toxicologist*, 66, 234, 2002.
118. Ruppert, P.H., Dean, K.F., and Reiter, L.W., Comparative developmental toxicity of triethyltin using split-litter and whole-litter dosing, *J. Toxicol. Environ. Health*, 12, 73, 1983.
119. Pearson, D.E., Teicher, M.H., Shaywitz, B.A., Cohen, D.J., Young, J.G., and Anderson, G.M., Environmental influences on body weight and behavior in developing rats after neonatal 6-hydroxydopamine, *Science*, 209, 715, 1980.
120. Pizzi, W.J., Rode, E.C., and Barnhart, J.E., Differential effects of methylphenidate on the growth of neonatal and adolescent rats, *Neurotoxicol. Teratol.*, 9, 107, 1987.
121. Pizzi, W.J., Rode, E.C., and Barnhart, J.E., Methylphenidate and growth: demonstration of a growth impairment and a growth-rebound phenomenon, *Dev. Pharmacol. Ther.*, 9, 361, 1986.
122. Greely, G.H. and Kizer, J.S., The effects of chronic methylphenidate treatment on growth and endocrine function in the developing rat, *J. Pharmacol. Exp. Ther.*, 215, 545, 1980.
123. Zoetis, T., Tassinari, M.S., Bagi, C., Walthall, K., and Hurtt, M.E., Species comparison of postnatal bone growth and development, *Birth Defects Res. Part B Dev. Reprod. Toxicol.*, 68, 86, 2003.
124. Padgett, E.L., Chengelis, C.P., Rhodes, K.K., Holbrook, D.L., Haas, M.C., Lehman, J.L., Bell, C.R., and Nemec, M.D., Postpartum developmental reference data from beagle dogs over four defined pediatric stages, *The Toxicologist*, 66, 236, 2002.
125. Roy, A.K., Schiop, M.J., and Dowbenko, D.J., Regulation of the hepatic synthesis of alpha-2 urinary globulin and its corresponding messenger RNA in maturing male rats, *FEBS Letters*, 70, 137, 1976.
126. Neuhaus, O.W. and Flory, W., Age-dependent changes in the excretion of urinary proteins by the rat, *Nephron.*, 22, 570, 1978.
127. Gopinath, C., Prentice, D.E. and Lewis, D.J., *Atlas of Experimental Toxicological Pathology*, Kluwer Academic Publishers, Boston, 1987, p. 78.
128. Gomez R.A., Sequeira Lopez, M.L., Fernandez, L., Chernavvksy, D.R., and Norwood, V.F., The maturing kidney: development and susceptibility, *Ren. Fail.*, 21, 283, 1999.
129. Organization for Economic Cooperation and Development (OECD), Guideline for the testing of chemicals, *Developmental Neurotoxicity Study, Proposal for a New Guideline 426*, September 2003 (Draft).
130. Lochry, E.A., Concurrent use of behavioral/functional testing in existing reproductive and developmental toxicity screens, *J. Am. Coll. Toxicol.*, 6, 433, 1987.
131. ICH Expert Working Group, ICH Harmonised Tripartite Guideline, *Detection of Toxicity to Reproduction for Medicinal Products*, S5A, International Conference on Harmonisation of Technical Requirements for Registration of Pharmaceuticals for Human Use (ICH), Geneva, Switzerland, Finalized June 1993.
132. Adams, J., Methods in behavioral teratology, in *Handbook of Behavioral Teratology*, Riley, E.P. and Vorhees, C.V., Eds, Plenum Press, New York, 1986, p. 67.
133. Altman, J. and Sudarshan, K., Postnatal development of locomotion in the laboratory rat, *Anim. Behav.*, 23, 896, 1975.

134. Rinaldi, P.C. and Thompson, R.R., Age, sex and strain comparison of habituation of the startle response in the rat, *Physiol. Behav.*, 35, 9, 1985.
135. Glass, D.C. and Singer, J.E., *Urban Stress: Experiments on Noise and Social Stressors*, Academic Press, New York, 1972, p. 10.
136. Thompson, R.F. and Spencer, W.A., Habituation: a model phenomenon for the study of neuronal substrates of behavior, *Psychol. Rev.*, 73, 16, 1966.
137. Clark, R.L., Endpoints of reproductive system development, in *An Evaluation and Interpretation of Reproductive Endpoints for Human Health Risk Assessment*, Daston, G. and Kimmel, C.A., Eds., ILSI Press, Washington DC, 1998, 10.
138. Korenbrot, C.C., Huhtaniemi, I.T., and Weiner, R.I., Preputial separation as an external sign of pubertal development in the male rat, *Biol. Reprod.*, 17, 298, 1977.
139. Cameron, J.L., Metabolic cues for the onset of puberty, *Horm. Res.*, 36, 97, 1991.
140. Ashby, J. and Lefevre, P.A., The peripubertal male rat assay as an alternative to the Hershberger castrated male rat assay for the detection of anti-androgens, oestrogens and metabolic modulators, *J. Appl. Toxicol.*, 20, 35, 2000.
141. Kelce, W.R., Stone, C.R., Laws, S.C., Gray, L.E., Kemppainen, J.A., and Wilson, E.M., Persistent DDT metabolite p,p'-DDE is a potent androgen receptor antagonist, *Nature*, 375, 581, 1995.
142. Hossain, M., Devi, P.U., and Bisht, K.S., Effect of prenatal gamma irradiation during the late fetal period on the postnatal development of the mouse, *Teratology*, 59, 133, 1999.
143. Devi, P.U. and Hossain, M., Effect of early fetal irradiation on the postnatal development of mouse, *Teratology*, 64, 45, 2001.
144. Torrente, M., Colomina, M.T., and Domingo, J.L., Effects of prenatal exposure to manganese on postnatal development and behavior in mice: influence of maternal restraint, *Neurotoxicol. Teratol.*, 24, 219, 2002.
145. Adams, J., Buelke-Sam, J., Kimmel, C.A., Nelson, C.J., Reiter, L.W., Sobotka, T.J., Tilson, H.A., and Nelson, B.K., Collaborative behavioral teratology study: protocol design and testing procedures, *Neurobehav. Toxicol. Teratol.*, 7, 579, 1985.
146. Gray, L.E. Jr. and Ostby, J.S., In utero 2,3,7,8-tetrachlorodibenzo-p-dioxin (TCDD) alters reproductive morphology and function in female rat offspring, *Toxicol. Appl. Pharmacol.*, 133, 285, 1995.
147. Kim, H.S., Shin, J.H., Moon, H.J., Kang, I.H., Kim, T.S., Kim, I.Y., Seok, J.H., Pyo, M.Y., and Han, S.Y., Comparative estrogenic effects of p-nonylphenol by 3-day uterotrophic assay and female pubertal onset assay, *Reprod. Toxicol.*, 16, 259, 2002.
148. Wood, S.L., Beyer, B.K., and Cappon, G.D., Species comparison of postnatal CNS development: functional measures, *Birth Defects Res. Part B Dev. Reprod. Toxicol.*, 68, 391, 2003.
149. Campbell, B.A., Lytle, L.D., and Fibiger, H.C., Ontogeny of adrenergic arousal and cholinergic inhibitory mechanisms in the rat, *Science*, 166, 635, 1969
150. Buelke-Sam, J., Kimmel, G.L., Webb, P.J., Slikker, W. Jr., Newport, G.D., Nelson, C.J., and Kimmel, C.A., Postnatal toxicity following prenatal reserpine exposure in rats: effects of dose and dosing schedule, *Fundam. Appl. Toxicol.*, 4, 983, 1984.
151. Buelke-Sam, J., Ali, S.F., Kimmel, G.L., Slikker, W. Jr., Newport, G.D., and Harmon, J.R., Postnatal function following prenatal reserpine exposure in rats: neurobehavioral toxicity, *Neurotoxicol. Teratol.*, 11, 515, 1989.
152. Hoffman, H.S. and Fleshler, M., Startle reaction: modification by background acoustic stimulation, *Science*, 141, 928, 1963.
153. Ison, J.R. and Hammond, G.R., Modification of the startle reflex in the rat by changes in the auditory and visual environments, *J. Comp. Physiol. Psychol.*, 75, 435, 1971.
154. Vorhees, C.V., Acuff-Smith, K.D., Schilling, M.A., Fisher, J.E., Moran, M.S., and Buelke-Sam, J., A developmental neurotoxicity evaluation of the effects of prenatal exposure to fluoxetine in rats, *Fundam.. Appl. Toxicol.*, 23, 194, 1994.
155. Biel, W.C., Early age differences in maze performance in the albino rat, *J. Gen. Psych.*, 56, 439, 1940.
156. Vorhees, C.V., Butcher, R.E., Brunner, R.L., and Sobotka, T.J., A developmental test battery for neurobehavioral toxicity in rats: a preliminary analysis using monosodium glutamate, calcium carrageenan and hydroxyurea, *Toxicol. Appl. Pharmacol.*, 50, 267, 1979.
157. Morris, R.G.M., Spatial localization does not require the presence of local cues, *Learn. Motivat.*, 12, 239, 1981.

158. Olton, D.S. and Wenk, G.L., Dementia: animal models of the cognitive impairments produced by degeneration of the basal forebrain cholinergic system, in *Psychopharmacology: The Third Generation of Progress*, Meltzer, H.Y., Ed., Raven Press, New York, 1987, p. 941.
159. Tilson, H.A. and Harry, G.J., Neurobehavioral toxicology, in *Neurotoxicology*, Abou-Donia, M.B., Ed., CRC Press, Boca Raton, FL, 1999, p. 527.
160. Barrett, R.J., Leith, N.J., and Ray, O.S., Permanent facilitation of avoidance behavior by d-amphetamine and scopolamine, *Psychopharmacologia*, 25, 321, 1972.
161. Barrett, R.J., Leith, N.J., and Ray, O.S., A behavioral and pharmacological analysis of variables mediating active-avoidance in rats, *J. Comp. Physiol. Psychol.*, 82, 489, 1973.
162. Barrett, R.J., Leith, N.J., and Ray, O.S., An analysis of the facilitation of avoidance acquisition produced by d-amphetamine and scopolamine, *Behav. Biol.*, 11, 189, 1974.
163. Barrett, R.J. and Steranka, L.R., An analysis of d-amphetamine produced facilitation of avoidance acquisition in rats and performance subsequent to drug termination, *Life Sci.*, 14, 163, 1974.
164. Caul, W.F. and Barrett, R.J., Shuttle-box vs. Y-maze avoidance: The value of multiple response measures in interpreting active avoidance performance, *J. Comp. Physiol. Psychol.*, 84, 572, 1973.
165. Moser, V.C., Applications of a neurobehavioral screening battery, *J. Am. Coll. Toxicol.*, 10, 661, 1991.
166. U.S. Environmental Protection Agency, *Health Effects Test Guideline, Neurotoxicity Screening Battery*, Office of Prevention, Pesticides and Toxic Substances (OPPTS), 870.6200, August 1998.
167. Organization for Economic Cooperation and Development (OECD), *Guideline for Testing of Chemicals, Guideline 424, Neurotoxicity Study in Rodents*, Paris, France, 1997.
168. Hurley, J.M., Losco, P.E., Hermansky, S.J., and Gill, M.W., Functional observational battery (FOB) and neuropathology in New Zealand White rabbits [Abstract], *The Toxicologist*, 15, 247, 1995.
169. Moser, V.C., Screening approaches to neurotoxicity: a functional observational battery, *J. Am. Coll. Toxicol.*, 8, 85, 1989.
170. Office of Food Additive Safety, IV.C.9.a., Guidelines for reproduction studies, *Redbook 2000—Toxicological Principles for the Safety Assessment of Food Ingredients, Office of Premarket Approval*, U.S. FDA, CFSAN, College Park, MD, 2000.
171. Nelson, C.J., Felton, R.P., Kimmel, C.A., Buelke-Sam, J., and Adams, J., Collaborative behavioral teratology study: statistical approach, *Neurobehav. Toxicol. Teratol.*, 7, 587, 1985.

CHAPTER 9

Significance, Reliability, and Interpretation of Developmental and Reproductive Toxicity Study Findings

Joseph F. Holson, Mark D. Nemec, Donald G. Stump, Lewis E. Kaufman,
Pia Lindström, and Bennett J. Varsho

CONTENTS

I. Introduction 330
   A. Basis of Chapter 330
   B. Overview of Chapter 331
II. Significance of Developmental and Reproductive Toxicology 333
   A. Historical Background on Predictive Power of Animal Studies
      with Regard to Human Developmental and Reproductive Outcomes 333
   B. Animal-to-Human Concordance for Reproductive Toxicity 337
III. Study Design Considerations 342
   A. History of Study Design 344
      1. Pharmaceuticals 344
      2. Agricultural and Industrial Chemicals 347
   B. Species 348
   C. Characterization of Dose-Response Curve 350
   D. Dose Selection and Maximum Tolerated Dose 352
   E. General Statistical Considerations 353
      1. Statistical Power 353
      2. The Litter Effect 353
IV. Dose Range–Characterization Studies vs. Screening Studies 354
   A. Dose Range–Characterization Studies 354
   B. Developmental and Reproductive Toxicity Screening Studies 358
   C. Converging Designs 361
V. Developmental Toxicity Studies 361
   A. Guideline Requirements 362
      1. Animal Models 362
      2. Timing and Duration of Treatment 362
   B. Interpretation of Developmental Toxicity Study Endpoints 364
      1. Maternal Toxicity and Its Interrelationship with Developmental Toxicity 365

      2. Endpoints of Maternal Toxicity 366
      3. Intrauterine Growth and Survival 368
      4. Fetal Morphology 371
VI. Reproductive Toxicity Studies 382
   A. Guideline Requirements 382
      1. Animal Models 382
      2. Timing and Duration of Treatment 385
   B. Interpretation of Reproductive Toxicity Study Endpoints 385
      1. Viable Litter Size/Live Birth Index 386
      2. Neonatal Growth 387
      3. Neonatal Survival Indices 387
      4. Prenatal Mortality 388
      5. Assessment of Sperm Quality 388
      6. Weight and Morphology (Macroscopic and Microscopic) of Reproductive Organs 390
      7. Estrus Cyclicity and Precoital Interval 391
      8. Mating and Fertility Indices and Reproductive Outcome 392
      9. Duration of Gestation and Process of Parturition 394
      10. Endpoints Assessed during Lactation, Nesting, and Nursing 395
      11. Sexual Behavior 398
      12. Landmarks of Sexual Maturity 398
      13. Sex Ratio in Progeny 400
      14. Functional Toxicities and CNS Maturation 401
      15. Oocyte Quantitation 401
VII. Evaluation of Rare (Low Incidence) Effects 401
   A. Introduction 401
   B. Scope of the Problem 402
   C. Criticality of Accurate and Consistent Historical Control Data 404
   D. Case Studies 404
      1. Dystocia 404
      2. ACE Fetopathy 405
      3. Retroesophageal Aortic Arch 406
      4. Lobular Agenesis of the Lung 407
   E. Approach to Evaluating Rare Findings 408
VIII. The Role of Expert Judgment 409
   A. Integrity of Database 410
   B. Biologic Dynamics and Dimensions 412
   C. Relevancy and Risk Analysis 413
IX. Conclusion 414
Acknowledgements 416
References 418

I. INTRODUCTION

A. Basis of Chapter

The impetus for this chapter was the editor's request for, and the present authors' goal to produce, the first in-depth publication on the interpretation of guideline developmental and reproductive toxicity study findings. In recent years, increasing numbers of individuals without training in either of these disciplines have begun working in areas requiring the oversight, execution, and interpretation of these types of toxicity studies. For this reason, the authors have supplemented the data interpretation theme with information concerning the significance and reliability of the studies.

While recognizing that developmental toxicology is a subdiscipline of reproductive toxicology, the authors have separated them for reasons of convenience when examining these complex topics. However, regulatory protocol designs may entail exposure during, and evaluation of, both reproduction and development. Therefore, these terms occasionally have been used interchangeably in this chapter.

The "significance" aspect of this chapter is presented initially to demonstrate the scientific basis for viewing these studies as strong signals of potential human hazard. The "reliability" components of this chapter are interspersed, as they are associated with the particular measures or procedures of import. The sections on interpretation constitute the remainder of the chapter.

The guidance provided in this chapter is in part based on the extensive database at WIL Research Laboratories, Inc. (WIL) and is augmented by selected reports from the open literature. For 15 years, WIL has been one of a small number of contract research organizations (CROs) conducting large numbers of reproductive and developmental toxicity studies for regulatory submission. The authors have been involved with the oversight, review, and/or interpretation of more than 1500 such studies over a composite of 80 person-experience years. The experience gained at our laboratory is based on consistent management, staffing, and methodologies, with studies conducted primarily at a single facility maintaining best husbandry practices, training, and progressive approaches. In addition, many additional nonguideline studies have been designed and conducted to verify and/or clarify findings or elucidate modes of action for a wide array of products and chemistries, thus providing additional information concerning variability and validity of certain endpoints.

Central themes of this chapter include the following: (1) validity and concordance of experimental animal studies to human outcome scenarios for both developmental and reproductive toxicity endpoints, (2) analyses of selected aspects of the guidelines and their effects on reliability of the data, (3) appropriate statistical paradigms, in simple language, with computational examples, (4) critique and guidance relative to experimental techniques, (5) new (including previously unpublished) data for selected anatomic variants and other key measures common to these studies, (6) guidance for interpretation of substatistical findings (including rare events), and (7) the role of kinetic data in the interpretation of study outcomes.

It is not possible to cover all contingencies or combinations of findings in the limited space of this chapter. For example, in a single developmental toxicity study, the possible number of permutations of the variables may exceed tens of thousands. The numerous values and "trigger levels" of various endpoints cited herein probably extrapolate to other laboratory venues, databases, and experimental scenarios. However, exceptions may occur because of differences in levels of control of husbandry, training, and consistency of applying state-of-the-science standard operating procedures. In the final analysis, however, recognition of the interrelatedness of various endpoints is the basis for study interpretation.

B. Overview of Chapter

This chapter begins with a brief review and discussion of animal-to-human concordance of developmental and reproductive toxicity. A historical perspective is presented, which includes earlier work conducted in the late 1970s and early 1980s. This work forms the basis for many of today's regulatory and scientific practices in reproductive toxicology. Much of that early work has not been published in the open literature; therefore, parts of it are presented here. Of equal importance is the significant role of this work in validating animal models as reliable predictors of potential hazards to human development and reproduction. The latter fact elevates the importance of best possible practices in the design, conduct, interpretation, and extrapolation (use) of these data. Those issues are of great importance because animal studies are, and will remain for the foreseeable

Table 9.1 Sources of regulatory agency guidance for interpretation of study data

Title	Agency	Date	Type	Status	Ref.
Guidelines for Developmental Toxicity Risk Assessment	EPA	1991	Developmental toxicity	Older[a]	39
Standard Evaluation Procedure: Developmental Toxicity Studies	EPA	1992	Developmental toxicity	Older[a]	147
Standard Evaluation Procedure: Reproductive Toxicity Studies	EPA	1993	Reproductive toxicity	Older[a]	147
Guidelines for Reproductive Toxicity Risk Assessment	EPA	1996	Reproductive toxicity	Older[a]	40
Reviewer Guidance: Interpretation of Study Results to Assess Concerns about Human Reproductive and Developmental Toxicities	FDA	2001	Reproductive and developmental toxicity	Draft	5

[a] "Older" refers to documents written prior to significant regulatory guideline–driven changes in study design, methodology, and/or endpoints required.

future, the linchpin between hazard identification and protection of human development and reproduction.

This chapter also presents a brief comparison of salient differences between the various regulatory guidelines and requirements under which developmental and reproductive toxicity studies are conducted. This is followed by a practical discussion on interpretation of the various data endpoints derived from these studies. Such interpretation has not been addressed consistently or sufficiently in-depth, although various regulatory agencies have produced guidance documents on the subject (see Table 9.1).

In addition to the regulatory agency documents listed in Table 9.1, DeSesso et al.[1] and Palmer and Ulrich[2] have produced general overviews of the subject. To date, a committee of the International Life Sciences Institute (ILSI) has compiled the most comprehensive treatment of the interpretation of reproductive toxicology data.[3] However, this report is oriented toward risk assessment and does not directly address the interpretive aspects of the relevant endpoints that would typically be required for people making direct use of these data sets. That effort had been initiated, in part, out of concern expressed by some segments of industry because the newly promulgated U.S. Environmental Protection Agency (EPA) guidelines contain requirements for many new endpoints. The most recent publication of guidance for the interpretation of reproductive toxicity data was produced in 2002 under the aegis of the European Centre for Ecotoxicology and Toxicology of Chemicals (ECETOC).[4] This document provides guidance in the form of a structured approach to evaluation of reproductive toxicity data, taking into account the possible role of maternal toxicity and presenting examples from several fertility and developmental toxicity studies.

The above-mentioned publications present varying structures in which developmental and reproductive toxicity data may be evaluated. For example, the reviewer guidance document produced by the U.S. Food and Drug Administration (FDA)[5] broadly classifies the various indicators of developmental toxicity (mortality, dysmorphogenesis, alterations to growth, and functional toxicities) and reproductive toxicity (effects on fertility, parturition, and lactation). It utilizes a numerical scheme to assign levels of concern to the various findings. Regulators use these levels of concern in the overall assessment of risk to human reproduction and development. This classification scheme, popularly known as "the Wedge" (due to its pyramidal depiction), is primarily used to judge signal strength and is useful as an overall framework in which to evaluate the data. No system, however, no matter how well intentioned, can replace the breadth of scientific experience (including technical competency) and judgment garnered from the conduct and reporting of many developmental and reproductive toxicity studies in the same laboratory. Experience remains the crux of this level of application of the scientific method. In fact, scientific interpretation of the data drives the application of such classification schemes. Hence, providing guidance to interpretation of these data, as contained in this chapter, is a crucial exercise.

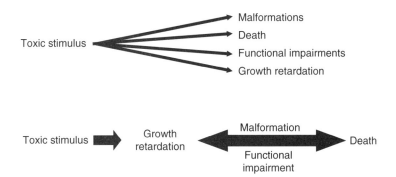

Figure 9.1 Relationship between toxic stimulus and adverse developmental outcome.

Table 9.2 Developmental toxicity following a single dose of mitomycin-C given to pregnant mice on gestational day 6

7½ days	Embryos smaller but morphologically normal
8½ days	Embryos smaller but morphologically retarded
10½ days	Embryos smaller but morphologically almost normal
13½ days	Embryos smaller, defects in one litter, reduced litter size
Birth	Size normal at birth, stillborn pups
Postnatal	Death (64% at 21 days), runts, motor defects, tremors
Maturity	Reduced fertility

Source: Snow, M.H.L. and Tam, P.P.L., *Nature*, 279, 555, 1979.

The appendix to this chapter contains the abbreviations used in this chapter.

II. SIGNIFICANCE OF DEVELOPMENTAL AND REPRODUCTIVE TOXICOLOGY

A. Historical Background on Predictive Power of Animal Studies with Regard to Human Developmental and Reproductive Outcomes

Prior to the early 1980s, numerous peer-reviewed articles were published comparing effects in laboratory animals and human beings.[6–9] Although animal studies to predict adverse human findings were federally mandated, there appeared to be a number of different views regarding the reliability of these animal models. The view at that time conceptualized the relationship between noxious insults to development as having four possible outcomes (malformation, death, functional impairment, and growth retardation), rather than constituting a continuum of response (see Figure 9.1). Experimental evidence to the contrary was found in work by Snow and Tam,[10] who utilized a longitudinal study design to demonstrate that the relationship between a toxic stimulus and adverse developmental outcomes is complex and exhibits a temporal dimension in both detection and expression. In their study, summarized in Table 9.2, they showed that a number of manifestations resulted from a single treatment of mice with mitomycin-C on gestational day (GD) 6 and that these manifestations varied dramatically, depending on the stage of development examined.

Further consideration of a number of known human teratogens led one of the present authors (J.F. Holson) to conclude that human manifestations of teratogenicity across exposure levels were most commonly multiple outcomes (see Table 9.3). Indeed, in these human cases, multiple manifestations of developmental toxicity (then referred to as teratogenicity) commonly resulted from exposure to the same external dose during the same period of development.

Table 9.3 Multiple manifestations of known human embryotoxins

Ionizing radiations	IUGR[a]
	Functional impairments
	Childhood cancer
	Malformations
	Death
Diethylstilbestrol	Malformations
	Clear-cell adenocarcinoma of cervix
	Functional changes
Alcohol	Malformations
	Functional impairment
	IUGR
	Lethality

[a] Intrauterine growth retardation.

Source: From U.S. Food and Drug Administration, National Center For Toxicological Research, *Reliability of Experimental Studies for Predicting Hazards to Human Development* (Kimmel, C.A., Holson, J.F., Hogue, C.I., and Carlo, G.L., Eds.), NCTR Technical Report for Experiment No. 6015, Jefferson, AR, 1984.

The above issues were recognized by J.F. Holson to be crucial for the use of animals in developmental and reproductive risk assessment, and became the basis for a concerted effort led by J.F. Holson, funded by the FDA and carried out at the National Center for Toxicological Research (NCTR) to evaluate the concordance (predictive reliability) of experimental animal findings to known human situations of xenobiotic-induced maldevelopment. In 1984, a final report was completed for this project, entitled *Reliability of Experimental Studies for Predicting Hazards to Human Development*.[11] This report was never published in the open literature. Selected highlights are presented here because they have direct bearing on the subject of significance of findings from regulatory reproductive toxicity studies. Also, the findings of that study were used to reinforce the value of experimental studies as animal models in EPA data assessment guidelines for developmental toxicology.

In Table 9.4, the five original assumptions that were made for assessing reliability of animal studies are presented. Unfortunately, in none of the other publications on this topic at the time of this research, nor since, were these assumptions considered in evaluating concordance.

In Table 9.5, a list of agents known to be human prenatal toxicants at the time of this study is presented. The list has not been updated from the original report. Thus, it reflects the original findings of this study of concordance and provides a historical perspective. This list was dictated by Assumption 1 in Table 9.4 that concordance between animals and humans could not be assessed in any other manner than by using established effects in the human. For each agent, the first year for which there was a report is presented in the second column. The species for which adverse effects on prenatal development were discovered are presented in the third column, with the first species presented in parentheses.

A closer evaluation of these data leads to several inferences. First, agents known to disrupt human development affect multiple species. Second, with the exception of thalidomide and polychlorinated biphenyls, all agents disrupting human development were first reported in a laboratory animal study. Third, in general, scientists, physicians, and regulators did not heed these initial reports of laboratory animal studies as adequate signals of hazard. Were one to update this list with current knowledge, similar relationships would be found, with the exception that much more credence is now given to findings from well-designed and interpreted animal developmental toxicity studies. That credence emanates from this NCTR study and experience gained during the intervening period.

The fifth assumption made in the NCTR study was that sensitivity of the model was based on comparability of "effect levels." These comparisons are presented in Table 9.6. At the time of the study, little pharmacokinetic (PK) or toxicokinetic (TK) data were available so only the "administered dose" could be used. However, with the exception of thalidomide, there was remarkable

Table 9.4 Original assumptions for assessing reliability of animal tests

1. Only agents with an established effect in humans and adequate information for both humans and animals could be evaluated for concordance of effects.
    - Compounds for which no effect was indicated may actually have been negative or may have been a false negative due to the inability to detect effects because of inadequate power of the studies.
2. Statistical power of the study designs had to be considered in order to evaluate adequacy of the data and apparent "species differences" in response.
    - Situations in which large animal studies may have been matched to a few case reports and a conclusion drawn as to the poor predictability of the animal studies were noted and reevaluated.
3. The multiplicity of endpoints in developmental toxicity comprise a continuum of response (i.e., dysmorphogenesis, prenatal death, intrauterine growth retardation, and functional impairment represent different degrees of a developmental toxicity response).
    - Although this assumption would be debated by some, the weight of experimental and epidemiological evidence tended to support rather than refute the assumption.
    - The examples of fetal alcohol syndrome, diethylstilbestrol, and methylmercury were discussed in support of this assumption.
4. Manifestations of prenatal toxicity were not presumed to be invariable among species (i.e., animal models were not expected to exactly mimic human response).
    - Also, the human population has exhibited an array of responses that were determined by the magnitude of the exposure, timing of the exposure, interindividual differences in sensitivities due to genotype, interaction with other types of exposure, and interaction among all of these factors.
    - Just as the human and rat are not the same, all human subjects are not identically responsive to exogenous influences.
5. Sensitivity was based on comparability of the "effect levels" among species.
    - This was necessary because for most established human developmental toxicants there was still not adequate dose-response information available to compare sensitivities among species.

Source: Summarized from U.S. Food and Drug Administration, National Center for Toxicological Research, *Reliability of Experimental Studies for Predicting Hazards to Human Development* (Kimmel, C.A., Holson, J.F., Hogue, C.I., and Carlo, G.L., Eds.), NCTR Technical Report for Experiment No. 6015, Jefferson, AR, 1984.

Table 9.5 Agents that cause developmental toxicity

Agent	Year	Species[a]	Ref.
Alcohol(ism)	1919	(rat), gp, ch, hu, mo	148
Aminopterin	1950	(mo & rat), ch, hu	149
Cigarette smoking	1941	(rab), hu, rat	150
Diethylstilbestrol	1939	(mo & rat), hu, mi	151
Heroin/morphine	1969	(rat), ha, hu, rab	152
Ionizing radiation	1950	(mo), ha, hu, rat, rab	153
Methylmercury	1965	(rat), ca, hu, mo	154
Polychlorinated biphenyls	1969	(hu), rat	155
Steroidal hormones	1943	(monk), ha, hu, mo, rat, rab	156
Thalidomide	1961	(hu), mo, monk, rab	157,158

[a] ca — cat, ch — chicken, ha — hamster, gp — guinea pig, hu — human, mi — mink, mo — mouse, monk — monkey, rat — rat, rab — rabbit.

Source: Adapted from U.S. Food and Drug Administration, National Center For Toxicological Research, *Reliability of Experimental Studies for Predicting Hazards to Human Development* (Kimmel, C.A., Holson, J.F., Hogue, C.I., and Carlo, G.L., Eds.), NCTR Technical Report for Experiment No. 6015, Jefferson, AR, 1984.

Table 9.6 Effect levels for selected developmental toxicants in humans and test species

Agent	Species	Dose	Response
Aminopterin	Human	0.1 mg/kg/day	Death and/or malformations
	Rat	0.1 mg/kg/day	Death and/or malformations
Diethylstilbestrol	Human	0.8–1.0 mg/kg	Genital tract abnormalities and/or death
	Mouse	1 mg/kg	Genital tract abnormalities and/or death
Ionizing radiation	Human	20 rd/day	Malformations
	Rat and mouse	10–20 rd/day	Malformations
Cigarette smoking	Human	20 cigarettes/day	Growth retardation
	Rat	>20 cigarettes/day	Growth retardation
Thalidomide	Human	0.8–1.7 mg/kg	Malformations
	Monkey	1.25–20 mg/kg	Malformations
	Rabbit	150 mg/kg	Malformations

Source: Adapted from U.S. Food and Drug Administration, National Center For Toxicological Research, *Reliability of Experimental Studies for Predicting Hazards to Human Development* (Kimmel, C.A., Holson, J.F., Hogue, C.I., and Carlo, G.L., Eds.), NCTR Technical Report for Experiment No. 6015, Jefferson, AR, 1984.

similarity in effect levels among animals and people for the examples chosen. With the dramatic increase in analytical capabilities and the more generalized use of metabolic, and PK and pharmacodynamic (PD) studies, comparison of adverse effects is now more routinely done based on the area under the curve (AUC) and the peak concentration (C_{max}) of the agent. Hence, differences in internal exposure among species are often demonstrated, and these reveal that most historical instances that were initially believed to be "species differences" in response actually resulted from differences in bioavailability or metabolism. Specific examples of true "biologically based differences" in embryonal response have not been demonstrated. This is understandable given the conserved nature of embryology and development and its great similarity among mammalian and submammalian species. For further discussion of this topic, the reader is referred to an excellent report prepared by the Committee on Developmental Toxicology of the National Research Council.[12]

Examples of compounds that are developmentally toxic in humans are presented in Table 9.7. As indicated in this table, the embryo or fetal toxicity of a substance may manifest at exposure levels that do not significantly affect the mother during pregnancy, at levels that cause maternal stress or alter the physiological state of the mother, or at or near levels that produce overt maternal toxicity. Animal developmental toxicity studies have revealed a similar range of responses. This striking similarity in the range of relationships between maternal toxicity (or absence thereof) and developmental outcome in humans is mirrored in laboratory animal studies. This point is important because it further confirms the similarities among species in regard to maternal toxicity and developmental outcome. It also indicates that the presence of maternal toxicity cannot *a priori* be assumed as the causative agent of maldevelopment in any particular instance.

The database from the epidemiology and animal studies in the NCTR study is sufficient in terms of relative power. Along with identification of dose-response relationships, it permits the deduction that, overall, toxic insults to the maternal and embryonal or fetal organisms are similar enough for all agents to presume concordance between animal models and humans, even when only external dose is considered. Inclusion of PK and TK studies has refined (and will continue to refine) the quantitative aspects of extrapolation of experimental findings to human risk assessment.

Table 9.8 presents a list of publications and reports that address the issue of animal-to-human concordance. These are presented along with key attributes for each because they represent excellent source material. Except for the first two reports, none contains critical analyses of the primary literature or use preestablished evaluation criteria.

As the findings from the NCTR study indicate, further bolstered by the other studies and publications listed in Table 9.8, the significance of experimental indications of developmental toxicity should garner great concern. Many scientists not working in the reproductive toxicology

Table 9.7 Relationship between maternal and fetal toxicity for established human developmental toxicants

Embryo and fetal toxicity produced at or near exposure levels that elicit overt maternal toxicity:

- Aminopterin
- Methylmercury
- Polychlorinated biphenyls

Embryo and fetal toxicity produced at exposure levels that cause maternal physiologic changes or stress:

- Cigarette smoking
- Steroidal hormones
- Ethanol consumption

Embryo and fetal toxicity produced at exposure levels that do not produce significant maternal effects during pregnancy:

- Ionizing radiation
- Diethylstilbestrol
- Thalidomide

Source: Adapted from Holson, J. F., Estimation and extrapolation of teratologic risk, in *Proceedings of NATO Conference on in vitro Toxicity Testing of Environmental Agents: Current and Future Possibilities*, NATO, Monte Carlo, Monaco, 1980.

area, corporate representatives, animal rights activists, and lay people are often confused or misguided by inconsistencies in technical reports or lay publications that allege differences between human experience and animal studies. Such allegations present a problem for the future of the discipline because they have contributed to past catastrophes. In Table 9.9, an attempt has been made to summarize the reasons for the high frequency of such misrepresentations. These reasons are self-explanatory, with the exception of the last entry, which the authors have included out of necessity because none can predict all possibilities for the future. No established or generally accepted example of an agent that has adversely affected human development but has not induced adverse developmental effects in an animal model currently exists. The reader is referred to the original assumptions made in the NCTR study to better reconcile the reasons for apparent discrepancies listed in Table 9.9.

Developmental toxicity studies in laboratory animals provide the only controlled and ethical means of identifying the potential risks of a large array of drugs, chemicals, and other agents to the developing human embryo and fetus. However, they cannot replace direct human evidence or experience. Often, the findings from the animal models will be supplemented by confirmation in pregnancy registries. These registries involve collection of information from human pregnancies through use of questionnaires and may be conducted after product approval and/or marketing. However, pregnancy registries are variably sensitive, and rarely serve as early predictors of developmental hazard. Therefore, the design and interpretation of results from animal developmental toxicity studies are critical to the hazard identification phase of risk assessment, and in the case of equivocal findings, such studies will guide the use of postmarketing registry analyses.

B. Animal-to-Human Concordance for Reproductive Toxicity

Reproductive toxicology is also a complex field of science, as many authors of textbooks and review articles have declared in their opening paragraphs. The reason for this complexity is that sexual reproduction involves the union of gametes from two genders, resulting in progeny that in turn experience development and maturation of the myriad processes that enable continuation of the "life cycle." That cycle comprises several life stages, each having a unique biology and temporally entrained acquisition of features. Reproductive physiology (and its dynamic anatomy) includes

Table 9.8 Studies and publications concerning animal-to-human concordance for developmental and reproductive toxicity

Authors	Ref.	Attributes
Holson, 1980 (Proceedings of NATO Conference) Holson et al., 1981 (Proceedings of Toxicology Forum) NCTR Report No. 6015, 1984	11, 95, 159	Interdisciplinary team (epidemiologists and developmental toxicologists) Critical analysis of primary literature Applied criteria for acceptance of data/conclusions in reports and included power considerations Established and applied concept of multiple developmental toxicity endpoints as representing signals of concordance Qualitative outcomes and external dose comparisons made No measures of internal dose Critical analysis of primary literature
Nisbet and Karch, 1983 (Report for the Council on Environmental Quality)	129	Many chemicals and agents addressed Limited review of primary literature Not a critical analysis of primary literature Relied on authors' conclusions No power analyses Limited use of internal dose information
Brown and Fabro, 1983 (journal article)	160	Not a critical analysis of primary literature Made use of findings from other reviews Excellent review and presentation of overall concordance issues
Hemminki and Vineis, 1985 (journal article)	130	Interspecies inhalation doses adjusted Relied on authors' conclusions Twenty three occupational chemicals and mixtures No measures of internal dose
Buelke-Sam and Mactutus, 1990 (journal article)	161	Small number of agents covered Critical review of the qualitative and quantitative comparability of human and animal developmental neurotoxicity Limited use of internal dose measures
Newman et al., 1993 (journal article)	132	Provided detailed information Only four drugs evaluated Emphasis on morphology Focus on NOAELs[a] No measures of internal dose
Shepard, 1995 (book, 8th ed.)	162	Computer-based annotated bibliography Catalog of teratogenic agents Not a critical analysis Limited comments regarding animal-to-human concordance for a limited number of agents No use of internal dose measures
Schardein, 2000 (book, 3rd ed.)	163	Extensive compilation of open literature Not a critical analysis Variably relied on authors' conclusions No measures of internal dose nor criteria for inclusion or exclusion of studies Only partially devoted to concordance issues

[a] No-observed-adverse-effect level.

many of biology's most complicated endocrinologic feedback and control mechanisms. It further encompasses the transient development (appearance and disappearance) of structures and functions, the roles of which are crucially timed to comprise the reproductive cycle. To our knowledge, and based on a recent search of the literature, there is no published rigorous and critical analysis–based study of concordance between humans and laboratory animal models for reproductive toxicity outcomes. Several publications in the early 1980s (refer to Table 9.10) attempted to address the issue, but were limited in scope because of data gaps (e.g., lack of exposure quantification, lack of animal and/or human data for key reproductive endpoints). Admittedly, because of the numerous endpoints typically evaluated in reproduction studies, in contrast to the more focused issues of

Table 9.9 Reasons for apparent discrepancies between animal and human effects

When it is reported or found that an agent causes developmental effects in test animals, but not in humans, this may be due to the following factors:

- The animal study is flawed or improperly interpreted.
- The chemical does not reach the marketplace so no human exposure occurs.
- Exposure occurs but is not high enough to cause an effect.
- Large enough populations with documented exposures are difficult to identify, and the effects may be confounded by other factors.
- Effects may occur but are below the level of detection for epidemiological studies.
- The agent actually does not affect development in humans.

Table 9.10 Publications concerning animal-to-human concordance for reproductive toxicity

Authors	Ref.	Attributes
Barlow and Sullivan, 1982 (book)	164	Many industrial chemicals addressed
		Critical analysis of primary literature
		Did not apply criteria for inclusion of chemicals reviewed based on extent of available data
		No power analysis
		Limited number of endpoints compared
		Limited internal dose information and human external exposure data lacking
U.S. Congress, Office of Technology Assessment, 1985 (chapter in book)	165	Eleven agents covered
		Not a critical analysis of primary literature
		Limited to one reproductive toxicity concordance table (created entirely from previously referenced Barlow and Sullivan (1982) and Nisbet and Karch (1983) reviews) and a very brief general discussion
		Only addressed concordance of effect (no assessment of concordance of dose)

developmental toxicity studies, development of a comprehensive review would be a daunting task. Thus a crucial need remains for an organized and critical analysis of the primary literature in reproductive toxicology to evaluate the concordance of regulatory reproductive toxicity studies to human reproductive outcomes. There have been attempts to compare effects in laboratory animals with those in humans for specific endpoints or by gender. One such study, Ulbrich and Palmer,[13] examined 117 substances or substance classes and proposed that histopathology and organ-weight analysis were the best general-purpose means for detecting substances that affect male fertility. These authors likewise concluded that sperm analysis was a realistic alternative to histopathology and organ-weight analysis when these proved to be impractical. Further, it has been reported that predictability may be observed in results from well-validated models,[14] and several publications have reviewed factors that are important in assessing risk to the male reproductive system.[14–17] Certainly, however, the most critical feature of an animal model, as with the use of models in all areas of toxicology, should be the use of TK data, especially biotransformation data.

In mammalian systems, perturbation of the reproductive organs or disturbances in the integrated control of the process of reproduction can originate from multiple sources. For example, cadmium disrupts testicular function secondarily to its main effect on vascular integrity to the vessels supplying the testes. The nematocide dibromochloropropane (DBCP) has several sites of action, including the testis and epididymis. Through conversion to α-chlorohydrin, DBCP causes epididymal vascular damage (progressing to sperm granulomas and spermatoceles) and decreases epididymal respiratory activity and motility, providing the basis of its putative action in the epididymis.[18] Both cadmium and DBCP ultimately induce infertility but act via divergent pathways. Standard hazard evaluation studies would not reveal the mode of action of either agent, and follow-up studies would be required to identify the ultimate target site.

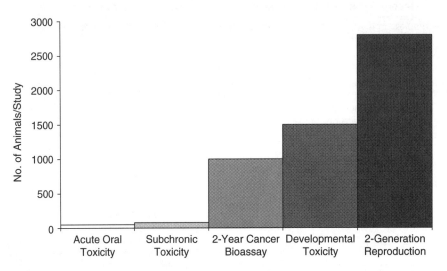

Figure 9.2 Comparison of the number of animals on study in various experimental designs.

Changes in a specific reproductive process in which the mode of action is well understood and conserved among species are considered a clear threat to human health. Command of the issues will afford the most enlightened use (or will prevent misuse) of the data for the intended purpose. Reproductive and developmental toxicity studies are large in scale and complex (Figure 9.2). No matter what systematic approach is applied to the data to formulate conclusions, the final interpretation must consider interrelated endpoints collectively if the risk assessment process is to proceed efficiently and effectively.

One potential problem related to extrapolation of reproductive hazard from animals to humans is embedded in the phenotypic diversity of humanity. Expansion of the human population over many years has given rise to a heterogeneous gene pool. Animals that are purpose-bred for experimental applications are genotypically more homogeneous than, and less predictive of, the outlier responders in the diverse human population. The logical assumption is that humans are more likely to be predisposed to idiosyncratic toxicologic responses. A second potential problem is that with reproductive toxicity, as compared to developmental toxicity, there are many more structures and processes in the maternal and paternal animals that may be affected or manifest toxicity. This is in contrast to prenatal development, with its highly conserved embryo and fetus, which has a much more limited phenotypic variety among species. While some phenotypic diversity may exist in embryos and fetuses among species, it is certainly far less than in the adult life stage, in which numerous structures reside in either or both of the parental sexes. Hence, caution must be exercised when extrapolating effects from animal models to humans for many aspects of reproductive toxicity. Nevertheless, the finding of reproductive toxicity in a well-conducted guideline study should be considered to constitute a strong signal, which currently could be negated only by appropriate information on mode of action or the existence of an adequate study in humans. The use of epidemiology has become increasingly important in establishing cause and effect relationships.[19]

It is an essential scientific need that an organized and critical analysis of the primary literature in reproductive toxicology be conducted to evaluate the concordance of regulatory reproductive toxicity studies to human reproductive outcomes. Such a study would need to be modeled after the FDA-NCTR study discussed in the beginning of this chapter. Assumptions for inclusion and exclusion of papers, consideration of power, establishment of exposure parameters, and criteria for concordance would need to be developed and used.

Figure 9.3 is an idealized depiction of the relationship between progression of time in development and the degree of phenotypic variability, both on an ontogenic and species basis. The earlier

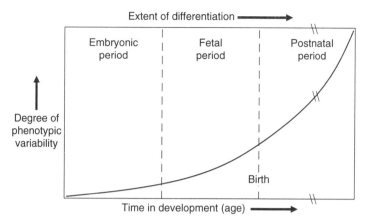

Figure 9.3 Relationship between progression of time in development and the degree of phenotypic variability.

in development, the more similar or conserved are the fundamental processes of development and their similarity among species. As development progresses, expression of individual and "species-specific" characteristics increases. In middle and late postnatal development, it would be expected that species-specific attributes that have evolved for adaptive success will be expressed and may represent vulnerable phenomena, not present in humans or having less importance to human health. Thus, on a theoretical basis, it might be expected that the later in development a model is used, the greater the number of species differences that might be encountered among models. It would be expected that for mammalian species these differences would be few and not so numerous as to thwart the rational use of laboratory models. With the advent of regulations for juvenile (pediatric) studies, experience over the next decade will be key to answering this question. The timing of onset of functions may be problematic, as the time of parturition in different species becomes a confounding factor for conducting the experimental studies. An example is maturation of the renin-angiotensin system, which occurs periparturitionally and postnatally in the rat and prenatally in the human.

DBCP is an insecticide that has been widely used on banana plantations and as a fumigant for grain and soil. Reported effects among workers at DBCP production sites in California and in Israel included prolonged azoospermia and oligospermia, with a strong correlation between the duration of employment and testicular function. Potashnik and Porath published data indicating that agricultural workers exposed to DBCP in Israel had impaired fertility or were sterile.[20] These investigators reported that while follicle stimulating hormone (FSH) and luteinizing hormone (LH) were significantly increased, testosterone levels were not significantly decreased in men who were severely affected by the DBCP exposure.[20] Similarly, increased gonadotropin levels were correlated with decreased sperm counts in agricultural applicators in California.[21] Meistrich et al. reported prolonged oligospermia in LBNF$_1$ rats injected with DBCP.[22] Within 6 to 20 weeks post-treatment, only 20% of the seminiferous tubules contained differentiating germ cells. A majority of the tubules (70%) had germinal epithelium and Sertoli cells but no differentiating germ cells. Morphologic alteration of the Sertoli cells was noted in the remaining 10% of the tubules. In contrast to the human hormone levels, both intratesticular testosterone and gonadotropin levels were increased in the DBCP-treated rats. The authors concluded that DBCP exerted its effects by inducing Type A spermatogonia to undergo apoptosis rather than differentiation. These data in humans and rodents demonstrate adequate concordance for hazard identification purposes.

Rats and humans exhibit other differences in the reproductive cycle, such as the intrinsic feature of rat reproduction in which the LH surge is synchronized with the onset of heat to ensure fertility. Agents that block norepinephrine binding to alpha-2 receptors, inhibitors of dopamine beta-hydroxylase activity, gamma-amino butyric acid (GABAnergic) receptor agonists, delta-9-tetrahydrocannabinol, and methanol attenuate or ablate the amplitude and alter the timing of the LH surge in the rat, thereby delaying or preventing ovulation.[23] Alteration of the LH surge in rats appears to be a common site of vulnerability for a wide array of agents and chemistries. Most often the results of such disruptions were initially observed as changes in fertility as well as changes in indices such as fertility, ovulation, implantation efficiency, and litter size in two-generation reproductive performance toxicity studies. Administration of a single dose of LH-disrupting agents on the day of proestrus can produce multiple adverse reproductive outcomes. By interfering with the hypothalamic regulation of the preovulatory surge of LH, these agents accelerate the normal loss of ovarian cycling in rats by inducing early onset of the normal age-associated impairment of central nervous system (CNS)–pituitary control of ovulation. The hormonal milieu present in the aging female rat is one of persistent estrogen secretion (from persistently developing follicles) and no secretion of progesterone. This milieu is the precise endocrine environment known to facilitate the development of mammary gland tumors in rats reported in a number of studies.[24–26] However, it is unlikely that this response bears relevance to humans because menopause in women results from depletion of primordial ovarian follicles and the subsequent decline in estrogen levels. Although inhibition of ovulation in the rat by these agents is relevant to human reproduction, effects on human fertility are generally considered threshold-driven responses. Therefore, the benefits of drugs inducing these changes can outweigh the risks if a sufficiently large margin of safety can be demonstrated. Most regulatory attention, however, focuses on the basis of carcinogenesis because it is viewed as obeying a linear dose-response, even when the mode of action is on a reproductive process.

From the previous discussion, the need for a well-conceived and conducted concordance study becomes evident. Likewise, the complexities of such an endeavor are also obvious.

III. STUDY DESIGN CONSIDERATIONS

Developmental toxicity studies are designed to assess effects on prenatal morphogenesis, growth, and survival. These studies are generally required for all regulated agents (e.g., drugs, biologics, pesticides, industrial chemicals, food additives, veterinary drugs, and vaccines). While the study designs are similar for the various compounds that are tested, customized guideline requirements are available to address unique properties and characteristics of each of the above categories. For example, the basic developmental toxicity study design for drugs suggests initiation of exposure either at implantation or immediately following mating. However, administration of vaccines takes place prior to mating to ensure that an immune response will occur during gestation. A pivotal consideration in the evaluation of biologics is the frequent lack of demonstrable maternal internal dose. In these cases, a key pharmacologic biomarker may serve as a proxy measure of exposure. The key features to all developmental toxicity designs are: (1) that maternal exposure is sustained throughout major organogenesis and (2) that the dam is necropsied 1 to 2 days prior to expected parturition so that all products of conception can be evaluated.

In contrast to developmental toxicity studies, reproductive toxicity studies are designed to assess fertility and reproductive outcome. Since the advent of the 1966 FDA guidelines,[27] the testing of medicinal products has used a segmented approach to reflect human exposure. Because most pharmaceuticals are administered for discrete courses of therapy, possess short half-lives, and minimally (if at all) bioaccumulate, the segmented approach using administration to a single generation is appropriate. Alternatively, reproductive toxicity testing of industrial chemicals, pesticides, and food additives is intended to assess effects of low-level exposure (intentional and/or unintentional) over a significant portion of the life span that may result in bioaccumulation. The test animals are dosed via routes of administration that mimic human exposure. Later sections of this chapter provide more in-depth discussion of specific aspects of the guideline developmental and reproductive toxicity study designs relating to data interpretation. Figure 9.4 presents a schematic comparison of the various developmental and reproductive toxicity study designs.

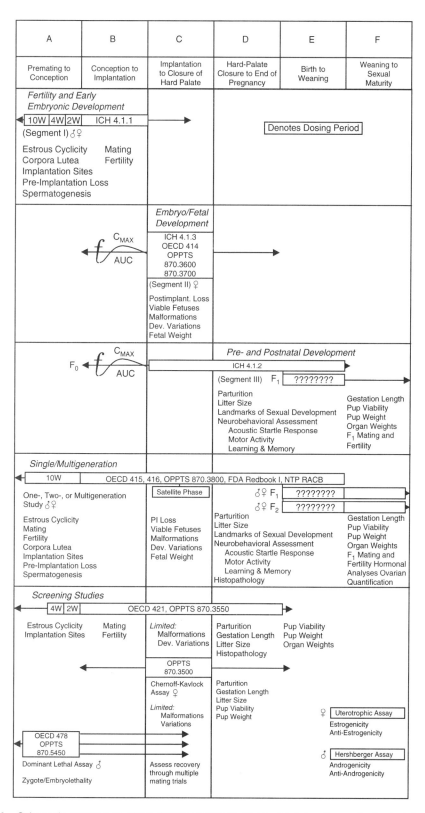

Figure 9.4 Schematic comparison of the design of various developmental and reproductive toxicity studies.

A. History of Study Design

Figure 9.5 depicts the major milestones in the evolution of developmental and reproductive toxicology (DART), including examples of various human tragedies that provided the impetus for improvement of study designs. In this section, pertinent events surrounding the historical development of DART study designs are addressed initially, followed by a similar treatment of the study designs relevant to agricultural and industrial chemicals.

1. Pharmaceuticals

Regulations governing the development, and particularly the safety assessment, of new drugs have changed dramatically since the establishment of the FDA. A variety of factors have driven the dynamics. These include human tragedies arising from the absence of (or deficiencies in) appropriate data, sociologic pressures, political pressures, governmental initiatives (including harmonization of testing guidelines to promote consistency among nations and to reduce redundant studies), economic pressures, and consumer advocacy. The net result has been continual advancement of the complexity and breadth of safety assessment protocols. Figure 9.5 highlights selected events in chronological order.

Prior to formation of the FDA in 1930, Congress passed the original Food and Drugs Act in 1906.[28] This action restricted the interstate commerce of misbranded and adulterated foods, drinks, and drugs. The Food, Drug and Cosmetic Act (FDC Act) of 1938[29] strengthened the provisions of the 1906 act. The most important provisions of this legislation were the introduction of the landmark concept of mandatory evaluation of drug safety in animal studies before marketing and the establishment of safe tolerances for unavoidable poisons. The FDC Act laid the foundations for the first FDA laboratory animal studies in the early 1940s. Just prior to passage of this law, diethylstilbestrol (DES) was synthesized as the first active estrogen mimic and was intended to be a therapeutic for maintenance of pregnancy in humans. Green et al. published the initial reports of adverse reproductive effects (intersexuality in rodents) in 1939,[30,31] but clinicians and scientists widely and mistakenly discounted the significance of the animal data. Approximately 35 years later, Herbst et al. established the causal and latent relationship of DES exposure *in utero* to vaginal clear-cell adenocarcinoma and reproductive tract dysplasia in progeny at about the onset of puberty.[32,33]

Procedures for the Appraisal of the Toxicity of Chemicals in Foods, published in 1949,[34] was among the earliest FDA guidance documents for the evaluation of specific effects of drugs on reproduction in rats. The default assumptions of risk focused on the following concepts: (1) that drugs were intended to be administered for short periods of time such that insults to the body would be limited and (2) that with properly designed toxicity studies, risk-benefit analysis could be used as a justification for introduction of the drug in humans. Further, the document suggested placement of more emphasis on the toxicologic responses resulting from acute and subacute exposures covering at least 10% of the life span of the species. Results of chronic bioassays were required during evaluation of chemical food additives because potential exposure represented a greater portion of the life span, human exposure was not volitional, and most additives were without direct benefit to health.

In the 1950s, reproductive toxicity tests gained more importance, especially when deleterious and selective responses were expected in sex organs. The studies were conducted concurrently with the chronic toxicity study. The experimental approach included the exposure of 16 males and 8 females per group for 100 days, followed by two mating trials (F_{1a} and F_{1b} litters). Selected offspring from the F_{1b} litters were exposed continuously after weaning until sexual maturity, and the assessment of functional reproductive outcome was repeated as for the first generation. The studies

DEVELOPMENTAL AND REPRODUCTIVE TOXICITY STUDY FINDINGS

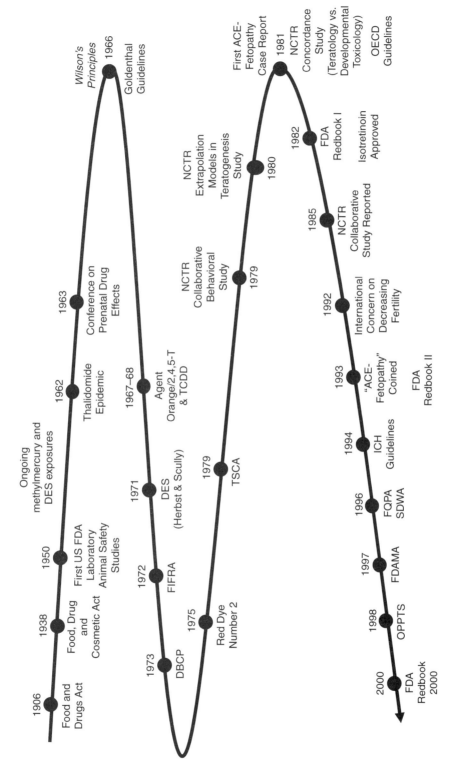

Figure 9.5 Major milestones in the evolution of developmental and reproductive toxicology.

continued through a third generation to produce F_{3a} and F_{3b} progeny. Offspring from the second mating of the F_2 generation were euthanized at weaning for histologic evaluation. Specific endpoints suggested by the guidance document were limited and included tracking success of mating and fertility, number of offspring in the litters at postnatal day (PND) 1, 5, and 21, and weight of the litter at 21 days postpartum. Paired-feeding studies were encouraged when issues of palatability arose or when large doses prevented normal nutritional intake. The Association of Food & Drug Officials of the United States endorsed these concepts of reproductive toxicity study design.[35]

In September 1960, the FDA received a New Drug Application (NDA) for the sedative-hypnotic thalidomide (under the brand name Kevadon®). The FDA refused to approve the NDA, largely because of the lingering concerns that there were insufficient data to support safe use in humans. By 1962, the causal relationship between thalidomide exposure during human pregnancy and limb reduction defects emerged, initially in Europe and Australia. At that time, testing for prenatal toxicity was not a mature field of science. Because of the thalidomide tragedy, the Pharmaceutical Manufacturer's Association formed the Commission on Drug Safety in 1962 and sponsored workshops with the FDA to advance the understanding of test methods for the discipline.[36] Hence, the discipline of teratology arose. At approximately the same time, Congress enacted the Kefauver-Harris Amendments (modifying the FDC Act of 1938) requiring drug manufacturers to provide the FDA with data that drugs were both safe and effective prior to marketing. Using the information gleaned from the workshops, the FDA launched the segmented three-phase testing paradigm,[27] which remains the essence of the modern safety assessment guidelines for developmental and reproductive toxicity hazard assessment.

The "fertility and general reproductive performance study," formerly termed the Segment I study, was devised to identify effects following the pairing of treated males and females. The exposure regime consists of a 10-week premating period for the males (encompassing an entire spermatogenic cycle) and a 2-week period (encompassing two to three estrous cycles) for females. The resultant offspring are evaluated for effects postnatally, through weaning. The study was also intended to screen for male-mediated "dominant lethal" effects. The "teratology study" (Segment II) was designed to reveal effects on morphogenesis, intrauterine growth, and intrauterine survival caused by prenatal exposures commencing after implantation and continuing through the period of major organogenesis. The "perinatal and postnatal study" (Segment III), also known as the "pre- and postnatal [PIP] study," originally employed exposure from late gestation through weaning, It was intended to characterize effects on late fetal development, the process of parturition, alterations in maternal behavior during lactation (nesting and nursing), and growth and salubrity of offspring during postnatal development. Recent emphasis on juvenile toxicity has increased awareness that adverse effects can be elicited during the late phase of gestation and early postnatal life despite treatment periods that are short relative to most treatment regimens for reproductive toxicity studies.

A major innovation to the 1966 guidelines included the addition of endpoints addressing functional toxicity. Wilson had described the four major manifestations of developmental toxicity as dysmorphogenesis, prenatal mortality, intrauterine growth retardation (IUGR), and functional deficit.[6] Although studies conducted under the 1966 guidelines provided experimental data for the first three aforementioned outcomes, **functional toxicity to the CNS and reproductive impairment following *in utero* and postnatal exposures were not widely studied.** The event responsible for changing this thinking stemmed from latent effects of the widespread methylmercury poisoning at Minamata Bay in Japan during the late 1950s and early 1960s. The resultant horrific and devastating neurobehavioral effects of this tragedy led to decades-long study of neurobehavioral teratogenicity, culminating in the execution and reporting of the NCTR Collaborative Behavioral Teratology Study (CBTS).[37] The CBTS, published in 1985, proved to be instrumental in formalizing neurobehavioral assessments in the United States and directly influenced changes in the European Union (EU) and Japanese Ministry of Health, Labour and Welfare (MHLW) guidelines.

Considerable country-to-country variation in the principles of study design plagued product development efforts globally during the 1980s and early 1990s. Pharmaceutical development programs utilized the 1966 guidelines to permit marketing in the United States. However, additional studies containing design modifications addressing the concern regarding functional toxicity were generally required prior to marketing a drug internationally.

Under the auspices of the International Conference on Harmonisation (ICH) in 1993, a memorandum of understanding standardized requirements across a broader range of regulatory agencies. Many of the concepts of modern study design emerged from this effort. Major innovations included the encouragement of internal dose determination to better quantify exposures across species, studies targeting effects on fertility and early embryonic development, and expansion of the treatment regimen in the study of pre- and postnatal development to better address functional toxicity.

In addition to providing guidance regarding study designs, the FDA has recently published a draft guidance document for integrated assessment of developmental and reproductive toxicity study results.[5] This document describes a suggested process for estimating the human developmental and reproductive risks resulting from drug exposure when only animal data are available.

2. Agricultural and Industrial Chemicals

Shortly after the formation of the EPA in 1970 and in response to the growing number of chemicals entering commerce, the Office of Pesticide Programs proposed guidelines for registering pesticides in the United States.[38] These guidelines were codified (CFR Vol. 43, No. 163) and included provisions for teratogenicity (163.83-3) and reproduction (163.83-4) studies.

Under the 1978 guidelines, EPA's pesticide evaluation was a modification of the FDA's historical three-generation approach to pesticides and food colorants. Representing a significant departure from classic FDA paradigms, the design was reduced to an assessment of two generations producing single litters in each generation. The EPA reasoned that the litters produced from the first mating cycle in each generation yielded highly variable results because of the relatively immature status of the mothers. To mitigate this confounding variable, the agency adopted the single litter per generation approach and delayed the onset of the mating cycle until the maternal animals were of sufficient age to produce uniform and stable litters. A rationale for eliminating the third generation was the introduction of the first requirements for mutagenicity studies in the 1978 proposed guidelines. The assumption was that a battery of mutagenicity studies was collectively more sensitive than a three-generation study for identifying genetic damage to germ cells. Another reason for reducing the number of generations was the agency's experience that for persistent chemicals, cumulative toxicity was rarely expressed for the first time in the third generation.

Requirements for the reproductive toxicity test were fundamentally unchanged when the guidelines were fully enacted in 1982. However, significant changes were introduced when the Office of Prevention, Pesticides and Toxic Substances (OPPTS) 870.3800 guidelines were promulgated in 1998. Differences in the 1998 guidelines arose from concerns over endocrine-active compounds. These concerns led to addition of new endpoints, such as landmarks of sexual development, semenology, estrous cyclicity, ovarian follicle quantification, and weanling organ weights, to improve overall sensitivity.

For studies designed to evaluate the effects of prenatal exposure on fetal outcome (teratogenicity studies and prenatal developmental toxicity studies), the concepts of study design have been remarkably constant since the advent of the 1978 guidelines proposed by the Federal Insecticide, Fungicide and Rodenticide Act (FIFRA). The major developments in content of those guidelines involve extending maternal exposure during the late fetal period, strengthening statistical power (group size) in the second species (rabbit), and requiring a more balanced evaluation of the products of conception for visceral and skeletal abnormalities in the rodent. The requirement for exposure during the late fetal period appeared initially in the 1978 guideline, was later deemphasized in the 1982 FIFRA and 1984 Toxic Substances Control Act (TSCA) guidelines, and was reintroduced under the OPPTS 870.3800 guidelines in 1998.

In addition to the guidelines for conduct of developmental and reproductive toxicity studies, the EPA has provided guidance documents to aid in interpretation and risk assessment of the data derived from these studies.[39,40] These risk assessment guidelines represent the best effort thus far at codifying and documenting a thorough approach to risk assessment for the discipline.

B. Species

Ideally, the principal factor in selecting the test species for developmental and reproductive toxicity studies would be that they respond to toxicity in the same manner as humans. That is not possible, so historically, animal models were not selected for any reasons other than size, availability, economics, fecundity, and (probably) not appearing too anthropomorphic. The rat, mouse, and rabbit have fortunately proven to be acceptable surrogates, and the advent of higher-quality PK and TK studies has improved the utility of these models. On a scientific basis, and assuming the aforementioned, several key attributes of these models include the following: (1) basic physiology and anatomy that are known or well-studied, (2) similar PK and TK profiles with humans, (3) comparable PD, and (4) in the absence of such information, apparently sufficient similarities in anatomical structure and reproductive physiology to permit comparison with humans. The prevailing assumption is that mammalian systems are most appropriate. The ideal test system would be easily maintained in a laboratory environment, possess significant fecundity, display stable reproductive indices, be polytocous, have a relatively low historical incidence of spontaneous structural malformations, and have other characteristics suitable for evaluating significant numbers of chemical entities cost effectively. The rat is often preferred because the endocrinology and reproductive physiology of this species have been thoroughly studied. In addition, general pharmacology models have been fully validated in the rat. Typically, animal models should be nulliparous because confirmation of pregnancy in previously gravid females is frequently problematic (because of implantation scars). However, in the case of the rabbit and dog, use of proven breeders can lend stability to the reproductive indices. For most product development efforts, the rat or mouse is selected initially, but another relevant and presumably susceptible species is typically required. Alternatives include the rabbit, guinea pig, hamster, dog, and nonhuman primate. Perceived advantages and disadvantages of each model are listed in Table 9.11 (enhanced from the ICH testing guidelines).

In assessment of developmental toxicity data, the investigator must be cognizant of a key difference in reproductive physiology among rodents (particularly rats), rabbits, and humans. Prolactin mediates maintenance of early pregnancy in the rat, unlike the rabbit and human. In the case of the rabbit, progesterone secreted from the corpus luteum sustains maintenance of pregnancy. The fundamental control in the human and rabbit changes with advancing pregnancy to control via the fetal-placental unit. This feature of endocrinology may predispose the rabbit to a higher risk of postimplantation loss and abortion when only a few implantations and limited hormonal signaling exist to maintain luteal function. This concept is illustrated by the data presented in Table 9.12. Of 60 control rabbits evaluated at laparohysterectomy that had a single implantation site, 26 (43.3%) had completely resorbed litters while 6 additional rabbits (10.0%) terminated pregnancy prematurely via abortion. The remaining 28 animals had normal reproductive outcomes. Animals with two implants also were at high risk for abnormal reproductive outcomes. There was less association between number of implantation sites and abnormal reproductive outcome in cases where the intrauterine contents consisted of three to five implants. Rabbits with greater than five implants usually have no difficulty maintaining pregnancy to term. Therefore, when interpreting the significance of abortion rates and total litter loss in compound-treated rabbits, the data must be evaluated in light of the number of embryos that were implanted in the uterus following fertilization.

Table 9.11 Comparison of attributes of species used in developmental and reproductive toxicity studies

Model	Perceived Advantages	Perceived Disadvantages
Rat	Studies of kinetics possible Large and well-characterized historical database Used extensively in repeated-dose toxicity studies Polytocous High fertility Processes of biotransformation, hormonal cascades, and organ system ontogeny well understood Short gestation period Cost-effective	Susceptible to dopamine agonists (dependence on prolactin for maintenance of early pregnancy) Prone to premature reproductive senescence following treatment with GABAnergic and other CNS-active agents Increased susceptibility to Leydig cell tumors Increased susceptibility to mammary tumors Inverted yolk sac placentation[a] Limited fetal period
Mouse	Studies of kinetics possible Require less test article Polytocous Processes of biotransformation, hormonal cascades, and organ system ontogeny well understood High fertility Inbred, knock-out and transgenic models available Cost-effective	High basal metabolic rate Visceral microdissection procedure hampered by small size of fetus Compressed developmental windows; prone to malformation clusters Limited blood sampling sites Inverted yolk sac placentation[a] Limited fetal period
Hamster	Alternative rodent model Polytocous Require less test article High fertility Very short gestation period Cost-effective Commonly used in biologic research	Not routinely used in repeated-dose toxicity studies Visceral microdissection procedure hampered by small size of fetus Cheek pouches (can conceal dosage) Inverted yolk sac placentation for all of gestation Cannibalism Fetal period occurs extrauterinally in part Poor depth of historical database Limited blood sampling sites
Guinea pig	Alternate rodent model Cost-effective True fetal phase	Small litter size Longer gestation period Sensitive to local gastrointestinal disturbances (e.g., antibiotics) Limited blood sampling sites Not routinely used in repeated-dose toxicity studies Poor depth of historical database Lower fertility
Rabbit	Studies of kinetics possible Polytocous Significant fetal period Large fetus is ideal for visceral microdissection procedure Semen can be obtained for longitudinal assessment Cost-effective	Consume diet inconsistently Prone to abortion and toxemia Induced ovulator Sensitive to local gastrointestinal disturbances (e.g., antibiotics) Not routinely used in repeated-dose toxicity studies Prone to resorption when few implantations are present Inverted yolk sac placentation[a]
Ferret	Nonrodent model Significant fetal period	Not routinely used in repeated-dose toxicity studies Poor depth of historical database Seasonal breeders Expensive

Table 9.11 Comparison of attributes of species used in developmental and reproductive toxicity studies (continued)

Model	Perceived Advantages	Perceived Disadvantages
Dog	Studies of kinetics possible Nonrodent model Offspring number sufficient Routinely used in repeated-dose toxicity studies Semen can be obtained for longitudinal assessment Radiographs possible to study fetal morphology Placentation more like humans (no inverted visceral yolk sac) Significant fetal period	Seasonal breeders Poor depth of historical database High inbreeding coefficient Expense may lead to small group sizes Limited history of use with some validation studies
Nonhuman primate	Size and multiple sampling sites Studies of kinetics possible Nonrodent model True fetal phase Routinely used in repeated-dose toxicity studies Phylogenetically close to humans Reproductive physiology more similar to that of humans Excellent model to assess biologics Semen can be obtained for longitudinal assessment Radiographs possible to study fetal morphology Placentation similar to humans (no inverted visceral yolk sac)	Kinetics often different than in humans Generally single offspring per pregnancy Long gestation period Prone to abortion Poor depth of historical database Studies of fertility and functional reproductive outcome problematic Interpretation dependent on adequate historical control data due to cost factors usually leading to small group sizes Expense and limited availability may lead to small group sizes Some species are seasonal breeders (e.g., Rhesus)

[a] Listed as a negative attribute because it is anatomically different than in higher orders of mammals; however, probably should only be considered a negative, that is, producing nonconcordant effects, for large, proteinaceous molecules or other agents that affect proteolysis and/or pinocytosis.

Source: Holson, J.F., Pearce, L.B., and Stump, D.G., *Birth Defects Res. (Part B)*, 68, 249, 2003.

C. Characterization of Dose-Response Curve

Broadly defined, the dose response of a particular chemical is the dynamic relationship between exposure and biological response(s) in the test system. In toxicity studies, plotting the effect (abscissae) versus a series of doses (ordinates) describes the dose-response relationship of an adverse finding for a population of animals. This model assumes that all members of the population are either positive responders or nonresponders, and the relationship is defined as a quantal dose-response relationship.[41] Dose-response curves are typically classified into one of three general models: (1) threshold model, (2) linear model, or (3) U-shaped model. The threshold model for adverse responses assumes that there is a dose at which no effect is produced. The linear model assumes that some effect occurs at any dose. With increasing dose, the shape of the dose-response curve will change until either a maximal response or maternal mortality occurs. For cancer endpoints, the linear model is typically assumed and applied. For noncancer endpoints, including developmental and reproductive toxicity, the threshold model is typically used. U-shaped models have been described for important nutritional substances required for homeostasis and for certain hormones.[42] For example, at low doses associated with dietary deficiency, adverse effects may manifest. However, as the dose increases through the essential range, the adverse response is diminished or ablated. As the dose is increased further, an adverse response manifests and increases with increasing dose similar to that described for the threshold and linear models.

Table 9.12 Modal distribution of spontaneous litter complications in control Hra:(NZW)SPF rabbits

Distribution of Implantation Sites		Females with Complete Litter Resorption		Females that Aborted		Total Abnormal Reproductive Outcomes	
No. Sites	No. Animals	No.	%	No.	%	No.	%
1	60	26	43.3	6	10.0	32	53.3
2	60	12	20.0	2	3.3	14	23.3
3	85	1	1.2	3	3.5	4	4.7
4	118	2	1.7	0	0.0	2	1.7
5	130	1	0.8	1	0.8	2	1.5
6	188	0	0.0	0	0.0	0	0.0
7	237	0	0.0	2	0.8	2	0.8
8	246	1	0.4	1	0.4	2	0.8
9	238	1	0.4	2	0.8	3	1.3
10	134	0	0.0	1	0.7	1	0.7
11	87	0	0.0	2	2.3	2	2.3
12	48	0	0.0	1	2.1	1	2.1
13	9	0	0.0	1	11.1	1	11.1
14	6	0	0.0	0	0.0	0	0.0
15	0	0	0.0	0	0.0	0	0.0
16	1	0	0.0	0	0.0	0	0.0

Data tabulated from 84 definitive developmental toxicity studies (1647 females) conducted at WIL Research Laboratories, Inc. (1992 to 2003). Average historical litter size = 6.5 viable fetuses.

In application, investigators typically attempt to use at least three graded dose levels to characterize a range of doses in toxicity studies that attempt to identify a no-effect level (NOEL), a threshold dose, and the maximum tolerated dose (MTD). Because the slope of the dose-response curve is often steeper in developmental toxicity studies than in other toxicity studies, large increments between dose levels should be avoided after achievement of the threshold. When a deleterious response increases in frequency with ascending dose, the dose-response relationship may be the critical factor in discriminating between effects that are treatment related and those that arise from biological variability. Because the dose-response curve characterizes the potency of an agent, it provides a common means to compare and contrast the attributes of various compounds relative to margin of safety, therapeutic index, potency, and efficacy.

One example of research aimed at investigating this important concept (i.e., presence or absence of threshold in developmental toxicity) is the work of Shuey et al. with the antineoplastic agent, 5-fluorouracil (5-FU).[43] Although this research effort was thorough and well conceived, it may not be directly applicable to all developmental toxicity scenarios. The agent, 5-FU, was selected for study because it is a teratogen in humans and animals, and the mode of action was known (inhibition of the enzyme thymidylate synthase [TS] that blocks DNA synthesis and cell proliferation). The process is saturable, and a threshold can be demonstrated by blocking the active sites for this enzyme. The method for analysis of TS in embryonic tissues was sufficiently sensitive to detect changes well below those required to induce IUGR and structural malformations. The investigators demonstrated that inhibition of TS in the fetus was measurable in the absence of associated changes in morphology or growth. Hence, a threshold model for 5-FU developmental toxicity was purported. However, the extent to which this phenomenon applies broadly to xenobiotics is not known. Additionally, whether the absence of morphologic effects in the presence of inhibition of TS constitutes reserve or repair remains unclear.

Many scientists believe mammalian embryos are capable of a high rate of repair. However, this remains largely unquantified and speculative. The subject has not been studied to the extent that any quantifiable correlation to a toxic insult can be made. Additionally, belief in the direct role of maternal homeostatic mechanisms in embryonal protection is based on knowledge of a limited number of endogenous elements and compounds that are preferentially taken up by the developing conceptus at the mother's expense (e.g., calcium). However, the role of these homeostatic mechanisms in protecting the conceptus at the time of insult by a xenobiotic is neither well studied nor understood, largely because of the limited assessment of maternal biochemistry and physiology in developmental and reproductive toxicity studies. Furthermore, given that the timing of development is so critical (e.g., movement of palatal shelves in concert with head growth), it would seem that repair mechanisms would have to be exquisite and rapid to protect against xenobiotic disruption of key temporal relationships.

Many homeostatic mechanisms, including active, facilitated, and pinocytotic transport processes, assure supply of essential nutrients to developing tissues at the expense of maternal levels. Although such protective mechanisms exist, there is ample evidence to indicate that severe perturbations in maternal homeostasis may have adverse effects on development. Nevertheless, it is often surprising that extensive effects on maternal physiology may occur in the absence of adverse developmental effects. Refer to Chapter 4 in this text and Section V.B.1 of this chapter for further discussion of the role of maternal toxicity in developmental outcome.

D. Dose Selection and Maximum Tolerated Dose

A fundamental tenet of well-designed toxicity studies is the characterization of toxicity at an MTD. Regulatory agencies, in general, require the highest dose in a hazard identification study to characterize toxicologic responses at the MTD. In practice, however, there is little agreement regarding the working definition of an MTD and its value and toxicologic relevance. In some instances, a dose that results in decrements in maternal or parental body weight gain of 5% to 10% (without inducing mortality) is of sufficient magnitude to define toxic effects at the upper range of the dose–response curve. Additional information such as TK data,[44,45] manifestations of developmental toxicity, and/or indications of reproductive dysfunction or failure may be used as evidence of an achieved MTD. In a typical dose-response study, a minimum of two additional treatment groups utilizing appropriate fractions of the MTD are evaluated, along with a control group. However, a more rigorous approach using more than three dose groups has been suggested,[46] and appropriately conducted dose–range-finding studies are essential in characterizing the lower segment of the dose–response curve.

The approach of using an MTD as a high-dose level in nonclinical safety studies is not universally accepted. The opposing viewpoint is founded principally on the premise that for human exposure, whether direct (as in the case of pharmaceutical entities or food additives) or indirect (through consumption of residues, as in the case of pesticides), evaluation of effects at the MTD grossly overpredicts risk. For example, inorganic arsenic injected intraperitoneally at high doses results in prenatal mortality and an array of structural fetal malformations in rodents, presumably through saturation of the methylation pathway in the liver.[47,48] Interestingly, doses administered orally at levels far exceeding relevant environmental exposures posed insignificant risk to the developing embryo, even when significant maternal toxicity occurred.[49,50] This example illustrates that for trace exposures, parenteral routes of delivery may create an exaggerated concern, even though it is physically not possible to deliver enough inorganic arsenic transplacentally under these conditions to elicit dysmorphogenesis. In this example, use of the MTD resulted in exaggerated doses that overwhelmed absorptive and metabolic pathways that have evolved to provide detoxification for semicontinuous environmental exposure. Barring situations of acute intoxication, there is little benefit to evaluating and committing resources to such an exposure scenario, especially compared with other routes of exposure more appropriate to humans. At the point when maternal toxicity manifests, pathways of elimination and metabolism may change dramatically, and saturation of plasma protein binding may occur, resulting in exaggerated exposure and less precise extrapolation between animal models and the human. Therefore, knowledge gained from PK and PD studies is essential in bridging this gap.

Prediction of hazard in large human populations from small-scale animal studies is difficult because of the limited statistical power of the latter. Regulatory decisions derived from animal studies may be based on the results from 20 to 25 pregnancies per dose level, whereas human exposure may involve thousands or millions. Studying toxicity at the MTD presumes maximization of response, which ameliorates the difference in power between the two situations. Thus, bioassays utilizing conventional numbers of sampling units and doses near the therapeutic dose may not indicate what would be the important toxicologic events in larger human populations. Likewise, for chemicals, use of the MTD in the absence of kinetic data indirectly, though crudely, establishes exposure and maximizes the ability to detect sensitive responders. Advocates of the MTD concept also contend that studying toxicity approaching the mortality portion of the dose-response curve is essential in the routine evaluation of pharmaceuticals to provide clinicians the types of adverse events that they may encounter at relevant therapeutic doses in clinical trial patients. Clearly, a negative data set in a preclinical study diminishes the ability of the clinical investigator to rapidly identify and manage dose-limiting toxicities. Data generated at the MTD may assist in identification of the symptomatology that is critical for assessment of human toxicity or overdose and for selecting antidotes or appropriate courses of therapy.

E. General Statistical Considerations

Statistical considerations play an especially important role in the evaluation of developmental and reproductive toxicity data. Both continuous (e.g., fetal body weight) and binary (e.g., postimplantation loss) measures are produced by these studies, and proper interpretation requires consideration of the statistical power of the study and the litter effect.

1. Statistical Power

Statistical power is the probability that a true effect will be detected if it occurs. It is formally defined as 1β, where β is the probability of committing a Type II error (false negative).[51] Power is dependent on the sample size, background incidence, and variability of the endpoint in question, and the significance (α) level of the analysis method. For example, the sample size (number of litters) needed to detect a 5% or 10% change in an endpoint is dramatically lower for a continuous measure with low variability, such as fetal body weight, than for a binary response with high variability, such as embryolethality (resorptions).[52] Consequently, significant changes in fetal weight are often detected at dose levels lower than those at which effects on embryo or fetal survival are observed. However, despite an adequate sample size, the large number of endpoints evaluated in a developmental or reproductive toxicity study dictates that several spurious statistically significant differences will likely occur because given a significance level of 0.05, a Type I error (false positive) will occur 5% of the time. For example, because a standard developmental toxicity study with ANOVA[53]/Dunnett's[54] and Kruskal-Wallis/Mann-Whitney[55] statistical analyses performed on all parametric and nonparametric data, respectively, may involve as many as 100 to 300 individual statistical hypothesis tests, the possibility exists for numerous spurious statistical findings.

A replicated or unbalanced study design may also augment the statistical power of a study. For example, a large-scale, replicated dose-response study of the herbicide 2,4,5-trichlorophenoxyacetic acid demonstrated that such a study design may assist in resolution of problems of interspecies variability determination, high- to low-dose response extrapolation, and reproducibility of low-level effects.[52]

2. The Litter Effect

The litter must be considered the experimental unit in developmental and reproductive toxicity studies because the litter is the unit that is randomized, and individual fetuses or pups within litters do not respond completely independently.[41,56] The propensity for fetuses of a given litter to exhibit similar responses to toxic insult, thereby artificially inflating the apparent group response, has been designated the "litter effect." Mathematically, using the litter as the experimental unit to account for this influence requires a determination of the percentage of embryos or fetuses within each litter that are affected. A grand mean is then calculated from the individual litter means. Using this approach, the variance among litters (standard deviation and standard error) can also be calculated.

While the litter proportion calculation is generally applied to embryo and fetal survival data, many investigators fail to use this approach to analyze fetal malformation data. These investigators simply determine a percentage of litters with at least one malformed fetus, failing to apply correct litter-based statistics. With this approach, the number of malformed fetuses within each litter is not taken into account (e.g., a litter with 1 of 12 fetuses malformed is given the same weighting as a litter with 6 of 12 fetuses malformed). Therefore, variance among litters cannot be determined. Initially determining the percentage of malformed fetuses within a litter, followed by calculating a litter grand mean, is the most appropriate way to analyze fetal malformation data.

Table 9.13 presents five comparative examples of calculations using the litter as the experimental unit. These examples include the following endpoints: numbers of resorptions (prenatal death), malformations, and fetal weight. For each example, an explanation of the derivation of the pertinent endpoint is provided. These examples contrast the incorrect and correct litter-based statistics and clearly demonstrate the different means of computation. Surprisingly, even though considerable research has been conducted and literature published on this topic in the 1960s and 1970s, few laboratories, and even fewer commercial software programs, properly calculate these values. Depending on a particular data set, the correct and incorrect means may differ little, but variance (sum-of-mean-square-derived) may not be obtained at all using the improper calculations. When the incorrect statistic is used, the N (number of fetuses) is exaggerated, increasing false positive results, and in the case of fetal weight, "within-litter variance components" are not determined or incorporated. The latter omission often fails to identify the within-litter response dimension (i.e., increase in number of affected fetuses per litter), which becomes lost in the overall among-litter value. Because of the incorrect statistics, litters with large or small numbers of fetuses will be disproportionately weighted in the analysis. In addition, the analysis will not be able to assess clustering of effect in limited numbers of litters. The precise calculations, which obey the litter unit, were first applied in the late 1970s at the NCTR.[56,57] Use of the appropriate method of calculation can make significant differences in interpretations. This becomes of paramount importance to postnatal data evaluations because litter influences persist well beyond weaning.[37]

IV. DOSE RANGE–CHARACTERIZATION STUDIES VS. SCREENING STUDIES

Dose range–characterization studies provide information to design a definitive study (one used for hazard identification in risk assessment). Under most circumstances, dose range–characterization studies do not eliminate a test article from development, although unexpected results can lead to that decision. Conversely, toxicologic screening studies may be used expressly for selecting product candidates. The following sections discuss the differences between dose range–characterization and screening studies, and where the two types of studies may converge.

A. Dose Range–Characterization Studies

Dose range–characterization studies (also referred to as dose range–finding, preliminary, pilot, or dose-finding studies) are an essential component of a valid research program. These studies provide investigators with information necessary to properly select dose levels for definitive developmental and reproductive toxicity studies. They represent a different level of interpretation and are pivotal

Table 9.13 Litter proportion calculation examples

Example 1: Resorptions (Prenatal Deaths)

Litter	Resorptions (No.)	Implantation Sites (No.)	Postimplantation Loss (%)
1	0	15	0%
2	0	14	0%
3	5	10	50%
4	0	13	0%
5	1	15	7%

Incorrect Statistics

Numerator	=	6	=	Total no. of resorptions
Denominator	=	67	=	Total no. of implantations
Incidence	=	**9.0%**	=	Total no. of resorptions/total no. of implantations
Among-litter variability	=	—	=	Incalculable

In Example 1, the numbers of resorptions are summed and divided by the total number of implantation sites. This calculation [(5 + 1)/67*100 = 9.0%] provides a simple incidence, without regard to weighting of effects by litter, as with correct litter-based statistics.

Correct Litter-Based Statistics

Numerator	=	57%	=	Sum of postimplantation loss (%) per litter
Denominator	=	5	=	Total no. of litters
True %PL	=	**11.3%**	=	Sum of postimplantation loss (%) per litter/total no. of litters
Among-litter variability	=	**21.8%**	=	Standard deviation of postimplantation loss

Using Example 1, correct litter-based statistics are calculated by summation of the percent per litter postimplantation loss and division by the number of litters, the randomized, true experimental unit [(50% + 7%)/5 = 11.3%].

Example 2: Resorptions (Prenatal Deaths)

Litter	Resorptions (No.)	Implantation Sites (No.)	Postimplantation Loss (%)
1	1	15	7%
2	1	14	7%
3	2	10	20%
4	1	13	8%
5	1	15	7%

Incorrect Statistics

Numerator	=	6	=	Total no. of resorptions
Denominator	=	67	=	Total no. of implantations
Incidence	=	**9.0%**	=	Total no. of resorptions/total no. of implantations
Among-litter variability	=	—	=	Incalculable

In Example 2, the numbers of resorptions are summed and divided by the total number of implantation sites, as in Example 1. Note that while the distribution of resorptions has changed dramatically from Example 1, the overall number of resorptions is the same, and the incorrect statistics do not change [(1 + 1 + 2 + 1 + 1)/67*100 = 9.0%].

Correct Litter-Based Statistics

Numerator	=	48%	=	Sum of postimplantation loss (%) per litter
Denominator	=	5	=	Total no. of litters
True %PL	=	**9.6%**	=	Sum of postimplantation loss (%) per litter/total no. of litters
Among-litter variability	=	**5.8%**	=	Standard deviation of postimplantation loss

Table 9.13 Litter proportion calculation examples (continued)

In Example 2, correct litter-based statistics are again calculated by summation of the percent per litter postimplantation loss and division by the true experimental unit [(7% + 7% + 20% + 8% + 7%)/5 = 9.6%]. Note that by use of the correct litter-based statistics, the change (from Example 1) in resorption distribution throughout the litters is reflected in both the percent per litter of resorptions and the among-litter variability (expressed here as standard deviation).

Example 3: Malformations

Litter	No. of Fetuses Malformed	Total No. of Fetuses	Percent Malformed
1	0	16	0%
2	0	14	0%
3	3	10	30%
4	0	18	0%
5	0	15	0%

Incorrect Statistics 1

Numerator	=	3	=	Total no. of fetuses malformed
Denominator	=	73	=	Total no. of fetuses
Incidence	=	**4.1%**	=	Total no. of fetuses malformed/total No. of fetuses
Among-litter variability	=	—	=	Incalculable

With incorrect statistics 1 in Example 3, the numbers of malformed fetuses are summed and divided by the total number of fetuses. This calculation [(3)/73*100 = 4.1%] provides a simple incidence, without regard to weighting of effects by litter, as with correct litter-based statistics.

Incorrect Statistics 2

Numerator	=	1	=	Total No. of litters with at least 1 malformed fetus
Denominator	=	5	=	Total No. of litters
False %	=	**20%**	=	Total No. of litters with at least 1 malformed fetus/total No. of litters
Among-litter variability	=	—	=	Incalculable

By applying incorrect statistics 2 to Example 3, the number of litters containing at least one malformed fetus is divided by the total number of litters [1/5*100 = 20%]. This approach may have some utility when comparing among studies where one is evaluating dispersion among litters for a specific-type finding (i.e., resorptions, a type of malformation, or a type of variant). This should only be used as a supplement to true litter-based statistics.

Correct Litter-Based Statistics

Numerator	=	30%	=	Sum of percent malformed per litter
Denominator	=	5	=	Total No. of litters
True %PL	=	**6.0%**	=	Sum of percent malformed per litter/total No. of litters
Among-litter variability	=	**13.4%**	=	Standard deviation of percent malformed per litter

Using Example 3, correct litter-based statistics are calculated by summation of the percent malformed fetuses per litter and division by the number of litters, the randomized, true experimental unit [(30%/5 = 6.0%].

Example 4: Malformations

Litter	No. of Fetuses Malformed	Total No. of Fetuses	Percent Malformed
1	0	16	0%
2	0	14	0%
3	1	10	10%
4	0	18	0%
5	2	15	13%

DEVELOPMENTAL AND REPRODUCTIVE TOXICITY STUDY FINDINGS 357

Table 9.13 Litter proportion calculation examples (continued)

Incorrect Statistics 1

Numerator	=	3	=	Total no. of fetuses malformed
Denominator	=	73	=	Total no. of fetuses
Incidence	=	**4.1%**	=	Total no. of fetuses malformed/total no. of fetuses
Among-litter variability	=	—	=	Incalculable

In Example 4, the numbers of malformed fetuses are summed and divided by the total number of fetuses, as in Example 3. Note that while the distribution of malformed fetuses has changed from Example 3, the overall number of malformed fetuses is the same, and the incorrect statistics 1 do not change [(3)/73*100 = 4.1%].

Incorrect Statistics 2

Numerator	=	2	=	Total no. of litters with at least 1 malformed fetus
Denominator	=	5	=	Total no. of litters
False %	=	**40%**	=	Total no. of litters with at least 1 malformed fetus/total no. of litters
Among-litter variability	=	—	=	Incalculable

In Example 4, incorrect statistics 2 are again calculated by dividing the number of litters containing at least 1 malformed fetus by the total number of litters [2/5*100 = 40%]. Note that by using incorrect statistics 2, the change (from Example 3) in malformation distribution is exaggerated; it ignores all of the unaffected fetuses in every litter, and it obligatorily produces a high incidence due to just spontaneous occurrences, potentially masking real treatment-related effects.

Correct Litter-Based Statistics

Numerator	=	23%	=	Sum of percent malformed per litter
Denominator	=	5	=	Total No. of litters
True %PL	=	**4.7%**	=	Sum of percent malformed per litter/total No. of litters
Among-litter variability	=	**6.5%**	=	Standard deviation of percent malformed per litter

In Example 4, correct litter-based statistics are again calculated by summation of the percent malformed fetuses per litter and division by the true experimental unit [(10% + 13%%/5 = 4.7%]. Note that by use of the correct litter-based statistics, the change (from Example 3) in malformation distribution throughout the litters is reflected in both the percent per litter of malformed fetuses and the among-litter variability (expressed here as standard deviation).

Example 5: Fetal weight

Litter	Litter mean	Uterine Position[a] and Weight (g)																
		1	2	3	4	5	6	7	8	9	10	11	12	13	14	15	16	17	18
1	3.6	3.8	3.6	3.5	3.7	3.4	3.8	3.6	3.5	3.7	3.4	3.8	3.5						
2	2.8	2.9	2.7	3.0	3.0	2.9	2.4	3.1	2.8	3.0	2.7	2.4	2.6	2.5	2.9	3.0	2.8	2.5	2.4
3	3.6	3.8	3.7	3.4	3.8	3.6	3.5	3.7	3.4	3.3	3.7	3.7	3.5	3.7	3.4				
4	3.6	3.8	3.8	3.7	3.6	3.5	3.7	3.3	3.7	3.4	3.8	3.8	3.6	3.7					
5	4.2	4.1	4.3																

[a] Uterine position begins at distal left uterine horn continuing through cervix and to distal right uterine horn.

Incorrect Statistics

Numerator		=	198.9	=	Sum of all fetal weights
Denominator		=	59	=	Total no. of fetuses
Incorrect group mean fetal weight (g)	=	**3.4**	=	Sum of all fetal weights/total no. of fetuses
Variability		=	**0.46**	=	Standard deviation of individual fetal weights

In Example 5, a continuous variable, fetal weight (g), is used to illustrate further the use of incorrect statistics. All individual fetal weights are summed and divided by the total number of fetuses in the group. By use of these incorrect statistics, litters with higher numbers of fetuses will be inappropriately weighted in the calculation. In addition, this approach further confounds hypothesis testing by artificially reducing the variance around the continuous data endpoint through inflation of the degrees of freedom.

Table 9.13 Litter proportion calculation examples (continued)

Correct Litter-Based Statistics

Numerator	=	17.8	=	Sum litter mean fetal weights
Denominator	=	5.0	=	Total no. of litters
True group mean fetal weight (g)	=	**3.6**	=	Sum litter mean fetal weights/total no. of litters
Among-litter variability	=	**0.52**	=	Standard deviation of litter mean fetal weights

By use of correct litter-based statistics [(3.6 + 2.8 + 3.6 + 3.6 + 4.2)/5 = 3.6 g], each litter is weighted similarly, and among-litter variability can be calculated. Example 5 demonstrates a 0.2-g difference in group mean fetal weight between the correct and incorrect statistics. Because a similar difference between a treated group and a concurrent control group of this relative magnitude in a continuous variable is often test article–related, the use of incorrect statistics can confound data interpretation.

---

to conducting an overall developmental and reproductive toxicity program. Interpretation of data collected from dose range–finding studies will not differ appreciably from interpretation of the definitive study results, with two important exceptions. First, the number of dams used in pilot studies is substantially smaller (usually 5 to 10 animals, versus 22 to 25 animals per group in the definitive study). Lower statistical power, because of the use of fewer animals, will potentially confound data interpretation, especially when pregnancy rates are less than 100%. Standard statistical analyses may be performed for a pilot data set; however, they must be interpreted with greater caution because of the reduced statistical power afforded by the low numbers of animals per group. The lower statistical power in these studies can increase both Type I and Type II errors. In addition, historical control data derived from definitive studies must be applied to dose range–characterization studies with caution. With diminished statistical power, the distribution of values around the central tendency will be much larger for the preliminary study. Furthermore, not all endpoints are evaluated in a range-finding study, making it unlikely to elicit strong conclusions from the data. Nevertheless, these studies are critical to conducting a successful definitive study. Specifically, they allow the investigator to develop a preliminary estimate of the threshold of response and the MTD, and they may provide early evidence of developmental or reproductive toxicity.

B. Developmental and Reproductive Toxicity Screening Studies

Toxicity screens are simplified studies or models designed to identify agents having a certain set of characteristics that will either exclude the agents from further investigation (e.g., drug candidates) or cause them to be assigned for further, more rigorous investigations (e.g., industrial and agricultural chemicals).

With respect to drug evaluation, screening studies can provide a number of advantages over performing guideline studies. These advantages include, but are not limited to, fewer animals and less test article required, more rapid execution, earlier attrition of candidate products, improved potency and selectivity evaluation, and economic savings. In the case of chemical testing, advantages of employing a screening regimen may include resource conservation, quantitative structure-activity relationships (QSARs) database creation and expansion, and increased ability to evaluate additional chemicals.

Pharmaceutical companies have historically used a number of developmental and reproductive toxicity screens for determining whether to halt or continue test article development (e.g., Hershberger and uterotrophic assays); however, the FDA has not published guidelines for screening studies. Conversely, the EPA and the Organisation for Economic Cooperation and Development (OECD) have promulgated a variety of *in vitro* and *in vivo* screening studies for evaluation of developmental and reproductive toxicity potential. Only the *in vivo* screens are discussed in this chapter; refer to Chapter 16 of this text and Brown et al.[58] for discussion of *in vitro* screens. The

Table 9.14 Guideline-driven developmental and reproductive toxicity screens, by agency

Type of Screen	Guideline Number	
	EPA/OPPTS	OECD
Preliminary developmental toxicity screen (Chernoff-Kavlock assay)	870.3500	—
Reproduction and developmental toxicity screening test	870.3550	421
Combined repeated dose toxicity study with the reproduction and developmental toxicity screening test	870.3650	422

Table 9.15 Selected features of developmental and reproductive toxicity screening studies

	OPPTS 870.3500	OPPTS 870.3550/ OECD 421	OPPTS 870.3650/ OECD 422
Species	Rat or mouse	Rat or mouse	Rat or mouse
Group size	15 females	10 per sex	10 per sex
Treatment regimen	GD 7 to 11 or to closure of hard palate	Male: 28 days Female: 14 days Premating through LD[a] 3	Male: 28 days Female: 14 days Premating through LD 3
Toxicokinetics	None	None	None
Mating period	Not applicable (bred prior to start of dosing)	Two weeks	Two weeks
Behavioral testing	None	None	FOB[b] motor activity
Organ weights	None	Testes, epididymides	Testes, epididymides, liver, kidneys, adrenals, thymus, spleen, brain, heart
Histopathology	None	Testes, epididymides, accessory sex organs, ovaries	Testes, epididymides, accessory sex organs, ovaries, uterus, gastrointestinal tract, urinary tract, lungs, central & peripheral nervous system, liver, kidneys, adrenals, thymus, spleen, heart, bone marrow
Pup necropsy	Dead pups only	External exam only	External exam only

Note: Additional subchronic endpoints: hematology, blood biochemistry
[a] Lactational day.
[b] Functional observational battery.

in vivo screening guidelines for chemical evaluation and their respective requirements are presented in Table 9.14 and Table 9.15. Another useful screen is the Reproductive Assessment by Continuous Breeding (RACB) screen used routinely by the National Toxicology Program.[59]

While screening studies will identify potent toxicants if appropriate endpoints are included and/or are correlated, Type I and Type II statistical errors can also result from the reduced sample sizes employed. The specificity of a screen refers to its capacity to produce false positives (Type I error), while the sensitivity of a screen speaks to its ability to produce false negatives (Type II errors). The authors' experiences with the Reproduction/Developmental Toxicity Screening Test (OECD 421/OPPTS 870.3550) and Combined Repeated Dose Toxicity Study with the Reproduction/Developmental Toxicity Screening Test (OECD 422/OPPTS 870.3650) confirm that these errors are not infrequent. In particular, these OECD and OPPTS screens are useful if employed as true screens. However, they are not useful if they are considered apical studies for hazard identification when there is significant potential for human exposure. An expert panel of the German Chemical Society's Advisory Committee on Existing Chemicals has concluded that OECD 421 and 422 screening tests are neither alternatives to definitive studies (i.e., those conducted under OECD guidelines 414, 415, and 416) nor replacements for these studies, particularly because the planned validation of these screening tests has not been completed.[60]

Table 9.16 Screens used as an antecedent to a two-generation reproductive toxicity study

Adult Endpoints	Screen (S) F_0	Two-Generation (2-G) F_0	F_1	Concordance (%) 2-G F_0 vs. S F_0	2-G F_1 vs. S F_0
Mortality	2/9	1/9	3/8	89	75
Body weight	5/9	8/9	7/8	67	63
Food consumption	5/9	8/9	6/8	67	50
Clinical observations	4/9	3/9	2/8	67	63
Fertility index	0/9	0/9	0/8	100	100
Mating index	0/9	0/9	0/8	100	100
Implantation index	3/9	2/9	1/8	89	88
Postimplantation loss	3/8	1/9	1/8	89	88
Live birth index	3/8	1/8	1/8	63	63
Organ weights	3/6	4/8	3/8	67	80
Gestation length	0/8	0/8	0/8	100	100
Dystocia	0/7	0/8	0/8	100	100
Neonatal endpoints	**(F_1)**	**(F_1)**	**(F_2)**	**2-G F_1 vs. S F_1**	**2-G F_2 vs. S F_1**
Survival	2/8	1/8	2/8	63	75
Body weight	3/8	5/8	5/8	75	75

Note: For endpoint-finding enumeration, the denominator represents the number of test articles for which both screening and two-generation studies were performed. The numerator reflects the number of studies for which an effect was noted. For this table, five of the nine studies used extrapolated dose levels, that is; dose levels in these studies that were close enough to have considered them comparable for the observed effects. Concordance reflects cumulative agreement (both positive and negative) between the screening results and the definitive results from each generation of the two-generation reproductive toxicity study, expressed as a percentage.

Source: Data from studies conducted at WIL Research Laboratories, Inc.

Table 9.17 Examples of failures in sensitivity of screens

Adult Endpoints	Screen (S) F_0	Two-Generation (2-G) F_0	F_1	Concordance (%) 2-G F_0 vs. S F_0	2-G F_1 vs. S F_0
Fertility index	0/3	0/3	3/3	100	0
Mating index	0/3	0/3	3/3	100	0
Live birth index	1/3	1/3	2/3	100	67
Organ weights	0/1	2/3	2/2	(33)	(0)
Gestation length	0/3	0/3	1/3	100	67
Dystocia	0/3	1/3	1/3	67	67
Neonatal endpoints	**(F_1)**	**(F_1)**	**(F_2)**	**2-G F_1 vs. S F_1**	**2-G F_2 vs. S F_1**
Sex ratio	0/3	1/3	2/3	67	33
Hypospadias	Not measured	1/3	Not measured	(0)	(0)

Note: The denominator represents the number of test articles for which both screening and full two-generation studies were performed. The numerator reflects the number of studies for which an effect was noted. For this table, dose levels were identical between the screening studies and the definitive two-generation reproductive toxicity studies. Concordance reflects cumulative agreement (both positive and negative) between the screening results and the definitive results from each generation of the two-generation reproductive toxicity study, expressed as a percentage.

Employing the guideline-minimum number of animals per group in these two general designs can often result in effects that appear to be nondose-responsive and/or inconclusive when faced with rare (low incidence) events. Table 9.16 presents selected endpoint-by-endpoint results for screens performed at the authors' laboratory when used as an antecedent to guideline two-generation studies (OECD 416/OPPTS 870.3800). Table 9.17 presents data from three additional case studies in which multiple failures of sensitivity (false negatives) occurred.

From the data presented in Table 9.16 and Table 9.17, the sensitivity and specificity limitations intrinsic to the referenced screens are apparent. Strong conclusions regarding developmental and reproductive hazards can be drawn only from definitive studies containing adequate sample sizes, multiple endpoints, appropriate treatment duration, and ideally, some assessment of internal exposure (TK).

The screening studies examined above are short-term experiments used to select and/or sort a series of molecules on the basis of developmental and reproductive toxicity potential. These studies may be used to predict potential hazard, but the results generally cannot be used for risk assessment. Heuristically, two axioms have arisen: (1) if the primary intent of the screen is to reduce the number of animals used, careful consideration must be given to the possible necessitation of subsequent studies because of poor characterization of the dose-response curve, and (2) if the intent of the screen is to reduce resource consumption, an analogy to the first point exists with the exception of the application to agents not developed for biologic activity, with limited human exposure and economic significance.

C. Converging Designs

Dose range-characterization studies can converge with screening studies for selected purposes. This convergence often occurs in development of drugs in particular pharmacologic classes. An example in the authors' laboratory is a hybridized dose range–characterization and definitive embryo and fetal development study that was employed to screen retinoic acid analogs for teratogenic potency. The hybridized design utilized more animals than are typically employed for dose range-characterization but fewer than would be necessary for a definitive study. Generally, several candidate analogs were evaluated concomitantly at high dose levels with one concurrent control group and one comparator group treated with a known potent analog (all-*trans*-retinoic acid). Teratogenic potency was determined through a postmortem examination of greater scope than is regularly employed for a dose range-characterization study, as it consisted of both visceral and skeletal components. Those analogs with the least expression of the classic terata were identified as potential candidates for further development. This type of specialized screening study can be adapted for other pharmacologic classes possessing known teratogenic potential. This, however, is only possible when the investigator is able to correlate the selected study endpoints with a known overall pattern of developmental effects. Furthermore, since teratogenic responses to certain classes of compounds are well known and expected, an abbreviated treatment regimen could be employed, targeting a critical, susceptible period of organogenesis.

V. DEVELOPMENTAL TOXICITY STUDIES

This section begins with a presentation of the key differences between various regulatory guideline requirements for developmental toxicity studies, including a brief discussion on the importance of animal models as well as the timing and duration of treatment. The interrelatedness of endpoints is discussed relative to both development (or delay thereof) and to the complexities of unraveling any interaction between developmental and maternal toxicity. A discussion of specific endpoints from these studies, within the context of the various interrelationships that may occur, is presented with historical control data to aid in determining toxicologic relevance. An in-depth evaluation of the challenges inherent in interpretation of fetal morphology data, including an exposition of experimental considerations relative to these critical examinations, is then presented. The section concludes with case studies from the authors' laboratory, illustrating the difficulties inherent in interpreting the significance of several fetal developmental variations.

A. Guideline Requirements

Table 9.18 presents selected differences in study design requirements between various agency guidelines. The OPPTS of the EPA has produced prenatal developmental toxicity testing guidelines for use in the testing of pesticides and toxic substances. This guidance document harmonized the previously separate testing requirements under TSCA and FIFRA. The OECD and the Japanese Ministry for Agriculture, Forestry and Fisheries (MAFF) have developed guidelines very similar to those of the EPA. For medicinal products, the ICH developed technical requirements for the conduct of developmental toxicity studies in support of pharmaceutical products registration for the European Union, Japan, and the United States. For food ingredients, the Center for Food Safety & Applied Nutrition (CFSAN) within the FDA has developed guidelines for developmental toxicity studies. Finally, the Maternal Immunization Working Group of the Center for Biologics Evaluation and Research (CBER) at the FDA has developed a draft guidance document for developmental toxicity studies of vaccines.

1. Animal Models

Relevance of the mode of action of the test agent, when known, is the most important factor in the choice of animal models for developmental toxicity testing, regardless of guideline requirements. Developmental toxicity studies are typically performed in two species, one rodent and one nonrodent. Historically, the rat and rabbit have been the preferred rodent and nonrodent species, respectively, for these types of studies. However, alternatives to these species exist (refer to Section III.B of this chapter for further details).

A recent publication has proposed the use of a tiered approach to developmental toxicity testing for veterinary pharmaceutical products for food-producing animals.[61] Using this approach, if teratogenicity were to be observed in the rodent, no testing in a second species would be required. However, if a negative or equivocal result for teratogenicity were to be observed in the rodent, then testing in a second species, preferably the rabbit, would be conducted.

In special cases, investigators may use the dog or nonhuman primate model in lieu of the rabbit. However, the cost of these models and the need for specialized techniques and experience to conduct a successful study necessitate a well-articulated, scientific rationale justifying their use.

An example of effective use of the canine model in developmental toxicology was a series of studies performed with a hemoglobin-based oxygen carrier (HBOC) intended to be an oxygen bridge in lieu of human blood transfusion. A preliminary developmental toxicity study in the rat revealed significant embryo toxicity (lethality, growth retardation, and structural malformations in multiple organ systems) following infusion of the test article on GD 9.[62] These effects coincided with the prechorioallantoic placental period of development when histiotrophic nutrition via the inverted visceral yolk sac placenta (InvYSP) is essential to development in rodents. The results of the preliminary study led to the hypothesis that the embryo toxicity in the rat was unique to the presence of the InvYSP during the time the malformations were occurring, and that these findings represented a false positive signal for human developmental toxicity.[63] Subsequent rat whole-embryo culture studies confirmed a pronounced effect on endocytotic and proteolytic functions in the visceral yolk sac placenta. Because the canine model and humans do not possess an InvYSP, a series of phased-exposure studies in the dog was conducted. The results of the canine studies revealed no evidence of developmental toxicity, substantiating the hypothesis.[62]

2. Timing and Duration of Treatment

Of particular note in the comparison of developmental toxicity study guidelines is the difference between EPA and ICH requirements for the timing and duration of treatment. The EPA guidelines require dose administration throughout the period of prenatal organogenesis (beginning at implantation

DEVELOPMENTAL AND REPRODUCTIVE TOXICITY STUDY FINDINGS

Table 9.18 Selected comparison points between various regulatory agency guideline requirements for developmental toxicity studies

	EPA (1998)	ICH (1994)	OECD (2001)	FDA/Redbook 2000 (2000 Draft)	FDA/Vaccines[a] (2000 Draft)	Japan/MAFF (1985)
Ref.	166	167	168	169	170	171
Choice of animal model	Rat and rabbit	Rat and rabbit	Rat and rabbit	Rat and rabbit	Vaccine should elicit immune response	Rat and rabbit
Single or multiple species	Multiple	Multiple	Multiple	Multiple	Single	Multiple
Randomization	Yes	Yes	Yes	Yes	Yes	Not mentioned
Group Size						
Rodents	Approx. 20 females with implants at necropsy	16–20 litters	Approx. 20 females with implants at necropsy	Approx. 20 females with implants at necropsy	16–20 litters per phase	At least 20 pregnant females
Nonrodents	Approx. 20 females with implants at necropsy	16–20 litters	Approx. 20 females with implants at necropsy	Approx. 20 females with implants at necropsy	16–20 litters per phase	At least 12 pregnant females
Timing and duration of treatment/Exposure	Period of major organogenesis[b]	Implantation to closure of hard palate	Implantation to day prior to scheduled laparohysterectomy[b]	Implantation to day prior to scheduled laparohysterectomy[b]	Prior to mating and implantation to birth	Period of major organogenesis
Morphologic examination of fetuses[c]						
Rodents	Approx. 1/2 of each litter for skeletal alterations (preferably bone and cartilage)	1/2 of each litter for skeletal alterations	Approx. 1/2 of each litter for skeletal alterations	Approx. 1/2 of each litter for skeletal alterations	1/2 of each litter for skeletal alterations	1/3 to 1/2 of each litter for skeletal alterations
	Remaining fetuses from each litter for soft tissue alterations (acceptable to examine all fetuses viscerally followed by skeletal examination)	Remaining fetuses from each litter for soft tissue alterations	Remaining fetuses from each litter for soft tissue alterations	Remaining fetuses from each litter for soft tissue alterations (acceptable to examine all fetuses viscerally followed by skeletal examination)	Remaining fetuses from each litter for soft tissue alterations	Remaining fetuses from each litter for soft tissue alterations
Nonrodents	All fetuses for both soft tissue and skeletal anomalies (capita of 1/2 removed and evaluated internally, thus precluding skeletal examination of those capita)	Each fetus for soft tissue (by dissection) and skeletal anomalies	All fetuses for both soft tissue and skeletal anomalies (capita of 1/2 removed and evaluated internally, thus precluding skeletal examination of those capita)	All fetuses for both soft tissue and skeletal anomalies (capita of 1/2 removed and evaluated internally, thus precluding skeletal examination of those capita)	Each fetus for soft tissue (by dissection) and skeletal anomalies	Each fetus for soft tissue (by dissection) and skeletal anomalies

[a] The FDA guidelines for developmental toxicity studies of vaccines also include a requirement for a follow-up phase from birth to weaning for evaluation of effects on preweaning development and growth, survival, developmental landmarks, and functional maturation.
[b] If preliminary studies do not indicate a high potential for preimplantation loss, treatment in the definitive study may include the entire period of gestation (typically gestational days 0–20 for rats and 0–29 for rabbits).
[c] ICH guidelines do not require examination of the low- and middose fetuses for soft tissue and skeletal alterations where evaluation of the control and high-dose animals did not reveal any relevant findings.

or alternatively at fertilization), whereas ICH guidelines require administration only during the period of major organogenesis (implantation through closure of the hard palate). This may appear to be a deficiency in the ICH guidelines, although the design was likely predicated on the presumption that pre- and postnatal studies would also be conducted, establishing exposure during the late gestational period. However, under regulatory regimens that do not include pre- and postnatal studies, as in the case of biologics, the absence of late gestational treatment may result in the inability to detect adverse effects on functional maturational changes and osteogenesis that occur during this period.

An understanding of the toxicokinetics of the test agent may also require broadening the duration of treatment to include the period prior to implantation. In studying the effects of agents that require protracted administration to achieve steady state in the maternal organism, it is advisable to begin dosing in advance of the processes under evaluation. An example of this understanding was reflected in a series of developmental toxicity studies of inorganic arsenic, in which freely circulating arsenic levels were maximized by administration prior to fertilization to compensate for sequestration of the test agent in erythrocytes. In one of these studies, females received the test substance for 14 days prior to mating, throughout mating, and continuing until GD 19, so that circulating levels of arsenic in the dams were high enough to ensure transplacental delivery to the conceptuses.[49] The timing and approach to these issues is compound specific and based on kinetics, pharmacodynamics, or other exposure information.

Alternatively, factors such as induction of detoxification enzymes may make beginning dose administration at implantation, rather than during the preimplantation phase, more appropriate to detect significant effects on early differentiation.

B. Interpretation of Developmental Toxicity Study Endpoints

Table 9.19 presents the endpoints typically evaluated in developmental toxicity studies, listed first in approximate chronological order of collection, and then ranked by approximate sensitivity of the endpoints. In this ranking, endpoints considered most sensitive are those that are most often affected in general or by lower doses of xenobiotics.

These endpoints may be either continuous or binary measures. Fetal body weight, one of the few endpoints that is continuous, is also the most sensitive because it is gravimetrically determined. Given good laboratory techniques, fetal body weight should exhibit the least variance because of the lack of subjectivity inherent in its collection. Because sensitivity of an endpoint may be methodologically or biologically dependent, the ranking in Table 9.19 is relatively arbitrary and should not be considered a definitive categorization. Mode of action may determine which measure is more sensitive, although in most cases of developmental toxicity established in humans and laboratory models, fetal weight is the most sensitive and is often associated with arrays of developmental disruption. Here, the critical importance of examining patterns of effect on multiple endpoints and reconciliation of the biological plausibility of the overall effect cannot be overemphasized.

Patterns of effect on multiple endpoints, arising often from depression of fetal body weight, are the key to understanding the arrays of developmental disruption associated with toxic insult to *in utero* progeny. Because events such as body cavity development (e.g., gut rotation and retraction) and midline closures (neural tube, hard palate, spine, thorax, and abdomen) proceed so systematically and harmoniously, retardation of growth may simply shift the timing of these discrete developmental events by a matter of hours or days in most species. For example, an insult that manifests as omphalocele or cleft palate may result from developmental delay due to intrauterine growth retardation rather than a direct effect on morphogenesis of those structures. This distinction may be important in risk assessment. Fraser[64] and Holson et al.[57] have presented examples involving palatal closure in mice. Furthermore, growth retardation may be associated with functional impairment and/or death. Therefore, all components of a developmental toxicity data set must be evaluated

Table 9.19 Endpoints of developmental toxicity studies

Approximate Chronological Order of Collection	Ranked by Sensitivity (Most Sensitive to Least Sensitive)
Maternal survival	Fetal weights
Maternal clinical observations	Male
Maternal body weight	Female
Maternal body weight change	Combined
Maternal gravid uterine weight	Postimplantation loss
Maternal net body weight	Early resorptions
Maternal net body weight change	Late resorptions
Maternal food consumption	Fetal viability
Maternal necropsy findings	Fetal malformations
Maternal clinical pathology[a]	Fetal developmental variations
Maternal organ weights[a]	Maternal body weight
Fetal viability	Maternal body weight change
Corpora lutea	Maternal gravid uterine weight
Implantation sites	Maternal net body weight
Preimplantation loss	Maternal net body weight change
Postimplantation loss	Maternal food consumption
Early resorptions	Maternal survival
Late resorptions	Maternal clinical observations
Fetal weights	Maternal necropsy findings
Male	Maternal clinical pathology[a]
Female	Maternal organ weights[a]
Combined	Corpora lutea
Fetal malformations	Implantation sites
Fetal developmental variations	Preimplantation loss

[a] Maternal clinical pathology and organ weight data can provide extremely useful information for resolving issues of maternal toxicity, but these endpoints are not required by regulatory agencies for developmental toxicity studies and therefore, unfortunately, typically are not collected.

holistically and under the premise that selective effects on discrete endpoints, although they may exist, are the exception to the rule of patterns of effect. Figure 9.6 illustrates the intrinsic interrelationships among developmental toxicity endpoints.

1. Maternal Toxicity and Its Interrelationship with Developmental Toxicity

Characterization of maternal toxicity is essential because of the nature of the developing organism within the maternal milieu. Progeny *in utero* are dependent upon the maternal animal for their physical environment, nutrients, oxygen, and metabolic waste disposal. The relationship of maternal toxicity to embryo and fetal toxicity is subject to various interpretations; therefore, due diligence must be given to the effects of the test agent on the dam, especially to determine whether the pregnant female is more sensitive to a given agent than is the conceptus. Embryo and fetal effects that occur in the presence of frank maternal toxicity sometimes, although perhaps not always logically, elicit less concern from a human risk assessment perspective than those effects on the conceptus that occur at maternally nontoxic doses.

In a developmental toxicity study, the dam and the products of conception coapt and are interdependent in many ways. Thus, it is biologically plausible that maternal toxicity may affect development of the progeny *in utero*. However, if toxic effects manifest in both the maternal and fetal organisms, there is no way to determine the relationship of the findings without, at minimum, measurement of concomitant plasma levels of the test agent in both organisms. In an acute dosing regimen, the dam would be expected have a higher C_{max} so fetal exposure, as judged from plasma levels of an agent, will lag behind that of the mother in time, and often in magnitude. Therefore, fetal effects that occur in the presence of maternal toxicity cannot necessarily be attributed to the

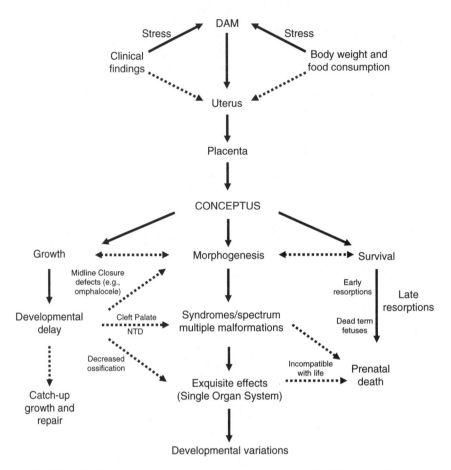

Figure 9.6 Depiction of intrinsic interrelationships among developmental toxicity endpoints.

effects on the mother. It has long been known that additional information on clinical pathology, organ weights, and measures of pharmacodynamics would further clarify this relationship. Unfortunately, the regulatory impetus to gather such information has not been forthcoming. Because this investment has not been made, the debate and uncertainty in this regard will continue.

Study of the ontogeny of physiologic regulation has revealed that mammalian maternal organisms are exquisitely adapted to protect their developing organisms through homeostatic protective mechanisms.[65–68] An understanding of this phenomenon will help prevent underestimation of developmental sensitivity that can occur when fetal effects are evaluated only in relation to maternal toxicity. In the final analysis, the debate over the relationship between maternal toxicity and developmental toxicity becomes more an issue of risk than one of causality. After all, the mother's health is still important, regardless of whether the progeny are affected. However, from a risk management perspective, maternal effects are not necessarily equivalent to adverse developmental outcomes. The reader is referred to Chapter 4 of this text for a more detailed discussion of maternally mediated developmental toxicity.

2. Endpoints of Maternal Toxicity

Data interpretation for a developmental toxicity study begins with an evaluation of the basic maternal data collected (survival, clinical observations, body weights, food consumption, and necropsy findings). Robust assessment of maternal toxicity includes measurements of organ weights

DEVELOPMENTAL AND REPRODUCTIVE TOXICITY STUDY FINDINGS

Table 9.20 Maternal gestation body weight gain ranges, by species/strain

Endpoint	Mouse Crl:CD®-1(ICR)BR	Rat Crl:CD®(SD)IGS BR	Rabbit Hra:(NZW)SPF
Mean total body weight gain during gestation[a] (g)	24–29	122–171	434–820
Mean net body weight gain during gestation[a,b] (g)	4–8	41–76	71–427
Mean percentage of total body weight gain attributed to the products of conception	75–82	55–67	47–93

Note: The day of observation of evidence of mating (rodents) or artificial insemination (rabbits) was designated gestational day 0. Approximate ages at GD 0 were 80-100 days, 12-15 weeks, and 5-7 months for mice, rats, and rabbits, respectively. Laparohysterectomies were conducted on GD 18 for mice, GD 20 for rats, and GD 29 for rabbits.

[a] Gestational days 0–18 for mice, GD 0–20 for rats, and GD 0–29 for rabbits.
[b] Lean body mass of dam/doe.

Source: Data tabulated from studies conducted at WIL Research Laboratories, Inc., including 10 mouse studies (1991 to 2001), 25 rat studies (1999 to 2003), and 25 rabbit studies (1998 to 2003).

and clinical pathology evaluations. The onset, relationship to dose, and timing of clinical findings of toxicity may be important factors to note, depending upon the effects observed in the developing embryos or fetuses. For example, an adverse effect on the gestating female that manifests clinically during a critical developmental window could be implicated in the subsequent morphologic alteration of the offspring.

In the absence of anatomic and clinical pathology evaluations and organ weight measurements, maternal body weight data (usually accompanied by food and/or water consumption data) provide the clearest measure of maternal toxicity in these studies. Thorough evaluation of maternal growth should include assessment of body weight and body weight gain at minimum intervals of 3 to 4 days throughout the treatment period, gravid uterine weight (weight of the uterus and contents), net body weight (the terminal body weight exclusive of the weight of the uterus and contents), and net body weight change (the overall body weight change during gestation exclusive of the weight of the uterus and contents). Body weight deficits of 5% or greater that are sustained over a period of several days are generally considered to be a signal of an adverse effect on maternal growth. Table 9.20 presents reference data from the authors' laboratory for mean total and net body weight gain during gestation for the most commonly used species, as well as the mean percentages of total weight gain during gestation that are due to the products of conception.

Maternal food consumption should be evaluated and presented for the corresponding intervals of body weight gain. Food consumption measurements in these studies are critical for monitoring of maternal homeostasis. Many studies have revealed an association between dietary restriction in mice, rats, and rabbits and adverse outcomes on the progeny during both gestation and lactation. Studies of dietary restriction have demonstrated that reductions in food consumption of as little as 10% of the normal dietary intake (approximately 7 to 8, 20 to 25, and 150 to 200 g/d for mice, rats, and rabbits, respectively) may be associated with increased prenatal death, dysmorphogenesis, and/or growth retardation.[69–75] Therefore, reduced maternal food intake of 10% or greater in a developmental toxicity study may be an indication not only of maternal toxicity but also of secondary insult to the developing progeny. However, the extent to which dietary restriction mimics maternal inanition, failed weight gain, or weight loss due to compound-related toxicity is not known. Regulatory developmental toxicity studies are generally not designed to separate maternal anorexic effects from other potential insults to the developing progeny.

Despite the general correlation in various species between restricted maternal food consumption or fasting and adverse developmental outcome, exceptions to this rule do exist. In a recent study conducted at the authors' laboratory, a proprietary compound that produced excessive maternal toxicity when tested in Crl:CD®(SD)IGS BR rats (extreme decrements in maternal food consumption and body weight gain) at the high-dose level had no resultant (or concomitant) effect on the developing progeny. At the high-dose level, mean food consumption over the treatment period (GD

6 to 18) was 14 g/animal/d, compared with the expected mean of 22 g/animal/d in the control group. Mean net (corrected) body weight in the high-dose group was 20% lower than the control group value. However, no adverse effects on fetal survival, weight, or morphology manifested at any dose level tested. This example illustrates the evolutionary principle of conservation of progeny, even at the expense of the mother, and also shows that frank maternal toxicity will not always cause adverse fetal outcome.

3. Intrauterine Growth and Survival

Basic gestational data collected for a developmental toxicity study at laparohysterectomy (mistakenly referred to as cesarean section) include parameters such as numbers of pregnant and nonpregnant females, corpora lutea, implantations, early and late resorptions, alive and dead fetuses, abortions, fetal body weight, and fetal sex ratio. From these data, the indices of pre- and postimplantation loss may be calculated on a proportional litter basis (refer to Section III.E.2 for a detailed discussion of the litter effect). Intrauterine parameters may provide important information with which to assess the relative concern of fetal morphologic findings. These endpoints are evaluated within the context of the health of the dam because the well-being of the gestating mother may influence fetal outcome.

a. Preimplantation Loss

Preimplantation loss is determined by comparing the number of corpora lutea produced with the number of successful implantations, as indicated below (presented on a proportional litter basis).

$$\text{Summation per group (\%)} = \frac{\Sigma \text{ preimplantion loss/litter (\%)}}{\text{Number litters/group}}$$

where

$$\text{Preimplantation loss/litter (\%)} = \frac{(\text{Number corpora lutea} - \text{Number implantation sites})/\text{litter}}{\text{Number corpora lutea/litter}} \times 100$$

In a study where females are treated prior to implantation, an increase in preimplantation loss may indicate an adverse effect on gamete transport, fertilization, the zygote or blastula, and/or the process of implantation itself. If treatment occurs at or after actual nidation, an increase in preimplantation loss probably reflects typical biological variability. However, in either case if the data reflect an apparent dose-response relationship and/or are remarkably lower than the historical control range, further studies may be necessary to elucidate the mode of action and extent of the effect. Table 9.21 presents, by species, typical ranges for numbers of corpora lutea and implantation sites, as well as mean litter proportions of preimplantation loss.

b. Postimplantation Loss

Effects on survival of the embryo/fetus are manifested by an increase in postimplantation loss. Postimplantation loss should be calculated on a proportional litter basis as indicated below.

$$\text{Summation per group (\%)} = \frac{\Sigma \text{ postimplantion loss/litter (\%)}}{\text{Number of litters/group}}$$

Table 9.21 Historically observed means (ranges) for preimplantation parameters, by species and strain

Endpoint	Mouse Crl:CD®-1(ICR)BR	Rat Crl:CD® (SD)BR	Rat Crl:CD®(SD)IGS BR	Rabbit Hra:(NZW)SPF
Corpora lutea (no./litter)	15.0 (13.1–17.3)	16.8 (14.4–20.5)	17.7 (16.0–19.4)	10.4 (7.9–13.6)
Implantations (no./litter)	13.0 (11.2–14.3)	15.0 (11.5–18.3)	15.9 (14.6–17.5)	7.0 (5.1–9.3)
Preimplantation loss (%/litter)	12.7 (5.7–19.2)	10.5 (2.8–25.4)	9.5 (4.0–15.7)	30.7 (6.6–52.0)

The day of observation of evidence of mating (rodents) or artificial insemination (rabbits) was designated gestational day 0. Laparohysterectomies were conducted on GD 18 for mice, GD 20 for rats, and GD 29 for rabbits.

Source: Data tabulated from studies conducted at WIL Research Laboratories, Inc., including 20 Crl:CD®-1 (ICR)BR mouse studies (1984-2000), 158 Crl:CD®(SD)BR rat studies (1982-1997), 61 Crl:CD®(SD)IGS BR rat studies (1998-2003), and 81 Hra:(NZW)SPF rabbit studies (1992-2003). These four databases contain information derived from 460, 3585, 1452, and 1608 pregnant control dams/does, respectively.

Table 9.22 Historically observed means (ranges) for postimplantation parameters by species and strain

Endpoint	Mouse Crl:CD®-1(ICR)BR	Rat Crl:CD®(SD)BR	Rat Crl:CD®(SD)IGS BR	Rabbit Hra:(NZW)SPF
Total postimplantation loss (%/litter)	7.4 (3.2–12.0)	5.7 (2.2–13.5)	4.4 (2.0–8.6)	9.1 (0.6–23.4)
Early resorptions (%/litter)	5.7 (3.2–10.4)	5.6 (1.8–13.5)	4.4 (1.5–8.6)	7.9 (0.6–22.7)
Late resorptions (%/litter)	1.4 (0.0–3.2)	0.1 (0.0–3.2)	0.1 (0.0–0.8)	1.2 (0.0–6.2)
Dead fetuses	0.3 (0.0–1.2)	0.0 (0.0–1.4)	0.0 (0.0–0.0)	0.0 (0.0–0.6)
Viable fetuses (%/litter)	92.6 (88.0–96.8)	94.3 (86.5–97.8)	95.6 (91.4–98.0)	90.9 (76.6–99.4)

The day of observation of evidence of mating (rodents) or artificial insemination (rabbits) was designated gestational day 0. Laparohysterectomies were conducted on GD 18 for mice, GD 20 for rats, and GD 29 for rabbits.

Source: Data tabulated from studies conducted at WIL Research Laboratories, Inc., including 20 Crl:CD®-1(ICR)BR mouse studies (1984 to 2000), 158 Crl:CD®(SD)BR rat studies (1982 to 1997), 61 Crl:CD®(SD)IGS BR rat studies (1998 to 2003), and 81 Hra:(NZW)SPF rabbit studies (1992 to 2003).

where

$$\text{Postimplantation loss/litter (\%)} = \frac{\text{Number dead fetuses + resorptions (early and late)/litter}}{\text{Number corpora lutea/litter}} \times 100$$

In rodents and rabbits, postimplantation loss may manifest as early resorptions, late resorptions, or dead fetuses. In both rodents and rabbits, the dead conceptus undergoes gradual degradation, followed by maternal reabsorption, and is referred to as a resorption; in rabbits, it may be aborted instead of being reabsorbed. Early or late resorptions are identified by the absence (early) or presence (late) of distinguishable features, such as the head or limbs. In guideline developmental toxicity studies, it is not possible to determine from evaluation of the products of conception whether intrauterine deaths were spontaneous in origin, the result of malformations, or the result of a direct toxic insult to the conceptus.

Table 9.22 provides historical ranges for postimplantation loss in mice, rats, and rabbits. As indicated in this table, there is typically an inverse relationship between mean litter proportions of total postimplantation loss and viable fetuses, except in the case of significant differences in the numbers of successful implantations between control and treatment groups. Because of the variability that is typically observed in these data, increased postimplantation loss in a test substance–treated group is typically considered an adverse effect only when it reaches a level that is at least double that observed in the concurrent controls. More confidence may be placed in the decision if a dose-response relationship is present and the concurrent control mean is within the historical control range.

Table 9.23 Modal distribution of resorption rates in Crl:CD®(SD)IGS BR rats

Resorptions (No.)	Females (No.)
0	788
1	447
2	161
3	35
4	9
5	2
6	2
7	3
8	0
9	1
10	1
11	1
Total	**1450**

Source: Data tabulated from 61 Crl:CD®(SD)IGS BR rat studies (1998–2003) conducted at WIL Research Laboratories, Inc.

An additional factor that must be considered for rat developmental toxicity studies is the number of females per treatment group with an increased number of resorptions. Table 9.23 presents the modal distribution of the total numbers of females with resorptions in the authors' Crl:CD®(SD)IGS BR rat historical control database. These data indicate that an increase in the number of females that have three or more resorptions is a signal of developmental toxicity.

c. Prenatal Growth

The most sensitive and reliable indicator of an alteration to intrauterine growth is a reduction in fetal body weight. Fetal body weight collection in developmental toxicity studies occurs on the day of laparohysterectomy, generally GD 18, 20, and 29 for mice, rats, and rabbits, respectively. These time points represent the day prior to the expected day of delivery for each species.

The consistency of fetal body weights, particularly for the rat model, has enabled investigators to censor those data that resulted from erroneous determination of evidence of mating. Incorrect assessment of the timing of mating in the Crl:CD®(SD)IGS BR rat, even if only displaced by 24 hours, may result in mean litter weights greater than 5.0 g (compared with the normal weight of 3.6 g on GD 20). When evaluated against the considerable historical control database compiled in the authors' laboratory, the conclusion is that the fetuses are actually 21 or 22 days old, as opposed to the intended age of 20 days. In these presumably rare cases, the heavier litters should not be included in the group mean.

A mature historical control database provides the best means of gauging the reasonableness of a group mean fetal body weight (presented on a litter basis). It has been the authors' observations that in mature historical control databases, the range of variation for control rat fetal body weights is only 0.3 to 0.6 g, depending upon the sex and strain evaluated (see Table 9.24). In rabbits this variation is greater, but it is still small enough to enable consistent interpretation of study results. Furthermore, male fetal body weights are typically greater than female fetal body weights for the most commonly used species.

d. Fetal Sex Ratio

The sex ratio per litter, evaluated in conjunction with mean fetal body weights for males and females, may reveal whether or not the test agent preferentially affects survival of a particular

Table 9.24 Historically observed litter means (ranges) for fetal body weight by species and strain

Endpoint	Mouse Crl:CD®-1(ICR)BR (g)	Rat Crl:CD®(SD)BR (g)	Rat Crl:CD®(SD)IGS BR (g)	Rabbit Hra:(NZW)SPF (g)
Combined sexes	1.31 (1.21–1.36)	3.5 (3.3–3.9)	3.6 (3.4–3.8)	46.9 (39.2–51.8)
Males	NA	NA	3.7 (3.5–3.9)	47.3 (39.9–51.7)
Females	NA	NA	3.5 (3.4–3.7)	45.8 (38.2–49.9)

Note: NA = Data not available. The day of observation of evidence of mating (rodents) or artificial insemination (rabbits) was designated gestational day 0. Laparohysterectomies were conducted on GD 18 for mice, GD 20 for rats, and GD 29 for rabbits.

Source: Data tabulated from studies conducted at WIL Research Laboratories, Inc., including 20 Crl:CD®-1(ICR)BR mouse studies (1984–2000), 158 Crl:CD®(SD)BR rat studies (1982-1997), 61 Crl:CD®(SD)IGS BR rat studies (1998–2003), and 81 Hra:(NZW)SPF rabbit studies (1992–2003). These four databases contain information derived from 460, 3585, 1452, and 1608 pregnant control dams/does, respectively. From these litters, 5552, 50,858, 22,047, and 10,278 viable fetuses, respectively, were available for evaluation.

gender. It has been hypothesized that the hormone levels of both parents around the time of conception may partially control the offspring sex ratio and manifestation of sex-biased malformations.[76,77] However, such effects are rarely observed at the embryo or fetal stage of development in hazard identification studies. Mild variations from an equal male to female sex distribution ratio frequently occur, and little significance is attached to values that fall within normally expected ranges. The historically observed mean percentage of males per litter for Crl:CD®-1(ICR)BR mice (1984 to 2000) and Crl:CD®(SD)IGS BR rats (1998 to 2003) is typically tighter (44.0% to 58.0% per litter) than the observed range for Hra:(NZW)SPF rabbit litters (37.9% to 74.1% per litter, 1982 to 2002) in the authors' laboratory. This difference is likely due to mathematical variability resulting from the smaller sample sizes for rabbits (average of 6.5 fetuses/litter) when compared with mice (12.1 fetuses/litter) or rats (15.2 fetuses/litter). For further discussion of this topic, the reader is referred to Section VI.B.13.

4. Fetal Morphology

a. Context and Interrelatedness of Findings

Measures of normalcy in morphogenesis are the most important endpoints in developmental toxicity studies. Beyond good study design and conduct, it is important to remember that assessment of effects on prenatal development occurs in a single snapshot (at the time of laparohysterectomy). Not all fetuses in these litters are at the same point in their developmental schedules. The discipline's current methods of assessment allow us to examine, at least macroscopically, every structure in these fetuses. Normal morphology may be confounded by disruptions in the temporal pattern of development. These insults may retard development by a direct effect on the conceptus (retardation of its growth), so that by the time the products of conception are evaluated, the natural timing and patterns of development are affected, including increasing cortical masses with decreasing cavity volumes, degree of cavitation of the brain ventricles and kidneys, closure of the neural tube, and development of the thorax and abdominal walls. More severe retardation of fetal growth may produce cardiac valve defects and altered histogenesis.

Actual diagnosis or designation of a defect may be confused by gradations of tissue types and appearances in affected structures, because microscopic evaluation procedures are not routinely employed in these studies. Examination of skeletal and cartilaginous structures may be confounded by imprecise staining procedures that do not enable quantification of the structures that are stained. Attempts have been made to semiquantify these procedures, but these methods have not been validated or generally applied.[78,79] Skeletal examination is further confounded because for the mouse and rat, the time of laparohysterectomy (usually 24 h prior to expected delivery) coincides with the period of peak osteogenesis. Ossification of rodent fetal bones occurs rapidly during the last

48 h of gestation. Because of these factors, the skeletal system in rodent fetuses is immature when it is evaluated.

Additional critical factors should be kept in mind when interpreting fetal morphology data. For example, the terminology and experimental basis for teratologic evaluation originated from the human medical field and human anatomy, for which no comparable detailed knowledge base exists (e.g., regarding human sternal and spinal cord development). Therefore, the certainty of many of these observations is not well founded relative to the human experience. Because of the incompleteness of the human database for newborn morphology (or lack of publication of these data), the interspecies differences and similarities for many of these body regions and structures are not well understood. Also problematic in these guideline studies is that species with short *in utero* developmental periods have been selected for use. In these species, the temporal spacing of complex developmental processes may make the model especially vulnerable to subtle disruptions in growth and timing. Finally, in species with short gestation periods, examination of specimens one day prior to birth fails to address a critical period in morphogenesis and histogenesis that would naturally occur *in utero* in human development.

Because of the numerous potential permutations of effects in developmental toxicity studies, the authors have chosen to summarize a critical approach to evaluation of the data in the form presented in Table 9.25. It should be remembered that each data set is unique, and therefore these questions are intended to serve as a basic framework (as opposed to a "cookbook") within which to evaluate developmental toxicity studies. Evaluation of data from a developmental toxicity study occurs at two levels. First, all data must be examined to differentiate between effects that manifest at statistically significant levels and those that manifest substatistically. This superficial review of the data is necessary for thorough reporting and regulatory Good Laboratory Practice (GLP) completeness. Most important is interpretation of the toxicologic significance of the affected endpoints. Table 9.26 presents selected examples of factors affecting interpretation of the data.

b. Experimental Considerations

1) Training of Prosectors — Macroscopic examination of fetal specimens requires extensive training and an understanding of developmental morphology. The potential impact to product development and, hence, human health may be substantial if any effects are unrecognized or erroneously created. This situation parallels that of a histopathologist's importance for the outcome of a carcinogenicity study, with one major difference. Professionals with veterinary degrees obtain board certification and learn both standardized nomenclature and a process for classification of histopathology. In contrast, there are no formally recognized training programs or certification boards for professionals who carry out reproductive and developmental toxicity studies. Investigators must rely upon on-the-job training to conduct morphologic evaluations of progeny, where one or two findings may be critical. In addition (or perhaps as a consequence), a universally accepted nomenclature and classification system for developmental morphologic effects has yet to be developed and adopted. With regard to standardized nomenclature for malformations in common laboratory animals, an FDA-sponsored effort in the mid-1990s (also supported by the International Federation of Teratology Societies) resulted in the publication "Terminology of Developmental Abnormalities in Common Laboratory Mammals (Version 1)."[80] Currently, the Terminology Committee of the Teratology Society is in the process of updating this glossary.

The following example illustrates the importance of a standardized nomenclature and classification scheme. Muscular ventricular septal defects have never been detected in among over 21,000 control Crl:CD®(SD)IGS BR rat fetuses evaluated at WIL. However, other laboratories have reported up to 200 ventricular septal defects (membranous and muscular) in similar databases. Realizing one practical aspect (i.e., degree of technical competence of personnel) of this discrepancy between laboratories is important. Are technicians not recognizing these defects because they lack training?

Table 9.25 Key considerations during assessment of developmental toxicity data

Questions	Implications and Comments
Are minor test article-treated group changes inappropriately weighted by uncommonly low/high control group values?	Although the data from test article-treated groups initially should be compared with the concurrent control group, use of the laboratory's historical control data will enable identification of atypical concurrent control values.
Is the pregnancy rate significantly reduced (either relative to the concurrent control group or in all groups)?	Reduced pregnancy rates caused by the test article are rare, considering the onset of treatment in most developmental toxicity studies (typically after implantation). The power of the study (ability to detect dose-response relationship) may be impaired.
Are clinical signs of toxicity present?	Severe clinical signs of toxicity that disrupt maternal homeostasis and nutritional status may result in subsequent insults to the products of conception. Conversely, frank increases in terata in the absence of significant maternal toxicity probably signal an exquisite effect on morphogenesis and suggest that the embryo or fetus is more sensitive.
Is there an increased incidence of abortion (in rabbits)?	Related maternal data must be examined carefully (females may have stopped eating), including the historical control rate of abortion and temporal distribution of events (abortions during GD 18 to 23 are rare, compared to those occurring later in gestation).
Does decreased maternal body weight gain correlate with similar changes in food consumption?	Concomitant reductions in body weight and food consumption usually indicate systemic toxicity, and in some cases signal a maternal CNS effect.
Do body weight deficits occur in a dose-related manner?	Dose-related effects on body weight generally indicate a compound-related effect. If maternal mortality is present at the high-dose level, effects on body weight gain may not manifest because of elimination of sensitive members of the group. In this case, body weight effects occurring only at lower doses may represent an adverse effect.
Does maternal body weight gain "rebound" after cessation of treatment?	A rebound in body weight gain following treatment may indicate recovery from the effects of the compound.
Do maternal body weight gain and food consumption decrease following cessation of treatment?	This may indicate a withdrawal effect during the fetal growth phase because of maternal CNS dependency (e.g., opioid compounds). However, it is difficult to correlate such an effect with fetal outcome.
Is maternal net body weight affected?	A decrease in maternal net body weight is most likely an indicator of systemic toxicity. However, reduced maternal body weight in the absence of an effect on maternal net body weight indicates an effect on intrauterine growth and/or survival, rather than maternal systemic toxicity.
Is food consumption reduced?	Changes in food consumption generally signal either palatability issues or systemic toxicity caused by the test substance. Reduced food consumption without a concomitant effect on body weight gain is likely a transient effect and probably not adverse. If food utilization is unaffected when food intake is reduced, the test article is probably affecting caloric intake (i.e., a test article or vehicle with high caloric content may cause an animal to consume less food with minimal or no net effect on body weight gain).
Is maternal body weight gain reduced during a specific period of gestation?	Decreased maternal body weight gain during the late gestational period may be associated with reduced fetal skeletal mineralization. In all cases, risk to the developing embryo or fetus is greater the longer the reduction in maternal body weight gain is sustained.
Is gravid uterine weight affected?	Reduced gravid uterine weights are usually due to reduced viable litter size and/or reduced fetal weights. In rare cases, effects on the placenta or the uterus may be initially recognized by nonfetus-related changes in gravid uterine weight.
Is the percentage of affected litters increased relative to controls?	This may indicate a maternal dimension if the proportion of fetuses in those litters is not similarly affected.
Is the percentage of affected fetuses per litter increased relative to controls?	This probably indicates a proximate fetal effect.
Is viable litter size decreased?	Decreased viable litter size generally represents either increased resorption rate (usually attributable to the test agent) or decreased numbers of corpora lutea and/or implantation sites (not likely to be a test substance–related effect unless treatment began at or prior to fertilization).

Table 9.25 Key considerations during assessment of developmental toxicity data (continued)

Questions	Implications and Comments
Is the resorption rate increased?	Complete litter resorptions are rare (3 in 1452 Crl:CD®(SD)IGS BR rat litters). Therefore, even one completely resorbed rat litter in the high-dose group merits attention. Furthermore, an increased incidence of control female rats with more than three resorptions usually constitutes a signal of developmental toxicity (refer to Table 9.23). Refer to Section III.B. for a discussion of the relationship between number of implantation sites and increased resorption and/or abortion rate in rabbits.
Is fetal weight subtly decreased in the high-dose group (e.g., mean rat fetal weights of 3.6, 3.6, 3.6, and 3.4 g in control, low-, mid-, and high-dose groups, respectively)?	If other signals of developmental toxicity are present, the decrease in fetal weight is probably adverse. If no other signals of developmental toxicity exist, and the low fetal weight is within the laboratory's historical control range, it is probably not an adverse effect. A robust, highly consistent historical control database in this instance would indicate that fetuses weighing 3.4 g are able to survive and thrive, and therefore the weight reduction is likely to be temporary. Such a conclusion may be corroborated with data from postnatal assessment studies.
Does a correlation exist between low-weight-for-age fetuses and dysmorphogenesis?	If malformations are limited to low-weight-for-age fetuses, the malformations may be due to generalized growth retardation (e.g., omphalocele in rabbits or cleft palate in rats).
Is a syndrome or spectrum of effects present?	A syndrome of effects indicates multiple insults on a specific organ system during development (e.g., tetralogy of Fallot, or a cascade of events during heart development, beginning with ventricular septal defects and including pulmonary stenosis and valvular defects). A spectrum of dysmorphogenic effects implies a less targeted, more generalized response (e.g., vertebral agenesis occurring with ocular field defects).
Is the total malformation rate per group affected?	Several organ systems may have slight malformation rate increases that may not be outside the historical control range when evaluated individually. However, in summation they may signal an increased generalized dysmorphogenic effect.
Are there qualitative differences in the relative functional utility of affected structures or systems?	Bent long bones (e.g., femur) are of greater concern than bent ribs. Unossified centra or sternebrae are not of great import if the underlying cartilaginous structures are present and properly articulated. Malformed caudal vertebrae or facial papillae in rats would not merit the concern that similar alterations to analogous human organs or structures would elicit.
Is fetal ossification generally retarded (evidenced by Alizarin Red[a] and/or Alcian Blue[b] uptake)?	Delayed ossification may be due to developmental delay resulting from intrauterine growth retardation or may stem from properties of the test agent expected to cause delayed mineralization of skeletal structures (e.g., calcium chelation, altering of blood urea nitrogen, alkaline phosphatase inhibition, or changes in parathyroid hormones or vitamin D).
Is a dose-related response observed when the endpoint of "affected implants" is evaluated?	For this endpoint, each implantation is counted once (tabulated as "affected"), whether the outcome is an early or late resorption, malformed fetus, or dead fetus. As the rate of malformation increases, the rate of late resorption declines. Likewise, as the number of early resorptions rises, the numbers of late resorptions and malformed fetuses decline. Refer to Figure 9.7 for an idealized graphical depiction of this interrelationship between most common fetal outcomes.
Are there similar effects in other studies?	Comparison to effects in a second species developmental toxicity study, malformations manifested as anatomical or functional changes in the pre- and postnatal development study, and maternal toxicity relative to systemic toxicity observed in subchronic toxicity studies may indicate that the pregnant animal is more susceptible to the test agent than the nonpregnant animal.

[a] Dawson, A. B., *Stain Technol.*, 1, 123, 1926.
[b] Inouye, M., *Congen. Anom.*, 16, 171, 1976.

Alternatively, is the laboratory confident (because of the highly trained technicians) that there are no such occurrences? A highly trained and stable group of professionals would indicate the latter. Such professionals would have been trained to recognize "normalcy" when judging fetal morphology, the most important criterion in these kinds of evaluations. In addition, they would be most likely to have developed and be using a consistent approach to nomenclature and definitions of malformations and variations.

Table 9.26 Factors affecting interpretation of developmental toxicity data

Procedural Artifacts
    Diagnosis
        Failure to recognize rare findings when present
    Fixation
        General tissue shrinkage
        Ocular opacities due to formaldehyde fixation
    Mishandling of specimens
        Deformed bone and cartilage (e.g., apparent arthrogryposis caused by uterine compression because fetuses were not removed from the uterus in a timely manner)
        Hematomas or petechial hemorrhages
        Imprecise heart cross-sections
Endpoint Interrelatedness (Refer to Table 9.25 for Further Details)
    Absence of dose-related response in malformation rate because of dose-related increase in postimplantation loss
    Fetal weight inversely related to litter size (intralitter competition)
    Late gestation maternal body weight gain relative to intrauterine survival
    Generalized intrauterine growth retardation vs. specific structural malformations (e.g., omphalocele, cleft palate, and increased renal cavitation)
Temporal Patterns
    Continuum of responses (effect of lumping versus splitting of findings)
        Overrepresenting an effect by tabulating both malformations and variations in the same body region, potentially confounding delineation of the NOAEL
        Insult to specific anatomic region manifested as variations at lower dose levels and as malformations at higher dose levels
        Shifts in the timing of intrauterine death (early to late resorption) relative to dose-response curve (refer to Figure 9.8 for an idealized graphical depiction)
    Lack of a dose-response relationship because of saturation of absorption, binding, and/or receptor capacity
    Symmetrical expression of the abnormality

Ideally, to minimize the bias introduced by subjectivity, the same individuals should perform the fetal evaluations across all dosage groups in a single study. To minimize unnecessary variability between outcomes, the same personnel should conduct the entire reproductive and developmental program for a particular test agent, if possible.

2) Methodology — Prenatal morphogenesis may be evaluated via one of two primary methods: (1) the older Wilson[81] method, involving serial freehand razor sections of decalcified specimens, or (2) microdissection of nonfixed fetuses (commonly referred to as fresh dissection[82] or the Staples method[83]). Though combinations of these two methods have been published (e.g., the Barrow and Taylor technique[84]), the Wilson and fresh dissection methods are the most commonly employed approaches. Following is a discussion of advantages and disadvantages of each method.

The primary advantage of the fresh dissection method is that all fetuses can be examined for both soft tissue and skeletal changes (see Figure 9.7), thus maintaining the greatest possible statistical power of a study. In addition, artifactual changes as a result of fixation can occur with the Wilson method (refer to Table 9.26), and visualization of spatial relationships in two dimensions (e.g., following structures that transcend multiple sections) is much more difficult to master than when they are visualized *in situ* (refer to Table 9.27 for additional discussion of the numerous practical and scientific advantages of fresh dissection of fetuses).

Another reason that the Wilson sectioning method, which allows the processed specimens to be stored and examined later, is not the best approach is that this method necessitates *a priori* "subsetting" of the fetuses into those that will be examined viscerally and those that will be examined skeletally. Such a practice is usually followed simply out of convenience; it is undesirable statistically and may diminish the value of the study. For example, from the external perspective, a fetus may manifest multiple malformations. Complete characterization of underlying internal morphologic defects would require both thorough visceral and skeletal examinations. However, the

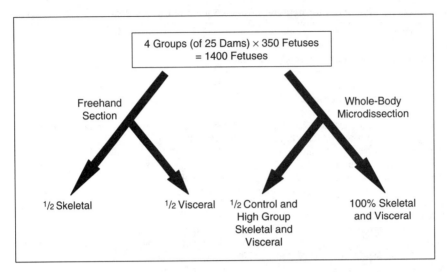

Figure 9.7 Comparison of the number of fetuses available for visceral and skeletal examination using the freehand section and whole-body microdissection techniques.

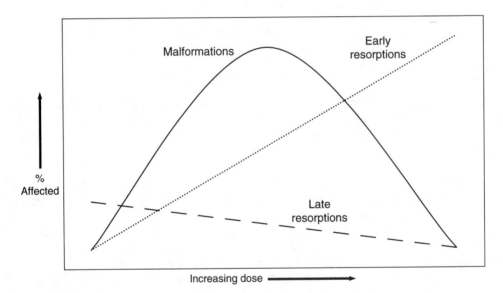

Figure 9.8 Idealized depiction of the relationship of dose and adverse developmental outcome.

Wilson sectioning method requires the investigator to commit to either a visceral or skeletal examination, but not both.

Some of the rationale behind *a priori* subsetting of the fetuses may be defensible if the investigator is using a high dose of a compound that will affect at least 15% to 20% of the fetuses. However, this rationale presumes that each compound tested will evoke such a response and that this would be known prior to the study, which is not usually the case. Therefore, adverse effects on fetal morphology are much more likely to be overlooked if a large number of compounds are tested by the subsetting approach.

Further compounding the visceral-skeletal subsetting problem is the ICH guideline–driven option of only evaluating control and high-dose group fetuses when no apparent effects are observed at the high-dose level. This approach assumes that any treatment-related findings that occur will

Table 9.27 Advantages of fresh dissection of fetuses

Prosectors are more easily trained.
Good visualization with less chance of missing a cleft, breach, or absence of structures (e.g., valve in heart), because of thickness and perpendicularity of cross-section of the sample.
Direct visualization of structures, as one would learn in anatomy, instead of having to rotate images mentally in three dimensions.
Coloration (organs, vessels, tissues).
Direct ascertainment of vessel patency by tactile manipulation of vessels to determine patency and if blood is moving.
Evaluation of vascular tissue perfusion by coloration.
All fetuses can be examined for both soft tissue and skeletal changes.
Avoids the difficulty of sectioning every fetus in exactly the same way as is needed to ensure comparability of views among fetuses.
Decreased probability of creating artifacts due to fixative tissue shrinkage

be discernible in the high-dose group (refer to Section VII for a further discussion of the dangers inherent in this assumption). The statistical power of these studies is already low relative to extrapolation to the human population, and when the dosage group has already been subsetted prior to evaluation, the probability of detecting treatment-related effects declines dramatically. Therefore, the authors conclude that morphologic evaluation of all fetuses by use of the fresh dissection method is the most powerful approach for developmental toxicity studies, yielding the most conclusive results for hazard identification.

c. Relative Severity of Findings (Malformations vs. Variations)

A critical first step in fetal morphologic evaluation is determining the relative severity of the alteration. This determination is complicated by the potential continuum of responses between normal and extremely deviant fetal morphology. Fetal dysmorphogenic findings have been reported as developmental deviations, structural changes, malformations, anomalies, congenital defects, anatomic alterations, terata, structural alterations, deformations, abnormalities, anatomic variants, or developmental variations, with little consistency across laboratories in definition of the terms. The current authors have chosen to report, and accordingly define, these external, visceral, and skeletal findings as either developmental variations or malformations. Variations are defined as alterations in anatomic structure that are considered to have no significant biological effect on animal health or body conformity and/or occur at high incidence, representing slight deviations from normal. Malformations are defined as those structural anomalies that alter general body conformity, disrupt or interfere with normal body function, or may be incompatible with life.

d. Presentation of Findings

Fetal morphologic findings should be summarized by: (1) presenting the incidence of a given finding both as the number of fetuses and the number of litters available for examination in the group and (2) considering the litter as the basic unit for comparison and calculating the number of affected fetuses in a litter on a proportional basis as follows:

$$\text{Summation per group (\%)} = \frac{\Sigma \text{ viable fetuses affected per litter (\%)}}{\text{Number of litters per group}}$$

where

$$\text{Viable fetuses affected per litter (\%)} = \frac{\text{Number of viable fetuses affected per litter}}{\text{Number of viable fetuses per litter}} \times 100$$

Table 9.28 Most commonly occurring developmental variations in control Crl:CD®(SD)IGS BR rats

Total Number Examined (1998–2003)	Fetuses	Litters	% per Litter
External	22,047	1447	—
None observed in control rats			
Visceral	21,853	1447	—
Major blood vessel variation[a]	15	15	0.0–0.4
Renal papilla(e) not developed and/or distended ureter(s)	5	4	0.0–0.8
Hemorrhagic ring around the iris	3	3	0.0–0.3
Hemorrhagic iris	1	1	0.0–0.3
Skeletal	21,843	1446	—
Cervical centrum no. 1 ossified	3,919	1065	6.6–32.1
Sternebra(e) no. 5 and/or no. 6 unossified	1,920	632	0.3–23.1
14th rudimentary rib(s)	1,328	565	1.4–15.1
Hyoid unossified	325	208	0.0–3.4
7th cervical rib(s)	121	91	0.0–2.7
Reduced ossification of the 13th rib(s)	119	74	0.0–3.0
Sternebra(e) no. 1, 2, 3, and/or 4 unossified	50	44	0.0–1.3
Sternebra(e) malaligned (slight or moderate)	32	31	0.0–0.8
25 presacral vertebrae	30	20	0.0–2.0
27 presacral vertebrae	29	21	0.0–1.8
7th sternebra	19	5	0.0–2.5
14th full rib(s)	18	14	0.0–0.9
Bent rib(s)	18	18	0.0–4.0

[a] The most commonly occurring manifestations of this finding are: (1) right carotid and right subclavian arteries arising independently from the aortic arch (no brachiocephalic trunk), (2) left carotid artery arising from the brachiocephalic trunk, and (3) retroesophageal right subclavian artery.

Source: Data collected at WIL Research Laboratories, Inc.

e. Significance of Developmental Variants

An important issue in interpreting developmental toxicity study data is the developmental (anatomic) variant. Table 9.28 and Table 9.29 present the most commonly occurring developmental variations observed in control Crl:CD®(SD)IGS BR rats and Hra:(NZW)SPF rabbits, respectively, in the authors' laboratory. The total numbers of descriptors presented in these tables (17 for rats and 23 for rabbits) represent approximately 42% of the total number of descriptors for each species when findings from both control and test substance-exposed animals are tabulated. At least 41 unique developmental variations have been observed over the past 6 years of investigations using the Crl:CD®(SD)IGS BR rat in the authors' laboratory (61 studies, 1447 litters with viable fetuses). In the case of the Hra:(NZW)SPF rabbit, at least 53 different developmental variations have been observed in 1529 litters (81 studies) from 1992 through 2003.

Developmental variants are of somewhat lesser concern to the investigator and regulatory reviewer than malformations; however, they often pose a great dilemma if not interpreted properly. In assessment of whether a fetal morphologic deviation represents a malformation or a variation, the factors listed in Table 9.30 must be considered.

Findings that are classified as developmental variants must then be assessed for their toxicologic significance and potential impact on hazard identification decisions. In the latter case, the EPA historically has been much more mindful of developmental variants, often differentiating the NOEL and NOAEL based upon the context of these variants, than has the FDA.[39] The toxicologic significance (human versus animal) of developmental variants generally will be dependent upon the combined weight of their impact on salubrity and their fate (the extent to which they are

Table 9.29 Most commonly occurring developmental variations in control Hra:(NZW)SPF rabbits

Total Number Examined (1992–2003)	Fetuses	Litters	% per Litter
External	10,278	1529	—
Twinning	1	1	0.0–0.8
Visceral	10,278	1529	—
Accessory spleen(s)	1,198	681	4.8–33.2
Major blood vessel variation[a]	565	329	0.0–17.5
Gallbladder absent or small	150	115	0.0–7.8
Retrocaval ureter	142	110	0.0–5.4
Hemorrhagic ring around the iris	46	33	0.0–3.6
Spleen — small	6	6	0.0–1.0
Hemorrhagic iris	4	4	0.0–0.8
Liver — pale	2	2	0.0–0.6
Eye(s) — opacity	2	1	0.0–1.0
Accessory adrenal(s)	1	1	0.0–0.7
Renal papilla(e) not developed and/or distended ureter(s)	1	1	0.0–1.2
Skeletal	10,278	1529	—
13th full rib(s)	4,082	1240	19.4–59.1
13th rudimentary rib(s)	1,982	1042	8.1–32.5
27 presacral vertebrae	1,724	766	4.5–32.1
Hyoid arch(es) bent	504	357	0.0–22.2
Sternebra(e) no. 5 and/or 6 unossified	448	274	0.0–11.4
Sternebra(e) with threadlike attachment	146	121	0.0–9.1
Sternebra(e) malaligned (slight or moderate)	117	108	0.0–5.0
Extra site of ossification anterior to sternebra no. 1	106	84	0.0–7.4
Accessory skull bone(s)	80	69	0.0–5.0
7th cervical rib(s)	73	59	0.0–7.7
25 presacral vertebrae	35	31	0.0–7.4

[a] The most commonly occurring manifestations of this finding are: (1) right carotid and right subclavian arteries arising independently from the aortic arch (no brachiocephalic trunk), (2) left carotid artery arising from the brachiocephalic trunk, and (3) retroesophageal right subclavian artery.

Source: Data collected at WIL Research Laboratories, Inc.

Table 9.30 Observational determinants of anatomic or functional deviations

Degree of deviation from average
   Magnitude of morphologic change
   Nature of functional impact, if any
Incidence (prevalence)[a]
Impact on salubrity
Cosmetic significance

[a] In humans, the convention has been to use an incidence of less than 4% and no medical or surgical significance. In experimental studies, no convention has been established.

reversible). The difficulty is that for animal models the impact on salubrity is generally unknown (unless adult bone scans are conducted to follow the fate of skeletal variations). However, if a clear dose-related response manifests, additional factors, including statistical power of the study and potential for temporal changes (due to breeding, penetrance, environmental factors, etc.), must be addressed.

Table 9.31 Determination of the significance of developmental variations

Learned scientific debate
Performing confirmatory studies if necessary
Determining if an effect is reversible with maturation or if significant remodeling is likely
Comparing the anatomy of mothers and progeny (e.g., vascular variations)
Assessing the prevalence of paternal contributions (e.g., gallbladder variations in the rabbit)
Taking advantage of newer technologies when possible (e.g., assessing potential expression of the variant in the adult phenotype via computed topography [CT] scan or bone densitometry)

A suggested approach to distinguishing the significance of developmental variants is outlined in Table 9.31. A combination of the steps may determine such significance.

Further discussion of related examples — the rabbit gallbladder issue and expression of 27 presacral vertebrae and supernumerary ribs in rats and rabbits — is presented in the following sections.

1) Gallbladder Variations in Rabbits — A fetal observation in the rabbit that has generated considerable regulatory discussion and discord is the "absent" or "small" gallbladder. The classic work by Sawin and Crary clearly demonstrated that the spontaneous incidence of "absent" gallbladder was significant and variable in several stocks of rabbits.[85] Sawin and Crary adroitly pointed out that most of the literature referenced the "presence" or "absence" of the gallbladder as a species characteristic, rather than an individual characteristic. Sawin and Crary studied genetic influences on the gallbladder in 25 unrelated stocks, several of which had been inbred for as many as 22 generations. Of particular note to the present day use of the Hra:(NZW)SPF stock rabbit were the present authors' studies demonstrating that some animals repeatedly produce offspring lacking a gallbladder, even when the authors' investigation of the phenomenon was extended to include other stocks. The "absence" of the gallbladder was part of a graded series of variations in its size and shape.

In the authors' experience, it is relatively rare to see a true and confirmed (by an additional study) dose-related increase in gallbladder absence. More often than not, these scenarios appear dose related but are not reproducible. Of course, there may be exceptions.

Over the course of hundreds of studies, the authors compiled a historical database on this observation and analyzed paternal contributions to size and/or presence of the gallbladder. These data indicate that there is a highly variable and sometimes dramatic trend that certain stock bucks sire litters with many more fetuses having either a "small" or "absent" gallbladder (refer to Table 9.32 for representative examples). Continued monitoring of the colony of breeders and maintaining appropriate records for use in interpreting related fetal findings is critical. However, purchasing date-mated females has become commonplace in some laboratories engaged in reproductive toxicology. Without sire records, these paternally influenced fetal observations and their relationship to treatment cannot be ascertained or appropriately analyzed.

2) Supernumerary Ribs and Presacral Vertebrae — Historically, the authors' laboratory has classified 27 presacral vertebrae (PV) and supernumerary ribs (SR) as developmental variations. These variants are routinely quantified and occur in control populations at well-characterized frequencies. It is the authors' opinion that 27 PV and 13th full/rudimentary rib, in the absence of other fetal effects, should not automatically be considered a finding of toxicologic concern or teratogenicity in developmental toxicity studies with rabbits. Furthermore, caution is warranted when considering whether an increased occurrence of 27 PV or 13th full or rudimentary rib is relevant when extrapolating to human development.

The thoracolumbar border that gives rise to 27 PV and SR is highly labile in rabbits, making its significance in developmental toxicity problematic. Results of 10,037 fetal skeletal evaluations

Table 9.32 Paternally influenced prevalence of absent or reduced gallbladder in New Zealand white rabbit fetuses

Stock Male No.	Percent Fetuses Affected per Litter[a]	No. of Litters[b]	No. with Absent Gallbladder/Litter[c]
Hra:(NZW)SPF Rabbits[d]			
3508	20.2	109	8/7
2876	20.0	40	5/4
8457	14.6	48	0/0
2877	11.1	166	5/4
2871	3.9	152	7/6
Lsf:NZW Rabbits[e]			
481	58.3	36	20/13
252	27.0	37	10/7
445	24.1	29	8/7
251	4.8	21	1/1

[a] Percent of fetuses per litter with absent or reduced gallbladder.
[b] Total litters sired.
[c] No. fetuses/No. litters.
[d] A limited number of bucks (5) is presented to demonstrate the range of values, representing more than 100 animals in the entire Hra:(NZW)SPF rabbit database at WIL Research.
[e] Four bucks were used to demonstrate the range of values in the Lsf:NZW rabbit database at WIL Research.

from 79 rabbit studies recorded in the WIL historical control database showed that 13th full rib, 13th rudimentary rib, and 27 PV are the first, second, and third most frequently noted skeletal variations, respectively, in control rabbits. The mean litter percent of fetal 27 PV was 17.4% (range of 5% to 32%; Q1–Q3 interquartile range of 13% to 21%). One-half of the control females carried at least one fetus with 27 PV, and over 1% of control females exhibited a 100% occurrence of fetal 27 PV. The mean litter percentages of fetuses with 13th full or 13th rudimentary rib were 39% (range of 19% to 59%; Q1–Q3 interquartile range of 35% to 44%) and 20% (range of 8% to 32%; Q1–Q3 interquartile range of 17% to 23%), respectively. There was also a strong positive correlation between SR and 27 PV among control populations. The presacral region is more labile in rabbits than in rats and mice, which exhibit mean litter percentages of 27 PV of 0.15% and 0.13%, respectively. The mean litter percent of supernumerary or rudimentary ribs is also lower than in rabbits, i.e., 6.1% in rats and 12.9% in mice.

A recent study examined the prevalence of 27 PV and SR in 62 adult male and 100 adult female rabbits by radiography.[86] The incidence of 27 PV among adult control rabbits was 22%. This incidence agreed well (especially given the difference in N) with the mean litter percent of 27 PV among fetal rabbits (17%), suggesting that there is no detriment or decrease in survival of progeny with this variant. These findings also provide evidence that the presence of 27 PV is without significant effect on salubrity and suggest little or no remodeling of these structures in the rabbit, as there is in rodents. This radiographic study further showed that 54.6% of adult control rabbits have some type of SR (e.g., full, rudimentary, bilateral, unilateral or in combination). Published reports have suggested that both chemically induced and control fetal SR can resolve during maturation in rats[87–90] and that both 12 and 13 pairs of ribs are wild-type phenotypes in adult rabbits.[91,92] The radiographic study found that 80% of adult rabbits with 27 PV also exhibited 13th full rib.[86] In other words, the occurrence of 13th full rib (considered wild type) in the rabbit occurs predominantly with additional vertebrae in the axial skeleton. Separate laboratories have reported similar correlations between PV and SR in adult rabbits.[88,90,93] It is also worth noting that approximately 5% of humans exhibit supernumerary (3%) or missing (2%) presacral vertebrae.[94]

In the absence of any other fetal effects (i.e., intrauterine growth retardation, major or minor malformations, fetal death, or functional impairment), an increased occurrence of 27 PV and/or 13th full ribs is not sufficient evidence of developmental toxicity. This is especially true in the rabbit model, where there is a high background incidence of these skeletal variations in controls, where 12 or 13 pairs of ribs is considered the wild-type phenotype, and where 13th full rib is commonly associated with 27 PV.

The usefulness of endpoints such as 27 PV and 13th full rib in predicting risk for the human population is complex and has not been well studied or validated. Reviews of known teratogenic agents (e.g., thalidomide, alcohol, ionizing radiation, steroidal hormones, heroin, or morphine) have been published, taking into account the endpoints evaluated, the relative power of the studies, the dose-response patterns, and overall toxicity to the maternal and fetal systems.[11,95–97] Based on the limited number of agents for which adequate published data were found, it was determined that the most reliable endpoints for predicting risk of developmental toxicity in humans were fetal growth retardation, malformations, functional impairment, and spontaneous abortion. There is currently inadequate information regarding skeletal developmental variations to make judgments concerning concordance or nonconcordance between human and animal data. The lack of concordance across animal species (e.g., rabbits versus rodents) for 27 PV and 13th full rib suggests that these findings might simply reflect species-specific anatomic variation, without detrimental effects on salubrity.

VI. REPRODUCTIVE TOXICITY STUDIES

This section begins with a discussion of the various regulatory guideline requirements for reproductive toxicity studies. Specific endpoints measured in these studies are then addressed, and guidance on their interpretation is provided, based on the authors' collective experience and historical data.

A. Guideline Requirements

Table 9.33 presents the salient differences between the various regulatory guidelines for reproductive toxicity study designs that may affect interpretation of the data. In the United States, OPPTS (within the EPA) has developed reproductive toxicity testing guidelines for use in the testing of pesticides and toxic substances. These guidelines harmonized the previously separate testing requirements under TSCA and FIFRA, and the OECD in Europe has developed guidelines very similar to those of the EPA. For medicinal products, ICH developed technical requirements for the conduct of reproductive toxicity studies for supporting the registration of pharmaceutical products for the EU, Japan, and the United States. Finally, for food ingredients, CFSAN (within the FDA) has developed guidelines for reproductive toxicity studies.

1. Animal Models

The rat is the recommended species for both the EPA and OECD multigeneration study and the ICH fertility and pre- and postnatal development studies. There are several advantages to using the rat, such as a high fertility rate, a large historical control database, and cost effectiveness. However, the rat is not always the most appropriate model for these studies. Based on mode of action, kinetics, nature of the compound (e.g., recombinant protein) or sensitivity, other species, such as the rabbit, dog, or nonhuman primate, have at times been used instead of the rat. However, because the duration

Table 9.33 Selected comparison points between various regulatory agency guideline requirements for fertility and reproductive toxicity studies

Comparison Point	FDA/Redbook 2000 (2000 Draft)	OECD (2001)	EPA (1998)	ICH (1994)
Reference	172	173	174	175,176
Species	Preferably rat	Preferably rat	Preferably rat	Preferably rat
Group size	30/sex/group (F_0) and 25/sex/group (maximum of 2/sex/litter) for F_1, to yield approx. 20 pregnant females/generation	Yield preferably not less than 20 pregnant females/generation	Yield approx. 20 pregnant females/generation	16–20 litters
Age at start of treatment	5–9 weeks	5–9 weeks	5–9 weeks	Young, mature adults at time of mating
Number of generations directly exposed	2, optional 3	2	2	1
Duration of treatment per generation	Males: 10 weeks prior to and throughout the mating period Females: 10 weeks prior to mating, throughout mating and pregnancy and up to weaning of the offspring	Males: 10 weeks prior to and throughout the mating period Females: 10 weeks prior to mating, throughout mating and pregnancy and up to weaning of the offspring	Males: 10 weeks prior to and throughout the mating period Females: 10 weeks prior to mating, throughout mating period and pregnancy and up to weaning of the offspring	Males: 4 weeks prior to and throughout the mating period Females: 2 weeks prior to mating, throughout mating period and through at least implantation
Estrous cycle determination	Minimum of 3 weeks prior to cohabitation and during cohabitation	Prior to cohabitation and optionally during cohabitation	Minimum of 3 weeks prior to cohabitation and throughout cohabitation	At least during the mating period
Mating period	1:1 until copulation occurs or 2 to 3 weeks have elapsed	1:1 until copulation occurs or 2 weeks have elapsed. In case pairing is unsuccessful, remating of females with proven male from same group could be considered	1:1 until copulation occurs or either 3 estrous periods or 2 weeks have elapsed. If mating has not occurred, animals should be separated without further opportunity for mating	1:1; most laboratories would use a mating period of between 2 and 3 weeks
Spermatogenesis assessment	If there is evidence of male-mediated effects on developing offspring, all (at least of the control and high dose) males per generation assessed for sperm motility and morphology and enumeration of homogenization-resistant spermatids and cauda epididymal sperm	At least 10 males per generation assessed for sperm motility and morphology and enumeration of homogenization-resistant spermatids and cauda epididymal sperm	All (at least of the control and high dose) males per generation assessed for sperm motility and morphology and enumeration of homogenization-resistant spermatids and cauda epididymal sperm	Sperm count in epididymides or testes, as well as sperm viability (later changed to optional)

Table 9.33 Selected comparison points between various regulatory agency guideline requirements for fertility and reproductive toxicity studies (continued)

Comparison Point	FDA/Redbook 2000 (2000 Draft)	OECD (2001)	EPA (1998)	ICH (1994)
Organ weights — adults	Uterus, ovaries, testes, epididymides (total and cauda), seminal vesicles, prostate, brain, liver, kidneys, adrenals, spleen, pituitary, known target organs	Uterus, ovaries, testes, epididymides (total and cauda), seminal vesicles, prostate, brain, liver, kidneys, adrenals, spleen, thyroid, pituitary, known target organs	Uterus, ovaries, testes, epididymides (total and cauda), seminal vesicles, prostate, brain, liver, kidneys, adrenals, spleen, pituitary, known target organs	Not required
Adult histopathology	10 control and high dose animals selected for mating	All control and high dose animals selected for mating	10 control and high dose animals selected for mating	Not required
Examination of offspring	From as soon as possible after birth until weaning	From as soon as possible after birth until weaning	From as soon as possible after birth until weaning	At any point after mid-pregnancy
Standardization of litters	Optional	Optional	Optional	Optional as part of a pre- and postnatal development study
Developmental landmarks	Age of vaginal opening and preputial separation	Age of vaginal opening and preputial separation	Age of vaginal opening and preputial separation	Assessment recommended as part of a pre- and postnatal development study
Offspring functional tests	Optional, excellent vehicle to screen for potential developmental neurotoxicants	Recommended when separate studies on neurodevelopment are not available	Not required	Assessment recommended as part of a pre- and postnatal development study
Organ weights — pups	Brain, spleen and thymus from at least two pups/sex/litter at weaning	Brain, spleen and thymus from one pup/sex/litter at weaning	Brain, spleen and thymus from one pup/sex/litter at weaning	Not required
Teratology phase	Optional as either F_{2b} or F_{3b} litter	Separate study	Separate study	Optional
Immunotoxicity screening	Optional for F_0, F_1, and F_2 generations	Not required	Separate study	Not required

and cost of the study are typically greatly increased, reproductive toxicity studies using such alternative species are only rarely performed.

2. Timing and Duration of Treatment

The EPA, OECD, and FDA food additives requirements for reproductive toxicity studies are designed to assess the effects of long-term exposure, while the treatment periods for drug studies are much shorter and less comprehensive. The rationale behind the long-term exposure regimens employed by the EPA and OECD is to mimic the ambient, low-level, long-term exposures to industrial chemicals in the workplace and to low-level residues of pesticides in the diet. Similarly, exposure to food additives may occur over a lifetime. In contrast, treatment during defined stages (phases) of reproduction in the ICH guidelines generally better reflects human exposure to medicinal products and allows more specific identification of stages at risk. In the EPA, OECD, and FDA multigeneration studies, both generations of parental animals are exposed prior to mating and throughout the mating, gestation, and lactation periods in the rat. The ICH studies are divided into three segments: (1) fertility assessment, (2) embryo-fetal development assessment, and (3) pre- and postnatal development assessment. Within each of the ICH segments (phases), the direct exposure period is also more limited than is required in the EPA and OECD study designs. For example, a 10-week premating period is recommended for the EPA, OECD, and FDA multigeneration studies, while the premating period in the ICH fertility study is generally only 4 and 2 weeks for the males and females, respectively. In a collaborative study of 16 compounds, the shorter premating male exposure period for ICH studies was found to be appropriate for detection of drug effects on male fertility.[98] In that collaborative study, the 4-week premating period was found to be as effective as a 9-week premating period for identification of male reproductive toxicants, albeit in a limited set of agents. In addition, exposure ends at the time of implantation in the ICH fertility study but continues throughout the entire gestation and lactation periods in the EPA, OECD, and FDA multigeneration study. Unlike the EPA, OECD, and food additive multigeneration studies, exposure of offspring following weaning cannot be assumed in the ICH pre- and postnatal study.

Because of the short premating treatment regimen in the ICH fertility study, compound administration does not begin until the rats are 8 to 9 weeks of age, which is after puberty has occurred. This is in contrast to the EPA, OECD, and food additive design in which direct F_1 exposure begins at weaning (prior to sexual maturation) and F_0 exposure begins typically at 5 to 9 weeks of age, during the process of sexual maturation in the females and prior to maturation in the males.

Another difference among the guidelines is that only the OECD requires adult thyroid weight evaluation. This is a surprising omission from the food additive and EPA guidelines because of the role of the thyroid in development and sexual maturation, recent welling concerns for endocrine modulation, and questions and concerns regarding the thyroid toxicity of ammonium perchlorate.[99,100]

B. Interpretation of Reproductive Toxicity Study Endpoints

In this section, the endpoints listed in Table 9.34 (male-female or coupled) for evaluating reproductive toxicity are discussed. These endpoints are listed first in approximate chronological order of collection and then ranked by approximate sensitivity of the endpoints (those endpoints that are bolded may be assessed in and/or appropriate for evaluation in humans). In this ranking, endpoints considered most sensitive are those that are most often affected in general or those most often affected by lower doses of xenobiotics. In the following sections, discussion of Crl:CD®SD(IGS) BR rat historical control data for reproductive toxicity endpoints is based on the database compiled in the authors' laboratory. Unless specifically noted otherwise, these data originate from 55 studies (92 separate matings) conducted during 1996 to 2002.

Table 9.34 Endpoints of reproductive toxicity studies

Approximate Chronological Order of Collection	Ranked by Sensitivity
Estrous cyclicity	Viable litter size/live birth index
Precoital interval	**Neonatal growth**[a]
Sexual behavior	**Neonatal survival indices**
Mating/fertility indices and reproductive outcome	**Prenatal mortality**
Duration of gestation	**Assessment of sperm quality**
Parturition	Weight and morphology (macroscopic and microscopic) of reproductive organs
Neonatal survival indices	
Prenatal mortality	**Estrous cyclicity**
Nesting and nursing behavior	Precoital interval
Assessment of sperm quality	**Mating and fertility indices and reproductive outcome**
Weight and morphology (macroscopic and microscopic) of reproductive organs	
Oocyte quantification	**Duration of gestation**
Qualitative and quantitative physiologic endpoints revealing unique toxicities of pregnancy and lactation	Parturition
	Landmarks of sexual maturity
	Functional toxicities and CNS maturation
	Learning and memory
Viable litter size/live birth index	Qualitative and quantitative physiologic endpoints revealing unique toxicities of pregnancy and lactation
Sex ratio in progeny	
Neonatal growth	Nesting and **nursing behavior**
Landmarks of sexual maturity	Sexual behavior
Functional toxicities and CNS maturation	Sex ratio in progeny
Learning and memory	Oocyte quantification

[a] Bolded endpoints may be assessed and/or appropriate for evaluation in humans.

Because sensitivity of an endpoint may be methodologically or biologically dependent, the ranking in Table 9.34 is relatively arbitrary and should not be considered a definitive ranking of endpoints. Mode of action may determine which measure is more sensitive.

1. Viable Litter Size/Live Birth Index

The calculation of live litter size and live birth index is as follows:

$$\text{Live litter size} = \frac{\text{Total viable pups on day 0}}{\text{Number litters with viable pups on day 0}}$$

$$\text{Live birth index} = \frac{\text{Number of live offspring}}{\text{Number live offspring delivered}} \times 100$$

As an apical measure, viable litter size is frequently the most sensitive indicator of reproductive toxicity and is a very stable index. Decreased numbers in this endpoint relative to the concurrent control group can result from a reduction in the ovulatory rate and timing, from shifts in the timing of tubal transport, and from reductions in implantation rate, postimplantation survival, or sperm parameters (motility or concentration). In assessing the number of viable pups, it is important to examine the litters as soon as possible following birth to ascertain the number of live pups versus the number of stillborn pups. In addition, cannibalism of the progeny by the dam may result in an inaccurate determination of the number of pups born. However, this concern has probably been overstated as a technical problem except in instances where offspring are born dead or malformed, or there is a specific effect on CNS function by the test agent. Problems with cannibalism have decreased because of improvements in husbandry and laboratory practices, coupled with better-defined stocks of animals. Optimal offspring quantification can be accomplished by examining the dam at least twice each day (in the morning and afternoon) during the period of expected parturition (beginning on GD 21). These frequent examinations are necessary to detect dystocia. Within the authors' laboratory, the historical control mean live litter size for Crl:CD®(SD)IGS BR rats is 14.1 pups per litter. When sufficient numbers of animals are included in the study to produce approximately 20 litters per group, compound-treated group mean reductions of as little as one pup per litter may be an indicator of an adverse effect. Decrements of 1.5 pups per litter and greater are strong signals of reproductive toxicity. This is especially true if differences from the control group occur in a dose-related manner.

2. Neonatal Growth

A change in neonatal growth compared to the concurrent control is another endpoint that often indicates toxicity in a reproductive toxicity study. Adverse effects can be manifested by reduced birth weights caused by growth retardation *in utero* and/or by reduced body weight gain following birth. Reductions in body weight gain following birth may be a result of indirect exposure to the test article through nursing or may be attributable to decreased milk production by the dam. Group mean male pup birth weights are generally approximately 0.5 g greater than group mean female birth weights. It is of critical importance to evaluate the neonatal growth curve in conjunction with litter size to control for the confounding effects of within-litter competition. This can be accomplished by using a nested analysis of variance, in which the covariate is live litter size.[56] Mean absolute pup body weight differences of 10% or more will typically result in statistical significance. In studies that use the dietary route of exposure, reductions in offspring body weight that are observed only during the week prior to weaning (postnatal days [PND] 14 to 21) are most likely a result of direct consumption of the feed by the pups. In a study using a fixed concentration of the test article in the diet, the milligram per kilogram exposure of these pups during the week prior to weaning and immediately following weaning is often much greater than that of the dam. This is because of the higher proportional consumption of food by the pups. It is also important to consider route of exposure when comparing pup body weights with historical control data. For example, in the typical EPA multigeneration study design via whole-body inhalation, the dams are removed from the pups and are exposed to the test material for 6 h/d. Because of this separation from the litter, pup body weights at weaning on PND 21 are generally approximately 20% less than the weights of pups from a study employing an oral (dietary, gavage, or drinking water) route of exposure.

3. Neonatal Survival Indices

Neonatal survival indices, in conjunction with offspring body weights, are used to gauge disturbances in postnatal salubrity, growth, and development. When evaluating postnatal survival, it is important to maintain the litter as the experimental unit. However, it is also useful to inspect the absolute numbers of deaths in each group. Standardization of litter size is an option in all guidelines. The United States convention is to standardize litters on PND 4, while litter standardization is often not performed in Europe. Neither approach is clearly right or wrong. The main advantage of litter size standardization is reduction of the variance in offspring weight due to litter size differences. This approach will improve sensitivity and make the historical control data range less variable. However, using this approach can decrease the likelihood of identifying outlier pup growth, because some pups are randomly euthanized. Assuming that litters are standardized on PND 4 (e.g., to 8 or 10 pups), the method for calculating the mean litter proportion of postnatal survival is as follows:

$$\text{Postnatal survival between birth and PND 0 or PND 4 (prior to selection) (\% per litter)} = \frac{\Sigma(\text{Viable pups per litter on PND 0 or PND 4/No. pups born per litter})}{\text{Number litters per group}} \times 100$$

$$\text{Postnatal survival for all other intervals (\% per litter)} = \frac{\Sigma(\text{Viable pups per litter at the end of interval N/viable pups per litter at the start of interval N})}{\text{Number of litters per group}} \times 100$$

where $N =$ PND 0 to 1, 1 to 4 (prior to selection), 4 (post-selection) to 7, 7 to 14, 14 to 21, or 4 (postselection) to 21.

Adverse effects on pup survival occur most frequently during the period prior to litter-size standardization (culling) on PND 4. Control group pup survival is generally high prior to culling, with group mean values greater than 90%. Following culling, survival is usually greater than 95%. Total litter losses occur infrequently in the rat. In the authors' laboratory, approximately 1% of dams that deliver have total litter loss. Therefore, a treatment group with just two total litter losses would constitute at least a fourfold increase over historical control. While total litter loss is occasionally observed in single litters in a control group, it is very rare to see two or more total litter losses in a control group in a guideline study (N of approximately 25 to 30 females per group). Moreover, relative to the authors' historical database, more than one total litter loss or mean litter postnatal survival values below 90% in compound-treated groups are typically indicators of toxicity in a guideline study.

4. Prenatal Mortality

Prenatal mortality is assessed by comparing the number of former implantation sites in the dam (determined at necropsy) with the number of pups born to that dam. This endpoint is not so precise as the viable litter size because of the uncertainty regarding the total number of pups born and because of the length of time passing from the implantation process to determination of the number of implantation sites (3 weeks following parturition). Prenatal mortality remains a useful parameter; however, in the authors' experience, there may be up to a 15% error in this parameter. Therefore, this endpoint must ultimately be evaluated relative to other reproductive parameters. A standard multigeneration study includes approximately 100 litters per generation, and it is not feasible to observe active parturition of each dam. As stated previously, parturition observations are generally performed two or three times per day for a study. Cannibalism can occur prior to identification of the total number of pups born within a litter. Because of the potential for cannibalism, the accuracy of prenatal mortality is not so great as the accuracy of the number of former implantation sites (determined by direct examination of the uterus of the dam following weaning). The term "unaccounted-for sites" should be used instead of postimplantation loss in reproduction studies. The number of unaccounted-for sites for a dam is calculated as follows:

$$\text{Unaccounted-for sites} = \frac{\text{Number of former implantation sites} - \text{number of pups born}}{\text{Number of former implantation sites}}$$

The number of unaccounted-for sites provides a good estimate of the prenatal mortality contributing to the perinatal mortality measured postnatally. Prenatal mortality generally occurs at a low incidence in the rat. Mean numbers of unaccounted-for sites greater than 1.5 per litter are usually an indicator of test article–induced prenatal mortality.

5. Assessment of Sperm Quality

Sperm quality is assessed by determination of motility, concentration, and morphology. For assessment of sperm motility (when collected from the epididymis or vas deferens, more aptly termed spermatozoon motility), the use of computer-assisted sperm analysis (CASA) systems has become the industry standard technology, although it is not required. The original methodology of manually counting cells by use of a hemocytometer and motility determination through video techniques will suffice, although the CASA system appears to be more accurate, precise, and cost-effective over the course of multiple studies. Based on user-defined criteria, the CASA system analyzes sperm motion and determines total motility (any sperm motion) and progressive motility (net forward sperm motion). The CASA system uses stroboscopic illumination at 662 nm (a series of light flashes of 1-msec duration at a rate of 60 Hz) to obtain precise optical images that are then converted into a digital format. The digital images are analyzed by processing algorithms to determine the properties of sperm motion.

A recent effort was undertaken to compare different techniques between multiple laboratories.[101] There are no standards for sperm motility assessment; therefore, procedures and criteria vary among laboratories. It is thus incumbent on each laboratory to validate its CASA system under conditions of use. The media used for dilution of the sperm sample and the incubation conditions must be validated to show that they do not affect sperm motility. In addition, gate, size, and intensity parameters on a CASA system must be adjusted to distinguish between motile and nonmotile sperm and between sperm and debris. Determination of the percentage of progressively motile sperm relies on user-defined thresholds for average path velocity and straightness or linear index.

Rat sperm can be obtained from the cauda epididymis or from the vas deferens. Mean control group values for total rat sperm motility are usually 80% to 90%. In the authors' laboratory, control rat progressive motility is typically 10% lower than total motility. When assessing rodent sperm motility, one must be cognizant that the sperm being analyzed are not fully mature. Sperm collected from the epididymis and vas deferens are not accompanied by the secretory fluids from the prostate, seminal vesicles, and bulbourethral gland that are added prior to ejaculation. Therefore, sperm motility and viability determined as part of the standard rodent evaluation may not precisely reflect the actual sperm motility and viability of an ejaculated semen sample. However, ejaculated semen can be obtained and analyzed from the rabbit, dog, and primate. The advantages of using these three species instead of the rodent for sperm assessment are several: (1) longitudinal assessment of sperm quality may be performed, (2) assessment of reversibility of adverse effects on sperm quality may be conducted in the same animal, (3) baseline (pretest) data may be obtained, and (4) an absolute number of ejaculated motile sperm may be determined. Sperm assessment in species other than rodents is not conducted frequently. However, following guidance from regulatory agencies, the authors' laboratory has measured sperm motility in nonrodent species by use of a CASA system.

Concentration of spermatozoa in the rodent is measured by homogenizing a testis or epididymis and then counting the number of homogenization-resistant sperm. The concentration values are normalized to the weight of the tissue. Separate enumeration of spermatozoa in the testis and epididymis is necessary to determine the site of insult. If a test material adversely affects spermatogenesis, then both testicular spermatid and epididymal sperm numbers will be reduced. However, if sperm maturation is affected, testicular spermatid counts will be normal, while epididymal sperm counts will be reduced. Frequently, reductions in sperm concentration are accompanied by reductions in sperm motility and reproductive organ weights. If large reductions are observed in any of these parameters, male fertility rates may be reduced.

Sperm morphology is determined qualitatively by light microscopy. Quantitative CASA measurements of sperm morphology (primarily the head) can be made. However, because there are no published sources of the range of normal values for these measurements and the significance of slight alterations in these measurements on fertility is not known, the majority of laboratories do not use them. The most common approach for morphologic assessment is to prepare a wet-mount slide and assess any abnormalities in the head, midpiece, and tail of the sperm.[102] Alterations include detachment of the head from the tail, dicephalic heads, irregularly shaped heads, and coiled tails. A certain degree of subjectivity is involved in the morphologic assessment. By use of the criteria

Table 9.35 Historically observed mean spermatogenesis data by species and strain[a]

Endpoint	Rat [Crl:CD®(SD)IGS BR]	Mouse [Crl:CD®-1(ICR)BR]	Rabbit [Hra:(NZW)SPF]
Sperm motility (%)	84.3	81.4	75.8
Sperm production rate (sperm/g testis/day)	14.3×10^6	28.9×10^6	16.2×10^6
Epididymal sperm concentration (sperm/g epididymis)	470×10^6	616×10^6	638×10^6
Morphologically abnormal sperm (%)	1.1	1.8	4.1

Source: Data from WIL Research Laboratories, Inc. historical control database.

described by Linder et al.,[102] morphologically abnormal control sperm are generally observed at a low frequency (<5%). Therefore, slight increases in abnormal sperm can be an indicator of sperm toxicity. For these reasons, the criteria used for determination of abnormal sperm must be well defined and consistently applied, which makes training of the observer of paramount importance.

Table 9.35 presents the mean historical control values in the authors' laboratory for sperm parameters for the most commonly used species for reproductive toxicity studies. Morphologically abnormal control sperm values much greater (1.5 times) than the values reported in Table 9.35 probably indicate poor technique and training and should be discarded.

6. Weight and Morphology (Macroscopic and Microscopic) of Reproductive Organs

Organ weights, macroscopic findings, and microscopic findings of reproductive tissues are often highly correlated, sensitive indicators of toxicity. The most commonly used designs for reproduction studies will expose both males and females to a test article. Therefore, the resulting data may provide essential information to determine if an adverse fertility effect is male-mediated, female-mediated, or due to effects in both sexes.

Male reproductive organs that are usually weighed include the testes, epididymides, ventral prostate, and seminal vesicles with coagulating gland. Nonreproductive organs often weighed include the brain, pituitary, and thyroid. It is generally thought that the reproductive tract is well conserved in the presence of weight loss. However, this is not always the case. A common interpretive trap is to discount decreases in absolute testis weight by alluding to a lack of change in testis weight relative to final body weight. However, in the authors' experience, if treatment of males begins when the animals are adults, tests weight will not vary with body weight. This conclusion is supported by the results of a recent study in which a dietary restriction regimen was employed to determine which reproductive tract organs were body-weight independent.[103] In that study, a 50% reduction in food consumption, with a decrease in body weight to approximately 75% of the control weight, demonstrated that absolute testes weights are completely insensitive to body weight changes. Moreover, epididymides are relatively insensitive, and the accessory sex gland unit (seminal vesicles with coagulating gland plus prostate) is quite sensitive. Therefore, a sophisticated approach must be employed when interpreting changes in reproductive tract organs. Decrements of 10% or greater in organ weight measures are usually of toxicologic significance. However, because of the remarkable consistency of testicular and epididymal weights, changes of as little as 5% may be indicators of toxicity, especially when they are accompanied by correlative macroscopic or microscopic findings.

Histopathologic examination of male reproductive organs provides information regarding severity of effects and selectivity of the organs affected. The most valuable histopathologic information is generally obtained in the testes, where staging can be performed. The process of spermatogenesis has been divided into 14 stages.[104] By examining individual seminiferous tubules, a trained pathologist can determine if a toxicologic insult targeted the germ cells (spermatogonia), more mature spermatids, the Leydig cells, and/or Sertoli cells. The type of cells affected may provide information regarding the potential reversibility of the effect. For example, a compound that destroys primary spermatogonia is more likely to produce a permanent deficit in fertility than a compound that causes spermatid retention. Interpretation of testicular histopathologic data should be conducted in conjunction with assessment of sperm parameters, testicular weights, fertility data, and macroscopic data. The testicular histopathologic examination may be the most sensitive indicator of toxicity, particularly when a lesion is observed to have been dose responsive. Presence of debris and sloughed cells in the epididymides can also provide indicators of adverse reproductive effects. Rarely are adverse reproductive effects manifested primarily in the prostate or seminal vesicles.

Female reproductive organs that are usually weighed include the uterus and the ovaries; nonreproductive organs include the brain, pituitary, and thyroid gland. When evaluating uterine weight, it is essential to determine the stage of estrus of the animal at the time of necropsy. This is best accomplished by histopathologic examination of the vagina and uterus. A lavage of vaginal contents may also be taken prior to necropsy and examined by light microscopy. However, the stage determined from the vaginal lavage is approximately one-half day behind the stage determined by histopathologic examination of the uterus and vagina. Therefore, results of histopathologic examinations may not be consistent with vaginal lavage data, particularly if the vaginal lavage is performed several hours prior to the necropsy. Fluid retention and cellular proliferation vary greatly in female rats, depending on the stage of estrus. Weights are highest during the estrus stage and lowest during the diestrus stage. Unless uterine weight is correlated to stage of estrus, false positive and false negative interpretations may result.

Reproductive organs subjected to histopathologic examination in the female include the ovaries, uterus, vagina, and mammary glands. Examination of these organs in combination with vaginal cytology data provides a determination as to whether the female is cycling normally, has entered an anovulatory stage, or has entered premature reproductive senescence.

7. *Estrus Cyclicity and Precoital Interval*

The normal rat estrous cycle is four or five days in duration. For the original and extensive treatment of this topic, refer to Long and Evans.[105] The techniques and principles for determining the estrous cycle have remained the same as in this original reference. However, the application and understanding of parameters of the estrous cycle have greatly improved in the last two decades. In the authors' laboratory, the estrous cycle is divided into the following four stages: estrus (E), metestrus (M), diestrus (D), and proestrus (P). During E, the majority of the cells present are cornified epithelial cells. These cells are irregularly shaped and may or may not have remnants of a nucleus. Some nucleated epithelial cells may be present; however, no leukocytes are present. Estrus usually lasts 10 to 15 h. During M, nucleated epithelial cells are surrounded by increasing numbers of leukocytes. Metestrus is the shortest stage, lasting approximately 6 h. At the beginning of D, cell types are equally distributed among leukocytes, cornified epithelial cells, and nucleated epithelial cells. During the later part of this stage, the nucleated epithelial cells become more spherical in appearance. Diestrus is the longest stage, lasting two to three days. During P, the majority of cells are nucleated epithelial cells, with some cornified epithelial cells that become more prevalent as the E stage approaches. The nucleated epithelial cells appear spherical in shape and are usually grouped in clusters. Proestrus usually lasts 8 to 12 h.

To determine the estrous cycle pattern in rats, vaginal lavages are performed once per day. Female rats will only be receptive for mating on the evening of the P stage. Therefore, animals that do not proceed through a cycle will not mate. It is not unusual for animals to occasionally have an extended cycle (six or more days); however, repeated extended cycles may be an indicator of reproductive toxicity.

A variety of methods are used to evaluate the estrous cycle. One approach is to determine the percentage of days that an animal is in each stage of the cycle (i.e., the percentage of days in E, M, D, or P). Another approach is to determine the number of animals that have one or more irregular cycles (i.e., a cycle of six or more days or a cycle of three or fewer days). Lavages are usually performed only once per day. Therefore, one or more of the shorter stages (P, E, and M) may be missed in a normal cycle, and this may confound evaluation of mean estrous cycle length. In the authors' laboratory, the average cycle length is calculated for complete estrous cycles (i.e., the total number of returns to M or D from E or P). Estrus cycle length is determined by counting the number of days from the first M or D in a cycle to the first M or D in a subsequent cycle. In this approach, it is assumed that if an animal has a cycle pattern of P followed by an M or a D, the animal actually went through E, but the estrus stage was missed because vaginal cytology was examined only once per day.

The composition of the cell types and the movement through the estrous cycle is hormonally controlled. During D, estradiol levels are low and gradually rise because of increased secretion by the growing follicles. This results in vaginal epithelial cell proliferation and the transition of the vaginal cytology into P. The elevated levels of estradiol trigger a surge in prolactin and luteinizing hormone. The surge in LH then triggers ovulation early in the morning of E. Serum estradiol levels decline dramatically on the evening of P as the follicle begins to luteinize and vaginal epithelial cell proliferation declines. Estrus cycle length is also dependent on progesterone secretion by the ovarian follicles and corpora lutea. In the rat, the hypothalamic regulation of the LH surge in response to elevated levels of estradiol is lost when an aging female enters reproductive senescence. As a result, ovulation does not occur, and the levels of estradiol remain persistently elevated. Evaluation of vaginal cytology in this condition reveals a persistent E state. The elevated estradiol levels stimulate prolactin release, producing mammary duct ectasia, and ultimately, galactoceles. If this normal aging process is accelerated in test article-treated groups, fertility would be affected, and this would be an indication of reproductive toxicity. An array of chemicals and drugs has been reported to cause this state of premature senescence through similar mechanisms. In a state of persistent D, estradiol levels are low, and follicular development and ovulation do not occur, again resulting in reduced fertility.

The precoital interval is defined as the number of days from the initiation of cohabitation of a breeding pair until evidence of copulation is observed. As previously stated, female rats will only be receptive to the male on the evening of P. Therefore, precoital interval data should be evaluated in conjunction with the estrous cycle data. For example, an increase or decrease in the precoital interval for an exposure group compared with the control group is not an adverse effect if the animals in the exposure group are cycling normally and mate on the evening of the first P following the start of the breeding period. However, an increase in the precoital interval can provide weight-of-evidence support for identifying slight effects on estrus cyclicity. Investigators also need to be cognizant that precoital intervals cannot be calculated for animals that do not cycle or mate. Therefore, this endpoint cannot be assessed in isolation without also evaluating estrous cycling and the mating index. Improved sensitivity of detecting effects on the precoital interval and conduct of studies to evaluate further such effects may be made more cost-effective by using the "shortened mating period" methodology.[56] In this method, the mating period may be reduced to a 2-h period by the use of perineal manipulation and its elicitation of the ear-quivering response.

8. Mating and Fertility Indices and Reproductive Outcome

Traditionally, assessment of functional aspects of reproduction, such as calculation of mating and fertility indices, has been an important cornerstone of well-designed studies and has served a key role in the identification and characterization of injury to the reproductive system.[40] Table 9.36 presents the historically observed range of values for male and female mating and fertility indices calculated from studies conducted in the authors' laboratory.

The mating index is a mathematical expression derived from *post facto* observations confirming that coitus has occurred. The index is generally considered a reliable measure of normal sexual behavior and libido. It also provides indirect information relative to the function and status of the hypothalamic-pituitary-gonadal axis. In other words, changes in this index may arise from disturbance

Table 9.36 Historically observed functional reproductive indices for [Crl:CD®(SD)IGS BR rats][a]

		Mean (%)	25th Quartile (%)	75th Quartile (%)
Male Mating Index				
$\dfrac{\text{No. of males with evidence of mating}}{\text{Total no. of males used for mating}}$	× 100	96.1	95.0	100.0
Female Mating Index				
$\dfrac{\text{No. of females with evidence of mating (or confirmed pregnancy)}}{\text{Total no. of females used for mating}}$	× 100	97.3	95.9	100.0
Male Fertility Index				
$\dfrac{\text{No. of males siring a litter}}{\text{Total no. of males used for mating}}$	× 100	89.6	86.7	95.0
Female Fertility Index				
$\dfrac{\text{No. of females with confirmed pregnancy}}{\text{Total no. of females used for mating}}$	× 100	90.9	88.0	96.0

Source: Data from WIL Research Laboratories, Inc. historical control database.

in the CNS or from imbalances in hormonal cascades. Indirect measures of mating performance are obtained by observing either the presence of sperm from a vaginal lavage (microscopically) or a copulatory plug in the vagina or below the mating cage. The latter observation technique may also be used for the mouse. However, results are more variable, and the endpoint becomes less sensitive because retention of the vaginal plug is longer and mistiming of pregnancy may be a consequence. The process of confirming that coitus has occurred is confounded because, even among the most technically skilled, the rate of proficiency of timing the event by vaginal lavage is only 90% to 93%, in the authors' experience. Thus, up to 10% of the data for a given breeding cycle may be unavailable for the final interpretation. For this reason, differences in the mating index that represent one or two fewer successful matings in a group should not cause an increased level of concern. However, decrements in this endpoint may also signal that abnormalities have been introduced. Such abnormalities may consist of changes in the duration and/or the orderly sequence of stages of the rodent estrous cycle, or they can also arise from deficits in sperm quality. These decrements may occur in the presence of disturbances in the elapsed time between the start of cohabitation and observed evidence of mating (referred to as the precoital interval). The precoital interval is usually reported as a fraction of the number of days required for mating. Most rodents would be expected to show evidence of mating within the first estrous cycle (4 or 5 days). It is noteworthy that normal precoital intervals may occur without successful impregnation if effects on intromission are observed. In addition, the females are rarely synchronized at the onset of the breeding period in terms of stage of estrus, and statistical transformations may be necessary to account for these variations. In essence, to determine biological significance of differences in group mean data of less than two days between treated groups and the control group, one must include a detailed assessment of the individual data.

On rare occasions, an unexpectedly high number of females may show evidence of copulation on the first morning after pairing. This is thought to be a result of forced copulation of females that are not in the receptive phase of the cycle (proestrus). A reduction in the number of successful pregnancies can be expected in this instance.[106]

The fertility index is the number of animals inducing pregnancy or becoming pregnant divided by the number of mating sets, and it is the primary measure describing the outcome of a mating trial. In the rodent, several intromissions must occur prior to successful conception. These intromissions trigger uterine contraction, enhanced sperm transport, and hormonal changes that prepare the uterus to accept the blastulae. When the fertility index is evaluated in isolation from other endpoints, it is generally considered a poor predictor of perturbations in the reproductive system. Therefore, it is essential to evaluate other interrelated endpoints (e.g., organ weights, macroscopic and microscopic tissue changes, estrous cycle data, and oocyte quantification) to provide initial insight into the processes of the reproductive system that have been compromised. Some reproductive toxicants are so selective in their actions that a full characterization of the response is only possible by means of specially designed studies in which "pulse doses" are administered acutely over discrete phases of the reproductive cycle. At the other extreme are reproductive toxicants for which a decline in the fertility index is the most sensitive endpoint. The best example of the latter can be found in a report produced by the Center for the Evaluation of Risks to Human Reproduction (CERHR) on the reproductive and developmental toxicity of 1-bromopropane (1-BP).[107] A study of the reproductive toxicity of 1-BP was conducted via the inhalation route at the authors' laboratory under the OPPTS 870.3800 guideline. Reproductive performance was highly compromised in this study over a broad range of exposure concentrations (100, 250, 500, and 750 ppm). In the first parental generation (F_0), the blockade of pregnancy was complete at 750 ppm (0% fertility index). At the higher exposure concentrations (over 250 ppm), numerous reproductive endpoints were affected (generally statistically significantly), and they included the following: increases in estrous cycle length and ovarian follicular cysts; and decreased ovarian weight, numbers of corpora lutea, fertility indices, and litter size; and/or changes in sperm quality. At the lowest exposure concentration, 100 ppm, the only alteration was a decrease in the male and female fertility indices (greater than 20%). The decreased fertility indices were interpreted to constitute the most sensitive indicator of reproductive toxicity for 1-BP, judged by the dose response expressed. As an apical measure, the fertility index may reflect cumulative, immeasurable decrements in many subordinate processes and unmeasured endpoints (e.g., sperm capacitation and tubal transport). The CERHR Expert Panel did not consider the diminution in this key functional endpoint to be an effect. Presumably, the final interpretation of the lowest-observed-adverse-effect level was arrived at merely on the basis of lack of statistical significance, rather than from comparisons to normally expected ranges (the laboratory's historical control data).

9. Duration of Gestation and Process of Parturition

The length of gestation is a derived value as determined through these regulatory study designs, which compares the elapsed time between fertilization and initiation of parturition. In the Crl:CD®(SD)IGS BR rat, the average length of gestation is 22 days, but it ranges from 21 to 23 days postcoitus. Near the period of expected parturition, the dams are observed several times daily to establish, as closely as possible, the onset of parturition. The mean data are generally reported as a fractional value, with units expressed in days. The duration of gestation is highly predictable in control populations and has been well characterized in a variety of laboratory animal models and stocks. Exposures that result in premature birth yield pups that have lower body mass because of the decreased gestation length. These low-weight litters are underdeveloped and at increased risk for mortality. The greater the prematurity, the more pronounced are the effects on development and survival. Conversely, prolongation of the gestation period may induce other adverse sequelae, including maternal death, increased pup mortality, and other confounded measures. In the most severe instances of delayed parturition, mortality rates among the dams may be excessive because of possible septicemia associated with the degenerating products of conception and other related major events such as circulatory collapse. The process of parturition is usually completed within 1 to 2 h in rats, with longer times expected for larger litters. Dystocia may be a manifestation of maternal toxicity resulting from exposures that cause prolonged labor, difficult labor, or termination of the contractions necessary to evacuate the uterus. Pharmaceutical agents, such as those antihypertensives that evoke blockade of calcium channels, may also directly cause dystocia.[108] Dystocia is a relatively rare event in the control population in standard laboratory strains; in the authors' laboratory, dystocia has been observed in less than 0.5% of Crl:CD®(SD)IGS BR rat litters (7 occurrences in 1905 litters from 1997 through 2003). Therefore, the appearance of dystocia in exposed animals should always trigger a heightened level of concern.

10. Endpoints Assessed during Lactation, Nesting, and Nursing

a) Endocrine Endpoints

Near the end of gestation, rapid changes in maternal endocrine levels occur to prepare the mammary gland for the functional demands of nurturing and supporting the offspring. The milk ejection reflex in the rodent is triggered by the suckling stimulus and involves the release of oxytocin from stores in the pituitary gland. This action, in turn, induces proliferative changes in the mammary tissue and stimulates milk production and release. Agents, or their metabolites, that cause disturbances in these pathways can also cause adverse changes in milk composition and/or milk quantity. Likewise, some chemical agents or their metabolites that enter the milk from the maternal circulation may suppress postnatal development and maturation. Adverse effects on the growth of nursing pups may also be caused by factors unrelated to the dam. For example, agents that disrupt the cascade of developmental events necessary for palatal closure can result in poor suckling and retarded growth or mortality in offspring. The support of lactation and nursing behavior may be dependent on audible cues emitted from the pups.[109,110] Agents that interfere with these cues may increase maternal neglect and subsequent postnatal wastage.

b) Blood and Tissue Dosimetry

Direct measurements of milk production, milk composition, and dispersion of parent drugs and/or metabolites are rarely conducted in standard guideline studies. Figure 9.9 amplifies key considerations in the design and interpretation of studies that include bioanalytical assays of various components of the blood and tissues following maternal exposure during the reproductive cycle. Blood and tissue dosimetry in the postnatal animal may also establish and clarify exposure of the progeny to the test compound. Ensuring exposure of the progeny to the test compound during the lactational period has received increasing interest and support, and is more frequently included in safety assessment studies. This approach also provides the strongest basis for designing studies with improved relevance to protection of children's health.

Figure 9.9 is a schematic depiction of the comparative modes of translocation of materials between the maternal systemic circulation, placenta, and embryo or fetus vs. the maternal systemic circulation, mammary glands, and neonate. This comparison demonstrates bidirectional fluxes of materials (endogenous and xenobiotic) between the maternal circulation, the placenta, and the embryo or fetus. On the other hand, unidirectional translocation of like materials occurs between the maternal circulation, the mammae, and the neonate. In the former case, flux between the embryo or fetus, placenta, and maternal circulation is intravenous (IV) in nature, with the only unidirectional absorptive activity occurring in the placental membranes, either chorioallantoic or yolk sac. In the latter case of maternal circulation to the mammae and neonate, translocation is secretive and entails gastrointestinal absorption by the neonate. Gastrointestinal absorption involves dynamic acquisition of specific absorptive functions and maturation of the overall gastrointestinal physiology over the period of postnatal development. Figure 9.9 illustrates that for estimation of prenatal exposure of

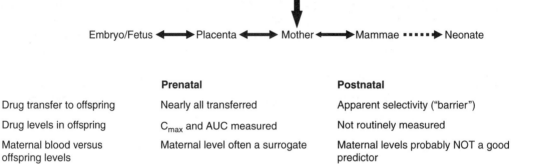

	Prenatal	**Postnatal**
Drug transfer to offspring	Nearly all transferred	Apparent selectivity ("barrier")
Drug levels in offspring	C_{max} and AUC measured	Not routinely measured
Maternal blood versus offspring levels	Maternal level often a surrogate	Maternal levels probably NOT a good predictor
Exposure route to offspring	Modulated IV exposure via placenta	Oral via immature gastrointestinal tract
Commentary	Timing of exposure is critical	Extent of transfer to milk and neonatal bioavailability is key to differentiating indirect (maternal) effects

Figure 9.9 Comparative modes of translocation of materials between the mother and the offspring.

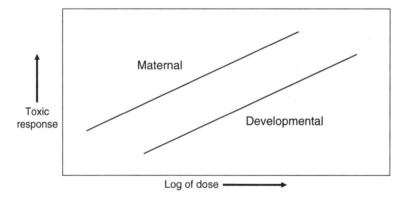

Figure 9.10 Hypothetical relationship of the log of dose and toxic response for maternal and developmental toxicity.

the products of conception, sampling of maternal blood and/or plasma concentrations of a xenobiotic is usually a good dosimeter. Hence, sampling the maternal blood for estimation of exposure of the embryo or fetus prenatally suffices as a surrogate for direct sampling of the embryo or fetus, which may not be feasible because of limitations of sample sizes and analytical sensitivity. Conversely, measuring levels of a xenobiotic in maternal milk is a poor dosimeter of exposure in postnatal progeny because of the secretory nature of milk production and the immaturity of the developing gastrointestinal tract. Therefore, the best estimate of postnatal exposure is made by sampling directly the blood and/or plasma of the progeny. In both instances, prenatal and postnatal, these toxicokinetic determinations are crucial to examining relative sensitivities of the adult vs. the developing offspring. The latter determination is crucial to differentiating relative biologic susceptibility from differences in exposure of the two life stages.

Without the toxicokinetic information, plotting and comparison of the dose-response curves for dams vs. progeny (prenatal or postnatal) are based solely on the less sensitive, more subjective, and less direct toxicologic observations routinely recorded in these reproductive toxicity studies. Figure 9.10 further depicts this very important relationship by displaying two hypothetical log-dose-response curves. The placement and slopes of the curves for maternal versus developmental (prenatal or postnatal) responses might suggest a biologically based or mediated difference in sensitivity of the two organisms. However, the difference between the curves may be merely due to differences in exposure. This is an important dimension to the overall issue of interpretation of individual parameters in these bioassay studies, as well as to the overall processes of risk assessment and risk management. Measurement of the levels of a xenobiotic in maternal milk is better than no information; however, it is not the best indicator of postnatal exposure of the progeny. For early embryonic development in which embryonic weight is in the microgram wet-weight range, maternal blood levels of a xenobiotic remain the most practical and reasonable estimate of prenatal exposure.

c) Milk Composition

Generally, the presence or absence of nursing is evaluated qualitatively by observing milk in the stomach of the pups. This is possible in early postnatal life because the abdominal tissues overlying the stomach are translucent at that time, and the absence of fur allows for visualization. Observations of decreased milk content in pups that die early in postnatal development may signal an adverse effect in nesting and nursing behavior. However, this observation provides little insight relative to the mode of action of the agent to which the pups are exposed. While quantitative evaluation of milk composition, milk quality, and xenobiotic kinetics by means of hormonal induction and manual expression techniques may be accomplished in guideline studies, these tests are not routinely carried out.

Unlike the breadth of knowledge that has been developed describing changes in the physiologic status of the prenatal rat, there is considerably less data characterizing the progressive changes in milk composition that occur from birth until weaning.[111] Therefore, quantitative evaluation of the following endpoints may be important considerations for further study: potassium, lactose, sodium, serum albumin, transferrin, and casein. Consistent with the human, increases in the concentration of potassium and lactose and decreases in sodium concentration begin from 12 h before birth to 48 h postpartum in the rat. These events correspond to the transition from the production of colostrum to that of milk. Serum albumin in the rat increases by a factor of two within a few days after delivery, but it does not increase in the human. Levels of transferrin in rat milk are low from parturition until lactational day (LD) 10, when marked elevations occur that are sustained through weaning. Casein is present in milk at term, and the concentration increases steadily until weaning. In response to the termination of nursing at the end of the postnatal period (LD 20), the levels of both lactose and potassium decline dramatically, while sodium content rises.

When conducting a reproduction study, it is important to recognize and observe innate behavioral characteristics of rodents as they care for their progeny because alterations in these behaviors may represent an overt change in the CNS status of the mother. Rodents, particularly mice, are expert nest builders and constantly strive to maintain an orderly arrangement of the pups to enhance suckling. Suckling occurs frequently over both the light and dark cycles. Rodents will also immediately retrieve pups removed from the nest.

d) Maternal Physiological Endpoints

It is important to compare physiologic endpoints in maternal animals, such as body weight during gestation and lactation, to those obtained during the premating phase. This can help to determine whether unique toxicities have occurred under the conditions of pregnancy and/or lactation. Food consumption will be elevated during gestation; this is due to elevations in circulating levels of progesterone. On the other hand, the rapidly changing physiologic status of the dams and the flux in hormones from the ovary and thyroid during pregnancy, coupled with rapid weight gain, tend to confound the utility of organ weight relative to body weight comparisons. Likewise, other confounders in reproductive toxicity studies can involve flaws in the features of experimental design relative to administered dose. For example, the traditional approach of correcting individual doses for gavage delivery based on current body weights should be refined for advancing pregnancy. If not, excessive overdosing of the maternal animals can induce profound effects during the period of expected parturition, when the mass of the products of conception (but not necessarily the lean body mass of the maternal animal) are maximal.

11. Sexual Behavior

While it is clear that normal sexual behavioral function is dependent on the complex integration of input from several organ systems, little direct evaluation is devoted to this important measure in animal studies. This void in the data is often created by practicality at the operational level because of the nocturnal breeding habits of rodents. In females, the most direct measure of sexual behavior involves the observation of lordosis to confirm receptivity, and this behavior is tied to the timing of late proestrus. Indirect measures of this endpoint are collected and evaluated as described in previous sections.

In males, periods of mounting and intromission can be quantified and may be included as experimental endpoints if class effects are anticipated. The timing and number of intromissions and release of ejaculatory plugs in the rat model can be used to assess potential effects on sexual behavior. The extent to which these endpoints relate to human sexual behavior is unclear, however, because the authors are unaware of any studies of the concordance or validity of such measures. From a basic hazard identification and/or screening perspective, these measures may have value and as such are commonly published in the neurobehavioral literature. Nevertheless, sexual behavior endpoints have been given little attention in the guidelines for reproductive toxicity studies. Indeed, the presence of ejaculatory plugs in cage pans may be a useful observation, but this parameter is not commonly reported in the toxicologic literature.

Rats have a specialized anatomical structure, termed the coagulating gland, which rapidly produces an ejaculatory plug after coitus. The function of this structure is to provide a physical barrier to prevent leakage of the seminal plasma from the vagina. Classically, little attention has been directed at quantifying changes in the frequency of these emissions in guideline reproductive toxicity studies. In the authors' laboratory, several guideline two-generation studies have revealed qualitative and dose-related increases in the number of ejaculatory plugs per animal in comparison with values from concurrent control animals. Interestingly, the response was expressed several weeks before the onset of the first generation mating cycle. This effect was initially encountered in inhalation studies using whole-body inhalation exposure regimens. More recently, it has been observed occasionally in studies utilizing more common routes of exposure. Several authors studying sexual behavior in rats have reported that rats have daily spontaneous ejaculations.[112–114] It has also been reported that rats normally lick their genitals at the time of ejaculation and consume the plugs.[115] It is conceivable that toxicants could be eliminated from the body in these structures, and poor palatability might preclude consumption. Thus, qualitative differences in the number of ejaculatory plugs observed in these studies may have no origin in neuroendocrine dysfunction. Nevertheless, the increased production of the ejaculatory plugs in the aforementioned studies had no impact on the fertility of the males, and the relevance of these findings to human risk assessment is unknown.

12. Landmarks of Sexual Maturity

The primary landmarks of sexual development are anogenital distance, balanopreputial separation (BPS) in males, and vaginal patency (VP) in females. In rodents, anogenital distance is greater for males than for females. Anogenital distance (AGD) is typically measured at birth, and reflects the linear distance between the genital tubercle and the anus. Mean values for males and females are not presented from the authors' laboratory because techniques for measuring AGD vary from one laboratory to another, based on anatomical reference points. It is critical that consistent procedures

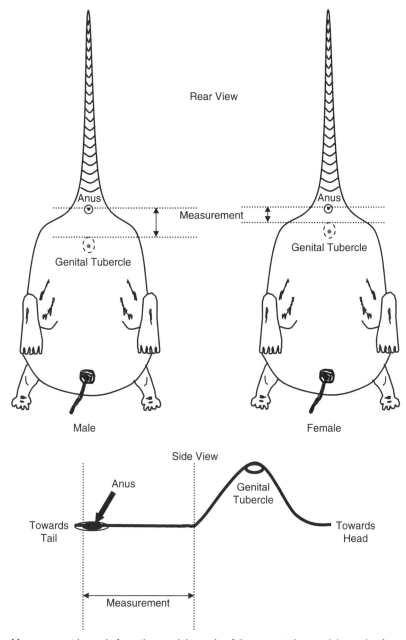

Measurement is made from the caudal margin of the anus to the caudal margin of the genital tubercle.

Figure 9.11 Measurement landmarks of anogenital distance.

be used within a single laboratory for this measurement so that observer bias is controlled. A diagram of the technique employed at the authors' laboratory is presented in Figure 9.11.

The skin should be stretched to facilitate consistency of the measurement. In addition, the technician should examine equal numbers of pups within each treatment group. The absolute AGD is directly correlated with body weight (i.e., pups with greater weight have larger AGDs). When statistically analyzed, the AGD should be normalized relative to the cube root of the body weight.[116]

The most frequently observed adverse effect on the AGD is a reduced distance in males in response to antiandrogenic agents or 5α-reductase inhibitors. When mean litter anogenital distances of approximately 20 litters are evaluated, differences of 5% or greater are generally indicators of reproductive toxicity.

BPS is evaluated by expressing the penis to determine if the prepuce can be retracted from the glans penis. An animal is not considered positive for this endpoint unless complete retraction of the prepuce can be performed. BPS is mediated by androgens, and males that do not achieve BPS will be infertile.[117] In control Crl:CD®(SD)IGS BR male rats in the authors' laboratory, the mean day of acquisition of BPS is PND 44.8 (the range is 41.6 to 49.0). Because body weight can influence the timing of BPS, it is important to record the weight of the animal on the day BPS is observed. When growth retardation (as measured by reduced body weight) is observed, the day of acquisition for BPS is likely to be delayed.[117,118] Preputial separation in males occurring in parallel with delays in vaginal patency in females indicates a generalized growth delay, as opposed to a selective endocrine-mediated mechanism. In the absence of an effect on body weight, a change in the day of acquisition of BPS of 2 days or greater is typically an indicator of toxicity.

VP is evaluated by examining the vagina to determine if the septum covering the vagina has ruptured. VP generally occurs around the time of first ovulation and occurs in response to an increase in serum estradiol levels as females enter puberty. A female is not considered positive for this endpoint unless the septum is completely absent. Occasionally, a thin strand of tissue will remain over the vagina. This threadlike strand would not preclude an animal from mating; as such, the animal would still be considered to have achieved VP. In control Crl:CD®(SD)IGS BR female rats in the authors' laboratory, the mean day of acquisition of VP is PND 33.5 (range is 31.7 to 38.8). As described for BPS, body weight can be a confounder in interpreting the timing of VP. In the absence of an effect on body weight, a change in the day of acquisition of VP of 2 days or greater is typically an indicator of toxicity.

13. Sex Ratio in Progeny

Several publications have reported sex-ratio effects of organochlorines (e.g., 2,3,7,8-tetrachlorodibenzo-*p*-dioxin [TCDD]) on human populations, although cross-study results have not been consistent. Lowered offspring sex ratio has been associated with paternal exposure to TCDD following an accidental exposure at Seveso in northern Italy in 1976 (0.38 male to female ratio),[119] and for two cohorts of pregnancies fathered by Russian workers at a pesticide-producing factory in the city of Ufa, Bashkortostan, Russia, from 1961 through 1988 (0.40 male to female ratio).[120] The Seveso data indicated a greater effect in men that were exposed before the age of 19; however, no similar correlation could be made for the Ufa cohorts. In contrast to these studies, conflicting results were found in a study of paternal occupational exposure to TCDD in factories producing Agent Orange in the United States. In this study, serum samples collected in 1987 revealed no differences in offspring sex ratio between pregnancies fathered by workers with high levels of TCDD exposure and those of nonexposed neighborhood referents.[121]

In contrast to the above-mentioned putative human examples, selective effects on survival of a particular gender are rarely observed in animal hazard identification studies. In the authors' experience with approximately 1500 studies, a confirmed test article–related shift in sex ratio has never been observed. Conceivably, alterations could arise from direct effects on the male germ cell populations that would favor the production or destruction of Y or X chromosome-bearing sperm. An alternate hypothesis would involve direct insults on the male or female conceptuses; however, the plausibility of this phenomenon is not fully established. Mild shifts in the male to female sex distribution ratio often occur, and little significance is focused on values that fall within normally expected ranges (44% to 56%, based on the authors' historical control data for Crl:CD®(SD)IGS BR rats).

14. Functional Toxicities and CNS Maturation

The study of immediate onset and latent behavioral expression of CNS damage associated with exposures during prenatal and early postnatal development has been accorded a higher level of concern in recent years. Tests to evaluate these aspects of development are routinely performed during the ICH pre- and postnatal study of development,[122] as well as under the more robust and focused EPA and OECD developmental neurotoxicity protocols.[123,124] A plethora of standard testing paradigms have been proposed, validated, and/or used in the process of identifying and characterizing agents that may represent a threat to the developing nervous system. The reader is referred to Chapter 8 of this text for additional exposition.

15. Oocyte Quantitation

In cases where female fertility is reduced, oocyte quantitation may provide information regarding the mode of action of an agent. Oocyte quantitation is used to assess whether the follicle maturation process in the ovary is occurring normally.[125,126] Three major types of follicles may be quantified: small or primordial, growing, and antral follicles. Small or primordial follicles contain isolated oocytes or oocytes with a single surrounding layer of granulosa cells. Growing follicles have an oocyte surrounded by multiple layers of granulosa cells. Antral follicles have a central oocyte within a fluid-filled space bordered by hundreds of layered granulosa cells. For quantifying the follicle numbers, serial sections of the ovary are usually cut approximately 100 µm apart. By cutting the serial sections 100 µm apart, the same primordial and growth follicles will not be counted multiple times. Many laboratories combine the primordial and growing follicles when reporting oocyte numbers. Counting antral follicles is more problematic. Because these follicles have cross sections of 100 to 550 µm, the antral follicles will be counted multiple times when sections taken 100 µm apart are examined. Therefore, quantification of antral follicles is typically not performed. There is another potential confounder in quantifying the small and growing follicles. For animals that are anovulatory (having fewer numbers of corpora lutea and antral follicles), increased numbers of primordial and growing follicles will often be recorded because the larger follicles are not present to obscure some of the small and growing follicles. In the authors' experience, changes in the numbers of small and growing follicles are rarely observed.

VII. EVALUATION OF RARE (LOW INCIDENCE) EFFECTS

A. Introduction

This section addresses the most critical, long existing, and often overlooked or misunderstood issue of rare event (low incidence) developmental and reproductive findings. Rare events may be caused by a xenobiotic or may be spontaneous. The difficulties in determining whether a small increase in response is related to treatment or is a chance occurrence are not straightforward but are commonly unappreciated. To the best of our knowledge, this is the first published attempt to fully characterize the attendant issues and recommend a paradigm for resolution.

The daunting task for most scientists working in this discipline is how to distinguish between those findings occurring spontaneously and those produced by a xenobiotic. The limited statistical power of the bioassays, relative to the large potential number of human beings exposed, dictates this to be a nonstatistical exercise and one that must rely on probabilistic theory, intuition for developmental processes and patterns, and arrays and timing of effects.

The material in this section was initially developed for a presentation at the 2002 Winter Toxicology Forum.[127] At that meeting of largely FDA and private-sector scientists, rare (low incidence) events and the difficulties they present for risk assessment, product licensure, and labeling were the topics of interest because of recent experience with the pharmaceutical class effects of inhibitors of the cyclooxygenase enzymes (COX inhibitors). Although there have been many previous examples of rare-effect issues, most have gone unnoticed or unpublicized, and not until this major class effect was identified was the topic given any focused regulatory attention. Because this issue is so important and intrinsic to developmental and reproductive toxicity bioassays, this section of the chapter was modified from the Toxicology Forum presentation in which both the topics of developmental and reproductive effects, and cancer were addressed. It is important to understand that in developmental and reproductive toxicity studies, most responses are likely to manifest at substatistical rates. For a review of the COX-2 rare events evaluation, see Cook et al.[128]

In this section, examples from the authors' historical control database illustrate the importance of accruing and managing consistent control data (over time and based on good animal husbandry and techniques) to assist in the interpretation of results. The advantages of having a highly skilled and stable work force are also discussed. A few case studies are included to demonstrate how data were interpreted in selected controversial study outcomes. Finally, a stepwise approach to evaluating rare findings and low incidence occurrences is presented.

B. Scope of the Problem

The goal for regulators is to gain knowledge of whether a condition, procedure, or chemical or drug exerts adverse effects on reproduction or development. Their mandate is to evaluate risk to humans with a sufficient degree of confidence to enable sound decision-making. This is accomplished by critically reviewing results of animal safety studies and then extrapolating those results to humans. A high degree of concordance has been shown between human adverse effect scenarios and what can be demonstrated in rodents (as well as in the rabbit).[11,129–132] However, when the incidence of a finding is too low for statistical identification or confirmation, it becomes more difficult to separate an adverse effect from normal biological variability. For example, what does a ventricular septal defect in one high-dose group fetus (out of 150 to 325 fetuses evaluated) signify in terms of potential human effects? Likewise, what is the human hazard of two to three rodents with total litter loss during the lactation period?

The FDA has defined *rare incidence* as "…an endpoint that occurs in less than 1% of the control animals in a study and in historical control animals."[5] This definition, from a statistical point of view, has an inherent problem. Animal safety studies are not (per guideline) designed with such a degree of statistical power (number of animals) that the statistical significance of 1 in 100 occurrences can be detected. Furthermore, upon evaluation of historical control data, responses for almost all of the critical developmental defects are seen to occur at a frequency of less than 1 in 100.

Table 9.37 provides a comparison of overall spontaneous malformation prevalence in different species.[133] The data demonstrate the challenge of detecting rare events in safety testing. In addition to the rare event–related statistical issues, incidences of malformations are much lower in laboratory animals than in humans, further complicating extrapolation. The background incidence of spontaneous malformations is particularly low in the rat. Because of this, the rat is a sensitive model for detecting rare events.

Concern for two to three malformations per group in animal studies is appropriate, even though there may not be statistically significant differences from the concurrent control group for the incidences of individual or total malformations. The challenge is how to differentiate the effect of a test agent from mere biological variation.

An example of the challenge inherent in evaluating low incidence findings occurred in a recent study in the authors' laboratory in which athyrosis was noted upon macroscopic examination in a high-dose group Crl:CD®(SD)IGS BR rat fetus (this malformation had not been seen previously in approximately 80,000 control and test fetuses of this rat strain). Several other fetuses distributed in multiple litters in the high-dose group had enlarged or rudimentary thyroids, and cleft palate was observed at 4.5% per litter in that same dosage group. Although athyrosis occurred in only

Table 9.37 Spontaneous malformation rates in various species[a]

Species	Mean (%)	Range (%)	Fetuses (N)
Rat	0.33	0–1.6	9643
Mouse	1.2	0–3	5207
Rabbit	3.2	0–10	4708
Dog	5.5	5.3–5.7	167
Human	4	3–9	Multiple surveys

Source: Data for rat, mouse, rabbit, and dog malformation rates originate from the historical control database at WIL Research Laboratories, Inc. Human data were obtained from the March of Dimes (2003) and personal communication from Ken Jones (2003).

Table 9.38 Selected malformation rates in control fetuses (means of study means)

Malformation	Rat Incidence (% per Litter)[a]	Rabbit Incidence (% per Litter)[b]
Ventricular septal defect	0.00	0.02
Cleft lip or palate	0.02	0.04
Abdominal wall defect (including gastroschisis)	0.04	0.06
Hydrocephaly	0.03	0.20
Spina bifida	0.00	0.17
Renal agenesis	0.01	0.02
Diaphragmatic hernia	0.00	0.04
Retroesophageal aortic arch	0.01	0.01

[a] 61 Crl:CD®(SD)IGS BR rat developmental toxicity studies conducted at WIL (1998 to 2003)
[b] 81 Hra:(NZW)SPF rabbit developmental toxicity studies conducted at WIL (1992 to 2003)

one fetus, this finding was considered an adverse effect in view of the following: (1) the rare nature of the finding, (2) other clear signals of structural malformation (cleft palate) at the high dose level, and (3) other alterations in fetal thyroid structure, revealing a pattern of insult. The relationship to treatment of a single case of athyrosis occurring independently would have been much less clear.

To further illustrate the challenge of discerning the significance of rare event findings, a few examples of malformations that have occurred as rare events or low incidence findings in numerous scenarios are listed in Table 9.38. Among the 21,000+ control Crl:CD®(SD)IGS BR rat fetuses in the authors' historical control database, ventricular septal defects, spina bifida, and diaphragmatic hernia have never been detected. However, spina bifida and diaphragmatic hernia have been observed in treated fetuses. Because any occurrence of a malformation is such a rare event, the relationship between the agent tested and the occurrence of a malformation is difficult to discern. On the other hand, the low background incidence of malformations makes it easier to distinguish subtle teratogenic events in those cases where the agent causes only a few malformations, given appropriately rigid and consistent historical control values. In the absence of such standards, the historical control database will be of limited use for identification of teratogens, and interpretation of the findings becomes even more confounded and without a reliable basis for comparison.

Further complicating matters is the interdependence of endpoints. The developing embryo or fetus, for example, is entirely reliant upon the maternal milieu for proper growth and development. Maternal stress has been shown to cause cleft palate in mice.[70] Antibiotics may produce embryo or fetal lethality and/or cause abortion in rabbits by adversely affecting the maternal gut flora,[75,134] and the rat embryo and fetus has been shown to depend upon maternal zinc for proper development.[135] Therefore, determining whether a fetal malformation is caused by the maternal condition

Table 9.39 Historical control data and values indicating a positive signal

Endpoint	Historical Control	Positive Signal Threshold
Viable litter size	Mean = 14.1 ± 0.94	Decrease of ≥1
Survival before/on PND 4	Mean = 95.9% Range = 91.3%–99.3%	≤91%
Total litter loss	Mean = 1.21% (N = 1905 litters)	1 is equivocal 2 is a stronger signal
PND 1 pup weights	Mean = 7.1 g ± 0.25 Range = 6.5–7.6 g (N = 1100 litters)	≤6.5 g is a strong signal

Source: Data from WIL Research Laboratories, Inc. Crl:CD®(SD)IGS BR rat reproductive historical control database (55 studies conducted during 1996–2002).

or is a direct effect of the test agent presents its own challenges and is discussed in detail in Chapter 4 of this text.

C. Criticality of Accurate and Consistent Historical Control Data

A comparison with the laboratory's historical control data should be a first step toward determining whether a small increase or decrease (not statistically significant) in an endpoint constitutes a treatment-related effect. To illustrate this point, control mean values and ranges from over 1300 litters for selected reproductive endpoints are shown in Table 9.39. These data originate from the Crl:CD®(SD)IGS BR rat historical control database at the authors' laboratory. Since inception of collecting these data for this animal model (1996), endpoints such as mean viable litter size, survival before or on PND 4, total litter loss, and newborn pup weights have been found to be very consistent when coupled with good animal husbandry. Occurrences above or below means and ranges of these endpoints may, therefore, indicate the threshold of a positive signal for a treatment-related effect.

Live litter size and early pup viability (before PND 4) are sensitive endpoints. Neonatal deaths may be small as a mean percentage, and a death rate exceeding approximately 5% per group should be closely evaluated for a potential treatment-related effect. Likewise, a decrease in PND 1 pup weight below a mean of 6.5 g is most likely past the threshold of a positive signal. Finally, in today's modern facilities with good animal husbandry and highly skilled technicians, total litter loss for one or two dams in the same group probably constitutes a treatment-related effect, even in the absence of statistical significance. These examples demonstrate the importance of creating a historical control database and using it consistently when interpreting overall results from individual studies.

D. Case Studies

The following discussion examines scenarios involving three different compounds that demonstrate the difficulties with rare event data interpretation and consequences that may occur if the signals are missed in safety testing.

1. Dystocia

The example shown in Table 9.40 refers to a two-generation reproductive toxicity and developmental neurotoxicity study with a second mating phase for the F_1 generation.[136] The agent tested, octamethylcyclotetrasiloxane (D4), was used industrially as well as in over-the-counter products and was administered via whole-body vapor inhalation. Five groups of 30 rats per sex received target

Table 9.40 Rates of dystocia in a two-generation reproductive toxicity study of octamethylcyclotetrasiloxane

Exposure (ppm)	0	70	300	500	700
No./sex	30	30	30	30	30
F_0	0	0	0	2/24	3/26
F_1-1st	0	0	0	0	1/17
F_1-2nd	0	0	1/21	1/18	0/12

exposures of 0, 70, 300, 500, or 700 ppm of D4. Exposure of the F_0 and F_1 males were conducted for 10 weeks before mating and through the mating period. Exposures of the F_0 and F_1 females were conducted for 10 weeks before mating, throughout mating and gestation (suspended from GD 21 through LD 4), and from LD 5 through weaning of the offspring.

One of the endpoints affected by D4 was parturition. To determine whether dystocia in this study was treatment related, comparisons were made to the laboratory's historical control rate of 0.6%. In the F_0 generation, there were five instances of dystocia in a total of 150 litters (3.33%); two in 24 litters (8.3%) at 500 ppm and three in 26 litters (11.5%) at 700 ppm. In the F_1 generation first mating, the single incidence (1/17) at 500 ppm was 5.9%. In the F_1 generation second mating, there were two occurrences, 1/21 (4.8%) at 300 ppm and 1/18 (5.6%) at 500 ppm. Because of the lack of evidence of dystocia in the concurrent control groups for either generation and the very low incidence of dystocia in the historical control data, the staff at the conducting laboratory viewed dystocia as a treatment-related effect.

This conclusion could be challenged because no statistically significant incidence of dystocia had occurred in any generation, there was no consistent pattern across generations and/or matings, and there was no apparent dose-response pattern. There were, however, plausible explanations for dystocia being a treatment-related effect in this study. Statistical significance was not (and would never have been) detected because the effect occurred at such a low incidence. In addition, the offspring of those animals exhibiting dystocia in the F_0 generation were not represented in the F_1 generation because of death. Therefore, an apparently decreased response in the first F_1 mating was thought to be the result of loss of the more sensitive animals from the second generation. When two more instances of dystocia were observed in the F_1 generation second mating (including one at a lower exposure level than previously observed), the laboratory's conclusion was strengthened, thereby confirming the generational effect.

This effect was initially not acknowledged by many toxicologists because of the low incidence at which it occurred. This case study demonstrates how a rare event finding may be misinterpreted if one ignores any comparison with historical data and only relies upon finding a dose-response relationship, consistency across generations, and statistical significance.

2. ACE Fetopathy

The fetal effects of angiotensin converting enzyme (ACE) inhibitors that cross the placenta were first recognized in the human. Subsequent reevaluations of animal data revealed that a positive signal (increased postnatal mortality) had been present in the original studies. However, the low (and not statistically significant) incidence above concurrent controls of this finding was deemed not treatment related. Current awareness of the real risks for birth defects when using ACE inhibitors that cross the placenta allows practitioners to minimize risk and gain therapeutic benefits for treating pregnancy-related hypertension. When ACE inhibitors that cross the placenta are used in the third trimester, these antihypertensive agents cause fetal hypotension. Additional fetal effects are illustrated in Figure 9.12.

Major organogenesis (defined as the time from implantation to closure of the hard palate) is unaffected by ACE inhibitors; however, other effects on the developing organism are severe,

Figure 9.12 Spectrum of fetal effects following maternal administration of ACE inhibitors.

including renal compromise resulting in anuria. This in turn leads to a reduced volume of amniotic fluid (oligohydramnios), hypoxia, and calvarial hypoplasia. Humans born with this syndrome are stunted because of the anuria, and many die at an early age.

What happened in the risk assessment of these ACE inhibitors? The original animal safety testing was extensive and of good quality.[137] In spite of this, the small increases in postnatal mortality, evident in the initial reproduction studies, were not identified as positive treatment-related effects, probably because these increases were marginally statistically significant.

An elegant series of postnatal studies (reported by Spence et al.), in which ACE inhibitors were directly administered to pups, revealed increased mortality (greater than 30%), growth retardation, and renal alterations (anatomical and functional).[138,139] There is a temporal difference in developmental schedules between humans and rodents. In humans, renal glomerular function matures *in utero*, while in the rat, it matures postnatally on day 17 or 18. Thus, initial lack of recognition of this difference led to an incorrect outcome in the extrapolation of results to humans. It was not until additional confirmatory studies were conducted that the complete picture emerged, and the ACE inhibitors became the first examples of non-CNS functional developmental toxicants.

3. Retroesophageal Aortic Arch

This case, involving a rat developmental toxicity study conducted in the authors' laboratory with an antibiotic to be used via the oral mucosal route, again demonstrates the criticality of maintaining a robust and up-to-date historical control database. Single occurrences of retroesophageal aortic arch in the mid- and high-dose groups of the study presented an interpretation dilemma. These malformations were initially considered to be a treatment-related rare event, but upon closer evaluation of the findings within the context of the historical control data were interpreted to represent increased penetrance of this trait in Crl:CD®(SD)IGS BR strain rats. However, reassessment of the data after two additional years of historical control data compilation pointed back to the original interpretation, that the findings were related to treatment with the test article.

The drug was tested for developmental effects in both the rat and rabbit during July 2000. In rats, it was administered orally by gavage as a single daily dose from GD 6 through 17 at a dose of 0, 5, 25, or 100 mg/kg/d. No maternal toxicity or effects on intrauterine growth and survival were observed. However, retroesophageal aortic arches were present in one fetus each at 25 and 100 mg/kg/d. Retroesophageal aortic arch is a malformation in which the great arch of the aorta from the heart develops behind the esophagus. It requires a major change in the early development of cardiac outflow, and it is a rarely occurring malformation. In spite of this malformation, both fetuses were apparently thriving based on their body weights, which were similar to their respective mean litter (3.6 and 3.5 g) and group mean litter (3.6 and 3.5 g) weights. Viscerally, these fetuses had no evidence of impaired peripheral perfusion. Furthermore, in the concurrently performed embryo and fetal development study in Hra:(NZW)SPF rabbits, no retroesophageal arches were observed in the 427 fetuses (63 litters) from does administered the test article orally at doses up to 60 mg/kg/d.

The appearance of retroesophageal aortic arch in two test article–treated groups was deemed unlikely to be a spontaneous event, considering that this finding was observed for only 2 of 9642 fetuses evaluated viscerally from 635 litters in 27 studies (1998 to 2001) in the WIL

Table 9.41 Retroesophageal aortic arch in Crl:CD®(SD)IGS BR rat developmental toxicity studies[a]

Dose (mg/kg/d)	Case Study[a] Group Incidences				Incidences in Historical Control Databases	
	0	5	25	100	1[b]	2[c]
Fetuses affected/total fetuses evaluated	0/386	0/368	1/353	1/351	2/9642	3/21,853[d]
Total litters evaluated	24	24	23	22	635	1447
Mean %/litter affected	0.0	0.0	0.3	0.3	0.0–0.3	0.0–0.3

[a] The case study was conducted during June 2000.
[b] Database 1 is the database existing at the time of initial interpretation of the case study data (27 studies conducted from April 1998 through March 2001).
[c] Database 2 is the current database (61 studies conducted from April 1998 through June 2003); this database is a continuation of Database 1.
[d] The initial two findings of retroesophageal aortic arch were from studies conducted in 1999, and the third instance of this finding occurred in a study conducted in 2002.

Source: All studies referenced were conducted at WIL Research Laboratories, Inc.

Crl:CD®(SD)IGS BR rat historical control database (refer to Table 9.41). While such a distribution (in the case study) arising merely from chance is statistically possible, the authors instead posited the explanation described below, based on the temporal placement of these findings in the historical control database at the time. Because the only instances of retroesophageal aortic arch in the historical control database at the time of the study were clustered in the year prior to conduct of the case study (1999), the occurrence of these malformations in this study was interpreted to reflect increased penetrance of this trait. Further evidence of increased penetrance at the time was derived from single occurrences of these findings in test article–treated groups (unrelated to test article treatment) of two studies that had been recently conducted for other sponsors.

The occurrence of retroesophageal aortic arches in the case study, while unusual, was not ascribed to treatment because the incidences of this finding in the mid- and high-dose groups were within the WIL historical control data range, the WIL historical control database suggested an increased penetrance of this trait, and no concordance was observed in the concurrently performed embryo and fetal development study in rabbits.

Further maturation of the WIL historical control database (addition of 2 years and 12,211 control fetuses; refer to Table 9.41) revealed only a single additional finding of retroesophageal aortic arch (in 2002). Since the time of the case study, another unrelated test article demonstrated a nearly singular effect on major blood vessels, with the predominant malformation being retroesophageal aortic arch (100% of the high-dose group litters had fetuses with this finding). A xenobiotic has subsequently been shown to exquisitely affect major blood vessel development, while no sustained increase in the historical control incidence of retroesophageal aortic arch has been observed. Therefore, the authors now consider the original two occurrences of this finding in the case study to likely have been associated with treatment.

4. Lobular Agenesis of the Lung

Although ultimately discounted in the aforementioned case, penetrance does remain a confounding factor in interpretation of rare event manifestation. As an example, Table 9.42 presents the temporal distribution of another malformation, lobular agenesis of the lung (right accessory lobe), in Hra:(NZW)SPF rabbits in the authors' historical control database.

During a 2-year period (1999 and 2000), lobular agenesis of the lung was observed in 13 of 2951 fetuses evaluated (average yearly incidence of 0.45%). In the previous and subsequent 3-year periods, only 2 of 3002 and 4 of 2763 fetuses, respectively, were observed with this finding (average yearly incidences of 0.05% and 0.17%, respectively). The increased incidence of lobular agenesis

Table 9.42 Temporal distribution of lobular agenesis of the lung in control Hra:(NZW)SPF rabbit fetuses[a]

Year	1996	1997	1998	1999	2000	2001	2002	2003
Total fetuses evaluated	1088	519	1395	1659	1292	1177	707	879
Lobular agenesis of the lung								
Fetuses affected (no.)	1	0	1	6	7	1	3	0
Fetuses affected (%)	0.09	0.00	0.07	0.36	0.54	0.08	0.42	0.00

[a] Data presented are from 67 Hra:(NZW)SPF rabbit developmental toxicity studies (1996 to 2003) in the historical control database at WIL Research Laboratories, Inc.

Table 9.43 Typical reaction to rare events and subsequent scenarios

Disbelief; rely on lack of statistical significance
Comparison to concurrent control
Comparison to laboratory's historical control database
Comparison to other historical control databases
Ask experience and opinions of others
Construct explanation to negate
Agency rejects
Repeat study or label appropriately

may indicate a heritable trait and, as such, may represent localized, time-bounded penetrance. This increase in the incidence of such a rare finding during a 2-year period may or may not be considered remarkable. However, given the temporal boundaries of the increase, the apparent penetrance of lobular agenesis may significantly affect interpretation of a study conducted during that time frame. Because the finding was still a rare event during 1999 to 2000, a spontaneously occurring distribution of findings solely in treated groups was certainly possible. Without systematic maintenance of historical control data and recognition of penetrance as it occurs, the hypothetical distribution of a few occurrences of lobular agenesis in the treated groups would likely be interpreted as a test article–related finding. Such an interpretation could possibly reduce the margin of safety of a candidate pharmaceutical, potentially keeping a needed therapy from patients.

E. Approach to Evaluating Rare Findings

Table 9.43 presents the typical reaction to the manifestation of low incidence findings and the subsequent scenario. The recommended paradigm for interpretation of rare events includes a number of steps that should be considered to determine if a finding is biologically relevant (even though it is not statistically significant) when compared with the concurrent control. These steps are listed in Table 9.44; however, their order may vary depending on the situation.

When evaluating and interpreting rare events, the best practice for resolution of the relationship between the agent tested and the findings observed is use of a large historical control database for comparison. Such a database should be developed at the laboratory performing the studies, using consistent methodology, staff, and animal husbandry, as well as appropriately designed confirmatory studies. The design of confirmatory studies may involve increasing the number of animals overall, as well as in the control group, to counteract the 3:1 probability of a rare event occurring in the treatment groups versus the control group (see Figure 9.13).

Evaluation of whether a signal is real should also include comparisons of results from other reproductive and developmental studies, as well as observations in other species.

With the current designs of animal safety testing protocols, it is not possible to detect rare events or small changes in endpoints by use of the guideline-recommended statistical methodologies. Other approaches (such as the most current version of the Monte-Carlo analysis) will have to be used.

DEVELOPMENTAL AND REPRODUCTIVE TOXICITY STUDY FINDINGS

Table 9.44 Approach to determining biological relevance of rare event findings

Stepwise Approach	Consideration and Caution
Dose response (including TK[a])	Possibility of any rare event occurring is 3:1 in favor of test article–treated groups
Historical control comparison	Mean and range values (laboratory's own data, then others)
	Values near or outside upper or lower limits of historical control range provide a strong signal that rare event is "real"
Additional statistical tests	Monte-Carlo analysis to examine sampling error and predict population estimates relative to sampling values
Second species comparison (including TK data)	Establishment of similar internal dose (C_{max} and/or AUC) critical
Other studies comparison	Concordant outcome from embryo, fetal, and pre- and postnatal studies (e.g., slight decrease in fetal BW in developmental study, along with slight decrease in PND 1 BW in reproductive study) is considered to be "real" event
Confirmatory study	Increased N per group; increased N in control group; increased dose; limited exposure regime (if developmental timing of organ or system affected is known, only administer doses to achieve same AUC on limited regime, which could illuminate whether the underlying embryology and the outcome make sense)
Mechanistic studies	Evaluate pharmacological action relative to ontogeny of receptors, signaling pathways, and other possible modes of action

[a] N = number, BW = body weight, TK = toxicokinetics, PND = postnatal day, AUC = area under the curve.

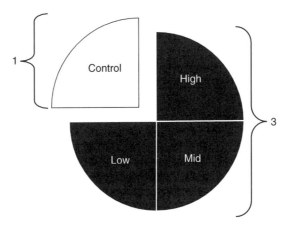

Figure 9.13 Probability of a rare event occurring in the treatment groups relative to the control group.

Because concurrent control groups are limited in size (N), they represent, at best, the current rate of high-incidence findings and/or continuous variable values. They are poor estimators of the true population statistic. To estimate the incidence of a developmental defect in the population of a given species for occurrences of less than 1 in 100 requires approximately 100 control data sets for reliability. Therefore, concurrent controls have little value in discerning treatment-related versus spontaneous occurrences for the majority of developmental toxicity endpoints (other than fetal weight and embryolethality). The best probabilistic approach to matching current findings to occurrences in a historical control population is a Monte-Carlo analysis.[140]

VIII. THE ROLE OF EXPERT JUDGMENT

This section will address additional factors that must be considered when interpreting developmental and reproductive toxicity data.

Table 9.45 Frequently asked key regulatory questions following data package submission

Integrity of database	Was an established, validated model used?
	Was a NOAEL (NOEL) demonstrated?
	Does increasing dose increase severity or incidence?
Biologic dynamics and dimensions	When was the effect exerted?
	Is the effect reversible?
	Do indications of sensitization (generational effects or imprinting) exist?
	Is the effect gender specific?
Relevancy and risk analysis	Were appropriate TK comparisons made between experimental studies and estimated human PK or exposure scenarios?
	Are there significant differences in the pattern, timing, or magnitude of exposure between guideline studies and human scenarios?
	Is concordance of effects among species evident?
	Is the mode of action known or deducible?
	Is the mechanism of action known?

The final report for a developmental or reproductive toxicity study is most often submitted to the appropriate regulatory agency. Following submission, careful regulators not only will question the study director's final interpretation but will also examine issues pertaining to the integrity of the database, key biological dynamics and dimensions, and the relevance and risk to the human exposure scenario. Table 9.45 presents 12 key questions frequently asked by representatives of regulatory agencies concerning the aforementioned issues.

The expert toxicologist must anticipate the regulators' concerns and ask these questions prior to data interpretation and regulatory submission, as well as considering them prior to study design. The following subsections will address these common questions pertaining to the integrity of the database, key biologic dynamics and dimensions, and the relevancy and risk related to the human exposure scenario.

A. Integrity of Database

Questions surrounding the integrity of the data generated from any toxicity study should be paramount in the mind of an expert toxicologist. These questions involve the ability of the study design to detect potential toxicologic changes under study, the training of personnel conducting a study, and the possible presence of dose-response relationships, among others. Without these underpinnings, the utility of toxicologic data, and any interpretations derived therefrom, is diminished.

1. Was an established, validated model used?

    While the rat and rabbit are the most commonly used species in studies of toxicity to development, and the rat is the primary test system for reproductive toxicity determinations, examples of the use of the dog and primate have been discussed in this chapter. Probably the single most important criterion of an appropriate model is metabolism that is similar to that of the human, especially when a metabolite is the developmentally toxic agent. Typically, this level of understanding of metabolism is not determined *a priori* except in certain pharmaceutical development situations.

2. Was a NOAEL (or NOEL) demonstrated?

    From numerous examples in this chapter, the potential problems that arise from relying solely on statistical significance and/or the presence of a dose-response relationship are evident. The absence of a dose response does not fully negate the findings, especially in the case of rare events or threshold responses. Regulators often will rely heavily on concurrent control data for comparison. Slight changes in response across treatment groups relative to the concurrent control, in the absence of statistical significance or a clear dose-response relationship, may be an indication of toxicity.

Table 9.46 How maldevelopment differs from tumorigenesis

Not more frequent with time
Myriad possible underlying mechanisms
    Amniotic banding, oligohydramnios-skull dysgenesis
    Interference with signaling pathways (e.g., TGF-β)
    Mutation
Maternal influences possible
Multiple endpoints interrelated
    Weight alterations causing cleft palate and neural tube defects
Occurs early in life and hence greater economic and social impact

Table 9.47 Selected differences in developmental toxicity vs. oncogenicity endpoint ascertainment

Smaller group sizes (25 litters vs. minimum 50/sex/group)
Macroscopic examination (histopathology very rare)
Physical limitations based on size
Less standardized nomenclature
No certification of personnel, controls over training, etc.
    This has potentially more of an affect because of the earlier inclusion of women in clinical trials
Involves coapt organisms (dam and fetuses)
    Always potential for maternal influence, but fetal effects may affect the dam as well
Dynamic morphology and function
    ACE fetopathy example
Animals evaluated in the midst of changing morphology
No two points in development are the same
    Exposure hourly and daily key to outcome
    An important aspect of human studies

In these instances, the use of historical control data may be the more appropriate comparison for determining relationship to treatment. However, some data sets will always appear uninterpretable. In these cases, additional studies with increased dosage levels, perhaps with an unbalanced design, can be useful to confirm or refute the initial findings.

3. Does increasing dose increase severity and/or incidence?

Despite the concerns raised above, the presence of a dose-response relationship can provide reasons for enhanced concern. Slightly increased responses (relative to the control group) at the lowest exposure level and greater responses with increasing exposure provide confidence that the effects observed at the lowest exposure level are likely to have been related to treatment, even in the absence of statistical significance. The proper spacing of dose levels is critical to discernment of dose-response relationships. Because of animal-to-animal variability, differences between dose levels of less than twofold may result in overlapping of responses with increasing exposure. Conversely, dose levels too widely spread (generally greater than fivefold) will hinder characterization of the dose-response relationship. A likely outcome of this scenario is identification of a no-effect dose and one response level; however, it is likely that no information regarding the slope of the dose-response curve will be obtained. In either case, improper spacing of dose levels can make identification of the threshold response level problematic.

It is interesting to contrast the dose-response relationships of developmental toxicity and carcinogenicity. The cancer endpoint (nonepigenetic) is considered to exhibit linear dose-response behavior, whereas most regulators consider reproductive or developmental toxicity to be a threshold phenomenon. Both developmental defects and cancer are generally irreversible endpoints, and certain reproductive lesions may also be irreversible. The basis for distinction between carcinogenicity and developmental toxicity is not totally rational, nor is it entirely consistent with our understanding of the possible mechanisms of carcinogenicity or with the dose-response outcomes of the two. Many differences exist between these two toxicologic endpoints (see Table 9.46 and Table 9.47) that contribute variably to this distinction (linear versus threshold). Nevertheless, it is more probable that both obey similar dose-response behavior at corresponding exposure levels, with the exception that developmental toxicity can be caused by nonstoichiometric phenomena.

B. Biologic Dynamics and Dimensions

A full command of the biologic dynamics and dimensions surrounding the possible manifestations of developmental and reproductive toxicity is essential for proper data interpretation. Questions under this umbrella address the discernment of the timing and reversibility of effects, the recognition of those effects that appear in a sensitization pattern, and the resolution of gender-specific effects.

4. When was the effect exerted?

    Identification of the timing of a developmental or reproductive insult is important for extrapolation of animal data to effects in humans for subsequent risk assessment. Intimate knowledge of developmental and reproductive processes will guide the investigator in designing subsequent studies to determine whether single or multiple exposures were necessary for the insult, to determine whether central or peripheral sites of action were affected, and to elucidate mode of action. Understanding the mode of action of the test substance in animal models is essential to determining relevance of the insult to humans, particularly for effects on fertility and other reproductive processes. There are numerous examples of agents that require only a limited window of exposure to elicit an adverse response. For example, agents that block the LH surge in female rats can adversely affect fertility following a single dose on the day of proestrus. These agents would be regulated quite differently than those agents requiring extended exposure to attain sufficient bioaccumulation to elicit an adverse response.

5. Is the effect reversible?

    Reversibility of developmental and reproductive insults may greatly influence regulation of the test agent. Assessment of reversibility for fertility toxicants may be accomplished through use of a recovery (nonexposure) period. Doses at or near the MTD may induce irreversible changes; however, those same effects may be reversible at lower dosage levels, indicating a threshold for reversibility. Interrelatedness of the endpoints in developmental toxicity studies makes it difficult to assess reversibility. Disruption of the schedule of *in utero* development may have consequences postnatally. In the case of intrauterine growth retardation, even though postnatal growth may normalize to the level of concurrent controls, developmental schedules may be disturbed, resulting in increased postnatal loss. Postweaning assessments beyond measurement of bodyweight, such as evaluation of neurobehavioral endpoints, may be important in revealing latent effects that signal lack of reversibility.

6. Do indications of sensitization (generational effects or imprinting) exist?

    Generational effects and imprinting represent some of the most insidious risks to human health. The multigeneration study is designed to assess the potential for increased risk in subsequent generations, relative to the initially exposed (F_0) generation. Specious decreases in adverse reproductive responses in subsequent generations must be recognized as such, as adverse effects can be masked by the loss of sensitive individuals from the F_0/F_1 parental animal populations (selection phenomenon). See Section VII.D.1 for a description of the D4 example from the authors' laboratory. Sensitizing phenomena can be separated from bioaccumulative effects by the use of toxicokinetic data to compare internal exposures in offspring with those of the parental generation.

7. Is the effect gender-specific?

    In determining if the effect observed is gender specific, one can identify the population at risk from the exposure. With agents not producing overt toxicity, discerning a gender basis may not always be straightforward. However, gender-specific effects can often be deduced from organ histopathology or other endpoints (e.g., reproductive hormone levels). In classic fertility and two-generation reproductive toxicity studies, in which both males and females are given the test article, additional studies generally will be required to determine if one sex is preferentially affected.

Alternatively, additional breeding phases can be introduced to the extant study designs, using naive males and/or females in an effort to minimize the number of animals required. When identified, gender-specific effects can give valuable guidance toward identification of possible modes of action.

C. Relevancy and Risk Analysis

Despite adequate integrity of the database and resolution of particular biologic dynamics and dimensions, consideration of the relevancy of the data and subsequent risk analysis must be addressed. These issues include interspecies and temporal comparisons of measures of dosimetry and knowledge of mode and mechanism of action.

8. Were appropriate toxicokinetic comparisons made between experimental studies and estimated human pharmacokinetics or exposure scenarios?

   The appropriate toxicokinetic assessments are key to standardizing internal dose, establishing bioavailability, and determining possible dose-dependent kinetics and/or potential changes in metabolism over the course of exposure. Adverse developmental and reproductive effects may be the result of sufficiently high C_{max} or AUC exposures or a combination thereof. Because examples of both cases exist (e.g., valproic acid for C_{max} and cyclophosphamide for AUC), the observed effects must be considered in terms of the proper internal exposure measure. In the case of at least one chemical (2-methoxyethanol), dysmorphogenesis can be attributed to both the AUC and C_{max}, depending on the critical period of development. In mice, 2-methoxyethanol exposure on GD 8 caused exencephaly, with a high correlation to the maternal plasma C_{max}, independent of the maternal AUC.[141] However, 2-methoxyethanol exposure of mice on GD 11 elicited paw malformations with a better correlation to the AUC than to the C_{max}.[142] Therefore, prenatal response to internal dose (as measured by C_{max} or AUC) at one specific point in development does not necessarily apply to all points in development.

   Further use of toxicokinetic data may help determine whether effects observed in neonates occur because of exposure during gestation or lactation. Elaborate schemes involving milk production, collection, consumption, and bioavailability may be constructed to address neonatal exposure questions. However, the use of simple fostering studies (mutual exchange of litters between treated and untreated dams) with direct toxicokinetic measurements from offspring blood may provide a more reliable and straightforward means of answering those questions.

9. Are there significant differences in the pattern, timing, or magnitude of the exposure between guideline studies and human scenarios?

   Standard experimental animal models employed in hazard identification have greatly compressed periods of organogenesis relative to the timing of human development. Single daily administrations of drugs with short half-lives may result in very low systemic exposures during critical developmental windows and may not be adequate for assessing developmental toxicity over the entire susceptible period. These cases may require multiple administrations each day or alternate exposure methods (e.g., continuous infusion or dietary administration) to maximize exposure duration. Multiple administrations per day have become more commonplace in the last 5 years, with the advent of more extensive use of toxicokinetics.

10. Is concordance of effects among species evident?

    Effects manifesting in multiple species signal enhanced concern for human risk, and the concern increases further when the number of species increases. However, the absence of such multiplicity does not necessarily diminish concern.[5] Response variability between species most often appears to indicate a mode of action difference that can be critical to appropriate animal model selection for human hazard identification. "Apparent" lack of concordance among species can arise via several different scenarios. A difference in sensitivity, implying a biological basis (i.e., not a maternally controlled or metabolic basis), is often cited as a primary reason for discordance between species. However, this claim has historically been made without comparing appropriate measures of internal dose (exposure) among the species in question. Examples where an apparent lack of concordance based on sensitivity issues was resolved through exhaustive measures of AUC and C_{max} are the cases of isotretinoin and valproic acid. For isotretinoin, comparisons of maternal and embryonic blood levels of 13-*cis*-retinoic acid (isotretinoin) and all-*trans*-retinoic acid, as well as their predominant metabolites, revealed significant differences between the more sensitive (primates and lagomorphs) and less sensitive (rodents) species in elimination, glucuronidation, and placental transport.[143–146]

Another common source of false discordance is lack of appropriate statistical power in a comparison species. Because of the prohibitive cost and animal welfare concerns of large group sizes, misleading discordance frequently occurs when the dog or nonhuman primate is used as the test species.

11. Is the mode of action known or deducible?

Insight into the mode of action of an agent under study is highly desirable when the toxicologist is faced with the interpretation of possible equivocal developmental and reproductive effects. Knowledge concerning the pharmacologic action or class effect may aid the investigator in targeting specialized study designs and obtaining a better assessment of the degree to which the findings represent a hazard to human reproduction.

12. Is the mechanism of action known?

If the mechanism of action is defined as the effect precipitating the initiation of pathogenesis, then the precise action of no known human developmental toxicant has been elucidated. However, increasing knowledge of mode and mechanism of actions is being gained through research and is probably most advanced in the case of the retinoids. The value of knowledge of mechanism of action is commonly extolled and obvious.

The adequacy of the current ICH, EPA OPPTS, and OECD developmental and reproductive toxicity guideline studies to answer these 12 questions is presented in Table 9.48.

IX. CONCLUSION

The goal of nonclinical reproductive toxicity studies is identification of potential human hazards. This chapter is meant to be a practical guide to the significance, reliability, and interpretation of reproductive and developmental toxicity study findings. When reviewing these data, one must be mindful that the validity and reliability of the study are based on many assumptions. These assumptions include the following:

1. Staff for study execution is well-trained and competent.
2. Animal husbandry follows best practices.
3. Animal model selected is appropriate.
4. Sample size is sufficient for statistical power.
5. Dosage levels are adequately spaced and selected.
6. Kinetics are well defined.

In addition, knowledge of the mode of action of test substances is becoming more available and will continue to be useful in the evaluation of reproductive and developmental toxicity study findings.

Once the investigator is confident that these assumptions have been met, the data set then can be presumed valid and reliable for extrapolation to the human situation. In order to draw a conclusion regarding reproductive toxicity, the investigator must understand the interrelatedness of the endpoints under study. Often, subtle reproductive insults can be recognized by identification of a pattern of effect across multiple endpoints. Historical control data developed at the performing laboratory under similar conditions is essential in assessment of the significance of rare event findings.

DEVELOPMENTAL AND REPRODUCTIVE TOXICITY STUDY FINDINGS

Table 9.48 Extent to which guideline studies answer key regulatory questions

What do regulators want to know?	ICH Fert. 4.1.1	ICH P/P 4.1.2	ICH DT 4.1.3	EPA OPPTS (8700.) DT 3700	EPA OPPTS 2-G 3800	EPA OPPTS DNT 6300	OECD DT 414	OECD 1-G 415	OECD 2-G 416	OECD Screen 421	OECD DNT 426
1 Validated model	Yes	Yes	Yes	Yes	Yes	Yes	Yes	Yes	Yes	Yes	Yes
2 NOAEL determined	Yes	Yes	Yes	Yes	Yes	Yes	Yes	Yes	Yes	Yes	Yes
3 Dose response	No	Yes	Yes	Yes	Yes	No	Yes	Some	Yes	Some	No
4 Insult timing elucidated	No	No	No	No	No	No	No	No	No	No	No
5 Reversibility	No	No	No	No	Yes	No	No	No	Yes	No	No
6 Imprinting phenomenon	No	No	No	No	No	No	No	No	No	No	No
7 Gender basis	No	Yes	Yes	Yes	No	Yes	Yes	No	No	No	Yes
8 TK profiled	Yes	Yes	Yes	No	No	No	No	No	No	No	No
9 Exposure mimicked	Some	Some	Some	?	?	?	?	?	?	?	?
10 Interspecies concordance	No	No	Yes	Yes	No	No	Yes	No	No	No	No
11 Mode of action	No	No	No	No	No	No	No	No	No	No	No
12 Mechanism of action	No	No	No	No	No	No	No	No	No	No	No

Note: Fert = fertility; P/P = pre and postnatal; DT = developmental toxicity; 2-G = two-generation; DNT = developmental neurotoxicity; 1-G = one-generation.

Despite confirmation of the aforementioned assumptions and the presence of a rich and accurate historical control database, the investigator may not be able to draw definitive conclusions because of equivocal findings. In this scenario, the only recourse is to perform additional studies, perhaps with increased sample size and/or increased dosage levels.

Important decisions must be made based on these data sets. In the case of drugs, many of the endpoints assessed in nonclinical reproductive safety studies may not be principal endpoints in clinical trials (e.g., functional assessments, hormone analyses, and intrauterine growth and development). For environmental chemicals, human exposure is often unknown and unmonitored, and could involve multiple compounds. As a result, determination of a proximate human reproductive toxicant is often impossible. Therefore, reproductive hazard identification in animal studies is paramount in protecting human health.

Human tragedies have led to the refinement and enhancement of regulatory guideline requirements. As a result, there exists a general confidence in the ability to detect reproductive hazards. This is especially true when effects are manifested across multiple well-validated animal models and/or studies with overlapping treatment regimes or endpoints. As our knowledge of the basic biology of animal models and humans advances, study designs and the strength of assumptions used for animal-to-human extrapolation will continue to improve. The challenge facing investigators is application of emerging scientific principles to hazard identification and ultimately risk assessment. Faulty study design and interpretation can keep a needed treatment from an affected human population or conversely expose those populations to unnecessary risk.

ACKNOWLEDGEMENTS

The authors would like to acknowledge with deep appreciation the efforts of various individuals who contributed in tangible ways to the completion of this project. Special thanks to Kerin Clevidence, Michelle Edwards, and Jeanette Howell, who labored diligently to compile and verify much of the WIL historical control data referenced in this chapter. The authors would also like to recognize the numerous staff members at WIL, too many to mention individually, who through their dedication, hard work, and meticulous data collection over the past 10 to 20 years have laid the foundation for this chapter.

REFERENCES

1. DeSesso, J.M., Harris, S.B., and Swain, S.M., The design, evaluation, and interpretation of developmental toxicity tests, in *Toxicology and Risk Assessment: Principles, Methods, and Applications*, Fan, A.M. and Chang, L.W., Eds., Marcel Dekker, New York, 1996, p. 227.
2. Palmer, A.K. and Ulbrich, B., Data presentation in developmental toxicity studies: an aid to evaluation? in *Handbook of Developmental Toxicology*, Hood. R.D., Ed., CRC Press, Boca Raton, FL, 1997, chap. 8.
3. Daston, G. and Kimmel, C.A., Eds., *An Evaluation and Interpretation of Reproductive Endpoints for Human Health Risk Assessment*, International Life Sciences Institute, Washington, DC, 1998.
4. European Centre for Ecotoxicology and Toxicology of Chemicals (ECETOC), *Guidance on Evaluation of Reproductive Toxicity Data*, Monograph No. 31, 2002.
5. U.S. Food and Drug Administration, *Reviewer Guidance: Integration of Study Results to Assess Concerns about Human Reproductive and Developmental Toxicities (Draft)*, Center for Drug Evaluation and Research, 2001.
6. Wilson, J.G., *Environment and Birth Defects*, Academic Press, New York, 1973.
7. Brent, R.L., Drug testing for teratogenicity: its implications, limitations and application to man, in *Drugs and Fetal Development*, Klingberg, M.A., Abramovici, A., and Cheneke, J., Eds., Plenum Press, New York, 1972, p. 31.

8. Nishimura, H. and Tanimura, T, *Chemical Aspects of the Teratogenicity of Drugs*, Elsevier, New York, 1976.
9. Fraser, F.C., Relation of animal studies to the problem in man, in *Handbook of Teratology*, Vol. 1, Wilson, J.G. and Fraser, F.C., Eds., Plenum Press, New York, 1979, p. 75.
10. Snow, M.H.L. and Tam, P.P.L., Is compensatory growth a complicating factor in mouse teratology?, *Nature*, 279, 555, 1979.
11. U.S. Food and Drug Administration, National Center for Toxicological Research, *Reliability of Experimental Studies for Predicting Hazards to Human Development* (Kimmel, C.A., Holson, J.F., Hogue, C.I., and Carlo, G.L., Eds.), NCTR Technical Report for Experiment No. 6015, Jefferson, AR, 1984.
12. National Research Council, *Scientific Frontiers in Developmental Toxicology and Risk Assessment*, National Academy Press, Washington, DC, 2000.
13. Ulbrich, B. and Palmer, A.K., Detection of effects on male reproduction — a literature survey, *J. Am. Coll. Toxicol.*, 14, 293, 1995.
14. Clegg, J.A., Adverse drug reactions, *Toxicol. Rev.*, 13, 235, 1994.
15. Buiatti, E, Barchielli, A., Geddes, M., Nastasi, L., Kriebel, D., Franchini, M., and Scarselli, G., Risk factors in male infertility, *Arch. Environ. Health*, 39, 266, 1984.
16. Paul, M. and Himmelstein J., Reproductive hazards in the workplace: what the practitioner needs to know about chemical exposures, *Obstet. Gynecol.*, 71, 921, 1988.
17. Thomas, M.J. and Thomas, J.A., Toxic responses of the reproductive system, in *Casarett & Doull's Toxicology, the Basic Science of Poisons*, Klassen, C.D., Ed., McGraw-Hill, New York, 2001, p. 673.
18. Waller, D.P., Killinger, J.M., Zaneveld, L.J.D., Physiology and toxicology of the male reproductive tract, in *Endocrine Toxicology*, Thomas, J.A., Korach, K.S., and McLachlan, J.A., Eds., Raven, New York, 1985, p. 269.
19. Scialli, A.R. and Lemasters, G.K., Epidemiologic aspects of reproductive toxicology, in *Reproductive Toxicology*, 2nd ed., Witorsch, R.J., Ed., Raven Press, 1995, p. 241.
20. Potashnik, G. and Porath, A., Dibromochloropropane (DBCP): a 17-year reassessment of testicular function and reproductive performance, *J. Occup. Environ. Med.*, 37, 1287, 1995.
21. Goldsmith, J.R., Dibromochloropropane: epidemiological findings and current questions, *Ann. N. Y. Acad. Sci.*, 26, 837, 1997.
22. Meistrich, M.L., Wilson, G., Shuttlesworth, G.A., and Porter, K.L., Dibromochloropropane inhibits spermatogonial development in rats, *Reprod. Toxicol.*, 17, 263, 2003.
23. Cooper, R.L., Neuroendocrine control of female reproduction, in *Reproductive and Endocrine Toxicology (Comprehensive Toxicology*, Vol. 10), Boekelheide, K., Chapin, R.E., Hoyer, P.B., and Harris, C., Eds., Elsevier, New York, 1997, p. 273.
24. Eldridge, J.C., Wetzel, L.T., Stevens, J.T., and Simpkins, J.W., The mammary tumor response in triazine-treated female rats: a threshold-mediated interaction with strain and species-specific reproductive senescence, *Steroids*, 64, 672, 1999.
25. Cooper, R.L., Goldman, J.M., and Rehnberg, G.L., Neuroendocrine control of reproductive function in the aging female rodent, *J. Am. Geriatric Sci.*, 34, 735, 1986.
26. Cooper, R.L., Stoker, T.E., Goldman, J.M., Parrish, M.B., and Tyrey, L., Effect of atrazine on ovarian function in the rat, *Reprod. Toxicol.*, 10(4), 257, 1996.
27. U.S. Food and Drug Administration, *Guidelines for Reproduction Studies for Safety Evaluation of Drugs for Human Use*, 1966.
28. U.S. Federal Food and Drugs Act of 1906, Public Law No. 59-384, 34 Stat. 768, 1906.
29. U.S. Federal Food, Drug and Cosmetic Act, 21 U.S.C. 301, 1938.
30. Green, R., Burrill, M., and Ivy, A., Experimental intersexuality: the paradoxical effects of estrogens on the sexual development of the female rat, *Anat. Rec.*, 74, 429, 1939.
31. Green, R., Burrill, M., and Ivy, A., Experimental intersexuality: modification of sexual development of the white rat with synthetic estrogen, *Proc. Soc. Exp. Biol. Med.*, 41, 169, 1939.
32. Herbst, A.L., Ulfelder, H., and Poskanzer, D.C., Adenocarcinoma of the vagina: association of maternal stilbestrol therapy with tumor appearance in young women, *N. Eng. J. Med.*, 284, 878, 1971.
33. Herbst, A.L., Cole, P., Colton, T., Robboy, S.J., and Scully, R.E., Age-incidence and risk of diethylstilbestrol-related clear cell adenocarcinoma of the vagina and cervix, *Am. J. Obstet. Gynecol.*, 128, 43, 1977.

34. Lehman, A.J., Laug, E.P., Woodard, G., Draize, J.H., Fitzhugh, O.G., and Nelson, A.A., Procedures for the appraisal of the toxicity of chemicals in foods, *Food Drug Cosmetic Law Quarterly,* September, 1949.
35. Mauer, I.,Ed., *Appraisal of the Safety of Chemicals in Foods, Drugs and Cosmetics*, The Association of Food and Drug Officials of the United States, Washington, D.C., 1959.
36. Hendrickx, A.G. and Binkerd, P.E., Nonhuman primates and teratological research, *J. Med. Primatol.*, 19, 81, 1990.
37. Buelke-Sam, J., Kimmel, C.A., Adams, J., Nelson, C.J., Vorhees, C.V., Wright, D.C., Omer, V.St., Korol, B.A., Butcher, R.E., Geyer, M.A., Holson, J.F., Kutscher, C.L., and Wayner, M.J., Collaborative behavioral teratology study: results, *Neurotoxicol. Teratol.*, 7, 591, 1985.
38. U.S. Environmental Protection Agency, Office of Pesticide Programs. Proposed guidelines for registering pesticides in the United States; Hazard evaluation: Humans and domestic animals, *Fed. Regist.*, 43, 37336, 1978.
39. U.S. Environmental Protection Agency, Guidelines for developmental toxicity risk assessment; notice, *Fed Regist.*, 56, 63798, 1991.
40. U.S. Environmental Protection Agency, Reproductive toxicity risk assessment guidelines; notice, *Fed. Regist.*, 61, 56274, 1996.
41. Eaton, D.L. and Klaassen, C.D., Principles of toxicology, in *Casarett & Doull's Toxicology: The Basic Science of Poisons*, 6th ed., Klaassen, C.D., Ed., McGraw-Hill, New York, 2001, p. 11.
42. vom Saal, F.S., Timms, B.G., Montano, M.M., Palanza, P., Thayer, K.A., Nagel, S.C., Dhar, M.D., Ganjam, V.K., Parmigiani, S., and Welshons, W.V., Prostate enlargement in mice due to fetal exposure to low doses of estradiol or diethylstilbestrol and opposite effects at high doses, *Proc. Natl. Acad. Sci. USA*, 94, 2056, 1997.
43. Shuey, D.L., Lau, C., Logsdon, T.R., Zucker, R.M., Elstein, K.H., Narotsky, M.G., Setxer, R.W., Kavlock, R.J., and Rogers, J.M., Biologically-based dose-response modeling in developmental toxicology: biochemical and cellular sequelae of 5-fluorouracil exposure in the developing rat, *Toxicol. Appl Pharmacol.*, 126, 129, 1994.
44. Young, J.F. and Holson, J.F., Utility of pharmacokinetics in designing toxicological protocols and improving interspecies extrapolation, *J. Environ. Pathol. Toxicol.*, 2, 169, 1978.
45. Scott, W.J. Jr., Collins, M.D., and Nau, H., Pharmacokinetic determinants of embryotoxicity in rats associated with organic acids, *Environ. Health Perspect.*, 102(Suppl. 11), 97, 1994.
46. Faustman, E.M., Allen, B.C., Kavlock, R.J., and Kimmel, C.A., Dose-response assessment for developmental toxicity. I. Characterization of database and determination of no observed adverse effect levels, *Fundam. Appl. Toxicol.*, 23, 478, 1994.
47. Hood, R.D., Effects of sodium arsenite on fetal development, *Bull. Environ. Contam. Toxicol.*, 7, 216, 1972.
48. DeSesso, J.M., Jacobson, C.F., Scialli, A.R., Farr, C.H., and Holson, J.F., An assessment of the developmental toxicity of inorganic arsenic, *Reprod. Toxicol.*, 12, 385, 1998.
49. Holson, J.F., Stump, D.G., Clevidence, K.J., Knapp, J.F., and Farr, C.H., Evaluation of the prenatal developmental toxicity of orally administered arsenic trioxide in rats, *Food Chem. Toxicol.*, 38, 459, 2000.
50. Stump, D.G., Holson, J.F., Fleeman, T.L., Nemec, M.D., and Farr, C.H., Comparative effects of single intraperitoneal or oral doses of sodium arsenate or arsenic trioxide during in utero development, *Teratology*, 60, 283, 1999.
51. National Academy of Sciences, Sample size determination, in *Guidelines for the Care and Use of Mammals in Neuroscience and Behavioral Research*, National Academies Press, Washington, DC, 2003, p. 176, Appendix A.
52. Nelson, C.J. and Holson, J.F., Statistical analysis of teratologic data: problems and advancements, *J. Environ. Pathol. Toxicol.*, 2, 187, 1978.
53. Steel, R.G.D. and Torrie, J.H., *Principles and Procedures of Statistics, A Biometrical Approach*, 2nd ed., McGraw, New York, 1980, pp. 504-506.
54. Dunnett, C.W., New tables for multiple comparisons with a control, *Biometrics*, 20, 482, 1964.
55. Kruskal, W.H. and Wallis, W.A., Use of ranks in one-criterion variance analysis, *J. Am. Statistical Assoc.*, 47, 583, 1952.
56. Holson, J.F., Scott, W.J., Gaylor, D.W., and Wilson, J.G., Reduced interlitter variability in rats resulting from a restricted mating period, and reassessment of the "litter effect", *Teratology*, 14, 135, 1976.

57. Holson, J.F., Gaines, T.B., Nelson, C.J., LaBorde, J.B., Gaylor, D.W., Sheehan, D.M., and Young, J.F., Developmental toxicity of 2,4,5-trichlorophenoxyacetic acid (2,4,5-T). I. Multireplicated dose-response studies in four inbred strains and one outbred stock of mice, *Fundam. Appl. Toxicol.*, 19, 286, 1992.
58. Brown, N.A., Spielmann, H., Bechter, R., Flint, O.P., Freeman, S.J., Jelinek, R.J., Koch, E., Nau, H., Newall, D.R., Palmer, A.K., Renault, J.Y., Repetto, M.F., Vogel, R., and Wiger, R., Screening chemicals for reproductive toxicity: the current alternatives; the report and recommendations of an ECVAM/ETS workshop (ECVAM workshop 12), *Altern. Lab. Anim.*, 23, 868, 1995.
59. Chapin, R.E. and Sloane, R.A., Reproductive assessment by continuous breeding: evolving study design and summaries of ninety studies, *Environ. Health Persp.*, 105(Suppl. 1), 199, 1997.
60. Gesellschaft Deutscher Chemiker (GDCh — German Chemical Society), *Evaluation of OECD Screening Tests 421 (Reproduction/Developmental Toxicity Screening Test) and 422 (Combined Repeated Dose Toxicity Study with the Reproduction/Developmental Toxicity Screening Test)*, edited by the GDCh Advisory Committee on Existing Chemicals (BUA), BUA Report 229, S. Hirzel Wissenschaftliche Verlagsgesellschaft, Stuttgart, 2002.
61. Hurtt, M.E., Cappon, G.D., and Browning, A., Proposal for a tiered approach to developmental toxicity testing for veterinary pharmaceutical products for food-producing animals, *Food Chem. Toxicol.*, 41, 611, 2003.
62. Stump, D.G., Holson, J.F., Pearce, L.B., Rentko, V.T., and Gawryl, M.S., Finding of developmental toxicity studies of HBOC-201 in rodent and canine models [abstract], *Birth Defects Res. (Part B)*, 68, 250, 2003.
63. Holson, J.F., Pearce, L.B., and Stump, D.G., A probable false positive finding of prenatal toxicity in the rodent model with a high molecular weight protein oxygen therapeutic: evidence and implications [abstract], *Birth Defects Res. (Part B)*, 68, 249, 2003.
64. Fraser, F.C., The multifactorial/threshold concept — use and misuses, *Teratology*, 14, 267, 1976.
65. Adolph, E.F., Ontogeny of physiological regulations in the rat, *Quart. Rev. Biol.*, 32, 89, 1957.
66. Gibson, K.J. and Lumbers, E.R., Mechanisms by which the pregnant ewe can sustain increased salt and water supply to the fetus, *J. Physiol.*, 445, 569, 1992.
67. Gibson, K.J. and Lumbers, E.R., The roles of arginine vasopressin in fetal sodium balance and as a mediator of the effects of fetal "stress," *J. Develop. Physiol.*, 19, 125, 1993.
68. Torchinsky, A., Fein, A., and Toder, V., Modulation of embryo sensitivity to teratogen by nonspecific intrauterine immunopotentiation, *Toxicol. Methods*, 5, 131, 1995.
69. Szabo, K.T. and Brent, R.L., Reduction of drug-induced cleft palate in mice, *Lancet*, 1, 1296, 1975.
70. Hemm, R.D., Arslanoglou, L., and Pollock, J.J., Cleft palate following prenatal food restriction in mice: association with elevated maternal corticosteroids, *Teratology*, 15, 243, 1977.
71. Ellington, S., In-vivo and in-vitro studies on the effects of maternal fasting during embryonic organogenesis in the rat, *J. Repro. Fertil.*, 60, 383, 1980.
72. Ikemi, N., Imada, J., Goto, T., Shimazu, H., and Yasuda, M., Effects of food restriction on the fetal development during major organogenesis in rats, *Cong. Anom.*, 33, 363, 1993.
73. Petrere, J.A., Rohn, W.R., Grantham II, L.E., and Anderson, J.A., Food restriction during organogenesis in rabbits: effects on reproduction and the offspring, *Fundam. Appl. Toxicol.*, 21, 517, 1993.
74. Matsuzawa, T., Nakata, M., Goto, I., and Tsushima, M., Dietary deprivation induces fetal loss and abortions in rabbits, *Toxicology*, 22, 255, 1981.
75. Clark, R.L., Robertson, R.T., Peter, C.P., Bland, J.A., Nolan, T.E., Oppenheimer, L., and Bokelman, D.L., Association between adverse maternal and embryo/fetal effects in norfloxacin-treated and food-deprived rabbits, *Fundam. Appl. Toxicol.*, 7, 272, 1986.
76. James, W.H., Evidence that mammalian sex ratios at birth are partially controlled by parental hormone levels at the time of conception, *J. Theor. Biol.*, 180, 271, 1996.
77. James, W.H., Exposure to chemicals, offspring sex ratios, and their relevance to teratology [letter], *Teratology*, 62, 75, 2000.
78. Menegola, E., Broccia, M.L., Di Renzo, F., and Giavini, E., Comparative study of sodium valproate-induced skeletal malformations using single or double staining methods, *Reprod. Toxicol.*, 16, 815, 2002.
79. Sucheston, M.E., Hayes, T.G., and Eluma, F.O., Relationship between ossification and body weight of the CD-1 mouse fetus exposed in utero to anticonvulsant drugs, *Terat. Carcinog. Mutagen.*, 6, 537, 1986.

80. Wise, L.D., Beck, S.L., Beltrame, D., Beyer, B.K., Chahoud, I., Clark, R.L., Clark, R., Druga, A.M., Feuston, M.H., Guittin, P., Henwood, S.M., Kimmel, C.A., Lindstrom, P., Palmer, A.K., Petrere, J.A., Solomon, H.M., Yasuda, M., and York, R.G., Terminology of developmental abnormalities in common laboratory animals, *Teratology*, 55, 249, 1997.
81. Wilson, J.G., Embryological considerations in teratology. In *Teratology: Principles and Techniques;* Wilson, J.G. and Warkany, J., Eds., The University of Chicago Press, Chicago, 1965, p. 251.
82. Stuckhardt, J.L. and Poppe, S.M., Fresh visceral examination of rat and rabbit fetuses used in teratogenicity testing, *Terat., Carcin. Mutagen.*, 4, 181, 1984.
83. Staples, R.E., Detection of visceral alterations in mammalian species [abstract], *Teratology*, 9, 37, 1974.
84. Barrow, M.V. and Taylor, W.J., A rapid method for detecting malformations in rat fetuses, *J. Morphol.*, 127, 291, 1969.
85. Sawin, P. and Crary, D., Morphogenetic studies of the rabbit. X. Racial variations in the gall bladder, *Anat. Rec.*, 110, 573, 1951.
86. Pitt, J.A., Snyder, L.T., Smith, B.L., Varsho, B.J., Nemec, M.D., and Holson, J.F., An x-ray study of the incidence of supernumary ribs and presacral vertebrae in adults rats, mice and rabbits, *Teratology*, 65, 328, 2002.
87. Chernoff, N., Rogers, J.M., Turner, C.I., and Francis, B.M., Significance of supernumerary ribs in rodent developmental toxicity studies: postnatal persistence in rats and mice, *Fundam. Appl. Toxicol.*, 17, 448, 1991.
88. Foulon, O., Girard, H., Pallen, C., Urtizberea, M., Repetto-Larsay, M., and Blacker, A.M., Induction of supernumerary ribs with sodium salicylate, *Reprod. Toxicol.*, 13, 369, 1999.
89. Wickramaratne, G.A. de S., The post-natal fate of supernumerary ribs in rat teratogenicity studies, *J. Appl. Toxicol.*, 8, 91, 1988.
90. Foulon, O., Jaussely, C., Repetto, M., Urtizberea, M., and Blacker, A.M., Postnatal evolution of supernumerary ribs in rats after a single administration of sodium salicylate, *J. Appl. Toxicol.*, 20, 205, 2000.
91. Green, E.L., The inheritance of a rib variation in the rabbit, *Anat. Rec.*, 74, 47, 1939.
92. Greenaway, J.G., Partlow, G.D., Gonsholt, N.L., and Fisher, R.S., Anatomy of the lumbosacral spinal cord in rabbits, *J. Amer. Animal Hosp. Assoc.*, 37, 27, 2001.
93. Beck, S.L., Assessment of adult skeletons to detect prenatal exposure to trypan blue in mice, *Teratology*, 28, 271, 1983.
94. Moore, K. and Persaud, T., Eds., *The Developing Human: Clinically Oriented Human Embryology*, 6th ed., W.B. Saunders, Philadelphia, 1998, p. 414.
95. Holson, J.F., Kimmel, C.A., Hogue, C.I., and Carlo, G.L., Suitability of experimental studies for predicting hazards to human development, in *Proceedings of the 1981 Annual Winter Meeting of the Toxicology Forum*, Arlington, VA, 1981.
96. Holson, J.F., DeSesso, J., Jacobson, C., and Farr, C., Appropriate use of animal models in the assessment of risk during prenatal development: an illustration using inorganic arsenic, *Teratology*, 62, 51, 2000.
97. National Academy of Sciences, *Evaluating Chemical and Other Agent Exposures for Reproductive and Developmental Toxicity*, National Academy Press, Washington, DC, 2001.
98. Takayama, S., Akaike, M., Kawashima, K., Takahashi, M., and Kurokawa, Y., A collaborative study in Japan on optimal treatment period and parameters for detection of male fertility disorders induced by drugs in rats, *J. Am. Coll. Toxicol.*, 14, 266, 1995.
99. Barton, H.A. and Andersen, M.E., Endocrine active compounds: from biology to dose response assessment, *Crit. Rev. Toxicol.*, 28, 363, 1998.
100. Sterner, T.R. and Mattie D.R., Perchlorate literature review and summary: developmental effects, metabolism, receptor kinetics and pharmacological uses, *NTIS Technical Report (NTIS/ADA367421)*, 1998.
101. Kato, M., Fukunishi, K., Ikegawa, S., Higuchi, H., Sato, M., Horimoto, M., and Ito, S., Overview of studies on rat sperm motion analysis using a Hamilton-Thorne Sperm Analyzer — collaborative working study, *J. Toxicol. Sci.*, 26, 285, 2001.
102. Linder, R.E., Strader, L.F., Slott, V.L., and Suarez, J.D., Endpoints of spermatotoxicity in the rat after short duration exposures to fourteen reproductive toxicants, *Repro. Toxicol.*, 6, 491, 1992.

103. O'Connor, J.C., Davis, L.G., Frame, S.R., and Cook, J.C., Evaluation of a Tier I screening battery for detecting endocrine-active compounds (EACs) using the positive controls testosterone, coumestrol, progesterone, and RU486, *Toxicol. Sci.*, 54, 338, 2000.
104. Russell, L.D., Ettlin, R., Sinha Hikim, A.P., and Clegg, E.D., *Histological and Histopathological Evaluation of the Testis*, Cache River Press, Clearwater, FL, 1990.
105. Long, J.A. and Evans, H.M., The oestrous cycle in the rat and its associated phenomena, *Mem. Univ. Calif.*, 6, 1, 1922.
106. Foster, P.M.D., Reproductive performance, in *An Evaluation and Interpretation of Reproductive Endpoints for Human Health Risk Assessment*, Daston, G. and Kimmel, C., Eds., International Life Sciences Institute, Washington, DC, 1998, p. 5.
107. *NTP-CERHR Expert Panel Report on the Reproductive and Developmental Toxicity of 1-Bromopropane*, National Toxicology Program (NTP), Center for the Evaluation of Risks to Human Reproduction (CERHR), 2002.
108. Horimoto, M., Takeuchi, K., Iijima, M., and Tachibana, M., Reproductive and developmental toxicity studies with amlodipine in rats and rabbits, *Oyo Yakuri*, 42, 167, 1991.
109. Zippelius, H.M. and Schleidt, W.M., Ultraschallante bei jungen Mausen [Ultrasounds in young mice], *Naturwissenschaften*, 43, 502, 1956.
110. Noirot, E. and Pye, D., Sound analysis of ultrasonic distress calls of mouse pups as a function of their age, *Animal Behavior*, 17, 340, 1969.
111. Nicholas, K.R. and Hartmann, P.E., The foetoplacental unit and the initiation of lactation in the rat, *Aust. J. Biol. Sci.*, 34, 455, 1981.
112. Orbach, J., Spontaneous ejaculation in rat, *Science*, 134, 1072, 1961.
113. Kihlström, J.E., Diurnal variation in the spontaneous ejaculations of the male albino rat, *Nature*, 209, 513, 1966.
114. Beach, F.A., Variables affecting "spontaneous" seminal emission in rats, *Physiol. Behav.*, 15(1), 91, 1975.
115. Orbach, J., Miller, M., Billimoria, A., and Solhkhah, N., Spontaneous seminal ejaculation and genital grooming in rats, *Brain Res.*, 5, 520, 1967.
116. Gallavan, Jr., R.H., Holson, J.F., Stump, D.G., Knapp, J.F., and Reynolds, V.L., Interpreting the toxicologic significance of alterations in anogenital distance: potential for confounding effects of progeny body weights, *Repro. Toxicol.*, 13, 383, 1999.
117. Clark, R.L., Endpoints of reproductive system development, in *An Evaluation and Interpretation of Reproductive Endpoints for Human Health Risk Assessment*, Daston, G. and Kimmel, C.A., Eds., International Life Sciences Institute, Washington, DC, 1998, p. 10.
118. Ashby, J. and Lefevre, P.A., The peripubertal male rat assay as an alternative to the Hershberger castrated male rat assay for the detection of anti-androgens, oestrogens and metabolic modulators, *J. Appl. Toxicol.*, 20, 35, 2000.
119. Mocarelli, P., Gerthoux, P.M., Ferrari, E., Patterson, Jr., D.G., Kieszak, S.M., Brambilla, P., Vincoli, N., Signorini, S., Tramacere, P., Carreri, V., Sampson, E.J., Turner, W.E., and Needham, L.L., Paternal concentrations of dioxin and sex ratio of offspring, *Lancet*, 355(9218), 1838, 2000.
120. Ryan, J.J., Amirova, Z., and Carrier, G., Sex ratios of Russian pesticide producers exposed to dioxin, *Environ. Health Persp.*, 110, A 699, 2002.
121. Schnorr, T.M., Lawson, C.C., Whelan, E.A., Dankovic, D.A., Deddens, J.A., Piacitelli, L.A., Reefhuis, J., Sweeney, M.H., Conally, L.B., and Fingerhut, M.A., Spontaneous abortion, sex ratio, and paternal occupational exposure to 2,3,7,8-tetrachlorodibenzo-*p*-dioxin, *Environ. Health Perspec.*, 109, 1127, 2001.
122. International Conference on Harmonisation (ICH), Tripartite Guideline on Detection of Toxicity to Reproduction for Medicinal Products, *Fed. Regist.*, Section 4.1.2, September 22, 1994.
123. U.S. Environmental Protection Agency, Health Effects Test Guideline, Developmental Neurotoxicity Study, OPPTS 870.6300, August 1998.
124. Organization for Economic Cooperation and Development, Guideline for the testing of chemicals, Developmental Neurotoxicity Study, Proposal for a New Guideline 426, September 2003 Draft.
125. Bolon, B., Bucci, J.J., Warbritton, A.R., Chen, J.J., Mattison, P.R., and Heindel, J.J., Differential follicle counts as a screen for chemically-induced ovarian toxicity in mice: results from continuous breeding bioassays, *Fundam. Appl. Toxicol.*, 39, 1, 1997.

126. Bucci, T.J., Bolon, B., Warbritton, A.R., Chen, J.J., and Heindel, J.J., Influence of sampling on the reproducibility of ovarian follicle counts in mouse toxicity studies, *Reprod. Toxicol.*, 11, 689, 1997.
127. Holson, J.F., Interpretation of low-incidence findings in developmental and reproductive toxicity studies, presented at the *2002 Annual Winter Meeting of the Toxicology Forum*, Washington, DC, 2002.
128. Cook, J.C., Jacobson, C.F., Gao, F., Tassinari, M.S., Hurtt, M.E., and DeSesso, J.M., Analysis of the nonsteroidal anti-inflammatory drug literature for potential developmental toxicity in rats and rabbits, *Birth Defects Res. Part B Dev. Reprod. Toxicol.*, 68, 5, 2003.
129. Nisbet, I.C.T. and Karch, N.J., *Chemical Hazards to Human Reproduction*, Noyes Data Corp., Park Ridge, IL, 1983.
130. Hemminki, K. and Vineis, P., Extrapolation of the evidence on teratogenicity of chemicals between humans and experimental animals: chemicals other than drugs, *Teratog. Carcinog. Mutagen.*, 5, 251, 1985.
131. Kimmel, C.A., Rees, D.C., and Francis, E.Z., Eds., Proceedings of the Workshop on the Qualitative and Quantitative Comparability of Human and Animal Developmental Neurotoxicity, *Neurotoxicol. Teratol.*, 12, 173, 1990.
132. Newman, L.M., Johnson, E.M., and Staples, R.E., Assessment of the effectiveness of animal developmental toxicity testing for human safety, *Reprod. Toxicol.*, 7, 359, 1993.
133. March of Dimes Birth Defects Foundation, 1996, *Birth Defects and Infant Mortality: A National and Regional Profile*, March of Dimes Birth Defects Foundation, New York.
134. Frohberg, H., Gleich, J., and Unkelbach, H.D., Reproduction toxicological studies on cefazedone, *Arzneim. Forsch.*, 29, 419, 1979.
135. Duffy, J.Y., Overmann, G.J., Keen, C.L., and Daston, G.P., Cardiac abnormalities induced by zinc deficiency are associated with alterations in the expression of genes regulated by zinc-finger transcription factors, *Teratology*, 59, 382, 1999.
136. Stump, D.G., Holson, J.F., Kirkpatrick, D.T., Reynolds, V.L., Siddiqui, W.H., and Meeks, R.G., Evaluation of octamethylcyclotetrasiloxane (D4) in a 2-generation reproductive toxicity study in rats, *The Toxicologist* 54, 370, 2000.
137. Youreneff, M.A., Singh, A.R., Hazelette, J.R., Yau, E.T., and Traina, V.M., Teratogenic evaluation of benazepril hydrochloride in mice, rats, and rabbits, *J. Am. Coll. Toxicol.*, 9, 647, 1990.
138. Spence, S.G., Allen, H.L., Cukierski, M.A., Manson, J.M., Robertson, R.T., and Eydelloth, R.S., Defining the susceptible period of developmental toxicity for the the AT_1-selective angiotensin II receptor antagonist Losartan in rats, *Teratology*, 51, 367, 1995.
139. Spence, S.G., Cukierski, M.A., Manson, J.M., Robertson, R.T., and Eydelloth, R.S., Evaluation of the reproductive and developmental toxicity of the AT_1-selective angiotensin II receptor antagonist Losartan in rats, *Teratology*, 51, 383, 1995.
140. Sweeney, L.M., Tyler, T.R., Kirman, C.R., Corley, R.A., Reitz, R.H., Paustenbach, D.J., Holson, J.F., Whorton, M.D., Thompson, K.M., and Gargas, M.L., Proposed occupational exposure limits for select ethylene glycol ethers using PBPK models and Monte Carlo simulations, *Toxicol. Sci.*, 62, 124, 2001.
141. Terry, K.K., Elswick, B.A., Stedman, D.B., and Welsch, F., Developmental phase alters dosimetry-teratogenicity relationship for 2-methoxyethanol in CD-1 Mice, *Teratology*, 49, 218, 1994.
142. Clarke, D.O., Elswick, B.A., Welsch, F., and Conolly, R.R., Pharmacokinetics of 2-methoxyethanol and 2-methoxyacetic acid in the pregnant mouse: a physiologically based mathematical model, *Toxicol. Appl. Pharm.*, 121, 239, 1993.
143. Nau, H., Teratogenicity of isotretinoin revisited: species variation and the role of all-trans-retinoic acid, *J. Am. Acad. Dermatol.*, 45, S183, 2001.
144. Tzimas, G., Thiel, R., Chahoud, I., and Nau, H., The area under the concentration-time curve of all-*trans*-retinoic acid is the most suitable pharmacokinetic correlate to the embryotoxicity of this retinoid in the rat, *Toxicol. Appl. Pharm.*, 143, 436, 1997.
145. Collins, M.D., Tzimas, G., Hummler, H., Bürgin, H., and Nau, H., Comparative teratology and transplacental pharmacokinetics of all-*trans*-retinoic acid, 13-*cis*-retinoic acid, and retinyl palmitate following daily administration in rats, *Toxicol. Appl. Pharm.*, 127, 132, 1994.
146. Nau, H., Embryotoxicity and teratogenicity of topical retinoic acid, *Skin Pharmacol.*, 6(Suppl.1), 35, 1993.
147. U.S. Environmental Protection Agency, *Standard Evaluation Procedure, Developmental Toxicity Studies*, Health Effects Division, Office of Pesticides Programs, Washington, DC, 1993.

148. Arlitt, A.H., The effect of alcohol on the intelligent behavior of the white rat and its progeny, *Psychol. Monogr.*, 26, 1, 1919.
149. Thiersch, J.B. and Phillips, F.S., Effect of 4-amino-pteroylglutamic acid (aminopterin) on early pregnancy, *Proc. Soc. Exp. Biol. Med.*, 74, 204, 1950.
150. Shoeneck, F.J., Cigarette smoking in pregnant women, *N.Y. State J. Med.*, 41, 1945, 1941.
151. Greene, R.R., Burrill, M.W., and Ivy, A.C., Experimental intersexuality. The effects of estrogens on the antenatal sexual development of the rat, *Am. J. Anat.*, 67, 305, 1940.
152. Geber, W.F. and Schramm, L.C., Comparative teratogenicity of morphine, heroin and methadone in the hamster, *Pharmacologist,* 11, 248, 1969.
153. Russell, L.B, X-ray induced developmental abnormalities in the mouse and their use in the analysis of embryological patterns, *J. Exp. Zool.*, 114, 545, 1950.
154. Matsumoto, H., Koya, G., and Takeuchi, T., Fetal Minamata disease: A neuropathological study of two cases of intrauterine intoxication by methyl mercury compound, *J. Neuropathol. Exp. Neurol.*, 24, 563, 1965.
155. Taki, I., Hisanaga, S., and Amagase, Y., Report on Yusno (chlorobiphenyls poisoning). Pregnant women and their fetuses, *Fukuoka Acta Med.*, 60, 471, 1969.
156. van Wagenen, G. and Hamilton, J.B., The experimental production of pseudohermaphroditism in the monkey, *Essays Biol.*, 583, 1943.
157. Lenz, W., Kindliche Missbildungen nach Medikament wahrend der Draviditat?, *Dtsch. Med. Wochenschr.*, 86, 2555, 1961.
158. McBride, W.G., Thalidomide and congenital abnormalities, *Lancet*, 2, 1358, 1961.
159. Holson, J.F., Estimation and extrapolation of teratologic risk, in *Proceedings of NATO Conference on* in vitro *Toxicity Testing of Environmental Agents: Current and Future Possibilities*, NATO, Monte Carlo, Monaco, 1980.
160. Brown, N.A. and Fabro, S.F., The value of animal teratogenicity testing for predicting human risk, *Clin. Obstet. Gynecol.*, 26(2), 467, 1983.
161. Buelke-Sam, J. and Mactutus, C., Workshop on the qualitative and quantitative comparability of human and animal developmental neurotoxicity: testing methods in developmental neurotoxicity for use in human risk assessment, *Neurotoxicol. Teratol.*, 12, 269, 1990.
162. Shepard, T.H., *Catalog of Teratogenic Agents*, 8th ed., The Johns Hopkins University Press, Baltimore, 1995.
163. Schardein, J.L., *Chemically Induced Birth Defects*, 3rd ed., Marcel Dekker, New York, 2000.
164. Barlow, S.M. and Sullivan, F.M., *Reproductive Hazards of Industrial Chemicals*, Academic Press, London, 1982.
165. U.S. Congress, Office of Technology Assessment, *Reproductive Health Hazards in the Workplace*, OTA-BA-266, U.S. Government Printing Office, Washington, DC, 1985, p. 161.
166. U.S. Environmental Protection Agency, *Health Effects Test Guidelines: Prenatal Developmental Toxicity Study,* Office of Prevention, Pesticides and Toxic Substances (OPPTS) 870.3700, 1998.
167. U.S. Food and Drug Administration, International Conference on Harmonisation; Guideline on detection of toxicity to reproduction for medicinal products. *Fed. Regist.*, 59(No. 183), 1994.
168. Organization for Economic Cooperation and Development, *Guidelines for Testing of Chemicals*. Section 4, No. 414: *Teratogenicity*, adopted 22 January 2001.
169. U.S. Food and Drug Administration, Office of Food Additive Safety, IV.C.9.b., Guidelines for developmental toxicity studies, *Redbook 2000—Toxicological Principles for the Safety Assessment of Food Ingredients*, College Park, MD, 2000.
170. U.S. Food and Drug Administration, Guidance for Industry—Considerations for Reproductive Toxicity Studies for Preventive Vaccines for Infectious Disease Indications, Rockville, MD, 2000 (Draft).
171. Japanese Ministry of Agriculture, Forestry and Fisheries (MAFF), *Developmental Toxicity Testing Guidelines*, 1985.
172. U.S. Food and Drug Administration, Office of Food Additive Safety, IV.C.9.a., Guidelines for reproduction studies, *Redbook 2000—Toxicological Principles for the Safety Assessment of Food Ingredients*, College Park, MD, 2000.
173. OECD, *OECD Guideline for Testing of Chemicals, Proposal for Updating Guideline 416, Two-Generation Reproduction Study*, OECD, Paris, France, 2001.

174. U.S. Environmental Protection Agency, *Health Effects Test Guidelines; Reproduction and Fertility Effects*, Office of Prevention, Pesticides and Toxic Substances (OPPTS) 870.3800, 1998b.
175. ICH Expert Working Group, *ICH Harmonised Tripartite Guideline, Detection of Toxicity to Reproduction for Medicinal Products, S5A*, International Conference on Harmonisation, Geneva, Switzerland, 1994.
176. ICH Expert Working Group, *Maintenance of the ICH Guidelines on Toxicity to Male Fertility, an Addendum to the ICH Harmonised Tripartite guideline, Detection of Toxicity to Reproduction for Medicinal Products, S5B(M)*, International Conference on Harmonisation, Geneva, Switzerland, 1994.

CHAPTER 10

Testing for Reproductive Toxicity*

Robert M. Parker

CONTENTS

- I. Introduction 428
  - A. Brief Overview of DART Study Guidelines 428
    1. Fertility Studies 429
    2. Developmental Toxicity Studies 430
    3. Reproduction Studies 430
    4. Reproductive Assessment by Continuous Breeding 431
    5. Developmental Neurotoxicity Study 432
    6. Combined Repeated Dose Toxicity Study with the Reproduction and Developmental Toxicity Screening Test 433
    7. Dominant Lethal Study 435
  - B. Endpoints for Reproductive Studies 436
    1. Couple-Mediated Endpoints 436
    2. Male-Specific Endpoints and Paternally Mediated Effects 436
    3. Female-Specific Endpoints and Maternally Mediated Effects 436
- II. Reproductive Performance 437
  - A. Overview 437
  - B. Treatment Duration for Males in Fertility and Reproduction Studies 437
  - C. Cohabitation and Evidence of Mating 438
    1. Vaginal Plugs 438
    2. Vaginal Smears 438
  - D. Reproductive Indices 439
    1. Male Mating and Fertility Index 439
    2. Female Fertility Index 440
    3. Mean Gestational Length 442
    4. Litter Size 442
    5. Live-Birth Index 443
    6. Survival Index 443
    7. Sex Ratio 444
- III. Male Reproductive Toxicology 444
  - A. Biology 444
  - B. Observations 445
  - C. Male Body Weight and Food Consumption 445
  - D. Male Necropsy 445

* The views expressed in this chapter are those of the author and do not necessarily reflect the views of Hoffmann-LaRoche, Inc.

- E. Male Reproductive Organ Weight 446
- F. Histopathological Evaluation of the Male Reproductive Organs 446
  1. Testis 447
  2. Epididymides, Accessory Sex Glands and Pituitary 448
- G. Sperm Evaluations 449
  1. Sperm Number or Sperm Concentration 449
  2. Computer-Assisted Sperm Analysis 450
  3. Cauda Epididymal Sperm Motility 454
  4. Sperm Morphology 456
- H. Seminiferous Epithelium Staging 457
  1. Spermatogenesis 457
  2. The Staging of Spermatogenesis 458

IV. Female Reproductive Toxicology 458
- A. Biology 458
- B. Observations 459
- C. Female Body Weight 459
- D. Female Necropsy 459
  1. Gestation Day 13 Uterine Examination (Segment I Studies) 459
  2. Gestation Day 20 or 21 Term Cesarean Section (Segment II Studies) 459
  3. End of Lactation (Segment III Studies) 460
- E. Female Reproductive Organ Weight 460
  1. Ovary 460
  2. Uterus, Oviducts, Vagina, and Pituitary Gland 460
- F. Histopathological Evaluation of the Female Reproductive Organs 461
  1. Ovaries 461
  2. Uterus 461
  3. Oviducts 461
  4. Vagina and External Genitalia 461
  5. Pituitary 462
  6. Mammary Gland and Lactation 462
  7. Reproductive Senescence 463
  8. Developmental and Pubertal Alterations 463
- G. Vaginal Cytology 463
  1. Description 463
  2. Vaginal Lavage Methodology 464
  3. Estrous Cycle Staging 465
  4. Estrous Cycle Evaluation 465
- H. Corpora Lutea Counts 466
- I. Uterine Evaluation 467
  1. Assessment of Resorptions 467
  2. Assessment of Implantation in "Apparently Nonpregnant" Dams 467
- J. Oocyte Quantitation 467

V. Parturition and Litter Evaluations 468
- A. Parturition 468
- B. Milk Collection 469
- C. Litter Evaluations 469
  1. Litter Observations 469
  2. Gender Determination 470
  3. Viability 470
  4. External Alterations 470
  5. Missing Pups 470

		6. Pups Found Dead 470
		7. Dam Dies before Scheduled Sacrifice 470
		8. All Pups in a Litter Die 471
		9. Gender Determination Errors 471
VI.	Pup Evaluations 471
	A.	Pup Body Weights 471
		1. Pup Birth Weight 471
		2. Postnatal Pup Body Weights 471
		3. Crown-Rump Length 471
		4. Anogenital Distance 472
	B.	Developmental Landmarks 472
		1. Balanopreputial Separation 472
		2. Vaginal Patency 473
		3. Day of First Estrus 473
		4. Pinna Unfolding 473
		5. Hair Growth 473
		6. Incisor Eruption 474
		7. Eye Opening 474
		8. Nipple Retention 474
		9. Testes Descent 474
	C.	Behavioral or Reflex Testing 474
		1. Surface-Righting Reflex 474
		2. Cliff Avoidance 474
		3. Forelimb Placing 475
		4. Negative Geotaxis Test 475
		5. Pinna Reflex 475
		6. Auditory Startle Reflex 475
		7. Hind Limb Placing 475
		8. Air Righting Reflex 475
		9. Forelimb Grip Test (Categorical) 476
		10. Pupil Constriction Reflex 476
	D.	Examination of Rodents for Hypospadias 476
		1. In-life Procedure for Male Rodents 476
		2. Postmortem Procedure for Male Rodents 477
		3. Postmortem Procedure for Female Rodents 477
VII.	Functional Observation Battery 477
	A.	Adult FOB 477
		1. In-Cage Observations 477
		2. Removing from Cage Observations 478
		3. Standard Open Field Arena Observations 478
		4. Manipulations in the Standard Open Field Arena 478
		5. Grip Strength Manipulations 478
		6. Foot Splay 478
		7. Pupillary Response 479
		8. Body Temperature 479
	B.	FOB in Preweanling Rats 479
		1. In Cage Observations 479
		2. Removing from Cage Observations 479
		3. Standard Open Field Arena Observations 479
		4. Manipulations in the Standard Open Field Arena 480
References 480

I. INTRODUCTION

The objective of this chapter is to provide the methodologies for evaluations used during the performance of reproductive toxicity studies. Traditional developmental and reproductive toxicity (DART) studies in rodents are a major source of data on the effects of potential developmental and reproductive toxicants. Reproductive toxicities include structural and functional alterations that may affect reproductive competence (fertility, parturition, and lactation). Evaluations of fertility, pregnancy, lactation, and maternal and paternal behaviors provide measures of the consequences of reproductive injury. These evaluations provide information concerning gonadal function, estrus cyclicity, mating behavior, conception, parturition, lactation, weaning, and the growth and development of the offspring. Developmental toxicities are generally those that affect the F_1 or F_2 generations. The four manifestations of developmental toxicity as described by Wilson[1] are mortality, dysmorphogenesis (structural alterations), alterations to growth, and functional alterations. Mortality due to developmental toxicity may occur at any time from conception to adulthood. Dysmorphogenic effects are generally seen as malformations or variations to the skeleton or viscera of the offspring. Alterations to growth are generally seen as growth retardation, although excessive growth or early maturation may also be seen. Functional alterations could include any persistent change in normal physiologic or biochemical function, but typically only reproductive function and developmental neurobehavioral effects are measured in these studies.

A. Brief Overview of DART Study Guidelines

The purposes of DART studies are to determine whether a test substance has the potential to cause adverse effects on the male and female reproductive system or the developing conceptus, and to determine a developmental and/or reproductive no observable adverse effect level (NOAEL) for the test substance. The developmental or reproductive NOAEL is the highest treatment level tested that shows no developmental or reproductive effects, respectively. It should be noted that DART study designs were not developed to cover every male and female reproductive parameter, developmental parameter, specific target organ, or mechanism of toxicity. For example, traditional DART studies do not provide information on maternal endocrine status, the levels of an agent or its metabolites in the milk and the fetus or pup, placental pathology, or reproductive senescence.

The reproductive cycle has historically been divided into three segments for DART studies. Segment I study (also known as reproduction and fertility studies) covers the period of premating, cohabitation and mating, and early pregnancy through implantation (gestation day [GD] 6 in the rat). Segment II study (also known as teratology or developmental toxicity studies) covers pregnancy from implantation through major organogenesis (closure of the hard palate; GD 15 in the rat) up to the day before delivery. Segment III study (perinatal and postnatal studies) covers late pregnancy and postnatal development (usually until weaning at postnatal day [PND] 21). Multigenerational studies (studies covering two to three generations) are required for chemicals that pose a low-level chronic exposure risk in humans. Combination studies such as Segment I and II and Segment II and III studies are acceptable as long as all required parameters and minimum study requirements for each segment are met. Another example of an acceptable combinatory study is the 90-Day Repeated Dose Subchronic Study with a One-Generation Reproduction Study.

DART study guidelines have been promulgated by the U.S. Food and Drug Administration (FDA),[2–7] the U.S. Environmental Protection Agency (EPA),[8–11] the Japan Ministry of Agriculture, Forestry and Fisheries (MAFF),[12,13] Canada,[14,15] Great Britain,[16] World Health Organization,[17] and the Organization for Economic Cooperation and Development (OECD).[18–22] The ICH (International Conference on Harmonization of the Technical Requirements for Registration of Pharmaceuticals for Human Use)[6,7,23] has been concerned with unifying the study guideline requirements for pharmaceutical safety assessment for the European Union, Japan, and the FDA. Guideline compliant reproductive and fertility studies must be performed in accordance with associated Good Laboratory

TESTING FOR REPRODUCTIVE TOXICITY

Practice Regulations,[24–26] and biosafety[27] and animal welfare guidelines.[28,29] A number of these guidelines and additional references for reproduction studies can be found in the appendixes for Chapter 19.

1. Fertility Studies

Fertility studies[4–7,10] (also known as Segment I studies) test for toxic effects resulting from treatment with the test substance before mating (males and/or females), during cohabitation and mating, and until implantation (GD 6). For females this study design detects effects on the libido, estrous cycle, ovulation, mating behavior, oviductal transport, development of preimplantation stages of the embryo, and implantation. For males this study design permits detection of functional effects on libido, mating, and sperm quality (motility, count, and morphology) that may not be detected by histological examinations of the male reproductive organs alone. Males are treated with the test substance 2, 4, or 10 weeks (depending upon guideline requirements) prior to cohabitation and mating, and until termination. Females are treated 2 weeks prior to cohabitation, through cohabitation and mating, and until implantation (usually GD 6). The dam is killed on either GD13 or 21 and necropsied, and the fetuses are evaluated for viability (see Figure 10.1). Reproductive endpoints evaluated include mating, pregnancy rate, implantation sites, conceptus viability, corpora lutea, pre- and postimplantation losses, and litter size. The Combined Male and Female Fertility Study (FDA Segment I and ICH s5 and s5a) guidelines are described in detail in Appendix 1 of Chapter 19. In male or female fertility studies, the test substance is administered only to males or females, respectively.

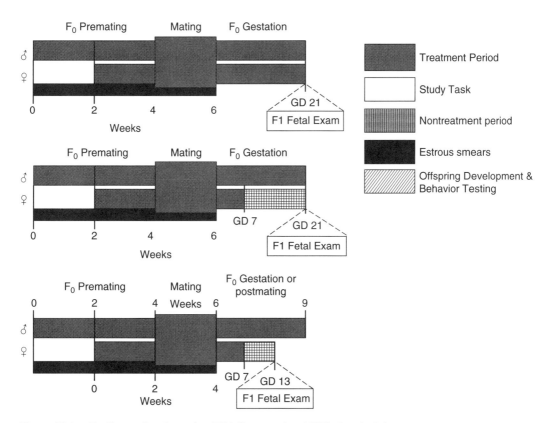

Figure 10.1 Fertility study schematics (FDA Segment I and ICH s5 and s5a).

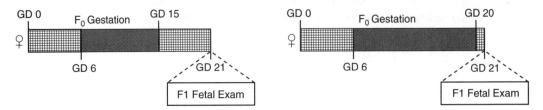

Figure 10.2 Developmental toxicity study schematics (FDA Segment II and EPA, OPPTS 870.3700 and OECD 414).

2. Developmental Toxicity Studies

Developmental toxicity[9,14,17] or teratology studies (also known as Segment II studies) are used to detect adverse effects of treatment on the pregnant female and development of the embryo and fetus from implantation (GD 6) to closure of the hard palate (GD 15); however, more recently the treatment period has been extended to GD 20 (see Figure 10.2), with necropsy and fetal evaluation on GD 21. Rabbit teratology studies dose on GD 6 through 18 with necropsy and fetal evaluations performed on GD 29. The endpoints assessed include maternal toxicity (body weight, clinical and necropsy observations, feed consumption), embryo-fetal death (early and late resorptions), growth retardation, and structural alterations (malformations and variations). Functional deficits cannot be determined using this paradigm. See Tyl and Marr [30] for detailed discussion of the developmental toxicity study. This study is also described in Appendixes 1 and 2 of Chapter 19.

3. Reproduction Studies

The perinatal and postnatal study[4–7] (also known as the Segment III study) design detects adverse effects on the pregnant and lactating female and on development of the conceptus and the offspring following treatment of the female from implantation (GD 6) or the end of organogenesis (GD 15) through lactation and weaning (usually PND 21). This study is described in detail in Appendix 1 and 2 of Chapter 19.

The one-generation reproduction testing guideline (OECD 415[18]) requires treatment of the P_1 males at least 10 weeks and females at least 2 weeks prior to cohabitation, during cohabitation and mating, until termination after the mating period for the males, and through gestation and lactation until weaning at PND 21 for the females. F_1 pups are killed on PND 21 and evaluated (see Figure 10.3). This study is designed to provide information concerning the effects of a test substance on male and female reproductive capacity, including gonadal function, the estrous cycle, mating behavior, conception, gestation, parturition, lactation, and weaning, and on the growth and development of the offspring. The study may also provide information about the effects of the test substance on neonatal morbidity, mortality, and target organs in the offspring (F_1 generation), and preliminary data on prenatal and postnatal developmental toxicity. It may even serve as a guide for subsequent tests.

The two-generation reproduction testing guideline (OPPTS 870.3800[10] and OECD 416[22]) is designed to provide information concerning the effects of a test substance on the integrity and performance of the male and female reproductive systems, as described above for the one-generation study, while including an F_2 generation. The study design is similar to the one-generation study for the P_1 and F_1 generations; however, the test substance is administered postweaning to selected F_1 offspring (usually one male and one female per litter) during their growth into adulthood, during cohabitation and mating, and during gestation, parturition, and birth, continuing until the F_2 generation is weaned (PND 21) (see Figure 10.4). For further description of this study, refer to the OPPTS 870.3800 guideline presented in Appendix 2 of Chapter 19.

TESTING FOR REPRODUCTIVE TOXICITY

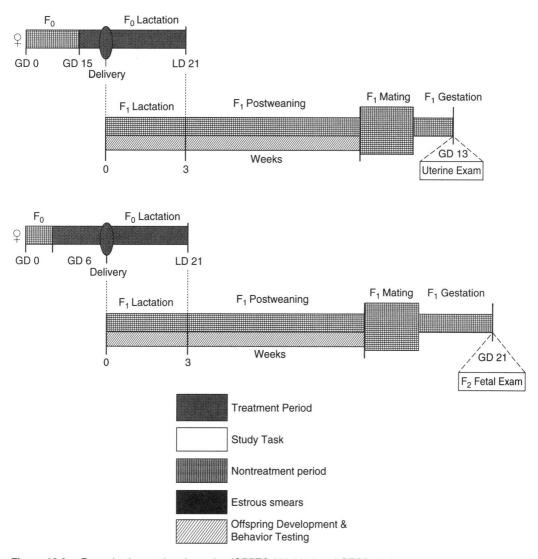

Figure 10.3 Reproduction study schematics (OPPTS 870.3800 and OECD 415).

4. Reproductive Assessment by Continuous Breeding

Data from reproductive assessment by continuous breeding (RACB) studies[31,32] identify hazards to reproduction, help to characterize their toxic effects, and indicate dose-response relationships. Each RACB study is separated into four tasks (see Figure 10.5).

Task 1 is a modified 4- to 5-week range-finding study consisting of a 1-week treatment period for five treatment levels and concurrent control group, followed by a 1-week cohabitation treatment period, a 3-week gestation treatment period, and treatment continuing until delivery.

Task 2 is the main study. A control plus three treatment levels are treated for 1 week prior to cohabitation, and for 18 weeks while the rats are cohoused as breeding pairs. Normally, four litters are delivered per adult pair during the 18-week cohabitation period. The last litter is weaned at PND 21, with one pup per gender per litter selected to continue in Task 4. If Task 2 results are negative, then Task 3 is not conducted and Task 4 is conducted with control and high treatment groups only. If Task 2 shows toxic effects, Task 3 is performed, and Task 4 requires all four groups (the control and three treatment groups).

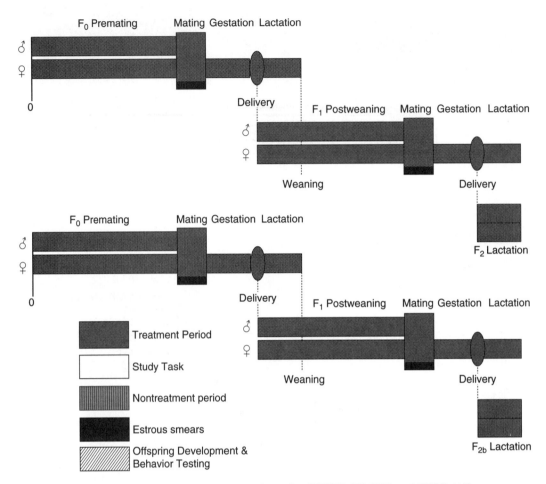

Figure 10.4 Two-generation reproduction study schematics (OPPTS 870.3800 and OECD 416).

Task 3, if required, is the crossover mating trial (no treatment required) that is performed to determine whether the effects are couple-mediated or single-sex mediated.

Task 4 is the F_1 generation test. Treatment with the test substance starts at weaning (PND 21), with each pup receiving the same treatment that the P_1 generation received. When 80 days of age, F_1 breeding pairs are cohabited for 1 week and separated when mating is confirmed. The F_1 female continues on treatment and delivers its litter (F_2), and the adult F_1 and neonatal F_2 generation animals are killed and necropsied.

5. Developmental Neurotoxicity Study

The developmental neurotoxicity study (OPPTS 870.6300[33] and OECD 426[34]) design provides information for use in evaluating the potential for neurotoxic effects in offspring after exposure to a test substance *in utero* and/or via maternal milk or by direct dosing of the pup during the lactation period. Female rats are administered the test substance once daily beginning on GD 6 and continuing through PND 21. If there is no lactational transfer of the test substance or active metabolites, then direct dosing of the pups beginning as early as PND 4 may be required. Female rats will be evaluated for adverse clinical signs observed during parturition, and for the duration of gestation, for litter size, live litter size, and pup viability at birth. F_1 generation males and females are tested as pups and as adults for spontaneous locomotor activity, learning and memory, reflex development, and time to sexual maturation. The functional observational battery (FOB) for pups and adults is

TESTING FOR REPRODUCTIVE TOXICITY

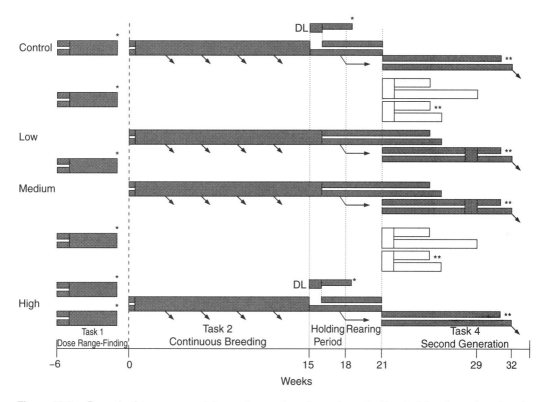

Figure 10.5 Reproductive assessment by continuous breeding schematic. See text for discussion of each task. The dark bars indicate treatment while the empty bars indicate no treatment. The two separate bars indicate single housing while the thicker bar indicates cohousing. The descending arrows indicate the birth of a litter. DL, dominant lethal; *, limited necropsy; **, full necropsy. (Adapted from Chapin and Sloan, *Environ. Health Perspect.*, 105(Suppl 1), 199, 1997.)

performed at specified intervals. Neurohistopathological examination and morphometry are performed on selected brain regions (see Figure 10.6).

6. Combined Repeated Dose Toxicity Study with the Reproduction and Developmental Toxicity Screening Test

The combined repeated dose toxicity study with the reproduction and developmental toxicity screening test (OECD 422[20] and OPPTS 870.3650[35]) is being required in the testing program of high production volume (HPV) chemicals. HPV chemicals are those chemicals that are produced in or imported to the United States in amounts over 1 million pounds per year. The Chemical Right-to-Know Initiative (1998) is the response to an EPA study that found that very little basic toxicity information is publicly available on most of the HPV commercial chemicals made and used in the United States. For further information concerning this program, go to http://www.epa.gov/opptintr/chemrtk/sidsappb.htm. This study design provides limited information on systemic toxicity and neurotoxicity following repeated treatments. It provides initial information on possible effects on male and female reproductive performance, such as gonadal function, mating behavior, conception, development of the conceptus, and parturition. It is not an alternative to, nor does it replace, the existing guidelines in OPPTS 870.3700, OPPTS 870.3800, OPPTS 870.6200, and OPPTS 870.7800, or the equivalent OECD studies. This study cannot be used as a definitive study for the setting of a developmental or reproductive NOAEL.

Males are treated for a minimum of 4 weeks (2 weeks precohabitation, during the mating period, and routinely 2 weeks postmating). Females are treated for 2 weeks precohabitation, during the

434 DEVELOPMENTAL REPRODUCTIVE TOXICOLOGY: A PRACTICAL APPROACH, SECOND EDITION

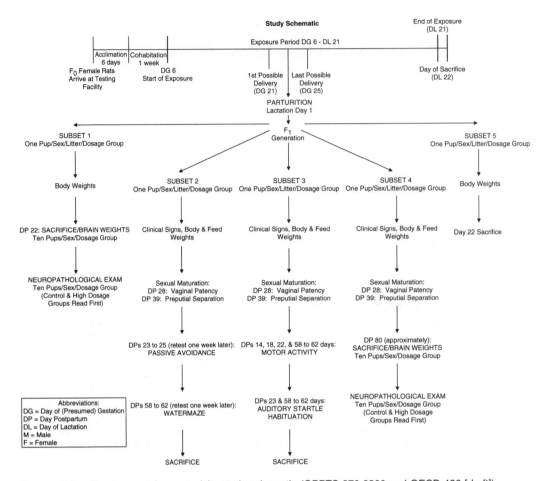

Figure 10.6 Developmental neurotoxicity study schematic (OPPTS 870.6300 and OECD 426 [draft]).

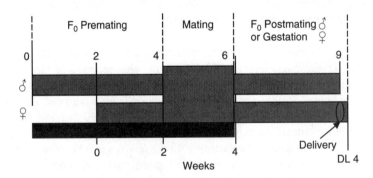

Figure 10.7 Combined repeated dose toxicity study with the reproduction and developmental toxicity screening test schematic (OPPTS 870.3650 and OECD 422).

mating period, during pregnancy, and at least 4 d after delivery (day before the scheduled kill). A FOB and a motor activity assessment are conducted pretest and after the fourth week of treatment. Clinical pathology data are collected in the fourth week. Tissues are collected at necropsy for a complete histopathological evaluation. Pups are sacrificed on day 4 of lactation (DL 4) (see Figure 10.7).

TESTING FOR REPRODUCTIVE TOXICITY

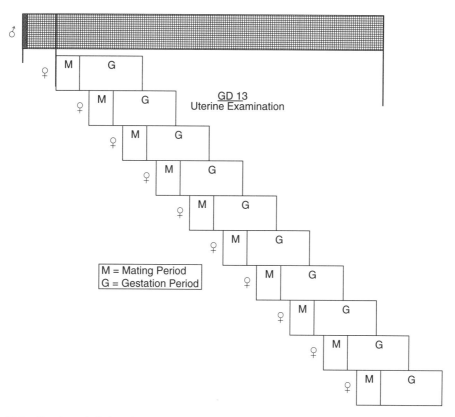

Figure 10.8 Dominant lethal study schematic (OPPTS 870.5450 and OECD 478).

7. Dominant Lethal Study

A dominant lethal mutation is a mutation occurring in a germ cell that does not cause dysfunction of the gamete but is lethal to the fertilized egg or developing embryo. Dominant lethal effects are generally thought to be the result of chromosomal damage (structural and numerical anomalies), but gene mutations and toxic effects cannot be excluded. In the dominant lethal study (OECD 478[36] and OPPTS 870.5450[37]), adult males receive a single treatment with the test substance (see Figure 10.8). Individual males are mated sequentially (usually at weekly intervals) to one or two virgin females per week. Females are cohoused with the males for at least the duration of one estrous cycle or until mating has been confirmed. The number of males (usually 25) in each treatment group should be sufficient to provide between 30 and 50 pregnant females per interval. A concurrent positive control (e.g., triethylenemelamine, cyclophosphamide, or ethyl methanesulfonate), a negative (vehicle) control, and three treatment levels should be used. The females are sacrificed in the second half of the pregnancy (usually around GD 13), and the uterine contents are examined for the numbers of implantation sites and numbers of live or dead conceptuses. The numbers of corpora lutea in the ovaries are also counted, and pre-and postimplantation losses are calculated. The mating index and the fertility index are also calculated (see below). The dominant lethal effect is based on a statistically significant decrease in the number of live embryos conceived with treated sperm compared with controls. A test substance that produces a statistically significant dose-related decrease in the number of live embryos at any one of the test points is considered mutagenic in the test system.

The number of mating periods should ensure that spermatogenesis is adequately covered. The duration of spermatogenesis is approximately 63 d in the rat; therefore the intervals need to extend over a total period of approximately 10 weeks. The weekly intervals (from the time of test substance administration to mating) permit sampling of germ cells exposed as mature sperm (effects seen in first week), spermatids (effects seen in the second through fourth weeks), spermatocytes (effects seen in the fifth through seventh weeks), and spermatogonial and stem cell stages of maturation (effects seen in the eight through tenth weeks).

B. Endpoints for Reproductive Studies

The endpoints of reproductive toxicity in test species can be divided into three categories: couple-mediated, male-specific, and female-specific endpoints.[38]

1. Couple-Mediated Endpoints

Couple-mediated endpoints are reproductive endpoints where both parents may have a contributory role. Couple-mediated endpoints from reproduction studies may include mating behaviors (mating rate, time to mating), pregnancy rate, pre- and postimplantation loss, number of implantation sites, gestation length, litter size (total and live), number of live and dead offspring, sex ratio, birth weight, external and internal malformations and variations, postnatal weight, postnatal structural and functional development, offspring survival, and offspring reproduction. Most regulatory reproduction and fertility guidelines specify cohabitation of treated male rats with treated female rats, complicating the resolution of gender-specific influences.

2. Male-Specific Endpoints and Paternally Mediated Effects

Data on the potential male reproductive toxicity of a test substance may be collected from many guideline studies including acute, subchronic, chronic, reproduction, and fertility studies. Male-specific endpoints that are determined in these studies include, but are not limited to: body weight; mating behavior (libido, mounting, erection, ejaculation); reproductive organ weights; histopathology of the testes, epididymides, seminal vesicles, prostate, and pituitary; sperm quality (count, viability, morphology, motility); sperm transport, maturation, and storage in the epididymis; production of seminal fluid; and production and secretion of hormones (LH, FSH, testosterone, estrogen, prolactin, etc.) in the pituitary-hypothalamus-gonadal axis.

Male-only treatment with a variety of agents has been shown to produce adverse effects in offspring including pre-and postimplantation loss, growth and behavioral deficits, and structural malformations.[39-41] Many of the chemicals reported to cause paternally mediated effects display genotoxic activity and may cause transmissible genetic alterations. However, other mechanisms of induction of male-mediated effects are also possible,[42] including nongenetic (e.g., presence of drug in seminal fluid) or epigenetic (e.g., an effect on imprinted gene expression from the paternal alleles, resulting in loss of gene function). Additional studies will be needed to clarify the mechanisms of paternal exposure associated with adverse effects on offspring. If a test substance is identified as causing a paternally mediated adverse effect on offspring in the test species, the effect should be considered an adverse reproductive effect.

3. Female-Specific Endpoints and Maternally Mediated Effects

The reproductive life cycle of the female may be divided into several phases (embryo-fetal, prepubertal, cycling adult, pregnant, lactating, and reproductively senescent). Detailed descriptions of these phases are beyond the scope of this chapter but are readily available.[43] Detailed descriptions of maternally mediated effects on development are beyond the scope of this chapter but are described by Hood and Miller.[44] Studies should be conducted to detect adverse effects occurring during any of these phases. Regulatory developmental and reproductive studies primarily detect adverse effects on the female's ability to become pregnant, on pregnancy outcome, and on offspring survival and development. Adverse alterations in the nonpregnant female reproductive system have been reported at treatment levels below those that result in reduced fertility or produce adverse effects on pregnancy or pregnancy outcomes.[45–47]

Female-specific endpoints of reproductive toxicity include, but are not limited to: organ weights, histopathology, estrous cycle onset, length, and other characteristics, ovarian follicular development, ovulation, mating behavior, production and secretion of hormones in the pituitary-hypothalamus-gonadal axis, fertility, gestation length, number of corpora lutea, pre- and postimplantation losses, parturition, lactation, nursing behavior, uterine decidualization and implantation, placentation, and senescence.

Unlike that of the male, the status of the adult female reproductive system fluctuates. In nonpregnant rats, the ovarian and uterine structures (and other reproductive organs) change throughout the estrous cycle. Although not cyclic, other changes normally occur during pregnancy, lactation, and return to cyclicity after lactation. These normal fluctuations may affect or confound the evaluation of female reproductive endpoints. It is important to be aware of the reproductive status of the female at necropsy, including estrous cycle stage. This facilitates interpretation of effects with endpoints, such as uterine weight and histopathology of the ovary, uterus, and vagina.[38]

II. REPRODUCTIVE PERFORMANCE

A. Overview

Although methods are available to quantify aspects of sexual performance in most test species, routinely no direct evaluation of sexual behavior is performed in most regulatory reproductive toxicology studies. Sexual behavior requires complex interactions between the neural, endocrine, and reproductive components; therefore, it is susceptible to disruption by a variety of toxic substances and pathologic conditions. Treatment-related alterations of sexual behavior in either sex of test species may represent potentially significant reproductive concern for humans.

B. Treatment Duration for Males in Fertility and Reproduction Studies

The fertility and reproduction study designs assume that data from previously conducted toxicity studies (e.g., histopathology and weights of reproductive organs from the acute and subchronic studies) will be used to select the appropriate treatment levels. The duration of treatment selected should be adequate for determining effects on spermatogenesis. Because of the duration of spermatogenesis, most fertility and reproductive studies require treatment of the males for approximately 10 weeks prior to mating.

Provided that no effects were found in a repeated dose toxicity study (of at least 4 weeks duration), a premating treatment interval of 2 weeks for females and 2 weeks for males should be used in ICH fertility studies.[6,7] The reduction of the premating treatment period by the ICH was based upon a reevaluation[48,49] of the basic research on the effects of male reproductive toxicants on the process of spermatogenesis. The ICH concluded: (1) that compounds that induce selective effects on male reproduction are rare; (2) that mating is an insensitive means of detecting effects on spermatogenesis in rodents; (3) that good pathological and histopathological examination of the male reproductive organs provides a more sensitive means of detecting effects on spermatogenesis; (4) that compounds that affect spermatogenesis almost invariably affect postmeiotic stages; and (5) that there are no conclusive examples of a male reproductive toxicant that could only be detected after treating the males for 9 to 10 weeks and then mating them. Studies conducted in Japan[50] indicate that a 2-week premating treatment period is adequate to detect toxicity in male reproductive organs. Because the 2-week repeated dose study was validated as being as effective as a 4-week repeated-dose study for determining male reproductive organ toxicity, treatment of the males for 2 weeks before mating is considered acceptable in ICH fertility studies. Justification for the length of the premating treatment period should be stated in the protocol. In fertility studies, treatment should continue throughout the cohabitation period until termination for males and at least through implantation for females. Multigenerational studies require a premating treatment period of at least 10 weeks for both P_1 and F_1 males and females.

C. Cohabitation and Evidence of Mating

Females are introduced into the male's cage in the afternoon, and they are checked for evidence of mating the following morning. When a receptive sexually competent female rat is placed with a sexually competent male rat, the male will attempt to mount within a relatively short time. The time from female placement into the cage until the first mount is defined as the *latency to first mount*. Prior to ejaculation, the male will attempt several (six to nine) mounts with intromission. In standard reproductive toxicology studies, the evidence of sexual receptivity and successful mating is based on the presence of copulatory plugs or sperm-positive vaginal lavage. However, these *post hoc* findings do not demonstrate that male sexual activity resulted in adequate sexual stimulation of the female rat.[51] Failure to adequately stimulate the female has been shown to impair sperm transport in the genital tract of female rats, thereby reducing the probability of successful fertilization.[52] The reduced fertility (as calculated in the fertility index) might be erroneously attributed to an effect on the spermatogenic process in the male or on fertility of the female, rather than on altered mating behavior. Effects on sexual behavior should be considered as adverse reproductive effects. However, impairment to sexual behavior that is secondary to more generalized systemic effects (e.g., severe acute intoxication, impaired motor activity, or general lethargy) should not be considered an adverse reproductive effect.

1. Vaginal Plugs

When the male rat intromits and ejaculates, the semen quickly coagulates, forming a copulatory plug. Secretions from the male accessory sex glands are responsible for copulatory plug formation, without which rodent fertility can be impaired. Copulatory plugs can remain *in situ* (in the vagina), or they can be dislodged from the vagina and fall into the cage pan. In male-treated fertility studies, fewer plugs found in the treated groups when compared with the control group indicates that male mating behavior may have been altered.

Mating (copulation) is confirmed by the presence of a copulatory plug or by the observation of spermatozoa in the vaginal smear ("sperm positive"). While in cohabitation, the females and their cages should be checked daily for the presence of copulatory plugs. (Note: The recording methodology presented here is an example only.) If a copulatory plug is observed *in situ* (in the vagina), a V is recorded on the cohabitation sheet. Those animals observed to have a copulatory plug *in situ* do not require a vaginal lavage. A V may also be recorded on the sample slide next to the animal number to show that no vaginal lavage sample was required for that animal. The presence of copulatory plug in the cage pan is recorded by writing a C on the cohabitation sheet. (Note: Copulatory plugs in the cage pan are not sufficient evidence that mating has occurred.) If only a cage pan plug is observed or when no evidence for mating is observed, then a vaginal lavage should be performed to detect the presence of sperm.

2. Vaginal Smears

Vaginal lavage for the presence of sperm can be performed at any time of day; however, morning is recommended. The samples should be collected at approximately the same time each day over the course of the study. Before any vaginal lavage samples are taken, the aliquot of fresh saline should be verified as contaminant-free (e.g., free of dirt, dust, spermatozoa, epithelial cells). The female rat is removed from the cage and its identification is verified. One or two drops (approximately 0.25 ml) of physiological (0.9%) saline are withdrawn from the aliquot into a new, clean dropping pipette. The tip of the dropping pipette is gently inserted into the vaginal canal. The pipette bulb is firmly but gently depressed to expel the saline from the pipette into the vagina. The saline is gently drawn back into the dropping pipette, and the pipette is removed from the vaginal canal. If the flush appears very clear, it is recommended that the lavage be repeated. Care must be taken not to insert the tip too far into the vaginal canal, because excessive stimulation of the cervix can result in pseudopregnancy (prolonged diestrus). The female is returned to her cage. The contents of the pipette are placed onto a clean glass slide. Each slide is numbered, a clear top and bottom of each slide is identified, and a data sheet is marked to indicate the location of each vaginal smear. All slides should be handled carefully to avoid mixing vaginal lavage samples from different animals. If samples become mixed, they should be discarded and the sampling procedure repeated for those animals. After vaginal lavage sampling has been completed for the day, the saline should be reevaluated for contamination that may have occurred during sample collection.

The vaginal lavage slides are allowed to air dry and are examined microscopically. When sperm are present in the vaginal smear, a positive (+) is recorded. When no sperm are present in the vaginal smear, a negative (−) is recorded. A second vaginal lavage sample may be required to more conclusively determine mating status (e.g., when the cage pan plugs are present but the vaginal lavage sample is not sperm positive or only a few sperm were present). When an additional vaginal lavage sample is collected and examined, an "a" for additional sample is noted, and the results of the second lavage are recorded on the cohabitation sheet.

D. Reproductive Indices

1. Male Mating and Fertility Index

a. Male Mating Index

The male mating index is reported in studies where the male is the test system. It is calculated using the following formula:

$$\text{Male mating index} = \frac{\text{number of males with confirmed mating}}{\text{number of males cohabitated}} \times 100$$

Historical control values for the male mating index are approximately 94% with a range of 80% to 100%. In fertility or reproduction studies where both sexes are treated, this index can be used as an indicator of reproductive toxicity, but not of specific male or female reproductive toxicity. Confirmation of male mating is usually based on the presence of a copulatory plug or sperm in the vaginal smear. Mating and pregnancy can also be confirmed by the unexpected delivery of a litter or by the discovery of resorbed uterine implantation sites in the postmortem uterine evaluation of presumed nonpregnant females (no confirmed mating and considered not pregnant). The male mating index can be influenced by many factors, including physical impairment, acute intoxication, alterations to the neuroendocrine-gonadal axis affecting libido, hormonal imbalance, etc. Other factors that might affect the male mating index include estrous cycle disruption (prolonged cycles or persistent estrus or diestrus) and lack of female receptivity.

b. Male Fertility Index

The male fertility index is calculated using the following formula:

$$\text{Male fertility index} = \frac{\text{number of males impregnating a female}}{\text{number of males cohabitated}} \times 100$$

This index measures the male's ability to produce sperm that are capable of impregnating a female. Because mating does not imply impregnation, this index supplies additional information. The male fertility index can be influenced by many of the same factors as those described above for the male mating index, as well as many other factors, such as sperm capacitation and transport, fertilization, oviductal milieu, development of the conceptus, and nidation. Historical control values for the male fertility index are approximately 85%, with a range between 70% and 100%. Again, in fertility or reproduction studies where both sexes are treated, the male mating index can be used as an indicator of reproductive toxicity but not of specific male or female reproductive toxicity.

2. Female Fertility Index

a. Fertilization Rate

Fertilization rate (the proportion of ova that were fertilized divided by the total number of ova available) is seldom reported in reproduction studies because the data collection must be done very early in gestation. Multiple factors are involved in the timing and integrity of gamete and zygote transport. These factors are quite susceptible to chemical perturbation and may affect the survival of the gametes, fertilization, and development of the conceptus.

b. Female Mating Index

The female mating index is calculated using the following formula:

$$\text{Female mating index} = \frac{\text{number of females mating}}{\text{number of cohabitated females}} \times 100$$

This index measures the female's ability to mate. Mating implies that the female was in a period of behavioral heat (estrus), was sexually attractive to the male (emitting chemicals that attract the male), and performed mating behaviors (solicitational behaviors [such as ear wiggling, hopping, and darting] and receptive behaviors [lordosis response]).[51] In routine reproductive toxicity testing, the mating is not actually observed, so unusual mating behaviors are not identified. Historical control values for the female mating index are approximately 94%, with a range between 80% and 100%. The female mating index can be influenced by many factors, such as physical impairment, acute intoxication, alteration of the neuroendocrine-gonadal axis, diminished libido, hormonal imbalance, and estrous cycle disruptions. When both sexes are treated, the female mating index can be used as an indicator of reproductive toxicity but not of specific male or female reproductive toxicity.

c. Female Fertility Index

The *Female Fertility Index* is calculated using the following formula:

$$\text{Female fertility index} = \frac{\text{number of pregnant females}}{\text{number of females cohabitated}} \times 100$$

TESTING FOR REPRODUCTIVE TOXICITY

This index measures the female's ability to achieve pregnancy. The female fertility index can be influenced by many of the same factors described above for the male fertility and female mating indices. Historical control values for the female fertility index are approximately 85%, with a minimum of 72% and a maximum of 100%. Again, when both sexes are treated, the female fertility index can be used as a general indicator of reproductive toxicity.

d. Fecundity or Pregnancy Index

The fecundity or pregnancy index is calculated using the following formula:

$$\text{Fecundity index} = \frac{\text{number of pregnant females}}{\text{number of females with confirmed mating}} \times 100$$

This index measures the female's ability to achieve pregnancy. Historically, the fecundity or pregnancy index for Sprague-Dawley rats averages 93.0% (SD ±9.88%).[53] The fecundity or pregnancy index can be influenced by many of the same factors described above in the male fertility and female mating indices. Again, the fecundity or pregnancy index can only be used as a general indicator of reproductive toxicity when both sexes are treated.

e. Implantation Index and Pre- and Postimplantation Losses

The implantation index is calculated by dividing the number of implantation sites by the number of corpora lutea. Historical control values for the implantation index are approximately 90%, with a range between 85% and 100%.

$$\text{Implantation index} = \frac{\text{number of implants}}{\text{number of corpora lutea}} \times 100$$

Disruption of early developmental processes may contribute to a reduction in fertilization rate and increased early embryonic death prior to implantation. This can be calculated as the *preimplantation loss*, which is the number of corpora lutea minus the number of implantation sites, divided by the number of corpora lutea.

$$\text{Preimplantation loss} = \frac{\text{number of corpora lutea} - \text{number of implants}}{\text{number of corpora lutea}} \times 100$$

Embryonic deaths after implantation plus fetal deaths are calculated as *postimplantation loss*, which, in fertility studies, is the total number of implantation sites minus the number of live conceptuses, divided by the number of implantation sites.

$$\text{Postimplantation loss} = \frac{\text{number of implants} - \text{number of viable fetuses}}{\text{number of implants}} \times 100$$

Postimplantation loss can be determined in reproduction studies following delivery of a litter. The uterus is examined (usually at necropsy after weaning of the litter), and the number of implantation sites is determined. Postimplantation loss in reproduction studies is the total number of implantation sites minus the number of full-term pups, divided by the number of implantation sites. The historical average preimplantation loss (in developmental toxicity studies) is 7.6%, while the average postimplantation loss in developmental toxicity and natural delivery studies is 6.0 and 8.8%, respectively.[53]

To identify the cause of increased preimplantation loss, additional studies using direct assessments of fertilized ova and early embryos would be necessary. Pre- and postimplantation loss occurs in untreated as well as treated rodents and contributes to the normal variation in litter size.

f. Gestation Index

The gestation index is calculated using the following formula:

$$\text{Gestation index} = \frac{\text{number of females with live born}}{\text{number of females with evidence of pregnancy}} \times 100$$

This index measures the female's ability to maintain pregnancy, based on having delivered at least one live pup. Historically, the gestation index for Sprague-Dawley rats averages 97.3% (SD ± 10.26%).[53] In RACB studies, this index is also known as the *pregnancy rate* (the proportion of pairs with confirmed mating that have produced at least one pregnancy within a fixed period). The gestation index is not a sensitive indicator of reproductive performance because all litters are treated as having equal biological significance regardless of their size.

g. Precoital Interval (Mean Time to Mating)

The precoital interval or mean time to mating (the average number of days after the initiation of cohabitation required for each pair to mate) is a useful indicator of altered reproductive function. Precohabitational estrous cycle data and cohabitational vaginal smears usually provide enough evidence to determine when mating has occurred. The normal precoital interval for rodents is 1 to 4 d, depending on the estrous cycle stage at the time of initiation of cohabitation. Historical control values for the precoital interval are approximately 3.3 ± 0.4 d, with a range between 2.8 and 3.8 d. If the stage of the estrous cycle at the time of cohabitation is known, this initial variance can be corrected during the data analysis. An increased precoital interval suggests impaired sexual behavior in one or both partners or abnormal estrous cyclicity.

3. Mean Gestational Length

Mean gestational length (duration of pregnancy) is calculated as the duration from GD 0 (day of positive evidence of mating) to the day of parturition. This information is easily derived from the breeding records and maternal records. Historically, the mean gestational length for Sprague-Dawley rats averages 22.3 d (SD = 0.44).[53] Decreased gestational length is associated with decreased pup weight and viability, while increased gestational length is associated with higher birth weights (due to the longer growth period in the uterus). However, increased gestational length may also be associated with dystocia (difficult labor and/or delivery), which may result in death or the physical impairment of the dam and/or the pups. Changes in gestational length can be due to several factors, including alteration in hormone levels and fetal growth retardation.

4. Litter Size

Litter size is the number of offspring delivered and should be determined at or as soon after birth as possible. Litter size is an important endpoint in the overall evaluation of reproductive performance in rodents and is frequently more sensitive than the other reproductive indices. For most reproduction studies, the mean litter size is calculated at standardized time points (PND 0, 4, 7, 14, and 21).

The day of birth is defined as PND 0. The litter size on PND 0 is calculated using the following formula:

$$\text{Litter size PND0} = \frac{\text{total number of pups delivered (live and stillborn)}}{\text{number of dams that delivered}}$$

Litter size should include dead as well as live offspring; therefore, numbers of live and dead offspring should be determined. Historically, the litter size PND 0 for Sprague-Dawley rats averages 14.75 pups (14.46 liveborn and 0.29 stillborn).[53] Dams that die or that resorb their entire litter are excluded from the litter size calculation. This value can be affected by maternal cannibalism, a common occurrence with stillborn or malformed pups. While cannibalism of healthy pups by untreated dams does occur, excessive cannibalism by treated dams should be considered an adverse treatment-related maternal behavior. The ovulation rate (number of ova available for fertilization), fertilization rate, implantation rate, and postimplantation survival rate affect litter size. A decreased litter size may indicate an adverse reproductive effect. Again, in fertility or reproduction studies where both sexes are treated, the litter size can be used as a nonspecific indicator of reproductive toxicity.

5. Live-Birth Index

The live birth index is calculated using the following formula:

$$\text{Live birth index} = \frac{\text{number of pups born alive}}{\text{total number of pups born}} \times 100$$

Normal delivery usually requires about 1 hour, depending on the number of pups in the litter. Litter viability checks are best done immediately after delivery; however, this usually is not possible because most litters are delivered late at night, when the technical staff is not present. All newborn pups should be examined for viability and for external abnormalities. It is difficult to determine visually whether the pup was stillborn (died *in utero*) or died shortly after birth. One way to distinguish the stillborn pups from the pups that died shortly after birth is by floating the lungs in saline (see below). Historically, the live birth index for Sprague-Dawley rats averages 98.0%.[53]

6. Survival Index

Survival indices are primary endpoints in a reproduction study, reflecting the offspring's ability to survive postnatally to weaning. Survival and viability indices are usually measured on PND 4, 7, 14, and 21. Survival indices are calculated using the following formula:

$$\text{Survival index} = \frac{\text{total number of live pups (at designated timepoint)}}{\text{number of pups born}} \times 100$$

If litter standardization or culling of the litter is performed on PND 4, then the PND 4 survival index should be based on the preculling number of offspring. Historically, the survival index for Sprague-Dawley rats at PND 4 averages 97.24% (derived from the percentage of dead pups between PND 0 and 4).[53] All subsequent survivability index calculations will use the number of offspring remaining after standardization as the denominator in the equation. If litter standardization was performed, the PND 21 calculation is known as the lactation index. If litter standardization was not performed, the PND 21 calculation is known as the weaning index.

Pup survival endpoints can be affected by the toxicity of the test substance, either by direct effects on the offspring (i.e., structural or functional developmental defects, low birth weight, impaired suckling ability), indirectly through effects on the dam (i.e., lactational ability of the dam, maternal neglect, acute intoxication during the treatment period), or by a combination of maternal and offspring effects. Pup survival may also be impaired by nontreatment-related factors, such as litter size.

7. Sex Ratio

Offspring gender in mammals is determined at fertilization of an ovum by sperm with an X or a Y chromosome. Altered sex ratios may be related to several factors, including selective loss of male or female offspring, sex-linked lethality (genetic germ cell abnormalities), abnormal production of X or Y chromosome–bearing sperm, or hormonal alterations that result in intersex conditions (masculinized females or feminized males). Determination of gender is generally done shortly after parturition, and the data are usually presented as a ratio, using the following format:

$$\text{Sex ratio} = \frac{\text{number of male offspring}}{\text{number of female offspring}}$$

These data are often presented as a percentage of male offspring, using the following formula:

$$\text{Percentage of male offspring} = \frac{\text{number of male offspring}}{\text{total number of offspring}} \times 100$$

Historically, the percentage of male offspring for Sprague-Dawley rats averages 50.0% (SD ±3.39%).[53] The calculations should include both live and dead pups because sex can easily be determined (except in the case of cannibalized pups or in some sexually ambiguous offspring). Treatment effects on sex ratio can indicate an adverse reproductive effect, especially if embryonic or fetal loss is observed. Because of the altered anogenital distance, visual determination of sex is often incorrect when the pups have an intersex condition. The correct diagnosis is usually determined later during lactation, when the condition becomes obvious, or at the postmortem evaluation, when the examination reveals male or female reproductive organs.

III. MALE REPRODUCTIVE TOXICOLOGY

A. Biology

Evaluation of the rodent male reproductive system usually consists of a gross evaluation (including organ weights) and histological evaluation of the testes, epididymides, vasa deferentia, seminal vesicles, coagulating gland, and prostate, and evaluation of the sperm. There are primarily four sperm measures required in the reproductive testing guidelines: epididymal sperm count, motility, and morphology, and testicular spermatid head count.

The male rat reproductive system consists of the testis (with scrotum), epididymis, vas deferens, ampullary gland, seminal vesicle, coagulating gland, bulbourethral gland, prostate, preputial gland, and penis. While the male genital protuberance is generally larger than the females', it is not a reliable method for determining gender of a pup. Determination of the sex of a newborn rat pup is based on the visual examination of the relative distance between the genital tubercle and the anus. This distance is larger in the male than female (approximately 3.5 mm versus 1.4 mm).[54] The testes are situated in the abdomen of the newborn male and do not descend into the scrotum until the rat is 4 to 6 weeks old. When viewed externally, the large testicles make the scrotal sacs protrude. The inguinal canal, unlike that of the human, remains open in the adult rat, so the testes may be withdrawn into the abdomen. Also visible externally is the aperture of the prepuce. The prepuce is a loose sheath that protects the penis. The penis and the external orifice of the urethra can be exposed for examination by gentle backward pressure at the sides of the prepuce.

B. Observations

There are three basic types of observations performed by trained technicians: a viability check, a clinical observation, and an extended or detailed clinical observation. A viability check is a daily cage-side observation performed once (at the start of the workday) or twice (at the start and end of the workday) to determine mortality, morbidity, early delivery, or abortion. During the viability check, food and water availability should also be monitored. The clinical observation is a more detailed evaluation of the physical or behavioral characteristics of the rat. The rat is removed from the cage and observed for any abnormalities, physical or behavioral. This observation is usually performed prior to treatment and often at a selected interval after treatment. The extended or detailed clinical observation is performed in an open field arena by a technician trained to administer the FOB. The extended or detailed clinical observation is essentially a FOB without including forelimb and hind limb grip strength, hind limb foot splay, and body temperature measurements (see below). Extended or detailed clinical observations are usually performed weekly or monthly depending upon the protocol.

C. Male Body Weight and Food Consumption

Body weights are recorded daily, twice weekly, or weekly, depending upon the guideline requirements. Food consumption is usually recorded at the same interval as body weight. Body weight and food consumption should be recorded at approximately the same time of day during the course of the study. Recording male body weight during treatment provides an indicator of the general health status that may be relevant in the interpretation of male reproductive effects. Body weight loss or reduction in body weight gain can be caused by several factors, including systemic toxicity, stress (such as restraint), treatment-induced anorexia, or decreased feed or water consumption due to altered palatability. Caloric restriction studies that caused small reductions in body weight have shown little effect on the male reproductive organs or function.[55] If a direct link between body weight change and the male reproductive effect cannot be established, then male reproductive system alterations should be considered adverse reproductive effects and not secondary to the occurrence of systemic toxicity. However, in the presence of severe body weight change or other significant systemic toxic effects, it should be stated that an adverse effect on a male reproductive endpoint occurred, but the effect may have resulted from a systemic toxic effect.

D. Male Necropsy

The male is usually terminated after the cohabitation period; however, it may be advisable to continue treatment and delay termination until after pregnancy outcome is known. In the event of an equivocal effect on fertility, the treated males can be mated with untreated females to determine if the effect was male mediated. The male is euthanized and subjected to a complete necropsy. The reproductive organs, and other tissues specified in the protocol, are weighed and saved for possible histopathological evaluation. At necropsy, sperm quality may also be assessed (see below).

E. Male Reproductive Organ Weight

Male reproductive organ weights are collected for the testes, epididymides, seminal vesicles (with coagulating glands), and prostate. The pituitary gland is also routinely harvested and weighed. The organ weight values should be reported as both absolute weights and relative weights (organ weight to body weight ratio). Organ weight to brain weight data is often reported because the weight of the brain in the adult rat usually remains quite stable.[56] Mean normal testis weight varies only slightly within a given experimental species,[57,58] suggesting that absolute testis weight should be a good indicator of testicular injury. However, reduction in testis weight is often detected only at treatment levels greater than those required to produce significant effects on other gonadal parameters.[59–61]

There can be a temporal delay between testicular cell death and testicular weight change. While cell death would be expected to reduce testicular weight, testicular edema and inflammation, cellular infiltration, and/or Leydig cell hyperplasia may actually increase testicular weight. Sloughed dead germinal epithelium cells can cause a blockage of the efferent ducts, causing increased testis weight as a result of fluid accumulation.[62,63] While testis weight may not be a good indicator for certain adverse testicular effects, a significant alteration in testicular weight should be considered an adverse male reproductive effect.

Epididymal and cauda epididymal weights are often required in reproductive studies, especially when sperm parameters are evaluated. Because sperm are within the epididymis and therefore contributes to its weight, reductions in sperm production can alter the epididymal weight. The presence of lesions such as sperm granulomas or leukocytic infiltration (inflammation) in the epididymal epithelium can also affect the epididymal weight.

Because the seminal vesicle and prostate are androgen-dependent organs, weight changes in these organs may reflect alterations in endocrine status or testicular function. The seminal vesicles and prostate can respond differently to a test substance; therefore, information will be improved if the weights were collected separately (with and/or without their secretory fluids). Differential loss of secretory fluids prior to weighing can cause artifactual organ weights or increased variation in organ weight values.

Pituitary weight can provide information on the reproductive status of an animal. However, the pituitary contains many other cell types that are responsible for the regulation of numerous physiologic functions unrelated to reproduction. Therefore, alterations in pituitary weight do not necessarily reflect altered reproductive function. Selective histological stains may be used to confirm alterations in regions of the pituitary associated with reproductive functions. If the pituitary weight change can be associated with changes in reproductive function of the pituitary, then the pituitary weight change should be considered an adverse reproductive effect.

Significant changes in male reproductive organ weights (absolute and/or relative) should be considered an adverse male reproductive effect. However, lack of reproductive organ weight effects should not be used to negate significant changes in other reproductive endpoints.

F. Histopathological Evaluation of the Male Reproductive Organs

Histopathological evaluations of the male reproductive organs have a prominent role in toxicity assessment. Detailed descriptions of the histology and pathology of the male reproductive system are beyond the scope of this chapter; however, brief discussions of the organs comprising the male reproductive system are presented. Routinely evaluated male reproductive organs include the testes, epididymides, prostate, seminal vesicles, coagulating glands, and pituitary. Histopathological evaluations can be especially useful by (1) providing location (including target cells) and severity of the lesion, (2) providing comparative data from a variety of reproductive study designs, (3) providing a relatively sensitive indicator of damage in short-term dosing studies, and (4) describing recovery after the cessation of treatment. When similar targets or mechanisms exist in humans, the basis for interspecies extrapolation is strengthened. Significant, biologically meaningful histopathological alterations of male reproductive organs should be considered adverse reproductive effects.

1. Testis

Numerous chapters[64–66] and books[67–69] have been written concerning the testis. While detailed descriptions of testicular histology and spermatogenesis are beyond the scope of this chapter, a review is presented.

The testis is covered by the tunica albuginea, a tough fibrous capsule. The rat testis contains very little intertubular connective tissue, in contrast to the human testis, which is partitioned by connective tissue septa. However, the rat testis is similar to the human testis in having two major compartments: the interstitial compartment and the seminiferous tubule compartment. The interstitial compartment contains the connective tissue, interstitial (Leydig) cells, nerves, blood vessels, and lymphatic vessels. Unlike human testicular capillaries, rat testicular capillaries are not fenestrated. The appearance of Leydig cells in the developing testes coincides with the production of testosterone. Macrophages are commonly observed in the interstitium and may account for up to 25% of the interstitial cells.

A limiting membrane or boundary tissue formed by lymphatic endothelial cells, myoid cells, and acellular components (basal lamina) bounds the seminiferous tubule compartment. The myoid cells are contractile and provide the major motive force for the movement of sperm and fluid through the seminiferous tubule. The basal lamina and the myoid cells provide the structural base on which the Sertoli cells and the basal compartment cells of the seminiferous epithelium rest. Sertoli cells extend around the periphery of the tubule through to the lumen, with invaginations surrounding germ cells at various stages of differentiation. Specialized junctions between the Sertoli cells form the blood-tubule barrier in the rat at approximately 18 days of age. After that, the Sertoli cells do not divide again. Sertoli cells maintain metabolic support for the seminiferous epithelium, maintain hormonal control of spermatogenesis and testicular function, and have a phagocytic function.

The seminiferous tubules are highly convoluted loops that have their two ends connected to short segments of straight tubules (tubuli recti). Although highly convoluted, the tubules straighten along the long axis of the testis. A transverse histological section through the long axis of the testis allows examination of cross-sectioned seminiferous tubules. The tubuli recti connect to the rete testis, which is connected to the cephalic portion of the epididymis by 10 to 20 long and tortuous efferent ducts (ductuli efferentes).

The most sensitive endpoint for the determination of testicular toxicity is histopathology. Testicular histology and sperm parameters are linked,[70] and alterations in testicular structure are usually accompanied by alterations in testicular function. However, some functional changes may occur in the absence of detectable structural changes in the testis. Alterations in the testicular structure and function are also linked to alterations in epididymal structure and function, but changes in epididymal function may occur in the absence of testicular or epididymal structural changes.

Proper fixation (using Bouin's, Davidson's, or a comparable fixative for at least 48 h) and careful trimming, embedding (paraffin), sectioning (5 to 6 μm), and staining (hematoxylin and eosin or periodic acid-Schiff's-hematoxylin stain) of testicular tissue[71,72] improves the quality of the histological evaluation of spermatogenesis. Care must be taken to minimize the compression of the testes during tissue dissection, as it can result in artifactual damage to the testes (such as sloughing of germ cells from the seminiferous epithelium).

Either a longitudinal or transverse section of the testis can be taken. The advantages of a longitudinal section are that there is more tissue to examine and that longitudinal sections of the seminiferous tubules allow determination of consecutive stages of the spermatogenic cycle. The advantage of a transverse section is a consistent cross-section of seminiferous tubules, allowing spermatogenic staging, quantification of cell numbers, and tubular measurements (such as seminiferous tubule diameters). Most technicians take a midline transverse section. However, the testis has asymmetrical structures (e.g., rete testis) that may be missed by a transverse section.

Histopathological evaluations of the testis can determine whether sloughed cells are present in the tubule lumen, whether the germinal epithelium is degenerating or severely depleted, or whether multinucleated giant cells are present. Evaluation of the seminiferous tubules should include the pachytene spermatocyte. This cell type has the longest life of the primary spermatocytes with the most RNA and protein synthesis. The pachytene spermatocyte appears to be uniquely susceptible to toxicants.[73] The rete testis should be examined for dilation due to obstruction or disturbances in fluid dynamics and for the presence of proliferative lesions and rete testis tumors. Both spontaneous and chemically induced lesions in the rete testis often appear as germ cell degeneration and depletion. More subtle lesions, such as missing germ cell types or retained spermatids, can reduce the number of sperm being released into the tubule lumen. Such effects may not be detected when less than adequate methods of tissue preparation have been used. Also, knowledge of the detailed testicular morphology and spermatogenesis assists in the identification of lesions that may accompany lower treatment levels or result from short-term exposure.[68]

Several methodologies for qualitative or quantitative assessment of testicular tissue are available that can assist in the identification of testicular lesions, including the use of "spermatogenic cycle staging."[68,74] To provide high resolution cellular morphologic detail required for spermatogenic cycle staging, semithin (2 μm) sections will have to be produced from testes embedded in glycol methacrylate resin. A detailed description of spermatogenesis and spermiogenesis required for staging of spermatogenesis is beyond the scope of this chapter; however, detailed morphological descriptions with light and electron microscopic photographs of these processes are available in an excellent book by Russell et al.[68]

Rodent males produce sperm in numbers that greatly exceed the minimum requirements for fertility, particularly as evaluated in reproductive studies that allow multiple matings.[75,76] In some strains of rats and mice, sperm production can be drastically reduced (by up to 90% or more) without affecting fertility.[77-79] Human male sperm production appears to be much closer to the infertility threshold; therefore, less severe sperm count reductions may cause human infertility. Negative results in rodent studies that are limited to only fertility and pregnancy outcomes provide insufficient information to conclude that the test substance poses no reproductive hazard in humans.

2. Epididymides, Accessory Sex Glands and Pituitary

The basic morphology of other male reproductive organs (epididymides, accessory sex glands, and pituitary) has been described[80,81] as well as the histopathological alterations that may accompany certain disease states.[82,83] While detailed descriptions of histology and pathology of these organs are beyond the scope of this chapter, a discussion is presented.

The epididymis, a single, long, highly tortuous tube surrounded by connective tissue, has three portions: the caput (head) located on the anterior pole of the testis, a narrow corpus (body) lying along the dorsomedial aspect of the testis, and the cauda (tail) located on the posterior pole of the testis. The cephalic portion of the epididymis is almost entirely encapsulated in fat. The ductus (vas) deferens is the continuation of the epididymis. The lumen of the vas deferens becomes less tortuous leading to the prostatic urethra. The secretions of the seminal vesicles, coagulating glands, ampullary glands, dorsal and ventral prostate glands, and bulbourethral glands are added to the sperm in the prostatic and penile urethra to produce semen. Preputial glands do not contribute to the semen, but rather secrete musklike compounds.

A longitudinal section of the epididymis that includes all epididymal segments should be included for histopathological examination. During dissection of the epididymis from the testis, care must be taken not to cut the testicular capsule and cause extrusion of the seminiferous tubules, as such damage may disrupt the testicular architecture. Histopathological evaluation of the epididymis should include information about the caput, corpus, and cauda segments. Presence of debris and/or sloughed cells in the epididymal lumen is an indicator of damage to the germinal epithelium or to the excurrent ducts of the epididymis. The presence of lesions such as leukocytic infiltration, sperm granulomas, or absence or paucity of clear cells in the cauda epididymal epithelium should be recorded.

The density of sperm and the presence of germ cells in the lumena of the three epididymal segments reflect time-dependent events in the testis. Sperm in the caput were released from the testis only a few days earlier, while the sperm in the cauda were released approximately 14 days earlier. A careful examination of the accessory sex glands is indicated when decreased copulatory plugs are observed during cohabitation.

Histopathologic evaluation of the pituitary should include cellular morphology of gonadotropin-producing cells.

G. Sperm Evaluations

The sperm quality parameters are sperm number, sperm motility, and sperm morphology. Although effects on sperm production can be reflected in other measures, such as testicular spermatid count or cauda epididymis weight, no other measures are adequate to reflect effects on sperm morphology or motility. The application of digital video technology to sperm motility analysis allows a detailed evaluation of sperm motion, including precise information on individual sperm tracks. Computer-assisted sperm analysis (CASA) also provides permanent storage of the actual sperm track images, which can be reanalyzed as necessary, either manually or computer assisted. With CASA technology, sperm velocity (rectilinear and curvilinear) and track amplitude and frequency can be obtained on large numbers of sperm. Chemically induced alterations in sperm motion have been detected by use of CASA technology,[84–86] and often these changes are correlated with the fertility of the exposed animals.[87–89] Studies indicate that significant reductions in sperm velocity are associated with infertility, even when the percentage of motile sperm is not affected. It is also important to be able to distinguish between the sperm with progressive motility and those showing other patterns of movement.[90]

While changes in spermatogenesis and sperm maturation endpoints have been related to fertility in several test species, the ability to predict infertility from these data (in the absence of actual fertility data) may not be reliable. Fertility is dependent not only on having adequate sperm count, but also on having normal sperm. When sperm quality is good, then a significant reduction in sperm count is required to affect fertility. Rat sperm counts have to be reduced substantially before fertility is affected.[91] In fertile and infertile men, the distribution of sperm count overlaps, with the mean sperm count for fertile men being greater than for infertile men.[92] However human fertility is usually reduced when sperm counts decrease below 20 million per milliliter.[93] If the sperm counts are normal in rodents, sperm motility or velocity must be greatly reduced before fertility is affected.[87] Reduction in sperm count or quality may not cause infertility in rats, but the reductions may be predictive of infertility.

1. Sperm Number or Sperm Concentration

Sperm number or sperm concentration for test species, as well as sperm motility, morphology, and viability, may be determined from ejaculated, epididymal, or testicular samples.[90] Rat sperm evaluations routinely use sperm from the cauda epididymis. Sperm count is reported as an absolute sperm count and relative to the cauda epididymal weight. However, because the number of sperm contributes to the weight of the epididymis, expression of the data as a ratio may actually mask a reduction in sperm count. As one would expect, epididymal sperm counts can be directly influenced by level of sexual activity.[75,94] Sperm production data may be derived by counting the number of homogenization-resistant testicular spermatids.[75,78,95,96] These counts are a measure of sperm production from the stem cells and their development through spermatocytogenesis and spermiogenesis.[78,97] The spermatid count may serve as a substitute for quantitative histological analysis of sperm production[68] if the duration of treatment is sufficient for a lesion to be fully expressed. However, spermatid counts may be misleading when the duration of treatment is too short.

2. Computer-Assisted Sperm Analysis

Technological advances in computer-assisted sperm analysis have made objective evaluation of the motile characteristics of rat sperm possible. CASA systems utilize a collection of components (VCR, microscope, multiple microscope objectives, optical disk drive, computer hardware, and CASA software) to determine sperm concentration, sperm motility and other motion parameters, and sperm morphology. CASA systems usually have a microprocessor-controlled stepping motor that can move a temperature-controlled microscope stage to exact positions. Light (of a selected wavelength) is provided by a strobe source that illuminates the image field with a series of short flashes (7 to 60 flashes/sec, each of 1- to 3-msec duration). The CASA system then collects a series of phase contrast or fluorescent images, with up to 20 image fields analyzed per specimen. Motile and stationary objects from each field of view are located and identified. Moving objects (sperm heads) are identified by use of background subtraction techniques, and their mean size and brightness are calculated. In the case of rat sperm, only the head is detected as motile.

After the motile cells are identified, the static sperm are then discriminated from detritus (typically leukocytes, erythrocytes, epithelial cells, spermatogonia, and cellular debris). The static sperm must be properly counted to allow percent motility to be calculated. The static rat sperm head and its connected midpiece can be identified and combined by the computer to form a single elongated object. The extreme elongation allows static sperm to be distinguished from other static cells and debris. Three modes of discrimination are employed to identify the static sperm: (1) size and intensity, (2) elongation (ratio of head width to head length), and (3) use of a DNA-specific fluorescent stain that stains the highly condensed DNA in the sperm head. Most of the other static cell types are basically round; therefore, elongation of the head and midpiece is a useful separator. Detritus does not routinely have significant DNA, and thus it does not stain or fluoresce.

Motile sperm are tracked by comparing successive frames by means of a nearest neighbor algorithm that selects the closest object from the next frame. The individual sperm track is calculated by joining the successive sperm positions with straight lines. Typically rat sperm move at a rate of 200 to 400 μm/sec or less, so a spermatozoon will usually move less than 3 to 6 μm between frames collected at 60 frames/sec.

After the motile and nonmotile sperm have been identified, counted, and tracked, the cell concentrations are calculated. Concentration calculations require the number of motile and nonmotile sperm, the area of the field of view, the dilution ratio, and the depth of the specimen chamber. The dilution factor is a variable and needs to be precisely controlled. The CASA system calculates sperm count and concentration, motility, progressive velocity, track speed, path velocity, straightness, linearity, cross beat frequency, and amplitude of lateral head displacement (see Figure 10.9).

The following list describes the various sperm motility endpoints:

Parameter	Definition
Total sperm counted	Number of static and motile sperm counted in all fields analyzed
Total concentration (millions/ml)	Calculated using concentration (M/ml) multiplied by the ejaculate volume
Concentration	The number of sperm per ml of sample determined by dividing the total sperm count by the total volume of the fields analyzed
Number motile	Population of sperm that are moving at or above a minimum defined speed
Number progressive	The number of cells moving with both the VAP greater than the medium average path velocity and STR greater than the threshold straightness
Path velocity (VAP)	Total distance along the smoothed average position of the sperm divided by the time elapsed

Parameter	Definition
Progressive velocity (VSL or rectilinear velocity)	Straight-line distance from the beginning to the end of the sperm track, divided by the time elapsed
Track speed (VCL or curvilinear velocity)	Total distance between the actual point-to-points for a given sperm during acquisition, divided by the time elapsed
Amplitude of lateral head displacement (ALH)	Mean width of the head oscillation as the sperm swims
Straightness (STR)	Average value of the ratio VSL/VAP (STR measures the departure of the sperm path from a straight line)
Linearity (LIN)	Average value of the ratio VSL/VCL (LIN measures the departure of the sperm track from a straight line)
Motility	The ratio of number of motile sperm to the total number of sperm
Sperm per gram	Number of sperm, in millions per gram of tissue

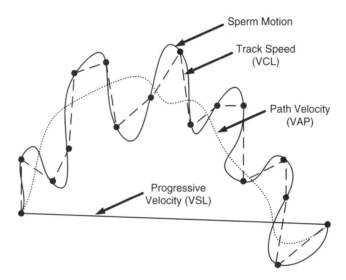

Figure 10.9 Illustration of sperm velocities and motion parameters.

a. Epididymal Sperm Counts

Samples should be processed and analyzed sequentially across treatment groups (i.e., first animal in each group followed by the second animal in each group). (Note: The Hamilton-Thorne Integrated Visual Optical System [IVOS] is used only as an example of a CASA system. Cryo Resources makes the CellSoft Computer Assisted Sperm Analyzer, the other CASA system available for rat sperm analysis at this time. No endorsement of either system is intended or implied.)

Routinely, the first printout of the summary report for each animal should be maintained in the study records and defined as raw data. An optical image of each field analyzed should be stored on an optical disk that can be maintained for future analysis. The following description of the use of a CASA system assumes that the system has been properly installed, the sperm parameters have been defined, and the system has been validated for use in a study. Select the study setup (e.g., Analysis Setup No. 7 [Rat Ident]). A unique filename for each study should be entered for printing ASCII files. The setup report for each day should be printed and maintained in the study records. Check each day of use to determine that the CASA system is correctly identifying sperm by selecting the "playback" option. Approximately 20 sperm per field should be counted.

1) Epididymal Sample Preparation — When the rat is euthanized, the epididymides are removed and weighed. If the selected epididymis cannot be evaluated immediately, it can be stored frozen until evaluation. To freeze the epididymis, place it in a Bitran® bag, flash freeze it in liquid nitrogen, and store it in an ultralow freezer (–70°C) until ready for analysis. When ready to evaluate the epididymis, thaw the specimen. For fresh tissue or after thawing the frozen tissue, excise the cauda epididymis from the caput and corpus. Trim off the fat, and cut the vas deferens close to the cauda. Weigh and record the cauda weight. (Note: If the cauda weight is less than expected, the volumes must be adjusted.) Place the cauda epididymis in a glass vial containing 15 ml of diluent, consisting of Triton X-100 [0.05% (v/v)] in 0.9% (w/v) NaCl (i.e., 1.8 g NaCl and 100 µl Triton X-100 in 200 ml deionized water). Homogenize (e.g., with a Polytron® homogenizer at setting 5) on ice for approximately 1 min. Bring the homogenate up to a final volume of 40 ml and vortex on high until the sample is mixed. Transfer 2 ml of homogenate to a scintillation vial. Sonicate (e.g., with a Sonifier® at setting 3) on ice for approximately 30 sec. Store vials on ice until analysis.

2) Epididymal Sperm Count Analysis — Place 200 µl of well vortexed homogenate into the IDENT stain tube (for DNA staining) and vortex to mix thoroughly. Incubate at 37° ± 2°C for 3 to 5 minutes. Using a standard pipette tip, transfer approximately 6 µl of the suspension to a clean 2X-CEL 20-µm slide with coverslip. (Note: The 2X-CEL 20-µm slides from the vendor may not be clean. Clean before use, if necessary.) Load the slide into the IVOS, and let the temperature equilibrate for approximately 20 sec. Move to the fifth or sixth field of cells using the Jog In/Jog Out button. Select the Ident Motile option and focus the sample on the screen. Visually examine the sample to determine if it is acceptable. If the image field is unacceptable (e.g., clumping is evident, too many or too few sperm are present, excessive debris or bubbles in the field), repeat the steps until an acceptable image field is found. Use Manual Select to select the first field to be analyzed. Move to the next field and repeat the selection process until a total of 10 fields have been selected. Run Start Scan. (Note: If a field is out of focus, select Start Scan, refocus, and repeat steps until 10 fields have been counted.) The summary page printout should be maintained in the study records.

b. Testicular Homogenization-Resistant Spermatid Counts

1) Testicular Sample Preparation — When the rat is euthanized, the testes are removed and weighed. (Note: If the testis weight is less than expected, adjusted volumes must be used.) If the testis cannot be evaluated immediately, it can be stored frozen until evaluation (as described above for the epididymis). For fresh or thawed testis, remove and weigh the tunica albuginea. Place the parenchyma in a glass vial containing 15 ml of diluent (see above for diluent composition). Homogenize on ice for approximately 30 sec. Bring the homogenate up to a final volume of 20 ml and vortex on high until the sample is mixed. Transfer 2 ml of homogenate to a scintillation vial. Sonicate on ice for approximately 45 sec. Store vials on ice (approximately 4°C) until analysis. This process varies in the duration of homogenization and sonication from the methods published by Meistrich and van Beek.[97]

2) Testicular Homogenization-Resistant Spermatid Count Analysis Using CASA — Use the same methodology described for the epididymal sperm count using the CASA system (above). (Note: If no stained spermatids are present, check that the stain was properly mixed. If the stain was not properly mixed, invert the stain kit vial two or three times, tapping it each time, and then revortex with homogenate and allow time to stain.)

c. Manual Epididymal and Testicular Sperm Counts

1) Epididymal Sample Preparation — Use the same methodology described for the epididymal sperm count (above), except homogenize the epididymis for 30 sec to 3 min to adequately disrupt the tissue. Bring the homogenate to the appropriate final volume (generally between 50 and 100 ml), using deionized water. Record the final volume in the study records.

2) Testicular Sample Preparation — For fresh tissue or after thawing the frozen tissue, weigh and record the testis weight. Remove and weigh the tunica albuginea. Place the parenchyma in a glass vial containing 30 ml of diluent. Homogenize the parenchyma for 30 sec to 3 min to adequately disrupt the tissue. Bring the homogenate up to appropriate final volume (generally between 50 and 100 ml), using deionized water. Record the final volume in the study records.

3) Manual Sperm and Spermatid Counts
Loading the Hemocytometer — Manual concentration analysis is conducted by producing a homogenate of standard volume from a cauda epididymis or testis sample of known weight and counting the number of sperm heads in a known volume of homogenate by use of a hemocytometer. Place the coverslip on the hemocytometer. Transfer 10 µl of the suspension into both chambers of the hemocytometer. Touch the center edge of the coverslip with the pipette tip and allow each chamber to fill by capillary action. To facilitate counting, place the loaded hemocytometer in a humid area (e.g., a Petri dish with moist gauze) for approximately 5 min so the sperm heads can settle to a common focal plane.

Counting using the Hemocytometer[96] — Count the four corner squares and the middle square in each chamber. Include cells touching the middle line at the top and the left side. Do not count cells touching the middle line at the bottom and on the right side. Obtain two sets of hemocytometer counts (total of four chambers counted). If there are 15 sperm or less per counting chamber, record all counts. If there are greater than 15 sperm, determine if a recount is necessary within each set of hemocytometer counts. Generally, if there is less than a 20% difference between hemocytometer counts, the sperm head counts are recorded and accepted. If there is a 20% to 50% difference between counts, the counts are recorded, the hemocytometer is cleaned and reloaded, and the new sample is counted and recorded. If greater than 50% difference exists between counts, the count is recorded but not used for tabulation. Then the hemocytometer is cleaned and reloaded, and a new sample is counted and recorded. (Note: Remember to always clean the hemocytometer and cover slip immediately after completion of counting by rinsing with deionized water and air drying or drying with lens tissue.)

4) Epididymal Calculations for Manual Sperm and Spermatid Counts
Sperm per Cauda Epididymis

$$\text{Sperm per cauda epididymis} = \frac{\text{mean count multiplied by dilution volume}}{\text{volume of counting chamber}}$$

For example, with a mean chamber count of 20, a dilution volume of 100 ml, and a chamber volume of 0.0001 ml, the calculation would be $(20 \times 100)/0.0001 = 2 \times 10^7$ sperm per cauda epididymis.

Sperm per Gram Cauda Epididymis

$$\text{Sperm per gram cauda epididymis} = \frac{\text{sperm count per cauda}}{\text{weight of cauda (grams)}}$$

For example, with a cauda sperm count of 2×10^7 and a cauda weight of 0.25 g, the calculation would be $2 \times 10^7/0.25 = 8 \times 10^7$ sperm per gram epididymis.

5) Calculations for Manual Testicular Sperm and Spermatid Counts

Spermatids per Testis

$$\text{Spermatids per testis} = \frac{\text{mean count multiplied by dilution volume}}{\text{volume of counting chamber}}$$

For example, with a mean chamber count of 20, the dilution volume of 100 ml, and a chamber volume (hemocytometer factor) of 0.0001 ml, the calculation would be $(20 \times 100)/0.0001 = 2 \times 10^7$ spermatids per testis.

Spermatids per Gram Testis

$$\text{Spermatids per gram testis} = \frac{\text{spermatid count per testis}}{\text{weight of parenchyma (in grams)}}$$

For example, with a testicular sperm count of 2×10^7 and a parenchyma weight of 0.25 g, the calculation would be $2 \times 10^7/0.25 = 8 \times 10^7$ spermatids per gram testis.

Spermatid Production — Spermatid production can also be expressed as number of spermatids produced per day per gram of tissue. This calculation uses the amount of time (in days) during spermatogenesis while these cells are resistant to homogenization (4.84 d for mice and 6.10 d for rats).

$$\text{Daily sperm production} = \frac{\text{spermatid count per gram tissue}}{\text{days resistant to homogenization}}$$

For example, with a value for spermatids per gram testis of 8×10^7 and a daily production factor of 6.10 (rat), the calculation would be 8×10^7 spermatids per gram testis/6.10 d = 1.3×10^7 spermatids per day per gram testis.

3. Cauda Epididymal Sperm Motility

Sperm motility estimates may be obtained from sperm samples collected from the cauda epididymis or vas deferens, or from ejaculated sperm samples. Sperm motility can be recorded and evaluated immediately, or the sperm images can be recorded, stored, and analyzed later, either by CASA or manually.[84,98–100] Sperm motility is a reproductive endpoint and an integral part of some reproductive toxicity tests.[101–103] Standardized procedures within the laboratory are essential. Sperm motility can be affected by many factors, including (1) method of sample collection and handling, (2) elapsed time between sampling and observation, (3) dilution medium characteristics (e.g., pH and concentration of energy sources), (4) temperature (of dilution medium, slides, and room), (5) sperm dilution, and (6) the microscopic chamber employed for the observations.[87,90,104–106]

The procedures described below can be used to collect, prepare, and analyze sperm motility in samples from the rat epididymis by means of a CASA system (e.g., IVOS). These procedures are generally chronological, but it is not absolutely necessary that the order presented be rigidly followed. The sample collection methods generally follow the procedures outlined in Toth et al..[107] An image of each field analyzed should be stored on an optical disk. These images can be maintained for future analysis. A printout of the summary report is routinely maintained in the study records and is often defined as the raw data.

a. Procedures

1) Set-up of the IVOS for Determining Sperm Motility — Turn the optical disk drive on first and off last. Check the magnification with a 100-µm grid prior to the start of the study or when changing from analyzing counts to motility. Document magnification in the study records. Calibrate the phase annulus daily before using the IVOS and document in the study records. Check the illumination daily before using the IVOS and document the low photometer reading (reading must be between 95 and 105) in the study records. Select the appropriate analysis setup. Assign a unique filename to each study for printing ASCII files. Print the Setup Report daily for each study and maintain in the study records.

2) Sample Collection — Prepare the appropriate volume of 10 mg/ml bovine serum albumin/phosphate buffered saline (BSA/PBS) solution daily. (Note: Use an unopened bottle of PBS for each day of solution preparation. Weigh 1.0 g BSA for every 100 ml of PBS required and mix until dissolved.) Place 10 ml of the BSA/PBS solution into Petri dishes for each sample and incubate at approximately 36°C until use. Document the incubator temperature in the study records each day of usage. Prepare test tubes with appropriate amounts of BSA/PBS diluent to be used for 1:9 dilution (i.e., 100 µl sample to 900 µl diluent), cover them with Parafilm®, and incubate them at approximately 36°C until use. Euthanize the male rat. (Note: Tissue harvesting should be done immediately. Do not allow the carcass to become chilled.) Open the abdominal cavity and excise the epididymis. Place the epididymis in a tared weigh boat, weigh it, and record the weight. Isolate the caudal portion of the epididymis with a scalpel blade. Place the cauda epididymis in a tared weigh boat, weigh it, and record the weight. Transfer the cauda epididymis to a prelabeled Petri dish containing 10 ml of BSA/PBS diluent prewarmed to approximately 36°C. Pierce the epididymis four to six times, using an 18-gauge needle. Cover the Petri dish and incubate it at approximately 36°C to allow for sperm swimout. Five minutes is the optimal amount of time for swimout; 5 to 8 min is an acceptable time. It is important that swimout times be kept uniform. Record incubation time in the study records. Upon removal of the dish from the incubator, gently swirl the Petri dish and its contents to disperse the sample. Place the appropriate amount of sperm solution into the dilution tube (e.g., 100 µl in a 1:9 dilution tube), re-cover it with Parafilm, and gently mix it by inverting the tube. Transfer approximately 15 µl of the diluted suspension to each chamber of an 80-µm 2X CEL slide.

3) Sample Analysis — Load the chamber into the IVOS. Perform a preliminary check to determine the suitability of the dilution used. Twenty to forty sperm per analysis field is considered a suitable dilution. If the sample appears too dilute, redilute (1:4) from the original sample. (Note: The Jog In/Jog Out button is used to move through the various sample fields.) Enter the appropriate diluent ratio (e.g., 1:9, 1:4, etc.), animal number, initials, and date in the information fields. Select the first sample field to be analyzed and press Manual Scan. (Note: The technician must check that each sample field is acceptable for analysis, i.e., no large debris or air bubbles, etc.) If the sample field is not acceptable, select the next field or prepare a new slide for analysis. Press the Jog In/Jog Out button to move to the next field, and press Manual Scan to select the next field to be analyzed. Press Start Scan after the desired sample fields have been selected. Review the count summary that appears on the screen to ensure that at least 200 sperm have been analyzed. If less than 200 sperm have been analyzed, move to a new field and select by pressing Add Scan. Keep adding new sample fields until at least 200 cells have been analyzed. Print the summary and maintain printouts in the study records.

4) Troubleshooting

    Sperm Are Present but Not Motile — If sperm are present but not motile, check the pH of the medium (should be 7.2), the temperature of the medium (approximately 36°C), the time passed since sample collection (approximately 15 min or less), the latency from death to harvesting tissue, and the necropsy observations concerning the left epididymis, the left testis, or the seminal vesicles.

    pH Is out of Range — If the pH is discovered to be less than 7.1, note the deviation in the comments section of the information screen and ensure that the correct pH is listed in the pH field. For subsequent samples, place the medium into the Petri dish earlier, giving it more time to shift toward the alkaline side. If placing the medium into the dish earlier does not allow the pH to shift back into range, then try adjusting the medium with 1 N NaOH.

If the pH is discovered to be greater than 7.3, note the deviation in the comments section of the information screen and ensure that the correct pH is listed in the pH field. For subsequent samples adjust the medium with 1N HCl.

- **Temperature Is out of Range** — Record the slide warmer temperature in the comments section of the information screen with the added comment that the medium temperature was out of range. Adjust the slide warmer temperature and wait until it is within range before preparing another sample.
- **Too Much Time Has Passed Since the Sample Collection** — Record that an excessive amount of time has passed in the comment section of the information screen. Reduce the time for sample collection by ensuring that the epididymis is collected, trimmed, and weighed as quickly as possible. Complete sample collection prior to all other steps of the necropsy, or assign the remaining necropsy steps to another technician.
- **Increased Latency from Death to Harvesting Tissue** — Ensure that no excess time passes between the time of death and the time of sample collection.
- **Check Necropsy Observations** — If there are necropsy findings for the epididymis, testis, or seminal vesicles, note relevant observations in the comments section of the information screen.
- **No Sperm Are Present** — First, prepare another slide. If no sperm are present, record the finding. Check to see if there are necropsy observations concerning the epididymis, the testis, or the seminal vesicles that might explain the lack of sperm.

4. Sperm Morphology

Sperm morphology refers to structural evaluation of the sperm. Sperm morphology can be evaluated from vas deferens or ejaculated samples, but it is routinely evaluated using cauda epididymal samples. Sperm morphology profiles are relatively stable in a normal individual (and a strain within a species) over time, which may enhance their use in the detection of spermatotoxic events.[108] The traditional sperm morphology methodology characterizes defects of the sperm head, midpiece, and flagellum in either stained preparations (by brightfield microscopy)[109] or fixed unstained preparations (by phase contrast microscopy).[90,110] Data should be collected that categorize the sperm abnormality by type and determine the frequency of occurrence. However, because rodents have a high proportion of normal sperm with little morphological variability, it has been recommended that the sperm should be classified as normal or abnormal for statistical analysis so that changes in the proportion of total abnormal sperm can detected.[90] Objective, quantitative approaches (e.g., automated sperm morphology analysis) are now available and should result in a higher level of confidence than more subjective measures.[111]

a. Sperm Morphology (from Samples Not Previously Used for Motility Analysis)

Prepare 1% Eosin Y solution in deionized water. Euthanize the male rat using CO_2 anesthesia and exsanguination. Make a midline abdominal incision, grasp the fat surrounding the caput epididymis, and pull the testis into the abdominal cavity. Excise the testis by cutting the vas deferens about 0.5 cm from the cauda epididymis, dissect any remaining tissues, and place the testis plus epididymis in a weighing pan. Carefully trim away surrounding fat and remove the intact epididymis from the testis. Weigh testis and epididymis (to the nearest 0.1 mg). The cauda region is separated from the corpus at the point just below where the vas deferens exits the epididymis. Isolate the cauda epididymis with a scalpel blade, and weigh it to the nearest 0.1 mg.

Transfer the cauda epididymis to a Petri dish containing 10 ml of Dulbecco's phosphate-buffered saline prewarmed to 37°C. (Note: As an alternate buffered saline, use 1% Medium 199 modified [M199] in Hank's Salts with 20 mM HEPES buffer and 5% BSA added, but without sodium bicarbonate and L-glutamine.) Pierce the epididymis 4 to 6 times, using an 18-gauge needle, and incubate it approximately 15 min at 37°C. Periodically gently swirl the Petri dish and its contents to facilitate release of sperm from the cauda. Pipette the suspension gently approximately 20 times, using a 200-µl pipette with wide-bore tip to disperse sperm aggregates while being careful not to create air bubbles in the suspension. Prepare a 1:1 dilution of sperm suspension in 1% Eosin Y stain (i.e., 200 µl sample to 200 µl Eosin Y stain). Gently tap the bottom of the test tube to mix. Incubate at room temperature for approximately 30 to 60 min to allow for staining. Gently remix the suspension with the 200 µl pipette. Prepare a smear (using a procedure similar to making a blood smear) on a labeled microscope slide. A Miniprep® Blood Smearing Instrument (Geometric Data, Wayne, PA) can be used to prepare uniform smears. Prepare two to four additional slides from each sample to use for confirmation or for additional evaluation. Air dry slides overnight. Samples may be fixed (using methanol or formalin) and a cover slip applied with Permount or other suitable medium for long-term storage.

b. Sperm Morphology (from Samples Previously Used for Motility Analysis)

Prepare 1% Eosin Y solution in deionized water. Incubate cauda epididymis sample at 37°C for approximately 15 min. Follow the same procedures described above. Remove the cauda epididymis from the dish and save it in an appropriate fixative for possible histological examination.

c. Evaluation of Sperm Morphology

A minimum of 200 to 1000 sperm per animal should be morphologically examined at 400× to 1000× magnification. Slides should be quickly scanned to determine the quality of the preparation. Common artifacts are clumping, slides that contain too many sperm, and improperly prepared sperm smears. Most sperm smears will contain at least 500 sperm. Sperm on the periphery of the slide are routinely excluded from examination because of the greater likelihood of preparation artifacts (broken sperm, clumping, etc.).

The assessment and classification of rat sperm morphology is a subjective technique. Abnormalities may affect the head, midpiece, or tail,[109] and some sperm may have multiple abnormalities. Values for normal sperm vary between laboratories, ranging from 94% to 99.7%.[70] In normal human males, up to 50% of the sperm may be misshapen. There are no standard classifications for rat sperm,[112] but many investigators use the sperm morphology classifications proposed by Linder et al.[110] Spermatozoa were classified as normal, normally shaped head separated from the flagellum, misshapen head separated from the flagellum, misshapen head with a normal flagellum, misshapen head with a normal flagellum, and normal head with degenerative flagellum. Other descriptions include banana-head, pinhead, two-head, two-tailed, bent tail, and blunt hook.

H. Seminiferous Epithelium Staging

1. Spermatogenesis

While no quantitative assessment of the spermatogenic cycle (e.g., frequency distribution of tubules in each stage) is required by current reproductive regulatory guidelines, qualitative knowledge of staging is needed to evaluate the normality of the cellular composition of the seminiferous tubule (e.g., which cells are missing or are inappropriately present). A detailed description of spermatogenesis and spermiogenesis is beyond the scope of this chapter; however, detailed morphological descriptions with light and electron microscopic photographs of these processes are available in a book by Russell et al.[68] The process of formation and development of the spermatozoon is known as spermatogenesis. Spermatogenesis is routinely divided into three phases: the proliferative, meiotic, and spermiogenic phases.

Flagellar development is a continuous process, continuing until sperm release. Sperm are essentially immotile when released into the lumina of the seminiferous tubules. Sperm motility is developed in the epididymis, but vigorous motility occurs only after deposition into the female reproductive tract.

2. The Staging of Spermatogenesis

While no guideline studies require staging of spermatogenesis, the FDA Redbook 2000 states "testicular tissue should be examined with a knowledge of testis structure, the process of spermatogenesis, and the classification of spermatogenesis." This statement implies that at least the pathologist should be aware of spermatogenesis staging. Other regulatory reproductive study guidelines have similar statements.

The classification of spermatogenesis is based on the examination of cross-sectioned seminiferous tubules, and the organization of their cells has been extensively studied in the rat.[113,114] A stage is "a defined grouping of germ cell types at particular phases of development" in cross-sectioned tubules. Each cell type of the cell association is morphologically integrated with others in its developmental processes,[115] and each stage has a defined germ cell composition. The rat spermatogenic cycle has been divided into 14 stages (Roman numerals I to XIV).

Single cell associations develop and remain in one region of the tubule, called a *segment*. Spermatids will develop from Stage I through Stage XIV. Although the cells do not move within the tubule, the cell associations are distinctly arranged along the length of a tubule. Starting at the rete testis and proceeding into both ends of the seminiferous tubule, segments will show a linked order of stages decreasing as they go further into the tubule. For example, if the initial segment next to the rete testis were a Stage IX, the next segment further away from the rete testis would be at Stage VIII, then Stage VII, VI, and so forth. The linking of the adjacent consecutive cell associations is called the *continuity of segmental order*. The pattern of decreasing segmental order in a direction away from the rete testis is called the *descent in segmental order*. The descent in segmental order continues from both ends of the seminiferous tubule (away from the rete testis) until the two patterns of descent meet at a point within the tubule. This point is called the *site of reversal*. The site of reversal is not equidistant from the rete testis. This decreasing segmental pattern is not perfect, as some variations, called *modulations*, do occur (e.g., Stage X, IX, VIII, IX, VIII, VII). A complete series of stages along a tubule (including modulations) is called a *wave of the seminiferous epithelium*. Many waves (usually about 24) are usually present in a seminiferous tubule. In rats, a wave averages 2.6 cm in length, but waves are not uniform in length (ranging from 1 to 5 cm). Spermatogenesis (from spermatogonia to spermatozoa) requires about 63 d in the Sprague-Dawley rat. The *cell cycle duration* is the time that it takes to go from Stage I to Stage XIV. The cell cycle duration is 12.9 to 13.3 d in the Sprague-Dawley rat.[116]

IV. FEMALE REPRODUCTIVE TOXICOLOGY

A. Biology

Evaluation of the rodent female reproductive system usually consists of a gross evaluation (including organ weights) plus a histological evaluation of the ovary, uterus, cervix, vagina, and mammary glands. Vaginal cytology is assessed for determination of the estrous cycle. Female reproductive function is regulated through complicated interactions involving the central nervous system (particularly the hypothalamus), pituitary, ovaries, the reproductive tract, and the secondary sexual organs. In addition, nongonadotrophic components of the endocrine system may also affect the female reproductive system. To understand the significance of effects on female reproductive endpoints, the relationships between the reproductive hormones and the reproductive organs must be understood.

B. Observations

There are three basic types of observations — a viability check, a clinical observation, and an extended or detailed clinical observation — which are performed as described above for males.

C. Female Body Weight

Body weights are recorded daily, twice weekly, or weekly, depending upon the guideline requirements and reproductive phase (premating, gestation, or lactation). During cohabitation and gestation, body weights are recorded daily or at protocol-specified intervals (e.g., GD 0, 6, 9, 12, 15, 18, and 20). Maternal and pup body weights in the lactation or postnatal period are usually recorded weekly (PND 0, 7, 14, 21). Food consumption during gestation is usually recorded at the same intervals as body weights. Body weights and food consumption should be recorded at approximately the same time of day during the course of the study. Monitoring female body weight during treatment provides an index of the general health status of the animal that may be relevant in the interpretation of reproductive effects. Female body weight fluctuates normally because estrogen and progesterone influence food intake, energy expenditure,[117] water retention, and fat deposition rates.[118,119] Reduction in food intake and body weight in the female rat is one of the most sensitive noninvasive indicators of an estrogenic compound. In addition to endocrine-related factors, body weight loss or reduction in body weight gain can be caused by other factors, including systemic toxicity, stress (such as restraint), treatment-induced anorexia, or decreased feed or water consumption due to altered palatability caused by the test substance.

Unless a direct causal link between the observed female reproductive effect and decreased body weight or body weight gain can be established, female reproductive system alterations should be considered an adverse effect and not secondary to the occurrence of systemic toxicity. Therefore, in the presence of mild to moderate body weight changes, alteration in a female reproductive measure should be considered an adverse female reproductive effect. In the presence of severe body weight change, it can be stated that an adverse effect on a female reproductive endpoint occurred, but the effect may have resulted from a more generalized toxic effect.

D. Female Necropsy

1. Gestation Day 13 Uterine Examination (Segment I Studies)

The dams are routinely necropsied around GD 13 (early in the second half of gestation). The dam is euthanized, and the uterine contents are examined (as described below). Uterine position and viability (live or dead) of each implant is recorded, as are the corpora lutea counts from each ovary. The dam is subjected to a complete necropsy, and the reproductive organs and other tissues specified in the protocol are weighed and saved for possible histopathological evaluation. Certain guidelines require termination on GD 20 or 21 (term Cesarean section).

2. Gestation Day 20 or 21 Term Cesarean Section (Segment II Studies)

The dam is euthanized, the uterus is removed and weighed (recorded as gravid uterine weight), and the uterine contents are exposed. The number and uterine position of each resorption (early or late) and/or fetus (viable or dead) are recorded. The number of corpora lutea from each ovary is counted and recorded. The individual term fetuses are removed from their membranes, and the umbilical cord is clamped and cut. Each placenta is examined and weighed, if required, or discarded. The empty uterus is weighed and the value recorded. Tag, box position, or other suitable means of identification individually identifies each fetus. The fetus is examined externally for any abnormalities, sexed, and weighed. Abnormalities are classified as either malformations or variations. A malformation is "a permanent structural change that may adversely affect survival, development or function" while a variation is a "divergence beyond the usual range of structural constitution that may not adversely affect survival or health."[11] In rat studies, half of the fetuses are subjected to visceral evaluation while either the remaining half or all of the fetuses are processed for skeletal evaluations. For a detailed discussion of necropsy, uterine evaluation, and the fetal gross, visceral, and skeletal evaluations used in the developmental toxicity study, see Tyl and Marr.[30]

3. End of Lactation (Segment III Studies)

In Segment III and multigenerational studies, the dam is killed after the weaning of the offspring on PND 21. The dam is euthanized, subjected to a complete necropsy, and the reproductive organs and other tissues specified in the protocol are weighed and saved for possible histopathological evaluation.

E. Female Reproductive Organ Weight

1. Ovary

Ovarian weight in the normal rat does not show significant fluctuations throughout the estrous cycle. However, persistent polycystic ovaries, oocyte depletion, decreased corpus luteum formation, luteal cysts, altered pituitary-hypothalamic function, and reproductive senescence may be associated with ovarian weight changes. Ovarian gross morphology and histology should be examined to detect alterations associated with ovarian weight changes.

2. Uterus, Oviducts, Vagina, and Pituitary Gland

Uterine weight fluctuates throughout the estrous cycle, peaking at proestrus, when the uterus is distended with watery fluid (sometimes misdiagnosed as hydrometra) in response to increased estrogen secretion. Compounds that inhibit steroidogenesis and cyclicity can cause the uterus to become small and atrophic, thereby decreasing the uterine weight. Uterine weight change has been used as a basis for comparing relative potency of estrogenic compounds in bioassays.[120]

The oviducts are not routinely evaluated or weighed in reproductive toxicity studies. The oviductal weight fluctuates depending upon the stage of the estrous cycle. The oviductal lumen contains a variable amount of fluid, which is greatest at metestrus, the period of ovulation.[121] Gross evaluations can be of value in detecting morphologic anomalies, such as agenesis and segmental aplasia.

Vaginal weight changes usually parallel uterine weight changes during the estrous cycle, although the magnitude is less. The vaginal and cervical epithelia show cyclic changes in morphology and thickness associated with stages of the estrous cycle.

Pituitary gland weight can provide information into the reproductive status of the female. Pituitary weight normally increases with age and also increases during pregnancy and lactation. Increased pituitary weight often precedes cancerous lesions induced by estrogenic compounds, and conversely, reduced pituitary weight may result from decreased estrogenic stimulation.[122] However, the pituitary contains other cell types that are responsible for the regulation of a variety of physiologic functions unrelated to reproduction. Therefore, changes in pituitary weight do not necessarily reflect reproductive impairment. Selective histological stains should be used to confirm changes in regions of the pituitary associated with reproductive functions. If the pituitary weight changes can be associated with changes in reproductive functions of the pituitary, then such pituitary weight changes should be considered as adverse female reproductive effects.

F. Histopathological Evaluation of the Female Reproductive Organs

1. Ovaries

Numerous articles,[123] chapters,[124–127] reviews,[128,129,131] and books[130] have been written concerning the ovary and ovulation. A detailed description of ovarian histology and ovulation is beyond the scope of this chapter. The ovary serves a number of functions that are critical to reproductive activity, including ovulation of oocytes and production of hormones. The developing ovarian follicles produce estrogen, while the corpora lutea formed after ovulation produce progesterone. Histopathological evaluation of the three major compartments of the ovary (i.e., follicular, luteal, and interstitial) plus the capsule and stroma should be performed to reveal possible toxicological conditions.[132–134] Methods described below can be used to quantify ovarian follicular development.

Delayed ovulation can alter oocyte viability and cause trisomy and polyploidy in the conceptus.[135–137] Altered follicular development, ovulation failure, or altered corpus luteum formation or function can result in disruption of cyclicity and reduced fertility. Therefore, significant increases in follicular atresia, evidence of oocyte toxicity, interference with ovulation, or altered corpus luteum formation or function should be considered adverse female reproductive effects. In addition, significant changes in folliculogenesis or luteinization (e.g., increased cystic follicles, luteinized follicles), altered puberty, and/or premature reproductive senescence should be considered adverse female reproductive effects.

2. Uterus

The histological appearance of the normal uterus fluctuates according to the stage of the estrous cycle and pregnancy. The endometrium consists of simple columnar epithelium with simple tubular glands with little or no branching. During proestrus, the epithelium becomes more cuboidal, possibly a result of compression and stretching of the dilated uterus. During proestrus and estrus, inflammatory cells (mostly neutrophils) infiltrate the myometrium and endometrial stroma. Inhibition of ovarian activity with the resultant reduction in steroid secretion can result in endometrial hypoplasia and atrophy, as well as altered vaginal cytology. Uterine hypertrophy and hyperplasia can also be caused by prolonged estrogen or progesterone treatment, respectively. Significant dose-related increases in endometrial hyperplasia, hypoplasia, or aplasia should be considered adverse female reproductive effects.

3. Oviducts

The oviducts are not routinely examined histologically in reproductive toxicity studies, although gross and histological evaluations of the oviduct for anomalies induced during development (agenesis, segmental aplasia, and hypoplasia) can be performed. Oviductal hypoplasia or hyperplasia can occur because of reduced estrogen stimulation or prolonged estrogenic stimulation, respectively. Significant dose-related increases in oviductal hyperplasia or hypoplasia should be considered adverse female reproductive effects.

4. Vagina and External Genitalia

Developmental abnormalities of the vagina (agenesis, hypoplasia, dysgenesis) can be due to genetic factors, prenatal toxicant exposure, or may have an unknown etiology. *In utero* exposure of females to agents that disrupt reproductive tract development can cause development of ambiguous external genitalia (e.g., masculinization of female rodents with increased anogenital distance).[138] Errors in gender determination can be made when observing newborn pups with ambiguous external genitalia, resulting in erroneous at-birth sex ratio observations. Offspring from sex steroid–treated rodent females commonly have malpositioned vaginal and urethral ducts. Agents with estrogenic activity (such as DES) can accelerate the occurrence of vaginal patency while other agents (such as phytoestrogens and TCDD) can delay vaginal opening. Infantile or malformed vagina or vulva (including masculinized vulva or increased anogenital distance) and/or altered timing of vaginal opening should be considered adverse developmental effects.

The estrous cycle can be monitored in rodents by observing changes in the vaginal smear cytology.[139,140] Repetitive occurrence of the estrous cycle stages at regular normal intervals suggests that neuroendocrine control of the cycle and ovarian responses to that control are normal. A more complete description of the vaginal cytology associated with the estrous cycle is presented below.

In reproduction and fertility studies, vaginal smear cytology should be examined daily for at least three normal estrous cycles (approximately 2 weeks) prior to treatment, after onset of treatment, and before necropsy.[141] In addition to the determination of estrous cycle length and the occurrence or persistence of estrus or diestrus, daily vaginal smear data provide the incidence of spontaneous pseudopregnancy, an altered endocrine state reflected by persistent diestrus. The daily vaginal smear data can be used to distinguish pseudopregnancy from pregnancy based on the number of days the smear remains leukocytic. The evaluation of vaginal cytology also can detect onset of reproductive senescence in rodents.[142] Generally, the estrous cycle will be lengthened or the females will become acyclic. Lengthening of the cycle may be a result of increased duration of estrus or diestrus. A significant alteration in the interval between occurrences of estrus for a treatment group when compared with the control group should be considered an adverse female reproductive effect.

5. Pituitary

In histological evaluations of the rodent anterior pituitary, the relative sizes of the acidophils and basophils have been reported to vary during the reproductive cycle and pregnancy.[143] Significant histopathological damage involving cells in the anterior pituitary that control gonadotropin or prolactin production should be considered an adverse reproductive effect.

6. Mammary Gland and Lactation

The development and growth, histology, and pathology of the rat mammary glands have been well documented.[144] Mammary tissue is highly endocrine dependent for development and function.[145–147] Development of the six pairs of mammary glands in rats occurs between birth and puberty.

Toxic agents can adversely affect mammary gland size, histological appearance, and milk production and release, and many exogenous chemicals and drugs or their metabolites can be transferred into milk.[46,148,149] Milk may contain lipophilic agents (mobilized from the adipose tissue) at concentrations greater than those present in the blood or organs of the dam, thereby exposing the suckling offspring to elevated levels of such agents.

During lactation, the mammary glands can be dissected, weighed, and processed for histological examination. Techniques have been developed for measuring mammary tissue development, milk production, and milk composition in rodents.[147] Milk production may be estimated by weighing milk-deprived litters before and after nursing or by measuring the milk collected from the stomachs of pups at necropsy. The DNA, RNA, and lipid content of the mammary gland and the composition of the milk can be measured as indicators of toxicity.

Reduced pup growth may be caused by altered milk palatability, reduced milk availability, or by ingestion of a toxic agent secreted into the milk. However, other factors unrelated to lactation can cause reduced pup growth (e.g., altered maternal behavior, stress, litter size, or poor sucking ability in the pup). Significant milk production decrements or alterations in milk quality, whether measured directly or reflected in altered development of the pup, should be considered adverse reproductive effects.[38]

7. Reproductive Senescence

There is a loss of the regular ovarian cycles and associated normal cyclical changes in the uterine and vaginal epithelium with advancing age in the female rat. Although the mechanisms causing reproductive senescence are not thoroughly understood, there are age-dependent alterations that occur within the hypothalamic-pituitary control of ovulation.[150,151] Cumulative estrogen exposure may play a role in the loss of ovarian function in rats since adulthood estrogen treatment can accelerate the loss of function.[152] In contrast, depletion of oocytes is the principal cause of the loss of ovarian cycling in humans.[153] Early onset of reproductive senescence in females (e.g., as evidenced by cessation of normal cycling, ovarian pathology, or an endocrine profile that is consistent with reproductive senescence) should be considered an adverse female reproductive effect.

8. Developmental and Pubertal Alterations

The Guidelines for Developmental Toxicity Risk Assessment[11] and the Guidelines for Reproductive Toxicity Risk Assessment[38] are available as more detailed references to determine whether effects induced or observed during the pre- or perinatal period should be considered adverse. Significant effects on sexual development (e.g., malformations of the internal or external genitalia, altered anogenital distance) or age at puberty, whether early or delayed, should be considered adverse. Other adverse effects for females include alterations in age for various developmental landmarks (nipple development, vaginal opening) or alterations in vaginal cyclicity.

G. Vaginal Cytology

1. Description

Evaluation of the estrous cycle is used to detect alterations in the normal female reproductive system. Beginning at puberty (approximately 45 days of age in the rat), distinctive cyclic changes in uterine cytology occur in response to the ovarian hormones, estradiol and progesterone. Female rats are polyestrous, with normal cycles ranging between 4 and 5 days. Estrous cycles continue throughout the reproductive lifespan of the female, except during pregnancy and lactation or pseudopregnancy. Vaginal lavage is a simple, noninvasive method of collecting vaginal cells for determination of the estrous cycle. Vaginal lavage can also be used to confirm mating (presence of sperm). Changes in vaginal cytology reflect changes in circulating ovarian hormone levels (estradiol and progesterone). Monitoring of the estrous cycle in rats over several weeks provides information about the length of the cycle and about alterations in cyclicity, such as the induction of prolonged or persistent estrus or diestrus (pseudopregnancy).

The estrous cycle is separated into three or four distinct phases: metestrus (also called Diestrus I), diestrus (also called Diestrus II), proestrus, and estrus.[154] The metestrus or diestrus (Diestrus I and II) duration is usually 2 to 3 d, the proestrus duration is usually 1 d, and the estrus duration is also 1 d. Metestrus (Diestrus I) vaginal smears contain leukocytes, few polynucleated epithelial cells, and a variable number of cornified epithelial cells (some of these cells are beginning to roll).[140] Diestrus (Diestrus II) vaginal smears contains a mixture of cell types, primarily leukocytes, a variable number of polynucleated epithelial cells, and few to no cornified epithelial cells. During this period, vaginal epithelial cells are proliferating and sloughing off into the vaginal lumen. The proestrus vaginal smear contains numerous, round, polynucleated epithelial cells (in clumps or strands), a variable number of leukocytes, and few to no cornified epithelial cells. Serum estradiol falls rapidly on the evening of proestrus as the ovarian follicles leutinize. Ovulation occurs early in the morning of vaginal estrus. Proliferation of the vaginal epithelium decreases, and cornified epithelial cells predominate. The estrus vaginal smear contains no

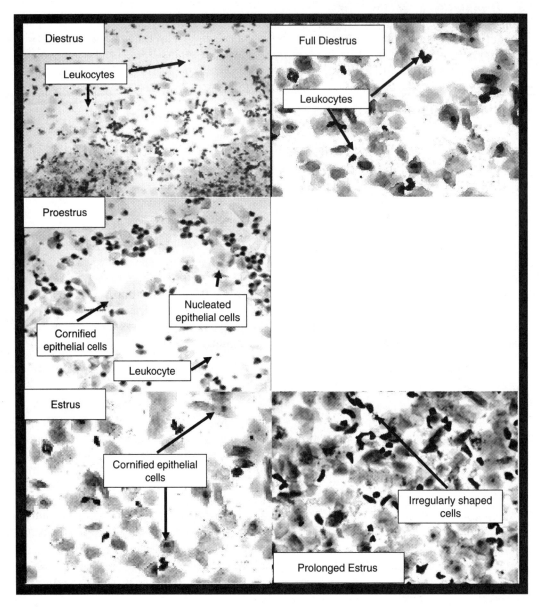

Figure 10.10 Estrous cycle evaluation.

leukocytes, few or no polynucleated epithelial cells, and many flat cornified cells. The changes in vaginal cytology are shown in Figure 10.10.

2. Vaginal Lavage Methodology

Vaginal lavage for the determination of the stage of estrous can be performed at any time of the day; however, it is important that the samples be collected at approximately the same time each day over the course of a study. A pre- and postcheck of the saline should be made. (Note: If glass pipettes are used, the rubber bulbs can be saved and washed for future use, while the glass stems should be discarded after each sampling. If plastic pipettes are used, the pipette should be discarded following each sampling. Dropping pipettes should be held with the tip pointing down whenever they contain a vaginal lavage or saline sample, to avoid contaminating the bulb. Contaminated bulbs should be discarded.)

The female rat is removed from the cage, and her identification is verified. One or two drops (approximately 0.25 ml) of physiological saline are drawn into a new clean dropping pipette, and the tip of the pipette is gently inserted into the vaginal canal. The pipette bulb is firmly but gently depressed to expel the saline into the vagina. The saline is gently drawn back into the dropping pipette, and the pipette is removed from the vaginal canal. If the flush is very clear, the lavage should be repeated. Care must be taken not to insert the tip too far into the vaginal canal because stimulation of the cervix can result in pseudopregnancy (prolonged diestrus). The animal is returned to its cage. The contents of the pipette are then delivered onto either (1) a clean glass slide or (2) a ring slide containing designated areas for the placement of each sample (vaginal smear). Each slide should be numbered, the top and bottom of each slide clearly identified, and a corresponding record made indicating the animal number and location for each vaginal smear. Technicians should record the date, the start and end times of sample collection, and their initials on appropriate forms. All slides should be handled carefully to avoid mixing vaginal lavage samples from different animals. If samples become mixed, they are discarded, and the sampling procedure is repeated for those animals.

3. *Estrous Cycle Staging*

Each vaginal lavage sample (fresh or stained) is examined microscopically (low power, 100×). The time and the technician's initials should be recorded on the estrous cycle evaluation data sheet when sample evaluations begin. Cellular characteristics of vaginal smears reflect structural changes of vaginal epithelium. The stage of the estrous cycle is determined by recognition of the predominant cell type present in the smear. Diestrus smears contain predominantly leukocytes, proestrus smears contain predominantly round, polynucleated epithelial cells (in clumps or strands), and estrus smears contain predominantly flat cornified cells. The stage of the estrous cycle is recorded for each animal on the appropriate data sheet. The end time and the technician's initials should be recorded on the estrous cycle evaluation data sheet.

4. *Estrous Cycle Evaluation*

Measurement of estrous cycle length is done by selecting a particular stage and counting until the recurrence of the same stage (usually day of first estrus, day of first diestrus, or day of proestrus). The estrous cycle lengths should be calculated for each female. It is often easier to visualize the estrous cycle by plotting the data (Figure 10.11). The group mean estrous cycle length should be calculated, and the number of females with a cycle length of 6 d or greater should be calculated. There is a general consensus that a single cycle with a diestrus period of 4 d or longer or an estrus period of 3 d or longer is aberrant. Cycles that have two or more days of estrus during most of the cycles are classified as showing *persistent* estrus. Cycles that have four or more days of diestrus during most of the cycles are classified as showing *persistent* or prolonged diestrus. Pseudopregnancy is a persistent diestrus (approximately 14 days in duration). Constant estrus and diestrus, if observed, should be reported. Persistent diestrus indicates at least temporary and possibly permanent cessation of follicular development and ovulation, and thus at least temporary infertility. Persistent diestrus, or anestrus, may be indicative of toxicants that interfere with follicular development, that deplete the pool of primordial follicles, or that perturb gonadotropin support of the ovary. The ovaries of anestrous females are atrophic, with few primary follicles and an unstimulated uterus, and as expected, serum estradiol and progesterone are abnormally low. The presence of regular estrous cycles after treatment does not necessarily indicate that ovulation occurred because luteal tissue may form in follicles that have not ruptured. However, that effect should be reflected in reduced fertility. Altered estrous cyclicity or complete cessation of vaginal cycling in response to

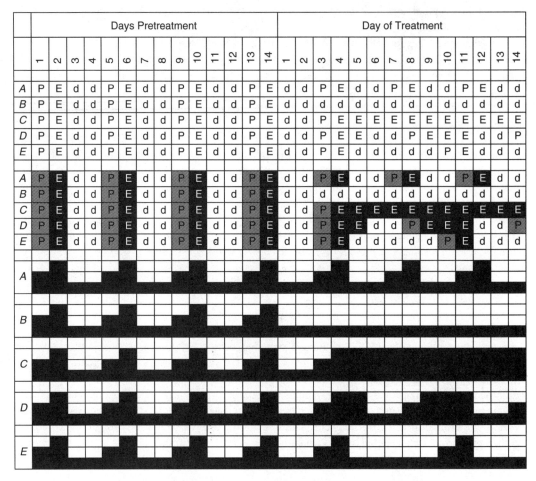

Figure 10.11 Graphical display of vaginal smear data. Three methods of depicting the estrous cycle data are presented. The first method presents the data on a spreadsheet with the estrus in bold. The second method presents the estrus and proestrus shaded. The third method uses stacked bars (estrus, 3 bars; proestrus, 2 bars; and diestrus, 1 bar). Cycle A represents a normal cycle. Cycle B shows the pattern in constant diestrus or pseudopregnancy. Cycle C shows constant estrus. Cycle D shows persistent or prolonged estrus. Cycle E shows persistent or prolonged diestrus. (Adapted from Cooper and Goldman in *An Evaluation and Interpretation of Reproductive Endpoints for Human Risk Assessment*, 1999.)

toxicants should be considered an adverse female reproductive effect. Subtle changes of cyclicity can occur at doses below those that alter fertility. However, subtle changes in cyclicity without associated changes in reproductive or hormonal endpoints would not be considered adverse.

H. Corpora Lutea Counts

The corpus luteum is formed from the thecal and granulosa cells of the postovulatory follicle and is a transitory endocrine organ.[155] Progesterone is the major hormone produced by the corpus luteum and is necessary for implantation and the maintenance of the pregnancy. Each corpus luteum appears as a round, slightly pink, discrete swelling on the surface of the ovary. Corpora lutea can be counted with the naked eye or under a dissecting microscope to determine the number of eggs ovulated. The number of corpora lutea should be recorded for each ovary. In a litter with no obvious signs of resorption, the number of corpora lutea should be equal to or exceed the number of total implantation sites. For partially resorbed litters, the number of corpora lutea may actually be less than the total number of implantations. When some of the corpora lutea are no longer required to support the remaining conceptuses, some of the corpora lutea involute to become corpora albicantia. However, the concordance between the number of corpora lutea that involute and the number of resorbed implants is not precise.

Corpora lutea can be either discounted when the count is less than the number of implants or the number of corpora lutea can be set to correspond to the number of implants to correct for involuting corpora lutea in partially resorbed litters.[30] The corpora lutea count is used in determining pre- and postimplantation losses. A reduction in the number of corpora lutea or an increase in the pre- and/or postimplantation loss should be considered an adverse reproductive effect.

I. Uterine Evaluation

1. Assessment of Resorptions

An implanted conceptus that is undergoing autolysis is termed a *resorption*. An early resorption is defined as a conceptus that has implanted but has no recognizable embryonic characteristics evident upon examination. An early resorption is usually visualized by ammonium sulfate staining or by observation through pressed glass plates. A late resorption is defined as fetal remains or tissues that have recognizable fetal characteristics (such as limb buds with no discernable digits present) and are undergoing autolysis. In some fertility studies where the necropsy is performed near term, fetuses are classified as live (fetus exhibits normal appearance, spontaneous movement, heart beat, etc.) or dead (nonliving full size fetus with discernable digits, with a body weight greater than 0.8 g for rats). In fertility studies where the pregnant female rodents are necropsied around GD 13 to 15, the conceptuses are categorized as live (normal appearance, heart beat, etc.) or dead (discolored, undergoing autolysis). The dead classification in these studies is analogous to resorption. The number and uterine positions of any early and late resorptions and viable and dead fetuses are recorded. Segment III and multigeneration studies routinely evaluate the reproductive organs of the dam at necropsy, postweaning.

2. Assessment of Implantation in "Apparently Nonpregnant" Dams

When dams appear not to be pregnant at necropsy, a thorough examination of the uterus must be performed. If no implantation sites are observed, the opened uterus is stained with a 10% aqueous solution of ammonium sulfide.[156] Record (under maternal gross findings) that the uterus was placed in stain. If implantation sites are found after staining, the pregnancy status must be updated.

J. Oocyte Quantitation

The U.S. EPA OPPTS 807.3800 Reproductive and Fertility Effects Guideline[10] requires a quantitative evaluation of the primordial follicle population of the ovary. The postlactational ovary should contain primordial and growing follicles, as well as the large corpora lutea of lactation. Ovarian histology, oocyte quantitation, and differential follicle count methodology have been described.[157] Ovarian follicles have been classified into 10 categories.[158] These 10 categories have been reduced into three classes for the purposes of quantification: primordial follicles, growing follicles, and antral follicles. The earliest primordial follicle is an oocyte with no surrounding granulosa cells (type 1), while the latest primordial follicle has a complete single layer of granulosa cells surrounding the oocyte. The earliest growing follicle has an oocyte that is starting to grow, and the surrounding cells have formed two layers (granulosa and theca), while the latest growing follicle has multiple layers of granulosa and thecal cells but no evidence of antrum formation. The earliest antral follicle has a large oocyte, multiple cell layers, and a developing antrum, while the latest antral (Graafian) follicle has mature multiple layers of granulosa and thecal cells and a large antrum.

Histopathological examination should detect qualitative depletion of the primordial follicle population; however, reproductive study guidelines require a quantitative evaluation of primordial follicles of the F_1 females. The number of animals, ovarian section selection method, and section sample size should be statistically appropriate for the evaluation procedure used. The guidelines cite ovarian follicular counting methodologies[159–161] that use a differential count of the primordial follicles, growing follicles, and antral follicles. The current EPA Multigeneration Protocol Modification[10] requires ovaries from 10 randomly selected females from highest treatment and control groups. Five 5-µm ovarian sections are made from the inner third of each ovary, at least 100 µm apart. The number of primordial follicles (and small growing follicles) is counted, and the presence or absence of growing follicles and corpora lutea is confirmed for comparison of highest treatment group and control group ovaries. If statistically significant differences occur, the next lower treatment group will be evaluated.

Differential follicular counts are associated with age of the female, number of animals evaluated, number and location of the ovarian sections, the number of sections per ovary, section thickness, differences between evaluators, and the criteria for defining primordial follicles. Ovarian follicle count appears to be more sensitive endpoint than ovarian weight. In 18 RACB studies using Sprague-Dawley rats,[161] the number of small follicles counted had a wide variation, ranging from 147 ± 57 to 556 ± 144, and another investigator reported 301 ± 13 small follicles.[131] A decrease in ovarian follicle count is usually considered a biomarker of an adverse reproductive effect because no recovery is possible. A recent review of follicle count historical control data in rats[162] suggests that: (1) with a large number of animals, there is relatively good replication within a generation (when evaluated by the same person); (2) there is great variability within individual ovaries and sections; (3) random selection of ovarian sections has a high probability of biasing effects; (4) ovarian follicle counts do not appear to be correlated with either ovarian weight or body weight at termination; and (5) ovarian follicle counts should not be the sole endpoint used in risk assessment. However, even with the great variability between reported values, a detectable decrease in follicle count in a treated group when compared with the control group is considered an adverse effect.

V. PARTURITION AND LITTER EVALUATIONS

A. Parturition

Female rats should be housed to litter in boxes with bedding material no later than day 20 of presumed gestation (2 d prior to the day of expected delivery; earlier if required by protocol). The dams are observed periodically throughout the day for stretching, visible uterine contractions, vaginal bleeding, and/or placentas in the nesting box. These signs are strong indicators of parturition onset, and indicate that dams should be monitored closely for delivery of the first pup.

As soon as a pup is found, parturition is considered initiated. If required by protocol, the time should be recorded. A marker (e.g., a tape "flag") may be placed on the outside of the nesting box, indicating a delivery in progress. Once delivery has begun, the dams are checked periodically during the day for evidence of difficulty in labor or delivery (e.g., a pup only partially delivered, a dam cold to the touch and pale). Any apparent difficulty is recorded, and a supervisor or veterinarian should be notified if the difficulty seems extreme or unusual.

The calendar date on which delivery of a litter appears to be completed is defined as lactation day 0 (LD 0), postpartum day 0 (PND 0), and day 0 of age. The following maternal behaviors offer evidence that parturition is complete: (1) removal of amniotic sacs, placentas, and umbilical cords, and grooming of the pups by the dam; (2) self-grooming by the dam; (3) nesting behavior by the dam; and (4) nursing. However, parturition may be complete without all of these behaviors present. Gentle palpation of the dam may also be used. If the time of completion of parturition is required, that information is recorded. Individual pup observations, such as appearance and viability, should not be recorded until delivery is completed. Apparently dead pups should be removed from the nesting box during parturition to preclude cannibalization by the dam. The pups should be retained in a container such as a weigh boat, labeled with the dam number, until pup statuses are recorded following completion of parturition. Pups delivered but not present at the end of parturition because of cannibalization should be recorded as "undetermined sex, undetermined viability."

In order to preclude disruption of delivery and/or maternal care, dams in the process of delivering pups should not be administered the test substance. If a dam has completed parturition by the time the last additional animal in the normal sequence has been dosed, this dam can be administered the test substance also. If the dam continues to actively deliver after daily test substance administration is completed, the dam's daily administration can be skipped. A delayed or missed daily test substance administration and the reason for its occurrence should be recorded on the appropriate form.

Occasionally, additional pups are born to a litter that has already been weighed and observed on PND 0. These pups are handled as follows:

> If delivery of a litter has been completed and it is within the same day, the additional pup(s) and dam are weighed on "LD 0/PND 0" (for example, a Monday).
> If an additional pup(s) is found on PND 1 (for example, Tuesday), when this litter is counted, the following comment (or a comment that conveys the same information) should be entered on the litter observation form for PND 1. "One additional (sex) pup present that was delivered after completion of weighing and recording of litter observations for all pups delivered on PND 0." The additional pup(s) is not weighed until PND 4; the next scheduled weighing interval for that litter (for example, Friday). The pup(s) is tattooed if required by protocol.

Edits will be required to change the number of pups in the litter. The date of the dam's DL 0 remains the same (it is *not* changed to the date that the additional pup was found). Pups are weighed on the days specified in the protocol. On those days when pup body weights are recorded, bedding is usually changed and each litter is evaluated for maternal and pup nesting behavior. Each litter should be checked for viability and counted every day, with the number and status of the pups recorded.

B. Milk Collection

Prior to milk collection, the pups are withheld from the dam for approximately 4 h. At least 5 min prior to milk collection, 0.05 ml of oxytocin (20 units/ml) is administered intravenously to the dam. The time that the oxytocin is administered should be documented on the appropriate form. The dam is then held in a position that facilitates access to the teats. Milk is expressed from one teat at a time by applying gentle digital pressure. Initially the pressure is applied at the base of the teat and then is directed distally. The milk is collected into a microtainer or other suitable container (labeled with the protocol and animal number, date, identification of sample, and test system, i.e., rat milk). The time of collection and volume of milk collected is documented on the appropriate form. If more milk is required, then another teat is selected and milked. When the milk collection is completed, the dam is returned to her litter.

C. Litter Evaluations

1. Litter Observations

The entire litter is observed for general appearance and maternal and pup nesting behavior. The following items should be checked: (1) that the pups are warm and clean, (2) that there is evidence of a nest, (3) that the pups are grouped together, and (4) that the pups are nursing or have milk present in their stomachs (milk spot present).

2. Gender Determination

The nesting box is removed from the rack, the dam's identification is verified, the dam is placed into the holding cage, the pups are removed from the nesting box, and the gender of each pup is determined. Anogenital distance (see below) is used to determine gender (the male anogenital distance is greater than the female anogenital distance). Pups that have been cannibalized to the extent that their gender cannot be determined are recorded as "sex unable to be determined due to degree of cannibalization." If a malformation precludes sex determination, this is recorded (for example, "tail absent, no apparent anus, sex could not be determined"). If the malformed pup is dead, the sex is determined by internal examination.

3. Viability

The viability of each pup is determined. Live pups are recorded as alive. Dead pups are necropsied, the trachea is tied off with string, and the lungs are removed. The lungs are then placed in a container of water. If the dead pup's lungs float, the pup is assumed to have taken a breath, and the viability is recorded as "born alive, but found dead." If the lungs sink, the pup is assumed to have never breathed, and the viability is recorded as "stillborn." Dead pups can be cannibalized to the point where viability cannot be determined. The sex and viability status are recorded for each pup (including dead, cannibalized, and malformed pups) in all litters.

4. External Alterations

Each pup in the litter is evaluated for external alterations. The following items are evaluated for each pup: the general shape of the head; presence and completeness of features; trunk continuity; absence of bruises, lesions, or depressions; length and shape of limbs (toes are counted); length and diameter of the tail; presence of anus; milk in stomach; and whether partial cannibalization has occurred. All live pups are weighed. If a live, partially cannibalized pup is weighed, its weight is footnoted in the raw data.

5. Missing Pups

If a pup cannot be found during the daily litter count, the pups are sorted by gender to determine the sex of the missing pup(s) by process of elimination. The missing pup is recorded as "missing, presumed cannibalized." If the pups have been tattooed or otherwise identified, the actual number of the missing pup should be recorded.

6. Pups Found Dead

All found dead pups are necropsied. Stomach contents (i.e., "stomach contains or does not contain milk") of all dead pups should be recorded. Unless otherwise dictated by the protocol, dead pups found on LD 0 to 4 with gross lesions are fixed in Bouin's fixative. Gross lesions from pups found dead on LD 5 through weaning are fixed in 10% neutral buffered formalin. Because of postmortem autolysis, dead pups are not routinely weighed.

7. Dam Dies before Scheduled Sacrifice

If a dam with a litter dies before her scheduled sacrifice, the pups are sacrificed and necropsied. The sacrifice and necropsy observations are recorded.

8. All Pups in a Litter Die

If all the pups in a litter die, the dam may remain in the study until her scheduled sacrifice date, or the dam may be sacrificed at the discretion of the study director.

9. Gender Determination Errors

If it is discovered that an error was made in the determination of pup gender, the appropriate edit should be written and the raw data (manual form or computer printout) corrected for that day and each preceding day. For example, if a sexing error is found at necropsy of a culled pup on LD 7, an edit is written to correct the gender, and the data from the previous lactation days are corrected.

After the litter observations, the bedding in the nesting box can be replaced with clean bedding or a new box with clean bedding can be obtained. The dam is weighed and observed for clinical signs, and the feed jar is weighed (if appropriate). The dam and pups are placed into the clean box, and the dam's cage tag and feed jar are transferred to the new box. The nesting box is returned to the rack.

VI. PUP EVALUATIONS

A. Pup Body Weights

1. Pup Birth Weight

Birth weight should always be measured for each pup on the day of parturition. Data from individual pups are reported as the entire litter weight and as weight by sex per litter. Growth rate *in utero* is influenced by the normality of the fetus, uterine milieu, uterine position, and fetal gender, with females tending to be smaller than males.[163] Maternal nutritional status, intrauterine growth rates, litter size, and gestation length influence birth weights. Individual pups in small litters tend to be larger than pups in larger litters. Therefore, reduced birth weights that can be related to large litter size or increased birth weights that can be related to small litter size should not be considered an adverse effect unless the altered litter size is considered treatment related and survivability and/or development of the offspring have been compromised.

2. Postnatal Pup Body Weights

Postnatal weights are dependent on litter size and on pup gender, birth weight, suckling ability, and normality, as well as on maternal milk production. With large litters, small or weak offspring may not thrive and may show further impairments in growth or development. Because one cannot determine whether growth retardation or decreased survival rate was due solely to the increased litter size, these effects are usually considered adverse developmental effects. Conversely, pup weights in very small litters may appear comparable to or greater than control weights and therefore may mask decreased postnatal weights in other litters.

3. Crown-Rump Length

Crown-rump lengths for fetuses are measured with a caliper. Fetuses may be euthanized before measuring the crown-rump length. Place each fetus on its side on a flat surface, and place the caliper with one prong touching the top of the head and the opposite prong touching the base of the tail. Do not stretch or compress the fetus. Record the measurements in millimeters.

4. Anogenital Distance

Anogenital distance is the length between the anus and the genital tubercle. In rodents and primates, the anogenital distance is greater for males than females.[54] For rodents, anogenital distances are greater for males than females beginning on GD 17 and continuing through PND 21. Mean anogenital distances at birth for males and females are approximately 3.5 and 1.4 mm, respectively. The increased growth of this region occurs in response to testosterone. The genders of rabbit fetuses or neonates cannot be reliably determined using anogenital distance but can be determined by examining the gonads.

Anogenital distances for rat fetuses (GD 20 or 21) or pups on PND 0 to 3 are measured with a micrometer and a stereomicroscope. Measurements taken on pups on PND 4 or later should be done using a caliper. The fetus or PND 0 to 3 pup is held in the technician's hand, and the tail is raised with the other hand to an 80° to 90° angle from the horizontal, exposing the anus. (Note: Be careful not to pull the tail, as this stretches the anogenital distance.) The anogenital area is brought into focus with a calibrated stereomicroscope. The anogenital distance is measured from the cranial (or anterior) edge of the anus, which comes to a point, to the base (or posterior edge) of the genital tubercle. The base of the tubercle is not clearly differentiated as it slopes into the anogenital area. A baseline between the distinct edges of the genital tubercle is visually estimated. It is very important that the anus and the base of the genital tubercle be kept in the same focal plane. The length from the base of the genital tubercle to the cranial edge of the anus is recorded.

Hold the PND 4 or older pup by its tail, keeping the tail at an 80° to 90° angle from the horizontal. The arms of the caliper should be aligned as follows: for males, the anogenital distance is measured from the cranial (or anterior) edge of the anus to the base (or posterior edge) of the anogenital aperture; and for females, the anogenital distance is measured from the cranial edge of the anus to the base of the urinary aperture (*not* the base of the vulva). The anogenital distance is recorded in millimeters. (For further description, see Chapter 9, Figures 9.9 through 9.11)

B. Developmental Landmarks

The onset of various developmental landmarks can be used to assess postnatal development. Sexual maturation landmarks (balanopreputial separation, vaginal patency) are required or recommended in perinatal-postnatal, multigenerational studies, and developmental neurotoxicity studies, where the offspring are raised to adulthood. Additional developmental landmarks can be assessed including pinna detachment, hair growth (pilation), incisor eruption, eye opening (pups are born with eyelids closed), nipple development, and testes descent. The development of the following reflexes can also be evaluated: surface righting reflex, cliff avoidance, forelimb placing, negative geotaxis, pinna reflex, auditory startle reflex (pups are born with the external auditory canal closed), hind limb placing, air righting reflex, forelimb grip test, and pupillary constriction reflex. The examinations must begin prior to the landmark' historical day of onset and continue daily until each animal in the litter meets the criterion.[164] Data for each day's testing should be expressed as the number of pups that have achieved the criterion for each developmental landmark, divided by the total number of pups tested in the litter. Forelimb grip test and pupillary constriction tests are conducted on PND 21 only. Data for the forelimb grip test and pupillary constriction tests should be expressed as the number of pups that have achieved the criterion, divided by the total number of pups tested in the litter. The postnatal days listed below for each developmental landmark were compiled from several sources and are subject to variability between laboratories and subtle differences in assessment of the criteria.[164–166]

1. Balanopreputial Separation

In rats, balanopreputial separation is considered to result from balanopreputial membrane cornification, which leads to the detachment of the prepuce from the glans penis in the rat. Preputial separation occurs dorsolaterally and then ventrally on the penis and down the shaft of the penis.[54] The prepuce remains attached to the glans penis on its ventral surface by the frenulum. (Note: The process of development of the prepuce in humans is different from that of the rat.) Male rodents are examined for balanopreputial separation beginning on PND 22 (mouse), PND 27 (hamster), or PND 35 to 40 (rat). Published preputial separation age ranges from PND 32 to 42 (Japanese studies)[167–169] or PND 41 to 46 (American studies),[54,170] depending on the observation criterion. The Japanese preputial separation criterion is based on the age that the prepuce separates from the glans penis, while the American criterion is met when the prepuce completely retracts from the glans penis and the foreskin is cuffed around the base of the penis. Therefore, it is important to define the criterion in the protocol.

Each male rodent is removed from its cage and held in a supine position. Because manipulation of the prepuce can accelerate the process of preputial separation, the males must be examined gently. Gentle digital pressure is applied to the sides of the prepuce, and the criterion is met when the prepuce completely retracts from the head of the penis (see above for alternate criterion). The foreskin can be attached along the shaft of the penis, but it cannot be attached to the opening of the urethra. Each male rodent is examined daily until acquisition or until PND 55, whichever is earlier. Body weight should be recorded on the day the criterion is met.

2. Vaginal Patency

After the canalization of the vagina occurs in rats, the vaginal opening remains covered by a septum or membrane. The age when the septum is broken or no longer evident is described as the age of vaginal patency, and vaginal patency is the most readily determined marker for puberty in rats.[54] Published Sprague-Dawley rat vaginal patency values range from PND 30.8 to 38.4.[169,171] Female rats are examined beginning a few days prior to expected age of maturation (e.g., PND 25 to 28), and the examinations are continued until the criterion for patency has been achieved or until PND 43, whichever comes first. The female is removed from the cage and held in a supine position, exposing the genital area. Pressure is gently applied to the side of the vaginal opening to see if the septum or membrane remains. When the membrane is present, the area has a slight "puckered" appearance; however, when the membrane has broken, the vaginal opening is about the size of a pinhead. The criterion has been met when the vagina is completely open. Body weight should be recorded on the day the criterion is met.

3. Day of First Estrus

After vaginal opening has been determined, vaginal smears can be taken daily to determine the day of first estrus (defined as the first day on which only cornified epithelial cells are observed on a vaginal smear). First estrus normally occurs within a few days after vaginal opening (approximately PNDs 40 to 45) in the Sprague-Dawley rat.[172]

4. Pinna Unfolding

Beginning on PND 1, each pup in a litter is examined individually until all pups in the litter meet the criterion for pinna unfolding (i.e., the first day that the point of a pinna [earflap] is detached from a rodent pup's head). Each pup is examined closely for detachment of either the right or left pinna from the side of the head, and criterion is met when the point of either pinna is detached. This usually occurs within PND 1[164] to 7[165] in Sprague-Dawley rats, with the average acquisition around PND 2 or 3.

5. Hair Growth

Beginning on PND 1, rodent pups are examined for the appearance of hair. All pups in each litter are examined daily until bristles appear on the dorsal surface of all pups in the litter. Criterion is met when there are bristles on the dorsal surface. This usually occurs within PND 1 to 5[164] in Sprague-Dawley rats.

6. Incisor Eruption

Rodent pups are examined for the appearance of incisor eruption beginning on PND 7. Each pup in every litter is carefully examined for eruption through the gum of either an upper or lower incisor. Criterion is met when an incisor is visible. This usually occurs within PND 8 to 16[164,165] in Sprague-Dawley rats, with the average acquisition around PND 11.[173]

7. Eye Opening

Beginning on PND 10, all pups in each litter are examined daily until all pups in the litter meet the criterion for eye opening. Each pup is carefully examined for any break in the membrane connecting the upper and lower eyelids, and criterion is met when there is a break in the membrane of either eye. This usually occurs within PND 11[164] to 18[165] in Sprague-Dawley rats, with the average acquisition around PND 13 or 14.

8. Nipple Retention

On PND 11 through 13, all males in each litter should be examined for the presence of areola and/or nipples, which should no longer be present by PND 12 or 13. Each pup is carefully examined for the presence of nipples and/or areola (usually only in the thoracic region) by brushing the hair coat against the nap. The data are usually presented as percent of males with areolae and percent of males with nipples.

9. Testes Descent

Male rodents are examined for testis descent beginning on PND 19. Each male rodent is removed from its cage and examined for the presence of one or both testes in the scrotum. This usually occurs within PND 20 to 29[164] in Sprague-Dawley rats, with the published average acquisition between PND 18[165,169] and 25.[166]

C. Behavioral or Reflex Testing

1. Surface-Righting Reflex

The surface-righting reflex is evaluated beginning on PND 1 for rat pups. Surface righting is regaining the normal position after the pup is placed on its back. This is a complex coordinated action requiring many different muscles in the neck, trunk, and limbs. Pups are placed on their backs on a flat surface and quickly released. The criterion is met when the pup regains its normal position on four paws on the floor within 5 sec. This usually occurs between PND 1 and 9[164,165] in Sprague-Dawley rats, with the average acquisition around PND 6[169] or 7.[166]

2. Cliff Avoidance

Cliff avoidance is evaluated beginning on PND 1 for all pups. Cliff avoidance is the behavior of crawling away from an edge of a flat surface edge (cliff). Pups are placed on a table or platform, with their front paws over the edge. The criterion is met when the pup attempts to crawl away from the edge within a 10-sec period. This usually occurs between PND 2[164] and 12[165] in Sprague-Dawley rats, with the average acquisition around PND 8 or 9.[166]

3. Forelimb Placing

Beginning on PND 7, rodent pups are tested for forelimb placing or lifting in response to tactile stimulus on the dorsal surface of the foot. The pup is suspended by holding the scruff of the neck so that one forelimb is in contact with a stainless-steel plate (e.g., cage tag holder) held horizontal to the working surface by a clamp and ring stand. The dorsum of the suspended foot is gently touched with a thin (approximately 2 to 3 mm in diameter) metal rod. The pup must immediately raise and place the suspended foot on the rod to meet the criterion. This usually occurs around PND 9 to 10.[165]

4. Negative Geotaxis Test

Beginning on PND 7, rodent pups are tested for the ability to change from a downward to an upward orientation on an inclined plane of approximately 30°. Each pup is placed facing "downhill" on a platform tilted at a 30° angle. The pup must turn 180° to face "uphill" within a 60-sec interval to meet the criterion (a pup with its body positioned sideways and its head facing "uphill" does not meet the criterion). This usually occurs between PND 7 and 14[165] in Sprague-Dawley rats, with the average acquisition around PND 7[169] or 8.[166]

5. Pinna Reflex

The pinna reflex tests the somatomotor component of the 7th cranial nerve in rats. The presence of the pinna reflex is evaluated daily, beginning on PND 13. The inner surface (near the concha) of the pinna is lightly touched with a filament or the tip of an artist's brush. The criterion is met with the presence of any movement of the pinna (sudden twitch or flattening of the ear) made in response to the applied stimulus. If the first ear tested does not respond to the stimulus, the opposite ear is tested. This reflex usually occurs around PND 14[165] in Sprague-Dawley rats.

6. Auditory Startle Reflex

Beginning on PND 10, all pups are examined daily for the auditory startle reflex. The auditory startle reflex is noted as a sudden flinch or cessation of ongoing movement following the auditory stimulus. Each nesting box is removed from the study room and placed in a quiet room. Littermates remain outside the testing room in order to mitigate habituation to the auditory stimulus. The pup is placed into a container, such as a beaker, with approximately 600 ml of bedding material and taken into the testing room. A clicker is held directly above the beaker, but not touching it, and the clicker stimulus is delivered. Any observable whole body response (e.g., flinching, jumping, and freezing of activity) meets the criterion for the startle response. This usually occurs between PNDs 12 and 13[164,165] in Sprague-Dawley rats.

7. Hind Limb Placing

Beginning on PND 14, rodent pups are tested for hind limb placing in response to a tactile stimulus on the dorsal surface of the foot. Each pup is held so that one hind limb is in contact with the metal plate (as in IV.C.3. above). The dorsum of the suspended foot is gently touched with a thin rod. The pup must immediately raise and place the suspended foot on the rod to meet the criterion. This usually occurs around PND 16.[165]

8. Air Righting Reflex

The air righting reflex is the ability of pups to land on all four paws when dropped from an inverted position. All pups in each litter are tested once each day beginning on PND 14 until all pups in the litter demonstrate the reflex. The pup is held in a supine position above a well-padded surface and then released. Rat and hamster pups are held approximately 38 cm above the surface; mouse pups are held 17 to 20 cm above the surface. The pup must land on all four limbs to meet the criterion. This usually occurs between PND 8[164,165] and 18 in Sprague-Dawley rats, with the average acquisition around PND 16[169] or 17.[166] The use of videotape to record the response will allow more information to be obtained, such as response speed, character, and progression.[174,175]

9. Forelimb Grip Test (Categorical)

Each pup is tested for forelimb grip on PND 21. A thin rod (approximately 2 to 3 mm in diameter) is supported by a ring stand suspended horizontally above the padded surface. The pup is held so it can grasp the thin rod with its forepaws and is then released. It must remain suspended for at least 1 sec to meet the criterion. The number of pups that met the criterion, divided by the total number of pups tested, is recorded.

10. Pupil Constriction Reflex

All pups in each litter are examined once on PND 21 for direct and consensual pupillary constriction of both eyes in response to the beam from a penlight. This test evaluates the autonomic component of cranial nerve reflexes. Each nesting box is transferred from the study room to a quiet isolated testing room, sufficiently dim to dilate the pupils of the eyes of the pups. Each pup is tested individually, with the following requirements to meet the criterion:

> The pup is removed from the nesting box and the penlight is directed into one eye. Immediate constriction of the pupil of the eye being tested is the initial requirement.
> Without turning the light off, it is immediately directed into the contralateral eye, the pupil of which should already be constricted.
> The light is turned off for a minimum of 5 sec. This sequence is then repeated for the other eye. Responses should be identical to meet the criterion.

D. Examination of Rodents for Hypospadias

In males, hypospadias is defined as an opening of the urethra at an abnormal location on the ventral surface of the penis. In normal males, the urethral opening is located at the tip of the penis. In affected males, the urethral opening extends a variable distance from the base of the penis. Other physical characteristics that may be seen in the affected males are inability to fully extend the penis, distended preputial glands, undescended testes, and decreased anogenital distance. In females, hypospadias is defined as an opening of the urethra into the vagina rather that at the urinary aperture adjacent to the vaginal opening.

1. In-life Procedure for Male Rodents

Male rodents are examined beginning on PND 39 (rat) or PND 22 (mouse). Each male is removed from its cage and held in a supine position. Gentle digital pressure is applied to the sides of the rodent's prepuce. The ventral surface of the penis is examined for the location of the urethral opening. The presence of hypospadias is documented as follows: If the prepuce has not separated and the location of the urethral opening is not apparent, the observation is documented as "hypospadias not determined" (this observation is continued daily until it is determined whether hypospadias is present); if the urethral opening is located at the tip of the penis, the observation is recorded as "hypospadias absent"; or if the urethral opening is located ventrally anywhere along the shaft of the penis, the observation is recorded as "hypospadias present." Severity scores, if required, are assigned. If the penis can be extended and the urethral opening is located approximately one-half to two-thirds the distance from the base of the penis, then the severity is recorded as "slight." If the penis cannot be fully extended and the urethral opening is located approximately one-quarter to one-half the distance from the base of the penis, the severity is recorded as "moderate." Moderate hypospadias may also be associated with distended preputial glands and undescended testes. If the penis cannot be extended and the urethral opening is located at the base of the penis, the severity is recorded as "extreme." Extreme hypospadias is usually associated with distended preputial glands, undescended testes, and visual presence of the os penis as the unattached flaps of penile tissue spread laterally.

2. Postmortem Procedure for Male Rodents

The male rodent is euthanized and then held in a supine position while gentle digital pressure is applied to the sides of its prepuce. Forceps are used to extend the penis, and the ventral surface of the penis is examined for the location of the urethral opening. The presence of hypospadias and severity scores are documented as described above.

3. Postmortem Procedure for Female Rodents

The female rodent is euthanized, its abdomen is cut open, and a dye (e.g., gentian violet) is injected into its bladder. Gentle pressure is applied to the distended bladder, and the location of the urethral opening is determined by watching for exiting dye. The presence of hypospadias is documented as follows: (1) if the dye exits from the urinary aperture adjacent to the vaginal opening, the observation is documented as "hypospadias absent," or (2) if the dye exits from the vagina, the observation is documented as "hypospadias present."

VII. FUNCTIONAL OBSERVATION BATTERY

Neurobehavioral screening methods, such as the FOB, have been routinely used to identify potential neurotoxicity of new and existing chemicals, primarily in adult rats. The U.S. EPA published guidelines for neurotoxicity screening tests, including the FOB and motor activity, in 1985[176] and subsequently revised them in 1991[177] and 1998.[178] These methods have been validated and large databases exist for effects of a wide range of chemicals. The FOB consists of approximately 20 to 30 endpoints that were selected to evaluate aspects of nervous system function. Most of the data collected is subjective. Therefore, the quality of the data depends primarily on the observer's ability to detect and describe changes in the animals' behaviors. The former American Industrial Health Council has developed a video and manual to train observers and to achieve consistency in the FOB across laboratories.[179]

Training of the technicians must be a continuous process, and variability between technicians should be minimized. The same observer should perform the FOB on all of the animals at a time point and, ideally, at all time points. When performing the FOB, the observer should be unaware of the animal's treatment group. Stable environmental conditions (e.g., light level, temperature, background noise) must be maintained or data may be difficult to interpret. An assessment of the various tests used in the FOB is presented in Ross.[175] A detailed or extended clinical observation, a modified FOB (without grip strength, foot splay, or body temperature measurements), is currently being used by many laboratories. FOB trained technicians have greater training in neurological assessment, and therefore, the quality of the information collected in the detailed or extended clinical observations using FOB trained technicians is increased.[180]

A. Adult FOB

1. In-Cage Observations

The adult rat is observed while it is in its home cage for the following characteristics: posture (e.g., sitting, lying on one side, limbs spread), palpebral (eyelid) closure (e.g., open, closed, drooping), writhing, circling, biting, and gait and coordination abnormalities. The observations and/or scores are recorded on the appropriate form.

2. Removing from Cage Observations

While the adult rat is being removed from its cage, it is observed for the following characteristics: ease of removal, ease of handling, vocalizations, muscle tone (e.g., limp, rigid), bite marks on tail and or paws, palpebral closure, piloerection, fur appearance (includes observation of skin, mucous membranes, and eyes), lacrimation, salivation, and exophthalmos. The observations and/or scores are recorded on the appropriate form.

3. Standard Open Field Arena Observations

The adult rat is placed in the standard open field arena for approximately 2 min. The technicians observes and assesses the following characteristics: presence of righting reflex, posture, respiration ease, rate of respiration, convulsions, tremors, muscle fasiculations, muscle spasm, grooming, gait and coordination abnormalities (e.g., unbalanced, ataxic, unable to move), level of arousal (e.g., stupor, excited), palpebral closure, diarrhea, polyuria, rearing (number of times both forepaws are lifted above cageboard), and vocalizations. The observations and/or scores are recorded on the appropriate form.

4. Manipulations in the Standard Open Field Arena

While the adult rat is in the open field arena, the technician observes and assesses the following characteristics: approach and touch response, auditory stimulus, and tail pinch. For the approach and touch response, the rat is rapidly approached toward the front of its head with a blunt probe (e.g., a finger) and then is touched between the eyes. The observer records the presence or absence of the approach and touch response. For auditory stimulus, an appropriate stimulus (e.g., finger cricket or snap) is presented directly above the rat's head and observer records the presence or absence of the flinch reflex or other responses. For the tail pinch, the observer grasps the middle portion of the tail firmly between the thumb and index finger and pinches the tail. The observer records the presence or absence of a tail pinch response. Additional observations are noted when present (e.g., sniffing, head weaving, Straub tail, syncope [loss of consciousness], stereotypic behavior).

5. Grip Strength Manipulations

The technician selects the appropriate setting for the force gauge (for example, a Chatillon Digital Force gauge) and tares it. The adult rat is grasped to provide support for the weight of the animal. The paws are brought into contact with the appropriate grip surface, ensuring that only forelimb paws come into contact with the forelimb grip surface and only the hind limb paws come into contact with the hind limb grip surface. The rat is allowed to grasp the surface with both paws being tested. The technician then pulls the animal with a fluid motion away from the gauge until the paws release. The technician performs at least three forelimb and three hind limb trials for each animal. The gauge should be tared between trials, and the gauge readouts are documented on the appropriate form.

6. Foot Splay

The technician applies ink to both heels of the hind paws of the adult rat. The rat is grasped by the base of the tail with one hand and around the thorax with the other hand and held approximately

32 cm above a sheet of paper. When the rat is still, the technician releases it, and it drops onto the paper. The technician marks the position of the first footprints made and measures and records the distance (in cm) between the centers of the backs of the heel prints.

The rationale for the use of the width of the foot splay measurement, as an indicator of neurological function, is not clear. Increased foot splay width is considered an adverse neurological effect. The anatomical basis for this reflex is unknown; see Ross[175] (p. 470) for a discussion of this parameter.

7. Pupillary Response

This test evaluates the autonomic component of cranial nerve reflexes. Light from a penlight is shined into one eye of the adult rat while the other eye is shielded from the light. The normal response to the light is pupillary constriction. This procedure is repeated for the other eye, and the responses are recorded on the appropriate form.

8. Body Temperature

The adult rat's rectal temperature is determined with a clean, calibrated temperature probe. A lubricant (such as K-Y Jelly®) is placed on the tip of probe. The animal is removed from its cage and gently restrained to expose the anus, while also limiting its movement. The tip of the lubricated probe is gently inserted into the rectum until the bulb end is no longer visible (approximately 0.6 to 0.95 cm). (Note: Handle only the cable and not the probe after insertion of the tip of the probe into the rectum because touching the probe may alter temperature readings.) The temperature is recorded when the digital display has stabilized (approximately 10 to 20 sec).

B. FOB in Preweanling Rats

The Developmental Neurotoxicity Study (OPPTS 870.6300) requires the assessment of the central nervous system of 10 male and 10 female offspring per dose group on PND 4, 11, 21, 35, 45, and 60. The standard FOB is appropriate for rats 21 days or older, but some of the assessments are inappropriate for preweanling rats because they cannot be performed with young rats (PND 4 or 11) or the data values obtained are too variable. The following points were recommended[181] to be excluded from the pup evaluation: pupillary response, ataxia, foot splay, home cage activity, ease of removal from the home cage, approach response, touch response, body temperature, and grip strength.

1. In Cage Observations

The preweanling rat is observed while it is in its home cage for the following characteristics: posture, palpebral closure, writhing, circling, biting, and gait and coordination abnormalities.

2. Removing from Cage Observations

While the rat pup is being removing from its cage, it is observed for the following characteristics: ease of removal, ease of handling, vocalizations, bite marks on tail and or paws, palpebral closure, piloerection, fur appearance (includes observation of skin and mucous membranes), lacrimation, salivation, and exophthalmos.

3. Standard Open Field Arena Observations

The preweanling rat is placed in the standard open field arena for approximately 2 min. The technicians observes and assesses the following characteristics: surface righting reflex, posture, respiration ease, rate of respiration, convulsions, tremors, grooming, gait and coordination abnormalities, arousal, palpebral closure, diarrhea, urination, and vocalizations.

4. Manipulations in the Standard Open Field Arena

While the preweanling rat is in the open field arena, the technician observes and assesses the auditory stimulus response. An auditory stimulus (e.g., finger cricket or snap) is presented directly above the preweanling rat's head and the observer records the presence or absence of the flinch reflex or other responses.

References

1. Wilson, J.G., *Environment and Birth Defects*, Academic Press, New York, 1973.
2. U.S. Food and Drug Administration, Bureau of Foods. *Guidelines for Reproduction Studies for Safety Evaluation of Drugs for Human Use*, Washington, DC, 1966.
3. U.S. Food and Drug Administration, Bureau of Foods, *Toxicological Principles for the Safety Assessment of Direct Food Additives and Color Additives Used in Foods*, Washington, DC, 1982.
4. U.S. Food and Drug Administration, Center for Food Safety and Applied Nutrition, *Toxicological Principles for the Safety Assessment of Direct Food Additives and Color Additives Used in Foods, "Redbook II"* (draft), Washington, DC, 1993.
5. U.S. Food and Drug Administration, International Conference on Harmonisation; Guideline on detection of toxicity to reproduction for medicinal products, *Fed. Regist.*, Part IX, 59 (183), 48746–48752, 1994.
6. U.S. Food and Drug Administration, International Conference on Harmonisation; Guideline on detection of toxicity to reproduction for medicinal products; Addendum on toxicity to male fertility, *Fed. Regist.*, April 5, 1996, Vol. 61, No. 67, 1996.
7. U.S. Food and Drug Administration, *International Conference on Harmonisation; Maintenance of the ICH guideline on toxicity to male fertility: An addendum to the ICH tripartite guideline on detection of toxicity to reproduction for medicinal products*, Amended November 9, 2000.
8. U.S. Environmental Protection Agency, *Reproductive and Fertility Effects. Pesticide Assessment Guidelines, Subdivision F. Hazard Evaluation: Human and Domestic Animals*, Office of Pesticides and Toxic Substances, Washington, DC, EPA 540/9-82-025, 1982.
9. U.S. Environmental Protection Agency, *Health Effects Test Guidelines OPPTS 870.3700, Prenatal Developmental Toxicity Study*, Washington, DC, EPA 712-C-96-207, 1996, p. 1.
10. U.S. Environmental Protection Agency, *Health Effects Test Guidelines OPPTS 870.3800, Reproduction and Fertility Effects,* Washington, DC, EPA 712-C-96-208, 1996, p. 1.
11. U.S. Environmental Protection Agency, Guidelines for developmental toxicity risk assessment, *Fed. Regist.* 56(234): 63798–63826, 1991.
12. Japan Ministry of Agriculture, Forestry and Fisheries, *Guidance on Toxicology Study Data for Application of Agricultural Chemical Registration*, 59 NohSan No. 4200, 1985, p. 45.
13. Japan Ministry of Agriculture, Forestry and Fisheries, *Guidelines for Screening Toxicity Testing of Chemicals*, 59 NohSan No. 4200, 1985, p. 209.
14. Canada, Health Protection Branch, *The Testing of Chemicals for Carcinogenicity, Mutagenicity and Teratogenicity*, Ministry of Health and Welfare, Canada, 1977.
15. Drug Directorate Guidelines, *Toxicological Evaluation 2.4, Reproductive Studies*, Ministry of National Health and Welfare, Health Protection Branch, Health and Welfare, Canada, 1990.
16. Committee on the Safety of Medicines (CSM), *Notes for Guidance on Reproduction Studies*, Department of Health and Social Security, Great Britain, 1974.
17. World Health Organization (WHO), Principles for the testing of drugs for teratogenicity, *WHO Tech. Rept. Series,* No. 364, Geneva, 1967.
18. Organization for Economic Cooperation and Development, *OECD 415: One-Generation Reproduction Toxicity Study* (Original Guideline, adopted 26th May 1983), OECD Paris, 1983.
19. Organization for Economic Cooperation and Development, *OECD 421: Reproduction/Developmental Toxicity Screening Test* (Original Guideline, adopted 27th July 1995), OECD Paris, 1995.

20. Organization for Economic Cooperation and Development, OECD 422: *Combined Repeated Dose Toxicity Study with the Reproduction/Developmental Toxicity Screening Test* (Original Guideline, adopted 22nd March 1996), OECD Paris, 1996.
21. Organization for Economic Cooperation and Development, *OECD 414: Prenatal Developmental Toxicity Study* (Updated Guideline, adopted 22nd January 2001), OECD Paris, 2001.
22. Organization for Economic Cooperation and Development. *OECD 416: Two-Generation Reproduction Toxicity Study* (Updated Guideline, adopted 22nd January 2001), OECD Paris, 2001.
23. International Conference on Harmonization of Technical Requirements of Pharmaceuticals for Human Use, Detection of toxicity to reproduction of medicinal products, *Fed. Regist.*, 59(1831), 48746–48752, 1994.
24. U.S. Food and Drug Administration, *Good Laboratory Practice Regulations; Final Rule*, 21 CFR Part 58, September 4, 1987.
25. U.S. Environmental Protection Agency, *Toxic Substance Control Act (TSCA), Good Laboratory Practice Standards, Final Rule*, 40 CFR Part 792, August 17, 1989.
26. U.S. Environmental Protection Agency. *Federal Insecticide, Fungicide and Rodenticide Act (FIFRA), Good Laboratory Practice Standards, Final Rule*, 40 CFR Part 160, August 17, 1989.
27. U.S. Department of Health and Human Services, *Biosafety in Microbiological and Biomedical Laboratories,* 4th ed., Richmond, J.Y. and McKinney, R.W., Eds., Public Health Service Centers for Disease Control and Prevention, and National Institutes of Health, Washington, D.C., 1999.
28. Institute for Laboratory Animal Research, National Research Council, *Guide for the Care and Use of Laboratory Animals*, Washington, DC, 1996.
29. National Institute of Mental Health, *Methods and Welfare Considerations in Behavioral Research with Animals: Report of the National Institute of Mental Health Workshop*, Morrison, A.R., Evans, H.L., Ator, N.A. and Nakamura, R.K., Eds., NIH Publication No. 02-5083, U.S. Government Printing Office, Washington, DC, 2002.
30. Tyl, R.W. and Marr, M.C., Developmental toxicity testing — methodology, in *Handbook of Developmental Toxicology*, Hood R.D., Ed., CRC Press, Boca Raton, FL, 1997, p. 175.
31. Gulati, D.K., Hope, E., Teague, J., and Chapin, R.E., Reproductive toxicity assessment by continuous breeding in Sprague-Dawley rats: a comparison of two study designs, *Fundam. Appl. Toxicol.*, 17, 270, 1991.
32. Chapin, R.E. and Sloane, R.A., Reproductive assessment by continuous breeding: evolving study design and summaries of ninety studies, *Environ. Health Perspect.*, 105(Suppl 1), 199, 1977.
33. U.S. Environmental Protection Agency, *Health Effects Test Guidelines; OPPTS 870.6300, Developmental Neurotoxicity Study*, Office of Prevention, Pesticides and Toxic Substances, August 1998.
34. Organization for Economic Cooperation and Development, *OECD 426: Developmental Neurotoxicity Study* (Revised Draft Guideline, October 1999), OECD Paris, 1999.
35. U.S. Environmental Protection Agency, *Health Effects Test Guidelines; OPPTS 870.3650, Combined Repeated Dose Toxicity Study with the Reproduction/Developmental Toxicity Screening Test*, Office of Prevention, Pesticides and Toxic Substances, July 2000.
36. Organization for Economic Cooperation and Development. *OECD 478: Genetic Toxicology: Rodent Dominant Lethal Test* (Updated Guideline, adopted 4 April 1984), OECD Paris, 1984.
37. U.S. Environmental Protection Agency. *Health Effects Test Guidelines; OPPTS 870.5450, Rodent Dominant Lethal Assay*, Office of Prevention, Pesticides and Toxic Substances, August 1998.
38. Risk Assessment Forum, U.S. Environmental Protection Agency. Guidelines for reproductive toxicity risk assessment, *Fed. Regist.*, 61(212), 56274–56322, October 31, 1996.
39. Davis, D.L., Friedler, G., Mattison, D., and Morris, R., Male-mediated teratogenesis and other reproductive effects: biologic and epidemiologic findings and a plea for clinical research, *Reprod. Toxicol.*, 6, 289, 1992.
40. Colie, C.F., Male mediated teratogenesis, *Reprod. Toxicol.*, 7, 3, 1993.
41. Savitz, D.A., Sonnenfeld, N.L., and Olshan, A.F., Review of epidemiologic studies of paternal occupational exposure and spontaneous abortion, *Am. J. Ind. Med.*, 25, 361, 1994.
42. Trasler, J.M. and Doerksen, T., Teratogen update: paternal exposures — reproductive risks, *Teratology*, 60, 161, 1999.
43. Knobil, E., Neill, J.D., Greenwald, G.S., Markert, C.L., and Pfaff, D.W., *The Physiology of Reproduction*, Raven Press, New York, 1994.

44. Hood, R.D. and Miller, D.B., Maternally mediated effects on development, in *Handbook of Developmental Toxicology*, Hood R.D., Ed., CRC Press, Boca Raton, FL, 1997, p. 61.
45. Le Vier, R.R. and Jankowiak, M.E. The hormonal and antifertility activity of 2,6-*cis*-diphenylhexamethylcyclotetrasiloxane in the female rat, *Biol. Reprod.*, 7, 260, 1972.
46. Sonawane, B.R. and Yaffe, S.J., Delayed effects of drug exposure during pregnancy: reproductive function, *Biol. Res. Pregnancy*, 4, 48, 1983.
47. Cummings, A.M. and Gray, L.E., Methoxychlor affects the decidual cell response of the uterus but not other progestational parameters in female rats, *Toxicol. Appl. Pharmacol.*, 90, 330, 1987.
48. Ulbrich, B. and Palmer, A.K., Detection of effects on male reproduction — a literature survey, *J. Am. Coll. Toxicol.*, 14, 293, 1995.
49. Takayama, S., Akaike, M., Kawashima, K., Takahashi, M., and Kurokawa, Y., Collaborative study in Japan on optimal treatment period and parameters for detection of male fertility disorders in rats induced by medicinal drugs, *J. Am. Coll. Toxicol.*, 14, 66, 1995.
50. Sakai, T., Takahashi, M., Mitsumori, K., Yasuhara, K., Kawashima, K., Mayahare, H., and Ohno, Y., Collaborative work to evaluate toxicity on male reproductive organs by 2-week repeated dose toxicity studies in rats. Overview of the studies, *J. Toxicol. Sci.*, 25, 1, 2000.
51. Cooper, R.L., Goldman, J.M., and Stoker, T.E., Measuring sexual behavior in the female rat, in *Methods in Toxicology, Volume 3, Part B. Female Reproductive Toxicology*, Heindel, J.J. and Chapin, R.E., Eds., Academic Press, San Diego, CA, 1993, p. 42.
52. Adler, N.T. and Toner, J.P., The effect of copulatory behavior on sperm transport and fertility in rats, in *Reproduction: Behavioral and Neuroendocrine Perspective*, Komisaruk, B.R., Siegel, H.I., Chang, M.F., and Feder, H.H., Eds., New York Academy of Science, New York, 1986, p. 21.
53. MARTA (Middle Atlantic Reproduction and Teratology Association) and MTA (Midwest Teratology Association), *Historical Control Data (1992–1994) for Developmental and Reproductive Toxicity Studies using the Crl:CD®(SD)BR Rat*, Charles Rivers Laboratories, March 1996. (Available on-line at www.crivers.com.)
54. Clark, R.L., Endpoints of reproductive system development, in *An Evaluation and Interpretation of Reproductive Endpoints for Human Risk Assessment*, Daston, G. and Kimmel, C., Eds., ILSI Press, Washington, DC, 1999, p. 28.
55. Chapin, R.E., Gulati, D.K., Barnes, L.H., and Teague, J.L., The effects of feed restriction on reproductive function in Sprague-Dawley rats, *Fundam. Appl. Toxicol.*, 20, 23, 1993.
56. Stevens, K.R. and Gallo, M.A., Practical considerations in the conduct of chronic toxicity studies, in *Principles and Methods of Toxicology*, Hayes, A.W., Ed., Raven Press, New York, 1989, p. 237.
57. Schwetz, B.A., Rao, K.S., and Park, C.N., Insensitivity of tests for reproductive problems, *J. Environ. Pathol. Toxicol.*, 3, 81, 1980.
58. Blazak, W.F., Ernst, T.L., and Stewart, B.E., Potential indicators of reproductive toxicity, testicular sperm production and epididymal sperm number, transit time and motility in Fischer 344 rats, *Fundam. Appl. Toxicol.*, 5, 1097, 1985.
59. Berndtson, W.E., Methods for quantifying mammalian spermatogenesis: a review, *J. Anim. Sci.*, 44, 818, 1977.
60. Foote, R.H. and Berndtson, W.E., The germinal cells, in *Reversibility in Testicular Toxicity Assessment*, Scialli, A.R. and Clegg, E.D., Eds., CRC Press, Boca Raton, FL, 1992, p. 1.
61. Ku, W.W., Chapin, R.E., Wine, R.N., and Gladen, B.C., Testicular toxicity of boric acid (BA): relationship of dose to lesion development and recovery in the F344 rat, *Reprod. Toxicol.*, 7, 305, 1993.
62. Hess, R.A., Moore, B.J., Forrer, J., Linder, R.E., and Abuel-Atta, A.A., The fungicide Benomyl (methyl 1-(butylcarbamoyl)-2-benzimidazolecarbamate) causes testicular dysfunction by inducing the sloughing of germ cells and occlusion of efferent ductules, *Fundam. Appl. Toxicol.*, 17, 733, 1991.
63. Nakai, M., Moore, B.J., and Hess, R.A., Epithelial reorganization and irregular growth following carbendazim induced injury of the efferent ductules of the rat testis, *Anat. Rec.*, 235, 51, 1993.
64. Boorman, G.A., Chapin, R.E., and Mitsumori, K., Testis and epididymis, in *Pathology of the Fischer Rat*, Boorman, G.A., Eustis, S.L., Elwell, M.R., Montgomery, C.A., Jr., and MacKenzie, W.F., Eds., Academic Press, San Diego, 1990, p. 405.
65. Russell, L.D., Normal testicular structure and methods of evaluation under experimental and disruptive conditions, in *Reproductive and Developmental Toxicity of Metals*, Clarkson, T.W., Nordberg, G.F., and Sager, P.R., Eds., Plenum Publishing Co., New York, 1983, p. 227.

66. De Kretser, D.M. and Kerr, J.B., The cytology of the testis, in *The Physiology of Reproduction*, Knobil, E. and Neill, J., Eds., Raven Press, New York, 1988, p. 837.
67. Johnson, A.D. and Gomes, W.R., The testis, in *Advances in Physiology, Biochemistry and Function*, Vol. IV, Academic Press, New York, 1977.
68. Russell, L.D., Ettlin, R., Sinha-Hikim, A.P., Clegg, E.D., Eds., *Histological and Histopathological Evaluation of the Testis*, Cache River Press, Clearwater, FL, 1990.
69. Scialli, A.R. and Clegg, E.D., *Reversibility in Testicular Toxicity Assessment*, CRC Press, Boca Raton, FL, 1992.
70. Chapin, R.E. and Conner, M.W., Testicular histology and sperm parameters, in *An Evaluation and Interpretation of Reproductive Endpoints for Human Risk Assessment*, Daston, G. and Kimmel, C., Eds., ILSI Press, Washington, DC, 1999, p. 28.
71. Chapin, R.E., Morphologic evaluation of seminiferous epithelium of the testis, in *Physiology and Toxicology of Male Reproduction*, Lamb, J.C. and Foster, P.M.D., Eds., Academic Press, New York, 1988, p. 155.
72. Hess, R.A. and Moore, B.J., Histological methods for evaluation of the testis, in *Methods in Toxicology: Male Reproductive Toxicology*, Chapin, R.E. and Heindel, J.J., Eds., Academic Press, San Diego, 1993, p. 52.
73. Mangelsdorf, I. and Buschman, J., *Extrapolation from Animal Studies to Humans for the Endpoint of Male Fertility*, Federal Institute for Occupational Safety and Health, Dortman/Berlin/Dresden, Germany, 2002.
74. Hess, R.A., Quantitative and qualitative characteristics of the stages and transitions in the cycle of the rat seminiferous epithelium: light microscopic observations of perfusion-fixed and plastic-embedded testes, *Biol. Reprod.*, 43, 525, 1990.
75. Amann, R.P., A critical review of methods for evaluation of spermatogenesis from seminal characteristics, *J. Androl.*, 2, 37, 1981.
76. Working, P.K., Male reproductive toxicity: comparison of the human to animal models, *Environ. Health*, 77, 37, 1988.
77. Aafjes, J.H., Vels, J.M., and Schenck, E., Fertility of rats with artificial oligozoospermia, *J. Reprod. Fertil.*, 58, 345, 1980.
78. Meistrich, M.L., Quantitative correlation between testicular stem cell survival, sperm production, and fertility in the mouse after treatment with different cytotoxic agents, *J. Androl.* 3, 58, 1982.
79. Robaire, B., Smith, S., and Hales, B.F., Suppression of spermatogenesis by testosterone in adult male rats: effect on fertility, pregnancy outcome and progeny, *Biol. Reprod.*, 31, 221, 1984.
80. Fawcett, D.W., *Bloom and Fawcett: A Textbook of Histology*, W.B. Saunders, Philadelphia, PA, 1986.
81. Boorman, G.A., Elwell, M.R., and Mitsumori, K., Male accessory sex glands, penis and scrotum, in *Pathology of the Fischer Rat*, Boorman, G.A., Eustis, S.L., Elwell, M.R., Montgomery, C.A., Jr., and MacKenzie, W.F., Eds., Academic Press, San Diego, 1990, p. 419.
82. Jones, T.C., Mohr, U., and Hunt, R.D., *Genital System*, Springer-Verlag, New York, 1987.
83. Haschek, W.M. and Rousseaux, C.G., *Handbook of Toxicologic Pathology*, Academic Press, New York, 1991.
84. Toth, G.P., Stober, J.A., Read, E.J., Zenick, H., and Smith, M.K, The automated analysis of rat sperm motility following subchronic epichlorohydrin administration: methodologic and statistical considerations, *J. Androl.*, 10, 401, 1989.
85. Slott, V.L., Suarez, J.D., Simmons, J.E., and Perreault, S.D., Acute inhalation exposure to epichlorohydrin transiently decreases rat sperm velocity, *Fundam. Appl. Toxicol.*, 15, 597, 1990.
86. Klinefelter, G.R., Laskey, J.W., Kelce, W.R., Ferrell, J., Roberts, N.L., Suarez, J.D., and Slott, V., Chloroethylmethanesulfonate-induced effects on the epididymis seem unrelated to altered Leydig cell function, *Biol. Reprod.*, 51, 82, 1994.
87. Toth, G.P., Stober, J.A., George, E.L., Read, E.J., and Smith, M.K., Sources of variation in the computer-assisted motion analysis of rat epididymal sperm, *Reprod. Toxicol.*, 5, 487, 1991.
88. Oberlander, G., Yeung, C.H., and Cooper, T.G., Induction of reversible infertility in male rats by oral ornidazole and its effects on sperm motility and epididymal secretions, *J. Reprod. Fertil.*, 100, 551, 1994.
89. Slott, V.L., Jeffay, S.C., Suarez, J.D., Barbee, R.R., and Perreault, S.D., Synchronous assessment of sperm motility and fertilizing ability in the hamster following treatment with alpha-chlorohydrin, *J. Androl.*, 16, 523, 1995.

90. Seed, J., Chapin, R.E., Clegg, E.D., Darney, S.P., Dostal, L., Foote, R.H., Hurtt, M.E., Klinefelter, G.R., Makris, S.L., Schrader, S., Seyler, D., Sprando, R., Treinen, K.A., Veeranachaneni, R., and Wise, L.D., Methods for assessing sperm motility, morphology, and counts in the rat, rabbit and dog: a consensus report, *Reprod. Toxicol.*, 10, 237, 1996.
91. Klinefelter, G.R., Laskey, J.W., Perreault, S.D., Ferrell, J., Jeffay, S., Suarez, J., and Roberts, N., The ethane dimethanesulfonate-induced decrease in the fertilizing ability of cauda epididymal sperm is independent of the testis, *J. Androl.* 15, 318, 1994.
92. Meistrich, M.L. and Brown, C.C., Estimation of the increased risk of human infertility from alterations in semen characteristics, *Fertil. Steril.*, 40, 220, 1983.
93. World Health Organization, *WHO Laboratory Manual for the Examination of Human Semen and Sperm-Cervical Mucus Interaction*, 3rd Ed., Cambridge University Press, Cambridge, 1992.
94. Hurtt, M.E. and Zenick, H., Decreasing epididymal sperm reserves enhances the detection of ethoxyethanol induced spermatotoxicity, *Fundam. Appl. Toxicol.*, 7, 348, 1986.
95. Cassidy, S.L., Dix, K.M., and Jenkins, T., Evaluation of a testicular sperm head counting technique using rats exposed to dimethoxyethyl phthalate (DMEP), glycerol alpha-monochlorohydrin (GMCH), epichlorohydrin (ECH), formaldehyde (FA), or methyl methanesulphonate (MMS), *Arch. Toxicol.* 53, 71, 1983.
96. Blazak, W.F., Treinen, K.A., and Juniewicz, P.E., Application of testicular sperm head counts in the assessment of male reproductive toxicity, in *Methods in Toxicology: Male Reproductive Toxicology*, Chapin, R.E. and Heindel, J.J., Eds., Academic Press, San Diego, 1993, p. 86.
97. Meistrich, M.L. and van Beek, M.E.A.B., Spermatogonial stem cells: assessing their survival and ability to produce differentiated cells, in *Methods in Toxicology: Male Reproductive Toxicology*, Chapin, R.E. and Heindel, J.J., Eds., Academic Press, San Diego, 1993, p. 106.
98. Boyers, S.P., Davis, R.O., and Katz, D.F., Automated semen analysis, *Curr. Probl. Obstet. Gynecol. Fertil.*, 12, 173, 1989.
99. Yeung, C.H., Oberlander, G., and Cooper, T.G., Characterization of the motility of maturing rat spermatozoa by computer-aided objective measurement, *J. Reprod. Fertil.*, 96, 427, 1992.
100. Slott, V.L. and Perreault, S.D., Computer-assisted sperm analysis of rodent epididymal sperm motility using the Hamilton-Thorne motility analyzer, in *Methods in Toxicology: Male Reproductive Toxicology*, Chapin, R.E. and Heindel, J.J., Eds., Academic Press, San Diego, 1993, p. 319.
101. Morrissey, R.E., Schwetz, B.A., Lamb, J.C., Ross, M.D., Teague, J.L., and Morris, R.W., Evaluation of rodent sperm, vaginal cytology, and reproductive organ weight data from National Toxicology Program 13-week studies, *Fundam. Appl. Toxicol.*, 11, 343, 1988.
102. Toth, G.P., Stober, J.A., Zenick, H., Read, E.J., Christ, S.A., and Smith, M.K., Correlation of sperm motion parameters with fertility in rats treated subchronically with epichlorohydrin, *J. Androl.*, 12, 54, 1991.
103. Morrissey, R.E., Lamb, J.C., Morris, R.W., Chapin, R.E., Gulati, D.K., and Heindel, J.J., Results and evaluations of 48 continuous breeding reproduction studies conducted in mice, *Fundam. Appl. Toxicol.*, 13, 747, 1989.
104. Slott, V.L., Suarez, J.D., and Perreault, S.D., Rat sperm motility analysis: methodologic considerations, *Reprod. Toxicol.*, 5, 449, 1991.
105. Chapin, R.E., Filler, R.S., Gulati, D., Heindel, J.J., Katz, D.F., Mebus, C.A., Obasaju, F., Perreault, S.D., Russell, S.R., Schrader, S., Slott, V., Sokol, R.Z., and Toth, G., Methods for assessing rat sperm motility, *Reprod. Toxicol.*, 6, 267, 1992.
106. Schrader, S.M., Chapin, R.E., Clegg, E.D., Davis, R.O., Fourcroy, J.L., Katz, D.F., Rothmann, S.A., Toth, G., Turner, T.W., and Zinaman, M., Laboratory methods for assessing human semen in epidemiologic studies: a consensus report, *Reprod. Toxicol.*, 6, 275, 1992.
107. Toth, G.P., Read, E.J., and Smith, M.K., Utilizing Cryo Resources CellSoft computer-assisted sperm analysis system for rat sperm motility studies, in *Methods in Toxicology: Male Reproductive Toxicology*, Chapin, R.E. and Heindel, J.J., Eds., Academic Press, San Diego, 1993, p. 303.
108. Zenick, H., Clegg, E.D., Perreault, S.D., Klinefelter, G.R., and Gray, L.E., Assessment of male reproductive toxicity: a risk assessment approach, in *Principles and Methods of Toxicology*, Hayes, A.W., Ed., Raven Press, New York, 1994, p. 937.
109. Filler, R., Methods for evaluation of rat epididymal sperm morphology, in *Methods in Toxicology: Male Reproductive Toxicology*, Chapin, R.E. and Heindel, J.J., Eds., Academic Press, San Diego, 1993, p. 334.

110. Linder, R.E., Strader, L.F., Slott, V.L., and Suarez, J.D., Endpoints of spermatotoxicity in the rat after short duration exposures to fourteen reproductive toxicants, *Reprod. Toxicol.*, 6, 491, 1992.
111. Davis, R.O., Gravance, C.G., Thal, D.M., and Miller, M.G., Automated analysis of toxicant-induced changes in rat sperm head morphometry, *Reprod. Toxicol.*, 8(6), 521, 1994.
112. UK Industrial Reproductive Toxicology Discussion Group (IRDG), Computer-Assisted Sperm Analysis (CASA) Group, *Rat Sperm Morphological Assessment, Guideline Document*, 1st ed., October 2000. (http://www.irdg.co.uk/IRDG_spermmorphology.doc)
113. Clermont, Y. and Perey, B., The stages of the cycle of the seminiferous epithelium of the rat: practical definitions in PA-Schiff-hematoxylin-eosin stained sections, *Rev. Canad. Biol.*, 16:451, 1957.
114. Leblond, C.P. and Clermont, Y., Definition of the stages of the cycle of the seminiferous epithelium in the rat, *Ann. N.Y. Acad. Sci.* 55:548, 1952.
115. Russell, L.D., Ettlin, R., Sinha Hikim, A.P., and Clegg, E.D., The classification and timing of spermatogenesis, in *Histological and Histopathological Evaluation of the Testis*, Russell, L.D., Ettlin, R., Sinha Hikim, A.P., and Clegg, E.D., Eds., Cache River Press, Clearwater, FL, 1990, chap. 2.
116. Keller, K.A., Developmental and reproductive toxicology, in *Toxicology Testing Handbook, Principles, Applications, and Data Interpretation*, Jacobson-Kram, D. and Keller, K.A., Eds., Marcel Dekker, Inc., New York, 2001, p. 209.
117. Wade, G.N., Gonadal hormones and behavioral regulation of body weight, *Physiol. Behav.*, 8, 523, 1972.
118. Galletti, F. and Klopper, A., The effect of progesterone on the quantity and distribution of body fat in the female rat, *Acta Endocrinol.*, 46, 379, 1964.
119. Hervey, E. and Hervey, G.R., The effects of progesterone on body weight and composition in the rat, *J. Endocrinol.*, 37, 361, 1967.
120. Kupfer, D., Critical evaluation of methods for detection and assessment of estrogenic compounds in mammals: strengths and limitations for application to risk assessment, *Reprod. Toxicol.*, 2, 147, 1987.
121. Leininger, J.R. and Jokinen, M.P., Oviduct, uterus and vagina, in *Pathology of the Fischer Rat*, Boorman, G.A., Eustis, S.L., Elwell, M.R., Montgomery, C.A., Jr., and MacKenzie, W.F., Eds., Academic Press, San Diego, 1990, p. 443.
122. Cooper, R.L., Chadwick, R.W., Rehnberg, G.L., Goldman, J.M., Booth, K.C., Hein, J.F., and McElroy, W.K., Effect of lindane on hormonal control of reproductive function in the female rat, *Toxicol. Appl. Pharmacol.*, 99, 384, 1989.
123. Hirschfield, A.N., Histological assessment of follicular development and its applicability to risk assessment, *Reprod. Toxicol.*, 1, 71, 1987.
124. Baker, T., Oogenesis and ovarian development, in *Reproductive Biology*, Balin, H. and Glasser, S., Eds., Exerpta Medica, Amsterdam, 1972.
125. Alison, R.H., Morgan, K.T., and Montgomery, C.A., Jr., Ovary, in *Pathology of the Fischer Rat*, Boorman, G.A., Eustis, S.L., Elwell, M.R., Montgomery, Jr., C.A., and MacKenzie, W.F., Eds., Academic Press, San Diego, 1990, p. 429.
126. Eppig, J., Mammalian oocyte development *in vivo* and *in vitro*, in *Elements of Mammalian Fertilization*, Vol. 1, Wassarman, P.M., Ed., CRC Press, Boca Raton, FL, 1991, p. 57.
127. Wassarman, P.M., The mammalian ovum, in *Physiology of Reproduction*, Vol. 1, Knobil, E. and Neill, J.D., Eds., Raven Press, New York, 1988, p. 69.
128. Crisp, T.M., Organization of the ovarian follicle and events in its biology: oogenesis, ovulation or atresia, *Mutat. Res.*, 296, 89, 1992.
129. Eppig, J.J., Oocyte control of ovarian follicular development and function in mammals, *Reproduction*, 122, 829, 2001.
130. Peters, H. and McNatty, K.P., *The Ovary*, Paul Elek, New York, 1980.
131. Plowchalk, D.R., Smith, B.J., and Mattison, D.R., Assessment of toxicity to the ovary using follicle quantitation and morphometrics, in *Methods in Toxicology: Female Reproductive Toxicology*, Heindel, J.J. and Chapin, R.E., Eds., Academic Press, San Diego, 1993, p. 57.
132. Kurman, R. and Norris, H.J., Germ cell tumors of the ovary, *Pathol. Annu.*, 13, 291, 1978.
133. Kwa, S.L. and Fine, L.J., The association between parental occupation and childhood malignancy, *J. Occup. Med.*, 22, 792, 1980.
134. Langley, F.A. and Fox, H., Ovarian tumors: classification, histogenesis, etiology, in *Haines and Taylor's Obstetrical and Gynaecologic Pathology*, Fox, H., Ed., Churchill Livingstone, Edinburgh, 1987, p. 542.

135. Fugo, N.W. and Butcher, R.L., Overripeness and the mammalian ova. I. Overripeness and early embryonic development, *Fertil. Steril.*, 17, 804, 1966.
136. Butcher, R.L. and Fugo, N.W., Overripeness and the mammalian ova. II. Delayed ovulation and chromosome anomalies, *Fertil. Steril.*, 18, 297, 1967.
137. Butcher, R.L., Blue, J.D., and Fugo, N.W., Overripeness and the mammalian ova. III. Fetal development at midgestation and at term, *Fertil. Steril.*, 20, 223, 1969.
138. Gray, L.E. and Ostby, J.S., In utero 2,3,7,8-tetrachlorodibenzo-p-dioxin (TCDD) alters reproductive morphology and function in female rat offspring, *Toxicol. Appl. Pharmacol.*, 133, 285, 1995.
139. Long, J.A. and Evans, H.M., The oestrous cycle in the rat and its associated phenomena, *Mem. Univ. Calif.*, 6, 1, 1922.
140. Cooper, R.L., Goldman, J.M., and Vandenbergh, J.G., Monitoring of the estrous cycle in the laboratory rodent by vaginal lavage, in *Methods in Toxicology: Female Reproductive Toxicology*, Heindel, J.J. and Chapin, R.E., Eds., Academic Press, San Diego, 1993, p. 45.
141. Kimmel, G.L., Clegg, E.D., and Crisp, T.M. Reproductive toxicity testing: a risk assessment perspective, in *Reproductive Toxicology*, Witorsch, R.J., Ed., Raven Press, New York, 1995, p. 75.
142. LeFevre, J. and McClintock, M.K., Reproductive senescence in female rats: a longitudinal study of individual differences in estrous cycles and behavior, *Biol. Reprod.*, 38, 780, 1988.
143. Holmes, R.L. and Ball, J.N., *The Pituitary Gland: A Comparative Account*, Cambridge University Press, Cambridge, 1974.
144. Boorman, G.A., Wilson, J.T., van Zwieten, M.J., and Eustis, S.L., Mammary gland, in *Pathology of the Fischer Rat*, Boorman, G.A., Eustis, S.L., Elwell, M.R., Montgomery, C.A., Jr., and MacKenzie, W.F., Eds., Academic Press, San Diego, 1990, p. 295.
145. Wolff, M.S., Lactation, in *Occupational and Environmental Reproductive Hazards*, Paul, M., Ed., Williams and Wilkins, Baltimore, 1993, p. 60.
146. Imagawa, W., Yang, J., Guzman, R., and Nandi, S., Control of mammary gland development, in *The Physiology of Reproduction*, Knobil, E. and O'Neill, J.D., Eds., Raven Press, New York, 1994, p. 1033.
147. Tucker, H.A., Lactation and its hormonal control, in *The Physiology of Reproduction*, Knobil, E. and O'Neill, J.D., Eds., Raven Press, New York, 1994, p. 1065.
148. American Academy of Pediatrics Committee on Drugs, The transfer of drugs and other chemicals into human milk, *Pediatrics*, 93, 137, 1994.
149. Oskarsson, A., Hallen, I.P., and Sundberg, J., Exposure to toxic elements via breast milk, *Analyst*, 120, 765, 1995.
150. Cooper, R.L., Conn, P.M., and Walker, R.F., Characterization of the LH surge in middle-aged female rats, *Biol. Reprod.*, 23, 611, 1980.
151. Finch, C.E., Felicio, L.S., and Mobbs, C.V., Ovarian and steroidal influences on neuroendocrine aging processes in female rodents, *Endocrinol. Rev.*, 5, 467, 1984.
152. Brawer, J.R. and Finch, C.E., Normal and experimentally altered aging processes in the rodent hypothalamus and pituitary, in *Experimental and Clinical Interventions in Aging*, Walker, R.F. and Cooper, R.L., Eds., Marcel Dekker, New York, 1983, p. 45.
153. Mattison, D.R., Clinical manifestations of ovarian toxicity, in *Reproductive Toxicology*, Dixon, R.L., Ed., Raven Press, New York, 1985, p. 109.
154. Cooper, R.L. and Goldman, J.M., Vaginal cytology, in *An Evaluation and Interpretation of Reproductive Endpoints for Human Risk Assessment*, Daston, G. and Kimmel, C., Eds., ILSI Press, Washington, DC, 1999, p. 42.
155. Nelson, S. and Gibori, G., Dispersion, separation, and culture of the different cell populations of the rat corpus luteum, in *Methods in Toxicology: Female Reproductive Toxicology*, Heindel, J.J. and Chapin, R.E., Eds., Academic Press, San Diego, 1993, p. 340.
156. Salewski, E., Farbemethode zum makroskopischen Nachweis von Implantationsstellan am Uterus der Ratte. Naunyn Schmiedebergs, *Arch. Exp. Pathol. Pharmakol.*, 247, 367, 1964.
157. Heindel, J.J., Oocyte quantitation and ovarian histology, in *An Evaluation and Interpretation of Reproductive Endpoints for Human Risk Assessment*, Daston, G. and Kimmel, C., Eds., ILSI Press, Washington, DC, 1999, p. 57.
158. Pedersen, T. and Peters, H.J., Proposal for the classification of oocytes and follicles in the mouse ovary, *Reprod. Fert.*, 17, 555, 1968.

159. Bolon, B., Bucci, T.J., Warbritton, A.R., Chen, J.J., Mattison, D.R., and Heindel, J.J., Differential follicle counts as a screen for chemically induced ovarian toxicity in mice: results from continuous breeding bioassays, *Fundam. Appl.Toxicol.*, 39, 1, 1997.
160. Bucci, T.J., Bolon, B., Warbritton, A.R., Chen, J.J., and Heindel, J.J., Influence of sampling on the reproducibility of ovarian follicle counts in mouse toxicity studies, *Reprod. Toxicol.*, 11(5), 689, 1997.
161. Heindel, J.J., Thomford, P.J., and Mattison, D.R., Histological assessment of ovarian follicle number in mice as a screen of ovarian toxicity, in *Growth Factors and the Ovary*, Hirshfield, A.N., Ed., Plenum, New York, 1989, p. 421.
162. Christian, M.S. and Brown, W.R., Control primordial follicle counts in multigeneration studies in Sprague-Dawley ("gold standard") rats, *The Toxicologist*, 66(1-S), March 2002.
163. Tyl, R.W., Developmental toxicity in toxicologic research and testing, in *Perspectives in Basic and Applied Toxicology*, Ballantyne, B., Ed., John Wright, Bristol, 1987, p. 203.
164. Bates, H.K., Cunny, H.C., and Kebede, G.A., Developmental neurotoxicity testing methodology, in *Handbook of Developmental Toxicology*, Hood, R.D., Ed., CRC Press, Boca Raton, FL, 1997, p. 291.
165. Henck, J.W., Developmental neurotoxicology: testing and interpretation, in *Handbook of Neurotoxicology*, Volume II, Massaro, E.J., Ed., Humana Press, Totowa, NJ, 2002, p. 461.
166. Iezhitsa, I.N., Spasov, A.A., and Bugaeva, L.I., Effects of bromantan on offspring maturation and development of reflexes, *Neurol. Toxiclo.*, 23, 213, 2001.
167. Kai, S. and Kohmura, H., Comparison of developmental indices between Crj:CD(SD)IGS rats and Crj:CD(SD) rats, in *Biological Reference Data on CD(SD)IGS-1999*, Matsuzawa, T. and Inoue, H., Eds., CD(SD)IGS Study Group, Yokohama, 1999, p. 157.
168. Ohkubo, Y., Furuya, K., Yamada, M., Akamatsu, H., Moritomo, A., Ishida, S., Watanabe, K., and Hayashi, Y., Reproductive and developmental data on Crj:CD(SD)IGS rats, in *Biological Reference Data on CD(SD)IGS-1999*, Matsuzawa, T. and Inoue, H., Eds., CD(SD)IGS Study Group, Yokohama, 1999, p. 153.
169. Ema, M., Fujii, S., Furukawa, M., Kiguchi, M., Ikka, T., and Harazano, A., Rat two-generation reproductive toxicity study of bisphenol A, *Repro. Toxicol.*, 15, 505, 2001.
170. Lewis, E.M., Christian, M.S., Barnett, J., Jr., and Hoberman, A.M., Control preputial separation data for F1 generation CRL Sprague-Dawley ("gold standard") rats in EPA developmental neurotoxicology, EPA multigeneration and FDA peri-postnatal studies, *The Toxicologist*, 66(1), 1149, 2002.
171. Hoberman, A.M., Christian, M.S., Lewis, E.M., and Barnett, J., Jr., Control vaginal opening data for F1 generation CRL Sprague-Dawley ("gold standard") rats in EPA developmental neurotoxicology, EPA multigeneration and FDA peri-postnatal studies, *The Toxicologist*, 66(1), 1137, 2002.
172. Tinwell, H., Haseman, J., Lefevre, P.A., Wallis, N., and Ashby, J., Normal sexual development of two strains of rat exposed in utero to low doses of Bisphenol A, *Toxicol. Sci.*, 68, 339, 2002.
173. Weisenburger, W.P., Neurotoxicology, in *Toxicology Testing Handbook*, Jacobson-Kram, D. and Keller, K.A., Eds., Marcel Dekker, New York, 2002, p. 255.
174. Vorhees, C.V., Acuff-Smith, K.D., Moran, M.S., and Minck, D.R., A new method for evaluating air-righting reflex ontogeny in rats using prenatal exposure to phenytoin to demonstrate delayed development, *Neurotoxicol. Teratol.*, 16(6), 563, 1994.
175. Ross, J.F., Tier 1 neurological assessment in regulated animal safety studies, in *Handbook of Neurotoxicology*, Volume II, Massaro, E.J., Ed., Humana Press, Totowa, NJ, 2002, p. 461.
176. U.S. Environmental Protection Agency, Toxic Substances Control Act Testing Guidelines, 40 CFR part 798 subpart G section 798.6050, *Fed. Regist.*, 50, 39458–39460, 1985.
177. U.S. Environmental Protection Agency, *Neurotoxicity Guidelines, Addendum 10, Pesticide Assessments Guidelines Subdivision F*, Publication number PB 91-154617, National Technical Information Services, Springfield, VA, 1991.
178. U.S. Environmental Protection Agency, *Health Effects Test Guidelines, OPPTS 870.6200, Neurotoxicity Screening Battery*, EPA 712-C-98-238, 1998.
179. Moser, V.C. and Ross, J.F., Eds., *Training Video and Reference Manual for a Functional Observational Battery*, American Industrial Health Council, Washington, DC, 1996.
180. Ross, J.F., Mattsson, J.L., and Fix, A.S., Expanded clinical observations in toxicity studies: historical perspectives and contemporary issues, *Regul. Toxicol. Pharmacol.*, 28, 17, 1998.
181. Moser, V.C., The functional observational battery in adult and developing rats, *Neurotoxicology*, 21(6), 989, 2000.

CHAPTER 11

The U.S. EPA Endocrine Disruptor Screening Program: *In Vitro* and *In Vivo* Mammalian Tier I Screening Assays

Susan C. Laws, Tammy E. Stoker, Jerome M. Goldman, Vickie Wilson, L. Earl Gray, Jr., and Ralph L. Cooper

CONTENTS

I. Introduction 489
II. *In Vitro* Assays for the Detection of Endocrine Disrupting Chemicals 490
    A. Estrogen and Androgen Receptor Competitive Binding Assays 490
    B. Androgen and Estrogen Receptor-Dependent Gene Transcription Assays 493
    C. Steroidogenesis 496
    D. Aromatase Assay 500
III. *In Vivo* Assays for the Detection of Endocrine Disrupting Chemicals 502
    A. Uterotrophic Assay 502
    B. Hershberger Assay 505
    C. Male and Female Pubertal Protocols 507
    D. In Utero-Lactational Exposure 510
IV. Considerations for Evaluation of Data from the TIS Battery 514
V. Summary 515
Acknowledgments 516
References 516

I. INTRODUCTION

In response to emerging concerns that environmental chemicals may have adverse effects on human health by altering the function of the endocrine system,[1] the Food Quality Protection Act mandated that the U.S. Environmental Protection Agency (U.S. EPA) develop and implement an endocrine disruptor screening program (EDSP). Working toward this goal, the U.S. EPA is currently implementing a proposed EDSP that is designed to detect chemicals that alter the estrogen, androgen, and thyroid systems in humans, fishes, and wildlife.[2] This program, based largely upon recommendations made in 1998 by the Endocrine Disruptor Screening and Testing Advisory Committee (EDSTAC),[3] will allow for: (1) the initial sorting and prioritization of the 80,000 plus chemicals under the purview of the EPA, (2) the identification of chemicals for further testing, using a Tier I screening (TIS) battery that includes both *in vitro* and *in vivo* mammalian and ecotoxicological assays, and (3) the characterization of adverse effects and establishment of dose-response relationships for hazard assessment, using a Tier II testing battery. In addition to developing a priority setting approach for chemical selection, the U.S. EPA is currently engaged in a process of optimization and validation of those Tier I and Tier II assays that have been proposed for inclusion in the final screening and testing program. To guide this process forward, in 2001 the U.S. EPA established the Endocrine Disruptor Methods Validation Subcommittee (EDMVS) made up of government and nongovernment scientists, along with stakeholder representatives from various interest groups, to advise and review ongoing and newly initiated methodological studies. The agency is also working closely with the Interagency Coordinating Committee for the Validation of Alternative Methods (ICCVAM)[4] and the Organisation for Economic Cooperation and Development (OECD),[5] who have agreed to assist with the validation of a number of the assays. Specifically, each assay that will be included in the final TIS or Tier II testing batteries will initially undergo a prevalidation process that will optimize the protocol and assure that the assay can be conducted by a qualified laboratory. The validation process will include a thorough evaluation to demonstrate the biological relevance of each assay, as well as the assay's ability to be replicated within a laboratory and among multiple laboratories. Performance criteria for each assay will be determined during the validation process and will include estimates of intra- and interlaboratory variance.

Here we review the mammalian *in vitro* and *in vivo* assays that will likely be included as components of the U.S. EPA's final TIS battery. An overview of each assay and the mode of action that each reflects are shown in Table 11.1. Each assay is discussed with respect to its purpose, technical issues, strengths, limitations, and interpretation of the data. Criteria for evaluating data from the TIS battery, as well as avenues for identifying new assays that will enhance and/or refine the first-generation screening and testing batteries, are discussed.

II. *IN VITRO* ASSAYS FOR THE DETECTION OF ENDOCRINE DISRUPTING CHEMICALS

A. Estrogen and Androgen Receptor Competitive Binding Assays

In vitro, cell-free or whole-cell mammalian androgen receptor (AR) and estrogen receptor (ER) competitive binding assays were proposed for the TIS battery to detect compounds that bind to these receptors and potentially affect normal hormone action. These assays can be used to determine the relative binding affinities of environmental chemicals for the receptor, as compared with those of a high affinity radiolabeled hormone, such as 17-β-estradiol for the ER and R1881 or dihydrotestosterone (DHT) for the AR. The ability of a steroid hormone to bind to its respective receptor plays a central role in steroid hormone action. After the receptor binds to its ligand in an intact cell, the protein undergoes a conformational change that facilitates the formation of receptor-ligand complexes. These complexes can then bind to specific sequences on DNA and initiate the transcription of target genes[6] (Figure 11.1). Environmental chemicals that compete with endogenous ligands for AR or ER binding have the potential to either induce hormone-dependent transcriptional activity on their own (agonist) or to block normal hormone function by preventing the endogenous hormone from binding to the receptor (antagonist). Although there are slight differences between the ER and AR competitive binding assays, the basic steps for each are the same. The assay measures the ability of a radiolabeled ligand to bind with its respective receptor in the presence of increasing concentrations of a test chemical (e.g., competitor). A test chemical with a high affinity for the receptor will compete with the radiolabeled ligand for binding to the receptor at a lower concentration than a chemical that does not have a high affinity for the receptor. Specifically, the steps of the assay are: (1) the receptor and radiolabeled ligand are incubated alone or in the presence of

Table 11.1 Mammalian assays under consideration for the TIS battery of the U.S. EPA's EDSP

Assay	Estrogen Agonist	Estrogen Antagonist	Androgen Agonist	Androgen Antagonist	Steroid Synthesis	HPG[g]	Thyroid Homeostasis
In Vitro Assays							
Estrogen Receptor							
Competitive binding[a]	X	X					
Gene transcription[b]	X	X					
Androgen Receptor							
Competitive binding[a]			X	X			
Gene transcription[b]			X	X			
Steroidogenesis[a]					X		
Aromatase[a,c]					X[d]		
In Vivo Assays							
Uterotrophic[e]	X	X					
Hershberger[e]			X	X			
Pubertal development							
Female[a]	X	X			X	X	X
Male[a,c]			X	X	X	X	X
In utero–lactational[f]	X	X	X	X	X	X	X

[a] Assay is currently undergoing validation by the EPA.
[b] Methods have been reviewed by an ICCVAM Expert Panel, but no specific assays have been selected for validation.
[c] Alternative assay recommended by EDSTAC.
[d] Assay detects inhibition of aromatase only.
[e] The OECD is directing the validation of this assay in collaboration with the EPA.
[f] Alternative assay under development by the EPA to evaluate the effects of pre- and perinatal exposures.
[g] Hypothalmic-pituitary-gonadal axis.

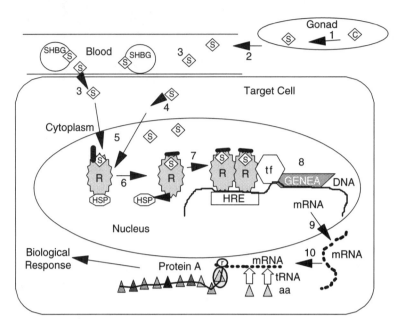

Figure 11.1 The steroid receptor pathway. Estrogen (S) and testosterone are synthesized in the gonads from cholesterol and enter the blood where they are transported by steroid hormone binding globulins (SHBG). S binds to its receptor, causing a transformational change in the macromolecule [e.g., loss of heat shock proteins (HSP)]. This activates the steroid-receptor complex and allows the receptors to bind in dimer formation to the S hormone response element (HRE). Gene transcription is initiated, and new mRNA is produced that results in the synthesis of a new protein.

increasing concentrations of a nonradioactive competitor; (2) once the reaction has reached a steady-state condition, the ligand-receptor complexes (bound) are separated from the unbound ligand (free); (3) the amount of radioactivity bound is quantitated; and (4) the results are analyzed and evaluated.[7]

Historically, the source of the receptor for ER and AR binding assays has been cytosolic extracts prepared from tissues (usually uterus for ER and prostate or epididymides for AR), and their use is well documented in the scientific literature.[8-14] However, other options are available, including whole-cell binding assays and the use of recombinant receptors.[15-19] Steroid-receptor binding assays are performed at various concentrations of test compound (usually over six to seven orders of magnitude) in competition with a fixed concentration of radiolabeled high-affinity reference ligand (such as [^{3}H]-17-β-estradiol for ER and [^{3}H]-R1881 or [^{3}H]-DHT for AR) to generate a dose-response curve. The test chemical is added to a tissue preparation or to a recombinant receptor preparation, and the amount of displacement of the radiolabeled ligand by the test chemical is determined. Assay tubes are also included to assess both total (no competitor present) and nonspecific binding (low-affinity binding not reflective of interaction with the ER or AR). Nonspecific binding is detected by measuring the amount of radioligand bound in the presence of a saturating concentration of unlabeled reference ligand (e.g., 100× the concentration of radioligand). The nonspecific binding is subtracted from the total binding for each concentration of test chemical to calculate the specific binding at that concentration. The IC_{50} (inhibitory concentration 50%) is the concentration of test chemical that inhibits the specific binding of the radioactive ligand by 50%.

Each type of *in vitro* binding assay has its strengths and limitations. These assays only assess the ability of the chemical to bind the receptor and are unable to differentiate agonist from antagonist responses. Binding assays using ER or AR preparations from tissues have been used extensively, are relatively simple to conduct, and produce consistent results with practice and proper handling. In addition, tissues from multiple species are amenable for use in these assays. Whenever tissue preparations are used, one must keep in mind that other steroid receptors are likely to be present as well as the receptor of interest. This is a particular concern when conducting assays for the androgen receptor because ligands for the androgen receptor may also interact with the glucocorticoid receptor (GR) or progesterone receptor (PR). In this case, a compound such as triamcinolone acetonide, which binds to and blocks PR and GR but not AR, can be added to the buffer to ensure specific AR binding.[20,21] Limitations of this type of assay include the use of animals to acquire the tissue, use of radioactive ligand, and the fact that they are not easily amenable to a high-throughput system.

The use of cultured cells and/or recombinant receptors eliminates the need to isolate receptors from animal tissues, can be performed with human cells or receptors, and has the possibility of adaptation for high throughput screening. With the use of recombinant receptors, the researcher also has more control over the receptors that are present in the preparation. For example, transient expression of AR in cells that do not express PR or GR eliminates the need for additional blocking components in the buffer.[16,18] Assays can also be conducted with purified recombinant receptor.[17,19] In this case, additional factors, such as heat shock proteins, may need to be added to help stabilize the receptor in its proper conformation for binding. Cell-based assays also require cell culture equipment, thorough training in cell culture techniques, and the use of radioactivity. Although purified recombinant receptor techniques have not been used as extensively as receptor preparations from tissues, they have the potential to provide an alternative method for the TIS battery.

Data interpretation is an important component of steroid-receptor binding studies. Typically, the data from each competitive binding assay are plotted on a semilog graph so that a sigmoid plot is obtained. The data can be plotted as counts per minute bound or as percent of total specific binding on the Y axis versus the concentration of the test chemical on the X axis. If the test chemical and the radiolabeled ligand compete for a single common binding site, the competitive binding curve will have a shape (Figure 11.2A) determined by the law of mass action.[22] Specifically, the curve should descend from 90% to 10% specific binding over approximately an 81-fold increase in the concentration of the test chemical (e.g., this portion of the curve will cover approximately two log units, Figure 11.2B). A binding curve that drops dramatically (i.e., from 70% to 0%) over one order of magnitude should be questioned, as should one that is U-shaped (i.e., percent bound at first decreases with increasing concentration of competitor but then begins to increase, Figure 11.2C). In the both cases, something has happened to the dynamics of the binding assay, and the reaction is no longer following the law of mass action (e.g., the test chemical may be binding to multiple binding sites or the test chemical may not be soluble at higher concentrations in the assay buffer and is precipitating from the solution). In these cases, an additional K_i experiment may be required to determine if the test chemical is truly a competitive inhibitor. While it is possible to calculate a K_i value from the IC_{50} by use of the Cheng-Prusoff equation,[23] the use of this equation is only valid if the competitive binding curve reflects a pure competition for a single binding site.[24] An experimentally derived K_i requires adding increasing concentrations of radiolabeled ligand in the presence of fixed concentrations of the test chemical and then plotting the data on a double reciprocal plot.[25,26] A pattern of lines that intersect on the ordinate axis is characteristic of competitive inhibition (Figure 11.3). Slopes obtained from each of the double reciprocal plots are then replotted as a function of the inhibitor concentration. The slope and intercept of the replot can be used to calculate a K_i value for the test chemical.[26]

Performance criteria for binding assays are being developed by the Office of Science, Coordination and Policy (OSCP) of the U.S. EPA through an agreement with the ICCVAM. Even though steroid receptor binding assays have been used for years, they have not been rigorously standardized and validated as required by the EDSP. Currently, protocols for ER and AR competitive binding assays are undergoing the standardization and validation process, under the direction of the OSCP.

B. Androgen and Estrogen Receptor-Dependent Gene Transcription Assays

Cell-based gene transcriptional activation assays were also proposed by EDSTAC, either to complement or as an alternative to the ER and AR competitive binding assays. These assays can be

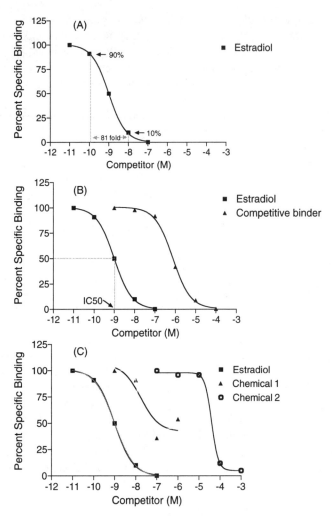

Figure 11.2 Examples of data from a steroid competitive binding assays. (A) Typical competitive binding curve, demonstrating that inert estradiol and [³H]-E₂ compete for a single common binding site (ER), following the law of mass action. (B) Comparison of competitive binding curves for inert estradiol and a competitive inhibitor. (C) Examples of data that do not follow the law of mass action. The U-shaped curve produced by Chemical 1 is indicative of a solubility problem. The dramatic drop from 100% to 12% binding over a one order of magnitude increase in Chemical 2 indicates a problem with the dynamics of the binding assay.

used to identify compounds that bind to the ER or AR and, unlike standard receptor binding assays, can also discriminate agonists from antagonists because they assess the ability of a chemical to affect hormone-dependent gene transcription. A vector carrying a hormone response element for the receptor of interest attached to a reporter gene is introduced into the cells. In some cases, a second vector that expresses the receptor (if not already endogenous to the cells) is also cotransfected into the cells. When the ligand-receptor complex binds to the hormone response element in dimer formation, the transcription of the reporter gene is initiated. The level of expression of the reporter can then be measured. The assays are conducted by diluting a range of concentrations of the test chemical in the cell culture medium. The test chemical is examined alone to assess its ability to activate hormone-dependent gene transcription (agonism) or in the presence of a high-affinity ligand (such as 17-β-estradiol for ER or DHT for AR) to assess its ability to interfere with or block hormone-dependent gene transcription (antagonism).

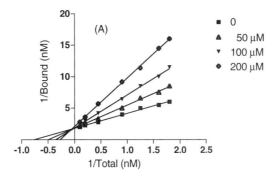
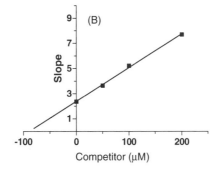

Figure 11.3 Experiment demonstrating competitive inhibition of steroid receptor binding by a test chemical. (A) Double-reciprocal plots of inhibition of [^{3}H]-ligand specific binding. (B) Slope-replot analysis for the determination of the inhibition constant (x intercept = K_i). These data represent a test chemical that is a competitive inhibitor of estrogen receptor binding with a K_i = 89 μM.

There are benefits as well as limitations to the use of mammalian versus yeast cells for these assays. Yeast-based assays are simple, sensitive, and generally less expensive to conduct than assays using mammalian cells. However, with yeasts there are rank-order differences in potencies of endocrine active test chemicals,[27–29] as well as differences in permeability of compounds through the yeast cell wall relative to the plasmalemma of mammalian cells. In this respect, assays utilizing mammalian cells are thought to be more representative of risk to humans.

A number of *in vitro* transcriptional activation assays using mammalian cells have been developed that assess either estrogen- or androgen-dependent gene transcription. Transient transfection assays in which both the receptor and a hormone-responsive reporter are cotransformed into cells[18,30–34] have been widely used. However, in transient transfection assays, target DNA sequences are over expressed and therefore do not reflect physiological levels of receptor. In addition, transfection efficiencies (i.e., percentage of cells expressing the receptor) may vary greatly from assay to assay. Unfortunately, the responsiveness is limited in time, as the transgenes are usually lost within 72 h, so transfections must be repeated with each new set of assays.

Adenoviral transduction, which uses a replication-defective virus to deliver biologically competent genes, has also been used to develop androgen receptor transcriptional activation assays.[35] Since the virus is replication defective, it represents no hazard of infection. Both the AR and a reporter gene are introduced into either CV-1 or MDA-MB-453 cells. As with transient transfection assays, transduction requires that the procedure be repeated for each new set of assays. However, viral infection appears to be much more efficient than transfection because induction is consistently high, with much lower variability. Transduction and transfection techniques require similar facilities, and both can be easily accomplished with basic cell culture supplies and equipment (culture hood and cell culture incubator). Stably transformed cells eliminate the need for repeated transient transfections and reduce the variability associated with this technique. However, stably transformed cell lines utilizing endogenous or transformed androgen receptor combined with a stably transformed hormone-responsive luciferase (or other) tagged reporter were not readily available until recently. These are genetically modified cell lines that are derived by transfection followed by antibiotic selection and clonal expansion. Ideally, this methodology produces a stable population of cells that will respond uniformly to stimuli.[36–41]

Transcriptional activation assays have several strengths. They utilize a well-defined mechanism of action to distinguish agonists from antagonists and can be performed using mammalian cells. Thus, the results are likely applicable to humans. Transient transformation and viral-transduction assays are highly responsive (i.e., give high levels of induction in comparison with vehicle controls) and provide more control over the specificity of response. For example, androgen receptors or glucocorticoid receptors can be transfected into cells separately, and the responses can be assessed independently. Both adenoviral-transduced and stable cell lines are amenable to a 96-well plate format and give a consistent response with low variability, thus making them attractive for adaptation to high throughput systems (HTS). Generally, *in vitro* transcriptional activation assays are cost effective and, with proper training and equipment, are rapid and easy to perform. Cell lines derived from human tissue can be used to develop these assays, so it is human receptors that are used in the assessments. In addition, only a small amount of test chemical is needed to run multiple assays.

Transcriptional activation assays also have some limitations. Compared with traditional receptor binding assays that use cytosolic preparations, there is a requirement for cell culture equipment, techniques, and training. The presence of more than one receptor in the cells (especially if using cells with endogenous receptors) that can activate the hormone response element of the reporter may be viewed as a limitation. This requires additional assays using selective competitors to differentiate individual receptor activity. Transient transfection assays require separate transformations for each set of assays conducted, can have high interassay variability, and are generally not suited for high throughput systems. Stable cell lines have many advantages, as discussed above, but require a considerable investment of time and resources to produce each cell line. Another concern is the possibility of false negatives and false positives. These systems assess the test chemical, but if a particular test chemical requires metabolic activation to produce its hormonally active metabolite, it could produce a false negative in the assay. It is also necessary to assess the issue of cytotoxicity when working with cell-based assays. In tests for antagonist activity, decreases in reporter activity could be due to either the test chemical or general cellular toxicity from the compound. The incorporation of cell toxicity tests into routine chemical testing is an important extension of the method. Otherwise, a test chemical could be falsely identified as an antagonist.

C. Steroidogenesis

The EDSTAC final report recommended two assays for the evaluation of effects of environmental chemicals on steroidogenesis. The first, a sliced testis protocol, was designed to measure testosterone production *in vitro* from testicular parenchymal fragments. The second, an aromatase assay designed to assess the potential effect of toxicant exposure on the conversion of testosterone to estradiol, was recommended as an alternate approach. Thus, the focus on steroidogenesis was on two major biologically active hormones, testosterone and estradiol, both generated in the later portion of the gonadal steroidogenic pathway (Figure 11.4).

The rationale for recommending the sliced testis assay was based on a number of previous studies that demonstrated the utility of the method for identifying toxicant-induced endocrine alterations.[42,43] In general, this method measures testosterone release *in vitro* under both nonstimulated and human chorionic gonadotropin (hCG)-stimulated conditions over a period of hours following testis decapsulation. The assay may be conducted using testes from animals following *in vivo* exposure, or it may evaluate *in vitro* testosterone production by using testes from animals with no prior exposure to the test chemical. Since a number of variations in the procedure have

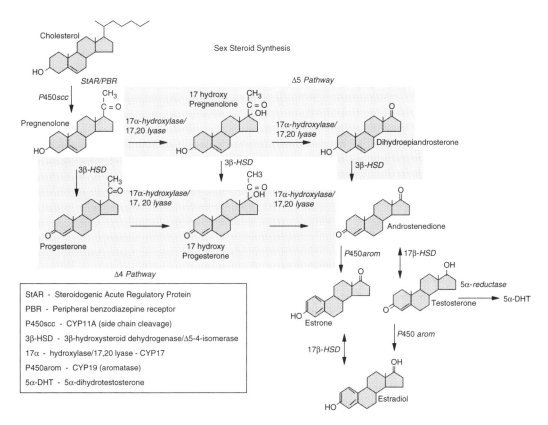

Figure 11.4 The sex steroid pathway. Chemical structures and enzymes for the synthesis of steroids.

been employed over the years, optimization and standardization of the protocol has been a prerequisite to using the method as a screen. Consequently, as part of the prevalidation process of the assay for possible use in the TIS battery, a number of experiments are being conducted to optimize the type of incubation medium, gaseous environment, temperature, parenchymal fragment size, time to achieve initial baseline testosterone secretion, appropriate hCG concentration, and measures of tissue viability. Moreover, the optimal age range of the animals from which the tissues are obtained is being established.

The basic sliced testis protocol is depicted in Figure 11.5. Following an initial replacement of medium to eliminate any hormone released by trauma, sampling from incubated fragments is performed at 1-h intervals. This establishes a baseline hormonal secretion and gauges the magnitude of the response to a stimulating concentration of hCG. However, some modifications in the protocol (Table 11.2) will likely emerge from the optimization process currently being conducted under the oversight of the U.S. EPA.

For use in characterization of the effects of environmental toxicants on steroid synthesis, the sliced testis approach can be compared with the evaluation of circulating steroid concentrations in response to *in vivo* exposures and to the assessment of steroid production *in vitro* by use of enriched or purified gonadal cell preparations. All have strengths and limitations when considered in the context of a screening application for endocrine disruptors (Table 11.3). Assessments of serum steroid concentrations are easily performed, whether done under terminal conditions or as part of a periodic sampling during an *in vivo* study. The serum samples can be obtained without extensive training, and characteristics of xenobiotic absorption, distribution, metabolism, and excretion (ADME) can be evaluated with an *in vivo* exposure paradigm. However, the specificity of the effect on steroidogenesis would be questionable. Any alteration in testosterone secretion observed following

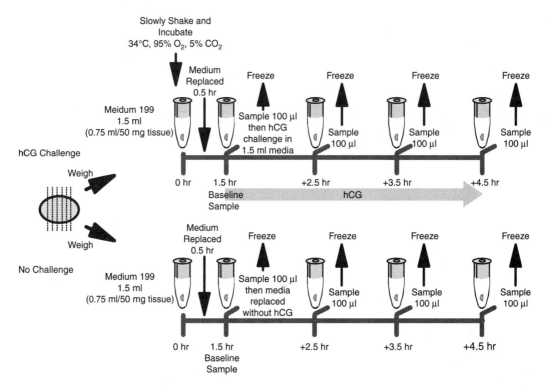

Figure 11.5 Example of the sliced testes protocol. Medium replacement after an initial 30-min incubation minimizes the influence of traumatic testosterone release on baseline secretion. Aliquots of medium are then sampled on an hourly basis under hCG-stimulated or nonstimulated conditions.

the *in vivo* exposure, for example, could reflect either a direct effect on steroidogenesis or a response secondary to an indirect gonadotropic stimulation. An alternative approach would be to administer the test chemical systemically and assess the gonadal secretory response *in vitro*, using testicular tissue taken at necropsy. In this case, unlike the case of *in vitro* only exposures, ADME conditions would still be maintained. Also, temporal controls over steroid sampling can be more strictly regulated under culture conditions where hormonal stimulation is artificially imposed. This type of stimulation, often using luteinizing hormone (LH) or hCG, is employed to evaluate the effects of a xenobiotic on the responsiveness of the gonadal preparation to gonadotropic stimulation in culture. Costs tend to be higher than for strictly *in vitro* exposures because the animal usage is greater and larger amounts of the compound under study are needed. A third alternative is to evaluate gonadal preparations exposed only *in vitro*. In this case, any measured alteration in steroid secretion would reflect a direct effect on the process of steroidogenesis.

The sliced testis protocol recommended by EDSTAC employs fragments of parenchymal tissue, which are relatively straightforward to prepare. Moreover, animal usage is reduced if the *in vitro*–only exposure paradigm is used. Multiple parenchymal tissue fragments from a single testis or testis pair can be randomly assigned to different treatment conditions, minimizing the influence of animal-to-animal variability. Nevertheless, one of the issues raised concerning the utility of a sliced (or minced) testis approach to the assessment of testosterone secretion is the question of variability of the obtained data in comparison to other types of incubated testicular preparations. For example, isolated Leydig cells have been used for the same purpose and have the advantage of increased cellular uniformity and sensitivity. Both crude enriched[44–47] and purified[48–50] Leydig cell preparations have been employed to measure testosterone secretion, with the latter (85% to 90% purity) exhibiting less variability in controls, as reflected by coefficients of variation (COV) calculated from reported control data (Table 11.4). On the other hand, crude Leydig cell preparations

Table 11.2 Potential modifications to sliced testes protocol currently under consideration

Fragment size: The protocol included in the EDSTAC final report suggested using a parenchymal fragment size of 50 mg. This would allow a significant number of samples from a single testis to be distributed across multiple chemical treatments for both nonstimulated and hCG-stimulated conditions, thereby reducing the variability inherent in using individual animals for exposures to different xenobiotics. At the same time, use of such smaller fragment sizes would minimize the number of animals required for testing. Ultimately, the determination of an appropriate amount of tissue will depend on the variability in hormonal secretion observed.

Incubation medium: Incubation times for the sliced testis protocol are short term and typically extend no longer than 5 h. While sterile-filtered Medium 199 has frequently been used for gonadal incubations, a variety of nutrient media (e.g., Krebs-Ringers bicarbonate, Dulbecco's modified Eagle's medium) should be acceptable. Nevertheless, modest disparities in secretion with different media may exist. There has been a trend away from using phenol red as a pH indicator in these media, since a growing number of publications have shown that it possesses estrogenic properties[a–c] and may have an adverse effect on Leydig cell testosterone production.[d]

Gaseous environment: Incubations of testicular fragments and enriched Leydig cell preparations have commonly been carried out under 95% air (approximately 19% oxygen) and 5% CO_2. Oxygenated media have also been employed,[e,f] and it is not likely that over the short term the increase in oxygen content in the incubation medium would have an adverse effect on tissue viability.

Tissue viability: For any type of *in vitro* system, it is essential that an assessment of cell and tissue viability be performed because reductions in steroid secretion could be attributable to a general toxicant-induced cell death and not a targeted alteration in steroidogenesis. The use of techniques relying on perturbations in cell membrane integrity (e.g., trypan blue, propidium iodide, lactic dehydrogenase) are later stage indices of cytotoxicity and may not identify earlier events in cell death. Alternatively, measures of mitochondrial integrity (i.e., tetrazolium-based MTT, MTS assays) are more likely to reflect a cytotoxic insult taking place within the 4 to 5 hour time frame of the sliced testis procedure. An estimation of impairments in mitochondrial membrane potential using specific dye probes appears to be a very sensitive index of early apoptotic changes, although such procedures are more appropriate for cell-based assays and their utility for assessments of tissue cytotoxicity would have to be determined.

[a] Hofland, L.J., van Koetsveld, P., Koper, J.W., den Holder, A., and Lamberts, S.W., Weak estrogenic activity of phenol red in the culture medium: its role in the study of the regulation of prolactin release in vitro, *Mol. Cell. Endocrinol.*, 54, 43, 1987.
[b] Welshons, W.V., Wolf, M.F., Murphy, C.S., and Jordan, V.C., Estrogenic activity of phenol red, *Mol. Cell. Endocrinol.*, 57, 169, 1988.
[c] Dumesic, D.A., Renk, M., and Kamel, F., Estrogenic effects of phenol red on rat pituitary cell responsiveness to gonadotropin-releasing hormone, *Life Sci.*, 44, 397, 1989.
[d] Abayasekara, D.R., Kurlak, L.O., Band, A.M., Sullivan, M.H., and Cooke, B.A., Effect of cell purity, cell concentration, and incubation conditions on rat testis Leydig cell steroidogenesis, *In Vitro Cell Dev. Biol.*, 27A, 253, 1991.
[e] Wilker, C.E., Welsh, T.H. Jr, Safe, S.H., Narasimhan, T.R., and Johnson, L., Human chorionic gonadotropin protects Leydig cell function against 2,3,7,8-tetrachlorodibenzo-*p*-dioxin in adult rats: role of Leydig cell cytoplasmic volume, *Toxicology*, 95, 93, 1995.
[f] Harris, G.C. and Nicholson, H.D., Characterization of the biological effects of neurohypophysial peptides on seminiferous tubules, *J. Endocrinol.*, 156, 35, 1998.

(approximately 12% to 15% purity) show COVs that are comparable with those obtained using sliced testicular tissue. As a screening approach, the increase in the technical demands of the isolation procedure, the costs, and the time necessary to complete the Leydig cell enrichment process are factors militating against this option.

One additional option for consideration is the use of stabilized steroidogenic cells. A growing number of studies in steroid toxicology have employed cells derived from mouse or rat Leydig cell tumors.[51–53] Depending on which portion of the sex steroid pathway is of concern, different cell lines can be used. The mouse MA-10 Leydig cell tumor line has been shown to be particularly useful for detection of alterations in steroidogenic acute regulatory (StAR) protein, cytochrome P450 side chain cleavage enzyme, and 3-hydroxysteroid dehydrogenase activity, all of which affect progesterone synthesis early in the steroidogenic pathway.[54–56] In contrast, the KGN human granulosa tumor cell line has been shown to be particularly high in P450 aromatase activity, making it potentially valuable for assessment of toxicant-induced insult to estradiol production.[57] However, this cell line shows low 17-hydroxylase and 17,20-lyase activity, which would make it unsuitable for detection of effects on androstenedione synthesis. One cell line that appears to hold promise

Table 11.3 Strengths and limitations of different assessments of testosterone production

Type of Assessment	Strength	Limitation
In vivo exposure and in vivo sampling	• Relatively easy to perform, extensive training not necessary • Can account for xenobiotic absorption, distribution, metabolism and excretion (ADME)	• Testicular specificity of response questionable • Animal usage and costs higher than for in vitro exposure conditions
In vivo exposure and in vitro sampling (sliced testis protocol)	• Not difficult to perform • ADME parameters maintained • Temporal conditions of stimulation more strictly controlled.	• Testicular specificity of response questionable • Animal usage and costs higher than for in vitro exposure conditions.
In vitro exposure and in vitro sampling (sliced testis protocol)	• Not difficult to perform • Testicular specificity of response • Animal usage and costs reduced • Multiple tissue fragments from single testis or testis pair can be randomly assigned to treatment groups, minimizing animal-to-animal variability	• Variability of data tends to be higher than for purified Leydig cell preparations • Sensitivity of response less than for more purified cell preparations • ADME factors absent
In vitro exposure and in vitro sampling (enriched and purified Leydig cells)	• Increased uniformity in cell population • Increased sensitivity of response • Purified Leydig cell preparations show reduced variability in response	• Technically more demanding, particularly for purified Leydig cells • Increase in cost over sliced testis approach • ADME factors absent

Table 11.4 Control Group Coefficients of Variation: Testosterone Secretion

Preparation	Non-stim.	LH*/hCG-stim.	Reference
Sliced Testis	10	27	Lau & Saksena (1981) Int. J. Androl 4:291.
	23	—	Gray et al. (1995) TAP 130:248.
	28	29	Powlin et al. (1998) Tox. Sci. 46:61
	23	22	Wilker et al. (1995) Toxicology 95:93.
	50	—	Banczerowski et al. (2001) Br.Res. 906:25.
Crude Leydig Cell Prep (~12–15%)	45	30	Chambon et al. (1985) Andrologia 17:172.
	26	27	Kan et al. (1985) J. Steroid Biochem. 23:1023.
	40	25*	Raji & Bolarinwa (1997) Life Sci. 61:1067.
	15	12	Laskey & Phelps (1991) TAP 108:296.
Purified Leydig Cell Prep (80–95%)	11	23	Ronco et al. (2001) Toxicology 159:99.
	9	—	Nagata et al. (1999) FEBS Lett. 444:160.
	6	8	Romanelli et al. (1997) Life Sci. 61:557.
	—	12	Klinefelter et al. (1991) TAP 107:460.
	13	12	Guillou et al. (1985) FEBS Lett. 184:6.

Incubation parameters: 10^5–10^6 cells/well; 3–4h collection period; 100mIU hCG or *50 ng/ml oLH stimulation

for evaluations of xenobiotic alterations in steroid production is the H295R line, which is derived from human adrenocortical tumor cells. This cell line appears to contain all of the enzymes of the steroidogenic pathway[58–60] and has already been used to a limited extent to evaluate effects on aromatase activity.[61,62] Since the use of a cell line would eliminate the necessity for animals and the line is available through the American Type Culture Collection (ATCC), this approach offers distinct advantages for studying the entire sex steroid pathway. For these reasons, its utility is currently under study by the U.S. EPA for possible substitution in the TIS battery.

D. Aromatase Assay

Aromatase (CYP19) is a cytochrome P450 enzyme complex that converts androgens to **estrogens** (Figure 11.4). The enzyme is present in multiple tissues, such as the ovary, placenta, uterus, testis, brain, and adipose tissue, and is essential for normal reproductive development and function in both males and females. As indicated in the previous section, an *in vitro* aromatase assay is currently included as an alternative assay in the TIS battery. This assay may be used if the final screening battery does not include the female pubertal assay, a protocol that can also detect alterations in the synthesis of estrogen. The need for an assay to screen specifically for changes in aromatase activity is justified by numerous reports that environmental contaminants can affect the activity of this enzyme. Several flavonoids and related phytoestrogens,[63–65] as well as certain pesticides,[66] have been shown to inhibit aromatase activity *in vitro*. Studies to evaluate the effects on reproductive function in mammalian and nonmammalian species following exposure to several pesticides, including fenarimol, imazalil, prochloraz, and triadimefon, have demonstrated reduced fertility as a result of decreased serum estrogens and altered gonadotropin concentrations.[67–69]

In vitro methods may be readily used to identify chemicals that inhibit aromatase activity by acting as suicide substrates. Generally, homogenates or microsomal preparations of aromatase-containing tissues, such as ovaries, testes, brain, or human placenta, are incubated for a designated period with the test chemical, radiolabeled androstenedione or testosterone, and essential cofactors. The estrogen product, estrone or estradiol, that forms during the enzymatic reaction can be isolated by use of thin-layer chromatography or high-pressure liquid chromatography (HPLC) and quantified by liquid scintillation spectroscopy. Alternatively, the tritiated water assay is a method that measures the amount of ^3H_2O released during incubation with [1β-^{3}H]-androstenedione. The basis of this method is that 1 Mol of ^3H_2O is released during the aromatization of 1 Mol of the A-ring of [1β-^{3}H]-androstenedione to estrone.[70] The advantage of the ^3H_2O method is that it does not require any chromatography to separate steroid substrate. Rather, the amount of ^3H_2O released is measured in the aqueous phase following extractions with organic solvents and dextran-coated charcoal. Results from the product isolation method and the ^3H_2O method have been shown to be comparable for human placental microsomes, rodent ovarian microsomes and tissues, and cell culture systems. Aromatase activity is typically reported as enzyme velocity or rate (nanomole product formed per milligram of protein per minute). The source of tissue used for these two methods can vary, depending upon the level of aromatase activity required for a particular study. Human placentas collected at term contain very high levels of aromatase and provide a good source for a microsomal preparation. Rodent ovarian microsomes and recombinant-human aromatase can also be used successfully with these methods although the level of aromatase activity is lower than that observed using human placental microsomes.

A number of *in vitro* cell culture systems have also been used to evaluate the effects of environmental chemicals on aromatase activity. These cell culture systems have the advantage of potentially detecting both inhibition and induction of enzyme activity. Additionally, some cell lines may provide metabolic activation of some test chemicals as well as providing information on the ability of the chemical to cross the cell membrane. Cell lines most commonly used include the human JEG-3 and JAR choriocarcinoma cells,[71] R2C rat Leydig carcinoma cells,[72] MCF-7 human mammary carcinoma cells,[73] and H295R human adrenocortical carcinoma cells.[61,62,74] The JEG-3 and JAR cell lines display a fairly high basal level of aromatase activity, a characteristic that is advantageous when evaluating chemicals that might have an inhibitory effect. In contrast, the H295R cell line has been used to detect induction of aromatase activity by a number of chlorotriazine herbicides (e.g., atrazine, simazine, and propazine) and the fungicide vinclozolin.[61,62,74] Elevated levels of CYP19 mRNA detected using RT-PCR indicate that the induction of enzyme activity in these cells is regulated at the level of gene transcription rather than by a change in the activation of the enzyme itself. Although induction of aromatase activity following *in vivo* exposure to these chemicals has not been currently documented, Stoker et al. have reported increased serum estrone and estradiol concentrations in male rats exposed to atrazine,[75] suggesting that aromatase activity may be induced. Another potential advantage of the H295R cell line is that it has been reported to possess multiple promoters that are responsible for the tissue-specific regulation of aromatase synthesis in mammals.[61] Thus, this cell line may provide an opportunity to determine whether a particular test chemical has an effect on aromatase activity through one of a number of promoter sites, as opposed to other cell lines that do not possess multiple promoters.

The approach employed to evaluate whether a particular chemical can disrupt the regulation of aromatase activity depends upon a number of factors. Chemicals that inhibit aromatase activity by acting as suicide substrates can be identified with fairly simple *in vitro* techniques. Since inhibition of aromatase activity has been demonstrated to be a common mechanism for a variety of environmental chemicals, *in vitro* assays using microsomal preparations and the production of ^3H_2O to evaluate aromatase activity provide a relatively easy and inexpensive screening method. However, chemicals that might induce aromatase activity by altering the regulation of enzyme synthesis or stability require the use of *in vitro* whole cell systems and/or *in vivo* studies that evaluate multiple aromatase-containing tissues. Finally, the design of *in vivo* studies is critical for understanding (1) whether a chemical has a direct effect on aromatase activity versus an indirect effect mediated through a change in the regulation of the hypothalamic-pituitary-gonadal axis and (2) whether any effect on aromatase is tissue dependent. For example, circulating estrogen concentrations are influenced by the levels and cyclicity of pituitary gonadotropins.[76] Thus, the evaluation of aromatase activity after longer exposures to a given chemical may not reflect its primary mode of action. Additionally, since aromatase expression is under the control of tissue-specific promoters regulated by different cohorts of transcription factors,[77] it is important to evaluate multiple tissues for aromatase activity in the *in vivo* studies. In some cases there could be local alterations in estrogen biosynthesis in a tissue that would not be reflected by an alteration of overall circulating hormone levels.[78]

III. *IN VIVO* ASSAYS FOR THE DETECTION OF ENDOCRINE DISRUPTING CHEMICALS

A. Uterotrophic Assay

The rodent uterotrophic assay, recommended by EDSTAC as one of the core *in vivo* assays for the TIS, is generally accepted as a robust, technically simple, and reliable method for detecting estrogenic activity. The assay was first developed during the mid-1930s for use in both mice and rats,[79–81] and continues to be routinely used today as the gold standard to detect both potent and weak estrogen agonists[82] as well as antagonists.[83,84] Specifically, the assay measures the ability of a test chemical to significantly increase the weight of the rodent uterus after three consecutive days of exposure, or in the case of an antagonist, to prevent an estrogen-dependent increase in uterine weight. The assay is based upon an estrogen mode of action that includes estrogen binding to its receptor, initiation of gene transcription, and induction of uterine growth.[85,86] A number of studies have correlated *in vitro* ER binding affinity with a uterotrophic response following *in vivo* exposure to a broad range of pharmaceuticals and environmental chemicals.[87–89]

In collaboration with the U.S. EPA, the OECD has agreed to direct the optimization and validation of the rodent uterotrophic assay for the TIS battery. The OECD has concentrated its international validation program on evaluating the performance of two versions of the uterotrophic assay, one using sexually immature female rats and the other using adult ovariectomized (OVX) rats. Both of these versions provide an animal model that is devoid of endogenous estrogen, thereby ensuring that any observed estrogenic response will be a direct influence of the test chemical. Owens and Ashby[82] have published an overview of the OECD protocol for the uterotrophic assay, along with an extensive review of the biological basis and use of the assay throughout the past 60 years. In brief, sexually immature female (less than 21 days of age) or adult OVX rats (approximately 60 days of age at ovariectomy with a minimum of 14 days' recovery) are dosed with a test chemical by oral gavage or subcutaneous injection for three consecutive days, at approximately 24-h intervals. Body weights are monitored daily, and the absolute uterine weights of both the wet uterus (includes imbibed fluid) and blotted uterus are recorded at necropsy, 24 h following the final dose. Uterine weights are evaluated by analysis of variance, with body weight at necropsy as a covariable.

Although there are numerous issues that must be considered during the optimization and validation process for the uterotrophic assay, the OECD has developed a strategy and begun to methodically move through the process.[90] Initially, three major variables that required consideration included species selection (e.g., mouse versus rat), age of the test animal (sexually immature versus adult OVX), and route of administration (oral gavage versus subcutaneous injection [s.c.]). Additional factors, such as animal strain, animal husbandry (influence of diet or vehicle), and length of time between last dose and necropsy (6 versus 24 h), that could possibly confound the results also required special consideration.

Drawing upon the vast amount of historical data and published literature on the uterotrophic assay, the OECD has based its approach to the validation process on scientific rationale and practical experience. The group chose to standardize the protocol with the laboratory rat as opposed to the mouse, as the former is the preferred rodent model in toxicology testing paradigms for regulatory purposes. It was also decided that studies would be designed to address questions regarding dosing route, strain variation, and animal husbandry during the multiple laboratory evaluation. Phase I of the OECD validation process was designed to test, refine, and standardize the sexually immature and the adult OVX rat uterotrophic assays using a high-potency estrogen, ethinyl estradiol, and the estrogen receptor antagonist, ZM 189,154. This phase was also designed to provide an initial determination of intra- and interlaboratory variability. Phase 2 was designed to demonstrate the capability of the standardized uterotrophic protocols to identify estrogenicity in a set of test chemicals composed of weak estrogens and a known negative, as well as to continue to document intra- and interlaboratory variations. The potential effects of low levels of isoflavones in laboratory animal chow were also evaluated to determine if the animal chows containing higher concentrations of phytoestrogens would hamper the detection of estrogenic activity during tests of weak agonists.

The results from Phases 1 and 2 of the validation process for the uterotrophic assay have recently been published. Phase I studies were conducted by 19 laboratories from Japan, Denmark, France, Germany, Korea, the Netherlands, the United Kingdom, and the United States.[90] The study evaluated four protocols: (1) sexually immature, female rat model, with oral dosing for 3 days, (2) sexually immature, female rat model, with dosing by s.c. injection for 3 days, (3) adult, OVX rat model, with dosing by s.c. injection for 3 days, and (4) adult, OVX rat model with extended dosing by s.c. injection for 7 days. Animals were exposed to ethinyl estradiol (0.03 to 3.0 µg/kg/d), or ZM 189,154 (0.1 and 1.0 µg/kg/d) coadministered with ethinyl estradiol (0.3 µg/kg/d). All animals were killed 24 h after the last dose. Other experimental conditions, such as animal strain, vehicle, diet, and bedding, varied between laboratories. Data from these studies demonstrated that all laboratories successfully conducted each of the four protocols, and that each protocol detected a significant increase in uterine growth following exposure to ethinyl estradiol. For each protocol there was good agreement among laboratories with respect to the lowest effective dose (LOEL) of ethinyl estradiol. Although the shape of the dose-response curve for ethinyl estradiol did vary between laboratories, there was general agreement in dose response when the 10% and 50% effective doses (ED_{10} and ED_{50}) were calculated. Given the differences in animal strain and husbandry, the robust nature of each of the protocols for detecting increases in uterine weight was demonstrated. In addition, preliminary evidence that the protocols are a feasible method for detecting antagonists was also provided. The recommendations resulting from these studies included: (1) moving forward with the validation of both the sexually immature and adult, OVX animal models, (2) refining the protocols to address minor technical issues, and (3) continuing to test the protocols, using several weak estrogen agonists.

Studies conducted under Phase 2 of the OECD validation process are reported by Kanno et al.[91,92] and Owens et al.[93] One study evaluated the same four protocols as employed in Phase I, using ethinyl estradiol, several weak estrogens (genistein, methoxychlor, nonylphenol, bisphenol A, and o,p'-DDT), and a negative control, dibutylphthalate. Twenty-one laboratories from nine nations participated in the study, although not all conducted each protocol or tested all of the chemicals. Overall, the uterotrophic assay provided reproducible results within the same laboratory and across the participating laboratories for all the test chemicals.[91] With the exception that the lowest effect levels (LOELs) differed for the test chemicals by dosing routes, no major differences were noted between the protocols. As a result, the sexually immature and adult, OVX animal models were judged to be qualitatively equivalent to one another, and both protocols will continue to move forward through the validation process. There were some false positives for dibutylphthalate in which 5 of 36 experiments resulted in a statistical increase in uterine weight as compared with the control (vehicle). Of these, three sets of data for the dibutylphthalate had uterine weights significantly higher than the vehicle control, yielding a false positive rate of 8% for this test chemical.[92] These data reinforce the importance of using a weight-of-the-evidence approach for data interpretation, using all the assays within the TIS battery. Finally, a comparison of the phytoestrogen content in the diets used in the participating laboratories showed that levels lower than 325 to 350 μg/g total genistein equivalents did not impair the responsiveness of the uterotrophic assay when testing the weak estrogen-receptor agonists, nonylphenol and bisphenol A.[93] However, recommendations to monitor dietary phytoestrogen content will most likely be included in the final OECD guidelines for the uterotrophic assay as well as recommendations to monitor control uterine weights to be certain they fall within a range consistent with historical data.

The advantages of using the uterotrophic assay as one of the *in vivo* components of the TIS battery are exemplified by its extensive use as an indicator of an estrogen receptor–mediated response. Historical data, as well as data recently collected during the OECD validation studies, have proven the assay to be both biologically relevant (e.g., it detected chemicals known to bind to the estrogen receptor) and reliable (in that results can be replicated within a laboratory and between laboratories). The robust nature of the assay has also been proven through the ability of multiple laboratories to replicate results, even with variations in experimental conditions, such as strain, dosing route, animal husbandry, and animal model (e.g., sexually immature versus adult, OVX female rats). Within the TIS battery, the uterotrophic assay is one of the most technically simple. It requires less time to complete and fewer animals per treatment group than the female pubertal protocol, and in some cases it appears to be more sensitive to weak environmental estrogens. For example, bisphenol A is known to possess a weak binding affinity for the estrogen receptor[89,94] and to increase uterine weight following oral or s.c. exposures.[89,95] However, this chemical did not alter the age of vaginal opening when administered by oral gavage from 21 to 35 d of age at doses two to three times higher than the dose required to stimulate uterine growth.[89] Ashby and Tinwell[95] also reported no change in the age of vaginal opening following a 3-d exposure to bisphenol A (600 and 800 mg/kg) by oral gavage, although uterine weights were significantly increased in the same animals. In the same study when bisphenol A was administered by s.c. injection, 4 of 7 (600 mg/kg) and 3 of 8 (800 mg/kg) animals displayed early vaginal opening. Thus, although uterotrophic responses were similar following oral or s.c. exposures to bisphenol A, only the s.c. route altered the age at vaginal opening.

While the uterotrophic assay has many strengths, the method is limited in that it can only detect estrogenic or antiestrogenic activity. The female pubertal protocol, for example, is capable of detecting a broader range of mechanisms of action. In addition, there are some technical caveats that may avert false negatives: Guidelines for the dose selection for the uterotrophic assay must be developed to ensure that test chemicals with short-term estrogenic activity will not be missed if the endpoints are measured 24 h after the last dose. Several reports have demonstrated that the length of the uterotrophic response following exposure to some chemicals is much shorter than for others. Thus, the dosing route and timing following exposure may dictate whether a significant change in uterine weight may still be detected at necropsy.[79,85,96,97] A comparison of the uterotrophic responses 6 and 24 h following oral exposures to 4-*tert*-octylphenol or bisphenol A showed significant increases in uterine weights at the 6 h but not at the 24 h time point.[89] Yamasaki et al.[98] also reported a higher increase in uterine weight at 6 h as compared with the 24 h time point following the last of three daily s.c. injections of bisphenol A. However, in the later study the increased uterine weight at 24 h was still significantly different from the control. The higher doses of bisphenol A used in the OECD validation studies (e.g., 600 mg/kg by oral gavage and 300 mg/kg for s.c. injection) did result in significant increases in uterine weight 24 h following the last dose.[91] Thus, it is imperative that the final protocol for the uterotrophic assay include dose selection criteria to allow a sufficient dose of the test chemical for each treatment route that will enable the detection of estrogenic activity at the time of necropsy.

B. Hershberger Assay

The Hershberger assay is an *in vivo* short-term test similar in concept to the rodent uterotrophic assay discussed above. Both assays measure changes in specific tissues that normally respond to endogenous steroid hormones. The assay conditions are designed to achieve low endogenous hormone levels and examine target tissues that are highly responsive to administration of exogenous hormones. The Hershberger assay provides a method to determine whether a compound possesses androgenic or antiandrogenic activity. It involves weighing tissues in castrated, sexually immature male rats following exposure to a test chemical with or without androgen supplementation.[99] The biological relevance of the assay is based upon the fact that accessory sex tissues and glands depend on androgen stimulation to gain and maintain weight during and after puberty.[100]

The Hershberger assay may also present the opportunity to detect inhibitors of 5-α-reductase by taking advantage of a differential response among the accessory sex glands and tissues. Testosterone is converted by 5-α-reductase to DHT, which has significantly higher binding affinity than testosterone for the androgen receptor. The levator ani and bulbocavernosus (LABC) muscles lack the enzyme 5-α-reductase, while in other tissues the enzyme is present (e.g., ventral prostate, seminal vesicles, and coagulating glands). It is plausible then that tissues with 5-α-reductase activity are more responsive to compounds that are metabolized to DHT than tissues without the enzyme.

Acknowledging the substantial amount of historical data and the long standing use of the Hershberger assay for the detection of androgens and antiandrogens,[99–104] the EDSTAC recommended that the Hershberger be a component of the TIS battery. As with the uterotrophic assay, the OECD agreed to assist the OSCP, U.S. EPA, by directing the validation process for the Hershberger assay. Using an approach similar to that used for the validation of the uterotrophic assay, the OECD designed a series of studies that will be conducted in multiple laboratories throughout the world. Phase I was designed to evaluate the standardized protocol, using a reference androgen and an antiandrogen. Phase II of the process will evaluate the ability of the assay to detect other androgens, several weak antiandrogens, and a 5-α-reductase inhibitor. Estimates of intra- and interlaboratory variance will be determined throughout the validation process.

The protocol being evaluated currently by the OECD uses castrate, sexually immature male rats. The animals are castrated after peripuberty (generally occurs after postnatal day [PND] 42) by removing both the testes and epididymides. In most rat strains, such as the Sprague-Dawley, Long-Evans, or Wistar, peripuberty is expected to take place at approximately 6 weeks of age, within an average age range of 5 to 7 weeks. Peripuberty is marked by prepuce separation from the glans penis (GP). Experimentally, preputial separation is necessary so that the GP can be properly dissected and accurately weighed. At the peripubertal stage of sexual development, the GP and other androgen-dependent sex accessory tissues are sensitive to androgens, having both androgen receptors and the appropriate steroidogenic enzymes. The advantage of using a rodent of this age is that the sex accessory tissues have a high sensitivity and small relative weight, which minimizes variation in responses between individual animals. In addition, castration removes the inhibitory feedback of testosterone on the hypothalamic-pituitary axis, resulting in a characteristic increase in the concentration of LH. Depending upon the action of the test substance in the hypothalamus, the pattern of LH secretion with increasing doses of the test substance may be altered.

Following a minimum of 7 days after castration (but no later than PND 60 because the age at necropsy should not be greater than PND 70), the males are assigned to treatment groups ($n = 6$), housed 2 to 3 per cage, and dosed by oral gavage for 10 consecutive days with one of several dosages of the test chemical. For detection of androgenic activity, only the test chemical is administered. To detect antiandrogenic activity, the test chemical is administered with and without the reference androgen agonist, testosterone propionate (TP) 0.1 mg/rat/d or 0.4 mg/kg/d, s.c. The males are killed approximately 24 hours later on the 11th day. The five sex accessory tissues, the ventral prostrate (VP), seminal vesicles with coagulating glands (SV), LABC muscles, Cowper's gland (or bulbourethral gland, COW), and GP are carefully dissected, trimmed of excess adhering tissue and fat, and weighed to the nearest 0.1 mg. Each tissue is handled with particular care to avoid loss of fluids or desiccation. In addition to the sex accessory tissues, it is useful to include weights of the liver, adrenals, and kidney, as they may provide supplementary information about the systemic toxicity, target organs, and other effects of the test substance. For example, possible induction of testosterone metabolizing enzymes in the liver or an impact on adrenal steroidogenesis could occur with exposure to some test chemicals. Optional endpoints, such as the measurement of serum testosterone and LH concentrations, may also provide possible insight into metabolism changes and hypothalamic responses. Considerations of the assessment of thyroid function (serum tetraiodothyronine [T_4], tri-iodothyronine [T_3], and thyroid-stimulating hormone [TSH]; and thyroid weight and histology) and adrenal steroidogenesis also seem reasonable as optional endpoints.

If the evaluation of each chemical requires necropsy of more animals than is reasonable for a single day, necropsy may be continued over two consecutive days. In this case, the work could be divided so that necropsy of three animals per treatment per day (one cage) takes place on the first day. Statistical comparisons in individual laboratories will be made for the different sex accessory tissues by analysis of variance. For an androgen agonist, the mean of each variable at each dose of the test chemical will be compared with the vehicle control. A statistically significant increase in tissue weight will be considered a positive androgen agonist result. For an androgen antagonist, the means from the test chemical coadministered with TP will be compared with the TP-only control. A statistically significant decrease in tissue weight will be considered a positive antagonist result. If more than one set of comparisons is required, all comparisons will be conducted separately for each test group against its control.

Several issues should be considered to adequately interpret the results from the Hershberger assay. Differences in body weight may be a source of variability in the weight of tissues of interest (especially the liver) within and among groups of animals. This variability could potentially reduce the assay sensitivity and possibly lead to false positives or false negatives. Therefore, variations in body weight should be both experimentally and statistically controlled. The statistical analysis should be conducted both with and without body weight as a covariate. Because toxicity may also affect body weight, the body weight on the first day of dosing, as opposed to the body weight at necropsy, should be used as the covariate in these cases.

Experimental control of body weight is accomplished in two steps. The first step involves selection of animals with relatively small variation in body weight for the study cohort from a larger population of animals. Use of unusually small or large animals should be avoided. A reasonable level of body weight variation within the study cohort can be tolerated (i.e., 20% of the mean body weight for the cohort population, e.g., 175 ± 35 g). The second step involves the assignment of animals to different treatment groups ($n = 6$) by a randomized complete block approach. Under this approach, animals are randomly assigned to treatment groups so that each group's body weights at the beginning of the study have the same mean and standard deviation.

Phase 1 of the OECD validation process has recently been completed and summarized.[105] In these studies, the androgenic response to TP and the antiandrogenic effects of flutamide (FLU) were evaluated in 17 laboratories. Data from these studies showed that the assay was robust and reproducible across several laboratories, even if the participating laboratory used a variation of the recommended protocol (e.g., strain of rat, diet, modest differences in the age of castration). All laboratories and all protocols were successful in detecting increases in the weights of the accessory sex organs and tissues in response to TP and in detecting the antiandrogenic effects of FLU. There was also good agreement among laboratories with regard to the dose responses obtained.

Phase II of the validation process, currently in progress, will evaluate the performance of the laboratories in using the Hershberger protocol to detect two additional agonists (methyl testosterone and trenbolone), weak AR antagonists (vinclozolin, procymidone, linuron, p,p´-DDE), and a 5-α-reductase inhibitor (finasteride). Each chemical will be examined in three to five laboratories to determine reproducibility and to provide estimates of intra- and interlaboratory variation. The OECD will use data from these studies to set performance criteria and to evaluate whether it is plausible to use the assay as a component of the TIS battery.

Finally, the strengths and limitations of the Hershberger assay should be addressed while evaluating the method for use in the TIS battery. As was noted for the uterotrophic assay, the Hershberger assay is supported by a wealth of historical data. The biological relevance is clearly documented and supported by the physiological dependence of the sex accessory glands and tissues on androgens during puberty. Phase I of the OECD validation process has demonstrated the robust nature of the assay in the ability of multiple laboratories to replicate the results, even with variations in animal strain. One disadvantage of the Hershberger assay is that it requires surgical castration, a potentially stressful procedure that requires technical skill. Another potential limitation of the assay is that although the androgens play a predominant role in the growth and maintenance of male reproductive structures, several other factors can influence organ weights. These factors include the thyroid and growth hormones, prolactin, estrogens, and EGF,[106] and could possibly lead to false positives. The fact that the Hershberger assay allows for the evaluation of optional adjunct measures that can facilitate the identification of the mechanism of action may help to alleviate this problem.

C. Male and Female Pubertal Protocols

Two *in vivo* assays currently undergoing validation for the TIS battery are the male and female pubertal protocols. These protocols provide methods for evaluating the effects of chemicals on pubertal development and thyroid function in the rat through alterations in the estrogen, androgen, or thyroid hormone systems. Additionally, the protocols are able to detect steroid-associated toxic insult and allow for an evaluation of additional apical effects on reproductive function, such as hypothalamic alterations, that may not be associated with an alteration in steroidogenesis or a steroid-receptor mechanism. Thus, while the pubertal protocols are less mechanistically targeted than the uterotrophic and Hershberger assays, they do allow for a broader casting of the endocrine disruptor net in the TIS battery.

The 21-day dosing period used in the female pubertal protocol encompasses the period of development in the rat during which the female brain begins to respond to the positive feedback of estrogen, resulting in the occurrence of the first estrous cycle and ovulation. The age of vaginal patency (vaginal opening) is correlated with the first estrus and provides a noninvasive endpoint to monitor the onset of puberty. Since pubertal development is dependent upon the presence of estrogen, this assay is capable of detecting chemicals with estrogenic or antiestrogenic (estrogen receptor or steroid-enzyme mediated) activity, as well as chemicals that alter pubertal development via changes in LH, follicle stimulating hormone (FSH), prolactin, or growth hormone secretion, or that alter hypothalamic neurotransmitter function. Finally, endpoints are included in the protocol to detect alterations in thyroid hormone homeostasis. Goldman et al.[107] have previously described this protocol and reviewed in detail the historical data that support the biological relevance of the endpoints (Figure 11.6). In brief, the assay is conducted using weanling Sprague-Dawley or Long-Evans female rats (culled to 8 to 10 per litter on PND 3). Females are gavaged with the test chemical in corn oil (2.5 to 5.0 ml/kg) between 0700 and 0900 (lights 14:10, on 0500 h) from PND 22 to

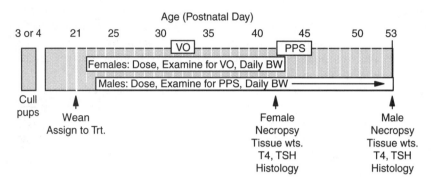

Figure 11.6 Overview of the male and female protocols for the assessment of pubertal development and thyroid function. Rats are weaned on PND 21, ranked by body weight and litter, and randomly assigned to treatment groups such that the mean body weights and variances are approximately equal for all groups. Females are dosed from PND 22 to 42, while males are dosed from PND 23 to 53. Endpoints of pubertal development are evaluated throughout the dosing period. Animals are killed on PND 42 (females) or 53 (males). At necropsy, tissues are weighed and serum is used for hormone assay.

42, with 15 females per treatment group. Required endpoints include growth, age at vaginal opening with vaginal cytology until necropsy, serum T4 and TSH, liver, kidney, pituitary, adrenal, uterine (with and without fluid) and ovarian weights, and thyroid histology. Optional endpoints include serum T_3, estradiol, prolactin, thyroid weight, systemic and vaginal tissue histology, *ex vivo* ovarian and pituitary hormone production, hypothalamic neurotransmitter concentrations, and extension of estrous cycle length.

The male pubertal protocol, an assay recommended as an alternate method by EDSTAC, is conducted over a 31-d period, during which a number of pubertal indices are measured. This assay is capable of detecting chemicals with antithyroid, estrogenic, androgenic, or antiandrogenic (AR or steroidogenic enzyme mediated) activity and agents that alter pubertal development via changes in FSH, LH, prolactin, growth hormone, or hypothalamic function. Weanling Sprague-Dawley or Long-Evans rats, housed two to three per cage, are dosed by oral gavage from PND 23 to 53.[108] Required endpoints include growth; age at preputial separation; serum T4 and TSH; seminal vesicle plus coagulating gland (with and without fluid), ventral prostate, and LABC weights; epididymal and testis weights and histology; and thyroid histology. Optional endpoints include serum T_3, testosterone, estradiol, LH, prolactin, and liver, kidney, adrenal, and pituitary weights, *ex vivo* testis and pituitary hormone production, and hypothalamic neurotransmitter concentrations.

From a historical perspective, the male and female pubertal protocols use a wealth of robust and sensitive endpoints capable of detecting a broad variety of endocrine disrupting chemicals.[107,108] However, prior to use in the U.S EPA's EDSP, the protocols will undergo additional testing to document the biological relevance of the endpoints, as well as the ability to produce data that are consistent within a single laboratory and across different laboratories for a broad range of chemicals. In addition, several technical issues are currently being addressed to facilitate the successful use of either of the protocols, to reduce intralaboratory variability, and to interpret the data correctly. For example, there have been reports that some laboratory rodent dietary formulations contain levels of phytoestrogens that are sufficient to induce alterations in uterine weight and histology.[109,110] As a result, there is some debate about the influence of the choice of diet on the pubertal endpoints, and whether a low phytoestrogen content in commercially prepared dietary formulations may present a confounding factor in the interpretation of test data. There are studies demonstrating no neonatal or pubertal alterations in the offspring of dams maintained on a phytoestrogen containing diet,[111] and the validation studies for the uterotrophic assay suggest that phytoestrogen contents of less than 325 to 350 µg/g total genistein equivalents did not impair assay responsiveness.[93] However, some consideration should be given to the selection of a diet with minimal or undetectable levels of estrogenic activity.

Another technical concern that has generated debate is whether a reduction in body weight gain during the study might have a confounding effect on the endpoints associated with pubertal development. While it is known that drastic reductions (greater than 25% to 50% as compared with controls)[112–115] can influence pubertal development, body weight decrements of 10% or less do not appear to cause significant changes in vaginal opening (VO) or preputial separation (PPS).[116–118] Thus, the dose selection criteria for the pubertal studies will remain at or just below the maximum tolerated dose (MTD), which is typically defined as a 10% lower body weight (as compared with the control) by the end of the dosing period. While it is generally recognized that doses of a test chemical at or below this level will have no adverse systemic toxicity that would interfere with the endpoint measures used in these protocols, this will be thoroughly evaluated during the validation process. Additionally, the fact that estrogenic compounds can decrease appetite and food consumption should be noted, and a drop in body weight following dosing with an estrogenic chemical should not automatically be considered indicative of general systemic toxicity. Differences in the magnitude of body weight decrement may differ by animal strain, a factor that should be considered when using an MTD that was determined in another strain.

The final pubertal protocols will likely contain recommendations that will help to minimize variability in the hormone and tissue weight endpoints. For example, variability in hormone concentrations can be reduced by carefully monitoring the time of day and the method by which the animals are euthanized. Carbon dioxide can be used as the method of euthanasia if only the thyroid hormones and TSH are to be measured, as a stress-related change in thyroid hormone does not generally occur in a short amount of time.[119] However, if additional endocrine endpoints, such as levels of the stress-responsive hormone, prolactin, are to be measured, decapitation would be recommended. In all cases, efforts should be made to prevent any stressors that might occur prior to euthanasia (animal transfer, cage changes, excessive noise) to avoid unwanted variability in the hormone data. In addition, because of the circadian rhythms in serum pituitary and steroid hormone secretions, necropsy should be performed within a narrow window of time following the last dose, although no time of day is specified for the necropsy of either females or males. For example, there is a circadian rhythm in the secretion of TSH in the rat, with a zenith at midday[120–122] and minimum secretion at the start of the dark period. Thus, a restricted period for the necropsy during a period of basal secretion would assure more consistent results. Finally, although the female pubertal protocol does not currently recommend necropsy on a specific day of the estrous cycle, it is important to understand the endocrine status of the animal at necropsy because serum hormone concentrations and the weights of the reproductive organs will vary throughout the estrous cycle. Either synchronizing the necropsy on a specific day of the estrous cycle or at least identifying the day of the cycle on which the females are necropsied will greatly facilitate proper interpretation of the hormone and tissue weight data. For example, uterine histology can be used to determine changes in uterine epithelium that correlate to changes in serum steroid and gonadotropin concentrations,[123] as is the uterine weight for detection of estrogenic compounds.[124,125] However, without considering the hormonal profile and the day of the estrous cycle, the use of these data is questionable. This is also true for ovarian histological assessments and the measurement of any pituitary peptide hormones, such as prolactin or luteinizing hormone.

Although both the male and female pubertal protocols are undergoing the validation process for the U.S. EPA's EDSP, only one of the protocols will likely be included in the final TIS battery. While both protocols are technically simple and provide numerous sensitive endpoints that have been used for decades to detect endocrine alterations, each protocol possesses its own unique set of strengths and weaknesses when used as part of the TIS battery. The endpoints in the female pubertal protocol are quite sensitive to estrogens and antiestrogens[126,127] but are not as useful for the detection of androgenic or antiandrogenic activity as compared with the endpoints of the male pubertal protocol. For example, dietary administration of nonylphenol affects VO but not PPS,[128] and methoxychlor accelerates VO at dosage levels nearly fivefold lower than those needed to delay PPS in the male.[128,129] The female pubertal protocol would also provide a better screen for chemicals that inhibit aromatase activity. Marty et al.[117] detected a delay in the age of VO with ketoconazole (100 mg/kg) and the aromatase inhibitor fadrozole (0.6 and 6 mg/kg), using the female pubertal protocol. Replication of results between laboratories has also been well documented for the pubertal endpoints.[118,130,131] Conversely, the male pubertal protocol is preferable for detecting androgenic and antiandrogenic activity. Although the Hershberger assay is also a sensitive assay for the detection of chemicals that alter androgen receptor function, the male pubertal assay more comprehensively assesses the endocrine system. The male pubertal protocol has been successfully used to identify chemicals that affect the hypothalamic-pituitary axis,[132–134] inhibit steroid hormone synthesis,[135] alter androgen receptor function,[136] or alter thyroid hormone homeostasis.[137–139] The effects of several toxicants on the timing of PPS have been replicated successfully by several laboratories. For example, the ability of vinclozolin to delay PPS at 100 mg/kg/d has been demonstrated in two laboratories by different investigators.[136,140] In addition, several laboratories using methoxychlor treatment have also delayed PPS at similar dosage levels.[129,141] A limitation of the male protocol is that it requires a longer dosing period than either the Hershberger assay or female pubertal protocol. Also, additional *in vivo* and *in vitro* tests would be required to identify the specific mechanism of action following a delay in PPS and altered sex accessory gland weights to determine whether the effect is mediated through an antiandrogen or a disruption of hypothalamic-pituitary function.[108,142]

D. *In Utero*-Lactational Exposure

The EDSTAC believed that the recommended T1S battery, if validated, would have the necessary breadth and depth to detect any currently known disruptors of estrogen, androgen, and thyroid hormone homeostasis. There was concern, however, that chemical substances or mixtures could produce effects from prenatal or prehatch exposure that would not be detected from pubertal or adult exposure.[3] Furthermore, there were differing views in the EDSTAC about whether there is scientific evidence of known endocrine disruptors or reproductive toxicants that can affect the prenatal stage of development without affecting the adult or prematuration stages, and whether effective doses and affected endpoints may differ among the three life stages.

The EDSTAC therefore recommended that the U.S. EPA take affirmative steps, in collaboration with industry and other interested parties, to develop a protocol for a full life cycle (i.e., with embryonic exposure and evaluation of the adult offspring) developmental exposure screening assay that can be subjected to standardization and validation. The EDSTAC believed such an assay must involve prenatal exposure and retention of offspring through puberty to adulthood and provide structural, functional, and reproductive assessment. Furthermore, the EDSTAC recommended that if such an assay was identified, standardized, and validated, the decision to include it in the final TIS battery would involve an evaluation of its potential to replace one or more of the originally recommended TIS assays and the overall impact on the cost effectiveness of the TIS battery. The proposed protocol has been identified by the EPA as the *in utero*/lactational exposure testing protocol (Figure 11.7) and has been assigned for development under the EDSP.[143] The objective of this bioassay is to detect effects mediated by alterations in the estrogen, androgen, and thyroid-signaling pathways as a consequence of exposure during pre- and postnatal development in the laboratory rat. The treatment paradigm allows for an evaluation of effects on organogenesis, sexual differentiation, and puberty. The U.S. EPA considered at least four variations of the *in utero*/lactational exposure protocol (Table 11.5). After deliberation, it was agreed that protocol D would be examined further for standardization. In this protocol, dosing of the dam begins on GD 6 and treatment is

THE U.S. EPA ENDOCRINE DISRUPTOR SCREENING PROGRAM

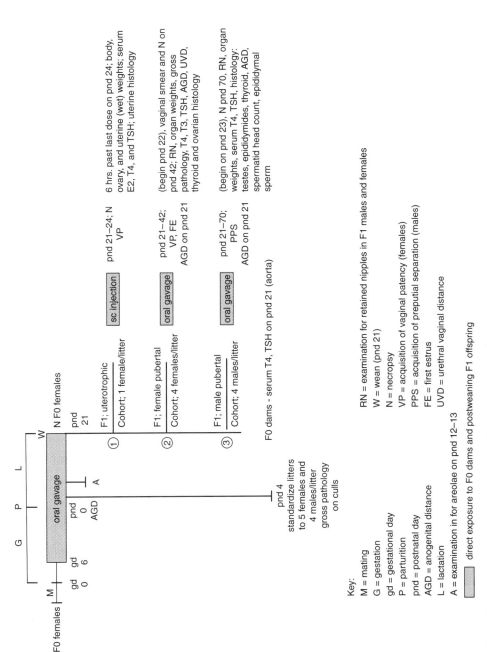

Figure 11.7 Overview of the *in utero*/lactational exposure protocol.

Table 11.5 Comparison of the parameters included in protocols A, B, C, and D

Parameter	Protocol A	Protocol B	Protocol C	Protocol D
Maternal Data (F_0)				
Dosing Period	GD 6–PND 21	GD 6–PND 21	GD 6–PND 21	GD 6–PND 21
Body weight	X	X	X	X
Food consumption	X	X	X	X
Gestational length	X	X	X	X
Mean litter size	X	X	X	X
Mating index	X	—	—	—
Fertility index	X	—	—	—
Preimplantation loss	X	X	X	X
Clinical signs and behavior	X	X	X	X
Gross necropsy (PND 21)	X	X	X	X
Organ weights	X	X	X	X
Histopathology	X	X	X	X
Hormone concentrations	—	X	X	X
Offspring Data (F_1)				
Dosing Period (preweaning)	GD 6-PND 21	GD 6-PND 21	GD 6-PND 18	GD 6-PND 21
Dosing Period (postweaning)	—	—	PND 18-21 (F) PND 18-42 (F) PND 18-52 (M)	PND 21-24 (F) PND 21-42 (F) PND 21-60 (M)
Litter size	X	X	X	X
Viability	X	X	X	X
Clinical signs	X	X	X	X
Body weight	X	X	X	X
Sex ratio	X	X	X	X
Gestational index	X	—	—	—
Life birth index	X	—	—	—
Survival index (preweaning)	X	X	X	X
Lactation index	X	X	X	X
Anogenital distance	—	X	X	X
PND-4 necropsy	—	X	X	X
Retained nipples	—	X	X (males only)	X (male and female)
Developmental Landmarks				
Pinna detachment	X	—	—	—
Eye opening	X	—	—	—
Vaginal patency	X	X	X	X
Preputial separation	X	X	X	X
Functional Assessment				
Eye examination	X	—	—	—
Auditory function	X	—	—	—
Motor activity	X	—	—	—
Learning and memory	X	—	—	—
Uterotrophic assay	—	—	X (PND 21)	X (PND 24)
Male weanling necropsy	—	X	X	—
Female pubertal evaluation	—	X (undosed)	X (direct dose)	X (direct dose)
Male pubertal evaluation	—	X (undosed)	X (direct dose)	X (direct dose)
F_1 gross necropsy	X (16 weeks of age)	X (at puberty)	X (PND 21 or puberty)	X (PND 24 or puberty)
Organ weights	X	X	X	X
Histopathology	X	X	X	X
Hormone concentrations	—	X	X	X
F_1 mating	X	—	—	—

Table 11.5 Comparison of the parameters included in protocols A, B, C, and D (continued)

Parameter	Protocol			
	A	B	C	D
F_2 evaluation (PND 4)	X	—	—	
Study duration[a]	21 weeks	12 weeks	12 weeks	12 weeks
Cost	$350,000[b]	$300,000[c]	$350,000[c]	$350,000[c]

Note: GD = gestation day, PND = postnatal day, X = parameter is included in the protocol, — = parameter is not included in the protocol, M = male, F = female.

[a] Study duration includes quarantine, and assumes in-house bred animals.
[b] Study cost does not include hormone assays, but includes chemistry and histopathology costs.
[c] Study cost includes all chemistry, histopathology, and hormone assay costs.

continued until PND 21. The dam is monitored for clinical signs daily throughout the treatment period. Body weight and food consumption are monitored during this time. Beginning on PND 20, the female is also observed twice daily for evidence of littering. On PND 21 (day of weaning), the dam is killed and blood is collected for subsequent T_4 and TSH determinations. The dam is then examined for gross morphological changes, and the uterus is examined for implantation scars, which are counted and recorded.

For the F_1 progeny, the *in utero*/lactational protocol requires the following: At birth, the pups are counted, sexed, and examined externally. Each live pup is then weighed individually and its anogenital distance is measured. On PND 4, the size of each litter is adjusted to nine pups (5 females and 4 males). Gross pathology is performed on the culls. There should be a maximum of 10 litters per dosage group. Pups are then weighed and examined daily. The presence or absence of nipples and areolae on the ventrum is recorded for all F_1 male and female offspring on PND11-13.

On PND21, the F_1 offspring are then divided into three different treatment cohorts, which include:

- The uterotrophic cohort, in which one female from each litter is selected and dosed (s.c.) from PND 22 to 24. On PND 24, the female is necropsied 6 h after the last dose and the body weight, ovarian weight, and uterine weight (wet) are recorded. Also, blood is collected for subsequent serum estradiol, T_4, and TSH determinations.
- The F_1 female pubertal cohort, comprising two dosed and two nondosed females from each litter. Dosing (oral gavage) begins on PND 21 and continues until PND 42 or 43. During this time the female is examined for vaginal patency, and her body weight is recorded. Clinical observations are made twice daily. Beginning on the day of VO, daily vaginal smears are obtained to determine the first estrous cycle and whether vaginal cycles are present prior to necropsy.
- The F_1 male pubertal cohort, composed of two dosed and two nondosed males from each litter. Dosing (oral gavage) begins on PND 21 and continues until PND 70. Body weights are recorded daily, and clinical observations are made twice daily. Beginning on PND 23, each F_1 study male is examined daily for PPS.

The *in utero*/lactational protocol lists a number of specific endpoints that must be carefully examined at necropsy for each cohort (Table 11.5). In using a developing system as the basis for the test, it is understood that modes of action other than those of the estrogen, androgen, and thyroid-signaling pathways may be involved in the induction of toxicity. As such, any observed effects must be interpreted in light of the overall weight of the evidence as to whether they are endocrine dependent. It does not appear that this assay would be suitable as an initial screen for endocrine disrupting chemicals because of the number of animals required and the length of the exposure. However, the assay might be useful as a bridge between the TIS battery and the Tier II testing assays (e.g., Tier 1.5), or as an alternative to the standard multigenerational tests in some limited cases.

IV. CONSIDERATIONS FOR EVALUATION OF DATA FROM THE TIS BATTERY

Within the conceptional framework of the EDSTAC's recommendations for the EDSP, the assays in the TIS battery were intended to identify chemicals capable of altering estrogen, androgen, and thyroid hormone systems. Criteria for selection of the assays to be incorporated in the TIS battery included the use of assays that would (1) maximize sensitivity to minimize false negatives, (2) include a range of organisms representing differences in metabolism, (3) detect a range of modes of action and of taxonomic groups, and (4) include enough diversity among the endpoints to permit weight-of-evidence conclusions. Thus, when using the TIS battery as a whole, the EDSTAC believed that the screening battery would possess the necessary breadth and depth to detect all currently known modes of action for endocrine disrupting chemicals. Those chemicals producing significant changes in the endpoints in both *in vitro* and *in vivo* TIS assays would be further evaluated by Tier II testing assays to characterize the dose response, relevance of the exposure route, sensitive life stages, and hazard potential.

Following the completion of the validation studies for the TIS assays, the data will be reviewed by a FIFRA Science Advisory Panel (SAP), which will provide recommendations for the selection of the TIS assays that will be used to screen the first set of chemicals.[144] It is likely that the SAP will also recommend an approach for evaluating the data from the TIS battery, and the scientific basis for moving forward with Tier II testing. In cases where there may be equivocal results, it may be prudent to replicate certain TIS assays before moving forward to the Tier II testing. Gray et al.[145] have suggested that replication of some TIS assays in such cases may almost entirely eliminate statistical false positives and minimize animal use by not moving directly to a multigenerational study. Alternatively, the *in utero*/lactational protocol might be employed as a Tier 1.5 study to provide additional data prior to engaging in a more extensive multigenerational study. Finally, the concept of tailoring the Tier II tests using results of the TIS battery will be considered. For example, if a test chemical produces androgenic or antiandrogenic effects in the TIS, then endpoints such as anogenital distance at birth and nipple and areola retention in infant rats should be included in the mammalian multigenerational test because these are sensitive, permanent biomarkers that are highly correlated with malformations and reproductive organ weight changes later in life.[145] Moreover, alterations in estrogen receptor–related assays or altered steroid hormone synthesis might lead one to include different endpoints in the Tier II tests.

Although the specific assays under consideration for the Tier II testing battery are not formally reviewed in this chapter, the EDSTAC used the following criteria for assay selection: assays that (1) include the most sensitive developmental life stage, (2) identify the specific hazard caused by exposure to the chemical and the dose response, and (3) cover a range of taxa. As such, the EDSTAC recommended a two-generation mammalian reproductive toxicity study, an avian reproduction test, fish and mysid life cycle tests, and an amphibian development and reproduction test as the initial candidates for the Tier II testing assays. While these assays were designed to work as a whole and to provide dose response data for any observed adverse effects, any mechanistic data obtained from the TIS battery will likely also be used to support the potential relevance to humans and wildlife during hazard identification and risk assessment.

The EPA updated its standard mammalian two-generation test in 1998 and included new endocrine sensitive endpoints that would be of value for EDSP Tier II testing.[146] More recently the agency has begun evaluating the utility of extending the observation period of the F_1 generation through puberty and adulthood, as compared with the standard protocol, to determine whether additional information could be obtained that would better detect endocrine disrupting chemicals. This study will use two known antiandrogens, dibutylphthalate and vinclozolin, and involve observation of male offspring survival, growth, and development through lactation, weaning, puberty, and adulthood following maternal exposure from GD 6 to PND 20. Secondary hypotheses to be tested include whether some effects might be detected more easily in the adult as compared with the weanling, and if increasing the number of males observed at either age would be beneficial. Data from these studies may ultimately increase the sensitivity of the mammalian multigeneration test for the detection of endocrine disrupting chemicals.

Potential alternatives to the standard mammalian two-generation test include several transgenerational protocols that have been developed recently. Many of these protocols typically use fewer litters (7 to 10 per dosage group) but examine all animals in the litter.[147–150] Even though the protocols employ fewer animals, they provide more statistical power to detect reproductive effects in the F_1 generation. In addition, the measurement of anogenital distance (AGD) at 1 to 2 days of age and areolae and nipples at 12 to 13 days of age in the F_1 generation are included. These endpoints are not only extremely useful "biomarkers" of an effect on androgen function, but also the changes in AGD and nipples can be permanent. Moreover, these endpoints are highly correlated with malformations and other alterations in androgen-dependent tissues.[148] Unfortunately neither of these markers is measured in the F_1 generation under the EPA standard two-generation test guidelines. In this regard, the proposal that the Tier II testing assays should be based upon the results of TIS remains an important consideration for obtaining the most relative data on the suspected mode of action. Another transgenerational protocol, described as alternative reproduction test (ART)[140,151] or alternative mammalian reproductive test (ARMT),[3] has been used to study xenoestrogens, such as the phthalates, the herbicide, linuron, and the fungicide, fenarimol. This protocol initiates oral exposure by daily gavage of P_0 male and female rats at weaning and continues through puberty, mating, gestation, and lactation. The F_1 are monitored throughout life but not dosed. While this protocol is longer than the *in utero*/lactational protocol, it may be appropriate for xenoestrogens and inhibitors of steroidogenesis that affect the P_0 as well as the F_1 generation in the lower dosage groups.

V. SUMMARY

In this chapter we have briefly discussed the *in vitro* and *in vivo* assays that are currently under consideration for use in the TIS battery of the EPA's EDSP. The current TIS battery represents the product of considerable scientific evaluation and deliberations by the EDSTAC, EDMVS, OECD, ICCVAM, and the EPA's EDSP. The strengths and limitations of these assays, as well as their usefulness for detecting known modes of action for the disruption of the mammalian endocrine system, have been described. The assays are currently in various stages of the validation process that will document the biological relevance and quantify intra- and interlaboratory variance as measures of assay reliability. The ultimate goal of the EPA is to provide a set of assays that will be capable of detecting chemicals with endocrine disrupting activity, while minimizing the occurrence of false negatives.

The EPA issued a notice of its approach for selecting the chemicals for the initial round of screening.[152] Of the approximately 87,000 chemicals under its regulatory purview, the agency is proposing to select and screen 50 to 100 chemicals drawn from pesticide active ingredients and high production volume chemicals, with some having pesticide inert uses. As previously recommended by the SAP subcommittee, data resulting from the screening process will be evaluated by an independent external panel of experts to determine whether the program needs to be improved or further optimized. In addition, as the agency moves forward with its EDSP, new methods and state-of-the-art approaches will continue to be evaluated for possible substitution in the TIS and Tier II testing batteries. New approaches that will reduce animal use, such as using QSAR models or HTS technology, will continued to be developed and evaluated to identify chemicals likely to disrupt the endocrine system. Thus, the assays used in the EDSP will continue to be refined and updated as the program evolves toward full implementation.

ACKNOWLEDGMENTS

We would like to acknowledge the efforts of our colleagues Dr. Joseph Merenda, Mr. Gary Timm, Mr. Jim Kariya, Mr. Greg Schweer, and the staff in OSCP, EPA, that have the direct responsibility for coordinating the implementation of the EDSP; Battelle Northwest Pacific Laboratories for providing the ER competitive binding curves shown in Figure 11.2; Janet Ferrell for assistance with the references; and Dr. Audrey Cummings and Dr. Ron Hood for their reviews and helpful comments on earlier drafts of the manuscript.

This manuscript has been reviewed in accordance with the policy of the National Health and Environmental Effects Research Laboratory, U.S. Environmental Protection Agency, and approved for publication. Approval does not signify that the contents necessarily reflect the views and policy of the Agency, nor does mention of trade names or commercial products constitute endorsement or recommendation for use.

References

1. The International Programme on Chemical Safety, http://www.who.int/pcs/.
2. U.S. EPA, Endocrine Disruptor Screening Program Web Site, http://www.epa.gov/scipoly/oscpendo/index.htm.
3. U.S. EPA, Office of Prevention, Pesticides, and Toxic Substances, EDSTAC Final Report, http://www.epa.gov/scipoly/oscpendo/edspoverview/finalrpt.htm.
4. ICCVAM, NICEATM, http://iccvam.niehs.nih.gov/methods/endocrine.htm.
5. OECD Endocrine Testing and Assessment. http://www.oecd.org/document/62/0,2340,en_2649_34377_2348606_1_1_1_1,00.html.
6. Tsai, M. and O'Malley, B.W., Molecular mechanisms of action of steroid/thyroid receptor superfamily members, *Ann. Rev. Biochem.*, 63, 451, 1994.
7. Bylund, D.B and Toews, M.L., Radioligand binding methods: practical guide and tips, *Am. J. Physiol.*, 265(5 pt 1), L421, 1993.
8. Evans, R.M., The steroid and thyroid hormone receptor superfamily, *Science*, 240, 889, 1988.
9. Nonneman, D.J., Ganjam, V.K., Welshons, W.V., and Vom Saal, F.S., Intrauterine position effects on steroid metabolism and steroid receptors of reproductive organs in male mice, *Biol. Reprod.*, 47, 723, 1992.
10. Tekpetey, F.R. and Amann, R.P., Regional and seasonal differences in concentrations of androgen and estrogen receptors in ram epididymal tissue, *Biol. Reprod.*, 38, 1051, 1988.
11. Traish, A.M., Muller, R.E., and Wotiz, H.H., Binding of 7 alpha, 17 alpha-dimethyl-19-nortestosterone (mibolerone) to androgen and progesterone receptors in human and animal tissues, *Endocrinology*, 118, 1327, 1986
12. Zava, D.T., Landrum, B., Horwitz, K.B., and McGuire, W.L., Androgen receptor assay with [3H]methyltrienolone (R1881) in the presence of progesterone receptors, *Endocrinology*, 104, 1007, 1979.
13. NIEHS, Interagency Coordinating Committee on the Validation of Alternative Methods (ICCVAM), NTP Interagency Center for the Evaluation of Alternative Toxicological Methods, ICCVAM Evaluation of In Vitro Test Methods for Detecting Potential Endocrine Disruptors: Estrogen Receptor and Androgen Receptor Binding and Transcriptional Activation Assays. http://iccvam.niehs.nih.gov/methods/endodocs/edfinrpt/edfinrpt.pdf.
14. NIEHS, Interagency Center for the Evaluation of Alternative Toxicological Methods (NICETM). Current Status of Test Methods for Detecting Endocrine Disruptors: *In Vitro* Androgen Receptor Binding Assays. http://iccvam.niehs.nih.gov/methods/endodocs/final/arbndbrd/arbndall.pdf
15. Andersen, H.R., Andersson, A., Arnold, S.F., Autrup, H., Barfoed, M., Beresford, N.A., Bjerregaard, P., Christensen, L.B., Gissel, B., Hummel, R., Jorgensen, E.B., Korsgaard, B., Le Guevel, R., Leffers, H., McLachlan, J., Moller, A., Nielsen, J.B., Olea, N., Oles-Karasko, A., Pakdel, F., Pedersen, K., Perez, P., Skakkeboek, N.E., Sonnenschein,C., Sots, A.M., Sumpter, J.P., Thorpe, S.M., and Grandjean, P., Comparison of short-term estrogenicity tests for identification of hormone-disrupting chemicals, *Environ. Health Perspect.*, 107, 89, 1999.

16. Wong, C., Kelce, W.R., Sar, M., and Wilson, E.M., Androgen receptor antagonist versus agonist activities of the fungicide vinclozolin relative to hydroxyflutamide, *J. Biol. Chem.*, 270, 1998, 1995.
17. Bauer, E.R.S., Daxenberger, A., Petri, T., Sauerwein, H., and Meyer, H.H.D., Characterization of the affinity of different anabolics and synthetic hormones to the human androgen receptor, human sex hormone binding globulin and to the bovine progestin receptor, *APMIS*, 108, 838, 2001.
18. Lambright, C.L., Ostby, J., Bobseine, K., Wilson, V.S., Hotchkiss, A.K., Mann, P.C., and Gray, L.E., Jr., Cellular and molecular mechanisms of action of linuron: an antiandrogenic herbicide that produces reproductive malformations in male rats, *Toxicol. Sci.*, 56, 389, 2000.
19. Kuiper, G.G., Lemmen, J.G., Carlsson, B., Corton, J.C., Safe, S.H., van der Saag, P.T., van der Burg, B., and Gustafsson, J., Interaction of estrogenic chemicals and phytoestrogens with estrogen receptor b, *Endocrinology*, 139, 4252, 1998.
20. Kyakumoto, S., Sato, N., Nemoto, T., Ohara-Nemoto, Y., and Ota, M., Binding of [^{3}H]methyltrienolone to androgen receptor in rat liver, *Biochim. Biophys. Acta.*, 800, 214, 1984.
21. Murthy, L.R., Johnson, M.P., Rowley, D.R., Young, C.Y., Scardino, P.T., and Tindall, D.J., Characterization of steroid receptors in human prostate using mibolerone, *Prostate*, 8, 241, 1986.
22. GraphPad Software, Inc., *Introduction to Radioligand Binding*, http://www.graphpad.com/curvefit/introduction9e.htm
23. Cheng, Y. and Prusoff, W.H., Relationship between the inhibition constant (Ki) and the concentration of inhibitor which causes 50 percent inhibition (I_{50}) of an enzymatic reaction, *Biochem. Pharmacol.*, 22, 3099, 1973.
24. Mailman, R.B. and Lawler, C.P., Toxicant-receptor interactions: fundamental principles, in *Introduction to Biochemical Toxicology*, Hodgson, E. and Levi, P., Eds., Appleton and Lange, Norwalk, CT, 1994, p. 319
25. Kelce, W.R., Monosson, E., Gamcsik, M.P., Laws, S.C., and Gray, L.E., Jr., Environmental hormone disruptors: evidence that vinclozolin developmental toxicity is mediated by antiandrogenic metabolites, *Toxicol. Appl. Pharmacol.*, 126, 276, 1994.
26. Segel, I.H., *Enzyme Kinetics*, John Wiley & Sons, New York, pp. 100–159.
27. Gaido, K.W., Leonard, L.S., Lovell, S., Gould, J.C., Babai, D., Portier, C.J., and McDonnell, D.P., Evaluation of chemicals with endocrine modulating activity in a yeast-based steroid hormone receptor gene transcription assay, *Toxicol. Appl. Pharmacol.*, 143, 205, 1997.
28. Lyttle, C.R., Damian-Matsumura, P., et al., Human estrogen receptor regulation in a yeast model system and studies on receptor agonists and antagonists, *J. Steroid Biochem. Mol. Biol.*, 42, 677, 1992.
29. Kohno, H., and Gandini, O., et al., Anti-estrogen activity in the yeast transcription system: Estrogen receptor mediated agonist response, *Steroids*, 59, 572, 1994.
30. Gaido, K.W., Leonard, L.S., et al., Differential interaction of the methoxychlor metabolite 2,2-bis-(*p*-hydroxyphenyl)-1,1,1-trichloroethane with estrogen receptors alpha and beta, *Endocrinology*, 140(12), 5746, 1999.
31. Kemppainen, J.A., Langley E., et al., Distinguishing androgen receptor agonists and antagonists: distinct mechanisms of activation by medroxyprogesterone acetate and dihydrotestosterone, *Mol. Endocrinol.*, 13, 440, 1999.
32. Lobaccaro, J.M., Poujol, N., Terouanne, B., Georget, V., Fabre, S., Lumbroso, S., and Sultan, C., Transcriptional interferences between normal or mutant androgen receptors and the activator protein 1 — dissection of the androgen receptor functional domains, *Endocrinology*, 140, 350, 1999.
33. Tremblay, A., Tremblay G.B., et al., EM-800, a novel antiestrogen, acts as a pure antagonist of the transcriptional functions of estrogen receptors alpha and beta, *Endocrinology*, 139, 111, 1998.
34. Vinggaard, A.M., Joergensen E.C., et al., Rapid and sensitive reporter gene assays for detection of antiandrogenic and estrogenic effects of environmental chemicals, *Toxicol. Appl. Pharmacol.*, 155, 150, 1999.
35. Hartig, P.C., Bobseine, K.L., et al., Development of two androgen receptor assays using adenoviral transduction of MMTV-*luc* reporter and/or hAR for endocrine screening, *Toxicol. Sci.*, 66, 82, 2002.
36. Blankvoort, B.M., de Groene, E.M., et al., Development of an androgen reporter gene assay (AR-LUX) utilizing a human cell line with an endogenously regulated androgen receptor, *Annal. Biochem.*, 298, 93, 2001.
37. Legler, J., van den Brink, C.E., et al., Development of a stably transfected estrogen receptor-mediated luciferase reporter gene assay in the human T47D breast cancer cell line, *Toxicol. Sci.*, 48, 55, 1999.

38. Rogers, J.M. and Denison, M.S., Recombinant cell bioassays for endocrine disruptors: development of a stably transfected human ovarian cell line for the detection of estrogenic and antiestrogenic chemicals, *In Vitro Mol. Toxicol.*, 13, 67, 2000.
39. Terouanne, B., Tahiri, B., et al., A stable prostatic bioluminescent cell line to investigate androgen and antiandrogen affects, *Mol. Cell. Endocrinol.*, 160, 39, 2000.
40. Wilson, V.S., Bobseine, K., et al., A novel cell line, MDA,kb2, that stably expresses an androgen- and glucocorticoid-responsive reporter for the detection of hormone receptor agonist and antagonist, *Toxicol. Sci.*, 66, 69, 2002.
41. Zacharefwski, T.R., Meek, M.D., et al., Examination of the in vitro and in vivo estrogenic activities of eight commercial phthalate esters, *Toxicol. Sci.*, 46, 282, 1998.
42. Rehnberg, G.L., Linder, R.E., Goldman, J.M., Hein, J.F., McElroy, W.K., and Cooper, R.L., Changes in testicular and serum hormone concentrations in the male rat following treatment with *m*-dinitrobenzene, *Toxicol. Appl. Pharmacol.*, 95, 255, 1988.
43. Gray, L.E., Klinefelter, G., Kelce, W., Laskey, J., Ostby, J., and Ewing, L., Hamster Leydig cells are less sensitive to ethane dimethanesulfonate when compared to rat Leydig cells both *in vivo* and *in vitro*, *Toxicol. Appl. Pharmacol.*, 130, 248, 1995.
44. Wilker, C.E., Welsh, T.H., Jr., Safe, S.H., Narasimha, T.R., and Johnson, L., Human chorionic gonadotropin protects Leydig cell function against 2,3,7,8-tetrachlorodibenzo-*p*-dioxin in adult rats: role of Leydig cell cytoplasmic volume, *Toxicology*, 95, 93, 1995.
45. Chambon, M., Grizard, G., and Boucher, D., Bromocryptine, a dopamine agonist, directly inhibits testosterone production by rat Leydig cells, *Andrologia*, 17, 172, 1985.
46. Kan, P.B., Hirst, M.A., and Feldman, D., Inhibition of steroidogenic cytochrome P-450 enzymes in rat testis by ketoconazole and related imidazole anti-fungal drugs, *J. Steroid Biochem.*, 23, 1023, 1985.
47. Laskey, J.W. and Phelps, P.V., Effect of cadmium and other metal cations on in vitro Leydig cell testosterone production, *Toxicol. Appl. Pharmacol.*, 108, 296, 1991.
48. Guillou, F., Martinat, N., and Combarnous, Y., Rapid in vitro desensitization of the testosterone response in rat Leydig cells by sub-active concentrations of porcine luteinizing hormone, *FEBS Lett.*, 184, 6, 1985.
49. Klinefelter, G.R., Laskey, J.W., and Roberts, N.L., In vitro/in vivo effects of ethane dimethanesulfonate on Leydig cells of adult rats, *Toxicol. Appl. Pharmacol.*, 107, 460, 1991.
50. Romanelli, F., Fillo, S., Isidori, A., and Conte, D., Stimulatory action of endothelin-1 on rat Leydig cells: involvement of endothelin-A subtype receptor and phospholipase A2-arachidonate metabolism system, *Life Sci.*, 61, 557, 1997.
51. Mack, S.O., Lorence, M.C., Andersson, S., and Mason, J.I., Expression of cytochrome P-450(17) alpha, 3 beta-hydroxysteroid dehydrogenase/delta 5,4-isomerase, and steroid 5 alpha-reductase in rat H540 Leydig tumor cells, *Mol. Cell. Endocrinol.*, 74R, 11, 1990.
52. Stocco, D.M, King, S., and Clark, B.J., Differential effects of dimethylsulfoxide on steroidogenesis in mouse MA-10 and rat R2C Leydig tumor cells, *Endocrinology*, 136, 2993, 1995.
53. Nikula, H., Talonpoika, T., Kaleva, M., and Toppari, J., Inhibition of hCG-stimulated steroidogenesis in cultured mouse Leydig tumor cells by bisphenol A and octylphenols, *Toxicol. Appl. Pharmacol.*, 157, 166, 1999.
54. Walsh, L.P., Webster, D.R., and Stocco, D.M., Dimethoate inhibits steroidogenesis by disrupting transcription of the steroidogenic acute regulatory (*StAR*) gene, *J. Endocrinol.*, 167, 253, 2000.
55. Dees, J.H., Gazouli, M., and Papadopoulos, V., Effect of mono-ethylhexyl phthalate on MA-10 Leydig tumor cells, *Reprod. Toxicol.*, 15, 171, 2001.
56. Chen, L.Y., Huang, Y.L., Liu, M.Y., Leu, S.F., and Huang, B.M., Effects of amphetamine on steroidogenesis in MA-10 mouse Leydig tumor cells, *Life Sci.*, 72, 1983, 2003.
57. Nishi, Y., Yanase, T., Mu, Y., Oba, K., Ichino, I., Saito, M., Nomura, M., Mukasa, C., Okabe, T., Goto, K., Takayanagi, R., Kashimura, Y., Haji, M., and Nawata, H., Establishment and characterization of a steroidogenic human granulose-like tumor cell line, KGN, that expresses functional follicle-stimulating hormone receptor, *Endocrinology*, 142, 437, 2001.
58. Logié, A., Boulle, N., Gaston, V., Perin, L., Boudou, P., Le Bouc, Y., and Gicquel, C., Autocrine role of IGF-II in proliferation of human adrenocortical carcinoma NCI H295R cell line, *J. Mol. Endocrinol.*, 23, 23, 1991.

59. Ohno, S., Shinoda, S., Toyoshima, S., Nakazawa, H., Makino, T., and Nakajin, S., Effects of flavonoid phytochemicals on cortisol production and on activities of steroidogenic enzymes in human adrenocortical H295R cells, *J. Steroid Biochem. Mol. Biol.*, 80, 355, 2002.
60. Harvey, P.W. and Everett, D.J., The adrenal cortex and steroidogenesis as cellular and molecular targets for toxicity: critical omissions from regulatory endocrine disruptor screening strategies for human health?, *J. Appl. Toxicol.*, 23, 81, 2003.
61. Sanderson, J.T., Boerma, J., Lansbergen, G.W., and van den Berg, M., Induction and inhibition of aromatase (CYP19) activity by various classes of pesticides in H295R human adrenocortical carcinoma cells, *Toxicol. Appl. Pharmacol.*, 182, 44, 2002.
62. Sanderson, J.T., Letcher, R.J., Heneweer, M., Giesy, J.P., and van den Berg, M., Effects of chloro-s-triazine herbicides and metabolites on aromatase activity in various human cell lines and on vitellogenin production in male carp hepatocytes, *Environ. Health Perspect.*, 109, 1027, 2001.
63. Kellis, J.T., Jr. and Vickery, L.E., Inhibition of human estrogen synthase (aromatase) by flavones, *Science*, 225, 1032, 1984.
64. Kellis, J.T., Jr., Nesnow, S., and Vickery, L.E., Inhibition of aromatase cytochrome P450 (estrogen synthase) by derivative of alpha-napthoflavone, *Biochem. Pharmacol.*, 35, 2887, 1986.
65. Ibrahim, A.R. and Abul-Hajj, Y.J., Aromatase inhibition by flavonoids, *J. Steroid Biochem. Mol. Biol.*, 37, 257, 1990.
66. Vinggaard, A.M., Hnida, C., Breinholt, V., and Larsen, J.C., Screening of selected pesticides for inhibition of CYP19 aromatase activity in vitro, *Toxicol. In Vitro*, 12, 227, 2000.
67. Hirsh, K.S., Weaver, D.E., Black, L.J., Falcone, J.F., and MacLusky, N.J., Inhibition of central nervous system aromatase activity: a mechanism for fenarimol-induced infertility in the male rat, *Toxicol. Appl. Pharm.*, 91, 235, 1987.
68. Kragie, L., Aromatase in primate pregnancy: a review, *Endocr. Res.*, 28, 121, 2002.
69. Vinggaard, A.M., Nellemann, C., Dalgaard, M., Jorgensen, E.B., and Andersen, H.R., Antiandrogenic effects in vitro and in vivo of the fungicide prochloraz, *Toxicol. Sci.*, 69, 344, 2002.
70. Lephart, E.D. and Simpson, E.R., Assay of aromatase activity, *Methods Enzymol.*, 206, 477, 1991.
71. Letcher, R.J., Van, H.I., Drenth, H.J., Norstrom, R.J., Bergman, A., Safe, S., Pieters, R.M., and van Den, B.M., Cytotoxicity and aromatase (CYP19) activity modulation by organochlorines in human placental JEG-3 and JAR choriocarcinoma cells, *Toxicol. Appl. Pharmacol.*, 160, 10, 1999.
72. Doody, K.M., Murry, B.A., and Mason, J.I., The use of rat Leydig tumor (R2C) and human hepatoma (HEPG2) cells to evaluate potential inhibitors of rat and human steroid aromatase, *J. Enzyme Inhib.*, 4, 153, 1990.
73. Catalano, S., Marsico, S., Giordano, C., Mauro, L., Rizza P, Panno, M.L., and Ando, S., Leptin enhances, via AP-1, expression of aromatase in MCF-7 cell line, *J. Biol. Chem.*, 279(19), 19908, 2004.
74. Sanderson, J.T., Seinen, W., Giesy, J.P., and van den Berg, M., 2-Chloro-s-triazine herbicides induce aromatase (CYP 19) activity in H295R human adrenocortical carcinoma cells; A novel mechanism for estrogenicity, *Toxicol. Sci.*, 54, 121, 2000.
75. Stoker, T.E., Laws, S.C., Guidici, D.L., and Cooper, R.L., The effect of atrazine on puberty in male Wistar rats: an evaluation in the protocol for the assessment of pubertal development and thyroid function, *Toxicol. Sci.*, 58, 50, 2000.
76. Ojeda, S.R. and Urbanski, H.F., Puberty in the rat, in *The Physiology of Reproduction*, 2nd Ed., Knobil, E. and Neill, J.D., Eds., Raven Press, New York, 1999.
77. Simpson, E.R., Clyne, C., Rubin, G., Boon, W.C., Robertson, K., Britt, K., Speed, C., and Jones, M., Aromatase — a brief overview, *Annual Rev. Physiol.*, 64, 93, 2002.
78. Miller, W.R. and O'Neill, J., The importance of local synthesis of estrogen within the breast, *Steroids*, 50, 537, 1987.
79. Levin, L. and Tyndale, H.H., The quantitative assay of follicle stimulating substances, *Endocrinology*, 21, 619, 1937.
80. Dorfman, R.I., Gallagher, T.F., and Koch, F.C., The nature of the estrogenic substance in human male urine and bull testis, *Endocrinology*, 19, 33, 1935.
81. Bulbring, E. and Burn, J.H., The estimation of oestrin and of male hormone in oily solution, *J. Physiol.*, 85, 320, 1935.

82. Owens, J.W. and Ashby, J., Critical review and evaluation of the uterotrophic bioassay for the identification of possible estrogen agonists and antagonists: in support of the validation of the OECD uterotrophic protocols for the laboratory rodent, *Critical Rev. Tox.*, 32, 445, 2002.
83. Wakeling, A.E. and Bowler, J., Novel antioestrogens without partial agonist activity, *J. Steroid. Biochem.*, 31, 645, 1988.
84. MacGregor J.I. and Jordan, V.C., Basic guide to the mechanisms of antiestrogen action, *Pharmacol. Rev.*, 50, 151, 1998.
85. Anderson J.N., Clark J.H., and Peck, E.J., Jr., The relationship between nuclear receptor-estrogen inding and uterotrophic responses, *Biochem. Biophys. Res. Comm.*, 48, 1460, 1972.
86. Clark, J.H. and Peck, E.J., Jr., *Female Sex Steroids, Receptors and Function*, Springer-Verlag, Berlin, 1979.
87. Zacharewski, T.R., Meek, M.D., Clemons, J.H., Wu, Z.F., Fielden, M.R., and Matthews, J.B., Examination of the *in vitro* and *in vivo* estrogenic activities of eight commercial phthalate esters, *Toxicol. Sci.*, 46, 282, 1998.
88. Ashby J., Tinwell, J., Pennie, W., Brooks, A., Lefevre, P.A., Beresford, N., and Sumpter, J.P., Partial and weak oestrogenicity of the red wine constituent resveratrol, consideration of its superagonist activity in MCF-7 cells and its suggested cardiovascular protective effects, *J. Appl. Toxicol.*, 19, 39, 1999.
89. Laws, S.C., Carey, S.A., Ferrell, J.M., Bodman, G.J., and Cooper, R.L., Estrogenic activity of octylphenol, nonylphenol, bisphenol A and methoxychlor in rats, *Toxiocol. Sci.*, 54, 154, 2000.
90. Kanno, J., Onyon, L., Haseman, J., Fenner-Crisp, P., Ashby, J., and Owens, J.W., The OECD program to validate the rat uterotrophic bioassay to screen compounds for in vivo estrogenic responses, Phase 1, *Environ. Health Perspect.*, 109, 785, 2001.
91. Kanno, J., Onyon, L., Peddada, S., Ashby, J., Jacob, E., and Owens, W., The OECD program to validate the rat uterotrophic bioassay, phase two: dose response studies, *Environ. Health Perspect.*, 111(12), 1530, 2003.
92. Kanno, J., Onyon L., Peddada, S., Ashby, J., Jacob, E., and Owens, W., The OECD program to validate the rat uterotrophic bioassay, phase two — coded single dose studies, *Environ. Health Perspect.*, 111(12), 1550, 2003.
93. Owens, W., Ashby, J., Odum, J., and Onyon, L., The OECD program to validate the rat uterotrophic bioassay, phase two: dietary phytoestrogen analyses, *Environ. Health Perspect.*, 111(12), 1559, 2003.
94. Gould, J.C., Leonard, L.S., Maness, S.C., Wagner, B.L., Conner, K., Zacharewski, T., Safe, S., McDonnell, D.P., and Gaido, K.W., Bisphenol A interacts with the estrogen receptor alpha in a distinct manner from estradiol, *Mol. Cell Endocrinol.*, 142, 203, 1998.
95. Ashby, J. and Tinwell, J., Uterotrophic activity of bisphenol A in the immature rat, *Environ. Health Perspect.* 106, 719, 1998.
96. Katzenellenbogen, B.S., Biology and receptor interactions of estriol and estriol derivatives in vitro and in vivo, *J. Steroid Biochem.*, 20, 1033, 1984.
97. Jordan, V.C., Mittal, S., Gosden, B., Koch, R., and Lieberman, M.E., Structure-activity relationships of estrogens, *Environ. Health Perspect.*, 61, 97, 1985.
98. Yamasaki, K., Sawaki, M., Noda, S., and Takatuki, M., Effects of olive, corn, sesame or peanut oil on the body weights and reproductive organ weights of immature male and female rats, *Exp. Animals,.* 50, 173, 2001.
99. Hershberger, L., Shipley, E., and Meyer, R., Myotrophic activity of 19-nortestosterone and other steroids determined by modified levator ani muscle method, *Proc. Soc. Exp. Biol. Med.*, 83, 175, 1953.
100. Ojeda, S.R., Andrews, W.W., Advis, J.P., and White, S.S., Recent advances in the endocrinology of puberty, *Endocr. Rev.*, 1, 228, 1980.
101. Wainman, P. and Shipounoff, G.C., The effects of castration and testosterone propionate on the striated perineal musculature in the rat, *Endocrinology*, 29, 975, 1941.
102. Eisenberg, E., Gordan, G.S., and Elliott, E.W., Testosterone and tissue respiration of the castrate male rat with a possible test for myotrophic activity, *Endocrinology*, 45, 113, 1949.
103. Korenchevsky, V., The assay of testicular hormone preparations, *Biochem. J.*, 26, 413, 1932.
104. Korenchevsky, V. and Dennison, M., The assay of transdehydroandrosterone and its effects on male and female gonadectomized rats, *Biochem. J.*, 29, 1514, 1935.

105. OECD, Final OECD report of the work towards the validation of the rat Hershberger assay: Phase-1, androgenic response to testosterone propionate and anti-androgenic effects of flutamide. ENV/JM/TG/EDTA(2002)1/REV2, October 2002, Paris.
106. Luke, M.C. and Coffey, D.S., Human androgen receptor binding to the androgen response element of prostate specific antigen, *J. Androl.*, 15, 41, 1994.
107. Goldman, J.M., Laws, S.C., Balchak, S.K., Cooper, R.L., and Kavlock, R.J., Endocrine-disrupting chemicals, prepubertal exposures and effects on sexual maturation and thyroid activity in the female rat. A focus on the EDSTAC recommendations, *Crit. Rev. Toxicol.*, 30, 135, 2000.
108. Stoker, T.E., Parks, L.G., Gray, L.E., and Cooper, R.L., Endocrine-disrupting chemicals, prepubertal exposures and effects on sexual maturation and thyroid function in the male rat. A focus on the EDSTAC recommendations, *Crit. Rev. Toxicol.*, 30, 197, 2000.
109. Thigpen, J.E., Setchell, K.D., Ahlmark, K.B., Locklear, J., Spahr, T., Caviness, G.F., Goelz, M.F., Haseman, J.K., Newbold, R.R., and Forsythe, D.B., Phytoestrogen content of purified, open- and closed-formula laboratory animal diets, *Lab. Anim. Sci.*, 49, 530, 1999.
110. Boettger-Tong, H., Murthy, L., Chiappetta, C., Kirkland, J.L., Goodwin, B., Adlercreutz, H., Stancel, G.M., and Makela, S., A case of a laboratory animal feed with high estrogenic activity and its impact on in vivo responses to exogenously administered estrogens, *Environ. Health Perspect.*, 106, 369, 1998.
111. Casanova, M., You, L., Gaido, K.W., Archibeque-Engle, S., Janszen, D.B., and Heck, H.A., Developmental effects of dietary phytoestrogens in Sprague-Dawley rats and interactions of genistein and daidzein with rat estrogen receptors and in vitro, *Toxicol. Sci.*, 51, 236, 1999.
112. Engelbregt, M.J.T., vanWeissenbruch, M.M., Pop-Snijders, D., Lips, P., and Delemarr-van de Waal, H.A., Body mass index, body composition, and leptin at onset of puberty in male and female rats after intrauterine growth retardation, and after early postnatal food restriction, *Ped. Res.*, 50, 474, 2001.
113. O'Connor, J.C., Davis, L.G., Frame, S.R., and Cook, J.C., Evaluation of a Tier I screening battery for detecting endocrine-active compounds (EACs) using the positive controls testosterone, coumestrol, progesterone, and RU486, *Toxicol. Sci.*, 54, 338, 2000.
114. Merry, B.J. and Holehan, A.M., Onset of puberty and duration of fertility in rats fed a restricted diet, *J. Reprod. Fert.*, 57, 253, 1979.
115. Kennedy, G.C. and Mitra, J., Body weight and food intake as initiating factors for puberty in the rat, *J. Physiol.*, 166, 408, 1979.
116. Cooper, R.L., Stoker, T.E., Ferrell, J., Leffler, K., Bremser, K., and Laws, S.C., Impact of body weight change on the EDSTAC Tier I male and female pubertal protocols, *Toxicologist*, 72(S-1), 129, 2003.
117. Marty, M.S., Crissman, J.W., and Carney, E.W., Evaluation of the EDSTAC female pubertal assay in CD rats using 17-beta-estradiol, steroid biosynthesis inhibitors, and a thyroid inhibitor, *Toxicol. Sci.*, 52, 269, 1999.
118. Laws, S.C., Ferrell, J.M., Stoker, T.E., Schmid, J., and Cooper, R.L., The effect of atrazine on puberty in female Wistar rats: an evaluation in the protocol for assessing pubertal development and thyroid function, *Toxicol. Sci.*, 58, 366, 2000.
119. Armario, A., Lopez-Calderon, A., Jolin, T., and Balasch, J., Response of anterior pituitary hormones to chronic stress. The specificity of adaptation, *Neurosci. Biobehav. Rev.*, 10, 245, 1986.
120. Jordan, D., Veisseire, M., Borson-Chazot, F., and Mornex, R., Postnatal development of TRH and TSH rhythms in the rat, *Hormone Res.*, 27, 216, 1987.
121. Selgas, L., Arce, A., Esquifino, A.I., and Cardinali, D.P., Twenty-four-hour rhythms of serum ACTH, prolactin, growth hormone, and thyroid-stimulating hormone, and of median-eminence norepinephrine, dopamine, and serotonin, in rats injected with Freund's adjuvant, *Chronobiol. Int.*, 14, 253, 1997.
122. Pallardo, L.F., Pericas, I., and Jolin, T., Thyroid iodine uptake, thyroid iodine secretion and plasma TSH levels in male rats during the day and night, *Acta Endocrinol.* (Copenhagen), 83, 517, 1976.
123. Spornitz, U.M., Socin, C.D., and Dravid, A.A., Estrus stage determination in rats by means of scanning electron microscopic images of uterine surface epithelium, *Anat. Rec.*, 254, 116, 1999.
124. Bulger, W.H., Feil, V.J., and Kupfer, D., Role of hepatic monooxygenases in generating estrogenic metabolites from methoxychlor and from its identified contaminants, *Mol. Pharmacol.*, 27, 115, 1985.
125. Grunert, G., Porcia, M., and Tchernitchin, A.N., Differential potency of oestradiol-17 beta and diethylstilboestrol on separate groups of responses in the rat uterus. *J. Endocrinol.*, 110, 103, 1986.

126. Kim, H.S., Shin, J.H., Moon, H.J., Kim, T.S., Kang, I.H., Seok, J.H., Kim, I.Y., Park, K.L., and Han, S.Y., Evaluation of the 20-day pubertal female assay in Sprague-Dawley rats treated with DES, tamoxifen, testosterone, and flutamide, *Toxicol. Sci.*, 67, 52, 2002.
127. Kim, H.S., Shin, J.H., Moon, H.J., Kang, I.H., Kim, T.S., Kim, I.Y., Seok, J.H., Pyo, M.Y., and Han, S.Y., Comparative estrogenic effects of *p*-nonylphenol by 3-day uterotrophic assay and female pubertal onset assay, *Reprod. Toxicol,,* 16, 259, 2002.
128. Chapin, R.E., Delaney, J., Wang, Y., Lanning, L., Davis, B., Collins, B., Mintz, N., and Wolfe, G., The effects of 4-nonylphenol in rats, a multigeneration reproductive study, *Toxicol. Sci.*, 52, 80, 1999.
129. Chapin, R.E., Harris, M.W., Davis, B.J., Ward, S.M., Wilson, R.E., Mauney, M.A., Lockhart, A.C., Smialowicz, R.J., Moser, V.C., Burka, L.T., and Collins, B.J., *Fundam. Appl. Toxicol.* 40, 138, 1997.
130. Ashby, J., Tinwell, H., Stevens, J., Pastoor, T., and Breckenridge, C.B., The effects of atrazine on the sexual maturation of female rats, *Regul. Toxicol. Pharmacol.*, 35, 468, 2002.
131. Gray, L.E., Jr, Ostby, J., Ferrell, J., Rehnberg, G., Liner, R., Cooper, R.L., Goldman, J.M., Slott, V., and Laskey, J., A dose-response analysis of methoxychlor-induced alterations in reproductive development and function in the rat, *Fundam. Appl. Toxicol.*, 12, 92, 1989.
132. Hostetter, M.W. and Piacsek, B.E., The effect of prolactin deficiency during sexual maturation in the male rat, *Biol. Reprod.*, 17, 574, 1977.
133. Ramaley, J.A. and Phares, C.K., Delay of puberty onset in males due to suppression of growth hormone, *Neuroendocrinology*, 36, 32, 1983.
134. Stoker, T.E., Laws, S.C., Guidici, D.L., and Cooper, R.L., The effect of atrazine on puberty in male Wistar rats, an evaluation in the protocol for the assessment of pubertal development and thyroid function, *Toxicol. Sci.*, 58, 50, 2000.
135. Marty, M.S., Crissman, J.W., and Carney, E.W., Evaluation of the male pubertal onset assay to detect testosterone and steroid biosynthesis inhibitors in CD rats, *Toxicol. Sci.*, 60, 285, 2001.
136. Monosson, E., Kelce, W.R., Lambright, C., Ostby, J., and Gray, L.E., Jr., Peripubertal exposure to the antiandrogenic fungicide, vinclozolin, delays puberty, inhibits the development of androgen-dependent tissues, and alters androgen receptor function the male rat, *Toxicol. Ind. Health*, 15, 65, 1999.
137. Stoker, T.E., Ferrell, J., Hedge, J.M., Crofton, K.M., Cooper, R.L., and Laws, S.C., Assessment of DE-71, a commercial polybrominated diphenyl ether (PBDE) mixture, in the EDSP male pubertal protocol, *Toxicologist*, 72(S-1), 135, 2003.
138. Marty, M.S., Crissman, J.W., and Carney, E.W., Evaluation of the male pubertal assay's ability to detect thyroid inhibitors and dopaminergic agents, *Toxicol. Sci.*, 60, 63, 2001.
139. Yamasaki, K., Tago, Y., Nagai, K., Sawaki, M., Noda, S., and Takatsuki, M., Comparison of toxicity studies based on the draft protocol for the 'Enhanced OECD Test Guideline No. 407' and the research protocol of 'Pubertal Development and Thyroid Function in Immature Male Rats' with 6-*n*-propyl-2-thiouracil, *Arch. Toxicol.*, 76, 495, 2002.
140. Anderson, S.A., Pearce, S.W., Fail, P.A., McTaggart, B.T., Tyle, R.W., and Gray, L.E. Jr., Validation of the Alternative Reproductive Test Protocol (ART) to assess toxicity of methoxychlor in rats, *Toxicologist*, 15, 1995.
141. Gray, L.E., Jr., Ostby, J., Sigmon, R., Ferrell, J., Rehnberg, G., Linder, R., Cooper, R., Goldman, J., and Laskey J., The development of a protocol to assess reproductive effect of toxicants in the rat, *Reprod. Toxicol.*, 2, 281, 1988.
142. Huhtaniemi, I., Nikula, H., Rannikko, S., and Clayton, R., Regulation of testicular steroidogenesis by gonadotrophin-releasing hormone agonists and antagonists, *J. Steroid Biochem.*, 24, 169, 1986.
143. In Utero/Lactational *Exposure Testing Protocol*, http://www.epa.gov/scipoly/oscpendo/meetings/2002/march/inuterolactationprotocol.pdf.
144. U.S. EPA, *About the Science Advisory Panel*, http://www.epa.gov/oscpmont/sap/about.htm.
145. Gray, L.E., Jr., Ostby, J., Wilson, V., Lambright, C., Bobseine, K., Hartig, P., Hotchkiss, A., Wolf, C., Furr, J., Price, M., Parks, L., Cooper, R.L., Stoker, T.E., Laws, S.C., Degitz, S.J., Jensen, K.M., Kahl, M.D., Kahl, J.J., Korte, E.A., Makynen, E.A., Tietge, J.E., and Ankley, G.T., Xenoendocrine disruptors-tiered screening and testing: filling the data gaps, *Toxicology*, 181, 371, 2002.
146. U.S. Environmental Protection Agency, Prevention, Pesticides and Toxic Substances. Health Effects Test Guidelines, OPPTS 870.3700, Prenatal Developmental Toxicity Study, EPA 712-C-98-207, August 1998.

147. McIntyre, B.S., Barlow, N.J., Wallace, D.G., Maness, S.C., Gaido, K.W., and Foster, P.M., Effects of in utero exposure to linuron on androgen-dependent reproductive development in the male Crl, CD(SD)BR rat, *Toxicol. Appl. Pharmacol.*, 167, 87, 2000.
148. McIntyre, B.S., Barlow, N.J., and Foster, P.M., Male rats exposed to linuron in utero exhibit permanent changes in anogenital distance, nipple retention, and epididymal malformations that result in subsequent testicular atrophy, *Toxicol. Sci.*, 6, 62, 2002.
149. McIntyre, B.S., Barlow, N.J., and Foster, P.M., Androgen-mediated development in male rat offspring exposed to flutamide in utero, permanence and correlation of early postnatal changes in anogenital distance and nipple retention with malformations in androgen-dependent tissues, *Toxicol. Sci.*, 62, 236, 2001.
150. Gray, L.E., Wolf, C., Lambright, C., Mann, P., Price, M., Cooper, R.L., and Ostby, J., Administration of potentially antiandrogenic pesticides (procymidone, linuron, iprodione, chlozolinate, p,p'-DDE, and ketoconazole) and toxic substances (dibutyl- and diethylhexyl phthalate, PCB 169, and ethane dimethane sulphonate) during sexual differentiation produces diverse profiles of reproductive malformations in the male rat, *Toxicol. Ind. Health*, 15, 94, 1999.
151. Anderson, S.A., Fail, P.A., McTaggart, B.J., Tyl, R.W., and Gray, L.E., Jr., Reproductive toxicity of methoxychlor in corn oil to male and female Long-Evans hooded rats using the alternative reproduction test protocol (ART). *Biol. Reprod.*, 50(Suppl. 1), 101, 1994.
152. U.S. EPA, Endocrine Disruptor Screening Program, Proposed Chemical Selection Approach for Initial Round of Screening (EPA Fact Sheet). http://www.epa.gov/scipoly/oscpendo/factsheet.htm.

CHAPTER 12

Role of Xenobiotic Metabolism in Developmental and Reproductive Toxicity

Arun P. Kulkarni

CONTENTS

I. Introduction 527
II. Overview of Xenobiotic Metabolism 527
    A. Phase I Reactions 528
        1. Cytochrome P450 528
        2. Flavin-Containing Monooxygenase 529
        3. Peroxidases and Prostaglandin Synthase 529
        4. Lipoxygenase 529
        5. Lipid Peroxidation 530
    B. Phase II Reactions 530
        1. UDP-Glucuronosyl Transferase 530
        2. Sulfotransferase 531
        3. Glutathione Transferase 531
III. Assay of Xenobiotic Metabolizing Enzymes 531
    A. General Comments 531
    B. Assay Methods 532
        1. Cytochrome P450 Estimation 532
        2. Cytochrome P450–Mediated Xenobiotic Oxidation 532
        3. Flavin-Containing Monooxygenase 533
        4. Lipoxygenase 533
        5. Prostaglandin Synthase 534
        6. Peroxidase 535
        7. Epoxide Hydrase 536
        8. Glutathione Transferase 536
        9. UDP-Glucuronosyl Transferase 537
       10. Sulfotransferase 537
IV. Metabolism of Developmental Toxicants 538
    A. Maternal Liver 538
        1. Cytochrome P450 538
        2. Flavin-Containing Monooxygenase 540
        3. Epoxide Hydrase 540

    4. UDP-Glucuronosyl Transferase 540
    5. Glutathione Transferase 540
  B. Uterus and Decidua 540
    1. Cytochrome P450 540
    2. Peroxidase 541
    3. Lipid Peroxidation 541
    4. Lipoxygenase and Prostaglandin Synthase 541
    5. Flavin-Containing Monooxygenase 542
  C. Intrauterine Conceptual Tissues 542
    1. Animal Species 542
    2. Humans 542
  D. Embryo 543
    1. Cytochrome P450 543
    2. Prostaglandin Synthase 544
    3. Lipoxygenase 545
    4. Peroxidase 545
  E. Fetus 546
    1. Cytochrome P450 546
    2. Flavin-Containing Monooxygenase 547
    3. Prostaglandin Synthase 547
    4. Peroxidase 547
    5. Lipoxygenase 547
    6. Epoxide Hydrase 548
    7. Glutathione Transferase 548
    8. UDP-Glucuronosyl Transferase 548
    9. Sulfotransferase 548
  F. Fetal Membranes 549
    1. Cytochrome P450 549
    2. Peroxidative Enzymes 549
  G. Placenta 549
    1. Cytochrome P450 549
    2. Flavin-Containing Monooxygenase 550
    3. Prostaglandin Synthase 550
    4. Peroxidase 550
    5. Lipoxygenase 551
    6. Lipid Peroxidation-Coupled Cooxidation of Xenobiotics 553
    7. Other Enzymes of Xenobiotic Oxidation 554
    8. Epoxide Hydrase 555
    9. UDP-Glucuronosyl Transferase 555
    10. Sulfotransferase 555
    11. Glutathione Transferase 555
V. Role of Xenobiotic Metabolism in Reproductive Toxicology 556
  A. General 556
  B. Male Reproductive System 556
    1. Testicular Enzymes of Xenobiotic Metabolism 556
    2. Male Reproductive Toxicants 556
  C. Female Reproductive System 557
    1. Ovarian Enzymes of Xenobiotic Metabolism 557
    2. Female Reproductive Toxicants 558
VI. Conclusions and Future Research Needs 559
References 560

I. INTRODUCTION

If it is a must to ingest food, drink water, and breathe, it is equally critical for survival to have the ability to defend ourselves from pernicious chemicals entering the body. Human exposure to potentially harmful natural or man-made xenobiotics is inevitable, and it is only a matter of extent. Evolutionary forces have endowed humans and animals with the necessary armor in the form of biochemical machinery composed of enzymes, each with unique ability to transform and neutralize lipophilic xenobiotics into hydrophilic metabolites suitable for exit from the body via feces and urine. Unfortunately, biochemical accidents do sometimes occur, leading to the generation of powerful ultimate toxicants. That process, instead of providing protection, predisposes body tissues and cells to demise or destruction. Xenobiotics interact with enzymes as substrates, inhibitors, activators, or inducers.

Knowledge of xenobiotic metabolism is vital to understanding the developmental and reproductive toxicity of chemical agents that require biotransformation. The bulk of the available information originates from studies conducted with adult rodent liver. However, extrahepatic tissues possess the ability to metabolize xenobiotics, and these tissues display major qualitative and quantitative differences in their metabolic capacity. Unfortunately, human data are very limited for conceptual tissues. Similarly, the evidence linking metabolism with developmental or reproductive toxicity of chemicals is not always transparent. Several excellent reviews covering earlier progress in developmental toxicology are available, providing lists of original publications. The reader is encouraged to consult these reviews covering xenobiotic metabolism during pregnancy[1,2] and in embryonic,[3-6] fetal,[6-8] and placental[6,8-10] tissues. This chapter mainly focuses on the data published in recent years. Methodology employed in the study of major enzymes involved in xenobiotic metabolism is also briefly described.

Since interpretation of *in vivo* xenobiotic metabolism data in terms of enzymes involved in a particular reaction is often difficult, many scientists focus their efforts on well-designed *in vitro* studies. This experimental approach permits investigators studying xenobiotic metabolism to: (1) establish the role of a specific enzyme, (2) examine the enzyme of interest under fully defined assay conditions, (3) make qualitative and quantitative estimates of the metabolites of a toxicant, (4) reveal the modulatory effects of endobiotics and other factors, and (5) study the effects of *in vivo* exposure to chemicals on different pathways. In light of this, only select publications primarily dealing with *in vitro* xenobiotic metabolism are considered here. No attempt is made to present a comprehensive review of all aspects.

II. OVERVIEW OF XENOBIOTIC METABOLISM

The process of transforming lipophilic xenobiotics to polar entities in the body is widely perceived to occur in two distinct metabolic phases. During Phase I, functionalization, an atom of oxygen is incorporated in the chemical, and functional groups, such as –OH or –COOH, are either introduced or exposed. In other cases, xenobiotic bioconversion involves reduction or hydrolysis. Although most reactions reduce chemical toxicity (detoxication), some Phase I metabolites are more reactive than their parent compounds (bioactivation). They interact with critical macromolecules, such as protein, RNA, and DNA, and initiate toxic sequelae. Phase II reactions consist of synthetic reactions in which functional or polar groups are conjugated with endobiotics and include sulfation, glucuronidation, acetylation, methylation, and conjugation with reduced glutathione (GSH) or amino acids. These reactions increase xenobiotic hydrophilicity and greatly promote excretion. In most cases the conjugates are biologically nonreactive. Bioactivation of xenobiotics during Phase II reactions can also occur, but it is less common.

The increased polarity induced by conjugation greatly retards the ability of compounds to diffuse across the lipid bilayer of the cell membrane. The exit of conjugates from cells is facilitated by members of the ATP-binding cassette superfamily of active transport proteins (e.g., multidrug resistance associated protein [MRP], multispecific organic anion transporter [MOAT], G-glycoproteins, and others). Active cellular efflux of xenobiotics by these transporters is often referred to as Phase III detoxication and elimination, because Phase I and/or Phase II reactions often precede or are required for Phase III processes.[11] A few investigators use the term "Phase III metabolism" to refer to a process that involves further metabolism of products emanating from Phase II reactions.[12] This is exemplified by enzymatic catalysis of GSH conjugates leading to mercapturic acid synthesis. Another example is bioactivation of cysteine conjugate after methylation or by a pyridoxal phosphate–dependent β-lyase.[13] A brief description of major enzymes involved in Phase I and Phase II is given below.

A. Phase I Reactions

Oxidation represents the Phase I reaction most, if not all, xenobiotics encounter inside the body. It is mediated by microsomal cytochrome P450, flavin-containing monooxygenase, peroxidase, prostaglandin synthase, lipoxygenase, and processes occurring during lipid peroxidation. Dehydrogenases and amine oxidases also contribute to oxidation of certain xenobiotics. Reduction of xenobiotics is mediated by azoreductases and nitroreductases, while epoxide hydrolases, esterases, and amidases affect hydrolytic cleavage of specific linkages in xenobiotics.

The liver was believed to be the chief site for the generation of ultimate developmental and reproductive toxicants, but that is not true for those chemicals in which the ultimate toxicant is a reactive species or a radical with a relatively short biological half-life. Also P450, an enzyme known for its broad substrate specificity and dominant role in xenobiotic metabolism in the liver, was presumed to be the sole catalyst responsible for the bioactivation of most developmental toxicants in conceptual tissues, especially placenta. The research efforts devoted to this enzyme system for the past several years resulted in unjustifiable neglect of alternate pathways for xenobiotic metabolism and hampered their exploration. Recent reports (discussed later) documenting the role of other enzymes capable of xenobiotic oxidation in conceptual tissues have changed the direction of research, and as a result, a new picture is emerging.

Of the various Phase I reactions, oxidation, which not only represents the first metabolic reaction for most xenobiotics, but in many cases also results in bioactivation, has received the most attention. In view of this, an overview of the major enzymes involved in xenobiotic oxidation is presented below.

1. Cytochrome P450

Of the enzymes involved in xenobiotic oxidation, cytochrome P450 (P450) is the most extensively studied. P450 is a heme thiolate protein (E.C. 1.14.14.1, nonspecific oxygenase). Multiple species of P450 enzymes (not isozymes, in a strict sense) are known to occur in different animal tissues. At least 154 gene products constitute this superfamily. The number of P450 genes is currently estimated to be 53 in humans alone. The total number of known P450 sequences exceeds 500.[14] Human liver microsomes contain more than 15 different P450 enzymes capable of oxidation of xenobiotics and endobiotics. The major human microsomal P450 enzymes involved in xenobiotic oxidation include CYP2A6, CYP2B6, CYP2C8, CYP2C9, CYP2C19, CYP2E1, CYP1A2, CYP2D6, and CYP3A4. Similar information is available for laboratory animal species. The titer of each form varies interindividually within a species, depending upon tissue, gender, genetics, hormonal status, exposure to chemicals, and other factors. The obligatory components for all P450s to function as monooxygenases include NADPH, O_2, and NADPH P450 reductase, while the need for phospholipids, NADH, cytochrome B_5, and NADH cytochrome B_5 reductase varies widely among different P450 species. The steady state concentration of NADPH in hepatocytes is estimated to be about 100 μM.

Xenobiotic catalysis by P450 involves multiple steps, some being rate limiting. The types of 2e oxidation of xenobiotics catalyzed by P450 mainly include hydroxylation of aliphatic or aromatic carbons, epoxidation of a C=C double bond, heteroatom (S– and N–) oxygenation and N-hydroxylation, and heteroatom (N–, S– and O–) dealkylation. A few examples of P450-mediated 1e reduction or xenobiotic oxidation are also known. Several excellent reviews on P450 cover mechanisms of catalysis,[15] purification and properties,[16] structure and function,[17] substrate specificity,[18–21] induction,[22] inhibition and stimulation,[23] and species differences.[24]

2. Flavin-Containing Monooxygenase

Flavin-containing monooxygenases (FMO; E.C. 1.14.13.8) are microsomal flavoproteins capable of one step 2e oxygenation of many heteroatom (N–, S–, and P–) containing chemicals, drugs, and pesticides.[25,26] FMO is primarily a detoxication enzyme, although a few examples of xenobiotic bioactivation are known. At least six forms (FMO1 to FMO6) of mammalian FMOs are described, and some may present in multiple tissues of a given species. NADPH is the preferred cofactor, while NADH supports xenobiotic oxidation by FMO with about 40% efficiency. Relatively long-lived reduced hydroperoxyflavin is the oxidant responsible for the oxygenation of xenobiotics. The release of a water molecule and NADP+ represent the rate limiting steps in the catalytic cycle of FMO.[25,26] In general, FMO is more thermolabile than P450 and also exhibits a higher pH optimum (pH 8–10) *in vitro*. It is noteworthy that antibodies raised against FMO do not inhibit the enzyme. FMO is refractive to the inductive effects of many chemicals.

3. Peroxidases and Prostaglandin Synthase

Peroxidases, e.g., lactoperoxidase, myeloperoxidase, and tissue specific peroxidases, are heme proteins. Among different peroxidases, protaglandin synthase (PGS; E.C. 1.14.99.1) is the most extensively studied enzyme.[27,28] PGS is a bifunctional enzyme with distinct cyclooxygenase and peroxidase activities. Cyclooxygenase is responsible for bis-dioxygenation of free arachidonic acid to produce cyclic endoperoxide hydroperoxide (PGG_2). PGG_2 is reduced to PGH_2 by the peroxidase activity of the enzyme. In most tissues, free arachidonate occurs at a concentration of about 20 μM.[27] Although PGS activity predominantly resides in endoplasmic reticulum, some activity can be noted in the nuclear envelope. There are two structurally related isoforms of PGS, called COX-1 and COX-2, encoded by separate genes. COX-1 is the constitutive form that synthesizes prostaglandins responsible for "housekeeping" functions, while COX-2 is induced by proinflammatory cytokines and growth factors and is believed to play a role in inflammation and cell growth. Despite differences in sensitivity to inhibitors, substrate specificity, and expression, both forms of PGS display a remarkable similarity in structure and catalytic function. Several mechanisms have been proposed to explain xenobiotic oxidation catalyzed by PGS.[27,28] Stated briefly, many reducing cosubstrates are directly oxidized to free radical species by the peroxidase activity of PGS. In certain cases, the free radical interacts with O_2 and generates a secondary oxidant capable of xenobiotic oxidation. Lipid peroxyl radicals, formed as intermediates, can serve as potent oxidants for epoxidation and other xenobiotic bioactivation reactions. Both COX-1 and COX-2 occur in many mammalian tissues, and several intrauterine tissues possess PGS activity.[6] Ram seminal vesicles express only COX-1 and represent a rich source of PGS activity. Because of its low content in the liver, PGS is believed not to contribute significantly to hepatic xenobiotic metabolism.[27,28]

4. Lipoxygenase

Lipoxygenases (LO) comprise a family of non–heme-iron containing cytosolic proteins that catalyze the insertion of a single molecule of O_2 into polyunsaturated fatty acid (**PUFA**) at defined positions. Accordingly, 5-, 12-, and 15-LOs, and other LOs have been described in the literature.[29] The stereospecific actions of these enzymes require a 1,4-*cis-cis*-pentadiene structure in the substrate molecule. This moiety is transformed into 1-hydroperoxy-2-trans, 4-*cis*-pentadiene. Subsequently, this relatively unstable hydroperoxy derivative of arachidonic acid undergoes catalysis, giving rise to a wide array of biologically active metabolites, such as leukotrienes, hydroxyeicosatetraenoic acids (HETEs), lipoxins, and hepoxilins.[29–31] The substrate requirements of LO are less stringent than those of PGS. Thus, certain LOs not only use free PUFAs but can also peroxidize di- and triglycerides, lipoproteins, and even membrane-bound lipids. LOs are ubiquitous in plants and animals, occurring in several mammalian organs, including intrauterine tissues.[6,29–31] Various arachidonate metabolites generated by LOs are essential for fertilization, implantation, normal growth and development of the conceptus, maintenance of pregnancy, and certain aspects of parturition.[6] A single LO protein not only exhibits dioxygenase activity toward PUFAs but also displays cooxidase activity toward a large number of xenobiotics.[29–31] The peroxidaselike (cooxidase) activity of different LOs is not only efficiently supported by PUFAs and their hydroperoxides but also by substitutes such as H_2O_2, *tert*-butyl hydroperoxide, and cumene hydroperoxide.[29–31] The principal mechanisms of xenobiotic oxidation include one-electron oxidation of reducing substances to free radical species by LO. Concurrently, the active ferric LO is converted into inactive ferrous LO. Ferrous iron oxidation by hydroperoxide regenerates the active enzyme for the next catalytic cycle. The intermediate peroxyl radicals of PUFA generated during dioxygenation also serve as oxidants and trigger epoxidation or sulfoxidation of xenobiotics. Additional mechanisms of LO-mediated xenobiotic catalysis have been described.[29–31]

5. Lipid Peroxidation

Lipid peroxidation is a normal biochemical process constantly occurring at a basal rate in essentially all cell types. It occurs in tissues as a metal-catalyzed nonenzymatic process, as well as processes mediated by LO and PGS. Lipid peroxidation products are essential for maintenance of pregnancy and for normal growth and development of the conceptus. They are involved in the induction of labor and in the parturition process at term. Recent data show that increased lipid peroxidation due to aberrant PGS and/or LO activities in reproductive tissues may be linked to various disorders, such as hypertension and preeclampsia in pregnancy.[32] Uncontrolled lipid peroxidation may also be responsible for the developmental or reproductive toxicities of some chemicals. The proposed mechanisms include the generation of reactive oxygen species (ROS), the toxic products of lipid peroxidation, and/or xenobiotic free radicals.[3,4] Lipid peroxidation-coupled cooxidation represents another pathway for oxidation of certain xenobiotics.[1,6]

B. Phase II Reactions

1. UDP-Glucuronosyl Transferase

Glucuronidation is a Phase II reaction catalyzed by two families of microsomal UDP-glucuronosyl transferases (UGTs: E.C. 2.4.17), with each family made up of several members.[33] The process involves transfer of glucuronic acid from UDP-glucuronic acid (UDPGA) to aglycone by UGT. The major site for glucuronidation is the liver. The concentration of UDPGA in the liver is estimated at about 350 μM. UGT-mediated hepatic glucuronidation of xenobiotics is considered a "low-affinity–high-capacity" process. UGT is inducible by xenobiotics. Operation of the enzyme *in vivo* requires at least two transporters. One is for the transport of UDPGA from cytoplasm to the lumen of endoplasmic reticulum, and the other exports glucuronide into the cytoplasm. The enzyme exhibits latency because of its luminal location, and prior membrane disruption by, e.g., brief sonication or detergent addition, is needed for full expression of activity. Many endobiotics and xenobiotics or their metabolites bearing –OH, –COOH, –NH$_2$, and –SH groups undergo glucuronidation. Although most conjugates are hydrophilic, biologically inactive products, many examples of activation of drugs, carcinogens, and other chemicals are known.[33]

2. Sulfotransferase

Sulfotransferases (SULTs: E.C. 2.8.2) are cytosolic enzymes that mediate the transfer of a sulfo (SO_3) moiety from 3′-phosphoadenosine-5′-phosphosulfate (PAPS) to a nucleophilic site of the acceptor molecule. Several excellent reviews provide detailed information on SULTs.[34] At least 10 SULTs are described in humans alone. Because of the rate limiting availability of PAPS in liver and SULT inhibition at high substrate concentrations, sulfoconjugation is considered a "high-affinity–low-capacity" pathway of xenobiotic metabolism. Under normal conditions, PAPS content is 30 to 70 nmol/g in the liver and 3 to 30 nmol/g in other tissues. Although many xenobiotics are detoxified by sulfoconjugation, in a few cases, bioactivation does occur.[34]

3. Glutathione Transferase

Glutathione transferases (GSTs; E.C. 2.5.1.18) are multifunctional proteins ubiquitously distributed in plants and animals. They effect thioether bond formation through the conjugation of GSH with chemicals containing an electrophilic center. GSH conjugation represents the first step in mercapturic acid synthesis. Additionally, certain GSTs isomerize unsaturated compounds, reduce hydroperoxides, serve as binding, storage, or transport proteins, and modulate signal transduction processes.[35] GSTs belong to two distinct superfamilies. The first comprised at least 16 cytosolic proteins in humans. The cytosolic GSTs have been assigned to eight families or classes, designated Alpha (α), Mu (μ), Pi (π), Sigma (σ), Theta (θ), Zeta (ζ), Omega (ω), and Kappa (κ). Four additional members, Beta (β), Delta (δ), Phi (ϕ), and Tau (τ) occur in bacteria, insects, and plants.[35] The second is composed of microsomal GSTs and contains at least six members in humans.[36] The cofactor, GSH, is the most abundant intracellular nonprotein thiol. It occurs in rodent liver at a concentration of about 10 mM.

Electrophilic chemicals serve as substrates for GST. The process of GSH conjugation of xenobiotics by GSTs primarily involves the displacement of an electron withdrawing group by GSH in a substrate containing a halide, sulfate, sulfonate, phosphate, or nitro group or the direct addition of GSH across a carbon-carbon double bond (Michael addition reaction). GST can open a strained epoxide ring during the GSH addition. In general, GSH conjugation leads to the loss of biological reactivity of the parent compound. However, GSH conjugation of geminal and vicinal dihaloalkanes generates mutagenic species.[13] With polyhalogenated alkenes, further metabolism of the initial GSH conjugates yields ultimate toxicants.[13,35]

III. ASSAY OF XENOBIOTIC METABOLIZING ENZYMES

A. General Comments

Observation of a specific xenobiotic metabolism reaction in a tissue may not necessarily identify the catalyst involved. Multiple enzymatic pathways should be considered. For example, aldrin epoxidation, long believed to be an indication specific for P450,[37] is also mediated by PGS[20,21,27,28] and LO.[29–31] To assess the importance of an enzyme, both qualitative and quantitative aspects should be considered. The presence of a bioactivation enzyme signals the potential of that pathway for the generation of an ultimate toxicant. However, confirmation of the pathway's biological significance requires more than mere demonstration of a barely detectable level of the enzyme in a tissue (e.g., P450 in nonsmoker's placenta or in rodent embryos) by ultrasensitive methods. Although of academic interest, such a finding may not be meaningful from a practical point of view because it may not provide an adequate explanation for the toxicity of the chemical. Since the quantity of product formed is determined by the amount of enzyme, it is important that the bioactivation enzyme occurs in the tissue in biologically significant amounts. Furthermore, the catalytic efficiency of the enzyme must be high enough to generate an amount of ultimate toxicant adequate to interact with target molecules and induce toxicity. This implies that the generation of ultimate toxicant must overwhelm the endogenous defense and repair mechanisms. That thresholds exist and the majority of babies born today are healthy and normal despite everyday low exposures of women to many chemicals throughout pregnancy attests to this contention. In this section, assay procedures are given for estimating activities of major xenobiotic metabolizing enzymes.

B. Assay Methods

1. Cytochrome P450 Estimation

The most commonly used method is that of Omura and Sato,[38] which is based on the reduced carbon monoxide (CO) difference spectrum.

a. Reagents

1. 0.1 M potassium phosphate buffer, pH 7.4
2. Sodium dithionite, reagent grade
3. CO gas, reagent grade (*use in fume hood*)

b. Procedure

Washed microsomes in the buffer (1 to 3 mg protein/ml) are placed in two matching cuvettes (1.0-cm light path). By use of a spectrophotometer equipped with a turbid sample accessory operated in split beam mode, a baseline of equal absorption is recorded between 400 and 500 nm. The contents of the sample cuvette are saturated with CO (one bubble per second for about 30 sec). A few crystals (about 1 mg) of sodium dithionite are added to both cuvettes. The cuvettes are covered with Parafilm®, and the contents are gently mixed by inverting the cuvettes three or four times. Difference spectra are then recorded (400 to 500 nm) until absorbance at the 450-nm peak reaches maximum. P450 content is determined using the following equation:

$$[(A450 \text{ to } 490 \text{ nm}) \text{ observed} (A450 \text{ to } 490 \text{ nm}) \text{ baseline}]/0.091 = \text{nmol P450/milliliter}$$

c. Remark

In the case of significant hemoglobin (Hb) contamination, the procedure of Matsubara et al.[39] is recommended.

2. Cytochrome P450-Mediated Xenobiotic Oxidation

Xenobiotic oxidation observed in a reconstituted system containing purified P450 enzyme and/or inhibition of the reaction by the inclusion of antibody to NADPH P450 reductase provide strong evidence for the involvement of P450. The results of experiments using preferred substrates in combination with selective inhibitor(s) and a specific antibody can provide useful data on the possible contribution of individual P450 enzymes to xenobiotic oxidation in microsomes. P450 can catalyze a wide spectrum of xenobiotic oxidation reactions. For each type of reaction, a choice of substrate is available. Typical assay conditions employed in such experiments are presented below for aldrin epoxidation. With appropriate modification(s), these assay conditions can be adapted for the study of other reactions.

a. Epoxidation

The highly sensitive and reproducible procedure of Wolff et al.,[37] which determines the rate of aldrin epoxidation, is commonly used.

b. Reagents

1. 0.1 M potassium phosphate or Tris-HCl buffer, pH 7.4
2. 0.5 M MgCl$_2$
3. 1.0 mM NADPH or NADPH generating system (1 IU glucose-6-phosphate dehydrogenase/ml, 0.5 mM NADP$^+$, 10 mM glucose-6-phosphate)
4. Aldrin solution in acetone or ethanol
5. Dieldrin standard solution in hexane

c. Procedure

After preincubation of microsomes in buffer with 0.25 mM aldrin at 37°C for 3 to 5 min, the reaction is initiated by the addition of NADPH (1.0 ml final volume). The flasks are incubated for 5 to 15 min at 37°C with constant shaking. The reaction is terminated by the addition of 250 µl of 10% trichloroacetic acid (TCA), followed by chilling in an ice bath. The dieldrin produced in the incubation medium is extracted by vortexing with 2.5 ml of hexane. The mixture is centrifuged at 2500 × g for 5 min to accelerate phase separation. Dieldrin is quantified by injecting 2 to 5 µl of hexane extract into a gas chromatograph equipped with FID or ^{63}Ni ECD. With ECD, levels of dieldrin as low as 5 to 10 pg can be detected.

3. Flavin-Containing Monooxygenase

The activity of FMO can be readily measured by following methimazole-dependent oxygen uptake or NADPH depletion.[40]

a. Reagents

1. 50 mM Tris-glycine buffer, pH 8.5
2. 2.0 mM GSH
3. 1.0 mM NADPH
4. 0.2 mM methimazole
5. 2.4 mM n-octylamine

b. Procedure

Mix microsomes with methimazole, GSH, n-octylamine, and the buffer in sample and reference cuvettes. Initiate the reaction by the addition of NADPH to the sample cuvette. A decrease in the absorbance at 340 nm is followed for 3-5 min. An extinction coefficient of 6.22 mM^{-1} cm^{-1} is used to calculate NADPH depletion rate, which equals to the rate of methimazole sulfoxidation.

4. Lipoxygenase

Dioxygenase activity of tissue LO can be assayed either spectrophotometrically by estimating the conjugated dienes formed or by measuring oxygen uptake (Section III.B.5). The identity of 5-, 12-, or 15-LO can be determined from reverse-phase HPLC of radioactive arachidonic acid metabolites. The spectrophotometric method[41] described below is simple, rapid, and sensitive for routine estimation of the rate of dioxygenase activity of LO. The protocol described in Section III.B.6 can be adapted with some modifications to measure the cooxidase (peroxidaselike) activity of LO. Hemoglobin (Hb) exhibits both pseudodioxygenase and pseudoperoxidase activities,[29-31] and so Hb contamination poses a serious problem for the estimation of tissue LO activity. The method devised in our laboratory for human term placental cytosolic LO (HTPLO) alleviates this problem. In brief, the protocol requires the treatment of cytosol with $ZnSO_4$ (0.5 mM).[42] The treatment selectively removes Hb (greater than 97%) without affecting the dioxygenase and cooxidase activities of HTPLO. After centrifugation, the supernatant is dialyzed thoroughly before assaying for dioxygenase or cooxidase activity.

a. Reagents

1. 50 mM Tris buffer, pH 9.0, ice cold, deaerated or argon bubbled
2. Substrate solution is prepared fresh daily by dissolving 466 µl pure linoleic acid in 500 µl absolute ethanol in a test tube on ice. A drop of Tween-80 is added, the volume is adjusted with buffer to 10 ml, and the mixture is vortexed briefly (5 sec). The substrate solution is protected from light and stored on ice under argon.

b. Procedure

A matched pair of quartz cuvettes is used. In the sample cuvette, a suitable amount (25 to 100 µl) of enzyme ($ZnSO_4$ treated, dialyzed 100,000 × g supernatant or purified LO) is mixed with buffer, while an equal volume of buffer is pipetted into the reference cuvette (no enzyme). After 2 min of preincubation at 37°C, the reaction is initiated by the addition of linoleic acid solution (5 mM or desired final concentration), first to the reference cuvette and then to the sample cuvette (1 to 3 ml final volume). The change in absorbance at 234 nm is recorded for 3 min (or longer if needed). The initial linear portion of the recording, reflecting conjugated diene formation, is used to compute the rate of reaction. Specific activity is calculated using a molar extinction coefficient of 25,000 for linoleic acid hydroperoxide.

c. Remarks

The optimum pH and substrate concentration vary, depending upon the tissue source of LO; therefore, it is essential to run pilot experiments if the optimum assay conditions are unknown. Arachidonic acid or another suitable PUFA can be substituted for linoleic acid. In case of a Hb contamination problem, the use of (partially) purified enzyme is strongly recommended. Within a narrow range, dioxygenase activity is proportional to the amount of enzyme protein used in the assay. Specific activity decreases when a high enzyme concentration is used, due primarily to self-inactivation by fatty acid hydroperoxide accumulation in the medium. Native enzyme contains most of its iron in the inactive (ferrous) state and commonly displays a lag phase, which may last up to several minutes. Especially in the case of crude enzyme preparations, the presence of endogenous chemicals (e.g., GSH, catecholamines) in the reaction medium also increases the lag period.[43] Prior dialysis or addition of a low (micromolar) concentration of H_2O_2,[44,45] or other lipid hydroperoxides is required to fully activate dioxygenase. Certain LOs, e.g., 5-LO, require calcium ions and/or ATP for maximum expression of activity, and this aspect should be considered.[29,30]

5. Prostaglandin Synthase

The measurement of cyclooxygenase activity can be followed by O_2 uptake and is described below. The assay procedure outlined in Section III.B.6 can be adapted with necessary modifications for the estimation of peroxidase activity of PGS.

a. Reagents

1. Arachidonic acid (substrate) solution can be prepared as described for linoleic acid in Section III.B.4. Arachidonate is relatively susceptible to autooxidation and therefore, arachidonate solutions should be stored under argon or nitrogen.
2. 0.1 M potassium phosphate buffer, pH 7.4.

b. Procedure

A microsomal suspension (0.5 to 2.0 mg protein) and the buffer are placed in a water-jacketed vessel maintained at 37°C and containing a Clark-type oxygen electrode connected to an oxygen monitor. The contents are allowed to equilibrate at 37°C, and a stable baseline is established (2 to 3 min). The reaction is started by adding the substrate solution (0.1 mM arachidonate) with a microsyringe (final volume 2 ml) though the vessel's outlet, and the oxygen uptake is recorded for 3 to 5 min. The ambient oxygen concentration in the aerated buffer at 37°C is approximately 220 µM.

c. Remarks

This procedure can be adapted with suitable modifications for the measurement of the dioxygenase activity of LO. Similar to LO, the cyclooxygenase activity of PGS exhibits a linear response within a narrow range of enzyme protein concentration. Low or no activity is observed when high enzyme concentration is used. When purified enzyme is reconstituted, the addition of heme (hematin), phenol, and GSH is required to observe maximal activity.

6. Peroxidase

A number of compounds can be used to monitor tissue peroxidase activity. The most commonly used method, described below for human placental peroxidase, is an adaptation of the spectrophotometric method reported for peroxidase-mediated tetraguaiacol formation from guaiacol in the presence of H_2O_2.[46]

a. Reagents

1. 50 mM Tris-HCl buffer, pH 7.2 (or appropriate buffer for the enzyme under study)
2. 2.6 M guaiacol (O-methoxyphenol) solution in absolute ethanol
3. 3.3 mM H_2O_2 (dilute 30% hydrogen peroxide with the buffer)

b. Procedure

Pipette and mix a suitable amount of buffer, 5 to 50 µl of enzyme (absent in the reference cuvette), and guaiacol (13 mM; 5 µl/ml) in the sample and reference cuvettes, and preincubate for 2 to 3 min at 37°C. Initiate the reaction by the addition of H_2O_2 solution (100 µl/ml or desirable volume) (1 or 3 ml final volume of reaction mixture). Record the increase in the absorbance at 470 nm for 3 to 5 min. Specific activity is calculated using an extinction coefficient of 26.6 mM^{-1} cm^{-1} for tetraguaiacol. The initial linear portion of the response curve is used to compute the enzyme velocity.

c. Remarks

Tetraguaiacol is the major product of the reaction. Small amounts of other metabolites are also formed. It is very important to consider the high peroxide specificity exhibited by different peroxidases to support xenobiotic oxidation. Thus, H_2O_2 serves as an efficient cofactor for xenobiotic oxidation catalyzed by human placental peroxidase,[47] but it is a very poor cofactor for PGS.[27]

Similarly, *tert*-butyl hydroperoxide or cumene hydroperoxide are less efficient in supporting peroxidase activity of PGS. The peroxidase activity of PGS requires the presence of PGG_2 or PUFA hydroperoxide in the medium.[27,28] Since these hydroperoxides are generated by the dioxygenase activity, cooxidation of xenobiotics by PGS and LO is routinely assayed using reaction media supplemented with PUFA alone. Peroxide specificity of LO is relatively broad. Not only H_2O_2[48] and PUFA hydroperoxides[29–31] but either *tert*-butyl hydroperoxide or cumene hydroperoxide[49] can also efficiently support the cooxidase activity of LO.

7. Epoxide Hydrase

The procedure of Oesch et al.,[50] which employs styrene oxide, is commonly used to assay epoxide hydrase (EH; E.C. 3.3.2.3) activity of tissue microsomes. The assay can also be performed using [^{3}H]BP-4,5-oxide.

a. Reagents

1. 7-[^{3}H]styrene oxide, 16 mM, in tetrahydrofuran containing 0.1% triethylamine
2. 0.5 M tris-HCl buffer, pH 8.7

b. Procedure

Add substrate solution (20 µl) to the suitable amount of enzyme in buffer (300 µl final volume) and incubate for 5 to 10 min at 37°C. The reaction is terminated by addition of 5 ml petroleum ether (bp 30°C to 60°C) to the test tube. The mixture is vortexed and then centrifuged for 5 min at 1000 × g. The tube is then cooled by immersion in an ice-acetone bath. After the aqueous phase is frozen, the upper organic phase is discarded. This extraction procedure is repeated twice. Finally, 2 ml of ethyl acetate is added, and the contents are mixed and centrifuged as before. An aliquot (300 µl) of the ethyl acetate extract containing styrene glycol is transferred to a scintillation vial for radioactivity counting. The incubation mixture containing boiled enzyme (100°C for 10 min) or no enzyme is used as the control.

c. Remarks

EH activity has been observed in rodent liver nuclei, mitochondria, and microsomes (mEH) as well as cytosol (sEH). Of these, the microsomal and cytosolic forms have received the most attention. According to some reports, selective substrates for mEH include phenanthrene oxide, BP-4,5-oxide, and *cis*-stilbene oxide while sEH hydrolyzes *trans*-stilbene oxide and *trans*-β-benzyl styrene oxide more efficiently. Most substrates are potentially carcinogenic; therefore, routine precautions for carcinogen and radioisotope handling should be taken.

8. Glutathione Transferase

A number of assay procedures are available for assaying GST activity. This includes estimation of depletion of GSH or test substrate from the medium or quantitation of specific product formation from the test substrate, either by radiometry, spectrophotometry, or HPLC. The most commonly used spectrophotometric method, described by Habig et al.[51] and modified by Radulovic and Kulkarni[52] for human placental GST (GSTP1-1), is described below.

a. Reagents

1. 0.2 M 1-chloro-2,4-dinitrobenzene (CDNB) solution in acetone or ethanol
2. 0.1 M potassium phosphate buffer, pH 6.5
3. 25 mM reduced glutathione solution in buffer
4. Enzyme: purified or cytosol (100,000 × g supernatant)

b. Procedure

The rate limiting amount of GST (5 to 50 µl/ml; absent in the reference cuvette), GSH (2.5 mM; 100 µl/ml), and buffer are pipetted into the sample and reference cuvettes and allowed to preincubate for 2 to 3 min at 37°C. The reaction is initiated by the addition of CDNB solution (1 mM, 5 µl/ml), and the increase in absorbance at 340 nm with time, reflecting the formation of the GSH adduct of CDNB, is recorded. The specific activity is estimated by use of the initial linear portion of the response curve (1 to 3 min) and an extinction coefficient of 9.6 mM^{-1} cm^{-1}.

c. Remarks

The inhibition of purified rat live and human placental GST by exogenous Hg is concentration dependent,[53] and contamination of tissue cytosol with Hb is commonly noted. Thus, for accurate determination of GST activity, the method devised by Sajan and Kulkarni[53] to remove Hb from mammalian tissue cytosol prior to the assay is strongly recommended. GST occurs in multiple forms.[35,36] CDNB serves as a substrate for most, if not all, isozymes of GST. When unpurified enzyme is used, the use of preferred substrates, along with selective inhibitors and specific antibodies, can provide useful information regarding individual isozymes of GST.

9. UDP-Glucuronosyl Transferase

Several spectrophotometric, radiometric, fluorometric, and HPLC methods are available in the literature to assay different forms of UGT activity in tissue microsomes. One of the routinely employed spectrophotometric methods measuring *p*-nitrophenol glucuronidation is described below.

a. Reagents

1. 1.0 mM UDP-glucuronate (UDPGA), triammonium salt, adjusted to pH 7.4 with KOH
2. 1.0 mM *p*-nitrophenol in 50 mM Tris-HCl buffer, pH 7.4, containing 10 mM MgCl$_2$
3. 50 mM Tris-HCl buffer, pH 7.4
4. 0.2 M glycine buffer containing 0.15 M NaCl, pH 10.4
5. Triton X-100 detergent

b. Procedure

Mix UDPGA (cofactor), *p*-nitrophenol (substrate), and Tris-HCl buffer (final volume 1.4 ml) in a tube and preincubate for 3 min at 37°C. Start the reaction by adding 0.25 to 1.0 mg Triton X-100 activated microsomes (100 µl). Microsomes are activated prior to incubation by mixing 0.2 µl Triton X-100/mg microsomal protein. After incubation for 10 min at 37°C, the reaction is terminated by the addition of 5 ml of glycine buffer. Measure *p*-nitrophenol concentration spectrophotometrically at 405 nm. Control incubation mixture is prepared by omitting UDPGA. The extinction coefficient of *p*-nitrophenol at pH greater than 10 is 18.1 mM^{-1} cm^{-1}.

10. Sulfotransferase

A variety of assay procedures are available to estimate SULT activity.[34] The general radiometric assay described below is versatile and convenient.

a. Reagents

1. 0.5 M sodium phosphate buffer, pH 6.5
2. 0.1 M 2-mercaptoethanol, prepared by adding 0.07 ml of the thiol to 10 ml H$_2$O

3. 5 mM 2-naphthol, dissolved in acetone
4. ^{35}S-labeled (0.1 to 1.0 µCi) plus cold PAPS
5. 0.1 M barium acetate
6. 0.1 M barium hydroxide
7. 0.1 M zinc sulfate

b. Procedure

Mix buffer, mercaptoethanol, 2-naphthol (substrate), and PAPS (cofactor) and preincubate for 3 min at 37°C. Start the reaction by adding a suitable amount of enzyme (final volume 1.0 ml). After 10 min incubation at 37°C, terminate the reaction by addition of the phosphate buffer and 0.2 ml barium acetate solution. Then add 0.2 ml barium hydroxide solution followed by 0.2 ml zinc sulfate solution. Remove the precipitate by low speed centrifugation (2500 × g for 10 min) and repeat the barium plus zinc step. The unreacted [^{35}S]PAPS remaining after the reaction is precipitated by barium, but most sulfoconjugates are water soluble and are not precipitated. Therefore, the radioactivity remaining in the supernatant after centrifugation is used to quantify the sulfoconjugate formed. Controls are assayed as above, omitting only the substrate.

c. Remarks

This procedure can be combined with other analytical procedures, such as thin layer chromatography, HPLC, or spectrophotometry using methylene blue reagent. If necessary, any other desirable substrate can substitute for the 2-naphthol. The optimum pH for the reaction varies by substrate and enzyme. This assay may not be suitable for some substrates if their sulfoconjugates are precipitated to various degrees by the barium treatment (e.g., sulfoconjugates containing a carboxyl group).

IV. METABOLISM OF DEVELOPMENTAL TOXICANTS

A. Maternal Liver

During pregnancy, several major shifts from the nonpregnant state occur in the maternal liver. With some xenobiotics these changes may not be of consequence to the conceptus if the chemical requires prior metabolism and the ultimate toxicant is a reactive species with a relatively short biological half-life. In these cases, xenobiotic metabolism in the maternal liver determines the amount of parent compound and its stable metabolite(s) reaching the conceptus. Logic dictates that the generation of ultimate toxicant must occur in the conceptus to exert toxicity on the embryo or fetus.

1. Cytochrome P450

a. General

The literature is inconsistent and either no change or pregnancy-related decreases in P450 content have been reported.[54–56] Turcan et al.[57] observed a significant decrease in the proportion of P450 in the high spin state in hepatic microsomes from pregnant rats. The spin state of the P450 is important in overall metabolism. A shift to a lower spin state may alter the redox potential of the NADPH cytochrome P450 reductase–hemoprotein coupling, causing a reduction in electron flow to the P450 and a decrease in the rate of substrate oxidation. Besides changes in the amount of P450, Lambert et al.[58] reported synthesis of a new major pregnancy-specific form, P450$_{gest}$, in the hepatic microsomes of two mouse strains. The elevation of this P450 was maximal by gestation day (GD) 6. Another pregnancy-induced P450 has been identified in the rabbit lung.[59]

In animals, a significant increase in maternal liver weight during pregnancy is noted and is a result of the proliferation of parenchymal cells.[54,60] This increases microsomal yield per liver.[54] However, according to Combes et al.,[61] one should keep in mind that the liver is not enlarged during pregnancy in humans. Several reports describe changes in the rates of hepatic microsomal P450-mediated xenobiotic oxidation during pregnancy.[54,55,62–65] For example, in the mouse, the declines in specific activity for N-demethylation of carbaryl and nicotine were 20% and 50%, respectively, on day 12 of gestation, and remained statistically lower on GD 18.[54] Similarly, the rates of 7-ethoxycoumarin O-dealkylase (ECOD) decreased 11% and 39% on GD 12 and 18, respectively.[54] A report[64] suggested no pregnancy-related change in the *N,N*-dimethylaniline demethylase activity in mouse hepatic microsomes while aminopyrine N-demethylase showed a modest decline. The rates of dearylation of EPN, but not parathion, declined significantly by day 12 of pregnancy but returned to nonpregnant levels by day 18.[54] Hepatic aldrin epoxidase activity continued to decline toward term[54,65] but posted no change in total capacity. Parathion undergoes dearylation and desulfuration catalyzed by hepatic P450. Although the dearylation of parathion exhibited no change in the rate, a 51% decline was noted in desulfurase activity in mouse liver during pregnancy.[66] Our observations suggest[54,64,65] that the total capacity of liver for dealkylation of nicotine, carbaryl, aminopyrine, *N,N*-dimethylaniline, and 7-ethoxycoumarin, and epoxidation of aldrin does not change as result of pregnancy. Similar observations have been noted for parathion.[54,67]

b. *Induction*

In pregnant golden hamsters, ethanol treatment resulted in a 50% increase in CYP2E protein.[56] Treatment of pregnant mice with polychlorinated biphenyls (PCBs) significantly increased the dimethylnitrosamine demethylase activity.[63] Pretreatment of pregnant rats with 2,3,7,8-tetrachlorodibenzo-*p*-dioxin (TCDD) on GD 10 has been reported to cause an approximate 15-fold induction of maternal liver AHH on GD 20.[62] Cytochrome P450 IIIA1, which is normally undetectable in the maternal liver, was significantly elevated at GD 15 to 21 following pregnenolone-16-carbonitrile (PCN) exposure of pregnant rats.[68]

Many reports have described the effect of chemical pretreatment of pregnant animals on the teratogenicity of test chemicals. Qualitative and quantitative changes in P450s were believed to be responsible for the observed results. Cyclophosphamide teratogenicity depends on its bioconversion to phosphoramide mustard and/or acrolein. Rodent embryos are incapable of this activation and require supplementation with adult liver S_9 fraction as an activation system in the culture medium. More effective activation was noted with a liver S_9 fraction prepared from adult male rats pretreated with phenobarbital or a mixture of PCBs other than that from untreated or 3-methylcholanthrene (3-MC) pretreated rats.[69,70] In contrast to these results, an S_9 fraction prepared from adult male rats pretreated with either 3-MC, Aroclor 1254, or isosafrole, but not from phenobarbital or PCN, was found to be more effective in enhancing 2-acetylaminofluorene (AAF) dysmorphogenesis in rat embryos.[70,71] The required P450s are not induced in the maternal liver by pretreatment of rats with 3-MC, Aroclor 1254, isosafrole, phenobarbital, or PCN. Transplacental exposure of GD 8 embryos to 3-MC resulted in a high percentage of malformed embryos following explantation on GD 10 in a medium containing AAF.[70,71]

For several reasons, the approach of using inducers in studies of chemical teratogenicity is seriously flawed, and considerable caution should be exercised in the interpretation of results of such models. Under *in vivo* conditions, it is difficult to identify the enzyme responsible for bioactivation of the test chemical. For example, cyclophosphamide oxidation is not only mediated by P450, but also by PGS[72] and LO,[73] and various chemicals also induce LOs.[29,30] The P450 inducers 3-MC, BP, or phenobarbital are themselves teratogenic and thus may weaken and predispose embryos to the teratogenicity of a test agent. Thus, the observed exaggerated response may not be due to the test agent alone but may reflect a potentiative or synergistic interaction between inducer and test chemical. Furthermore, the residues of inducers and/or their toxic metabolites occur as contaminants in the liver microsomes or S_9 fractions of pretreated animals. Therefore, their use for *in vitro* embryo culture models may alter the response of embryos to test chemicals.

2. Flavin-Containing Monooxygenase

Relatively little published information exists describing the dynamics of hepatic FMO-catalyzed xenobiotic metabolism during pregnancy. Although the hepatic dimethylaniline N-oxidase activity did not change, lung FMO increased by 57% in 28-day pregnant rabbits.[74] In mice, no pregnancy-related changes were noted in N-oxidase activity in liver, lung, or kidney microsomes[64] or for the hepatic S-oxidation of phorate.[54]

3. Epoxide Hydrase

The microsomal epoxide hydrase activity in guinea pigs toward styrene oxide declined 30% to 49% in intestinal preparations at GD 54 and 67,[75] while a spurt of activity (a 20% to 67% rise) was noted in the lung, intestine, and kidney microsomes of 20-day pregnant rabbits.[75] No such increase in epoxide hydrolase activity was noted for the liver.

4. UDP-Glucuronosyl Transferase

In rats, pregnancy appears to decrease *p*-nitrophenol glucuronidation by 50% by GD 20, as compared to the activity 8 days after parturition.[62] The maternal enzyme is inducible and increased about 15-fold at term when rats were treated with 5 μg of TCDD on the tenth day of pregnancy.

5. Glutathione Transferase

a. General

Pregnancy did not alter maternal liver GST in mice when CDNB was used as the substrate.[76] Similar results were reported for rat liver GST activity toward 3,4-dichloro-1-nitrobenzene (DCNB) at midpregnancy.[77] However, the liver cytosol from 19- or 20-day pregnant rats exhibited a significant increase in CDNB and 1,2-epoxy-3-(*p*-nitrophenoxy) propane conjugation and a decrease in DCNB conjugation activity. The authors also noted no change in GST activity toward *p*-nitrobenzylchloride.[77] GSH conjugation of sulfobromophthalein in rats[78] and of styrene oxide in rabbits[75] is depressed in rats because of pregnancy. Mitra et al.[79] demonstrated that in rats, the addition of purified maternal liver GST and GSH to the culture medium markedly enhances 1,2-dibromoethane (DBE) toxicity to embryos. In the absence of exogenous GST, the same concentrations of DBE and GSH caused only marginal toxicity, suggesting the absence of GST in rat embryos. Since the maternally generated toxic episulfonium ions of DBE do not escape the liver, *in vivo* bioactivation of DBE is expected to produce maternal liver toxicity but not embryotoxicity.

b. Induction

Hepatic GST activity was noninducible by treatment of pregnant rats with phenobarbital.[80] However, induction of hepatic GST was reported in pregnant rats and mice exposed to 5,6-benzoflavone.[81] Kulkarni et al.[76] observed a threefold induction of hepatic GST activity toward CDNB in 18-day pregnant mice exposed to *trans*-stilbene oxide, a potent inducer of the enzyme.

B. Uterus and Decidua

1. Cytochrome P450

During GD 7 to 10, the weight of the rat embryo amounts to less than 1% of the whole implantation site (embryo plus decidua).[82] No AHH activity could be detected in homogenates of whole implantation sites (GDs 7 through 10) or of the decidua (GD 11) of control rats. However, induction occurred when pregnant rats were pretreated with BP 24 h before sacrifice.[82]

2. Peroxidase

a. General

Rat uterine peroxidase has been purified and characterized.[83] Nelson and Kulkarni[84] noted that membrane-bound uterine peroxidase activity was very high in nonpregnant mice but declined rapidly with the onset of pregnancy, and only 0.2% remained by GD 18. An earlier study[85] reported that total (membrane-bound and luminal-soluble) uterine peroxidase activity is initially low but increases on GD 7 in rats and mice, and this high level is maintained throughout pregnancy. Also, uterine peroxidase has been shown to oxidize estradiol to polymeric products capable of binding to protein.[86] Byczkowski and Kulkarni[87] noted the generation of BP-7,8-dihydrodiol-9,10-epoxide, a postulated ultimate teratogen in animals,[88] from BP-7,8-dihydrodiol (BP-diol) by rat uterine peroxidase.

b. Induction

Jellinck et al.[89] studied the characteristics of estrogen-induced peroxidase in mouse uterine luminal fluid. The peroxidase activity was stimulated by tyrosine and 2,4-dichlorophenol, and evidence was presented establishing uterine peroxidase-catalyzed oxidation of diethylstilbestrol (DES).[90] A semiquinone metabolite was presumed to be the primary intermediate produced during one-electron oxidation of DES by the peroxidase. In contrast to the widely known inducibility of peroxidase by estrogenic compounds, treatment of young female rats with 2,3,4,7,8-pentachlorodibenzofuran or TCDD was found to significantly lower the constitutive uterine peroxidase activity in a dose-dependent manner.[91] In cotreatment studies with 17-estradiol, TCDD antagonized the increase in uterine peroxidase activity.

3. Lipid Peroxidation

In general, lipid peroxidation in the rat uterus is decreased during gestation. To eliminate any contribution from conceptual tissues, investigators studied lipid peroxidation in the rat uterus following deciduoma induction by intraluminal injection of arachis oil on GD 5.[92] The endogenous content of thiobarbiturate-reactive substances decreased significantly in uterine tissue during decidua induction. The lowest values were observed on GD 11 and 14, and the values increased thereafter. Ascorbate-supported lipid peroxidation decreased during the progressive phase of induction and increased later, during the regressive phase. Cumene hydroperoxide–triggered lipid peroxidation was low, while that supported by NADPH did not change. Low levels of peroxidized lipids were noted in human uterine microsomes in the absence of any exogenous cofactors.[93] The presence of this basal level of thiobarbiturate-reactive substances in the microsomes suggests *in vivo* lipid peroxidation. An increase in malondialdehyde production was observed when increasing amounts of ascorbate, ferrous iron, and arachidonate were added to the incubation medium.

4. Lipoxygenase and Prostaglandin Synthase

The results of radioimmunoassays[94] indicated that leukotriene generation in rats remained unaltered during days 1 to 3 but exhibited a marked increase on day 4, with a peak at noon. This was followed by a sharp decline on day 5, prior to implantation. Prostaglandin biosynthesis displayed an essentially similar pattern. At midpregnancy in rats, myometrium incubated with the attached decidua released more prostanoids into the incubation medium than when incubated without decidua.[95] As pregnancy progressed to GD 21, more release of prostanoids was noted. In animals, uterine prostaglandin release is pulsatile and can be stimulated by oxytocin or estradiol.[96] However, this hormonal stimulation is not sustained and diminishes even when the stimulus is applied continuously. This decline may be related to the liberation and availability of the substrate, arachidonic acid, from the cell membranes.

In the nonpregnant human uterus, both endometrium and myometrium exhibit cyclooxygenase as well as 5- and 12-LO activities toward arachidonate.[97] Prostanoid synthesis has been studied in human endometrium obtained at different phases of the ovarian cycle and in decidua in early pregnancy.[98] Endometrial prostaglandin production was maximal during the midsecretory phase, while it was markedly suppressed in decidua. Essentially similar conclusions have been reached by others. The production of arachidonic acid metabolites via cyclooxygenase and LO pathways occurs in decidua vera at term.[99] Rees et al.[100] observed the release of different leukotrienes when decidua samples obtained at term with different modes of delivery were examined by use of a short-term incubation technique. Recently, Lei and Rao[101] noted the presence of immunoreactive 15-LO in rough endoplasmic reticulum and found 15-HETE in myofilaments. Quite unexpectedly, both were also present in nuclear chromatin. Flatman et al.[102] purified and characterized 12-LO from human uterine cervix and found the highest specific activity to be associated with microsomes.

5. Flavin-Containing Monooxygenase

So far only one report has described the changes in microsomal FMO activity during pregnancy. Osimitz and Kulkarni[64] reported a decline in FMO activity in the mouse uterus during pregnancy. The activity decreased by 56% and 28% in 12- and 18-day pregnant mice, respectively, as compared with that of nonpregnant animals.

C. Intrauterine Conceptual Tissues

1. Animal Species

Whole rodent embryo culture is one of the most popular *in vitro* test systems employed to investigate the developmental toxicity of xenobiotics. Even though the use of only the embryo proper is implied, in reality, preparations usually not only include embryo, but also ectoplacental cone and chorioallantoic placenta, visceral yolk sac, and amniotic membranes. According to Schlede and Merker,[82] decidua cannot be excised from 7- to 10-day-old rat embryos. Although a mixture of tissues is commonly used, there appears to be no strong rationale behind the common practice of not including other tissues of close proximity, such as the parietal yolk sac and Reichert's membrane. In any case, one should realize that the biological response observed following exposure to the test chemical is a net outcome of metabolic interactions of the xenobiotic with all the tissues involved. It is noteworthy that drug metabolizing enzymes, such as P450, are relatively more active and abundant in the extraembryonic tissues (especially visceral yolk sac) than in the embryo proper.[5] A similar argument can be made in the case of the chick embryo *in ovo* test system. Availability of information on the collective ability of analogous preparations of HICT (human intrauterine conceptual tissues: embryo, together with amnion, chorion, placenta, decidua, and umbilical cord) to metabolize xenobiotics would be of great value in understanding the developmental toxicity of chemicals in humans. Limited data collected in our laboratory are available on this aspect (detailed later).

2. Humans

a. Peroxidase

Membrane-bound peroxidase was isolated and purified from pooled samples of HICT of 4 to 6 weeks' gestation.[103] The multistep procedure involved $CaCl_2$ extraction of membranes and affinity chromatography on Concanavalin A sepharose 4B, followed by hydrophobic chromatography on phenylsepharose CL 4B. This yielded preparations of HICT peroxidase with relatively high specific activity (15 µmol tetraguaiacol formed/min/mg protein), reflecting a 44% yield of activity and 95-fold purification. Besides guaiacol, the purified embryonic enzyme was able to oxidize all of the tested compounds (pyrogallol, o-dianisidine, p-phenylenediamine, N,N,N,N-tetramethyl phenylenediamine [TMPD], benzidine, 3,3′,5,5′-tetramethyl benzidine [TMBZ], epinephrine, bilirubin, and thiobenzamide) at a significant rate in the presence of H_2O_2.[103] Qualitatively similar results were observed with peroxidase isolated and purified from pooled samples of HICT of 10 weeks' gestation.[104] Further study[105] has shown that the teratogenic arylamine 2-aminofluorene (2-AF) is easily oxidized by the purified HICT peroxidase. Kinetic data obtained under optimal assay conditions yielded a K_m value of 41 µM for 2-AF and a V_{max} value of 1.2 µmol/min/mg protein. The reactive intermediates generated during the oxidation of 2-AF were capable of binding to calf thymus DNA and protein (bovine serum albumin, BSA). Covalent binding of radioactivity (nmol of [^{3}H]2AF equivalent bound/min/mg enzyme protein/mg of DNA or BSA) occurred at the rate of approximately 1.90 and 3.75 to DNA and to protein, respectively.

b. Lipoxygenase

LO activity has been partially purified from the cytosol of HICT at 4[106] or 8 to 10[104,107] weeks of gestation by affinity chromatography. Besides model compounds,[104] the purified HICT LO was capable of activation of teratogens in the presence of PUFAs. Thus, the enzyme easily epoxidized aflatoxin B_1 (ATB$_1$) in the incubation media supplemented with linoleic acid.[107] It is interesting to note that the rate of epoxidation was greater than that noted with purified term placental LO. The purified LO isolated from HICT of 4 weeks of gestation was also capable of oxidation of the proximal animal teratogen, BP-diol.[106] Both formation of the putative ultimate toxicant BP-diol-9,10-epoxide and significant covalent binding of radioactivity to protein were observed during the reaction supported by linoleic acid.[106]

c. Glutathione Transferase

Datta et al.[108] purified GST activity present in cytosol of HICT of 6 to 10 weeks of gestation using affinity chromatography and examined its ability to form GSH conjugates with xenobiotics. A 200-fold purification was achieved when GST activity toward CDNB was assayed. There was no significant difference in the specific activity toward CDNB among the HICT GST preparations of 6-, 8-, and 10-weeks' gestation. Besides CDNB, the enzyme was found to be active toward DCNB, ethacrynic acid, p-nitrobenzyl chloride, and 4-nitropyridine-oxide. Additionally, the enzyme also was capable of DBE activation. The rate of GSH conjugation of DBE was about twofold greater with HICT-GST of early gestation (6 weeks) as compared with those from later periods (8 and 10 weeks). The covalent binding of [^{14}C]DBE to protein was approximately 10-fold higher when the reaction was mediated by HICT-GST from 6 weeks gestation compared with that from 10 weeks gestation. Taken together, these results strongly suggest that DBE may be a developmental toxicant in humans.

D. Embryo

1. Cytochrome P450

a. General

Flint and Orton[109] evaluated the ability of 46 compounds to inhibit differentiation of rat midbrain and limb bud cells in culture. No increase in the extent of inhibition was observed with 24 out of

27 active compounds in the presence of rat liver S_9 fraction, suggesting that rat embryonic cells possess enzymes capable of xenobiotic activation. Several attempts by other investigators to definitively identify P450s in rodent, human, and chick embryos and extraembryonic tissues have met with limited success. Recent reviews[2,5] of the published data (most of which are indirect, conflicting, and/or negative) on this subject have finally concluded that constitutive functional P450 enzymes capable of xenobiotic oxidation have not been definitely identified in rodent or human embryos during organogenesis. It appears that, with a possible exception of P4503A7, the constitutive forms of P4501A1, P4501A2, and probably P4502E1 and P4502B are either absent or their signal is so weak that it can not be detected by ultrasensitive probes such as reverse transcriptase–polymerase chain reaction (RT-PCR). Thus, the marginally positive results claiming P450 involvement reported by some investigators appear to be of questionable significance and relevance to the bioactivation of teratogens during human organogenesis.

b. Induction

Although a few studies[5] have reported the inductive effects of 3-MC, β-naphthoflavone, and PCN on rodent embryonic xenobiotic oxidation, published data are too scant for other inducers. P450s in rodent embryos are refractive to induction following exposure to phenobarbital.[70,71] However, those from MC-pretreated dams were capable of measurable oxidative biotransformation of AAF, suggesting P450 induction.[70,71] In preimplantation stage mice, activation of BP, possibly by CYP1A1, can be detected on GD 4.[110] Mouse embryos preexposed to 3-MC exhibit a higher incidence of BP-induced fetal abnormality, suggesting possible induction of CYP1A1.[111] Interestingly, the incidence of fetal anomalies was much lower in noninducible littermates. The signals for P4501A1 mRNA were detectable in conceptal tissues from 3-MC–treated rats only on GD 12.[112] However, the intensity of the signal was lowest in tissues of the embryo proper. This casts doubts regarding the presumed activation of teratogenic chemicals by embryonic P450s. In light of other known pathways, Wells and Winn[3] have recently questioned whether the elevated levels of embryonic P450 are sufficient to mediate teratologically relevant bioactivation of xenobiotics. Treatment of eggs with 3,3′,4,4′-tetrachloro-biphenyl has been shown to result in a 2- and 14-fold increase in ECOD and AHH activities, respectively, in livers from 5-day-old chick embryos.[113]

2. Prostaglandin Synthase

PGS activity occurs in mouse[114] and rat[115] blastocysts before implantation. The liberation of prostaglandin 6-keto-PGF_1, PGE, and PGF from endogenous arachidonate was noted in GD 11 rat embryo homogenates.[116] As expected, nonsteroidal antiinflammatory drugs, such as aspirin, indomethacin, salicylate, meclofenamic acid, ibuprofen, 5,8,11-eicosatriynoic acid (ETI), and 5,8,11,14-eicosatetraynoic acid (ETYA), caused dose-dependent inhibition of prostaglandin synthesis. Nevertheless, the authors opined that the inhibition of PGS is not likely to be the cause of aspirin-induced limb defects in rats. Recent reports[117,118] have described the occurrence of PGS activity in mouse embryos. Even though, the presence of PGS in human embryos is expected, it appears that relevant published data on this subject are not currently available.

A series of investigations reported by Wells and coworkers provide compelling evidence implicating, in part, embryonic PGS as one of the major enzymes responsible for activation of phenytoin and other teratogenic compounds. The results supporting the hypothesis that phenytoin teratogenicity involves a free radical mechanism include oxidation of DNA and covalent binding to protein,[118,119] oxidation of GSH to GSSG,[118,120] and partial protection from phenytoin-induced cleft palate formation in mice when dams were pretreated with either EYTA,[120] caffeic acid (a dual inhibitor of PGS and LO), and aspirin (a cyclooxygenase inhibitor) or with α-phenyl-N-t-butyl nitrone (PBN, a spin trap agent for free radicals).[121] Purified PGS in the presence of arachidonate generates reactive metabolites from phenytoin that bind covalently to proteins and can be detected by ESR spectrometry.[122] Further ESR studies[123] revealed that the putative unstable nitrogen-centered free radical and a stable carbon-centered radical are formed from phenytoin by PGS. A correlation was noted between the amounts of free radical adducts formed by PGS and the teratogenicity of several chemicals. Thus, highly teratogenic dimethadione produced substantially more free radicals than its minimally teratogenic parent compound, trimethadione. Phenytoin and its *p*-hydroxylated metabolite, 5-(*p*-hydroxyphenyl)-5-diphenylhydantoin, yielded similar amounts of free radicals. Both D- and L-isomers of mephenytoin and its N-demethylated metabolite generated different amounts of free radicals. Among the chemicals that are less potent teratogens in humans, phenobarbital yielded only minimal amounts of free radicals, while no free radical formation was detectable from carbamazepine.[123] Incubation of mouse embryos in culture with BP decreases anterior neuropore closure, turning, yolk sac diameter, and somite number, and similar to phenytoin, triggers DNA and protein oxidation.[118] Superoxide dismutase addition provides complete protection, suggesting that BP embryotoxicity in mice may be mediated by ROS generated by free radicals produced during peroxidative metabolism of BP. Similarly, free radicals generated by peroxidase from thalidomide, the most potent known human teratogen, initiate teratogenic insult and DNA oxidation in rabbit embryos. These effects are abolished by the pretreatment with either PBN[124] or aspirin.[125]

3. Lipoxygenase

Two laboratories have independently documented the existence of arachidonate LO activity in rat embryos. Earlier, Vanderhoek and Klein[126] presented evidence for the existence of LO activity in rat embryos. Our laboratory has not only confirmed the occurrence of LO in 9- and 10-day-old rat embryos but has also established the importance of this enzyme system in xenobiotic metabolism.[127] The enzyme was found to oxidize benzidine, *O*-dianisidine, TMPD, and TMBZ in the presence of linoleic acid. Although direct evidence for the presence of a LO pathway in the human embryo proper is still lacking, our earlier studies have shown an abundance of LO activity in HICT during organogenesis. The enzyme preparations were capable of oxidation of xenobiotics at biologically significant rates (see Section IV.C.2.b for details).

The hypothesis was tested that phenytoin teratogenicity in mouse embryos may, in part, result from peroxidative bioactivation of phenytoin by LO to a toxic reactive free radical intermediate.[3,120,128,129] Embryoprotective effects were observed when mouse embryos were cultured *in vitro* in the presence of ETYA, a dual inhibitor of the PGS and LO pathways. The combined treatment resulted in a significant increase in yolk sac diameter, somite development, and crown-rump length, relative to phenytoin alone. These results suggest that embryonic PGS and/or LO play a major role in phenytoin embryotoxicity. Direct evidence supporting phenytoin activation by LO is also available.[128,129] It was shown that both arachidonate and linoleic acid support phenytoin oxidation by soybean LO, leading to covalent binding of reactive intermediates to protein. As expected, the magnitude of covalent binding was diminished by the addition of LO inhibitors, such as NDGA, quercetin, BW755, or ETYA.

4. Peroxidase

By means of the diaminobenzidine staining reaction, Balakier[130] was able to detect endogenous peroxidase activity in early mouse embryos. The bulk of the activity was localized inside or close to the numerous apical vacuoles in the endoderm. Ectoderm, mesoderm, ectoplacental cone, and trophoblast cells did not contain peroxidase activity. Using guaiacol as a model substrate and H_2O_2, Nelson and Kulkarni[84] observed easily measurable peroxidase activity in calcium extracts of mouse embryo membranes. The activity increased about sixfold between GD 12 and 14. Although published data are not available for the human embryo proper, HICT isolates were found to possess considerable peroxidase activity at 4 to 6 weeks gestation.[103] Preparations of partially purified enzyme, free of Hb contamination, were capable of oxidizing several xenobiotics, such as guaiacol, thiobenzamide, 3,3′-dimethoxybenzidine, TMPD, TMBZ, 2,2′-azinobis(3-ethylbenzothiazoline-6-sulfonic acid) (ABTS), benzidine, pyrogallol, epinephrine, bilirubin, *p*-phenylenediamine, and 2-AF.

E. Fetus

The subject of xenobiotic metabolism in mammalian fetal tissues has been reviewed periodically.[6–8,131–133] The following is a brief synopsis of the literature.

1. Cytochrome P450

a. General

In rats, the total P450 content in fetal liver microsomes is approximately 10% of that in the maternal liver.[134] Several studies have suggested that the capacity of rodent fetal liver to mediate xenobiotic oxidation is much lower (approximately 5%) than that observed in adults, and the activity becomes more easily detectable toward term. However, information regarding the identity of P450s involved in these reactions is still unclear. In the case of human fetal liver, early efforts were primarily focused on revealing the ability of various preparations to mediate xenobiotic oxidation. By use of cell cultures, homogenates, postmitochondrial supernatant, or microsomes, several drug substrates were reported to be oxidized by human fetal liver.[7,8,131,132,135] The total P450 content in the human fetal liver varies from 20% to 70% of the level in adult liver.[131,132] Compared with rates in adult human liver, the reported rates of aromatic and aliphatic hydroxylation and N-demethylation of different xenobiotics in human fetal liver are generally very low. However, in the case of ethylmorphine demethylation, fetal liver microsomal activity may approach the levels noted in adult liver.[136,137]

Although the ability of human fetal liver microsomes to oxidize xenobiotics has been known for some time, very little was clear about the biochemical characteristics of P450s until purification attempts by Kitada et al.[135] and Komari et al.[138] Now, the presence of several P450 enzymes is known. Very low levels (about one-tenth the adult level) of CYP1A1 protein are expressed before 20 weeks' gestation. The enzyme is associated with ECOD and hydroxylation of AAF, and the activities are inhibited by antibodies to CYP1A1 and 7,8-benzoflavone.[139,140] CYP1A2 and CYP1B1 appear to be expressed at negligible levels in the human fetus.[141] As regards the expression of CYP2C enzymes, CYP2C8 may be present as a major form,[139] while the concentration of 2C9, and 2C19 may be extremely low or undetectable.[142] Although the expression of CYP2D6 is detectable at 11 weeks of pregnancy by RT-PCR,[139] functional protein is found later after 20 weeks of gestation.[143] CYP2E1 mRNA was detectable in fetal liver after 16 weeks gestation.[144] Both protein level and ethanol oxidation rate are elevated in cultured fetal hepatocytes following exposure to ethanol or clofibrate. The results of RT-PCR probe and RNase protection assays suggest the presence of CYP3A4[145] and possibly CYP3A5.[146] CYP3A7 is the major constituent that accounts for about one-third of human fetal liver P450. It has been characterized and purified.[147,148] This form is shown to be capable of metabolism of about a dozen chemicals and plays a major role in fetal xenobiotic dealkylation or activation.

b. Induction

In general, P450 responsible for xenobiotic oxidation in animal fetuses is not readily inducible by chemical exposure of the mother during pregnancy. This appears to be particularly true for phenobarbital-type inducers, such as *p,p*-DDT.[149] Some elevation in certain monooxygenase activities can be detected if the maternal exposure to specific inducers (e.g., 3-MC) occurs near term. Simultaneous exposure to both types of inducers usually produces additive effects. A study reported an approximate fourfold increase in P450 content exclusively in fetal liver on GD 21 when 3-MC was administered to pregnant rats on GD 18 and 19.[134] However, the reduced CO difference spectrum exhibited a peak at 450 nm, while the maternal liver microsomes displayed a peak at 448 nm, as expected.

AHH in fetal rat liver is induced following treatment of pregnant rats with TCDD.[62] Induction was evident 3 days prior to birth when pregnant rats were administered TCDD on GD 10 or 16 but not on GD 5. TCDD treatment did not change the fetal liver microsomal P450 content. Dimethylnitrosamine demethylase activity in fetal mouse liver was low but readily detectable on GD 16.[63] When pregnant mice were treated with PCBs on GD 14, the fetal livers showed a significant increase in enzyme activity on GD 16. Induction was not detectable in fetal liver when assayed on GD 19, after treatment of mothers on GD 15 or 17.

Sherratt et al.[150] investigated induction of CYP1A1 in cultured rat fetal hepatocytes and found that exposure to 1,2-benzanthrecene alone resulted in a 60-fold increase. The inductive effect was potentiated about threefold when dexamethasone was included in the culture medium. CYPIIIA1 in fetal rat liver is inducible by PCN as early as GD 15.[68] Ethanol treatment of pregnant hamsters was found to result in a twofold increase in the CYP2E protein in fetal livers.[56] Recently, Srivastava et al.[151] reported a significant decrease in fetal hepatic microsomal P450 content and aminopyrine N-demethylase, aniline hydroxylase, and AHH activities on GD 20 following daily exposure of pregnant rats to styrene.

Negative findings about inducibility of P450 in the human fetus should not be construed as absence of a mechanism of induction, considering the fact that *in vivo* exposures to chemicals during pregnancy are generally low. At midgestation, phenobarbital exposure of mothers or cigarette smoking failed to induce fetal AHH or *N*–methyl aniline demethylase activities.[131] On the other hand, induction of prazepam metabolism was indicated when cultured fetal liver explants were exposed to phenobarbital.[152] AHH activity is induced when cells are exposed to BP, 3-MC, and benz(a)anthracene.[131] In monolayer culture of human fetal hepatocytes, phenobarbital and α-naphthoflavone exposures caused up to eightfold induction of AHH activity, while *t*-stilbene oxide or benzanthracene treatment had no effect.

2. Flavin-Containing Monooxygenase

In human fetal liver, N-oxidation of *N,N*-dimethylaniline by FMO can be detected in microsomes as early as 13 weeks of gestation.[153] There appears to be no correlation between the rate of N-oxidation and the smoking habit of the mother.

3. Prostaglandin Synthase

Although systematic studies on the role played by PGS in xenobiotic metabolism have not appeared in the literature, available data strongly suggest that this pathway is operative in human[154–156] and animal[157] fetal tissues.

4. Peroxidase

In the mouse, peroxidase activity toward guaiacol in the presence of H_2O_2 was found to increase logarithmically between GD 12 and 18, resulting in a 50-fold increase.[84] A 400-fold purification of peroxidase from whole human fetuses obtained at gestation weeks 8 to 10 has been reported.[158]

5. Lipoxygenase

Analyses of arachidonate metabolites suggested functional 5-, 12-, and 15-LO pathways in the rabbit fetus.[157] Human fetal tissues exhibiting LO activity include lung,[154] adrenal, brain, heart, kidney, and liver.[156] Partially purified LO from the cytosol of HICT at 8 to 10 weeks of gestation.[104,107] oxidized model compounds,[104] and epoxidized ATB_1 in the presence of linoleic acid.[107]

6. Epoxide Hydrase

A study on prenatal changes in microsomal epoxide hydrase activity toward styrene oxide revealed that the enzyme specific activity in rabbit fetal liver at GD 20 is less than 30% of the adult value.[75] The activity was low in intestine, lung, and kidney on GD 28. In guinea pigs, fetal hepatic and intestinal epoxide hydrase activities were less than 10% of adult values, and levels of the enzyme in lung and kidney were too low to measure. Phenobarbital caused dose-dependent induction (1.2 to 3.1-fold) in human fetal hepatocytes obtained at 17 to 20 weeks' gestation.[159]

7. Glutathione Transferase

Hepatic GST titer is virtually undetectable in rodent embryos and early fetal liver, low near term, and reaches adult levels within a few weeks postpartum.[75,78,80,149,160] It appears that fetal GST activity is not induced when pregnant rats are treated with either DCNB or *p,p*-DDT.[149,160] On the other hand, daily exposure of pregnant rats to styrene was found to be associated with a significant decline in GST activity in the fetal liver on GD 20.[151]

GST is abundant in human fetal liver. However, controversy surrounds the presence of acidic, basic, and neutral forms.[161–163] Mitra et al.[164] purified five GST isozymes from fetal liver (16 to 18 weeks' gestation) cytosol by affinity and ion-exchange HPLC. Two anionic forms with pI values of 5.5 (P-2) and 4.5 (P-3) and one basic form with the pI value of 8.7 (P-6) were clearly separated, and small amounts of two near neutral forms (P-4 and P-5) were also identified. Earlier, Radulovic et al.[165] studied the metabolism of methyl parathion by GST purified from human fetal livers at 14 to 23 weeks of gestation. Methyl parathion was detoxified by O-dealkylation. HPLC, TLC, and radiometry indicated desmethyl parathion as the sole metabolite. Kulkarni et al.[166] reported the GSH conjugation of DBE by human fetal liver GST. Subsequently,[164] a novel model was devised to assess the likelihood of developmental toxicity of DBE in humans. In the proposed *in vitro* model, rat embryos (lacking GST) served as passive targets, and DBE bioactivation was mediated by purified isozymes of human fetal liver GST. All of the tested human fetal liver GST isozymes were found capable of bioactivating DBE. Covalent binding of radioactivity from DBE to DNA and protein suggested DBE bioactivation by GST, which was 144% and 212% higher, respectively, with the P-3 anionic form compared with the P-6 basic isoform. DBE bioactivation by the GST P-3 isoform resulted in developmental toxicity to cultured rat embryos. The results of this study suggest that DBE may be a human developmental toxicant.

8. UDP-Glucuronosyl Transferase

Glucuronide formation in the liver does not occur until after birth in guinea pigs, mice, and humans,[167,168] and is barely measurable with *o*-aminophenol in rat fetuses near term.[160] In rats treated with TCDD on GD 5, fetal liver UGT activity toward *p*-nitrophenol and testosterone was first evident at GD 18 and 22, respectively.[62] This activity increased rapidly, and a 20-fold increase in activity was observed by GD 22. In humans, UGT activity is measurable toward morphine, eostriol, and 2-naphthol in fetal liver microsomes.[169] UGT activity in human fetal liver was low, about 0.1% of adult values with bilirubin,[170] less than 5% with harmol,[171] and approximately 10% with 2-naphthol and morphine.

9. Sulfotransferase

Cappiello et al.[172] conducted a comparative study of human SULT activity in the cytosolic fractions of adult and fetal liver and placenta, using 2-naphthol as a substrate. Activity in fetal liver was approximately 15% of the value for adult liver. Activity of the enzyme, rather than the availability of PAPS, was suggested as the rate-limiting factor in the sulfation reaction. Both SULT and PAPS were higher in fetal liver than in placenta. The rate of sulfation correlated with the rate of 2-naphthol glucuronidation; however, no relationship with gestational age was observed.[169]

F. Fetal Membranes

1. Cytochrome P450

In the rat, the fetal membranes mainly consist of the visceral yolk sac and a fine layer of amnion, and these are difficult to separate.[82] Homogenates of fetal membranes from untreated rats occasionally exhibit detectable levels of AHH activity. Treatment of pregnant rats with BP 24 h prior to sacrifice resulted in a three- to eightfold higher induction of AHH activity in fetal membranes than in placenta.[82]

2. Peroxidative Enzymes

When stimulated by calcium ions, rabbit amnion and splanchnopleure produced a mixture of 11-, 12-, and 15-HETEs via the LO pathway.[173] Metabolite production was greater on GD 20 to 22 than on GD 28 to 30. The cyclooxygenase products of arachidonate were more abundantly produced by these tissues in late pregnancy. Earlier studies suggested the operation of the cyclooxygenase pathway in human amnion[174] under the influence of estrogens.[175] Subsequent studies[99,100] showed the existence of both the cyclooxygenase and LO pathways in human amnion and chorion laeve. The production of leukotrienes by amnion showed no difference when samples were collected after labor at term, after preterm labor, or after cesarean section at term.[100] In contrast to this, another study noted a threefold greater release of PGE_2 by amnion following spontaneous delivery compared with cesarean section at term.[176] The opposite was true for PGE_2 in decidua, while no difference in prostaglandin production was observed for chorion.

Price et al.[177] studied immunohistochemical localization of PGS in human fetal membranes and decidua. Decidualized stromal cells were stained most intensely and consisted of all cell types. Cytotrophoblast, as well as early placental villi and syncytiotrophoblast of all gestational ages, demonstrated lighter, more variable staining patterns. Regardless of gestational age, amnion stained in a heterogeneous fashion, with some cells demonstrating intense staining and others having no staining. A more recent study[178] indicated that tissues obtained after vaginal delivery released significantly higher quantities of leukotrienes than did tissues obtained following cesarean section. The leukotriene release was stimulated by oxytocin. Despite the presence of both PGS and LO, there is no published report on xenobiotic oxidation by these enzymes from human fetal membranes.

G. Placenta

1. Cytochrome P450

a. Nonsmoker's Placenta

Numerous studies on oxidative metabolism of xenobiotics in various preparations of placenta have been reported in the literature, and the information has been critically reviewed.[1–4,8–10,131,132] Based on the results of over 25 years of research, Juchau[10] concluded that the microsomal P450 responsible for xenobiotic oxidation is either undetectable or occurs in extremely small amounts, in the placentas of nonsmoking women. Similar opinions have been echoed by other investigators.[1,3,6,8] Low but measurable deethylation of 7-ethoxycoumarin or 7-ethoxyresorufin, which is routinely observed in the placentas of nonsmokers, may be an attribute of CYP19 responsible for steroid aromatization.[179]

The supportive evidence is weak and inconclusive for the suspected presence of CYP2E1, while the occurrence of other functional forms of P450 enzymes (CYP3A, detectable in first trimester, and CYP4B1) in human term placentas is debatable.[8]

b. Induction

In general, phenobarbital and other barbiturates, as well as ethanol, do not induce placental xenobiotic enzymes at midgestation or at term.[131] The concomitant induction of P450 and xenobiotic monooxygenation activities in human placental microsomes by the components of cigarette smoke has been known for quite some time.[8–10] There is compelling evidence that only CYP1-1 is induced as a result of maternal cigarette smoking. Although several enzyme activities are co-induced to some extent by cigarette smoking, AHH and deethylation of ethoxycoumarin and ethoxyresorufin have received the most attention.[8] Although a satisfactory explanation is still lacking, marked AHH induction due to cigarette smoking during pregnancy is observed at term but not during the first or second trimester.[8–10] Apparently, there is no report on the influence of cigarette smoking during pregnancy on the terminal epoxidation of BP-diol by human placental microsomes. Wong et al.[180] and Lucier et al.[181] studied inducibility of the P450 system in placentas of women exposed to PCBs and dibenzofurans. The results revealed a dramatic elevation (about 100-fold) of AHH activity. This increase in enzyme activity paralleled an increase (150-fold) in the placental microsomal protein related to P450 form 6 (CYP1A1). No other P450 forms were detected. There was an excellent correlation between induction of 7-ethoxyresorufin deethylase and AHH activities. Induction of CYP1A1 and associated enzyme activities has also been noted in human chorionic carcinoma cell line JEG-3 after treatment with 3-MC, BP, or TCDD.[8]

2. Flavin-Containing Monooxygenase

By use of *N,N*-dimethylaniline as a model substrate, the presence of FMO activity was observed in mouse placental microsomes.[64] The activity increased nearly fivefold when measured at GD 18, in comparison with GD 12. Subsequent experiments indicated that *N,N*-dimethylaniline N-oxidase activity is undetectable or absent in microsomes isolated from human term placentas of nonsmokers. Further evaluations with other substrates are needed before definitive conclusions can be reached regarding the existence of this enzyme in human placenta.

3. Prostaglandin Synthase

Earlier reports suggested the presence of PGS activity, albeit at very low levels, in human and rabbit placentas.[6] It was suggested that the low PGS activity in human placenta may, in part, be related to circulating endogenous inhibitors during pregnancy.[182] Although the role played by this enzyme in xenobiotic oxidation has never been rigorously tested in placental microsomes of any animal species, the results of a recent study indicate that neither benzidine oxidation nor the PGS antigen can be detected in microsomal preparations from term placentas.[183]

4. Peroxidase

At least two studies have shown the presence of peroxidase activity in the placentas of mice and rats, measured with guaiacol as the model substrate.[84,85] The results of the mouse study indicated that peroxidase activity does not change during GD 7 through 20.[84]

Earlier peroxidase activity was described in crude human placental homogenates.[184] This activity was detectable in human placenta as early as the sixth week of gestation and increased toward term. Nelson and Kulkarni[47] confirmed the abundance of peroxidase activity in membranes of human term placentas from nonsmokers. They described a procedure for the isolation, solubilization, and partial purification of placental peroxidase free from contaminating heme proteins. The protocol was refined further to obtain the enzyme with electrophoretic homogeneity.[185,186] The modified procedure yields final enzyme preparations displaying over 110-fold purification compared to the initial calcium extract of membranes, with an overall recovery of about 55% of enzyme activity. At present it is not known whether exposure to chemicals present in diet, cigarette smoke, etc., modulates peroxidase activity.

Several studies have now established peroxidase as one of the major pathways of xenobiotic metabolism in human placenta. Available evidence suggests that both membrane-bound and partially purified preparations of the enzyme display the ability to catalyze oxidation of several xenobiotics in the presence of H_2O_2. Hydrogen donors, such as guaiacol, pyrogallol, benzidine, TMBZ, 3,3′-dimethoxybenzidine, TMPD, ABTS, and endogenous chemicals (e.g., bilirubin, epinephrine) are easily oxidized by the enzyme.[47,185] Thiobenzamide, a hepatotoxicant, appears to be one of the best substrates, and its oxidation to a sulfoxide by purified peroxidase was found to proceed at the rate of about 18 μmol/min/mg protein.[186] Spectrophotometric and radiometric studies have yielded data suggesting that 2-AF also serves as an excellent substrate and undergoes peroxidase-mediated oxidation at an impressive rate.[187] It is reported that BP-diol, a proximate human developmental toxicant, is epoxidized to generate the putative ultimate toxicant BP-diol-9,10-epoxide by placental peroxidase.[188,189] The presence of phenylbutazone or amino acids, e.g., tyrosine, was required to observe a maximal rate of the reaction.[188,189]

A few studies[190-192] have reported a positive correlation between the use of phenothiazine drugs during early pregnancy and induction of malformations, while a suggestive relationship of borderline significance in terms of cardiovascular effects was noted in another study.[193] Sobel[194] opined that chlorpromazine induces abortion but causes no birth defects in the babies of mothers who received the drug. It is widely accepted that prior oxidation is essential for the expression of the pharmacological and toxicological effects of phenothiazines. Yang and Kulkarni[195] studied *in vitro* metabolism of seven phenothiazines by purified placental peroxidase. All the phenothiazines tested were found to undergo one-electron oxidation to their respective cation radicals. These relatively stable species can be detected spectrophotometrically. The highest rate of cation radical formation was noted with promazine (approximately 350 nmol/min/mg protein), while the lowest rate was observed with trifluoperazine. More recently, eugenol, a naturally occurring plant compound widely used as a food flavoring agent, an additive to fragrances, cosmetics, and tobacco products, and in medicine, was found to be oxidized to a toxic quinone methide by purified placental peroxidase.[196]

5. Lipoxygenase

The concentration of free arachidonic acid in most tissues is about 20 μM.[27] Therefore, the availability of substrate, arachidonic acid or other PUFAs can be a rate limiting factor for tissue LO activity. However, this may not be an issue for the placenta because free arachidonic acid alone is abundant (about 0.7 mM) in human term placenta.[197] Considering the questionable activities of PGS and P450, it appears that arachidonic acid metabolism in human placenta occurs exclusively via the LO pathway. The studies on leukotriene biosynthesis in human and rabbit placentas[6,100,155,198,199] lend strong support to this contention. Human term placental LO (HTP-LO) has been purified and partially characterized.[199] At present, there is no published report detailing the effect of dietary substances, ethanol consumption, or cigarette smoking during pregnancy on HTP-LO activity.

Aberrant activities of LO and/or PGS in reproductive tissues have been implicated in various pregnancy related hypertension disorders and preeclampsia.[6,32] The liberation of leukotrienes suggests *in vivo* operation of the 5-LO pathway in human placenta. The production of LTB_4 is high in early pregnancy (7 to 12 weeks of gestation) but remains low during the third trimester, with a significant increase during spontaneous labor at term.[198] The observations that plenty of LO activity occurs in HICT at 4 weeks gestation[106] and midpregnancy[104] are in line with this report. No difference in leukotriene release was observed in placentas examined after labor at term, after preterm labor, or after cesarean section.[100] Leukotriene output is low in uncomplicated preterm labor but markedly increases in chorioamnionitis-associated preterm labor. The fact that 5- and 12-HETE are generated when homogenates were incubated with ^{14}C-arachidonate signals that functional 5- and 12-LO in human placenta is present at spontaneous vaginal delivery or after cesarean section.[155,156] In subsequent studies,[100,200] the predominance of the 12-LO pathway was observed. Decreased formation of 5- and 12-HETE from arachidonic acid was noted in preeclampsic placentas.[200]

Recently, cellular oxidative stress was proposed as one of the underlying mechanisms of developmental toxicity of many chemicals.[3,4] Concurrent depletion of cellular GSH plays a major contributory role in this process. GSH provides the first line of defense against excessive production of xenobiotic free radicals within cells. While performing the sacrificial role as an intracellular nucleophilic trap, GSH is extensively consumed. In the absence of GSH participation, an interaction of xenobiotic free radicals with molecular oxygen leads to the liberation of ROS.[3,4] Other investigators have also stressed the role played by ROS in the developmental toxicity and teratogenicity of various chemicals. In this regard, it is noteworthy that although heme proteins, such as P450 and various peroxidases, lack the ability to oxidize GSH, LO has been found to trigger its direct oxidation in the absence of xenobiotics. This phenomenon, first noted with soybean LO in the presence of linoleic acid,[201,202] is also displayed by HTP-LO in reaction media supplemented with different PUFAs.[203] GSH oxidation by LO is accompanied by superoxide anion production.[201] More intense superoxide anion radical generation can be noted during LO-catalyzed linoleate-coupled cooxidation of NAD(P)H,[204] but not of ascorbate,[205] in the absence of oxidizable xenobiotics.

The importance of these findings is further heightened by the demonstration that GS• generated during PUFA-supported GSH oxidation by LO can attack a C=C bond in xenobiotics, and the process ultimately leads to the formation of a GSH-xenobiotic conjugate. This novel mechanism of GSH conjugate formation from xenobiotics has been documented for ethacrynic acid with soybean LO and HTP-LO in the presence of various PUFAs.[202,203] It is noteworthy that the rate of soybean LO-catalyzed ethacrynic acid conjugation with GSH was up to 1650-fold greater than the rates reported in the literature for various preparations of purified GST.[202] This novel reaction with HTP-LO proceeds at a biologically significant rate (about 60% of the maximal velocity noted under optimum assay conditions) under physiologically relevant conditions of pH and concentrations of GSH and arachidonic acid.[203] Further studies have shown that *p*-aminophenol, a developmental toxicant in animals, also rapidly undergoes LO-mediated GSH conjugation to generate an ultimate toxic species.[206]

The study by Joseph et al.[199] has shown that purified HTP-LO cooxidizes pyrogallol, ABTS, benzidine, 3,3′-dimethoxybenzidine, TMPD, TMBZ, and guaiacol in the presence of linoleic acid. The oxidation rate was highest with TMPD and lowest with guaiacol, the most commonly used peroxidase substrate. Hypervitaminosis-A has been linked with birth defects in animals and humans. Several studies have suggested that the parent molecule *per se* may not be teratogenic and bioactivation is essential. The study conducted by Datta and Kulkarni[207] suggested that all-*trans* retinol acetate is an excellent substrate for cooxidation by soybean LO, as well as HTP-LO, in the presence of linoleic acid. The dioxygenase-generated peroxyl radical of linoleic acid was proposed to attack the π-electrons of the C=C bond and bring about epoxidation of the all-*trans* retinol acetate. Earlier, HTP-LO-mediated epoxidation was also shown with the developmental toxicant BP-diol.[106] Significant covalent binding of radiocarbon from BP-diol and a preponderance of *trans*-anti-7,8,9,10-tetrahydrotetrol suggested bioactivation of BP-diol by purified LO isolated from nonsmokers' placenta. These observations clearly explain the literature reports of detection of BP adducts in human term placentas from nonsmokers.[208] ATB_1, a substituted coumarin, is a known animal and human teratogen. Many reports have documented biologically significant levels of ATB_1 in maternal and cord blood,[209,210] ATB_1-DNA adducts in placentas and cord blood,[211] and transplacental carcinogenicity.[212] It is widely accepted that ATB_1 requires prior epoxidation to cause toxicity. The notion that P450 is the sole catalyst for the bioactivation of this toxicant has been questioned.[6] The postulate that LO activates ATB_1 finds strong support in the demonstration that purified adult human liver LO[213] can efficiently perform this reaction. Datta and Kulkarni[107] observed that not only HTP-LO but also LO purified from HICT at early gestation can rapidly oxidize ATB_1 to an electrophilic species, the ATB_1-8,9-epoxide, in the presence of PUFAs. Such findings explain, in part, the developmental toxicity associated with *in utero* exposure to this fungal toxin.

Occupational exposure of workers in the dye industry to arylamines, such as 4-aminobiphenyl (4-ABP) and benzidine, can result in bladder cancer. Currently, there is increasing concern over possible transplacental carcinogenicity associated with gestational active or passive exposure to 4-ABP present in environmental tobacco smoke. Interestingly, the emission of 4-ABP is 30 times greater in sidestream smoke than in mainstream smoke. It crosses the placenta and binds to fetal Hb,[214,215] and 4-ABP adducts have been detected in the fetuses of both smoking and nonsmoking mothers. In rats, the presence of detectable levels of DNA adducts has been noted in all fetal tissues examined following maternal dosing with 4-ABP.[216]

It is widely presumed that prior metabolism is obligatory to observation of 4-ABP toxicity. The known absence of P450, FMO, and PGS triggered interest in examining the possible metabolism of 4-ABP by HTP-LO.[217] Both soybean LO and HTP-LO oxidized 4-ABP to a stable metabolite that was identified by gas chromatography–mass spectrometry (GC-MS) as 4,4′-azobis(biphenyl) when the reaction media were supplemented with either PUFAs or H_2O_2. HTP-LO was superior to soybean LO as a catalyst of the reaction. Datta et al.[217] hypothesized that 4-ABP undergoes one-electron oxidation by peroxy radicals generated by LO from PUFA to produce a free radical species that dimerizes to the coupling azo product. Similar metabolic conversion of 4-ABP by LO from HICT is expected. Collectively, the results suggest that the cooxidation catalyzed by LO may be the underlying biochemical mechanism responsible for the transplacental toxicity of 4-ABP in humans.

As with other xenobiotics, N-dealkylation of drugs is one of the major detoxication reactions. PUFAs support HTP-LO-mediated N-demethylation of phenothiazines, and thus far, the reaction has been noted with promazine, chlorpromazine, promethazine, and trimeprazine.[218] Both soybean LO and HTP-LO can mediate the reaction in the presence of H_2O_2.[219] It is noteworthy that besides linoleic acid and H_2O_2, both cumene hydroperoxide and *tert*-butyl hydroperoxide efficiently couple LO-mediated N-demethylation of phenothiazines.[49] In addition to studies of phenothiazines, animal data are available for other drugs taken during pregnancy. Other drugs that undergo H_2O_2-dependent N-demethylation under the influence of soybean LO include aminopyrine,[220] imipramine, and related tricyclic antidepressants.[221] In addition, model compounds, such as, *N*-methyl aniline, *N,N*-dimethyl aniline, *N,N,N,N*-TMBZ, *N,N*-dimethyl-*p*-phenylenediamine, *N,N*-dimethyl-3-nitroaniline, and *N,N*-dimethyl-*p*-toluidine are demethylated by HTP-LO in the presence of H_2O_2.[222]

Several organophosphate and carbamate pesticides have been reported to possess teratogenic properties, and metabolism plays a major role in determining their developmental toxicity.[6] Currently, the information on this subject is scant for human conceptual tissues, especially placenta. Hu and Kulkarn[223] demonstrated that soybean LO detoxifies several pesticides by mediating N-demethylation in the presence of H_2O_2. In a subsequent study, purified HTP-LO was shown to support cooxidative N-demethylation of aminocarb, chlordimeform, dicrotofos, and zectran at a biologically significant level in the presence of linoleic acid.[218] It was proposed that the initial step in xenobiotic N-dealkylation by LO involves one-electron oxidation of the substrate to produce a corresponding nitrogen-centered cation free radical. The radical undergoes a rapid decay to form an iminium cation, either by deprotonation or hydrogen atom abstraction. Subsequent hydrolysis of the iminium cation finally yields formaldehyde and a monodemethylated product.

6. Lipid Peroxidation-Coupled Cooxidation of Xenobiotics

a. Lipid Peroxidation

Many studies have documented that pregnancy in human is associated with increased *in vivo* lipid peroxidation. Uotila et al.[224] found a 45% increase in serum levels of conjugated dienes when pregnancy advanced from the first to the second trimester. An earlier study[225] reported conjugated diene levels nearly doubled by 36 weeks of gestation, and the values were about 40% greater in women with preeclampsia. The evidence suggests that placenta (and/or other conceptual tissues) may be the major source of lipid peroxidation products during pregnancy, as the values for conjugated dienes return to normal within 24 h of delivery.

In contrast to the compelling *in vivo* evidence, the reports on *in vitro* lipid peroxidation in placental microsomes are inconsistent. Although the occurrence of ascorbate-supported nonenzymatic lipid peroxidation has been repeatedly demonstrated in human placenta,[1,6,93] anomalies exist in the reports of NADPH-supported reactions. Thus Juchau and Zachariah[226] could not detect lipid peroxidation when human term placental microsomes were incubated with NADPH. Similar negative findings were reported by Arvela et al.[227] with placental microsomes from rats, rabbits, and guinea pigs. Kulkarni and Kenel[228] pointed out that previous negative findings[226] may have been due to suboptimal assay conditions employed to evaluate NADPH-dependent lipid peroxidation in human placental microsomes. The presence of chelated iron was found to be essential to the process. In fact, the results suggested a high susceptibility of human placental microsomes to lipid peroxidation because of an apparent lack of endogenous antioxidants in the microsomes. Several reports from other labs[229,230] have now confirmed the occurrence of NADPH-dependent lipid peroxidation in human term placental microsomes and mitochondria.

Many environmental pollutants initiate, promote, or undergo one-electron oxidation under aerobic conditions, and in the presence of lipids they promote lipid peroxidation. Using linoleic acid, the occurrence of the process has been demonstrated for copollutants such as reduced vanadium (vanadyl), hydrated sulfur dioxide (bisulfite), and asbestos fibers (Canadian chrysotile).[231–233] Vanadium accumulates in human placenta at a concentration of about 3 ng/g. When added to placental microsomes in the presence of NADPH, vanadium triggers intense lipid peroxidation.[234,235]

b. Effect of Redox Cycling Agents

It has been proposed that redox cycling of certain nitroheterocycles, quinones, and pesticides, such as nitrophen and paraquat, coupled with NADPH-cytochrome P450 reductase, may represent a serious embryotoxic or fetotoxic threat.[4,6] Mounting evidence points out that lipid peroxidation may be one of the major contributing factors in the developmental toxicity of the agents listed above. This contention finds support in the strong stimulatory effects exerted by paraquat on NADPH-supported lipid peroxidation in human placental microsomes or mitochondria.[230,236,237] NADPH-dependent redox cycling of menadione in placental microsomes also leads to extensive lipid peroxidation.[237]

c. Cooxidation of Xenobiotics

Earlier reports[1,6] have shown that rat liver microsomes undergoing ascorbate- or NADPH-supported lipid peroxidation are capable of cooxidation of xenobiotics. In light of these reports, the possibility of lipid peroxidation-coupled cooxidation of xenobiotics in human term placenta was examined. NADPH- and iron-dependent lipid peroxidation of human term placental microsomes supported cooxidation of both BP and BP-diol to several metabolites.[238] Cooxidation of BP-diol was also observed during placental microsomal lipid peroxidation triggered by vanadium in the presence of NADPH.[235]

7. Other Enzymes of Xenobiotic Oxidation

It should be borne in mind that not all the enzymes responsible for xenobiotic bioactivation of developmental toxicants in human placenta are known. Similarly, the presumption that each teratogen requiring oxidative bioactivation must serve as a substrate for one or more of the known enzymes described above may not always be true. The reported presence of indanol dehydrogenase activity in human term placenta of nonsmokers[239] represents a case in point. Earlier, the results of an *in vitro* test suggested indan and 1-indanone to be teratogenic.[240] Thus, the noted rapid oxidation of 1-indanol to 1-indanone by human placental microsomes in the presence of either NAD⁺ or NADP⁺ represents one step in bioactivation.[239]

8. Epoxide Hydrase

In rabbit and guinea pig placentas, microsomal epoxide hydrase activity toward styrene oxide is near the limit of detection and exhibits a downward trend as pregnancy advances.[75] In humans, such activity is detectable in placenta as early as 8 weeks gestation. Apparently, the enzyme matures early during the first trimester since no gestational increase in epoxide hydrase activity was observed later.[241]

9. UDP-Glucuronosyl Transferase

Lucier et al.[62] measured placental UGT activity toward *p*-nitrophenol on GD 20 in rats. The activity was induced threefold when pregnant rats were pretreated with TCDD on GD16, but not when animals were treated on GD 5 or 10. The ability of rabbit placenta to form oxazepam glucuronide *in vitro* was nearly doubled on GD 14 by pretreatment with phenobarbital but was not affected by oxazepam.[242]

10. Sulfotransferase

Cappiello et al.[172] measured SULT activity in human placentas at 11 to 27 weeks of gestation. The values obtained represent about 6% of the activity observed in fetal liver. Placental PAPS content was estimated at 3.6 nmol/g tissue. The authors opined that the availability of SULT and not PAPS is the rate-limiting factor for sulfation of xenobiotics in human placenta.

11. Glutathione Transferase

a. General

The occurrence of GST activity has been noted in the placentas of rabbits, guinea pigs, mice, rats, and monkeys.[1,6] In general, the activity measured with different substrates was found to decline gradually toward term. Earlier, conventional methods to purify human placental GST met with limited success. Radulovic and Kulkarni[52] were the first to introduce a novel, rapid method combining affinity chromatography and HPLC for the purification of this GST. Their procedure resulted in high yields of electrophoretically homogeneous enzyme, with about 1500-fold purification. The enzyme, now identified as GSTP1-1, can exist in two charge isomers. These forms can be separated by HPLC and are interconvertible, depending upon the concentration of GSH in the surrounding medium.[243] Since the tissue concentration of GSH was estimated as 0.11 mM, it was postulated that the enzyme predominantly exists as a high-affinity–low-activity conformer *in vivo*. Preterm placental GST activity toward CDNB has been found to exhibit a twofold variation during weeks 13 to 23 of gestation.[244] GST activity toward styrene oxide remains fairly constant between 8 and 25 weeks of gestation, followed by a decline of about 50% at full term.[245] A significant inhibition of the enzyme's activity toward CDNB and other model substrates can be observed at relatively low concentrations of Hb,[53] PUFAs, and their esters,[246] a variety of retinoids,[247] and nitrobenzenes.[248]

Besides being able to metabolize CDNB, DCNB, *p*-nitrobenzyl chloride, 4-nitropyridine-*N*-oxide, and ethacrynic acid,[243–248] the purified GST1-1 can metabolize methyl parathion, an animal embryotoxicant and teratogen.[244] HPLC analyses revealed the presence of desmethyl parathion as the sole metabolite from O-dealkylation. The metabolism of DBE by purified GST1-1 generates reactive species that bind covalently to protein and DNA.[249] Rat embryos exhibit an inhibition of growth and development when DBE is bioactivated in the culture medium by purified human placental GST.

b. Induction

It appears that human placental GST activity is not altered following exposure of women to the constituents of tobacco smoke.[245,250] Studies with laboratory animals have also indicated the refractive nature of placental GST to the effects of such inducers as *trans*-stilbene oxide,[76] phenobarbital, AAF, butylated hydroxyanisole, and 3-methyl-4-dimethylaminobenzene.[251]

V. ROLE OF XENOBIOTIC METABOLISM IN REPRODUCTIVE TOXICOLOGY

A. General

Concerns regarding possible ill effects of chemicals present in the working and living environment on human reproduction have markedly increased over the years. The pathways of hormone biosynthesis, gametogenesis, and secretory functions of accessory sex glands represent some of the obvious targets for reproductive toxicants. A multitude of mechanisms are expected to be responsible for reproductive failure. Unfortunately, in most cases, cause and effect relationships are not firmly established. Although biotransformation of toxicants is fairly well understood, especially in animals, there is paucity of information on the metabolism of toxicants in the affected cell types and the identity of the ultimate toxicants responsible for toxic insults.

B. Male Reproductive System

1. Testicular Enzymes of Xenobiotic Metabolism

The presence of P450, AHH activity, epoxide hydrase, and glutathione transferase activities in the testes of laboratory animals has been documented.[252–254] These enzymes are induced following exposure to TCDD. Seminal vesicle microsomes represent a rich source of PGS,[27] and LO activity has been detected in testis, vesicular gland, prostate, Ledig cells, and spermatozoa.[6]

2. Male Reproductive Toxicants

Drugs, pesticides, environmental chemicals and metals have been identified as potential testicular toxicants. Affected men exhibit loss of libido, testicular atrophy, abnormal sperm morphology, low numbers of competent spermatozoa, or dysfunctions related to erection, ejaculation, and secretion by accessory sex glands.

Although the biological significance is currently unknown, low levels of residues of currently used organophosphorus and carbamate pesticides and those of now abandoned organochlorine insecticides have been detected in food and water, human tissues, blood, and milk.[255] Organochlorines (e.g., DDT, dieldrin, heptachlor), carbamates (e.g., carbaryl), organophosphates (e.g., parathion), and other pesticides (e.g., DBE and 1,2-dibromo-3-chloropropane, DBCP) have been suspected of causing testicular injury in animals, although supportive evidence for humans is available for only a few. The literature on P450-dependent oxidation of pesticides has been reviewed.[19,21,26] The nematocide DBCP is probably the first example linking chemical exposure under an occupational setting to testicular toxicity in workers.[256] Oligospermia, azoospermia, and increased FSH and LH levels were noted in exposed men. Following discontinuation of exposure, reversal of DBCP toxicity was noted in oligospermic but not in azoospermic men. In rats, GSH conjugates are formed from DBCP in liver, kidney, and testis cytosol.[254] The identification of the ultimate toxicant has been illusive,[257] but an Ames assay suggested that oxidative metabolism is needed to generate genotoxic metabolite(s) from DBCP.[257]

Chlordecone (kepone) causes reversible testicular toxicity and low sperm count, decreased sperm motility, abnormal morphology, and arrest in sperm maturation in exposed men.[258] In humans, as in animals, a small portion of kepone in the body is reduced to its alcohol or converted to glucuronides and/or GSH conjugates before fecal excretion. Whether kepone itself or one of its metabolites is the reproductive toxicant is not yet established. Among pesticides, linking exposures to Agent Orange (a 1:1 mixture of 2,4-D and 2,4,5-T herbicides) sprayed in Vietnam to systemic illnesses in soldiers and increased incidences of miscarriages, stillbirths, and birth defects has been frustrating, illusive, and hotly debated for the past 30 years. Many conflicting reports have appeared in the literature. It has been suspected that TCDD contamination, which occurred during the manufacture of 2,4,5-T, may be a culprit.

Testicular cancer due to exposure to soot in chimney sweeps was documented long ago. Since then, several polycyclic aromatic hydrocarbons (PAHs) are suspected to cause testicular toxicity. Many studies have documented that smoking adversely affects testicular function in men. A significant decrease in sperm count and motility and an increase in abnormal sperm morphology have been documented. Although over 3000 different chemicals occur in the cigarette smoke, PAHs and a few other compounds are the suspected toxicants. The blood-testis barrier does not prevent exposure of the testis to PAHs, such as BP. P450-catalyzes epoxidation of BP to the proximate BP-diol. Further oxidation forms the ultimate 7,8-diol-9,10-epoxide, which is believed to exert toxicity. Evidence gathered in rats by use of perfusion of the testis and testis homogenates indicates that testicular enzymes can activate BP as well as detoxify toxic reactive intermediates.[252,253] The plasticizer diethylhexyl phthalate and its metabolite monoethylhexyl phthalate, resulting from dealkylation, are testicular toxicants in rodents.[259] They cause sloughing of spermatids and spermatocytes and vacuolization of Sertoli cell cytoplasm. 2,5-Hexanedione, a metabolite of *n*-hexane, is a testicular toxicant,[260] but it is not known whether testicular P450 mediates this reaction. Ethylene glycol monoethyl ether and its metabolite, 2-methoxy ethanol (2-ME), are testicular toxicants.[261] 2-Methoxyacetic acid, the oxidation product of 2-ME generated by sequential oxidation by alcohol dehydrogenase and acetaldehyde dehydrogenase, is believed to be the ultimate testicular toxicant.

Among drugs, several agonists or antagonists of natural hormones, antineoplastic agents, anesthetics, and antibiotics are known to produce adverse effects on the male reproductive system. Cyclophosphamide is a therapeutic agent commonly used for the treatment of boys with nephrotic syndrome and Hodgkin's disease. The results of a few studies have suggested that high drug exposure, especially later in the puberty, is associated with elevated risks of decreased spermatogenesis. Although the enzymes responsible for cyclophosphamide bioactivation in testis are not known, published data suggest that P450, PGS, and LO are capable of its conversion to acrolein and phosphoramide mustard.[72,73] Among antipsychotic agents, chlorpromazine has been shown to increases the rate of prolactin secretion that may be associated with breast engorgement in men. Chlorpromazine is metabolized by several heme containing proteins and LO. Therapeutic use of imipramine and amitriptyline causes delayed orgasm in some men, while other antidepressants have been shown to suppress spermatogenesis. Imipramine and related tricyclic compounds are known to be metabolized by P450 and flavin-containing monooxygenase, as well as by LO.[219]

C. Female Reproductive System

1. Ovarian Enzymes of Xenobiotic Metabolism

The evidence suggests that the ovary has the full complement of enzymes necessary for metabolism of xenobiotics. Thus, the demonstration of AHH activity and BP metabolism indirectly suggest the presence of P450 and epoxide hydrase activities in the murine ovary.[253] Various reports have shown active prostaglandin synthesis in animal ovaries, while the LO pathway is operational in mammalian ovary, preovulatory follicles, and oocytes.[6] Other metabolically competent reproductive tissues in close proximity to the embryo, such as uterus, placenta, fetal membranes, etc., are also expected to contribute to biotransformation of reproductive toxicants.

2. Female Reproductive Toxicants

In women, destruction of ova; interference with ovulation, fertilization, or implantation; early, late, or missed abortions, etc., are some of the lead indicators of unfavorable reproductive outcomes. Pesticide exposure is essentially unavoidable for women engaged in the agriculture industry. Epidemiological surveys have noted a positive correlation between the residues of organochlorine pesticides in maternal or cord blood, placenta, and fetal tissues and the incidence of spontaneous abortion, missed abortion, fetal prematurity, induction of early labor, premature delivery, and stillbirth.[262] Recent recognition that methoxychlor, dieldrin, o,p-DDT, toxaphene, and endosulfan may act as endocrine disruptors has renewed interest in the reproductive toxicology of pesticides. Some data for other pesticides are available.[263] Many of these pesticides act as estrogen agonists and/or androgen antagonists. Prochloraz reacts as both an estrogen and an androgen antagonist.[263]

Epoxidation of aldrin to dieldrin has long been employed as a model reaction catalyzed by P450[19–21] and has been noted with PGS.[21] The demonstration that soybean LO can mediate this reaction suggests LO as an additional pathway for this metabolic step.[264] Heptachlor, isodrin, and chlordene also follow this path of metabolism.[19] Methoxychlor itself is not estrogenic, but its metabolites are.[265] Human liver microsomes catalyze primarily mono-O-demethylation of methoxychlor, and CYP2C19, CYP2C18, and CYP1A2 are more active in this catalysis. Further metabolism occurs during longer incubation. The second O-demethylation yields bis-hydroxy methoxychlor, which is a more potent estrogen than mono-hydroxy methoxychlor. CYP2B6 and CYP1A2 mediate ortho-hydroxylation, whereas CYP3A4 is most active in ortho-hydroxylation of mono- and bis-demethylated methoxychlor.[265]

Chronic exposure to organophosphates causes abnormal menstruation, amenorrhea, and induction of early menopause.[266] Animal studies have revealed reproductive toxicity and teratogenicity of several organophosphorous and carbamate insecticides. Metabolic catalysis of these pesticides has been extensively studied in animals; however, data are very limited for human tissues.[19–21] Desulfuration catalyzed by P450, FMO,[19–21,26] and LO[267] forms their oxons, which are more potent anticholinesterases. The pesticides lose their potency when metabolized by other pathways. Detoxication by dealkylation occurs via P450[19–21] and LO.[218,223]

Many PAHs can cause ovarian toxicity.[253] Thus, BP, 3-MC, and DMBA all cause destruction of primordial oocytes in DBA/2N and C57BL/6N inbred mice. In rodents, the process is species, strain, age, dose, and metabolism dependent. However, oocyte destruction does not appear to be linked to the *Ah* locus or AHH activity. For example, conversion of BP to BP-diol epoxide is essential for toxicity. Although P450 was believed to be the catalyst responsible for this bioactivation,[253] possible involvement of other enzymes cannot be ruled out at present because PGS,[27,28] soybean LO, and animal tissue LOs[106,268,269] can perform this final activation. It is highly likely that ovarian PGS and/or LO also bioactivates BP and other PAHs.

Several epidemiological studies have found cigarette smoking in women to decrease the age of spontaneous menopause, suggesting oocyte destruction. BP has been the prime suspect; however, involvement of other chemicals is likely. Although specific details are sketchy, arylamines, such as 4-ABP, benzidine, and 2-AF, may also contribute to female reproductive toxicity. These chemicals are oxidized by P450, PGS, and LO.[27,28,217,270] Despite the obvious importance of this subject, studies on these toxicants with human ovarian tissue enzymes are still lacking.

Acrylonitrile exposure in polymer industries seems to be associated with abnormal menses. Animal studies have revealed that acrylonitrile is a developmental toxicant and a teratogen.[271] They also point out that the ultimate toxicant may be the epoxide and/or cyanide released during acrylonitrile oxidation. Earlier suggestions on the role played by P450 in acrylonitrile bioactivation have been questioned.[272] The data point out the ability of LO to convert acrylonitrile to its epoxide and cyanide.[272] A few reports have similarly indicated reproductive toxicity for styrene and vinyl chloride. Metabolic data for these chemicals are not currently available for human female reproductive tissues.

DES is a synthetic estrogen that was used to prevent miscarriage. DES is carcinogenic in animals, and its human use has been linked to the occurrence of vaginal and cervical clear cell adenocarcinoma in young women. *In utero* exposure to DES has been associated with a high incidence of vaginal adenosis, cervical ectropion, and structural abnormalities of the uterus and vagina in girls, and with epididymal cysts, hypoplastic testes, varicoceles, and spermatozoal defects in boys.[273] The results of several studies have led to the conclusion that estrogenic activity alone is insufficient to explain DES toxicity. It is now established that DES oxidation to Z,Z-dienestrol is required. Horseradish peroxidase, uterine peroxidase, and PGS, but not P450, were shown to be capable of this reaction.[27,28,273] Recently, soybean LO has been found to activate DES by a one-electron oxidation process.[274] It was proposed that two molecules of the initial DES oxidation product, DES-semiquinone, undergo dismutation to generate DES-quinone and DES. The quinone metabolites can bind irreversibly to cellular macromolecules and may be the ultimate toxicants. At present, there are no reports on human female reproductive tissue enzymes catalyzing DES activation to explain the toxicity reported in boys and girls.

Although intake of phenothiazines during pregnancy appears to be relatively safe, conflicting reports on toxicities do exist in the literature. Sobel[194] suggested that chlorpromazine induces abortions but causes no birth defects, while Slone et al.[193] noted a positive relationship between antenatal phenothiazine use and intelligence quotient score, birth defects, and perinatal mortality rate. Phenothiazines are teratogenic in laboratory animals. Several oxidative enzymes can metabolize these compounds. Recent studies with soybean LO have revealed another metabolic complexity with possible unfavorable pharmacological or toxicological outcomes.[275,276] It was demonstrated that the chlorpromazine cation radical, the product of one-electron oxidation of chlorpromazine by LO, can serve as a shuttle oxidant in a secondary reaction that stimulates oxidation of benzidine and other xenobiotics several-fold. The shuttle oxidant production was also noted with other phenothiazines.[275,276] Antiepileptic drug therapy is associated with reproductive dysfunction. For example, the anticonvulsant drug phenytoin has been reported to cause hirsutism in young women. As pointed out earlier, phenytoin is metabolized by P450 and PGS, as well as by LO.[3] It remains to be established whether phenytoin itself or a metabolite generated by human female reproductive tissues is responsible for the outcome.

VI. CONCLUSIONS AND FUTURE RESEARCH NEEDS

Because of ethical and legal reasons, few controlled *in vitro* xenobiotic metabolism studies have been conducted with human reproductive tissues. Unfortunately, the paucity of human data dictates continued heavy reliance on information generated from laboratory animals. The thalidomide tragedy has taught us a lesson by clearly demonstrating that no animal species can be a complete surrogate for humans. It is therefore extremely important to consider both qualitative and quantitative species-specific differences in xenobiotic metabolizing enzymes between the laboratory animal species commonly employed in developmental and reproductive toxicity testing and humans. Notable among these are P450 and GST. The specific P450 content and the xenobiotic oxidizing ability of adult human liver are significantly lower than is true of rodents. However, the reverse is true when one considers fetal liver. Significant amounts of this hemoprotein occur in human fetal liver, while it is barely detectable in rodent fetal liver late in gestation. Further, the GST content of human fetal liver equals or even exceeds adult levels, while this enzyme is barely detectable in rodent fetal liver, and the same is true for placental GST. Significant commonality resides in the existence of peroxidative enzymes in the conceptal tissues of both humans and rodents. In the placentas of nonsmokers, functional PGS and P450 are essentially absent. A relative abundance of peroxidase and LO strongly suggest that they may be primarily responsible for oxidative xenobiotic biotransformation in human placenta. Although our understanding of the *in vivo* metabolism of developmental and reproductive toxicants provides some comfort, a wide gap remains with regard to the cause-and effect-relationships involved in toxicity, the chemical identities of the ultimate toxicants, and the subcellular sites of their biosynthesis in human target tissues.

References

1. Kulkarni, A.P., Role of xenobiotic metabolism in developmental toxicity, in *Handbook of Developmental Toxicology*, Hood, R.D., Ed., CRC Press, Boca Raton, FL, 1997, chap.12.
2. Miller, M.S., et al., Drug metabolic enzymes in developmental toxicology, *Fundam. Appl. Toxicol.*, 34, 165, 1996.
3. Wells, P.G. and Winn, L.M., Biochemical toxicology of chemical teratogenesis, *Crit. Rev. Biochem. Mol. Biol.*, 31, 1, 1996.
4. Wells, P.G., et al., Oxidative damage in chemical teratogenesis, *Mut. Res.*, 396, 65, 1997.
5. Juchau, M.R., Boutelet-Bochan, H., and Huang, Y., Cytochrome-P450-dependent biotransformation of xenobiotics in human and rodent embryonic tissues, *Drug. Met. Rev.*, 30, 541, 1998.
6. Kulkarni, A.P., Role of biotransformation in conceptal toxicity of drugs and other chemicals, *Curr. Pharmaceut. Designs*, 7, 833, 2001.
7. Kitada, M. and Kamataki, T., Cytochrome P450 in human fetal liver: Significance and fetal-specific expression, *Drug Met. Rev.*, 26, 305, 1994.
8. Hakkola, J., et al., Xenobiotic-metabolizing cytochrome P450 enzymes in the human feto-placental unit: role in intrauterine toxicity, *Crit. Rev. Toxicol.*, 28, 35, 1998.
9. Pasanen, M. and Pelkonen, O., The expression and environmental regulation of P450 enzymes in human placenta, *Crit. Rev. Toxicol.*, 24, 211, 1994.
10. Juchau, M.R., Placental enzymes: cytochrome P450s and their significance, in *Placental Toxicology*, Rama Sastry, B.V., Ed., CRC Press, Boca Raton, FL, 1995, p. 197.
11. LeBlanc, G.A. and Dauterman, W.C., Conjugation and elimination of toxicants, in *Introduction to Biochemical Toxicology*, 3rd Ed., Hodgson, E. and Smart, R.C., Eds., Wiley Interscience, New York, 2001, chap. 6.
12. Ioannides, C., Xenobiotic metabolism. An overview, in *Enzyme Systems that Metabolise Drugs and Other Xenobiotics*, Ioannides, C., Ed., Wiley, New York, 2002, chap.1.
13. Anders, M.W. and Dekant, W., Glutathione-dependent bioactivation of haloalkenes, *Annu. Rev. Pharmacol. Toxicol.*, 38, 501, 1998.
14. Nelson, D.R., et al., P450 superfamily: update on new sequences, gene mapping, accession numbers and nomenclature, *Pharmacogenetics*, 6, 1, 1996.
15. Guengerich, F.P., Comparisons of catalytic selectivity of cytochrome P450 subfamily members from different species, *Chem.-Biol. Interact.*, 106, 161, 1997.
16. Ryan, D.E. and Levin, W., Purification and characterization of hepatic microsomal cytochrome P-450, *Pharmacol. Ther.*, 45, 153, 1990.
17. Waterman, M.R., Cytochrome P450: cellular and structural considerations, *Curr. Opinions Struct. Biol.*, 2, 384, 1992.
18. Juchau, M.R., Substrate specificities and functions of the P540 cytochromes, *Life Sci.*, 47, 2385, 1990.
19. Kulkarni, A.P. and Hodgson, E., The metabolism of insecticides: the role of monooxygenase enzymes, *Annu. Rev. Pharmacol. Toxicol.*, 24, 19, 1984.
20. Kulkarni, A.P. and Hodgson, E., Metabolism of insecticides by mixed function oxidase systems, in *Differential Toxicities of Insecticides and Halogenated Aromatics, International Encylopedia of Pharmacology and Therapeutics, Section 113*, Matsumura, F., Ed., Pergamon Press, New York, 1984, chap. 2.
21. Kulkarni, A.P., Naidu, A.K., and Dauterman, W.C., Oxidative metabolism of insecticides, in *Recent Advances in Insect Physiology and Toxicology*, Gujar, G.T., Ed., Agricole Publishing Acad., New Delhi, India, 1993, chap. 13.

22. Batt, A.M., Enzyme induction by drugs and toxins, *Clin. Chim. Acta*, 209, 109, 1992.
23. Halpert, J.R. and Guengerich, F.P., Enzyme inhibition and stimulation, in *Biotransformation, Vol. 3, Comprehensive Toxicology*, Guengerich, F.P., Ed., Elsevier Science, Oxford, 1997, chap.4.
24. Smith, D.A., Species differences in metabolism and pharmacokinetics: Are we close to an understanding? *Drug Met. Rev.*, 23, 355, 1991.
25. Cashman, J.R., Human flavin-containing monooxygenase: substrate specificity and role in drug metabolism, *Curr. Drug Met.*, 1, 1037, 2000.
26. Hodgson, E., et al., Flavin-containing monooxygenase and cytochrome P450-mediated metabolism of pesticides: from mouse to human, *Rev. Toxicol.*, 6, 231, 1998.
27. Smith, W.L. and Marnett, L.J., Prostaglandin endoperoxide synthase: structure and catalysis, *Biochim. Biophys. Acta*, 1083, 1, 1991.
28. Degan, G.H., Vogel, C., and Abel, J., Prostaglandin synthases, in *Enzyme Systems that Metabolise Drugs and Other Xenobiotics*, Ioannides, C., Ed., John Wiley, Chichester, UK, 2002, chap.6.
29. Kulkarni, A.P., Lipoxygenases, in *Enzyme Systems that Metabolise Drugs and Other Xenobiotics*, Ioannides, C., Ed., John Wiley, Chichester, UK, 2002, chap.7.
30. Kulkarni, A.P., Lipoxygenase — a versatile biocatalyst for biotransformation of endobiotics and xenobiotics, *Cell. Mol. Life Sci.*, 58, 1805, 2001.
31. Kulkarni, A.P., Lipoxygenase-mediated metabolism of xenobiotics: relevance to toxicology and pharmacology, *Recent Res. Dev. Drug Met. Disp.*, in press, 2002.
32. Zahradnik, H.P., et al., Hypertensive disorders in pregnancy. The role of eicosanoids, *Eicosanoids*, 4, 213, 1991.
33. Bock, K.W., UDP-glucuronosyltransferases, in *Enzyme Systems that Metabolise Drugs and Other Xenobiotics*, Ioannides, C., Ed., John Wiley, Chichester, UK, 2002, chap.8.
34. Glatt, H., Sulfotransferases, in *Enzyme Systems that Metabolise Drugs and Other Xenobiotics*, Ioannides, C., Ed., John Wiley, Chichester, UK, 2002, chap.10.
35. Sherratt, P.J. and Hayes, J.D., Glutathione transferases, in *Enzyme Systems that Metabolise Drugs and Other Xenobiotics*, Ioannides, C., Ed., John Wiley, Chichester, UK, 2002, chap. 9.
36. Jakobsson, P.J., et al., Membrane associated proteins in eicosanoid and glutathione metabolism (MAPEG). A widespread superfamily, *Am. J. Resp. Crit. Care Med.*, 161, S20, 2000.
37. Wolff, T., Erhard, D., and Wanders, H., Aldrin epoxidation, a highly sensitive indicator specific for cytochrome P450-dependent monooxygenase activities, *Drug Met. Disp.*, 7, 301, 1979.
38. Omura, T. and Sato, R., The carbon monoxide binding pigment of liver microsomes. I. Evidence for its hemoprotein nature, *J. Biol. Chem.*, 239, 2370, 1964.
39. Matsubara, T., et al., Quantitative determination of cytochrome P450 in rat liver homogenate, *Anal. Biochem.*, 75, 596, 1976.
40. Ziegler, D.M., Microsomal flavin-containing monooxygenase: oxygenation of nucleophilic nitrogen and sulfur compounds, in *Enzymatic Basis of Detoxication*, Vol. I, Jakoby, W.B., Ed., Academic Press, New York, chap. 9.
41. Kulkarni, A.P. and Cook, D.C., Hydroperoxidase activity of lipoxygenase: a potential pathway for xenobiotic metabolism in the presence of linoleic acid. *Res. Comm. Chem. Pathol. Pharmacol.*, 61, 305, 1988.
42. Hover, C. and Kulkarni A.P., A simple and efficient method for hemoglobin removal from mammalian tissue cytosol by zinc sulfate and its application to the study of lipoxygenase, *Prost. Leuk. Essential Fatty Acids*, 62, 97, 2000.
43. Byczkowski, J.Z., Ramgoolie, J., and Kulkarni, A.P., Some aspects of activation and inhibition of rat brain lipoxygenase, *Int. J. Biochem.*, 24, 1691, 1992.
44. Kulkarni, A.P., et al., Hydrogen peroxide: a potent activator of dioxygenase activity of soybean lipoxygenase, *Biochem. Biophys. Res. Commun.*, 166, 417, 1990.
45. Kulkarni, A.P., et al., Dioxygenase and peroxidase activities of soybean lipoxygenase: Synergistic interaction between linoleic acid and hydrogen peroxide, *Res. Commun. Chem. Pathol. Pharmacol.*, 66, 287, 1989.
46. Maehly, A.C. and Chance, B., Assay of catalases and peroxidases, *Methods Biochem. Anal.*, 1, 385, 1954.
47. Nelson, J.L. and Kulkarni, A.P., Partial purification and characterization of a peroxidase activity from human placenta, *Biochem. J.*, 268, 739, 1990.

48. Kulkarni, A.P. and Cook, D.C., Hydroperoxidase activity of lipoxygenase: hydrogen peroxide-dependent oxidation of xenobiotics, *Biochem. Biophys. Res. Commun.*, 155, 1075, 1988.
49. Hover, C.G. and Kulkarni, A.P., Hydroperoxide specificity of plant and human tissue lipoxygenase: an in vitro evaluation using N-demethylation of phenothiazines. *Biochim. Biophys. Acta*, 1475, 256, 2000.
50. Oesch, F., Jerina, D.M., and Daly, J.A., A radiometric assay for hepatic epoxide hydrase activity with [7–3H]styrene oxide, *Biochim. Biophys. Acta*, 227, 685, 1971.
51. Habig, W.H., Pabst, M.J., and Jakoby, W.B., Glutathione S-transferase, the first enzymatic step in mercapturic acid formation, *J. Biol. Chem.*, 249, 7130, 1974.
52. Radulovic, L.L. and Kulkarni, A.P., A rapid, novel high performance liquid chromatography method for the purification of glutathione S-transferase: An application to the human placental enzyme, *Biochem. Biophys. Res. Commun.*, 128, 75, 1985.
53. Sajan, M. and Kulkarni, A.P., A simple and rapid method for hemoglobin removal from mammalian tissue cytosol by zinc sulfate and its application to the study of glutathione transferase, *Toxicol. Methods*, 7, 55, 1997.
54. Osimitz, T.G. and Kulkarni, A.P., Hepatic microsomal oxidative metabolism of pesticides and other xenobiotics in pregnant CD1 mice, *Pestic. Biochem. Physiol.*, 23, 328, 1985.
55. Symons, A.M., Turcan, R.C., and Parke, D.V., Hepatic drug metabolism in pregnant rat, *Xenobiotica*, 12, 365, 1982.
56. Miller, M.S., et al., Expression of the cytochrome P-4502E and 2B gene families in the lungs and livers of nonpregnant, pregnant and fetal hamsters, *Biochem. Pharmacol.*, 44, 797, 1992.
57. Turcan, R.G., et al., Drug metabolism, cytochrome P-450 spin state and phospholipid changes during pregnancy in the rat, *Biochem. Pharmacol.*, 30, 1223, 1981.
58. Lambert, G.H., Lietz, H.W., and Kotake, A.N., Effect of pregnancy on the cytochrome P-450 system in mice, *Biochem. Pharmacol.*, 36, 1965,1987.
59. Williams, D.E., et al., A prostaglandin w-hydroxylase cytochrome P450 (P-450$_{PG-w}$) purified from lungs of pregnant rabbits, *J. Biol. Chem.*, 259, 14600, 1984.
60. Cambell, R.M., Fell, B.F., and Mackie, W.S., Ornithine decarboxylase activity, nucleic acids, and cell turnover in the livers of pregnant rats, *J. Physiol. (London)*, 241, 699, 1974.
61. Combes, B., et al., Alterations in sulphobromophthalein sodium removal mechanism from blood during normal pregnancy, *J. Clin. Invest.*, 42, 1431, 1963.
62. Lucier, G.W., et al., Postnatal stimulation of hepatic microsomal enzymes following administration of TCDD to pregnant rats, *Chem.-Biol. Interact.*, 11, 15, 1975.
63. Jannettl, R.A. and Anderson, L.M., Dimethylnitrosamine demethylase activity in fetal, suckling, and maternal mouse liver and its transplacental and transmammary induction by polychlorinated biphenyls, *J. Natl. Cancer Inst.*, 67, 461, 1981.
64. Osimitz, T.G. and Kulkarni, A.P., Oxidative metabolism of xenobiotics during pregnancy: significance of flavin-containing monooxygenase, *Biochem. Biophys. Res. Commun.*, 109, 1164, 1982.
65. Osimitz, T.G. and Kulkarni, A.P., Polyamine effects on cytochrome P-450 and flavin containing monooxygenase-mediated oxidation of xenobiotics, *Drug Met. Disp.*, 13, 197, 1985.
66. Weitman, S.D., Vidicnik, M.J., and Lech, J.J., Role of pregnancy on the activation and detoxication of parathion, *Pharmacologist*, 24, 93, 1982.
67. Weitman, S.D. and Lech, J.J., Influence of pregnancy on the hepatic metabolism of parathion, *Toxicologist*, 3, A78, 1983.
68. Hulla, J.E. and Juchau, M.R., Occurrence and inducibility of cytochrome P450IIIA in maternal and fetal rats during prenatal development, *Biochemistry*, 28, 4871, 1989.
69. Greenaway, J.C., et al., The *in vitro* teratogenicity of cyclophosphamide in rat embryos, *Teratology*, 25, 335, 1982.
70. Juchau, M.R., et al., Effect of 3-methylcholanthrene and phenobarbital on the capacity of embryos to bioactivate teratogens during organogenesis, *Toxicol. Appl. Pharmacol.*, 80, 137, 1985.
71. Juchau, M.R., et al., Generation of reactive dysmorphogenic intermediates by rat embryo in culture: effects of cytochrome P450 inducers, *Toxicol. Appl. Pharmacol.*, 81, 533, 1985.
72. Kanekal, S. and Kehrer, J.P., Metabolism of cyclophosphamide by lipoxygenases, *Drug Met. Disp.*, 22, 74, 1994.
73. Kanekal, S. and Kehrer, J.P., Evidence for peroxidase mediated metabolism of cyclophosphamide, *Drug Met. Disp.*, 21, 37, 1993.

74. Devereux, T.R. and Fouts, J.R., Effect of pregnancy or treatment with certain steroids on N,N-dimethylaniline demethylation and N-oxidation by rabbit liver and lung microsomes, *Drug Met. Disp.*, 3, 254, 1975.
75. James, M.O., et al., The perinatal development of epoxide metabolizing enzymes activities in liver and extrahepatic organs of guinea pigs and rabbits, *Drug Met. Disp.*, 5, 19, 1977.
76. Kulkarni, A.P., et al., Glutathione S-transferase activity during pregnancy in the mouse: effects of trans-stilbene oxide pretreatment, *Biochem. Pharmacol.*, 33, 3301,1984.
77. Polidoro, G., et al., Effect of pregnancy on hepatic glutathione S-transferase activities in the rat, *Biochem. Pharmacol.*, 30, 1859, 1981.
78. Combes, B. and Stakelum, G.S., Maturation of sulfobromophthalein sodium-glutathione conjugating system in rat liver, *J. Clin. Invest.*, 41, 750, 1962.
79. Mitra, A.K., et al., Rat hepatic glutathione S-transferase mediated embryotoxic bioactivation of ethylenedibromide, *Teratology*, 46, 439, 1992.
80. Bell, J.V., Hansell, M.M., and Ecobichon, D.J., The influence of phenobarbitone on maternal and perinatal hepatic drug metabolizing enzymes in the rat, *Can. J. Physiol. Pharmacol.*, 53, 1147, 1975.
81. Rouet, P., et al., Metabolism of benzo(a)pyrene by brain microsomes of fetal and adult rats and mice. Induction by 5,6-benzoflavone, comparison with liver microsomal activities, *Carcinogenesis*, 2, 919, 1981.
82. Schlede, E. and Merker, H.J., Benzo(a)pyrene hydroxylase activity in the whole implantation site, decidua, placenta and fetal membranes of the rat, *Naunyn Schmiedebergs Arch. Pharmacol.*, 282, 59, 1974.
83. Keeping, H.S. and Lyttle, C.R., Monoclonal antibody to rat uterine peroxidase and its use in identification of the peroxidase as being of eosinophil origin, *Biochim. Biophys. Acta*, 802, 399, 1984.
84. Nelson, J.L. and Kulkarni, A.P., Peroxidase activity in the mouse uterus, placenta and fetus during pregnancy, *Biochem. Internat.*, 13, 131, 1986.
85. Agrawal, P. and Laloraya, M.M., Ascorbate and peroxidase changes during pregnancy in albino rat and Swiss mouse, *Am. J. Physiol.*, 236, E386, 1979.
86. McNabb, T., Sproul, J., and Jellinck, P.H., Effects of phenols on the oxidation of estradiol by uterine peroxidase, *Can. J. Biochem.*, 53, 855, 1975.
87. Byczkowski, J.Z. and Kulkarni, A.P., Activation of benzo(a)pyrene-7,8-dihydrodiol in rat uterus: an in vitro study, *J. Biochem. Toxicol.*, 5, 139, 1990.
88. Shum, S., Jenson, N.M., and Nebert, D.W., The murine ah locus: in utero toxicity and teratogenesis associated with genetic differences in benzo(a)pyrene metabolism, *Teratology*, 20, 365, 1979.
89. Jellinck, P.H., Newbold, R.R., and McLachlan, J.A., Characteristics of estrogen-induced peroxidase in mouse uterine luminal fluid, *Steroids*, 56, 162, 1991.
90. Metzer, M. and Jellinck, P.H., Peroxidase-mediated oxidation, a possible pathway for metabolic activation of diethylstilbestrol, *Biochem. Biophys. Res. Commun.*, 85, 874, 1978.
91. Astroff, B. and Safe, S., 2,3,7,8-Tetrachlorodibenzo-*p*-dioxin as an antiestrogen: effect on rat uterine peroxidase activity, *Biochem. Pharmacol.*, 39, 485, 1990.
92. Devasagayam, T.P.A., Sivabalan, R., and Tarachand, U., Lipid peroxidation in the rat uterus during decidua induced cell proliferation, *Biochem. Int.*, 21, 27, 1990.
93. Falkay, G., Herczeg, J., and Sas, M., Microsomal lipid peroxidation in human uterus and placenta, *Biochem. Biophys. Res. Commun.*, 79, 843, 1977.
94. Malathy, P.V., Cheng, H.C., and Dey, S.K., Production of leukotrienes and prostaglandins in the rat uterus in endometrium during preimplantation period, *Prostaglandins*, 32, 605, 1986.
95. Gu, W. and Rice, G.E., Arachidonic acid metabolites in pregnant rat uterus, *Prost. Leukot. Essent. Fatty Acids*, 42, 15, 1991.
96. Poyser, N.L., A possible explanation for the refractiveness of uterine prostaglandin production, *J. Reprod. Fertil.*, 91, 371, 1991.
97. Demers, L.M., Rees, M.C.P., and Turnbull, A.C., Arachidonic acid metabolism by the non-pregnant human uterus, *Prost. Leukotrienes Med.*, 14, 175, 1984.
98. Ishihara, O., et al., Metabolism of arachidonic acid and synthesis of prostanoids in human endometrium and decidua, *Prost. Leukotrienes Med.*, 24, 93, 1986.
99. Mitchell, M.D. and Grzyboski, C.F., Arachidonic acid metabolism by lipoxygenase pathways in intrauterine tissues of women at term of pregnancy, *Prost. Leukotrienes Med.*, 28, 303, 1987.

100. Rees, M.C.P., et al., Leukotriene release by human fetal membranes, placenta and decidua in relation to parturition, *J. Endocrinol.*, 118, 497, 1988.
101. Lei, Z.M. and Rao, C.V., The expression of 15-lipoxygenase gene and the presence of functional enzyme in cytoplasm and nuclei of pregnant human myometria, *Endocrinology*, 130, 861, 1992.
102. Flatman, S., et al., Biochemical studies on a 12-lipoxygenase in human uterine cervix, *Biochim. Biophys. Acta*, 883, 7, 1986.
103. Joseph, P., Srinivasan, S., and Kulkarni, A.P., Xenobiotic oxidation during early pregnancy in human: peroxidase catalyzed chemical oxidation in conceptal tissues, *Xenobiotica*, 24, 583, 1994.
104. Datta, K., et al., Peroxidative xenobiotic oxidation by partially purified peroxidase and lipoxygenase from human fetal tissues at 10 weeks of gestation, *Gen. Pharmacol.*, 26, 107, 1995.
105. Kulkarni, A.P. and Murthy, K.R., Xenobiotic metabolism in humans during early pregnancy: peroxidase-mediated oxidation and bioactivation of 2-aminofluorene, *Xenobiotica*, 25, 799, 1995.
106. Joseph, P., et al., Bioactivation of benzo(a)pyrene-7,8-dihydrodiol catalyzed by lipoxygenase purified from human term placenta and conceptal tissues, *Reprod. Toxicol.*, 8, 307, 1994.
107. Datta, K. and Kulkarni, A.P., Oxidative metabolism of aflatoxin B_1 by lipoxygenase purified from human term placenta and intrauterine conceptal tissues, *Teratology*, 50, 311, 1994.
108. Datta, K., et al., Glutathione S-transferase-mediated detoxification and bioactivation of xenobiotics during early human pregnancy, *Early Human Dev.*, 37, 167, 1994.
109. Flint, O.P. and Orton, T.C., An *in vitro* assay for teratogens with cultures of rat embryo midbrain and limb bud cells, *Toxicol. Appl. Pharmacol.*, 76, 383, 1984.
110. Filler, R. and Lew, K.J., Developmental onset of mixed function oxidase activity in preimplantation mouse embryos, *Proc. Natl. Acad. Sci. USA.*, 78, 6991, 1981.
111. Nebert, D.W., Genetic differences in drug metabolism: proposed relationship to human birth defects, in *Handbook of Experimental Pharmacology, Vol. 65, Teratogenesis and Reproductive Toxicology*, Johnson, E.M. and Kochhar, D.M., Eds., Springer-Verlag, New York, 1983, p. 49.
112. Omiecinski, C.J., Redlich, C.A., and Costa, P., Induction and developmental expression of cytochrome P450iA1 messenger RNA in rat and human tissues: detection by the polymerase chain reaction, *Cancer Res.*, 50, 4315, 1990.
113. Brunstrom, B., Activities in chick embryos of 7-ethoxycoumarin O-deethylase and aryl hydrocarbon (benzo[a]pyrene) hydroxylase and their induction by 3,3′,4,4′-tetrachlorobiphenyl in early embryos, *Xenobiotics*, 16, 865, 1986.
114. Nimura, S. and Ishida, K.J., Immunohistochemical detection demonstration of prostaglandin E_2 in preimplantation mouse embryos, *J. Reprod. Fert.*, 80, 505, 1987.
115. Parr, M.B., et al., Immunohistochemical localization of prostaglandin synthase in the rat uterus and embryo during the implantation period, *Biol. Reprod.*, 38, 333, 1988.
116. Klein, K.L., Scott, W.J., and Clark, K.E., Prostaglandin synthesis in rat embryo tissue: the effect of non-steroidal anti-inflammatory drugs *in vivo* and *ex vivo*, *Prostaglandins*, 27, 659, 1984.
117. Liu, L. and Wells, P.G., Potential molecular targets mediating chemical teratogenesis: *in vitro* peroxidase-catalyzed phenytoin bioactivation and oxidative damage to proteins and lipids in murine maternal and embryonic tissues, *Toxicol. Appl. Pharmacol.*, 134, 71, 1995.
118. Winn, L.M. and Wells, P.G., Evidence for embryonic prostaglandin H synthase-catalyzed bioactivation and reactive oxygen species-mediated oxidation of cellular macromolecules in phenytoin and benzo(a)pyrene teratogenesis, *Free Rad. Biol. Med.*, 22, 607, 1997.
119. Winn, L.M. and Wells, P.G., Phenytoin-initiated DNA oxidation in murine embryo culture, and embryoprotection by antioxidative enzymes superoxide dismutase and catalase: evidence for reactive oxygen species mediated DNA oxidation in the molecular mechanism of phenytoin teratogenicity, *Mol. Pharmacol.*, 48, 112, 1995.
120. Miranda, A.F., Wiley, M.J., and Wells, P.G., Evidence for embryonic peroxidase-catalyzed bioactivation and glutathione-dependent cytoprotection in phenytoin teratogenicity: modulation by eicosatetraynoic acid and buthionine sulfoximine in murine embryo culture, *Toxicol. Appl. Pharmacol.*, 124, 230, 1994.
121. Wells, P.G., et al., Modulation of phenytoin teratogenicity and embryonic covalent binding by acetylsalicylic acid, caffeic acid, and alpha-phenyl-N-t-butylnitrone: implications for bioactivation by prostaglandin synthetase, *Toxicol. Appl. Pharmacol.*, 97, 192, 1989.

122. Kubow, S. and Wells, P.G., In vitro bioactivation of phenytoin to a reactive free radical intermediate by prostaglandin synthase, horseradish peroxidase, and thyroid peroxidase, *Mol. Pharmacol.*, 35, 504, 1989.
123. Parman, T., Chen, G., and Wells, P.G., Free radical intermediates of phenytoin and related teratogens. Prostaglandin H synthase-catalyzed bioactivation electron paramagnetic resonance spectrometry, and photochemical product analysis, *J. Biol. Chem.*, 273, 25079, 1998.
124. Parman T., Wiley, M.J., and Wells, P.G., Free radical mediated oxidative DNA damage in the mechanism of thalidomide teratogenicity, *Nature Med.*, 5, 582, 1999.
125. Arlen, R.R. and Wells, P.G., Inhibition of thalidomide teratogenicity by acetylsalicylic acid: evidence for prostaglandin H synthase-catalyzed bioactivation of thalidomide to a reactive intermediate, *J. Pharmacol. Exp. Ther.*, 277, 1649, 1996.
126. Vanderhoek, J.Y. and Klein, K.L., Lipoxygenases in rat embryo tissue, *Proc. Soc. Exp. Biol. Med.*, 188, 370, 1988.
127. Roy, S.K., et al., Lipoxygenase activity in rat embryo and its potential for xenobiotic oxidation, *Biol. Neonate*, 63, 297, 1993.
128. Kubow, S. and Wells, P.G., *In vitro* evidence for lipoxygenase-catalyzed bioactivation of phenytoin, *Pharmacologist*, 30, A74, 1988.
129. Yu, W.K. and Wells, P.G., Evidence for lipoxygenase-catalyzed bioactivation of phenytoin to a teratogenic reactive intermediate: *in vitro* studies using linoleic acid-dependent soybean lipoxygenase, and *in vivo* studies using pregnant CD-1 mice, *Toxicol. Appl. Pharmacol.*, 131, 1, 1995.
130. Balakier, H., Endogenous peroxidase in the visceral endoderm of early mouse embryos, *J. Exp. Zool.*, 231, 243, 1984.
131. Pelkonen, O., Environmental influences on human foetal and placental xenobiotic metabolism, *Eur. J. Clin. Pharmacol.*, 18, 17, 1980.
132. Pelkonen, O., Biotransformation of xenobiotics in the fetus, *Pharmacol. Ther.*, 10, 261, 1980.
133. Farrar, H.C. and Blumer, J.L., Fetal effects of maternal drug exposure, *Ann. Rev. Pharmacol. Toxicol.*, 31, 525, 1991.
134. Mizokami, K., et al., 3-Methylcholanthrene induces phenobarbital-induced cytochrome P450 hemoprotein in fetal liver and not cytochrome 448 hemoproteins induced in maternal liver of rats, *Biochem. Biophys. Res. Commun.*, 107, 6, 1982.
135. Kitada, M., et al., Four forms of cytochrome P450 in human fetal livers: purification and their capacity to activate promutagens, *Jpn. J. Cancer Res.*, 82, 426, 1991.
136. Rane, A. and Ackermann, E., Metabolism of ethylmorphine and aniline in human fetal liver, *Clin. Pharmacol. Ther.*, 13, 663, 1972.
137. Ladona, M.G., et al., Human fetal and adult liver metabolism of ethylmorphine, *Biochem. Pharmacol.*, 38, 3147, 1989
138. Komari, M., et al., Fetus specific expression of a form of cytochrome P450 in human livers, *Biochemistry*, 29, 4430, 1990
139. Hakkola, J., et al., Expression of xenobiotic metabolizing cytochrome P450 forms in human adult and fetal liver, *Biochem. Pharmacol.*, 48, 59, 1994.
140. Shimada, T., et al., Characterization of microsomal cytochrome P450 enzymes involved in oxidation of xenobiotic chemicals in human fetal livers and adult lungs, *Drug Met. Disp.*, 24, 515, 1996.
141. Hakkola, J., et al., Expression of CYP1B1 in human adult and fetal tissues and differential inducibility of CYP1B1 and CYP1A1 by Ah receptor ligands in human placenta and cultured cells, *Carcinogenesis*, 18, 391, 1997.
142. Shimada, T., Misono, K.S., and Guengerich, F.P., Human liver microsomal cytochrome P450 mephenytoin 4-hydroxylase, a prototype of genetic polymorphism in oxidative drug metabolism, *J. Biol. Chem.*, 261, 909, 1986.
143. Jacqz-Aigrain, E. and Cresteil, T., Cytochrome P450-dependent metabolism of dextromethorphan: fetal and adult studies, *Dev. Pharmacol. Ther.*, 18, 161, 1992.
144. Carpenter, S.J., Lasker, J.M., and Raucy, J.L., Expression, induction, and catalytic activity of the ethanol-inducible cytochrome P450 (CYP2E1) in human fetal liver and hepatocytes, *Mol. Pharmacol.*, 49, 260, 1996.
145. Greuet, J., et al., The fetal specific gene CYP3A7 is inducible by rafampicin in adult hepatocytes in primary culture, *Biochem. Biophys. Res. Commun.*, 225, 689, 1996.

146. Schuetz, J.D., Beach, D.L., and Guzelian, P.S., Selective expression of cytochrome P4450 CYP3A mRNA in embryonic and adult human liver, *Pharmacogenetics*, 4, 11, 1994.
147. Wrighton, S.A. and VandenBranden, M., Isolation and characterization of human fetal cytochrome P450 P450HLp2: a third member of the P450III gene family, *Arch. Biochem. Biophys.*, 268, 144, 1989.
148. Komari, M., et al., Molecular cloning and sequence analysis of cDNA containing the genetic encoding region for human fetal liver cytochrome P450, *J. Biochem.*, 105, 161,1989.
149. Bell, J.U., Hansell, M.M., and Ecobichon, D.J., The influence of DDT on the ontogenesis of drug metabolizing enzymes in the perinatal rats, *Toxicol. Appl. Pharmacol.*, 35, 165, 1976.
150. Sherratt, A.J., Banet, D.E., and Prough, R.A., Glucocorticoid regulation of polycyclic aromatic hydrocarbon induction of cytochrome P450IAi, glutathione S-transferase, and NAD(P)H:quinone oxidoreductase in cultured fetal rat hepatocytes, *Mol. Pharmacol.*, 37, 198, 1990.
151. Srivastava, S., Seth, P.K., and Srivastava, S.P., Altered activity of hepatic mixed function oxidase, cytochrome P450, and glutathione S-transferase by styrene in rat fetal liver, *Drug Chem. Toxicol.*, 15, 233, 1992.
152. Nau, H., et al., Preparation, morphology and drug metabolism of isolated hepatocytes and liver organ cultures from human fetus, *Life Sci.*, 23, 2361, 1978.
153. Rane, A., N-oxidation of a tertiary amine (N,N-dimethylaniline) by human fetal liver microsomes, *Clin. Pharmacol. Ther.*, 15, 32, 1973.
154. Saeed, S.A. and Mitchell, M.D., Arachidonate lipoxygenase activity in human fetal lung, *Eur. J. Pharmacol.*, 78, 389, 1982.
155. Saeed, S.A. and Mitchell, M.D., Formation of arachidonate lipoxygenase metabolites by human fetal membranes, uterine decidua vera, and placenta, *Prost. Leukotrienes Med.*, 8, 635, 1882.
156. Saeed, S.A. and Mitchell, M.D., Conversion of arachidonic acid to lipoxygenase products by human fetal tissues, *Biochem. Med.*, 30, 322, 1983.
157. Morykawa, M.J., et al. Arachidonic acid metabolites: effect on inflammation of fetal rabbit excisional wounds, *Inflammation*, 16, 251, 1992.
158. Dimitrijevic, L., et al., Purification and characterization of an estrogen-binding peroxidase from human fetus, *Biochimie*, 61, 535, 1979.
159. Peng, D., Pacifici, G.M., and Rane, A., Human fetal liver cultures: basal activities and inducibility of epoxide hydrolases and aryl hydrocarbon hydroxylase, *Biochem. Pharmacol.*, 33, 71, 1984.
160. Stevens, L.A., A comparative study of enzymes in foetal, young and adult rat, *Comp. Biochem. Physiol.*, 6, 129, 1962.
161. Pacifici, G.M., Norlin, A., and Rane, A., Glutathione S-transferases in human fetal liver, *Biochem. Pharmacol.*, 30, 3367, 1981.
162. Guthenberg, C., et al., Two distinct forms of glutathione S-transferase from human fetal liver, *Biochem. J.*, 235, 741, 1986.
163. Kashiwada, M., et al., Purification and characterization of acidic form of glutathione S-transferase in human fetal livers: high similarity to placental form, *J. Biochem.*, 110, 743, 1991.
164. Mitra, A.K., et al., A novel model to assess developmental toxicity of dihaloalkanes in human: embryotoxic bioactivation of 1,2-dibromoethane by isozymes of human fetal liver glutathione S-transferase, *Teratogen. Carcinog. Mutagen.*, 12, 113, 1992.
165. Radulovic, L.L., Dauterman, W.C., and Kulkarni, A.P., Biotransformation of methyl parathion by human fetal liver glutathione S-transferases: an in vitro study, *Xenobiotica*, 17, 105, 1987.
166. Kulkarni, A.P., Edwards, J., and Richards, I., Metabolism of 1,2-dibromoethane in the human fetal liver, *Gen. Pharmacol.*, 23, 1, 1992.
167. Dutton, G.J., Glucuronide synthesis in foetal liver and other tissues, *Biochem. J.*, 71, 141, 1959.
168. Burchell, B., UDP-glucuronyltransferase activity towards eostriol in fresh and cultured foetal tissues from man and other species, *J. Steroid Biochem.*, 5, 261, 1974.
169. Pacifici, G.M., et al., Development of glucuronyltransferase and sulfotransferase toward 2-naphthol in human fetus, *Dev. Pharmacol. Ther.*, 14, 108, 1990.
170. Kawade, N. and Onishi, S., The prenatal and postnatal development of UDP-glucuronyltransferase activity towards bilirubin and the effect of premature birth on this activity in human liver, *Biochem. J.*, 196, 257, 1981.
171. Tan, T.M.C., Sit, K.H., and Wong, K.P., UDP-glucuronyltransferase activity toward harmol in human liver and human fetal liver cells in culture, *Anal. Biochem.*, 185, 44, 1990.

172. Cappiello, M., et al., Sulfotransferase and its substrate: adenosine-3′-phosphate-5′-phosphosulfate in human fetal liver and placenta, *Dev. Pharmacol. Ther.*, 14, 62, 1990.
173. Elliott, W.J., et al., Arachidonic acid metabolism by rabbit fetal membranes of various gestational ages, *Prostaglandins*, 27, 27, 1984.
174. Kinoshita, K. and Green, K., Bioconversion of arachidonic acid to prostaglandins and related compounds in human amnion, *Biochem. Med.*, 23, 185, 1980.
175. Olson, D.M., Skinner, K, and Challis, J.R.G., Estradiol-17 and 2-hydroxyestradiol-17-induced differential production of prostaglandins by cells dispersed from human intrauterine tissues at parturition, *Prostaglandins*, 25, 639, 1983.
176. Cheung, P.Y.C. and Challis, J.R.G., Prostaglandin E_2 metabolism in the human fetal membranes, *Am. J. Obstet. Gynecol.*, 161, 1580, 1989.
177. Price, T.M., et al., Immunohistochemical localization of prostaglandin endoperoxide synthase in human fetal membranes and decidua, *Biol. Reprod.* 41, 701, 1989.
178. Pasetto, N., et al., Influence of labor and oxytocin on *in vitro* leukotriene release by human fetal membranes and uterine decidua at term gestation, *Am. J. Obstet. Gynecol.*, 166, 1500, 1992.
179. Meigs, R.A., The constitutive 7-ethoxycoumarin O-deethylase activity of human placental microsomes: relationship to aromatase, *Biochem. Biophys. Res. Commun.*, 145, 1012, 1987.
180. Wong, T.K., et al., Correlation of placental microsomal activities with protein detected by antibodies to rabbit cytochrome P450 isozyme 6 in preparations from humans exposed to polychlorinated biphenyls, quaterphenyls, and dibenzofurans, *Cancer Res.*, 46, 999, 1986.
181. Lucier, G.W., et al., Placental markers of human exposure to polychlorinated biphenyls and polychlorinated dibenzofurans, *Environ. Hlth. Persp.*, 76, 79, 1987.
182. Mortimer, G., Mackay, M.M., and Stimson, W.H., The distribution of pregnancy associated prostaglandin synthase inhibitor in the human placenta, *J. Pathol.*, 159, 239, 1989.
183. Flammang, T.J., et al., Arachidonic acid dependent peroxidative activation of carcinogenic arylamines by extrahepatic human tissue microsomes, *Cancer Res.*, 49, 1977, 1989.
184. Matkovics, B., Fodor, I., and Kovacs, K., Properties of enzymes III. A comparative study of peroxide-decomposing enzymes and superoxide dismutase of human placenta of different ages, *Enzyme*, 19, 285, 1975.
185. Joseph, P., et al., Peroxidase: a novel pathway for chemical oxidation in human term placenta, *Placenta*, 13, 545, 1992.
186. Joseph, P., Srinivasan, S., and Kulkarni, A.P., Placental peroxidase — further purification of the enzyme and oxidation of thiobenzamide, *Placenta*, 14, 309, 1993.
187. Murthy, K.R., Joseph, P., and Kulkarni, A.P., 2-Aminofluorene bioactivation by human term placental peroxidase, *Terat. Carcinog. Mut.*, 15, 115, 1995.
188. Kulkarni, A.P., Joseph, P., and Byczkowski, J.Z., Effects of phenylbutazone and amino acids on benzo(a)pyrene-7,8-dihydrodiol co-oxygenation by partially purified placental peroxidase, *Toxicologist*, 12, 197, 1992.
189. Madhavan, N.D. and Naidu, K.A., Purification and partial characterization of peroxidase from human placenta of non-smokers: metabolism of benzo(a)pyrene-7,8-dihydrodiol, *Placenta*, 21, 501, 2000.
190. Pumeau-Rouquette, C., Goujard, J., and Hual, G., Possible teratogenic effect of phenothiazines in human beings, *Teratology*, 15, 57, 1977.
191. Edmund, M.J. and Craig, T.J., Antipsychotic drug use and birth defects: an epidemiological reassessment, *Comp. Psychiatry*, 25, 32, 1984.
192. Altshuler, L.L., et al., Pharmacologic management of psychiatric illness during pregnancy: dilemmas and guidelines, *Am. J. Psychiatry*, 153, 592, 1996.
193. Slone, D., et al., Antenatal exposure to phenothiazines in relation to congenital malformations perinatal mortality rate, birth weight and intelligence quotient score, *Am. J. Obstet. Gynecol.*, 128, 486, 1977.
194. Sobel, D.E., Fetal damage due to ECT, insulin coma, chlorpromazine, or reserpine, *Arch. Gen. Psychiatry*, 2, 606, 1960.
195. Yang, X. and Kulkarni, A.P., Oxidation of phenothiazines by human term placental peroxidase in non-smokers. *Teratogen. Carcinog. Mutagen.*, 17, 139, 1997.
196. Zhang, R. Kulkarni, K.A., and Kulkarni, A.P., Oxidation of eugenol by human term placental peroxidase, *Placenta*, 21, 234, 2000.

197. Ogburn, P.L., et al., Leukotriene B_4, 6-keto-prostaglandin F_1, thromboxane B_2, and free arachidonic acid levels in human placental tissue, *Ann. N. Y. Acad. Sci.*, 524, 434, 1988.
198. Lopez Bernal, A., et al., Placental leukotriene B_4 release in early pregnancy and in term and preterm labour, *Early Hum. Dev.*, 23, 93, 1990.
199. Joseph, P., Srinivasan, S.N., and Kulkarni, A.P., Purification and partial characterization of lipoxygenase with dual catalytic activities from human term placenta, *Biochem. J.*, 293, 83, 1993.
200. Walsh, W.S. and Parisi, V.M., The role of arachidonic metabolites in preeclampsia, *Semin. Perinatol.*, 10, 334, 1986.
201. Roy, P., Sajan, M., and Kulkarni, A.P., Lipoxygenase mediated glutathione oxidation and superoxide generation, *J. Biochem. Toxicol.*, 10, 111, 1995.
202. Kulkarni, A.P. and Sajan, M., A novel mechanism of glutathione conjugation formation by lipoxygenase: a study with ethacrynic acid, *Toxicol. Appl. Pharmacol.*, 143, 179, 1997.
203. Kulkarni, A.P. and Sajan, M., Glutathione conjugation of ethacrynic acid by human term placental lipoxygenase, *Arch. Biochem. Biophys.*, 371, 220, 1999.
204. Roy, P., et al., Superoxide generation by lipoxygenase in the presence of NADH and NADPH, *Biochim. Biophys. Acta*, 1214, 171, 1994.
205. Roy, P. and Kulkarni, A.P., Oxidation of ascorbic acid by lipoxygenase: effect of selected chemicals, *Food Chem. Toxicol.*, 34, 563, 1996.
206. Yang, X. and Kulkarni, A.P., Lipoxygenase-mediated biotransformation of p-aminophenol in the presence of glutathione. Possible conjugate formation, *Toxicol. Lett.*, 111, 253, 2000.
207. Datta, K. and Kulkarni, A.P., Co-oxidation of t-retinol acetate by human term placental lipoxygenase and soybean lipoxygenase, *Repr. Toxicol.*, 10, 105, 1996.
208. Manchester, D.K., et al., Detection of benzo(a)pyrene diol epoxide-DNA adducts in human placenta, *Proc. Natl. Acad. Sci. USA.*, 85, 9243, 1988.
209. Denning, D.W., et al., Transplacental transfer of aflatoxins in humans, *Carcinogenesis*, 11, 1033, 1990.
210. Wild, C.P., et al., In-utero exposure to aflatoxin in West Africa, *Lancet*, 337, 1602, 1991.
211. Heish, L.L. and Heish, T.T., Detection of aflatoxin B_1-DNA adducts in human placenta and cord blood, *Cancer Res.*, 53, 1278, 1993.
212. Cameron, H.M. and Warwick, G.P., Primary cancer of the liver in Kenyan children, *Brit. J. Cancer*, 36, 793, 1977.
213. Roy, S. and Kulkarni, A.P., Aflatoxin B_1 epoxidation catalyzed by partially purified adult human liver lipoxygenase, *Xenobiotica*, 27, 231, 1996.
214. Coghlin, P.H., et al., 4-aminobiphenyl-hemoglobin adducts in fetuses exposed to tobacco smoke carcinogen *in utero*, *J. Natl. Cancer Inst.*, 83, 274, 1991.
215. Hammond, S.K., et al., Relationship between environmental tobacco smoke exposure and carcinogen-hemoglobin adduct levels in nonsmokers, *J. Natl. Cancer Inst.*, 85, 474, 1993.
216. Lu, L.J.W., et al., ^{32}P-postlabelling assay in mice of transplacental DNA damage induced by the environmental carcinogens safrole, 4-aminobiphenyl and benzo(a)pyrene, *Cancer Res.*, 46, 3046, 1986.
217. Datta, K., Sherblom, P.M., and Kulkarni, A.P., Co-oxidative metabolism of 4-aminobiphenyl by lipoxygenase from soybean and human term placenta, *Drug Met. Disp.*, 25, 196, 1997.
218. Hover, C.G. and Kulkarni, A.P., Human term placental lipoxygenase-mediated — demethylation of phenothiazines and pesticides in the presence of polyunsaturated fatty acids, *Placenta*, 21, 647, 2000.
219. Rajadhyaksha, A., et al., N-demethylation of phenothiazines by lipoxygenase from soybean and human term placenta in the presence of hydrogen peroxide, *Teratogen. Carcinogen. Mutagen.*, 19, 211, 1999.
220. Yang, X. and Kulkarni, A.P., N-dealkylation of aminopyrine catalyzed by soybean lipoxygenase in the presence of hydrogen peroxide, *J. Biochem. Toxicol.*, 12, 175, 1997.
221. Hu, J., Sajan, M., and Kulkarni, A.P., Lipoxygenase-mediated N-demethylation of imipramine and related tricyclic antidepressants in the presence of hydrogen peroxide, *Int. J. Toxicol.*, 18, 251, 1999.
222. Hover, C. and Kulkarni, A.P., Lipoxygenase-mediated hydrogen peroxide-dependent N-demethylation of N,N'-dimethylaniline and related compounds, *Chem.-Biol. Interact.*, 124, 191, 2000.
223. Hu, J. and Kulkarni, A.P., Soybean lipoxygenase-catalyzed demethylation of pesticides, *Pestic. Biochem. Physiol.*, 61, 145, 1998.
224. Uotila, J., et al., Lipid peroxidation products, selenium-dependent glutathione peroxidase and vitamin E in normal pregnancy, *Eur. J. Obstet. Gynecol. Reprod. Biol.*, 42, 95, 1991.
225. Wickens, D., et al., Free radical oxidation (peroxidation) products in plasma in normal and abnormal pregnancy, *Ann. Clin. Biochem.*, 18, 158, 1981.

226. Juchau, M.R. and Zachariah, P.K., Comparative studies on the oxidation and reduction of drug substances in human placenta versus rat hepatic microsomes, *Biochem. Pharmacol.*, 24, 227, 1975.
227. Arvela, P., Karki, N.T., and Pelkonen, R.O., Lipoperoxidation rates and drug oxidizing enzyme activities in the liver and placenta of some animal species during the perinatal period, *Experientia*, 32, 1311, 1976.
228. Kulkarni, A.P. and Kenel, M.F., Human placental lipid peroxidation. I. Some characteristics of the NADPH-supported microsomal reaction, *Gen. Pharmacol.*, 18, 491, 1987.
229. Diamant, S., Kisselevitz, R., and Diamant, Y., Lipid peroxidation system in human placental tissue: general properties and the influence of age, *Biol. Reprod.*, 23, 776, 1980.
230. Klimek, J., Influence of lipid peroxidation on the progesterone biosynthesis in human placental mitochondria, *J. Steroid Biochem. Mol. Biol.*, 42, 729, 1992.
231. Byczkowski, J.Z. and Kulkarni, A.P., Lipid peroxidation and benzo(a)pyrene derivative co-oxygenation by environmental pollutants, *Bull. Environ. Contam. Toxicol.*, 45, 633, 1990.
232. Kulkarni, A.P. and Byczkowski, J.Z., Effect of transition metals on biological oxidations, in *Environmental Oxidants*, Nriagu, J.O. and Simmons, M.S., Eds., John Wiley, New York, 1994, chap. 16.
233. Byczkowski, J.Z. and Kulkarni, A.P., Oxidative stress and asbestos, in *Environmental Oxidants*, Nriagu, J.O. and Simmons, M.S., Eds., John Wiley, New York, 1994, chap.15.
234. Byczkowski, J.Z., Wan, B., and Kulkarni, A.P., Vanadium-mediated lipid peroxidation in microsomes from human term placenta, *Bull. Environ. Contam. Toxicol.*, 41, 696, 1988.
235. Byczkowski, J.Z. and Kulkarni, A.P., Vanadium redox cycling and lipid peroxidation, *Biochim. Biophys. Acta*, 1125, 134, 1992.
236. Kenel, M.F., Bestervelt, L.L., and Kulkarni, A.P., Human placental lipid peroxidation. II. NADPH and iron dependent stimulation of microsomal lipid peroxidation by paraquat, *Gen. Pharmacol.*, 18, 373, 1987.
237. Byczkowski, J.Z. and Kulkarni, A.P., NADPH-dependent drug redox cycling and lipid peroxidation in microsomes from human term placenta, *Int. J. Biochem.*, 21, 183, 1989.
238. Byczkowski, J.Z. and Kulkarni, A.P., Lipid peroxidation-coupled co-oxygenation of benzo(a)pyrene and benzo(a)pyrene-7,8-diol in human term placental microsomes, *Placenta*, 11, 17, 1990.
239. Kulkarni, A.P., Strohm, B., and Houser, W.H., Human placental indanol dehydrogenase: some properties of the microsomal enzyme, *Xenobiotica*, 15, 513, 1985.
240. Braun, A.G. and Weinreb, S.L., Metabolism of thalidomide analogues inhibit attachment of cells to concanavalin A coated surfaces, *Teratology*, 27, 33A, 1983.
241. Pacifici, G.M. and Rane, A., Epoxide hydrase in human placenta at different stages of pregnancy, *Dev. Pharmacol. Ther.*, 6, 83, 1983.
242. Berte, F., et al., Ability of placenta to metabolize oxazepam and aminopyrine before and after drug stimulation, *Arch. Int. Pharmacodyn.*, 182, 182, 1969.
243. Radulovic, L.L. and Kulkarni, A.P., High performance liquid chromatographic separation and study of the charge isomers of human placental glutathione S-transferase, *Biochem. J.*, 239, 53, 1986.
244. Radulovic, L.L., LaFerla, J.J., and Kulkarni, A.P., Human placental glutathione S-transferase mediated metabolism of methyl parathion, *Biochem. Pharmacol.*, 35, 3473, 1986.
245. Pacifici, G.M. and Rane, A., Glutathione S-transferase in the human placenta at different stages of pregnancy, *Drug Met. Disp.*, 9, 472, 1981.
246. Mitra, A., et al., Inhibition of human term placental and fetal liver glutathione S-transferases by fatty acids and fatty acid esters, *Toxicol. Lett.*, 60, 281, 1992.
247. Kulkarni, A.A. and Kulkarni, A.P., Retinoids inhibit mammalian glutathione transferases, *Cancer Lett.*, 91, 185, 1995.
248. Sajan, M., Reddy, G., and Kulkarni, A.P., In vitro inhibition of mammalian glutathione transferases by selected nitrobenzenes, *Int. J. Toxicol.*, 19, 285, 2000.
249. Mitra, A.K., et al., In vitro embryotoxicity of ethylene dibromide bioactivated by purified human placental glutathione S-transferase, *Int. J. Occup. Med. Toxicol.*, 2, 101, 1993.
250. Manchester, D.K. and Jacoby, E.H., Glutathione S-transferase activities in placentas from smoking and non-smoking women, *Xenobiotica*, 12, 543, 1982.
251. Kitahara, A., et al., Changes in molecular forms of rat hepatic glutathione S-transferase during chemical hepatocarcinogenesis, *Cancer Res.*, 44, 2698, 1984.
252. Dixon, R.L., Aspects of male reproductive toxicology, in *Occupational Hazards and Reproduction*, Hemminki, K, Sorsa, M., and Vainio, H., Eds., Hemisphere, New York, 1985, chap. 4.

253. Mattison, D.R., Shiromizu, K., and Nightingale, M.S., The role of metabolic activation in gonadal and gamete toxicity, in *Occupational Hazards and Reproduction*, Hemminki, K, Sorsa, M., and Vainio, H., Eds., Hemisphere, New York, 1985, chap.7.
254. Miller, G.E. and Kulkarni, A.P., 1,2-dibromo-3-chloropropane as a substrate for hepatic, renal and testicular glutathione S-transferases in the rat, *Toxicologist*, 3, 416A, 1983.
255. Kulkarni, A.P. and Mitra, A., Pesticide contamination of food, in *Food Contamination from Environmental Sources*, Simons, M.S. and Nriagu, J., Eds., Adv. Environ. Sci. Technol., Vol. 23, John Wiley, New York, 1990, chap. 9.
256. Babich, H., Davis, D.L., and Stotzky, G., Dibromochloropropane (DBCP): a review, *Sci. Total Environ.*, 17, 207, 198.
257. Miller, G.E., Brabec, M.J., and Kulkarni, A.P., Mutagenic activation of 1,2-dibromo-3-chloro-propane by cytosolic glutathione S-transferases and microsomal enzymes, *J. Toxicol. Environ. Hlth.*, 19, 503, 1986.
258. Guzelian, P.S., Comparative toxicology of chlordecone (kepone) in humans and experimental animals, *Ann. Rev. Pharmacol. Toxicol.*, 22, 89, 1982.
259. Gray, T.J.B. and Beamand, J.A., Effect of some phthalate esters and other testicular toxins on primary cultures of testicular cells, *Food Cosmet. Toxicol.*, 22, 123, 1984.
260. Boekelheide, K., 2,5-Hexanedione alters microtubule assembly: I. Testicular atrophy, not nervous system toxicity, correlates with enhanced tubulin polymerization, *Toxic. Appl. Pharmacol.*, 88, 370, 1987.
261. Wang, W. and Chapin, R.E., Differential gene expression detected by suppression substractive hybridization in ethylene glycol monomethyl ether-induced testicular lesion, *Toxicol. Sci.*, 56, 165, 2000.
262. Kulkarni, A.P., Treinen, K.A., and Bestervelt, L.L., Human placental Ca^{2+}-ATPase: target for organochlorine pesticides? *Trophoblast Res.*, 2, 329, 1987.
263. Andersen, H.R., et al., Effects of currently used pesticides in assays for estrogenicity, adrogenicity, and aromatase activity in vitro, *Toxicol. Appl. Pharmacol.*, 179, 1, 2002.
264. Naidu, A.K., Naidu, A.K., and Kulkarni, A.P., Aldrin epoxidation: catalytic potential of lipoxygenase coupled with linoleic acid oxidation, *Drug Metab. Disp.*, 19, 758, 1991.
265. Stresser, D.M. and Kupfer, D., Human cytochrome P450-catalyzed conversion of the proestrogenic pesticide methoxychlor into an estrogen. Role of CYP2C19 and CYP1A2 in O-demethylation, *Drug Met. Disp.*, 26, 868, 1998.
266. Nicoise, T., Chronic organophosphorus intoxication in women, *J. Jpn. Assoc. Rural Med.*, 22, 756, 1974.
267. Naidu, A.K., Naidu, A.K., and Kulkarni, A.P., Role of lipoxygenase in xenobiotic oxidation: parathion metabolism catalyzed by highly purified soybean lipoxygenase, *Pestic. Biochem. Physiol.*, 41, 150, 1991.
268. Byczkowski, J.Z. and Kulkarni, A.P., Lipoxygenase-catalyzed epoxidation of benzo(a) pyrene-7,8-dihydrodiol, *Biochem. Biophys. Res. Commun.*, 159, 1199, 1989.
269. Byczkowski, J.Z. and Kulkarni, A.P., Linoleate-dependent co-oxygenation of benzo(a)pyrene-7,8-dihydrodiol by rat cytosolic lipoxygenase, *Xenobiotica*, 22, 609, 1992.
270. Roy, S. and Kulkarni, A.P., Lipoxygenase: a new pathway for 2-aminofluorene bioactivation. *Cancer Lett.*, 60, 33, 1991.
271. Saillenfait, A.M., et al., Modulation of acrylonitrile-induced embryotoxicity *in vitro* by glutathione depletion, *Arch. Toxicol.*, 67, 164, 1993.
272. Roy, P. and Kulkarni, A.P., Co-oxidation of acrylonitrile by soybean lipoxygenase and partially purified human lung lipoxygenase, *Xenobiotica*, 29, 511, 1999.
273. Metzler, M., Biochemical toxicology of diethylstilbestrol, *Rev. Biochem. Toxicol.*, 6, 191, 1984.
274. Nunez-Delicado, E., Sanchez-Ferrer, A., and Garcia-Carmona, F., Hydroperoxidative oxidation of diethylstilbestrol by lipoxygenase, *Arch. Biochem. Biophys.*, 348, 411, 1997.
275. Hu, J. and Kulkarni, A.P., Metabolic fate of chemical mixtures. I. Shuttle oxidant effect of lipoxygenase-generated radical of chlorpromazine and related phenothiazines on the oxidation of benzidine and other xenobiotics, *Teratogen, Carcinogen. Mut.*, 20, 195, 2000.
276. Persad, A.S., Stedeford, T.J., and Kulkarni, A.P., The hyperoxidation of L-dopa by chlorpromazine in a lipoxygenase catalyzed reaction, *Toxicologist*, 54, 1746A, 2000.

CHAPTER 13

Use of Toxicokinetics in Developmental and Reproductive Toxicology

Patrick J. Wier

CONTENTS

- I. Introduction 572
- II. Principles of Absorption, Distribution, Biotransformation, and Elimination 572
  - A. Movement across Biological Membranes 572
  - B. Plasma Protein Binding 573
  - C. Volume of Distribution 573
  - D. Blood and Seminiferous Tubule Distribution 574
  - E. Placental Transfer 574
  - F. Biotransformation 575
  - G. Elimination 576
  - H. Physiological Changes in Pregnancy 576
  - I. Physiological Changes in Neonates 577
- III. Classical Pharmacokinetics 578
  - A. Simple Compartmental Models 578
  - B. Basic Pharmacokinetic Parameters 578
- IV. Physiologically Based Pharmacokinetics 582
  - A. Physiologically Based Pharmacokinetic Models of Pregnancy 582
- V. Sampling Strategies 583
- VI. Regulatory Expectations 585
  - A. ICH Reproduction Test Guideline 585
  - B. ICH Toxicokinetic Guideline 587
- VII. Toxicokinetics in Developmental and Reproductive Toxicology 587
  - A. When Are Toxicokinetics *Not* Essential? 588
  - B. Use of Toxicokinetics in Study Design 588
  - C. Use of Toxicokinetics in Study Interpretation 588
  - D. Use of Toxicokinetics in Risk Assessment 590
- VIII. Conclusions 592
- References 592

I. INTRODUCTION

It has long been recognized that access of a nontopical drug to its ultimate site of action requires the agent to enter and be distributed through the systemic circulation. Although a drug may be bound to plasma protein or blood cells, the aqueous phase of blood plasma is a vehicle for distribution of drugs to tissues throughout the body. Tissues can sequester, metabolize, or eliminate the drug from the body. Pharmacokinetics quantitatively defines the absorption, distribution, metabolism, and elimination of drugs in the context of understanding how these factors influence a drug's efficacy. Quantitative expressions (pharmacokinetic parameters) are used to define the extent of drug absorption, the magnitude of drug binding to plasma proteins, and the rate of drug elimination from blood plasma, to name a few examples.

A logical extension of the principles of pharmacokinetics is their application toward understanding how the absorption, distribution, metabolism, and elimination of a compound influence its toxicity. If an orally ingested compound is poorly absorbed from the gastrointestinal tract, extensively bound to plasma proteins, rapidly metabolized to a nonreactive moiety, or quickly eliminated by urinary excretion, the relative potential for toxicity is diminished. Conversely, greater absorption of a xenobiotic, formation of reactive metabolites, and/or prolonged retention of the agent in the body generally leads to a relatively higher potential for toxicity. Toxicokinetics is not simply the collection of pharmacokinetic parameters in a toxicology study; it means *quantitatively* establishing how such parameters relate to the toxicity potential of the test article.

Toxicokinetics is practically useful in reproductive and developmental toxicology for: (1) experimental study design, (2) interpretation of toxicology data, and (3) risk assessment, especially in extrapolation of findings between species or between different exposure conditions. Each of these is detailed by example in this chapter. To provide the necessary background, this chapter begins with an overview of how chemicals are absorbed, distributed, biotransformed, and eliminated in the body, followed by definition of classical pharmacokinetic parameters.

II. PRINCIPLES OF ABSORPTION, DISTRIBUTION, BIOTRANSFORMATION, AND ELIMINATION

Xenobiotics may enter the body by a variety of routes, classically divided into two broad categories, enteral (via the gastrointestinal tract) and parenteral (not via the gastrointestinal tract). The former includes sublingual, oral (by diet or intubation), and rectal administration. Parenteral routes include dermal, intranasal, ocular, inhalation, subcutaneous, intramuscular, and intravascular. In experimental studies, the route of administration can be based on a variety of considerations, for example, the route of administration in humans or physical characteristics of the test article. As subsequently described in this chapter, toxicokinetics can be an important part of the rationale for route of administration in experimental studies, for example, when striving to optimize the level of systemic exposure in animals.

A. Movement across Biological Membranes

By any route of administration, fundamental processes govern the rate and extent of xenobiotic transit across biological membranes. These processes are important determinants of xenobiotic absorption and distribution. Many agents cross biological membranes by simple passive diffusion, which depends on the characteristics of the biological membrane, the concentration gradient of unbound xenobiotic, and the molecular properties of the xenobiotic, particularly molecular radius, degree of ionization, and lipid-water parturition coefficient. Most xenobiotics with molecular weight less than 800 Da transfer across the placenta by passive diffusion.[1] In facilitated diffusion (e.g., placental transfer of D-glucose or L-lactate),[2] the xenobiotic binds to a "carrier" molecule, forming a complex that diffuses across membranes, depending on the concentration gradient. Active transport is a cellular-energy requiring process for movement of molecules against a concentration gradient. Several essential nutrients (e.g., amino acids, calcium, vitamin B_{12}) cross the placenta by active transport.[2] Xenobiotics may also be subject to carrier-mediated transport. The p-glycoproteins (also known as multidrug-resistant [mdr] proteins), organic-anion transporters, and organic-cation transporters can actively "pump" certain xenobiotics out of cells. Gastrointestinal absorption of several drugs is limited by mdr activity in intestinal enterocytes.[3] Placental mdr has been shown to modulate the teratogenic potential of avermectin, as mdr null-mutant mice are more susceptible to the induction of cleft palate by that drug.[4] Macromolecules can move across cellular membranes by cellular pinocytosis (engulfing of fluid) or endocytosis (receptor mediated internalization, e.g., movement of iron-transferrin across hemochorial placentas).[2]

When movement of a substance (e.g., anesthetic gases such as nitrous oxide or halothane[1]) across cell membranes is very rapid relative to the rate of blood flow to the tissue, uptake into that tissue is termed "flow-limited." Conversely, when uptake is rate limited by movement across cell membranes (e.g., antibiotics such as penicillin or streptomycin[5]), uptake is considered to be "membrane limited" (or "diffusion limited" when that movement occurs by passive diffusion).

B. Plasma Protein Binding

The binding of a xenobiotic to blood plasma proteins can be an important determinant of its distribution and elimination. Many drugs noncovalently bind to albumins, globulins, or lipoproteins.[5] The large molecular radius of the drug-protein complex can effectively retard passive diffusion across membranes and thereby limit drug distribution into tissues and limit glomerular filtration, both of which can influence toxicity. The extent of plasma protein binding by a compound can be assessed by ultrafiltration or equilibrium dialysis and is commonly expressed as %fu, the fraction of drug unbound by plasma protein, i.e., the "free fraction." It is not uncommon for there to be species differences in %fu, and small differences can be important. For example, consider the case of a drug with %fu = 0.5% (99.5% bound) in the rat and %fu = 1.5% (98.5% bound) in humans. At the same total concentration in plasma, the "free" drug concentration, which is freely diffusible across cellular membranes into target organs, is three times higher in humans than in rats. Such a difference could be associated with a greater potential for toxicity. At the same time, urinary excretion of the drug could be greater in humans and might lower the potential for toxicity. It should also be appreciated that %fu can vary according to total drug concentration, given a finite number of binding sites on plasma protein.[5]

Unexpected drug interactions can occur when there is competition for binding to plasma proteins. If a drug with a low %fu, such as digitoxin (%fu = 2%), is only 10% displaced from plasma protein binding, the "free" drug concentration increases sixfold (%fu increases from 2% to 12%).[6] Such interactions can have adverse consequences. An example is kernicterus in infants resulting from displacement of bilirubin from plasma proteins by sulfonamide drugs, which compete for the bilirubin binding site, leading to more "free" bilirubin diffusion into the brain.[6]

C. Volume of Distribution

An empirical indication of how extensively a xenobiotic is distributed throughout the body is given by the apparent volume of distribution (V_d), the virtual fluid volume in which the total amount of drug in the body appears to be dissolved to account for the measured drug concentration. Conceptually, a drug restricted to blood plasma (and assuming that the drug is not bound, metabolized, or eliminated) has a V_d equivalent to the total blood plasma volume (about 3 liters for a 70-kg human); a drug widely distributed throughout all intracellular and extracellular (interstitial) space has a V_d equivalent to the total body water volume (about 41 liters). The V_d can be determined experimentally by giving the test article as a bolus intravenous dose, in which case:

$$V_d \text{ (l)} = \text{intravenous dose (mg)}/C_{(0)} \text{ (mg/liter)} \quad (13.1)$$

where $C_{(0)}$ is the unbound plasma drug concentration determined by extrapolation back to the time of injection. A large V_d can be due to extensive distribution throughout body water and/or extensive binding to tissues. Xenobiotics that partition into body fat can have a V_d many times greater than the total body water volume. In the absence of having tissue concentrations of xenobiotics to compare, differences in V_d can sometimes help explain differences in the toxicity potential among drugs with comparable absorption and elimination rates.

D. Blood and Seminiferous Tubule Distribution

Distribution of xenobiotics between blood plasma and the seminiferous tubule lumen is limited by several cells and surrounding layers, including capillary endothelium and basal lamina, lymphatic endothelium, peritubular myocytes, seminiferous tubule basal lamina, and Sertoli cell–to–Sertoli cell junctions.[7] These layers can prevent the access of certain molecules (e.g., immunoglobulins, albumin, inulin, and certain lower molecular weight compounds) to the seminiferous tubule lumen and hence limit exposure of the germinal epithelium and its progeny.[8,9] Certain p-glycoproteins that can serve as efflux transporters and further control access of xenobiotics to germ cells have been detected in the testes. Testicular concentrations of verapamil are significantly higher in *mdr1a* (-/-) mice relative to *mdr1a* (+/+) mice given the same dose,[10] and the administration of p-glycoprotein inhibitors has been shown to increases testicular penetration of several drugs.[10–12]

E. Placental Transfer

There are multiple pathways by which the embryo or fetus can be exposed to xenobiotics during development. Initially, the preimplantation blastocyst is freely exposed to solutes within the uterus. Embryonic exposure can be rapid (thiopental reaches the blastocyst within minutes of maternal administration) and extensive (the blastocyst attains concentration of nicotine that are approximately 10-fold greater than the maternal plasma concentration).[13,14] In rodents and rabbits, the predominant interface between the maternal and embryonic circulations during very early organogenesis is the yolk-sac placenta, a saclike cavity of extraembryonic endoderm. Subsequently during gestation, the predominant interface in these species is the chorioallantoic placenta, beginning at the 20 somite stage of embryonic development (about day 11 and 10 postfertilization in the rat and rabbit, respectively).[15,16]

There are species differences in the histology of the chorioallantoic placenta, including different numbers of cell layers separating the maternal and fetal blood. Table 13.1 gives a simplified description of the maternal-fetal interface in common species during the fetal period of development.

Histological features of placentas vary during gestation. For example, the rabbit chorioallantoic placenta is epitheliochorial on day 8 postcoitus (pc), endotheliochorial on day 9.5 pc, and hemochorial from day 10 pc.[16] Furthermore, placental histology is not entirely uniform. For example, human chorionic epithelium variably consists of one layer (syncytiotrophoblast) or two layers (cytotrophoblast and syncytiotrophoblast) over the villous surface.

Most importantly, the placenta should not be regarded as a barrier nor as a sieve. It is both active and selective in the transport of nutrients and xenobiotics. While placental permeability to most proteins is low, immunoglobulin G (IgG) is actively (via receptor-mediated endocytosis) and selectively (all subclasses of IgG, but not other immunoglobulins) transferred across the placenta.[2] But also there are gaps in the intervening cell layers, so virtually any substance might to some extent cross the placenta; for example, even erythrocytes pass in both directions across the placenta. Therefore, the salient questions are the rate and extent of placental transfer, which can vary during gestation as a result of the dynamic nature of placentation and can vary between species having

Table 13.1 Histological features of the chorioallantoic placenta in common species

	Type of Placentation and Representative Animal Species			
	Epitheliochorial (Pig, Horse)	Syndesmochorial (Sheep, Cow)	Endotheliochorial (Dog, Cat, Ferret)	Hemochorial[a] (e.g., Human, Monkey, Rabbit, Rat, Mouse)
	Maternal Blood			
Maternal capillary endothelium	√	√	√	
Uterine connective tissue	√	√		
Uterine epithelium	√			
Chorionic epithelium	√	√	√	√
Fetal connective tissue	√	√	√	√
Fetal capillary endothelium	√	√	√	√
	Fetal Blood			

[a] The number (1 to 3) of trophoblast cell layers that form the chorionic epithelium varies by species and stage of gestation.

Source: Amoroso, E.C., Placentation, in *Marshall's Physiology of Reproduction*, 3rd ed., Parkes, A.S., Ed., Longmans, London, 1952, Vol. II, chap. 15.

different types of placentation. For example, placental transfer of digoxin is much greater in humans and rodents (hemochorial placenta) than in sheep (syndesmochorial placenta).[17]

F. Biotransformation

The term *metabolism* is properly used to describe the absorption, distribution, biotransformation, and elimination of a xenobiotic, but for the present discussion it can be understood to refer to chemical transformations catalyzed by enzymes. These transformations are broadly categorized as Phase I or Phase II reactions.[18] Phase I reactions include chemical oxidations, reductions, or hydrolyses that generally increase polarity (hydrophilicity) but may also result in a more reactive metabolite.[19] Phase II reactions can further increase hydrophilicity (which generally enhances elimination), typically by conjugation with polar moieties (e.g., via acetylation, sulfation, glucuronidation, and glutathione or amino acid conjugation). Biotransforming enzymes are widely distributed throughout the body, including the gonads and placenta,[20–26] where "local" formation of reactive metabolites can feature in a compound's reproductive or developmental toxicity. The human fetal liver has significant capability to metabolize xenobiotics (much more so than those of rodents) using cytochrome P450 enzymes, including CYP1A1, CYP2D6, and CYP3A7, which account for up to half the total fetal liver P450 content.[27] There is active sulfation but limited glucuronidation enzyme activity in human fetal liver.[27]

In some cases metabolism can be the rate limiting determinant of xenobiotic elimination, but metabolism of a given agent may differ markedly between species or over time (e.g., because of cytochrome P450 induction) within a given species. This will generally influence and be reflected by changes in the pharmacokinetic parameters described below. However, attempts to correlate pharmacokinetic parameters with a xenobiotic's toxicity potential are confounded if only the parent molecule is considered and a metabolite mediates the toxicity (e.g., testicular toxicity of diethyl hexyl phthalate,[28] ethylene glycol monomethyl ether,[29] and *n*-hexane[30]; the teratogenicity of 2-methoxyethanol is discussed in later sections).

G. Elimination

The principal routes of elimination from the body are urinary excretion (typically for polar xenobiotics and metabolites), fecal excretion (for nonabsorbed material and via biliary excretion for products of hepatic metabolism), and exhalation (for gases and volatile compounds). As a result of physiological changes, the rate of compound elimination can vary markedly during pregnancy and postnatal development, as discussed below.

Xenobiotics can also be excreted in semen. Basic xenobiotics tend to achieve semen:blood plasma concentration (S:P) ratios greater than unity (e.g., methadone[31,32] and naltrexone[33] attain S:P ratios as high as 14), neutral drugs (e.g., caffeine[34,35] acetaminophen[36]) have a ratio near 1, and acidic drugs tend to have a low S:P ratio (e.g., acetylsalicylic acid, 0.12; valproic acid, 0.07).[37,38] This apparent relationship between drug disposition in semen and drug pK_a has been attributed to the slightly acidic (pH 6.6)[39,40] prostatic fluid component of seminal plasma and the phenomenon of ion trapping.[37,41,42] Given equilibration of the nonionized form of a drug by passive diffusion, a transmembrane difference in total drug concentration depends on the pH gradient and drug pK_a. Valproic acid (pK_a = 4.95) is ionized to a lesser extent in prostatic fluid (pH 6.6) than in blood plasma (pH 7.4), where ion trapping occurs. Although seminal vesicle fluid is slightly basic (pH 7.4–8.0) and constitutes a major percentage of semen volume, partitioning into prostatic fluid appears to be a more important determinant (protein binding aside) of drug disposition in semen. On a mass balance basis, semen is not a major route of elimination (the average human ejaculate volume is about 3 ml), but xenobiotics in semen can potentially alter spermatozoa and be transferred during intercourse. This transfer can present toxicity directly (e.g., vincristine in semen may cause vaginitis[43]) or after absorption into the partner's blood. Seminal transmission to a pregnant female presents a potential concern for a teratogenic agent, but the relatively small ejaculate volume (3 ml) delimits the dose. For example, a drug with a maximum plasma concentration of 10 µg/ml that equilibrates with semen gives a body weight adjusted dose less than 1 µg/kg to a 55-kg partner. Nonetheless, malformations were produced in naive female rabbits mated to male rabbits treated with thalidomide prior to mating,[44] so implications of seminal transmission for potent teratogens should be considered.

Xenobiotics may also be excreted in milk. The pH of milk is slightly acidic relative to that of plasma, and hence weak acids, like phenytoin and valproic acid, appear in low concentration in milk because of the phenomenon of ion trapping in plasma.[45] Milk contains 3% to 5% emulsified fat that favors excretion of lipophilic drugs, such as diazepam and chlorpromazine.[45] Drugs may transfer from plasma to milk by passive diffusion or by carrier-mediated transport (e.g., cimetidine by the organic cation–transport system).[46] Most drugs have a milk to plasma ratio of 1.0 or less, and only about 15% of drugs studied achieve a milk to plasma ratio greater than twofold.[47] Movement of drug between blood plasma and milk is bidirectional, and retrograde diffusion from milk into plasma can occur as the plasma concentration declines.[45] Hence, excretion of drugs in milk ultimately depends on the rate and extent of milk secretion and not simply the milk to blood concentration ratio. Consequently, the percentage of dose excreted in milk is generally less than 1%.[48] The milk to blood plasma ratio *per se* also does not predict infant systemic exposure, which is determined by the absolute concentration in milk, quantity of milk intake, oral bioavailability, and rate of drug clearance by the infant.[45,47,49] Furthermore, oral bioavailability is influenced by the composition of the milk. For example, calcium in milk seems to inhibit the absorption of tetracyclines.[50]

H. Physiological Changes in Pregnancy

Physiological changes in pregnancy can affect absorption, distribution, metabolism, and elimination of xenobiotics.[51–53] Increased gastric emptying time prolongs residence time in the stomach and small intestine that could enhance oral bioavailability. Pulmonary tidal volume increases while interstitial changes in the lung decrease gas diffusion rates. There is increased blood flow to the skin that could enhance dermal absorption of chemicals. Plasma volume increases nearly 50% during human pregnancy,[54] along with an increase in total body water, which can result in a higher V_d. At the same time, plasma albumin levels decrease as much as 38% while most lipid fractions rise,[54] which can change %fu. There is no major sustained change in liver blood flow or xenobiotic metabolizing enzyme activities during pregnancy.[53] However, a xenobiotic with low hepatic extraction due to high protein binding in nonpregnant females may be subject to enhanced hepatic elimination during pregnancy if %fu increases.[53] Human renal blood flow and glomerular filtration rate increase up to 33% and 66%, respectively,[54] which promotes more rapid clearance of xenobiotics eliminated predominantly unchanged by the kidneys. There can be significant species differences in gestational physiology. For example, contrary to the situation in humans, there is no increase in renal blood flow in pregnant rabbits.[55]

I. Physiological Changes in Neonates

Reproductive toxicology studies can include indirect (via milk) or direct administration of test articles to immature animals. Neonates, particularly rodents because of their relatively short period of fetal development, and infants have physiological differences from adults that can result in qualitative or quantitative differences in compound absorption, distribution, biotransformation, or elimination including:

- An immature gastrointestinal tract (e.g., fewer layers of intestinal epithelium, longer gastric emptying time, lower peptic activity),[56] which may result in enhanced absorption of substances ranging from lead[57] to insulin[58]
- Exponential development of lung alveoli from birth to 4 years of age in children[59]
- Greater total body water and extracellular fluid volume as a percentage of body weight that can affect xenobiotic distribution[60]
- Generally lower plasma protein binding of drugs in newborns[61]
- Incompletely developed blood-brain barrier with greater hematoencephalic exchange[62]
- Minimal or low levels of xenobiotic-metabolizing enzymes[60,63–65]
- Reduced biliary excretion of xenobiotics[66,67]
- Low glomerular filtration rate (about one-third the adult rate) and renal tubular secretion rates[68,69] which can result in a slower rate of renal excretion of xenobiotics[60]

These differences can sometimes account for the greater toxicity of certain compounds in neonates. The use of analgesics during labor and delivery can result in transplacental exposure followed by clinical toxicity in the neonate because of the slow rate of drug elimination. For example, meperidine readily equilibrates across the placenta, but its elimination half-life in the neonate is two- to sevenfold longer than in the adult.[70] Moreover the incomplete blood-brain barrier potentially increases the risk of respiratory depression in the infant.[71] A second classical example is that of chloramphenicol, an anti-infective that produced cases of "gray-baby syndrome" (cyanosis) in infants because of hematotoxicity (blood dyscrasia).[72,73] About 90% of chloramphenicol is excreted as a monoglucuronide adduct in adult humans. Premature and newborn infants have low rates of both glucuronide conjugation and glomerular filtration that result in a longer half-life (10 to 48+ h in newborns vs. 4 h in adults)[74] with elevated and prolonged plasma concentrations of chloramphenicol. Greater sensitivity to toxicants is not always simply a matter of less drug metabolism in the infant. Theophylline is extensively metabolized in adults, with little free drug excreted in urine. The half-life of theophylline is extended in infants because of an undeveloped rate of N-demethylation,[75] but at the same time there is active methylation of theophylline to produce caffeine,[76] itself having a half-life nearly 20 times greater in infants (100 h)[77] vs. adults (4.9 h).[78]

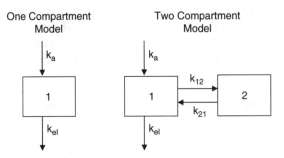

Figure 13.1 One- and two-compartment pharmacokinetic models.

III. CLASSICAL PHARMACOKINETICS

In classical pharmacokinetics, the body is represented by theoretical compartments of drug distribution.[79–81] A complete model would require many compartments to account for the amount of compound distributed among intestinal lumen, blood, various tissues, bile, feces, urine, etc. In practice, data on chemical concentration in each "compartment" are rarely collected, but blood plasma concentration data are readily obtained and generally provide a reliable indication of systemic exposure. However, in some cases (e.g., placental transfer studies), tissue samples must be collected to meet the objectives of the toxicokinetic study.

A. Simple Compartmental Models

In basic modeling, the body is represented by one or two "compartments." The simplest model is a one-compartment model in which blood plasma (or whole blood) is assumed to be in rapid equilibrium with all other compartments, including those where compound absorption (k_a) and elimination (k_{el}) occur. A two-compartment model adds a peripheral compartment representing all tissues that do not rapidly equilibrate with blood plasma (Figure 13.1). The latter model does not imply equivalent compound concentrations in those tissues, but only that changes in compound concentration in those tissues occur proportionally to changes in compound concentration in the peripheral compartment.

In a typical toxicokinetic study, blood samples are collected at various times after compound administration (oral in this example), compound concentration in the blood plasma samples is measured, and the data are presented graphically (Figure 13.2). Note how the aforementioned processes of absorption, distribution, and elimination conceptually relate to the shape of the curve.

B. Basic Pharmacokinetic Parameters

The fundamental parameter used to describe systemic exposure is $C_{(time)}$, compound concentration at a given time. $C_{(time)}$ is given in units of concentration (e.g., μg/ml). $C_{(0)}$ indicates initial concentration after bolus intravenous administration (determined by extrapolation back to the moment of injection). For intravenous infusion studies when the rate of compound administration equals the rate of elimination, C_{ss} refers to the steady-state concentration. The basic parameters that describe the blood plasma drug concentration vs. time curve are:

C_{max}: the peak concentration
t_{max}: the time (min or h) at which C_{max} occurs
AUC_{0-t}: area-under concentration vs. time curve from time (zero) to time (t), in the units of the product of time and concentration (e.g, min × μg/ml)
$AUC_{0-\infty}$: pertains to AUC from time zero until the concentration is virtually zero.

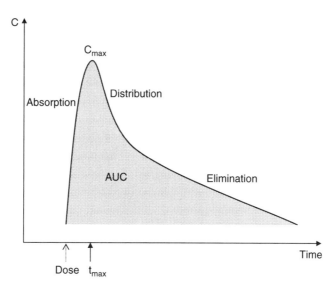

Figure 13.2 Typical blood plasma drug concentration vs. time curve following a single oral dose, depicting the maximum plasma drug concentration (C_{max}) achieved at time t_{max}. Shading indicates the area under the concentration vs. time curve (AUC).

C_{max} and AUC can be determined graphically or computationally, with the AUC calculated as the sum area of trapezoids defined by the sampling time points on the horizontal axis. C_{max} indicates the highest systemic concentration achieved, whereas the AUC gives an indication of the totality of systemic exposure. Depending upon route of administration and the rates of absorption, distribution, and elimination, a wide variety of blood plasma concentration vs. time profiles may be observed. As will be discussed subsequently, a key application of toxicokinetics is to identify the exposure metric (e.g., C_{max}, C_{ss}, or AUC) that best correlates with toxicity potential. Experimentally, different routes of administration can be used to produce profiles having comparable AUC while C_{max} is significantly different, or vice versa (Figure 13.3).

The net rate of compound elimination from the central compartment (blood plasma in these examples) is calculated from the semilogarithmic plot of compound concentration vs. time (Figure 13.4). If such a plot is linear, the data are well described by the one-compartment model, and the slope of the resulting straight line gives the elimination rate constant (k_{el}) in the reciprocal unit of time (e.g., min^{-1} or h^{-1}):

$$k_{el} = 2.303 \times \text{slope} \qquad (13.2)$$

Another useful pharmacokinetic parameter is the apparent elimination half-life ($t_{1/2}$), the time for the compound concentration in the central compartment to decrease by one-half, which can be determined graphically (Figure 13.4) or computationally. The units of half-life are time (e.g., min or h). In a one compartment model:

$$t_{1/2} = 0.693 / k_{el} \qquad (13.3)$$

Typically in a two-compartment model, the terminal elimination half-life ($t_{1/2}$) is quoted. More precisely, there are two composite rate constants (k_α and k_β) reflecting distribution (k_{12} and k_{21}) and elimination (k_{el}). These can be used to calculate $t_{1/2\alpha}$ and $t_{1/2\beta}$, respectively, and one can compute an "effective" half-life for the central compartment from the AUC-weighted average of $t_{1/2\alpha}$ and $t_{1/2\beta}$, i.e., effective half-life = $(t_{1/2\alpha})(fAUC_\alpha) + (t_{1/2\beta})(fAUC_\beta)$, where $fAUC_\alpha$ and $fAUC_\beta$ represent the fraction of the total AUC marked by lines α and β, respectively (see Figure 13.4).

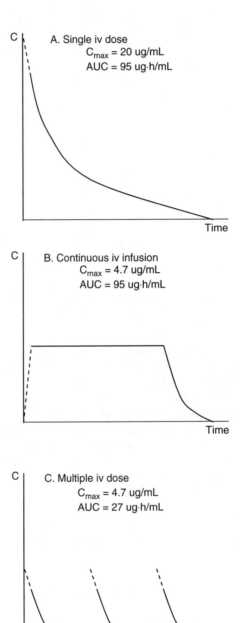

Figure 13.3 Theoretical blood plasma compound concentration vs. time curves following various dosing regimens. In panels 3a (single bolus intravenous injection) and 3b (continuous intravenous infusion), the AUC_{0-t} is comparable while the C_{max} differs. In panel 3c (divided bolus intravenous injections), the C_{max} is comparable while the AUC_{0-t} differs from that in panel 3b.

Clearance (Cl) is a concept related to half-life. Clearance from the body depends on biotransformation and excretion. It is the virtual volume of the central compartment that is completely "cleared" of the xenobiotic in a given unit of time, hence the units are in terms of flow (e.g., liter/min). Just as the virtual volume V_d can be experimentally determined following intravenous administration:

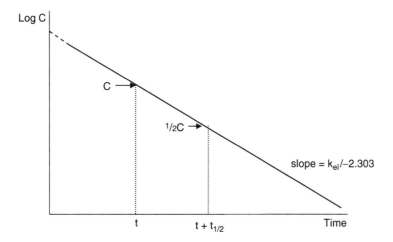

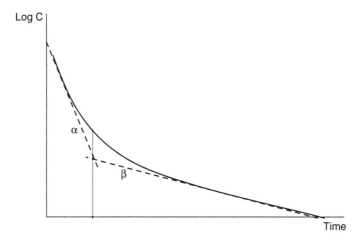

Figure 13.4 Semilogarithmic plots depicting one-compartment (top) and two-compartment (bottom) models. In the one-compartment model, first-order kinetics follow, and the plot reveals a straight line with decreasing slope proportional to the elimination rate and inversely proportional to $t_{1/2}$. In the two-compartment model, the plot shows a decreasing exponential curve that can be resolved into two component straight lines (α and β). Their slopes reflect hybrid rate constants (α and β) and that are inversely proportional to $t_{1/2\alpha}$ and $t_{1/2\beta}$, respectively.

$$V_d \text{ (l)} = \text{intravenous dose (mg)} / C(0) \text{ (mg/l)} \quad (13.4)$$

Cl can be computed from:

$$\text{Cl (l/min)} = \text{intravenous dose (mg)} / \text{AUC}_{(0-\infty)} \text{ (min} \times \text{mg/l)} \quad (13.5)$$

V_d and Cl are related by the expression:

$$\text{Cl (l/min)} = [0.693 \, V_d \text{ (l)}] / t_{1/2} \text{ (min)} \quad (13.6)$$

Thus, more rapid elimination results in a shorter half-life and more rapid clearance, but clearance also depends on distribution from the central compartment. Hence, a larger V_d yields a higher rate of clearance from the central compartment.

The fraction of the dose reaching the systemic circulation (the central compartment in these examples) is referred to as bioavailability (F). Oral bioavailability indicates the extent to which a compound is absorbed into blood plasma following oral administration. Experimentally, oral bioavailability can be determined by comparing AUC following oral (AUC_{po}) and intravenous (AUC_{iv}) administration of the identical dose:

$$F = AUC_{po} / AUC_{iv} \tag{13.7}$$

where the same expression (0–t or 0–∞) of AUC is used for both.

Some useful interrelationships among the foregoing pharmacokinetic parameters are:

$$AUC_{iv} = dose(i.v.) \times t_{\frac{1}{2}} / 0.693 \, V_d \tag{13.8}$$

$$AUC_{po} = F \times dose(p.o.) \times t_{\frac{1}{2}} / 0.693 \, V_d \tag{13.9}$$

These expressions illustrate that low bioavailability, a shorter half-life, and a greater volume of distribution lower the blood plasma AUC. Conversely, high bioavailability, slow elimination (long half-life), and a limited volume of distribution increase the blood plasma AUC. Because these relationships reflect the underlying, fundamental processes of compound absorption, distribution, biotransformation, and elimination, they provide a rationale for explaining differences in toxicokinetics among different compounds or among different conditions (e.g., pregnant vs. nonpregnant) for the same compound.

IV. PHYSIOLOGICALLY BASED PHARMACOKINETICS

Physiologically based pharmacokinetic (PBPK) models depict individual or grouped organs (or organ systems) as compartments with interconnections representing pathways for compound uptake and disposition. Unlike classical pharmacokinetic models, PBPK models are not generic but are based mechanistically on anatomical, physiological, biochemical, and physicochemical parameters. These parameters include, for example, body weight, organ weight or volume, cardiac output, regional blood flow, tissue permeability and tissue to blood partition coefficients, and metabolic rate constants. Mass balance differential equations are used to calculate rates of change in the amount of xenobiotic in each compartment. Consequently, PBPK models require more data and more sophisticated computational methods than classical pharmacokinetic models. PBPK models offer a unique benefit, in that they allow extrapolation beyond the exposure paradigm, physiological state, and species used to generate the original model. This is possible because the PBPK parameters fundamentally relate to processes of absorption, distribution, biotransformation, and elimination. After defining a PBPK model for one set of conditions, the model parameters can be scaled to another set of conditions (e.g., another species, another time in pregnancy) to predict blood and tissue concentrations. This application of PBPK models is particularly useful in studies of compound disposition in pregnancy, including placental transfer, where the cost of performing experiments on multiple days of gestation and/or in multiple species can be prohibitive. An overview of selected PBPK models of pregnancy is presented below, and the reader is referred to additional sources for more information on PBPK models and their applications in toxicology.[82–85]

A. Physiologically Based Pharmacokinetic Models of Pregnancy

Early PBPK models of pregnancy described the disposition of drugs (tetracycline,[86] morphine,[87] theophylline,[88] methadone,[89] and pethidine[90]) among blood and various maternal and fetal tissues. These models successfully described drug distribution, identifying major reservoirs and the impact of physiological changes (e.g., blood flow) on drug elimination. In some cases, they were used to simulate human tissue concentration vs. time curves in the mother and fetus.

A PBPK model described by Fisher et al.[91] was among the first to incorporate gestation time-dependent physiological parameters, including maternal body weight, maternal body fat content, blood flow, and placental and fetal growth (the last two were not modeled prior to gestation day [GD]14). In addition the model employed different partition coefficients and metabolic constants for naive versus pregnant rats. Pregnant rats were exposed throughout gestation to trichloroethylene (TCE) by inhalation, bolus gavage or drinking water, and the maternal and fetal blood levels of TCE and plasma levels of its oxidative metabolite trichloroacetic acid (TCA) were measured on GD 20 or 21. Values generated by computer simulation with the PBPK model compared favorably with actual data generated at isolated time points near term. Although the "validation" data are scant, in principle this model could be used to simulate maternal and fetal exposure to TCE and TCA at any point from GD 14 to term following either inhalation or oral exposure.

Further evolution of the rat pregnancy PBPK model was evident in the work of O'Flaherty et al.[92] in their pharmacokinetic studies of dimethyloxazolidine-2,4-dione (DMO). This model includes a more complete (entire period of gestation) and more extensive description, for both rat and mouse, of gestation time-dependent changes in maternal, placental, and embryo or fetal tissue volumes, blood flow, and pH. The disposition of DMO, a weak organic acid with high %fu, was primarily influenced by compartment pH and volume. A variation of this PBPK model has been used to study the disposition of 2-methoxyethanol (2-ME) and its metabolite 2-methoxyacetic acid (2-MAA),[93] a rodent teratogen. Model simulation agreed well with actual concentrations in maternal plasma (2-ME, 2-MAA), the embryo (2-MAA), and extraembryonic fluid (2-MAA) following bolus gavage or subcutaneous injection of mice on GD 11. PBPK modeling was subsequently applied to study the maternal-fetal disposition of 2-MAA in mice on GD 8 and 13 following intravenous administration of 2-ME,[94] and in pregnant rats on GD 11 to 15 following oral, intravenous,[95] or inhalation exposure to 2-ME.[96] Furthermore, a PBPK model for human pregnancy was developed using the general structure of the rat model but with partition coefficients and physiologic and metabolic constants for humans.[96] PBPK modeling provided a rational basis for predicting systemic levels of 2-MAA arising from 2-ME inhalation in humans and rats, taking into account several key interspecies differences (e.g., hepatic clearance of 2-ME, half-life of 2-MAA, and ventilation rate). This model was used to predict the 2-ME inhalation exposure concentration estimated to yield maternal plasma C_{max} and AUC values in women corresponding to those associated with developmental toxicity in rats.

The foregoing provides a foundation for PBPK pregnancy model representation and parameterization, including established values for gestation time-dependent anatomical and physiological parameters. Application of such models would require consideration of a given compound's absorption, distribution, metabolism, and elimination, including the generation of chemical-specific parameters (e.g., partition coefficients, uptake, metabolism, and excretion rate constants). Given the magnitude of experimentation required to perform classical pharmacokinetics on multiple days of gestation by multiple routes (and bearing in mind that the collection of fetal tissue requires terminal sampling procedures), there is high potential value for the application of PBPK modeling in developmental toxicology and risk assessment.[97,98] This is further enabled by the recent development of a comprehensive (including 27 maternal and 16 embryo-fetal compartments) PBPK model for pregnancy that can be fitted with either rat or human data, permitting comparisons across species and at multiple times of gestation.[99,100]

V. SAMPLING STRATEGIES

Generally six to eight time points over a 24-h period are required to produce an accurate plasma concentration vs. time curve. A serial sampling strategy obtains a sample from treated subjects at

Table 13.2 Example of a composite sampling strategy

Time (h)	Subject Identification Number											
	1	2	3	4	5	6	7	8	9	10	11	12
0	✓	✓	✓									
0.5				✓	✓	✓						
1.0							✓	✓	✓			
2.0										✓	✓	✓
4.0	✓	✓	✓									
8.0				✓	✓	✓						
12							✓	✓	✓			
24										✓	✓	✓

The general relationship is A = B, where A = (number of animals sampled) × (number of time points sampled per animal) and B = (number of time points) × (samples/time point)

Table 13.3 Sampling strategies for adult animals

Species	Status[a]	Site	Frequency (per 24 h)	Sample Blood Volume (ml)[b]
Rat	Live	Tail vein	1 to 8	0.2
Rat	Terminal	Vena cava	1	5.0
Mouse	Terminal	Vena cava	1	0.5
Rabbit	Live	Ear vein	1 to 8	0.5

[a] Where status = live, either a serial or composite sampling strategy may be used.
[b] Approximate blood volumes offered for general reference only. Typically within a 24-h period, the maximum blood withdrawn from adult animals ranges from 7.5% to 20% of total blood volume, depending upon the recovery period; for further guidance refer to Diehl, K.-H., et al., *J. Appl. Toxicol.*, 21, 15, 2001.

each time point. This affords individual subject concentration vs. time curves, and pharmacokinetic parameters can be calculated for individual subjects. These parameters can be compared with individual subject toxicity data or can be averaged to obtain group mean pharmacokinetic parameters. This strategy can be used when the total sampling volume required to obtain all time points is compatible with the test conditions (species, test article, analytical sensitivity). In some situations, there can be a concern for the influence of subject restraint and/or blood or milk withdrawal on toxicology study endpoints. In this case, a separate ("satellite") group of animals, given the same treatment regimen as the toxicology study animals, can be serially sampled to obtain toxicokinetic data only. Serially sampling a minimum of three animals is typically needed to characterize systemic exposure for a given treatment regimen.

Some test articles have an effect on blood pressure or blood volume that limits the volume of blood samples obtainable. Also, for some species or ages of animals, it may not be appropriate to remove the required amount of blood to cover all the desired time points (see Table 13.2, Table 13.3, and Table 13.4). Both of these situations require a composite sampling strategy. A composite sampling strategy obtains samples from individual subjects at only one or a subset of the desired time points. For example, 12 subjects per group, with each sampled at 2 time points, yields 3 independent blood samples at each of 8 time points (Table 13.2).

The composite sampling strategy can be applied to toxicology study animals to minimize blood sampling influences on study endpoints. Composite sampling can also be applied to satellite study animals when terminal sampling is required; for example when a relatively large volume of blood per sample is needed or when taking tissue (target organs, embryos, fetuses) samples.

Table 13.4 Sampling strategies for embryos and fetuses

Species	Day Postcoitus	Sample	Sample Blood Volume (ml)[a]
Mouse	10 to 18	Tissue	—
Rat	9 to 20	Tissue	—
Rat	21	Blood (decapitation)	0.2
Rabbit	10 to 18	Tissue	—
Rabbit	29	Blood (decapitation)	1.0

[a] Approximate blood volumes offered for general reference.

Table 13.5 Sampling strategies for neonatal rats

Age (days)	Status	Site	Sample Blood Volume (ml)[a]
2	Terminal	Decapitate	0.4
12	Terminal	Vena cava	0.8

[a] Approximate blood volumes offered for general reference.

By use of a vacuum-assisted apparatus,[102] it is generally possible to collect 1 to 10 ml of milk from lactating rats injected with oxytocin. An alternative approach is to euthanize the suckling, neonatal animal, from which one can collect mother's milk from the neonate's stomach. Neonatal blood can then be obtained to assess neonatal systemic exposure.

VI. REGULATORY EXPECTATIONS

The only comprehensive regulatory guidance relating to toxicokinetics in reproductive and developmental toxicology was written under the auspices of the International Conference on Harmonization (ICH). The topic is addressed both in the *Guideline on Detection of Toxicity to Reproduction for Medicinal Products* (ICH Topic S5A)[103] and in the guideline entitled *The Assessment of Systemic Exposure in Toxicity Studies* (ICH Topic S3A).[104] These guidelines are intended to apply to nonclinical toxicity studies of human medicinal products.

A. ICH Reproduction Test Guideline

This guideline uses the term "kinetics" equivalently to the term "toxicokinetics" as used in this chapter. The guideline discusses kinetics as it relates to study design and as it relates to study interpretation. Relevant passages of the guideline are quoted verbatim, along with this author's comments reflecting practical experience.

Kinetic information may be obtained in a variety of toxicology studies preceding the conduct of reproductive or developmental toxicity studies. Such information can be extremely helpful in the design of the latter studies. "It is preferable to have some information on kinetics before initiating reproduction studies since this may suggest the need to adjust choice of species, study design, and dosing schedules. At this time the information need not be sophisticated nor derived from pregnant or lactating animals."

The following paragraphs describe the application of kinetics to species selection, study design, and dosing rationale.

According to ICH guidelines, kinetic information could result in a justification for a single, relevant (to humans) species or for the use of two rodent species for embryo-fetal development studies (normally both a rodent and a nonrodent species would be required). The rationale to justify a single species would include kinetic (including test compound metabolism) and pharmacological data showing that the species is uniquely relevant to humans. Deselection of a nonrodent species may be justified by kinetic data, for example, if the maximum practical or tolerated dose results in a systemic exposure level less than would be seen in humans.

By ICH guidelines, a rodent embryo-fetal development study may be conducted as a stand-alone study (embryo-fetal development study, 4.1.3) or as an extension of the rodent fertility study (fertility, early embryonic, *and* embryo-fetal development study, 4.1.1 *and* 4.1.3). Furthermore, ICH Guidelines indicate: "For drugs where alterations in plasma kinetics are seen following repeated administration, the potential for adverse effects ... may not be fully evaluated ...." It stands to reason that for drugs that induce their own metabolism (so-called "autoinducers"), the systemic exposure level during organogenesis might be much lower if dosing were initiated before mating (as in study 4.1.1 *and* 4.1.3) than if dosing began at the time of implantation (study 4.1.3). Other compounds may reach a higher level of systemic exposure upon repeated administration (accumulate), in which case if dosing was initiated at the time of implantation (study 4.1.3), the systemic exposure level during organogenesis might be much lower than if dosing began before mating (study design 4.1.1 *and* 4.1.3).

Kinetics often plays a decisive role in the rationale for dose selection. For poorly bioavailable compounds, kinetic data can show that a plateau in systemic exposure is reached at a certain dose. That dose may be justifiable as a high dose because there is, "little point in increasing administered dosage if it does not result in increased plasma or tissue concentration." Selection of lower dosages is made, "depending on kinetic and other toxicity factors." Such kinetic factors include the anticipated level of systemic exposure in humans. The lowest animal test dose is generally selected to provide a small multiple of the C_{max} and AUC measured in humans at therapeutic dosages. Animal test doses should also be selected to achieve good separation in the systemic exposure levels among the groups. Use of kinetics data can point to a dose selection surprisingly different from the typical proportional spacing of administered doses because compounds frequently exhibit nonproportional increases in systemic exposure with increasing administered dose. For a compound with greater than dose-proportional systemic exposure, the AUC at the usual middose might be undesirably close to that at the high dose if kinetic information were not considered.

Generally the route of administration to animals is selected to match the clinical route of administration. ICH guidelines indicate: "If it can be shown that one route provides a greater body burden, e.g., ... (AUC), there seems little reason to investigate routes that would provide a lesser body burden ...." This can be a practical advantage for the conduct of animal studies when the clinical route of administration is difficult to use in animals (e.g., intranasal) or when the clinical route of administration results in poor systemic exposure in the test animals.

Some medicines are used clinically by multiple routes of administration. For the supporting animal toxicology studies, "One route ... may be acceptable if ... a similar distribution (kinetic profile) results from different routes." Therefore, a single route of administration in nonclinical studies may support multiple clinical routes if the selected route of administration results in animal systemic exposure that sufficiently exceeds human C_{max} and AUC by the clinical routes.

According to ICH guidelines, "The usual frequency of administration is once daily but consideration should be given to use either more frequent or less frequent administration taking kinetic variables into account." This can apply if the drug's half-life in the preclinical species is much shorter than in humans, especially considering that embryonic development is more rapid in rodents than in humans. For example, if a drug with an elimination half-life less than 2 h is given once daily, there would be minimal systemic exposure for more than 12 h each day, and multiple daily administration (e.g., b.i.d. or t.i.d.) should be considered. As an alternative to more frequent administration, another method of drug administration (e.g., continuous infusion) could be selected to prolong the duration of exposure or a means (e.g., dietary or subcutaneous depot administration) that prolongs the absorption phase could be employed.

In addition to the use of kinetics to assist with study design, kinetics is important for study interpretation: "At the time of study evaluation further information on kinetics in pregnant or lactating animals may be required according to the results obtained." Also, kinetics in pregnant animals "may pose problems due to the rapid changes in physiology. It is best to consider this as a two or three phase approach." This may refer to the fact that kinetics evaluated at an *a priori* selected time during gestation (e.g., near term) may not provide relevant toxicokinetic information for developmental toxicity induced at a different time in gestation (e.g., during organogenesis) because of changes in maternal physiology and placentation that occur throughout pregnancy.

B. ICH Toxicokinetic Guideline

According to this guideline, toxicokinetics are not an absolute requirement in reproduction studies: "The limitation of exposure in reproductive toxicity is usually governed by maternal toxicity. Thus, while toxicokinetic monitoring in reproductive toxicity studies may be valuable in some instances, especially with compounds with low toxicity, such data are not generally needed for all compounds."

The interpretation of these statements could be that toxicokinetics in reproduction studies may not add much value if: (1) dose-response relationships for toxicity are the same in nonpregnant and pregnant animals, (2) the high dose could be no higher due to maternal toxicity, and (3) preexisting toxicokinetic data are available for the same species, albeit in nonpregnant animals. However, "Where adequate systemic exposure might be questioned because of absence of pharmacological response or toxic effects, toxicokinetic principles could usefully be applied ...," and "Consideration should be given to the possibility that the kinetics will differ in pregnant and nonpregnant animals." So, when toxicity data from nonpregnant animals are used to select doses for pregnant animals and the expected maternal effects do not occur (rationale for dose selection did not hold true, possibly because of a difference in kinetics during pregnancy), toxicokinetic data may be required to determine if the high dose was associated with an adequate level of systemic exposure.

ICH guidelines do not absolutely require placental or milk transfer studies, but "In some situations, additional studies may be necessary or appropriate in order to study embryo/foetal transfer and secretion in milk." Also, "Consideration should be given to the interpretation of reproductive toxicity tests in species in which placental transfer of the substance cannot be demonstrated."

The second point could be raised when there is no effect on the embryo or fetus. It is not a concern, however, if a maternally toxic dose was used and the maternal systemic exposure level was sufficiently higher than that observed in the clinic, where it is unlikely the extent of placental transfer would be known. Nonetheless, the point should be considered when presenting negative developmental toxicity data for a compound in a therapeutic class previously associated with teratogenicity.

ICH guidelines state, "Secretion in milk may be assessed to define its role in the exposure of newborns." However, drug concentration in milk needs to be considered with respect to its bioavailability in milk. It is otherwise difficult to interpret the toxicological significance of the observed concentration (see Section II.G.). In terms of validating the exposure of the neonate in a postnatal study, measurement of the drug in pup's blood is more relevant than measurement of drug in milk. Finally, regardless of the preclinical data, product labeling will typically indicate (1) a drug has the potential to be secreted in breast milk or (2) the (more important) existence of human data on secretion of the drug in milk.

VII. TOXICOKINETICS IN DEVELOPMENTAL AND REPRODUCTIVE TOXICOLOGY

This section of the chapter combines scientific, regulatory, and practical aspects of toxicokinetics in developmental and reproductive toxicology to define how it can support nonclinical safety assessment.

A. When Are Toxicokinetics *Not* Essential?

When systemic exposure data at similar doses are available in the same species from preexisting studies, it is generally not necessary

- To generate exposure data in a rodent fertility study because preexisting data exist.
- To generate exposure data in developmental toxicity studies with animals treated up to a dose producing toxicity in the dams, given toxicokinetic data generated in nonpregnant animals up to the same doses and the high dose resulted in similar toxicity.
- To see if a compound crosses the placenta or to measure how much compound is in the embryo without regard to any specific biological question. (Because human data for placental transfer or embryonic drug level are very rare, there is no value in generating placental transfer data in animals that cannot be interpreted in the absence of a specific question.)
- To see if a drug is present in milk (see the preceding section).

B. Use of Toxicokinetics in Study Design

As mentioned previously, toxicokinetics can support study design with regard to selection of species, dose, and route of test material administration. For a species not commonly used in general toxicology studies but frequently used for developmental toxicity studies (e.g., rabbit), a preliminary toxicokinetic evaluation (not necessarily in pregnant animals) provides assurance that an adequate level of systemic exposure can be achieved at a maximally tolerated dose. A preliminary study can also look for evidence of major metabolites produced in humans. Such data may result in deselection of a conventional species in favor of a more clinically relevant species. The use of toxicokinetics in dose selection was discussed in the preceding section.

An alternative route of administration to animals can be selected to enhance systemic exposure relative to that resulting from dosing by the intended clinical route. Drugs with low oral bioavailability in animals can be associated with C_{max} and AUC values not substantially different from those seen in humans despite orally administered doses in animals that are high multiples of the clinical dose on a milligram per kilogram basis. However, an alternate route of administration (e.g., intravenous) could provide a much higher C_{max} and AUC in animals at a lower dose.

C. Use of Toxicokinetics in Study Interpretation

Often the doses for developmental toxicity studies are selected based on preceding data, usually including systemic exposure data collected in toxicology studies with nonpregnant animals. If a dose producing clear evidence of toxicity in nonpregnant animals does not do so in the maternal animals of the developmental toxicity study, it is useful to obtain toxicokinetic data in pregnant animals to determine if the lack of maternal toxicity is attributable to lower than expected systemic exposure. Despite the absence of maternal toxicity, the additional data may indicate the study is adequate if the maternal exposure level is a sufficient multiple of that observed in humans. It is also possible that an unexpectedly high degree of maternal toxicity can be observed relative to that expected from studies in nonpregnant animals. Here again, a comparison of systemic exposure between pregnant and nonpregnant animals given the same dose can help determine if the difference in toxicity is attributable to differences in kinetics or inherent susceptibility.

The outcome of developmental and reproductive toxicity studies may include: species-specific developmental toxicity, unique sensitivity of neonatal animals, and/or absence of a predicted class effect. While it is important to consider mechanisms of toxicity in these cases, an evaluation of toxicokinetics is the most practical first step and can aid in the design of mechanistic studies. A general approach for investigating such outcomes involves

Table 13.6 Administered doses of 13-*cis*-retinoic acid (13-*cis*-RA) and systemic levels of 13-*cis*-RA and 13-*cis*-4-oxoretinoic acid (13-*cis*-4-oxoRA) reported to be teratogenic in various species

Species	Dose (mg/kg)	Maternal Plasma AUC (ng·h/ml)		Embryo AUC (ng·h/g)	
		13-*cis*-RA	13-*cis*-4-oxoRA	13-*cis*-RA	13-*cis*-4-oxoRA
Mouse[a]	100	21205	N/A	449	N/A
Rat[b]	75	9226	10907	789	259
Rabbit[c]	15	49100	58500	2390	1830
Monkey[d]	2.5	2325	1784	956	590

Note: AUCs determined on last day of treatment.

[a] NMRI mice treated on day 11 postcoitus (pc) (Kochar, D.M., Kraft, J., and Nau, H., in *Pharmacokinetics in Teratogenesis, Vol. 2*, Nau, H. and Scott, W.J., Eds., chap. 16.); N/A indicates data not available.

[b] Teratology data in Sprague-Dawley rats treated on days 8–10 pc (Agnish, N.D., Rusin, G., and DiNardo, B., *Fund. Appl. Toxicol.*, 15, 249, 1990), and kinetic data from Wistar rats treated on days 7–12 pc (Collins, M.D., et al., *Toxicol. Appl. Pharmacol.*, 127, 132, 1994)

[c] Rabbits treated on days 8–11 pc (Eckhoff, C., et al., *Toxicol. Appl. Pharmacol.*, 125, 34, 1994).

[d] Cynomolgus monkey, 2.5 mg/kg once daily on days 16–26 of pregnancy then 2.5 mg/kg twice daily on days 27-31 of pregnancy (Hummler, H., Korte, R., and Hendrickx, A.G., *Teratology*, 42, 263, 1990; Hummler, H., Hendrickx, A.G., and Nau, H., *Teratology*, 50, 184, 1994).

1. Identifying the period of susceptibility
2. Evaluating maternal systemic exposure and compound metabolism
3. Evaluating embryo-fetus, pup, or target organ (e.g., testes, ovary) exposure level, including major metabolites

Because of the potential for differences in toxicokinetics according to the period of gestation or lactation, it can be important to establish the sensitive period for toxicity before conducting a toxicokinetic evaluation. So-called critical window studies may be conducted to define the most appropriate day of gestation or lactation for subsequent investigation.

If developmental toxicity occurs in one species and not another, an initial evaluation of maternal systemic exposure levels (parent molecule and major metabolites) in both species can help explain the results. If maternal systemic exposure levels are comparable between sensitive and nonresponsive species, then studies of placental or lactational transfer in both species at the relevant times of development are warranted. Because terminal sampling is required for such investigations and therefore more animals must be used, it is logical to first determine that the developmental toxicity results are not explainable simply in terms of the maternal systemic exposure data, which typically requires many fewer animals to collect.

The compound 13-*cis*-retinoic acid provides an illustration of how maternal systemic exposure and metabolism, as well as placental transfer, can be used to understand species differences in teratogenicity. The reported teratogenic oral dose of 13-*cis*-retinoic acid varies 40-fold among species (Table 13.6).

Toxicokinetic evaluations in these species revealed significant differences in metabolism. Beta-glucuronidation is the major pathway in rodents[107,111] ($t_{1/2}$ less than 1 h)[112,113] but not in rabbits or monkeys ($t_{1/2}$ approximately 10 to 13 h).[108,110] Given the relationship between half-life and AUC, it is understandable that at teratogenic doses, the maternal AUCs for 13-*cis*-retinoic acid and the active metabolite 13-*cis*-4-oxo-retinoic acid are more comparable (4- to 33-fold different) than administered dose (6- to 40-fold different). Moreover, placental transfer studies revealed significantly less placental transfer of 13-*cis*-retinoic acid and 13-*cis*-4-oxo-retinoic acid in mice, rats, or rabbits (embryo to maternal AUC ratios range from 0.02 to 0.09) vs. monkeys (embryo to maternal concentration AUC ratios of 0.41 and 0.33 for 13-*cis*-retinoic acid and 13-*cis*-4-oxo-retinoic acid, respectively).[105–108,110,111] Hence embryonic exposure levels at the teratogenic dose are much more

Table 13.7 Maternal plasma systemic exposure levels of valproic acid producing approximately equal teratogenicity in mice[116]

Dosing Regimen	Dose (mg/kg)	C_{max} (µg/ml)	AUC (µg·h/ml)
One injection[a]	356	445	652
Four injections[b]	181	225	1369
Infusion[c]	3809	248	5972

[a] Single subcutaneous injection on day 8 postcoitus (pc).
[b] Multiple subcutaneous injections every 6 h on days 7–8.5 pc.
[c] Infusion subcutaneously by implanted minipump on days 7.5–8.5 pc. $C_{max} = C_{ss}$.

comparable (0.5 to 5-fold different) than administered dose. Indeed, given that 10 mg/kg is teratogenic in rabbits,[114] and assuming dose-proportional AUC values, the embryonic AUC values associated with teratogenicity are quite similar. Taken together, these toxicokinetic data provide a rational basis for at least partly understanding species differences in teratogenic doses of 13-*cis*-retinoic acid. These include humans, for which metabolism, half-life, and low teratogenic dose threshold (0.5 to 1.5 mg/kg/day) are comparable with those of the monkey.[109,110]

To address unique sensitivity of neonatal animals (for example, unexpected pup deaths in a pre- and postnatal study), it can be useful to determine if the pups received an unusually high dose of compound via milk. It might be assumed that measurement of drug concentration in milk can provide an explanation for such events, but, as mentioned previously, drug concentration in milk (or milk to plasma concentration ratio) does not predict pup systemic exposure without taking into consideration the quantity of milk intake, oral bioavailability, and rate of drug clearance by the infant. Assessment of the plasma drug level in pups can be more useful than assessment of excretion in milk for understanding marked toxicity to the neonate because such toxicity could have resulted in part from prenatal drug accumulation. Additionally, measurement of drug concentration in the pup's blood in natural vs. cross-fostered litters can be used to determine the relative amount of a pup's drug body burden arising from prenatal versus lactational exposure.

D. Use of Toxicokinetics in Risk Assessment

The comparison of systemic exposure between test species and humans is an important element of risk assessment. It is well established that interspecies extrapolation can be grossly inaccurate if made solely on the basis of administered dose (e.g., 13-*cis*-retinoic acid) because of significant species differences in the absorption, distribution, metabolism, and/or excretion of xenobiotics. Pharmacokinetic parameters can be used quantitatively in risk assessment, but this begs the question as to which parameters are most influential for expression of toxicity. Experimentally, this can be explored by measuring changes in the incidence or severity of toxicity in response to changes in the mode of compound administration that dramatically alter systemic exposure. This approach is illustrated below by way of two examples.

Valproic acid is an anticonvulsant drug for the treatment of certain seizure disorders. It produces neural tube defects in the mouse (exencephaly) but not in the rat, rabbit, or monkey.[115] Manipulation of dose and route was used by Nau[116] to investigate relationships among C_{max}, AUC, and teratogenic potential in the mouse. Analyses of equally teratogenic regimens suggested that peak drug level (C_{max}), but not administered dose or AUC (which varied by approximately 10-fold), was the most predictive dose metric, i.e., a threshold concentration must be exceeded (Table 13.7).

The mouse C_{max} associated with teratogenicity is not much beyond the range of plasma drug concentration (up to 160 µg/ml) reported in the clinic following daily doses ranging up to 2400 mg,[117] although serum protein binding of valproic acid is more extensive in humans than in the mouse

(%fu = 8 and 65, respectively).[118] Nonetheless, prenatal exposure to valproic acid in human pregnancy has been associated with an increased risk of spina bifida,[119–121] and the studies by Nau[115,116] suggest that antiepileptic therapy that minimizes high peak concentrations of valproic acid could reduce the risk of teratogenicity.

Another example is provided by studies with cyclophosphamide, a cytotoxic agent that produces a range of defects in all species tested.[122] Reiners and coworkers[123] studied developmental toxicity in mice administered cyclophosphamide subcutaneously (s.c.) by single injection vs. continuous infusion by implanted minipumps. A relatively low maternal plasma cyclophosphamide concentration (0.09 μg/ml) maintained for 24 h was estimated to be comparably teratogenic to a single injection achieving a maternal plasma concentration of 13 μg/ml, suggesting that AUC (estimated to be 2 to 3 μg·h/ml), rather than C_{max} or administered dose, is the most predictive dose metric. Note that although a cyclophosphamide metabolite, phosphoramide mustard, may be the proximate teratogen,[124] it is expected to be present in the same proportion to the parent drug after injection or infusion because the same route (s.c.) was used. The plasma half-life of cyclophosphamide in the mouse is about 10 min[124] and in humans is about 9 h.[125] For a patient given a clinical dose (6.4 mg/kg, i.v.), the observed C_{max} was approximately 13 μg/ml, and, owing to the prolonged half-life in humans, the cyclophosphamide AUC is greater than 400 μg·h/ml and that for phosphoramide mustard is about 9 μg·h/ml,[125] exposure levels significantly above those associated with developmental toxicity in mice.

Whereas the teratogenic potential for some agents seemingly best correlates with C_{max} (e.g., valproic acid, caffeine) and for other agents it is AUC (e.g., cyclophosphamide, retinoids),[126,127] it is possible that different manifestations of developmental toxicity induced by the same agent can have different predictive dose metrics. Following 2-ME exposure of mice on GD 8, there appears to be a correlation between peak concentrations of 2-MAA and the occurrence of exencephaly,[128] but on GD 11 there is a correlation between the AUC for 2-MAA and the incidence of digit malformations.[129] The latter relationship holds true for the rat despite a very different plasma elimination half-life for 2-MAA in rats (approximately 20 h)[130] compared with mice (approximately 6 h).[129]

The litter incidence of dexamethasone-induced developmental toxicity (resorptions, death, cleft palate) among rats given an identical maternal dose (0.8 mg/kg/day on GD 9 through 14) was not closely correlated with either maternal plasma C_{max} or AUC values.[131] Instead, the maternal plasma $t_{1/2\alpha}$ was positively correlated with the percentage of the litter affected, suggesting that elapsed time during which the dexamethasone concentration exceeded a threshold value was the key metric.

Similar AUCs can be achieved with very different concentration vs. time profiles, for which the elapsed time during which a given concentration is exceeded can be markedly different (Figure 13.5). Hence, an AUC value is a generic concentration-time product that *per se* does not precisely define a toxicity threshold in terms of magnitude or duration of a critical plasma concentration. In this context, a C_{max} can be viewed as achieving a given concentration for a very brief period. Most precisely toxicokinetics should define both the critical concentration and time thresholds, which are only approximated by C_{max} or AUC.

As the foregoing examples illustrate, a more accurate assessment of risk is made possible by characterization of maternal systemic exposure than by extrapolation of administered dose. In principle, the risk of direct (as opposed to maternally mediated) developmental toxicity depends on the toxicants' concentration in the embryo or fetus. However, such data are generally not available for humans, whereas dose-systemic exposure relationships for the mother can be obtained or extrapolated from similar data in men or nonpregnant women. Establishing correlations between maternal systemic exposure and the incidence of animal teratogenicity is further substantiated when embryonic and maternal concentrations have a similar profile, as has been shown for 2-MAA in rats following maternal 2-ME treatment.[132]

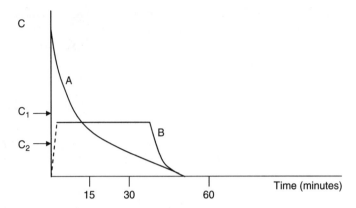

Figure 13.5 Theoretical blood plasma compound concentration (C) vs. time curves following a single bolus intravenous injection (A) or continuous intravenous infusion (B). The AUCs are comparable, whereas the elapsed time exceeding a certain concentration (C_1 or C_2) is different. The higher of these concentrations (C_1) is only exceeded following the bolus dose, while the duration of time during which C_2 is exceeded is greater following infusion.

VIII. CONCLUSIONS

The rate and extent of absorption, distribution, biotransformation, and elimination influence the toxicity potential of xenobiotics. Pharmacokinetic parameters reflecting these events are used in developmental and reproductive toxicology study design, in species selection, and in the rationale for route and dose selection. Although general assessments of systemic exposure and metabolism can be collected in a relatively standardized manner, toxicokinetic assessments are ideally made in the context of having first identified the nature of the toxicity, critical "windows" of exposure, and sensitive vs. nonresponsive species. This is because toxicokinetics is not merely the assessment of pharmacokinetics in the toxicology laboratory but is the determination of the relationships between the compound's absorption, distribution, biotransformation, and elimination, and the expression of toxicity. These relationships, based on classical pharmacokinetic parameters and physiologically based models, can be used to explain experimental results and to extrapolate among different exposure conditions or between species. In that regard, toxicokinetics plays an integral role in the risk assessment of developmental and reproductive toxicity.

REFERENCES

1. Crawford, J.S., The placenta, drugs, and the fetus, in *Principles and Practice of Obstetric Anesthesia*, 5th ed., Blackwell Scientific Publications, Oxford, 1984, p. 132.
2. Morriss, Jr., F.H. and Boyd, R.D.H., Placental Transport, in *The Physiology of Reproduction*, Knobil, E. and Neill, J., Eds., Raven Press Ltd., New York, 1988, chap. 50.
3. Rozman, K.K. and Klaassen, C.D., Absorption, distribution, and excretion of toxicants, in *Casarett and Doull's Toxicology: The Basic Science of Poisons*, 6th ed., Klaassen, C.D., Ed., McGraw-Hill, New York, 2001, chap. 5.
4. Landas, G.R., et al., Placental p-glycoprotein deficiency enhances susceptibility to chemically induced birth defects in mice, *Reprod. Toxicol.*, 12, 457, 1998.
5. Goldstein, A., Aronow, L., and Kalman, S.M., *Principles of Drug Action: The Basics of Pharmacology*, 2nd ed., John Wiley & Sons, New York, 1974, chap 2.
6. Goldstein, A., Aronow, L., and Kalman, S.M., *Principles of Drug Action: The Basics of Pharmacology*, 2nd ed., John Wiley & Sons, New York, 1974, p. 821.

7. Dym, M. and Fawcett, D.W., The blood-testis barrier in the rat and the physiological compartmentation of the seminiferous epithelium, *Biol. Reprod.*, 3, 300, 1970.
8. Setchell, B.P., Vogimayr, J.K., and Waites, G.M.H., A blood-testis barrier restricting passage from blood lymph into rete testis fluid but not into lymph, *J. Physiol.*, 200, 73, 1969.
9. Okumura, K., Lee, I.P., and Dixon, R.L., Permeability of selected drugs and chemicals across the blood-testis barrier of the rat, *J. Pharmacol. Exp. Ther.*, 194, 89, 1975.
10. Hendrikse, N.H., et al., Complete *in vivo* reversal of p-glycoprotein pump function in the blood-brain barrier visualized with positron emission tomography, *Br. J. Pharmacol.*, 124, 1413, 1998.
11. Hughes, C.S., et al., Modulation of doxorubicin concentration by cyclosporin A in brain and testicular barrier tissues expressing p-glycoprotein in rats, *J. Neuro-Oncology*, 37, 45, 1998.
12. Choo, E.F., et al., Pharmacological inhibition of p-glycoprotein transport enhances the distribution of HIV-1 protease inhibitors into brain and testes, *Drug Metabol. Disposition*, 28, 655, 2000.
13. Sieber, S.M. and Fabro, S., Identification of drugs in the preimplantation blastocyst and in the plasma, uterine secretion and urine of the pregnant rabbit, *J. Pharmacol. Exp. Ther.*, 176, 65, 1971.
14. McLachlan, J.A., et al., Accumulation of nicotine in the uterine fluid of the six-day pregnant rabbit, *Fertil. Steril.*, 27, 1204, 1967.
15. Beck, F., Comparative placental morphology and function, in *Developmental Toxicology*, Kimmel, C.A. and Buelke-Sam, J., Eds., Raven Press, New York, 1981, p. 35.
16. Amoroso, E.C., Placentation, in *Marshall's Physiology of Reproduction*, 3rd ed., Parkes, A.S., Ed., Longmans, London, 1952, Vol. II, chap. 15.
17. Mirkin, B.L., Perinatal pharmacology: placental transfer, fetal localization, and neonatal disposition of drugs, *Anesthesiology*, 43, 156, 1975.
18. Williams, R.T., *Detoxification Mechanisms*, 2nd ed., Wiley, New York, 1971.
19. Parkinson, A., Biotransformation of xenobiotics, in *Casarett and Doull's Toxicology: The Basic Science of Poisons*, 6th Ed., Klaassen, C.D., Ed., McGraw-Hill, New York, 2001, chap. 6.
20. Juchau, M.R., Mechanisms of drug biotransformation reactions in the placenta, *Fed. Proc.*, 31, 48, 1972.
21. Juchau, M.R., Drug biotransformation in the placenta, *Pharmacol. Ther.*, 8, 501, 1980.
22. Heinrichs, W.L. and Juchau, M.R., Extrahepatic drug metabolism: the gonads, in *Extrahepatic Metabolism of Drugs and Other Foreign Compounds*, Gram, T.E., Ed., SP Medical and Scientific Books, New York, 1980, p. 313.
23. Ishimura, K., et al., Ultrastructural and immunohistochemical studies on steroid-secreting cells of testis and ovary of normal and 3-methylcholanthrene-treated mice, *Cell & Tissue Research*, 245, 681, 1986.
24. Batt, A.M., et al., Enzyme induction by drugs and toxins, *Clinica Chimica Acta*, 209, 109, 1992.
25. Krishna, D.R. and Klotz, U., Extrahepatic metabolism of drugs in humans, *Clin. Pharmacokinet.*, 26, 144, 1994.
26. Slikker, W. and Miller, R.K., Placental metabolism and transfer: role in developmental toxicology, in *Developmental Toxicology*, 2nd ed., Kimmel, C.A. and Buelke-Sam, J., Eds., Raven Press, New York, 1994, p. 245.
27. Ring, J.A., et al., Fetal hepatic drug elimination, *Pharmacol. Ther.*, 84, 429, 1999.
28. Thomas, J.A., Curto, K.A., and Thomas, M.J., MEHP/DEHP gonadal toxicity and effects on rodent accessory sex organs, *Environ. Health Perspect.*, 45, 85, 1982.
29. Foster, P.M.D., et al., Testicular toxicity of 2-methoxyacetaldehyde, a possible metabolite of ethylene glycol monomethyl ether in the rat, *Toxicol. Lett.*, 32, 73, 1986.
30. Boekelheide, K., 2,5-Hexanedione alters microtubule assembly: I. Testicular atrophy, not nervous system toxicity, correlates with enhanced tubulin polymerization, *Toxicol. Appl. Pharmacol.*, 88, 370, 1987.
31. Gerber, N. and Lynn, R.K., Excretion of methadone in semen from methadone addicts; comparison with blood levels, *Life Sci.*, 19, 787, 1976.
32. Swanson, B.N., et al., Seminal excretion, vaginal absorption, distribution and whole blood kinetics of *d*-methadone in the rabbit, *J. Pharmacol. Exp. Ther.*, 206, 507, 1978.
33. Taylor, S.M., et al., The seminal excretion, plasma elimination, tissue distribution and metabolism of naltrexone in the rabbit, *J. Pharmacol. Exp. Ther.*, 213, 289, 1980.

34. Beach, C.A., Bianchine, J.R., and Gerber, N., The excretion of caffeine in the semen of men: pharmacokinetics and comparison of the concentrations in blood and semen, *J. Clin. Pharmacol.*, 24, 120, 1984.
35. Beach, C.A., et al., Metabolism, distribution, seminal excretion and pharmacokinetics of caffeine in the rabbit, *J. Pharmacol. Exp. Ther.*, 233, 18, 1985.
36. Kershaw, R.A., et al., Excretion of acetaminophen in semen of men: comparison with concentration in plasma, *J. Am. Coll. Toxicol.*, 1, 113, 1982.
37. Kershaw, R.A., et al., Disposition of aspirin and its metabolites in the semen of man, *J. Clin. Pharmacol.*, 27, 304, 1987.
38. Swanson, B.N., et al., Excretion of valproic acid into semen of rabbits and man, *Epilepsia*, 19, 541, 1978.
39. Lundquist, F., Aspects of biochemistry of human semen, *Acta Physiol. Scand.*, 19, 66, 1949.
40. White, M.A., Changes in pH of expressed prostatic secretion during the course of prostatitis, *Proc. R. Soc. Med.*, 68, 511, 1975.
41. Winningham, D.G., Nemoy, N.J., and Stamey, T.A., Diffusion of antibiotics from plasma into prostatic fluid, *Nature*, 219, 139, 1968.
42. Winningham, D.G. and Stamey, T.A., Diffusion of sulfonamides from plasma into prostatic fluid, *J. Urol.*, 104, 559, 1970.
43. Paladine, W.J., et al., Possible sensitivity to vinblastine in prostatic fluid. *N. Engl. J. Med.*, 292, 52, 1975.
44. Lutwak-Mann, C., Schmid, K., and Keberle, H., Observations of the progeny of thalidomide treated male rabbits, *Br. Med. J.*, 1, 1090, 1964.
45. Anderson, P.O., Drug use during breast feeding, *Clin. Pharm.*, 10, 594, 1991.
46. Oo, C.Y., et al., Active transport of cimetidine into human milk, *Clin. Pharmacol. Ther.*, 58, 548, 1995.
47. Ito S. and Koren G., A novel index for expressing exposure of the infant to drug in breast milk, *Br. J. Clin. Pharmacol.*, 38, 99, 1994.
48. Atkinson, H.C., Begg, E.J., and Darlow, B.A., Drugs in human milk: clinical pharmacokinetic considerations, *Clin. Pharmacokinet.*, 14, 217, 1988.
49. Ito, S., Drug therapy for breast-feeding women, *New Engl. J. Med.*, 343, 118, 2000.
50. Matsuda, S., Transfer of antibiotics into maternal milk, *Biol. Res. Pregnancy*, 5, 57, 1984.
51. Mattison, D.R., Physiological variations in pharmacokinetics during pregnancy, in *Drug and Chemical Action in Pregnancy*, Fabro, S. and Scialli, A.R., Eds., Dekker, New York, 1986, p. 37.
52. Mattison, D.R., Blann, E., and Malek, A., Physiological alterations during pregnancy: impact on toxicokinetics, *Fund. Appl. Toxicol.*, 16, 215, 1991.
53. Krauer, B., Physiological changes and drug disposition during pregnancy, in *Pharmacokinetics in Teratogenesis, Vol. 1*, Nau, H. and Scott, W.J., Eds., CRC Press, Boca Raton, 1987, chap. 1.
54. Hytten, F.E. and Thomson, A.M., Maternal physiological adjustments, in *Biology of Gestation, Vol. 1, The Maternal Organism*, Assali, N.S., Ed., Academic Press, New York, 1968, chap. 8.
55. Johnson, R.L., et al., Cardiac output distribution and uteroplacental blood flow in the pregnant rabbit: a comparative study, *Am. J. Obstet. Gynecol.*, 151, 682, 1985.
56. Koldovský, O., Digestion and absorption, in *Perinatal Physiology*, Stave, U., Ed., Plenum Medical Book Company, New York, 1978, chap. 15.
57. Calabrese, E.J., *Age and Susceptibility to Toxic Substances*, John Wiley & Sons, New York, 1986.
58. Znamenacek, K. and Pibylová, H., The effect of glucose and insulin administration on the blood glucose levels of the newborn, *Cesk. Pediatr. (Prague)*, 18, 104, 1963.
59. Angus, G.E. and Thurlbeck, W.M., Number of alveoli in the human lung, *J. Appl. Physiol.*, 32, 483, 1972.
60. Besunder, J.B., Reed, M.D., and Blumer, J.L., Principles of drug biodisposition in the neonate: a critical evaluation of the pharmacokinetic-pharmacodynamic interface (part I), *Clin. Pharmacokinet.*, 14, 189, 1988.
61. Kurz, H., Michels, H., and Stickel, H.H., Differences in the binding of drugs to plasma proteins from newborn and adult man (part II), *Eur. J. Clin. Pharmacol.*, 11, 469, 1977.
62. Himwich, W.A., Physiology and pharmacology of the central nervous system, in *Perinatal Physiology*, Stave, U., Ed., Plenum Medical Book Company, New York, 1978, chap. 29.
63. Neims, A.H., et al., Developmental aspects of the hepatic cytochrome P-450 monooxygenase system, *Ann. Rev. Pharmacol. Toxicol.*, 16, 427, 1976.

64. Klinger, W., Biotransformation of drugs and other xenobiotics during postnatal development, *Pharmacol. Ther.*, 16, 377, 1982.
65. Juchau, M.R., Fetal and neonatal drug biotransformation, in *Drug Toxicity and Metabolism in Pediatrics*, Kacew, S., Ed., CRC Press, Boca Raton, FL, 1990, p.15.
66. Klaassen, C.D., Immaturity of the newborn rat's hepatic excretory function for ouabain, *J. Pharmacol. Exp. Ther.*, 183, 520, 1972.
67. Ballatori, N. and Clarkson, T.W., Developmental changes in the biliary excretion of methylmercury and glutathione, *Science*, 216, 61, 1982.
68. Fetterman, G.H., et al., The growth and maturation of human glomeruli and proximal convolutions from term to adulthood, *J. Pediat.*, 35, 601, 1965.
69. Gomez, R.A., et al., The maturing kidney: development and susceptibility, *Renal Failure*, 21, 283, 1999.
70. Morselli, P.L. and Provei, V., Placental transfer of pethidine and norpethidine and their pharmacokinetics in the newborn, *Eur. J. Clin. Pharmacol.*, 18, 25, 1980.
71. Rajchgot, P. and MacLeod, S., Perinatal pharmacology, in *Developmental Pharmacology*, MacLeod, S., Okey, A.B., and Spielberg, S.P., Eds., Alan R. Liss, New York, 1983, p. 14.
72. Kauffman, R.E., The role of chloramphenicol in pediatric therapy, in *Developmental Pharmacology*, MacLeod, S., Okey, A.B., and Spielberg, S.P., Eds., Alan R. Liss, New York, 1983, p. 315.
73. Goldstein, A., Aronow, L., and Kalman, S.M., *Principles of Drug Action: The Basics of Pharmacology*, 2nd Ed., John Wiley & Sons, New York, 1974, pp. 291.
74. Ambrose, P.J., Clinical pharmacokinetics of chloramphenicol and chloramphenicol succinate, *Clin. Pharmacokinet.*, 9, 222, 1984.
75. Aranda, J.V., et al., Pharmacokinetic aspects of theophylline in premature newborns, *New Engl. J. Med.*, 295, 413, 1976.
76. Boutroy, M.J., et al., Caffeine, a metabolite of theophylline during the treatment of apnea in the premature infant, *J. Pediat.*, 94, 996, 1979.
77. Aranda, J.V., Turmen, T., and Sasyniuk, B.I., Pharmacokinetics of diuretics and methylxanthines in the neonate, *Eur. J. Clin. Pharmacol.*, 18, 55, 1980.
78. Busto, U., Bendayan, R., and Sellers, E.M., Clinical pharmacokinetics of non-opiate abused drugs, *Clin. Pharmacokinet.*, 16, 1, 1989.
79. Wagner, J.G., Do you need a pharmacokinetic model and, if so, which one? *J. Pharmacokinet. Biopharm.*, 3, 457, 1975.
80. Rowland, M. and Tozer, T.N., *Clinical Pharmacokinetics: Concepts and Applications*, Lea and Febiger, Philadelphia, 1980.
81. Gibaldi, M. and Perrier, D., *Pharmacokinetics*, 2nd ed., Marcel Dekker, New York, 1982.
82. Himmelstein, K.J. and Lutz, R.J., A review of the applications of physiologically based pharmacokinetic modeling, *J. Pharmacokinet. Biopharm.*, 7, 127, 1979.
83. Gerlowski, L.E. and Jain, R.K., Physiologically based pharmacokinetic modeling: principles and applications, *J. Pharm. Sci.*, 72, 1103, 1983.
84. Conolly, R.B. and Andersen, M.E., Biologically based pharmacodynamic models: tools for toxicological research and risk assessment, *Annu. Rev. Pharmacol. Toxicol.*, 31, 503, 1991.
85. Krishnan, K. and Andersen, M.E., Physiologically based pharmacokinetic modeling in toxicology, in *Principles and Methods of Toxicology*, 4th ed., Hayes, A.W., Ed., Taylor and Francis, Philadelphia, 2001, chap. 5.
86. Olanoff, L.S. and Anderson, J.M., Controlled release of tetracycline – III: A physiological pharmacokinetic model of the pregnant rat, *J. Pharmacokinet. Biopharm.*, 8, 599, 1980.
87. Gabrielsson, J.L. and Paalzow, L.K., A physiological pharmacokinetic model for morphine disposition in the pregnant rat, *J. Pharmacokinet. Biopharm.*, 11, 147, 1983.
88. Gabrielsson, J.L., Paalzow, L.K., and Nordstrom, L., A physiologically based pharmacokinetic model for theophylline disposition in the pregnant and nonpregnant rat, *J. Pharmacokinet. Biopharm.*, 12, 149, 1984.
89. Gabrielsson, J.L., et al., Analysis of methadone disposition in the pregnant rat by means of a physiological flow model, *J. Pharmacokinet. Biopharm.*, 13, 355, 1985.
90. Gabrielsson, J.L., et al., Analysis of pethidine disposition in the pregnant rat by means of a physiological flow model, *J. Pharmacokinet. Biopharm.*, 14, 381, 1986.

91. Fisher, J.W., et al., Physiologically based pharmacokinetic modeling of the pregnant rat: a multiroute exposure model for trichloroethylene and its metabolite, trichloroacetic acid, *Toxicol. Appl. Pharmacol.*, 99, 395, 1989.
92. O'Flaherty, E.J., et al., A physiologically based kinetic model of rat and mouse gestation: disposition of a weak acid, *Toxicol. Appl. Pharmacol.*, 112, 245, 1992.
93. Clarke, D.O., et al., Pharmacokinetics of 2-methoxyethanol and 2-methoxyacetic acid in the pregnant mouse: a physiologically based mathematical model, *Toxicol. Appl. Pharmacol.*, 121, 239, 1993.
94. Terry, K.K., et al., Development of a physiologically based pharmacokinetic model describing 2-methoxyacetic acid disposition in the pregnant mouse, *Toxicol. Appl. Pharmacol.*, 132, 103, 1995.
95. Hays, S.M., et al., Development of a physiologically based pharmacokinetic model of 2-methoxyethanol and 2-methoxyacetic acid disposition in pregnant rats, *Toxicol. Appl. Pharmacol.*, 163, 67, 2000.
96. Gargas, M.L., et al., A toxicokinetic study of inhaled ethylene glycol monomethyl ether (2-ME) and validation of a physiologically based pharmacokinetic model for the pregnant rat and human, *Toxicol. Appl. Pharmacol.*, 165, 53, 2000.
97. Gabrielsson, J.L. and Larsson, K.S., Proposals for improving risk assessment in reproductive toxicology, *Pharmacol. Toxicol.*, 66, 10, 1990.
98. Welsch, F., Blumenthal, G.M., and Conolly, R.B., Physiologically based pharmacokinetic models applicable to organogenesis: extrapolation between species and potential use in prenatal toxicity risk assessments, *Toxicol. Letters*, 82–83, 539, 1995.
99. Luecke, R.H., et al., A physiologically based pharmacokinetic computer model for human pregnancy, *Teratology*, 49, 90, 1994.
100. Luecke, R.H., et al., A computer model and program for xenobiotic disposition during pregnancy, *Computer Methods Programs Biomed.*, 53, 201, 1997.
101. Diehl, K.-H., et al., A good practice guide to the administration of substances and removal of blood, including routes and volumes, *J. Appl. Toxicol.*, 21, 15, 2001.
102. Smith Kline and French Laboratories, Philadelphia, *Small Animal Milking Apparatus*, U.S. Patent 3,580,220.
103. ICH Guideline S5A: Reproductive Toxicology: Detection of Toxicity to Reproduction for Medicinal Products (CPMP/ICH/386/95).
104. ICH Guideline S3A: Toxicokinetics: The Assessment of Systemic Exposure in Toxicity Studies (CPMP/ICH/384/95).
105. Kochar, D.M., Kraft, J., and Nau, H., Teratogenicity and disposition of various retinoids in vivo and in vitro, in *Pharmacokinetics in Teratogenesis, Vol. 2*, Nau, H. and Scott, W.J., Eds., CRC Press, Boca Raton, 1987, chap. 16.
106. Agnish, N.D., Rusin, G., and DiNardo, B., Taurine failed to protect against the embryotoxic effects of isotretinoin in the rat, *Fund. Appl. Toxicol.*, 15, 249, 1990.
107. Collins, M.D., et al., Comparative teratology and transplacental pharmacokinetics of all-*trans*-retinoic acid, 13-*cis*-retinoic acid, and retinyl palmitate following daily administration in rats, *Toxicol. Appl. Pharmacol.*, 127, 132, 1994.
108. Eckhoff, C., et al., Teratogenicity and transplacental pharmacokinetics of 13-*cis*-retinoic acid in rabbits, *Toxicol. Appl. Pharmacol.*, 125, 34, 1994.
109. Hummler, H., Korte, R., and Hendrickx, A.G., Induction of malformations in the cynomolgus monkey with 13-cis retinoic acid, *Teratology*, 42, 263, 1990.
110. Hummler, H., Hendrickx, A.G., and Nau, H., Maternal toxicokinetics, metabolism, and embryo exposure following a teratogenic dosing regimen with 13-*cis*-retinoic acid (Isotretinoin) in the cynomolgus monkey, *Teratology*, 50, 184, 1994.
111. Creech Kraft, J., et al., Teratogenicity and placental transfer of all-*trans*-13-*cis*, 4-oxo-all-*trans*- and 4-oxo-13-*cis*-retinoic acid after administration of a low oral dose during organogenesis in mice, *Toxicol. Appl. Pharmacol.*, 100, 162, 1989.
112. Kalin, J.R., Wells, J., and Hill, D.L., Disposition of 13-*cis*-retinoic acid and N-(2-hydroxyethyl) retinamide in mice after oral doses, *Drug Metab. Dispos.*, 10, 391, 1982.
113. Shelly, R.S., et al., Blood level studies of all-*trans*- and 13-*cis*-retinoic acid in rats using different formulations, *J. Pharm. Sci.*, 71, 904, 1982.
114. Kamm, J.J., Toxicology, carcinogenicity, and teratogenicity of some orally administered retinoids, *J. Am. Acad. Dermatol.*, 6, 652, 1982.

115. Nau, H., Species differences in pharmacokinetics, drug metabolism, and teratogenesis, in *Pharmacokinetics in Teratogenesis, Vol. 1*, Nau, H. and Scott, W.J., Eds., CRC Press, Boca Raton, 1987, chap. 6.
116. Nau, H., Teratogenic valproic acid concentrations: infusion by implanted minipumps vs. conventional injection regimen in the mouse, *Toxicol. Appl. Pharmacol.*, 80, 243, 1985.
117. Meijer, J.W.A. and Hessing-Brand, L., Determination of lower fatty acid, particularly the antiepileptic dipropylacetic acid, in biological material by means of microdiffusion and gas chromatography, *Clinica Chimica Acta*, 43, 215, 1973.
118. Nau, H. and Löscher, W., Valproic acid and metabolites: pharmacological and toxicological studies, *Epilepsia*, 25 (Suppl 1), S14, 1984.
119. Robert, E. and Guibaud, P., Maternal valproic acid and congenital neural tube defects, *Lancet*, 2, 937, 1982.
120. Robert, E. and Rosa, F., Valproate and birth defects, *Lancet*, 2, 1142, 1983.
121. Lindhout, D. and Meinardi, H., Spina bifida and in-utero exposure to valproate, *Lancet*, 2, 396, 1984.
122. Mirkes, P.E., Cyclophosphamide teratogenesis: a review, *Teratogenesis, Carcinogenesis, Mutagenesis*, 5, 75, 1985.
123. Reiners, J., et al., Teratogenesis and pharmacokinetics of cyclophosphamide after drug infusion as compared to injection in the mouse during day 10 of gestation, in *Pharmacokinetics in Teratogenesis, Vol. 2*, Nau, H. and Scott, W.J., Eds., CRC Press, Boca Raton, 1987, chap. 4.
124. Nau, H., et al., Mutagenic, teratogenic and pharmacokinetic properties of cyclophosphamide and some of its deuterated derivatives, *Mutation Res.*, 95, 105, 1982.
125. Juma, F.D., Rogers, H.J., and Trounce, J.R., The pharmacokinetics of cyclophosphamide, phosphoramide mustard and nor-nitrogen mustard studied by gas chromatography in patients receiving cyclophosphamide therapy, *Br. J. Clin. Pharmacol.*, 10, 327, 1980.
126. Nau, H., Species differences in pharmacokinetics and drug teratogenesis, *Environ. Health Perspect.*, 70, 113, 1986.
127. Nau, H., Pharmacokinetic considerations in the design and interpretation of developmental toxicity studies, *Fund. Appl. Toxicol.*, 16, 219, 1991.
128. Terry, K.K., et al., Developmental phase alters dosimetry-teratogenicity relationship for 2-methoxyethanol in CD-1 mice, *Teratology*, 49, 218, 1994.
129. Clarke, D.O., Duignan, J.M., and Welsch, F., 2-Methoxyacetic acid dosimetry-teratogenicity relationships in CD-1 mice exposed to 2-methoxyethanol, *Toxicol. Appl. Pharmacol.*, 114, 77, 1992.
130. Sleet, R.B., et al., Developmental phase specificity and dose-response effects of 2-methoxyethanol teratogenicity in rats, *Fund. Appl. Toxicol.*, 29, 131, 1996.
131. Hansen, D.K., et al., Pharmacokinetic considerations of dexamethasone-induced developmental toxicity in rats, *Toxicological Sciences*, 48, 230, 1999.
132. Hays, S.M., et al., Development of a physiologically based pharmacokinetic model of 2-methoxyethanol and 2-methoxyacetic acid disposition in pregnant rats, *Toxicol. Appl. Pharmacol.*, 163, 67, 2000.

CHAPTER 14

Cellular, Biochemical, and Molecular Techniques in Developmental Toxicology*

Barbara D. Abbott, Mitchell B. Rosen, and Gary Held

CONTENTS

I. Introduction 600
   A. Overview and Introduction 600
   B. General Approach to the Study of Mechanisms of Developmental Toxicity 600
II. Descriptive Approaches: Evaluating Protein and mRNA Expression Patterns 601
   A. Immunohistochemistry 602
   B. In Situ Hybridization 604
III. Approaches for Quantifying mRNA Expression...607
   A. Semiquantitative Methods 607
      1. Northern Blot...607
      2. Dot Blot or Slot Blot 608
   B. PCR and RT-PCR Quantitative Methods 609
      1. Real-Time RT-PCR 609
      2. Quantitative RT-PCR Based on Image Analysis of PCR Products in Electrophoretic Gels 611
   C. Solution Hybridization: RNase Protection Assay 611
   D. RNA Transcription *In Vitro*: Nuclear Run-on Assay 612
IV. Characterizing the Role of Specific Genes and Pathways of Response...........................613
   A. Transgenic Animals 613
   B. *In Vitro* Antisense RNA 615
   C. DNA Array...615
   D. Genetic Library Screening..615
   E. Reverse Southern Blots and RT-PCR *In Situ* Transcription 616
References 617

---

* Disclaimer: This chapter has been reviewed by NHEERL and by U.S. Environmental Protection Agency, and has been approved for publication. Mention of trade names of commercial products does not constitute endorsement or recommendation for use.

I. INTRODUCTION

A. Overview and Introduction

Cellular, molecular, and biochemical approaches vastly expand the possibilities for revealing the underlying mechanisms of developmental toxicity. The increasing interest in embryonic development as a model system for the study of gene expression has resulted in a cornucopia of information and approaches available to the toxicologist. The typical teratology screening test examines near-term fetuses after exposure throughout organogenesis and evaluates the potential for an exposure to disrupt development. However, this assay can also provide insight regarding the organs and/or cellular processes that might be targeted. More specific information can be derived from dose response studies, where exposure is restricted to one or only a few days during development. Examination of the target organs or cells at critical stages of morphogenesis, particularly at early time points after exposure, is important for identifying specific genes or proteins that may be involved in the response. This chapter presents an overview of some of these cellular, biochemical, and molecular biological methods and discuss their applicability to developmental toxicology, providing examples of applications of each method.

The molecular biological techniques presented in this chapter may appear to be routine and straightforward. However, most or all of these methods require a considerable investment of effort, time, patience, and laboratory resources to allow their introduction into a research program. This cautionary note is intended only to alert the investigator that these are not trivial undertakings. Even for experienced investigators, successful application of some of these techniques will not always be straightforward but may require intuitive adjustment of specific protocol conditions to suit the particular experimental situation. An investigator who has no previous experience with cellular and molecular biological methods should consider enrollment in a class that provides basic information regarding the principles and protocols. If possible, the best option is to learn a new technology in the laboratory of a mentor or colleague who is currently applying the method with success. It is important to gain an understanding of the principles and pitfalls of each assay, and there are a number of excellent manuals available, some general and others specifically addressing a single technique. Most of the major suppliers of molecular biology reagents and equipment have detailed laboratory manuals and periodic newsletters available on request. New investigators should search the literature and compile and compare variations of each technique, particularly with regard to applications of the assay under conditions similar to those for their intended study. Contacting more experienced researchers directly can also provide invaluable assistance, as most are very willing to share their methods and expertise.

Although it is beyond the scope of this chapter to provide detailed methods and protocols, several sources of such information will be cited. There are a great many excellent books and publications available, including general molecular cell biology texts[1] and manuals for molecular biological methods.[2-4] Because each technique is discussed in this chapter, examples of applications in developmental biology and toxicology are briefly described. These represent only a few of the many excellent studies that have been published, and readers are encouraged to seek out other examples from the literature that are relevant to their particular special interests.

B. General Approach to the Study of Mechanisms of Developmental Toxicity

The developmental toxicity of a particular chemical or class of chemicals is generally characterized initially in a standard screening protocol that involves exposure throughout organogenesis and evaluation late in gestation. This protocol will identify particular patterns of developmental toxicity, malformations, or other effects of interest. To identify the mechanisms leading to these effects, it is generally important to narrow the exposure to a specific developmental period (one of only a few days of exposure) and to determine the dose-response for the effect within the narrow exposure period. Identifying the target tissue and the most sensitive developmental stage allows some inferences to be made regarding cellular processes or regulatory pathways that could be the target of the toxicant. The etiology of these effects becomes the subject of detailed study at early time points following exposure, as well as throughout subsequent development and differentiation.

In general, toxicants may disrupt development through maternally mediated effects, direct effects on the conceptus (placenta, extraembryonic membranes, or the embryo), or a combination of effects on mother and conceptus. The mechanisms may involve disruption of gene expression, signal transduction, protein function, enzyme activity, membrane integrity, ionic balance, or other cellular activities critical to development. The cellular responses to these insults can include changes in cell cycle regulation, cell differentiation, cell-cell and cell-extracellular matrix interactions, cell migration, cell survival, or cell death. Depending on the toxicant and the level and timing of exposure, these biochemical and cellular alterations have the potential to affect pattern formation, organogenesis, and tissue and organ functions in later gestation and postnatally, and may result in prenatal lethality.

An overview of the potential mechanisms of developmental toxicity and a discussion of 18 signaling pathways that are considered critical to normal embryonic development was recently published by the National Academy of Sciences.[5] Study of these complex, interactive signal transduction pathways offers an approach for revealing the mechanisms involved in abnormal development and response to toxicants. The effects on gene transcription and protein synthesis could involve altered spatial and temporal patterns of expression, changed rates of synthesis or degradation, altered posttranscriptional and translational modifications, or altered activity and function in the cell. The methods described in this chapter can be applied to localize and/or quantitate the expression of specific mRNAs and proteins in these critical regulatory pathways and to search for novel protein and mRNA partners in the response pathways.

The assays used to study embryos and/or small specimens must be highly sensitive to detect subtle changes from a limited number of cells. It is also important to have methods that can evaluate gene or protein expression *in situ*; because the specimens are histologically complex, it may be necessary to identify specific cells or regions that are affected. This chapter will discuss descriptive approaches to evaluate patterns and levels of protein and mRNA expression *in situ*, quantitative and semiquantitative methods to evaluate mRNA transcription, and methods to characterize the role of specific genes and pathways or to identify novel genes that may be involved. Approaches described in the chapter include evaluating the patterns of protein expression with immunohistochemistry in sectioned tissues or in whole mount preparations and localizing mRNA through *in situ* hybridization in sections or whole mounts. Additional approaches to determine effects on mRNA expression include Northern blot and dot blot, RNase (S1) nuclease protection assay, reverse transcriptase polymerase chain reaction (RT-PCR) amplification, and nuclear run-on assays for detection of newly transcribed RNA. The role of specific genes in mediating a developmental response can be studied using transgenic mice that have constitutive expression of a gene of interest, that "report" activation of transcription of a gene, or that fail to express the gene of interest ("knockout mice"). Such mice can provide insight into the function or role of a gene in the mechanism of toxicity, as well as demonstrating redundancy of function within the embryo (if no effects of gene knockout are detectable). Effects of reducing or eliminating mRNA translation can also be studied *in vitro* by use of antisense RNA hybridization in the whole embryo, the organ, or cell culture.

II. DESCRIPTIVE APPROACHES: EVALUATING PROTEIN AND mRNA EXPRESSION PATTERNS

Methods of particular value in mechanistic studies of developmental toxicology are those that localize the effect within the embryo and have the resolution to detect effects on a particular subset of cells within the developing tissue. Two such techniques that are frequently applied in developmental biology and are valuable in mechanistic developmental toxicology are immunohistochemical localization of proteins and *in situ* hybridization to detect specific mRNAs. Both approaches localize the target molecules within the embryo to specific regions, and some have the resolution to detect expression in specific cell types. There are considerable advantages to applying these methods in tandem to describe effects on both gene and protein expression to evaluate responses to a toxicant.

A. Immunohistochemistry

For immunohistochemical methods to be used, an antibody that specifically recognizes the protein of interest must be available. There are numerous commercial antibody suppliers, and antisera are now available to study peptides involved in signal transduction, proliferation, growth factors, cytokines, and cytoskeletal and membrane peptides, hormones, and receptors. A search of the Internet can reveal useful sources of information (e.g., http://www.antibodyresource.com or http://www.abcam.com) for contract antibody suppliers, books, educational resources about antibodies, online databanks and databases, online journals, and immunology and biotechnology sites. A useful reference for finding antibodies is Linscott's Directory, which contains a list of more than 17,000 monoclonal antibodies and the companies that sell them. Contacting investigators who have previously worked with the peptide of interest may also provide leads in the search for antisera. Preparation of antibody in your own facility or contracting for preparation by an outside company may also be a possibility if resources and/or experienced personnel are available. Protocols for immunohistochemical analysis are relatively simple and are available in handbooks[6] and from companies providing antisera for these applications.

Selection of fixative and frozen or paraffin embedded sections will depend to some extent on the properties of a particular antibody. Some will give good results on both Western blots and sectioned tissues, but not all antisera are compatible with both protocols. The detection methods are also diverse, and choice of method will depend on the desired application. Fluorescent labels (rhodamine 123 or fluorescein are commonly used) are sensitive and give outstanding color photographs, but quantitation is not generally an option because of the fading of the fluorescent signal that occurs during illumination. Enzymatic detection systems are also commonly used and include horseradish peroxidase or alkaline phosphatase to generate colored products. Some but not all of the substrates produce precipitates that are stable in the alcohols and solvents used to prepare permanently coverslipped slides. The enzyme may also be incorporated as part of a complex in which a biotinylated secondary antibody complexes with avidin-biotin-peroxidase (ABC) or phosphatase. This method is useful for amplification of low levels of bound primary antibody.

Immunostaining of intact tissues or small embryos provides a three-dimensional view of protein expression and can be very helpful in evaluating changes over time during development or in response to treatment. The size of the tissue or embryo that can be successfully stained in this whole mount approach is limited, and generally only small specimens have been used (e.g., presomite mouse embryos and stages just after neural tube closure, as well as some dissected tissues, e.g., limb bud or palate).[7] The method typically involves fixation of the specimen followed by permeabilization with either a detergent, enzymatic digestion (with proteinase), or a salt solution. Incubation with hydrogen peroxide serves to inhibit endogenous peroxide and clears (makes more transparent) the tissues. The incubation with primary antibody can require a prolonged period for penetration (perhaps 4 to 5 days). The primary antibody is localized with secondary antibody complexed with peroxidase enzyme that then reacts with a substrate (e.g., 3,3′-diaminobenzidine [DAB]) to give a colored product or with a substrate such as 4-chloro-1-naphthol followed by ethanolic incubations to give a fluorescent product. The immunostained whole mount specimens are likely to degrade over time and the color of the label can change, so it is important to obtain images of the outcomes as soon as possible.

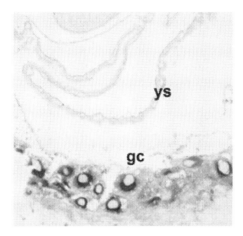

Figure 14.1 Vascular endothelial cell growth factor (VEGF) expression was detected immunohistochemically in trophoblastic giant cells of the GD 9.5 conceptus using a primary antibody complexed with a biotinylated secondary antibody that complexed with avidin-biotin-peroxidase. The substrate for the enzymatic reaction was 3,3′-diaminobenzidine (DAB), which reacted to produce a red-brown precipitate.

A specialized version of the whole mount application exists for detection of apoptosis, i.e., programmed cell death. Nuclear DNA is fragmented during programmed cell death through the apoptotic pathway, and that is diagnostic of this form of cell death. The TUNEL (terminal deoxynucleotidyl transferase [TdT]–mediated dUTP-biotin nick end labeling) method involves enzymatic labeling of the ends of the DNA fragments and localization of these tagged ends with either fluorescent label or biotin-antibody complex methods (such as the avidin-biotin-peroxidase approach). This localization in early stage, whole embryos can be very informative regarding specific sites of action and mechanisms involving cell death. The method is also applicable to sectioned tissues, and in either case there are a number of commercially available kits and reagents (Roche Diagnostics Corp, Indianapolis, IN; Gibco/BRL, Gaithersburg, MD; Oncor, Gaithersburg, MD, and others). Tissues are incubated with the enzyme to label the DNA and then processed through the detection step (fluorescent imaging or typical peroxidase enzymatic immunolabeling steps).[8,9]

For developmental studies, the immunolocalization is generally not quantitated but is described as patterns of expression (within specific cell types or certain organs) in the embryo. Many examples of this application are available in the literature, and recent studies in our laboratory include immunolocalization of members of the vascular endothelial growth factor (VEGF) pathway in placenta, extraembryonic membranes, and embryo.[10] The VEGF (vascular endothelial cell growth factor) localization in trophoblastic giant cells of a gestation day (GD) 9.5 conceptus shown in Figure 14.1 illustrates the sensitivity and resolution that can be achieved using the ABC technique with DAB substrate. Other good examples include studies localizing expression of transforming growth factor (TGF)-β isoforms in the developing embryo[11–14] or in embryonic cultured palatal mesenchymal cells.[15] Mahmood et al.[13] examined the effects of retinoic acid on expression of TGFβ in sections of early stage mouse embryos (GD 8.5 to 10.5) with the avidin-biotin-peroxidase localization method. Studies of the role of serotonin as a regulator of craniofacial and cardiac morphogenesis used whole embryo culture to evaluate the effects of inhibitors of 5-hydroxytryptophan(5-HT) uptake *in vitro*, and immunohistochemical analysis of inhibition provided valuable insight into the role of serotonin metabolism in morphogenesis and teratogenesis.[16,17]

In toxicology studies, the detection of differences between levels of expression in control versus treated tissues is an important goal, and measures of the intensity of immunohistochemical staining can provide some insight into the effects of a treatment. Semiquantitative analysis of stained sections can be performed with densitometry when the substrate is permanent (e.g., in our studies densitometry was performed using avidin-biotin-peroxidase complex with diaminobenzidine). The sections are not counterstained with histological stains because staining interferes with the densitometric analysis. For quantitative applications, extreme care must be exercised to expose all specimens to identical solutions for precisely timed incubation periods, as well as to include appropriate controls for endogenous peroxidase and nonspecific binding of either primary or secondary antibodies. Also, substrate development should be monitored on test slides to determine the reaction time that produces a density allowing detection of either greater or lesser densities in the specimens. The intensity of immunostaining can be ranked by the investigator based on viewing under the microscope (typically three levels are assigned: low = 1, medium = 2, high = 3). However, this method should be considered semiquantitative at best and susceptible to observer bias (the scorer should be "blinded" to specimen identity).

There are several computer software products available for densitometric analysis (e.g., Image-Pro Plus, NIH Image), and these can generate density values from images of immunostained sections. Capture of images through a digital camera mounted on the microscope offers a most convenient method, although print or film images are also analyzed with these programs (generally this requires scanning of the film or print to generate a digital file). In all of the densitometric methods, it is critical to acquire images under tightly controlled, uniform illumination conditions for consistency across specimens. The darkest and lightest regions must not be "off scale" (not maximal values). All images intended for densitometric analysis should be acquired with a monochrome (not color) camera, and sections should not be counterstained. The densitometric data are relative values for a specific protein, and treatment effects are most appropriately expressed as comparisons between the same cell type in the same region of treated and control sections. Such data may be statistically analyzed by ANOVA (analysis of variance), but consultation with a statistician is recommended because these data sets may require transformation or other approaches.

B. *In Situ* Hybridization

With this technique, mRNA transcribed from the gene of interest can be localized on either sectioned tissue or in whole mount preparations. These protocols require preparation of antisense probes that hybridize to the sense strand (mRNA) in the tissue. Before this method can be used, it is therefore necessary to have a gene sequence from which to synthesize the antisense probe. Through a computer search of a gene database, it may be possible to determine if the gene of interest has been sequenced and entered in the database. A literature search may identify a research group that has the cDNA (complementary DNA) for the gene of interest, and most investigators will make the material available on request after publication of its sequence.

Alternatively, the American Type Culture Collection (ATCC) maintains an NIH repository of human and mouse DNA probes and libraries.[18] cDNAs are placed in the repository by investigators and can be ordered from the catalog for a small handling fee. cDNAs from ATCC or private sources are not usually ready to use for making *in situ* hybridization probes. To prepare a quantity of the cDNA for use in probe preparation, it may be feasible to amplify the sequence with polymerase chain reaction (PCR). Alternatively, the cDNA can be cloned into a plasmid vector. Preparation of RNA antisense–labeled probes for *in situ* hybridization will involve *in vitro* transcription of the cDNA sequence, which requires a promoter sequence for the polymerase (typically T3 or T7 polymerase). Cloning or insertion of the cDNA into an appropriate vector and production of large quantities of that plasmid in bacterial culture is a standard, classical method. When received as purified isolated cDNA or in an unsuitable vector, the cDNA must be cloned into a suitable vector by standard protocols (as described in molecular biology reference texts). It is also possible to use synthesized oligonucleotide probes that require no cloning.

The choice of probe type (RNA, double strand DNA, single strand DNA, oligonucleotide), labeling method (*in vitro* transcription, end labeling, nick translation, random primer synthesis), and detection system (radioactive — ^{32}P, ^{35}S, ^{125}I, ^{3}H — or nonradioactive tag on nucleotide precursor) will affect the sensitivity and resolution of the probe. The methods selected may be determined not only by the properties desired in a probe but also by individual experience, access to necessary facilities, and level of resources and funding. There are several general reviews and sources of information that may be helpful in making these choices, in particular reviews by Angerer and Angerer[19] and by Singer et al.,[20] which give good overviews of the advantages and disadvantages of the different approaches.

A general consensus protocol has emerged based on experiences of many investigators, and this version was used in our laboratory[11] and in most of the studies cited in this section. Briefly, the protocol for examination of embryonic gene expression uses 4% paraformaldehyde-fixed tissues embedded in paraffin. In all the protocol steps, extreme care must be taken to prevent degradation of the RNA by RNase enzymes. The sections are placed on triethoxysilane coated (TESPA) slides. Probes are synthesized with [^{35}S]CTP incorporation, using *in vitro* transcription from an RNA polymerase promoter (T3, T7, or Sp6) on the linearized vector. Probe length may be adjusted to 100 to 200 bases by alkaline hydrolysis at 60°C.

Prehybridization washes prepare the sections with proteinase K digestion, and additional fixation and acetylation steps. Hybridization is performed in humid chambers in an oven or incubator overnight. Posthybridization treatment generally includes washes of various stringencies at elevated temperatures, with 50% formamide in some steps, as well as an RNase A digestion step to remove unhybridized probe (single stranded RNA). Slides are dipped in liquid photographic emulsion (Kodak NTB2), dried, and stored (4°C) for days to weeks. After the emulsion is developed and coverslipped, the slides can be viewed under the microscope by darkfield or (if they contain abundant label) brightfield illumination. This method has been successfully applied to embryonic tissues to characterize expression of specific genes, generating time, and tissue specific patterns of expression. An example from our laboratory is shown in Figure 14.2A, in which the expression of the aryl hydrocarbon receptor (AhR) was characterized in the developing mouse embryo.

Whole mount *in situ* hybridization offers the potential to visualize the expression of genes in three dimensions in small specimens.[12,21] The power of this approach is illustrated in Figure 14.2B, in which *bmp4* mRNA is detected in a GD 10 mouse embryo by use of a digoxigenin-labeled probe. As was discussed for whole mount immunohistochemical staining, the size of the specimen is a limiting factor because the reagents and probe need to penetrate the tissue. This also mandates the use of a treatment to permeabilize the tissues, which is most often done by pretreating the specimen with proteinase K. The length of pretreatment must be adjusted based on the age of the embryo and the activity of the enzyme. Following proteinase K digestion, the embryos are usually refixed in an aldehyde fixative. The probe is synthesized as before, using *in vitro* transcription of antisense RNA with a nonradioactive label. Probe length is generally not reduced by most laboratories, and experiments using riboprobes of 1 kilobase (kb) or greater are reported in the literature. Some protocols for whole mount *in situ* include incubations with methanol and hydrogen peroxide as a bleaching step. This incubation will also inhibit endogenous peroxidase and thus prevent any potential for background with a peroxidase-based enzymatic detection method. The posthybridization washes may include RNase digestion and 50% formamide in low salt buffer at elevated temperature. The specimens may also be heated to 70°C for 20 min prior to probe detection in order to inhibit endogenous alkaline phosphatases. The hybridized probe is then detected with antidigoxigenin antisera conjugated to alkaline phosphatase. Alkaline phosphatase reacts with substrate to the enzyme, producing a colored precipitate. Alternatively, the antisera may carry a fluorescent tag that can be visualized with a fluorescent dissecting microscope or by confocal laser scanning microscopy. Posthybridization incubations to render the specimen more transparent, thus allowing visualization of the label within the tissue and not just on the surface, may also be utilized.

Other examples of applications of *in situ* hybridization to localize gene expression in the developing embryo include studies of the effects of retinoic acid. Retinoic acid (RA) is an endogenous compound important to embryonic development, but it also has the ability to disrupt morphogenesis

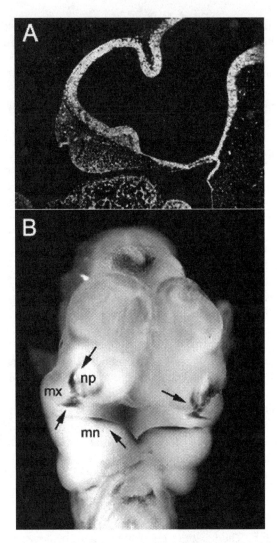

Figure 14.2 (A) Expression of the aryl hydrocarbon receptor (AhR) was examined in the GD 10 C57BL/6N mouse embryo by *in situ* hybridization. Tissues were fixed in paraformaldehyde and embedded in paraffin, and sections were incubated with radiolabeled RNA riboprobes. At this development stage, AhR mRNA was expressed at high levels in the neuroepithelial cells of the developing brain. (B) Whole mount *in situ* hybridization in a GD 10 mouse embryo by use of a digoxigenin-labeled antisense riboprobe for *bmp4*. Arrows indicate regions of the nasal prominence and the maxillary and mandibular processes, which express *bmp4*. *Note:* np = nasal prominence, mx = maxillary process, mn = mandibular process. (Probe courtesy of Dr. Brigid Hogan.)

when present inappropriately during development. A variety of molecular approaches have been utilized in the study of this compound in the embryo. The differential distribution of the cellular retinoic acid binding proteins (CRABP I and II), the cellular retinol binding protein (CRBP), and the retinoic acid receptors (multiple isoforms of RAR and RXR) provide excellent examples of the localization of mRNA spatially and temporally in developing embryos.[21-23] *Hox* gene expression has also been extensively studied with this technique, both in the normal embryo and in determination of the effects of retinoic acid treatment on *Hox* and *Krox* gene expression.[24,25] Whole mount detection was able to demonstrate that exogenous RA-induced expression of *Hox*-2 genes occurred in anterior regions of the embryo in a stage dependent manner.[26] The embryonic expression of several TGFβs and growth and differentiation factors was also studied by several investigators[27,28] using *in situ* hybridization. The role of growth factors in cardiac development was investigated using whole mount hybridization, sectioned tissue *in situ*, and immunohistochemistry,[12] and in our laboratory the expression of growth factors, including TGFβs, was examined in embryonic palatal shelves after exposure to chemicals that induce cleft palate.[11,29]

III. APPROACHES FOR QUANTIFYING MRNA EXPRESSION

A. Semiquantitative Methods

1. Northern Blot

Northern blot assays are used to examine mRNA expression and can provide quantitative information concerning levels of expression. However, unlike the case with *in situ* hybridization, information regarding localized expression within a tissue cannot be obtained because the samples are homogenized for preparation of total RNA. Protocols for Northern blots are available in the general references provided in the introduction,[2-4] and a brief outline of the basics follows. Investigators should bear in mind that modifications and adjustments may be required to optimize conditions for a particular gene. In order to learn the method, it may be advisable to probe for a gene with well characterized expression and that has been previously examined by Northern blot.

Total or polyA$^+$RNA is prepared from specific tissues or the whole embryo, and this RNA is separated by electrophoresis on a denaturing gel. Northern blots allow multiple samples to be examined simultaneously on a single gel. The method is reasonably sensitive, detecting mRNAs representing as little as 0.003% of total mRNA. Nevertheless, the small size of most embryonic samples generally requires pooling of multiple embryos (and in some cases pooling entire litters) to obtain sufficient total RNA. Also, when the target mRNA has very low expression, the treatment decreases expression, or relatively few cells in the tissue express the gene, this method may not be sensitive enough to detect or quantitate the mRNA. After electrophoresis in the gel, the RNA is blotted onto a filter of nylon or nitrocellulose and bound to the filter by cross-linking with UV irradiation or baking in a vacuum oven. An antisense radiolabeled probe (RNA or DNA) is prepared against the specific mRNA being examined, and filter, probe, and hybridization buffer are incubated overnight. After the excess probe is washed away, the filter is placed on x-ray film. After incubation at 80°C, the exposed film is developed, and the bands are quantified by densitometry.

The amount of probe hybridizing to the RNA is proportional to the amount of target RNA on the blot. Thus, the intensity of the band on the film can be used to estimate differences between samples. Typically, the amount of RNA loaded in each lane is also estimated by hybridization of a probe to a constitutively expressed gene (e.g., glyceraldehyde 3-phosphate dehydrogenase [GAPDH] or β-actin). This allows a correction between lanes for any variability in the amount loaded in different lanes. Hybridization to the mRNA of a constitutive gene can be performed on the blot by either stripping off the first probe (this requires that the blot never be allowed to dry during posthybridization manipulations, including exposure of the film) or by allowing the radioactivity to diminish to the point that the first probe no longer exposes the film (only a matter of weeks for ^{32}P-labeled probes). The option to reprobe can also be very useful for screening for the expression of several genes in the same samples. The size of the bands to which the probe hybridizes can be estimated by comparing the migration of the band to the migration of known molecular weight standards or by use of the 28S and 18S ribosomal RNA in total RNA samples. Ethidium bromide staining allows the migration distances to be observed with a UV transilluminator and photographed, either in the gel or on the blot.

Titration on slot blots or RNase protection, as described in the following sections, is most often used for quantitation in conjunction with Northern blot data. Northern blots are more often used for semiquantitative, qualitative, or temporal studies rather than for quantitation. Quantitation with Northern blots alone can be relatively difficult, and some additional considerations are important.

As is the case with protein staining on acrylamide gels, x-ray film exposure is linearly related to band (or spot) quantity only up to a point. As a result of a combination of intense signal and "flaring," linearity is lost at high levels of signal. Therefore, it is generally advisable to use multiple exposure times and multiple amounts of standards to assure that the band intensities are in the linear range.

Examples of Northern blot analysis of embryonic gene expression include studies of the expression of retinoic acid receptor-β (RARβ) and the growth factor, TGFβ. Chick embryos were examined for expression of RARβ mRNA at developmental stages 22, 24, and 25.[30] The effects of retinoic acid on the expression of RARβ were examined in stage-28 embryos. In this study, 10 μg total RNA was loaded per lane, and type I collagen expression was used to control for variability in loading. The embryos were dissected into frontonasal mass, mandibular primordia, head, limb buds, and trunk, and each region was homogenized for total RNA preparation. In this study, the Northern blots were not quantitated but were used to demonstrate that the mRNA is expressed in these tissues. RARβ transcripts were localized with *in situ* hybridization, and the trends for increased expression found with Northern analysis were demonstrated over sections that also provided information on regional expression patterns.

There have been numerous studies localizing the expression of the TGFβ isoforms in the mouse embryo with *in situ* hybridization. In Schmid et al.,[31] Northern analysis was used to confirm specificity of the probes and to determine transcript size in GD 13, 14, and 16 mouse embryos and in neonates.

Northern blot analysis of cultured embryonic cells can also provide insight into the mechanism of action of a teratogen. Assays of the effects of retinoic acid on gene expression in cultured murine embryonic palate mesenchymal (MEPM) cells included Northern blot analysis of effects on TGFβ3, type III TGFβ receptor, CRABP I, histone H3, and tenascin.[32,33] This approach detected a differential time course of effects on transcription of these genes, and this information should be useful in designing further studies. In a similar application, Northern blot analysis was used extensively to determine the regulation of early immediate genes (*c-jun, c-fos, junB*) in developing lung and brain of mouse embryos.[34] In conjunction with the blots, immunohistochemistry localized the expression of these transcription factors.

There are nonradioactive detection methods for hybridization assays that can be used, but these methods may be unable to detect very low abundance mRNA. Northern blotting is less sensitive than the RNase protection assay but may be useful for genes that are widely expressed or have high transcription rates.

2. Dot Blot or Slot Blot

These blots are similar to Northern blots as binding of sample to a membrane support is followed by hybridization with a probe. This is a faster and easier method that may be useful for screening for specific target genes, but it is less accurate and sensitive than the Northern blot. The RNA or DNA sample is applied directly to restricted regions of the filter, often using a special vacuum apparatus. The hybridization, film exposure, and quantitation are similar to those described for Northern blots. This method will not detect degradation of the RNA, and nonspecific hybridization also cannot be determined. This is because the range of mRNA band sizes within the sample cannot be observed.

In a study of 3-hydroxy-3-methylglutaryl coenzyme A (HGM-CoA) reductase mRNA in the rat embryo, investigators localized expression with *in situ* hybridization, confirmed the specificity of their probe with Northern blots, and used dot blots to compare stimulated and unstimulated cellular expression of the transcript.[35] A titration of total RNA from 5 to 0.1 μg on the dot blot demonstrated the sensitivity of the response, as well as enhanced transcription in the stimulated cell sample. The application of multiple techniques was not redundant in this study, as each provided slightly different information, and the combination proved advantageous in defining the response to the treatment.

B. PCR and RT-PCR Quantitative Methods

Polymerase chain reaction amplification is an enzymatic method to increase the amount of a specific target DNA or RNA available for study. Short oligonucleotide sequences (primers) are prepared that bracket the DNA sequence that is to be amplified, i.e., they specifically recognize and bind to bases at the 5' and 3' ends of the target region. A thermal cycler is used to alternatively denature the double-stranded DNA (higher temperature) and promote primer hybridization and strand synthesis (lower setting), with repeated, automated cycling. Cycles typically require 5 min, and a full run may involve 30+ cycles. The template to which the primers bind must be DNA, so that the amplification of RNA requires the additional step of reverse transcription of the RNA to synthesize cDNA (RT-PCR). This is an extremely valuable technique for the study of embryos as qualitative and quantitative information can be obtained from very small tissue samples, potentially even those consisting of only a few cells. The thermostable enzymes required for these reactions are expensive, and large scale assays could become major components of a research budget. General information and protocols are available in the three volume manual *Molecular Cloning: A Laboratory Manual*[2] and in *PCR Protocols, A Guide to Methods and Applications*.[36]

RT-PCR can be used to detect gene expression in small specimens, and some studies in developmental biology included detection of gene transcripts in developing chick limb bud, during mouse mandibular and tooth bud morphogenesis, and in developing mouse lung.[37-40] Detection of the presence of RXRγ mRNA in chick wing buds was achieved by reverse transcription and amplification of total RNA; β-actin was amplified as a positive control, and in some tubes the enzyme reverse transcriptase was omitted to confirm that the reaction was not contaminated by genomic DNA. These are useful controls when attempting to exhibit the presence of a low-level transcript in embryonic tissues. In similar studies in the mouse, dissected tooth buds, mandibles, or lung buds were prepared from different stages of development. In the tooth bud, epithelial growth factor (EGF) mRNA was detected by use of as little as 1 ng of RNA. The relative levels of expression in each tissue were compared between developmental stages, and the findings were supplemented by immunolocalization of translated protein for the genes being studied.[38-40]

1. Real-Time RT-PCR

Real-time RT-PCR[41] is rapidly becoming the method of choice for determining the levels of mRNA expression. Real time RT-PCR involves measuring the amount of PCR product produced after each cycle as the reactions proceed. Several companies produce dedicated instruments to accomplish these measurements. Although, these instruments differ widely in detail, they all are essentially thermal cyclers combined with a fluorometer. The simplest approach to doing real time RT-PCR consists of adding an intercalating dye, such as SYBR Green 1 (that fluoresces when bound to DNA) to the PCR reaction mixture.[42] The increase in fluorescence as the PCR product accumulates is measured and is proportional to the amount of product. A disadvantage to this method is that the intercalating dyes will bind to all DNA synthesized, including primer dimers and any other nonspecific products produced.

To improve specificity several related approaches have been developed. These methods rely on synthesizing two probes, a reporter that will hybridize with the PCR product and is also labeled with a fluorescent dye and a quencher, a dye that will absorb the light emitted by the fluorescent dye when they are in close proximity. Hydrolysis probes (i.e., Taqman™) are end-labeled with the fluorescent dye and quencher on opposite ends of the probe.[43,44] Because the probes are short (typically 20 to 40 base pairs), the signal from the reporter is effectively quenched. This probe is designed to anneal to the specific amplicon (DNA product of the polymerase chain reaction) between the forward and reverse primers. During the extension phase, the 5-exonuclease activity of Taq polymerase degrades the probe, releasing the reporter and allowing it to fluoresce. Common alternatives to hydrolysis probes are hybridization probes (i.e., molecular beacons).[45,46] Like the

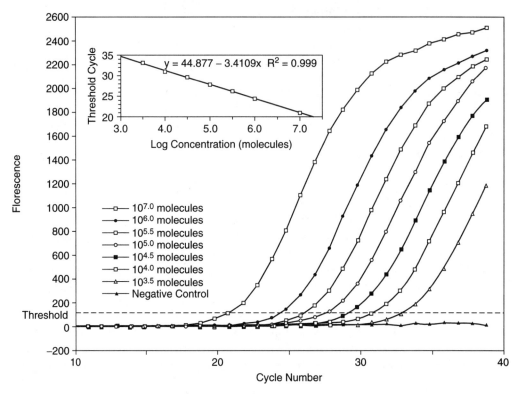

Figure 14.3 The plot illustrates a typical real-time RT-PCR experimental run in which the increase in fluorescence is tracked by cycle number. The samples are DNA standards with 10^7 to $10^{3.5}$ molecules per reaction tube at ½ log intervals, except for the $10^{6.5}$ interval, which was not run. The inset graph is a plot of the initial concentration of DNA vs. the threshold cycle, the cycle at which each tube exhibited the same level of amplification.

hydrolysis probes, these are also end-labeled with a reporter and quencher. However, the ends of the probe contain short complementary sequences designed to allow the probe to form a stem-loop structure when free in solution. This results in the reporter and quencher being in very close proximity and having low background fluorescence when compared with hydrolysis probes. When the probe hybridizes to the specific amplicon, the stem-loop structure opens, the distance between the reporter and quencher increases, and the reporter fluoresces. As the PCR primers are extended, the probe is displaced rather than being degraded.

Figure 14.3 illustrates a typical real-time RT-PCR experiment that measured the increase in fluorescence tracked by cycle number. The curve for each sample is characterized by a region where insufficient product has been synthesized to allow detection, followed by an exponential increase in the accumulation of product, and finally by the appearance of a plateau. The relative amount of mRNA in each sample is determined by comparing the cycle number when the fluorescence reaches a threshold level. The threshold level is selected at a point sufficiently above the noise level of the instrument to obtain accurate fluorescence measurements while amplification is still occurring exponentially. The amount of mRNA in the original sample is inversely proportional to the threshold cycle number. Reproducibility between real-time RT-PCR runs is sufficient that for many purposes the use of standards is not required.[47] Standards can be used to reduce run-to-run variability. These standards can be either cDNA or RNA samples. If it is necessary to determine the actual number of copies of a specific mRNA contained in a given sample, it is important to control for the efficiency of the reverse transcription reaction. This can be accomplished by including a standard curve derived from a pure RNA sample of the sequence being amplified with each run.

2. Quantitative RT-PCR Based on Image Analysis of PCR Products in Electrophoretic Gels

Quantitative RT-PCR in gel-based applications requires that a method be used to standardize the density values between specimens and/or lanes on the gel. Typical approaches are to spike each sample with an internal standard (synthesized DNA oligo or reverse transcribed RNA sequence) or to amplify a housekeeping gene for comparison purposes. One of these approaches is to compare the RT-PCR product to an artificial mRNA internal standard or reference gene transcript that uses the same primers as the target gene.[48] This involves adding increasing amounts of "competitor" (internal standard) to fixed amounts of RT-PCR product in a series of reaction tubes, with simultaneous amplification. The relative amounts of amplified internal standard and target cDNA can be identified by electrophoretic separation on a gel and visualization with ethidium bromide. Densitometric analysis produces a ratio of standard to target, and regression analysis identifies the point at which both standard and target were equally amplified. By inference, the known amount of standard predicts an equimolar amount of target present in the tube. Other methods spike the PCR reactions with a mutant, modified, or shortened cDNA or oligo sequence of the target gene. Some studies express mRNA levels relative to a housekeeping gene amplified from the same samples. Typically a constitutively expressed gene is amplified in a separate PCR reaction from the gene of interest. These approaches should be used with caution. Variation between the quality or quantitation of the original total RNA samples, as well as differences in amplification efficiency between different genes, may contribute to errors when comparisons are made to housekeeping genes or sequences that do not amplify using the same primers and in the same reaction tube as the target gene of interest. Methods using a recombinant mRNA internal standard (rcRNA) eliminate these problems as the internal standard is reverse transcribed using primer sequences of the target gene, and this includes a T7 DNA polymerase promoter site and a polyA sequence. The internal standard would be added to the total RNA sample prior to the reverse transcriptase reaction, and the amplified products vary in length from the target gene (50 bases approximately) to allow separation on the gel.[48] This protocol gives extremely sensitive detection using very low amounts of total RNA (100 ng total RNA per tube).

This approach was used in our study of the expression of AhR, ARNT, and CYP1A1 in human and mouse embryonic palatal tissues.[49,50] The mRNA levels were expressed in units of molecules per 100 fg of total RNA. This common unit of expression allowed comparison of the expression of these genes and the effects of treatment in the mouse after an *in vivo* exposure to 2,3,7,8-tetrachlorodibenzo-*p*-dioxin (TCDD) to expression in mouse and human palates exposed *in vitro*. This method was also used to examine the dose-response relationship for induction by TCDD of a hepatic enzyme (CYP1A1) in adult rats, along with effects on PAI2 and TGFα.[51]

Another application of RT-PCR involves *in situ* transcription of mRNA from sectioned tissues or single cells, followed by amplification in the presence of radiolabeled nucleotides. These transcripts are used as probes on reverse Southern blots to which cDNAs of suspected target genes are bound.[52] This interesting method is discussed in more detail in the discussion of reverse Southern blots and RT-PCR.

C. Solution Hybridization: RNase Protection Assay

The RNase protection assay is a solution hybridization–based method. The target RNA and probe (radiolabeled RNA or DNA) are allowed to hybridize in solution and are then exposed to S1 nuclease (RNA-DNA hybrids) or RNase (RNA-RNA hybrids), which degrades all single stranded (nonhybridized) RNA. Electrophoresis on a denaturing polyacrylamide gel separates the samples, and the dry gel can be placed with film for autoradiography. As with Northern blots, the signal intensity is proportional to the amount of target mRNA. This assay is often performed as a titration, where the amount of sample increases while a constant known amount of probe is added to each tube. This allows estimation of the amount of target RNA by means of linear regression of the hybridization densities for each ratio of target to probe. Solution hybridization is more sensitive than Northern blots as the hybridization reactions go to completion (on blots a proportion of the target will be unavailable because of binding to the membrane).

This method is particularly valuable for quantitation of mRNA in embryonic samples and has been used effectively in studies of retinoic acid effects on RAR-β and CRABP-I.[53,54] It has also been employed in developmental biology studies of glucocorticoid receptor (GR) regulation during lung development[55] and to characterize expression of PDGF A (platelet derived growth factor A) and its receptor in early embryogenesis.[56] In the first example, the effects of retinoic acid exposure on various isoforms of RAR mRNA were determined in GD 11 mouse embryos (whole embryo or dissected limb buds, head, and body) and on GD 14 in whole embryo, dissected limb bud, brain, liver, body, and craniofacial region (including brain).[53] Treated embryos showed tissue and stage specific increases (2- to 12-fold) in mRNA for various isoforms of RAR. The sensitivity of the assay allowed quantitation of the treatment effects by use of ^{32}P-labeled antisense cRNA probes and only 1 to 10 µg of total RNA.

In a study of GR regulation, both Northern blots and solution hybridization were used to detect mRNA transcription during lung development in fetal rats on GD 16 to 21.[55] The probe for Northern blot analysis was ^{32}P-labeled antisense cRNA, with 20 µg total RNA loaded per lane. The solution hybridization probe was a ^{35}S-labeled antisense cRNA, and β-actin was used as a control gene in both assays. The protection assay confirmed qualitative information from the Northern blot analysis and provided statistically significant evidence for down-regulation of the gene in response to glucocorticoid exposure. The PDGF study also used a [^{32}P]cRNA probe with β-actin as a control gene, and more detail was provided regarding the necessity to pool embryos to obtain sufficient total RNA to perform the analysis.[56] In this case, embryonic mice from GD 6.5 to 8.5 were examined, and 20 µg RNA per lane was used for each hybridization reaction. Whole embryos were pooled to prepare total RNA as follows: GD 6.5, n=10; GD 7.5, n=5; GD 8.5, n=3. This protocol indicates that even with a sensitive assay, the size of the embryos mandated pooling, and data were not obtained from individual embryos. In order to study the responses of single embryos, or even more importantly, the effects on particular target tissues, the analytical method of choice would involve PCR amplification of reverse transcribed mRNA (i.e., RT-PCR).

D. RNA Transcription *In Vitro*: Nuclear Run-on Assay

The nuclear run-on assay measures the rate of RNA transcription (transcripts synthesized per gene per minute). The Northern blot, dot blot, solution hybridization, and *in situ* hybridization assays all determine amounts of RNA present. Those techniques may not reflect transcription rates, however, because RNA levels may be affected by posttranslational processing and half-life in the cytoplasm. In the nuclear run-on assay, RNA is transcribed *in vitro*, by use of isolated nuclei and radiolabeled nucleotides. The labeled RNA transcripts are isolated and hybridized to cDNAs bound on filters (reverse Northern blot). The intensity of the autoradiographic signal is then measured and used to estimate the relative rates of transcription between different samples. Care must be taken to have equal levels of radiolabeled precursor in each reaction tube. This method may not be sensitive enough for poorly expressed mRNAs, and the experimental examples given below both utilized cultured cells to define the levels at which genes were regulated.

The nuclear run-on approach was used to define the up-regulation by retinoic acid of *Hox*-2[57] and the PDGF α-receptor gene[58] in cultured F9 embryonal carcinoma cells. In the *Hox*-2 study, embryonic stem cells (which are more closely related to cells of the early embryo than the F9 cells are) and *Xenopus* embryos were also examined. Exposure to RA increased levels of *Hox*-2.1 mRNA, as determined by Northern blots, and nuclear run-on experiments with the F9 cells demonstrated that the increase in steady-state levels of *Hox*-2.1 was not accompanied by an increased transcription rate. Examination of the transcription rate by this method suggested that the effects of the chemical exposure were at a posttranscriptional level.[57] The embryonal carcinoma cells also express PDGF α-receptor, and the combination of retinoic acid and dibutyryl cyclic AMP stimulated expression of mRNA for that receptor.[58] The PDGF growth factors are expressed in embryos, and a lethal mutation in the *patch* locus correlates with failure to express the PDGF α-receptor. Exploring the regulation of this receptor with Northern blots and nuclear-run-on analyses demonstrated strong induction by the RA-cAMP treatment. The study also showed that transcriptional rates increased proportionally with mRNA levels, indicating transcriptional regulation was involved in the response.

IV. CHARACTERIZING THE ROLE OF SPECIFIC GENES AND PATHWAYS OF RESPONSE

Molecular techniques can be helpful in identifying potential biomolecules involved in response to a teratogen. This approach is conceptually the inverse of that discussed previously. In this alternative case, the approach is to screen the genome of the embryo. Differences between treated and control embryonic gene expression are determined, and a survey across the developmental period of interest may be involved. The scope of these studies must of necessity be wider, and thus there is the potential to identify contributors to a response that would otherwise go unrecognized. One approach involves screening of genetic libraries derived from embryos or specific organs in various stages of development. Restriction fragment length polymorphism (RFLP) analysis has also been valuable in studies of human diseases and birth defects and has identified links between specific genes and malformations.[59,60] Mutational analysis and targeted gene disruption allow the phenocopy approach to be used, in which a defect induced by a teratogenic agent can be compared to similar defects caused by mutation or targeted gene disruption. Although potentially valuable for developmental toxicology studies, the RFLP and phenocopy approaches are not presented here. The following discussion focuses on two methods, subtractive screening and *in situ* transcription RT-PCR reverse Southern blot, which have the potential to detect differential gene expression across development and between experimental treatments.

A. Transgenic Animals

Transgenic mice are currently being generated to examine the regulation, function, and expression patterns of a wide range of genes that are implicated in controlling embryonic development. These transgenic animals can be "reporter mice" that allow the expression of a particular gene to be monitored during various stages of development. Reporter mice typically use the regulatory sequences from a gene of interest to drive the expression of a reporter gene, such as *lacZ*, to which the regulatory sequences are fused. This reporter transgene is activated in the regions that have endogenous gene expression and is detected by the appearance of dark blue staining caused by β-galactosidase activity.

Other transgenic mice are generated to "knock out" expression of the gene of interest, generally through targeted mutation that results in a homozygous inactivation. These loss-of-function mutants are produced by disruption of the gene by homologous recombination in embryonic stem (ES) cells. The mutant ES cells are then introduced into a mouse blastocyst to generate chimeric embryos that are then placed in the hormonally primed uterus of a host, pseudopregnant female. The neonates will be chimeric, and coat color is usually employed (host blastocyst being from the albino and ES cells being from the pigmented strain) as a means of readily identifying the transgenic animals (ES cell homologous recombination was reviewed by Capecchi[61]). The chimeric mice are bred to generate mice homozygous for the disrupted gene, and the phenotype of the resulting embryos can be informative concerning the role of that gene in development.

Similarly, gain of function transgenic mice can be produced that inappropriately express the gene during development, either constitutively or under the regulation of an inducible promoter. These are typically dominant mutations that are produced in a manner similar to that just described for producing ES cell chimeric transgenic individuals, or they may be generated by directly injecting the cloned DNA into fertilized mouse eggs. The presence of the transgene is detected in the resulting pups by Southern blot analysis of a tail-clip tissue sample.

Analysis of a gene of interest, generation of either activating or inactivating mutations, and insertion of these mutations or reporter constructs into the embryonic cells to produce transgenic lines of mice are not trivial endeavors and require a considerable investment of time and resources. There are increasing numbers of these transgenic lines available, and the developmental toxicologist may be in a position to benefit from the wide interest of developmental biologists in this area. These mice are frequently obtainable from the laboratory that produced them if the researchers involved are contacted with an interesting proposal to examine effects of exogenous agents. This may be an option for projects in which the developmental toxicant is suspected to act through the gene modified in the transgenic mouse (or reported in the *lacZ* transgenic form) or in which the phenotype of the transgenic animal is homologous to that produced by the agent.

The study of retinoic acid has taken advantage of transgenic technology to examine the effects of RA exposure on *Hox* gene expression patterns, RA activation of retinoic acid receptors, and consequences of constitutive expression of RAR in the lens of the developing eye. *Hox* genes are expressed in the mouse embryo with spatially restricted but overlapping domains, and the combination of expression of sets of *Hox* genes may represent a system for determining regional morphological specificity.[62] *Hox-lacZ* reporter transgenic mice have provided spatial and temporal three-dimensional patterning of *Hox* expression during development,[63] and these transgenic mice were used to characterize the effects of RA on this pattern.[64] In this and other studies, *Hox* and *Krox* genes were shown to be responsive to RA and to exposure induced homeotic transformation of hindbrain rhombomere identity.

A similar approach was applied to the characterization of retinoic acid receptor gene expression using transgenic reporter mice to localize spatially and temporally the distribution of RARs and the inducibility of the gene by RA.[65,66] The teratogenic RA exposure correlated with abnormal RARβ2 gene expression in the CNS, a region of the embryo sensitive to RA induced dysmorphology. A transgenic line was created in which constitutively expressed RAR-*LacZ*-fusion protein was targeted to the ocular lens by linking the transgene to the α-crystallin promoter. The transgenic mice overexpressed RAR in the developing lens and exhibited eye defects similar to those observed after RA exposure.[67] That particular application demonstrated an ability to target a particular system for overexpression of a gene suspected to have a role in the pathogenesis of teratogenic exposure.

Growth factor genes are also being studied using transgenic mutations, and examples are available for overexpression or knockout of TGFα and TGFβ1.[68–72] TGFβ1 null mutants exhibit a rapid wasting syndrome 2 to 3 weeks after birth, excessive inflammatory responses, and early death.[71] The *int-2* gene, a member of the FGF growth factor family, was disrupted in transgenic individuals, and these mutants showed inner ear and tail defects and generally did not survive to adulthood.[72] TGFα appeared to promote tumor progression in overexpressing transgenic individuals, while embryogenesis seemed to progress normally. The transgenic mice with a knockout mutation for TGFα also develop without major effects on organogenesis but have anomalies of the hair and skin.[69] Bryant et al.[73] used TGFα and EGF knockout mice to study the mechanism through which TCDD produces cleft palate and hydronephrosis in embryonic mice. TCDD disrupted control of cell proliferation and differentiation in the embryonic epithelial cells, and the responses in the palate and ureter correlated with tissue specific altered expression of EGF and/or TGFα. Using the transgenic knockouts revealed that EGF expression is important in mediating the responses in palatal cells and that the ureteric cells were more responsive in the absence of EGF.

A number of transgenic gene knockouts have minimal effects on the developing embryo, and it has been concluded that other closely related proteins are able to compensate functionally. Other hypotheses suggest that the proteins are expressed superfluously, where they have no function; eliminating these proteins thus has no discernible consequences.[70] On the other hand, knockout of genes involved in response to hypoxia or vascularization has proven to be lethal *in utero* (reviewed by Hanahan[74]), and studies of the etiology of the lethality have given new insight into the development of the placenta and vascularization of the yolk sac and the embryo.[10,75,76] Interpretation of the developmental consequences of gene knockout or constitutive expression in transgenic mice appears to be a complex issue, and use of these models in developmental toxicology would also require qualified interpretations.

B. *In Vitro* Antisense RNA

Antisense oligo probes can be designed to be reasonably stable in culture medium, to enter cultured embryos, organs, or cells, and to hybridize specifically with a target mRNA to inhibit translation.[77] This approach is similar to transgenic knockout of genes for determining effects on development of failure to express a specific gene. The effectiveness of this approach depends on the ability of the oligo probe to reach the specific target in sufficient concentration. Size and stability in culture are important factors, along with concentration and incubation temperature. These characteristics may vary with oligonucleotide class, modification of the oligos, and type of culture (e.g., whole embryo vs. cell monolayer); the outcome may differ between different cell types.

Use of antisense oligonucleotides directed against endogenous expression of the growth factor EGF in tooth bud organ culture resulted in abnormal tooth morphogenesis and inhibited growth of the cultures.[39] The relative levels of endogenous EGF and EGF receptor expressed in developing teeth were also determined in this study by use of RT-PCR and immunolocalization of expression patterns. The application of multiple methodologies provided strong support for the role of EGF in odontogenesis. The same research group applied this multitechnique approach (using RT-PCR, antisense *in vitro*, and immunolocalization) to determine the role of TGFβs in the morphogenesis of the mandible and tooth.[40,78] The antisense oligo inhibition of TGFβ1 resulted in dysmorphology of Meckel's cartilage, while treatment with antisense TGFβ2 had no such effect.

In whole embryo culture, the role of *wnt* proto-oncogene family members in development was examined by injecting an antisense oligo into embryonic amniotic cavities.[79] PCR analysis detected decreased *wnt* mRNA, and malformations were induced in the treated embryos. This approach may provide an alternative to transgenic analysis when models are available that allow developmental events of interest to proceed in cultured cells, organ culture, or whole embryo culture. This may allow the role of several genes to be investigated in culture with less investment of resources and time than are required for generating transgenic animals. However, as with the data from knockout transgenic mice, interpretation of these experiments should be approached with caution as multiple factors can influence their outcomes.

C. DNA Array

This is a powerful technique for assessing gene expression and has extensive applications in the area of developmental and reproductive toxicity. For this reason, the full discussion of this method is presented in a separate chapter (Chapter 15).

D. Genetic Library Screening

This approach is directed toward detection of specific effects on regulation of gene expression and allows one to survey the entire genome to identify target genes. The method has been used to examine gene expression at different stages of development or variable expression in different tissues and to detect altered gene expression after treatment during development. Differential and subtractive screening methods have been developed for these applications. The differential screen is generally not as likely to detect transcripts from rarely expressed genes or to discern differences between low-level expression mRNAs from a contrasted pair of samples. Subtractive hybridization cross-hybridizes two mRNA or DNA samples that have been isolated from different experimental samples and then isolates the nonhybridized sequences. All nucleic acid sequences occurring in both samples hybridize and are removed, leaving behind the unmatched mRNA or DNA. This isolated material represents the genes expressed in one sample and not the other. Using a higher concentration of one sample than the other results in detection of either increased or decreased expression depending on which sample (control or treated, or different developmental stages) was present in excess.

Branch et al.[80] adapted this approach to examine the effects of restraint of pregnant mice on gene expression in the embryos. Maternal restraint significantly increases the incidence of lumbar ribs and encephalocele in mice. GD 8 embryonic mRNA was isolated from the treated group and reverse transcribed to give single stranded cDNA with c-tailed 3′ ends. Control mRNA was biotinylated and then cross-hybridized to the cDNA. Streptavidin binding removes all biotinylated mRNA (hybridized and single stranded), subtracting or leaving behind the unmatched cDNA of the treated embryos. After second strand synthesis, this cDNA can be PCR amplified and cloned into vectors for transformation of *E. coli*. The bacterial colonies with vector containing the cDNA insert are isolated and sequenced, with the intent to identify the gene by comparing the sequence to those in a gene data bank. In the stress-restraint study, seven cDNAs were isolated. One was identified as encoding a protein important in cell cycle regulation, and the others are under further study.

Subtractive hybridization was also used to isolate mRNAs associated with programmed cell death induced by exposure of thymoctyes to glucocorticoids.[81] One of the isolated genes was sequenced and found to contain a zinc-finger domain, suggesting involvement in DNA regulation, while a second gene appeared to encode a membrane-bound protein.

When the isolated sequence is compared with known gene sequences, the isolation of a sequence that is differentially expressed may lead to identification of the gene involved. The cDNA clone may also be used to construct fusion proteins that can be synthesized in *E. coli*, purified, and used to prepare antibodies. These antibodies may then be useful in isolating the protein that is encoded by the gene. Alternatively, the clone sequence may be used to synthesize a small peptide against which to raise antibody.

E. Reverse Southern Blots and RT-PCR *In Situ* Transcription

In situ transcription of mRNAs on a sectioned embryo coupled with RT-PCR and reverse Southern blotting has recently been proposed as a screen of differential gene activity.[52,82–84] With this method, specific regions of a tissue may be examined and compared with like regions from other developmental stages, or controls can be compared with treated embryos. The *in situ* transcription uses oligo primers incubated directly on the sectioned tissue (or selected portion of the section following removal of unwanted regions) to synthesize cDNA directly from mRNA. Following separation of the mRNA-cDNA hybrids and second strand synthesis to generate double stranded cDNA, this material can be amplified by PCR for use as a probe or it can be cloned into a vector. The riboprobes prepared from this cDNA can be used to screen Southern dot or slot blots, in which cDNA sequences of genes suspected to be involved in the mechanism of toxicity or the developmental pathway for that tissue are bound to the membrane. Hybridization intensity reflects the relative abundance of the mRNAs of interest. Gene expression in the Splotch mutant mouse, a model for neural tube defects, was examined for a range of growth factors, transcription factors, *Hox* genes, and extracellular proteins.[82] This approach has detected effects on candidate genes after exposure to valproic acid, retinoic acid, or hyperthermia in SWV mice.[83] Similar studies examined changes in gene expression after exposure of SWV mice to phenytoin.[84]

REFERENCES

1. Lodish, H., Berk, A., Zipursky, L., Matsudaira, P., Baltimore, D., and Darnell, J., *Molecular Cell Biology*, 4th ed., W.H. Freeman, New York, 2000.
2. Sambrook, J., MacCallum, P., and Russell, D., *Molecular Cloning: A Laboratory Manual*, 3rd ed., Cold Spring Harbor Laboratory, Cold Spring Harbor, NY, 2001.
3. Perbal, B., *A Practical Guide to Molecular Cloning*, 2nd ed., John Wiley, New York, 1988.
4. Ausubel, F.M., Brent, R., Kingston, R.E., Moore, D.D., Seidman, J.G., Smith, J.A., and Struhl, K., *Short Protocols in Molecular Biology*, 4th ed., John Wiley, New York, 1999.
5. *Scientific Frontiers in Developmental Toxicology and Risk Assessment*, National Academy Press, Washington, DC, 2000.
6. Harlow, E. and Lane, D., *Using Antibodies: A Laboratory Manual*, Cold Spring Harbor Laboratory, Cold Spring Harbor, NY, 1999.
7. Mark, M., Lufkin, T., Vonesch, J.L., Ruberte, E., Olivo, J.C., Dolle, P., Gorry, P., Lumsden, A., and Chambon, P., Two rhombomeres are altered in Hoxa-1 mutant mice, *Development*, 119, 319, 1993.
8. Stadler, B., Phillips, J., Toyoda, Y., Federman, M., Levitsky, S., and McCully, J.D., Adenosine-enhanced ischemic preconditioning modulates necrosis and apoptosis: effects of stunning and ischemia-reperfusion, *Ann. Thorac. Surg.*, 72, 555, 2001.
9. Takagi, T.N., Matsui, K.A., Yamashita, K., Ohmori, H., and Yasuda, M., Pathogenesis of cleft palate in mouse embryos exposed to 2,3,7, 8-tetrachlorodibenzo-p-dioxin (TCDD), *Teratog. Carcinog. Mutagen.*, 20, 73, 2000.
10. Abbott, B.D. and Buckalew, A.R., Placental defects in ARNT-knockout conceptus correlate with localized decreases in VEGF-R2, Ang-1, and Tie-2, *Dev. Dyn.*, 219, 526, 2000.
11. Abbott, B.D., Harris, M.W., and Birnbaum, L.S., Comparisons of the effects of TCDD and hydrocortisone on growth factor expression provide insight into their interaction in the embryonic mouse palate, *Teratology*, 45, 35, 1992.
12. Dickson, M.C., Slager, H.G., Duffie, E., Mummery, C.L., and Akhurst, R.J., RNA and protein localisations of TGF beta 2 in the early mouse embryo suggest an involvement in cardiac development, *Development*, 117, 625, 1993.
13. Mahmood, R., Flanders, K.C., and Morriss-Kay, G.M., Interactions between retinoids and TGF beta s in mouse morphogenesis, *Development*, 115, 67, 1992.
14. Pelton, R.W., Saxena, B., Jones, M., Moses, H.L., and Gold, L.I., Immunohistochemical localization of TGF beta 1, TGF beta 2, and TGF beta 3 in the mouse embryo: expression patterns suggest multiple roles during embryonic development, *J. Cell Biol.*, 115, 1091, 1991.
15. Linask, K.K., D'Angelo, M., Gehris, A.L., and Greene, R.M., Transforming growth factor-beta receptor profiles of human and murine embryonic palate mesenchymal cells, *Exp. Cell Res.*, 192, 1, 1991.
16. Shuey, D.L., Sadler, T.W., and Lauder, J.M., Serotonin as a regulator of craniofacial morphogenesis: site specific malformations following exposure to serotonin uptake inhibitors, *Teratology*, 46, 367, 1992.
17. Yavarone, M.S., Shuey, D.L., Tamir, H., Sadler, T.W., and Lauder, J.M., Serotonin and cardiac morphogenesis in the mouse embryo, *Teratology*, 47, 573, 1993.
18. Nierman, W.C. and Maglott, D.R., American Type Culture Collection NIH Repository of Human and Mouse DNA Probes and Libraries, American Type Culture Collection, Rockville, MD, 1992.
19. Angerer, L.M. and Angerer, R.C., Localization of mRNAs by in situ hybridization, *Methods Cell. Biol.*, 35, 37, 1991.
20. Singer, R.H., Lawrence, J.B., and Villnave, C., Optimization of in situ hybridization using isotopic and non-isotopic detection methods., *Biotechniques*, 4, 230, 1986.
21. Ruberte, E., Friederich, V., Morriss-Kay, G., and Chambon, P., Differential distribution patterns of CRABP I and CRABP II transcripts during mouse embryogenesis, *Development*, 115, 973, 1992.
22. Ruberte, E., Friederich, V., Chambon, P., and Morriss-Kay, G., Retinoic acid receptors and cellular retinoid binding proteins. III. Their differential transcript distribution during mouse nervous system development, *Development*, 118, 267, 1993.
23. Perez-Castro, A.V., Toth-Rogler, L.E., Wei, L.N., and Nguyen-Huu, M.C., Spatial and temporal pattern of expression of the cellular retinoic acid-binding protein and the cellular retinol-binding protein during mouse embryogenesis, *Proc. Natl. Acad. Sci. USA*, 86, 8813, 1989.

24. Hunt, P., Wilkinson, D., and Krumlauf, R., Patterning the vertebrate head: murine Hox 2 genes mark distinct subpopulations of premigratory and migrating cranial neural crest, *Development*, 112, 43, 1991.
25. Morriss-Kay, G.M., Murphy, P., Hill, R.E., and Davidson, D.R., Effects of retinoic acid excess on expression of Hox-2.9 and Krox-20 and on morphological segmentation in the hindbrain of mouse embryos, *EMBO J.*, 10, 2985, 1991.
26. Conlon, R.A. and Rossant, J., Exogenous retinoic acid rapidly induces anterior ectopic expression of murine Hox-2 genes in vivo, *Development*, 116, 357, 1992.
27. Lehnert, S.A. and Akhurst, R.J., Embryonic expression pattern of TGF beta type-1 RNA suggests both paracrine and autocrine mechanisms of action, *Development*, 104, 263, 1988.
28. Millan, F.A., Denhez, F., Kondaiah, P., and Akhurst, R.J., Embryonic gene expression patterns of TGF beta 1, beta 2 and beta 3 suggest different developmental functions in vivo, *Development*, 111, 131, 1991.
29. Abbott, B.D. and Birnbaum, L.S., TCDD alters medial epithelial cell differentiation during palatogenesis, *Toxicol. Appl. Pharmacol.*, 99, 276, 1989.
30. Rowe, A., Richman, J.M., and Brickell, P.M., Retinoic acid treatment alters the distribution of retinoic acid receptor-beta transcripts in the embryonic chick face, *Development*, 111, 1007, 1991.
31. Schmid, P., Cox, D., Bilbe, G., Maier, R., and McMaster, G.K., Differential expression of TGF beta 1, beta 2 and beta 3 genes during mouse embryogenesis, *Development*, 111, 117, 1991.
32. Nugent, P. and Greene, R.M., Interactions between the transforming growth factor beta (TGF beta) and retinoic acid signal transduction pathways in murine embryonic palatal cells, *Differentiation*, 58, 149, 1994.
33. Nugent, P., Sucov, H.M., Pisano, M.M., and Greene, R.M., The role of RXR-alpha in retinoic acid-induced cleft palate as assessed with the RXR-alpha knockout mouse, *Int. J. Dev. Biol.*, 43, 567, 1999.
34. Molinar-Rode, R., Smeyne, R.J., Curran, T., and Morgan, J.I., Regulation of proto-oncogene expression in adult and developing lungs, *Mol. Cell Biol.*, 13, 3213, 1993.
35. Brewer, L.M., Sheardown, S.A., and Brown, N.A., HMG-CoA reductase mRNA in the post-implantation rat embryo studied by in situ hybridization, *Teratology*, 47, 137, 1993.
36. Innis, M.A., Gelfand, D.H., Sninsky, J.J., and White, T.J., *PCR Protocols, A Guide to Methods and Applications*, Academic Press, New York, 1989.
37. Thaller, C., Hofmann, C., and Eichele, G., 9-cis-retinoic acid, a potent inducer of digit pattern duplications in the chick wing bud, *Development*, 118, 957, 1993.
38. Hu, C.C., Sakakura, Y., Sasano, Y., Shum, L., Bringas, P., Jr., Werb, Z., and Slavkin, H.C., Endogenous epidermal growth factor regulates the timing and pattern of embryonic mouse molar tooth morphogenesis, *Int. J. Dev. Biol.*, 36, 505, 1992.
39. Warburton, D., Seth, R., Shum, L., Horcher, P.G., Hall, F.L., Werb, Z., and Slavkin, H.C., Epigenetic role of epidermal growth factor expression and signalling in embryonic mouse lung morphogenesis, *Dev. Biol.*, 149, 123, 1992.
40. Chai, Y., Mah, A., Crohin, C., Groff, S., Bringas, P., Jr., Le, T., Santos, V., and Slavkin, H.C., Specific transforming growth factor-beta subtypes regulate embryonic mouse Meckel's cartilage and tooth development, *Dev. Biol.*, 162, 85, 1994.
41. Heid, C.A., Stevens, J., Livak, K.J., and Williams, P.M., Real time quantitative PCR, *Genome Res.*, 6, 986, 1996.
42. Gentle, A., Anastasopoulos, F., and McBrien, N.A., High-resolution semi-quantitative real-time PCR without the use of a standard curve, *Biotechniques*, 31, 502, 2001.
43. Giulietti, A., Overbergh, L., Valckx, D., Decallonne, B., Bouillon, R., and Mathieu, C., An overview of real-time quantitative PCR: applications to quantify cytokine gene expression, *Methods*, 25, 386, 2001.
44. Holland, P.M., Abramson, R.D., Watson, R., and Gelfand, D.H., Detection of specific polymerase chain reaction product by utilizing the 5'-3' exonuclease activity of Thermus aquaticus DNA polymerase, *Proc. Natl. Acad. Sci. USA*, 88, 7276, 1991.
45. Piatek, A.S., Tyagi, S., Pol, A.C., Telenti, A., Miller, L.P., Kramer, F.R., and Alland, D., Molecular beacon sequence analysis for detecting drug resistance in Mycobacterium tuberculosis, *Nat. Biotechnol.*, 16, 359, 1998.
46. Tyagi, S. and Kramer, F.R., Molecular beacons: probes that fluoresce upon hybridization, *Nat. Biotechnol.*, 14, 303, 1996.

47. Pfaffl, M.W., A new mathematical model for relative quantification in real-time RT-PCR, *Nucleic Acids Res.*, 29, E45, 2001.
48. Vanden Heuvel, J.P., Tyson, F.L., and Bell, D.A., Construction of recombinant RNA templates for use as internal standards in quantitative RT-PCR, *Biotechniques*, 14, 395, 1993.
49. Abbott, B.D., Held, G.A., Wood, C.R., Buckalew, A.R., Brown, J.G., and Schmid, J., AhR, ARNT, and CYP1A1 mRNA quantitation in cultured human embryonic palates exposed to TCDD and comparison with mouse palate in vivo and in culture, *Toxicol. Sci.*, 47, 62, 1999.
50. Abbott, B.D., Schmid, J.E., Brown, J.G., Wood, C.R., White, R.D., Buckalew, A.R., and Held, G.A., RT-PCR quantification of AHR, ARNT, GR, and CYP1A1 mRNA in craniofacial tissues of embryonic mice exposed to 2,3,7,8-tetrachlorodibenzo-*p*-dioxin and hydrocortisone, *Toxicol. Sci.*, 47, 76, 1999.
51. Vanden Heuvel, J.P., Clark, G.C., Kohn, M.C., Tritscher, A.M., Greenlee, W.F., Lucier, G.W., and Bell, D.A., Dioxin-responsive genes: examination of dose-response relationships using reverse-transcription polymerase chain reaction, *Cancer Res.*, 54, 62, 1994.
52. Eberwine, J., Spencer, C., Miyashiro, K., Mackler, S., and Finnell, R., Complementary DNA synthesis in situ: methods and applications, *Methods Enzymol.*, 216, 80, 1992.
53. Harnish, D.C., Jiang, H., Soprano, K.J., Kochhar, D.M., and Soprano, D.R., Retinoic acid receptor beta 2 mRNA is elevated by retinoic acid in vivo in susceptible regions of mid-gestation mouse embryos, *Dev. Dyn.*, 194, 239, 1992.
54. Harnish, D.C., Soprano, K.J., and Soprano, D.R., Mouse conceptuses have a limited capacity to elevate the mRNA level of cellular retinoid binding proteins in response to teratogenic doses of retinoic acid, *Teratology*, 46, 137, 1992.
55. Bronnegard, M. and Okret, S., Regulation of the glucocorticoid receptor in fetal rat lung during development, *J. Steroid Biochem. Mol. Biol.*, 39, 13, 1991.
56. Mercola, M., Wang, C.Y., Kelly, J., Brownlee, C., Jackson-Grusby, L., Stiles, C., and Bowen-Pope, D., Selective expression of PDGF A and its receptor during early mouse embryogenesis, *Dev. Biol.*, 138, 114, 1990.
57. Papalopulu, N., Lovell-Badge, R., and Krumlauf, R., The expression of murine Hox-2 genes is dependent on the differentiation pathway and displays a collinear sensitivity to retinoic acid in F9 cells and Xenopus embryos, *Nucleic Acids Res.*, 19, 5497, 1991.
58. Wang, C., Kelly, J., Bowen-Pope, D.F., and Stiles, C.D., Retinoic acid promotes transcription of the platelet-derived growth factor alpha-receptor gene, *Mol. Cell Biol.*, 10, 6781, 1990.
59. Sassani, R., Bartlett, S.P., Feng, H., Goldner-Sauve, A., Haq, A.K., Buetow, K.H., and Gasser, D.L., Association between alleles of the transforming growth factor-alpha locus and the occurrence of cleft lip, *Am. J. Med. Genet.*, 45, 565, 1993.
60. Chenevix-Trench, G., Jones, K., Green, A.C., Duffy, D.L., and Martin, N.G., Cleft lip with or without cleft palate: associations with transforming growth factor alpha and retinoic acid receptor loci, *Am. J. Hum. Genet.*, 51, 1377, 1992.
61. Capecchi, M.R., The new mouse genetics: altering the genome by gene targeting, *Trends Genet.*, 5, 70, 1989.
62. Hunt, P., Gulisano, M., Cook, M., Sham, M.H., Faiella, A., Wilkinson, D., Boncinelli, E., and Krumlauf, R., A distinct Hox code for the branchial region of the vertebrate head, *Nature*, 353, 861, 1991.
63. Sham, M.H., Hunt, P., Nonchev, S., Papalopulu, N., Graham, A., Boncinelli, E., and Krumlauf, R., Analysis of the murine Hox-2.7 gene: conserved alternative transcripts with differential distributions in the nervous system and the potential for shared regulatory regions, *EMBO J.*, 11, 1825, 1992.
64. Marshall, H., Nonchev, S., Sham, M.H., Muchamore, I., Lumsden, A., and Krumlauf, R., Retinoic acid alters hindbrain *Hox* code and induces transformation of rhombomeres 2/3 into a 4/5 identity, *Nature*, 360, 737, 1992.
65. Balkan, W., Colbert, M., Bock, C., and Linney, E., Transgenic indicator mice for studying activated retinoic acid receptors during development, *Proc. Natl. Acad. Sci. USA*, 89, 3347, 1992.
66. Zimmer, A., Induction of a RAR beta 2-*lacZ* transgene by retinoic acid reflects the neuromeric organization of the central nervous system, *Development*, 116, 977, 1992.
67. Balkan, W., Klintworth, G.K., Bock, C.B., and Linney, E., Transgenic mice expressing a constitutively active retinoic acid receptor in the lens exhibit ocular defects, *Dev. Biol.*, 151, 622, 1992.
68. Lee, G.H., Merlino, G., and Fausto, N., Development of liver tumors in transforming growth factor alpha transgenic mice, *Cancer Res.*, 52, 5162, 1992.

69. Mann, G.B., Fowler, K.J., Gabriel, A., Nice, E.C., Williams, R.L., and Dunn, A.R., Mice with a null mutation of the TGF alpha gene have abnormal skin architecture, wavy hair, and curly whiskers and often develop corneal inflammation, *Cell*, 73, 249, 1993.
70. Erickson, H.P., Gene knockouts of *c-src*, transforming growth factor beta 1, and tenascin suggest superfluous, nonfunctional expression of proteins, *J. Cell Biol.*, 120, 1079, 1993.
71. Kulkarni, A.B., Huh, C.G., Becker, D., Geiser, A., Lyght, M., Flanders, K.C., Roberts, A.B., Sporn, M.B., Ward, J.M., and Karlsson, S., Transforming growth factor beta 1 null mutation in mice causes excessive inflammatory response and early death, *Proc. Natl. Acad. Sci. USA*, 90, 770, 1993.
72. Mansour, S.L., Goddard, J.M., and Capecchi, M.R., Mice homozygous for a targeted disruption of the proto-oncogene int-2 have developmental defects in the tail and inner ear, *Development*, 117, 13, 1993.
73. Bryant, P.L., Schmid, J.E., Fenton, S.E., Buckalew, A.R., and Abbott, B.D., Teratogenicity of 2,3,7,8-tetrachlorodibenzo-p-dioxin (TCDD) in mice lacking the expression of EGF and/or TGF-alpha, *Toxicol. Sci.*, 62, 103, 2001.
74. Hanahan, D., Signaling vascular morphogenesis and maintenance, *Science*, 277, 48, 1997.
75. Maltepe, E., Keith, B., Arsham, A.M., Brorson, J.R., and Simon, M.C., The role of ARNT2 in tumor angiogenesis and the neural response to hypoxia, *Biochem. Biophys. Res. Commun.*, 273, 231, 2000.
76. Kozak, K.R., Abbott, B., and Hankinson, O., ARNT-deficient mice and placental differentiation, *Dev. Biol.*, 191, 297, 1997.
77. Crooke, R.M., In vitro toxicology and pharmacokinetics of antisense oligonucleotides, *Anticancer Drug Des.*, 6, 609, 1991.
78. Slavkin, H.C., Sasano, Y., Kikunaga, S., Bessem, C., Bringas, P., Jr., Mayo, M., Luo, W., Mak, G., Rall, L., and Snead, M.L., Cartilage, bone and tooth induction during early embryonic mouse mandibular morphogenesis using serumless, chemically-defined medium, *Connect. Tissue Res.*, 24, 41, 1990.
79. Augustine, K., Liu, E.T., and Sadler, T.W., Antisense attenuation of Wnt-1 and Wnt-3a expression in whole embryo culture reveals roles for these genes in craniofacial, spinal cord, and cardiac morphogenesis, *Dev. Genet.*, 14, 500, 1993.
80. Branch, S., Francis, B.M., Rosen, M.B., Brownie, C.F., Held, G.A., and Chernoff, N., Differentially expressed genes associated with 5-aza-2′-deoxycytidine-induced hindlimb defects in the Swiss Webster mouse, *J. Biochem. Mol. Toxicol.*, 12, 135, 1998.
81. Owens, G.P., Hahn, W.E., and Cohen, J.J., Identification of mRNAs associated with programmed cell death in immature thymocytes, *Mol. Cell Biol.*, 11, 4177, 1991.
82. Bennett, G.D., An, J., Craig, J.C., Gefrides, L.A., Calvin, J.A., and Finnell, R.H., Neurulation abnormalities secondary to altered gene expression in neural tube defect susceptible Splotch embryos, *Teratology*, 57, 17, 1998.
83. Finnell, R.H., Van Waes, M., Bennett, G.D., and Eberwine, J.H., Lack of concordance between heat shock proteins and the development of tolerance to teratogen-induced neural tube defects, *Dev. Genet.*, 14, 137, 1993.
84. Musselman, A.C., Bennett, G.D., Greer, K.A., Eberwine, J.H., and Finnell, R.H., Preliminary evidence of phenytoin-induced alterations in embryonic gene expression in a mouse model, *Reprod. Toxicol.*, 8, 383, 1994.

CHAPTER 15

Functional Genomics and Proteomics in Developmental and Reproductive Toxicology

Ofer Spiegelstein, Robert M. Cabrera, and Richard H. Finnell

CONTENTS

I. Introduction 622
II. Gene Expression 622
    A. Background 622
    B. Microarrays 623
        1. Complementary DNA (cDNA) Arrays 623
        2. Oligonucleotide Microarrays 628
        3. Data Analysis and Bioinformatics 629
        4. Experimental Considerations 631
    C. Serial Analysis of Gene Expression 633
        1. General 633
        2. Methodology 633
        3. Limitations 635
        4. SAGE and Developmental Biology 635
III. Proteomics 636
    A. Background 636
    B. Protein Expression Analysis 636
        1. Protein Isolation 636
        2. Protein Identification 638
        3. Protein Arrays 640
IV. Predictive Values 640
V. Relevant Websites and Databases 641
    A. Microarrays 641
    B. SAGE 641
    C. Proteomics 641
References 642

I. INTRODUCTION

Ongoing worldwide efforts continue to produce sequences for genomic assembly and single nucleotide polymorphism (SNP) typing of DNA, from model organisms to humans, with the initial draft of the human genome having been published in 2001.[1,2] The significance of genetic information is derived from the transcription of genomic DNA into messenger ribonucleic acids (mRNA) that encode the amino acid sequences of translated proteins. The process culminates in posttranslational modifications that are required to form biologically active and functional proteins that participate as enzymes in metabolic pathways. To understand the factors involved in dynamics of biological activity, the intricate relationships between DNA sequences (genomics), RNA levels (functional genomics), and expressed proteins (functional proteomics) must be examined.[3,4] Developmental and reproductive toxicological phenomena in model organisms can also be discerned by careful examination at these three fundamental levels. Environmental and pharmaceutical agents can induce abnormal expression of genes, which, should it occur at specific gestational time points of high susceptibility, may result in abnormal developmental processes that culminate in the birth of an infant with congenital defects. Such deleterious effects may be examined by techniques that are described in this chapter. Another level of topical organization to study involves protein-toxicant interactions. This level of inquiry seeks to elucidate altered protein presence, altered binding, or conformational changes that interfere with or damage developmental biological processes.

This chapter provides an overview of information and procedures used by contemporary investigators to study functional genomics and proteomics. These new cutting edge tools provide the means to explore the etiology of developmental and reproductive toxicants. Each topic is presented in such a way as to provide the reader not only with the background information but also with the theory and assay limitations that can help make practical decisions about the application of these new techniques.

II. GENE EXPRESSION

A. Background

Gene expression studies may be approached using several different RNA-based technologies:

- *In situ* hybridization (ISH)
- Northern blotting
- Ribonuclease protection assays (RPA)
- Gene reporter assay
- Branched DNA amplification
- Reverse transcription–polymerase chain reaction (RT-PCR)
- Real-time quantitative RT-PCR
- Differential display
- Scintillation proximity assay
- Rapid analysis of gene expression (RAGE)
- Serial analysis of gene expression (SAGE)
- Microarrays

Each of these methods provides the means to gather information on gene expression levels. Specific experiments must be designed around the limitations of each assay and the type of data needed to fulfill the aims of a given study. To study complex biological processes at the molecular level, these approaches are best used in combination. For instance, microarrays can initially be used to screen thousands of both previously characterized and unknown genes for changes in expression following exposure to a model compound. Once genes of interest have been identified, the data should be confirmed and validated by additional approaches, such as real time-PCR or Northern blotting.

With these tools now readily available to the scientific community, gene expression profiles that are characteristic of modifications caused by developmental or reproductive toxicants can and should be identified. Experimental designs presently adhere to traditional teratogen testing methods, i.e., comparing patterns of gene expression in experimental samples vs. those observed in controls. This section specifically reviews two recent methods for global gene expression profiling, namely microarrays and SAGE, that may be used in toxicology studies. With these tools, global changes in gene expression profiles can be identified and characterized. These are critical first steps for understanding the mechanisms of action of reproductive or developmental toxicants or for predictive purposes in the identification of potential teratogens.

B. Microarrays

The platform typically associated with gene expression profiling is the microarray, also known as the "DNA chip" or "gene chip."[5] Microarrays come on different platforms and can be custom manufactured in-house or purchased from selected biotechnology companies. The common theme concerning microarrays is that scientists are able to study the expression of a large number of genes simultaneously in a single experiment. Experiments that in the past would have taken several years to complete can now be performed within a matter of days. Aside from the high throughput nature of the microarray technology, one of the most prominent features is the ability to study novel and previously uncharacterized genes. Thereby previously unknown genes or genes never before linked to the experimental questions being pursued may be implicated. This, in turn, provides opportunities to target these newly implicated genes as part of drug discovery processes or to study their function, regulation, and cellular networks. In toxicological studies, discovery of novel genes that are perturbed in response to toxicological stimuli may enhance understanding of the underlying mechanisms of toxicity associated with a specific compound.

In retrospect, microarray methodology is merely a recent extension of well-established methods to assess gene expression by hybridization. Such methods have been successfully applied for many years. The first hybridization-based method, Southern blotting,[6] was used to study only a limited set of genes. Subsequent methodologies such as "dot-blots," enabled investigators to explore a larger set of genes that are deposited on porous nylon membranes.[7] What distinguishes microarrays from the aforementioned methods is primarily the use of glass as the solid support. Contrary to nylon membranes, glass (being a nonporous material) enables high-density and spatially reproducible deposition of target DNA, resulting in the ability to screen thousands of genes simultaneously. Using a nonporous support has also simplified hybridization kinetics and improved image acquisition and processing procedures.[8] Concurrent to developments in fluorescent labeling of nucleic acids, microarrays enable scientists to evaluate gene expression from both experimental and control samples on the same slide simultaneously. There are two major types of DNA arrays: complementary DNA arrays and oligonucleotide arrays. Several features, such as the length of target DNA, the method for DNA deposition, and the slide chemistry, distinguish the two types of arrays.

1. Complementary DNA (cDNA) Arrays

Complementary DNA arrays, better known as cDNA arrays, were initially developed in the laboratory of Patrick Brown and colleagues from Stanford University. Their first publication describing the use of this technology with a small subset of *Arabidopsis* genes appeared in 1995.[9] Since then, cDNA arrays have become the most popular microarray platform, and it is now quite common for institutions and departments to have their own core microarray manufacturing facilities. The main reasons for the popularity of cDNA arrays are:

1. The DNA deposited on the solid support can be easily prepared by investigators using standard PCR amplification.
2. For producers of their own arrays, sets of cloned genes can be purchased from biotechnology companies, obtained from the National Institutes of Health (NIH), or prepared in-house. Thus, researchers have a great deal of control over the content of the arrays they make and use.
3. The cost of preparing cDNA arrays in-house is substantially lower than the cost of commercially available arrays. Once the basic equipment for array fabrication has been obtained and the target DNA prepared, hundreds of microarray slides can be produced in a relatively short time.

a. Array Fabrication

The choice of which genes or expressed sequence tags (ESTs) to use for cDNA array fabrication depends on their availability; however, all deposited target DNA is prepared by standard PCR amplification of cDNA cloned into bacterial plasmids. Individually cloned genes and ESTs come in sets of several hundred to several thousand. They are either commercially available, obtained free of charge (e.g., from the NIH), or they may even be developed within the institution. See Figure 15.1 for an outline of cDNA microarray fabrication.

Comprehensive collections of genes and ESTs are particularly useful for applications such as gene discovery, when the goal is to associate specific genes with a specific toxicological process. Several cDNA clone sets are commercially available.[10] These include: the BMAP clone set (Res-Gen™, Carlsbad, CA), which was derived from adult C57BL6J mouse brain, spinal cord, and retinal tissues; arrayTAG™ (Lion Biosciences, San Diego, CA), which is a family of cDNA clones (mouse, rat, and dog) that were specifically designed for microarray applications by use of several unique methodologies; the full-length mouse cDNA clones from the RIKEN Genomic Sciences Center (Yokohama, Japan); and more.

Two of the most useful clone sets for developmental biologists and developmental toxicologists alike, the mouse 15K and 7.4K sets, were assembled at the National Institute of Aging. The 15K clone set was constructed to specifically represent developmentally important mammalian genes.[11,12] The approximately15,000 genes in this set were derived from pre- and peri-implantation embryos and from placenta. Between a third to a half of the 15,000 set contains known and previously characterized genes, whereas the reminder is composed of ESTs of currently unknown function.[11] The second clone set, released in 2002, is the 7400 clone set that contains approximately 7400 genes with no redundancy within the set or with the original 15,000 set. The 7400 clone set is especially valuable for developmental biologists and toxicologists as it includes genes expressed in a variety of stem cells, early embryos, and newborn organs, such as the heart and brain.

Preparing in-house clone sets requires considerable molecular biology expertise and resources; however, it enables researchers to construct microarray slides with a tissue-specific set of genes or with genes that are expressed during specific developmental processes, such as palatal shelf fusion. Using the in-house approach, investigators studying eye development created a library of approximately 15,000 ESTs that were obtained from mouse retina at embryonic day 14.5.[13] By use of microarray slides that were prepared from this clone set, a potential regulatory association between the transcription factor Brn-3b and GAP-43, a protein associated with axonal growth, has been identified.

Similarly, at the National Center for Toxicogenomics (NCT), which is part of the National Institute of Environmental Health Sciences (NIEHS), a toxicological human cDNA chip called the ToxChip has been developed.[14,15] The 2000 genes arrayed on the early version of the ToxChip (version 1.0) have been previously implicated in mechanisms of toxicity and cellular processes in response to different types of toxic insults. The latter include genes involved in apoptosis, DNA repair, and drug metabolism, transcription factors, tumor suppressor genes, and more. A similar approach was used to evaluate the effects of the environmental toxicant β-naphthoflavone on mice by use of a toxicological gene array and Northern blotting.[16] The authors compared the two methods

FUNCTIONAL GENOMICS AND PROTEOMICS

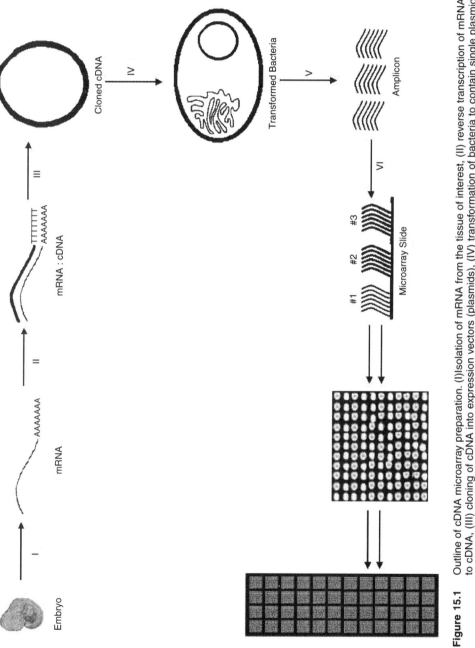

Figure 15.1 Outline of cDNA microarray preparation. (I) Isolation of mRNA from the tissue of interest, (II) reverse transcription of mRNA to cDNA, (III) cloning of cDNA into expression vectors (plasmids), (IV) transformation of bacteria to contain single plasmid inserts, (V) amplification of cDNA from plasmids by PCR, and (VI) high density deposition of purified target DNA on glass slides.

and concluded that the sensitivity of microarrays is comparable to that of Northern blotting. The data variability introduced by the printing process was less than the interanimal (biological) variability observed in the study.

After the target DNA has been amplified and purified by use of standard protocols, the DNA is dissolved in an arraying buffer. Several types of arraying buffers are commonly used, usually based on saline sodium citrate (SSC) and sodium dodecyl sulfate (SDS). However, some prefer to use printing buffers based on dimethyl sulfoxide (DMSO); evaporation from both the printed glass surface and the microtiter plates is greatly reduced, thereby improving spot homogeneity.[17–19] The target DNA solution, stored in either 96- or 384-well microtiter plates, is printed (arrayed) by microarray printers on specially treated glass microscope slides. Commonly used slide-coating materials are poly-L-lysine, epoxysilane, aminosilane, and aldehyde.[17,18] All types of pretreated microarray slides are commercially available, but they can also be prepared in-house with regular glass microscope slides. These special slide coatings are important for optimizing the slide's surface area to promote the adhesion of target DNA, thereby preventing lateral dispersion or detachment during hybridization.

Microarray printers are equipped with specialty printing tips that have the capability to pick up a submicroliter volume of target DNA solution and deposit nanoliter droplets with exceptionally high accuracy and precision. A typical slide may contain several thousands of target DNA spots, 100 to 200 μm in diameter, spaced 100 to 250 μm apart. When a large (usually exceeding 10,000) set of genes is to be printed, each target DNA can be arrayed once to maximize the total number of genes represented on the slide. However, when a relatively small set of genes is to be printed, each target DNA can be arrayed multiple times at separate locations of the slide. This facilitates statistical analysis of the data, thereby increasing sensitivity and confidence in the results.[20]

After the slides have been printed, several additional steps are necessary prior to their use in experiments. cDNA slides are often rehydrated to distribute the DNA more evenly on the glass surface. Slides are also irradiated by UV light to cause cross-linking of DNA so that it remains attached to the surface during hybridization and washing. In the final step of array fabrication, the slide's surface is deactivated (blocked) in order to minimize nonspecific binding of the probe DNA, thereby decreasing background signal levels, increasing the signal-to-noise ratio (SNR), and improving the overall sensitivity of detection.

b. Probe Preparation

From the RNA sample, a labeled cDNA probe is prepared by standard reverse transcription methodology. A radiolabeled probe is often used with microarrays printed on membranes because the probe enables high detection sensitivity. However, membrane microarrays are not as popular as glass microarrays in which the probes are usually labeled with a fluorophore (fluorescent tag). Use of fluorescently labeled dNTPs is the most popular approach, primarily because of the ability to cohybridize two samples simultaneously, which is not possible with radiolabeled targets. The ability to hybridize two samples simultaneously is a tremendous advantage. For example, one could label a sample from the developing neural tube of a drug-treated mouse embryo with one fluorophore and label a control sample with a different fluorophore. Thus, on a single microarray slide, a developmental toxicologist can examine the isolated effects of *in utero* exposure to a compound and distinguish the altered patterns of gene expression that may result. See Figure 15.2 for an outline of probe preparation.

The two most widely used fluorescent dNTPs are the CyDyes (Amersham Biosciences Corp, Piscataway, NJ), namely, Cy-3-dUTP™ (Cy-3) and Cy-5-dUTP™ (Cy-5). In both dyes, the fluorophores attached to the dUTP backbone have unique chemical structures that are spectrally distinct from one another. This enables the simultaneous detection of both dyes without cross interference, as it allows the concomitant measurement of signal from both the control and experimental samples.

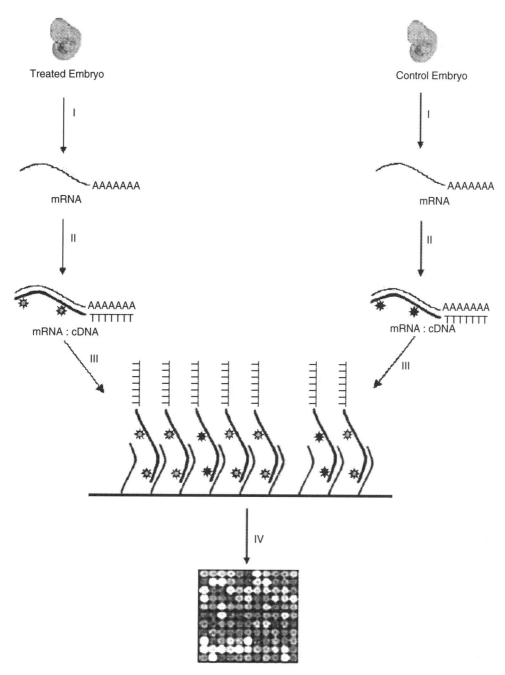

Figure 15.2 Outline of probe preparation and hybridization to microarrays. (I) Extraction of RNA from control and drug-treated embryos, (II) reverse transcription of mRNA to fluorescent cDNA probes, (III) probes are mixed and cohybridized to a microarray slide, and (IV) differential gene expression is visualized after the slide is processed and scanned.

The relative spectral intensities of the two samples are then obtained, and the transcriptional changes that occurred in response to the experimental treatment are determined by calculating the signal ratios for each gene. Other fluorescently tagged dNTPs are also available, such as the Alexa Fluor® series of dyes (Molecular Probes, Eugene, OR). The Alexa dyes seem to be somewhat superior to the CyDyes because of their higher SNRs and better chemical stability.[21,22]

These dyes can be incorporated using "direct labeling" of cDNA, i.e., the fluorescently tagged dUTPs are directly incorporated into the probe DNA as it is being synthesized from RNA. Both of these tagged nucleotides have the ability to cause premature termination of target DNA synthesis. That is because the large size of the fluorophore may cause steric hindrance, resulting in detachment of the reverse transcriptase enzyme. An additional problem with the direct labeling technique is that the differential incorporation ratios of each dye have the potential to introduce bias in data analysis.

Since the direct labeling methodology has several pitfalls, an alternative method called "indirect labeling" has been developed. In this technique, the fluorescent tag is conjugated after the probe DNA has been synthesized. Indirect labeling is achieved by the use of aminoallyl-dUTP in the initial reverse transcription step. This chemical modification creates an activated dUTP molecule to which the fluorescent dye can later be covalently attached. The chemical modification of the dUTP does not interfere with reverse transcription and nor does it cause premature termination of cDNA synthesis. In addition, the problem of differential incorporation ratios into cDNA is eliminated.[23] After the probe DNA has been synthesized, monoreactive Alexa or CyDyes are conjugated to the aminoallyl-dUTP groups to make the fluorescent probe.

c. Hybridization, Washing, and Image Acquisition

After the two probe samples have been prepared from both an experimental and control sample, each labeled with a different fluorescent dye, they are mixed together in a hybridization buffer. The hybridization buffer usually contains SSC or formamide with blocking agents. Blocking agents that are typically used are *Cot-1* DNA, poly-dA oligos, and/or salmon sperm DNA. The purpose of these blocking agents is to minimize nonspecific binding of probe DNA to the slide. Hybridization is performed at elevated temperatures, ranging from 40°C to 70°C depending on the hybridization buffer's composition, the average length of the probe DNA, and the stringency requirements. In addition, hybridization is best accomplished in specially designed chambers that maintain optimal humidity to prevent evaporation of water from the hybridization solution. Typically, the volume of the hybridization solution is 10 to 40 μl; therefore, even a small loss of volume may result in relatively large changes in salt concentration, which may adversely affect the quality of the hybridization. Collectively, buffer composition and hybridization temperature will determine the stringency of hybridization and the resulting signal and background fluorescence intensities.

Hybridization time can range from a few hours to overnight, after which the slide is washed to remove excess hybridization buffer and unattached probe DNA. Buffers of different stringencies are used in this step to reduce background signal. After the slide is dried, it is scanned by special laser scanners that detect the emitted signals from two separate channels (532 and 635 nm in the case of Cy-3-dUTP and Cy-5-dUTP, respectively), where each channel represents the signal from one of the fluorescent dyes.

2. Oligonucleotide Microarrays

The second form of DNA microarrays is the oligonucleotide array, of which two types currently exist. Affymetrix® (Santa Clara, CA) pioneered the development of the first type of oligonucleotide array, called the GeneChip®, which is made by photolithography technology and solid-phase DNA synthetic chemistry.[24] These combined methods are similar to techniques used in combinatorial chemistry synthesis. They enable the simultaneous synthesis of approximately 300,000 different 25-mer oligos at high density and high spatial resolution, covalently attached to a solid glass support.[25] The oligomers' sequences are the target DNA and are designed exclusively on DNA sequence information. They do not require any clone sets or PCR procedures, as is necessary with cDNA arrays. For each gene represented on the GeneChip, 20 different 25-mer sequences are selected. These represent different areas along the coding sequence of the gene that serve as sensitive sequence-specific sensors of expression. These sequences are optimized so that there is minimal overlap with other genes, gene-family members, or interfering genomic elements, such as repetitive sequences. The rational for using this probe redundancy (multiple sequences within a single gene) is to create improved accuracy and SNRs, minimizing false positives, false negatives, and nonspecific binding.[25] Another important feature of the GeneChip is the use of 20 "mismatch control" sequences. They are identical to their corresponding "true sequences" with the exception of a single base difference. These mismatch sequences act as specificity controls for the true sequences, thereby allowing reduction of background and cross-hybridization signals and allowing increased confidence in the results. It is important to note that while the GeneChip technology offers an exciting microarray platform with which to work, researchers must purchase proprietary hardware, software, and the GeneChips from Affymetrix. As previously mentioned, cDNA arrays offer the advantage of being amenable to customized, in-house design and the production of multiple slides, whereas using GeneChips is restrictive in the gene content and is substantially more expensive.

The second type of oligonucleotide microarray is a hybrid between cDNA and the previously described GeneChip oligonucleotide array platform. The target DNA oligomers are prepared using standard DNA synthetic methods, and the considerations for designing the oligomers' sequences are essentially the same as with the GeneChip. The oligos are then printed in-house on pretreated glass slides, similar to cDNA arrays. The oligo array method offers the advantages of specificity inherent in preselected short DNA sequences, plus the ability to prepare multiple slides containing the same set of DNA sequences, thus substantially reducing costs. However, this type of array platform does not quite have the highly controlled and reproducible nature of the GeneChip and nor does it have the probe redundancy and specificity control sequences.

3. Data Analysis and Bioinformatics

a. Normalization

Normalization of data can be performed in numerous ways, depending on the experimental design. The goal is to extract meaningful information and expand the investigator's ability to make comparisons between data sets. The immediate effect of normalization is to correct systematic hybridization variation so that the data more closely represent the true biological variation. Generally, normalization methods can be broken into three categories: global normalization, single gene (housekeeping gene) normalization, and external control normalization.[26]

Global normalization assumes that the expression of only a small proportion of the genes will vary significantly between the experimental and control samples.[27] Global normalization is the most widely used and is based on the overall median or mean of spot signal intensity.[28-30] It is worth noting that global normalization is not effective when working with transcripts that are expressed at low levels.[31] There is also evidence of spatially or intensity dependent dye biases in numerous experiments that are not considered and go uncorrected in global normalization. It is important to understand that while global normalization relies on the generally correct assumption that the two fluorescent dye intensities are related by a constant factor, there can be spatial variations in signal and background intensities that are not displayed evenly by both dyes. Furthermore, the dyes do not necessarily share common emission curves; therefore, the factor that relates the two dyes can change according to their intensity. This results in a dye-bias that often becomes apparent in the upper and lower bounds of dye emission intensities.

A set of housekeeping genes (e.g., β-actin, glyceraldehyde-3-phosphate dehydrogenase, ubiquitin) that are believed to maintain constant expression across a variety of conditions can also be used in normalization.[26] Although using single gene normalization does not require the assumption that only a small proportion of the genes varies significantly, it does require the assumption that the expression of the housekeeping gene used for normalization is not significantly altered by the experimental conditions. However, not all housekeeping genes are equally expressed under all conditions, and experiment-specific genes thus need to be identified and evaluated. Only after a gene has been shown to have constant expression under the relevant experimental conditions can it be used to normalize microarray data. Another limitation of using housekeeping genes is that they tend to be highly expressed, and their normalization parameters, when applied to other genes, may not be representative of true variation.[31]

An alternative method for normalizing gene expression data is the use of spiked-in controls.[26,32] These externally added genes could be used either as positive or negative controls or as a titration series, thereby enabling normalization to account for intensity-based variation. To incorporate these controls into a microarray experiment, a cDNA or oligonucleotide target DNA must be added to the array at the printing step. The spiked controls and their corresponding capture targets must be obtained from a different species than the organism of study so that they are composed of nonhomologous sequences. Common sources of spike control DNA are *Saccharomyces cerevisiae* and *Arabidopsis thaliana*. Their use minimizes cross hybridization with, for example, DNA of the mouse and rat, which are conventional species used in developmental and reproductive toxicology studies. Generating external foreign target DNA often entails PCR amplification from genomic samples by use of primers that contain a T7-RNA polymerase promoter sequence. That enables the subsequent synthesis of complementary RNA (cRNA) in an *in vitro* transcription (IVT) reaction. An additional advantage to using this method is that it introduces a basis for quantification of the absolute abundance of individual mRNAs. That is not possible with any of the other normalization methods.[27,33,34] This approach has the added benefit of allowing the comparison of multiple array experiments, if investigators use the same control sets. However, the use of spiked controls is not problem-free, and it has been shown that there are a number of uncertainties when extrapolating from the spiked control to sample derived mRNA.[27]

Debate on the most effective method or "hybrid" method of normalization continues. After considerable testing, it is currently accepted as being both plausible and feasible to incorporate more than a single method to normalize microarray data.[26,35] Despite the debate, the primary goal of normalization remains the same. Assessment of alternatives to the widely used global normalization procedure depends on whether they expand the investigator's ability to compare data sets. Ideally, a procedure should account for or correct systematic hybridization variation so that array data more closely represents actual biological activity.

b. Clustering

After data have been generated and normalized, an appropriate analytic method must be selected. Presently, the acceptable basis for analyzing and organizing (clustering) gene expression data is by grouping genes with similar patterns of expression, similar function, or similar regulatory elements.[36–39] These grouping methods address different key questions, such as which genes share common expression patterns, which are expressed or controlled in a coordinated manner across different sets of conditions, and which share related biological pathways. The goal of clustering culminates in the latter question: what are the dynamics of the genetic network?

There are several different clustering methods that can be used to organize gene expression data: hierarchical clustering,[36,40] k-means clustering,[41] principal components analysis,[42] mixture model methodology,[43] multidimensional scaling,[44] self-organizing maps,[45] graph-theoretic techniques,[46] and a few additional less-well known approaches.[47–49] These methods often provide graphical representations of the data in dendrograms, which are branched treelike graphical structures used to represent a hierarchy. Ratio-based coloring is also used in dendrograms to quantitatively and qualitatively represent the experimental observations, the observed changes in gene expression. Such clustering methods are capable of demonstrating expression profiles of developmental time points, tissue specific expression, or the changes in expression that occur following exposure to a drug. The basic format of the various clustering methods assembles genes in matrices according to criteria applied by an algorithm. For example, rows may represent the genes spotted on the microarray, while columns represent the samples used to hybridize to the array. The signal ratios, as measures of relative changes in gene expression between experimental and control samples, are depicted according to a color-based scale. By convention, up-regulation of genes in the experimental samples appears in red, whereas down-regulation is green. No change in gene expression is represented by yellow, whereas gray or black cells indicate missing or excluded data.[36,50]

4. Experimental Considerations

a. Limited RNA Availability

Typically, running a single sample on a microarray slide requires 20 to 100 µg of total RNA. For those working with large tissue samples or cell cultures, these RNA requirements are usually easy to meet. However, if the biological sample of interest is derived from a very small tissue sample, such as a portion of the neural tube of a neurulation-stage mouse embryo, the amount of total RNA that can be obtained is limited and will most likely be less than 1 µg. For those using laser-capture microdissection, single-sample RNA yields may be in the low nanogram range. To overcome such problems, it may be possible to pool a large number of individual samples so that the combined RNA content is sufficient for labeling and hybridization. While appropriate under specific experimental situations, this approach is likely to mask small and subtle changes in gene expression (i.e., have reduced sensitivity) due to a dilution effect. Most importantly, pooling of samples eliminates the ability to measure the biological variability that is a particularly important experimental parameter when working with microarrays.

An alternative approach to overcome the problem is to use labeling methods that increase the specific activity of the probes, thus compensating for the limited quantities of RNA. Examples of such methods are the Micromax® Tyramide Signal Amplification (TSA) technique (NEN Life Science Products, Zaventem, Belgium), the HiLight Array Detection System (QIAGEN, Valencia, CA), and the 3DNA Dendrimer technology (Genisphere®, Hatfield, PA), One must keep in mind that none of these techniques has emerged over time as the single most powerful and reliable approach.[23,51–53]

Another option to deal with limited RNA availability in gene expression experiments has been developed by James Eberwine at the University of Pennsylvania. His method is based on linear amplification of mRNA by IVT from cDNA.[54,55] The cDNA used for IVT is made by reverse transcription of mRNA by use of a poly-dT oligonucleotide with a T7 RNA polymerase promoter sequence. Thus, by utilizing the well-known ability of T7 RNA polymerase to bind to its promoter and synthesize RNA at a constant rate of approximately 100 bases/sec, linear amplification of RNA is achieved with high fidelity. The extent of amplification depends on the amount of template used. It is typically 40 to 200-fold with relatively large amounts of template and up to 2000-fold when the starting amount of template is very small, such as that obtained from single cells.[54,56]

This method has been successfully used for determining changes in gene expression in mouse embryos following administration of several known teratogens. In these studies, dot blots containing up to 50 genes of interest were prepared by immobilizing cDNA on nylon membranes. The probes used in these experiments were prepared by the T7-based IVT method and labeled with a radioactive nucleotide. With this approach, it has been shown that there is a significant down-regulation in the expression of several folate pathway genes (e.g., folate binding protein-1 and methylene tetrahydrofolate reductase) in neurulation stage neural tube defect–sensitive SWV mouse embryos following *in utero* exposure to valproic acid.[57] In contrast, the more resistant LM/Bc strain embryos had an up-regulation of these same genes in response to the teratogenic valproate exposure. These observations provided a plausible mechanism for relative resistance to valproate-induced neural tube defects. Similar experimental designs were used to study additional teratogen-induced changes in expression of growth and transcription factors as well as cell cycle checkpoint genes.[58–61]

As previously described, microarray technology has overtaken the laborious low-throughput dot-blot technology; however, the basic experimental principles and technical considerations remain the same. Thus, T7 RNA polymerase–based IVT can be integrated into microarray experiments to examine changes in gene expression following *in utero* exposure to a variety of known and suspected teratogens. Currently, only a handful of published studies used T7 RNA polymerase amplification in conjunction with microarrays, and all were nonreproductive toxicology studies; however, they serve to emphasize the feasibility of coupling the two techniques to approach these critically important issues.[62–64]

b. Importance of Replicates

The data generated from microarrays are subject to the variability inherent in all experimental systems. One source of this variability is introduced by biological factors, such as the differences in response between individual animals. Additional variables, such as slide-surface inconsistencies and probe preparation variations, can arise from any one of the numerous steps in manufacturing microarray slides. Therefore, considering the variability that can be generated from microarray data, replication is absolutely necessary to allow confidence in the results. In the early days of using microarrays, it was not uncommon for studies to include data from a single slide per experimental group. This was largely due to the high costs involved and the limited availability of microarray slides. However, this has changed in recent years, and it is no longer acceptable to perform microarray studies without adequate replicates.

The primary method of introducing replication in microarrays involves repeated measurements of expression. The most common way to achieve this is to print the same target DNA at multiple locations on the same slide, thereby creating intra-array replicates. This provides the means to assess the average expression level of every gene on the array, as well as the variability of the measurement, thus providing a much more reliable estimate of the measurement itself. An advantage of this type of replication is that spatial problems on the array surface can be minimized. If, for example, a particular area of the slide is defective and meaningful data cannot be extracted from it, expression data can be obtained from the other replicates. When spotting replicate targets, one must take into account the limited surface area available for printing because printing replicate spots on the same slide will reduce the overall number of genes that can be arrayed on a given slide. This may not pose any problem if the gene collection is relatively small or if the microarray printer is capable of spotting in very high densities. However, in cases where the gene collection is large and each gene is spotted in replicates, multiple slides may be necessary to print the entire collection. This adds to both cost and inconvenience. An additional way to obtain repeated measurements is to hybridize probes derived from the same original RNA sample on multiple arrays, thereby creating interarray replicates. This is in many ways similar to intra-array replication, but it also provides information on the variability between slides.

It is most important to account for the biological variability within experiments. This type of variability is present in all experimental models and is likely the single major contributor to the system's "noise." The only way to overcome the problem of biological variability is to include sufficient biological samples in each experimental group. Thus, in very noisy models, such as outbred mice or heterogeneous cell populations with highly variable responses, a larger number of samples per experimental group will be necessary to achieve the experimental goals. However, if the model is relatively "clean," as is the case when working with inbred mice strains or stable cell lines, experimental questions may be properly addressed with a smaller number of slides per group (e.g., three to four slides). An advantage of performing sufficient biological replicates is that interpretation of the results can go beyond the experimental groups and can be extrapolated to the general population.

Applying the most basic statistical consideration that "more is better" implies that the reliability of data is increased as the number of replicates (regardless of which type) is increased.[65] The specific number of replicates to be used in any given experiment varies according to the aforementioned sources of variability. Therefore, these parameters must be assessed in each experiment.[66] And it would seem most logical to print as many replicate target DNA spots onto the microarray slide as is practical and to perform several replicate experiments whenever possible. However, performing replicate hybridizations is relatively expensive, and the appropriate number of replications should be determined by proper statistical considerations.[65,66]

c. Data Validation

Considering the limitations in reliability of data generated from microarrays, it is absolutely vital to use other experimental methods to validate the data generated by both types of microarray platforms. One such technology considered by many to be the 'gold-standard' for gene expression data is RT-PCR, which is even now giving way to the more state-of-the-art and informative method of quantitative (also known as real time) PCR.[67-69] This method is used to examine a limited number of genes of interest that have been identified by other methods, such as microarrays. Quantitative PCR is based on the detection of fluorescence produced during the amplification of gene-specific PCR products, and it can even determine the copy numbers of specific genes within a tissue sample. Since microarrays tend to have a relatively small (three- to fourfold) dynamic range, their use can lead to under-representation of changes that occur. Real time PCR has a greater dynamic range and is able to validate observed trends identified by microarray studies.[13,70-73] A more detailed description of the use of RT-PCR, real time PCR, and other more conventional methods for validating microarray data can be found in Chapter 14.

C. Serial Analysis of Gene Expression

1. General

Concomitant to the development of microarrays, SAGE methodology has been developed as another technique for analyzing and screening thousands of expressed genes. In contrast to microarrays, which totally depend on genes that have already been cloned and at least partially sequenced, SAGE is independent of such limitations. SAGE enables the simultaneous detection and quantification of thousands of expressed genes in tissues or cells without any prior knowledge of their identities or sequences.[74]

SAGE relies upon short, unique 9- or 10-base pair (bp) sequences that serve as distinctive markers (tags) to distinguish between different gene transcripts within tissues or cells of interest.[74] According to the mathematical probabilities, using 9- or 10-bp tags allows the generation of 262,144 or 1,048,576 different sequences, respectively (i.e., 4^9 and 4^{10}, respectively). Thus, these short tags provide random coverage throughout the genome, representing each expressed gene in a discriminative manner. The final stage of SAGE involves sequencing the cloned tags; however, in contrast to conventional sequencing of cloned genes, a process that requires considerable time and effort, sequencing in SAGE is extremely efficient. Each clone used in SAGE sequencing is composed of many different tags serially aligned (concatenated) and cloned in a single plasmid vector. See Figure 15.3 for an outline of the SAGE process.

2. Methodology

The SAGE procedure starts by isolation of mRNA from the tissue of interest and the synthesis of double stranded cDNA by use of a biotinylated oligo-dT primer. The double stranded cDNA is subsequently cleaved with a restriction endonuclease ("anchoring enzyme") that has a 4-bp recognition sequence. Generally, a 4-bp recognition site will cleave DNA on average every 256 base pairs, which is significantly shorter than the average size of mRNAs. The most commonly used

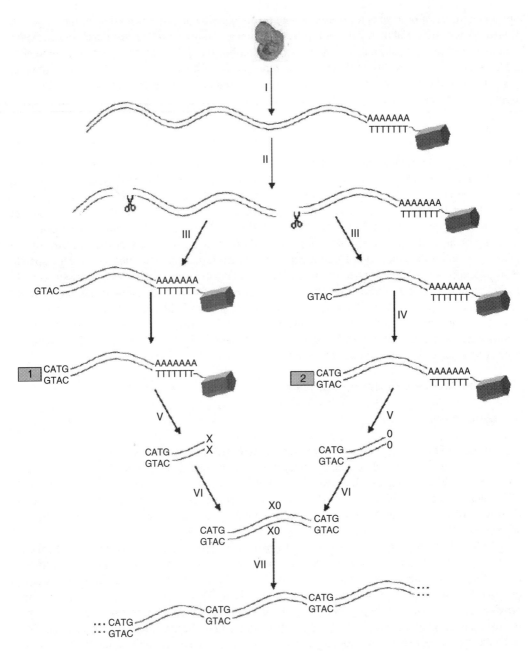

Figure 15.3 Outline of the SAGE methodology. (I) RNA extraction from tissue of interest and reverse transcription to double stranded cDNA with a biotin tag, (II) cleavage of cDNA with anchoring enzyme, (III) affinity purification of biotinylated cDNA and division of sample, (IV) ligation of the two linkers, (V) cleavage with tagging enzyme and blunt ending, (VI) ligation of two cDNA samples to form the ditag, and (VII) concatenate ditags, clone, and sequence

enzyme is *Nla*III, although in actuality, almost any restriction enzyme with a 4-bp recognition site can be used for the procedure.

Following cleavage, the resulting cDNA that contains a 3′-cohesive end is affinity purified with streptavidin-coated magnetic beads. The purified cDNA is then divided into two portions, and two separate linkers are ligated to the cohesive ends. These linkers are designed to contain a type IIS restriction endonuclease recognition site near the 3′-*Nla*III sequence. These two restriction sequences enable digestion by their respective restriction enzymes (usually *Fok*I and *Bsm*FI), also known as tagging enzymes. These restriction endonucleases cleave the cDNA at a distance of 9 to 10 bp from the recognition sequence, resulting in small tags of a linker sequence. Blunt ending is performed on the two populations of cDNAs, which are then mixed and ligated. The resulting product, a ditag with endonuclease recognition sites at either end, is PCR amplified and separated on a polyacrylamide gel. The endonuclease-anchoring site is cleaved, and concatamers are cloned into plasmid vectors, which are then used in standard sequencing procedures. The resulting sequence data provides a gene expression profile of the original mRNA population. For detailed SAGE procedures, consult Yamamoto et al.[75]

3. Limitations

Although SAGE offers an exciting and unique way to obtain global gene expression profiles of any type of eukaryotic cell, it has several drawbacks that have thus far limited its use and popularity. Aside from several technical issues, the major disadvantage of this technology is that it is difficult to perform SAGE successfully. The problem is mainly due to the inherent difficulty in creating tag libraries. Several steps of the SAGE protocol are relatively inefficient and introduce contaminating sequences and bias in the data.

One of the first drawbacks to be recognized was the need for relatively large amounts of starting RNA; however, in the past few years several modifications have been made to enable the use of SAGE with considerably less RNA. MicroSAGE requires 500 to 5000 times less RNA than the original SAGE protocols demanded and has been simplified by combining several steps within a single tube.[76] Similarly, SAGE-lite was developed to allow analysis from as little as 100 ng of RNA.[77] Another technical difficulty with the SAGE was that contaminating linker molecules generated from PCR were cloned into the plasmid vector, thus introducing confusing and erroneous data. By using biotinylated PCR primers, the ditags synthesized in PCR can be removed by affinity to streptavidin beads.[78] Another major problem of SAGE is the difficulty of further analyzing the unknown tags. Originally, the way to isolate and clone specific genes based on tag sequences was done by oligo-based plaque lifting. However, even though this method was used successfully in several cases, it appears to produce a high rate of false positives.[74] By using oligo-dT primers with the identified tag sequences, this problem was partially resolved, mainly as a result of truncation at the 5' ends of the genes.[79]

4. SAGE and Developmental Biology

The majority of the published literature on the use of SAGE technology has been in the cancer and immunology fields, whereas only a handful of developmental biology studies utilized this technology for global gene expression profiling. One such study has addressed the issue of limb-specific gene expression in the developing embryo.[80] By using SAGE on the forelimbs and hindlimbs of developing mouse embryos, Margulies and colleagues have been able to identify novel genes involved in the determination of limb identity. In order to obtain a refined and more relevant tag collection for limb identity, these investigators developed a new algorithm for sequence analysis. They also have improved the method for tag-identity determination, thereby increasing the specificity of the tag-UniGene match. By using these novel approaches, they observed that the most differentially expressed gene was *Pitx1*, a transcription factor known to play a vital role in limb development.[81] The differential expression of additional transcription factors critical to limb development, such as *Tbx4*, *Tbx5*, and limb-specific *Hox* genes was also detected by SAGE.[82] The authors have further refined their tag library by performing a "virtual subtraction" with nonlimb specific tag libraries. The number of candidate genes was thereby reduced by 74%, yet the previously identified regulators of limb identity were preserved.

To date, there is no published literature on the use of SAGE to study developmental or reproductive toxicology issues, although several studies have addressed toxicological questions in general, such as identification of genes involved in dioxin-induced hepatotoxicity and carcinogenesis.[83] Given the advantages of SAGE, especially with the recent developments making the technology more accessible to the scientific community and the data more reliable and informative, we expect SAGE to become increasingly popular among developmental and reproductive biologists and toxicologists.

III. PROTEOMICS

A. Background

The term 'proteomics' was first introduced in 1995 as the study of "the total protein complement of a genome", and it focuses on the isolation, quantification, and identification of the biologically active molecules coded by genes.[84] Typical questions in this field of study include queries into proteins' functions and how they relate to the larger scientific questions being pursued. One must remember that proteins are the moieties that carry out the majority of the cells' functions. Thus, the processing of DNA to RNA, followed by the synthesis of proteins, places proteomics downstream of functional genomics in the study of biological systems. Conceptually, proteomics is in many ways similar to gene expression profiling with microarrays or SAGE since it seeks to describe and study protein expression patterns during different stages of development or in response to exposure to toxicants. In essence, proteomics is a necessary complement to gene expression analysis. Although it presently lacks well developed and high-throughput platforms, such as microarrays, new technologies are constantly emerging that will soon overcome these notable deficits and enable high-throughput and sensitive profiling of protein expression.

Protein expression analysis can be used to analyze the myriad of proteins expressed by cells during embryonic and fetal development, whereas a toxicology-oriented study can also utilize proteomics to compare expression profiles of treated versus control samples. Similar to microarrays, where gene expression of an experimental sample is compared with that of a matched control, the underlying assumption in proteomics is that the altered state induced by exposure to a toxicant will be characterized by either the up- or down-regulation of certain proteins. These changes in protein content will reflect various aspects of the toxic insult. Such changes may involve proteins that have been inactivated by a reactive metabolite, proteins that transduce cellular signals that produce toxic end points, or even proteins that participate in detoxification. Once these indicators have been identified, they can be screened against expression profiles induced by exposure to other drugs or toxicants, thereby allowing the deduction of mechanisms of action or providing predictive toxicological assessments.[85,86]

B. Protein Expression Analysis

1. Protein Isolation

a. Two-Dimensional Gel Electrophoresis

The first method that was developed for quantitatively studying global changes in protein expression was two-dimensional gel electrophoresis (2DE).[87] In spite of the many years that have passed since it was originally developed, 2DE is still the predominant technology used in proteomics to separate and observe changes in expression of specific proteins within biological samples. This approach can only provide separation between proteins, while providing a gross visual representation. 2DE

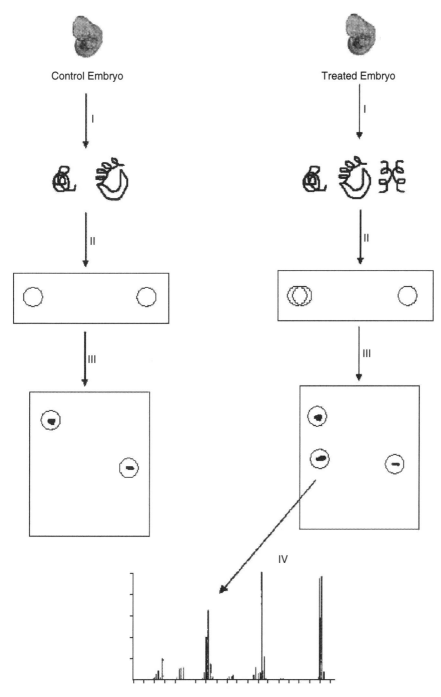

Figure 15.4 Outline of protein expression analysis with two-dimensional gel electrophoresis. (a) Extraction of proteins from tissues of interest, (b) separation of proteins by isoelectric focusing, (c) separation of proteins by mass, and (d) extraction and purification of differentially expressed proteins, followed by mass-spectrometry analysis.

is not able to provide additional information on the identity of the proteins, and thus, supplemental technologies for protein identification, such as mass spectrometry, must be used in conjunction with 2DE. See Figure 15.4 for an outline of the 2DE process.

The basis for the separation of the hundreds of proteins within a sample is primarily the protein's isoelectric point (IP), but separation is also based on molecular weight, solubility, and abundance.[88,89] Briefly, the IP of a protein is the pH at which no net charge is exhibited by the protein. Protein extracts are placed onto an immobilized IP gradient (IIPG) strip, and an electric field is applied. The IP of a protein is based upon the amino acid content; therefore, according to the pH range used, different proteins can be resolved. After this primary one-dimensional separation, the IIPG strips are loaded onto an SDS polyacrylamide gel for electrophoresis. This second dimensional separation is thus based on the molecular weight of each protein, so that those proteins with similar IP's now become resolved according to their molecular weights.[90–92] Following 2DE, the proteins are visualized by the traditional Coomassie Blue or silver staining, although the dynamic range of these detection methods is limited. Recently, new methods of detection with fluorescent dyes, which increase the dynamic range in which proteins can be distinguished and quantified, have been introduced. For instance, the use of SYPRO orange and SYPRO red (Molecular Probes, Eugene, OR) enable increased sensitivity and reproducibility.[93–96]

Although 2DE is regarded as a standard method in proteomics, there are several aspects of its application that limit its use. Proteins expressed at low levels often cannot be visualized because of the detection limits of this technology.[97] The difficulty in detecting low abundance proteins is further exacerbated because of masking by other, more highly expressed proteins.[98] Additional technological limitations include the minimal solubility of membrane proteins, which often have important roles in transporting chemicals across the cell membrane.[97,99] It is estimated that membrane proteins make up approximately 30% of the total cellular protein content, and resolving this particular problem is still a major challenge facing proteomics research.

In order to better facilitate the use of 2DE and address some of the previously mentioned limitations, fractionation is now often performed prior to isoelectric focusing. Fractionations are performed with respect to different biochemical properties such as isolation according to cellular compartments and gradients of aqueous solubility. Chromatography and free flow electrophoresis devices have also been applied, not only to reduce protein mixture complexity but also to enrich the presence of low expression proteins.[100]

b. Multidimensional Liquid Chromatography

Although the majority of proteomic studies utilize 2DE, alternatives have been under development. One such method incorporates the use of multidimensional liquid chromatography to separate the proteins prior to employing protein identification steps, such as mass spectrometry.[101,102] Liquid chromatography is more adaptable to high throughput and robotics, and more importantly, it offers several advantages in terms of protein separation. As previously discussed, some of the limitations of 2DE relate to the inability to detect some proteins. Liquid chromatography, on the other hand, has better capabilities to separate protein mixtures and thus may be more informative than 2DE. Separation may be performed based on the molecular charge by use of an ion-exchange column. It is also possible to separate proteins on the basis of their hydrophobic interactions with the stationary phase or by specific affinity. The eluted protein fractions are subsequently identified, primarily by mass spectrometry.

2. Protein Identification

a. Mass Spectrometry

Once differentially expressed proteins have been visualized, the next step involves protein identification. This is accomplished with mass spectrometry (MS), an approach that has lead to major advancements in proteomics research. Briefly, MS is a method that identifies molecules based on their molecular mass and charge signature (m/z) after they have been ionized and fragmented. The fragmentation pattern of each molecule is unique and generates a spectrum that distinguishes it from others. As a result, proteins and peptides can be identified and quantified by comparison against existing databases. By coupling Edman degradation (removal of single amino acids from the N-terminal of the protein for sequencing) along with MS, it is possible to identify the amino acid sequences of the protein by comparison with established protein databases. Although this latter method is highly precise, it is also regrettably slow. There is clearly a need for new approaches that offer the increased throughput needed to analyze the numerous resulting spots from 2DE.[103]

1) MALDI and MALDI-TOF — Matrix-assisted laser desorption ionization (MALDI), coupled with MS, is accomplished by laser excitation of a protein that has been digested with a site specific protease, such as trypsin. The protein digest is deposited on a special matrix, and upon laser excitation, the digested product becomes ionized and vaporizes to a gaseous state.[91,104] In the gaseous state, the protein fragments can be detected and analyzed by MS and their identity elucidated. Currently, MALDI is often coupled with time-of-flight (TOF), which separates peptides and proteins according to the time it takes them to travel from the MALDI matrix to the MS detector. After the peptides have been vaporized and ionized by the laser, they travel across a chamber toward the MS detector. Analysis is then carried out on the ionized molecules as they enter the detector according to their TOFs, and similar to all other MS methods, the m/z signature is used for identification of the protein of interest.

2) MS/MS and ESI-MS/MS — Tandem mass spectrometry (MS/MS) has also emerged as a common approach for the identification of proteins after extraction from 2-DE.[105] Once extracted, the protein of interest is digested and undergoes electrospray ionization (ESI) that allows ionization of peptides directly from the liquid phase. The sample next enters into a tandem MS consisting of two mass analyzers. The first MS is used to isolate peptides ionized from the original sample. The fragments analyzed by the first MS are subsequently sent into a collision chamber, where peptides are further fragmented by collision with an inert gas, such as argon. The second MS analyzes the secondarily formed ionized fragments, and a unique m/z signature is obtained. This method allows for the partial determination of amino acid sequences within the protein (also called peptide sequence tags), which can be used to search protein databases and identify the protein of interest.

b. Isotope-Coded Affinity Tag

The recently developed isotope-coded affinity tag (ICAT) methodology is conceptually similar to microarray technology in the sense that it enables the simultaneous identification and quantification of peptides derived from two biological samples of interest.[106–108] ICAT is an exciting technology that will certainly have a tremendous affect on our ability to examine proteomewide changes in protein expression.

ICAT is based upon MS combined with site-specific, stable isotopic tagging of proteins, thereby enabling distinct and high throughput analysis of protein mixtures. The ICAT reagent contains several components: a protein reactive group, a biotin tag, and a linker. The protein reactive group is specific for the thiol groups of cysteine residues, where it covalently attaches to the reduced thiol form. The biotin tag is used for isolation by affinity chromatography of tagged-peptides that can be used for analysis. The linker group is composed of an ethylene glycol region that comes in two forms, light or heavy. The heavy form has eight deuterium atoms, whereas the light reagent has eight hydrogen atoms; therefore, they are identical and differ only in mass. As a result of the ability of MAS to distinguish the 8-Da difference in mass, two distinct populations of proteins may be compared against each other.

For example, let us consider an experiment where the mechanism of 2,3,7,8-tetrachloro dibenzodioxin (TCDD)-induced palatal clefting in mice is being investigated. Pregnant dams at gestational day (GD) 12 are treated with either a relevant oral dose of TCDD or the vehicle. Palatal tissues are removed from fetuses 24 to 48 h after TCDD exposure, and the proteins are isolated. Proteins from vehicle-treated fetuses are labeled with the light reagent, whereas proteins from TCDD-treated fetuses are labeled with the heavy reagent. Following labeling, the protein lysates are subjected to digestion by trypsin, and the resulting peptide fragments are purified by avidin affinity. Only those peptides with modified cysteine residues that have reacted with the ICAT reagents will possess the biotin group necessary for retention. The complexity of the affinity purified peptide mixture is substantially reduced in comparison with the original sample because it contains approximately 10% the amount of peptides. If desired, the purified peptide mixture can be further separated and purified by liquid chromatography. Eventually, peptides from both TCDD and control samples are pooled and analyzed simultaneously by MS. Tagged peptides from both groups are comparable and will fragment and ionize similarly in the mass spectrometer; however, because of the differences in mass, the detector can easily distinguish between the two forms of identical peptides. In other words, proteins from both samples can be identified and comparatively quantified by the heavy to light labeling shifts. Hence, TCDD-specific effects on the palatal protein content could be identified, thereby providing unique insight into potential mechanisms of TCDD-induced palatal clefting in mice.

3. Protein Arrays

High bandwidth expression assays, such as microarrays and SAGE, which enable the screening of numerous genes simultaneously, have revolutionized genomics. Yet, it is the protein that is the functional motif of biological reactions in mammals. Thus, the next logical step is to implement protein microarrays, so that large numbers of proteins may be coanalyzed on a single platform. Protein arrays are currently not commonly used; however, exploratory and revolutionizing technologies and platforms are being developed and tested. Protein arrays can be used to detect protein-antibody interactions, protein–nucleic acid interactions, protein-protein interactions, protein modifications, enzymatic activities, and more.[109–111] Several technical obstacles complicate the development of protein arrays. These include the tendency of proteins to denature and interact in a three-dimensional fashion, thus complicating the maintenance of relevant interactions on a surface. Emerging technologies that will affect this field in the upcoming years may be found in the following references.[109–113]

IV. PREDICTIVE VALUES

One of the areas that is expected to benefit substantially from the development of toxicogenomics and proteomics is the ability to predict pathological processes induced by xenobiotics or other potentially harmful agents. Microarrays have already been shown to have predictive merit in tumor classification, thereby improving the prognostic capabilities of clinicians as well as the optimization of chemotherapeutic regimens.[69,114] Similar efforts are being made to increase the ability to predict a novel compound's toxicity based on similarities in gene expression.[115] Such efforts are currently being led by Phase-1 (Santa Fe, NM), a biotechnology company dedicated to developing predictive toxicological tools based on gene expression profiles. The company has created TOXbank™, a database made up of toxicology (blood chemistry, histopathology, and more) and gene expression data. All gene expression data were generated with the company's proprietary toxicology-specific rat arrays, from animals that had been treated with classical toxicants. Thus, gene expression profiles of new investigational compounds can be compared against the database, and any significant match in gene expression profile and histopathology is suggestive of potential toxicity that may be elicited by the chemical in question.[116]

Gene expression profiling studies in developmental toxicology have focused on implantation,[117] early-stage differentiation of embryonic stem cells,[118] and the development of the nervous system[37] and the retina.[13] Given the ethical and practical complexities of obtaining relevant samples from human embryos or fetuses, no studies on their gene expression have as yet been published. Even developmental toxicology studies in laboratory animals are often difficult to perform and interpret given the complexity of experimental design, routes of exposure, and differences in metabolism between the different animal species and humans.

Developmental and reproductive toxicity studies are designed to assess the potential of agents to induce toxicity in both a qualitative and quantitative manner. Most studies encompass the use of several animal species under different exposure paradigms. When there are specific, cross-species, and dose-dependent developmental abnormalities, extrapolation to humans is considered relatively reliable. On the other hand, when toxicity is species specific, extrapolation of adverse outcomes to humans might be established by more mechanistic approaches.[119] Therefore, using global gene expression or protein profiling data of developmental and reproductive toxicants and newly developed pharmaceuticals to establish potential hazards in humans is one such approach that is constantly gaining support from both the academic and industrial sectors. However, it is important that the limitations and advantages of each the technologies described be kept in mind so that when planning such experiments, the most appropriate technique is chosen. Similar to the advancements recently made in the functional genomics of cancer biology, increasing knowledge and understanding of the mechanisms of developmental and reproductive toxicity will enable us to predict those compounds that posses the greatest potential for harm.[120-122]

V. RELEVANT WEBSITES AND DATABASES

A. Microarrays

- The National Center for Biotechnology Information (NCBI), Gene Expression Omnibus, the microarray data repository
  - http://www.ncbi.nlm.nih.gov/geo
- The National Institute of Aging (NIA) mouse cDNA project home page
  - http://lgsun.grc.nia.nih.gov/cDNA/cDNA.html
- Microarray user group
  - http://tango01.cit.nih.gov/sig/home.taf?_function=main&SIGInfo_SIGID=58
- Basic information and relevant links
  - http://www.gene-chips.com
  - A public source for microarray protocols and software http://www.microarrays.org/index.html
- The microarrays Yahoo eGroup
  - http://groups.yahoo.com/group/microarray

B. SAGE

- NCBI's SAGEmap
  - http://www.ncbi.nlm.nih.gov/SAGE
- The Cancer Genome Anatomy Project
  - http://cgap.nci.nih.gov/SAGE
- SAGE home page
  - http://www.sagenet.org/

C. Proteomics

- NCBI's Structure Group
  - http://www.ncbi.nlm.nih.gov/Structure
- Protein Data Bank
  - http://www.rcsb.org/pdb/

REFERENCES

1. Lander, E.S., et al., Initial sequencing and analysis of the human genome, *Nature*, 409, 860, 2001.
2. Venter, J.C., et al., The sequence of the human genome, *Science*, 291, 1304, 2001.
3. Freeman, W.M., Robertson, D.J., and Vrana, K.E., Fundamentals of DNA hybridization arrays for gene expression analysis, *Biotechniques*, 29, 1042, 2000.
4. Brown, P.O. and Botstein, D., Exploring the new world of the genome with DNA microarrays, *Nat. Genet.*, 21, 33, 1999.
5. Gerhold, D., Rushmore, T., and Caskey, C.T., DNA chips: promising toys have become powerful tools, *Trends Biochem. Sci.*, 24, 168, 1999.
6. Southern, E.M., Detection of specific sequences among DNA fragments separated by gel electrophoresis, *J. Mol. Biol.*, 98, 503, 1975.
7. Kafatos, F.C., Jones, C.W., and Efstratiadis, A., Determination of nucleic acid sequence homologies and relative concentrations by a dot hybridization procedure, *Nucleic Acids Res..*, 7, 1541, 1979.
8. Southern, E., Mir, K., and Shchepinov, M., Molecular interactions on microarrays, *Nat. Genet.*, 21, 5, 1999.
9. Schena, M., et al., Quantitative monitoring of gene expression patterns with a complementary DNA microarray, *Science*, 270, 467, 1995.
10. Ko, M.S., Embryogenomics: developmental biology meets genomics, *Trends Biotechnol.*, 19, 511, 2001.
11. Tanaka, T.S., et al., Genome-wide expression profiling of mid-gestation placenta and embryo using a 15,000 mouse developmental cDNA microarray, *Proc. Natl. Acad. Sci. USA*, 97, 9127, 2000.
12. Kargul, G.J., et al., Verification and initial annotation of the NIA mouse 15K cDNA clone set, *Nat. Genet.*, 28, 17, 2001.
13. Mu, X., et al., Gene expression in the developing mouse retina by EST sequencing and microarray analysis, *Nucleic Acids Res.*, 29, 4983, 2001.
14. Lovett, R.A., Toxicogenomics. Toxicologists brace for genomics revolution, *Science*, 289, 536, 2000.
15. Nuwaysir, E.F., et al., Microarrays and toxicology: the advent of toxicogenomics, *Mol. Carcinog.*, 24, 153, 1999.
16. Bartosiewicz, M., et al., Development of a toxicological gene array and quantitative assessment of this technology, *Arch. Biochem. Biophys.*, 376, 66, 2000.
17. Hegde, P., et al., A concise guide to cDNA microarray analysis, *Biotechniques*, 29, 548, 2000.
18. Eisen, M.B. and Brown, P.O., DNA arrays for analysis of gene expression, *Methods Enzymol.*, 303, 179, 1999.
19. Diehl, F., et al., Manufacturing DNA microarrays of high spot homogeneity and reduced background signal, *Nucleic Acids Res.*, 29, E38, 2001.
20. Wildsmith, S.E. and Elcock, F.J., Microarrays under the microscope, *Mol. Pathol.*, 54, 8, 2001.
21. Wildsmith, S.E., et al., Maximization of signal derived from cDNA microarrays, *Biotechniques*, 30, 202, 2001.
22. Panchuk-Voloshina, N., et al., Alexa dyes, a series of new fluorescent dyes that yield exceptionally bright, photostable conjugates, *J. Histochem. Cytochem.*, 47, 1179, 1999.
23. Manduchi, E., et al., Comparison of different labeling methods for two-channel high-density microarray experiments, *Physiol. Genomics*, 10, 169, 2002.
24. Fodor, S.P., et al., Light-directed, spatially addressable parallel chemical synthesis, *Science*, 251, 767, 1991.
25. Lipshutz, R.J., et al., High density synthetic oligonucleotide arrays, *Nat. Genet.*, 21, 20, 1999.
26. Yang, Y.H., et al., Normalization for cDNA microarray data: a robust composite method addressing single and multiple slide systematic variation, *Nucleic Acids Res.*, 30, e15, 2002.
27. Hill, A.A., et al., Evaluation of normalization procedures for oligonucleotide array data based on spiked cRNA controls, *Genome Biol.*, 2, RESEARCH0055, 2001.
28. Alon, U., et al., Broad patterns of gene expression revealed by clustering analysis of tumor and normal colon tissues probed by oligonucleotide arrays, *Proc. Natl. Acad. Sci. USA*, 96, 6745, 1999.
29. Selinger, D.W., et al., RNA expression analysis using a 30 base pair resolution Escherichia coli genome array, *Nat. Biotechnol.*, 18, 1262, 2000.
30. Cho, R.J., et al., A genome-wide transcriptional analysis of the mitotic cell cycle, *Mol. Cell*, 2, 65, 1998.

31. Yang, M.C., et al., A statistical method for flagging weak spots improves normalization and ratio estimates in microarrays, *Physio. Genomics*, 7, 45, 2001.
32. Li, C. and Wong, W.H., Model-based analysis of oligonucleotide arrays: expression index computation and outlier detection, *Proc. Natl. Acad. Sci. USA*, 98, 31, 2001.
33. Lockhart, D.J., et al., Expression monitoring by hybridization to high-density oligonucleotide arrays, *Nat. Biotechnol.*, 14, 1675, 1996.
34. Ishii, M., et al., Direct comparison of GeneChip and SAGE on the quantitative accuracy in transcript profiling analysis, *Genomics*, 68, 136, 2000.
35. Schuchhardt, J., et al., Normalization strategies for cDNA microarrays, *Nucleic Acids Res.*, 28, E47, 2000.
36. Eisen, M.B., et al., Cluster analysis and display of genome-wide expression patterns, *Proc. Natl. Acad. Sci. USA*, 95, 14863, 1998.
37. Wen, X., et al., Large-scale temporal gene expression mapping of central nervous system development, *Proc. Natl. Acad. Sci. USA*, 95, 334, 1998.
38. Sherlock, G., Analysis of large-scale gene expression data, *Curr. Opinion Immunol.*, 12, 201, 2000.
39. Zhang, K. and Zhao, H., Assessing reliability of gene clusters from gene expression data, *Funct. Integr. Genomics*, 1, 156, 2000.
40. Heyer, L.J., Kruglyak, S., and Yooseph, S., Exploring expression data: identification and analysis of coexpressed genes, *Genome Res.*, 9, 1106, 1999.
41. Tavazoie, S., et al., Systematic determination of genetic network architecture, *Nat. Genet.*, 22, 281, 1999.
42. Hilsenbeck, S.G., et al., Statistical analysis of array expression data as applied to the problem of tamoxifen resistance, *J Natl. Cancer Inst.*, 91, 453, 1999.
43. Ghosh, D. and Chinnaiyan, A.M., Mixture modelling of gene expression data from microarray experiments, *Bioinformatics*, 18, 275, 2002.
44. Shmulevich, I. and Zhang, W., Binary analysis and optimization-based normalization of gene expression data, *Bioinformatics*, 18, 555, 2002.
45. Tamayo, P., et al., Interpreting patterns of gene expression with self-organizing maps: methods and application to hematopoietic differentiation, *Proc. Natl. Acad. Sci. USA*, 96, 2907, 1999.
46. Ben-Dor, A., Shamir, R., and Yakhini, Z., Clustering gene expression patterns, *J. Comput. Biol.*, 6, 281, 1999.
47. Lukashin, A.V. and Fuchs, R., Analysis of temporal gene expression profiles: clustering by simulated annealing and determining the optimal number of clusters, *Bioinformatics*, 17, 405, 2001.
48. Herrero, J., Valencia, A., and Dopazo, J., A hierarchical unsupervised growing neural network for clustering gene expression patterns, *Bioinformatics*, 17, 126, 2001.
49. Chiang, D.Y., Brown, P.O., and Eisen, M.B., Visualizing associations between genome sequences and gene expression data using genome-mean expression profiles, *Bioinformatics*, 17(Suppl. 1), S49, 2001.
50. Bittner, M., Meltzer, P., and Trent, J., Data analysis and integration: of steps and arrows, *Nat. Genet.*, 22, 213, 1999.
51. Wong, K.K., Cheng, R.S., and Mok, S.C., Identification of differentially expressed genes from ovarian cancer cells by MICROMAX cDNA microarray system, *Biotechniques*, 30, 670, 2001.
52. Stears, R.L., Getts, R.C., and Gullans, S.R., A novel, sensitive detection system for high-density microarrays using dendrimer technology, *Physiol. Genomics*, 3, 93, 2000.
53. Shchepinov, M.S., et al., Oligonucleotide dendrimers: stable nano-structures, *Nucleic Acids Res.*, 27, 3035, 1999.
54. Van Gelder, R.N., et al., Amplified RNA synthesized from limited quantities of heterogeneous cDNA, *Proc. Natl. Acad. Sci. USA*, 87, 1663, 1990.
55. Kacharmina, J.E., Crino, P.B., and Eberwine, J., Preparation of cDNA from single cells and subcellular regions, *Methods Enzymol.*, 303, 3, 1999.
56. Baugh, L.R., et al., Quantitative analysis of mRNA amplification by in vitro transcription, *Nucleic Acids Res.*, 29, E29, 2001.
57. Finnell, R.H., et al., Strain-dependent alterations in the expression of folate pathway genes following teratogenic exposure to valproic acid in a mouse model, *Am. J. Med. Genet.*, 70, 303, 1997.
58. Bennett, G.D., et al., Valproic acid-induced alterations in growth and neurotrophic factor gene expression in murine embryos [corrected], *Reprod. Toxicol.*, 14, 1, 2000.

59. Wlodarczyk, B.C., et al., Valproic acid-induced changes in gene expression during neurulation in a mouse model, *Teratology*, 54, 284, 1996.
60. Wlodraczyk, B., et al., Arsenic-induced alterations in embryonic transcription factor gene expression: implications for abnormal neural development, *Dev. Genet.*, 18, 306, 1996.
61. Wlodarczyk, B.J., et al., Arsenic-induced neural tube defects in mice: alterations in cell cycle gene expression, *Reprod. Toxicol*, 10, 447, 1996.
62. Wang, E., et al., High-fidelity mRNA amplification for gene profiling, *Nat. Biotechnol.*, 18, 457, 2000.
63. Ohyama, H., et al., Laser capture microdissection-generated target sample for high-density oligonucleotide array hybridization, *Biotechniques*, 29, 530, 2000.
64. Luo, L., et al., Gene expression profiles of laser-captured adjacent neuronal subtypes, *Nat. Med.*, 5, 117, 1999.
65. Lee, M.L., et al., Importance of replication in microarray gene expression studies: statistical methods and evidence from repetitive cDNA hybridizations, *Proc. Natl. Acad. Sci. USA*, 97, 9834, 2000.
66. Pan, W., Lin, J., and Le, C.T., How many replicates of arrays are required to detect gene expression changes in microarray experiments? A mixture model approach, *Genome Biol.*, 3, research0022, 2002.
67. Ginzinger, D.G., Gene quantification using real-time quantitative PCR: an emerging technology hits the mainstream, *Exp. Hematol.*, 30, 503, 2002.
68. Salonga, D.S., et al., Relative gene expression in normal and tumor tissue by quantitative RT-PCR, *Methods Mol. Biol.*, 191, 83, 2002.
69. Snider, J.V., Wechser, M.A., and Lossos, I.S., Human disease characterization: real-time quantitative PCR analysis of gene expression, *Drug Discov. Today*, 6, 1062, 2001.
70. Mayanil, C.S., et al., Microarray analysis detects novel Pax3 downstream target genes, *J. Biol. Chem.*, 276, 49299, 2001.
71. Wurmbach, E., et al., Gonadotropin-releasing hormone receptor-coupled gene network organization, *J. Biol. Chem.*, 276, 47195, 2001.
72. Han, E., et al., cDNA expression arrays reveal incomplete reversal of age-related changes in gene expression by calorie restriction, *Mech. Ageing Dev.*, 115, 157, 2000.
73. Choi, J.D., et al., Microarray expression profiling of tissues from mice with uniparental duplications of chromosomes 7 and 11 to identify imprinted genes, *Mamm. Genome*, 12, 758, 2001.
74. Velculescu, V.E., et al., Serial analysis of gene expression, *Science*, 270, 484, 1995.
75. Yamamoto, M., et al., Use of serial analysis of gene expression (SAGE) technology, *J. Immunol. Methods*, 250, 45, 2001.
76. Datson, N.A., et al., MicroSAGE: a modified procedure for serial analysis of gene expression in limited amounts of tissue, *Nucleic Acids Res.*, 27, 1300, 1999.
77. Peters, D.G., et al., Comprehensive transcript analysis in small quantities of mRNA by SAGE-lite, *Nucleic Acids Res.*, 27, e39, 1999.
78. Powell, J., Enhanced concatemer cloning-a modification to the SAGE (Serial Analysis of Gene Expression) technique, *Nucleic Acids Res.*, 26, 3445, 1998.
79. van den Berg, A., van der Leij, J., and Poppema, S., Serial analysis of gene expression: rapid RT-PCR analysis of unknown SAGE tags, *Nucleic Acids Res.*, 27, e17, 1999.
80. Margulies, E.H., Kardia, S.L., and Innis, J.W., A comparative molecular analysis of developing mouse forelimbs and hindlimbs using serial analysis of gene expression (SAGE), *Genome Res.*, 11, 1686, 2001.
81. Lanctot, C., et al., Hindlimb patterning and mandible development require the Ptx1 gene, *Development*, 126, 1805, 1999.
82. Takeuchi, J.K., et al., Tbx5 and Tbx4 genes determine the wing/leg identity of limb buds, *Nature*, 398, 810, 1999.
83. Kurachi, M., et al., Identification of 2,3,7,8-tetrachlorodibenzo-p-dioxin-responsive genes in mouse liver by serial analysis of gene expression, *Biochem. Biophys. Res. Commun.*, 292, 368, 2002.
84. Wasinger, V.C., et al., Progress with gene-product mapping of the Mollicutes: Mycoplasma genitalium, *Electrophoresis*, 16, 1090, 1995.
85. Hellmold, H., et al., Identification of end points relevant to detection of potentially adverse drug reactions, *Toxicol. Lett.*, 127, 239, 2002.
86. Aardema, M.J. and MacGregor, J.T., Toxicology and genetic toxicology in the new era of "toxicogenomics": impact of "-omics" technologies, *Mutat. Res.*, 499, 13, 2002.

87. O'Farrell, P.H., High resolution two-dimensional electrophoresis of proteins, *J. Biol. Chem.*, 250, 4007, 1975.
88. Righetti, P.G., Gianazza, E., and Bjellqvist, B., Modern aspects of isoelectric focusing: two-dimensional maps and immobilized pH gradients, *J. Biochem. Biophys. Methods*, 8, 89, 1983.
89. Gorg, A., Postel, W., and Gunther, S., The current state of two-dimensional electrophoresis with immobilized pH gradients, *Electrophoresis*, 9, 531, 1988.
90. Aebersold, R. and Leavitt, J., Sequence analysis of proteins separated by polyacrylamide gel electrophoresis: towards an integrated protein database, *Electrophoresis*, 11, 517, 1990.
91. Patterson, S.D., and Aebersold, R., Mass spectrometric approaches for the identification of gel-separated proteins, *Electrophoresis*, 16, 1791, 1995.
92. Molloy, M.P., Two-dimensional electrophoresis of membrane proteins using immobilized pH gradients, *Anal. Biochem.*, 280, 1, 2000.
93. Steinberg, T.H., Haugland, R.P., and Singer, V.L., Applications of SYPRO orange and SYPRO red protein gel stains, *Anal. Biochem.*, 239, 238, 1996.
94. Steinberg, T.H., et al., SYPRO orange and SYPRO red protein gel stains: one-step fluorescent staining of denaturing gels for detection of nanogram levels of protein, *Anal. Biochem.*, 239, 223, 1996.
95. Lopez, M.F., et al., A comparison of silver stain and SYPRO Ruby Protein Gel Stain with respect to protein detection in two-dimensional gels and identification by peptide mass profiling, *Electrophoresis*, 21, 3673, 2000.
96. Yan, J.X., et al., Postelectrophoretic staining of proteins separated by two-dimensional gel electrophoresis using SYPRO dyes, *Electrophoresis*, 21, 3657, 2000.
97. Celis, J.E. and Gromov, P., 2D protein electrophoresis: Can it be perfected? *Curr. Opin. Biotechnol.*, 10, 16, 1999.
98. Fey, S.J. and Larsen, P.M., 2D or not 2D. Two-dimensional gel electrophoresis, *Curr. Opin. Chem. Biol.*, 5, 26, 2001.
99. Lognonne, J.L., 2D-page analysis: a practical guide to principle critical parameters, *Cell Mol. Biol. (Noisy-le-grand)*, 40, 41, 1994.
100. Madoz-Gurpide, J., et al., Protein based microarrays: a tool for probing the proteome of cancer cells and tissues, *Proteomics*, 1, 1279, 2001.
101. Issaq, H.J., The role of separation science in proteomics research, *Electrophoresis*, 22, 3629, 2001.
102. Issaq, H.J., et al., Multidimensional high performance liquid chromatography—capillary electrophoresis separation of a protein digest: an update, *Electrophoresis*, 22, 1133, 2001.
103. Hille, J.M., Freed, A.L., and Watzig, H., Possibilities to improve automation, speed and precision of proteome analysis: a comparison of two-dimensional electrophoresis and alternatives, *Electrophoresis*, 22, 4035, 2001.
104. Karas, M. and Hillenkamp, F., Laser desorption ionization of proteins with molecular masses exceeding 10,000 daltons, *Anal. Chem.*, 60, 2299, 1988.
105. Lahm, H.W. and Langen, H., Mass spectrometry: a tool for the identification of proteins separated by gels, *Electrophoresis*, 21, 2105, 2000.
106. Gygi, S.P., et al., Quantitative analysis of complex protein mixtures using isotope-coded affinity tags, *Nat. Biotechnol.*, 17, 994, 1999.
107. Smolka, M.B., et al., Optimization of the isotope-coded affinity tag-labeling procedure for quantitative proteome analysis, *Anal. Biochem.*, 297, 25, 2001.
108. Smolka, M., Zhou, H., and Aebersold, R., Quantitative protein profiling using two-dimensional gel electrophoresis, isotope-coded affinity tag labeling, and mass spectrometry, *Mol. Cell Proteomics*, 1, 19, 2002.
109. Zhu, H., et al., Global analysis of protein activities using proteome chips, *Science*, 293, 2101, 2001.
110. Zhu, H., et al., Analysis of yeast protein kinases using protein chips, *Nat. Genet.*, 26, 283, 2000.
111. Lueking, A., et al., Protein microarrays for gene expression and antibody screening, *Anal. Biochem.*, 270, 103, 1999.
112. MacBeath, G. and Schreiber, S.L., Printing proteins as microarrays for high-throughput function determination, *Science*, 289, 1760, 2000.
113. Zhang, H.T., et al., Protein quantification from complex protein mixtures using a proteomics methodology with single-cell resolution, *Proc. Natl. Acad. Sci. USA*, 98, 5497, 2001.

114. Golub, T.R., et al., Molecular classification of cancer: class discovery and class prediction by gene expression monitoring, *Science*, 286, 531, 1999.
115. Burczynski, M.E., et al., Toxicogenomics-based discrimination of toxic mechanism in HepG2 human hepatoma cells, *Toxicol. Sci.*, 58, 399, 2000.
116. Gant, T.W., Classifying toxicity and pathology by gene-expression profile--taking a lead from studies in neoplasia, *Trends Pharmacol. Sci.*, 23, 388, 2002.
117. Hines, R.S., Molecular analysis of implantation, *Semin Reprod. Med.*, 18, 91, 2000.
118. Guo, X., et al., Proteomic characterization of early-stage differentiation of mouse embryonic stem cells into neural cells induced by all-trans retinoic acid in vitro, *Electrophoresis*, 22, 3067, 2001.
119. Guittina, P., Elefant, E., and Saint-Salvi, B., Hierarchization of animal teratology findings for improving the human risk evaluation of drugs, *Reprod. Toxicol.*, 14, 369, 2000.
120. Pennie, W.D., et al., The principles and practice of toxigenomics: applications and opportunities, *Toxicol. Sci.*, 54, 277, 2000.
121. Gross, S.J., et al., Gene expression profile of trisomy 21 placentas: a potential approach for designing noninvasive techniques of prenatal diagnosis, *Am. J. Obstet. Gynecol.*, 187, 457, 2002.
122. Miertus, J., and Amoroso, A., Microarray-based genetics of cardiac malformations, *Ital. Heart J.*, 2, 565, 2001.

CHAPTER 16

In Vitro Methods for the Study of Mechanisms of Developmental Toxicology

Craig Harris and Jason M. Hansen

CONTENTS

I. Introduction 648
II. Use of *In Vitro* Developmental Systems in Studies of Mechanisms 649
    A. Mammalian Embryo Culture 649
        1. Preimplantation Embryo Culture 649
        2. Peri-implantation Embryo Culture 655
        3. Postimplantation Embryo Culture 655
    B. Other Systems and Approaches Employing Whole Embryo Culture 666
        1. Transgenic Animals 666
        2. Proteomics and Genomics 668
    C. Nonmammalian Whole Embryo Cultures 670
        1. Caenorhabditis elegans Cultures 670
        2. Drosophila Cultures 670
        3. Amphibian Embryo Culture 670
        4. Fish Embryo Culture 671
        5. Avian Embryo Culture 671
    D. Organ and Tissue Culture 672
        1. Limb Bud and Related Perinatal Bone Cultures 673
        2. Palatal Cultures 676
        3. Visceral Yolk Sac Cultures 677
        4. Sex Organs 678
        5. Heart 678
        6. Eye and Lens 678
        7. Embryonic Tooth Germ 678
        8. Spinal Cord 678
        9. Lung Buds 678
    E. Cell Culture and Clonal Analysis 678
III. Summary 683
References 684

I. INTRODUCTION

Current applications of *in vitro* developmental systems can be broadly divided into two areas: (1) prescreening for developmental toxicants and (2) testing for elucidation of mechanisms of normal and abnormal embryogenesis. Manifestations of abnormal development are, herein, described as dysmorphogenesis or developmental toxicity to identify the functional or morphological abnormalities produced by chemical or environmental insults *in vitro*. The relative merits of these models for their respective purposes have been discussed previously at great length.[1-9] It is generally agreed that *in vitro* models, such as the rodent whole embryo culture system, may require additional validation before they can be used universally as predictive teratogen prescreens. Nevertheless, clear advantages support the continued development of the various methods as *in vitro* tools to study mechanisms of developmental toxicity. This is especially true for developmental investigations in viviparous species. Major obstacles encountered in understanding mechanisms of abnormal development and embryotoxicity include experimental inaccessibility, a paucity of material, and the numerous confounding maternal, nutritional and physiological influences present *in utero*. Most of these limitations can now be overcome through systematic study of the whole conceptus or its dissociated component parts *in vitro* at various stages of gestation. A complete understanding of any mechanism, however, is unlikely to emerge from analyses using only a single approach. Ultimately, several *in vivo* and *in vitro* methods, utilizing different levels of biological organization and complexity, must be combined in a systematic manner to elucidate mechanisms of developmental toxicity. In addition, the likelihood that a single universal mechanism will be found to explain embryotoxic manifestations caused by environmental extremes and chemical exposure is highly improbable.

Various manifestations of embryotoxicity are not expressed uniformly throughout the complex, dynamic and interactive continuum of development and are exhibited during the species' period of sensitivity for that specific tissue. Identical maternal exposures to a given chemical agent at different times in gestation may, for example, produce a spectrum of deleterious effects characteristic only of disturbances in specific developmental processes particularly vulnerable at the time of exposure. Lesions produced by chemical agents and environmental extremes will, therefore, be manifest both temporally and spatially at all levels of biological organization. These range from effects on the entire conceptus to highly selective alterations at the cellular and molecular levels. In the discussion to follow, emphasis is placed on the use of mammalian cells, tissues, organs, and embryos, although nonmammalian species are highlighted as deemed appropriate.

The study of mechanisms of developmental toxicity bears the close and expected resemblance to similar *in vitro* methods used for investigating a broad spectrum of toxicological effects in mature organisms. Systems employing decreasing levels of biological complexity are commonly used to isolate specific processes and probe for specific biochemical and molecular mechanisms that may be involved in developmental toxicology. Endpoints for analysis can be developed and customized, based on the needs of a particular study. The range of possibilities for analytical endpoints may include (1) an evaluation of gross malformations using visual examination, light microscopy, scanning and transmission electron microscopy, scanning laser confocal microscopy, and computer assisted morphometric analysis; (2) functional and physiological alterations in organs or tissues, including studies of transport, metabolism, biosynthesis, turnover, electrical and metabolic coupling, and maintenance of cellular redox status, and (3) molecular events, including studies of time and tissue-specific gene expression, differentiation, pattern formation, induction, receptor binding and signal transduction, programmed cell death, and imprinting. Experimental models with higher levels of biological organization and complexity are useful in the evaluation of mechanisms involving organ-organ interactions, pharmacokinetics, and systemic metabolic regulation. Recent advances in molecular and developmental biology have created new possibilities for evaluation of the dynamic molecular and biochemical processes of selected cell populations within an intact embryo. The techniques involved include the use of specific fluorescent antibody probes and ribonucleotide hybridization markers.

In vitro approaches to the study of mechanisms of developmental toxicity discussed in this chapter are classified according to a descending order of biological organization beginning with whole embryo culture, followed by organ, cell, and tissue culture. Specific details of various procedures to be used as examples have been described in several excellent reviews and in hundreds of methodological texts to which the reader is referred. Only a small portion of the available literature has been cited as examples in this chapter and the omission of otherwise excellent work is, regrettably, unavoidable. The focus of the following discussion is to describe selected examples of *in vitro* approaches for the study of mechanisms of developmental toxicity, to describe some advantages and disadvantages of each, and to suggest where additional applications may be appropriate for future studies. Many current advances in the fields of developmental and molecular biology have yet to be exploited for studies of chemical embryotoxicity.

II. USE OF *IN VITRO* DEVELOPMENTAL SYSTEMS IN STUDIES OF MECHANISMS

A. Mammalian Embryo Culture

In vitro approaches used for mechanistic studies of developmental toxicity employ whole embryos taken from various stages of the developmental spectrum. Embryos for culture range from pronuclear and early cleavage stage embryos (preimplantation) to conceptuses grown from the postegg cylinder-head fold stage to late limb bud stages (postimplantation). Each of the approaches described below provides certain advantages for the experimental isolation and study of a potential chemical interaction and its accompanying mechanism. Not the least of these advantages is the ability to isolate and focus on single stage–specific processes. The limitations of each method, however, should be borne in mind. Not the least of these is the restricted duration for which the cultured tissue retains growth and differentiation characteristics *in vitro* that are comparable with those of littermates still growing *in vivo*.

As growth, differentiation, and complexity of function increase during development, the ability to maintain normal growth rate is generally the first parameter to be compromised *in vitro*. Conceptuses of very young or advancing gestational age are likely to undergo additional morphological and functional differentiation, but they universally lag with respect to overall growth as determined by overall size and macromolecule (protein and DNA) content. The loss of optimal growth is a more important consideration in the whole embryo than with cell or organ cultures because of the more specific need in the latter instances to preserve only function and the capacity to differentiate. Isolated tissues and organs grown in culture can lose important physical and chemical stimuli normally found in their embryonic microenvironment while still maintaining function. For these and other reasons pertaining to development in general, the effective length of culture has also been shown to decrease with advancing gestational age as has been delineated by New.[10] Major *in vitro* systems that utilize whole embryos are divided into three categories, namely, those that include starting material from preimplantation, peri-implantation, and postimplantation stages of development. Each of these requires a different methodological approach for the isolation and maintenance of the embryo, and they do not allow for a continuous culture from fertilization through to the close of organogenesis.

1. Preimplantation Embryo Culture

Evaluation of the teratogenic and embryotoxic effects of chemical during fertilization and the preimplantation phase of development has relied primarily on the premise set forth by Austin[11] that chemical exposure of the preimplantation conceptus has one of two consequences: (1) the embryotoxic agent will destroy all or a significant number of the existing embryonic cells, resulting in early embryonic death, i.e., reproductive failure, or (2) chemical exposure will result in the destruction of a lesser portion of the embryonic cells, and from those that remain, the embryo can recover and grow normally without malformation or growth retardation. The latter result is due to the demonstrated totipotency and general plasticity of early cleavage stage embryonic cells, where a single viable cell is capable of regenerating an entire normal conceptus. Although generally correct, the universality of this "all-or-none" theory has recently been challenged, partly through work using preimplantation culture systems that show that persistent functional and dysmorphogenic lesions may persist without embryonic death.

Initial studies employing preimplantation whole embryo cultures were initiated as early as 1912, utilizing the rabbit blastocyst.[12] The subsequent development and use of preimplantation embryo cultures for the study of developmental mechanisms has been recently reviewed by Spielmann and Vogel[13] and Lawitts and Biggers.[14] Fertilized or unfertilized preimplantation embryos can be obtained after synchronization and superovulation of the test animal, the most common experimental species used are the rabbit, rat, mouse, and hamster, and the procedures follow the general outline depicted in Figure 16.1. The collection of sufficient numbers of eggs or fertilized embryos for use in studies of mechanisms often requires superovulation of the female and some means to control gestational age and stage. Successful superovulation can depend on a number of factors, including the age, weight, and strain of the animal used, as well as the time of injection.

Superovulation and synchronization can be accomplished in the mouse by use of an initial i.p. injection of 5 IU pregnant mare's serum gonadotropin (PMSG), followed 44 to 48 h later by an i.p. injection of human chorionic gonadotropin (hCG, 5 IU)[14,15] These methods may be used to obtain fertilized and nonfertilized ova, and early cleavage and later preimplantation stage embryos depending on the time at which material is removed from the oviduct and whether mating is allowed prior to recovery. For *in vitro* experimentation involving the culture of fertilized preimplantation embryos, dams are placed with proven stud males immediately after hCG injection. Successful mating is determined by the appearance of a vaginal plug on the following morning. Pregnant dams are sacrificed at 17 to 28 h after hCG administration, and their oviducts are removed and flushed to release pronuclear- or two-cell–stage embryos. Flushing of embryos fertilized *in vivo* from the oviduct at later cleavage stages (two- to eight-cell; 48 h) also offers some advantages, in that certain complicating factors related to *in vitro* culture are eliminated. The second polar body (Figure 16.1B), which appears within hours after fertilization, allows fertilized eggs to be distinguished from those that have not been fertilized (Figure 16.1A). More accurate determination of fertilization can be made by observation of the formation of a pronucleus or of the decondensing sperm head.

Once collected, embryos are staged and cultured individually in microdrops of M-16 or T-6 culture medium and bovine serum albumin. The embryos are cultured under paraffin or silicone oil as specified by the *in vitro* fertilization (IVF) protocol, or as groups in rotating cultures containing various media.[15] Paraffin oil is used to prevent unnecessary drying and cross-contamination of embryos during culture. Its use can present additional problems for studies involving exposures, especially when the chemical exposure involves addition of highly water-soluble compounds that may precipitate in the oil prior to reaching the embryo. To avoid this problem, preimplantation embryos may also be successfully cultured in 96-well microplates.

During the preimplantation period of development, several properties of the conceptus provide advantages for experimental manipulation and mechanistic studies. Two-cell to four-cell embryos can be easily disaggregated and the cells stained with fluorescein isothiocyanate (FITC) (Figure 16.1G). Reaggregation of cells into control and experimental chimeras allows the fate of treated and untreated cells to be easily followed and identified during periods of exposure and response. This "chimera assay" approach has been used to evaluate both direct and indirect effects on embryogenesis. The direct effects of 2,3,7,8-tetrachlorodibenzo-*p*-dioxin (TCDD) on cell proliferation and differentiation were reported by Blankenship and coworkers.[16] They used FITC-labeled chimeras to show that TCDD exposure of two-cell embryos did not appear to cause subsequent alterations

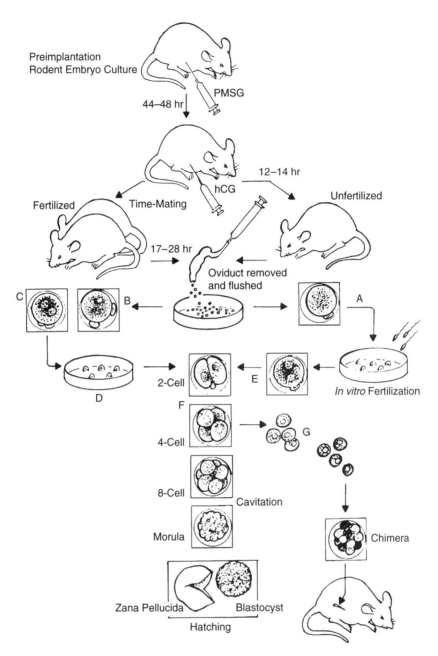

Figure 16.1 Schematic diagram of the general strategy for preimplantation rodent embryo culture. Superovulation is initiated by the successive injection of pregnant mare's serum gonadotropin (PMSG) and human chorionic gonadotropin (hCG). Unfertilized oocytes are collected 12 to 14 h after hCG administration by flushing them from the oviduct. (A) Oocytes at this stage have extruded the first polar body prior to ovulation and fertilization and can be used for tests involving *in vitro* fertilization (IVF). (B) Mating of superovulated females after hCG administration allows for collection of fertilized eggs. They are identified by extrusion of the second polar body. (C to E) Sperm nucleus decondensation, formation of the pronucleus, and migration of pronuclei to the center of the egg occur in succession as cells prepare for the first cleavage division, which can occur either *in vivo* or *in vitro*. Depending on the length of time after mating, embryos flushed from the oviduct may be at the two- or the four-cell stage. This would circumvent the two-cell block typically seen in embryos cultured *in vitro*. (F) Compaction of the embryo begins at the eight-cell stage. This is followed by cavitation at the 32-cell stage, as development proceeds through the morula stage to hatching of the late blastocyst from the zona pellucida. (G) Embryos can be dissociated at the four-cell stage and reassociated with other labeled cells; this procedure produces chimeras to be evaluated *in vitro* at the blastocyst stage or reintroduced into a receptive female, allowing development to continue to term.

in the rate of cell proliferation but did accelerate the rate of differentiation. This was manifested by an increase in the rate of cavitation, which serves as a measure of trophectoderm differentiation.

In a study by Oudiz et al.,[17] male mice were treated with ethylene glycol monomethyl ether (EGME) or remained untreated and were mated with superovulated females. Cells from the resulting embryos were disaggregated; those from embryos sired by control males were labeled with FITC, and chimeras were formed from the cells obtained from the control and experimental matings. Proliferation ratios calculated using this approach showed that embryos from EGME male matings had significant proliferation disadvantages compared with embryos obtained entirely from control mating.

Large quantities of unfertilized eggs collected as cumulus masses after superovulation of the nonpregnant animal can also be used for experiments looking at direct toxicity to the gamete or mechanisms involving the process of IVF. The latter approach provides the distinct added advantage of being able to synchronize development for better dose and response fidelity in evaluation of chemical effects. An example of the use of this approach in developmental toxicity studies was reported by Kholkute et al.[18] using B6D2F1 mice. They describe the effects of polychlorinated biphenyl (PCB) mixtures (Arochlor 1221, 1254, and 1268) and the individual congener (3,3′,4,4′-tetrachlorobiphenyl) added directly to the IVF medium containing capacitated sperm. Direct effects on fertilization and increased incidence of oocyte degeneration and abnormalities were reported. Additional modifications of this approach are possible but have not been thoroughly exploited for the evaluation of developmental toxicity mechanisms.

Success rates of IVF are sometimes low but can be improved considerably through coculture with oviductal epithelial cells and/or by supplementation from appropriate growth factors. This procedure will yield higher rates of fertilization and more robust embryos because of the expected contribution of natural growth factors provided by the epithelial cells. Such an approach may provide important insights into the toxicological significance of oviduct response and regulation.

The combination of these two *in vitro* methods may provide an even more relevant environment for mechanistic studies of early development. For example, the direct and interactive consequences of chemical exposure in the oviduct may be addressed by use of primary cultures of oviduct epithelium with early preimplantation embryos under IVF conditions.

In vitro evaluations of the response of cultured oviduct epithelial cells to modulators of glutathione (GSH) status and a pesticide (lindane) have been completed. They show that oxidative stress and altered cellular thiol status in the oviduct may lead to altered function and subsequently influence embryonic growth and development.[19,20] This approach has the mechanistic advantage of allowing evaluations of chemical effects on the developmental environment, including the nutritive and signaling functions of the oviduct epithelium. Investigations are underway to elucidate functional changes elicited in oviduct function and subsequent effects on the process of fertilization *in vitro*. These studies should provide valuable mechanistic information regarding the manner in which toxicants exert direct effects on fertilization and early embryogenesis and also regarding the environment in which these events take place.

Eggs and embryos during the one- and two-cell stages are particularly sensitive to environmental conditions, especially ambient temperature and oxygen concentration. Several species differ with respect to the ease with which their embryos undergo cleavage, compaction, and blastocyst formation, and progress through development. Because of a much higher success rate for culturing two-cell embryos through to the blastocyst stage, the mouse is generally the species of choice for these types of experiments. A characteristic difficulty in culturing embryos of several species, including mouse and human, is a developmental two- or four-cell "block," beyond which the fertilized embryo is unable to progress. Usually considered an artifact of *in vitro* growth and closely associated with oxidative stress, this condition may actually hold relevance to some mechanisms of early reproductive failure elicited by chemical agents. Indeed, recent analyses have provided evidence that reactive oxygen species generated *in vitro* are responsible for the developmental arrest.[21,22]

Understanding the mechanistic interactions of the preimplantation embryo with its environment and possible chemical insults could provide some explanations for currently unknown causes of infertility. It is also conceivable, but untested, that xenobiotic exposures within the oviduct could alter redox status and antioxidant functions and lead to similar consequences *in vivo*. In relation to possible disturbances of cellular homeostasis following chemical insult, Gardiner and Reed[23] have begun to characterize the GSH status of the preimplantation embryo. These authors have provided evidence that this antioxidant and embryoprotectant may be extensively involved in the mechanism of numerous chemical toxicants. They have shown that the oxidative stress produced by exposure to *t*-butylhydroperoxide (tBH) significantly alters GSH status, the overall manifestations of which change during the course of development. Very little systematic mechanistic or descriptive information exists regarding the exposure, toxicokinetics, biotransformation, and/or manifestations of cellular embryotoxicity for the developing embryo while in the oviduct, including possible roles played by the oviduct in regulating chemical exposure, biotransformation, and detoxication. A great deal of information may be gained by the use of *in vitro* methods to elucidate developmental effects at various stages of early gestation.

From another perspective, preimplantation embryo cultures have been used extensively for mechanistic studies designed to understand the biochemical and physiological processes that regulate normal development. Much of this early developmental work with preimplantation embryo cultures to date has evaluated the uptake and incorporation of labeled precursors into embryonic macromolecules and the functional and morphological effects of inhibitors of metabolism.[24–28] These procedures provide valuable endpoints for use in the evaluation of chemical embryotoxicity and have, in fact, been employed in studies designed to describe the embryotoxicity of agents such as cyclophosphamide and trypan blue.[13,29] Procedures for the estimation of embryonic cell number, repair capacity, biotransformation, and drug uptake have also been devised.[13,15,29–31]

Conditions for successful culture vary with respect to species and gestational age. They depend, to a large degree, on the presence of specific nutritive and trophic factors to ensure progression of the embryogenesis beyond the characteristic *in vitro* block of development. *In vitro* studies have been instrumental in describing amino acid uptake and utilization, particularly for use in protein synthesis, in the preimplantation embryo.[26–28,32] Borland and Tasca[24] demonstrated the inhibition of L-methionine uptake and incorporation into four-cell stage mouse embryos. The inhibition appeared to be a function of alterations in active methionine influx and not a result of accelerated efflux or inhibition of protein synthesis. Physiological and biochemical functions, such as growth factor interactions,[33,34] prostaglandins,[35] nutritive functions,[36,37] glucose uptake and utilization,[38–40] intermediary metabolism,[41,42] steroid hormone function and metabolism,[43–45] receptor-mediated cellular functions,[33,34] and amino acid uptake and transport functions,[42,46,47] have all been evaluated in preimplantation cultures. Most have not been exploited for use in studies of developmental embryotoxicity.

More realistic evaluation of the maternal, embryonic, and chemical factors that affect preimplantation growth and development requires the ability to assess the influence of the environment and maternal factors on the process. Extensions of the *in vitro* method to utilize other *in vivo* and *in vitro* approaches can greatly enhance the power of this approach. The use of IVF, coculture (with oviductal epithelial, stromal, or decidual cells), and transplantation have also been demonstrated in developmental and physiological studies, but these paradigms have not yet been thoroughly exploited for use in mechanistic studies of embryotoxicity. Heterogenic cultures and *in vivo–in vitro* procedures can be used to test whether a xenobiotic is interfering directly with embryonic development or whether a maternal effect (e.g., decidualization in the uterus) is responsible. The factors that contribute to the "two-cell block" of rodent and rabbit conceptuses cultured *in vitro* have been associated with bursts of reactive oxygen species that may be controlled by intracellular protective capacities and the soluble constituents found in culture media or oviductal fluid.[21–23] The results of these studies are reinforced by work involving inclusion of cysteine, glycine, hypotaurine, and other amino acids or antioxidants in the culture medium. Such supplementation can significantly improve viability and reproductive success under IVF conditions and where embryos are subsequently reintroduced into a receptive female for continuation of development.[48] A major disadvantage of the transplantation approach is that success rates among laboratories doing the same procedures differ greatly. Also, such methods can be quite time consuming and costly, especially if it is necessary to establish a dose response for several different chemicals or modes of exposure.

Preimplantation embryo culture is possible using a number of species and is not dependent on extraembryonic membrane structure, as is postimplantation embryo culture. A major disadvantage of this system is the inability to grow embryos optimally beyond the late blastocyst stages. Attempts to extend cultures beyond the normal time of implantation and through organogenesis have, however, been described.[49,50] Although differentiation continues in a near normal fashion *in vitro*, the rates of growth and differentiation are considerably slower in comparison with patterns of growth and differentiation *in vivo*. In general, embryos explanted before the egg cylinder stage are not likely to develop into the somite stages *in vitro*, and most do not establish an active heartbeat or circulation. A great deal of differentiation is possible with regard to function, establishment of primary germ layers, and microstructural and morphological form, although growth lags and the failure rate is very high.

The direct use of *in vitro* preimplantation culture methods for the study of teratogenicity mechanisms has been reviewed by Spielmann and Eibs,[29] with particular reference to *in vivo* exposure to chemicals and/or therapeutic agents and subsequent culture and evaluation of embryos *in vitro*. Many early studies on the sensitivity of preimplantation embryos to external insults were conducted *in vitro* and focused on the effects of UV and x-ray irradiation. It was found that the rate of development to the blastocyst stage is actually a less sensitive indicator of embryotoxicity than is cell number in the blastocyst.[29]

Kaufmann and Armant[51] used mouse preimplantation cultures to evaluate the *in vitro* embryotoxicity of cocaine and its metabolite, benzoylecognine. They showed that these agents are both able to directly block embryogenesis, particularly at the one and two-cell stages, and that metabolism of cocaine seems to abate the embryotoxic potential. Other approaches to evaluate the potential contribution of biotransformation to chemical embryotoxicity have utilized the inclusion of an exogenous biotransformation system to generate reactive intermediates *in vitro*. A rat hepatic S-9 fraction and appropriate cofactors was used by Iyer et al.[31] to show that day-3 blastocysts had altered hatching, culture dish attachment, and trophoblastic outgrowth when exposed to naphthalene *in vitro*. The use of the exogenous bioactivation source demonstrated the need for bioactivation of naphthalene to produce embryotoxicity and helped to define the mechanism of action of this and possibly other polycyclic aromatic hydrocarbons. Direct exposure to 5-bromodeoxyuridine (BrdU) in preimplantation cultures results in alterations in both cell division and differentiation.[52]

Preimplantation cultures have been used extensively for the study of mechanisms of development in general and the endogenous and exogenous factors that influence early cell cleavage and differentiation. Endpoints for evaluation include morphological, physiological, biochemical, and molecular events, each of which can be used in conjunction with other determinations to probe mechanisms of development and the chemical disturbances that may influence the normal process. The processes of regulated gene expression, including the events governing the transition from maternal to embryonic genomic control, can also be investigated. A more detailed discussion of the use of molecular techniques in the study of developmental toxicology can be found in Chapters 14 and 15 of this volume. Gene expression, posttranscriptional modifications, and protein synthesis have been studied and found to be critical components in the regulation of normal development as well as sensitive targets for chemical agents that interfere with normal development.[53–55]

An example of a recent molecular approach to integrate the various expression events and understand their mechanistic controls is described in an evaluation of HSP 70.1 expression concurrent with the activation of the zygotic genome.[56] This approach used transgenic mice from C57BL6/CBA F1 embryos that contained two to three copies of the HSP70.1 *Luc* plasmid/haploid genome. Transgenic offspring were identified by slot blot hybridization, using polymerase chain reaction (PCR) primers specific to the firefly luciferase gene. One-cell zygotes obtained from transgenic crosses were typed by the luciferase assay when the HSP70.1 gene was expressed. The study concluded that this product was expressed only during the one-cell stage and did not continue into the two-cell stage. Considering the expected relevance of HSP 70.1 to environmental insult and chemical exposure, it is easy to suppose that this methodology might be appropriately applied to the study of the effects of pre- and postfertilization chemical insult on the same parameters. Similar molecular approaches could be used to evaluate the roles and potential perturbations of mechanisms related to cell division and/or cleavage, gap junctions, second messengers systems, growth factors, receptors, and the expression and function of many other components.

2. Peri-implantation Embryo Culture

In conjunction with concerns of mechanisms of preimplantation development that involve direct effects and consequences to the developing conceptus itself, chemically elicited effects on the process of uterine implantation are also of considerable importance in developmental toxicity. The overall process of implantation, especially in the rodent and human (where the blastocyst becomes completely imbedded into the uterine endometrial mucosa) has been the least experimentally accessible process in the developmental continuum. Recent advances, however, have produced an *in vitro* primary culture system that can address the mechanistic bases of toxic events that can affect implantation.[57]

The system that has been developed utilizes a primary cultured monolayer of endometrial cells grown on a Cellagen™ membrane matrix impregnated with stromal cells. This approach not only allows recognition of the trophoblast by the endometrium but results in the ingress of the conceptus into the stromal layer, accompanied by displacement of resident cells. In addition, membranous cell culture dish inserts may be used to provide a proper environment for the attachment and normal polar orientation of columnar epithelial cells. Unlike other approaches, this method preserves a large number of functional and structural characteristics seen in the same cells *in vivo*. These include absorptive and excretory functions, appropriate orientation of polarized cell surface markers, polar orientation of intracellular organellar structures (such as the Golgi apparatus and endoplasmic reticulum), preservation of estrogen and progesterone responsiveness, and synthesis and release of proteins. While not ideal in every sense of functional integrity, this system has been shown to promote the recognition and implantation of competent blastocysts. It may provide the best available model for the mechanistic study of early embryonic growth as related to implantation and toxicity factors that may affect the embryo during the peri-implantation period. Attachment of human, bovine, and mouse blastocysts has been compared by the use of this system,[58] but few, if any, systematic studies have been conducted using this approach to assess the direct effects of chemicals on the process of implantation. Utilization of such an *in vitro* approach could lead to important mechanistic insights regarding the reproductive toxicity of environmental and therapeutic chemicals during implantation.

3. Postimplantation Embryo Culture

Anatomical peculiarities of the developing rodent conceptus make it particularly well suited for whole embryo culture during the teratogen-sensitive period of organogenesis. Unlike a majority of mammalian species, including human beings, rodents (rats and mice) possess a visceral yolk sac (VYS) that completely surrounds the amnion and developing embryo through inversion and antimesometrial attachment via the presumptive ectoplacental cone (Figure 16.2). In this arrangement, the VYS serves as a physical protective barrier encompassing the embryo; it also serves in numerous vital detoxication, metabolic, and nutritional roles. This compact arrangement results in conceptuses that are easy to observe, manipulate, and perturb as needed experimentally and represents the

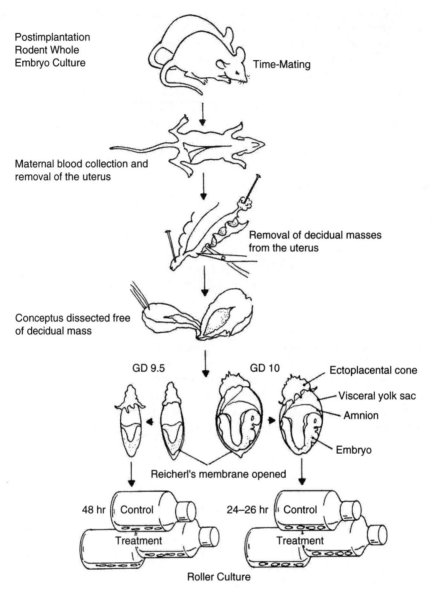

Figure 16.2 Schematic diagram of rat whole embryo culture. Time-mated pregnant females are anesthetized on GD 9.5 or 10, prior to opening of the abdomen and removal of maternal blood (to be used for preparation of culture media). The uterus is removed with the ovaries and cervix attached. Uteruses containing the conceptuses are pinned to a silicone-filled dissecting dish, and decidual masses are cut free from the uterus with iridectomy scissors. Conceptuses are teased free of the decidual mass by use of fine watchmaker's forceps, and the Reichert's membrane is removed to complete the preparation for culture. Embryos are then placed for 24 to 72 h at 37°C in roller bottles containing rat serum in the appropriate gas mixture for optimal growth at the given developmental stage. Morphological, biochemical, physiological, and molecular endpoints can be evaluated throughout the culture period by removing selected embryos from the culture. Relative size, degree of development, and identification of major structures for the GD 9.5 and GD 10 conceptuses are shown.

highest order of *in vitro* biological organization. Because of its versatility, the rodent whole embryo culture system has been used to evaluate the embryotoxicity of several hundred compounds. It continues to grow in popularity as a method of choice for mechanistic evaluations of embryotoxicants, as well as for investigations of mechanisms of development in general.

The historical evolution of the rodent whole embryo culture technique has been described by New.[10] It encompasses a review of the first successful attempts to transfer and culture rabbit preimplantation embryos, including the attempts to devise an effective artificial uterus. In the initial approach, a rodent embryo encased in an intact VYS is suspended in a 5-cm diameter watch glass containing heat-inactivated rat serum. The watch glass is placed in a petri dish lined with wet cotton to maintain humid conditions, and the dish is kept in a constant gas environment of 60% to 95% O_2 and 3% to 5% CO_2. This method is effective for growing conceptuses from early somite to early limb bud stages but has not been used extensively for mechanistic studies. This is a result in part of the difficulties of culturing large numbers of conceptuses.

By far the most popular and universally applicable method for the culture and study of whole embryos for investigation of mechanisms is the rotator culture system, first described by New[59] in 1971, and summarized in Figure 16.2. The specific details of embryo explantation and preparation for culture have been described in detail in a number of excellent texts.[60–62] Briefly, a pregnant dam is anesthetized or euthanized, and its abdomen is opened to expose the gravid uterus. Maternal blood is withdrawn from the abdominal aorta and further processed for heat-inactivated serum to be used as a component of the culture medium. The uterus, with ovaries and cervix attached, is removed, pinned to a dissecting surface submerged in Hanks' balanced salt solution (HBSS) and the decidual swellings are exposed by use of iridectomy scissors. The decidua is separated with fine watchmaker's forceps, and intact conceptuses are removed. Reichert's membrane is then torn away to expose the VYS and the conceptus is ready for culture.

Several types of culture media have been described, although culture for more than a few hours appears to have an absolute requirement for 35% to100% heat inactivated rat serum. The composition of the remainder of the medium is usually a balanced salt solution or other defined medium (e.g., Waymouth's) in various combinations. Conceptuses can be cultured in serum-free defined media, such as HBSS, for several hours if the conceptus is then returned to the usual serum-replete medium. Subsequent growth in this type of culture is without apparent deleterious long-term effects. Media are warmed to 37°C and equilibrated with stage-appropriate O_2 and CO_2 concentrations. The conceptuses are submerged in glass vials (approximately 3 ml) that can be inserted into a rotating culture apparatus capable of continuous gassing and maintenance of constant temperature or grown in sealed roller bottles placed on a roller in an incubator and gassed every 24 h.[63] The rotating culture apparatus and roller bottle apparatus can each be used and have distinct advantages. Roller bottles generally contain 60 to 175 ml of media and support growth of multiple conceptuses up to 24 h following gassing (1 ml media per conceptus is usually considered sufficient). However, a continuous gassing rotator apparatus requires smaller volumes of media and better facilitates the culture of individual conceptuses. In either case, these two approaches are able to maintain the correct stage specific O_2 and CO_2 concentrations required for optimal growth.

The direct exposure of cultured conceptuses to chemical agents can usually be accomplished by addition of the agent to the culture medium. This is especially simple for water-soluble chemicals, whereas, lipophilic agents require prior solubilization in an appropriate vehicle. Acetone, dimethyl sulfoxide (DMSO), and ethanol have been used successfully for this purpose and do not appear to adversely affect development as long as concentrations are kept reasonably low (less than 0.1%).

There is a range of stages of embryonic growth during which conceptuses can be easily manipulated and for which *in vitro* growth closely parallels that seen in the intact uterus. These range from the egg cylinder-head fold and early somite stages (at which time the anterior neural tube has not yet closed, the heart has not yet begun to beat, and the embryo has yet to rotate to the ventral orientation) to the later stages of organogenesis (characterized by complete closure of the anterior neural tube, establishment of a functional heart and circulatory system, and complete rotation of the embryo).[64] Vascularization of the VYS becomes much more evident over this period of development and is used as a marker of development *in vitro*. It is also during this critical juncture of development that the embryo undergoes a number of important metabolic and functional

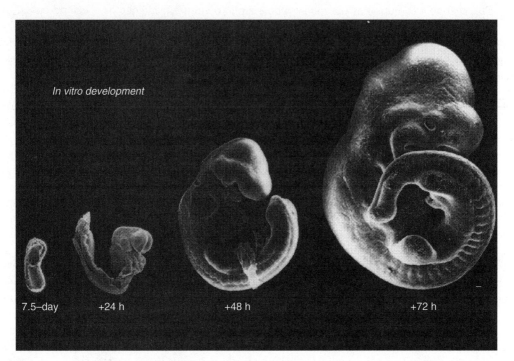

Figure 16.3 Scanning electron micrographs of the best mouse embryos cultured in one experiment, showing the extent of growth and morphogenesis that may be achieved *in vitro* (bar: 100 μm). Day 8.5 embryos (plug day = day 0) were cultured in rotating bottles containing DR50 medium (changed after 24 h of culture) and gassed intermittently with 5% O_2, 5% CO_2, and 90% N_2 for the first 38 h; they were then cultured in fresh DR50 medium in a gas phase with 20% O_2, 5% CO_2 and 75% N_2 until 48 h. Embryos were then cultured in rotating bottles gassed continuously with 20% to 40% O_2 until 72 h *in vitro*. (From Sturm, K. and Tam, P.P.L., *Methods Enzymol.*, 225, 164, 1993. With permission.)

transformations, including conversion from an essentially anaerobic, glycolysis-based metabolism to an aerobic, citric acid cycle–based dependence. In the mouse, the window of optimal *in vitro* growth and differentiation begins on gestational day (GD) 8 (plug day = GD 0). In the rat, this period begins on GD 9.5. In each case, a period of 48 h in culture is believed to encompass the optimal period wherein the *in vitro* conceptus will grow at almost the same rate as if it had remained in the uterus, although earlier and later stages have also been grown successfully. The entire spectrum of *in vitro* growth, spanning a period of 72 h, is shown in Figure 16.3.[61]

While most of the early efforts to perfect whole embryo culture have utilized mouse and rat conceptuses, whole embryo culture of other species, such as the rabbit, is becoming more common. Rabbit whole embryo culture has not been widely used in toxicological investigations but has great potential. In the past decade, procedures for rabbit whole embryo culture have been developed largely by the efforts of Naya and coworkers,[65] Ninomiya and coworkers,[66] and Pitt and Carney.[67,68] Criteria for uniform evaluation of rabbit conceptuses have also been developed and used as a standard of reference during *in vitro* development.[67,68] While the logistics of rabbit whole embryo culture are similar to those for rodent embryo culture (maximum culture time, timing of explantation, periods of organogenesis, gassing conditions, etc.), some physiologic differences require special attention and make rabbit embryo culture unique (Figure 16.4).

Rabbit conceptuses require that a portion of the bidiscoid placenta remain attached throughout the culture period. Although not critical for conceptual viability, the placental bed serves the function of keeping the conceptus intact during the course of incubation. A striking difference between the rabbit and rodent conceptus is that the rabbit VYS only surrounds the caudal portion of the partially rotated embryo, unlike mouse and rat where the VYS is inverted and completely surrounds the

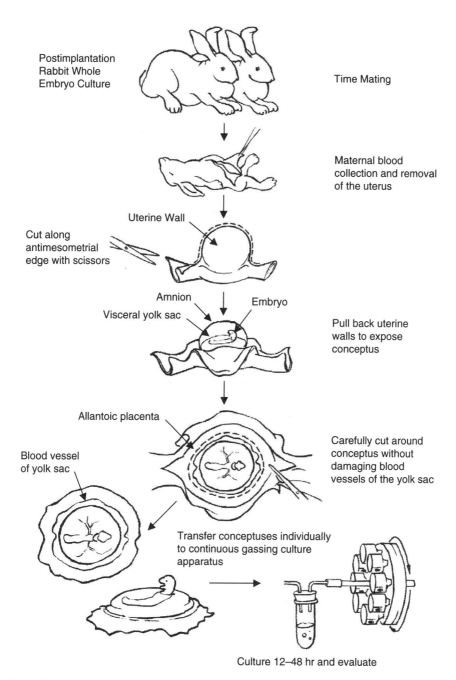

Figure 16.4 Schematic diagram of rabbit whole embryo culture. Time-mated pregnant females are anesthetized on GD 9 prior to opening of the abdomen and removal of maternal blood (to be used for preparation of culture media). The uterus is removed with the ovaries and cervix attached. Each conceptual swelling is individually removed and pinned to a silicone filled dish for dissection. A single cut is made along the antimesometrial edge of the uterus, and the overlying uterine tissue is peeled back to expose the embryo. To free the embryo, it is necessary to cut through the placenta, leaving a portion of the placenta attached to maintain the integrity of the conceptus for culture. Special care must be taken not to damage the visceral yolk sac or any related blood vessels. Embryos are then placed for 24 to 48 h at 37°C in a continuous gassing incubator for optimal growth at the given developmental stage. Morphological, biochemical, physiological, and molecular endpoints can be evaluated throughout the culture period by removing selected embryos from the culture. Relative size, degree of development, and identification of major structures for the GD 9.5 and GD 10 conceptuses are shown.

embryo. It has been suggested that the VYS in the rat and mouse has a protective effect, as is evident by higher GSH concentrations and more biotransforming enzymes in the VYS than in the embryo proper.[69,70] The distinctive structure of the rabbit VYS is more closely related to human VYS in that it does not envelop the embryo. While this is a distinct advantage of rabbit whole embryo culture, it can prove to be an inconvenience during culture, and extra care must be taken to avoid physically damaging the conceptus.

The combined use of both rat and rabbit whole embryo cultures allow investigators to compare common developmental processes, and biochemical and molecular parameters that may contribute to mechanisms of toxicity and species sensitivity, and to resistance to chemical agents. This approach was used to provide insight into thalidomide's species selectivity and its mechanism of action. Hansen et al.[71] have shown that GSH and cysteine, two major contributors to intracellular redox status, are preferentially depleted in the rabbit conceptus by thalidomide, while rat conceptal GSH and cysteine remain relatively replete. This suggests that thalidomide-induced oxidative stress and possible misregulation of redox status occur only in the rabbit conceptus and may contribute to the mechanisms of teratogenesis. This approach, the comparison of both sensitive and resistant species *in vitro*, provides a means to determine species-specific differences and should prove useful in the elucidation of specific developmental toxicity mechanisms.

Whole embryo cultures have been used successfully to investigate the mechanisms of a number of relevant toxicological endpoints. Some of the first and most significant studies attempting to describe mechanisms of chemical embryotoxicity were investigations that dealt with conceptal biotransformation of chemical teratogens. Inherent in these studies were attempts to understand the relationships between maternal chemical biotransformation and the biotransformation capacities of the conceptal tissues. Some known *in vivo* teratogens, such as cyclophosphamide (CP) and 2-acetylaminofluorene (2-AAF), were found to produce little relevant dysmorphogenesis when added directly to cultures of early somite-stage rat conceptuses.[72,73] That maternal bioactivation is a prerequisite for producing dysmorphogenesis was demonstrated *in vitro* by the addition of rat hepatic 9000 × g supernatant fractions (S-9) and the appropriate cofactors (i.e., an NADPH generating system) for supporting cytochrome P450 monooxygenation activities. Because of the experimental accessibility of the *in vitro* system and the ease with which environmental conditions could be controlled, this approach was found to be useful in characterization of the maternal and conceptal biotransformation necessary to produce embryotoxicity. Some of the agents tested include cyclophosphamide,[72] 2-AAF,[73] acetylsalicylic acid,[74] adriamycin,[75] acetaminophen,[76] retinoic acid (RA),[77,78] valproic acid,[79,80] niridazole,[81] cytochalasin D,[82] and rifampin.[83]

Hepatic S-9 fractions prepared from male rats exposed to chemical inducers of selected subclasses of cytochrome P-450 isozymes have helped to identify the requisite biotransformations necessary for making the chemicals in question dysmorphogenic. For example, it was demonstrated that major phenobarbital (PB)-inducible isoforms were required to metabolize cyclophosphamide to its reactive and teratogenic forms; other classes of inducible enzymes were ineffective.[73,84] Conversely, the 3-methylcholanthrene (3-MC)-inducible isoforms (predominantly CYP 1A1)[58] were shown to be necessary for the conversion of 2-AAF to a variety of embryotoxic and dysmorphogenic products, including the putative proximal dysmorphogen, 7-hydroxy-acetylaminofluorene (7-OH-AAF). The latter metabolite was found to be dysmorphogenic without need for an exogenous bioactivation system. It was suggested to be the proximal teratogen because it produced neural tube defects not altogether unlike those elicited when the parent 2-AAF was administered to the pregnant animal.[73,76] Further studies revealed that, unlike PB-inducible isoforms, the major MC-inducible isoform can also be induced *in utero* in conceptal tissues through maternal exposure to MC. These studies further showed that conceptuses explanted from MC-exposed dams and grown in culture were able to constitutively bioactivate AAF without an exogenous bioactivation system.[84] Subsequent *in vitro* studies using acetaminophen (APAP), a compound physically and chemically similar to 7-OH-AAF, allowed for direct testing of the prediction that similar manifestations of dysmorphogenesis would result from both agents because of the common physical and chemical properties.

This was found to be the case, leading to additional studies. These studies concluded that the embryotoxic mechanism of both 7-OH-AAF and APAP was likely to involve additional metabolism by an endogenous bioactivation source to form more chemically reactive species, such as their respective catechols.[76]

The mechanistic implications from these studies were that constitutive and inducible cytochrome P450s exist in the embryo and VYS of the organogenesis-stage rodent conceptus. Cytochromes P450 have been partially characterized in the VYS, using a series of resorufin ether analogs that have been shown to be selectively metabolized by various cytochrome P450 isozymes. This approach provides evidence that P450 activities are present and differentially distributed in various tissue compartments of the conceptus. Total P450 specific activity is approximately 10-fold higher in the VYS than in the embryo proper. At least four constitutive isoforms and one inducible isoform were postulated from these studies. Specific antibodies raised against purified adult P450 isozymes were able to identify only the inducible form as CYP 1A1 by use of Western blot analysis,[85,86] suggesting that the others may be uniquely expressed only in the conceptus.

Subsequent work has sought to identify and further characterize the functions and metabolic capacities of these enzymes.[87] Such studies using the whole embryo culture system have been critical in the characterization of P450 function in the organogenesis-stage rat conceptus. Specific mechanistic studies into the molecular regulation of these activities and their consequences are underway but may need to be combined with investigations of cell culture and cell free systems to clearly define complex cellular mechanisms. *In vitro* models have been extremely useful in elucidating these pathways. They provide a means to systematically evaluate the mechanisms of conceptual biotransformation in an experimentally friendly system that has been removed from potentially confounding maternal factors.

Other enzymes capable of the bioactivation and detoxication of chemical agents in the conceptus have also recently been characterized by use of the whole embryo culture system. Recent studies of the biotransformation of phenytoin have provided compelling mechanistic evidence that the previously postulated generation of an arene oxide metabolite by the actions of cytochrome P450 is most likely not responsible for the resultant embryotoxicity and dysmorphogenesis.[88,89] The data now suggest that the majority of phenytoin's embryotoxic and teratogenic effects are mediated through the formation of organic free radicals or reactive oxygen species, generated through the activity of endogenous lipoxygenases or prostaglandin H synthetase.[90,91] Significant lipoxygenase activities have been reported in the rodent conceptus.[92] Evidence for free radical–mediated embryotoxicity is supported by a concomitant reactive oxygen-mediated oxidation of DNA (8-OH-2′-deoxyguanosine) and findings that the embryonic lesions can be eliminated by addition of the antioxidants catalase and/or superoxide dismutase (SOD) to the culture medium.[93] Activities of these and related enzymes of the arachidonic acid pathway have been shown to be high in the developing conceptus. They may well account for the bioactivation and subsequent embryotoxicity of a number of other known teratogens and *in vitro* dysmorphogens for which cooxidation to free radicals and reactive intermediates has already been described.

To use *in vitro* methods for reliable prediction of mechanisms operating *in vivo*, one must be able to address important pharmacokinetic factors that would determine the actual amount of chemical reaching the conceptus *in utero*. Whole embryo cultures have been used for comparative *in vivo–in vitro* studies to define mechanisms related to the distribution, metabolism, and elicited effects in the whole animal.[94] Studies into the mechanisms of 2-methoxyethanol embryotoxicity have utilized the whole embryo culture system to clarify the site and mode of metabolism of (methoxy-^{14}C)-2-methoxyacetic acid (2-MAA) to $^{14}CO_2$, suggesting that the compound is metabolized in the dam and not in the conceptus.[95] Short term *in vitro* cultures of CD-1 mouse embryos were also used to demonstrate the inhibition of DNA synthesis by 2-MAA and the attenuation of inhibition and embryotoxicity by compounds such as formate, acetate, and sarcosine.[96] Other investigators have also integrated *in vitro* and *in vivo* approaches to study the pharmacokinetics and biotransformation of the teratogen, valproic acid.[79,94] These authors point out that in using *in vitro* models, care must be taken to consider relevant pharmacokinetic behavior so as to avoid possible misinterpretation of data. One significant property of the developing rodent conceptus that also should not be overlooked in pharmacokinetic and mechanistic considerations is the discovery that the conceptal milieu during early organogenesis is significantly more alkaline than the surrounding maternal fluids and tissues. This can result in the selective accumulation of weak acids, such as valproic acid, within the conceptus.[97,98] The cellular regulation of conceptal pH has not been adequately studied, especially from the perspective of possible deleterious alterations elicited by chemical agents. Preliminary studies also indicate that the viable, intact conceptus may be used *in vitro* with a customized perfusion apparatus and minute microfiberoptic (MFOP) sensors capable of monitoring real-time alterations in pH caused by environmental changes and chemical exposure.[99,100] The ability to isolate the embryo *in vitro* and conduct evaluations of the specific effects of individual chemical modulators and inhibitors provides a means to probe the mechanisms of complex processes that are extremely difficult to unravel in the whole organism.

An understanding of the entire spectrum of events leading to persistent anatomical and/or functional defects in the developing organism will not only require elucidation of the means by which chemicals are taken up and metabolized within the conceptus. It will also demand knowledge of how the conceptal response to that insult affects dysmorphogenic or functional alterations. The rodent whole embryo culture system is an excellent model for the study of antioxidant, detoxication, and other protective systems. In particular, the studies described in the next sections have taken advantage of the ability, in whole embryo culture, to selectively modulate several biochemical and physiological pathways and to isolate and study specific effects caused by chemicals.

Early reports indicated that organogenesis-stage rat conceptuses have very low or undetectable activities of enzymes such as UDP-glucuronosyltransferase, sulfotransferase and glutathione-S-transferases, the predominant enzymes used by the adult for detoxication of reactive chemical intermediates.[101] While these conclusions still appear to be valid, enzyme activities may be present but highly localized.

Recent investigations also suggest that the conceptus maintains significant concentrations of GSH and has ample activities of enzymes involved in its synthesis, turnover, and metabolism.[69,102,103] The important role of GSH in embryo protection has been well established through both *in vitro*[104–109] and *in vivo* studies. These have provided the basis for use of whole embryo culture to investigate the spatial and temporal mechanisms that control GSH synthesis, turnover, and metabolism. Chemical modulators used in the manipulation of GSH status for mechanistic and descriptive studies include L-buthionine-*S,R*-sulfoximine (BSO), a selective and irreversible inhibitor of GSH synthesis,[109] resulting in GSH depletion as a function of the intrinsic cellular rates of GSH turnover;[102] diethyl maleate (DEM), an, α,β-unsaturated carbonyl compound that readily and nonezymatically forms GSH adducts, resulting in rapid GSH depletion;[69] a cysteine pro-drug, 2-oxothiazolidine-4-carboxylate (OTC), which is capable of supporting higher rates of GSH synthesis through increasing intracellular levels of the rate-limiting precursor cysteine;[104] glutathione monoethyl ester (GME), which produces selective elevations of intracellular GSH without the need for active GSH synthesis;[110] and acivicin, which selectively inhibits the activity of γ-glutamyltranspeptidase (GGT), a plasma membrane glutathionase enzyme also believed to function in amino acid transport and metabolism of GSH conjugates.[111] Microinjection of these and other selective modulators has also been shown to result in selective spatial alteration of GSH within the conceptus.[110,112] Such abilities to selectively alter GSH status will be of considerable advantage in dissecting the tissue selective effects of chemicals and the focal biotransformations that may define the basis for differential tissue sensitivity and resistance and the ability to respond to specific chemical agents. The ability to synthesize new GSH is an important consideration in this context. *In vitro* studies have shown that depletion of conceptal GSH (to 30% to 40% of control in embryo and VYS) by addition of DEM directly to the culture medium is followed by a differential recovery in the two compartments. Visceral yolk sac recovery begins immediately after reaching the 30-min nadir and is replenished at a rate to exceed initial levels within 3 h.[69] The embryo, however, remains refractory until VYS concentrations are restored to preexposure levels. This suggests that repletion of the VYS may be involved in the regulation of embryonic GSH synthesis through control of the supply of cysteine (the rate limiting precursor for GSH *de novo* synthesis) to the embryo or through inhibitory signals.

Specific rates of GSH synthesis have been determined by measurement of the specific incorporation of [^{35}S]cysteine into GSH during periods of normal embryogenesis, and the parallel use of [^{35}S]methionine to show that no cystathionine pathway exists in the conceptus during this phase of development.[69] Although the embryo has the enzymatic capacity to replenish GSH, embryonic synthesis is not activated until the VYS is GSH-replete, perhaps due to regulation of the precursor amino acid supply. Investigations using whole embryo culture are currently underway to understand the regulation of GSH precursors, their supply to the embryo, and possible interruptions resulting from xenobiotic exposure. The mechanisms of amino acid uptake and utilization have already been addressed,[113-115] but a lack of information still exists regarding the perturbation of these processes by chemical agents. The *in vitro* whole embryo culture model will also serve well for studies designed to help understand the mechanisms of GSH stimulation, induction, and response during chemical and oxidative stress.

The ability to control conditions such as the time and concentration of exposure, oxygen tension, temperature, pH, and other environmental factors makes whole embryo culture a very useful system for evaluation of certain endpoints. These include effects of oxidative stress, production of reactive oxygen species (ROS),[116] and the antioxidant functions that protect the conceptus from these conditions. Direct additions of the enzymes superoxide dismutase, catalase, and glutathione peroxidase have all been shown to affect outward manifestations of embryotoxicity in cases of exposure to agents such as glucose (hyperglycemia),[117] arsenic,[118] and phenytoin.[119] It is not yet understood, however, what the mechanisms of action are and/or whether the enzymes are exerting their effects intra- or extracellularly. Superoxide dismutase, catalase, glutathione disulfide reductase, glutathione peroxidase, γ-glutamyl transpeptidase, and other related protective enzymes have been postulated to be expressed differentially in space and time throughout the continuum of development. This may well be responsible for some of the selective sensitivities and resistance seen at different junctures in gestation. Mechanistic biochemical and molecular investigations are currently underway to probe the possibility that the regulation of programmed cell death is a function of ROS production and the cells ability to maintain the proper redox and antioxidant status.[120]

Closely related to all of the foregoing discussions of biotransformation, oxidative stress, cell death, DNA repair, and maintenance of the proper intracellular environment are the roles of extracellular calcium,[121] calcium channel blockers,[122,123] and the potential susceptibility of altered calcium homeostasis in the dysmorphogenesis of neural tube closure.[124] Regulation and maintenance of pyridine nucleotide [NAD(H) and NADP(H)] status during development and the critical alterations that occur following chemical insult have also been identified as important factors in mechanisms of teratogenesis and embryonic development, but have not yet been systematically studied in the developing conceptus. The status and turnover of NAD(H) in aminothiadiazole teratogenicity,[125] possibly related to DNA repair initiated through poly-ADP ribosylation and hypoxia,[126] have been addressed using *in vitro* systems. An important role of the major cellular reducing equivalents, NADP(H) and NAD(H), has been postulated for a number of critical biochemical pathways and reactions involving normal developmental metabolism and their response to xenobiotic insult, including the endogenous production of antimetabolites.[126-128] Because of the need to monitor very rapid changes in redox status that may be altered due to chemical or environmental disruption of cellular function, Thorsrud and Harris[126] have used the whole embryo culture system and noninvasive, real-time MFOP fluorescence probes. These probes were used to measure pyridine nucleotide fluxes from the VYS surface of intact, viable rat and mouse conceptuses maintained in a customized perfusion chamber (Figure 16.5). Chemical modulation of antioxidant functions and concurrent determinations of GSH status have shown that measured changes of

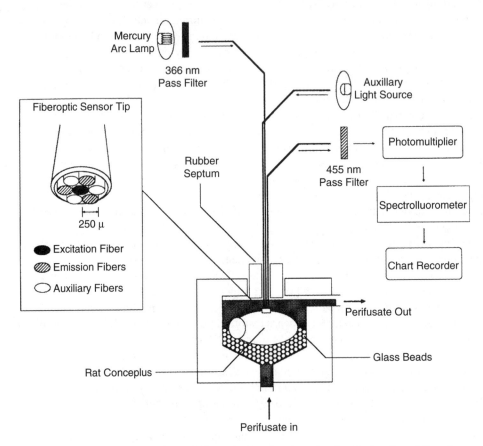

Figure 16.5 Schematic diagram of the MFOP monitoring chamber, showing the location of the rat conceptus in the chamber and the placement of the sensor on the surface of the visceral yolk sac. Insert shows an amplified view of the microfiberoptic sensor tip. (From Thorsrud, B.A. and Harris, C., *Teratology*, 48, 347, 1993. With permission.)

surface fluorescence in the VYS can involve changes in different pyridine nucleotide pools.[129] These included the net reduction of NAD^+ following cyanide exposure and hypoxia. Another example was the net oxidation of NADPH following GSH oxidation and utilization by glutathione disulfide reductase after exposure to the diabetogenic agent, alloxan, and the glutathione oxidant, diamide. Further development of this and other similar technologies will aid in the study of other potential mechanisms of embryotoxicity involving biotransformation, biosynthesis, mitochondrial function, intermediary metabolism, and macromolecular repair mechanisms.

Since we are able to grow the intact conceptus (embryo proper plus associated extraembryonic membranes) during critical periods of organogenesis *in vitro*, both spatial and temporal investigations into toxicological mechanisms have been possible. Whole conceptuses maintain the structural integrity and important physiological and functional interrelationships between embryo proper and extraembryonic membranes, which may be lost in disrupted systems. Careful selection of tissue-specific agents allows studies of how damage to one tissue may affect the function of another. Elegant studies[130–138] have demonstrated that several known teratogenic agents appear to act through disruption of the nutritional pathways of the rat VYS. The relatively unique placement of the inverted rodent VYS provides a physical as well as functional barrier between the embryo and the maternal milieu. During organogenesis, this organ serves an indispensable function in taking up maternal proteins. This is accomplished via active fluid phase pinocytosis and adsorptive endocytosis, followed by enzymatic degradation in VYS lysosomes, thus providing the major source of precursor amino acids for new protein synthesis. These proteins, incorporated into acidic lysosomal vacuoles, are rapidly degraded, principally by cathepsins D and L. The constituent amino acids are then transported to the embryo for use in protein synthesis and as precursors for numerous other biosynthetic and regulatory processes. These studies showed that virtually all of the amino acids incorporated into new protein synthesized in the developing conceptus were derived from VYS proteolytic digestion of maternal (or serum) proteins, hence the designation of this process as *histiotrophic nutrition*.

Chemical teratogens and embryotoxins known to specifically interfere with either pinocytosis (trypan blue) or proteolysis (leupeptin, chloroquine) are believed to exert their effects by these principal means. Careful control of dose, exposure, and environmental conditions have allowed for the study and further characterization of these processes as they relate to the mechanism of action of a number of chemical toxicants.[130,131,133,134,137] *In vitro* methods focused on the function of the VYS have also shown that hyperglycemic embryopathies are accompanied by sequential yolk sac failure.[138]

The development of new biochemical tools, such as selective fluorescent probes, has provided additional insights into the mechanisms and systems that control proteolysis and their relationships to chemical embryotoxicity and dysmorphogenesis. Studies have also been conducted using intact conceptuses complemented by the culture of VYSs and the use of organ cultures, such as "giant" yolk sacs.[139] Gaining an understanding of which mechanisms of developmental toxicity involve the VYS and are directly dependent on the unique relationship between embryo and extraembryonic membranes in the rodent is important for possible extrapolation to the human. That is because the human does not have an inverted VYS encompassing the embryo and entirely different functions and responses to chemical insult may result.

Other potential targets for developmental embryotoxicants related to VYS function that can be effectively studied by use of the whole embryo culture system include the processes of intermediary metabolism and energy production. Frienkel et al.[140] utilized the whole embryo culture system in their evaluation of the "honeybee syndrome." They proposed that the rodent embryo undergoes a dramatic transition from an essentially anaerobic, glycolysis-dependent organism during early organogenesis to an aerobically metabolizing, Krebs cycle–dependent organism following closure of the anterior neuropore and onset of an active cardiovascular system.[141–143] Addition of mannose and other nutrient substrates to the culture medium allowed these and other investigators to eliminate confounding maternal factors and determine the direct capabilities of the conceptus to metabolize and respond to glucose.[144] Hiranruengchok and Harris[146–148] have incorporated methods used to assess intermediary metabolism and energy production into a chemical exposure paradigm in which several events and responses can be correlated to help define the mechanisms of toxicity. Exposure to the thiol oxidant, diamide, revealed that GSH oxidation, protein mixed-disulfide formation, and altered glycolytic and pentose phosphate pathway enzymatic activities all correlate temporally. That finding suggests a role for disrupted metabolism in the mechanism of diamide embryotoxicity.

A considerable amount of important work has also been conducted in the laboratories of Sadler, Eriksson, and others. Those workers described potential mechanisms involved in glucose metabolism[148–150] and toxicity[151] and factors that may relate to maternal insulin-dependent diabetes mellitus (IDDM) and diabetic embryopathy,[142–144] as well the consequences of other maternal metabolic disorders, such as phenylketonuria (PKU).[152] These studies provide considerable important information as to the role of maternal and conceptal biotransformation and the need for specific nutrients in fuel processing. *In vitro* systems have been beneficial in determining that some diabetic factors, such as ketosis and hypoglycemia, probably affect the embryo directly; others, such as somatomedin inhibitors, likely alter embryonic nutritional pathways by interference with the function of the VYS.[130–135] It has also been suggested by Eriksson and Borg[151] that mechanisms of embryotoxicity related to hyperglycemia include the production of reactive oxygen species and may also involve GSH status and the antioxidant systems described above.

B. Other Systems and Approaches Employing Whole Embryo Culture

The examples cited above represent only a small part of the body of work that has in common the use of rodent whole embryo culture systems as a means to probe mechanisms of developmental toxicity. The application of these techniques and approaches has also been covered in some detail in other sections of this book in conjunction with specific subject areas. Numerous additional applications and techniques employing rodent embryo culture will surely be reported in the future. Several mechanistic theories not explicitly covered in the foregoing sections, but which have benefited by current use of whole embryo culture or will benefit from its use in the future include:

- Morphological events such as neural tube closure, craniofacial anomalies, limb defects, and ocular development[153–158]
- Functional/physiological alterations involving growth factors, pH regulation, angiogenesis, and hemopoiesis[159–162]
- Macromolecule synthesis and turnover, including protein synthesis; DNA synthesis, damage, and repair; cell death; and stress protein responses[163–166]
- Receptor binding and function (described in detail in Chapter 14)
- Gene expression, including gene product regulation, genetic alterations associated with pattern formation, gene expression, sense-antisense manipulations, *in situ* hybridizations, and a myriad of other molecular approaches, several which are covered in detail in Chapters 14 and 15

Some examples of the novel approaches being employed *in vitro* for studies of mechanisms of developmental embryotoxicity involve use of the rodent whole embryo culture system. For example, this system has been used to evaluate the causes and repercussions of programmed cell death produced by such chemical agents as *N*-acetoxy-acetylaminofluorene. Thayer and Mirkes[166] (Figure 16.6) have adapted novel *in situ* procedures to identify and characterize cells undergoing programmed cell death. An *in situ* procedure identified as the TUNEL (TdT-mediated dUTP-biotin nick end labeling) method pinpoints apoptotic cells in histological sections of any region of treated and control conceptuses. A vital dye, Nile blue sulfate, is used for identification of changing regions of cell death at the gross light microscopic level. The morphological and cytological characteristic of the dying cell populations can be further characterized by transmission electron microscopy (TEM, Figure 16.6b). Such methods are valuable for the study of mechanisms because they allow investigators to focus on the specific cell populations that seem to be affected by chemical exposure.

A second example of new approaches to study mechanisms of developmental toxicology involves the incorporation of molecular biology techniques into studies of normal and abnormal mechanisms of embryogenesis. Sadler et al.[167] have microinjected antisense oligonucleotides that are specifically targeted toward the *engrailed*-1 (*En-1*) gene into GD 9 mouse embryos. These authors have demonstrated that *En-1* is involved in developmentally determined pattern formation in the mouse and that dysmorphogenesis produced by the microinjection of antisense oligonucleotides (Figure 16.7) is most likely related to elicited reductions in En-1 protein levels. These experimental models may allow direct evaluation of the interactions between chemicals and genes in multifactorial causes of developmental embryotoxicity.

1. Transgenic Animals

Over the past 25 years, advances in the manipulation of the mouse genome have opened doors to understanding specific developmental pathways and their relationship to mechanisms of developmental toxicity. Transgenic approaches provide a more in depth insight into the roles of individual

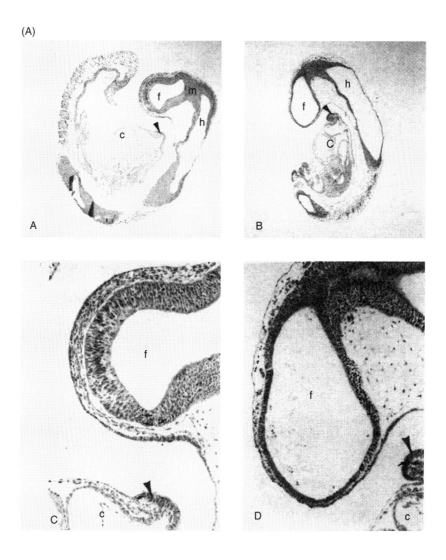

Figure 16.6 (A) The TUNEL method was used on tissue sections of control embryos and embryos exposed to 50 µg/ml N-Ac-AAF, at the 10-h time point. Methyl green was used as a counterstain. A: Nearly midsagittal section through a control embryo. Three regions of the brain are shown: forebrain (f), midbrain (m) and hindbrain (h). Between the heart (c) and brain a small portion of one of the visceral arches can be seen (arrowhead). B: Sagittal section of treated embryo. C: Higher magnification of head of embryo shown in A. Arrow indicates DAB-positive (brown-stained) cells or cellular particles. D: Higher magnification of section shown in B. (B) Transmission electron microscopy (TEM) of tissues from a control embryo and an embryo exposed to N-Ac-AAF, 5 h after exposure. A: Typical appearance of mesenchymal cells in a control embryo. Cellular processes were abundant, but there were no obvious apoptotic cells or debris. B: Embryo exposed to 50 µg/ml N-Ac-AAF. Occasional apoptotic bodies were noted (arrowhead), and electron-dense, intracellular inclusions were common (arrows). C: Cell from embryo shown in B, clearly undergoing apoptosis (arrowheads). D: Embryo exposed to 200 µg/ml N-Ac-AAF. Occasional apoptotic cells (arrowheads) were observed, and extracellular, heterogeneous, electron-dense bodies were common (open arrows). E: Section from embryo shown in D. In addition to apoptotic cells (arrowhead), some cells had discontinuous cell membranes and small electron-dense nuclei (large arrows), suggestive of necrosis. (Thayer, J.M. and Mirkes, P.E., *Teratology*, 51, 418, 1995. With permission.)

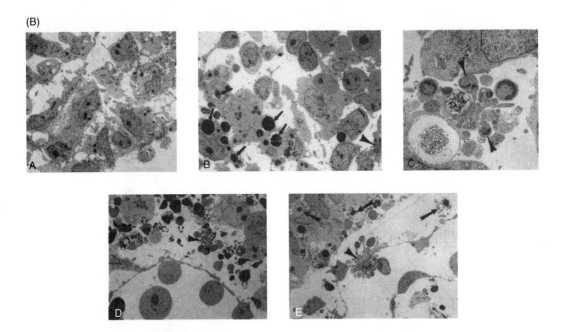

Figure 16.6 (continued)

genes during the development of specific tissues. There are numerous methods by which transgenic approaches can modify eventual developmental outcomes, including the overexpression or misexpression of a specific gene in an ectopic tissue, expression of antisense transgenes resulting in hybridization with endogenous mRNA and effectively silencing gene expression, and gene knockout technologies. In developmental toxicology, the use of transgenic animals is still in its infancy, but the technique is being employed ever more frequently. An example is the use of antioxidant enzyme (SOD, GSH-peroxidase, etc.) deletions and overexpressions to assess their effects on chemical sensitivity and dysmorphogenesis. Clearly, the use of transgenic mice will aid in understanding the relationships between specific gene expression and toxicant exposures and responses. These approaches are covered in greater detail in Chapter 14.

2. *Proteomics and Genomics*

As proteomic and genomic technologies are being perfected, microarray analyses of conceptuses, embryos, or developing tissues can potentially yield vast quantities of descriptive data about responses to teratogenic insult. *In vitro* manipulation allows for precise culture and treatment conditions and would allow for more exact, teratogen-specific determinations of gene and protein expression. Many teratogens presumably produce abnormalities by inhibiting or hyperstimulating specific tissues. In all likelihood, this is a result of misregulation of gene expression. Microarray analysis could provide information very useful for elucidating the initiation and manifestation of birth defects.

Understanding the structural modification of specific conceptal macromolecules following chemical exposure can aid greatly in elucidating the biochemical and molecular mechanisms of toxicant action. Application of new time-of flight mass spectrometry techniques such as SELDI and MALDI offer the potential of identifying and characterizing alterations to target molecules and the pathways responsible for toxicity. As the proteomic databases are expanded and developed, this significant knowledge can be applied to mechanisms of developmental toxicology.

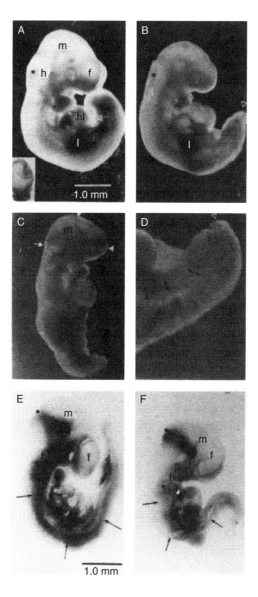

Figure 16.7 Phenotypes of mouse embryos injected with *En*-1 sense (A) or antisense (B) oligonucleotides at the 4- to 6-somite stage (inset) and cultured for 48 h. (A) Embryo injected with *En*-1 sense oligonucleotides (100 nl, 25 µM) showing normal morphology, including expansion of the 4th ventricle (large asterisk) and establishment of the 1st and 2nd pharyngeal arches (small asterisks). (B) Embryo exposed to *En*-1 antisense oligonucleotides (100 nl, 25µM), exhibiting abnormal development of the 4th ventricle and reduction in caudal extension (arrow). (C) Embryo exposed to *En*-1 antisense probes (100 nl, 60 µM), exhibiting craniofacial defects (white arrowheads) and hypoplasia of the 4th ventricle (arrow). Reduction in caudal segments is evident from the forelimb-bud to the tip of the tail. (D) Higher magnification of the tail region of the embryo in (C). Kinks are apparent in the neural tube (black arrows), and caudal segments are missing (white arrow). (E and F) Embryos exposed to sense (E) and antisense (F) *En*-1 oligonucleotides at the early somite stage, cultured for 48 h, and stained with anti-*En* antibodies. Note that staining is decreased (arrows) in the neural tube and somites of the antisense-treated embryo, but that a band of staining remains at the midbrain-hindbrain junction (asterisk). (A through F: f, forebrain; h, hindbrain; ht, heart; l, limb bud; m, midbrain.) (From Sadler, T.W., Liu, E.T., and Augustine, K.A., *Teratology*, 51, 292, 1995. With permission.)

C. Nonmammalian Whole Embryo Cultures

1. Caenorhabditis elegans Cultures

There are numerous advantages to the use of the relatively ubiquitous nematode, *Caenorhabditis elegans* (*C. elegans*). Because *C. elegans* can be hermaphroditic, self-fertilization is possible. Cultures are relatively easy to grow and maintain, with a generation time of only 3 days. The worms are also transparent throughout their entire developmental period, making observations of internal development possible with the use of a light microscope. The *C. elegans* genome is very small (97 megabases) and has been completely mapped. Another advantage is that the mature *C. elegans* has relatively few somatic cells, approximately 1000 in adults, and embryonic cell lineages have been carefully defined,[168,169] so each mature cell can be traced back to its embryonic precursor. This advantage was exploited in laser ablation experiments, where specific cells were killed and development was followed to adulthood. This approach allows for understanding cell-cell interactions during development, an approach that is very difficult, if not impossible, in other animal models without further genetic manipulations.

While many developmental pathways that may be relevant to developmental toxicology can be studied in *C. elegans*, this model has primarily been utilized in developmental biology. Pathways studied in *C. elegans* have concentrated on receptor tyrosine kinases,[170,171] transforming-growth factor β,[172–174] *Notch* and *Delta*,[175,176] *Wnt*,[177] and apoptosis[178,180] pathways. The use of *C. elegans* in developmental toxicology has predominantly been limited to embryo toxicity screens. The organism has not been widely used as a model for the evaluation of mechanisms of mammalian developmental toxicity, despite its obvious advantages.

2. Drosophila Cultures

Drosophila larvae have been used extensively in studies of the control and regulation of gene expression and pattern formation during embryogenesis. The virtues of mechanistic toxicity studies in *Drosophila* mirror those of *C. elegans* toxicity testing. *Drosophila* are inexpensive and generally easy to maintain and grow, but the greatest advantage to their use is the wealth of information about them that has accrued since the 1920s. *Drosophila* genetics were first elucidated by T.H. Morgan at the University of Columbia, and in the 1970s *Drosophila* studies led to the discovery of early egg organization and body patterning. Among the genes that were discovered during these initial experiments, the homeobox genes were found to be highly conserved in species ranging from flies to mammals. Later, it was found that many other *Drosphila* genes had mammalian homologues, providing an easily manipulable nonmammalian model for developmental biology. Such knowledge and relatively well-understood developmental genes and specific pathways should render *Drosophila* valuable for furthering an awareness of developmental toxicology mechanisms. However, much like *C. elegans*, *Drosophila* has not been used to elucidate developmental toxicology mechanisms but rather has been primarily employed as a potential developmental toxicity screening test organism.

3. Amphibian Embryo Culture

Some of the earliest studies of mechanisms of development were conducted in amphibians. Amphibians are likely have somewhat more relevance to humans than *C. elegans* or *Drosophila* because they are vertebrates and thus develop more like humans. These simple organisms have been used to study neural crest cells, osteogenesis, vasculogenesis, cardiogenesis, and development of gut derivatives (e.g., liver and pancreas), all of which have much in common with human developmental processes.

In particular, the African clawed frog, *Xenopus laevis*, has figured prominently in developmental studies, as it has served as a model system for events such as fertilization, gene expression, pattern formation, organogenesis, morphogenesis, and metabolism. Recent use has been made of the ability to incubate and manipulate *Xenopus* eggs for developmental toxicity studies. The approach, identified as the FETAX system (Frog Embryo Teratogenesis Assay Xenopus) is intended for use as a rapid screen for chemical teratogens but has recently been applied for use in more mechanistic studies. Fort and colleagues have demonstrated thalidomide-induced limb reduction defects with the FETAX system,[181] although, no mechanistic findings have come from these initial studies. Many other chemicals have been tested using the FETAX system, including the environmental contaminant mercury,[182] the pesticide and redox cycler paraquat,[183,184] cadmium,[185] the carcinogen benzo[a]pyrene,[186] and caffeine.[187] One of the potential advantages to the FETAX system, as with avian cultures, is that it allows xenobiotic effects to be followed throughout development from fertilization to hatching and senescence because of the relatively short lifespan of *X. laevis*. Chemical-induced behavioral deficits can be determined in a shorter time span that would be the case with mammals. Another advantage to the FETAX system, like *C. elegans* and *Drosophila*, is the relatively low cost and the high output, as numerous concentrations of test chemicals can be assessed concurrently on many developing eggs, giving enormous amounts of data. In its current form, however, the FETAX test system has a number of weaknesses, including inadequate inter-laboratory reproducibility, which must be addressed if it is ever to become a reliable mainstream developmental toxicity assay.

4. Fish Embryo Culture

The use of zebrafish (*Danio rerio*) in developmental and developmental toxicity studies is relatively new but may have advantages for modeling vertebrate development, including low cost and high output. Toxicology testing can be simplified, in that embryos are extremely permeable to toxicants dissolved in water. A handful of toxicology studies have been performed on zebrafish, including evaluations of cadmium,[188] TCDD,[189] and 4-chloroaniline.[190] Some genetic manipulations of zebrafish have also been performed with excellent results, which have encouraged their increased use in toxicological studies. One unique benefit of zebrafish in toxicological assays is that during development they are transparent, allowing specific organs and tissues to be visualized and more easily studied (i.e., with light microscopy). Zebrafish manipulation with reporter gene constructs coupled with this transparency during development allow for detailed study of a specific toxicant's effects on certain tissues. The green fluorescent protein (GFP) transgenic fish was first constructed in 1995. Since then it has been utilized in studies analyzing gene expression patterns, and tissue and organ development, dissection of promoters and enhancers, cell lineage, axonal pathfinding, cellular localization of protein products, chimeric embryo and nuclear transplantation, and cell sorting.[191] The use of zebrafish in studies of mechanisms of developmental toxicology has, thus far, been limited. Mattingly and coworkers[195] developed an aryl hydrocarbon receptor specific GFP construct and injected it into developing zebrafish. Following fertilization, zebrafish were exposed to TCDD (10 n*M*). Early in development (72 h), GFP was detected in the eye, nose, and vertebrae, while no GFP expression was detected in DMSO-treated controls. In longer-term examinations (5 days), terata coincided with the previously demonstrated regions of GFP expression and included craniofacial and vertebral defects.

5. Avian Embryo Culture

Classical studies in developmental biology have exploited the chick as a model for which a large number of developmental processes and mechanisms have been defined. Accessibility for experimental manipulation and observation have contributed to the success of these experiments. The chick has also been used as a model for classic studies in teratology. Incubation of eggs with 6-amino-nicotinamide,[192] retinoic acid,[193,194] and other agents has provided fundamental knowledge of developmental mechanisms and their perturbation with chemical agents. More basic studies, including exposure of chick embryos to ethyl alcohol and retinoic acid derivatives, followed by an evaluation of the function and behavior of cranial neural crest cells, have looked at mechanisms involving specific cell types and related processes. Coupled with use of *in vitro* cranial neural crest cell cultures, this approach has provided considerable information showing that the neural crest may be sensitive to chemical agents that produce or increase the production of reactive oxygen species. This sensitivity may be due to deficiencies in intrinsic detoxication capacity, particularly involving superoxide dismutase. In many respects, analogous studies can be conducted for early postimplantation embryogenesis in both avian and rodent species. The chick model has the distinct advantage, however, of being able to continue the developmental process through hatching and subsequent maturation.

In general, most nonmammalian culture systems have numerous advantages, giving them great potential, which, unfortunately, has not been fully utilized in the study of toxicological mechanisms. When the genomes of these organisms are completely or nearly completely mapped, their utilization for mechanistic studies should increase. They may eventually be fashioned into screening systems of choice, providing useful information regarding developmental toxicity mechanisms.

D. Organ and Tissue Culture

Organ culture and *tissue culture* are terms often used interchangeably to describe the *in vitro* growth of structurally and functionally intact subsets of a complete organism. The levels of biological organization encompassed by this definition range from intact, functional, normally vascularized complex organs to essentially avascular, relatively homogeneous tissue fragments. Organ culture provides numerous advantages for the study of mechanisms in developmental toxicity. It allows for the maintenance of a functional unit of an organism in a controlled environment, while retaining an intact architectural structure and without the need of perfusion via an intact vasculature.[196,197] Specific functional and morphological processes carried out by many organ systems are influenced by chemical and physiological factors produced endogenously and also those provided by other organs and tissues. The roles and functions of these factors can be evaluated by their selective addition or removal from the culture medium.

Successful elucidation of mechanisms of toxicity produced through chemical interactions ultimately requires a system in which the temporal and spatial relationships of morphological, biochemical, and molecular alterations can be evaluated. This approach necessitates the removal of tissues and organs from confounding influences, so that their interactions can be identified and characterized. Several embryonic organs can now be dissected free and grown successfully in culture, still meeting a number of selected experimental criteria. These include maintenance of similarities in growth and differentiation, compared with *in vivo* patterns, and retention of the timed sequence of production of organ-specific products. These techniques are useful in manipulating the system to probe mechanisms and are aided by the ability to determine the exact time and duration of exposure, the precise stage of development at the time of exposure, and the degree of chemical metabolism prior to reaching organ targets.

Organs and tissues that have been used successfully in *in vitro* culture applications for the study of mechanisms of developmental toxicity include limb buds, limbs, bone, heart, palatal shelves, visceral yolk sac, eyes, sex organs, and embryonic tooth germs. Many of the initial culture techniques were based on Trowell's method,[198] where the organ is grown attached to a membrane in a humidified atmosphere. Many of these techniques have been subsequently modified to use submerged cultures and a roller apparatus in defined media to improve morphological and biochemical fidelity in comparison with parallel *in vivo* developmental events (Figure 16.8). Cocultures with other organs or cells, dissociation of specific cell types from the organ, and the use of nontoxic

IN VITRO METHODS FOR THE STUDY OF MECHANISMS OF DEVELOPMENTAL TOXICOLOGY

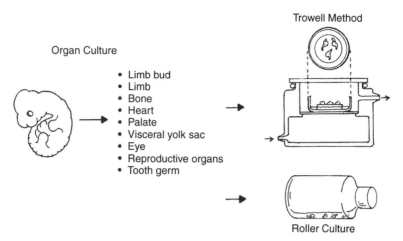

Figure 16.8 Organ culture of numerous structures can be initiated from embryos obtained during a broad range of developmental stages, depending on experimental needs. Organs can be grown in culture on static grids or filters using the Trowell method (*Exp. Cell Res.*, 16, 118, 1959), as depicted in the upper part of this illustration, or in roller cultures, as shown in the bottom portion. Growth and differentiation are generally superior in roller cultures.

bioactivation systems extend the utility of these experimental models. They allow for the study of organ-organ interactions and direct xenobiotic effects on interrelated cells and tissues. Endpoints for evaluation of organ cultures range from gross morphogenetic changes to cellular effects, biochemical alterations, metabolic transitions, and physiological responses to the molecular bases of differentiation, as well as the genetic events that regulate or are regulated by xenobiotics.

1. Limb Bud and Related Perinatal Bone Cultures

A number of toxicants produce persistent malformations involving the developing limbs. This suggests that these organs may be particularly sensitive to chemicals during development and may, therefore, be good systems in which to study mechanisms of developmental toxicology.[199,200] Undifferentiated limb buds from the chick were first successfully grown in culture by use of a plasma clot environment.[201] Later, chick, rat, and mouse limb buds were cultured by use of the Trowell method,[198] where limbs were grown attached to Millipore® filters in a humidified environment. Plasma clot techniques are difficult to work with in studies involving the addition of chemical agents or radioisotopes because the clot material interferes with chemical distribution and quantitation. Initial studies by Aydelotte and Kochhar[202] established the limb bud as an important early tool for the mechanistic study of chemical induced limb malformations. Limb buds removed from 40 to 45 somite rodent embryos provided the greatest degree of growth and differentiation and could be cultured successfully for up to 6 d, when they reached optimal differentiation (Figure 16.9). Organ cultures of this type produce the distinct patterns of chondrogenesis and establishment of the skeletal anlage seen with normal limb growth *in vivo*, but they require a longer period to complete differentiation and undergo developmental growth. The expected sequence of development and the programmed production of unique products of differentiation, however, show that the cultured limb buds bear close morphological and biochemical resemblance to the intact limb *in vivo*. Subsequent modifications to the Trowell method by Neubert and Barrach[203] utilized a submerged roller bottle culture system. Their modification effectively improved the cultured limbs' morphological appearance, decreased the time needed for optimal growth and differentiation, and provided greater accessibility of organs and tissues to growth factors, nutrients, oxygen, and test chemicals. This approach effectively overcomes many of the morphological infidelities associated with growing organs attached to a solid filter substrate and produces differentiated structures in

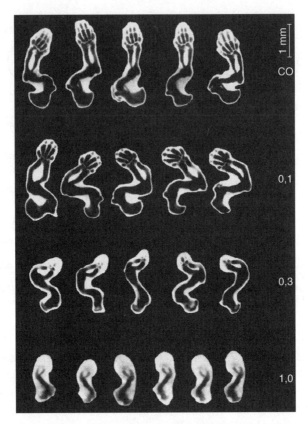

Figure 16.9 Effects of (bis[tri-n-butyltin])oxide (TBTO) on the differentiation of mouse forelimb buds in organ culture. The cultures were initiated with explants from 12-d-old mouse embryos, and the culture period was 6 d. TBTO was present in the culture medium for the entire culture period. (A) Control limbs (CO); all cartilaginous bone anlagen are well developed. (B) 0.1 µg/ml TBTO in the medium; in comparison with controls, development of the phalanges was found to be inhibited, and the scapula was less well developed. (C) 0.3 µg/ml TBTO in the medium; drastic impairment of differentiation of the paw skeleton and the scapula; in some cases ulna and radius were also affected. (D) 1.0 µg/ml TBTO in the medium; almost no morphogenetic differentiation occurs; only the cartilage of the humerus and parts of the ulna and radius are recognizable. (From Krowke, R.R., Bluth, U., and Neubert, D., *Arch. Toxicol.*, 58, 125, 1986. With permission.)

about half the time (3 d) when compared with the Trowell cultures.[198] An example of the degree of growth and differentiation that can be obtained is shown in Figure 16.9. Controls (CO) have undergone significant morphogenesis *in vitro*, while (bis[tri-*n*-butyltin])oxide (TBTO)-treated organs failed to grow or differentiate properly.[204]

Endpoints used for the effective study of mechanisms of limb malformation are intimately related to morphometric structure and to the patterns of gene expression and biochemical synthesis that result in the formation of the cartilaginous anlage of long bones and digits. Investigations into mechanisms of limb teratogenesis have focused on the evaluation of two main processes important for normal and abnormal limb growth. First, the morphological and functional differentiations of cells characteristic of the mature limb bud occur developmentally as a consequence of temporally and spatially controlled gene expression. The patterned production of new enzymes, structural proteins, and regulatory elements results in accumulation of the structural and functional elements necessary for altering, organizing, and establishing morphological form. Second, as a result of programmed gene expression, the actual spatial and functional organization of the elements into cartilaginous anlage is accomplished. These events are accompanied by genetic regulation of programmed cell death, resulting in the final sculpting of the finished limb. Selective alteration in any one or a combination of these processes as a result of exposure to chemical agents may represent a relevant endpoint to be used as a probe for specific mechanisms of action of chemical embryotoxicity. In this context, the limb culture technique has been used to investigate the effects of over 60 agents, such as retinoic acid,[205] retinol,[206] cyclophosphamide,[207,208] 6-mercaptopurine riboside,[209] sodium salicylate, cytosine arabinoside, eserine sulfate,[210] thiabendazole,[211] and many others, as reviewed by Friedman.[200] Several clues to the etiology of the toxicity were derived from these studies, including effects on mitochondrial function, maternal electrolyte and water balance, cartilage and extracellular matrix biochemistry, gene expression and regulation, receptor function and pattern formation, programmed cell death, and others, although no common mechanism was identified. It is highly unlikely that a common mechanism will be found. Chondrogenesis and the tissue localization of various forms of collagen,[212] glycosaminoglycans,[213] chondroitin sulfate, and other products,[199,207] specifically associated with cartilage formation have been studied. The advantages of organ culture for quantitation and characterization of these biochemical products can now be effectively combined with other *in vivo* and *in vitro* techniques by which temporal and spatial expression of selected genes can be related to functional outcome. Immunohistochemical localization, gene expression (pattern formation), *in situ* hybridization, and other molecular approaches can now be used in attempts to understand the effects of chemicals on limb development in organ culture.

The possible mechanisms to be evaluated and related endpoints to be measured in studies of limb dysmorphogenesis are too numerous to be covered exhaustively in this chapter. The reader is referred to several reviews and texts on this topic.[199,200,203] An overview of possibilities can be best illustrated by drawing from a selected few examples from the vast variety of studies conducted with this system using retinoic acid and its related metabolites. Earlier studies in mice established the role of retinoic acid as a limb teratogen and defined gross morphological effects with respect to patterns of progressive differentiation.[202,214] Single maternal doses of retinoids produced phocomelia or partial phocomelia in patterns that followed differential gradients established in fore- and hindlimbs, depending on gestational age and time of exposure. The role of retinoids in development and the endpoints and processes that may be investigated for studies of mechanisms of developmental toxicity have been reviewed in detail.[215] Experiments *in vitro* using mouse and chick limb culture systems determined that grafting a portion of the ZPA (zone of polarizing activity) of a region from the posterior limb bud to the anterior margin of another limb bud altered the gradient of RA distribution within the limb and resulted in a mirror-symmetrical duplication of digits. The relationship with RA distribution in the limb and a causal effect on the observed pattern of development was confirmed by obtaining a similar result through grafting a bead coated with all-*trans*-retinoic acid to the anterior site normally reserved for the ZPA graft. These studies suggested that RA was acting as a morphogen, a diffusible chemical distributed in a discrete concentration gradient that dictates the final form of the organ through concentration-dependent interactions. Such results also indicated that excess RA could be exerting its teratogenic effects through disruption of the normal chemical gradient in the limb.

Antibodies directed against a RA-bovine serum albumin antigen have allowed the immunohistochemical localization of endogenous and exogenously added RA *in vitro*. Subsequent studies have shown, however, that although patterns of RA distribution correlate well with various aspects of pattern formation, RA cannot be a complete morphogen. That is because of the requirement for a number of other factors essential for pattern formation and differentiation. Considerable interest and insight have been generated from the observations that exogenous application of retinoids to the anterior margin of the chick wing bud has the same effect as transplantation of the ZPA or coated beads. *In vitro* limb culture methods can be used in these ways for testing the effects of chemical exposure on other morphogenetic and cellular processes.

Other aspects of RA effects on the altered morphology and differentiation of the limb bud *in vitro* have focused on characterization of the role and distribution of binding proteins and specific classes of retinoic acid receptors, and the mechanisms that regulate them.[215] Specific changes in the types and distribution of retinoic acid receptors have been identified by *in vitro* methods. Approaches for the study of mechanisms of receptor and binding protein involvement in teratogenesis have included the histochemical and immunohistochemical localization of receptors and binding proteins. They have also involved the use of *in situ* hybridization techniques to determine the specific pattern-related distribution of receptor and cytosolic retinoic acid binding protein mRNA expression under different exposure conditions. A suspected role for RA as a tissue morphogen and the probable role of this naturally occurring compound in regulating gene expression responsible for chondrogenesis and pattern formation have led to a number of studies aimed at understanding the effects of RA on gene expression. Observations that embryotoxic agents can, indeed, act through regulation or disruption of programmed gene expression have fueled investigations of these events as mechanisms for the action of teratogenic agents. These studies have provided considerable motivation for the continued study of mechanisms for understanding RA teratogenicity. The ability to coordinate specific enzyme activities, macromolecule synthesis, gene expression, and pharmacological effects with observed patterns of differentiations *in vitro* makes limb culture a useful research tool for studying mechanisms of limb development.

Sequential expression of gene products directs the patterning of the limb and the delineation of the chondrogenic anlage. This must be followed by the events associated with programmed cell death that result in the final sculpting of the digits to result in the functional hand, wing, or foot. *In vivo* and *in vitro* observations of incomplete or abnormal sculpting as a result of chemical exposure make the mechanistic understanding of this process extremely interesting. Differentiation of specific cellular and molecular characteristics results in programmed cell death and altered cellular function. Several important studies have been conducted *in vivo* and utilizing the cultured limb bud to help delineate the changes that occur as cells commit to die.[216]

2. Palatal Cultures

Incomplete closure of the primary and secondary palates is frequently seen in clinical manifestations of malformation (cleft palate and cleft lip), both in humans and in experimental animal models. A number of chemical agents, including TCDD,[217] glucocorticoids,[218] phenytoin,[219] and retinoic acid,[220] have been shown to selectively induce incomplete palatal closure. Excision of the intact palatal shelves and their subsequent development *in vitro* have aided greatly in understanding the mechanisms that regulate palatal growth and differentiation *in vivo*. Palates were first successfully grown in culture by use of the Trowell method, where palatal shelves were placed intact on a Millipore membrane and allowed to undergo fusion.[221] Several drawbacks of this method became apparent in that only differentiation and morphology directly related to shelf fusion could be evaluated. The more global processes of shelf elevation, cell movement and related differentiation, and functional interactions could not be evaluated. The modified serum-free cultures of Saxen and Saxen,[221] Saxen,[222,223] and Abbot and Buckalew[224] improved growth and differentiation but fell short of the more important advances that were realized by Shiota et al.,[225] who grew whole palatal explants in submerged, rotating cultures.

Because exposure to dioxins *in vivo*, particularly 2,3,7,8-tetrachlorodibenzo-*p*-dioxin resulted in an increased incidence of cleft palate, an *in vitro* technique was developed to investigate the mechanisms of palatal closure and biochemical, molecular, and morphological disruption by chemical agents. Intact palatal shelves, removed from rodent embryos on GD 10 to12, have been shown to elevate and fuse in *in vitro* cultures in much the same manner that they do *in vivo*. Influences on the palatal closure process have been characterized and have been found to include induction through glucocorticoids, *Ah*-receptor responsive chemicals, and other metabolic inhibitors.[226,227] The *Ah* responsiveness was determined by parallel experiments using selected strains of mice that characteristically have different *Ah* receptor phenotypes. Retinoic acids have also been evaluated with this approach to elucidate the role of growth factor expression (EGF, TGF-α, and TGF-β) in altered palatogenesis.[228,229] These studies could not demonstrate that perturbation of growth factor receptors was the sole cause of the altered palatal growth and differentiation, but did indicate that their role is important. The perturbation of neural crest cell migration to the appropriate mesenchymal environments has also been investigated, suggesting additional possible mechanisms by which palatal development can be altered.[230] Molecular studies are easily conducted by use of this system but are not covered in this section.

3. Visceral Yolk Sac Cultures

The visceral yolk sac plays a vital role in the nourishment, metabolism, and protection of the developing conceptus, especially during the sensitive period of early organogenesis. The inverted VYS epithelium is structured to have the brush border facing the decidual blood lumen. This facilitates the uptake by pinocytosis and the enzymatic degradation of maternal histiotroph (proteins and other macromolecules derived from the decidual cell mass and from a transudate of the maternal serum) and the transport of carbohydrates, amino acids, and other nutrients. The rodent VYS is distinct from those of other mammalian species because it retains its form and function until term.

A number of embryotoxins and teratogens have been reported to alter VYS function, and their effects on this organ have been suggested as central to their mechanisms of toxicity. The teratogen trypan blue was first shown to elicit its deleterious effects via inhibition of pinocytosis and/or proteolysis by the VYS.[231] Since these reports, other agents (e.g., leupeptin, anti-VYS antibody, chloroquine, diazomethyl ketones, E-64) have been shown to elicit their effects through inhibition of pinocytosis or through inhibition of lysosomal proteolysis.[130–137,232–234]

Because of its direct contact with the culture environment, a number of direct exposures and manipulations can be performed on the VYS in whole embryo culture without compromising conceptal viability and function. Many of the characterizations of VYS function, however, used a detached, free-floating VYS in a roller culture system. These studies were used successfully to determine the extent of proteolytic and pinocytotic activity in the VYS and to characterize changes in these activities elicited through chemical exposure or anti–yolk sac antibody recognition.[114,115,233,235] Similar approaches were taken to evaluate the mechanisms of uptake of other macromolecules, such as antibodies, ferritin, and enzymes. Relatively few studies have been conducted to determine specific effects of chemicals on the VYS epithelium, endothelium, or stem cell populations, even though considerable evidence exists to support an extensive role of the VYS in several nutritional, metabolic, and physiologic functions.

Organ culture of the VYS can also be accomplished at a level of biological organization intermediate between whole embryo culture and the isolated VYS. These investigations utilize "giant yolk sacs," in which conceptuses are explanted at an early somite stage (GD 9 in the rat), and the early or presomite-stage embryos are surgically ablated using dissecting needles.[139] Return of the severed conceptus to culture results in the continued growth of only the VYS. Such preparations have been shown to retain numerous vital functional and structural characteristics without interference from the embryo proper or interactions associated with the vitelline circulation. Because of its ability to trap and sequester products in the exoceloemic fluid, the closed yolk sac that results from this procedure may have a number of advantages in studies of mechanism involving directional transport across the VYS epithelial barrier. It has not been clearly delineated, however, how a lack of vitelline circulation and possible hydrodynamic factors might affect VYS function or response. Nevertheless, such preparations could be of use in determining the mechanisms and effects of chemical induction of avascular yolk sacs.[161]

Organ cultures similar to the examples cited above have been used to investigate probable mechanisms that are specific to the functions of other organ systems. Some examples are briefly mentioned in the following sections.

4. Sex Organs

Investigations have looked into mechanisms of the normal and abnormal morphogenesis of the primary sex organs. These have included *in vitro* studies of differentiation of the Wolffian and Müllerian ducts of avian and mammalian species.[236]

5. Heart

Differentiation and functional responses in heart that may be perturbed by chemical embryotoxicants include coordinated muscular contractions, gap junctional communication, and electrophysiological coupling. The latter aspect has been reviewed[237] in terms of reaggregation and electric coupling of dissociated cells, although the direct effects of chemical embryotoxicants have not been systematically evaluated.

6. Eye and Lens

The *in vitro* approaches taken to study the effects of toxicants on this organ can be coordinated with other studies of neural tube closure and central nervous system development.[153] Sensitivity of the avian and human lens with regard to cataract formation following exposure to rubella virus was evaluated *in vitro* by use of eyes collected from human embryos obtained from therapeutic abortions.[196] Verification of direct viral involvement in the mechanism of cataract formation was possible through use of the organ culture model.

7. Embryonic Tooth Germ

Early tooth germ formation and dentition can be studied in whole organ cultures *in vitro*. This can be done on solid supports or in flotation cultures.[238,239]

8. Spinal Cord

Postmeiotic growth and differentiation of neuronal cells and structures was examined in cultures of spinal cord from the chick.[240] And with *in vitro* glutamate treatment, neurite outgrowth and growth cone motility from the mouse spinal cord were significantly decreased, implicating glutamate as a regulator of neuronal differentiation, development, and function.[241]

9. Lung Buds

Branching morphogenesis and N-*myc* expression were inhibited with TGFβ1 in cultures from these organs from the day 11.5 mouse.[242] Dihydrotestosterone (DHT) significantly increased lung bud branching, a phenomenon that was reduced by flutamide, an androgen receptor competitive inhibitor.[243]

E. Cell Culture and Clonal Analysis

Morphological, histological, and ultrastructural analyses have demonstrated that teratogenicity or embryotoxicity may be elicited because of the particular sensitivity of a single cell type within complex tissues of the conceptus. The ultimate description of biochemical and molecular mechanisms in the complex organism may depend on an understanding of the specific functions and behaviors of individual cell types and how they are induced to change in response to endogenous and exogenous cues. The utility of employing isolated primary cell cultures and clonal cell lines to meet this end will depend on the direct demonstration that they behave *in vitro* in a manner similar to their behavior in the intact organism. Some of the first successful attempts at clonal analysis to demonstrate similar behavior *in vivo* and *in vitro* involved the culture of quail myoblasts for which various degrees of success were obtained, depending on the substrate and the constituents of the culture medium used.[244] A number of cell types can now be dissociated from the intact organism, conceptus, organ, or tissue, and successfully grown in culture, where the proper choices of medium and substrate are still the critical components in determining success. Techniques for primary culture have been established for heart myoblasts,[245] chondrocytes,[246] cartilage,[247] pigmented retina,[248] lens,[249] and cranial neural crest.[250,251]

Use of primary embryonic cell cultures and established cell lines in toxicology usually focuses on proliferation or on the progression of differentiation into a specific cell type, but few have addressed the effects of teratogens on embryonic stem (ES) cells. Since ES cells are pluripotent, effects of specific toxicants may perturb signal transduction pathways involved in specific cell fates, resulting in the promotion of cell death, decreased proliferation, or faulty differentiation. In any case, such effects could result in severe birth defects and embryotoxicity. ES cells have been used to study PCB toxicity in neurogenesis. Endosulfan and 2,2′,4,4′,5,5′-hexachlorobiphenyl (2,4,5-HCB) both caused an inhibition of neurite formation in a dose-dependent manner and decreased intracellular gap junctions and connexin 43 expression, implicating intracellular communication in early neurogenesis.[252] The use of ES cells in mechanistic studies may serve as a more appropriate *in vitro* model of teratogenesis than other established cell lines. Measurement could be assessed by morphologic, biochemical, or genetic markers following toxicant exposure. These approaches could yield a great deal of information about sensitive pathways of toxicant-induced dysmorphogenesis as well as highlight the role of these pathways in development.

Hypotheses relating to the proposed mechanisms of several known teratogens have implied that alterations in cranial neural crest cell (CNC) migration, differentiation, and function are probable causes for observed developmental anomalies. The clonal primary culture of CNC was originally described in Japanese quail, using procedures such as those reported by Loring et al.[250] and Glimelius and Weston.[251] With this method, the distal portion of the trunk (containing the last 8 to 10 somites) is dissected free. It is then exposed to mild enzymatic digestion (pancreatin) to loosen tissue adhesions and allow the neural tube to be physically pulled free from the notochord and ectoderm. Additional treatments (trypsin, collagenase) may then be needed to remove other adherent tissues before placing the neural tubes in media-filled plastic or collagen-coated petri dishes for culture. Various culture media have been used successfully, although, all require an appropriate amount of serum, 10-d chick embryo extract, and glutamine supplementation for optimal growth and differentiation. The relative proportions of serum and chick embryo extract affect both cell proliferation and cell cluster formation. This methodology takes advantage of the inherent high proliferative and rapid migratory properties of CNC cells and produces a very homogeneous population that will respond to a number of chemical and environmental stimuli (Figure 16.10).

Chemical agents, such as the tumor promoter 12-*O*-tetradecanoyl phorbol-13-acetate (TPA), have been shown to promote cell proliferation while causing significant delays in differentiation, as manifested by alterations in melanocyte production.[251] Davis et al.[253–255] utilized the avian CNC culture system to investigate the role of free radical production and oxidative stress in the cellular pathogenesis of ethanol (EtOH) and retinoid (isotretinoin and 4-oxo-isotretinoin)-induced craniofacial abnormalities. They showed that treatment of CNC with these agents *in vitro* is accompanied by the significant generation of reactive oxygen species, producing cellular blebs, loss of normal Ca^{2+} homeostasis, and decreased viability.

This system allows investigators the flexibility to measure multiple biochemical and morphological endpoints and to alter several environmental conditions to probe mechanisms of ROS generation and cellular response. Using these approaches, Davis et al.[254] determined the CNC to be devoid of superoxide dismutase and catalase, two important antioxidant enzymes critical for elimination of ROS. Addition of these enzymes directly to the culture medium provided significant protection against ROS-mediated damage. In addition, those investigators were able to characterize

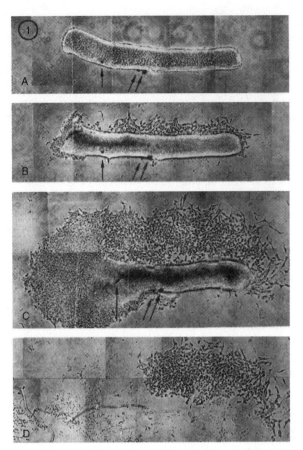

Figure 16.10 Migration of neural crest cells from the neural tube is shown in this series of phase contrast micrographs of a single living neural tube explanted *in vitro*. To indicate polarity, blood charcoal (arrows) was applied to the ventral (notochordal) surface of the neural tube before it was explanted. One of the lateral surfaces of the neural tube is in contact with the collagen substratum of the petri plate. (A) No cells migrate from the explant during the first 2 h. (B) After 8 h, mesenchymal cells being to migrate from the dorsal aspect of the tube. (C) Within 20 h, hundreds of cells have emigrated from the neural tube, primarily from the dorsal surface. The few mesenchymal cells seen along the ventral surface arose from an area damaged while the tube was being explanted. In addition, one end of the tube has flattened to form an epithelial sheet. (D) The mesenchymal outgrowth was isolated by carefully dissecting away the neural tube and its flattened epithelial end. The remaining primary outgrowth is grown for an additional 2 d until enough cells are available for routine cloning procedures. (From Cohen, A.M. and Konigsberg, I.R., *Devel. Biol.*, 46, 262, 1975. With permission.)

related alterations in intracellular Ca^{2+} homeostasis. This was done by use of ionophores and sulfhydryl reagents, and it provided detailed ultrastructural descriptions of the accompanying morphological changes.

The primary culture of cerebrocortical neural cells has also been used to evaluate mechanisms of methyl iodide–induced neurotoxicity, as reported by Bonnefoi.[256] Endpoints for evaluation in these studies focused on the role of intracellular cytosolic and mitochondrial GSH status in the mechanism of methyl iodide toxicity. Addition of a series of antioxidants and correlations to cell injury, as indicated by lactate dehydrogenase leakage, showed that manifestations of toxicity were correlated with a minimal 50% reduction in mitochondrial GSH and were not altered by cytosolic GSH depletion. Similar approaches and an expanded palette of measurable endpoints make the isolation and culture of clonal cell lines from the developing embryo a valuable tool for the study of mechanisms of developmental toxicity, especially when effects can be correlated with the findings of companion *in vivo* studies.

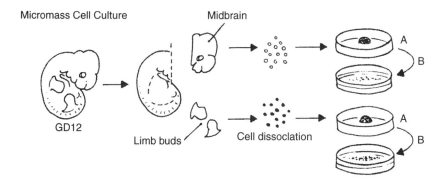

Figure 16.11 Schematic diagram showing the procedure for midbrain or limb bud micromass cultures. Limb buds or midbrains are dissected from GD 13 rat embryos or GD 12 rabbit embryos and dissociated to a suspension of single cells in medium containing trypsin and EDTA. (A) High-density cell droplets (micromasses) are cultured in small petri dishes or 24-well culture plates in a humidified incubator at 37°C with a 5% CO_2 atmosphere for 1 to 1.5 h. (B) After cells have attached, cultures are flooded with medium, to which can be added test chemicals, inhibitors, modulators, bioactivation systems, and growth factors. (Adapted from Kistler, A. and Howard, W.B., *Methods Enzymol.*, 190, 427, 1990. With permission.)

A more recent advancement in the culture of mammalian embryonic cells for the study of mechanisms has been the use of micromass cultures. This system, as with many other *in vitro* approaches, has been touted as a possible rapid screen for teratogenic agents[257–259] but may find greater utility in mechanistic studies. This method involves the removal of midbrains (CNS) and limb buds (LB) of 34 to 36 somite rat or mouse embryos by microdissection, followed by successive washing in calcium- and magnesium-free phosphate buffered saline solution and trypsin to weaken cell adhesions (Figure 16.11). The cells are resuspended in an appropriate culture medium, dissociated into individual cells by repeated titration through a Pasteur pipette, and filtered on nylon mesh to remove clumps and provided a suitable starting cell suspension. Cell suspensions are plated on plastic petri dishes at concentrations of 1×10^5 to 4×10^5 cells per 20 µl in culture medium drops. Ninety minutes after initial plating, additional culture medium containing the test compound (and possibly an S-9 microsomal bioactivating system and additional modulators) is added. However, the S-9 mix can be toxic and therefore must be limited to a specific period of culture.[257] At the end of a 5-d culture period, the medium is removed, and the cells are fixed and stained for morphometric analysis.

While midbrain cultures generally provide an excellent means to gather qualitative data, limb bud micromass cultures are typically stained with Alcian Blue to visualize regions of active chondrogenesis. Stained chondrogenic foci can be eluted and then measured spectrophotometrically to provide quantitative data. In regions of initial plating of the LB micromass, massed mesenchymal cells generally differentiate in place, but other cell types will proliferate at the edge of the mass and spread outward (Figure 16.12). This finding provides information about two distinct populations in the same culture — a proliferating population and a differentiating population — which may prove useful for various toxicological assays. When stained with mercury orange to assess small biothiol content (i.e., GSH and cysteine), cells in the differentiation zone are seen to contain much less GSH than cells in the proliferative zone (Figure 16.13). This may predispose differentiating cells to greater damage during oxidative stress.[260] Other biochemical analyses can be performed on these two populations and may be extrapolated to *in vivo* exposures.

Numerous agents have now been tested with the micromass approach, but mostly during screening for teratogens, the intended use of the assay.[257–259] Validation of these procedures, especially the limb bud micromass technique, was attempted by Flint and Orton.[257] Eighteen different known teratogens were injected into time-mated rats. Sixteen hours later, the embryos treated *in utero* were removed and their limb bud cells were plated as micromass cultures. Of the 18 teratogens

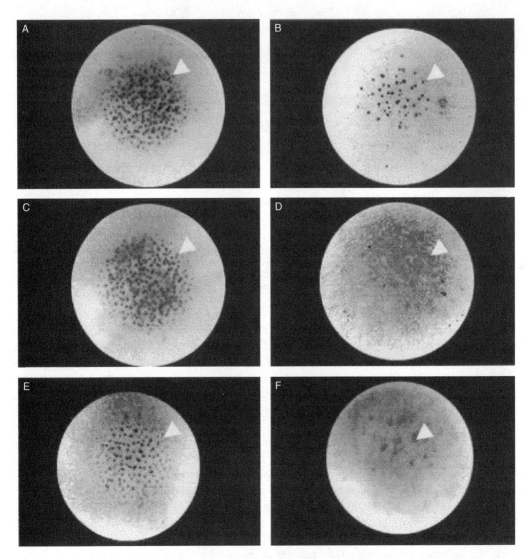

Figure 16.12 Photomicrographs of rat and rabbit limb bud micromass cultures treated with the glutathione depleting agents buthionine sulfoximine (BSO) and diethyl maleate (DEM). Limb bud micromasses treated with 100 μM BSO [(C) rat and (D) rabbit] or 25 μM DEM [(E) rat and (F) rabbit] showed a decrease in chondrogenesis in comparison with control micromasses from the respective species (A) rat and (B) rabbit. In addition, a complete inhibition of chondrogenic focus formation was observed in rabbit micromasses, indicating that rabbit limb bud tissues may be more sensitive to glutathione depletion and modulation. (Hansen, J.M., Carney, E.W., and Harris C., *Free Rad. Biol. Med.*, 31, 1582, 2001. With permission.)

tested, 17 tested positive (6% false negative) for the inhibition of chondrogenesis. In a follow-up study, 27 known teratogens and 19 nonteratogens were used to treat rat limb bud micromass cultures via addition to the culture medium.[261] Of the 27 known teratogens, only two (7% false negative) did not have inhibitory effects on chondrogenesis (2,4-dichlorophenoxyacetic acid and thalidomide). Conversely, only two of the nonteratogens (11% false positive) showed a decrease in chondrogenic focus formation (glutethimide and dimenhydrinate). Other *in vitro* prescreens used to test chemicals have shown some power to estimate the teratogenicity of specific chemicals, but the assessments described above clearly support the micromass system as a prescreen for teratogenic potential.

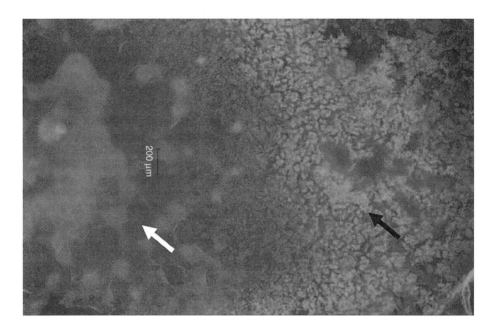

Figure 16.13 Limb bud micromass stained with mercury orange (HgO) for small biothiols, mainly cysteine and reduced glutathione. Regions of high chondrogenic differentiation near the center of the micromass (white arrow) contain much less glutathione than does the highly proliferative region (black arrow) surrounding the differentiative zone. (Hansen, J.M., Carney, E.W., and Harris C., *Free Rad. Biol. Med.*, 31, 1582, 2001. With permission.)

While this procedure has been touted and primarily used as a screen for teratogenic potential, the utility of these assays have been exploited in developmental biology as well. Plating at high-density allows for the formation of an artificial "slice" of the limb and can be easily manipulated across different scientific disciplines. Much work utilizing the micromass technique has revolved around the function of the different bone morphogenetic proteins (BMP-2 and -6) and sonic hedgehog (*Shh*) and their role in chondrogenesis and limb differentiation.[262–265] Similarly, midbrain cultures have been utilized in comparable fashion but to a much more limited manner.[266] More specific to developmental toxicology, LB micromass cultures treated with GSH depleting agents, such as buthionine sulfoximine (BSO) and DEM, have demonstrated a substantial decrease in chondrogenic focus formation (Figure 16.12). Such findings implicate GSH and oxidative stress as possible modulators of early cartilage development.[260] Although not employed as a developmental toxicology tool to the same extent as the limb bud micromass approach, midbrain micromass cultures may potentially provide similar information in regard to neurogenesis and developmental neurotoxicology.

III. SUMMARY

A number of useful culture methods have been described for application in studies designed to probe mechanisms of developmental toxicity. The systems used for these purposes span the entire spectrum of development, from fertilization to parturition (and, in some cases, beyond). They encompass virtually all levels of biological organization, from the intact, viable embryo or fetus to cellular organelles and individual cells. The ability to select a single methodological approach that can be used in the examination of a potential mechanism and that covers the entire developmental spectrum is not currently possible. Rather, experimental protocols will likely require numerous methodologies. Inconsistent rates of growth and differentiation between different systems, even at similar developmental stages and levels of biological complexity, also make absolute conclusions regarding mechanisms difficult. Multiple experimental models must still be created in order to probe various aspects of the mechanisms of developmental toxicity. Cocultures and combined approaches utilizing different *in vitro* methods and *in vivo–in vitro* models will further extend the utility of these approaches as research tools. Continued refinement of *in vitro* procedures and the additional acquired knowledge regarding basic developmental mechanisms will continue to stimulate research in areas of developmental toxicity by providing a more complete understanding of endpoints and their relevance to environmental insults.

REFERENCES

1. Schwetz, B.A., Morrissey, R.E., Welsch, F., and Kavlock, R.A., In vitro teratology, *Environ. Health Perspect.*, 94, 265, 1991.
2. Welsch, F., In vitro approaches to the elucidation of mechanisms of chemical teratogenesis, *Teratology*, 46, 3, 1992.
3. Flynn, T.J., Teratological research using *in vitro* systems. I. Mammalian whole embryo culture. *Environ. Health Perspect.*, 72, 203, 1987.
4. Kimmel, G.L. and Kochhar, D.M., *In Vitro Methods in Developmental Toxicology*, CRC Press, Boca Raton, 1990.
5. Sadler, T.W., Horton, W.E., Jr., and Hunter, E.S., Mammalian embryos in culture: a new approach to investigating normal and abnormal developmental mechanisms, *Prog. Clin. Biol. Res.*, 171, 227, 1985.
6. Sadler, T.W., and Warner, C.W., Use of whole embryo culture for evaluating toxicity and teratogenicity, *Pharmacol. Rev.*, 36, 145S, 1984.
7. Steele, C.E., Whole embryo culture and teratogenesis, *Human Reprod.*, 6, 144, 1991.
8. New, D.A.T., Whole embryo culture, teratogenesis, and the estimation of teratologic risk, *Teratology*, 42, 635, 1990.
9. Wilson, J.G., Survey of *in vitro* systems: their potential use in teratogenicity screening, in *Handbook of Teratology*, Wilson, J.G. and Fraser, F.C., Eds., Plenum Press, New York, 1978, p. 135.
10. New, D.A.T., Introduction, in *Postimplantation Mammalian Embryos: A Practical Approach*, Copp, A.J. and Cockroft, D.L., Eds., IRL Press, Oxford, 1990, p. 1.
11. Austin, C.R., Embryo transfer and sensitivity to teratogenesis, *Nature*, 244, 333, 1973.
12. Brachet, A., Dévelopment *in vitro* de blastomères et jeunes embryon de mammifères, *C.R. Acad. Sci. Paris*, 155, 1191, 1912.
13. Spielmann, H. and Vogel, R., Unique role of studies on preimplantation embryos to understand mechanisms of embryotoxicity in early pregnancy, *Crit. Rev. Toxicol.*, 20, 51, 1989
14. Lawitts, J.A. and Biggers, J.D., Culture of preimplantation embryos, *Methods Enzymol.*, 225, 153, 1993.
15. Pratt, H.P.M., Isolation, culture and manipulation of pre-implantation mouse embryos, in *Mammalian Development: A Practical Approach*, Monk, M., Ed., IRL Press, Washington, D.C., 1987, p. 13.
16. Blakenship, A.L., Suffia, M.C., Matsumura, F., Walsh, K.J., and Wiley, L.M., 2,3,7,8-tetrachlorod-ibenzo-p-dioxin (TCDD) accelerates differentiation of murine preimplantation embryos in vitro, *Reprod. Toxicol.*, 7, 255, 1993.
17. Oudiz, D.J., Walsh, K., and Wiley, L.M., Ethylene glycol monomethyl ether (EGME) exposure of male mice produces a decrease in cell proliferation of preimplantation embryos, *Reprod. Toxicol.*, 7, 101, 1993.
18. Kholkute, S.D., Rodriguez, J., Dukelow, W.R., Effects of polychlorinated biphenyls (PCBs) on in vitro fertilization in the mouse, *Reprod. Toxicol.*, 8, 69, 1994.
19. McNutt, T.L. and Harris, C., Glutathione and cysteine modulation in bovine oviduct epithelium cells cultured in vitro, *Teratology*, 49, 413, 1994.
20. McNutt, T.L. and Harris, C., Influence of lindane on cysteine (CYS) and glutathione (GSH) levels in bovine oviduct epithelium cells cultured in vitro, *The Toxicologist*, 15, 251, 1995.
21. Goto, Y., Noda, Y., Mori, T., and Nakano, M., Increased generation of reactive oxygen species in embryos cultured in vitro, *Free Rad. Biol. Med.*, 15, 69, 1993.

22. Goto, Y., Noda, Y., Narimoto, K., Umaoka, Y., and Mori, T., Oxidative stress on mouse embryo development in vitro, *Free Rad. Biol. Med.*, 13, 47, 1992.
23. Gardiner, C.S. and Reed D.J., Status of glutathione during oxidant-induced oxidative stress in the preimplantation mouse embryo. *Biol. Reprod.*, 51, 1307, 1994.
24. Borland, R.M. and Tasca, R.J., Inhibition of L-methionine uptake and incorporation by chlorpromazine in preimplantation mouse embryos, *J. Reprod. Fert.*, 42, 473, 1975.
25. Heyner, S., Rao, L.V., Jarett, L., and Smith, R.M., Preimplantation mouse embryos internalize maternal insulin via receptor-mediated endocytosis: pattern of uptake and functional correlations, *Dev. Biol.*, 134, 48, 1989.
26. Sellens, M.H., Stein, S., and Sherman, M.I., Protein and free amino acid content in preimplantation mouse embryos and in blastocysts under various culture conditions, *J. Reprod. Fert.*, 61, 307, 1981.
27. Poueymirou, W.T. and Schultz, R.M., Regulation of mouse preimplantation development: inhibition of synthesis of proteins in the two-cell embryo that require transcription by inhibitors of cAMP-dependent protein kinase, *Dev. Biol.*, 133, 588, 1989.
28. Murach, K., Frei, M., Gerhauser, D., and Illmensee, K., Protein synthesis in embryonic tissues during mouse postimplantation development, *J. Cell. Biol.*, 44, 19, 1990.
29. Spielman, H. and Eibs, H.-G., Recent progress in teratology: a survey of methods for the study of drug actions during the preimplantation period, *Arzneim.-Forsch.*, 28, 1733, 1978.
30. Dumoulin, J.C.M., Evers, J.L.H., Bakker, J.A., Bras, M., Pieters, M.H.E.C., and Geraedts, J.P.M., Temporal effects of taurine on mouse preimplantation development in vitro, *Human Reprod.*, 7, 403, 1992.
31. Iyer, P., Martin, J.E., and Irvin, R.T., Role of biotransformation in the in vitro preimplantation embryotoxicity of naphthalene, *Toxicology*, 66, 257, 1991.
32. Winkle, L.J.V., Amino acid transport in developing animal oocytes and early conceptuses, *Biochim. Biophys. Acta*, 947, 173, 1988.
33. Mattson, B.A., Rosenblum, I.Y., Smith, R.M. and Heyner, S. Autoradiographic evidence for insulin and insulin-like growth factor binding to early mouse embryos, *Diabetes*, 37, 585, 1988.
34. Rao, L.V., Wikarczuk, M.L., and Heyner, S., Functional roles of insulin and insulinlike growth factors in preimplantation mouse embryo development, *In Vitro Cell. Dev. Biol.*, 26, 1043, 1990.
35. Niimura, S. and Ishida, K., Immunohistochemical demonstration of prostaglandin E-2 in preimplantation mouse embryos, *J. Reprod. Fert.*, 80, 505, 1987.
36. Biggers, J.D. and Borland, R.M. Physiological aspects of growth and development in the preimplantation mammalian embryo. *Ann. Rev. Physiol.* 38, 95, 1976.
37. Kane, M.T. and Bavister, B.D., Vitamin requirements for development of eight-cell hamster embryos to hatching blastocysts in vitro, *Biol. Reprod.*, 39, 1137, 1988.
38. Butler, J.E., Lechene, C., and Biggers, J.D., Noninvasive measurement of glucose uptake by two populations of murine embryos, *Biol. Reprod.*, 39, 779, 1988.
39. Seshagiri, P.B. and Bavister, B.D., Glucose inhibits development of hamster 8-cell embryos in vitro, *Biol. Reprod.*, 40, 599, 1989
40. McKiernan, S.H., Bavister, B.D., and Tasca, R.J., Energy substrate requirements for in vitro development of hamster 1- and 2-cell embryos to the blastocyst stage, *Human Reprod.*, 6, 64, 1991.
41. Brinster, R.L., Studies on the development of mouse embryos in vitro. II. The effect of energy source. *J. Exp. Zool.*, 158, 59, 1965.
42. Barbehenn, E.K., Wales, R.G., and Lowry, O.H., Measure of metabolites in single preimplantation embryos; a new means to study metabolic control in early embryos, *J. Embryol. Exp. Morph.*, 43, 29, 1978.
43. Gupta, J., Dey, S.K., and Dickman, Z., Evidence that "embryonic estrogen" is a factor which controls the development of the mouse preimplantation embryo, *Steroids*, 29, 363, 1977.
44. Wu, J.T., Metabolism of progesterone by preimplantation mouse blastocysts in culture, *Biol. Reprod.*, 36, 549, 1987.
45. Garris, D.R., Role of the preimplantation embryo in the timing of LH-dependent progesterone secretion from the rat corpus luteum, *Proc. Soc. Exp. Biol. Med.*, 170, 443, 1982.
46. Lamb, V.K. and Leese, H.J., Uptake of a mixture of amino acids by mouse blastocysts. *J. Reprod. Fert.*, 102, 169, 1994.
47. Epstein, C.J. and Smith, S.A., Amino acid uptake and protein synthesis in preimplantation embryonic development, *Biol. Reprod.*, 12, 82, 1973.

48. Bavister, B.D. and McKiernan, S.H., Regulation of hamster embryo development in vitro by amino acids, in *Preimplantation Embryo Development*, Bavister, B.D., Ed., Springer-Verlag, New York, 1993, p. 57.
49. Hsu, Y., Baskar, J., Stevens, L. and Rash, J., Development *in vitro* of mouse embryos from the two-cell egg stage to the early somite stage, *J. Embryol. Exp. Morphol.*, 31, 235, 1974.
50. Hsu, Y.-C. *In vitro* development of individually cultured whole mouse embryos from blastocyst to early somite stage, *Dev. Biol.* 68, 453, 1979.
51. Kaufman, R.A., and Armant, D.R., *In vitro* exposure of preimplantation mouse embryos to cocaine and benzoylecgonine inhibits subsequent development, *Teratology*, 46, 85, 1992
52. Pollard, D.R., Baran, M.M. and Bachvarova, R., The effect of 5-bromodeoxyuridine on cell division and differentiation of preimplantation mouse embryos, *J. Embryol. Exp. Morph.*, 35, 169, 1976.
53. Rothstein, J.L., Johnson, D., DeLoia, J.A., Skowronski, J., Solter, D., and Knowles, B., Gene expression during preimplantation mouse development, *Genes Dev.*, 6, 1190, 1992.
54. Bolton, V.N., Oades, P.J. and Johnson, M.H. The relationship between cleavage, DNA replication and gene expression in the mouse 2-cell embryo, *J. Embryol. Exp. Morphol.*, 79, 139, 1984.
55. Conover, J.C., Temeles, G.L., Zimmerman, J.W., Burke, B., and Schultz, R.M., Stage-specific expression of a family of proteins that are major products of zygotic gene activation in the mouse embryo, *Dev. Biol.*, 144, 392, 1991
56. Christians, E., Campion, E., Thompson, E.M., and Renard, J.-P., Expression of the HSP 70.1 gene, a landmark of early zygotic activity in the mouse embryo, is restricted to the first burst of transcription, *Development*, 121, 113, 1995.
57. Kimber, S.J., Waterhouse, R., and Lindenberg, S., In vitro models for implantation of the mammalian embryo, in *Preimplantation Embryo Development*, Bavister, B.A., Ed., Springer-Verlag, New York, 1993, 244-263.
58. Lindenberg, S., Hyttel, P., Sjgren, A., and Greve, T., A comparative study of attachment of human, bovine and mouse blastocysts to uterine epithelial monolayer, *Human Reprod.*, 4, 446, 1989.
59. New, D.A.T., Methods for the culture of postimplantation embryos of rodents, in *Methods in Mammalian Embryology*, Daniel, J.C., Jr., Ed., W.H. Freeman & Co., San Francisco, 1971, p. 305.
60. Cockroft, D.L., Dissection and culture of postimplantation embryos, in *Postimplantation Mammalian Embryos: A Practical Approach*, Copp, A.J. and Cockroft, D.L., Eds., IRL Press, Oxford, 1990, p. 15.
61. Sturm, K. and Tam, P.P.L., Isolation and culture of whole postimplantation embryos and germ layer derivatives, *Methods Enzymol.*, 225, 164, 1993.
62. Freeman, S.J., Coakley, M.E., and Brown, N.A., Post-implantation embryo culture for studies of teratogenesis, in *Biochemical Toxicology: A Practical Approach*, Snell, K. and Mullock, B., Eds., IRL Press, Oxford, 1987, p. 4.
63. Tarlatzis, B.C., Sanyal, M.K., Biggers, W.J. and Naftolin, F., Continuous culture of the postimplantation rat conceptus, *Biol. Reprod.*, 31, 415, 1984.
64. Kaufmann, M.H., Morphological stages of postimplantation embryonic development, in *Postimplantation Mammalian Embryos: A Practical Approach*, Copp, A.J. and Cockroft, D.L., Eds., IRL Press, Oxford, 1990, p. 81.
65. Naya M., Kito Y., Eto K., and Deguchi T., Development of rabbit whole embryo culture during organogenesis, *Cong. Anom.*, 31, 153, 1991.
66. Ninomiya H., Kishida K., Ohno Y., Tsurumi K., and Eto K., Effects of trypan blue on rat and rabbit embryos cultured in vitro, *Toxicol. In Vitro*, 7, 707, 1993.
67. Pitt J. and Carney, E.W., Development of a morphologically-based scoring system for postimplantation New Zealand White rabbit embryos, *Teratology*, 59, 88, 1999.
68. Pitt J. and Carney, E.W. Evaluation of various toxicants in rabbit whole-embryo culture using a new morphologically-based evaluation system, *Teratology*, 59, 102, 1999.
69. Harris, C., Glutathione biosynthesis in the postimplantation rat conceptus in vitro, *Toxicol. Appl. Pharmacol.*, 120, 247, 1993.
70. Hansen, J., Choe, Y., Carney, E.W., and Harris, C., Differential antioxidant enzyme activities and glutathione content between rat and rabbit conceptuses, *Free Rad. Biol. Med.*, 30, 1078, 2001.
71. Hansen, J., Carney, E.W., and Harris, C., Differential alteration by thalidomide of the glutathione content of rat vs. rabbit conceptuses in vitro, *Reprod. Toxicol.* 13, 547, 1999.

72. Fantel, A.G., Greenaway, J.C., Juchau, M.R., and Shepard, T.H., Teratogenic bioactivation of cyclophosphamide in vitro, *Life Sci.*, 25, 67, 1979.
73. Faustman-Watts, E.M., Greenaway, J.C., Namkung, M.J., Fantel, A.G., and Juchau, M.R., Teratogenicity in vitro of 2-acetylaminofluorene: role of biotransformation in the rat, *Teratology*, 27, 19, 1983.
74. Greenaway, J.C., Bark, D.H., and Juchau, M.R., Embryotoxic effects of salicylates: role of biotransformation, *Toxicol. Appl. Pharmacol.*, 74, 141, 1984.
75. Fantel, A.G., Greenaway, J.C. and Juchau, M.R., The embryotoxicity of adriamycin in rat embryos in vitro, *Toxicol. Appl. Pharmacol.*, 80, 155, 1985.
76. Harris, C., Stark, K.L., Luchtel, D.L. and Juchau, M.R., Abnormal neurulation induced by 7-hydroxy-2-acetylaminofluorene and acetaminophen: evidence for catechol metabolites as proximate dysmorphogens, *Toxicol. Appl. Pharmacol.*, 101, 432, 1989.
77. Steele, C.E., Marlow, R., Turton, J., and Hicks, R.M., In vitro teratogenicity of retinoids, *Br. J. Exp. Path.*, 68, 215, 1987.
78. Creech Kraft, J. and Juchau, M.R., Conceptal biotransformation of 4-oxo-all-*trans*-retinoic acid, 4-oxo-13-*cis*-retinoic acid and all-*trans*-retinoyl-beta-glucuronide in rat whole embryo culture, *Biochem. Pharmacol.*, 43, 2289, 1992.
79. Klug, S., Lewandowski, C., Zappel, F., merker, H.-J., Nau, H. and Neubert, D., Effects of valproic acid, some of its metabolites and analogues on prenatal development of rats *in vitro* and comparison with effects *in vivo*, *Arch. Toxicol.* 64, 545, 1990.
80. Bruckner, A., Lee, Y.J., O'Shea, K.S., and Henneberry, R.C., Teratogenic effects of valproic acid and diphenylhydantoin on mouse embryos in culture, *Teratology*, 27, 29, 1983.
81. Fantel, A.G., Greenaway, J.C., Walker, E., and Juchau, M.R., The toxicity of niridazole in rat embryos in vitro, *Teratology*, 33, 105, 1986.
82. Fantel, A.G., Greenaway, J.C. Shepard, T.H., Juchau, M.R., and Selleck, S.B., The teratogenicity of cytochalasin D and its inhibition by drug metabolism, *Teratology*, 23, 223, 1981.
83. Greenaway, J.C. and Fantel, A.G., Enhancement of rifampin teratogenicity in cultured rat embryos. *Toxicol. Appl. Pharmacol.*, 69, 81, 1983.
84. Faustman-Watts, E.M., Namkung, M.J., Greenaway, J.C. and Juchau, M.R., Analyses of metabolites of 2-acetylaminofluorene generated in an embryo culture system: relationship of biotransformation to teratogenicity in vitro, *Biochem. Pharmacol.*, 34, 2853, 1985.
85. Yang, H.-Y.L., Namkung, M.J., and Juchau, M.R. Cytochrome P-450-dependent biotransformation of a series of phenoxazone ethers in the rat conceptus during early organogenesis: evidence for multiple P-450 isozymes. *Mol. Pharmacol.*, 34, 67, 1988.
86. Yang, H.-Y.L., Zelus, B.D. and Juchau, M.R., Organogenesis-stage cytochrome P450 isoforms: utilization of PCR for detection of CYP1A1 mRNA in rat conceptal tissues, *Biochem. Biophys. Res. Commun.*, 178, 236, 1991.
87. Chapman, D.E., Yang, H.Y., Watters, J.J., and Juchau, M.R., Induction in vitro and complete coding region sequence of cytochrome P4501A1 cDNA from cultured whole rat conceptuses during early organogenesis, *Biochem. Pharmacol.*, 48, 1807, 1994.
88. Martz, F., Failinger, C., III, and Blake, D.A., Phenytoin teratogenesis: correlation between embryopathic effect and covalent binding on putative arene oxide metabolite in gestational tissue, *J. Pharmacol. Exp. Ther.*, 203, 231, 1986.
89. Wells, P.G., Zubovits, J.T., Wong, S.T., Molinari, L.M., and Ali, S., Modulation of phenytoin teratogenicity and embryonic covalent binding by the cyclooxygenase inhibitor acetylsalicylic acid, caffeic acid, and α-phenyl-N-*t*-butylnitrone: implications for bioactivation by prostaglandin H synthease, *Toxicol. Appl. Pharmacol.*, 97, 192, 1989.
90. Kubow, S. and Wells, P.G., In vitro bioactivation of phenytoin to a reactive free radical intermediate by prostaglandin H synthase, horseradish peroxidase, and thyroid peroxidase, *Mol. Pharmacol.*, 35, 504, 1989.
91. Yu, W.K. and Wells, P.G., Evidence for lipoxygenase-catalyzed bioactivation of phenytoin to a teratogenic reactive intermediate: *in vitro* studies using linoleic acid-dependent soybean lipoxygenase, and *in vivo* studies using pregnant CD-1 mice, *Toxicol. Appl. Pharmacol.*, 131, 1, 1995.
92. Roy, S.K., Mitra, A.K., Hilbelink, D.R., Dwornik, J.J., and Kulkarni, A.P., Lipoxygenase activity in rat embryos and its potential for xenobiotic oxidation, *Biol. Neonate*, 63, 297, 1993.

93. Liu, L. and Wells, P.G., DNA oxidation as a potential molecular mechanism mediating drug-induced birth defects: Phenytoin and structurally related teratogens initiate the formation of 8-hydroxy-2′-deoxyguanosine in vitro and in vivo murine maternal hepatic and embryonic tissues, *Free Rad. Biol. Med.*, 19, 639, 1995.
94. Nau, H., Pharmacokinetic aspects of in vitro teratogenicity studies: comparison to in vivo, in *In Vitro Methods in Developmental Toxicology: Use in Defining Mechanisms and Risk Parameters*, Kimmel, G.L. and Kochhar, D.M., Eds., CRC Press, Boca Raton, 1990, p. 29.
95. Mebus, C.A., Clarke, D.O., Stedman, D.B. and Welsch, F., 2-Methoxyethanol metabolism in pregnant CD-1 mice and embryos, *Toxicol. Appl. Pharmacol.*, 112, 87, 1992.
96. Stedman D.B. and Welsch, F., Inhibition of DNA synthesis in mouse whole embryo culture by 2-methoxyacetic acid and attenuation of the effects by simple physiological compounds, *Toxicol. Lett.* 45, 111, 1989.
97. Scott, W.J., Duggan, C.A., Schreiner, C.M., Collins, M.D., and Nau, H., Intracellular pH of rodent embryos and its association with teratogenic response, in *Approaches to Elucidate Mechanisms in Teratogenesis*, Welsch, F., Ed., Hemisphere, New York, 1987, p. 99.
98. Nau, H. and Scott, W.J., Jr., Teratogenicity of valproic acid and related substances in the mouse: drug accumulation and pHi in the embryo during organogenesis and structure-activity considerations, *Arch. Toxicol. Suppl.*, 11, 128, 1987.
99. Harris, C., Tan, W., Thorsrud, B.A., and Kopelman, R., Ultramicrofiberoptic pH sensors for measurement of real time responses to chemical insult in intact, organogenesis stage rat conceptuses in vitro, *Teratology*, 49, 413, 1994.
100. Terlouw, G.D.C. and Bechter, R., Comparison of the metabolic activity of yolk sac tissue in the whole embryo and isolated yolk sac culture, *Reprod. Toxicol.*, 6, 85, 1992.
101. Lucier, G.W., Lui, E.M.K., and Lemartiniere, C.A., Metabolic activation/deactivation reactions during perinatal development, *Environ. Health Perspect.*, 29, 7, 1979.
102. Harris, C., Fantel, A.G., and Juchau, M.R. Differential glutathione depletion in rat embryo vs. visceral yolk sac in vivo and in vitro by L-buthionine-S,R-sulfoximine, *Biochem. Pharmacol.*, 35, 4437, 1986.
103. Harris, C., Stark, K.L., and Juchau, M.R. Glutathione status and the incidence of neural tube defects elicited by direct acting teratogens in vitro, *Teratology*, 37, 577, 1988.
104. Harris, C., Namkung, M.J., and Juchau, M.R., Regulation of intracellular glutathione in rat embryos and visceral yolk sacs and its effects on 2-nitrosofluorene-induced malformations in the whole embryo culture system, *Toxicol. Appl. Pharmacol.*, 88, 141, 1987.
105. Harris, C., Juchau, M.R., and Mirkes, P.E. The role of glutathione and Hsp70 in the acquisition of thermotolerance in postimplantation rat embryos, *Teratology*, 43, 229, 1991.
106. Slott, V.L. and Hales, B.F., Enhancement of the embryotoxicity of acrolein, but not phosporamide mustard, by glutathione depletion in rat embryos in vitro, *Biochem. Pharmacol.*, 36, 2091, 1987.
107. Wong, M., Helston, L.M.J., and Wells, P.G., Enhancement of murine phenytoin teratogenicity by the γ-glutamylcysteine synthetase inhibitor L-buthionine-(S,R)-sulfoximine and by the glutathione depletor diethyl maleate, *Teratology*, 40, 127, 1089.
108. Naya, M., Mataki, Y., Takanira, H. Deguchi, T., and Yasuda, M., Effects of phorone and/or buthionine sulfoximine on teratogenicity of 5-fluorouracil in mice, *Teratology*, 41, 275, 1990.
109. Griffith, O.W. and Meister, A., Potent and specific inhibition of glutathione synthesis by buthionine sulfoximine (S-*n*-butylhomocysteine sulfoximine), *J. Biol. Chem.*, 254, 7558, 1979.
110. Lee, E. and Harris, C., Differential chemical modulation of glutathione and cysteine status in rat embryo and visceral yolk sac in vitro: microinjection and addition to the culture medium, *In Vitro Toxicol.* (in press), 1995.
111. Stark, K.L., Harris, C., and Juchau, M.R., Embryotoxicity elicited by inhibition of gamma-glutamyl transferase by acivicin and transferase antibodies in cultured rat embryos, *Toxicol. Appl. Pharmacol.*, 89, 88, 1987.
112. Stark, K.L., Lee, Q.P., Namkung, M.J., Harris, C., and Juchau, M.R., Dysmorphogenesis elicited by microinjected acetaminophen analogs and metabolites in rat embryos cultured in vitro, *J. Pharmacol. Exp. Ther.*, 255, 74, 1990.
113. Rowe, P.B. and Kalaizis, A., Serine metabolism in rat embryos undergoing organogenesis, *J. Embryo. Exp. Morphol.*, 87, 137, 1985.

114. Beckman, D.A., Pugarelli, J.E., Jensen, M., Koszalka, T.R., Brent, R.L., and Lloyd, J.B., Sources of amino acid for protein synthesis during early organogenesis in the rat. I. Relative contributions of free amino acids and of proteins, *Placenta*, 11, 109, 1990.
115. Beckman, D.A., Pugarelli, J.E., Jensen, M., Koszalka, T.R., Brent, R.L., and Lloyd, J.B., Sources of amino acid for protein synthesis during early organogenesis in the rat. II. Exchange with amino acid and protein pools in embryo and yolk sac, *Placenta*, 12, 37, 1991.
116. Jenkinson, P.C., Anderson, D., and Gangolli, S.D., Malformations induced in cultured rat embryos by enzymatically generated active oxygen species, *Teratogen. Carcinogen. Mutagen.*, 6, 547, 1986.
117. Eriksson, U.J. and Borg, L.A.H., Protection by free radical scavenging enzymes against glucose induced embryonic malformations in vitro, *Diabetologia*, 34, 325, 1991.
118. Tabacova, S. and Hunter, E.S., Modulation of arsenic teratogenicity by free radical scavengers, *Teratology*, 47, 390, 1993.
119. Wells, P.G., Leeder, J.S., and Winn, L.M., Phenytoin-initiated protein oxidation in murine embryo culture: a potential molecular mechanism mediating phenytoin teratogenicity, *The Toxicologist*, 15, 276, 1995.
120. Hockenberry, D.M., Oltvai, Z.N., Yin, X.M., Millman, C.L., and Korsmeyer, S.J., Bcl-2 functions in an antioxidant pathway to prevent apoptosis, *Cell*, 75, 241,1993.
121. Smedley, M.J. and Stanisstreet, M., Calcium and neurulation in mammalian embryos, *J. Embryol. Exp. Morph.*, 89, 1, 1985.
122. Smedley, M.J. and Stanisstreet, M., Calcium and neurulation in mammalian embryos. II. Effects of cytoskeletal inhibitors and calcium antagonists on the neural folds of rat embryos, *J. Embryol. Exp. Morph.*, 93, 167, 1986.
123. Stein, G., Srivastava, M.K., Merker, H.-J., and Neubert, D., Effects of calcium channel blockers on the development of early rat postimplantation embryos in culture, *Arch. Toxicol.*, 64, 623, 1990.
124. Lee, H. and Nagele, R.G., Toxic and teratologic effects of verapamil on early chick embryos: evidence for the involvement of calcium in neural tube closure, *Teratology*, 33, 203, 1986.
125. Beaudoin, A.R., NAD precursors as antiteratogens against aminothiadiazole, *Teratology*, 13, 95, 1976.
126. Thorsrud, B.A. and Harris, C., Real time, micro-fiberoptic monitoring of endogenous fluorescence in the rat conceptus during hypoxia, *Teratology*, 48, 343, 1993.
127. Krowke, R. and Neubert, D., Embryonic intermediary metabolism under normal and pathological conditions, in *Handbook of Teratology*, Wilson, J.G. and Fraser, F.C., Eds., Plenum Press, New York, 1978, p. 117.
128. Kitchin, K.T., Sanyal, M.K., and Schmid, B.P., A coupled microsomal-activating/embryo culture system: toxicity of reduced β-nicotinamide adenine dinucleotide phosphate(NADPH), *Biochem. Pharmac.*, 30, 985, 1981.
129. Thorsrud, B.A. and Harris, C., Real time microfiberoptic redox fluorometry: modulation of the pyridine nucleotide status of the organogenesis-stage rat visceral yolk sac with cyanide and alloxan, *Toxicol. Appl. Pharmacol.*, 135, 237, 1995.
130. Freeman, S.J. and Lloyd, J.B., Evidence that suramin and sodium aurothiomalate are teratogenic in rat by disturbing yolk sac-mediated embryonic protein nutrition, *Chem. Biol. Interact.*, 58, 149, 1986.
131. Freeman, S.J. and Lloyd, J.B., Inhibition of proteolysis in rat yolk sac as a cause of teratogenesis. Effects of leupeptin in vitro and in vivo, *J. Embryol. Exp. Morphol.*, 78, 183, 1983.
132. Ambroso, J.L. and Harris, C., Chloroquine embryotoxicity in the postimplantation rat conceptus in vitro, *Teratology*, 48, 213, 1993.
133. Ambroso, J.A. and Harris, C., In vitro embryotoxicity of the cysteine proteinase inhibitors benzyloxycarbonyl-phenylalanine-alanine-diazomethane (Z-Phe-Ala-CHN2) and benzyloxycarbonyl-phenylalanine-phenylalanine-diazomethane (Z-Phe-Phe-CHN2), *Teratology*, 50, 214, 1994.
134. Ambroso, J.L. and Harris, C., Chloroquine accumulation and alterations of proteolysis and pinocytosis in the rat conceptus in vitro, *Biochem. Pharmacol.*, 47, 679, 1994.
135. Hunter, E.S., III, Phillips, L.S., Goldstein, S., and Sadler, T.W., Altered visceral yolk sac function produced by a low-molecular-weight somatomedin inhibitor, *Teratology*, 43, 331, 1991.
136. Lerman, S., Koszalka, T.R., Jensen, M., Andrew, C. L., Beckman, D.A. and Brent, R.L., In vitro studies on the effect of yolk sac antisera on functions of the viscera yolk sac. I. Pinocytosis and transport of small molecules, *Teratology*, 34, 335, 1986.

137. Marlow, R. and Freeman, S.J., Differential effect of zinc on teratogen-induced inhibition of pinocytosis by cultured rat yolk sac, *Life Sci.*, 40, 1717, 1987.
138. Pinter, E., Reece, E.A., Leranth, C.Z., Sanyal, M.K., Hobbins, J.C., Mahoney, M.J., and Naftolin, F., Yolk sac failure in embryopathy due to hyperglycemia: ultrastructural analysis of yolk sac differentiation associated with embryopathy in rat conceptuses under hyperglycemic conditions, *Teratology*, 33, 73, 1986.
139. Dunton, A., Al-Alousi, L.A., Pratten, M.K., and Beck, F., The giant yolk sac: a model for studying early placental transport, *J. Anat.*, 145, 189, 1986.
140. Freinkel, N., Lewis, N.J., Akawawa, S., Roth, S.I., and Gorman, L., The honeybee syndrome — implications of the teratogenicity of mannose in rat-embryo culture, *N. Engl. J. Med.*, 310, 223, 1984.
141. Tanimura, T. and Shepard, T.H., Glucose metabolism by rat embryos in vitro, *Proc. Soc. Exp. Biol. Med.*, 135, 51, 1970.
142. Ellington, S.K.L., In vitro analysis of glucose metabolism and embryonic growth in postimplantation rat embryos, *Development*, 100, 431, 1987.
143. Clough, J.R. and Whittingham, D.G., Metabolism of [14C]glucose by postimplantation mouse embryos in vitro, *J. Embryol. Exp. Morph.*, 74, 133, 1983.
144. Moore, D.C.P., Stanistreet, M., Beck, F., and Clarke, C.A., The effects of mannose on rat embryos grown in vitro, *Life Sci.*, 41, 1885, 1987.
145. Hiranruengchok, R. and Harris, C., Formation of protein-glutathione mixed disulfides in the developing rat conceptus following diamide treatment in vitro, *Teratology*, 52, 196, 1995.
146. Hiranruengchok, R. and Harris, C., Glutathione oxidation and embryotoxicity elicited by diamide in the developing rat conceptus in vitro, *Toxicol. Appl. Pharmacol.*, 120, 62, 1993.
147. Hiranruengchok, R. and Harris, C., Diamide-induced alterations of intracellular thiol status and the regulation of glucose metabolism in the developing rat conceptus in vitro, *Teratology*, 52, 205, 1995.
148. Horton, W.E., Jr. and Sadler, T.W., Effects on maternal diabetes on early embryogenesis. Alterations in morphogenesis produced by the ketone body, β-hydroxybutyrate, *Diabetes*, 32, 610, 1983.
149. Sadler, T.W., Hunter, E.S., III, Balkan, W., and Horton, W.E., Jr., Effects of maternal diabetes on embryogenesis, *Am. J. Perinatol.*, 5, 319, 1988.
150. Sadler, T.W., Denno, K.M. and Hunter, E.S., III, Effects of altered glucose metabolism during gastrulation and neurulation stages of embryogenesis, *Ann. New York Acad. Sci.*, 678, 48, 1993.
151. Eriksson, U.J. and Borg, L.A.H., Diabetes and embryonic malformations: role of substrate-induced free-oxygen radical production for dysmorphogenesis in cultured rat embryos, *Diabetes*, 42, 411, 1993.
152. Denno, K.M. and Sadler T.W., Phenylalanine and its metabolites induce embryopathies in mouse embryos in culture, *Teratology*, 42, 565, 1990.
153. Sulik, K.K. and Sadler, T.W., Postulated mechanisms underlying the development of neural tube defects. Insights from in vitro and in vivo studies, *Ann. New York Acad. Sci.*, 678, 8, 1993.
154. Bolon, B., Welsch, F., and Morgan, K.T., Methanol-induced neural tube defects in mice: pathogenesis during neurulation, *Teratology*, 49, 497, 1994.
155. Cuthbertson, R.A. and Beck, F., Postimplantation whole embryo culture: a new method for studying ocular development, *Invest. Ophthalmol. Vis. Sci.*, 31, 1653, 1990.
156. Seegmiller, R.E., Harris C., Luchtel, D.L., and Juchau, M.R., Morphological differences elicited by two weak acids, retinoic and valproic, in rat embryos grown *in vitro*, *Teratology*, 43, 133, 1991.
157. Alles, A.J. and Sulik, K.K., Retinoic-acid-induced limb-reduction defects: perturbation of zones of programmed cell death as a pathogenic mechanism, *Teratology*, 40, 163, 1989.
158. Eto, K. and Takakubo, F., The role of the yolk sac in craniofacial development of cultured rat embryos, *J. Craniofac. Genet. Dev. Biol.*, 5, 357, 1985.
159. Baltz, J.M., Biggers, J.D., and Lechene, C., Intracellular pH regulation by the preimplantation embryo, in *Preimplantation Embryo Development*, Bavister, B.A., Ed., Springer-Verlag, New York, 1993, p. 97.
160. Gressens, P., Hill, J.M., Gozes, I., Fridkin, M., and Brenneman, D.E., Growth factor function of vasoactive intestinal peptide in whole cultured mouse embryos, *Nature*, 362, 155, 1993.
161. Yasuda, Y., Nishi, N., Takahashi, J.A., Konishi, H., Ohara, I., Fujita, H., Ohta, M., Itoh, N., Hatanaka, M., and Tanimura, T., Induction of avascular yolk sac due to reduction of basic fibroblast growth factor by retinoic acid in mice, *Dev. Biol.*, 150, 397, 1992.
162. Chen, L.T. and Hsu, Y.C., Hemopoesis of the cultured whole mouse embryo, *Exp. Hemat.*, 7, 232, 1979.

163. Mirkes, P.E., Cyclophosphamide teratogenesis: a review, *Teratogen. Carcinogen. Mutagen.*, 5, 75, 1985.
164. Mirkes, P.E. and Cornel, L., A comparison of sodium arsenite- and hyperthermia-induced stress responses and abnormal development in cultured postimplantation rat embryos, *Teratology*, 46, 251, 1992.
165. Mirkes, P.E., Grace, R.H., and Little, S.A., Developmental regulation of heat shock protein synthesis and HSP 70 RNA accumulation during postimplantation rat embryogenesis, *Teratology*, 44, 77, 1991.
166. Thayer, J. and Mirkes, P.E., Programmed cell death and N-acetoxy-2-acetylaminofluorene-induced apoptosis in the rat embryo, *Teratology*, 51, 418, 1995.
167. Sadler, T.W., Liu, E.T., and Augustine, K.A., Antisense targeting of *engrailed*-1 causes abnormal axis formation in mouse embryos, *Teratology*, 51, 292, 1995.
168. Sulston, J.E. and Horvitz, H.R., Post-embryonic cell lineages of the nematode, *Caenorhabditis elegans*, *Dev. Biol.*, 56, 110, 1977.
169. Sulston, J.E., Schierenberg, E., White, J.G., and Thomson, J.N. The embryonic cell lineage of the nematode *Caenorhabditis elegans*, *Dev. Biol.*, 100, 64, 1983.
170. Kokel, M., Borland, C.Z., DeLong, L., Horvitz, H.R., and Stern, M.J., clr-1 encodes a receptor tyrosine phosphatase that negatively regulates an FGF receptor signaling pathway in *Caenorhabditis elegans*, *Genes Dev.*, 12, 1425, 1998.
171. Gleason, J.E., Korswagen, H.C., and Eisenmann, D.M., Activation of Wnt signaling bypasses the requirement for RTK/Ras signaling during *C. elegans* vulval induction, *Genes Dev.*, 16, 1281, 2002.
172. Krishna, S., Maduzia, L.L., and Padgett, R.W., Specificity of TGF-beta signaling is conferred by distinct type I receptors and their associated SMAD proteins in *Caenorhabditis elegans*, *Development*, 126, 251, 1999.
173. Morita, K., Flemming, A.J., Sugihara, Y., Mochii, M., Suzuki, Y., Yoshida, S., Wood, W.B., Kohara, Y., Leroi, A.M., and Ueno, N., A *Caenorhabditis elegans* TGF-beta, DBL-1, controls the expression of LON-1, a PR-related protein, that regulates polyploidization and body length, *EMBO J.*, 21, 1063, 2002.
174. Yoshida, S., Morita, K., Mochii, M., and Ueno, N. Hypodermal expression of *Caenorhabditis elegans* TGF-beta type I receptor SMA-6 is essential for the growth and maintenance of body length, *Dev. Biol.*, 240, 32, 2001.
175. Berset, T., Hoier, E.F., Battu, G., Canevascini, S., and Hajnal, A., Notch inhibition of RAS signaling through MAP kinase phosphatase LIP-1 during *C. elegans* vulval development, *Science*, 291, 2255, 2001.
176. Hermann, G.J., Leung, B., and Priess, J.R., Left-right asymmetry in *C. elegans* intestine organogenesis involves a LIN-12/*Notch* signaling pathway, *Development*, 127, 3429, 2000.
177. Herman M., *C. elegans* POP-1/TCF functions in a canonical *Wnt* pathway that controls cell migration and in a noncanonical *Wnt* pathway that controls cell polarity, *Development*, 128, 581, 2001.
178. Liu, Q.A. and Hengartner, M.O., The molecular mechanism of programmed cell death in *C. elegans*. *Ann. N. Y. Acad. Sci.*, 887, 92, 1999.
179. Tzur, Y.B., Hersh, B.M., Horvitz, H.R., and Gruenbaum, Y., Fate of the nuclear lamina during *Caenorhabditis elegans* apoptosis. *J. Struct. Biol.* 137, 146, 2002.
180. Schumacher, B., Hofmann, K., Boulton, S., and Gartner, A., The *C. elegans* homolog of the *p53* tumor suppressor is required for DNA damage-induced apoptosis, *Curr. Biol.*, 11, 1722, 2001.
181. Fort, D.J., Stover, E.L., Bantle, J.A., and Finch, R.A., Evaluation of the developmental toxicity of thalidomide using frog embryo teratogenesis assay-*Xenopus* (FETAX): biotransformation and detoxification, *Teratog. Carcinog. Mutagen.*, 20, 35, 2000.
182. Prati, M., Gornati, R., Boracchi, P., Biganzoli, E., Fortaner, S., Pietra, R., Sabbioni, E. and Bernardini, G., A comparative study of the toxicity of mercury dichloride and methylmercury, assayed by the frog embryo teratogenesis assay — *Xenopus* (FETAX), *Altern. Lab Anim.*, 30, 23, 2002.
183. Vismara, C., Bacchetta, R., Cacciatore, B., Vailati, G. and Fascio, U., Paraquat embryotoxicity in the *Xenopus laevis* cleavage phase, *Aquat. Toxicol.*, 55, 85, 2001.
184. Vismara, C., Battista, V.V., Vailati, G., and Bacchetta, R., Paraquat induced embryotoxicity on *Xenopus laevis* development. *Aquat. Toxicol.*, 49, 171, 2000.
185. Kotyzova, D. and Sundeman, F.W., Maternal exposure to Cd(II) causes malformations of *Xenopus laevis* embryos, *Ann. Clin. Lab. Sci.*, 28, 224, 1998.

186. Propst, T.L., Fort, D.J., Stover, E.L., Schrock, B., and Bantle, J.A., Evaluation of the developmental toxicity of benzo[a]pyrene and 2-acetylaminofluorene using *Xenopus*: modes of biotransformation. *Drug Chem Toxicol.*, 20, 45, 1997.
187. Fort, D.J., Stover, E.L., Propst, T.L., Faulkner, B.C., Vollmuth, T.A., and Murray, F.J., Evaluation of the developmental toxicity of caffeine and caffeine metabolites using the frog embryo teratogenesis assay — *Xenopus* (FETAX), *Food Chem. Toxicol.*, 36, 591, 1998.
188. Van den Belt, K., Van Puymbroeck, S., and Witters, H., Toxicity of cadmium-contaminated clay to the zebrafish *Danio rerio*, *Arch. Environ. Contam. Toxicol.*, 38, 191, 2000.
189. Henry, T.R., Spitsbergen, J.M., Hornung, M.W., Abnet, C.C., and Peterson, R.E., Early life stage toxicity of 2,3,7,8-tetrachlorodibenzo-*p*-dioxin in zebrafish (*Danio rerio*), *Toxicol. Appl. Pharmacol.*, 142, 56, 1997.
190. Bresch, H., Beck, H., Ehlermann, D., Schlaszus, H., and Urbanek, M., A long-term toxicity test comprising reproduction and growth of zebrafish with 4-chloroaniline, *Arch. Environ. Contam. Toxicol.*, 19, 419, 1990.
191. Gong, Z., Ju, B., and Wan, H., Green fluorescent protein (GFP) transgenic fish and their applications, *Genetica*, 111, 213, 2001.
192. Iwama, M., Honda, A., Mori, Y., Alterations of glycosaminoglycans synthesized by chick embryo cartilage treated with 6-amino-nicotinamide, *J. Pharmacobiodyn.*, 6, 613, 1983.
193. Shiga, T., Gaur, V.P., Yamaguchi, K., and Oppenheim, R.W., The development of interneurons in the chick embryo spinal cord following in vivo treatment with retinoic acid, *J. Comp. Neurol.*, 360, 463, 1995.
194. Richman, J.M. and Delgado, J.L., Locally released retinoic acid leads to facial clefts in the chick embryo but does not alter the expression of receptors for fibroblast growth factor, *J. Craniofac. Genet. Dev. Biol.*, 15, 190, 1995.
195. Mattingly, C.J., McLachlan, J.A., and Toscano, W.A., Green fluorescent protein (GFP) as a marker of aryl hydrocarbon receptor (AhR) function in developing zebrafish (*Danio rerio*), *Environ. Health Perspect.*, 109, 845, 2001.
196. Karkinen-Jaaskelainen, M. and Saxen, L. Advantages of organ culture techniques in teratology, in *Tests of Teratogenicity In Vitro*, Marois, M., Ed. North Holland Publishers, Amsterdam, 1976, p. 275.
197. Shepard, T.H. and Pious, D., Cell, tissue and organ culture as teratologic tools, in *Handbook of Teratology*, Wilson, J.G. and Fraser, F.C., Eds. Plenum Press, New York, 1978, p. 71.
198. Trowell. O.A., The culture of mature organs in a synthetic medium, *Exp. Cell Res.*, 16, 118, 1959.
199. Lessmollmann, U., Neubert, D., and Merker, H.-J., Mammalian limb buds differentiating in vitro as a test system for the evaluation of embryotoxic effects, in *New Approaches to the Evaluation of Abnormal Embryonic Development*, Neubert, D., Merker, H.J., Bedürftig, A., and Kreft, R., Eds., Georg Thieme Verlag, Stuttgart, 1975, p. 99.
200. Friedman, L., Teratological research using in vitro systems. II. Rodent limb bud culture system, *Environ. Health Perspect.*, 72, 211, 1987.
201. Strangeways, T.S.P. and Fell, H.B., Experimental studies on the differentiation of embryonic tissues growing in vivo and in vitro. I. The development of the undifferentiated limb-bud (a) when subcutaneously grafted into the post-embryonic chick and (b) when cultured in vitro, *Proc. Roy. Soc. London B*, 99, 340, 1926.
202. Aydelotte, M.B. and Kochhar, D.M., Development of mouse limb buds in organ culture: chondrogenesis in the presence of a proline analog, 1-azetidine-2-carboxylic acid, *Dev. Biol.*, 28, 191, 1972.
203. Neubert, D. and Barrach, H.J., Techniques applicable to study morphogenic differentiation of limb buds in organ culture, in *Methods in Prenatal Toxicology*, Neubert, D., Merker, H.-J., and Kwasigroch, T.E., Eds., George Thieme Publishers, Stuttgart, 241, 1977.
204. Krowke, R.R., Bluth, U., and Neubert, D., In vitro studies on the embryotoxic potential of (bis[tri-n-butyltin])oxide in a limb bud organ culture system, *Arch Toxicol.*, 58, 125, 1986.
205. Kochar, D.M. and Aydelotte, M.B., Susceptible stages and abnormal morphogenesis in the developing mouse limb, analyzed in organ culture after transplacental exposure to vitamin A (retinoic acid), *J. Embryol. Exp. Morphol.*, 31, 721, 1974.
206. Nakamura, H., Analysis of limb anomalies induced in vitro by vitamin A (retinol), *Teratology*, 12, 61, 1975.
207. Manson, J.M. and Smith, C. C., Influence of cyclophosphamide and 4-ketocyclophosphamide on mouse limb development, *Teratology*, 15, 291, 1977.

208. Stahlmann, R., Bluth, U., and Neubert, D., Effects of the cyclophosphamide metabolite acrolein in mammalian limb bud cultures, *Arch. Toxicol.*, 57, 163, 1985.
209. Neubert, D., Lessmollmann, U., Hinz, N., Dillman, I., and Fuchs, G., Interference of 6-metcaotopurine riboside, 6-methylmercaptopurine riboside and azathioprine with the morphogenic differentiation of mouse extremities in vivo and in an organ culture, *Naunyn-Schmiedebergs Arch. Pharmakol.*, 289, 1093, 1977.
210. Flint, O.P., The effects of sodium salicylate, cytosine arabinoside, and eserine sulphate on rat limb buds in culture, in *Teratology of the Limbs*, Merker, H.-J., Hau, H., and Neubert, D., Eds., Walter de Gruyter, Berlin, 1980, p. 325.
211. Tsuchiya, T. and Tanaka, A., Effects of thiabendazole on the mouse limb-bud organ and cell in culture, *Toxicol. Lett.*, 30, 19, 1986.
212. Barrach, H.J., Mark, K.V.D., and Gay, S., Location of type I and type II collagen in developing limb buds of mammalian, in *New Approaches to the Elucidation of Abnormal Embryonic Development*, Neubert, D. and Merker, H.-J., Eds., Georg Thieme Publishers, Stuttgart, 1975, p. 145.
213. Lilja, S. and Schimmelpfennig, K., Studies on glycosaminoglycans in cultured mouse limb buds, in *New Approaches to the Elucidation of Abnormal Embryonic Development*, Neubert, D. and Merker, H.-J., Eds., Georg Thieme Publishers, Stuttgart, 1975, p. 241.
214. Kwasigroch, T.E. and Kochhar, D.M., Locomotory behavior of limb bud cells: effects of excess vitamin A in vivo and in vitro, *Exp. Cell. Res.*, 95,269, 1975.
215. Hoffman, C. and Eichele, G., Retinoids in development, in *The Retinoids: Biology, Chemistry, and Medicine*, 2nd Ed., Sporn, M.B., Roberts, A.B., and Goodman, D. S., Eds., Raven Press, Ltd., New York, 1994, p. 387.
216. Knudsen, T.B., In vitro approaches to the study of embryonic cell death in developmental toxicity, in *In Vitro Methods in Developmental Toxicology: Use in Defining Mechanisms and Risk Parameters*, Kimmel, G.L. and Kochhar, D.M., Eds., CRC Press, Boca Raton, 1990, p. 129.
217. Pratt, R.M., Dencker, L., and Diewert, V.M., 2,3,7,8-tetrachlorodibenzo-pdioxin-induced cleft palate in the mouse: evidence for alterations in palatal shelf fusion, *Teratogen. Carcinogen. Mutagen.*, 4, 427, 1984.
218. Pratt, R.M., Receptor-dependent mechanisms of glucocorticoid and dioxin-induced cleft palate, *Environ. Health Perspect.*, 61, 35, 1985.
219. Mino, Y., Mizusawa, H., and Shiota, K., Effects of anticonvulsant drugs on fetal mouse palates cultured in vitro, *Reprod. Toxicol.*, 8, 225, 1994.
220. Abbott, B.D., Harris, M.W., and Birnbaum, L.S., Etiology of retinoic acid-induced cleft palate varies with the embryonic stage, *Teratology*, 40, 533, 1989.
221. Saxen, I. and Saxen, L., Organ culture in teratology: closure of the palatal shelves as a model system, in *New Approaches to the Evaluation of Abnormal Embryonic Development*, Neubert, D., Merker, H.J., Bedürftig, A., and Kreft, R., Eds., Georg Thieme Verlag, Stuttgart, 1975, p. 84.
222. Saxen, I., The effect of tetracycline on osteogenesis in vitro, *J. Exp. Zool.*, 162, 269, 1966.
223. Saxen, I., Effects of hydrocortisone on the development in vitro of the secondary palate in two inbred strains of mice, *Arch. Oral Biol.*, 18, 1469, 1973.
224. Abbott, B.D. and Buckalew, A.R., Embryonic palatal responses to teratogens in serum-free organ culture, *Teratology*, 45, 369, 1992.
225. Shiota, K., Kosazuma, T., Klug, S., and Neubert, D., Development of the fetal mouse palate in suspension organ culture, *Acta Anat.*, 137, 59, 1990.
226. Pratt, R.M., Receptor-dependent mechanisms of glucocorticoid and dioxin-induced cleft palate, *Environ. Health Perspect.*, 61, 434, 1985.
227. Pratt, R.M., Receptor-dependent Mechanisms of craniofacial malformations, in *Approaches to Elucidate Mechanisms in Teratogenesis*, Welsch, F., Ed., Hemisphere, New York, 1987, p. 149.
228. Abbot, B.D. and Pratt, R.M., Retinoids and epidermal growth factor alter embryonic mouse palatal epithelial and mesenchymal cell differentiation in organ culture, *J. Craniofac. Gen. Dev. Biol.*, 7, 219, 1987.
229. Abbott, B.D. and Pratt, R.M., Human embryonic palatal epithelial differentiation is altered by retinoic acid and epidermal growth factor in organ culture, *J. Craniofac. Gen. Dev. Biol.*, 7, 241, 1987.
230. Pratt, R.M., Goulding. E.H., and Abbott, B.D., Retinoic acid inhibits migration of cranial neural crest cells in the cultured mouse embryo, *J. Craniofac. Gen. Dev. Biol.*, 7, 205, 1987.

231. Beck, F., Lloyd, J.B., and Griffiths, A., Lysosomal enzyme inhibition by trypan blue: a theory of teratogenesis, *Science*, 157, 1180, 1967.
232. Beck, F. and Lowy, A. The effect of cathepsin inhibitor on rat embryos grown in vitro, *J. Embryol. Exp. Morphol.*, 71, 1, 1982.
233. Duncan, R. and Lloyd, J.B., Pinocytosis in the visceral yolk sac. Effects of temperature, metabolic inhibitors and some other modifiers, *Biochim. Biophys. Acta*, 544, 647, 1978.
234. Stevenson, G.B. and Williams, K.E., Ethanol-induced inhibition of pinocytosis and proteolysis in rat yolk sac in vitro, *Development*, 99, 247, 1987.
235. Lerman, S., Koszalka, T.R., Jensen, M., Andrew, C.L., Beckman, D.A., and Brent, R.L., In vitro studies on the effect of yolk sac antisera on functions of the visceral yolk sac. I. Pinocytosis and transport of small molecules, *Teratology*, 34, 335, 1986.
236. Haffen, K., Organ culture and the determination of sex organs, in *Tests of Teratogenicity In Vitro*, Marois, M., Ed., North Holland Publishers, Amsterdam, 1976, p. 285.
237. DeHaan, R.L., Cell coupling and electrophysiological differentiation of embryonic heart cells, in *Tests of Teratogenicity In Vitro*, Marois, M., Ed. North Holland Publishers, Amsterdam, 1976, p. 225.
238. Kollar, E. J., The use of organ cultures of embryonic tooth germs for teratological studies, in *Tests of Teratogenicity In Vitro*, Marois, M., Ed. North Holland Publishers, Amsterdam, 1976, p. 303.
239. Lasnitzki, I., Organ culture, In *Animal Cell Culture: A Practical Approach*, Freshney, R.I., Ed., IRL Press, Oxford, 1992, p. 213.
240. Fisher, K.R.S. and Fedoroff, S., The development of chick spinal cord in tissue culture, *In vitro*, 14, 878, 1978.
241. Bird, M.M. and Owen, A., The effect of calcium ionophore A23187 on neurites from embryonic mouse spinal cord explants in culture, *J. Electron Microsc.* (Tokyo), 49, 379, 2000.
242. Serra, R., Pelton, R.W., and Moses, H.L., TGFb1 inhibits branching morphogenesis and N-*myc* expression in lung bud organ cultures, *Development*, 120, 2153, 1994.
243. Levesque, B.M., Vosatka, R.J., and Nielsen, H.C., Dihydrotestosterone stimulates branching morphogenesis, cell proliferation, and programmed cell death in mouse embryonic lung explants, *Pediatr. Res.*, 47, 481, 2000.
244. Konigsberg, I.R., Clonal analysis of development, in *Tests of Teratogenicity In Vitro*, Marois, M., Ed. North Holland Publishers, Amsterdam, 1976, p. 275.
245. Kasten, F.H., Mammalian myocardial cells, in *Tissue Culture Methods and Applications*, Kruse, Jr., P.F., and Patterson, M.K., Jr., Eds., Academic Press, New York, 1973, p. 72.
246. Zimmerman, B., Effects of matrix influencing substances on chondrocyte in monolayer culture, in *Teratology of the Limbs*, Merker, H.-J., Hau, H., and Neubert, D., Eds., Walter de Gruyter, Berlin, 1980, p. 207.
247. Coon, H.G., Clonal stability and phenotypic expression of chick cartilage cells in vitro, *Proc. Nat. Acad. Sci. U.S.A.*, 55, 66, 1966.
248. Cahn, R.D. and Cahn, M.B., Heritability of cellular differentiation: clonal growth and expression of differentiation in retinal pigment cells in vitro, *Proc. Nat. Acad. Sci. U.S.A.*, 55, 106, 1966.
249. Eguchi, G. and Okada, T.S., Differentiation of lens tissue from the progeny of chick retinal pigment cells cultured in vitro: a demonstration of a switch of cell types in clonal cell culture, *Proc. Nat. Acad. Sci. U.S.A.*, 70, 1495, 1973.
250. Loring, J., Glimelius, B., Erickson, C., and Weston, J. A., Analysis of developmentally homogeneous neural crest cell populations in vitro, *Dev. Biol.*, 82, 86, 1981.
251. Glimelius, B. and Weston, J. A., Analysis of developmentally homogeneous neural crest cell populations in vitro. II. A tumor-promoter (TPA) delays differentiation and promotes cell proliferation, *Dev. Biol.*, 82, 95, 1981.
252. Kang, K.S., Park, J.E., Ryu, D.Y., and Lee, Y.S., Effects and neuro-toxic mechanisms of 2,2′,4,4′,5,5′-hexachlorobiphenyl and endosulfan in neuronal stem cells, *J. Vet. Med. Sci.*, 63, 1183, 2001
253. Davis, W.L., Crawford, L.A., Cooper, O.J., Farmer, G.R., Thomas, D.L., and Freeman, B.L., Ethanol induces the generation of reactive free radicals by neural crest cells in vitro, *J. Craniofac. Genet. Dev. Biol.*, 10, 277, 1990.
254. Davis, W.L., Crawford, L.A., Cooper, O.J., Farmer, G.R., Thomas, D.L., and Freeman, B.L., Generation of radical oxygen species by neural crest cells treated in vitro with isotretinoin and 4-oxo-isotretinoin, *J. Craniofac. Genet. Dev. Biol.*, 10, 295, 1990.

255. Davis, W.L., Jacoby, B.H., Farmer, G.R., and Cooper, O.J., Changes in cytosolic calcium, bleb formation, and cell death in neural crest cells treated with isotretinoin and 4-oxo-isotretinoin, *J. Craniofac. Genet. Dev. Biol.*, 10, 295, 1990.
256. Bonnefoi, M.S., Mitochondrial glutathione and methyl iodide-induced neurotoxicity in primary neural cell cultures, *Neuro. Toxicol.*, 13, 401, 1992.
257. Flint, O.P. and Orton, T.C., An in vitro assay for teratogens with cultures of rat embryo midbrain and limb bud cells, *Toxicol. Appl. Pharmacol.*, 76, 383, 1984.
258. Kistler, A. and Howard, W.B., Testing of retinoids for teratogenicity in vitro: use of micromass limb bud culture, *Methods Enzymol.*, 190, 427, 1990.
259. Kistler, A., Limb bud cell cultures for estimating the teratogenic potential of compounds. Validation of the test system with retinoids, *Arch. Toxicol.*, 60, 403, 1987.
260. Hansen, J.M., Carney, E.W., Harris, C., Altered differentiation in rat and rabbit limb bud micromass cultures by glutathione modulating agents, *Free Radic. Biol. Med.*, 31, 1582, 2001
261. Flint, O.P., Orton, T.C., and Ferguson, R.A., Differentiation of rat embryo cells in culture: response following acute maternal exposure to teratogens and non-teratogens, *J. Appl. Toxicol.*, 4, 109, 1984.
262. Roark, E.F. and Greer, K., Transforming growth factor-beta and bone morphogenetic protein-2 act by distinct mechanisms to promote chick limb cartilage differentiation in vitro, *Dev. Dyn.*, 200, 103, 1994.
263. Lee, Y.S. and Chuong, C.M., Activation of protein kinase A is a pivotal step involved in both BMP-2- and cyclic AMP-induced chondrogenesis, *J. Cell Physiol.*, 170, 153, 1997.
264. Boskey, A.L., Paschalis, E.P., Binderman, I., and Doty, S.B., BMP-6 accelerates both chondrogenesis and mineral maturation in differentiating chick limb-bud mesenchymal cell cultures, *J. Cell Biochem.*, 84, 509, 2002.
265. Kruger, M., Mennerich, D., Fees, S., Schafer, R., Mundlos, S. and Braun T., Sonic hedgehog is a survival factor for hypaxial muscles during mouse development, *Development*, 128, 743, 2001.
266. Whittaker, S.G., Wroble, J.T., Silbernagel, S.M. and Faustman, E.M. Characterization of cytoskeletal and neuronal markers in micromass cultures of rat embryonic midbrain cells, *Cell. Biol. Toxicol.*, 9, 359, 1993.

CHAPTER 17

Statistical Analysis for Developmental and Reproductive Toxicologists

James J. Chen

CONTENTS

I. Introduction 697
II. Analysis of Continuous Response Data 699
    A. Litter-Based ANOVA Method 699
    B. Linear Mixed-Effects Models 700
    C. Example-Fetal Body Weight Data 700
III. Analysis of Binary Response Data 701
    A. Independent Binary Response Data 701
    B. Correlated Binary Response Data 703
        1. Litter-Based ANOVA Method 703
        2. Likelihood and Quasi-Likelihood Approaches 703
        3. Example — Hydroxyurea Data 704
IV. Analysis of Count Data 705
    A. Likelihood and Quasi-likelihood Methods 706
    B. Example 706
V. Quantitative Risk Assessment and Benchmark Dose 707
    A. Dose-Response Model and Benchmark Dose 707
    B. Example 709
VI. Discussion 709
References 710

I. INTRODUCTION

Reproductive and developmental toxicity experiments are conducted in laboratory animals for the evaluation of potential adverse effects of chemical compounds on fertility, reproduction, and fetal development. Depending on the study design, a test compound is administered to either parent prior to conception, to the female during gestation, or postnatally. Various reproductive and developmental endpoints are measured, recorded, and analyzed. These endpoints can be divided into two categories: (1) parental and (2) embryonic or fetal endpoints. Parental endpoints used to assess reproductive effects include: body weight and weight gain, mating index, fertility index, changes in gestation length, sperm count, numbers of corpora lutea, implantation sites, dead or resorbed implants, viable fetuses, and target organ weights, as well as food and water consumption. Embryonic or fetal endpoints include: individual malformations (external or gross, visceral, and skeletal) and variations in viable fetuses, number of normal fetuses, fetal weight, fetal length, sex ratio, preweaning and postweaning survival and body weight, physical development, and functional test outcomes (see Chapters 7, 8, 10, and 19).

Mice, rats, and rabbits are the most commonly used species for reproductive and developmental toxicology studies. An experiment typically consists of an untreated control and three dosage groups. The highest dose is chosen to produce minimal maternal toxicity, ranging from marginal body weight reduction to not more than 10% mortality. The low dose should be a no-observed adverse-effect level for both maternal and developmental toxicity. The midlevel dose usually is halfway between the low and high doses. A fourth dosage group may be added to avoid excessive dosage intervals and to assess dose-response relationships. Ideally the dose should be the amount of active substance at the affected tissue site. The estimate of the internal tissue-site dose requires data from toxicokinetic and/or toxicodynamic experiments. In the absence of this information, the amount of the internal dose is assumed proportional to the administered or exposure dose. Typically, dosage is measured in milligrams per kilogram body weight per day, and animals are randomly assigned to dosage groups. Each animal is monitored throughout the experiment to detect toxic and pharmacologic effects of the test compounds, and all adult animals are necropsied at terminal sacrifice.

Sample size determination is an important issue in detecting a dose effect. The probability of detecting a statistically significant effect increases with the increased magnitude of effect and increased sample size. It is customary to specify a range of the magnitudes of the effect of interest; an optimal sample size then can be calculated to provide a reasonable power of detecting the effect. Sample size calculations and references for various tests are provided in Desu and Raghavarao.[1] The U.S. regulatory guidelines generally recommend about 20 pregnant rodents and 15 nonrodent animals per dosage group. The ICH guideline recommends use of 16 to 20 pregnant animals. Typical litter sizes (number of viable offspring) for the control animals range from 8 for rabbits to 12 and 14 for mice and rats, respectively. However, it should be emphasized that sample size determination is preferable to the arbitrary requirements for testing 10 or 20 animals as indicated in various guidelines (cf., Chapter 19).

In the analysis of animal developmental endpoints, the experimental unit is the entire litter rather than an individual fetus. This is because the test compound is administered to the adult animal, so the effect of the test compound occurs in the female that receives the compound or is mated to a male that receives the compound. The treatment thus affects the embryo or fetus indirectly, via the dam. Thus, the development of individual offspring in a dam is not independent, and the fetal responses of individuals from the same litter are expected to be more alike than responses of individuals from different litters. This phenomenon is referred to as the "litter effect." The proper experimental unit in the analysis should be the litter, with the fetal responses representing multiple observations from a single experimental unit. Failure to account for the intralitter correlations by using each fetus as the experimental unit can inflate the Type I error and reduce the validity of the test. The classical approach to the analysis of reproductive data is based on the litter mean. An alternative approach is to model the fetal endpoints in a litter as correlated outcomes (clustered data). These two approaches are described in this chapter.

Statistical analyses of various toxicity endpoints have been concerned with two aspects: qualitative testing for adverse effects and quantitative estimation for risk assessment. The qualitative testing is to determine if there is a statistically significant difference among the groups. Three statistical tests are generally conducted:

1. Homogeneity tests for detecting significant differences among groups
2. Pairwise comparisons between two specific groups (usually control versus a dosage group)
3. Trend tests to identify dose-related increases or decreases

Trend tests are more sensitive (powerful) for detecting dose effects than are pairwise comparisons because trend tests use information from all dosage groups. The quantitative estimation is to determine a safe level of human exposure from an assessment of the dose-response relationship. However, the quantitative risk estimation is typically used for environmental or workplace exposures and does not apply to pharmaceutical drugs.

The reproductive and developmental toxicity endpoints described above can generally be divided into three categories: (1) binary (e.g., presence or absence of a malformation) or proportion (of malformation among the viable progeny), (2) count (e.g., number of implants), and (3) continuous (e.g., fetal body weight). This chapter describes statistical methods for the analysis of the three data types.

II. ANALYSIS OF CONTINUOUS RESPONSE DATA

Consider a typical experiment of g groups, a control ($d_1 = 0$) and $g - 1$ dose groups ($d_2, \cdots, d_g$). Assume that the ith group contains n_i female animals. Let y_{ijk} be the response from the kth fetus out of n_{ij} examined or tested for a particular developmental outcome, $1 \leq i \leq g$, $1 \leq j \leq n_i$, and $1 \leq k \leq n_{ij}$. For example, y_{ijk} may be a fetal body weight or behavioral measurement conducted on offspring following birth, and n_{ij} is the number of viable fetuses from the jth female in the ith group. The continuous endpoints can also be measured in an adult animal (e.g., maternal body or organ weight) denoted as y_{ij}. The data can be summarized as the mean with a standard error of the mean (SE) for each dosage group, using the litter as the experimental unit.

A. Litter-Based ANOVA Method

Analysis of variance (ANOVA) is the most commonly used procedure for the analysis of continuous response data. The ANOVA method assumes that data are independently and normally distributed with homogeneous variance. Prior to ANOVA, Bartlett's test[2] is conducted to test for homogeneity of variance. Transformations, such as the logarithmic, are often applied to satisfy the normality assumption and stabilize the variance across groups. Transformations of the data are acceptable because the scale of measurement of variables is often arbitrary. Graphical normal plots and Box plots[3] can be used to determine if the transformed data satisfy the normality and variance homogeneity assumptions. The SAS PROC UNIVARIATE[4] procedure provides various statistical tests and graphical plots for checking the underlying assumptions for ANOVA. The litter-based analysis is based on the per-litter response. For maternal endpoints, a simple one-way ANOVA on the observed response y_{ij} is used for testing homogeneity among groups. Developmental endpoints are analyzed similarly but in terms of the average within each litter,

$$y_{ij} = \sum_k y_{ijk} / n_{ij}$$

A wide variety of *post hoc* tests is available for pairwise comparisons or trend analysis once a significant result is obtained in an ANOVA. The Dunnett[5] and Williams[6] tests are the two mostly used multiple test procedures for comparison of each dosage group with the control. The objectives of interest often are to identify the largest dose that is not statistically significantly different from the control (NOEL) or the lowest dose that is significantly different from the control. The Williams test assumes an increasing (or decreasing) dose-response trend. When this assumption is valid, the Williams' test is the more powerful to detect differences between control and dosage groups. Linear regression is the simplest method to assess dose-response trends. But contrast tests, with a proper choice of contrast coefficients in conjunction with ANOVA, are more commonly used for dose-response

Table 17.1 Mean fetal weights (per litter) and dosage group means with their standard errors (SEs)

Dose	Mean fetal weight/litter									Mean (SE)
Control	6.845	6.296	7.355	6.289	6.885	6.966	6.235	5.894	5.365	6.535	6.467 (.183)
Low	5.725	7.310	5.796	5.820	6.076	6.008	5.451	6.145	6.164	6.043	6.054 (.156)
High	5.243	5.407	7.237	5.873	6.564	6.298	6.001				6.089 (.259)

trend testing. A contrast test performs custom hypothesis tests by specifying coefficients for testing a univariate hypothesis or multivariate hypothesis. The trend analysis is simplified by examining orthogonal contrasts among the treatment factor levels that measure the linear, quadratic, and higher level of polynomial effects. Table values of the orthogonal polynomial coefficients can be found in experimental design textbooks.[2] Other trend tests are Tukey's test,[7] which has been proposed to identify the highest dose at which there is no statistically significant difference from the control trend, and Bartholomew's test [8] of ordered alternatives.

Nonparametric methods are used when the assumption of normality fails. The nonparametric analysis is initiated by ranking all observations of the combined groups. The nonparametric method is then applied to ranked data. Further details may be found in general texts addressing nonparametric methods.[9]

B. Linear Mixed-Effects Models

A general approach to modeling fetal data can be carried out in terms of a mixed-effects model. Dempster et al.[10] proposed a normal mixed-effects model with two levels of variance in which litter effect is modeled by a nested random factor and dose by a fixed factor. The response y_{ijk} in a litter is modeled with two sources of variation: the between litter γ_{ij} and within litter variation e_{ijk},

$$y_{ijk} = \mu_{ij} + \gamma_{ij} + e_{ijk}$$

The random components γ_{ij} and e_{ijk} are independently normally distributed with $E(\gamma_{ij}) = 0$, $Var(\gamma_{ij}) = \sigma_f^2$, and $E(e_{ijk}) = 0$, $Var(e_{ijk}) = \sigma^2$. Thus, the mean and variance of y_{ijk} are $E(y_{ijk}) = \mu_i$ and $Var(y_{ijk}) = \sigma_f^2 + \sigma^2$. The intralitter correlation between y_{ijk} and $y_{ijk'}$ for $k \neq k'$ is $\phi = \sigma_f^2 / (\sigma^2 + \sigma_f^2)$. The mean parameter μ_{ij} is often modeled as a linear function of dose $\mu_{ij} = \beta_0 + \beta_1 d$ for trend tests $H_o : \beta_1 (= 0)$. The maximum likelihood estimates of the parameters of linear mixed effects models can be obtained using the PROC MIXED procedure of SAS.[11] Alternatively, the generalized estimating equations (GEE) approach[12] can be used to estimate the fixed effects parameters. The software for the GEE method is available publicly[13] or is found in the SAS PROC GENMOD procedure.

C. Example-Fetal Body Weight Data

An analysis of fetal body weight data from a reproductive study by Dempster et al.[10] is used for illustration. The experiment consisted of three dosage groups: control, low, and high. The numbers of animals in the three groups were 10, 10, and 7, respectively. The litter means y_{ij}s are given in Table 17.1. (The individual fetal weight data are listed in Table 17.2 of Dempster et al.[10]) The three hypothesis tests, homogeneity, pairwise comparisons of dosage groups with control, and trend, were conducted using the litter-based analysis and linear mixed-effects model approaches. Table 17.2 contains the summary of the analysis. The SAS PROC GLM and PROC MIXED procedures were used for the litter-based analysis and the linear mixed-effects model, respectively.

Using the method of litter-based ANOVA on the litter means, the homogeneity test of no difference among the three means yields an F value of 1.50 ($p = 0.244$) with 2 and 24 degrees of

Table 17.2 *p*-values for testing homogeneity, control versus dosage groups, and trend, using the litter-based analysis and mixed-effects model approaches

Approach	Homogeneity	Control vs. Dose				Trend
		LSD (t-test)		Dunnett		
		$\mu_1 = \mu_2$	$\mu_1 = \mu_3$	$\mu_1 = \mu_2$	$\mu_1 = \mu_3$	
Litter-based analysis	.244	.138	.198	.215	.332	.198
Mixed-effect model	.185	.093	.161	.163	.276	.161

freedom. The *p* value is greater than the 5% significance level. The comparison could stop here, but for the purpose of illustration, the dosage group comparisons and trend test are performed. For the mixed-effects model, the maximum likelihood estimates of the variance parameters are $\hat{\sigma}^2 = 0.196$ and $\hat{\sigma}_f^2 = 0.277$. The estimated intralitter correlation is $\hat{\phi} = 0.277/(0.196+0.277) = 0.586$, indicating the presence of litter effect. The two methods give very similar results; however, the mixed-effect modeling approach appears more powerful because it uses individual fetal data that accounts for litter effects in the analysis.

III. ANALYSIS OF BINARY RESPONSE DATA

Binary endpoints can also be measured either at the parent level, such as success or failure of pregnancy, or at the individual fetal level, such as presence or absence of a particular malformation type. The binary responses from individual dams are independent, while the binary responses from the fetuses in a litter are correlated. Statistical methods for the analyses of the parental and fetal responses are different.

A. Independent Binary Response Data

Given a typical reproductive experiment, if n_i represents the total number of animals from the *i*th group and y_i is the number of animals found to present a particular response of interest, the data can be summarized as

Dose	d_1	d_2	...	d_g	Total
No. of responses	y_1	y_2	...	y_g	y
Total	n_1	n_2	...	n_g	n
Proportions	p_1	p_2	...	p_g	p

y and n are the row marginal totals, $p = y/n$, and $p_i = y_i/n_i$ is the observed incidence rate in the *i*th group.

An asymptotic chi-square test for homogeneity is commonly used to compare incidence rates (proportions) among several groups:

$$\chi^2_{g-1} = \sum_{i=1}^{g} \frac{(y_i - n_i p)^2}{n_i p(1-p)}$$

For pairwise comparisons, the chi-square test is often used with the continuity correction. It can be simplified as

$$\chi^2_1 = \frac{(|y_i - n_1 p| - 0.5)^2}{n_1 p(1-p)}$$

The Cochran-Armitage (CA) test[14,15] is used to test for dose-related trend,

$$Z = \frac{\sum_i y_i d_i - p \sum_i n_i d_i}{\sqrt{p(1-p)[\sum_i n_i d_i^2 - (\sum_i n_i d_i)^2 / n]}}$$

Alternatively, logistic regression provides a simple approach for analyzing binary data. The probability of response is modeled as

$$P(d_i) = \frac{\exp(\beta_0 + \beta_1 d_i)}{1 + \exp(\beta_0 + \beta_1 d_i)}$$

Logistic regression can be used for group comparisons and trend test. The model can include any additional explanatory variables (e.g., fetal weight, litter size).

When the total number of occurrences is small, the asymptotic test may not be reliable, and an exact permutation test is recommended. The Fisher's exact test is the best known permutation test for comparing two groups. We describe a generalization of the Fisher's exact test to the $2 \times g$ table for trend test.

Conditional on $y_.$ and $n_.$, under the null hypothesis of no difference among groups, the distribution of $(y_1, y_2, ..., y_g)$ is the multivariate hypergeometric distribution

$$P(y_1, y_2, ..., y_g) = \frac{\binom{n_1}{y_1}\binom{n_2}{y_2} \cdots \binom{n_g}{y_g}}{\binom{n_.}{y_.}}$$

The trend score associated with $(y_1, y_2, ..., r_g)$ is defined by $S = \sum_{i=1}^{g} d_i y_i$. The probability distribution for the trend score statistic S is

$$P(S = s) = \sum_\Omega P(y_1, y_2, ..., y_g)$$

where Ω consists of all possible permutations of y_i such that $\sum_{i=1}^{g} y_i = y$ and $\sum_{i=1}^{g} d_i y_i = s$. Let s^* denote the trend score associated with the observed outcome. The exact one-tailed p-value for a positive dose-related trend is

$$p\text{-value} = \sum_{s \geq s^*} P(S = s)$$

An example of a hypothetical animal experiment with four dosage groups, $d_1 = 0, d_2 = 1, d_3 = 2$, and $d_4 = 4$, with 50 animals in each group is used for illustration. Assume that the observed outcome for the number of responses in the four groups is (0, 0, 3, 3), which corresponds to a trend score of $S = 18$. Note that two other permutations (0, 2, 0, 4) and (1, 0, 1, 4) also give the score 18. The probability for all possible permutations such that $S >= 18$ are given below:

Response Pattern				S	Prob.	Cumul. Prob.
0	0	0	6	=24	0.00019	0.00019
0	0	1	5	=22	0.00129	0.00148
0	1	0	5	=21	0.00129	0.00277
1	0	0	5	=20	0.00129	0.00406
0	0	2	4	=20	0.00341	0.00747
0	1	1	4	=19	0.00699	0.01446
0	2	0	4	=18	0.00342	0.01788
0	0	3	3	=18	0.00466	0.02254
1	0	1	4	=18	0.00699	0.02953

The one-sided exact *p*-value is *p*-value = 0.00019 + 0.00129 + ... + 0.00699 = 0.02953. To compare with the asymptotic test, the *p*-value is 0.01778 from the CA test and is 0.02735 from the logistic regression Wald test. The exact permutation test generally gives a larger *p*-value (less likely to be significant) than an asymptotic test. In a two group comparison, this method becomes the Fisher's exact test. That is, the *p*-value calculated from the Fisher's exact test is usually larger than the *p*-value obtained from an asymptotic chi-square (Wald) test. But, the *p*-value from a chi-square test with the continuity correction and the Fisher's exact *p*-value are comparable.

The chi-square test and Fisher's exact test are available in the SAS[11] PROC FREQ procedure. The SAS PROC LOGISTIC procedure provides the *p*-values of the CA test (score test) and Wald test, and the PROC MULTEST procedure provides the exact *p*-value for the permutation test. The exact *p*-value can also be obtained using StatXact software distributed by Cytel Software Corporation.[16]

B. Correlated Binary Response Data

As discussed, the fetal responses within a litter are not independent. The proper experimental unit in the analysis should thus be the litter. As was the case for the analysis of continuous response data, two methods can be described.

1. Litter-Based ANOVA Method

The conventional approach for the analysis of fetal binary endpoints is to consider the proportion per litter, such as the proportion of live fetuses with a certain type of malformation or the proportion of deaths and/or resorptions among the number of implants. If y_{ijk} is the binary response from a fetus out of n_{ij} examined or tested for a particular developmental outcome, the total number of responses in a litter is $y_{ij} = \sum_k y_{ijk}$. The litter-based analysis is based on the litter proportion $p_{ij} = y_{ij}/n_{ij}$. Typically, the proportions are transformed by an arcsine transformation, $\arcsin\sqrt{p_{ij}}$. The transformed data are then analyzed as continuous data by use of the parametric ANOVA methods described in Section II. Nonparametric methods, such as the Kruskal-Wallis[9] ANOVA test for homogeneity, the Mann-Whitney[9] test for pairwise comparisons, and the Jonckheere[17] trend test, can be used to analyze litter proportions by ranks. The nonparametric methods are available in the SAS PROC NPR1WAY procedure.[11]

The litter-based approach does not use the data effectively because it does not account for litter size. For example, 1 out of 2 is treated as the same as 5 out of 10. Thus, the beta-binomial likelihood and quasi-likelihood approaches have been developed recently to analyze correlated binary data.

2. Likelihood and Quasi-Likelihood Approaches

Williams[18] proposed the beta-binomial model for the analysis of developmental data. This model assumes that responses within the same litter occur according to a binomial distribution, and the probability of responses is assumed to vary among litters according to a beta distribution. The distribution for the number of responses in a litter y_{ij} is

Table 17.3 The observed numbers of adverse developmental effects (death/resorption/malformation) (y) per litter from exposure to hydroxyurea and the numbers of implantations per litter (n)

Control	(y)	0	0	0	0	2	0	1	1	0	0	1	0	0	0	0	0	1
	(n)	10	10	12	13	11	9	9	12	11	8	11	13	12	6	8	11	9
150 ppm	(y)	1	0	0	5	1	2	2	4	1	0	6	1	0				
	(n)	14	11	11	8	12	8	11	10	12	14	13	7	9				
200 ppm	(y)	2	0	13	9	12	10	4	3									
	(n)	14	11	16	11	13	13	13	11									
250 ppm	(y)	8	11	7	10	9	4	4	9	8	5	11	9	10	10	6	9	9
	(n)	9	11	11	11	11	11	9	10	8	5	12	12	12	11	11	13	12
	(y)	8	11	6	11	3												
	(c)	9	12	10	12	9												

$$P(y_{ij}) = \binom{n_{ij}}{y_{ij}} \frac{B(a_i + y_{ij}, b_i + n_{ij} - y_{ij})}{B(a_i, b_i)}$$

where $B(a_i, b_i) = \Gamma(a_i)\Gamma(b_i)/\Gamma(a_i + b_i)$, where $\Gamma(\cdot)$ is the gamma function, $a_i > 0$, and $b_i > 0$. Under the reparameterization $\mu_i = a_i/(a_i + b_i)$ and $\phi_i = (a_i + b_i + 1)^{-1}$, the parameters μ_i and ϕ_i are, respectively, the mean and the intralitter correlation parameters in the ith group. The mean of y_{ij} is $E(y_{ij}) = n_{ij}\mu_i$, and the intralitter correlation is $Corr(y_{ijk}, y_{ijk'}) = \phi_i$. The variance of y_{ij} is $Var(y_{ij}) = [\phi_i(n_{ij} - 1) + 1][n_{ij}\mu_i(1 - \mu_i)]$. (The intralitter correlation coefficient is positive ($\phi_i > 0$) for most developmental toxicity endpoints. Thus, the variance of the total response y_{ij} in a litter is greater than the nominal binomial variance $n_{ij}\mu_i(1 - \mu_i)$ (here the mean parameter μ_i corresponds to the p_i in the binomial model). The distribution of y_{ij} is known as an overdispersion binomial, **and when** $\phi_i = 0$, then y_{ij} becomes a binomial.

As for the binomial model, logistic regression can be applied to the beta-binomial model for group comparisons and trend tests. The parameters βs and ϕs are estimated by the maximum likelihood method. The likelihood ratio or Wald test is often used to test for the significance of parameters. However, the likelihood ratio test is preferable. One problem with the beta-binomial model is the bias and instability of the maximum likelihood estimates (MLEs) of the coefficients in fitting the logistic dose-response model. Williams[19] proposed using the quasi-likelihood method as an alternative approach to the beta-binomial model.

In the quasi-likelihood approach, only assumptions regarding the mean and variance are required: $E(y_{ij}|n_{ij}) = n_{ij}\mu_i$ and $Var(y_{ij}|n_{ij}) = n_{ij}\mu_i(1 - \mu_i)[\phi(n_{ij} - 1) + 1]$. Note that in the quasi-likelihood estimation, the intralitter correlations typically are modeled as constant across groups. The coefficients of the βs can be obtained by solving the quasi-likelihood estimating (score) equations. The intralitter correlation coefficient is calculated by equating with the mean of Pearson chi-square statistics. Estimation details are given by Williams.[18] The Wald test is used to test for the significance of βs in terms of the logistic regression model. Chen[20] and Chen et al.[21] compared several trend tests and homogeneity tests for the analysis of overdispersion binomial data from developmental studies.

3. Example – Hydroxyurea Data

A study of the effect of maternal exposure to hydroxyurea in mice is used for illustration. The experiment consisted of four treatment groups: 0, 150, 200, and 250 ppm. Table 17.3 shows the frequency table of litters according to number of fetuses affected (including death, resorption, and malformation) (y_{ij}) and number of implants (n_{ij}). Table 17.4 contains the means and standard errors from the litter-based analysis, beta-binomial model, and binomial model. Note that the binomial model assumes that all fetuses are independent samples. Therefore, the total number of responses

Table 17.4 The mean (with standard error) from the litter-based analysis and the maximum likelihood estimates (with standard errors) from the beta-binomial and binomial models

Model	Parameter	0	100	200	250
Litter-based	μ	0.0340(.014)	0.1777(.056)	0.5058(.128)	0.7733(.044)
Beta-binomial	μ	0.0343(.014)	0.1720(.053)	0.4895(.111)	0.7700(.040)
	ϕ	0.0017(.040)	0.1758(.097)	0.3574(.125)	0.1136(.057)
Binomial	μ	0.0342(.014)	0.1642(.031)	0.5196(.049)	0.7768(.028)

Table 17.5 *p*-values for comparisons between the control and dosage groups, using the litter-based analysis, beta-binomial, quasi-likelihood, and binomial approaches

	Control vs. Dose					
	LSD (t-test)			Dunnett		
Approach	$\mu_1 = \mu_2$	$\mu_1 = \mu_3$	$\mu_1 = \mu_4$	$\mu_1 = \mu_2$	$\mu_1 = \mu_3$	$\mu_1 = \mu_4$
Litter-based analysis	.010	<.001	<.001	.028	<.001	<.001
Beta-binomial	.003	<.001	<.001	.009	<.001	<.001
Quasi-likelihood	.019	<.001	<.001	.051	<.001	<.001

($\sum_j \sum_k y_{ijk}$) out of the total number of fetuses ($\sum_j \sum_k n_{ijk}$) in each dosage group is a binomial. The mean estimates from the three procedures are close, but the standard errors based on the binomial model are much smaller than those based on the litter-based analysis and beta-binomial model, except for the 0 ppm group. The standard errors from the litter-based method and beta-binomial model are close. The estimated intralitter correlations for the four groups were 0.002, 0.176, 0.357, and 0.114. All three dosed groups showed evidence of overbinomial variation. Note that the binomial model does not account for the intralitter correlation and will underestimate the variation of responses among dosage groups (Table 17.4). Therefore, an analysis based on the binomial model will result in large false positives.

The three hypotheses were tested using the litter-based analysis, beta-binomial, and quasi-likelihood approaches. The likelihood ratio test was used in the beta-binomial approach and the Wald test was used in the quasi-likelihood method. Table 17.5 contains the summary of the analyses. The *p*-values from the homogeneity and trend tests are all less than 0.0001; those results are not shown. The three approaches give similar results except that the (two-sided) *p*-value from the comparison between control and 150 ppm groups is 0.051 (greater than 0.05) from the quasi-likelihood approach. The quasi-likelihood analysis was run using the SAS PROC LOGISTIC procedure with the Williams' option, and the beta-binomial analysis was run from the program developed by the author at the National Center for Toxicological Research.

IV. ANALYSIS OF COUNT DATA

Statistical methods developed for the evaluation of reproductive and developmental toxicity endpoints generally focus on the analysis of continuous and proportional data. Methods for the analysis of count data have seldom been discussed. However, a number of primary reproductive endpoints are measured as counts. In a dominant lethal assay, male mice are treated with a suspect mutagen and then are mated with females. The numbers of corpora lutea, implantations, and live or dead conceptuses are counted to assess reproductive effects of the test compound.

In the litter-based analyses, count data are often normalized by the square root transformation. The transformed data are then analyzed as continuous data by use of parametric ANOVA methods. Count data can also be analyzed by nonparametric methods or by use of likelihood or quasi-likelihood methods.

A. Likelihood and Quasi-likelihood Methods

Let n_{ij} be an observed count from the jth animal in the ith group $1 < i < g$ and $1 < j < m_i$. The counts n_{ij} are often modeled by a Poisson distribution,

$$p(n_{ij}) = \frac{\mu_i^{n_{ij}} e^{-\mu_i}}{n_{ij}!}, \quad n_{ij} = 0, 1, 2, \ldots,$$

the mean and variance of n_{ij} are $E(n_{ij}) = \text{Var}(n_{ij}) = \mu_i$. A common complication in the analysis of count data is that the observed variation often exceeds or falls behind the variation that is predicted from a Poisson model. A generalized variance for n_{ij} is of the form $\text{Var}(n_{ij}) = \mu_i(1 + \phi_i\mu_i)$. If $\phi_i > 0$, then n_{ij} has an extra-Poisson variation; if $\phi_i < 0$, then n_{ij} has a sub-Poisson variation; and if $\phi_i = 0$, then n_{ij} becomes a Poisson. In the Poisson model, the mean function is often modeled by a log-linear function. The dose-response model for trend test is $\mu_i = \exp(\beta_0 + \beta_1 d_i)$. This analysis is referred to as the Poisson regression.

The classical extra-Poisson modeling assumes that the mean of the Poisson has a gamma distribution, which leads to a negative binomial (gamma-Poisson) distribution for the observed data,

$$p(n_{ij}) = \frac{\Gamma(n_{ij} + \phi_i^{-1})}{\Gamma(n_{ij} + 1)\Gamma(\phi_i^{-1})} \left(\frac{\phi_i\mu_i}{1 + \phi_i\mu_i}\right)^{n_{ij}} \left(\frac{1}{1 + \phi_i\mu_i}\right)^{\frac{1}{\phi_i}}$$

where $\phi_i > 0$. The maximum likelihood estimation of the negative binomial model was described in detail by Lawless.[22] As in the beta-binomial model, the restriction on $\phi \geq 0$ can limit its application. For example, the number of implantations or the number of corpora lutea may exhibit a sub-Poisson variation. The quasi-likelihood approach[23,24] provides a method to model both extra- or sub-Poisson variation data. The quasi-likelihood approach assumes that the mean and variance of count data are of a negative binomial form, $E(n_{ij}) = \mu_i$ and $\text{Var}(n_{ij}) = \mu_i(1 + \phi_i\mu_i)$.

B. Example

The example is the hydroxyurea data considered in Section III. In this example, the effect of hydroxyurea exposure on the number of implants was analyzed. Because the exposure occurred after implantation, the numbers of implants would not be expected to differ among the four groups. The means and variances for the four groups are

$$\hat{\mu}_1 = 10.29 \, (3.72), \quad \hat{\mu}_2 = 10.76 \, (5.19), \quad \hat{\mu}_3 = 12.75 \, (3.07), \quad \hat{\mu}_4 = 10.50 \, (3.21)$$

The means are much larger than the variance in all four groups. The data exhibited sub-Poisson variation, so the negative binomial model is not appropriate for the analysis.

Table 17.6 contains the summary of the analyses using the litter-based analysis, Poisson model (Poisson regression), and quasi-likelihood approach. In the litter-based analysis, the numbers of implants were transformed by a square root transformation for data normalization. The litter-based and quasi-likelihood approaches gave similar results in all three tests at the 5% significance level. The Poisson model shows larger p-values in all tests because it does not adjust for sub-Poisson variation. The SAS PROC GLM and PROC GENMOD procedures were used in the analysis.

Table 17.6 p-values for testing homogeneity, control versus dosage groups, and trend using the litter-based analysis, Poisson, and quasi-likelihood approaches

Approach	Homogeneity	Control vs. Dose						Trend
		LSD (t-test)			Dunnett			
		$\mu_1 = \mu_2$	$\mu_1 = \mu_3$	$\mu_1 = \mu_4$	$\mu_1 = \mu_2$	$\mu_1 = \mu_3$	$\mu_1 = \mu_4$	
Litter-based	.044	.543	.007	.743	.881	.020	.978	.130
Poisson	.671	.691	.086	.843	.960	.206	.995	.587
Quasi-likelihood	.006	.502	.001	.743	.841	.003	.977	.404

V. QUANTITATIVE RISK ASSESSMENT AND BENCHMARK DOSE

A fundamental goal of reproductive and developmental experiments is to determine safe levels of human exposure to a substance. Such estimated levels are commonly referred to as reference doses (RfDs) (see Chapter 21). The RfDs are presumed to cause no or negligible health risks in humans. Risk assessment for toxic effects other than cancer has traditionally been based on the concept of a dosage "threshold." At doses below such a threshold, adverse effects are not expected to occur. RfDs for reproductive and developmental effects are commonly set by dividing experimental no-observed-adverse-effect-levels (NOAEL) by appropriate safety (uncertainty) factors. Various limitations for the NOAEL plus uncertainty factor approach have been documented.[25,26] Crump[27] proposed that the NOAEL be replaced by a benchmark dose estimated to be associated with an excess risk in the range of 1% to 10%. This section describes a procedure for constructing RfDs for correlated binary developmental endpoints.

A. Dose-Response Model and Benchmark Dose

Dose-response models are usually formulated to establish a mathematical relationship between the observed response and the dose. In addition to providing a mathematical framework for testing hypotheses as shown in the previous sections, an important use of dose-response models is to determine RfDs for risk assessment. The standard approach involves fitting a dose-response function to the experimental data and then using the fitted model to estimate the RfD. Because relatively small numbers of animals are exposed to only a few dose levels, the dose level used to produce a response is often much higher than the environmental exposure level of interest. It is well known that various dose response functions that fit the experimental data adequately can yield extrapolated doses differing by several orders of magnitude. In general, one should aim for a model that is both consistent with the available biological knowledge and fits the observations reasonably well.

For binary endpoints, such as malformation or death, the dose response function $P(d)$ is the probability of occurrence of the adverse effect associated with the dose level d. The quantity of interest from the risk assessment viewpoint is

$$\pi(d) = P(d) - P(0)$$

which measures the excess risk (above the background risk) due to the added dose. A "safe" dose is defined as the dose for which the excess risk satisfies $\pi(d) = \pi^*$, where π^* is a predetermined small number (e.g., 10^{-5}).

Krewski and Van Ryzin[28] demonstrated that the risk estimates based on a variety of dose-response models do not differ significantly above an incidence of 1%, although large relative differences can be derived at doses corresponding to lower risks. In general, the dose corresponding to a risk of 10% can be estimated with adequate precision and is not particularly dependent on the dose-response model used to fit the data. The doses corresponding to an excess risk between 1% (ED_{01}) and 10% (ED_{10}) have been suggested as a point of departure (an anchor point) for risk assessment.[27,29,30] Thus, a "benchmark" dose is defined as a dose that is estimated to correspond to an excess risk in the range of 1% to 10%. With this benchmark dose as a starting point, uncertainty factors (or linear extrapolation) are then employed to establish RfDs. In order to accommodate experimental variation, it is recommended that the lower confidence limit for the benchmark dose be used.[26,28] For developmental effects, fetal death and malformation rates often exceeds 1% in typical developmental studies, and replacing the NOAEL with the ED_{01} (1% benchmark dose) will often result in a lower human exposure guideline. On the other hand, the ED_{10} can be regarded as the LOAEL (lowest-observed-adverse-effect-level) that can be detected to show a significant dose effect. Thus, the lower confidence limit dose corresponding to excess risk between 1% and 10% (LED_p) is used as a benchmark dose. By use of an uncertainty factor of 10,000 (linear extrapolation) from a lower bound on the benchmark dose with an estimated risk of 10% (LED_{10}), the dose corresponding to a risk of less than 10^{-5} is estimated by $LED_{10}/10,000$. Calculation of a benchmark dose is illustrated below.

As illustrated in Section III, the distribution of the total number of (adverse) responses in a litter corresponds to a beta-binomial model,

$$P(y_{ij}) = \binom{n_{ij}}{y_{ij}} \frac{B(a_i + y_{ij}, b_i + n_{ij} - y_{ij})}{B(a_i, b_i)}$$

The probability of the adverse response at the dose d_i is $P(d_i) = \mu_i = a_i/(a_i + b_i)$. A parametric dose-response function is used to describe the relationship between the probability of response and the exposure level. Chen and Kodell[31] proposed a Weibull dose response model,

$$P(d_i) = \mu_i = 1 - \exp(-\beta_0 - \beta_1 d_i^w)$$

The power parameter on dose, $w > 0$, increases the flexibility in the shape of the dose response function. The benchmark dose corresponding to 10% excess risk (ED_{10}) is the dose that satisfies the equation

$$0.10 = \exp(-\beta_0) - \exp(-\beta_0 - \beta_1 d_i^w)$$

The estimate of ED_{10} is computed from the coefficient estimates of the fitted Weibull dose-response function. To estimate a lower confidence limit on the ED_{10}, Chen and Kodell[31] proposed using the likelihood confidence limit approach by solving this problem:

$$\text{minimize} \frac{\log[1 - 0.1\exp(\beta_0)]}{-\beta_1}$$

subject to the constraint

$$2(L_{max} - L_1) = 1.645^2$$

where 1.645 is the 95th percentile of the standard normal distribution (for a 95% confidence limit), L_{max} is the maximum value of the log-likelihood subject to no constraint, and L_1 is the maximum value of the log-likelihood subject to the 0.10 constraint.

B. Example

We use the hydroxyurea data to illustrate the procedure. The maximum likelihood estimates of the Weibull dose-response model are $\hat{\alpha} = 3.46 \times 10^{-2}, \hat{\beta} = 9.88 \times 10^{-11}$, and $\hat{w} = 4.24$. The point estimates and the 95% lower confidence limits of benchmark doses are

Benchmark dose	1%	5%	10%
MLE estimate	78.05	114.67	135.92
Confidence limit	49.15	84.62	107.51

Based on the result from the quasi-likelihood method in Table 17.4, the NOAEL is 100 ppm. This number is very close to the LED_{10} of 107.51 ppm. Benchmark doses can be calculated with software such as TOX_RISK[32] and ToxTools.[33]

VI. DISCUSSION

Statistical analysis of developmental endpoints presents a number of important issues. Developmental endpoints consist of both continuous and discrete outcomes. The analysis should account for the correlations induced by the clustering effects within litters. The litter-based ANOVA approach is the simplest method for data analysis. Once the data are normalized, the ANOVA can be applied directly as described. The ANOVA is also used in more complex experiments involving crossed (e.g., dose levels, replicates) and nested (e.g., litter effects) factors. The repeated measures ANOVA is used for the analysis of postnatal behavioral data. The ANOVA and conventional *post hoc* test procedures such as the Dunnett test and linear regression analysis are all available in standard statistical software packages.

The parametric likelihood-based and quasi-likelihood (generalization estimating equations) approaches are two alternative methods. The mixed-effects model is a general parametric approach to modeling continuous outcomes. The parametric model for the binary outcomes from fetal responses is described in terms of the sum of correlated binary variables resulting in a generalized binomial distribution. The distribution for the common developmental endpoints, such as the number of malformations per litter, is often an overdispersed binomial, but the distribution of sex combinations can be an underdispersed binomial. For count data, the negative binomial distribution has been used to model overdispersed count data, such as results from an ovarian toxicity experiment. However, the parametric approach has disadvantages in that it does not allow for underdispersed variations and its maximum likelihood estimates can be biased or numerically unstable. The quasi-likelihood approach is a semiparametric approach. It can be applied to modeling both over- and undervariation data. The estimates derived are generally more stable than the parametric maximum likelihood estimates. The quasi-likelihood approach does not provide inference on the dispersion parameter.

In a typical reproductive and developmental study, various endpoints are collected in order to obtain the maximum information from the study. As described, the standard approach for risk assessment of a compound has been based on the analysis of each reproductive and developmental endpoint separately. Statistical analysis often involves a large number of tests, both in terms of multiple comparisons among groups and multiple tests on many endpoints. Multiple testing procedures, e.g., Dunnett's test, are commonly applied to group comparisons, rather than to the analysis of test endpoints. In a single experiment, statistical tests are performed on 10 to 20 reproductive and developmental endpoints. Because of the conduct of a large number of statistical tests, the chance of false positive findings is considerably increased. For example, the overall false positive rate is about $0.40 \approx 1-(1-0.05)^{10}$ for tests of 10 independent developmental and reproductive endpoints, all at $\alpha = 0.05$.

Interpreting results of a reproductive or developmental experiment is a complex process, and there are risks of both false negative and false positive results. The relatively small number of animals used and the low incidence rates can cause reproductive or developmental effects of a compound not to be detected. Because of the large number of comparisons involved, there is also a great potential for finding statistically significant positive trends or differences in some endpoints that are due to chance alone. Furthermore, a significant result for an endpoint tested may be statistical, biological, or both. The issues of false positive and false negative error rates in testing multiple reproductive and developmental endpoints have not been addressed. The main consideration should be to maximize the power of identifying every significant endpoint, while controlling the overall Type I or familywise error rate. Several authors[34–38] have proposed joint modeling of multiple developmental outcomes. However, development of risk assessment methods for combined analysis of several reproductive and developmental outcomes will require further statistical research.

REFERENCES

1. Desu, M.M. and Raghavarao, D., *Sample Size Methodology*, Academic Press, New York, 1990.
2. Winer, B.J., Brown, D.R., and Michels, K.M., *Statistical Principles in Experimental Design*, 3rd ed., McGraw-Hill, New York, 1991.
3. Hamilton, L.C., *Regression with Graphics: A Second Course in Applied Statistics*, Duxbury Press, Belmont, CA, 1991.
4. SAS Institute Inc., *Procedures Guide, Version 8*, Cary, NC, 1999.
5. Dunnett, C.W., A multiple comparison procedure for comparing several treatments with a control, *J. Amer. Stat. Assoc.*, 50, 1096, 1955.
6. Williams, D.A., The comparison of several dose levels with a zero dose control, *Biometrics*, 28, 519, 1972.
7. Tukey, J.W., Ciminera, J.L., and Heyse, J.F., Testing the statistical certainty of a response to increasing doses of a drug, *Biometrics*, 41, 295, 1985.
8. Barlow, R.E., et al., *Statistical Inference Under Order Restrictions*, John Wiley, New York, 1972.
9. Lehmann, E.L., *Nonparametrics: Statistical Methods Based on Ranks*, Holden-Day, San Francisco, 1975.
10. Dempster, A.P., et al., Statistical and computational aspects of mixed model analysis, *Appl. Stat.*, 33, 203, 1984.
11. SAS Institute Inc., *SAS/STAT User's Guide, Version 8*, Cary, NC, 1999.
12. Liang, K.Y. and Zeger, S.L., Longitudinal data analysis using generalized linear models, *Biometrika*, 73, 13, 1986.
13. Carey, V. and McDermott, A., gee() function, StatLib Archives, Carnegie Mellon University, Pittsburgh, PA. (S-PLUS function obtainable via anonymous FTP from lib.stat.cmu.edu., 1996.)
14. Cochran, W.G., Some methods for strengthening the common χ^2 tests, *Biometrics*, 10, 417, 1954.
15. Armitage, P., Tests for linear trends in proportions and frequencies, *Biometrics*, 11, 375, 1955.
16. StatXact, Cytel Software Corporation, Cambridge, MA, 1992.
17. Jonckheere, A.R., A distribution-free k-sample test against ordered alternatives, *Biometrika*, 41, 133, 1954.
18. Williams, D.A., The analysis of binary responses from toxicological experiments involving reproduction and teratogenicity, *Biometrics*, 31, 949, 1975.
19. Williams, D.A., Extra-binomial variation in logistic linear models, *Appl. Stat.*, 31, 144, 1982.
20. Chen, J.J., Trend test for overdispersed proportions, *Biometrical J.*, 35, 949, 1993.
21. Chen, J.J., Ahn, H., and Cheng, K.F., Comparison of some homogeneity tests in analysis of overdispersed binomial data, *Environ. Ecolo. Stat.*, 1, 315, 1996.
22. Lawless, J.F., Negative binomial and mixed Poisson regression. *Canadian J. Stat.*, 15, 209, 1987.
23. Breslow, N.E., Extra-Poisson variation in log-linear model, *Appl. Stat.*, 33, 38, 1984.
24. Chen, J.J. and Ahn, H., Fitting mixed Poisson regression models using quasi-likelihood methods, *Biometrical J.*, 38, 81, 1996.

25. Munro, I. and Krewski, D., Risk assessment and regulatory decision making, *Food Cosmet. Toxicol.*, 19, 549. 1981.
26. Kimmel, C.A. and Gaylor, D.W., Issues in qualitative and quantitative risk analysis for developmental toxicology, *Risk Analysis*, 8, 15, 1988.
27. Crump, K.S., A new method for determining allowable daily intakes, *Fundam. Appl. Toxicol.*, 854, 1984.
28. Krewski, D. and Van Ryzin, J., Dose response models for quantal response toxicity data, in *Statistics and Related Topics,* George, M., Dawaon, D., Rao, J.N.K., and Saleh, E., Eds., North-Holland, Amsterdam, 1981, p. 201.
29. U.S. Environmental Protection Agency, Proposed Guidelines for Carcinogen Risk Assessment: Notice, *Fed. Regist.*, 56, 17995, 1996.
30. Gaylor, D.W., et al., A unified approach to risk assessment for cancer and noncancer endpoints based on benchmark doses and uncertainty/safety factors, *Regulat. Toxicol. Pharmacol.,* 29, 151, 1999.
31. Chen, J.J. and Kodell, R.L., Quantitative risk assessment for teratological effects, *J. Am. Stat. Assoc.*, 84, 966, 1989.
32. ICF/Clement, TOX_RISK, Ruston, LA. 1989.
33. ToxTools, Cytel Software Corporation, Cambridge, MA, 1999.
34. Lefkopoulou M, Moore D., and Ryan L.M., The analysis of multiple correlated binary outcomes: application to rodent teratology experiments, *J Am. Stat. Assoc.*, 84, 810, 1989.
35. Chen, J.J., et al., Analysis of trinomial responses from reproductive and developmental toxicity experiments, *Biometrics*, 41, 47, 1991.
36. Catalano, P.J. and Ryan, L.M., Bivariate latent variable models for clustered discrete and continuous outcomes, *J. Am. Stat. Assoc.*, 87, 651, 1992.
37. Ryan, L.M., Quantitative risk assessment for developmental toxicity, *Biometrics*, 48, 163, 1992.
38. Chen, J.J., A malformation incidence dose response model incorporating fetal weight and/or litter size as covariates, *Risk Anal.*, 13, 559, 1993.

CHAPTER 18

Quality Concerns for Developmental and Reproductive Toxicologists

Kathleen D. Barrowclough and Kathleen L. Reed

CONTENTS

I. Introduction 714
II. Good Laboratory Practice Standards 714
    A. Background 714
    B. Regulations 714
    C. Regulations Specific to Electronic Records and Electronic Signatures 715
    D. Meeting 21 CFR Part 11 Requirements 715
    E. Current Trends 716
III. The GLP Compliance Program 716
    A. Overview 716
    B. Roles and Responsibilities 717
        1. Management 717
        2. Quality Assurance 718
        3. Study Directors 719
IV. Auditing Developmental and Reproductive Toxicity Studies 719
    A. Overview 719
    B. Best Practices 720
        1. Protocol Review 720
        2. In-Life Phase Audits 720
        3. Records and Report Audit 725
V. Quality Concerns and Quality Control Measures 727
    A. Overview 727
    B. Prestudy 727
    C. In-Life 727
VI. Conclusions 730
Acknowledgments 730
References 731

I. INTRODUCTION

In today's business and regulatory climate the ability to generate quality products in the most efficient manner is critical to maintaining competitiveness. Regulators must rely on the scientific community to provide the best possible data for input into their decision-making process. Those responsible for submitting applications to the regulatory agencies must continually strive for the most efficient way to develop a quality product. This chapter will

- Review the Good Laboratory Practice Standards (GLPs), which describe minimal standards for conducting nonclinical laboratory studies that support or are intended to support applications for research or marketing permits for products regulated by the Food and Drug Administration (FDA) or the Environmental Protection Agency (EPA).
- Provide guidance for developing and managing an effective GLP compliance program.
- Discuss specific procedures for auditing developmental and reproductive studies.
- Summarize quality concerns as solicited from a number of toxicologists, study personnel, and quality assurance unit (QAU) personnel from multiple organizations.
- Discuss recommendations for addressing the quality concerns.

We hope that the information presented herein will provide the developmental and reproductive toxicologist with useful tools to consider during all aspects of a study, from protocol preparation to final report issue.

II. GOOD LABORATORY PRACTICE STANDARDS

A. Background

In the mid-1970s, inconsistencies and unacceptable as well as fraudulent practices were uncovered in several studies that had been submitted to the FDA.[1] The kinds of issues that surfaced had the potential to have a detrimental, if not harmful, effect on the consumer.

Because of the FDA's commitment to and responsibility for protecting the consumer, the agency promulgated the FDA Good Laboratory Practice for Nonclinical Laboratory Studies, finalized in 1979 and revised in 1984 and again in 1987.[2] By 1983, the EPA had promulgated the EPA Toxic Substance Control Act (TSCA) and Federal Insecticide, Fungicide, Rodenticide Act (FIFRA) GLPs, both of which have since been revised.[3,4] The EPA's TSCA and FIFRA GLPs were intended to conform as closely with the FDA GLPs as possible.[5]

Be aware that the issues prompting these responses by the federal regulatory agencies were not simply minor judgmental oversights with little consequence. There were occurrences of underqualified personnel and supervision, selective reporting, poor animal care procedures, and inadequate recordkeeping and missing data, sometimes a result of fraudulent activity.[3]

Over the years several international agencies have developed regulations or guidelines that emulate the U.S. GLPs. For example, the Organization of Economic Cooperation and Development (OECD) Principles of GLP are based on the FDA GLPs and are similar though not identical to them in content.[1]

B. Regulations

While the FDA and the EPA TSCA and FIFRA GLPs are not identical, there are enough similarities among all three that if one adheres to the principles articulated by one GLP standard, very little adjustment will be required to be compliant with another.

If one considers the issues that prompted promulgation of GLPs, it is not difficult to envision the major thread running through the regulations – *documentation*:

- Plan it well — protocol requirements
- Conduct it well — standard operating procedures, documentation requirements
- Report it well — reporting requirements

Other GLP requirements include:

- Personnel training
- Study director responsibilities
- QAU responsibilities
- Adequacy and operation of facilities
- Calibration and maintenance of equipment
- Handling and characterization of test, control, and reference substances
- Retention of records

C. Regulations Specific to Electronic Records and Electronic Signatures

With the coming of the "computer age" came a desire to take advantage of the increased efficiencies and reduced opportunities for error in the generation and manipulation of data. What better way to accomplish this than by computerizing one's processes?

Unfortunately, in the early 1990s when this proliferation of electronic record-keeping was occurring, there were no regulations to provide guidance for the industry or to provide assurance to regulatory agencies that electronic records were as reliable as manually created records or that electronic signatures were irrefutable. Along came the Good Automated Laboratory Practices (GALP)[6] issued by the EPA in an effort to provide such guidance to regulated industry. Unlike GLPs, the GALPs are guidance for industry, rather than regulation.

In 1997, FDA promulgated the Electronic Records, Electronic Signatures Rule (21 CFR Part 11).[7] This rule has not been without its controversies and issues. The regulated pharmaceutical industry, while striving to achieve compliance with 21 CFR Part 11, is not considered by most to have reached this goal. The onset of 21 CFR Part 11, coupled with the increasing automation found in testing entities, raises a new issue for the developmental and reproductive toxicologist: Are all the systems validated, and are there appropriate controls to assure integrity of data in automated data collection systems?

FDA is not alone in its concern for the integrity of electronic records. The EPA has proposed a Cross-Media Electronic Reporting and Record-Keeping Rule (CROMERRR),[8] which at the time of this writing has not been published as a final rule.

D. Meeting 21 CFR Part 11 Requirements

The pharmaceutical industry struggled with developing 21 CFR Part 11 compliant programs. On the one hand, there was a lack of available technology to meet the demands of the rule. However, the primary focus for industry in 1999 was the potential for significant glitches in computer systems when the new year arrived. It was believed that internal system clocks based on two digit years (e.g., 97, 98, 99) might not identify "00" in a manner that allowed computer operations dependent on dates to function normally. This concern, felt around the world, was labeled "Y2K." The pharmaceutical industry, like most industry was devoting significant resources to assuring Y2K effects would not halt or disable business practices.

Once the Y2K dilemma had come and gone, the pharmaceutical industry, as well as the FDA, began to tackle 21 CFR Part 11 in earnest. Recognizing the need to provide guidance for industry,

FDA issued a Compliance Policy Guide in 1999. The purpose of the guide was to provide guidance for both industry and regulators regarding what criteria would be used in making decisions on whether to pursue regulatory actions.[9]

FDA decisions would be made based on the following:

- Nature and extent of Part 11 deviations
- Effect on product quality and data integrity
- Adequacy and timeliness of planned corrective measures
- Compliance history of the establishment[9]

Industry soon discovered that what FDA wanted to see was a good-faith plan, describing what steps would be taken to become compliant and establishing milestones that would measure progress. Some potential tasks were:

- Develop an overall strategy and action plan
- Develop policies and procedures that address 21 CFR Part 11 requirements, including the requirement for validation
- Train personnel on the meaning of electronic signatures and hold them accountable
- Conduct an inventory of existing systems (referred to as legacy systems)
- Prioritize the systems
- Conduct evaluations to determine the Part 11 gaps in those existing systems
- Develop plans to address the gaps
- Revalidate systems, if warranted

There is no "one way" to establish a 21 CFR Part 11 compliance program; however, a methodical approach that identifies and prioritizes the tasks may make this overwhelming project more manageable.

E. Current Trends

It is our belief that GLPs have played a significant role in improving consistency in the quality of data submitted to regulatory agencies. Not only have they established standards by which the scientist must conduct research, standards beyond scientific principles, but also they have provided a mechanism for research facilities to establish procedures for monitoring themselves. GLPs assign responsibility for overall study conduct to the study director (SD). He or she must assure that the work is conducted properly, meets the GLP requirements, is documented clearly, is conducted with well-maintained and appropriately calibrated equipment, and results in reliable, trustworthy data.

III. THE GLP COMPLIANCE PROGRAM

A. Overview

An effective GLP compliance program needs a proactive QAU and strong management support. Assuring management that studies have been conducted in compliance with the appropriate GLP regulations is but one facet of the QAU's responsibility. "The Quality Assurance Unit is the only unit in the laboratory that has a total view of the personnel, facilities, equipment, and processes that are occurring. From this perspective, the QAU can analyze the work and paper flow, identify defective practices and procedures, and recommend suitable corrections."[10] In order to realize the greatest benefit, the GLP compliance program incorporates GLP regulations and draws from the organization's experiences (metrics) to define processes that build quality and efficiency into all phases of a study.

B. Roles and Responsibilities

1. *Management*

Management must clearly define its expectations for the overall organization. The GLPs provide minimal requirements but leave it up to an organization to determine how it will set the standards. Minimally, testing facility management must

- Designate a study director before the study is initiated
- Replace the study director promptly when necessary
- Assure the presence of a QAU
- Assure that test, control, and reference substances or mixtures have been appropriately tested for identity, strength, purity, stability, and uniformity, as applicable
- Assure that personnel, resources, facilities, equipment, materials, and methodologies are available as scheduled
- Assure that personnel clearly understand the functions they are to perform
- Assure that any deviations from the GLP regulations reported by the QAU are communicated to the study director and corrective actions are taken and documented[2–4]

One approach in setting quality expectations that are in alignment with the GLP regulations is to develop high-level policies, which may be referred to as "quality directives." These policy statements present the goals, expectations, requirements and standards of practice. They provide direction and guidance for meeting regulatory requirements as well as for supporting good business practices and are signed by the highest level of management in an organization. In addition, quality directives provide a mechanism for assuring that the procedures used throughout the organization are consistent. An added benefit is that personnel may now move from one assignment to another without the additional burden of learning new policies. Policies that may be addressed in quality directives are:

- **Creating, Revising, and Retiring Standard Operating Procedures (SOPs):** "A testing facility must have standard operating procedures in writing setting forth study methods that management is satisfied are adequate to ensure the quality and integrity of the data generated in the course of a study."[2–4] This quality directive provides consistent guidance for the creation, revision, and retirement of SOPs, including but not limited to formatting, numbering, and approval structure.
- **Best Practices for Meeting Electronic Records and Signature Requirements:** Electronic records and signature requirements may be too extensive to address in one policy document. One approach is to issue two or three separate documents:
  - Define the organization's policy on electronic signatures used in the electronic records subject to 21 CFR 11, other applicable regulations, and internal policies. Explain the requirements, components, and controls specific to electronic signatures.
  - Address why validation is required, what needs to be validated in the GLP environment, when validation needs to occur, and who is responsible for the validation process. The directives do not address how to validate, but do stress the importance of documentation for the process and maintenance of the documentation.
- **Study Director and Principle Investigator Review of Data:** Define expectations for regular review of data for completeness, accuracy, and organization to ensure that the records demonstrate proper study conduct, adherence to the protocol, and compliance with applicable SOPs and GLPs.
- **Training Records:** GLPs require that each individual engaged in the conduct of or responsible for supervision of a study shall have the education, training, and experience, or combination thereof, to enable that individual to perform the assigned functions.[2–4] This Directive provides the guidance needed to assure that the appropriate documentation exists to demonstrate compliance with the GLP requirement.

As quality programs are developed and a need for management expectations is identified, other quality directives may be created. Additional topics that may be addressed by policy documents include data handling, security, and a metrology program.

While some individuals are concerned about yet another set of documents to be managed and audited against, the benefits from having an organization that clearly understands management expectations far outweigh the perceived inconvenience of additional paperwork. In addition, having management approved policies that set expectations for a quality-driven, GLP compliant organization goes a long way toward assuring customers of management's commitment to assuring the quality of the work product.

2. Quality Assurance

GLPs require that management "assure the presence of a QAU."[2-4] The responsibilities of the QAU include:

- Assure management that facilities, personnel, practices, and records are in compliance with regulations
- Maintain a master schedule sheet of studies
- Inspect each nonclinical study at intervals to assure compliance
- Report findings to the study director and management
- Review the final report to assure that it accurately reflects the raw data
- Prepare and sign a QA statement in the final report

The regulations also require that procedures used by the QAU to fulfill its responsibilities be in writing. An added benefit of having these written procedures is assuring consistency among the auditors. One way to document the QAU procedures is to utilize a mixture of QAU SOPs and guidance policies. SOPs define the practices that must be adhered to for each study reviewed by the QAU. Potential SOP topics include:

- Activities and responsibilities of the QAU
- Maintenance of the QAU master schedule
- Maintenance of QAU files
- Routine audit functions (protocol, conduct, records, reports)
- Internal facility inspections
- Distribution of audit reports to principal investigators, study directors, and management

Guidance policies may contain information on how to use a specific program (i.e., user's manual for the QAU master schedule) or how to conduct regulatory or customer inspections. Or they may provide guidance on auditor training. Guidance policies are not as prescriptive as SOPs and afford greater flexibility.

GLPs define the role of the QAU as "monitoring each study to assure management that the facilities, equipment, personnel, methods, practices, records, and controls are in conformance with the regulations in this part."[2-4] The QAU can and should play a larger role in the success of a business organization. Regulators, technical management, and the scientific community have recognized the enormous potential of the QAU. "The role of the quality assurance (QA) auditor should transcend simple verification of whether columns of numbers are added correctly or all of the cross-outs have been duly initialed and dated. The QA professional should be capable of bringing to light considerations of scientific quality as well as data integrity."[11] This sentiment is echoed by Paul Lepore, formerly of the FDA.[10] Consequently, many organizations today include the following QAU roles and responsibilities in the GLP compliance pogram:

- Liaison program
- Compilation, analysis, and report back of metrics
- GLP training for the technical staff (introductory and ongoing)
- Consultation
- Mentoring

QAU auditors interact on a day-to-day basis with the individuals involved in study work, and while it is important for auditors to pay attention to what is happening with the current study, they should be continually looking at the work processes from a much broader perspective. As reported by Caulfield, "Management expects the QAU to serve as the primary conduit through which they receive information regarding organizational needs like training, equipment quality and availability, inadequacies in facility, or gaps in study administration."[12]

Though the QAU is primarily responsible for monitoring studies and facilities for GLP compliance, "it is clearly the QAU who is expected to provide much of the informational raw materials needed for them [management] to make informed decisions and to take any needed action."[12] The most effective organization will expect the QAU to identify and recommend changes to existing processes that will result in increased efficiency and effectiveness.

3. Study Directors

The study director is defined as "the individual responsible for the overall conduct of a study" and is responsible "for the interpretation, analysis, documentation and reporting of results, and represents the single point of control."[2-4] The study director:

- Assures the protocol is approved and followed. All changes in or revisions of an approved protocol and the reasons for the changes and revisions shall be documented, signed by the study director, dated, and maintained with the protocol.
- Assures all data are accurately recorded and verified.
- Assures unforeseen circumstances that may affect the quality and integrity of the study are noted when they occur, and corrective action is taken and documented.
- Assures test systems are as specified in the protocol.
- Assures all applicable GLPs are followed.
- Assures all raw data, documentation, protocols, specimens and final reports are archived.
- Assures all deviations from SOPs in a study are authorized by the study director and documented in the raw data.[2-4]

While the list above may not seem exhaustive, it must be pointed out that the study director must assure that *all applicable GLPs are followed*. This is a daunting task, and one that the study director cannot hope to successfully accomplish without assistance. "It is incumbent upon the QAU to assist them [study directors] in meeting the formidable responsibilities that have been placed on them by the regulations."[12] For instance, the study director needs to know that those personnel assigned to a study are adequately trained; however, management is responsible for assuring personnel clearly understand the functions they are to perform.[2-4] Additionally, the QAU can assist by reviewing training documentation for personnel observed working on a particular study. Management may designate supervision to be responsible for an annual review of training documentation and assignment of qualified personnel to a particular study function.

Additionally, procedures should be in place to assure the study director that the facilities are appropriate, equipment functions as required, reagents and solutions are appropriately labeled, and that appropriate SOPs are in place. The QAU should utilize its audit program to assist the study director in determining the compliance status of the resources to be used on his or her study and to provide recommendations for meeting the GLP requirements.

IV. AUDITING DEVELOPMENTAL AND REPRODUCTIVE TOXICITY STUDIES

A. Overview

The QAU is required to:

- Inspect each study at intervals adequate to ensure the integrity of the study.
- Review the final report to assure that such report accurately describes the methods and standard operating procedures, and that the reported results accurately reflect the raw data of the study.[2-4]

Nowhere in the regulations does it indicate how the QAU is to accomplish these tasks. Therefore, the QAU should:

- Define audit functions in SOPs.
- Determine an audit plan.
- Decide who the responsible individuals are and how audit findings will be communicated and acted upon.

The following sections provide one way of managing the audit function. The information presented incorporates best practices from various organizations. Keep in mind that as changes occur in testing guidelines and technology, the most effective QAU continually looks for ways to improve its audit practices — always looking for the "biggest bang for the buck."

B. Best Practices

1. Protocol Review

Ideally, the protocol is provided to the QAU as a draft document. The auditor reviews the protocol to assure that all items required by the GLPs have been incorporated and addressed. While responsibility for the content of the protocol is clearly the responsibility of the study director, some organizations expect the QAU to verify also that the protocol incorporates testing guideline requirements. In either case, use of a standard checklist may facilitate the review. An example of a protocol checklist for GLP compliance is provided in Table 18.1.

Once the review is complete, the auditor works with the study director to resolve any issues. A formal audit report requiring the study director to respond to findings may be issued if warranted. However, GLPs do not specifically require that an audit report be issued for the protocol review. Since the signed protocol needs to be submitted to the QAU, a review of that document provides an indication of whether issues have been addressed appropriately. Should the QAU not have an opportunity to review the draft document and should changes be required to the issued protocol, a protocol amendment will be requested.

There may be instances where a protocol needs to be issued prior to having all the requisite information available. For instance, a range finding study to determine dosage levels is ongoing, but the protocol for the main study must be issued to ensure that the resources necessary for the study are available. It is a requirement that the protocol include "each dosage level, expressed in milligrams per kilogram of body or test system weight or other appropriate units, of the test, control, or reference standard to be administered and the method and frequency of administration."[2-4] In this example, it is appropriate to indicate in the protocol that an amendment will be issued to address dosage levels once the range finding study has been completed.

Another example of a protocol required item that may not be known at the time of protocol development or issue is the source of supply for the test system. In this case, it is acceptable to indicate that the exact source of supply will be documented in the study records. Care should be taken against using this approach indiscriminately. It is not acceptable to have a protocol riddled with the "to be addressed in the study records" statement.

2. In-Life Phase Audits

The QAU "inspects each study at intervals adequate to ensure the integrity of the study."[2-4] Specifics of what constitutes adequate intervals are not defined in the GLPs. The GLPs also do not define

Table 18.1 Protocol review checklist

I.	General	Yes	No	Comment
1.	Descriptive title of the study			
2.	Statement of purpose of the study			
3.	Name and address of the sponsor			
4.	Name and address of the testing facility			
5.	Proposed experimental start and termination dates			
6.	Records to be maintained			
7.	Area for the signature and date of the study director			
8.	Sponsors date of approval and/or area for sponsor signature			
II.	**Test System (w/a[a])**	**Yes**	**No**	**Comment**
9.	Species (w/a; animal-rat, plant-apple tree, or soil- silt loam)			
10.	Strain/substrain (w/a; Sprague-Dawley, McIntosh, or Keyport)			
11.	Source of supply (w/a)			
12.	Number (w/a)			
13.	Sex, age, body weight range (w/a)			
14.	Procedure for identification of the test system (pot, sample ID)			
15.	Justification for selection of the test system			
III.	**Test and Control Substance**	**Yes**	**No**	**Comment**
16.	Identification of the test/control/reference substance			
17.	Identification of solvents/emulsifiers/suspensions and carriers to be used with the test, control or reference substance(s)			
18.	Specifications for acceptable levels of contaminants			
19.	Route of administration and reason for its choice			
IV.	**Methods (Study Conduct)**	**Yes**	**No**	**Comment**
20.	Description of the experimental design, including methods for control of bias			
21.	Each dosage level of the test, reference or control substance to be administered, expressed in the appropriate units of measure			
22.	Method and frequency of the administration of the test, control, or reference substance			
23.	Type and frequency of tests, analyses, and measurements			
24.	Proposed statistical methods			
V.	**Feed and Water for Plant and Animal Studies (w/a)**	**Yes**	**No**	**Comment**
25.	Description and/or identification of diet used (w/a)			
26.	Specifications for acceptable levels of contaminants (w/a)			

[a] When applicable.

Table 18.2 Critical phases

Dose and diet preparation and sampling for analysis
Dose and diet analysis
Dose administration and exposure (test system identification and housing)
Data collection (body weights, clinical observations, food consumption)
Mating
Culling procedures
Blood collection
Vaginal lavage and slide preparation (estrous cycle evaluation)
Epididymal and testicular sperm counts, morphology, motility
Pup data collection (weighing and sexing)
Vaginal patency, balanopreputial separation, anogenital distance
Necropsy and fetal examination (visceral and skeletal)
Neurotoxicology endpoints (observe one or more):
   FOB* and motor activity
   Auditory startle
   Water maze
   Brain collection and measurement of weights
   Neuropathology

* Functional observation battery.

"critical phases." Therefore, the QAU works with the technical staff to determine adequate intervals and critical phases for each study type. Critical phases essential to the quality and integrity of the study are identified through dialogue between the study director and the QAU. The technical staff and QAU routinely review this list to assure that changes in testing guidelines or study design have not affected the phases considered critical to a particular study. Table 18.2 identifies critical phases typical for developmental or reproductive studies. Whether a phase relates to a particular study is driven by the protocol.

When determining adequate intervals for critical phase inspections, the following should be considered:

- Duration of experimental phase of the study
- Complexity of the study
- Frequency with which the study is conducted
- Experience of the personnel working on the study

One critical phase per study is audited for studies that last less than 28 days and are routine in nature. The procedure for assigning the critical phase to be inspected ensures that all critical phases of a particular study type are subject to inspection within a reasonable time frame (e.g., each critical phase for a particular study type is reviewed at least once per year). The identity of the critical phase, date of inspection, identity of auditor, and dates the findings are reported to the study director and management are entered into the audit log for the subject study. In addition to the critical phase inspection, the protocol, records, and final report are also reviewed. Additional critical phases may be reviewed at the discretion of the study director. The auditor may choose to review additional phases if the personnel working on the study are fairly inexperienced or if the auditor is looking to increase his or her understanding of the particular study type.

Studies with duration of 28 days or greater and those studies that are nonroutine in nature follow a different auditing scheme. Each critical phase is inspected at least once during the course of the study. Inspections should be spread throughout the course of the study. For example, having conducted all critical phase inspections during the first 3 months of a 2-year study without any additional monitoring would not be considered monitoring at "adequate intervals."

Auditors review the protocol and any pertinent SOPs prior to conducting the phase inspection. Checklists are used to provide guidance for the auditor and to ensure consistency between auditors. Table 18.3 is an example of an inspection checklist for test material preparation and administration.

Table 18.3 Phase inspection checklist for test material preparation and administration

Training Records		Yes	N/A	No	Comment
1.	Personnel have been trained; training is documented.	☐	☐	☐	☐
2.	Personnel are wearing the appropriate protective clothing.	☐	☐	☐	☐
SOPs, Protocol, and Amendment(s)		**Yes**	**N/A**	**No**	**Comment**
3.	SOPs are readily available; the protocol and SOPs are followed.	☐	☐	☐	☐
4.	Deviations from protocol or SOP are documented and authorized by the study director. Protocol amendments are issued as necessary.	☐	☐	☐	☐
5.	Unforeseen circumstances are noted when they occur; corrective action is taken and documented.	☐	☐	☐	☐
Equipment		**Yes**	**N/A**	**No**	**Comment**
6.	Equipment is adequately calibrated and maintained according to the appropriate SOP. Documentation is present.	☐	☐	☐	☐
7.	Appropriate instrument qualification and software validation has been conducted for electronic data collection systems.	☐	☐	☐	☐
8.	Users have been trained in the use of the equipment and software; training is documented; users know whom to contact if there is a system problem.	☐	☐	☐	☐
9.	Correct computer operation procedures are followed (login, security, etc).	☐	☐	☐	☐
General		**Yes**	**N/A**	**No**	**Comment**
10.	All data are recorded accurately, directly, promptly, and legibly, signed and dated. Audit trails accompany any changes to the data.	☐	☐	☐	☐
11.	All reagents and solutions are labeled to indicate identity, titer, or concentration, storage requirements, and expiration date.	☐	☐	☐	☐
12.	Beakers, vials, and storage containers of dose and diet formulations are properly identified.	☐	☐	☐	☐
13.	The storage container for the test or control material is appropriately labeled.	☐	☐	☐	☐
14.	Test material preparation forms and usage logs are properly completed; all required information is recorded correctly.	☐	☐	☐	☐
15.	Each test system is uniquely identified.	☐	☐	☐	☐
16.	Test system housing units are appropriately labeled.	☐	☐	☐	☐
17.	Feed bin stored in the animal room is properly labeled; feed is within its expiration date.	☐	☐	☐	☐
18.	The animal room is clean and well maintained. Accountability, rack change, and stock records are adequately completed.	☐	☐	☐	☐
19.	Fish culture, feeding records, and purchase information are available.	☐	☐	☐	☐

During the critical phase inspection, study personnel, including the study director, should be able to discuss the procedures that ensure the following:

- Is there evidence that personnel have been trained and that the training is documented?
- Are SOPs readily available?
- Are the protocol and applicable SOPs being followed?

- Have deviations from the protocol or SOPs been documented and authorized by the study director?
- Are protocol amendments issued as they are necessary?
- Are unforeseen circumstances noted when they occur, and is corrective action taken and documented?
- Has equipment been adequately calibrated and maintained according to the appropriate SOP? Is documentation present?
- Has the appropriate instrument qualification or software validation been conducted for electronic data collection systems, and is there documentation of user training?
- Are all data recorded accurately, directly, promptly, and legibly; are entries signed and dated; do audit trails accompany any changes to the data?

"The QAU role is to assist the team with ensuring that sufficient documentation exists to allow for full study reconstruction and to help study areas prepare for potential regulatory inspections."[13] Clear, concise documentation is essential for providing the tools needed to accurately reconstruct a study. Regulators will expect study personnel to demonstrate proficiency in the tasks they are performing as well as exhibit comprehensive understanding of the reasons for performing the tasks.

Where appropriate, the items listed above are reviewed during each critical phase inspection, regardless of the phase assigned. Examples of items specific to a particular critical phase are discussed below.

Critical Phase	Audit Items
Dose or diet preparation and sampling for analysis	Labeling of containers adheres to GLP requirements.
Dose or diet analysis	Expired reagents and solutions are not in use.
	Appropriate equipment (e.g., balances, pipettes) is clean, calibrated, and appropriately maintained; documentation (maintenance and calibration logs) to support routine and nonroutine maintenance, and calibration requirements is readily available.
	Measurements are accurately recorded in the study records, including units of measure.
	Mixing times, where appropriate, are accurately recorded in the study records.
Dose administration and exposure (includes test system identification and housing)	Test system is appropriately and uniquely identified.
	Housing units are uniquely identified.
	Test system identification is verified upon removal from and return to a housing unit, and prior to administering the dosage and diet.
	Feed containers (if appropriate) are identified (minimally) by dose level.
	Amount of dose or diet is accurately recorded.
Data collection (body weights, clinical observations, food or water consumption)	Balance is calibrated, and calibration is documented.
	Test system identification is verified upon removal from and return to a housing unit.
Mating, breeding, and culling procedures	Appropriate terminology is used for clinical observations.
	Care is taken not to mate siblings.
	Appropriate procedures for determining the gestation period are followed.
Blood collection	Tubes are labeled with the appropriate information (e.g., animal ID, dose level, date of collection, etc.).
	Injection site and size of needle and syringe are documented.
	Where required, blood is mixed in tubes appropriately.
	Appropriate chain-of-custody documentation for sample tracking is available.
	Animal identification is verified prior to blood collection.
Vaginal lavage and slide preparation	Slide is labeled uniquely with appropriate study information.
Necropsy and fetal examination	Animal identification is verified against record.
	Balance is calibrated, and calibration is documented.
	Fetuses are identified with dam code.
Neurotoxicology endpoints (functional observation battery [FOB] and motor activity, auditory startle, water maze, brain collection and weight measurement, neuropathology)	Test is conducted in a way that precludes tester from knowing dose level.
	Appropriate controls are in place to minimize disturbance (white noise for motor activity).
	Equipment is appropriately calibrated, and calibration is documented.
	Animal identification is verified.

Critical Phase	Audit Items
Determination of epididymal and testicular sperm counts and morphology	Slides are labeled appropriately. Timing and temperature are consistent with protocol and/or SOP requirements. Equipment is appropriately calibrated, and calibration is documented. Expired reagents and solutions are not in use.
Pup data collection (determination of sex, weighing)	Balance is calibrated, and calibration is documented. Dam identification is verified prior to data collection.

3. Records and Report Audit

Auditors are required to "review the final study report to assure that such report accurately reflects the methods and standard operating procedures, and that the reported results accurately reflect the raw data of the study."[2-4]

For the purposes of this chapter, we assume the report is formatted in the following manner:

Text — multiple sections:
    GLP compliance statement
    Summary
    Introduction
    Methods and materials
    Results and discussion
    Conclusion
Tables — generally a summary of the data presented in the appendixes
Appendixes — consisting of individual animal data (e.g., body weights, clinical observations, food consumption, etc.)

In order to review the report in the most efficient manner, an organized approach is recommended (see Table 18.4 for a final report checklist example):

Review the study records (raw data) to verify that all endpoints specified in the protocol have been documented clearly and concisely, and that the date of entry and identity of the individual making the entry are known.

If individual data are reported in appendixes, utilize a predetermined process to select and verify values from the report back to the raw data. For instance, when using validated, automated data collection systems, it might be sufficient to spot check a random sampling of data points and review the data to ensure that any out-of-the ordinary or unexpected notes and comments are captured in footnotes. However, if the data were collected manually, a much higher level of review would be required. Many organizations require that as much as 100% of the data is verified for manually collected data. The QAU SOPs should indicate the process used to audit the data.

Verify summary tables to the appendixes and/or study records, as appropriate. Again, the appropriate level of data verification is defined in the QAU SOPs.

Verify any values mentioned in the final report against summary tables, appendixes, and/or raw data, as appropriate.

Check the following:

- Is the introduction consistent with the protocol?
- Are the methods as stated in the final report supported by the protocol, amendments, SOPs, and raw data?
- Is the information presented in the results and discussion section supported by the summary tables, appendixes, and raw data?
- Does the results and discussion section support the conclusion section?

Table 18.4 Final report checklist

Final Report		Yes	N/A	No	Comment
1.	Name and address of test facility	☐	☐	☐	☐
2.	Dates on which the study was initiated and completed	☐	☐	☐	☐
3.	Objectives and procedures as stated in the protocol and amendments	☐	☐	☐	☐
4.	Statistical methods employed for analyzing the data	☐	☐	☐	☐
5.	Identification of test and control materials (chemical name, unique sample no., chemical abstracts no. or code no., strength, purity, and composition)	☐	☐	☐	☐
6.	Stability of the test and control materials under the conditions of administration	☐	☐	☐	☐
7.	Methods used (e.g., generation in inhalation studies)	☐	☐	☐	☐
8.	Test system used (no. of animals, species, strain, and substrain, source, age, sex, body weight range)	☐	☐	☐	☐
9.	Methods of identification of test system (e.g., tattoo, cage card, etc.)	☐	☐	☐	☐
10.	Dosage, dosage regimen, route and duration of administration	☐	☐	☐	☐
11.	All circumstances that may have affected study quality or integrity	☐	☐	☐	☐
12.	Name of study director, names of other scientists or professionals, and names of all supervisory personnel involved in the study	☐	☐	☐	☐
13.	Description of the transformations, calculations, or operations performed on the raw data, a summary and analysis of the data, and a statement of conclusions drawn from the analysis	☐	☐	☐	☐
14.	Signed and dated reports of other scientists or professionals involved in the study (pathologist, clinical pathologist, statistician, chemist, ophthalmologist)	☐	☐	☐	☐
15.	Storage location of specimens, raw data, and the final report	☐	☐	☐	☐
16.	Dated signature of the study director	☐	☐	☐	☐
17.	GLP compliance page signed by the study director	☐	☐	☐	☐

- Is the information reported in the summary, results and discussion, and conclusion sections consistent?
- Are all items required by GLPs included in the final report?
- Is the GLP compliance statement accurate?

The time required to audit a final report depends upon a number of factors:

- Complexity of the study
- Experience of the auditor
- Experience of the study personnel
- Data collection — manual or automated
- Quality control (QC) processes

Use of validated, automated data collection systems has the potential to significantly reduce not only the time spent during the experimental portion of the study, but also the amount of time an auditor needs to review the data. Likewise, having personnel responsible for QC of the data and report before the package is sent to the QAU will have a positive affect on the time spent in review.

V. QUALITY CONCERNS AND QUALITY CONTROL MEASURES

A. Overview

Drawing from the experiences of scientists, technical support personnel, and quality assurance professionals, critical quality issues encountered during the conduct of developmental and reproductive toxicity studies have been identified. Based on "what makes sense" and "what has worked," several alternatives for diminishing the potential of these issues to affect data integrity are presented. In no way should these approaches be considered the only means for correcting deficiencies. While most scientists would agree that a well-designed study conducted by a well-trained staff has high potential for success, experience shows that it requires much more.

B. Prestudy

Concern	Approach
Lack of understanding of study protocol and SOPs by study personnel and supporting scientists.	Hold a prestudy meeting with study personnel to review protocol and SOPs. Set clear expectations around communication. Have a QC process that requires data checks on a regular basis and includes follow-through when consistent errors are observed.
Inaccurate database-driven collection intervals for weights, food consumption, and clinical observations.	Establish adequate software validation procedures. Involve the study director, statistician, primary study technician, and application administrator in the prestudy meeting. Study director should issue and communicate amendments or changes to the protocol in a timely fashion.

C. In-Life

Concern	Approach
Dosing or diet preparation errors.	SD should review and sign documentation for calculations. Use dated version control measures to assure that the most current procedure is used in diet or dosing preparation. Create a specialized, well-trained diet preparation or formulations group. Observe diet preparation to assure that all procedures are followed. Develop, follow, and document procedures for equipment cleaning that minimize potential for cross-contamination, e.g., mixer, polytron, mortar and pestle. Develop and follow appropriate measuring and pipetting techniques. Use of disposable equipment (e.g., pipettes) is another option. Analyze each batch of test substance (GLP requirement) and recalculate diet or dosing formulas. Include a process for checking animal identification before and after dosing. Develop a process for comparing test substance identification in protocol, study records, documentation accompanying the test substance, and the final report. Verify that all test substance storage containers are labeled according to GLP requirements — name, CAS number, batch number, expiration date, if any, and, where appropriate, storage conditions necessary to maintain the identity, strength, purity, and composition.[2-4] This can prevent use of expired, incorrect, or improperly stored test substances. Follow well-documented test substance accountability practices. Assure that apparatus used in continuous i.v. dosing has been adequately calibrated so that the correct dose is injected over the correct timeframe.

Concern	Approach
Gavage misdosings.	Study director and QC reviewers should be on the alert for increased incidences of misdosings and investigate to determine their cause. Determine whether it is a training issue or if there is a possible systemic effect causing hyperactivity that leads to misdosing. Pregnant animals can be more susceptible to toxic effects of the test substance or the dosing solution. Ensure that dosing solutions are neither too hot nor too cold. Provide specific instructions and assure that they are followed. Severe discrepancies in temperature of the dosing solution can cause pain to the animal. Inappropriate temperature of the test substance may cause a change in the physical or chemical properties of the test substance, affecting delivered dose or making the delivery difficult. Training about proper rate of administration for gavage feedings is critical. Adherence to the protocol and SOPs must be verified. Recertification training, especially when there has been a time lapse, is one way to assure retention of the needed skill.
Blood collection — inadequate handling of specimens and insufficient sample volume.	Personnel training is critical. Create a blood collection form that includes collection requirements, such as volume, handling conditions (e.g., dry ice), spin time, and spin temperature. Assure that the appropriate preservatives are in the tubes by including a checkpoint in the procedure. Properly maintain and calibrate the equipment. Document the appropriate size needle and syringe to reduce opportunity for animal injury due to trauma.
Equipment malfunctions, operator error, and improperly maintained equipment.	Develop adequate calibration and preventive maintenance procedures. There is a growing trend toward establishing metrology programs to manage instrument calibration and maintenance. These programs can be a cost-effective means of maintaining an inventory, leveraging preventive maintenance programs with vendors, and providing appropriate maintenance routines that reduce equipment downtime and enable consistent equipment performance. Develop a cleaning schedule that includes follow-up swipe tests for equipment such as diet mixers and blenders that have a high cross-contamination potential.
Clinical observations, especially during FOB testing.	To reduce variability in classifying observations, several trainees may observe the same animals and document their findings. Follow this by reconciling the differences over several trials until consistency among the trainees is reached. By recertifying at regular intervals, a high level of proficiency is maintained. Standardize terms used in describing clinical observations. Design the database using controlled vocabulary to ensure consistency. Minimize injury and skewed results in the water maze battery by documenting water temperature prior to the run. Prior to FOB testing, designate primary and secondary observers so that the observations are verified. Prior to water maze tests, verify that the maze has been set up using the correct goals, i.e., learning retention vs. latency.
Observation and reporting of weight loss.	When weight loss is observed, one should review the previous clinical observations, previous weight, and veterinarian and study director notes to determine whether anything has been documented that supports the potential for the weight loss. One way of correlating weight loss with clinical observations is to conduct the activities simultaneously. Check the feed and water supply systems for indications of malfunction. Check food consumption data and feed bowl to evaluate the test system's eating habits; diet aversion can be a real issue, generally in higher dose groups, but potentially in any group. Check the animal for injury or signs of illness. To assure appropriate measures are initiated, it is critical for the observer to bring all issues to the attention of the study director immediately.
Transcription errors.	Personnel training that includes acceptable standards for data documentation and built-in QC checks can reduce the number of transcription and inputting errors. Use of validated automated data collection systems can significantly reduce data errors, calculation discrepancies, and protocol deviations.

Concern	Approach
Inaccurate recording of gestational onset.	Sometimes a state of "pseudo" pregnancy is provoked as a result of rough handling during vaginal lavage. Personnel training and awareness and observation of personnel during the procedure reduce the incidence of such problems.
Because there is a potential for vaginal plugs (sign of successful mating) to be formed without a successful mating, other means to confirm mating should be considered. One can conduct a wet reading of the vaginal lavage slide to determine the presence of sperm, which is especially important if the plug was found in the cage rather than *in situ*.
Processing vaginal lavage samples for estrous cycle monitoring is one way to manage a study. Accurate labeling of slides is essential. Some organizations create a grid slide to capture samples over several days. One way to assure correlation between the estrous cycle dates for gestational onset (as determined from slide samples) and the date that was recorded in the database or notebook is to clearly mark the grid on the estrous cycle slide. For example, one could mark the slide with a "0" on the day that the plug was observed. Note: Verify that the mark will withstand contact with staining chemicals.
When checking cages for plugs in combination with conducting vaginal lavage for estrous cycle data, it is critical to check for plugs prior to conducting the lavage. Otherwise, the estrous cycle slide may be processed and result in a disparity between the time to mating data gathered via slides and the day "0" of gestation date entered when the plug was observed.
For rodents — confusion over which day the male should be removed from cohabitation.	Clearly define in the SOPs and/or protocol what constitutes the maximum duration of the cohabitation period, the criteria for evidence of mating, and when the male rodent should be removed from cohabitation. For example, the first day of cohabitation usually begins in the afternoon. The following morning, each female is examined for evidence of mating (a plug *in situ* or a sperm-positive vaginal lavage). When evidence of mating is observed, cohabitation is ended for that pair. Checking continues each morning until evidence of mating is observed. If no evidence of mating is evident, the cohabitation is continued (up to the maximum duration designated). For example, if the cohabitation period was defined as no more than a 7-d duration, and for seven consecutive morning checks no evidence of mating is observed, the male is removed from cohabitation.
Personnel training and study director involvement in the study will minimize errors.
Misdiagnosed male fertility effects.	Carefully identify and document the correct size restraint used for animals undergoing nose-only inhalation exposures. A too tight fit in the restraint can cause increased temperature, especially near the testes, resulting in hyperthermia with possible reduced fertility.
Artifacts on slides can cause inaccurate sperm count readings. Personnel training in appropriate techniques, maintaining slides in a manner that reduces potential for artifacts, and QC checks of slides combine to improve data quality.
When sperm sample slides are not read in a timely fashion, erroneous sperm counts may result. See that personnel are trained and focused when conducting the study. Temperature and timing are critical issues in sperm motility.
Test system issues.	If males are used over several matings, review the mating history to assure that the male has successfully mated and is not introducing bias into the study.
Confirm that the breeder colony has not gone beyond the age prescribed in the protocol.
Inaccurate pup litter weights, pup mix-ups, and disproportionate weights for culled litters.	Define the randomization process and follow it.
Check the weigh bucket after weighing litters. Occasionally, pups may adhere to the inside of the weigh bucket. If the pup is not noticed, there is potential for intermingling with the next weighed litter.
If pups were mis-sexed, assure that appropriate editing techniques are used. Sex must be changed in all applicable locations, e.g., database, notebooks.
Errors during fetal observation and dissection.	Personnel training and recertification, along with the previously discussed approach (whereby trainees observe the same specimen but record their findings separately and then reconcile them), is an excellent way to reduce variability across workers.

VI. CONCLUSIONS

Developmental and reproductive toxicity studies can be fragile. We have described specific approaches to handling known issues that are critical to the success of the study. However, it all comes down to having sound scientific practices, which in essence are tantamount to adhering to GLPs.

As we mentioned earlier, there is no *one way* to conduct a successful developmental or reproductive toxicity study. Success has been achieved by following GLPs and by implementing procedures that facilitate adherence to quality standards. Being open to new techniques and benchmarking with other scientists are valuable tools for the developmental and reproductive toxicologist.

Pendergast has described the rewards of a quality study as follows: "We promise that sponsors who produce high quality data with the highest integrity will be rewarded with reduced reviewing time, quicker approvals, and a reduced threat of private lawsuits. In the end, good quality data serves the interest of both the industry and the public."[14]

The role of the study director is to manage the study in a way that results in a high quality product that can be used in making the right decisions regarding substances with a potential to affect human health or the environment. One of the questions we have posed to study directors, study personnel, and quality assurance auditors is: "What do you believe are the most significant factors impacting study quality?" The top three answers:

1. Study director involvement in the study
2. An effective quality assurance unit
3. A quality control process

Please do not confuse "quality assurance" with "quality control." Quality assurance includes monitoring to assure GLP compliance. It is not merely a data checking exercise but rather a program that integrates all facets of GLPs. The QAU plays a resource and advisory role. Part of that role is to assure that adequate training, facilities, equipment, procedures, protocol, study conduct, and reporting are combined to produce a GLP-compliant study.

On the other hand, the role of the quality control reviewer is to assure that data at a given point in time has been accurately recorded or transcribed according to procedures. When quality control inspections are conducted at frequent intervals (typically at the end of the day or week), resolution of errors is more expeditious.

Reliable, trustworthy data should result when:

Study personnel are knowledgeable and well trained.
Study director involvement is adequate.
Prestudy design has been given proper emphasis.
Study protocol and procedures are followed.
Facilities are adequate and well maintained.
Equipment is adequately calibrated and maintained, and functions as intended.
Quality control reviews are established and followed periodically during the course of the study.

ACKNOWLEDGMENTS

We would like to express our appreciation to the following experienced individuals who provided valuable insight into issues encountered when conducting and monitoring GLP studies: Kimberly B. Brebner, DuPont; Joyce L. Henry, B.S., Dow Corning Corporation; Deborah Little, RQAP-GLP, DLL Quality Consulting, LLC; Susan M. Munley, M.A., Independent Consultant; Eve Mylchreest, Ph.D., DuPont; Robert M. Parker, Ph.D., DABT, DuPont; Barbara J. Patterson, Argus Research; Patricia O'Brien Pomerleau, M.S., RQAP-GLP, CIIT Centers for Health Research; Ronald L. Poore, DuPont; and Deborah L. Tyler, DuPont.

REFERENCES

1. Huber, L., *Good Laboratory Practice*, Hewlett-Packard Co., 1993.
2. FDA GLP Regulations; Final Rule, 21 CFR Part 58, 33768, 1987.
3. EPA TSCA GLP Standards, Final Rule, 40 CFR Part 792,34034, 1989.
4. EPA FIFRA GLP Standards, Final Rule, 40 CFR Part 160,34052, 1989.
5. Keener, S.K. and Hoover, B.K., Good laboratory practices: a comparison of the regulations, *J. Am. Coll. Toxicol.*, 4 (6), 339, 1985.
6. *EPA Good Automated Laboratory Practices*, Office of Information Resources Management, 1995.
7. FDA, 21 CFR Part 11, *Electronic Records: Electronic Signatures*, Final Rule, 62, 54, 13430, 1997.
8. *EPA Establishment of Electronic Reporting: Electronic Records*, Proposed Rule, 40 CFR Parts 3, 51, et al., 66, 170, 46162, 2001.
9. FDA, Enforcement Policy: 21 CFR Part 11, *Electronic Records: Electronic Signatures* (CPG 7153.17), 1999.
10. Lepore, P., The 10 essential activities of the quality assurance unit, *Quality Assurance: Good Practice, Regulation, and Law*, 4(1), 29, 1995.
11. Greenspan, B.J., Quality considerations in inhalation toxicology, *Quality Assurance: Good Practice, Regulation, and Law*, 2(1,2), 105, 1993.
12. Caulfield, M., Customer focus and satisfaction, *Quality Assurance: Good Practice, Regulations, and Law*, 3(1), 15, 1994.
13. Usher, R.W., Developing an effective toxicology/QA partnership, *Quality Assurance: Good Practice, Regulations, and Law*, 4(4), 308, 1995.
14. Pendergast, M. K., U. S. FDA. Integrity, accuracy, and quality of scientific information, *Quality Assurance: Good Practice, Regulation, and Law*, 2(½), 57, 1993.

CHAPTER 19

Perspectives on the Developmental and Reproductive Toxicity Guidelines

Mildred S. Christian, Alan M. Hoberman, and Elise M. Lewis

CONTENTS

I. Introduction 734
II. Regulatory History 734
III. Roles of Various Regulatory Groups and Evolution of Testing Concerns 738
    A. IRLG Guidelines..738
    B. ICH Guidelines 738
    C. OECD Guidelines..739
    D. EPA Guidelines..739
IV. Putting the Guidelines into the Final Protocol Design 740
    A. Fertility and General Reproductive Performance 740
    B. Embryo and Fetal Development
       (Developmental Toxicology or Teratology Studies) 741
    C. Perinatal and Postnatal Evaluations 742
    D. ICH Stages...743
       1. ICH Stage A — Premating to Conception..743
       2. ICH Stage B — Conception to Implantation 743
       3. ICH Stage C — Implantation to Closure of the Hard Palate.........................744
       4. ICH Stage D — Closure of the Hard Palate to the End of Pregnancy..............744
       5. ICH Stage E — Birth to Weaning 744
       6. ICH Stage F — Weaning to Sexual Maturity...744
V. Differences in the Guidelines for Testing Pharmaceutical Products and the Effect
of the ICH Harmonization Process..744
    A. Compliance with Good Laboratory Practice (GLP) Regulations...........................744
    B. Species Selection 744
    C. Numbers of Dosage Groups and Number of Animals per Dosage Group 745
    D. Dosage Selection 745
    E. Timing of Gestation and Postnatal Development 746
       1. Exposure Intervals 746
    F. Reproductive Performance 747
       1. Evaluation of the Ovaries..752

G. Maternal and Pup Interactions and Lactation 752
H. Growth and Development of Conceptus through Puberty 752
   1. Fetal Evaluations 752
   2. Postnatal Evaluations for Gender and Sexual Maturation 752
I. Endocrine-Mediated Findings 753
J. Comparison of FDA and EPA Requirements and Methods for Postnatal Behavioral and Functional Evaluations 754
K. Developmental Neurotoxicity 754
References 754
Appendix 1 759
Appendix 2 779
Appendix 3 792

I. INTRODUCTION

The purpose of this chapter is to describe the various regulatory guidelines specifically designed to address safety concerns regarding reproduction and development and to identify commonalities and differences in these guidelines. Regulatory guidelines issued for conduct of developmental and reproductive toxicology studies are essentially categorized into three sectors: pharmaceuticals, food and food additives, and chemicals. Because the guidelines for development of pharmaceuticals are flexible, *The Guidelines for Toxicity to Reproduction of Medicinal Products*, an effort of the International Conference for Harmonization (ICH),[1] addresses all elements of reproductive and developmental toxicology studies. The ICH guidelines are accepted by the U.S. Food and Drug Administration (FDA),[2-4] the European Union (EU), and Japan. The ICH guidelines also provide an overview of the various study designs and endpoints evaluated in the regulatory testing of pharmaceuticals, biotechnology products,[5] indirect food additives and devices,[6] chemicals,[7-10] and pesticides, fungicides, and rodenticides.[7,11-13]

As the harmonization processes continue, it is likely that the currently available guidelines will become more, rather than less, similar. For instance, the ICH guidelines[2-4] replace the former guidelines issued by the U.S. FDA Bureau of Drugs,[5] EU countries,[14,15] and Japan.[16-18] Those issued by the U.S. Environmental Protection Agency (EPA) Office of Prevention, Pesticides and Toxic Substances (EPA-OPPTS) replaced the U.S. EPA Toxic Substances Control Act (TSCA) guidelines[7-10] and the Federal Insecticide, Fungicide and Rodenticide (FIFRA) guidelines.[7,11-13] International harmonization efforts are also ongoing between the EPA and the Organization for Economic and Cooperative Development (OECD).

Table 19.1 lists the principal regulatory guidelines currently in use. Additional information can also be obtained from various websites, identified in Table 19.2 and Table 19.3, many of which provide updates on regulatory issues, information on professional society activities and projects, and access to historical databases.

II. REGULATORY HISTORY

The reproduction and developmental toxicity regulatory guidelines were developed in response to human health concerns. Three human tragedies that resulted from *in utero* exposure to a drug provided the momentum for initial development of the guidelines. These were: (1) 1961 — congenital malformations associated with the use of thalidomide;[32,33] (2) early 1970s — cancer associated with maternal use of diethylstilbesterol;[34] and (3) 1976 — behavioral and functional alterations associated with exposure to methyl mercury.[35] For a detailed history of the development of the guidelines associated with these issues, see Christian and Hoberman[36] in the previous edition of this book.

Table 19.1 Current regulatory guidelines — ICH, FDA, OECD, and EPA

Regulatory Agency	Ref.	Guideline
Medical Agents		
ICH	1	Detection of toxicity to reproduction for medicinal products
ICH	19	Detection of toxicity to reproduction for medicinal products
FDA	2	International Conference on Harmonisation: guideline on detection of toxicity to reproduction for medicinal products
FDA	3	International Conference on Harmonisation: guideline on detection of toxicity to reproduction for medicinal products: Addendum on toxicity to male fertility
FDA	4	International Conference on Harmonisation: maintenance of the ICH guideline on toxicity to male fertility; An addendum to the ICH tripartite guideline on detection of toxicity to reproduction for medicinal products, amended November 9, 2000
Indirect Food Additives		
FDA, Bureau of Foods	20	Toxicological principles and procedures for priority based assessment of food additives (Red Book), guidelines for reproduction testing with a teratology phase
FDA, Center for Food Safety and Applied Nutrition	6	Toxicological principles for the safety assessment of direct food additives and color additives used in food (Red Book II), guidelines for reproduction and developmental toxicity studies
FDA, Center for Food Safety and Applied Nutrition	21	Toxicological principles for the safety of food ingredients (Red Book 2000)
Chemicals		
EPA, OPPTS	22	Health effects test guidelines: prenatal developmental toxicity study
EPA, OPPTS	23	Health effects test guidelines: reproduction and fertility effects
EPA, OPPTS	24	Health effects test guidelines: developmental neurotoxicity study
EPA, Risk Assessment Forum	25	Guidelines for developmental toxicity risk assessment
EPA	26	Guidelines for developmental toxicity risk assessment
OECD	27	Guidelines for testing of chemicals: teratogenicity
OECD	28	Guidelines for testing of chemicals: one-generation reproduction toxicity
OECD	29	Guidelines for testing of chemicals: two-generation reproduction toxicity study
OECD	30	Guidelines for testing of chemicals: combined repeated dose toxicity study with the reproduction and developmental toxicity screening test
OECD	31	Guidelines for testing of chemicals: developmental neurotoxicity study

Table 19.2 List of useful websites for obtaining regulatory documents

Federal Register	http://www.access.gpo.gov/su_docs/aces/aces140.html
FDA guidance documents	http://www.fda.gov/cder/guidance/index.htm
ICH guidelines	http://www.ifpma.org/ich5s.html
Red Book 2000	http://vm.cfsan.fda.gov/~redbook/red-toct.html
EPA (public drafts)	http://www.epa.gov/OPPTS_Harmonized/870_Health_Effects_Test_Guidelines/Drafts
EPA (final guidelines)	http://www.epa.gov/docs/OPPTS_Harmonized/870_Health_Effects_Test_Guidelines/Series/
EPA guidelines for reproductive toxicity risk assessment	http://www.epa.gov/ordntrnt/ORD/WebPubs/repro/index.html
EPA's EDSTAC	http://www.epa.gov/scipoly/oscpendo/index.htm
EPA Food Quality Protection Act of 1996	http://www.epa.gov/opppsps1/fqpa
OECD guidelines	http://www.oecd.org/ehs/test/testlist.htm
OECD endocrine disrupter testing and assessment	http://www.oecd.org/EN/document/0,,EN-document-524-14-no-24-6685-0,00.html
U.S. National Toxicology Program (NTP) testing information and study results	http://ntp-server.niehs.nih.gov/main_pages/NTP_ALL_STDY_PG.html
California Proposition 65 documents	http://www.oehha.org/prop65.html

Table 19.3 List of useful research organizations

Reproductive and Developmental Toxicology

Teratology Society	http://www.teratology.org/
Middle Atlantic Reproduction and Teratology Association (MARTA)	http://www.teratology.org/members/MARTAtext.htm
Midwest Teratology Association (MTA)	http://www.midwest-teratology.org/
Society of Toxicology Reproductive and Developmental Specialty Section	http://www.toxicology.org/memberservices/specsection/specsection.html
Australian Teratology Society (ATS)	http://www.cchs.usyd.edu.au/bio/ats/atshome.htm
European Teratology Society (ETS)	http://www.etsoc.com/
Japanese Teratology Society	http://www.med.hiroshima-u.ac.jp/med/med/kiso/kaibo1/CongAnom/eWelcome.html

Toxicology

Society of Toxicology (SOT)	http://www.toxicology.org/
American College of Toxicology (ACT)	http://www.landaus.com/toxicology/
Mid-Atlantic Chapter of Society of Toxicology (MASOT)	http://www.masot.org/
Italian Society of Toxicology	http://users.unimi.it/~spharm/sit/engSIThome.html
British Toxicology Society	http://www.thebts.org/
Eurotox	http://www.eurotox.com/
American Board of Toxicology (ABT)	http://www.abtox.org/
Academy of Toxicological Sciences (ATS)	http://www.acadtox.com/
Society of Toxicology of Canada	http://meds.queensu.ca/stcweb/index.html

Neurotoxicology

Neurobehavioral Teratology Society (NBTS)	http://nbts.bsbe.umn.edu/
Behavioral Toxicology Society	http://www.behavioraltoxicology.org/BTS.html
The Society for Neuroscience (SFN)	http://www.sfn.org
The International Neurotoxicology Association	http://www.neurotoxicology.org/

Additional Resources

Reproductive Toxicology ($850/year online)	http://www.elsevier.nl/locate/reprotox
International Federation of Teratology Societies (IFTS) Atlas of Developmental Abnormalities	http://www.ifts-atlas.org/
National Institute of Environmental Health Sciences (NIEHS)	http://www.niehs.nih.gov/
Teratogen Information System (TERIS, $1000/year)	http://depts.washington.edu/~terisweb/

Recently, new guidelines have been developed to address specific political and economic concerns, particularly regarding potential hazards to children. These guidelines are related to concerns regarding treatment of neonatal and pediatric patients with pharmaceuticals and chronic and inadvertent exposure to food additives and other environmental chemicals, particularly late in gestation and/or before puberty, as described in the following information.

The current FDA and ICH guidelines[2] for pharmaceuticals include postnatal evaluations in animals for behavioral and functional alterations, in particular those concerning potential exposure of the central nervous system and other systems that continue to develop postnatally (e.g., pulmonary, immune, renal, osseous, and gastrointestinal) and thus may be potentially affected during late gestation and the early neonatal period. The FDA currently has a pediatric study requirement (1998 Final Rule: Pediatric Studies).[37] To conform with this requirement, all new drugs, generally prescription drugs (including biologics or drugs derived from living organisms), that are intended for common use in children or are important in medical treatment of children must include labeling information on safe pediatric use. Similar efforts are ongoing in the EU.[15]

More extensive increases in requirements for pediatric testing have been enacted or are under consideration by the EPA and the OECD. In 1997, the EPA issued entirely new guidelines in response to children's health concerns and established the Office of Children's Health Protection (OCHP).[38] This federal agency identified four major environmental threats to children: (1) lead poisoning; (2) pesticides in household chemicals and food, (3) asthma, and (4) drinking water contaminants (principally microbial), polluted water, toxic waste dumps, PCBs, second-hand tobacco smoke, UV light, and endocrine disruptors, and potential effects from particulate matter air pollution. These relatively new EPA guidelines are based on the perceived greater relative vulnerability of children to environmental threats. This perception is based on children having immature body organs and tissues, rapidly changing as they develop from infants to adolescents. They have weaker immune systems during infancy and are potentially exposed to more environmental threats while being least able to protect themselves from such hazards. One response to these concerns was the issue of the EPA guidelines for developmental neurotoxicity.[24]

Another important response of the EPA is its endocrine disruption program, which was mandated by the Food Quality Protection Act[39] and authorized by the amended Safe Water Act.[40] These two acts resulted in formation of the Endocrine Disruptor Screening and Testing Advisory Committee (EDSTAC), which has completed its mandate. EDSTAC's purpose was to develop recommendations for an endocrine disruptor screening program.[41] One of EDSTAC's recommendations was expansion of the endocrine disruptor screening program to include not only pesticides but also chemicals regulated under the TSCA, nutritional supplements, food additives, and cosmetics. EDSTAC also recommended that screens be performed to evaluate endocrine-mediated effects on the androgen and thyroid systems and other effects on fish and wildlife.[42,43]

The EDSTAC recommendations were used to design the EPA Endocrine Disruptor Screening Program (EDSP). The Endocrine Disruptor Method Validation Subcommittee (EDMVS), which is under a Federal Advisory Committee — the National Advisory Council for Environmental Policy and Technology (NACEPT)[44] — is one essential part of implementation of the EDSP. The EDMVS is responsible for evaluating the methods and the results of validation efforts for the new protocols recommended by EDSTAC. New protocols currently being considered for validation or in the validation process include a one-generation extension study[45] and a 20-day pubertal development and thyroid function study in juvenile male and female rats.[46,47]

Internationally, an OECD Task Force on Endocrine Disruptor Testing and Assessment (EDTA) was instituted in December, 1997 in an effort to: (1) consider and recommend priorities for the development of methods to detect estrogenic and androgenic effects and (2) harmonize test methods and assessments of endocrine disruptors. Under the direction of the EDTA, OECD Validation Management groups were established to manage the technical conduct of validation work for screening and testing for endocrine-mediated mammalian and ecotoxicological effects. The validation testing phase has been completed for the rodent uterotrophic assay.[48] Revisions of the OECD Two Generation Reproductive Toxicity[29] and Teratogenicity[27] guidelines have been issued as OECD Test Guidelines 416 and 414, respectively. Additional projects currently being addressed by the OECD include validation of the rodent Hershberger assay[48] and of studies to assess the feasibility of enhancing the current OECD TG 407 (Repeated Dose Toxicity) guideline.[49] The EPA and OECD are working in close cooperation regarding the validation programs.

In toto, the concepts regarding the ability of agents to affect *in utero* and early postnatal development, including sexual maturation and function, have changed remarkably since 1966, when the FDA issued its first guidelines.[5] In 1966, the focus was on identifying "teratogens." Thus the concerns and ways in which multiple endpoints, ranging from delayed or advanced sexual maturation to frank malformation, are to be addressed have evolved. As a result, it is now recognized that it is important to screen for adverse effects on the entire reproductive process as well as for *in utero* development and maturation.

III. ROLES OF VARIOUS REGULATORY GROUPS AND EVOLUTION OF TESTING CONCERNS

The pharmaceutical and chemical industries are extensively regulated internationally. Remarkable international and local concern exists for both industries regarding regulatory requirements for reproductive and developmental toxicity testing. As a result, harmonization of guidelines and ever increasing appreciation of the multiplicity of the endpoints to be evaluated in reproductive and developmental toxicity studies has been in progress for the last 20 years. In general, current guidelines for use in the development of medicines for humans are the ICH guidelines.[2-4] FDA guidelines relevant to development of medicines for animals, foods, and food additives ("FDA Red Book") are moving in the direction of the ICH guidelines, but have some differences. OECD and EPA guidelines relevant to the testing of chemicals are less flexible and often have different technical requirements, although the same endpoints are evaluated. Some information regarding the various organizations involved in guideline development and harmonization are provided below. Copies of the relevant ICH and EPA guidelines (with the exception of the EPA developmental neurotoxicity guideline, addressed in another chapter) are provided in Appendixes 1 and 2 of this chapter.

A. IRLG Guidelines

The Interagency Regulatory Liaison Group (IRLG), which is no longer active, was composed of five U.S. regulatory agencies: (1) Consumer Product Safety Commission; (2) Environmental Protection Agency; (3) Food and Drug Administration (U.S. Department of Human Health and Safety); (4) Food Safety and Quality Service (U.S. Department of Agriculture), and (5) the Occupational Safety and Health Administration (U.S. Department of Labor). In 1980, the IRLG issued harmonized guidelines for the conduct of developmental toxicity studies that were acceptable in these U.S. agencies.[50] Those guidelines became the basis for the early international OECD guidelines.

B. ICH Guidelines

The European Economic Community (EEC) guideline update[15] provided the basis for the ICH process. Beginning in 1990,[51,52] the International Federation of Pharmaceutical Manufacturers Associations (IFPMA) began to bring together the regulatory authorities of Europe, Japan, and the United States, as well as experts from the pharmaceutical industry in these three major geographical regions. Their task was to identify ways to eliminate redundant technical requirements. The objective was to expedite global development and availability of new medicines without sacrificing safeguards on quality, safety, or efficacy; the process continues to this day.

The ICH effort had six cosponsors: the Commission of the European Communities (CEC), the European Federation of Pharmaceutical Industries Associations (EFPIA), the Ministry of Health and Welfare (MHW, Japan), the Japan Pharmaceutical Manufacturers Association (JPMA), the FDA, and the Pharmaceutical Manufacturers Association (PMA, United States), which has since become PhRMA (Pharmaceutical Research and Manufacturers of America). The result of these efforts was the international acceptance of the regulatory requirements of the ICH guidelines.

The ICH guideline on *Detection of Toxicity to Reproduction for Medicinal Products* was the first to be approved.[2] Since 1994, it has been amended twice[3,4] to clarify testing procedures for the male fertility portion of the studies. A history of the regulatory activities and the ICH program regarding these guidelines can be found in the forerunner to this book.[36]

C. OECD Guidelines

The Organisation for European Economic Co-operation (OEEC) was formed to administer American and Canadian aid under the Marshall Plan for the reconstruction of Europe following World War II. At that time it consisted of its European founder countries plus the United States and Canada. The OECD took over from the OEEC in 1961. The OECD currently includes 30 member countries in Europe, North America, Asia, and the Pacific.

In 1979 to 1980, OECD assisted in updating and standardizing testing procedures for chemicals.[53] More recently the OECD Pesticide Working Group, part of OECD's Environment Program, was established as the first forum for national pesticide regulators from developed countries to discuss common issues. The OECD Pesticide Working Group was established in 1994 as the Pesticide Forum.[54] The purpose of the Pesticide Working Group is to assist countries in managing their task of conducting new risk assessments for hundreds of established pesticides, as well as assessments of the active ingredients in newer pesticides. The major activity areas of the Pesticide Working Group include conventional, biological, and antimicrobial pesticides.[54]

D. EPA Guidelines

The EPA regulates pesticides and other chemicals for reproductive and developmental hazard and risk under multiple federal statutes, the most important of which are the Federal Insecticide, Fungicide, and Rodenticide Act[55] and the Federal Food, Drug, and Cosmetic Act (FFDCA).[56] Under the authority of FIFRA, the EPA reviews and registers pesticides for specific uses. It can also suspend or cancel pesticide registration if it deems the pesticide to be hazardous. Section 408 of the Federal Food, Drug, and Cosmetic Act[56] allows the EPA to establish maximum residue levels, or tolerances, for pesticides used in or on foods consumed by humans or animals. In 1996, the Food Quality Protection Act (FQPA) was enacted, amending FIFRA and FFDCA. The FQPA established higher safety standards for all pesticides used on foods, required uniformity in food processing, and changed pesticide regulations by requiring greater health and environmental protection for infants and children.[57] The Office of Pollution Prevention and Toxics (OPPT), formed in 1977,[58] has the primary responsibility for administering the Toxic Substances Control Act.[59] This law covers production and distribution of commercial and industrial chemicals produced by or imported into the United States.

The EPA OPPTS has developed uniform guidelines for testing pesticides and other substances to meet the EPA's data requirements.[60] These data requirements are regulated by two laws, the Toxic Substances Control Act[59] and the Federal Insecticide, Fungicide, and Rodenticide Act.[55] Before the harmonization process, slightly different testing procedures and requirements existed in separate publications by the OPPT, the Office of Pesticide Programs (OPP), and the OECD.[60]

The EPA works closely with the OECD to develop and carry out OECD Pesticide Program activities. OPP is involved in cooperative work to harmonize pesticide data requirements, focus test guidelines on pesticide regulatory needs, and harmonize industry data submissions and government data review formats and contents. OECD encourages the exchange of review reports and has developed an electronic database to facilitate such exchanges and collaboration on reviews. OPP is also working with OECD countries on electronic data submission for industry. As the result of international agreements, the OECD guidelines may be applied to EPA-regulated chemicals in development.

IV. PUTTING THE GUIDELINES INTO THE FINAL PROTOCOL DESIGN

As previously mentioned, a copy of the ICH guideline and amendments[2-4] are provided as Appendix 1 of this chapter. References for guidelines and other information generally considered in the development of drugs, food additives, pesticides, and other chemicals are provided as Appendix 3. To understand how to appropriately select and combine the various "ICH Stages" of the reproductive cycle,[2] it is necessary to understand the endpoints evaluated and some of the technical concerns regarding conduct of these studies. The following information cites the relevant endpoints and how they are addressed in the various guidelines currently in use.

A. Fertility and General Reproductive Performance

The evaluation of fertility and general reproductive performance includes examination of potential adverse effects on gonadal function and mating behavior in male and female animals, conception rates, early and late stages of gestation, parturition, lactation, and development of the offspring. Thus, the study design is an abbreviated multigeneration study, usually ending at weaning of the second generation.

For a full evaluation of male fertility and reproductive performance (i.e., a multigeneration study), the agent is given to sexually mature males for 60 to 70 days before mating, i.e., a complete cycle of spermatogenesis, and then continues until mating occurs. If there is no evidence of impaired male reproductive performance from other studies, males may be treated for 28-day or 14-day intervals before mating with cohort females. Although not mandated, evaluation of the histopathology of the testes and epididymides is generally performed, as well as evaluation of sperm for count, concentration, and morphology.

To study female reproductive performance and fertility, sexually mature females are evaluated for estrous cycling for 14 days before treatment, for the first 14 days of treatment before cohabitation, and until mating occurs. Treatment continues through cohabitation and mating, and in the female animals until they are necropsied on day 13 of gestation or at completion of a 21-day lactation period. Parturition, lactation, maternal-pup interaction, and pup growth and development until weaning are monitored.

Adult reproductive function may be evaluated by treating animals of only one sex and mating these animals with untreated animals of the opposite sex, or by treatment of both males and females. A common refinement of the test is to use double sized control and high dose groups during the premating period, with cross-mating of one-half of the animals in these two groups. This is done to determine whether administration of the test agent to animals of only one sex causes different effects from those observed when both sexes are treated.[61]

Alternate procedures include:

1. Extension of the treatment period through the remainder of gestation (e.g., for rats, until day 17 or 19 of gestation, with caesarean-sectioning on day 20 or 21 of gestation), and evaluation of the fetuses for viability and gross (external), soft tissue, and skeletal alterations — a design addressing both male and female fertility and reproductive performance and developmental toxicity.
2. Extension of the treatment period through the remainder of gestation and then through parturition and lactation (i.e., for rats until day 21 of lactation), and assessing the pups for developmental landmarks and postnatal behavior and function (pups are selected at weaning for continued evaluation through mating and gestation and examined for general sensory [auditory, visual] and behavioral development and fertility. Additional functional evaluations are sometimes considered for inclusion[62,63]).

Canada is not one of the countries participating directly in the ICH process; however, it is usually considered in international marketing plans. The proposed guidelines for Canada[64,65] differ from the ICH guidelines[2] on the following points: (1) an 80-day dosage period before mating is suggested for male rodents, and (2) it was suggested that consideration be given to use of the United Kingdom (UK) protocol,[14] which is a design in which treatment of rats of both sexes continues from before cohabitation until sacrifice after mating (males) or completion of a 21-day lactation period (females) and postnatal evaluation of the offspring through gestation. Because the UK is part of ICH, Canada now generally accepts ICH guideline compliant studies.

The guidelines in the FDA Red Book[21] require both "teratology" and multigeneration studies. However, the teratology study may have an exposure period extending from before mating of the parental generation through gestation and can be conducted separately or as part of the multigeneration study. The EPA[23] and OECD[28,29] guidelines also require multigeneration studies.

B. Embryo and Fetal Development (Developmental Toxicology or Teratology Studies)

The study design formerly identified as a "Segment II" or FDA "teratology" study is currently generally described as a "developmental toxicity study," i.e., a study of potential embryo and fetal toxicity or embryotoxicity. Developmental toxicity studies are usually conducted in two species, one rodent and one nonrodent. Two species are tested because: (1) adverse effects on development, especially in the absence of maternal effects, remain the most sensitive regulatory concern; and (2) thalidomide did not cause phocomelia in the rat, although it can produce phocomelia in the rabbit.[66] Despite Wilson[67] noting that one cannot assume mimicry of responses across species, and Palmer[68] identifying that appropriate interpretation of rat data from a two-litter test would have identified that thalidomide was selectively toxic to the development of the conceptuses (i.e., it increased resorption and reduced live litter sizes at dosages that were not toxic to the dams), the rabbit continues to be the nonrodent species most commonly tested in this assay. However, rats and rabbits should not automatically be selected as the two most appropriate species. ICH guidelines emphasize that the agent should have pharmacologic activity in the species tested. The increased consideration of the metabolism of test agents has resulted in the animal species whose metabolism of an agent is most like that of humans generally being considered the most appropriate species. Concerns about pharmacological metabolic comparability and the ability to justify species choice have increased the use of nonhuman primate testing for biologics and the development of homologues of human proteins for testing of biologics in rabbits, rats, or mice.

Despite increased concern and appreciation of other endpoints, the developmental toxicity study continues to be commonly perceived as an evaluation of the potential of an agent to malform the conceptus. However, this study is actually an evaluation of embryo and fetal toxicity and differs from a true teratology study, in which a "pulse" exposure is typically the most effective treatment method.[69] In the developmental toxicity study design, pregnant females are treated throughout the period of major organogenesis (embryogenesis), generally defined as the interval between implantation and palate closure, and necropsied shortly before expected parturition. The ovaries are examined for the number of corpora lutea. The uterus is examined for implantation sites, live and dead fetuses, early and late resorptions, and placental abnormalities. The fetuses are weighed, the fetal sex is identified, and if indicated, anogenital distances are determined. Occasionally, crown-rump lengths for the fetuses are recorded. In rodent studies, approximately one-half of the fetuses in each litter are examined for soft tissue alterations by using either Wilson's sectioning technique[70] or Staples' visceral examination technique[71] (fully described in Christian,[72] possibly as adapted by Stuckhardt and Poppe[73]). The remaining fetuses are stained with alizarin red-S,[74] and if desired, also with alcian blue,[75–77] and examined for skeletal alterations. In nonrodent studies, all fetuses are examined for soft tissue and skeletal alterations.

Day 0 of presumed gestation is generally recognized as the day spermatozoa, a vaginal or copulatory plug (rodents) or insemination (rabbit) occurs. The period of organogenesis or embryogenesis is identified in the ICH guidelines[2] as occurring between implantation and palate closure. This period was extended from that in the FDA 1966 guidelines to include the fetal intervals when sexual differentiation and initial renal development occur, i.e., gestation days (GD) 15 through 18 or 19 in rodents. The fetuses are caesarean-delivered one or two days before parturition is expected to occur, i.e., usually on GD 18, 21, and 29 in mice, rats and rabbits, respectively.

The ICH guidelines[2] recommend that there be 16 to 20 animals per dosage group, with a control and at least three dosage groups administered the test material. Generally, 20 rodents and 16 nonrodents per dosage group have been considered appropriately sized populations for testing. The FDA Red Book,[21] OECD,[27] and EPA guidelines,[22] however, recommend 20 pregnant animals per group, regardless of species. A frequent modification of this design is to use double-sized groups, with one-half of the pregnant animals caesarean-sectioned and the remaining pregnant animals allowed to deliver and to nurse their litters until weaned. The pups may be examined for postnatal maturation and physical development, including mating of the F_1 generation as adults and examination of uterine contents of the female rats, using the same methods described for the first generation animals.

The FDA Red Book[21] allows conduct of a teratology study in rodents either within the multigeneration reproduction study, in which case exposure is continual in the parental generation before mating and in the dams throughout gestation, or as a separate study. These guidelines also require conduct of a rabbit study. FDA Red Book,[21] OECD,[27] and EPA[22] requirements differ from the ICH[2] guidelines in terms of the numbers of animals per dosage group, as previously described, and also in the duration of exposure. These guidelines now require treatment to continue from implantation (i.e., GD 6 or 7) until the day before parturition is expected, i.e., through GD 17, 20, and 28 in mice, rats and rabbits, respectively. The EPA guidelines consider it appropriate to treat throughout the entire gestation period, although studies performed with treatment initiated at implantation are generally acceptable. The exception is when the percutaneous route is used, for which preimplantation exposure is often recommended so that appropriate blood levels can be achieved before embryogenesis begins. The EPA guidelines also recommend double-staining of fetuses. In addition, while the OECD[30] and EPA-OPPTS[78] guidelines require testing of two species, the two species need not necessarily be a rodent and nonrodent species (i.e., two rodent species may be tested, when necessary). The FDA Red Book,[21] OECD,[27] and EPA[22] guidelines request uterine and maternal carcass weights. On all other points, the EPA-FIFRA guidelines[11] allow performance of essentially the same evaluations as those recommended by ICH.[2]

C. Perinatal and Postnatal Evaluations

The FDA Segment III study,[5] or perinatal and postnatal evaluation, was designed to evaluate the effects of an agent on the late (fetal) stages of gestation and on parturition and lactation. This study design may still be used if it is warranted by the toxicity of an agent or a specific finding of interest, provided both embryo, fetal, and pup exposures are evaluated for postnatal effects. However, the peri-postnatal study is generally performed with treatment of the pregnant female rats beginning on GD 7 and continuing through gestation, parturition, and a 21-day lactation period, combining embryo, fetal, and peri-postnatal exposures in one study. It is recognized that postnatal changes may be induced by insult at any time during gestation. Thus, unless the developmental toxicity study includes a postnatal phase, potential adverse postnatal findings associated with embryo exposure could not be identified. Parturition and pup survival and development are monitored, including observations for prolonged gestation, dystocia, impaired maternal nesting or nursing behavior, and postnatal mortality. If indicated by the results of this study or other studies, a cross-fostering procedure may be included to demonstrate whether findings in the pups are direct effects of exposure during gestation and/or lactation or if they are associated with abnormalities in maternal reproductive behavior or health. The peri-postnatal evaluation is not required by the FDA Red Book, OECD, or EPA guidelines. Rather, these evaluations are part of the multigeneration studies performed to meet the needs of these regulatory bodies that consider this design an appropriate overall evaluation of reproductive capacity.[79,80]

The ICH guideline[2] acknowledges the overlap in the methods used to test chemicals and pharmaceuticals for adverse effects on the reproductive process, but recommends that segmented study designs generally be used because humans usually take a pharmaceutical product for limited intervals, rather than an entire lifetime. However, the ICH document[2] also notes that lifetime exposure to a drug sometimes occurs, and that there may be drugs that can be more appropriately tested by exposure throughout the entire reproductive life, i.e., through use of a multigeneration study design.

Rather than a standard checklist approach, the ICH guideline[2] emphasizes flexibility in developing the testing strategy. Although the most probable options are identified in the guideline, the development of the testing strategy is to be based on the following points:

- Anticipated drug use, especially in relation to reproduction
- The form of the substance and route(s) of administration intended for humans
- Making use of any existing data on toxicity, pharmacodynamics, kinetics, and similarity to other compounds in structure and activity

The notes in the ICH guideline[2] describe some methods appropriate for use in specific portions of the tests; however, it emphasizes that the individual investigator is at liberty to identify the tests used and that the ICH guideline[2] is a guideline and not a set of rules. The overall aim of the described reproductive toxicity studies is to identify any effect of an active substance on mammalian reproduction, with subsequent comparison of this effect with all other pharmacologic and toxicologic data. The objective is to determine whether human risk of effects on the reproductive process is the same as, increased, or reduced, in comparison with the risks of other toxic effects of the agent. The document clearly states that for extrapolation of results of the animal studies, other pertinent information should be used, including human exposure considerations, comparative kinetics, and the mechanism of the toxic effect.

Because the objective of these tests is to assess all stages of reproduction, the total exposure period includes mature adults and all stages of development of the offspring, from conception to weaning. Observations should be made over one complete life cycle (from conception in the first generation through conception and pregnancy in the second generation). The reproductive cycle is segmented into six ICH stages (ICH Stages A through F). To assure that there are no gaps in exposure when ICH stages are tested separately or in combination, it was suggested that there be a one-day overlap of treatment. The six ICH stages are described below.

D. ICH Stages

1. ICH Stage A — Premating to Conception

This ICH stage evaluates reproductive functions in adult males and females, including development and maturation of gametes, mating behavior, and fertilization. By convention, it was agreed that pregnancy would be timed on the basis of when spermatozoa are identified in a vaginal smear or a copulatory plug is observed, and that these events would identify GD 0, even should mating occur overnight.

2. ICH Stage B — Conception to Implantation

This ICH stage examines reproductive functions in the adult female and preimplantation and implantation stages of the conceptus. Unless data are provided that prove otherwise, it is assumed that implantation occurs in rats, mice, and rabbits on days 6 to 7 of pregnancy.

3. ICH Stage C — Implantation to Closure of the Hard Palate

This ICH stage evaluates adult female reproductive functions, embryonic development, and major organ formation. The period of embryogenesis is completed at closure of the hard palate, which by convention is considered to occur on days 15 through 18 of pregnancy in rats, mice, and hamsters, respectively.

4. ICH Stage D — Closure of the Hard Palate to the End of Pregnancy

Adult female reproductive function continues to be examined in this ICH stage, as well as fetal development and growth and organ development and growth.

5. ICH Stage E — Birth to Weaning

Again adult female reproductive function is examined; this ICH stage also evaluates adaptation of the neonate to extrauterine life, including preweaning development and growth of the neonate. By convention, the day of birth is considered postnatal day 0, unless specified otherwise. It is also noted that when there are delays or accelerations of pregnancy, postnatal age may be optimally based on postcoital age.

6. ICH Stage F — Weaning to Sexual Maturity

This ICH stage is generally treated only when the intended use of the pharmaceutical product is in children. The ICH guideline[2] recognizes that it may sometimes be appropriate to also treat the neonate during this ICH stage, in addition to the required evaluations of the offspring. This ICH stage provides observations of postweaning development and growth, adaptation to independent life, and attainment of full sexual development.

V. DIFFERENCES IN THE GUIDELINES FOR TESTING PHARMACEUTICAL PRODUCTS AND THE EFFECT OF THE ICH HARMONIZATION PROCESS

A. Compliance with Good Laboratory Practice (GLP) Regulations[81–85]

In general, all studies submitted in support of the safety of a pharmaceutical or chemical product should be conducted in conformance with the GLPs of one or more regulatory organizations. Although preliminary dosage range-finding studies need not be conducted in complete conformance with GLPs, the results of such studies should be submitted.

B. Species Selection

The various guidelines generally identify testing in a rodent and nonrodent species, usually the rat and rabbit. To avoid additional studies to the extent possible, the ICH guideline[2] recommends use of the same species and strain of animals tested in other toxicology studies and in kinetic studies. It is also recommended that kinetic, pharmacological, and toxicological data be considered to demonstrate that the selected species provides a model relevant for the human. As mentioned earlier, the OECD[30] and EPA-OPPTS[78] combined guidelines allow use of two rodent species.

Animals should have comparable ages, weights, and parity when assigned to the study. Historical incidences of abnormalities in fertility, fecundity, fetal morphology, and resorption, and information regarding consistency from study to study should be available. For EPA studies, positive control data are often requested to ensure that the laboratory has appropriate experience in the conduct of the studies. In addition, for EPA regulated studies, it is preferred that historical data be restricted to the 2-year period during which the study is conducted. For pharmaceuticals the same species and strain should be used to the extent possible in all toxicologic and kinetic studies. Although this guideline is not consistently followed in EPA regulated studies, this would assist in producing meaningful data.

Because of the increased interest in identifying specific genes that may be affected by an agent, transgenic models and knock-out models of certain disease states are available. There are times when it may be appropriate to test an agent in an animal expressing the disease state to be treated; such tests would usually supplement studies in outbred laboratory species. For example, testing an antidiabetic agent in a normal animal may result in unexpected and irrelevant findings for human diabetics, as the result of abnormally lowering blood sugar. Testing the same agent in a diabetic animal has the potential to demonstrate whether any developmental abnormalities produced were the result of abnormal blood sugar levels in normal animals and whether such states would be attained in a diabetic animal.

C. Numbers of Dosage Groups and Number of Animals per Dosage Group

The various guidelines require different numbers of dosage groups and animals per dosage group. The ICH guideline[2] identifies that all studies should include a control and at least three groups given the drug, although additional groups may be used if considered scientifically necessary. It recommends use of 16 to 20 animals per dosage group for most studies (pregnant with litters, when appropriate), based on the observations that biological responses can generally be identified by using these numbers of animals and that use of additional animals does not greatly enhance the ability to detect changes, except in the case of infrequent events. The general practice is to use 20 pregnant animals per group, regardless of species, with the exception of studies performed in nonhuman primates or other large animal species. When these are used, it is recommended that the protocol design be discussed with the regulatory agencies before conduct of the study to ensure that there is consensus regarding the sensitivity of assay.

D. Dosage Selection

The basis for identifying an appropriate high dosage is not clearly defined in the various existing guidelines. In general, for pharmaceuticals, a 10% reduction in maternal body weight gain was the criterion used, unless the drug had so little toxicity that a very high multiple of the clinical dosage was considered appropriate. For chemicals, a dose that did not result in greater than 10% mortality was the usual criterion. Obviously, this subject is one of great concern, especially in consideration of labeling issues. The ICH guideline[2] recognizes that the high dosage should produce minimal maternal (adult) toxicity and be selected on the basis of data from all available studies, including pharmacologic, acute, and subchronic toxicity and kinetic studies. Dosage-range studies are recommended when other studies do not provide sufficient information for selection of an appropriate high dosage. Issues of toxicity in the high dosage group dam may include:

- Reduction in body weight gain.
- Increased body weight gain, particularly when related to perturbation of homeostatic mechanisms.
- Toxicity to specific target organs.
- Changes in hematology and clinical chemistry results.
- Exaggerated pharmacological response, which may or may not be reflected as marked clinical reactions (e.g., sedation, convulsions).
- Practical limitations in the dose that can be administered because of the physico-chemical properties of the test substance, dosage formulation, and/or the route of administration. Under most circumstances, 1 g/kg/day should be an adequate "limit dose."

Table 19.4 ICH treated and evaluated stages

	FDA	UK	EEC	JAPAN
Segment I				
Treated	A–E[a]	A–E	A–E	A, B
Evaluated	A–E	A–F	A–F	A–D
Segment II				
Treated	C	C	C	C, ½ D
Evaluated	C, D	C, D	C, D	C–F
Segment III				
Treated	D, E	D, E	D, E	½ D, E
Evaluated	D, E	D, E	D, E	½ D, E, F

[a] A = premating to conception, B = conception to implantation, C = implantaion to closure of the hard palate, D = closure of the hard palate to the end of pregnancy, E = birth to weaning, F = weaning to sexual maturity.

- Kinetics can be useful in determining high dose exposure for low toxicity compounds. There is, however, little point in increasing the administered dosage if it does not result in increased plasma or tissue concentrations.
- Marked increase in embryo-fetal lethality in preliminary studies

Selection of the appropriate high dosage for studies regulated by OECD and EPA remains problematic because unless required as the result of negotiations with the regulatory body, parameters commonly studied in development of pharmaceuticals are omitted from developmental and reproductive toxicology studies, and toxicokinetics are not usually assessed unless problems are identified. Thus, interaction with the regulatory agency is recommended in this activity.

E. Timing of Gestation and Postnatal Development

Unless otherwise identified, GD 0 is defined as the day spermatozoa are observed in a smear of the vaginal contents and/or a copulatory plug is observed *in situ*, and day 0 of lactation is defined as the day of birth. Should alternate timing be used, such as GD 1 or day 1 of lactation, such a change should be clearly identified in the data and reports.

1. Exposure Intervals

As previously mentioned, the ICH guideline[2] segments, the entire reproductive process into six ICH stages, each of which has specific endpoints evaluated. As shown in Table 19.4, earlier guidelines that used a three-segment approach combined these ICH segments in various ways, just as did the various OECD and EPA approaches. For example, in the FDA,[5] UK,[14] and EEC[15] Segment I studies, treatment of the animals occurred in ICH stages A through E. The UK[14] and EEC[15] studies also evaluated the animals in ICH Stage F. The Japanese Segment I study[16,17] treated the animals in ICH Stages A and B but included evaluations of ICH Stages C and D. The FDA,[5] UK,[14] and EEC[15] Segment II studies treated the animals in ICH Stage C and also evaluated ICH Stage D. The Japanese Segment II study[16,17] treated the animals in ICH Stage C and part of ICH Stage D, with evaluations continued into ICH Stage F. The FDA,[5] UK,[14] and EEC[15] Segment III studies treated and evaluated the animals in ICH Stages D and E. The Japanese Segment III study[16,17] treated the animals beginning near the end of ICH Stage D and through ICH Stage E, with evaluations continuing through ICH Stage F.

Table 19.5 Comparison of current guidelines for the assessment of fertility and early embryonic development

Guideline	ICH 4.1.3[a]	EPA OPPTS[b]	FDA Red Book[c]	OECD[d]
Species (No.)	At least one species, preferably rats (16 to 20 litters)	—	—	—
Mating	1:1 advisable; mating period of 2–3 weeks or until observation of plug	—	—	—
Route of test compound administration	Similar to the route used by humans	—	—	—
Exposure to test compound	Premating treatment interval of 2 weeks for females and 2, 4, or 10 weeks for males; treatment should continue throughout mating to termination of males and at least through implantation for females	—	—	—
Procedures on test animals	Sacrifice females after midgestation and males any time after mating. Dams: clinical signs; body weight; feed; gross necropsy; vaginal smears; preserve ovaries, uteri and organs with macroscopic findings; count corpora lutea and implantation sites (live and dead conceptuses). Males: gross necropsy; preserve testes, epididymides, and organs with macroscopic findings; assess sperm count and viability	—	—	—

[a] FDA, *Fed. Regist.*, 59(183), 1994; FDA, *Fed. Regist.*, 61(67), April 5, 1996.
[b] Reproduction and fertility assessments are covered in the EPA OPPTS 870.3800 multigenerational guideline.
[c] Reproduction and fertility assessments are covered in the FDA "Redbook" two generation guideline.
[d] Reproduction and fertility assessments are covered in the OECD 415 and 416 multigeneration guidelines.

The current ICH guideline[2] allows any combination of intervals A through F, provided all the intervals are evaluated. This procedure eliminates differences in durations of the various segments and allows evaluation of the conceptuses at any time in gestation. Additional flexibility was added to the fertility assessments by allowing the duration of the premating treatment period in male animals to be reduced from 60 to 28 days (4 weeks)[3] and later to 14 days[4] when the results of a 4-week subchronic study in the same strain and species do not identify adverse effects on male reproductive organs. Most studies continue to be performed using the 28-day exposure period. Should adverse effects be present in the subchronic (multidose) study, the 60-day premating treatment period may be used, or other studies may be performed to characterize the effects. In contrast, the FDA Red Book[21] and the OECD[28,29] and EPA[23] approaches are to provide more fixed protocol designs to evaluate the same endpoints as those in studies performed for evaluation of pharmaceuticals. A full comparison of the ICH endpoints and the multiple protocols that evaluate these endpoints is provided in Table 19.5, Table 19.6, Table 19.7, and Table 19.8.

F. Reproductive Performance

Mating and fertility are essential endpoints of reproductive studies; the techniques used in these evaluations are the same as those used for impregnating animals to be used in developmental

Table 19.6 Comparison of current guidelines for the assessment of embryo and fetal development

Guideline	ICH 4.1.3[2]	EPA OPPTS 870.3700[22]	FDA Red Book[21]	OECD[27]
Species (No.)	One rodent, preferably rats; one nonrodent, preferably rabbits	Rodent, e.g., rat, and nonrodent, e.g., rabbit (approximately 20 animals with implantation sites at necropsy)	Rats, rabbits, hamsters, or mice (minimum 20 pregnant)	Rats, rabbits, mice, or hamsters (20 pregnant)
Mating	Plug, sperm, or insemination is day 0	Mate females with males of the same strain and species; plug or sperm is day 0	In-house recommended; one male to one or two females; plug or sperm is day 0	Mate females with males of the same strain and species; plug or sperm is day 0
Route of test compound administration	Similar to the route used by humans	Orally by intubation or potential human exposure (need justification)	Intended human exposure (diet or drinking water)	Orally by intubation; alternate route with justification
Exposure to test compound	From implantation to the closure of the hard palate	From at least the day of implantation to day before parturition, or from fertilization to day before termination	From implantation to day before parturition, or from fertilization to termination	From preimplantation through entire gestation period to the day before caesarean-sectioning
Procedures on test animals	Dam: clinical signs, body weight, feed; necropsy, preserve organs with macroscopic findings; count corpora lutea; implantations (live and dead conceptuses) Fetal: gender, body weight, fetal abnormalities (gross, soft tissue, and skeletal; rodents: half skeletal and half soft tissue; rabbits: all fetuses, gross evaluation of placentas	Dam: clinical signs, body weight, feed; terminal weight, gross necropsy; examine uterine contents (live and dead conceptuses), examine nongravid uteruses, weigh gravid uterus (with cervix), count corpora lutea Fetal: gender, body weight, fetal abnormalities (external, skeletal, and soft tissue anomalies; rodents: half skeletal and half soft tissue; rabbits: all fetuses)	Dam: clinical signs, body weight, feed; terminal weight, gross necropsy; gravid uterine weight, examine uterine contents (live and dead conceptuses), count corpora lutea Fetal: gender, body weight, fetal abnormalities (external, skeletal, and soft tissue anomalies; rodents: half for skeletal and half for soft tissue; rabbits: all fetuses)	Dam: clinical signs, body weight, feed, terminal weight, gross necropsy; examine nongravid uteri, gravid uterine weight (with cervix), examine uterine contents (live or dead conceptuses), count corpora lutea Fetal: gender, body weight, fetal abnormalities (external, skeletal, and soft tissue anomalies; rodents: half for skeletal and half for soft tissue; rabbits: all fetuses)

Table 19.7 Comparison of current guidelines for the postnatal and behavioral assessments

Guideline	ICH 4.1.2[a]	EPA OPPTS 870.6300[b]	FDA Red Book[c]	OECD[d]
Species (No.)	At least one species, preferably rats	Approximately 20 pregnant rats in control group; similar numbers for mating pairs in treatment groups	—	—
Mating	Plug, sperm, or insemination is day 0	Mate females with males of the same strain and species (plug or sperm is day 0)	—	—
Route of test compound administration	Similar to the route used by humans	Oral (diet, drinking water, or gavage)	—	—
Exposure to test compound	From implantation to the end of lactation	From implantation through day 10 postnatally	—	—
Procedures on test animals	Females deliver and rear offspring to weaning; select one weanling/sex/litter to assess reproductive competence Maternal observations: clinical signs, body weight, and feed consumption Necropsy: all adults, preserve organs with macroscopic findings for possible histological evaluation; implantations, abnormalities in offspring, live or dead offspring at birth, body weight at birth, pre- and postweaning survival and growth and body weight, maturation and fertility, physical development, sensory functions, reflexes, and behavior	Dams: detailed clinical observations, including assessment of signs of autonomic function; description of abnormal movements, posture, gait, and behavior, etc.; body weights (at least weekly, day of delivery, and PDs 11 and 21) Observation of offspring: daily cageside observations (PDs 4, 11, 21, 35, 45, and 60) for gross signs of mortality or morbidity Developmental landmarks: count and weigh live pups (PDs 0, 4, 7, 14, and 21, then every 2 weeks) and assess sexual maturation; cull on PD 4 Behavior: learning and memory (weaning and adulthood), motor activity (PDs 13, 17, 21, and 60 ± 2), auditory startle (PDs 22 and approximately PD 60) Neuropathology: on PND 11 and at termination, remove and weigh brains of 1 pup/sex/litter (total 10/sex/dosage group), then select 6 pups/sex/dose for neuropathological evaluation	—	—

[a] FDA, *Fed. Regist.*, 59(183), 1994.
[b] Postnatal assessments are covered in the EPA OPPTS 870.6300 Development Neurotoxicity study and EPA OPPTS 870.3800 multigeneration guideline.
[c] Postnatal assessments are covered in the FDA "Redbook" two generation guideline.
[d] Postnatal assessments are covered in the OECD 415 and 416 multigeneration guidelines PND(s) = day(s) postnatal.

toxicity studies. As noted for male animals, reduced fertility may have many causes, and the production of a litter is not sufficient to assume the absence of an effect on reproductive performance.

Rats and mice generally have multiple intromissions in one copulatory interval, and such activity generally is associated with improved fertility in the females. Should it appear that aberrant copulatory behavior is present, videotaping the animals and grading the numbers of copulations (matings) and intromissions, duration of intromission, and expected female receptivity (lordosis) is recommended.

The ICH guidelines do not identify specific methods for performing the sperm evaluations, while EPA protocols require evaluation of specific parameters by recommended methodologies, generally including automated sperm evaluations. There are two generally used automated techniques, the Cellsoft System and the Hamilton Thorne System. While the Cellsoft system had the most advanced technical capacity, and hand methods are sufficient to meet the guideline requirements,[86,87] the Hamilton Thorne System has become that most widely used.

Table 19.8 Comparison of current guidelines for evaluating reproductive function

Guideline	EPA OPPTS 870.3800[23]	FDA Red Book[21]
Species (No.)	Rats, both males and females 5 to 9 weeks in age (approximately 20 pregnant females in the control group); similar numbers for mating pairs in treatment groups	Minimum of 20 rats/sex; begin F_0 with 30 rats/sex/group; begin F_1 with 25 rats/sex/group
Mating	Mate females with males of the same strain and species (plug or sperm is day 0)	1:1 from the same dose until pregnant or 2–3 weeks elapsed (plug or sperm is day 0)
Route of test compound administration	Oral (diet, drinking water, or gavage)	Diet, drinking water, or gavage
Exposure to test compound	P: at least 10 weeks before mating and through termination F_1: begin at weaning and continue into adulthood, mating, and production of F_2 and until the F_2 offspring are weaned	F_0 males: at least 10 weeks before mating and during mating F_0 females: at least 10 weeks before mating and during mating and pregnancy to weaning of F_1 litter F_1 litter: prenatal period through postnatal life
Procedures on test animals	Parental: clinical observations; body weights (first day of treatment, then weekly, GDs 0, 7, 14, and 21, and during lactation on the same days as the weighing of litters); food and/or water consumption; estrous cycle length (vaginal smears for a minimum of 3 weeks prior to mating and throughout cohabitation) Sperm evaluation: (all P and F_1 males) sperm from one testis and one epididymis for enumeration of homogenization-resistant spermatids and cauda epididymal sperm reserves, respectively; sperm motility, and sperm morphology Offspring: number and sex of pups, stillbirths, live births, and the presence of gross anomalies; count and sex live pups, weigh (days 0, 4, 7, 14, and 21 of lactation, at sexual maturation, and at termination) and assess sexual maturation; cull on day 4 (optional) Termination: all P and F_1 adult males and females when they are no longer needed for assessment of reproductive effects; all unselected F_1 offspring and all F_2 offspring at comparable ages after weaning Gross necropsy on all P and F_1 parental animals and at least three pups per sex per litter from unselected F_1 weanlings and F_2 weanlings; observe reproductive organs; vaginal smears to determine stage of estrous; examination of uteruses of all cohabited females (for presence and number of implantation sites) Organ weights on all P and F_1 parental animals: weigh uterus (with oviducts and cervix), ovaries; testes, epididymides (total weights for both and cauda weight for either one or both), seminal vesicles (with coagulating glands and their fluids) and prostate; brain, pituitary, liver, kidneys, adrenal glands, spleen, and known target organs Organ weights on all F_1 and F_2 weanlings: brain, spleen, and thymus Tissue preservation and histopathology on P and F_1 parental animals: vagina, uterus with oviducts, cervix, and ovaries; one testis, one epididymis, seminal vesicles, prostate, and coagulating gland; pituitary and adrenal glands; target organs, grossly abnormal tissue; for F_1 and F_2 weanlings selected for macroscopic examination: grossly abnormal tissue and target organs; primordial follicle count	Recommendation: two generations; randomly select F_1 generation by taking at least one rat pup/sex/litter to mate with a pup from the same dosage group but a different litter Parental: clinical observations at least twice daily; record behavioral changes, signs of toxicity, and mortality; evaluate all F_0 and F_1 females for estrous cycle length by daily vaginal smears (minimum of 3 weeks prior to mating and during cohabitation); body weights before first dosage, once weekly thereafter, and at necropsy; feed consumption weekly, water consumption (if the test substance is given in water) Offspring: number and sex of pups, stillbirths, live births, and the presence of gross anomalies; evaluate dead pups; body weights (PD 0, 4, 7, 14, and 21); measure anogenital distance of F_2 pups (PD 0); and pubertal evaluations for F_1 weanlings (i.e., age and weight at vaginal opening or balanopreputial separation) Necropsy on all F_0 and F_1 control and treated parental animals: observe and weigh brain, pituitary, liver, kidneys, adrenal glands, spleen, known target organs, and reproductive organs (i.e., uterus and ovaries, both testes, seminal vesicles with coagulating glands, and the prostate) Tissue preservation: (females) vagina, uterus with cervix, ovaries with oviducts, adrenal and pituitary glands, target organs, and grossly abnormal tissue; (male) one testis, one epididymis, seminal vesicles, coagulating glands, prostate, and adrenal and pituitary glands, target organs, and grossly abnormal tissue Histopathology: male (e.g., corpus, cauda, and caput epididymis) and female (e.g., primordial follicles) reproductive organs Additional evaluations: testicular spermatid numbers and sperm motility, morphology, and count; fertility, gestation, live-born, weaning, and viability indices; percentage by sex Options: culling; a third generation (i.e., if overt reproductive, morphologic, and/or toxic effects of a test substance are observed) and a teratology phase; neurotoxicity and immunotoxicity screens

Guideline	OECD 415 (One Generation)[28]	OECD 416 (Two Generation)[29]
Species (No.)	Rats 5 to 9 weeks in age (approximately 20 pregnant/group)	Rats 5 to 9 weeks of age (approximately 20 pregnant/group)
Mating	1:1 or 1:2 male to female; plug or sperm is day 0	Parental: 1:1 until mating occurs or 2 weeks elapse; plug or sperm is day 0 F_1; one weanling/sex/litter with other pups of the same dose level but from a different litter
Route of test compound administration	Oral (diet, drinking water or gavage)	Oral (diet, drinking water or gavage)
Exposure to test compound	Parental males: at least 10 weeks before mating Parental females: at least 2 weeks before mating; continuing during 3-week mating period, pregnancy, and until weaning the F_1 offspring	Parental males: daily beginning at 5 to 9 weeks of age, 10 weeks before mating Parental females: 2 weeks before mating; continuing during mating (3 weeks), pregnancy, and through weaning of F_1 offspring F_1: begin at weaning and continue until termination
Procedures on test animals	Parental: clinical observations (at least once daily); body weights (first day of treatment, then weekly); weekly feed consumption (premating and mating) and during lactation on the days litters are weighed Offspring: number and sex of pups, stillbirths, live births, and the presence of gross anomalies in the offspring; count and weigh live pups (days 0, 4, and 7, then weekly until termination); culling optional Gross necropsy: all parental animals, with emphasis on reproductive organs Histopathology: ovaries, uterus, vagina, cervix, testes, epididymides, seminal vesicles, prostate, coagulating gland, pituitary, and target organs of all P parental animals (i.e., in control and high dosage group P and F_1 parental animals selected for mating, and rats that die)	Parental: clinical observations (at least once daily); body weights (first day of dosing, then weekly); weekly feed consumption (premating and mating) and during lactation on the days litters are weighed Offspring: number and sex of pups, stillbirths, live births, and the presence of gross anomalies in the offspring; count and weigh live pups (days 0, 4, and 7, then weekly until term); culling optional Gross necropsy: all parental animals (all P and F_1), with emphasis on reproductive organs Histopathology: ovaries, uterus, vagina, cervix, testes, epididymides, seminal vesicles, prostate, coagulating gland, pituitary, and target organs of all control and high dose group P and F_1 parental animals selected for mating, and rats that die

Note: P = parental generation, F_0 = parental generation, F_1 = first-generation offspring, F_2 = second-generation offspring, PD = postnatal day, GD = gestational day.

1. Evaluation of the Ovaries

Ovarian weights may be recorded close to the time the ovaries are removed and trimmed to avoid dehydration, or after trimming and fixation. These same techniques as those generally used for recording organ weights.

Although follicle number and size can be identified by histological evaluation of cross-sections of the ovary,[88,89] this procedure is extremely time consuming and expensive (30 to 60 sections per ovary for mice and rats, resulting in 300 and 600 sections per group of 10 animals, respectively). A shorter screening method, described by Plowchalk et al.[90] for potential use when indicated by observations in companion studies, e.g., pharmacologic action or reduced ovarian weight or atresia, has the potential to save time and expense. However, when based only on primordial follicles, it should not be used as the sole endpoint to identify reproductive toxicity.[91] This statement conforms with the harmonization efforts between the EPA and OECD. The EPA draft guidance for reproduction and fertility effects[92] was modified to remove "triggers" that would be inappropriate to identify chemicals requiring further evaluation and to include an assessment of F_1 females for every study. The modifications also added flexibility in methodological criteria (e.g., laboratory-based selection of ovarian section, number of females to be evaluated, and methods to increase the speed in counting and the statistical power of the sample[93]).

G. Maternal and Pup Interactions and Lactation

Lactation, maternal-pup interaction, and pup growth and development until weaning are monitored during ICH Stage E and in FDA Red Book[21], OECD, and EPA multigeneration studies,[23,28,29] as well as in the OECD/EPA Developmental Neurotoxicity Study[24,31] and the EPA One Generation Extension[45] and Pubertal Assays.[46,47]

H. Growth and Development of Conceptus through Puberty

1. Fetal Evaluations

The various guidelines have differences in terms of the numbers of litters and fetuses evaluated and the technical procedures recommended. The ICH guideline[2] requires evaluation of all nonrodent fetuses for gross external, soft tissue, and skeletal alterations. Although it is possible to similarly evaluate all rodent fetuses, for practical reasons (the litters are large) only external alterations are noted for all rodent fetuses. Subsequently rodent litters are assigned to soft tissue or skeletal evaluations, approximately one-half of the fetuses to each evaluation, regardless of the method used to evaluate the fetal soft tissue. When a finding is of special interest, the full set of examinations may be performed, although such is not common practice.

The ICH guideline[2] recognizes each of the four possible outcomes[67] of adverse exposure of the conceptus as important. Regardless of differences in nomenclature, these endpoints remain death (preimplantation loss, early or late resorption, and fetal death, or postimplantation loss, and in some cases, abortion resulting in postimplantation loss), reduced body weight, dysmorphology (malformation, variations associated with retarded growth or with the genetic background of the species), and functional alterations (e.g., behavioral teratology, developmental neurotoxicity, functional teratology, functional alterations of organs, transplacental carcinogenicity).

2. Postnatal Evaluations for Gender and Sexual Maturation

a. Culling of Litters

In the United States and Japan, litters are commonly culled to a standard number of pups of each sex. Some investigators believe that this practice can bias study outcome.[94] The ICH guideline[2] allows use of either procedure, provided justification is given for use of the selected procedure. It is expected that additional study in this area may clarify whether and when culling should be performed.

b. Juvenile Evaluations, including Postnatal Behavioral and Functional Evaluations

The ICH guideline[2] requires that behavioral and functional evaluations be performed after *in utero* exposure of ICH Stages C, D, and E, i.e., embryogenesis, fetogenesis, and lactation. Thus, when the treatment interval includes these three ICH stages, postnatal evaluations need be performed only once; when these ICH stages are tested separately, postnatal evaluations are necessary in each study. This position was supported by the observation that for many agents, fetal and pup body weights were as sensitive or more sensitive indicators of developmental toxicity than behavioral and functional evaluations,[95] regardless of whether treatment occurred continually during gestation or was limited to either embryogenesis or fetogenesis and lactation.[96] It must also be noted that assessment of body weight and other developmental parameters can be confounded by live litter size, sex ratio, and gestational age.

Although the specific postnatal evaluations to be performed are not provided by the ICH guideline,[2] the testing facility must justify the appropriateness of the tests to be used and demonstrate their historical experience and the value of the tests. Because developmental landmark tests highly correlate with body weights and gestational ages,[95] such factors should be considered when these tests are performed. In general, the only physical landmarks considered to be necessary observations in the behavioral and functional evaluations were those related to sexual maturation, i.e., age at vaginal patency or balanopreputial separation. It was recommended that tests evaluating learning, memory, and activity be included in the behavioral and functional evaluations, and that these parameters be evaluated by using tests that are not subject to investigator bias (i.e., to the extent possible, automated tests should be used). Because behavioral and functional testing procedures are described elsewhere in this book, only those relevant to sexual maturation are described (see Table 19.8).

c. Balanopreputial Separation and Vaginal Patency

Pups should be evaluated daily for balanopreputial separation and vaginal patency, beginning on approximately postnatal days (PD) 39 and 28, respectively. Sexual maturation assessments are best performed on at least three randomly selected weanling rats per sex, although common practice has been to evaluate only one per sex. The criteria for identifying these endpoints are achieved when complete preputial separation or vaginal opening are observed, i.e., when the prepuce completely retracts from the head of the penis or when there is any visible break in the membranous sheath covering the vaginal orifice, respectively. Detailed descriptions of these procedures and diagrams are available in Lewis et al.[97]

I. Endocrine-Mediated Findings

As previously noted, the EPA, in combination with the OECD, has an ongoing program to develop and validate new protocols for identifying endocrine-mediated effects. These new protocols were first cited in the Final Report of the Endocrine Disruptor Screening and Testing Advisory Committee.[98] They are variations of the ICH peri-postnatal and juvenile studies and, in general, assess the same endpoints. However, because they will be EPA-regulated studies, the designs identified after validation of these protocols should be used. In other words, the flexibility and scientific digression acceptable in ICH guideline-compliant studies will not be present in these OECD/EPA study designs. The three study designs currently undergoing the validation process are:

1. One-generation extension study.[45]
2. Assessments of pubertal development and thyroid function in juvenile female rats on PD 22 through 42 or 43.[47] The purpose of this assay is to evaluate the effects of estrogens on vaginal patency. The theoretical advantage of this assay over the uterotrophic assay is that both estrogen agonists and antagonists are detected by the same test. Additionally, this pubertal assay detects xenoestrogens, aromatase inhibitors, and alterations in hypothalamic-pituitary-gonadal function.
3. Assessment of pubertal development and thyroid function in juvenile male rats on PD 23 through 52 or 53.[46] The purpose of this pubertal protocol is to assess the effects of androgens and antiandrogens on preputial separation and the entire endocrine system after approximately 3 weeks of treatment.

J. Comparison of FDA and EPA Requirements and Methods for Postnatal Behavioral and Functional Evaluations

The revised EPA multigeneration (two-generation) protocol has multiple new endpoints that have been historically performed in the development of pharmaceuticals. However, as previously noted, EPA protocols differ from those developed for compliance with the ICH guidelines[2–4] in that the EPA protocols are set procedures. Changes must be negotiated, rather than based solely on scientific judgment and justification regarding design. Endpoints not previously assessed in EPA-compliant protocols include:

Estrous cyclicity data (vaginal smears) for 3 weeks prior to mating, during mating, and at termination
Sperm parameters (count, motility, and morphology)
Developmental milestones (age at vaginal opening and preputial separation for the F_1 generation, anogenital distance for the F_2 generation, if triggered by a treatment-related effect on the sexual maturation of the F_1 generation)
Gross pathology and selected histopathology of weanlings and adult organ weight data (uterus, ovaries, testes, one epididymis, including the total weight and cauda epididymal weight, seminal vesicles with coagulating glands and fluids, prostate, brain, liver, kidneys, adrenal glands, spleen, thymus, and all known target organs)

K. Developmental Neurotoxicity

The EPA Developmental Neurotoxicity Test (DNT) is an extension of the procedures used for behavioral and functional testing during postnatal development. In this test, the vehicle is administered to a control group, and at least three groups of pregnant animals are administered the test substance. Exposure of the mated dams occurs during gestation and early lactation; pups are randomly selected from within litters and assessed for neurotoxicity. Observations are made to detect gross neurologic and behavioral abnormalities, as well as to assess motor activity, response to auditory startle, and learning. A neuropathological evaluation is performed, and brain weights and morphometry are determined. The DNT can be performed as a separate study, as a follow up to a standard developmental toxicity and/or adult neurotoxicity study, or as part of a two-generation reproduction study. Assessment of the offspring is conducted on the second (F_2) generation.

References

1. International Conference on Harmonisation (ICH) Harmonised Tripartite Guideline, Detection of toxicity to reproduction for medicinal products (proposed rule endorsed by the ICH Steering Committee at Step 4 of the ICH Process, 24 June 1993), in *Proceedings of the Second International Conference on Harmonization*, D'Arcy, P.F. and Harron, D.W.G., Eds., Greystone Books, Ltd. Antrim, N. Ireland, 1993, p. 557.
2. U.S. Food and Drug Administration, International Conference on Harmonisation, Guideline on detection of toxicity to reproduction for medicinal products, *Fed. Regist.*, 59(183), 1994.

3. U.S. Food and Drug Administration, International Conference on Harmonisation, Guideline on detection of toxicity to reproduction for medicinal products, Addendum on toxicity to male fertility, *Fed. Regist.*, 61(67), April 5, 1996.
4. U.S. Food and Drug Administration, International Conference on Harmonisation, *Maintenance of the ICH Guideline on Toxicity to Male Fertility*, An addendum to the ICH tripartite guideline on detection of toxicity to reproduction for medicinal products, amended November 9, 2000.
5. U.S. Food and Drug Administration, *Guidelines for Reproduction Studies for Safety Evaluation of Drugs for Human Use*, 1966.
6. U.S. Food and Drug Administration, *Draft: Toxicological Principles for the Safety Assessment of Direct Food Additives and Color Additives Used in Food* (Red Book II), *Guidelines for Reproduction and Developmental Toxicity Studies*, Center for Food Safety and Applied Nutrition, Washington, DC, 1993, p. 123.
7. Christian, M.S. and Voytek, P.E., In vivo reproductive and mutagenicity tests, in *A Guide to General Toxicology*, Homburger, F., Hayes, J.A., and Pelikan, E.W., Eds., Karger, Basel, Switzerland, 1982, p. 294.
8. U.S. Environmental Protection Agency, Subpart E — specific organ/tissue toxicity, No. 798.4900: developmental toxicity study, 40 CFR Part 798, Toxic Substances Control Act test guidelines, final rules, *Fed. Regist.* 50:39433-39434, 1985a.
9. U.S. Environmental Protection Agency, Subpart E — specific organ/tissue toxicity, No. 798.4700: reproduction and fertility effects, 40 CFR Part 798, Toxic Substances Control Act test guidelines, final rules, *Fed. Regist.*,, 50:39432-39433, 1985b.
10. U.S. Environmental Protection Agency, *Toxic Substances Control Act Test Guidelines* (TSCA), (OPPTS-42193: FRL-5719-5), Final rule, 40 CFR Part 799, 1997.
11. U.S. Environmental Protection Agency (EPA-FIFRA), *Hazard Evaluation Division Standard Evaluation Procedure: Teratology Studies*, Office of Pesticide Programs, Washington, DC, EPA-540/9-85-018, 1985.
12. U.S. Environmental Protection Agency (EPA-FIFRA), Pesticide Assessment Guideline, *Subdivision F — Hazard Evaluation: Human and Domestic Animals, Addendum 10: Neurotoxicity, Health Effects Division*, Office of Pesticide Programs, March 1991.
13. U.S. Environmental Protection Agency (EPA-FIFRA), *Health Effects Division Draft Standard Evaluation Procedure, Developmental Toxicity Studies*, Office of Pesticides Programs, Washington, DC, 1993.
14. Committee on the Safety of Medicines, *Notes for Guidance on Reproduction Studies*, Department of Health and Social Security, Great Britain, 1974.
15. European Economic Community, The rules of governing medicinal products in the European Community, in *Guidelines on the Quality, Safety and Efficacy of Medicinal Products for Human Use*, Vol. III, Office of Official Publications of the European Communities, Brussels, 1988.
16. Ministry of Health and Welfare, Japan, *On Animal Experimental Methods for Testing the Effects of Drugs on Reproduction*, Notification No. 529 of the Pharmaceutical Affairs Bureau, Ministry of Health and Welfare, March 31, 1975.
17. Ministry of Health and Welfare, Japan, *Information on the Guidelines of Toxicity Studies Required for Applications for Approval to Manufacture (Import) Drugs*, Notification No. 118 of the Pharmaceutical Affairs Bureau, Ministry of Health and Welfare, February 15, 1984.
18. Tanimura, T., Kameyama, Y., Shiota, K., Tanaka, S., Matsumoto, N., and Mizutani, M., *Report on the Review of the Guidelines for Studies of the Effect of Drugs on Reproduction*, Notification No. 118, Pharmaceutical Affairs Bureau, Ministry of Health and Welfare, Japan, 1989.
19. International Conference on Harmonisation (ICH) Harmonised Tripartite Guideline, Male fertility studies in reproductive toxicology, in *Proceedings of the Third International Conference on Harmonization*, D'Arcy, P.F. and Harron, D.W.G., Eds., Greystone Books, Ltd., Antrim, N. Ireland, 1995, p. 245.
20. U.S. Food and Drug Administration, *Toxicological Principles and Procedures for Priority Based Assessment of Food Additives (Red Book), Guidelines for Reproduction Testing with a Teratology Phase*, Bureau of Foods, Washington, DC, 1982, p. 80.
21. U.S. Food and Drug Administration, *Toxicological Principles for the Safety of Food Ingredients (Red Book 2000)*, Sections issued electronically July 7, 2000 at http:// vm.cfscan.fda.gov/~Redbook/Redtcct.html, 2000.

22. U.S. Environmental Protection Agency, *Health Effects Test Guidelines: Prenatal Developmental Toxicity Study*, Office of Prevention, Pesticides and Toxic Substances (OPPTS) 870.3700, 1998.
23. U.S. Environmental Protection Agency, *Health Effects Test Guidelines: Reproduction and Fertility Effects*, Office of Prevention, Pesticides and Toxic Substances (OPPTS) 870.3800, 1998.
24. U.S. Environmental Protection Agency, *Health Effects Test Guidelines: Developmental Neurotoxicity Study*, Office of Prevention, Pesticides and Toxic Substances (OPPTS) 870.6300, 1998.
25. U.S. Environmental Protection Agency (EPA-Risk Assessment Forum), Guidelines for developmental toxicity risk assessment, *Fed. Regist.*, 56(234): 63798-63826, Dec. 5, 1991.
26. U.S. Environmental Protection Agency, *Guidelines for Reproductive Toxicity Risk Assessment*, NTIS PB No. PB97-100098, 1996a.
27. Organization for Economic Cooperation and Development, *Guidelines for Testing of Chemicals, Section 4, No. 414: Teratogenicity*, adopted 22 January 2001.
28. Organization for Economic Cooperation and Development, *OECD Guidelines for Testing of Chemicals, Section 4, No. 415: One-Generation Reproduction Toxicity*, adopted 26 May 1983.
29. Organization for Economic Cooperation and Development, *OECD Guidelines for Testing of Chemicals, Section 4, No. 415: Two-Generation Reproduction Toxicity*, adopted 22 January 2001(b).
30. Organization for Economic Cooperation and Development, *OECD Guidelines for Testing of Chemicals, Section 4, No. 422: Combined Repeated Dose Toxicity Study with the Reproduction/Developmental Toxicity Screening Test*, adopted 22 March 1996.
31. Organization for Economic Cooperation and Development, *OECD Guidelines for Testing of Chemicals, Section 4, No. 426: Developmental Neurotoxicity Study*, drafted October 1999.
32. Lenz, W. Kindliche micebildungen nach medikament wahrend der draviditat? *Deutsch. Med. Wochenschr.*, 86, 2555, 1961.
33. McBride, W.G., Thalidomide and congenital abnormalities, *Lancet*, 2, 1358, 1961.
34. Herbst, A.L., Ulfelder, H., Poskanzer, D.C., Adenocarcinoma of the vagina: association of maternal stilbestrol therapy with tumor appearance in young women, *New Engl. J. Med.*, 284, 878, 1971.
35. Koos, B.J. and Longo, L., Mercury toxicity in pregnant women, fetus and newborn infant, *Am. J. Obstet. Gynecol.*, 126, 390, 1976.
36. Christian M.S. and Hoberman A.M., Perspectives on the U.S., EEC, and Japanese developmental toxicity testing guidelines, in *Handbook of Developmental Toxicology*, Hood, R.D., Ed., CRL Press, Inc., New York, 1997, p. 551.
37. U.S. Food and Drug Administration, *1998 Final Rule: Pediatric Studies*, adopted April 1, 1999.
38. U.S. Environmental Protection Agency, *Executive Order: Protection of Children from Environmental Health Risks and Safety Risks*, available at http://www.epa.gov/children/whatwe/executiv.htm (verified 11 June 2002).
39. U.S. Food and Drug Administration, Food Quality Protection Act (FQPA): Public Law 104-170, 1996a.
40. U.S. Environmental Protection Agency, Safe Drinking Water Act. (SDWA): Public Law 104-182, 1996b.
41. U.S. Environmental Protection Agency, *Endocrine Disruptor Screening Program*, available at http://www.epa.gov/scipoly/oscpendo/index.htm (verified 11 June 2002).
42. Endocrine Disruptor Screening and Testing Advisory Committee (EDSTAC), *EDSTAC Final Report, Compilation of EDSTAC Recommendations*, chap. 7, available at http://www.epa.gov/scipoly/oscpendo/history/chap7v14.pdf (verified 11 June 2002).
43. U.S. Environmental Protection Agency, Endocrine disruptor screening program: statement of policy notice, *Fed. Regist.*, 63(248): 71541-71568, available at http://www.epa.gov/scipoly/oscpendo/fr122898_1.pdf (verified 11 June 2002).
44. U.S. Environmental Protection Agency, Endocrine Disruptor Method Validation Subcommittee under the National Advisory Council for Environmental Policy and Technology, *Fed. Regist.*, 66(197): 51951–51952, available at http://www.epa.gov/oscpmont/oscpendo/fedregnotice.htm (verified 11 June 2002).
45. Research Triangle Institute, *Draft Protocol: One-Generation Extension Study of Vinclozolin and di-n-butyl Phthalate Administered by Gavage on Gestational Day 6 to Postnatal Day 21 in CD® (Sprague-Dawley) Rats, Study No. RTA-ED01*, Research Triangle Institute, Research Triangle Park, NC, for the US EPA, 2001.
46. Endocrine Disruptor Screening and Testing Advisory Committee, Male 20-day thyroid/pubertal assay in rodents, *EDSTAC Final Report* 1:5–30, 1998.
47. Endocrine Disruptor Screening and Testing Advisory Committee, Female 20-day thyroid/pubertal assay in rodents, *EDSTAC Final Report* 1:5–26, 1998.

48. Endocrine Disruptor Screening and Testing Advisory Committee, *EDSTAC Final Report*, Chap. 5, Appendix L, Protocols for Tier 1 screening assays, available at http://www.epa.gov/scipoly/oscpendo/history/app-lv14.pdf (verified 11 June 2002).
49. Organization for Economic Cooperation and Development, *OECD Guidelines for Testing of Chemicals, Section 4, No. 407: Repeated Dose 28-Day Oral Toxicity Study in Rodents*, adopted 27 July 1995.
50. Interagency Regulatory Group, *Recommended Guideline for Teratogenicity Studies in the Rat, Mouse, Hamster or Rabbit*, 1980.
51. Christian, M.S., Harmonization of reproductive guidelines: a perspective from the International Federation of Teratology Societies, *J. Am. Coll. Toxicol.*, 11(3), 299, 1993.
52. D'Arcy, P.F. and Harron, D.W.G., Eds., Topic 2: Reproductive toxicology, in *Proceedings of the First International Conference on Harmonisation*, The Queen's University of Belfast, Belfast, Ireland, 1992, p. 255.
53. Organization for Economic Cooperation and Development (OECD), Final report — OECD short-term and long-term toxicology groups, *Teratogenicity*, 110, 1979.
54. U.S. Environmental Protection Agency, *Harmonization and Regulatory Coordination: OECD Pesticide Working Group*, available at http://www.epa.gov/oppfead1/international/harmonization.html (verified 18 June 2002).
55. U.S. Environmental Protection Agency, Federal Insecticide, Fungicide, and Rodenticide Act: 7 U.S.C. s/s 136 et seq., available at http://www4.law.cornell.edu/uscode/7/ch6.html (verified 18 June 2002).
56. U.S. Food and Drug Administration, The Federal Food, Drug and Cosmetic Act: §408, 21 U.S.C. 348a.
57. U.S. Environmental Protection Agency, *About the Office of Pesticide Programs (OPP)*, available at http://www.epa.gov/pesticides/about.htm (verified 18 June 2002).
58. U.S. Environmental Protection Agency, *About OPPT*, available at http://www.epa.gov/oppt/oppt-abt.htm (verified 18 June, 2002).
59. U.S. Environmental Protection Agency, Toxic Substance Control Act (TSCA): 15 U.S.C. s/s 2601 et seq., available at http://www4.law.cornell.edu/uscode/15/ch53.html (verified 18 June 2002).
60. U.S. Environmental Protection Agency, *About Harmonized Test Guidelines*, available at http://epa.gov/docs/OPPTS_Harmonized/abguide.txt.html (verified 18 June 2002).
61. PMA guidelines, in *Reproduction, Teratology and Pediatrics*. Christian, M., Diener, R., Hoar, R., and Staples, R, Eds., 1981 (revised).
62. Sullivan, F.M., Behavioral teratology, in *Proc. 11th Conf. of the European Teratology Society*, Paris, 1983.
63. Christian, M.S., Postnatal alterations of gastrointestinal physiology, hematology, clinical chemistry, and other non-CNS parameters, in *Handbook of Experimental Pharmacology: Teratogenesis and Reproductive Toxicology*, Vol. 65, Springer, Berlin, 1983, p. 263.
64. Canada, Bureau of Drugs, Health Protection Branch, Health and Welfare, *Draft of Preclinical Txicologic Guidelines,* Section 2.4, 35, 1979.
65. Drug directorate guidelines, *Toxicological Evaluation 2.4, Reproductive Studies*, Ministry of National Health and Welfare, Health Protection Branch, Health and Welfare, Canada, 1990.
66. Somers, G.F, Letter to the editor, *Lancet*, 1, 912, 1962.
67. Wilson, J.G., Principles of teratology, in *Environment and Birth Defects*, Academic Press, New York, 1973, p. 11.
68. Palmer, A.K., Some thoughts on reproductive studies for safety evaluations, *Proc. Europ. Soc. Study Drug Tox.*, 14, 79, 1973.
69. Johnson, E.M. and Christian, M.S., When is a teratology study not an evaluation of teratogenicity? *J. Am. Coll. Toxicol.*, 3, 431, 1984.
70. Wilson, J.G., Methods for administering agents and detecting malformations in experimental animals, *Teratology, Principles and Techniques*, University of Chicago Press, Chicago, 1965, p. 262.
71. Staples, R.E., Detection of visceral alterations in mammalian fetuses, *Teratology*, 9, 37, 1974.
72. Christian, M.S., Test methods for assessing female reproductive and developmental toxicology, in *Principles and Methods of Toxicology*, Hayes, A.W., Ed., Taylor and Francis, Philadelphia, 2001, p. 1301.
73. Stuckhardt and Poppe, Fresh visceral examination of rat and rabbit fetuses used in teratogenicity testing, *Teratogenesis, Carcinogenesis and Mutagenesis*, 4, 181, 1984.
74. Staples, R.E. and Schnell, J.L., Refinement in rapid clearing technique in the KOH-alizarin red S method for fetal bone, *Stain. Technol.*, 39, 61, 1963.

75. Peters, P.W.J., Double staining of fetal skeletons for cartilage and bone, in *Methods in Prenatal Toxicology*, Neubert, D., Merker, W.G., and Kwasigroch, T.E., Eds., Georg Thieme Publ., Stuttgart, 1977, p.153.
76. Whitaker, J. and Dix, K.M., Double staining technique for rat foetus skeletons in teratological studies, *Lab. Animals*, 13:309, 1979.
77. Webb G.N. and Byrd R.A., Simultaneous differential staining of cartilage and bone in rodent fetuses: an alcian blue and alizarin red S procedure without glacial acetic acid, *Biotech Histochem.*, 69:181, 1994.
78. U.S. Environmental Protection Agency, *Health Effects Test Guidelines: Combined Repeated Dose Toxicity with the Reproduction/Developmental Toxicity Screening Test*, Office of Prevention, Pesticides and Toxic Substances (OPPTS) 870.3650, 2000.
79. Johnson, E.M., The scientific basis for multigeneration safety evaluation, *J. Am. Coll. Toxicol.*, 5(4), 197, 1986.
80. Christian, M.S., A critical review of multigeneration studies, *J. Am. Coll. Toxicol.*, 5(2), 161, 1986.
81. U.S. Food and Drug Administration, *Good Laboratory Practice Regulations, Final Rule 21*, Code of Federal Regulations, Part 58.
82. Organization for Economic Cooperation and Development, *Laboratory Practice in the Testing of Chemicals, Final Rule of the OCED Expert Group on Laboratory Practices* [C(81) 3.0 (Final), Annex 2], 1981.
83. Japanese Ministry of Agriculture, Forestry and Fisheries (MAFF), Good Laboratory Practice Regulations, 59 Noshan No. 3850, 1984.
84. U.S. Environmental Protection Agency, Federal Insecticide, Fungicide and Rodenticide Act (EPA-FIFRA), *Good Laboratory Practice Standards, Final Rule*, 40 CFR Part 160.
85. U.S. Environmental Protection Agency, Toxic Substances Control Act, *Good Laboratory Practice Standards, Final Rule*, 40 CFR Part 792.
86. Christian, M.S., The use of CASA systems for reproductive toxicology (CRYO Resources Ltd. CellSoft Series 4000 Tox. System vs the Hamilton Thorne HTM-IVOS), presented at the *82nd Meeting of the Mid-Atlantic Reproduction and Teratology Association*, Lambertville, PA, October 14–15, 1993.
87. Christian, M.S., Compliance with tripartite guideline for male reproductive performance, presented at the *Pharmaceutical Manufacturers Association Drug Safety Subsection Annual Meeting*, Austin, TX, October 17–20, 1993c.
88. Plowchalk, D.R. and Mattison, D.R., Phosphoramide mustard is responsible for the ovarian toxicity of cyclophosphamide, *Toxicol. Appl. Pharmacol.*, 107, 472, 1991.
89. Smith, B.J., Plowchalk, D.R., Sipes, I.G., and Mattison, D.R., Comparison of random and serial sections in assessment of ovarian toxicity, *Reprod. Toxicol.*, 5, 379, 1991.
90. Plowchalk, D.R., Smith, B.J., and Mattison, D.R., Assessment of toxicity to the ovary using follicle quantitation and morphometrics, in *Methods In Toxicology*, Vol. 3B, *Female Reproductive Toxicology*, Heindel, J.J. and Chapin, R.E., Eds., Academic Press, San Diego, 1993, p. 57.
91. Christian, M.S. and Brown W.R., Control primordial follicle counts in multigeneration studies in CRL Sprague-Dawley ("Gold Standard") rats, *The Toxicologist*, 66(1S), A1140, 2002.
92. U.S. Environmental Protection Agency, *Proposed Testing Guidelines, Notice of Availability and Request for Comments*, 61 FR 31522, 1996c.
93. U.S. Environmental Protection Agency, *Overview and Summary of Changes Made in the Harmonization of OPPTS 870 Toxicology Guidelines with OECD Guidelines*, available at http://www.epa.gov/docs/OPPTS_Harmonized/870_Health_Effects_Test_Guidelines/Summary/summchge.htm (verified 18 June 2002).
94. Palmer, A.K. and Ulbrich, B.C., The cult of culling, *Fund. Applied Toxicol.*, 38, 7, 1997.
95. Diener, R.M., Timing and utility of behavioral studies in developmental toxicology, in *Proceedings of the First International Conference on Harmonisation*, D'Arcy P.F. and Harron, D.W.G., Eds., The Queen's University of Belfast, Antrim, Northern Ireland,, 1993, p. 273.
96. Lochry, E.A. and Christian, M.S., Behavioral evaluations in reproductive safety assessments, *J. Am. Coll. Toxicol.*, 10(5), 585, 1991.
97. Lewis, E.M., Barnett Jr., J.F., Freshwater, L., Hoberman, A.M., and Christian, M.S., Sexual maturation data for Crl Sprague-Dawley rats: criteria and confounding factors, *Drug Chem. Toxicol.*, 2002.
98. Endocrine Disruptor Screening and Testing Advisory Committee, EDSTAC final report, available at http:// http://www.epa.gov/scipoly/oscpendo/history/finalrpt.htm (verified 11 June 2002).

APPENDIX 1

International Conference on Harmonisation of Technical Requirements for Registration of Pharmaceuticals for Human Use

ICH HARMONISED TRIPARTITE GUIDELINE
DETECTION OF TOXICITY TO REPRODUCTION FOR MEDICINAL PRODUCTS

Recommended for Adoption at Step 4 of the ICH Process on 24 June 1993 by the ICH Steering Committee

This Guideline has been developed by the appropriate ICH Expert Working Group and has been subject to consultation by the regulatory parties, in accordance with the ICH Process. At Step 4 of the Process the final draft is recommended for adoption to the regulatory bodies of the European Union, Japan and USA.

CONTENTS

1. Introduction
    1.1 Purpose of the Guideline
    1.2 Aim of Studies
    1.3 Choice of Studies
2. Animal Criteria
    2.1 Selection and Number of Species
    2.2 Other Test Systems
3. General Recommendations Concerning Treatment
    3.1 Dosages
    3.2 Route and Frequency of Administration
    3.3 Kinetics
    3.4 Control Groups
4. Proposed Study Designs — Combination of Studies
    4.1 The Most Probable Option
        4.1.1 Study of Fertility and Early Embryonic Development to Implantation
        4.1.2 Study For Effects on Pre- and Postnatal Development, including Maternal Function
        4.1.3 Study for Effects on Embryo-Fetal Development
    4.2 Single Study Design (Rodents)
    4.3 Two Study Design (Rodents)
5. Statistics
6. Data Presentation
7. Terminology

1. INTRODUCTION

1.1 Purpose of the Guideline

There is a considerable overlap in the methodology that could be used to test chemicals and medicinal products for potential reproductive toxicity. As a first step to using this wider methodology for efficient testing, this guideline attempts to consolidate a strategy based on study designs currently in use for testing of medicinal products; it should encourage the full assessment on the safety of chemicals on the development of the offspring. It is perceived that tests in which animals are treated during defined stages of reproduction better reflect human exposure to medicinal products and allow more specific identification of stages at risk. While this approach may be useful for most medicines, long term exposure to low doses does occur and may be represented better by a one or two generation study approach.

The actual testing strategy should be determined by:

> anticipated drug use especially in relation to reproduction,
> the form of the substance and route(s) of administration intended for humans and making use of any existing data on toxicity, pharmaco-dynamics, kinetics, and similarity to other compounds in structure/activity.

To employ this concept successfully, flexibility is needed (Note 1). No guideline can provide sufficient information to cover all possible cases, all persons involved should be willing to discuss and consider variations in test strategy according to the state of the art and ethical standards in human and animal experimentation. Areas where more basic research would be useful for optimization of test designs are male fertility assessment, and kinetic and metabolism in pregnant/lactating animals.

1.2 Aim of Studies

The aim of reproduction toxicity studies is to reveal any effect of one or more active substance(s) on mammalian reproduction. For this purpose both the investigations and the interpretation of the results should be related to all other pharmacological and toxicological data available to determine whether potential reproductive risks to humans are greater, lesser or equal to those posed by other toxicological manifestations. Further, repeated dose toxicity studies can provide important information regarding potential effects on reproduction, particularly male fertility. To extrapolate the results to humans (assess the relevance), data on likely human exposures, comparative kinetics, and mechanisms of reproductive toxicity may be helpful. The combination of studies selected should allow exposure of mature adults and all stages of development from conception to sexual maturity. To allow detection of immediate and latent effects of exposure, observations should be continued through one complete life cycle, i.e., from conception in one generation through conception in the following generation. For convenience of testing this integrated sequence can be subdivided into the following stages.

> A. Premating to conception (adult male and female reproductive functions, development and maturation of gametes, mating behavior, fertilisation).
> B. Conception to implantation (adult female reproductive functions, preimplantation development, implantation).
> C. Implantation to closure of the hard palate (adult female reproductive functions, embryonic development, major organ formation).
> D. Closure of the hard palate to the end of pregnancy (adult female reproductive functions, fetal development and growth, organ development and growth).
> E. Birth to weaning (adult female reproductive functions, neonate adaptation to extrauterine life, preweaning development and growth).
> F. Weaning to sexual maturity (postweaning development and growth, adaptation to independent life, attainment of full sexual function). For timing conventions, see Note 2.

1.3 Choice of Studies

The guideline addresses the design of studies primarily for detection of effects on reproduction. When an effect is detected, further studies to characterise fully the nature of the response have to be designed on a case by case basis (Note 3). The rationale for the set of studies chosen should be given and should include an explanation for the choice of dosages.

Studies should be planned according to the "state of the art", and take into account preexisting knowledge of class-related effects on reproduction. They should avoid suffering and should use the minimum number of animals necessary to achieve the overall objectives. If a preliminary study is performed the results should be considered and discussed in the overall evaluation (Note 4).

2.. ANIMAL CRITERIA

The animals used must be well defined with respect to their health, fertility, fecundity, prevalence of abnormalities, embryofetal deaths and the consistency they display from study to study. Within and between studies animals should be of comparable age, weight and parity at the start; the easiest way to fulfill these criteria is to use animals that are young, mature adults at the time of mating with the females being virgin.

2.1 Selection and Number of Species

Studies should be conducted in mammalian species. It is generally desirable to use the same species and strain as in other toxicological studies. Reasons for using rats as the predominant rodent species are practicality, comparability with other results obtained in this species and the large amount of background knowledge accumulated.

In embryotoxicity studies only, a second mammalian species traditionally has been required, the rabbit being the preferred choice as a "non-rodent". Reasons for using rabbits in embryotoxicity studies include the extensive background knowledge that has accumulated, as well as availability and practicality. Where the rabbit is unsuitable, an alternative non-rodent or a second rodent species may be acceptable and should be considered on a case by case basis (Note 5).

2.2 Other Test Systems

Other test systems are considered to be any developing mammalian and non-mammalian cell systems, tissues, organs, or organism cultures developing independently in vitro or in vivo. Integrated with whole animal studies either for priority selection within homologous series or as secondary investigations to elucidate mechanisms of action, these systems can provide invaluable information and, indirectly, reduce the numbers of animals used in experimentation. However, they lack the complexity of the developmental processes and the dynamic interchange between the maternal and the developing organisms. These systems cannot provide assurance of the absence of effect nor provide perspective in respect of risk/exposure. In short, there are no alternative test systems to whole animals currently available for reproduction toxicity testing with the aims set out in the introduction (Note 6).

3. GENERAL RECOMMENDATIONS CONCERNING TREATMENT

3.1 Dosages

Selection of dosages is one of the most critical issues in design of the reproductive toxicity study. The choice of the high dose should be based on data from all available studies (pharmacology, acute and chronic toxicity and kinetic studies, Note 7). A repeated dose toxicity study of about 2 to 4 weeks duration provides a close approximation to the duration of treatment in segmental designs of reproductive studies. When sufficient information is not available preliminary studies are advisable (see Note 4).

Having determined the high dosage, lower dosages should be selected in a descending sequence, the intervals depending on kinetic and other toxicity factors. Whilst it is desirable to be able to determine a "no observed adverse effect level", priority should be given to setting dosage intervals close enough to reveal any dosage related trends that may be present (Note 8).

3.2 Route and Frequency of Administration

In general the route or routes of administration should be similar to those intended for human usage. One route of substance administration may be acceptable if it can be shown that a similar distribution (kinetic profile) results from different routes (Note 9). The usual frequency of administration is once daily but consideration should be given to use either more frequent or less frequent administration taking kinetic variables into account (see also Note 10).

3.3 Kinetics

It is preferable to have some information on kinetics before initiating reproduction studies since this may suggest the need to adjust choice of species, study design and dosing schedules. At this time the information need not be sophisticated nor derived from pregnant or lactating animals. At the time of study evaluation further information on kinetics in pregnant or lactating animals may be required according to the results obtained (Note 10).

3.4 Control Groups

It is recommended that control animals be dosed with the vehicle at the same rate as test group animals. When the vehicle may cause effects or affect the action of the test substance, a second (sham- or untreated) control group should be considered.

4. PROPOSED STUDY DESIGNS — COMBINATION OF STUDIES

All available pharmacological, kinetic, and toxicological data for the test compound and similar substances should be considered in deciding the most appropriate strategy and choice of study design. It is anticipated that, initially, preference will be given to designs that do not differ too radically from those of established guidelines for medicinal products (The most probable option). For most medicinal products the 3-study design will usually be adequate. Other strategies, combinations of studies and study designs could be as valid or more valid as the "most probable option" according to circumstances. The key factor is that, in total, they leave no gaps between stages and allow direct or indirect evaluation of all stages of the reproductive process (Note 11). Designs should be justified.

4.1 The Most Probable Option

The most probable option can be equated to a combination of studies for effects on

Fertility and early embryonic development
Pre- and postnatal development, including maternal function
Embryo-fetal development

4.1.1 Study of Fertility and Early Embryonic Development to Implantation

AIM

To test for toxic effects/disturbances resulting from treatment from before mating (males/females) through mating and implantation. This comprises evaluation of stages A and B of the reproductive process (see 1.2). For females this should detect effects on the oestrous cycle, tubal transport, implantation, and development of preimplantation stages of the embryo. For males it will permit detection of functional effects (e.g., on libido, epididymal sperm maturation) that may not be detected by histological examinations of the male reproductive organs (Note 12).

ASSESSMENT OF

   maturation of gametes,
   mating behavior,
   fertility,
   preimplantation stages of the embryo,
   implantation.

ANIMALS

At least one species, preferably rats.

NUMBER OF ANIMALS

The number of animals per sex per group should be sufficient to allow meaningful interpretation of the data (Note 13).

ADMINISTRATION PERIOD

The design assumes that, especially for effects on spermatogenesis, use will be made of data from repeated dose toxicity studies of at least one month duration. Provided no effects have been found that preclude this, a premating treatment interval of 2 weeks for females and 4 weeks for males can be used (Note 12). Selection of the length of the premating administration period should be stated and justified (see also chapter 1.1, pointing out the need for research). Treatment should continue throughout mating to termination of males and at least through implantation for females. This will permit evaluation of functional effects on male fertility that cannot be detected by histologic examination in repeated dose toxicity studies and effects on mating behavior in both sexes. If data from other studies show there are effects on weight or histologic appearance of reproductive organs in males or females, or if the quality of examinations is dubious or if there are no data from other studies, then a more comprehensive study should be designed (Note 12).

MATING

A mating ratio of 1:1 is advisable and procedures should allow identification of both parents of a litter (Note 14).

TERMINAL SACRIFICE

Females may be sacrificed at any point after mid pregnancy. Males may be sacrificed at any time after mating but it is advisable to ensure successful induction of pregnancy before taking such an irrevocable step (Note 15).

OBSERVATIONS

During study:

   signs and mortalities at least once daily,
   body weight and body weight change at least twice weekly (Note 16),
   food intake at least once weekly (except during mating),

record vaginal smears daily, at least during the mating period, to determine whether there are effects on mating or precoital time
observations that have proved of value in other toxicity studies.

At terminal examination:

necropsy (macroscopic examination) of all adults,
preserve organs with macroscopic findings for possible histological evaluation; keep corresponding organs of sufficient controls for comparison,
preserve testes, epididymides, ovaries and uteri from all animals for possible histological examination and evaluation on a case by case basis. Tissues can be discarded after completion and reporting of the study.
sperm count in epididymides or testes, as well as sperm viability
count corpora lutea, implantation sites (Note 16),
live and dead conceptuses.

4.1.2 Study for Effects on Pre- and Postnatal Development, Including Maternal Function

AIM
To detect adverse effects on the pregnant/lactating female and on development of the conceptus and the offspring following exposure of the female from implantation through weaning. Since manifestations of effect induced during this period may be delayed, observations should be continued through sexual maturity (i.e., stages C to F listed in 1.2) (Notes 17, 18).

ADVERSE EFFECTS TO BE ASSESSED

enhanced toxicity relative to that in non-pregnant females,
pre- and postnatal death of offspring,
altered growth and development,
functional deficits in offspring, including behavior, maturation (puberty) and reproduction (F1).

ANIMALS
At least one species, preferably rats;

NUMBER OF ANIMALS
The number of animals per sex per group should be sufficient to allow meaningful interpretation of the data (Note 13).

ADMINISTRATION PERIOD
Females are exposed to the test substance from implantation to the end of lactation (i.e., stages C to E listed in 1.2).

EXPERIMENTAL PROCEDURE
The females are allowed to deliver and rear their offspring to weaning at which time one male and one female offspring per litter should be selected (document method used) for rearing to adulthood and mating to assess reproductive competence (Note 19).

OBSERVATIONS
During study (for maternal animals):

signs and mortalities at least once daily,
body weight and body weight change at least twice weekly (Note 16),

food intake at least once weekly at least until delivery,
observations that have proved of value in other toxicity studies,
duration of pregnancy,
parturition.

At terminal examination (for maternal animals and where applicable for offspring)

necropsy (macroscopic examination) of all adults,
preservation and possibly histological evaluation of organs with macroscopic findings; keep corresponding organs of sufficient controls for comparison,
implantations (Note 16),
abnormalities,
live offspring at birth,
dead offspring at birth,
body weight at birth,
pre- and postweaning survival and growth/body weight (Note 20), maturation and fertility,
physical development (Note 21),
sensory functions and reflexes (Note 21),
behavior (Note 21).

4.1.3 Study for Effects on Embryo-Fetal Development

AIM
To detect adverse effects on the pregnant female and development of the embryo and fetus consequent to exposure of the female from implantation to closure of the hard palate (i.e., stages C to D listed in 1.2).

ADVERSE EFFECTS TO BE ASSESSED

enhanced toxicity relative to that in non-pregnant females,
embryofetal death,
altered growth
structural changes.

ANIMALS
Usually, two species: one rodent, preferably rat; one non-rodent, preferably rabbit (Note 5). Justification should be provided when using one species.

NUMBER OF ANIMALS
The number of animals should be sufficient to allow meaningful interpretation of the data (Note 13).

ADMINISTRATION PERIOD
The treatment period extends from implantation to the closure of the hard palate (i.e., end of C) — see 1.2.

EXPERIMENTAL PROCEDURE
Females should be killed and examined about one day prior to parturition. All fetuses should be examined for viability and abnormalities. To allow subsequent assessment of the relationship between observations made by different techniques fetuses should be individually identified (Note 22).

When using techniques requiring allocation to separate examination for soft tissue or skeletal changes, it is preferable that 50% of fetuses from each litter be allocated for skeletal examination. A minimum of 50% rat fetuses should be examined for visceral alterations, regardless of the technique used. When using fresh microdissection techniques for soft tissue alterations — which is the strongly preferred method for rabbits — 100% of rabbit fetuses should be examined for soft tissue and skeletal abnormalities.

OBSERVATIONS
During study (for maternal animals):

> signs and mortalities at least once daily,
> body weight and body weight change at least twice weekly (Note 16),
> food intake at least once weekly, and
> observations that have proved of value in other toxicity studies.

At terminal examination:

> necropsy (macroscopic examination) of all adults,
> preserve organs with macroscopic findings for possible histological evaluation; keep correspondingorgans of sufficient controls for comparison,
> count corpora lutea, numbers of live and dead implantations (Note 16),
> individual fetal body weight,
> fetal abnormalities (Note 22),
> gross evaluation of placenta.

4.2 Single Study Design (Rodents)

If the dosing period of the fertility study and pre- and postnatal study are combined into a single investigation, this comprises evaluation of stages A to F of the reproductive process (see 1.2). If such a study, if it includes fetal examinations, provided clearly negative results at sufficiently high exposure no further reproduction studies in rodents should be required. Fetal examinations for structural abnormalities can also be supplemented with an embryo-fetal development study (or studies) to make a 2-study approach (Note 3, 11).

Results from a study for effects on embryo-fetal development in a second species are expected (see also 4.1.3).

4.3 Two Study Design (Rodents)

The simplest two segment design would consist of the fertility study and the pre- and postnatal development study, if it includes fetal examinations. It can be assumed, however, that if the pre- and postnatal development study provided no indication of prenatal effects at adequate margins above human exposure, the additional fetal examinations (4.1.3.) are most unlikely to provide a major change in the assessment of risk.

Alternatively, female treatment in the fertility study (4.1.1.) could be continued until closure of the hard palate and fetuses examined according to the procedures of the embryo-fetal development study (4.1.3.). This, combined with the pre- and postnatal study (4.1.2.), would provide all the examinations required in "the most probable option" but use considerably less animals (Note 3, 11).

Results from a study for effects on embryo-fetal development in a second species are expected (see also 4.1.3).

5. STATISTICS

Analysis of the statistics of a study is the means by which results are interpreted. The most important part of this analysis is to establish the relationship between the different variables and their distribution (descriptive statistics), since these determine how groups should be compared. The distributions of the endpoints observed in reproductive tests are usually non-normal and extend from almost continuous to the extreme categorical. When employing inferential statistics (determination of statistical significance) the mating pair or litter, not the foetus or neonate, should be used as the basic unit of comparison. The tests used should be justified (Note 23).

6. DATA PRESENTATION

The key to good reporting is the tabulation of individual values in a clear concise manner to account for every animal that was entered into the study. A reader should be able to follow the history of any individual animal from initiation to termination and should be able to deduce with ease the contribution that the individual has made to any group summary values. Group summary values should be presented in a form that is biologically plausible (i.e., avoid false precision) and that reflects the distribution of the variable. Appendices or tabulations of individual values such as bodyweight, food consumption, litter values should be concise and, as far as possible, consist of absolute rather than calculated values; unnecessary duplication should be avoided.

For tabulation of low frequency observations such as clinical signs, autopsy findings, abnormalities etc. it is advisable to group together the (few) individuals with a positive recording. Especially in the presentation of data on structural changes (fetal abnormalities) the primary listing (tabulation) should clearly identify the litters containing abnormal fetuses, identify the affected fetuses in the litter and report all the changes observed in the affected fetus. Secondary listings by type of change can be derived from this, if necessary.

7. TERMINOLOGY

Besides effects on the reproductive competence of adult animals toxicity to reproduction includes:

> Developmental toxicity: Any adverse effect induced prior to attainment of adult life. It includes effects induced or manifested in the embryonic or fetal period and those induced or manifested postnatally.
> Embryotoxicity, fetotoxicity, embryo-fetal toxicity: Any adverse effect on the conceptus resulting from prenatal exposure, including structural or functional abnormalities or postnatal manifestations of such effects. Terms like "embryotoxicity", "fetotoxicity" relate to the timepoint/-period of induction of adverse effects, irrespective of the time of detection.
> One, two or three generation studies: are defined according to the number of adult breeding generations directly exposed to the test material. For example, in a one generation study there is direct exposure of the F0 generation and indirect exposure (via the mother) of the F1 generation and the study is usually terminated at weaning of the F1 generation. In a two generation study as used for agrochemicals and industrial chemicals there is direct exposure of the F0 generation, indirect and direct exposure of the F1 generation and indirect exposure of the F2 generation. A three generation study is defined accordingly.
> Body burden: The total internal dosage of an individual arising from the administration of a substance, comprising parent compound and metabolites, taking distribution and accumulation into account.
> Kinetics: The term "kinetics" is used consistently throughout this guideline, irrespective of intending to mean pharmaco- and/or toxicokinetics. No better single term was available.

NOTES
Note 1 (1.1) Scientific flexibility
These guidelines are not mandatory rules, they are a starting point rather than an end point. They provide a basis from which an investigator can devise a strategy for testing according to available knowledge of the test material and the state of the art. For encouragement some alternative test

designs have been mentioned in this document but there are others that can be sought out or devised. In devising a strategy, the primary objective should be to detect and bring to light any indication of toxicity to reproduction. Fine details of study design and technical procedures have been omitted from the text. Such decisions rightly belong in the field of the investigator since a technique that may be suitable for one laboratory may not be suitable in another. The investigator needs to utilize staff and resources to do the best he or she can achieve and should know how to do this better than any outsider; human attributes of attitude, ability, and consistency are more important than material facilities. For necessary compliance to GLP, reference is made to such regulations.

Note 2 (1.2) Timing conventions
In this guideline the convention for timing of pregnancy is to refer to the day that a sperm positive vaginal smear and/or plug is observed as day 0 of pregnancy even if mating occurs overnight. Unless shown otherwise it is assumed that, for rats, mice and rabbits implantation occurs on day 6-7 of pregnancy, and closure of the hard palate on day 15-18 of pregnancy. Other conventions are equally acceptable but MUST be defined in reports. Also, the investigator must be consistent in different studies to assure that no gaps in treatment occur. It is an advisable precaution to provide an overlap of at least one day in the exposure period of related studies.

The accuracy of the time of mating should be specified since this will affect the variability of fetal and neonatal parameters. Similarly, for reared litters, the day offspring are born will be considered as postnatal or lactation day 0 unless otherwise specified. However, particularly with regard to delays in, or prolongation of parturition, reference to a postcoital time frame may be useful.

Note 3 (1.3) First pass and secondary testing
To a greater or lesser degree all first pass (guideline) tests are apical in nature, i.e., an effect on one endpoint may have several different origins. A reduced litter size at birth may be due to a reduced ovulation rate (corpora lutea count), higher rate of preimplantation deaths, higher rate of postimplantation deaths or immediate postnatal deaths. In turn these deaths may be the consequence of an earlier physical malformation that can no longer be observed due to subsequent secondary changes and so on. Particularly for effects with a natural low frequency among controls, discrimination between treatment induced and coincidental occurrence is dependent upon association with other types of effects.

A toxicant usually induces more than one type of effect in a dose dependent manner. For example, induction of malformation is almost invariably associated with increased embryonic death and an increased incidence of less severe structural changes. Given an effect on one endpoint secondary investigations for possible associations should be considered, i.e., the nature, scope and origins of the substance's toxicity should be characterized. Characterization should also include identification of dose-response relationships to facilitate risk assessment; this is different from the situation in first pass tests where the presence or absence of a dose-response assists discrimination between treatment related and coincidental differences.

Note 4 (1.3) Preliminary studies
At the time most reproduction studies are planned or initiated there is usually information available from acute and repeated dose toxicity studies of at least one month's duration. This information can be expected to be sufficient in identifying doses for reproductive studies. If adequate preliminary studies are performed, they are part of the justification of the choice of dose for the main study. Such studies should be submitted regardless of their GLP-status in principle. This may avoid unnecessary use of animals.

Note 5 (2.1) Selection of species and strains
In choosing an animal species and strain for reproductive toxicity testing care should be given to select a relevant model. Selection of the species and strain used in other toxicology studies may

avoid the need for additional preliminary studies. If it can be shown — by means of kinetic, pharmacological and toxicological data — that the species selected is a relevant model for the human, a single species can be sufficient. There is little value in using a second species if it does not show the same similarities to humans. Advantages and disadvantages of species (strains) should be considered in relation to the substance to be tested, the selected study design and in the subsequent interpretation of the results.

All species have their advantages. Rats, and to a lesser extent mice, are good, general purpose models; the rabbit has been somewhat neglected as a "non-rodent" species for repeated dose toxicity and other reproduction studies than embryotoxicity testing. It has attributes that would make it a useful model for fertility studies, especially male fertility. For both rabbits and dogs (which are often used as a second species for chronic toxicity studies) it is feasible to obtain semen samples without resorting to painful techniques (electro ejaculation) for longitudinal semen analysis. Most of the other species are not good, general purpose models and probably are best used for very specific investigations only. All species have their disadvantages, for example:

Rats: sensitivity to sexual hormones, unsuitable for dopamine agonists due to dependence on prolactin as the primary hormone for establishment and maintenance of early pregnancy, highly susceptible to non-steroidal anti-inflammatory drugs in late pregnancy.
Mice: fast metabolic rate, stress sensitivity, malformation clusters (which occur in all species) particularly evident, small fetus.
Rabbits: often lack of kinetic and toxicity data, susceptibility to some antibiotics and to disturbance of the alimentary tract, clinical signs can be difficult to interpret.
Guinea pigs: often lack of kinetic and toxicity data, susceptibility to some antibiotics and to disturbance of the alimentary tract, long fetal period, insufficient historical background data.
Domestic and/or mini pigs: malformation clusters with variable background rate, large amounts of compound required, large housing necessary, insufficient historical background data.
Ferrets: seasonal breeder unless special management systems used (success highly dependent on human/animal interaction), insufficient historical background data.
Hamsters: intravenous route difficult if not impossible, can hide doses in the cheek pouches and can be very aggressive, sensitive to intestinal disturbance, overly sensitive teratogenic response to many chemicals, small foetus.
Dogs: seasonal breeders, inbreeding factors, insufficient historical background data.
Non-human primates: kinetically they can differ from humans as much as other species, insufficient historical background data, often numbers too low for detection of risk. They are best used when the objective of the study is to characterize a relatively certain reproductive toxicant, rather than detect a hazard.

Note 6 (2.2) Uses of other test systems than whole animals
Other test systems have been developed and used in preliminary investigations ("pre-screening" or priority selection) and secondary testing. For preliminary investigation of a range of analogue series of substances it is essential that the potential outcome in whole animals is known for at least one member of the series to be studied (by inference, effects are expected). With this strategy substances can be selected for higher level testing. For secondary testing or further substance characterisation other test systems offer the possibility to study some of the observable developmental processes in detail, e.g., to reveal specific mechanisms of toxicity, to establish concentration-response relationships, to select "sensitive periods," or to detect effects of defined metabolites.

Note 7 (3.1) Selection of dosages
Using similar doses in the reproductive toxicity studies as in the repeated dose toxicity studies will allow interpretation of any potential effects on fertility in context with general systemic toxicity. Some minimal toxicity is expected to be induced in the high dose dams.

According to the specific compound, factors limiting the high dosage determined from repeat dose toxicity studies or from preliminary reproduction studies could include:

reduction in bodyweight gain

increased bodyweight gain, particularly when related to perturbation of homeostatic mechanisms

specific target organ toxicity

haematology, clinical chemistry

exaggerated pharmacological response, which may or may not be reflected as marked clinical reactions (e.g., sedation, convulsions)

the physico-chemical properties of the test substance or dosage formulation which, allied to the route of administration, may impose practical limitations in the amount that can be administered. Under most circumstances 1 g/kg/day should be an adequate limit dose.

kinetics, they can be useful in determining high dose exposure for low toxicity compounds. There is, however, little point in increasing administered dosage if it does not result in increased plasma or tissue concentration

marked increase in embryo-fetal lethality in preliminary studies

Note 8 (3.1) Determination of dose-response relationships

For many of the variables in reproduction studies the power to discriminate between random variation and treatment effect is poor and the presence or absence of a dosage related trend can be a critical means of determining the probability of a treatment effect. It has to be kept in mind that in these studies dose-responses may be steep, and wide intervals between doses would be inadvisable. If an analysis of dose-response relationships for the effects observed is attempted in a single study, it is recommended to use at least three dose levels and appropriate control groups. If in doubt, a fourth dose group should be added to avoid excessive dosage intervals. Such a strategy should provide a "no observed adverse effect level" for reproductive aspects. If not, the implication is that the test substance merits a greater depth of investigation and further studies.

Note 9 (3.2) Exposure by different routes of administration

If it can be shown that one route provides a greater body burden, e.g., area under the curve (AUC), there seems little reason to investigate routes that would provide a lesser body burden or which present severe practical difficulties (e.g., inhalation). Before designing new studies for a new route of administration existing data on kinetics should be used to determine the necessity of another study.

Note 10 (3.3) Kinetics in pregnant animals

Kinetic investigations in pregnant and lactating animals may pose some problems due to the rapid changes in physiology. It is best to consider this as a two or three phase approach. In planning studies kinetic data (often from non-pregnant animals) provide information on the general suitability of the species, and can assist in deciding study designs and choice of dosage. During a study kinetic investigations can provide assurance of accurate dosing or indicate marked deviations from expected patterns.

Note 11 (4) Examples for choosing other options

For compounds causing no lethality at 2 g/kg and no evidence of repeated dose toxicity at 1 g/kg, conduct of a single 2 generation study with one control and 2 test groups (0.5 and 1.0 g/kg) would seem sufficient. However, it might pose the question as to whether the correct species had been chosen or whether the compound was an effective medicine.

For compounds that may be given as a single dose, once in a life time, (e.g., diagnostics, medicines used in operations) it may be impossible to administer repeated dosages more than twice the human therapeutic dosage for any length of time. A reduced period of treatment allowing a higher dose would seem more appropriate. For females considerations of human exposure suggest little or no need for exposures beyond the embryonic period. For dopamine agonists or compounds reducing circulating prolactin levels female rats are poor models; the rabbit would probably make a better choice for all the reproductive toxicity studies, but it does not appear to have been attempted. This also applies to other types of compound when the rabbit shows a pattern of metabolism considerably closer to humans than the rat.

For drugs where alterations in plasma kinetics are seen following repeated administration, the potential for adverse effects on embryo-fetal development may not be fully evaluated in studies according to 4.1.3. In such cases it may be desirable to extend the period of drug administration to females in a 4.1.1 study to day 17. With sacrifice at term, both fertility and embryo-fetal development can be assessed.

Note 12 (4.1.1) Premating treatment
The design of the fertility study, especially the reduction in the premating period for males, is based on evidence accumulated and re-appraisal of the basic research on the process of spermatogenesis that originally prompted the demand for a prolonged premating treatment period. Compounds inducing selective effects on male reproduction are rare; mating with females is an insensitive means of detecting effects on spermatogenesis; good pathological and histopathological examination (e.g., by employing Bouin's fixation, paraffine embedding, transverse sections of 2-4 microns for testes, longitudinal sections for epididymides, PAS and haematoxylin staining) of the male reproductive organs provides a more sensitive and quicker means of detecting effects on spermatogenesis; compounds affecting spermatogenesis almost invariably affect post meiotic stages; there is no conclusive example of a male reproductive toxicant the effects of which could be detected only by dosing males for 9-10 weeks and mating them with females.

Information on potential effects on spermatogenesis can be derived from repeated dose toxicity studies. This allows the investigations in the fertility study to be concentrated on other, more immediate, causes of effect. It is noted that the full sequence of spermatogenesis (including sperm maturation) in rats lasts 63 days. When the available evidence, or lack of it, suggests that the scope of investigations in the fertility study should be increased, or extended from detection to characterization, appropriate studies should be designed to further characterise the effects.

Note 13 (4.1.1, 4.1.2, 4.1.3) Number of animals
There is very little scientific basis underlying specified group sizes in past and existing guidelines nor in this one. The numbers specified are educated guesses governed by the maximum study size that can be managed without undue loss of overall study control. This is indicated by the fact that the more expensive the animal is to obtain or keep, the smaller the group size proposed. Ideally, at least the same group size should be required for all species and there is a case for using larger group sizes for less frequently used species such as primates. It should also be made clear that the numbers required depend on whether or not the group is expected to demonstrate an effect. For a high frequency effect few animals are required, to presume the absence of an effect the number required varies according to the variable (endpoint) being considered, its prevalence in control populations (rare or categorical events) or dispersion around the central tendency (continuous or semi-continuous variables). See also Note 23.

For all but the rarest events (such as malformations, abortions, total litter loss), evaluation of between 16 to 20 litters for rodents and rabbits tends to provide a degree of consistency between studies. Below 16 litters per evaluation, between study results become inconsistent, above 20–24 litters per group consistency and precision is not greatly enhanced. These numbers relate to evaluation. If groups are subdivided for different evaluations the number of animals starting the study should be doubled. Similarly, in studies with 2 breeding generations, 16-20 litters would be required for the final evaluation of the litters of the F1 generation. To allow for natural wastage, the starting group size of the F0 generation must be larger.

Note 14 (4.1.1) Mating
Mating ratios: When both sexes are being dosed or are of equal consideration in separate male and female studies, the preferred mating ratio is 1:1 since this is the safest option in respect of obtaining good pregnancy rates and avoiding incorrect analysis and interpretation of results.

Mating period and practices: most laboratories would use a mating period of between 2-3 weeks, some remove females as soon as a positive vaginal smear or plug is observed whilst others leave the pairs together. Most rats will mate within the first 5 days of cohabitation (i.e., at the first available estrus), but in some cases females may become pseudopregnant. Leaving the female with the male for about 20 days allows these females to restart estrus cycles and become pregnant.

Note 15 (4.1.1) Terminal sacrifice

FEMALES
When exposure of the females ceases at implantation, termination of females between days 13-15 of pregnancy in general is adequate to assess effects on fertility or reproductive function, e.g., to differentiate between implantation and resorption sites. In general, for detection of adverse effects, it is not thought necessary, in a fertility study, to sacrifice females at day 20/21 of pregnancy in order to gain information on late embryo loss, fetal death, and structural abnormalities.

MALES
It would be advisable to delay sacrifice of the males until the outcome of mating is known. In the event of an equivocal result, males could be mated with untreated females to ascertain their fertility or infertility. The males treated as part of study 4.1.1 may also be used for evaluation of toxicity to the male reproductive system if dosing is continued beyond mating and sacrifice delayed.

Note 16 (4.1.1, 4.1.2, 4.1.3) Observations
Daily weighing of pregnant females during treatment can provide useful information. Weighing an animal more frequently than twice weekly during periods other than pregnancy (premating, mating, lactation) may also be advisable for some compounds. For apparently non-pregnant rats or mice (but not rabbits), ammonium sulphide staining of the uterus might be useful to identify peri-implantation death of embryos.

Note 17 (4.1.2) Treatment of offspring
Consequent to derivation from existing guidelines for medicines this guideline does not fully cover exposures from weaning through puberty, nor does it deal with the possibility of reduced reproductive life span. To detect adverse effects for medicinal products that may be used in infants and juveniles, special studies (case by case designs) involving direct treatment of offspring, at ages to be specified, should be considered.

Note 18 (4.1.2) Separate embryotoxicity and peri-postnatal studies
If a pre- and postnatal study is separated into two studies, one covering the embryonic period the other the fetal period, parturition, and lactation, postnatal evaluation of offspring is required in both studies.

Note 19 (4.1.2) F1-animals
The guideline suggests selection of 1 male and 1 female per litter on the evidence that it is feasible to conduct behavioral and other functional tests on the same F1 individuals that will be used for assessment of reproductive function. This has the advantage of allowing cross referencing of performance in different tests at the individual level. It is recognised, however, that some laboratories prefer to select separate sets of animals for behavior testing and for assessment of reproductive function. Which is the most suitable for an individual laboratory will depend upon the combination of tests used and the resources available.

Note 20 (4.1.2) Reduction of litter size
The value of culling or not culling for detection of effects on reproduction is still under discussion. Whether or not culling is performed, it should be explained by the investigator.

Note 21 (4.1.2) Physical development, sensory functions, reflexes, and behavior
The best indicator of physical development is bodyweight. Achievement of preweaning landmarks of development such as pinna unfolding, coat growth, incisor eruption etc., is highly correlated with pup bodyweight. This weight is better related to postcoital time than postnatal time, at least when significant differences in gestation length occur. Reflexes, surface righting, auditory startle, air righting, and response to light are also dependent on physical development.

Two postweaning landmarks of development that are advised are vaginal opening of females and cleavage of the balanopreputial gland of males. The latter is associated with increasing testosterone levels whereas testis descent is not. These landmarks indicate the onset of sexual maturity and it is advised that bodyweight be recorded at the time of attainment to determine whether any differences from control are specific or related to general growth.

Functional tests: To date, functional tests have been directed almost exclusively to behavior. Even though a great deal of effort has been expended in this direction it is not possible to recommend specific test methods. Investigators are encouraged to find methods that will assess sensory functions, motor activity, learning and memory.

Note 22 (4.1.3) Individual identification and evaluation of fetuses
It must be possible to relate all findings by different techniques (i.e., body weight, external inspection, visceral and/or skeletal examinations) to single specimen in order to detect patterns of abnormalities. The examination of mid and low dose fetuses for visceral and/or skeletal abnormalities may not be necessary where the evaluation of the high dose and the control groups did not reveal any relevant differences. It is advisable, however, to store the fixed specimen for possible later examination. If fresh dissection techniques are normally used, difficulties with later comparisons involving fixed fetuses should be anticipated.

Note 23 (5.) Inferential statistics
"Significance" tests (inferential statistics) can be used only as a support for the interpretation of results. The interpretation itself must be based on biological plausibility. It is unwise to assume that a difference from control values is not biologically relevant simply because it is not "statistically significant". To a lesser extent it can be unwise to assume that a "statistically significant" difference must be biologically relevant. Particularly for low frequency events (e.g., embryonic death, malformations) with one sided distributions, the statistical power of studies is low.

Confidence intervals for relevant quantities can indicate the likely size of the effect. When using statistical procedures, experimental units of comparison should be considered: the litter, not the individual conceptus, the mating pair, when both sexes are treated, the mating pair of the parent generation in a two generation study.

DETECTION OF TOXICITY TO REPRODUCTION FOR MEDICINAL PRODUCTS: ADDENDUM ON TOXICITY TO MALE FERTILITY

TABLE OF CONTENTS
I. Introduction (1)
II. Amendments (2)
   A. Introduction (Last Paragraph) (I.A)

B. Study of Fertility and Early Embryonic Development to Implantation (4.1.1)
  1. Administration period (IV.A.1.e. (4.1.1))
  2. Observations (IV.A.1.h (4.1.1))
C. Note 12 (IV.A.1 (4.1.1)) Premating Treatment

This guideline was developed within the Expert Working Group (Safety) of the International Conference on Harmonisation of Technical Requirements for Registration of Pharmaceuticals for Human Use (ICH) and has been subject to consultation by the regulatory parties, in accordance with the ICH process. This document has been endorsed by the ICH Steering Committee at *Step 4* of the ICH process, November 29, 1995. At *Step 4* of the process, the final draft is recommended for adoption to the regulatory bodies of the European Union, Japan and the USA. This guideline was published in the *Federal Register* on April 5, 1996 (61 FR 15360) and is applicable to drug and biological products. Although this guideline does not create or confer any rights for or on any person and does not operate to bind FDA or the industry, it does represent the agency's current thinking on the detection of toxicity to reproduction for medicinal products. For additional copies of this guideline, contact the Drug Information Branch, HFD-210, CDER, FDA, 5600 Fishers Lane, Rockville, MD 20857 (Phone: 301-827-4573) or the Manufacturers Assistance and Communication Staff (HFM-42), CBER, FDA, 1401 Rockville Pike, Rockville, MD 20852-1448. Send one self-addressed adhesive label to assist the offices in processing your request. An electronic version of this guidance is also available via Internet using the World Wide Web (WWW) (connect to the CDER Home Page at http://www.fda.gov/cder and go to the "Regulatory Guidance" section).

I. INTRODUCTION (1)

This text is an addendum to the ICH Tripartite Guideline on Detection of Toxicity to Reproduction for Medicinal Products and provides amendments to the published text. At the time of adoption, it was accepted that the male fertility investigation, as included in the currently harmonized guideline, would need scientific and regulatory improvement and optimization of test designs.

The amendments are intended to provide a better description of the testing concept and recommendations, especially those addressing:

Flexibility
Premating treatment duration
Observations

The general principles and background were contained in two papers published in the *Journal of American College of Toxicology*. These papers contain the necessary experimental data (prospective and retrospective) for reaching consensus and have been commented on. The individual data from the Japanese collaborative study were also published in the *Journal of Toxicological Science*.

II. AMENDMENTS (2)

A. Introduction (Last Paragraph) (I.A)

To employ this concept successfully, flexibility is needed (Note 1). No guideline can provide sufficient information to cover all possible cases. All persons involved should be willing to discuss and consider variations in test strategy according to the state-of-the-art and ethical standards in human and animal experimentation.

B. Study of Fertility and Early Embryonic Development to Implantation (4.1.1)

1. Administration period (IV.A.1.e. (4.1.1))

The design assumes that, especially for effects on spermatogenesis, use will be made of available data from toxicity studies (e.g., histopathology, weight of reproductive organs, in some cases hormone assays and genotoxicity data). Provided no effects have been found in repeated dose toxicity studies of at least 4 weeks duration that preclude this, a premating treatment interval of 2 weeks for females and 4 weeks for males can be used (Note 12). Selection of the length of the premating administration period should be stated and justified. Treatment should continue throughout mating to termination for males and at least through implantation for females. This will permit evaluation of functional effects on male fertility that cannot be detected by histopathologic examination in repeated dose toxicity studies and effects on mating behavior in both sexes. If data from other studies show there are effects on weight or histology of reproductive organs in males or females, or if the quality of examinations is dubious, or if there are no data from other studies, the need for a more comprehensive study should be considered (Note 12).

2. Observations (IV.A.1.h (4.1.1))

At terminal examination, the following should be done:

- perform necropsy (macroscopic examination) of all adults;
- preserve organs with macroscopic findings for possible histopathological evaluation; keep corresponding organs of sufficient controls for comparison;
- preserve testes, epididymides, ovaries, and uteri from all animals for possible histopathological examination and evaluation on a case by case basis;
- count corpora lutea,
- implantation sites (note 16);
- count live and dead conceptuses; and
- sperm analysis can be used as an optional procedure for confirmation or better characterization of the effects observed (Note 12).

C. Note 12 (IV.A.1 (4.1.1)) Premating Treatment

The design of the fertility study, especially the reduction in the premating period for males, is based on evidence accumulated and on re-appraisal of the basic research on the process of spermatogenesis. Compounds inducing selective effects on male reproduction are rare; compounds affecting spermatogenesis almost invariably affect postmeiotic states and weight of testis; mating with females is an insensitive means of detecting effects on spermatogenesis. Histopathology of the testis has been shown to be the most sensitive method for the detection of effects of spermatogenesis. Good pathological and histopathological examination (e.g., by employing Bouin's fixation, paraffin embedding, transverse section of 2094 microns for testes, longitudinal section for epididymides, PAS and hematoxylin staining) of the male reproductive organs provides a direct means of detection. Sperm analysis (sperm counts, sperm motility, sperm morphology) can be used as an optional method to confirm findings by other methods and to further characterize effects. Sperm analysis data are considered more relevant for fertility assessment when samples from vas deferens or from cauda epididymis are used. Information on potential effects on spermatogenesis (and female reproductive organs) can be derived from repeated dose toxicity studies or reproductive toxicity studies.

For detection of effects not detectable by histopathology of male reproductive organs and sperm analysis, mating with females after a premating treatment of 4 weeks has been shown to be at least as efficient as mating after a longer duration of treatment (2 weeks may be acceptable in some cases). However, when 2 weeks treatment period is selected, more convincing justification should be provided. When the available evidence suggests that the scope of investigations in the fertility study should be increased, appropriate studies should be designed to characterize the effects further.

INTERNATIONAL CONFERENCE ON HARMONISATION OF TECHNICAL REQUIREMENTS FOR REGISTRATION OF PHARMACEUTICALS FOR HUMAN USE

ICH HARMONISED TRIPARTITE GUIDELINE MAINTENANCE OF THE ICH GUIDELINE ON TOXICITY TO MALE FERTILITY

An Addendum to the ICH Tripartite Guideline on
DETECTION OF TOXICITY TO REPRODUCTION FOR
MEDICINAL PRODUCTS

Recommended for Adoption at Step 4 of the ICH Process on 29 November 1995
and amended on 9 November 2000 by the ICH Steering Committee

This Guideline has been developed by the appropriate ICH Expert Working Group and has been subject to consultation by the regulatory parties, in accordance with the ICH Process. At Step 4 of the Process the final draft is recommended for adoption to the regulatory bodies of the European Union, Japan and USA.

Background Note (not part of the guideline)
In June 1993 an ICH Harmonised Tripartite Guideline on Detection of Toxicity to Reproduction for Medicinal Products was adopted by the ICH Steering Committee and has since been implemented in the three ICH regions. Additional data on the duration of the premating treatment and on the parameters suited for the detection of effects on the male reproductive system and on fertility have been collected and are presented in two papers that were published in J. Amer. Coll. Toxicol. in August 1995 and also in J. Toxicol. Sci. as Special Issue in September 1995. The Japanese collaborative study which was conducted came to the conclusion that histopathology of reproductive organs is the most sensitive method for detecting effects on spermatogenesis. Sperm analysis gives similar information and could be a useful method where histopathology is not feasible, as well as for confirmative purposes and further characterisation. Mating trials were found to be sensitive for the detection of effects of sperm maturation, sperm motility and of behavioural changes (e.g., libido). The Japanese parties to ICH compared premating treatments of 4 and 9 weeks duration and found no differences in detection rate, i.e., at least 4 weeks is necessary. Data from the European literature survey show equal detection efficiency with premating treatments of 2 and 4 weeks duration. Validation studies conducted in Japan recently indicated that the 2 week study with an appropriate dose setting has equal potential in detecting the toxicity on male reproductive organs (Sakai et al 2000).

References

1. Takayama S, Akaike M, Kawashima K, Takahashi M, Kurokawa Y; A Collaborative Study in Japan on Optimal Treatment Period and Parameters for Detection of Male Fertility Disorders in Rats Induced by Medical Drugs. *J. Amer. Coll. Toxicol.*, 14: 266-292, 1995.
2. Ulbrich B, Palmer AK; Detection of Effects on Male Reproduction – A Literature Survey. J. Amer. Coll. Toxicol., 14; 293-327, 1995 3. *J. Toxicol. Sci.,* Special Issue, Testicular Toxicity, 20: 173-374, 1995.
3. Sakai T, Takahashi M, Mitsumori K, Yasuhara K, Kawashima K, Mayahara H, Ohno Y; Collaborative work to evaluate toxicity on male reproductive organs by 2-week repeated dose toxicity studies in rats. Overview of the studies. *J. Toxicol. Sci.* 25, 1-21 (2000)

TOXICITY TO MALE FERTILITY
An Addendum to the ICH Tripartite Guideline on
DETECTION OF TOXICITY TO REPRODUCTION FOR MEDICINAL PRODUCTS

Introduction

This text is an addendum to the ICH Tripartite Guideline on Detection of Toxicity to Reproduction for Medicinal Products 1 and provides amendments to the published text. At the time of adoption, it was accepted that the male fertility investigation, as included in the currently harmonised guideline, would require scientific and regulatory improvement and optimisation of test designs. The amendments are intended to provide a better description of the testing concept and recommendations, especially those addressing:

- flexibility
- premating treatment duration
- observations

The general principles and background in two papers were published in the *J. Amer. Coll. Toxicol.* These papers contain the necessary experimental data (prospective and retrospective) for reaching consensus, and have been commentated. The individual data from the Japanese collaborative study were also published in the *J. Toxicol. Sci.*

AMENDMENTS

INTRODUCTION (Last Paragraph)

To employ this concept successfully, flexibility is needed (Note 1). No guideline can provide sufficient information to cover all possible cases. All persons involved should be willing to discuss and consider variations in test strategy according to the state of the art and ethical standards in human and animal experimentation.

4.1.1. Study of Fertility and Early Embryonic Development to Implantation
ADMINISTRATION PERIOD

It is assumed that, especially for effects on spermatogenesis, use will be made of available data from toxicity studies (e.g., histopathology, weight of reproductive organs, in some cases hormone assays and genotoxicity data). Provided no effects have been found in repeated dose toxicity studies of at least 2 weeks duration that preclude this, a premating treatment interval of 2 weeks for females and 2 weeks for males can be used (Note 12). Selection of the length of the premating administration period should be stated and justified. Treatment should continue throughout mating to termination of males and at least through implantation for females. This will permit evaluation of functional effects on male fertility that cannot be detected by histopathological examination in repeated dose toxicity studies and effects on mating behaviour in both sexes. If data from other studies show there are effects on weight or histology of reproductive organs in males or females, or if the quality of examinations is dubious or if there are no data from other studies, then the need for a more comprehensive study should be considered (Note 12).

OBSERVATIONS

At terminal examination:

- Perform necropsy (macroscopic examination) of all adults,
- Preserve organs with macroscopic findings for possible histopathological evaluation; keep corresponding organs of sufficient controls for comparison,
- Preserve testes, epididymides, ovaries and uteri from all animals for possible histopathological examination and evaluation on a case-by-case basis,
- Count corpora lutea, implantation sites (Note 16),

- Count live and dead conceptuses.
- Sperm analysis can be used as an optional procedure for confirmation or better characterisation of the effects observed (Note 12).

Note 12 (4.1.1) Premating treatment
The design of the fertility study, especially the reduction in the premating period for males, is based on evidence accumulated and on re-appraisal of the basic research on the process of spermatogenesis. Compounds inducing selective effects on male reproduction are rare; compounds affecting spermatogenesis almost invariably affect post meiotic stages and weight of testis; mating with females is an insensitive means of detecting effects on spermatogenesis. Histopathology of the testis has been shown to be the most sensitive method for the detection of effects on spermatogenesis. Good pathological and histopathological examination (e.g., by employing Bouin's fixation, paraffin embedding, transverse section of 2-4 microns for testes, longitudinal section for epididymides, PAS and haematoxylin staining) of the male reproductive organs provides a direct means of detection. Sperm analysis (sperm counts, sperm motility, sperm morphology) can be used as an optional method to confirm findings by other methods and to characterise effects further. Sperm analysis data are considered more relevant for fertility assessment when samples from vas deferens or from cauda epididymis are used. Information on potential effects on spermatogenesis (and female reproductive organs) can be derived from repeated dose toxicity studies or reproductive toxicity studies. For detection of effects not detectable by histopathology of male reproductive organs and sperm analysis, mating with females after a premating treatment of 4 weeks has been shown to be at least as efficient as mating after a longer duration of treatment. Since 2 week study was validated to be as effective as a 4 week study, 2 weeks treatment before mating is also acceptable. When the available evidence suggests that the scope of investigations in the fertility study should be increased, appropriate studies should be designed to characterise the effects further.

APPENDIX 2

United States Environmental Protection Agency
Prevention, Pesticides and Toxic Substances
(7101)
EPA 712–C–98–207
August 1998

Health Effects Test Guidelines
OPPTS 870.3700

Prenatal Developmental Toxicity Study

Introduction

This guideline is one of a series of test guidelines that have been developed by the Office of Prevention, Pesticides and Toxic Substances, United States Environmental Protection Agency for use in the testing of pesticides and toxic substances, and the development of test data that must be submitted to the Agency for review under Federal regulations.

The Office of Prevention, Pesticides and Toxic Substances (OPPTS) has developed this guideline through a process of harmonization that blended the testing guidance and requirements that existed in the Office of Pollution Prevention and Toxics (OPPT) and appeared in Title 40, Chapter I, Subchapter R of the Code of Federal Regulations (CFR), the Office of Pesticide Programs (OPP) which appeared in publications of the National Technical Information Service (NTIS) and the guidelines published by the Organization for Economic Cooperation and Development (OECD).

The purpose of harmonizing these guidelines into a single set of OPPTS guidelines is to minimize variations among the testing procedures that must be performed to meet the data requirements of the U.S. Environmental Protection Agency under the Toxic Substances Control Act (15 U.S.C. 2601) and the Federal Insecticide, Fungicide and Rodenticide Act (7 U.S.C. 136, *et seq.*).

Final Guideline Release: This guideline is available from the U.S. Government Printing Office, Washington, DC 20402 on disks or paper copies: call (202) 512–0132. This guideline is also available electronically in ASCII and PDF (portable document format) from EPA's World Wide Web site (http://www.epa.gov/epahome/research.htm) under the heading "Researchers and Scientists/Test Methods and Guidelines/OPPTS Harmonized Test Guidelines."

OPPTS 870.3700 Prenatal developmental toxicity study.

(a). Scope
  (1) Applicability. This guideline is intended to meet testing requirements of both the Federal Insecticide, Fungicide, and Rodenticide Act (FIFRA) (7 U.S.C. 136, *et seq.*), as amended by the Food Quality Protection Act (FQPA)(Pub. L.104–170), and the Toxic Substances Control Act (TSCA) (15 U.S.C. 2601).
  (2) Background. The source material used in developing this harmonized OPPTS test guideline is the OPPT guideline under 40 CFR 798.4900, OPP guideline 83–3, and OECD guideline 414.
(b). Purpose. This guideline for developmental toxicity testing is designed to provide general information concerning the effects of exposure of the pregnant test animal on the developing organism; this may include death, structural abnormalities, or altered growth and an assessment of maternal effects. For information on testing for functional deficiencies and other postnatal effects, the guidelines for the two-generation reproductive toxicity study and the developmental neurotoxicity study should be consulted.
(c). Good laboratory practice standards. The study should be conducted in accordance with the laboratory practices stipulated in 40 CFR Part 160 (FIFRA) and 40 CFR Part 792 (TSCA)—Good Laboratory Practice Standards.

(d). Principle of the test method. The test substance is administered to pregnant animals at least from implantation to one day prior to the expected day of parturition. Shortly before the expected date of delivery, the pregnant females are terminated, the uterine contents are examined, and the fetuses are processed for visceral and skeletal evaluation.

(e). Test procedures
  (1) Animal selection
    (i) Species and strain. It is recommended that testing be performed in the most relevant species, and that laboratory species and strains which are commonly used in prenatal developmental toxicity testing be employed. The preferred rodent species is the rat and the preferred non-rodent species is the rabbit.
    (ii) Age. Young adult animals should be used.
    (iii) Sex. Nulliparous female animals should be used at each dose level. Animals should be mated with males of the same species and strain, avoiding the mating of siblings, if parentage is known. Day 0 in the test is the day on which a vaginal plug and/or sperm are observed in the rodent or that insemination is performed or observed in the rabbit.
    (iv) Animal care. Animal care and housing should be in accordance with the recommendations contained in the DHHS/PHS NIH Publication No. 86–23, 1985, *Guidelines for the Care and Housing of Laboratory Animals*, or other appropriate guidelines.
    (v) Number of animals. Each test and control group should contain a sufficient number of animals to yield approximately 20 animals with implantation sites at necropsy.
  (2) Administration of test and control substances
    (i) Dose levels and dose selection.
      (A) At least three-dose levels and a concurrent control should be used. Healthy animals should be randomly assigned to the control and treatment groups, in a manner which results in comparable mean body weight values among all groups. The dose levels should be spaced to produce a gradation of toxic effects. Unless limited by the physical/chemical nature or biological properties of the test substance, the highest dose should be chosen with the aim to induce some developmental and/or maternal toxicity but not death or severe suffering. In the case of maternal mortality, this should not be more than approximately 10 percent. The intermediate dose levels should produce minimal observable toxic effects. The lowest dose level should not produce any evidence of either maternal or developmental toxicity (i.e., the no-observed-adverse-effect level, NOAEL) or should be at or near the limit of detection for the most sensitive endpoint. Two- or four-fold intervals are frequently optimal for spacing the dose levels, and the addition of a fourth test group is often preferable to using very large intervals (e.g., more than a factor of 10) between dosages.
      (B) It is desirable that additional information on metabolism and pharmacokinetics of the test substance be available to demonstrate the adequacy of the dosing regimen. This information should be available prior to testing.
      (C) The highest dose tested need not exceed 1,000 mg/kg/day by oral or dermal administration, or 2 mg/L (or the maximum attainable concentration) by inhalation, unless potential human exposure data indicate the need for higher doses. If a test performed at the limit dose level, using the procedures described for this study, produces no observable toxicity and if an effect would not be expected based upon data from structurally related compounds, then a full study using three-dose levels may not be considered necessary.
    (ii) Control group.
      (A) A concurrent control group should be used. This group should be a sham-treated control group or a vehicle control group if a vehicle is used in administering the test substance.
      (B) The vehicle control group should receive the vehicle in the highest volume used.
      (C) If a vehicle or other additive is used to facilitate dosing, consideration should be given to the following characteristics: Effects on the absorption, distribution, metabolism, or retention of the test substance; effects on the chemical properties of the test substance which may alter its toxic characteristics; and effects on the food or water consumption or the nutritional status of the animals.

(iii) Route of administration.
    (A) The test substance or vehicle is usually administered orally by intubation.
    (B) If another route of administration is used, for example, when the route of administration is based upon the principal route of potential human exposure, the tester should provide justification and reasoning for its selection, and appropriate modifications may be necessary. Further information on dermal or inhalation exposure is provided under paragraphs (h)(12), (h)(28), and (h)(29) of this guideline. Care should be taken to minimize stress on the maternal animals. For materials administered by inhalation, whole-body exposure is preferable to nose-only exposure due to the stress of restraint required for nose-only exposure.
    (C) The test substance should be administered at approximately the same time each day.
    (D) When administered by gavage or dermal application, the dose to each animal should be based on the most recent individual body weight determination.
(iv) Dosing schedule. At minimum, the test substance should be administered daily from implantation to the day before cesarean section on the day prior to the expected day of parturition. Alternatively, if preliminary studies do not indicate a high potential for preimplantation loss, treatment may be extended to include the entire period of gestation, from fertilization to approximately 1 day prior to the expected day of termination.

(f) Observation of animals
  (1) Maternal.
    (i) Mortality, moribundity, pertinent behavioral changes, and all signs of overt toxicity should be recorded at this cageside observation. In addition, thorough physical examinations should be conducted at the same time maternal body weights are recorded.
    (ii) Animals should be weighed on day 0, at termination, and at least at 3-day intervals during the dosing period.
    (iii) Food consumption should be recorded on at least 3-day intervals, preferably on days when body weights are recorded.
    (iv) Termination schedule.
      (A) Females should be terminated immediately prior to the expected day of delivery.
      (B) Females showing signs of abortion or premature delivery prior to scheduled termination should be killed and subjected to a thorough macroscopic examination.
    (v) Gross necropsy. At the time of termination or death during the study, the dam should be examined macroscopically for any structural abnormalities or pathological changes which may have influenced the pregnancy. Evaluation of the dams during cesarean section and subsequent fetal analyses should be conducted without knowledge of treatment group in order to minimize bias.
    (vi) Examination of uterine contents.
      (A) Immediately after termination or as soon as possible after death, the uteri should be removed and the pregnancy status of the animals ascertained. Uteri that appear non-gravid should be further examined (e.g., by ammonium sulfide staining) to confirm the non-pregnant status.
      (B) Each gravid uterus (with cervix) should be weighed. Gravid uterine weights should not be obtained from dead animals if autolysis or decomposition has occurred.
      (C) The number of corpora lutea should be determined for pregnant animals.
      (D) The uterine contents should be examined for embryonic or fetal deaths and the number of viable fetuses. The degree of resorption should be described in order to help estimate the relative time of death of the conceptus.
  (2) Fetal.
    (i) The sex and body weight of each fetus should be determined.
    (ii) Each fetus should be examined for external anomalies.
    (iii) Fetuses should be examined for skeletal and soft tissue anomalies (e.g., variations and malformations or other categories of anomalies as defined by the performing laboratory).
      (A) For rodents, approximately one-half of each litter should be prepared by standard techniques and examined for skeletal alterations, preferably bone and cartilage. The remainder

should be prepared and examined for soft tissue anomalies, using appropriate serial sectioning or gross dissection techniques. It is also acceptable to examine all fetuses by careful dissection for soft tissue anomalies followed by an examination for skeletal anomalies.
    (B) For rabbits, all fetuses should be examined for both soft tissue and skeletal alterations. The bodies of these fetuses should be evaluated by careful dissection for soft-tissue anomalies, followed by preparation and examination for skeletal anomalies. An adequate evaluation of the internal structures of the head, including the eyes, brain, nasal passages, and tongue, should be conducted for at least half of the fetuses.
(g) Data and reporting.
    (1) Treatment of results. Data should be reported individually and summarized in tabular form, showing for each test group the types of change and the number of dams, fetuses, and litters displaying each type of change.
    (2) Evaluation of study results. The following should be provided:
        (i) Maternal and fetal test results, including an evaluation of the relationship, or lack thereof, between the exposure of the animals to the test substance and the incidence and severity of all findings.
        (ii) Criteria used for categorizing fetal external, soft tissue, and skeletal anomalies.
        (iii) When appropriate, historical control data to enhance interpretation of study results. Historical data (on litter incidence and fetal incidence within litter), when used, should be compiled, presented, and analyzed in an appropriate and relevant manner. In order to justify its use as an analytical tool, information such as the dates of study conduct, the strain and source of the animals, and the vehicle and route of administration should be included.
        (iv) Statistical analysis of the study findings should include sufficient information on the method of analysis, so that an independent reviewer/statistician can reevaluate and reconstruct the analysis. In the evaluation of study data, the litter should be considered the basic unit of analysis.
        (v) In any study which demonstrates an absence of toxic effects, further investigation to establish absorption and bioavailability of the test substance should be considered.
    (3) Test report. In addition to the reporting requirements as specified under 40 CFR part 792, subpart J, and 40 CFR part 160, subpart J, the following specific information should be reported. Both individual and summary data should be presented.
        (i) Species and strain.
        (ii) Maternal toxic response data by dose, including but not limited to:
            (A) The number of animals at the start of the test, the number of animals surviving, the number pregnant, and the number aborting.
            (B) Day of death during the study or whether animals survived to termination.
            (C) Day of observation of each abnormal clinical sign and its subsequent course.
            (D) Body weight and body weight change data, including body weight change adjusted for gravid uterine weight.
            (E) Food consumption and, if applicable, water consumption data.
            (F) Necropsy findings, including gravid uterine weight.
        (iii) Developmental endpoints by dose for litters with implants, including:
            (A) Corpora lutea counts.
            (B) Implantation data, number and percent of live and dead fetuses, and resorptions (early and late).
            (C) Pre- and postimplantation loss calculations.
        (iv) Developmental endpoints by dose for litters with live fetuses, including:
            (A) Number and percent of live offspring.
            (B) Sex ratio.
            (C) Fetal body weight data, preferably by sex and with sexes combined.
            (D) External, soft tissue, and skeletal malformation and variation data. The total number and percent of fetuses and litters with any external, soft tissue, or skeletal alteration, as well as the types and incidences of individual anomalies, should be reported.
        (v) The numbers used in calculating all percentages or indices.

(vi) Adequate statistical treatment of results.
(vii) A copy of the study protocol and any amendments should be included.
(h) References. The following references should be consulted for additional background information on this test guideline:
(1) Aliverti, V.L. et al. The extent of fetal ossification as an index of delayed development in teratogenicity studies in the rat. *Teratology* 20:237–242 (1979).
(2) Barrow, M.V. and W.J. Taylor. A rapid method for detecting malformations in rat fetuses. *Journal of Morphology* 127:291–306 (1969).
(3) Burdi, A.R. Toluidine blue-alizarin red S staining of cartilage and bone in whole-mount skeltons in vitro. *Stain Technolology* 40:45–48 (1965).
(4) Edwards, J.A. The external development of the rabbit and rat embryo. In *Advances in Teratology* (ed. D.H.M. Woolam) Vol. 3. Academic, NY (1968).
(5) Fritz, H. Prenatal ossification in rabbits as indicative of fetal maturity. *Teratology* 11:313–320 (1974).
(6) Fritz, H. and R. Hess. Ossification of the rat and mouse skeleton in the perinatal period. *Teratology* 3:331–338 (1970).
(7) Gibson, J.P. et al. Use of the rabbit in teratogenicity studies. *Toxicology and Applied Pharmacology* 9:398–408 (1966).
(8) Inouye, M. Differential staining of cartilage and bone in fetal mouse skeleton by alcian blue and alizarin red S. *Congenital Anomalies* 16(3):171–173 (1976).
(9) Igarashi, E. et al. Frequence of spontaneous axial skeletal variations detected by the double staining technique for ossified and cartilaginous skeleton in rat fetuses. *Congenital Anomalies* 32:381–391 (1992).
(10) Kaufman, M. (ed.) *The Atlas of Mouse Development*. Academic Press, London (1993).
(11) Kimmel, C.A. et al. Skeletal development following heat exposure in the rat. *Teratology* 47:229–242 (1993).
(12) Kimmel, C.A. and E.Z. Francis. Proceedings of the workshop on the acceptability and interpretation of dermal developmental toxicity studies. *Fundamental and Applied Toxicology* 14:386–398 (1990).
(13) Kimmel, C.A. and C. Trammell. A rapid procedure for routine double staining of cartilage and bone in fetal and adult animals. *Stain Technology* 56:271–273 (1981).
(14) Kimmel, C.A. and J.G. Wilson. Skeletal deviation in rats: malformations or variations? *Teratology* 8:309–316 (1973).
(15) Marr, M.C. et al. Comparison of single and double staining for evaluation of skeletal development: the effects of ethylene glycol (EG) in CD rats. *Teratology* 37:476 (1988).
(16) Marr, M.C. et al. Developmental stages of the CD (Sprague-Dawley) rat skeleton after maternal exposure to ethylene glycol. *Teratology* 46:169–181 (1992).
(17) McLeod, M.J. Differential staining of cartilage and bone in whole mouse fetuses by Alcian blue and alizarin red S. *Teratology* 22:299–301 (1980).
(18) Monie, I.W. et al. *Dissection procedures for rat fetuses permitting alizarin red staining of skeleton and histological study of viscera.* Supplement to Teratology Workshop Manual, pp. 163–173 (1965).
(19) Organization for Economic Cooperation and Development, No. 414: Teratogenicity, Guidelines for Testing of Chemicals. [C(83)44(Final)] (1983).
(20) Salewski (Koeln), V.E. Faerbermethode zum makroskopischen nachweis von implantations stellen am uterus der ratte. *Naunyn-Schmeidebergs Archiv fü" r Pharmakologie und Experimentelle Pathologie* 247:367 (1964).
(21) Spark, C. and A.B. Dawson. The order and time of appearance of centers of ossification in the fore and hind limbs of the albino rat, with special reference to the possible influence of the sex factor. *American Journal of Anatomy* 41:411–445 (1928).
(22) Staples, R.E. Detection of visceral alterations in mammalian fetuses. *Teratology* 9(3): A37–A38 (1974).
(23) Staples, R.E. and V.L. Schnell. Refinements in rapid clearing technique in the KOH—alizarin red S method for fetal bone. *Stain Technology* 39:61–63 (1964).
(24) Strong, R.M. The order time and rate of ossification of the albino rat *(mus norvegicus albinus)* skeleton. *American Journal of Anatomy* 36: 313–355 (1928).

(25) Stuckhardt, J.L. and S.M. Poppe. Fresh visceral examination of rat and rabbit fetuses used in teratogenicity testing. *Teratogenesis, Carcinogenesis, and Mutagenesis* 4:181–188 (1984).
(26) U.S. Environmental Protection Agency. Guideline 83-3: Teratogencity Study. Pesticide Assessment Guidelines, Subdivision F. Hazard Evaluation: Human and Domestic Animals. Office of Pesticides and Toxic Substances, Washington, DC, EPA–540/9–82–025 (1982).
(27) U.S. Environmental Protection Agency. Subpart E—Specific Organ/Tissue Toxicity, 40 CFR 798.4900: Developmental Toxicity Study.
(28) U.S. Environmental Protection Agency. Health Effects Test Guidelines, OPPTS 870.3200, 21/28-Day Dermal Toxicity, July 1998.
(29) U.S. Environmental Protection Agency. Health Effects Test Guidelines, OPPTS 870.3465, 90-Day Inhalation Toxicity, July 1998.
(30) U.S. Environmental Protection Agency Guidelines for Developmental Toxicity Risk Assessment. FEDERAL REGISTER (56 FR 63798–63826, December 5, 1991).
(31) Van Julsingha, E.B. and C.G. Bennett. A dissecting procedure for the detection of anomalies in the rabbit foetal head. In: *Methods in Prenatal Toxicology* (eds. D. Neubert, H.J. Merker, and T.E. Kwasigroch). University of Chicago, Chicago, IL, pp. 126–144 (1977).
(32) Walker, D.G. and Z.T. Wirtschafter. *The Genesis of the Rat Skeleton*. Thomas, Springfield, IL (1957).
(33) Whitaker, J. and D.M. Dix. Double-staining for rat foetus skeletons in teratological studies. *Laboratory Animals* 13:309–310 (1979).
(34) Wilson, J.G. Embryological considerations in teratology. In "Teratology: Principles and Techniques" (ed. J.G. Wilson and J. Warkany). University of Chicago, Chicago, IL, pp 251–277 (1965).
(35) Wilson, J.G. and F.C. Fraser, ed. *Handbook of Teratology*, Vol. 4. Plenum, NY (1977).
(36) Yasuda, M. and T. Yuki. *Color Atlas of Fetal Skeleton of the Mouse, Rat, and Rabbit*. Ace Art Co., Osaka, Japan (1996).

United States Environmental Protection Agency
Prevention, Pesticides and Toxic Substances
(7101)
EPA 712–C–98–208
August 1998

Health Effects Test Guidelines
OPPTS 870.3800
Reproduction and Fertility Effects

Introduction

This guideline is one of a series of test guidelines that have been developed by the Office of Prevention, Pesticides and Toxic Substances, United States Environmental Protection Agency for use in the testing of pesticides and toxic substances, and the development of test data that must be submitted to the Agency for review under Federal regulations.

The Office of Prevention, Pesticides and Toxic Substances (OPPTS) has developed this guideline through a process of harmonization that blended the testing guidance and requirements that existed in the Office of Pollution Prevention and Toxics (OPPT) and appeared in Title 40, Chapter I, Subchapter R of the Code of Federal Regulations (CFR), the Office of Pesticide Programs (OPP) which appeared in publications of the National Technical Information Service (NTIS) and the guidelines published by the Organization for Economic Cooperation and Development (OECD).

The purpose of harmonizing these guidelines into a single set of OPPTS guidelines is to minimize variations among the testing procedures that must be performed to meet the data requirements of the U. S. Environmental Protection Agency under the Toxic Substances Control Act (15 U.S.C. 2601) and the Federal Insecticide, Fungicide and Rodenticide Act (7 U.S.C. 136, *et seq.*).

Final Guideline Release: This guideline is available from the U.S. Government Printing Office, Washington, DC 20402 on disks or paper copies: call (202) 512–0132. This guideline is also available electronically in ASCII and PDF (portable document format) from EPA's World Wide Web site (http://www.epa.gov/epahome/research.htm) under the heading "Researchers and Scientists/Test Methods and Guidelines/OPPTS Harmonized Test Guidelines."

OPPTS 870.3800 Reproduction and fertility effects.
(a) Scope
  (1) Applicability. This guideline is intended to meet testing requirements of both the Federal Insecticide, Fungicide, and Rodenticide Act (FIFRA) (7 U.S.C. 136, *et seq.*), as amended by the Food Quality Protection Act (FQPA)(Pub. L. 104–170) and the Toxic Substances Control Act (TSCA) (15 U.S.C. 2601).
  (2) Background. The source material used in developing this harmonized OPPTS test guideline is the OPPT guideline under 40 CFR 798.4700, OPP guideline 83–4, and OECD guideline 416.
(b) Purpose. This guideline for two-generation reproduction testing is designed to provide general information concerning the effects of a test substance on the integrity and performance of the male and female reproductive systems, including gonadal function, the estrous cycle, mating behavior, conception, gestation, parturition, lactation, and weaning, and on the growth and development of the offspring. The study may also provide information about the effects of the test substance on neonatal morbidity, mortality, target organs in the offspring, and preliminary data on prenatal and postnatal developmental toxicity and serve as a guide for subsequent tests. Additionally, since the study design includes *in utero* as well as post-natal exposure, this study provides the opportunity to examine the susceptibility of the immature/neonatal animal. For further information on functional deficiencies and developmental effects, additional study segments can be incorporated into the protocol, utilizing the guidelines for developmental toxicity or developmental neurotoxicity.
(c) Good laboratory practice standards. The study should be conducted in accordance with the laboratory practices stipulated in 40 CFR Part 160 (FIFRA) and 40 CFR Part 792 (TSCA)—Good Laboratory Practice Standards.
(d) Principle of the test method. The test substance is administered to parental (P) animals prior to and during their mating, during the resultant pregnancies, and through the weaning of their F1 offspring. The sub-stance is then administered to selected F1 offspring during their growth into adulthood, mating, and production of an F2 generation, until the F2 generation is weaned.
(e) Test procedures
  (1) Animal selection
    (i) Species and strain. The rat is the most commonly used species for testing. If another mammalian species is used, the tester should provide justification/reasoning for its selection, and appropriate modifications will be necessary. Healthy parental animals, which have been acclimated to laboratory conditions for at least 5 days and have not been subjected to previous experimental procedures, should be used. Strains of low fecundity should not be used.
    (ii) Age. Parental (P) animals should be 5 to 9 weeks old at the start of dosing. The animals of all test groups should be of uniform weight, age, and parity as nearly as practicable, and should be representative of the species and strain under study.
    (iii) Sex.
      (A) For an adequate assessment of fertility, both males and females should be studied.
      (B) The females should be nulliparous and nonpregnant.
    (iv) Animal care. Animal care and housing should be in accordance with the recommendations contained in DHHS/PHS NIH Publication No. 86–23, 1985, *Guidelines for the Care and Use of Laboratory Animals*, or other appropriate guidelines.
    (v) Number of animals. Each control group should contain a sufficient number of mating pairs to yield approximately 20 pregnant females. Each test group should contain a similar number of mating pairs.
    (vi) Identification of animals. Each animal should be assigned a unique identification number. For the P generation, this should be done before dosing starts. For the F1 generation, this should be done for animals selected for mating; in addition, records indicating the litter of origin should be maintained for all selected F1 animals.

(2) Administration of test and control substances
   (i) Dose levels and dose selection.
      (A) At least three-dose levels and a concurrent control should be used. Healthy animals should be randomly assigned to the control and treatment groups, in a manner which results in comparable mean body weight values among all groups. The dose levels should be spaced to produce a gradation of toxic effects. Unless limited by the physical/chemical nature or biological properties of the test substance, the highest dose should be chosen with the aim to induce some reproductive and/or systemic toxicity but not death or severe suffering. In the case of parental mortality, this should not be more than approximately 10 percent. The intermediate dose levels should produce minimal observable toxic effects. The lowest dose level should not produce any evidence of either systemic or reproductive toxicity (i.e., the no-observed-adverse-effect level, NOAEL) or should be at or near the limit of detection for the most sensitive endpoint. Two- or four-fold intervals are frequently optimal for spacing the dose levels, and the addition of a fourth test group is often preferable to using very large intervals (e.g., more than a factor of 10) between dosages.
      (B) It is desirable that additional information on metabolism and pharmacokinetics of the test substance be available to demonstrate the adequacy of the dosing regimen. This information should be available prior to testing.
      (C) The highest dose tested should not exceed 1,000 mg/kg/day (or 20,000 ppm in the diet), unless potential human exposure data indicate the need for higher doses. If a test performed at the limit dose level, using the procedures described for this study, produces no observable toxicity and if an effect would not be expected based upon data from structurally related compounds, then a full study using three dose levels may not be considered necessary.
   (ii) Control group.
      (A) A concurrent control group should be used. This group should be an untreated or sham treated group or a vehicle-control group if a vehicle is used in administering the test substance.
      (B) If a vehicle is used in administering the test substance, the control group should receive the vehicle in the highest volume used.
      (C) If a vehicle or other additive is used to facilitate dosing, consideration should be given to the following characteristics: Effects on the absorption, distribution, metabolism, or retention of the test substance; effects on the chemical properties of the test substance which may alter its toxic characteristics; and effects on the food or water consumption or the nutritional status of the animals.
      (D) If a test substance is administered in the diet and causes reduced dietary intake or utilization, the use of a pair-fed control group may be considered necessary.
   (iii) Route of administration.
      (A) The test substance is usually administered by the oral route (diet, drinking water, or gavage).
      (B) If administered by gavage or dermal application, the dosage administered to each animal prior to mating and during gestation and lactation should be based on the individual animal body weight and adjusted weekly at a minimum.
      (C) If another route of administration is used, for example, when the route of administration is based upon the principal route of potential human exposure, the tester should provide justification and reasoning for its selection, and appropriate modifications may be necessary. Further in-formation on dermal or inhalation exposure is provided under paragraphs (g)(18) and (g)(19) of this guideline. Care should be taken to minimize stress on the maternal animals and their litters during gestation and lactation.
      (D) All animals should be dosed by the same method during the appropriate experimental period.
   (iv) Dosing schedule.
      (A) The animals should be dosed with the test substance on a 7–days–a–week basis.
      (B) Daily dosing of the parental (P) males and females should begin when they are 5 to 9

weeks old. Daily dosing of the F1 males and females should begin at weaning. For both sexes (P and F1), dosing should be continued for at least 10 weeks before the mating period.

(C) Daily dosing of the P and F1 males and females should continue until termination.

(3) Mating procedure
   (i) Parental.
      (A) For each mating, each female should be placed with a single randomly selected male from the same dose level (1:1 mating) until evidence of copulation is observed or either 3 estrous periods or 2 weeks has elapsed. Animals should be separated as soon as possible after evidence of copulation is observed. If mating has not occurred after 2 weeks or 3 estrous periods, the animals should be separated without further opportunity for mating. Mating pairs should be clearly identified in the data.
      (B) Vaginal smears should be collected daily and examined for all females during mating, until evidence of copulation is observed.
      (C) Each day, the females should be examined for presence of sperm or vaginal plugs. Day 0 of pregnancy is defined as the day a vaginal plug or sperm are found.
   (ii) F1 mating. For mating the F1 offspring, at least one male and one female should be randomly selected from each litter for mating with another pup of the same dose level but different litter, to produce the F2 generation.
   (iii) Second mating. In certain instances, such as poor reproductive performance in the controls, or in the event of treatment-related alterations in litter size, the adults may be re-mated to produce an F1b or F2b litter. If production of a second litter is deemed necessary in either generation, the dams should be remated approximately 1–2 weeks following weaning of the last F1a or F2a litter.
   (iv) Special housing. After evidence of copulation, animals that are presumed to be pregnant should be caged separately in delivery or maternity cages. Pregnant animals should be provided with nesting materials when parturition is near.
   (v) Standardization of litter sizes.
      (A) Animals should be allowed to litter normally and rear their offspring to weaning. Standardization of litter sizes is optional.
      (B) If standardization is performed, the following procedure should be used. On day 4 after birth, the size of each litter may be adjusted by eliminating extra pups by random selection to yield, as nearly as possible, four males and four females per litter or five males and five females per litter. Selective elimination of pups, i.e., based upon body weight, is not appropriate. Whenever the number of male or female pups prevents having four (or five) of each sex per litter, partial adjustment (for example, five males and three females, or four males and six females) is acceptable. Adjustments are not appropriate for litters of eight pups or less.

(4) Observation of animals
   (i) Parental.
      (A) Throughout the test period, each animal should be observed at least once daily, considering the peak period of anticipated effects after dosing. Mortality, moribundity, pertinent behavioral changes, signs of difficult or prolonged parturition, and all signs of overt toxicity should be recorded at this cageside examination. In addition, thorough physical examinations should be conducted weekly on each animal.
      (B) Parental animals (P and F1) should be weighed on the first day of dosing and weekly thereafter. Parental females (P and F1) should be weighed at a minimum on approximately gestation days 0, 7, 14 and 21, and during lactation on the same days as the weighing of litters.
      (C) During the premating and gestation periods, food consumption should be measured weekly at a minimum. Water consumption should be measured weekly at a minimum if the test substance is administered in the water.
      (D) Estrous cycle length and pattern should be evaluated by vaginal smears for all P and F1 females during a minimum of 3 weeks prior to mating and throughout cohabitation; care should be taken to prevent the induction of pseudopregnancy.

(E) For all P and F1 males at termination, sperm from one testis and one epididymis should be collected for enumeration of homogenization-resistant spermatids and cauda epididymal sperm reserves, respectively. In addition, sperm from the cauda epididymis (or vas deferens) should be collected for evaluation of sperm motility and sperm morphology.

    (1) The total number of homogenization-resistant testicular sperm and cauda epididymal sperm should be enumerated (see paragraphs (g)(3) and (g)(13) of this guideline). Cauda sperm reserves can be derived from the concentration and volume of sperm in the suspension used to complete the qualitative evaluations, and the number of sperm recovered by subsequent mincing and/or homogenizing of the remaining cauda tissue. Enumeration in only control and high-dose P and F1 males may be performed unless treatment-related effects are observed; in that case, the lower dose groups should also be evaluated.

    (2) An evaluation of epididymal (or vas deferens) sperm motility should be performed. Sperm should be recovered while minimizing damage (refer to paragraph (g)(13) of this guideline), and the percentage of progressively motile sperm should be determined either subjectively or objectively. For objective evaluations, an acceptable counting chamber of sufficient depth can be used to effectively combine the assessment of motility with sperm count and sperm morphology. When computer-assisted motion analysis is performed (refer to paragraph (g)(13) of this guideline), the derivation of progressive motility relies on user-defined thresholds for average path velocity and straightness or linear index. If samples are videotaped, or images otherwise recorded, at the time of necropsy, subsequent analysis of only control and high-dose P and F1 males may be performed unless treatment-related effects are observed; in that case, the lower dose groups should also be evaluated. In the absence of a video or digital image, all samples in all treatment groups should be analyzed at necropsy.

    (3) A morphological evaluation of an epididymal (or vas deferens) sperm sample should be performed. Sperm (at least 200 per sample) should be examined as fixed, wet preparations (refer to paragraphs (g)(7) and (g)(13) of this guideline) and classified as either normal (both head and midpiece/tail appear normal) or abnormal. Examples of morphologic sperm abnormalities would include fusion, isolated heads, and misshapen heads and/or tails. Evaluation of only control and high-dose P and F1 males may be performed unless treatment-related effects are observed; in that case, the lower dose groups should also be evaluated.

(ii) Offspring.

    (A) Each litter should be examined as soon as possible after delivery (lactation day 0) to establish the number and sex of pups, stillbirths, live births, and the presence of gross anomalies. Pups found dead on day 0 should be examined for possible defects and cause of death.

    (B) Live pups should be counted, sexed, and weighed individually at birth, or soon thereafter, at least on days 4, 7, 14, and 21 of lactation, at the time of vaginal patency or balanopreputial separation, and at termination.

    (C) The age of vaginal opening and preputial separation should be determined for F1 weanlings selected for mating. If there is a treatment-related effect in F1 sex ratio or sexual maturation, anogenital distance should be measured on day 0 for all F2 pups.

(5) Termination schedule.

    (i) All P and F1 adult males and females should be terminated when they are no longer needed for assessment of reproductive effects.

    (ii) F1 offspring not selected for mating and all F2 offspring should be terminated at comparable ages after weaning.

(6) Gross necropsy.

    (i) At the time of termination or death during the study, all parental animals (P and F1) and when litter size permits at least three pups per sex per litter from the unselected F1 weanlings and the F2 weanlings should be examined macroscopically for any structural abnormalities or pathological changes. Special attention should be paid to the organs of the reproductive system.

(ii) Dead pups or pups that are terminated in a moribund condition should be examined for possible defects and/or cause of death.
(iii) At the time of necropsy, a vaginal smear should be examined to determine the stage of the estrous cycle. The uteri of all cohabited females should be examined, in a manner which does not compromise histopathological evaluation, for the presence and number of implantation sites.

(7) Organ weights.
  (i) At the time of termination, the following organs of all P and F1 parental animals should be weighed:
    (A) Uterus (with oviducts and cervix), ovaries.
    (B) Testes, epididymides (total weights for both and cauda weight for either one or both), seminal vesicles (with coagulating glands and their fluids), and prostate.
    (C) Brain, pituitary, liver, kidneys, adrenal glands, spleen, and known target organs.
  (ii) For F1 and F2 weanlings that are examined macroscopically, the following organs should be weighed for one randomly selected pup per sex per litter.
    (A) Brain.
    (B) Spleen and thymus.

(8) Tissue preservation. The following organs and tissues, or representative samples thereof, should be fixed and stored in a suitable medium for histopathological examination.
  (i) For the parental (P and F1) animals:
    (A) Vagina, uterus with oviducts, cervix, and ovaries.
    (B) One testis (preserved in Bouins fixative or comparable preservative), one epididymis, seminal vesicles, prostate, and coagulating gland.
    (C) Pituitary and adrenal glands.
    (D) Target organs, when previously identified, from all P and F1 animals selected for mating.
    (E) Grossly abnormal tissue.
  (ii) For F1 and F2 weanlings selected for macroscopic examination: Grossly abnormal tissue and target organs, when known.

(9) Histopathlogy
  (i) Parental animals. Full histopathology of the organs listed in paragraph (e)(8)(i) of this guideline should be performed for ten randomly chosen high dose and control P and F1 animals per sex, for those animals that were selected for mating. Organs demonstrating treatment-related changes should also be examined for the remainder of the high-dose and control animals and for all parental animals in the low- and mid-dose groups. Additionally, reproductive organs of the low- and mid-dose animals suspected of reduced fertility, e.g., those that failed to mate, conceive, sire, or deliver healthy offspring, or for which estrous cyclicity or sperm number, motility, or morphology were affected, should be subjected to histopathological evaluation. Besides gross lesions such as atrophy or tumors, testicular histopathological examination should be conducted in order to identify treatment-related effects such as retained spermatids, missing germ cell layers or types, multinucleated giant cells, or sloughing of spermatogenic cells into the lumen (refer to paragraph (g)(11) of this guideline). Examination of the intact epididymis should include the caput, corpus, and cauda, which can be accomplished by evaluation of a longitudinal section, and should be conducted in order to identify such lesions as sperm granulomas, leukocytic infiltration (inflammation), aberrant cell types within the lumen, or the absence of clear cells in the cauda epididymal epithelium. The postlactational ovary should contain primordial and growing follicles as well as the large corpora lutea of lactation. Histopathological examination should detect qualitative depletion of the primordial follicle population. A quantitative evaluation of primordial follicles should be conducted for F1 females; the number of animals, ovarian section selection, and section sample size should be statistically appropriate for the evaluation procedure used. Examination should include enumeration of the number of primordial follicles, which can be combined with small growing follicles (see paragraphs (g)(1) and (g)(2) of this guideline), for comparison of treated and control ovaries.

(ii) Weanlings. For F1 and F2 weanlings, histopathological examination of treatment-related abnormalities noted at macroscopic examination should be considered, if such evaluation were deemed appropriate and would contribute to the interpretation of the study data.

(f) Data and reporting
  (1) Treatment of results. Data should be reported individually and summarized in tabular form, showing for each test group the types of change and the number of animals displaying each type of change.
  (2) Evaluation of study results.
    (i) An evaluation of test results, including the statistical analysis, should be provided. This should include an evaluation of the relationship, or lack thereof, between the exposure of the animals to the test substance and the incidence and severity of all abnormalities.
    (ii) When appropriate, historical control data should be used to enhance interpretation of study results. Historical data, when used, should be compiled, presented, and analyzed in an appropriate and relevant manner. In order to justify its use as an analytical tool, information such as the dates of study conduct, the strain and source of the animals, and the vehicle and route of administration should be included.
    (iii) Statistical analysis of the study findings should include sufficient information on the method of analysis, so that an independent reviewer/statistician can reevaluate and reconstruct the analysis.
    (iv) In any study which demonstrates an absence of toxic effects, further investigation to establish absorption and bioavailability of the test substance should be considered.
  (3) Test report. In addition to the reporting requirements as specified under 40 CFR part 792, subpart J and 40 CFR part 160, subpart J, the following specific information should be reported. Both individual and summary data should be presented.
    (i) Species and strain.
    (ii) Toxic response data by sex and dose, including indices of mating, fertility, gestation, birth, viability, and lactation; offspring sex ratio; precoital interval, including the number of days until mating and the number of estrous periods until mating; and duration of gestation calculated from day 0 of pregnancy. The report should provide the numbers used in calculating all indices.
    (iii) Day (week) of death during the study or whether animals survived to termination; date (age) of litter termination.
    (iv) Toxic or other effects on reproduction, offspring, or postnatal growth.
    (v) Developmental milestone data (mean age of vaginal opening and preputial separation, and mean anogenital distance, when measured).
    (vi) An analysis of P and F1 females cycle pattern and mean estrous cycle length.
    (vii) Day (week) of observation of each abnormal sign and its subsequent course.
    (viii) Body weight and body weight change data by sex for P, F1, and F2 animals.
    (ix) Food (and water, if applicable) consumption, food efficiency (body weight gain per gram of food consumed), and test material consumption for P and F1 animals, except for the period of cohabitation.
    (x) Total cauda epididymal sperm number, homogenization-resistant testis spermatid number, number and percent of progressively motile sperm, number and percent of morphologically normal sperm, and number and percent of sperm with each identified anomaly.
    (xi) Stage of the estrous cycle at the time of termination for P and F1 parental females.
    (xii) Necropsy findings.
    (xiii) Implantation data and postimplantation loss calculations for P and F1 parental females.
    (xiv) Absolute and adjusted organ weight data.
    (xv) Detailed description of all histopathological findings.
    (xvi) Adequate statistical treatment of results.
    (xvii) A copy of the study protocol and any amendments should be included.
(g) References. The following references should be consulted for additional background information on this test guideline:
  (1) Bolon, B. et al. Differential follicle counts as a screen for chemically induced ovarian toxicity in mice: results from continuous breeding bioassays. *Fundamental and Applied Toxicology* 39:1-10 (1997).

(2) Bucci, T.J. et al. The effect of sampling procedure on differential ovarian follicle counts. *Reproductive Toxicology* 11(5): 689-696 (1997).

(3) Gray, L.E. *et al.* A dose-response analysis of methoxychlor-induced alterations of reproductive development and function in the rat. *Fundamental and Applied Toxicology* 12:92–108 (1989).

(4) Heindel, J.J. and R.E. Chapin, (eds.). Part B. Female Reproductive Systems, *Methods in Toxicology*, Academic, Orlando, FL (1993).

(5) Heindel, J.J. *et al.* Histological assessment of ovarian follicle number in mice as a screen of ovarian toxicity. In: *Growth Factors and the Ovary*, A.N. Hirshfield (ed.), Plenum, NY, pp. 421–426 (1989).

(6) Korenbrot, C.C. *et al.* Preputial separation as an external sign of pubertal development in the male rat. *Biology of Reproduction* 17:298–303 (1977).

(7) Linder, R.E. *et al.* Endpoints of spermatoxicity in the rat after short duration exposures to fourteen reproductive toxicants. *Reproductive Toxicology* 6:491–505 (1992).

(8) Manson, J.M. and Y.J. Kang. Test methods for assessing female reproductive and developmental toxicology. In: *Principles and Methods of Toxicology*, A.W. Hayes (ed.), Raven, New York (1989).

(9) Organization for Economic Cooperation and Development, No. 416: Two Generation Reproduction Toxicity Study, Guidelines for Testing of Chemicals. [C(83)44 (Final)] (1983).

(10) Pederson, T. and H. Peters. Proposal for classification of oocytes and follicles in the mouse ovary. *Journal of Reproduction and Fertility* 17:555–557 (1988).

(11) Russell, L.D. *et al. Histological and Histopathological Evaluation of the Testis*, Cache River, Clearwater, FL (1990).

(12) Sadleir, R.M.F.S. Cycles and seasons, In: *Reproduction in Mammals*: I. Germ Cells and Fertilization, C.R. Auston and R.V. Short (eds.), Cambridge, NY (1979).

(13) Seed, J., R.E. Chapin, E.D. Clegg, L.A. Dostal, R.H. Foote, M.E. Hurtt, G.R. Klinefelter, S.L. Makris, S.D. Perreault, S. Schrader, D. Seyler, R. Sprando, K.A. Treinen, D.N.R. Veeramachaneni, and L.D. Wise. Methods for assessing sperm motility, morphology, and counts in the rat, rabbit, and dog: a consensus report. *Reproductive Toxicology* 10(3):237–244 (1996).

(14) Smith, B.J. *et al.* Comparison of random and serial sections in assessment of ovarian toxicity. *Reproductive Toxicology* 5:379–383 (1991).

(15) Thomas, J.A. Toxic responses of the reproductive system. In: *Casarett and Doull's Toxicology*, M.O. Amdur, J. Doull, and C.D. Klaassen (eds.), Pergamon, NY (1991).

(16) U.S. Environmental Protection Agency. OPP Guideline 83–4: Reproductive and Fertility Effects. Pesticide Assessment Guidelines, Subdivision F, Hazard Evaluation: Human and Domestic Animals. Office of Pesticides and Toxic Substances, Washington, DC, EPA–540/9–82–025 (1982).

(17) U.S. Environmental Protection Agency. Subpart E—Specific Organ/Tissue Toxicity, 40 CFR 798.4700: Reproduction and Fertility Effects.

(18) U.S. Environmental Protection Agency. Health Effects Test Guidelines, OPPTS 870.3250, 90-Day Dermal Toxicity, July 1998.

(19) U.S. Environmental Protection Agency. Health Effects Test Guidelines, OPPTS 870.3465, 90-Day Inhalation Toxicity, July 1998.

(20) U.S. Environmental Protection Agency. Reproductive Toxicity Risk Assessment Guidelines. Federal Register 61 FR 56274–56322 (1996)

(21) Working, P.K. and M. Hurtt. Computerized videomicrographic analysis of rat sperm motility. *Journal of Andrology* 8:330–337 (1987).

(22) Zenick, H. et al. Assessment of male reproductive toxicity: a risk assessment approach. In: *Principles and Methods of Toxicology*, A.W. Hayes (ed.), Raven, NY (1994).

APPENDIX 3

REFERENCES FOR GUIDELINES AND OTHER INFORMATION PHARMACEUTICALS INFORMATION INTENDED FOR INTERNATIONAL USE INTERNATIONAL CONFERENCE ON HARMONIZATION (ICH)

International Conference on Harmonisation of Technical Requirements for the Registration of Pharmaceuticals for Human Use. *ICH Harmonised Tripartite Guideline; Detection of Toxicity to Reproduction for Medicinal Products*, June 24, 1993, Presented at the Second International Conference on Harmonisation, Orlando, Florida, October 27-29, 1993 (Provided in Appendix 1).

Bass, R., Ulbrich, B., Hildebrandt, A., Weissinger, J., Doi, O., Baeder, C., Fumero, S., Harada, Y., Lehmann, H., Manson, J., Neubert, D., Omori, Y., Palmer, A., Sullivan, F., Takayama, S., Tanimura, T., Draft guideline on detection of toxicity to reproduction for medical products, *Adverse Drug React. Toxicol.*, Rev.9, 127, 1991.

World Health Organization (WHO)

World Health Organization (WHO), Principles for the Testing of Drugs for Teratogenicity, *WHO Tech. Rept. Series,* No. 364, Geneva, 1967.

Information Intended for Local Use

Canada

Canada, Health Protection Branch, *The Testing of Chemicals for Carcinogenicity, Mutagenicity and Teratogenicity*, Ministry of Health and Welfare, Canada, 1977.

Canada, Bureau of Drugs, Health Protection Branch, Health and Welfare, *Draft of Preclinical Toxicologic Guidelines*, Section 2.4, 35, 1979.

Drug Dictorate Guidelines, Toxicological Evaluation 2.4, Reproductive Studies, Ministry of National Health and Welfare, Health Protection Branch, Health and Welfare, Canada, 1990.

European Economic Council (EEC)

European Economic Community. Council recommendation of 26 October 1983 concerning tests relating to the placing on the market of proprietary medicinal products. *Official Journal of the European Communities*, No. L 332 (83/571/EEC), November 28, 1983.

European Economic Community, The Rules Governing Medicinal Products in the European Community, Vol. III, Guidelines on the Quality, Safety and Efficacy of Medicinal Products for Human Use. Office of Official Publications of the European Communities, Brussels, 1988.

Great Britain/United Kingdom (U.K.) — Committee on the Safety of Medicines (CSM)

Committee on the Safety of Medicines (CSM), *Notes for Guidance on Reproduction Studies*, Department of Health and Social Security, Great Britain, 1974.

Japan — Ministry of Health and Welfare (MHW)

Ministry of Health and Welfare, Japan (MHW), *On Animal Experimental Methods for Testing the Effects of Drugs on Reproduction*, Notification No. 529 of the Pharmaceutical Affairs Bureau, Ministry of Health and Welfare, March 31, 1975.

Ministry of Health and Welfare, Japan (MHW), *Information on the Guidelines of Toxicity Studies Required for Applications for Approval to Manufacture (Import) Drugs*, Notification No. 118 of the Pharmaceutical Affairs Bureau, Ministry of Health and Welfare, February 15, 1984.

Tanimura, T., Kameyama, Y., Shiota, K., Tanaka, S., Matsumoto, N. and Mizutani, M., Report on the review of the guidelines for studies of the effect of drugs on reproduction. Notification No. 118, Pharmaceutical Affairs Bureau, Ministry of Health and Welfare, Japan, 1989.

Pharmaceutical Manufacturers Association (PMA)

Pharmaceutical Manufacturers Association (PMA), *Second Workshop in Teratology. Supplement to the Teratology Workshop Manual* (A collection of lectures and demonstrations from the Second Workshop in Teratology held January 25-30, 1965, Berkeley, California), Pharmaceutical Manufacturers Association, Washington, D.C., 1965.

PMA Guidelines: *Reproduction, teratology and pediatrics*. Contributors: Christian, M., Diener, R., Hoar, R. and Staples, R., 1981 (revised)

USA — Food and Drug Administration (FDA)

Reproductive Toxicity

U.S. Food and Drug Administration (1959) *Appraisal of the Safety of Chemicals in Foods, Drugs, and Cosmetics,* third printing: 1975, the Association of Food and Drug Officials of the United States, Washington, DC.

U.S. Food and Drug Administration (FDA), Guidelines for Reproduction Studies for Safety Evaluation of Drugs for Human Use, Washington, D.C., 1966.

U.S. Food and Drug Administration (FDA). International Conference on Harmonisation; Guideline on detection of toxicity to reproduction for medicinal products. Federal Register, Vol. 59, No. 183, 1994.

U.S. Food and Drug Administration (FDA). International Conference on Harmonisation; Guideline on detection of toxicity to reproduction for medicinal products; Addendum on Toxicity to male fertility. *Federal Register*, April 5, 1996, Vol. 61, No. 67, 1996.

U.S. Food and Drug Administration (FDA). International Conference on Harmonisation; Maintenance of the ICH guideline on toxicity to male fertility; An addendum to the ICH tripartite guideline on detection of toxicity to reproduction for medicinal products. Amended November 9, 2000.

Good Laboratory Practices (GLPS)

U.S. Food and Drug Administration. Good Laboratory Practice Regulations; Final Rule. 21 CFR Part 58.

USA — Interagency Regulatory Group (IRLG)

Interagency Regulatory Group (IRLG): Recommended guideline for teratogenicity studies in the rat, mouse, hamster or rabbit, April, 1980.

Recommended Guideline for Teratogenicity Studies in Rat, Mouse, Hamster or Rabbit. Interagency Regulatory Liaison Group Testing Standards and Guidelines Work Group, US Consumer product safety commission, US Environmental Protection Agency, Food and Drug Administration (US Dept. of HHS), Food Safety and Quality Service (US Dept. of Agriculture) and Occupational Safety and Health Administration (US Dept. of Labor), 2, 1981.

Foods — Information Intended for Local Use

USA — Food and Drug Administration (FDA)

U.S. Food and Drug Administration (1959) Appraisal of the Safety of Chemicals in Foods, Drugs, and Cosmetics, third printing: 1975, the Association of Food and Drug Officials of the United States, Washington, DC.

U.S. Food and Drug Administration (FDA), Advisory Committee on Protocols for Safety Evaluations: Panel on Reproduction. Report on reproduction studies in the safety evaluation of food additives and pesticide residues, *Toxicol. Applied Pharmacol.*, 16, 264, 1970.

U.S. Food and Drug Administration (FDA). *Toxicological Principles and Procedures for Priority Based Assessment of Food Additives. (Red Book)*, Guidelines for Reproduction Testing with a Teratology Phase, Bureau of Foods, U.S. Food and Drug Administration, Washington, D.C., 1982, pp. 80-117.

U.S. Food and Drug Administration (FDA): Toxicological principles for the safety assessment of direct food additives and color additives uses in food. National Technical Information Services, PB83-170696, 1982, pp. 80-119.

U.S. Food and Drug Administration (FDA). Draft — Toxicological Principles for the Safety Assessment of Direct Food Additives and Color Additives Used in Food. (Red Book II), Guidelines for Reproduction and Developmental Toxicity Studies, Center for Food Safety and Applied Nutrition, U.S. Food and Drug Administration, Washington, D.C., 1993, pp. 123-134.

U.S. Food and Drug Administration (FDA). *Toxicological Principles for the Safety of Food Ingredients (Redbook 2000)*: Sections issued electronically July 7, 2000 at http:// vm.cfscan.fda.gov/~Redbook/Red-tcct.html, 2000.

USA — National Academy of Sciences (NAS)

National Academy of Sciences, Food Protection Committee, *Evaluating the Safety of Food Chemicals*, National Academy of Sciences, Washington, D.C., 1970.

USA — National Toxicology Program (NTP)

Collins, T.F.X., Reproduction and teratology guidelines: Review of deliberations by the National Toxicology Advisory Committee's Reproduction Panel, *J. Environ. Path. Toxicol.*, 2, 141, 1978.

Chemicals — Information Intended for International Use

Organization on Economic Cooperation and Development (OECD)

Reproductive Toxicity

Organization for Economic Cooperation and Development (OECD): Final Report- OECD short-term and long-term toxicology groups, Teratogenicity, December 31, 1979, 110.

Organization for Economic Cooperation and Development. *Guidelines for Testing of Chemicals*. Section 4, No. 414: Teratogenicity, 22 January 2001.

Organization for Economic Cooperation and Development. *First Addendum to OECD Guidelines for Testing of Chemicals*. Section 4, No. 415: One-Generation Reproduction Toxicity, 1983.

Organization for Economic Cooperation and Development. *Guidelines for Testing of Chemicals.* Section 4, No. 416: Two Generation Reproduction Toxicity Study, 22 January 2001.

Organization for Economic and Co-operative Development (OECD) Screening Information Data Set (SIDS). *Guideline for Testing of Chemicals*, Section 4, No. 421: Reproduction/Developmental Toxicity Screening Test, 27 July 1995.

Good Laboratory Practices

Organization for Economic Cooperation and Development. *Good Laboratory Practice in the Testing of Chemicals*, Final Report of the OECD Expert Group on Laboratory Practices [C(81) 3.0 (Final), Annex 2], 1981.

World Health Organization (WHO)

Food and Agriculture Organization of the United Nations (FAO), Working Party on Pesticide Residues. Pesticide Residues in Food, World Health Organization, Geneva, 1990.

World Health Organization (WHO), Monograph of the Joint Meeting on Pesticide Residues, Geneva, 1975.

INFORMATION INTENDED FOR LOCAL USE

Japan — Ministry of Agriculture, Forestry and Fisheries (MAFF)

Reproductive Toxicity

Japanese Ministry of Agriculture, Forestry and Fisheries: Procedure for performing toxicity tests on chemicals, 1982.

Japanese Ministry of Agriculture, Forestry and Fisheries. *Guidance on Toxicology Study Data for Application of Agricultural Chemical Registration.* Notification No. 4200, January 28, 1985.

Good Laboratory Practices (GLPS)

Japanese Ministry of Agriculture, Forestry and Fisheries. *Good Laboratory Practice Regulations.* 59 Nohsan No. 3850, 1984.

USA — Environmental Protection Agency (EPA)

Reproductive Toxicity

Christian, M.S. and Voytek, P.E., *In Vivo Reproductive and Mutagenicity Tests.* Environmental Protection Agency, Washington, D.C. National Technical Information Service, U.S. Department of Commerce, Springfield, VA 22161, 1982.

U.S. Environmental Protection Agency: Guidelines for the Health Assessment of Suspect Developmental Toxicants. Federal Register, Vol. 51, No. 185 34027-34040. Sept. 14, 1986.

U.S. Environmental Protection Agency Workshop on One vs. Two-Generation Reproductive Effects Studies, Sponsored by the USEPA Risk Assessment Forum, Washington, D.C., 1987.

U.S. Environmental Protection Agency Proposed Guidelines for Assessing Male Reproductive Risk. Fed. Reg. 53:24850-24869, 1988.

U.S. Environmental Protection Agency Proposed Guidelines for Assessing Female Reproductive Risk. Fed. Reg. 53:24834-24847, 1988.

U.S. Environmental Protection Agency Health Effects Division Peer Review of Procymidone, Office of Pesticide Programs, October 31, 1990.

U.S. Environmental Protection Agency (EPA-FIFRA), Pesticide Assessment Guideline. Subdivision F — Hazard Evaluation: Human and Domestic Animals, Addendum 10, Neurotoxicity, Health Effects Division, Office of Pesticide Programs, 540/09-91-123, PB 91-154617, March, 1991.

U.S. Environmental Protection Agency Guidelines for Developmental Toxicity Risk Assessment. Fed. Reg. 56(234):63798-63826, 1991.

U.S. Environmental Protection Agency Memorandum: Atrazine Two-Generation Study, HED Ad hoc Committee for Atrazine Reproductive Issues, for a Discussion of the Setting of a NOEL in a Reproduction Study with Varying Mean Litter Sizes, September 23, 1992.

U.S. Environmental Protection Agency Draft Guidelines for Reproductive Toxicity Risk Assessment, 1993.

USA — Environmental Protection Agency (EPA) — Office of Pesticides and Toxic Substances (OPTS) or Federal Insecticide, Fungicide, and Rodenticide Act (FIFRA)

Reproductive Toxicity

U.S. Environmental Protection Agency, OPP Guideline 83-4: Reproductive and Fertility Effects. Pesticide Assessment Guidelines, Subdivision F, Hazard Evaluation: Human and Domestic Animals. Office of Pesticides and Toxic Substances, Washington D.C., EPA-540/9-82-025, 1982.

U.S. Environmental Protection Agency, *Pesticide Assessment Guidelines.* Subdivision F. Hazard Evaluation: Humans and Domestic Animals, Nov. 1984.

U.S. Environmental Protection Agency, *Pesticide Assessment Guidelines.* Subdivision F — Hazard Evaluation: Human and Domestic Animals, November, 1984 (Revised Edition), 1984.

U.S. Environmental Protection Agency, Hazard Evaluation Division Standard Evaluation Procedure: Teratology Studies. Office of Pesticide Programs, Washington, DC. EPA-540/9-85-018, 1985.

U.S. Environmental Protection Agency, PR Notice 86-5, Standard Format For Data Submitted Under the Federal Insecticide, Fungicide and Rodenticide Act (FIFRA) and Certain Provisions of the Federal Food, Drug and Cosmetic Act (FFDCA), Issued July 29, 1986.

U.S. Environmental Protection Agency, Health Effects Division Standard Evaluation Procedure. Inhalation Toxicity Testing. Office of Pesticide Programs, Washington, DC, EPA-540/098-88-101, 1988.

U.S. Environmental Protection Agency, FIFRA Accelerated Reregistration Phase 3 Technical Guidance, Office of Pesticides and Toxic Substances, Washington, DC, EPA No. 540/09-90-078, 1988.

U.S. Environmental Protection Agency, (EPA-FIFRA), Pesticide Assessment Guideline. Subdivision F — Hazard Evaluation: Human and Domestic Animals, Addendum 10, Neurotoxicity, Health Effects Division, Office of Pesticide Programs, March, 1991.

U.S. Environmental Protection Agency, Pesticide Assessment Guidelines, Subdivision F, hazard Evaluation: Human and Domestic Animals, Addendum 10, Neurotoxicity, Series 81, 82 and 83, March, EPA 540/09-91-123, PB 91-154617, 1991.

U.S. Environmental Protection Agency, *Code of Federal Regulations*, Title 40 (40CFR) — Protection of Environment, Subchapter E — Pesticide Programs, Part 158, Data Requirements for Registration, 88, 1992.

U.S. Environmental Protection Agency, Health Effects Division Draft Standard Evaluation Procedure, Developmental Toxicity Studies, Office of Pesticide Programs, Washington, DC, 1993.

Good Laboratory Practices (GLPS)

U.S. Environmental Protection Agency, Federal Insecticide, Fungicide and Rodenticide Act (FIFRA); Good Laboratory Practice Standards; Final Rule. 40 CFR Part 160.

USA — Environmental Protection Agency (EPA) — Toxic Substances Control Act (TSCA)

Reproductive Toxicology

U.S. Environmental Protection Agency, Subpart E — Specific Organ/Tissue Toxicity, No. 798.4900: Developmental Toxicity Study. 40 CFR Part 798 — Toxic Substances Control Act Test Guidelines; Final Rules. *Fed. Regist.* 50:39433-39434, 1985.

U.S. Environmental Protection Agency, Toxic Substances Control Act Test Guidelines, Final Rules. CFR 40 Part 798, 1985.

U.S. Environmental Protection Agency, Subpart E — Specific Organ/Tissue Toxicity, No. 798.4700: Reproduction and Fertility Effects. 40 CFR Part 798 — Toxic Substances Control Act Test Guidelines; Final Rules. *Fed. Regist.* 50:39432-39433, 1985.

U.S. Environmental Protection Agency, Toxic Substances Control Act, Part 798 — Health Effects Testing Guidelines. Subpart E — Specific Organ/Tissue Toxicity, 798.4700, Reproduction and Fertility Effects. Fed. Regist. 50(188):3942-39433, 1985.

U.S. Environmental Protection Agency, Toxic Substances Control Act Test Guidelines; Final Rule. *Fed. Regist.*, September 27, 1985. Part II, Vol. 50, No. 188, 1985.

Good Laboratory Practices (GLPS)

U.S. Environmental Protection Agency. Toxic Substances Control Act (TSCA); Good Laboratory Practice Standards; Final Rule. 40 CFR Part 792.

USA — Environmental Protection Agency (EPA) — Office of Drinking Water (ODW)

U.S. Environmental Protection Agency Office of Drinking Water Health Advisories, Drinking Water Health Advisory: Pesticides, Lewis Publishers, Michigan, 1989.

USA — National Research Council (NRC)

National Research Council (NRC) Reproduction and Teratogenicity Tests. In *Principles and Procedures for Evaluating the Toxicity of Household Substances*, National Academy Press, Washington, D.C., 1977.

National Research Council (NRC), Nutrient Requirements of Laboratory Animals, Vol. 10, National Academy Press, Washington, D.C., 1978.

National Research Council (NRC), Biologic Markers in Reproductive Toxicology: Subcommittee on Reproductive and Neurodevelopmental Toxicology Committee on Biological Markers, National Academy Press, Washington, D.C., 1989.

National Research Council (NRC), Pharmacokinetics: Defining Dosimetry for Risk Assessment, National Academy of Sciences Workshop, Washington, D.C., March 4-5, 1992.

National Research Council (NRC), Risk Assessment in the Federal Government: Managing the Process. National Academy Press, Washington, DC, p 159. ISBN 0-309-03979-7, 1983.

USA — Office of Technical Assessment (OTA)

U.S. Congress, Office of Technical Assessment (OTA), Reproductive Health Hazards in the Workplace, Chapter 4, Evidence of Workplace Hazards to Reproductive Function, J.B. Lippincott Company, Philadelphia, PA, 1988.

USA — President's Science Advisory Committee (PSAC)

President's Science Advisory Committee, *Use of Pesticides*, Government Printing Office, Washington, D.C., 1963.

CHAPTER 20

Human Studies — Epidemiologic Techniques in Developmental and Reproductive Toxicology

Bengt Källén

CONTENTS

I. Introduction	801
II. Adverse Reproductive Outcomes to Study	802
A. Infertility and Subfertility	802
1. Definition of Infertility and Subfertility	802
2. Time to Pregnancy	802
3. Clinical Infertility and Subfertility	803
4. Study Design	803
B. Spontaneous Abortion (Miscarriage)	804
1. Definition	804
2. Rates and Confounding	805
3. Study Design	806
C. Infant and Child Death	806
1. Definition	806
2. Rates and Confounding	807
3. Classification	807
4. Statistical Analysis	808
D. Birth Weight, Pregnancy Length, and Growth Retardation	808
1. Definition	808
2. Rates and Confounding	810
3. Analysis of Prenatal Growth	810
E. Congenital Malformations	810
1. Definition	810
2. Classification and Grouping	811
3. Mutagenesis and Teratogenesis	811
4. Source of Information on Congenital Malformations	812
5. Rates and Confounding	813
F. Mental Retardation and Behavioral Problems	813
1. Definition	813
2. Sources of Information	813
3. Confounding	814

- G. Childhood Cancer 815
  - 1. Definition and Classification 815
  - 2. Sources of Information 815
- III. Methods to Record Exposures 815
  - A. Embryonic and Maternal Exposure 815
  - B. Estimates of Maternal Exposure 815
    - 1. Concerns about Exposure Data 815
    - 2. Information on Drug Exposure 817
    - 3. Information on Occupational Exposure 818
    - 4. Information on General Environmental Exposure 818
    - 5. Misclassification of Exposure 819
  - C. Prospective and Retrospective Exposure Information 819
- IV. Methods to Record Outcomes 819
  - A. Questionnaire and Interview Information 819
  - B. Medical Record Studies 820
  - C. Register Studies 820
- V. Case Control Studies 820
  - A. Principles 820
  - B. Selection of Cases 821
  - C. Control Definition 821
  - D. Selection of Controls 822
  - E. Statistical Analysis of Case Control Materials 823
    - 1. Unmatched Studies 823
    - 2. Stratified Analysis — A Method to Control for Confounding 825
    - 3. Logistic Multiple Regression Analysis 826
    - 4. Analysis of Matched Pairs or Triplets 826
    - 5. The Multiple Testing Problem 828
- VI. Cohort Studies 828
  - A. Principles 828
  - B. The Formation of Cohorts 829
  - C. Comparison Material 829
  - D. Statistical Analysis of Cohort Data 830
- VII. Nested Case Control Studies 831
- VIII. Confounding 832
  - A. General Considerations 832
  - B. Maternal Age and Reproductive History 832
  - C. Maternal Diseases 832
  - D. Social Characteristics 833
  - E. Biological Characteristics 833
  - F. Incomplete Compensation of a Confounder 833
- IX. Surveillance 834
  - A. Birth Defects Surveillance in the Population 834
  - B. Surveillance after Specific Exposures 835
  - C. Outcome-Exposure Surveillance 835
  - D. Statistical Considerations 835
- X. Cluster Analysis 836
  - A. Definition of a Cluster 836
  - B. Clusters as Chance Phenomena or Artifacts 836
  - C. Exposure Analysis of Clusters 836
  - D. Clusters and Statistics 836
- XI. Concluding Words 837
- References 838

I. INTRODUCTION

Epidemiological studies of human reproduction are important sources of information on developmental toxicity for the human. Because of species variability in developmental processes and in responses to different harmful factors, the ultimate answer to questions on risks to humans can only be obtained by information on humans. Controlled experiments are usually out of the question for practical and ethical reasons. The most important source of information will therefore be epidemiology: studies of possible statistical associations between specific exposures and reproductive outcome. Such associations do not necessarily prove causality because they may be due to unidentified or uncontrolled confounders, to bias, or to chance.

In rare situations, when a very high risk exists, formal epidemiological studies may not be necessary in order to identify a risk. Examples are thalidomide and rubella embryopathy. Both of these important human teratogens were detected by alert clinicians. For such effects to be found, however, the risk increase associated with the exposure must be huge, which means that the effect must be highly unusual. In most instances, however, we are dealing with small or moderate risk increases (e.g., anticonvulsant drugs, maternal smoking). It is then necessary to make strict epidemiological analyses, but it is easy to draw wrong conclusions because most such studies have some imperfection and are burdened with uncertainty.

High risks are cause for concern for the individual — examples are thalidomide, isotretinoin, rubella, and maternal alcoholism. Low risks may be of little concern for any one exposed individual, but if many women are exposed, the number of reproductive failures caused by the factor may be large anyway. Maternal smoking is an example. It has been shown to cause reproductive anomalies, including an increased risk for perinatal deaths. But the risk increase is moderate and does not noticeably affect the probability that a specific pregnancy will end with a perinatal death (perhaps an increase from 0.6% to 0.9% in Sweden). Nevertheless, it has been estimated that some 10% of all perinatal deaths are related to maternal smoking.[1]

From the individual's point of view, absence of a demonstrable risk may be enough. Epidemiological studies can never prove a lack of reproductive toxicity; they can only show that the risk is below a certain level. A low risk, perhaps not detectable even with a number of well-performed epidemiological studies, may nevertheless represent a major cause of birth defects or other reproductive failures in the population. This also means that society's decision about possible risks from specific exposures must sometimes be based not only on epidemiological data (notably when these are negative), but also on other types of data, including animal toxicological tests.

The basic question which the epidemiological study tries to answer is this: Is there an association between a specific exposure (in the widest sense of the word) and an outcome; if so, can this association be explained by chance, by the action of confounding factors (for a discussion, see Section VIII), or by bias? Such associations can be estimated with different techniques, the two classical ones being the case control (or case referent) and cohort approaches. The first compares exposures in "cases" (the abnormal reproductive outcome under study) and "controls" (lacking that abnormal reproductive outcome). With this design, many exposures can be studied simultaneously in relation to the selected outcome. The cohort approach studies reproductive outcome in parents who were exposed in a specific way and the outcome is compared with that of unexposed cohorts or the total population; with this approach, many different abnormal outcomes can be studied after a specific exposure. In both methods, two variables must be defined: reproductive outcome and exposure.

In the analysis and interpretation of all epidemiological studies, two main problems exist: confounding and bias. Confounding means that a third factor covaries with both exposure and outcome, resulting in a statistical association between the two. Bias means that the information on exposure is influenced by the outcome or vice versa. These problems are discussed later in the text.

II. ADVERSE REPRODUCTIVE OUTCOMES TO STUDY

A disturbance of reproductive function can result in a multitude of different adverse outcomes or endpoints for study. It is not necessary that a specific exposure causes all types of adverse outcomes. Some result in a very broad spectrum of disturbances, e.g., maternal alcoholism, which causes embryonic, fetal, and perinatal death, growth retardation, major and minor congenital malformations, and mental disturbances, including mental retardation.[2] Other agents may have a rather specific effect on only one or a few types of congenital malformations, e.g., the possible effect of lithium in increasing the risk for a congenital heart defect.[3] The full evaluation of the reproductive toxicology of an agent therefore necessitates study of many different aspects of reproductive outcome. Even after such studies, it is always possible that the exposure results in a late, unexpected effect, as was the case with DES (diethylstilbestrol).[4]

A. Infertility and Subfertility

1. Definition of Infertility and Subfertility

The fertility of a couple will be a function of both male and female fertility. The fertility rate in a population will be affected by voluntary reduction of family size with the aid of contraceptive practice and induced abortions.

Female fertility depends on normal ovulation of a normal egg, its unimpaired passage through the fallopian tube to the uterus, thinning of the mucus surrounding the mouth of the cervix enabling the sperm to pass into the uterine cavity, and changes in the uterine lining enabling the fertilized egg to implant. These processes are under endocrine regulation from the pituitary gland and the ovaries.

Male fertility depends on the production of adequate numbers of healthy sperm in the testes and on the ability to achieve erection and ejaculation. Sperm production is under endocrine control from the pituitary gland, and testicular hormone production is necessary for normal male genital function.

Infertility occurs when a couple is unable (without medical intervention) to have a child; this is a rather frequent phenomenon. Ten percent or more of all couples trying to conceive will seek medical help. In about half of the couples, male infertility is the cause and in about half, it is female infertility.

Modern medical technology often makes it possible to help these couples have children. Treatment will depend on the cause of the infertility. Examples are surgical opening of occluded fallopian tubes, fertility drug stimulation of the ovary and/or ovulation, and various types of *in vitro* fertilization (IVF). Notably for male infertility, a special form of IVF called ICSI (intracytoplasmic sperm injection) can be used. The egg is then fertilized by one sperm cell, introduced into the egg *in vitro* with micromanipulation.

Subfertile couples have a reduced capability to achieve a pregnancy. They may have a pregnancy after a longer delay than is usually the case. It is a matter of clinical judgement when a subfertile couple is judged to be infertile—often 2 to 3 years of involuntary childlessness is needed before IVF procedures, for instance, are recommended.

Exogenous causes may harm either the male or the female reproductive organs. Classical examples are male occupational exposure to dibromochloropropane, which resulted in a (temporary) arrest of spermatogenesis,[5] and prenatal female exposure to progestational agents or diethylstilbestrol, which has been said to affect future fertility.[6]

2. Time to Pregnancy

A popular way to study an effect on fertility is to use the time taken to achieve pregnancy (TTP). This means the time (usually measured in months or menstrual cycles) it takes for a woman who is actively trying to conceive to become pregnant. Usually this means that she becomes aware of her pregnancy. We will later see that a pregnancy may cease as a very early miscarriage before the woman realizes she is pregnant. Such pregnancies will not be detected, but the delay in conception will be registered as an extended TTP.

Usually estimates of TTP are based on prospective or retrospective interview or questionnaire information. A large number of variables, some of which are difficult to control for, will affect TTP measurement, and great care has to be used in the interpretation of such data.[7-9]

3. Clinical Infertility and Subfertility

Studies on infertility are usually made on populations identified in fertility clinics. Information on clinical subfertility (e.g., having experienced involuntary childlessness for at least 1 year) is usually based on interview or questionnaire data, even though such information is available in sources such as the Swedish Medical Birth Registry.

Related to this problem is the use of fertility drugs and IVF. Couples treated in this way represent selected subfertility groups. That is because they did not achieve pregnancy without medical help, and access to these methods may differ according to socioeconomic characteristics. Identification of women who underwent IVF has been made in different ways, e.g., from existing registries of all IVF pregnancies.[10]

4. Study Design

Fertility problems can be of interest both as outcome measures and as risk factors. Does the existence of a fertility problem affect the outcome of a pregnancy that has occurred, with or without medical intervention? Does the outcome of pregnancies occurring after fertility drug or IVF treatments differ from the expected outcome?

The presence of information on length of involuntary childlessness in the Swedish Medical Birth Registry made it possible to describe epidemiological characteristics of affected women. The period of involuntary childlessness increases with maternal age and (of course) decreases with parity, and it is associated with maternal smoking, previous spontaneous abortions, extrauterine pregnancies, and stillbirths. Previous induced abortions appears as a "protective factor" because subfertile couples are less likely to have an unwanted pregnancy.[11]

When subfertility was looked upon as a risk factor and delivery outcome was studied, an increased rate of preterm births, low birth weight infants, and intrauterine growth retardation was noticed among singleton infants. Perinatal mortality was increased, but this was mainly explainable by the maternal age and parity distribution.[11] An association between the risk for spontaneous abortion and previous subfertility has also been described.[12]

Studies of infants born after IVF show clear-cut deviations from the expected outcome. The most obvious effect is an increase in multiple births (because of more than one embryo being transferred), but effects were also seen on preterm birth, low birth weight, intrauterine growth retardation, and presence of congenital malformations. These effects are explained mainly and perhaps exclusively by the subfertility factor itself and not by the IVF procedure.[10]

A theoretically interesting finding is an increased risk for infant hypospadias after ICSI but not after standard IVF.[13] The link is probably male subfertility due to a genetic defect resulting in both testicular anomalies and an increased risk for infant hypospadias.

One study from the Swedish Medical Birth Registry used data on the rate of women who reported involuntary childlessness to demonstrate that an increase in fertility had occurred during one decade (1983 to 1993).[14] Great care must be used in the interpretation of this type of information, however, as recording of involuntary childlessness is incomplete and may well vary with time.

When the possible effects of various exposures have been studied, sperm counts or morphology[15,16] or TTP[17] have sometimes been assessed. Other investigators have studied exposures

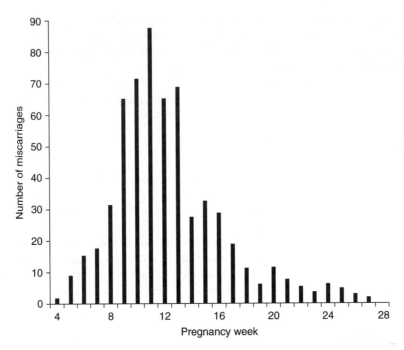

Figure 20.1 The number of miscarriages (from a total of 610) during each gestational week. Data from one hospital. (Sandahl, B., *Europ. J. Obst. Gynecol. Reprod. Biol.*, 31, 23, 1989.)

of men attending an infertility clinic and showing poor sperm production. Men attending the clinic with normal sperm counts are used as controls, thereby avoiding possible bias from the fact that an infertility clinic had been contacted.[18]

B. Spontaneous Abortion (Miscarriage)

1. Definition

A spontaneous abortion is the death (and expulsion) of an embryo or fetus before it has reached the legal status of a child, which is usually defined by gestational age and/or birth weight. The limit differs among countries. In some the upper age limit for a fetus is 28 completed gestational weeks (calculated from the last menstrual period, LMP), but in many populations, lower limits are used, e.g., 20 or 22 weeks of gestation. To some extent, the upper limit will affect estimates of spontaneous abortion rates, but rather few fetuses die between weeks 20 through 28. A dead fetus expelled after the upper time limit of spontaneous abortion is referred to as a stillborn infant. If born alive before that limit, it will be considered an infant.

Figure 20.1 shows the gestational week distribution of spontaneous abortions treated in one hospital.[19] Early miscarriages that did not make the woman seek medical advice and miscarriages that occurred before the woman realized she was pregnant were not included. Some studies[20] have indicated that in a high percentage of pregnancies, the embryo dies before or just after implantation, and the woman will never realize that she was pregnant. Such very early pregnancy losses are extremely difficult to identify and are therefore not very suitable for epidemiological studies. A few such studies have been made[21] however. Various exposures were studied in pregnancies diagnosed with sensitive pregnancy tests, and the possible association with very early pregnancy loss was determined.

As an indirect method, TTP can be used (see above). If a pregnancy loss occurs briefly after conception, a new pregnancy may occur in the next menstrual cycle, but it will result in a lengthening

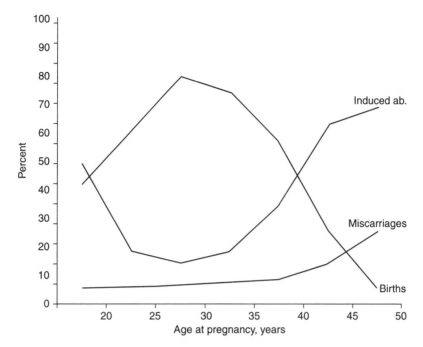

Figure 20.2 Percent distribution of miscarriages, induced abortions, and deliveries among Swedish women according to age at pregnancy. (Reprinted with permission from Källén, B., *Epidemiology of Human Reproduction*, CRC Press, Boca Raton, FL, 1988.)

of the time from when the couple begins to try to achieve pregnancy until a pregnancy becomes clinically apparent. A delay caused by a very early pregnancy loss cannot, however, be distinguished from a delay due to reduced fecundity. The measurement of time to pregnancy is therefore a summary of at least two effects.

Because of the considerations described above, most studies assess clinically overt miscarriages. Information on the occurrence of a miscarriage can be obtained by interview or questionnaire, but experience has shown that these data may be uncertain and sometimes biased.[22] Such data can also be obtained from medical records or registries based on such records. The study will then probably be restricted to true miscarriages, but early events will be underrepresented. Furthermore, confounding factors (e.g., education and social class) may affect the probability that a woman with an early miscarriage will seek medical help and may have an even greater effect on the likelihood that she will be hospitalized (many registries on miscarriages are based on discharge diagnoses from hospitals).[23]

2. Rates and Confounding

The actual rate of recorded clinically identified miscarriages varies with the method of ascertainment. Often figures around 10% to 15% are mentioned, but these are very crude and are influenced by many factors, including maternal age (Figure 20.2). The relatively high rate makes miscarriages suitable for statistical analysis, but this is counterbalanced by the fact that spontaneous abortions occur for a variety of reasons. In individual cases, little is known about the etiology and pathogenesis. In early miscarriages, a high percentage of the conceptuses show gross chromosomal anomalies that made continued development impossible. Maternal infections may cause a miscarriage as may certain immunological relationships between the parents, uterine anomalies, etc. Fetuses with some structural malformations are aborted spontaneously at a high rate, but the aborted specimen is seldom investigated from this point of view. An increase in fetal anomalies caused by an exogenous agent will result in an increase in miscarriage rate, but it will often be moderate in size and difficult to discern from the "background noise" of miscarriages from other causes.

3. Study Design

There are thus some inherent difficulties in counting the number of miscarriages occurring in a group of women, and it is still more difficult to use the result for an estimate of miscarriage risk. To become meaningful, the number of miscarriages has to be related to the total number of pregnancies occurring. Very often, the denominator is restricted to the number of deliveries, and a large proportion of pregnancies are left out, i.e., those ending in a voluntarily induced abortion, legal or (in some populations) illegal. An induced abortion will prevent a delivery but will not prevent a spontaneous abortion if it occurs before the time of the induced abortion. The later that abortions are induced, the stronger will be their effect on miscarriage rate. This phenomenon can easily be demonstrated in pregnancies occurring in girls below the age of 15 in Sweden. Very few such pregnancies end as deliveries. If they do not abort spontaneously, an abortion will be induced. If the spontaneous abortion rate is calculated in girls of this age class, without consideration of induced abortions, the risk for a miscarriage appears very high, while it is really quite low. This is an extreme situation, but less dramatic differences can be seen between different socioeconomic groups of women, for example. Information on induced abortions is often missing or incomplete, making rates of miscarriage uncertain.

In a case control approach, the same problem also arises. The control of a woman who miscarries should not be a woman who delivered a normal infant. It should be a woman with an intrauterine pregnancy (at the stage of pregnancy when the case woman miscarried) who did not miscarry any time during the pregnancy. Her pregnancy may end with the birth of a normal or a malformed infant or as a preterm birth, or it may end as an induced abortion. Representatives of all these reproductive outcomes should be eligible as controls but are seldom included. Therefore, any exposures that are somehow related to preterm birth or to induced abortion (e.g., maternal smoking) will appear as risk factors for spontaneous abortions if only normal infants are used as controls.[24,25] The perfect analysis of miscarriage rates takes this into consideration and is based on life-table principles,[19] but this necessitates access to data that are often unavailable. If the analysis comes short of this standard, due caution is needed in its interpretation.

All these factors make miscarriage less suitable as an endpoint for developmental toxicity studies than would be thought from its relatively high occurrence rate. Especially when small effects on miscarriage rates (even when statistically significant) are described, a sound scepticism is needed in their interpretation.

C. Infant and Child Death

1. Definition

For practical reasons, child death is usually divided into different groups:

Stillbirth — the birth of an infant (see definition above) who shows no signs of life.
Early neonatal death — the death of a live born infant within the first 7 days (or 168 h) of postnatal life.
Perinatal death — the sum of stillbirth and early neonatal death.
Neonatal death — the death of a live born infant before the age of 1 month (or 28 days).
Infant death — the death of a live born infant before the age of 1 year.

The limits between the groups are often but not always sharp. An infant giving a few gasps after birth before dying may be classified either as stillborn or live born. The lower limit of stillbirth (even when strictly defined by gestational duration) may also be somewhat arbitrary, and a classification into

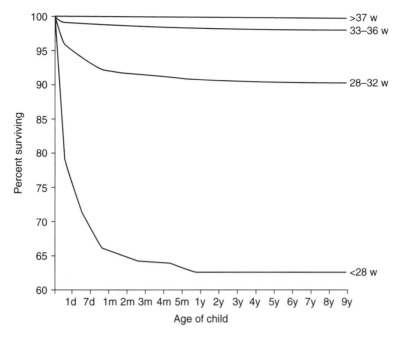

Figure 20.3 Survival rate of infants according to gestational duration. (w = week, d = day, m = month, y = year).

stillbirth or late fetal death may be uncertain. By and large, however, infant death is a definite and certain variable to study.

Events during pregnancy and around delivery can have a strong impact on perinatal deaths. They may also influence later death, even though it then often becomes difficult to differentiate effects of agents encountered during pregnancy from those acting after delivery (e.g., the effect of maternal smoking on sudden infant death syndrome[26]).

2. Rates and Confounding

The drawback to using infant death in epidemiological studies is the low numbers encountered, at least in developed countries. In Sweden, the perinatal death rate is below 7 per 1000 births, and in order to see effects on that rate (or a subfraction of it), either very large study populations are needed or very strong effects. It should also be remembered that medical care affects the perinatal death rate and may also shift early neonatal deaths into a later age period, thereby artificially reducing the perinatal death rate. When infant death is used in epidemiological studies of reproductive hazards, the two main confounders that must be taken into consideration are socioeconomic level and medical care quality. In many societies, social factors have a strong impact on perinatal death rate; in welfare states, this phenomenon may still exist but is more difficult to demonstrate.[27] Similarly, medical care quality can differ considerably. Even within a country with a generally high level of medical care, a differential that can confound comparisons of infant death rates between groups may exist between the quality of care of different delivery and neonatal units.

The infant death rate is strongly influenced by gestational duration (Figure 20.3) or birth weight and also by the presence of severe congenital malformations.

3. Classification

As was the case with miscarriages, perinatal deaths may have many different explanations. A large group is related to immaturity and preterm birth, and another large group is related to congenital malformations, but in many, no clear-cut cause of death is apparent (especially in the case of stillbirths). In contrast to miscarriages, perinatal (and later) deaths can be divided into subgroups according to the main cause of death. Precision in the definition of the adverse outcome can therefore be rather high. For instance, it is possible to see whether an increased perinatal death risk is due to a shift in gestational age–birth weight distribution, if it is due to an increased rate of lethal congenital malformations, or if it is due to other causes, e.g., to a high rate of abruptio placentae.

Various classification systems of perinatal or neonatal deaths are available. Some are built on cause of death information from death certificates, a source of information which is notoriously uncertain.[28] Various hierarchic systems have been constructed, and Wigglesworth published the most well known.[29] A classification that can be used in studies based on medical birth registries has also been published.[30] However, for one large group, intrauterine deaths before the onset of delivery, the cause of death is often not known.

4. Statistical Analysis

For the statistical analysis of infant and child survival, life table analysis is often suitable. Figure 20.3 shows the survival rate of term and preterm infants (dividing lines 29, 33, and 37 completed weeks of gestation). For each period, these rates are calculated based on the number of infants who could have died then. The denominator for the calculation of the death rate during the second week is therefore the mean number of infants surviving the first week and the number of infants also surviving the second week.

In a certain period i, the number of infants entering that period is n_i, and the number of infants living at the end of that period is n_{i+1}. The number of infants dead during the period is d_i, and the death rate during the period is

$$p = \frac{d_i}{0.5(n_i + n_{i+1})}$$

The standard error of that estimate is

$$s_p = \sqrt{\frac{p(1-p)}{0.5(n_i + n_{i+1})}}$$

assuming that a normal approximation can be made (d_i is at least 5).

D. Birth Weight, Pregnancy Length, and Growth Retardation

1. Definition

Birth weights can be analyzed in different ways. They can be described as continuous variables, with means, standard deviations, and standard errors of the means. They can also be described as discontinuous variables, either as the percentage of very low birth weight (VLBW) infants, that is, with a weight below 1500 g, or as the percentage of low birth weight (LBW) infants, below 2500 g. There are advantages and disadvantages with both methods. From a clinical point of view, the VLBW infants are the most important group because they represent the highest risk for perinatal and later death and sequelae. However, they make up a very small proportion of live births (some 0.5%), and marked shifts may occur in this fraction without any major change in the mean birth weight. On the other hand, most environmental factors affecting birth weight do so by increasing the proportion of infants in the range of 1500 to 2499 g.

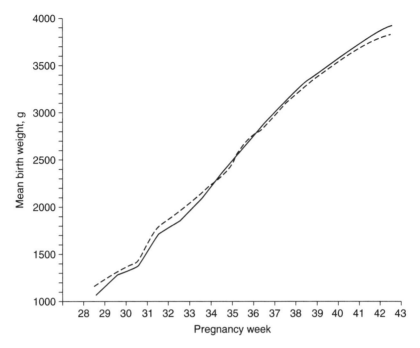

Figure 20.4 Mean birth weight at different gestational ages in two groups of infants. Infants of mothers with only compulsory school (- - - - -); and infants of mothers with higher education (———).

The most important determinant of birth weight is gestational duration, and this variable is often difficult to adequately ascertain. Information based on LMP is often inaccurate, and ultrasound determination of gestational duration is often not available. An infant is regarded as born at term if gestational duration exceeds 37 completed weeks, counted from the LMP, and preterm if born before that time. Very preterm infants are born before 32 completed weeks — this group and still more infants born before 28 completed weeks represent the group with the highest demand on neonatal intensive care.

The interaction between gestational duration and birth weight is twofold. A preterm birth of a normal baby will result in low birth weight. Alternatively, birth weight may be reduced at a given gestational duration. This reflects growth retardation during development, resulting in an infant who is small for gestational age (SGA). In order to identify infants that are small or large for gestational age, standard birth weight graphs that are determined in different ways are used. In principle, such graphs should be based on the same population as that studied, but when two or more groups of infants are compared, the details of the growth curve are of less importance.

It is likely that most environmental factors affecting birth weight will do so by retarding growth, but growth retardation in itself increases the risk for preterm birth. This phenomenon gives rise to great problems in the determination of normal growth weight distribution curves for different gestational weeks.[31] Much of the published data underestimate the "normal" weight in a given week and overestimate the "normal" dispersion. Typically in epidemiological studies, however, the important matter is not to classify an individual infant as growth retarded or not but to compare mean birth weights at different gestational weeks in two or more groups of infants.

Figure 20.4 shows an example: the mean birth weight at different gestational durations in two groups of Swedish infants selected by maternal educational level. This figure also illustrates a paradoxical finding that is quite common; even though at term infants of mothers with low education weighed less than the infants of women with high education do (as expected), the opposite is the case in preterm births. The explanation is that in the former group a higher proportion of preterm infants are normal infants born too early, while in the latter group the majority of preterm infants are preterm because they are growth retarded. This phenomenon gives rise to unexpected effects if standardization is made for gestational duration in analyses of birth weight.[32]

2. Rates and Confounding

Birth weight information is usually very reliable, although in large birth registries errors will occur that, even though rare, may be important. For example, some 9% of infants with a birth weight below 1500 g in the Swedish registry were misrepresentations of infants with normal birth weight.[33] This falsely reduced perinatal mortality in infants with a birth weight below 1500 g. The incidence of preterm and LBW infants varies between populations and races within populations. In Sweden, approximately 5% of singleton infants are born preterm, and 3.5% have a birth weight below 2.5 kg.

The high accuracy of birth weight information and the large numbers available (practically all infants have a birth weight recorded) offer good opportunities for statistical analysis. It is also well known that many environmental factors, e.g., maternal smoking,[34] affect birth weight. The problem is that an infant's birth weight is influenced by a large number of variables. Some of them, such as maternal age and parity (these two variables have to be crossed because of a strong interaction between them)[35] and infant sex are usually easily controlled. Others, including maternal weight and weight increase during pregnancy, parental height, social class, and smoking habits, are more difficult to control.

Similar problems exist in the analysis of gestational duration, but with the added uncertainty of the accuracy with which duration may be determined.

3. Analysis of Prenatal Growth

Crude comparisons of birth weight or gestational duration are easily made with standard statistical techniques, e.g., t-tests of means with their standard errors. When confounding factors are considered, a multiple regression analysis is often useful (although it is based on a supposition of near-normality in data distribution). Frequencies of preterm or LBW infants are also easily compared using suitable statistical tests, e.g., chi-square tests. When confounders are considered, a stratified Mantel-Haenszel or a logistic multiple regression method can be employed. Details of these techniques are discussed later.

Intrauterine growth retardation can also be analyzed either as continuous or dichotomous data. In the former situation, the deviation in birth weight for each infant is expressed as standard deviations (SD) from the value of the relevant gestational week in the normal growth curve. As a rule of thumb, SD can be approximated to 12% of the mean weight for a specific week. This measurement (SDS, standard deviation score) can be used in the analysis in a similar way as, for instance, birth weight. If a dichotomous analysis is preferred, the usual cutoff is more than ±2 SD from the mean (large and small for gestational age, respectively), but obviously other cutoffs (e.g., 2.5 or 3 SD) can be used.

Easy access to gestational duration and birth weight information, and the large amount of data available make statistical analysis simple. They also make it simple to demonstrate differences between infants born, for example, in different geographical areas.[35] The main problem lies in the interpretation of the findings because there are so many confounding factors, many of which are difficult or impossible to control.

E. Congenital Malformations

1. Definition

There is no strict definition of a congenital malformation. Very loosely, it can be defined as a congenital structural anomaly of significance for the individual. Its severity may, however, vary from monstrous, usually lethal conditions, through severe and handicapping conditions, to conditions that can be repaired to more or less complete normality. Even conditions that only have cosmetic significance or that are so insignificant that the individual can live a completely normal life are also sometimes included.

We can illustrate this with various results of a disturbance of the closure of the neural folds to form a neural tube, the embryonic process that leads to the formation of the central nervous system. In the most severe instances, an infant with *anencephaly* is born; remnants of a heavily malformed brain lie exposed, and the infant is stillborn or dies during the first few days after birth. A disturbance occurring at a slightly later time and hitting more caudal parts of the nervous system may lead to a *spina bifida aperta*, with an open spine and herniation of the spinal cord and meninges. When severe, this leads to pareses and hydrocephaly, with severe brain damage and perhaps lifelong handicap. In less severe forms, neurosurgery can correct the conditions, sometimes with little or no remaining handicap and normal mental development. In still less pronounced instances, a *spina bifida occulta* may develop, with an opening of the spine but with a normal or near normal spinal cord. In most instances, this gives no problems for the carrier, although it sometimes results in disturbances of bladder innervation.

2. Classification and Grouping

The classification of congenital malformations is usually based on the ICD (International Classification of Diseases) codes, possibly with extensions. It should be realized that the ICD code is constructed for clinical use and is often not specific enough for detailed studies of congenital malformations. Thus, specially designed coding systems that enable a greater specification than is permitted by the ICD code have sometimes been used.

Congenital malformations are a very heterogeneous group of conditions from the point of view of aetiology and pathogenesis. It is therefore rather useless to analyze this outcome as the total malformation rate, and some division into categories must be made. How such a division should be made can be debated. A common method is to use organ system subdivisions, but the group "limb malformations," for example, contains many very different subgroups. These could include limb reductions, polydactyly, syndactyly, or positional defects, and a change in the rate of one type (e.g., radial reduction defects) can easily be hidden by the large numbers of more common defects.

A problem is that the pathogenesis of specific defects is not always agreed upon and may even vary for the same malformation. A small bowel atresia, for instance, may be the result of a very early disturbance in gut development, the result of faulty secondary canalization of the gut rudiment, or the result of an ischemic insult perhaps occurring rather late in development. A transverse limb reduction defect may be caused by an early disturbance in limb development, may be the secondary consequence of a vascular accident, or may occur as the result of a so-called amniotic constriction band. It is therefore not possible to set up a strict categorization of congenital malformations that mirrors pathogenesis, but it must be done in a more intuitive way. Nevertheless, if the subdivision is carried too far, the number of infants with malformations in each subgroup becomes so low that analysis is meaningless.

3. Mutagenesis and Teratogenesis

An environmental factor may cause congenital malformations by a mutagenic process affecting germ cells, resulting either in DNA point mutations or in chromosomal anomalies (e.g., nondisjunction resulting in trisomy or breaks resulting in translocation). To cause congenital malformations, mutagenesis usually has to affect the germ cell genome. This means that an effect can be seen from paternal or maternal exposures and that most likely the event has occurred before conception. It could also occur during very early embryonic cell divisions. Mutagenic events during embryonic or fetal development are more likely to cause malignancy rather than malformation. In studies of mutagenic events, therefore, exposures before pregnancy should be studied, and exposures of both prospective parents are of interest. On the other hand, a mutagenic exposure is probably relatively nonspecific, and a limited number of easily identified conditions can be used as indicators. For DNA point mutations, so-called sentinel phenotypes[36] can be used, i.e., conditions caused by dominant autosomal mutations (e.g., achondroplasia). Such conditions are rare and either very large study populations or very strong effects are needed in order to detect a mutagenic effect of an environmental factor.

It has been demonstrated that obtaining very precise diagnoses and identifying actual dominant mutations is perhaps not so important. Equal power can be obtained by using large data sets with less diagnostic accuracy.[37] In a similar way, trisomy 21 (Down syndrome) is often used as the "sentinel phenotype" for chromosome anomalies (or at least trisomies). This condition is rather easily identified and occurs at a reasonable rate (some 1 in 700 births), but it must be remembered that its rate is highly sensitive to maternal age distribution. Intensive prenatal diagnosis (especially among older women) further affects the rate of infants born, and this is also a problem in the case of other conditions which are readily diagnosed prenatally.

The second way in which an environmental factor can cause a malformation is by teratogenesis. *Teratogenesis* is used here to mean a direct effect on embryonic development, acting during the actual development. In practice this occurs by exposure of the pregnant woman, and only exposures during pregnancy (or even during a short part of the pregnancy) are of interest.

Both from a practical and theoretical point of view, mutagenic and teratogenic processes differ, and in the design and interpretation of epidemiological studies of environmental factors causing malformations, this has to be taken into consideration. So far, most convincing evidence of a harmful effect of environment (in its widest sense) is suggestive of teratogenic processes.

This discussion is also relevant for investigations of the possible role of paternal exposures and congenital malformations. As stated above, mutagenic events can occur at sperm production, and the most sensitive period is perhaps 3 months before conception. There is indirect evidence that the majority of new dominant DNA mutations causing birth defects are actually paternal in origin, but few, if any, specific risk factors have been identified, except for paternal age. On the other hand, the vast majority of nondisjunctional processes leading to trisomy 21 occur at maternal meiosis (usually the first meiosis, which takes place before the time of conception). Teratogenic effects due to paternal exposure may occur, notably by secondary exposure of the embryo via the maternal organism; examples are passive smoking and "carry-home" exposures, e.g., of toxic substances from the working place.

4. Source of Information on Congenital Malformations

Information on congenital malformations in infants is best retrieved from medical records because information from parents is often incomplete and nonspecific. It should be realized that only a proportion of congenital malformations in infants are recorded at the time of delivery. Many are detected during the first week of life, and others may give symptoms later, sometimes even years or decades after birth. The follow-up time is therefore important for the recorded rate, especially for mild or internal malformations, but even severe cardiac defects may go undetected for some time after birth.

In many areas of the world, a special opportunity is offered by the existence of registries of congenital malformations. Such registries were usually set up for monitoring purposes but are also very useful as a basis for epidemiological investigations. Complete ascertainment is seldom achieved, and the possibility of a biased ascertainment must be kept in mind. The major strength lies in the possibility of retrieving data for large numbers of infants with rare specific defects. Most such registries are limited to recording malformations detected during the newborn period, but some have an extended follow-up time, e.g., to the age of 1 year.

A distinction is sometimes made between registries that are based on data collection by staff scrutiny of hospital records in a limited area ("active" programs) and registries that rely on the direct reporting of the congenital malformations, often from a large area such as a whole country ("passive" programs). I think the terms are misnomers because the activity of a registry is best evaluated by the intensity in the analysis of the incoming data, not the way in which data are obtained.

5. Rates and Confounding

The inclusion or exclusion of especially mild forms will give very different total rates of congenital malformations. In the literature you can find total rates ranging from less than 1% to nearly 10%. Relatively severe malformations that result in infant death, handicap, or major medical intervention occur at an approximate rate of 2%, but this rate is also sensitive to inclusion criteria.

Rates recorded in a specific study related to one problem can never be compared with a generally applicable malformation rate; the rate used for comparison must be defined in the same way as the rate in the study group. Specifically, consideration must be given to the hospital of birth because recording of congenital malformations varies among hospitals, especially for minor or variable conditions such as less severe congenital heart defects or positional foot defects.

Compared with studies on birth weight, for example, studies on congenital malformations have fewer problems with confounding. Maternal age effects exist for some conditions (e.g., increasing risk for Down syndrome with maternal age, high risk of gastroschisis at low maternal age), while parity effects usually are weak. Maternal smoking can slightly increase the risk for some specific malformations (e.g., orofacial clefts) but is by and large a weak confounder. Rather few drugs show a teratogenic effect of any significance. An effect of socioeconomic level exists in some countries for some malformations. In most studies, the strongest confounder is actually the recording hospital, which makes analyses of the geographical distribution of malformations difficult.

F. Mental Retardation and Behavioral Problems

1. Definition

For the individual, the family, and society, exposures capable of harmful effects on fetal brain development are probably more important than most other hazards. They may result in a reduction of the intelligence quotient (IQ), perhaps leading to mental retardation and a need of social support. At the least, the intellectual capacity of the child, and therefore its social success, may be reduced below the capability defined by its genetic makeup. Mild mental retardation is usually defined as an IQ of 50 to 69, moderate mental retardation as an IQ of 35 to 49, severe mental retardation as an IQ of 20 to 34, and profound mental retardation as an IQ under 20.

Other effects of brain disturbance may be less dramatic but nevertheless result in social consequences; minimum brain damage may result in lack of concentration and in behavioral disturbances followed by difficulties in school and in social adaptation. It has been suggested that an increased rate of non–right-handedness in a population can be an indicator of slight brain damage.[38]

2. Sources of Information

Effects on mental development are important but also difficult to study, most notably the less marked deviations from normality. Limited groups of exposed infants can be followed and studied with various psychological tests. For example, this has been done for maternal alcoholism[2] and lead exposure.[39] It is much more difficult to conduct such studies on large groups of infants in situations where weak or moderate effects are expected, e.g., after workplace exposures, environment pollution,

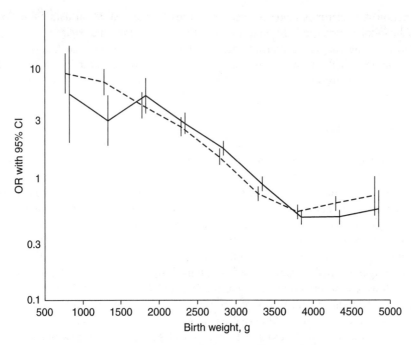

Figure 20.5 Odds ratio for mental retardation as a function of birth weight in singletons. Infants with mental retardation identified from special register of mentally handicapped infants (———); and infants with a diagnosis of mental retardation identified from a hospital discharge registry (- - - - -).

or food contamination. Nevertheless, toxic levels of organic mercury (e.g., the Minamata catastrophe[40]) have been shown to cause major brain damage.

In some countries or areas, registries of infants with mental retardation exist and can be used for large scale studies. Efforts have also been made to make use of hospital discharge registries, even though it must be realized that only a portion of children with mental retardation need hospitalization. Figure 20.5 compares the odds ratios for mental retardation according to birth weight, based on two different populations: a specific register of infants in social care because of mental retardation and infants hospitalized with a diagnosis of mental retardation. As this figure shows, the results are relatively similar.

In order to study mild forms of brain damage, central registries are of little help. Registries of medical characteristics of young men at the time of conscription have been used to demonstrate an association between late prenatal ultrasound exposure and an increased risk for non–right-handedness.[38]

3. Confounding

Perinatal complications, including short gestational duration, low birth weight, and intrauterine growth retardation, are major causes of brain damage. Obviously, such variables may be intermediary. Because IVF procedures increase the risk for preterm birth among singletons and for multiple births, they are likely to increase the risk of brain damage, including effects on mental development. However, detailed studies of small groups of children have a low power to demonstrate such effects.[41]

A more difficult confounder is parental mental characteristics because genetics play a large role in the mental development of the child. A confounding of socioeconomic conditions is therefore to be expected, most notably for mild effects. When registries are based on hospitalization, other parental characteristics, including maternal age, parity, and subfertility, may affect the probability of hospitalization.[42]

G. Childhood Cancer

1. Definition and Classification

Many instances of childhood cancer are thought to be initiated prenatally, perhaps as a result of mutations in somatic cells of the embryo or fetus. The peak incidence of childhood cancer (that is, a malignancy developing before the age of 15) occurs at 3 to 4 years of age, and it is probable that malignancies with an early onset are especially likely to result from a disturbance during intrauterine development. In spite of this, rather few instances of factors resulting in intrauterine carcinogenesis are known. The best known example is intrauterine exposure to diethylstilbestrol (DES) resulting in vaginal cancer, which, however, was not observed until after the age of 15.[4]

In a study of childhood cancer in Sweden, about 40% were leukemias (mainly acute lymphoblastic leukemia), 28% were tumors of the central nervous system, and 9% were kidney tumors.[43]

2. Sources of Information

Childhood cancer is a relatively rare occurrence, with a rate of about 2 per 1000 children. Individual follow-up of children is therefore tedious, and large groups are needed for such studies. Usually, case identification is through central cancer registries. This necessitates that each individual can be identified with certainty, e.g., by the use of personal identification numbers, as is the case in Nordic countries. The completeness of cancer registries is often good, although not perfect. One should also realize that when a group of children is followed from birth, detection of the cancer necessitates that the child survive and also that the child does not emigrate from the area covered by the cancer registry. It is thus easy to demonstrate that children born in Sweden of immigrant parents have a lower probability of appearing in the cancer registry than children born of Swedish parents. This is a result of a higher emigration rate among the former.

III. METHODS TO RECORD EXPOSURES

A. Embryonic and Maternal Exposure

In most instances, the embryonic (or fetal) exposure is the critical information we seek. It is seldom known for certain, even though such measurements as lead content in deciduous teeth[44] or cotinine in infant hair (to estimate exposure due to parental smoking)[45] have been used. Instead, maternal exposure is usually used as a proxy for embryonic exposure, but maternal exposure is seldom known exactly. In drug teratogenesis studies, for example, the use of a drug may be ascertained (although even this may be difficult), but actual maternal serum levels of the drug during those few days or weeks when a specific malformation is formed are seldom known or estimable. Nevertheless, such information can sometimes be obtained for chronic medication, where drug concentrations are routinely monitored (e.g., anticonvulsant drugs, lithium).

B. Estimates of Maternal Exposure

1. Concerns about Exposure Data

In studies on occupational exposures or exposures from environmental pollution, actual exposure data are usually rather unspecific. Very often exposure data mainly indicate with a reasonable probability that the woman has been exposed, but the actual level is not known. This will result in a bias toward null effects if a large proportion of women regarded as exposed were not actually exposed or were exposed only to a very low degree.

In practice, occupational exposure usually means that the woman has had a reasonable probability of having been exposed during pregnancy. Such exposure information can have different levels of precision: occupation information (e.g., nursing), actual working situations (e.g., working in anesthesiology), types of exposures (e.g., description of working place, including types of volatile anesthetics handled), or measurements of air content of the volatile anesthetics used combined with exact information on working hours. Moving along this list, better and better approximations of the actual embryonic exposure will be obtained; at the same time, most likely fewer and fewer pregnancies will have that level of exposure information.

The increased precision in exposure information is often strongly counterbalanced by a decreasing power of analysis due to the lower number of cases. This will be especially important in the study of rare events, such as perinatal deaths or congenital malformations. It is not very informative to say that among 10 women with an exposure of so many ppm of halothane 6 h a day during 5 working days during weeks 4 through 5 of pregnancy, none had a baby with spina bifida. This estimate can actually mean anything from 0% to 37% risk of such a pregnancy outcome. It may be more informative to get an idea of the risk (with its confidence interval) of having an infant with spina bifida for any pregnant woman working in anesthesiology.

On the other hand, exposure information may be so crude that it is meaningless. For example, a study of all women working in industrial production will probably give more information on the effect of social factors on reproductive outcome than on the effect of specific chemical or physical exposures.

Similar concerns are valid for studies of exposures from environmental pollution. Very often, the only available information is the place of residence of the woman (quite often only the place of residence at delivery, which does not necessarily mean the place of residence in early pregnancy, when most malformations are initiated), and estimates of actual pollution (e.g., in air) are very rough. Furthermore, exposure occurs only to some extent at the place of residence. Most women spend a considerable part of their time in other places, such as their workplace, and sometimes the place of residence is a better estimate of social situation than of actual exposure to pollution.

Before starting data collection, it is useful to make a power analysis to see whether the design of the study may permit the identification of a specific risk. The exact structure of such an analysis differs according to the design of the study, but we can exemplify it with a typical situation: one wants to compare congenital malformation rates in two groups of infants, one exposed and one not exposed, for a suspected harmful factor.

The power calculation is based on two concepts, α and β, where α is the probability that the finding is false (due to chance), and β is the probability that the study reveals an existing effect. Very often in power calculations α is set to 0.05 and β to 0.80 (but other values can be chosen).

The following parameters are of interest: the number of exposed infants (ne), the number of unexposed infants (nc), the expected rate of the outcome (e.g., congenital malformations) (p), and the risk increase caused by the exposure (r). The following equation shows the relationship but is based on normal approximations, which means that the expected number of outcomes in the group with the smallest number should exceed 5.

The probability of a malformation in the exposed group is then $P_1 = rp$. The probability for not having a malformation is then

$$Q_1 = 1 - P_1$$

The probability for a malformation in the unexposed group is $P_2 = p$, the probability of not being malformed is $Q_2 = 1 - p$. The probabilities in the two groups together are called P_m and Q_m, respectively. The fraction of the total that is exposed $ne/(ne + nc)$ is called f.

$C_{1-\beta}$ is then

$$\frac{(P_1 - P_2)\sqrt{f(1-f)(ne + nc)} - c_{1-\alpha/2}\sqrt{P_m Q_m}}{\sqrt{(1-f)P_1 Q_1 + f P_2 Q_2}}$$

If p is small (e.g., malformations, perinatal deaths), $1-p$ approximates 1, and the equation can be simplified.

The equation can be solved in different ways, e.g., how large is α (the probability of finding an effect) at other given parameters, what number of ne and nc are needed in order to detect the risk r, or what risk increase is likely to be detectable given the other parameters? Computer programs for power calculations are available in most statistical software packages.

Let us apply this to the specific study of nurses exposed to anesthetic gases. We want to be able to demonstrate a doubling of the congenital malformation risk (put as 2% in the unexposed population), and we want to compare two groups of infants born of nurses. The mothers of one group worked as anesthesiology nurses, and the others did not. How many cases in each group (if the two groups are equal in size) are needed to have an 80% probability of identifying a doubled risk in the exposed group with a 95% certainty? The equation above gives the answer: 1140. A study of only 50 women in each group, for example, would require a nearly 10-fold increase in malformation risk (18% occurrence of malformations) to have a reasonable chance of detection. Such an increase is not a very likely event, even if exposure criteria are very strict.

A doubling of the risk in a group of women who may have a 50% chance to have had a real exposure means that the actual risk in the exposed group is fourfold. It is thus easier to detect the risk in a large group of women where exposure is present in only half than to detect it in a small group of women where exposure has been individually verified.

2. Information on Drug Exposure

Since the occurrence of the thalidomide-induced epidemic of severe congenital malformations, interest has been paid to the possible hazards associated with maternal use of drugs during early pregnancy. With the continuous addition of new drugs to the therapeutic arsenal, this problem is always relevant. Even though new drugs are extensively tested in animals, the possibility of a human-specific teratogenic effect exists. This necessitates continued studies on the association between drug use and birth defects.

In most classical studies on drug teratogenesis, exposure information is obtained by retrospective interview or questionnaire studies, usually in a case control design. Sometimes the study has been specifically designed for the purpose of studying the association between a specific malformation and maternal use of drugs. In other studies, data are obtained from ongoing recording of drug use by mothers of malformed infants and control mothers without any restriction to malformation type or drug type.[46] Sometimes data have been extracted from data retrieved for other purposes, and the study of drugs has just been a spin-off of another study, e.g., the Vietnam Veteran Study.[47]

Other studies have tried to record all drugs used and all birth defects present in a specific population.[48] In these studies, the recording of drug use has usually been prospective, but no specific hypothesis is tested. Instead, all drugs have been crossed with all groups of malformations to identify combinations showing an excess. Such an approach is good for hypothesis generation, but the fact that multiple associations are tested makes it nearly impossible to distinguish causal associations in a given study from random occurrences. Most of these studies are research projects of a limited size.

Groups of women who have used a specific drug or a group of drugs can be identified in different ways, and their pregnancy outcomes can be studied. An example is women with epilepsy who are using anticonvulsants. During the last decades, such information has been published from Teratology Information Services (TIS). Women and doctors can approach such organizations to ask for advice when a drug has been used in early pregnancy, and by following up such exposed pregnancies, information on possible hazards can be obtained. The main drawbacks are that only relatively small numbers of exposures are usually identified, that pregnancy outcome is not always known, and that it is sometimes difficult to get control groups for comparison. A positive characteristic is that exposure data with respect to timing and drug dosage are often precise.

During the past few years, there have been further opportunities. Prospective recording of drug use in early pregnancy has become part of the Swedish Medical Birth Registry. This provides a growing database of prospectively recorded drug information with outcome data that can easily be compared with that of all births.[49] The main drawback is that there are few data on time of exposure and amounts of drug used. Other programs have used linkage between registries of drug prescription and birth registries.[50] Exposure information will be crude because it will not be known whether the woman actually used the drug during pregnancy, especially if it is not a chronic medication. In one system, prospectively recorded drug exposure data are combined with retrospective data, obtained by interview.[51]

In studies on drug teratogenesis, various confounders, such as maternal age and parity, subfertility, and smoking habits, should be taken into consideration. The most important confounder, the disease or complaint for which the drug was used, is difficult to control. One way is to compare different drug categories used for the same condition.

3. Information on Occupational Exposure

In many studies on occupational hazards to reproduction, maternal occupation has been used as a proxy for exposure.[52] This will usually give rather crude exposure information, and within the studied group a subgroup with markedly different exposure conditions that may represent a risk might be hidden. The more detailed the occupational characterization, the better the study, but usually it is not possible to get accurate actual exposure levels. In some studies (usually designed as nested case control studies), the actual exposure for each involved woman has been evaluated blindly by an occupational hygienist. In such a situation, the relevant period when the malformation in question could have been formed should also be considered. Direct measurements will further increase exposure precision, but as stressed above, the necessary small number of cases will often reduce the study power too greatly.

In other studies, exposure information is obtained by retrospective interviews or questionnaires. Except for the risks involved in all retrospective studies (see Section III.C), lack of knowledge among the women of their real exposure is a problem.

4. Information on General Environmental Exposure

In studies of the impact of the environment on pregnancy outcome, one is usually dealing with rather generalized exposures of large populations (e.g., air pollution, drinking water contamination, radioactive fallout) where very strong effects are not expected. In studies of congenital malformations, the environment for the woman during her early pregnancy is of interest. Often only the mother's address at the time of delivery is available, and a change of address is not uncommon during pregnancy, especially the first pregnancy. In most studies, a rather crude geographical localization is made, e.g., parish or postal code. In more sophisticated studies, the actual coordinates of the address can be used and perhaps linked to air pollution measurements or information on drinking water composition.

When a source of environmental pollution is studied, the distance between the address of the woman and the source of pollution is often used as a proxy for exposure. In some studies of airborne pollution, the dominating wind direction has been taken into consideration.

5. Misclassification of Exposure

Except for small studies with very exact exposure information, some misclassification is expected to occur. There may be two types of misclassification. One is due to biased exposure information, e.g., obtained in retrospective case control studies. This can result in wrong and usually exaggerated risk estimates. The second type of misclassification is obtained as a more or less random phenomenon among prospectively recorded exposure information. For example, in a study of maternal smoking, some smoking women may deny that they smoke. As such a behavior probably has nothing to do with the outcome, the misclassification will be nondirectional, i.e., it will be equal among mothers of infants with birth defects and mothers of normal infants. Such a nondirectional misclassification will in principle result in a reduction of the risk estimate, i.e., a bias toward unity. It will therefore be of greatest importance in studies where the outcome of the study is that the exposure in question has no harmful effect — this could be the effect of an underestimate of an actual risk. A woman who reported the use of a drug but used it before or after the critical period when a malformation can arise or a woman working in a specific occupational setting who had a holiday during the critical period would be additional examples of this type of misclassification.

If the actual risk increase after a specific exposure is a doubling of the population rate of congenital malformations and the outcome is studied in a group of women where only 75% were actually exposed, the estimated risk increase will amount to 1.8, and if 50% were exposed, the estimate would be only 1.5.

C. Prospective and Retrospective Exposure Information

In general, exposure information is obtained by questionnaires or interviews. This can be done *prospectively*, that is, before the outcome of the pregnancy is known, or *retrospectively*, after the outcome is known. The latter is the most common method. The drawback is that the woman who has had an adverse reproductive outcome is sometimes more careful in the reporting of exposures. She may even imagine that exposures have occurred or that they occurred during the relevant part of the pregnancy instead of earlier or later. This has been called "memory bias" or "recall bias." Its existence has been denied[53,54] or has been thought to be insignificant, but there is ample evidence that a recall bias may exist and may give misleading results.[55] When information is retrieved by interview, the interviewer's knowledge of the pregnancy outcome may also affect the thoroughness of the interview and the recording of the answers, thus introducing an "interviewer bias."

Both of the foregoing types of bias have a tendency to increase the estimate of a risk, while a reduction of the risk estimate will be obtained if the woman understates the exposure. This is often the case when information on smoking, alcohol use, etc., is sought. As most women know that these factors are harmful for the pregnancy, they have a tendency to underreport them. This may also, however, be influenced by the outcome of the pregnancy.

In some circumstances, exposure information can be obtained from independent sources to reduce information bias. At least crude exposure information may be obtained from central registries (e.g., concerning maternal and paternal occupation)[56] and more refined information may be received from independent sources, such as industrial hygienists.[57] Measurements of environmental pollution may give an unbiased estimate of exposure.[58]

IV. METHODS TO RECORD OUTCOMES

A. Questionnaire and Interview Information

Pregnancy outcome can be identified in different ways. One is by questionnaire or interview studies of women who have had a specific exposure. Some care should be applied because experience has shown that some types of data are well recalled and reported (e.g., birth weight), while others are not. The self-reporting of miscarriages, for instance, has been shown to be relatively inexact and may also be biased by the woman's apprehension regarding various exposures she has had.[22] Congenital malformations are often underreported, most notably lethal and minor ones. Self-reported data may be useful in studies of long-term follow-up of children, e.g., for mental disturbances.

B. Medical Record Studies

Another possible source is medical record information. Often information retrieval is restricted to delivery hospital files, which may give a serious underestimate of congenital malformations and will usually say nothing about survival beyond the first few days of life. A complete search of other relevant clinical records (pediatric, pediatric surgery, plastic surgery, etc.) will increase ascertainment but will obviously be very tedious. The data obtained are reasonably unbiased.

C. Register Studies

In many parts of the world, central registries are available. These include medical birth registries (which can sometimes be linked to death registries), special registries of congenital malformations, and even central registries of spontaneous and induced abortions in a few countries.[23,59] These sources give very easy access to large amounts of data, and are of great value for the selection of large number of infants with a specific congenital malformation. They are also valuable in the selection of controls in case control studies and as the basis for cohort studies. When it is possible to link the medical birth registries with other outcome registries (e.g., hospital discharge registries), further opportunities exist for long term follow-up of specific groups of children.

On the other hand, it must be realized that errors usually exist in large computerized registries, and a thorough knowledge of the registry and its shortcomings is necessary for its proper usage. When used in a correct way, the registries offer good opportunities for conducting epidemiological studies; when misused, the opportunity to obtain false results is also great.

V. CASE CONTROL STUDIES

A. Principles

In case control or case referent studies, exposure rates are compared among "cases" (in this context, women with an adverse reproductive outcome) and "controls" or "referents" (women without that outcome). This study design has both advantages and disadvantages. Among the advantages is that the number of cases to study can be relatively high, and therefore the exposure rate in cases can be determined reasonably well if exposure is not too rare. The exposure rate is compared with that among individuals where the reproductive outcome did not contain the anomaly under study. If information is available on the total population (which may be the case when population registries are available), the exposure rate among controls can be accurately estimated. Often such information is not available, but control exposure rates are estimated from a set of control individuals. The higher the number of controls per case, the more exact the estimate of the control exposure rate, but the uncertainty in the case exposure rate of course remains. Therefore, only a small increase in power is obtained by increasing the number of controls per case above two or three.

The case control approach makes it possible to use the same set of cases for a study of many different exposures. A typical example is a case control study of maternal occupation and a certain malformation. Each occupation can be studied separately, but it should be remembered that if many such tests are made, false positive associations can be obtained for one or even a few occupations just by chance.[34]

Table 20.1 Maternal smoking and odds ratios for different types of facial clefts[60]

Facial Cleft Category	Odds Ratio	95% Confidence Interval
Cleft lip and /or palate, isolated	1.11	0.99–1.29
Associated	1.41	0.98–2.09
Median cleft palate, isolated	1.35	1.12–1.63
Associated	0.88	0.51–1.54
All facial clefts, isolated	1.18	1.06–1.31
Associated	1.19	0.88–1.62

B. Selection of Cases

It may seem easy to define a case in studies of reproductive epidemiology, but it sometimes offers difficulties. If we are interested in the effect of environmental factors, obviously it would be good to exclude all infants where the malformation has a nonenvironmental cause, e.g., a monogenic condition or a chromosome anomaly. Even though environment may play a role in the etiology of a specific malformation (e.g., duodenal atresia) in a Down infant, the major cause of the malformation in that infant is the nondisjunction process leading to trisomy 21. In many conditions, e.g., cleft lip and palate, a genetic component occurs. It can then be debated whether infants with a known family history should be excluded. The ideal is to treat them as a separate group, but unfortunately such splitting results in a reduced number of cases available for analysis. It can also be argued that the genetic background may act by increasing the susceptibility to environmental agents, and an exclusion of infants with a known family history may reduce the proportion of "susceptible" cases.

Case selection can be more or less strict. Should all neural tube defects be lumped together, or should anencephaly and spina bifida be kept apart, and should the latter perhaps be divided into high and low forms? It is possible that only a subset of the group is associated with the environmental exposure. If all are analyzed together, the risk is that this subgroup will be drowned among the remaining ones that are not associated with the exposure. Again, the ideal is to subdivide the material and look for heterogeneity in it, but the risk is that numbers and statistical power are diminished. A balance must be kept between large enough numbers to permit statistical analysis and groupings specific enough to prevent dilution of possible associations, thereby reducing power.

One important distinction that is often made is between infants with isolated congenital malformations and infants with two or more associated malformations. There is some good evidence that these may be different etiological and pathogenetic entities that should be analyzed separately.

In some circumstances, separating cases into subgroups that make biological sense does not seem to affect the association with an exposure. Table 20.1 exemplifies this from a study on the possible association between maternal smoking and facial clefts.[60] The material was divided into cleft lip with or without cleft palate and isolated cleft palate; these are regarded as very different entities. Cases involving known chromosomal anomalies were deleted, and those remaining were categorized as "isolated" and "associated." The association with maternal smoking was seen in three of the four subgroups, and the estimated odds ratios did not differ significantly. Statistical significance was not reached in all groups separately but was attained when the groups were totalled.

C. Control Definition

Control definition can also offer problems. It is important to realize the definition of a control. It is not necessarily a delivery of a normal infant; it is a pregnancy that could have ended as a case but did not. A spontaneous abortion can never be a control to an infant with a congenital malformation, but an infant with a congenital malformation can be a control to a spontaneous abortion.

The control pregnancy for a case of spontaneous abortion is an intrauterine pregnancy that did not miscarry, but it can end as an induced abortion, a preterm infant, a malformed infant, a stillbirth, etc. If normal surviving infants are used as controls, all factors associated with other types of reproductive outcome will appear as risk factors for spontaneous abortions, and studies based on such comparisons are invalid.

Similarly, the appropriate control for a case with one malformation (e.g., isolated spina bifida) is a nonmalformed infant, while the control for a case with two or more malformations can be either a normal infant or an infant with one malformation. The suitable control for an infant developing mental retardation or childhood cancer is an infant alive at the time of diagnosis but without these characteristics. If dead, the control could not have received the diagnosis under study! If, for example, the birth weight distribution of infants developing mental retardation is compared with that of all infants born, very low birth weight will appear as a protective factor for mental retardation. In a way it is, because the risk of death is high and a perinatally dead infant cannot develop mental retardation. But if adequate controls are selected (surviving infants), one finds that very low birth weight is a risk factor for mental retardation.

D. Selection of Controls

In principle there are two ways to select a control: randomly or by matching. The random control is drawn by a randomization process, e.g., from a registry of all births or a subset of that registry. If no registries are available, some other type of randomization must be applied.

For control matching, one looks for controls that share some characteristics with the case, e.g., maternal age, parity, or geographical location. Matching can be so-called frequency matching, which means that in the set of controls, the variable in question (e.g., maternal age) occurs with the same distribution as among the cases. If, for example, 10% of the cases have a maternal age between 15 and 19, then 10% of the controls are selected from that maternal age stratum. This ensures that the cases and controls resemble each other as groups. Individual matching selects controls that have the same values as the cases for the matching variables. For example, for a specific case with a certain maternal age and parity and born at a certain hospital, one or more controls are selected with the same or similar (e.g., the same 5-year age group) maternal age and the same parity, born at the same hospital.

The purpose of matching is to reduce the influence of confounding variables (see further on). In other circumstances, matching is more a question of easy control selection. A typical example is the "next baby born" control, perhaps with restrictions concerning life or sex. This will give a very heavy matching for date of birth (which is usually not very important) and for hospital of birth, but no further matching. This type of control is standard in some registries for congenital malformations, but it should be realized that the matching is limited in extent.

Matching makes impossible an analysis of the variables for which matching is made. Many investigators therefore prefer to use random controls and control for confounding in other ways (see further on). When some variables are very skewed (e.g., maternal age when the infant has gastroschisis or Down syndrome), a lack of matching may reduce the information on controls available for extreme groups (very young women in the case of gastroschisis and older women in the case of Down syndrome), thus reducing power in the analysis. One should remember that the purpose of the controls is to estimate exposure rate in the population, and if we are interested in exposure rate in very young women (who had infants with gastroschisis), we must use the exposure rate in control women of that age stratum for comparison. If few such women are included in the sample of controls, the estimate of the exposure rate will be uncertain, regardless of the statistical method used.

Quite often, one chooses an infant with a congenital malformation (sometimes called a "sick control") as a control for an infant with another congenital malformation. The beauty of this approach is that the estimate of exposure rates may be less biased than when "healthy" controls

Table 20.2 Information on smoking in early pregnancy in women who had an infant with Down syndrome and in controls

	Smoker	Nonsmoker	Total
Down infant	183	569	752
Control infant	444	1060	1504
Total	627	1629	2256

Generalized Form of a 2×2 Table

	Exposed	Nonexposed	Total
Cases	a	b	n_1
Controls	c	d	n_2
Total	n_3	n_4	n

Source: Data from the Swedish Medical Birth Registry, 1983–1990.

are used, but the problem is that the risk factor under study may affect both malformations. In most instances, this is not a major problem because teratogenic effects usually have a rather high degree of specificity, but it may pose a problem.

E. Statistical Analysis of Case Control Materials

1. Unmatched Studies

In this situation, two groups of individuals are compared: cases and controls. Table 20.2 shows an example, the smoking rate among women who had infants with Down syndrome and those who had infants without a chromosome anomaly. The latter group was randomly selected among all delivered women.

Smokers and nonsmokers in the two groups form a 2×2 table with marginal sums (see bottom half of Table 20.2). For each of the four cells, the expected numbers can be calculated from the marginal sums. For example, the expected number for cell a is $n_1 n_3/n$. The distribution is really hypergeometric, but if numbers are large enough, it can be approximated as normal. The smallest expected number in each cell should exceed 5. If a normal approximation is made, the χ^2-distribution can be applied as a test.

The χ^2 value calculated from the 2×2 table will be $\chi^2 = \Sigma(o - e)^2/e$, where o is the observed and e the expected value in each cell. There is an easier and quicker way to reach the same result:

$$\chi^2 = (a*d - b*c)^2 * n / (n_1 * n_2 * n_3 * n_4)$$

If this formula is applied to the data from Table 20.2, the resulting $\chi^2 = 6.72$, and from a statistical table, it is found that the probability of this occurring by chance is only 0.0095. Mothers of infants with Down syndrome were less likely to be smokers (24%) than mothers of chromosomally normal infants (30%), and this difference is unlikely to have been due to chance.

As stated above, this method supposes that numbers (as they were in the example) are reasonably large. The smallest expected number in the table (smokers among mothers with Down syndrome infants) was 627*752/2256=209. Even so, the *p* value is not quite correct, but it is a reasonably good estimate (the exact *p* value is 0.010 instead of 0.0095, which matters little). Let us suppose, however, that Table 20.2 had been based on only 10% of the sample. There would perhaps have been 18 smokers among 75 mothers of infants with Down syndrome and 44 among 150 controls. The proportion of smokers in the two groups would be the same, but $\chi^2=0.71$, which gives $p = 0.40$. Even though the mothers of infants with Down syndrome smoke less than control mothers,

Table 20.3 Smoking in early pregnancy among women who had infants with Down syndrome and age-matched controls, divided according to maternal age class

Maternal Age Class	Down Infants		Controls	
	Smoking	Nonsmoking	Smoking	Nonsmoking
15–19	5	8	11	15
20–24	43	67	72	148
25–29	44	144	94	282
30–34	45	156	103	299
35–39	35	137	72	272
40–44	11	52	31	95
45–49	0	5	1	9

Source: Data from the Swedish Medical Birth Registry, 1983–1990.

the restricted study could not exclude the possibility that this was due to chance. Note that a statistically nonsignificant result does not mean that the effect is lacking, only that it can be due to chance, perhaps because the sample was too small.

When we are dealing with very small numbers, the normal approximation, which is a prerequisite for the χ^2 test, is not permissible. In a study, 4 cases among 10 were exposed to a specific drug, with only two cases among 30 controls (3 controls per case); $\chi^2 = 5.68$, which is statistically significant ($p = 0.02$). The lowest expected number is 10*6/40=1.5, which is below 5, and the normal approximation is not permissible. It is, however, possible to calculate the exact probability that this distribution is random with the Fisher test.

Using the symbols of Table 20.2, the exact p value for this distribution is:

$$p = n_1!*n_2!*n_3!*n_4!/(n!*a!*b!*c!*d!)$$

where $n!$ means n factorial, that is, $1*2*3*...*(n1)*n$

Such a p value must be calculated for this distribution and for all distributions that are still more deviant (in our case 5-5 vs. 1-29 and 6-4 vs. 0-30), and the sum of these p values tells the exact probability that the observed or still more deviant distributions were obtained by chance. The p value obtained is one-tailed, and in order to compare this value of p with that obtained in a χ^2 test (which is two-tailed), it must be doubled. One-tailed tests are permissible only when it is possible to say, *a priori*, that if a deviation is obtained, it must be in one direction only (e.g., an increased smoking rate). In the example above, no such prior statement can be made; smoking could conceivably increase as well as decrease the risk to have an infant with Down syndrome.

Exact tests are available in most statistical computer packages. When applied to the observations just described, a p value of 0.03 is obtained, which is also statistically significant.

The p value tells how likely it is that the finding was obtained by chance but tells nothing about the strength of the association between the exposure the outcome studied. This association can be expressed as a ratio between the two groups. In case control studies, the odds for being exposed are usually compared between two groups, cases and controls. The odds are calculated as the ratio between exposed and unexposed individuals. Using the designations in Table 20.2, the odds for cases to be exposed are a/b and those for controls are c/d. These two odds are compared as an odds ratio (OR), which will be OR = a/b:c/d. If the OR = 1, cases and controls had the same exposure rate, an OR > 1 means that cases were more often exposed than controls, and the opposite is true for OR < 1. The OR in Table 20.2 is 0.77, which means that smoking occurs less often among case women than among control women.

The odds ratio has a confidence interval. Its 95% confidence interval (CI) means that with 95% probability the true OR lies within that interval. If the lower limit of the interval is above 1, it means that the OR is significantly increased, while if the upper limit is below 1, the OR is significantly decreased. The value of the CI is such that one can evaluate not only the significance but also the magnitude of an effect. The 95% confidence interval of the odds ratio can be calculated by using Miettinen's method (test based estimate) according to the following formula, although other methods are available:

$$95\% \text{ CI of OR} = \text{OR}^{\left(1 \pm 1.96/\sqrt{\chi^2}\right)}$$

Let us take the example in Table 20.2 — the OR is 0.77 and the 95% CI is 0.63 to 0.94, while in the smaller material the OR is nearly the same (0.76), but the 95% CI is much larger (0.40 to 1.44). From the former estimate, it can be said that most likely the OR is reduced by at least 6% and possibly by as much as 37%, while from the latter estimate, the OR could be increased by up to 44%.

It should be observed that the odds ratio is not a direct estimate of the risk or the risk decrease for a smoking woman to have an infant with Down syndrome, but when very rare occurrences (such as Down syndrome) are studied, the estimate will be quite adequate. If we suppose that the population risk of having an infant with Down syndrome is 1 in 700, it means that the 752 cases were drawn from a total of some 526,400 births. Among them (judged from the smoking rate among controls), 155,400 smoked and 371,000 did not. The risk of having an infant with Down syndrome among smokers is thus 1.18 per 1000, and among nonsmokers it is 1.53 per 1000. The risk ratio will be 0.77, exactly the odds ratio that was estimated.

2. Stratified Analysis — A Method to Control for Confounding

In the example in Table 20.2, we saw a difference in smoking rate among case and control women. Is this difference causal? Does smoking really prevent the birth of a Down infant, or is it a secondary effect of a so-called confounder, that is, a variable that affects both smoking rate and the risk of having a Down infant?

Table 20.3 addresses this problem. Here the material has been divided into 5-year maternal age classes. From the control data it is apparent that smoking declines with age, and it is well known that the risk of having an infant with Down syndrome increases with age. The seemingly protective effect of smoking may therefore be secondary to these differences in smoking distribution and Down infant risk between age classes.

There are different methods to control for such a confounder (or a set of confounders). One is to match cases and controls, so each case and control will form a pair with the same maternal age. Another is to make a stratified analysis. The data in Table 20.3 can be looked upon as a series of 2×2 tables, one for each age class or stratum. The Mantel-Haenszel method makes it possible to sum up the differences over the strata and to calculate a χ^2 (based on one degree of freedom, df) common for all strata.

The method is as follows, using the designations of Table 20.2 for each 2×2 table, with $E(a)$ designating the expected value of a (= $n_1 n_3 / n$). The subscript i refers to the ith 2×2 table:
$s = \Sigma[a_i - E(a_i)]$

The variance estimate of a_i is:

$$Va_i = \frac{n_1 * n_2 / (n_3 * n_4)}{n_i^2 * (n_1 - 1)}$$

$v = \Sigma(Va_i)$

$R_1 = \Sigma(a_i * d_i / n_i)$

$R_2 = \Sigma(c_i * b_i / n_i)$

$\chi^2 = s^2/v$. This χ^2 is based on only 1 df.

The odds ratio is OR = R_1/R_2

The analysis thus gives an odds ratio and a χ^2. Note that the χ^2 can be calculated even if, in individual 2×2 tables, the expected number for one of the cells is small.

If we apply this method to Table 20.3, we find that the OR approximates 1, and its statistical significance disappears: OR = 0.94 with a 95% CI of 0.76 to 1.13.

It should be noted that the conclusion supposes that all 2×2 tables estimate the same risk. This may not be so: The effect of smoking may be stronger at young age or vice versa. In order to test this possibility, one can test for homogeneity between the strata. If numbers in each stratum are large enough, this can be done with Breslow and Day's χ^2 test. If a normal approximation is not permissible, there are also exact tests available (e.g., Zelen's test). In the present example, the p value for homogeneity is $p = 0.71$, which indicates homogeneity, that is, the different age strata behave in a similar way with respect to the effect of smoking. We thus have reason to believe that the "protective" effect of smoking was due to the confounding effect of maternal age.

3. Logistic Multiple Regression Analysis

The most popular current method to control for confounders is to use logistic multiple regression. This is a complicated technique that necessitates the use of computer programs, and many such programs are available. Here we can only indicate the principle of the basic type of analysis.

A linear logistic model can be set up from an equation of the following type. Let p be the rate of occurrence of a specific event in the studied material (e.g., maternal smoking), then

$$\ln(p/(1 - p)) = \alpha + \beta_1 * X_1 + \beta_2 X_2 + \ldots + \beta_n X_n$$

for n different variables, where X_1 can be the control or case status (0,1). Using an iterative technique, the best fit of the equation to the available data can be made, and one can also get estimates (with errors) of the coefficients for each term, independent of the effects of the other terms. Each coefficient can be statistically tested against 0, and if it differs from 0, the variable has an effect. An estimate of an OR for each term (including the case control property) can also be obtained.

Two variables may interact. For example, the risk of having a Down syndrome infant might be affected differently by smoking in different age classes (even though we did not find this in the homogeneity study above). This can be controlled by adding a term for interaction, consisting of the product of the two variables (age and smoking). If the interaction term becomes statistically significant, it shows that smoking does have different effects in different age classes.

The method has many advantages. Its greatest value compared with the stratified analysis is that exact values for the variables can be used, e.g., exact maternal ages instead of 5-year classes. Classification of the data always carries the risk that the distribution within the class differs between cases and controls. In each maternal age class, mothers of infants with Down syndrome may average a few years older than control mothers. The standard analysis is, however, based on first-degree linear regressions, and these may not be valid. The effect of maternal age on birth weight is an example. This effect is U-shaped. When analyzed in a linear regression model, no maternal age effect is found because the linear distribution forced on a U-shaped function becomes practically horizontal. Obviously, regressions of higher order can be constructed to take care of such phenomena, or the variable distribution can be transformed to linearity.

4. Analysis of Matched Pairs or Triplets

When data have been collected from matched cases and controls, those confounders that were matched for will have been eliminated (if matching has been effective). Sometimes the matching

Table 20.4 Smoking among women who had an infant with Down syndrome and age-matched controls

Case	Control	Number of Pairs
Smoking	Smoking	$n_1 = 38$
Not smoking	Not smoking	$n_2 = 411$
Smoking	Not smoking	$n_3 = 145$
Not smoking	Smoking	$n_4 = 158$

Source: Data from the Swedish Medical Birth Registry, 1983–1990.

is "broken," and the two sets of data (cases and controls) are analyzed as described above. This usually reduces the efficiency (if the matching has been meaningful). If the matching is done correctly, the variation between individuals within the strata used for matching is less than the total variation in the population. Differences between cases and controls can be revealed against the background of the low variation within the stratum but may not be observable when compared with the large variation in the total population. Effectively matched materials should therefore be analyzed without breaking the matching.

There are many methods to analyze such data sets. Table 20.4 shows a matched sample of Down syndrome infants and maternal age-matched controls, with information on maternal smoking. The material can be divided into four groups: (1) both case and control mothers smoked (++), (2) neither smoked (--), (3) the case mother but not the control mother smoked (+-), and (4) the control but not the case mother smoked (-+). If smoking was unrelated to the risk of having a Down infant, $n_3 = n_4$, and the probability can be estimated that the observed values of n_3 and n_4 are taken from the binomial distribution of 50% of the total, $n_3 + n_4$ (which can be expressed as Bin($n_3 + n_4$, 0.5)]. This probability can easily be evaluated (e.g., from a binomial table).

The exact formula is: $p = n!*p^x*(1 - p)^{n-x}/(x!*(n - x)!)$, where n is the total number of pairs, $p = 1 - p = 0.50$, and x is the observed number of pairs with characteristics +-. Note that the p value also has to be calculated for still more skewed distributions ($x + 1/n - x - 1$; $x + 2/n - x - 2$, etc., if $x > n/2$). It should be noted that the two concordant groups (++ and --) do not contribute to the evaluation of the effect studied.

A similar exact analysis can be made with triplets (or a mixture of pairs and triplets). When numbers are large, approximate methods can be used instead. One much used technique is the McNemar test, which can be used both for pairs and triplets. The formula for pairs is: If there are n_3 +- pairs and n_4 -+ pairs, then $\chi^2 = (n_3 - n_4)^2/(n_3 + n_4)$. The odds ratio is then OR = n_3/n_4. From the example in Table 20.4, we see that $\chi^2 = (145\ 158)^2/(145 + 158) = 0.50$, which is far from significant, and the odds ratio is 0.92. Its 95% CI, calculated as described above (see p. 827), is 0.72 to 1.16.

With triplets, the following will be informative (triplets with all exposed or all nonexposed are not informative): +-- (n_1), ++- (n_2), -++ (n_3), -+- (n_4). Then $\chi^2 = (2*n_1 - n_4 + n_2 - 2*n_3)^2/(2*n_1 + n_2 + n_3 + n_4)$. The estimate of the OR is complicated but can be made.

Another possibility is to use the Mantel-Haenszel test described above. There is no minimum size limit for a stratum in that technique, and in principle a stratum can be a pair (or a triplet or a quadruplet). If pairs are analyzed, a 2×2 table for a +- pair (see above) will be:

	Exposed	Unexposed	Total
Case	1	0	1
Control	0	1	1
Total	1	1	2

For a triplet with one case and one control exposed, the table will look like:

	Exposed	Unexposed	Total
Case	1	0	1
Control	1	1	2
Total	2	1	3

There are also methods available to apply logistic multiple regression techniques to matched materials.

5. The Multiple Testing Problem

Case control studies are rarely made with one specific question in mind. Quite often, information is collected on a number of exposures. This may be because some of these exposures may be important confounders to the one of real interest, but very often many different exposures are studied simultaneously. For example, one may wish to know if there is any occupation or maternal drug use that is associated with a specific outcome.

It is important to realize that if many different statistical tests are used on the same material, "significant" results may be obtained merely by chance. Statistical significance merely means that such a deviant outcome is likely to be obtained by chance only once in 20 tests ($\alpha = 0.05$). For example, if 20 different drugs are studied, it is expected that one will deviate (either as a risk or a protective factor), and more than one may well do so. Great care must be taken to avoid overinterpretation of statistical tests that are not made on prior hypotheses. Ideally, the project plan states which test or tests will be made. All "significant" findings obtained by the use of any additional testing should be regarded as "fishing expedition" results that may suggest new studies on fresh material.

A similar situation will occur when 20 different studies of one drug-malformation relation are performed. Very likely one of them will end up with a "statistical significance" merely by chance.

VI. COHORT STUDIES

A. Principles

A cohort study can be thought of as a study of a group of individuals characterized by a specific situation. They may be born in the same year (a birth cohort), they may have been similarly exposed (e.g., have worked in a specific occupation or have taken a specific drug), or they may even be characterized by having had an infant with a specific malformation (e.g., mothers of Down syndrome infants). The question which is asked is whether the reproductive outcome of these women differs from the expected one. For instance, will women who have taken an antiepileptic drug have an increased risk of having a malformed infant? Or will mothers of infants with Down syndrome have an increased risk of having an infant with a malformation in *future pregnancies*, compared with the risk of "normal" women?

A cohort study can thus tell us about many different reproductive outcomes, e.g., infertility, spontaneous abortion, low birth weight, congenital malformations, mental retardation, or childhood cancer. The multiple testing situation, commented upon in the section for case control studies, will also be relevant in this situation. For example, if 20 different types of malformations are studied, it is rather likely that one or more will show a "significantly" high or low incidence.

B. The Formation of Cohorts

In the classical cohort study, a group of persons is identified because of an exposure and is then followed with respect to the development of a disease, but this is seldom the case in reproductive epidemiology. As exposures during pregnancy are of major interest, this approach necessitates that women are followed into pregnancy with respect to their exposures. Even when this approach makes it possible to obtain rather exact exposure information, it will usually result in small numbers of subjects; therefore, it remains difficult to study rare outcomes, such as congenital malformations.

Very often, retrospective cohorts are formed instead, e.g., women who work at a specific workplace, and information on previous pregnancies and their outcome is identified (ideally, this should include determination of whether exposure occurred during those pregnancies). There is a clear danger in this approach, notably with respect to occupational exposures. If women have a tendency to stop working after childbirth (which is often the case and was previously even more prevalent), women remaining in the workforce will be selected for previous poor reproductive outcome or subfertility. Any such cohort will thus have a tendency to show an abnormally high incidence of reproductive failures. This may also be true when a specific geographic area is studied; if the area is less suitable for families with small children, a differential emigration from the area may occur after successful pregnancies, leaving couples with poor reproductive outcome.

A third possibility is that the cohort is defined from recorded exposures during the pregnancies to be studied, e.g., drugs taken, occupations held, place of living. The problem lies in identifying such cohorts. Sometimes this can be done by use of prospective or retrospective records obtained for all pregnant women (e.g., by interviews in early pregnancy or after delivery), and sometimes it can be achieved by linkage of different registries, e.g., census information linked to medical birth registry information.

C. Comparison Material

The pregnancy outcome in the defined cohort must be compared with that of another group. Quite often, a comparison is made with the expected outcomes known from the entire population or a reasonable sample of the population. If, for instance, the perinatal death rate is known in the population, the rate in a specific cohort can be evaluated. The problem resembles that in case control studies; the cohort under study may have characteristics that by themselves influence pregnancy outcome, e.g., maternal age distribution or socioeconomic level. This may be avoided by a method similar to that used in the matched case control study, i.e., the comparison with another cohort should be selected to ensure as many similarities with the study cohort as possible except for the exposure under study. If a specific industrial occupational exposure is of interest, women with other industrial jobs may form a suitable cohort for comparison. If nurses working in anesthesiology are studied, nurses working in wards of internal medicine might be chosen as controls.[61] As when sick controls are used in case control studies, it must be remembered that the cohort of comparison may also be exposed to reproductive hazards.

There are two main problems with the cohort approach: the difficulty in definition and selection of cohorts, and the problem with size. One needs very large numbers to study rare outcomes, such as specific congenital malformations, and even large cohorts may be insufficient. We made a study of reproductive outcome among women working as dentists, dental assistants, or dental technicians, because of an alleged increased risk of having an infant with spina bifida.[62] We studied more than 8000 such births and found 3 infants with spina bifida when the expected number was 4.2. We could thus not confirm the alleged risk, but in spite of the large size of the study group, it is possible that an increased risk exists. This is true because the observed number of 3 may be a randomly low sample estimate of a true expected number of 8.8 (i.e., a doubling of the risk).

Table 20.5 Infant mortality when the mother worked in the chemical industry with a comparison group of all women working in industry

Outcome	Mother Worked in Chemical Industry	Mother Worked in Any Industry
Number of infants	221	21,605
Stillbirths	3	123
Early neonatal deaths	5	82
Later deaths	2	70
Total deaths	10 (4.5%)	275 (1.3%)

Source: Swedish data, infants born in 1971, 1981, or 1986. Källén, B. and Landgren, O., J. Occup. Med., 36, 563, 1994.

D. Statistical Analysis of Cohort Data

In principle, the statistical analysis is similar to that for unmatched case control studies, but risks are often expressed as risk ratios instead. In the exposed cohort, the risk is measured as the number of adverse outcomes divided by the total number of pregnancies or infants. This risk can be compared with the risk in the total population and expressed as a risk ratio (the statistical uncertainty of the risk ratio will be nearly completely due to the estimate from the exposed cohort); alternatively, a risk ratio can be calculated between the risk in the exposed cohort and that in a cohort of comparison. Table 20.5 shows an example: infant and child deaths recorded in two cohorts of women working in industry, the first in the chemical industry and the second (control cohort) in other industries.[63] The table shows that among 221 infants born to women working in the chemical industry, 10 died during the observation period (4.5%), while in the comparison cohort of 21,605 infants, 1.3% died. The risk ratio is thus 3.5.

In all such comparisons it is important to be certain that outcome information is of similar quality in the cohorts or in the cohort and the population. If detailed follow-up has been made of each child in the cohort, risks from that cohort cannot be compared with risks derived from sources such as birth registries or congenital malformation registries, where ascertainment is often incomplete and risks therefore underestimated.

The statistical significance of risk differences is evaluated with χ^2 tests or exact tests, as described in the section on case control studies. The effect of confounding must again be taken into consideration. If the study cohort, for example, differs in maternal age or parity distribution from the material used for comparison, this must be compensated for. This can be done in a similar way as in the case control study, using the Mantel-Haenszel procedure or a regression method. If the comparison group is very large (e.g., the total population or as in Table 20.5), it is possible instead to calculate the expected number of outcomes for the study population, taking maternal age, parity, or other confounders into consideration. The observed number can then be compared with the expected number as a ratio that will represent the risk ratio.

The observed-to-expected ratio is sometimes expressed multiplied by 100 as an SMR (standard morbidity or mortality rate). In this situation, the confidence interval is most easily estimated from the observed number of reproductive outcomes: if this is a rare event (less than 10% of all), it will behave as a Poisson distributed variable, and the 95% confidence interval can be determined as $\pm 2\sqrt{n}$, where n is the observed number. These two values can be divided by the expected number to obtain the confidence interval of the risk ratio. If n is small (less than 5), an exact calculation should be made instead (or a Poisson table can be used). In the example in Table 20.5, the expected number of deaths is 2.8, and a normal approximation is not allowed. The 95% CI of the observed value of 10 is 4.8 to 18.4, and that of the risk ratio therefore is 1.7 to 6.6.

Table 20.6 Perinatal death rate according to maternal smoking in early pregnancy

Smoking Status	Total Births (No.)	Perinatal Deaths (No.)	%
Not smoking	62,993	361	0.57
Smoking < 10 cigs/day	16,571	130	0.78
Smoking ≥ 10 cigs/day	10,606	88	0.83

Source: Swedish infants born in 1985; data from the Medical Birth Registry.

Sometimes cohorts with different amounts of exposure are compared. An example is nonsmoking women, women smoking less than 10 cigarettes per day, and women smoking 10 or more cigarettes per day. Table 20.6 shows an example, stating the number of perinatally dead infants in each cohort. In order to see whether a trend exists in such a table, the Armitage-Cochran test for a trend in a frequency table can be used. In principle, this test compares the χ^2 in the total material with that around a straight line fitted to the data, and the difference is due to the linear regression in the material. For low numbers, exact trend tests should be used. The trend analysis of Table 20.6 gives $\chi^2 = 14.6$ at 1 df, $p < 0.001$.

VII. NESTED CASE CONTROL STUDIES

We have seen advantages and disadvantages with the two main epidemiological techniques: case control and cohort studies. They can sometimes be combined in a rather efficient way in what may be called a nested case control study. The principle is that by making a crude cohort with a high probability of exposure and then making a case control study within the cohort, the size of the material needed for study can be considerably reduced. All types of reproductive outcomes of relevance to the exposure can be studied.

The definition of the crude cohort can, for example, be an occupation, perhaps identified from registries or by linkage of different registries. Such a cohort may not be "true," in that all women identified actually had the occupation during pregnancy, but they are relatively likely to have had it. Furthermore, only some of the working women may have the exposure of interest. For example, one may be interested in women exposed to plastic monomers during pregnancy. One can then form a crude cohort of all women working in the plastic industry, and some but not all will have been exposed to the monomers.[64]

This crude cohort is then analyzed, and adverse outcomes are identified. Cases are selected among these outcomes (perhaps after exclusion of inherited conditions or other causes that obviously have no relationship to the workplace), and one or more controls is selected for each case from the crude cohort (using the control definition that was discussed above). Then a regular case control study is made, retrieving exposure information in the best possible way. This leads to two results: (1) that the control group estimates the proportion of the crude cohort that was actually exposed and (2) a comparison of exposure rates can be made between the case and control groups. If close to 100% of controls were exposed, obviously no difference from the case group can be expected (unless the exposure was protective), but then it can be concluded that the outcome of the total crude cohort is representative for the exposure under study. If (as is usually the case) it appears that only a proportion of the controls were exposed, an odds ratio can be calculated in the standard way, comparing case and control exposures. In this situation, the study can be regarded as any case control study, but with the advantage that primary selection of the crude cohort has markedly increased the exposure rate in comparison with that of the total population.

VIII. CONFOUNDING

A. General Considerations

We have already mentioned the phenomenon of confounding: a variable that affects both the exposure rate and the outcome rate. An example mentioned was maternal age, which affects both smoking rate and the risk of having an infant with Down syndrome. Many possibly important confounders can exist in reproductive epidemiology; some thought should be given to the biological meaning of the statistical relationships, otherwise completely misleading results may be drawn. A classical example is maternal smoking and perinatal death rate, with birth weight as a confounder. Smoking causes a decrease in birth weight and an increase in perinatal mortality. Low birth weight is an important determinant of perinatal death risk and could therefore be regarded as a confounder. By stratifying for birth weight, not only does the effect of smoking on perinatal death disappear, but also an apparent protective effect appears. This does not mean that maternal smoking does not affect perinatal death risk, but it means that it does so to a large extent by changing the birth weight distribution. After birth weight stratification, one will compare small but otherwise healthy infants of smoking women with small but sick (growth retarded) infants of nonsmoking women. This will naturally result in a seemingly protective effect of smoking on perinatal death rate. Similar effects were seen in a study on socioeconomic characteristics and infant mortality.[27]

B. Maternal Age and Reproductive History

Both of these variables are important confounders in reproductive epidemiology, especially in studies of spontaneous abortion rate, birth weight distribution, and perinatal mortality. They can, however, have a complex effect on outcome, and it is not absolutely clear that a compensation for them should always be made. Let us look at the group of relatively old primiparous women (say, above 35 years of age). This group contains two main subgroups: women who have had difficulty becoming pregnant and therefore give birth for the first time late in reproductive life, and women who have voluntarily postponed pregnancy because of considerations such as education or career. These two groups will have different distributions in, for example, different occupational groups. Differences recorded in reproductive outcome will not be due to actual differences in exposures related to occupation, but to the fact that incomparable groups are compared. If the presence of long-standing subfertility is recorded in an adequate way, this can be entered into the model, but often such information is not available.

Previous reproductive history, including subfertility, may be an important confounder in studies on occupational exposures, as was pointed out above. It may also affect drug usage, smoking, and other possible exposures. It can be shown that there is a positive association between the use of antidepressive drugs and the occurrence of hypospadias. This is completely explained by the excess use of these drugs by subfertile women and an association between subfertility and hypospadias.[65]

C. Maternal Diseases

Maternal diseases are obvious confounders in the study of the reproductive effects of maternal use of drugs during pregnancy. If the disease directly or indirectly (e.g., by genetic linkage) increases the risk for a congenital malformation, any drug used for that disease will appear to be a teratogen. A classical example is insulin and maternal diabetes. It is generally thought that the increased risk for malformations seen in infants of diabetic mothers[66] is due to the disease and not to insulin; actually, good control of the disease with insulin may reduce the risk. Similarly, epilepsy as such may (perhaps mainly because of linked genes) increase the risk for facial clefts,[67] even though maternal anticonvulsant drug use has an effect of its own.

When different drugs are used for the same disease or when drugs are not always used, it is often possible to control for the effect of the disease in an analysis of drug effects. Sometimes studies involving different populations with different therapeutic traditions can help.

In the study of other environmental conditions, drugs usually appear as a very weak confounder. Very few drugs have any marked effect on the teratogenic risk, and in these few instances, typical outcomes that can be identified are usually seen. In studies of occupational risks or the risk of exposures from the general environment, it is generally not worthwhile to record drug exposures. Some chronic diseases (e.g., diabetes and epilepsy) have distinct teratogenic effects, but they occur at low frequency. In studies of occupational exposures, such chronic diseases may give a "healthy worker" effect if the disease is not compatible with the occupation under study, but this effect is usually weak.

D. Social Characteristics

Social characteristics may have an effect on gestational duration, birth weight, stillbirth rate, and infant death rate. In a few circumstances, an effect on specific congenital malformations is seen (e.g., neural tube defects in some populations).[68] Highly skewed social distributions are commonly observed in studies of occupational exposures and general environmental exposures, and it is often difficult to distinguish between the effect of an occupation and the consequences of the social characteristics of that occupation. When very specific exposures are studied, this can be done by selecting women with a similar but unexposed occupation as controls, but even then differences may remain. There are also difficulties in the characterization of the social level, as most such groupings are based on occupation classifications. One always has to consider the possibility that differences demonstrated between groups, such as occupational groups, are not due to actual differences in exposure but to social characteristics, including lifestyle (e.g., smoking, alcohol and drug abuse).

As a proxy for socioeconomic level, parental education is sometimes used. In many societies, paternal education seems to have the strongest effect; in others (e.g., the Nordic) it appears that maternal education is the stronger determinant of, for instance, birth weight. Especially for young women, who give birth before finishing their training, education level at the time of delivery may be less informative than final education as a proxy of social level. When registries are available of the education level of the members of the population, it may sometimes be more effective to use "final" education, even if it is attained many years after the delivery.

E. Biological Characteristics

Some biological characteristics of the women may appear as confounders. As pointed out above, subfertility not only can affect reproductive outcome, but also can affect the probability of staying in the workforce or living in a specific geographical area. Physical fitness may be a prerequisite for a specific job (e.g., physiotherapist), and this may result in a positive reproductive outcome compared with the general population.[69] Exposures within the job may nevertheless carry a risk that is difficult to identify when comparisons are made between the exposed group and the population.

F. Incomplete Compensation of a Confounder

It is often thought that the possible effect of a confounder can be disregarded if the confounding variable is entered into the statistical model, but this is not always true. In a study on socioeconomic level and perinatal deaths, a standardization was made for maternal smoking as recorded in early pregnancy.[32] Standardization for smoking did reduce the risk ratio, but only partly. There are at least three possible explanations. The obvious one is that socioeconomic level plays a role, irrespective of smoking. Another is that women of different social classes have different tendencies to stop smoking during pregnancy. The third is that women of different social classes have different tendencies to underreport their smoking when answering an interviewer in early pregnancy. The statistical compensation of the confounder will be incomplete if the registration of the confounder is incomplete or biased! Further, it is most important to note that there is always the possibility that confounders exist that have not been realized or for which one cannot control.

IX. SURVEILLANCE

A. Birth Defects Surveillance in the Population

Population surveillance of reproductive outcomes has concentrated on congenital malformations. Most such activities began after the thalidomide tragedy. The introduction of a new drug, thought to be low in toxicity, caused a large number of infants with previously very rare malformations, especially absent or strongly reduced limbs (amelia, phocomelia). It took some time before this "epidemic" was detected, and thousands of children were damaged. The congenital malformation registries that have been built up usually had surveillance as a primary purpose, but — as pointed out above — are also excellent sources for case finding for epidemiological studies. A crude surveillance of perinatal or neonatal deaths occurs in most developed countries, but more elaborate systems aiming at the detection of changes in specific subgroups are rare. Such an effort was published from Sweden, based on data in computerized perinatal registries.[30]

A change in the prevalence at birth of a specific congenital malformation can occur as a result of the introduction of a new teratogenic agent in the environment. If such an agent causes a relatively common malformation, it is extremely difficult to detect the change in rate.[70] Only if the agent has a very strong teratogenic effect or exposure to it becomes widespread in the population can effects on the population rates of common malformations be detected. Minor changes in the rates will be masked by the random fluctuations in the number of malformed babies born, and this masking effect is exacerbated by changes in ascertainment of malformations. Prenatal diagnosis followed by selective abortion will also complicate an analysis. It is rather unlikely that population surveillance of common malformations will ever be able to detect a new teratogenic agent causing a moderate increase in the rate of common malformations.

What population surveillance can detect is the appearance of previously rare congenital malformations or combinations of malformations. Many of the detected human teratogens had such effects, e.g., amelia or phocomelia seen after thalidomide, congenital cataract seen after rubella, and the malformation syndrome typically seen after isotretinoin. This, of course, means that the risk increase caused by these agents is enormous. For amelia or phocomelia the "normal" population rate is perhaps 3 per 100,000 births, while after thalidomide exposure something like 6% of infants had such limb defects, a 2000-fold increase.

Congenital malformation surveillance is therefore less a matter of statistical analysis of the recorded rates of common malformations than it is the search for previously unusual malformations or combinations of malformations. It must be realized that the absence of a noticeable change in the occurrence of a specific malformation does not show that no new teratogen has been introduced. This only implies that if there was such an introduction, its impact on the total malformation rate was small. We can illustrate this with valproic acid, an anticonvulsant that causes spina bifida in some 1% to 2% of exposed embryos.[71] In the Swedish population, spina bifida occurs at an approximate rate of 1 in 2000 (50 to 60 cases a year). If valproic acid were suddenly introduced, and 20% of all epileptic women used the new drug (a rather unlikely supposition), among 400 epileptic women giving birth each year, the new drug would cause one new case of spina bifida. Such a modest increase would be completely impossible to detect because of the random variability of the number of spina bifida cases born each year (with 55 as a mean and a 95% CI of 41 to 72). On the other hand, if thalidomide had been used by 80 pregnant women, perhaps 5 would have borne infants with amelia or phocomelia, more than doubling the rate of these very unusual malformations.

B. Surveillance after Specific Exposures

The effectiveness of the surveillance process can be markedly increased if it is restricted to specific groups of women characterized by exposure. Surveillance of the reproductive outcomes of women with specific diseases (e.g., epilepsy), with specific occupations (e.g., working in the plastics industry), or living in areas with specific pollution (e.g., exposed to radioactive fallout after the Chernobyl accident) has a much greater chance to detect changes and identify the effect of a new drug, a changed work environment, or a specific pollutant.

Such specific surveillance can be made, but it should be realized that a very large number of possible objects of surveillance exist, and surveillance can only give a signal that something *may* be wrong. Suppose, for example, that during the surveillance of infants born of epileptic women, the rate of infants with spina bifida suddenly doubled, and half of these cases had been exposed to a specific anticonvulsant drug; this could be due to a causal association, but specific investigation of the subject should be made on an independent population.

C. Outcome-Exposure Surveillance

Another way to increase the power of routine surveillance is to make an on-going surveillance of outcome-exposure associations, e.g., maternal drug usage and congenital malformations in the offspring. It may be based on the reporting of adverse drug reactions or similar systems, but it can also be based on a systematic data collection, e.g., the MADRE system run by the International Clearinghouse for Birth Defects Monitoring Systems.[72,73] The Swedish Medical Birth Registry provides an opportunity for an ongoing surveillance similar to that just described.

D. Statistical Considerations

There is a statistical problem in all surveillance processes because no prior hypothesis exists. Many different types of malformations or other adverse outcomes are monitored during an extended time, and an increase in the rate of any malformation will give a signal. It is therefore not possible to run statistical tests on the observed changes. For example, during a 3-month period 10 infants are observed with omphalocele, and the expected number of such infants is only 3.7. One cannot then say that the probability for that increase being due to chance is only $p = 0.01$ (based on a two-tailed Poisson distribution). There are dozens of malformations that can show such a change, monitoring may have occurred for many years, and there may be many quarters in which such a change could occur. Therefore, such a large difference between the observed and expected numbers of infants with omphalocele can very well be due to chance, in spite of the "statistical significance."

Any such observation is only an indication that something may be wrong and that perhaps a specific study should be conducted. Personally, I am less alarmed by a finding of this type than a finding of the birth of two or three infants with cyclopia during the period of a few weeks, even though this also may have been due to chance.

A similar situation exists in surveillance of drug-malformation combinations. Many drugs are surveyed, and many types of malformation are studied; thus, a seemingly strong association observed between a specific drug and a specific malformation may well be random. Such an observation can only serve as a signal, to be supported or rejected by other studies. Continued surveillance is one way to get new data, although at the price of a delay.

X. CLUSTER ANALYSIS

A. Definition of a Cluster

A cluster is a time-space aggregation of adverse outcomes, e.g., infertility, congenital malformations, perinatal deaths, or miscarriages. It is sometimes observed at surveillance (see the preceding section) but usually is noted in the "periphery" where the cluster occurs.

B. Clusters as Chance Phenomena or Artifacts

A cluster can be quite impressive but nevertheless be due to chance. According to the same reasoning as that presented above, even strong clusters not only may arise by chance but *should* arise due to the random fluctuation of numbers. Technical explanations for a cluster may also exist, such as local peculiarities in the diagnosis or registration of congenital malformations. Other clusters may be caused by local teratogenic factors; that is the obvious reason for studying clusters at all. It is seldom possible to directly say whether a cluster is random or actually caused by a local factor. In many circumstances possible harmful factors can be suggested, irrespective of whether they caused the cluster, something else did, or the cluster merely occurred by chance.

Clusters are of interest in the study of the affect of the environment on reproduction but must be handled with great care. The first task is to see whether the cluster can be explained by some technical anomaly. If not, one should analyze the cases entering the cluster in order to look for similarities. A cluster consisting of infants with very different types of malformations is less impressive than a cluster where the malformations of the infants seem to have something in common. The analysis is less a statistical exercise than an evaluation of the biological plausibility of the observed events.

C. Exposure Analysis of Clusters

If one decides to follow up on a cluster, exposure information becomes important. Sometimes a tentative cause can be identified, such as emission from an industry. If so, a case control study can be carried out in the area where the cluster appeared. The study might compare the distance to the source of emission from the homes of the cluster cases and those of controls selected from the same geographical area. Other examples of tentative exposure sources are local water supply or occupational exposure in a local industry. No tentative explanation may exist, or the explanation is not supported by exposure data. In order to get new hypotheses to explain the cluster, detailed interviews of the (usually few) cluster cases can be conducted. If an exogenous factor of moderate strength caused the cluster, a majority of the cluster cases should have been exposed to it, and formal control materials are often not necessary. If a possible explanation is found, the hypothesis has to be tested on a new population, if that is possible.

The most difficult situation is when the suspected teratogenic factor is ubiquitous, e.g., a widespread air or water pollutant. All pregnancies will then be exposed, and the resulting cluster can be due to a relatively low teratogenic property. Such a teratogen can never be detected in a case control study within the area. If it is possible to identify other areas with the same type of exposure, follow-up studies can be made in them to support or reject the hypothesis, but it can always be argued that the exposure situation was not identical.

D. Clusters and Statistics

Very often one has tested the "statistical significance" of a cluster, e.g., by applying a Poisson distribution test. Suppose, for example, that in a specific geographic area, eight infants were born with spina bifida and only 2.8 were expected from the country rate of the malformation. A Poisson model gives $p = 0.008$, which seems rather convincing. This is the probability that eight cases occurred (with 2.8 expected) if this area had been selected for analysis for some specific reason, e.g., that an industry emitted something that was thought to cause spina bifida.

In cluster analysis the situation is reversed; one observes an accumulation of cases and secondarily tries to fit that to an exposure situation. There are numerous areas where an aggregation of spina bifida cases could have occurred, and there are many different types of malformations or other adverse reproductive outcomes that could have been clustered. Thus, it is not unlikely that an aggregation of the magnitude described will occur in one of the areas for one of the malformations. The p value reached therefore does not have the common meaning of a p value, estimating the probability for the occurrence to be due to chance.

We once conducted a study of the distribution per municipality of infants with Down syndrome in Sweden. The distribution exactly followed what could be expected from a Poisson distribution, but nevertheless when specific municipalities and time periods were scrutinized, rather impressive clusters were found. During the years 1978 to 1986 in one such municipality, 11 cases were found when the expected number was 4.0 ($p = 0.003$); in another municipality, 7 infants with Down syndrome were born in 1978 against the expected number of 1.5 ($p = 0.0009$). These strong aggregations can be due to chance, but obviously they can also have a specific cause, and the latter cluster resulted in an intense follow-up (without any reasonable cause being identified).

XI. CONCLUDING WORDS

Epidemiological studies of human reproduction are burdened with many problems and difficulties. It is rare that single epidemiological studies give a final answer on hazards associated with specific exposure. Each epidemiological study should be regarded as a piece in a jigsaw puzzle. When many pieces have been put together, perhaps together with information from other sources, a picture of the truth develops. It is not necessarily the truth, but hopefully resembles the truth as much as possible. If too much importance is paid to the statistical significance levels in a single study, there is a definite risk that the importance of the findings will be overestimated. One must be careful in drawing practical conclusions from single studies if they do not unequivocally show very high risks as well as biological plausibility. One phenomenon should be realized — mass significance also exists with respect to scientific studies. If 20 different researchers investigate the same problem (e.g., a possible role between exposure to electromagnetic fields and child leukemia), it is quite probable that in one of the studies a "statistically significant" effect will be seen, even if no causal relationship exists between exposure and outcome. To this is added the "publication bias" problem. Studies that show an association usually have an advantage in competing for publication in scientific journals. Thus, there is a tendency for the first reports appearing in the literature to overestimate risks, sometimes even when the risk does not actually exist.

To the above is added the information problem, namely, how to convey to society and to the public the digested interpretation of the available evidence. There are special problems in informing the public, who often cannot accurately appreciate the concept of risk. Individual risks may be small, but the attributable risk for the outcome under study may be large in the total population. Individual risks may be high (and may influence decisions about termination of pregnancy), but the total number of damaged infants born may be negligible. Epidemiological studies may be "negative," but they do not prove that the exposure under study is harmless. To transform the scientific information into recommendations for society and the public is a delicate matter, and the scientists who conducted the studies may not be the best persons to interpret their findings in an unbiased way. Indeed, probably one of the strongest biases in the field is to favor one's own study!

REFERENCES

1. Cnattingius, S., Haglund, B., and Meirik, O., Cigarette smoking as a risk factor for late fetal death and early neonatal death, *Brit. Med. J.*, 297, 258, 1988.
2. Jones, K.L., Smith, D.W., Streissguth, A.P., and Myrianthopoulos, N.C., Outcome in offspring of chronic alcoholic women, *Lancet*, 1, 1076, 1974.
3. Källén, B., Lithium therapy and congenital malformations, in *Lithium in Biology and Medicine*, Schrauzer, G.N. and Klippel, K.-F., Eds., Weinheim, New York, 1991, p. 121.
4. Herbst, A.L., Ulfelder, H., and Poskanzer, D.C., Adenocarcinoma of the vagina. Association of maternal stilbestrol therapy and tumor appearance in young women, *New Engl. J. Med.*, 284, 878, 1971.
5. Whorton D., Krauss R., Marshall, S., and Milby, T., Infertility in male pesticide workers, *Lancet*, 2, 1259, 1977.
6. Palmer, J.R., Hatch, E.E., Rao, R.S., Kaufman, R.H., Herbst, A.L., Noller, K.L., Titusernstoff, L., and Hoover, R.N. Infertility among women exposed prenatally to diethylstilbestrol, *Am. J. Epidemiol.*, 154, 316, 2001.
7. Joffe, M. and Barnes, I. Do parental factors affect male and female fertility? *Epidemiology*, 11, 6, 2000.
8. Jensen, T.K., Scheike, T., Keiding, N., Schaumburg, I., and Grandjean, P., Selection bias in determining the age dependence of waiting time, *Am. J. Epidemiol.*, 152, 565, 2000.
9. Basso, O., Juul, A., and Olsen, J., Time to pregnancy as a correlate of fecundity: differential persistence in trying to become pregnant as a source of bias, *Int. J. Epidemiol.*, 29, 856, 2000.
10. Bergh, T., Ericson, A., Hillensjö, T., Nygren, K-G., and Wennerholm, U.-B., Deliveries and children born after in-vitro fertilisation in Sweden 1982–1995: a retrospective cohort study, *Lancet*, 354, 1579, 1999.
11. Ghazi, H.A., Spielberger, C., and Källén, B. Delivery outcome after infertility — a registry study, *Fert. Ster.*, 55, 726, 1991.
12. Gray, R.H. and Wu, L., Subfertility and risk of spontaneous abortion, *Am J Publ Health*, 90, 1452, 2000.
13. Ericson, A. and Källén, B., Congenital malformations in infants born after IVF: a population-based study, *Human Reprod.*, 16, 504, 2001.
14. Akre, O., Cnattingius, S., Bergström, R., Kvist, U., Trichopoulos, D., and Ekbom, A., Human fertility does not decline: evidence from Sweden, *Fert. Steril.*, 71, 1066, 1999.
15. Tielemans, E. Burdorf, A. Velde, E.R.T., Weber, R.F.A., Vankooij, R.J., Veulemans, H., and Heederik, D.J.J., Occupationally related exposures and reduced semen quality; a case-control study, *Fert. Steril.*, 71, 690, 1999.
16. Kolstad, H.A., Bonde, J.P., Spano, M., Giwercman, A., and Zschiesche,Y., Subfertility and risk of spontaneous abortion, W., Kaae, D, Larsen, S.B., and Roeleveld, N., Change in semen quality and sperm chromatin structure following occupational styrene exposure, *Int. Arch. Occup. Environ. Health*, 72, 135, 1999.
17. Wennborg, H., Bodin, L., Vainio, H., and Axelsson, G., Solvent use and time to pregnancy among female personnel in biomedical laboratories in Sweden, *Occup. Environ. Med.*, 58, 225, 2001.
18. Kurinczuk, J.J. and Clarke, M., Case-control study of leatherwork and male infertility, *Occup. Environ. Med.*, 58, 217, 2001.
19. Sandahl, B., Smoking habits and spontaneous abortion, *Europ. J. Obst. Gynecol. Reprod. Biol.*, 31, 23, 1989.
20. Miller, J.F., Williamson, E., Glue, J., Gordon, Y.B., Grudzinskas, J.G., and Sykes, A., Fetal loss after implantation. A prospective study, *Lancet*, 2, 554, 1980.
21. Juutilainen, J., Matilainen, P., Saarikoski, S., Läärä, E., and Suonio, S., Early pregnancy loss and exposure to 50 Hz magnetic fields, *Bioelectromagnetics*, 14, 229, 1993.
22. Axelsson, G. and Rylander, R., Validation of questionnaire reported miscarriage, malformation and birth weight, *Int. J. Epidemiol.*, 13, 94, 1984.
23. Hemminki, K., Niemi, M.-L., Soloniemi, I., Vainio, H., and Hemminki, E., Spontaneous abortion by occupation and social class in Finland, *Int. J. Epidemiol.*, 9, 199, 1980.
24. Kline, J., Stein, Y.A., and Susser, M., Smoking: a risk factor for spontaneous abortion, *New Engl. J. Med.*, 297, 793, 1977.
25. Goldhaber, M.K., Pole, M.R., and Hiatt, R.A., The risk of miscarriage and birth defects among women who use visual display terminals during pregnancy, *Amer. J. Industr. Med.*, 13, 695, 1988.

26. Haglund, B. and Cnattingius, S., Cigarette smoking as risk factor for sudden infant death syndrome: a population-based study, *Amer. J. Pub. Health*, 80, 29, 1990.
27. Ericson, A., Eriksson, M., Källén, B., and Zetterström, R., Socio-economic variables and pregnancy outcome. 2. Infant and child survival, *Acta Paediatr. Scand.*, 79, 1009, 1990.
28. Winbo, I.G.B., Serenius, F.H., and Källén, B.A.J., Lack of precision in neonatal death classifications based on the underlying cause of death stated on death certificates, *Acta Paediatr.*, 87, 1167, 1998.
29. Wigglesworth, J.S., Monitoring perinatal mortality. A pathophysiological approach, *Lancet*, 2, 684, 1980.
30. Winbo, I.G.B., Serenius, F.H., Dahlquist, G.G., and Källén, B., NICE, a new cause of death classification for stillbirths and neonatal deaths, *Int. J. Epidemiol.*, 27, 499, 1998.
31. Källén, B., A birth weight for gestational age standard based on data in the Swedish Medical Birth Registry, 1985–1989, *Europ. J. Epidemiol.*, 11, 601, 1993.
32. Ericson, A., Eriksson, M., Källén, B., and Zetterström, R., Secular trends in the effect of socio-economic factors on birth weight and infant survival in Sweden, *Scand. J. Soc. Med.*, 21, 10, 1993.
33. Ericson, A., Gunnarskog, J., Källén, B., and Otterblad Olausson, P., A registry study of very low birthweight liveborn infants in Sweden, 1973–1988, *Acta Obst. Gynecol. Scand.*, 71, 104, 1992.
34. English, P.B. and Eskenazi, B., Reinterpreting the effects of maternal smoking on infant birthweight and perinatal mortality: a multivariate approach to birthweight standardizationm *Int. J. Epidemiol.*, 21, 1097, 1992.
35. Ericson, A., Eriksson, M., Källén, B., and Meirik, O., Birth weight distribution as an indicator of environmental effects on fetal development, *Scand. J. Soc. Med.*, 15, 11, 1987.
36. Mulvihill, J.J., and Czeizel, A.A., 1983 view of sentinel phenotypes, *Mutation Res.*, 123, 345, 1983.
37. Källén, B., Knudsen, L.B., Mutchinick, O., Mastroiacovo, P., Lancaster, P., Castilla, E., and Robert, E., Monitoring dominant germ cell mutations using skeletal dysplasias registered in malformation registries. An international feasibility study, *Int. J. Epidemiol.*, 22, 107, 1993.
38. Kieler, H., Cnattingius, S., Haglund, B., Palmgren, J., and Axelsson, O., Sinistrality — a side effect of prenatal sonography: a comparative study of young men, *Epidemiology*, 47, 618, 2001.
39. Feldman, R.G. and White, R.F., Lead neurotoxicity and disorders of learning, *J. Child Neurol.*, 7, 354, 1992.
40. Harada, M., Congenital Minimata disease: intrauterine methylmercury poisoning, *Teratology*, 18, 285, 1978.
41. Sutcliffe, A.G., Dsouza, S.W., Cadman, J., Richards, B., McKinlay, I.A., and Lieberman, B., Minor congenital anomalies, major congenital malformations and development in children conceived from cryopreserved embryos, *Human Reprod.*, 10, 3332, 1995.
42. Ericson, A., Nugren, K.G., Otterblad Olausson, P., and Källén, B., Hospital care utilization of infants born after in vitro fertilization. *Human Reprod.*, 17, 929, 2002.
43. Forsberg, J.-G. and Källén, B., Pregnancy and delivery characteristics of women whose infants develop child cancer, *APMIS*, 98, 37, 1990.
44. Needleman, H.L., Gunnoe, C.G., Leviton, A., Reed, R., Peresie, H., Maher, C., and Barrett, P., Deficits in psychologic and classroom performance of children with elevated dentine lead levels, *New Engl. J. Med.*, 300, 689, 1979.
45. Oryszczyb, M.P., Godin, J., Annesi, I., Hellier, G., and Kauffman, F., In utero exposure to parental smoking. Cotinine measurements and cord blood IgE, *J. Allergy Clin. Immunol.*, 87, 6, 1991.
46. Martínez-Frías, M.L. and Rodrigues-Pinilla, E., Epidemiologic analysis of prenatal exposure to cough medicines containing dextromethorphan: no evidence of human teratogenicity, *Teratology*, 63, 38, 2001.
47. Boneva, R.S., Moore, C.A., Botto, L., Wong, L.Y., and Erickson, J.D. Nausea during pregnancy and congenital heart defects: a population-based case-control study, *Am. J. Epidemiol.*, 149, 717, 1999.
48. Heinonen, O.P., Slone, D., and Shapiro, S., *Birth Defects and Drugs in Pregnancy*, Publishing Sciences Group, Inc., Littleton, MA, 1977.
49. Källén, B., Use of omeprazole during pregnancy — no hazard demonstrated in 955 infants exposed during pregnancy, *Obst. Gynecol.*, 96, 63, 2000.
50. Nielsen, G.L., Sorensen, H.T., Larsen, H., and Pedersen, L., Risk of adverse birth outcome and miscarriage in pregnant users of non-steroidal anti-inflammatory drugs: population based observational study and case-control study, *Brit. Med. J.*, 322, 266, 2001.

51. Czeizel, A.E., Rockenbauer, M., Sorensen, H.T., and Olsen, J., The teratogenic risk of trimethoprim-sulfonamides: a population-based case-control study, *Reprod. Toxicol.*, 15, 637, 2001.
52. Hemminki, K., Niemi, M.-L., Saloniemi, I., Vainio, H., and Hemminki, E. Spontaneous abortions by occupation and social class in Finland. *Int. J. Epidemiol.*, 9, 149, 1980.
53. Feldman, Y., Koren, G., Mattice, D., Shear, H., Pellegrini, E., and MacLeod, S.M., Determinants of recall and recall bias in studying drugs and chemical exposure in pregnancy, *Teratology*, 40, 37, 1989.
54. Mackenzie, S.G., and Lippman, A., An investigation of report bias in a case-control study of pregnancy outcome, *Amer. J. Epidemiol.*, 129, 65, 1989.
55. Källén, B., Castilla, E.E., Robert, E., Lancaster, P.A.L., Kringelbach, M., Mutchinick, O., Martínez-Frías, M.L., and Mastroiacovo, P., An international case-control study on hypospadias. The problem with variability and the beauty of diversity, *Europ. J. Epidemiol.*, 8, 256, 1992.
56. Ericson, A., Eriksson, M., Källén, B., and Zetterström, R., Maternal occupation and delivery outcome: a study using central registry data, *Acta Paediatr. Scand.*, 76, 512, 1987.
57. Kurppa, K., Holmberg, P.C., Hernberg, S., Rantala, K., Riala, R., and Nurminen, T., Screening for occupational exposures and congenital malformations. Preliminary results from a nationwide case-referent study, *Scand. J. Work. Environ. Health*, 9, 89, 1983.
58. Ericson, A., Källén, B., and Löfkvist, E., Environmental factors in the etiology of neural tube defects: a negative study, *Environ. Res.*, 45, 38, 1988.
59. Knudsen, L.B., Possibilities in the study of adverse reproductive outcome in Denmark, based on registries, in *Proc. 1st Int. Meet. Genetic and Reproductive Research Society (GRERS)*, Mastroiacovo, O., Källén, B., and Castilla, E.E., Eds., Ghedini Editori, Milano, Italy, 1995, p. 35.
60. Källén, K., Maternal smoking and orofacial clefts, *Cleft Palate-Cran.fac. J.*, 34, 11, 1997.
61. Ericson, A. and Källén, B., Hospitalization for miscarriage and delivery outcome among Swedish nurses working in operating rooms 1973-1978, *Anesth. Analg.*, 64, 981, 1985.
62. Ericson, A. and Källén, B., Pregnancy outcome in women working as dentists, dental assistants or dental technicians, *Int. Arch. Occup. Environ. Health*, 61, 329, 1989.
63. Källén, B. and Landgren, O., Delivery outcome in pregnancies when either parent worked in chemical industry. A study with central registries, *J. Occup. Med.*, 36, 563, 1994.
64. Ahlborg, G., Jr., Bjerkedal, T., and Egenaes, J., Delivery outcome among women employed in the plastic industry in Sweden and Norway, *Amer. J. Industr. Med.*, 12, 507, 1987.
65. Källén, K., Role of maternal smoking and maternal reproductive history in the etiology of hypospadias in the offspring, *Teratology*, 66, 185, 2002.
66. Åberg, A., Westbom, L., and Källén, B. Congenital malformations among infants whose mothers had gestational diabetes or preexisting diabetes, *Early Human Dev.*, 61, 85, 2001.
67. Friis, M.L., Broeng-Nielsen, B., Sindrup, E.H., Lund, M., Fogh-Andersen, P., and Hauge, M., Increased prevalence of cleft lip and/or cleft palate among epileptic patients, in *Epilepsy, Pregnancy, and the Child*, Janz, D., Dam, M., Richens, A., Bossi, L., Helge, H., and Schmidt, D., Eds., Raven Press, New York, 1982, p. 297.
68. Jones, W.H., Myelomeningocele and social class, *Dev. Med. Child. Neurol.*, 22, 244, 1980.
69. Källén, B., Malmquist, G., and Moritz, U., Delivery outcome among physio-therapists in Sweden: Is non-ionizing radiation a fetal hazard? *Arch. Environ. Health*, 37, 81, 1982.
70. Källén, B., Population surveillance of congenital malformations. Possibilities and limitations. Review article, *Acta Paediatr. Scand.*, 78, 657, 1989.
71. Lammer, E.J., Sever, L.E., and Oakley, G.P., Jr., Teratogen update: valproic acid, *Teratology*, 36, 465, 1987.
72. Robert, E., Handling surveillance types of data on birth defects and exposures during pregnancy. Reproductive toxicology review, *Reprod. Toxicol.*, 6, 205, 1992.
73. Arpino, C., Brescianini, S., Robert, E., Castilla, E.E., Cocchi, G., Cornel, M.C., de Vigan, C., Lancaster, P.A., Merlob, P., Sumiyoshi, Y., Zampino, G., Renzi, C., Rosano, A., and Mastroiacovo, P., Teratogenic effect of antiepileptic drugs; use of an international database on malformations and drug exposure, *Epilepsia*, 41, 1436, 2000.

CHAPTER 21

Developmental and Reproductive Toxicity Risk Assessment for Environmental Agents*

Carole A. Kimmel, Gary L. Kimmel, and Susan Y. Euling

CONTENTS

I. Introduction 841
    A. Definitions..842
    B. Historical Perspective on Risk Assessment Approaches: 1983 to 2002 842
II. Framework for Developmental and Reproductive Toxicity Risk Assessment..................843
    A. Planning and Problem Formulation 844
    B. The Risk Assessment Process 845
        1. Hazard Characterization 845
        2. Dose-Response Assessment..856
        3. Exposure Assessment 859
        4. Risk Characterization 860
III. Emerging Issues for Risk Assessment...864
    A. Research Efforts to Develop Human Data on Developmental and Reproductive Effects of Environmental Exposures — Contributions to Risk Assessment............864
    B. Using Mode of Action and Pharmacokinetic Data in Risk Assessment..................865
    C. Harmonization of Approaches for Human Health Risk Assessment 866
    D. Use of Genomics, Proteomics, and Other "Futures Technologies" 866
    E. Implications of Benefits Analysis for Risk Assessment 867
References 867

I. INTRODUCTION

The approach to risk assessment for environmental agents has progressed significantly over the last two decades since the publication of the National Academy of Sciences landmark report: *Risk Assessment in the Federal Government: Managing the Process*.[1] Although risk assessment approaches can be applied to all classes of chemical or physical agents, this chapter only deals with the practice of risk assessment for environmental agents. Likewise, rather than deal with risk

---

* The views expressed in this chapter are those of the authors and do not necessarily reflect the views or policies of the U.S. Environmental Protection Agency. Mention of trade names of commercial products does not constitute endorsement or recommendation for use.

assessment for all types of health effects, this chapter focuses on the process as it relates to reproductive and developmental toxicity, recognizing that other types of health effects, including mutagenicity, neurotoxicity, carcinogenicity, and immunotoxicity, may also affect reproductive and developmental toxicity.

This chapter updates and expands on much of the material covered in the first edition of this book[2] but does not go into detail on many of the aspects of hazard characterization covered there. Instead, issues of interest and importance in the rapidly changing area of environmental risk assessment are discussed here.

A. Definitions

The following definitions are used in this chapter. Other relevant definitions can be found in the appropriate risk assessment guidelines[3-5]

- Developmental toxicology — The study of adverse effects on the developing organism that may result from exposure prior to conception (either parent), during prenatal development, or postnatally to the time of sexual maturation. Adverse developmental effects may be detected at any point in the lifespan of the organism. The major manifestations of developmental toxicity include: (1) death of the developing organism, (2) structural abnormality, (3) altered growth, and (4) functional deficiency.
- Reproductive toxicology — The study of the occurrence of biologically adverse effects on the reproductive systems of females or males that may result from exposure to environmental agents. The toxicity may be expressed as alterations to the female or male reproductive organs, to the related endocrine system, or to pregnancy outcomes.

B. Historical Perspective on Risk Assessment Approaches: 1983 to 2002

Developmental and reproductive toxicity testing procedures have been available for the past 40 to 50 years.[6] However, it was not until the development of a risk assessment paradigm by the National Research Council (NRC)[1] that approaches for assessing the test data with a focus on potential human health risks began to be formalized.

Using the NRC paradigm, the U.S. Environmental Protection Agency (EPA) was instrumental in developing risk assessment guidance in a number of health- and exposure-related areas. For developmental toxicity risk assessment, the *Guidelines for the Health Assessment of Suspect Developmental Toxicants* were published in 1986,[7] and later updated as *Guidelines for Developmental Toxicity Risk Assessment*.[3] Following this, the *Guidelines for Reproductive Toxicity Risk Assessment* were published in 1996.[4] In addition, *Guidelines for Neurotoxicity Risk Assessment*, published in 1998,[5] addressed developmental effects on the nervous system.

In the process of developing guidance for risk assessment, issues important to the consistent assessment of developmental and reproductive toxicity data were discussed, and some were specifically incorporated into the default assumptions. For example, the guidance made it clear that all manifestations of developmental toxicity are of concern in assessing potential risk. This broadened the traditional focus on fetal death and malformations (teratology) to other, sometimes more sensitive, indicators of growth and functional alteration. Likewise, the reproductive endpoints included not only effects on primary and secondary sexual organs and their function but also cycle normality, sexual behavior, puberty, and reproductive senescence.

In 1993, the NRC published an influential report *Pesticides in the Diets of Infants and Children*.[8] While the focus of this report was specifically on childhood pesticide exposure, the effect of the report in the general area of childhood risk assessment was dramatic. In 1996, the Food Quality Protection Act (FQPA) was passed and included a number of the recommendations from the NRC report. FQPA amended the Federal Insecticide, Fungicide, and Rodenticide Act (FIFRA) and the

Federal Food, Drug, and Cosmetic Act (FFDCA) and included a number of special provisions for infants and children. These provisions included the requirement for determining whether pesticide tolerances are safe for children, the use of an additional safety factor of up to 10-fold (the FQPA 10X factor) to account for data uncertainty with regard to toxicity and exposure to children, and the collection of improved pesticide exposure data for infants and children. In addition, amendments to the Safe Drinking Water Act (SDWA) in 1996 mandated research for new water standards on the health effects of subpopulations at greater risk and thus had an effect on children's health risk assessment. The FQPA and the SDWA amendment requirements are the drivers of a number of current developmental and reproductive toxicology risk assessment issues and activities. These include the development of a screening and testing program for potential endocrine disrupting chemicals (EDCs) and the development of approaches for cumulative risk assessment for exposure to multiple chemicals having a similar mode of action.

As a result of these and other events, there has been increasing attention paid to the potential for environmental exposures to have an effect on development and reproductive health. Schmidt[9] recently reported on the "growth spurt" in children's health executive and legislative initiatives. These include Executive Order (EO) 13045 (issued in 1997, amended in 2001 by EO 13229), which mandated the consideration of children in the development of health policy and created the Task Force on Environmental Health Risks and Safety Risks to Children. Also of importance is the Children's Health Act of 2000 (PL 106-310). Among other things, this act established the planning for the National Children's Study, a large prospective longitudinal cohort study of environmental factors on children's health. These and several other efforts will be discussed in greater detail below. The efforts surrounding endocrine disrupting chemicals (EDCs) have focused a great deal of attention on issues of developmental and reproductive effects related to such environmental agents.[10-13]

Finally, the field of risk assessment is moving toward harmonization of approaches for health risk assessment. This trend results in large part from the NRC report *Science and Judgment in Risk Assessment*,[14] which pointed out the importance of a risk assessment approach that is less fragmented, more consistent in application of similar concepts, and more holistic than the current endpoint-specific guidelines. Given that different approaches are used for various types of effects (cancer versus other health endpoints, inhalation versus oral exposures, children versus adults, single chemicals versus mixtures), a consistent set of principles and guidelines is needed for drawing inferences from scientific information. Coupled with the explosion of new technology and data collection at the cellular and molecular level, the fields of developmental and reproductive toxicology will be challenged to evolve further. For example, a recent EPA review of the reference dose (RfD) and reference concentration (RfC) processes[15] promoted the development of a broader view of risk assessment. It contained recommendations for derivation of acute, short-term, and longer-term reference values, as well as the chronic RfD and RfC. The review also considered potentially susceptible populations (including life stages), mode of action, pharmacokinetics, aggregate exposure and cumulative risk issues, and harmonization of approaches for all health effects.

II. FRAMEWORK FOR DEVELOPMENTAL AND REPRODUCTIVE TOXICITY RISK ASSESSMENT

Hazard identification and characterization of reproductive and developmental effects must be viewed as part of the larger framework of information considered for the regulatory decision-making process. Thus, a framework for risk assessment and risk management includes: (1) the planning and problem formulation phase, (2) the risk assessment itself, which comprises hazard characterization, dose-response assessment, exposure assessment, and risk characterization, and (3) risk management and risk communication (Figure 21.1). The first two components of this framework are discussed in the following sections.

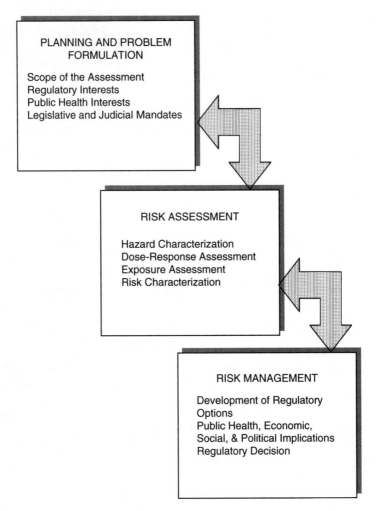

Figure 21.1 Framework for risk assessment and risk management.

A. Planning and Problem Formulation

At the same time that risk assessment is moving toward the goal of harmonization and consistency, it is also true that the application of the risk assessment paradigm cannot be "one size fits all." Assessments vary in scope, depending on the regulatory and public health interests, and have ranged from covering all possible exposures and effects to focusing on a specific exposure and health endpoint. Moreover, legislative and judicial mandates often direct the type of assessment that may be appropriate. The planning and problem formulation process allows the risk assessor to understand the needs of the risk manager, the interests of the stakeholders, and the extent and limitations of the available database and analytical tools. Most important, the process should result in clearly defined questions and goals for the assessment.

Specific to developmental and reproductive toxicity risk assessments, the planning and problem formulation process identifies which exposure scenarios are being considered. As defined, developmental and reproductive toxicity may include exposure of almost any segment of the population (except perhaps nonreproducing adults). But identifying the exposure scenario(s) to be considered can help identify specific populations of interest (e.g., school children, pregnant women, migrant workers), as well as exposure characteristics with respect to medium (e.g., air, water, soil), route, duration, frequency, and life stage.

The planning and problem formulation process should also help define what information is available and applicable to the risk assessment process, as well as appropriate analytical methods. Over the years, the type of toxicity test data available, especially for pesticides and some high profile agents, has not only increased in size but also in depth. For example, developmental and reproductive toxicity testing now includes such measurements as sperm motility, estrous cycling, and endocrine function, and may sometimes include measurements of behavior and specific organ and system development. In addition, technological advances permit measurements at the cellular and molecular levels, and this has prompted discussion on how to incorporate these data into risk assessment. Coupled with improved methods of statistical analysis and mathematical model development, it is important to consider what will be included in the risk assessment and to discuss how it will be analyzed in the initial planning and problem formulation process.

B. The Risk Assessment Process

The process of risk assessment for reproductive and developmental toxicity is based on general and specific scientific principles but also includes a great deal of scientific judgment. As previously mentioned, risk assessment consists of four major components, hazard characterization, dose-response assessment, exposure assessment, and risk characterization, each of which is discussed briefly in the following sections.

1. Hazard Characterization

Hazard characterization involves determining whether and under what conditions exposure to an environmental agent causes adverse effects. This information often comes from animal data, although infrequently human data are available. *Guidelines for Developmental Toxicity Risk Assessment*,[3] *Guidelines for Reproductive Toxicity Risk Assessment*,[4] *Guidelines for Neurotoxicity Risk Assessment*,[5] and *Guidelines for Carcinogen Risk Assessment*[16] all address some aspect of reproductive and developmental toxicity risk assessment. Endocrine disrupting chemicals (EDCs) comprise a subclass of agents that alter the normal functioning of the endocrine system. The mode of action for endocrine disruption may be receptor dependent, e.g., through interaction with a particular receptor, or receptor independent, e.g., through alteration in the synthesis or metabolism of a hormone. Reproductive and/or developmental endpoints are almost always affected. In particular, the development of the reproductive system is often affected by early developmental EDC exposure. *Guidelines for Reproductive Toxicity Risk Assessment*[4] addresses many of the types of endpoints that may be affected by EDCs. The general evaluation and interpretation of data are discussed here.

a. Default Assumptions

The process of interpreting toxicity data requires a great deal of scientific judgment because a complete database to conduct a risk assessment is rarely if ever available. Assumptions must be made because of uncertainties due to limited or nonexistent information on toxicokinetics, mechanism of action, low-dose response relationships, and human exposure patterns. Several of these assumptions are basic to the extrapolation of toxicity data from animals to humans, while others are specific to reproductive and developmental toxicity. The general assumptions (in the following list), which are used in the absence of data to the contrary, are based on the assumptions discussed in the developmental toxicity risk assessment guidelines[3] as expanded upon in the reproductive toxicity guidelines.[4]

- An agent that produces an adverse reproductive and/or developmental effect in experimental animal studies is assumed to pose a potential reproductive and/or developmental threat to humans following sufficient exposure at the right stages of reproduction and development.
    - This assumption is based on the evaluation of data for agents known to cause human reproductive and/or developmental toxicity; these data indicate that, in general, adverse reproductive and/or developmental effects seen in experimental animals are also seen in humans.
- Effects of xenobiotics on male and female reproductive processes and/or developmental outcomes are assumed generally to be predictive of similar effects across species, although outcomes seen in experimental animal species may not necessarily be the same as those seen in humans.
    - For example, the differences between species in developmental and reproductive outcomes may occur because of species-specific differences in timing of exposure relative to critical periods of development or reproductive system function, pharmacokinetics (including metabolism), developmental patterns, placentation, or modes of action. Because of this, all of the four major manifestations of developmental toxicity (death, structural abnormalities, growth alterations, and functional deficits) are considered to be of concern for hazard characterization. This is in contrast to the approach in the 1960s and 1970s, when there was a tendency to consider only malformations and death as endpoints of concern. More recently, a biologically significant increase in any of the four manifestations of altered development is considered indicative of an agent's potential for disrupting development and producing a developmental hazard.
- When sufficient data (e.g., pharmacokinetic, mode of action) are available to allow a decision, the most appropriate species for estimating risk to humans should be used for hazard characterization. In the absence of such data, it is assumed that the most sensitive species should be used.
    - This is based on observations that humans are as at least as sensitive or more so than the most sensitive animal species tested for the majority of agents known to cause human reproductive and/or developmental toxicity.
- In the absence of specific information to the contrary, it is assumed that a chemical that affects reproductive function or development of offspring as a result of exposure of one sex may also adversely affect reproductive function or offspring development as a result of exposure of the other sex.
    - This assumption is based on three considerations: (1) for most agents, the nature of the testing and the data available are limited, reducing confidence that the potential for toxicity to both sexes and their offspring has been adequately examined, (2) exposure of either males or females has been shown to result in developmental toxicity, and (3) many of the mechanisms controlling important aspects of reproductive system function are similar in males and females, and therefore they could be susceptible to the same agents.
- When detailed information on mode of action and pharmacokinetics is available, it should be used to predict the shape of the dose-response curve at low doses. In the absence of such information, a threshold is generally assumed for the dose-response curve for agents that produce reproductive and/or developmental toxicity.
    - This assumption is based on known homeostatic, compensatory, or adaptive mechanisms that must be overcome before a toxic endpoint is manifested and on the rationale that cells and organs of the reproductive system and the developing organism are known to have some capacity to repair damage. In addition, because of the multipotency of cells at certain stages of development, multiple insults at the molecular or cellular level may be required to produce an effect on the whole organism. It should be recognized that background levels of toxic agents, the lifestage at exposure, preexisting disease conditions, and other factors may increase the sensitivity of some individuals in the population. Thus, exposure to a toxic agent may result in an increased risk of adverse effects for some, but not necessarily all, individuals within the population. Although a threshold may exist, it is usually not feasible to distinguish empirically between a true threshold and a nonlinear low-dose response relationship.

b. *Types of Data Used for Hazard Characterization*

Data used for hazard characterization include both laboratory animal and human data. Adverse reproductive endpoints include altered reproductive organs, endocrine system function, or pregnancy

Table 21.1 Common types of human studies

Epidemiologic

Cohort	Groups defined by exposure
	Follow groups over time
Case control	Groups defined by selected health effects
	Exposure history determined
Ecologic	Study groups defined by general characteristic
	Exposure history assumed
Cross-sectional	Exposure and outcome examined at same time

Other Human Data

Clusters	Concern generated because of perceived excess of health effect
Case reports/series	Typically based on clinical observation
	May be endpoint or exposure based

outcomes, such as effects on ovulation and menstruation, sexual behavior, fertility, onset of puberty, gestation, heritable reproductive effects, and premature reproductive senescence.[4] Adverse developmental endpoints include structural malformations, altered growth, premature mortality, and functional deficits, such as mental retardation, sensory loss, or immunological effects.[3]

1) Human Studies — Adequate human data are the most appropriate sources of information for determining an agent's potential for causing human reproductive or developmental toxicity. However, ethical and practical reasons often limit collection of human exposure and effects data, especially on reproductive and developmental toxicity because of potential harm to the developing fetus or child or because of ethical issues or social taboos on collecting such types of data. Nevertheless, guidance for data evaluation has been developed for cases where human data are available. The human studies category includes a variety of data sources, ranging from epidemiologic studies to individual case reports (Table 21.1). The general design characteristics of human studies are covered in detail in Chapter 20 and were also reviewed by Adams and Selevan.[17]

Well-designed epidemiologic studies provide the most relevant information for assessing human risk. There are several epidemiologic study designs, each having certain advantages and limitations. Cohort studies define study populations by exposure and then examine outcomes associated with differing exposures. This study design can examine multiple health effects and is most powerful when more common developmental effects are the endpoints. Prospective cohort studies, which examine exposures before the outcomes are known, avoid many of the problems with recall bias. In the case referent design, populations are defined by a selected health effect and exposure history is determined retrospectively. This design can be used to examine multiple exposures and is useful when rare outcomes are being considered. Both the cohort and the case referent study design can have significant power or ability to detect a true effect, depending on population size, outcome frequency, and the level of excess risk to be identified. But both can be fairly expensive and resource intensive.

The ecologic study is an epidemiologic study that defines the study population by some general characteristic (e.g., zip code, county of residence), and the exposure history is assumed, based on that characteristic. Detailed information on individuals is most often not collected. This makes the ecologic study a less expensive method of data collection, and it can be very useful in developing hypotheses for further testing. However, the lack of individual exposure and outcome data limit the usefulness of this design in establishing clear associations.

The cross-sectional study combines some of the features of other study designs and examines both exposure and outcome at the same time point. Exposure data are generally more reliable than those obtained from an ecologic study, since individual exposures are assessed through interviews or monitoring. Exposures are assumed to be similar to historical exposure levels that may have occurred at the critical period of development when the effect was induced. Since outcome is measured at the same time, the choice of an appropriate outcome can be influenced by such things as time-to-appearance and its transience or persistence. The risk assessment must consider these factors in drawing associations between exposure and effect from this type of study.

Other data that can provide indications of a possible effect are reports of clusters and case reports or series. Reports of clusters indicate a concentration of an outcome in a particular geographical area or population. Case reports or series generally come from reports of individuals or their physicians and are recognized as a potential pattern only after the data are brought together. Rare events are much more likely to be reported as an outcome than are more common events. For example, the thalidomide tragedy was recognized by physicians in large part because of the rare malformation (phocomelia) that the drug produced. Had exposure been associated with a more common effect, recognition of a possible association between the effect and the exposure would have been more difficult, if not impossible, from case reports. Tabacova et al.[18] have developed an approach for combining multiple case reports as a case series for a particular exposure and pattern of associated outcomes for use in generating hypotheses for further study. Case reports were defined by a particular exposure (e.g., enalapril, a hypotensive agent) during pregnancy. They were then evaluated using several criteria to determine the strength of the association between an exposure (including the timing of exposure) and various pregnancy outcomes. Using this approach, the authors were able to confirm previous reports of enalapril-related effects following second and third trimester pregnancy exposures.

The differences between human and laboratory animal developmental toxicity studies are considerable and must be recognized by the risk assessor. Human studies rarely, if ever, provide the controlled study population and exposures characteristic of the laboratory study. Often, the size of the human study population is small, limiting the statistical power of the study. The potential for bias in data collection is more likely a concern in human studies because the study population is almost always heterogeneous and exposures are seldom uniform and often are not specific enough to provide actual dose information. Initial and continued participation in a study can be influenced by an individual's perceived exposure, concern over a health effect, verbal skills, education, etc. Health records and personal recall can differ considerably and can be influenced by a physician's awareness or a patient's memory.

In spite of these and other issues, it is important to reiterate that sufficient human data are the most appropriate source of information for determining an agent's potential for causing human developmental toxicity. Moreover, there have been advances in the collection of both exposure and health data and in data analysis that have strengthened the utility of human data in the overall toxicological profile. In addition, integration of human and animal data often lends strength to the database that neither alone can provide. Studies to evaluate the comparability of animal and human data have provided a good deal of information on the bases for similarities and differences in species-specific responses and have pointed the way toward additional animal or human studies needed to further define or clarify exposure-related effects.[19–21]

2) Laboratory Animal Studies—Standard Testing Protocols — Three testing protocols are commonly used in the regulatory setting for assessing the reproductive and developmental toxicity of environmental agents in laboratory animals: (1) the prenatal developmental toxicity study, (2) the reproduction and fertility effects study, and (3) the developmental neurotoxicity (DNT) study[22] (Table 21.2). The first two studies are routinely required for pesticides and may be required for industrial chemicals if there is cause for concern. The DNT study has been required on a case-by-case basis and was included as part of a data call-in for organophosphate insecticides.[23] The prenatal developmental toxicity study involves exposure of pregnant females during the period of organogenesis through the end of pregnancy (typically gestation days [GD] 6 to 20 in the rat) and evaluation of fetuses at term. This study focuses on general reproductive indices of the maternal animals, survival of the fetuses to term, fetal weight at term, and a detailed structural evaluation of the fetuses. The reproduction and fertility study involves dosing the parental animals from 10 weeks

Table 21.2 OPPTS harmonized guideline protocols for reproductive and developmental toxicity testing

Protocol	Guideline	Brief Description
Prenatal developmental toxicity study	870.3700[a]	Exposure from implantation to termination just prior to parturition — gestation day (GD) 18 in mice, 20 in rats, and 29–30 in rabbits. Detailed evaluation of offspring morphology.
Reproduction and fertility effects study	870.3800[a]	Exposure of males and females for 10 weeks prior to mating and during mating, exposure of females throughout gestation and lactation, and exposure of offspring after weaning — typically conducted in rats. Estrous cyclicity in parental (P) and F_1 females; semen quality in P and F_1 males; survival, growth, and reproductive development of offspring (age at vaginal opening and preputial separation, anogenital distance in F_2 pups if an effect is seen on F_1 sex ratio or sexual maturation); organ weights (reproductive, brain, spleen, and thymus, other target organs); histology, including primordial follicle counts in F_1 female ovaries.
Developmental neurotoxicity study	870.6300	Exposure from implantation to postnatal day (PND) 10 or 21. Offspring evaluated pre- and postweaning for general growth and development, timing of puberty, clinical signs (modified functional observation battery), motor activity, auditory startle habituation, cognitive function, and neurohistology.

[a] These complete guidelines are presented in Chapter 19, Appendix II.

before mating and during pregnancy and lactation, with continued exposure of selected F_1 offspring that are mated and allowed to deliver and maintain their (F_2) pups to weaning. This study focuses on the evaluation of reproductive development (both structure and function), as well as on adult reproductive structure and function and the developmental effects of continuous exposure. The developmental neurotoxicity study protocol involves exposure during organogenesis, through the end of gestation, and into the postnatal period, either to midlactation or to weaning. The study focuses on the evaluation of the developing nervous system and includes measures of physical and reflex development, motor activity, auditory startle habituation, learning and memory, and neuropathology in young animals and adults. This is the only study that evaluates aspects of latency to response and reversibility of effects following termination of exposure.

These guideline animal testing protocols and their use in regulatory testing are described in detail by Kimmel and Makris[24] and in Chapters 7 to 9 and 19 of the current volume. From such studies, a number of endpoints of general adult toxicity and reproductive and developmental toxicity are used to determine the types of effects and the effective dose levels for a given agent. These data are used in hazard characterization and dose-response assessment.

3) Other Types of Data

a) **Data on Functional Endpoints** — Functional developmental and reproductive toxicity refers to the effects of agents on the functioning of essentially any organ system. The two-generation reproduction study discussed above provides information on reproductive function in the form of data on fertility, semen and sperm quality, estrous cycling, and ovarian function (follicle numbers, number of ova produced and fertilized). However, it should be recognized that there are limitations in assessing reproductive function by means of the two-generation protocol. For example, exposure is continuous, and effects induced during development may not be separable from those produced at later ages. Since both males and females are exposed, it is not clear which sex is primarily affected, although this can be ascertained by evaluating endpoints in individual sexes or by cross-mating control and treated animals. Other aspects of the interpretation of reproductive function induced in males and females during development are discussed in Chapter 9 and in the reproductive toxicity risk assessment guidelines.[4] More detailed evaluations of male and female reproductive function[25] were added to the guideline protocol in 1994[26] and were republished as harmonized guidelines for pesticides and toxic substances in 1998.[22]

Standard testing approaches to evaluate a wide range of functional effects have not been developed, with the exception of the developmental neurotoxicity study discussed in the preceding sections. Moreover, this study has not been required on a routine basis; yet developmental neurotoxicity is one of the hallmarks of several environmental agents in humans, e.g., lead, methylmercury, and polychlorinated biphenyls (PCBs). Recent data on lead and neurological effects suggest that the developing nervous system may be sensitive at blood lead levels as low as 5 µg/dl.[27]

Strong support for using animal models to evaluate the potential for developmental neurotoxicity in humans came from a workshop on the qualitative and quantitative comparability of human and animal developmental neurotoxicity.[28] Newer more sensitive measures that can be used in both animals and humans are becoming available and should be considered for incorporation into testing protocols (see Sharbaugh et al.[29]).

Effects of chemicals on the developing immune system have been reviewed by several authors.[30-34] Although there are no testing guidelines for the evaluation of such effects, a few endpoints (spleen and thymus weights) were added to the fertility and reproduction study guideline[22] to screen for possible effects on the immune system and to trigger further study. Development of a testing guideline is needed and has been discussed in recent EPA reports.[15,35]

Lau and Kavlock[36] summarized the literature on functional developmental toxicity of the heart, lungs, and kidneys. Their review revealed a plethora of information on a wide variety of chemicals and endogenous agents (hormones and neurotransmitters), but they found a lack of uniform treatment paradigms or observations among studies, although protocols have been developed for evaluating renal function. Interpretation of data on these organ systems is hindered by the limited understanding of mechanisms controlling normal development in these organs and by the lack of toxicological evaluation of these organs in an integrated manner, despite their interdependence during development. The authors were able to summarize some common features emerging from studies on functional developmental toxicity of these three organs that are probably applicable to the functional development of other organ systems. These include

1. Assessment of multiple endpoints is needed to adequately characterize the spectrum of toxicity.
2. Compensatory mechanisms may mask functional deficiencies, necessitating the use of tests that evaluate maximal responses of organ systems.
3. Because of physiological interactions among functional systems, examples of organ-specific toxicants are rare.
4. Functional systems display broad periods of developmental susceptibility to toxic insult, with accelerations or delays having potential long term consequences.
5. Within the limits of the database on functional developmental toxicity in humans, there is generally good concordance of effects across species.
6. Although toxicologic mechanisms are unknown, disruption of endogenous trophic factors and altered cellular and molecular interactions are likely common pathways for the induction of functional developmental toxicity (adapted from Lau and Kavlock[36]).

A number of publications have linked prenatal undernutrition and reduced birth weight in humans and animals with later development of heart disease,[37-40] diabetes and obesity,[41-47] and osteoporosis.[48] In addition, several studies have shown a positive association between birth weight and later development of breast cancer (for example, see Michels et al.,[49] Innes et al.,[50] and Mellemkjaer et al.[51]). Together, these studies strongly suggest that prenatal factors can play an important role in the function of adult systems and in the aging process and that environmental chemicals that affect prenatal nutrition and growth may also affect later organ function and occurrence of disease. Thus, developmental factors should be considered in the etiology of disease states that are not recognized until adulthood or old age.

b) **Data on Endocrine Disrupting Chemicals** — EDCs can be defined as chemicals that positively or negatively alter hormone activity and thereby may affect reproduction or development.

For example, in rodent studies, prenatal exposure to the antiandrogen pesticide vinclozolin can lead to effects on the development of the male reproductive system, including femalelike anogenital distance, nipple retention, cleft phallus with hypospadias, suprainguinal ectopic scrota and/or testes, a vaginal pouch, epididymal granulomas, and small to absent sex accessory glands.[52]

Given the potential for developmental and reproductive effects in humans and wildlife, EDCs have emerged as a major international environmental issue. Some human population studies of accidental exposures have found associations between high EDC exposure and developmental and/or reproductive outcomes. For example, in the Michigan exposures to polybrominated biphenyls (PBBs), girls who experienced a high (greater than 7 ppb) *in utero* exposure and were breastfed had a significantly earlier age of menarche and more advanced pubic hair development than either girls that were not breastfed or had a lower *in utero* exposure (less than 7 ppb).[53] Among women who were exposed before menarche to high dioxin levels from the Seveso explosion, a tenfold increase in serum dioxin concentration was associated with a lengthened menstrual cycle.[54]

Findings from laboratory animal as well as human studies suggest that the timing of puberty is a sensitive biomarker of effect. For example, in rodent studies, females showed delayed vaginal opening in the absence of body weight decrements after developmental lead exposures.[55–57] Two recent epidemiologic studies found an association between higher concurrent blood lead levels and a later timing of puberty in girls, as indicated by one or more delayed puberty measures.[58,59] Additionally, racial and ethnic background modified the association, e.g., the association was found to be statistically significant for Tanner stages in Mexican-American girls and for Tanner stages as well as onset of menarche in African-American girls; no measures were significant in Caucasian girls.[58] The associations observed between blood lead level and puberty timing were present after controlling for body size,[58] suggesting that lead affects puberty via other pathways, such as the hypothalamic-pituitary-gonadal axis.

How to define EDCs, what to call them (e.g., "hormonally active agent," "endocrine modulator"), whether environmental exposure levels are high enough to affect the health of humans in the general population, and whether there are "low dose" effects are all hotly debated questions. Multiple EDCs with the same mode of action are of concern because individual exposures to each EDC may be "low" but the total exposure to a mode-of-action class may be "high." A recent study found that multiple "no effect" level exposures to EDCs with an estrogenic mode of action were dose additive, leading to observable effects.[60] Furthermore, the fetus and child are likely to be more susceptible to EDC exposure than the adult because of critical windows of exposure in the endocrine, neural, and reproductive systems.[61]

The FQPA and the SDWA amendments passed by Congress required the EPA to assess the estrogenic potential of pesticides and water contaminants that may affect human health. As a result of this legislation, the EPA formed the Endocrine Disruptor Screening and Testing Committee (EDSTAC), a multistakeholder federal advisory committee, to develop a screening program. EDSTAC expanded the scope of their charge to include testing as a second tier, and androgen and thyroid activity effects in addition to estrogenic effects. The EDSTAC broadened the chemicals to be assessed to include industrial chemicals, pesticide "inert" ingredients, pharmaceutical chemicals, and environmental contaminants. A final consensus report of their recommendations was published in August 1998.[62]

A number of endpoints in the current EPA testing guidelines for reproduction and fertility effects[22] are sensitive to altered estrogen, androgen, or thyroid hormone activity. For example, the timing of preputial separation (PPS), a marker of male puberty, can be affected by altered androgen activity. A number of additional test endpoints sensitive to hormone activity were recommended by EDSTAC, and these include brain histology and weight, some motor activity and memory endpoints, TSH and T4 levels, and testis descent.[62] See Chapter 11 for detailed coverage of the EDSTAC test protocols.

One of the challenges of developing such a screening and testing program was to include assays that would cover the numerous modes of action for EDCs. Similar to carcinogens, the EDC classification focuses on mode of action data in determining whether a chemical is an EDC. The assays recommended by EDSTAC were designed to detect estrogenic, antiestrogenic, androgenic, antiandrogenic, thyroid, and antithyroid activity. Two tiers of testing batteries, one screening (T1S) and one testing (T2T), were recommended. The intention of T1S is to determine whether a chemical is capable of affecting the activity of estrogens, androgens, or thyroid hormones. T1S *in vitro* screening assays include an estrogen receptor binding or reporter gene assay, an androgen receptor binding or reporter gene assay, and a steroidogenesis assay that uses minced testis. T1S *in vivo* screening assays include a rodent 3-d uterotrophic assay, a rodent 20-d pubertal female assay with enhanced thyroid endpoints, a rodent 5- to 7-d Hershberger assay, a frog metamorphosis assay, and a fish reproductive screening assay. The analysis of the data from T1S will either conclude that the chemical be moved into T2T or that no further screening or testing is currently needed (placed "on hold").

The T2T battery was designed to determine whether a chemical may be an EDC and to quantify the effects on estrogen, androgen, and thyroid hormone activity. The tests include dosing during development and a larger dose range; most include testing in two generations. T2T *in vivo* testing assays include a two-generation mammalian reproductive toxicity study or a less comprehensive alternative mammalian reproductive toxicity test, an avian reproduction toxicity test, a fish life cycle toxicity test, an opossum shrimp (*Mysidacea*) or other invertebrate life cycle toxicity test, and an amphibian development and reproduction test.* Information from the T2T battery is of the type necessary for hazard assessments for human health, and since the outcome of tests is more definitive than that of screens, a negative outcome in T2T will supplant a positive finding from T1S.

Interpreting the data from the EDC screening and testing program and then developing risk assessment methods are the next challenges. While the screening and testing tiers were not designed expressly for risk assessment purposes, the data from the screens and tests will be available for use in risk assessment. Specific guidance for the risk assessment of EDCs has not been developed. The attention on EDCs has raised a number of important issues for developmental and reproductive toxicology risk assessment that will likely apply to other types of effects and chemical classes.

One of the most contentious issues has been the potential for effects at "low doses" or "environmental levels" (i.e., likely levels of human exposure). A National Toxicology Program (NTP) Workshop (Endocrine Disruptors Low Dose Peer Review) evaluated the existing evidence for low dose reproductive and developmental effects for a number of EDCs.[63] The peer review defined "low dose" as equal to human exposure levels, below those used in EPA's standard reproductive and developmental toxicity testing, or those below the current no effect level (NOEL). Based on these definitions, the panel concluded that there is credible evidence for low dose effects on certain endpoints for estradiol, diethylstilbestrol, nonylphenol, methoxychlor, and genistein.[63] Other EDC-related issues of interest are: (1) the effective addition and subtraction (depending on whether the agent acts as an agonist or antagonist) from endogenous steroid hormone levels,[64] (2) critical and sensitive windows of EDC exposure that correspond to developmental events regulated by steroid hormones,[65] and (3) cumulative effects of exposure to chemicals with the same or opposing modes of action.[60] There is also interest in understanding the significance of nonmonotonic dose response curves, hormesis, and U-, J-, or inverted U-shaped dose response curves,[66-71] which may reflect multiple modes of action for one chemical at different doses.[68-70]

4) Toxicokinetics and Toxicodynamics — Toxicokinetic data provide information on the absorption, metabolism, distribution (including placental transfer), and/or excretion (including via breast milk) of an agent. When available, toxicokinetic data can be extremely useful in determining target dosimetry and critical levels or duration of exposure at the target site (e.g., peak concentration, area under the curve) and can be related to expression of the toxic effect (toxicodynamics). In addition, toxicokinetic information provides support for defining similarities and differences among

---

* Further information is available at http://www.epa.gov/scipoly/oscpendo/index.htm.

routes of exposure and among species, or in accurately extrapolating information from one species to another.

At a minimum, absorption data may assist in the interpretation of developmental and reproductive toxicity studies in animal models and will aid in the extrapolation of data from animal models to humans if human absorption data are available. Absorption data are essential for studies in which dermal exposure is used, whether or not any developmental or reproductive effects are produced.[71] A study showing no adverse effects without evidence of dermal absorption is considered insufficient for risk assessment and is potentially misleading, especially if interpreted as a "negative" study.

Toxicokinetic studies related to developmental toxicity are most useful if conducted in animals at the stage when developmental insults occur, rather than at the end of gestation. In this way, specific toxicokinetic patterns can be correlated with developmental outcomes to define the parameters that contribute most to the outcomes measured.[72–74] Certain drugs and chemicals have been classified according to the correlation between the malformations produced and either the integrated exposure area under the curve (AUC) or the peak concentration (C_{max}).[75] A study by Terry et al.,[76] however, showed that for the same agent (2-methoxyacetic acid, a metabolite of 2-methoxyethanol [2-ME]), the toxicokinetic pattern that correlated best with outcome differed depending on the stage of development at which treatment occurred. In this case, C_{max} correlated better with exencephaly produced by a dose of the parent compound, 2-ME, on gestation day (GD) 8, while AUC correlated better with limb defects produced on GD 11.[73] Thus, developmental stage–specificity may be as important as or more important than the toxicokinetic characteristics of a particular agent. If true, this becomes especially important in extrapolation of data to humans, where the rate of development is much slower than that of most laboratory animal species used.

The ontogeny of metabolic capabilities in the developing fetus and child is important for understanding potential susceptibility of children to chemical exposures.[77] Several recent studies have contributed significantly to our understanding of these processes. For example, Cresteil[78] published data on the activity of several P450 enzymes in human liver from birth to 10 years of age. Koukouritaki et al.[79] showed the developmental ontogeny of the flavin-containing monooxygenase (FMO) 1 and 3 enzymes in human liver samples ranging from 8 weeks gestation to 18 years of age.

Several physiologically based pharmacokinetic (PBPK) models have been developed that take into account the changes occurring during pregnancy in rodents and nonhuman primates[80–82] and in humans.[83–85] Other efforts have been made to model lactational transfer[86,87] and pharmacokinetics in young children.[88,89] Ginsberg et al.[90] evaluated the published literature for data on child and adult pharmacokinetic differences for pharmaceutical agents, and Hattis et al.[91] used the database to analyze the variability in children's drug elimination half-lives for the purpose of building a PBPK model. These approaches are likely to improve the ability of researchers to extrapolate information across species because the approaches can accommodate the anatomic and physiological changes that occur as pregnancy advances in different species.[92] An in-depth review of information available on perinatal pharmacokinetics was developed recently.[93]

Physiologically based models for reproductive systems have also been developed. For example, Keys et al.[94] developed a PBPK model for di(2-ethylhexyl) phthalate (DEHP) and mono(2-ethylhexyl) phthalate (MEHP) to quantify the testis dose of MEHP. Plowchalk and Teeguarden[95] developed a PBPK model for estradiol in rats and humans to explore the importance of various physiological parameters on circulating levels of estrogen in reproductive tissues and responses to endocrine-active chemicals. Several researchers have studied 4-vinylcyclohexene to determine the basis for species differences in ovarian toxicity in rats and mice.[96,97] For a more extensive discussion of pharmacokinetics, see Chapter 13.

Understanding toxicodynamics (mechanism or mode of action) is an important aspect of hazard characterization and may influence significantly many aspects of the approach taken to risk assessment. Such data may come from a variety of short-term *in vitro* or *in vivo* studies or comparative studies in different species. Not a great deal is known as yet about the mechanisms of action of reproductive and developmental toxicants, although there has been a strong research effort investigating EDCs, especially those that act by binding and activating the estrogen or androgen receptor. Other endocrine modes of action, such as effects on steroidogenesis and thyroid hormone antagonism, have been identified for some chemicals, and more mode of action data are expected from the Endocrine Disruptor Screening and Testing Program.[98] The National Research Council Report,[99] *Scientific Frontiers in Developmental Toxicology and Risk Assessment*, suggested areas of research for gaining a better mechanistic understanding of developmental toxicity.

As toxicokinetic and mechanistic information becomes available for reproductive and developmental toxicity risk assessment, these types of information can be incorporated into biologically based dose-response models that can provide a more accurate assessment of risk. For example, information from studies on enzyme function, receptor-ligand interaction, metabolic and synthetic pathways, and cell growth, function, and control were incorporated into mathematical models to define the overall dose-response relationship for 5-fluorouracil.[100–102] Such information can provide a conceptual framework for defining the underlying events associated with cellular control and the relationship between exposure parameters (e.g., dose, frequency, duration, and timing) and the ultimate health outcome.[103–106]

c. Characterization of the Database

Once the toxicity data in both humans and animals have been thoroughly reviewed, the database is characterized as being sufficient or insufficient to proceed further in the risk assessment process. This characterization of hazard considers the context of exposure, e.g., dose, route, duration, and timing, relative to the life stage(s) during which exposure occurred. Characterizing the hazard potential in this manner allows for better determination of the strengths and weaknesses, as well as the uncertainties, of the data associated with a particular exposure. This does not, however, address the nature and magnitude of the human health risks, which are determined after the exposure assessment is completed and the final characterization of risk is conducted. A great deal of scientific judgment and expertise in the area of developmental and reproductive toxicity, including laboratory studies, epidemiology, and statistics, may be necessary to characterize the hazard for a given database.

The EPA's *Guidelines for Developmental Toxicity Risk Assessment*[3] and *Guidelines for Reproductive Toxicity Risk Assessment*[4] present similar categorization schemes for characterizing the database for developmental and reproductive toxicity. The scheme from the latter document is presented in Table 21.3. Two broad categories are included: sufficient evidence and insufficient evidence. Within the sufficient evidence category, there is a further breakdown, depending on whether human data are available and the strength of those data. Human data are considered primary in determining a potential hazard but are often limited or lacking, so that few agents can be classified as having sufficient human evidence. Often, however, limited human data and sufficient animal data can be used together in characterizing the hazard of a particular agent.

The text in the categories in Table 21.3 defines the minimum evidence necessary to determine sufficiency, and the strength of evidence for a database increases with replication of the findings and with additional human studies or additional animal species tested. All data, whether indicative of a hazard potential or not, are to be considered in a weight-of-evidence approach for making judgments about the strength of evidence available to support a complete risk assessment for reproductive and developmental toxicity. Thus, within each category, minimum criteria are detailed for determining whether data are sufficient to judge whether an agent does or does not have the potential to cause reproductive or developmental toxicity.

Information on pharmacokinetics and/or pharmacodynamics (mechanisms of action), especially at the developmental stage or in the reproductive tissue of concern, may reduce uncertainties in determining relevance to humans. More evidence (e.g., more studies, species, or endpoints) is

Table 21.3 Categorization of the health-related database

Sufficient Evidence

The Sufficient Evidence category includes data that collectively provide enough information to judge whether a reproductive hazard exists within the context of effect as well as dose, duration, timing, and route of exposure. This category may include both human and experimental animal evidence.

Sufficient Human Evidence

This category includes agents for which there is convincing evidence from epidemiologic studies (e.g., case control and cohort) to judge whether exposure is causally related to reproductive toxicity. A case series in conjunction with other supporting evidence also may be judged as Sufficient Evidence. An evaluation of epidemiologic and clinical case studies should discuss whether the observed effects can be considered biologically plausible in relation to chemical exposure.

Sufficient Experimental Animal Evidence — Limited Human Data

This category includes agents for which there is sufficient evidence from experimental animal studies and/or limited human data to judge if a potential reproductive hazard exists. Generally, agents that have been tested according to EPA's two-generation reproductive effects test guidelines (but not limited to such designs) would be included in this category. The minimum evidence necessary to determine if a potential hazard exists would be data demonstrating an adverse reproductive effect in a single appropriate, well-executed study in a single test species. The minimum evidence needed to determine that a potential hazard does not exist would include data on an adequate array of endpoints from more than one study with two species that showed no adverse reproductive effects at doses that were minimally toxic in terms of inducing an adverse effect. Information on pharmacokinetics, mechanisms, or known properties of the chemical class may also strengthen the evidence.

Insufficient Evidence

This category includes agents for which there is less than the minimum sufficient evidence necessary for assessing the potential for reproductive toxicity. Included are situations such as when no data are available on reproductive toxicity as well as for databases from studies on test animals or humans that have a limited study design or conduct (e.g., small numbers of test animals or human subjects, inappropriate dose selection or exposure information, other uncontrolled factors); data from studies that examined only a limited number of endpoints and reported no adverse reproductive effects; or databases that were limited to information on structure-activity relationships, short-term or in vitro tests, pharmacokinetic data, or metabolic precursors.

Source: From EPA, 1996.

necessary to judge that an agent is unlikely to pose a hazard than is required to support the finding of a potential hazard. This is because it is more difficult, both biologically and statistically, to support a finding of no apparent adverse effect than to support a positive effect.

The insufficient evidence category may include situations in which no data are available or situations in which data are available from studies with a limited study design or of a type (e.g., some short-term studies) that are not yet considered sufficient to use for human risk assessment. Such data are often useful in recommending the need for additional studies. In some cases, a substantial database may exist in which no one study is considered sufficient but the body of data may be judged as meeting the sufficient evidence category.

When the database is considered sufficient for risk assessment, the hazard potential along with information on other relevant health effects and dose-response, pharmacokinetic, pharmacodynamic, or other supporting information can be used to calculate reference values (e.g., the chronic RfD or RfC).

A recent review of the RfD and RfC processes urges expansion of the data considered in hazard characterization and discusses criteria to be considered in a weight-of-evidence approach.[15] For example, a robust database would include evaluation of a variety of health endpoints (both structural and functional), durations of exposure, timing of exposure, life stages, and susceptible subpopulations. In the absence of complete human data, mechanistic and other data that show the relevance of the animal data for predicting human response would provide a strong basis for extrapolation of animal data to humans. The database for a reference value of less-than-chronic duration should also address the issue of reversibility of effects and latency to response, taking into consideration the possibility that less-than-chronic exposure may lead to effects some time after exposure. Biological and chemical characteristics of the exposure and outcomes, as well as known limits on reserve capacities and repair of damage, may provide helpful information in determining whether the length of follow-up is sufficient to detect latent effects or the degree of reversibility.

2. Dose-Response Assessment

The evaluation of the dose-response relationship for developmental toxicity includes an evaluation of both laboratory animal and human data. Unfortunately, adequate exposure information (and less frequently, actual dose data) are rarely available from human studies, so that laboratory animal data are most frequently relied upon. Evidence for a dose-response relationship is an important criterion for identifying an agent as causing developmental or reproductive toxicity.

The lowest effective doses in adults and young are often similar or may be the same, but the type and severity of effects may be very different, e.g., the developmental effects may be permanent while the maternal effects may be reversible. The difference between the maternally toxic dose and the developmentally toxic dose may at times be related to the relative thoroughness with which endpoints are evaluated in dams and offspring. It is also important to reiterate a point made previously[2] regarding the relationship of maternal and developmental toxicity. From a risk assessment perspective, developmental toxicity in the presence of maternal toxicity cannot be simply considered "secondary to maternal toxicity" and discounted. This presumption was never considered appropriate in evaluating data for risk assessment purposes,[3,7] is still not considered a valid argument for ignoring developmental toxicity in an animal study, and should have been put to rest long ago.

In most cases, the lowest dose at which adverse developmental or reproductive effects are seen (lowest observed adverse effect level — LOAEL), and the dose at which no adverse effects can be detected (no observed adverse effect level — NOAEL) are determined. These doses are often identified based on statistical differences from controls, but may be determined by examining the trend in response and certain biological considerations, such as rarity of the effect. When sufficient data are available to do mathematical modeling of the dose-response relationship, this approach is preferable to the NOAEL approach. Mathematical modeling may be performed to determine a quantitative estimate of response at low doses in the experimental range. This approach can be used to determine the benchmark dose (BMD) and its lower confidence limit (BMDL), which may be used in place of the NOAEL.[107,108]

a. Duration Adjustment and Derivation of Human Equivalent Concentrations and Human Equivalent Doses

Prior to derivation of BMDs or NOAELs, data from toxicity studies are adjusted to convert the duration of exposure in a particular study to a continuous exposure scenario. For inhalation exposures, this adjustment also involves the application of a concentration × time ($c \times t$) adjustment, while for oral studies, the adjustment is to a daily exposure (e.g., a 5-d/week exposure is converted to 7 d/week). For inhalation exposures, this adjustment has not traditionally been done in the case of developmental toxicity studies as it has for other types of health effects. That is because of concerns about peak versus integrated exposure and the likelihood of a threshold for effects.[3] However, the EPA, in a review of the RfD and RfC processes, examined additional newer data as well as the basis for the adjustment of other endpoints and recommended that inhalation developmental toxicity studies be adjusted in the same way as for other endpoints.[15]

A number of approaches have been developed for establishing short-duration (less than lifetime) exposure limits.[109,110] Each approach addresses specific exposure scenarios, but a common feature of most of these approaches is the assumption that the probability of response depends on the cumulative exposure, i.e., the product of exposure concentration and duration. The concept of a constant relationship between concentration and duration is generally referred to as Haber's

Law.[111,112] And even though the approach was not intended for exposures of long durations and/or low concentrations, it has come to be more generally applied and is used in extrapolating to inhalation exposures that are not covered by the experimental data. The reliability of this approach in estimating the effects of a particular exposure at various concentrations and durations is relatively undefined. However, recent reviews of this area have begun to question its general applicability over wide ranges of concentration and duration.[113,114]

The development of new and more appropriate models should define the relationship between exposure concentration and duration. Initial application of such models will require broad assumptions about the mode of action. Ultimately, more refined models should permit interpolation of data to untested $c \times t$ combinations, reducing the uncertainty in assessing the data and increasing confidence in the prediction of potential human risk. The duration of exposure, even for agents with a short half-life, can influence the developmental effects that are produced, as demonstrated by recent studies on all-*trans*-retinoic acid[115] and ethylene oxide.[116] The effects of exposure to hyperthermia during a brief period of development (GD 10 in the rat) have also been shown to be a function of the level and duration of temperature elevation.[117] Thus, the default adjustment of inhalation developmental toxicity studies for duration of exposure will likely be more protective with adjustment from shorter to longer durations of exposure.[116,117]

Derivation of the human equivalent concentration (HEC) for inhalation exposures is intended to account for pharmacokinetic differences between humans and animals, while the potential pharmacodynamic differences are usually accounted for by a portion of the interspecies uncertainty factor (typically $10^{0.5}$). The application of dosimetric adjustment factors (DAFs) to derive the HEC is discussed in detail in the EPA's guidance for RfC derivation.[118]

Currently, there is no similar procedure available to account for pharmacokinetic differences in deriving the human equivalent dose (HED), and the basis for deriving the dose metric for oral exposure differs between cancer and noncancer risk assessment. In the case of noncancer risk assessment, oral dose is expressed on the basis of body weight[1] (e.g., milligrams per kilogram per day), while for cancer risk assessment, dose is expressed on the basis of body weight to the 0.75 power (e.g., milligrams per kilogram$^{0.75}$ per day). Harmonization of these approaches is needed for more consistent dose-response assessment.

b. Benchmark Dose Modeling

The benchmark dose modeling (BMD) approach has several advantages over the NOAEL approach. For example, dose-response modeling with derivation of a BMD uses all the data in fitting a model, accounts for the variability in the data, and does not limit the BMD to one of the experimental doses. The BMD is defined as a dose that corresponds to a particular level of response — the benchmark response (BMR). The BMR is defined as a particular response rate for a quantal response, e.g., 1%, 5%, 10%, or as a specified change from control values for a continuous response, e.g., 1 SD change from the control mean. The BMR must be selected for the derivation of a BMD, and this is not straightforward, particularly in the case of continuous data, because the decision is based on selecting the magnitude of change from controls that is considered adverse.

Studies have been conducted to evaluate several approaches to calculating BMDs for standard prenatal developmental toxicity data.[119–122] These studies applied several dose-response models, both generic and developmental toxicity–specific, to a large number of standard developmental toxicity studies in animals (rats, mice, rabbits) that were dosed throughout the period of major organogenesis (or in some cases throughout pregnancy). These studies showed that such models can be used successfully with standard developmental toxicity data. The BMD for a 5% excess risk above controls for binomial endpoints corresponded on average to the NOAEL. Incorporation of variables, such as intralitter correlation and litter size, appeared to enhance the fit of the developmental toxicity–specific models, whereas inclusion of a threshold did not. Threshold in this case refers to a model-derived estimate of the point at which the model can no longer distinguish treatment effects from background control levels and has nothing to do with a biological threshold. Application of various models to fetal weight data, a continuous endpoint, also was successful.[122] Several definitions of change from control values were found to give BMDs similar to the NOAEL, including a change of 1 SD from control values.

The EPA has published a draft technical guidance document for BMD derivation,[108] which recommends deriving, as a default, a 10% BMD and lower confidence limit (BMDL) for quantal data, and a BMD and BMDL for 1 SD from the control mean for continuous data. These default values allow comparison between endpoints and between chemicals. The BMDL is used as the point of departure (POD), i.e., the point on the dose-response curve used for deriving the reference values or for linear low-dose extrapolation and estimation of risk.

c. Application of Uncertainty Factors and Derivation of Reference Values (RfDs and RfCs)

In the absence of pharmacokinetic or mode of action data to the contrary, developmental and reproductive effects (as well as many other types of noncancer health effects) are generally assumed to have a threshold or nonlinear dose-response relationship at low dose levels. Because of the nonlinear or threshold assumption, extrapolation to low dose levels by use of mathematical models is not typically done. Instead, reference values, such as the RfD and RfC, are derived.

RfDs and RfCs are estimates of a daily exposure to the human population assumed to be without appreciable risk of deleterious effects over a lifetime. A major disadvantage of the RfD and RfC approach is that it does not give an estimate of risk at particular dose levels, so when exposure occurs at levels above the RfD or RfC, the risk is unknown.

RfDs and RfCs are derived by dividing the POD by uncertainty factors that account for such things as animal to human extrapolation (UF_A), variations in sensitivity within the human population (UF_H), LOAEL to NOAEL extrapolation when a NOAEL has not been determined (UF_L; not applied when a BMDL has been derived), subchronic to chronic extrapolation (UF_S), and deficiencies in the database (UF_D). RfDs and RfCs are set for chronic exposures. In some cases, less than lifetime reference values are set. For example, acute, short-term, longer-term, and chronic health advisories are set for drinking water contaminants, and acute and short-term RfDs are sometimes set for pesticide exposures, e.g., for acute dietary and short-term dermal exposures. The EPA review of the RfD and RfC process recommended that less than lifetime values, in addition to the chronic RfD and RfC, be set for all chemicals where appropriate data are available.[15] In setting less than lifetime values, it is important to consider developmental and reproductive effects along with other types of toxicity. The EPA's guidelines for developmental toxicity risk assessment[3] had recommended setting RfD_{DT}s to account for the fact that a lifetime exposure is not necessary to produce a developmental effect. However, this approach will not be necessary if shorter-term reference values are generally derived to be protective of all potential health effects for a given exposure scenario and to cover all lifestages.

The default value for uncertainty factors is 10, but this may be reduced (often to $10^{0.5}$) depending on the confidence in the data or information that provides assurance of reduced intra- or interspecies variability (see Moore et al.,[123,124] for examples). The UF_A and UF_H are considered to cover both pharmacokinetic and pharmacodynamic components of inter- and intraspecies differences. Chemical specific data on pharmacokinetics and pharmacodynamics (i.e., mode of action) may be used to replace part or all of these UFs. The methodology for deriving RfCs uses physiological and pharmacokinetic data to replace the PK component of the UF_A, reducing it to $10^{0.5}$.[118] A number of studies have been done to evaluate the adequacy of the various 10-fold uncertainty factors as a default and to determine whether these factors adequately protect susceptible subpopulations.[125–129] Peleki et al.[130] used a physiological model–based approach to examine the magnitude of the pharmacokinetic part of the intraspecies UF for VOCs for adults and children. They reported that no adult-children differences in the parent chemical concentrations were likely to be seen with inhalation exposures.

Models that are biologically based should provide a more accurate estimate of low-dose risk to humans. The development of biologically based dose-response models has been limited by intra- and interspecies differences in developmental and reproductive effects, a lack of understanding of the biological mechanisms underlying developmental and reproductive toxicity, and other factors. The advent of the use of molecular techniques for understanding developmental and reproductive biology gives promise for understanding mechanisms of toxicity[99] and will support the construction of biologically based models.

3. Exposure Assessment

The exposure assessment describes the magnitude, duration, schedule, and route of human exposures from various sources for comparison with the toxicity data. An important point to remember is that based on the definition of developmental and reproductive toxicity used in the EPA guidelines,[3,4] almost any segment of the population may be at risk for developmental or reproductive toxicity. The Guidelines for Estimating Exposures[131,132] discusses the various approaches that can be taken for estimating exposures through actual monitoring or by modeling various sources and exposure scenarios. Default values have been published[133] for use in estimating exposures, e.g., from food and water consumption in adults and children, soil ingestion in children, and respiration rates in children and adults. In some cases, behavior patterns of importance in children, e.g., crawling on the floor, pica, and food preferences or patterns, should be taken into account.[8] Recently, the Child-Specific Exposure Factors Handbook has been published.[134] This discussion does not attempt to provide definitive guidance for an exposure assessment but focuses on those issues of specific importance to developmental and reproductive toxicity risk assessment.

Several exposure conditions are unique for developmental toxicity. For example, during pregnancy a woman is exposed directly and her developing conceptus is exposed indirectly. For physical agents, exposure may be direct to the embryo or fetus, e.g., x-irradiation and heat, with only a minimal modification of the exposure by the maternal system. Exposure of the conceptus to chemical agents is dependent on maternal absorption, distribution, metabolism, and excretion, as well as placental metabolism and transfer. Transit time in the conceptus also may depend on its ability to metabolize and/or excrete the chemical. In a few cases, a chemical agent may have its primary effect on the maternal system,[135] and the effect on the conceptus does not depend on exposure to the agent and/or its metabolites, but on some factor induced in the mother. Another unique exposure for infants is via breast milk, particularly for agents that are fat soluble and stored in maternal fat. Data suggest that lead may be mobilized along with calcium from bone stores in women during pregnancy and lactation[136] and be excreted in milk,[137-139] although the levels absorbed by the infant may be low and blood levels in the baby may actually decrease during nursing.[89] In addition, continuous bone remodeling in the child, in addition to lead exposure from house dust, paint, etc., may contribute to blood lead.[88] Thus, exposure of the conceptus and child may not be the same as for the pregnant or lactating mother, and measurement of the agent in cord blood and in breast milk may give a better estimate of exposure.

Considerations unique to exposures that affect the male reproductive system are related to the stages of spermatogenesis. An acute exposure may result in different effects on the germ cells that are possible to detect only at certain periods after the exposure occurs. For example, germ cells that were spermatozoa, spermatids, spermatocytes, or spermatogonia at the time of acute exposure take 1 to 2, 3 to 5, 5 to 8, or 8 to 12 weeks to appear in the ejaculate. Thus, an acute exposure will not be detected if the affected endpoint is looked for too early or too late. With longer, more continuous exposures, the effects may be more apparent, but timing of exposure and expected outcomes must still be considered.

Exposures that affect the female reproductive system are also important with respect to the stage of reproductive life or different physiological states (e.g., nonpregnant, pregnant, lactating, perimenopausal, postmenopausal). Exposure of pregnant women to diethylstilbestrol (DES) at a time when critical stages of reproductive system development were occurring resulted in a number of cases of reproductive tract abnormalities and a rare form of cancer in young women exposed *in utero*.[140,141] Compounds that cause destruction of primordial follicles or primary oocytes or that increase the rate of apoptosis in the germ cells may result in infertility or alter time to pregnancy, disrupt follicular recruitment, ovulation, and ovum transport, and lead to changes in viability or fertilizability of the egg.

Children and adults are exposed directly to contaminants via water, food, air, and soil. Dust and paint or anything that can be mouthed by a young child should also be considered a potential source of exposure. The period and duration of exposure should be related to life stage of development or reproduction. For example, exposures should be characterized as preconception, first, second, or third trimester of pregnancy (or more specific periods, if possible), infancy, early, middle, and late childhood (specific ages, if possible), adolescence (stage of puberty and development), adulthood (by age), pre- or postmenopausal, older adult, and the aged. These life stages may have different sensitivities to an agent, and exposure estimates should be characterized for as many as possible. In addition, exposure to either parent prior to conception should be considered as a possibility in the production of adverse developmental effects.[142–144]

Exposures that can cause developmental and reproductive toxicity may be for short or long durations. Many experimental studies have confirmed for a number of agents that a single exposure is sufficient to cause developmental or reproductive toxicity, depending on the life stage, i.e., repeated exposure is not a necessary prerequisite for adverse outcomes. Thus, agents with short half-lives, as well as those that show accumulation with repeated exposures or that have long half-lives, may cause effects. The pattern of exposure, whether intermittent or continuous, as well as peak exposures and average exposure levels over the period of exposure, should be examined.

Exposures affecting development and reproduction will often result in latent effects. For example, effects of exposure during pregnancy may not be manifest until after birth in the form of structural malformations, growth retardation, cancer, or mental retardation or other functional defects. Likewise, effects of exposures that enhance follicular degeneration early in development may not become apparent until an early onset of menopause is noted. Other types of effects that may not become apparent until long after exposure during early life stages include cancer, heart disease, neurodegenerative disease, and shortened lifespan.

4. Risk Characterization

Risk characterization is the final step in the risk assessment process and involves integration of the hazard characterization and dose-response assessment with the exposure assessment. Integral to risk characterization is information on the quality of the data for the hazard evaluation and the exposure assessment, relevancy to humans, strengths and weaknesses of the database, assumptions made, scientific judgments, and the level of confidence in determining potential risks to humans. All of these factors are part of the weight-of-evidence determination. Table 21.4 gives a list of issues to be considered in each of the components of risk assessment and can serve as a guide for the questions to be addressed in risk characterization.

The weight-of-evidence information is communicated to the risk manager along with the quantitative information on the range of effective exposure levels, dose-response modeling estimates (e.g., the BMD and BMDL), appropriate uncertainty factors for each duration reference value, as well as the reference values. Information on developmental and reproductive toxicity is considered an important part of the database used in calculating chronic RfDs and RfCs and must be weighed appropriately relative to other adverse outcomes. When developmental and reproductive toxicity data are not available, the database is considered to be deficient, and an additional database uncertainty factor may be applied. In the case of pesticides, the Food Quality Protection Act requires the addition of a 10-fold margin of safety to take into account the potential for pre- and postnatal toxicity and the completeness of the toxicology and exposure databases. The EPA's Office of Pesticide Programs has published a guidance document[35] that provides direction for applying the additional 10-fold FQPA safety factor and clarifies that there is overlap between this tenfold factor and the traditional uncertainty factors, particularly the database deficiency factor. It presumes that in most cases the traditional uncertainty factors will be sufficient to be protective of children's health. Nevertheless, there may be residual uncertainties about health effects not captured in the RfD process or in the exposure assessment that provide a basis for maintaining part or all of the FQPA factor.

Three types of descriptors of human risk are especially useful and important. The first of these is related to interindividual variability, i.e., the range of variability in population response to an agent and the potential for highly sensitive or susceptible subpopulations. A default assumption is made that the most sensitive individual in the population will be no more than 10-fold more sensitive than the average individual; thus, a default 10-fold uncertainty factor is often applied in calculating reference values to account for this potential difference. When data are available on highly sensitive or susceptible subpopulations, the risk characterization can be done on them separately or by using more accurate factors to account for the differences. When data are not available to indicate differential susceptibility between developmental stages and adulthood, all stages of development and reproduction may be assumed to be highly sensitive or susceptible. Certain age subpopulations can sometimes be identified as more sensitive because of critical periods for exposure; for example, pregnant or lactating women, infants, children, adolescents, adults, or the elderly. In general, not enough is understood about how to identify sensitive subpopulations without obtaining specific data on each agent, although it is known that factors such as nutrition, personal habits (e.g., smoking, alcohol consumption, drug abuse), quality of life (e.g., socioeconomic factors), preexisting disease (e.g., diabetes), race, ethnic background or other genetic factors may predispose some individuals to be more sensitive than others to the developmental effects of various agents.

The second descriptor of importance is for highly exposed individuals. These are individuals who are more highly exposed because of occupation, residential location, behavior, or other factors. For example, children are more likely than adults to be exposed to agents deposited in dust or soil, either indoors or outdoors, both because of the time spent crawling or playing on the floor or ground and the mouthing behavior of young children. The inherent sensitivity of children may also vary with age, so that both sensitivity and exposure must be considered in the risk characterization. If population data are absent, various scenarios representing high end exposures may be assumed by use of upper percentile or judgment based values. This approach must be used with caution, however, to avoid overestimation of exposure.

The third descriptor that is sometimes used to characterize risk is the margin of exposure (MOE). The MOE is the ratio of the NOAEL (or BMDL) from the most appropriate or sensitive species to the estimated human exposure level from all potential sources. Considerations for the acceptability of the MOE are similar to those for the uncertainty factors applied to the NOAEL or BMDL to calculate the reference values. Examples of the calculation of an MOE based on developmental toxicity data have been described for dinoseb,[145] lithium,[123] and boric acid and borax.[124] In the case of dinoseb, the MOEs were very small, in some cases less than 1, indicating toxicity in the animal studies at levels to which people were being exposed. The analysis of the data on dinoseb led to an emergency suspension of this pesticide in 1986 and ultimately to its removal from the market. The risk characterization is communicated to the risk manager, who uses the results along with technological factors (e.g., exposure reduction measures), as well as social and economic considerations, in reaching a regulatory decision. Depending on the statute involved and other considerations, risk management decisions are usually made on a case-by-case basis. This may result in different, but appropriate, regulation of an agent under different statutes.

Table 21.4 Guide for developing chemical-specific risk characterizations for developmental and reproductive effects[a]

Part One: Summarizing Major Conclusions in Risk Characterization

I. Hazard Characterization
  A. What are the key toxicological studies that provide the basis for health concerns for developmental and reproductive effects?
   - How good are the key studies?
   - Are the data from laboratory or field studies? In one or several species?
   - What adverse developmental and reproductive endpoints were observed, and what is the basis for the effects at the lowest exposure levels?
   - Describe other studies that support these findings.
   - Discuss any valid studies that conflict with these findings.
  B. Besides the developmental and reproductive effects observed in the key studies, are there other health endpoints of concern? What are the significant data gaps?
  C. What epidemiological or clinical data are available? For epidemiological studies:
   - What types of studies were available (e.g., human ecologic, case control, or cohort studies, or case reports or series)?
   - Describe the degree to which exposures were described.
   - Describe the degree to which confounding factors were considered.
   - Describe the major demographic and other personal factors examined (e.g., age, sex, ethnic group, socioeconomic status, smoking status, occupational exposure).
   - Describe the degree to which other causal factors were excluded.
   - Describe the degree to which the findings were examined for biological plausibility, internal and external consistency of the findings, and the influence of limitations of the design, data sources, and analytic methods.
  D. How much is known about how (through what biological mechanism) the chemical produces adverse developmental and reproductive effects?
   - Discuss relevant studies of mechanisms of action or metabolism.
   - Does this information aid in the interpretation of the hazard data?
   - What are the implications for potential adverse developmental and reproductive effects?
  E. Comment on any nonpositive data in animals or humans and whether these data were considered in the hazard characterization.
  F. If adverse health effects have been observed in wildlife species, characterize such effects by discussing the relevant issues as in A through E above.
  G. Summarize the hazard characterization and discuss the significance of each of the following:
   - Confidence in the conclusions
   - Alternative conclusions that are also supported by the data
   - Significant data gaps
   - Highlights of major assumptions

II. Characterization of Dose Response
  A. What data were used to develop the dose response curve? Would the result have been significantly different if based on a different data set?
   - If laboratory animal data were used:
     - Which species were used?
     - Most sensitive, average of all species, or other?
     - Were any studies excluded? Why?
   - If human data were used:
     - Which studies were used?
     - Only positive studies, all studies, or some other combination?
     - Were any studies excluded? Why?
     - Was a meta-analysis performed to combine the epidemiological studies? What approach was used?
     - Were studies excluded? Why?
  B. Was a model used to develop the dose response curve and, if so, which one? What rationale supports this choice? Is chemical-specific information available to support this approach? What other models were considered?
   - How were the reference values (or the acceptable range) calculated?
   - What assumptions and uncertainty factors were used?
   - What is the confidence in the estimates?
  C. Discuss the route, level, timing (i.e., life stage), and duration of exposure observed, as compared to expected human exposures.
   - Are the available data from the same route of exposure as the expected human exposures? If not, are pharmacokinetic data available to extrapolate across route of exposure?
   - What information was used to support duration adjustment and to calculate the human equivalent dose or concentration (HED or HEC)?
   - How far does one need to extrapolate from the observed data to environmental exposures? One to two orders of magnitude? Multiple orders of magnitude? What is the impact of such an extrapolation?
  D. If adverse health effects have been observed in wildlife species, characterize dose response information using the process outlined in A through C above.

Table 21.4 Guide for developing chemical-specific risk characterizations for developmental and reproductive effects[a] (continued)

III. Characterization of Exposure
   A. What are the most significant sources of environmental exposure?
      Are there data on sources of exposure from different media?
      What is the relative contribution of different sources of exposure?
      What are the most significant environmental pathways for exposure?
   B. Describe the populations that were assessed, including the general population, highly exposed groups, and highly susceptible groups, and including different life stages, e.g., pregnant women, children, adolescents, adults, elderly.
   C. Describe the basis for the exposure assessment, including any monitoring, modeling, or other analyses of exposure distributions, such as Monte Carlo or krieging.
   D. What are the key descriptors of exposure?
      Describe the (range of) exposures to: "average" individuals, "high-end" individuals, general population, high exposure group(s), children, susceptible populations, males, females (nonpregnant, pregnant, lactating).
      How was the central tendency estimate developed?
      What factors and/or methods were used in developing this estimate?
      How was the high-end estimate developed?
      Is there information on highly exposed subgroups?
      Who are they?
      What are their levels of exposure?
      How are they accounted for in the assessment?
   E. Is there reason to be concerned about cumulative or multiple exposures to classes of agents with a similar mechanism of action?
      Are there biological, behavioral, ethnic, racial, or socioeconomic factors that may affect exposures?
   F. If adverse developmental and reproductive effects have been observed in wildlife species, characterize the wildlife exposure by discussing the relevant issues in A through E above.
   G. Summarize exposure conclusions and discuss the following:
      Results of different approaches, i.e., modeling, monitoring, probability distributions
      Limitations of each, and the range of most reasonable values
      Confidence in the results obtained, and the limitations to the results

Part Two: Risk Conclusions and Comparisons
   I. Risk Conclusions
      A. What is the overall picture of risk, based on the hazard, quantitative dose response, and exposure characterizations?
      B. What are the major conclusions and strengths of the assessment in each of the three main analyses (i.e., hazard characterization, quantitative dose-response, and exposure assessment)?
      C. What are the major limitations and uncertainties in the three main analyses?
      D. What are the science policy options in each of the three major analyses?
         What are the alternative approaches evaluated?
         What are the reasons for the choices made?
   II. Risk Context
      A. What are the qualitative characteristics of the developmental and reproductive hazard (e.g., voluntary vs. involuntary, technological vs. natural, etc.)? Comment on findings, if any, from studies of risk perception that relate to this hazard or similar hazards.
      B. What are the alternatives to this developmental or reproductive hazard? How do the risks compare?
      C. How does this developmental or reproductive risk compare to other risks?
         How does this risk compare to other risks in this regulatory program or other similar risks that the EPA has made decisions about?
         Where appropriate, can this risk be compared with past agency decisions, decisions by other federal or state agencies, or common risks with which people may be familiar?
         Describe the limitations of making these comparisons.
      D. Comment on significant community concerns that influence public perception of risk.
   III. Existing Risk Information
      A. Comment on other reproductive risk assessments that have been done on this chemical by the EPA, other federal agencies, or other organizations. Are there significantly different conclusions that merit discussion?
   IV. Other Information
      A. Is there other information that would be useful to the risk manager or the public in this situation that has not been described above?

Source: Modified from EPA, 1996.

III. EMERGING ISSUES FOR RISK ASSESSMENT

A. Research Efforts to Develop Human Data on Developmental and Reproductive Effects of Environmental Exposures — Contributions to Risk Assessment

Several large research efforts are underway or being planned to address data gaps on environmental exposures and links with children's health effects. These include the NIEHS, EPA, and CDC Centers of Excellence in Children's Environmental Health and Disease Prevention Research.* These centers (currently 12 in number) are focusing on research to address possible links between environmental exposures and health effects of great concern for children, e.g., neurobehavioral effects, including autism, and respiratory diseases, including asthma.

Another large effort currently being planned is the National Children's Study, a national cohort study of environmental effects on child health and development (www.nationalchildrensstudy.gov). The study is being organized by the U.S. Department of Health and Human Services (DHHS) and the EPA to follow children from early gestation through infancy, childhood, and adolescence, into adulthood. The National Institute of Child Health and Human Development (NICHD), together with a consortium of federal agencies, was charged with conducting the study by the Children's Health Act of 2000. The EPA, together with NICHD, NIEHS, and CDC, has been engaged in planning the study, which is scheduled to be launched in late 2005. The NCS will evaluate the effects of environment, defined broadly to include chemical, physical, biological, and social factors, on growth and development of children. Because data on health and environment will be collected at times throughout the study, there will be an opportunity to evaluate links between health and environment in a way that has not been possible in smaller, more limited duration studies. In addition, genetic factors will be analyzed to allow the evaluation of gene-environment interactions. Biological and environmental samples will also be collected and stored to serve as a national resource. This will allow the investigation of hypotheses in addition to those posed for the core study, including questions that may become important in the future.[146]

These data, which will help us understand the effects of environmental factors on human health, has a number of implications for risk assessment. For example, the issue of relative susceptibility of adults and children to environmental factors can be more fully explored, as well as the question of when and to what extent children are more highly exposed to environmental agents. Because of the longitudinal nature and large size of the National Children's Study, a number of important risk assessment questions can be addressed. For example:

> What is the role of various environmental factors in accounting for children's health problems? Does exposure to environmental agents increase the burden of disease in children?
> Are there long-term health effects, e.g., asthma, cancer, cardiovascular disease, obesity, diabetes, or neurological diseases, from early exposures of children to environmental agents?
> What are the factors that increase or decrease the susceptibility of children from exposure to environmental agents?
> Are there disparities in children's environmental health because of race, ethnicity, poverty, housing, income, nutrition, location near heavy industry or toxic waste sites, etc?
> What genetic factors can increase the susceptibility of children to environmental exposures?
> What is the variability in the response of children at different ages or life stages to environmental exposures?
> Are the default risk assessment approaches used to account for children's health in both the toxicity and exposure assessments sufficient to protect children?

---

* Information on each of the centers and their research focus can be found at http://es.epa.gov/ncer/centers/.

B. Using Mode of Action and Pharmacokinetic Data in Risk Assessment

The mode of action for a chemical can be defined as the key steps after interaction with the target site upon which all other changes are dependent.[147] Incorporating mode of action (MOA) information into risk assessment will reduce uncertainties, because changes in the key steps can be measured, and endpoints relevant to the MOA, as opposed to biomarkers, can be assessed.

There are several examples of developmental and reproductive toxic agents for which MOA data exist. One example is the androgen antagonist pesticide vinclozolin (reviewed by Euling and Kimmel[148]). Another example is the use of MOA information to classify chemicals into equivalency groups. For example, inclusion in the dioxin and dioxinlike chemical grouping was done using the toxic equivalency factor approach, based on binding to the aryl hydrocarbon receptor. In individual risk assessments, information supporting a particular MOA may be available for a chemical but is not always considered conclusive. Even in cases where the MOA has been relatively well established, such as for perchlorate or ethylene oxide, chemical-specific or biologically based dose-response models have yet to be established. For atrazine, a triazine herbicide, the MOA was a critical factor in eliminating the finding of cancer in the rodent as relevant to humans, leaving reproductive effects as the most sensitive endpoint.[149]

The concept that chemicals with the same mode of toxicity can be considered dose additive has broad implications for risk assessment (e.g., for NOAEL and RfD determination). Cumulative risk for endocrine disruptors has been shown to be of concern for estrogenic chemicals.[60] Specifically, these authors found that combinations of weak estrogen agonists, when combined at doses below their no-observed-effect concentration (NOEC) or EC01, were dose additive (considering potency differences).

Developmental stage sensitivity information for a developmentally toxic agent can help inform MOA (and vice versa). For example, if a critical window of exposure for a chemical corresponds to a period of male sexual differentiation, then the chemical may be affecting androgen action at some level. Improving the incorporation of developmental stage information into risk assessment addresses the changes in sensitivity to exposures that occur during development. An example of this for carcinogenesis was recently published by Ginsberg,[150] and the concept has been incorporated into the recent draft guidance for assessing risks of carcinogenesis to children.[151] Efforts to compile and model developmental stage sensitivity information are ongoing.[148,152,153] Information on critical windows during development for different organ systems in humans and model organisms was compiled and compared at the Workshop to Identify Critical Windows of Exposure for Children's Health.[61]

FQPA requires that cumulative effects of chemicals with the same mechanism of toxicity be incorporated into the assessment when determining the adequacy of tolerances for infants and children. Thus, mode and mechanism of action data became the basis for determining the cumulative risk of multiple chemicals. The U.S. EPA Office of Pesticide Programs (OPP) developed guidance that relies on mechanism of action for cumulative risk assessment of pesticides[154] and a policy on common mechanism of toxicity, defined as the same sequence of major biochemical events (also considered the "mode of action").[155] These guidance documents were informed by a case study and a classification system for common mechanism and mode of toxicity (based on acetylcholinesterase activity) developed for organophosphate pesticides.[156,157] These and other efforts culminated in the development of the integrated, multichemical, multipathway risk assessment for organophosphate pesticides.[158]

FQPA also requires that tolerances on food can only be established if they do not result in harm from aggregate exposure to the pesticide residue, including all anticipated dietary exposure as well as all other exposures. This means that all nonoccupational exposures must be aggregated for a single pesticide from all routes and pathways. OPP has developed operating guidance for conducting aggregate exposure assessments.[159]

Finally, as MOA is defined either in general for a particular chemical or for specific biological parameters and responses, we will be able to examine dose-dependent transitions from homeostatic situations to toxicologic responses and the critical elements in pathogenesis at levels relevant to likely human exposure conditions.[160] This is an area of toxicology that is of growing interest, especially as the field becomes more focused on the events that occur at the lower end of the exposure range.[70]

C. Harmonization of Approaches for Human Health Risk Assessment

Harmonization of risk assessment refers to developing a consistent set of principles and guidelines for drawing inferences from scientific information. While it is unlikely that a single method will be developed for the assessment of all toxicities and exposures, it is likely that endpoint-specific risk assessment guidance will give way to guidance that is applicable to many different endpoints, relies more on our understanding of mechanism (mode) of action, and models potential response to a stressor at several levels of biological complexity.

There is progress in this area with respect to environmental health risk assessment. Following the NRC report on Science and Judgment in Risk Assessment,[14] the EPA organized two internal colloquia to discuss harmonization within the context of the agency.[161,162] Following these colloquia, the agency cosponsored a Society of Toxicology workshop, Harmonization of Cancer and Noncancer Risk Assessment.[163] These efforts saw consensus develop around the importance of moving away from traditional risk assessment dichotomies, of better defining the mode of action and toxicity, and of assessing the entire available database in the context of exposure and response. More recently, the agency has completed a review of the RfD and RfC process that identified a number of specific elements where the consistency of the current approach needs to be further evaluated.[15] Among these were the modification of study protocols to provide more comprehensive coverage of life stages for both exposures and outcomes, the collection of more information from less-than-lifetime exposures to evaluate latency to effect and reversibility of effect, and further evaluation of appropriate adjustment of doses for duration of exposure. Of particular importance to a more unified approach is the recommendation that health effects no longer be categorized as "cancer" or "noncancer" for the purposes of hazard characterization and dose-response analysis. The agency's Risk Assessment Forum has established a Harmonization Steering Committee that is beginning to address many of these issues.

Another level of harmonization was addressed by the EPA's Science Policy Council in its *Guidance on Cumulative Risk Assessment, Part 1, Planning and Scoping*.[164] This document supports a global shift toward aggregate and cumulative approaches to risk assessment, attempting to "harmonize" and integrate areas of risk assessment that are now considered separately (e.g., ecological and human health assessments, as well as cancer and noncancer assessments).

How do these issues relate to reproductive and developmental toxicity? Initially, a broader perspective of the relationship of these fields to other areas of toxicology must be developed. To a large extent, this is consistent with reproductive and developmental toxicology methods because the issues addressed are not limited to a particular organ system but include a wide variety of endpoint parameters (e.g., many specific organ systems that include growth, structural, and functional manifestations). However, our knowledge of basic MOA in reproduction and development must greatly expand. As risk assessment moves to a more unified approach, current testing and assessment approaches may have to be modified, and the biological reasoning behind endpoint-specific approaches must be made consistent within and across all exposures and responses.

D. Use of Genomics, Proteomics, and Other "Futures Technologies"

DNA microarrays can be used to measure global gene expression changes in tissues after toxic agent exposure. This technology has been used to illustrate that chemicals sharing a MOA have similar gene expression profiles.[165,166] Hamadeh et al.[167] suggest that a gene expression profile database for existing chemicals will be very useful for screening purposes. For example, the database could be queried about a chemical with DNA microarray data but other missing information (e.g., unknown MOA). While this technology generates a lot of detailed information on gene expression, it is critical to establish the links between the microarray expression profile and the adverse effects of a given chemical before this technique can be used routinely in risk assessment. EPA's Science Policy Council Interim Policy on Genomics[168] states that the incorporation of genomics data into risk assessment will be on a case-by-case basis.

Toxicogenomic data will become increasingly available. Thus, it will be important to establish whether observed gene expression changes are required for the adverse effects, are part of a compensatory mechanism, or are related to effects seen only at high doses. In the future, data on proteomics and metabonomics may ultimately be found more useful than gene microarray data. High-throughput techniques for these assays are becoming available, but there will continue to be issues with analysis and interpretation of the data for some time to come, as there have been with genomics. A more detailed coverage of gene expression profiling can be found in Chapter 15.

E. Implications of Benefits Analysis for Risk Assessment

Benefits analysis (assessing the benefits of reducing potential health effects) is an area that is currently being explored. This approach typically has been used for cancer, where a linear low-dose response relationship is assumed and modeling of data can be used to estimate risks at low levels. Estimating the benefit of reducing exposure is relatively straightforward in this case. Benefits analysis for other types of health effects for which a nonlinear low dose response relationship is assumed also requires the development of quantitative estimates of risk at low levels.[169-175] However, an approach for low dose quantitative risk estimation has not been adopted, so the total benefit of reducing exposures cannot be determined and is likely underestimated. Because the draft *Guidelines for Carcinogen Risk Assessment*[16] introduced the possibility of a nonlinear dose response relationship for some types of cancers (e.g., thyroid tumors resulting from long-term stimulation of the thyroid), efforts are underway to develop dose response models for extrapolating nonlinear dose response relationships to low levels of environmental exposure. This is an area that holds a good deal of promise for moving forward with dose response modeling of all types of health effects and more accurate assessment of the financial impact of health effects on society.

REFERENCES

1. National Research Council, *Risk Assessment in the Federal Government: Managing the Process*, Committee on the Institutional Means for the Assessment of Risks to Public Health, Commission on Life Sciences, National Research Council, National Academy Press, Washington, D.C., 1983, p. 17.
2. Kimmel, C.A. and Kimmel, G.L., Principles of developmental toxicity risk assessment, in *Handbook of Developmental Toxicology*, Hood, R.D., Ed., CRC Press, Boca Raton, 1996, p. 667.
3. US Environmental Protection Agency, Guidelines for developmental toxicity risk assessment, *Fed. Reg.*, 56, 63798, December 5, 1991.
4. US Environmental Protection Agency, Guidelines for reproductive toxicity risk assessment, *Fed. Reg.*, 61, 56274, October 31, 1996.
5. U.S. Environmental Protection Agency, Guidelines for neurotoxicity risk assessment, *Fed. Reg.*, 63, 26926, May 14, 1998.
6. Palmer, A.K., Regulatory requirements for reproductive toxicology: theory and practice, in *Developmental Toxicology*, Kimmel, C.A. and Buelke-Sam, J., Eds., Raven Press, New York, 1981, p. 259.
7. U.S. Environmental Protection Agency, Guidelines for the health assessment of suspect developmental toxicants, *Fed. Reg.*, 51, 34028, 1986.
8. National Research Council, *Pesticides in the Diets of Infants and Children*, Committee on Pesticides in the Diets of Infants and Children, National Research Council, 408, p. 1993.

9. Schmidt, C. W., A growth spurt in children's health laws, *Environ. Health Perspect.*, 109, A270, 2001.
10. Colborn, T. and Clement, C., Eds., *Chemically-Induced Alterations in Sexual and Functional Development -- The Wildlife/Human Connection*, Princeton Scientific, Princeton, New Jersey, 1992, p. 403.
11. Colborn, T., Dumanoski, D., Myers, J.P. *Our Stolen Future: Are We Threatening Our Fertility, Intelligence, and Survival?* Penguin, New York, 1997, p. 316.
12. National Research Council, *Hormonally Active Agents in the Environment*, Committee on Hormonally Active Agents in the Environment, National Academy Press, Washington, D.C., 2000, p. 452.
13. U.S. Environmental Protection Agency, *Special Report on Environmental Endocrine Disruption: An Effects Assessment and Analysis*, EPA/630/R-96/012, February 1997.
14. National Research Council, *Science and Judgment in Risk Assessment*, National Academy Press, Washington, D.C., 1994, p. 672.
15. U.S. Environmental Protection Agency, *A Review of the Reference Dose and Reference Concentration Processes*, EPA/630/P-02/002F, 01 Dec 2002, Risk Assessment Forum, Washington, D.C., 2002, p. 192; available at http://cfpub.epa.gov/ncea/cfm/recordisplay.cfm?deid=55365.
16. U.S. Environmental Protection Agency, *Guidelines for Carcinogen Risk Assessment (March 2005)*, EPA/630/P-03/001F, April 7, 2005, Risk Assessment Forum, Washington, D.C., 2003, p. 125; available at http://cfpub.epa.gov/ncea/raf/recordisplay.cfm?deid=116283.
17. Adams, J. and Selevan, S.G., Issues and approaches in human developmental toxicology, in *Developmental Toxicology*, 2nd ed., Kimmel, C.A. and Buelke-Sam, J., Eds., Raven Press, New York, 1994, p. 287.
18. Tabacova, S., Little, R., Tsong, Y., Vega, A., and Kimmel, C.A., Adverse pregnancy outcomes associated with maternal enalapril antihypertensive treatment, *Pharmacoepidem. Drug Safety*, published online January 3, 2003, DOI: 10.1002/pds.796.
19. Kimmel, C.A., Holson, J.F., Hogue, C.J., and Carlo, G., *Reliability of Experimental Studies for Predicting Hazards to Human Development*, National Center for Toxicological Research, Jefferson, AR, NCTR Technical Report for Experiment No. 6015, 1984.
20. Tabacova, S.A. and Kimmel, C.A., Enalapril: pharmacokinetic/dynamic inferences for comparative developmental toxicity, *Reprod. Toxicol.*, 15, 467, 2001.
21. Tabacova, S.A. and Kimmel, C.A., Atenolol: pharmacokinetic/dynamic aspects of comparative developmental toxicity, *Reprod. Toxicol.*, 16, 1, 2002.
22. U.S. Environmental Protection Agency, *OPPTS Harmonized Test Guidelines, Health Effects, 870 Series*, August, 1998; available at http://www.epa.gov/opptsfrs/OPPTS_Harmonized/870_Health_Effects_Test_Guidelines/Series/.
23. U.S. Environmental Protection Agency, *Data Call-in Notice for Cholinesterase-Inhibiting Organophosphate Pesticides*, letter from Lois A. Rossi to registrants, September 10, 1999.
24. Kimmel, C.A. and Makris, S.L., Recent developments in regulatory requirements for developmental toxicology, *Toxicol. Lett.*, 120, 73, 2001.
25. Daston, G. and Kimmel C., *An Evaluation and Interpretation of Reproductive Endpoints for Human Health Risk Assessment*, ILSI Press, Washington, D.C., 1999, p. 103.
26. U.S. Environmental Protection Agency, *OPPTS Health Effects Test Guidelines*, 870 Series, EPA 712-C-94-208, July 1994.
27. Canfield, R.L., Henderson, C.R. Jr., Cory-Slechta, D.A., Cox, C., Jusko, T.A., and Lanphear, B.P., Intellectual impairment in children with blood lead concentrations below 10 microg per deciliter, *New Engl. J. Med.*, 348, 1517, 2003.
28. Kimmel, C.A., Rees, D.C., and Francis, E.Z., Eds., Proceedings of the Workshop on the Qualitative and Quantitative Comparability of Human and Animal Developmental Neurotoxicity, *Neurotoxicol. Teratol.*, 12, 173, 1990.
29. Sharbaugh, C., Viet, S.M., Fraser, A., and McMaster, S.B., Comparable measures of cognitive function in human infants and laboratory animals to identify environmental health risks to children, *Environ. Health Perspect.*, 111, 1630, 2003, DOI:10.1289/ehp.6205, published online July 2, 2003.
30. Dostal, M., Developmental immunotoxicology, *Reprod. Toxicol.* 6, 179, 1992.
31. Holladay, S.D. and Luster, M.I., Developmental immunotoxicology, in *Developmental Toxicology*, 2nd ed., Kimmel, C.A. and Buelke-Sam, J., Eds., Raven Press, New York, 1994, p. 93.
32. Holladay, S.D. and Smialowicz, R.J., Development of the murine and human immune system: differential effects of immunotoxicants depend on time of exposure. *Environ. Health Perspect.*, 108(Suppl. 3), 463, 2000.

33. Holsapple, M.P., Developmental immunotoxicology and risk assessment: a workshop summary. *Hum. Exp. Toxicol.*, 21, 473, 2002.
34. Holsapple, M.P., Developmental immunotoxicity testing: a review, *Toxicology,* 185, 193, 2003.
35. U.S. Environmental Protection Agency, *Determination of the Appropriate FQPA Safety Factor(s) for Use in the Tolerance-Setting Process,* Office of Pesticide Programs, 2002; available from: http://www.epa.gov/oppfead1/trac/science/#10_fold.
36. Lau, C. and Kavlock, R.J., Functional toxicity in the developing heart, lung, and kidney, in *Developmental Toxicology*, 2nd ed., Kimmel, C.A. and Buelke-Sam, J., Eds., Raven Press, New York, 1994, p. 119.
37. Osmond, C., Barker, D. J.P., Winter, P.D., Fall, C.H.D., and Simmonds, S.J., Early growth and death from cardiovascular disease in women, *Br. Med. J.*, 307, 1519, 1993.
38. Langley-Evans, S.C., Phillips, G.J., Benediktsson, R., Gardner, D.S., Edwards, C.R., Jackson, A.A., and Seckl, J.R., Protein intake in pregnancy, placental glucocorticoid metabolism and the programming of hypertension in the rat, *Placenta*, 17, 169, 1996.
39. Barker, D.J., Fetal programming of coronary heart disease, *Trends Endocrinol. Metab.*, 13, 364, 2002.
40. Eriksson, J.G. and Forsen, T.J., Childhood growth and coronary heart disease in later life, *Ann. Med.*, 34, 157, 2002.
41. Snoeck, A., Remacle, C., Reusens, B., and Hoet, J.J., Effect of low protein diet during pregnancy on the fetal rat endocrine pancreas, *Biol. Neonate*, 57, 107, 1990.
42. Gauguier, D., Bihoreau, M.T., Ktorza, A., Berthault, M.F., and Picon, L., Inheritance of diabetes mellitus as consequence of gestational hyperglycemia in rats, *Diabetes*, 39, 734, 1990.
43. Van Assche, F.A., Aerts, L., and Holemans, K., Metabolic alterations in adulthood after intrauterine development in mothers with mild diabetes, *Diabetes*, 40, 106, 1991.
44. Dahri, S., Snoeck A., Reusens-Billen B., Remacle, C., and Hoet, J.J., Islet function in offspring of mothers on low-protein diet during gestation, *Diabetes*, 40, 115, 1991.
45. Hoet, J.J., Ozanne, S., and Reusens, B., Influences of pre- and postnatal nutritional exposures on vascular/endocrine systems in animals, *Environ. Health Perspect.*, 108(Suppl. 3), 563, 2000.
46. Eriksson, J., Forsen, T., Tuomilehto, J., Osmond, C., and Barker, D., Size at birth, childhood growth and obesity in adult life, *Int. J. Obes. Relat. Metab. Disord.*, 25, 735, 2001.
47. Eriksson, J.G., Forsen, T., Tuomilehto, J., Osmond, C., and Barker, D.J., Early adiposity rebound in childhood and risk of Type 2 diabetes in adult life, *Diabetologia*, 46, 190, 2003. [Epub 2003 Jan 08].
48. Cooper, C., Javaid, M.K., Taylor, P., Walker-Bone, K., Dennison, E., and Arden N., The fetal origins of osteoporotic fracture, *Calcif. Tissue Int.*, 70, 391, 2002. [Epub 2002 Apr 19].
49. Michels, K.B., Trichopoulos, D., Robins, J.M., Rosner, B.A., Manson, J.E., Hunter, D.J., Colditz, G.A., Hankinson, S.E., Speizer, F.E., and Willett, W.C., Birthweight as a risk factor for breast cancer, *Lancet*, 348, 542, 1996.
50. Innes, K., Byers, T., and Schymura, M., Birth characteristics and subsequent risk for breast cancer in very young women, *Am. J. Epidemiol.*, 152, 1121, 2000.
51. Mellemkjaer, L., Olsen, M.L., Sorensen, H.T., Thulstrup, A.M., Olsen, J., and Olsen, J.H., Birth weight and risk of early-onset breast cancer (Denmark), *Cancer Causes Control*, 14, 61, 2003.
52. Gray, Jr. L.E., Ostby, J.S., and Kelce, W.R., Developmental effects of an environmental antiandrogen: the fungicide vinclozolin alters sex differentiation of the male rat, *Toxicol. Appl. Pharmacol.*, 129, 46, 1994.
53. Blanck, H.M., Marcus, M., Tolbert, P.E., Rubin, C., Henderson, A.K., Hertzberg, V.S., Zhang, R.H., and Cameron, L., Age at menarche and tanner stage in girls exposed in utero and postnatally to polybrominated biphenyl, *Epidemiology*, 11, 641, 2000.
54. Eskenazi, B., Warner, M., Mocarelli, P., Samuels, S., Needham, L.L., Patterson, D.G., Jr., Lippman, S., Vercellini, P., Gerthoux, P.M., Brambilla, P., and Olive, D., Serum dioxin concentrations and menstrual cycle characteristics, *Am. J. Epidemiol.*, 156, 383, 2002.
55. Kimmel, C.A., Grant, L.D., Sloan, C.S., and Gladen, B.C., Chronic low-level lead toxicity in the rat. I. Maternal toxicity and perinatal effects, *Toxicol. Appl. Pharmacol.*, 56, 28, 1980.
56. Grant, L.D., Kimmel, C.A., West, G.L., Martinez-Vargas, C.M., and Howard, J.L., Chronic low-level lead toxicity in the rat. II. Effects on postnatal physical and behavioral development, *Toxicol. Appl. Pharmacol.*, 56, 42, 1980.
57. Dearth, R.K., Hiney, J.K., Srivastava, V., Burdick, S.B., Bratton, G.R., and Dees, W.L., Effects of lead (Pb) exposure during gestation and lactation on female pubertal development in the rat, *Reprod. Toxicol.*, 16, 343, 2002.

58. Selevan, S.G., Rice, D.C., Hogan, K.A., Euling, S.Y., Pfahles-Hutchens, A., and Bethel, J., Blood lead concentration and delayed puberty in girls, *New Engl. J. Med.*, 348, 1527, 2003.
59. Wu, T., Buck, G.M., and Mendola, P., Blood lead levels and sexual maturation in U.S. girls: The Third National Health and Nutrition Examination Survey, 1988–1994, *Environ. Health Perspect.*, 111, 737, 2003.
60. Silva, E., Rajapakse, N., and Kortenkamp, A., Something from "nothing" — eight weak estrogenic chemicals combined at concentrations below NOECs produce significant mixture effects, *Environ. Sci. Technol.*, 36, 1751, 2002.
61. Selevan, S.G., Kimmel, C.A., and Mendola, P., Eds., Identifying critical windows of exposure for children's health, *Environ. Health Perspect.*, 108(Suppl 3), 449, 2000.
62. U.S. Environmental Protection Agency, *Endocrine Disruptor Screening and Testing Advisory Committee Final Report*, EPA/743/R-98/003, August 1998; available at http://www.epa.gov/scipoly/oscpendo/history/finalrpt.htm.
63. National Toxicology Program, *Report of the Endocrine Disruptors Low-Dose Peer Review*, August 2001; available at http://ntp-server.niehs.nih.gov/htdocs/liason/LowDosePeerFinalRpt.pdf.
64. Euling, S.Y., Gennings, C., Wilson, E.M., Kemppainen, J.A., Kelce, W.R., and Kimmel, C.A., Response-surface modeling of the effect of 5-dihydrotestosterone and androgen receptor levels on the response to the androgen antagonist vinclozolin, *Toxicol. Sci.*, 69, 332, 2002.
65. Wolf, C.J., LeBlanc, G.A., Ostby, J.S., and Gray, L.E., Jr., Characterization of the period of sensitivity of fetal male sexual development to vinclozolin, *Toxicol. Sci.*, 55, 152, 2000.
66. vom Saal, F.S., Timms, B.G., Montano, M.M., Palanza, P., Thayer, K.A., Nagel, S.C., Dhar, M.D., Ganjam, V.K., Parmigiani, S., and Welshons, W.V., Prostate enlargement in mice due to fetal exposure to low doses of estradiol or diethylstilbestrol and opposite effects at high doses, *Proc. Natl. Acad. Sci. USA*, 94, 2056, 1997.
67. Kohn, M.C. and Melnick, R.L., Biochemical origins of the non-monotonic receptor-mediated dose-response, *J. Mol. Endocrinol.*, 29, 113, 2002.
68. Almstrup, K., Fernandez, M.F., Petersen, J.H., Olea, N., Skakkebaek, N.E., and Leffers, H., Dual effects of phytoestrogens result in u-shaped dose-response curves, *Environ. Health Perspect.*, 110, 743, 2002.
69. Gaido, K.W., Maness, S.C., McDonnell, D.P., Dehal, S.S., Kupfer, D., and Safe, S., Interaction of methoxychlor and related compounds with estrogen receptor alpha and beta, and androgen receptor: structure-activity studies, *Mol. Pharmacol.*, 58, 852, 2000.
70. Slikker, W. Jr., Anderson, M.E., Bogdanffy, M.S., Bus, J.S., Cohen, S.D., Conolly, R.B., David, R.M., Doerrer, N.G., Dorman, D.C., Gaylor, D.W., Hattis, D., Rogers, J.M., Setzer, R.W., Swenberg, J.A., and Wallace, K., Dose-dependent transitions in mechanisms of toxicity, *Toxicol. Appl. Pharmacol.*, 15, 203, 2004.
71. Kimmel, C.A. and Francis, E.Z., Proceedings of the workshop on the acceptability and interpretation of dermal developmental toxicity studies, *Fundam. Appl. Toxicol.*, 14, 386, 1990.
72. Kimmel, C.A. and Young, J.F., Correlating pharmacokinetics and teratogenic endpoints, *Fundam. Appl. Toxicol.*, 3, 250, 1983.
73. Clarke, D.O., Duignan, J.M., and Welsch, F., 2-Methoxyacetic acid dosimetry-teratogenicity relationships in CD-1 mice exposed to 2-methoxyethanol, *Toxicol. Appl. Pharmacol.*, 114, 77, 1992.
74. Nau, H., Teratogenic valproic acid concentrations: infusion by implanted minipumps vs conventional injection regimen in the mouse, *Toxicol. Appl. Pharmacol.*, 80, 243, 1985.
75. Nau, H., Species differences in pharmacokinetics, drug metabolism and teratogenesis, in *Pharmacokinetics in Teratogenesis*, Vol. 1, Nau, H. and Scott, W. J., Eds., CRC Press, Boca Raton, 1987, p. 81.
76. Terry, K.K., Elswick, B.A., Stedman, D.B., and Welsch, F., Developmental phase alters dosimetry-teratogenicity relationship for 2-methoxyethanol in CD-1 mice, *Teratology*, 49, 218, 1994.
77. McCarver, D.G. and Hines, R.N., The ontogeny of human drug-metabolizing enzymes: phase II conjugation enzymes and regulatory mechanisms, *J. Pharmacol. Exp. Ther.*, 300, 361, 2002.
78. Cresteil, T., Onset of xenobiotic metabolism in children: toxicological implications, *Food Addit. Contam.*, 15(Suppl), 45, 1998.
79. Koukouritaki, S.B., Simpson, P., Yeung, C.K., Rettie, A.E., and Hines, R.N., Human hepatic flavin-containing monooxygenases 1 (FMO1) and 3 (FMO3) developmental expression, *Pediatr. Res.*, 51, 236, 2002.

80. Gabrielsson, J.L., Johansson, P., Bondesson, U., and Paalzow, L.K., Analysis of methadone disposition in the pregnant rat by means of a physiological flow model, *J. Pharmacokinet. Biopharm.*, 13, 355, 1985.
81. O'Flaherty, E.J., Scott, W.J., Nau, H., and Beliles, R.P., Simulation of valproic acid kinetics in primate and rodent pregnancy by means of physiologically-based models, *Teratology*, 43, 457, 1992.
82. Clarke, D.O., Elswick, B.A., Welsch, F., and Conolly, R.B., Pharmacokinetics of 2-methoxyethanol and 2-methoxyacetic acid in the pregnant mouse: a physiologically-based mathematical model, *Toxicol. Appl. Pharmacol.*, 121, 239, 1993.
83. Mattison, D.R., Physiologic variations in pharmacokinetics during pregnancy, in *Drug and Chemical Action in Pregnancy: Pharmacologic and Toxicologic Principles*, Fabro, S. and Scialli, A.R., Eds., Marcel Dekker, New York, 1986, p. 37.
84. Androit, M.D. and O'Flaherty, E.J., *Development of a Physiologically-Based Pharmacokinetic Model for Lead during Pregnancy*, Report prepared for the U.S. Environmental Protection Agency under Cooperative Agreement No. 820303-01-0, 1995.
85. Luecke, R.H., Wosilait, W.D., Pearce, B.A., and Young, J.F., A physiologically based pharmacokinetic computer model for human pregnancy, *Teratology*, 49, 90, 1994.
86. Byczkowski, J.Z., Gearhart, J.M., and Fisher, J.W., "Occupational" exposure of infants to toxic chemicals via breast milk, *Nutrition*, 10, 43, 1994.
87. Byczkowski, J.Z. and Lipscomb, J.C., Physiologically based pharmacokinetic modeling of the lactational transfer of methylmercury, *Risk Anal.*, 21, 869, 2001.
88. O'Flaherty, E.J., Physiologically based models for bone-seeking elements. V. Lead absorption and disposition in childhood, *Toxicol. Appl. Pharmacol.*, 131, 297, 1995.
89. Manton, W.I., Angle, C.R., Stanek, K.L., Reese, Y.R., and Kuehnemann, T.J., Acquisition and retention of lead by young children, *Environ. Res.*, 82, 60, 2000.
90. Ginsberg, G., Hattis, D., Sonawane, B., Russ, A., Banati, P., Kozlak, M., Smolenski, S., and Goble, R., Evaluation of child/adult pharmacokinetic differences from a database derived from the therapeutic drug literature, *Toxicol. Sci.*, 66, 185, 2002.
91. Hattis, D., Ginsberg, G., Sonawane, B., Smolenski, S., Russ, A., Kozlak, M., Goble, R., Differences in pharmacokinetics between children and adults--II. Children's variability in drug elimination half-lives and in some parameters needed for physiologically-based pharmacokinetic modeling, *Risk Anal.*, 23, 117, 2003.
92. O'Flaherty, E.J. and Clarke, D.O. Pharmacokinetic/pharmacodynamic approaches for developmental toxicity, in *Developmental Toxicology*, 2nd ed., Kimmel, C.A. and Buelke-Sam, J., Eds., Raven Press, New York, 1994, p. 215.
93. U.S. Environmental Protection Agency, *Exploration of Perinatal Pharmacokinetic Issues*, EPA/630/R-01/004. May 10, 2001, Risk Assessment Forum, Washington, D.C.; available at http://cfpub.epa.gov/ncea/cfm/recordisplay.cfm?deid=29420.
94. Keys, D.A., Wallace, D.G., Kepler, T.B., and Conolly, R.B., Quantitative evaluation of alternative mechanisms of blood and testes disposition of di(2-ethylhexyl) phthalate and mono(2-ethylhexyl) phthalate in rats, *Toxicol. Sci.*, 49, 172, 1999.
95. Plowchalk, D.R. and Teeguarden, J., Development of a physiologically based pharmacokinetic model for estradiol in rats and humans: a biologically motivated quantitative framework for evaluating responses to estradiol and other endocrine-active compounds, *Toxicol. Sci.*, 69, 60, 2002.
96. Smith, B.J., Carter, D.E., and Sipes, I.G., Comparison of the disposition and in vitro metabolism of 4-vinylcyclohexene in the female mouse and rat, *Toxicol. Appl. Pharmacol.*, 105, 364, 1990.
97. Fontaine, S.M., Hoyer, P.B., and Sipes, I.G., Evaluation of hepatic cytochrome P4502E1 in the species-dependent bioactivation of 4-vinylcyclohexene, *Life Sci.*, 69, 923, 2001.
98. U.S. Environmental Protection Agency, Endocrine Disruptor Screening Program Web Site, http://www.epa.gov/scipoly/oscpendo/index.htm, 2003.
99. National Research Council, *Scientific Frontiers in Developmental Toxicology and Risk Assessment*, National Academy Press, Washington, D.C., 2000, p. 354.
100. Shuey, D.L., Lau, C., Logsdon, T.R., Zucker, R.M., Elstein, K.H., Narotsky, M.G., Setzer, R.W., Kavlock, R.J., and Rogers, J.M., Biologically-based dose-response modeling in developmental toxicology: biochemical and cellular sequelae of 5-fluorouracil exposure in the developing rat, *Toxicol. Appl. Pharmacol.*, 126, 129, 1994.

101. Lau, C., Mole, M.L., Copeland, M.F., Rogers, J.M., Kavlock, R.J., Shuey, D.L., Cameron, A.M., Ellis, D.H., Logsdon, T.R., Merriman, J., and Setzer, R.W., Toward a biologically based dose-response model for developmental toxicity of 5-fluorouracil in the rat: acquisition of experimental data, *Toxicol. Sci.*, 59, 37, 2001.
102. Setzer, R.W., Lau, C., Mole, M.L., Copeland, M.F., Rogers, J.M., and Kavlock, R.J., Toward a biologically based dose-response model for developmental toxicity of 5-fluorouracil in the rat: a mathematical construct, *Toxicol. Sci.*, 59, 49, 2001.
103. Barton, H.A., Andersen, M.E., and Clewell, J.H., Harmonization: developing consistent guidelines for applying mode of action and dosimetry information to cancer and noncancer risk assessment, *Human Ecol. Risk Assess.*, 4, 75, 1998.
104. Barton, H.A., Deisinger, P.J., English, J.C., Gearhart, J.M., Faber, W.D., Tyler, T.R., Banton, M.I., Teeguarden, J., and Andersen, M.E., Family approach for estimating reference concentrations/doses for series of related organic chemicals, *Toxicol. Sci.*, 54, 251, 2000.
105. Barton, H.A. and Clewell, J.H., Evaluating effects of trichloroethylene: dosimetry, mode of action, and risk assessment, *Environ. Health Perspect.*, 108(Suppl. 2), 323, 2000.
106. Clewell, J.H., Andersen, M.E., and Barton, H.A., A consistent approach for the application of pharmacokinetic modeling in cancer and noncancer risk assessment, *Environ. Health Perspect.*, 110, 85, 2002.
107. Crump, K.S., A new method for determining allowable daily intakes, *Fund. Appl. Toxicol.*, 4, 854, 1984.
108. U.S. Environmental Protection Agency, *Benchmark Dose Technical Guidance Document*, External Review Draft, EPA/630/R-00/001, October 2000; available at http://cfpub.epa.gov/ncea/raf/recordisplay.cfm?deid=20871.
109. Jarabek, A.M., Consideration of temporal toxicity challenges current default assumptions, *Inhal. Toxicol.*, 7, 927, 1995.
110. Kimmel, G.L., Exposure-duration relationships: the risk assessment process for health effects other than cancer, *Inhal. Toxicol.*, 7, 873, 1995.
111. Witschi, H., The story of the man who gave us Haber's Law, *Inhal. Toxicol.*, 9, 201, 1997.
112. Witschi, H., Some notes on the history of Haber's Law, *Toxicol. Sci.*, 50, 164, 1999.
113. Eastern Research Group, *Summary of the U.S. EPA Workshop on the Relationship Between Exposure Duration and Toxicity*, Prepared for the U. S. Environmental Protection Agency, Risk Assessment Forum, Washington, D.C., under Contract No. 68-D5-0028, 1998; available from: Technical Information Staff (202-564-3261).
114. Pierano, W.B., Mattie, D., and Smith, P., Proceedings of the Conference on Temporal Aspects in Risk Assessment for Noncancer Endpoints, *Inhal. Toxicol.*, 7, 837, 1995.
115. Tzimas, G., Thiel, R., Chahoud, I., and Nau, H., The area under the concentration-time curve of all-*trans*-retinoic acid is the most suitable pharmacokinetic correlate to the embryotoxicity of this retinoid in the rat, *Toxicol. Appl. Pharmacol.*, 143, 436, 1997.
116. Weller, E., Long, N., Smith, A., Williams, P., Ravi, S., Gill, J., Henessey, R., Skornik, W., Brain, J., Kimmel, C., Kimmel, G., Holmes, L. and Ryan, L., Dose-rate effects of ethylene oxide exposure on developmental toxicity, *Toxicol. Sci.*, 50, 259, 1999.
117. Kimmel, G.L., Williams, P.L., Claggett, T.W., and Kimmel, C.A., Response-surface analysis of exposure-duration relationships: the effects of hyperthermia on embryonic development of the rat in vitro, *Toxicol. Sci.*, 69, 391, 2002.
118. U.S. Environmental Protection Agency, *Methods for Derivation of Inhalation Reference Concentrations and Application of Inhalation Dosimetry*, EPA/600/8-90/066F, 1994; available from: National Technical Information Service, Springfield, VA.
119. Faustman, E.M., Allen, B.C., Kavlock, R.J., and Kimmel, C.A., Dose-response assessment for developmental toxicity: I. Characterization of data base and determination of NOAELs, *Fundam. Appl. Toxicol.*, 23, 478, 1994.
120. Allen, B.C., Kavlock, R.J., Kimmel, C.A., and Faustman, E.M., Dose-response assessment for developmental toxicity: II. Comparison of generic benchmark dose estimates with NOAELs, *Fundam. Appl. Toxicol.*, 23, 487, 1994.
121. Allen, B.C., Kavlock, R.J., Kimmel, C.A., and Faustman, E.M., Dose-response assessment for development toxicity: III. Statistical models, *Fundam. Appl. Toxicol.*, 23, 496, 1994.

122. Kavlock, R.J., Allen, B.C., Kimmel, C.A. and Faustman, E.M., Dose-response assessment for developmental toxicity. IV. Benchmark doses for fetal weight changes, *Fundam. Appl. Toxicol.*, 26, 211, 1995.
123. Moore, J.A. and an IEHR Expert Scientific Committee, An assessment of lithium using the IEHR evaluative process for assessing human developmental and reproductive toxicity of agents, *Reprod. Toxicol.*, 9, 175, 1995.
124. Moore, J.A. and an Expert Scientific Committee, An assessment of boric acid and borax using the IEHR Evaluative Process for assessing human developmental and reproductive toxicity of agents, *Reprod. Toxicol.*, 11, 123, 1997.
125. Dourson, M.L., Felter, S.P., and Robinson, D., Evolution of science-based uncertainty factors in noncancer risk assessment, *Regul. Toxicol. Pharmacol.*, 24, 108, 1996.
126. Renwick, AG., Toxicokinetics in infants and children in relation to the ADI and TDI, *Food Addit. Contam.*, 15(Suppl), 17, 1998.
127. Renwick, A.G. and Lazarus, N.R., Human variability and noncancer risk assessment — an analysis of the default uncertainty factor, *Regul. Toxicol. Pharmacol.*, 27, 3, 1998.
128. Calabrese, E.J., Assessing the default assumption that children are always at risk, *Human Ecol. Risk Assess.*, 7, 37, 2001.
129. Schilter, B., Renwick, A.G., and Huggett, A.C., Limits for pesticide residues in infant food: a safety-based approach, *Regul. Toxicol. Pharmacol.*, 24, 126, 1996.
130. Peleki, M., Gephart, L.A., and Lerman, S.E., Physiological-model-based derivation of the adult and child pharmacokinetic intraspecies uncertainty factors for volatile organic compounds, *Regul. Toxicol. Pharmacol.*, 33, 12, 2001.
131. U.S. Environmental Protection Agency, Guidelines for estimating exposures, *Fed. Reg.*, 51, 34042, 1986.
132. U.S. Environmental Protection Agency, Guidelines for exposure assessment, *Fed. Reg.*, 57, 22888, May 29, 1992.
133. U.S. Environmental Protection Agency, *Exposure Factors Handbook*, Office of Health and Environmental Assessment, Washington, D.C., EPA/600/8-98/043, 1990.
134. U.S. Environmental Protection Agency, *Child-Specific Exposure Factors Handbook (Interim Report)*, EPA-600-P-00-002B, 01 Sep 2002, Office of Research and Development, National Center for Environmental Assessment, Washington Office, Washington, D.C., 2002, p. 448; available at http://cfpub.epa.gov/ncea/cfm/recordisplay.cfm?deid=55145
135. Daston, G.P., Relationships between maternal and developmental toxicity, in *Developmental Toxicology*, 2nd ed., Kimmel, C.A. and Buelke-Sam, J., Eds., Raven Press, New York, 1994, p. 189.
136. Manton, W.I., Angle C.R., Stanek, K.L., Kuntzelman, D., Reese, Y.R., and Kuehnemann, T.J., Release of lead from bone in pregnancy and lactation, *Environ. Res.*, 92, 139, 2003.
137. Silbergeld, E.K., Lead in bone: implications for toxicology during pregnancy and lactation, *Environ. Health Perspect.*, 91, 63, 1991.
138. Silbergeld, E.K., Schwartz, J., and Mahaffey, K., Lead and osteoporosis: mobilization of lead from bone in postmenopausal women, *Environ. Res.*, 47, 79, 1988.
139. Watson, L.R., Gandley, R., and Silbergeld, E.K., Lead: interactions of pregnancy and lactation with lead exposure in rats, *Toxicologist*, 13, 349, 1993.
140. Herbst, A.L., Ulfelder, H., and Poskanzer, D.C., Adenocarcinoma of the vagina. Association of maternal stilbestrol therapy with tumor appearance in young women, *New Engl. J. Med.*, 284, 878, 1971.
141. Herbst, A.L., Hubby, M.M., Blough, R.R., and Azizi, F., A comparison of pregnancy experience in DES-exposed and DES-unexposed daughters, *J. Reprod. Med.*, 24, 62, 1980.
142. Olshan, A.F. and Mattison, D.R., Eds., *Male-Mediated Developmental Toxicity*, Plenum Press, New York, 1994.
143. Robaire, B. and Hales, B.F., Eds., *Advances in Male Mediated Developmental Toxicity, Advances in Experimental Medicine and Biology*, Vol. 518, Kluwer Academic, New York, 2003, p. 298.
144. Selevan, S.G., Kimmel, C.A., and Mendola, P., Identifying critical windows of exposure for children's health, *Environ. Health Perspect.*, 108(Suppl 3), 451, 2000.
145. U.S. Environmental Protection Agency, Pesticide products containing dinoseb; notices, *Fed. Reg.*, 51, 36645, 1986.

146. The National Children's Study Interagency Coordinating Committee (Branum, A.M., Collman, G.W., Correa, A., Keim, S.A., Kessel, W., Kimmel, C.A., Klebanoff, M.A., Longnecker, M.P., Mendola, P., Rigas, M., Selevan, S.G., Scheidt, P.C., Schoendorf, K., Smith-Khuri, E., and Yeargin-Alsopp, M.), The National Children's Study of environmental effects on child health and development, *Environ. Health Perspect.*, 111, 642, 2003.
147. International Programme on Chemical Safety. *IPCS Workshop on Developing a Conceptual Framework for Cancer Risk Assessment*, 16-18 February, Lyon, France, IPCS/99.6, final draft, 1999.
148. Euling, S.Y. and Kimmel, C.A., Developmental stage sensitivity and mode of action information for androgen agonists and antagonists, *Sci. Total Environ.*, 274, 103, 2001.
149. Stoker, T.E., Guidici, D.L., Laws, S.C., and Cooper, R.L., The effects of atrazine metabolites on puberty and thyroid function in the male Wistar rat, *Toxicol. Sci.*, 67, 198, 2002.
150. Ginsberg, G.L., Assessing cancer risks from short-term exposures in children, *Risk Anal.*, 23, 19, 2003.
151. U.S. Environmental Protection Agency, *Supplemental Guidance for Assessing Cancer Susceptibility from Early-Life Exposure to Carcinogens*, EPA/630/R-03/003F. April 7, 2005, Risk Assessment Forum, Washington, D.C., 2005, p. 126; available at http://cfpub.epa.gov/ncea/raf/recordisplay.cfm?deid=116283.
152. Clewell, H.J., Teeguarden, J., McDonald, T., Sarangapani, R., Lawrence, G., Covington, T., Gentry, P.R., and Shipp, A.M., Review and evaluation of the potential impact of age and gender-specific pharmacokinetic differences on tissue dosimetry, *Crit. Rev. Toxicol.*, 32, 329, 2002.
153. Price, K., Haddad, S., and Krishnan, K., Physiological modeling of age-specific changes in the pharmacokinetics of organic chemicals in children, *J. Toxicol. Environ. Health*, 14, 66, 417, 2003.
154. U.S. Environmental Protection Agency, *Guidance on Cumulative Risk Assessment of Pesticide Chemicals That Have a Common Mechanism of Toxicity*, January 16, 2002, Docket Number: OPP-00658B; available at http://www.epa.gov/pesticides/trac/science/cumulative_guidance.pdf.
155. U.S. Environmental Protection Agency, *Guidance for Identifying Pesticide Chemicals and Other Substances That Have a Common Mechanism of Toxicity (Revised)*, February 5, 1999, Docket Number: OPP-00542; available at http://www.epa.gov/pesticides/trac/science/cumulative_guidance.pdf.
156. Mileson, B.E., Chambers, J.E., Chen, W.L., Dettbarn, W., Ehrich, M., Eldefrawi, A.T., Gaylor, D.W., Hamernik, K., Hodgson, E., Karczmar, A.G., Padilla, S., Pope, C.N., Richardson, R.J., Saunders, D.R., Sheets, L.P., Sultatos, L.G., and Wallace, K.B., Common mechanism of toxicity: a case study of organophosphorus pesticides, *Toxicol. Sci.*, 41, 8, 1998.
157. ILSI Risk Science Institute, *A Framework for Cumulative Risk Assessment*, Report of an ILSI Risk Science Institute workshop, 2001.
158. US Environmental Protection Agency, *Organophosphate Pesticides: Revised OP Cumulative Risk Assessment*, June 10, 2002; available at http://www.epa.gov/pesticides/cumulative/rra_op/.
159. U.S. Environmental Protection Agency, Guidance for performing aggregate exposure and risk assessments, *Fed. Reg.*, 66, 59428, November 28, 2001.
160. Slikker, W. Jr., Anderson, M.E., Bogdanffy, M.S., Bus, J.S., Cohen, S.D., Conolly, R.B., David, R.M., Doerrer, N.G., Dorman, D.C., Gaylor, D.W., Hattis, D., Rogers, J.M., Setzer, R.W., Swenberg, J.A., and Wallace, K., Dose-dependent transitions in mechanisms of toxicity: case studies, *Toxicol. Appl. Pharmacol.*, 15, 226, 2004.
161. U.S. Environmental Protection Agency, *Summary of the U.S. EPA Colloquia on a Framework for Human Health Risk Assessment*, Vol. 1; available at http://www.Cfpub.epa.gov/ncea/raf/index.cfm, 1997.
162. U.S. Environmental Protection Agency, *Summary of the U.S. EPA Colloquia on a Framework for Human Health Risk Assessment*, Vol. 2; available at http://www.Cfpub.epa.gov/ncea/raf/index.cfm, 1998.
163. Bogdanffy, M.S., Daston, G., Faustman, E.M., Kimmel, C.A., Kimmel, G.L., Seed, J., and Vu, V., Harmonization of cancer and non-cancer risk assessment: proceedings of a consensus-building workshop, *Toxicol. Sci.*, 61, 18, 2001.
164. US Environmental Protection Agency, *Guidance on Cumulative Risk Assessment, Part 1, Planning and Scoping*, Science Policy Council, July 3, 1997; available at http://www.epa.gov/osp/spc/cumrisk2.htm.
165. Hamadeh, H.K., Bushel, P.R., Jayadev, S., Martin, K., DiSorbo, O., Sieber, S., Bennett, L., Tennant, R., Stoll, R., Barrett, J.C., Blanchard, K., Paules, R.S., and Afshari, C.A., Gene expression analysis reveals chemical-specific profiles, *Toxicol. Sci.*, 67, 219, 2002.

166. Waring, J.F., Jolly, R.A., Ciurlionis, R., Lum, P.Y., Praestgaard, J.T., Morfitt, D.C., Buratto, B., Roberts, C., Schadt, E., and Ulrich, R.G., Clustering of hepatotoxins based on mechanism of toxicity using gene expression profiles, *Toxicol. Appl. Pharmacol.*, 175, 28, 2001.
167. Hamadeh, H.K., Bushel, P.R., Jayadev, S., DiSorbo, O., Bennett, L., Li, L., Tennant, R., Stoll, R., Barrett, J.C., Paules, R.S., Blanchard, K., and Afshari, C.A., Prediction of compound signature using high density gene expression profiling, *Toxicol. Sci.*, 67, 232, 2002.
168. U.S. Environmental Protection Agency, *Science Policy Council Interim Policy on Genomics*, July 25, 2002; available at http://www.epa.gov/osp/spc/genomics.htm.
169. Baird, S.J.S., Cohen, J.T., Graham, J.D., Shlyakhter, A.I., and Evans, J.S., Noncancer risk assessment: a probabilistic alternative to current practice, *Human Ecol. Risk Assess.*, 2, 79, 1996.
170. Brand, K.P., Rhomberg, L., and Evans, J.S., Estimating noncancer uncertainty factors: are ratios NOAELs informative? *Risk Anal.*, 19, 295, 1999.
171. Evans, J.S., Rhomberg, L.R., Williams, P.L., Wilson, A.M., and Baird, S.J.S., Reproductive and developmental risks from ethylene oxide: a probabilistic characterization of possible regulatory thresholds, *Risk Anal.*, 21, 697, 2001.
172. Gaylor, D.W. and Kodell, R.L., Percentiles of the product of uncertainty factors for establishing probabilistic reference doses, *Risk Anal.*, 20, 245, 2000.
173 Maull, E.A., Cogliano, V.J., Scott, C.S., Barton, H.A., Fisher, J.W., Greenberg, M., Rhomberg, L., and Sorgen, S.P., Trichloroethylene health risk assessment: a new and improved process. *Drug Chem. Toxicol.*, 20, 427, 1997.
174 Slob, W. and Pieters, M.N., A probabilistic approach for deriving acceptable human intake limits and human health risks from toxicological studies: general framework, *Risk Anal.*, 18, 787, 1998.
175. Swartout, J.C., Price, P.S., Dourson, M.L., Carlson-Lynch, H.L., and Keenan, R.E., A probabilistic framework for the reference dose (probabilistic RfD), *Risk Anal.*, 18, 271, 1998.

CHAPTER 22

Principles of Risk Assessment — FDA Perspective

Thomas F. X. Collins, Robert L. Sprando, Mary E. Shackelford, Marion F. Gruber, and David E. Morse

CONTENTS

I. Introduction and Background 878
II. Food Safety 878
   A. Perspective 878
   B. The Statutory Safety Standard for Additives 879
      1. Guidance for Food Additives 880
      2. Estimating Exposure to a Direct Food Additive 880
      3. Safety Testing of a Food or Color Additive 881
III. Vaccine Safety 890
   A. Perspective 890
   B. General Considerations 891
   C. Specific Considerations 891
      1. Animal Models 891
      2. Dose 892
      3. Immunization Interval 892
      4. Exposure Period and Follow-Up Period 892
      5. Study Endpoints 892
   D. Summary Remarks 893
IV. Drug Safety 893
   A. Perspective 893
   B. Product Labels 893
   C. Pregnancy Labeling — Practical Considerations...894
   D. Assessment of Reproductive and Developmental Toxicity 894
      1. Nonclinical Reproductive Toxicity Data Assessment — Future 895
      2. Data Needed for the Integration Process 895
      3. Types of Reproductive and Developmental Toxicity Evaluated........................895
      4. The Integration Process 896
      5. Section A. Overall Decision Tree Applicable to All Data-Sets.........................896
      6. Section B. No Signal (Applicable to Data-sets without Evidence
         of Reproductive or Developmental Toxicity)..898
      7. Section C. Applicable to Data-sets with One or More Positive Indications
         of Reproductive or Developmental Toxicity 900

8. Factor 1. Signal Strength, Part I 900
9. Factor 2. Signal Strength, Part II 902
10. Factor 3. Pharmacodynamics 903
11. Factor 4. Concordance between the Test Species and Humans 904
12. Factor 5. Relative Exposures 905
13. Factor 6. Class Alerts 906
14. Summary and Integration of Positive Findings 906
Acknowledgments 907
References 907

I. INTRODUCTION AND BACKGROUND

The mission of the Food and Drug Administration (FDA) is "to promote and protect the public health by helping safe and effective products reach the market in a timely way, and monitoring products for continued safety after they are in use" (http://www.fda.gov). The FDA thus regulates a multitude of diverse consumer products.

Five centers within the FDA are responsible for regulating the use of a wide range of products for humans and animals. For humans, the safety of food is regulated by the Center for Food Safety and Applied Nutrition (CFSAN) (excluding meat and poultry products regulated by the U.S. Department of Agriculture). Human drugs used for the treatment and prevention of disease are regulated by the Center for Drug Evaluation and Research (CDER). Biologics (a category that includes vaccines, blood products, biotechnology products, and gene therapy) are regulated by the Center for Biologics Evaluation and Research (CBER). Medical devices (a category that includes items that range from simple tongue depressors to complex items such as pacemakers and dialysis machines) are regulated by the Center for Devices and Radiological Health. The Center for Veterinary Medicine at FDA regulates drugs and devices used for animals (pets and also animals that produce food). Before manufacturers can market animal drugs, they must obtain FDA approval by providing proof of their safety and effectiveness. Livestock drugs are also evaluated for their safety to the environment and to the people who eat the animal products. Of the many products regulated by the FDA, only human food additives (regulated by CFSAN), drugs (regulated by CDER), and vaccines (regulated by CBER) are discussed here. The emphasis is on how reproductive and developmental toxicity studies are utilized in risk assessment at the FDA. The chapter reflects the complexity of the utilization of these studies within the agency.

II. FOOD SAFETY

A. Perspective

Food safety in the United States is intricately interwoven with legal decisions. From the mid-1800s to the early 1900s, use of chemical preservatives and toxic colors in foods was virtually uncontrolled. In 1938, Congress passed the Federal Food, Drug, and Cosmetic Act (FFDCA). Although imperfect, the law provided some measure of food safety. Some of the provisions were: poisonous substances could no longer be added to foods except where unavoidable or required in production; factories could be inspected; food standards were required to be set up when needed in the interest of consumers; legal remedies against violations included Federal court injunctions.

During the 1940s and early 1950s, with the increased needs of World War II and rapid advances in chemical and food technology, many new food additives were introduced by the food industry. Congress and the general public began to be concerned and to ask questions about the safety of some of the additives. The Pesticide Amendments of 1956 were the first Congressional actions requiring the preclearance of safety studies for chemicals added to food and granted the FDA authority to set tolerances for pesticides in foods (Section 408).

In 1958, Congress enacted the Food Additives Amendment (Section 409) to the FFDCA. This amendment requires that a food additive be shown to be safe under its intended conditions of use before it is allowed in food. The amendment requires premarket safety evaluation of food additives and provides the FDA with authority to issue tolerances and regulations approving the use of food additives. In establishing the standard for demonstrating the safety of a food additive, Congress recognized that it is not possible to determine with absolute certainty that no harm will result from the intended use of an additive. Specifically, therefore, the legislation requires that the petitioner provide proof of reasonable certainty that no harm will result from the intended use of the additive.

Prior to 1958, upon request of the industry, FDA periodically would review the safety of various chemicals and render letters of advisory opinion on their acceptability, under certain conditions, as safe additives in food.[1] These written FDA Advisory opinions were recognized by Congress in the 1958 Food Additives Amendment as "prior sanctions." In the food additives regulations, these are categorized in 21 CFR Part 181.

Under Section 409 of the act as originally established, food additives require a food additive petition, premarket approval by FDA, and publication of a regulation authorizing their intended use. The Food Additives Amendment applies to both "direct" and "indirect" food additives. Direct additives (21 CFR Part 172) are substances added deliberately in finite amounts, for whatever technical purpose served, and remain in the food as consumed. Secondary food additives (21 CFR Part 173) represent substances added to food or food components during manufacture or processing but removed before the final product is consumed. Indirect food additives (21 CFR Parts 175 and 179) are not intentionally added to food; they are substances used as articles or components of articles that are intended for use in packaging, transporting, or holding food. Because indirect food additives are not intended to become components of the food, their presence in food may be the result of migration or inadvertent extraction from the food contact surface.[1]

Section 309 of the Food and Drug Administration Modernization Act of 1997 (FDAMA) amends Section 409 of the FFDCA (21 USC 348) to establish a notification procedure as the primary method by which the FDA regulates food additives that are food contact substances (FCS). Food contact substances include all substances that are intended for use as components of materials used in manufacturing, packaging, transporting, or holding food if the use is not intended to have any technical effect in the food.

The Delaney Clause, part of the 1958 Food Additives Amendment (Section 409) to the FFDCA, sets a zero-tolerance cancer-risk standard for substances added directly or indirectly to food. This far-reaching standard is still the priority test for food additives, and any other risks or risk assessments are based on knowledge of the noncarcinogenic properties of the additive.

In 1960, Congress passed the Color Additive Amendments (Section 721) of the FFDCA, which required that color additives be shown to be safe for use in or on food, drugs, or cosmetics prior to approval. The amendments created an analogous premarket approval process for color additives used in foods, drugs, and cosmetics, or medical devices. Approved color additives are listed in 21 CFR Parts 73 and 74.

The act also recognizes GRAS (Generally Recognized as Safe) status for certain food ingredients. GRAS substances are generally recognized by experts as safe, based either on their extensive history of use in food before 1958 or on generally available and accepted scientific evidence.[1] The available data and information must establish that the intended use of the substance is safe.

B. The Statutory Safety Standard for Additives

The Food Additives Amendment of the 1938 Act requires that a food additive be shown to be safe under the intended conditions of use before it is allowed in food. This safety requirement is often referred to as the general safety clause for food additives. Although what is meant by "safe" is not explained in the general safety clause for food additives, FDA regulations define "safety" as a reasonable certainty in the minds of competent scientists that the substance is not harmful under the intended conditions of use. It does not require proof beyond any possible doubt that no harm will result under any conceivable circumstance. This concept of safety requires the consideration of available data on the additive in question, as well as additional relevant information including: (1) the probable consumption of the additive, (2) the cumulative effect of such additive in the diet of humans or animals, and (3) safety factors that qualified experts consider appropriate.

1. Guidance for Food Additives

The act requires that chemicals added to food be tested and/or evaluated for safety. This is a general requirement for food additives and color additives. Safety evaluation usually involves toxicology testing by a battery of studies in animals. Protocols for safety testing reflect the increased sophistication of the science of toxicology as well as an increased concern about food safety.[1] Despite the increased sophistication of knowledge and techniques, the four principles of toxicity testing for the evaluation of food additives published by the National Research Council (NRC)[2] remain essentially the same. These principles are:

- Animal test endpoints are appropriate for identifying human risk.
- Extrapolation of safety information from animal test data to humans requires an understanding of species differences in metabolism and disposition, sensitivity of target tissue, and rates of injury and repair.
- The principle of dose response relationship, that intensity, time of onset, and duration of biological effects are related to dose, is central to toxicological evaluation. Generally, the greater the dose, the greater the likelihood of observing a toxic effect.
- The potential exposure from the intended use must be assessed in the evaluation of the safety of food additives.

Shortly after the NRC publication, the FDA issued *Toxicological Principles for the Assessment of Direct Food Additives and Color Additives Used in Food*.[3] The publication was intended to serve several purposes: to assess a substance's potential for causing toxicity, to determine if safe conditions of use can be established for food additives, and to provide testing guidance, including a priority-based ranking scheme for food additives that ranks compounds according to their level of health concern (based on the extent of human exposure and the compound's known or presumed toxicological effects). The 1982 guidelines provided general toxicity guidelines and promulgated individual tests to be used for determining a compound's potential for causing specific types of toxicity, such as reproductive and teratological effects. Tests for reproductive and teratological effects had been available before this time, but this was the first time they were issued together with other tests as part of a set of guidelines. Further expansion and refinement of the tests was done in the 1993 Draft Redbook II.[4] As the individual guidelines are reviewed and updated, they are made available online on FDA's Web site.[5] The testing guidelines for female reproduction and teratology studies are posted online (http://www.cfsan.fda.gov/~redbook/redivc9a.html for reproduction studies and http://www.cfsan.fda.gov/~redbook/redivc9a.html for developmental toxicity studies) and are readily available, so they will not be repeated here.

2. Estimating Exposure to a Direct Food Additive

Section 409 of the act describes the statutory requirements that are required to establish safety of a food additive. The chemical and technological data encompass four general areas of information: (1) identity of the additive, (2) proposed use of the additive, (3) intended technical effect of the additive, and (4) a method of analysis for the additive in food. Section 409 also requires that probable consumption of an additive and of any substance formed in or on food because of its use be considered in determining whether the proposed use is safe. An estimate of probable human exposure to food additives or any of its by-products, resulting from the ingestion of food containing the substance, is most often referred to as an estimated daily intake (EDI). A cumulative estimated daily intake (CEDI) is calculated if a new use of a regulated additive results in an increase in consumption. A detailed evaluation of the consumer exposure to direct food additives is posted at FDA's Web site in the FDA Guidance for Industry *Recommendations for Submissions of Chemical and Technological Data for Direct Food Additives and GRAS Food Ingredient Petitions*.[6]

3. Safety Testing of a Food or Color Additive

Under the Food Drug and Cosmetic Act, the safety of food additives must be established prior to marketing by evaluating the probable exposure to the new substance and appropriate toxicological and other scientific information. Thus, the approval of any new food additive used in food depends in part upon the outcome of toxicity tests that are performed and evaluated prior to marketing. Through the years, the FDA has and continues to adjust testing recommendations for food and color additives used in food as necessary to reflect the steady progress of science and current information about population exposure.[7] Protocols for safety testing reflect the increased sophistication of the science of toxicology as well as an increased concern about food safety.

a. Direct Food Additives and Color Additives

The FDA's *Toxicological Principles for the Assessment of Direct Food Additives and Color Additives Used in Food*[3] recommended a series of individual tests to be used to determine a compound's potential for causing specific types of toxicity. The document sets out specific principles to determine the likelihood that the test substance poses a potential for health risk to the public. The document proposed that the development of information needed for the safety assessment process should be done in a tiered procedure whereby the small number of additives with the highest probable risk would receive the greatest amount of scrutiny and the large number of substances with minimal potential toxicity will receive a less comprehensive and detailed evaluation. This concept was referred to as the "concept of concern," and the safety evaluation was determined by the "levels of concern." This tiered approach to testing has been the subject of several reviews.[1,8-10]

Three concern levels are generally used, based on molecular structures combined with exposure information for the entire population. As the duration of exposure increases, the types of effects are determined with greater sensitivity. Tests for Concern Level 1 are short-term tests and are the least sensitive. Tests for Concern Level 2 are of intermediate sensitivity, and consist of subchronic feeding studies, a reproduction study with a teratology phase, and short-term tests for carcinogenicity. Tests for Concern Level 3 are the most demanding. In addition to chronic studies, a reproduction study with a teratology phase is required, with the possibility of testing additional generations.

Expansion and refinement of the tests was done in the 1993 Draft Redbook II.[4] The FDA continues to propose the use of Concern Levels as the basis for determining minimum toxicology testing recommendations. The tests for Concern Levels 2 and 3 still require multigeneration reproduction studies, each with a teratology phase. As the individual guidelines are reviewed and updated, they are made available on FDA's Web site.

b. Food Contact Substances

Section 309 of the Food and Drug Administration Modernization Act of 1997[11] amended Section 409 of the FFDCA to establish a notification procedure as the primary method by which the FDA regulates food additives that are food contact substances. The most recent FDA Guidance for Industry is *Preparation of Food Contact Notifications for Food Contact Substances: Toxicology Recommendations*.[12] These recommendations are consistent with the general principle that the potential risk of a substance is likely to increase as exposure increases. Food contact substances with a dietary concentration at or greater than 0.5 ppb are now regulated under the food contact notification process for food contact substances.[13]

Because the safety standard is the same for all food additives, the chemical and technological information needed for food contact substances is comparable to that recommended for a direct food additive. The probable consumer exposure may include the food contact substance itself and the degradation products and impurities in the food contact substance. The cumulative estimated daily intake from all regulated uses and effective food contact substance notifications are used by the FDA in the safety evaluation of a food contact substance. The most recent information on the estimation of consumer exposure to food contact substances is posted at FDA's Web site in FDA Guidance for Industry, *Preparation of Premarket Notifications for Food Contact Substances: Chemistry Recommendations.*[14]

The FDA Guidance for Industry recommends safety testing to assess the safety of the food contact substance and/or various constituents of the food contact substance. If the probable human dietary exposure for a single use is at or less than 0.5 ppb (1.5 µg/person/day), then no new safety studies are recommended. If the cumulative human dietary exposure is greater than 0.5 ppb (1.5 µg/person/day) but not exceeding 50 ppb (150 µg/person/day), then a battery of genotoxicity tests is recommended to evaluate the potential for carcinogenicity of the food contact substance and/or constituent. If the cumulative human dietary exposure is between 50 ppb (150 µg/person/day) and 1 ppm (3 mg/person/day), then the potential toxicity of the food contact substance and/or constituent should be evaluated in a battery of genotoxicity studies as well as subchronic oral toxicity studies in rodents and nonrodents. These studies are used to determine if longer term or specialized toxicity tests are needed to assess safety. Additional specialized toxicity tests include metabolism studies, teratology studies, reproductive toxicity studies, neurotoxicity studies, and immunotoxicity studies.

Food contact substances that are used primarily for their antimicrobial or fungicidal effects are toxic by design. The minimum safety testing recommendation is one-fifth the value of the cumulative estimated daily intake (CEDI) used to determine the appropriate safety testing for other types of food contact substances.[15]

c. *Determination of the ADI*

The FDA's 1982 guidelines[3] defined a tiered system of tests to be conducted to determine the effects of food additives. The goal of the testing is to assess the substance's potential for causing toxicity and determine if safe conditions of use can be established, provided the additive serves its intended function. Not only do the guidelines provide testing guidance, they also set out a priority-based ranking scheme for additives in food that ranks compounds according to their level of health concern. Health concern is, in turn, based on the extent of human exposure and the assessed or presumptive toxicological effects.

A fundamental principle used in setting tolerances (maximum acceptable limits of intake) for chemicals in food is that toxicity is a result of dose response and that there is no toxic effect at some low dose. In ensuring safe conditions for use, the FDA must consider whether a tolerance limit for a food additive should be specified by regulation. For this, the likely level of consumer exposure to an additive (from the estimated daily intake) and the maximum level that can be considered safe or noninjurious to health (from the maximum acceptable daily intake, or ADI) must be compared. The EDI is the likely level of consumer exposure to an additive from the daily use of a food containing it. The ADI is the daily intake of a chemical that, during an entire lifetime, appears to be without appreciable risk on the basis of all the known facts at the time. The definition refers to average lifetime exposure, and thus it can be inferred that slightly exceeding the ADI for a period may be compensated for by periods when consumption is below the ADI.[16] The ADI is an estimate that depends on a number of statistically calculated levels, some of which may be quite primitive. It is a yardstick that should be used with care to set permitted levels for additives in food.

Both chemical and biological data are needed to establish an ADI, and all the toxicological data are taken into account in deriving an ADI. The chemical information helps to determine the relevance and importance of a study in estimating human risk. Chemical information and data required include identity and purity of the chemical, impurities, stability, metabolic products, products formed in combination with other food components, absorption, distribution, metabolism, etc.

In deriving an ADI, the no effect level (NOEL) is used as determined in the most susceptible but appropriate animal and with the most sensitive indicator of toxicity. Determination of the appropriateness of a species requires in-depth knowledge of comparative physiology, biochemistry, and metabolism in laboratory animals and humans. Acute toxicity studies and short-term toxicity studies are valuable, but the most sensitive indicator has historically been found in long-term studies. For substances that could affect pregnant women, special studies should be conducted, including tests for reproductive function and developmental toxicity.

If the probable level of consumer exposure to the additive (i.e., the EDI) is determined to be well below the maximum safe level of consumption (i.e., the ADI) and typical use levels are unlikely to be exceeded, the use of the additive could be considered self-limiting and would not require a quantitative tolerance. Use of the additive under conditions of good manufacturing practice is considered acceptable where there is a wide margin of safety (i.e., the EDI is well within the ADI) and the use is self-limiting. If, however, the exposure may exceed safe limits under certain conditions of use, a quantitative tolerance may be required. The tolerance may be set to limit the amount of additive to be used in all foods, in a range of foods, in specific foods, or foods in certain food categories.[1]

d. Interpretation and Evaluation of Data from Developmental and Reproductive Toxicity Tests

There are difficulties in interpreting animal studies, in defining a strict cutoff point between what can be regarded as adverse and extrapolating from this point to give acceptable levels for humans, as the regulatory process requires. To allow for these uncertainties, a safety factor is used to extrapolate from the no-effect level obtained in animal studies to an ADI. The safety factor is intended to allow for differences in sensitivity between the animal species and humans, variations in susceptibility among the human population, and the fact that the number of animals tested is small compared with the size of the human population that may be exposed. A safety factor of 100 was proposed by WHO in 1958,[17] and this has been accepted as adequate by the FDA for some substances. The safety factor is influenced by the nature of the observed toxicity and the adequacy of the data. Larger factors are used to compensate for slight deficiencies in toxicity data, such as small numbers of animals. The safety factor of 1000, as proposed by Jackson,[18] linked dosage with the severity of anomalies seen in developmental toxicity studies.

The interpretation of data from developmental toxicity studies is similar to the interpretation of data from other toxicity tests in that the animal tests must be well designed and properly conducted. After the data have been collected, it must be determined whether any observed differences between treated and control groups are real, whether they can be reproduced, and whether they can be attributed to treatment with the test chemical.[18]

Observed differences that are spontaneous must be distinguished from those that are induced by exposure to the test chemical. Spontaneous changes may result from hereditary factors, disease, or environmental factors. If proper randomization has been done, these changes are as likely to occur with the same frequency in treated and untreated animals. The use of concurrent experimental controls provides the primary evidence of whether a particular change can be regarded as treatment related. Colony control data and strain data collected from other laboratories can be useful when

Table 22.1 Types of adverse developmental effects

Types	Changes	Developmental Toxicity Examples
I	Permanent (irreversible), possibly life-threatening changes, frequently associated with gross malformations	Decreased number live fetuses (litter size) Increased number resorptions Increased number fetuses with malformations, structural changes
II	Nonpermanent (reversible) changes, not life-threatening, usually not associated with gross malformations	Increased number fetuses with retarded development of skeletal or soft-tissues

Source: Jackson, B.A., 1988, p. 207.

evaluating the significance of the occurrence of low incidence or rare malformations. Statistical analyses are employed for identifying changes that are unlikely to be chance occurrences.[18]

The endpoints observed in the animal test system represent various manifestations of outcomes of altered development. These endpoints fit three general and well-known responses: (1) altered rate of development, (2) abnormal development, and (3) death.

In the past, attention was focused on the occurrence of developmental malformations. Now, neurological, immunological, and carcinogenic effects are recognized as possible manifestations of developmental toxicity. In reproduction studies, it is sometimes more difficult to determine whether intrauterine or postnatal lethality was a result of a lethal malformation or a manifestation of maternal or fetal toxicity. Examination of dead fetuses for structural changes or malformations in many cases reveals only partial information regarding whether malformations preceded or accompanied death.

A multigeneration test with two generations and one litter per generation, as currently recommended by FDA, can help evaluate the adverse effects of agents on the reproductive systems of both males and females, as well as postnatal maturation and reproductive capacity of the offspring. Also, and perhaps most importantly, the cumulative effects of a substance through several generations can be evaluated.

Of the multitude of developmental and reproductive effects, some changes are more serious than others. Jackson[18] proposed classifying observations into two types, according to severity and reversibility (Table 22.1). Type I changes are perceived as permanent and irreversible, possibly life threatening, and frequently associated with gross malformations. Type II changes are perceived as nonpermanent and reversible, nonlife threatening, and not necessarily associated with induced malformations. Type I changes include those associated with death and abnormal development, such as decreased litter size, increased resorptions, and gross malformations. Type II changes include retarded skeletal and soft-tissue development, decreased fetal weight, and decreased postnatal growth. In using the latter category, emphasis should be placed on acquiring sufficient information to rule out the possibility that the test chemical presents a special and unique hazard to the developing organism.[18]

A stepwise evaluation of developmental toxic effects should be done. Any combination of Type I and Type II changes would result in the provisional classification of the test chemical as a Type I compound. Disparities between the outcomes of two test systems might indicate metabolic, mechanistic, or other differences that could be used in clarifying its hazard to humans.[18]

Appropriate procedures are necessary to translate the information on developmental toxicity and reproductive effects into regulatory action. Such procedures must consider all the information on a particular chemical, including the results from other toxicological evaluations. For example, if metabolic data indicate that humans detoxify a compound more rapidly than the test species that exhibited an adverse effect, then this can be reflected in the safety assessment.

Any attempt to predict the conditions under which a compound can be safely used brings sharply into focus those areas in which our knowledge is less well established. Two such areas can be identified, and they merit some discussion here. One area that has been much studied but which still contains deficiencies is how to separate the range of abnormalities that are manifestations of interference with development. In the Type I and Type II classification scheme proposed above, the two extremes of the range of deficiencies are considered. Between the extremes, a determination of safety can present difficulties. One possible link is the dose response relationship. Materials that induce frank malformations can be expected to induce other effects at the same and lower doses. Another problem area is whether a distinction can be made between substances that produce malformations at dosages that are toxic to the mother and those substances that produce these changes at doses that are not maternally toxic. The classic example of this phenomenon is thalidomide's adverse effects on the fetus at levels that are not maternally toxic.

Safety factors applied to the results of animal toxicity studies are employed to establish the acceptable exposure levels for the use of compounds by humans. A 1000-fold safety factor would be applied in those cases where there was evidence of serious, irreversible developmental effects (Type I effects), and a 100-fold safety factor would be applied to a Type II hazard. In those instances where the results in two test systems are discordant, consideration could be given to information on maternal toxicity, metabolic differences, and mechanism as mitigating factors.

Two-generation reproduction studies with a developmental toxicity phase are used as the basic screening tool for identifying compounds with the potential for producing adverse developmental effects. Testing needs are identified in a stepwise fashion from the results of the initial screening and information such as exposure and relation to known developmental toxicities. The combination of endpoints is used to maximize test sensitivity.

For developmental and reproductive effects, the primary measures of toxicity are generally the proportion of dead and resorbed implants per litter and the proportion of various types of malformations among the live fetuses in a litter. The variability of these proportions among litters within dose groups must be considered in testing for toxicity. Litter size and fetal weight are examined as well and may be used as covariates or intermediate variables for dose response relationships to estimate the risk.[19]

Measurements of biological effects, such as body weight, organ weight, hormone concentrations, enzyme concentrations, etc., are measured on a continuous scale. For a biological effect measured on a continuous scale, there generally is not a sharp point of demarcation between a normal and an adverse response. For such cases, a normal range can be established in unexposed control animals or individuals, and the risk can be based upon the magnitude of the abnormal response.

e. *Interpretation and Evaluation of Data from Male Reproductive Toxicity Tests*

Between the FDA guidelines published in 1982 and those published in1993, a major revision involved the increased emphasis on reproductive effects in males. Compared with other animal species, the human male is relatively infertile and therefore is at greater reproductive risk from exposure to various toxicants than laboratory models commonly used in testing protocols. Unfortunately, in the past, the ability of a compound to induce damage to the male reproductive system was assessed by examining fertility. It is now recognized that fertility, which measures the end result of the reproductive process, is not a sensitive predictor of male reproductive toxicity in the rodent. A comprehensive assessment of male reproductive toxicity, which requires information from multiple endpoints, has been included in testing guidelines from national and international organizations (EPA, OECD, ICH, etc.).

Data obtained from male reproductive toxicity studies must be evaluated in light of the interspecies variation between the human and the test species. At present, the extrapolation of animal data to the human is troublesome, especially with respect to the male because of a lack of knowledge of the physiological variation between the human male and males from various test species. Therefore, it is very important that regulatory animal testing protocols for regulatory purposes be well designed and properly conducted. Testing protocols utilized in the regulatory process should be designed to simulate the factors that affect the test results for reproductive risk assessment (i.e., route of exposure, potential for bioaccumulation, and the comparability of the test species to the human) for the compound in question. Several important parameters that must be considered in designing a study to test for effects on the male reproductive system are the duration of exposure, appropriate routes of exposure, level of exposure associated with adverse risks in humans, choice of the test species, and the endpoints selected for study. Other parameters that should be considered include appropriate fixatives for the preservation of reproductive tissues and the most appropriate methods to detect changes in the selected endpoints.

The endpoints commonly observed in male reproductive toxicity testing include animal body weight, fertility, reproductive organ weights, sperm quality (morphology and motility), sperm quantity (cauda epididymal sperm numbers and testicular homogenization-resistant spermatid numbers), histopathological observations on reproductive organs (and in particular the testis), anogenital distance, and when necessary, reproductive endocrine profiles. Many of the endpoints obtained from these studies are interrelated. The strength of collecting this type of data lies in the fact that data obtained on one endpoint can help corroborate effects observed on other endpoints. It is therefore suggested that a weight-of-evidence approach be taken when interpreting male reproductive toxicity data obtained in regulatory testing protocols.

The endpoints observed in the animal test system represent various manifestations of outcomes of altered sperm production, sperm quality, and sperm quantity. A multitude of reproductive effects can be observed from exposure to a toxicant, but the effects are unequal. Effects observed should not only be classified as to their severity but also as to whether they are reversible. For example, compounds that destroy stem cells will have a long-term irreversible effect on the testis, whereas another compound may have only a short-term effect on a different cell population.

The following presents a brief discussion of some endpoints of male reproductive toxicity commonly collected and their use in the risk assessment process. Among the endpoints of male reproductive toxicity commonly assessed, two endpoints have received the most attention in the last 10 years: evaluation of testicular histopathology and assessment of sperm motility. These endpoints are discussed first and are followed by a discussion of other selected endpoints.

1) Testicular Histopathology — Careful histopathological examination of the testis is recognized as a sensitive method to identify effects on spermatogenesis. The sensitivity of this method can be increased or decreased depending on the quality of the tissue fixation and the embedment of the testicular tissues. In the past, histopathological evaluation of the testis identified only major changes between control and treated groups. Subtle lesions (retained spermatids, missing germ cell layers) may not have been observed because the tissue preservation methodology was substandard and pathologists were not familiar with testicular morphology and the kinetics of spermatogenesis of the test species examined. Testicular morphological changes can be used to identify injury that is not severe enough to cause a reduction in fertility or sperm production in the rat. Morphological changes that occur in the absence of functional changes (i.e., libido, sperm motility, fertility) should be interpreted as an indication of the potential for the compound to affect spermatogenesis adversely in the human. Although histopathological changes may fail to reveal treatment related effects, consideration should be given to the possible presence of other testicular or epididymal effects not detected histologically that may affect reproductive function.

Excellent articles and books have been written describing methods that can be used to preserve testicular tissue as well as to evaluate testicular histopathology.[20,21] These resources provide various approaches for the qualitative and/or quantitative assessment of testicular lesions. Included in the books are detailed descriptions of the various stages of the cycle of the seminiferous epithelium and extensive information on tissue preparation and examination, and interpretation of observations from normal microscopic and ultrastructural histology of many experimental animals.

The histological picture seen after toxic insult may provide insight into the site and/or mechanism of action of the toxicant on a particular reproductive organ. If similar targets or mechanisms exist in humans and other test species, the data can be utilized for extrapolation between species and may allow a prediction of the extent of damage and the chances for recovery.[22]

2) Sperm Motility — Sperm motility may be assessed either from the vas deferens or the cauda epididymis (rat and mouse) or from an ejaculate (rabbit and dog). In rats and mice, placing the vas or cauda epididymis in an appropriate medium and allowing sperm to swim out is a popular method for collecting sperm for analysis.[23] Sperm motion may be captured on videotape or digitally formatted and analyzed at a later date, either manually or by computer-assisted sperm analysis (CASA) systems.

Although there is no standard for the percentage of motile sperm, the general consensus is that in a control sample the number of "progressively" motile sperm should be at least 70%.[24,25] Visual estimates of counts of moving sperm are straightforward. CASA systems have given us the ability to quantify a number of sperm motion parameters. For the CASA systems, a variety of parameters must be defined correctly for accurate image acquisition and analysis. It is important to remember that the information obtained from a CASA system is dependent upon the instrument settings. Two different laboratories can use the same instrument but use different settings and obtain different results for various motion parameters. Therefore, data obtained by these instruments cannot be compared across laboratories, and all values obtained should be compared to concurrent controls, not historical controls.

The use of the CASA system allows for the rapid and efficient collection of motility information, including velocity (straight-line and curvilinear) measurements, beat cross-frequency, and amplitude of lateral head displacement measurements, on a large number of sperm. Recently, using CASA systems, compound-induced changes in motility have been detected[26-29] and associated with changes in fertility[30-32] in experimental animals. Preliminary results have suggested that significant reductions in sperm velocity are associated with reduced fertility, even when the percentage of motile sperm is not affected. It is important to distinguish motility from progressive motility. Sperm may be motile but not demonstrate a directed forward movement. A change in motility may be independent of changes in testicular histopathology because the agent may adversely affect epididymal function. Motility is also influenced by abstinence, the time between obtaining the sample and the evaluation of the sample, medium pH, sample chamber depth, and temperature. The investigator should be careful to avoid artifactual cell death during specimen preparation so that the percentage of motile sperm from control animals is consistently high. A reduction in the velocity or pattern of motility of treated animals when compared with controls may indicate a need for further testing. Of all the sperm motion parameters, the percent motility is clearly related to fertility. Therefore, any change in the percent motility should be considered an adverse effect, especially if it is dose related.

3) *Other Selected Endpoints of Male Reproductive Toxicity*

a) Organ Weights — Other endpoints commonly evaluated are reproductive organ weights and histopathological effects observed on male secondary sex organs. The male reproductive organs for which weights may be useful for reproductive risk assessment include the testis, epididymis, pituitary, seminal vesicle with coagulating gland, and the prostate. Reproductive organ weight changes may occur irrespective of a decrease in body weight; however, it should be remembered that changes in other endpoints related to reproductive function (i.e., motility) might be altered without a concomitant change in reproductive organ weight. Organ weights should be reported as absolute weights or relative weights (weights per gram of body weight; weights per gram of brain weight).[33] Care should be taken to ensure that deleterious reproductive effects are not masked because the organ weight and body weight are both reduced by treatment. Testicular weights are relatively stable and vary only modestly within a given test species.[34,35] Increases or decreases in testis weight indicate that a compound has significantly altered testis structure but do not help to characterize the effect.

Depending on the treatment period, not all testicular effects will result in a decrease or an increase in testicular weight. For example, short treatment periods may produce increased germ cell degeneration that may not immediately be reflected in testicular weights. Additionally, blockage of the efferent ducts by cells sloughed from the seminiferous epithelium may result in increased testis weights, or slight increases in testis weight due to increased intratesticular fluid might offset the weight change induced by germ cell depletion.[36,37] Testicular effects may also be observed at dose levels higher than those required to affect other measures of gonadal status.[38-40] Although a change in testicular weight may not reflect certain adverse testicular effects and does not indicate the nature of an effect, a significant increase or decrease is indicative of an effect.

In conclusion, statistically or biologically significant changes in absolute or relative male reproductive organ weights may indicate an adverse reproductive effect and provide a basis for obtaining additional information on toxicity of an agent. It is important to note that significant alterations in other endpoints related to reproductive function may not be reflected in organ weight data. Therefore, lack of a change in organ weight does not indicate a lack of effect in other reproductive endpoints.

b) Histopathology — The histopathological evaluation of the testis and other reproductive organs (epididymis, prostate, seminal vesicles with coagulating gland, and pituitary) plays a very important role in reproductive risk assessment. Histopathological observation can provide information on the toxicity of the compound, the site and extent of toxicity, and the potential for recovery. It can be useful as a tool for identifying low-dose effects. Significant histopathological damage in excess of the level seen in the control tissue of any male reproductive organ should be considered an adverse reproductive effect. Of paramount importance in any histological examination is the proper fixation and embedding of the tissue. Appropriate fixatives and embedding materials that minimize the extent of fixation artifact should be utilized.

c) Sperm Morphology — Sperm morphology profiles are relatively stable and characteristic in a normal individual over time. Sperm morphology is one of the least variable sperm measures in normal individuals, which may enhance its use in the detection of spermatotoxic events.[41] During spermiogenesis there is a progressive condensation of chromosomal material in the spermatid during successive stages of differentiation, and this results in the tight packaging of DNA in the heads of the spermatozoa. Although temporal changes in nuclear basic proteins are involved, the chemical basis for changes in chromatin packing is not well understood. In eutherian mammals, somatic histones present in cells at early stages of spermatogenesis are replaced entirely by protamine. In the rat it has been shown that three histonelike proteins first appear in primary spermatocytes but are not retained in late spermatids and that new proteins are synthesized just prior to chromatin condensations in elongating spermatid nuclei. Thus, chromatin condensation is a highly complex process that involves the appearance and disappearance of various nucleoproteins throughout the condensation process. Therefore, changes in sperm head morphology indicate that the agent has in some way affected spermatogenesis and that the quality of the sperm produced has been in some way affected. A number of studies in test species and humans have suggested that abnormally shaped sperm may not reach the oviduct or participate in fertilization,[42,43] suggesting that the greater the number of abnormal sperm in the ejaculate, the greater the probability of reduced fertility. Care should be taken in evaluating abnormal sperm morphology because altered nuclear morphology should not be interpreted as indicating genetic damage, that is, not all dysmorphogens are mutagens.

Traditionally the characterization of abnormalities in sperm morphology is based on the appearance of the sperm head, midpiece, and tail. The sperm may be visualized as unfixed samples by bright field microscopy,[44] or fixed, unstained preparations can be examined by phase contrast microscopy.[24,45] This approach is useful in species in which the sperm population is homogeneous (i.e., rats and mice); however, in species in which sperm morphology is heterogeneous (human), sperm morphological alterations are difficult to assess. Data that categorize the types of abnormalities observed and quantify the frequencies of their occurrences are preferred to estimation of the overall proportion of abnormal sperm. Objective, quantitative approaches that are done properly should result in a higher level of confidence than more subjective measures. Recently CASA systems have been developed to assess sperm morphological alterations in both the head and tail for humans and other test species.

At the present time it is not clear how alterations in sperm morphology influence fertility. Sperm shape abnormalities have been correlated with poor fertility in humans and laboratory species. Therefore, because the events underlying the process of nuclear condensation in all species examined to date are similar, it is believed that changes in sperm morphology in test species can be directly related to the same endpoint in humans. An increase in abnormal sperm morphology has been considered evidence that the agent has gained access to the germ cells.[46] The implication is that the greater the number of abnormal sperm in the ejaculate, the greater the probability of reduced fertility. Changes in sperm nuclear morphology should be considered as an adverse reproductive effect.

d) Sperm and Spermatid Numbers — One frequently reported semen variable for the human is sperm concentration.[47] The quantitation of sperm numbers can be assessed from either the epididymis (cauda epididymal sperm counts), the testis (homogenization resistant spermatid numbers), or the ejaculate.

Epididymal counts reflect both the production of sperm by the testis and the ability of the epididymis to store sperm. Epididymal sperm counts are directly related to fertility, and any statistical reduction in epididymal sperm counts should be considered adverse. The ability of the cauda epididymis to store sperm is assessed by counting cauda epididymal sperm, which are commonly used for both sperm morphology and motility assessments. Epididymal sperm numbers can be determined by performing counts on the sperm found in the suspension used to evaluate sperm motility and sperm morphology or the sperm obtained by the subsequent mincing and/or homogenization of the remaining cauda tissue. Epididymal sperm numbers can be counted using either a hemocytometer or CASA systems. Typically, epididymal sperm counts are expressed in relation to the weight of the cauda epididymis; however, this can be misleading because sperm contribute to the epididymal weight and the expression of the data as a ratio may mask a reduction in sperm numbers. Therefore, absolute cauda epididymal sperm numbers should be included with the data obtained from toxicity studies. It should be remembered that cauda epididymal sperm counts are influenced by the level of sexual activity.[48,49]

The quantitation of testicular spermatid numbers is a rapid, inexpensive method commonly used to evaluate the production of sperm by stem cells and to evaluate the ability of the developing germinal cells to survive through the proliferative, meiotic, and spermiogenic phases of spermatogenesis. The quantitation of testicular spermatid numbers will identify agents that have a toxic effect on spermatogenesis but will not provide information on the mechanism of action or the specific cell types sensitive to the agent. The enumeration of spermatid numbers should primarily be used in chronic studies where spermatid numbers have stabilized, but not in short term studies where treatment may not have affected the late spermatid population. If the evaluation was conducted when the effect of a lesion would be reflected adequately in the spermatid counts, then spermatid count may serve as a substitute for quantitative histologic analysis of sperm production.[21] However, spermatid counts may be misleading when the duration of exposure is shorter than the time required for a lesion to be fully expressed in the count. A reduction in homogenization resistant spermatid head counts does not indicate that the agent is toxic to the germ cells. The agent may have adversely affected the Sertoli cell, Leydig cell, or other elements such as the hypothalamo-pituitary-gonadal axis, and therefore a reduction in homogenization resistant spermatid counts would be a secondary response. Testicular spermatid counts may rise if spermatids are not being released from the Sertoli cell. Small increases in spermatid retention will not be detected in the homogenization resistant spermatid counts or in the epididymal sperm counts but should be picked up in the histopathological evaluation of the testis. A total inhibition of sperm release will severely reduce epididymal sperm counts. Remember that a normal sperm count does not provide information on motility, fertilizing potential, or the genetic integrity of the cells that are important for fertility. Homogenization resistant spermatid count data may be reported as numbers of homogenization resistant spermatids per testis, efficiency of sperm production, or daily sperm production rates.

Sperm count determinations from ejaculates in dogs, rabbits, and other species are affected by the frequency of collection. Therefore, care should be taken to collect these counts at appropriate intervals, to ensure that the same male is used, and to ensure that the entire ejaculate is collected. Baseline information from these species (humans and test species) is also valuable, as changes during exposure and recovery can be better defined.

In summary, sperm quantity and quality are both related to fertility. Although the precise relationship between sperm quantity and fertility may not be the same for all species, there is nevertheless a distinct relationship. Fertility can be affected if sperm count is significantly reduced in the presence of normal sperm quality.[30,32]

e) Fertility — Fertility can be affected if sperm quality is significantly reduced in the presence of adequate numbers of sperm.[50] Sperm shape has also been linked to infertility in the human and all species of laboratory animals examined. Because morphological effects are also observed in the presence of decreased sperm count and decreased motility, it is not easy to separate effects of abnormal sperm morphology from effects related to decreased sperm quantity or quality. Nevertheless, the underlying processes for nuclear condensation are similar in both test species and the human, and therefore data on sperm morphology are important in the risk assessment process. Finally, fertility can be impaired in the presence of small decreases in sperm quality and quantity. It is important to remember that small changes in either sperm quantity or quality may not impair fertility but can be predictive of infertility in the human.

III. VACCINE SAFETY

A. Perspective

The FDA's Center for Biologics Evaluation and Research (CBER) reviews a broad spectrum of investigational vaccines. For the purpose of the following discussion a vaccine is a product, the administration of which is intended to elicit an immune response that can prevent and/or lessen the severity of one or more infectious diseases. A vaccine may be a live, attenuated preparation of bacteria, viruses, or parasites, inactivated (killed) whole organisms, living irradiated cells, crude fractions, or purified immunogens, including those derived from recombinant DNA in a host cell, conjugates formed by covalent linkage of components, synthetic antigens, polynucleotides (such as plasmid DNA vaccines), living vectored cells expressing specific heterologous immunogens, or cells pulsed with immunogen. It may also be a combination of vaccines listed above. Therapeutic vaccines, e.g., viral vector–based gene therapy and tumor vaccines, are not considered here.

Although advances in analytical technologies have improved biological product characterization, these products are usually composed of, or extracted from, living organisms and tend to be more complex than conventional drug products. Therefore, for many biotherapeutics and vaccines, unique issues arise that frequently require modifications to standard preclinical testing requirements and designs that are used to evaluate drugs.

There are numerous vaccines in clinical development for the prevention of infectious diseases in adolescents and adults. Thus, the target population for vaccines often includes females in their reproductive years. In addition, there are a number of vaccines in clinical development specifically intended for maternal immunization, with the goal of preventing infectious disease in the pregnant woman and/or young infant through passive antibody transfer from mother to fetus. There are special considerations, such as the potential for adverse effects of the vaccine on normal fetal development, that need to be employed in the assessment of the risks versus the benefits of immunization programs for pregnant women and/or females of childbearing potential.

Currently, unless the vaccine is specifically indicated for maternal immunization, no data are collected regarding the vaccine's safety in pregnant women during the prelicensure phase. Thus, in most cases, preclinical reproductive toxicity studies provide the only data source upon which to base estimations of risk to the developing fetus. However, there is a scarcity of scientific literature on animal reproductive toxicity testing for preventive vaccines. The following outlines considerations for reproductive toxicity study designs for preventive vaccines. It should be noted that approaches to reproductive toxicity assessment for these products may evolve as new methods and test systems become available.

B. General Considerations

If the vaccine is indicated for females of childbearing potential, subjects may be included in clinical trials without reproductive toxicity studies, provided appropriate precautions are taken to exclude pregnant women by the use of pregnancy testing and appropriate contraceptive methods. However, reproductive toxicity studies should be included in the initial Biologics License Application (BLA) when the vaccine is indicated for or may have the potential to be indicated for females of childbearing potential. If the vaccine is indicated for maternal immunization and if pregnant women are enrolled in a clinical trial, data from preclinical reproductive toxicity studies should be available prior to the initiation of the clinical trial.

The International Conference on Harmonization (ICH) has published *Detection of Toxicity to Reproduction for Medicinal Products* (ICHS5a), which should be consulted as a point of reference to assist in the design of reproductive toxicity studies of a preventive vaccine.[51] However, as preventive vaccines present a diverse class of biologics, product specific issues frequently arise that may necessitate modifications to standard study designs. It is, therefore, recommended that the vaccine manufacturer establish an early dialogue with CBER to reach agreement on specific design issues and study endpoints prior to the conduct of such studies.

All data generated from prior preclinical toxicity studies should be reviewed for their possible contribution to the interpretation of any adverse developmental effects that appear in the reproductive toxicology studies, e.g., fetal toxicity secondary to maternal toxicity.[52] In addition, available clinical experience in pregnant women should be considered for any potential application to the design of reproductive toxicity studies in animals. Clinical experience derived from immunization of pregnant women may be helpful in the evaluation of the potential for any adverse outcome on the viability and development of offspring.

C. Specific Considerations

1. Animal Models

Appropriate rationales should be provided in the selection of a relevant animal species for developmental toxicity testing for preventive vaccines. The primary criterion used to define a relevant animal model is the ability of the vaccine to elicit an adequate immune response in the animal model. Species specific differences in the immune response and frequent lack of availability of a "standard species" model can be major obstacles that are encountered. It is recognized that responses induced in a particular animal species may not always be predictive of the expected human response. However, in the absence of alternative test methods, animal models remain the best option in the attempt to predict toxicities for humans. It is important to realize that such studies do not necessarily need to be conducted in the traditional species, i.e., rat and rabbit, nor is the use of two species mandatory. It is the responsibility of the sponsor to provide a scientific rationale for either the animal model used or the lack thereof.

2. Dose

Traditional approaches, i.e., the selection of low, intermediate, and high doses to identify potential dose response relationships with respect to adverse effects may not always be applicable to the reproductive toxicity studies for preventive vaccines because immune-based reactions frequently do not follow dose response relationships. The dose level(s) selected should elicit an immune response in the animal model and, when feasible, should include a full human dose equivalent (e.g., 1 human dose equals 1 rabbit dose). Injection volume is one consideration for feasibility. Data from other available preclinical toxicology studies may guide in the selection of the doses for the reproductive toxicity studies. Control animals should be dosed with the vehicle at the same frequency as test group animals.

3. Immunization Interval

The immunization interval and frequency of immunization(s) should be based on the clinically proposed immunization interval. Similarly, the route or routes of administration should mimic the intended clinical routes, if possible. Since preventive vaccines indicated for females of childbearing potential or pregnant females are administered infrequently, episodic dosing of pregnant animals is likely to be more relevant than daily dosing. Thus, in order to mimic potential exposure in the clinical situation, several groups of pregnant animals should be dosed only at defined intervals during sensitive periods of organogenesis. Modifications to the timing and frequency of dosing may be necessary, depending on the kinetics of the antibody response induced in the animal species. Thus, in addition to immunizing the female animal during gestation, it may be necessary to also administer a priming dose prior to conception to allow for an immune response to occur during gestation.

4. Exposure Period and Follow-Up Period

Of primary concern are potential adverse effects of the vaccine on embryo-fetal development, as unintentional clinical exposure to the vaccine will likely occur during early stages of pregnancy. Thus, the pregnant animal should be exposed to the vaccine during the period of organogenesis, i.e., from implantation to birth, defined as stages C to D of the ICHS5A document. In addition, some females should be allowed to deliver and the offspring should be followed until weaning to evaluate the health of the neonates as well as look for evidence of maternal antibody transfer. Data on reproductive performance (i.e., fertility for males and females) are usually not necessary. Additional data regarding the integrity of the reproductive organs for both sexes can be obtained from histopathology data generated in the toxicology studies in nonpregnant animals.

5. Study Endpoints

The parameters and endpoints for assessing the potential reproductive toxicity of a preventive vaccine will be based on the theoretical risk and concern associated with the particular product under study. For selection of standard developmental toxicity endpoints, the ICHS5a document should be consulted. Briefly, parameters that should be included are maternal weight gain, clinical observations, implantation number, corpora lutea number, litter size, number of live fetuses, fetal and embryonic deaths, resorptions, fetal weight, and crown-rump length, as well as incidence and description of external, visceral, and skeletal malformations. Other parameters may also be included.

Postnatal evaluations may include maternal-newborn relationship, neonate adaptation to extrauterine life, preweaning development and growth, survival incidence, developmental landmarks, and functional testing. Again, other parameters may also be included.

Since the immune system represents the primary target organ for preventive vaccines, the pharmacodynamic effects of the vaccine antigen may potentially result in unwanted findings in the mother and/or the offspring. Therefore, it is important to measure various immune response parameters that may be induced by the vaccine. At a minimum, serum samples collected from pregnant animals should be assessed for antibody production at designated time points. In addition, cord blood as well as blood samples from newborn animals should be evaluated for the presence of antibodies (i.e., antibody transfer from the mother to the fetus or neonate). If adverse effects on embryo-fetal development are observed, additional studies, including studies for assessments of potential immunotoxicity (i.e., lymphocyte subset tests, immune function tests) may be necessary to further evaluate the mechanism for the toxicity.

For certain vaccines, there may be concerns that immunization of pregnant females may interfere with the ability of the offspring to mount an active immune response to either the same or a related vaccine antigen. Such concerns may need to be addressed on a case-by-case basis in clinical immunogenicity studies in infants born to mothers that have been immunized with the vaccine during pregnancy.

D. Summary Remarks

CBER has formulated guidance for developmental toxicity studies for preventive vaccines in a draft guidance document for industry *Considerations for Reproductive Toxicity Studies for Preventive Vaccines for Infectious Disease Indications*.[52] However, regulatory policy with regard to reproductive and developmental toxicity assessment of preventive vaccines will need to be viewed in the context of the evolving field of biotechnology and may change as new test systems and information become available. It is important, therefore, to initiate a dialogue with CBER to discuss proposed protocol designs for preclinical reproductive toxicity studies well in advance of implementation. Preclinical protocols for these products should follow general principles set forth in the CBER draft guidance document, applicable ICH documents, and other available guidance. However, the most useful approach for reproductive and developmental toxicity testing of preventive vaccines should be based on rational science.

IV. DRUG SAFETY

A. Perspective

CDER is currently involved in an attempt to change the way animal data are translated into risk assessment for humans. The primary focus is on molecular entities new to the market that have been tested in a relatively small series of clinical trials, sometimes 100 patients or less. When a drug first goes to market, it is likely that there will be very little information about its effects on mother or offspring when given during pregnancy, unless it is very specifically intended to be used during pregnancy. The most common situation with drugs at the FDA is that the molecular entity is new, animal data is available, and essentially no human experience data is available to predict what will happen to the general population. This is a very different situation from one in which a product has been on the market for several years, and product registries, behaviors, etc., exist for it. In a study, the use of a drug is tightly controlled, whereas drugs in the general population can be taken in combination with other prescription medications, over-the-counter medications, and herbal and food supplements.

B. Product Labels

At the present time, the FDA uses established pregnancy labeling categories for drugs, but many complaints have been made against the use of the categories. The categories are defined by the availability of human data (positive or negative), the availability of animal reproductive toxicity

Table 22.2 Current pregnancy categories

Category	Effects	Criteria
A	No adverse effects demonstrated in human clinical trials	Most stringent criteria
B	No effect in humans with evidence of adverse effects in animals or no effects in animals without adequate human data	Less stringent criteria
C	Adverse effects in animals in the absence of human data or no data available for animals or humans	Assumption of a level of risk without data
D	Adverse effects demonstrated in humans or adverse effects in animals with strong mechanistic expectation of effects in humans	Expectation of translation of effects in humans based on conservative genetic properties
X	Adverse effects in humans or animals without indication for use during pregnancy	May indicate high risk for adverse reproductive effects

data (positive or negative), and the indication for use (Table 22.2). The "limits" or "extremes" of pregnancy labeling are generally defined by clinical data. Risk and benefit are codeterminants of the drug product category. Based on the *1996 Physicians' Desk Reference* (PDR), approximately 25% of the products in the PDR have no information on reproduction.

There is an ongoing attempt to provide the clinician with more information than is currently available with the five pregnancy categories. It is an attempt to deal with some of the inconsistencies that have occurred in the evaluation of some drugs. For example, historically, all oncology drugs have been given a pregnancy category of "D," and some of the assignments may not be justified. It is time to improve the system. The changes being proposed are designed to reduce some of the assumptions that have been made up to now and to evaluate the risk of reproductive effects of drugs with a more objective series of questions. The changes will involve primarily the evaluation of new products.

C. Pregnancy Labeling — Practical Considerations

The important aspects of pregnancy category assignment and risk assessment are that nonclinical study data generally define the pregnancy labeling for new products and systematically collected clinical data are unlikely for agents with "significant" or "meaningful" positive effects in nonclinical studies.

One issue to be addressed is that risk and benefit are codeterminants of the pregnancy category for a drug or therapeutic. Risk ranges from none or minimal to significant, and benefit ranges from minimal to significant. When risk is minimal and benefit significant, or risk is significant and benefit is minimal, the outcome of the codetermination process is easily discerned. However, when risk and benefit are roughly equivalent, the outcome of the analysis becomes less certain and the final decision is more difficult. When a particularly difficult situation arises in the evaluation process, the product may be referred to the FDA's Reproductive Toxicity Committee and to a policy group that oversees the committee. One issue to be addressed in the new process is whether risk and benefit should be addressed separately in drugs and therapeutics used in pregnancy or should they continue to be combined. This issue has been extensively debated within the agency.

D. Assessment of Reproductive and Developmental Toxicity

There are three types of underlying animal studies: (1) Segment I (ICH Stages A and B): In this type of study, the effects of a compound are evaluated from premating to conception or from conception to implantation. In the first type, male and female reproductive function are tested, as well as gamete maturation, mating, behavior, and fertilization. In the second type, female reproductive function, preimplantation development, and implantation are tested. (2) Segment II (ICH stage C): In this type of study, the effects of a compound are evaluated from implantation through closure of the hard palate. Female reproductive function is tested, as well as embryonic development and major organ formation during the period of organogenesis. (3) Segment III (ICH stages D, E, and F): The effects of a compound are evaluated perinatally or postnatally; from hard palate closure to parturition, from parturition to weaning, or from weaning to sexual maturity.

Clearly, the aim of the underlying reproductive and developmental toxicity studies in animals is to reveal possible effects of an agent on human reproduction. Both latent and immediate effects should be known, and multigeneration reproductive studies in animals are particularly valuable because it is unlikely that multigeneration effects in humans will be measurable.

1. Nonclinical Reproductive Toxicity Data Assessment — Future

A Pregnancy Data Integration Working Group has been formed and an integration process is being developed to serve a threefold purpose: (1) to assist in interpretation and integration of reproductive and developmental toxicity data, (2) to promote consistency in the interpretation of reproductive toxicity study findings, and (3) to provide a common framework for the review, interpretation, and discussion of findings. The Working Group is developing a "process" or "tool" that reflects the consensus thought processes of experts in reproductive toxicity and the regulatory sciences as applied to the interpretation of findings from studies of reproductive and developmental toxicity.

A draft guidance document was published in October 2001 by CDER to describe a process for estimating the increase in human developmental and reproductive risks as a result of drug exposure when definitive human data are unavailable.[53] The integration process would be used on drug labels. The draft document reflects current thinking on the topic; it is not final. The overall approach integrates nonclinical information from a variety of sources (i.e., reproductive toxicology, general toxicology, and toxicokinetic and pharmacokinetic information, including absorption, distribution, metabolism, and elimination findings) and available clinical information to evaluate a drug's potential to increase the risk of an adverse developmental or reproductive outcome in humans. The integration process focuses on the likelihood that a drug will increase the risk of adverse human developmental or reproductive effects.

2. Data Needed for the Integration Process

The integration process should be based on an evaluation of a complete set of the expected general toxicology, reproductive toxicology, and pharmacokinetics studies. This evaluation should include an assessment of the ability of the drug to produce a positive finding in relevant animal studies (e.g., whether doses used were large enough to induce toxicity of some kind). The evaluation should also compare animal and human pharmacodynamic effects, animal and human metabolism and disposition, animal and human pharmacological and toxic effects, and drug exposures in animal studies in relation to the highest proposed dose in humans. The type and extent of available toxicology data may vary, depending on the biological actions of the product, test systems available for studying the compound, and other factors.

3. Types of Reproductive and Developmental Toxicity Evaluated

For purposes of the integrative process, two broad categories of toxicity are considered: reproductive and developmental toxicity. In the reproductive toxicity category, three classes of toxicity are evaluated: effects on fertility, parturition, and lactation. In the developmental toxicity category, four classes of toxicity are evaluated: mortality, dysmorphogenesis (structural alterations), alterations to growth, and functional toxicities. For a given drug, each class of toxicity should be assessed. A positive signal in any class of reproductive or developmental toxicity should be evaluated to estimate the likelihood of increased reproductive or developmental risk for humans.

a. Reproductive Toxicities

Reproductive toxicities include structural and functional alterations that may affect reproductive competence in the F_0 generation.

1. **Fertility:** Male reproductive toxicity associated with administration of a drug may be due to degeneration or necrosis of the cells in the reproductive organs, and may be assessed by several endpoints, including reduced sperm count, altered sperm motility or morphology, aberrant mating behavior, altered ability to mate, altered endocrine function, and overall reduced fertility.
   Female reproductive toxicity may be due to damage to the reproductive organs or alterations to endocrine regulation of gamete maturation and release, and may manifest as aberrant mating behavior, altered ability to mate, or overall reduction in fertility. Diminished fertility in female animals is typically detected by reductions in the fertility index, the number of implantation sites, time to mating, or fecundity.
2. **Parturition:** Changes in the duration of parturition are frequently reported as mean time elapsed per pup or as total duration of parturition.
3. **Lactation:** Drugs administered to lactating animals may be a source of unwanted exposure in the nursing neonate, may alter the process of lactation in the nursing mother (e.g., the quality or quantity of milk), or may alter maternal behavior toward the nursing offspring.

b. Developmental Toxicities

Developmental toxicities, generally those that affect the F_1 generation, include:

1. **Mortality:** Mortality due to developmental toxicity may occur at any time, from early conception to postweaning. A positive signal may appear as pre- or peri-implantation loss, early or late resorption, abortion, stillbirth, neonatal death, or periweaning loss.
2. **Dysmorphogenesis:** Dysmorphogenic effects (structural alterations) are observed as malformations or variations to the skeleton or soft tissues of the offspring.
3. **Alterations to growth:** Growth retardation, excessive growth, and early maturation are considered alterations to growth. Body weight is the most common measurement for assessing growth rate, but crown-rump length and anogenital distance are also frequently measured.
4. **Functional toxicities:** Any persistent alteration of normal physiological or biochemical function is considered a functional toxicity, but typically only effects on neurobehavior and reproductive function are measured. Common assessments include locomotor activity, learning and memory, reflex development, time to sexual maturation, mating behavior, and fertility.

4. The Integration Process

According to the October 2001 CDER Guidance Document, the data should be passed through the integration process in a stepped procedure consisting of three sections that are applicable to all data sets, applicable only to data sets without evidence of reproductive or developmental toxicity, or applicable to data sets with positive indications of reproductive or developmental toxicity.

5. Section A. Overall Decision Tree Applicable to All Data-Sets (Figure 22.1)

The decision tree process outlined here should be used for each class of reproductive or developmental toxicity. A single study may address several classes of toxicity. Each study should be evaluated individually. A sequential series of decisions should be made in evaluating the various situations that may be encountered and the next steps that should be taken where there are studies that can be evaluated.

PRINCIPLES OF RISK ASSESSMENT — FDA PERSPECTIVE

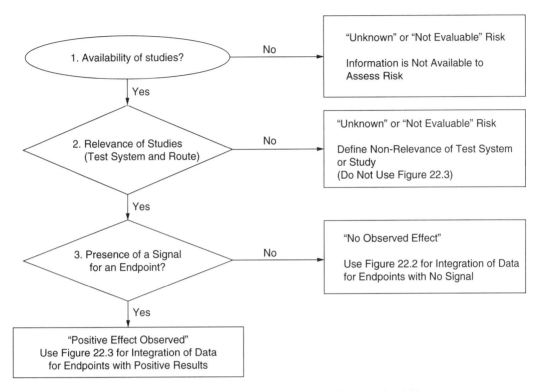

Figure 22.1 Overall decision tree for evaluation of reproductive/developmental toxicities.

a. Availability of Studies

If adequate studies were not done or the results are not available for comprehensive evaluation, risk to humans is considered unknown or not possible to evaluate, and the product labeling should reflect that conclusion. If studies were conducted and are available for comprehensive evaluation, the assessment should continue.

b. Relevance of Studies

If the test system was not relevant (e.g., nonrelevant route of administration) or not appropriate (e.g., inappropriate test protocol or species), the risk to humans is considered unknown or not amenable to evaluation, and the product labeling should reflect that conclusion. If the studies are relevant to evaluating the risk to humans, the risk integration process should continue with the next question.

c. Presence or Absence of a Signal

If the test system is relevant and appropriate for assessing the risk of toxicity in humans, the next question that should be asked is: "Was there a positive signal (suggesting toxicity)?" If no positive signal (suggesting toxicity) was seen, the evaluation process should continue to Section B. If a positive signal was seen, the evaluation process should continue to Section C.

6. Section B. No Signal (Applicable to Data-sets without Evidence of Reproductive or Developmental Toxicity) (Figure 22.2)

Where there is no positive signal for one of the seven classes of reproductive or developmental toxicity, the risk assessment should be a stepwise process leading to a recommendation about the relevance of the nonfinding in humans. If multiple studies are available and all the results are negative, the process in Section B should be used. If any study has a positive signal, the process in Section C should be used. The following factors should be considered during the evaluation of each class of reproductive or developmental toxicity for which there was no signal:

a. The Model or Test Species Predictive Adequacy

The following questions bear on the determination of a model's predictive adequacy. Do any of the model or test species (or systems) demonstrate or have the capability of responding to the pharmacodynamic effects of the drug? Do any of the model or test species (or systems) demonstrate an overall toxicity profile that is relevant to the human toxicity profile? Do any of the model or test species (or systems) demonstrate pharmacokinetic (including absorption, distribution, metabolism, and elimination) profiles for the drug that are qualitatively similar to those in humans? If the responses to these questions suggest that the response of the test species is of little relevance to humans, the review should explain why the test may have low predictive value. Even if the test system is determined to be of limited relevance, the review should consider the remaining factors 2 to 4 (see following sections) and describe any additional uncertainties.

b. Adequacy of Study Doses and Exposure

The following elements should be considered in assessing the relevance to humans of the drug doses and exposure in the test system: Were adequate doses (concentrations) of the drug administered to the test species or test systems (e.g., maximum tolerated dose [MTD])? Were the exposures (based on the area under the curve of plasma concentration time course [AUC], maximum achieved concentration [C_{max}], or other appropriate systemic exposure metric) achieved in the test species or test systems adequate relative to those demonstrated in humans at the maximum recommended human dose? If the answer to either of these questions is no, the evaluation should state that the animal studies are inadequate and explain why. The evaluation should proceed to Section C below.

c. Class Alert

Class alerts should be based on adverse reproductive or developmental effects previously demonstrated in humans by closely related chemical entities or compounds with similar pharmacodynamic effects. If there is a class alert for the drug, based on a related chemical structure of parent or metabolite or related pharmacologic effect, the class-specific information relevant to the class of toxicity reported to be negative should be included in the risk evaluation and discussion of the drug. The basis for the class alert for adverse effects in humans should be reasonably applicable to the drug being evaluated.

d. Signals for Related Types of Reproductive and Developmental Toxicity

The next step in evaluating the relevance is to assess findings for related reproductive and developmental toxicities. A positive signal for one class of toxicity may suggest some risk in humans for other toxicities in the same category for which there were no findings in animals. The issue of related toxicities is most relevant for developmental toxicities. For example, if there is no signal for fetal mortality but there is a positive signal for alterations to growth or dysmorphogenesis in

PRINCIPLES OF RISK ASSESSMENT — FDA PERSPECTIVE

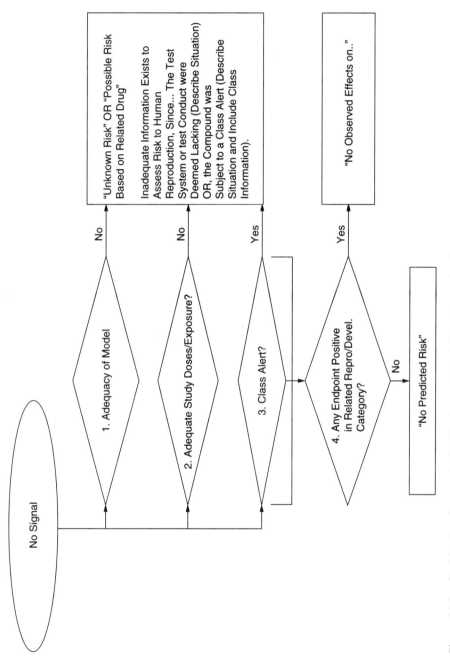

Figure 22.2 Decision tree for reproductive/developmental toxicities with no signal.

one or more animal species, it may be inappropriate to conclude there is no risk of fetal mortality for humans. Related toxicities may also be relevant for reproductive toxicities where a hormonal mechanism is identified and the mechanism is relevant to humans.

If positive signals for related classes of toxicity were observed in the animal studies, the evaluation should state that there was no observed effect on the type of toxicity being assessed, but positive signals were seen for related toxicities. If there is no positive signal for any class of reproductive or developmental toxicity, the evaluation should state that there is no expected increase in risk for reproductive or developmental toxicity in humans based on the results of animal studies.

7. Section C. Applicable to Data-sets with One or More Positive Indications of Reproductive or Developmental Toxicity (Figure 22.3)

There are six factors that may affect the level of concern for a positive signal: (1) signal strength with respect to cross-species concordance (where more than one species has been studied), presence of multiple effects, and presence of adverse effects at several times; (2) signal strength analyzed with respect to the coexistence of maternal toxicity, the presence of a dose response relationship, and the observation of rare events; (3) pharmacodynamics, (4) concordance (metabolic and toxicologic concordance with the human); (5) relative exposure; and (6) class alerts.

The positive signal may be from a reproductive or developmental toxicology study or an effect observed on a reproductive tissue, system, or behavior in a general toxicology study. Each factor should be analyzed independently. Guidance is provided on what types of observations for each of the six factors might increase, decrease, or leave unchanged the level of concern for that factor. These analyses should be a weighted integration that takes into account the quality and nature of the data under consideration. The assessments of concern for each of the six factors should be assigned values of +1, −1 or 0, respectively, if the factor is perceived as increasing, decreasing, or leaving unchanged the level of concern for a class of reproductive or developmental toxicity. Conclusions from the six analyses should be summed to arrive at a comprehensive evaluation of the potential increase in risk for each class of the seven reproductive or developmental classes for which there was a positive signal.

Intra- and interspecies concordance of adverse effects in animals deserves some special consideration in this risk integration process. Positive signals for related types of reproductive or developmental toxicity within the same species indicate intraspecies concordance of effects (e.g., a reduction in normal growth and an increase in developmental mortality). Positive signals for the same or a related type of toxicity across species indicate interspecies concordance. In general, findings for which there is intra- or interspecies concordance are more convincing than a positive signal in only one toxicity class in only a single species.

In evaluating potential human risk for adverse reproductive or developmental outcomes, if there is interspecies concordance for a single adverse effect, it may be reasonable to conclude that a similar effect is the most likely adverse event to be seen in humans treated with the drug. If different but related adverse effects are seen in multiple test species (e.g., alterations to growth in one species and developmental mortality in another, or parturition effects in one species with lactation effects in the second), it may be reasonable to assume there is some level of risk for categorically related endpoints in humans.

A detailed discussion of the overall integrative analysis, the six individual factors, the contributory elements for each factor, and the assignment of the level of concern for each factor is presented here.

8. Factor 1. Signal Strength, Part I

A positive signal in any reproductive or developmental toxicity class should be analyzed with respect to whether the finding is present in more than one setting: cross-species concordance (where

PRINCIPLES OF RISK ASSESSMENT — FDA PERSPECTIVE

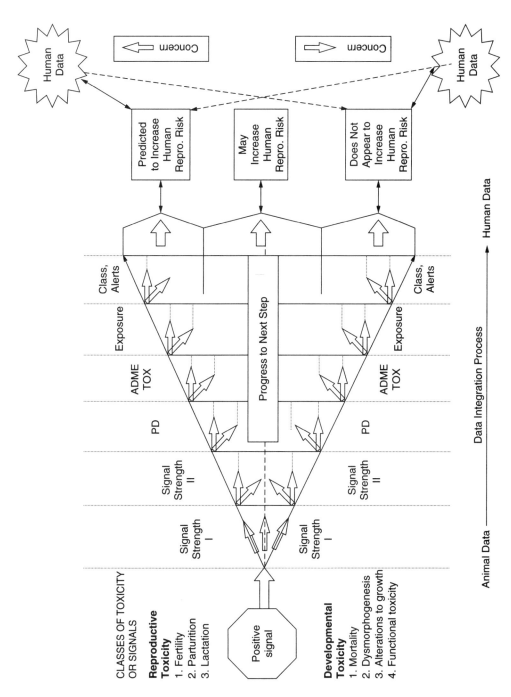

Figure 22.3 Integration tool for reproductive or developmental toxicites with a positive signal.

more than one species has been studied), multiplicity of effects, and adverse effects seen at more than one time.

a. Cross-Species Concordance

The defining characteristic of cross-species concordance is a positive signal in the same class of toxicity in more than one species. Cross-species concordance is most likely to be identified for structural abnormalities (dysmorphogenesis) or developmental mortality because these toxicities are frequently detected in the organogenesis testing paradigm, in which multiple species are typically evaluated. In addition, alterations to endocrine function or gonadal histopathology (which may alter fertility) may be indirectly detected in subchronic and chronic toxicity studies in rodents and nonrodents. When cross-species concordance is observed, there is increased concern for reproductive or developmental toxicity in humans. In contrast, there is decreased concern when a signal is detected in only one species (with the proviso that the negative species is an appropriate animal model and the studies were adequate in design, dosing, and implementation).

b. Multiplicity of Effects

Multiplicity of effects refers to observation, in a single species or animal model, of two or more positive signals within one of the two general categories of toxicity (reproductive or developmental) or within one of the seven classes of reproductive or developmental toxicities. The observation of increased embryo-fetal death and structural abnormalities in an animal test species is an example of multiple positive signals within a general category. The observation of two or more positive signals for structural abnormalities in tissues of multiple embryonic origin (e.g., defects affecting soft tissue, skeletal tissue, and/or neural tissue) is an example of multiple positive signals in a toxicity class.

If all species examined demonstrate multiplicity of effects, there is increased concern for reproductive or developmental toxicity in humans. If there are positive signals in two or more species, but multiplicity of effects is observed in only one species, concern is unchanged for this element. If no species studied exhibits multiplicity of effects, there is decreased concern.

c. Adverse Effects at Different Stages of the Reproductive or Developmental Process

Evidence of toxicity may arise during any stage of the reproductive or developmental process. If a positive signal in animals is observed in multiple stages of development, there is generally greater concern for adverse human reproductive outcomes. If a positive signal is observed only during a single, discrete interval, the level of concern is unchanged. If the positive signal occurs only during processes that are of limited relevance to humans, there would be less concern for adverse human reproductive outcomes. In addition to its relevance to this evaluation process, it is also important to define the timing of the period of susceptibility for the observed positive signal to provide a context for the human risk.

9. Factor 2. Signal Strength, Part II

A positive signal should also be analyzed with respect to the coexistence of maternal toxicity, presence of a dose response relationship, and the observation of rare events.

a. Maternal Toxicity

In weighing a signal of toxicity, the magnitude of adverse effects in the offspring versus the severity of maternal (and, for fertility studies, paternal) toxicity should be considered when drawing a conclusion about the relevance of the F_0 toxicity to effects observed in the offspring. A positive signal occurring at doses that are not maternally toxic increases concern for human reproductive or developmental toxicity. If a positive signal is observed only in the presence of frank maternal toxicity, there is decreased concern, provided that the positive signal may be reasonably attributed to maternal toxicity.

When evaluating a positive signal in two or more species, assessment of the implications of maternal or paternal toxicity should be based on a composite analysis of the data from all adequately studied species. If a positive signal is seen in two or more species in the absence of maternal toxicity, there is increased concern for adverse human reproductive outcomes. If a positive signal is seen only in the presence of clear relevant maternal toxicity in multiple species, there is decreased concern. If there is nonconcordance between test species as to the presence and relevance of maternal toxicity, there may be no change in the overall level of concern for this contributory element.

b. Dose Response Relationship

Concern for human reproductive or developmental toxicity is increased when a positive signal is characterized by any of the following: increased severity of adverse effects with an increase in dose, increased incidence of adverse effects with an increase in dose, or a high incidence of adverse effects across all dosed groups. Conversely, the absence of all three of these indicators of dose response would be cause for unchanged or decreased concern.

If multiple species are evaluated, a clear dose response across all tested species increases concern. If a positive signal occurs in more than one species, only one of which demonstrates one of the dose response relationships described above, the level of concern will generally be unchanged. If there is no dose response in any species, concern is decreased.

c. Rare Events

Developmental toxicity studies usually lack the statistical power to detect subtle increases in rare events. Thus, an increased frequency of positive signals for rare events in drug-exposed animals increases concern for reproductive or developmental toxicity in humans. The absence of an increased frequency of rare events, however, does not decrease concern. When multiple species are studied, an increased frequency of positive signals for rare events in more than one species increases concern for adverse outcomes in humans.

10. Factor 3. Pharmacodynamics

A positive signal should be analyzed with respect to the therapeutic index, biomarkers as a benchmark, and similarity between the pharmacologic and toxicologic mechanisms.

a. Therapeutic Index

The therapeutic index (TI) is used to identify the extent to which there is overlap between therapeutic doses and doses that cause reproductive or developmental toxicity. It is unusual to obtain well-defined dose–response curves for toxicity and efficacy from a single species. Thus, the use of estimations or surrogate endpoints (related to the therapeutic mechanism) for this evaluation may be warranted. To reduce the impact of variation in the slope of the dose response curves, estimation of the TI should generally be based on comparison of the TD10 and the ED90 concentrations, and resulting in the composite score called TI10/90.

The TD10 (toxic dose or concentration) should be defined by the C_{max} (or other appropriate exposure metric) that produced the toxic reproductive or developmental response in 10% of a responsive or sensitive species, whereas the ED90 (efficacious dose or concentration) should be defined by the C_{max} (or other appropriate exposure metric) that produced the desired effect in 90%

of the test species. These parameters can be estimated. Preferably, both the TD10 and ED90 would be defined in the same species. In some instances, estimation of the ED90 can be based on *in vitro* cell inhibition studies (frequently seen for antibiotics and antineoplastic agents). Although less desirable, efficacy data can be derived from another species, but caution should be exercised in such situations. The same exposure metric should be used in the estimation of the TD10 and ED90 values. Scientific justification for the drug exposure metrics used for comparison should be provided.

If the TI10/90 is less than 5 (obtained by dividing TD10 by ED90), there is increased concern for reproductive or developmental toxicity in humans, as there is limited separation in the doses causing adverse effects from those responsible for efficacy. If the TI10/90 ratio falls between 5 and 20, the level of concern is unchanged. If the TI10/90 ratio is greater than 20, there is decreased concern because of the wide separation in doses causing adverse effects from those resulting in efficacy.

If there are data available to determine the TI10/90 ratio in multiple species, assessment of the level of concern for this element should be based on an integrated analysis of data from all adequately studied species. The extent of concordance in the size of the TI10/90 between species may increase, decrease, or leave unchanged the level of concern (i.e., the greater the concordance, the more likely concern will be increased). In the event of nonconcordance of the TI ratios between multiple test species, the nature of the positive signals observed and the relevance of the endpoint and test species to the human condition should be considered before making an assessment. In the event that one species is considered inappropriate to the analysis, the evaluation should be performed without reference to that species.

b. Biomarkers as a Benchmark

There may be circumstances in which an effect on a biomarker is consistently seen in multiple species at doses lower than the NOEL for demonstrable reproductive and developmental toxicity. If there is an effect on this biomarker at or below the therapeutic dose in humans, there is increased concern for reproductive or developmental toxicity in humans. If this biomarker is responsive to the drug in humans, can be monitored, and is not affected at the therapeutic dose, there may be decreased concern.

c. Similarity between Pharmacologic and Reproductive Developmental Toxicologic Mechanisms

If a positive signal is an extension of, progression of, or related response to the intended pharmacologic effect of the drug (e.g., delay of parturition by drugs known to suppress uterine smooth muscle contractility or hypotension in the offspring of dams treated during late gestation with a drug known to lower blood pressure), there is increased concern for reproductive or developmental toxicity in humans. There is less concern if the positive signal is attributed to an animal-specific pharmacological response, even though it may be an extension of the pharmacologic effect of the drug (e.g., pregnancy loss in rats due to hypoprolactinemia).

11. Factor 4. Concordance between the Test Species and Humans

Concordance between the test species and humans should be evaluated with respect to the metabolic and drug distribution profiles, the general toxicity profiles, and biomarker profiles.

a. Metabolic and Drug Distribution Profiles

Drug distribution, elimination, and biotransformation (pathways and metabolites) in the test species and in humans should be compared. Quantitative differences in metabolic and drug distribution profiles between the test species and humans are often seen, may not have important implications, and should not be overemphasized. Reproductive and developmental toxicities induced by compounds whose metabolic and distribution profiles are very similar in animals and humans increases concern for reproductive or developmental toxicity in humans. For compounds with highly dissimilar metabolic or tissue distribution profiles in animals and humans, there is less concern if the toxic effect seen in the test species can be attributed to a metabolite or tissue distribution profile not seen in humans. For any other scenario, the concern is unchanged.

When there are significant differences in drug distribution or metabolic profiles between several species, yet each test species demonstrates a positive signal for a reproductive or developmental toxicity, the toxicity is assumed to be attributable to the parent drug or a common biotransformed product, and concern is increased.

b. General Toxicity Profiles

If the overall toxicity profile of a drug in one or more test species with a positive signal is similar to that in humans, there is increased concern for reproductive or developmental toxicity in humans. If the overall toxicity profiles are dissimilar, there may be decreased concern. When general toxicology data are available for more than one species, the determination of the level of concern (increased, decreased, or unchanged) should be based on an assessment of each test species' ability to indicate human adverse effects in response to the drug.

c. Biomarker Profiles

When biomarker profiles are available for comparison, an approach similar to that described in the previous section on assessment of general toxicity profiles may be useful.

12. Factor 5. Relative Exposures

When considering the relative exposure comparisons discussed in the next section, more emphasis should be placed on a parameter within this factor when there is a scientifically plausible link between the exposure metric (or biomarker) and the adverse reproductive (or developmental) effect.

a. Kinetic Comparison of Relative Exposure

Comparison of systemic drug exposure at the NOEL for the reproductive or toxicity class in the test species to that in humans at the maximum recommended dose is a critical determination. This comparison should be based on the most relevant metric (e.g., AUC, C_{max}, C_{min}, BSA [body surface area] adjusted dose). In general, there is increased concern for reproductive or developmental toxicity in humans for relative exposure ratios (animal:human) that are 10 or less, decreased concern for exposure ratios that are 25 or more, and no change in concern for ratios between 10 and 25. When applicable, the relative exposure ratio should consider both the parent compound and its metabolites. For example, it is appropriate to combine parent compound and metabolite when both are pharmacologically active and the activity relates to the reproductive or developmental toxicity.

Where there are exposure data for multiple test species, the NOEL exposure for each should be compared to human exposure at the maximum recommended dose. If the exposure ratios are low (10-fold or less) in multiple species with a positive signal, there is increased concern. If the exposure ratios are high (25-fold or more), there is decreased concern. In the event a significant difference in relative exposures is observed between multiple test species, the appropriateness of the metric (e.g., AUC, C_{max}) being used to define the interspecies exposure comparisons should be reassessed. If an alternative metric fails to reduce the disparity between the species, the assessment of concern should be based on the lowest ratio (i.e., in the most sensitive species).

Relative interspecies exposure data should be evaluated in light of species differences in protein binding (free drug concentration), receptor affinity (if related to the positive signal), or site-specific drug concentrations. In the absence of meaningful differences between the test species and humans in these parameters, the interspecies comparisons should be based on total drug exposure.

b. Biomarkers as a Measure of Relative Exposure

The purpose of this relative exposure metric is to compare the dose causing a reproductive toxic effect in the test species to the therapeutic dose in humans, normalized to the doses causing a response common to both species. In practice, this is done by taking the NOEL for the adverse reproductive or developmental effect and dividing by the dose at which the biomarker response is seen in the test species. This is compared with the human therapeutic dose divided by the dose at which the biomarker response is seen in the human. The ratio calculated for animals is then divided by the ratio calculated for humans. When this ratio of relative biomarker exposure (animal to human) is 10 or less, there is increased concern for human reproductive or developmental toxicity. When this ratio is 25 or more, there is less concern. When this ratio falls between 10 and 25, the level of concern is unchanged.

Where there are data to compute relative biomarker exposure ratios for multiple species, the level of concern assessment should be based on an integrated analysis of species. Where there are nonconcordant biomarker ratios between multiple test species, the relevance of the biomarker as expressed in the various species should be considered before making an assessment. If there is no scientific rationale for the disparity between species, the biomarker, as a measure of exposure, will be of questionable utility.

13. Factor 6. Class Alerts

Consideration of a class associated effect should be based on prior human experience for a drug with related chemical structure (parent or metabolite) or related pharmacologic effect, and with known reproductive or developmental outcomes in humans. There is increased concern for reproductive or developmental toxicity in humans when the drug is from a class of compounds known to produce adverse effects in the same toxicity class in humans and animals. There is decreased concern only in circumstances in which a class of compounds, although demonstrating adverse effects in animals, has been previously shown to have no related adverse effects on human reproduction or development. In the absence of adequate human reproduction or developmental data for a class, the level of concern is unchanged.

14. Summary and Integration of Positive Findings

The factors discussed here are derived from a limited sample of pharmaceuticals where the clinical outcomes are reasonably well defined. CDER believes that using specific factors and benchmark values to assess the potential to increase risk to humans for adverse reproductive and developmental outcomes will result in a more unbiased and uniform evaluation. CDER also believes this approach will help identify specific areas of additional information about a pharmaceutical that would be useful in more fully defining risk and will allow specific analysis of areas of disagreement that influence the risk evaluation.

- Where there is a positive finding in nonclinical or general toxicology studies for one of the seven classes of reproductive or developmental toxicity, there is a potential for increased human risk. In evaluating the level of increased risk, positive findings from each of the seven classes of reproductive and developmental toxicity should be assessed separately. All relevant information should be considered.

- In evaluating the level of concern for each of the six factors in the overall assessment, the analysis should reflect the weight of evidence, taking into account the quality and type of data under consideration for each factor (i.e., should not be merely an arithmetic summation of the contributory elements for each factor). For each factor there should be a determination of increased (+1), decreased (–1), or no change (0) in the level of concern.
- The values for the six factors should then be summed to arrive at one of the following overall conclusions for each class of reproductive or developmental toxicity: (1) the drug is predicted to increase risk, (2) the drug may increase risk, or (3) the drug does not appear to increase risk of that class of reproductive or developmental toxicity in humans. Where there is sufficient information about the drug to assess each of the six factors within Figure 22.3, a net value of +3 or more suggests that a drug is predicted to increase risk for that class of toxicity in humans, a value between +2 and 2 suggests that the drug may increase the risk, and a value of –3 or less suggests the drug does not appear to increase the risk.

The summary risk conclusions for the outcomes of analyses using Figure 22.3 are:

- **Does not appear to increase risk:** The drug is not anticipated to increase the incidence of adverse reproductive (or developmental) effects above the background incidence discussed in humans when used in accordance with dosing information in the product label.
- **May increase risk:** The drug may increase the incidence of adverse reproductive (or developmental) events above the background incidence in humans when used in accordance with the dosing information in the product label.
- **Predicted to increase risk:** The drug is expected to increase the incidence of adverse reproductive (or developmental) events above the background incidence in humans when used in accordance with the dosing information in the product label.

ACKNOWLEDGMENTS

Many thanks are due to the following persons from the various FDA centers who reviewed or helped expedite the manuscript: Mitchell Cheeseman, Karen Goldenthal, David Hattan, Karen Midthun, Robert Osterberg, George Pauli, Mercedes Serabian, and John Welsh.

REFERENCES

1. Kokoski, C.J., et al., Methods used in safety evaluation, in *Food Additives*, Branen, A.L., Davidson, P.M., and Salminen, S., Eds., Marcel Dekker, New York, 1990, p. 579.
2. Food and Nutrition Board, National Research Council, *Risk Assessment/Safety Evaluation of Food Chemicals*, NAS Press, Washington, D.C., 1980.
3. FDA, Toxicological Principles for the Safety Assessment of Direct Food Additives and Color Additives Used in Food, U.S. Food and Drug Administration, Bureau of Foods, Washington, DC, 1982.
4. FDA, Toxicological Principles for the Safety Assessment of Direct Food Additives and Color Additives Used in Food, "Draft Redbook II," U.S. Food and Drug Administration, Center for Food Safety and Applied Nutrition, Washington, D.C., 1993.
5. FDA, *Toxicological Principles for the Safety of Food Ingredients*, U.S. Food and Drug Administration, Center for Food Safety and Applied Nutrition (http://www.cfsan.fda.gov/~redbook/red-toca.html), 2000.
6. FDA, Recommendations for Submission of Chemical and Technological Data for Direct Food Additive and GRAS Food Ingredient Petitions, U.S. Food and Drug Administration, Center for Food Safety and Applied Nutrition, Washington, D.C., 1993.
7. FDA, Color Additive Petitions: FDA Recommendations for Submission of Chemical and Technological Data on Color Additives for Food, Drugs, or Cosmetics, U.S. Food and Drug Administration, Center for Food Safety and Applied Nutrition (http://www.cfsan.fda.gov/~dms/opa-col1.html), 1997.

8. Kokoski, C.J. and Flam, W.G., Establishment of acceptable limits of intake, *Proceedings of the Second National Conference for Food Protection*, Department of Health and Human Services, Washington, DC, 1984, p. 61.
9. Kokoski C.J., Regulatory toxicology and food flavors, in *Thermal Generation of Aromas*, Parliament, T.H., McGorrin, R.J., and Ho, C.T., Eds. American Chemical Society, Washington, DC, 1989, p. 23.
10. Rulis, A.M., Safety assurance margins for food additives currently in use, *Regul. Toxicol. Pharmacol.*, 7, 160, 1987.
11. FDA, The Food and Drug Modernization Act, P.L. 105-115, 105th Congress, 1997.
12. FDA, Guidance for Industry. Preparation of Food Contact Notifications for Food Contact Substances: Toxicology Recommendations, Final Guidance, February 2002 (http://intranet.cfsan.fda.gov/ofas/internetprep/opa-fcnt.html), 2002.
13. FDA, Food additives: food contact substance notification system, *Fed. Regist.*, 67, 35724, May 21, 2002.
14. FDA, Guidance for Industry. Preparation of Premarket Notifications for Food Contact Substances: Chemistry Recommendations (http://www.cfsan.fda.gov/~dms/opa-pmnc.html), 2000.
15. FDA, Guidance for Industry, Preparation of Premarket Notifications for Food Contact Substances: Toxicology Recommendations (http://www.cfsan.fda.gov/~dms/opa-pmnt.html), 1999.
16. Copestake, P., The acceptable daily intake, *Food Chem. Toxicol.*, 27, 273, 1989.
17. World Health Organization (WHO), *Procedures for the Testing of Intentional Food Additives to Establish Their Safety for Use, Second Report*, FAO Nutrition Meetings Report Series, 1958, No. 17; WHO Technical Report Series 1958, No. 144.
18. Jackson, B.A., Regulatory implications of reproductive toxicity, in *The Toxicology Forum 1988 Annual Winter Meeting*, Toxicology Forum, Washington, D.C., 1988, p. 207.
19. Gaylor, D.W., et al., Health risk assessment practices in the U.S. Food and Drug Administration, *Regulat. Toxicol. Pharmacol.*, 26, 307, 1997.
20. Hess, R.A. and Moore, B.J., Histological methods for evaluation of the testis, in *Male Reproductive Toxicology*, Chapin, R.E. and Heindel, J., Eds., Academic Press, San Diego, CA, 1993, p. 52.
21. Russell, L.D. et al., *Histological and Histopathological Evaluation of the Testis*, Cache River Press, Clearwater, FL, 1990.
22. Russell, L.D., Normal testicular structure and methods of evaluation under experimental and disruptive condition, in *Reproductive Toxicity of Metals*, Clarkson, T.W., Nordberg, G.F., and Sager, P.R., Eds., Plenum, New York, 1983, p. 227.
23. Klinefelter, G., Gray, E.L., and Suarez, J.D., The method of sperm collection significantly influences sperm motion parameters following ethane dimethanesulphonate administration in the rat, *Reprod. Toxicol.*, 5, 39, 1991.
24. Seed, J., et al., Consensus report: methods for assessing sperm motility, morphology, and counts in the rat, rabbit and dog, *Reprod. Toxicol.*, 10, 237, 1996.
25. Chapin, R.E., et al., Methods for assessing rat sperm motility, *Reprod. Toxicol.*, 6, 267, 1992.
26. Toth, G.P., et al., The automated analysis of rat sperm motility following subchronic epichlorhydrin administration: methodologic and statistical consideration, *J. Androl.*, 10, 401, 1989.
27. Toth, G.P., et al., Effects of three male reproductive toxicants on rat cauda epididymal sperm motion, *Reprod. Toxicol.*, 6, 507, 1992.
28. Slott, V.L., et al., Acute inhalation exposure to epichlorhydrin transiently decreases rat sperm velocity, *Fundam. Appl. Toxicol.*, 115, 597, 1990.
29. Klinefelter, G.R., et al., Chlorethylmethanesulfonate-induced effects on the epididymis seem unrelated to altered Leydig cell function, *Biol. Reprod.*, 51, 82, 1994.
30. Toth, G.P., et al., Sources of variation in the computer-assisted motion analysis of rat epididymal sperm, *Reprod. Toxicol.*, 5, 487, 1991.
31. Oberlander, G., Yeung, C.H., and Cooper, T.G., Indication of reversible infertility in male rats by oral ornidazole and its effects on sperm motility and epididymal secretions, *J. Reprod. Fertil.*, 100, 551, 1994.
32. Slott, V.L. et al., Synchronous assessment of sperm motility and fertilizing ability in the hamster following treatment with alpha-chlorhydrin, *J. Androl.*, 16, 523, 1995.
33. Stevens, K.R. and Gallo, M.A., Practical considerations in the conduct of chronic toxicity studies, in *Principles and Methods of Toxicology*, Hays, A.W., Ed., Raven Press, New York, 1989, p. 237.

34. Schwetz, B.A., Rao, K.S., and Park, C.N., Insensitivity of tests for reproductive problems, *J. Environ. Pathol. Toxicol.*, 3, 81, 1980.
35. Blazak, W.F., Ernst, T.L., and Stuart B.E., Potential indicators of reproductive toxicity, testicular sperm production and epididymis sperm numbers, transit time and motility in Fisher 144 rats, *Fundam. Appl. Toxicol.*, 5, 1097, 1985.
36. Hess, R.A., et al., The fungicide benomyl methyl-1-(butylcarbamoyl)-2-(benzimidazolecarbamate) causes testicular dysfunction by inducing the sloughing of germ cells and occlusion of efferent ductules, *Fundam. Appl. Toxicol.*, 17, 733, 1991.
37. Nakai, M., Moore, B.J., and Hess, R.A., Epithelial reorganization and irregular growth following carbendazim-induced injury of the efferent ductules of the rat testis, *Anat. Rec.*, 235, 51, 1993.
38. Berndtson, W.E., Methods for quantifying mammalian spermatogenesis: a review, *J. Anim. Sci.*, 44, 818, 1977.
39. Foote, R.H., Schermerhorn, E.C., and Simken, M.E., Measurement of semen quality, fertility and reproductive hormones to assess dibromochloropropane (DBCP) effects in live rabbits, *Fundam. Appl. Toxicol.*, 6, 628, 1986.
40. Ku, W.W., et al., Testicular toxicity of boric acid (BA): relationship of dose to lesion development and recovery in F344 rat, *Reprod. Toxicol.*, 7, 305, 1993.
41. Zenick, H., et al., Assessment of male reproductive toxicity: a risk assessment approach, in *Principles and Methods of Toxicology*, Hays, A.W., Ed., Raven Press, New York, 1994, p. 937.
42. Nestor, A. and Handel, M., The transport of morphologically abnormal sperm in the female reproductive tract of mice, *Gamete Res.*, 10, 119, 1984.
43. Redi, C.A. et al., Spermatozoa of chromosomally heterozygous mice and their fate in male and female genital tracts, *Gamete Res.*, 9, 273, 1984.
44. Filler, R., Methods for evaluation of rat epididymal sperm morphology, in *Methods in Toxicology: Male Reproductive Toxicology*, Chapin, R.E. and Heindel, J.J., Eds., Academic Press, San Diego, CA, 1993, p. 334.
45. Linder, R.E., et al., Endpoints of spermatotoxicity in the rat after short duration exposures to 14 reproductive toxicants, *Reprod. Toxicol.*, 6, 491, 1992.
46. U.S. Environmental Protection Agency (EPA), Guidelines for mutagenicity risk assessment, *Fed. Regist.*, 51, 34006, 1986.
47. Wyrobek, A.J., et al., An evaluation of the mouse sperm morphology test and other sperm tests in non-human mammals, *Mutat. Res.*, 115, 1, 1983.
48. Amann, R.P., A critical review of methods for the evaluation of spermatogenesis from seminal characteristics, *J. Androl.*, 2, 37, 1981.
49. Hurtt, M.E. and Zenick, H., Decreasing epididymal sperm reserves enhances the detection of ethoxyethanol-induced spermatotoxicity, *Fundam. Appl. Toxicol.*, 7, 348, 1986.
50. Klinefelter, G.R. et al., The ethane dimethanesulfonate-induced decrease in the fertilizing ability of cauda epididymal sperm is independent of the testis, *J. Androl.*, 15, 318, 1994.
51. International Conference on Harmonization (ICH) Harmonized Tripartite Guideline, Detection of toxicity to reproduction for medicinal products, *Fed. Regist.*, 59, 48746, 1994.
52. FDA, Considerations for reproductive toxicity studies for preventive vaccines for infectious disease indications, *Fed. Regist.*, 65, 54534, 2000.
53. FDA, Reviewer Guidance. Integration of Study Results to Assess Concerns about Human Reproductive and Developmental Toxicities (http://www.fda.gov/cder/guidance/4625dft.pdf), October 2001.

APPENDIX A

Terminology of Anatomical Defects

GLOSSARY OF DEVELOPMENTAL DEFECTS

Ronald D. Hood and Kit Keller

ablepharia   reduction or absence of the eyelids, with continuous skin covering the eyes
abrachia   absence of the arms (forelimbs)
acampsia   rigidity or inflexibility of a joint [ankylosis]
acardia   absence of the heart
acaudia, acaudate   without a tail [anury]
acephaly   agenesis of the head [acephalia]
acheiria   congenital absence of one or both hands (forepaws)
achondroplasia   cartilage defect causing inadequate bone formation and resulting in short limbs and other defects
acorea   absence of the pupil of the eye
acrania   partial or complete absence of the cranium
acystia   absence of the urinary bladder
adactyly   absence of digits
agenesis   lack of development of an organ
aglossia   absence of the tongue
agnathia   absence of the lower jaw (mandible)
agyria   characterized by a small brain, lacking the normal convolutions of the cerebral cortex [lissencephaly]
amastia   absence of the mammae (breasts)
amelia   absence of a limb or limbs (see also ectromelia)
ametria   absence of the uterus
anasarca   generalized edema
anencephaly   absence of the cranial vault, with the brain missing or greatly reduced
anephrogenesis   absence of kidney(s)
aniridia   absence of the iris
anisomelia   inequality between paired limbs
ankyloglossia   partial or complete adhesion of the tongue to the floor of the mouth

ankylosis   abnormal fixation of a joint; implies bone fusion
anodontia   absence of some or all of the teeth
anonchia   absence of some or all of the nails
anophthalmia   absent or vestigial eye(s)
anorchism   uni- or bilateral absence of the testes [**anorchia**]
anostosis   defective development of bone; failure to ossify
anotia   absence of the external ear(s) (i.e., pinnae, auricles)
anovarism   absence of the ovaries [**anovaria**]
anury   see *acaudia*
aphakia   absence of the eye lens
aphalangia   absence of a digit or of one or more phalanges
aplasia   see *agenesis*
apodia   absence of one or both of the feet (paws)
aprosopia   partial or complete absence of the face
arachnodactyly   abnormal length and slenderness of the digits
arrhinia   absence of the nose
arthrogryposis   persistent flexure or contracture of a joint
arthrogryposis multiplex congenita   syndrome distinguished by congenital fixation of the joints and muscle hypoplasia
asplenia   absence of the spleen
astomia   absence of the opening of the mouth
atelectasis   incomplete expansion of a fetal lung or a portion of the lung at birth
athelia   absence of the nipple(s)
athymism   absence of the thymus gland [**athymia**]
atresia   congenital absence of a normally patent lumen or closure of a normal body opening
atresia ani   agenesis or closure of the anal opening [**imperforate anus**]
atrial septal defect   postnatal communication between the atria
bifid tongue   cleft tongue
bipartite   division of what is normally a single structure into two parts; usually refers to areas of skeletal ossification
brachydactyly   abnormal shortness of digits
brachygnathia   abnormal shortness of the mandible
brachyury   abnormally short tail
buphthalmos   enlargement and distension of the fibrous coats of the eye; congenital glaucoma [**buphthalmia, buphthalmus**]
camptodactyly   permanent flexion of one or more digits [**camptodactylia**]
carpal flexure   abnormal flexion of the fetal carpus (wrist); most often seen in the rabbit; may be transient or permanent
celoschisis   congenital fissure of the abdominal wall
celosomia   fissure or absence of the sternum, with visceral herniation
cephalocele   protrusion of part of the brain through the cranium
cleft face   clefting due to incomplete fusion of the embryonic facial primordia
cleft lip   cleft or defect in the upper lip [**hare lip**]
cleft palate   fissure or cleft in the bony palate [**palatoschisis**]
clinodactyly   permanent lateral or medial deviation of one or more fingers
club foot   see *talipes*
coarctation   stricture or stenosis, usually of the aorta

coloboma fissure or incomplete development of the eye
conceptus everything that develops from the fertilized egg, including the embryo or fetus and the extraembryonic membranes
craniorachischisis fissure of the skull and vertebral column
cranioschisis congenital cranial fissure
craniosynostosis premature ossification of cranial sutures
craniostenosis premature cessation of cranial growth because of craniosynostosis
cryptorchidism failure of one or both testes to enter the scrotum [**cryptorchism**]
cyclopia fusion of the orbits into a single orbit, with the nose absent or present as a tubular appendage (proboscis) above the orbit
delayed ossification incomplete mineralization of an otherwise normal ossification center
dextrocardia abnormal displacement of the heart to the right
diaphragmatic hernia protrusion of abdominal viscera through a defect in the diaphragm
dicephalus conjoined twins with two heads and one body
diplomyelia complete or incomplete doubling of the spinal cord due to a longitudinal fissure
diplopagus symmetrically duplicated conjoined twins with largely complete bodies that may share some internal organs
diprosopus fetus with partial duplication of the face
dysarthrosis malformation of a joint
dysgenesis defective development
dysplasia abnormal tissue development
dysraphism failure of fusion, especially of the neural folds [**dysraphia**]
dystocia prolonged, abnormal, or difficult delivery
ectopia displacement or malposition of an organ or body part
ectopia cordis displacement of the heart outside of the thoracic cavity
ectopic kidney abnormal position of one or both kidneys
ectopic pregnancy pregnancy located outside the uterine cavity
ectrodactyly absence of all or only part of one or more digits
ectromelia hypoplasia or absence of one or more limbs
ectropion abnormal eversion of the margin of the eyelid
encephalocele herniation of part of the brain, encased in meninges, through an opening in the skull [**encephalomeningocele, meningoencephalocele**]
endocardial cushion defect heart defects resulting from incomplete fusion of the embryonic endocardial cushions
entropion abnormal inversion of the margin of the eyelid
epispadias absence of the upper wall of the urethra; more common in males, where the urethra opens on the dorsal surface of the penis
ethmocephalus form of cyclopia in which the eyes are closely set and the nose is hypoplastic and may be displaced upward [**ethmocephaly**]
eventration protrusion of bowels through the abdominal wall
exencephaly absence of part or the entire cranium, leaving the brain exposed
exomphalos see *omphalocele* [**umbilical hernia**]
exophthalmos abnormal protrusion of the eyeball [**exophthalmus**]
exostosis abnormal bony growth projecting outward from the surface of a bone
exstrophy congenital eversion of a hollow organ, e.g., of the bladder
fetal wastage postimplantation death of embryo or fetus
fetus developing mammal from the completion of major organogenesis to birth

gastroschisis fissure of the abdominal wall, not involving the umbilicus and usually involving protrusion of viscera

gonadal dysgenesis general term for various abnormalities of gonadal development, e.g., gonadal aplasia, hermaphroditism, "streak gonads"

hamartoma benign nodular or tumorlike mass resulting from faulty embryonal development of cells and tissues natural to the part

hemimelia absence of all or part of the distal half of a limb

hemivertebra incomplete development of one side of a vertebra

hermaphrodite individual with both male and female gonadal tissue (see also pseudohermaphrodite)

heterotopia development of a normal tissue in an abnormal location

holocardius grossly defective monozygotic twin whose circulation is dependent upon the heart of a more perfect twin

holoprosencephaly failure of division of the prosencephalon, resulting in a deficit in midline facial development, with hypotelorism; cyclopia occurs in the severe form

hydramnion, hydramnios see *polyhydramnios*

hydranencephaly complete or almost complete absence of cerebral hemispheres, which have been replaced by cerebrospinal fluid

hydrocele collection of fluid in the tunica vaginalis of the testis or along the spermatic cord

hydrocephalus marked dilation of the cerebral ventricles with excessive fluid, usually accompanied by a vaulted or dome-shaped head **[hydrocephaly, hydrencephaly, hydrencephalus]**

hydronephrosis distension of the pelvis and calyces of the kidney with fluid, as a result of obstruction of urinary outflow

hydroureter distension of the ureter with urine because of obstruction of the ureter

hypermastia presence of one or more supernumerary mammary glands **[polymastia]**

hypertelorism abnormally great distance between paired parts or organs, e.g., between the eyes

hypospadias abnormal opening of the urethra on the underside of the penis, on the perineum in males, or into the vagina in females

ichthyosis developmental skin disorders characterized by excessive or abnormal keratinization, with dryness and scaling

iniencephaly anomaly of the brain and neck characterized by an occipital bone defect, spina bifida of the cervical vertebrae, and fixed retroflexion of the head on the cervical spine

kyphosis abnormal dorsal convexity in the curvature of the thoracic spine (humpback, hunchback)

lissencephaly see *agyria*

lordosis abnormal anterior convexity in the curvature of the spine (swayback)

macroglossia abnormally large tongue, often protruding

macrophthalmia abnormally large eye(s)

macrosomia abnormally large body

macrostomia abnormally wide mouth

malformation permanent structural deviation that generally is incompatible with or severely detrimental to normal postnatal development or survival

meningocele herniation of the meninges through a defect in the cranium **[cranial m.]** or spinal column **[spinal m.]**

meningoencephalocele herniation of meninges and brain tissue through a defect in the cranium **[encephalocele, encephalomeningocele]**

meningomyelocele herniation of meninges and spinal cord through a defect in the spinal column [**myelomeningocele**]
meromelia absence of part of a limb
micrencephaly abnormally small brain
microcephaly abnormally small head [microcephalia; microcephalus]
microglossia tongue hypoplasia
micrognathia abnormally small jaw (usually the mandible)
micromelia having abnormally small or short limb(s)
microphthalmos abnormal smallness of one or both eyes [**microphthalmia**]
microstomia hypoplasia of the mouth
microtia hypoplasia of the pinna, with an absent or atretic external auditory meatus
misaligned sternebrae when the two ossification centers of each sternebra are not aligned in the transverse plane
myelocele herniation of spinal cord through a vertebral defect
myelomeningocele (see *meningomyelocele*)
myeloschisis cleft spinal cord caused by failure of neural tube closure
neural tube defect any malformation (e.g., anencephaly, exencephaly, spina bifida) resulting from failure of closure of the neural tube during embryonic development
nevus circumscribed skin malformation, usually hyperpigmented or with abnormal vascularization
oligodactyly having fewer than the normal number of digits
oligohydramnios abnormally reduced quantity of amniotic fluid
omphalocele herniation of intestine covered with peritoneum and amnion through a defect in the abdominal wall at the umbilicus [**exomphalos, umbilical hernia**]
otocephaly extreme underdevelopment of the mandible, allowing close approximation or union of the ears on the anterior aspect of the neck
overriding aorta displacement of the aorta to the right, so that it appears to arise from both ventricles
pagus combining form (suffix) indicating conjoined twins
palatine rugae alteration misaligned or otherwise abnormal palatal ridges
patent ductus arteriosus an open communication between the pulmonary trunk and aorta, persisting postnatally
patent foramen ovale failure of adequate postnatal closure of the foramen ovale, the atrial septal defect allowing interchange of blood between the atria
phocomelia absence of the proximal part of a limb(s), the distal part attached to the trunk by a small, irregularly shaped bone
plagiocephaly asymmetrical condition of the cranium resulting from premature closure of the cranial sutures on one side
polydactyly supernumerary digits
polyhydramnios abnormally increased quantity of amniotic fluid
polymastia see *hypermastia*
proboscis cylindrical protuberance of the face
pseudohermaphroditism partial masculinization or partial feminization, with gonadal tissue of only one sex present
ptosis drooping of the upper eyelid because of abnormal muscle or nerve development
rachischisis congenital fissure of the vertebral column

resorption conceptus that died after implantation and is being, or has been, reabsorbed into the mother's bloodstream

rhinocephaly possessing a proboscis-like nose above partially or completely fused eyes [**rhinocephalus**]

runt normally developed fetus or newborn significantly smaller than the remainder of a litter

scoliosis lateral deviation of the vertebral column

sirenomelia any of several degrees of side-to-side fusion of the lower extremities and concomitant midline reduction of the pelvis; soft tissues and long bones, lower paws (feet), and viscera of the pelvis tend to be reduced or absent; anus and external genitalia are often absent [**symmelia**]

situs inversus lateral transposition of the viscera

spina bifida localized defective closure of the vertebral arches, through which the spinal cord and/or meninges may protrude

spina bifida aperta spina bifida in which the neural tissue is exposed

spina bifida cystica spina bifida with herniation of a cystic swelling containing the meninges (meningocele), spinal cord (myelocele), or both (myelomeningocele)

spina bifida occulta spina bifida with intact skin and no herniation

stillbirth birth of a dead fetus

sympodia fusion of the feet

syndactyly webbing or fusion between adjacent digits

talipes congenital deformity of the foot, which is twisted out of shape or position [**clubfoot**]

talipes equinovalgus talipes in which the heel is elevated and turned outward from the midline

talipes equinovarus talipes in which the foot is plantarflexed and turned inward from the midline; the most common talipes

teratogen agent that can cause abnormal development of the embryo or fetus

teratogenicity ability to cause defective development of the embryo or fetus

tetralogy of Fallot combination of cardiac defects consisting of pulmonary or infundibular stenosis, interventricular septal defect, overriding aorta, and right ventricular hypertrophy

tracheoesophageal fistula abnormal connection between trachea and esophagus

umbilical hernia see *omphalocele*

ventricular septal defect persistent communication between the ventricles

wavy ribs extra bends in one or more ribs

[Adapted in part, with modifications, from the following sources: Anderson, D. M., Keith, J., Novak, P. D., and Elliott, M.A., *Dorland's Illustrated Medical Dictionary*, 28th ed., W. B. Saunders, Philadelphia, 1983; MARTA Committee, Glossary of fetal anomalies, in *Handbook of Developmental Toxicology*, R. Hood, Ed., CRC Press, Boca Raton, 1997, 697 pp.; Keller, K. unpublished glossary; Moore, K.L. and Persaud, T. V. N., *The Developing Human*, 6th ed., W. B. Saunders, Philadelphia, 1998; O'Rahilly, R. and Müller, F., *Human Embryology & Teratology*, 2nd ed., Wiley-Liss, New York, 1996; Schardein, J.L., unpublished glossary (1995); Spraycar, M., Ed., *Stedman's Medical Dictionary*, 26th ed. Williams & Wilkins, Baltimore, 1995; U.S. EPA, Health Effects Division, *Standard Evaluation Procedure, Developmental Toxicity Studies*. Health Effects Division, Office of Pesticide Programs, Washington, DC, 1993; Wise, L.D., Beck, S.L., Beltrame, D., Beyer, B.K., et al. Terminology of developmental abnormalities in common laboratory mammals (version 1), *Teratology* 55, 249, 1997.]

TERMINOLOGY OF DEVELOPMENTAL ABNORMALITIES IN COMMON LABORATORY MAMMALS*

L. David Wise,[1,19,20,21,]* Sidney L. Beck,[2,22] Diana Beltrame,[3,19,†] Bruce K. Beyer,[4,22] Ibrahim Chahoud,[5,19] Robert L. Clark,[6,20] Ruth Clark,[7,19,††] Alice M. Druga,[8,19] Maureen H. Feuston,[9,20] Pierre Guittin,[10,19,†††] Susan M. Henwood,[11,22] Carole A. Kimmel,[12,19] Pia Lindström[13, 21, 23], Anthony K. Palmer[14, 19], Judith A. Petrere[15, 22], Howard M. Solomon,[16,20,21] Mineo Yasuda,[17,19] and Raymond G. York[18,20]

[1] Merck Research Laboratories, Safety Assessment, West Point, Pennsylvania, 19486
[2] DePaul University, Department Biological Sciences, Chicago, Illinois, 60614
[3] Pharmacia & Upjohn, Worldwide Toxicology, Milan, Italy
[4] Bristol-Myers Squibb, Pharmaceutical Research Institute, Evansville, Indiana, 47721
[5] Freie Universitat, Institut fur Toxikologie and Embryo Pharmakologie, 14195 Berlin 33, Germany
[6] Rhône-Poulenc Rorer Research and Development, Department of Toxicology, Collegeville, Pennsylvania, 19426-0994
[7] Ruth Clark Associates Ltd., North Lincolnshire, DN17 3JB, United Kingdom
[8] Institute for Drug Research Ltd., Division of Safety Studies Budapest, Hungary, H-1325
[9] Sanofi Research, Division of Sanofi Pharmaceuticals, Inc., Malvern, Pennsylvania, 19355
[10] Rhône-Poulenc Rorer, Drug Safety, General and Reproductive Toxicology Department, 94403, Vitry-sur-Seine, France
[11] Covance Laboratories Inc., Madison, Wisconsin, 53707
[12] U.S. Environmental Protection Agency, NCEA, Washingtom, DC, 20460; currently on detail to the Food and Drug Administration, NCTR, Rockville, Maryland, 20857
[13] Quintiles, Inc., Regulatory Affairs, Research Triangle Park, North Carolina, 27709
[14] Huntingdon, PE18 7XJ, United Kingdom
[15] Warner-Lambert/Parke-Davis Pharmaceutical Research Division, Pathology/Experimental Toxicology Department, Ann Arbor, Michigan, 48105
[16] SmithKline Beecham Pharmaceuticals, Research & Development Division, King of Prussia, Pennsylvania, 19406-0939
[17] Hiroshima University School of Medicine, 734, Hiroshima, Japan
[18] Argus Research Laboratories, Horsham, Pennsylvania, 19044
[19] International Federation of Teratology Societies Committee on International Harmonization of Nomenclature in Developmental Toxicology
[20] Middle Atlantic Reproduction and Teratology Association Nomenclature Committee
[21] Teratology Society Terminology Committee
[22] Midwest Teratology Association Nomenclature Committee
[23] Food and Drug Administration Standardized Nomenclature Program

[†] Representing the Italian Nomenclature Working Group. Other members include Dr. Rita Bussi, RBM, Ivrea; Prof. Erminio Giavini, University of Milan; and Dr. Alberto Mantovani, Instituto Superiore di Sanità, Rome.
[††] Representing the UK Foetal Pathology Terminology Group. Other members include Mr. Robert Bramley, Zeneca Pharmaceuticals, Macclesfield; Mrs. Elizabeth J. Davidson, Medeva Group Development, Surrey; Mr. Keith P. Hazelden, Inveresk Research Ltd., Edinburgh (currently Huntingdon Life Sciences, Eye); Mr. David M. John, Huntingdon Life Sciences, Eye; Mrs. Mary Moxon, Zeneca Central Toxicology Laboratory, Macclesfield; Ms. Meg M. Parkinson, Glaxo Wellcome Research and Development, Ware; and Mrs. Sheila A. Tesh, Tesh Consultants International, Saxmundham.
[†††] Representing the French Teratology Association's Nomenclature Working Group. Other members include Dr. Paul Barrow, Chrysalis International, L'Arbresle; Dr. Catherine Boutemy, Synthelabo Recherche, Gargenville; Dr. Didier Hiss, Ruffey les Echirey; Dr. Andre Morin, Faculte de Medecine, Lyon; Dr. Charles Roux, CHU St-Antoine, Paris; and Dr. Jeanne Stadler, Pfizer, Amboise.

* Correspondence to: L. David Wise, Merck Research Laboratories, W45-1, West Point, PA 19486

* This article was previously published as: Wise, L.D. et al., Terminology of Developmental Abnormalities in Common Laboratory Mammals (Version 1). *Teratology* 55:249-292, 1997. With permission.

ABSTRACT

This paper presents the first version of an internationally-developed glossary of terms for structural developmental abnormalities in common laboratory animals. The glossary is put forward by the International Federation of Teratology Societies (IFTS) Committee on International Harmonization of Nomenclature in Developmental Toxicology, and represents considerable progress toward harmonization of terminology in this area. The purpose of this effort is to provide a common vocabulary that will reduce confusion and ambiguity in the description of developmental effects, particularly in submissions to regulatory agencies world-wide. The glossary contains a primary term or phrase, a definition of the abnormality, and notes, where appropriate. Selected synonyms or related terms, which reflect a similar or closely related concept, are noted. Nonpreferred terms are indicated where their usage may be incorrect. Modifying terms used repeatedly in the glossary (e.g., absent, branched) are listed and defined separately, instead of repeating their definitions for each observation. Syndrome names are generally excluded from the glossary, but are listed separately in an appendix. The glossary is organized into broad sections for external, visceral, and skeletal observations, then subdivided into regions, structures, or organs in a general overall head to tail sequence. Numbering is sequential, and not in any regional or hierarchical order. Uses and misuses of the glossary are discussed. Comments, questions, suggestions, and additions from practitioners in the field of developmental toxicology are welcomed on the organization of the glossary as well on as the specific terms and definitions. Updates of the glossary are planned based on the comments received.

INTRODUCTION

Nomenclature used to describe observations of fetal and neonatal morphology often varies considerably among laboratories, investigators, and textbooks in the fields of teratology and developmental toxicology. Standard medical, veterinary, and anatomical dictionaries often differ in the definitions of commonly used terms. This lack of a common vocabulary leads to confusion and uncertainty when communicating scientific findings, and can be a particular problem in submissions to regulatory agencies worldwide. Thus, there is a recognized need for a common nomenclature to use when describing such observations.

Over the years numerous efforts have been undertaken to provide a common nomenclature. Various groups in different countries have made attempts to provide solutions, and international organizations, such as the International Programme on Chemical Safety (IPCS) and the Organization for Economic Cooperation and Development (OECD) have shown interest in promoting international harmonization. For various reasons, previous efforts have not been adopted on a widespread basis. In an attempt to provide an internationally acceptable common vocabulary, the International Federation of Teratology Societies (IFTS*) in May 1995 appointed a Committee on International Harmonization of Nomenclature in Developmental Toxicology. The first priority for this committee was to collect and review existing glossaries and ongoing efforts in several countries aimed at developing atlases of fetal abnormalities, and to attempt to develop a glossary and atlas that reflected international harmonization. Several groups, notably in France, Germany, Japan, and the United Kingdom (UK) have begun developing atlases of normal and/or abnormal fetal morphology either as books, CD-ROM, or both. Although these image-based atlases promise to provide the best means of portraying specific observations with photographs and drawings, they are time-consuming to prepare. The IFTS committee soon decided to concentrate first on harmonization of the terminology. In this way the image-based products could incorporate this harmonized terminology, avoid confusion from one atlas to another, and be used in further refinement of the terminology.

---

* The IFTS is composed of members of the teratology societies of Australia, Europe, Japan, and North America

The most convenient material available for initiating the process at the time the IFTS committee began its work was the glossary previously developed and distributed in the U.S.A. by the Middle Atlantic Reproduction and Teratology Association (MARTA). A revision of this glossary was underway, in part, due to a request by the U.S. Food and Drug Administration for assistance in their project to standardize submissions of data on new drugs. The effort to revise the MARTA glossary was initiated with the involvement of the Midwest Teratology Association (MTA). After a review by the IFTS Nomenclature Committee, the three organizations agreed to work jointly on developing an international glossary. The glossary was revised in response to hundreds of comments received from the initial request as well as comments from the IFTS committee and associated nomenclature committees in the UK, France, and Italy. The document presented herein represents considerable progress towards harmonization. Although not the product of a more formal international harmonization process such as ICH guideline development, this document is the first version of an international glossary that will continue to evolve over the coming years with input from worldwide IFTS members as well as other users of the glossary.

We invite and would welcome questions, comments, additions, and suggestions of either a specific or general nature. Based on these comments and proposed modifications, an update of the glossary will be initiated by the nomenclature committees. Specific or detailed comments and proposed changes to the glossary may be directed to any author listed above, or, if not convenient, to Dr. L. David Wise (Merck Research Laboratories, W45-1, West Point, PA 19446; phone 215-652-6974; fax 215-652-3423; e-mail david_wise@merck.com). Comments or information regarding the overall IFTS initiative may be addressed to Dr. Carole A. Kimmel (U.S. Food and Drug Administration, NCTR [HFT-10] Rm. 16B-06, 5600 Fishers Lane, Rockville, MD 20857; phone 301-827-3403; fax 301-443-3019; e-mail ckimmel@nctr.fda.gov).

Organization of the Glossary

The terms included in the glossary describe morphological changes observed grossly or with the aid of a dissecting microscope in the most common laboratory animals used for developmental studies (mainly rats, mice, and rabbits). The term 'abnormalities' is used here as a collective term to mean differences from normal in the specimen under examination relative to the perceived norm of control specimens for a particular species. It is perhaps not an ideal term since normality is relative and may change over time, but is considered an acceptable term for this list.

Terms are included which describe observations in fetal as well as neonatal animals. Thus, for example, hypospadias and ocular coloboma, which are not readily detectable at birth in these animals, but are observable at some later time, are included in the glossary. On the other hand, cryptorchidism is not included as an observation, since the applicable event (descent of the testes into the scrotum) does not occur until after weaning in these species. We sought to describe observations of abnormalities with mainly descriptive terms and to avoid or limit terms that were more diagnostic or perhaps implied mechanisms. Some observations may require follow-up studies or examinations to determine more detail.

The glossary is organized into broad sections for external, visceral, and skeletal observations, which reflect how the data are typically collected in developmental toxicity studies. The observations are subdivided into regions, structures, or organs in a general overall head to tail sequence; the observations within each are listed in alphabetical order. Some external observations are better defined with a subsequent visceral and/or skeletal examination, and these terms should then be associated to reflect a single observation. The notes for external observations sometimes indicate where this is the case.

Each term is sequentially numbered in this version in order to facilitate tracking of future revisions. These code numbers are not intended to provide any regional or hierarchical information.

The first digit of all terms in this version is "1", and corresponds to the present version number. Observation entries that are revised in subsequent versions will be indicated by changing the first digit to correspond to the revision number of the document. Totally new observations added in subsequent versions will have a first digit corresponding to that version number, and will be numbered sequentially from the preceding version. Thus, the first new term added in version 2 will be numbered 20869.

An observation is represented by a primary term or phrase, and definition of the observation is included along with notes, where appropriate. Among other uses, the notes indicate when an alteration is part of a common syndrome. Selected synonyms or related terms, which reflect a similar or closely related concept, are then noted. Nonpreferred terms are indicated where their usage may be incorrect. In some cases, the nonpreferred term may be synonymous with the observation but it is not recommended for general use since other terms are more descriptive or more widely accepted.

Certain modifying terms used repeatedly in the glossary (e.g., absent, branched) are listed and defined separately in Appendix I, instead of repeating their definitions for each observation. In addition, syndrome names are generally excluded from the glossary even if the name of the syndrome is common. For example, there is no entry for ectrosyndactyly, since this can be described by using the separate entries of ectrodactyly and syndactyly. Common syndromes omitted from the glossary are listed in a separate section (Appendix II). Various options for the organization of the glossary were considered so comments from users would be very welcomed.

The list of terms is formatted so that it can be handled electronically. One of the designated objectives of the list was to facilitate transfer of data within and between computer databases. Also, the IFTS intends to make the glossary with an associated atlas of images accessible on the internet for easy access by the wider scientific community.

Uses and Misuses of the Glossary

During the course of creating and revising this glossary a number of issues arose regarding its purpose and uses.

1. Not all observations in this relatively extensive list are necessarily intended to be used in the summarization of data for the purpose of reporting and comparing group incidences. Initially, the list of terms may be used so that separate but coincident abnormalities are enumerated as separate findings. Then, multiple coincidental abnormalities can be described using a single combining term. As a simple example, domed head may be described externally, but upon visceral examination, if hydrocephaly is found, only this term is used on a summary table. Another example would be where multiple fetuses have different digits that are variably short. Instead of enumerating the values for absent and short phalanges, the summary table could use brachydactyly as the general combining term.
2. To reiterate, the terms are to be used to describe morphological observations outside the normal range. Thus, for example, it is intended that the use of the term "small" is understood to describe a structure that is morphologically similar to the norm but small compared to control specimens of the same general age.
3. A ranking or classification of terms into categories, such as malformation and variation, does not appear in the glossary since a given observation may be a malformation in one species but a variation in another species, or the classification may change depending on the gestational day of examination. In addition, there is no consensus at present as to which classification scheme is most relevant. Classification of each observation into a certain category is a common practice, and must be left as an option to the user.
4. It is beyond the scope of this glossary to specify the normal number of skeletal elements for each skeletal region as separate observations; although, for general reference, these numbers have been listed in the notes for various vertebral regions. This omission does NOT negate the common

practice of reporting number of ossified elements in specific regions (e.g., sacral and/or caudal vertebrae), for this approach does offer certain advantages.
5. In general, observations are represented by simple descriptive terms or phrases, except in cases where a common, generally accepted medical term is available.
6. Terminology in the glossary typically reflects the basic observation. For example, open eye includes observations of partially open eyelid and completely open eyelid, and cervical rib includes unilateral and bilateral. Modifiers, such as those relating to symmetry (e.g., right, left, medial, unilateral, bilateral) and degree (e.g., slight, moderate, marked), are not included in the glossary, except where sidedness is an integral part of the specific observation (e.g., right-sided aortic arch). Some users may wish to further define abnormalities using these modifiers (e.g., "slightly dilated ureter" and "markedly dilated ureter").
7. Analysis of data on treatment-related effects may be more useful when glossary terms are combined. For example, many of the individual observations could be combined into general categories (e.g., tail abnormalities or thoracic vertebral abnormalities). Alternatively, some terms may be subdivided into relevant categories based on modifying terms (e.g., median versus unilateral cleft lip). Thus, tables used to present the individual or summary results of the examinations of fetuses or offspring may use the individual terms, names for groups of terms, or subsets of the terms.

Finally, this glossary represents the start of a process that is working toward harmonization of terminology among scientists from many countries. The intent is for it to be a basic support for describing developmental abnormalities. However, scientific judgment is required to use the observations appropriately to allow an accurate interpretation of the data.

The following dictionary was used as the primary reference for the glossary when inconsistencies were found in other reference sources: *International Dictionary of Medicine and Biology* (3 volumes), 1986, Wiley Medical Publications, John Wiley & Sons, NY (ISBN 0-471-01849-X). Terms covering the field of teratology within this dictionary were authored by the late Dr. James G. Wilson, an internationally recognized expert in the field of teratology. Modifications to definitions may have been made in some cases in order to reach an agreed upon, workable definition.

Terminology of developmental abnormalities in common laboratory mammals (version 1): Extrernal abnormalities[1]

Region/Organ/Structure	Code Number	Observation	Synonym or Related Term	Non-preferred Term	Definition	Note
General	10001	Anasarca	Generalized edema		Generalized edema	
	10002	Conjoined twins	Double monster		Monozygotic twins with variable incomplete separation into two during cleavage or early stages of embryogenesis	Site and extent of fusion may be described
	10003	Cutis aplasia			Localized region of no skin development	
	10004	Hemorrhage	Petechia, Purpura, Ecchymosis, Hematoma		An accumulation of extravasated blood	
	10005	Local edema			Localized accumulation of fluid	
	10006	Oligohydramnios			Less than normal amount of amniotic fluid	
	10007	Polyhydramnios			Excessive amount of amniotic fluid	
Cranium	10008	Acephalostomia		Acrania	Absence of head but with the presence of mouth-like orifice in the neck region	Acrania is a skeletal term
	10009	Acephaly			Absence of the head	
	10010	Anencephaly		Acrania	Absence of the cranial vault, with the brain missing or reduced to small mass(es)	
	10011	Cranioschisis			Fissure of the cranium	[Craniorachischisis]
	10012	Domed head			Cranium appears more elevated and rounded than normal	May or may not be associated with hydrocephaly
	10013	Exencephaly		Acrania	The brain protrudes outside the skull due to absence of the cranial vault	Erosion of brain tissue has not occurred as in Anencephaly
	10014	Iniencephaly			Exposure of occipital brain and upper spinal cord tissue. Involves extreme retroflection of the head.	
	10015	Macrocephaly			Disproportionate largeness of head	
	10016	Meningocele			Herniation of meninges through defect in skull	
	10017	Meningoencephalocele	Encephalomeningocele	Cephalocele, Craniocele, Encephalocele	Herniation of brain and meninges through a cranial opening	May or may not be covered by skin
	10018	Microcephaly	Leptocephaly, Nanocephaly		Small cranium	

TERMINOLOGY OF ANATOMICAL DEFECTS

Region	Code	Term	Synonym	Colloquial	Definition	Comments
Ear	10019	Anotia	Agenesis, Aplastic		Absence of pinna	
	10020	Macrotia			Large pinna	
	10021	Malpositioned pinna	Displaced, Ectopic, Low set			
	10022	Microtia			Small pinna	
	10023	Misshapen pinna	Abnormally shaped, Irregularly shaped			
	10024	Synotia			Fusion or abnormal approximation of pinnae below the face	[Otocephaly]
Eye	10025	Absent eye bulge				May indicate micro- or anophthalmia
	10026	Cryptophthalmia	Cryptophthalmos		Skin continuous over eye(s) without formation of eyelid(s)	Usually associated with microphthalmia
	10027	Cyclopia	Monophthalmia, Synophthalmia		Single orbit; eye(s) can be absent, completely or incompletely fused	Nose may be absent or appear as a frontonasal appendage (proboscis) above the orbit
	10028	Enlarged eye bulge				
	10029	Exophthalmos	Proptosis	Pop-eye	Excessive protrusion of the eyeball	
	10030	Malpositioned eye	Displaced, Ectopic			
	10031	Microblepharia			Short vertical dimension of eyelid	
	10032	Ocular coloboma			A family of eye defects generally involving a cleft or fissure of an eye structure	Not readily apparent in fetuses or soon after birth
	10033	Open eye	Ablepharia		Partial or complete deficiency of the eyelid	Eyeball usually visible
	10034	Palpebral coloboma			A notch or fissure of the eyelid	
	10035	Small eye bulge	Reduced			May indicate microphthalmia
Face	10036	Cleft face	Prosoposchisis		Fissure of the face and jaw	[Cheilognathoschisis, Cheilognathopalatoschisis]
Nose	10037	Short snout	Rudimentary			
	10038	Arhinia	Agenesis, Aplastic		Absence of the nose	
	10039	Malpositioned naris	Displaced, Ectopic			
	10040	Malpositioned nose	Displaced, Ectopic			
	10041	Misshapen nose	Abnormally shaped, Irregularly shaped			
	10042	Naris atresia	Imperforate			
	10043	Proboscis			Tubular projection replaces the nose	[Rhinocephaly]
	10044	Single naris			One naris instead of two nares	

Terminology of developmental abnormalities in common laboratory mammals (version 1): External abnormalities[1] (continued)

Region/ Organ/ Structure	Code Number	Observation	Synonym or Related Term	Non-preferred Term	Definition	Note
	10045	Small nose	Hypoplastic, Reduced, Rudimentary			
Mouth/Jaw	10046	Aglossia	Agenesis, Aplastic		Absence of the tongue	
	10047	Agnathia	Agenesis, Aplastic		Absence of the lower jaw (mandible)	
	10048	Ankyloglossia			Shortness or absence of the frenum of the tongue; tongue fused to the floor of the mouth	
	10049	Anodontia	Edentia		Partial or complete absence of one or more teeth	
	10050	Astomia	Agenesis, Aplastic		Absence of the mouth	
	10051	Cleft lip	Cheiloschisis	Harelip	Fissure of the upper lip	
	10052	Cleft palate	Palatoschisis, Uranoschisis		Fissure of the palate	
	10053	High-arched palate			A higher than normal palatal arch	
	10054	Macroglossia			Large tongue	
	10055	Macrostomia			Abnormal elongation of the mouth	
	10056	Mandibular cleft				
	10057	Mandibular macrognathia	Prognathism	Exognathia	Enlarged or protruding lower jaw (mandible)	May be protruding
	10058	Mandibular micrognathia	Brachygnathia, Micromandible		Small lower jaw (mandible)	
	10059	Maxillary macrognathia	Prognathism	Exognathia	Enlarged or protruding upper jaw (premaxilla/maxilla)	
	10060	Maxillary micrognathia	Micromaxilla		Small upper jaw (premaxilla/maxilla)	
	10061	Microglossia			Small tongue	
	10062	Microstomia			Small mouth	
	10063	Misaligned palate rugae	Asymmetric			
	10064	Misshapen palate rugae	Abnormally shaped, Irregularly shaped			
	10065	Protruding tongue				
Limb	10066	Amelia	Agenesis, Aplastic, Ectromelia		Complete absence of one or more limbs	Fleshy tab may be present
	10067	Bowed limb				Usually seen as an outward bending of the limb

TERMINOLOGY OF ANATOMICAL DEFECTS

	Code	Term	Synonyms	Definition	Comments
	10068	Hemimelia	Ectromelia	Absence or shortening of the distal two segments of limb	May be further defined as being fibular, radial, tibial, or ulnar
	10069	Limb hyperextension	Arthrogryposis	The excessive extension or straightening of a limb or a joint	
	10070	Limb hyperflexion	Flexed	The excessive flexion or bending of a limb or a joint	
	10071	Macromelia		An excessive size of one or more limbs	
	10072	Malrotated limb		Limb turned toward the center (i.e., inward rotation) or the periphery (i.e., outward rotation)	
	10073	Micromelia	Nanomelia	Disproportionately short or small limb(s)	
	10074	Phocomelia		Reduction or absence of proximal portion of limb, with the paws being attached to the trunk of the body	
	10075	Sirenomelia	Symelia	Fusion of lower limbs	May involve lower torso
	10076	Absent claw	Agenesis, Aplastic		Refers to distal-most tip, nail
Paw/Digit	10077	Adactyly	Agenesis, Aplastic	Absence of all digits	
	10078	Apodia	Acheiria, Acheiropodia	Absence of one or more paws	
	10079	Brachydactyly	Microdactyly	Shortened digit(s)	Expected skeletal alterations include absence or shortening of phalanx(ges)
	10080	Ectrodactyly	Agenesis, Aplastic, Oligodactyly	Absence of one or more digit(s)	Expected skeletal alterations include absence of all phalanges in each affected digit
	10081	Enlarged digit	Dactylomegaly, Macrodactyly		
	10082	Malpositioned claw	Displaced, Ectopic		Refers to distal-most tip, nail
	10083	Malpositioned digit	Clinodactyly, Camptodactyly, Displaced, Ectopic	Deflection of digit(s) from the central axis	Includes fixed flexion deformity of digit(s)
	10084	Malrotated paw	Clubbed paw, Talipes	Paw turned toward the center (i.e., inward) or the periphery (i.e., outward)	
	10085	Misshapen digit	Abnormally shaped, Irregularly shaped		
	10086	Paw hyperextension	Arthrogryposis	The excessive extension or straightening of a paw	

Terminology of developmental abnormalities in common laboratory mammals (version 1): Extrernal abnormalities[1] (continued)

Region/Organ/Structure	Code Number	Observation	Synonym or Related Term	Non-preferred Term	Definition	Note
	10087	Paw hyperflexion			The excessive flexion or bending of a paw	
	10088	Polydactyly	Supernumerary digit(s)		Supernumerary digit(s)	
	10089	Small claw	Hypoplastic, Reduced, Rudimentary			Refers to distal-most tip, nail
	10090	Small paw	Hypoplastic, Microcheiria, Reduced, Rudimentary		Smallness of paw(s)	
	10091	Syndactyly	Ankylodactyly		Partial or complete fusion of, or webbing between, digits	Includes bony, cartilaginous, and/or soft tissue
Tail	10092	Sympodia			Fusion of the hindlimb paws	
	10093	Acaudate	Agenesis, Anury, Aplastic	Anurous	Absence of tail	
	10094	Bent tail	Angulated, Bowed		Shaped like an angle	
	10095	Blunt-tipped tail			Rounded or flat at the end, not tapered	
	10096	Curled tail	Curly		Curved into nearly a full circle	
	10097	Double-tipped tail	Forked		Duplication of the tail at the end	
	10098	Fleshy tail tab				
	10099	Hooked tail			Approximately 180 degree bend or curve of the tail	
	10100	Kinked tail	Kinky		Localized undulation(s) of the tail	
	10101	Malpositioned tail				
	10102	Narrowed tail	Constricted	Ring tail		
	10103	Short tail	Brachyury, Rudimentary			
	10104	Thread-like tail	Filamentous, Filiform			
Trunk	10105	Anal atresia	Aproctia, Imperforate		Absence or closure of the anal opening	
	10106	Absent genital tubercle	Agenesis, Aplastic			
	10107	Decreased anogenital distance	Reduced perineum		Shortened anogenital distance (AGD)	
	10108	Ectopia cordis			Heart displaced outside thoracic cavity due to failure of ventral closure	
	10109	Gastroschisis	Laparoschisis, Schistocelia		Fissure of abdominal wall, not involving the umbilicus, and usually accompanied by protrusion of viscera which may or may not be covered by a membranous sac	May be further defined as medial (gastroschisis) or lateral fissure (laparoschisis)

TERMINOLOGY OF ANATOMICAL DEFECTS

10110	Holorachischisis		Fissure of the entire spinal column	[Craniorachischisis]
10111	Hypospadias		Urethra opening on the underside of the penis or on the perineum	Not readily apparent in fetuses or soon after birth
10112	Increased anogenital distance		Increased anogenital distance (AGD)	
10113	Kyphosis	Humpback, Hunchback	Increased convexity in the curvature of the thoracic spinal column as viewed from the side	
10114	Lordosis	Hollowback, Swayback, Saddleback	Anterior concavity in the curvature of the lumbar and cervical spinal column as viewed from the side	
10115	Omphalocele	Exomphalos	Protrusion of intra-abdominal viscera into the umbilical cord with viscera contained in a thin translucent sac of peritoneum and amnion	The condition may present with ruptured sac
10116	Scoliosis		Lateral curvature of the spinal column	
10117	Short trunk			
10118	Small anus			
10119	Small genital tubercle	Hypoplastic, Reduced, Rudimentary		
10120	Spina bifida	Meningocele, Meningomyelocele, Myelocele, Myelomeningocele, Rachischisis	A family of defects in the closure of the spinal column	May be covered with skin (spina bifida oculta); may involve protrusion of spinal cord and/or meninges
10121	Thoracogastroschisis	Thoracoceloschisis	Failure of closure leaving the thoracic and abdominal cavities, or major parts thereof, exposed ventrally	
10122	Thoracoschisis		Fissure of thoracic wall	
10123	Thoracostenosis		Narrowness of the thoracic region	
10124	Umbilical hernia		Protrusion of a skin-covered segment of the gastrointestinal tract and/or greater omentum through a defect in the abdominal wall at the umbilicus, the herniated mass being circumscribed and covered with skin; or protrusion of skin-covered viscera through the umbilical ring with prominence of the navel	Lung(s) may be herniated

[1]For terms in brackets, see Appendix 1-B below.

Terminology of developmental abnormalities in common laboratory mammals (version 1): Visceral abnormalities[1]

Region/Organ/Structure	Code Number	Observation	Synonym or Related Term	Non-preferred Term	Definition	Note
General	10125	Aneurysm			Localized sac formed by distention of an artery or vein that is filled with blood	
	10126	Fluid-filled abdomen	Ascites, Hemorrhage		Effusion and accumulation of watery fluid (ascites) or blood (hemorrhage)	
	10127	Situs inversus			Mirror-image transposition of the abdominal and/or thoracic viscera	
Brain	10128	Dilated cerebral ventricle	Bulbous		Slight to moderate dilatation of cerebral ventricles, with no evidence of decrease in cortical thickness	
	10129	Hemorrhagic cerebellum				
	10130	Hemorrhagic cerebrum				
	10131	Hydrocephaly			Excessive cerebrospinal fluid within the skull	May be further defined by location
	10132	Misshapen cerebellum	Abnormally shaped, Irregularly shaped			
	10133	Misshapen cerebrum	Abnormally shaped, Irregularly shaped			Includes absence of interhemispheric fissure
	10134	Small cerebellum	Hypoplastic, Reduced, Rudimentary			
	10135	Small cerebrum	Hypoplastic, Reduced, Rudimentary			
Ear	10136	Misshapen inner ear	Abnormally shaped, Irregularly shaped			
Eye	10137	Anophthalmia	Agenesis, Aplastic		Absence of eye(s)	
	10138	Aphakia	Agenesis, Aplastic		Absence of lens	
	10139	Cataract			Opacity of the crystalline lens	
	10140	Hemorrhagic eye				
	10141	Macrophthalmia	Megalophthalmia (-mus,-mos)		Large eye, eyeball	
	10142	Malpositioned eye	Displaced, Ectopic			
	10143	Microphthalmia	Hypoplastic, Reduced, Rudimentary		Small eye	
	10144	Misshapen lens	Abnormally shaped, Irregularly shaped			

TERMINOLOGY OF ANATOMICAL DEFECTS

	10145	Ocular coloboma		A family of eye defects generally involving a cleft or fissure of an eye structure
	10146	Retina fold		Undulation of retinal layers
	10147	Small lens	Hypoplastic, Reduced, Rudimentary	May be due to processing artifact
Nose	10148	Absent nasal conchae	Agenesis, Aplastic	
	10149	Absent nasal septum	Agenesis, Aplastic	
	10150	Enlarged nasal cavity		Enlarged nasal cavity
	10151	Malpositioned nasal conchae	Displaced, Ectopic	
	10152	Malpositioned nasal septum	Displaced, Ectopic	
	10153	Small nasal cavity	Reduced	Reduced nasal cavity
	10154	Small nasal conchae	Hypoplastic, Reduced, Rudimentary	
	10155	Small nasal septum	Hypoplastic, Reduced, Rudimentary	
Thyroid gland	10156	Absent thyroid gland	Agenesis, Aplastic	
	10157	Malpositioned thyroid gland	Displaced, Ectopic	
Thymus	10158	Absent thymus	Agenesis, Aplastic, Athymia, Athymism	
	10159	Hemorrhagic thymus		
	10160	Malpositioned thymus	Displaced, Ectopic	
	10161	Misshapen thymus	Abnormally shaped, Irregularly shaped	
	10162	Small thymus	Hypoplastic, Reduced, Rudimentary, Trace, Vestigial	Reduced size, imperfect form, or remnant of thymus
	10163	Split thymus	Bifid, Bipartite, Cleft	Consisting of two separate parts
	10164	Supernumerary thymus	Additional, Extra	Includes reduced number of convolutions
Heart	10165	Absent aortic valve	Agenesis, Aplastic	
	10166	Absent cordae tendinae	Agenesis, Aplastic	
	10167	Absent left A-V valve	Agenesis, Aplastic	
	10168	Absent papillary muscle	Agenesis, Aplastic	
	10169	Absent pulmonic valve	Agenesis, Aplastic	A-V = atrioventricular

Terminology of developmental abnormalities in common laboratory mammals (version 1): Visceral abnormalities[1] (continued)

Region/Organ/Structure	Code Number	Observation	Synonym or Related Term	Non-preferred Term	Definition	Note
	10170	Absent right A-V valve	Agenesis, Aplastic			A-V = atrioventricular
	10171	Atrial septum defect			Postnatal communication between atria; includes defects of septa primum and secundum; premature closure of foramen ovale apertum	
	10172	Cardiomegaly			Large heart	
	10173	Cor biloculare			Two-chambered heart with an atrium and a ventricle	
	10174	Cor triloculare			Three-chambered heart with two atria and a ventricle or one atrium and two ventricles	
	10175	Defect of A-V septum			Inappropriate communication between atrium and ventricle	A-V = atrioventricular
	10176	Dextrocardia			Heart located on the right side of the thoracic cavity	Can occur as isolated alteration or in association with situs inversus
	10177	Enlarged aortic valve				
	10178	Enlarged atrial chamber				
	10179	Enlarged A-V ostium			Enlargement of an atrioventricular orifice	A-V = atrioventricular
	10180	Enlarged left A-V valve				A-V = atrioventricular
	10181	Enlarged pulmonic valve				
	10182	Enlarged right A-V valve				A-V = atrioventricular
	10183	Enlarged ventricular chamber				
	10184	Enlarged ventricular wall				
	10185	Hydropericardium			Accumulation of fluid in the sac that envelopes the heart	
	10186	Malpositioned heart	Displaced, Ectopic			Levocardia abnormal only in situs inversus
	10187	Membranous ventricular septum defect	Membranous VSD		An opening in the membranous septum between the ventricles	[Tetralogy of Fallot]

TERMINOLOGY OF ANATOMICAL DEFECTS

	ID	Term	Synonyms	Notes
	10188	Microcardia	Hypoplastic, Reduced, Rudimentary	Small heart
	10189	Misshapen aortic valve	Abnormally shaped, Irregularly shaped	Includes alterations in number of cusps
	10190	Misshapen left A-V valve	Abnormally shaped, Irregularly shaped	A-V = atrioventricular
	10191	Misshapen pulmonic valve	Abnormally shaped, Irregularly shaped	Includes alterations in number of cusps
	10192	Misshapen right A-V valve	Abnormally shaped, Irregularly shaped	Includes alterations in number of cusps. A-V=atrioventricular
	10193	Muscular ventricular septum defect	Muscular VSD	An opening in the muscular septum between the ventricles [Tetralogy of Fallot]
	10194	Persistent A-V canal		Defects of endocardial cushions resulting in low atrial and high ventricular septal defects A-V = atrioventricular
	10195	Small aortic valve	Hypoplastic, Reduced, Rudimentary	
	10196	Small left A-V valve	Hypoplastic, Reduced, Rudimentary	A-V = atrioventricular
	10197	Small pulmonic valve	Hypoplastic, Reduced, Rudimentary	
	10198	Small right A-V valve	Hypoplastic, Reduced, Rudimentary	A-V = atrioventricular
	10199	Small ventricular chamber	Hypoplastic, Reduced, Rudimentary	
Aorta	10200	Aortic atresia	Imperforate	
	10201	Dilated aorta		
	10202	Double aorta	Bulbous	
	10203	Malpositioned aorta	Displaced, Ectopic	
	10204	Narrowed aorta	Coarctation, Constricted, Stenosis, Stricture	
	10205	Overriding aorta		Biventricular origin of aorta
	10206	Aortic arch atresia	Imperforate	
	10207	Dilated aortic arch	Bulbous	
	10208	Interrupted aortic arch		Ascending aorta not connected to descending aorta [Tetralogy of Fallot]
Aortic arch	10209	Narrowed aortic arch	Coarctation, Constricted, Stenosis, Stricture	

Terminology of developmental abnormalities in common laboratory mammals (version 1): Visceral abnormalities[1] (continued)

Region/Organ/Structure	Code Number	Observation	Synonym or Related Term	Non-preferred Term	Definition	Note
	10210	Retroesophageal aortic arch				
	10211	Right-sided aortic arch				
Carotid artery	10212	Absent carotid	Agenesis, Aplastic			
	10213	Dilated carotid	Bulbous			
	10214	Malpositioned carotid branch	Displaced, Ectopic			Includes right carotid originating from aorta (i.e., no innominate)
	10215	Narrowed carotid	Coarctation, Constricted, Stenosis, Stricture			
	10216	Retroesophageal carotid				
Ductus arteriosus	10217	Absent ductus arteriosus	Agenesis, Aplastic			
	10218	Dilated ductus arteriosus	Bulbous			
	10219	Malpositioned ductus arteriosus	Displaced, Ectopic			
	10220	Narrowed ductus arteriosus	Coarctation, Constricted, Stenosis, Stricture			
	10221	Patent ductus arteriosus	Persistent		Open and unobstructed ductus arteriosus postnatally	Normal occurrence after birth Normal only in fetal period
Great vessels	10222	A-P septum defect			Communication between ascending aorta and pulmonary trunk	A-P = Aorticopulmonary
	10223	Persistent truncus arteriosus	Truncus arteriosus communis		A common aortic and pulmonary truncus	
	10224	Transposition of great vessels			Origin of aorta from right ventricle and pulmonary trunk from left ventricle	
Innominate artery	10225	Dilated innominate	Bulbous			Also known as brachiocephalic trunk
	10226	Elongated innominate				Also known as brachiocephalic trunk
	10227	Malpositioned innominate	Displaced, Ectopic			Also known as brachiocephalic trunk
	10228	Narrowed innominate	Coarctation, Constricted, Stenosis, Stricture			Also known as brachiocephalic trunk

TERMINOLOGY OF ANATOMICAL DEFECTS

	Code	Term	Modifier	Notes
Pulmonary artery	10229	Short innominate	Rudimentary	Also known as brachiocephalic trunk
	10230	Malpositioned pulmonary artery branch	Displaced, Ectopic	
Pulmonary trunk	10231	Dilated pulmonary trunk	Bulbous	
	10232	Narrowed pulmonary trunk	Coarctation, Constricted, Stenosis, Stricture	
	10233	Pulmonary trunk atresia	Imperforate	[Tetratology of Fallot]
	10234	Retroesophageal pulmonary trunk		
	10235	Right-sided pulmonary trunk		
Subclavian artery	10236	Absent subclavian	Agenesis, Aplastic	
	10237	Dilated subclavian	Bulbous	
	10238	Malpositioned subclavian branch	Displaced, Ectopic	Includes right subclavian artery originating from aorta
	10239	Narrowed subclavian	Coarctation, Constricted, Stenosis, Stricture	
	10240	Retroesophageal subclavian		
Trachea	10241	Malpositioned trachea	Displaced, Ectopic	
	10242	Narrowed trachea	Coarctation, Constricted, Stenosis, Stricture	
	10243	Tracheoesophageal fistula		
Esophagus	10244	Absent esophagus	Agenesis, Aplastic	
	10245	Atresia of esophagus	Imperforate	
	10246	Diverticulum of esophagus		
	10247	Malpositioned esophagus	Displaced, Ectopic	
	10248	Narrowed esophagus	Coarctation, Constricted, Esophagostenosis, Stenosis, Stricture	
Lung	10249	Abnormal lung lobation	Fused, Absent fissure	Fusion of lung lobes

Terminology of developmental abnormalities in common laboratory mammals (version 1): Visceral abnormalities[1] (continued)

Region/ Organ/ Structure	Code Number	Observation	Synonym or Related Term	Non-preferred Term	Definition	Note
	10250	Absent lung	Agenesis, Aplastic, Apulmonism		Absence of one or more lobes	Refers to one or more lobes. In rabbits most often the caudate lobe; see notes for code 10260
	10251	Atelectasis			Incomplete expansion of lungs or portions of lungs, due to collapse of pulmonary alveoli during postnatal life	
	10252	Discolored lung	Mottled			
	10253	Enlarged lung				
	10254	Hemorrhagic lung				
	10255	Infarct of lung			Well demarcated macroscopic area of tissue discoloration due to insufficiency of blood supply	Usually whitish in color and the result of necrosis, which should be confirmed histologically
	10256	Malpositioned lung	Displaced, Ectopic			
	10257	Misshapen lung	Abnormally shaped, Irregularly shaped			
	10258	Pale lung				
	10259	Small lung	Hypoplastic, Reduced, Rudimentary			
	10260	Supernumerary lung lobe	Additional, Extra			Normal lobation number for rabbits: 3 lobes on right (diaphrag. lobe further divided into large lateral and small caudate), and 2 on left. Normal number for rodents: 3 lobes on right, one medial lobe (from right bronchus), and 1 on left
Veins	10261	Absent anterior vena cava	Agenesis, Aplastic			May also be called cranial or superior vena cava
	10262	Absent azygos				Right only normal in rabbit, left only normal in rat and mouse

TERMINOLOGY OF ANATOMICAL DEFECTS

	10263	Absent posterior vena cava	Agenesis, Aplastic	May also be called caudal or inferior vena cava
	10264	Dilated anterior vena cava	Bulbous	May also be called cranial or superior vena cava
	10265	Dilated posterior vena cava	Bulbous	May also be called caudal or inferior vena cava
	10266	Interrupted anterior vena cava		May also be called cranial or superior vena cava
	10267	Interrupted posterior vena cava		May also be called caudal or inferior vena cava
	10268	Malpositioned anterior vena cava	Displaced, Ectopic	May also be called cranial or superior vena cava
	10269	Malpositioned posterior vena cava	Displaced, Ectopic	May also be called caudal or inferior vena cava
	10270	Narrowed anterior vena cava	Coarctation, Constricted, Stenosis, Stricture	May also be called cranial or superior vena cava
	10271	Narrowed posterior vena cava	Coarctation, Constricted, Stenosis, Stricture	May also be called caudal or inferior vena cava
	10272	Supernumerary azygos	Additional, Extra	Azygos veins on both sides. Right only normal in rabbit, left only normal in rat and mouse
	10273	Transposed azygos		Azygos vein on side other than normal. Right only normal in rabbit, left only normal in rat and mouse
Diaphragm	10274	Absent diaphragm	Agenesis, Aplastic	
	10275	Diaphragmatic hernia		Absence of portion of the diaphragm with protrusion of some abdominal viscera
	10276	Eventration of diaphragm		Abnormal anterior protrusion of a part of the diaphragm which is thin and covers variably displaced abdominal viscera
Liver	10277	Abnormal liver lobation	Fused, Absent fissure	Fusion of liver lobes
	10278	Absent liver	Agenesis, Aplastic	Absence of one or more lobes
	10279	Discolored liver	Mottled	
	10280	Hemorrhagic liver	Hepatorrhagia	
	10281	Hepatomegaly		Enlarged liver
	10282	Infarct of liver		Well demarcated, macroscopic area of tissue discoloration. Usually whitish in color and is the result of cell destruction which should be confirmed histologically
	10283	Malpositioned liver	Displaced, Ectopic	

Terminology of developmental abnormalities in common laboratory mammals (version 1): Visceral abnormalities[1] (continued)

Region/ Organ/ Structure	Code Number	Observation	Synonym or Related Term	Non-preferred Term	Definition	Note
	10284	Misshapen liver	Abnormally shaped, Irregularly shaped			
	10285	Pale liver				
	10286	Small liver	Hypoplastic, Microhepatia, Reduced, Rudimentary			
	10287	Supernumerary liver lobe	Additional, Extra			Refers to one or more extra lobes
Gallbladder	10288	Absent bile duct	Agenesis, Aplastic			Not present in rats
	10289	Absent gallbladder	Agenesis, Aplastic			Not present in rats
	10290	Bilobed gallbladder				Not present in rats
	10291	Elongated bile duct				Not present in rats
	10292	Enlarged gallbladder				Not present in rats
	10293	Malpositioned gallbladder	Displaced, Ectopic			Not present in rats
	10294	Misshapen gallbladder	Abnormally shaped, Irregularly shaped Rudimentary			Not present in rats
	10295	Short bile duct				Not present in rats
	10296	Small gallbladder	Hypoplastic, Reduced, Rudimentary			Not present in rats
	10297	Supernumerary gallbladder	Additional, Extra			Not present in rats
Stomach	10298	Absent stomach	Agastria, Agenesis, Aplastic			
	10299	Atresia of stomach	Atretogastria, Imperforate			
	10300	Distended stomach				
	10301	Diverticulum of stomach				
	10302	Enlarged stomach	Gastromegaly			
	10303	Malpositioned stomach	Displaced, Ectopic, Gastroptosis	Dextrogastria		
	10304	Narrowed stomach	Coarctation, Constricted, Stenosis, Stricture			Includes pylorus
	10305	Small stomach	Hypoplastic, Microgastria, Reduced, Rudimentary			

TERMINOLOGY OF ANATOMICAL DEFECTS

Pancreas	10306	Absent pancreas	Agenesis, Aplastic	
	10307	Small pancreas	Hypoplastic, Reduced, Rudimentary	
	10308	Supernumerary pancreas	Additional, Extra	
Spleen	10309	Asplenia	Agenesis, Aplastic	Absence of the spleen
	10310	Cyst on spleen		
	10311	Discolored spleen	Mottled	
	10312	Malpositioned spleen	Displaced, Ectopic	
	10313	Misshapen spleen	Abnormally shaped, Irregularly shaped	
	10314	Pale spleen		
	10315	Small spleen	Hypoplastic, Microsplenia, Reduced, Rudimentary	
	10316	Splenomegaly		Enlarged spleen
	10317	Supernumerary spleen	Additional, Extra	
Intestines	10318	Absent intestine	Agenesis, Aplastic	Includes any region from duodenum to rectum
	10319	Atresia of intestine	Imperforate	Includes any region from duodenum to rectum
	10320	Diverticulum of intestine		Includes any region from duodenum to rectum
	10321	Enlarged intestine	Enteromegaly	Includes any region from duodenum to rectum
	10322	Fistula of intestine		Includes any region from duodenum to rectum
	10323	Malpositioned intestine	Displaced, Ectopic	Includes any region from duodenum to rectum
	10324	Narrowed intestine	Coarctation, Constricted, Stenosis, Stricture	Includes any region from duodenum to rectum
	10325	Short intestine	Rudimentary	Includes any region from duodenum to rectum
Kidney	10326	Absent kidney	Agenesis, Aplastic	
	10327	Absent renal papilla		
	10328	Cyst on kidney		
	10329	Dilated renal pelvis	Bulbous	Slight or moderate dilatation of renal pelvis
	10330	Discolored kidney	Mottled	
	10331	Enlarged kidney		
	10332	Fused kidney		Horseshoe

Terminology of developmental abnormalities in common laboratory mammals (version 1): Visceral abnormalities[1] (continued)

Region/ Organ/ Structure	Code Number	Observation	Synonym or Related Term	Non-preferred Term	Definition	Note
	10333	Hemorrhagic kidney				
	10334	Hydronephrosis			Marked dilatation of renal pelvis and calices secondary to obstruction of urine flow, usually combined with destruction of the renal parenchyma	Destruction should be confirmed histologically. Often associated with dilated ureter
	10335	Infarct of kidney				
	10336	Malpositioned kidney	Displaced, Ectopic			
	10337	Misshapen kidney	Abnormally shaped, Irregularly shaped			
	10338	Pale kidney				
	10339	Small kidney	Hypoplastic, Reduced, Rudimentary			
	10340	Small renal papilla	Hypoplastic, Reduced, Rudimentary			
	10341	Small renal pelvis	Hypoplastic, Reduced, Rudimentary			
	10342	Supernumerary kidney	Additional, Extra			
Adrenal gland	10343	Absent adrenal	Agenesis, Aplastic			
	10344	Enlarged adrenal				
	10345	Extracapsular adrenal tissue				
	10346	Fused adrenal				
	10347	Hemorrhagic adrenal				
	10348	Malpositioned adrenal	Displaced, Ectopic			
	10349	Misshapen adrenal	Abnormally shaped, Irregularly shaped			
	10350	Small adrenal	Hypoplastic, Reduced, Rudimentary			
	10351	Supernumerary adrenal	Additional, Extra			
Bladder	10352	Acystia	Agenesis, Aplastic		Absence of the bladder	
	10353	Cyst on bladder				
	10354	Distended bladder				

TERMINOLOGY OF ANATOMICAL DEFECTS

Category	Code	Term	Description	Notes
	10355	Small bladder	Hypoplastic, Reduced, Rudimentary	
Ureter	10356	Absent ureter	Agenesis, Aplastic	
	10357	Convoluted ureter	Coiled, Folded, Kinked, Twisted	Folded, curved, and/or tortuous windings
	10358	Dilated ureter	Bulbous, Distended	Slight or moderate distention of ureter(s)
	10359	Doubled ureter		
	10360	Hydroureter		Markedly dilated ureter(s). Compare to Dilated; may accompany hydronephrosis
	10361	Retrocaval ureter		Passing dorsally to the vena cava
Gonad	10362	Absent gonads	Agenesis, Agonad, Aplastic	Used when sex can not be determined
	10363	Hermaphroditism	Hermaphrodism	Presence of both male and female gonadal tissue
	10364	Small gonad	Hypoplastic, Reduced, Rudimentary	Used when sex can not be determined
	10365	Supernumerary gonad	Additional, Extra	Used when sex can not be determined
Testis	10366	Anorchia	Agenesis, Aplastic	Absence of a testis. If both gonads are absent use "Absent gonad"
	10367	Enlarged testis		
	10368	Hemorrhagic testis		
	10369	Malpositioned testis	Displaced, Ectopic	Cryptorchidism = testis(es) never descended into scrotum
	10370	Misshapen testis	Abnormally shaped, Irregularly shaped	
	10371	Small testis	Hypoplastic, Microrchidia, Reduced, Rudimentary	
	10372	Supernumerary testis	Additional, Extra	
Epididymis	10373	Absent epididymis	Agenesis, Aplastic	
	10374	Hemorrhagic epididymis		
	10375	Malpositioned epididymis	Displaced, Ectopic	
	10376	Short epididymis	Rudimentary	
	10377	Small epididymis	Hypoplastic, Reduced, Rudimentary	
Vas deferens	10378	Absent vas deferens	Agenesis, Aplastic	

Terminology of developmental abnormalities in common laboratory mammals (version 1): Visceral abnormalities[1] (continued)

Region/ Organ/ Structure	Code Number	Observation	Synonym or Related Term	Non-preferred Term	Definition	Note
	10379	Atresia of vas deferens	Imperforate			
	10380	Malpositioned vas deferens	Displaced, Ectopic			
	10381	Short vas deferens	Rudimentary			
Ovary	10382	Absent ovary	Agenesis, Aplastic			If both gonads are absent use Absent gonads
	10383	Cyst on ovary				
	10384	Enlarged ovary				
	10385	Hemorrhagic ovary	Oophorrhagia			
	10386	Malpositioned ovary	Displaced, Ectopic			
	10387	Misshapen ovary	Abnormally shaped, Irregularly shaped			
	10388	Small ovary	Hypoplastic, Reduced, Rudimentary			
	10389	Supernumerary ovary	Additional, Extra			
Oviduct	10390	Absent oviduct	Agenesis, Aplastic			
	10391	Cyst on oviduct				
	10392	Enlarged oviduct				
	10393	Hemorrhagic oviduct				
	10394	Malpositioned oviduct	Displaced, Ectopic			
	10395	Misshapen oviduct	Abnormally shaped, Irregularly shaped			
	10396	Short oviduct	Rudimentary			
Uterus	10397	Absent uterine horn	Agenesis, Ametria, Aplastic		Absence of one uterine horn	Use Absent uterus if both horns are absent Specify total or partial
	10398	Atresia of uterus	Hysteratresia, Imperforate			
	10399	Hemorrhagic uterus				
	10400	Misshapen uterus	Abnormally shaped, Irregularly shaped			
	10401	Small uterus	Hypoplastic, Reduced, Rudimentary			

[1] For terms in brackets, see Appendix 1-B.

Terminology of developmental abnormalities in common laboratory mammals (version 1): Skeletal abnormalities[1]

Region/Organ/Structure	Code Number	Observation	Synonym or Related Term	Non-preferred Term	Definition	Note
Skull, General	10402	Acrania			Absence of the calvarium and variably other bones comprising the braincase	
	10403	Craniofenestria			The incomplete ossification of the bones of the cranial vault such that the calvarium appears fenestrated	
	10404	Craniostenosis			Premature closure of cranial sutures and fontanels resulting in small maldeveloped skull	Used to describe multiple skull bone fusions
	10405	Enlarged fontanel				Specify anterior or posterior
	10406	Extra ossification site	Additional, Bone island, Fontanellar bone, Suture bone		Isolated site of ossification	Specify location
Alisphenoid	10407	Absent alisphenoid	Agenesis, Aplastic			
	10408	Alisphenoid hole(s)				
	10409	Fused alisphenoid				Specify structures that are fused
	10410	Incomplete ossification of alisphenoid	Reduced ossification			
	10411	Misshapen alisphenoid	Abnormally shaped, Irregularly shaped			
	10412	Small alisphenoid	Hypoplastic, Reduced, Rudimentary			
	10413	Unossified alisphenoid				
Auditory ossicles	10414	Absent auditory ossicles	Agenesis, Aplastic			Includes incus, malleus, and stapes
	10415	Fused auditory ossicles				Specify structures that are fused. Includes incus, malleus, and stapes
	10416	Misshapen auditory ossicles	Abnormally shaped, Irregularly shaped			Includes incus, malleus, and stapes
	10417	Unossified auditory ossicles				Includes incus, malleus, and stapes
Basioccipital	10418	Absent basioccipital	Agenesis, Aplastic			
	10419	Basioccipital hole(s)				
	10420	Fused basioccipital				

Terminology of developmental abnormalities in common laboratory mammals (version 1): Skeletal abnormalities[1] (continued)

Region/ Organ/ Structure	Code Number	Observation	Synonym or Related Term	Non-preferred Term	Definition	Note
	10421	Incomplete ossification of basioccipital	Reduced ossification			
	10422	Misshapen basioccipital	Abnormally shaped, Asymmetric, Irregularly shaped			
	10423	Small basioccipital	Hypoplastic, Reduced, Rudimentary			
	10424	Unossified basioccipital				
Basisphenoid	10425	Absent basisphenoid	Agenesis, Aplastic			
	10426	Basisphenoid hole(s)				
	10427	Fused basisphenoid				Specify structures that are fused
	10428	Incomplete ossification of basisphenoid	Reduced ossification			
	10429	Misshapen basisphenoid	Abnormally shaped, Asymmetric, Irregularly shaped			
	10430	Small basisphenoid	Hypoplastic, Reduced, Rudimentary			
	10431	Unossified basisphenoid				
Exoccipital	10432	Absent exoccipital	Agenesis, Aplastic			
	10433	Exoccipital hole(s)				
	10434	Fused exoccipital				Specify structures that are fused
	10435	Incomplete ossification of exoccipital	Reduced ossification			
	10436	Misshapen exoccipital	Abnormally shaped, Irregularly shaped			
	10437	Small exoccipital	Hypoplastic, Reduced, Rudimentary			
	10438	Unossified exoccipital				
Frontal	10439	Absent frontal	Agenesis, Aplastic			
	10440	Frontal hole(s)				
	10441	Fused frontal				Specify structures that are fused

TERMINOLOGY OF ANATOMICAL DEFECTS

	10442	Incomplete ossification of frontal	Reduced ossification
	10443	Misshapen frontal	Abnormally shaped, Asymmetric, Irregularly shaped
	10444	Small frontal	Hypoplastic, Reduced, Rudimentary
Hyoid	10445	Unossified frontal	
	10446	Absent hyoid	Agenesis, Aplastic
	10447	Bent hyoid	Angulated, Bowed
	10448	Incomplete ossification of hyoid	Reduced ossification
	10449	Misshapen hyoid	Abnormally shaped, Asymmetric, Irregularly shaped
	10450	Small hyoid	Hypoplastic, Reduced, Rudimentary
Interparietal	10451	Unossified hyoid	
	10452	Absent interparietal	Agenesis, Aplastic
	10453	Bipartite ossification of interparietal	Bifid ossification
	10454	Fused interparietal	Specify structures that are fused
	10455	Incomplete ossification of interparietal	Reduced ossification
	10456	Interparietal hole(s)	
	10457	Misshapen interparietal	Abnormally shaped, Asymmetric, Irregularly shaped
	10458	Small interparietal	Hypoplastic, Reduced, Rudimentary
Lacrimal	10459	Unossified interparietal	
	10460	Absent lacrimal	Agenesis, Aplastic
	10461	Fused lacrimal	Specify structures that are fused
	10462	Incomplete ossification of lacrimal	Reduced ossification
	10463	Misshapen lacrimal	Abnormally shaped, Irregularly shaped
	10464	Small lacrimal	Hypoplastic, Reduced, Rudimentary

Terminology of developmental abnormalities in common laboratory mammals (version 1): Skeletal abnormalities[1] (continued)

Region/ Organ/ Structure	Code Number	Observation	Synonym or Related Term	Non-preferred Term	Definition	Note
Mandible	10465	Unossified lacrimal				
	10466	Absent mandible	Agenesis, Aplastic			
	10467	Fused mandible				Specify structures that are fused
	10468	Incomplete ossification of mandible	Reduced ossification			
	10469	Misshapen mandible	Abnormally shaped, Asymmetric, Irregularly shaped			
	10470	Small mandible	Hypoplastic, Reduced, Rudimentary			
Maxilla	10471	Unossified mandible				
	10472	Absent maxilla	Agenesis, Aplastic			
	10473	Fused maxilla				Specify structures that are fused
	10474	Incomplete ossification of maxilla	Reduced ossification			
	10475	Misshapen maxilla	Abnormally shaped, Irregularly shaped			
	10476	Small maxilla	Hypoplastic, Reduced, Rudimentary			
Nasal	10477	Unossified maxilla				
	10478	Absent nasal	Agenesis, Aplastic			
	10479	Fused nasal				Specify structures that are fused
	10480	Incomplete ossification of nasal	Reduced ossification			
	10481	Misshapen nasal	Abnormally shaped, Asymmetric, Irregularly shaped			
	10482	Nasal hole(s)				
	10483	Small nasal	Hypoplastic, Reduced, Rudimentary			
Palatine	10484	Unossified nasal				
	10485	Absent palatine	Agenesis, Aplastic			

TERMINOLOGY OF ANATOMICAL DEFECTS

	10486	Fused palatine	Specify structures that are fused
	10487	Incomplete ossification of palatine	Reduced ossification
	10488	Misshapen palatine	Abnormally shaped, Asymmetric, Irregularly shaped
	10489	Small palatine	Hypoplastic, Reduced, Rudimentary
	10490	Split palatine	Bifid, Bipartite, Cleft palate
Parietal	10491	Unossified palatine	
	10492	Absent parietal	Agenesis, Aplastic
	10493	Fused parietal	Specify structures that are fused
	10494	Incomplete ossification of parietal	Reduced ossification
	10495	Misshapen parietal	Abnormally shaped, Asymmetric, Irregularly shaped
	10496	Parietal hole(s)	
	10497	Small parietal	Hypoplastic, Reduced, Rudimentary
Premaxilla	10498	Unossified parietal	
	10499	Absent premaxilla	Agenesis, Aplastic
	10500	Fused premaxilla	Specify structures that are fused
	10501	Incomplete ossification of premaxilla	Reduced ossification
	10502	Misshapen premaxilla	Abnormally shaped, Asymmetric, Irregularly shaped
	10503	Premaxilla hole(s)	
	10504	Small premaxilla	Hypoplastic, Reduced, Rudimentary
Presphenoid	10505	Unossified premaxilla	
	10506	Absent presphenoid	Agenesis, Aplastic
	10507	Fused presphenoid	Specify structures that are fused
	10508	Incomplete ossification of presphenoid	Reduced ossification

Terminology of developmental abnormalities in common laboratory mammals (version 1): Skeletal abnormalities[1] (continued)

Region/ Organ/ Structure	Code Number	Observation	Synonym or Related Term	Non-preferred Term	Definition	Note
	10509	Misshapen presphenoid	Abnormally shaped, Asymmetric, Irregularly shaped			
	10510	Presphenoid hole(s)				
	10511	Small presphenoid	Hypoplastic, Reduced, Rudimentary			
	10512	Unossified presphenoid				
Squamosal	10513	Absent squamosal	Agenesis, Aplastic			Bone may be called temporal
	10514	Fused squamosal				Specify structures that are fused Bone may be called temporal
	10515	Incomplete ossification of squamosal	Reduced ossification			Bone may be called temporal
	10516	Misshapen squamosal	Abnormally shaped, Irregularly shaped			Bone may be called temporal
	10517	Small squamosal	Hypoplastic, Reduced, Rudimentary			Bone may be called temporal
	10518	Squamosal hole(s)				
	10519	Unossified squamosal				Bone may be called temporal
Supraoccipital	10520	Absent supraoccipital	Agenesis, Aplastic			
	10521	Bipartite ossification of supraoccipital	Bifid ossification			
	10522	Fused supraoccipital				Specify structures that are fused
	10523	Incomplete ossification of supraoccipital	Reduced ossification			
	10524	Misshapen supraoccipital	Abnormally shaped, Asymmetric, Irregularly shaped			
	10525	Small supraoccipital	Hypoplastic, Reduced, Rudimentary			
	10526	Supraoccipital hole(s)				

Tympanic annulus	10527	Unossified supraoccipital	
	10528	Absent tympanic annulus	Agenesis, Aplastic
	10529	Fused tympanic annulus		Specify structures that are fused
	10530	Incomplete ossification of tympanic annulus	Reduced ossification
	10531	Misshapen tympanic annulus	Abnormally shaped, Irregularly shaped
	10532	Small tympanic annulus	Hypoplastic, Reduced, Rudimentary
	10533	Unossified tympanic annulus	
Vomer	10534	Absent vomer	Agenesis, Aplastic
	10535	Fused vomer		Specify structures that are fused
	10536	Incomplete ossification of vomer	Reduced ossification
	10537	Misshapen vomer	Abnormally shaped, Asymmetric, Irregularly shaped
	10538	Small vomer	Hypoplastic, Reduced, Rudimentary
	10539	Unossified vomer	
Zygomatic	10540	Absent zygomatic	Agenesis, Aplastic	Bone may be called jugal or malar
	10541	Fused zygomatic		Specify structures that are fused Bone may be called jugal or malar
	10542	Incomplete ossification of zygomatic	Reduced ossification	Bone may be called jugal or malar
	10543	Misshapen zygomatic	Abnormally shaped, Irregularly shaped
	10544	Small zygomatic	Hypoplastic, Reduced, Rudimentary	Bone may be called jugal or malar
	10545	Unossified zygomatic		Bone may be called jugal or malar
Clavicle	10546	Absent clavicle	Agenesis, Aplastic
	10547	Bent clavicle	Angulated, Bowed
	10548	Incomplete ossification of clavicle	Reduced ossification

Terminology of developmental abnormalities in common laboratory mammals (version 1): Skeletal abnormalities[1] (continued)

Region/ Organ/ Structure	Code Number	Observation	Synonym or Related Term	Non-preferred Term	Definition	Note
	10549	Misshapen clavicle	Abnormally shaped, Irregularly shaped			
	10550	Small clavicle	Hypoplastic, Reduced, Rudimentary			
	10551	Thickened clavicle				
	10552	Unossified clavicle				
Scapula	10553	Absent scapula	Agenesis, Aplastic			
	10554	Bent scapula	Angulated, Bowed			
	10555	Incomplete ossification of scapula	Reduced ossification			
	10556	Misshapen scapula	Abnormally shaped, Forked spine, Irregularly shaped			
	10557	Thickened scapula				
	10558	Unossified scapula				
Humerus	10559	Absent humerus	Agenesis, Aplastic			
	10560	Bent humerus	Angulated, Bowed			
	10561	Fused humerus				Specify structures that are fused
	10562	Incomplete ossification of humerus	Reduced ossification			
	10563	Malpositioned humerus	Displaced, Ectopic			
	10564	Misshapen humerus	Abnormally shaped, Irregularly shaped			
	10565	Short humerus	Rudimentary			
	10566	Thickened humerus				
	10567	Unossified humerus				
Radius	10568	Absent radius	Agenesis, Aplastic, Radial hemimelia			
	10569	Bent radius	Angulated, Bowed			
	10570	Fused radius				Specify structures that are fused
	10571	Incomplete ossification of radius	Reduced ossification			
	10572	Malpositioned radius	Displaced, Ectopic			

TERMINOLOGY OF ANATOMICAL DEFECTS

Category	Code	Term	Modifier	Comment
	10573	Misshapen radius	Abnormally shaped, Irregularly shaped	
	10574	Short radius	Rudimentary	
	10575	Thickened radius		
	10576	Unossified radius		
Ulna	10577	Absent ulna	Agenesis, Aplastic, Ulnar hemimelia	
	10578	Bent ulna	Angulated, Bowed	
	10579	Fused ulna		Specify structures that are fused
	10580	Incomplete ossification of ulna	Reduced ossification	
	10581	Malpositioned ulna	Displaced, Ectopic	
	10582	Misshapen ulna	Abnormally shaped, Irregularly shaped	
	10583	Short ulna	Rudimentary	
	10584	Thickened ulna		
	10585	Unossified ulna		
Carpal bone	10586	Absent carpal bone	Agenesis, Aplastic	
	10587	Fused carpal bone		Specify structures that are fused
	10588	Incomplete ossification of carpal bone		
	10589	Malpositioned carpal bone	Displaced, Ectopic	
	10590	Misshapen carpal bone	Abnormally shaped, Irregularly shaped	
	10591	Small carpal bone	Hypoplastic, Reduced, Rudimentary	
	10592	Supernumerary carpal bone	Additional, Extra	
	10593	Unossified carpal bone		
Metacarpal	10594	Absent metacarpal	Agenesis, Aplastic	
	10595	Fused metacarpal		Specify structures that are fused
	10596	Incomplete ossification of metacarpal	Reduced ossification	
	10597	Malpositioned metacarpal	Displaced, Ectopic	
	10598	Misshapen metacarpal	Abnormally shaped, Irregularly shaped	

Terminology of developmental abnormalities in common laboratory mammals (version 1): Skeletal abnormalities[1] (continued)

Region/ Organ/ Structure	Code Number	Observation	Synonym or Related Term	Non-preferred Term	Definition	Note
	10599	Small metacarpal	Hypoplastic, Reduced, Rudimentary			
	10600	Supernumerary metacarpal	Additional, Extra			
	10601	Unossified metacarpal				
Forepaw phalanx	10602	Absent phalanx	Agenesis, Aphalangia, Aplastic		Absence of one or more phalanges in one or more digits	
	10603	Fused phalanx				Specify structures that are fused
	10604	Incomplete ossification of phalanx	Reduced ossification			
	10605	Malpositioned phalanx	Displaced, Ectopic			
	10606	Misshapen phalanx	Abnormally shaped, Irregularly shaped			
	10607	Small phalanx	Hypoplastic, Reduced, Rudimentary			
	10608	Supernumerary phalanx	Additional, Extra			
	10609	Thickened phalanx				
	10610	Unossified phalanx				
Sternebra	10611	Absent sternebra	Agenesis, Aplastic			
	10612	Bipartite ossification of sternebra	Bifid ossification			
	10613	Extra sternebral ossification site	Additional	Supernumerary		Refers to advanced ossification between sternebral centers
	10614	Fused sternebra				Specify structures that are fused
	10615	Incomplete ossification of sternebra	Reduced ossification			
	10616	Malpositioned sternebra	Displaced, Ectopic			
	10617	Misaligned sternebra	Offset	Checkerboard		
	10618	Misshapen sternebra	Abnormally shaped, Asymmetric, Irregularly shaped			

TERMINOLOGY OF ANATOMICAL DEFECTS

	ID	Term	Synonyms	Description	Notes
Rib	10619	Sternoschisis	Bifid, Bipartite, Cleft sternum	Split sternum involving the cartilage and one or more sternebral centers	
	10620	Unossified sternebra			
	10621	Absent rib	Agenesis, Aplastic		Generally the normal number of rib pairs is 13 in rodent and 12 or 13 in rabbit; but must be related to normal number for given species, strain, and time in each laboratory.
	10622	Bent rib	Angulated, Bowed	Shaped like an angle	
	10623	Branched rib	Bifurcated, Forked	A rib that divides distally into two ribs	
	10624	Branched rib cartilage	Bifurcated, Forked	A rib cartilage that divides distally into two	
	10625	Cervical rib	Supernumerary cervical	Ossification lateral to a cervical arch	Usually seen on last cervical arch
	10626	Detached rib		Rib with no attachment to the vertebral column	
	10627	Discontinuous rib	Interrupted ossification	Absence of cartilage and Alizarin Red uptake in a central region of a rib	
	10628	Full supernumerary rib	Extra full, Long supernumerary thoracolumbar	An extra rib that is distally joined by a cartilaginous portion	Generally the normal number of rib pairs is 13 in rodent and 12 or 13 in rabbit; but must be related to normal number for given species, strain, and time in each laboratory.
	10629	Fused rib			Specify structures that are fused
	10630	Fused rib cartilage			
	10631	Incomplete ossification of rib	Reduced ossification		
	10632	Intercostal rib		A non-articulating rib-like structure located between two other ribs	
	10633	Knobby rib	Focal enlargement, Nodulated	A rounded protuberance on a rib	
	10634	Malpositioned rib	Displaced, Ectopic		
	10635	Misaligned rib			
	10636	Misshapen rib	Abnormally shaped, Asymmetric, Irregularly shaped		

Terminology of developmental abnormalities in common laboratory mammals (version 1): Skeletal abnormalities[1] (continued)

Region/ Organ/ Structure	Code Number	Observation	Synonym or Related Term	Non-preferred Term	Definition	Note
	10637	Short rib	Rudimentary			Not used for supernumerary rib; generally less than half normal length
	10638	Short supernumerary rib	Extra rudimentary, Rudimentary supernumerary thoracolumbar, Short supernumerary thoracolumbar	Lumbar rib	An extra rib that is usually rounded distally and without cartilaginous extensions	Generally the normal number of rib pairs is 13 in rodent and 12 or 13 in rabbit; but must be related to normal number for given species, strain, and time in each laboratory
	10639	Thickened rib				
	10640	Unossified rib				
	10641	Wavy rib		Clubbed	Undulation(s) along the length of a rib	
Vertebra, General	10642	Absent vertebra	Agenesis, Aplastic			Used when subregion not specified
	10643	Supernumerary vertebra	Additional, Extra			Used when subregion not specified
Cervical arch	10644	Absent cervical arch	Agenesis, Aplastic			Atlas = #1, Axis = #2
	10645	Fused cervical arch				Specify structures that are fused Atlas = #1, Axis = #2
	10646	Incomplete ossification of cervical arch	Reduced ossification			Atlas = #1, Axis = #2
	10647	Malpositioned cervical arch	Displaced, Ectopic			Atlas = #1, Axis = #2
	10648	Misaligned cervical arch				Atlas = #1, Axis = #2
	10649	Misshapen cervical arch	Abnormally shaped, Asymmetric, Irregularly shaped			Atlas = #1, Axis = #2
	10650	Small cervical arch	Hypoplastic, Reduced, Rudimentary			Atlas = #1, Axis = #2
	10651	Supernumerary cervical arch	Additional, Extra			Atlas = #1, Axis = #2

TERMINOLOGY OF ANATOMICAL DEFECTS

Cervical centrum	10652	Unossified cervical arch		Atlas = #1, Axis = #2
	10653	Absent cervical centrum	Agenesis, Aplastic	Atlas = #1, Axis = #2
	10654	Bipartite ossification of cervical centrum	Bifid ossification	Atlas = #1, Axis = #2
	10655	Dumbbell ossification of cervical centrum	Dumbbell-shaped	Atlas = #1, Axis = #2
	10656	Dumbbell-shaped cartilage of cervical centrum		Atlas = #1, Axis = #2
	10657	Fused cervical centrum		Specify structures that are fused Atlas = #1, Axis = #2
	10658	Fused cervical centrum cartilage		Specify structures that are fused Atlas = #1, Axis = #2
	10659	Hemicentric cervical centrum		Atlas = #1, Axis = #2
	10660	Incomplete ossification of cervical centrum	Reduced ossification	Atlas = #1, Axis = #2
	10661	Misaligned cervical centrum		Atlas = #1, Axis = #2
	10662	Misshapen cervical centrum	Abnormally shaped, Asymmetric, Irregularly shaped	Atlas = #1, Axis = #2
	10663	Split cartilage of cervical centrum	Bifid, Bipartite, Cleft	Atlas = #1, Axis = #2
	10664	Supernumerary cervical centrum	Additional, Extra	Atlas = #1, Axis = #2
	10665	Unilateral cervical centrum cartilage		Atlas = #1, Axis = #2
	10666	Unossified cervical centrum		Atlas = #1, Axis = #2
Cervical vertebra	10667	Absent cervical vertebra	Agenesis, Aplastic	Normal: 7 in rodent and rabbit
	10668	Cervical hemivertebra		
	10669	Malpositioned cervical vertebra	Displaced, Ectopic	
	10670	Supernumerary cervical vertebra	Additional, Extra	
Thoracic arch	10671	Absent thoracic arch	Agenesis, Aplastic	Normal: 7 in rodent and rabbit

Terminology of developmental abnormalities in common laboratory mammals (version 1): Skeletal abnormalities[1] (continued)

Region/ Organ/ Structure	Code Number	Observation	Synonym or Related Term	Non-preferred Term	Definition	Note
	10672	Fused thoracic arch				Specify structures that are fused
	10673	Incomplete ossification of thoracic arch	Reduced ossification			
	10674	Malpositioned thoracic arch				
	10675	Misaligned thoracic arch				
	10676	Misshapen thoracic arch	Abnormally shaped, Asymmetric, Irregularly shaped			
	10677	Small thoracic arch	Hypoplastic, Reduced, Rudimentary			
	10678	Supernumerary thoracic arch	Additional, Extra			
	10679	Unossified thoracic arch				
Thoracic centrum	10680	Absent thoracic centrum	Agenesis, Aplastic			
	10681	Bipartite ossification of thoracic centrum	Bifid ossification			
	10682	Dumbbell ossification of thoracic centrum	Dumbbell-shaped			
	10683	Dumbbell-shaped cartilage of thoracic centrum				
	10684	Fused thoracic centrum				Specify structures that are fused
	10685	Fused thoracic centrum cartilage				Specify structures that are fused
	10686	Hemicentric thoracic centrum				
	10687	Incomplete ossification of thoracic centrum	Reduced ossification			
	10688	Misaligned thoracic centrum				

TERMINOLOGY OF ANATOMICAL DEFECTS

	10689	Misshapen thoracic centrum	Abnormally shaped, Asymmetric, Irregularly shaped
	10690	Split cartilage of thoracic centrum	Bifid, Bipartite, Cleft
	10691	Supernumerary thoracic centrum	Additional, Extra
	10692	Unilateral thoracic centrum cartilage	
	10693	Unossified thoracic centrum	
Thoracic vertebra	10694	Absent thoracic vertebra	Agenesis, Aplastic
	10695	Malpositioned thoracic vertebra	Displaced, Ectopic
	10696	Thoracic hemivertebra	
	10697	Supernumerary thoracic vertebra	Additional, Extra — Normal: 13 in rodent, 12 or 13 in rabbit
Lumbar arch	10698	Absent lumbar arch	Agenesis, Aplastic
	10699	Fused lumbar arch	Specify structures that are fused
	10700	Incomplete ossification of lumbar arch	Reduced ossification
	10701	Malpositioned lumbar arch	Displaced, Ectopic
	10702	Misaligned lumbar arch	
	10703	Misshapen lumbar arch	Abnormally shaped, Asymmetric, Irregularly shaped
	10704	Small lumbar arch	Hypoplastic, Reduced, Rudimentary
	10705	Supernumerary lumbar arch	Additional, Extra
	10706	Unossified lumbar arch	
Lumbar centrum	10707	Absent lumbar centrum	Agenesis, Aplastic
	10708	Bipartite ossification of lumbar centrum	Bifid ossification
	10709	Dumbbell ossification of lumbar centrum	Dumbbell-shaped
	10710	**Dumbbell-shaped cartilage of lumbar centrum**	

Normal: 13 in rodent, 12 or 13 in rabbit (10697)

Terminology of developmental abnormalities in common laboratory mammals (version 1): Skeletal abnormalities[1] (continued)

Region/Organ/Structure	Code Number	Observation	Synonym or Related Term	Non-preferred Term	Definition	Note
	10711	Fused lumbar centrum				Specify structures that are fused
	10712	Fused lumbar centrum cartilage				Specify structures that are fused
	10713	Hemicentric lumbar centrum				
	10714	Incomplete ossification of lumbar centrum	Reduced ossification			
	10715	Misaligned lumbar centrum				
	10716	Misshapen lumbar centrum	Abnormally shaped, Asymmetric, Irregularly shaped			
	10717	Split cartilage of lumbar centrum	Bifid, Bipartite, Cleft			
	10718	Supernumerary lumbar centrum	Additional, Extra			
	10719	Unilateral lumbar centrum cartilage				
	10720	Unossified lumbar centrum				
Lumbar vertebra	10721	Absent lumbar vertebra	Agenesis, Aplastic			Normal: 6 in rodent, 6 or 7 in rabbit
	10722	Lumbar hemivertebra				
	10723	Malpositioned lumbar vertebra	Displaced, Ectopic			
	10724	Supernumerary lumbar vertebra	Additional, Extra			Normal: 6 in rodent, 6 or 7 in rabbit
Sacral arch	10725	Absent sacral arch	Agenesis, Aplastic			
	10726	Fused sacral arch				Specify structures that are fused
	10727	Incomplete ossification of sacral arch	Reduced ossification			
	10728	Malpositioned sacral arch	Displaced, Ectopic			
	10729	Misaligned sacral arch				

TERMINOLOGY OF ANATOMICAL DEFECTS

	ID	Term	Description	Notes
	10730	Misshapen sacral arch	Abnormally shaped, Asymmetric, Irregularly shaped	
	10731	Small sacral arch	Hypoplastic, Reduced, Rudimentary	
	10732	Supernumerary sacral arch	Additional, Extra	
	10733	Unossified sacral arch		
	10734	Absent sacral centrum	Agenesis, Aplastic	
	10735	Bipartite ossification of sacral centrum	Bifid ossification	
Sacral centrum	10736	Dumbbell ossification of sacral centrum	Dumbbell-shaped	
	10737	Dumbbell-shaped cartilage of sacral centrum		
	10738	Fused sacral centrum		Specify structures that are fused
	10739	Fused sacral centrum cartilage		Specify structures that are fused
	10740	Hemicentric sacral centrum		
	10741	Incomplete ossification of sacral centrum	Reduced ossification	
	10742	Misaligned sacral centrum		
	10743	Misshapen sacral centrum	Abnormally shaped, Asymmetric, Irregularly shaped	
	10744	Split cartilage of sacral centrum	Bifid, Bipartite, Cleft	
	10745	Supernumerary sacral centrum	Additional, Extra	
	10746	Unilateral sacral centrum cartilage		
	10747	Unossified sacral centrum		
Sacral vertebra	10748	Absent sacral vertebra	Agenesis, Aplastic	Normal: 4 in rodent and rabbit
	10749	Malpositioned sacral vertebra	Displaced, Ectopic	

Terminology of developmental abnormalities in common laboratory mammals (version 1): Skeletal abnormalities[1] (continued)

Region/ Organ/ Structure	Code Number	Observation	Synonym or Related Term	Non-preferred Term	Definition	Note
	10750	Sacral hemivertebra				
	10751	Supernumerary sacral vertebra	Additional, Extra			Normal: 4 in rodent and rabbit
Caudal arch	10752	Absent caudal arch	Agenesis, Aplastic			
	10753	Fused caudal arch				Specify structures that are fused
	10754	Incomplete ossification of caudal arch	Reduced ossification			
	10755	Malpositioned caudal arch	Displaced, Ectopic			
	10756	Misaligned caudal arch				
	10757	Misshapen caudal arch	Abnormally shaped, Asymmetric, Irregularly shaped			
	10758	Small caudal arch	Hypoplastic, Reduced, Rudimentary			
	10759	Unossified caudal arch				
Caudal centrum	10760	Absent caudal centrum	Agenesis, Aplastic			
	10761	Bipartite ossification of caudal centrum	Bifid ossification			
	10762	Dumbbell ossification of caudal centrum	Dumbbell-shaped			
	10763	Fused caudal centrum				Specify structures that are fused
	10764	Hemicentric caudal centrum				
	10765	Incomplete ossification of caudal centrum	Reduced ossification			
	10766	Misaligned caudal centrum				
	10767	Misshapen caudal centrum	Abnormally shaped, Asymmetric, Irregularly shaped			
	10768	Unossified caudal centrum				

TERMINOLOGY OF ANATOMICAL DEFECTS

Caudal vertebra	10769	Absent caudal vertebra	Agenesis, Aplastic
	10770	Caudal hemivertebra	
	10771	Malpositioned caudal vertebra	Displaced, Ectopic
	10772	Supernumerary caudal vertebra	Additional, Extra
Ilium	10773	Absent ilium	Agenesis, Aplastic
	10774	Bent ilium	Angulated, Bowed
	10775	Fused ilium	Specify structures that are fused
	10776	Incomplete ossification of ilium	Reduced ossification
	10777	Malpositioned ilium	
	10778	Misshapen ilium	Abnormally shaped, Irregularly shaped
	10779	Small ilium	Hypoplastic, Reduced, Rudimentary
	10780	Thickened ilium	
	10781	Unossified ilium	
Ischium	10782	Absent ischium	Agenesis, Aplastic
	10783	Bent ischium	Angulated, Bowed
	10784	Fused ischium	Specify structures that are fused
	10785	Incomplete ossification of ischium	Reduced ossification
	10786	Malpositioned ischium	
	10787	Misshapen ischium	Abnormally shaped, Irregularly shaped
	10788	Small ischium	Hypoplastic, Reduced, Rudimentary
	10789	Thickened ischium	
	10790	Unossified ischium	
Pubis	10791	Absent pubis	Agenesis, Aplastic
	10792	Bent pubis	Angulated, Bowed
	10793	Incomplete ossification of pubis	Reduced ossification
	10794	Malpositioned pubis	
	10795	Misaligned pubis	
	10796	Misshapen pubis	Abnormally shaped, Irregularly shaped

Terminology of developmental abnormalities in common laboratory mammals (version 1): Skeletal abnormalities[1] (continued)

Region/Organ/Structure	Code Number	Observation	Synonym or Related Term	Non-preferred Term	Definition	Note
	10797	Small pubis	Hypoplastic, Reduced, Rudimentary			
	10798	Thickened pubis				
	10799	Unossified pubis				
Femur	10800	Absent femur	Agenesis, Aplastic			
	10801	Bent femur	Angulated, Bowed			
	10802	Fused femur				Specify structures that are fused
	10803	Incomplete ossification of femur	Reduced ossification			
	10804	Malpositioned femur	Displaced, Ectopic			
	10805	Misshapen femur	Abnormally shaped, Irregularly shaped			
	10806	Short femur	Rudimentary			
	10807	Thickened femur				
	10808	Unossified femur				
Fibula	10809	Absent fibula	Agenesis, Aplastic			
	10810	Bent fibula	Angulated, Bowed			
	10811	Fused fibula				Specify structures that are fused
	10812	Incomplete ossification of fibula	Reduced ossification			
	10813	Malpositioned fibula	Displaced, Ectopic			
	10814	Misshapen fibula	Abnormally shaped, Irregularly shaped			
	10815	Short fibula	Rudimentary			
	10816	Thickened fibula				
	10817	Unossified fibula				
Tibia	10818	Absent tibia	Agenesis, Aplastic, Tibial hemimelia			
	10819	Bent tibia	Angulated, Bowed			
	10820	Fused tibia				Specify structures that are fused
	10821	Incomplete ossification of tibia	Reduced ossification			

TERMINOLOGY OF ANATOMICAL DEFECTS

Category	Code	Term	Modifier	Notes
	10822	Malpositioned tibia	Displaced, Ectopic	
	10823	Misshapen tibia	Abnormally shaped, Irregularly shaped	
	10824	Short tibia	Rudimentary	
	10825	Thickened tibia		
	10826	Unossified tibia		
Calcaneus	10827	Absent calcaneus	Agenesis, Aplastic	
	10828	Fused calcaneus		Specify structures that are fused
	10829	Incomplete ossification of calcaneus	Reduced ossification	
	10830	Malpositioned calcaneus	Displaced, Ectopic	
	10831	Misshapen calcaneus	Abnormally shaped, Irregularly shaped	
	10832	Small calcaneus	Hypoplastic, Reduced, Rudimentary	
	10833	Supernumerary calcaneus	Additional, Extra	
	10834	Unossified calcaneus		
Talus	10835	Absent talus	Agenesis, Aplastic	Bone may be called astragalus
	10836	Fused talus		Specify structures that are fused Bone may be called astragalus
	10837	Incomplete ossification of talus	Reduced ossification	Bone may be called astragalus
	10838	Malpositioned talus	Displaced, Ectopic	Bone may be called astragalus
	10839	Misshapen talus	Abnormally shaped, Irregularly shaped	Bone may be called astragalus
	10840	Small talus	Hypoplastic, Reduced, Rudimentary	Bone may be called astragalus
	10841	Thickened talus		Bone may be called astragalus
	10842	Unossified talus		Bone may be called astragalus
Tarsal bone	10843	Absent tarsal bone	Agenesis, Aplastic	Excludes calcaneus and talus

Terminology of developmental abnormalities in common laboratory mammals (version 1): Skeletal abnormalities[1] (continued)

Region/ Organ/ Structure	Code Number	Observation	Synonym or Related Term	Non-preferred Term	Definition	Note
	10844	Fused tarsal bone				Specify structures that are fused Excludes calcaneus and talus
	10845	Incomplete ossification of tarsal bone				
	10846	Malpositioned tarsal bone	Displaced, Ectopic			Excludes calcaneus and talus
	10847	Misshapen tarsal bone	Abnormally shaped, Irregularly shaped			Excludes calcaneus and talus
	10848	Small tarsal bone	Hypoplastic, Reduced, Rudimentary			Excludes calcaneus and talus
	10849	Supernumerary tarsal bone	Additional, Extra			Excludes calcaneus and talus
	10850	Thickened tarsal bone				Excludes calcaneus and talus
	10851	Unossified tarsal bone				
Metatarsal	10852	Absent metatarsal	Agenesis, Aplastic			
	10853	Fused metatarsal				Specify structures that are fused
	10854	Incomplete ossification of metatarsal	Reduced ossification			
	10855	Malpositioned metatarsal	Displaced, Ectopic			
	10856	Misshapen metatarsal	Abnormally shaped, Irregularly shaped			
	10857	Small metatarsal	Hypoplastic, Reduced, Rudimentary			
	10858	Supernumerary metatarsal	Additional, Extra			
	10859	Unossified metatarsal				
Hindpaw phalanx	10860	Absent phalanx	Agenesis, Aphalangia, Aplastic		Absence of one or more phalanges in one or more digits	
	10861	Fused phalanx				Specify structures that are fused
	10862	Incomplete ossification of phalanx	Reduced ossification			

10863	Malpositioned phalanx	Displaced, Ectopic
10864	Misshapen phalanx	Abnormally shaped, Irregularly shaped
10865	Small phalanx	Hypoplastic, Reduced, Rudimentary
10866	Supernumerary phalanx	Additional, Extra
10867	Thickened phalanx	
10868	Unossified phalanx	

[1]For terms in brackets, see Appendix 1-B.

Appendix 1-A Descriptive terminology used more than once in the glossary

Term	Definition	Synonym or Related term	Non-preferred Term
Absent	An absolute failure of development of an organ or part; in the case of bone, no cartilage model is present	Agenesis, Aplastic	Missing
Atresia	Absence or closure of a normal body orifice or tubular organ	Imperforate	
Bipartite ossification	Having two separate ossification sites	Bifid ossification	Split
Bent	Abnormal curvature of a normally straight structure	Angulated, Bowed	
Branched	Deviation from the predominant pattern due to displacement and/or duplication of one or more normally occurring structures	Bifurcated, Forked	
Cyst	Any abnormal sac; usually containing fluid or other material		
Dilated	Widened or expanded orifice or vessel	Bulbous	Swollen
Discolored	Not the normal color	Mottled, Pale	
Distended	Enlarged or expanded organ		Swollen
Diverticulum	A localized sac or pouch formed in the wall of a hollow viscus and opening into its lumen		
Dumbbell ossification	Two roughly spherical ossification sites attached at or near the mid-line by an ossified bridge	Dumbbell-shaped ossification	Bilobate, Bilobed
Enlarged	Larger than normal		Hyperplastic
Fistula	Abnormal passage or communication between two normally unconnected structures, body cavities, or the surface of the body		
Fused	Joined or blended together		
Hemicentric	Absence of either hemicentrum of a centrum		
Hemivertebra	Absence of a lateral half (arch + hemicentrum), or major part of a lateral half of a vertebra		
Hemorrhagic	Descriptive of any tissue into which abnormal bleeding is observed (may be graded as petechia, purpura, ecchymosis, or hematoma)		
Incomplete ossification	Partial ossification (as assessed by Alizarin Red uptake) at a site which, in controls of the same age, usually has a more advanced degree of ossification	Reduced ossification	Delayed ossification, Retarded ossification
Malpositioned	Not occurring in the proper position and/or orientation	Displaced, Ectopic	Misdirected, Dislocated
Misaligned	Abnormal relative position of structures on opposite sides of a dividing line or about the center or axis		Unaligned
Misshapen	Abnormally shaped [Not to be used to describe sites of incomplete ossification]	Abnormally shaped, Asymmetric, Irregularly shaped	
Narrowed	Constriction of a cylindrical structure, such as the aorta or tail, or of a lumen	Coarctation, Constricted, Stenosis, Stricture	
Retro-esophageal	Passing dorsal to the esophagus		
Retrotracheal	Passing dorsal to the trachea		

Appendix 1-A Descriptive terminology used more than once in the glossary (continued)

Term	Definition	Synonym or Related term	Non-preferred Term
Short	Less than the normal or expected length	Rudimentary	
Small	Incompletely developed structure, or less than normal in size	Aplastic, Hypoplastic, Reduced, Rudimentary	Underdeveloped
Split	Division of a single structure (usually into two parts) with no intervening structure between the parts	Bifid, Bipartite, Cleft	
Supernumerary	More than the usual or expected number	Additional, Extra	Accessory
Thickened	A widening of a skeletal element relative to normal		Bulbous, Clubbed
Unossified	Absence of ossification (as assessed by Alizarin Red uptake) at a site which, in controls of the same age, is usually at least partially ossified		Nonossified

Appendix 1-B Syndromes and combining terms

Caudal dysplasia	Severe reduction of caudal structures, including reduction of or absence of hindlimbs, tail, and/or sacral area.
Cheilognathopalatoschisis	Cleft lip, jaw, and palate; also called Cheilognathouranoschisis
Cheilognathoschisis	Cleft lip and jaw
Craniorachischisis	Failure of the neural tube to close in regions of both the brain and spinal cord.
Ethmocephaly	Some degree of cyclopia in which the eyes may be closely set but the nose is hypoplastic (rudimentary).
Otocephaly	Extreme underdevelopment of the lower jaw, permitting close approximation or union of the ears on the anterior aspect of the neck.
Rhinocephaly	Proboscis-like nose above partially of completely fused eyes.
Sirenomelia	Any of several degrees of side-to-side fusion of lower extremities and concomitant midline reduction of the pelvis. Soft tissues and long bones, lower paw (feet), and viscera of the pelvis tend to be reduced or absent; anus and external genitalia are often absent.
Tetralogy of Fallot	Defect of the heart which includes all of the following: pulmonary stenosis, interventricular septal defect, dextraposed aorta overriding the ventricular septum, and enlarged right ventricular wall.

APPENDIX B

Books Related to Developmental and Reproductive Toxicology

Ronald D. Hood

Please note: It is hoped that the following list will be useful and informative, but it contains only those references that I could readily discover. Therefore, it is virtually certain that potentially useful books have inadvertently been omitted. Further, the list contains a number of books that I have not reviewed. Thus, I cannot vouch for their accuracy or merit. Only books published in 1989 or later have been included, and several are no longer in print, although they may still be available in limited quantities.

Abel, E. L., *Behavioral Teratogenesis and Behavioral Mutagenesis: A Primer in Abnormal Development*, Plenum, New York, 1989.
Abel, E. L., Ed., *Fetal Alcohol Syndrome: From Mechanism to Prevention*, CRC Press, Boca Raton, FL, 1996.
Boekelheide, K., Gandolfi, A.J., Harris, C., Hoyer, P.B., McQueen, C.A., Sipes, I. G., and Chapin, R., Eds., *Comprehensive Toxicology*, vol. 10: *Reproductive and Endocrine Toxicology*, Pergamon Press, New York, 1997.
Briggs, G. G., Freeman, R. K., and Yaffe, S. J., Eds., *Drugs in Pregnancy and Lactation*, 7th ed., Lippincott Williams & Wilkins, Philadelphia, PA, 2005.
Chapin, R. E., Heindel, J. J., Tyson, C. A., Witschi, H. R., Eds., *Methods in Toxicology, Part A: Male Reproductive Toxicology*, Academic Press, San Diego, CA, 1993.
Daston, G. P., *Molecular and Cellular Methods in Developmental Toxicology*, CRC Press, Boca Raton, FL, 1996.
Frazier, L. M. and Hage, M. L., *Reproductive Hazards of the Workplace*, John Wiley & Sons, New York, 1997.
Golub, M. S., *Metals, Fertility, and Reproductive Toxicity*, CRC Press, Boca Raton, FL 2005.
Harry, G. J., Ed., *Developmental Neurotoxicology*, CRC Press, Boca Raton, FL, 1994.
Haschek, W. M., Rousseaux, C. G., and Wallig, M. A., Eds., *Handbook of Toxicologic Pathology*, 2nd ed., Vol. 2, Academic Press, San Diego, CA, 2001.
Holladay, S. D., Ed., *Developmental Immunotoxicology*, CRC Press, Boca Raton, FL, 2004.
Hood, R. D., Ed., *Developmental Toxicology — Risk Assessment and the Future*, Van Nostrand Reinhold, New York, 1990.
Hood, R. D., Ed., *Handbook of Developmental Toxicology*, CRC Press, Boca Raton, FL, 1997.
Hoyer, P. B., Ed., *Ovarian Toxicology* (Target Organ Toxicology Series), CRC Press, Boca Raton, FL, 2004.
Kavlock, R. J. and Daston, G. P., Eds., *Drug Toxicity in Embryonic Development. I: Advances in Understanding Mechanisms of Birth Defects: Morphogenesis and Processes at Risk* (Handbook of Experimental Pharmacology, vol. 142/2), Springer Verlag, Berlin, 1997.

Kimmel, C. A. and Buelke-Sam, J., Eds., *Developmental Toxicology* (Target Organ Toxicology Series), 2nd ed., Raven Press, New York, 1994.

Kochhar, D. M., *In Vitro Methods in Developmental Toxicology: Use in Defining Mechanisms and Risk Parameters*, CRC Press, Boca Raton, FL, 1989.

Korach, K. S., *Reproductive and Developmental Toxicology*, Marcel Dekker, New York, 1998.

Koren, G., *Maternal-Fetal Toxicology: A Clinicians' Guide*, 2nd ed., Marcel Dekker, New York, 1994.

Lumley, C. E. and Walker, S. R., Eds., *Current Issues in Reproductive and Developmental Toxicology: Can an International Guideline Be Achieved?* (CMR Workshop Series), Quay Publishing, Lancaster, 1991.

McLachlan, J. A., Guillette, L. J., and Iguchi, T., Eds., *Environmental Hormones: The Scientific Basis of Endocrine Disruption* (Annals of the New York Academy of Sciences, V. 948), New York Academy of Sciences, New York, 2001.

NAS Committee on Hormonally Active Agents in the Environment, National Research Council, *Hormonally Active Agents in the Environment*, National Academy Press, Washington, DC, 1999.

National Research Council, Committee on Developmental Toxicology, Board on Environmental Studies and Toxicology, Commission on Life Sciences, *Scientific Frontiers in Developmental Toxicology and Risk Assessment*, National Academy Press, Washington, DC, 2000.

National Research Council, Subcommittee on Reproductive and Developmental Toxicology, Committee on Toxicology, Board on Environmental Studies and Toxicology, Commission on Life Sciences, *Evaluating Chemical and Other Agent Exposures For Reproductive and Developmental Toxicity*, National Academy Press, Washington, DC, 2001.

Needleman, H. L. and Bellinger, D., *Prenatal Exposure to Toxicants: Developmental Consequences*, Johns Hopkins University Press, Baltimore, MD, 1994.

Neubert, D., *Risk Assessment of Prenatally-Induced Adverse Health Effects*, Springer Verlag, New York, 1993.

Olshan, A. F. and Mattison, D. R., Eds., *Male-Mediated Developmental Toxicity*, Plenum, New York, 1994.

Paul, M. and Paul, S., Eds., *Occupational and Environmental Reproductive Hazards: A Guide for Clinicians*, Lippincott, Williams & Wilkins, Philadelphia, PA, 1993.

Rama Sastry, B.V., Ed., *Placental Toxicology*, CRC Press, Boca Raton, FL, 1995.

Richardson, M., Ed., *Reproductive Toxicology*, VCH Publishing, Weinheim, 1993.

Robaire, B. and Hales, B. F., *Advances in Male Mediated Developmental Toxicity* (Advances in Experimental Medicine and Biology, V.), Plenum, New York, 2003.

Schardein, J. L., *Chemically Induced Birth Defects*, 3rd ed., Marcel Dekker, New York, 2000.

Scialli, A. R. and Clegg, E. D., *Reversibility in Testicular Toxicity Assessment*, CRC Press, Boca Raton, FL, 1992.

Scialli, A. R. and Zimaman, M. J., *Reproductive Toxicology and Infertility*, McGraw Hill, New York, 1993.

Scialli, A. R., *A Clinical Guide to Reproductive and Developmental Toxicology*, CRC Press, Boca Raton, FL, 1992.

Scialli, A. R., Lione, A., and Padgett, G. K. B., *Reproductive Effects of Chemical, Physical, and Biologic Agents: Reprotox*, Johns Hopkins University Press, Baltimore, MD, 1995.

Slikker, W. and Chang, L. W., Eds., *Handbook of Developmental Neurotoxicology*, Academic Press, San Diego, CA, 1998.

Sullivan, F. M., Watkins, W. J., and van der Venne, M. Th., Eds., *The Toxicology of Chemicals: Series 2 - Reproductive Toxicology (Industrial Health and Safety)*, European Communities/Union, Office for Official Publications of the EC, Luxembourg, 1993.

Thiel, R. and Klug, S., *Methods in Developmental Toxicology and Biology*, Blackwell Publishers, Boston, 1997.

Thorogood, P., Embryos, Genes and Birth Defects, John Wiley & Sons, Chichester, 1997.

Tyson, C. A., Heindel, J. J., and Chapin, R. E., Eds., *Methods in Toxicology, Part B: Female Reproductive Toxicology,* Academic Press, San Diego, CA, 1993.

Witorsch, R. J., Ed., *Reproductive Toxicology*, 2nd ed., Raven Press, New York, 1995.

World Health Organization, *Principles for Evaluating Health Risks to Reproduction Associated with Exposure to Chemicals (Environmental Health Criteria)*, World Health Organization (WHO), Geneva, 2001.

APPENDIX C

Postnatal Developmental Milestones

APPENDIX C-1*
SPECIES COMPARISON OF POSTNATAL BONE GROWTH AND DEVELOPMENT

Tracey Zoetis,[1] Melissa S. Tassinari,[2] Cedo Bagi,[2] Karen Walthall,[1] and Mark E. Hurtt[2,3]

---

[1] Milestone Biomedical Associates, Frederick, MD 21701, USA
[2] Pfizer Global Research & Development, Groton, CT 06340, USA
[3] Correspondence to: Dr. Mark E. Hurtt, Pfizer Global Research & Development, Drug Safety Evaluation, Eastern Point Road, Mailstop 8274-1306, Groton, CT 06340. 860-715-3118. Fax (860) 715-3577. Email: mark_e_hurtt@groton.pfizer.com

1 Introduction

The purpose of this review is to identify the critical events in human postnatal bone growth and to compare human bone development to that of other species. Bone is a dynamic connective tissue whose postnatal development in all species reflects its multiple tasks within the body. The two major functions of the skeleton in all mammals are 1) to provide mechanical support as a part of the locomotor apparatus, and 2) to ensure the safe environment of the vital organs of the body. In addition, bones are involved in metabolic pathways associated with mineral homeostasis serving as a reservoir for calcium and phosphorus. Also, bone matrix is a repository for many growth factors and cytokines that are released locally and systemically during the process of bone resorption (autocrine, paracrine and endocrine function). Finally, bones provide safe and steady environment for bone marrow.

Important characteristics of postnatal bone growth are expressed in different types of bone. Each bone in our body has a unique shape that reflects a specific function of that bone and time when this function starts to develop. Growth patterns of specific bones that exhibit important characteristics of postnatal bone development are used to illustrate these characteristics. Thus, this discussion focuses on the postnatal growth of one bone of the upper limb (humerus – non weight bearing bone), one bone of the lower limb (femur – weight bearing bone) and one bone from the craniofacial region (mandible – non weight bearing bone with intermittent mechanical load).

---

* Source: Zoetis, T., Tassinari, M. S., Bagi, C., Walthall, K., and Hurtt, M. E., Species comparison of postnatal bone growth and development, *Birth Defects Research, Part B: Developmental and Reproductive Toxicology*, 68, 86-110, 2003.

The following events are critical milestones in postnatal bone growth and development:

- The appearance of secondary ossification centers
- Longitudinal bone growth at the epiphyseal growth plate
- The fusion of secondary ossification centers
- Diametric bone growth
- Bone vascularity

This review begins with a general discussion of the composition and structure of bone, discusses the appearance and fusion of secondary ossification centers, details the role of the epiphyseal growth plate and metaphysis in bone growth, discusses diametric increases in the diaphysis as well as formation and remodeling of osteons in human compact bone, and emphasizes the importance of bone vascularity in postnatal growth and development.

1.1 Overview of Postnatal Bone Growth

The earliest evidence of bone development in humans is marked by the formation of mesenchymal cell clusters during gestation weeks 4 and 5 (1). In general, bone is formed either by endochondral bone formation or ossification (arising from cartilage) and by intramembranous bone formation or ossification (arising from membranous tissue) (2).

For example, cranial bones such as the mandible are formed by intramembranous bone formation. During mandibular development, soft connective tissue is replaced directly by bone tissue. Similarly, during prenatal development, long bones develop by intramembranous formation; however, the soft connective tissue is first replaced by cartilage, which is replaced by bone through endochondral bone formation during postnatal development (2).

Long bones grow by means of a structure called the epiphyseal or cartilagenous growth plate that is formed after the appearance of secondary ossification centers. The growth plate is a thin disk of cartilage that develops between the metaphysis and epiphysis. Under normal conditions, the growth plate facilitates lengthening of long bones throughout childhood and adolescence and stops functioning when adult height is attained. At the growth plate, endochondral bone formation occurs, i.e. layers of cartilage are successively replaced by new layers of bone. As a result, the growth plate and epiphysis grow away from the central portion of the bone depositing calcified tissue into the transitional metaphyseal region which gradually increases the length of the diaphyseal shaft (2).

As bones grow in length, they must also grow in diameter to support the increased load. Diametric increases in bone occur by periosteal apposition. Unlike longitudinal bone growth, periosteal apposition can occur at any time during development and is a form of intramembranous bone formation that involves the osteogenic fibrous membrane called the periosteum.

Remodeling of immature bone also occurs during postnatal development. Initially, the bone formed in newborns consists primarily of woven or immature bone. As development continues, the woven bone is remodeled into mature adult bone. In compact bone of humans and other large mammals, primary osteons are characteristically found in immature bone whereas secondary osteons are found in maturing bone.

Normal postnatal bone growth and development can only occur with an adequate vascular supply. Bone vascularity plays a vital role in the successful growth and development of both human and animal bones. Thus, to ensure proper nutrition of bone, specialized blood vessels supply various areas of a growing bone.

2 Bone Structure and Composition

Bone is a highly organized structure at both the macroscopic and microscopic levels. Macroscopic observation of bone reveals two different forms of tissue, cortical or compact bone and spongy or

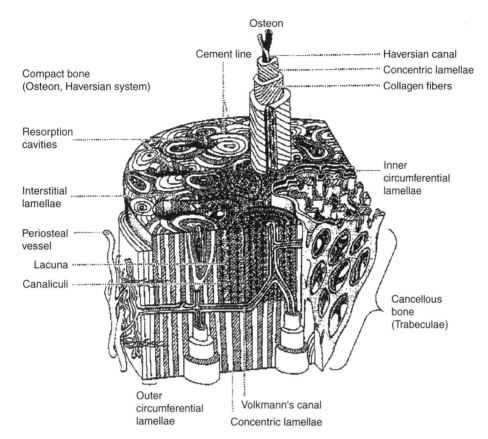

Figure 1 (6) Schematic view of a part of the proximal shaft of a long bone containing cortical (left) and trabecular (right) bone. The cortical bone is composed mainly of Haversian systems or osteons of concentric lamellae with inner and outer circumferential lamellae at the periphery. Many completed osteons can be seen, each with its central Haversian canal, cement line, concentric lamellae, canaliculi, and lacunae. Note the orientation of the collagen fibers in successive lamellae in the protruding osteon. Cross connections between osteons are established by Volkmann's canals. Volkmann's canals connect with periosteal vessels and the marrow cavity and, unlike the Haversian canals, lack a covering of concentric lamellae. Interstitial lamellae are seen between completed Haversian systems. Three resorption cavities are also depicted.

cancellous bone (3, 4, 5, 6). In general, cancellous bone is composed of a lattice of thin bony rods, plates, and arches that form the spaces occupied by bone marrow. Macroscopically, compact bone appears as a solid continuous mass; however, microscopic examination reveals that compact bone is made up of densely packed tubular structures called osteons (3). In the adult human skeleton, compact bone makes up about 80% of total bone mass.

Mature compact and cancellous bone are both made up of lamellae, or stacks of parallel or concentrically curved sheets (Figure 1). Each lamella is about 3 to 7 μm thick and is composed of highly organized, densely packed collagen fibers (6, 4). There are 4 general types of lamellar bone (trabecular lamellae, inner and outer circumferential lamellae, and lamellae of osteons), three of which are found only in compact bone (4).

Cancellous or spongy bone is made up of trabecular lamellae. Trabecular lamellae form the spaces in spongy bone that are filled with bone marrow (Figure 1). Also housed in trabeculae are osteocytes that are connected by tiny branching tubular passages called canaliculi. The canaliculi provide a channel for the exchange of metabolites between osteocytes and the extracellular space.

Outer and inner circumferential lamellae are found only in compact bone (Figure 1). Outer circumferential lamellae are located directly underneath the periosteum extending around the entire circumference of the bone shaft. Inner circumferential lamellae are located on the inner surface of

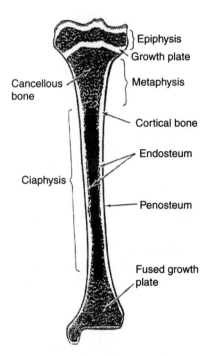

Figure 2 (6) Schematic diagram of a tibia. The interior of a typical long bone showing the growing proximal end with a growth plate and a distal end with the epiphysis fused to the metaphysis.

the shaft adjacent to the endosteum (6). Finally, in mature compact bone circular rings of lamellae are arranged concentrically around longitudinal vascular channels called Haversian canals. The vascular channel and the concentric lamellae form an osteon or Haversian system. The outer limits of osteons are defined by distinct borders called cement lines. Osteons communicate with one another, the periosteum, and the bone marrow via oblique channels called Volkman's canals (3, 6).

Compact and cancellous bone also contain regularly spaced small cavities called lacunae that each contain an osteocyte (Figure 1). Extending from each lacuna in a 3 dimensional fashion are branching tubular passages called canaliculi that connect with adjacent lacuna (6, 3).

2.1 Long Bones

The cylindrical shaft of a long bone is called the diaphysis (Figure 2). The diaphysis is made up of a thick wall of compact bone surrounding a central medullary cavity made up of cancellous bone (3). The bulbous ends of long bones in growing humans and animals are called the epiphyses. In mature long bones, the epiphysis is primarily composed of cancellous bone covered by a thin layer of compact bone. In addition, articular cartilage can be found covering the compact bone at the uppermost portion of the epiphysis. During postnatal bone growth, epiphyses are separated from the diaphysis by a cartilaginous epiphyseal plate that is connected to the diaphysis by an area of spongy bone called the metaphysis. Growing bones increase in length in the growth zone that is formed by the epiphyseal growth plate and the metaphysis.

In general, the external surfaces of bone are covered by a layer of fibrous connective tissue called periosteum that has the ability to produce bone (Figure 3). The periosteum is made up of 2 layers, an outer dense fibrous layer, and an inner vascular cellular layer (4). The inner layer contains osteoblasts and osteoclasts. The periosteum is anchored to bone by bundles of collagen fibers called Sharpey's fibers.

The inner cavities of bone are lined with a layer of osteogenic connective tissue called endosteum (Figure 3). Endosteum is made up of a single layer of flattened osteoprogenitor cells and vascular

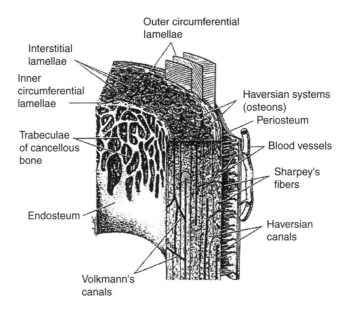

Figure 3 (3) Diagram of a sector of the shaft of a long bone, showing the disposition of the lamellae in the osteons, the interstitial lamellae, and the outer and inner circumferential lamellae.

connective tissue and is thinner than the periosteum. All of the cavities of bone, including the medullary cavities and haversian canals are lined with endosteum.

The long bones include the clavicle, humerus, radius, ulna, femur, tibia, fibula, metacarpals, metatarsals, and phalanges (7).

2.2 Short Bones

Unlike long bones, short bones do not consist of a diaphysis, metaphysis, and epiphysis. Short bones are made up of a core of cancellous bone that is surrounded by a thin layer of compact bone that is covered with periosteum.

The short bones include the carpus, tarsus and sometimes the patellae (7).

2.3 Flat Bones

Many of the flat bones are found in the skull, these include the occipital, parietal, frontal, nasal, lacrimal, and vomer (7). The flat bones of the skull are made up of two layers of compact bone called inner and outer plates or tables. A layer of cancellous bone called the diploe separates these inner and outer plates from one another (8, 3). These flat bones are lined by connective tissue that is similar to the periosteum and endosteum of long bones. The outer surfaces of the flat bones of the skull are covered with a layer of vascular and osteogenic connective tissue called the pericranium, whereas the inner surface of the skull flat bones are lined with a vascular connective tissue called the dura mater (3).

Other flat bones include the scapula, os coxae, sternum, and ribs (7). These bones are also made up of two thin layers of compact bone separated by cancellous bone and are covered with periosteum.

2.4 Irregular Bones

The irregular bones of the skeleton are those that due to their shape cannot be grouped with the long, short, or flat bones. These bones are made up of cancellous tissue that is enclosed in a thin layer of compact bone covered by the periosteum. Irregular bones include the vertebrae, sacrum, coccyx, temporal, sphenoid, ethmoid, zygomatic, maxilla, mandible, palatine, inferior nasal concha, and hyoid (7).

2.5 The Cells of Bone

Microscopically, actively growing bone is composed of intercellular calcified material called bone matrix, and four types of cells (osteoprogenitor cells, osteoblasts, osteocytes, and osteoclasts) (3, 9, 10, 11, 12). The functions of each of these cells are discussed below.

Osteoprogenitor cells are found on or near all surfaces of bone. These cells have the potential to differentiate into chondroblasts (cartilage cells) or osteoblasts (bone cells). Osteoprogenitor cells are most active in the postnatal period during the growth of bones, but are also active in adult life during repair of bone injury (3).

Osteoblasts originate from local mesenchymal stem cells. Although the primary function of osteoblasts is the production of bone, these cells have been shown to have a variety of other roles. For example, osteoblasts are involved in aiding bone resorption, organizing collagen fibrils, and synthesizing matrix components such as type I collagen, proteoglycans, and glycoproteins (3, 10). During embryonic and postnatal growth the periosteum contains an inner layer of osteoblasts, that are in direct contact with bone.

Osteocytes have an important role as a mechanosensors and are associated with local activation of bone turnover. Osteocytes are formed from osteoblasts that become confined within the newly formed bone matrix during bone formation (10, 11, 9). The exact function of osteocytes is unknown (9). In mature bone, osteocytes reside within cavities called lacunae that are found throughout the interstitial spaces of bone. Osteocytes have numerous cytoplasmic projections that extend from canaliculi into the bone matrix and contact adjacent cells (3, 11).

Osteoclasts arise from hematopoietic mononuclear cells in the bone marrow. Osteoclasts are large mutinucleated cells that adhere to the surface of bone for the sole purpose of resorption (13, 3). Osteoclasts secrete products that lower the pH of the bone surface to aid in resorption of mineralized bone and cartilage (13).

2.6 Bone Matrix

The matrix of bone, occupies more volume than the cells. Bone matrix is composed of approximately 65% inorganic material (calcium, phosphorus, sodium, magnesium), > 20% organic material (type I, V and XII collagen, glycoproteins, bone proteoglycans), and > 10% water (3, 4).

Bone matrix is formed in 2 stages, deposition and mineralization. During matrix deposition, osteoblasts secrete the initial matrix called osteoid, that is made up of type I collagen, various proteins, and sulfated glycoaminoglycans. During mineralization, calcium phosphates and carbonates that were previously stored in vessicles of cells are released into the matrix where they are deposited onto collagen fibrils with the help of glycoproteins. Osteoblasts also secrete alkaline phosphatase that is important for mineralization of osteoid.

3 Ossification

The process by which bone tissue replaces its precursor connective tissue is called ossification. Ossification can either be intramembranous or endochondral. Intramembranous ossification involves the replacement of membranous fibrous tissue by bone, whereas endochondral ossification involves the replacement of cartilage by bone (2).

In general, ossification begins and is concentrated in a focal area that subsequently expands in size until the previously existing tissue is completely replaced by bone. The initial site of ossification is called the primary ossification center (2). Most primary ossification centers develop during the embryonic and early fetal period; however, a few also develop postnatally. In some bones (e.g.,

Table 1 Age of Appearance and Fusion of Secondary Ossification Centers in the Humerus and Femur

	Age			
	Appearance		Fusion	
	Proximal Epiphysis	Distal Epiphysis	Proximal Epiphysis	Distal Epiphysis
Humerus				
Human	Gestation week 36–4 years	6 months–10 years	12–20 years	11–19 years
Monkey	Birth	Birth–1 month	4–6 years	1.75–4.5 years
Dog	1–2 weeks	2–9 weeks	10–12 months	6–8 months
Rabbit	1 day	1 day	32 weeks	32 weeks
Rat	8 days	8–30 days	52–181 weeks	31–158 days
Mouse	5–10 days	5–19 days	6–7 weeks	3 weeks
Femur				
Human	1–12 years	Gestation week 36–40	11–19 years	14–19 years
Monkey	Birth–6 months	Birth	2.25–6 years	3.25–5.75 years
Dog	1 week–4 months	2–4 weeks	6–13 months	8–11 months
Rabbit	1–5 days	1 day	16 weeks	32 weeks
Rat	21–30 days	8–14 days	78–156 weeks	15–162 weeks
Mouse	14–15 days	7–9 days	13–15 weeks	12–13 weeks

long bones), the primary ossification center does not expand into the entire precursor tissue area. Secondary centers of ossification develop in regions where primary centers do not extend. These secondary ossification centers are generally formed postnatally (2).

Shortly after secondary centers of ossification appear in the epiphyseal region of long bones, the center grows in all directions and results in development of the epiphyseal growth plate. The epiphyseal growth plate creates a barrier between the epiphysis and the diaphysis. As bone is formed at the growth plate, the epiphysis moves away from the diaphysis and bone is deposited in the transitional region called the metaphysis located directly below the epiphyseal growth plate (Figure 2). As bone is formed at the growth plate, lengthening of the diaphysis occurs. When the long bones have reached their adult length, the epiphyseal growth plate fuses with the diaphysis thus removing the barrier between these two regions of the bone. Once fusion of the epiphyseal growth plate occurs, longitudinal bone growth can no longer occur. Union of the epiphysis with the diaphysis is also called fusion of secondary ossification centers, or fusion of the epiphyseal growth plate.

The number, location, and time of appearance of secondary centers of ossification varies between species. Also, within species, there is some variation regarding the timing of ossification (14).

The following sections include information regarding the ossification of the humerus, femur, and mandible in humans, monkeys, dogs, rabbits, rats, and mice. Table 1 provides an overview of the ages at which appearance and fusion of secondary ossification centers occur in the humerus and the femur.

3.1 Human

Ossification centers that develop prenatally in humans include those in the skull, vertebral column, ribs, sternum, the primary centers in the diaphysis of major long bones, their girdles, and the phalanges of the hands and feet (2). In addition, some primary centers in the ankle and secondary centers around the knee also develop during the last few weeks of gestation.

In humans, postnatal ossification centers develop over a timespan ranging from immediately after birth through early adulthood. Ossification centers that fuse during adolescence include the

Table 2 Timing of Appearance and Fusion of Secondary Ossification Centers in the Human Humerus (2, 19, 22)

Region of Humerus	Time of Appearance	Time of Fusion
Proximal Epiphysis		
Head	Gestation week 36–40 or 2–6 months after birth	2–7 yrs: composite proximal epiphysis formed
Greater Tubercle	3 months–3 years	12–19 yrs: proximal epiphysis fused with diaphysis in females
Lesser Tubercle	4+ years	15.75–20 yrs: proximal epiphysis fuses with diaphysis in males
Distal Epiphysis		
Capitulum	6 months–2 years	10–12 yrs: composite distal epiphysis is formed
Lateral Epicondyle	10 years	11–15 yrs: distal epiphysis fuses with diaphysis in females
Trochlea	By 8 years	12–17 yrs: distal epiphysis fuses with diaphysis in males
Medial Epicondyle	4+ years	11–16 yrs: fusion with diaphysis in females 14–19 yrs: fusion with diaphysis in males

epiphyses of the major long bones of the limbs, those of the hands and feet, and the spheno-occipital synchondrosis of the skull. Following adolescence, fusion occurs in the jugular growth plate of the skull, and in the secondary ossification centers of the vertebrae, scapula, clavicle, sacrum, and pelvis (2).

The appearance and fusion of ossification centers in the humerus and femur are detailed below. Pre- and postnatal growth of the mandible is also discussed below.

3.1.1 Humerus

Primary ossification centers appear in the diaphysis of the humerus during gestation. At birth, 79% of the human humerus is made up of an ossified diaphyseal shaft and 21% is made up of non-ossified cartilaginous material found primarily in the proximal and distal epiphyses (2).

In the newborn infant, the humerus has a rounded proximal end, and a triangular distal region, separated from each other by the diaphyseal shaft. The proximal epiphysis of the humerus develops from 3 separate secondary ossification centers: one in the head, one in the greater tubercle, and one in the lesser tubercle. The time of appearance for each of the secondary ossification centers of the human humerus is listed in Table 2. Studies have shown that the ossification center in the head usually appears by 6 months after birth, but sometimes develops during weeks 36-40 of gestation (15, 16, 2). Early appearance of the ossification center in the head of the humerus has been shown to be related to birth weight, sex, nationality, maternal history, size, and maturity (2, 15, 16, 17, 18). The timing of appearance for the secondary ossification center of the greater tubercle ranged from 3 months of age to 3 years of age. In general, it appears earlier in girls than in boys. The existence of a third center in the lesser tubercle has been debated in the literature. Some studies note that only 2 centers of ossification develop in the proximal epiphysis (head and greater tubercle), others note a third at the lesser tubercle (19, 20). Appearance of the lesser tubercle was noted to occur between the ages of 4 and 5 years (2).

The ages reported for unification of the secondary ossification centers in the proximal epiphysis of the humerus range from 2 to 7 years. Radiologic studies report that each of the secondary ossification centers of the proximal epiphysis join to form a single compound proximal epiphysis between the ages of 5 and 7 years (2). Histological studies demonstrate that a compound proximal epiphysis forms as early as 2 or 3 years of age (19, 2).

The proximal epiphysis of the humerus is responsible for up to 80% of the growth in length of the diaphyseal shaft (21). When diaphyseal growth is complete, fusion of the proximal epiphysis occurs. Fusion of the proximal epiphysis has been reported to occur at ages ranging from 12 to 19 years in females and 15.75 to 20 years in males (2).

Ossification in the distal epiphysis of the humerus occurs via 4 separate secondary centers that develop in the capitulum, medial epicondyle, lateral epicondyle, and trochlea. By the age of 2 years, the secondary ossification center of the capitulum is evident; however, this center may also appear as early as 6 months after birth (2). The secondary ossification center of the medial epicondyle is usually visible by age 4 but develops slowly thereafter (2). Development of the secondary ossification center in the trochlea begins with the appearance of multiple foci at the age of 8 years. Soon after appearance, the trochlear epiphysis becomes joined to the capitulum. Ossification in the lateral epicondyle is evident by the age of 10 years.

The secondary centers of the capitulum, trochlea, and lateral epicondyle join with each other between the ages of 10 and 12 years. Fusion of these structures with the diaphyseal shaft begins posteriorly and leaves an open line above the capitulum, lateral trochlea, and proximal lateral epicondyle which becomes fused at approximately the age of 15 years.

The ossification center in the medial epicondyle does not fuse with the capitulum, trochlea, and lateral epicondyle prior to uniting with the diaphyseal shaft. In females and males, fusion has been seen to occur at ages ranging from 11 to 16 and 14 to 19 years, respectively (22, 2).

3.1.2 Femur

In the femur, primary ossification centers of the diaphysis appear during the prenatal period.

The epiphysis that is found at the distal end of the femur is the largest and fastest growing epiphysis in the body (2). This secondary ossification center is the first long bone epiphysis that appears during skeletal development and one of the last to fuse (2). This distal epiphysis that develops from a single center of ossification usually appears during prenatal weeks 36-40; however, some variation in the time of appearance exists. For example, in premature infants, this center is not always present and some studies have found that the distal femoral epiphysis is sometimes visible as early as prenatal week 31. Even so, the distal epiphysis is always visible by the postnatal age of 3 months (15, 20, 23, 22).

At birth, the distal femoral epiphysis of females is about 2 weeks more advanced than that of males. By the time of puberty, girls are about 2 years more developmentally advanced than boys (2).

During postnatal months 6-12, the distal epiphyseal plate begins to develop and the epiphysis takes on an ovoid shape (2). During postnatal years 1 to 3, the width of the epiphysis increases rapidly as ossification spreads throughout the epiphyseal region. When females and males reach ages 7 and 9, respectively, the epiphysis is as wide as the metaphysis (2).

The distal epiphysis is responsible for approximately 70% of the longitudinal growth of the femur (24, 25, 2). When growth of the femur is complete, fusion of the distal femoral epiphysis occurs. Fusion of the distal femoral epiphysis occurs in females and males between the ages of 14 to 18 and 16 to 19 years, respectively (2).

At the proximal end of the femur, 3 to 4 separate secondary centers of ossification develop. Unlike those in the proximal epiphysis of the humerus, these centers develop and fuse independently with either the neck or shaft of the femur. At birth, the proximal epiphyseal growth plate is divided into 3 sections, medial, subcapital, and lateral subtrochanteric portions. By the age of 2 years, the neck has grown and divided the epiphyseal region into the head and the greater trochanter. The lesser trochanter lies below and medial to the epiphyseal region. The head, greater trochanter and lesser trochanter each develop a separate secondary ossification center. The center in the head of

Table 3 Timing of Appearance and Fusion of Secondary Ossification Centers in the Human Femur (2, 22)

Region of Femur	Time of Appearance	Time of Fusion
Proximal Epiphysis		
Head	By year 1	11–16 years: females 14–19 years: males
Greater Trochanter	2–5 years	14–16 years: females 16–18 years: males
Lesser Trochanter	7–12 years	16–17 years
Distal Epiphysis	Gestation weeks 36–40	14–18 years: females 16–19 years: males

the femur appears in most infants between 6 months to 1 year after birth (15, 16, 2). Fusion of the femoral head has been reported to occur in females and males between the ages of 11 to 16 and 14 to 19 years, respectively (22, 2). The secondary center of ossification of the greater trochanter appears between the ages of 2 and 5 years. Fusion occurs at about 14 to 16 years in girls and 16 to 18 years in boys (2). Appearance of a secondary ossification center in the lesser trochanter of the femur is reported to range from age 7 to age 11. Fusion of the lesser trochanter with the femoral shaft is seen at the age of 16 to 17 years.

A summary of the ages at which secondary ossification centers in the femur appear and fuse is provided in Table 3.

3.1.3 Mandible

The mandible, or lower jaw, is the second bone in the body to begin ossification. Ossification of the human mandible primarily takes place during the prenatal period. The mandible is made up of a curved, horizontal portion called the body, and two perpendicular portions called the rami (7). In general, the body of the mandible includes the alveolar region which includes deep pockets that house teeth later in development, and the symphyseal region that connects the 2 two portions of the bone early in development. Each ramus is made up of a condylar process and a coronoid process.

Postnatal growth and development of the human mandible — Of all the facial bones, the mandible undergoes the most postnatal variation in size and shape. As a result, the perinatal mandible differs greatly in size and shape from the mature mandible.

At birth, the mandible consists of two halves that are connected by a fibrous region called the symphysis. During the first year of life (usually by 6 months), the right and left halves of the mandibular body fuse at the midline in the symphyseal region (2). At birth, both the mandible and maxilla bone are equal in size, but the mandible is located posterior to the maxilla. As rapid postnatal growth continues, the mandible assumes its normal position in line with the maxilla (achieving normal occlusion) (2).

Two types of bone growth occur in the mandible, appositional/resorptive and condylar growth. The posterior border of the ramus is an active site of bone apposition and the anterior border of the mandibular body is an active site of bone resorption (26). Deposition of bone on the posterior end of the ramus and resorption of bone on the anterior portion allows the mandibular body to lengthen and make space for developing teeth (2).

The condyle has been shown to play a primary role in development of the mandible. Growth of the condyle essentially leads to a downward and forward "displacement of the mandible." This growth occurs as cartilage in the condyle is replaced by bone. Condylar growth plays a role in decreasing the mandibular or gonial angle. During the perinatal period, the mandibular angle ranges from 135° to 150°; however, soon after birth, it decreases to 130° to 140° (2). In the adult mandible, the gonial angle measures between 110° and 120° (2, 7).

Table 4 Growth and Ossification of the Human Mandible (2)

Age	Event
Prenatal week 6	Intramembranous ossification center develops lateral to Meckel's cartilage
Prenatal week 7	Coronoid process begins differentiating
Prenatal week 8	Coronoid process fuses with main mandibular mass
About Prenatal week 10	Condylar and coronoid processes are recognizable. The anterior portion of Meckel's cartilage begins to ossify
Prenatal weeks 12–14	Secondary cartilages for the condyle, coronoid, and symphysis appear
Prenatal weeks 14–16	Deciduous tooth germs start to form
Birth	Mandible consists of separate right and left halves
Postnatal year 1	Fusion of the right and left halves of the mandible at the symphysis
Infancy and childhood	Increase in size and shape of the mandible
	Eruption and replacement of teeth
By 12–14 years	All permanent teeth emerged except third molars

In the perinatal mandible, the condyle is essentially at the same level as the superior border of the mandibular body. After birth a rapid increase in the height of the ramus occurs. This increase in ramus height results in the condyle being situated on a higher plane than the alveolar surface (the top portion of the mandibular body that is hollowed into cavities for teeth) of the mandibular body. In addition, during growth, an increase in the height of the mandibular body occurs as a result of alveolar bone growth (26).

As the cranial width increases, resorption and deposition of bone also results in widening of the mandibular body. Other changes occurring in the mandible during growth include changes in the horizontal and vertical position of the mental foramen. Initially, the mental foramen is positioned below and between the canine and first molar. Once dentition (cutting of teeth) begins, the foramen moves under the first molar and is later is positioned between the first and second molar. The foramen also changes vertical position as the depth of the alveolar process increases (2).

The mental proturbance or chin region also changes during postnatal development. Growth of the mental proturbance has been shown to occur rapidly from birth until the age of 4 years old. During this time, the depth of the chin increases to make room for the developing roots of the incisor teeth (27, 2).

The sequence of growth and ossification in the human mandible is shown in Table 4.

3.2 Monkey

When the timing of appearance and fusion of secondary ossification centers in rhesus monkeys was studied, it was noted that more ossification centers were present in monkeys at birth than in man. In general, bone ossification in the newborn rhesus monkey, resembles that of a 5 to 6 year old human (28). Ossification in the limbs of male and female rhesus monkeys is complete by 5.25 and 6.5 years, respectively.

In the chimpanzee, the average onset of postnatal ossification in long and short bone centers occurs about 12 to 20 months earlier, than ossification in man (28).

3.2.1 Humerus

At birth in male rhesus monkeys, two separate ossification centers are present in the proximal epiphysis (29). In the epiphyseal region of the distal humerus, one secondary ossification center is present at birth (29). Another 2 centers develop in the distal humerus during the first postnatal month. At 9 months after birth, the separate centers in the proximal and distal humeral epiphyses unite to form single secondary ossification centers in their respective regions (29).

Fusion of the proximal epiphysis to the diaphysis in male and female cynomolgus monkeys occurs at approximately 6 and 4.75 to 5.5 years, respectively (30, 31). Union of the distal epiphysis

Table 5a Timing of Secondary Ossification Center Appearance and Epiphyseal Fusion in the Rhesus Monkey (29)

Bone	Male		Female	
	Time of Appearance	Time of Fusion	Time of Appearance	Time of Fusion
Humerus				
Proximal Epiphysis	Birth	4–6 years	Birth	4–5.25 years
Distal Epiphysis	Birth–1 month	2 years	Birth	1.75–2 years
Femur				
Proximal Epiphysis: Lesser Trochanter	6 months	2.75–3.75 years	6 months	Information not available
Head	Birth	3–3.75 years	Birth	2.25–3.25 years
Medial Epicondyle	Information not available	3–3.75 years	Information not available	2.25–2.5 years
Distal Epiphysis	Birth	4–5.75 years	Birth	3.25–4.25 years

Table 5b Timing of Secondary Ossification Center Appearance and Epiphyseal Fusion in the Cynomologus Monkey (30, 31)

Bone	Male		Female	
	Time of Appearance	Time of Fusion	Time of Appearance	Time of Fusion
Humerus				
Proximal Epiphysis	< 5 months	6 years	< 4 months	4.75–5.5 years
Distal Epiphysis	< 5 months	3.4–4.5 years	< 4 months	2.25 years
Femur				
Proximal Epiphysis	< 5 months	6 years	< 4 months	4.75 years
Distal Epiphysis	< 5 months	5.25 years	< 4 months	4.75 years

to the diaphysis is said to occur in male monkeys between 3.4 and 4.5 years after birth (30, 31). Fusion of the distal epiphysis occurs in female monkeys at approximately 2 years after birth (31).

The timing of appearance and fusion of the epiphyses in the humerus of rhesus monkeys and cynomolgus monkeys is shown in Tables 5a and 5b, respectively.

3.2.2 Femur

In rhesus monkeys at birth, an epiphyseal ossification center is present in the femoral head (29). Epiphyseal ossification centers in the distal femur are also present at birth. In addition, at 6 months after birth, secondary ossification centers develop in the lesser trochanter of the femur (29).

When the femur of female cynomolgus monkeys was examined at 4 months after birth, ossification centers were already present in the proximal and distal epiphyses. In male monkeys, ossification centers were present in the proximal and distal epiphyses when examined at 5 months after birth (31).

In male cynomolgus monkeys, fusion the proximal and distal epiphyses to the respective diaphysis occurred in the femur at 6 and 5.25 years of age, respectively. Fusion of both the proximal and distal epiphyses occurred in females at 4.75 years of age (31).

The timing of appearance and fusion of the epiphyses in the femur of rhesus monkeys and cynomolgus monkeys is shown in Tables 5a and 5b, respectively.

3.2.3 Mandible

During postnatal growth in the mandible of monkeys, a large increase in overall size of the mandible, occurs as a result of extensive growth and remodeling processes that take place from infancy to adulthood (32). Essentially, bone growth occurs on all surfaces of the primate mandible during postnatal development (33, 34).

The following discussion includes information taken primarily from a study conducted by McNamara and Graber (32) to investigate the postnatal development of the mandible in rhesus monkeys. In this study, 4 groups of monkeys (at various ages) were evaluated for 6 consecutive months. The groups included infant monkeys ages 5.5 to 7 months, juvenile monkeys ages 18 to 24 months, adolescent monkeys ages 45 to 54 months, and young adult monkeys over the age of 72 months (32).

In rhesus monkeys, the mandible grew most rapidly during infancy (32). Growth at the mandibular condyle was most rapid in infant monkeys and successively slowed as animals aged. Specifically, during the 6-month observation period, growth of the condyle in the infant, juvenile, adolescent, and young adult monkeys was 5.92, 4.47, 3.00, and 1.07 mm, respectively.

During postnatal development, deposition of bone occurs along the posterior border of the ramus and resorption of bone occurs along the anterior border of the ramus. The largest increase in the width of the ramus was seen in infant monkeys (5.5 to 13 months). When compared to animals over the age of 18 months, the mandibular ramus of infant monkeys had 4 times as much bone deposited along the posterior border than was resorbed from the anterior border. Bone deposition on the posterior border of the ramus decreased as the animals aged. Changes in the width of the mandibular ramus in the adult monkey were slight (32). In addition, bone deposition occurred along the posterior border of the condyle (32).

In the infant, juvenile, and adolescent animals, remodeling of the mandibular angle was noted; however, remodeling of the mandibular angle was not seen in the adult monkeys (32). During the 6-month observation period, the largest change in the mandibular angle occurred in infant monkeys. The average decrease in mandibular angle over the 6 month observation period in infant, juvenile, and adolescent monkeys was 6.2, 2.4, 1.7 degrees, respectively.

3.3 Dog

Appearance of secondary ossification centers and fusion of the proximal and distal epiphyses with the diaphysis is summarized in Table 6.

3.3.1 Humerus

The humerus of the dog develops from five secondary centers of ossification: one found in the diaphysis that is ossified at birth, one in the head and tubercles of the proximal epiphysis, and three in the distal epiphysis (one each for the medial condyle, lateral condyle, and medial epicondyle) (35, 36).

A secondary ossification center appears in the head of the humerus at 1 to 2 weeks after birth (37, 35, 38). At 5 months after birth, partial fusion of the epiphysis to the diaphysis begins. Closure of the epiphyseal plate occurs at approximately 10 - 12 months of age; however, in some dogs, the fusion line may still be evident at 12 months, but usually disappears completely by 14 months (35, 37, 30).

In the distal epiphysis of the humerus, the ossification centers in the medial and lateral condyles appear during the 2nd, 3rd, or 4th postnatal week (35, 37, 36). The center in the medial epicondyle appears during postnatal weeks 5, 6, 7, 8, or 9 (36, 38, 35). The center of the medial epicondyle unites with the medial condyle during postnatal months 4, 5, or 6, and the whole distal extremity fuses with the diaphyseal shaft during the 6th, 7th, or 8th month after birth (35, 37, 36). At approximately 10 months of age, the fusion line in the distal epiphysis of the humerus is barely visible (37).

Table 6 Appearance and Fusion of Ossification Centers in the Dog (36, 35, 38, 25, 37)

Bone	Age at Appearance	Age at Fusion
Humerus		
Proximal Epiphysis		
Head	1 to 2 weeks[1,2,3,4]	10 to 12 months [1,3]
Distal Epiphysis		
Medial condyle	2, 3 or 4 weeks [1,2,3]	To lateral condyle 6 to 9 weeks[1]
Lateral condyle	2 to 3 weeks[1,2,3]	See above
Epicondyle (medial and lateral)	5 to 9 weeks[1,2,3]	To condyles 4 to 6 months[1,3] complete to shaft 6 to 8 months[1,3]
Femur		
Proximal Epiphysis		
Head	2 to 3 weeks;[1] 1 to 3 weeks[2]	6 to 10 months[1]
Greater Trochanter	9 to 10 weeks;[1] 2 months[2]	8 to 11 months[1]
Lessor Trochanter	3 to 4 months;[1] 7 to 11 weeks[2]	10 to 13 months[1]
Distal Epiphysis		
Trochlea	2 weeks[1]	Trochlea to condyles 3 to 4 months[1]
Medial condyle	3 weeks;[1,5] 2 to 4 weeks[2]	Complete to diaphysis at 8 to 11 months;[1] 250 to 325 days[5]
Lateral condyle	3 weeks;[1] 2 to 4 weeks[2]	

[1] 36
[2] 38
[3] 35
[4] 37
[5] 25

3.3.2 Femur

A slight amount of variation exists in the scientific literature regarding the number of ossification centers in the femur of the dog. Andersen and Floyd (25) reports that the canine femur develops from the six centers of ossification: one in the diaphysis that is ossified at birth, one each in the head, greater trochanter, and lessor trochanter of proximal region, and one each in the medial and lateral condyle of the distal epiphysis. Parcher and Williams (36) lists seven centers of ossification by an additional center in the trochlea noting one in the trochlea. Hare (38) reports five centers of ossification in the femur of the dog, listing only one center in the distal epiphysis.

In spite of the variation in the number of ossification centers reported in the literature, reasonable consistency exists concerning the age at which secondary ossification centers appear. At 7-10 days after birth, ossification is evident in the diaphysis of the femur (25). By postnatal week 2 and 3, ossification centers develop in the head of the femur (36, 38). In the greater trochanter and lessor trochanter, secondary ossification centers do not develop until 8 to 10 weeks and 2, 3 or 4 months after birth, respectively (36, 38). Parcher and Williams (36) list that fusion of the epiphyseal plates in the femoral head, greater trochanter, and lessor trochanter occurs at 6 to 10, 8 to 11, and 10 to 13 months after birth, respectively.

By 21 days after birth in the distal region of the femur, epiphyseal growth plates develop in the medial and lateral epicondyle (25, 36). Similarly, Hare (38) reports appearance of the epiphyseal growth plate in the distal epiphysis of the femur at 2, 3, or 4 weeks. Parcher and Williams (36) report that the secondary ossification center of the trochlea appeared one week earlier at 2 weeks after birth. By 3 and 4 months after birth, it is reported that the epiphyseal growth plate in the

Table 7 Appearance and Fusion of Ossification Centers in the Rabbit (41, 42)

Bone	Age at Appearance	Age at Fusion
Humerus		
Proximal Epiphysis	1 day	32 weeks
Distal Epiphysis	1 day	32 weeks
Femur		
Proximal Epiphysis	1–5 days	16 weeks
Distal Epiphysis	1 day	32 weeks

trochlea begins to fuse with those of the medial and lateral condyles (36). By 8 to 11 months, the epiphyseal growth plate in the distal region of the femur fuses with the diaphysis (36, 25).

3.3.3 Mandible

During prenatal development, the mandible of the dog, is formed by intramembranous ossification. During postnatal development, endochondral bone formation occurs at the condyle; however, elsewhere in the mandible, bone growth occurs by the apposition of bone along the mandibular surfaces (39).

The rate at which bone is added to the mandible is most rapid during the early postnatal period. After dogs reach about 40 to 60 days of age, the rate of bone formation in the mandible plateaus (39, 40). By the age of 6 to 7 months, even though bone apposition and resorption continues in the mandible, the volume of the mandible remains constant because of a balance between the two processes (39).

During postnatal development, the mandible increases in overall size, length, width, and height. The increase in the length that occurs in the mandible of the dog is primarily due to bone formation on the rear portions of both the ramus and the mandibular body (39). The mandibular ramus increases in width as more bone is deposited at the caudal border of the mandible than is being resorbed at the rostral border. Resorption at the rostral border of the mandible occurs to makes room for eruption of the molars. The height of the mandible increases by apposition of bone on the ventral surface of the mandible, and by growth of alveolar bone (39). A change in the mandibular angle occurs as bone is laid down on the caudo-ventral portion of the mandible (39). Growth of the mandible in a downward and backward direction is caused by endochondral bone formation at the mandibular condyle (39).

3.4 Rabbit

A summary of the appearance of secondary ossification centers and the fusion of the epiphyses of the humerus and femur is provided in Table 7.

3.4.1 Humerus

When maturation of secondary ossification centers was studied in the Japanese white rabbit, no significant differences between male and female rabbits were noted (41). As early as postnatal Day 1, secondary ossification centers were present in the proximal and distal epiphyses of the humerus (41). By 1 week after birth, all secondary ossification centers had appeared in the long bones with the exception of those in the proximal epiphysis of the fibula, which appeared at 2 weeks of age (41).

At 32 weeks of age, the proximal and distal epiphyses of the humerus fused with the diaphyses (41).

3.4.2 Femur

According to Heikel (42), secondary ossification centers in the head of the femur appear in the rabbit by the 4th or 5th postnatal day. However, Fukuda and Matsuoka reported that secondary ossification centers were present in the head and distal epiphysis of the femur at 1 day after birth (41).

Growth in the rabbit femur occurs rapidly until 3 months after birth at which time growth slows until completely stopping at about 6 months (43). This suggests that fusion of the distal and proximal epiphyseal growth plates occur at approximately 6 months after birth. However, Fukuda and Matsuoka reports fusion of the femoral head and distal epiphysis of the femur occur at approximately 16 and 32 weeks after birth, respectively (41).

3.4.3 Mandible

Unlike the human mandible, the rabbit mandible does not contain a prominent coronoid process, nor do the right and left half of the mandibular body fuse during development (44). The two portions of the mandibular body remain separated at the junction between the right and left half.

In the rabbit, postnatal craniofacial growth occurs most rapidly immediately after birth. The length of the rabbit mandible increases rapidly from birth to 16 weeks at which time approximately 90% of the adult length is achieved. The largest increase in mandibular length occurs during the first 2-6 weeks after birth (45).

During postnatal growth in the rabbit mandible, several combined growth processes work together to move the ramus in a upward and backward direction. First, the ramus grows in an upward and backward direction via endochondral bone formation that occurs within the condylar cartilage (44). Second, the combination of resorption and deposition on opposite sides of the mandible contribute to movement of the ramus in a backward and upward direction (44). For example, the lingual side of the ramus (area facing the tongue or oral cavity) mainly exhibits bone deposition whereas the buccal surface (area nearest to the check) primarily experiences resorption of bone (44). Thirdly, as bone is being deposited on the posterior border of the ramus, resorption is taking place on the anterior border of the ramus. This combination of resorption and deposition also contributes to posterior growth of the mandibular ramus (44).

As the ramus moves posteriorly via the growth mechanisms discussed above, elongation of the mandibular body occurs. Areas of bone that were once part of the ramus are remodeled and become part of the mandibular body (44). In addition, as bone is deposited along the lower border of the mandibular angle the angular process enlarges (44).

3.5 Rat

3.5.1 Humerus

Johnson (46) reported that secondary ossification centers appear in the head and greater tubercle of the humerus on postnatal day 8. This account corresponds with that of Fukuda and Matsuoka (47) who reported that secondary ossification centers in the head of the humerus appear during postnatal week 2. However, considerable variation exists regarding the timing of epiphyseal fusion in the proximal region of the humerus. Fukuda and Matsuoka (47) reported that fusion of the secondary ossification centers in the proximal epiphysis of the humerus is complete by 52 weeks after birth; however, Dawson (48) noted that the epiphysis in the head of the humerus did not unite with its diaphyseal shaft until 162-181 weeks after birth.

Fukuda and Matsuoka (30, 47) acknowledge the existence of other reports in the literature regarding incomplete epiphyseal fusion in aged mice, rats, and rabbits. This incomplete fusion leaves the epiphyseal line visible for quite some time after fusion has begun and progressed far enough to ensure that longitudinal growth of the bone is no longer occurring. Fukuda and Matsuoka

Table 8 Appearance and Fusion of Postnatal Secondary Ossification Centers in the Rat (47, 48, 46)

Bone	Time of Appearance (day/week after birth)	Time of Fusion
Humerus:		
Proximal Epiphysis		
Greater Tubercle	8 days[1]	52 weeks[2]
Head	8 days[1], 4 weeks[2]	52 weeks[2], 162–181 weeks[3]
Distal Epiphysis		
Capitellum	8 days[1]	4 weeks[2], 31–42 days[3]
Trochlea	8 days[1]	4 weeks[2], 31–42 days[3]
Medial Condyle	30 days[1]	4 weeks[2], 31–42 days[3]
Lateral Condyle	30 days[1]	4 weeks[2], 31–42 days[3]
Epicondyle	Information not available	130–158 days[3]
Femur		
Proximal Epiphysis		
Greater Trochanter	30 days[1], 4 weeks[2]	78 weeks[2], 143–156 weeks[3]
Lesser Trochanter	21 days[1]	Information not available
Head	21 days[1], 4 weeks[2]	104 weeks[2], 143–156 weeks[3]
Distal Epiphysis		
Medial Condyle	8 days[1], 2 weeks[2]	15–17 weeks[2], 162 weeks[3]
Lateral Condyle	8 days[1], 2 weeks[2]	15–17 weeks[2], 162 weeks[3]

[1] 46
[2] 47
[3] 48

(41) noted that the incomplete fusion seen in some long bones of aged mice, rats, and rabbits might be a common occurrence. He notes that in these cases, if the incomplete fusion remains for an extended period of time, then the earlier timepoint of partial fusion would be accepted as the "complete" fusion of the epiphysis (47, 30).

In the distal epiphysis of the rat humerus, secondary ossification centers appear in the capitulum, and trochlea during week 2 after birth, specifically, on postnatal day 8 (46, 47). Dawson (48) reports that by postnatal day 15, a single well-defined secondary center of ossification can be seen in the capitulum and trochlea. Fukuda and Matsuoka (47) report that fusion is noted in the distal epiphysis of the humerus by week 4. The epiphysis that develops in the distal region of the humerus is the first long bone epiphysis to unite with its corresponding diaphyseal shaft (48). By day 42, the epiphyseal growth plate of the distal humerus is no longer visible, indicating that epiphyseal fusion has occurred. Dawson (48) reported that in the epiphysis of the capitulum and trochlea, the characteristic columnar arrangement of cartilage cells within the growth plate was not seen. As a result it was postulated that in this case, the typical sequence of events is not followed due to the rapid rate of epiphyseal fusion (48).

The appearance and fusion of secondary ossification centers in the humerus is summarized in Table 8.

3.5.2 Femur

Secondary ossification centers appear in the greater trochanter and head of the femur at 4 weeks after birth (47). According to Fukuda and Matsuoka (47) and Dawson (48), fusion occurred in the greater trochanter at 78 and 143-156 weeks respectively, and in the head of the femur at 104 and 143-156 weeks, respectively.

At the distal end of the femur, secondary ossification centers appear at 2 weeks after birth (47). Fusion of the distal epiphysis occurs between 15 and 17 weeks as reported by Fukuda and Matsuoka (47) and at 162 weeks as reported by Dawson (48). The discrepancy noted here regarding the time of fusion could be due to the fact that by 15 weeks, all of the secondary ossification centers in the fore and hind limbs of the rat had begun fusion, but all had not completed the fusion process (47). For example, it was also reported that fusion of secondary ossification centers in the distal epiphysis of the radius, both epiphysis of the ulna, the proximal epiphyses of the tibia and fibula had not completed fusion even at 134 weeks of age (47).

The appearance and fusion of secondary ossification centers in the femur is summarized in Table 8.

3.5.3 Mandible

Prenatal — Ossification of the rat mandible begins in the fetus during prenatal week 15 or 16. The mandible of the rat becomes ossified from a single center of ossification that allows bone growth in various directions and results in formation of the mandible. This ossification centers appears lateral to Meckel's cartilage, in the same general area as the first molar tooth germ (49).

Postnatal — Bone deposition and resorption occurs rapidly in the rat immediately after birth and throughout the early postnatal period. These activities play an important role in the growth and development of the mandible. At one week after birth, bone is rapidly being deposited in various areas of the rat mandible (50). For example, resorption is occurring at the incisor socket making way for the incisor tooth that will erupt at 2 weeks.

At birth, the mandibular body of the rat consists of 2 separate halves joined by the symphysis. At one day after birth a wedged shaped cartilaginous bar is present in the symphyseal region of the rat mandible (51). At seven days after birth, the cartilage elongates posteriorly. At 14 days after birth, the symphyseal cartilage is still present; however, endochondral bone has begun to form in this region (50). By 15 days after birth, sealing of the cartilaginous border begins. By 21 days after birth, the cartilaginous border of the symphysis is completely sealed (51).

During the first postnatal week, bone begins to be remodeled in the developing mandibular ramus (50). Growth that occurs at the mandibular condyle contributes to the normal growth of the mandible, particularly to the growth of the ramus (52). Specifically, increases in mandibular length and height that occur during postnatal growth are primarily due to the directional growth of the mandibular condyle and the angular process (50). At birth, the condyle is positioned above and slightly lateral to the angular process (50). During the first 3 days after birth, both the condylar and angular processes grow primarily in a backward direction (49). This backward growth contributes to the lengthening of the mandibular ramus. However, growth of the condylar cartilage has a slight dorsal and lateral component and angular cartilage growth also has a slight ventral component. These components not only facilitate increases in length, but also cause the divergence of the condyle and angular process from one another in the vertical and horizontal planes. As a result of this divergence, from postnatal day 3 forward, the condylar cartilage changes its direction of growth, and begins to primarily grow upward, though also continuing backward and lateral growth (49).

Up to about 12 days after birth, the increased in the width of the ramus is due to progressive increases in the circumference of the condylar and angular cartilages. From postnatal day 12 to 15, the portion of the ramus that lies just behind the molar region increases rapidly in width due to the enlargement of the incisal socket and the posterior growth of the incisor. From 15 to 30 days after birth, growth continues ventromedially and backward (49).

At approximately 2 weeks after birth, the coronoid process of the rat mandible is very thin and bone begins to be deposited on its surfaces. As bone is deposited, the coronoid process grows

Table 9 Appearance and Fusion of Postnatal Secondary Ossification Centers in the Mouse (30, 46)

Bone	Appearance	Fusion
Humerus		
Proximal Epiphysis		
Greater Tubercle	5 days	6–7 weeks
Head	7–10 days	
Distal Epiphysis		
Capitellum	5 days	
Trochlear	5 days	3 weeks
Medial Condyle	9 days	
Lateral Condyle	19 days	
Femur		
Proximal Epiphysis		
Greater Trochanter	14 days	
Lesser Trochanter	14 days	13, 14, or 15 weeks
Head	15 days	
Distal Epiphysis		
Medial Condyle	7 days	12 or 13 weeks
Lateral Condyle	9 days	

backward, and increases both in size and height. By 10 weeks after birth, the coronoid process is composed of compact bone and exhibits little bone forming activity (50).

By three months after birth, bone is no longer being deposited on the surfaces of the mandible but internal bone remodeling continues. Specifically, bone growth at the condylar cartilage decreases substantially at 6 weeks after birth and by 3 months after birth the angular process is composed of dense inactive bone (50).

By 14 weeks after birth, growth of the rat mandible has essentially ceased (50).

3.6 Mouse

When compared to the rat or human, ossification in the mouse skeleton occurs over a relatively short period of time (14). A summary of the maturation of secondary ossification centers in the humerus and femur of mice is provided in Table 9.

3.6.1 Humerus

In the proximal epiphysis of the humerus, secondary ossification centers develop in the greater tubercle, on postnatal day 5 (46). Shortly thereafter, these secondary ossification centers unite and form a single broad plate of bone in the epiphysis (46). On postnatal day 7, an ossification center appears in the head of the humerus. By postnatal day 17, the centers in the head and in the greater tubercle unite and form a single secondary ossification center in the proximal epiphysis (46). Fusion of the proximal epiphysis to the diaphysis occurs at approximately 6 to 7 weeks after birth (30).

In the distal epiphysis, the medial and lateral condyles develop separate centers of ossification at 9 and 19 days after birth, respectively. In addition, a secondary center of ossification first appears in the trochlea and capitulum at 5 days after birth (46). Union of the distal epiphysis of the humerus with the diaphysis usually occurs in the mouse at 3 weeks after birth (30).

3.6.2 Femur

In the proximal region of the femur, by day 14 after birth ossification centers develop in the greater and lesser trochanters. The head of the femur exhibits a small ossification center on the 15 day after birth (46).

In the femur, at 7 days after birth, a secondary ossification center appears in the medial condyle at the distal portion of the femur. At 9 days after birth, the lateral condyle develops a secondary ossification center. By postnatal day 13, these centers form a single broad plate of bone in the distal epiphysis (46).

At the end of the first postnatal month, the proximal and distal epiphyses of the femur are almost completely ossified; however, they have not yet fused with the diaphysis (46). Fusion of the proximal epiphysis occurs approximately between 13 and 15 weeks after birth (30). Fusion of the distal epiphysis occurs prior to postnatal week 15, approximately between week 12 and 13 (30).

3.6.3 Mandible

Ossification — In the mouse, a mandibular ossification center develops by day 15 of gestation (46). By gestation day 17, the alveolar process of the mandible is relatively well developed and the mental foramen is fully developed (46). At birth, the coronoid process, the condyloid process, and the mandibular (gonial) angle are apparent. As late as postnatal week 5, cartilage can still be seen in the condylar and coronoid processes and fusion of the anterior end of the mandible has not yet occurred in the symphysis (46).

Postnatal growth — During the first 3 postnatal weeks, morphological and histochemical changes are noted in the condylar cartilage of the mouse (53). In addition, the height of the condylar cartilage increases dramatically during the first 2 postnatal weeks. As a result, the height of the mandibular ramus also increases during this period. The width of the condylar cartilage also increases, especially during the 2nd and 3rd weeks of life (53).

The cartilagenous growth center in the mouse condyle reaches skeletal maturity by postnatal week 8 and a well-developed region of primary spongiosa is visible below the growth center (53). However, by 8 weeks after birth, the primary spongiosa in the condyle no longer exists, instead lamellar bone is now in direct contact with the undersurface of the mature cartilage (53).

Livne et al., (54) noted that the period of most rapid and active growth of the mouse mandible occurs from 0 to 8 weeks after birth and results in appositional growth of cartilage followed by endochondral ossification (54). These results were consistent with those of Silberman and Livne (53) who reported that the cartilagenous growth center in the mouse condyle reaches skeletal maturity by postnatal week 8. As maturity progresses a well-developed area of primary spongiosa is visible below the condylar growth center (53). From this point forward, the mature condylar cartilage no longer serves as a growth center but begins to serve as an articulating surface for the squamoso-mandibular joint (54).

4 Epiphyseal Growth Plate

As briefly discussed in Section 3, soon after the appearance of the secondary ossification centers in the epiphyses of developing long bones, a region containing cartilage, bone, and fibrous tissue develops between the epiphysis and the diaphysis. This region is called the epiphyseal growth plate and is responsible for increases in diaphyseal length that occur during postnatal development (55, 6). As bone is produced at the epiphyseal growth plate a transitional region between the diaphysis and the epiphysis is created called the metaphysis (6, 2).

In mammals, during postnatal development, longitudinal growth of long bones occurs at the epiphyseal growth plate by the process of endochondral ossification (56, 57, 58, 59, 60, 61). In general, this process involves production of cartilagenous matrix by proliferating chondrocytes, mineralization of the cartilage matrix, removal of calcified and uncalcified matrix, and the replacement

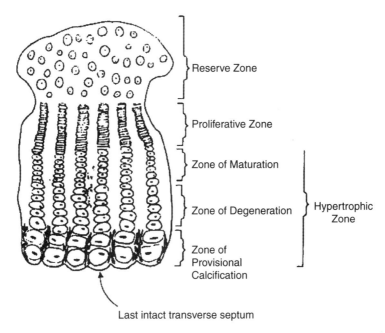

Figure 4a (70) Drawing depicting the various zones of the cartilaginous portions of the growth plate.

of matrix by bone (61). When the epiphyseal growth plate is active, the above-mentioned processes are continually occurring in the growth plate and results in deposition of bone in the metaphysis. When the rate of bone formation begins to exceed the rate of cartilage proliferation, the epiphyseal plate begins to narrow, and epiphyseal fusion occurs and results in the disappearance of the epiphyseal growth plate and its replacement with bone and marrow. In both humans and animals, epiphyseal fusion marks the end of longitudinal bone growth (62, 2).

The epiphyseal growth plates of human bone are structurally and functionally similar to the epiphyseal growth plates that develop in the bones of other mammals (63, 64, 60, 65, 57, 66, 67, 68, 69). Thus, the following description of the structure of the growth plate is applicable to mammals in general.

The growth plate is divided into several functional zones, the names of which have not been standardized throughout the scientific literature (Figure 4a-b). The zone closest to the epiphysis and farthest from the diaphysis is called the resting, germinal, or reserve zone (2, 70). Chondrocytes in this area are small, quiescent, and randomly distributed. It is thought that the cells in the resting zone store lipids, glycogen, and other material (70, 71). Adjacent to the germinal zone is the proliferative zone. This zone is dedicated to rapid, ordered cell division and matrix production. In the proliferative zone, chondrocytes increase in size (accumulating glycogen in their cytoplasm), exhibit mitotic division, synthesize and secrete matrix, and become arranged in columns. Directly beneath the proliferative zone is the zone of cartilage transformation, which is sometimes further divided into upper and lower hypertrophic or maturation and degeneration zones. In the zone of cartilage transformation, chondrocytes hypertrophy (swell), preparing for their replacement by bone, and continuing their synthesis of matrix (70, 71, 2). Many of the chondrocytes in the zone of cartilage transformation progressively degenerate and their intracellular connections are removed by the advancing metaphyseal sinusoidal loops (2). In the zone of ossification also called the zone of provisional calcification, hyaline matrix becomes calcified. Calcification of the matrix results in restricted diffusion of nutrients and the eventual death of hypertrophic chondrocytes. In this final region of the epiphyseal growth plate, osteoblasts form a layer of bone on the remaining mineralized cartilage (2, 70, 71). Additionally, invading metaphyseal blood vessels bring nutrients and osteogenic cells into the zone of ossification to aid in the formation of bone.

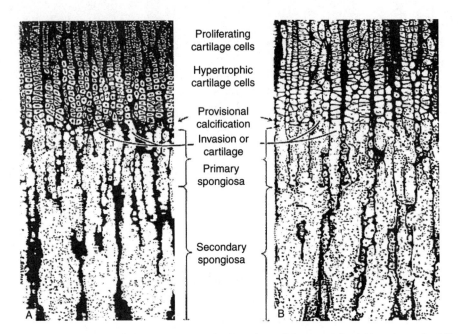

Figure 4b (3) Endochondral ossification in longitudinal sections of the zone of epiphyseal growth of the distal end of the radius in a puppy. (A) Neutral formalin fixation, no decalcification, von Kossa and hematoxylin-eosin stains. All deposits of bone salt are stained black; thus, bone and calcified cartilage matrix stain alike. (B) Zenker-formol fixation, specimen decalcified, and stained with hematoxylin-eosin-azure II.

Discussed below is the rate of longitudinal growth in both humans and animals during postnatal development as a result of cell proliferation at the epiphyseal growth plate.

4.1 Human

In humans, skeletal growth is initiated by the formation of a cartilage template that is subsequently replaced by bone (72). Linear growth, particularly of the long bones, commences with the proliferation of epiphyseal cartilage cells and ceases with epiphyseal closure at puberty. In general an increase in rate of growth of the long bones occurs between the ages of 9 and 14 (43).

4.2 Dog

In dogs, during postnatal development, increases in diaphyseal length occur by proliferation and maturation of chondrocytes at the epiphyseal growth plate followed by calcification of matrix and endochondral bone formation (69, 60).

In general, rapid growth of the limb bones has been shown to cease by approximately 5 months after birth (37).

4.3 Rabbit

The structure and function of the epiphyseal growth plate in the rabbit is similar to that of humans (73, 74, 75). Thus, in the rabbit, increases in the length of long bones also occur as a result of cellular activity at the epiphyseal growth plate.

Fukuda and Matsuoka (41) report that in the rabbit, increases in limb length occur rapidly from 0 to 8 weeks after birth. Once animals reach 8 weeks of age, longitudinal growth of the limbs slows until 32 weeks after birth at which time growth ceases completely (41). Khermosh et al., (43) report that a steady increase in growth rate of the femur was seen from a few days after birth until 3 months. After 3 months, there is a gradual decrease in growth rate up to 6 months when the growth ceases. In later studies, Rudicel et al., (76) report that the length of the rabbit femur rapidly increases between 2 and 4 weeks of age. Growth rate begins to plateau at 8 weeks of age and by 10 to 14 weeks of age growth essentially ceases.

As is true in all mammals, the growth rate of long bones in the rabbit decreases with increasing age (77). When compared to rats and humans, rabbits have the highest longitudinal growth rate/day (77). The rate of longitudinal growth in the rabbit ages 20-60 days is greater than that seen in the rat ages 20-60 days, and than that seen in humans ages 2-8 years (77). At 20 days old, the growth rate in rabbits and rats were 554 and 375 μm/day, respectively, when measured in the proximal tibias (77). The growth rate in humans (femur) at 2 years of age was 55 μm/day. By 60 days old the growth rate in the rabbit and rat had decreased to 378 and 159 μm/day, respectively. The growth rate in humans at age 5-8 years decreased to 38 μm/day (77).

4.4 Rat

The structure of the epiphyseal growth plate in the rat is similar to that of the human. In the rat, at about the time of weaning, the epiphyseal growth plate has been formed (67, 68).

A rapid increase in length of each long bone of the fore and hind limb has been seen up until the rats reach the age of 8 weeks (47). The growth of the long bones in male and female rats slows almost to the point of stopping between the age of 15 to 17 weeks (47).

5 Metaphysis

The metaphysis is a transitional region of bone that develops between the epiphyseal growth plate and the diaphysis during postnatal growth of the long bones. The structure and function of the metaphysis is similar among mammals (60, 78, 79, 64, 80). The metaphyseal region is one of the active sites of bone turnover in the long bone of growing mammals (79, 55).

In long bones, the metaphyseal region consists of an area of spongy bone directly beneath the epiphyseal growth plate. Together, the epiphyseal growth plate and the metaphysis form the growth zone (3). During postnatal bone growth and development, the composition of the metaphysis changes significantly. Early in development, calcified cartilage is the predominant hard tissue in the metaphysis; however, as development progresses the metaphyseal tissue is converted to bone with minimal amounts of cartilage (79).

During postnatal development, the cancellous bone of the metaphysis is divided into two different regions, the primary spongiosa and the secondary spongiosa (81, 79, 6). As a long bone grows in length, the primary spongiosa fills the areas that were previously occupied by the growth plate (6). The primary spongiosa contains many osteoblasts and osteoprogenitor cells and is characterized by thin trabeculae made up of bone covered calcified cartilage. Capillary loops at the front of the primary spongiosa, play a role in introducing osteoprogenitor cells that later form osteoblasts and osteoclasts (60).

The tissue outside the primary spongiosa is called the secondary spongiosa. The secondary spongiosa is mainly composed of bone with small amounts of calcified cartilage (79). As bone formation proceeds at the growth plate, most of the primary spongiosa is converted into secondary spongiosa (6).

Debates have been raised regarding whether rat long bone metaphysis is an appropriate model to simulate human cancellous bone remodeling. In humans the two processes, mini-modeling and remodeling, are responsible for developing and maintaining cancellous bone mass and architecture. Mini modeling changes trabeculae by removing old bone from one surface and adding new bone on the opposite surface. Remodeling removes pieces of old lamellar bone and refills the space (lacuna) with new lamellar bone (78). Mini-modeling is the mostly occurs before skeletal maturity while remodeling occurs mostly after skeletal maturity has occurred.

Similarly, in the metaphysis of rats, the remodeling of primary bone results in secondary lamellar bone (78). Specifically, in the rat proximal tibial metaphysis cancellous bone remodeling is similar to that described in humans once a rat reaches the age when longitudinal bone growth plateaus (78).

6 Diaphysis

During postnatal growth, the diaphysis of long bones increase in length by the process of endochondral bone formation whereas the diaphyseal cortex increases in diameter by intramembranous bone formation (55). In mammals, longitudinal growth must occur prior to fusion of the epiphysis with the diaphysis; however, bones can increase or decrease in width at any time during life by appositional growth.

6.1 Growth in Diameter

During postnatal growth, in both humans and animals, as bones grow in length, they also grow circumferentially to accommodate the increase in load (82). Long bones increase in diameter by appositional bone growth i.e. deposition of new bone on the periosteal surface (i.e. beneath the periosteum) (55, 3). During appositional growth, osteoblasts under the periosteum secrete matrix on the outside surfaces of bone. As bone is deposited on the periosteal surface, osteoclastic resorption on the endosteal surface regulates the thickness of the bone forming the cortex of the diaphyseal shaft. The amount of bone in the cortex remains fairly constant throughout growth due this coupled deposition and resorption. Thus, the overall diameter of the diaphysis increases rapidly whereas the thickness of the cortex increases slowly (3).

During postnatal growth, the shape of a growing bone is maintained by continuous remodeling of the bone surface. This involves bone deposition on some periosteal surfaces and absorption on other sometimes-adjacent periosteal surfaces (3).

6.1.1 Human

At birth, the diameter of the diaphysis in long bones is rapidly increasing in diameter. Kerly (83) reports that in humans at birth resorption by osteoclasts is highest in the mid region of the diaphyseal cortex. By four years of age, osteoclastic resorption is occurring throughout the cortex in preparation for the formation of secondary osteons. Between the ages of 10 and 17 years, as the long bones undergo rapid growth, osteoclastic resorption and osteoblastic apposition occur extensively at the periosteal surface of bone to preserve bone shape (83, 3). Once the bone achieves adult size, resorption is greatly decreased and is balanced by osteoblastic deposition of bone in the form of osteons.

Parfitt et al., (84) report that when appositional bone growth was studied in the iliac bone of humans ages 1.5 to 23 years, significant increases in the diameter of the diaphysis was seen to occur with age. The results indicated that between 2 and 20 years of age, the ilium increased in width by 3.8 mm by periosteal apposition, and 1.0 mm by endosteal bone formation. During this time period, resorption at the endosteal and periosteal surfaces accounted for removal of 3.2 and 0.4 mm of bone, respectively. As a result of coupled apposition and resorption at the periosteal and endosteal surfaces, the width of the diaphyseal cortex increased from 0.52 mm at 2 years of age to 1.14 mm by the age of 20 years (84).

6.1.2 Dog

In the beagle dog at approximately 100 days of age, the medullary cavity of the diaphysis reaches its adult diameter; however, the cortex or outside diameter of the shaft does not stop increasing until the animal is approximately 180 days of age (25).

6.1.3 Rat

In the femur of Norwegian male rats, growth of the long bones occurred by both diametric and longitudinal growth coupled with remodeling (85). During the early stages of growth (at approximately 30 days old), widening of the marrow cavity occurred by resorption of bone at the endosteal surface of the diaphysis, and an increase in cortical thickness occurred by apposition of new bone at the periosteal surface. Endosteal resorption was highest in the posterior wall of the femur where the cortex is visibly curved.

6.2 Osteon Formation and Remodeling

6.2.1 Human

Bone that is added to the cortex of the diaphysis during appositional bone growth is immature or woven bone (3). Much of the diaphyseal bone of newborns is made up of this nonlamellar woven bone (86). During development, as the diaphysis expands diametrically, blood vessels become trapped in the cavities of the bone matrix. In these cavities, woven bone fills the spaces around the blood vessel forming primary osteons or Haversian systems (55).

In humans, remodeling of osteons begins shortly after birth and continues throughout life (55, 3). Immediately after birth, the substantia compacta of growing bones consist only of primary (immature woven) bone. As growth progresses, the primary bone is completely or partly replaced by secondary, Haversian bone by the process of remodeling (87).

During bone growth and remodeling, primary osteons are resorbed and replaced by secondary osteons in which lamellar bone is formed around central vascular canals (86, 3). The formation of secondary osteons begins when osteoclasts form a cutting cone that advances through the bone creating a resorption cavity (Figure 5). Osteoblasts lining the resorption cavity lay down osteoid filling the resorption cavity with concentric lamellae from the outside of the cavity inwards (6). Osteoblastic apposition of concentric lamellae continues in the resorption cavity until a normal Haversian canal diameter has been achieved. The end result of this process is completion or closure of a new secondary Haversian system or osteon (6).

Fawcett (3) reports that in humans aged 1 year and older, only organized lamellar bone is deposited in the long bone shaft and that all primary bone is eventually replaced by secondary Haversian bone. In, addition Fawcett (3) notes that internal bone resorption and reconstruction does not cease once primary bone is replaced by secondary bone. Resorption and remodeling take place throughout life. Thus, in adult bone the following can be seen; mature osteons, forming osteons, and new absorption cavities. In addition, portions of former osteons that are not destroyed become interstitial lamellae and are situated between mature osteons (55).

In humans, approximately 1 µm of bone is deposited in developing Haversian systems per day (3). As osteons near completion, the rate of appositional bone formation slows (3). Osteon completion or closure occurs in 4 to 5 weeks.

6.2.2 Animals

The fact that rodent models of human skeletal disorders lack Haversian system like those of larger mammals, and that systemic internal remodeling of cortical and cancellous bone is dissimilar to one in humans was the reason for use of so called "second species" (88). Skeletal maturity (sealed growth plates) as well as presence of Haversian system and intra-cortical type of bone remodeling is required in order to more completely assess the effect of novel therapies or procedures aimed to ameliorate bone disorders of various etiologies. It is worth mentioning here that quadrupeds, non-primates have different bone biomechanical characteristics than bipeds, the fact to keep in mind when animal bone studies are extrapolated to humans (89, 90).

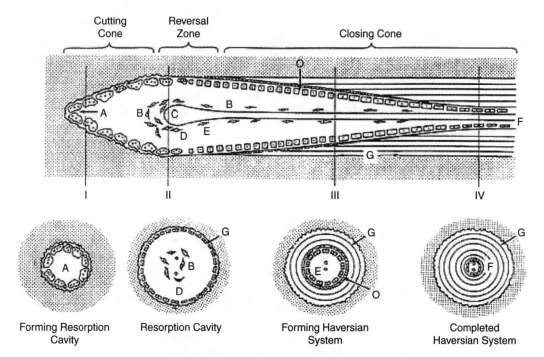

Figure 5 (6) Diagram showing a longitudinal section through a cortical remodeling unit with corresponding transverse sections below. **A.** Multinucleated osteoclasts in Howship's lacunae advancing longitudinally from right to left and radially to enlarge a resorption cavity. **B.** Perivascular spindle-shaped precursor cells. **C.** Capillary loop. **D.** Mononuclear cells lining reversal zone. **E.** Osteoblasts apposing bone centripetally in radial closure and its perivascular precursor cells. **F.** Flattened cells lining Haversian canal of completed Haversian system. Transverse sections at different stages of development: **(I)** resorption cavities lined with osteoclasts; **(II)** completed resorption cavities lined by mononuclear cells, the reversal zone; **(III)** forming Haversian system or osteons lined with osteoblasts that had recently apposed three lamellae; and **(IV)** completed Haversian system with flattened bone cells lining canal. Cement line **(G)**; osteoid **(stippled)** between osteoblast **(O)** and mineralized bone.

As reported by Hert et al., (91), the diaphysis of long bones in adult humans and large adult mammals (longer living) such as dogs and sheep is composed of Haversian, secondary bone that replaces the original primary bone. However, the compact bone of smaller adult mammals such as mice and rats is not made up of Haversian systems (87). In fact, Bellino (92) notes that the rabbit is the smallest species known to undergo Haversian bone remodeling.

Jowsey (93) reports that secondary osteons are not present in all species with the same degree of frequency that is seen in mature human compact bone. He states that in some animals, primary bone is replaced at an early age, while in others, primary bone persists for quite some time. For example, it was noted that in the 2-year-old human, cortical bone of the femur consists mostly of secondary osteons whereas the cortical bone in the femur of a 7-year-old rabbit only has "occasional" osteons (93).

When the tibia and fibula of rats, hamsters, guinea pigs, and rabbits were investigated at ages ranging from 6 months to 5 years, it was noted that in varying degrees, primary bone is present in compact bone throughout the life of each of these animals (87). Specifically, in guinea pigs and rats, secondary bone in the form of osteons was found in the compact bone of only a few "very old" animals. In the adult rabbit, however, a greater degree of "Haversian transformation" i.e. remodeling of primary osteons to secondary osteons was seen to occur throughout the life of the animal (87).

Enlow (94) also noted that there was a characteristic absence of Haversian systems in the compact bone of many vertebrate species, specifically naming the white rat. However, Enlow (94) reported that the appearance of primary osteons is quite common in the young growing dog and monkey. Albu et al., (95) noted that Haversian systems do exist in the compact diaphyseal bone of larger mammals such as the dog, pig, cow, and horse. In fact, Georgia et al., (96) reports that in dogs, the density of Haversian canals in the compact bone of the femoral diaphysis is higher than that of humans whereas the diameter of the Haversian canals are smaller in dogs than they are in man.

The rate of osteoblastic apposition of bone in the resorption cavity of a forming secondary osteon has been extensively studied in dogs (97, 98, 99). Lee (97) noted that age plays an important role in the rate of formation of osteons. In dogs, as age increased, the rate of appositional bone growth in osteons decreased (97). The mean growth rate of appositional growth in forming osteons in a 3 month old, 1 to 2 year old, and an adult dog of unknown age was 2.0 μ/day, 1.5 μ/day, and 1.0 μ/day, respectively (97). In addition, osteons in the dog have been shown to grow to maturity within 4 to 8 weeks (99).

In dogs, the appositional rate of bone formation was highest during the early stages of osteon formation when compared to the later stages. It was postulated that the rates of appositional bone growth were most rapid initially to smoothe over absorption lacunae and to convert the resorption cavity into a concentric canal. As the Haversian canal cavity grew smaller due to addition of concentric lamellae by osteoblasts, the appositional bone formation decreased (97, 100). Thus, the rate of appositional bone formation decreased as the cavity resorption cavity was closing and neared completion (100).

7 Vascularity

Successful growth and development of human bone is vitally dependent upon a vascular blood supply (2). The vascular supply to a typical long bone is strikingly similar among mammals (55).

In general, long bones, as well as many flat and irregular bones are vascularized by nutrient arteries and veins that pass through the compact bone acting as channels for the entry and exit of blood (101). These vessels gain entry to the compact bone via openings called nutrient foramens and canals. When a nutrient artery reaches the marrow cavity, it divides and branches proximally and distally (101).

During postnatal development, the diaphysis, metaphysis, and epiphysis of long and short bones are separated by a cartilagenous growth plate. While the growth plate persists, the terminal branches of nutrient arteries do not cross through the growth plate into the epiphysis (101, 102). Instead, the capillary ends of these branches terminate just below the growth plate. In the adult, the growth plate has fused which allows the terminal branches of nutrient arteries to extend unhindered into the epiphyseal region where they connect with arteries arising from the periosteum (101). It is important to note that the functional importance of the vessels that cross into the epiphysis during adulthood is uncertain as they cannot supply sufficient blood to the entire epiphysis in the event of epiphyseal artery damage (4). It is also important to note that while the growth plate persists, blood is supplied to the region by 3 different sources.

In addition to the nutrient artery, the blood supply of many flat bones is also derived from periosteal vessels. With age, long bones can sometimes also become increasingly dependent on periosteal vessels (103).

The blood supply of the metaphysis, diaphysis, epiphysis, and epiphyseal growth plate (Section 7.2) is discussed below.

7.1 Vessels of Long Bones

Specific groups of arterial networks are responsible for supplying blood to the various portions of long bones (Figure 6a). These networks include the diaphyseal arteries, the metaphyseal arteries, and the epiphyseal arteries (2). In addition, vessels arising from the periosteum, can also provide a source of vasculature to bones (55).

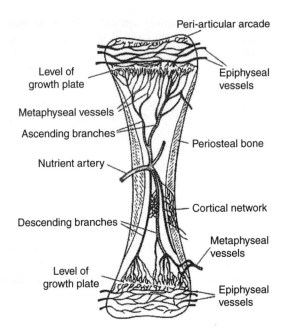

Figure 6a (2) A summary of the main arterial supply to a developing long bone.

Postnatally, nutrient arteries supply the diaphysis and central portion of the metaphysis with blood while the epiphyseal arteries supply only the epiphyses. The metaphyseal arteries supply the peripheral regions of the metaphysis (103).

In addition, during early postnatal development in the epiphyseal region, cartilage canals persisting from the fetal period are a source of nutrition for the epiphyses. Cartilage canals may also play a role in the appearance of the secondary center of ossification (55, 103).

7.1.1 Diaphyseal Arteries

Diaphyseal arteries gain entry to the diaphysis of long bones through nutrient foramens that lead into nutrient canals (2). Once inside of the medullary cavity, the diaphyseal arteries branch proximally and distally (2, 104). The ascending and descending branches of the vessel travel throughout the marrow cavity and end in helical loops near the metaphysis. In the metaphyseal region these vessels connect with both the metaphyseal and periosteal arteries (104).

The main nutrient artery is surrounded by several venules that merge into 1 or 2 large nutrient veins. In fact, in bone, many openings or foramina give exit to veins alone so that the number of nutrient veins draining a bone usually exceeds the number of the nutrient arteries supplying it (103).

Long bones are sometimes supplied by more than one nutrient diaphyseal artery (103). For example, the rabbit tibia has 2 diaphyseal arteries and the artery of the trochanteric fossa in the rat femur can develop into a second afferent vessel to the diaphysis. On occasion, the diaphysis of the human humerus may have 3 nutrient arteries and the diaphysis of the human femur may have up to 2 nutrient arteries (103).

In the scientific literature conflicting views exist regarding the direction and source of arterial blood flow in the cortex of the diaphysis. Historically, scientists believed that the compact bone of the diaphyseal cortex was fed by periosteal vessels, and the inner portion of the shaft received its blood supply from the nutrient artery (102, 2). However, modern research has demonstrated that in young bones (< 35 years old), the diaphyseal cortex primarily receives its blood supply from the nutrient artery of the diaphysis, but as bones age, periosteal arteries may begin to supply blood to the cortex (2).

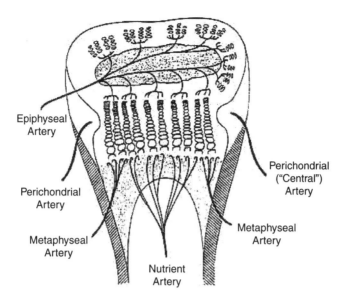

Figure 6b (107) Drawing showing the blood supply of a typical growth plate

7.1.2 Metaphyseal Arteries

The blood supply of the metaphysis comes from both the diaphyseal nutrient artery and the metaphyseal arteries (Figure 6a-b).

In early fetal development, the blood supply to the metaphysis comes only from branches of the diaphyseal nutrient artery (103, 70). However, as development continues, the metaphyseal arteries become an additional source of nutrition (103). As a result, during postnatal bone growth, the active metaphysis receives its blood supply from both the branches of the diaphyseal nutrient artery and from the metaphyseal arteries that pierces the substantia compacta in the vicinity of the metaphysis (103, 105). When these 2 nutrient sources are present, the central region of the metaphysis receives blood from the diaphyseal arteries, while the peripheral portions of the metaphysis are supplied by metaphyseal arteries (102, 103). In the adult, the metaphyseal arteries branch into the entire (non-growing) metaphysis and supply blood to the whole region (103).

The metaphyseal arteries enter bone from the periosteum next to the metaphysis and join with branches of the nutrient artery (106). In the absence of a diaphyseal artery, the metaphyseal arteries can supply blood to the entire diaphyseal shaft (103).

7.1.3 Epiphyseal Arteries

Early in development (fetal and early postnatal periods), cartilage canals provide nutrition to the cartilage of the epiphysis; however, during postnatal bone growth, it is the epiphyseal arteries take the lead role in providing a vascular supply to the growing epiphysis.

In humans and large vertebrates such as dogs and rabbits, cartilage canals are present in the epiphyses of developing bone months before the formation of secondary ossification centers. These cartilage canals contain arterioles, venules, and intervening capillaries in a connective tissue matrix (108, 103, 55). The purpose of these cartilage canals is to provide nutrition to the cartilage. Later in development cartilage canals may play a role in the formation of secondary ossification centers (108, 55). In the late embryonic and early postnatal periods, cartilage canals can been seen crossing partly or completely through the region where the growth plate later develops (108). Postnatally, the vessels that cross through the region where the epiphyseal growth plate develops are destroyed leaving this region avascular. These changes occur in the rabbit a few weeks after birth, and in humans several months after birth (108).

The role of cartilage canals in the formation of secondary ossification centers has not been clearly established. However, it is believed that during postnatal development, osteoprogenitor cells from the connective tissue covering the canals differentiate into the osteoblasts that begin ossification (55).

Epiphyseal arteries enter long bones through nutrient foramens in the non-articular region of the epiphysis between the articular cartilage and the cartilaginous growth plate. During postnatal bone growth, epiphyseal arteries are responsible for supplying blood to the epiphysis and to the zone of resting cells at the top of the growth plate (see Section 7.2, Vascularity of the Epiphyseal Growth Plate) (103, 104).

While the growth plate persists, diaphyseal nutrient arteries do not cross the epiphyseal growth plate and thus the epiphysis relies solely on epiphyseal arteries for its blood supply (104, 2).

7.1.4 Periosteal Vessels

The periosteum, which covers the external surface of bone is an extremely vascular connective tissue. This tissue provides an additional source of vasculature to bones. Specifically, blood vessels from the periosteum form vascular networks that penetrate the outer surface of the bone and provide nutrition to the outer portions of the compact bone. If the blood supply to the medullary cavity is destroyed, the periosteum can provide blood to the entire diaphyseal cortex through Volkmanns and Haversian canals (55).

The periosteum is also a major source of vasculature in flat bones (103).

7.2 Vascularity of the Epiphyseal Growth Plate

The vascular supply of the growth plate has been shown to come from 3 sources: epiphyseal arteries, metaphyseal arteries, and perichondrial arteries (Figure 6b) (105, 107, 70). Perichondrial arteries are derived from the fibrous connective tissue called the perichondrium the covers the peripheral regions of the cartilaginous growth plate (107, 70).

From the epiphyseal side, terminal branches of the epiphyseal arteries provide blood to the uppermost regions of the cartilaginous growth plate. Once the inside the epiphysis, smaller arterioles branch from the main epiphyseal artery and pass through small cartilage canals in the reserve or resting zone to end at the top of the columns of cells in the proliferative zone (70). As a result, the proliferative zone is well supplied with blood from the epiphyseal arteries. It is important to note that these branching epiphyseal arteries do not penetrate into the hypertrophic zone.

From the metaphyseal side, the surface of the growth plate is supplied by metaphyseal arteries (105, 107, 70). Terminal branches from the nutrient and metaphyseal arteries enter the zone of ossification and end in vascular loops at the base of the cartilage portion of growth plate (55). It is here that the vessels loop back so that the venous branches can descend to drain (70).

Perichondrial arteries supply the peripheral regions of the growth plate (65, 107, 70, 103).

7.3 Vascularity of the Flat Bones of the Skull

Like the long bones, growth of flat bones in the skull (calvaria) is dependent upon a vascular supply. The vascularity of the calvaria is derived from pericranial vessels, dural vessels, sutural vessels, meningeal vessels, and calvaria veins (103).

7.3.1 Pericranial Vessels

The outer surface of the brain is covered with a layer of connective tissue that is called the pericranium (3). The pericranium has a vascular network that is similar to that of long bone periosteum. The vessels of the pericranium are connected to the internal vessels of the cranial bones by arteries and veins that pass through the juxtasutural foramina (103).

In addition, capillaries of the outer table connect the diploic veins with those of the pericranium (103).

7.3.2 Dural Vessels

The dura mater is the outer and most fibrous of the 3 membranes that surround the brain. The dura mater is made up of two fused layers; an endosteal outer layer called the endocranium and the inner meningeal layer. Located between these 2 layers are the venous sinuses (103).

Two distinct vascular networks are found within the dura mater. The inner vascular plexus, and the outer vascular plexus. The inner vascular plexus is made up of arteries, veins, and a capillary network called the primary dural plexus. The arteries are derived from the meningeal vessels, and the veins communicate with those of the outer vascular plexus (103).

Nutrient arteries derived from dural vessels penetrate the inner tables of flat bones. Branches of these arteries extend peripherally from the ossification centers the develop during the fetal period to the sutural boundaries of the frontal, parietal, and occipital bones (103).

7.3.3 Sutural Vessels

In the skull, many bones such as the frontal, parietals, squamous occipital, squamous temporal, and greater wing of sphenoid are connected together by junctions called sutures (2). During postnatal development, sutures are composed of dense connective tissue which later becomes ossified (109).

During postnatal skull growth, a plethora of small foramina that contain arteries and veins can be seen near the sutural borders (103). These "juxta-sutural vessels" play an important role in aiding the growth in surface area of individual flat bones (103). In addition, the fibrous tissue of the suture contains sutural veins. Trans-sutural connections of veins can be found which unite the veins of one bone with those of the neighboring bones. The sutures also connect the blood vessels of the pericranium (the connective tissue membrane that surrounds the skull) with the vessels of the dura (the outermost and most fibrous of 3 membranes surrounding the brain). The edges of bones connected by sutures contain many holes for the passage of blood vessels (103).

After ossification and closure of sutures, the skull maintains vascularity as the surface of bones are covered with small foramina filled with capillaries that serve to join the capillaries of the diploe with those of the pericranium (103). Blood vessels can also be seen passing through the inner table joining with those of the diploe and the dura mater (103).

7.3.4 Meningeal Arteries

The cranial meningeal arteries supply the dura mater and the 2 other membranes covering the brain (leptomeninges) with blood; however, the main function of the meningeal arteries is to supply blood to the cranial bones. (103).

7.3.5 Calvarial Veins

In the developing calvaria small arteries and large venules join together without capillary mediation (arteriovenous anastomoses). In addition, in the adult skull, venous sinuses of the diploe are connected to vessels of the pericranium and dura mater by capillaries from the inner and outer tables. Large diploic sinuses (the veins of Breschet) are proximal and parallel to the sutures. The diploic sinuses exit the outer table of the skull through the frontal, anterior temporal, posterior temporal, and occipital diploic foramina (103).

Conclusions

Growth and weight gain within normal ranges is viewed as part of the assessment of the overall health of young children and animals. Bone growth and development occurs in a similar fashion in all mammalian species studied during the course of this literature review. Bone growth has been assessed in animal studies by direct measure of selected bones at necropsy (110) and by noninvasive techniques such as dual energy X-ray absorptiometry (DEXA) (111, 112) and autoradiography. Using direct measurements of body and femur and cranial length, Schunior et al. (110) were able to study the relationship between postnatal growth and body weight. Bone growth kinetics over a matter of hours has been studied using tetracycline as a marker followed by histopathology (113). Bone quality is also assessed bone mineral density (BMD) and bone mineral content (BMC) analyses.

This review summarizes available literature regarding important processes in postnatal bone growth. The timing of important developmental milestones was compared between humans and laboratory animal species. These milestones included ossification, epiphyseal growth plate, metaphysis, as well as diaphysis development, osteon formation, and vascularization.

In conclusion, postnatal skeletal growth can be assessed in appropriately designed animal studies since the growth and developmental patterns are similar for both laboratory animals and humans.

REFERENCES

(1) Moore KL and Persaud TVN. *The Developing Human: Clinically Oriented Embryology* (Philadelphia: W.B. Saunders Company, 1998), 405-424
(2) Scheuer, Louise and Black, Sue, *Developmental Juvenile Osteology* (San Diego: Academic Press, 2000).
(3) Fawcett DW. *Bloom and Fawcett A Textbook of Histology*, Twelfth Edition. (New York: Chapman & Hall, 1994), 194-259
(4) Buckwalter JA, Glimcher MJ, Cooper RR and Recker R. "Bone Biology I: Structure, Blood Supply, Cells, Matrix, and Mineralization." In: *Instructional Course Lectures*. Ed. Pritchard, DJ, vol. 45. (Rosemont, IL: American Academy of Orthopaedic Surgeons, 1996), 371-86
(5) Price JS, Oyajobi BO and Russell RGG. The Cell Biology of Bone Growth. *Eur. J. Clin. Nutr.* 48(Suppl. 1):S131-149 (1994)
(6) Jee WSS. "The Skeletal Tissues." In: *Cell and Tissue Biology: A Textbook of Histology*, Sixth Edition. Ed. Leon Weiss, M.D., with 35 contributors. Baltimore: Urban & Schwarzenberg, Inc., 1988.
(7) Gray, H. Anatomy of the Human Body. Philadelphia: Lea & Febiger, 1918; Bartleby.com, 2000. www.bartleby.com/107/. [10-12-01]
(8) Junqueira LC, Carneiro J and Kelley RO. "Bone." In: *a LANGE Medical Book Basic Histology*, Ninth Edition. (Stamford, CT: Appleton & Lange, 1998), 134-151 (1998)
(9) Zerwekh JE. Bone Metabolism. *Seminars in Nephrology.* 12(2):79-90 (1992)
(10) Freemont AJ. Basic Bone Cell Biology. Review. *Int. J. Exp. Pathol.* 74(4):411-6 (1993)
(11) Marks SC and Hermey DC. The Structure and Development of Bone. In *Principles of Bone Biology*. J.P. Bilezekian, L.G. Raiz, G.A. Rodan (Eds.). Academic Press,1996, 3-14.
(12) Boskey AL and Posner AS. Bone Structure, Composition, and Mineralization. *Orthop. Clin. North Am.* 15(4):597-612 (1984)
(13) Marks SC and Popoff SN. Bone Cell Biology: The Regulation of Development Structure and Function in the Skeleton. *Am. J. Anat.* 183:1-44 (1988)
(14) Patton JT and Kaufman MH. The Timing of Ossification of the Limb Bones, and Growth Rates of Various Long Bones of the Fore and Hind Limbs of the Prenatal and Early Postnatal Laboratory Mouse. *J. Anat.* 186(Pt. 1):175-85 (1995)
(15) Davies DA and Parsons FG. The Age Order of the Appearance and Union of the Normal Epiphyses as Seen by X-rays. *J. of Anat.* 62:58-71 (1927)

(16) Menees TO and Holly LE. The Ossification in the Extremities of the New-Born. *Am. J. Roentgenology.* 28(3):389-90 (1932)
(17) Kuhns LR, Sherman MP, Poznanski AK and Holt JF. Humeral-Head and Coracoid Ossification in the Newborn. *Radiology.* 107:145-149 (1973)
(18) Kuhns LR and Finnstrom O. New Standards of Ossification of the Newborn. *Radiology.* 119:655-660 (1976)
(19) Ogden JA, Conlogue GJ and Jensen P. Radiology of Postnatal Development: The Proximal Humerus. *Skeletal Radiol.* 2:153-160 (1978)
(20) Paterson RS. A Radiological Investigation of the Epiphyses of the Long Bones. *J. Anat.* 64:28-46 (1929)
(21) Pritchett JW. Growth-plate Activity in the Upper Extremity. *Clin. Orthop. and Related Res.* 268:235-242 (1991)
(22) Hansman CF. Appearance and Fusion of Ossification Centres in the Human Skeleton. *Am. J. Roentgenology.* 88(3):476-482 (1962)
(23) Christie A. Prevalence and Distribution of Ossification Centers in the Newborn Infant. *Am. J. Dis. Child.* 77:355-361 (1949)
(24) Gill GG and Abbott LC. Practical Method of Predicting the Growth of the Femur and Tibia in the Child. *Arch. Surg.* 45:286-315 (1942)
(25) Anderson AC and Floyd M. Growth and Development of the Femur in the Beagle. *Am. J. Vet. Res.* 24:348-351 (1963)
(26) Sarnat BG. Growth Pattern of the Mandible: Some Reflections. *Am. J. Orthod. Dentofac. Orthop.* 90(3):221-33 (1986)
(27) Brodie AG. On the Growth Pattern of the Human Head from the Third Month to the Eighth Year of Life. *Am. J. Anat.* 68:209-262 (1941)
(28) Michejda M. Growth Standards in the Skeletal Age of Rhesus Monkey (*M. mulatta*) Chimpanzee (Pan tryglodytes) and Man. *Dev. Biol. Stand.* 45(2):45-50 (1980)
(29) van Wagenen G and Asling CW. Roentgenographic Estimation of Bone Age in the Rhesus Monkey (Macaca Mulatta). *Am. J. Anat.* 103:163-186 (1958)
(30) Fukuda S and Matsuoka O. Comparative Studies on Maturation Process of Secondary Ossification Centers of Long Bones in the Mouse, Rat, Dog, and Monkey. *Exp. Anim.* 29(3):317-26 (1980)
(31) Fukuda S, Cho F and Honjo S. Bone Growth and Development of Secondary Ossification Centers of Extremities in the Cynomolgus Monkey (*Macaca fascicularis*). *Exp. Anim.* 27(4):387-97 (1978)
(32) McNamara JA Jr and Graber LW. Mandibular Growth in the Rhesus Monkey (*Macaca mulatta*). *Am. J. Phys. Anthrop.* 42:15-24 (1975)
(33) Baume LJ. The Postnatal Growth of the Mandible in Macaca Mulatta. A Metric, Roentgenographic and Histologic Study. *Am. J. Orthod.* 39:228-229 (1953)
(34) Turpin DL. Growth and Remodeling of the Mandible in the Macava Mulatta Monkey. *Am. J. Orthod.* 54:251-271 (1968)
(35) Hare WCD. Radiographic Anatomy of the Canine Pectoral Limb. Part II. Developing Limb. *J.A.V.M.A.* 305-310 (1959)
(36) Parcher JW and Williams JR. "Ossification." In: *The Beagle as an Experimental Dog.* Ed., Andersen AC, Technical Ed., Good LS (Ames: The Iowa State University Press Inc., 1970), 158-161
(37) Yonamine H, Ogi N, Ishikawa T and Ichiki H. Radiographic Studies on Skeletal Growth of the Pectoral Limb of the Beagle. *Jpn. J. Vet. Sci.* 42(4):417-25 (1980)
(38) Hare WCD. The Ages at Which the Centers of Ossification Appear Roentgenographically in the Limb Bones of the Dog. *Am J of Vet. Res.* 22(90):825-835 (1961)
(39) Hennet PR and Harvey CE. Craniofacial Development and Growth in the Dog. Reviewed Article – Anatomy. *J Vet Dent.* 9(2):11-18 (1992)
(40) Sullivan PG. On Growth of the Canine Mandible. *Brit. J. Orthod.* 2(3):159-163 (1975)
(41) Fukuda S and Matsuoka O. Radiographic Studies on Maturation Process of Secondary Ossification Centers in Long Bones of the Japanese White Rabbit. *Exp. Anim.* 30(4):497-501 (1981)
(42) Heikel HVA. On Ossification and Growth of Certain Bones of the Rabbit; with a Comparison of the Skeletal Age in the Rabbit and in Man. *Acta Orthop. Scand.* 29(39):171-184 (1959-1960)
(43) Khermosh O, Tadmor A, Weissman SL, Michels CH and Chen R. Growth of the Femur in the Rabbit. *Am. J. Vet. Res.* 33(5):1079-82 (1972)

(44) Bang S and Enlow DH. Postnatal Growth of the Rabbit Mandible. *Arch. Oral. Biol.* 12(8):993-8 (1967)
(45) Masoud I, Shapiro F and Moses A. Longitudinal Roentgencephalometric Study of the Growth of the New Zealand White Rabbit: Cumulative and Biweekly Incremental Growth Rates for Skull and Mandible. *J. Craniofac. Genet. & Dev. Biol.* 6:259-87 (1986)
(46) Johnson ML. The Time and Order of Appearance of Ossification Centers in Albino Mouse. *Am. J. Anat,* 52(2):241-271 (1933)
(47) Fukuda S and Matsuoka O. Maturation Process of Secondary Ossification Centers in the Rat and Assessment of Bone Age. *Exp. Anim.* 28(1):1-9 (1979)
(48) Dawson AB. The Age Order of Epiphyseal Union in the Long Bones of the Albino Rat. *Anat. Rec.* 31(1):1-17 (1925)
(49) Bhaskar SN. Growth Pattern of the Rat Mandible from 13 Days Insemination Age to 30 Days after Birth. *Am. J. Anat.* 92(1):1-53 (1953)
(50) Manson JD. *A Comparative Study of the Postnatal Growth of the Mandible.* (London: Henry Kimpton, 1968)
(51) Bernick S and Patek PQ. Postnatal Development of the Rat Mandible. *J. Dent. Res.* 48(6):1258-63 (1969)
(52) Meikle MC. The Role of the Condyle in the Postnatal Growth of the Mandible. *Am. J. Orthod.* 64(1):50-62 (1973)
(53) Silbermann M and Livne E. Skeletal Changes in the Condylar Cartilage of the Neonate Mouse Mandible. *Biol Neonate.* 35:95-105 (1979)
(54) Livne E, Weiss A and Silbermann M. Changes in Growth Patterns in Mouse Condylar Cartilage Associated with Skeletal Maturation and Senescence. Growth, *Dev. & Aging.* 54(4):183-93 (1990)
(55) Kincaid SA and Van Sickle DC. Bone Morphology and Postnatal Osteogenesis. Potential for Disease. *Vet. Clin. North Am.: Small Anim. Pract.* 13(1):3-17 (1983)
(56) Sissons HA. Experimental Determination of Rats of Longitudinal Bone Growth. *J Anat., Lond.* 87:228-236 (1953)
(57) Hert J. Growth of the Epiphyseal Plate in Circumference. *Acta Anat.* 82:420-36 (1972)
(58) Hansson LI. Determination of Endochondral Bone Growth in Rabbit by Means of Oxytetracycline. *Acta Univ. Lund.* II(1):2-10 (1964)
(59) Wallis GA. Bone Growth: Coordinating Chondrocyte Differentiation. *Curr. Biol.* 6(12):1577-80 (1996)
(60) Nap RC and Hazewinkel HAW. Growth and Skeletal Development in the Dog in Relation to Nutrition; a Review. *Vet. Quart.* 16(1):50-9 (1994)
(61) Rosati R, Horan GSB, Pinero GJ, Garofalo S, Keene DR, Horton WA, Vuorio E, de Crombrugghe B and Behringer RR. Normal Long Bone Growth and Development in Type X Collagen-null Mice. *Nat. Genet.* 8(2):129-35 (1994)
(62) Haines RW. The Histology of Epiphyseal Union in Mammals. *J. Anat.* 120(1):1-25 (1975)
(63) Thyberg J and Friberg U. Ultrastructure of the Epiphyseal Plate of the Normal Guinea Pig. *Z. Zellforsch.* 122:254-272 (1971)
(64) Ingalls TH. Epiphyseal growth: Normal Sequence of Events at the Epiphyseal Plate. *Endocrinology.* 29:710-719 (1941)
(65) Hansson LI. Daily Growth in Length of Diaphysis Measure by Oxytetracycline in Rabbit Normally and after Medullary Plugging. *Acta Orthop. Scand.* Suppl 101:9-199 (1967)
(66) Miralles-Flores C, Delgado-Baeza E. Histomorphometric Differences Between the Lateral Region and Central Region of the Growth Plate in Fifteen-Day-Old Rats. *Acta Anat.* 139:209-13 (1990)
(67) Kember NF. Comparative Patterns of Cell Division in Epiphyseal Cartilage Plates in the Rat. *J. Anat.* 111(1):137-142 (1972)
(68) Moss-Salentijn L. Studies of Long Bone Growth. I. Determination of Differential Elongation in Paired Growth Plates of the Rat. *Acta Anat.* 90:145-60 (1974)
(69) Guthrie S, Plummer JM, Vaughan LC. Post natal development of the canine elbow joint: a light and electron microscopical study. *Res Vet Sci.* 52(1):67-71 (1992)
(70) Brighton CT. The growth Plate. *Orthop. Clin. North Am.* 15(4):571-95 (1984)
(71) Zaleske DJ. Cartilage and Bone Development. AAOS Instr. Course Lect. 47:461-8 (1998)
(72) Raisz LG and Kream BE. Hormonal control of skeletal growth. *Ann. Rev. Physiol.* 43:225 (1981)
(73) Ring PA. Ossification and Growth of the Distal Ulnar Epiphysis of the Rabbit. *J Anat., Lond.* 89:457-464 (1955)

(74) Seinsheimer F III and Sledge CB. Parameters of longitudinal growth rate in rabbit epiphyseal growth plates. *J. Bone Joint Surg.* AM63-A(4):627-30 (1981)
(75) McCormick MJ, Lowe PJ and Ashworth MA. Analysis of the relative contributions of the proximal and distal epiphyseal plate to the growth in length of the tibia in the New Zealand white rabbit. *Growth.* 36(2):133-44 (1972)
(76) Rudicel S, Lee KE and Pelker RR. Dimensions of the rabbit femur during growth. *Am. J. Vet. Res.* 46(1):268-9 (1985)
(77) Thorngren K-G and Hansson LI. Cell Production of Different Growth Plates in the Rabbit. *Acta. Anat.* 110:121-7 (1981)
(78) Li XJ, Jee WSS, Ke HZ, Mori S and Akamine T. Age-related Changes of Cancellous and Cortical Bone Histomorphometry in Female Sprague-Dawley Rats. *Cells and Materials.* Suppl 1:25-35 (1991)
(79) Kimmel DB and Jee WS. A Quantitative Histologic Analysis of the Growing Long Bone Metaphysis. *Calcif. Tissue Int.* 32(2):113-22 (1980)
(80) Lozupone E. A Quantitative Analysis of Bone Tissue Formation in Different Regions of the Spongiosa in the Dog Skeleton. *Anat. Anz.* 145(5):425-52 (1979)
(81) Stump CW. The Histogenesis of Bone. *J. Anat.* 59:136-154 (1925)
(82) Parfitt AM. The Two Faces of Growth: Benefits and Risks to Bone Integrity. *Osteoporosis Int.* 4:382-98 (1994)
(83) Kerley ER. Age Determination of Bone Fragments. *J. Forensic Sci.* 14(1):59-67 (1969)
(84) Parfitt AM, Travers R, Rauch F and Glorieux FH. Structural and Cellular Changes During Bone Growth in Healthy Children. *Bone.* 27(4):487-94 (2000)
(85) Tomlin DH, Henry KM and Kon SK. Autoradiographic Study of Growth and Calcium Metabolism in the Long Bones of the Rat. *Brit. J. Nutr.* 7:235-253 (1953)
(86) Cooper RR, Milgram JW and Robinson RA. Morphology of the Osteon. An Electron Microscopic Study. *J. Bone Jt. Surg.* 48-A(7):1239-1271 (1966)
(87) Fiala P. Age-related Changes in the Substantial compacta of the Long Limb Bones. *Folia Morphologica.* 26(4):316-20 (1978)
(88) Rogers, MJ, Nicholson, PHF, Davie, MWJ, Lohmann, G, and Turner, AS. The pathophysiology of osteoporous – species differences. In *Current Research in Osteoporosis and Bone Mineral Measurement* 2. Eds: Ring, EFJ, Elvins, DM and Bhalla, AK, pp. 50-57 (1993)
(89) Biewener, AA., Biomechanics of mammalian terresterial locomotion. *Science* 250: 1097-1103 (1990)
(90) Einhorn, TA., Bone strength: The bottom line. *Calcif Tissue Int* 51: 333-339 (1992)
(91) Hert J, Fiala P and Petrtyl M. Osteon Orientation of the Diaphysis of the Long Bones in Man. *Bone.* 15(3):269-77 (1994)
(92) Bellino FL. Nonprimate animal models of menopause: workshop report. Menopause: *J. North Am. Menopause Soc.* 7(1):14-24 (2000)
(93) Jowsey J. Studies of the Haversian Septums in Man and Some Animals. *J. Anat.* 100(4):857-864 (1966)
(94) Enlow DH. Functions of the Haversian System. *Amer. J. Anat.* 110: 269-305 (1962)
(95) Albu I, Georgia R and Georoceanu M. The Canal System in the Diaphysial Compacta of the Femur in Some Mammals1. *Anat. Anz., Jena.* 170:181-187 (1990)
(96) Georgia R, Albu I, Sicoe M and Georoceanu M. Comparative Aspects of the Density and Diameter of Haversian Canals in the Diaphyseal Compact Bone of Man and Dog. *Morphol.-Embryol.* 28(1):11-4 (1982)
(97) Lee WR. Appositional Bone Formation in Canine Bone: A Quantitative Microscopic Study Using Tetracycline Markers. *J Anat., Lond.* 98(4):665-677 (1964)
(98) Marotti G. The Dynamics of Osteon Formation in Inert Bones. *Calcif Tissue Res.* 2(Suppl):86-88 (1968)
(99) Vanderhoeft PJ, Kelly PJ and Peterson LFA. Determination of Growth Rates in Canine Bone by Means of Tetracycline-labeled Patterns. *Lab. Invest.* 11(9):714-726 (1962)
(100) Anderson C and Danylchuck KD. Appositional Bone Formation Rates in the Beagle. *Am. J. Vet. Res.* 40(7):907-10 (1979)
(101) Miller ME, Christensen GE and Evans HE. *Anatomy of the Dog* (Philadelphia: W.B. Saunders Company, 1965, 1964), 1-94
(102) Lewis OJ. The Blood Supply of Developing Long Bones with Special Reference to the Metaphyses. *J. Bone Joint Surg.* 38(4):928-933 (1956)

(103) Brookes M and Revell WJ. "Nutrient Vessels in Long Bones," In: *Blood Supply of Bone. Scientific Aspects*. (Great Britain: Springer-Verlag London Limited, 1998)

(104) Buckwalter JA and Cooper RR. "Bone structure and function." In: *Instructional Course Lectures*. American Academy of Orthopaedic Surgeons, Ed. Griffin PP, vol. 36. Easton, PA: Mack Printing Company, 1987, 27-48

(105) Irving MH. The Blood Supply of the Growth Cartilage in Young Rats. *J Anat., London.* 98(4):631-9 (1964)

(106) Spira E and Farin I. The Vascular Supply to the Epiphyseal Plate Under Normal and Pathological Conditions. *Acta Orthop. Scandinav.* 38: 1-22 (1967)

(107) Brighton CT. Structure and Function of the Growth Plate. *Clin. Orthop.* 136:22-32 (1978)

(108) Shapiro F. Epiphyseal and Physeal Cartilage Vascularization: A Light Microscopic and Tritiated Thymidine Autoradiographic Study of Cartilage Canals in Newborn and Young Postnatal Rabbit Bone. *Anat. Rec.* 252(1):140-8 (1998)

(109) Zimmerman B, Moegelin A, de Souza P and Bier J. Morphology of the Development of the Sagittal Suture of Mice. *Anat. Embryol.* 197:155-65 (1998)

(110) Schunior A, Zengel AE, Mullenix PJ, Tarbell NJ, Howes A and Tassinari MS. An Animal Model to Study Toxicity of Central Nervous System Therapy for Childhood Acute Lymphoblastic Leukemia: Effects on Growth and Craniofacial Proportion. *Cancer Research.* 50:6455-6460 (1990)

(111) Arikoske, P., Komulainen, J., Riikonen, P., Voutilainen, R., Knip, M., and Kroger, H. Alterations in bone turnover and impaired development of bone mineral density in newly diagnosed children with cancer: A 1-year prospective study. *J. Clin. Endocrinol. Metabolism* 84(9):3174-3181 (1999)

(112) Juhn A., Weiss A., Mendes D., and Silbermann M. Non-invasive assessment of bone mineral density during maturation and aging of wistar female rats. *Cells and Materials.* Suppl 1: 19-24 (1991)

(113) Tam CS, Reed R, Campbell JE and Cruickshank B. Bone Growth Kinetics II. Short-term Observations on Bone Growth in Sprague-Dawley Rats. *J. Pathol.* 113(1):39-46 (1974)

APPENDIX C-2*
SPECIES COMPARISON OF ANATOMICAL AND FUNCTIONAL RENAL DEVELOPMENT

Tracey Zoetis[1], and Mark E. Hurtt[2,3]

[1] Milestone Biomedical Associates, Frederick, MD 21701, USA
[2] Pfizer Global Research & Development, Groton, CT 06340, USA
[3] Correspondence to: Dr. Mark E. Hurtt, Pfizer Global Research & Development, Drug Safety Evaluation, Eastern Point Road, Mailstop 8274-1306, Groton, CT 06340. 860-715-3118. Fax (860) 715-3577. Email: mark_e_hurtt@groton.pfizer.com

Introduction

Renal development involves both anatomical and functional aspects that occur in predictable timeframes. Xenobiotics can interrupt either anatomic or functional development or both. The purpose of this paper is to identify critical time frames for anatomical and functional development of the kidney in the humans and to compare these events to other species. This comparison should result in data that will be useful in designing and interpreting studies of the possible prenatal and/or postnatal developmental effects of chemicals on the maturing kidney.

Kidney maturation has been assessed by the size and distribution of the tubules and by the histologic appearance of glomeruli (1). Major renal anatomic developmental events occur in humans prenatally, with functional development continuing into the postnatal periods. A brief description of human anatomical renal development is presented followed by a description of comparative species. Similarly, a description of human functional renal maturation is presented using the major renal functions of glomerular filtration, urine concentration, acid base equilibrium, and urine volume control, followed by a description of comparative species for each function.

Anatomical Development

Humans

The human kidney begins to develop through a process of reciprocal inductive interaction during week 5 of gestation. The interaction occurs between the primitive ureteric bud and the metanephric mesenchyme (a "temporary" kidney derived from the pronephros). Beginning at gestation week 5, the diverticulum of the mesonephric duct (primitive ureter) comes into contact with the caudal measenchyme of the nephrogenic cord and induces it to undergo epithelial transformation that gives rise to the upper urinary tract or ureteric (or metanephric) bud. The mesenchyme then induces the ureteric bud to grow, differentiate, and branch into it. The mesenchymal cells condense around the tip of each branch of the developing ureter. These condensates develop into vesicles, and then into a comma-shaped body that develops into an S-shaped glomerulus. Endothelial cells collect in the S-shaped body and form a glomerular capillary loop. Nephrogenesis and kidney vasculature develop during the same timeframe and are complete about week 35 of gestation (2, 3, 4, 5, 6). A diagram of this process is presented below.

---

* Source: Zoetis, T. and Hurtt, M. E., Species comparison of anatomical and functional renal development, *Birth Defects Research, Part B: Developmental and Reproductive Toxicology*, 68, 111-120, 2003.

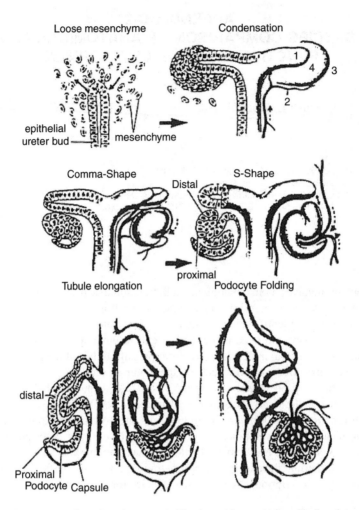

Figure 1 Schematic representation of nephrogenesis. The branching ureteric epithelium interaction with loose metanephric mesenchyme (A) results in condensation of the mesenchyme (B). Cell lineages shown include: 1) ureteric epithelium, 2) vasculature, 3) undifferentiated mesenchyme, and 4) condensed mesenchyme differentiating into epithelia. These stages are followed by infolding of the primitive glomerular epithelium to form comma- and S-shaped bodies (C and D). Elongation of the proximal and distal tubular elements subsequently occur (E) along with further infolding of the glomerular epithelium and vascular structures to form the mature glomerular capillary network (F). The initial phases of glomerular vascularization are believed to occur during the early stages of glomerular differentiation (C and D). From Ekblom, (1984), with permission.

By the fifth month there are 10 to 12 branchings (7). Approximately 20% of nephrons are formed by 3 months of gestation, 30% by 5 months, and nephrogenesis is complete with a total of approximately 800,000 nephrons by 34 weeks of gestation (1, 7, 8). Juxtamedullary nephrons are formed earliest (by 5 months gestation) and superficial cortical nephrons form later (by 34 weeks gestation). Postnatal maturation of the nephrons and elongation of the tubules continues during the first year of life (7). An illustration of the branching process is presented below.

Three developmental periods have been defined based on the rate of glomerular proliferation (9). The first period begins during gestation week 10 and is marked by a slow increase in the number of glomeruli. The second period begins during gestation week 17 or 18 when glomerular proliferation increases abruptly and occurs rapidly until gestation week 32. The period between gestation weeks 18 – 32 is a critical time point in renal development; it is at this time that nephrogenetic development reaches its peak (9). The third period lasts from gestation week 32 on,

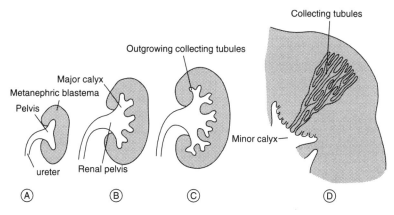

Figure 11-5. Schematic drawings showing the development of the renal pelvis, calyces and collecting tubules of the metanephros. A, At 6 weeks; B, end of sixth week; C, 7 weeks; D, newborn. Note the pyramid form of the collecting tubules entering the minor calyx. From: Langman, 1975. (4)

when no increase in the number of glomeruli is observed, the nephrogenic blastema disappears, and the nephrogenetic process is complete.

Low birth weight premature infants exhibit nephrogenesis at a level commensurate with age rather than body weight, and maturation continues during early postnatal weeks (1). The number of nephrons in a given species is constant; however, cellular growth can be influenced by exposure to environmental factors, renal blood flow and glomerular filtration, capacity for excretion of sodium and water, renal prostaglandin production and urinary excretion of calcium (5, 7).

Comparative Species

Mammalian kidneys follow similar developmental pathways; however, the time frame with regard to birth varies between species. Cell culture studies using glomeruli from monkey, sheep, dog, rabbit, and rat kidneys demonstrate that the pattern of growth and morphologic features of each cell type, including numbers present and rates of division were the same for all species studied (10). Cell culture studies using nephrons from various species have demonstrated similar genetic expression of patterns that can be related to time scales in morphogenesis (3). An electron microscopic study of granulated glomerular epithelial cells from 17 mammalian species demonstrated similarities in morphology between all species studies (11). Thus, animal studies provide critical information that aids in the understanding of renal development across species, including primates such as humans. The following table marks the gestation day of the first appearance of metanephros cells that differentiate and proliferate to form the kidney for several species.

Onset of Kidney Development as Evidenced by Metanephros

Species	Metanephros (gestation day)	Total gestation period (days)
Man	35–37	267
Macaque	38–39	167
Guinea Pig	23	67
Rabbit	14	32
Rat	12.5	22
Mouse	11	19
Hamster	10	16
Chick	6	21

From: Evan, et. al., 1984 (12)

The following table illustrates the timing of completion nephrogenesis for various species, as summarized in Kleinman (13), Fouser and Avner (14), and Gomez et. al, (2). More detailed descriptions of individual species follow.

Species	Timing of Nephrogenesis Completion
Man	35 weeks gestation
Sheep	Before birth
Guinea Pig	Before birth
Dog	Postnatal week 2
Pig	Postnatal week 3
Mouse	Before birth
Rat	Postnatal week 4–6

Rats

In the rat, nephrogenesis was observed to occur at a rapid pace between birth and 8 days and is complete by 11 days of age (15), tubular differentiation continues until the time of weaning, and functional maturity occurs even later (16).

One of the factors involved in nephrogenesis is the nutritional status of the mother. In rats fed low protein diets during days 8 – 14 or 15 – 22 of gestation, offspring were observed to have lower numbers of nephrons and low renal size. This observation persisted until 19 weeks of age, at which time glomerular filtration rate (GFR) was normal (17).

Maturation of the loop of Henle has been studied in rats, with regard to succinic dehydrogenase and acid phosphatase activity (18). Neonatal loops of Henle are relatively short loops without thin ascending limbs. As maturation occurs, apoptic deletion of thick ascending cells and transformation into thin ascending limb cells yields a well-defined boundary between inner and outer medullas by postnatal day 21 in the rat (18).

Mice

Mice and humans share, among renal histological characteristics, timing of onset of nephrogenesis (19). This process occurs prenatally in both mice and humans. In mice nephrogenesis begins on gestation day 11 and is complete by birth (14).

Dogs

The puppy kidney is immature in both structure and function at birth (20). Functionally, the puppy has a lower glomerular filtration rate, renal plasma flow, and filtration fraction compared to the adult dog (20). In a study of casts of renal vessels from puppies age 1 – 21 days after birth, investigators found evidence of immaturity of the intrarenal vascular system and the proximal tubule when compared to adult dog. Nephrogenesis continues for at least 2 weeks postnatally in the dog (20).

Glomerular blood flow and thus maturation was studied using microsphere injection in 26 puppies ranging in age from 5h to 42 days and in 5 adult dogs (21). The renal cortex was divided in to four equal zones progressing from the outer to inner cores, and glomeruli were counted in each. Results confirmed that nephrogenesis is still underway at the time of birth. Positive identification of the differentiation of the primitive metanephric vesicle from the vascularized glomeruli was not always possible since the glomeruli become more separated as the dog ages and tubules elongate. In newborns, the glomeruli of the inner zone are larger than those in the outer zone, but this difference gradually disappears as the kidney grows (21). A similar study showed that plasma flow is an important factor in this process in maturing puppies (22).

Sheep

Nephrogenesis is complete in sheep at birth (23, 24).

Primates

Nephrogenesis is complete in monkeys at birth (23, 24).

Rabbits

Nephrogenesis is completed 2 – 3 weeks postnatally in rabbits (24). Maturation of the loop of Henle has been studied in rabbits, with regard to succinic dehydrogenase and acid phosphatase activity, and carbonic anhydrase IV activity (18). Succinic dehydrogenase and acid phosphatase have high levels of activity in the neonate and decrease to adult levels by about postnatal day 28 in the rabbit. Neonatal loops of Henle are relatively short loops without thin ascending limbs. As maturation occurs, apoptic deletion of thick ascending cells and transformation into thin ascending limb cells yields a well-defined boundary between inner and outer medullas. The expression of carbonic anhydrase IV has been correlated with this maturation process in the rabbit (18).

Functional Development

Compared to the adult renal function, the human infant has decreased renal blood flow, glomerular filtration rate, tubular secretion, and a more acidic urinary pH (7). Urine production begins during the 10th week of gestation (6, 2). Maturation of glomerular filtration, concentrating ability, acid-base equilibrium, and urine volume control are described below.

Glomerular Filtration

Humans

The primary function of the glomeruli is to act as a filter before plasma reaches the proximal tubule. Glomerular filtration rate (GFR) continues to increase after birth and reaches adult levels at 1 – 2 years of age. After one month of age, creatinine clearance is used to measure GFR; however, GFR is commonly estimated in units of ml/min/1.73m^2, based on the formula K L/S$_{Cr}$, where K is a constant*, L is length (or height in cm) and S$_{Cr}$ is serum creatinine (mg/dl). Serum creatinine levels during the postnatal days 1 and 2 reflect maternal values and decrease to 0.2 – 0.4 mg/dl by 3 months. An increase in creatinine level indicates a decrease in GFR regardless of gestational age (2). Normal values for GFR are presented in the following table.

---

* The K constant is 0.55 in children and adolescent girls, 0.70 in adolescent boys, 0.45 in term neonates, and 0.33 in preterm infants.

Normal Values of Glomerular Filtration Rate for Humans by Age

Age	GFR (ml/min/1.73m^2)
Preterm infant (gestation weeks 25–28)	
1 week	11.0 ± 5.4
2–8 weeks	15.5 ± 6.2
Preterm infant (gestation weeks 29–34)	
1 week	15.3 ± 5.6
2–8 weeks	28.7 ± 13.8
Term infant	
5–7 days	50.6 ± 5.8
1–2 months	64.6 ± 5.8
3–4 months	85.8 ± 4.8
5–8 months	87.7 ± 11.9
9–12 months	86.9 ± 8.4
2–12 years	133 ± 27

From: Gomez et al. 1999 (2)

The kidney is a well-perfused organ, receiving approximately 20% of cardiac output at rest. Glomerular filtration delivers filtered plasma to the proximal tubule. Small molecules (~5000 daltons) pass through the glomerular barrier and restriction increases as molecular mass increases to where molecules the size of albumin (~68,000 daltons) do not pass the barrier. Glomerular filtration rate (GFR) is determined using clearance studies comparing urine volume excreted with amount of excreted inulin or some other metabolically inert material not reabsorbed or secreted by the renal tubules. A normal GFR for newborn infants is less than 50 mL/min/1.73m^2, rises to 100 to 140 mL/min/1.73m^2 in humans over 1 year of age, and reaches an adult level by the age of 2 years (1, 7, 8). After birth the rapid rise in glomerular filtration rate is attributed to several factors including increased mean arterial blood pressure and glomerular hydraulic pressure, a sharp decline in renal vascular resistance, with a redistribution of intrarenal blood flow from the juxtamedullary to the superficial cortical nephrons (25).

Immature glomeruli are present for months after birth and glomerular maturation and filtration increases during early infancy. Glomerular size follows a regular growth curve in childhood, with an average diameter of 100 μm at birth, progressing to 300 μm (1). However, glomerular filtration rate does not increase proportionally with body size. Prior to gestation week 34, body size increases but glomerular filtration rate does not; after that time glomerular filtration rate increases more rapidly than body weight (26). Immature glomeruli are functional once capillary function has been established (1).

Comparative Species

Three stages of glomerular filtration rate development have been identified by studying various species (13). The first stage is characterized by equivalent rates of increase in GFR and kidney mass. The second stage is characterized by a greater rate of increase in GFR than kidney growth. In the final stage, GFR increases at the same rate as that of kidney mass. These three stages do not necessarily correlate with anatomical development across species. In addition to empirical measurements, allometric scaling has been used to predict GFR for various species (27).

Changes in glomerular filtration rate parallel those for renal blood flow. Intrarenal blood flow occurs at different rates for the inner and outer cortexes. Thus glomerular filtration development occurs during different time periods as intra renal blood flow distributes between inner and outer cortexes (13).

Dogs — Inner cortical blood flow begins to increase with age about postnatal Day 12 in dogs. This results in a sharp decline in the ratio between inner and outer cortical blood flow, that has been reported as ~0.95 on Day 1 to 0.3 on Day 12 and throughout adulthood (13). In the case of the neonatal dog the developmental change in intra renal blood flow distribution correlates with the period of nephrogenesis. An illustration of the timing of maturation of the inner cortical and outer cortical blood flow in dogs is presented below.

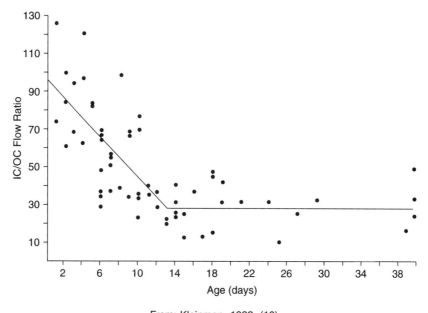

From: Kleinman, 1982. (13)

Glomerular filtration rate increases with age postnatally in dogs. Based on clearance studies using small to large molecules, an increase in glomerular capillary surface area and pore density occurs between postnatal weeks 1 and 6 in dogs (28).

Rats — Glomerular filtration rate sharply increases in rats during the first six weeks of postnatal life (29). A comparison of early postnatal glomerular filtration rates in rats and dogs is presented below.

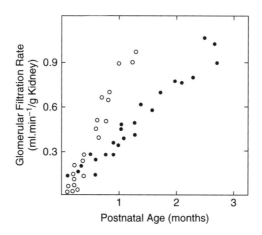

Glomerular filtration rate in rat (○) and canine (●) kidney during postnatal maturation. From: Horster, 1977. (29)

Sheep — The first stage of GFR development has been observed in fetal lambs. Blood pressure and renal vascular resistance effect renal blood flow. In the early postnatal period, there is little or no change in renal blood flow per gram kidney (13).

In fetal sheep, renin-angiotensin type 1 receptor antagonist (10-mg/kg losartan) was dosed on gestation days 125 – 132. Although renal blood flow increased, glomerular filtration rate decreased, resulting in decreased filtration fraction. However, glomarulotubular balance was maintained since there was no change in sodium reabsorption (30).

Rabbit — Rabbit studies indicate that body temperature of the neonate can to have an effect on glomerular filtration rate. In newborn rabbits, a 2°C decrease in body temperature causes renal vasoconstriction accompanied by a decrease in glomerular filtration rate (25).

Concentrating Ability

Humans

The newborn human infant is incapable of excreting concentrated urine at birth and this function reaches maturity by the first year of life.

The ability of the kidney to concentrate urine is controlled by mechanisms that control water balance, including antidiuretic hormone, short loops of Henle, low NaCl transport in the thick ascending limb, and decreased tubular response to arginine vasopressin. Normal osmolality is regulated by antidiuretic hormone (ADH) controlled by the hypothalamus. The mechanism that regulates ADH begins to operate 3 days after birth and the kidney becomes responsive to ADH at that time (8). A series of active transports and passive diffusions occur along the tubule, with chloride actively transported out of the ascending tubule and remaining water is transported by diffusion. The primary function of the tubules is to reabsorb glomerular filtrate and the loop of Henle functions to concentrate and dilute urine. Tubular length and volume increases postnatally, allowing for increased capacity for transport and metabolism (1). The loop of Henle forms a hairpin turn as it enters and exits the medulla. At a specific point in the loop of Henle, urea enters into the descending limb and plays an important role in generating a high sodium concentration in the fluid. Importantly for the neonate, quantities of urea in breast milk or formula are not sufficient for maximal urine concentration. Thus the neonate cannot excrete highly concentrated urine, but has no difficulty in excreting dilute urine (1). Fluid from a higher level is progressively concentrated as water diffuses out of the descending limb (1). Maturational changes in the ability to concentrate urine are illustrated in the following table.

Maximal Urine Osmolality by Age

Postnatal Age	mOsm/L
3 days	151 ± 172
6 days	663 ± 133
10–30 days	896 ± 179
10–12 months	1118 ± 154
14–18 years	1362 ± 109

From: Gomez et. al., 1999 (2)

Sodium excretion also changes in the maturing infant, as the ability to respond to aldosterone matures. Premature infants excrete more sodium than do full term infants. At gestation week 31, the fractional excretion of sodium is about 5% and declines to about 1% by the second postnatal month (2). The term infant retains about 30% of dietary sodium, which is necessary to support normal growth; however, they have difficulty excreting an acute load of sodium and water, which can lead to edema. Low GFR and enhanced distal tubular sodium reabsorbtion is responsible for sodium retention. The ability to excrete a sodium load is fully developed by the end of the first year of life (2).

Comparative Species

Rats — Neonatal rats do not excrete concentrated urine at birth, but the concentration increases dramatically with age (15). As single-nephron GFR increases with maturation, proximal tubular sodium reabsorption increases proportionally (13).

Dogs — The fractional reabsorption of water in dogs is constant during postnatal maturation and similar in adults (13). Sodium excretion was studied in 1, 2, 3, and 6 week old puppies by expanding intravascular volume with either isotonic saline or isoncotic albumin in saline (31). Glomerular filtration rate, sodium excretion, fractional excretion of sodium, and plasma volume measurements were made. Of these parameters the sodium excretion was increased over controls at all ages. The highest level of sodium excretion was observed in 3-week-old puppies. The authors state that the mechanism underlying the difference between the response to isotonic saline and isoncotic albumin in saline is already operative at birth in dogs (31). A graph depicting the timing of urine concentrating ability in dogs is presented below.

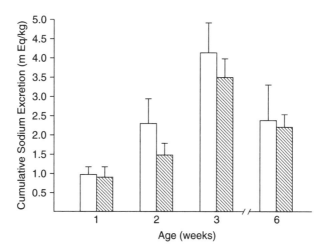

Cumulative sodium excretion in developing animals sustaining volume expansion with either saline (*open bars*) or isoncotic albumin (*hatched bars*). From: Aladjem, et al. 1982. (31)

Results from other sodium loading studies in neonatal and adult dogs indicate that the pressure natriuresis occurs in the proximal tubules and the newborn proximal tubule is more sensitive to renal arterial blood pressure changes when compared to the adult (32).

Rabbits — Using a polyclonal antibody directed at rabbit carbonic anhydrase IV, Schwartz et al. (18) noted the maturation pattern for the expression of the enzyme in the medulla of the maturing kidney paralleled that of the urine concentrating system. These investigators note that the localization of the carbonic anhydrase IV expression within the kidney is important to its function. For example, in rabbits carbonic anhydrase IV is expressed in the outer medullar collecting ducts, but not in rats. With the regard to function, the rabbit medulla matures after 21 days of age, and clear distinction between inner an outer medulla is not visible at 3 weeks of age, but is at 5 weeks. Further investigation demonstrated that the maturational pattern observed in the inner medulla was similar to carbonic anhydrase IV expression that was approximately one-fourth adult levels at 2 weeks of age and surged during postnatal weeks 3 and 4. The ability to concentrate urine developed along this same timeline.

Sheep — Renal water and electrolyte reabsorbtion have been studied in fetal sheep. The percentage of water, sodium and chloride reabsorbed increased throughout the later stages of gestation. However, electrolyte reabsorbtion exceeded water reabsorbtion to a degree that resulted in hypotonic urine until the last 2 weeks of gestation, at which time hypertonic urine was produced (8).

Guinea Pigs — The fractional reabsorption of water in guinea pigs is constant during postnatal maturation and similar in adults (13).

Acid-Base Equilibrium

Humans

Growth rate and the composition of intake determine renal acid-base control in infants. Excreted phosphate levels prior to birth and for the first two days after birth are very low, resulting in low titratable acidity (8). The intake of the fetus is relatively constant and regulated by the placenta, which limits the renal ability to contribute to acid-base equilibrium (8). As newborn feeding begins, the acidity varies directly in relation to the amount of protein, sulfate, and phosphate in the diet, and inversely with the rate of body growth. Walker (8) reports disturbances in equilibrium can be produced by changes in intake in the two to three week old child, where these same changes would not affect a 2-month-old child.

Comparative Species

Studies in animals have demonstrated the role of enzymes in establishing and maintaining acid base equilibrium. Carbonic anhydrase is a zinc metalloenzyme that catalyzes the hydration of CO_2 and the dehydration of carbonic acid. Carbonic anhydrase activity has been detected in rat proximal convoluted tubules and inner medullary collecting duct, and rabbit outer medullary collecting duct (18). Inhibition of luminal carbonic anhydrase has been found to diminish renal acid excretion and reduce HCO_3^- reabsorption (H^+ excretion) in the proximal tubule, suggesting an important role for carbonic anhydrase IV in maintaining acid-base equilibrium. The authors hypothesized that the low levels of carbonic anhydrase in the neonatal kidney may help to explain the difficulty in maintaining acid-base homeostasis.

Rats — In the rat carbonic anhydrase IV mRNA is expressed in the 20-day rat fetal kidney and increases dramatically by postnatal day 17 (18).

Rabbits — In rabbits, the expression of carbonic anhydrase was one-fourth adult levels at 2 weeks postnatally and surged to adult levels during postnatal weeks 3 and 4 (18).

Dogs — In a study using mongrel dogs, postnatal excretion of uric acid decreased from 83% at birth to 51% at 90 days of age (33). A direct correlation was observed between uric acid and sodium clearance during early development. This study indicates acid-base homeostasis develops postnatally in dogs.

Urine Volume Control

Urine volume control in response to water diuresis is demonstrable in the human infant on Day 3 after birth, and this capacity increases over a period of weeks (8). The diuretic response in infants differs from that observed in adults in that it is controlled by an increased GFR accompanied with a minimal reduction in specific gravity in the infant, whereas the adult GFR remains constant. Normal values for neonatal renal function are presented in the following table.

Normal Values of Neonatal Renal Function

	Premature First 3 days	Term infant First 3 days	2 Weeks
Daily Excretion (ml/kg/24 hr.)	15–75	20–75	25–125
Max. osmolality (mosmos/kg H_2O)	400–500	600–800	800–900
GFR (ml/min x 1.73m^2)	10–15	15–20	35–45

Hentschel et al. 1996 (34)

Renin-Angiotensin System

The role of angiotensin converting enzyme (ACE) activity differs between mature and immature systems. ACEs play a role in renal anatomical and functional development and maturation. An illustration of the differences of angiotensin effects between the neonate and adult is presented below.

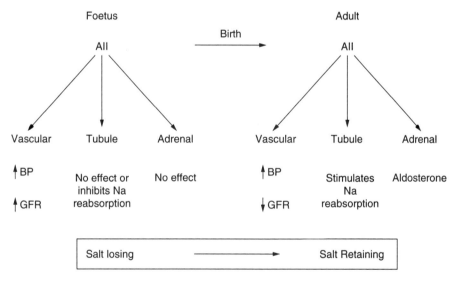

Differences between the renal actions of fetal and adult renin-angiotensin systems. From: Lumbers, 1995. (23)

The developing kidney has different periods of susceptibility to either functional or anatomical injury. The normal development of the kidney can be further understood by studying abnormal development that has occurred in the presence of known xenobiotics.

An example of an agent that interferes with normal renal development is the **ACE inhibitor**. The effects of this drug class on renal development are briefly reviewed as a case study below.

ACE is a peptidyl dipeptidase that catalyzes the conversion of angiotensin I to angiotensin II, which in turn, acts as a vasoconstrictor. Angiotensin II also stimulates aldosterone secretion by the adrenal cortex. Inhibition of ACE results in decreased plasma levels of angiotensin II, and subsequent vasopressor activity and decreased aldosterone secretion. ACE inhibitors are prescribed as antihypertensives.

Humans

The role of ACE inhibitors in infant renal failure has been well documented in the literature (35, 36, 37, 38). Udwadia-Hegde et al. (38) present a review of case histories of mothers taking ACE inhibitors during the second and third trimesters of pregnancy. Oligohydramnios was observed which resulted in preterm delivery of neonates with intrauterine growth retardation, severe hypotension, and anuria. Biopsy of the kidneys generally showed renal tubular dysplaisia. Major anomalies induced by ACE inhibitors in humans include oligohydramnios, neonatal anuria/renal tubular dysgenesis, pulmonary hypoplasia, intrauterine growth retardation, persistent patent ductus arteriosus, calvarial hypoplasia/acalvaria, fetal or neonatal death (35). Since 1992, ACE inhibitors marketed in the United States carry a black box warning in the label regarding the use of this drug class during pregnancy, as in the following label for Lotrel.

> **When used in pregnancy during the second and third trimesters, angiotensin converting enzyme inhibitors can cause injury and even death to the developing fetus.** When pregnancy is detected, Lotrel should be discontinued as soon as possible (Physician's Desk Reference, 2000).

Comparative Species

Mice — Renin angiotensin is essential for the development of the mammalian kidney and urinary tract (19). Using mutant mice carrying a targeted null mutation of the angiotensin I or II receptor, Miyazaki and Ichikawa (19) demonstrated that both mutants have distinct phenotype in the kidney and urinary tract system. Angiotensin II is involved in multiple aspects of the early morphogenesis of the kidney and urinary tract. Angiotensin I receptor induces the development of the renal pelvis, which promotes the removal of urine from the renal parenchyma. Failure of the angiotensin receptors to operate properly during specific developmental stages results in congenital anomalies of the kidney and urinary tract in utero and hydronephrosis ex utero (19).

Other investigators have noted the important role of renin-angiotensin in nephrogenesis, vascularization, and architectural and functional development of the kidney (39). Renal development was studied using ACE inhibitors in neonatal rats and ACE mutant mice and similar renal pathology was observed in both cases. The authors hypothesized that the primary lesion was a disturbance in renal vessel development, with tubular pathology due to the close temporal and spatial relationship of the tubules to the vessels during late stages of development (39).

Rats — Rat studies illustrate the importance of the timing of exposure to ACE inhibitors as a critical factor in the development of altered renal morphology. The rat is susceptible to altered renal morphology mainly during the last 5 days of pregnancy and the first two weeks of life (2, 39). Treatment of newborn rats with an ACE inhibitor for the first 12 days of postnatal life resulted in marked renal abnormalities (2). Microscopic findings included relatively few and immature glomeruli, distorted and dilated tubules, relatively few and short thick arterioles that resulted in less branching and arrested maturation. The changes did not resolve after termination of treatment after 23 days of age (2). Similarly, when rats were treated with enalapril on postnatal days 3 – 13, abnormalities in renal morphology were correlated with functional abnormalities (16). Functional abnormalities included impairment of urinary concentrating ability, which correlated with the degree of papillary atrophy. When losartan treatment began in 21-day-old rats or enalapril treatment began in 14-day-old rats, no changes in renal morphology were observed (16, 36, 39). The timeframe in which rats are susceptible to renal injury induced by ACE inhibitor exposure correlates with the critical period for nephrogenesis and marked tubular growth and differentiation.

The human kidney is more mature than the rat kidney at birth, and therefore the adverse effects of renin-angiotensin blockers on human kidneys are more likely to occur when exposure occurs during the last few weeks of pregnancy (39).

Rabbits — A single oral dose of 30-mg/kg enalapril to pregnant rabbits on Day 26 resulted in 100% fetal death (34). At doses within the human therapeutic range, fetal deaths occurred in middle to late gestation with a peak effect on gestational day 26 (of 31 in the rabbit) (36).

The mechanism of fetal death has been postulated as decreased fetal-placental blood flow in sheep and rabbits (35).

Sheep — The fetal renin-angiotensin system in sheep (gestation days 125 – 132) helps to regulate fetal renal blood flow and is essential in the maintenance of fetal glomerular function (30, 40). Differences between fetal and adult renin-angiotensin systems have been documented using sheep (23). During gestation, the fetal renin-angiotensin system maintains glomerular filtration rate, thus the author proposed that the renal excretion of sodium and water into the amniotic cavity, ensuring adequate amniotic fluid volume to support normal growth and development, was impaired. However, in the adult, the glomerular filtration rate is not regulated by angiotensin II. Angiotensin stimulates tubule sodium reabsorption in the adult but not in the fetus (23).

Baboons — Baboons were treated with enalapril at a level comparable to a clinical dose used to achieve moderate but sustained ACE inhibition, and below doses generally selected for toxicologic study (41). Treatment began prior to mating and continued throughout pregnancy. Eight out of 13 had adverse outcomes (fetal death or intrauterine growth retardation) compared with 0 of 13 in the placebo group; no histopathologic evaluation was performed to determine the cause of death and no fetal malformations were observed (41). A direct effect on the fetal renin-angiotensin system and placental ischemia has been postulated to be the mechanism of toxicity in baboons (35).

Conclusion

Both anatomical and functional development of the kidney must be considered when making comparisons between species. This literature review has shown that the end of the anatomical development of the kidney is marked by completion of nephrogenesis. Nephrogenesis is complete prior to birth in humans, monkeys, mice, sheep, and guinea pigs, and after birth in rats, dogs, and pigs. Maturation of renal function occurs during different time frames for different species. The focus of this paper was on the maturation of major renal functions including glomerular filtration, concentrating ability, acid-base equilibrium, and urine volume control. Glomerular filtration can be detected as early as the first trimester of pregnancy and is important in tubular reabsorption of Na^+ and Cl that helps to maintain sodium balance in amniotic fluid. Concentrating ability develops postnatally in humans, rats, rabbits, and sheep, and prenatally in dogs and guinea pigs. Acid base equilibrium develops postnatally in all species reported which includes humans, rats, rabbits, and dogs. Control of urine volume also develops postnatally.

In conclusion, design and interpretation of studies in prenatal and juvenile animals regarding renal development should include careful consideration of the variability in time points of maturation of both anatomical and functional developmental milestones between species.

REFERENCES

1. Bernstein J. "Morphologic Development and Anatomy." In: *Rudolph's Pediatrics*. Eds. Rudolph, AM; coeditors, Hoffman, JIE, Rudolph, CD; assistant editor, Sagan P; associate editor, Travis LB, Chapter. 25. *The Kidneys and Urinary Tract*. (Norwalk, CT: Appleton & Lange, 1991), pp.1223-1224.
2. Gomez, R.A, Maria Luisa S. Sequeira Lopez, Lucas Fernadez, Daniel R. Chernavvksy, and Victoria F. Norwood, "The Maturing Kidney: Development and Susceptibility," *Renal Failure, vol.21, no.3&4* 1999: 283-91.
3. Horster, MF, Gerald S. Braun, and Stephan M. Huber, "Embryonic Renal Epithelia: Induction, Nephrogenesis, and Cell Differentiation," *Physiological Reviews,* vol.79, no.4 Oct. 1999: pp.1157-91.

4. Langman J. "Urogenital Systems." Chapt. 11 in *Medical Embryology; Human Development-Normal and Abnormal*. Baltimore: Williams & Wilkins, 1975.
5. Larsson SH, and Aperia A. "Renal Growth in infancy and childhood – experimental studies of regulatory mechanisms," *Pediatr Nephrol, vol.5* 1991: pp.439-42.
6. Norwood, Victoria F., Scott G. Morham, and Oliver Smithies, "Postnatal development and progression of renal dysplasia in cyclooxygenase-2 null mice," *Kidney International, vol.58* 2000: pp2291-2300.
7. Witte MK, Stork JE, Blumer JL. Diuretic Therapeutics in the Pediatric Patient. Am J Cardiol. 1986; 57: 44A-53A.
8. Walker DG. "Functional Differentiation of the Kidney." In: *Intra-Uterine Development*. Ed. Barnes, AC, Chapt. 13 (Philadelphia, PA: Lea & Febiger, 1968), 245-252.
9. Gasser, B., Y. Mauss, J.P. Ghnassia, R. Favre, M. Kohler, O. Yu, J.L. Vonesch, "A Quantitative Study of Normal Nephrogenesis in the Human Fetus: Its Implication in the Natural History of Kidney Changes due to Low Obstructive Uropathies," *Fetal Diagn Ther, vol.8* 1993: pp.371-84.
10. Holdsworth, SR, Glasgow EF, Atkins RC and Thomson NM, "Cell Characteristics of Cultured Glomeruli from Different Animal Species," *Nephron, vol.22, no.4-6* 1978: pp.454-9.
11. Gall, JA, Daine Alcorn, Aldona Butkus, John P. Goghlan, and Graeme B. Ryan, "Distribution of glomerular peripolar cells in different mammalian species," *Cell Tissue Res, vol.244, no.1* 1986: pp.203-8.
12. Evan, AP, Vincent H. Gattone, II, and Philip M. Blomgren, "Application of scanning electron microscopy to kidney development and nephron maturation," *Scan Electron Microsc, pt.1* 1984: pp.455-73.
13. Kleinman LI. "Developmental Renal Physiology," *Physiologist*. 1982 Apr; 25(2): 104-10.
14. Fouser, M.D., Laurie and Ellis D. Avner, M.D., "Normal and Abnormal Nephrogenesis," *American Journal of Kidney Disease, vol.21, no.1* Jan. 1993: pp.64-70.
15. Kavlock, RJ, and Jacqueline A. Gray, "Evaluation of Renal Function in Neonatal Rats," *Biol Neonate, vol.41* 1982: pp.279-88.
16. Guron, Gregor, Niels Marcussen, Annika Nilsson, Birgitta Sundelin, and Peter Friberg, "Postnatal Time Frame for Renal Vulnerability to Enalapril in Rats," *J Am Soc Nephrol, vol. 10* 1999: pp.1550-60.
17. Langley-Evans, S.C., Simon JM Welham, and Alan A. Jackson, "Fetal exposure to a maternal low protein diet impairs nephrogenesis and promotes hypertension in the rat," *Life Sciences, vol.64, no.11* 1999: pp.965-74.
18. Schwartz GL, Olson J, Kittelberger AM, Matsumoto T, Waheed A, Sly WS. Postnatal development of carbonic anhydrase IV expression in rabbit kidney. Am J Physiol. 1999 Apr; 276(4 Pt 2): F510-20.
19. Miyazaki Y, Ichikawa I. Role of the angiotensin receptor in the development of the mammalian kidney and urinary tract. *Comp Biochem Physiol A Mol Integr Physiol*. 2001 Jan; 128(1): 89-97.
20. Evan, AP, James A. Stoeckel, Vickie Loemaker, and Jeffrey T. Baker, "Development of the vascular system of the puppy kidney," *Anat Rec, vol.194, no.2* Jun 1979: pp.187-99.
21. Olbing H, M. Donald Blaufox, Lorenzo C. Aschinberg, Geraldine I. Silkalns, Jay Bernstein, Adrian Spitzer, and Chester M. Edelmann, Jr., "Postnatal Changes in Renal Glomerular Blood Flow Distribution in Puppies," *J Clin Invest, vol.52, no.11* Nov. 1973: pp.2885-95.
22. Tavani, N Jr, Philip Calcagno, Steve Zimmet, Walter Flamenbaum, Gilbert Eisner, and Pedro Jose, "Ontogeny of Single Nephron Filtration Distribution in Canine Puppies," *Pediatr Res, vol.14, no.6* Jun 1980: pp.799-802.
23. Lumbers ER. Functions of the renin-angiotensin system during development. *Clin Exp Pharmacol Physiol*. 1995 Aug; 22(8): 499-505.
24. Seikaly MG, Billy S. Arant, Jr., "Development of Renal Hemodynamics: Glomerular Filtration and Renal Blood Flow," *Clin Perinatol, vol.19, no.1* 1992: pp.1-13.
25. Toth-Heyn P, Drukker A, Guignard JP. The stressed neonatal kidney: from pathophysiology to clinical management of neonatal vasomotor nephropathy. Pediatr. Nephrol. 2000 Mar; 14(3): 227-39.
26. Arant, BS. Jr., The Newborn Kidney. In *Rudolph's Pediatrics*. Eds. Rudolph, AM; coeditors, Hoffman, JIE, Rudolph, CD; assistant editor, Sagan, P; associate editor,
27. Singer MA. Of Mice and Men and Elephants: Metabolic Rate Sets Glomerular Filtration Rate. *Am J Kidney Dis*. 2001 Jan; 37(1): 164-178.
28. Goldsmith, DI, Roberto A. Jodorkovsky, Julius Sherwinter, Stuart R. Kleeman, and Adrian Spitzer, "Glomerular capillary permeability in developing canines," *Am J Physiol, vol.251, no.3, pt.2* 1986: pp.F528-31.

29. Horster, M, "Nephron Function and Perinatal Homeostatis," *Ann Rech Vet., vol.8, no.4* 1977: pp.468-82.
30. Stevenson KM, Gibson KJ, Lumbers ER. Effects of losartan on the cardiovascular system, renal haemodynamics and function and lung liquid flow in fetal sheep. *Clin Exp Pharmacol Physiol.* 1996 Feb; 23(2): 125-33.
31. Aladjem, Mordechai, Adrain Spitzer, and David I. Goldsmith, "The Relationship between Intravascular Volume Expansion and Natriuresis in Developing Puppies," *Pediatr Res, vol.16, no.10* Oct 1982: pp.840-5.
32. Kleinman LI, and Robert O. Banks, "Pressure natriuresis during saline expansion in newborn and adult dogs," *Am J Physiol, vol.246, no.6, pt.2* Jun 1984: pp.F828-34.
33. Stapleton FB, and Billy S. Arant, Jr., "Ontogeny of Renal Uric Acid Excretion in the Mongrel Puppy," *Ped Res, vol.15, no.12* Dec 1981: pp.1513-6.
34. Hentschel R, Lodige B, Bulla M. Renal insufficiency in the neonatal period. *Clin Nephrol.* 1996 Jul; 46(1): 54-8. Review.
35. Buttar HS. An overview of the influence of ACE inhibitors on fetal-placental circulation and perinatal development. *Mol Cell Biochem.* 1997 Nov; 176(1-2): 61-71.
36. Sedman AB, Kershaw DB, Bunchman TE. Recognition and management of angiotensin converting enzyme inhibitor fetopathy. *Pediatr Nephrol.* 1995 Jun; 9(3): 382-5.
37. Sorensen AM, Christensen S, Jonassen TE, Andersen D, Petersen JS. [Teratogenic effects of ACE-inhibitors and angiotensin II receptor antagonists]. Ugeskr Laeger. 1998 Mar 2; 160(10): 1460-4. Danish.
38. Udwadia-Hegde A, Parekji S, Ali US, Mehta KP. "Angiotensin converting enzyme inhibitor fetopathy," *Indian Pediatr.* 1999 Jan; 36(1): 79-82.
39. Hilgers KF, Norwood VF, Gomez RA. Angiotensin's role in renal development. *Semin Nephrol.* 1997 Sep; 17(5): 492-501.
40. Lumbers ER, Bernasconi C, Burrell JH. Effects of inhibition of the maternal renin-angiotensin system on maternal and fetal responses to drainage of fetal fluids. *Can J Physiol Pharmacol.* 1996 Aug; 74(8): 973-82.
41. Harewood, W.J., Andrew F. Pippard, Geoffrey G. Duggin, John S. Horvath, and David J. Tiller, "Fetotoxicity of angiotensin-converting enzyme inhibition in primate pregnancy: A prospective, placebo-controlled study in baboons (*Papio hamadryas*)," *Am J Obstet Gynecol*, vol.171, no.3 Sep. 1994: pp.633-42

ADDITIONAL RELATED REFERENCES

Airede A, Bello M, Weerasinghe HD. Acute renal failure in the newborn: incidence and outcome. *J Pediatric Child Health.* 1997 Jun; 33(3): 246-9.

Bernardini N, Mattii L, Bianchi F, Da Prato I, Dolfi A. TGF-Alpha mRNA Expression in Renal Organogenesis: A Study in Rat and Human Embryos. *Exp Nephrol.* 2001 Mar; 9(2): 90-98.

Capulong MC, Kimura K, Sakaguchi N, Kawahara H, Matsubara K, Likura Y. Hypoalbuminemia, oliguria and peripheral cyanosis in an infant with severe atopic dermatitis. *Pediatr Allergy Immunol.* 1996 May; 7(2): 100-2.

Casellas D, Bouriquest N, Artuso A, Walcott B, Moore LC. New method for imaging innervation of the renal preglomerular vasculature. Alterations in hypertensive rats. *Microcirculation.* 2000 Dec; 7(6 Pt 1): 429-37.

Dawson R Jr., Liu S, Jung B, Messina S, Eppler B. Effects of high salt diets and taurine on the development of hypertension in the stroke-prone spontaneously hypertensive rat. *Amino Acids.* 2000; 19(3-4): 643-65.

De Heer E, Sijpkens YW, Verkade M, den Dulk M, Langers A, Schutrups J, Bruijn JA, van Es La. Morphometry of interstitial fibrosis. *Nephrol Dial Transplant.* 2000; 15 Suppl 6: 72-3.

Forhead AJ, Gillespie CE, Fowden AL. Role of cortisol in the ontogenic control of pulmonary and renal angiotensin-converting enzyme in fetal sheep near term. *J Physiol* 2000 Jul 15; 526 Pt. 2: 409-16.

Gomez, M.D., R.A., and Victoria F. Norwood, M.D., "Recent advances in renal development," *Current Opinion in Pediatrics, vol.11* 1999: pp.135-40.

Griffet J, Bastiani-Griffet F, Jund S, Moreigne M, Zabjek KF. Duplication of the leg-renal agenesis: congenital malformation syndrome. *J Pediatr Orthop B.* 2000 Oct; 9(4): 306-8.

Hamar P, Peti-Peterdi J, Szabo A, Becker G, Flach R, Rosivall L, Heemann U. Interleukin-2-Dependent Mechanisms are involved in the development of the glomerulosclerosis after partial renal ablation in rats. *Exp Nephrol.* 2001 Mar; 9(2): 133-141.

Hegde AU, Parekji S, Ali US, Mehta KP. Angiotensin converting enzyme inhibitor fetopathy. *Indian Pediatr.* 1999 Jan; 36(1): 79-82.

Ibrahim SH, Bhutta ZA, Khan IA. Haemolytic uraemic syndrome in childhood: an experience of 7 years at the Aga Khan University. *JPMA J Pak Med Assoc.* 1998 Apr; 48(4): 100-3.

Jones, MD, DP, and Russell W. Chesney, MD, "Development of Tubular Function," *Clin Perinatol, vol.19, no.1* Mar 1992: pp.33-57.

Karnak I, Muftuoglu S, Cakar N, Tanyel FC. Organ growth and lung maturation in rabbit fetuses. *Res Exp Med* (Berl). 1999 Mar; 198(5): 277-87.

Kleinman LI, and John H. Reuter, "Maturation of Glomerular Blood Flow Distribution in the New-Born Dog," *J. Physiol, vol.228* 1973: pp.91-103.

Kleinman, LI, "Developmental Renal Physiology," *Physiologist, vol.25, no.2* Apr 1982: pp.104-10.

Komhoff, Martin, Jun-Ling Wang, Hui-Fang Cheng, Robert Langenbach, James A. McKanna, Raymond C. Harris, and Matthew D. Breyer, "Cyclooxygenase-2-selective inhibitors impair glomerulogenesis and renal cortical development," *Kidney International, vol.57* 2000: pp.414-22.

Kusuda S, Kim TJ, Miyagi N, Shishida N, Litani H, Tanaka Y, Yamairi T. Postnatal change of renal artery blood flow velocity and its relationship with urine volume in very low birth weight infants during the first month of life. *J Perinat Med.* 1999; 27(2): 107-11.

Landau D, Shelef I, Polacheck H, Marks K, Holcberg G. Perinatal vasoconstrictive renal insufficiency associated with material nimesulide use. *Am J Perinatol.* 1999; 16(9): 441-4.

Lankin VZ, Sherensheva NI, Konovalova GG, Tikhaze AK. Beta-carotene-containing preparation carinat inhibits lipid peroxidation and development of renal tumors in rats treated with chemical carcinogen. *Bull Exp Biol Med.* 2000 Jul; 130(7): 694-6.

Lavoratti G, Seracini D, Fiorini P, Cocchi C, Materassi M, Donzelli G, Pela I. Neonatal anuria by ACE inhibitors during pregnancy. *Nephron* 1997; 76(2): 235-6.

Ludders JW, G.F. Grauer, R.R. Dubielzig, G.A., Ribble, J.W. Wilson, "Renal microcirculatory and correlated histologic changes associated with dirofilariasis in dogs," *Am J Vet Res, vol.49, no.6* Jun 1988: pp.826-30.

McCracken GH, Ginsberg C, Chrane DF, Thomas ML, Horton LJ. Clinical pharmacology of penicillin in new born infants. *J Ped.* 1973 April; 82(4): 692-698.

McDonald MC, Mota-Filipe H, Paul A, Cuzzocrea S, Abdelrahman M, Harwood S, Plevin R, Chatterjee PK, Yaqoob MM, Thiemermann C. Calpain inhibitor I reduces the activation of nuclear factor-{kappa}B and organ injury/dysfunction in hemorrhagic shock. *FASEB J.* 2001 Jan 1; 15(1): 171-186.

Neuhuber WL, Eichhorn U, Worl J. Enteric co-innervation of striated muscle fibers in the esophagus: Just a "hangover"? *Anat Rec.* 2001 Jan 1; 262(1): 41-46.

O'Brien KL, Selanikio JD, Hecdivert C, Placide MF, Louis M, Barr BD, Barr JR, Hospedales CJ, Lewis MJ, Schwartz B, Philen RM, St. Victor S. Espindola J, Needham, LL, Denerville K. Epidemic of pediatric deaths from acute renal failure caused by diethylene glycol poisoning. Acute Renal Failure Investigation Team. *JAMA.* 1998 Apr 15; 279(15): 1175-80.

Okada T, Iwamoto A, Kusakabe K, Mukamot M, Kiso Y, Morioka H, Kodama H, Sasaki F, Morikawa Y. Perinatal Development of the Rat Kidney: Proliferative Activity and Epidermal Growth Factor. *Biol Neonate.* 2001 Jan; 79(1): 46-53.

O'Rourke, Dawn A., Hiroyuki Sakurai, Katherine Spokes, Crystal Kjelsberg, Masahide Takahashi, Sanjay Nigam, and Lloyd Cantley, "Expression of c-ret promotes morphogenesis and cell survival in mIMCD-3 cells," *American Physiological Society* 1999: pp.F581-88.

Peneyra RS, and Roger S. Jaenke, "Functional and Morphologic Damage in the Neonatally Irradiated Canine Kidney," *Radiat Res, vol.104, no.2, pt.1* Nov 1985: pp.166-77.

Peruzzi, Licia, Bruno Gianoglio, Maria Gabriella Porcellini, and Rosanna Coppo, "Neonatal end-stage renal failure associated with maternal ingestion of cyclo-oxygenase-type-1 selective inhibitor nimesulide as tocolytic," *The Lancelet, vol.354, no.9190* Nov. 6, 1999: pp.1615.

Querfeld U, Ortmann M, Vierzig A, Roth B. Renal tubular dysgenesis: a report of two cases. *J Perinatol.* 1996 Nov-Dec; 16(6): 498-500.

Ramirez O, Jimenez E. Opposite transitions of chick brain catalytically active cytosolic creatine kinase isoenzymes during development. *Int J Dev Neurosci.* 2000 Dec; 18(8): 815-23.

Sandberg K, Ji H. Kidney angiotensin receptors and their role in renal pathophysiology. *Semin Nephrol.* 2000 Sep; 20(5): 402-16.

Sitdikov FG, Gil'mutdinova RI, Minnakhmetov RR, Zefirov TL. Asymmetrical effects of vagus nerves on functional parameters of rat heart in postnatal ontogeny. *Bull Exp Biol Med.* 2000 Jul; 130(7): 620-3.

Strehl, R., Will W. Minuth, "Nephron induction-the epithelial mesenchymal interface revisited," *Pediatr Nephrol, vol.16* 2001: pp.38-40.

Suzuki T, Kimura M, Asano M, Fujigaki Y and Hishida A. Role of Atrophich Tubules in Development of Interstitial Fibrosis in Microembolism-Induced Renal Failure in Rat. *Am J Pathol.* 2001 Jan; 158(1): 75-85.

Taylor HA, Delany ME. Ontogeny of telomerase in chicken: impact of downregulation of pre- and postnatal telomere length in vivo. *Dev Growth Differ.* 2000 Dec' 42(6): 613-21.

Traebert, Martin, Marius Lotscher, Ralph Aschwanden, Theresia Ritthaler, Jurg Biber, Heini Murer, and Brigitte Kaissling, "Distribution of the Sodium/Phosphate Transporter during Postnatal Ontogeny of the Rat Kidney," *J Am Soc Nephrol, vol.10* 1999: pp.1407-15.

Travis LB. *The Kidneys and Urinary Tract.* Rudolph's Pediatrics. 1991; 19th Ed., Chapt. 25: 1223-1236.

Tucker LB, Stehouwer DJ. L-DOPA-induced air-stepping in the preweaning rat: electromyographic and kinematic analyses. *Behav Neurosci.* 2000 Dec; 114(6): 1174-82.

Vesna, Lackovic, and Mujovic Spomenka, "Postnatal development of the kidney juxtaglomerular appartus in rats," *Acta anat., vol.108* 1980: pp.281-87.

White, DVM JV, D.R. Finco, DVM, Ph.D., W.A. Crowell, DVM, S.A. Brown, DVM, Ph.D., D.A. Hirakawa, Ph.D., "Effect of dietary protein on functional, morphologic, and histolgic changes of the kidney during compensatory renal growth in dogs," *Am J Vet Res,* vol.52, no.8 Aug 1991: pp.1357-65.

Zhang G, Oldroyd SD, Huang LH, Yang B, Li Y, Ye R, El Nahas AM. Role of Apoptosis and Bcl-2/Bax in the Development of Tubulointerstitial Fibrosis during Experimental Obstructive Nephropathy. *Exp Nephrol.* 2001 Mar; 9(2): 71-80.

Zhang SL, To C, Chen X, Filep JG, Tang SS, Ingelfinger JR, Carriere S, Chan JS. Effect of Renin-Angiotensis System Blockade on the Expression of the Angiotensinogen Gene and Induction of Hypertrophy in Rat Kidney Proximal Tubular Cells. *Exp Nephrol.* 2001 Mar; 9(2): 109-117.

APPENDIX C-3*
SPECIES COMPARISON OF LUNG DEVELOPMENT

Tracey Zoetis[1], and Mark E. Hurtt[2,3]

[1] Milestone Biomedical Associates, Frederick, MD 21701, USA
[2] Pfizer Global Research & Development, Groton, CT 06340, USA
[3] Correspondence to: Dr. Mark E. Hurtt, Pfizer Global Research & Development, Drug Safety Evaluation, Eastern Point Road, Mailstop 8274-1306, Groton, CT 06340. 860-715-3118. Fax (860) 715-3577. Email: mark_e_hurtt@groton.pfizer.com

Introduction

The purpose of this paper is to identify critical time frames for development of the lung in humans and to compare these events to other species. This comparison should result in data that will be useful in designing and interpreting studies of the possible prenatal and/or postnatal developmental effects of chemicals and drugs on the developing lung. A brief description of growth and development of the human lung is presented as a baseline for comparison with other species. This is followed by an inter-species comparison of lung development.

Growth and Development of the Human Lung

The development of the human lung is a relatively steady and continuous process, arbitrarily divided into the following 6 stages: embryonic, pseudo-glandular, canalicular, saccular, alveolar, and vascular maturation (1). The first 4 developmental stages are complete during fetal development. At birth, the human neonate has entered the alveolar stage of development. Over 80% of alveoli are formed after birth by a process of air space septation (1). The developmental stages are illustrated in Figure 1.

Structural changes occur on a continuum with increases in the length of the respiratory tract and in number of alveoli as the child ages (3). Lung surface area increases as a result of the increase in the alveoli numbers. Lung surface area increases during late developmental stages may simply represent an expansion of airspace of the lung. A schematic representation of human lung development at various stages was developed by Reid (3) as is presented in Figure 2.

Alveoli increase in number and surface area with increasing age and begin to level off between the ages of 2 to 4 years (4, 3). This is accompanied by decreasing interstitial tissue. The following graph illustrates the rate of development. Data are inconclusive regarding the timing that alveolar formation is complete in humans, and range from 2 years (1, 4) to 8 years (3).

Accurate interpretation of this graph requires knowledge of the criteria used to determine the completion of alveolization. Although all authors did not explain this, interpretation of the endpoint could range from the age of detection of the last known immature alveolus to determination of the critical timepoint for lung function. Another confounding factor in the interpretation of this graph is the fact that some of the data were obtained from post mortem examinations of patients whose alveolar development may have been compromised by disease. Dietert et al. (2000)(6) report the alveolar proliferation takes place during the first 2 years of life while expansion takes place during the ages of 2 – 8 years.

---

* Source: Zoetis, T. and Hurtt, M. E., Species comparison of Lung development, *Birth Defects Research, Part B: Developmental and Reproductive Toxicology*, 68, 121-124, 2003.

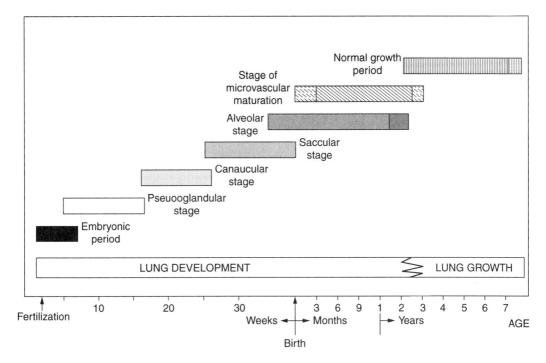

Figure 1 Timing of stages of human lung development. (Reproduced from Zeltner and Burri, 1987(1)).

Inter-species Comparison of Lung Growth and Development

Developmental stages identified for the human lung are also found in other mammalian species reported in the literature. In this section, the growth and development of the pulmonary system in common laboratory species will be discussed and compared with that of humans.

The timing of each developmental stage and the degree of lung development at birth varies widely between species (3). As an example, at birth the opossum lung is very primitive; rat and mouse lungs have no alveoli; kittens, calf and humans have relatively few alveoli; and the lamb lungs are quite well developed. Postnatal development of the lung has also been studied in other species: guinea pig, hamster, dog, monkey and humans. A comparative table of the timing of each stage for several species is presented below (7).

Species	Glandular	Canalicular	Saccular	Alveolar
Mouse	14–16	16.5–17.4	17.4–	PD 5–
Rat	13–18	19–20	21–PD	PD 7–21
Rabbit	19–24	24–27	27–	—
Sheep	–95	95–120	120–	—
Human	42–112	112–196	196–252	252–childhood

Rat Lung Growth and Development

Rats are perhaps the most extensively studied laboratory species for lung development (2). Many hypotheses regarding human lung development are based on research conducted in rats. Postnatal changes in the rat lung occur in 3 phases: short expansion; proliferation; and equilibration. The first phase lasts from birth to day 4 of postnatal age. During this period, the lung transforms from the saccular stage into the alveolar stage. During the proliferating phase, alveolar numbers increases more rapidly than body weight. So does the lung volume and lung weight. During the last phase,

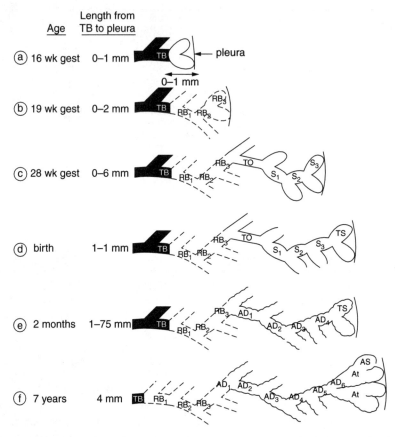

Figure 2 A schematic presentation of human lung development at various ages. TB = Terminal bronchiole; Rb$_i$ = respiratory bronchioles; TD = transitional duct; Ad$_i$ = Alveolar duct; At = Atrium; AS = alveolar sac. From Thurlbeck (3).

the equilibrated growth phase, growth of the lung parallels overall body growth. This phase may last until very late in age.

The rat appears to be an acceptable model for study of juvenile populations because of similarities between rats and humans regarding the stages of lung development. The following table is a comparison of lung parameters between rats and humans.

Lung Growth in Humans and Rats (Numbers Represent the Fold Change from Newborn to Adult)

Pulmonary Parameter	Human	Rat
Lung volume (ml)	23.4	23.5
Parenchyma airspace volume (ml)	30.2	26.9
Septal volume (ml)	13.5	13.6
Alveolar surface area (m^2)	21.4	20.5
Capillary surface area (m^2)	23.3	19.2

Source: Zeltner et al (1987)(4).

Limitations exist, but can be overcome to some degree with careful planning for timing and duration of dosing. Nevertheless, there are differences between rats and humans in lung development. For example, rats are born during the saccular developmental stage and humans are born during the alveolar developmental stage. Rat lungs reach alveolar stage at the age of 4 days. Rat

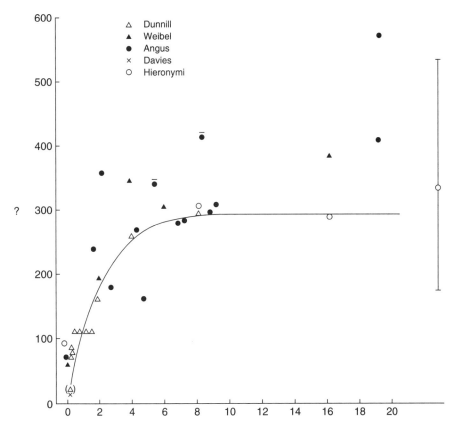

Figure 3 Postnatal development of lung alveoli in humans. Data with a wide range on the right side was from Angus and Thurlbeck (1972) (5). The solid (regression) line was calculated from Dunnill's data only. Source: W.M. Thurlbeck, 1975(3).

lungs develop rapidly, with most lung development complete within the first 2 weeks after birth. Furthermore, the rat lung continues to proliferate at a slow rate through out its life span. By comparison, human lung development continues until about age 3 to 8, with functional parameters peaking at the age of 18 – 25 years (8).

Dog Lung Growth and Development

The few studies that specifically address postnatal lung development in dogs present conflicting findings. Boyden and Tompsett (1961)(9) reported that morphologically, dog lungs are slightly less mature at birth than human lungs. Mansell and colleagues (1995)(10) report that lung development is more mature at birth in dogs than in humans. Biological variation may contribute significantly to this difference. The degree of lung maturity at postnatal day 11 in dogs appears to be comparable to humans at birth (9). Evidence of lung development and growth activity was noted at the age of 8 weeks. One investigator estimates the maximal functional efficiencies of the lung are reached at about the age of 1 year in dogs compared to 20 years in humans (8). The dog is considered an acceptable species for testing the safety of inhaled drugs intended for pediatric populations.

Other Species Considered

Other species considered to date include the rabbit, sheep, pig, and monkey. These species were not considered to be acceptable models of postnatal lung development because of their advanced development at birth when compared to humans (3, 10, 11, 12, 13, 14, 15, 16 and 17).

Conclusions

The literature does not provide definitive evidence for the timing of the completion of alveolar development, and a weight-of-evidence approach to address the safety of inhaled drugs intended for pediatric populations is needed. Inhaled drugs intended to treat children over 2 years of age generally do not require extensive testing in juvenile animals. For children over the age of 2 years, the pulmonary safety of the drug is demonstrated in clinical trials by measuring lung function parameters during the course of the trial. The issue of local postnatal developmental toxicity for inhaled drugs is most important in children under 2 years of age. For the purposes of testing the safety of inhaled agents in pediatric populations under 2 years of age, the rat and dog appear to be acceptable models.

REFERENCES

1. Zeltner, T.B. and Burri, P.H. (1987). The postnatal development and growth of the human lung. II. Morphology. *Respirat. Physiol.* 67: 269-282.
2. Burri, P.H. (1996). Structural Aspects of Prenatal and Postnatal Development and Growth of the Lung.
3. Thurlbeck, W.M. (1975). Postnatal Growth and Development of the Lung American Review of Respiratory Disease. Vol. 111.
4. Zeltner, T.B., Cauduff, J.H., Gehr, P., Pfenninger, J. and Burri, P.H. (1987). The postnatal development and growth of the human lung. I. Morphometry. *Respiration Physiology.* 67:247-267.
5. Angus, G.E., and Thurlbeck, W.M. 1972. Number of alveoli in the human lung. *J Appl Physiol* 32(4): 483-5.
6. Dietert, R.R., Etzel, R.A., Chen, D., Halonen, M., Holladay, S.D., Jarabek, A.M., Landreth, K., Peden, D.B., Pinkerton, K., Smialowicz, R.J., Zoetis, T. (2000). Workshop to Identify Critical Windows of Exposure for Children's Health: Immune and Respiratory Systems Work Group Summary. Environmental Health Perspectives. Vol. 108, Supplement 3:483-90.
7. Lau, C. and Kavlock, R.J. (1994). *Functional Toxicity in the Development Heart, Lung, and Kidney. Developmental Toxicity,* 2nd ed. 119-188.
8. Mauderly, J.L., Effect of age on pulmonary structure and function of immature and adult animals and man. (1979), *Fed. Proc.* Vol. 38, No. 2. February 173-177.
9. Boyden, E.A., and Tompsett, D.H. (1961). The Postnatal Growth of Lung in the Dog. *Acta Anat.* Vol. 47, No. 3: 185 – 215
10. Mansell, A.L., Collins, M.H., Johnson, E., Jr. and Gil, J. (1995) Postnatal Growth of Lung Parenchyma in the Piglet: Morphometry Correlated With Mechanics. *Anat. Rec.* 241: 99-104.
11. Winkler, G.C. and Cheville, N.F. (1985). Morphometry of Postnatal Development in the Porcine Lung. *Anat. Rec.* 211(4): 427-433.
12. Mills, A.N., Lopez-Vidriero, M.T., and Haworth, S.G. 1986. Development of the airway epithelium and submucosal glands in the pig lung: changes in epithelial glycoprotein profiles. *Br J Exp Pathol* 67(6): 821-9.
13. Zeilder, R.B. and Kim, H.D. (1985). Phagocytosis, chemiluminescence, and Cell volume of Alveolar Macrophages From Neonatal and Adult Pigs. *J Leukoc Biol.* 37(1): 29-43.
14. Rendas, A., Branthwaite, M. and Reid, L., (1978). Growth of pulmonary circulation in normal pig – structural analysis and cardiopulmonary function. *J. Appl. Physiol.* 45(5): 806-817.
15. Kerr, G.R., Couture, J., and Allen, J.R. (1975). Growth and Development of the Fetal Rhesus Monkey. VI. Morphometric Analysis of the Development Lung. *Growth.* 39:67-84.
16. Boyden, E.A. (1977). Development and Growth of the Airways. In: *Development of the Lung.* W.A. Hodson, Ed. Marcel Decker, New York, pp. 3-35.
17. Hislop, A., Howard, S. and Fairweather, D.V.I. (1984). Morphometric studies on the structural development of the lung in Macaca fascicularis during fetal and postnatal life. *J. Anat.* 138(1):95-112.

ADDITIONAL RELEVANT REFERENCES

Adamson, I.Y.R. and King, G.M. (1986). Epithelial-Interstitial Cell Interactions in Fetal Rat Lung Development Accelerated by Steroids. *Laboratory Investigations.* Vol. 55, No. 2, 145.

Barry, B.E., Mercer, R.R., Miller, F.J., and Crapo, J.D., (1988). Effects of Inhalation of 0.25 ppm Ozone on the Terminal Bronchioles of Juvenile and Adult Rats. *Experimental Lung Research.* 14:225-245.

Bisgaard, H., Munck, S.L., Nielsen, J.P., Petersen, W., and Ohlsson, S.V. (1990). Inhaled budesonide of treatment of recurrent wheezing in early childhood. *The Lancet.* Vol. 336:649-651.

Blanco, L.N. and Frank, L. (1993). *The Formation of Alveoli in Rat Lung during the Third and Fourth Postnatal Weeks: Effect of Hyperoxia, Dexamethasone, and Deferoxamine.* International Pediatric Research Foundation, Inc. Vol. 34, No. 3.

Boyden, E.A., (1977). The Development of the Lung in the Pig-tail Monkey (Macaca nemestrina, L.). *Anat. Rec.* 186:15-38.

Burri, P.H. (1974). The Postnatal Growth of the Rat Lung. III. Morphology. *Anat. Rec.* 180:77-98.

Burri, P.H., Dbaly, J., and Weibel, E.R. (1973). The Postnatal Growth of the Rat Lung. I. Morphometry. *Anat. Rec.* 278:711-730.

Carson, S.H., Taeusch, H.W, Jr., Avery, M.E., Inhibition of Lung cell division after hydrocortisone injection into fetal rabbits. *Journal of Applied Physiology.* Vol. 34, No. 5, May 1973.

Chang, L-Y., Graham, J.A., Miller, F.J., Ospital, J.J., Crapo, J.D. (1986). Effects of Subchronic Inhalation of Low Concentrations of Nitrogen Dioxide. *Toxicology and Applied Pharmacology.* 83: 46-61.

Ellington, B., McBride, J.T., and Stokes, D.C. (1990). Effects of corticosteriods on postnatal lung and airway growth in the ferret. *J. Appl. Physiol.* 68(5): 2029-2033.

Hyde, D.M., Bolender, R.P., Harkema, J.R., Plopper, C.G. (1994). Morphometric Approaches for Evaluating Pulmonary Toxicity in Mammals: Implications for Risk Assessment. *Risk Analysis.* Vol 14, No. 3: 293-302.

Kamada, A.K., Szefler, S.J., Martin, R.J., Boushey, H.A., Chinchilli, V.M. Drazen, J.M., Fish, J.E., Israel, E., Lazarus, S.C., Lemanske, R.F. (1996) Issue in the Use of Inhaled Glucocorticoids. *Am. J. Respir. Crit. Care Med.* Vol. 153. 1739-1748.

Massaro, D., Teich, N., Maxwell, S., Massaro, G.D., Whitney, P. Postnatal development of Alveoli; Regulation and Evidence of a Critical Period in Rats. *J. Clin. Invest.* Vol. 76, October 1985, 1297-1305.

Merkus, P.J.F.M., Have-Opbroek, A.A.W., and Quanjer, P.H. (1996). Human Lung Growth: A Review. *Pediatric Pulmonology* 21:383-397.

Moraga, F.A., Riquelme, R.A., PharmD, López, A.A., Moya, F.R., Llanos, A.J. (1994). Maternal administration of glucocorticoid and thyrotropin-releasing hormone enhances fetal lung maturation in undisturbed preterm lambs. *Am J. Obstet. Gynecol.* 171(3): 729-734.

Morishige, W.K. (1982). Influence of Glucocorticoids on Postnatal Lung Development in the Rat: Possible Modulation by Thyroid Hormone. *Endocrinology* Vol. 111, No. 5:1587-1594.

Murphy, S. and Kelly, H.W. (1992). Evolution of Therapy for Childhood Asthma. *Am. Rev. Respir. Dis.* 146:544-575.

Odom, M.W., Ballard, P.L. Developmental and Hormonal Regulation of the Surfactant System. 495-575.

Ogasawara, Y., Kuroki, Y., Tsuzuke, A., Ueda, S., Misake, H., Akino, T. Pre-and Postnatal Stimulation Surfactant Protein D by *In vivo* Dexamethasone Treatment of Rats. *Life Sciences,* Vol. 50:1761-1767.

Picken, J., Lurie, M. and Kleinerman, J. (1974). Mechanical and Morphologic Effects of Long-Term Corticosteroid Administration on the Rat Lung. *Am Rev. Respa. Dis.* Vol. 110:746-753.

Robinson, D.S., Geddes, D.M. (1996). Review Article Inhaled Corticosteroids: Benefits and Risks. *J. Asthma.* 33(1): 5-16.

Rooney, S.A., Dynia, D.W., Smart, D.A., Chu, A.J., Ingleson, L.D., Wilson, C.M. and Gross, I. (1986). Glucocorticoid stimulation of choline-phosphate cytidylyltransferase activity in fetal rat lung: receptor-response relationships. *Biochem. Biophys. Acta.* 888(2): 208-216.

Sahebjami, H. and Domino, M. (1989). Effects of Postnatal Dexamethasone Treatment of Development of Alveoli in Adult Rats. *Exp. Lung Res.* 15(6): 961-973.

Schellenberg, J-C., Liggins, G.C., (1987). New approaches to hormonal acceleration of fetal lung maturation. *J. Perinat. Med.* 15(5): 447-451.

Sindhu, R.K., Rasmussen, R.E. and Kikkawa, Y. (1996). Exposure to Environmental Tobacco Smoke Results in an Increased Production of (+)-*anti*-Benzo[a]pyrene-7,8-Dihydrodiol-9,10-Eposide in Juvenile Ferret Lung Homogenates. *J. Toxicol. Environ. Health* 46 (6): 523-534.

Tabor, B.L., Lewis, J.F., Ikegami, M., Polk, D. and Jobe, A.H. (1994). Corticosteroids and Fetal Intervention Interact to Alter Lung Maturation in Preterm Lambs. *Pediatr Res.* 35(4): 479-483.

Tough, S.C., Green, F.H.Y., Paul, J.E., Wigle, D.T., Butt, J.C. (1996). Sudden Death from Asthma in 108 Children and Young Adults. *J. Asthma.* 33(3): 179-188.

Van Essen-Zandvliet, E.E., Hughes, M.D., Waalens, H.J., Duiverman, E.J., Pocock, S.J., Kerrebijn, K.F. and The Dutch Chronic Non-Specific Lung Disease Study Group. (1992). *Am. Rev. Respir. Dis.* 146: 547-554.

Wallkens, H.J., Vanessen-Zandvliet, E.E., Hughes, M.D., Gerritsen, L., Duiverman, E.J., Knol, K., Kerrebijn, K.F., and The Dutch CNSLD Study Group (1993). *Am. Rev. Respir. Dis.* 148:1252-1257.

Ward, R.M. (1994). Pharmaoclogic Enhancement of Fetal Lung Maturation. *Clin Perinatol.* 21(3): 523-542.

Weiss, S.T., Tosteson, T.D., Segal, M.R., Tager, I.B., Redline, S. and Speizer, F. (1992). Effects of Asthma of Pulmonary Function in Children. *Am. Rev. Respir. Dis.* 145: 58-64.

Yeh, H-C., Hulbert, A.J., Phalen, R.F., Velasques, D.J., Harris, T.D. (1975). A Stereoradiographic Technique and Its Application to the Evaluation of Lung Casts. *Investigative Radiology.* Vol. 10, July-August:351-357.

APPENDIX C-4*
DEVELOPMENT AND MATURATION
OF THE MALE REPRODUCTIVE SYSTEM

M. Sue Marty,[1,5] Robert E. Chapin,[2] Louise G. Parks,[3] and Bjorn A. Thorsrud[4]

[1] Dow Chemical Company, Midland, MI 48674, USA
[2] Pfizer Global Research & Development, Groton, CT 06340, USA
[3] Merck & Company, West Point, PA 19486, USA
[4] Springborn Laboratories, Inc., Spencerville, OH 45887, USA
[5] Correspondence to: Dr. M. Sue Marty, Dow Chemical Company, Toxicology Research Laboratory, 1803 Building, Midland, MI 48674, 517-636-6653, Fax 517-638-9863, Email: martym@dow.com

Introduction

This review briefly describes some of the key events in the postnatal development of the male reproductive system in humans, non-human primates, rats and dogs. Topics discussed include development of the testes, epididymides, the blood-testis barrier, anogenital distance, testicular descent, preputial separation, accessory sex glands (prostate and seminal vesicles), and the neuroendocrine control of the reproductive system. The objective of this work is merely to allow the reader to make initial comparisons of the developmental processes and timing of these events in human versus animal models. This review is not intended to be comprehensive, but merely provides an initial overview of these processes. In some cases, information was not available for all species. Available information is summarized in Table 1.

1 Reproductive Organs

1.1 Testes

1.1.1 Human

In the human, spermatogenesis does not begin until puberty; however, the prenatal, early postnatal and prepubescent testis plays a critical role in hormone production. In the early postnatal testis, immature Sertoli cells are the most common cell type[1,2] with limited numbers of germ cells in a relatively undifferentiated state. According to Cortes *et al.*[2] total Sertoli cell number increases from the fetal period through childhood, puberty and early adulthood. In contrast, Lemasters *et al.*[3] reported that Sertoli cells proliferate after birth, ceasing at 6 months when the adult number of Sertoli cells are achieved. Sertoli cells secrete inhibin B until 2-4 years of age and anti-Müllerian hormone during the entire prepubescent period.[4] In humans, there are three known testosterone surges, one from 4-6 weeks of gestation, one from 4 months of gestation to 3 months of age, and the last from 12 to 14 years of age.[5] Consistent with the early postnatal increase in testosterone, there is a biphasic increase in Leydig cell number, which includes an early increase in Leydig cells, followed by a decrease to the lowest level at 1.5 years of age,[6] then a continuous increase in Leydig cells until their numbers plateau in adulthood. In infant boys, an increase in total germ cell number occurs with peak numbers achieved between 50 and 150 days of age, followed by a decrease in older boys.[7] During the early proliferative period, there is an overall decrease in germ cell density

---

* Source: Marty, M. S., Chapin, R. E., Parks, L. G., and Thorsrud, B. A., Development and maturation of the male reproductive system, *Birth Defects Research, Part B: Developmental and Reproductive Toxicology*, 68, 125-136, 2003.

Table 1

	Human	Non-human Primate	Rat	Dog	Mouse
Developmental Stages	0-1 month Neonatal 1 m – 2 yr Infantile 2 –12 yr children 12-16 yr Adolescents[127]		0-7 days Neonatal 8-21 days Infantile 21-35 days juvenile 35- (55-60) days peripubertal depending on sex [20]	3 phases of testis growth[52]: 1: 0-22 wks 2: 22-36 wks 3: 36-46 wks	
Anogenital Distance (AGD)		Similar to rat[57]	2.5 X > in males compared to females[49]		
Preputial Separation (PPS)	Begins during late gestation[66] Complete from 9 months to 3 years of age[67,68,69]. Androgens play key role[70]		Sprague Dawley PND 42-46[58] Androgens play key role		
Puberty	12-14 yrs[9,10,11,13]	2.5 years of age[19]	Early puberty begins at PPS (~PND 43)[58]	34-36 weeks, as defined by presence of ejaculated sperm[41, 52]	PND 35
Prostate Structure	Not sharply demarcated and appears as a single gland with several zones. It is the middle lobe that obstructs the urethra in men with enlarged prostate. Lateral lobe, dorsal (or posterior) lobe, and median (or middle) lobe[74]		Discrete lobes (ventral, lateral and the paired anterior lobes, also known as the coagulating glands),. Dorsal and lateral lose distinct borders at adulthood Form lobes between PND 1-7, tubular lumen between PND 7-14, secretory granules PND 14-21 and shows adult cytology PND 28-35 Secretory activity approaches adult levels ca. PND 43-46[73,88]	The only well-developed accessory sex gland in the dog. Completely surrounds the urethra[73] and divided into right and left lobes with middle lobe either poorly developed or absent. Dogs are the only lab species that spontaneously develops benign prostatic hypertrophy (BPH). There are differences between dogs and humans in their response to antiandrogen treatment.[84,85] Prostate secretion starts at ~4 months	Ventral prostate and dorsolateral prostate increase dramatically between PND 1-15, reach adult secretory levels by ~PND 30. For more details see Sugimura et al[89,90,91]

POSTNATAL DEVELOPMENTAL MILESTONES

Seminal Vesicles (SV)	The SV are present by gestational month 6 and have attained adult form by the 7th month of gestation. Development of the muscular wall is mediated by estrogen stimulation. Growth of the SV continues slowly until puberty.[66] Secretory activity of the SV is androgen dependent.[95]		The basic pattern of the SV are present at PND 10, with lumen formation from PND 2-15 and markedly increasing in size between PND 11-24 with adult appearance and secretory properties between PND 40-50.[84] Secretory granules are evident at PND 16.	The dog has neither SV nor bulbourethral glands[73]
Spermatozoa	Spermatogonia increase 6-fold between birth and 10 years of age, then increase at puberty. Mean age at spermarche is 13.4 years; ejaculation possible during middle to late puberty.[1,8,13,16]	In the testis, spermatogonia become more numerous by the end of the 1st year. Spermatozoa in the testis appear as early as 3 yrs with fertility starting at approximately 3.5 yrs.[17,19]	In seminiferous tubules PND 45,[31] in vas deferens PND 58-59.[32] Spermatogenesis is stimulated by increased testosterone and gonadotrophin production.	In testis at 26-28 wks of age. First visible in the epididymis 26-28 wks (beagles[41]); first in ejaculate 32-34 weeks (fox terriers[52])
Leydig Cells	Testosterone (T) producing cells, Leydig cells (LC), begin producing T ~7-8 weeks of gestation[96] and ultimately come under control of placental gonadotrophin (human chorionic gonadotrophin, hCG).[97] Pituitary gonadotrophin synthesis begins at ~12 weeks of gestation after T production begins.[98]	LC are prominent in fetal life but decrease in number during the 1st yr and dedifferentiate. By the end of 3 yrs the LCs redifferentiate. T levels in immature males is ~30-250 ng/100 ml rising to 230-1211 ng/100ml in mature males.[18]	Testosterone production begins during late gestation and decreases just prior to birth. During the infantile-juvenile period (PND 8-35) the primary androgens produced include androstenedione, 5 alpha-androstanediol, and dihydrotestosterone NOT testosterone[105,106] but by ~PND 25 and onward testosterone becomes the primary androgen.[105,107] Different androgen levels are primarily due to changes in steroidogenic enzyme levels or activity.	First visible histologically GD 36-46.[39] Testicular LH receptors begin to increase at 2 mos., reach max. 12-24 mos. Testic. T and DHT levels begin to increase at 6 mos., plateau 12-24 mos (beagles[128])	LCs proliferation is dependent on gonadotrophin and ends ~ PND 21-33[33]

Table 1 (continued)

	Human	Non-human Primate	Rat	Dog	Mouse
Sertoli Cells	Sertoli cell (SC) differentiation and production of antimullerian hormone (MIS) begins during the end of the first trimester[18] and is triggered by an unknown mechanism mediated by the Y chromosome.		Increase in number at ~ GD 16 with peak division at GD19 and cease division at ~PND 14-16 [23,24,25,26]. Follicle stimulating hormone (FSH) receptors on SCs increases markedly before birth[27,28,22] and peaks ~PND 18 with a decline until adult levels met ~PND 40-50 [29]	First visible GD 36-46 (Beagles[39]) Divide up until 8 wks post-partum; stable thereafter[41]	Division is under the influence of pituitary gonadotrophin and ends at PND-17.[36,37,38]
Testis Descent	Testis descent occurs prenatally[62,63]	The testes descend at birth but soon after birth ascend into the inguinal canal (postnatal regression). At ~3 yrs the testes descend while increasing in size[64,65]	Testes attached to internal inguinal ring Gestational day 20-21 Testes into scrotum ~ PND 15[60]	Passage of testes through inguinal canal begins PND 3 or 4. Descent complete ca. PND 35-42 [61]	
Epididymal Ontogeny			Undifferentiated PND 0-15, differentiation PND 16-44 and expansion > PND 44 [46]	Postnatally: low columnar epithelium in all segments from birth to 20 wks post-partum. Diameter of lumina increase slowly until week 20, after which there is a burst of growth, which levels off ca. week 48 (caput), 36 (corpus) or >48 (cauda).[41]	

partially due to increased testicular volume. Spermatogonia increase in number 6-fold between birth and 10 years of age[1,8] and this number increases exponentially at puberty along with an increase in testicular volume.

Puberty signals the trigger for the daily production of millions of spermatozoa. Spermarche, the age at which spermatogenesis begins, occurs early in puberty and can be verified by the appearance of sperm in the urine. Increasing levels of testosterone play a key role in the initiation of spermatogenesis at puberty. With the initiation of the spermatogenic cycle, the development of the tubular lumen and the formation of the blood-testis barrier are critical events. The start of pubertal testicular growth in healthy boys (one-sided testicular volume 3-4 ml) occurs between 11.8 and 12.2 years.[9,10,11] Precocious puberty is before the age of 9 years.[12] Nielson et al.[13] reported the mean age at spermarche as 13.4 years (range: 11.7 to 15.3 years). Approximately two years after spermarche, adult levels of testosterone are achieved.[14] From puberty onset, 3.2 ± 1.8 years are required to attain adult testicular volume.[15] Although spermatozoa are produced early in puberty, ejaculation is not possible until middle or late puberty. This signifies the onset of male fertility. Unlike spermatogonia, Sertoli cells and Leydig cells do not proliferate in the adult.[16]

1.1.2 Non-Human Primate (Rhesus; Macaca mulatta)

The steroid hormone producing cells of the testes are the leydig cells (LC) and are prominent during fetal life. During the first year the LCs decrease in number and dedifferentiate entering a period of suspended development. By the end of the third year the LCs redifferentiate and begin to produce the primary steroid hormone testosterone.[17] The testosterone levels in immature males is approximately 30 to 250 ng/100ml and in the mature male range from 230 to 1211 ng/100ml.[18]

In the Yale colony, puberty began at approximately 2.5 years of age.[19] Spermatogonia in the testis become more numerous by the end of the first year but the earliest appearance of spermatozoa is approximately 3 years of age.

1.1.3 Rat

In contrast to humans and primates, rats do not exhibit a period of testicular quiescence in which there is a sustained interruption in gonadotrophin secretion. In rats, postnatal testicular development begins early and maturation steadily progresses. Ojeda and colleagues described male rat postnatal sexual development in 4 stages:[20] a neonatal period from birth to 7 days of age, an infantile period from postnatal days 8-21, a juvenile period which extends to approximately 35 days of age, and a peripubertal period until 55 to 60 days of age, ending when mature spermatozoa are seen in the vas deferens. Spermatocytes, which arise from gonocytes in the fetal testis, cease dividing on gestation day 18 and many degenerate between postnatal days 3 and 7. The remainder (often as few as 25% of those present at birth) begin to divide to form the first spermatogonia.[21] Testicular luteinizing hormone (LH) receptors, interstitial cells and testicular testosterone content increase during gestation, reaching a maximum at about the time of birth.[22] Concentrations of testosterone decline soon after birth. Sertoli cells within the seminiferous tubules begin to increase in number at approximately GD 16, reach a peak of division near GD19, and cease at approximately postnatal day 14-16.[23,24,25,26] Similarly, Sertoli cell follicle stimulating hormone (FSH) receptors increase markedly before birth.[27,28,22] Sertoli cells do not proliferate after postnatal day 16.[16] FSH responsiveness peaks postnatally about PND 18, and declines thereafter until it reaches adult levels at about PND 40-50.[29] There is a rapid proliferation of Leydig cells between days 14 and 28 in the rat with a second cell division between days 28 and 56. Similarly, androgen production increases slowly until 28 days of age, then increases rapidly until 56 days of age.[30] Leydig cells typically do not proliferate in adult rats, although this can occur to replace damaged or destroyed cells.

In male rats, the first spermatozoa appear in the lumen of the seminiferous tubules at 45 days of age[31] and transit through the epididymis to the vas deferens, where they can be detected at 58-59 days of age.[32] Testicular sperm reach a plateau by 77 days of age.[30]

In contrast to these values, Maeda *et al.* cited the following postnatal developmental time lines for Wistar rats:[33] testicular spermatids first appear in the testis at day 20-30 after birth and 100% of animals contain testicular sperm at day 70. In the tail of the epididymis, some males contain sperm at approximately day 40 and almost all animals do so by day 90. The relative weights of the testis and epididymis reach their peak value at around day 70. Testicular weight is around 1% of body weight at 50 days after birth, the time of puberty.

It is important to note that the first wave or two of spermatogenesis in rats is quite inefficient, and there is much greater cell loss peripubertally than there is in adult rats.[34] Thus, studies evaluating animals peripubertally should expect to see greater levels of cell death and structural abnormalities in the controls, which will complicate the identification of such effects in treated animals. The testes continue to increase in size after puberty due to increased total sperm production, although production per unit weight of the testis rapidly reaches a maximal value.[21]

1.1.4 Mouse

Prenatally, germ cells and Sertoli cells migrate from the mesonephric ridge much as in the rat.[35] Postnatally, the earliest Type A spermatogonia can be observed at PND 3, and these are proliferating. Sertoli cells proliferate until PND17 under the influence of pituitary gonadotrophins.[36,37,38] Leydig cells proliferate much later, closer to puberty (PND 21-33),[15] and this is dependent on gonadotrophin in serum.[38] Puberty in the mouse is approximately PND 35.

1.1.5 Dog

Testis differentiation was observed at GD 36 in Schnauzers and beagles;[39] this was followed closely by regression of the Müllerian ducts. Postnatally, the seminiferous tubules are composed of immature Sertoli cells and gonocytes at 2 weeks of age. During the first 3 postnatal weeks, the Leydig cells appear to be mature, then they appear to regress from weeks 4 to 7, although there is no appreciable change in androgen levels. At 8 weeks of age, Leydig cells appear to be active and mitotic figures can be found in both Sertoli cells and spermatogonia. Between 16 and 20 weeks, the number of germ cells per cross section of cord decreases and the amount of lipid increases. At 16 weeks, evidence for the leptotene stage of meiotic prophase is evident as condensed spermatogonial chromatin.[40] At wk 18-20, the germ cells begin their rapid division, and tubule cellularity and diameter increase rapidly.[41] Round spermatids begin to appear at wk 22, long spermatids at wk 26, and by wk 28 in beagles, diameter is nearly at adult values, when all cell types are represented.

1.2 Epididymides

1.2.1 Human

As with other mammals, development of the epididymis is dependent upon androgens from the testicular Leydig cells. Furthermore, differentiation of the epididymal epithelium is dependent upon constituents of luminal fluid from the testis or proximal epididymis.[42]

1.2.2 Rat

Like the testis, the epididymis arises embryologically from the middle of three nephroic regions in the fetus. The most caudal region gives rise to the kidney. The cells in the mesonephros, in contrast, migrate caudally, and forms a diffuse networks of ducts. In the presence of Müllerian inhibiting substance, the cranial portion of the Wolffian duct forms the epididymis. These cells develop and proliferate under the influence of testosterone, and is formed into a single tubule by late gestation.[43] Prenatal exposure to compounds that reduce testosterone synthesis androgenic signal interfere with epididymal development, frequently resulting in the absence of whole sections of the organ.[44,45]

Sun and Flickinger[46] name three periods of rat epididymal ontogeny: an undifferentiated period (birth to PND15), a period of differentiation (PND16-44), and a period of expansion (>PND44). DeLarminat et al[47] found the time of greatest cell division was PND25. This picture becomes more complex when cell division is considered by region,[48] but still, the vast majority of cell division occurs prior to PND30.[49] This is reasonably consistent with the picture presented by Limanowski et al.[50] Both testosterone and the arrival of germ cells and their bathing fluid from the rete testis are believed to contribute to epididymal differentiation.[16] Spermatozoa appear in the epididymis at 49 days of age and reach their highest levels in the epididymis at 91 days of age. This period corresponds with the intervals with the greatest increases in epididymal weight between days 49-63 and 77-91.[30]

Another critical element in epididymal development is the formation of the blood-epididymis barrier, which is complete in rats by postnatal day 21, prior to the appearance of spermatozoa in the epididymal lumen.[51]

1.2.3 Dog

Kawakami[41] found that beagle epididymal epithelial cell height and duct diameter both rose very slowly postnatally until ca. postnatal wk 22, whereupon both measures increased sharply and then leveled off. Visible epididymal sperm apparent density was zero at wk 24, "a small number" at wk 26, more at wk 28, and apparent sperm density in the epididymis plateaued at and after wk 30. Mailot[52] (using fox terriers) followed ejaculated sperm measures from postnatal wks 30-57, and found a continual increase in count; obvious influences on this number could be the underlying maturation of spermatogenesis, confounded by the dog's accommodation to the sample collection process.

1.3 The Blood-Testis Barrier

1.3.1 Human

At 5 years of age, some junctional particles are visible in freeze-fractured preparations of the human testis. By 8 years of age, rows and plaques of communication junctions are present. Lanthanum readily penetrates into the intercellular space. The blood-testis barrier is completed at puberty, when a continuous belt of junctional particles can be seen at the onset of spermatogenesis.[53]

1.3.2 Rat

Permeability of the blood-testis barrier decreases markedly between 15 and 25 days of age.[54,55] The Sertoli cell tight junctions form between postnatal days 14 and 19, concurrent with the cessation of Sertoli cell divisions, and the movement of the first early spermatocytes up to the luminal side of the barrier. The blood-testis barrier in adults excludes even smaller molecules than those restricted from entering the testis in prepubescent animals.[56]

1.3.3 Dog

At 8 weeks of age, there are no occluding junctions between Sertoli cells in the beagle testis, but septate junctions are present that partially limit the penetration of lanthanum. At 13 to 17 weeks of age, Sertoli cells exhibit incompletely formed tight junctions, which appear in patches. By 20 weeks of age, these tight junctions appear linearly arranged and connected with the septate junctions. At this time, lanthanum can no longer penetrate into the adluminal compartment.[40]

1.4 Anogenital Distance

1.4.1 Rat

In rats, gender is often determined by external examination of anogenital distance (AGD) during early neonatal periods. AGD, defined as the distance between the genital tubercle and the anus, is approximately 2.5X greater for males than females. A similar sex difference in AGD has been reported for rhesus monkey fetuses.[57] It is not clear whether a change in AGD in laboratory animals corresponds to an adverse effect in humans. There are no reports of decreased AGD in pseudohermaphroditic men, who lack 5-reductase, and finasteride, a 5-reductase inhibitor, failed to alter the AGD of monkeys at doses causing hypospadias.[57] Age-related changes in AGD are cited in Clark.[29] Mean AGD for control groups of Sprague-Dawley rats on postnatal day 0 was 3.51 (3.27-3.83) mm for males and 1.42 (1.29-1.51) mm for females. Absolute AGD is affected by body size.[58,59] Note that there may be some interlaboratory variation in AGD measurements.

1.5 Testicular Descent

The process of testicular descent is similar in most species, although there are some timing and topographical differences (e.g., gubernacular outgrowth, mesenchymal regression and development of the cremaster muscle[60]). Basically, the testicles are attached to the mesenephros intra-abdominally during gestation. During testicular descent, the mesenephros degenerates and the gubernaculum proprium increases in size, thereby dilating the inguinal canal and allowing the testicle to pass through. Once testicular descent is complete, the gubernaculum shortens, allowing the testicle to move into the scrotum.[61]

1.5.1 Humans

Hogan et al.[62] cites gestation weeks 7 to 28 as the period for testicular descent in humans whereas Gondos[63] identifies the seventh month of gestation as the time the processus vaginalis grows and the inguinal canal increases in diameter for passage of the testis. Descent occurs due to degeneration of part of the gubernaculum. Thus, Gondos[63] cites the latter part of gestation as the period when the testis completes its descent into the scrotum.

1.5.2 Non-Human Primate

Testicular descent occurs at birth.[64] However, soon after birth testes ascend into the inguinal canal and precedes into a postnatal regression.[65] At approximately 3 years of age the testes descend again while also increasing in size.

1.5.3 Dog

Testicular descent is not complete in dogs at the time of birth. Most of the mesenephros degeneration is complete prior to initiation of testicular descent and there is little increase in testicular size while the testes remain in the abdomen. Passage of the testicle through the inguinal canal occurs on postnatal day 3 or 4. Regression of the gubernaculum begins after testicular passage and is complete by 5-6 weeks of age.[61]

1.5.4 Rat

On gestation day 20-21, the testicle in the rat is attached to the internal inguinal ring with its caudal pole and the cauda epididymis located within the canal. Enlargement of the gubernaculum occurs postnatally with descent of the testicles occurring on approximately postnatal day 15. Unlike many species, the inguinal canal remains wide in the adult rat, allowing the male rat to lift its testicles into its abdomen.[60]

1.6 Preputial Separation

1.6.1 Human

In humans, PPS begins during late gestation[66] and is generally completed postnatally between 9 months and 3 years of age.[67,68,69] Androgens are assumed to play a role in PPS in humans.[70]

1.6.2 Rat

An external sign of puberty onset in the male rat is balano-preputial separation (PPS) when the prepuce separates from the glans penis.[71] Initially, the rat penis looks similar to the clitoris of the female and the glans of the penis is difficult to expose by 30 days after birth.[33] The shape of the tip of the penis changes from a V-shape to a W-shape (20-30 days after birth) to the U-shape, which is seen in 100% of animals by day 70 after birth.[33] Mean age at PPS for Sprague-Dawley rats was 43.6 ± 0.95 (X + SD) days of age (range: 41.8-45.9[58]). There is a complex relationship between the onset of puberty and body weight.[72]

2 Accessory Sex Glands

2.1 Prostate

Although serving a similar function, the prostate gland varies among mammals, making selection of an animal model problematic. For example, latent cancer of the prostate is a high incidence disease in older men, yet it is rare among animals. In animal models, neither prostate cancer (spontaneous or induced) nor prostatic hypertrophy parallels exactly the human disease in morphology, biochemistry, response to hormonal manipulation and metastatic spread.[73]

2.1.1 Human

Although present, the lobes of the human prostate are not sharply demarcated; the prostate appears as a single gland with several zones. During the fetal period, the prostate appears as a few widely-spaced tubules supported by stromal cells.[74] Prior to birth, the number and proximity of prostatic tubules increases markedly, and proliferation and secretion of the tubule epithelium occurs. The middle and lateral prostate lobes exhibit prominent squamous metaplasia of tubular epithelium. Eventually, desquamation occurs and epithelial cells in the prostatic tubules are shed into the lumen. Thus, for a short period after birth (~1 month), the histology of the prostate does not change appreciably; however, metaplastic changes that occurred during the fetal period, regress after birth leaving empty tubules with only a few remnants of the previous metaplasia by 3 months of age.[74] The tubule epithelium also regresses.[74] During the period of regression, the state of the prostatic tubules may vary somewhat from being filled with metaplastic cells to being predominantly empty with some debris from desquamated cells.[74] It should be noted that squamous metaplasia is limited to the fetal period, except for recurrence during pathological conditions in the adult.[74] The remaining lobes (anterior and posterior lobes) exhibit little or no metaplastic changes.[74]

Prostatic secretion is present at the time of birth. Androgen-stimulated secretion from the prostate gland initially occurs at approximately gestation week 14[74] and increases in the incidence and degree of secretion throughout the remainder of the fetal period. Up to the last month of gestation, secretion is limited to small tubules at the periphery of the lateral and anterior lobes of the fetal prostate. During this time, an intermediate zone separates the secreting tubules from the more central parts, which are still undergoing varying degrees of squamous metaplasia. Once squamous metaplasia has subsided and the metaplastic epithelial cells have been shed, secretory activity may begin in these areas too.[74] After birth, when squamous metaplasia has entirely subsided, secretion will continue for some time. Strong secretion can still be noted 1 month after birth, but becomes more variable thereafter.[74,75]

The transition of the prostate from metaplasia to desquamation to secretion is a hormonally-mediated process. *In utero* estrogen stimulation induces metaplasia in the prostatic utricle and the surrounding glands. As gestation progresses, the metaplastic cells become squamous and are eventually shed, thus leading to lumen formation and its sac-like structure. After birth, metaplastic changes regress and estrogen stimulation ends, resulting in a cessation of the metaplastic process and a decrease in utricle distension.[42] As estrogen levels decline, the changing hormonal balance favors increased androgen levels and stimulation of prostatic secretion. Because squamous metaplasia and secretion are controlled by different hormones, these processes are localized to different areas of the prostate; thus, secretion is initiated in the periphery and extends into the central areas of the prostate once squamous metaplasia has been completed in these areas.[74]

2.1.2 Dog

The prostate is the only well-developed accessory sex gland in the dog. It is relatively large, completely surrounding the urethra.[73] The prostate is divided into a right and a left lobe with the middle lobe either poorly developed or absent. In contrast, it is the middle lobe that obstructs the urethra in men with enlarged prostate.[76] In the canine prostate, branching secretory acini and ducts radiate from each side of the urethra. According to O'Shea,[77] the size and weight of the adult canine prostate are as follows: 1.9-2.8 cm long, 1.9-2.7 cm wide, 1.4-2.5 cm high and 4.0-14.5 g (0.21-0.57 g/kg) in weight.

In newborn puppies, the prostate is comprised primarily of stroma with some discernable parenchyma. During adolescence, the parenchyma proliferate more quickly than the stroma, making parenchymal cells the primary cell type in the sexually mature adult dog. Parenchyma are not equally distributed; the stroma still predominates in the central area of the prostate with little parenchyma visible anterior to the colliculus seminalis and dorsal to the urethra, even in the adult dog.[78]

The prostate of the adult dog passes through 3 stages: normal growth in the young animal, hyperplasia during the middle of adult life, and senile involution.[73] Similar to other species, prostate development and function are controlled by androgens; however, the histological structure of the canine prostate varies with age. Up to 1 year of age, prostatic growth is slow until puberty approaches, then growth is rapid and associated with the development of structural and functional maturity.[73] As androgen levels rise during puberty, androgens reach a sufficient level to complete normal prostatic growth and maturation. As androgenic stimulation continues, a phase of hyperplastic growth occurs which is manifested as a loss of normal histology and onset of glandular hyperplasia. Cysts may form during this period.[79,73] Growth of the prostate in the adult dog appears to proceed at a steady rate up to about 11 years of age. Senile involution occurs from ~11 years onward. During this stage, there is a steady decline in prostate weight, probably due to decreased androgen production. The prostate may or may not exhibit histological evidence of atrophy.[73]

Evidence for prostatic secretion by the alveolar epithelium can be detected at ~4 months of age in dogs. Under normal conditions, secretory epithelial cells originate from differentiated basal reserve cells rather than through a process involving metaplasia. Similar to the parenchyma in the

Table 2

Rat	Man
Ventral prostate	—
Lateral prostate	Lateral lobes
Dorsal prostate	Dorsal (or posterior) lobe
Anterior prostate	Median (or middle) lobe

periphery of the prostate, some glandular tissue is present in the suburethral submucosa and in the colliculus seminalis and surrounding tissue at 9 months of age. This glandular tissue does not seem to increase significantly in number or size throughout life.[73]

Aside from man, dogs are the only common mammals that spontaneously develop benign prostatic hypertrophy (BPH).[80,81] BPH in these two species is similar with respect to older age at onset (5 years or 5th decade), requirement for normal testicular function and prevention by early In both species, dihydrotestosterone levels are elevated in hyperplastic prostate tissue.[82,83] However, human and canine BPH differ in histology,[76,77,79,81] symptomatology and magnitude of their response to antiandrogen treatment.[84,85]

Canine prostate glands can respond to both androgenic and estrogenic signals, having receptors for both steroids with a prevalence of estrogen receptors.[86] Estradiol increases cytosolic androgen-binding protein and thereby, stimulates androgen-mediated prostate growth.[87] Within the prostate, dihydrotestosterone is the predominant hormone at various ages, including in immature, mature and hypertrophic prostate glands.[82]

2.1.3 Rat

Unlike the human prostate, the rat prostate is comprised of discrete lobes (ventral, lateral, dorsal and the paired anterior lobes that are also known as the coagulating glands). In immature and young rats, separation of the lobes of the prostate, particularly the dorsal and lateral lobes, is possible, whereas in the adult rat, the dorsolateral prostate is often examined due to the lack of a distinct border between these glands.[73] Table 2 lists probable homologies between the rat and human prostates.[73,88] Note that there is no embryological corresponding structure in the adult man to the rat ventral prostate.

Much of rat prostate development occurs postnatally. The prostatic lobes form between postnatal days 1-7 with prostatic tubular lumen forming between postnatal days 7-14. Secretory granules are evident in the developing prostate between postnatal days 14-21. The prostate has attained its adult appearance by postnatal days 28-35, which parallels the postnatal increase in testosterone levels.

Several other distinctions appear between rat and human prostates. The rat prostate lacks a strong, well developed fibromuscular stroma.[73] Furthermore, there are differences in enzyme activities, zinc uptake and concentration, and antibacterial factor distribution (present throughout the prostate in humans, but only in dorsolateral prostate in the rat).[73]

2.1.4 Mouse

The postnatal development of the murine prostate was evaluated in a series of papers by Sugimura et al.[89,90,91] These studies found that the number of branch points and tips in both the ventral prostate and the dorsolateral prostate increased dramatically in the first 15 days postpartum. There were significant differences between lobes in terms of the number of branch points, and morphologies (both gross and microscopic). The numbers of main ducts and tips of ducts reached adult levels by ≈ PND30. The differences between lobes noted here have been reported in numerous other studies, mostly in rats.[92,93,94]

2.2 Seminal Vesicles

2.2.1 Human

The seminal vesicles and the ductus deferens are present by the 6th month of fetal development in an arrangement similar to that seen in adults. The seminal vesicles have a larger lumen and thicker, stronger muscular wall than the ductus deferens. Prenatal development of the muscular wall is mediated by estrogenic stimulation. By the 7th month of gestation, the seminal vesicles have attained their adult form, although it is not until term that the mucous membrane surrounding the lumen begins to arrange itself in folds.[42] The seminal vesicles continue to grow slowly until puberty.[42]

Secretory activity is present in the seminal vesicles by the 7th month of gestation and slowly increases thereafter, persisting for a considerable time after birth (detected at 17 months). At 4 years of age, seminal vesicle secretion was no longer detected.[74] Secretion by the seminal vesicles is androgen dependent.[95]

2.2.2 Dog

The dog has neither seminal vesicles nor bulbourethral glands.[73]

2.2.3 Rat

In rats, the basic pattern for seminal vesicle formation is present at postnatal day 10. Lumen formation occurs over a relatively protracted period from postnatal days 2-15. Secretory granules become evident at postnatal day 16. The seminal vesicles markedly increase in size between postnatal days 11-24, and continues to grow until adult appearance and secretory properties are attained between PND 40 and 50.[84] Thus, proliferation and differentiation of the seminal vesicles parallels the postnatal increase in testosterone levels.

3 Neuroendocrine Control of the Reproductive System

In mammals, the pituitary and gonads are capable of supporting gametogenesis prior to puberty; however, events in the brain are required to change the hypothalamic-pituitary-gonadal (HPG) axis and trigger maturational changes.

3.1 Human

In humans, much of the process controlling testicular androgen production is present at the time of birth. In the fetal testis, production of testosterone and antimüllerian hormone begins at the end of the first trimester of pregnancy.[18] Sertoli cells, triggered through an unknown mechanism mediated by the Y chromosome, differentiate and produce antimüllerian hormone. Subsequently, Leydig cells differentiate at approximately 7-8 weeks of gestation and begin producing androgens[96] that ultimately come under the control of the placental gonadotrophin, human chorionic gonadotrophin (hCG).[97] Pituitary gonadotrophin synthesis begins around week 12 of gestation with initially high levels that decline towards the end of gestation, the likely period for the onset of negative feedback regulation.[98] Thus, the hypothalamic-pituitary-gonadal axis is fully functional in the fetus and neonate. Neonatal exposure of the hypothalamus to androgen is need for sexual differentiation of the LH release mechanism, allowing LH secretion to be modified by either androgen or estrogen.[16] Thus, transient elevations of FSH, LH and testosterone have been noted in boys during the first 6 months of life,[7] then pulsatile secretion of gonadotrophin declines, reaching its lowest point at 6 years of age.[99] Thereafter, pulsatile gonadotrophin secretion begins to increase.

Both LH and FSH are involved in the initiation of spermatogenesis. Pulses of LH elicit increases in androgen concentrations. As age increases, pulsatile gonadotrophin secretion increases in frequency and amplitude in response to pulsatile GnRH secretion. During puberty, inhibin from Sertoli cells is the primary negative feedback agent to control FSH release. Testosterone regulates both FSH and LH at the level of the hypothalamus. From puberty onward, androgen increases libido.

3.2 Rat

A comprehensive review of rat neuroendocrine development for the control of reproduction has been compiled by Ojeda and Urbanski.[100]

Pituitary-gonadal maturation occurs later in rats than in humans, although the sequence of events is similar.[98] Gonadotrophin secretion and testicular androgen production begin during the last third of gestation and continue to gradually decline during the first two weeks of postnatal life.[101] In neonates, testicular androgen production is needed during the first few days of life to imprint male sexual behavior. Unless exposure to steroids occurs, the rat hypothalamus will show a female pattern of discharge exhibiting cyclical activity.[21]

Postnatal development of the reproductive system requires hormonal signals from the hypothalamic-pituitary (HP) axis, a subsequent testicular response and feedback from the testis to the HP axis to modulate gonadotrophin release. Initially during the neonatal period, serum gonadotrophins are high in male rats, but decline rapidly within a few days.[102] Leydig cells undergo rapid proliferation from postnatal days 14 to 28 and another cell division between 28 and 56. Similarly, androgen production increases slowly until 28 days of age with a pronounced increase between 50 and 60 days of age.[102,103,104] During this peripubertal period, hormone production by the testis changes due to changes in steroidogenic enzyme levels. During the prepubescent period, androstenedione, 5-androstanediol, and dihydrotestosterone are the primary hormones produced by the testis;[105,106] however, after 40 days of age, testosterone becomes the primary testicular androgen.[107] Adult testosterone levels are achieved at approximately 56 days of age.[30] Increased testosterone production, coupled with gonadotrophins, stimulates spermatogenesis and development and maintenance of the accessory sex organs.

Maturation of the reproductive system is initiated by a centrally-mediated process. Gonadotrophin release is negatively controlled by testosterone and inhibin. As early as the neonatal period, testosterone-induced negative feedback to the HP is present.[108,109,110] During postnatal life and into adulthood, concentrations of GnRH in the hypothalamus continue to increase.[111,72,112] Coincident with this, LH and FSH levels in the pituitary rise postnatally and the pituitary response to GnRH stimulation also is enhanced.[113,114] FSH supports spermatogenesis within the seminiferous tubules and stimulates the formation of gonadotrophin receptors in the testis,[115,116] while LH triggers interstitial cells to produce and secrete testosterone. Increases in serum FSH promote the production of steroidogenic enzymes[117] and overall testicular growth.[118] As the animal matures, the HP becomes progressively less sensitive to negative feedback allowing for puberty onset. At this time, pulsatile GnRH release is enhanced, resulting in increased circulating levels of LH and FSH from the pituitary.[119,120,121] Thus, testicular maturation and puberty onset occur secondary to changes in the secretion of pituitary gonadotrophins.

After puberty has occurred, gonadotrophin levels stabilize. Thus, after maximum serum FSH levels have been achieved (30-40 days), serum testosterone rises and FSH decreases to relatively low adult levels.[102,122,120,123] The maximum responses to FSH and LH occur between 25-35 and 35-45 days of age, respectively,[124,125] then decline to adult levels. Adult responses are achieved between 60 and 80 days of age.[126,80]

Conclusion

Overall, there is evidence of similar patterns of male postnatal reproductive system development between humans and experimental animal models. However, a more detailed examination of the developmental processes reveals pertinent cross-species differences. As research progresses, more thorough comparisons on the cellular and molecular level will become possible and the specific pathways targeted by chemicals can be identified. This knowledge will assist in the selection of sensitive and predictive animal models and further reduce uncertainty when extrapolating animal data to human risk.

REFERENCES

1. Müller J. and Skakkebaek, N.E. (1992). The prenatal and postnatal development of the testis. *Bailliere's Clin. Endo. Metab.* 6, 251-271.
2. Cortes, D., Müller, J. and Skakkebaek, N.E. (1987). Proliferation of Sertoli cells during development of the human testis assessed by stereological methods. *Int. J. Androl.* 10, 589-596.
3. Lemasters, G.K., Perreault, S.D., Hales, B.F., Hatch, M., Hirshfield, A.N., Hughes, C.L., Kimmel, G.L., Lamb, J.C., Pryor, J.L., Rubin, C. and Seed, J.G. (2000). Workshop to identify critical windows of exposure for children's health: reproductive health in children and adolescents work group summary. *Environ. Health Perspect.* 108 (Suppl. 3), 505-509.
4. Rey, R. (1999). The prepubertal testis: a quiescent or a silently active organ? *Histol. Histopathol.* 14, 991-1000.
5. Claudio, L., Bearer, C.F. and Wallinga, D. (1999). Assessment of the U.S. Environmental Protection Agency methods for identification of ;hazards to developing organisms, part I: the reproduction and fertility testing guidelines. *Am. J. Ind. Med.* 35, 543-553.
6. Clements, J.A., Reyes, F.I. Winter, J.S.D. and Faiman, C. (1976). Studies on human sexual development. III: fetal pituitary and serum and amniotic fluid concentrations of LH, CG, and FSH. *J. Clin. Endocrinol. Metab.* 42, 9-19.
7. Müller, J. and Skakkebaek, N.E. (1984). Fluctuations in the number of germ cells during the late foetal and early postnatal periods in boys. *Acta Endocrinol.* 105, 271-274.
8. Müller, J. and Skakkebaek, N.E. (1983). Quantification of germ cells and seminiferous tubules by stereological examination of testicles from 50 boys who suffered from sudden death. *J. Androl.* 6, 143-156.
9. Largo, R.H. and Prader, A. (1983). Pubertal development in Swiss boys. *Helv. Paediatr. Acta* 38, 211-228.
10. Biro, F.M., Lucky, A.W. and Huster, G.A. (1995). Pubertal staging in boys. *J. Pediatr.* 127, 100-102.
11. Nysom, K., Pedersen, J.L., Jorgensen, M., Nielsen, C.T., Müller, J., Keiding, N. and Skakkebaek, N.E. (1994). Spermaturia in two normal boys without other signs of puberty. *Acta Paediatr.* 83, 520-521.
12. Partsch, C.-J. and Sippell, W.G. (2001). Pathogenesis and epidemiology of precocious puberty. Effects of exogenous oestrogens. *Hum. Reprod. Update* 7, 292-302.
13. Nielsen, C.T., Skakkebaek, N.E., Richardson, D.W., Darling, J.A., Hunter, W.M., Jorgensen, M., Nielsen, A., Ingerslev, O., Keiding, N. and Müller, J. (1986). Onset of the release of spermatozoa (spermarche) in boys in relation to age, testicular growth, pubic hair, and height. *J. Clin. Endocrinol. Metab.* 62, 532-535.
14. Nielsen, C.T., Skakkebaek, N.E., Darling, J.A.B., Hunter, W.M., Richardson, D.W., Jorgensen, M. and Keiding, N. (1986). Longitudinal study of testosterone and luteinizing hormone (LH) in relation to spermarche, pubic hair, height and sitting height in normal boys. *Acta Endocrinol. Suppl.* 279, 98-106.
15. Buchanan, C.R. (2000). Abnormalities of growth and development in puberty. *J. R. Coll. Physicians Lond.* 34, 141-145.
16. Pryor, J.L., Hughes, C., Foster, W., Hales, B.F. and Robaire, B. (2000). Critical windows of exposure for children's health: the reproductive system in animals and humans. *Environ. Health Perspect.* 108 (Suppl. 3), 491-503.

17. Catchpole, H.R. and Van Wagenen, G., Reproduction in the Rhesus Monkey, Macaca mulatta. In: The Rhesus Monkey Vol. II; Management, Reproduction and Pathology (ed. Geoffrey H. Bourne), Academic Press, New York, Pgs. 132-138, 1975.
18. Bennett, W.I, Dufau, M.L., Catt, K.J. and Tullner, W.W. (1973). Effect of human menopausal gonadotropin upon spermatogenesis and testosterone production in juvenile rhesus monkeys. *Endocrinology* 92, 813-812.
19. Van Wagenen, G. and Catchpole, H.R. (1956). *Amer. J. Phys. Anthropol.* 14, 245-274.
20. Ojeda, S.R., Andrews, W.W., Advis, J.P., and Smith-White, S. (1980). Recent advances in the endocrinology of puberty. *Endocr. Rev.* 1, 228-257.
21. Setchell, B.P. (1978). Spermatogenesis. *In: The Mammalian Testis.* Cornell University Press, Ithaca, New York, pp. 181-232.
22. Warren, D.W., Huhtaniemi, I.T., Tapanainen, J., Dufau, M.L., and Catt, K.J. (1984). Ontogeny of gonadotropin receptors in the fetal and neonatal rat testis. *Endocrinology* 114, 470-476.
23. Steinberger, A., and Steinberger, E. (1971). Replication pattern of Sertoli cells in maturing rat testis *in vivo* and in organ culture. *Biol. Reprod.* 4, 84-87.
24. Orth, J. (1982). Proliferation of Sertoli cells in fetal and postnatal rats: A quantitative autoradiographic study. *Anat. Rec.* 203, 485-492.
25. Wang, Z.X., Wreford, N.G.M., and de Kretser, D.M. (1989). Determination of Sertoli cell numbers in the developing rat testis by stereological methods. *Int. J. Androl.* 12, 58-64.
26. Van den Dungen, H.M., van Dieten, J.A.M.J., van Rees, G.P., and Schoemaker, J. (1990). Testicular weight, tubular diameter and number of Sertoli cells in rats are decreased after early prepubertal administration of an LHRH-antagonist; the quality of spermatozoa is not impaired. *Life Sci.* 46, 1081-1089.
27. Steinberger, A., Thanki, K.H., and Siegal, B. (1974). FSH binding in rat testes during maturation and following hypophysectomy. Cellular localization of FSH receptors. *In: Current Topics in Molecular Endocrinology. Vol. 1.* (M.L. Dufau, and A.R. Means, eds.) Plenum Press, New York, pp. 177-192.
28. Hodgson, Y., Robertson, D.M., and de Kretser, D.M. (1983). The regulation of testicular function. *Int. Rev. Physiol.* 27, 275-327.
29. Waites, G.M.H., Speight, A.C., and Jenkins, N. (1985). The functional maturation of the Sertoli cell and Leydig cell in the mammalian testis. *J. Reprod. Fertil.* 75, 317-326.
30. Scheer, H. and Robaire, B. (1980). Steroid ^{4}-5-reductase and 3-hydroxysteroid dehydrogenase in the rat epididymis during development. *Endocrinology* 107, 948-953.
31. Clermont, L., and Perey, B. (1957). Quantitative study of the cell population of the seminiferous tubules in immature rats. *Am. J. Anat.* 100, 241-267.
32. Clegg, E.J. (1960). The age at which male rats become fertile. *J. Reprod. Fertil.* 1, 119-120.
33. Maeda, K., Ohkura, S., and Tsukamura, H. (2000). Physiology of Reproduction. *In: The Laboratory Rat.* (G. J. Krinke, ed.) Academic Press, New York, pp. 145-176.
34. Russell, L.D., Alger, L.E., and Nequin, L.G. (1987). Hormonal control of pubertal spermatogenesis. *Endocrinol.* 120, 1615-1632.
35. Anderson, R., Garcia-Castro, M., Heasman, J., and Wylie, C. (1998). Early stages of male germ cell differentiation in the mouse. *APMIS* 106, 127-133.
36. Vergouwen, R.P.F.A., Jacobs, S.G.P.M., Huiskamp, R., Davids, J.A.G., and de Rooij, D.G. (1991) Proliferative activity of gonocytes, Sertoli cells, and interstitial cells during testicular development in mice. *J. Reprod. Fertil.* 93, 233-243.
37. Vergouwen, R.P.F.A., Huiskamp, R., Bas, R.J., Roepers-Gajadien, H.L., Davids, J.A.G., and de Rooij, D.G. (1993) Postnatal development of testicular cell populations in mice. *J. Reprod. Fertil.* 99, 479-485.
38. Baker, P.J. and O'Shaughnessy, P.J. (2001) Role of gonadotrophins in regulating numbers of Leydig and Sertoli cells during fetal and postnatal development in mice. *Reproduction* 122, 227-234.
39. Myers-Wallen, V.N., Manganaro, T.F., Kuroda, T., Concannon, P.W., MacLaughlin, D.T., and Donahoe, P.K. (1991) The critical period for Mullerian duct regression in the dog embryo. *Biol. Reprod.* 45, 626-633.
40. Connell, C.J. (1980). Blood-testis barrier formation and the initiation of meiosis in the dog. *In: Testicular Development, Structure, and Function.* (A. Steinberger and E. Steinberger, eds.) Raven Press, New York, pp. 71-78.

41. Kawakami, E., Tsutsui, T, and Ogasa, A (1991) Histological observations of the reproductive organs of the male dog from birth to sexual maturity. *J. Vet. Med. Sci.* 53, 241-248.
42. Rogriguez, C.M., Kirby, J.L. and Hinton, B.T. (2002). The development of the epididymis. *In: The Epididymis. From Molecules to Clinical Practice.* (B. Robaire and B.T. Hinton, eds.), Kluwer Academic/Plenum Publishers, New York, pp. 251-267.
43. Büskow, A.G., and Hoyer, P.E. (1988) *The Physiology of Reproduction.* Vol. 1. Raven Press, New York, pp. 265-302.
44. Mylchreest, E., D. G. Wallace, R. C. Cattley and P. M. D. Foster: Dose-Dependent Alterations in Androgen-Regulated Male Reproductive Development in Rats Exposed to Di(n-butyl) Phthalate during Late Gestation. *Toxicol. Sci.* 2000, 55, 143-151.
45. McIntyre, B. S., N. J. Barlow and P. M. D. Foster: Androgen-mediated development in male rat offspring exposed to flutamide in utero: Permanence and correlation of early postnatal changes in anogenital distance and nipple retention with malformations in androgen-dependent tissues. *Toxicol. Sci.* 2001, 62, 236-249.
46. Sun, E.L. and Flickinger, C.J. (1979). Development of cell types and of regional differences in the postnatal rat epididymis. *Am. J. Anat.* 154, 27-56.
47. DeLarminat, M.A., Cuasnicu,P.S., and Blaquier, J.A. (1981). Changes in trophic and functional parameters of the rat epididymis during sexual maturation. *Biol. Reprod.* 25, 813-819.
48. Sun, E.L. and Flickinger, C.J. (1982). Proliferative activity in the rat epididymis during postnatal development. *Anat. Record.*203, 273-284.
49. Ramirez, R., Martin, R., Martin, J.J., Ramirez, J.R., Paniagua, R., and Santamaria, L (1999). Changes in the number, proliferation rates, and bcl-2 protein immunoexpression of epithelial and periductal cells from rat epididymis during postnatal development. *J. Androl.* 20, 702-712.
50. Limanowski, A., Miskowiak, B., Otulakowski, B., Partyka, M., and Konwerska, A. (2001). Morphometric studies on rat epididymis in the course of postnatal development (computerised image analysis). *Folia Histochemica et Cytobiol.* 39, 201-202
51. Agarwal, A. and Hoffer, A.P. (1989). Ultrastructural studies on the development of the blood-epididymis barrier in immature rats. *J. Androl.* 10, 425-431.
52. Mailot, J.P., Guerin, C, and Begon, D (1985) Growth, testicular development, and sperm output in the dog from birth to post-pubertal period. *Andrologia* 17, 450-460.
53. Camatini, M., Franchi, E. and DeCurtis, I. (1981). Differentiation of inter-Sertoli junctions in human testis. *Cell Biol. Int. Rep.* 5, 109.
54. Vitale, R., Fawcett, D.W., and Dym, M. (1973). The normal development of the blood-testis barrier and the effects of clomiphene and estrogen treatment. *Anat. Rec.* 176, 333-344.
55. Setchell, B.P., Laurie, M.S., and Fritz, I.B. (1980). Development of the function of the blood-testis barrier in rats and mice. *In: Testicular Development, Structure, and Function.* (A. Steinberger and E. Steinberger, eds.) Raven Press, New York, pp. 65-69.
56. Setchell, B.P., Voglmayr, J.K., and Waites, G.M.H. (1969). A blood-testis barrier restricting passage from blood into rete testis fluid but not into lymph. *J. Physiol.* 200, 73-85.
57. Prahalada, S., Tarantal, A.F., Harris, G.S., Ellsworth, K.P., Clarke, A.P., Skiles, G.L., MacKenzie, K.I., Kruk, L.F., Ablin, D.S., Cukierski, M.A., Peter, C.P., vanZwieten, M.J., and Hendrickx, A.G. (1997). Effects of finasteride, a type 2 5-alpha reductase inhibitor, on fetal development in the rhesus monkey (*Macaca mulatta*). *Teratology* 55, 119-131.
58. Clark, R.L. (1999). Endpoints of reproductive system development. *In: An Evaluation and Interpretation of Reproductive Endpoints for Human Risk Assessment.* International Life Sciences Institute/Health and Environmental Science Institute, Washington, D.C. pp. 27-62.
59. Wise, L.D., Vetter, C., Anderson, C.A., Antonello, J.M., and Clark, R.L. (1991). Reversible effects of triamcinolone and lack of effects with aspirin or L-656,224 on external genitalia of male Sprague-Dawley rats exposed *in utero*. *Teratology* 44, 507-520.
60. Wensing, C.J. (1986). Testicular descent in the rat and comparison of this process in the rat with that in the pig. *Zoology* 423-434.
61. Wensing, C.J.G. (1973). Testicular descent in some domestic mammals. I. Anatomical aspect of testicular descent. *Zoology* 423-434.

62. Hogan, M.D., Newbold, R.R. and McLachlan, J.A. (1987). Extrapolation of teratogenic responses observed in laboratory animals to humans: DES as an illustrative example. Cold Spring Harbor Laboratory, Cold Spring Harbor, New York.
63. Gondos, B. (1985). Development of the reproductive organs. *Ann. Clin. Lab. Sci.* 15, 363-373.
64. Wislocki, G.B. (1933). *Anat. Rec.* 57, 133-148.
65. Catchpole, H.R. and Van Wagenen, G., Reproduction in the Rhesus Monkey, Macaca mulatta. In: The Rhesus Monkey Vol. II; Management, Reproduction and Pathology (ed. Geoffrey H. Bourne), Academic Press, New York, Pgs. 132-138, 1975.
66. Deibert, G.A. (1933). The separation of the prepuce in the human penis. *Anat. Rec.* 57, 387-393.
67. Ben-Ari, J., Merlob, P., Mimouni, F., and Reisner, S.H. (1985). Characteristics of the male genitalia in the newborn: penis. *J. Urol.* 134, 521-522.
68. Oster, J. (1968). Further fate of the foreskin. Incidence of preputial adhesions, phimosis, and smegma among Danish schoolboys. *Arch. Dis. Child* 43, 200-203.
69. Gairdner, D. (1949). The fate of the foreskin: a study of circumcision. *Br. Med. J.* 2, 1433-1437.
70. Burrows, H. (1944). The union and separation of living tissue as influenced by cellular differentiation. *Yale J. Biol. Med.* 17, 397-407.
71. Korenbrott, C.C., Huhtaniemi, I.T., and Weiner, R.L. (1977). Preputial separation as an external sign of pubertal development in the male rat. *Biol. Reprod.* 17, 298-303.
72. Cameron, J.L. (1991). Metabolic cues for the onset of puberty. *Horm. Res.* 36, 97-103.
73. Sandberg, A.A., Karr, J.P., and Müntzing, J. (1980). The prostates of dog, baboon and rat. *In: Male Accessory Sex Glands.* Biology and Pathology. (E. Spring-Mills and E.S.E. Hafez, eds.). Elsevier/North Holland Biomedical Press, New York, p. 565-608.
74. Zondek, L.H., and Zondek, T. (1980). Congenital malformations of the male accessory sex glands in the fetus and neonate. *In: Male Accessory Sex Glands. Biology and Pathology.* (E. Spring-Mills and E.S.E. Hafez, eds.). Elsevier/North Holland Biomedical Press, New York, p. 17-37.
75. Zondek, L.H., and Zondek, T. (1979). Effects of hormones on the human fetal prostate. *In: Fetal Endocrinology.* (L.H. Zondek and T. Zondek, eds.) Karger, New York.
76. Schlotthauer, C.F. (1932). Observations on the prostate gland of the dog. *J. Amer. Vet. Med. Assoc.* 81, 645.
77. O'Shea, J.D. (1962). Studies on the canine prostate gland. I. Factors influencing its size and weight. *J. Comp. Path.* 72, 321.
78. Berg, O.A. (1958a). The normal prostate gland of the dog. *Acta Endocrinol.* 27, 129.
79. Berg, O.A. (1958b). Parenchymatous hypertrophy of the canine prostate gland. *Acta Endocrinol.* 27, 140.
80. Huggins, C., and Moulder, P.V. (1945). Estrogen production by Sertoli cell tumors of the testis. *Cancer Res.* 5, 510.
81. Ofner, P. (1968). Effects and metabolism of hormones in normal and neoplastic prostate tissue. *Vitam. Horm.* 26, 237-291.
82. Gloyna, R.E., Siiteri, P.K., and Wilson, J.D. (1970). Dihydrotestosterone in prostatic hypertrophy. II. The formation and content of dihydrotestosterone in the hypertrophic canine prostate and the effect of dihydrotestosterone on prostate growth in the dog. *J. Clin. Invest.* 49, 1746-1753.
83. Siiteri, P.K., and Wilson, J.D. (1970). Dihydrotestosterone in prostatic hypertrophy. I. The formation and content of dihydrotestosterone in the hypertrophic prostate of man. *J. Clin. Invest.* 49, 1737-1745.
84. Brooks, J.E., Busch, R.D., Patanelli, D.J., and Steelman, S.L. (1973). A study of the effects of a new anti-androgen on the hyperplastic dog prostate. *Proc. Soc. Exp. Biol. Med.* 143, 647-655.
85. Scott, W.W., and Wade, J.C. (1969). Medical treatment of benign nodular prostatic hyperplasia with cyproterone acetate. *J. Urol.* 101, 81-85.
86. Robinette, C.L., Blume, C.D., and Mawhinney, M.G. (1978). Androphilic and estrophilic molecules in canine prostate glands. *Invest. Urol.* 15, 425-431.
87. Moore, R.J., Gazak, J.M., and Wilson, J.D. (1979). Regulation of cytoplasmic dihydrotestosterone binding in dog prostate by 17-estradiol. *J. Clin. Invest.* 63, 351-357.
88. Price, D. (1963). Comparative aspects of development and structure in the prostate. *Nat. Cancer Inst. Monogr.* 12, 1.
89. Sugimura, Y., Cunha, G.R., and Doncajour, A.A. (1986). Morphogenesis of ductal networks in the mouse prostate. *Biol. Reprod.* 34, 961-971.

90. Sugimura, Y., Cunha, G.R., and Doncajour, A.A. (1986). Morphological and histological study of castration-induced degeneration and androgen-induced regeneration in the mouse prostate. *Biol. Reprod.* 34, 973-983.
91. Sugimura, Y., Cunha, G.R., Doncajour, A.A., Bigsby, R.M., and Brody, J.R. (1986). Whole-mount autoradiography study of DNA synthetic activity during postnatal development and androgen-induced regeneration in the mouse prostate. *Biol. Reprod.* 34, 985-995.
92. Lee, C., Sensibar, J.S., Dudek, S.M., Hiipakka, R.A., and Liao, S. (1990). Prostatic ductal system in rats: regional variation in morphological and functional activities. *Biol. Reprod.* 43, 1079-1086.
93. Prins, G.S. (1989). Differential regulation of androgen receptors in the separate rat prostate lobes: androgen-independent expression in the lateral lobe. *J. Steroid Biochem.* 33, 319-326.
94. Higgins, S.J., Smith, S.E., and Wilson, J. (1982). Development of secretory protein synthesis in the seminal vesicles and ventral prostate of the male rat. *Mol. Cell. Endocrinol.* 27, 55-65.
95. Mann, T., and Lutwak-Mann, C. (1976). Evaluation of the functional state of male accessory glands by the analysis of seminal plasma. *Andrologia* 8, 237-242.
96. Wilson, J.D., George, F.W., and Griffin, J.E. (1981). The hormonal control of sexual development. *Science* 211, 1278-1284.
97. Huhtaniemi, I.T., Korenbrot, C.C., and Jaffe, R.B. (1977). HCG binding and stimulation of testosterone biosynthesis in the human fetal testis. *J. Clin. Endocrinol. Metab.* 44(5), 963-967
98. Huhtaniemi, I., Pakarinen, P., Sokka, T., and Kolho, K.-L. (1989). Pituitary-gonadal function in the fetus and neonate. In: Control of the Onset of Puberty III. (H.A. Delemarre-van de Waal, ed.) Elsevier Science Publishers, New York, pp. 101-109.
99. Brook, C.G. (1999). Mechanism of puberty. *Horm. Res.* 51 (Suppl. 3), 52-54.
100. Ojeda, S.R. and Urbanski, H.F. (1994). Puberty in the Rat. *In: The Physiology of Reproduction. Second Edition. Vol. 2.* (E. Knobil and J.D. Neill, eds.), Raven Press, New York, pp.363-409.
101. Tapanainen, J., Kuopio, T., Pelliniemi, L.J., and Huhtaniemi, I. (1984). Rat testicular endogenous steroids and number of Leydig cells between the fetal period and sexual maturity. *Biol. Reprod.* 31, 1027-1035.
102. Döhler, K.D., and Wuttke, W. (1974). Serum LH, FSH, prolactin and progesterone from birth to puberty in female and male rats. *Endocrinology* 94, 1003-1008.
103. Piacsek, B.E., and Goodspeed, M.P. (1978). Maturation of the pituitary-gonadal system in the male rat. *J. Reprod. Fertil.* 52, 29-35.
104. .Resko, J.A., Feder, H.H., and Goy, R.W. (1968). Androgen concentrations in plasma and testis of developing rats. *J. Endocrinol.* 40, 485-491.
105. Podesta, E.J., and Rivarola, M.A. (1974). Concentration of androgens in whole testes, seminiferous tubules and interstitial tissue of rats at different stages of development. *Endocrinology* 95, 455-461.
106. Wiebe, J.P. (1976). Steroidogenesis in rat Leydig cells: Changes in activity of 5-ane and 5-ene 3-hydroxysteroid dehydrogenase during sexual maturation. *Endocrinology* 98, 505-513.
107. Gupta, D., Zarzycki, J., and Rager, K. (1975b). Plasma testosterone and dihydrotestosterone in male rats during sexual maturation and following orchidectomy and experimental bilateral cryptorchidism. *Steroids* 25, 33-42.
108. Goldman, B.D., Grazia, Y.R., Kamberi, I.A., and Porter, J.C. (1971). Serum gonadotropin concentrations in intact and castrated neonatal rats. *Endocrinology* 88, 771-776.
109. Ramirez, V.D., and McCann, S.M. (1965). Inhibitory effect of testosterone on luteinizing hormone secretion in immature and adult rats. *Endocrinology* 76, 412-417.
110. Negro-Vilar, A., Ojeda, S.R., and McCann, S.M. (1973). Evidence for changes in sensitivity to testosterone negative feedback on gonadotropin release during sexual development in the male rat. *Endocrinology* 93, 729-735.
111. Desjardin, C. (1981). Endocrine signaling and male reproduction. *Biol. Reprod.* 24, 1-21.
112. Brabant, G., Ray, A.K., Wagner, T.O.F., Krech, R., and von zur Mühlen, A. (1985). Peripubertal variation in GnRH-release from *in vitro* superfused hypothalami of male rats. *Neuroendocrinol. Lett.* 7, 197-202.
113. Chiappa, S.A., and Fink,G. (1977). Releasing factor and hormonal changes in the hypothalamic-pituitary-gonadotrophin and adrenocorticotrophin systems before and after birth and puberty in male, female and androgenized female rats. *J. Endocrinol.* 72, 211-224.

114. Chappel, S.C., and Ramaley, J.A. (1985). Changes in the isoelectric focusing profile of pituitary follicle-stimulating hormone in the developing male rat. *Biol. Reprod.* 32, 567-573.
115. Odell, W.D., and Swerdloff, R.S. (1976). Etiologies of sexual maturation: A model system based on the sexually maturing rat. *Recent Prog. Horm. Res.* 32, 245-288.
116. Huhtaniemi, I.T., Nozu, K., Warren, D.W., Dufau, M.L., and Catt, K.J. (1982). Acquisition of regulatory mechanisms for gonadotropin receptors and steroidogenesis in the maturing rat testis. *Endocrinology* 111, 1711-1720.
117. Murono, E.P., and Payne, A.H. (1979). Testicular maturation in the rat. *In vivo* effects of gonadotropins on steroidogenic enzymes in the hypophysectomized immature rat. *Biol. Reprod.* 20, 911-917.
118. Ojeda, S.R., and Ramirez, V.D. (1972). Plasma levels of LH and FSH in maturing rats: response to hemigonadectomy. *Endocrinology* 90, 466-472.
119. Chowen-Breed, J.A., and Steiner, R.A. (1989). Neuropeptide gene regulation during pubertal development in the male rat. *In: Control of the Onset of Puberty III.* (H.A. Delemarre-van de Waal, ed.) Elsevier Science Publishers, New York, pp. 63-77.
120. Payne, A.H., Kelch, R.P., Murono, E.P., and Kerlan, J.T. (1977). Hypothalamic, pituitary and testicular function during sexual maturation of the male rat. *J. Endocrinol.* 72, 17-26.
121. Araki, S., Toran-Allerand, C.D., Ferin, M., and Vande Wiele, R.L. (1975). Immunoreactive gonadotropin-releasing hormone (Gn-RH) during maturation in the rat: ontogeny of regional hypothalamic differences. *Endocrinology* 97, 636-697.
122. Döhler, K.D., and Wuttke, W. (1975). Changes with age in levels of serum gonadotropins, prolactin, and gonadal steroids in prepubertal male and female rats. *Endocrinology* 97, 898-907.
123. Nazian, S.J., and Cameron, D.F. (1992). Termination of the peripubertal FSH increase in male rats. *Am. J. Physiol.* 262, E179-E184.
124. Dullaart, J. (1977). Immature rat pituitary glands in vitro: Age- and sex-related changes in luteinizing hormone releasing hormone-stimulated gonadotrophin release. *J. Endocrinol.* 73, 309-319.
125. Debeljuk, L., Arimura, A., and Schally, A.V. (1972). Studies on the pituitary responsiveness to luteinizing hormone-releasing hormone (LH-RH) in intact male rats of different ages. *Endocrinology* 90, 585-588.
126. Dalkin, A.C., Bourne, G.A., Pieper, D.R., Regiani, S., and Marshall, J.C. (1981). Pituitary and gonadal gonadotropin-releasing hormone receptors during sexual maturation in the rat. *Endocrinology* 108, 1658-1664.
127. Federal Register, Part Two, Vol 63, no 231, Wednesday, Dec 2, 1998 Rules and regulations, Department of Health and Human Services, Food and Drug Administration 21 CFR Parts 201, 312, 314 and 601, pg 66650.
128. Inaba, T., Matsuoka, S., Kawate, N., and Mori, J. (1994). Developmental changes in testicular luteinizing hormone receptors and androgens in the dog. *Res. Vet. Sci.* 57, 305-309.

APPENDIX C-5*
LANDMARKS IN THE DEVELOPMENT
OF THE FEMALE REPRODUCTIVE SYSTEM†

David A. Beckman, PhD, and Maureen Feuston, PhD

Author affiliations:

David A. Beckman, PhD, Novartis Pharmaceuticals Corporation, Preclinical Safety, Toxicology, 59 Route 10, Building 406-142, East Hanover, NJ 07936 USA

Maureen H. Feuston, PhD, Sanofi-Synthelabo Research, Toxicology, 9 Great Valley Parkway, P.O. Box 3026, Malvern, PA 19355 USA

Address correspondence to:

David A. Beckman, PhD, Novartis Pharmaceuticals Corporation, Preclinical Safety, Toxicology, 59 Route 10, Building 406-142, East Hanover, NJ 07936 USA, E-mail: david.beckman@pharma.novartis.com, Tel: 862-778-3490, Fax: 862-778-5489

Drugs and environmental chemicals have the potential to adversely affect the developing female reproductive system. The consequences of an exposure are influenced by the magnitude and duration of the exposure, the mechanism of action for the drug/chemical, the sensitivity of target tissue, and the critical windows in the development of the reproductive system. For the purposes of this review, Table 1 is presented to define the postnatal age for phases of sexual development in the female organism and to indicate equivalent ages in selected common laboratory species, specifically the rat, Beagle dog, and non-human primates.‡ It is important to point out that the range of ages for a particular phase can vary somewhat depending on the organ systems under consideration and the species selected for the comparison. This follows from the observation that tissues and organs do not necessarily mature in the same sequence across all species or by the same overall stage of development.

Table 1 Summary of Locomotor Development

	Human	Rat	Dog	Non-Human Primate (Rhesus)
Crawling	≈ PND 270 (9 months)	PND 3-12	PND 4-20	PND 4-49
Walking	≈ PND 396 (13 months)	PND 12-16	PND 20-28	PND 49[a]

PND = Postnatal Day
[a] Bipedal locomotion

Although the number of oocytes is determined by birth, the postnatal maturation of the reproductive tissues, steroid hormone production, external genitalia, sexual behavior and cyclic signaling events enables the female to reach her reproductive potential. Furthermore, although puberty may be defined as the time at which the generative organs mature and reproduction may occur, it does not signify full or normal reproductive capacity.

Focusing on the rat first because it is the most widely used species in pharmaceutical/chemical industry research for toxicity assessments, plasma follicle-stimulating hormone (FSH) levels start

---

* Source: Beckman, D. A. and Feuston, M., Landmarks in the development of the female reproductive system, *Birth Defects Research, Part B: Developmental and Reproductive Toxicology*, 68, 137-143, 2003.

† The categories presented by the U.S. Department of Health and Human Services (95) to aid in the design of clinical investigations of medicinal products are not included in Table 1 since they are not consistent with those used for the laboratory species.

to increase shortly after birth, reaching maximum concentrations on day 12, then declining gradually to approximately 20% of the day-12 values by the end of the juvenile period (6,7,8). Plasma luteinizing hormone (LH) levels are also higher in neonatal-infantile rats than in juvenile rats and are characterized by sporadic surges (6,7,8,10). These surges disappear completely by the juvenile phase and LH levels remain low during this period.

During the infantile period, estradiol exerts relatively little negative feedback, due to high-affinity binding to the high levels of alpha-fetoprotein (11), and aromatizable androgens play a leading role in the steroid negative feedback control of gonadotropin secretion. This changes from a predominantly androgenic control to a dual estrogenic-androgenic control during the juvenile period. During the juvenile period, there is an increase in secretion of dehydroepiandrosterone (DHEA) and DHEA sulfate associated with the maturation of a prominent adrenal zona reticularis. Occurring at day 20 in the rat (12), adrenarche, the maturation of a prominent zona reticularis, is characterized by a prepubertal rise in adrenal secretion of DHEA and DHEA sulfate that is independent of the gonads or gonadotropins. The rise in DHEA and DHEA sulfate production results from the increased presence of 17alpha-hydroxylase/17,20-lyase and DHEA sulfotransferase and the decreased presence of 3beta-hydroxysteroid dehydrogenase (13,14).

Also during the juvenile period, morphological maturation of neurons in the hypothalamus coincides with changes in the diurnal pattern of LH release and the time at which the hypothalamic-pituitary unit becomes fully responsive to estradiol positive feedback (15).

The ovary is relatively insensitive to gonadotropin stimulation during the first week after birth, coming under strong gonadotropin control during the second week (16,17). Waves of follicular development and atresia occur during the juvenile period, but these follicles are not ovulated. At the end of the juvenile phase, the mode of LH release begins to change (9,10), animals are older than 30 days of age, their uteri are small (wet weight less than 100 mg), no intrauterine fluid can be detected and vaginal patency is not yet achieved (18). Animals progressing to the next phase have larger uteri with intrauterine fluid and the vagina remains closed (19). On the day of the first proestrus, there is a large amount of intrauterine fluid, the wet weight of the uterus is greater than 200 mg, the ovaries have large follicles, and the vagina is closed in most animals (18). On the day of the first ovulation, uterine fluid is no longer present, fresh corpora lutea have been formed, the vagina is open and vaginal cytology shows a predominance of cornified cells. A common parameter of sexual maturity in rats is vaginal opening. Vaginal opening occurs on the day after the first preovulatory surge of gonadotropins, i.e. day 36-37 (20,21,22,13) but can range from 32 to 109 days (24). Female rats may maintain their full reproductive potential to 300 days (1), however, this is influenced by the strain of rat.

As in the rat, there is an early postnatal rise in gonadotropin secretion in the human and monkey, followed by a reduction lasting from late infancy to the end of the prepubertal period (25,26,27,28). Until about 12-24 months in girls, the ovaries respond to the increased follicle-stimulating hormone (FSH) by secreting estradiol reaching levels that are not again achieved before the onset of puberty (29,30). The secretion of estradiol then decreases with a nadir, in the human, at about 6 years of age (29). During this time, there is a reduction in the pulsatile release of gonadotrophin releasing hormone (GnRh) which appears to involve both gonadal and central nervous system restraints. In the human, the timing of adrenarche may vary from 5 to 8 years (31,32). However, several reports suggest that adrenarche may be a gradual maturational process beginning at 7-8 years that continues to 13-15 years and is not the result of sudden rapid changes in adrenal enzyme activities or adrenal androgen concentrations (10,17,33,34). Interestingly, adrenarche appears to be absent in non-human primates (5,35,36). Although adrenal androgen concentrations are similar to those in humans, adrenal androgen levels throughout development in rhesus and cynomolgus monkeys and baboons are similar to those in adulthood (37,38).

Menarche, the onset of menstrual cyclicity, is regarded as an overt sign of the initiation of puberty, although endocrine profiles indicate significant hormonal changes years/months prior. The first ovulation occurs sometime after the first cycle. This supports the "adolescent sterility" hypothesis (the interval between menarche and first ovulation) proposed by Young and Yerkes (39). In most young women, ovulation does not occur until 6 or more months after menarche and a regular recurrence of ovulatory menstrual cycles does not appear to become established until several years later (40,41,42,43). The postmenarcheal phase of development in the rhesus monkey is the result of a high incidence of anovulatory and short luteal-phase cycles (44,45).

Estradiol concentrations increase at initial stages of puberty at 2.5 to 3 years in the non-human primate (46,47) and at 8 to 10 years in the human (48,49,50,51). Gonadotropes acquire the capacity to respond to the stimulatory action of estradiol on LH secretion after menarche (25,26,27). The onset of puberty in the human is marked by an increase in the amplitude of LH pulses, which is taken to indicate an increase in the amplitude of GnRH pulses. Studies in the rhesus monkey (28,46,52,53) suggest that the central nervous system mechanisms responsible for the inhibition of the GnRH pulse generator during childhood involve primarily gamma-aminobutyric acid (GABA) and GABAergic neurons. With the onset of puberty, the reactivation of the GnRH pulse generator is associated with a fall in GABAergic neurotransmission and an associated increase in the input of excitatory amino acid neurotransmitters (including glutamate) and possibly astroglial-derived growth factors (30,54).

It appears that adequate levels of leptin and a leptin signal are required to achieve puberty and to maintain cyclicity and reproductive function in the rodent (55,56,57,58,59,60,61), non-human primate and human (30,62,63,64). However, leptin does not appear to have a direct effect on the central control of the pulsatile release of GnRh (53). The available evidence suggests that leptin acts as one of several permissive factors but alone is not sufficient to initiate sexual maturation.

Prolactin has a luteotropic role in rats. The prolactin content and prolactin-containing cells of the anterior pituitary increase with postnatal age (65,66) but prolactin levels remain low until the prepubertal period (67,68). Prolactin is not leuteotropic in monkeys or humans (69).

In order to aid in pinpointing the timing of landmarks in the postnatal development of the female reproductive tract and for cross-species comparisons, Table 2 presents the timing of specific landmarks for the rat, dog, non-human primate and human, including endocrine status, follicle maturation, ovulation, reproductive tract development, estrous cyclicity/menarche, adrenarche, puberty and fertility.

The information in this review may be useful in the interpretation of results from investigations into the potential effects of chemical/compound exposures during the development of young animals. Although this information may also aid in the selection of the most appropriate species for an investigation, it cannot be separated from other issues in the design of a study that impact the final species selection, including:

Study purpose: Is the investigation a primary or screening study, with the primary purpose to identify potential adverse effects in juvenile animals that are not seen in adults of the same species, or is it secondary/focused study, with the primary purpose of investigating, for example, the scope, severity, potential for recovery, etc., for adverse effects on a specific organ system during postnatal development? What regulatory guidelines must be followed?

Exposure window and duration: Based on the purpose of the study, will the exposure be limited to a defined stage in postnatal development or should it include several stages, possibly from neonatal to sexual maturity? Also, do results from previous studies in adults or in young animals, if available, demonstrate changes in no-observed-effect levels over time and/or progression of target organ toxicities with chronic administration? Do considerations of enzyme induction, immature metabolism, and the desired stage(s) of development for exposure limit the species selection?

Compound specific considerations: Are there species-specific characteristics for the compound with respect to the mechanism of action, pharmacokinetics, target organ toxicity, etc., that limit potential species for consideration?

Concordance between animal and human toxicity: Are results available (most probably in the adult) that support concordance of target organ toxicity for the compound from animal and human studies?

While some of the framing of the preceding questions may be more appropriate for pharmaceutical development, they nevertheless provide useful discussion for other types of investigations.

Table 2 Landmarks in the development of the female reproductive tract for the rat, dog, non-human primate and human

Parameter	Rat (Wistar or Sprague Dawley, if possible)	Dog (Beagle, if possible)	Non-human primate (Macaca mulatta, if possible)	Human
Endocrine status				
Estrogen/estradiol	Range of free estradiol concentration during early neonatal period is similar to that during cycling adult (70).	Not available	2.5 to 3 years, increased estradiol concentrations at initial stages of puberty (46,47).	8 to 10 years, increased estradiol concentrations at initial stages of puberty (48,49,50,51).
Luteinizing hormone, LH	LH levels increase shortly after birth to a maximum on day 12 then declines to about one-fifth of the day 12 values by the end of the juvenile period (6,7,8). Prepubertal increase in LH 8-9 days before the expected day of first proestrus (9,10).	Gonadotropes acquire the capacity to respond to the stimulatory action of estradiol on LH secretion by day 20 and full capacity by day 28 (77). By 4 months, FSH concentrations are similar to those in adult anestrus (3).	Gonadotropes acquire the capacity to respond to the stimulatory action of estradiol on LH and FSH secretion after menarche (25,26,27). GnRH neurosecretory system is active in neonatal period then enters a dormant state. An increase in pulsatile release is essential for onset of puberty (28).	Gonadotropes acquire the capacity to respond to the stimulatory action of estradiol on LH and FSH secretion, probably after menarche (25,26,27). GnRH neurosecretory system is active in neonatal period then enters a dormant state. An increase in pulsatile release is essential for onset of puberty (28). Also see reference 51.
Follicle-stimulating hormone, FSH	FSH levels increase shortly after birth to a maximum on day 12 then declines to about one-fifth of the day 12 values by the end of the juvenile period (6,7,8).	By 4 months, FSH concentrations are similar to those in adult anestrus (3).	Gonadotropes acquire the capacity to respond to the stimulatory action of estradiol on LH and FSH secretion after menarche (25,26,27,46).	Gonadotropes acquire the capacity to respond to the stimulatory action of estradiol on LH and FSH secretion probably after menarche (25,26,27). Also see reference 51.
Prolactin	Luteotropic in rats and mice Anterior pituitary prolactin content and prolactin-containing cells increase with postnatal age (65,66) but prolactin levels remain low until the prepubertal period (67,68).	Luteotropic role is questionable (78,79).	No luteotropic effect in monkeys.	No luteotropic effect in humans. Also see reference 69.

Table 2 (continued) Landmarks in the development of the female reproductive tract for the rat, dog, non-human primate and human (continued)

Parameter	Rat (Wistar or Sprague Dawley, if possible)	Dog (Beagle, if possible)	Non-human primate (Macaca mulatta, if possible)	Human
Leptin	Necessary for maintenance of estrous cyclicity (60). Adequate levels essential but not sufficient for onset of puberty (57). Adequate levels essential for onset of puberty and maintenance of cyclicity and reproductive function (59).	Present (80) but role in reproduction not defined.	In "monkey" adequate levels essential but not sufficient for onset of puberty (57). Adequate levels essential for onset of puberty and maintenance of cyclicity and reproductive function (59). Adequate levels essential for onset of puberty (53).	Adequate levels essential for onset of puberty and maintenance of cyclicity and reproductive function (59).
Follicle maturation	Ovarian follicles become subjected to strong gonadotropin control during the second week of postnatal life (16,17).	5-6 months: primary follicles show antrum formation (81).	Not available	Follicles were seen in 86% of prepubertal girls and in 99% of pubertal girls (85).
Ovulation	1st ovulation on about day 38 (1). 1st ovulation on day 29 (19). 1st ovulation on day 37 (71).	8-12 months (3). 9-14 months (81).	The postmenarcheal phase of development in the rhesus monkey is the result of a high incidence of anovulatory and short luteal-phase cycles (44,45).	In most young women, ovulation does not occur until 6 or more months after menarche and a regular recurrence of ovulatory menstrual cycles does not appear to become established until several years later (40,41,42,43,86,87). 12-24 months post menarche (38). Ovarian function last ~30 years (51).

Reproductive tract maturation

Uterine maturation	Differentiation of the muscular and glandular epithelium: uterine glands first appear on postnatal day 9 and increased until day 15 (71). Prepubertal: Small uterus, wet weight less than 100mg with no intrauterine fluid (18). Postpubertal: Larger uterus, wet weight greater than 200 mg (18). Increase in uterine weight on day 27 (19).	Puberty (81).	Not available	Uterine growth begins before puberty (85).
Anogenital distance	Anogenital distance is 3.5 mm for males and approx one half that for females and (73). GD21: females, 1.29-1.41 mm; 3.15-4.44 mm for males (74). PND 0: females, 1.29-1.51 mm; males, 3.27-3.83 mm (74). Anogenital distance varies with body weight – normalization (75).	Not determined	Not determined	Not determined
Vaginal opening	32-109 days (24). Wistar 35.6 days ± 0.8 SE (76). Occurs on the day after the first preovulatory surge of gonadotropins, i.e. day 36-37 (20,21,22,23). Crl SD: 31.6-35.1 days; Mean, 33.4 days (74).	Not determined	Not determined	Not determined

Table 2 (continued) Landmarks in the development of the female reproductive tract for the rat, dog, non-human primate and human (continued)

Parameter	Rat (Wistar or Sprague Dawley, if possible)	Dog (Beagle, if possible)	Non-human primate (Macaca mulatta, if possible)	Human
Adrenarche	Day 20 (12).	11 weeks of age (82).	Absence of adrenarche in non-human primates (5,35,36). "In primates, adrenal androgen concentrations are similar to humans. However, rhesus and cynomegalus monkeys and baboons exhibit adrenal androgen levels throughout development that are no different than those during adulthood" (37,38).	5 years (32). 6-8 years (31). Onset 7-8 years (6-8 years skeletal age) that continues to 13-15 years (10,17,34).
Puberty	Vaginal opening and first ovulation at ~5 weeks (1,70).	8-12 months (3). 5-13 months (mean, 9 months) in 8-15 kg beagles (83).	2.5-3 years, nipple growth, increase in perineal swelling and coloration (46,47).	8-10 years: appearance of labial hair, initiation of breast enlargement (48,49,50,88). Menarche at 8-14 yr (89,90,91,92,93,94) Menses at 12-13 years (88) 8-13 years (34)
Estrous cyclicity/Menarche (1st estrus/mensus)	~5 weeks (24). 36.4 days ± 1.2 SE (74).	8-12 months (3).	2-3 years (84).	~13 years (90,91,92,94). 12-13 years (93). 8-14 years (89). 10-16.5 years; Mean, 13.4 years (34).
Sexual maturity/Fertility	50 ± 10 days (24).	6-12 months (2). 8-12 months (3).	2.6-3.5 years (84). 3-4 years (2).	11-16 years (2).

REFERENCES

1. Ojeda SR, Andrews WW, Advis JP, Smith-White S. Recent advances in the endocrinology of puberty. *Endocr Rev* 1980;1:228-57.
2. Altman PL, Dittmer DS. *Biology Data Book,* 2nd ed., Vol. 1. Bethesda: Fed Am Soc Exp Biol, 1972
3. Concannon PW. Biology of gonadotrophin secretion in adult and prepubertal female dogs. *J Reprod Fertil Suppl* 1993;47:3-27.
4. Andersen AC. Reproduction. In: Andersen AC, editor. *The Beagle as an Experimental Dog.* Ames, Iowa: The Iowa State University Press, 1970:31-39.
5. Meusy-Dessolle N, Dang DC. Plasma concentrations of testosterone, dihydrotestosterone, delta 4-androstenedione, dehydroepiandrosterone and oestradiol-17 beta in the crab-eating monkey (Macaca fascicularis) from birth to adulthood. *J Reprod Fertil* 1985;74:347-59.
6. Dohler KD, Wuttke W. Serum LH, FSH, prolactin and progesterone from birth to puberty in female and male rats. *Endocrinology* 1974;94:1003-8.
7. Kragt CL, Dahlgen J. Development of neural regulation of follicle-stimulating hormone (FSH) secretion. *Neuroendocrinology* 1972;9:30-40
8. Ojeda SR, Ramirez VD. Plasma levels of LH and FSH in maturing rats: response to hemigonadectomy. *Endocrinology* 1972;9066-72.
9. Andrews WW, Ojeda SR. A detailed analysis of the serum LH secretory profiles of conscious free-moving female rats during the time of puberty. *Endocrinology* 1981;109:2032-9.
10. MacDonald J, Dees WL, Ahmed CE, Ojeda SR. Biochemical and immunocytochemical characterization of neuropeptide Y in the immature rat ovary. *Endocrinology* 1987;120:1703-10.
11. Meijs-Roelofs HMA, Kramer P. Maturation of the inhibitory feedback action of oestrogen on follicle-stimulating hormone secretion in the immature female rat: a role for alphafoetoprotein. *J Endocrinol* 1979;81:199-208.
12. Pignatelli D, Bento MJ, Maia M, Magalhaes MM, Magalhaes MC, Masaon JI. Ontogeny of 3beta-hydroxysteroid dehydrogenase expression in the rat adrenal gland as studied by immunohistochemistry. *Endocr Res* 1999;25:21-7.
13. Gell JS, Carr BR, Sasano H, Atkins B, Margraf L, Mason JI, Rainey WE. Adrenarche results from development of a 3beta-hydroxysteroid dehydrogenase-deficient adrenal reticularis. *J Clin Endocrinol Metab* 1998;83:3695-701.
14. Parker CR Jr. Dehydroepiandrosterone and dehydroepiandrosterone sulfate production in the human adrenal during development and aging. *Steroids* 1999;64:640-7.
15. Urbanski HF, Ojeda SR. The juvenile-peripubertal transition period in the female rat: establishment of a diurnal pattern of pulsatile luteinizing hormone secretion. *Endocrinology* 1985;117:644-9.
16. Hage AJ, Groen-Klevant AC, Welschen R. Follicle growth in the immature rat ovary. *Acta Endocrinol* 1978;88:375-82.
17. Lunenfeld B, Kraiem Z, Eshkol A. The function of the growing follicle. *J Reprod Fertil* 1975;45:567-74.
18. Advis JP, Andrews WW, Ojeda SR. Changes in ovarian steroidal and prostaglandin E responsiveness to gonadotropins during the onset of puberty in the female rat. *Endocrinology* 1979;104:653-8.
19. Tamura K, Abe Y, Kogo H. Phenytoin inhibits both the first ovulation and uterine development in gonadotropin-primed immature rats. *Eur J Pharmacol* 2000;398:317-22.
20. Critchlow V, Elswers Bar-Sela M. Control of the onset of puberty. In: Martini L, Ganong WF, editors. *Neuroendocrinology,* Vol. II. New York: Academic Press, Inc., 1967:101-62
21. Meijs-Roelofs HMA, Uilenbroek JThJ, de Greef WJ, de Jong FH, Kramer D. Gonadotropin and steroid levels around the time of first ovulation in the rat. *J Endocrinol* 1975;67:275-82.
22. Oheda SR, Wheaton JE, Jameson HE, McCann SM. The onset of puberty in the female rat. I. Changes in plasma prolactin, gonadotropins, LHRH levels and hypothalamic LHRH content. *Endocrinology* 1976;98:630-8.
23. Ramirez VD. Endocrinology of puberty. In: Greep RO, Astwood EB, editors. *Handbook of Physiology,* Vol. II, Section 7. Washington, DC: American Physiological Society, 1973:1-28.
24. Kohn DF, Barthold SW. Biology and diseases of rats. In: Fox JG, Cohen, BJ, Loew, FM, editors. *Laboratory Animal Medicine,* Chapt. 4. San Diego: Academic Press, Inc., 1984:91-122.

25. Dierschke DJ, Weiss G, Knobil E. Sexual maturation in the female rhesus monkey and the development of estrogen-induced gonadotropic hormone release. *Endocrinology* 1974;94:198-206.
26. Reiter EO, Kulin HE, Hamwood SM. The absence of positive feedback between estrogen and luteinizing hormone in sexually immature girls. *Pediatr Res* 1974;8:740-5.
27. Presl J, Horejsi J, Stroufova A, Herzmann J. Sexual maturation in girls and the development of estrogen-induced gonadotropic hormone release. *Ann Biol Anim Biochem Biophys* 1976;16:377-83.
28. Terasawa E, Fernandez DL. Neurological mechanisms of the onset of puberty in primates. *Endocr Rev* 2001;22:111-51
29. Brook CG. Mechanism of puberty. *Horm Res* 1999;51 Suppl 3:52-4.
30. Grumbach MM. The neuroendocrinology of human puberty revisited *Horm Res* 2002;57 Suppl 2:2-14.
31. Dardis A, Saraco N, Rivarola MA, Belgorosky A. Decrease in the expression of the 3beta-hydroxysteroid dehydrogenase gene in human adrenal tissue during prepuberty and early puberty: implications for the mechanism of adrenarche. *Pediatr Res* 1999;45:384-8.
32. Suzuki T, Sasano H, Takeyama J, Kaneko C, Freije WA, Carr BR, Rainey WE. Developmental changes in steroidogenic enzymes in human postnatal adrenal cortex: imunohistochemical studies. *Clin Endocrinol* 2000;53:739-47.
33. Palmert MR, Hayden DL, Mansfield MJ, Crigler JF, Crowley WF, Chandler DW, Boepple PA. The longitudinal study of adrenal maturation during gonadal suppression: evidence that adrenarche is a gradual process. *J Clin Endocrinol Metab* 2001;86:4536-42.
34. Styne, DM, and MM Grumbach. Puberty in the male and female: Its physiology and disorders. In: Yen SS, Jaffe RB, editors. *Reproductive Endocrinology,* Chapter 12. Philadelphia: WB Saunders Company, 1986:313-384.
35. Crawford BA, Harewood WJ, Handelsman DJ. Growth and hormone characteristics of pubertal development in the hamadryas baboon. *J Med Primatol* 1997;26:153-63.
36. Mann DR, Castracane VD, McLaughlin F, Gould KG, Collins DC. Developmental patterns of serum luteinizing hormone, gonadal and adrenal steroids in the sooty mangabey (Cercocebus atys). *Biol Reprod* 1983; 28:279-84.
37. Gonzalez, F. Adrenarch. In: Knobil E, Neill JD, editors. *Encyclopedia of Reproduction,* Volume 1. San Diego: Academic Press, 1998:51-61.
38. Hatasaka, H. Menarche. In: Knobil E, Neill JD, editors. *Encyclopedia of Reproduction,* Volume 3. San Diego: Academic Press, 1998:177-83.
39. Young WC, Yerkes RM. Factors influencing the reproductive cycle in the chimpanzee: the period of adolescent sterility and related problems. *Endocrinology* 1943;33:121-54
40. Treloar AE, Boynton RE, Behn BG, Brown BW. Variation of the human menstrual cycle through reproductive life. *Int J Fertil* 1967;12:77-126.
41. Doring GK. The incidence of anovular cycles in women. *J Reprod Fertil* 1969;6(Suppl.):77-81.
42. Apter D, Viinikka L, Vihko R. Hormonal pattern of adolescent menstrual cycles. *J Clin Endocrinol Metab* 1978;47:944-54.
43. Metcalf, MG, Skidmore DS, Lowry GF, Mackenzie JA. Incidence of ovulation in the years after the menarche. *J Endocrinol* 1983;97:213-9.
44. Foster DL. Luteininzing hormone and progesterone secretion during sexual maturation of the rhesus monkey: short luteal phases during the initial menstrual cycles. *Biol Reprod* 1977;17:584-90.
45. Resko JA, Goy RW, Robinson JA, Norman RL. The pubescent rhesus monkey: some characteristics of the menstrual cycle. *Biol Reprod* 1982;27:354-61.
46. Terasawa E, Nass TE, Yeoman RR, Loose MD, Schultz NJ. Hypothalamic control of puberty in the female rhesus macaque. In: Norman RL, editor. *Neuroendocrine Aspects of Reproduction.* New York: Academic Press Inc., 1983:149-82.
47. Wilson ME, Gordon TP, Collins DC. Ontogeny of luteinizing hormone secretion and first ovulation in seasonal breeding rhesus monkeys. *Endocrinology* 1986;118:293-301.
48. Bidlingmaier F, Wagner-Barnach M, Butenandt O, Kron D. Plasma estrogen in childhood and puberty under physiologic and pathological conditions. *Pediatr Res* 1973;7:901-7.
49. Jenner MR, Kelch RP, Kaplan SL, Grumbach MM. Hormonal changes in puberty: IV. plasma estradiol, LH, and FSH in prepubertal children, pubertal females, and in precocious puberty, premature thelarche, hypogonadism and in a child with a feminizing ovarian tumor. *J Clin Endocrinol Metab* 1972;34:521-30.

50 Winter JSD, Faiman C. Pituitary-gonadal relations in female children and adolescents. *Pediatr Res* 1973;7:948-53.
51 Yen SS. The Human Menstrual Cycle. In: Yen SS, Jaffe RB, editors. *Reproductive Endocrinology*, Chapter 8. Philadelphia: WB Saunders Company, 1986:200-36.
52 Kasuya E, Nyberg CL, Mogi K, Terasawa E. A role of gamma-amino butyric acid (GABA) and glutamate in control of puberty in female rhesus monkeys: effect of an antisense oligodeoxynucleotide for GAD67 messenger ribonucleic acid and MK801 on luteinizing hormone-releasing hormone release. *Endocrinology* 1999;140:705-12.
53 Plant TM. Neurological bases underlying the control of the onset of puberty in the rhesus monkey: a representative higher primate. *Front Neuroendocrinol* 2001;22:107-39.
54 Ma YJ, Ojeda SR. Neuroendocrine control of female puberty: glial and neuronal interactions. *J Investig Dermatol Symp Proc* 1997;2:19-22.
55 Cheung CC, Thornton JE, Kuijper JL, Weigle DS, Clifton DK, Steiner RA. Leptin is a metabolic gate for the onset of puberty in the female rat. *Endocrinology* 1997;138:855-8.
56 Cheung CC, Thornton JE, Nurani SD, Clifton DK, Steiner RA. A reassessment of leptin's role in triggering the onset of puberty in the rat and mouse. *Neuroendocrinology* 2001;74:12-21.
57 Cunningham MJ, Clifton DK, Steiner RA. Leptin's actions on the reproductive axis: perspectives and mechanisms. *Biol Reprod* 1999;60:216-22.
58 Urbanski HF. Leptin and puberty. *Trends Endocrinol Metab* 2001;12:428-9.
59 Mantzoros CS. Role of leptin in reproduction. *Ann N Y Acad Sci* 2000;900:174-83
60 Carro E, Pinilla L, Seoane LM, Considine RV, Aguilar E, Casanueva FF, Dieguez C. Influence of endogenous leptin tone on the estrous cycle and luteinizing hormone pulsatility in female rats. *Neuroendocrinology* 1997; 66:375-7.
61 Watanobe H, Schioth HB. Postnatal profile of plasma leptin concentrations in male and female rats: relation with the maturation of the pituitary-gonadal axis. *Regul Pept* 2002;105:23-8.
62 Chan JL, Mantzoros CS. Leptin and the hypothalamic-pituitary regulation of the gonadotropin-gonadal axis. *Pituitary* 2001;4:87-92.
63 Farooqi IS. Leptin and the onset of puberty: insights from rodent and human genetics. *Semin Reprod Med* 2002;20:139-44
64 Bouvattier C, Lahlou N, Roger M, Bougneres P. Hyperleptinaemia is associated with impaired gonadotrophin response to GnRH during late puberty in obese girls, not boys. *Eur J Endocrinol* 1998;138:653-8.
65 Watanabe H, Takebe K. Involvement of postnatal gonads in the maturation of dopaminergic regulation of prolactin secretion in female rats. *Endocrinology* 1987;120:2212-9.
66 Yamamuro Y, Aoki T, Yukawa M, Sensui N. Developmental profile of prolactin cells in the anterior pituitary of postnatal rats during the lactational stage. *Proc Soc Exp Biol Med* 1999;220:94-9.
67 Natagani S, Guthikonada P, Foster DL. Appearance of a nocturnal peak of leptin secretion in the pubertal rat. *Horm Behav* 2000; 37:345-52.
68 Ojeda SR, Advis JP, Andrews WW. Neuroendocrine control of the onset of puberty in the rat. *Fed Proc* 1980;39:2365-71.
69 Yen SS. Prolactin in human reproduction. In: Yen SS, Jaffe RB, editors. *Reproductive Endocrinology*, Chapter 9. Philadelphia: WB Saunders Company, 1986:237-63.
70 Montano MM, Welshons WV, vom Saal FS. Free estradiol in serum and brain uptake of estradiol during fetal and neonatal sexual differentiation in female rats. *Biol Reprod* 1995; 53:1198-207.
71 Rice VM, Limback SD, Roby KF, Terranova PF. Changes in circulating and ovarian concentrations of bioactive tumour necrosis factor alpha during the first ovulation at puberty in rats and in gonadotrophin-treated immature rats. *J Reprod Fertil* 1998;113:337-41.
72 Branham WS, Sheehan DM, Zehr DR, Ridlon E, Nelson CJ. The postnatal ontogeny of rat uterine glands and age-related effects of 17 beta-estradiol. *Endocrinology* 1985;117:2229-37.
73 Tyl RW, Marr MC. Developmental toxicity testing – methodology, p. 204. In: Hood RD, editor. *Handbook of Developmental Toxicology*. Boca Raton: CRC Press, 1996:175-225.
74 Clark, RL. Endpoints of reproductive system development. In: G Daston and C Kimmel, editors. *An Evaluation and Interpretation of Reproductive Endpoints for Human Health Risk Assessment*. Washington, D.C.: ILSI Press, 1999.

75. Gallavan RH Jr, Holson JF, Stump DG, Knapp JF, Reynolds VL. Interpreting the toxicological significance of alterations in anogenital distance: potential for confounding effects of progeny body weights. *Reprod Toxicol* 1999; 13(5):383-90.
76. Eckstein B, Golan R, Shani J. Onset of puberty in the immature female rat induced by 5α-androstane-3B,17B-diol. *Endocrinology* 1973;92:941-5.
77. Andrews WW, Mizejewski GJ, Ojeda SR. Development of estradiol positive feedback on LH release in the female rat: a quantitative study. *Endocrinology* 1981;109:1404-13.
78. Okkens AC, Kooistra HS, Dieleman SJ, Bevers MM. Dopamine agonistic effects as opposed to prolactin concentration in plasma as the influencing factor on the duration of anoestrus in bitches. *J Reprod Fetil Suppl* 1997;51:55-8.
79. Onclin K, Verstegen JP. Secretion patterns of plasma prolactin and progesterone in pregnant compared with nonpregnant dioestrous beagle bitches. *J Reprod Fertil Suppl* 1997;51:203-8.
80. Le Bel C, Bourdeau A, Lau D, Hunt P. Biologic response to peripheral and central administration of recombinant human leptin in dogs. *Obes Res* 1999;7:577-85.
81. *The Beagle as an Experimental Dog.* AC Andersen, Editor. Iowa State University Press. 1970.
82. Perez-Fernandez R, Facchinetti F, Beira A, Lima L, Gaudiero GJ, Genazzani AR, Devesa J. Morphological and functional stimulation of adrenal reticularis zone by dopaminergic blockade in dogs. *J Steroid Biochem* 1987; 28:465-70.
83. Concannon, PW. Reproduction in dog and cat. In: *Reproduction in Domestic Animals.* PT Cupps, Editor. Academic Press, 1991.
84. Hendrickx AG, Dukelow WR. Reproductive biology. In: Bennett BT, Abee CR, Henrickson R, editors. *Nonhuman Primates in Biomedical Research.* San Diego: Academic Press, 1995:147-191.
85. Holm K, Laursen EM, Brocks V, Muller J. Pubertal maturation of the internal genitalia: an ultrasound evaluation of 166 healthy girls. *Ultrasound Obstet Gynecol* 1995;6:175-81.
86. Chabbert Buffet N, Djakoure C, Maitre SC, Bouchard, P. Regulation of the human menstrual cycle. *Front Neuroendocrinol* 1998;19:151-86.
87. Hartman CG. On the relative sterility of the adolescent organism. *Science* 1931;74:226-7.
88. Herman-Giddens ME, Slora EJ, Wasserman RC, Bourdony CJ, Bhapkar MV, Koch GG, Hasemeier CM. Secondary sexual characteristics and menses in young girls seen in office practice: a study from the Pediatric Research in Office Settings network. *Pediatrics* 1997;99:505-12.
89. Blondell RD, Foster MB, Dave KC. Disorders of puberty. *Am Fam Physician* 1999;60:209-4.
90. Chowdhury S, Shafabuddin AK, Seal AJ, Talukder KK, Hassan Q, Begum RA, Rahman Q, Tomkins A, Costello A, Talukder MQ. Nutritional status and age at menarche in a rural area of Bangladesh. *Ann Hum Biol* 2000; 27:249-56.
91. Graham MJ, Larsen U, Xu X. Secular trend in age at menarche in China: a case study of two rural counties in Anhui Province. *J Biosoc Sci* 1999;31:257-67.
92. Marrodan MD, Mesa MS, Arechiga J, Perez-Magdaleno A. Trend in menarcheal age in Spain: rural and urban comparison during a recent period. *Ann Hum Biol* 2000;27:313-9.
93. Marshall WA. Puberty. In: Falkner F, Tanner JM, editors. *Human Growth,* Volume 2: Post-natal Growth. New York: Plenum Press, 1978:141-81.
94. Pasquet P, Biyong AM, Rikong-Adie H, Befidi-Mengue R, Garba MT, Froment A. Age at menarche and urbanization in Cameroon: current status and secular trends. *Ann Hum Biol* 1999;26:89-97.
95. U.S. Department of Health and Human Services, Food and Drug Administration. Guidance for industry. E11. Clinical investigation of medicinal products in the pediatric population. www.fda.gov/cber/guidelines.htm, December 2000.

APPENDIX C-6*
POSTNATAL ANATOMICAL AND FUNCTIONAL DEVELOPMENT OF THE HEART: A SPECIES COMPARISON

Kok Wah Hew[1,3] and Kit A. Keller[2]

[1] Purdue Pharma L.P., Nonclinical Drug Safety Evaluation, Ardsley, New York
[2] Consultant, Washington DC
[3] Correspondence to: Kok Wah Hew, PhD, Purdue Pharma L.P., Nonclinical Drug Safety Evaluation, 444 Saw Mill River Road, Ardsley, NY 10502. E-mail: kok-wah.hew@pharma.com

Introduction

The purpose of this review is to summarize the postnatal development, growth and maturation of the human heart and its vascular bed, and explain how the infant/juvenile systems differ, both morphologically and functionally, from that seen in the adult. A species comparison with common laboratory animals, where available, is also included to aid in the extrapolation to human risk assessment. The available data is mostly obtained from smaller laboratory animals. The postnatal changes observed in the heart are relatively similar across the mammalian species, but do differ in the timing of events. As a general rule, the functional and morphological changes in the heart occur faster, according to the developmental rate, in small laboratory animals compared to humans due to the faster growth and maturation rate in these animals. This review is by no means a comprehensive compilation of all of the literature available on the human and animal postnatal heart development, but rather a basic outline of key events.

Postnatal Anatomical Growth and Maturation of the Heart

Immediate Postnatal Changes in Cardiac Morphology and Vascularization

The transition from prenatal to postnatal circulation involves three primary steps: 1) cessation of umbilical circulation, 2) the transfer of gas exchange function from the placenta to the lungs, and 3) closure of the prenatal shunts, at first functionally and then structurally (Rakusan, 1980, 1984; Friedman and Fahey, 1993; Smolich, 1995).

The switch from fetal to adult type of circulation is not immediate. During the neonatal period, a transient intermediate period occurs, characterized by decreasing pulmonary vascular resistance, increasing pulmonary blood flow, increasing left atrium blood volume, and increasing systemic vascular resistance. During this period there is constriction of the ductus arteriosus and closure of the foramen ovale. In addition, the heart function changes from acting as two parallel pumps (right and left) to acting as two pumps performing in series.

Human

Umbilical Vasculature — Elimination of the vascular lumina of the umbilical vessels takes from several weeks to months after birth (O'Rahilly and Muller, 1996). The umbilical vein between the

---

* Source: Hew, K. W. and Keller, K., Postnatal anatomical and functional development of the heart: A species comparison, *Birth Defects Research, Part B: Developmental and Reproductive Toxicology*, 68, 309-320, 2003.

umbilicus and liver becomes the ligamentum teres in the falciform ligament. The proximal parts of the umbilical arteries remain as the internal iliac arteries.

Closure of the Foramen Ovale — Functional closure of the foramen ovale occurs very rapidly in association with the first breath and occurs as a result of the hemodynamic changes that hold the valve of the foramen ovale closed (Walsh et al., 1974). Anatomical closure of the foramen ovale is a slow process, which normally does not occur before the end of the first year (Walsh et al., 1974). Anatomical closure is complete when the valve becomes fixed to the interatrial septum.

Closure of the Ductus Arteriosus — Following birth, functional closure of the ductus arteriosus, which is a direct connection of the pulmonary trunk into the dorsal aorta, is brought about by contraction of the smooth muscle in the wall of the ductus arteriosus. Anatomical closure is produced by structural changes and necrosis of the inner wall of the ductus arteriosus, followed by connective tissue formation, fibrosis, and permanent sealing of the lumen (Broccoli and Carinci, 1973; Clyman, 1987). In full-term infants, the ductus arteriosus begins to constrict soon after the first breath. Functional closure is complete in about 15 hours after birth (Heymann and Rudolph, 1975). Anatomical closure occurs in about one-third of infants by 2-3 weeks after birth, in nearly 90% by 2 months of age and in 99% by 1 year of age (Broccoli and Carinci, 1973; O'Rahilly and Muller, 1996). The remaining fibrous cord following anatomical closure is known in adult anatomy as the ligamentum arteriosum.

The biochemistry behind the initial functional closure of the ductus arteriosus is believed to involve a balance between the opposing actions of oxygen and prostaglandin E_2 (PGE_2). PGE_2 has a dilating effect and acts to keep the duct open *in utero*, while increased oxygen concentration after birth alters pulmonary and systemic arterial pressure causing the duct to constrict (Clyman, 1987). This is supported by reports that closure of the ductus arteriosus is delayed or absent in premature infants (who have high PGE_2 levels) and in neonates exposed to low oxygen environments, including high altitudes (Moss et al., 1964; Penaloza et al., 1964).

Closure of the Ductus Venosus — The ductus venosus, which carries portal and umbilical blood to the inferior vena cava, is functionally closed within hours of birth. Permanent closure begins a few days after birth and is complete by 18-20 days of age (Meyer and Lind, 1966; Fugelseth et al., 1997; Kondo et al., 2001). The duct becomes the ligamentum venosum in a fissure on the back of the liver.

Comparative Species

Closure of the Foramen Ovale — In the rat, the foramen ovale is diminished in the first two days after birth and is completely closed three days after birth (Momma et al., 1992a).

Closure of the Ductus Arteriosus — In the rat and mouse, functional closure of the ductus arteriosus is complete by 2-5 hours after birth (Jarkovska et al., 1989; Tada and Kishimoto, 1990). Functional closure is complete in a few minutes after birth in guinea pigs and rabbits (Clyman, 1987). In rats, rabbits, guinea pigs and lambs, anatomical closure is complete in 1-5 days after birth (Jones et al., 1969; Fay and Cooke, 1972; Heymann and Rudolph, 1975). In dogs, anatomical closure of the ductus arteriosus occurs at about 7-8 days after birth (House and Enderstrom, 1968).

As in the human infant, the biochemistry behind the initial functional closure of the ductus arteriosus is believed to involve a balance between the opposing actions of oxygen and PGE_2 (Clyman, 1987). Studies in guinea pigs and lambs suggest that this contractile response to oxygen is the results of interaction between oxygen and a cytochrome P_{450} - catalyzed enzymatic process, that can be inhibited by carbon monoxide (Fay and Jobsis, 1972; Coceani et al., 1984). Specifically, cytochrome a_3 becomes oxidized with subsequent generation of adenosine triphosphate (ATP) and muscle cell contraction. In full-term animals, the loss of responsiveness to PGE_2 shortly after birth prevents the duct from reopening once it has constricted (Clyman, et al., 1985a). This change in responsiveness may involve the presence of thyroid hormones (Clyman, et al., 1985b).

Closure of the Ductus Venosus — In the rat, the ductus venosus narrows rapidly after birth and closes completely in two days (Momma and Ando, 1992b). In the dog and lamb, functional closure of the ductus venosus occurs between 2-3 days after birth (Lohse and Suter, 1977; Zink and Van Petten, 1980).

A study in lambs suggests that thromboxane may play a role in the closure of the ductus venosus after birth (Adeagbo et al., 1985).

Shape, Position and Size of the Heart During Postnatal Development

The rate of cardiac growth during the postnatal period is species dependent (Hudlicka and Brown, 1996). In humans, the relative heart weight does not significantly change from infancy to adulthood and the highest variability in weights occurs in newborns. In contrast, the relative heart weights in most experimental animals decrease significantly with age and the highest variability in heart weight occurs in adults. In humans and laboratory animals, the right ventricular weight fraction is higher at birth than in the adult and this value changes over to a left ventricular dominance during the postnatal period. Postnatal changes in heart shape and dimensions vary according to species.

Human

In humans, the heart is relatively large in proportion to the chest at birth and is usually oval or globular in shape. At birth the length is approximately 75% of its width (Rakusan, 1984). By the end of the first year, the length increases to approximately 80% of its width. By adulthood, the length and width are almost the same.

At birth, the heart is positioned higher and more "transverse" than in the adult. The characteristic adult oblique lie is attained between 2 and 6 years of age (O'Rahilly and Muller, 1996).

The greatest rate of increase in absolute heart weight occurs during the first postnatal year. The heart doubles its birth weight by 6 months of age and triples by 1 year (Smith, 1928; Rakusan, 1980). It should be noted that heart weights in infants and small children are highly variable and available mean data show a deviation factor of 2 to 3 (Rakusan, 1980). After a brief surge during the early postnatal period, the increase in cardiac weight is proportional to the increase in body weight, resulting in a constant relative cardiac weight throughout the first half of life.

Human Heart Weight and Heart Weight as % of Body Weight (Smith, 1928)

Age in years	Mean Heart Weight (g)		% Body Weight	
	Male	Female	Male	Female
Birth to 1 year	45	35	0.62	0.55
2–4	69	55.5	0.43	0.37
5–7	87.7	94	0.35	0.37
8–12	149	120	0.44	0.42
13–17	225	180	0.43	0.38
18–21	308	245	0.42	0.42

Part of this growth in mass in early development involves a modest increase in right ventricular wall thickness and a marked increase in the left ventricular wall thickness (Eckner et al., 1969; Graham et al., 1971; Oberhansli et al., 1981). At birth, volumes of the right and left ventricular cavities are approximately the same. In infants, the right ventricular volume is about twice that of the newborns, whereas the left ventricular volume does not change. In the second year, the ratio of right ventricle:left ventricle volumes reaches 2:1 and does not change with age (Kyrieleis, 1963). Studies on the relation of the right and left ventricular weight at birth show high variability (Rakusan, 1980).

Comparative Species

The basic geometry of the heart does not change with age in the rabbit and guinea pig (Lee et al., 1975). In the rat and dog, the left ventricle is less spherical in the postnatal period, compared to adults, due to growth-related changes in ventricular dimensions (House and Ederstrom, 1968; Grimm et al., 1973; Lee et al., 1975). Heart weight, in relation to body weight, is greater in the newborn than the adult in most common laboratory animals (House and Ederstrom, 1968; Lee et al., 1975; Rakusan, 1980). Subsequently, the heart:body weight ratio decreases with age.

Heart Weight as % of Body Weight (Rakusan, 1980)

	Newborn	Adult
Rat	0.52–0.57	0.17–0.22
Mouse	0.46	0.44
Rabbit	0.49	0.20
Guinea Pig	0.39–0.51	0.23–0.19
Dog	0.93	0.76

In the rat, the heart grows in proportion to the weight of the body during the early postnatal period (first month of age). Later the cardiac growth rate is slower than that of the body, resulting in a gradual decrease in relative cardiac weight (Addis and Gray, 1950; Rakusan *et al.*, 1963; Mattfeldt and Mall, 1987). In the Wistar rat, the rate of increase of heart weight, compared to that of the body weight, is greater from 1-5 postnatal days and less from 5-11 postnatal days (Anversa *et al.*, 1980). Throughout the first 11 days of growth, the left ventricular weight increases at a faster rate than that of the right ventricle, being 16%, 51%, 113% greater at 1, 5, and 11 days, respectively. The 6.2-fold increase in left ventricular weight is accompanied by a 2.7-fold increase in wall thickness. In contrast, no significant change in right ventricular wall thickness is observed despite a 3.4-fold weight gain. A 400% increase in the thickness of the primum septum occurs in the first 2 postnatal days (Momma *et al.*, 1992a). The septum secundum also grows rapidly during this time, with a 69% increase in length and width.

Postnatal Heart Growth in Wistar Rats (Anversa *et al.*, 1980; Rakusan *et al.*, 1965)

Mean Values	Age (postnatal day)						
	1	5	11	23	40	68	120
Body weight, g	5.35	11.75	26.34	36	81	216	328
Heart weight, mg	21.44	60.6	104.9	181	318	641	778
Left ventricle	8.86	28.3	55.1	NA	NA	NA	NA
Right ventricle	7.64	18.7	25.9	NA	NA	NA	NA
Left ventricle wall thickness, mm	0.470	0.869	1.256	NA	NA	NA	NA
Right ventricle wall thickness, mm	0.383	0.335	0.354	NA	NA	NA	NA

Heart growth rate in the mouse is proportional to body growth from birth to 24 weeks of age (Wiesmann *et al.*, 2000). Left ventricular mass is reported to increases with age.

Reported relative heart weights in the guinea pig range from 0.39% to 0.51% at birth to 0.19% to 0.23% in adults, with decreasing relative fraction of right ventricular weight from 57% to 55% of the left ventricle (Webster and Liljegren, 1949; Lee *et al.*, 1975).

Relative heart weight in the rabbit is reported to decrease from 0.49% at birth to 0.20% in adults (Lee *et al.*, 1975; Boucek, 1982; Fried *et al.*, 1987). Investigators also reported a relative decrease in right ventricular weight from 86% of the left ventricular at birth to 54% in adults, with preferential growth of the left ventricle between postnatal days 1 and 7.

For the beagle dog, Deavers et al. (1972) reported a small but significant decrease in relative heart weights from birth to adulthood. The ratio of left ventricular volume and weight did not change with age (Lee et al., 1975).

Postnatal Development of Heart Cellular Constituents

At the cellular level, cardiac growth is due to both cellular proliferation and an increase in cellular volume (Hudlicka and Brown, 1996). Myocyte hypertrophy is accompanied by an increase in cellular DNA. Cardiac myocytes at birth contain one diploid nucleus. The number of binucleated cells rapidly increases during the early postnatal stages. In human myocytes, the majority of these nuclei subsequently fuse, leading to an increase in polyploidy. On the other hand, binucleated cardiac myocytes remain predominant throughout life in rodents. Hearts from other experimental animals contain both binucleated and polyploid myocytes. Other significant intracellular maturational changes include a progressive increase in sarcoplasmic reticulum and myofibrils.

Human

The newborn heart contains about half the total number of myocytes present in the adult heart. Adult values are reached probably before the age of 4 months (Linzbach, 1950, 1952; Hort, 1953). At birth, the majority of cardiac myocytes are mononucleated. During postnatal development, the percentage of binucleated cells increases up to 33% during late infancy or early childhood, with a subsequent decrease to adult values [5-13% of cells in the left ventricle are binucleated and 7% of the cells in the right ventricle are binucleated] (Schneider and Pfitzer, 1973). In infants, almost all nuclei of the cardiac myocyte are diploid. In adults, 60% are diploid, 30% are tetraploid and 10% are octoploid (Eisentein and Wied, 1970). Myocyte size is also reported to increase with age from approximately 5 µm at birth, 8 µm at 6 weeks, 11 µm at 3 years, 13 µm at 15 years to 14 µm in adults (Ashley, 1945; Rakusan, 1980).

An increase in myocardial contractile function during the early postnatal period is accompanied by intracellular changes such as increased sarcoplasmic reticulum and myofibrils, the organelles that regulate and utilize calcium to produce cardiac contraction (Fisher and Towbin, 1988).

Comparative Species

In the rat, as in humans, myocyte cell volume increases nearly 25-fold and myocyte cell number 3-4 fold from birth to 2 months of age, (Anversa et al., 1986; Englemann et al., 1986; Mattfeldt and Mall, 1987;Vliegen et al., 1987). The increase in cardiac mass is due to cell proliferation until postnatal day 3-6, hyperplasia and hypertrophy until postnatal day 14, and solely by hypertrophy thereafter (Chubb and Bishop, 1984; Anversa et al., 1986, 1992; Batra and Rakusan, 1992; Li et al., 1996). Most of the muscle cell nuclei are diploid in the rat heart (Korecky and Rakusan, 1978; Rakusan, 1984). The frequency of polyploidy, which is rather low at birth, even decreases slightly with age in the rat heart (Grove et al., 1969). Mast cells, which are postulated to play a role in capillary angiogenesis, are present in very low numbers in early rat pups, but start to increase in the postnatal weeks 3 and 4 to maximum numbers by 100 days of age (Rakusan et al., 1990).

In mice, the volume of myocytes remains relatively constant despite a concomitant increase in heart weight, indicating growth due to cell division during the first four postnatal days (Leu et al., 2001). After postnatal day 5, the volume of myocytes increases markedly until postnatal day 14, when the increase slows down. Myocytes reach their adult volume at around 3 months of age.

In rabbits, myocyte diameter increases progressively (from ~ 5 µm) towards adult values (14 µm), starting on postnatal day 8 (Hoerter et al., 1981). Rabbit neonate studies show that, as in the case of human infants, cellular sarcoplasmic reticulum changes dramatically during the perinatal period (Seguchi et al., 1986). Such changes suggest that the early perinatal hearts are more dependent on transsarcolemmal Ca^{2+} influx in excitation-contraction coupling, when compared to the adult heart, which depends on Ca^{2+} release and uptake from sarcoplasmic reticulum. This may be due to the immaturity in early neonates of the transverse tubular system that interconnects with the sarcoplasmic reticulum (Nakamura et al., 1986).

In dogs, the number of myocytes remains relatively constant as the heart enlarges with increasing age, but myocyte shape and dimensions and intercellular connections changed dramatically (Legato, 1979).

Average Diameter of Cardiac Myocytes (Rakusan, 1980; Hoerter et al., 1981)

	Age	Diameter of Myocytes (µm)
Human	Birth	5
	6 weeks	8
	3 years	11
	15 years	13
	Adult	14
Rat	5–14 days	5.5
	30 days	10.5–11.8
	Young adult	15.0–16.0
Rabbit	Day 8	5
	Adult	14
Dog	< 100 days	6.5
	0.5–0.9 year	13.3
	1–4.2 years	13.8

An increase in left ventricular mass, due to an increase in myocyte size rather than proliferation, is also observed in the neonatal pig (Satoh et al., 2001).

In addition to changes in myocyte diameter, a progressive change in the organization and pattern of association between gap junctions and cell adhesion junctions are also observed within the first 20 postnatal days in rats and dogs (Gourdie et al., 1992; Angst et al., 1997).

Postnatal Cardiac Vasculature

There are marked species differences in the development of the coronary vascular bed, depending on the degree of maturity of the species at birth (Hudlicka and Brown, 1996). Rapid developmental heart growth is accompanied by a proportional growth of capillaries but not always of larger vessels, and thus coronary vascular resistance gradually increases. Growth of adult hearts can be enhanced by thyroid hormones, catecholamines and the renin-angiotensin system hormones, but these do not always stimulate growth of coronary vessels. Likewise, chronic exposure to hypoxia leads to growth, mainly of the right ventricle and its vessels but without vascular growth elsewhere in the heart. On the other hand, ischaemia is a potent stimulus for the release of various growth factors involved in the development of collateral circulation. In humans, the coronary vascular bed is well established at birth, although some capillary angiogenesis and vessel maturation occurs postnatally. In comparison, rats show a greater immaturity in the development of the coronary vascular bed at birth, with marked development of the coronary arteries and capillaries postnatally. In all species, capillary density decreases with age into adulthood.

Human

The coronary vascular bed in humans is established well before birth, but some capillary angiogenesis occurs postnatally. Rakusan et al. (1994) report similar arteriolar density and vessel thickness in hearts from infants, children and adults. While larger branches are already formed at birth, there are reports that smaller arteriole vessels continue to proliferate postnatally (Ehrich et al., 1931; Rakusan et al., 1992). The rate of cardiac growth is higher than the growth of capillaries. The number of capillaries per arteriole decreases from 500 in infant hearts to 397 in children and 347 in adult hearts. During postnatal development, final vessel maturation is characterized by the establishment of definitive morphological and biochemical features in the vascular walls and changes in the spatial orientation of the capillaries (Rakusan et al., 1994). In newborns, the average diameter of coronary arteries is 1 mm. It doubles during the first postnatal year, and it reaches maximal values around the age of 30 years (Vogelberg, 1957; Rabe, 1973).

Comparative Species

In rats, development of both the coronary arteries and capillaries continues after birth (Olivetti et al., 1980a; Mattfeldt and Mall, 1987; Batra and Rakusan, 1992; Rakusan et al., 1994; Tomanek et al., 1996; Heron et al., 1999). Formation and maturation of coronary arterioles, including thickening of the media due to intensive production of connective tissue and hypertrophy, is complete within the first postnatal month (Looker and Berry, 1972; Olivetti et al., 1980b; Heron et al., 1999). Following birth, the coronary capillaries also continue to proliferate and mature, with close to half of all capillaries in the adult heart forming during the early postnatal period (Rakusan et al., 1994). During the first few weeks, capillaries grow 2-3 times more rapidly than heart mass. The volume fraction of capillaries in the rat heart increases from values around 4% at birth to 16% by postnatal day 28, with subsequent decrease to adult values of 8-9%. Recent studies of early postnatal vascularization in the rat report that both vascular endothelial growth factor (VEGF) and beta-fibroblast growth factor (bFGF) modulate capillary growth and bFGF facilitates arteriolar growth (Tomanek et al., 2001).

In rabbits, there is a rapid growth of the heart capillaries during the first postnatal week, but no growth was detected in adult (Rakusan et al., 1967).

During the postnatal period in dogs, the degree of myocyte surface area in close contact to capillary walls increases dramatically (Legato, 1979).

Postnatal Cardiac Innervation

There are some species differences in the development of cardiac innervation, depending on the degree of maturity of the heart of the species at birth, but the general processes appear to be similar.

Human

Innervation of the human heart is both morphologically and functionally immature at birth. The number of neurons gradually increases and reaches a maximum density in childhood, at which time an adult pattern in innervation is achieved (Chow et al., 2001).

Comparative Species

In rats, adrenergic innervation patterns in cardiac tissues and the vasculature reach adult levels by the third postnatal week, while the thickness and density of nerve fibers reach adult levels by the fifth postnatal week (De Champlain et al., 1970; Scott and Pang, 1983; Schiebler and Heene, 1986; Horackova et al., 2000). Histochemical studies suggest that ventricular innervation develops later than that of the atria in the rat (De Champlain et al., 1970). The cholinergic innervation is also structurally and functionally immature at birth in the rat heart (Truex et al., 1955; Winckler, 1969; Vlk and Vincenzi, 1977; Horackova et al., 2000). Neuropeptide Y and tyrosine hydroxylase positive nerve fibers are reported to increase rapidly during the first two weeks postnatally and reach adult levels by the third week (Nyquist-Battie et al., 1994).

Development and maturation in adrenergic innervation patterns are reported to occur earlier in the guinea pig, compared to rats (De Champlain et al., 1970; Friedman, 1972).

Friedman et al. (1968) reported comparable cholinergic density in neonate and adult rabbits.

In the dog, the adrenergic and cholinergic systems are incomplete at birth, and their development continues during the first 2 months of age (Ursell et al., 1990; Cua et al., 1997). The quantities of monoaminergic nerve endings are similar to adult levels in the atria by 6 weeks of age and in the ventricles by 4 months of age (Dolezel et al., 1981).

Postnatal Functional and Physiological Development of the Heart

Biochemistry of the Heart during Postnatal Development

A large amount of information, primarily in experimental animals, is available on the developmental changes in cardiac metabolism that is beyond the scope of this review (see Rakusan, 1980 for review). The major characteristic change during early development is the change from primarily carbohydrate energy metabolism to lipid energy metabolism, increasing concentrations of cytochromes, creatinine, and phosphocreatinine and increasing activity in mitochondrial enzymes activity.

Human

Water and Mineral Content — Widdowson and Dickerson (1960) reported a small but significant reduction in water content in the heart during postnatal development. No significant changes in calcium, magnesium, sodium and potassium were observed in heart autopsies in subjects ranging from newborn to 90 years of age (Eisenstein and Wied, 1970).

Energy Metabolism — The switching energy metabolism from carbohydrate (glucose) to fatty acids involves changes in phosphofructokinase isozymes. In humans, this isozyme switch occurs before birth (Davidson et al., 1983). In addition, the concentrations of cytochromes and mitochondrial protein in cardiac cells increase during the early postnatal period (Rakusan, 1980).

Comparative Species

Water and Mineral Content — Solomon et al. (1976) reported a rapid postnatal decline in water content of the rat heart up to 23 days of age. Thereafter, a smaller decline toward adult levels was observed. Sodium (Na) and potassium (K) levels also increased to the maximum value by 16 days of age with a subsequent marked decrease by 40 days of age (Hazlewood and Nichols, 1970; Solomon et al., 1976). The ratio of Na/K is very high at birth and declines to its minimum at 40 days of age. Chloride and calcium content declined with age.

Lipid and Protein Content — Immature rats are reported to have lower myocardial myoglobin concentrations compared to adults (Rakusan et al., 1965; Dhindsa et al., 1981). A slight increase in heart phospholipid levels was reported in the early postnatal rat (Carlson et al., 1968).

In hamsters, heart lipid and phospholipids increase during the first 4 months postnatally (Barakat et al., 1976).

Energy Metabolism — The heart in postnatal animals shows a higher dependence on glucose as an energy source than the adult heart (see Rakusan, 1980 for further review). Unlike humans, in which phosphofructokinase isozyme changes responsible for the switching of myocardial fuels from carbohydrate (glucose) to fatty acids occur before birth, rat isozyme changes occur during the first 2 postnatal weeks (Davidson et al., 1983). Carbohydrate dependence may also be partly due to the limited ability to oxidize long chain fatty acids, compared to adults.

Adenylate cyclase and guanylate cyclase activities, as well as cyclic GMP levels, are higher in heart homogenates of weanling compared to adult rats (Dowell, 1984). Cyclic AMP levels were similar in weanlings and adult rats.

The concentrations of cytochromes and mitochondrial proteins as well as enzyme activity in cardiac cells increase during the early postnatal period in rats and dogs (Rakusan, 1980; Schonfeld *et al.*, 1996). In comparison, the heart of the guinea pig, which is more mature at birth, showed no differences in cardiac mitochondrial enzyme activity between newborns and adults (Barrie and Harris, 1977).

Creatinine and phosphocreatine increased postnatally in the dog, rabbit and guinea pig during the early postnatal period (Rakusan, 1980). In the rat and mouse, creatinine phosphokinase levels in the heart increase after birth and reach adult values and isozyme types by 25 - 30 days of age (Hall and De Luca, 1975; Baldwin *et al.*, 1977).

Postnatal Electrophysiology of the Heart

In general, cardiac innervation, especially the adrenergic component, is functionally immature at birth. There are age-dependent alterations in the myocardial alpha 1-adrenergic, beta-adrenergic and muscarinic signal transduction cascades during postnatal development (Rakusan, 1980, 1984; Robinson, 1996; Garofolo *et al.*, 2002). An inhibitory alpha 1-adrenergic response appears in early maturation, which differs from the pre-existing excitatory response both with respect to the specific receptor subtype involved and its G protein coupling. Likewise, sympathetic innervation appears to be involved in the loss of an excitatory muscarinic response during development. The role of innervation in developmental regulation of the beta-adrenergic response is not fully understood. Functional innervation during the postnatal period favors excitation (chronotropic and/or inotropic) over inhibition. In addition, hearts from younger animals have higher stimulation thresholds than those from adults.

Human

Nerve Functional Development — Innervation of the human heart is both morphologically and functionally immature at birth (Rakusan, 1980, 1984). Human infants show a paucity of neurons with acetylcholinesterase (AChE), but a substantial amount of pseudocholinesterase activity, compared to adults (Chow *et al.*, 1993, 2001). During maturation into adulthood, a gradual loss in pseudocholinesterase activity occurs with increasing numbers of AChE-positive nerves. This coincides with initial sympathetic dominance in the neural supply to the human heart in infancy, and its gradual transition into a sympathetic and parasympathetic codominance in adulthood.

Electrocardiograpic and Vectorcardiographic Measurements — In general, the duration of ECG deflection and intervals increase with age, reflecting a postnatal decline in basal heart rate. The amplitudes of ECG waves are also age dependent, but vary widely. Early changes in QRS-T relationships from birth to 4 days of age reflect the immediate transition from *in utero* to postnatal life (Rautaharju *et al.*, 1979). Over a number of years, the horizontal QRS loops change from a complete clockwise rotation in newborns into a figure-of-eight loop that eventually unwinds into a counterclockwise loop orientation towards the left, as normally observed in adults (Namin and Miller, 1966; Rautaharju *et al.*, 1979). The magnitude of the ventricular gradient vector changes increases from 3 weeks of age until about 7 years of age (Rautaharju *et al.*, 1979). The spatial angle between QRS and STT vectors reaches its minimum at 1.5 to 4.5 years of age.

Comparative Species

Nerve Functional Development — There are marked species differences in the development of functional innervation of the heart in the perinatal period (Vlk and Vincenzi, 1977). At birth, the rat heart is quite immature functionally and requires several weeks to mature. On the other hand, guinea pig and rabbit heart innervation functions are relatively mature at birth, with cholinergic function more mature than adrenergic function.

In the rat, final functional sympathetic and parasympathetic innervation of the heart occurs after birth, with the parasympathetic system maturing first (Lipp and Rudolph, 1972; Mackenzie and Standen, 1980; Quigley et al., 1996). Adrenergic - and -receptors are present in the myocardium at birth in the rat; however, positive inotropic responses to sympathetic nerve stimulation do not develop until between 2 and 3 weeks of age (Mackenzie and Standen, 1980; Metz et al., 1996). Inotropic responsiveness in the rat atria shows significant levels at one week of age and reach 70% of adult values at two weeks of age. Receptor expression during this period appears to be regulated by thyroid hormone (Metz et al., 1996). Concomitantly, opposing A_1 adenosine receptor density is reported to be twice as numerous than adults and decrease to adult levels by two weeks of age (Cothran et al., 1995). It is believed that catecholaminergic stimulation plays an important role in maintaining basal heat rate during this early neonatal period in the rat (Tucker, 1985). After birth, sensitivity of the heart to catecholamines increases with postnatal age and generally shows higher sensitivity to catecholamines, compared to adult hearts (Penefsky, 1985; Shigenobu et al., 1988). Finally, it has been suggested that the cardiac renin-angiotensin system may also play a role in heart growth and in the adaptation of the heart to postnatal circulatory conditions (Sechi et al., 1993; Charbit et al., 1997).

Seven-day old rabbits and guinea pigs display adult cholinergic responsiveness to stimulation (Vlk and Vincenzi, 1977). However, neonatal rabbits appear to display some functional deficiency of adrenergic innervation as manifested by lack of blood pressure increase and heart rate decrease after asphyxiation.

Cardiac adrenergic and cholinergic innervation is also structurally and functionally immature at birth in the dog (Truex et al., 1955; Winckler, 1969; Vlk and Vincenzi, 1977; Haddad and Armour, 1991; Pickoff and Stolfi, 1996). Both the sympathetic and parasympathetic systems become fully functional by about 7 weeks of age (Haddad and Armour, 1991).

Cardiac Neurotransmitters — The concentration of norepinephrine in the heart is very low at birth but increases rapidly during the first postnatal weeks in the rat, rabbit, hamster, pig and sheep (Friedman et al. 1968; De Champlain et al., 1970; Heggenes et al., 1970; Friedman, 1972; Sole et al., 1975).

In the rat and rabbit, concentrations and uptake of catecholamine in the heart are very low at birth but increase rapidly during the first postnatal weeks (Glowinski et al., 1964; Friedman et al., 1968).

Tissue levels of acetylcholine in the atria of 3-4 day old rats are reported to be only one-sixth the levels found in the adult (Vlk and Vincenzi, 1977). Cholinesterase first appears in neurons at 4 days of age and is present in nerve fibers of the sinus node by 15 days of age.

Electrocardiographic and Vectorcardiographic Measurements — In the rat, the PQ interval decreases between 6 and 14 days of age and reaches adult values at 53 days of age (Yamori et al., 1976; Diez and Schwartze, 1991). The QT interval also decreases during the first 3 week of life and reaches adult values later than 53 days of age.

As in the case in humans, shorter QRS intervals are noted and horizontal QRS loops change from clockwise, figure-of-eight to counterclockwise loop oriented towards the left, as observed in the adult, during the early postnatal period in rats (Diez and Schwartze, 1991). QRS intervals are considerably shorter than in the adult, even at day 53. At this age, rats have attained the adult configuration of the QRS loop. From day 30 onward, QRS frequency merges with the T wave.

Similar findings are reported in the guinea pig where the QRS interval increases from 15.5 ms in the newborn to 32 ms in the adult (Yamori et al., 1976). The lengthening of the conduction course over the ventricles in adults, compared to infant and juvenile animals, might explain this increasing excitation period.

In mice, the entire J junction-S-T segment-T wave complex, R-R interval and Q-T intervals were shorter during early postnatal development and matured by postnatal days 7 to 14 (Wang *et al.*, 2000). Age-dependent ECG responses to K^+ channel blockers were also reported in mice.

Postnatal Cardiac Output and Hemodynamics

At birth, the cardiovascular system changes dramatically; arterial blood pressure, heart rate, and cardiac output increase, and blood flow distribution undergoes regional changes. During the early postnatal period, increasing heart rate, rather than stroke volume, is the primary mechanism by which cardiac output is increased during the neonatal period.

Human

The mean basal heart rate at birth is approximately 120 to 130 beats/minute. A steep increase in heart rate occurs between the 5th and 10th postnatal day, reaching maximum levels of ~140-150 beats/minute, which then gradually decrease to ~120 beats/minute over the first 100 postnatal days as parasympathetic restraints develop. Adult heart beat rates (55-85 beats/min) are achieved around 12 years of age (Oberhansli *et al.*, 1981; Mrowka *et al.*, 1996).

Mean Values for Heart Functional Parameters in Human Infants (Oberhansli *et al.*, 1981)

Heart Parameter	Age						
	1 day	3 days	6 days	1 month	2 month	6–11 months	12–14 months
Heart Rate (beats/min)	133	129	135	155	150	140	124
RV Pre-ejection Period (msec)	71	63	59	51	55	55	61
RV Ejection Time (msec)	199	203	203	193	204	232	243
LV Pre-ejection Period (msec)	65	61	59	55	59	59	65
LV Ejection Time (msec)	197	193	192	184	192	200	228
Mean Velocity	1.67	1.72	1.75	1.79	1.73	1.97	1.51
Ejection Fraction	70	70	71	70	70	77	71

RV- right ventricle; LV- left ventricle

Infants and young children are reported to have significantly smaller ventricular end-diastolic volumes and stroke indexes than adults, slightly smaller ejection fractions, but no differences in cardiac index (Graham *et al.*, 1971; Mathew *et al.*, 1976). Oh *et al.* (1966) reported marked increases in the distribution of cardiac output to the kidney during the first days after birth, but still a small fraction of output, when compared to adults.

The average blood pressure values (systolic/diastolic) increased rapidly from 62.1/39.7 mm Hg at age day one to up to 72.7/46.9 mm Hg at age one week and up to 85.0/47.4 mmHg at age two months (Hwang and Chu, 1995). Blood pressure changes relatively little between the ages of 6 months and 10 years. Systolic blood pressure rose from a mean of 88.5 mm Hg at age 6 months to 96.2 mm Hg at 8 years, and diastolic blood pressure rose from 57.8 mm Hg at 5 years to 61.8 mm Hg at 10 years (de Swiet *et al.*, 1992). Blood pressure was correlated with weight, weight adjusted for height, height, and arm circumference, at all ages studied. Blood pressure values (systolic/diastolic) increased rapidly from 68.0/43.7 mm Hg in children weighing less than 5 kg up to 87.6/41.8 mm Hg in those weighing 5 to 10 kg. Subsequently, these values increased gradually with body weight (Hwang and Chu, 1995).

Associations between birth weight, childhood growth and blood pressure are reported by numerous investigators (Georgieff et al., 1996; Uiterwaal et al., 1997; Walker et al., 2001). There was a significant correlation between systolic and diastolic blood pressure and both weight and length at birth to 4 months of age (Georgieff et al., 1996). Greater weight gain between ages 7 and 11 was associated with a greater increase in systolic blood pressure. The relation between growth and blood pressure is complex and has prenatal and postnatal components (Walker et al., 2001). Systolic blood pressure increased, with height, from 111 to 120 mm Hg in girls (body height 120 to 180 cm), as well as in boys from 112 to 124 mm Hg (Steiss and Rascher, 1996). However, mean diastolic blood pressure did not change during maturation and was 72 to 74 mm Hg, irrespective of height and sex.

The correlation coefficient between systolic and diastolic blood pressure increased steadily with age from 0.28 at 2 years to 0.59 at 10 years, but do not reach adult levels during this period (Levine et al., 1979; de Swiet et al., 1992).

In adults, arterial baroreceptors are the major sensing elements of the cardiovascular system. In the majority of species, including humans, baroreceptor reflexes at birth exhibit depressed sensitivity with a gradual postnatal maturation to adult levels (Gootman et al., 1979; Vatner and Manders, 1979).

Comparative Species

The heart rate in neonatal rats increases during the early postnatal period and then remains relatively constant into adulthood (Kyrieleis, 1963; Diez and Schwartze, 1991; Quigley et al., 1996). This increase in rate appears to be inversely proportional to the decrease in QT interval seen during this period, and may be explained by the increasing influence of the sympathetic system.

Heart Rate of Postnatal Wistar Rats (Diez and Schwartze, 1991)

Age (days)	Heart Rate (beats/min)
6	296 ± 60
14	405 ± 33
21	433 ± 48
30	436 ± 36
38	444 ± 24
53	435 ± 27

Hemodynamics in neonatal rats is characterized by a high cardiac output and low systemic resistance (Prewitt and Dowell, 1979). Shortly before puberty, there is a rise of systemic resistance, accompanied by a decrease in cardiac output (Smith and Hutchins, 1979). This is primarily due to structural maturation of the vasculature. Developmental changes in vascular sensitivity to vasoactive agents play a less important role for the maturational rise in systemic resistance. Marked maturation of the baroreceptor reflex also occurs during the postnatal period (Andresen et al., 1980). Normal systolic blood pressure in the young rat is more than double that of the neonate and reaches adult levels by 10 weeks of age (Litchfield, 1958; Gray, 1984; Rakusan et al., 1994).

Blood Pressure in Rats

Age	Blood Pressure (mm Hg)
Newborn	15–25
3-4 Weeks of Age	80–95

Estimation of coronary blood flow indicated no significant differences between immature and adult rats (Vizek and Albretch, 1973; Rakusan and Marcinek, 1973). However, marked changes have been reported in the organ distribution of cardiac output throughout postnatal development, especially during the first four weeks (Rakusan and Marcinek, 1973). The most prominent observation is the very low renal blood flow, which gradually increases during the postnatal period. Wiesmann et al. (2000) reported increases in cardiac output and stroke volume in mice with age, but no significant change in heart rate, ejection fraction, and cardiac index with age. On the other hand, Tiemann et al. (2003) recently reported an increase in heart rate during maturation from 396 beats/min at 21 days of age to 551 beats/min at 50 days of age. Mean arterial blood pressure also increased in parallel from 86 to 110 mm Hg and remained constant thereafter. Baroreflex heart rate control matures at around 2 weeks after birth in the mouse (Ishii, 2001).

Cardiac Output and the Heart Rate of C57bl/6 mice (Wiesmann et al., 2000)

	Age						
	3 days	10 days	3 weeks	4 weeks	5 weeks	10 weeks	16 weeks
LV cardiac output (ml/min)	1.1	5.3	8.7	9.3	11.2	15.7	14.3
LV stroke volume (µl)	3.2	15.0	20.8	23.9	30.5	35.6	40.2
Heart rate (beats/min)	372	418	422	390	366	442	360

LV-left ventricle

Cardiac output fractions to the kidneys and intestines are markedly lower in newborn rabbits, when compared to adults (Boda et al., 1971).

In dogs, a significant increase in blood pressure and a decrease in heart rate occurred with growth from 1 week to 6 months (Adelman and Wright, 1985). These changes are qualitatively similar to those observed in young infants and children. Baroreceptor reflex control in the neonate is less developed than in the adult heart (Hageman et al., 1986).

The heart rate in the pig does not change significantly with age (Satoh et al., 2001). Systolic and diastolic blood pressures increase (from 57 to 94 mm Hg and from 38 to 55 mm Hg, respectively) during the first 6 days postnatally in the pig (Satoh et al., 2001).

In the sheep, heart rate, cardiac output, ventricular output, and stoke volume are higher in neonates, when compared to adults (Klopfenstein and Rudolph, 1978; Berman and Musselman, 1979). Cardiac output in the newborn lamb is four times greater than in the adult. At birth, baroreceptor reflexes exhibit depressed sensitivity, with a gradual postnatal maturation to adult levels (Dawes et al., 1980; Segar et al., 1992).

Cardiac Parameters in Sheep

Heart Parameter	Term Fetus	Newborn	Adult
Heart Rate (beats/min)	150	200	100
Cardiac Output (ml/min/kg)	450	400	100
Ventricular Output (ml/min/kg) - Left	150	400	100
Ventricular Output (ml/min/kg) - Right	300	400	100
Stoke Volume (ml/kg) - Left	1	2	1
Stoke Volume (ml/kg) - Right	2	2	1

Summary Table – Postnatal Heart Development: A Species Comparison

Parameter	Human	Rat	Dog
Heart Size/Shape	Heart position higher and more transverse; oblique lie attained between 2 and 6 years of age. At birth, ventricular volumes are equal. Right ventricular volume doubles by 12 months, left ventricular volume unchanged. R/L ratio reaches 2:1 at 2 years.	Age-related changes in ventricular dimensions. Becomes more spherical with age.	Age-related changes in ventricular dimension. Becomes more spherical with age.
Heart Weight	Relative heart weight constant. Absolute weight doubled by 6 months and tripled by 1 year; achieved adult relative to body weight by about 21 years.	Decreasing relative heart weight with age. High weight increases on postnatal days 1-5 and slower growth afterwards.	Decreasing relative heart weight with age.
Cardiac Cells	Myocyte count at birth is 50% of adult, with proliferation, adult value reached by 4 months. Growth thereafter due to myocyte hypertrophy: 5 µm at birth, 8 µm at 6 wks, 11 µm at 3 yrs., 13 µm at 15 yrs and 14 µm in adults.	Myocyte proliferation (3-4 fold increase) and hypertrophy from birth to 2 months of age. Myocyte diameter 5.5 µm at 14 days, 10.5-11.8 µm at 30 days and 15-16 µm in adults.	Myocardial cell numbers relatively constant. Growth primarily through myocyte hypertrophy: 7 µm at 100 days, 13 µm at 0.5-0.9 year and 14 µm at 1-4.2 years.
	Myocytes diploid at birth, compared to 60% diploid, 40% polyploid in adults.	Myocytes primarily diploid in both infant and adult.	No information available
Coronary Vasculature	Primary arteries established before birth, diameter of coronary arteries doubled at 1st year reaching maximum at 30 years of age. Some capillary angiogenesis occurs postnatally. Capillary density decreases with age.	More immature at birth. Capillary and arteriole angiogenesis occurs postnatally. Arterial maturation by one month. Volume fraction of capillaries reaches maximum of 16% by day 28. Capillary density subsequently decreases with age.	Minimal information available. Capillary angiogenesis occurs postnatally.
Cardiac Innervation	Morphologically and functionally immature at birth. Number of neurons gradually increase and reach maximum density and adult patterns in childhood.	Morphologically and functionally immature at birth. Adrenergic patterns complete by 3 weeks and nerve density complete by 5 weeks. Cholinergic and other nerve types also matured postnatally.	Morphologically and functionally immature at birth and continues development during the first 2–4 months of age.

Summary Table – Postnatal Heart Development: A Species Comparison

Parameter	Human	Rat	Dog
Electrophysiology	Shorter QRS intervals in neonates with horizontal QRS loops changing from a complete clockwise rotation into a figure-of-eight loop that eventually unwinds into a counterclockwise loop orientation towards the left as normally observed in adults. The magnitude of the ventricular gradient vector changes increases from age 3 weeks until about 7 years of age. The spatial angle between QRS and STT vectors reaches its minimum at 1.5 to 4.5 years of age.	Shorter QRS intervals in neonates with horizontal QRS loops changing from clockwise, figure-of-eight to counterclockwise loop oriented towards the left. PQ interval decreases between 6 and 14 days and reaches adult values at 53 days of age. The QT interval also decreases during the first 3 weeks. From day 30 onward, QRS frequency merges with the T wave.	No information available.
Cardiac Biochemistry	Small decrease in water content but electrolyte concentrations relatively constant.	Rapid decline in water content up to 23 days of age. Increased Na and K with maximum levels at 16 days. Chloride and Ca levels decrease with age.	No information available.
	Concentrations of cytochromes and mitochondrial proteins increase during early postnatal period.	Concentrations of cytochromes, mitochondrial proteins and enzymes increase during early postnatal period.	Concentrations of cytochromes, mitochondrial proteins and enzymes increase during early postnatal period.
	Phosphofructokinase isozyme switch occurs before birth.	Phosphofructokinase isozyme switch occurs during first 2 postnatal weeks.	
Cardiac Output and Hemodynamics	Decreasing basal heart rate: 138 beats/min at birth to 55-85 beats/min in adults. Infants and young children have smaller ventricular volumes, stroke index and ejection fractions, but no differences in cardiac index compared to adults.	Early increase in heart rate, then remains relatively constant into adulthood. High cardiac output and low systemic resistance postnatally.	Significant increase in blood pressure and decrease in heart rate from 1 week to 6 months.
	Rapid increase in blood pressure from birth to two months (systolic/diastolic – 62/40 to 85/47); relatively constant from 6 months to 8 years of age (diastolic 58 to 62).	Systolic blood pressure doubles from neonate to young adults and reaches adult levels by 10 weeks of age.	No information available.

REFERENCES

Addis, T, Gray, H. 1950. Body size and organ weight. *Growth* 14:49-80.

Adeagbo AS, Bishai I, Lees J, Olley PM, Coceani F. 1985. Evidence for a role of prostaglandin I2 and thromboxane A2 in the ductus venosus of the lamb. *Can J Physiol Pharmacol* 63:1101-1105.

Adelman RD, Wright J. 1985. Systolic blood pressure and heart rate in the growing beagle puppy. *Dev Pharmacol Ther* 8:396-401.

Andresen MC, Kuraoka S, Brown AM. 1980. Baroreceptor function and changes in strain sensitivity in normotensive and spontaneously hypertensive rats. *Circ Res* 47:821-828.

Angst BD, Khan LU, Severs NJ, Whitely K, Rothery S, Thompson RP, Magee AI, Gourdie RG. 1997. Dissociated spatial patterning of gap junctions and cell adhesion junctions during postnatal differentiation of ventricular myocardium. *Circ Res* 80:88-94.

Anversa P, Capasso JM, Olivetti G, Sonnenblick EH. 1992. Monocyte cellular hypertrophy and hyperplasia in the heart during postnatal maturation and ageing. *Acta Paediatr* 383:29-31.

Anversa P, Olivetti G, Loud AV. 1980. Morphometric study of early postnatal development in the left and right ventricular myocardium of the rat. I. Hypertrophy, hyperplasia, and binucleation of myocytes. *Circ Res* 46:495-502.

Anversa P, Ricci R, Olivetti G. 1986. Quantitative structural analysis of the myocardium during physiologic growth and induced cardiac hypertrophy: a review. *J Am Coll Cardiol* 7:1140-1149.

Ashley LM. 1945. A determination of the diameters of myocardial fibers in man and other mammals. *J Anat* 77:325-347.

Baldwin KM, Cooke DA, Cheadle WG. 1977. Enzyme alterations in neonatal heart muscle during development. *J Mol Cell Cardiol* 9:651-60.

Barakat HA, Dohm GL, Loesche P, Tapscott EB, Smith C. 1976. Lipid content and fatty acid composition of heart and muscle of the BIO 82.62 cardiomyopathic hamster. *Lipids* 11:747-51.

Barrie SE, Harris P. 1977. Myocardial enzyme activities in guinea pigs during development. *Am J Physiol* 233:H707-H710.

Batra S, Rakusan K. 1992. Capillary network geometry during postnatal growth in rat hearts. *Am J Physiol* 262: H635-H640.

Berman W, Musselman J. 1979. Myocardial performance in the newborn lamb. *Am J Physiol* 237:H66-H70.

Boda D, Belay M, Eck E, Csernay L. 1971. Blood distribution of the organs examined by ^{86}Rb uptake under intrauterine conditions and in the newborn, in normal and hypoxic rabbits. *Biol Neonate* 18:71-77.

Boucek RJ Jr. 1982. Polyamine metabolism during the perinatal development of the rabbit right and left ventricle. *Pediatr Res* 16:721-727.

Broccoli F, Carinci P. 1973. Histological and histochemical analysis of the obliteration processes of ductus arteriosus Botalli. *Acta Anat* 85:69-83.

Carlson LA, Froberg SO, Nye ER. 1968. Effect of age on blood and tissue lipid levels in the male rat. *Gerontologia* 14:65-79.

Charbit M, Blazy I, Gogusev J, et al. 1997. Nitric oxide and the renin angiotensin system: contributions to blood pressure in the young rat. *Pediatr Nephrol* 11:617-22.

Chow LT, Chow SS, Anderson RH, Gosling JA. 1993. Innervation of the human cardiac conduction system at birth. *Br Heart J* 69:430-5.

Chow LT, Chow SS, Anderson RH, Gosling JA. 2001. Autonomic innervation of the human cardiac conduction system: changes from infancy to senility--an immunohistochemical and histochemical analysis. *Anat Rec* 264:169-82.

Chubb FJ, Bishop SP. 1984. Formation of binucleated cells in the neonatal rat. *Lab Invest* 50:571-577.

Clyman RI. 1987. Ductus arteriosus: current theories of prenatal and postnatal regulation. *Seminars in Perinatology* 11:64-71.

Clyman RJ, Campbell D, Heymann M., et al. 1985a. Persistent responsiveness of the neonate ductus arteriosus in immature lambs: A possible cause for reopening of patent ductus arteriosus after indomethacin-induced closure. *Circulation* 71:141-145.

Clyman RJ, Breall J, Maher P., et al. 1985b. Thyroid hormones and the ductus arteriosus. *Pediatr Res* 19:338A (abstr).

Coceani F, Hamilton NC, Labuc J., et al. 1984. Cytochrome P 450-linked monooxygenase: involvement in the lamb ductus arteriosus. *Am J Physiol* 246: H640-H643.

Cothran DL, Lloyd TR, Taylor H, Linden J, Matherne GP. 1995. Ontogeny of rat myocardial A_1 adenosine receptors. *Biol Neonate* 68:111-118.

Cua M, Shvilkin A, Danilo P, Rosen MR. 1997. Developmental changes in modulation of cardiac repolarization by sympathetic stimulation: The role of beta- and alpha-adrenergic receptors. *J Cardiovasc Electrophysiol* 8:865-871.

Dawes GS, Johnston BM, Walker DW. 1980. Relationship of arterial pressure and heart rate in fetal, newborn and adult sheep. *J Physiol* 309:405-417.

Davidson M, Collins M, Byrne J, Vora S. 1983. Alterations in phosphofructokinase isoenzymes during early human development. Establishment of adult organ-specific patterns. *Biochem J* 214:703-10.

Deavers S, Huggins RA, Smith EL. 1972. Absolute and relative organ weights of the growing beagle. *Growth* 36:195-208.

De Champlain J, Malmfors T, Olson L, Sachs C. 1970. Ontogenesis of peripheral adrenergic neurons in the rat: pre- and postnatal observations. *Acta Physiol Scand* 80: 276-288.

de Swiet M, Fayers P, Shinebourne EA. 1992. Blood pressure in first 10 years of life: The Brompton study. *Brit Med J* 304:23-26.

Diez U, Schwartze H. 1991. Quantitative electrocardiography and vectorcardiography in postnatally developing rats. *J Electrocardiol* 24:53-62.

Dhindsa DS, Metcalf J, Blackmore DW, Koler RD. 1981. Postnatal changes in oxygen affinity of rat blood. *Comp Biochem Physiol* 69A:279-283.

Dolezel S, Gervoa M, Gero J, Vasku J. 1981. Development of sympathetic innervation of the coronary arteries and the myocardium in the dog. *Morphologica* 29:189-191.

Dowell RT. 1984. Metabolic and cyclic nucleotide enzyme activities in muscle and non-muscle cells of rat heart during perinatal development. *Can J Physiol Pharmaco* 63:78-81.

Eckner FAO, Brown BW, Davidson DL, Glagov S. 1969. Dimension of normal human hearts. *Arch Pathol* 88:497-507.

Ehrich W, De La Chapelle C, Cohn AE. 1931-1932. Anatomical ontogeny. B. Man (a study of the coronary arteries). *Am J Anat* 49:241-282.

Eisenstein R, Wied GL. 1970. Myocardial DNA and protein in maturity and hypertrophied hearts. *Proc Soc Exp Biol Med* 133:176-179.

Englemann GL, Vitullo JC, Gerrity RG. 1986. Morphometric analysis of cardiac hypertrophy during development, maturation, and senescence in spontaneously hypertensive rats. *Circ Res* 60:487-494.

Fay FS, Cooke PH. 1972. Guinea pig ductus arteriosus. II. Irreversible closure after birth. *Am J Physiol* 222:841-849.

Fay FS, Jobsis FF. 1972. Guinea pig ductus arteriosus. III. Light absorption changes during response to O_2. *Am J Physiol* 223: 588-595.

Fisher DJ, Towbin J. 1988. Maturation of the heart. *Clin Perinat* 15:421-446.

Friedman AH, Fahey JT. 1993. The transition from fetal to neonatal circulation: Normal responses and implications for infants with heart disease. *Sem Perinat* 17:106-121.

Friedman WF. 1972. The intrinsic physiologic properties of the developing heart. *Prog Cardiovasc Dis* 15:87-111.

Friedman WF, Pool PE, Jacobowitz D, Seagren SC, Braunwald E. 1968. Sympathetic innervation of the developing rabbit heart. Biochemical and histochemical comparisons of fetal, neonatal, and adult myocardium. *Circ Res* 23:25-32.

Fried R, Jolesz FA, Lorenzo AV, Francis H, Adams DF. 1987. Developmental changes in proton magnetic resonance relaxation times of cardiac and skeletal muscle. *Invest Radiol* 23:289-293.

Fugelseth D, Lindemann R, Liestol K, Kiserud T, Langslet A. 1997. Ultrasonographic study of ductus venosus in healthy neonates. *Arch Dis Child Fetal Neonatal Ed* 77:F131-F134.

Garofolo MC, Seidler FJ, Auman JT, Slotkin TA. 2002. beta-Adrenergic modulation of muscarinic cholinergic receptor expression and function in developing heart. *Am J Physiol Regul Integr Comp Physiol* 282:R1356-63.

Georgieff MK, Mills MM, Gomez-Marin O, Sinaiko AR. 1996. Rate of change of blood pressure in premature and full term infants from birth to 4 months. *Pediatr Nephrol* 10:152-155.

Glowinski J, Axelrod J, Kopin IJ, Wurtman RG. 1964. Physiological disposition of H^3-norepinephrine in the developing rat. *J Pharm Exp Ther* 146: 48-53.

Gootman PM, Buckley NM, Gootman N. 1979. Postnatal maturation of neural control of the circulation. *Rev Perinat Med* 3:1-72.

Gourdie RG, Green CR, Severs NJ, Thompson RP. 1992. Immunolabelling patterns of gap junction connexins in the developing and mature rat heart. *Anat Embryol* 185:363-378.

Graham TP, Jarmakani JM, Canent RV, Morrow MN. 1971. Left heart volume estimation in infancy and childhood. Reevaluation of methodology and normal values. *Circulation* 43:895-904.

Gray SD. 1984. Pressure profiles in neonatal spontaneously hypertensive rats. *Biol Neonate* 45:25-32.

Grimm AF, Katele KV, Klein SA, Lin HL. 1973. Growth of the rat heart: left ventricular morphology and sarcomere lengths. *Growth* 37:189-201.

Grove D, Nair KG, Zak R. 1969. Biochemical correlates of cardiac hypertrophy. III. Changes in DNA content; the relative contributions of polyploidy and mitotic activity. *Circ Res* 25:463-471.

Haddad C, Armour JA. 1991. Ontogeny of canine intrathoracic cardiac nervous system. *Reg Integ Comp Physiol* 30:R920-R927.

Hageman GR, Neely BH, Urthaler F. 1986. Cardiac autonomic efferent activity during baroreflex in puppies and adult dogs. *Am J Physiol* 251:H443-H447.

Hall N, De Luca M. 1975. Developmental changes in creatine phosphokinase isoenzymes in neonatal mouse hearts. *Biochem Biophys Res Commun* 66:988-94.

Hazlewood CF, Nichols BL. 1970. *Johns Hopkins Med J* 127:136-145 (as cited in Rakusan, 1980).

Heggeness FW, Diliberto J, Distefano V. 1970. Effect of growth velocity on cardiac norepinephrine content in infant rats. *Proc Soc Exp Biol Med* 133:1413-1416.

Heron MI, Kuo C, Rakusan K. 1999. Arteriole growth in the postnatal rat heart. *Microvasc Res* 58:183-186.

Heymann MA, Rudolph AM. 1975. Control of the ductus arteriosus. *Physiol Review* 55:62-78.

Hoerter J, Mazet F, Vassort G. 1981. Perinatal growth of the rabbit cardiac cell: Possible implications for the mechanism of relaxation. *J Mol Cell Cardiol* 13:725-740.

Horackova M, Slavikova J, Byczko Z. 2000. Postnatal development of the rat intrinsic cardiac nervous system: A confocal laser scanning microscopy study in whole-mount atria. *Tissue Cell* 32:377-388.

Hort W. 1953. Quantitative Histologische Untersuchungen an Wachsenden Herzen. *Virchows Arch* [Pathol Anat] 323:223-242.

House EW, Ederstrom HE. 1968. Anatomical changes with age in the heart and ductus arteriosus in the dog after birth. *Anat Rec* 160:289-296.

Hudlicka O, Brown MD. 1996. Postnatal growth of the heart and its blood vessels. *J Vasc Res* 33:266-87.

Hwang B, Chu NW. 1995. Normal oscillometric blood pressure values in Chinese children during their first six years. *Zhonghua Min Guo Xiao Er Ke Yi Xue Hui Za Zhi* 36:108-12.

Ishii T, Kuwaki T, Masuda Y, Fukuda Y. 2001. Postnatal development of blood pressure and baroreflex in mice. *Auton Neurosci* 94:34-41.

Jarkovska D, Janatova T, Hruda J, Ostadal B, Samanek M. 1989. The physiological closure of ductus arteriosus in the rat, an ultrastructural study. *Anat Embryol* 180:497-504.

Jones M, Barrow MV, Wheat MW. 1969. An ultrastructural evaluation of the closure of the ductus arteriosus in rats. *Surgery* 66:891-898.

Kondo M, Itoh S, Kunikata T, Kusaka T, Ozaki T, Isobe K, Onishi S. 2001. Time of closure of ductus venosus in term and preterm neonates. *Arch Dis Child Fetal Neonatal Ed* 85:F57-F59.

Klopfenstein MS, Rudolph AM. 1978. Postnatal changes in the circulation and responses to volume loading in sheep. *Circ Res* 42:839-845.

Korecky B, Rakusan K. 1978. Normal and hypertrophic growth of the rat heart: Changes in cell dimensions and number. *Am J Physiol* 234:H123-H128.

Kyrieleis C. 1963. Die Formveränderungen des menschlichen Herzens nach der Geburt. *Virchows Arch* [Pathol Anat] 337:142-163.

Lee JC, Taylor JFN, Downing SE. 1975. A comparison of ventricular weights an geometry in newborn, young and adult mammals. *J Appl Physiol* 38:147-150.

Legato MJ. 1979. Cellular mechanisms of normal growth in the mammalian heart. I. Qualitative and quantitative features of ventricular architecture in the dog from birth to five months of age. *Circ Res* 44:250-262.

Leu M, Ehler E, Perriard JC. 2001. Characterisation of postnatal growth of the murine heart. *Anat Embryol* 204:217-24.

Levine RS, Hennekens CH, Klein B, et al., 1979. A longitudinal evaluation of blood pressure in children. *Am J Public Health* 69:1175-1177.

Li F, Wang X, Capasso JM, Gerdes AM. 1996. Rapid transition of cardiac myocytes from hyperplasia to hypertrophy during postnatal development. *J Mol Cell Cardiol* 28:1737-1746.

Linzbach AJ. 1950. Die Muskelfaserkonstante und das Wachstumgesetz der menschlichen Herzkammer. *Virchows Arch* [Pathol Anat] 318:575-618.

Linzbach AJ. 1952. Die Anzahl der Herzmuskelkerne in normalen, überlasteten, atrophen und Corhormon behandelten Herzkammer. *Z Kreislaufforsch* 41:641-658.

Lipp J, Rudolph A. 1972. Sympathetic nerve development in the rat and guinea-pig heart. *Biol Neonate* 21:76-82.

Litchfield JB. 1958. Blood pressure in infant rats. *Physiol Zool* 31:1-6.

Looker T, Berry CL. 1972. The growth and development of the rat aorta. II. Changes in nucleic acid and scleroprotein content. *J Anat* 113:17-34.

Lohse CL, Suter PF. 1977. Functional closure of the ductus venosus during early postnatal life in the dog. *Am J Vet Res* 38:839-844.

Mackenzie E, Standen NB. 1980. The postnatal development of adrenoreceptor responses in isolated papillary muscles from rat. *Pfluegers Arch* 383:185-187.

Mathew R, Thilenius OG, Arcilla RA. 1976. Comparative response of right and left ventricles to volume overload. *Am J Cardiol* 38:209-17.

Mattfeldt T, Mall G. 1987. Growth of capillaries and myocardial cells in the normal rat heart. *J Mol Cell Cardiol* 19:1237-46.

Metz LD, Seidler FJ, McCook EC, Slotkin TA. 1996. Cardiac -adrenergic receptor expression is regulated by thyroid hormone during a critical developmental period. *J Mol Cell Cardiol.* 28:1033-1044.

Meyer WW, Lind J. 1966. The ductus venosus and the mechanism of its closure. *Arch Dis Child* 41:597-605.

Momma K, Ito T, Ando M. 1992a. In situ morphology of the foramen ovale in the fetal and neonatal rat. *Pediatr Res* 32:669-672.

Momma K, Ito T, Ando M. 1992b. In situ morphology of the ductus venosus and related vessels in the fetal and neonatal rat. *Pediatr Res* 32:386-389.

Moss AJ, Emmanouilides GC, Adams FH, et al. 1964. Response of ductus arteriosus and pulmonary and systemic arterial pressure to changes in oxygen environment in newborn infants. *Pediatrics* 33:937-944.

Mrowka R, Patzak A, Schubert E, Persson PB. 1996. Linear and non-linear properties of heart rate in postnatal maturation. *Cardio Res* 31:447-454.

Nakamura S, Asai J, Hama K. 1986. The transverse tubular system of rat myocardium: Its morphology and morphometry in the developing and adult animal. *Anat Embryol* 173:307-315.

Namin EP, Miller RA. 1966. The normal electrocardiogram and vectorcardiogram in children. In: Cassetls DE and Ziegler RF (Eds) *Electrocardiography in Infants and Children*. Grune & Stratton, Orlando, Florida.

Nyquist-Battie C, Cochran PK, Sands SA, Chronwall BM. 1994. Development of neuropeptide Y and tyrosine hydroxylase immunoreactive innervation in postnatal rat heart. *Peptides* 15:1461-9.

Oberhansli I, Brandon G, Friedli B. 1981. Echocardiographic growth patterns of intracardiac dimensions and determination of function indices during the first year of life. *Helv Paedatr Acta* 36:325-340.

Oh W, Oh MA, Lind J. 1966. Renal function and blood volume in newborn infants related to placental transfusion. *Acta Paediat* 56:197-210.

Olivetti G, Anversa P, Loud AD. 1980a. Morphometric study of early postnatal development in the left and right ventricular myocardium of the rat. II. Tissue composition, capillary growth, and sarcoplasmic alterations. *Circ Res* 46:503-512.

Olivetti G, Anversa P, Melissari M, Loud AV. 1980b. Morphometry of medial hypertrophy in the rat thoracic aorta. *Lab Invest* 42:559-565.

O'Rahilly R, Müller F. 1996. *Human Embryology and Teratology.* 2nd Edition, Wiley-Liss, New York, pp. 159-206.

Penaloza D, Arias-Stella J, Sime F, et al. 1964. The heart and pulmonary circulation in children at high altitudes. *Pediatrics* 34:568-582.

Penefsky ZJ. 1985. Regulation of contractility in developing heart. In: Legato MJ (Ed) *The Developing Heart.* Martinus Nijhoff Publishing, Boston, p. 113.

Pickoff AS, Stolfi A. 1996. Postnatal maturation of autonomic modulation of heart rate. Assessments of parasympathetic and sympathetic efferent function in the developing canine heart. *J Electrocardiol* 29:215-222.

Prewitt RL, Dowell RF. 1979. Structural vascular adaptations during the developmental stages of hypertension in the spontaneously hypertensive rat. *Bibl Anal* 18:169-173.

Quigley KS, Shair HN, Myers MM. 1996. Parasympathetic control of heart during early postnatal development in the rat. *J Auto Nervous Sys* 59:75-82.

Rabe D. 1973. Kalibermessungen an den Herzkranzarterien und dem Sinus coronarius. *Basic Res Cardiol* 68:356-379.

Rakusan, K. 1980. Postnatal development of the heart. In: *Hearts and Heart-Like Organs*, Vol. 1, Academic Press, New York, pp. 301-348.

Rakusan, K. 1984. Cardiac growth, maturation, and aging. In: Zak R (Ed) *Growth of the Heart in Health and Disease*, Raven Press, New York, pp. 131-164.

Rakusan K, Cicutti N, Flanagan MF. 1994. Changes in the microvascular network during cardiac growth, development and aging. *Cell Mol Biol Res* 40:117-122.

Rakusan K, Du Mesnil De Rochemont W, Braasch W, Tschopp H, Bing RJ. 1967. Capacity of the terminal vascular bed during normal growth, in cardiomegaly and in cardiac atrophy. *Circ Res* 21:209-214.

Rakusan K, Flanagan MF, Geva T, et al. 1992. Morphometry of human coronary capillaries during normal growth and the effect of age in left ventricular pressure-overload hypertrophy. *Circulation* 86:38-46.

Rakusan K, Korecky B, Roth Z, Poupa O. 1963. Development of the ventricular weight of the rat heart with special reference to the early phases of postnatal ontogenesis. *Physiol Bohemoslov* 12:518-525.

Rakusan K, Marcinek H. 1973. Postnatal development of the cardiac output distribution in rat. *Biol Neonate* 22:58-63.

Rakusan K, Radl J, Poupa O. 1965. The distribution and content of myoglobins in the heart of rat during postnatal development. *Physiol Bohem* 14:317-319.

Rakusan K, Sarkar K, Turek Z, Wicker P. 1990. Mast cells in the rat heart during normal growth and in cardiac hypertrophy. *Circ Res* 66:511-516.

Rautaharju PM, Davignon A, Soumis F, Boiselle E, Choquette A. 1979. Evolution of QRS-T relationship from birth to adolescence in Frank-lead orthogonal electrocardiograms of 1492 normal children. *Circ* 60:196-204.

Robinson RB. 1996. Autonomic receptor--effector coupling during post-natal development. *Cardiovasc Res* 31:E68-E76.

Satoh K, Shirota F, Tsunajima T, Beinlich CJ, Morgan HE, Ichihara K. 2001. Limitation of heart growth in neonatal piglets by simvastatin and atorvastatin: Comparison with pravastatin. *Am J Physiol Heart Circ Physiol* 280:H2746-H2751.

Schiebler TH, Heene R. 1986. *Histochemie* 14: 28-334 (as cited in Rakusan, 1980).

Schneider R, Pfitzer P. 1973. Die Zahl der Kerne in isolierten Zellen des menschlichen Myokards. *Virchows Arch* [Pathol Anat] 12:238-258.

Schonfeld P, Schild L, Bohnensack R. 1996. Expression of the ADP/ATP carrier and expansion of the mitochondrial (ATP + ADP) pool contribute to postnatal maturation of the rat heart. *Eur J Biochem* 241:895-900.

Scott TM, Pang SC. 1983. The correlation between the development of sympathetic innervation and the development of medial hypertrophy in jejunal arteries in normotensive and spontaneously hypertensive rats. *J Auton Nerv Syst* 8: 25-32.

Sechi LA, Sechi G, De Carli S, Griffin CA, Schambelan M, Bartoli E. 1993. Angiotensin receptors in the rat myocardium during pre- and postnatal development. [Studio dei recettori dell'angiotensina nel miocardio di ratto durante lo sviluppo pre- e post-natale.] *Cardiologia* 38:471-476.

Segar JL, Hajduczok G, Smith BA, et al. 1992. Ontogeny of baroreflex control of renal sympathetic nerve activity and heart rate. *Am J Physiol* 263:H1819-H1826.

Seguchi M, Harding JA, Jarmakani JM. 1986. Developmental change in the function of sarcoplasmic reticulum. *J Mol Cell Cardiol* 18:189-195.

Shigenobu K, Tanaka H, Kasuya Y. 1988. Changes in sensitivity of rat heart to norepinephrine and isoproterenol during pre- and postnatal development and its relation to sympathetic innervation. *Dev Pharmacol Ther* 11:226-236.

Smith HL. 1928. The relation of the weight of the heart to the weight of the body and of the weight of the heart to age. *Am Heart J* 4:79-93.

Smith TL, Hutchins PM. 1979. Central hemodynamics in the developing stage of spontaneous hypertensive in the unanesthetized rat. *Hypertension* 1:508-517.

Smolich JJ. 1995. Ultrastructural and functional features of the developing mammalian heart: a brief overview. *Reprod Fertil Dev* 7:451-61.
Sole MJ, Lo CM, Laird CW, et al. 1975. Norepinephrine turnover in the heart and spleen of the cardiomyopathic Syrian hamster. *Circ Res* 37:855-862.
Solomon S, Wise P, Ratner A. 1976. *Proc Soc Exp Biol Med* 153:359-362 (as cited in Rakusan, 1980).
Steiss JO, Rascher W. 1996. Automated blood pressure measurement in children. *Z Kardiol* 85 (Suppl 3):81-84.
Tada T, Kishimoto H. 1990. Ultrastructural and histological studies on closure of the mouse ductus arteriosus. *Acta Anat* (Basel) 139:326-34.
Tiemann K, Weyer D, Djoufack PC, et al. 2003. Increasing myocardial contraction and blood pressure in C57BL/6 mice during early postnatal development. *Am J Physiol Heart Circ Physiol* 284:H464-H474.
Tomanek RJ, Haung L, Suvarna PR, O'Brien LC, Ratajska A, Sandra A. 1996. Coronary vascularization during development in the rat and its relationship to basic fibroblast growth factor. *Cardiovasc Res* 31:E116-E126.
Tomanek RJ, Sandra A, Zheng W, Brock T, Bjercke RJ, Holifield JS. 2001. Vascular endothelial growth factor and basic fibroblast growth factor differentially modulate early postnatal coronary angiogenesis. *Circ Res* 88:1135-41.
Truex RC, Scott JC, Ong DM, Smythe MQ. 1955. *Anat Rec* 123:201-225 (as cited in Rakusan, 1980).
Tucker DC. 1985. Components of functional sympathetic control of heart rate in neonatal rats. *Am J Physiol* 248:R601-R610.
Ursell PC, Ren CL, Danilo P. 1990. Anatomic distribution of autonomic neural tissue in the developing dog heart: I. Sympathetic innervation. *Anat Rec* 226:71-80.
Uiterwaal CS, Anthony S, Launer LJ, et al. 1997. Birth weight, growth, and blood pressure: An annual follow-up study of children aged 5 through 21 years. *Hypertension* 30:267-271.
Vatner SF, Manders WT. 1979. Depressed responsiveness of the carotid sinus reflex in conscious newborn animals. *Am J Physiol* 237:H40-H43.
Vizek M, Albretch I. 1973. Development of cardiac output in male rats. *Physiol Bohemoslov* 22:573-580.
Vliegen HW, van der Laarse S, Huysman JAN, et al. 1987. Morphometric quantification of myocyte dimensions validated in normal growing rat hearts and applied to hypertrophic human hearts. *Cardiovasc Res* 21:352-357.
Vlk J, Vincenzi FF. 1977. Functional autonomic innervation of mammalian cardiac pacemaker during the perinatal period. *Biol Neonate* 31:19-26.
Vogelberg K. 1957. Die Lichtungsweite der Koronarostien an normalen und hypertrophen Herzen. *Z Kreislaufforsch* 46:101-115.
Walker SP, Gaskin P, Powell CA, et al. 2001. The effects of birth weight and postnatal linear growth retardation on blood pressure at age 11-12 years. *J Epidemiol Community Health* 55:394-398.
Walsh SZ, Myer WW, Lind J. 1974. *The Human Fetal and Neonatal Circulation.* Charles C. Thomas, Springfield, Illinois.
Wang L, Swirp S, Duff H. 2000. Age-dependent response of the electrocardiogram to K+ channel blockers in mice. *Am J Physiol Cell Physiol* 278:C73-C80.
Webster SH, Liljegren EJ. 1949. *Am J Anat* 85:199-230 (as cited in Rakusan, 1980).
Winckler J. 1969. *Z Zellforsch Mikosk Anat* 98:106-121 (as cited in Rakusan, 1980).
Widdowson EM, Dickerson JWT. 1960. *J Mol Cell Cardiol* 9:859-866 (as cited in Rakusan, 1980).
Wiesmann F, Ruff J, Hiller KH, Rommel E, Haase A, Neubauer S. 2000. Developmental changes of cardiac function and mass assessed with MRI in neonatal, juvenile, and adult mice. *Am J Physiol Heart Circ Physiol* 278:H652-H657.
Yamori Y, Ohtaka U, Nara Y. 1976. Vectorcardiographic study on left ventricular hypertrophy in spontaneously hypertensive rats. *Jpn Circ J* 40:315.
Zink J, Van Petten GR. 1980. Time course of closure of the ductus venosus in the newborn lamb. *Pediatr Res* 14:1-3.

APPENDIX C-7*
SPECIES COMPARISON OF ANATOMICAL AND FUNCTIONAL IMMUNE SYSTEM DEVELOPMENT

Michael P. Holsapple[1,4], Lori J. West[2] and Kenneth S. Landreth[3]

[1] ILSI Health and Environmental Sciences Institute (HESI), Washington, DC;
[2] The Hospital for Sick Children, University of Toronto, Toronto, ON, Canada;
[3] Department of Microbiology, Immunology and Cell Biology, West Virginia University, Morgantown, WV.
[4] Address All Correspondence to: Michael P. Holsapple, Ph.D., Executive Director, ILSI Health and Environmental Sciences Institute, One Thomas Circle, NW, Ninth Floor, Washington, DC 20005-5802, PH: 202-659-3306, FAX: 202-659-3617. Email: mholsapple@ilsi.org

Introduction

In recent years, there has been increasing regulatory pressure to protect children's health because it is suspected that immature populations may be at greater risk for chemically induced toxicity. Congress enacted two statutes in 1996, the FOOD QUALITY PROTECTION ACT *(FQPA)* and the SAFE DRINKING WATER ACT *(SDWA)*. FQPA stated that "when establishing, . . . a tolerance . . . of a pesticide residue on food, EPA must perform s separate assessment for infants and children . . ."; and the SDWA "requires that EPA conduct studies to identify subpopulations, such as infants, children, pregnant women . . . that may be more susceptible than the general population . . .". In 1997, presidential Executive Order #13045 was issued which emphasized the following, "each Federal Agency shall make it a high priority to identify and assess environmental health risks and safety risks that may disproportionately affect children and ensure that its policies, programs, activities and standards address disproportionate risk to children that result from environmental health risks or safety risks . . . ".

Interestingly, a review of the available literature in virtually any area of health-related toxicology indicated that an overwhelming proportion of previous research and testing had been directed toward exposure of adults as opposed to children (Dietert et al., 2000). As the number of studies in young animals has begun to increase to address this regulatory pressure, it is already clear that there are many challenges to the design, conduct and interpretation of these new data.

Components of the immune system have not traditionally been emphasized as potential target organs in standard developmental and reproductive toxicity (DART) protocols. Moreover, although immunotoxicology has evolved as a science to the point where several regulatory agencies have crafted guidelines to address the immunotoxic potential of both drugs and non-drug chemicals, these protocols are generally performed in young adult animals, principally either rats or mice. Developmental immunotoxicology is predicated around the possibility that the immune system may exhibit unique susceptibility during its ontological development that may not be seen if data is only acquired in adults. It is important to emphasize at the onset, that there are currently no validated or widely accepted methods for evaluating the effects of a drug or chemical on the developing immune system. Nonetheless, because of concerns over children's health issues, specifically the possibility that the very young are uniquely susceptible to chemical perturbation, governmental regulators are beginning to ask for information about potential effects on the developing immune system.

---

* Source: Holsapple, M. P., West, L. J., and Landreth, K. S., Species comparison of anatomical and functional immune system development, *Birth Defects Research, Part B: Developmental and Reproductive Toxicology,* 68, 321-334, 2003.

POSTNATAL DEVELOPMENTAL MILESTONES

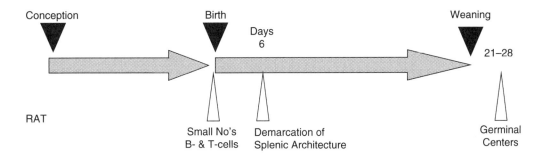

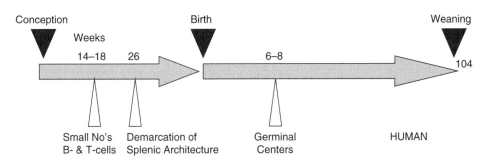

Figure 1

A number of workshops were recently organized to examine scientific questions that underlie developmental immunotoxicity tests, and the interpretation of results as they relate to human risk assessment. One of the most important questions considered in the workshops was how to determine the most appropriate species and strains to model the developing human immune system. A workshop organized by ILSI HESI in 2001 considered the following animal models in a series of plenary presentations: mice (Landreth, 2002; Herzyk et al., 2002 and Holladay and Blalock, 2002), rats (Smialowicz, 2002; Chapin, 2002), pigs (Rothkotter et al., 2002), dogs (Felsburg, 2002) and nonhuman primates (Hendrickx et al., 2002; Neubert et al., 2002), in addition to humans (West, 2002; Neubert et al., 2002). A subsequent panel discussion, with input from all participants, offered a number of important conclusions (Holsapple, 2002a). Although rabbits were emphasized as a preferred animal model for developmental toxicity studies, they have been rarely used in immunotoxicity studies. Although pigs and mini-pigs may offer advantages for mechanistic questions in developmental immunotoxicology, the consensus of the participants was that these species were not appropriate for screening. Although there are known differences between the development of the immune system in mice/rats and humans, rodents were judged to be the most appropriate model for screening for developmental immunotoxicology. The importance of the differences between rats and humans are highlighted in Figure 1, which illustrates three developmental landmarks. First, while small numbers of B- and T-lymphocytes can be detected in the spleen at the beginning of the second trimester of pregnancy in humans, these cells are only detectable in rats at birth. Second, the demarcation of the spleen into recognizable red and white pulp areas also occurs *in utero* for humans, but not until after parturition in rats. Finally, while germinal centers can be detected in humans very early during postnatal development, this landmark does not occur in rats until after weaning. These differences indicate that the maturation of the immune system in rats is delayed relative to humans and will be discussed in greater detail below.

The objective of this review is to compare the anatomical and functional development of the immune system in a number of species important to either preclinical studies for drug development and /or safety assessments for chemicals, with what is known in humans. Our current understanding of the development of the immune system in rodents has come almost exclusively from mechanistic experiments in mice. We know far less about the development of immunocompetent cells in rats. However, the general outline and timing of immune cell development in rats appear to closely parallel studies in mice where data are available. These points will be highlighted below. There is little doubt that we know the most about the immune systems of rodents and humans; but this review will also include what is known about the developing immune systems in dogs and subhuman primates. Dogs play an important role in the investigation of new drugs since they are one of the major species used in preclinical studies, including the effects of investigational drugs on the immune system (Felsberg, 2002). In the past, the major limitation of the use of dogs as an experimental model in immunologic research has been the paucity of reagents available to dissect the canine immune system. While still limited when compared to reagents for the murine immune system, great strides have been made in recent years to develop reagents to study dogs. Nonhuman primates have played an increasingly important role as a test species in preclinical testing due to the phylogenetic and physiologic similarities to man, especially in assessing the effects of immunomodulatory agents (Hendrickx et al., 2000).

Background — Immune System Development

The establishment of a functional immune system in all mammals, including humans, requires a sequential series of carefully timed and coordinated developmental events, beginning early in embryonic/fetal life and continuing through the early postnatal period. The immune system develops from a population of pluripotential hematopoietic stem cells that are generated early in gestation from uncommitted mesenchymal stem cells in the intraembryonic splanchnopleure surrounding the heart. This early population of hematopoietic stem cells gives rise to all circulating blood cell lineages, including cells of the immune system, via migration through an orderly series of tissues, and a dynamic process that involves continual differentiation of lineage restricted stem cells. Establishment of these populations of lymphoid-hematopoietic progenitor cells involves the migration of these cells from intraembryonic mesenchyme to fetal liver and fetal spleen, and ultimately, the relocation of these cells in late gestation to bone marrow and thymus. The latter two organs are the primary sites of lymphopoiesis and appear to be unique in providing the microenvironmental factors necessary for the development of functionally immunocompetent cells. The lineage-restricted stem cells expand to form a pool of highly proliferative progenitor cells that are capable of a continual renewal of short-lived functional immunocompetent cells, and that ultimately provide the necessary cellular capacity for effective immune responsiveness and the necessary breadth of the immune repertoire (Good, 1995). It is important to realize that immune system development does not cease at birth, and that immunocompetent cells continue to be produced from proliferating progenitor cells in the bone marrow and thymus. Mature immunocompetent cells leave these primary immune organs and migrate via the blood to the secondary immune organs - spleen, lymph nodes, and mucosal lymphoid tissues. The onset of functional competence depends on the specific parameter being measured and is also species-specific. Senescence of the immune responses is not well understood, but it is clear that both innate and acquired immune responses to antigens are different in the last quartile of life. This failure of the immune response is due, in part, to a continual reduction in the production of newly formed cells, and to the decreased survival of long-lived cells in lymphoid tissues.

The concept that any of a number of dynamic changes associated with the developing immune system may provide periods of unique susceptibility to chemical perturbation has been previously reviewed (Barnett, 1996; Dietert et al., 2000; Holladay and Smialowicz, 2000). In particular, an understanding of these developmental landmarks has prompted some to speculate about the existence of five critical windows of vulnerability in the development of the immune system (Dietert et al., 2000). The first 'window' encompasses a period of hematopoietic stem cell formation from undifferentiated mesenchymal cells. Exposure of the embryo to toxic chemicals during this period could result in failures of stem cell formation, abnormalities in production of all hematopoietic lineages, and immune failure. The second 'window' is characterized by migration of hematopoietic cells to the fetal liver and thymus, differentiation of lineage-restricted stem cells, and expansion of progenitor cells for each leukocyte lineage. This developmental window is likely to be particularly sensitive to agents that interrupt cell migration, adhesion, and proliferation. The critical developmental events during the third 'window' are the establishment of bone marrow as the primary hematopoietic site and the establishment of the bone marrow and the thymus as the primary lymphopoietic sites for B-cells and T-cells, respectively. The fourth 'window' addresses the critical periods of immune system functional development, including the initial period of perinatal immunodeficiency, and the maturation of the immune system to adult levels of competence. The final 'window', addresses the subsequent period during which mature immune responses are manifest, and functional pools of protective memory cells are established.

Emergence of Hematopoietic Stem Cells

Hematopoietic stem cells (HSC) are a population of multipotential stem cells that retain the capacity to self-renew and which have the capacity to differentiate to form all subclasses of leukocytes that participate in immune responses, as well as megakaryocytic and erythrocytic cells (Weissman, 2000).

Humans

Immune system development in the human fetus generally begins with HSC formation in the yolk sac, which first appear to migrate at approximately 5 weeks, and is followed by seeding of lymphoid and myeloid lineage progenitor cells (Haynes et al., 1988; Migliaccio et al., 1986).

Mice

HSC first appear developmentally in intraembryonic splanchnopleuric mesenchyme surrounding the heart (or the aorto-gonadomesonephros, AGM) at approximately 8 days of gestation in mice (Cumano and Godin, 2001). These cells are found at essentially the same developmental stage in the extraembryonic blood islands of the yolk sac. Embryonic circulation is established by gestational day 8.5 in the mouse and it remains unclear to what extent there is exchange of cells from intraembryonic hematopoietic tissues to extraembryonic sites in any rodent species. However, recent evidence clearly demonstrates that the population of intraembryonic stem cells, but not those which appear in the yolk sac, contribute to sustained intraembryonic blood cell development and the emergence of the immune system in postnatal rodents (Cumano et al., 2001).

Rats

No information available.

Dogs

No information available.

Primates

No information available.

Fetal Liver Hematopoiesis

The fetal liver serves as the initial hematopoietic tissue in early gestation (in some species it serves as the primary hematopoietic tissue throughout gestation) and is characterized by the emergence and rapid expansion of lineage restricted progenitor cells for all types of leukocytes.

Humans

Lymphocyte progenitors appear in human fetal liver at approximately the 7-8th week following conception (Migliaccio et al., 1986). Thus, an early period of susceptibility occurs during cell migration and early lymphohematopoiesis during which stem cells and lineage-committed progenitors are at risk (~7-10 weeks post-conception).

Mice

HSC migrate to the developing fetal liver by gestation day 10 in mice (Cumano and Godin, 2001 and Cumano et al., 2001). Differentiation of HSC to form lineage-restricted subpopulations of stem cells for the lymphoid and myeloid cell lineages has not been demonstrated prior to gestational day 10 in mice (Godin et al., 1999). Interestingly, the period of rapid hematopoietic progenitor cell expansion in fetal liver is not accompanied by wholesale differentiation to mature immunocompetent cell phenotypes (Godin et al., 1999). In fact, mature lymphocytes are not found in the developing liver until day 18 of gestation in the mouse, and after the initiation of hematopoiesis in embryonic bone marrow (Kincade, 1981).

Rats

No information available.

Dogs

No information available.

Primates

No information available.

Development of the Spleen

The role of the spleen in the immune system varies both across the timeline of development and according to a distinct species-specificity. The development of other relevant immune organs (e.g., lymph nodes, Peyer's patches, etc.) will be discussed as appropriate. The development of the thymus is discussed below.

Humans

Sites of lymphopoiesis begin to develop in the human fetus over the last half of the first trimester, starting with thymic stroma, which forms during the 6th week post-conception (Haynes et al., 1988). Stem cells and T-cell progenitors begin migration from fetal liver to thymus during the 9th week, while B-cell progenitors appear in blood by about week 12 (Loke, 1978; vonGaudecker, 1991; Kendall, 1991; Royo et al., 1987). Gut-associated lymphoid tissue develops from week 8 onward, beginning with the lamina propria during weeks 8-10, followed by Peyer's patches and the appendix from weeks 11-15. Lymph nodes begin development from 8-12 weeks post-conception, followed by tonsils and spleen from 10-14 weeks. Smith (1968) reported a similar sequence in that lymphocytes appear first in the fetal thymus, then in peripheral blood shortly after their appearance in the thymus and then in the fetal spleen. Although the spleen never completely loses hematopoietic function in the mouse, as described below, the human spleen has largely ceased hematopoiesis by the time of birth (Tavassoli, 1995).

Mice

HSC and lineage restricted hematopoietic progentior cells are found in fetal spleen at approximately gestational day 13 (Landreth, 2002). The spleen continues to contain limited reserves of hematopoietic cells throughout gestation and into postnatal life, and to support myeloid and erythroid cell development, particularly if the bone marrow is damaged. However, lymphopoiesis does not occur in the spleen of postnatal mice under any experimental conditions tested (Paige et al., 1981), suggesting that the bone marrow hematopoietic microenvironment is unique and required for lymphocyte production in mice.

Rats

As depicted in Figure 1, only small numbers of B- and T-lymphocytes are found in the spleens of newborn rats (Marshall-Clarke et al., 2000). The demarcation of the spleen into recognizable white and red pulp areas occurs at around postnatal day 6. B-cell follicles are not seen until about two weeks of age and the ability to form germinal centers does not develop until weeks three to four (Figure 1; Dijkstra and Dopp, 1983). For comparative purposes in Figure 1, although T- and B-cells can be found as early as week 14-18 of gestation and white pulp areas of the spleen are clearly demarcated by week 26, germinal centers are not observed until several weeks after birth (Namikawa et al., 1986). In addition, as discussed further below, the immune system of the neonate is immature and the phenotype of the cells that are found in the marginal zone of the spleen remains immature until around two years of age (Timens et al., 1989).

Dogs

Between days 27 and 28 of gestation, the primordia of the spleen is evident (Kelly, 1963; Felsberg, 1998). Lymphocytic infiltration of the spleen and lymph nodes with evidence of T-cell dependent zones is evident between days 45 to 52 (Bryant and Shifrine, 1972; Felsberg, 1998). Peyer's patches are present in the small intestine at about the same time, gestation days 45 to 55 (HogenEsch et al., 1987). Germinal centers and plasma cells appear in the spleen and lymph nodes shortly after birth (Yang and Gawlack, 1989; Felsberg, 1998).

Primates

Substantial lymphocyte differentiation occurs in fetal macaques by gestational day 65 and is accompanied by increasing lymphoid tissue organization (e.g., demarcation) into specific T-cell and B-cell areas that are apparent in peripheral lymphoid organs by gestational day 80 (Hendrickx et al, 2002). In the fetal spleen at gestational day 75, a large proportion of the lymphocytes were $CD20^+$ B-cells and there were low numbers of T-cells. By gestational day 145, the ratio of white pulp to red pulp was 1:1, similar to that seen in the mature primate spleen. At gestational day 80, large numbers of B- and T-cells were scattered throughout the parenchyma of lymph nodes, with early evidence of organization into the characteristic compartments. In the last stage examined (gd 145), both the follicle areas containing B-cells and the paracortex containing T-cells had expanded. In the small intestine at gestational day 80, both $CD20^+$ B-cells and $CD3^+$ T-cells were frequently encountered in the lamina propia. In the latter stages of fetal development (between gd 100 and gd 145), lymphoid aggregates were well defined within the lamina propia with B-cells oriented toward the luminal side and T-cells common on the muscularis side.

Development of the Thymus

Under the influence of thymic epithelial cells, developing thymic lymphocytes initiate expression of the T-cell receptor (TcR), undergo a series of selection events which remove autoreactive cells, and ultimately migrate out of that tissue expressing the TcR and a set of linage specific membrane glycoproteins that define phenotype and predict function of these cells (e.g., the so-called 'cluster designation' or CD markers). Interference during these important stages of T-cell development can alter the evolution of 'self' vs. 'non-self' recognition, leading either to autoimmunity or to immunodeficiencies resulting in increased susceptibility to infections (Jenkins et al., 1988). Thymic cell production wanes rapidly following sexual maturity in all vertebrates.

Humans

The seeding of the immune microenvironment during human fetal thymus development has been reviewed (Lobach and Haynes, 1987; Haynes and Hale, 1998). As noted above, the thymic stroma forms in the human fetus during the 6th week post-conception (Haynes et al., 1988), and T-cell progenitors begin migration from fetal liver to thymus during the 9th week (Kay et al., 1962; Royo et al., 1987; vonGaudecker, 1991; Kendall, 1991). T-lymphocyte development proceeds in the thymus, which divides into cortex and medulla at 10-12 weeks post-conception (Loke, 1978). The thymic medulla is fully formed by 15-16 weeks, after which there is an orderly progression of T-cell development beginning in the cortex as thymocytes proceed from the cortex towards the medulla (Royo et al., 1987; vonGaudecker, 1991; Kendall, 1991). Gene re-arrangement in developing thymocytes begins at approximately week 11, leading first to expression of the $\gamma\delta$-TcR, then the $\alpha\beta$-TcR (Royo et al., 1987; Penit and Vasseur, 1989; Kendall, 1991). Sequential expression of co-receptors follows: CD3, CD4 and CD8 (corresponding to expression of major histocompatibility complex class I and II antigens by thymic epithelial and dendritic cells) (Royo et al., 1987; Penit and Vasseur, 1989; vonGaudecker, 1991; Kendall, 1991). Lobach and Haynes (1987) reported that human fetal thymocytes express the T-cell markers, CD3, CD4, CD5 and CD8, at 10 weeks (Lobach and Haynes, 1987). The complex process of 'thymic education' continues to progress throughout the mid-trimester of gestation, with positive and then negative selection of 'double positive' T-lymphocytes co-expressing the CD4 and CD8 surface molecules (Kay et al., 1970; Loke, 1978; Gale, 1987). This selection process culminates eventually in an enormous reduction in total lymphocyte populations emerging from the thymus. Export of 'single positive' CD4+ and CD8+ T-lymphocytes begins after week 13 (Berry et al., 1992). Single positive T-cells are detectable in spleen by about week 14 and in cord blood by week 20 (Peakman et al., 1992). Thereafter follows an intense expansion of T-cell numbers occurring during weeks 14-26 (Kay et al., 1970; Loke, 1978; Royo et al., 1987; Gale, 1987; Berry et al., 1992; Peakman et al., 1992). Exposures during this period that alter the emerging T-cell repertoire, such that T-cells are more promiscuous in terms of antigen recognition, may result in wider 'holes' in the available T-cell response to certain antigens encountered later in life.

Mice

In mice, the thymus anlage develops from the 3rd and 4th pharyngeal pouches on gestational day 10 and is colonized by immigrating HSC by gestational day 11 (Shortman et al., 1998). Pluripotent HSC continue to be detectable in the thymus throughout gestation, however, the majority of immigrating cells differentiate within the thymic microenvironment to form immature proliferating lymphoid thymocytes which express the TcR and the TcR-specific cell surface proteins, CD3, CD4 and CD8. Thymic cell production wanes rapidly following sexual maturity resulting in the virtual absence of thymocyte production at one year of life in mice (Shortman et al., 1998).

Rats

Less is known about thymus development in the rat. The thymus anlage in rats is invaded by hematopoietic stem cells by day 13 of gestation, and these cells have differentiated to express both CD4 and CD8 (e.g., double positive cells) by day 18 (Aspinall et al., 1991). Mature single positive thymocytes are not detected until day 21 in the rat.

Dogs

The primordia of the thymus is evident in dogs between days 27 and 28, and it descends from the cervical region into the anterior thoracic cavity on day 35 (Kelly, 1963; Felsberg, 1998). At this time, the thymus is composed of epithelial lobules and mesenchymal stroma only. Between days 35-40, the fetal thymus becomes actively lymphopoietic and shows corticomedullary demarcation; by day 45 the thymic microenvironment has assumed its normal postnatal histologic appearance (Miller and Benjamin, 1985; Snyder et al., 1993; Felsberg, 1998). Fetal and postnatal thymopoiesis has been recently evaluated in dogs and the results indicate that normal thymopoiesis is occurring by day 45 of gestation with a distribution of thymocyte subsets virtually identical to that seen in the postnatal thymus, although with a markedly reduced total cellularity (Somberg et al., 1994; Felsberg, 1998). The thymus undergoes rapid postnatal growth and reaches maximum size at 1 to 2 months of age, as the percentage of body weight, and at 6 months of age in absolute terms (Yang and Gawlak, 1989).

Primates

Fetal thymic histogenesis in primates (rhesus) has been demonstrated to be gradual and continuous, as is the case in humans (Tanimura and Tanioka, 1975). The thymic anlage separates from the pharyngeal pouches and engages in early stages of proliferation at gestation days 37 – 48 (Hendrickx et al., 1975). These same studies showed that the differentiation of lymphoid elements occurred within the thymus at gestation days 50 – 73; and the maturation of the fetal lymphoid system occurred at gestation days 100 – 133. Later studies by the same laboratory using another species of primates (cynomolgus) indicated that the cortex and medulla of the thymus were distinguishable by gestational day 65, and most of the thymocytes were $CD3^+$ and localized throughout both regions of the fetal thymus, while $CD20^+$ B-cells were dispersed in the corticomedullary junction. In older fetuses (gd 100 and gd 145), $CD20^+$ B-cells increased in the medulla and corticomedullary junction, while the number of $CD3^+$ thymocytes remained similar throughout gestation (Hendrickx et al., 2002). In the same study, flow cytometric analysis of lymphocyte subsets indicated a slight increase in the single positive ($CD4^+/CD8^-$ and $CD4^-/CD8^+$) thymocytes and a slight decrease (86% to 76%) in double positive ($CD4^+/CD8^+$) thymocytes between gestational days 80 and 145.

Bone Marrow Hematopoiesis

The final critical stage in the embryonic development of the mammalian immune system is the establishment of the bone marrow as the primary hematopoietic organ.

Humans

Bone marrow lymphopoiesis begins in the human fetus at approximately week 12 (West, 2002). B-lymphocyte development begins at approximately the same time as T-cells in humans. B-cells bearing sIgM and sIgG are first found in the liver at 8 weeks of gestation and in the spleen at about 12 weeks. IgD- and IgA-bearing cells are also found at 12 weeks of gestation, and Anderson et al. (1981) reported that adult levels of B-cells bearing sIg of all classes are reached by 14-15 weeks.

Mice

The formation of the bone marrow cavity as long bones are mineralized late in gestation, and the immigration of hematopoietic cells into that tissue site, occurs on gestational day 17.5 in mice (Kincade et al., 1981). The bone marrow rapidly assumes primary hematopoietic function after gestational day 18 in mice, and this persists throughout postnatal life. The relationship of bone formation and emergence of hematopoietic tissue is particularly interesting, and it is unclear whether the migration of hematopoietic cells into this tissue actually initiates the process of bone ossification. The evacuated bone marrow cavity is colonized by HSC and these pluripotential cells expand to establish the primary hematopoietic reserves of stem cells for postnatal life. In fetal mice, B-cells expressing surface immunoglobulin (sIg) appear in the liver, spleen and bone marrow on approximately gestational day 17 (Verlarde and Cooper, 1984).

Rats

No information available.

Dogs

Bone marrow in fetal dogs becomes heavily cellular, including abundant hematopoietic stem cells between days 45 and 52 (Bryant and Shifrine, 1972; Felsberg, 1998).

Primates

No information available.

Postnatal Development of the Immune System

Because birth occurs at various stages of fetal maturity, the significance of parturition as a landmark in the development of the immune system can vary from species to species. As such, direct comparison of immune functional development between humans and animals is complicated by differences in the maturity of the immune system before and after parturition. This difference has been linked to the length of gestation (Holladay and Smialowicz, 2000) in that animals with short gestation periods (e.g., mice, rats, rabbits and hamsters) have relatively immature immune systems at birth compared to humans. However, this speculation is not absolute, because, as noted by Felsburg (2001), dogs are like humans in that their immune systems are essentially fully developed at birth even though their gestation period is markedly shorter than humans.

Certainly, the event of birth marks an emergence from intra-uterine maternal influences, including the maternal immune 'suppression' associated with pregnancy (Oldstone and Tishon, 1977; Barrett et al., 1982; Papdogiannakis et al., 1985; Papadogiannakis and Johnsen, 1988; Loke and King, 1991; Sargent et al., 1993). It is important to keep in mind that maternal influences continue to contribute throughout lactation and other maternal behaviors. After birth, infectious and similar antigenic exposures become more significant, including colonization with microbes through the gastro-intestinal tract and other sites. The importance of the early postnatal exposure to 'antigens' is clearly demonstrated in studies where there is a delay in lymphoid development in animals raised from birth in a germ-free environment (Thorbecke, 1959). More recent studies by Anderson et al. (1981) indicated that germ-free animals have significantly reduced levels of Ig, due to a 2-5-fold lower rate of Ig synthesis, with IgA and IgM more affected than IgG, and have approximately 10-fold lower numbers of plasma cells.

Humans

At the time of human birth, the proportion of lymphocytes represented by T-cells is lowest and increases over time (Semenzato et al., 1980; Series et al., 1991; Berry et al., 1992; Erkellar-Yuksel et al., 1992; Hannet et al., 1992; Plebani et al., 1993; Hulstaert et al., 1994). Specifically, absolute numbers of both CD4$^+$ and CD8$^+$ subsets increase, while the CD4/CD8 ratio has been variably reported to differ, or not, with time after birth (Foa et al., 1984; Hannet et al., 1992; Plebani et al., 1993). Usage of TcR Vβ subsets likewise has been variably reported to differ, or not, between newborn and adult $\alpha\beta$-T-cells (Foa et al., 1984; Hayward and Cosyns, 1994). The proportion of T cells expressing the $\gamma\delta$-TCR is high during fetal life, decreases at term, but remains higher in newborns than later in life (Peakman et al., 1992). Some reports have described circulating 'double-positive (CD4$^+$CD8$^+$) immature' T-cells; however generally positive and negative selection are probably complete by birth (Foa et al., 1984; Griffiths-Chu et al., 1984; Solinger, 1985; Reason et al., 1990; Hayward and Cosyns, 1994). Nonetheless, T-cells do have phenotypic and functional features of 'naïve' cells. In particular, markers of 'immature' or antigenically 'naïve' cells such as CD45RO$^-$/RA$^+$ on CD4$^+$ cells are much higher than in adults (Clement et al., 1990; Denny et al., 1992; Erkellar et al., 1992; Hannet et al., 1992; Hulstaert et al., 1994; Igegbu et al., 1994; Jennings et al., 1994). Expression of some surface molecules (e.g., integrins, adhesion molecules) is reported to be low at birth and increases post-natally, while expression of others is high and decreases (Clement et al., 1990; Hannet et al., 1992; Hayward and Cosyns, 1994).

Mice

Following birth, there is an immediate disappearance of hematopoietic cells from the liver as that organ assumes postnatal function. In the postnatal mouse, leukocytes are produced in the bone marrow and, except for T lymphocytes, complete their maturation in that tissue. It is known that splenic hematopoiesis persists for several weeks after birth in rodents (Marshall-Clarke et al., 2000). Spear et al. (1973) observed an increase in B-cells and a decrease in the T-cell to B-cell ratio in mice between 2 and 3 weeks of age, which coincided, with the onset of antigen responsiveness in their studies. As discussed below, similar results were presented in 21-day old rats by Ladics et al. (2000).

One of the clearest indications that the immune system continues to develop in mice is the fact that perinatal (<36 hours of age) thymectomy results in a wasting syndrome due to the failure of T-cell development (Barnett, 1996). This observation is contrasted with that observed in newborn infants. West (2002) described a not infrequent scenario in which newborn infants undergo heart transplantation within the first days of life. Thymectomy is routinely performed during these transplant procedures. The grave consequences described for the mouse are not seen, even after clinical immunosuppression is initiated (often using a polyclonal anti-lymphocyte antibody preparation and high-dose steroids, a calcineurin inhibitor, an antimetabolite, or in some cases prednisone following transplantation surgery).. One is left to conclude that perinatal thymectomy causes a wasting syndrome and death in the mouse, but is tolerated in humans even when requisite immunosuppressive therapy is initiated.

Rats

As seen with the mouse, a limited number of studies in the rat have determined that its immune system is very immature at birth and that it must undergo much maturation during the early postnatal period. Phenotypic analysis in 10-day old rats pups indicated the absence of lymphocyte subsets

(Ladics et al., 2000). Subsequent histological analysis of the spleens from 10-day old pups indicated that there were no germinal centers present either in untreated rats or in rats immunized with the T-dependent antigen, sheep erythrocytes (SRBC). As discussed below, these same studies indicated that no anti-SRBC antibody response could be detected in 10-day old rat pups. Taken together, these results suggested that the immune system of a 10-day rat pup is too immature to elicit a primary response. The phenotypic analysis of the spleen from a 21-day old rat indicated that the number of B-cells was comparable to adults, and that the number of T-cells and T-cell subsets was reduced relative to an adult (Ladics et al., 2000). As discussed below, a number of studies have demonstrated that rats at approximately 21 days of age are capable of an immune response, but that it is always less than that seen in an adult animal.

Dogs

Shortly after birth in dogs, germinal centers and plasma cells appear in the spleen and lymph nodes (Yang and Gawlack, 1989; Felsberg, 1998), and the thymus undergoes rapid postnatal growth (Yang and Gawlak, 1989; Somberg et al., 1994). The phenotype of the lymphocyte subpopulations in neonatal dogs differs significantly from that observed in adult dogs (Felsberg, 1998, Somberg et al., 1994; Felsberg, 2002). During the first 16 weeks, there is a gradual decline in the proportion of peripheral B-cells and an increase in the proportion of peripheral T-cells to normal adult levels, which occurs between 2 and 4 months of age. Thereafter, the proportion of B- and T-cells remains fairly constant throughout the life of the dog. Other age-related differences in lymphocyte subsets include the following: considerably higher proportions of CD4+ T-cells during the first six months, a resultant high CD4:CD8 ratios during the same time period, a decline in the proportion of CD4+ at 10-12 months; CD8+ T-cells increasing to normal adult levels during the same time with a resultant normalization of the CD4:CD8 ratios (1.5 – 2.0). Finally, during the neonatal and immediate postnatal period, greater than 90% of the peripheral T-cells are CD45RA+ (naïve) T-cells (Somberg et al., 1996; Felsberg, 2002). After 4 months of age, the relative frequency of CD45RA+ T-cells declines such that only 40-50% of the peripheral T-cells in adults are CD45RA+. Interestingly, very similar age-related changes were observed in healthy children and adults (Denny et al., 1992; Erkellar et al., 1992), an observation consistent with the recognition that the immune systems of dogs and humans are at a similar stage of development at birth.

Although most indicators suggest that the development of the immune systems of dogs and humans are very comparable, there is one important distinction that does impact on the respective postnatal development of the immune systems (Felsberg, 2002). Humans possess a hemochorial placenta in which the blood of the mother is in direct contact with the trophoblast, thereby permitting direct entry of maternal IgG into the fetal bloodstream. In contrast, dogs have an endotheliochorial placenta with four structures separating the maternal and fetal blood – the endothelium of the uterine vessels, the chorion, mesenchyme (connective tissue) and the endothelium of fetal tissues. These four layers of tissue separating the maternal and fetal circulation in the dog limit the *in utero* transfer of maternal IgG to the fetus. Ultimately, only 5 to 10% of maternal antibody in the dog is obtained *in utero* through the placenta with the majority being obtained via the colostrums during the first 24 hours after birth (Felsberg, 1998). The levels of serum IgG in newborn puppies that receive colostrums approach those levels found in adults. It is important to emphasize that the half-life of maternal antibody in the dog is approximately 8.4 days, so the average protection from maternal antibody in the neonate is 8 to 16 weeks. Following the decline in maternal IgG, there is a gradual increase of all three major classes of immunoglobulin, with IgM and IgG reaching normal adult levels by 2-3 months of age, and 6 to 9 months of age, respectively. As in other species, the synthesis of IgA lags behind the other isotypes and doesn't reach normal adult levels until approximately one year of age.

Primates

A study by Neubert et al. (1996) compared a number of phenotypic markers in peripheral blood lymphocytes between adults and newborns in both humans and primates (marmosets). While peripheral blood was used for adult humans, adult marmosets and newborn marmosets, umbilical cord blood was used as a surrogate for newborn humans. Their results indicated no significant differences between newborns and adults in either humans or primates and only trends will be discussed.

* % lymphocytes:	humans: umbilical = adult
	marmosets: newborn ≤ adult
* CD4+ (helper):	humans: umbilical = adult
	marmosets: newborn = adult
* CD4+/CD29+ (memory):	humans: umbilical < adult
	marmosets: newborn < adult
* CD4+/CD45RA (naïve):	humans: umbilical > adult
	marmosets: newborn >> adult
* CD8+ (suppressor):	humans: umbilical ≤ adult
	marmosets: newborn ≥ adult
* CD 56+ (NK cells):	humans: umbilical = adult
	marmosets: newborn = adult
* CD 20+ (B-cells):	humans: umbilical = adult
	armosets: newborn = adult

The biggest differences between newborns and adults observed in this study were for memory T-cells (e.g., CD4+/CD29+), which were lower in newborns, and for naïve T-cells (CD4+/CD45RA), which were higher in newborns. Both of these observations are consistent with, and do not detract from the interpretation that the immune systems of humans and primates are devoid of any antigenic exposure, but are effectively mature at birth. Neubert et al. (2002) emphasized that the immune system of newborns (e.g., marmosets and humans) is rather 'immature'; that these deficiencies are not so much an indication of a lack of important components of the immune system, but rather are the result of little intrauterine contact with environmental antigens. The biggest difference between humans and marmosets was in the number of white blood cells (e.g., measured either as total WBCs or as % lymphocytes) being considerably higher in the primates.

Acquisition of Functional Immunocompetence

As noted above, the onset of functional immunocompetence varies across species and is strikingly different between rodents and humans. Exposure to a specific antigen during the perinatal period results in a rapidly expanding accumulation of lymphocyte specificities in the pool of memory cells in secondary lymphoid tissues. As thymic function wanes and thymocytes are no longer produced in that tissue, it is this pool of memory B- and T-cells that maintains immunocompetence for the life of the individual.

Humans

Thymocytes derived from human fetal tissue of less than 11 weeks gestation show no demonstrable response to mitogen stimulation (Kay et al., 1970; Sites et al., 1974; Royo et al., 1987). Functional capability of human fetal T-lymphocytes begins to develop between the end of the first trimester and the end of the second trimester (~14-26 weeks). At 12-14 weeks, fetal thymocytes respond to PHA only, while by 13-14 weeks (after thymic colonization with stem cells), fetal thymocytes respond to most mitogens (Kay et al., 1970; Sites et al., 1974; Royo et al., 1987). Mitogen-responsive

T-cells can be demonstrated in spleen and peripheral blood by 16-18 weeks gestation (Kay et al., 1970). Similar results were reported by Mumford et al. (1978), who observed responses to Con A with thymus cells at 13-14 weeks of gestation, and with spleen cells at approximately 18 weeks of gestation; while August et al. (1971) observed that fetal thymocytes responded to mitogenic stimulation at 12 weeks of gestation and splenic T-cells responded between 14 and 16 weeks of gestation. Reactivity of fetal lymphocytes in the MLR has been variably reported as beginning during weeks 10-14 (Ohama and Kaji, 1974; Loke, 1978; Royo et al., 1987), followed by functional effector reactivity in cell-mediated lympholysis (CML) assays subsequently, proceeding in variable fashion depending on the lymphoid tissue tested (Granberg and Hirvonen, 1980; Murgita and Wigzell, 1981). Fetal thymocytes have been shown to demonstrate a positive MLR by 12 weeks gestation, and fetal splenocytes by about 19 weeks, although to a consistently-low degree until 23 weeks (Granberg and Hirvonen, 1980). Lymphocytes derived from cord blood have been demonstrated to generate a positive CML response at 18-22 weeks gestation; however high individual variability was found (Rayfield et al., 1980). The functionality of T cells from human neonates has been demonstrated to be reasonably well-developed (Hayward, 1983). Thus, neonatal T cells can proliferate in MLR and in response to most mitogens, and show weak proliferation and cytokine production with stimulation by endogenous antigen presenting cells or anti-CD3 monoclonal antibody (representative 'physiologic' stimuli) (Granberg and Hirvonen, 1980; Rayfield et al., 1980; Loke and King, 1991). Furthermore, with TcR-independent stimulation, neonatal T-cells show equivalent proliferation and cytokine production to adult T cells (Splawski and Lipsky, 1991; Demeure et al., 1994; Tsuji et al., 1994). However, the cytokine profiles of the human neonate show a deficient and/or delayed production of interleukin (IL)-2, interferon-γ, IL-4 and IL-6, thus possibly skewed toward a Th2 phenotype (Splawski and Lipsky, 1991; Demeure et al., 1994; Tsuji et al., 1994). In some reports, the cytokine networks appear less efficient, requiring increased stimulation and/or receptor maturation (Cairo et al., 1991; Tucci et al., 1991). T-cell reactivity can be demonstrated in CML assays, but at lower levels than adult T cells (Granberg and Hirvonen, 1980; Rayfield et al., 1980; Barret et al., 1980; Loke and King, 1991). Demonstrating *in vivo* functionality, newborns can reject organ and tissue allografts, however, they remain very sensitive to clinical immunosuppression (Demeure et al., 1994; Webber, 1996; Pietra and Boucek, 2000). Indeed, it has been demonstrated that a survival advantage of approximately 10-12% is maintained more than 10 years post-transplant if heart transplantation is performed within the first 30 days of life compared to 1-6 months of age, and that this survival advantage is due to decreased immune-related events (Pietra and Boucek, 2000).

The development of functional NK cells in the human fetus occurs at 28 weeks of gestation, with full-term newborns displaying peripheral blood NK activity at approximately 60% of adult levels (Toliven et al., 1981). NK cells are fewer in number at birth than later, and have been reported to be less active and less responsive to stimulation, with diminished cytotoxicity capacity (Kohl et al., 1984; Rabatic et al., 1990). However, levels of NK cells during early fetal life have been demonstrated to be significantly higher than during neonatal life, suggesting that certain aspects of innate immunity may play a more important role in the fetal immune response than adaptive immunity (Erkellar et al., 1992; Hulstaert et al., 1994).

As noted above, adult levels of B-cells bearing sIg of all classes are reached by 14-15 weeks gestation in humans (Anderson et al., 1981). Circulating B lymphocytes are generally at high levels in the neonate, with immature markers demonstrable, and these decline with age (Tucci et al., 1991; Erkeller-Yuksle et al., 1992; Hannet et al., 1992; Peakman et al., 1992; Plebani et al., 1993; Hulstaert et al., 1994; Nahmias et al., 1994). The development of mature plasma cells in the bone marrow is incomplete at birth (Loke, 1978; Durandy et al., 1990). Isotype switching is defective, with simultaneous surface expression of different isotypes, and immunoglobulin production is low (Loke, 1978; Gathings et al., 1981; Hayward, 1983; Nahmias et al., 1994). Human neonatal B-cells are functionally defective in their capacity to generate antibody-producing cells *in vitro,* compared to

B-cells from adults, and in general, human B-cells are assumed to be inherently immature at birth. As described above for neonatal T-cells, deficiencies in cytokine networks likely play an important role in diminished functionality of the humoral response (Splawsky and Lipsky, 1991; Tsuji et al., 1994). Although a small number of IgM-producing cells are detected, no IgG- or IgA-producing cells can be identified. The ability of human B-cells to produce either IgG or IgA antibodies increases with age, with adult levels being reached by 5 and 12 years of age, respectively (Miyawaki et al., 1981). These results prompted speculation that the greater susceptibility of human newborns to certain bacterial infections is due to deficient B-cell function, especially the delay in production of IgG and IgA antibodies. In contrast with what is observed in rodent, infants appear to respond well to T-dependent antigens primarily through adequate production of IgM. However, they respond poorly or not at all to polysaccharide antigens such as the antigens found on the cell walls of several infectious bacteria, which also contributed to increased susceptibility to infections (Garthings et al., 1981). Decreased responsiveness to antigen stimulation, particularly to non-protein 'T cell-independent' antigens, continues well into the second year of life. Examples of this deficiency are well-recognized clinically, such as the inability of the newborn to mount an immune response to *Streptococcus pneumoniae* and *Haemophilus influenza* - leading to increased susceptibility of newborns to infection with these pathogens - and to respond effectively to vaccines (Cadoz, 1998; Ahmad and Chapnick, 1999). As discussed below, it is interesting to note the distinction between humans and mice/rats in terms of the maturation of antibody responses to T-dependent and T-independent antigens. To date, there is no explanation for this dichotomy.

After the early neonatal period, there is continued acquisition of immune competence concomitant with increased antigen exposure throughout years one and two of childhood. It is important to consider the fact that attainment of immunocompetence does not necessarily mean closure of 'windows'. Lactation and other nutritional modalities make variable contributions depending on the immune compartment and on continued risks of particular exposures. During late childhood and adolescence, continuous growth processes play an important role in susceptibility to toxic exposures, as do the influences of hormonal fluctuations and the physiologic changes of adolescence. Furthermore, ongoing changes in social activities and behavioral modifications can increase susceptibility during these years. Importantly, there is need for continued tracking during childhood and adolescence of the effects of exposures occurring earlier in life.

Mice

Mouse thymocytes begin to respond to phytohemagglutinin (PHA), Concanavalin A (Con A) and in a mixed lymphocyte reaction (MLR) by gestation day 17 (Mosier, 1977). Although fetal thymocytes respond to PHA and in the MLR similar to adults, the response to Con A does not reach adult levels until 2-3 weeks after birth. It is also during the immediate postnatal period that acquired immune function is first detectable in mice (Ghia et al., 1998). Functional B- and T-lymphocytes are produced in the bone marrow and thymus, respectively, and migrate to the spleen, lymph nodes, and mucosal associated lymphoid tissues (MALT). However, a mature pattern of immune response to antigen is not achieved until approximately one month of age in rodents. During the first month of postnatal life in mice, the immune system remains immature, fails to produce antibodies to carbohydrate antigens, and is predominated by IgM mediated immune responses to antigen (Landreth, 2002). Holladay and Smialowicz (2000) reported that in mice, antibody responses to T-independent antigens occurs soon after birth and reach near-adult levels by 2 or 3 weeks. In contrast, T-cell dependent antibody responses in mice begin after about two weeks and do not reach adult levels until 6-8 weeks of age. Similarly, natural killer (NK) cell activity is absent in mice at birth and does not begin to appear until about three weeks of age (Santoni et al., 1982).

During the first six months of life in rodents, there is enormous production of T- and B-lymphocytes from primary lymphoid tissues (Kincade, 1981).

Rats

As discussed above, a recent study by Ladics et al. (2000) compared the antibody response to SRBC in 10-day and 21-day old rat pups. There was no antibody response to SRBC in 10-day rat pups. The results with the antibody response in 21-day old weanlings were mixed. The use of an ELISA to measure antibody titer was precluded due to a high background. No high background was seen when the response was measured as the number of antibody-producing B-cells; and the magnitude of the response was at the low end of the historical range of responses seen in adult rats. The latter results were consistent with histological analysis of the spleens in that prominent germinal centers were observed in 21-day old rat pups immunized with SRBC. Similar results were seen when another T-dependent antigen, keyhole limpet hemocyanin (KLH) was used in that while the immune status of rat weanlings could be measured, the response was less than that seen in adults (Bunn et al., 2001). The observed profile of age-dependent antibody responses in rat pups and weanlings is also consistent with previous studies by other laboratories. For example, Spear et al. (1973) found that consistent antibody responses to SRBC could not be detected until after 2 weeks of age. Kimura et al. (1985) reported a steady age-related increase in the antibody response to SRBC beginning around postnatal day 12. Interestingly, this same investigation showed a similar age-related kinetics in the antibody response to the T-independent antigens, TNP-Ficoll and TNP-dextran, but a markedly accelerated antibody response to another T-independent antigen, TNP-*Brucella abortus* (e.g., measurable at birth; and nearly at adult levels by postnatal day 8).

Dogs

Our knowledge of the ontogeny of immune responses in dogs is limited (Felsberg, 2002). Fetal dogs are capable of responding to various antigens, including an antibody response to bacteriophage, a T-dependent antigen, on gestational day 40, proliferation to PHA by lymphocytes from fetal spleen and lymph node on gestational day 45, an antibody response to RBC on gestational day 48, proliferation to PHA by fetal thymocytes and antibody response to vaccination with Brucella canis on gestational day 50 (Bryant et al., 1973; Shifrine et al., 1971; Klein et al., 1983). Nonetheless, it is generally considered that dogs become immunologically mature close to, or at parturition (Felsberg, 2002). Jacoby et al. (1969) demonstrated that colostrums-deprived, gnotobiotic puppies developed both primary and secondary specific antibody responses when immunized with bacteriophage within the first 24 hours after birth. These results indicated that neonatal dogs possess a functional B-cell and T-cell system at birth. However, it is important to note that these results did indicate that the humoral immune response matured with age, in that the titers were <1%, ~3% and ~13% of the adult titers, at fetal day 40, fetal day 45 and postnatal day 1, respectively (Felsberg, 2002). Similar results were observed when the lymphoproliferative response was measured in that the capability at <one week was ~66% of the adult responses which are manifest as early as one month (Felsberg, 2002).

Primates

Coincident with the demarcation of lymphoid organs into distinct B- and T-cell areas by gestational day 80, as described above, results also indicate that immunoglobulin- and cytokine-secreting cells are also detected at this time point in the development of the macaque (Hendrickx et al., 2002). Taken together, these results indicate that the fetal rhesus macaque has the potential to generate immune responses early in the second trimester. Although it is possible to measure a number of different types of immune parameters in nonhuman primates, there does not appear to have been any comparative studies across different stages of development. Consistent with what is described above for humans, a study by Watts et al. (1999) indicated that fetal immunization of baboons with the Recombinax HB vaccine resulted in a fetal-specific IgG antibody response.

Table 1 Summary of Immune System Development across Multiple Species

	Parameter			
Gestation	Mouse 20 days	Dog 60–63 d[1]	Primate 155–165 d[2]	Human 40 weeks
Stem cells appear	gd 8 (40)[3]	—	—	gw 5 (12.5)
Fetal liver hematopoiesis	gd 10 (50)	—	—	gw 7 (17.5)
Splenic primordia	gd 13 (65)	gd 28 (45)	—	gw 10 (25)
Spleen demarcation[4]	—	gd 45 (72)	gd 80 (50)	gw 26 (65)
Thymic primordia	gd 10 (50)	gd 28 (45)	gd 35 (22)	gw 6 (15)
Bone marrow hemat.	gd 18 (90)	gd 45 (72)	—	gw 12 (30)
Mitogen proliferation[5]	gd 17 (85)	gd 50 (81)	—	gw 12 (30)

[1] Assume a gestational period of 62 days for later calculations.
[2] Length of gestation affected by strain of primate; use 160 days for calculation.
[3] Number in parentheses is the % of total gestational length.
[4] Demarcation = organization of spleen into distinct red and white pulp areas.
[5] Proliferation in response to mitogens (PHA, Con A) by fetal thymocytes.

Summary

An attempt has been made here to capture comparisons of the development of the immune system across five mammalian species: mouse, rat, dog, primate and human (Table 1). Because it is generally accepted that the development of the immune system in mice and rats is comparable and because much more information is available for the mouse, only the results for one rodent species are depicted in the summary table. Developmental landmarks that are described in greater detail in the text above, are shown in the context of the known length of gestation for each species. This data base is incomplete and it is important to avoid over-interpreting these results. However, they do allow speculation about this comparison. The results clearly depict that we know the most about the immune systems of the mouse and human. Moreover, across every developmental landmark depicted, the mouse is developmentally delayed relative to humans. As emphasized above, this delay is ultimately manifested as a dramatic difference in the functional capabilities of the immune systems of mice and humans at birth. A comparison between rats and humans for some select developmental landmarks is shown in Figure 1 and indicates a similar trend in that the development in this rodent species is delayed. It is generally assumed that the development of the immune systems of rats and mice are largely comparable. Depicting the comparative development along a time-line illustrates major differences that must be taken into account in developmental immunotoxicology studies – e.g., some phases of the development of the human immune system that occur *in utero*, occur after birth in rodents; and some phases of the development of the human immune system that occur during the lactational neonatal period, occur after weaning in rodents. It is important to emphasize that these differences do not necessarily obviate the selection of rodents as a test species; but they must be factored into the interpretations of any results, especially in the context of what is known about transplacental transfer and/or transfer via the mother's milk of the drug or chemical being studied.

As emphasized above, the dog is like the human in that it is born with an essentially functional immune system. In that regard, the results in Table 1 are quite interesting. With an admittedly very limited database for the comparison, the results seem to suggest that the *in utero* development of the immune system in the dog is delayed, as compared to humans, and only slightly more advanced than what is seen with rodents. This observation may become important to the design of a developmental immunotoxicology study in the dog where exposure to pregnant bitches would occur at different times during gestation. Another major consideration in such studies relates to the difference between human and dog placentas. This distinction was discussed above in the context of the limited transfer of maternal antibodies to the fetus and the possible impact on the postnatal development of the immune system in dogs. However, the endotheliochorial placenta of the dog must be factored into preclinical studies for potential impacts on the transplacental transfer of a given chemical. As with the distinctions noted for the rat, although these differences should not obviate the selection of the dog as the test species, they must be considered in the interpretation of any results.

Unfortunately, there was an insufficient number of studies characterizing the *in utero* development of the immune system in primates to make comparisons to the other species. Defining with more precision the developmental landmarks in the various test species is essential to maximizing the interpretation of preclinical studies, and ensuring that the parameters we extrapolate from animal models are physiologically and clinically relevant to the developing human. As emphasized above, the sequence of developmental landmarks described here has led some investigators to think in terms of 'windows' of vulnerability (Holladay and Smialowicz, 2000). In fact, one of the cornerstones of the evolution of developmental immunotoxicology is that it is assumed that some chemicals may have a greater effect during one or more of these 'windows' than another. Stated another way, the premise is that the developing immune system may be more or less sensitive to toxic insult than the mature adult immune system. However, this premise should consider the physiology of the immune system during the various stages of development. As noted above, both the developing immune system and the mature immune system are characterized by a number of physiological processes, including cellular proliferation, cell migration, cell-to-cell interactions, expression of surface markers (e.g., for activation, differentiation, adhesion, etc.). As such, a chemical with a mode of action that targets one of these processes (e.g., such as an antiproliferative agent) would be expected to affect all 'windows' where that process (e.g., proliferation) played an important role.

One feature of the developing immune system that clearly distinguishes it from the mature immune system, especially during gestation, is the role played by organogenesis. Defects in the development of the immune system due to heritable changes in the lymphoid elements have provided clinical and experimental examples of the devastating consequences of impaired immune development (Rosen et al., 1995). Therefore, the effects of chemicals on the genesis of critical components of the immune system in the developing fetus may be more important than effects on these tissues after they have been populated by hematopoietic and lymphoid cells. However, a chemical that induces the formation of this kind of developmental abnormality in the fetus would legitimately be classified as a 'teratogen'. Interestingly, studies by Hendrickx et al. (2000) characterized the effects of several well-known teratogenic pharmaceutical agents on the developing immune systems in primates. Treatment with triamcinolone acetonide, 13-cis-retinoic acid and experimentally-induced zinc deficiency all were shown to trigger an adverse effect on fetal thymus in macaques and baboons. Moreover, one of the most widely studied environment contaminants, 2,3,7,8-tetrachlorodibenzo-p-dioxin (TCDD or 'dioxin'), that has been characterized by multiple laboratories as a developmental immunotoxicant (Barnett, 1996; Holladay and Smialowicz, 2000; Holladay and Blalock, 2002; Neubert et al., 1996; Neubert et al., 2002; and Smialowicz, 2002) is also a known animal teratogen (Couture et al., 1990). These results prompt the question as to how many known teratogens would have an impact on the thymus or other critical immune organs and whether these types of agents should be more appropriately classified as 'immunoteratogens', as opposed to developmental immunotoxicants.

Results using clinical immunosuppressants, which are devoid of teratogenic effects, in pregnant women are relevant to this discussion. Recent advances in organ transplant success have resulted in increasing numbers of women becoming pregnant after an organ or tissue transplant procedure. These women require therapeutic immunosuppression throughout pregnancy to prevent allograft rejection. Studies of pregnant women receiving the required polytherapy (e.g., prednisone plus either azathioprine or cyclosporine) following renal, heart or liver transplants indicate that the major risk is fetal growth restriction and that there is little evidence that exposure is associated with an increase in congenital anomalies (Hou, 1999). In contrast, another study reported that a transient but severe B-cell depletion was seen in neonates born to renal transplant mothers receiving cyclosporine, azathioprine and methylprednisolone (Takahashi et al., 1994). One report also showed an association between azathioprine and a slightly increased risk of nonspecific congenital anomalies as well as prematurity (Penn et al., 1980), and another study described neonatal lymphopenia and thrombocytopenia in children following maternal azathioprine therapy (Davidson, 1994). No malformations have been noted (Hou, 1999); but there have been reports of reduced birth weight and transient neonatal immunocompromise (Jain, 1993; Laifer et al., 1994) following the use of tacrolimus, a cyclosporine-like immunosuppressant, during pregnancy. Taken together, the available epidemiologic reports to date provide little evidence of increased malformations associated with the use of clinical immunosuppressants during pregnancy, and the major risks appear to be adverse effects on fetal growth and the fetal immune system following exposure to these agents during the second and third trimesters (Hendrickx et al., 2000).

This discussion is relevant to subsequent decisions about how to test for adverse effects on the developing immune system. Organs essential to the normal functioning of the immune system, like the thymus or spleen or bone marrow, are not typically assessed in developmental and reproductive toxicology (DART) studies, and this failure to assess developmental damage to the immune system should be reevaluated. The EPA has suggested that immune organs be weighed in rat pups being culled from a standard two-generation reproductive toxicity study (OPPTS 870-3800). While this is a logical first step, there is little rationale to support the predictive value of weighing neonatal immune organs.

In order to address this issue properly, it will be necessary to establish a more appropriate panel of tests to evaluate the effect of chemical toxicity on the developing immune system and to establish whether or not critical windows of vulnerability to potentially immunotoxic compounds exist at specific developmental stages in mammals. The primary conclusion from a recent workshop organized by ILSI HESI was that the best approach to developmental immunotoxicology testing was to address all of the 'windows' at once (e.g., using an exposure regimen that extended from early gestation through lactation and weaning and into young adulthood), and then go back and dissect specific windows if an effect is seen (Sandler, 2002; Holsapple, 2002b). Importantly, this approach has the potential to distinguish between highly transient effects and those that are genuinely persistent. This approach would also address the fact that the immune system continues to develop postnatally. This point is particularly relevant when rodents are used in studies because their immune system undergoes such considerable anatomical development after birth. However, an approach that addresses all of the critical windows is also important in species where it is recognized that the immune system is essentially functional at birth. Even in humans, there is considerable development of functional competence during the neonatal period, and virtually the entire scope of immunological memory is established after birth.

REFERENCES

Ahmad H. and Chapnick E.K. 1999. Conjugated polysaccharide vaccines. *Infectious Disease Clinics of North America.* 13:113.

Anderson, U., Bird, A.G., Britton, S. and Palacios, R. 1981. Humoral and cellular immunity in humans studied at the cell level from birth to two years of age. *Immunol. Rev.,* 57:1.

Aspinall, R., Kampinga, J. and Bogaerde, J. 1991. T cell development in the fetus and the invariant hypothesis. Immunol. *Today,* 12:7.

August, C.S., Berkel, A.I., Driscoll, S. and Merler, E. 1971. Onset of lymphocyte function in the developing human fetus. *Ped. Res.,* 5:539.

Barnett, J.B. 1996. Developmental Immunotoxicology. In: *Experimental Immunotoxicology.* (Smialowicz, R.J. and Holsapple, M.P., eds.) CRC Press, Boca Raton, FL. Pg. 47.

Barrett D.S., Rayfield L.S. and Brent L. 1982. Suppression of natural cell-mediated cytotoxicity in man by maternal and neonatal serum. *Clin. Exp. Immunol.,* 47: 742.

Berry S.M., Fine N., Bichalski J.A., Cotton D.B, Dombrowski M.P. and Kaplan J. 1992. Circulating lymphocyte subsets in second- and third-trimester fetuses: comparison with newborns and adults. *Am. J. Obstet. Gynecol.,* 167: 895.

Bryant, B.J. and Shifrine, M. 1972. Histiogenesis of lymph nodes during the development of the dog. *J. Ret. Soc.,* 12:96.
Bryant, B.J., Shifrine, M. and McNeil, C. 1973. Cell-mediated immune response in the developing dog. *Int. Arch. All. Appl. Immunol.,* 45:937.
Bunn, T.L., Dietert, R.R., Ladics, G.S. and Holsapple, M.P. 2001. Developmental immunotoxicology assessment in the rat: Age, gender and strain comparisons after exposure to lead. *Toxicol. Methods,* 11:41.
Cadoz, M. 1998. Potential and limitations of polysaccharide vaccines in infancy. *Vaccine* 16:1391.
Cairo, M.S., Suen Y., Knoppel E., VandeVen C., Nguyen A. and Sender L. 1991. Decreased stimulated GM-CSF production and GM-CSF gene expression but normal numbers of GM-CSF receptors in human term newborns compared with adults. *Pediatr. Res.,* 30:362.
Chapin, R.E. 2002. The use of the rat in developmental immunotoxicology studies. *Hum. Exp. Toxicol.* 21:521.
Chen, S., Golemboski, K.A., Sanders, F.S. and Dietert, R.R. 1999. Persistent effect of in utero meso-2,3-dimercaptosuccininic acid (DMSA) on immune function and lead-induced immunotoxicity. *Toxicology,* 132:67.
Clement, L.T., Vink P.E. and Bradley G.E. 1990. Novel immunoregulatory functions of phenotypically distinct subpopulations of $CD4^+$ cells in the human neonate. *J. Immunol.,* 145 (1):102.
Couture, L.A., Abbott, B.D. and Birnbaum, L.S. 1990. A critical review of the developmental toxicity and teratogenicity of 2,3,7,8-tertachlorodibenzo-p-dioxin: recent advances toward understanding the mechanism. *Teratology,* 42:619.
Cumano, A, and I Godin. 2001. Pluripotent hematopoietic stem cell development during embryogenesis. *Curr Opin Immunol* 13:166.
Cumano, A, C Ferraz, M Klaine, J Santo, I Godin. 2001. Intraembryonic, but not yolk sac hematopoietic precursors, isolated before circulation, provide long-term multilineage reconstitution. *Immunity* 15:477.
Davidson, J.M. 1994. Pregnancy in renal allograft recipients: problems, prognosis and practicalities. *Balliere's Clin. Obst. Gynecol.,* 8:501.
Demeure C.E., Wu C.Y., Shu U., et al. 1994. In vitro maturation of human neonatal CD4 T lymphocytes: II. Cytokines present at priming modulate the development of lymphokine production. *J. Immunol.,* 152:4775.
Denny, T., Yogev, R., Gellman, R., Skuza, C., Oleske, J., Chadwick, E., Cheng, S.C. and Connor, E. 1992. Lymphocyte subsets in healthy children during the first 5 years of life. *J. Am. Med. Assoc.,* 267:1484.
Dietert, R.R., Etzel, R.A., Chen, D., Halonen, M., Holladay, S.D., Jarabek, A.M., Landreth, K., Peden, D.B., Pinkerton, K., Smialowicz, R.J. and Zoetis, T. 2000. Workshop to identify critical windows of exposure for children's health: Immune and Respiratory Systems Work Group Summary. *Environ. Health Perspec.* 108, Suppl. 3:483.
Dietert, R.R., Lee, J.-E. and Bunn, T.L. 2002. Developmental immunotoxicology: Emerging issues. *Hum. Exp. Toxicol.* 21:479.
Dijkstra, C. and Dopp, E. 1983. Ontogenetic development of T- and B-lymphocytes and non-lymphoid cells in the white pulp of the rat. *Cell Tissue Res.,* 229:351.
Durandy, A., Thuillier, L., Forveille, M. and Fischer, A. 1990. Phenotypic and functional characteristics of human neonatal B lymphocytes. *J. Immunol.,* 144 (1):60.
Erkeller-Yuksel, F.M., Deneys, V., Yuksel, B., Hannet, I., Hulstaert, F., Hamilton, C., Mackinnon, H., Stokes, L.T., Munhyeshuli, V., Vanlangendonck, F., DeBruyere, M., Bach, B.A. and Lydyard, P.M. 1992. Age-related changes in human blood lymphocyte subpopulations. *J. Pediatr.,* 120 (2): 216.
Felsberg, P.J. 1998. Immunology of the dog. In: *Handbook of Vertebrate Immunology.* (Pastoret, P.P., Griebel, P., Bazin, H. and Govearts, A. eds.) Academic Press, New York, pg. 261.
Felsburg, P.J. 2002. Overview of immune system development in the dog: Comparison with humans. *Hum. Exp. Toxicol.* 21:487.
Foa, R., Giubellino, M.C., Fierro, M.T., Lusso, P. and Ferrando, M.L. 1984. Immature T lymphocytes in human cord blood identified by monoclonal antibodies: A model for the study of the differentiation pathway of T cells in humans. *Cell. Immunol.,* 89:194.
Gale, R.P. 1987. Development of the immune system in human fetal liver. *Thymus,* 10: 45.
Gathings, W.E., Kubagawa, H. and Cooper, M.D. 1981. A distinctive pattern of B-cell immaturity in perinatal humans. *Immunol. Rev.,* 57:107.
Ghia, P, E Boekel, A Rolink, F Melchers. 1998. B-cell development: a comparison between mouse and man. *Immunol Today* 19(10):480.

Godin, I, J Garcia-Porrero, F Dieterlen-Lievre, A Cumano. 1999. Stem cell emergence and hemopoietic activity are incompatible in mouse intraembryonic sites. *J. Exp. Med.,* 190(1):43.

Good, R A. 1995. Organization and development of the immune system. Relation to its reconstruction. *NY Acad. Sci.,* 77:8.

Granberg, C. and Hirvonen, T. 1980. Cell-mediated lympholysis by fetal and neonatal lymphocytes in sheep and man. *Cell. Immunol.,* 51:13.

Griffiths-Chu, S., Patterson, J.A.K., Berger, C.L., Edelson, R.L. and Chu, A.C. 1984. Characterization of immature T cell subpopulations in neonatal blood. *Blood,* 64(1):296.

Hannet, I., Erkeller-Yuksel, F., Lydyard, P., Deneys, V. and DeBruyere, M. 1992. Developmental and maturational changes in human blood lymphocyte subpopulations. *Immunol. Today,* 13(6):215.

Haynes, B.F., Martin, M.E., Kay, H.H. and Kurtzberg, J. 1988. Early events in human T cell ontogeny. *J. Exp. Med.,* 168:1061.

Haynes, B.F. and Hale, L.O. 1998. The human thymus. A chimeric organ comprised of central and peripheral lymphoid components. *Imm. Res.,* 18:61.

Hayward, A. and Cosyns, M. 1994. Proliferative and cytokine responses by human newborn T cells stimulated with staphylococcal enterotoxin B. *Pediatr. Res.,* 35(1):293.

Hayward, A.R. 1983. Development of immunity mechanism. In: *Pædiatric Immunology.* (Soothill, J.F., Hayward, A.R. and Wood, C.B.S., eds.) Oxford, Blackwell, pg. 48.

Hendrickx et al. 1975. Teratogenic effects of triamcinolone on the skeletal and lymphoid systems in nonhuman primates. *Fed. Proc.,* 34:1661.

Hendrickx, A.G., Makori, N. and Peterson, P. 2000. Nonhuman primates: their role in assessing developmental effects of immunomodulatory agents. *Hum. Exp. Toxicol.,* 19:219.

Hendrickx, A.G., Makori, N. and Peterson, P. 2002. The nonhuman primate as a model of developmental immunotoxicology. *Hum. Exp. Toxicol.* 21:537.

Herzyk, D.J., Bugelski, P.J., Hart, T.K. and Wier, P.J. 2002. Practical aspects of including functional endpoints in developmental toxicity studies. Case study: Immune function in HuCD4 transgenic mice exposed to anti-CD4 Mab *in utero. Hum. Exp. Toxicol.* 21:507.

HogenEsch, H., Housman, J.M. and Felsberg, P.J. 1987. Canine Peyer's patches: macroscopic, light microscopic, scanning electron microscopic and immunohistochemical investigations. *Adv. Exp. Med. Biol.,* 216A:249.

Holladay, S.D. and Smialowicz, R.J. 2000. Development of the murine and human immune system: Differential effects of immunotoxicants depend on time of exposure. *Environ. Health Perspec.* 108, Suppl. 3:463.

Holladay, S.D. and Blaylock, B.L. 2002. The mouse as a model for developmental immunotoxicology. *Hum. Exp. Toxicol.* 21:525.

Holsapple, M.P. 2002a. Developmental immunotoxicology and risk assessment: A workshop summary. *Hum. Exp. Toxicol.* 21:473.

Holsapple, M.P. 2002b. Developmental immunotoxicity testing: A review. *Toxicology* 185:193.

Hou, S. 1999. Pregnancy in chronic renal insufficiency and end-stage renal disease. *Am. J. Kid. Dis.,* 33:235.

Hulstaert, F., Hannet, I., Deneys, V., *et al.* 1994. Age-related changes in human blood lymphocyte subpopulations. *Clin. Immunol. Immunopath.,* 70:152.

Ibegbu, C., Spira, T.J., Nesheim, S., et al. 1994. Subpopulations of T and B cells in perinatally HIV-infected and noninfected age-matched children compared with those in adults. *Clin. Immunol. Immunopath.,* 71:27.

Jacoby, R.O., Dennis, R.A. and Griesemer, R.A. 1969. Development of immunity in fetal dogs: humoral responses. *Am. J. Vet. Res.,* 30:1503.

Jain, A. 1993. FK506 and pregnancy in liver transplant patients. *Transplantation,* 56:751.

Jenkins, M.K., Schwartz, R.H. and Pardoll, D.M. 1988. Effects of cyclosporine A on T cell development and clonal deletion. *Science,* 241:1655.

Jennings, C., Rich, K., Siegel, J.N. and Landay, A. 1994. A phenotypic study of CD8[+] lymphocyte subsets in infants using three-color flow cytometry. *Clin. Immunol. Immunopath.,* 71:8.

Kay, H.E.M., Playfair, J.H., Wolfenade, M. and Hopper, P.K. 1962. Development of the human thymus during fetal development. *Nature,* 196:238.

Kay, H.E.M., Doe, J. and Hockley, A. 1970. Response of human fœtal thymocytes to phytohæmagglutinin (PHA). *Immunology,* 18:393.

Kelly, W.D. 1963. The thymus and lymphoid morphogenesis in the dog. *Fed. Proc.,* 20:600.

Kendall, M.D. 1991. Functional anatomy of the thymic microenvironment. *J. Anat.,* 177:1.

Kimmel, C. 2001. Developmental immunotoxicity considerations in testing and risk assessment for children's health (abstract). *The Toxicologist,* 60:65.

Kimurs, S., Eldridge, J.H., Michalek, S.M., Morisaki, I., Hamada, S. and McGhee, J.R. 1985. Immunoregulation in the rat: Ontogeny of B cell responses to types 1, 2 and T-dependent antigens. *J. Immunol.,* 134:2839.

Kincade, P.W. 1981. Formation of lymphocytes in fetal and adult life. *Adv. Immunol.,* 31: 177.

Klein, A.K., Dyck, J.A., Stitzel, K.A., Shimizu, M.J., Fox, L.A. and Taylor, N. 1983. Characterization of canine lymphohematopoiesis. *Exp Hematol.,* 11:263.

Kohl, S., Loo, L.S. and Gonik, B. 1984. Analysis in human neonates of defective antibody-dependent cytotoxicity and natural killer cytotoxicity to herpes simplex virus infected cells. *J. Infect. Dis.,* 150:14.

Ladics, G.S., Smith, C., Bunn, T.L., Dietert, R.R., Anderson, P.K., Wiescinski, C.M. and Holsapple, M.P. 2000. Characterization of an approach to developmental immunotoxicology assessment in the rat using SRBC as the antigen. *Toxicol. Methods,* 10:283.

Laifer, S.A., Yeagley, C.J., Armitage, J.M. 1994. Pregnancy after cardiac transplantation. *Am. J. Perinatol.,* 11:217.

Landreth, K.S. 2002. Critical windows in development of the rodent immune system. *Hum. Exp. Toxicol.* 21:493.

Lobach, D.F. and Haynes, B.F. 1987. Ontogeny of the human thymus during fetal development. *J. Clin. Immunol.* 7:81.

Loke, Y.W. and King, A. 1991. Immunology of human pregnancy. In: *Scientific Foundation of Obstetrics and Gynæcology.* (Philipp, E., Setchell, M. and Ginsburg, J., eds.) Oxford, Butterworth-Heinemann Ltd., pg. 55.

Loke, Y.W. 1978. Development of immunocompetence in the human foetus. In: *Immunology and Immunopathology of the Human Foetal-Maternal Interaction.* (Loke, Y.W., ed.) Amsterdam, Elsevier North-Holland Biomedical Press, pg. 51.

Marshall-Clarke, S., Reen, D., Tasker, L. and Hassan, J. 2000. Neonatal immunity: How well has it grown up? *Immunol. Today,* 21:35.

Migliaccio, G., Migliaccio, A.R., Petti, S., *et al.* 1986. Human embryonic hemopoiesis - Kinetics of progenitors and precursors underlying the yolk sac --> liver transition. *J. Clin. Invest.,* 78:51.

Miller, G.K. and Benjamin, S.A. 1985. Radiation-induced quantitative alterations in prenatal thymic development in the beagle dog. *Lab. Invest.,* 52:224.

Miyawaki, T., Moriya, N., Nagaoki, T. and Tonigachi, N. 1981. Maturation of B-cell differentiation ability and T-cell regulatory function in infancy and childhood. *Immunol. Rev.,* 57:69.

Mosier, D.E. 1977. Ontogeny of T-cell function in the neonatal mouse. In: *Development of Host Defenses.* (Cooper, M.D. and Dayton, D.H. eds.). New York, Raven Press, pg. 115.

Mumford, D.M., Sung, J.S., Wallis, J.O. and Kaufmann, R.H. 1978. The lymphocyte transformation response of fetal hemolymphatic tissue to mitogens and antigens. *Pediatr. Res.* 12:171.

Murgita, R.A. and Wigzell, H. 1981. Regulation of immune functions in the fetus and newborn. *Prog. Allergy,* 29:54.

Nahmias, A., Ibegbu, C., Lee, F. and Spira, T. 1994. The development of the immune system - importance in the ascertainment of immunophenotypic changes in perinatal HIV infection. *Clin. Immunol. Immunopath.,* 71:2.

Namikawa, R., et al. 1986. Ontogenetic development of T- and B-cells and non-lymphoid cells in the white pulp of the human spleen. *Immunology,* 57:61.

Neubert, R., Helge, H. and Neubert, D. 1996. Nonhuman primates as models for evaluating substance-induced changes in the immune system with relevance for man. In: *Experimental Immunotoxicology.* (Smialowicz, R.J. and Holsapple, M.P., eds.) CRC Press, Boca Raton, FL. pg. 63.

Neubert, R.T., Webb, J.R. and Neubert, D. 2002. Feasibility of human trials to assess developmental immunotoxicity, and some comparison with data on New World monkeys. *Hum. Exp. Toxicol.* 21:543.

Ohama, K. and Kaji, T. 1974. Mixed cultures of fetal and adult lymphocytes. *Am. J. Obstet. Gynecol.,* 119:552.

Oldstone, M.B.A. and Tishon, A. 1977. Active thymus-derived suppressor lymphocytes in human cord blood. *Nature,* 269:333.

Paige, C.J., P.W. Kincade, L.A. Shinefield, and V.L. Satio. 1981. Precursors of murine B lymphocytes: physical and functional characterization and distinctions from myeloid stem cells. *J. Exp. Med.,* 153:154.

Papadogiannakis, N., Johnsen, S.A. and Olding, L.B. 1985. Strong prostaglandin associated suppression of the proliferation of human maternal lymphocytes by neonatal lymphocytes linked to T *versus* T cell interactions and differential PGE2 sensitivity. *Clin. Exp. Immunol.,* 61:125.

Papadogiannakis, N. and Johnsen, S.A. 1988. Distinct mitogens reveal different mechanisms of suppressor activity in human cord blood. *J. Clin. Lab. Immunol.,* 26:37.

Peakman, M., Buggins, A.G.S., Nicolaides, K.H., Layton, D.M. and Vergani, D. 1992. Analysis of lymphocyte phenotypes in cord blood from early gestation fetuses. *Clin. Exp. Immunol.,* 90:345.

Penit, C. and Vasseur, F. 1989. Cell proliferation and differentiation in the fetal and early postnatal mouse thymus. *J. Immunol.,* 142:3369.

Penn, I., Makowski, E.L. and Harris, P. 1980. Parenthood following renal transplantation. *Kidney International,* 18:221.

Pietra, B.A. and Boucek, M.M. 2000. Immunosuppression for pediatric cardiac transplantation in the modern era. *Progress in Pediatric Cardiology,* 11:115.

Plebani, A., Proserpio, A.R., Guarneri, D., Buscaglia, M., Cattoretti, G. 1993. B and T lymphocyte subsets in fetal and cord blood: age-related modulation of CD1c expression. *Biol. Neonate,* 63:1.

Rabatic, S., Benic, L., Mazuran, R. and Dekaris, D. 1990. Polymorphonuclear leucocyte and natural killer cell functions in mature and premature newborns. *Biol. Neonate,* 58:252.

Rayfield, L.S., Brent, L. and Rodeck, C.H. 1980. Development of cell-mediated lympholysis in human fœtal blood lymphocytes. *Clin. Exp. Immunol.,* 42:561.

Reason, D.C., Ebisawa, M., Saito, H., Nagakura, T. and Iikura, Y. 1990. Human cord blood lymphocytes do not simultaneously express CD4 and CD8 cell surface markers. *Biol. Neonate,* 58:87.

Rosen, F.S., Cooper, M.D. and Wedgwood, R.J.P. 1995. The primary immunodeficiencies. *N. Engl. J. Med.,* 333:431.

Rothkotter, H.J., Sowa, E. and Pabst, R. 2002. The pig as a model of developmental immunotoxicololgy. *Hum. Exp. Toxicol.* 21:533.

Royo, C., Touraine, J.-L. and DeBouteiller, O. 1987. Ontogeny of T lymphocyte differentiation in the human fetus: Acquisition of phenotype and function. *Thymus,* 10:57.

Sandler, J.D. 2002. Executive summary. *Hum. Exp. Toxicol.* 21:469.

Santoni, A., Riccardi, C., Barlozzari, T. and Herberman, R.B. 1982. Natural suppressor cells for murine NK activity. In: *NK Cells and Other Natural Effector Cells.* (Herberman, R.B., ed.). New York, Academic Press, pg. 443.

Sargent, I.L., Redman, C.W.G. and Starkey, P.M. 1993. The placenta as a graft. In: *The Human Placenta.* (Redman, C.W.G., Sargent, I.L. and Starkey, P.M., eds.) Oxford, Blackwell Scientific Publications, pg. 334.

Semenzato, G., Piovesan, A., Amadori, G., Colombatti, M., Gasparotto, G. and Rubaltelli, F.F. 1991. T cell immune function in newborn infants. *Biol. Neonate,* 37:8.

Series, I.M., Pichette, J., Carrier, C., et al. 1991. Quantitative analysis of T and B cell subsets in healthy and sick premature infants. *Early Hum. Devel.,* 26:143.

Shifrine, M.N., Smith, J.B., Bulgin, M.S., Bryant, J.B., Zec, Y.C. and Osborn, B.I. 1971. Response of canine fetuses and neonates to antigenic stimulation. *J. Immunol.,* 107:965.

Shortman K, Vremec D. Cocoran IM, Georgopolos K, Lukas K, Wu L. 1998. The linkage between T-cell and dendritic cell development in the mouse thymus. *Immuno. Rev.* 165:39.

Smialowicz, R.J. 2002. The rat as a model in developmental immunotoxicology. Hum. Exp. Toxicol. 21:513.

Smith, R.T. 1968. Development of fetal and neonatal immunological function. In: *Biology of Gestation. Volume II: The Fetus and Neonate.* (Asal, N.S., ed.). New York Academic Press, pg. 321.

Snyder, P.W., Kazacos, E.A. and Felsberg, P.J. 1993. Histologic characterization of the thymus in canine X-linked severe combined immunodeficiency. *Clin. Imm. Immunpath.,* 67:55.

Solinger, A.M. 1985. Immature T lymphocytes in human neonatal blood. Cell. Immunol., 92:115.

Somberg, R.L., Robinson, J. and Felsberg, P.J. 1994. T-lymphocyte development and function in dogs with X-linked severe combined immunodeficiency. *J. Immunol.,* 153:4006.

Somberg, R.L., Tipold, A., Hartnett, B.J., Moore, P.F., Henthorn, P.S. and Felsberg, P.J. 1996. Postnatal development of T-cells in dogs with X-linked severe combined immunodeficiency. *J. Immunol.,* 156:1431.

Spear, P.G., Wang, A., Rutishauser, U. and Edelman, G.M. 1973. Characterization of splenic lymphoid cells in fetal and newborn mice. *J. Exp. Med.,* 138:557.

Splawski, J.B. and Lipsky, P.E. 1991. Cytokine regulation of immunoglobulin secretion by neonatal lymphocytes. *J. Clin. Invest.,* 88(3):967.

Stites, D.P., Carr, M.C. and Fudenberg, H.H. 1974. Ontogeny of cellular immunity in the human fetus: development of responses to phytohemagglutinin. *Cell. Immunol.,* 11:257.

Takahashi, N., Nishida, H. and Hoshi, J. 1994. Severe B-cell depletion in newborns from renal transplant mothers taking immunosuppressive agents. *Transplantation,* 57:1617.

Tanimura, T. and Tanioka, Y. 1975. Comparison of embryonic and fetal development in man and rhesus monkey. In: *Breeding Simians for Developmental Biology, Laboratory Animal Handbooks.* Laboratory Animals Limited, London.

Tavassoli, M. 1995. Ontogeny of hemapoiesis. In: *Handbook of Human Growth and Development, Vol. III.* (Meisami, S., ed.) New York-CRC Press.

Thorbecke, G.J. 1959. Some histological and functional aspects of lymphoid tissue in germfree animals. I. Morphological studies. *Ann. NY Acad. Sci.,* 78:107.

Timens, W. et al. 1989. Immaturity of the human splenic marginal zone in infancy. Possible contribution to the deficient infant immune response. *J. Immunol.,* 143:3200.

Toivanen, P., Uksila, J., Leino, A., Lassila, O., Hirnonen, T. and Ruuskanen, O. 1981. Development of mitogen responding T cells and natural killer cells in the human fetus. *Immunol. Rev.,* 57:89.

Tsuji, T., Nibu, R., Iwai, K., et al. 1994. Efficient induction of immunoglobulin production in neonatal naive B cells by memory CD4$^+$ T cell subset expressing homing receptor L-selectin. *J. Immunol.,* 152:4417.

Tucci, A., Mouzaki, A., James, H., Bonnefoy, J.-Y. and Zubler, R.H. 1991. Are cord blood B cells functionally mature? *Clin. Exp. Immunol.,* 84:389.

Verlarde, A. and Cooper, M.D. 1984. An immunofluorescence analysis of the ontogeny of myeloid, T, and B lineage cells in mouse hematopoietic tissues. *J. Immunol.,* 133:672.

VonGaudecker, B. 1991. Functional histology of the human thymus. *Anat. Embryol.,* 183:1.

Watts, A.M., Stanley, J.R., Shearer, M.H., Hefty, P.S. and Kennedy, R.C. 1999. Fetal immunization of baboons induces a fetal-specific antibody response. *Nat. Med.,* 5:427.

Webber, S.A. 1996. Newborn and infant heart transplantation. *Current Opinion in Cardiology,* 11:68.

Weissman, I. 2000. Stem cells: Units of development, units of regeneration and units in evolution. *Cell* 100:157.

West, L.J. 2002. Defining critical windows in the development of the human immune system. *Hum. Exp. Toxicol.* 21:499.

Yang, T.J. and Gawlack, S.L. 1989. Lymphoid organ weights and organ:body weight ratios of growing beagles. *Laboratory Animals,* 23:143.

APPENDIX C-8*
SPECIES COMPARISON OF POSTNATAL CNS DEVELOPMENT: FUNCTIONAL MEASURES

Sandra L. Wood[1,4], Bruce K. Beyer[2] and Gregg D. Cappon[3]

[1] Merck Research Laboratories, West Point, PA 19486, USA
[2] Sanofi-Synthelabo Research, Malvern, PA 19355, USA
[3] Pfizer Global Research & Development, Groton, CT 06340, USA
[4] Correspondence to: Dr. Sandra L. Wood, Merck Research Laboratories, WP45-103, Safety Assessment, West Point, PA 19486. Phone (215) 652-6334. Fax (215) 652-7758. Email: sandra_wood2@merck.com

Introduction

The purpose of this review is to identify key events in human postnatal CNS development and compare them, when possible, to other species. Specifically, this review will focus on the behavioral measures of CNS development, but will not include reviews of neuroanatomical or neurochemical maturation. This review is part of the initial phase of a project undertaken by the Developmental and Reproductive Toxicology Technical Committee of the ILSI Health and Environmental Sciences Institute (HESI) to bring together information on a selected number of organ systems and compare their postnatal development across several species (Hurtt and Sandler, 2003). This review is not intended to propose effective study designs and testing strategies since this is the final objective of the HESI project. It is rather to be used as a resource to aid researchers when beginning to plan the necessary nonclinical safety evaluations for pediatric drugs (Department of Health and Human Services (FDA), 1998; FDA, 1998).

There is general recognition that the developing nervous system is qualitatively different from the adult system. With respect to the central nervous system (CNS), the human is an altricial species, i.e., born in a relatively immature state and has a prolonged postnatal dependency. The human CNS must undergo considerable postnatal development in order to reach its adult state of maturation. A large percentage of laboratory animals, such as mice, rats, cats, dogs and rabbits, are like humans and must undergo a prolonged period of neurologic development postnatally. However, some species, like guinea pigs and sheep, are born with very mature nervous systems, which undergo little CNS development postnatally. Therefore, these precocial species may be of little practical use as a comparative model to the human CNS.

Few would argue with the premise that the ultimate manifestation of a neural process is behavior. However, the links between the biological processes occurring during CNS development and the behavioral capacities exhibited are not easily established. Because behavior is the aspect of postnatal CNS development that is clinically monitored in humans as well as preclinically in many laboratory species, behavioral measures of neurologic function will be the focus of this review. The measures selected have no special status, but it is the hope of the authors to provide a broad overview of the developing features that are under CNS control. The maturation of the CNS has been organized in eight categories: 1) Locomotor Development, 2) Ontogeny of Fine Motor Development and Dexterity, 3) Sensory and Reflex Development, 4) Cognitive Development, 5) Communication, 6) Social Play, 7) Development of the Fear Response and 8) Development of Sleep Cycles.

It should be noted that the review will not include detailed descriptions of all behavioral paradigms mentioned. The reader is encouraged to refer to the cited text for the specific details of

---

* Source: Wood, S.L., Beyer, B.K., Cappon, G.D., Species comparison of postnatal CNS development: functional measures. *Birth Defects Research, Part B: Developmental and Reproductive Toxicology*, 68, 321-334, 2003.

a behavioral assessment and Chapter 4 of "The Handbook of Behavioral Teratology" for a general description of the methods most often utilized in behavioral evaluations (Adams, 1986).

Ontogeny of Locomotor Development

Comparisons of locomotor development among humans, non-human primates, rats and dogs are complicated since these later two species never walk upright. In addition, human infants cannot readily swim like rats or dogs. However, a comparison of motor develop in all these species is beneficial since humans, rats and dogs all gradually development from a rather immature state of locomotion (i.e., crawling) to a mature state of walking. Non-human primates also develop locomotor abilities postnatally, but at a faster pace than their human counterparts. This section of the review will compare the early ontogeny of locomotion between humans and common laboratory species.

Humans

By 12 to 14 weeks, an infant can hold her chin and shoulders off the floor with her forearms (Illingworth, 1983). By 24 weeks, an infant can hold her chest and upper part of her abdomen off the floor and maintain her weight by her hands with extended arms. By 28 weeks she can bear her weight on one hand, and by 9 months she can usually crawl. Interestingly, the first stage of crawling is typically progression backward. Around 10 months, she can creep on her hands and knees with her abdomen off the floor.

The last stage before walking around 13 months of age typically involves intermittently placing one foot flat on the floor and creeping like a bear on hands and feet. Interestingly, an infant can bear her full weight (i.e., stand) while being held by her hands by 24 to 28 weeks and can walk while holding onto a piece of furniture by 48 weeks. When an infant does begin walking on her own, her arms are held outward, elbows are flexed and her base is broad. In addition, the steps often vary in length and direction and are by no means completely adult-like. By 18 to 24 months, the toddler learns to walk with legs closer together, arms moved inward and a heel-to-toe gait develops. By 2.5 years, she can jump with both feet and walk on her tiptoes. By 3 years, the child begins to display a more adult-like motor pattern.

Rats

Locomotion in the rat matures during the first few weeks after birth (Altman and Sudarshan, 1975; Westerga and Gramsbergen, 1990). Rat pups are able to ambulate through the use of their forelimbs, upper torso and head beginning around Postnatal Day (PND) 3-4. This "crawling" behavior peaks around PND 7 and disappears around PND 15 (Altman and Sudarshan, 1975; Westerga and Gramsbergen, 1990). It is not until around PND 8-10 that rat pups can stand with their abdomens completely off the floor (Altman and Sudarshan, 1975; Bolles and Woods, 1964). Around PND 12-13, rat pups can walk while supporting their full weight, but the hindlimbs are typically rotated outward. Adult-like locomotion appears around PND 15-16 which is shortly after eye-opening in rats (Westerga and Gramsbergen, 1990). Fast running appears around PND 16 (Altman and Sudarshan, 1975). Beginning around PND 16, rats can stand on their hind-legs while leaning their forelegs against a wall support (Altman and Sudarshan, 1975; Bolles and Woods, 1964). Around PND 18, rats can rear for short periods of time without foreleg support.

An additional issue regarding rat locomotion is a peak period of hyperactivity in rats observed around PND 15-20, during which pups change from one activity to another very quickly (Bolles and Woods, 1964), e.g., switching from grooming to running to fighting in a matter of seconds. After this period, there is a slight decrease in their activity level then a gradual increase to adult activity levels around PND 50-60.

Swimming

Another type of rat "ambulation" that should be mentioned because of its importance in many behavioral paradigms (e.g., Morris Water maze, Biel Water maze) is the ontogeny of swimming. The stages of swimming have been well documented (Bekoff and Trainer, 1979; Butcher and Vorhees, 1979; Schapiro et al., 1970; Vorhees, 1986). When PND 6 rats are placed in water, they often display an asynchronous movement of the limbs which results in little forward motion but some small turns. Over the next few days, rats will begin to propel themselves forward by paddling in a coordinated fashion with all four limbs. Around PND16, paddling with all limbs begins to disappear, and what emerges is the more adult-like swimming style in which the hindlimbs are used to propel forward. The forelimbs are held stationary under the jaw, except during turning.

Dogs

Locomotion in the canine develops during the first month or so after birth (Fox, 1964a; Fox, 1971; Thomas, 1940). Crawling forward (or a rooting reflex) occurs when there is bilateral stimulation of the head with the hand cupped around the muzzle (Fox, 1964a). This reflex is demonstrated beginning around PND 4. This reflex becomes weaker and more variable around PND 14 and onward. Between PND 6 and 10, supporting of the forelimbs is first observed. Supporting of the hindlimbs begins around PND 11 to 15. Sitting upright on the forelimbs is seen around PND 20 while standing is observed around PND 21. The hindlimbs appear to tremor at this time indicating a weakness in muscle tone. There is a marked transition from variable locomotor responses to adult-like postures and equilibrium from PND 28 onward. The primitive reflexes that were an integral part of neonatal behavior disappear around PND 28, and the ability of the dog to remain upright when subjected to lateral tilting is well developed by PND 35-42.

Non-human Primates

It is normal for the young macaque to cling to its mother for the first 2 to 3 days of life (Hines, 1942). Shortly after this period, the infant can raise her chest off the surface of the floor by extending her arms and head. Surprisingly, the infant monkey begins to stand on all fours by 4 days after birth (Golub, 1990; Hines, 1942). Around approximately 3 weeks (in some cases as early as the end of first week), the position of the trunk becomes more horizontal to the floor surface (i.e., more adult-like), and between the 5th and 11th weeks, the extremities become more extended (Hines, 1942). Bipedal standing is observed as early as 7 weeks of age.

The infant macaque develops locomotor capabilities in both the horizontal plane (i.e., the floor surface) and the vertical plane (i.e., climbing and jumping). During the first two weeks of life, forward progression in the horizontal plane begins with reaching and grasping of the extremities in a rather abducted position (i.e., extremities extended away from the midline of the body). The extension of the extremities becomes more exaggerated between the 5th and 11th weeks of life, and the knee is rather extended compared to the adult form. Easy flexion-extension of the knee and adduction of the extremities (i.e., extremities extended toward the midline of the body) begins around 7 weeks, but is not seen with consistency until approximately 12 weeks of age.

The infant macaque begins to develop climbing capabilities as early as 2 days of age. However, the ability to return either by backing down or turning around does not begin to emerge until the end of the 1st week and the 5th week, respectively. Jumping across a space was not attempted by even the most adventurous macaque until 5 weeks of age.

It is interesting that adult standing is obtained by approximately the 3rd month, but an adult sitting posture is not obtained until 9 months of age. Specifically, by the 4th week of life, the infant macaque can squat on two feet with initially some aid from the tail. The infant begins to sporadically sit erect by 7 weeks, but the consistent use of the adult posture of sitting was not observed routinely until approximately 9 months of age.

Conclusions for Locomotor Development

In humans, rats, dogs and non-human primates, locomotor capabilities develop postnatally. In rats, dogs and humans, there appears to be a more gradual rostrocaudal maturation that leads to the shoulders in a raised position prior to the pelvis. In contrast, non-human primates develop motor abilities rather quickly after birth compared to humans. Therefore, assessments of motor development in juvenile toxicity studies would be beneficial in any of these species. However, studies in non-humans need to be carefully designed in order to accommodate the rapid development of locomotion in this species.

Table 1 Summary of Locomotor Development

	Human	Rat	Dog	Non-Human Primate (Rhesus)
Crawling	≈ PND 270 (9 months)	PND 3–12	PND 4–20	PND 4–49
Walking	≈ PND 396 (13 months)	PND 12–16	PND 20–28	PND 49[a]

PND = Postnatal Day
[a] Bipedal locomotion

Ontogeny of Fine Motor Development and Dexterity

Fine motor skills and dexterity are generally associated with primates, especially humans. However, rats possess dexterity in both their fore- and hindlimbs. Although a reflexive ability to grasp objects may be present early in the postnatal life of many species, fine motor skills and other coordinated motor movements develop over time.

Humans

The reflex grasping of an object placed in the palm of the newborn (i.e., the palmar grasp reflex) is present at birth and begins to disappear around 2 months of age (Herschkowitz et al., 1997; Rosenblith and Sims-Knight, 1904; Volpe, 1995). The amount of pre-reaching, i.e., opening of hand while approaching an object, decreases at around 7 weeks of age. The infant does not seem to lose interest in the object at this age, but attending to the object inhibits pre-reaching activity in some way. The form of pre-reaching also changes at this age. Initially, instead of opening the hand during the forward extension of the arm, the hand will be fisted. After this age, the amount of pre-reaching activity increases and the hand starts opening during the forward extension, but only when the infant looks at the object (von Hofsten and Fazel-Zandy, 1984).

Onset of reaching occurs in the fifth postnatal month, and infants reliably grasp for objects within their reach 3-4 months after this milestone (Konczak and Dichgans, 1997; Lantz et al., 1996). However, most kinematic grasping parameters do not reach adult-like levels before age 2 years. Between ages 2 and 3, improvements in hand- or joint-related movements are marginal. The formation of a consistent interjoint synergy between shoulder and elbow motion is not achieved within the first year of life. Stable patterns of temporal coordination across arm segments begin to emerge at 12-15 months and continue to develop to age 3 (Konczak and Dichgans, 1997).

Younger children (4-5 years) had relatively wider open grips than older (12 years), thus grasping with a higher safety margin. Movement duration and peak spatial velocity of the reaching hand was similar for children 4-12 years of age. In addition, hand trajectory straightened, and coordination between hand transport and grip formation improved by 12 years of age. Only the oldest children (12 years) were able to accurately adjust their grip according to the various sizes of target objects. Thus development of prehensile skills during childhood lasts until the end of first decade of life (Kuhtz-Buschbeck et al., 1998).

Sensorimotor memory coordinates the forces used to grip objects, and maturation of this system is believed to underlay the development of precision grip. Young children have an immature capacity to adapt to the frictional condition, e.g., silk vs. sandpaper. During the first 5 years of age, children show improvement in their ability to apply the correct grip to various frictional conditions. At 18 months, infants could adjust to the frictional condition in a series of lifts when the same surface structure was presented in blocks of trials, but failed when the surface was changed unpredictably between subsequent lifts (Forssberg et al., 1995). This suggests a poor capacity to form sensorimotor memory. Older children (12-13 years) required a few lifts and adults only one lift to update their force coordination to a new friction. Therefore, young children use excessive grip force, a strategy to avoid frictional slips to compensate for an immature tactile control of the precision grip (Forssberg et al., 1995).

Rats

Fine motor and dexterity development is difficult to measure in the rat, and there is little data available describing its development. When rats are suspended from their forepaws on a wire, their hindlimbs provide support to prevent falling and aid progression along the wire. Wistar rats demonstrate hindlimb grasping abilities as early as PND2 and can support their weight for about 10-20 seconds up until PND 16 (Petrosini et al., 1990). Beginning around PND 16, Wistar rats can remain suspended on the wire for 70-120 seconds with 100% of the animals mastering this task by PND 32.

Dogs

No information found.

Non-human primates

Primates demonstrate considerable postnatal development of fine motor movements and dexterity. The ontogeny of relatively independent finger movements in the newborn macaque monkey requires the development of functional cortico-motorneuronal (CM) connections (Lemon, 1999). Although corticospinal fibers reach all levels of the spinal cord before birth, and there may be substantial innervation of the spinal grey matter of the intermediate zone, significant CM connections to the motor nuclei innervating hand muscles are not present at birth.

The first appearance of relatively independent finger movements occurs around 3 months of age in monkeys (Lemon, 1999). Thereafter, both the anatomical and physiological development of the corticospinal system follows a protracted time course which, in general, parallels the maturation of motor skill. It is believed that fine finger movements are not observed before functional CM connections are well established, and that rapid changes in the physiological properties of the corticospinal system coincide with the period in which precision grip is known to mature (3-6 months) (Lemon, 1999). However, corticospinal development continues long after simple measures of dexterity indicate functional maturity, and these changes may contribute to the improved speed and coordination of skilled hand tasks (Olivier et al., 1997). Striking differences have been found between the strong CM projections to the hand motorneuron pools in the dextrous capuchin monkey compared with the much weaker projection in the far less-dextrous squirrel monkey. Similar findings have been reported for the macaque, which is also very dextrous (Lemon, 1999).

By 6 months of age, all basic forms of manipulation seen in adult capuchin monkeys have appeared (Adams-Curtis and Fragaszy, 1994). Actions that occur frequently in the first 8 weeks are gentle and involve sustained visual orientation and aimed reaching. Later actions are more vigorous and involve grasping. Large increases in the rate of activity are evident over the 6-month period of development studied. The increase from the first 8 weeks to the second 8 weeks may be due to a) an increase in the amount of time spent alert and active, b) a decrease in the amount of time spent in a ventral position on the mother, c) improvements in postural control and stamina, and d) the onset of independent locomotion. Changes in form can be attributed primarily to postural factors and to neuromuscular development, resulting in precisely aimed and controlled movements appearing in the 5th and 6th months.

Fine motor abilities develop rapidly in infant monkeys. By 14 weeks of age, infant macaques demonstrate well developed fine motor abilities including hand-to-mouth coordination, bimanual coordination (evaluated using a task which involves lifting a transparent inverted beaker with one hand while retrieving a reward from under it with the other hand) and finger-thumb opposition (Golub et al., 1991b). These results may not be readily generalized to humans because of differences in maturity of motor systems at birth. Monkey neonates are more advanced in motor function than human newborns. For example, monkeys crawl within a few days of birth.

Conclusions for Fine Motor and Dexterity Development

Fine motor and dexterity development is generally protracted. Human neonates do not possess the capacity to perform fine finger movements, and these may not make their appearance for many months after birth. For example, precise coordination involved with grasping an object and lifting may not be fully mature until the second decade of life. In contrast, much cruder grasp functions are present at birth. Postnatal development of fine motor activity in humans and non-human primates follows a similar pattern, although primates develop much more rapidly than humans. However, there is a paucity of data in rats and dogs. Therefore, use of this measure in routine studies is not likely to provide any benefit for risk assessment.

Table 2 Summary of Fine Motor Development and Dexterity

Fine Motor Development	Approximate Postnatal Age Ability Emerges
Human	
Palmar Grasping Reflex	< 2 months
Prehension	5–9 months
Precision Grip	1.5–13 years
Rats	
Fine Motor Development	PND 2–32
Dog	
No Information Found	No Information Found
Non-Human Primate	
Fine Motor Development	11 weeks
Precision Grip	3–6 months
Manipulations	6 months

Ontogeny of Sensory and Reflex Development

The various sensory systems develop in a sequential manner. Across a wide range of species, the functional onset of the auditory system precedes that of the visual system. Furthermore, the functional onset of tactile and olfaction usually precedes that of the auditory system (Alberts, 1984). Conditioned responses to various sensory modalities develop in the same sequence (Richardson et al., 2000), i.e., conditioned responses usually occur first to olfactory, then to auditory, and finally to visual cues.

While electrophysiological and histological research tools are very useful in determining the onset of functional sensory systems, the significance of sensory function is ultimately linked to behaviors that are measured as sensorimotor reflexes. Unfortunately, a limitation of this approach is that the development of sensorimotor system is dependent on the maturation of both the motor and sensory systems. Therefore, it is not always possible to completely separate the development of a given sensory system from the development of the motor system.

One of the most common ways to assess sensory and reflex functionality in a developing animal, especially the rodent, is through the use of test batteries (e.g., Functional Observational Battery (FOB)) (Carlson, 2000; Geyer and Reiter, 1985; Spear, 1990). Test batteries have also been developed to study behavioral functionality in monkeys and children (Paule, 2001).

Humans

There are several standardized clinical tests for assessing the intactness of the nervous system of infants and young children (Winneke, 1995). One of the most commonly used batteries is the Neonatal Behavioral Assessment Scale (NBAS) (Connolly, 1985; Stengel, 1980; Stewart et al., 2000). During early infancy (0-2 months), reflexive development is most commonly used as a developmental milestone. Those reflexes that develop during the first month of postnatal life include: 1) a response to touch [e.g., such as the in curving to the side of skin stimulation (Galant response)], 2) grasping responses (Palmar finger grasp), and 3) stepping and placing reflexes (automatic walking when held upright) (Connolly, 1985). The infantile oral reflexes, consisting of rooting, biting, mouth opening, lateral tongue movement and Babkin reflexes (Sheppard and Mysak, 1984) are commonly evaluated in humans. These reflexes are generally apparent at birth and are active through 4-6 months of age (Sheppard and Mysak, 1984). Onset of chewing behavior occurs around 5-8 months, and in normal development occurs during a waning of the infantile oral reflexes (Sheppard and Mysak, 1984). In general, persistence of infantile oral reflexes beyond the age at which they are reported to occur in normal infants has been viewed as pathological.

Somatosensory (touch/tactile) systems are fully developed during gestation. For example, at birth the human neonate should respond to touch by directing attention to the body part that is being touched (Bower, 1979). Auditory function also develops prenatally. Physiologic evidence of cochlear function after the 20th week of gestation supports findings of fetal responses to sound. At birth, neonates should respond to loud sounds, but not to pure auditory stimulus (Eisenberg, 1970).

Newborns contain a functional taste system; however, gradual transitions in electrical responses to taste stimuli are shown from fetal to adult stage (Rao et al., 1997). The behavioral measure of taste development is intake behavior. The ability to discriminate degrees of sweetness is well developed at birth (Desor et al., 1975), bitter aversion develops within a few days of birth (Rao et al., 1997), but salt discrimination may not develop until 4 years of age (Desor et al., 1975; Kajiura et al., 1992).

The human visual system is functional at birth but undergoes significant postnatal development, and most components are fully operative by the end of the second year of life (Suchoff, 1979). Different functions emerge in different age ranges, for example, flicker sensitivity is mature by 3 months postnatal, binocular functions emerge from 3-7 months of age and contrast sensitivity is not adult-like until several years after birth (Boothe et al., 1985). Visual acuity develops from about 20/600 at birth to 20/100 by 2 months and 20/20 by 4-6 months (Suchoff, 1979). Prior to the second month of life, there is little evidence indicating selective attention. Following the second month of life, human infants demonstrate clear fixation responses indicating that the perception of the infant is qualitatively similar to adults (Bond, 1972). Neonate are able to track slow linear movements but do not track fast rotary movements. This may indicate that while the visual-vestibular response is functional very early after birth, it continues to develop postnatally.

The eye blink response, a protective eye reflex that can be provoked by numerous stimuli, is a generalized phenomenon in mammals (Esteban, 1999). In humans, the glabellar reflex refers to the bilateral eye blink elicited by gentle stimulation of the glabellar region (Zametkin et al., 1979). Following repeated stimulation, the glabellar eye blink response diminishes in adults. The mean rate of spontaneous eye blink is less than 2 per minute in early infancy and increases steadily up until age 14 or 15 when adult levels of 10-16 blinks per minute are achieved (Zametkin et al., 1979).

Rats

Sensorimotor behaviors (e.g., visual and tactile orientation) and righting reflexes develop in a rostral-caudal pattern and achieve mature characteristics during the postnatal period (Almli and Fisher, 1977). The most commonly evaluated reflexes during development of rats are righting (vestibular) and brainstem stimulus-response reflexes (e.g., auditory startle, pupillary light).

The surface-righting reflex emerges soon after birth in the rat (PND1-3), and matures during the first week of development (Bignall, 1974; Smart and Dobbing, 1971). The organization of this reflex emerges a few days postnatally in the low pons and medulla, and during development it becomes dependent on, but not completely replaced by, the more rostral systems in the mescecephalon (Bignall, 1974). Soon after onset of the surface-righting reflex, rat pups achieve the ability to orient themselves upward on an inclined plane, a response known as negative geotaxis (achieved on PND 9-11) (Smart and Dobbing, 1971).

The air-righting reflex, also known as free-fall righting, has been used extensively to assess reflex integrity in rodents (Cripps and Nash, 1983; Smart and Dobbing, 1971; Vorhees et al., 1994). In rats, and presumably other rodents, triggering of the reflex is not dependent on vision (Pellis et al., 1989) while in cats vision is required. The air righting reflex in rats can be detected as early as PND 9, and the response is fully present by PND 18 (Smart and Dobbing, 1971; Vorhees et al., 1994) with certain rat strains showing a fully functional air-righting response by PND 15 (Unis et al., 1991).

Brainstem stimulus-response reflexes consist of motor responses to various sensory stimuli, such as smell, taste, touch or visual. Olfaction and taste are among the first sensory systems to develop in rats with expression of olfactory receptor and transduction genes occurring just prior to birth (Margalit and Lancet, 1993). While rats are likely born with a functional taste sense, they show increased sugar discrimination from birth through PND 15 and an adult-like salt preference by PND 21 (Hill and Mistretta, 1990).

Development of nociceptive and tactile sensation varies based on the specific organ; however, in general the sensory nerve supply to the developing organ occurs immediately prior to partus in the rat (Sisask et al., 1995). Visceral nociceptive receptors are absent at PND 5 and an adult-like response does not develop until weaning (Bronstein-Goral et al., 1986). The tail flick response (a spinal reflex to heat stimulus on the tail) is present at PND 10 but does not mature to adult levels until PND 25 (Ba and Seri, 1993). Vibrissa placing, a tactile response to a stimulus of the stiff hairs located about the nostrils that serve as a tactile organ, is developed by PND 12.5 in rats (Smart and Dobbing, 1971).

At birth, rats are non-responsive to sound (Uziel and Marot, 1981). The acoustic or auditory startle reflex (ASR) is a response elicited by an intense auditory stimulus. The ASR is functional in rats by PND 12 (Smart and Dobbing, 1971). Inhibition of the ASR by brief auditory, cutaneous or visual stimulus (e.g., pre-pulse inhibition) is a brainstem-mediated mechanism (Davis and Gendelman, 1977) that develops gradually beginning around the 3rd postnatal week and continues through the 6th postnatal week, with tactile inhibition developing to an adult-like response at a slightly earlier age (Parisi and Ison, 1979; Parisi and Ison, 1981). The difference between the times of onset of inhibition mediated by the different sensory modalities suggests that the shared central mechanisms that modulate the ASR mature during the 3rd week of postnatal development in rats.

As with most mammals, the visual system is the last sensory modality to develop in rats (Gottlieb, 1971). Rats are born with their eyes closed, and eyelid opening does not occur until approximately PND 14-16 (Crozier and Pincus, 1937). However, phototaxic reactions are present in rats as young as PND 5 (Crozier and Pincus, 1937; Routtenberg et al., 1979). Visual evoked potentials can be obtained as early as PND 11 (Rose, 1968), and rat pups are able to perform a visual placing task by PND 17 (Smart and Dobbing, 1971). This suggests that although rats are born with their eyes closed, the visual system of the rat may be at a developmental level roughly equivalent to species born with their eyes open, including humans.

Dogs

The age when positive conditioned responses can first be seen is dependent on the modality of the stimulus, i.e., a conditioned response involving integration at a higher sensory level occurs later in life than those at lower levels (e.g., rhinencephalic). The majority of the data available on the ontogeny of sensory and reflexive development in dogs can be found in "Integrative Development of Brain and Behavior in the Dog" by Michael W. Fox (Fox, 1964a).

Non-Human Primates

A trend exists among primates for those species that reach higher levels of functioning to have prolonged juvenile periods such that the duration for juvenile growth ranges from a period of months in tree shrews to about 18 months in prosimians (e.g., slow loris), 2-3 years in monkeys and 8 years in great apes (Schultz, 1969).

Primate infants demonstrate a rooting reflex at birth with a maximum response at PND 4, which disappears by PND 10 (Mowbray and Cadell, 1962). Other reflexes present at birth include hand and foot grasping and clasping (Mowbray and Cadell, 1962). From birth, a rhesus monkey placed dorsal side down will quickly "right" by turning to achieve ventral contact (Hines, 1942). Squirrel monkeys have the righting reflex during the first week post birth (Schusterman and Sjoberg, 1969), to orient up by 3 weeks of age (Elias, 1977) and to air-right at week 5 (King et al., 1974).

Somatosensory systems are functional at birth; infant rhesus monkeys have been shown to respond to a pinprick at birth (Golub et al., 1991a). While these systems are functional at birth, they may still be refined during development. For example, macaques have been shown to discriminate texture and size differences at the level of adult thresholds by postnatal week 10, indicating that tactile sensory systems, while functional at birth, continue to develop postnatally (Carlson, 1984).

In non-human primates, auditory responsiveness is observed at birth, indicating that a functional auditory sensory system develops prenatally (Winter et al., 1973), and infant monkeys show a startle response to an abrupt sound on PND 10 (Mowbray and Cadell, 1962).

Taste buds in the soft palate of the common marmoset increase in number after birth, reaching a maximum number at 2 months and then decrease until 9 years of age (Yamaguchi et al., 2001).

Nonhuman primates can visually track their mothers at birth (Dolhinow and Murphy, 1982), can track a moving object within the first few days of birth (Mendelson, 21982), and orient to a small object by PND 13 (Mowbray and Cadell, 1962). Optical quality (optical line spread function) for well-focused retinal images in nonhuman primates is very good at birth and improves rapidly to adult levels by postnatal week 9 (Williams and Boothe, 1981). Visual acuity (argued to be a function of maturation of neural elements) is low at birth and improves slowly during the first postnatal months (Neuringer et al., 1984) and reaches adult levels by about 1 year (Williams and Boothe, 1981).

Sensory and Reflex Conclusions

The sequence of development of sensory and reflex systems in rats, dogs and non-human primates correlate rather closely with those of the human infant. For example, the gustatory system in newborn animals is apparently important for selective ingestion, and olfaction and taste are among the first sensory systems to develop. Behavioral responses in humans and rats suggest the presence of functional taste and olfaction systems at birth (Hudson and Distel, 1983; Teicher and Blass,

1977; Yamaguchi et al., 2001), and olfaction may even be functional *in utero* (Pedersen et al., 1983; Stickord et al., 1982). However, while the sequence of development of sensory and reflex systems may be similar, timing of development relative to birth varies for some systems. In general, rats, dogs and humans tend to develop slowly postnatally and have a clearly defined neonatal period during which they are dependent on the mother for shelter and food. In contrast, non-human primates tend to be more developed at birth. For example, in monkeys tactile, auditory and visual systems are more developed at birth than in humans or rats. This likely reflects the need for rapid postnatal development in non-human primates. Within 3 months of birth the infant macaque develops from a state of complete dependence on its mother to one of relative independence.

Even though primates are precocial for reflexive and sensory behaviors, all of the common laboratory species are relatively developed at birth, with the exception of vision and taste. Therefore, evaluations of somatosensory development in juvenile toxicity studies are not likely to provide significant insight because many of these systems are fully developed during gestation in humans (Table 3).

Ontogeny of Cognitive Development

There are some learning abilities that appear extremely early in life in many species, e.g., classical conditioning tasks. However, the development of more complex learning abilities (e.g., delayed-response or spatial learning) is gradual and progressively improves with the developing animal. Learned responses are also instrumental in the ontogeny of social interactions across many species, including the development of play behavior and fear of strangers. In addition, normal sleep-wake rhythms also mature during the early periods of postnatal development.

Humans

General Understanding

Illingworth believed the first sign of understanding in a neonate can be observed in the first few days of life (Illingworth, 1983). A neonate will attentively watch her mother speak and will respond by opening and closing her mouth and/or bobbing her head up and down. By 4 to 6 weeks, an infant will begin to smile, and about 2 weeks later will begin to vocalize. Between 12 and 16 weeks, an infant opens her mouth when a bottle approaches, and by 24 weeks, stretches her arms out when she sees her mother is going to lift her. By 40 weeks, a toddler will pull her mother's clothes to attract attention. Around 44 weeks, a toddler realizes she has to extend her arms in order to put her coat on, and by 1 year may understand a phrase such as 'where is your shoe?' After the 1st birthday, her understanding is developing in numerous ways. A child can execute simple requests, has increased interest in toys and books, and is developing speech. By approximately 2 years of age, she can put words together or form a simple sentence. The complexity of her language continues to mature, and she will begin to count to 10 and name colors by 5 years of age.

Specific Understanding: Learning & Memory

Neonates — A classical conditioning paradigm was administered to newborns by presenting them with 50 conditioned-stimuli (CS) (i.e., a tone) that overlapped and terminated with an air-puff (Unconditioned Stimulus) which caused the neonate to blink (Little et al., 1984; Rovee-Collier and Gerhardstein, 1997). Neonates were trained at 10, 20, or 30 days of age. Ten days later they received a retention session. The difference between the number of responses during the retention test and during the initial training was the measure of retention. Significant retention was displayed in neonates trained at 20 and 30 days of age, but not those initially trained at 10-days of age.

Table 3 Postnatal Sensory and Reflexive Development of Rat, Dog, Non-human Primate and Human (Approximate age at Demonstration of Characteristic)

	Human	Rat	Dog	Non-human Primate
Eating/Rooting	Birth	Birth	Birth	Birth
Surface righting	No Information Found	PND 1–11	No Information Found	Birth
Grasping/clasping	Birth–2 months	No Information Found	No Information Found	Birth
Orientation	No Information Found	PND 9–11	No Information Found	3 weeks
Air righting	No Information Found	PND 15–18	No Information Found	5 weeks
Olfaction	No Information Found	Birth	Birth	No Information Found
Taste	Birth	Birth	Birth	Birth
Tactile	Discrimination Birth–4 yr	Discrimination; Birth–PND 21 Vibrissa placing: PND 12.5 Nociceptive: PND 21 Tail flick: PND 10–25	Discrimination Birth–PND14 Proprioceptive: PND1 Air puff: PND 10 Nociceptive: PND 1–18 Adult-like Nociceptive Response: PND 21	Pinprick: Birth Mature: 10 weeks
Auditory	Birth	ASR: PND 12 Inhibition of ASR: PND 21–42	ASR: PND 19–25	Birth ASR: PND 14–25
Visual	Visual tracking: Birth Perception: mature in 2 months Acuity: mature in 4–6 months	Phototaxis response: PND 5 Eyes open: PND 14–16 Visual placing: PND 17	Visual orientation: PND 21–25	Visual tracking: Birth Visual orientation: PND 13 Acuity: improves during first 2 months of life; mature: 1yr

ASR = Auditory Startle Response; PND = Postnatal Day; Mature: Age when animal demonstrates an adult like response.

Infants (1 month-2 years) — An operant conditioning task (i.e., a mobile conjugate reinforcement paradigm) has been used to assess "long-term" retention in infants 3 to 7 months of age (Rovee-Collier and Gerhardstein, 1997). In this task, the intensity of the reinforcement (mobile movement) is proportional to the rate and vigor of the infant's kick since a soft ribbon is connected from the infant's ankle to a mobile (acquisition) or an empty stand (nonreinforcement) (Rovee-Collier et al., 1978; Rovee and Rovee, 1969). All infants quickly learn that when the mobile is visually present movement of their ankle will elicit a reward (i.e., a moving mobile) (Hill et al., 1988). However, the rate at which they forgot this association is different depending on the age of the infant during the initial learning of this task. Vander Linde et al. (Vander Linde et al., 1985) found that forgetting occurred within 3 days of training at 2 months of age. Sullivan et al. (Sullivan et al., 1979) found forgetting within 6-8 days of training at 3 months of age, and Hill et al. (Hill et al., 1988) observed a gradual forgetting over the first 2 weeks after training at 6 months of age.

Learning and memory in infants between the ages of 6 and 18 months of age were evaluated in a paradigm very similar to the mobile paradigm. Instead of their ankle moving a mobile, infants learned to press a lever to move a miniature train around a circular track (Campos-de-Carvalho et al., 1993). Infants tested in this paradigm remember progressively longer between 6 and 18 months of age. Infants tested at 6 months of age gradually forgot the task over 2 weeks, and infants 18 months of age remembered the association between the lever press and train movement for nearly 12 weeks after the task was first administered.

Fivuch and Hammond (Fivush and Hammond, 1989) engaged 24- and 28-month old toddlers in a series of play events involving 4 different toy animals (zebra, duck, monkey and giraffe). One group of children re-enacted the play event 2 weeks after the first session while another group did not. All children were then tested for their retention 14 weeks later. Children who had repeated experiences in the play event recalled significantly more items than children in the single experience group.

Children/Juveniles (2 to 12 years) — Over the 1st and 2nd year of life, there is a steady increase in the length of the sequence that can be recalled accurately (Bauer, 1997). By 20 months of age, children were capable of reproducing three-act events (Bauer and Hertsgaard, 1993; Bauer et al., 1994; Bauer and Thal, 1990). At 24 months, children can reliably reproduce events five steps in length (Bauer and Travis, 1993). And by 30 months, they can reliably reproduce an event involving as many as 8 steps, such as "building a house" (Bauer and Fivush, 1992).

There are few studies that have evaluated long-term recall in infancy and early childhood. One study by Bauer et al. (Bauer et al., 1994) tested children 21, 24 and 29 months of age for their recall of a novel multi-step event sequence that they had experienced 8 months earlier, as 13-, 16- and 20-month-olds. The performance of the experienced children was significantly better compared to age- and gender-matched control children who had never experienced the test event. These results suggest that even very young children can maintain organized representations of novel events over long periods of time. Even though the long-term memory is present at 1 year of age, it does not imply that it is fully mature.

Fivush (Fivush, 1991) tested the recall ability of children who visited Disney World around 3 or 4 years of age. Half of the subjects were interviewed 6 months after the trip, the other half 12 months later. All children recounted a great amount of information about their Disney World experience, but there was no difference between retention interval groups. However, it was noted that the older children recalled their trip with more detailed information, and the recall was more rapid than the younger children's recall.

The ontogenetic development of explicit (declarative) and implicit (procedural) learning were studied in children between 5-10 years of age and compared to young adults (23 years of age) (Homberg et al., 1993). Explicit memory is the ability to recall words, dates, visual images or life-events; whereas, implicit memory is associated with a variety of nonconscious abilities, including the capacity for learning habits and skills and some forms of classical conditioning (Zola-Morgan and Squire, 1993). Two tasks, story recall and pictorial memory, were used to assess explicit memory

(Homberg et al., 1993). In the story recall task, individuals were read a story that comprised 11 significant items. In the pictorial memory task, the subjects were presented with a series of 6 pictures. In both explicit memory tasks, the number of recalled items gradually increased with age. Adult performance was not seen before 9 years of age in the picture recall task. Performance in the story recall task was superior in the young adults compared to the oldest children tested (11 years age). There were no differences in either implicit learning task between young adults and children (not described). These results demonstrated that implicit and explicit memory follow different maturational courses.

The first signs of understanding can be observed within the first few days of life, and infants very quickly learn to associate certain responses with learned behaviors. At all ages, humans have memories that forget gradually, but the rate at which they forget becomes slower with increasing age. In addition, as age increases the ability to recall more complex sequences and detailed information continues to improve.

Rats

The learning capabilities of the newborn rat are impressive. For example, Johanson and Hall (Johanson and Hall, 1979) demonstrated appetitive learning in 1-day-old rat pups. Specifically, 1-day-old pups were capable of discriminating between two paddles on the basis of odor and position when they were provided milk as a reward for the correct choice. Rudy and Cheatle (Rudy and Cheatle, 1977) demonstrated that when 2-day-old rats were exposed to a novel odor and injected with an illness-inducing drug, i.e., lithium chloride, these same pups, when retested at 8-days of age, avoided the odor, whereas control pups did not. These data suggest rat pups are capable of associative learning at a very early age.

Several investigators have suggested that immature animals have an enhanced pattern of forgetting compared to adults (Campbell and Spear, 1972; Coulter et al., 1976). Neonatal rats 7, 9 and 12 days of age were tested in a nondirectional active-avoidance task (Spear and Smith, 1978). Animals in the 12-day old group required fewer trials to reached criterion performance during the learning phase and had superior retention 24-hrs later compared to the 9-day old rats. The 7-day old rats never achieved criterion performance (5 out 6 avoidances) during the learning phase. Reactivation treatment (i.e., presentation of unconditioned stimulus (footshock) shortly before the retention test) was found to enhance retention over the 24-hr period for 12-day old rats, but only somewhat for 9-day old rats. The reactivation treatment seemed completely ineffective for rats 7 days of age. These experiments supported the authors' hypothesis that substantial deficits in young animals are due, in part, to a deficiency in memory retrieval.

The ontogeny of step-down passive avoidance was studied in Wistar rats 2-, 3-, 4-, 6-, 8- and 13-weeks of age (Myslivecek and Hassmannova, 1991). During the retention trial, the step-down latency was longest in 3-week old rats, but the most efficient learning was seen in rats 6 weeks of age. In addition, passive avoidance learning was present in rats 10 days of age, but retention improved considerably in rats 15 days of age (Stehouwer and Campbell, 1980). These data suggest that passive avoidance learning exists in early postnatal life, but the maturity of the storage and/or retrieval systems improves with age in the rat.

Biel (Biel, 1940) trained white rats at various ages in a multiple T-maze, i.e., the Biel maze. Group A, B, C and D rats were trained in the Biel maze between the ages of 17 - 23, 20 - 26, 23 – 29, and 30 - 36 days of age, respectively. His results showed that Group D rats were superior to and made fewer errors than Group C; Group C was better than Group B; and in turn, Group B was more proficient than Group A. In addition, Group A never reached the level of accuracy, i.e., as few errors, as the other groups, and Group D learned the task most rapidly. Furthermore, the average time it took to complete the maze was shorter with the increasing age of the groups. The results of this study indicated that there are maturational differences that influence a rat's ability to perform the Biel maze, and there is an improvement in maze performance with age.

Rudy et al. (Rudy et al., 1987) demonstrated a distinctive ontogeny of spatial navigation in Long-Evans Hooded rats. Using the Morris water maze, rats as young as 17 days of age were capable of using proximal cues to locate a safe platform. However, rats 20-days of age began to display only minimal evidence of distal-cue abilities. In another study investigating the ontogeny of spatial navigation (or place navigation), Long-Evans Hooded rats were given a total of 28 trials over 5 consecutive days beginning at 21, 28, 35 or 64 days of age (Schenk, 1985). Rats in the 21- and 28-day groups had slower escape latencies and took more circuitous routes to locate the hidden platform compared to older rats. In addition, the 21- and 28-day groups had poor spatial bias towards the location of the platform during the probe trial. The 35-day group showed adult-like learning during the training trials, but failed to demonstrate a proper search behavior in the probe trials. Only rats 40 days and older displayed typical adult behavior, e.g., swimming directly toward the platform from any start position or localized searching around the location of the platform during the probe trial.

Rats are capable of learning several types of behavioral tasks early in postnatal development, but adult efficiency and accuracy of learning do not fully emerge until later. Of course, some of the constraints on early learning are imposed by the status of several sensory systems, e.g., the auditory and visual systems do not function fully until late in the second week of postnatal development.

Dogs

The period between 18-28 days of age is considered a critical period in the development of canine learning (Fox, 1964b; Fox, 1966). Fox (Fox, 1971) did some outstanding work on the ontogeny of learning in puppies in which he evaluated the development of the following behaviors in young puppies: olfactory habituation, visual discrimination (including reversal learning), and delayed response. The results of these experiments will be briefly summarized.

During the first 5 days of life, puppies were exposed to anise oil that was liberally applied to the mammary region of the bitch. When tested on PND 6, puppies in the anise group oriented toward the Q-tip soaked in anise oil while the controls avoided it. These findings suggest that the neonatal dog can learn to respond differently to odors, and that habituation to an odor can develop at this early age. It also suggests that olfactory stimuli may play an important role in the neonatal life of puppies.

In another experiment by Fox (Fox, 1971), visual discrimination was evaluated in puppies at 5, 8, 12 and 16 weeks of age. The puppies were placed on 23-hour food deprivation and fed for 1 hour after testing. Puppies were trained to discriminate a white cube from 2 black cubes (initial discrimination task) and a black cube from 2 white cubes (reversal learning). The undersides of all cubes were smeared with meat, but the correct cube contained a small piece of meat. In the initial discrimination task, 15 trials per day were given until puppies reached 2 consecutive days with 85% or better correct responses (i.e., criterion). Reversal discrimination began the day after criterion for the initial task was reached, and the same criterion was used. Puppies in the 5- and 8-week old groups required significantly more trials to reach criterion performance on the initial discrimination task compared to the 12- and 16-week puppies. All groups had difficulty in the reversal discrimination task, but the best results were obtained in the oldest group tested. In both discrimination tasks, variability in the number of errors decreased with increasing age.

During the initial training for the delayed-response task, puppies 4, 8, 12 and 16-weeks of age were trained to select between three boxes (i.e., one correct, one incorrect and one neutral). As the puppy watched from behind a Plexiglas start box, an experimenter placed food in the box designated as the correct choice in an 8' X 8' arena. The Plexiglas was raised, and the puppy was then allowed to make a selection. Puppies where fed only one meal 1-2 hours after testing.

During the actual delayed-response task, after the puppies observed which box was the correct choice, the boxes were blocked from view for increasing lengths of time (i.e., the delay), and the experimenter left the arena. The delays were 3, 5, and 10 seconds, and increased by 5 seconds every time the dog made 4 consecutive correct choices. Puppies 4- and 8-weeks of age made more incorrect choices (of both the neutral and incorrect box) than the 12-week puppies. The 16-week old puppies had an unexpected increase in the number of errors to the incorrect box. However, both the 12- and 16-week groups were capable of increasingly longer delayed responses. After successfully achieving a delay time of 15 seconds, the 12-week group continued to increase their delay in a linear fashion (i.e., a 50 sec delay was possible on day 13 of testing), whereas the 16-week puppies were much slower (i.e., only a 35 sec delay was possible on day 13 of testing). Fox noted that the apparent inferior abilities of the puppies at 16 weeks compared to 12 weeks might be due to the development of inhibition reflexes that appear around 16 weeks of age. This experiment demonstrated that puppies 12- and 16-weeks of age can proficiently use both long-term memory (learned food-motor approach pattern) and short-term memory (remembering for a short time which box contained the reward) to solve a delayed response task. It is likely that the learning capabilities of dogs continue to improve with age based on the fact that in adult dogs the delayed response task can successfully be solved with as much as 120-sec delays.

A spatial delayed nonmatching-to-sample task was evaluated in mongrel dogs divided into three groups: 5 young (8 months-2.5 years), 3 middle-aged (5-8 years), and 5 aged dogs (Ω10 years) (Head et al., 1995). All ages performed equally well when a short delay was present, i.e., 20 seconds, but the young dogs had fewer errors compared to the older dogs when the delay was increased, i.e., 70 and 110 second delays. Identical results were seen by Milgram et al. (Milgram et al., 1994) in a delayed-nonmatching-to-sample paradigm using beagle dogs. In contrast, Fox (Fox, 1971) reported poorer performance of very young dogs relative to adults in a spatial delayed-response task. These studies emphasize the importance of selecting an appropriately aged animal when assessing the impact of a compound on early cognitive development. It also points out the significant maturation the CNS undergoes during early postnatal development in the canine.

Non-Human Primates

Harlow (Harlow, 1959) evaluated the development of learning in the rhesus monkey. A conditioned avoidance task was initiated on the third day of life. Infants showed improvement (i.e., learning) over the 5 consecutive days of testing. Interestingly, when a retention test was administered at 23 days of age, it revealed considerable learning loss ranging from no indication of retention to conditioned responses on half of the test trials.

Two groups of newborn macaques (one group started training at birth and the other on day 11) were trained on a black-white discrimination task, i.e., object discrimination. Learning of this task by the neonatal macaques was rapid. In the group trained from birth, an accuracy of 90% was attained beginning at 9 days of age. The second group was rewarded for climbing up either of two gray ramps for the first ten days of life. Then on day 11, the black and white ramps were substituted. By the second test day (i.e., day 12), these monkeys could solve the discrimination problem. Interestingly, during the reversal discrimination task, the infants made many errors when the problem was first reversed and showed severe emotional disturbances (e.g., balking, vocalization, urination, and defecation).

Even though 11-day old monkeys can solve a black-white discrimination task, the ability to solve a triangle-circle discrimination problem requires greater CNS maturation. Monkeys 20 to 30 days of age may require from 150 to 200 trials to solve a triangle-circle discrimination task. The development of more complex object discrimination (i.e., three dimensional objects differing in color, form, size and material) was evaluated in rhesus monkeys 30, 60, 90, 120, 150 and 366 days of age. The animals were given 25 trials a day, 5-days a week for 4 weeks. The number of trials required to have 10 consecutive correct responses decreased with increasing age and approached adult performance around 120 to 150 days of age.

In a delayed-response task, an animal was first shown a food reward, which was then concealed under one of two identical containers during the delay period. The rhesus monkeys began testing at 60, 90, 120, or 150 of age. Over time, all four groups demonstrated an increasing ability to solve the delayed response task. However, Harlow did note that the capacity to solve delayed-response problems was better in animals 125 to 135 days of age compared to 60-90 days of age. In addition, half of the monkeys at 200 to 250 days of age had essentially flawless performance in this task. It is clear that the capacity to solve delayed-response problems matures later than the capacity to solve object-discrimination tasks.

Two broad conclusions can be derived from the monkey data presented thus far. First, certain learning abilities appear extremely early in life in the rhesus monkey, e.g., conditioned avoidance or black-white discrimination. The ability to solve a problem may be present as early as 1 day of age, and within a matter of days, performance will change from chance levels to nearly 100% accuracy. Second, the development of certain, more complex, learning abilities is more gradual and progressively improves, e.g., delay-response learning and some discrimination learning tasks (e.g., color, pattern or shape discrimination). Interestingly, extensive early training on complex tasks (e.g., delayed response task) did not facilitate the solution, but rather success was clearly a function of age, i.e., even when young animals were trained more extensively than the older animals, their performance did not equal that of older monkeys (Zimmerman and Torrey, 1965).

Conclusions for Cognitive Development

There is evidence that rats, dogs, monkeys and human infants have learning capacities that change during development. In all species evaluated, the capacity to mediate reflexive behaviors emerged before learned behavioral reactions. Vogt and Rudy (Vogt and Rudy, 1984) suggested that the sequence of behavioral development reflects a sequential caudal to rostral maturation of components. That is, the lower centers of a sensory system are sufficient to support primitive reflexive behaviors, and it is well established that subcortical structures are sufficient to support many instances of Pavlovian conditioning (Girden et al., 1936; LeDoux et al., 1984). However, higher centers of learning (e.g., hippocampus and prefrontal cortex) are necessary for the acquisition of more complex behaviors.

Therefore, in all common laboratory species, evaluations of cognitive function during juvenile toxicity studies can provide insight into the ontogeny of the maturation of these systems. The time required to learn a particular task is often greater in neonates and infants compared to adults. In addition, the accuracy of learning that task is often not equal to an adult. This implies that even though a task can be completed at a fairly young age, it is by no means functionally mature, i.e., at adult levels of performance. Finally, it is important to remember that learning and memory are intimately associated with the maturation of many of the sensory systems described above (Table 4).

Ontogeny of Communication

Early in postnatal development, many species use some form of communication to express fundamental needs (e.g., hunger and fear). However, the complexity of the communication often undergoes a period of prolonged postnatal development.

Humans

Infants have many methods of preverbal communication, e.g., crying, smiling, clinging, and kissing (Illingworth, 1983; Morley, 1972; Murray and Murray, 1980; Nelson, 2000). At about 2 months, the infant listens to voices and coos in response, and around 3 months, she begins to make sounds (e.g., "aah", "ngah"). At 4 months, laughing out loud occurs, and the infant may show displeasure if social contact is broken or excitement at the sight of food. At 8 months, preference for the mother emerges, the infant begins to babble with new complexity and inflections that mimic the native language. Around 9 to 10 months a critical step occurs in speech development, an infant begins to

Table 4 Summary of Cognitive Development

Learning Task	Approximate Postnatal Age Ability Emerges
Humans	
Classical Conditioning (air puff)	20–30 days old
Operant Conditioning (task-dependent)	2 months–18 months
Long-term recall	≥13 months–20 months
Implicit Learning	<5 years
Explicit Learning	5–9 years
Rats	
Appetitive Learning	PND 1
Associative Learning (Olfactory)	PND 2–8
Active Avoidance	PND 9–12
Passive Avoidance	PND 10–21
Sequential Learning (Biel Maze)	PND 17–23
Proximal Cued Learning (Morris Water Maze)	PND 17
Spatial Learning (Morris Water Maze)	PND 35–40
Dogs	
Associative Learning (Olfactory)	PND 1–6
Visual Discrimination	5–16 weeks
Delayed Response Task	≥12–16 weeks
Non-Human Primates	
Conditioned Avoidance	PND 3–8
Black-White Discrimination Task	(with minimal retention on PND23)
Color, Pattern or Shape Discrimination	PND 9–12
Delayed-Response Task	PND 20–150
	PND 60 (rudimentary performance)
	PND 125–135 (improved performance)
	PND 200–250 (flawless performance)

imitate sound and responds to the sound of her name (Nelson, 2000). By 10 months, she can say one word with meaning, respond to NO and obey orders. By 12 months, she can imitate the sound of a dog or cat and can say 2 or 3 words with meaning. Between 15 and 18 months, a child begins to speak in her own jargon, i.e., speaking expressively with modulation and dramatic inflection in her voice with an occasional intelligible word. It is not until 21 to 24 months, that a child will begin to join words together.

Rats

Hofer et al. (Hofer et al., 1998) evaluated the ontogeny of ultrasonic vocalizations (USV) in preweanling rats when isolated immediately after brief periods of maternal interaction (potentiation). Pups on PND 3 vocalized, but it is thought to be predominantly under thermal control (Carden and Hofer, 1992). Consistent USV potentiation did not develop until PND 7-9 (Hofer et al., 1998). Potentiation reached maximum around PND 13 (i.e., just prior to eye-opening), and then declined until all USV responses ceased after PND 21.

Dogs

Distress vocalization begins around 3 to 3.5 weeks of age, and whining vocalization (distinct from "distress") begins around 4 to 5 weeks (Fox, 1971). The repertoire of vocal patterns improves during weeks 3 to 5. It is also believed that socialization processes are very instrumental in developing and reinforcing vocal behavior and distress vocalization.

Non-Human Primates

Three different emotional vocalizations have been described in the macaque: play cry, fear-pain cry and food-contentment cry (Foley Jr., 1934; Hines, 1942; Lashley and Watson, 1913). Emotional vocalizations predominate during the first 6 months of life and were uttered as single syllables or phrases. After 6 months of age, the number of distinct monosyllabic and bisyllabic sounds increased, and the sounds varied in pitch, timbre and inflection. Sounds of similar pitch were combined differently so that the phrase changed in response to definite situations in the laboratory. The number of phrases increased to 4 or 5 during the first year, but never to 7 or 8 like was heard in the adult.

Conclusions for the Development of Communication

Evaluations of the development of communication in juvenile toxicity studies with rats and dogs (and to some degree in non-human primates) are not likely to provide significant insight because communication in these species appears to function primarily as a means of expressing distress. In humans, the development of speech and language is a critical component in CNS maturation.

Table 5 Summary of Communication

Communication	Approximate Postnatal Age Ability Emerges
Human	
"Coos" in response to voices	2 months
Babble with inflection	8 months
Imitates sound	9–10 months
Can understand or say 2 or 3 words with meaning	12 months
Begins to join words together	21–24 months
Rats	
Ultrasonic vocalizations	PND 7–9
USV (maximum)	PND 13
Dog	
Distress vocalizations	3–3.5 weeks
Whining vocalizations	4–5 weeks
Non-Human Primate	
Emotional vocalizations (1 or 2 syllables)	0–6 months
Emotional vocalizations (3, 4 or 5 syllables)	6 months–1 year

Ontogeny of Social Play

Social play is one of the means by which an individual obtains information about his social environment. By playing with peers, the individual learns norms of social communication, social dominance, and social integration in a context where errors in performance usually have neither aversive nor lethal consequences.

Humans

At approximately 40 weeks of age, babies will begin to play simple games such as peek-a-boo and pat-a-cake; and beginning around their 1st birthday, they will play simple ball games (Nelson, 2000). Around 2.5 years of age, children begin to pretend play. During the third year of life, children develop a greater interest in others and will play simple games in parallel with other children. Around 4 years of age, children will play with several children and show the beginnings of social interactions, including role-playing. During the elementary school years (5-12), continued peer group relationships are essential (Kaye et al., 1988). These relationships help the child develop the capacity for socially acceptable aggressiveness, group cooperation and self-identity. During the early teenage years, the adolescent may appear to behave with no rhyme or reason as she searches for independence from her parents while looking for peers groups to achieve a sense of belonging. Closeness to the peer group is achieved by conformity of dress and frequent telephone communications.

Around 15 to 20 years of age, the adolescent begins to solidify her own ideals and values. She also begins to function as an independent thinking individual with a reasonable state of equilibrium among instinctual drives, self-needs and the outside world.

Rats

In rats, the earliest forms of social interactions are at birth with the competition to nurse and cluster together to sleep (Bolles and Woods, 1964). After PND 14, a new interaction begins to emerge where one pup may appear to spontaneously jump, and a chain reaction occurs involving all the littermates. Panksepp (Panksepp, 1981) evaluated the development of play behavior in young rats by observing the frequency and duration of pinning behavior. Play increased from 18-28 days, peaked during 32-40 days of age, and then gradually declined. He also found that social isolation of rats resulted in a marked increase in play behavior. One function of play may be to establish stable social relationships and help rats find their place in the existing social structure of a group.

Grooming in rats may be thought of as a social behavior because it results in the intimate interaction of siblings. Grooming consists primarily of three different kinds of behavior (Bolles and Woods, 1964). Face-washing and scratching in rudimentary forms can be observed as early as PND 2. By PND 10-12 both face-washing and scratching resemble the adult pattern and include the characteristic licking of the paw. Around PND 13, the third component, fur-licking, begins. In addition, grooming of its littermates also begins around PND 13 and is quite common in PND 40 rats. However, sibling grooming declines somewhat in adulthood. The younger animals spend a considerable amount of time washing their front paws and face and scratching with the hindpaws. However, licking the fur is the prevalent component of grooming in the adult. Grooming is the only activity other than eating and sleeping in which young animals will persist at for 5 or more minutes.

Dogs

Around 3 to 3.5 weeks of age, pups begin the earliest forms of playful social interactions (Fox, 1971). They will chew on each other's ears and lick faces of one another, but will either avoid or elicit a distress vocalization if the stimulation becomes too intense. Between 4 to 5 weeks of age, play-fighting, scruff-holding and "prey-killing" (i.e., head shaking movements) appear, along with pouncing, snapping and aggressive vocalizations (e.g., growling and snarling with teeth bared) (James, 1955; James, 1961). Submissive postures and play-soliciting gestures (e.g., ears pricked, tail wagging, vocalization) also appear around this time. By 4-months of age, dominant-subordinate relations are established between littermates (Fox, 1971; Pawlowski and Scott, 1956; Scott and Marston, 1948). In general, male pups also indulge in more rough-and-tumble play and biting than females, but there are differences in aggressiveness depending on the breed of dog (Fuller and DuBuis, 1962).

Non-Human Primates

Play behavior in the macaque has two forms: movement play and social play (Hines, 1942). As soon as the infant monkey is able to move easily (i.e., about 3 weeks of age), she will repeat movements such as jumping, running and climbing. Around 4 weeks of age, bouncing on a wire mesh is possible. Social play begins around 5 weeks of age and is described as dodging and chasing games with older monkeys that were allowed to run loose in the room. Around 6 weeks of age, the infant monkey begins stealing food from another monkey's cage in spite of the presence of an abundance of food in her own cage. Stealing food from another monkey's cage is a well-established behavior in the adult monkey. Around 7 to 9 weeks, infants will approach each other with arms extended and then lightly slap each other. This appeared to be a signal to initiate rough-and-tumble rolling and biting. In infant rhesus monkeys, males generally indulge in more vigorous rough-and-tumble wrestling and sham biting than do females (Harlow, 1962).

Conclusions for Social Play Development

Evaluations of the development of social play in juvenile toxicity studies may provide some insight because it appears to develop postnatally in all common laboratory species. However, evaluations of the development of the social play would not be straightforward in routine screening studies since methodologies across species are not easily comparable.

Table 6 Summary of Social Play

Social Play	Approximate Postnatal Age Ability Begins to Emerges
Human	40 weeks–1 year
Rat	PND 14–28
Dog	3–5 weeks
Non-Human Primate	5–9 weeks

Development of Fear Response

The development of a fear response to a novel environment or stranger is found in many species, including rats, dogs, rhesus monkeys, and human infants.

Humans

Two of the most prominent events that provoke fear in infants are the approach of unfamiliar people and temporary separation from the caregiver. Distress to both these events, even in different cultures, emerges between 7-9 months (Kaye et al., 1988; Nelson, 2000). While infants do not cry every time they encounter a stranger or experience separation, the more unfamiliar the stranger or the context, the more likely behavioral signs of fear will occur. Since 3-month olds can discriminate a stranger from a parent with no fear of strangers, improvement in working memory at 7-9 months is thought to play a role in these reactions.

The maturation of the dorsolateral prefrontal cortex and the hippocampus, as well as the integration of limbic structures and the cortical network, are considered important in the development of the fear response process (Herschkowitz et al., 1997). Not surprisingly in humans, the amygdala is thought to participate in fear reactions to unfamiliarity like non-human primates. The amygdala is linked reciprocally to the cerebral cortex by axons within the capsula interna, which develops mature myelin at 10 months in human infants, a time when stranger and separation fears are prominent (Herschkowitz et al., 1997).

Rats

It was found that fearful behavior in response to strange situations begins to increase in rats around 20 and 30 days of age (Candland and Campbell, 1962).

Dogs

At approximately 3 weeks of age (i.e., beginning of the socialization stage), young puppies will begin to bark or whine when placed in strange places. Elliot and Scott (Elliot and Scott, 1961) showed that this reaction to isolation peaks around 6 to 7 weeks of age, and declines thereafter.

Puppies that had only social contact to other dogs during the critical periods of socialization (i.e., 5-7 weeks of age) exhibited extreme fear when introduced to human contact (Freedman et al., 1961).

Non-Human Primates

Around 2.5-3 months of age, but not before, monkeys separated from their mothers become agitated, emit frequent distress vocalizations, and show a distinct increase in adrenocorticotropic hormone response (Harlow, 1959). Rhesus infants also begin to demonstrate fear reactions to threatening pictures, strange objects, and novel situations around this age. If a human enters a room with caged monkeys, the animals become silent and freeze, which are both signs of fear to unfamiliarity. Rhesus infants raised with an inanimate object do not show signs of fear to a novel location until they are 3-4 months old (Harlow, 1959).

Much work has been done to investigate the CNS systems involved in development of fear. The freezing response appears to be influenced by gamma-aminobutyric acid systems (Herschkowitz et al., 1997), while the distress cries are influenced by opiate systems. Both the anterior cingulated cortex and the amygdala, which have high densities of opiate receptors, participate in the production of distress calls. Bilateral ablation of the amygdala at 2.5 months produced a significant decrease in fear reactions to novel situations. Control animals who showed fear to an unfamiliar place took three times longer than lesioned animals to leave a familiar cage and enter an unfamiliar one. Amygdalectomized monkeys also showed significantly fewer fear reactions to pictures of monkey faces expressing fear.

The amygdala has been implicated in the mediation of emotional and species-specific social behavior. The amygdala in adults is involved in determining whether an object or organism is potentially dangerous. If danger is detected, it coordinates a variety of other brain regions to produce a species-typical response to avoid the danger. Therefore, the amygdala contributes to learning what is dangerous in the environment. Selective amygdala lesions in 2-week old macaques resulted in less fear of novel objects (e.g., rubber snakes) when the animals were tested at 6-8 months of age. However, the lesioned animals displayed substantially more fear behavior than controls during social interactions. These results suggest that neonatal amygdala lesions dissociate a system that mediates social fear from one that mediates fear of inanimate objects. Thus, amygdala lesions early in development have different effects on social behavior than lesions produced in adulthood, which decrease fear of inanimate objects and increase fear during social behavior (Prather et al., 2001).

Conclusions for Fear Response Development

Evaluations of the development of a fear response in juvenile toxicity studies may provide some insight because it appears to develop postnatally in all common laboratory species. However, methods to evaluate the development of the fear response are not widely applicable to screening toxicity studies.

Table 7 Summary of Fear Response

Fear Response	Approximate Postnatal Age Ability Begins to Emerges
Human	7–9 months
Rat	PND 20–30
Dog	3–7 weeks
Non-Human Primate	2.5–4 months

Development of Sleep Cycles

The sleep-wake cycle is developed postnatally in many species and is thought to be critical in the development of cognitive abilities.

Humans

In humans, the periodicity of wakefulness and sleep is observed by between 10-11 weeks of age (de Roquefeuil et al., 1993; Hellbrügge, 1960). Between the ages of 18 to 21 weeks, the longest duration of sleep occurred between 11 pm and 5 am (Kleitman and Engelmann, 1953) (Moore and Ucko, 1957). De Roquefeuil et al. (de Roquefeuil et al., 1993) examined the sleep-wake rhythms of 12 children admitted to a day-care at 4-5 months of age until approximately 15 months of age. The circadian rhythm of wakefulness and sleep is well established by 4 months of age. In this same study, the results indicated that at 7, 11 and 15 months of age the number of sleep spans continued to decrease in a 24-hr period (4, 3 and 2, respectively), and the duration of an individual sleep span progressively increased (200, 210 and 260 minutes, respectively).

Rats

Up until PND12, nursing and burrowing for preferred sleeping and nursing positions are the dominant activities in rats (Bolles and Woods, 1964). Any movement by one pup is likely to disturb the entire litter. Gradually after PND 12, activities such as exploring and grooming occur with increasing frequency. By PND 15, some rats prefer to bury their heads under the litter to sleep while others prefer to climb on top. By PND 19, sleeping becomes quieter with less twitching and moving around. By PND 27, an individual rat can sleep soundly while littermates are playing on top of him. Interestingly during the first 3 weeks of life, rat pups do not exhibit the normal diurnal sleep cycle of adult animals, i.e., sleep during daylight, and activity at night (Bolles and Woods, 1964). Alföldi et al. (Alföldi et al., 1990) demonstrated that by PND 23 waking activities predominated during the 12-h dark period.

Dogs

The development of electrocortical activity as measured by EEG in the dog has been well described by Fox (Fox, 1971). The results of the EEG analysis indicated changes in behavioral state (e.g., sleep, alertness) were not accompanied by significant changes in the EEG during the first two weeks of life. Up until 7 days of age, sleep in the puppy consisted of mainly body jerks and twitches including occasional vocalizations, sucking, crawling and scratching movements (which could easily be mistaken for an alert state). After the second week of life, slow-wave activity began to emerge and was associated with the development of quiet sleep. Three weeks of age marks a transitional point when the relative durations of wakefulness (or attentiveness) and activated sleep are similar, and the duration of quiet sleep is increasing. The onset of the mature EEG occurs between 3 and 4 weeks of age, which also marks a critical period of socialization (Scott, 1962). By 4-5 weeks of age, the EEG has developed the full range of activity that is present in the adult.

Non-Human Primates

Infant monkeys have a less well-defined sleep/wake pattern than adult animals. By 2 to 8 days of age, neonatal monkeys already tend to sleep more frequently at night than during the day (Bowman et al., 1970).

Conclusions for Sleep-Wake Cycle Development

While development of adult-like sleep-wake cycles occurs postnatally, utilizing sleep-wake cycles as a measure in juvenile toxicity studies will provide limited insight because its development is not well understood.

Table 8 Summary of Sleep-Wake Cycle

"Semi-Adult-like" Sleep Cycle	Approximate Postnatal Age Ability Begins to Emerges
Human	3–15 months
Rat	PND 15–23
Dog	2–4 weeks
Non-Human Primate	2–8 days

Overall Conclusions

There is no species that provides a direct correlation to human neurological development for all functional measures discussed. However, when development is tracked within particular functional domains (e.g., sensory or locomotor development), the common laboratory species are born in similar states of CNS development as the human, and there is an extended period of postnatal life where these species develop similar to humans. Both the rat and dog are appropriate models for comparative CNS development, with neither showing a clear advantage. Non-human primates are more advanced at birth for several of the most commonly measured endpoints of CNS development. Because of their advanced cognitive evolution monkeys are more similar to humans in cognitive development. However, for most applications, the rat is likely to be the preferred species for evaluation of CNS function in juvenile toxicity testing, but should not be chosen as the default species for all studies. It is important to select the species that most appropriately correlates to humans within the functional domain of concern.

Additional Research Needs

The ontogeny of memory is less well understood than adult memory systems for several reasons. First, unlike adults, infants do not typically suffer discrete brain lesions, which have helped researchers make associations between different types of memory and specific brain regions. Second, cognitive and linguistic limitations in infants provide additional challenges, which make studies of the development of early memory challenging.

Also, there is a strong need have additional reviews on postnatal development of the CNS which include both neuroanatomical and neurochemical maturation. Further research on the ontogeny of neurological development in all species is necessary in order to design better cognitive assessments of juvenile toxicity.

REFERENCES

Adams-Curtis LE, and Fragaszy DM (1994) Development of manipulation in capuchin monkeys during the first 6 months. *Developmental Psychobiology* 27: 123-136

Adams J (1986) Methods of Behavioral Teratology. In: Riley EP, Vorhees CV (eds) *Handbook of Behavioral Teratology.* Plenum Press, New York, pp 67-97

Alberts JA (1984) Sensory-perceptual development in the Norway rat: A view toward comparative studies. In: R.Kail, Spear NE (eds) *Comparative Perspectives on the Development of Memory.* Erlbaum, Hillsdale, NJ, pp 65-101

Alföldi P, Tobler, and Borbely A (1990) Sleep regulation in rats during early development. *American Journal of Physiology* 258: R634-R644

Almli CR, and Fisher RS (1977) Infant rats: sensorimotor ontogeny and effects of substantia nigra destruction. *Brain Res. Bull.* 2: 425-459

Altman J, and Sudarshan K (1975) Postnatal development of locomotion in the laboratory rat. *Anim. Behav.* 23: 896-920

Ba A, and Seri BV (1993) Functional development of central nervous system in the rat: ontogeny of nociceptive thresholds. *Physiology & Behavior* 54: 403-405

Bauer PJ (1997) Development of memory in early childhood. In: Cowan N (ed) *The Development of Memory in Childhood.* Psychology Press, pp 83-111

Bauer PJ, and Fivush R (1992) Constructing event representations: Building on a foundation of variation and enabling relations. *Cognitive Development* 7: 381-401

Bauer PJ, and Hertsgaard LA (1993) Increasing steps in recall of events: factors facilitating immediate and long-term memory in 13.5- and 16.5-month-old children. *Child Development* 64: 1204-1223

Bauer PJ, Hertsgaard LA, and Dow GA (1994) After 8 months have passed: long-term recall of events by 1- to 2-year-old children. *Memory* 2: 353-382

Bauer PJ, and Thal DJ (1990) Scripts or scraps: reconsidering the development of sequential understanding. *Journal of Experimental Child Psychology* 50: 287-304

Bauer PJ, and Travis LL (1993) The fabric of an event: Different sources of temporal invariance differentially affect 24-month-olds' recall. *Cognitive Development* 8: 319-341

Bekoff A, and Trainer W (1979) The development of interlimb co-ordination during swimming in postnatal rats. *The Journal of Experimental Biology* 83: 1-11

Biel WC (1940) Early Age Differences in Maze Performance in the Albino Rat. *The Journal of Genetic Psychology* 56: 439-453

Bignall KE (1974) Ontogeny of levels of neural organization: the righting reflex as a model. *Exp. Neurol.* 42: 566-573

Bolles RC, and Woods PJ (1964) The ontogeny of behavior in the albino rat. *Anim. Behav.* XII: 427-441

Bond E (1972) Perception of form by the human infant. *Psychol Bull* 77: 225-245

Boothe RG, Dobson V, and Teller DY (1985) Postnatal development of vision in humans and nonhuman primates. *Ann. Rev. Neurosci.* 8: 495-545

Bower T (1979) *Human Development.* W.H. Freeman & Co., New York,

Bowman RE, Wolf RC, and Sackett GP (1970) Circadian rhythms of plasma 17-hydroxycorticosteroids in the infant monkey. *Proceedings of the Society for Experimental Biology and Medicine* 133: 342-344

Bronstein-Goral H, Mitteldorf P, Sadeghi MM, Kirby K, and Lytle LD (1986) Visceral nociception in developing rats. *Developmental Psychobiology* 19: 473-487

Butcher RE, and Vorhees CV (1979) A preliminary test battery for the investigation of behavioral teratology of selected psychotropic drugs. *Neurobehav. Toxicol.* 1 Suppl 1: 207-212

Campbell BA, and Spear NE (1972) Ontogeny of memory. *Psychological Review* 79: 215-236

Campos-de-Carvalho M, Bhatt R, Wondoloski T, Klein P, and Rovee-Collier C (1993) Learning and Memory at nine months. Paper presented at the meeting of the International Society for the Study of Behavioral Development, Recife, Brazil

Candland, D. K. and Campbell, B. A. Private Communication . 1962.

Carden SE, and Hofer MA (1992) Effect of a social companion on the ultrasonic vocalizations and contact responses of 3-day-old rat pups. *Behavioral Neuroscience* 106: 421-426

Carlson M (1984) Development of tactile discrimination capacity in Macaca mulatta. Brain Res. 318: 69-82

Carlson SE (2000) Behavioral methods used in the study of long-chain polyunsaturated fatty acid nutrition in primate infants. [Review] [45 refs]. *Am. J. Clin. Nutr.* 71: 268S-274S

Connolly BH (1985) Neonatal Assessment: an overview. *Physical Therapy* 65: 1505-1513

Coulter X, Collier AC, and Campbell BA (1976) Long-term retention of early Pavlovian fear conditioning in infant rats. *J Exp Psychol Anim Behav Process* 2: 48-56

Cripps MM, and Nash DJ (1983) Ontogeny and adult behavior of mice with congenital neural tube defects. *Behavioral & Neural Biology* 38: 127-132

Crozier WJ, and Pincus G (1937) Photic stimulation of young rats. *J Gen Psychol* 17: 105-111

Davis M, and Gendelman PM (1977) Plasticity of the acoustic startle reflex in acutely decerebate rats. *J. Comp. Physiol. Psychol.* 91: 549-563

de Roquefeuil G, Djakovic M, and Montagner H (1993) New Data on the Ontogeny of the Child's Sleep-Wake Rhythm. *Chronobiology International* 10: 43-53

Department of Health and Human Services (FDA) (1998) 21 CFR Parts 201, 312, 314 and 601. *Federal Register* 63: 66632-66672

Desor JA, Maller O, and Andrews K (1975) Ingestive responses of human newborns to salty, sour and bitter stimuli. *J. Comp. Physiol. Psychol.* 89: 966-970

Dolhinow P, and Murphy G (1982) Langur monkey (Presbytis entellus) development: the first 3 months of life. *Folia Primatol* 39: 305-331

Eisenberg R (1970) Development of hearing in man: an assessment of current status. *J. Amer. Speech Hear. Assoc.* 12: 199-223

Elias MF (1977) Relative maturity of cebus and squirrel monkeys at birth and during infancy. *Developmental Psychobiology* 10: 519-528

Elliot O, and Scott JP (1961) *Journal of Genetic Psychology* 99:

Esteban A (1999) A neurophysiological approach to brainstem reflexes. Blink reflex. [Review] [162 refs]. *Neurophysiologie Clinique* 29: 7-38

FDA. FDA, Federal Register Part II, Vol 63, No. 231. Wednesday, December 2, 1998; Rules and Regulations. 12-2-1998.

Fivush R (1991) The social construction of personal narratives. *Merrill-Palmer Quarterly* 37: 59-81

Fivush R, and Hammond NR (1989) Time and Again: Effects of repetition and retention interval on 2-year-olds' event recall. *Journal of Experimental Child Psychology* 47: 259-273

Foley Jr. JP (1934) First year development of a rhesus monkey, reared in isolation. *Journal of Genetic Psychology* 45: 39-105

Forssberg H, Eliasson AC, Kinoshita H, Westling G, and Johansson RS (1995) Development of human precision grip. IV. Tactile adaptation of isometric finger forces to the frictional condition. *Experimental Brain Research* 104: 323-330

Fox MW (1964a) The ontogeny of behaviour and neurologic responses in the dog. In: Worden AN (ed) *Animal Behaviour*. Bailliere, Tindall and Cox, Ltd., London, pp 301-310

Fox MW (1964b) The Postnatal Growth of the Canine Brain and Correlated Anatomical and Behavioral Changes during Neuro-ontogenesis. *Growth* 28: 135-141

Fox MW (1966) Neuro-behavioral ontogeny. A Synthesis of ethological and neurophysiological concepts. *Brain Res.* 2: 3-20

Fox MW (1971) *Integrative Development of Brain and Behavior in the Dog*. The University of Chicago Press, Chicago,

Freedman DG, King JA, and Elliot O (1961) *Science* 133:

Fuller JL, DuBuis EM (1962) The behaviour of dogs. In: Hafez ESE (ed) *The Behavior of Domesticated Animals*. Bailliere, Tindall and Cox, London, pp 415-452

Geyer MA, and Reiter LW (1985) Strategies for the selection of test methods. *Neurobehav. Toxicol. Teratol.* 7: 661-662

Girden E, Mettler FA, Finch G, and Culler E (1936) Conditioned responses in a decorticate dog to acoustic, thermal, and tactile stimulation. *Journal of Comparative Psychology* 21: 367-385

Golub MS (1990) Use of monkey neonatal neurobehavioral test batteries in safety testing protocols. *Neurotoxicology & Teratology* 12: 537-541

Golub MS, Eisele JH, and Donald JM (1991a) Effect of intrapartum meperidine on the behavioral consequences of neonatal oxygen deprivation in Rhesus monkey infants. *Developmental Pharmacology & Therapeutics* 4: 231-240

Golub MS, Eisele LM, and Donald JM (1991b) Effect of intrapartum meperidine on the behavioral consecquences of neonatal oxygen deprivation in rhesus monkey infants. *Developmental Pharmacology & Therapeutics* 16: 231-240

Gottlieb G (1971) Ontogenesis of sensory functionin birds and mammals. In: Tobach E, Aronson LR, Shaw E (eds) *The Biopsychology of Development*. Academic Press, New York, pp 67-129

Harlow HF (1959) The Development of Learning in the Rhesus Monkey. *American Scientist Winter:* 459-479

Harlow HF (1962) Development of affection in primates. In: Bliss EL (ed) *Roots of Behavior*. Harper, New York,

Head E, Mehta R, Hartley J, Kameka M, Cummings BJ, Cotman CW, Ruehl WW, and Milgram NW (1995) Spatial learning and memory as a function of age in the dog. *Behavioral Neuroscience* 109: 851-858

Hellbrügge T (1960) The Development of circadian rhythms in infants. In: Anonymous *Biological Clocks*. Cold Spring Harbor Symp Quant Biol. pp 311-323

Herschkowitz N, Kagan J, and Zilles K (1997) Neurobiological bases of behavioral development in the first year. *Neuropediatrics* 28: 296-306

Hill DL, and Mistretta CM (1990) Developmental neurobiology of salt taste sensation. *TINS* 13: 188-195

Hill WL, Borovsky D, and Rovee-Collier C (1988) Continuities in infant memory development. *Developmental Psychobiology* 21: 43-62

Hines M (1942) The development and regression of reflexes, postures and progression in the young macaque. *Contributions to Embryology* 196: 155-209

Hofer MA, Masmela JR, Brunelli SA, and Shair HN (1998) The ontogeny of maternal potentiation of the infant rats' isolation call. *Developmental Psychobiology* 33: 189-201

Homberg V, Bickmann U, and Muller K (1993) Ontogeny is different for explicit and implicit memory in humans. *Neurosci Lett* 150: 187-190

Hudson R, and Distel H (1983) Nipple location by newborn rabbits: behavioral evidence for pheromonal guidance. *Behavior* 85: 260-275

Hurtt ME, and Sandler JD (2003) Comparative Organ System Development: Introduction. *Birth Defects Research* (Part B) 68: 85-85

Illingworth RS (1983) Normal Development. In: *Anonymous Development of the Infant and Young Child*. Churchhill Livingstone, New York, pp 120-162

James WT (1955) Behaviors involved in expression of dominance among puppies. *Psychological Reports* 1: 299-301

James WT (1961) Preliminary observations on play behaviour in puppies. *Journal of Genetic Psychology* 98: 273-277

Johanson IB, and Hall WG (1979) Appetitive learning in 1-day-old rat pups. *Science* 205: 419-421

Kajiura H, Cowhart BJ, and Beauchamp GK (1992) Early developmental changes in bitter taste responses in human infants. *Developmental Psychobiology* 25: 375-386

Kaye R, Oski F, Barness L, (editors) (1988) *Core Textbook of Pediatrics*. J.B. Lippincott Company, Philadelphia,

King JE, Fobes JT, and Fobes JL (1974) Development of early behaviors in neonatal squirrel monkeys and cotton-top tamarins. *Developmental Psychobiology* 7: 97-109

Kleitman N, and Engelmann TG (1953) Sleep characteristics in infants. *J. Appl. Physiol.* 6: 269-282

Konczak J, and Dichgans J (1997) The development toward stereotypic arm kinematics during reaching in the first 3 years of life. *Experimental Brain Research* 117: 346-354

Kuhtz-Buschbeck JP, Stolze H, Johnk J, Boczek-Funcke A, and Illert M (1998) Development of prehension movements in children: a kinematic study. *Experimental Brain Research* 122: 424-432

Lantz C, Melen K, and Forssberg H (1996) Early infant grasping involves radial fingers. *Developmental-medicine-and-child-neurology* 38: 668-674

Lashley KS, and Watson JB (1913) Notes on the development of a young monkey. *Journal of Animal Behavior* 3: 114-139

LeDoux JE, Sakaguchi A, and Reis DJ (1984) Subcortical efferent projections of the medial geniculate nucleus mediate emotional responses conditioned to acoustic stimuli. *J. Neurosc. Meth.* 4: 383-398

Lemon RN (1999) Neural control of dexterity: what has been acheived? *Exp. Brain Res.* 128: 6-12

Little AH, Lipsitt LP, and Rovee-Collier C (1984) Classical conditioning and retention of the infant's eyelid response: effects of age and interstimulus interval. *Journal of Experimental Child Psychology* 37: 512-524

Margalit T, and Lancet D (1993) Expression of olfactory receptor and transduction genes during rat development. Brain Res. *Developmental Brain Research*. 73: 7-16

Mendelson MJ (21982) Clinical examination of visual and social responses in infant rhesus monkeys. *Dev. Psychol.* 18: 658-664

Milgram NW, Head E, Weiner E, and Thomas E (1994) Cognitive functions and aging in the dog: acquisition of nonspatial visual tasks. *Behavioral Neuroscience* 108: 57-68

Moore T, and Ucko C (1957) Night Waking in Early Infancy. *Archives of disease in childhood* 33: 333-342

Morley ME (1972) *The Development and Disorders of Speech in Childhood*. Churchill Livingstone, Edinburgh,

Mowbray JB, and Cadell TE (1962) Early Behavior patterns in rhesus monkeys. *J. Comp. Physiol. Psychol.* 55: 350-357

Murray T, Murray I (1980) *Infant Communication, Cry and Early Speech*. College Hill Press, Houston,

Myslivecek J, and Hassmannova J (1991) Step-down passive avoidance in the rat ontogeny. *Acta Neurobiol Exp* (Warsz) 51: 89-96

Nelson WE (2000) The Newborn. In: Nelson WE (ed) *Nelson Textbook of Pediatrics*. W.B. Saunders Co, Philadelphia, pp 36-38

Neuringer M, Coner WE, Petten CV, and Barstad L (1984) Dietary omega-3 fatty acid deficiency and visual loss ininfant rhesus monkeys. *J. Clin. Invest.* 73: 272-276

Olivier E, Edgley SA, Armand J, and Lemon RN (1997) An electrophysiological study of the postnatal development of the corticospinal system in the macaque monkey. *J. Neurosc.* 17: 267-276

Panksepp J (1981) The ontogeny of play in rats. *Developmental Psychobiology* 14: 327-332

Parisi T, and Ison J (1979) Development of the acoustic startle response in the rat: ontogenetic changes in magnitude of inhibition by pre-pulse stimulation. *Developmental Psychobiology* 12: 219-230

Parisi T, and Ison JR (1981) Ontogeny of control over the acoustic startle reflex by visual prestimulation in the rat. *Developmental Psychobiology* 14: 311-316

Paule MG (2001) Using Identical Behavioral Tasks in Children, Monkeys and Rats to Study the Effects in Drugs. *Curr. Therap. Res.* 62: 820-833

Pawlowski AA, and Scott JP (1956) Hereditary differences in the development of dominance in litters of puppies. *J. Comp. Physiol. Psychol.* 49: 353-358

Pedersen PE, Stewart WB, Green AG, and Shepherd GM (1983) Evidence of olfactory function in *utero*. *Science* 221: 478-480

Pellis SM, Pellis VC, Morrissey TK, and Teitelbaum P (1989) Visual modification of vestibularly triggered air-righting in the rat. *Behav. Brain Res.* 35: 23-26

Petrosini, L. M., Molinari, M., and Gremoli, T. Hemicerebellectomy and motor behaviour in rats. I. Development of motor function after neonatal lesion. *Experimental Brain Research* 82(3), 472-482. 1990.

Prather MD, Lavenex P, Mauldin-Jourdain M.L., Mason W.A., Capitanio JP, Mendoza SP, and Amaral DG (2001) Increased social fear and decreased fear of objects in monkeys with neonatal amygdala lesions. *Neuroscience* 106: 653-658

Rao BS, Shankar N, and Sharma KN (1997) Ontogebny of taste sense. *Indian Journal of Physiology and Pharmacology* 41: 193-203

Richardson R, Paxinos G, and Lee J (2000) The ontogeny of conditioned odor potentiation of startle. *Behavioral Neuroscience* 114: 1167-1173

Rose GH (1968) The development of visually evoked electro-cortical responses in the rat. *Developmental Psychobiology* 1: 35-40

Rosenblith JF, Sims-Knight JE (1904) *In the Beginning*. Brooks Cole, Monterey, CA,

Routtenberg A, Strop M, and Jordan J (1979) Response of the infant rat to light prior to eyelid opening: mediation by the superior colliculus. *Developmental Psychobiology* 11: 469-478

Rovee-Collier C, Gerhardstein P (1997) The development of infant memory. In: Cowan N (ed) *The Development of Memory in Childhood*. Psychology Press, pp 5-39

Rovee-Collier C, Morrongiello BA, Aron M, and Kupersmidt J (1978) Topographical response differentiation in three-month-old infants. *Infant Behavior and Devlopment* 1: 323-333

Rovee C, and Rovee DT (1969) Conjugate reinforcement of infant exploratory behavior. *Journal of Experimental Child Psychology* 8: 33-39

Rudy JW, and Cheatle MD (1977) Odor-aversion learning in neonatal rats. *Science* 198: 845-846

Rudy JW, Stadler-Morris S, and Albert P (1987) Ontogeny of spatial navigation behaviors in the rat: dissociation of "proximal"- and "distal"-cue-based behaviors. *Behavioral Neuroscience* 101: 62-73

Schapiro S, Salas M, and Vukovich K (1970) Hormonal effects on ontogeny of swimming ability in the rat: assessment of central nervous system development. *Science* 168: 147-150

Schenk F (1985) Development of place navigation in rats from weaning to puberty. *Behavioral Neural Biology* 43: 69-85

Schultz AH (1969) *The Life of Primates*. Weidenfeld and Nicolson, London,

Schusterman RJ, Sjoberg A (1969) Early behavior patterns of squirrel monkeys (saimiri sciureus). In: Carpenter CR (ed) *Behavior*. Karger, Basel, pp 149-203

Scott JP (1962) Critical Periods in Behavioral Development. *Science* 138: 949-958

Scott JP, and Marston MV (1948) The development of dominance in litters of puppies. *Anatom. Rec.* 101: 696-696

Sheppard JJ, and Mysak ED (1984) Ontogeny of infantile oral reflexes and emerging chewing. *Child Development* 55: 831-843

Sisask G, Bjurholm A, Ahmed M, and Kreicbergs A (1995) Ontogeny of sensory nerves in the developing skeleton. *Anatom. Rec.* 243: 234-240

Smart JL, and Dobbing J (1971) Vulnerability of developing brain. VI. Relative effects of foetal and early postnatal undernutrition on reflex ontogeny and development of behaviour in the rat. *Brain Res.* 33: 303-314

Spear LP (1990) Neurobehavioral assessment during the early postnatal period. *Neurotoxicology & Teratology* 12: 489-495

Spear NE, and Smith GJ (1978) Alleviation of forgetting in preweanling rats. *Dev Psychobiol* 11: 513-529

Stehouwer DJ, and Campbell BA (1980) Ontogeny of passive avoidance: role of task demands and development of species-typical behaviors. *Dev Psychobiol* 13: 385-398

Stengel TJ (1980) The neonatal bhavioral assessment scale: description, clinical uses and research indications. *Physical and Occupational Therapy in Pediatrics* 1: 39-57

Stewart P, Reihman J, Lonky E, Darvill T, and Pagano J (2000) Prenatal PCB exposure and neonatal behavioral asessment scale (NBAS) performance. *Neurotoxicol. Teratol.* 22: 21-29

Stickord G, Kimble DP, and Smotherman WP (1982) *In utero* taste/odor aversion conditioning in the rat. *Physiol Behav* 28: 5-7

Suchoff IB (1979) Visual development. *J. Amer. Optom. Assoc.* 50: 1129-1135

Sullivan MH, Rovee-Collier C, and Tynes DM (1979) A conditioning analysis of infant long-term memory. *Child Development* 50: 162

Teicher MH, and Blass EM (1977) First suckling response of the newborn albino rat: the roles of olfaction and amniotic fluid. *Science* 196: 635-636

Thomas A (1940) *Equilibre et equilibration*. Masson, Paris,

Unis AS, Petracca F, and Diaz J (1991) Somatic and behavioral ontogeny in three rat strains: preliminary observations of dopamine-mediated behaviors and brain D-1 receptors. *Progress in Neuro-Psychopharmacology & Biological Psychiatry* 15: 129-138

Uziel ARR, and Marot M (1981) Development of cochlear potentials in rats. *Audiology* 20: 89-100

Vander Linde E, Morrongiello BA, and Rovee-Collier C (1985) Determinants of retention in 8-week-old infants. *Developmental Psychobiology* 21: 601-613

Vogt MB, and Rudy JW (1984) Ontogenesis of learning: IV. Dissociation of memory and perceptual-altering processes mediating taste neophobia in the rat. *Developmental Psychobiology* 17: 601-611

Volpe J (1995) The neurological examination normal and abnormal features. In: *Anonymous Neurology of the Newborn*. W.B. Saunders Co., Philadelphia, pp 101-101

von Hofsten C, and Fazel-Zandy S (1984) Development of visually guided hand orientation in reaching. *J Exp Child Psychol* 38: 208-219

Vorhees CV (1986) Methods for assessing the adverse effects of foods and other chemicals on animal behavior. *Nutrition Reviews* 44 Suppl: 185-193

Vorhees CV, Acuff-Smith KD, Moran MS, and Minck DR (1994) A new method for evaluating air-righting reflex ontogeny in rats using prenatal exposure to phenytoin to demonstrate delayed development. *Neurotoxicology & Teratology* 16: 563-573

Westerga J, and Gramsbergen A (1990) The development of locomotion in the rat. *Dev. Brain Res.* 57: 163-174

Williams R, and Boothe R (1981) Development of optical quality in the infant monkey (*Macaca nemestrina*) eye. *Invest. Opthalmol. Visual Sci.* 21: 728-736

Winneke G (1995) Endpoints of developmental neurotoxicity in enviromentally exposed children. *Toxicol. Lett.* 77: 127-136

Winter P, Handley P, Ploog D, and Schott D (1973) Ontogeny of squirrel monkey calls under normal conditions and under acoustic isolation. *Behaviour* 47: 131-147

Yamaguchi K, Harada S, Kanemaru N, and Kasahara Y (2001) Age-related alteration of taste bud distribution in the common marmoset. *Chemical Senses* 26: 1-6

Zametkin AJ, Stevens JR, and Pittman R (1979) Ontogeny of spontaneous blinking and of habituation of the blink reflex. *Ann. Neurol.* 5: 453-457

Zimmerman RR, Torrey CC (1965) Ontogeny of Learning. In: Schrier AM, Harlow HF, Stollnitz F (eds) *Behavior of Nonhuman Primates: Modern Research Trends*. Academic Press, New York and London, pp 405-447

Zola-Morgan S, and Squire LR (1993) Neuroanatomy of memory. *Annual Review of Neuroscience* 16: 547-563

Index*

A

Abdominal viscera, 239–240
Abnormal development, pathogenesis
  alterations, cell-signaling/cell-cell interaction, 71–73
  associations, 63
  basics, 61–62, 80–81
  cellular level, *67,* 67–75
  cleft palate, 76–77
  death of cells, 68–70, *69*
  deformations, 62–63
  developmental toxicity manifestations, 62–63
  development stage, 63–64
  differentiation of cells, 74–75, *75*
  different morphology, 78–80
  disruptions, 62–63
  dose and magnitude of exposure, 65–67, *66*
  examples, 75–80, *76*
  influential factors, 63–67
  malformations, 62–63
  migration of cells, 74–75, *75*
  multiple defects, 63
  neural tube defects, 77–78
  proliferation of cells, 70–71, *71*
  sequences, 63
  similar morphology, 76–78
  structural anomalies, 62–63
  syndromes, 63
  tissue specificity, 64–65
Absorption principles, 572–577
Accessory sex glands, 448–449, 1037–1040
Action mode study and data, 283–284, 865–866
Adapted adult toxicity study, 281–283, *283*
Additives statutory safety standard, 879–890
Administration duration, 226–228, *230–234*
Administration routes, 286–290, *287–289*
Advanced/late organogenesis, 7
Adverse events, 267, *268*
Adverse reproductive outcomes, 802–805
Agency intervention, 267, *268,* 269–271
Agricultural chemicals, 347–348
Air righting reflex, 475–476
Alkylating agents, 38–39

Allocation, mated females, 220–221
Alterations, cell-signaling/cell-cell interaction, 71–73
Altered energy sources, 32–33
Alternative categorization, stressors, 97
Amphibian embryo culture, 670–671
Anatomical defects, terminology, 911–965
Androgen receptors, 490, *492,* 492–496, *494–495*
Anesthesia, 236
Angiotensin converting enzyme fetopathy, 405–406, *406*
Animals, *see also specific animals;* Transgenic animals
  age, 211–213
  cellular, biochemical, and molecular techniques, 613–615
  dosage group, 745
  husbandry, 213–219
  ICH guidelines, 761
  interpretation of study findings, 362, 382
  intrauterine conceptual issues, 542
  maternally mediated effects, 110–111
  osteon formation and remodeling, 993–995
  paternally mediated effects, 127–128
Animal-to-human concordance, 337–342, *339–341*
Anogenital distance and disease, 472, 1036
ANOVA method, litter-based, 699–700
Antitubulin agents, 30
Aromatase assay, *497,* 500–502
Aromatic amines, 39
Array fabrication, cDNA arrays, 624, *625,* 626
Assay methods
  epoxide hydrase, 536
  flavin-containing monooxygenase, 533
  glutathione transferase, 536–537
  lipoxygenase, 533–534
  peroxidases, 535–536
  prostaglandin synthase, 534–535
  sulfotransferase, 537–538
  UDP-Glucuronosyl transferase, 537
  xenobiotic metabolism, role, 532–538
Associations, abnormal development pathogenesis, 63
Auditing studies, 719–726
Auditory startle reflex/response, 312–313, 475
Avian embryo culture, 671–672

---

* Numbers in italics represent figures and tables.

B

Baboons, 1017, *see also* Primates
Balanopreputial separation, 472–473, 753
Behavioral assessment and testing, 301, 306–316, 318, 474–476
Behavioral measures, 108–109
Benchmark dose, 707–709
Benefit analysis, risk assessment, 867
Best Practices, 717, 720–726, *721–726*
Between-litter design, *292,* 293, *294*
Binary response data, 701–705
Biochemical measures, 104–108
Biochemical techniques, *see* Cellular, biochemical, and molecular techniques
Bioinformatics, 629–631
Biological membranes, movement across, 572–573
Biologic dynamics and dimensions, 412–413
Biology, 444–445, 458
Biotransformation, 572–577
Birth weight, 471, 808–810
Blood dosimetry, 395–397, *396*
Blood plasma, 574
Blood-testis barrier, 1035–1036
Body temperature, 479
Body weight, 229, 445, 459, 471–472
Bone marrow meatopoiesis, 1087–1088
Bone structure and composition, 970–974, *971*
Book listing, 967–968

C

Caenorhabditis elegans cultures, 670
Calvarial veins, 999
Cardiac neurotransmitters, 1068
Case controls, human studies, 820–828
Case studies, interpretation of findings, 404–408
Catecholamines, 106
Cauda epididymal sperm mobility, 454–456
cDNA arrays, *see* Complementary DNA arrays
Cell-based gene transcription assays, 493–496
Cell-cell communication, *26,* 27–28
Cell culture, 678–683, *680–683*
Cell differentiation, 74–75
Cells, bone, 974
Cell-signaling pathways, *19, 23, 26,* 26–27
Cellular, biochemical, and molecular techniques
    basics, 600
    descriptive approaches, 601–606
    DNA array, 615
    dot blots, 608
    electrophoretic cells, 611
    genes, role, 613–616
    genetic library screening, 615–616
    immunohistochemistry, 602–604, *603*
    mRNAs, 601–613
    Northern blots, 607–608
    nuclear run-on assay, 612–613
    PCR quantitative method, 609–611
    proteins, 601–606
    real time RT-PCR quantitative method, 609–610, *610*
    response pathways, role, 613–616
    reverse Southern blots, 616
    RNase protection assay, 611–612
    RNA transcription *in vitro,* 612–613
    RT-PCR *in situ* transcription, 616
    RT-PCR quantitative method, 609–611
    semiquantitative methods, 607–608
    in situ hybridization, 604–606, *606*
    slot blots, 608
    solution hybridization, 611–612
    study approach, 600–601
    transgenic animals, 613–615
    in vitro antisense RNA, 615
Cellular level abnormal development pathogenesis, *67,* 67–75
Central nervous system, *see also* Nervous system
    basics, 1103–1104, 1125
    cognitive development, 1112, *1113,* 1114–1118, *1119*
    communication, 1118–1120, *1120*
    fear response, 1122–1123, *1123*
    fine motor development and dexterity, 1106–1108, *1108*
    interpretation of study findings, 401
    locomotor development, 1104–1106, *1106*
    reflex and sensory development, 1108–1112
    research needs, 1125
    sleep cycles, 1123–1124, *1125*
    social play, 1120–1122, *1122*
21 CFR Part 11, 715–716
Chemicals
    developmental toxicity testing, 215
    safety and handling, 209
    teratogenesis potentiation, 102–103
Child death, 806–808
Childhood cancer, 815
Children, 1114, *see also* Humans
Chlorpyrifos, 34
Circulatory system, *164–167*
Cleft palate, 76–77
Cliff avoidance, 474
Clinical observations, 228–229, *235*
Clonal analysis, 678–683
Cluster analysis, 836–837
Clustering, data analysis and bioinformatics, 630–631
Cocaine, 33
Cognitive development, 1112, *1113,* 1114–1118, *1119*
Cohabitation, 438–439
Cohort studies, 828–831
Combined repeated dose toxicity study, 433–434, *434*
Communication, 1118–1120, *1120*
Comparison material, human studies, 829
Competitive binding assays, 490, *492,* 492–493, *494–495*
Complementary DNA arrays, 623–628
Compliance with regulations, 254
Computer-assisted sperm analysis, 450–454, *451*
Conceptus exposure, 9
Confounding, 805, 807, 810, 813–814, 832–834
Congenital malformations, 810–813
Context and interrelatedness, 371–372, *373–375*
Continuous breeding, 431–432, *433*

Continuous response data, 699–701
Controls, maternally mediated effects, 111
Converging designs, 361
Cooxidation, 553–554
Corpora lutea counts, 466–467
Correlated binary response data, 703–705
Count data analysis, 705–706
Couple-mediated endpoints, 436
Craniofacial examination, 241–242, *242*
Critical events evaluation, *18–19,* 18–20
Crown-rump length, 471
Current trends, quality concerns, 716
Cytochrome P450
    basics, 528–529
    decidua, 540–541
    embryo, 543–544
    estimation, 532
    fetal membranes, 549
    fetus, 546–547
    maternal liver, 538–539
    placenta, 549–550
    uterus, 540–541
    xenobiotic oxidation, 532–533
Cytosine arabinoside, 36–37

D

Dams, 443, 470, *see also* Maternal issues
DART study guidelines, 428–436
Data, 252–253, 629–631, 633, 767
Databases, 410–411, *411,* 641
Death of cells, 68–70, *69*
Death of dams/pups, 470–471
Decidua
    cytochrome P450, 540–541
    flavin-containing monooxygenase, 542
    lipid peroxidation, 541
    lipoxygenase, 541–542
    peroxidases, 541
    prostaglandin synthase, 541–542
    xenobiotic metabolism, role, 540–542
Defects, terminology, 911–965
Deformations, 62–63, *see also* Malformations
Descriptive approaches, 601–606
Developing *vs.* developed systems, 272–273
Development
    abnormal development pathogenesis, 63–64
    defects, terminology, 911–965
    interpretation of study findings, 361–382
    landmarks, reproductive toxicity testing, 472–473
    maternally mediated effects, 100–102
    neurotoxicity studies, 432–433
    nonclinical juvenile toxicity testing, 295–297, *296–298*
    reproductive toxicity testing, 430, *430,* 463
    toxicants metabolism, 538–556
    toxicity manifestations, 62–63
Developmental biology, 635–636
Developmental toxicity testing
    age of animals, 211–213
    allocation, mated females, 220–221
    animal husbandry, 213–219
    basics, 202, *203–208,* 204–207
    body weight, 229
    chemicals, 215
    chemical safety and handling, 209
    clinical observations, 228–229, *235*
    compliance with regulations, 254
    data collection, 252–253
    discomfort limitation, 215–216
    dose formulation and analyses, 209–210
    dose selection, 220
    duration of administration, 226–228, *230–234*
    environmental conditions, 214–215
    experimental design, 219–253
    feed consumption, 229, 232
    fetal necropsy and postmortem examination, 236–252
    final reports, 254–255
    food, *213,* 213–214
    housing, *213,* 213–214
    identification, 215
    justification for live animals, 211
    light cycles, 214
    maintenance of records, 255–256
    materials and methods, 208–219
    maternal body weight, 229
    maternal feed consumption, 229, 232
    maternal necropsy and postmortem examination, 234–236
    mating, 216–219
    necropsy, 234–252
    noise, 215
    observation, mated females, 228–229, 232
    personnel, 255
    postmortem examination, 234–252
    rabbit mating, 217–219
    record storage, 253, *253*
    regulatory compliance, 254
    relative humidity, *213,* 214
    reports, 254–255
    rodent mating, 216–217
    routes of administration, 221–226, *225*
    species of animals, 210, *211*
    statistics, 252
    status reports, 254
    study design, 219–220
    supplier of animals, 210, *211*
    temperature, *213,* 214
    test animals, 210–213
    test substance, 208–210
    total number, animals, 211–213
    treatment, mated females, 221–228
    ventilation, 215
    water, *213,* 213–214
    weight of animals, 211–213
Developmental toxicology
    advanced/late organogenesis, 7
    basics, 3–5
    conceptus exposure, 9
    conducting tests, 11–12
    early differentiation/organogenesis, 7

interpretation of results, 11–12
placental transfer, 8–9
predifferentiation, 7
Wilson's principles, 6–11
Diaphyseal arteries, 996
Diaphysis, 992–995
Differentiation of cells, 74–75, *75*
Digestive system, *168–170*
Dioxin, 41, 43
Direct effects comparison, 99–100
Discomfort limitation, 215–216
Disruptions, abnormal development pathogenesis, 62–63
Distribution, toxicokinetics, 572–577
DNA array, 615
Dogs
    acid-base equilibrium, 1014
    blood-testis barrier, 1035–1036
    bone marrow meatopoiesis, 1088
    cellular constituents, heart, 1064
    cognitive development, 1116–1117
    communication, 1119
    concentrating ability, 1013
    diaphysis, 992
    ductus venosus, 1061
    epididymides, 1035
    epiphyseal growth plate, 990
    fear response, 1122–1123
    fetal liver hematopoiesis, 1084
    fine motor development and dexterity, 1107
    functional immunocompetence, 1094
    glomerular filtration, 1010–1011
    heart development. postnatal, 1059–1063, *1072–1073*
    hematopoietic stem cells, 1083
    innervation, heart, 1066
    locomotor development, 1105
    lung development, 1025
    nerve functional development, 1068
    ossification, 981–983, *982*
    output and hemodynamics, heart, 1071
    postnatal immune system development, 1090
    prostate, 1038–1039
    reflex and sensory development, 1111
    renal development, 1008–1009
    seminal vesicles, 1040
    shape, position, size of heart, 1062, *1062*, 1063, *1064*, *1072–1073*
    sleep cycles, 1124
    social play, 1121
    spleen development, 1085
    testes, 1034
    testicular descent, 1036
    thymus, 1087
    vasculature, heart, 1065
Dominant lethal study, *435*, 435–436
Dose-response model, 707–708
Dosing
    abnormal development pathogenesis, 65–67, *66*
    formulation and analyses, 209–210
    group number, 745
    guideline perspectives, 745–746
    range-characterization studies, 354, 358

response curve characterization, 350–352
risk assessment, 856–859, 892
selection, 220, 352–353
selection, nonclinical juvenile toxicity testing, 290–292
Dot blots, 608
Double staining, ossified bone, 243–244
Drosophila cultures, 670
Drugs, *see also* Pharmaceuticals
    exposure, 817–818
    paternally mediated effects, 128
    safety, 893–897
    usage, 275–276, *275–276*
Ductus arteriosus, 1060–1061
Ductus venosus, 1060–1061
Dural vessels, 999
Dystocia, 404–405, *405*

E

Early differentiation/organogenesis, 7
EDSP, *see* Endocrine disruptor screening program (EDSP)
Electrocardiographic measurements, 1067–1069
Electronic records and signatures, 715
Electrophoretic cells, 611, *see also* Gel electrophoresis
Electrospray ionization tandem mass spectrometry (ESI-MS/MS), 639
Elimination, toxicokinetics, 572–577
Embryo
    cytochrome P450, 543–544
    exposure, human studies, 815
    guideline perspectives, 741–742
    lipoxygenase, 545
    peroxidases, 545–546
    prostaglandin synthase, 544–545
    xenobiotic metabolism, role, 543–546
Embryological processes
    basics, 154–156, *157*, 158–159
    circulatory system, *164–167*
    digestive system, *168–170*
    embryological processes, *162–168*
    endocrine system, *171–173*
    excretory system, *185–186*
    gestational milestones, *164–167*
    integumentary system, *181–184*
    milestones, *162–188*
    muscular system, *181–184*
    nervous system, *175–177*
    reproductive system, *187–188*
    respiratory system, *174*
    sense organs, *178–180*
    skeletal system, *181–184*
Endocrine disruptor screening program (EDSP)
    androgen receptors, 490, *492*, 492–496, *494–495*
    aromatase assay, *497*, 500–502
    basics, 489–490, *491*, 515
    estrogen receptors, 490, *492*, 492–496, *494–495*
    Hershberger assay, 505–507
    pubertal protocols, 507–510, *508*
    steroidogenesis, 496–500, *497–500*

TIS battery, 514–515
in utero-lactational exposure, 510, *511–513,* 513
uterotrophic assay, 502–505
in vitro assays, 490, 492–502
in vivo assays, 502–514
Endocrine endpoints, 395
Endocrine-mediated findings, 753–754
Endocrine system, *171–173*
Endpoints
    interpretation of study findings, 364–382, *365–366,* 385–401, *386,* 395–398
    reproductive toxicity testing, 436–437
    risk assessment, 892–893
Energy metabolism, 1066–1067
Environmental conditions and exposures, 214–215, 818
Enzyme inhibition, 33–35
Enzymes assay, 531–538
EPA, *see* U.S. EPA
EPA EDSP, *see* Endocrine Disruptor Screening Program (EDSP)
Epidemiological studies, *126,* 126–127
Epididymides, 448–449, 1034–1035
Epigenetic changes, 138
Epiphyseal arteries, 997–998
Epiphyseal growth plate, 988–991, *989–990*
Epoxide hydrase
    assay methods, 536
    fetus, 548
    maternal liver, 540
    placenta, 555
ESI-MS/MS, *see* Electrospray ionization tandem mass spectrometry (ESI-MS/MS)
Estimation, cytochrome P450, 532
Estrogen receptors, 490, *492,* 492–496, *494–495*
Estrous cycle
    interpretation of study findings, 391–392
    reproductive toxicity testing, 473
    staging, 465–466, *466*
Ethanol, 44–46
Euthanasia, 236, *see also* Necropsy
Excretory system, *185–186*
Experimental considerations
    developmental toxicity testing, 219–253
    functional genomics and proteomics, 631–633
    interpretation of study findings, 372–377
    maternally mediated effects, 110–114
Expert judgment role, 409–414, *410*
Exposures
    environmental, 818
    intervals, guideline perspectives, 746–747, *746–747*
    magnitude, 65–67
    nonclinical juvenile toxicity testing, 286–292
    recording, human studies, 815–819
    risk assessment, 859–860, 892
Extant adult nonclinical animal toxicity data, 280
Extant data, 277–280
External alterations, 470
External examinations, 237–238, *238*
External genitalia, 461–462
Eyes, 474, 678

F

Fear response, 1122–1123, *1123*
Fecundity index, 441
Feed consumption, 229, 232, *see also* Food
Female mating index, 440
Female reproductive system
    development landmarks, *1048,* 1048–1054, *1051–1054*
    toxicants, 558–559
    xenobiotic metabolism, role, 557–559
Females, 436–437, 507–510, *508, see also* Maternal issues
Fertility, guideline perspectives, 740–741
Fertility indices, 392–394, *393,* 439–442
Fertility studies, 429, *429*
Fertilization rate, 440
Fetus
    body weight data example, 700–701, *700–701*
    cytochrome P450, 546–547
    development, guideline perspectives, 741–742
    epoxide hydrase, 548
    evaluations, 752
    flavin-containing monooxygenase, 547
    glutathione transferase, 548
    lipoxygenase, 547–548
    liver hematopoiesis, 1084
    membranes, 549
    morphology, 371–382
    necropsy and postmortem examination, 236–252
    peroxidases, 547
    prostaglandin synthase, 547
    sex ratio, 370–371
    sulfotransferase, 548–549
    UDP-Glucuronosyl transferase, 548
    xenobiotic metabolism, role, 546–549
Final reports, 254–255
Fine motor development and dexterity, 1106–1108, *1108*
Fish embryo culture, 671
Flat bones, 973, 998–999
Flavin-containing monooxygenase
    assay methods, 533
    basics, 529
    decidua, 542
    fetus, 547
    maternal liver, 540
    placenta, 550
    uterus, 542
5-Fluorouracil, 29–30, 35
Food, *213,* 213–214, 299, *300,* 445, 878–890, *see also* Feed consumption
Food and Drug Administration, 754
Foot splay, 478–479
Foramen ovale, 1060
Forelimb grip test, 476
Forelimb placing, 475
Fostering design, 294–295, *295*
Framework, risk assessment, 843–861, *844*
Frequency of exposure, 290
Functional alterations, 666
Functional evaluations, guideline perspectives, 753–754
Functional genomics and proteomics
    array fabrication, 624, *625,* 626

basics, 622
bioinformatics, 629–631
clustering, 630–631
complementary DNA arrays, 623–628
data analysis, 629–631
databases, 641
data validation, 633
developmental biology, 635–636
experimental considerations, 631–633
gene expression, 622–636, *634*
hybridization, 628
image acquisition, 628
isotope-coded affinity tag, 639–640
limitations, 635
limited RNA availability, 631–632
mass spectrometry, 638–639
microarrays, 623–633, 641
multidimensional liquid chromatography, 638
normalization, 629–630
oligonucleotide microarrays, 628–629
predictive values, 640–641
probe preparation, 625–628, *627*
protein arrays, 639–640
protein expression analysis, 636–640
protein identification, 638–640
protein isolation, 636–638
proteomics, 636–641
replicates, 632–633
SAGE technology, 635–636, 641
serial analysis, gene expression, 633–636
two-dimensional gel electrophoresis, 636–638, *637*
washing, 628
Websites, 641
Functional immunocompetence, 1091–1094
Functional observation battery, 316–318, *317*, 477–480
Functional toxicities, 401
Future research and technologies, 559–560, 866–867

G

Gallbladder variations, rabbits, 380, *381*
Gel electrophoresis, 636–638, *see also* Electrophoretic cells
Gender determination, 470–471, *see also* Sex ratio
Gene expression, 40–42, 622–636, *634*, 666
Genes, role, 613–616
Genetic library screening, 615–616
Gene transcription assays, cell-based, 493–496
Genitalia, reproductive toxicity testing, 461–462
Genomic conservation, *21–22*, 21–23, *24–25*
Genomics, 668, 866–867, *see also* Functional genomics and proteomics
Gestation
    day 20/21 term Cesarean section, 459–460
    day 13 uterine examination, 459
    duration, interpretation of study findings, 394–395
    guideline perspectives, 746–747
    index, 442
    milestones, *164–167*
Glossary, developmental defects, 911–965

Glutathione transferase
    assay methods, 536–537
    basics, 531
    fetus, 548
    intrauterine conceptual tissues, 543
    maternal liver, 540
    placenta, 555–556
Good Laboratory Practices (GLPs), 714–719, 744
Grip strength manipulations, 478
Growth
    inhibition, maternally mediated effects, 99
    nonclinical juvenile toxicity testing, 295–297, *296–298*
    retardation, human studies, 808–810
Guideline perspectives
    animals per dosage group, 745
    basics, 734, *735–736*
    conceptus through puberty, 752–753
    differences in guidelines, 744–754
    dosage group number, 745
    dosage selection, 745–746
    embryo development, 741–742
    endocrine-mediated findings, 753–754
    EPA guidelines, 739
    exposure intervals, 746–747, *746–747*
    FDA *vs.* EPA requirements and methods, 754
    fertility, 740–741
    fetal development, 741–742
    fetal evaluations, 752
    functional evaluations, 753–754
    gestation, 746–747
    Good Laboratory Practices, 744
    ICH guidelines and stages, 738–739, 743–744
    IRLG guidelines, 738
    lactation, 752
    maternal-pup interactions, 752
    neurotoxicity, 754
    OECD guidelines, 739
    ovaries evaluation, 752
    perinatal evaluations, 742–743
    postnatal development, 746–747
    postnatal evaluations, 742–743, 752–754
    protocol design, 740–744
    regulatory history, 734, 736–738
    reproductive performance, 740–741, 747, 749, 752
    roles of groups, 738–739
    sexual maturation, 752–753
    species selection, 744–745
    testing concerns evolution, 738–739
Guidelines
    differences, 744–754
    interpretation of study findings, 362–364, *363*, 382, *383–384*, 385
Guinea pigs
    cardiac innervation, 1066
    concentrating ability, 1014
    ductus arteriosus, 1061
    electrocardiographic measurements, 1068
    energy metabolism, 1067
    nerve functional development, 1068
    shape, position, size of heart, 1062, *1062*
    vectorcardiographic measurements, 1068

H

Hair growth, 473–474
Hamsters, 1068
Harmonization of approaches, 866
Hazard characterization, 845–856
Heart, 678, 1059
Heart, postnatal development
    basics, 1059, *1072–1073*
    biochemistry, 1066–1067
    cellular constituents, 1063–1064
    electrophysiology, 1067–1069
    hemodynamics, 1069–1071
    immediate postnatal changes, 1059–1061
    innervation, 1065–1066
    output, 1069–1071
    position, 1061–1063
    shape, 1061–1063
    size, 1061–1063
    vasculature, 1064–1065
Hematology, *285,* 299, 301, *301–305*
Hematopoietic stem cells, 1083
Hershberger assay, 505–507
Hind limb placing, 475
Histopathological evaluations, 446–447, 460, 462–463
Historical background, 94–95, 266–267
Historical control data, 404, *404*
Housing, *213,* 213–214
Humans
    acid-base equilibrium, 1014
    blood-testis barrier, 1035
    bone marrow meatopoiesis, 1087–1088
    cardiac innervation, 1065
    cardiac vasculature, 1064–1065
    cellular constituents, heart, 1063
    cognitive development, 1112, 1114–1115
    communication, 1118–1119
    concentrating ability, 1012–1013
    diaphysis, 992
    ductus arteriosus, 1060
    ductus venosus, 1060
    epididymides, 1034
    epiphyseal growth plate, 990
    fear response, 1122
    fetal liver hematopoiesis, 1084
    fine motor development and dexterity, 1106–1107
    foramen ovale, 1060
    functional immunocompetence, 1091–1093
    glomerular filtration, 1009–1010
    heart development, postnatal, 1059–1063, *1072–1073*
    hematopoietic stem cells, 1083
    intrauterine conceptual issues, 542–543
    locomotor development, 1104
    male reproductive system, 1029, *1030–1032,* 1033–1037
    neuroendocrine control, 1040–1041
    ossification, 975–979, *976, 978–979*
    osteon formation and remodeling, 993
    postnatal immune system development, 1089
    pregnancy outcomes, maternally mediated effects, 103
    preputial separation, 1037
    prostate, 1037–1038
    reflex and sensory development, 1109
    renal development, 1005–1007
    renin-angiotensin system, 1016
    seminal vesicles, 1040
    sleep cycles, 1124
    social play, 1120–1121
    spleen development, 1084–1085
    testes, 1029, 1033
    testicular descent, 1036
    thymus, 1086
    umbilical vasculature, 1059–1060
    urine volume control, 1014–1015
Human studies
    adverse reproductive outcomes, 802–805
    basics, 801, 837
    birth weight, 808–810
    case control studies, 820–828
    child death, 806–808
    childhood cancer, 815
    cluster analysis, 836–837
    cohort studies, 828–831
    comparison material, 829
    confounding, 805, 810, 813–814, 832–834
    congenital malformations, 810–813
    drug exposure, 817–818
    embryonic exposure, 815
    environmental exposure, 818
    exposures, recording, 815–819
    growth retardation, 808–810
    infant death, 806–808
    infertility, 802–804
    interviews, 819–820
    matched pairs or triplets, 826–828, *827*
    maternal exposure, 815–819
    medical record studies, 820
    mental retardation, 813–814
    miscarriage, *804,* 804–806
    misclassification of exposure, 819
    multiple regression analysis, 826
    multiple testing problem, 828
    nested case control studies, 831
    occupational exposure, 818
    outcomes, recording, 819–820
    pregnancy length, 808–810
    prospective exposure information, 819
    questionnaires, 819–820
    recording exposures, method, 815–819
    register studies, 820
    retrospective exposure information, 819
    spontaneous abortion, *804,* 804–806
    statistical analysis, 823–828, 830–831, *830–831,* 835
    stratified analysis, *823–824,* 825–826
    study design, 803–804, 806
    subfertility, 802–804
    surveillance, 834–835
    time to pregnancy, 802–803
    unmatched studies, *823,* 823–825
Hybridization, cDNA arrays, 628
Hydroxyurea, 35–36, 704–705, *704–705*
Hypospadias, 476–477

I

Identification, developmental toxicity testing, 215
Image acquisition, cDNA arrays, 628
Immune system development
    basics, 1080–1083, *1081, 1095,* 1095–1097
    bone marrow meatopoiesis, 1087–1088
    fetal liver hematopoiesis, 1084
    functional immunocompetence, 1091–1094
    hematopoietic stem cells, 1083
    postnatal immune system development, 1088–1091
    spleen development, 1084–1086
    thymus, 1086–1087
Immunization intervals, 892
Immunohistochemistry, 602–604, *603*
Implantation, 236–237
Implantation index, 441–442, 467
In-cage observations, 477–479
Incisor eruption, 474
Indanol dehydrogenase, 555
Independent binary response data, 701–703
Industrial chemicals, 347–348
Infants, 806–808, 1114, *see also* Humans
Infertility, human studies, 802–804
Injury levels, mechanistic approaches, 20, *20*
In-life concerns, 720, 722, *722–723,* 724, 727–729
In situ hybridization, 604–606, *606*
Integumentary system, *181–184*
Interagency Regulatory Liaison Group (IRLG), 738
International Conference of Harmonisation (ICH) guidelines
    animal criteria, 761
    basics, 759–761
    data presentation, 767
    male fertility addendum, 773–778
    perspectives, 739
    pharmaceuticals information, 793–798
    stages, 743–744
    statistics, 766–767
    study designs, 762–766
    terminology, 767–773
    treatment recommendations, 761–762
Interpretation of study findings
    agricultural chemicals, 347–348
    angiotensin converting enzyme fetopathy, 405–406, *406*
    animal models, 362, 382
    animal-to-human concordance, 337–342, *339–341*
    basics, 330–333, *332,* 414, 416
    biologic dynamics and dimensions, 412–413
    blood dosimetry, 395–397, *396*
    case studies, 404–408
    central nervous system maturation, 401
    context and interrelatedness, 371–372, *373–375*
    converging designs, 361
    database integrity, 410–411, *411*
    developmental toxicity studies, 361–382
    developmental toxicology, 11–12
    dose range-characterization studies, 354, 358
    dose-response curve characterization, 350–352
    dose selection, 352–353
    duration of treatment, 362, 364, 385
    dystocia, 404–405, *405*
    endocrine endpoints, 395
    endpoints, 364–382, *365–366,* 385–401, *386,* 395–398
    estrous cyclicity, 391–392
    experimental considerations, 372–377
    expert judgment role, 409–414, *410*
    fertility indices, 392–394, *393*
    fetal morphology, 371–382
    fetal sex ratio, 370–371
    functional toxicities, 401
    gallbladder variations, rabbits, 380, *381*
    gestation duration, 394–395
    guideline requirements, 362–364, *363,* 382, *383–384,* 385
    historical control data, 404, *404*
    industrial chemicals, 347–348
    intrauterine growth and survival, 368–371
    lactation, 395–398
    litter effect, 353–354, *355–358,* 386–387
    live birth index, 386–387
    lobular agenesis, lung, 407–408, *408*
    malformations, 377
    maternally mediated effects, 112–113
    maternal physiological endpoints, 397–398
    maternal toxicity, 365–368, *367*
    mating indices, 392–394, *393*
    maximum tolerated dose, 352–353
    milestones, 344–348, *345*
    milk composition, 397
    neonatal growth, 387
    neonatal survival indices, 387–388
    nesting, 395–398
    nursing, 395–398
    oocyte quantitation, 401
    organs weights and morphology, 390–391
    parturition process, 394–395
    pharmaceuticals, 344, 346–347
    physiological endpoints, 397–398
    postimplantation loss, 368–370, *369–370*
    precoital interval, 391–392
    predictive power background, 333–337, *333–339*
    preimplantation loss, 368, *369*
    prenatal growth, 370, *371*
    prenatal mortality, 388
    presacral vertebrae, 380–382
    presentation of findings, 377
    progeny sex ratio, 400
    rare effects evaluation, 401–409
    relative severity of findings, 377
    relevancy and risk analysis, 413–414, *415*
    reproductive toxicity studies, 382–401
    retroesophageal aortic arch, 406–407, *407*
    screening studies, 354, 358–361, *359–360*
    sex ratio, 370–371, 400
    sexual behavior, 398
    sexual maturity landmarks, 398–400, *399*
    significance, 333–342
    species, 348, *349–351*
    sperm quality assessment, 388–390, *390*
    statistical considerations, 353–354
    study design considerations, 342–354, *343*
    supernumerary ribs, 380–382

timing of treatment, 362, 364, 385
tissue dosimetry, 395–397, *396*
training of prospectors, 372, 374–375
treatment, timing and duration, 362, 364, 385
variants significance, *378–380,* 378–382
variations, 377
Interrelatedness and context, 371–372
Interviews, human studies, 819–820
Intrauterine conceptual tissues, 542–543
Intrauterine growth and survival, 368–371
In utero-lactational exposure, 510, *511–513,* 513
In vitro antisense RNA, 615
In vitro assays, 490, 492–502
In vitro methods, mechanisms study
    amphibian embryo culture, 670–671
    avian embryo culture, 671–672
    basics, 648–649, 683–684
    Caenorhabditis elegans cultures, 670
    cell culture, 678–683, *680–683*
    clonal analysis, 678–683
    Drosophila cultures, 670
    eyes, 678
    fish embryo culture, 671
    functional alterations, 666
    gene expression, 666
    genomics, 668
    heart, 678
    lens (eyes), 678
    limb bud cultures, 673–676, *674*
    lung buds, 678
    macromolecule synthesis and turnover, 666
    mammalian embryo culture, 648–665
    morphological events, 666
    nonmammalian whole embryo cultures, 670–672
    organ culture, 672–678, *673*
    palatal cultures, 676–677
    peri-implantation embryo culture, 655
    perinatal bone cultures, 673–676
    physiological alterations, 666
    postimplantation embryo culture, 655–658, *656, 658–659,* 660–665, *664*
    preimplantation embryo culture, 649–650, *651,* 652–655
    proteomics, 668
    receptor binding and function, 666
    sex organs, 678
    spinal cord, 678
    tissue culture, 672–678
    tooth germ, 678
    transgenic animals, 666, 668
    visceral yolk sac cultures, 677
In vivo assays, 502–514
Irregular bones, 973–974
Isotope-coded affinity tag, 639–640

J

Justification for live animals, 211
Juveniles, 753, 1114–1115, *see also* Humans; Nonclinical juvenile toxicity testing

L

Lactation, 395–398, 460, 462, 752, *see also* Milk
Lambs, 1061, 1071, *see also* Sheep
Late organogenesis, 7
Learning, 313–316, 1112, 1114–1115
Lens (eyes), 678
Levels of injury, mechanistic approaches, 20, *20*
Light cycles, 214
Likelihood and quasi-likelihood approaches, 703–704, 706
Limb bud cultures, 673–676, *674*
Limited RNA availability, 631–632
Linear mixed-effects models, 700
Lipid levels, 1066
Lipid peroxidation, 530, 541, 553–554
Lipoxygenase
    assay methods, 533–534
    basics, 529–530
    decidua, 541–542
    embryo, 545
    fetus, 547–548
    intrauterine conceptual tissues, 543
    placenta, 551–553
    uterus, 541–542
Liquid chromatography, multidimensional, 638
Litter-based ANOVA method, 699–700, 703
Litters
    culling, guideline perspectives, 752–753
    interpretation of study findings, 353–354, *355–358,* 386–387
    reproductive toxicity testing, 442–443, 469–471
Live-birth index, 386–387, 443
Lobular agenesis, lung, 407–408, *408*
Locomotor activity and development, 318, 1104–1106, *1106*
Long bones, 972–973, *972–973,* 995–998, *996*
Lung, lobular agenesis, 407–408
Lung buds, 678
Lung development, 1022–1026, *1023–1025,* 1026

M

Macaca mulatta, 1033, *see also* Primates
Macromolecule synthesis and turnover, 666
Macroscopic and microscopic evaluations, 301
Magnitude of exposure, 65–67
Maintenance of records, 255–256
MALDI/MALDI-TOF, *see* Matrix-assisted laser desorption ionization (MALDI)
Male fertility, 773–778
Male fertility index, 439
Male mating index, 439
Male reproductive system
    accessory sex glands, 1037–1040
    anogenital distance, 1036
    basics, 1029, *1030–1032*
    blood-testis barrier, 1035–1036
    epididymides, 1034–1035
    neuroendocrine control, 1040–1041

preputial separation, 1037
prostate, 1037–1039
reproductive organs, 1029, 1033–1037
seminal vesicles, 1040
testes, 1029, 1033–1034
testicular descent, 1036–1037
toxicants, 556–557
Males
   germ cells, paternally mediated effects, 128–132, *130–131*
   paternally mediated effects, 125–128
   pubertal protocols, Endocrine Disruptor Screening Program, 507–510, *508*
   specific endpoints, reproductive toxicity testing, 436
Malformations, *see also* Deformations
   abnormal development pathogenesis, 62–63
   experimental findings, 250
   interpretation of study findings, 377
   maternally mediated effects, 98–99
Mammalian embryo culture, 648–665
Mammary glands, 462
Management, quality concerns, 717–718
Mass spectrometry, 638–639
Matched pairs or triplets, 826–828, *827*
Maternal issues
   body weight, 229
   exposure, human studies, 815–819
   feed consumption, 229, 232
   liver, 538–540
   necropsy and postmortem examination, 234–236
   physiological endpoints, 397–398
   pup interactions, 752
   toxicity, 365–368, *367*
Maternally mediated effects, *see also* Paternally mediated effects
   alternative categorization, stressors, 97
   animal's environment control, 110–111
   basics, 94, 115
   behavioral measures, 108–109
   biochemical measures, 104–108
   catecholamines, 106
   chemical teratogenesis potentiation, 102–103
   controls, 111
   developmental toxicity test outcomes, 100–102
   direct effects comparison, 99–100
   experimental assessment, 110–114
   growth inhibition, 99
   historical background, 94–95
   human pregnancy outcomes, 103
   interpretation of results, 112–113
   malformations, 98–99
   mechanisms, 103–104
   methodology consistency, 112
   model systems, 113–114
   neurogenic stress, 96–97
   organ weights, 108
   outcome assessment, 112–113
   prenatal mortality, 99
   reproductive toxicity testing, 436–437
   restraint stress, 113–114
   stress, *95–96,* 95–99, 104–111
   stress cascade, 104–106
   stress proteins, 106–107
   supernumerary ribs, 98
   systemic stress, 97
   timing, 111–112
   toxicity, 97–99, *109,* 109–110
   zinc metabolism, 107–108
Mating
   developmental toxicity testing, 216–219
   evidence, 438–439
   indices, 439–442
   interpretation of study findings, 392–394, *393*
Matrix-assisted laser desorption ionization (MALDI), 639
Maximum tolerated dose, 352–353
Mean gestational length, 442
Mean time to mating, 442
Mechanistic approaches
   alkylating agents, 38–39
   altered energy sources, 32–33
   antitubulin agents, 30
   aromatic amines, 39
   basics, 16–18, *17,* 46–47
   cell-cell communication, *26,* 27–28
   cell-signaling pathways, *19,* 23, *26,* 26–27
   chlorpyrifos, 34
   cocaine, 33
   critical events evaluation, *18–19,* 18–20
   cytosine arabinoside, 36–37
   dioxin, 41, 43
   enzyme inhibition, 33–35
   ethanol, 44–46
   examples, 28–46
   5-fluorouracil, 29–30, 35
   gene expression alterations, 40–42
   genomic conservation, *21–22,* 21–23, *24–25*
   hydoxyurea, 35–36
   injury levels, 20, *20*
   methotrexate, 33–34
   methylmercury, *17,* 30–32, *31*
   mevinolin, 35
   mitochrondrial respiration inhibitors, 32–33
   mitotic interference, 29–32
   mutations, 37–40
   nucleic acids, 35–37
   omic research, *21,* 42
   organogenesis processes, 23, 26–28
   programmed cell death, 42–46
   radiation, 30, 39–40
   retinoic acid, 40–41, 43
   rhodamine dyes, 33
   three-dimensional evaluation context, 23, 26–28
Medical record studies, 820
Memory, 313–316, 1112, 1114–1115
Meningeal arteries, 999
Mental retardation, human studies, 813–814
Metaphyseal arteries, *996–997,* 997
Metaphysis, 991–992
Methotrexate, 33–34
Methylmercury, *17,* 30–32, *31*

Mevinolin, 35
Mice
    bone marrow meatopoiesis, 1088
    cellular constituents, heart, 1063
    ductus arteriosus, 1061
    electrocardiographic measurements, 1069
    energy metabolism, 1067
    fetal liver hematopoiesis, 1084
    functional immunocompetence, 1093
    hematopoietic stem cells, 1083
    ossification, *987,* 987–988
    postnatal immune system development, 1089
    prostate, 1039
    renal development, 1008
    renin-angiotensin system, 1016
    shape, position, size of heart, 1062, *1062, 1071*
    spleen development, 1085
    testes, 1034
    thymus, 1086–1087
    vectorcardiographic measurements, 1069
Microarrays, 623–633, 641
Migration of cells, 74–75, *75*
Milestones
    interpretation of study findings, 344–348, *345*
    vertebrate embryology, 159–161, *162–188*
Milestones, postnatal
    basics, 969–970, 1000
    bone structure and composition, 970–974, *971*
    diaphysis, 992–995
    epiphyseal growth plate, 988–991, *989–990*
    metaphysis, 991–992
    ossification, *972,* 974–988, *975*
    vascularity, 995–999
Milk, 397, 469, *see also* Lactation
Mineral content, 1066
Miscarriage, human studies, *804,* 804–806
Misclassification of exposure, 819
Missing pups, 470
Mitochndrial respiration inhibitors, 32–33
Mitotic interference, 29–32
Models
    interpretation of study findings, 362
    maternally mediated effects, 113–114
    nonclinical juvenile toxicity testing, 284–286
    toxicokinetics, 578, *578–579*
Mode of action data and study, 283–284, 865–866
Molecular techniques, *see* Cellular, biochemical, and molecular techniques
Monkeys, 979–981, *980,* 1025, *see also* Primates
Morphological events, 666
Morphology, 76–80
MRNAs, 601–613
MS/MS, *see* Tandem mass spectrometry (MS/MS)
Multidimensional liquid chromatography, 638
Multiple defects, 63
Multiple regression analysis, 826
Multiple testing problem, 828
Muscular system, *181–184*
Mutations, 37–40

N

Necropsy, 234–252, 445, 459–460
Negative geotaxis test, 475
Neonatal growth, interpretation of study findings, 387
Neonates, 387–388, 577, 1112, *see also* Humans
Nerve functional development, 1067–1068
Nervous system, *175–177, see also* Central nervous system
Nested case control studies, 831
Nesting, interpretation of study findings, 395–398
Neural tube defects, 77–78
Neuroendocrine control, 1040–1041
Neurogenic stress, 96–97
Neurotoxicity, 432–433, 754
Neurotransmitters, cardiac, 1068
Nipple retention, 474
Noise, 215
Nomenclature, 911–965
Nonclinical juvenile toxicity testing
    adapted adult toxicity study, 281–283, *283*
    administration route, 286–290, *287–289*
    adverse events, 267, *268*
    agency intervention, 267, *268,* 269–271
    basics, 264–266, *265,* 319
    behavioral assessment, 301, 306–316, 318
    between-litter design, *292,* 293, *294*
    clinical considerations, 273
    design considerations, 274–280
    developing *vs.* developed systems, 272–273
    development, 295–297, *296–298*
    dose selection, 290–292
    drug usage, 275–276, *275–276*
    duration of exposure, 290
    exposure, 286–292
    extant adult nonclinical animal toxicity data, 280
    extant data, 277–280
    food consumption, 299, *300*
    fostering design, 294–295, *295*
    frequency of exposure, 290
    growth, 295–297, *296–298*
    hematology, *285,* 299, 301, *301–305*
    historical perspectives, 266–267
    importance, 272–274
    macroscopic and microscopic evaluations, 301
    model selection, 284–286
    mode of action study, 283–284
    one pup per sex per litter design, 295
    parameters for evaluation, 295–318
    pharmacokinetic data, adult animals, 277–280, *278–289*
    pharmacological profile differences, 276–277, *277*
    physical developmental landmarks, 301, 306–316, 318
    predictive value, 274
    regulatory guidelines, 267, *268,* 269–271
    reproductive assessment, 318
    route of administration, 286–290, *287–289*
    sample size, 284–286
    screening, 281, *282*
    serum chemistry, *285,* 299, 301, *301–305*
    sexual developmental landmarks, 301, 306–316, 318
    species, 284, *285*

stand-alone studies, 280
statistical design, 318–319
study design, 280–319
target population, 275–276, *275–276*
test group organization, 292–295
timing of exposure, 275–276, *275–276*
toxicokinetic data, adult animals, 277–280, *278–289*
toxicological profile differences, 276–277, *277*
types of studies, 280–284
utility of studies, 273–274
within-litter design, *282, 292,* 292–293
Nonessential situations, 588
Non-human primates, 1033, 1036, *see also* Primates
Nonmammalian whole embryo cultures, 670–672
Nonpregnant dams, 467
Nonsmoker's placenta, 549–550
Normalization, data analysis and bioinformatics, 629–630
Northern blots, 607–608
Nuclear run-on assay, 612–613
Nucleic acids, 35–37
Nursing, 395–398

O

Observations, 228–229, 232, 445
Occupational exposure, human studies, 818
OECD, *see* Organisation for European Economic Eo-operation (OEEC/OECD)
OEEC/OECD, *see* Organisation for European Economic Eo-operation (OEEC/OECD)
Oligonucleotide microarrays, 628–629
Omic research, *21,* 42
One pup per sex per litter design, 295
Oocyte quantitation, 401, 467–468
Organ culture, 672–678, *673*
Organisation for European Economic Eo-operation (OEEC/OECD), 739
Organogenesis processes, 23, 26–28
Organ weights, 108, 390–391
Ossification, *972,* 974–988, *975*
Ossified bone, staining, 243
Outcomes, assessment and recording, 112–113, 819–820
Ovaries, 461, 557–558, 752
Oviducts, 460–461

P

Palatal cultures, 676–677
Parameters for evaluation, 295–318
Parturition, 394–395, 468–469
Paternally mediated effects, *see also* Maternally mediated effects
  animal experimentation, 127–128
  basics, 138
  drugs, 128
  epidemiological studies, *126,* 126–127
  epigenetic changes, 138

heritability, 132
male germ cells, 128–132, *130–131*
mechanisms involved, 128–132
methodological approaches, 132–138
posttesticular excurrent duct system, 129–131
reproductive toxicity testing, 436
reversibility, 131–132
role of male, 125–128
semen, 128
seminal fluid, 132
sperm, 132–138
testis, 129–131
toxicants, 128, 132
PCR quantitative method, 609–611
Pericranical vessels, 998–999
Peri-implantation embryo culture, 655
Perinatal bone cultures, 673–676
Perinatal evaluations, 742–743
Periosteal vessels, 997–998
Peroxidases
  assay methods, 535–536
  basics, 529
  decidua, 541
  embryo, 545–546
  fetal membranes, 549
  fetus, 547
  intrauterine conceptual tissues, 542–543
  placenta, 550–551
  uterus, 541
Personnel, developmental toxicity testing, 255
Pharmaceuticals, 344, 346–347, 793–798, *see also* Drugs
Pharmacokinetics
  nonclinical juvenile toxicity testing, 277–280, *278–289*
  risk assessment, 865–866
  toxicokinetics, *578–581,* 578–582
Pharmacological profile differences, 276–277, *277*
Phase I reactions, 528–530
Phase II reactions, 530–531
Physical developmental landmarks, 301, 306–316, 318
Physiological alterations, 666
Physiological changes, 576–577
Physiological endpoints, 397–398
Physiologically based pharmacokinetics, 582–583
Pigs, 1025, 1068, 1071
Pinna reflex and unfolding, 473, 475
Pituitary, 448–449, 460, 462
Placenta
  cytochrome P450, 549–550
  epoxide hydrase, 555
  flavin-containing monooxygenase, 550
  glutathione transferase, 555–556
  lipoxygenase, 551–553
  peroxidases, 550–551
  prostaglandin synthase, 550
  sulfotransferase, 555
  transfer, toxicokinetics, 574–575, *575*
  UDP-Glucuronosyl transferase, 555
  vertebrate embryology, 148–151, *149–150, 152–153,* 154
  xenobiotic metabolism, role, 549–556
Placental transfer, 8–9

Planning, risk assessment, 844–845
Plasma protein binding, 573
Postimplantation embryo culture, 655–658, *656, 658–659,* 660–665, *664*
Postimplantation losses, 368–370, *369–370,* 441–442
Postmortem examination, 234–252
Postnatal issues, 742–743, 746–747, 752–754, 1088–1091
Posttesticular excurrent duct system, 129–131
Precoital interval, 391–392, 442
Predictive power background, 333–337, *333–339*
Predictive values, 274, 640–641
Predifferentiation, 7
Pregnancy
    index, 441
    labeling, risk assessment, 894
    length, human studies, 808–810
    toxicokinetics, 576–577, 582–583
Preimplantation
    embryo culture, 649–650, *651,* 652–655
    interpretation of study findings, 368, *369*
    reproductive toxicity testing, 441–442
Prenatal issues, 99, 370, *371,* 388, 779–783
Preputial separation, 1037
Presacral vertebrae, 380–382
Presentation of findings, 377
Prestudy, quality concerns, *727*
Preweanling rats, 479
Primates
    bone marrow meatopoiesis, 1088
    cognitive development, 1117–1118
    communication, 1119–1120
    fear response, 1123
    fetal liver hematopoiesis, 1084
    fine motor development and dexterity, 1107
    functional immunocompetence, 1094
    hematopoietic stem cells, 1083
    locomotor development, 1105
    ossification, 979–981, *980*
    postnatal immune system development, 1091
    reflex and sensory development, 1111
    renal development, 1009
    renin-angiotensin system, 1017
    sleep cycles, 1124
    social play, 1121–1122
    spleen development, 1085
    testes, 1033
    testicular descent, 1036
    thymus, 1087
Principle Investigator Review of Data, 717
Probe preparation, cDNA arrays, 625–628, *627*
Problem formulation, risk assessment, 844–845
Process, risk assessment, 845–861
Product labels, 893–894, *894*
Progeny sex ratio, 400
Programmed cell death, 42–46
Proliferation of cells, 70–71, *71*
Prospective exposure information, 819
Prostaglandin synthase
    assay methods, 534–535
    basics, 529
    decidua, 541–542
    embryo, 544–545
    fetus, 547
    placenta, 550
    uterus, 541–542
Prostate, 1037–1039
Proteins, 601–606, 636–640, 1066
Proteomics, 636–641, 668, 866–867, *see also* Functional genomics and proteomics
Protocol design, 740–744
Pubertal alterations, 463
Pubertal protocols, 507–510, *508*
Pup evaluations, 471–477
Pupil constriction reflex, 476
Pupillary response, 479

Q

Quality assurance, 718–719
Quality concerns
    auditing studies, 719–726
    basics, 714, 730
    best practices, 720–726, *721–726*
    21 CFR Part 11, 715–716
    current trends, 716
    electronic records and signatures, 715
    GLP compliance program, 716–719
    good laboratory practice standards, 714–716
    in-life concerns, 720, 722, *722–723,* 724, *727–729*
    management, 717–718
    prestudy, *727*
    quality assurance, 718–719
    quality control measures, 727
    records and report audit, 725–726, *726*
    regulations, 714–715
    roles and responsibilities, 717–719
    study directors, 719
Quality control measures, 727
Quantitative risk assessment, 707–709
Questionnaires, human studies, 819–820

R

Rabbits
    acid-base equilibrium, 1014
    cellular constituents, heart, 1063–1064
    concentrating ability, 1013–1014
    developmental toxicity testing, 217–219
    double staining, ossified bone, 243–244
    ductus arteriosus, 1061
    epiphyseal growth plate, 990–991
    gallbladder variations, 380
    glomerular filtration, 1012
    lung development, 1025
    nerve functional development, 1068
    neurotransmitters, heart, 1068
    ossification, *983,* 983–984
    output and hemodynamics, heart, 1071

renal development, 1009
renin-angiotensin system, 1017
shape, position, size of heart, 1062, *1062, 1064*
vasculature, heart, 1065
Radiation, 30, 39–40
Rare effects evaluation, 401–409
Rats
    acid-base equilibrium, 1014
    anogenital distance, 1036
    blood-testis barrier, 1035
    bone marrow meatopoiesis, 1088
    cellular constituents, heart, 1063
    cognitive development, 1115–1116
    communication, 1119
    concentrating ability, 1013
    diaphysis, 993
    ductus arteriosus, 1061
    ductus venosus, 1061
    electrocardiographic measurements, 1068
    energy metabolism, 1066–1067
    epididymides, 1034–1035
    epiphyseal growth plate, 991
    fear response, 1122
    fetal liver hematopoiesis, 1084
    fine motor development and dexterity, 1107
    foramen ovale, 1061
    functional immunocompetence, 1094
    glomerular filtration, 1011
    heart development, postnatal, 1059–1063, *1062, 1064, 1070–1073*
    hematopoietic stem cells, 1083
    innervation, heart, 1065
    lipid levels, 1066
    locomotor development, 1104–1105
    lung development, 1023–1025
    mineral content, 1066
    nerve functional development, 1068
    neuroendocrine control, 1041
    neurotransmitters, heart, 1068
    ossification, 984–987, *985*
    output and hemodynamics, heart, 1070–1071
    postnatal immune system development, 1089–1090
    preputial separation, 1037
    preweanling, 479
    prostate, 1039, *1039*
    protein content, 1066
    reflex and sensory development, 1110–1111
    renal development, 1008
    renin-angiotensin system, 1016
    seminal vesicles, 1040
    shape, position, size of heart, 1062
    sleep cycles, 1124
    social play, 1121
    spleen development, 1085
    testes, 1033–1034
    testicular descent, 1037
    thymus, 1087
    vasculature, heart, 1065
    vectorcardiographic measurements, 1068
    water content, 1066
Real time RT-PCR quantitative method, 609–610, *610*

Receptor binding and function, 666
Receptor-dependent gene transcription assays, 493–496
Recording exposures, method, 815–819
Records, 253, *253,* 255–256, 725–726, *726*
Redox cycling agents, 554
Reflex and sensory development, 1108–1112
Reflex testing, 474–476
Regulations, quality concerns, 714–715
Regulatory issues
    compliance, 254
    expectations, 585–587
    guidelines, 267, *268,* 269–271
    history, 734, 736–738
Relative humidity, *213,* 214
Relative severity of findings, 377
Relevancy, 413–414, *415*
Removed-from-cage observations, 478–479
Renal development
    acid-base equilibrium, 1014
    anatomical development, 1005–1009
    basics, 1005
    concentrating ability, 1012–1014
    functional development, 1009–1017
    glomerular filtration, 1009–1012
    renin-angiotensin system, 1015–1017
    urine volume control, 1014–1015
Replicates, 632–633
Reports, 254–255, 725–726
Reproduction and fertility effects, 784–790
Reproduction studies, 430, *431–432*
Reproduction test guideline, ICH, 585–587
Reproductive assessment, 318
Reproductive indices, 439–444
Reproductive organs
    male reproductive system, 1029, 1033–1037
    weights and morphology, 390–391, 446, 460
Reproductive performance
    guideline perspectives, 740–741, 747, 749, 752
    reproductive toxicity testing, 437–444
Reproductive senescence, 463
Reproductive system, *187–188, see also* Female reproductive system; Male reproductive system
Reproductive toxicity studies, 382–401
Reproductive toxicity testing
    accessory sex glands, 448–449
    air righting reflex, 475–476
    anogenital distance and disease, 472
    auditory startle reflex, 475
    balanopreputial separation, 472–473
    basics, 428
    behavioral testing, 474–476
    biology, 444–445, 458
    birth weight, 471
    body temperature, 479
    body weight, 445, 459, 471–472
    cauda epididymal sperm mobility, 454–456
    cliff avoidance, 474
    cohabitation, 438–439
    combined repeated dose toxicity study, 433–434, *434*
    computer-assisted sperm analysis, 450–454, *451*
    continuous breeding, 431–432, *433*

corpora lutea counts, 466–467
couple-mediated endpoints, 436
crown-rump length, 471
dams, 443, 470
DART study guidelines, 428–436
death of dam, 470
death of pups, 470–471
developmental alterations, 463
developmental landmarks, 472–473
developmental neurotoxicity studies, 432–433
developmental toxicity studies, 430, *430*
dominant lethal study, *435,* 435–436
endpoints, 436–437
epididymides, 448–449
estrous, 473
estrous cycle, 465–466, *466*
estrous cycle staging, 465
external alterations, 470
external genitalia, 461–462
eye opening, 474
fecundity index, 441
female-specific endpoints, 436–437
fertility index, 439–442
fertility mating index, 440
fertility studies, 429, *429*
fertilization rate, 440
food consumption, 445
foot splay, 478–479
forelimb grip test, 476
forelimb placing, 475
functional observation battery, 477–480
gender determination, 470–471
genitalia, 461–462
gestation day 20/21 term Cesarean section, 459–460
gestation day 13 uterine examination, 459
gestation index, 442
grip strength manipulations, 478
hair growth, 473–474
hind limb placing, 475
histopathological evaluations, 446–447, 460, 462–463
hypospadias, 476–477
implantation, 441–442, 467
in-cage observations, 477–479
incisor eruption, 474
lactation, 460, 462
landmarks, 472–473
litters, 442–443, 469–471
live-birth index, 443
male-specific endpoints, 436
mammary glands, 462
maternally mediated effects, 436–437
mating evidence, 438–439
mating index, 439
mean gestational length, 442
mean time to mating, 442
milk collection, 469
missing pups, 470
necropsy, 445, 459–460
negative geotaxis test, 475
neurotoxicity studies, 432–433
nipple retention, 474

nonpregnant dams, 467
observations, 445
oocyte quantitation, 467–468
ovaries, 461
oviducts, 460–461
parturition, 468–469
paternally mediated effects, 436
pinna reflex, 475
pinna unfolding, 473
pituitary, 448–449, 460, 462
postimplantation losses, 441–442
precoital interval, 442
pregnancy index, 441
preimplantation losses, 441–442
preweanling rats, 479
pubertal alterations, 463
pup evaluations, 471–477
pupil constriction reflex, 476
pupillary response, 479
reflex testing, 474–476
removed-from-cage observations, 478–479
reproduction studies, 430, *431–432*
reproductive indices, 439–444
reproductive organ weight, 446, 460
reproductive performance, 437–444
reproductive senescence, 463
reproductive toxicology, 444–468
resorptions assessment, 467
segment I - III studies, 459–460
seminiferous epithelium staging, 457–458
sex ratio, 444
spermatogenesis, 457–458
sperm evaluations, 449
sperm morphology, 456–457
sperm number/concentration, 449–450
standard open field arena, 478–480
surface-righting reflex, 474
survival index, 443–444
temperature, 479
term Cesarean section, 459–460
testes descent, 474
testis, 447–448
toxicity screening test, 433–434, *434*
toxicity studies, 430, *430*
treatment duration, 437–438
uterus, 459–461, 467
vagina, 460–461
vaginal cytology, 463–466, *464*
vaginal lavage methodology, 464–465
vaginal patency, 473–474
vaginal plugs, 438
vaginal smears, 438–439
viability, 470
Reproductive toxicology, 444–468
Research needs, 1125
Resorptions assessment, 467
Respiratory system, *174*
Response pathways, role, 613–616
Responsibilities and roles, 717–719
Restraint stress, 113–114

Results interpretation, 11–12, *see also* Interpretation of study findings
Retinoic acid, 40–41, 43
Retroesophageal aortic arch, 406–407, *407*
Retrospective exposure information, 819
Reverse Southern blots, 616
Reversibility, paternally mediated effects, 131–132
Rhesus, 1033, *see also* Primates
Rhodamine dyes, 33
Risk analysis, 413–414
Risk assessment, environmental agents
    basics, 841–842
    benefit analysis, 867
    dose-response assessment, 856–859
    emerging issues, 864–867
    exposure assessment, 859–860
    framework, 843–861, *844*
    future technologies, 866–867
    genomics, 866–867
    harmonization of approaches, 866
    hazard characterization, 845–856
    historical perspectives, 842–843
    mode of action data, 865–866
    pharmacokinetic data, 865–866
    planning, 844–845
    problem formulation, 844–845
    process, 845–861
    proteomics, 866–867
    risk characterization, 860–861, *862–863*
Risk assessment, FDA perspective
    additives statutory safety standard, 879–890
    basics, 878
    dosing, 892
    drug safety, 893–897
    endpoints, 892–893
    exposures, 892
    food safety, 878–890
    immunization intervals, 892
    pregnancy labeling, 894
    product labels, 893–894, *894*
    toxicity assessment, 894–907
    vaccine safety, 890–893
Risk assessment use, *590,* 590–591, *592*
Risk characterization, 860–861, *862–863*
RNA availability, limited, 631–632
RNase protection assay, 611–612
RNA transcription *in vitro*, 612–613
Rodents, 216–217, 243, *see also* specific type of rodent
Roles and responsibilities, 717–719
Roles of groups, 738–739
Routes of administration
    developmental toxicity testing, 221–226, *225*
    nonclinical juvenile toxicity testing, 286–290, *287–289*
RT-PCR *in situ* transcription, 616
RT-PCR quantitative method, 609–611

S

SAGE technology, 635–636, 641
Sample size, 284–286
Sampling strategies, 583–585, *584–585*
Screening, 281, *282,* 354, 358–361, *359–360*
Segment I - III studies, 459–460
Semen, 128, *see also* Sperm
Seminal fluid, 132
Seminal vesicles, 1040
Seminiferous epithelium staging, 457–458
Seminiferous tubule distribution, 574
Semiquantitative methods, 607–608
Sense organs, *178–180*
Sequences, 63
Serial analysis, gene expression, 633–636
Serum chemistry, *285,* 299, 301, *301–305*
Sex organs, 678
Sex ratio, 370–371, 400, 444, *see also* Gender determination
Sexual behavior, 398
Sexual developmental landmarks, 301, 306–316, 318
Sexual maturation, 752–753
Sexual maturity landmarks, 398–400, *399*
Sheep, *see also* Lambs
    concentrating ability, 1014
    glomerular filtration, 1012
    lung development, 1025
    neurotransmitters, heart, 1068
    output and hemodynamics, heart, 1071
    renal development, 1009
    renin-angiotensin system, 1017
Short bones, 973
Single staining, ossified bone, 243
Skeletal examination and evaluation, 244–250, *245–250*
Skeletal system, *181–184*
Sleep cycles, 1123–1124, *1125*
Slot blots, 608
Social play, 1120–1122, *1122*
Solution hybridization, 611–612
Species, *see also* specific species
    developmental toxicity testing, 210, *211*
    interpretation of study findings, 348, *349–351*
    nonclinical juvenile toxicity testing, 284, *285*
    selection, guideline perspectives, 744–745
Sperm
    cauda epididymal sperm mobility, 454–456
    characteristics, 132–134
    chromatin packaging and function, 134–136, *135*
    evaluations, reproductive toxicity testing, 449
    genetic integrity, 136–137
    morphology, reproductive toxicity testing, 456–457
    number/concentration, reproductive toxicity testing, 449–450
    paternally mediated effects, 132–138, *135*
    quality, 134–138
    quality assessment, 388–390, *390*
    quantity, 132–134
Spermatogenesis, 457–458
Spinal cord, 678
Spleen development, 1084–1086
Spontaneous abortion, *804,* 804–806
Stages, ICH guidelines, 743–744
Stand-alone studies, 280
Standard open field arena, 478–480

Standard operating procedures (SOPs), 717
Staples's technique, 238–240
Statistical analysis
  basics, 697–699, 709–710
  benchmark dose, 707–709
  binary response data, 701–705
  continuous response data, 699–701
  correlated binary response data, 703–705
  count data analysis, 705–706
  dose-response model, 707–708
  examples, 700–701, *700–701,* 704–706, 709
  fetal body weight data example, 700–701, *700–701*
  human studies, 823–828, 830–831, *830–831,* 835
  hydroxyurea data, 704–705, *704–705*
  independent binary response data, 701–703
  likelihood and quasi-likelihood approaches, 703–704, 706
  linear mixed-effects models, 700
  litter-based ANOVA method, 699–700, 703
  quantitative risk assessment, 707–709
Statistics
  developmental toxicity testing, 252
  ICH guidelines, 766–767
  interpretation of study findings, 353–354
  nonclinical juvenile toxicity testing, 318–319
Status reports, 254
Steroidogenesis, 496–500, *497–500*
Stratified analysis, *823–824,* 825–826
Stress, *95–96,* 95–99, 104–111
Stress cascade, 104–106
Stress proteins, 106–107
Structural anomalies, 62–63
Study approach and design
  cellular, biochemical, and molecular techniques, 600–601
  developmental toxicity testing, 219–220
  human studies, 803–804, 806
  ICH guidelines, 762–766
  interpretation of study findings, 342–354, *343*
  nonclinical juvenile toxicity testing, 280–319
Study directors, 717, 719
Study interpretation, *see* Interpretation of study findings
Study use, toxicokinetics, 588–590, *589*
Subfertility, 802–804
Sulfotransferase
  assay methods, 537–538
  basics, 531
  fetus, 548–549
  placenta, 555
Supernumerary ribs, 98, 380–382
Supplier of animals, 210, *211*
Surface-righting reflex, 474
Surveillance, human studies, 834–835
Survival index, 443–444
Sutural vessels, 999
Swimming, 1105
Syndromes, 63
Systemic stress, 97

T

Tandem mass spectrometry (MS/MS), 639
Target population, 275–276, *275–276*
Temperature, *213,* 214, 479
Term Cesarean section, 459–460
Terminology, 767–773, 911–965
Test animals, 210–213
Testes, 474, 1029, 1033–1034
Test group organization, 292–295
Testicular descent, 1036–1037
Testicular enzymes, 556
Testing, *see* Developmental toxicity testing; Reproductive toxicity testing
Testing concerns evolution, 738–739
Testis, 129–131, 447–448
Test substance, 208–210
Thoracic viscera, 239
Three-dimensional evaluation context, 23, 26–28
Thymus, 1086–1087
Time to pregnancy, 802–803
Timing
  exposure, 275–276, *275–276*
  interpretation of study findings, 362, 364, 385
  maternally mediated effects, 111–112
TIS battery, 514–515
Tissue culture, 672–678
Tissue dosimetry, 395–397, *396*
Tissue specificity, 64–65
Tooth germ, 678
Total number, animals, 211–213
Toxicants
  female reproductive system, 558–559
  male reproductive system, 556–557
  paternally mediated effects, 128, 132
Toxicity
  maternally mediated effects, 97–99, *109,* 109–110
  reproductive toxicity testing, *430*
  risk assessment, 894–907
  screening test, 433–434, *434*
Toxicogenetic guideline, ICH, 587
Toxicokinetics
  absorption principles, 572–577
  adult animals, 277–280, *278–289*
  basics, 572, 587–592
  biological membranes, movement across, 572–573
  biotransformation, 572–577
  blood plasma, 574
  distribution, 572–577
  elimination, 572–577
  ICH reproduction test guideline, 585–587
  ICH toxicogenetic guideline, 587
  models, 578, *578–579*
  neonates, physiological changes, 577
  nonessential situations, 588
  pharmacokinetics, *578–581,* 578–582
  physiological changes, 576–577
  physiologically based pharmacokinetics, 582–583
  placental transfer, 574–575, *575*
  plasma protein binding, 573
  pregnancy, 576–577, 582–583

regulatory expectations, 585–587
risk assessment use, *590,* 590–591, *592*
sampling strategies, 583–585, *584–585*
seminiferous tubule distribution, 574
study use, 588–590, *589*
Toxicological profile differences, 276–277, *277*
Training of prospectors, 372, 374–375
Training records, 717
Transgenic animals, 613–615, 666, 668, *see also* Animals; *specific animals*
Transitional findings, 251
Treatment
  interpretation of study findings, 362, 364, 385
  mated females, 221–228
  recommendations, 761–762
  timing and duration, 362, 364, 385, 437–438
Two-dimensional gel electrophoresis, 636–638, *637*

U

UDP-Glucuronosyl transferase
  assay methods, 537
  basics, 530
  fetus, 548
  maternal liver, 540
  placenta, 555
Umbilical vasculature, 1059–1060
Unmatched studies, *823,* 823–825
U.S. EPA, 739, 754, 779–790
U.S. EPA EDSP, *see* Endocrine Disruptor Screening Program (EDSP)
Uterotrophic assay, 502–505
Uterus
  cytochrome P450, 540–541
  flavin-containing monooxygenase, 542
  lipid peroxidation, 541
  lipoxygenase, 541–542
  peroxidases, 541
  prostaglandin synthase, 541–542
  reproductive toxicity testing, 459–461, 467
  xenobiotic metabolism, role, 540–542
Utility of studies, 273–274

V

Vaccine safety, 890–893
Vagina, 460–461
Vaginal cytology, 463–466, *464*
Vaginal lavage methodology, 464–465
Vaginal patency, 473–474, 753
Vaginal plugs, 438
Vaginal smears, 438–439
Variants significance, *378–380,* 378–382
Variations, 251–252, 377
Vascularity, 995–999
Vectorcardiographic measurements, 1067–1069
Ventilation, 215

Vertebrate embryology, comparative features
  basics, 147–148, *148*
  circulatory system, *164–167*
  digestive system, *168–170*
  embryological processes, 154–156, *157,* 158–159, *162–168*
  endocrine system, *171–173*
  excretory system, *185–186*
  gestational milestones, *164–167*
  integumentary system, *181–184*
  milestones, 159–161, *162–188*
  muscular system, *181–184*
  nervous system, *175–177*
  placental characteristics, 148–151, *149–150, 152–153,* 154
  reproductive system, *187–188*
  respiratory system, *174*
  sense organs, *178–180*
  skeletal system, *181–184*
Viability, 470
Visceral examinations, 238–241
Visceral yolk sac cultures, 677

W

Washing, cDNA arrays, 628
Water, *213,* 213–214
Water content, 1066
Websites, 641
Weight of animals, 211–213
Wilson's principles, 6–11
Wilson's technique, 240–241
Within-litter design, *282, 292,* 292–293

X

Xenobiotic metabolism, role
  assay methods, 532–538
  basics, 527–531, 556–560
  cooxidation, 553–554
  cytochrome P450, 528–529, 532–533, 538–541, 543–544, 546–547, 549–550
  decidua, 540–542
  developmental toxicants metabolism, 538–556
  embryo, 543–546
  enzymes assay, 531–538
  epoxide hydrase, 536, 540, 548, 555
  female reproductive system, 557–559
  fetal membranes, 549
  fetus, 546–549
  flavin-containing monooxygenase, 529, 533, 540, 542, 547, 550
  future research, 559–560
  glutathione transferase, 531, 536–537, 540, 543, 548, 555–556
  indanol dehydrogenase, 555
  intrauterine conceptual tissues, 542–543

lipid peroxidation, 530, 541, 553–554
lipoxygenase, 529–530, 533–534, 541–543, 545, 547–548, 551–553
male reproductive system, 556–557
maternal liver, 538–540
nonsmoker's placenta, 549–550
ovarian enzymes, 557–558
peroxidases, 529, 535–536, 541–543, 545–547, 549–551
Phase I reactions, 528–530
Phase II reactions, 530–531
placenta, 549–556
prostaglandin synthase, 529, 534–535, 541–542, 544–545, 547, 550
sulfotransferase, 531, 537–538, 548–549, 555
testicular enzymes, 556
toxicants, 556–559
UDP-Glucuronosyl transferase, 530, 537, 540, 548, 555
uterus, 540–542

Z

Zinc metabolism, 107–108

JAN